Frequently Used Formulas

D0164592

Slope of a Line Containing the Points (x_1, y_1) and (x_2, y_2):

$$m = \frac{y_2 - y_1}{x_2 - x_1}, \text{ where } x_2 - x_1 \neq 0.$$

Slope—Intercept Form of a Line: $y = mx + b$

Standard Form of a Line:

$Ax + By = C$ where A, B, and C are integers and A is positive.

Point-Slope Formula: If (x_1, y_1) is a point on line L and m is the slope of line L, then the equation of L is given by

$$y - y_1 = m(x - x_1).$$

Factoring Formulas

1. Difference of Two Squares: $a^2 - b^2 = (a + b)(a - b)$
2. Sum and Difference of Two Cubes:

$$a^3 + b^3 = (a + b)(a^2 - ab + b^2)$$
$$a^3 - b^3 = (a - b)(a^2 + ab + b^2)$$

Distance Formula: The distance, d, between the points (x_1, y_1) and (x_2, y_2) is given by $d = \sqrt{(x_2 - x_1)^2 + (y_2 - y_1)^2}$.

Quadratic Formula: The solutions of any quadratic equation of the form $ax^2 + bx + c = 0$ $(a \neq 0)$ are

$$x = \frac{-b \pm \sqrt{b^2 - 4ac}}{2a}.$$

Standard Form for the Equation of a Parabola That Opens Vertically:

$y = a(x - h)^2 + k$, where the vertex is at (h, k)

Standard Form for the Equation of a Parabola That Opens Horizontally:

$x = a(y - k)^2 + h$, where the vertex is at (h, k)

Standard Form for the Equation of a Circle:

$(x - h)^2 + (y - k)^2 = r^2$, where the center is (h, k) and the radius is r

Standard Form for the Equation of an Ellipse

$$\frac{(x - h)^2}{a^2} + \frac{(y - k)^2}{b^2} = 1, \text{ where the center is } (h, k)$$

Standard Form for the Equation of a Hyperbola

$$\frac{(x - h)^2}{a^2} - \frac{(y - k)^2}{b^2} = 1 \text{ or } \frac{(y - k)^2}{b^2} - \frac{(x - h)^2}{a^2} = 1$$

where the center is (h, k)

Intermediate Algebra

Sherri Messersmith
College of DuPage

Connect
Learn
Succeed™

INTERMEDIATE ALGEBRA

Published by McGraw-Hill, a business unit of The McGraw-Hill Companies, Inc., 1221 Avenue of the Americas, New York, NY 10020. Copyright © 2012 by The McGraw-Hill Companies, Inc. All rights reserved. No part of this publication may be reproduced or distributed in any form or by any means, or stored in a database or retrieval system, without the prior written consent of The McGraw-Hill Companies, Inc., including, but not limited to, in any network or other electronic storage or transmission, or broadcast for distance learning.

Some ancillaries, including electronic and print components, may not be available to customers outside the United States.

This book is printed on acid-free paper.

1 2 3 4 5 6 7 8 9 0 DOW/DOW 1 0 9 8 7 6 5 4 3 2 1

ISBN 978–0–07–340617–6
MHID 0–07–340617–1

ISBN 978–0–07–329718–7 (Annotated Instructor's Edition)
MHID 0–07–329718–6

Vice President, Editor-in-Chief: *Marty Lange*
Vice President, EDP: *Kimberly Meriwether David*
Vice-President New Product Launches: *Michael Lange*
Editorial Director: *Stewart K. Mattson*
Executive Editor: *Dawn R. Bercier*
Developmental Editor: *Emily Williams*
Director of Digital Content Development: *Emilie J. Berglund*
Marketing Manager: *Peter A. Vanaria*
Lead Project Manager: *Peggy J. Selle*
Buyer II: *Sherry L. Kane*
Senior Media Project Manager: *Sandra M. Schnee*
Senior Designer: *David W. Hash*
Cover Designer: *Greg Nettles/Squarecrow Creative*
Cover Image: *©Jeff Urban*
Lead Photo Research Coordinator: *Carrie K. Burger*
Compositor: *Aptara, Inc.*
Typeface: *10.5/12 Times New Roman*
Printer: *R. R. Donnelley*

All credits appearing on page or at the end of the book are considered to be an extension of the copyright page.

Library of Congress Cataloging-in-Publication Data

Messersmith, Sherri.
   Intermediate algebra / Sherri Messersmith. — 1st ed.
      p. cm.
   Includes index.
   ISBN 978-0-07-340617-6—ISBN 0-07-340617-1 (hard copy : alk. paper) 1. Algebra—Textbooks. I. Title.
QA154.3.M47 2012
512.9—dc22

                                                          2010044513

www.mhhe.com

Building a Better Path to Success

Intermediate Algebra, helps students build a better path to success by providing the tools and building blocks necessary for success in their mathematics course. The author, Sherri Messersmith, learned in her many years of teaching that students had a better rate of success when they were connecting their knowledge of arithmetic with their study of algebra. By **making these connections between arithmetic and algebra** and **presenting the concepts in more manageable, "bite-size" pieces,** Sherri better equips her students to learn new concepts and strengthen their skills. In this process, students practice and build on what they already know so that they can understand and **master new concepts** more easily. These practices are integrated throughout the text and the supplemental materials, as evidenced below:

Connecting Knowledge

- **Examples** draw upon students' current knowledge while connecting to concepts they are about to learn with the positioning of arithmetic examples before corresponding algebraic examples.

- The very popular author-created **worksheets** that accompany the text provide instructors with additional exercises to assist with overcoming potential stumbling blocks in student knowledge and to help students see the connections among multiple mathematical concepts. The worksheets fall into three categories:
  - Worksheets to Strengthen Basic Skills
  - Worksheets to Help Teach New Concepts
  - Worksheets to Tie Concepts Together

Presenting Concepts in Bite-Size Pieces

- The **chapter organization** help break down algebraic concepts into more easily learned, more manageable pieces.

- New **Fill-It-In exercises** take a student through the process of working a problem step-by-step so that students have to provide the reason for each mathematical step to solve the problem, much like a geometry proof.

- New **Guided Student Notes** are an amazing resource for instructors to help their students become better note-takers. They contain in-class examples provided in the margin of the text along with additional examples not found in the book to emphasize the given topic so that students spend less time copying down information and more time engaging within the classroom.

- **In-Class Examples** give instructors an additional tool that exactly mirrors the corresponding examples in the book for classroom use.

Mastering Concepts:

- **You Try** problems follow almost every example in the text to provide students the opportunity to immediately apply their knowledge of the concept being presented.

- **Putting It All Together** sections allow the students to synthesize important concepts presented in the chapter sooner rather than later, which helps in their overall mastery of the material.

- **Connect Math hosted by ALEKS** is the combination of an online homework manager with an artificial-intelligent, diagnostic assessment. It allows students to identify their strengths and weaknesses and to take the necessary steps to master those concepts. Instructors will have a platform that was designed through a comprehensive market development process involving full-time and adjunct math faculty to better meet their needs.

Connecting Knowledge

Examples   The **examples** in each section begin with an arithmetic equation that mirrors the algebraic equation for the concept being presented. This positioning allows students to apply their knowledge of arithmetic to the algebraic problem, making the concept more easily understandable.

"Messersmith does a great job of addressing students' abilities with the examples and explanations provided, and the thoroughness with which the topic is addressed is excellent." Tina Evans, Shelton State Community College

Example 1

In-Class Example 1
Find the LCD of each pair of rational expressions.
a) $\dfrac{3}{14}, \dfrac{5}{20}$ b) $\dfrac{8}{15w^4}, \dfrac{6}{5w^2}$
answer: a) 140 b) 15w⁴

Find the LCD of each pair of rational expressions.

a) $\dfrac{7}{18}, \dfrac{11}{24}$ b) $\dfrac{4}{9t^3}, \dfrac{5}{6t^2}$

Solution
a) Follow the steps for finding the least common denominator.

   Step 1: Factor the denominators.
   $$18 = 2 \cdot 3 \cdot 3 = 2 \cdot 3^2$$
   $$24 = 2 \cdot 2 \cdot 2 \cdot 3 = 2^3 \cdot 3$$

   Step 2: The LCD will contain each unique factor the *greatest* number of times it appears in any factorization. *The LCD will contain 2^3 and 3^2.*

   Step 3: The LCD is the *product* of the factors in step 2.
   $$LCD = 2^3 \cdot 3^2 = 8 \cdot 9 = 72$$

b) To find the LCD of $\dfrac{4}{9t^3}$ and $\dfrac{5}{6t^2}$,

   Step 1: Factor the denominators.
   $$9t^3 = 3 \cdot 3 \cdot t^3 = 3^2 \cdot t^3$$
   $$6t^2 = 2 \cdot 3 \cdot t^2$$

   Step 2: The LCD will contain each unique factor the *greatest* number of times it appears in any factorization. *It will contain 2, 3^2, and t^3.*

   Step 3: The LCD is the *product* of the factors in Step 2.
   $$LCD = 2 \cdot 3^2 \cdot t^3 = 18t^3$$

"The author is straightforward, using language that is accessible to students of all levels of ability. The author does an excellent job explaining difficult concepts and working from easier to more difficult problems." Lisa Christman, University of Central Arkansas

Example 6

In-Class Example 6
Multiply and simplify.
a) $(\sqrt{5} + 2)^2$
b) $(2\sqrt{t} - 7)^2$
answer: a) $9 + 4\sqrt{5}$
b) $4t - 28\sqrt{t} + 49$

Multiply and simplify.

a) $(\sqrt{10} + 3)^2$ b) $(2\sqrt{x} - 6)^2$

Solution

a) Use $(a + b)^2 = a^2 + 2ab + b^2$.

$(\sqrt{10} + 3)^2 = (\sqrt{10})^2 + 2(\sqrt{10})(3) + (3)^2$ Substitute $\sqrt{10}$ for a and 3 for b.
$= 10 + 6\sqrt{10} + 9$ Multiply.
$= 19 + 6\sqrt{10}$ Combine like terms.

b) Use $(a - b)^2 = a^2 - 2ab + b^2$.

$(2\sqrt{x} - 6)^2 = (2\sqrt{x})^2 - 2(2\sqrt{x})(6) + (6)^2$ Substitute $2\sqrt{x}$ for a and 6 for b.
$= (4 \cdot x) - (4\sqrt{x})(6) + 36$ Multiply.
$= 4x - 24\sqrt{x} + 36$ Multiply.

You Try 6

Multiply and simplify.

a) $(\sqrt{6} + 5)^2$ b) $(3\sqrt{2} - 4)^2$ c) $(\sqrt{w} + \sqrt{11})^2$

About the Cover

In order to be successful, a cyclist must follow a strict training regimen. Instead of attempting to compete immediately, the athlete must practice furiously in smaller intervals to build up endurance, skill, and speed. A true competitor sees the connection between the smaller steps of training and final accomplishment. Similarly, after years of teaching, it became clear to Sherri Messersmith that mastering math for most students is less about the memorization of facts and more of a journey of studying and understanding what may seem to be complex topics. Like a cyclist training for a long race, as pictured on the front cover, students must build their knowledge of mathematical concepts by connecting and applying concepts they already know to more challenging ones, just as a cyclist uses training and hard work to successfully work up to longer, more challenging rides. After following the methodology applied in this text, like a cyclist following a training program, students will be able to succeed in their course.

Brief Contents

Message from the Author

Dear Colleagues,

Students constantly change—and over the last 10 years, they have changed a lot, therefore this book was written for today's students. I have adapted much of what I had been doing in the past to what is more appropriate for today's students. This textbook has evolved from the notes, worksheets, and teaching strategies I have developed throughout my 25-year teaching career in the hopes of sharing with others the successful approach I have developed.

To help my students learn algebra, I meet them where they are by helping them improve their basic skills and then showing them the connections between arithmetic and algebra. Only then can they learn the algebra that is the course. Throughout the book, concepts are presented in **bite-size pieces** because developmental students learn best when they have to digest fewer new concepts at one time. The **Basic Skills Worksheets** are quick, effective tools that can be used in the classroom to help strengthen students' arithmetic skills, and most of these worksheets can be done in 5 minutes or less. The **You Try** exercises follow the examples in the book so that students can practice concepts immediately. The **Fill-It-In exercises** take students through a step-by-step process of working multistep problems, asking them to fill in a step or a reason for a given step to prepare them to work through exercises on their own and to reinforce mathematical vocabulary. **Modern applications** are written with student interests in mind; students frequently comment that they have never seen "fun" word problems in a math book before this one. **Connect Mathematics hosted by ALEKS** is an online homework manager that will identify students' strengths and weaknesses and take the necessary steps to help them master key concepts.

The **writing style** is friendlier than that of most textbooks. Without sacrificing mathematical rigor, this book uses language that is mathematically sound yet easy for students to understand. Instructors and students appreciate its conversational tone because it sounds like a teacher talking to a class. The **use of questions** throughout the prose contributes to the conversational style while teaching students how to ask themselves the questions we ask ourselves when solving a problem. This friendly, less intimidating writing style is especially important because many of today's developmental math students are enrolled in developmental reading as well.

Intermediate Algebra, is a compilation of what I have learned in the classroom, from faculty members nationwide, from the national conferences and faculty forums I have attended, and from the extensive review process. Thank you to everyone who has helped me to develop this textbook. My commitment has been to write the most mathematically sound, readable, student-friendly, and up-to-date text with unparalleled resources available for both students and instructors. To share your comments, ideas, or suggestions for making the text even better, please contact me at sherri.messersmith@gmail.com. I would love to hear from you.

Sherri Messersmith

About the Author

Sherri Messersmith has been teaching at College of DuPage in Glen Ellyn, Illinois, since 1994. She has over 25 years of experience teaching many different courses from developmental mathematics through calculus. She earned a bachelor of science degree in the teaching of mathematics at the University of Illinois at Urbana-Champaign and went on to teach at the high school level for two years. Sherri returned to UIUC and earned a master of science in applied mathematics and stayed on at the university to teach and coordinate large sections of undergraduate math courses. Sherri has authored several textbooks, and she has also appeared in videos accompanying several McGraw-Hill texts.

Sherri lives outside of Chicago with her husband, Phil, and their daughters, Alex and Cailen. In her precious free time, she likes to read, play the guitar, and travel—the manuscripts for this and her previous books have accompanied her from Spain to Greece and many points in between.

Worksheets Supplemental **worksheets** for *every* section are available online through Connect. They fall into three categories: worksheets to strengthen basic skills, worksheets to help teach new concepts, and worksheets to tie concepts together. These worksheets provide a quick, engaging way for students to work on key concepts. They save instructors from having to create their own supplemental material, and they address potential stumbling blocks. They are also a great resource for standardizing instruction across a mathematics department.

Worksheet 3.1   Name: _____

Messersmith–Intermediate Algebra

1) Graph the following ordered pairs on the same coordinate system.

   A(3, 5)   B(−1, −2)   C$\left(0, \frac{4}{3}\right)$ D(−4, 4)   E(−7, 0)   F(3, −4)

2) Determine whether each ordered pair is a solution of $3x - 5y = 15$.

   a) (10, 3)

   b) (−5, 0)

   c) $\left(\frac{5}{3}, -2\right)$

Complete the table of values and graph the equation.

3) $y = 4x - 3$

x	y
0	
1	
2	
−1	

5) $x = 5$

x	y
	0
	−1
	3
	2

4) $2x + y = 4$

x	y
0	
1	
	0
	5

6) $y = -1$

x	y
0	
	2
	−3
1	

"I really like it and many topics are covered the way I would teach them in my classroom."
Pamela Harden,
Tennessee Tech University

Worksheet 3C   Name: _____

Messersmith–Intermediate Algebra

Find 2 numbers that . . .

MULTIPLY TO	*and* ADD TO	ANSWER
−27	−6	−9 and 3
72	18	
24	−11	
−4	3	
10	−7	
121	22	
−54	−3	
54	29	
16	−10	
30	17	
9	−6	
−8	−2	
21	10	
60	−19	
56	15	
−28	3	
−72	−6	
100	25	
−40	6	
11	−12	
20	12	
−35	−2	
77	18	
108	21	
−3	−2	

"Messersmith has a very simple and clear approach to each objective. Messersmith tends to think where students have most difficulties and provide examples and explanation on those areas." Avi Kar Abraham Baldwin, Agricultural College

Presenting Concepts in Bite-Size Pieces

Chapter Organization   The **chapter organization** is designed to present the context in "bite-size" pieces, focusing not only on the mathematical concepts but also on the "why" behind those concepts. By breaking down the sections into manageable chunks, the author has identified the core places where students traditionally struggle.

"The material is presented in a very understandable manner, in that it approaches all topics in bite-sized pieces and explains each step thoroughly as it proceeds through the examples."
Lee Ann Spahr,
Durham Technical Community College

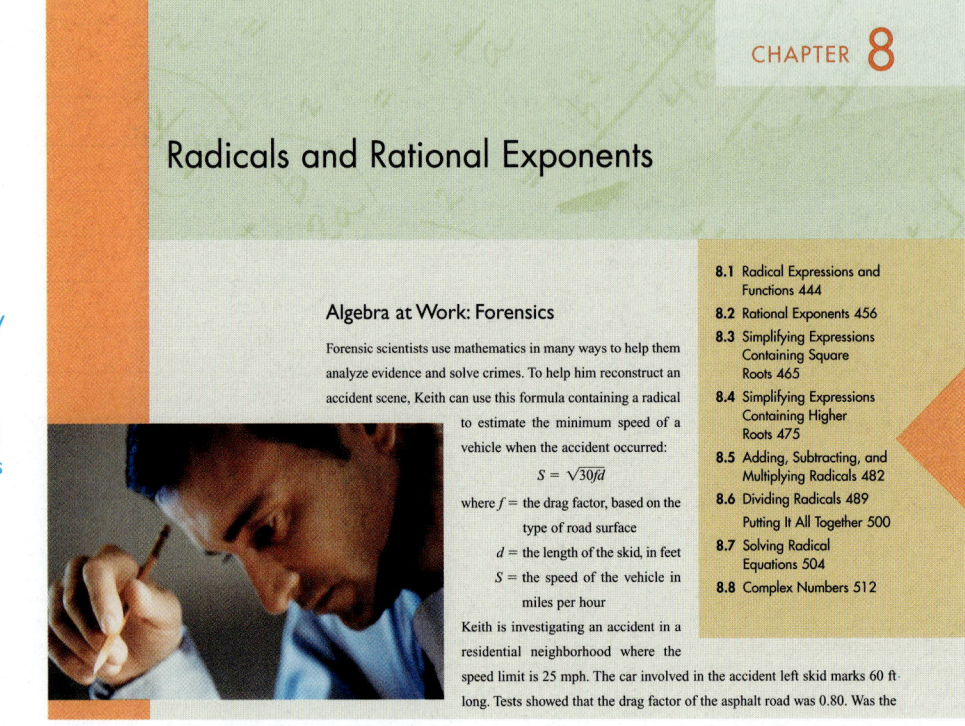

CHAPTER **8**

Radicals and Rational Exponents

Algebra at Work: Forensics

Forensic scientists use mathematics in many ways to help them analyze evidence and solve crimes. To help him reconstruct an accident scene, Keith can use this formula containing a radical to estimate the minimum speed of a vehicle when the accident occurred:

$$S = \sqrt{30fd}$$

where f = the drag factor, based on the type of road surface

d = the length of the skid, in feet

S = the speed of the vehicle in miles per hour

Keith is investigating an accident in a residential neighborhood where the speed limit is 25 mph. The car involved in the accident left skid marks 60 ft long. Tests showed that the drag factor of the asphalt road was 0.80. Was the

In-Class Examples   To give instructors additional material to use in the classroom, a matching ***In-Class Example*** is provided in the margin of the Annotated Instructor's Edition for every example in the book. The more examples a student reviews, the better chance he or she will have to understand the related concept.

Example 10

In-Class Example 10
Refer to Example 10. If Jenny is traveling at 50 mph and Alex is traveling at 65 mph, how long will it take Alex to catch Jenny?
answer: $1\frac{1}{3}$ hr

Alex and Jenny are taking a cross-country road-trip on their motorcycles. Jenny leaves a rest area first traveling at 60 mph. Alex leaves 30 min later, traveling on the same highway, at 70 mph. How long will it take Alex to catch Jenny?

Solution

Step 1:   **Read** the problem carefully, and identify what we are being asked to find.

We must determine how long it takes Alex to catch Jenny.

We will draw a picture to help us see what is happening in this problem.

Since both girls leave the same rest area and travel on the same highway, when Alex catches Jenny they have driven the *same* distance.

"I like that the teacher's edition gives in-class examples to use so the teacher doesn't have to spend prep time looking for good examples or using potential homework/exam questions for in-class examples."
Judith Atkinson,
University of Alaska–Fairbanks

x

Fill-It-In **Fill-It-In exercises** take a student through the process of working a problem step-by-step so that students either have to provide the reason for each mathematical step or fill in a mathematical step when the reason is given. These types of exercises are unique to this text and appear throughout.

Fill It In

Fill in the blanks with either the missing mathematical step or reason for the given step.

39) $64^{-1/2} = \left(\quad\right)^{1/2}$ The reciprocal of 64 is ___.

$= \sqrt{\dfrac{1}{64}}$ _____

$=$ ___   Simplify.

40) $\left(\dfrac{1}{1000}\right)^{-1/3} = (\_\_)^{1/3}$ The reciprocal of $\dfrac{1}{1000}$ is ___.

$= \sqrt[3]{1000}$ _____

$=$ ___   Simplify.

"I love how your problems are set up throughout each section and how they build upon the problems before them. It has great coverage of all types as well." Keith Pachlhofer, University of Central Arkansas

"I would describe this text as being student centered, one that offers readability for the student in both explanations and definitions, nice worked examples, and a wide variety of practice exercises to enhance the students', understanding of the concepts being discussed." Kim Cain, Miami University–Hamilton

Fill It In

Fill in the blanks with either the missing mathematical step or reason for the given step.

39) $64^{-1/2} = \left(\dfrac{1}{64}\right)^{1/2}$ The reciprocal of 64 is $\dfrac{1}{64}$.

$= \sqrt{\dfrac{1}{64}}$ The denominator of the fractional exponent is the index of the radical.

$= \dfrac{1}{8}$ Simplify.

40) $\left(\dfrac{1}{1000}\right)^{-1/3} = (1000)^{1/3}$ The reciprocal of $\dfrac{1}{1000}$ is 1000.

$= \sqrt[3]{1000}$ The denominator of the fractional exponent is the index of the radical.

$= 10$ Simplify.

Guided Student Notes   Name: _____

Messersmith–Intermediate Algebra

3.5 Introduction to Functions

Definition of a Relation   **Identifying a Domain**

Definition of a Function   **Identifying a Range**

Identify the domain and range of each relation, and determine whether each relation is a function.

1) $\{(3, 1), (-4, 2), (3, -7), (6, 4)\}$

2) $\left\{(4, -2), (11, 6), \left(\frac{2}{3}, 4\right), \left(0, -\frac{3}{4}\right)\right\}$

3)
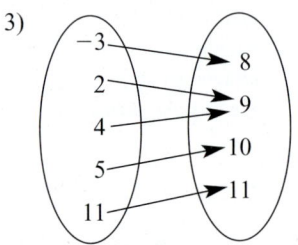

Guided Student Notes   New **Guided Student Notes** are an amazing resource for instructors to help their students become better note-takers. They contain in-class examples provided in the margin of the text along with additional examples not found in the book to emphasize the given topic so that students spend less time copying down information and more time engaging within the classroom. A note will be available for each section of the text.

"This text is easier to read than most texts, using questions to prompt the student's thinking. In general, it breaks the concepts down into smaller bite-size pieces, with exercises that build progressively from just-in-time review problems to more difficult exercises."
Cindy Bond, Butler Community College

Guided Student Notes   Name: _____

Messersmith–Intermediate Algebra

10.4 Properties of Logarithms

Product Rule for Logs   **Other Properties of Logs**

$$\log_a a^x = x$$

Quotient Rule for Logs $a^{\log_a x} = x$

Power Rule for Logs

Write as the sum of logarithms and simplify, if possible. Assume the variables represent positive real numbers.

1) $\log_a(5 \cdot 2)$ 3) $\log_8 w^4$

2) $\log_3 9b$ 4) $\log 7wx$

"This author has written one of the best books for this level in the past 15 years that I have been teaching. It is one of the top three books I have seen in my teaching career."
Edward Koslowska, Southwest Texas Junior College–Uvalde

Mastering Concepts

You Try After almost every example, there is a *You Try* problem that mirrors the example. This provides students the opportunity to practice a problem similar to what the instructor has presented before moving on to the next concept.

"Comparing Martin-Gay to Messersmith is like comparing Tin to Gold." Mark Pavitch, California State University–Los Angeles LA

"Sherri Messersmith weaves well-chosen examples with frequently occurring, student-directed *You Try* problems to reinforce concepts in a clear and concise manner." Jon Becker, Indiana University Northwest–Gary

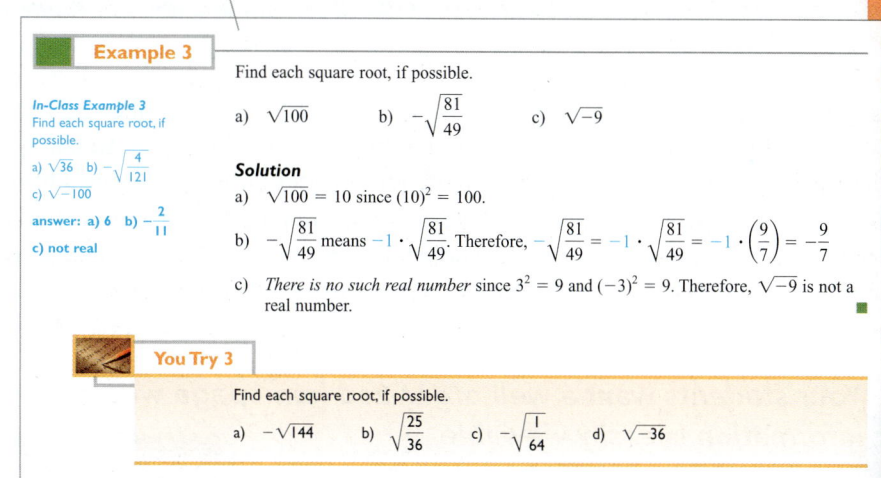

Example 3

Find each square root, if possible.

a) $\sqrt{100}$ b) $-\sqrt{\dfrac{81}{49}}$ c) $\sqrt{-9}$

In-Class Example 3
Find each square root, if possible.

a) $\sqrt{36}$ b) $-\sqrt{\dfrac{4}{121}}$

c) $\sqrt{-100}$

answer: a) 6   b) $-\dfrac{2}{11}$

c) not real

Solution

a) $\sqrt{100} = 10$ since $(10)^2 = 100$.

b) $-\sqrt{\dfrac{81}{49}}$ means $-1 \cdot \sqrt{\dfrac{81}{49}}$. Therefore, $-\sqrt{\dfrac{81}{49}} = -1 \cdot \sqrt{\dfrac{81}{49}} = -1 \cdot \left(\dfrac{9}{7}\right) = -\dfrac{9}{7}$

c) *There is no such real number* since $3^2 = 9$ and $(-3)^2 = 9$. Therefore, $\sqrt{-9}$ is not a real number.

You Try 3

Find each square root, if possible.

a) $-\sqrt{144}$ b) $\sqrt{\dfrac{25}{36}}$ c) $-\sqrt{\dfrac{1}{64}}$ d) $\sqrt{-36}$

Putting It All Together Several chapter include a *Putting It All Together* section, in keeping with the author's philosophy of breaking sections into manageable chunks to increase student comprehension. These sections help students synthesize key concepts before moving on to the rest of the chapter.

"I like the Putting It All Together midchapter summary. Any opportunity to spiral back around to previous concepts gives students the opportunity to make important connections with these properties rather than just leaving students with the impression that each section of the chapter is independent." Steve Felzer, Lenoir Community College

Putting It All Together

Objective
1. Learn Strategies for Factoring a Given Polynomial

1. Learn Strategies for Factoring a Given Polynomial

In this chapter, we have discussed several different types of factoring problems:

1) Factoring out a GCF (Section 6.1)
2) Factoring by grouping (Section 6.1)
3) Factoring a trinomial of the form $x^2 + bx + c$ (Section 6.2)
4) Factoring a trinomial of the form $ax^2 + bx + c$ (Section 6.2)
5) Factoring a perfect square trinomial (Section 6.3)
6) Factoring the difference of two squares (Section 6.3)
7) Factoring the sum and difference of two cubes (Section 6.3)

We have practiced the factoring methods separately in each section, but how do we know which factoring method to use given many different types of polynomials together?

Connect MATH

Hosted by **ALEKS Corp.**

Connect Math Hosted by ALEKS Corporation is an exciting, new assignment and assessment platform combining the strengths of McGraw-Hill Higher Education and ALEKS Corporation. Connect Math Hosted by ALEKS is the first platform on the market to combine an artificially-intelligent, diagnostic assessment with an intuitive ehomework platform designed to meet your needs.

Connect Math Hosted by ALEKS Corporation is the culmination of a one-of-a-kind market development process involving math full-time and adjunct Math faculty at every step of the process. This process enables us to provide you with a solution that best meets your needs.

Connect Math Hosted by ALEKS Corporation is built by Math educators for Math educators!

1 *Your students want a well-organized homepage where key information is easily viewable.*

Modern Student Homepage

▶ This homepage provides a dashboard for students to immediately view their assignments, grades, and announcements for their course. (Assignments include HW, quizzes, and tests.)

▶ Students can access their assignments through the course Calendar to stay up-to-date and organized for their class.

Modern, intuitive, and simple interface.

2 *You want a way to identify the strengths and weaknesses of your class at the beginning of the term rather than after the first exam.*

Integrated ALEKS® Assessment

▶ This artificially-intelligent (AI), diagnostic assessment identifies precisely what a student knows and is ready to learn next.

▶ Detailed assessment reports provide instructors with specific information about where students are struggling most.

▶ This AI-driven assessment is the only one of its kind in an online homework platform.

Recommended to be used as the first assignment in any course.

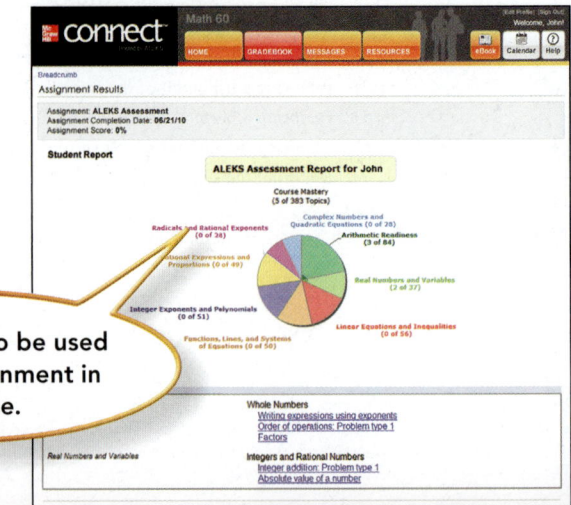

ALEKS is a registered trademark of ALEKS Corporation.

Built by Math Educators for Math Educators

③ *Your students want an assignment page that is easy to use and includes lots of extra help resources.*

Efficient Assignment Navigation

▶ Students have access to immediate feedback and help while working through assignments.

▶ Students have direct access to a media-rich eBook for easy referencing.

▶ Students can view detailed, step-by-step solutions written by instructors who teach the course, providing a unique solution to each and every exercise.

Students can easily monitor and track their progress on a given assignment.

④ *You want a more intuitive and efficient assignment creation process because of your busy schedule.*

Assignment Creation Process

▶ Instructors can select textbook-specific questions organized by chapter, section, and objective.

▶ Drag-and-drop functionality makes creating an assignment quick and easy.

▶ Instructors can preview their assignments for efficient editing.

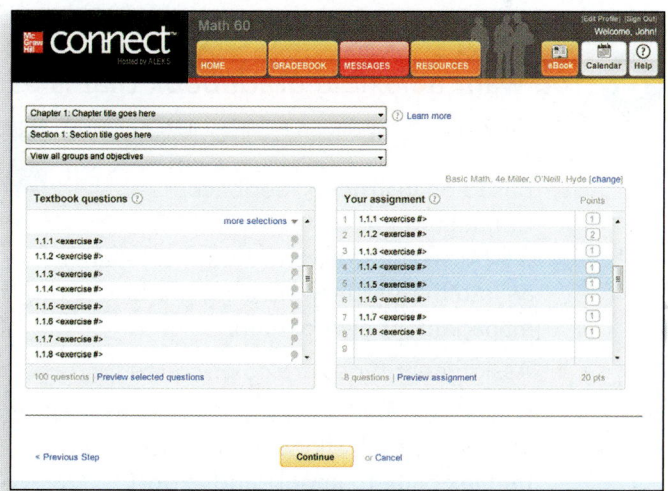

Connect
Learn
Succeed™

www.connectmath.com

5 *Your students want an interactive eBook with rich functionality integrated into the product.*

Hosted by **ALEKS Corp.**

Integrated Media-Rich eBook

▶ A Web-optimized eBook is seamlessly integrated within ConnectPlus Math Hosted by ALEKS Corp for ease of use.

▶ Students can access videos, images, and other media in context within each chapter or subject area to enhance their learning experience.

▶ Students can highlight, take notes, or even access shared instructor highlights/notes to learn the course material.

▶ The integrated eBook provides students with a cost-saving alternative to traditional textbooks.

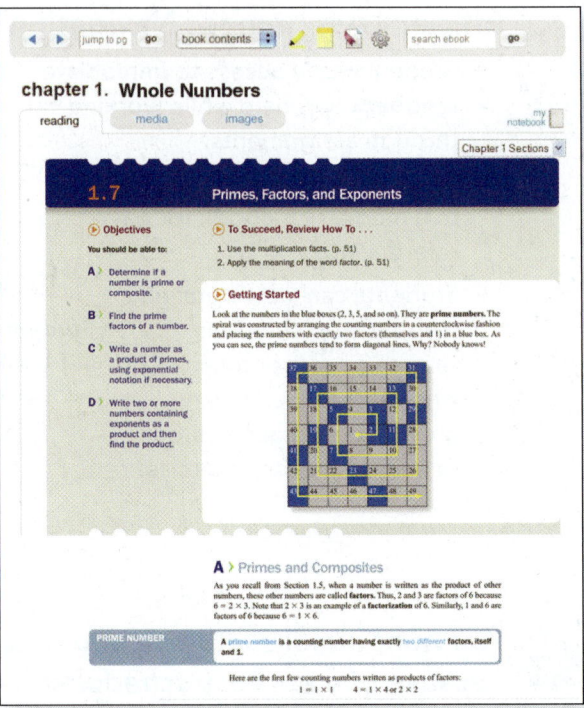

6 *You want a flexible gradebook that is easy to use.*

Flexible Instructor Gradebook

▶ Based on instructor feedback, Connect Math Hosted by ALEKS Corp's straightforward design creates an intuitive, visually pleasing grade management environment.

▶ Assignment types are color-coded for easy viewing.

▶ The gradebook allows instructors the flexibility to import and export additional grades.

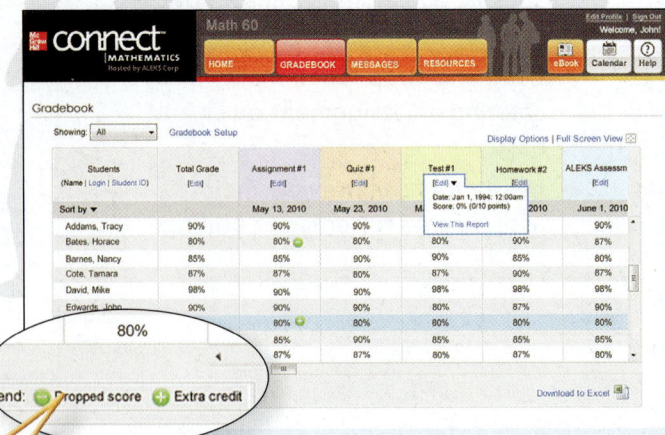

Instructors have the ability to drop grades as well as assign extra credit.

Built by Math Educators for Math Educators

 7 *You want algorithmic content that was developed by math faculty to ensure the content is pedagogically sound and accurate.*

Digital Content Development Story

The development of McGraw-Hill's Connect Math Hosted by ALEKS Corp. content involved collaboration between McGraw-Hill, experienced instructors, and ALEKS, a company known for its high-quality digital content. The result of this process, outlined below, is accurate content created with your students in mind. It is available in a simple-to-use interface with all the functionality tools needed to manage your course.

1. McGraw-Hill selected experienced instructors to work as Digital Contributors.
2. The Digital Contributors selected the textbook exercises to be included in the algorithmic content to ensure appropriate coverage of the textbook content.
3. The Digital Contributors created detailed, stepped-out solutions for use in the Guided Solution and Show Me features.
4. The Digital Contributors provided detailed instructions for authoring the algorithm specific to each exercise to maintain the original intent and integrity of each unique exercise.
5. Each algorithm was reviewed by the Contributor, went through a detailed quality control process by ALEKS Corporation, and was copyedited prior to being posted live.

Connect Math Hosted by ALEKS Corp.
Built by Math Educators for Math Educators

Lead Digital Contributors

Tim Chappell
Metropolitan Community College, Penn Valley

Jeremy Coffelt
Blinn College

Nancy Ikeda
Fullerton College

Amy Naughten

Digital Contributors

Al Bluman, *Community College of Allegheny County*

John Coburn, *St. Louis Community College, Florissant Valley*

Vanessa Coffelt, *Blinn College*

Donna Gerken, *Miami-Dade College*

Kimberly Graham

J.D. Herdlick, *St. Louis Community College, Meramec*

Vickie Flanders, *Baton Rouge Community College*

Nic LaHue, *Metropolitan Community College, Penn Valley*

Nicole Lloyd, *Lansing Community College*

Jackie Miller, *The Ohio State University*

Anne Marie Mosher, *St. Louis Community College, Florissant Valley*

Reva Narasimhan, *Kean University*

David Ray, *University of Tennessee, Martin*

Kristin Stoley, *Blinn College*

Stephen Toner, *Victor Valley College*

Paul Vroman, *St. Louis Community College, Florissant Valley*

Michelle Whitmer, *Lansing Community College*

www.connectmath.com

Build a Better Path to Success with ALEKS®

Experience Student Success!

ALEKS Aleks is a unique online math tool that uses adaptive questioning and artificial intelligence to correctly place, prepare, and remediate students . . . all in one product! Institutional case studies have shown that **ALEKS has improved pass rates by over 20% versus traditional online homework, and by over 30% compared to using a text alone.**

By offering each student an individualized learning path, ALEKS directs students to work on the math topics that they are ready to learn, Also, to help students keep pace in their course, instructors can correlate ALEKS to their textbook or syllabus in seconds.

To learn more about how ALEKS can be used to boost student performance, please visit www.aleks.com/highered/math or contact your McGraw-Hill representative.

ALEKS Pie
Each student is given her or his own individualized learning path.

Easy Graphing Utility!
Students can answer graphing problems with ease!

Course Calendar
Instructors can schedule assignments and reminders for students.

New ALEKS Instructor Module

Enhanced Functionality and Streamlined Interface Help to Save Instructor Time

ALEKS The new ALEKS Instructor Module features enhanced functionality and a streamlined interface based on research with ALEKS instructors and homework management instructors. Paired with powerful assignment-driven features, textbook integration, and extensive content flexibility, the new ALEKS Instructor Module simplifies administrative tasks and makes ALEKS more powerful than ever.

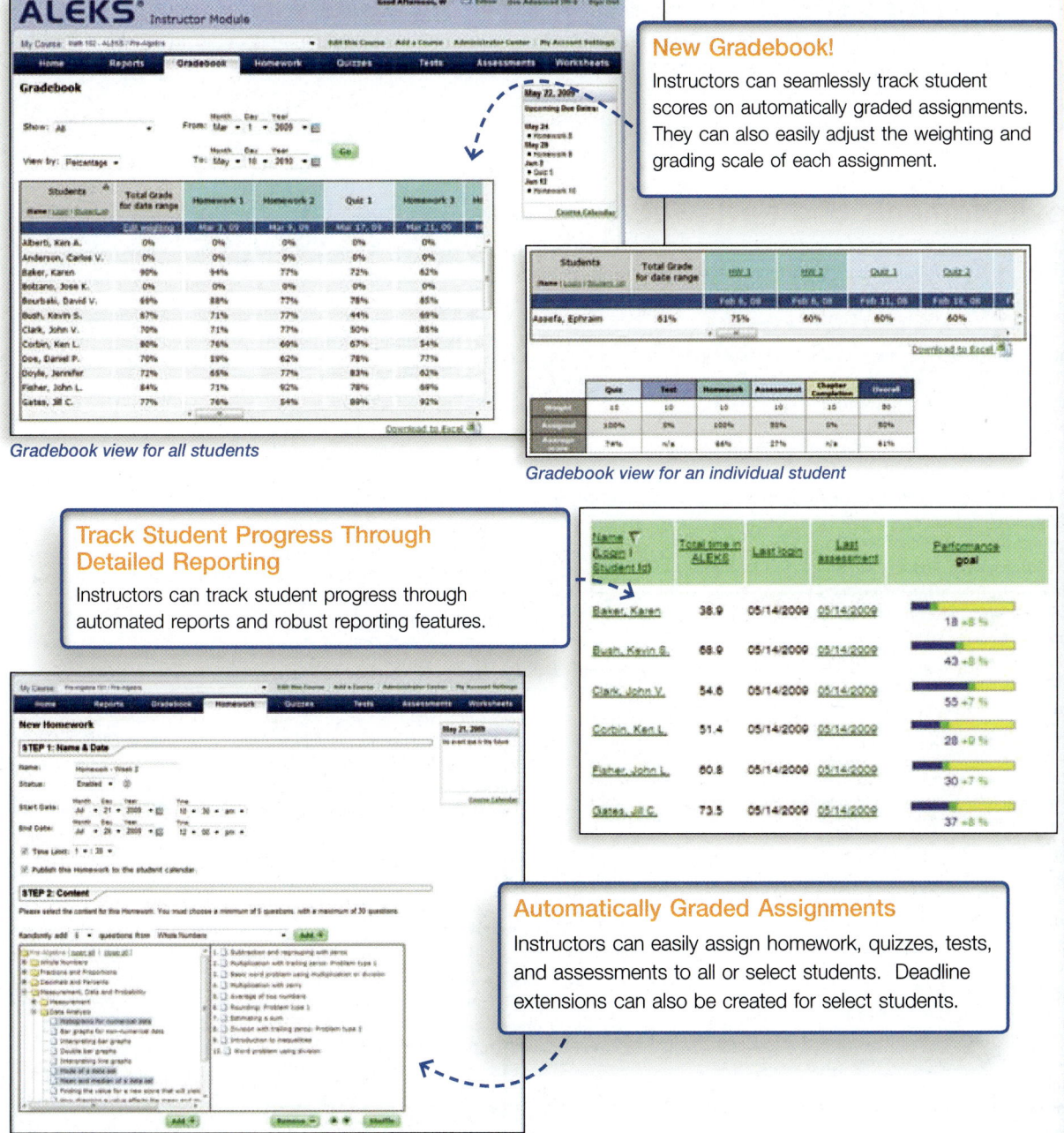

New Gradebook!

Instructors can seamlessly track student scores on automatically graded assignments. They can also easily adjust the weighting and grading scale of each assignment.

Gradebook view for all students

Gradebook view for an individual student

Track Student Progress Through Detailed Reporting

Instructors can track student progress through automated reports and robust reporting features.

Automatically Graded Assignments

Instructors can easily assign homework, quizzes, tests, and assessments to all or select students. Deadline extensions can also be created for select students.

Select topics for each assignment

Learn more about ALEKS by visiting www.aleks.com/highered/math **or contact your McGraw-Hill representative.**

360° Development Process

McGraw-Hill's 360° Development Process is an ongoing, never ending, market-oriented approach to building accurate and innovative print and digital products. It is dedicated to continual large scale and incremental improvement driven by multiple customer feedback loops and checkpoints. This is initiated during the early planning stages of our new products, and intensifies during the development and production stages, then begins again upon publication, in anticipation of the next edition.

A key principle in the development of any mathematics text is its ability to adapt to teaching specifications in a universal way. The only way to do so is by contacting those universal voices—and learning from their suggestions. We are confident that our book has the most current content the industry has to offer, thus pushing our desire for accuracy to the highest standard possible. In order to accomplish this, we have moved through an arduous road to production. Extensive and open-minded advice is critical in the production of a superior text.

Acknowledgments and Reviewers

The development of this textbook series would never have been possible without the creative ideas and feedback offered by many reviewers. We are especially thankful to the following instructors for their careful review of the manuscript.

Manuscript Reviewers

Kent Aeschliman, *Oakland Community College Highland Lakes*
Froozan Afiat, *College of Southern Nevada–Henderson*
Carlos Amaya, *California State University–Los Angeles*
Judy Atkinson, *University of Alaska–Fairbanks*
Rajalakshmi Baradwai, *University of Maryland Baltimore County*
Carlos Barron, *Mountain View College*
Jon Becker, *Indiana University–Northwest–Gary*
Abraham Biggs, *Broward College–South*
Donald Bigwood, *Bismarck State College*
Lee Brendel, *Southwestern Illinois College*
Joan Brown, *Eastern New Mexico University*
Shirley Brown, *Weatherford College*
Debra Bryant, *Tennessee Tech University*
Gail Butler, *Erie Community College North Campus–Williamsville*
Kim Cain, *Miami University–Hamilton*
Ernest Canonigo, *California State University–Los Angeles*
Douglas Carbone, *Central Maine Community College*
Randall Casleton, *Columbus State University*
Jose Castillo, *Broward College–South*
Dwane Christensen, *California State University–Los Angeles*
Lisa Christman, *University of Central Arkansas*
William Clarke, *Pikes Peak Community College*
Delaine Cochran, *Indiana University Southeast*
Wendy Conway, *Oakland Community College Highland Lakes*
Charyl Craddock, *University of Tennessee–Martin*
Greg Cripe, *Spokane Falls Community College*
Joseph De Guzman, *Riverside Community College*
Robert Diaz, *Fullerton College*
Paul Diehl, *Indiana University Southeast*
Deborah Doucette, *Erie Community College North Campus–Williamsville*
Scott Dunn, *Central Michigan University*
Angela Earnhart, *North Idaho College*
Hussain Elalaoui-Talibi, *Tuskegee University*
Joseph Estephan, *California State University–Dominguez Hills*
Tina Evans, *Shelton State Community College*
Angela Everett, *Chattanooga State Tech Community College (West)*
Christopher Farmer, *Southwestern Illinois College*
Steve Felzer, *Lenoir Community College*
Angela Fisher, *Durham Tech Community College*

Marion Foster, *Houston Community College–Southeast College*
Mitzi Fulwood, *Broward College–North*
Scott Garvey, *Suny Agriculture & Tech College–Cobleskille*
Antonnette Gibbs, *Broward College–North*
Sharon Giles, *Grossmont College*
Susan Grody, *Broward College–North*
Kathy Gross, *Cayuga Community College*
Margaret Gruenwald, *University of Southern Indiana*
Kelli Hammer, *Broward College–South*
Pamela Harden, *Tennessee Tech University*
Jody Harris, *Broward College–Central*
Terri Hightower Martin, *Elgin Community College*
Michelle Hollis, *Bowling Green Community College at Western Kentucky University*
Joe Howe, *Saint Charles County Community College*
Barbara Hughes, *San Jacinto College-Pasadena*
Michelle Jackson, *Bowling Green Community College at Western Kentucky University*
Pamela Jackson, *Oakland Community College–Farmington Hills*
Nancy Johnson, *Broward College–North*
Tina Johnson, *Midwestern State University*
Maryann Justinger, *Erie Community College South Campus–Orchard Park*
Cheryl Kane, *University of Nebraska–Lincoln*
Avi Kar, *Abraham Baldwin Agricultural College*
Ryan Kasha, *Valencia Community College–West Campus*
Joe Kemble, *Lamar University–Beaumont*
Pat Kier, *Southwest Texas Junior College–Uvalde*
Heidi Kilthau-Kiley, *Suffolk County Community College*
Jong Kim, *Long Beach City College*
Lynette King, *Gadsden State Community College*
Edward Koslowska, *Southwest Texas Junior College–Uvalde*
Randa Kress, *Idaho State University*
Debra Landre, *San Joaquin Delta Community College*
Cynthia Landrigan, *Erie Community College South Campus–Orchard Park*
Richard Leedy, *Polk Community College*
Janna Liberant, *Rockland Community College*
Shawna Mahan, *Pikes Peak Community College*
Aldo Maldonado, *Park University–Parkville*
Rogers Martin, *Louisiana State University–Shreveport*
Carol Mcavoy, *South Puget Sound Community College*
Peter Mccandless, *Park University–Parkville*

Gary Mccracken, *Shelton State Community College*
Margaret Messinger, *Southwest Texas Junior College–Uvalde*
Kris Mudunuri, *Long Beach City College*
Amy Naughten, *Middle Georgia College*
Elsie Newman, *Owens Community College*
Paulette Nicholson, *South Carolina State University*
Ken Nickels, *Black Hawk College*
Rhoda Oden, *Gadsden State Community College*
Charles Odion, *Houston Community College–Southwest*
Karen Orr, *Roane State Community College*
Keith Pachlhofer, *University of Central Arkansas*
Charles Patterson, *Louisiana Tech University*
Mark Pavitch, *California State University–Los Angeles*
Jean Peterson, *University of Wisconsin–Oshkosh*
Novita Phua, *California State University–Los Angeles*
Mohammed Qazi, *Tuskegee University*
L. Gail Queen, *Shelton State Community College*
William Radulovich, *Florida Community College*
Kumars Ranjbaran, *Mountain View College*
Gary Rattray, *Central Maine Community College*
David Ray, *University of Tennessee–Martin*
Janice Reach, *University of Nebraska at Omaha*
Tracy Romesser, *Erie Community College North Campus–Williamsville*
Steve Rummel, *Heartland Community College*
John Rusnak, *Central Michigan University*
E. Jennell Sargent, *Tennessee State University*
Jane Serbousek, *Noth Virginia Community College–Loudoun Campus*
Brian Shay, *Grossmont College*
Azzam Shihabi, *Long Beach City College*
Mohsen Shirani, *Tennessee State University*
Joy Shurley, *Abraham Baldwin Agricultural College*
Nirmal Sohi, *San Joaquin Delta Community College*
Lee Ann Spahr, *Durham Technical Community College*
Joel Spring, *Broward College–South*
Sean Stewart, *Owens Community College*
David Stumpf, *Lakeland Community College*
Sara Taylor, *Dutchess Community College*
Roland Trevino, *San Antonio College*
Bill Tusang, *Suny Agriculture & Technical College–Cobleskille*
Mildred Vernia, *Indiana University Southeast*
Laura Villarreal, *University of Texas at Brownsville*
James Wan, *Long Beach City College*
Terrence Ward, *Mohawk Valley Community College*
Robert White, *Allan Hancock College*
Darren White, *Kennedy-King College*
Marjorie Whitmore, *Northwest Arkansas Community College*
John Wilkins, *California State University–Dominguez Hills*
Henry Wyzinski, *Indiana University–Northwest-Gary*
Mina Yavari, *Allan Hancock College*
Diane Zych, *Erie Community College North Campus–Williamsville*

Instructor Focus Groups

Lane Andrew, *Arapahoe Community College*
Chris Bendixen, *Lake Michigan College*
Terry Bordewick, *John Wood Community College*
Jim Bradley, *College of DuPage*
Jan Butler, *Colorado Community Colleges Online*
Robert Cappetta, *College of DuPage*
Margaret Colucci, *College of DuPage*

Anne Conte, *College of DuPage*
Gudryn Doherty, *Community College of Denver*
Eric Egizio, *Joliet Junior College*
Mimi Elwell, *Lake Michigan College*
Vicki Garringer, *College of DuPage*
Margaret Gruenwald, *University of Southern Indiana*
Patricia Hearn, *College of DuPage*
Mary Hill, *College of DuPage*
Maryann Justinger, *Eric Community College–South*
Donna Katula, *Joliet Junior College*
Elizabeth Kiedaisch, *College of DuPage*
Geoffrey Krader, *Morton College*
Riki Kucheck, *Orange Coast College*
James Larson, *Lake Michigan College*
Gail Laurent, *College of DuPage*
Richard Leedy, *Polk State College*
Anthony Lenard, *College of DuPage*
Zia Mahmood, *College of DuPage*
Christopher Mansfield, *Durham Technical Community College*
Terri Martin, *Elgin Community College*
Paul Mccombs, *Rock Valley College*
Kathleen Michalski, *College of DuPage*
Kris Mudunuri, *Long Beach City College*
Michael Neill, *Carl Sandburg College*
Catherine Pellish, *Front Range Community College*
Larry Perez, *Saddleback College*
Christy Peterson, *College of DuPage*
David Platt, *Front Range Community College*
Jack Pripusich, *College of DuPage*
Patrick Quigley, *Saddleback College*
Eleanor Storey, *Front Range Community College*
Greg Wheaton, *Kishwaukee College*
Steve Zuro, *Joliet Junior College*
Carol Schmidt *Lincoln Land Community College*
James Carr *Normandale Community College*
Kay Cornelius *Sinclair Community College*
Thomas Pulver *Waubonsee Community College*
Angie Matthews *Broward Community College*
Sondra Braesek *Broward Community College*
Katerina Vishnyakova *Colin County Community College*
Eileen Dahl *Hennepin Technical College*
Stacy Jurgens *Mesabi Range Community and Technical College*
John Collado *South Suburban College*
Barry Trippett *Sinclair Community College*
Abbas Meigooni *Lincoln Land Community College*
Thomas Sundquist *Normandale Community College*
Diane Krasnewich *Muskegon Community College*
Marshall Dean *El Paso Community College*
Elsa Lopez *El Paso Community College*
Bruce Folmar *El Paso Community College*
Pilar Gimbel *El Paso Community College*
Ivette Chuca *El Paso Community College*
Kaat Higham *Bergen Community College*
Joanne Peeoples *El Paso Community College*
Diana Orrantia *El Paso Community College*
Andrew Stephan *Saint Charles County Community College*
Joe Howe *Saint Charles County Community College*
Wanda Long *Saint Charles County Community College*
Staci Osborn *Cuyahoga Community College*
Kristine Glasener *Cuyahoga Community College*
Derek Hiley *Cuyahoga Community College*
Penny Morries *Polk State College*
Nerissa Felder *Polk State College*

Digital Contributors

Donna Gerken, Miami–Dade College
Kimberly Graham
Nicole Lloyd, Lansing Community College
Reva Narasimhan, Kean University
Amy Naughten
Michelle Whitmer, Lansing Community College

Additionally, I would like to thank my husband, Phil, and my daughters, Alex and Cailen, for their patience, support, and understanding when things get crazy. A big high five goes out to Sue Xander and Mary Hill for their great friendship and support.

Thank you to all of my colleagues at College of DuPage, especially Betsy Kiedaisch, Christy Peterson, Caroline Soo, and Vicki Garringer for their contributions on supplements. And to the best Associate Dean ever, Jerry Krusinski: your support made it possible for me to teach and write at the same time.

Thanks to Larry Perez for his video work and to David Platt for his work on the Using Technology boxes. Thanks also go out to Kris Mudunuri, Diana Orrantia, Denise Lujan, K.S. Ravindhran, Susan Reiland, Pat Steele, and Lenore Parens for their contributions.

To all of the baristas at my two favorite Starbucks: thanks for having my high-maintenance drink ready before I even get to the register and for letting me sit at the same table for hours on end.

There are so many people to thank at McGraw-Hill: my fellow Bengal Rich Kolasa, Michelle Flomenhoft, Torie Anderson, Emilie Berglund, Emily Williams, Pete Vanaria, and Dawn Bercier. I would also like to thank Stewart Mattson, Marty Lange, and Peggy Selle for everything they have done.

To Bill Mulford, who has been with me from the beginning: thanks for your hard work, ability to multitask and organize everything we do, and for your sense of humor through it all. You are the best.

Sherri Messersmith

Sherri Messersmith
College of DuPage

Supplements for the Student

Connect

Connect Math hosted by ALEKS is an exciting, new assignment and assessment ehomework platform. Starting with an easily viewable, intuitive interface, students will be able to access key information, complete homework assignments, and utilize an integrated, media–rich eBook.

ALEKS Prep for Developmental Mathematics   ALEKS Prep for Beginning Algebra and Prep for Intermediate Algebra focus on prerequisite and introductory material for Beginning Algebra and Intermediate Algebra. These prep products can be used during the first 3 weeks of a course to prepare students for future success in the course and to increase retention and pass rates. Backed by two decades of National Science Foundation funded research, ALEKS interacts with student much like a human tutor, with the ability to precisely assess a student preparedness and provide instruction on the topics the student is most likely to learn.

ALEKS Prep Course Products Feature:

Artificial intelligence targets gaps in individual student knowledge

Assessment and learning directed toward individual students' needs

Open response environment with realistic input tools Unlimited online access—PC and Mac compatible Free trial at www.aleks.com/free_trial/instructor

Student Solution Manual   The student's solution manual provides comprehensive, worked-out solutions to the odd-numbered exercises in the section exercises, summary exercises, self-test, and the cumulative review. The steps shown in the solutions match the style of solved examples in the textbook.

Online Videos   In the online exercise videos, the author, Sherri Messersmith, works through selected exercises using the solution methodology employed in her text. Each video is available online as part of Connect and is indicated by an icon next to a corresponding exercise in the text. Other supplemental videos include eProfessor videos, which are animations based on examples in the book, exercise videos by Larry Perez, and Connect2Developmental Mathematics videos, which use 3D animations and lectures to teach algebra concepts by placing them in a real-world setting. The videos are closed-captioned for the hearing impaired, are subtitled in Spanish, and meet the Americans with Disabilities Act Standards for Accessible Design.

Supplements for the Instructor

Connect

Connect Math hosted by ALEKS is an exciting, new assignment and assessment ehomework platform. Instructors can assign an AI-driven assessment from the ALEKS corporation to identify the strengths and weaknesses of the class at the beginning of the term rather than after the first exam. Assignment creation and navigation is efficient and intuitive. The gradebook, based on instructor feedback, has a straightforward design and allows flexibility to import and export additional grades.

Instructor's Testing and Resource Online   Provides wealth of resources for the instructor. Among the supplements is a computerized test bank utilizing Brownstone Diploma algorithm-based testing software to create customized exams quickly. This user-friendly program enables instructors to search for questions by topic, format, or difficulty level; to edit existing questions, or to add new ones; and to scramble questions and answer keys for multiple versions of a single test. Hundreds of text-specific, open-ended, and multiple-choice questions are included in the question bank. Sample chapter tests are also provided. CD available upon request.

Annotated Instructor's Edition   In the Annotated Instructor's Edition (AIE), answers to exercises and tests appear adjacent to each exercise set, in a color used only for annotations. Also found in the AIE are icons with the practice exercises that serve to guide instructors in their preparation of homework assignments and lessons.

Instructor's Solution Manual   The instructor's solution manual provides comprehensive, worked-out solutions to all exercises in the section exercises, summary exercises, self-test, and the cumulative review. The steps shown in the solutions match the style of solved examples in the textbook.

Worksheets   The very popular author-created worksheets that accompany the text provide instructors with additional exercises to assist with overcoming potential stumbling blocks in student knowledge and to help students see the connection among multiple mathematical concepts. The worksheets fall into three categories: Worksheets to Strengthen Basic Skills, Worksheets to Help Teach New Concepts, Worksheets to Tie Concepts Together.

Guided Student Notes   Guided Student Notes are an amazing resource for instructors to help their students become better note-takers. They are similar to "fill-in-the-blank" notes where certain topics or definitions are prompted and spaces are left for the students to write down what the instructor says or writes on the board. The Guided Student Notes contain in-class examples provided in the margin of the text along with additional examples not found in the book to emphasize the given topic. This allows students to spend less time copying down from the board and more time thinking and learning about the solutions and concepts. The notes are specific to the Messersmith textbook and offer ready-made lesson plans for teachers to either "print and go" or require students to print and bring to class.

Powerpoints   These powerpoints will present key concepts and definitions with fully editable slides that follow the textbook. Project in class or post to a website in an online course.

Table of Contents

the future of custom publishing is here.

Introducing McGraw-Hill Create™ –a new, self-service website that allows you to create custom course materials by drawing upon McGraw-Hill's comprehensive, cross-disciplinary content and other third party resources. Select, then arrange the content in a way that makes the most sense for your course. Even personalize your book with your course information and choose the best format for your students–color print, black-and-white print, or eBook.

- **Build custom print and eBooks easily**
- **Combine material from different sources and even upload your own content**
- **Receive a PDF review copy in minutes**

begin creating now:
www.mcgrawhillcreate.com

Start by registering for a free Create account. If you already have a McGraw-Hill account with one of our other products, you will not need to register on Create. Simply Sign In and enter your username and password to begin using the site.

Sign In

Please sign in to access your Create projects. If you already have a McGraw-Hill Account, you can use your existing username and password.

* Email
Your email address here

* Password
●●●●●●●●

Forgot Your Password?

* Required Field

Sign In

New to Create?

Register online to set up a new account. Then you can take full advantage of Create to build custom books tailored to your course.

Register Now

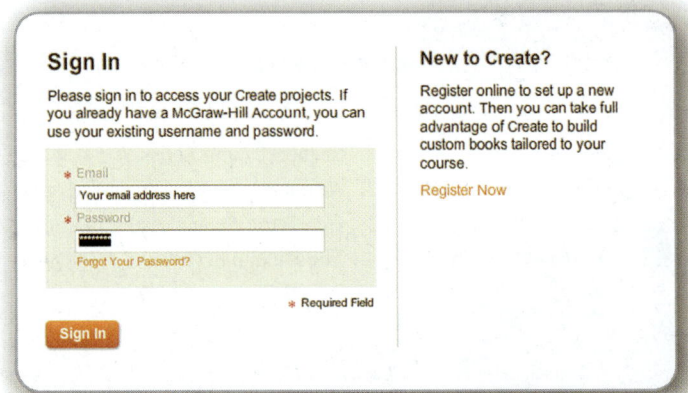

what you've only imagined.

edit, share and approve like never before.

After you've completed your project, simply go to the My Projects tab at the top right corner of the page to access and manage all of your McGraw-Hill Create™ projects. Here you will be able to edit your projects and share them with colleagues. An ISBN will be assigned once your review copy has been ordered. Click Approve to make your eBook available for student purchase or print book available for your school or bookstore to order. At any time you can modify your projects and are free to create as many projects as needed.

receive your pdf review copy in minutes!

Request an eBook review copy and receive a free PDF sample in minutes! Print review copies are also available and arrive in just a few days.

Finally—a way to quickly and easily create the course materials you've always wanted. *Imagine that.*

questions?

Please visit www.**mcgrawhill**create.com/createhelp for more information.

Mc Graw Hill

Connect
Learn
Succeed™

Applications Index

Animals
aquarium measurements, 355
cockroach reproduction rates, 643
deer and rabbit ratios, 423
dog food, 167, 693
dog house manufacturer's profit, 280
dog licenses issued, 655
dog run lengths, 364
dog weights, 53
fish tank measurements, 61, 564
holding pen measurements, 352–353
porcupine quills, 303
three toed sloth rate of movement, 271
veterinary care, 109, 249

Applied Problems and Solutions
absolute value inequality, 105
angle calculations, 55–56, 62–63, 109,
    110, 234, 236, 249–251
area calculations, 55, 62, 113, 270, 289,
    290, 304, 355, 567–568, A–12
arithmetic sequences and series, 770, 773
exponents, 270–271, 639, 670
factoring polynomials, 309
geometric sequences, 779
infinite series, 783, 787
involving addition and subtraction, 14
length, width and height calculations, 299,
    305, 349, 353, 355, 356, 362, 364,
    456, 474, 545, 570, 612, 614
linear equations, 42, 62, 151–152
linear equations in three variables,
    217–220
linear equations in two variables, 122,
    214–216
linear inequalities, 84–85, 87
logarithmic equations, 652
minimum and maximum value,
    586–588
perimeters, 280, 289, 304
proportions, 75, 417
Pythagorean Theorem, 351, 356, 361, 363,
    545
quadratic equations and formulas, 350,
    551–553, 568
radius calculations, 456, 481
rational expressions, 367
scientific notation, 270–271
sequences, 758
signed numbers, 9
slope, 130
surface area calculations, 431
unknown numbers, 38–39, 45–46, 53,
    55, 109, 233, 355, 356, 594, 739,
    749
volume, 290, 304, 355, 429, 431, 440,
    456, 481, 512, 566–567, 715

Arts and Crafts
fabric measurements, 363, 570
frame dimensions, 750
photo printer ink measurements, 271
picture dimensions, 234, 570
pillow sham constructions, 612
stained glass window measurements, 355

Biology
bacteria growth, 675, 681–682, 685, 694,
    787, 802
blood alcohol levels, 128
hair growth calculations, 122–124
pH calculations, 675, 694

Business and Manufacturing
advertising budget changes, 786
assembly line for filling soup cans,
    101–102
bicycle manufacturer's profits, 280
billboard advertising costs, 155
book publisher profits, 277
bookstore sales by genre, 418
break-even points, 739
calculator manufacturer's profits, 280
CD sales comparisons, 109
computer company profits, 195, 605
discount calculations, 48–49, 53–55, 695
employee increases/decreases, 770, 801
exports to Mexico, 125
factory machine calibrations, 167
jewelry prices, 228
labor force projections, 143
laptop computer manufacturer's profits,
    280
online sales for consumer electronics,
    116–117
press hole punch power requirements, 253
real estate values, 642
revenue calculations, 109
soccer ball manufacturing costs, 437
spiral notebook manufacturing costs, 431
Starbucks store sites, 54
toaster manufacturer's profits, 280

Construction and Home Improvement
board lengths, 44–45, 55
building lot dimensions, 226–227
chain lengths, 45
countertop measurements, 363
decorative trim installation, 53
door dimensions, 234
driveway slope design, 141
elliptical room/structure equations, 721,
    750
fencing measurements, 112, 234,
    587–588, 594, 613

floor built using Greek golden rectangle,
    423
flooring costs, A–13
fork lift rental costs, 112
home construction growth, 655
housing starts, 15
ladder heights, 545
Leaning Tower of Pisa reinforcement,
    127–128
nail and screw prices, 251
painting rates, 420–423
paint mixtures, 423
pipe lengths, 53
pipe stack totals, 803
playhouse construction boards required, 252
ramp measurements, 363, 570
roof pitch calculations, 140
room area calculations, 55–56
table top area, A–13
weight calculations for beam support, 431
wheel chair ramp slopes, 141
window dimensions, 749
wire lengths, 52, 55

Consumer Goods
backpack production, 614
battery purchases, 235
camera sales, 615
carpet costs, 426
carpet/rug dimensions, 511
CD measurements, 714
flower bouquet demands, 612
iPod Mini dimensions, 234
lawn mower production, 167
lip balm containers, A–13
paper towel market percentages by brand,
    236
personal consumption expenditures, 192
postage costs, 235, 703, 707, 748
purse productions, 605
purse sales and profits, 410
shoe size calculations, 155
shovel demand, 571
television depreciation, 764
wireless communication subscribers, 193

Demographics
babies born to teenage girls, 141, 594
city population calculations, 685
female work force percentages in
    Belgium, 156
Maine population calculations, 155
North Dakota population calculations, 155
Oakland, California population
    calculations, 15
obese population calculations, 77
suburban growth rates, 694

Section 1.1 Sets of Numbers

Objectives
1. **Identify Numbers and Graph Them on a Number Line**
2. **Compare Numbers Using Inequality Symbols**
3. **Find the Absolute Value of a Number**

1. Identify Numbers and Graph Them on a Number Line

Why should we review sets of numbers and arithmetic skills? Because the manipulations done in arithmetic are precisely the same set of skills needed to learn algebra. Let's begin by defining some numbers used in arithmetic:

The set of **natural numbers** is $\{1, 2, 3, 4, 5, \ldots\}$.

The set of **whole numbers** is $\{0, 1, 2, 3, 4, 5, \ldots\}$.

Natural numbers are often thought of as the counting numbers. Whole numbers consist of the natural numbers and zero. Let's look at other sets of numbers. We begin with integers. Remember that, on a number line, positive numbers are to the right of zero, and negative numbers are to the left of zero.

Definition
The set of **integers** includes the set of natural numbers, their negatives, and zero. The set of *integers* is $\{\ldots, -3, -2, -1, 0, 1, 2, 3, \ldots\}$.

Example 1

Graph each number on a number line: $5, 1, -2, 0, -4$.

Solution

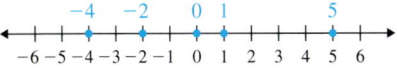

5 and 1 are to the right of zero since they are positive. -2 is two units to the left of zero, and -4 is four units to the left of zero.

In-Class Example 1
Graph each number on a number line.
$3, 6, -1, -4, 0$
answer:

You Try 1

Graph each number on a number line. $3, -1, 6, -5, -3$

Positive and negative numbers are also called *signed numbers*.

Example 2

Given the set of numbers $\left\{-11, 0, 9, -5, -1, \dfrac{2}{3}, 6\right\}$, list the

a) whole numbers   b) natural numbers   c) integers

Solution

a) whole numbers: 0, 6, 9   b) natural numbers: 6, 9

c) integers: $-11, -5, -1, 0, 6, 9$

In-Class Example 2
Given this set of numbers $\left\{-12, 6, 7, 0, -1, 1, \dfrac{1}{4}\right\}$, list the
a) whole numbers   b) natural numbers   c) integers
answer: a) 0, 1, 6, 7   b) 1, 6, 7
c) $-12, -1, 0, 1, 6, 7$

You Try 2

Given the set of numbers $\left\{3, -2, -9, 4, 0, \dfrac{5}{8}, -\dfrac{1}{3}\right\}$, list the

a) whole numbers   b) natural numbers   c) integers

Real Numbers and Algebraic Expressions

Algebra at Work: Landscape Architecture

Jill is a landscape architect and uses multiplication, division, and geometry formulas on a daily basis. Here is an example of the type of landscaping she designs. When Jill is asked to create the landscape for a new house, her first job is to draw the plans.

The ground in front of the house will be dug out into shapes that include rectangles and circles, shrubs and flowers will be planted, and mulch will cover the ground. To determine the volume of mulch that will be needed, Jill must use the formulas for the area of a rectangle and a circle and then multiply by the depth of the mulch. She will calculate the total cost of this landscaping job only after determining the cost of the plants, the mulch, and the labor. Her calculations involve whole numbers, fractions, and decimals, so Jill must know how to perform operations with all of these types of numbers.

It is important that her calculations are accurate. Her company and her clients must have an accurate estimate of the cost of the job. If the estimate is too high, the customer might choose another, less expensive landscaper to do the job. If the estimate is too low, either the client will have to pay more money at the end or the company will not earn as much profit on the job.

In this chapter, we will review operations and concepts from arithmetic.

Notice in Example 2 that $\frac{2}{3}$ did not belong to any of these sets. That is because the

whole numbers, natural numbers, and integers do not contain any fractional parts. $\frac{2}{3}$ is a
rational number.

Definition

A **rational number** is any number of the form $\frac{p}{q}$, where p and q are integers and $q \neq 0$.

A rational number is any number that can be written as a fraction where the numerator and denominator are integers and the denominator does not equal zero.

Rational numbers include much more than numbers like $\frac{2}{3}$, which are already in fractional form.

Example 3

Explain why each of the following numbers is rational.

a) 3   b) 0.5   c) -7

d) $2\frac{1}{8}$ e) $0.\overline{4}$ f) $\sqrt{9}$

In-Class Example 3
Explain why each of the following numbers is rational.

a) 9 b) 0.02 c) -4 d) $3\frac{4}{5}$
e) $0.\overline{6}$ f) $\sqrt{16}$

answer:

a) 9 can be written as $\frac{9}{1}$.

b) 0.02 can be written as $\frac{2}{100}$.

c) -4 can be written as $\frac{-4}{1}$.

d) $3\frac{4}{5}$ can be written as $\frac{19}{5}$.

e) $0.\overline{6}$ can be written as $\frac{2}{3}$.

f) $\sqrt{16} = 4$ and $4 = \frac{4}{1}$.

Solution

Rational Number	Reason
3	3 can be written as $\frac{3}{1}$.
0.5	0.5 can be written as $\frac{5}{10}$.
-7	-7 can be written as $\frac{-7}{1}$.
$2\frac{1}{8}$	$2\frac{1}{8}$ can be written as $\frac{17}{8}$.
$0.\overline{4}$	$0.\overline{4}$ can be written as $\frac{4}{9}$.
$\sqrt{9}$	$\sqrt{9} = 3$ and $3 = \frac{3}{1}$.

$\sqrt{9}$ is read as "the square root of 9." This means, "What number times itself equals 9?" That number is 3.

You Try 3

Explain why each of the following numbers is rational.

a) 6   b) 0.3   c) -2 d) $3\frac{2}{5}$ e) $0.\overline{3}$ f) $\sqrt{25}$

To summarize, the set of rational numbers includes

1) integers, whole numbers, and natural numbers.
2) repeating decimals.
3) terminating decimals.
4) fractions and mixed numbers.

The set of rational numbers does *not* include nonrepeating, nonterminating decimals. These decimals cannot be written as the quotient of two integers. Numbers such as these are called *irrational numbers*.

> **Definition**
>
> The set of numbers that cannot be written as the quotient of two integers is called the set of **irrational numbers**. Written in decimal form, an *irrational number* is a nonrepeating, nonterminating decimal.

Example 4

In-Class Example 4
Explain why each of the following numbers is irrational.
a) 1.459678512 ... b) $\sqrt{8}$
**answer: a) It is a nonrepeating, nonterminating decimal.
b) "8" is not a perfect square, and the decimal equivalent of the square root of a nonperfect square is a nonrepeating, nonterminating decimal. Here,
$\sqrt{8} \approx 2.828427....$**

Explain why each of the following numbers is irrational.

a) 0.21598354 . . .   b) π c) $\sqrt{5}$

Solution

Irrational Number	Reason
0.21598354. . .	It is a nonrepeating, nonterminating decimal.
π	$\pi \approx 3.14159265...$ It is a nonrepeating, nonterminating decimal.
$\sqrt{5}$	"5" is not a perfect square, and the decimal equivalent of the square root of a nonperfect square is a nonrepeating, nonterminating decimal. Here, $\sqrt{5} \approx 2.236067....$

You Try 4

Explain why each of the following numbers is irrational.

a) 0.7593614 . . .   b) $\sqrt{7}$

If we put together the sets of numbers we have discussed up to this point, we get the *real numbers*.

> **Definition**
>
> The set of **real numbers** consists of the rational and irrational numbers.

We summarize the information next with examples of the different sets of numbers:

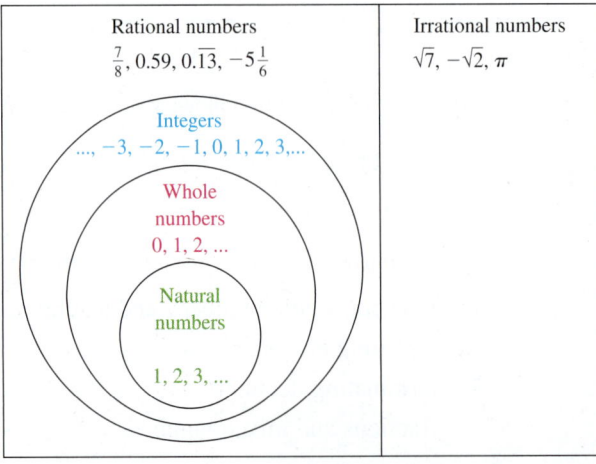

From the figure we can see, for example, that all whole numbers {0, 1, 2, 3, . . .} are integers, but not all integers are whole numbers (−3, for example).

Example 5

In-Class Example 5
Given the set of numbers
$\left\{ 17, -5, \dfrac{1}{4}, 9.178, \sqrt{20}, \right.$

$\left. 3.135117..., -0.\overline{7}, \pi \right\}$, list the

a) integers   b) natural numbers
c) whole numbers   d) rational
numbers   e) irrational numbers
f) real numbers
answer:
a) −5, 17   b) 17   c) 17
d) $17, -5, \dfrac{1}{4}, 9.178, -0.\overline{7}$
e) $\sqrt{20}, 3.135117..., \pi$
f) $17, -5, \dfrac{1}{4}, 9.178, \sqrt{20},$

   $3.135117, -0.\overline{7}, \pi$

Given the set of numbers $\left\{ 0.\overline{2}, 37, -\dfrac{4}{15}, \sqrt{11}, -19, 8.51, 0, 6.149235... \right\}$, list the

a)   integers   b)   natural numbers   c)   whole numbers

d)   rational numbers   e)   irrational numbers   f)   real numbers

Solution

a)   integers: $-19, 0, 37$

b)   natural numbers: 37

c)   whole numbers: $0, 37$

d)   rational numbers: $0.\overline{2}, 37, -\dfrac{4}{15}, -19, 8.51, 0$ Each of these numbers can be written

   as the quotient of two integers.

e)   irrational numbers: $\sqrt{11}, 6.149235...$

f)   real numbers: All of the numbers in this set are real.

$$\left\{ 0.\overline{2}, 37, -\dfrac{4}{15}, \sqrt{11}, -19, 8.51, 0, 6.149235... \right\}$$

You Try 5

Given the set of numbers $\left\{ -38, 0, \sqrt{15}, 6, \dfrac{3}{2}, 5.4, 0.\overline{8}, 4.981162... \right\}$, list the

a)   whole numbers   b)   integers   c)   rational numbers   d)   irrational numbers

2. Compare Numbers Using Inequality Symbols

Let's review the inequality symbols.

$<$ less than	$\leq$ less than or equal to
$>$ greater than	$\geq$ greater than or equal to
$\neq$ not equal to	$\approx$ approximately equal to

We use these symbols to compare numbers as in $5 > 2$, $6 \leq 17$, $4 \neq 9$, and so on.
How do we compare negative numbers?

Note

As we move to the *left* on the number line, the numbers get smaller. As we move to the *right* on the number line, the numbers get larger.

Example 6

In-Class Example 6
Insert > or < to make the statement true. Look at the number line if necessary.

−5 −4 −3 −2 −1 0 1 2 3 4 5

a) 3 __ −1   b) −4 __ −1
c) 3 __ 0   d) 1 __ −2
answer:
a) 3 > −1   b) −4 < −1
c) 3 > 0   d) 1 > −2

Insert > or < to make the statement true. Look at the number line, if necessary.

−5 −4 −3 −2 −1 0 1 2 3 4 5

a) 5 __ 1   b) −4 __ 3   c) −1 __ −5   d) −5 __ −2

Solution

a) $5 \geq 1$ 5 is to the right of 1.   b) $-4 \leq 3$ −4 is to the left of 3.

c) $-1 \geq -5$ −1 is to the right of −5.   d) $-5 \leq -2$ −5 is to the left of −2.   ■

You Try 6

Insert > or < to make the statement true.

a) 4 __ 9   b) 6 __ −8   c) −3 __ −10

Applications of Signed Numbers

Example 7

In-Class Example 7
Use a signed number to represent the change in each situation.
a) After driving down the mountain, the temperature rose 10°.
b) Sarah lost her iPod and had to use her old iPod. Her storage capacity dropped by 5000 songs.
answer: a) 10°   b) −5000

Use a signed number to represent the change in each situation.

a) After a storm passed through Kansas City, the temperature dropped 18°.

b) Between 1995 and 2000, the number of "Generation Xers" moving to the Seattle, Washington, area increased by 27,201 people.
(*American Demographics,* May 2004, Vol. 26, Issue 4, p. 18)

Solution

a) −18°   The negative number represents a decrease in temperature.

b) 27,201   The positive number represents an increase in population.   ■

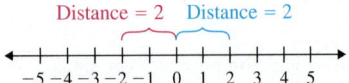

You Try 7

Use a signed number to represent the change.
After taking his last test, Julio raised his average by 3.5%.

3. Find the Absolute Value of a Number

Before we discuss absolute value, we will define *additive inverses*.

Distance = 2   Distance = 2

−5 −4 −3 −2 −1 0 1 2 3 4 5

Notice that both −2 and 2 are a distance of 2 units from 0 but are on opposite sides of 0. We say that 2 and −2 are *additive inverses*.

Definition

Two numbers are **additive inverses** if they are the same distance from 0 on the number line but on the opposite sides of 0. Therefore, if *a* is any real number, then −*a* is its additive inverse.

Furthermore, $-(-a) = a$. We can see this on the number line.

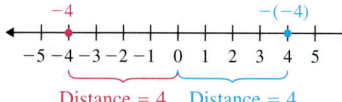

Example 8

In-Class Example 8
Find $-(-7)$.
answer: 7

Find $-(-4)$.

Solution

$$\begin{array}{c}
-4 \qquad\qquad\qquad -(-4)\\
\leftarrow\!\!+\!\!+\!\!+\!\!+\!\!+\!\!+\!\!+\!\!+\!\!+\!\!+\!\!+\!\!\rightarrow\\
-5\ -4\ -3\ -2\ -1\ \ 0\ \ 1\ \ 2\ \ 3\ \ 4\ \ 5
\end{array}$$
Distance = 4   Distance = 4

Beginning with -4, the number on the opposite side of zero and 4 units away from zero is 4. So, $-(-4) = 4$

You Try 8

Find $-(-11)$.

This idea of "distance from zero" can be explained in another way: *absolute value*.

Definition

If a is any real number, then the **absolute value of a**, denoted by $|a|$, is

    i) a if $a \geq 0$

    ii) $-a$ if $a < 0$

Remember, $|a|$ is never negative.

Note

The absolute value of a number is the distance between that number and 0 on the number line. It just describes the distance, *not* what side of zero the number is on. Therefore, the absolute value of a number is always positive or zero.

Example 9

In-Class Example 9
Evaluate each.
a) $|8|$ b) $|-3|$ c) $|0|$
d) $-|4|$ e) $|4 - 7|$
**answer: a) 8   b) 3   c) 0
d) -4 e) 3**

Evaluate each.

a) $|9|$ b) $|-7|$ c) $|0|$ d) $-|5|$ e) $|11 - 5|$

Solution

a) $|9| = 9$ 9 is 9 units from 0.

b) $|-7| = 7$ -7 is 7 units from 0.

c) $|0| = 0$

d) $-|5| = -5$ First, evaluate $|5|$. $|5| = 5$. Then, apply the negative symbol to get -5.

e) $|11 - 5| = |6|$ The absolute value symbols work like parentheses. First, evaluate what is inside: $11 - 5 = 6$.

            $= 6$ Find the absolute value.

You Try 9

Evaluate each.

a) $|16|$ b) $|-4|$ c) $-|6|$ d) $|14 - 9|$

Answers to You Try Exercises

1)   2) a) $0, 3, 4$ b) $3, 4$ c) $-9, -2, 0, 3, 4$

3) a) $6 = \dfrac{6}{1}$ b) $0.3 = \dfrac{3}{10}$ c) $-2 = \dfrac{-2}{1}$ d) $3\dfrac{2}{5} = \dfrac{17}{5}$ e) $0.\overline{3} = \dfrac{1}{3}$ f) $\sqrt{25} = 5 = \dfrac{5}{1}$

4) a) It is a nonrepeating, nonterminating decimal. b) 7 is not a perfect square, so the decimal equivalent of $\sqrt{7}$ is a nonrepeating, nonterminating decimal.

5) a) $0, 6$ b) $-38, 0, 6$ c) $-38, 0, 6, \dfrac{3}{2}, 5.4, 0.\overline{8}$ d) $\sqrt{15}, 4.981162...$

6) a) $<$ b) $>$ c) $>$ 7) 3.5% 8) 11 9) a) 16 b) 4 c) -6 d) 5

1.1 Exercises

*Additional answers can be found in the Answers to Exercises appendix.

Objective 1: Identify Numbers and Graph Them on a Number Line

1) Given the set of numbers

$$\left\{ -14, 6, \frac{2}{5}, \sqrt{19}, 0, 3.\overline{28}, -1\frac{3}{7}, 0.95 \right\},$$

VIDEO list the

a) whole numbers. $6, 0$

b) integers. $-14, 6, 0$

c) irrational numbers. $\sqrt{19}$

d) natural numbers. 6

e) rational numbers. $-14, 6, \frac{2}{5}, 0, 3.\overline{28}, -1\frac{3}{7}, 0.95$

f) real numbers. $-14, 6, \frac{2}{5}, \sqrt{19}, 0, 3.\overline{28}, -1\frac{3}{7}, 0.95$

2) Given the set of numbers

$$\left\{ 5.2, 34, -\frac{9}{4}, -18, 0, 0.\overline{7}, \frac{5}{6}, \sqrt{6}, 4.3811275... \right\},$$

list the

a) integers. $34, -18, 0$

b) natural numbers. 34

c) rational numbers. $5.2, 34, -\frac{9}{4}, -18, 0, 0.\overline{7}, \frac{5}{6}$

d) whole numbers. $34, 0$

e) irrational numbers. $\sqrt{6}, 4.3811275...$

f) real numbers. $5.2, 34, -\frac{9}{4}, -18, 0, 0.\overline{7}, \frac{5}{6}, \sqrt{6}, 4.3811275...$

Determine if each statement is true or false.

3) Every whole number is an integer.   true

4) Every rational number is a whole number.   false

5) Every real number is an integer.   false

6) Every natural number is a whole number.   true

7) Every integer is a rational number.   true

8) Every whole number is a real number.   true

Graph the numbers on a number line. Label each.

9) $6, -4, \dfrac{3}{4}, 0, -1\dfrac{1}{2}$

10) $5\dfrac{2}{3}, 1, -3, -4\dfrac{5}{6}, 2\dfrac{1}{4}$

11) $-5, 6\dfrac{1}{8}, 2\dfrac{3}{4}, -2\dfrac{5}{7}, 4.3$

12) $1.7, -\dfrac{4}{5}, 3\dfrac{1}{5}, -5, -2\dfrac{1}{2}$

VIDEO 13) $-6.8, -\dfrac{3}{8}, 0.2, 1\dfrac{8}{9}, -4\dfrac{1}{3}$

14) $-1, 5.9, 1\dfrac{7}{10}, -\dfrac{2}{3}, 0.61$

Objective 3: Find the Absolute Value of a Number

Evaluate.

15) $|-13|$ 13

16) $|8|$ 8

17) $\left| \dfrac{3}{2} \right|$ $\dfrac{3}{2}$

18) $|-23|$ 23

19) $-|10|$ -10

20) $-|6|$ -6

21) $-|-19|$ -19

22) $-\left| -1\dfrac{3}{5} \right|$ $-1\dfrac{3}{5}$

Find the additive inverse of each.

23) 11 -11

24) 5 -5

25) -7 7

26) $-\dfrac{1}{2}$ $\dfrac{1}{2}$

27) -4.2 4.2

28) 2.9 -2.9

Mixed Exercises: Objectives 2 and 3

Write each group of numbers from smallest to largest.

29) $7, -2, 3.8, -10, 0, \dfrac{9}{10}$ $-10, -2, 0, \dfrac{9}{10}, 3.8, 7$

30) $-6, -7, 5.2, 5.9, 6, -1$ $-7, -6, -1, 5.2, 5.9, 6$

31) $-4\dfrac{1}{2}, \dfrac{5}{8}, \dfrac{1}{4}, -0.3, -9, 1$ $-9, -4\dfrac{1}{2}, -0.3, \dfrac{1}{4}, \dfrac{5}{8}, 1$

32) $14, -5, 13.6, -5\dfrac{2}{3}, 1, \dfrac{6}{7}$ $-5\dfrac{2}{3}, -5, \dfrac{6}{7}, 1, 13.6, 14$

Decide if each statement is true or false.

33) $9 \geq -2$ true

34) $-6 > 3$ false

35) $-7 \leq -4$ true

36) $10.8 \geq 10.2$ true

37) $\dfrac{1}{6} \leq \dfrac{1}{8}$ false

38) $-8.1 > -8.5$ true

39) $-5\dfrac{3}{10} < -5\dfrac{3}{4}$ false

40) $\dfrac{4}{5} \neq 0.8$ false

41) $|-9| \geq 9$ true

42) $-|-31| = 31$ false

Use a signed number to represent the change in each situation.

43) In 2001, Barry Bonds set the all-time home run record with 73. In 2002, he hit 46. That was a decrease of 27 home runs. (Source: *Total Baseball: The Ultimate Baseball Encyclopedia*, 8th edition by John Thorn, Phil Birnbaum, and Bill Deane, © 2004, SPORT Media Pub., Inc.) -27

44) In 2007, the median income for households in Texas was $46,248, and in 2008 it increased by $2830 to $49,078. (Source: U.S. Census Bureau)   2830

45) In 2005, the U.S. unemployment rate was 5.1% and in 2006 it decreased by 0.5% to 4.6%. (Source: U.S. Census Bureau) -0.5%

46) During the 2005–2006 season, Kobe Bryant average 35.4 points per game (ppg). The following season he averaged 31.6 ppg, a decrease of 3.8 over the previous year. (Source: nba.com/playerfile/kobe_bryant/career_stats.html) -3.8

47) In Michael Jordan's last season with the Chicago Bulls (1997–1998) the average attendance at the United Center was 23,988. During the 2002–2003 season, the average attendance fell to 19,617. This was a decrease of 4371 people per game. (Source: *Total Basketball: The Ultimate Basketball Encyclopedia*, by Ken Shouler, Bob Ryan, Sam Smith, Leonard Koppett, and Bob Bellotti, © 2003, SPORT Media Pub., Inc.) -4371

48) The 2004 Indianapolis 500 was won by Buddy Rice with an average speed of 138.518 mph. In 2006, Sam Hornish Jr. won it with an average speed of 157.085 or 18.567 mph faster than in 2004. (Source: www.indy500.com)   18.567

49) According to the 2000 census, the population of North Dakota was 642,200. In 2006, it decreased by 6333 to 635,867. (Source: U.S. Census Bureau) -6333

50) The 1999 Hennessey Viper Venom can go from 0 to 60 mph in 3.3 sec. The 2000 model goes from 0 to 60 mph in 2.7 sec, which is a decrease of 0.6 sec to go from 0 to 60 mph. (Source: www.supercars.net) -0.6

Section 1.2 Operations on Real Numbers

Objectives

1. Add Real Numbers
2. Subtract Real Numbers
3. Solve Applied Problems Involving Addition and Subtraction
4. Multiply Real Numbers
5. Divide Real Numbers

In the previous section, we defined real numbers. In this section, we will discuss adding, subtracting, multiplying, and dividing real numbers.

1. Add Real Numbers

Recall that when we add two numbers with the same sign, the result has the same sign as the numbers being added.

$$2 + 5 = 7 \qquad -1 + (-4) = -5$$

Procedure Adding Numbers with the Same Sign

To add numbers with the same sign, find the absolute value of each number and add them. The sum will have the same sign as the numbers being added.

Example 1

In-Class Example 1
Add.
a) $-1 + (-7)$
b) $-15 + (-28)$
answer: a) -8 b) -43

Add.

a) $-8 + (-2)$ b) $-35 + (-71)$

Solution

a) $-8 + (-2) = -(|-8| + |-2|) = -(8 + 2) = -10$

b) $-35 + (-71) = -(|-35| + |-71|) = -(35 + 71) = -106$ ■

You Try 1

Add.

a) $-2 + (-9)$ b) $-48 + (-67)$

Let's review how to add numbers with different signs.

$$2 + (-5) = -3 \text{ and } -8 + 12 = 4$$

> **Procedure**   Adding Numbers with Different Signs
>
> To add two numbers with different signs, find the absolute value of each number. Subtract the smaller absolute value from the larger. The sum will have the sign of the number with the larger absolute value.

Let's apply this to $2 + (-5)$ and $-8 + 12$.

$2 + (-5)$: $|2| = 2$ $|-5| = 5$

Since $2 < 5$, subtract $5 - 2$ to get 3. Since $|-5| > |2|$, the sum will be negative.

$$2 + (-5) = -3$$

$-8 + 12$: $|-8| = 8$ $|12| = 12$

Subtract $12 - 8$ to get 4. Since $|12| > |-8|$, the sum will be positive.

$$-8 + 12 = 4$$

Example 2

In-Class Example 2
Add.
a) $-32 + 11$ b) $8.9 + (-3.5)$
c) $\dfrac{3}{8} + \left(-\dfrac{2}{3}\right)$ d) $-10 + 10$
answer: a) -21 b) 5.4
c) $-\dfrac{7}{24}$ d) 0

Add.

a) $-19 + 4$ b) $10.3 + (-4.1)$ c) $\dfrac{1}{4} + \left(-\dfrac{5}{9}\right)$ d) $-7 + 7$

Solution

a) $-19 + 4 = -15$ The sum will be negative since the number with the larger absolute value, $|-19|$, is negative.

b) $10.3 + (-4.1) = 6.2$ The sum will be positive since the number with the larger absolute value, $|10.3|$, is positive.

c) $\dfrac{1}{4} + \left(-\dfrac{5}{9}\right) = \dfrac{9}{36} + \left(-\dfrac{20}{36}\right)$ Get a common denominator.

$$= -\dfrac{11}{36}$$

The sum will be negative since the number with the larger absolute value, $\left|-\dfrac{20}{36}\right|$, is negative.

d) $-7 + 7 = 0$ ■

Note

The sum of a number and its additive inverse is always 0. That is, if a is a real number, then $a + (-a) = 0$. Notice in part d) of Example 2 that -7 and 7 are additive inverses.

You Try 2

Add.

a) $11 + (-8)$ b) $-17 + (-5)$ c) $-\dfrac{5}{8} + \dfrac{2}{3}$ d) $59 + (-59)$

2. Subtract Real Numbers

Subtraction of numbers can be defined in terms of the additive inverse. We'll begin by looking at a basic subtraction problem.

Represent $6 - 4$ on a number line.

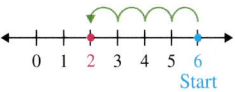

Begin at 6. To subtract 4, move 4 units to the left to get 2: $6 - 4 = 2$.
We use the same process to find $6 + (-4)$. This leads us to a definition of subtraction:

Definition

If a and b are real numbers, then $a - b = a + (-b)$.

The definition tells us that to subtract $a - b$,

1) change subtraction to addition.
2) find the additive inverse of b.
3) add a and the additive inverse of b.

Example 3

In-Class Example 3
Subtract.
a) $9 - 24$ b) $-4 - 34$
c) $13 - 4$ d) $43 - (-12)$
answer: a) -15 **b)** -38
c) 9 **d)** 55

Subtract.
a) $5 - 16$ b) $-19 - 4$ c) $20 - 5$ d) $7 - (-11)$

Solution

a) $5 - 16 = 5 + (-16) = -11$

           Change to   Additive inverse
           addition   of 16

b) $-19 - 4 = -19 + (-4) = -23$

           Change to   Additive inverse
           addition   of 4

c) $20 - 5 = 20 + (-5)$
             $= 15$

d) $7 - (-11) = 7 + 11 = 18$

           Change to   Additive inverse
           addition   of -11

You Try 3

Subtract.

a) $3 - 10$ b) $-6 - 12$ c) $18 - 7$ d) $9 - (-16)$

In part d) of Example 3, $7 - (-11)$ changed to $7 + 11$. This shows that *subtracting a negative number is equivalent to adding a positive number.* Therefore, $-2 - (-6) = -2 + 6 = 4$.

3. Solve Applied Problems Involving Addition and Subtraction

Sometimes, we use signed numbers to solve real-life problems.

Example 4

In-Class Example 4
The lowest temperature ever recorded in Chicago, Illinois, was $-27°$F. The highest temperature in Chicago was $105°$F. What is the difference between these two temperatures?
answer: 132°F

The lowest temperature ever recorded was $-129°$F in Vostok, Antarctica. The highest temperature on record is $136°$F in Al'Aziziyah, Libya. What is the difference between these two temperatures? (Source: *Encyclopedia Britannica Almanac* 2004)

Solution

$$\text{Difference} = \text{Highest temperature} - \text{Lowest temperature}$$
$$= \quad\quad\quad 136 \quad\quad\quad - \quad\quad (-129)$$
$$= 136 + 129$$
$$= 265$$

The difference between the temperatures is $265°$F.

You Try 4

The best score in a golf tournament was -12, and the worst score was $+17$. What is the difference between these two scores?

4. Multiply Real Numbers

Let's begin by reviewing the rules for multiplying real numbers.

> **Procedure** Multiplying Real Numbers
>
> 1) The product of two positive numbers is positive.
> 2) The product of two negative numbers is positive.
> 3) The product of a positive number and a negative number is negative.
> 4) The product of any real number and zero is zero.

Example 5

In-Class Example 5
Multiply.
a) $-11 \cdot (-12)$ b) $-0.5 \cdot 20$
c) $-\dfrac{5}{8} \cdot \left(-\dfrac{2}{3}\right)$
d) $-2 \cdot \left(-\dfrac{1}{5}\right) \cdot (-20)$
answer: a) 132   b) -10
c) $\dfrac{5}{12}$ d) -8

Multiply.

a) $-7 \cdot (-3)$ b) $-2.5 \cdot 8$ c) $-\dfrac{4}{5} \cdot \left(-\dfrac{1}{6}\right)$ d) $-3 \cdot (-4) \cdot (-5)$

Solution

a) $-7 \cdot (-3) = 21$ The product of two negative numbers is positive.

b) $-2.5 \cdot 8 = -20$ The product of a negative number and a positive number is negative.

c) $-\dfrac{4}{5} \cdot \left(-\dfrac{1}{6}\right) = -\dfrac{\overset{2}{\cancel{4}}}{5} \cdot \left(-\dfrac{1}{\underset{3}{\cancel{6}}}\right) = \dfrac{2}{15}$ The product of two negatives is positive.

d) $\underbrace{-3 \cdot (-4)}_{12} \cdot (-5) = 12 \cdot (-5) = -60$ Multiply from left to right.

You Try 5

Multiply.

a) $-2 \cdot 8$ b) $-\dfrac{10}{21} \cdot \dfrac{14}{15}$ c) $-2 \cdot (-3) \cdot (-1) \cdot (-4)$

Note

It is helpful to know that

1) an **even number** of negative factors in a product gives a positive result.

$$-3 \cdot 1 \cdot (-2) \cdot (-1) \cdot (-4) = 24 \qquad \text{Four negative factors}$$

2) an **odd number** of negative factors in a product gives a negative result.

$$5 \cdot (-3) \cdot (-1) \cdot (-2) \cdot (3) = -90 \qquad \text{Three negative factors}$$

5. Divide Real Numbers

Next we will review these rules for dividing signed numbers.

Procedure   Dividing Real Numbers

1) The quotient of two positive numbers is a positive number.
2) The quotient of two negative numbers is a positive number.
3) The quotient of a positive and a negative number is a negative number.

Example 6

In-Class Example 6
Divide.

a) $-30 \div 5$ b) $-\dfrac{4}{5} \div \left(-\dfrac{8}{7}\right)$

c) $\dfrac{-11}{-1}$ d) $\dfrac{10}{-100}$

answer: a) -6 b) $\dfrac{7}{10}$

c) 11 d) $-\dfrac{1}{10}$

Divide.

a) $-48 \div 6$ b) $-\dfrac{1}{12} \div \left(-\dfrac{4}{3}\right)$ c) $\dfrac{-6}{-1}$ d) $\dfrac{-27}{72}$

Solution

a) $-48 \div 6 = -8$

b) $-\dfrac{1}{12} \div \left(-\dfrac{4}{3}\right) = -\dfrac{1}{12} \cdot \left(-\dfrac{3}{4}\right)$ When dividing by a fraction, multiply by the reciprocal.

$$= -\dfrac{1}{\overset{}{\underset{4}{12}}} \cdot \left(-\dfrac{\overset{1}{3}}{4}\right)$$

$$= \dfrac{1}{16}$$

c) $\dfrac{-6}{-1} = 6$ The quotient of two negative numbers is positive, and $\dfrac{6}{1} = 6$.

d) $\dfrac{-27}{72} = -\dfrac{27}{72}$ The quotient of a negative number and a positive number is negative, so reduce $\dfrac{27}{72}$.

$$= -\dfrac{3}{8}$$ 27 and 72 each divide by 9.

It is important to note here that there are three ways to write the answer: $-\frac{3}{8}, \frac{-3}{8},$ or $\frac{3}{-8}$. These are equivalent. However, we usually write the negative sign in front of the entire fraction as in $-\frac{3}{8}$.

You Try 6

Divide.

a) $-\dfrac{4}{21} \div \left(-\dfrac{2}{7}\right)$ b) $\dfrac{72}{-8}$ c) $\dfrac{-19}{-1}$

Answers to You Try Exercises

1) a) -11 b) -115 2) a) 3 b) -22 c) $\dfrac{1}{24}$ d) 0   3) a) -7 b) -18 c) 11 d) 25

4) 29   5) a) -16 b) $-\dfrac{4}{9}$ c) 24   6) a) $\dfrac{2}{3}$ b) -9 c) 19

1.2 Exercises

*Additional answers can be found in the Answers to Exercises appendix.

Mixed Exercises: Objectives 1 and 2

1) Explain, in your own words, how to add two negative numbers.   Answers may vary.

2) Explain, in your own words, how to add a positive and a negative number.   Answers may vary.

3) Explain, in your own words, how to subtract two negative numbers.   Answers may vary.

4) Explain, in your own words, how to add two positive numbers.   Answers may vary.

Add or subtract as indicated.

5) $9 + (-13)$ -4

6) $-7 + (-5)$ -12

7) $-2 - 12$ -14

8) $-4 + 11$ 7

9) $-25 + 38$ 13

10) $10 - (-17)$ 27

11) $-1 - (-19)$ 18

12) $-40 - (-6)$ -34

13) $-794 - 657$ -1451

14) $380 + (-192)$ 188

15) $-\dfrac{3}{10} + \dfrac{4}{5}$ $\dfrac{1}{2}$

16) $\dfrac{2}{9} - \dfrac{5}{6}$ $-\dfrac{11}{18}$

17) $-\dfrac{5}{8} - \dfrac{2}{3}$ $-\dfrac{31}{24}$ or $-1\dfrac{7}{24}$

18) $\dfrac{3}{7} - \left(-\dfrac{1}{8}\right)$ $\dfrac{31}{56}$

19) $-\dfrac{11}{12} - \left(-\dfrac{5}{9}\right)$ $-\dfrac{13}{36}$

20) $-\dfrac{3}{4} + \left(-\dfrac{1}{6}\right)$ $-\dfrac{11}{12}$

21) $7.3 - 11.2$ -3.9

22) $-14.51 + 20.6$ 6.09

23) $-5.09 - (-12.4)$ 7.31

24) $8.8 - 19.2$ -10.4

25) $-1 - 4.2$ -5.2

26) $288.11 - 1.367$ 286.743

27) $18 - |-12|$ 6

28) $-14 + |-11|$ -3

29) $|13| + 9$ 22

30) $|-7| - 19$ -12

31) $|-5.2| - 4.8$ 0.4

32) $\left|-\dfrac{1}{4}\right| + \dfrac{1}{4}$ $\dfrac{1}{2}$

Determine whether each statement is true or false. For any real numbers a and b,

33) $|a - b| = |b - a|$ true

34) $|a + b| = a + b$ false

35) $|a| + |b| = a + b$ false

36) $|a + b| = |a| + |b|$ false

37) $a + (-a) = 0$ true

38) $-b - (-b) = 0$ true

Objective 3: Solve Applied Problems Involving Addition and Subtraction

Applications of Signed Numbers: Write an expression for each and simplify.

39) The world's tallest mountain, Mt. Everest, reaches an elevation of 29,028 ft. The Mariana Trench in the Pacific Ocean has a maximum depth of 36,201 ft below sea level. What is the difference between these two elevations? (Source: *Encyclopedia Britannica Almanac*, 2004)

40) In 2002, the total attendance at Major League Baseball games was 67,859,176. In 2003, the figure was 67,630,052. What was the difference in the number of people who went to ballparks from 2002 to 2003? (Source: *Total Baseball: The Ultimate Baseball Encyclopedia*, 8th edition © 2004)

41) The median income for a male with a bachelor's degree in 2004 was $51,801. In 2005, the median income was $51,700. What was the difference in the median income from 2004 to 2005? (Source: U.S. Census Bureau)

42) Mt. Washington, New Hampshire, rises to an elevation of 6288 ft. New Orleans, Louisiana, lies 6296 ft below this. What is the elevation of New Orleans? (Source: www.infoplease.com)
$6288 - 6296 = -8$. New Orleans lies 8 ft below sea level.

43) The lowest temperature ever recorded in the United States was −79.8°F in the Endicott Mountains of Alaska. The highest U.S. temperature on record was 213.8° more than the lowest and was recorded in Death Valley, California. What is the highest temperature on record in the United States? (Source: www.infoplease.com)

44) The lowest temperature ever recorded in Hawaii was 12°F while the lowest temperature in Colorado was 73° less than that. What was the coldest temperature ever recorded in Colorado?

$12 - 73 = -61$. The coldest temperature recorded in Colorado was −61°F.

45) During one offensive drive in the first quarter of Super Bowl XXXVIII, the New England Patriots ran for 7 yd, gained 4 yd on a pass play, gained 1 yd on a running play, gained another 6 yd on a pass by Tom Brady, then lost 10 yd on a running play. What was the Patriots' net yardage on this offensive drive? (Source: www.superbowl.com)

46) The bar graph shows Dale Earnhardt Jr.'s total winnings from races for the years 2004–2007. Use a signed number to represent the change in his winnings over the given years. (Source: www.dalejrpitstop.com)

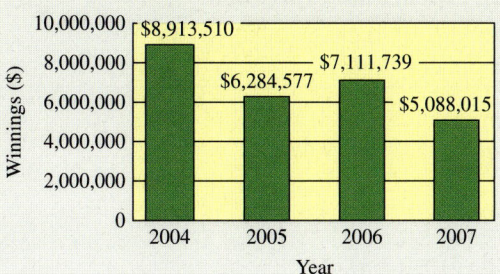

a) 2004–2005   −$2,628,933   b) 2005–2006 $827,162

c) 2006–2007   −$2,023,724

47) The bar graph shows the number of housing starts (in thousands) during five months in 2003 in the Northeastern United States. Use a signed number to represent the change over the given months. (Source: www.census.gov)

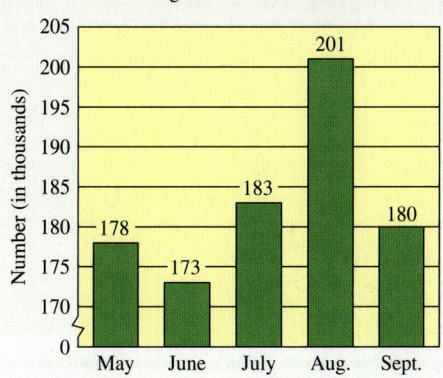

a) May–June   −5000   b) June–July   10,000

c) July–August   18,000   d) August–September   −21,000

48) The bar graph shows the population of Oakland, California from 2003 to 2006. Use a signed number to

represent the change in Oakland's population over the given years. (Source: U.S. Census Bureau)

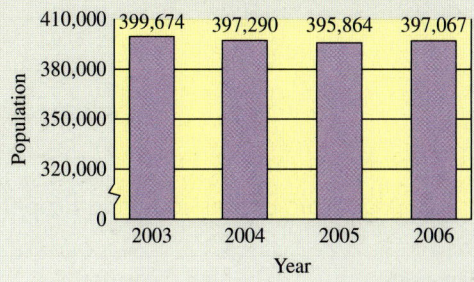

a) 2003–2004   −2384   b) 2004–2005   −1426

c) 2005–2006   1203

Objective 4: Multiply Real Numbers

Fill in the blank with *positive* or *negative*.

49) The product of two negative numbers is _____.   positive

50) The product of a positive number and a negative number is _____.   negative

Multiply.

51) $-5 \cdot 9$ −45   52) $3 \cdot (-11)$ −33

53) $-14 \cdot (-3)$ 42   54) $-16 \cdot (-31)$ 496

55) $-2 \cdot 5 \cdot (-3)$ 30   56) $-1 \cdot (-6) \cdot (-7)$ −42

57) $\frac{7}{9} \cdot \left(-\frac{6}{5}\right)$ $-\frac{14}{15}$ 58) $-\frac{15}{32} \cdot \left(-\frac{8}{25}\right)$ $\frac{3}{20}$

59) $(-0.25)(1.2)$ −0.3   60) $(-3.8)(-7.1)$ 26.98

61) $8 \cdot (-2) \cdot (-4) \cdot (-1)$ −64

62) $-5 \cdot (3) \cdot (-2) \cdot (-1) \cdot (-4)$ 120

63) $(-8) \cdot (-9) \cdot 0 \cdot \left(-\frac{1}{4}\right) \cdot (-2)$ 0

64) $(-6) \cdot \left(-\frac{2}{3}\right) \cdot 2 \cdot (-5)$ −40

Objective 5: Divide Real Numbers

65) The quotient of a negative number and a positive number is _____.   negative

66) The quotient of two negative numbers is _____.   positive

Divide.

67) $-42 \div (-6)$ 7   68) $-108 \div 9$ −12

69) $\frac{56}{-7}$ −8   70) $\frac{-32}{-4}$ 8

71) $\frac{-3.6}{0.9}$ −4   72) $\frac{12}{-0.5}$ −24

73) $-\frac{12}{13} \div \left(-\frac{6}{5}\right)$ $\frac{10}{13}$ 74) $-14 \div \left(-\frac{10}{3}\right)$ $\frac{21}{5}$ or $4\frac{1}{5}$

75) $\dfrac{0}{-4}$ 0

76) $-\dfrac{0}{9}$ 0

79) $\dfrac{\frac{20}{21}}{\frac{5}{7}}$ $\dfrac{4}{3}$ or $1\dfrac{1}{3}$

80) $\dfrac{-\frac{3}{5}}{\frac{3}{4}}$ $-\dfrac{4}{5}$

77) $\dfrac{360}{-280}$ $-\dfrac{9}{7}$ or $-1\dfrac{2}{7}$

78) $\dfrac{-84}{-210}$ $\dfrac{2}{5}$

81) $\dfrac{-0.5}{10}$ -0.05

82) $\dfrac{-11}{0.11}$ -100

Section 1.3 Algebraic Expressions and Properties of Real Numbers

Objectives

1. Use Exponents
2. Find Square Roots
3. Use the Order of Operations
4. Evaluate Algebraic Expressions
5. Properties of Real Numbers

1. Use Exponents

Remember that exponents can be used to represent repeated multiplication. For example,

$$2 \cdot 2 \cdot 2 \cdot 2 \cdot 2 = 2^5 \longleftarrow \text{exponent (or power)}$$

$$\uparrow$$

base

The *base* is 2, and 2 is a *factor* that appears five times. 5 is the *exponent* or *power*. We read 2^5 as "2 to the fifth power." 2^5 is called an **exponential expression.**

Example 1

In-Class Example 1
Rewrite each product in exponential form.
a) $8 \cdot 8 \cdot 8$ b) $(-5) \cdot (-5)$
answer: a) 8^3 **b)** $(-5)^2$

Rewrite each product in exponential form.

a) $6 \cdot 6 \cdot 6 \cdot 6$ b) $(-3) \cdot (-3)$

Solution

a) $6 \cdot 6 \cdot 6 \cdot 6 = 6^4$ b) $(-3) \cdot (-3) = (-3)^2$

 ### You Try 1

Rewrite each product in exponential form.

a) $4 \cdot 4 \cdot 4 \cdot 4 \cdot 4 \cdot 4$ b) $\dfrac{3}{8} \cdot \dfrac{3}{8} \cdot \dfrac{3}{8} \cdot \dfrac{3}{8}$ c) $(-7) \cdot (-7) \cdot (-7) \cdot (-7) \cdot (-7)$

We can also evaluate an exponential expression.

Example 2

In-Class Example 2
Evaluate.
a) 2^4 b) 6^2 c) $\left(\dfrac{3}{5}\right)^3$ d) -3^2
answer: a) 16 **b)** 36 **c)** $\dfrac{27}{125}$
d) -9

Evaluate.

a) 3^2 b) $\left(-\dfrac{4}{5}\right)^3$ c) -2^4

Solution

a) $3^2 = 3 \cdot 3 = 9$

b) $\left(-\dfrac{4}{5}\right)^3 = \left(-\dfrac{4}{5}\right) \cdot \left(-\dfrac{4}{5}\right) \cdot \left(-\dfrac{4}{5}\right)$ The negative sign is included in the parentheses.

$-\dfrac{4}{5}$ appears as a factor 3 times.

$\qquad = -\dfrac{64}{125}$

c) $-2^4 = -1 \cdot 2 \cdot 2 \cdot 2 \cdot 2$ The negative sign is not included in the parentheses.

$\qquad = -16$ The base is 2, so the exponent of 4 applies only to the 2.

Note

1 raised to any natural number power is 1 since 1 multiplied by itself equals 1.

You Try 2

Evaluate.

a) -3^4 b) 7^2 c) $\left(\dfrac{2}{5}\right)^3$ d) $\left(-\dfrac{1}{2}\right)^5$ e) -2^3

It is generally agreed that there are some skills in arithmetic that everyone should have in order to be able to acquire other math skills. Knowing the basic multiplication facts, for example, is essential for learning how to add, subtract, multiply, and divide fractions as well as how to perform many other operations in arithmetic and algebra. Similarly, memorizing powers of certain bases is necessary for learning how to apply the rules of exponents (Chapter 5) and for working with radicals (Chapter 8). Therefore, the powers listed here must be memorized in order to be successful in the previously mentioned, as well as other, topics. Throughout this book, it is assumed that students know these powers:

Powers to Memorize						
$2^1 = 2$	$3^1 = 3$	$4^1 = 4$	$5^1 = 5$	$6^1 = 6$	$8^1 = 8$	$10^1 = 10$
$2^2 = 4$	$3^2 = 9$	$4^2 = 16$	$5^2 = 25$	$6^2 = 36$	$8^2 = 64$	$10^2 = 100$
$2^3 = 8$	$3^3 = 27$	$4^3 = 64$	$5^3 = 125$			$10^3 = 1000$
$2^4 = 16$	$3^4 = 81$					
$2^5 = 32$				$7^1 = 7$	$9^1 = 9$	$11^1 = 11$
$2^6 = 64$				$7^2 = 49$	$9^2 = 81$	$11^2 = 121$
						$12^1 = 12$
						$12^2 = 144$
						$13^1 = 13$
						$13^2 = 169$

(Hint: Making flashcards might help you learn these facts.)

2. Find Square Roots

In Example 2, we saw that $3^2 = 3 \cdot 3 = 9$. Here we will list and use the rules for the opposite procedure, finding **roots** of numbers.

The $\sqrt{}$ symbol represents the *positive* square root, or **principal square root,** of a number. For example,

$$\sqrt{9} = 3.$$

BE CAREFUL

$\sqrt{9} = 3$ but $\sqrt{9} \neq -3$. The $\sqrt{}$ symbol represents *only* the positive square root.

To find the negative square root of a number we must put a negative symbol in front of the $\sqrt{}$. For example, $-\sqrt{9} = -3$.

$\sqrt{}$ is the **square root symbol** or the **radical sign.** The number under the radical sign is the **radicand.**

radical sign $\rightarrow \sqrt{9}$
↑
radicand

The entire expression, $\sqrt{9}$, is called a **radical.**

Property   Radicands and Square Roots

1)   If the radicand is a perfect square, the *square* root is a *rational* number.

   Example: $\sqrt{16} = 4$ 16 is a perfect square.

   $\sqrt{\dfrac{100}{49}} = \dfrac{10}{7}$ $\dfrac{100}{49}$ is a perfect square.

2)   If the radicand is a negative number, the square root is *not* a real number.

   Example: $\sqrt{-25}$ is not a real number.

3)   If the radicand is positive and *not* a perfect square, then the square root is an *irrational* number.

   Example: $\sqrt{13}$ is irrational.   13 is not a perfect square.

   The square root of such a number is a real number that is a nonrepeating, nonterminating decimal. It is important to be able to approximate such square roots because sometimes it is necessary to estimate their places on a number line or on a Cartesian coordinate system when graphing.

For the purposes of graphing, approximating a radical to the nearest tenth is sufficient. A calculator with a $\sqrt{}$ key will give a better approximation of the radical.

Example 3

In-Class Example 3
Find each square root, if possible.

a) $\sqrt{36}$ b) $-\sqrt{\dfrac{4}{121}}$
c) $\sqrt{-100}$

answer: a) 6   b) $-\dfrac{2}{11}$
c) not real

Find each square root, if possible.

a) $\sqrt{100}$ b) $-\sqrt{\dfrac{81}{49}}$ c) $\sqrt{-9}$

Solution

a) $\sqrt{100} = 10$ since $(10)^2 = 100$.

b) $-\sqrt{\dfrac{81}{49}}$ means $-1 \cdot \sqrt{\dfrac{81}{49}}$. Therefore, $-\sqrt{\dfrac{81}{49}} = -1 \cdot \sqrt{\dfrac{81}{49}} = -1 \cdot \left(\dfrac{9}{7}\right) = -\dfrac{9}{7}$

c)   *There is no such real number* since $3^2 = 9$ and $(-3)^2 = 9$. Therefore, $\sqrt{-9}$ is not a real number.

    You Try 3

Find each square root, if possible.

a) $-\sqrt{144}$ b) $\sqrt{\dfrac{25}{36}}$ c) $-\sqrt{\dfrac{1}{64}}$ d) $\sqrt{-36}$

3. Use the Order of Operations

We will begin this topic with a problem for the student:

You Try 4

Evaluate $36 - 12 \div 4 + (6 - 1)^2$.

What answer did you get? 31? or 58? or 8? Or, did you get another result?

Most likely you obtained one of the three answers just given. Only one is correct, however. If we do not have rules to guide us in evaluating expressions, it is easy to get the incorrect answer. Here are the rules we follow. This is called the **order of operations.**

Procedure   The Order of Operations

Simplify expressions in the following order:

1) If parentheses or other grouping symbols appear in an expression, simplify what is in these grouping symbols first.

2) Simplify expressions with exponents.

3) Perform multiplication and division from left to right.

4) Perform addition and subtraction from left to right.

Think about the "You Try" problem. Did you evaluate it using the order of operations? Let's look at that expression:

Example 4

Evaluate $36 - 12 \div 4 + (6 - 1)^2$.

In-Class Example 4
Evaluate.
$9 \cdot 5 - (12 \div 4) + 8 \cdot 4 - 32$.
answer: 42

Solution

$36 - 12 \div 4 + (6 - 1)^2$	First, perform the operation in the parentheses.
$36 - 12 \div 4 + 5^2$	
$36 - 12 \div 4 + 25$	Exponents are done before division, addition, and subtraction.
$36 - 3 + 25$	Perform division before addition and subtraction.
$33 + 25$	When an expression contains only addition and subtraction, perform
58	the operations starting at the left and moving to the right.

A good way to remember the order of operations is to remember the sentence, "**P**lease **E**xcuse **M**y **D**ear **A**unt **S**ally." (**P**arentheses, **E**xponents, **M**ultiplication, and **D**ivision from left to right, **A**ddition and **S**ubtraction from left to right) Don't forget that multiplication and division are at the same "level" in the process of performing operations and that addition and subtraction are at the same "level."

Example 5

Evaluate.

In-Class Example 5
Evaluate.
a) $3[15 - (48 \div 12)] + 8$
b) $\dfrac{(11 - 6)^2 \cdot 4}{7 + 13 - 4^2}$
answer: a) 41   b) 25

a) $-2[20 - (40 \div 5)] + 7$ 　　　 b) $\dfrac{(7 - 5)^3 \cdot 6}{30 - 9 \cdot 2}$

Solution

a) $-2[20 - (40 \div 5)] + 7$

This expression contains two sets of grouping symbols: **brackets** [ ] and **parentheses** ( ). Perform the operation in the **innermost** grouping symbol first which is the parentheses in this case.

$-2[20 - (40 \div 5)] + 7 = -2[20 - 8] + 7$	Innermost grouping symbol
$= -2[12] + 7$	Brackets
$= -24 + 7$	Perform multiplication before addition.
$= -17$	Add.

b) $\dfrac{(7 - 5)^3 \cdot 6}{30 - 9 \cdot 2}$

The fraction bar in this expression acts as a grouping symbol. Therefore, simplify the numerator, simplify the denominator, then simplify the resulting fraction, if possible.

$$\dfrac{(7 - 5)^3 \cdot 6}{30 - 9 \cdot 2} = \dfrac{2^3 \cdot 6}{30 - 18} \qquad \text{Parentheses}$$
$$\qquad\qquad \text{Multiply.}$$
$$= \dfrac{8 \cdot 6}{12} \qquad \text{Exponent}$$
$$\qquad\qquad \text{Subtract.}$$
$$= \dfrac{48}{12} \qquad \text{Multiply.}$$
$$= 4$$

You Try 5

Evaluate:

a) $5 + 3[15 - 2(3 + 1)]$

b) $\dfrac{7^2 - 3 \cdot 3}{5(12 - 8)}$

4. Evaluate Algebraic Expressions

Here is an algebraic expression: $5x^3 - 9x^2 + \frac{1}{4}x + 7$. The *variable* is x. A **variable** is a symbol, usually a letter, used to represent an unknown number. The *terms* of this algebraic expression are $5x^3$, $-9x^2$, $\frac{1}{4}x$, and 7. A **term** is a number or a variable or a product or quotient of numbers and variables. 7 is the **constant** or **constant term.** The value of a constant does not change. Each term has a **coefficient.**

Definition

An **algebraic expression** is a collection of numbers, variables, and grouping symbols connected by operation symbols such as $+, -, \times,$ and $\div$.

Here are more examples of algebraic expressions:

$$10k + 9, \qquad 3(2t^2 + t - 4), \qquad 6a^2b^2 - 13ab - 2a + 5$$

We can **evaluate** an algebraic expression by substituting a value for a variable and simplifying. The value of an algebraic expression changes depending on the value that is substituted.

Example 6

Evaluate $5x - 3$ when (a) $x = 4$ and (b) $x = -2$.

In-Class Example 6
Evaluate $8f - 3$ when
a) $f = 5$ and b) $f = -3$.
answer: a) 37   b) −27

Solution

a) $5x - 3$ when $x = 4$ Substitute 4 for x.
    $= 5(4) - 3$ Use parentheses when substituting a value for a variable.
    $= 20 - 3$ Multiply.
    $= 17$ Subtract.

b) $5x - 3$ when $x = -2$ Substitute -2 for x.
    $= 5(-2) - 3$ Use parentheses when substituting a value for a variable.
    $= -10 - 3$ Multiply.
    $= -13$

You Try 6

Evaluate $10x + 7$ when $x = -4$.

Example 7

Evaluate $4c^2 - 2cd + 1$ when $c = -3$ and $d = 5$.

In-Class Example 7
Evaluate $n^3 - 3mn - 11$ when $m = 4$ and $n = -2$.
answer: 5

Solution

$4c^2 - 2cd + 1$ when $c = -3$ and $d = 5$ Substitute -3 for c and 5 for d.
$\quad = 4(-3)^2 - 2(-3)(5) + 1$ Use parentheses when substituting.
$\quad = 4(9) - 2(-15) + 1$ Evaluate exponent and multiply.
$\quad = 36 - (-30) + 1$ Multiply.
$\quad = 36 + 30 + 1$
$\quad = 67$

You Try 7

Evaluate $b^2 + 7ab - 4a - 5$ when $a = \dfrac{1}{2}$ and $b = -6$.

5. Properties of Real Numbers

Like the order of operations, the properties of real numbers guide us in our work with numbers and variables.

Summary   Properties of Real Numbers

If a, b, and c are real numbers, then

Commutative Properties:	$a + b = b + a$ and $ab = ba$
Associative Properties:	$(a + b) + c = a + (b + c)$ and $(ab)c = a(bc)$
Identity Properties:	$a + 0 = 0 + a = a$
	$a \cdot 1 = 1 \cdot a = a$
Inverse Properties:	$a + (-a) = -a + a = 0$
	$b \cdot \dfrac{1}{b} = \dfrac{1}{b} \cdot b = 1 \,(b \neq 0)$
Distributive Properties:	$a(b + c) = ab + ac$ and $(b + c)a = ba + ca$
	$a(b - c) = ab - ac$ and $(b - c)a = ba - ca$

Example 8

Use the stated property to rewrite each expression. Simplify, if possible.

a) $x \cdot 9$; commutative

b) $5 + (2 + n)$; associative

c) $-(-10m + 3n - 8)$; distributive

In-Class Example 8
Use the stated property to rewrite each expression. Simplify, if possible.

a) $\left(-\dfrac{4}{9} \cdot \dfrac{8}{3}\right)\dfrac{3}{8}$; associative

b) $9 + 7$; commutative

c) $-2(8 - 3b)$; distributive

answer:

a) $\left(-\dfrac{4}{9} \cdot \dfrac{8}{3}\right)\dfrac{3}{8} = -\dfrac{4}{9}\left(\dfrac{8}{3} \cdot \dfrac{3}{8}\right) =$
$-\dfrac{4}{9}(1) = -\dfrac{4}{9}$

b) $9 + 7 = 7 + 9 = 16$

c) $-2(8 - 3b) = -2 \cdot 8 +$
$(-2)(-3b) = -16 + 6b =$
$6b - 16$

Solution

a) $x \cdot 9 = 9 \cdot x$ or $9x$ Commutative property

b) $5 + (2 + n) = (5 + 2) + n = 7n$ Associative property

c) $-(-10m + 3n - 8) = -1(-10m + 3n - 8)$
$\qquad\qquad\qquad\qquad = -1(-10m) + (-1)(3n) - (-1)(8)$ Apply the distributive property.

$\qquad\qquad\qquad\qquad = 10m + (-3n) - (-8)$ Multiply.
$\qquad\qquad\qquad\qquad = 10m - 3n + 8$ Simplify.

You Try 8

Use the stated property to rewrite each expression. Simplify, if possible.

a) $y \cdot 8$; commutative   b) $\left(-\dfrac{5}{12} \cdot \dfrac{4}{7}\right)\dfrac{7}{4}$; associative   c) $3(8x - 5y + 11z)$; distributive

---

Example 9

In-Class Example 9
Which property is illustrated by
each statement?
a) $-5 + 5 = 0$
b) $1.4 + 0 = 1.4$
c) $(-20)\left(-\dfrac{1}{20}\right) = 1$
d) $-8(1) = -8$
**answer: a) inverse property
b) identity property
c) inverse property
d) identity property**

Which property is illustrated by each statement?

a) $0 + 9 = 9$ b) $-1.3 + 1.3 = 0$

c) $\dfrac{1}{12} \cdot 12 = 1$ d) $7(1) = 7$

Solution

a) $0 + 9 = 9$ Identity property

b) $-1.3 + 1.3 = 0$ Inverse property

c) $\dfrac{1}{12} \cdot 12 = 1$ Inverse property

d) $7(1) = 7$ Identity property

You Try 9

Which property is illustrated by each statement?

a) $4 \cdot \dfrac{1}{4} = 1$ b) $-8 + 8 = 0$ c) $-7.4(1) = -7.4$ d) $5 + 0 = 5$

Answers to You Try Exercises

1) a) 4^6 b) $\left(\dfrac{3}{8}\right)^4$ c) $(-7)^5$ 2) a) -81 b) 49 c) $\dfrac{8}{125}$ d) $-\dfrac{1}{32}$ e) -8 3) a) -12 b) $\dfrac{5}{6}$

c) $-\dfrac{1}{8}$ d) not a real number   4) 58   5) a) 26 b) 2   6) -33 7) 8   8) a) $8y$ b) $-\dfrac{5}{12}$

c) $24x - 15y + 33z$ 9) a) inverse property   b) inverse property   c) identity property
d) identity property

---

*Additional answers can be found in the Answers to Exercises appendix.

Objective 1: Use Exponents

Write in exponential form.

1) $9 \cdot 9 \cdot 9 \cdot 9$ 9^4

2) $2 \cdot 2 \cdot 2 \cdot 2 \cdot 2 \cdot 2 \cdot 2 \cdot 2$ 2^8

Fill in the blank with *positive* or *negative*.

3) If a is a positive number, then $-a^6$ is _____.   negative

4) If a is a positive number, then $(-a)^6$ is _____.   positive

5) If a is a negative number, then $-a^5$ is _____.   positive

6) Explain the difference between how you would evaluate -3^4 and $(-3)^4$. Then, evaluate each.

Evaluate

7) 2^5 32

8) 9^2 81

9) $(11)^2$ 121

10) 4^3 64

11) $(-2)^4$ 16

12) $(-5)^3$ -125

13) -7^2 -49

14) -6^2 -36

15) -2^3 -8

16) -3^4 -81

17) $\left(\dfrac{1}{5}\right)^3$ $\dfrac{1}{125}$

18) $\left(\dfrac{3}{2}\right)^4$ $\dfrac{81}{16}$

19) Evaluate $(0.5)^2$ two different ways.   0.25 or $\dfrac{1}{4}$

20) Explain why $1^{200} = 1$ 1 raised to any natural number power equals 1.

Objective 2: Find Square Roots

Decide if each statement is true or false. If it is false, explain why.

21) $\sqrt{49} = 7$ and $\sqrt{49} = -7$ False; the $\sqrt{\ }$ symbol means to find only the positive square root of 49.

22) $\sqrt{121} = 11$ true

23) $-\sqrt{4} = -2$ true

24) The square root of a negative number is a negative number.
False; the square root of a negative number is not a real number.

Find all square roots of each number.

25) 64   8 and −8

26) 25   5 and −5

27) 400   20 and −20

28) 8100   90 and −90

29) $\dfrac{25}{16}$ $\dfrac{5}{4}$ and $-\dfrac{5}{4}$

30) $\dfrac{49}{144}$ $\dfrac{7}{12}$ and $-\dfrac{7}{12}$

Find each square root, if possible.

31) $\sqrt{36}$ 6

32) $\sqrt{169}$ 13

33) $-\sqrt{1}$ −1

34) $-\sqrt{900}$ −30

35) $\sqrt{-25}$ not real

36) $\sqrt{-36}$ not real

37) $\sqrt{\dfrac{100}{121}}$ $\dfrac{10}{11}$

38) $\sqrt{\dfrac{4}{9}}$ $\dfrac{2}{3}$

39) $-\sqrt{\dfrac{1}{64}}$ $-\dfrac{1}{8}$

40) $-\sqrt{\dfrac{1}{25}}$ $-\dfrac{1}{5}$

Objective 3: Use the Order of Operations

41) In your own words, summarize the order of operations.
Answers may vary.

Evaluate.

42) $35 - 7 + 8 - 3$ 33

43) $-50 \div 10 + 15$ 10

44) $6 \cdot (-4) - 2$ −26

45) $20 - 3 \cdot 2 + 9$ 23

46) $22 + 10 \div 2 - 1$ 26

47) $\dfrac{1}{2} \cdot \dfrac{4}{5} - \dfrac{2}{5} \cdot \dfrac{3}{10}$ $\dfrac{7}{25}$

48) $\left(\dfrac{3}{2}\right)^2 - \left(-\dfrac{5}{4}\right)^2$ $\dfrac{11}{16}$

49) $15 - 3(6 - 4)^2$ 3

50) $7 + 2(9 - 5)^2$ 39

51) $-6[21 \div (3 + 4)] - 9$ −27

52) $2[23 + (9 - 11)^3] + 3$ 33

53) $4 + 3[(3 - 7)^3 \div (10 - 2)]$ −20

54) $(8 + 2)^2 - 5[9 \cdot (3 + 1) - 5^2]$ 45

55) $\dfrac{12(5 + 1)}{2 \cdot 5 - 1}$ 8

56) $\dfrac{(14 - 4)^2 - 4^3}{4 \cdot 9 - 3 \cdot 11}$ 12

57) $\dfrac{4(7 - 2)^2}{(12)^2 - 8 \cdot 3}$ $\dfrac{5}{6}$

58) $\dfrac{6(3 - 5)^2}{10 + 12 \div 2 + 4}$ $\dfrac{6}{5}$

Objective 4: Evaluate Algebraic Expressions

59) Evaluate $2j^2 + 3j - 7$ when
   a) $j = 4$ 37
   b) $j = -5$ 28

60) Evaluate $6 - t^3$ when
   a) $t = -3$ 33
   b) $t = 3$ −21

Evaluate each expression when $x = -2$, $y = 7$, and $z = -3$.

61) $8x + y$ −9

62) $10z - 3x$ −24

63) $x^2 + xy + 10$ 0

64) $x^2 - 2z^2 + 4xy$ −70

65) $\dfrac{2x}{y + z}$ −1

66) $\dfrac{x - y}{3z}$ 1

67) $\dfrac{x^2 - y^2}{2z^2 + y}$ $-\dfrac{9}{5}$

68) $\dfrac{5 + 3(y + 4z)}{x^2 - z^2}$ 2

Objective 5: Properties of Real Numbers

69) What is the identity element for addition?   0

70) What is the identity element for multiplication?   1

71) What is the multiplicative inverse of 6? $\dfrac{1}{6}$

72) What is the additive inverse of −9?   9

Which property of real numbers is illustrated by each example? Choose from the commutative, associative, identity, inverse, or distributive property.

73) $(-11 + 4) + 9 = -11 + (4 + 9)$ associative

74) $5 \cdot 7 = 7 \cdot 5$ commutative

75) $20 + 8 = 8 + 20$ commutative

76) $16 + (-16) = 0$ inverse

77) $3(8 \cdot 4) = (3 \cdot 8) \cdot 4$ associative

78) $5(3 + 7) = 5 \cdot 3 + 5 \cdot 7$ distributive

79) $(10 + 2)6 = 10 \cdot 6 + 2 \cdot 6$ distributive

80) $\dfrac{3}{4} \cdot 1 = \dfrac{3}{4}$ identity

81) $-24 + 0 = -24$ identity

82) $\left(\dfrac{8}{15}\right)\left(\dfrac{15}{8}\right) = 1$ inverse

83) $9(a - b) = 9a - 9b$ distributive

84) $-3(c + d) = -3c - 3d$ distributive

Rewrite each expression using the indicated property.

85) $7(u + v)$; distributive $7u + 7v$

86) $12 + (3 + 7)$; associative $(12 + 3) + 7$

87) $k + 4$; commutative $4 + k$

88) $-8(c + 5)$; distributive $-8c - 40$

89) $-4z + 0$; identity $-4z$

90) $9 + 11r$; commutative $11r + 9$

91) Is $10c - 3$ equivalent to $3 - 10c$? Why or why not?
No. Subtraction is not commutative.

92) Is $8 + 5n$ equivalent to $5n + 8$? Why or why not?
Yes. Addition is commutative.

Rewrite each expression using the distributive property. Simplify if possible.

93) $5(4 + 3)$ $5 \cdot 4 + 5 \cdot 3 = 20 + 15 = 35$

94) $8(1 + 5)$ $8 \cdot 1 + 8 \cdot 5 = 8 + 40 = 48$

VIDEO 95) $-2(5 + 7)$ $(-2) \cdot 5 + (-2) \cdot 7 = -10 + (-14) = -24$

96) $6(9 - 4)$ $6 \cdot 9 + 6 \cdot (-4) = 54 + (-24) = 30$

97) $-7(2 - 6)$ $(-7) \cdot 2 + (-7) \cdot (-6) = -14 + 42 = 28$

98) $-(9 - 5)$ $-9 + 5 = -4$

99) $-(6 + 1)$ $-6 - 1 = -7$

100) $(8 - 2)4$ $8 \cdot 4 + (-2) \cdot 4 = 32 + (-8) = 24$

101) $(-10 + 3)5$ $(-10) \cdot 5 + 3 \cdot 5 = -50 + 15 = -35$

102) $2(-6 + 5 + 3)$ $2 \cdot (-6) + 2 \cdot 5 + 2 \cdot 3 = -12 + 10 + 6 = 4$

103) $9(g + 6)$ $9g + 9 \cdot 6 = 9g + 54$

104) $4(t - 5)$ $4t + 4(-5) = 4t - 20$

105) $-5(z + 3)$ $-5z + (-5) \cdot 3 = -5z - 15$

106) $-2(m + 11)$ $-2m + (-2) \cdot 11 = -2m - 22$

107) $-8(u - 4)$ $-8u + (-8) \cdot (-4) = -8u + 32$

108) $-3(h - 9)$ $-3h + (-3) \cdot (-9) = -3h + 27$

109) $-(v - 6)$ $-v + 6$

110) $-(y - 13)$ $-y + 13$

111) $10(m + 5n - 3)$ $10m + 10 \cdot 5n + 10 \cdot (-3) = 10m + 50n - 30$

112) $12(2a - 3b + c)$ $12 \cdot 2a + 12 \cdot (-3b) + 12c = 24a - 36b + 12c$

VIDEO 113) $-(-8c + 9d - 14)$ $8c - 9d + 14$

114) $-(x - 4y + 10z)$ $-x + 4y - 10z$

Definition/Procedure	Example

1.1 Sets of Numbers

Natural numbers: $\{1, 2, 3, 4, \ldots\}$
Whole numbers: $\{0, 1, 2, 3, 4, \ldots\}$
Integers: $\{\ldots, -3, -2, -1, 0, 1, 2, 3, \ldots\}$

Definition/Procedure	Example
A **rational number** is any number of the form $\frac{p}{q}$, where p and q are integers and $q \neq 0$. **(p. 2)**	The following numbers are rational: $-1, 2, \frac{3}{4}, 3.\bar{6}, 4.5$
An **irrational number** cannot be written as the quotient of two integers. **(p. 4)**	The following numbers are irrational: $\sqrt{7}, 5.1948\ldots$
The set of **real numbers** includes the rational and irrational numbers. **(p. 4)**	Any number that can be represented on the number line is a real number.
The **additive inverse** of a is $-a$. **(p. 6)**	The additive inverse of 13 is -13.

Absolute Value
Definition/Procedure	Example
$\lvert a \rvert$ is the distance of a from zero. **(p. 7)**	$\lvert -9 \rvert = 9$

1.2 Operations on Real Numbers

Adding Real Numbers
Definition/Procedure	Example
To add numbers with the *same sign*, add the absolute value of each number. The sum will have the same sign as the numbers being added. **(p. 9)**	$-7 + (-4) = -11$
To add two numbers with *different signs* subtract the smaller absolute value from the larger. The sum will have the sign of the number with the larger absolute value. **(p. 10)**	$-16 + 10 = -6$

Subtracting Real Numbers
Definition/Procedure	Example
To subtract $a - b$, change subtraction to addition and add the additive inverse of b: $a - b = a + (-b)$. **(p. 11)**	$5 - 9 = 5 + (-9) = -4$ $-14 - (-6) = -14 + 6 = -8$ $11 - 4 = 11 + (-4) = 7$

Multiplying Real Numbers
Definition/Procedure	Example
The product of two real numbers with the *same* sign is *positive*.	$9 \cdot 4 = 36 \qquad -6 \cdot (-5) = 30$
The product of a positive number and a negative number is *negative*.	$-3 \cdot 7 = -21 \qquad 8 \cdot (-1) = -8$
An *even number* of negative factors in a product gives a *positive* result.	$\underbrace{(-2)(-1)(3)(-4)(-5)}_{\text{4 negative factors}} = 120$
An *odd number* of negative factors in a product gives a *negative* result. **(p. 12)**	$\underbrace{(4)(-3)(-2)(-1)(3)}_{\text{3 negative factors}} = -72$

Dividing Real Numbers
Definition/Procedure	Example
The quotient of two numbers with the *same* sign is positive.	$\frac{100}{4} = 25 \qquad -63 \div (-7) = 9$
The quotient of two numbers with *different* signs is negative. **(p. 13)**	$\frac{-20}{4} = -5 \qquad 32 \div (-8) = -4$

Definition/Procedure	Example

1.3 Algebraic Expressions and Properties of Real Numbers

Exponents

An **exponent** represents repeated multiplication. **(p. 16)**

Write $8 \cdot 8 \cdot 8 \cdot 8 \cdot 8$ in exponential form.
$$8 \cdot 8 \cdot 8 \cdot 8 \cdot 8 = 8^5$$

Evaluating Exponential Expressions (p. 16)

Evaluate $(-2)^4$.
$$(-2)^4 = (-2)(-2)(-2)(-2) = 16$$
Evaluate -2^4.
$$-2^4 = -1 \cdot 2^4 = -1 \cdot 2 \cdot 2 \cdot 2 \cdot 2 = -16$$

Finding Roots

If the *radicand* is a perfect square, then the square root is a *rational* number.

$\sqrt{49} = 7$ since $7^2 = 49$.

If the *radicand* is a negative number, then the square root is *not* a real number.

$\sqrt{-36}$ is not a real number.

If the *radicand* is positive and not a perfect square, then the square root is an *irrational* number. **(p. 18)**

$\sqrt{7}$ is irrational because 7 is not a perfect square.

Order of Operations

Parentheses, **E**xponents, **M**ultiplication, **D**ivision, **A**ddition, **S**ubtraction.

Remember that multiplication and division are at the same "level" when performing operations and that addition and subtraction are at the same level. **(p. 19)**

Evaluate $10 + (2 + 3)^2 - 8 \cdot 4$

$$
\begin{aligned}
10 + (2 + 3)^2 - 8 \cdot 4 & \\
= 10 + 5^2 - 8 \cdot 4 \quad & \text{Parentheses} \\
= 10 + 25 - 8 \cdot 4 \quad & \text{Exponents} \\
= 10 + 25 - 32 \quad & \text{Multiply.} \\
= 35 - 32 \quad & \text{Add.} \\
= 3 \quad & \text{Subtract.}
\end{aligned}
$$

An **algebraic expression** is a collection of numbers, variables, and grouping symbols connected by operation symbols such as $+, -, \times,$ and $\div$. **(p. 20)**

$4y^2 - 7y + \dfrac{3}{5}$

Important terms

      Variable   Constant

      Term   Coefficient

We can evaluate expressions for different values of the variables. **(p. 20)**

Evaluate $2xy - 5y + 1$ when $x = -3$ and $y = 4$.

Substitute -3 for x and 4 for y and simplify.

$$
\begin{aligned}
2xy - 5y + 1 &= 2(-3)(4) - 5(4) + 1 \\
&= -24 - 20 + 1 \\
&= -24 + (-20) + 1 \\
&= -43
\end{aligned}
$$

Properties of Real Numbers

If $a, b,$ and c are real numbers, then the following properties hold.

Commutative Properties:

$$a + b = b + a$$
$$ab = ba$$

$$9 + 2 = 2 + 9$$
$$(-4)(7) = (7)(-4)$$

Associative Properties:

$$(a + b) + c = a + (b + c)$$
$$(ab)c = a(bc)$$

$$(4 + 1) + 7 = 4 + (1 + 7)$$
$$(2 \cdot 3)10 = 2(3 \cdot 10)$$

Definition/Procedure	Example

Identity Properties:
$$a + 0 = 0 + a = a$$
$$a \cdot 1 = 1 \cdot a = a$$

$$\frac{3}{4} + 0 = \frac{3}{4}, \qquad 5 \cdot 1 = 5$$

Inverse Properties:
$$a + (-a) = -a + a = 0$$
$$b \cdot \frac{1}{b} = \frac{1}{b} \cdot b = 1$$

$$6 + (-6) = 0, \qquad 8 \cdot \frac{1}{8} = 1$$

Distributive Properties:
$$a(b + c) = ab + ac \text{ and } (b + c)a = ba + ca$$
$$a(b - c) = ab - ac \text{ and } (b - c)a = ba - ca \textbf{ (p. 21)}$$

$$9(3 + 4) = 9 \cdot 3 + 9 \cdot 4$$
$$= 27 + 36$$
$$= 63$$
$$4(n - 7) = 4n - 4 \cdot 7$$
$$= 4n - 28$$

Additional answers can be found in the Answers to Exercises appendix.

(1.1)

1) Given the set of numbers,

$$\left\{ \sqrt{23}, -6, 14.38, \frac{3}{11}, 2, 5.\overline{7}, 0, 9.21743819\ldots \right\},$$

list the

a) whole numbers.   0, 2

b) natural numbers.   2

c) integers. $-6, 0, 2$

d) rational numbers. $-6, 14.38, \dfrac{3}{11}, 2, 5.\overline{7}, 0$

e) irrational numbers. $\sqrt{23}, 9.21743819\ldots$

2) Graph and label these numbers on a number line.

$$-2, 5\frac{1}{3}, 0.8, -4.5, 3, -\frac{3}{4}$$

3) Evaluate $|-10|$.   10

(1.2) Add or subtract as indicated.

4) $-18 + 4$ -14

5) $60 - (-15)$ 75

6) $-\dfrac{5}{8} + \left(-\dfrac{2}{3}\right)$ $-\dfrac{31}{24}$ or $-1\dfrac{7}{24}$

7) $0.8 - 5.9$ -5.1

8) The lowest temperature on record in the state of Wyoming is $-66°F$. Georgia's record low is $49°$ higher than Wyoming's. What is the lowest temperature ever recorded in Georgia? (Source: www.infoplease.com) $-17°F$

Multiply or divide as indicated.

9) $(-10)(-7)$ 70

10) $\left(-\dfrac{2}{3}\right)(15)$ -10

11) $(3.7)(-2.1)$ -7.77

12) $(-3)(-5)(-2)$ -30

13) $(-1)(6)(-4)\left(-\dfrac{1}{2}\right)(-5)$ 60

14) $-54 \div 6$ -9

15) $\dfrac{-24}{-12}$ 2

16) $\dfrac{38}{-44}$ $-\dfrac{19}{22}$

17) $-\dfrac{20}{27} \div \dfrac{8}{15}$ $-\dfrac{25}{18}$ or $-1\dfrac{7}{18}$

18) $-\dfrac{8}{9} \div (-4)$ $\dfrac{2}{9}$

(1.3) Evaluate.

19) -5^2 -25

20) $(-5)^2$ 25

21) $(-3)^4$ 81

22) $(-1)^9$ -1

23) -2^6 -64

24) $\sqrt{16}$ 4

25) $\sqrt{49}$ 7

26) $-\sqrt{4}$ -2

27) $-\sqrt{36}$ -6

28) $\sqrt{-64}$ not a real number

Use the order of operations to simplify.

29) $64 \div (-8) + 6$ -2

30) $15 - (3 - 7)^3$ 79

31) $-11 - 3 \cdot 9 + (-2)^1$ -40

32) $\dfrac{6 - 2(5 - 1)}{(-3)(-4) + 7 - 3}$ $-\dfrac{1}{8}$

33) $\dfrac{3}{4} \cdot \left|-\dfrac{5}{7}\right|$ $\dfrac{15}{28}$

34) $2[3 - 4 - (-2)^2] \div 5$ -2

35) $\dfrac{4^2 - (3 \cdot 5)}{|-4 - 2|}$ $\dfrac{1}{6}$

36) $\dfrac{-\left|-\dfrac{2}{3}\right|}{\dfrac{19}{9}}$ $-\dfrac{6}{19}$

37) $12 + \sqrt{16} - 2^3 + 1$ 9

38) $-\sqrt{4} + (-5) - |-9|$ -16

39) List the terms and coefficients of

$$c^4 + 12c^3 - c^2 - 3.8c + 11.$$

40) Evaluate $-3m + 7n$ when $m = 6$ and $n = -2$. -32

41) Evaluate $\dfrac{t - 6s}{s^2 - t^2}$ when $s = -4$ and $t = 5$. $-\dfrac{29}{9}$ or $-3\dfrac{2}{9}$

Which property of real numbers is illustrated by each example? Choose from the commutative, associative, identity, inverse, or distributive property.

42) $0 + 12 = 12$ identity

43) $(8 + 1) + 5 = 8 + (1 + 5)$ associative

44) $\left(\dfrac{4}{7}\right)\left(\dfrac{7}{4}\right) = 1$ inverse

45) $35 + 16 = 16 + 35$ commutative

46) $-6(3 + 8) = (-6)(3) + (-6)(8)$ distributive

Rewrite each expression using the distributive property. Simplify if possible.

47) $3(10 - 6)$ $3 \cdot 10 - 3 \cdot 6 = 30 - 18 = 12$

48) $(3 + 9)2$ $3 \cdot 2 + 9 \cdot 2 = 6 + 18 = 24$

49) $-(12 + 5)$ $-12 - 5 = -17$

50) $-7(2c - d + 4)$
$(-7) \cdot 2c + (-7)(-d) + (-7) \cdot 4 = -14c + 7d - 28$

*Additional answers can be found in the Answers to Exercises appendix.

1) Given the set of numbers,

$$\left\{41, -8, 0, 2.\overline{83}, \sqrt{75}, 6.5, 4\frac{5}{8}, 6.37528861\ldots\right\},$$

list the

a) integers. 41, −8, 0

b) irrational numbers. $\sqrt{75}$, 6.37528861...

c) natural numbers. 41

d) rational numbers. 41, −8, 0, 2.$\overline{83}$, 6.5, 4$\frac{5}{8}$

e) whole numbers. 41, 0

2) Graph the numbers on a number line. Label each.

$$6, \frac{7}{8}, -4, -1.2, 4\frac{3}{4}, -\frac{2}{3}$$

Perform the indicated operation(s). Write all answers in lowest terms.

3) $\frac{9}{14} \cdot \frac{7}{24}$ $\frac{3}{16}$

4) $\frac{1}{3} + \frac{4}{15}$ $\frac{3}{5}$

5) $5\frac{1}{4} - 2\frac{1}{6}$ $3\frac{1}{12}$

6) $\frac{12}{13} \div 6$ $\frac{2}{13}$

7) $\frac{4}{7} - \frac{5}{6}$ $-\frac{11}{42}$

8) $-11 - |-19|$ −30

9) $25 + 15 \div 5$ 28

10) $\frac{9}{10} \cdot \left(-\frac{2}{5}\right)$ $-\frac{9}{25}$

11) $-8 \cdot (-6)$ 48

12) $-13.4 + 6.9$ −6.5

13) $30 - 5[-10 + (2 - 6)^2]$ 0

14) $\frac{(50 - 26) \div 3}{3 \cdot 5 - 7}$ 1

Evaluate.

15) 2^5 32

16) -3^4 −81

17) $|-92|$ 92

18) $|2 - 12| - 4|6 - 1|$ −10

19) $-2(7 + \sqrt{25})$ −24

20) $4(3 - 2)^2 + 19$ 23

21) $\frac{(-\sqrt{36} \div 2)^3 + 7}{-|-11 + 3| \cdot (-4)}$ $-\frac{5}{8}$

Determine whether each statement is true or false.

22) $|b - a| = |a - b|$ true

23) $|a + b| = |a| + |b|$ false

24) If a is a positive number, then $-a^2$ is positive. false

25) The square root of a negative number is a negative number. false

26) Both the highest and lowest points in the continental United States are in California. Badwater, Death Valley, is 282 ft below sea level while, 76 mi away, Mount Whitney reaches an elevation of 14,505 ft. What is the difference between these two elevations? 14,787 ft

27) Evaluate $9g^2 + 3g - 6$ when $g = -1$. 0

28) Which property of real numbers is illustrated by each example? Choose from the commutative, associative, identity, inverse, or distributive property.

a) $9(5 - 7) = 9 \cdot 5 - 9 \cdot 7$ distributive

b) $-6 + 6 = 0$ inverse

c) $8 \cdot 3 = 3 \cdot 8$ commutative

29) Rewrite each expression using the distributive property. Simplify if possible.

a) $-2(5 + 3)$ $(-2) \cdot 5 + (-2) \cdot 3 = -10 + (-6) = -16$

b) $5(t + 9u + 1)$ $5t + 5 \cdot 9u + 5 \cdot 1 = 5t + 45u + 5$

30) Is $x - 8$ equivalent to $8 - x$? Why or why not? No. Subtraction is not commutative.

Linear Equations and Inequalities in One Variable

Algebra at Work: Landscape Architecture

A landscape architect must have excellent problem-solving skills.

Matthew is designing the driveway, patio, and walkway for this new home. The village has a building code that states that, at most, 70% of the lot can be covered with an impervious surface such as the house, driveway, patio, and walkway leading up to the front door. So, he cannot design just anything.

To begin, Matthew must determine the area of the land and find 70% of that num-

ber to determine how much land can be covered with these hard surfaces. He must subtract the area covered by the house to determine how much land he has left for the driveway, patio, and walkway. Using his design experience and problem-solving skills, he must come up with a plan for building the driveway, patio, and walkway that will not only please his client but will meet building codes as well.

In this chapter, we will learn different strategies for solving many different types of problems.

Section 2.1 Linear Equations in One Variable

Objectives

1. Define a Linear Equation in One Variable
2. Use the Properties of Equality
3. Combine Like Terms to Solve a Linear Equation
4. Solve Equations Containing Fractions or Decimals
5. Solve Equations with No Solution or an Infinite Number of Solutions
6. Use the Five Steps for Solving Applied Problems

1. Define a Linear Equation in One Variable

What is an equation? It is a mathematical statement that two expressions are equal. $5 + 1 = 6$ is an equation.

> An equation contains an "$=$" sign and an expression does not.

$$5x + 4 = 7 \rightarrow \text{equation}$$
$$9y + 2y \rightarrow \text{expression}$$

We can **solve** equations, and we can **simplify** expressions.

In this section, we will begin our study of solving algebraic equations. Examples of algebraic equations include

$$a + 6 = 11, \qquad t^2 + 7t + 12 = 0, \qquad \sqrt{n - 4} = 16 - n$$

The first equation is an example of a linear equation; the second is a quadratic equation, and the third is a radical equation. In this section we will learn how to solve a linear equation. We will work with the other equations later in this book.

To **solve an equation** means to find the value or values of the variable that make the equation true.

$a + 6 = 11$:   The **solution** is $a = 5$ since we can substitute 5 for the variable and the equation is true:

$$a + 6 = 11$$
$$5 + 6 = 11 \quad \checkmark$$

> **Definition**
>
> A **linear equation in one variable** is an equation that can be written in the form
>
> $$ax + b = 0$$
>
> where a and b are real numbers and $a \neq 0$.

Notice that the exponent of the variable, x, is 1 in a linear equation. For this reason, these equations are also known as first-degree equations. Here are other examples of linear equations in one variable:

$$5y - 8 = 19, \qquad 2(c + 7) - 3 = 4c + 1, \qquad \frac{2}{3}k + \frac{1}{4} = k - 5$$

2. Use the Properties of Equality

The properties of equality will help us solve equations.

> **Property**   The Properties of Equality
>
> Let $a, b,$ and c be expressions representing real numbers. Then,
>
> 1) If $a = b$, then $a + c = b + c$ Addition property of equality
> 2) If $a = b$, then $a - c = b - c$ Subtraction property of equality
> 3) If $a = b$, then $ac = bc$ Multiplication property of equality
> 4) If $a = b$, then $\dfrac{a}{c} = \dfrac{b}{c} \, (c \neq 0)$ Division property of equality
>
> These properties tell us that we can add, subtract, multiply, or divide both sides of an equation by the same real number without changing the solutions to the equation.

| Example 1 |

Solve and check each equation.

In-Class Example 1
Solve and check each equation.
a) $a + 14 = 7$ b) $8v = 96$
c) $\frac{2}{3}c - 4 = 9$

answer: a) $\{-7\}$ b) $\{12\}$
c) $\left\{\frac{39}{2}\right\}$

a) $w + 9 = 2$ b) $4k = -24$ c) $\frac{4}{7}a - 5 = 1$

Solution

Remember, to solve the equation means to find the value of the variable that makes the statement true. To do this, we want to *isolate* the variable; that is, we need to get the variable by itself.

a) $w + 9 = 2$:   Here, 9 is being *added to* w. To get the w by itself, *subtract* 9 from each side.

$$w + 9 = 2$$
$$w + 9 - 9 = 2 - 9 \qquad \text{Subtract 9.}$$
$$w = -7$$

Check: $w + 9 = 2$
$-7 + 9 = 2$
$2 = 2$ ✓

The solution set is $\{-7\}$.

b) $4k = -24$:   On the left-hand side of the equation, the k is being *multiplied* by 4. So, we will perform the "opposite" operation and *divide* each side by 4.

$$4k = -24$$
$$\frac{4k}{4} = \frac{-24}{4} \qquad \text{Divide by 4.}$$
$$k = -6$$

Check: $4k = -24$
$4(-6) = -24$
$-24 = -24$ ✓

The solution set is $\{-6\}$.

c) $\frac{4}{7}a - 5 = 1$:   On the left-hand side, the a is being multiplied by $\frac{4}{7}$, and 5 is being *subtracted* from the a-term. To solve the equation, begin by eliminating the number being subtracted from the a-term.

$$\frac{4}{7}a - 5 = 1$$
$$\frac{4}{7}a - 5 + 5 = 1 + 5 \qquad \text{Add 5 to each side.}$$
$$\frac{4}{7}a = 6 \qquad \text{Combine like terms.}$$
$$\frac{7}{4} \cdot \frac{4}{7}a = \frac{7}{4} \cdot 6 \qquad \text{Multiply each side by the reciprocal of } \frac{4}{7}.$$
$$1a = \frac{7}{\overset{}{\underset{2}{4}}} \cdot \overset{3}{6} \qquad \text{Simplify.}$$
$$a = \frac{21}{2}$$

Check: $\dfrac{4}{7}a - 5 = 1$

$$\dfrac{\overset{2}{\cancel{4}}}{\underset{1}{\cancel{7}}}\left(\dfrac{\overset{3}{\cancel{21}}}{\underset{1}{2}}\right) - 5 = 1$$

$$6 - 5 = 1 \quad \checkmark$$

The solution set is $\left\{\dfrac{21}{2}\right\}$.

You Try 1

Solve and check each equation.

a) $b - 8 = 5$ b) $-9w = 36$ c) $20 = 13 + \dfrac{1}{6}n$

---

3. Combine Like Terms to Solve a Linear Equation

Sometimes it is necessary to combine like terms before we apply the properties of equality. Here are the steps we use to solve a linear equation in one variable.

> **Procedure**   How to Solve a Linear Equation
>
> **Step 1:**   **Clear parentheses** and **combine like terms** on each side of the equation.
>
> **Step 2:**   **Get the variable on one side of the equal sign and the constant on the other side of the equal sign** (isolate the variable) using the addition or subtraction property of equality.
>
> **Step 3:**   **Solve for the variable** using the multiplication or division property of equality.
>
> **Step 4:**   **Check the solution** in the original equation.

There are some cases where we will not follow these steps exactly. We will see this with a couple of examples later in the section.

Example 2

Solve $8n + 5 - 3(2n + 9) = n + 2(2n + 7)$.

In-Class Example 2
Solve $3u - 4 + 6(2 - 2u) = -2u + 3(u - 4)$.
answer: $\{2\}$

Solution

We will follow the steps to solve this equation.

Step 1:   Clear the parentheses and combine like terms.

$$8n + 5 - 6n - 27 = n + 4n + 14$$
$$2n - 22 = 5n + 14$$

Step 2:   Isolate the variable. (The variable can be on either side of the equal sign.)

$$2n - 22 = 5n + 14$$
$$2n - 2n - 22 = 5n - 2n + 14 \qquad \text{Get the variable on the right side of the } = \text{ sign}$$
$$\text{by subtracting } 2n.$$

$$-22 = 3n + 14$$
$$-22 - 14 = 3n + 14 - 14 \qquad \text{Get the constants on the left side of the } = \text{ sign}$$
$$\text{by subtracting } 14.$$

$$-36 = 3n$$

Step 3: Solve for *n* using the division property of equality.

$$\frac{-36}{3} = \frac{3n}{3} \qquad \text{Divide each side by 3.}$$

$$-12 = n$$

Step 4: Check $n = -12$ in the original equation.

$$8n + 5 - 3(2n + 9) = n + 2(2n + 7)$$
$$8(-12) + 5 - 3[2(-12) + 9] = -12 + 2[2(-12) + 7]$$
$$-96 + 5 - 3(-24 + 9) = -12 + 2(-24 + 7)$$
$$-91 - 3(-15) = -12 + 2(-17)$$
$$-91 + 45 = -12 + (-34)$$
$$-46 = -46 \quad \checkmark$$

The solution set is $\{-12\}$.

You Try 2

Solve $5(3 - 2a) + 7a - 2 = 2(9 - 2a) + 15$.

4. Solve Equations Containing Fractions or Decimals

Some equations contain several fractions or decimals which make them appear more difficult to solve. Here are two examples:

$$\frac{2}{9}x - \frac{1}{2} = \frac{1}{18}x + \frac{2}{3} \qquad \text{and} \qquad 0.05c + 0.4(c - 3) = -0.3$$

Before applying the steps for solving a linear equation, we can eliminate the fractions and decimals from the equations.

> **Procedure**   Eliminating Fractions from an Equation
>
> To eliminate the fractions, determine the least common denominator (LCD) for all of the fractions in the equation. Then, multiply both sides of the equation by the LCD.

Example 3

In-Class Example 3
Solve $\frac{3}{8}s - \frac{1}{8} = \frac{1}{2}s + \frac{5}{8}$.
answer: $\{-6\}$

Solve $\frac{2}{9}x - \frac{1}{2} = \frac{1}{18}x + \frac{2}{3}$.

Solution

The least common denominator of all of the fractions in the equation is 18. Multiply both sides of the equation by 18 to eliminate the fractions.

$$18\left(\frac{2}{9}x - \frac{1}{2}\right) = 18\left(\frac{1}{18}x + \frac{2}{3}\right) \qquad \text{Multiply by 18 to eliminate the denominators.}$$

$$18 \cdot \frac{2}{9}x - 18 \cdot \frac{1}{2} = 18 \cdot \frac{1}{18}x + 18 \cdot \frac{2}{3} \qquad \text{Distribute.}$$

$$4x - 9 = x + 12$$
$$4x - x - 9 = x - x + 12 \qquad \text{Get the variable on the left side of the } = \text{ sign by subtracting } x.$$
$$3x - 9 = 12$$
$$3x - 9 + 9 = 12 + 9 \qquad \text{Get the constants on the right side of the } = \text{ sign by adding 9.}$$
$$3x = 21$$
$$\frac{3x}{3} = \frac{21}{3} \qquad \text{Divide by 3.}$$
$$x = 7 \qquad \text{Simplify.}$$

The check is left to the student. The solution set is $\{7\}$.

You Try 3

Solve $\dfrac{1}{5}y + 1 = \dfrac{3}{10}y + \dfrac{1}{4}$.

Just as we can eliminate the fractions from an equation to make it easier to solve, we can eliminate decimals from an equation to make it easier to solve.

> **Procedure**   Eliminating Decimals from an Equation
>
> To eliminate the decimals from an equation, multiply both sides of the equation by the smallest power of 10 that will eliminate all decimals from the problem.

Example 4

Solve $0.05c + 0.4(c - 3) = -0.3$.

In-Class Example 4
Solve
$0.09x - 0.3(x + 4) = -0.15$.
answer: $\{-5\}$

Solution

We want to eliminate the decimals. The number containing a decimal place farthest to the right is 0.05. The 5 is in the hundredths place. Therefore, multiply both sides of the equation by 100 to eliminate all decimals in the equation.

$$100\,[0.05c + 0.4(c - 3)] = 100(-0.3)$$
$$100 \cdot (0.05c) + 100 \cdot [0.4(c - 3)] = 100(-0.3) \qquad \text{Distribute.}$$

Now we will distribute the 100 eliminate the decimals.

$$5c + 40(c - 3) = -30$$
$$5c + 40c - 120 = -30 \qquad \text{Distribute.}$$
$$45c - 120 = -30$$
$$45c - 120 + 120 = -30 + 120 \qquad \text{Get the constants on the right side of the = sign.}$$
$$45c = 90$$
$$\frac{45c}{45} = \frac{90}{45} \qquad \text{Divide.}$$
$$c = 2$$

The check is left to the student. The solution set is $\{2\}$.

You Try 4

Solve $0.1d = 0.5 - 0.02(d - 5)$.

5. Solve Equations with No Solution or an Infinite Number of Solutions

Does every equation have a solution? Consider the next example:

Example 5

Solve $9a + 2 = 6a + 3(a - 5)$.

In-Class Example 5
Solve
$11d - 8 = 3(5d + 6) - 4d$.
answer: $\varnothing$

Solution

$$9a + 2 = 6a + 3(a - 5)$$
$$9a + 2 = 6a + 3a - 15 \qquad \text{Distribute.}$$
$$9a + 2 = 9a - 15 \qquad \text{Combine like terms.}$$
$$9a - 9a + 2 = 9a - 9a - 15 \qquad \text{Subtract } 9a.$$
$$2 = -15 \qquad \text{False}$$

Notice that the variable has "dropped out." Is $2 = -15$ a true statement? No! This means that the equation has *no solution*. We can say that the solution set is the **empty set,** or **null set,** denoted by $\varnothing$.

We have seen that a linear equation may have one solution or no solution. There is a third possibility—a linear equation may have an infinite number of solutions.

Example 6

In-Class Example 6
Solve
$8s + 6(3 - 2s) = 18 - 4s.$
answer: {all real numbers}

Solve $p - 3p + 8 = 8 - 2p$.

Solution

$$
\begin{aligned}
p - 3p + 8 &= 8 - 2p \\
-2p + 8 &= 8 - 2p \qquad \text{Combine like terms.} \\
-2p + 2p + 8 &= 8 - 2p + 2p \qquad \text{Add } 2p. \\
8 &= 8 \qquad \text{True}
\end{aligned}
$$

Here, the variable has "dropped out," and we are left with an equation, $8 = 8$, that is true. This means that any real number we substitute for p will make the original equation true. Therefore, this equation has an *infinite number of solutions*. The solution set is {**all real numbers**}.

Summary   Outcomes When Solving Linear Equations

There are three possible outcomes when solving a linear equation. The equation may have

1) **one solution.** Solution set: {a real number}. An equation that is true for some values and not for others is called a **conditional equation.**

or

2) **no solution.** In this case, the variable will drop out, and there will be a false statement such as $2 = -15$. Solution set: $\varnothing$. An equation that has no solution is called a **contradiction.**

or

3) **an infinite number of solutions.** In this case, the variable will drop out, and there will be a true statement such as $8 = 8$. Solution set: {all real numbers}. An equation that has all real numbers as its solution set is called an **identity.**

You Try 5

Solve.

a) $6 + 5x - 4 = 3x + 2 + 2x$ b) $3x - 4x + 9 = 5 - x$

6. Use the Five Steps for Solving Applied Problems

Equations can be used to describe events that occur in the real world. Therefore, we need to learn how to translate information presented in English into an algebraic equation. We will begin slowly, then throughout the chapter we will work our way up to more challenging problems. Yes, it may be difficult at first, but with patience and persistence, you can do it!

While no single method will work for solving all applied problems, the following approach is suggested to help in the problem-solving process.

> **Procedure**   Steps for Solving Applied Problems
>
> **Step 1:**   **Read** the problem carefully, more than once if necessary, until you understand it. Draw a picture, if applicable. Identify what you are being asked to find.
>
> **Step 2:**   **Choose a variable** to represent an unknown quantity. If there are any other unknowns, define them in terms of the variable.
>
> **Step 3:**   **Translate** the problem from English into an equation using the chosen variable. Some suggestions for doing so are:
>
> - Restate the problem in your own words.
> - Read and think of the problem in "small parts."
> - Make a chart to separate these "small parts" of the problem to help you translate to mathematical terms.
> - Write an equation in English, then translate it to an algebraic equation.
>
> **Step 4:**   **Solve** the equation.
>
> **Step 5:**   **Check** the answer in the original problem, and **interpret** the solution as it relates to the problem. Be sure your answer makes sense in the context of the problem.

Example 7

In-Class Example 7
Write an equation and solve. Seven more than twice a number is fifteen. Find the number.
answer: 2x + 7 = 15; 4

Write an equation and solve.
Five more than twice a number is nineteen. Find the number.

Solution

How should we begin?

Step 1:   **Read** the problem carefully. We must find an unknown number.

Step 2:   **Choose a variable** to represent the unknown.

$$\text{Let } x = \text{the number}$$

Step 3:   **Translate** the information that appears in English into an algebraic equation by rereading the problem slowly and "in parts."

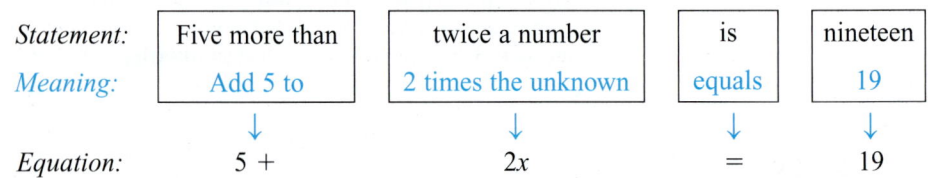

Statement:	Five more than	twice a number	is	nineteen
Meaning:	Add 5 to	2 times the unknown	equals	19
	↓	↓	↓	↓
Equation:	$5 +$	$2x$	$=$	19

The equation is $5 + 2x = 19$.

Step 4:   **Solve** the equation.

$$5 + 2x = 19$$
$$5 - 5 + 2x = 19 - 5 \qquad \text{Subtract 5 from each side.}$$
$$2x = 14 \qquad \text{Combine like terms.}$$
$$x = 7 \qquad \text{Divide each side by 2.}$$

Step 5:   **Check** the answer. Does the answer make sense? Five more than twice seven is $5 + 2(7) = 19$. The answer is correct. The number is 7. ∎

You Try 6

Write an equation and solve.
Nine more than twice a number is seventeen.

Sometimes, dealing with subtraction in a word problem can be confusing. So, let's look at an arithmetic problem first.

Example 8

In-Class Example 8
What is nine less than
seventeen?
answer: 8

What is four less than ten?

Solution

To solve this problem, do we subtract $10 - 4$ or $4 - 10$? "Four less than ten" is written as $10 - 4$, and $10 - 4 = 6$. Six is four less than ten. The 4 is *subtracted from* the 10. Keep this problem in mind as you read the next example. ■

You Try 7

Write an equation and solve.
A number decreased by eight is twelve.

Example 9

In-Class Example 9
Seven less than four times a
number is the same as the number increased by twenty.
Find the number.
answer: $4x - 7 = x + 20$; 9

Write the following statement as an equation, and find the number.

Eleven less than three times a number is the same as the number increased by nine. Find the number.

Solution

Step 1:   **Read** the problem carefully. We must find an unknown number.

Step 2:   **Choose a variable** to represent the unknown.

Let x = the number

Step 3:   **Translate** the information that appears in English into an algebraic equation by rereading the problem slowly and "in parts."

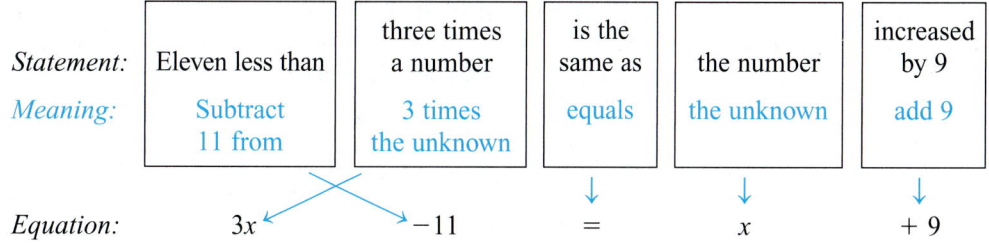

The equation is $3x - 11 = x + 9$.

Step 4:   **Solve** the equation.

$$3x - 11 = x + 9$$
$$3x - x - 11 = x - x + 9 \qquad \text{Subtract } x \text{ from each side.}$$
$$2x - 11 = 9 \qquad \text{Combine like terms.}$$
$$2x - 11 + 11 = 9 + 11 \qquad \text{Add 11 to each side.}$$
$$2x = 20 \qquad \text{Combine like terms.}$$
$$x = 10 \qquad \text{Divide each side by 2.}$$

Step 5:   **Check** the answer. Does the answer make sense? Eleven less than three times 10 is $3(10) - 11 = 19$. The number increased by nine is $10 + 9 = 19$. The answer is correct. The number is 10. ■

You Try 8

Write the following statement as an equation, and find the number.

Three less than five times a number is the same as the number increased by thirteen.

Using Technology

We can use a graphing calculator to solve a linear equation in one variable. First enter the left side of the equation in Y_1 and the right side of the equation in Y_2. Then graph the equations. The x-coordinate of the point of intersection is the solution to the equation.

We will solve $3x + 5 = 6$ algebraically and by using a graphing calculator, and then compare the results. First, use algebra to solve $3x + 5 = 6$. You should get $\left\{\dfrac{1}{3}\right\}$.

Next, use a graphing calculator to solve $3x + 5 = 6$.

1) Enter $3x + 5$ in Y_1 by pressing the [Y=] key and typing $3x + 5$ to the right of $\backslash Y_1 =$. Then press [ENTER].

2) Enter 6 in Y_2 by pressing the [Y=] key and typing 6 to the right of $\backslash Y_2 =$. Then press [ENTER].

3) Press [ZOOM] and select 6:ZStandard to graph the equations.

4) To find the intersection point, press [2nd] [TRACE] and select 5:intersect. Press [ENTER] three times. The x-coordinate of the intersection point is shown on the left side of the screen shown to the right, and is stored in the variable x.

5) Return to the home screen by pressing [2nd] [MODE]. Enter [X,T,Θ,n] [ENTER] to display the solution. Since the result in this case is a decimal value, we can convert it to a fraction by entering [X,T,Θ,n] [MATH] [ENTER] to convert the result to a fraction as shown on the screen to the right.

The calculator then gives us a solution set of $\left\{\dfrac{1}{3}\right\}$.

Solve each equation algebraically, and then verify your answer using a graphing calculator.

1) $2x - 3 = 5$ 2) $2x - 6 = 5$ 3) $3x - 6 = 7$

4) $4x - 8 = 3$ 5) $4x - 5 = x + 2$ 6) $3x + 7 = 2 - 5x$

Answers to You Try Exercises

1) a) $\{13\}$ b) $\{-4\}$ c) $\{42\}$ 2) $\{20\}$ 3) $\left\{\dfrac{15}{2}\right\}$ 4) $\{5\}$ 5) a) {all real numbers}

b) $\varnothing$ 6) $9 + 2x = 17; 4$ 7) $x - 8 = 12; 20$ 8) $5x - 3 = x + 13; 4$

Answers to Technology Exercises

1) $\{4\}$ 2) $\left\{\dfrac{11}{2}\right\}$ 3) $\left\{\dfrac{13}{3}\right\}$ 4) $\left\{\dfrac{11}{4}\right\}$ 5) $\left\{\dfrac{7}{3}\right\}$ 6) $\left\{-\dfrac{5}{8}\right\}$

2.1 Exercises

*Additional answers can be found in the Answers to Exercises appendix.

Objective 1: Define a Linear Equation in One Variable

Identify each as an expression or an equation.

1) $7t - 2 = 11$ equation

2) $\frac{3}{4}k + 5(k - 6) = 2$ equation

3) $8 - 10p + 4p + 5$ expression

4) $9(2z - 7) + 3z$ expression

5) Can we solve $3(c + 2) + 5(2c - 5)$? Why or why not?
No, it is an expression.

6) Can we solve $3(c + 2) + 5(2c - 5) = -6$? Why or why not?   Yes, it is an equation.

7) Which of the following are linear equations in one variable?   b, d

   a) $y^2 + 8y + 15 = 0$ b) $\frac{1}{2}w - 5(3w + 1) = 6$

   c) $8m - 7 + 2m + 1$ d) $0.3z + 0.2 = 1.5$

8) Which of the following are linear equations in one variable?   a, b

   a) $-7p = 0$

   b) $-2 = 5g - 4 + g + 10 + 3g - 1$

   c) $9x + 4y = 3$

   d) $10 - 6(4n - 1) + 7$

Determine if the given value is a solution to the equation.

9) $-8p = 12; p = -\frac{3}{2}$ yes

10) $2d + 1 = 13; d = -6$ no

11) $2(t - 5) + 7 = 3(2t - 9) - 2; t = 4$ no

12) $5 - 3(2k + 1) = 4k - 3; k = \frac{1}{2}$ yes

Objective 2: Use the Properties of Equality

Solve and check each equation.

13) $r - 6 = 11$ {17}

14) $c + 2 = -5$ {−7}

15) $-16 = k - 12$ {−4}

16) $8 = t + 1$ {7}

17) $a + \frac{5}{8} = \frac{1}{2}$ $\left\{-\frac{1}{8}\right\}$

18) $w - \frac{3}{4} = -\frac{1}{6}$ $\left\{\frac{7}{12}\right\}$

19) $3y = 30$ {10}

20) $-56 = -7v$ {8}

21) $-6 = \frac{k}{8}$ {−48}

22) $30 = -\frac{x}{2}$ {−60}

23) $\frac{2}{3}g = -10$ {−15}

24) $\frac{7}{4}r = 42$ {24}

25) $-\frac{5}{3}d = -30$ {18}

26) $-\frac{5}{6} = -\frac{4}{9}x$ $\left\{\frac{15}{8}\right\}$

27) $0.5q = 6$ {12}

28) $0.3t = 3$ {10}

29) $3x - 7 = 17$ {8}

30) $5g + 19 = 4$ {−3}

31) $8d - 15 = -15$ {0}

32) $4 = 7j - 8$ $\left\{\frac{12}{7}\right\}$

33) $\frac{4}{9}w - 11 = 1$ {27}

34) $\frac{5}{3}a + 6 = 41$ {21}

35) $\frac{10}{7}m + 3 = 1$ $\left\{-\frac{7}{5}\right\}$

36) $\frac{9}{10}x - 4 = 11$ $\left\{\frac{50}{3}\right\}$

37) $5 - 0.4p = 2.6$ {6}

38) $1.8 = 1.2n - 7.8$ {8}

Mixed Exercises: Objectives 3 and 5

Solve and check each equation.

39) $10v + 9 - 2v + 16 = 1$ {−3}

40) $-8g - 7 + 6g + 1 = 20$ {−13}

41) $5 = -12p + 7 + 4p - 12$ $\left\{-\frac{5}{4}\right\}$

42) $12 = 9y + 11 - 3y - 7$ $\left\{\frac{4}{3}\right\}$

43) $-12 = 7(2a - 3) - (8a - 9)$ {0}

44) $20 = 5r - 3 + 2(9 - 3r)$ {−5}

45) $2y + 7 = 5y - 2$ {3}

46) $8n - 21 = 3n - 1$ {4}

47) $6 - 7p = 2p + 33$ {−3}

48) $z + 19 = 5 - z$ {−7}

49) $-8x + 6 - 2x + 11 = 3 + 3x - 7x$ $\left\{\frac{7}{3}\right\}$

50) $10 - 13a + 2a - 16 = -5 + 7a + 11$ $\left\{-\frac{2}{3}\right\}$

51) $4(2t + 5) - 7 = 5(t + 5)$ {4}

52) $3(2m + 10) = 6(m + 4) - 8m$ $\left\{-\frac{3}{4}\right\}$

53) $-9r + 4r - 11 + 2 = 3r + 7 - 8r + 9$ ∅

54) $3(4b - 7) + 8 = 6(2b + 5)$ ∅

55) $j - 15j + 8 = -3(4j - 3) - 2j - 1$ {all real numbers}

56) $n - 16 + 10n + 4 = 2(7n - 6) - 3n$ {all real numbers}

57) $8(3t + 4) = 10t - 3 + 7(2t + 5)$ {all real numbers}

58) $2(9z - 1) + 7 = 10z - 14 + 8z + 2$ ∅

59) $8 - 7(2 - 3w) - 9w = 4(5w - 1) - 3w - 2$ {0}

60) $4m - (6m + 5) + 2 = 8m + 3(4 - 3m)$ {−15}

61) $7y + 2(1 - 4y) = 8y - 5(y + 4)$ $\left\{\frac{11}{2}\right\}$

Objective 4: Solve Equations Containing Fractions or Decimals

62) How can you eliminate the fractions from the equation $\frac{1}{6}x + \frac{5}{4} = \frac{1}{2}x - \frac{5}{12}$?   *Multiply the equation by 12.*

Solve each equation by first clearing fractions or decimals.

63) $\frac{1}{6}x + \frac{5}{4} = \frac{1}{2}x - \frac{5}{12}$ {5}   64) $\frac{3}{4}n + \frac{1}{2} = \frac{1}{2}n + \frac{1}{4}$ {−1}

65) $\frac{2}{3}d - 1 = \frac{1}{5}d + \frac{2}{5}$ {3}   66) $\frac{1}{5}c + \frac{2}{7} = 2 - \frac{1}{7}c$ {5}

67) $\frac{m}{3} + \frac{1}{2} = \frac{2m}{3} + 3$ $\left\{-\frac{15}{2}\right\}$ 68) $\frac{a}{8} - 1 = \frac{a}{3} - \frac{7}{12}$ {−2}

69) $\frac{1}{3} + \frac{1}{9}(k + 5) - \frac{k}{4} = 2$ {−8}

70) $\frac{1}{2} = \frac{2}{9}(3x - 2) - \frac{x}{9} - \frac{x}{6}$ $\left\{\frac{17}{7}\right\}$

71) $0.05(t + 8) - 0.01t = 0.6$ {5}

72) $0.2(y - 3) + 0.05(y - 10) = -0.1$ {4}

73) $0.1x + 0.15(8 - x) = 0.125(8)$ {4}

74) $0.2(12) + 0.08z = 0.12(z + 12)$ {24}

75) $0.04s + 0.03(s + 200) = 27$ {300}

76) $0.06x + 0.1(x - 300) = 98$ {800}

Objective 6: Use the Five Steps for Solving Applied Problems

Write each statement as an equation, and find the number.

77) Four more than a number is fifteen. $x + 4 = 15; 11$

78) Thirteen more than a number is eight. $x + 13 = 8; -5$

79) Seven less than a number is twenty-two. $x - 7 = 22; 29$

80) Nine less than a number is eleven. $x - 9 = 11; 20$

81) Twice a number is -16. $2x = -16; -8$

82) The product of six and a number is fifty-four. $6x = 54; 9$

83) Seven more than twice a number is thirty-five. $2x + 7 = 35; 14$

84) Five more than twice a number is fifty-three. $2x + 5 = 53; 24$

85) Three times a number decreased by eight is forty. $3x - 8 = 40; 16$

86) Twice a number decreased by seven is -13. $2x - 7 = -13; -3$

87) Half of a number increased by ten is three. $\frac{1}{2}x + 10 = 3; -14$

88) One-third of a number increased by four is one. $\frac{1}{3}x + 4 = 1; -9$

89) Three less than twice a number is the same as the number increased by eight. $2x - 3 = x + 8; 11$

90) Twelve less than five times a number is the same as the number increased by sixteen. $5x - 12 = x + 16; 7$

91) Ten more than one-third of a number is the same as the number decreased by two. $\frac{1}{3}x + 10 = x - 2; 18$

92) A number decreased by nine is the same as seven more than half the number. $x - 9 = \frac{1}{2}x + 7; 32$

93) Twice the sum of a number and five is sixteen. $2(x + 5) = 16; 3$

94) Twice the sum of a number and -8 is four. $2[x + (-8)] = 4; 10$

95) Three times a number is fifteen more than half the number. $3x = 15 + \frac{1}{2}x; 6$

96) A number increased by fourteen is nine less than twice the number. $x + 14 = 2x - 9; 23$

97) A number decreased by six is five more than twice the number. $x - 6 = 5 + 2x; -11$

98) A number divided by four is nine less than the number. $\frac{x}{4} = x - 9; 12$

Section 2.2 Applications of Linear Equations

Objectives

1. Solve Problems Involving General Quantities
2. Solve Problems Involving Lengths
3. Solve Consecutive Integer Problems
4. Solve Problems Involving Percent Change
5. Solve Problems Involving Simple Interest

In the previous section, you were introduced to solving applied problems. In this section, we will build on these skills so that we can solve other types of problems.

1. Solve Problems Involving General Quantities

Example 1

Mrs. Ramirez has 26 students in her third-grade class. There are four more boys than girls. How many boys and girls are in her class?

In-Class Example 1
Sarah has 112 marbles in a jar.
Some are red and some are
blue. If there are 16 more red
than blue, how many of the marbles in the jar are red and how
many are blue?
**answer: There are 64 red
marbles and 48 blue
marbles in the jar.**

Solution

Step 1:   **Read** the problem carefully, and identify what we are being asked to find.

We must find the number of boys and number of girls in the class.

Step 2:   **Choose a variable** to represent an unknown, and define the other unknown in terms of this variable.

In the statement "There are four more boys than girls," the number of boys is expressed *in terms of* the number of girls. Therefore, let

$$x = \text{the number girls}$$

Define the other unknown (the number of boys) in terms of x. Since there are four *more* boys than girls,

$$\text{The number of girls} + 4 = \text{the number of boys}$$
$$x + 4 = \text{the number of boys}$$

Step 3:   **Translate** the information that appears in English into an algebraic equation. One approach is to restate the problem in your own words.

We can think of the situation in this problem as:

The number of girls plus the number of boys is 26.

Let's write this as an equation.

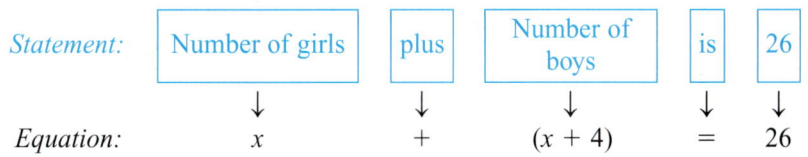

The equation is $x + (x + 4) = 26$.

Step 4:   **Solve** the equation.

$$x + (x + 4) = 26$$
$$2x + 4 = 26$$
$$2x + 4 - 4 = 26 - 4 \qquad \text{Subtract 4 from each side.}$$
$$2x = 22 \qquad \text{Combine like terms.}$$
$$\frac{2x}{2} = \frac{22}{2} \qquad \text{Divide each side by 2.}$$
$$x = 11 \qquad \text{Simplify.}$$

Step 5:   **Check** the answer and **interpret** the meaning as it relates to the problem.

Since x represents the number of girls, there are 11 girls in the class.

The expression $x + 4$ represents the number of boys, so there are $x + 4 = 11 + 4 = 15$ boys.

The answer makes sense because the total number of children in this third-grade class is $11 + 15 = 26$. ■

 You Try 1

The record low temperature for the month of January in Anchorage, Alaska, was (in °F) 84° less than January's record high temperature. The sum of the record low and record high is 16°. Find the lowest and highest temperatures ever recorded in Anchorage, Alaska, in January. (www.weather.com)

2. Solve Problems Involving Lengths

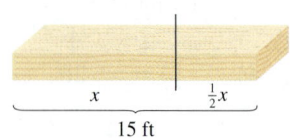

Example 2

In-Class Example 2
An electrician has a 60-ft. wire. He needs to cut the wire so that one piece is one-third as long as the other. What will be the length of each piece?
answer: One wire is 45 ft long and the other is 15 ft long.

A carpenter has a board that is 15 ft long. He needs to cut it into two pieces so that one piece is half as long as the other. What will be the length of each piece?

Solution

Step 1: **Read** the problem carefully, and identify what we are being asked to find.

We must find the length of each of two pieces of a board.

A picture will be very helpful in this problem.

Step 2: **Choose a variable** to represent an unknown, and define the other unknown in terms of this variable.

One piece of the board must be half the length of the other piece. Therefore, let

$$x = \text{the length of one piece}$$

Define the other unknown in terms of x.

$$\frac{1}{2}x = \text{the length of the second piece}$$

Step 3: **Translate** the information that appears in English into an algebraic equation. Let's label the picture with the expressions representing the unknowns and then restate the problem in our own words.

From the picture we can see that the

length of one piece plus the length of the second piece equals 15 ft.

Let's write this as an equation.

Statement:	Length of one piece	plus	Length of second piece	equals	15 ft
	↓	↓	↓	↓	↓
Equation:	x	$+$	$\frac{1}{2}x$	$=$	15

The equation is $x + \frac{1}{2}x = 15$.

Step 4: **Solve** the equation.

$$x + \frac{1}{2}x = 15$$

$$\frac{3}{2}x = 15 \qquad \text{Add like terms.}$$

$$\frac{2}{3} \cdot \frac{3}{2}x = \frac{2}{3} \cdot 15 \qquad \text{Multiply by the reciprocal of } \frac{3}{2}.$$

$$x = 10$$

Step 5:   **Check** the answer and **interpret** the solution as it relates to the problem.

One piece of board is 10 ft long.

The expression $\frac{1}{2}x$ represents the length of the other piece of board, so the length of the other piece is $\frac{1}{2}x = \frac{1}{2}(10) = 5$ ft.

The answer makes sense because the length of the original board was
10 ft + 5 ft = 15 ft. ■

You Try 2

A 24-ft chain must be cut into two pieces so that one piece is twice as long as the other. Find the length of each piece of chain.

3. Solve Consecutive Integer Problems

What *are* consecutive integers? "Consecutive" means one after the other, in order. In this section, we will look at consecutive integers, consecutive even integers, and consecutive odd integers.

Consecutive integers differ by 1. For example, look at the consecutive integers 5, 6, 7, and 8. If $x = 5$, then $x + 1 = 6$, $x + 2 = 7$, and $x + 3 = 8$. Therefore, to define the unknowns for consecutive integers, let

$$x = \text{first integer}$$
$$x + 1 = \text{second integer}$$
$$x + 2 = \text{third integer}$$
$$x + 3 = \text{fourth integer}$$

and so on.

Example 3

The sum of three consecutive integers is 126. Find the integers.

In-Class Example 3
The sum of three consecutive integers is 183. Find the numbers.
answer: The integers are 60, 61, and 62.

Solution

Step 1:   **Read** the problem carefully, and identify what we are being asked to find.

We must find three consecutive integers with a sum of 126.

Step 2:   **Choose a variable** to represent an unknown, and define the other unknowns in terms of this variable.

There are three unknowns. We will let x represent the first consecutive integer and then define the other unknowns in terms of x.

$$x = \text{the first integer}$$

Define the other unknowns in terms of x.

$$x + 1 = \text{the second integer}$$
$$x + 2 = \text{the third integer}$$

Step 3: **Translate** the information that appears in English into an algebraic equation. What does the original statement mean?

"The sum of three consecutive integers is 126" means that when the three numbers are *added* together the sum is 126.

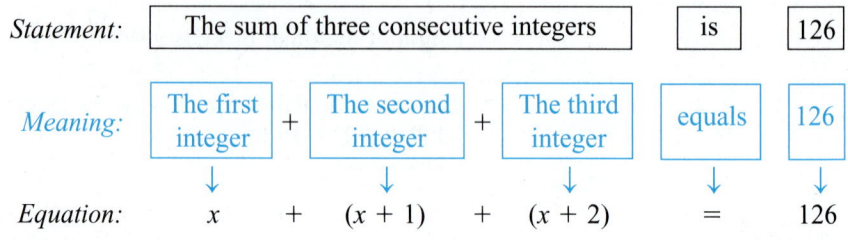

Statement: | The sum of three consecutive integers | | | is | 126

Meaning: | The first integer | + | The second integer | + | The third integer | equals | 126

Equation: x + $(x + 1)$ + $(x + 2)$ =   126

The equation is $x + (x + 1) + (x + 2) = 126$.

Step 4: **Solve** the equation.

$$x + (x + 1) + (x + 2) = 126$$
$$3x + 3 = 126$$
$$3x + 3 - 3 = 126 - 3 \qquad \text{Subtract 3 from each side.}$$
$$3x = 123 \qquad \text{Combine like terms.}$$
$$\frac{3x}{3} = \frac{123}{3} \qquad \text{Divide each side by 3.}$$
$$x = 41 \qquad \text{Simplify.}$$

Step 5: **Check** the answer and **interpret** the solution as it relates to the problem.

The first integer is 41. The second integer is 42 since $x + 1 = 41 + 1 = 42$, and the third integer is 43 since $x + 2 = 41 + 2 = 43$.

The answer makes sense because their sum is $41 + 42 + 43 = 126$.   ■

You Try 3

The sum of three consecutive integers is 177. Find the integers.

Next, let's look at **consecutive even integers,** which are even numbers that differ by two such as -10, -8, -6, and -4. If x is the first even integer, we have

$$\begin{array}{cccc} -10 & -8 & -6 & -4 \\ x & x+2 & x+4 & x+6 \end{array}$$

Therefore, to define the unknowns for consecutive even integers, let

$$x = \text{the first even integer}$$
$$x + 2 = \text{the second even integer}$$
$$x + 4 = \text{the third even integer}$$
$$x + 6 = \text{the fourth even integer}$$

and so on.

Will the expressions for **consecutive odd numbers** be any different? No! When we count by consecutive odds, we are still counting by twos. Look at the numbers 9, 11, 13, and 15 for example. If x is the first odd integer, we have

$$\begin{array}{cccc} 9 & 11 & 13 & 15 \\ x & x+2 & x+4 & x+6 \end{array}$$

To define the unknowns for consecutive odd integers, let

$$x = \text{the first odd integer}$$
$$x + 2 = \text{the second odd integer}$$
$$x + 4 = \text{the third odd integer}$$
$$x + 6 = \text{the fourth odd integer.}$$

Example 4

In-Class Example 4
The sum of two consecutive odd integers is 16 more than four times the smaller integer. Find the integers.
answer: The integers are −7 and −5.

The sum of two consecutive odd integers is 47 more than three times the larger integer. Find the integers.

Solution

Step 1:   **Read** the problem carefully, and identify what we are being asked to find.

We must find two consecutive odd integers.

Step 2:   **Choose a variable** to represent an unknown, and define the other unknowns in terms of this variable.

There are two unknowns. We will let x represent the first consecutive odd integer and then define the other unknown in terms of x.

$$x = \text{the first odd integer}$$
$$x + 2 = \text{the second odd integer}$$

Step 3:   **Translate** the information that appears in English into an algebraic equation. Read the problem slowly and carefully, breaking it into small parts.

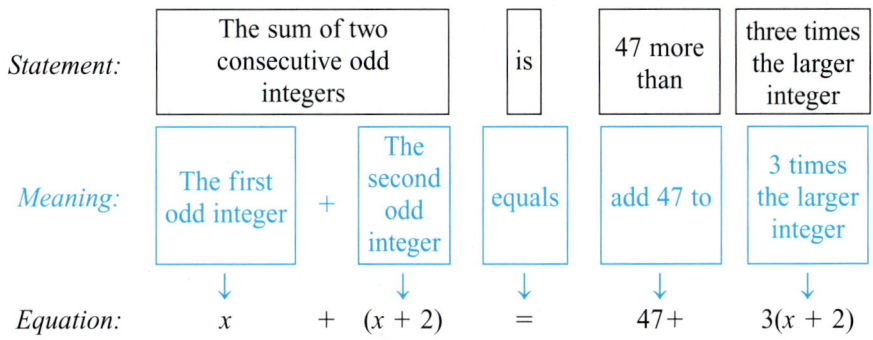

Statement:	The sum of two consecutive odd integers		is	47 more than	three times the larger integer
Meaning:	The first odd integer	+ The second odd integer	equals	add 47 to	3 times the larger integer
	↓	↓	↓	↓	↓
Equation:	x	$+ \ (x + 2)$	$=$	$47+$	$3(x + 2)$

The equation is $x + (x + 2) = 47 + 3(x + 2)$.

Step 4:   **Solve** the equation.

$$x + (x + 2) = 47 + 3(x + 2)$$

$2x + 2 = 47 + 3x + 6$	Combine like terms; distribute.
$2x + 2 = 3x + 53$	Combine like terms.
$2x + 2 - 2 = 3x + 53 - 2$	Subtract 2 from each side.
$2x = 3x + 51$	Combine like terms.
$2x - 3x = 3x - 3x + 51$	Subtract $3x$ from each side.
$-x = 51$	Combine like terms.
$x = -51$	Divide each side by −1.

Step 5:   **Check** the answer and **interpret** the solution as it relates to the problem.

The first odd integer is -51. The second integer is -49 since $x + 2 = -51 + 2 = -49$.

Check these numbers in the original statement of the problem. The sum of -51 and -49 is -100. Then, 47 more than three times the larger integer is $47 + 3(-49) = 47 + (-147) = -100$. The numbers are -51 and -49. ∎

You Try 4

Twice the sum of three consecutive even integers is 18 more than five times the largest number. Find the integers.

4. Solve Problems Involving Percent Change

Percents pop up everywhere. "Earn 4% simple interest on your savings account." "The unemployment rate increased 1.2% this year." We've seen these statements online, or we have heard about them on television. In this section, we will introduce applications involving percents.

Here's another type of percent problem we might see in a store: "Everything in the store is marked down 30%." Before tackling an algebraic percent problem, let's look at an arithmetic problem. Relating an algebra problem to an arithmetic problem can make it easier to solve an application that requires the use of algebra.

Example 5

In-Class Example 5
Shoes that normally sell for $80.00 are marked down 20%. What is the sale price?
answer: $64.00

Jeans that normally sell for $28.00 are marked down 30%. What is the sale price?

Solution

Concentrate on the *procedure* used to obtain the answer. This is the same procedure we will use to solve algebra problems with percent increase and percent decrease.

$$\text{Sale price} = \text{Original price} - \text{Amount of discount}$$

How much is the discount? It is 30% of $28.00.
Change the percent to a decimal. The amount of the discount is calculated by multiplying:

$$\text{Amount of discount} = (\text{Rate of discount})(\text{Original price})$$
$$\text{Amount of discount} = (0.30) \cdot (\$28.00) = \$8.40$$

$$
\begin{aligned}
\text{Sale price} &= \text{Original price} - \text{Amount of discount}\\
&= \$28.00 - (0.30)(\$28.00)\\
&= \$28.00 - \$8.40\\
&= \$19.60
\end{aligned}
$$

The sale price is $19.60.

You Try 5

A dress shirt that normally sells for **$39.00** is marked down 25%. What is the sale price?

Next, let's solve an algebra problem involving a markdown or percent decrease.

Example 6

In-Class Example 6
The sale price of a waffle iron is $15.60 after a 35% discount. What was the original price of the waffle iron?
answer: $24.00

The sale price of a Lil Wayne CD is $14.80 after a 20% discount. What was the original price of the CD?

Solution

Step 1: **Read** the problem carefully, and identify what we are being asked to find.

We must find the original price of the CD.

Step 2: **Choose a variable** to represent the unknown.

x = the original price of the CD

Step 3: **Translate** the information that appears in English into an algebraic equation. One way to figure out how to write an algebraic equation is to relate this problem to the arithmetic problem in Example 5. To find the sale price of the jeans in Example 5 we found that

$$\text{Sale price} = \text{Original price} - \text{Amount of discount}$$

where we found the amount of the discount by multiplying the rate of the discount by the original price. We will write an algebraic equation using the same procedure.

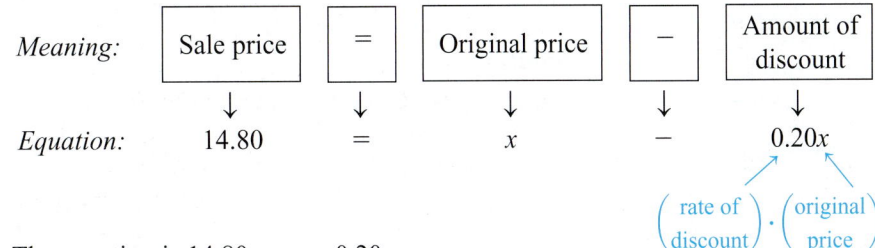

The equation is $14.80 = x - 0.20x$.

Step 4: **Solve** the equation.

$$14.80 = x - 0.20x$$
$$14.80 = 0.80x \qquad \text{Combine like terms.}$$
$$\frac{14.80}{0.80} = \frac{0.80x}{0.80} \qquad \text{Divide each side by 0.80.}$$
$$x = 18.5 \qquad \text{Simplify.}$$

Step 5: **Check** the answer and **interpret** the solution as it relates to the problem.

The original price of the CD was $18.50.

The answer makes sense because the amount of the discount is $(0.20)(\$18.50) = \$3.70$ which makes the sale price $\$18.50 - \$3.70 = \$14.80$.

You Try 6

A video game is on sale for $35.00 after a 30% discount. What was the original price of the video game?

5. Solve Problems Involving Simple Interest

When customers invest their money in bank accounts, their accounts earn interest. Why? Paying interest is a way for financial institutions to get people to deposit money.

There are different ways to calculate the amount of interest earned from an investment. In this section, we will discuss *simple interest*. **Simple interest** calculations are based on the initial amount of money deposited in an account. This initial amount of money is known as the **principal**.

The formula used to calculate simple interest is $I = PRT$, where

$$I = \text{interest earned (simple)}$$
$$P = \text{principal (initial amount invested)}$$
$$R = \text{annual interest rate (expressed as a decimal)}$$
$$T = \text{amount of time the money is invested (in years)}$$

We will begin with two arithmetic problems. The procedures used will help you understand more clearly how we arrive at the algebraic equation in Example 9.

Example 7

In-Class Example 7
If $375 is invested for 2 years in an account earning 7.5% simple interest, how much interest will be earned?
answer: $56.25

If $500 is invested for 1 year in an account earning 6% simple interest, how much interest will be earned?

Solution

Use $I = PRT$ to find I, the interest earned.

$$P = \$500, R = 0.06, T = 1$$
$$I = PRT$$
$$I = (500)(0.06)(1)$$
$$I = 30$$

The interest earned will be $30.

You Try 7

If $3500 is invested for 2 years in an account earning 4.5% simple interest, how much interest will be earned?

Example 8

In-Class Example 8
Dick invests $700 in an account earning 6% interest and $3000 in an account earning 4% interest. How much interest will Dick have earned after 1 year?
answer: $162.00

Tom invests $2000 in an account earning 7% interest and $9000 in an account earning 5% interest. After 1 year, how much interest will he have earned?

Solution

Tom will earn interest from two accounts. Therefore,

Total interest earned = Interest from 7% account + Interest from 5% account

$$
\begin{array}{ccc}
& P \quad R \quad T & P \quad R \quad T \\
\text{Total interest earned} = & (2000)(0.07)(1) \quad + & (9000)(0.05)(1) \\
\text{Total interest earned} = & 140 \qquad + & 450 \\
& = \$590 &
\end{array}
$$

Tom will earn a total of $590 in interest from the two accounts.

You Try 8

Donna invests $1500 in an account earning 4% interest and $4000 in an account earning 6.7% interest. After 1 year, how much interest will she have earned?

Note

When money is invested for 1 year, $T = 1$. Therefore, the formula $I = PRT$ can be written as $I = PR$.

This idea of earning interest from different accounts is one we will use in Example 9.

Example 9

Last year, Neema Reddy had $10,000 to invest. She invested some of it in a savings account that paid 3% simple interest, and she invested the rest in a certificate of deposit that paid 5% simple interest. In 1 year, she earned a total of $380 in interest. How much did Neema invest in each account?

In-Class Example 9
Olivia had $13,000 to invest. She
invested some of it in a savings
account that paid 4% simple
interest, and she invested the
rest in a certificate of deposit
that paid 6% simple interest. In
1 year, she earned a total of
$640 in interest. How much did
Olivia invest in each account?
**answer: Olivia invested
$7000 in the savings account
earning 4% interest and
$6000 in the certificate of
deposit earning 6% interest.**

Solution

Step 1:   **Read** the problem carefully, and identify what we are being asked to find.

We must find the amounts Neema invested in the 3% account and in the 5% account.

Step 2:   **Choose a variable** to represent an unknown, and define the other unknown in terms of this variable.

Let x = Amount Neema invested in the 3% account.

How do we write an expression, in terms of x, for the amount invested in the 5% account?

Total invested   Amount invested in 3% account

$\downarrow$ $\downarrow$

10,000 $-$ x = amount invested in the 5% account

We define the unknowns as

x = amount Neema invested in the 3% account
$10,000 - x$ = amount Neema invested in the 5% account

Step 3:   **Translate** the information that appears in English into an algebraic equation. Use the "English equation" we used in Example 8. Since $T = 1$, we can compute the interest using $I = PR$.

Total interest earned = Interest from 3% account + Interest from 5% account

$\qquad\qquad\qquad\qquad\qquad\quad P \quad R \qquad\qquad\qquad\qquad P \qquad\quad R$

$\qquad 380 \qquad\quad = \qquad (x)(0.03) \qquad + \qquad (10,000 - x)(0.05)$

The equation is $380 = 0.03x + 0.05(10,000 - x)$.

We can also get the equation by organizing the information in a table:

	Amount Invested (in dollars) P	Interest Rate R	Interest Earned After 1 Year I
3% account	x	0.03	$0.03x$
5% account	$10,000 - x$	0.05	$0.05(10,000 - x)$

Total interest earned = Interest from 3% account + Interest from 5% account
$\qquad 380 \qquad\quad = \qquad\quad 0.03x \qquad + \quad 0.05(10,000 - x)$

The equation is $380 = 0.03x + 0.05(10,000 - x)$.

Either way of organizing the information will lead us to the correct equation.

Step 4:   **Solve** the equation. Begin by multiplying both sides of the equation by 100 to eliminate the decimals.

$$380 = 0.03x + 0.05(10,000 - x)$$
$$100(380) = 100[0.03x + 0.05(10,000 - x)]$$
$$38,000 = 3x + 5(10,000 - x) \qquad \text{Multiply by 100.}$$
$$38,000 = 3x + 50,000 - 5x \qquad \text{Distribute.}$$
$$38,000 = -2x + 50,000 \qquad \text{Combine like terms.}$$
$$-12,000 = -2x \qquad\qquad\qquad \text{Subtract 50,000.}$$
$$6000 = x \qquad\qquad\qquad\quad \text{Divide by } -2.$$

Step 5:   **Check** the answer and **interpret** the solution as it relates to the problem.

Neema invested $6000 at 3% interest. The amount invested at 5% is $10,000 − x or $10,000 − $6000 = $4000.

Check:

Total interest earned = Interest from 3% account + Interest from 5% account
380 = 6000(0.03) + 4000(0.05)
= 180 + 200
= 380

You Try 9

Christine received an $8000 bonus from work. She invested part of it at 6% simple interest and the rest at 4% simple interest. Christine earned a total of $420 in interest after 1 year. How much did she deposit in each account?

Answers to You Try Exercises

1) record low: −34°F, record high: 50°F   2) 8 ft and 16 ft   3) 58, 59, 60
4) 26, 28, 30   5) $29.25   6) $50.00   7) $315   8) $328
9) $5000 at 6% and $3000 at 4%

2.2 Exercises

*Additional answers can be found in the Answers to Exercises appendix.

1) On a baseball team, there are 5 more pitchers than catchers. If there are c catchers on the team, write an expression for the number of pitchers. $c + 5$

2) On Wednesday, the Snack Shack sold 23 more hamburgers than hot dogs. Write an expression for the number of hamburgers sold if h hot dogs were sold. $h + 23$

3) There were 31 fewer people on a flight from Chicago to New York than from Chicago to Los Angeles. Write an expression for the number of people on the flight to New York if there were p people on the flight to L.A. $p - 31$

4) The test average in Mr. Muscari's second-period class was 3.8 points lower than in his first-period class. If the average test score in the first-period class was a, write an expression for the test average in Mr. Muscari's second-period class. $a - 3.8$

5) A survey of adults aged 20–29 revealed that, of those who exercise regularly, three times as many men run as women. If w women run on a regular basis, write an expression for the number of male runners. $3w$

6) At Roundtree Elementary School, s students walk to school. One-third of that number ride their bikes. Write an expression for the number of students who ride their bikes to class. $\frac{1}{3}s$

7) An electrician cuts a 14-ft wire into two pieces. If one is x ft long, how long is the other piece?

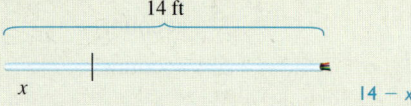

8) Jorie drives along a 142-mi stretch of highway from Oregon to California. If she drove m mi in Oregon, how far did she drive in California? $142 - m$ mi

Objective 1: Solve Problems Involving General Quantities

Solve using the five "Steps for Solving Applied Problems." See Example 1.

9) A 12-oz serving of Pepsi has 6.5 more teaspoons of sugar than a 12-oz serving of Gatorade. Together they contain 13.1 teaspoons of sugar. How much sugar is in each 12-oz drink? (www.dentalgentlecare.com)
Pepsi = 9.8 tsp, Gatorade = 3.3 tsp

10) Two of the smallest countries in the world are the Marshall Islands and Liechtenstein. The Marshall Islands covers 8 mi² more than Liechtenstein. Find the area of each country if together they encompass 132 mi². (www.infoplease.com)   MI = 70 mi², Lich = 62 mi²

11) In the 2004 Summer Olympics in Athens, Thailand won half as many medals as Greece. If they won a total of 24 medals, how many were won by each country? (www.olympics.org)   Greece: 16, Thailand: 8

12) Latisha's Golden Retriever weighs twice as much as Janessa's Border Collie. Find the weight of each dog if they weigh 96 lb all together.   Golden = 64 lb, BC = 32 lb

13) The Columbia River is 70 miles shorter than the Ohio River. Determine the length of each river if together they span 2550 miles. (ga.water.usgs.gov)
Columbia: 1240 mi, Ohio: 1310 mi

14) In the 2002 Alabama gubernatorial election, Don Siegelman had 3120 fewer votes than Bob Riley. If they received a total of 1,341,330 votes, how many people voted for each man? (www.sos.state.al.us)
Siegelman: 669,105, Riley: 672,225

Objective 2: Solve Problems Involving Lengths

Solve using the five "Steps for Solving Applied Problems." See Example 2.

15) A plumber has a 36-in.-long pipe. He must cut it into two pieces so that one piece is 14 in. longer than the other. How long is each piece?   11 in. and 25 in.

16) A builder has to install some decorative trim on the outside of a house. The piece she has is 75 in. long, but she needs to cut it so that one piece is 27 in. shorter than the other. Find the length of each piece.   51 in. and 24 in.

17) Calida's mom found an 18-ft-long rope in the garage. She will cut it into two pieces so that one piece can be used for a long jump rope and the other for a short one. If the long rope is to be twice as long as the short one, find the length of each jump rope.   12 ft and 6 ft

18) A 55-ft-long drainage pipe must be cut into two pieces before installation. One piece is two-thirds as long as the other. Find the length of each piece.   33 ft and 22 ft

Objective 3: Solve Consecutive Integer Problems

Solve using the five "Steps for Solving Applied Problems." See Examples 3 and 4.

19) The sum of three consecutive integers is 195. Find the integers.   64, 65, 66

20) The sum of two consecutive integers is 77. Find the integers.   38, 39

21) Find two consecutive even integers such that twice the smaller is 10 more than the larger.   12, 14

22) Find three consecutive odd integers such that four times the smallest is 56 more than the sum of the other two.   31, 33, 35

23) Find three consecutive odd integers such that their sum is five more than four times the largest integer.   −15, −13, −11

24) The sum of two consecutive even integers is 52 less than three times the larger integer. Find the integers.   48, 50

25) Two consecutive page numbers in a book add up to 345. Find the page numbers.   172, 173

26) The addresses on the east side of Arthur Ave. are consecutive odd numbers. Two consecutive house numbers add up to 36. Find the addresses of these two houses.
17 Arthur Ave. and 19 Arthur Ave.

Objective 4: Solve Problems Involving Percent Change

Find the sale price of each item.

27) A cell phone that regularly sells for $75.00 is marked down 15%. $63.75

28) A baby stroller that retails for $69.00 is on sale at 20% off. $55.20

29) A sign reads, "Take 30% off the original price of all DVDs." The original price on the DVD you want to buy is $16.50. $11.55

30) The $120.00 basketball shoes James wants are now on sale at 15% off. $102.00

31) At the end of the summer, the bathing suit that sold for $29.00 is marked down 60%. $11.60

32) An advertisement states that a TV that regularly sells for $399.00 is being discounted 25%. $299.25

Solve using the five "Steps for Solving Applied Problems." See Example 6.

33) A digital camera is on sale for $119 after a 15% discount. What was the original price of the camera? $140.00

34) Marie paid $15.13 for a hardcover book that was marked down 15%. What was the original selling price of the book? $17.80

35) In February, a store discounted all of its calendars by 60%. If Ramon paid $4.38 for a calendar, what was its original price? $10.95

36) An appliance store advertises 20% off all of its refrigerators. Mr. Kotaris paid $399.20 for the fridge. Find its original price. $499.00

37) The sale price of a coffeemaker is $22.75. This is 30% off of the original price. What was the original price of the coffeemaker? $32.50

38) Katrina paid $20.40 for a backpack that was marked down 15%. Find the original retail price of the backpack. $24.00

39) One hundred forty countries participated in the 1984 Summer Olympics in Los Angeles. This was 75% more than the number of countries that took part in the Summer Olympics in Moscow 4 years earlier. How many countries participated in the 1980 Olympics in Moscow? (www.olympics.org)   80

40) In 2010 there were about 1224 acres of farmland in Custer County. This is 32% less than the number of acres of farmland in 2000. Calculate the number of acres of farmland in Custer County in 2000.   1800 acres

41) In 2004, there were 7569 Starbucks stores worldwide. This is approximately 1681% more stores than 10 years earlier. How many Starbucks stores were there in 1994? (Round to the nearest whole number.) (www.starbucks.com)   425

42) Liu Fan's salary this year is 14% higher than it was 3 years ago. If he earns $37,050 this year, what did he earn 3 years ago? $32,500

Objective 5: Solve Problems Involving Simple Interest

Solve. See Examples 7 and 8.

43) Jenna invests $800 in an account for 1 year earning 4% simple interest. How much interest was earned from this account? $32

44) Last year, Mr. Jaworski deposited $11,000 in an account earning 7% simple interest for 1 year. How much interest was earned? $770

45) Sven Andersson invested $6500 in an account earning 6% simple interest. How much money will be in the account 1 year later? $6890

46) If $3000 is deposited into an account for 1 year earning 5.5% simple interest, how much money will be in the account after 1 year? $3165

47) Rachel Levin has a total of $5500 to invest for 1 year. She deposits $4000 into an account earning 6.5% annual simple interest and the rest into an account earning 8% annual simple interest. How much interest did Rachel earn? $380

48) Maurice plans to invest a total of $9000 for 1 year. In the account earning 5.2% simple interest he will deposit $6000, and in an account earning 7% simple interest he will deposit the rest. How much interest will Maurice earn? $522

Solve using the five "Steps for Solving Applied Problems." See Example 9.

49) Amir Sadat receives a $15,000 signing bonus on accepting his new job. He plans to invest some of it at 6% annual simple interest and the rest at 7% annual simple interest. If he will earn $960 in interest after 1 year, how much will Amir invest in each account? $9000 at 6% and $6000 at 7%

50) Lisa Jenkins invested part of her $8000 inheritance in an account earning 5% simple interest and the rest in an account earning 4% simple interest. How much did Lisa invest in each account if she earned $365 in total interest after 1 year? $3500 at 4% and $4500 at 5%

51) Enrique's money earned $164 in interest after 1 year. He invested some of his money in an account earning 6% simple interest and $200 more than that amount in an account earning 5% simple interest. Find the amount Enrique invested in each account. $1400 at 6% and $1600 at 5%

52) Saori Yamachi invested some money in an account earning 7.4% simple interest and twice that amount in an account earning 9% simple interest. She earned $1016 in interest after 1 year. How much did Saori invest in each account? $4000 at 7.4% and $8000 at 9%

53) Last year, Clarissa invested a total of $7000 in two accounts earning simple interest. Some of it she invested at 9.5%, and the rest was invested at 7%. How much did she invest in each account if she earned a total of $560 in interest last year? $2800 at 9.5% and $4200 at 7%

54) Ted has $3000 to invest. He deposits a portion of it in an account earning 5% simple interest and the rest at 6.5% simple interest. After 1 year he has earned $175.50 in interest. How much did Ted deposit in each account? $1300 at 5% and $1700 at 6.5%

Mixed Exercises

Write an equation and solve. Use the five "Steps for Solving Applied Problems."

55) Irina was riding her bike when she got a flat tire. Then, she walked her bike 1 mi more than half the distance she rode it. How far did she ride her bike and how far did she walk if the total distance she traveled was 7 mi?   ride: 4 mi, walk: 3mi

56) It is estimated that in 2003 the number of Internet users in Slovakia was 40% more than the number of users in Kenya. If Slovakia had 700,000 Internet users in 2003, how many people used the Internet in Kenya that year? (www.theodora.com)   500,000

57) On his 21st birthday, Jerry received $20,000 from a trust fund. He invested some of the money in an account earning 5% simple interest and put the rest in a high-risk investment paying 9% simple interest. After 1 year, he earned a total of $1560 in interest. How much did Jerry invest in each account? $6000 at 5% and $14,000 at 9%

58) A book is open to two pages numbered consecutively. The sum of the page numbers is 373. What are the page numbers?   186, 187

59) In August 2010, there were 2600 entering freshmen at a state university. This is 4% higher than the number of freshmen entering the school in August of 2009. How many freshmen were enrolled in August 2009?   2500

60) In 2002, Mia Hamm played in eight fewer U.S. National Team soccer matches than in 2003. During those 2 years she appeared in a total of 26 games. In how many games did she play in 2002 and in 2003? (www.ussoccer.com)   2002: 9 matches, 2003: 17 matches

61) A 53-in. board is to be cut into three pieces so that one piece is 5 in. longer than the shortest piece, and the other piece is twice as long as the shortest piece. Find the length of each piece of board.   12 in., 17 in., and 24 in.

62) Ivan has $8500 to invest. He will invest some of it in a long-term IRA paying 4% simple interest and the rest in a short-term CD earning 2.5% simple interest. After 1 year, Ivan's investments have earned $250 in interest. How much did Ivan invest in each account?   IRA: $2500, CD: $6000

VIDEO 63) One-sixth of the smallest of three consecutive even integers is three less than one-tenth the sum of the other even integers. Find the integers.   72, 74, 76

64) A 58-ft-long cable must be cut into three pieces. One piece will be 6 ft longer than the shortest piece, and the longest portion must be twice as long as the shortest. Find the length of the three pieces of cable.   13 ft, 19 ft, and 26 ft

65) In 2001, *Harry Potter* grossed $42 million more in theaters than *Shrek*. Together they earned $577.4 million. How much did each movie earn in theaters in 2001? (www.boxofficereport.com)   *Shrek*: $267.7 million, *Harry Potter*: $309.7 million

66) Henry marks up the prices of his fishing poles by 50%. Determine what Henry paid his supplier for his best-selling fishing pole if Henry charges his customers $33.75. $22.50

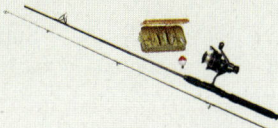

67) Tamara invests some money in three different accounts. She puts some of it in a CD earning 3% simple interest and twice as much in an IRA paying 4% simple interest. She also decides to invest $1000 more than what she's invested in the CD into a mutual fund earning 5% simple interest. Determine how much money Tamara invested in each account if she earned $290 in interest after 1 year.   CD: $1500, IRA: $3000, mutual fund: $2500

68) The three top-selling albums of 2009 were by Taylor Swift, Susan Boyle, and Michael Jackson. Susan Boyle sold 0.8 million more albums than Michael Jackson while Taylor Swift's album sold 0.9 million more than Michael's. Find the number of albums sold by each artist if together they sold 8.6 million copies. (www.billboard.com)   Taylor Swift: 3.2 mil; Susan Boyle: 3.1 mil; Michael Jackson: 2.3 mil

VIDEO 69) Zoe's current salary is $40,144. This is 4% higher than last year's salary. What was Zoe's salary last year? $38,600

70) Find the original price of a cell phone if it costs $63.20 after a 20% discount. $79.00

Section 2.3 Geometry Applications and Solving Formulas

Objectives

1. Solve Problems Using Formulas from Geometry
2. Solve Problems Involving Angle Measures
3. Solve Problems Involving Complementary and Supplementary Angles
4. Solve a Formula for a Specific Variable

In this section, we will make use of concepts and formulas from geometry to solve applied problems. We will also learn how to solve a formula for a specific variable. We begin with geometry.

1. Solve Problems Using Formulas from Geometry

Example 1

The area of a rectangular room is 180 ft^2. Its width is 12 ft. What is the length of the room?

Solution

Step 1: **Read** the problem carefully, and identify what we are being asked to find.

We must find the length of the room.

A picture will be very helpful in this problem.

In-Class Example 1
The area of a rectangular field is 208 yd^2. If the length is 16 yd, how wide is the field?
answer: The width is 13 yd.

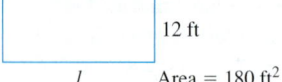

12 ft

l Area = 180 ft^2

Step 2:   **Choose a variable** to represent the unknown.

l = the length of the room

Label the picture with the width, 12 ft, and the length l.

Step 3:   **Translate** the information that appears in English into an algebraic equation. We will use a known geometry formula. How do we know which formula to use? List the information we are given and what we want to find:

The room is in the shape of a rectangle, its area = 180 ft^2, and its width = 12 ft. We must find the width. Which formula involves the area, length, and width of a rectangle?

$$A = lw$$

Substitute the known values into the formula for the area of a rectangle, and solve for l.

$$A = lw$$
$$180 = l(12) \qquad \text{Substitute the known values.}$$

Step 4:   **Solve** the equation.

$$180 = 12l$$
$$\frac{180}{12} = \frac{12l}{12} \qquad \text{Divide by 12.}$$
$$15 = l \qquad \text{Simplify.}$$

Step 5:   **Check** the answer and **interpret** the solution as it relates to the problem.

If $l = 15$ ft, then $l \cdot w = 15$ ft $\cdot$ 12 ft = 180 ft^2. Therefore, the length of the room is 15 ft. ∎

Note

Remember to include the correct units in your answer!

   You Try 1

The area of a rectangular ping-pong table is 45 ft^2. Its length is 9 ft. What is the width of the ping-pong table?

2. Solve Problems Involving Angle Measures

Example 2

Find the missing angle measures.

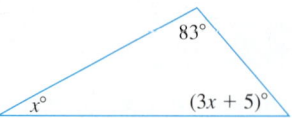

In-Class Example 2
Find the missing angle measures.
answer: The measure of one angle is 14° and the other is 102°.

64°

$(7x + 4)°$ $x°$

Solution

Step 1:   **Read** the problem carefully, and identify what we are being asked to find.

Find the missing angle measures.

Step 2:   The unknowns are already defined. We must find x, the measure of one angle, and then $3x + 5$, the measure of the other angle.

Step 3:   **Translate** the information into an algebraic equation. Since the sum of the angles in a triangle is 180°, we can write

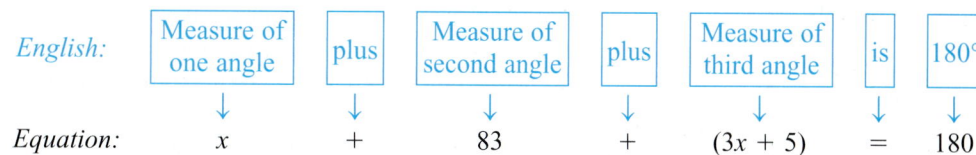

English:	Measure of one angle	plus	Measure of second angle	plus	Measure of third angle	is	180°
	↓	↓	↓	↓	↓	↓	↓
Equation:	x	$+$	83	$+$	$(3x + 5)$	$=$	180

The equation is $x + 83 + (3x + 5) = 180$.

Step 4:   **Solve** the equation.

$$x + 83 + (3x + 5) = 180$$
$$4x + 88 = 180 \qquad \text{Combine like terms.}$$
$$4x + 88 - 88 = 180 - 88 \qquad \text{Subtract 88 from each side.}$$
$$4x = 92 \qquad \text{Combine like terms.}$$
$$x = 23 \qquad \text{Divide each side by 4.}$$

Step 5:   **Check** the answer and **interpret** the solution as it relates to the problem.

One angle, x, has a measure of 23°. The other unknown angle measure is $3x + 5 = 3(23) + 5 = 74°$.

The answer makes sense because the sum of the angle measures is $23° + 83° + 74° = 180°$.

You Try 2

Find the missing angle measures.

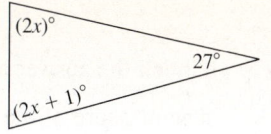

$(2x)°$
$27°$
$(2x + 1)°$

3. Solve Problems Involving Complementary and Supplementary Angles

Recall that two angles are **complementary** if the sum of their angles is 90°, and two angles are **supplementary** if the sum of their angles is 180°.

For example, if the measure of $\angle A$ is 52°, then

   a) the measure of its complement is $90° - 52° = 38°$.

   b) the measure of its supplement is $180° - 52° = 128°$.

Now, let's say an angle has a measure, $x°$. Using the same reasoning as above

   a) the measure of its complement is $90 - x$.

   b) the measure of its supplement is $180 - x$.

We will use these ideas to solve Example 3.

Example 3

Twice the complement of an angle is 18° less than the supplement of the angle. Find the measure of the angle.

In-Class Example 3
Four times the complement of an angle is 30° greater than the supplement of the angle. Find the measure of the angle.
answer: The measure of the angle is 50°.

Solution

Step 1:   **Read** the problem carefully, and identify what we are being asked to find.

We must find the measure of the angle.

Step 2: **Choose a variable** to represent an unknown, and define the other unknown in terms of this variable.

This problem has three unknowns: the measures of the angle, its complement, and its supplement. Choose a variable to represent the original angle, then define the other unknowns in terms of this variable.

$$x = \text{the measure of the angle}$$

Define the other unknowns in terms of x.

$$90 - x = \text{the measure of the complement}$$
$$180 - x = \text{the measure of the supplement}$$

Step 3: **Translate** the information that appears in English into an algebraic equation.

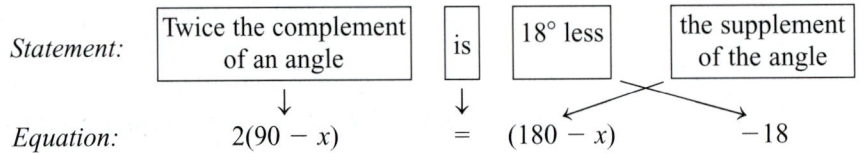

Statement:	Twice the complement of an angle	is	18° less	the supplement of the angle
Equation:	$2(90 - x)$	$=$	$(180 - x)$	-18

The equation is $2(90 - x) = (180 - x) - 18$.

Step 4: **Solve** the equation.

$$2(90 - x) = (180 - x) - 18$$
$$180 - 2x = 180 - x - 18 \qquad \text{Distribute.}$$
$$180 - 2x = 162 - x \qquad \text{Combine like terms.}$$
$$180 - 162 - 2x = 162 - 162 - x \qquad \text{Subtract 162 from each side.}$$
$$18 - 2x = -x \qquad \text{Combine like terms.}$$
$$18 - 2x + 2x = -x + 2x \qquad \text{Add 2x to each side.}$$
$$18 = x \qquad \text{Simplify.}$$

Step 5: **Check** the answer and **interpret** the solution as it relates to the problem.

The measure of the angle is 18°.

To check the answer, we first need to find its complement and supplement. The complement is $90° - 18° = 72°$, and its supplement is $180° - 18° = 162°$. Now we can check these values in the original statement: twice the complement is $2(72°) = 144°$. Eighteen degrees less than the supplement is $162° - 18° = 144°$. ■

You Try 3

Ten times the complement of an angle is 9° less than the supplement of the angle. Find the measure of the angle.

4. Solve a Formula for a Specific Variable

The formula $P = 2l + 2w$ allows us to find the perimeter of a rectangle when we know its length, l, and width, w. But, what if we were solving problems where we repeatedly needed to find the value of w? Then, we could rewrite $P = 2l + 2w$ so that it is solved for w:

$$w = \frac{P - 2l}{2}$$

Doing this means that we have *solved the formula $P = 2l + 2w$ for the specific variable, w.*

Solving a formula for a specific variable may seem confusing at first because the formula contains more than one letter. Keep in mind that we will solve for a specific variable the same way we have been solving equations up to this point.

We'll start by solving $2x + 5 = 13$ step by step for x and then applying the same procedure to solving $mx + b = y$ for x.

Example 4

In-Class Example 4
Solve $5x - 4 = 16$ and
$ax - b = c$ for x.

answer: $x = 4$; $x = \dfrac{c + b}{a}$

Solve $2x + 5 = 13$ and $mx + b = y$ for x.

Solution

Look at these equations carefully, and notice that they have the same form. Read the following steps in order.

Part 1   Solve $2x + 5 = 13$

Don't quickly run through the solution of this equation. *The emphasis here is on the steps used to solve the equation and why we use those steps!*

$$2\boxed{x} + 5 = 13$$

We are solving for x. We'll put a box around it. What is the first step? "Get rid of" what is being added to the $2x$, that is "get rid of" the 5 on the left. Subtract 5 from each side.

$$2\boxed{x} + 5 - 5 = 13 - 5$$

Combine like terms.

$$2\boxed{x} = 8$$

Part 2   Solve $mx + b = y$ for x.

Since we are solving for x, we'll put a box around it.

$$m\boxed{x} + b = y$$

The goal is to get the "x" on a side by itself. What do we do first? As in Part 1, "get rid of" what is being added to the "mx" term, that is "get rid of" the "b" on the left. Since b is being added to "mx," we will subtract it from each side. (We are performing the same steps as in Part 1!)

$$m\boxed{x} + b - b = y - b$$

Combine like terms.

$$m\boxed{x} = y - b$$

[We cannot combine the terms on the right, so the right remains $y - b$.]

Part 3   We left off needing to solve $2\boxed{x} = 8$ for x. We need to eliminate the "2" on the left. Since x is being multiplied by 2, we will *divide* each side by 2.

$$\frac{2\boxed{x}}{2} = \frac{8}{2}$$

Simplify.

$$x = 4$$

Part 4   Now, we have to solve $m\boxed{x} = y - b$ for x. We need to eliminate the "m" on the left. Since x is being multiplied by m, we will *divide* each side by m.

$$\frac{m\boxed{x}}{m} = \frac{y - b}{m}$$

These are the same steps used in Part 3!

Simplify.

$$\frac{\cancel{m}\boxed{x}}{\cancel{m}} = \frac{y - b}{m}$$

$$x = \frac{y - b}{m} \quad \text{or} \quad x = \frac{y}{m} - \frac{b}{m} \qquad \blacksquare$$

Note

To obtain the result $x = \dfrac{y}{m} - \dfrac{b}{m}$, we distributed the m in the denominator to each term in the numerator. Either form of the answer is correct.

When you are solving a formula for a specific variable, think about the steps you use to solve an equation in one variable.

You Try 4

Solve $an - pr = w$ for n.

Example 5

In-Class Example 5
$S = \dfrac{2\pi R}{T}$ is a formula used for circular motion. Solve this equation for R.

answer: $R = \dfrac{ST}{2\pi}$

$R = \dfrac{\rho L}{A}$ is a formula used in physics. Solve this equation for L.

Solution

$R = \dfrac{\rho \boxed{L}}{A}$ Solve for L. Put it in a box.

$AR = A \cdot \dfrac{\rho \boxed{L}}{A}$ Begin by eliminating A on the right.

Since ρL is being divided by A, we will *multiply* by A on each side.

$AR = \rho \boxed{L}$ Next, eliminate ρ.

$\dfrac{AR}{\rho} = \dfrac{\rho \boxed{L}}{\rho}$ Divide each side by ρ.

$\dfrac{AR}{\rho} = L$

Example 6

In-Class Example 6
$C = \dfrac{3}{8}k(a + b)$. Solve for b.

answer: $b = \dfrac{8C}{3k} - a$

$A = \dfrac{1}{2}h(b_1 + b_2)$ is the formula for the area of a trapezoid. Solve it for b_1.

Solution

$A = \dfrac{1}{2}h(\boxed{b_1} + b_2)$ There are two valid ways to solve this for b_1. We'll look at both of them.

Method I
We will put b_1 in a box to remind us that this is what we must solve for. In Method 1, we will start by eliminating the fraction. Multiply both sides by 2.

$$2A = 2 \cdot \dfrac{1}{2}h(\boxed{b_1} + b_2)$$
$$2A = h(\boxed{b_1} + b_2)$$

We are solving for b_1, which is in the parentheses. The quantity in parentheses is being multiplied by h, so we can *divide both sides by h* to eliminate it on the right.

$$\dfrac{2A}{h} = \dfrac{h(\boxed{b_1} + b_2)}{h}$$

$$\dfrac{2A}{h} = \boxed{b_1} + b_2$$

$$\dfrac{2A}{h} - b_2 = \boxed{b_1} + b_2 - b_2 \qquad \text{Subtract } b_2 \text{ from each side.}$$

$$\dfrac{2A}{h} - b_2 = b_1$$

Method 2

To solve $A = \dfrac{1}{2}h(b_1 + b_2)$ for b_1, we can begin by distributing $\dfrac{1}{2}h$ on the right.

We will put b_1 in a box to remind us that this is what we must solve for.

$$A = \frac{1}{2}h\left(\boxed{b_1} + b_2\right)$$

$$A = \frac{1}{2}h\boxed{b_1} + \frac{1}{2}hb_2$$

$$2A = 2\left(\frac{1}{2}h\boxed{b_1} + \frac{1}{2}hb_2\right) \qquad \text{Multiply by 2 to eliminate the fractions.}$$

$$2A = h\boxed{b_1} + hb_2$$

$$2A - hb_2 = h\boxed{b_1} + hb_2 - hb_2 \qquad \text{Since } hb_2 \text{ is being added to } hb_1, \text{ subtract } hb_2 \text{ from each side.}$$

$$2A - hb_2 = h\boxed{b_1}$$

$$\frac{2A - hb_2}{h} = \frac{h\boxed{b_1}}{h} \qquad \text{Divide by } h.$$

$$\frac{2A - hb_2}{h} = b_1$$

Or, b_1 can be rewritten as $b_1 = \dfrac{2A}{h} - \dfrac{hb_2}{h} = \dfrac{2A}{h} - b_2$.

So, $b_1 = \dfrac{2A - hb_2}{h}$ or $b_1 = \dfrac{2A}{h} - b_2$. The forms are equivalent.

You Try 5

Solve for the indicated variable.

a) $t = \dfrac{qr}{s}$ for q b) $R = t(k - c)$ for c

Answers to You Try Exercises

1) 5 ft   2) 76°, 77°   3) 81°   4) $n = \dfrac{w + pr}{a}$ 5) a) $q = \dfrac{st}{r}$ b) $c = \dfrac{kt - R}{t}$ or $c = k - \dfrac{R}{t}$

2.3 Exercises

*Additional answers can be found in the Answers to Exercises appendix.

Objective 1: Solve Problems Using Formulas from Geometry

Use a formula from geometry to solve each problem. See Example 1.

1) The Torrence family has a rectangular, in-ground pool in their yard. It holds 1700 ft^3 of water. If it is 17 ft wide and 4 ft deep, what is the length of the pool?   25 ft

2) A rectangular fish tank holds 2304 in^3 of water. Find the height of the tank if it is 24 in. long and 8 in. wide.   12 in.

3) A computer printer is stocked with paper measuring 8.5 in. × 11 in. But, the printing covers only 48 in^2 of the page. If the length of the printed area is 8 in., what is the width?   6 in.

4) The rectangular glass surface to produce copies on a copy machine has an area of 135 in^2. It is 9 in. wide. What is the longest length of paper that will fit on this copier?   15 in.

5) The face of the clock on Big Ben in London has a radius of 11.5 ft. What is the area of this circular clock face? Use 3.14 for π. Round the answer to the nearest square foot. (www.bigben.freeservers.com)   415 ft^2

6) A lawn sprinkler sprays water in a circle of radius 6 ft. Find the area of grass watered by this sprinkler. Use 3.14 for π. Round the answer to the nearest square foot.   113 ft^2

7) The "lane" on a basketball court is a rectangle of length 19 ft. The area of the lane is 228 ft^2. What is the width of the lane?   12 ft

8) A trapezoidal plot of land has the dimensions pictured below. If the area is 9000 ft^2,
   a) find the length of the missing side, x.   90 ft
   b) how much fencing would be needed to completely enclose the plot?   440 ft

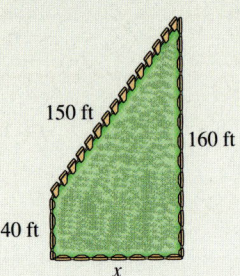

9) A large can of tomato sauce is in the shape of a right circular cylinder. If its radius is 2 in. and its volume is 24π in^3, what is the height of the can?   6 in.

10) A coffee can in the shape of a right circular cylinder has a volume of 50π in^3. Find the height of the can if its diameter is 5 in.   8 in.

Objective 2: Solve Problems Involving Angle Measures

Find the missing angle measures. See Example 2.

11)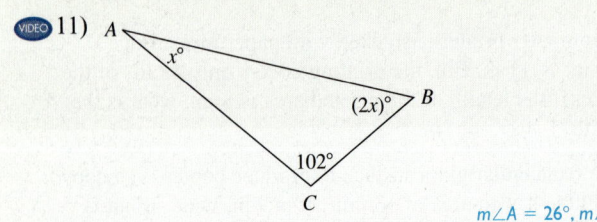

$m\angle A = 26°, m\angle B = 52°$

12)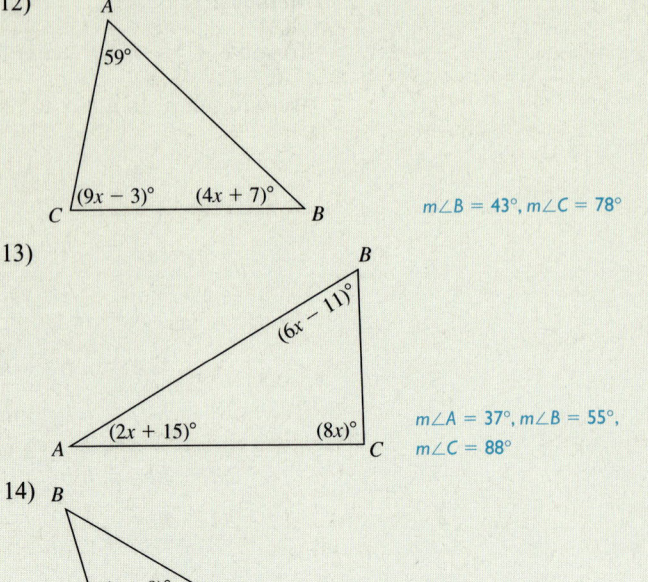

$m\angle B = 43°, m\angle C = 78°$

13) $m\angle A = 37°, m\angle B = 55°, m\angle C = 88°$

14) $m\angle A = 107°, m\angle B = 39°, m\angle C = 34°$

Find the measure of each indicated angle.

15)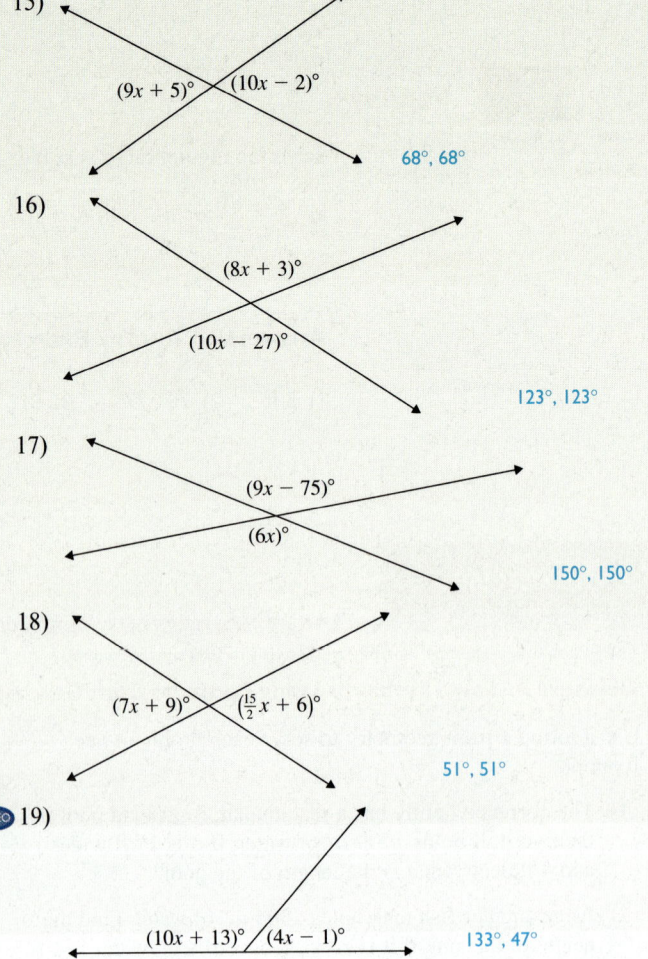

$(9x + 5)°$ $(10x - 2)°$ 68°, 68°

16) $(8x + 3)°$ $(10x - 27)°$ 123°, 123°

17) $(9x - 75)°$ $(6x)°$ 150°, 150°

18) $(7x + 9)°$ $\left(\frac{15}{2}x + 6\right)°$ 51°, 51°

19) $(10x + 13)°$ $(4x - 1)°$ 133°, 47°

20)

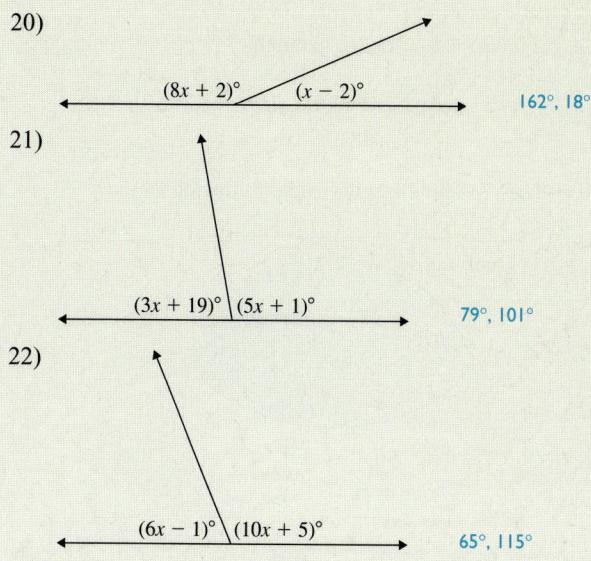

$(8x + 2)°$ $(x - 2)°$ 162°, 18°

21)

$(3x + 19)°$ $(5x + 1)°$ 79°, 101°

22)

$(6x - 1)°$ $(10x + 5)°$ 65°, 115°

Objective 3: Solve Problems Involving Complementary and Supplementary Angles

23) If x = the measure of an angle, write an expression for its supplement. $180 - x$

24) If x = the measure of an angle, write an expression for its complement. $90 - x$

Solve each problem. See Example 3.

25) Ten times the measure of an angle is 7° more than the measure of its supplement. Find the measure of the angle.   17°

26) The measure of an angle is 12° more than twice its complement. Find the measure of the angle.   64°

27) Four times the complement of an angle is 40° less than twice the angle's supplement. Find the angle, its complement, and its supplement.   angle: 20°, comp.: 70°, supp.: 160°

28) Twice the supplement of an angle is 24° more than four times its complement. Find the angle, its complement, and its supplement.   angle: 12°, comp.: 78°, supp.: 168°

29) The sum of an angle and three times its complement is 55° more than its supplement. Find the measure of the angle.   35°

30) The sum of twice an angle and its supplement is 24° less than 5 times its complement. Find the measure of the angle.   41°

31) The sum of 3 times an angle and twice its supplement is 400°. Find the angle.   40°

32) The sum of twice an angle and its supplement is 253°. Find the angle.   73°

Objective 4: Solve a Formula for a Specific Variable

Substitute the given values into the formula. Then, solve for the remaining variable.

33) $I = Prt$ (simple interest); if $I = 240$ when $P = 3000$ and $r = 0.04$, find t.   2

34) $I = Prt$ (simple interest); if $I = 156$ when $P = 650$ and $t = 3$, find r.   0.08

35) $V = lwh$ (volume of a rectangular box); if $V = 96$ when $l = 8$ and $h = 3$, find w.   4

36) $V = \frac{1}{3}Ah$ (volume of a pyramid); if $V = 60$ when $h = 9$, find A.   20

37) $P = 2l + 2w$ (perimeter of a rectangle); if $P = 50$ when $w = 7$, find l.   18

38) $P = s_1 + s_2 + s_3$ (perimeter of a triangle); if $P = 37$ when $s_1 = 17$ and $s_3 = 8$, find s_2.   12

39) $V = \frac{1}{3}\pi r^2 h$ (volume of a cone); if $V = 54\pi$ when $r = 9$, find h.   2

40) $V = \frac{1}{3}\pi r^2 h$ (volume of a cone); if $V = 32\pi$ when $r = 4$, find h.   6

41) $S = 2\pi r^2 + 2\pi rh$ (surface area of a right circular cylinder); if $S = 120\pi$ when $r = 5$, find h.   7

42) $S = 2\pi r^2 + 2\pi rh$ (surface area of a right circular cylinder); if $S = 66\pi$ when $r = 3$, find h.   8

43) $A = \frac{1}{2}h(b_1 + b_2)$ (area of a trapezoid); if $A = 790$ when $b_1 = 29$ and $b_2 = 50$, find h.   20

44) $A = \frac{1}{2}h(b_1 + b_2)$ (area of a trapezoid); if $A = 246.5$ when $h = 17$ and $b_2 = 16$, find b_1.   13

45) Solve for x.

   a) $x + 12 = 35$ $x = 23$ b) $x + n = p$ $x = p - n$

   c) $x + q = v$ $x = v - q$

46) Solve for t.

   a) $t - 3 = 19$ $t = 22$ b) $t - w = m$ $t = m + w$

   c) $t - v = j$ $t = j + v$

47) Solve for n.

   a) $5n = 30$ $n = 6$ b) $yn = c$ $n = \frac{c}{y}$

   c) $wn = d$ $n = \frac{d}{w}$

48) Solve for y.

   a) $4y = 36$ $y = 9$ b) $ay = x$ $y = \frac{x}{a}$

   c) $py = r$ $y = \frac{r}{p}$

49) Solve for c.

   a) $\frac{c}{3} = 7$ $c = 21$ b) $\frac{c}{u} = r$ $c = ur$

   c) $\frac{c}{x} = t$ $c = xt$

50) Solve for m.

   a) $\dfrac{m}{8} = 2$ $m = 16$ b) $\dfrac{m}{z} = p$ $m = zp$

   c) $\dfrac{m}{q} = f$ $m = qf$

VIDEO 51) Solve for d.

   a) $8d - 7 = 17$ $d = 3$ b) $kd - a = z$ $d = \dfrac{z+a}{k}$

52) Solve for g.

   a) $3g + 23 = 2$ $g = -7$ b) $cg + k = \pi$ $g = \dfrac{\pi - k}{c}$

53) Solve for z.

   a) $6z + 19 = 4$ $z = -\dfrac{5}{2}$ b) $yz + t = w$ $z = \dfrac{w-t}{y}$

54) Solve for p.

   a) $10p - 3 = 19$ $p = \dfrac{11}{5}$ b) $np - r = d$ $p = \dfrac{d+r}{n}$

Solve each formula for the indicated variable.

55) $F = ma$ for m (physics) $m = \dfrac{F}{a}$

56) $C = 2\pi r$ for r (geometry) $r = \dfrac{C}{2\pi}$

57) $n = \dfrac{c}{v}$ for c (physics) $c = nv$

58) $f = \dfrac{R}{2}$ for R (physics) $R = 2f$

59) $E = \sigma T^4$ for σ (meteorology) $\sigma = \dfrac{E}{T^4}$

60) $p = \rho gy$ for ρ (geology) $\rho = \dfrac{p}{gy}$

61) $V = \dfrac{1}{3}\pi r^2 h$ for h (geometry) $h = \dfrac{3V}{\pi r^2}$

62) $U = \dfrac{1}{2}LI^2$ for L (physics) $L = \dfrac{2U}{I^2}$

63) $R = \dfrac{E}{I}$ for E (electronics) $E = IR$

64) $A = \dfrac{1}{2}bh$ for b (geometry) $b = \dfrac{2A}{h}$

65) $I = PRT$ for R (finance) $R = \dfrac{I}{PT}$

66) $I = PRT$ for P (finance) $P = \dfrac{I}{RT}$

67) $P = 2l + 2w$ for l (geometry) $l = \dfrac{P - 2w}{2}$ or $l = \dfrac{P}{2} - w$

68) $A = P + PRT$ for T (finance) $T = \dfrac{A - P}{PR}$ or $T = \dfrac{A}{PR} - \dfrac{1}{R}$

69) $H = \dfrac{D^2 N}{2.5}$ for N (auto mechanics) $N = \dfrac{2.5H}{D^2}$

70) $V = \dfrac{AH}{3}$ for A (geometry) $A = \dfrac{3V}{H}$

71) $A = \dfrac{1}{2}h(b_1 + b_2)$ for b_2 (geometry) $b_2 = \dfrac{2A}{h} - b_1$ or $b_2 = \dfrac{2A - hb_1}{h}$

72) $A = \pi(R^2 - r^2)$ for r^2 (geometry) $r^2 = R^2 - \dfrac{A}{\pi}$ or $r^2 = \dfrac{\pi R^2 - A}{\pi}$

For Exercises 73 and 74, refer to the following figure.

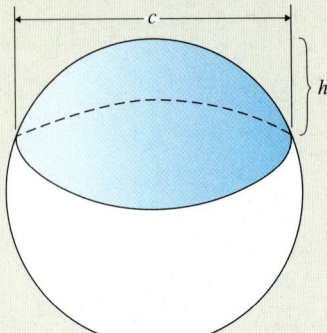

The surface area, S, of the spherical segment shown is given by $S = \dfrac{\pi}{4}(4h^2 + c^2)$, where h is the height of the segment and c is the diameter of the segment's base.

73) Solve the formula for h^2. $h^2 = \dfrac{S}{\pi} - \dfrac{c^2}{4}$ or $h^2 = \dfrac{1}{4}\left(\dfrac{4S}{\pi} - c^2\right)$

74) Solve the formula for c^2. $c^2 = \dfrac{4S}{\pi} - 4h^2$ or $c^2 = \dfrac{4S - 4\pi h^2}{\pi}$

VIDEO 75) The perimeter, P, of a rectangle is $P = 2l + 2w$, where $l =$ length and $w =$ width.

   a) Solve $P = 2l + 2w$ for w. $w = \dfrac{P - 2l}{2}$ or $w = \dfrac{P}{2} - l$

   b) Find the width of the rectangle with perimeter 28 cm and length 11 cm.   3 cm

76) The area, A, of a triangle is $A = \dfrac{1}{2}bh$, where $b =$ length of the base and $h =$ height.

   a) Solve $A = \dfrac{1}{2}bh$ for h. $h = \dfrac{2A}{b}$

   b) Find the height of the triangle that has an area of 30 in^2 and a base of length 12 in.   5 in.

77) $C = \dfrac{5}{9}(F - 32)$ is a formula that can be used to convert from degrees Fahrenheit, F, to degrees Celsius, C.

   a) Solve this formula for F. $F = \dfrac{9}{5}C + 32$

   b) The average high temperature in Mexico City, Mexico, in April is 25°C. Use the result in part a) to find the equivalent temperature in degrees Fahrenheit. (www.bbc.co.uk)   77°F

78) The average low temperature in Stockholm, Sweden, in January is −5°C. Use the result in Exercise 77a) to find the equivalent temperature in degrees Fahrenheit. (www.bbc.co.uk)   23°F

Section 2.4 More Applications of Linear Equations

Objectives

1. **Define Ratio and Proportion**
2. **Solve a Proportion**
3. **Solve Problems Involving a Proportion**
4. **Solve Problems Involving Money**
5. **Solve Mixture Problems**
6. **Solve Problems Involving Distance, Rate, and Time**

1. Define Ratio and Proportion

A **ratio** is a quotient of two quantities. For example, if a survey revealed that 28 people said that their favorite candy was licorice, while 40 people said that their favorite candy was chocolate, then the ratio of people who prefer licorice *to* people who prefer chocolate is

$$\frac{\text{Number of licorice lovers}}{\text{Number of chocolate lovers}} = \frac{28}{40} \text{ or } \frac{7}{10}$$

A *ratio* is a way to compare two quantities. If two ratios are equivalent, like $\frac{28}{40}$ and $\frac{7}{10}$, we can set them equal to make a *proportion*. A **proportion** is a statement that two ratios are equal. $\frac{28}{40} = \frac{7}{10}$ is an example of a proportion. How can we be certain that a proportion is true?

2. Solve a Proportion

Example 1

In-Class Example 1

Is the proportion $\frac{4}{5} = \frac{16}{20}$ true?

answer: yes

Is the proportion $\frac{5}{9} = \frac{10}{18}$ true?

Solution

One way to decide is to find the **cross products**. If the cross products are equal, then the proportion is true:

$$\frac{5}{9} \diagup\!\!\!\!= \frac{10}{18}$$

Multiply.   Multiply.

$$5 \cdot 18 = 9 \cdot 10$$
$$90 = 90 \quad \text{True}$$

The cross products are equal, so $\frac{5}{9} = \frac{10}{18}$ is a true proportion.

If the cross products are not equal, then the proportion is not true.

You Try 1

Decide if the following proportions are true.

a) $\frac{7}{3} = \frac{35}{15}$ b) $\frac{3}{4} = \frac{7}{10}$

We can make a general statement about cross products:

Property Cross Products

If $\frac{a}{b} = \frac{c}{d}$, then $ad = bc$ provided that $b \neq 0$ and $d \neq 0$.

This allows us to use cross products to solve equations. We will see later in the book that finding the cross products is the same as multiplying both sides of the equation by the least common denominator of the fractions.

Example 2

In-Class Example 2
Solve each proportion.
a) $\dfrac{2}{3} = \dfrac{x}{27}$ b) $\dfrac{b-6}{12} = \dfrac{b+2}{20}$
answer: a) {18}   b) {18}

Solve each proportion.

a) $\dfrac{x}{6} = \dfrac{12}{9}$ b) $\dfrac{a+4}{3} = \dfrac{a-1}{4}$

Solution

a) $\dfrac{x}{6} \diagdown \dfrac{12}{9}$

Multiply.   Multiply.
$$9x = 72$$
$$x = 8$$

b) $\dfrac{a+4}{3} \diagdown \dfrac{a-1}{4}$

Multiply.   Multiply.
$$4(a+4) = 3(a-1)$$
$$4a + 16 = 3a - 3$$
$$a + 16 = -3$$
$$a = -19$$

You Try 2

Solve each proportion.

a) $\dfrac{10}{14} = \dfrac{n}{21}$ b) $\dfrac{2k-3}{5} = \dfrac{3k+1}{10}$

3. Solve Problems Involving a Proportion

One application of proportions is in comparing prices.

Example 3

In-Class Example 3
If 4 lb of coffee cost $27.80, how much would 6 lb cost?
answer: The cost of 6 lb of coffee is $41.70.

If 3 lb of potatoes cost $1.77, how much would 5 lb of potatoes cost?

Solution

Step 1: **Read** the problem carefully, and identify what we are being asked to find.

We must find the cost of 5 lb of potatoes.

Step 2: **Choose a variable** to represent the unknown.

x = the cost of 5 lb of potatoes

Step 3: **Translate** the information that appears in English into an algebraic equation.

Write a proportion. We will write our ratios in the form of $\dfrac{\text{pounds of potatoes}}{\text{cost of potatoes}}$ so that the numerators contain the same quantities and the denominators contain the same quantities.

$$\text{pounds of potatoes} \rightarrow \dfrac{3}{1.77} = \dfrac{5}{x} \leftarrow \text{pounds of potatoes}$$
$$\text{cost of potatoes} \rightarrow \qquad\qquad \leftarrow \text{cost of potatoes}$$

The equation is $\dfrac{3}{1.77} = \dfrac{5}{x}$.

Step 4:   Solve the equation.

$$\frac{3}{1.77} = \frac{5}{x}$$

$$3x = 1.77 \cdot 5 \qquad \text{Set the cross products equal.}$$
$$3x = 8.85 \qquad \text{Multiply.}$$
$$x = 2.95 \qquad \text{Divide by 3.}$$

Step 5:   **Check** the answer and **interpret** the solution as it relates to the problem.

The cost of 5 lb of potatoes is $2.95. The check is left to the student.   ■

You Try 3

How much would a customer pay for 16 gal of unleaded gasoline if another customer paid $23.76 for 12 gal at the same gas station?

Another application of proportions is for solving similar triangles.

Similar Triangles

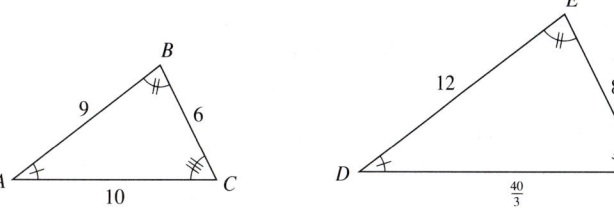

$$m\angle A = m\angle D, \quad m\angle B = m\angle E, \quad \text{and} \quad m\angle C = m\angle F$$

$\triangle ABC$ and $\triangle DEF$ are *similar triangles*. Two triangles are **similar** if they have the same shape. The corresponding angles have the same measure, and the corresponding sides are proportional.

The ratio of each pair of corresponding sides is $\frac{3}{4}$: $\frac{9}{12} = \frac{3}{4}$; $\frac{6}{8} = \frac{3}{4}$; $\frac{10}{\frac{40}{3}} = 10 \cdot \frac{3}{40} = \frac{3}{4}$.

We can set up and solve a proportion to find the length of an unknown side in two similar triangles.

Example 4

In-Class Example 4
Given the following similar triangles, find x.

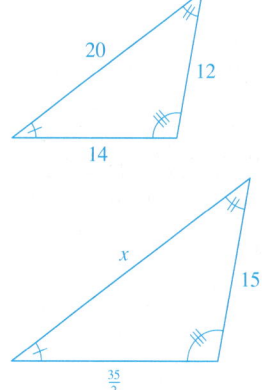

answer: 25

Given the following similar triangles, find x.

Solution

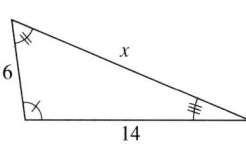 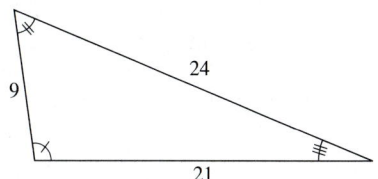

$$\frac{6}{9} = \frac{x}{24} \qquad \text{Set the ratios of two corresponding sides equal to each other. (Set up a proportion.)}$$
$$9x = 6 \cdot 24 \qquad \text{Solve the proportion.}$$
$$9x = 144 \qquad \text{Multiply.}$$
$$x = 16 \qquad \text{Divide by 9.}$$

   ■

You Try 4

Given the following similar triangles, find *x*.

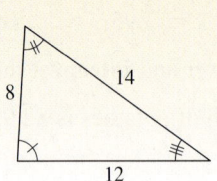
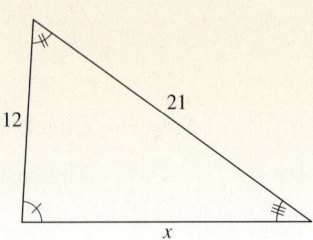

4. Solve Problems Involving Money

Many application problems involve thinking about the number of coins or bills and their values. Let's look at how arithmetic and algebra problems involving these ideas are related.

Example 5

In-Class Example 5
Determine the amount of
money you have in cents *and* in
dollars if you have
a) 12 nickels   b) 9 quarters
c) 12 nickels and 9 quarters
answer: a) 60¢; $0.60
b) 225¢; $2.25
c) 285¢; $2.85

Determine the amount of money you have in cents *and* in dollars if you have

a)   8 nickels   b)   7 quarters   c)   8 nickels and 7 quarters

Solution

You may be able to do these problems "in your head," but it is very important that we understand the *procedure* that is used to do this arithmetic problem so that we can apply the same procedure to algebra. So, read this carefully!

Parts a) and b):   Let's begin with part a), finding the value of 8 nickels.

Value in Cents   Value in Dollars

$$5 \cdot 8 = 40¢$$ $$0.05 \cdot 8 = \$0.40$$

Value of a nickel ↗   Number of nickels ↑   Value of 8 nickels ↖   Value of a nickel ↗   Number of nickels ↑   Value of 8 nickels ↖

Here's how we find the value of 7 quarters:

Value in Cents   Value in Dollars

$$25 \cdot 7 = 175¢$$ $$0.25 \cdot 7 = \$1.75$$

Value of a quarter ↗   Number of quarters ↑   Value of 7 quarters ↖   Value of a quarter ↗   Number of quarters ↑   Value of 7 quarters ↖

A table can help us organize the information, so let's put both part a) and part b) in a table so that we can see a pattern.

Value of the Coins (in cents)   **Value of the Coins (in dollars)**

	Value of the Coin	Number of Coins	Total Value of the Coins		Value of the Coin	Number of Coins	Total Value of the Coins
Nickels	5	8	$5 \cdot 8 = 40$	Nickels	0.05	8	$0.05 \cdot 8 = 0.40$
Quarters	25	7	$25 \cdot 7 = 175$	Quarters	0.25	7	$0.25 \cdot 7 = 1.75$

Notice that each time we want to find the total value of the coins we find it by multiplying.

$$\text{Value of the coin} \cdot \text{Number of coins} = \text{Total value of the coins}$$

c)   Now let's write an equation in English to find the total value of the 8 nickels and 7 quarters.

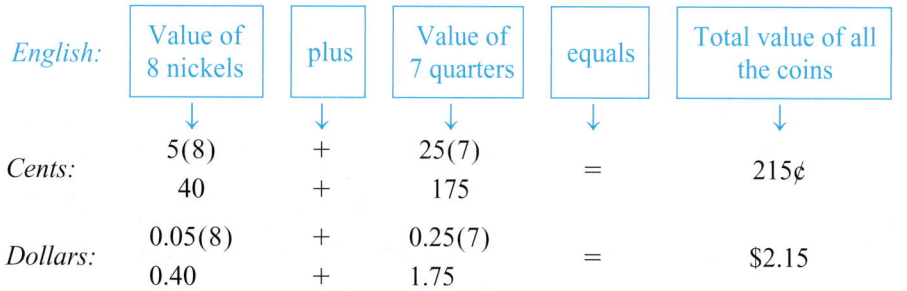

We will use the same procedure that we just used to solve these arithmetic problems to write algebraic expressions to represent the value of a collection of coins.

Example 6

Write expressions for the amount of money you have in cents *and* in dollars if you have

a)   *n* nickels   b)   *q* quarters   c)   *n* nickels and *q* quarters

In-Class Example 6
Write expressions for the amount of money you have in cents and in dollars if you have
a) *d* dimes   b) *q* quarters
c) *d* dimes and *q* quarters
answer: a) 10*d*; 0.10*d*
b) 25*q*; 0.25*q*
c) 10*d* + 25*q*; 0.10*d* + 0.25*q*

Solution

Parts a) and b):   Let's use tables just like we did in Example 5. We will put parts a) and b) in the same table.

Value of the Coins (in cents)

	Value of the Coin	Number of Coins	Total Value of the Coins
Nickels	5	*n*	$5 \cdot n = 5n$
Quarters	25	*q*	$25 \cdot q = 25q$

Value of the Coins (in dollars)

	Value of the Coin	Number of Coins	Total Value of the Coins
Nickels	0.05	*n*	$0.05 \cdot n = 0.05n$
Quarters	0.25	*q*	$0.25 \cdot q = 0.25q$

If you have *n* nickels, then the expression for the amount of money in cents is $5n$. The amount of money in dollars is $0.05n$. If you have *q* quarters, then the expression for the amount of money in cents is $25q$. The amount of money in dollars is $0.25q$.

c)   Write an equation in English to find the total value of *n* nickels and *q* quarters. It is based on the same idea that we used in Example 5.

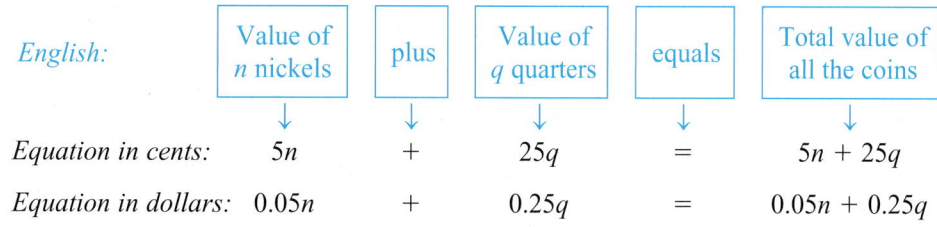

The expression in cents is $5n + 25q$. The expression in dollars is $0.05n + 0.25q$.

You Try 5

Determine the amount of money you have in cents *and* in dollars if you have

a)   11 dimes   b)   20 pennies   c)   8 dimes and 46 pennies

d)   *d* dimes   e)   *p* pennies   f)   *d* dimes and *p* pennies

Next, we'll apply this idea of the value of different denominations of money to an application problem.

Example 7

In-Class Example 7
A jar contains $14.35 in change. If there are only nickels and dimes in the jar, how many nickels and dimes are there if there are three times as many dimes than nickels?
answer: 41 nickels, 123 dimes

Jamaal has only dimes and quarters in his piggy bank. When he counts the change, he finds that he has $18.60 and that there are twice as many quarters as dimes. How many dimes and quarters are in his bank?

Solution

Step 1: **Read** the problem carefully, and identify what we are being asked to find.

We must find the number of dimes and quarters in the bank.

Step 2: **Choose a variable** to represent an unknown, and define the other unknown in terms of this variable.

In the statement "there are twice as many quarters as dimes," the number of quarters is expressed *in terms of* the number of dimes. Therefore, let

$$d = \text{the number of dimes}$$

Define the other unknown (the number of quarters) in terms of d:

$$2d = \text{the number of quarters}$$

Step 3: **Translate** the information that appears in English into an algebraic equation.

Let's begin by making a table to write an expression for the value of the dimes and the value of the quarters. We will write the expression in terms of dollars because the total value of the coins, $18.60, is given in dollars.

	Value of the Coin	Number of Coins	Total Value of the Coins
Dimes	0.10	d	$0.10d$
Quarters	0.25	$2d$	$0.25 \cdot (2d)$

Write an equation in English and substitute the expressions we found in the table and the total value of the coins to get an algebraic equation.

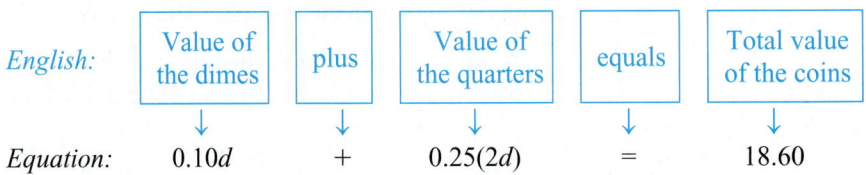

English:	Value of the dimes	plus	Value of the quarters	equals	Total value of the coins
	↓	↓	↓	↓	↓
Equation:	$0.10d$	$+$	$0.25(2d)$	$=$	18.60

Step 4: **Solve** the equation.

$$0.10d + 0.25(2d) = 18.60$$
$$100[0.10d + 0.25(2d)] = 100(18.60) \qquad \text{Multiply by 100 to eliminate the decimals.}$$
$$10d + 25(2d) = 1860 \qquad \text{Distribute.}$$
$$10d + 50d = 1860 \qquad \text{Multiply.}$$
$$60d = 1860 \qquad \text{Combine like terms.}$$
$$\frac{60d}{60} = \frac{1860}{60} \qquad \text{Divide each side by 60.}$$
$$d = 31 \qquad \text{Simplify.}$$

Step 5: **Check** the answer and **interpret** the meaning of the solution as it relates to the problem.

There were 31 dimes and $2(31) = 62$ quarters in the bank.

Check: The value of the dimes is $0.10(31) = \$3.10$, and the value of the quarters is $0.25(62) = \$15.50$. Their total is $\$3.10 + \$15.50 = \$18.60$. ■

You Try 6

A collection of coins consists of pennies and nickels. There are 5 fewer nickels than there are pennies. If the coins are worth a total of $4.97, how many of each type of coin is in the collection?

5. Solve Mixture Problems

Mixture problems involve combining two or more substances to make a mixture of them. We will begin with an example from arithmetic then extend this concept to be used with algebra.

Example 8

In-Class Example 8
Refer to Example 8. If the customer purchased 20 gal of gasoline, how many gallons of ethanol are in the 20 gal of gasoline?
answer: 2 gal of ethanol

The state of Illinois mixes ethanol (made from corn) in its gasoline to reduce pollution. If a customer purchases 15 gal of gasoline and it has a 10% ethanol content, how many gallons of ethanol are in the 15 gal of gasoline?

Solution

Write an equation in English first:

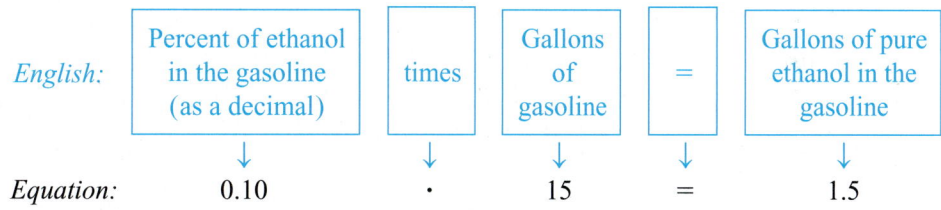

We *multiply* the percent of ethanol by the number of gallons of gasoline to get the number of gallons of ethanol in the gasoline.

The equation is $0.10(15) = 1.5$.

We can also organize the information in a table:

Percent of Ethanol in the Gasoline (as a decimal)	Gallons of Gasoline	Gallons of Ethanol in the Gasoline
0.10	15	$0.10(15) = 1.5$

Either way we find that there are 1.5 gallons of ethanol in 15 gallons of gasoline.

The idea above will be used to help us solve the next mixture problem.

Example 9

In-Class Example 9
A chemist needs to make 15 L of a 7% acid solution. It will be made from a mixture of some 1% acid solution and some 10% acid solution. How much of the 1% solution and 10% solution should be used?
answer: 5 L of 1% solution, and 10 L of 10% solution

A chemist needs to make 30 liters (L) of a 6% acid solution. She will make it from some 4% acid solution and some 10% acid solution that is in the storeroom. How much of the 4% solution and 10% solution should she use?

Solution

Step 1: **Read** the problem carefully, and identify what we are being asked to find.

We must find the amount of 4% acid solution and 10% acid solution she should use.

Step 2: **Choose a variable** to represent an unknown, and define the other unknown in terms of this variable. Let

x = the number of liters of 4% acid solution needed

Define the other unknown (the amount of 10% acid solution needed) in terms of x. Since she wants to make a total of 30 L of acid solution,

$30 - x$ = the number of liters of 10% acid solution needed

Step 3: **Translate** the information that appears in English into an algebraic equation.

Let's begin by arranging the information in a table. *Remember, to obtain the expression in the last column, multiply the percent of acid in the solution by the number of liters of solution to get the number of liters of acid in the solution.*

	Percent of Acid in Solution (as a decimal)	Liters of Solution	Liters of Acid in Solution
Mix these	0.04	x	$0.04x$
	0.10	$30 - x$	$0.10(30 - x)$
to make →	0.06	30	$0.06(30)$

Now, write an equation in English. Since we obtain the 6% solution by mixing the 4% and 10% solutions,

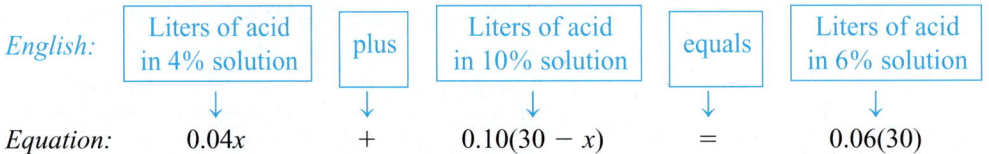

English: | Liters of acid in 4% solution | plus | Liters of acid in 10% solution | equals | Liters of acid in 6% solution |

Equation: $0.04x$ $+$ $0.10(30 - x)$ $=$ $0.06(30)$

The equation is $0.04x + 0.10(30 - x) = 0.06(30)$.

Step 4: **Solve** the equation.

$$0.04x + 0.10(30 - x) = 0.06(30)$$
$$100[0.04x + 0.10(30 - x)] = 100[0.06(30)]$$ Multiply by 100 to eliminate decimals.

$$4x + 10(30 - x) = 6(30)$$
$$4x + 300 - 10x = 180$$ Distribute.
$$-6x + 300 = 180$$ Combine like terms.
$$-6x = -120$$ Subtract 300 from each side.
$$x = 20$$ Divide by -6.

Step 5: **Check** the answer and **interpret** the meaning of the solution as it relates to the problem.

The chemist needs 20 L of the 4% solution.

Find the other unknown, the amount of 10% solution needed.

$$30 - x = 30 - 20 = 10 \text{ L of 10% solution}$$

Check:

Acid in 4% solution + Acid in 10% solution = Acid in 6% solution
$$0.04(20)   +   0.10(10)   =   0.06(30)$$
$$0.80   +   1.00   =   1.80$$
$$1.80 = 1.80$$ ∎

You Try 7

How many milliliters (mL) of a 7% alcohol solution and how many milliliters of a 15% alcohol solution must be mixed to make 20 mL of a 9% alcohol solution?

6. Solve Problems Involving Distance, Rate, and Time

If you drive at 50 mph for 4 hr, how far will you drive? One way to get the answer is to use the formula

$$\text{Distance} = \text{Rate} \times \text{Time}$$
$$\text{or}$$
$$d = rt$$
$$d = (50 \text{ mph}) \cdot (4 \text{ hr})$$
$$\text{Distance traveled} = 200 \text{ mi}$$

Notice that the rate is in miles per *hour* and the time is in *hours*. The units must be consistent in this way. If the time in this problem had been expressed in minutes, it would have been necessary to convert minutes to hours. The formula $d = rt$ will be used in Example 10.

Example 10

In-Class Example 10
Refer to Example 10. If Jenny is traveling at 50 mph and Alex is traveling at 65 mph, how long will it take Alex to catch Jenny?
answer: $1\frac{1}{8}$ hr

Alex and Jenny are taking a cross-country road-trip on their motorcycles. Jenny leaves a rest area first traveling at 60 mph. Alex leaves 30 min later, traveling on the same highway, at 70 mph. How long will it take Alex to catch Jenny?

Solution

Step 1: Read the problem carefully, and identify what we are being asked to find.

We must determine how long it takes Alex to catch Jenny.

We will draw a picture to help us see what is happening in this problem.

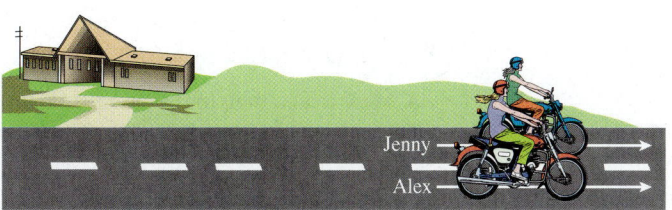

Since both girls leave the same rest area and travel on the same highway, when Alex catches Jenny they have driven the *same* distance.

Step 2: Choose a variable to represent an unknown, and define the other unknown in terms of this variable.

Alex's time is in terms of Jenny's time, so let

t = the number of hours Jenny has been riding when Alex catches her

Alex leaves 30 minutes ($\frac{1}{2}$ hour) after Jenny, so Alex travels $\frac{1}{2}$ hour *less than* Jenny.

$t - \frac{1}{2}$ = the number of hours it takes Alex to catch Jenny

Step 3: Translate the information that appears is English into an algebraic equation.

Let's make a table using the equation $d = rt$. Fill in the time and the rates first, then multiply those together to fill in the value for the distance.

	d	r	t
Jenny	$60t$	60	t
Alex	$70(t - \frac{1}{2})$	70	$t - \frac{1}{2}$

We will write an equation in English to help us write an algebraic equation. The picture shows that

English:

Jenny's distance	is the same as	Alex's distance
↓	↓	↓

Equation: $60t$ $=$ $70\left(t - \dfrac{1}{2}\right)$

The equation is $60t = 70\left(t - \dfrac{1}{2}\right)$.

Step 4:   **Solve** the equation.

$$60t = 70\left(t - \frac{1}{2}\right)$$

$$
\begin{aligned}
60t &= 70t - 35 &&\text{Distribute.}\\
-10t &= -35 &&\text{Subtract } 70t.\\
\frac{-10t}{-10} &= \frac{-35}{-10} &&\text{Divide each side by } -10.\\
t &= 3.5 &&\text{Simplify.}
\end{aligned}
$$

Step 5:   **Check** the answer and **interpret** the meaning of the solution as it relates to the problem.

Remember, Jenny's time is t. Alex's time is $t - \dfrac{1}{2} = 3\dfrac{1}{2} - \dfrac{1}{2} = 3$ hr.

It will take Alex 3 hr to catch Jenny.

Check to see that Jenny travels 60 mph · (3.5 hr) = 210 miles, and Alex travels 70 mph · (3 hr) = 210 miles. The girls travel the same distance.   ■

You Try 8

The Kansas towns of Topeka and Voda are 230 mi apart. Alberto left Topeka driving west on Interstate 70 at the same time Ramon left Voda driving east toward Topeka on I-70. Alberto and Ramon meet after 2 hr. If Alberto's speed was 5 mph faster than Ramon's speed, how fast was each of them driving?

Answers to You Try Exercises

1) a) true   b) false   2) a) $n = 15$ b) $k = 7$ 3) $31.68   4) 18   5) a) 110¢, $1.10
b) 20¢, $0.20   c) 126¢, $1.26   d) 10$d$ cents, 0.10d dollars   e) 1p cents, 0.01p dollars
f) 10d + 1p cents, 0.10d + 0.01p dollars   6) 87 pennies, 82 nickels
7) 15 ml of 7%, 5 ml of 15%   8) Ramon: 55 mph, Alberto: 60 mph

2.4 Exercises

*Additional answers can be found in the Answers to Exercises appendix.

Objective I: Define Ratio and Proportion

Write as a ratio in lowest terms.

1) 15 tea drinkers to 25 coffee drinkers $\dfrac{3}{5}$

2) 4 mi to 6 mi $\dfrac{2}{3}$

3) 60 ft to 45 ft $\dfrac{4}{3}$

4) 19 girls to 16 boys $\dfrac{19}{16}$

5) What is the difference between a ratio and a proportion?
   A ratio is a quotient of two quantities. A proportion is a statement that two ratios are equal.

6) Is 0.45 equivalent to the ratio 9 to 20? Explain.
   Yes, the ratio "9 to 20" can be written as $\dfrac{9}{20}$. This reduces to 0.45.

Determine if each proportion is true or false. See Example 1.

7) $\dfrac{2}{15} = \dfrac{8}{45}$ false

8) $\dfrac{50}{35} = \dfrac{10}{7}$ true

9) $\dfrac{42}{77} = \dfrac{6}{11}$ true

10) $\dfrac{20}{30} = \dfrac{4}{3}$ false

Objective 2: Solve a Proportion

Solve each proportion. See Example 2.

11) $\dfrac{m}{9} = \dfrac{10}{45}$ {2}

12) $\dfrac{c}{3} = \dfrac{16}{12}$ {4}

13) $\dfrac{120}{50} = \dfrac{x}{4}$ $\left\{\dfrac{48}{5}\right\}$

14) $\dfrac{2}{27} = \dfrac{r}{36}$ $\left\{\dfrac{8}{3}\right\}$

15) $\dfrac{2a + 3}{8} = \dfrac{2}{16}$ {−1}

16) $\dfrac{5d - 1}{2} = \dfrac{6d + 3}{3}$ {3}

17) $\dfrac{n - 4}{5} = \dfrac{5n - 2}{10}$ {−2}

18) $\dfrac{2w - 1}{7} = \dfrac{3 - 6w}{4}$ $\left\{\dfrac{1}{2}\right\}$

Objective 3: Solve Problems Involving a Proportion

For each problem, set up a proportion and solve. See Example 3.

19) David's hobby is archery. Usually, he can hit the bulls-eye 9 times out of 15 tries. If he shot 50 arrows, how many bulls-eyes could he expect to make?   30

20) Hector buys batteries for his Gameboy. He knows that 3 packs of batteries cost $4.26. How much should Hector expect to pay for 5 packs of batteries? $7.10

21) A 12-oz serving of Mountain Dew contains 55 mg of caffeine. How much caffeine is in an 18-oz serving of Mountain Dew? (www.nsda.org)   82.5 mg

22) An 8-oz serving of Coca-Cola Classic contains 23 mg of caffeine. How much caffeine is in a 12-oz serving of Coke? (www.nsda.org)   34.5 mg

23) The national divorce rate is about 4.8 divorces per 1000 people. How many divorces are expected in a town of 35,000 people if that town followed the national rate? (www.cobras.org)   168

24) If the exchange rate between the American dollar and the Norwegian krone is such that $2.00 = 13.60 krone, how many krone could be exchanged for $15.00? (moneycentral.msn.com)   102 krone

Find the length of the indicated side, x, by setting up a proportion. See Example 4.

25)

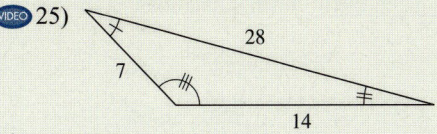

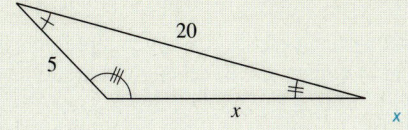

$x = 10$

26)

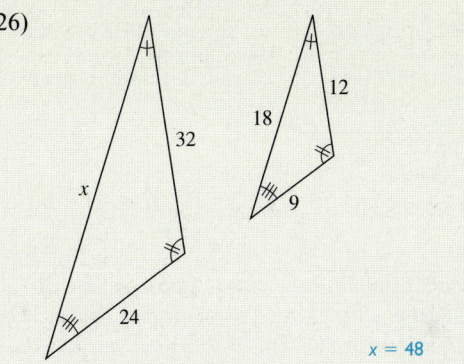

$x = 48$

27)

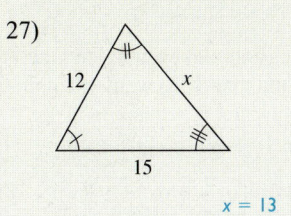

$x = 13$

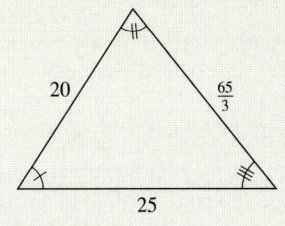

28)

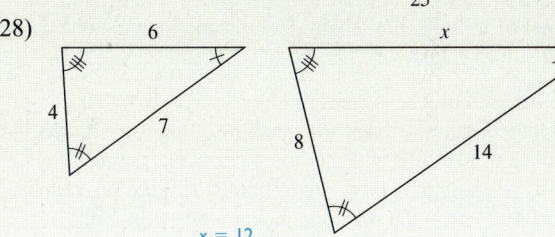

$x = 12$

29)

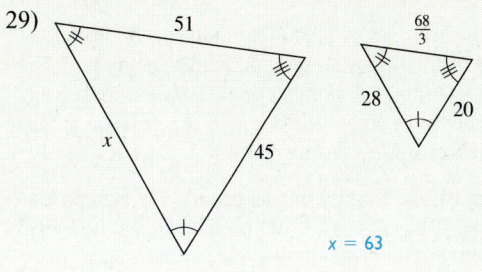

$x = 63$

30)

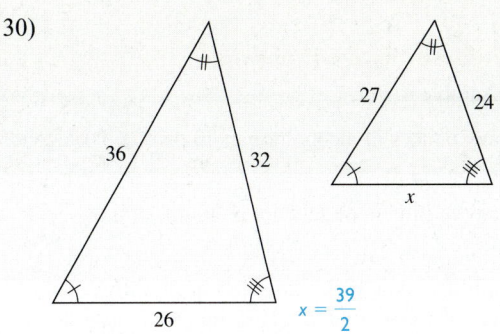

$x = \dfrac{39}{2}$

Objective 4: Solve Problems Involving Money

For Exercises 31–36, determine the amount of money a) in dollars and b) in cents given the following quantities.

31) 8 dimes
    a) $0.80   b) 80¢
32) 32 nickels
    a) $1.60   b) 160¢
33) 217 pennies
    a) $2.17   b) 217¢
34) 12 quarters
    a) $3.00   b) 300¢
35) 9 quarters and 7 dimes
    a) $2.95   b) 295¢
36) 89 pennies and 14 nickels VIDEO
    a) $1.59   b) 159¢

For Exercises 37–42 write an expression which represents the amount of money in a) dollars and b) cents given the following quanities.

37) q quarters
    a) 0.25q dollars   b) 25q cents
38) p pennies
    a) 0.01p dollars   b) p cents
39) d dimes
    a) 0.10d dollars   b) 10d cents
40) n nickels
    a) 0.05n dollars   b) 5n cents
41) p pennies and q quarters
    a) 0.01p + 0.25q dollars   b) p + 25q cents
42) n nickels and d dimes
    a) 0.05n + 0.10d dollars   b) 5n + 10d cents

Solve using the five "Steps for Solving Applied Problems." See Example 7.

43) Gino and Vince combine their coins to find they have all nickels and quarters. They have 8 more quarters than nickels, and the coins are worth a total of $4.70. How many nickels and quarters do they have?   9 nickels, 17 quarters

44) Danika saves all of her pennies and nickels in a jar. One day she counted them and found that there were 131 coins worth $3.43. How many pennies and how many nickels were in the jar?   78 pennies and 53 nickels

45) Kyung Soo has been saving her babysitting money. She has $69.00 consisting of $5 bills and $1 bills. If she has a total of 25 bills, how many $5 bills and how many $1 bills does she have?   11 $5 bills, 14 $1 bills

46) A bank employee is servicing the ATM after a busy Friday night. She finds the machine contains only $20 bills and $10 bills and that there are twice as many $20 bills remaining as there are $10 bills. If there is a total of $550.00 left in the machine, how many of the bills are twenties, and how many are tens?   22 $20 bills, 11 $10 bills

VIDEO 47) A movie theater charges $9.00 for adults and $7.00 for children. The total revenue for a particular movie is $475.00. Determine the number of each type of ticket sold if the number of children's tickets sold was half the number of adult tickets sold.   38 adult tickets, 19 children's tickets

48) At the post office, Ronald buys 12 more 44¢ stamps than 28¢ stamps. If he spends $11.04 on the stamps, how many of each type did he buy?   20 44¢ stamps, 8 28¢ stamps

Objective 5: Solve Mixture Problems

Solve. See Example 8.

49) How many ounces of alcohol are in 40 oz of a 5% alcohol solution?   2 oz

50) How many milliliters of acid are in 30 mL of a 6% acid solution?   1.8 mL

51) Sixty milliliters of a 10% acid solution are mixed with 40 mL of a 4% acid solution. How much acid is in the mixture?   7.6 mL

52) Fifty ounces of an 8% alcohol solution are mixed with 20 ounces of a 6% alcohol solution. How much alcohol is in the mixture?   5.2 oz

Solve using the five "Steps for Solving Applied Problems." See Example 9.

53) How many ounces of a 4% acid solution and how many ounces of a 10% acid solution must be mixed to make 24 oz of a 6% acid solution?   16 oz of the 4% acid solution, 8 oz of the 10% acid solution

54) How many milliliters of a 15% alcohol solution must be added to 80 mL of an 8% alcohol solution to make a 12% alcohol solution? $106\frac{2}{3}$ mL

55) How many liters of a 40% antifreeze solution must be mixed with 5 L of a 70% antifreeze solution to make a mixture that is 60% antifreeze? $2\frac{1}{2}$ L

56) How many milliliters of a 7% hydrogen peroxide solution and how many milliliters of a 1% hydrogen peroxide solution should be mixed to get 400 mL of a 3% hydrogen peroxide solution? $133\frac{1}{3}$ mL of the 7% solution, $266\frac{2}{3}$ mL of the 1% solution.

57) Custom Coffees blends its coffees for customers. How much of the Aztec coffee, which sells for $6.00 per pound, and how much of the Cinnamon coffee, which sells for $8.00 per pound, should be mixed to make 5 lb of the Winterfest blend to be sold at $7.20 per pound?   2 lb of Aztec and 3 lb of Cinnamon

58) All-Mixed-Up Nut Shop sells a mix consisting of cashews and pistachios. How many pounds of cashews, which sell for $7.00 per pound, should be mixed with 4 lb of pistachios, which sell for $4.00 per pound, to get a mix worth $5.00 per pound?   2 lb

Objective 6: Solve Problems Involving Distance, Rate, and Time

Solve using the five "Steps for Solving Applied Problems." See Example 10.

59) Two cars leave Indianapolis, one driving east and the other driving west. The eastbound car travels 8 mph slower than the westbound car, and after 3 hr they are 414 mi apart. Find the speed of each car.
    eastbound: 65 mph; westbound: 73 mph

60) Two planes leave San Francisco, one flying north and the other flying south. The southbound plane travels 50 mph faster than the northbound plane, and after 2 hours they are 900 miles apart. Find the speed of each plane.
    northbound: 200 mph; southbound: 250 mph

61) Maureen and Yvette leave the gym to go to work traveling the same route, but Maureen leaves 10 min after Yvette. If Yvette drives 60 mph and Maureen drives 72 mph, how long will it take Maureen to catch Yvette? $\frac{5}{6}$ hr

62) Vinay and Sadiva leave the same location traveling the same route, but Sadiva leaves 20 minutes after Vinay. If Vinay drives 30 mph and Sadiva drives 36 mph, how long will it take Sadiva to catch Vinay? $1\frac{2}{3}$ hr

63) A passenger train and a freight train leave cities 400 mi apart and travel toward each other. The passenger train is traveling 20 mph faster than the freight train. Find the speed of each train if they pass each other after 5 hr.
passenger train: 50 mph; freight train: 30 mph

64) A freight train passes the Old Towne train station at 11:00 A.M. going 30 mph. Ten minutes later a passenger train, headed in the same direction on an adjacent track, passes the same station at 45 mph. At what time will the passenger train catch the freight train?   11:30 A.M.

65) A truck and a car leave the same intersection traveling in the same direction. The truck is traveling at 35 mph, and the car is traveling at 45 mph. In how many minutes will they be 6 mi apart?   36 min

66) At noon, a truck and a car leave the same intersection traveling in the same direction. The truck is traveling at 30 mph, and the car is traveling at 42 mph. At what time will they be 9 mi apart?   12:45 P.M.

67) Ajay is traveling north on a road while Rohan is traveling south on the same road. They pass by each other at 3 P.M., Ajay driving 30 mph and Rohan driving 40 mph. At what time will they be 105 miles apart?   4.30 P.M.

68) When Lance and Jan pass each other on their bikes going in opposite directions, Lance is riding at 22 mph, and Jan is pedaling at 18 mph. If they continue at those speeds, after how long will they be 100 mi apart? $2\frac{1}{2}$ hr

69) At noon, a cargo van crosses an intersection at 30 mph. At 12:30 P.M., a car crosses the same intersection traveling in the opposite direction. At 1 P.M., the van and car are 54 miles apart. How fast is the car traveling?   48 mph

70) A freight train passes the Naperville train station at 9:00 A.M. going 30 mph. Ten minutes later a passenger train, headed in the same direction on an adjacent track, passes the same station at 45 mph. At what time will the passenger train catch the freight train?   9:30 A.M.

Mixed Exercises: Objectives 3–6

Solve.

71) At the end of her shift, a cashier has a total of $6.30 in dimes and quarters. There are 7 more dimes than quarters. How many of each of these coins does she have?
23 dimes, 16 quarters

72) An alloy that is 30% silver is mixed with 200 g of a 10% silver alloy. How much of the 30% alloy must be used to obtain an alloy that is 24% silver? $466\frac{2}{3}$ g

73) One-half cup of Ben and Jerry's Cherry Garcia Ice Cream has 260 calories. How many calories are in 3 cups of Cherry Garcia? (www.benjerry.com)   1560 calories

74) According to national statistics, 41 out of 200 Americans aged 28–39 are obese. In a group of 800 Americans in this age group, how many would be expected to be obese? (0-web10.epnet.com)   164

75) A jet flying at an altitude of 35,000 ft passes over a small plane flying at 10,000 ft headed in the same direction. The jet is flying twice as fast as the small plane, and thirty minutes later they are 100 mi apart. Find the speed of each plane.   jet: 400 mph, small plane: 200 mph

76) Tickets for a high school play cost $3.00 each for children and $5.00 each for adults. The revenue from one performance was $663, and 145 tickets were sold. How many adult tickets and how many children's tickets were sold?   114 adult tickets, 31 children's tickets

77) A pharmacist needs to make 20 cubic centimeters (cc) of a 0.05% steroid solution to treat allergic rhinitis. How much of a 0.08% solution and a 0.03% solution should she use?

8 cc of the 0.08% solution and 12 cc of the 0.03% solution

78) Geri is riding her bike at 10 mph when Erin passes her going in the opposite direction at 14 mph. How long will it take before the distance between them is 6 mi? $\frac{1}{4}$ hr or 15 min

79) How much pure acid must be added to 6 gal of a 4% acid solution to make a 20% acid solution? $1\frac{1}{5}$ gal

80) If the exchange rate between the American dollar and the Japanese Yen is such that $4.00 = 442 yen, how many yen could be exchanged for $30.00? (moneycentral.msn.com)   3315 yen

Section 2.5 Linear Inequalities in One Variable

Objectives

1. **Use Graphs and Set and Interval Notations**
2. **Solve Inequalities Using the Addition and Subtraction Properties of Inequality**
3. **Solve Inequalities Using the Multiplication Property of Inequality**
4. **Solve Inequalities Using a Combination of the Properties**
5. **Solve Compound Inequalities Containing Three Parts**
6. **Solve Applications Involving Linear Inequalities**

Recall the inequality symbols

$<$ "is less than" $\leq$ "is less than or equal to"

$>$ "is greater than" $\geq$ "is greater than or equal to"

We will use the symbols to form *linear inequalities in one variable*.

While an equation states that two expressions are equal, an *inequality* states that two expressions are not necessarily equal. Here is a comparison of an equation and an inequality:

Equation	**Inequality**
$3x - 8 = 13$	$3x - 8 \leq 13$

> **Definition**
>
> A **linear inequality in one variable** can be written in the form $ax + b < c$, $ax + b \leq c$, $ax + b > c$, or $ax + b \geq c$ where $a, b,$ and c are real numbers and $a \neq 0$.

The solution to a linear inequality is a set of numbers that can be represented in one of three ways:

1) On a graph

2) In *set notation*

3) In *interval notation*

In this section, we will learn how to solve linear inequalities in one variable and how to represent the solution in each of those three ways.

1. Use Graphs and Set and Interval Notations

Example 1

In-Class Example 1
Graph each inequality and express the solution in set notation and interval notation.
a) $x < 4$ b) $w \geq 0$
answer: a)

$\{x|x < 4\}; (-\infty, 4)$
b)

$\{w|w \geq 0\}; [0, \infty)$

Graph each inequality and express the solution in set notation and interval notation.

a) $x \leq -1$ b) $t > 4$

Solution

a) $x \leq -1$:

Graphing $x \leq -1$ means that we are finding the solution set of $x \leq -1$. What value(s) of x will make the inequality true? The largest solution is -1. Then, any number *less than* -1 will make $x \leq -1$ true. We represent this **on the number line** as follows:

The graph illustrates that the solution is the set of all numbers less than and including -1.

Notice that the dot on -1 is shaded. This tells us that -1 is included in the solution set. The shading to the left of -1 indicates that *any* real number (not just integers) in this region is a solution.

We can express the solution set in **set notation** this way: $\{x \mid x \leq -1\}$

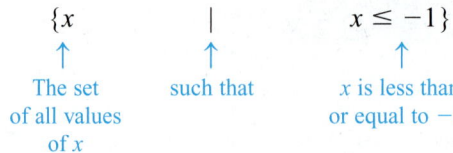

$\{x$	$\mid$	$x \leq -1\}$
↑	↑	↑
The set of all values of x	such that	x is less than or equal to -1.

In **interval notation** we write $(-\infty, -1]$

$-\infty$ is not a number. x gets infinitely more negative without bound. Use a "(" instead of a bracket

The bracket indicates the -1 is included in the interval.

Note

The variable does not appear anywhere in interval notation.

---

b) $t > 4$:

We will plot 4 as an *open circle* on the number line because the symbol is ">" and *not* "≥." The inequality $t > 4$ means that we must find the set of all numbers, t, *greater than* (but *not* equal to) 4. Shade to the right of 4.

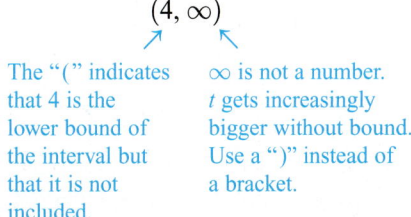
$$-4\ -3\ -2\ -1\ \ 0\ \ 1\ \ 2\ \ 3\ \ 4\ \ 5\ \ 6$$

The graph illustrates that the solution is the set of all numbers greater than 4 but not including 4.

We can express the solution set in *set notation* this way: $\{t \mid t > 4\}$
In *interval notation* we write

$$(4, \infty)$$

The "(" indicates that 4 is the lower bound of the interval but that it is not included.

∞ is not a number. t gets increasingly bigger without bound. Use a ")" instead of a bracket.

Hints for using interval notation:

1)   The variable never appears in interval notation.

2)   A number *included* in the solution set gets a bracket: $x \le -1 \rightarrow (-\infty, -1]$

3)   A number *not included* in the solution set gets a parenthesis: $t > 4 \rightarrow (4, \infty)$

4)   The symbols $-\infty$ and ∞ *always* get parentheses.

5)   The smaller number is always placed to the left. The larger number is placed to the right.

6)   Even if we are not asked to graph the solution set, the graph may be helpful in writing the interval notation correctly.   ■

You Try 1

Graph each inequality and express the solution in interval notation.

a) $k \ge -7$ b) $c < 5$

---

2. Solve Inequalities Using the Addition and Subtraction Properties of Inequality

The addition and subtraction properties of equality help us to solve equations. Similar properties hold for inequalities as well.

> **Property**   Addition and Subtraction Properties of Inequality
>
> Let a, b, and c be real numbers. Then,
>
> 1) $a < b$ and $a + c < b + c$ are equivalent
>
> *and*
>
> 2) $a < b$ and $a - c < b - c$ are equivalent.
>
> Adding the same number to both sides of an inequality or subtracting the same number from both sides of an inequality will not change the solution.

Note

The above properties hold for any of the inequality symbols.

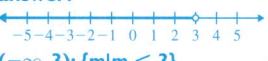

 **Example 2**

Solve $y - 8 \geq -5$. Graph the solution set and write the answer in interval and set notations.

In-Class Example 2
Solve $m - 4 < -1$. Graph the solution set and write the answer in interval and set notations.
answer:

```
←+++++++◇+++→
 −5−4−3−2−1 0 1 2 3 4 5
```

$(-\infty, 3)$; $\{m \mid m < 3\}$

Solution

$$y - 8 \geq -5$$
$$y - 8 + 8 \geq -5 + 8 \qquad \text{Add 8 to each side.}$$
$$y \geq 3$$

```
←++++++++++→
 −4−3−2−1 0 1 2 3 4
```

The solution set in interval notation is $[3, \infty)$. In set notation we write $\{y \mid y \geq 3\}$.   ■

You Try 2

Solve $k - 10 \geq -4$. Graph the solution set and write the answer in interval and set notations.

---

3. Solve Inequalities Using the Multiplication Property of Inequality

While the addition and subtraction properties for solving equations and inequalities work the same way, this is not true for multiplication and division. Let's see why.

Begin with an inequality we know is true: $2 < 5$. Multiply both sides by a *positive* number, say 3.

$$2 < 5 \qquad \text{True}$$
$$3(2) < 3(5) \qquad \text{Multiply by 3.}$$
$$6 < 15 \qquad \text{True}$$

Begin again with $2 < 5$. Multiply both sides by a *negative* number, say -3.

$$2 < 5 \qquad \text{True}$$
$$-3(2) < -3(5) \qquad \text{Multiply by } -3.$$
$$-6 < -15 \qquad \text{False}$$

To make $-6 < -15$ into a *true* statement, we must *reverse the direction of the inequality symbol.*

$$-6 > -15 \qquad \text{True}$$

If you begin with a true inequality and *divide* by a positive number or by a negative number, the results will be the same as above since division can be defined in terms of multiplication. This leads us to the multiplication property of inequality.

> **Property**   Multiplication Property of Inequality
>
> Let a, b, and c be real numbers.
>
> 1) If c is a *positive* number, then $a < b$ and $ac < bc$ are equivalent inequalities and have the same solutions.
>
> 2) If c is a *negative* number, then $a < b$ and $ac > bc$ are equivalent inequalities and have the same solutions.

It is also true that if $c > 0$ and $a < b$, then $\dfrac{a}{c} < \dfrac{b}{c}$. If $c < 0$ and $a < b$, then $\dfrac{a}{c} > \dfrac{b}{c}$.

For the most part, the procedures used to solve linear inequalities are the same as those for solving linear equations **except** *when you multiply or divide an inequality by a negative number, you must reverse the direction of the inequality symbol.*

Example 3

In-Class Example 3
Solve each inequality. Graph the solution set and write the answer in interval and set notations.
a) $3r > -12$ b) $-3r > 12$
answer: a)

$\leftarrow\!+\!\!+\!\!\diamond\!\!+\!\!+\!\!+\!\!+\!\!+\!\!+\!\!+\!\!+\!\!+\!\!+\!\!\rightarrow$
$-5\,-4\,-3\,-2\,-1\ \ 0\ \ 1\ \ 2\ \ 3\ \ 4\ \ 5$
$(-4, \infty);\ \{r\,|\,r > -4\}$
b)

$\leftarrow\!+\!\!+\!\!\diamond\!\!+\!\!+\!\!+\!\!+\!\!+\!\!+\!\!+\!\!+\!\!+\!\!+\!\!\rightarrow$
$-5\,-4\,-3\,-2\,-1\ \ 0\ \ 1\ \ 2\ \ 3\ \ 4\ \ 5$
$(-\infty, -4);\ \{r\,|\,r < -4\}$

Solve each inequality. Graph the solution set and write the answer in interval and set notations.

a) $-5w \le 20$ b) $5w \le -20$

Solution

a) $-5w \le 20$

First, divide each side by -5. *Since we are dividing by a negative number, we must remember to reverse the direction of the inequality symbol.*

$$-5w \le 20$$

$$\frac{-5w}{-5} \ge \frac{20}{-5} \qquad \text{Divide by } -5, \text{ so reverse the inequality symbol.}$$

$$w \ge -4$$

Interval notation: $[-4, \infty)$. Set notation: $\{w\,|\,w \ge -4\}$

b) $5w \le -20$

First, divide by 5. Since we are dividing by a *positive* number, the inequality symbol remains the same.

$$5w \le -20$$

$$\frac{5w}{5} \le \frac{-20}{5} \qquad \text{Divide by 5. Do } not \text{ reverse the inequality symbol.}$$

$$w \le -4$$

Interval notation: $(-\infty, -4]$.
Set notation: $\{w\,|\,w \le -4\}$

 You Try 3

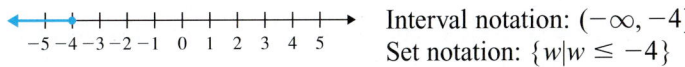

Solve $-\dfrac{1}{4}m < 3$. Graph the solution set and write the answer in interval and set notations.

4. Solve Inequalities Using a Combination of the Properties

Often it is necessary to combine the properties to solve an inequality.

Example 4

In-Class Example 4
Solve
$2(1 - 3f) + 4f \leq 2(7 + 2f)$.
Graph the solution set and write
the answer in interval and set
notations.
answer:

$-5\ -4\ -3\ -2\ -1\ \ 0\ \ 1\ \ 2\ \ 3\ \ 4\ \ 5$

$[-2, \infty)$; $\{f | f \geq -2\}$

Solve $4(5 - 2d) + 11 < 2(d + 3)$. Graph the solution set and write the answer in interval and set notations.

Solution

$$4(5 - 2d) + 11 < 2(d + 3)$$
$$20 - 8d + 11 < 2d + 6 \qquad \text{Distribute.}$$
$$31 - 8d < 2d + 6 \qquad \text{Combine like terms.}$$
$$31 - 8d - 2d < 2d - 2d + 6 \qquad \text{Subtract } 2d \text{ from each side.}$$
$$31 - 10d < 6$$
$$31 - 31 - 10d < 6 - 31 \qquad \text{Subtract 31 from each side.}$$
$$-10d < -25$$
$$\frac{-10d}{-10} > \frac{-25}{-10} \qquad \begin{array}{l}\text{Divide both sides by } -10.\\ \text{Reverse the inequality symbol.}\end{array}$$
$$d > \frac{5}{2} \qquad \text{Simplify.}$$

To graph the inequality, think of $\frac{5}{2}$ as $2\frac{1}{2}$.

$\overset{\frac{5}{2}}{\longleftarrow\!+\!+\!+\!+\!+\!+\!\circ\!+\!\longrightarrow}$
$-4\ -3\ -2\ -1\ \ 0\ \ 1\ \ 2\ \ 3\ \ 4$
Interval notation: $\left(\frac{5}{2}, \infty\right)$. Set notation: $\left\{ d \,\middle|\, d > \frac{5}{2} \right\}$ ■

You Try 4

Solve $4(p + 2) + 1 > 2(3p + 10)$. Graph the solution set and write the answer in interval and set notations.

5. Solve Compound Inequalities Containing Three Parts

A **compound inequality** contains more than one inequality symbol. Some types of compound inequalities are

$$-5 < b + 4 < 1, \quad t \leq \frac{1}{2} \text{ or } t \geq 3, \quad \text{and} \quad 2z + 9 < 5 \text{ and } z - 1 > 6$$

In this section, we will learn how to solve the first type of compound inequality. In Section 2.6 we will discuss the last two.

Consider the inequality $-2 \leq x \leq 3$. We can think of this in two ways:

1) x is *between* -2 and 3, and -2 and 3 are included in the interval.

or

2) We can break up $-2 \leq x \leq 3$ into the two inequalities $-2 \leq x$ *and* $x \leq 3$.

Either way we think about $-2 \leq x \leq 3$, the meaning is the same. On a number line, the inequality would be represented as

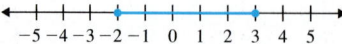
$-5\ -4\ -3\ -2\ -1\ \ 0\ \ 1\ \ 2\ \ 3\ \ 4\ \ 5$

Notice that the **lower bound** of the interval on the number line is -2 (including -2), and the **upper bound** is 3 (including 3). Therefore, we can write the interval notation as

$$[-2, 3]$$

The endpoint, -2,   The endpoint, 3,
is included in the   is included in the
interval, so use   interval, so use
a bracket.   a bracket.

The set notation to represent $-2 \le x \le 3$ is $\{x|-2 \le x \le 3\}$.

Next, we will solve the inequality $-5 < b + 4 < 1$. To solve this type of compound inequality you must remember that *whatever operation you perform on one part of the inequality must be performed on all three parts.* All properties of inequalities apply.

Example 5

In-Class Example 5
Solve $0 \le h + 3 \le 5$. Graph the solution set, and write the answer in interval notation.
answer: $[-3, 2]$

Solve $-5 < b + 4 < 1$. Graph the solution set, and write the answer in interval notation.

Solution

$$-5 < b + 4 < 1$$
$$-5 - 4 < b + 4 - 4 < 1 - 4 \qquad \text{To get the } b \text{ by itself subtract 4}$$
$$-9 < b < -3 \qquad \qquad \text{from each part of the inequality.}$$

Interval notation: $(-9, -3)$.

Note

Use parentheses in the interval notation since -9 and -3 are not included in the solution set.

 **You Try 5**

Solve $-2 \le 7k - 9 \le 19$. Graph the solution set, and write the answer in interval notation.

We can eliminate fractions in an inequality by multiplying by the LCD of all of the fractions.

Example 6

In-Class Example 6
Solve $-\frac{1}{2} < \frac{2}{3}p + \frac{3}{2} \le \frac{11}{6}$.
Graph the solution set, and write the answer in interval notation.
answer: $\left(-3, \frac{1}{2}\right]$

Solve $-\frac{7}{3} < \frac{1}{2}y - \frac{1}{3} \le \frac{1}{2}$. Graph the solution set, and write the answer in interval notation.

Solution

The LCD of the fractions is 6. Multiply by 6 to eliminate the fractions.

$$-\frac{7}{3} < \frac{1}{2}y - \frac{1}{3} \le \frac{1}{2}$$

$$6\left(-\frac{7}{3}\right) < 6\left(\frac{1}{2}y - \frac{1}{3}\right) \le 6\left(\frac{1}{2}\right) \qquad \begin{array}{l}\text{Multiply all parts of the} \\ \text{inequality by 6.}\end{array}$$

$$-14 < 3y - 2 \le 3$$

$$-14 + 2 < 3y - 2 + 2 \le 3 + 2 \qquad \text{Add 2 to each part.}$$

$$-12 < 3y \le 5 \qquad \qquad \text{Combine like terms.}$$

$$-\frac{12}{3} < \frac{3y}{3} \le \frac{5}{3} \qquad \qquad \text{Divide each part by 3.}$$

$$-4 < y \le \frac{5}{3} \qquad \qquad \text{Simplify.}$$

Interval notation: $\left(-4, \frac{5}{3}\right]$.

You Try 6

Solve $-\dfrac{3}{4} < \dfrac{1}{3}z - \dfrac{3}{4} \le \dfrac{5}{4}$. Graph the solution set, and write the answer in interval notation.

Remember, if we multiply or divide an inequality by a negative number, we reverse the direction of the inequality symbol. When solving a compound inequality like these, reverse *both* symbols.

Example 7

In-Class Example 7
Solve $-7 < -2x - 5 < 3$.
Graph the solution set, and
write the solution in interval
notation.
answer: (−4, 1)

Solve $11 < -3x + 2 < 17$. Graph the solution set, and write the answer in interval notation.

Solution

$$11 < -3x + 2 < 17$$

$$11 - 2 < -3x + 2 - 2 < 17 - 2 \qquad \text{Subtract 2 from each part.}$$

$$9 < -3x < 15$$

$$\dfrac{9}{-3} > \dfrac{-3x}{-3} > \dfrac{15}{-3} \qquad \text{When we divide by a negative number,}$$
$$\qquad\qquad\qquad\qquad \text{reverse the direction of the inequality symbol.}$$

$$-3 > x > -5 \qquad \text{Simplify.}$$

Think carefully about what $-3 > x > -5$ means. It means "x is less than -3 *and* x is greater than -5." This is especially important to understand when writing the correct interval notation.

The graph of the solution set is

Even though we got $-3 > x > -5$ as our result, -5 is actually the lower bound of the solution set and -3 is the upper bound. The inequality $-3 > x > -5$ can also be written as $-5 < x < -3$.

The solution in interval notation is $(-5, -3)$.

Lower bound on the left   Upper bound on the right

You Try 7

Solve $4 < -2x - 4 < 10$. Graph the solution set, and write the answer in interval notation.

6. Solve Applications Involving Linear Inequalities

Certain phrases in applied problems indicate the use of inequality symbols:

at least:	$\ge$	no less than:	$\ge$
at most:	$\le$	no more than:	$\le$

There are others. Next, we will look at an example of a problem involving the use of an inequality symbol. We will use the same steps that were used to solve applications involving equations.

Example 8

In-Class Example 8
Refer to Example 8. If the cost of the party is $150 for the first 8 children plus $3.50 for each additional child, how many children could attend the party if Joe can spend at most $215?
answer: 26 children

Joe Amici wants to have his son's birthday party at Kiddie Fun Factory. The cost of a party is $175 for the first 10 children plus $3.50 for each additional child. If Joe can spend at most $200, find the greatest number of children who can attend the party.

Solution

Step 1:   **Read** the problem carefully. We must find the greatest number of children who can attend the party.

Step 2:   **Choose a variable** to represent the unknown quantity. We know that the first 10 children will cost $175, but we do not know how may *additional* guests Joe can afford to invite.

x = number of children **over** the first 10 who attend the party

Step 3:   **Translate** from English to an algebraic inequality.

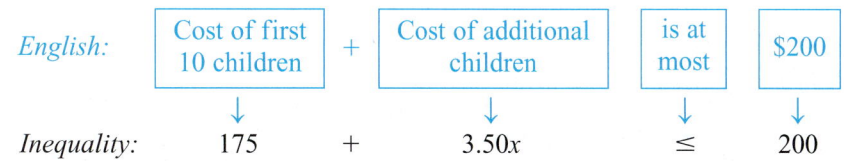

English:	Cost of first 10 children	+	Cost of additional children	is at most	$200
	↓		↓	↓	↓
Inequality:	175	+	3.50x	≤	200

The inequality is $175 + 3.50x \le 200$.

Step 4:   **Solve** the inequality.

$$175 + 3.50x \le 200$$
$$3.50x \le 25 \qquad \text{Subtract 175.}$$
$$x \le 7.142\ldots \qquad \text{Divide by 3.50.}$$

Step 5:   **Check** the answer and **interpret** the solution as it relates to the problem.

The result was $x \le 7.142\ldots$, where x represents the number of additional children who can attend the party. Since it is not possible to have $7.142\ldots$ people and $x \le 7.142\ldots$, in order to stay within budget, Joe can afford to pay for at most 7 additional guests *over* the initial 10.

Therefore, the greatest number of people who can attend the party is

The first 10 + additional = total
↓ ↓ ↓
10 + 7 = 17

At most, 17 children can attend the birthday party. Does the answer make sense?

$$\text{Total Cost of Party} = \$175 + \$3.50(7)$$
$$= \$175 + \$24.50$$
$$= \$199.50$$

We can see that one more guest (at a cost of $3.50) would put Joe over budget. ■

 You Try 8

For $4.00 per month, Van can send or receive 200 text messages. Each additional message costs $0.05. If Van can spend at most $9.00 per month on text messages, find the greatest number he can send or receive each month.

Answers to You Try Exercises

1) a) $[-7, \infty)$

b) $(-\infty, 5)$

2) interval: $[6, \infty)$, set: $\{k | k \geq 6\}$

3) interval: $(-12, \infty)$, set: $\{m | m > -12\}$

4) interval: $\left(-\infty, -\dfrac{11}{2}\right)$, set: $\left\{ p \left| p < -\dfrac{11}{2} \right. \right\}$

5) $[1, 4]$ 6) $(0, 6]$

7) $(-7, -4)$ 8) 300

2.5 Exercises

*Additional answers can be found in the Answers to Exercises appendix.

Objective 1: Use Graphs and Set and Interval Notations

1) When do you use parentheses when writing a solution set in interval notation?

2) When do you use brackets when writing a solution set in interval notation?

Graph the inequality. Express the solution in a) set notation and b) interval notation.

3) $x \geq 3$ 4) $t \geq -4$

5) $c < -1$ 6) $r < \dfrac{5}{2}$

7) $w > -\dfrac{11}{3}$ 8) $p \leq 2$

9) $1 \leq n \leq 4$ 10) $-3 \leq g \leq 2$

11) $-2 < a < 1$ 12) $-4 < d < 0$

13) $\dfrac{1}{2} < z \leq 3$ 14) $-2 \leq y < 3$

Mixed Exercises: Objectives 2 and 3

Solve each inequality. Graph the solution set and write the answer in a) set notation and b) interval notation. See Examples 2 and 3.

15) $n - 8 \leq -3$ 16) $p + 6 \geq 4$

17) $y + 5 \geq 1$ 18) $r - 9 \leq -5$

19) $3c > 12$ 20) $8v > 24$

21) $15k < -55$ 22) $16m < -28$

23) $-4b \leq 32$ 24) $-9a \geq 27$

25) $-14w > -42$ 26) $-30t < -18$

27) $\dfrac{1}{3}x < -2$ 28) $\dfrac{1}{5}z \geq -3$

29) $-\dfrac{2}{5}p \geq 4$ 30) $-\dfrac{9}{4}y < -18$

Objective 4: Solve Inequalities Using a Combination of the Properties

Solve each inequality. Graph the solution set and write the answer in interval notation. See Example 4.

31) $8z + 19 > 11$ 32) $5x - 2 \leq 18$

33) $12 - 7t \geq 15$ 34) $-1 - 4p < 5$

35) $-23 - w < -20$ 36) $16 - h \geq 9$

37) $7a + 4(5 - a) \leq 4 - 5a$

38) $6(7y + 4) - 10 > 2(10y + 13)$

39) $9c + 17 > 14c - 3$ 40) $-11n + 6 \leq 16 - n$

41) $\dfrac{8}{3}(2k + 1) > \dfrac{1}{6}k + \dfrac{8}{3}$

42) $\dfrac{1}{2}(c - 3) + \dfrac{3}{4}c \geq \dfrac{1}{2}(2c + 3) + \dfrac{3}{8}$

43) $0.04x + 0.12(10 - x) \geq 0.08(10)$

44) $0.09m + 0.05(8) \leq 0.07(m + 8)$

Objective 5: Solve Compound Inequalities Containing Three Parts

Solve each inequality. Graph the solution set and write the answer in interval notation. See Examples 5–7.

45) $-8 \leq a - 5 \leq -4$ 46) $1 \leq t + 3 \leq 7$

47) $9 < 6n < 18$ 48) $-10 < 2x < 7$

49) $-19 \leq 7p + 9 \leq 2$ 50) $-5 \leq 3k - 11 \leq 4$

51) $-6 \leq 4c - 13 < -1$ 52) $-11 < 6m + 1 \leq -3$

53) $2 < \dfrac{3}{4}u + 8 < 11$ 54) $2 \leq \dfrac{5}{2}y - 3 \leq 7$

55) $-\dfrac{1}{2} \leq \dfrac{5d + 2}{6} \leq 0$ 56) $2 < \dfrac{2b + 7}{3} < 5$

57) $-13 \leq 14 - 9h < 5$ 58) $3 < 19 - 2j \leq 9$

59) $0 \leq 4 - 3w \leq 7$ 60) $-6 < -5 - z < 0$

Mixed Exercises: Objectives 2–5

Solve each inequality. Write the answer in interval notation.

61) $k + 11 > 4$ $(-7, \infty)$ 62) $5 < x + 9 < 12$ $(-4, 3)$

63) $-12p \geq -16$ $\left(-\infty, \dfrac{4}{3}\right]$ 64) $2w + 7 \geq 13$ $[3, \infty)$

65) $5(2b - 3) - 7b > 5b + 9$
$(-\infty, -12)$ 66) $8 - m < 14$ $(-6, \infty)$

67) $-12 < \dfrac{8}{5}t + 12 \leq 6$ $\left(-15, -\dfrac{15}{4}\right]$

68) $0.29 \geq 0.04a + 0.05$ $(-\infty, 6]$

69) $\dfrac{5}{4}(k + 4) + \dfrac{1}{4} \geq \dfrac{5}{6}(k + 3) - 1$ $[-9, \infty)$

70) $-3 \leq 6c - 1 \leq 5$ $\left[-\dfrac{1}{3}, 1\right]$

71) $4 < 4 - 7y \leq 18$ $[-2, 0)$

72) $9z \leq -18$ $(-\infty, -2]$

Objective 6: Solve Applications Involving Linear Inequalities

Write an inequality for each problem and solve. See Example 8.

73) Oscar makes a large purchase at Home Depot and plans to rent one of its trucks to take his supplies home. The most he wants to spend on the truck rental is $50.00. If Home Depot charges $19.00 for the first 75 min and $5.00 for each additional 15 min, for how long can Oscar keep the truck and remain within his budget? (www.homedepot.com) He can rent the truck for at most 2 hr 45 min.

74) Carson's Parking Garage charges $4.00 for the first 3 hr plus $1.50 for each additional half-hour. Ted has only $11.50 for parking. For how long can Ted park his car in this garage? at most $5\dfrac{1}{2}$ hr

75) A taxi charges $2.00 plus $0.25 for every $\dfrac{1}{5}$ of a mile. How many miles can you go if you have $12.00? at most 8 mi

76) A taxi charges $2.50 plus $0.20 for every $\dfrac{1}{4}$ of a mile. How many miles can you go if you have $12.50? at most $12\dfrac{1}{2}$ mi

77) Melinda's first two test grades in Psychology were 87 and 94. What does she need to make on the third test to maintain an average of at least 90? 89 or higher

78) Russell's first three test scores in Geography were 86, 72, and 81. What does he need to make on the fourth test to maintain an average of at least 80? 81 or higher

---

Section 2.6 Compound Inequalities in One Variable

Objectives

1. Find the Intersection and Union of Two Sets
2. Solve Compound Inequalities Containing the Word And
3. Solve Compound Inequalities Containing the Word Or
4. Solve Special Compound Inequalities
5. Application of Intersection and Union

In Section 2.5, we learned how to solve a compound inequality like $-8 \leq 3x + 4 \leq 13$. In this section, we will discuss how to solve compound inequalities like these:

$$t \leq \dfrac{1}{2} \text{ or } t \geq 3 \qquad \text{and} \qquad 2z + 9 < 5 \text{ and } z - 1 > 6$$

But first, we must talk about set notation and operations.

1. Find the Intersection and Union of Two Sets

 Example 1

Let $A = \{1, 2, 3, 4, 5, 6\}$ and $B = \{3, 5, 7, 9, 11\}$.

The **intersection** of sets A and B is the set of numbers that are elements of A **and** of B. The *intersection* of A and B is denoted by $A \cap B$.

$A \cap B = \{3, 5\}$ since 3 and 5 are found in both A and B.

The **union** of sets A and B is the set of numbers that are elements of A **or** of B. The *union* of A and B is denoted by $A \cup B$. The set $A \cup B$ consists of the elements in A or in B or in both.

$$A \cup B = \{1, 2, 3, 4, 5, 6, 7, 9, 11\}$$

In-Class Example 1
Let $A = \{0, 2, 4, 6, 8\}$ and $B = \{1, 2, 3, 4, 5\}$.
Find a) $A \cap B$ b) $A \cup B$
answer: a) $A \cap B = \{2, 4\}$
b) $A \cup B =$
$\{0, 1, 2, 3, 4, 5, 6, 8\}$

Note

Although the elements 3 and 5 appear in both set A and in set B, we do not write them twice in the set $A \cup B$.

You Try 1

Let $A = \{2, 4, 6, 8, 10\}$ and $B = \{1, 2, 5, 6, 9, 10\}$. Find $A \cap B$ and $A \cup B$.

Note

The word "*and*" indicates *intersection*, while the word "*or*" indicates *union*. This same principle holds when solving compound inequalities involving "*and*" or "*or*."

2. Solve Compound Inequalities Containing the Word *And*

Example 2

In-Class Example 2
Solve the compound inequality $4q \geq -16$ and $q + 3 < 7$. Graph the solution set, and write the answer in interval notation.
answer:

$[-4, 4)$

Solve the compound inequality $c + 5 \geq 3$ and $8c \leq 32$. Graph the solution set, and write the answer in interval notation.

Solution

Step 1: Identify the inequality as "*and*" or "*or*" and understand what that means. These two inequalities are connected by "*and*." That means the solution set will consist of the values of c that make *both* inequalities true. The solution set will be the *intersection* of the solution sets of $c + 5 \geq 3$ and $8c \leq 32$.

Step 2: Solve each inequality separately.

$$c + 5 \geq 3 \quad \text{and} \quad 8c \leq 32$$
$$c \geq -2 \quad \text{and} \quad c \leq 4$$

Step 3: Graph the solution set to each inequality on its own number line even if the problem does not require you to graph the solution set. This will help you visualize the solution set of the compound inequality.

$c \geq -2$:

$c \leq 4$:

Step 4: Look at the number lines and think about where the solution set for the compound inequality would be graphed.

Since this is an "*and*" inequality, the solution set of $c + 5 \geq 3$ and $8c \leq 32$ consists of the numbers that are solutions to *both* inequalities. We can visualize it this way: if we take the number line above representing $c \geq -2$ and place it on top of the number line representing $c \leq 4$, what shaded areas would overlap (intersect)?

$c \geq -2$ and $c \leq 4$:

They intersect between -2 and 4, *including* those endpoints.

Step 5: Write the answer in interval notation.

The final number line illustrates that the solution to $c + 5 \geq 3$ and $8c \leq 32$ is $[-2, 4]$. The graph of the solution set is the final number line above. ■

Here are the steps to follow when solving a compound inequality.

Procedure   Steps for Solving a Compound Inequality

1) Identify the inequality as "*and*" or "*or*" and understand what that means.

2) Solve each inequality separately.

3) Graph the solution set to each inequality on its own number line even if the problem does not explicitly tell you to graph the solution set. This will help you to visualize the solution to the compound inequality.

4) Use the separate number lines to graph the solution set of the compound inequality.

   a) If it is an "*and*" inequality, the solution set consists of the regions on the separate number lines that would *overlap* (intersect) if one number line was placed on top of the other.

   b) If it is an "*or*" inequality, the solution set consists of the *total* (union) of what would be shaded if you took the separate number lines and put one on top of the other.

5) Use the graph of the solution set to write the answer in interval notation.

 You Try 2

Solve the compound inequality $y - 2 \le 1$ and $7y > -28$. Graph the solution set, and write the answer in interval notation.

Example 3

In-Class Example 3
Solve the compound inequality
$6u - 10 < -4$ and $\frac{2}{3} - 2u < 4$.
Write the solution set in interval notation.

answer: $\left(-\frac{5}{3}, 1\right)$

Solve the compound inequality $7y + 2 > 37$ and $5 - \frac{1}{3}y < 6$. Write the solution set in interval notation.

Solution

Step 1:   This is an "*and*" inequality. The solution set will be the *intersection* of the solution sets of the separate inequalities $7y + 2 > 37$ and $5 - \frac{1}{3}y < 6$.

Step 2:   We must solve each inequality separately.

$$7y + 2 > 37 \quad \text{and} \quad 5 - \frac{1}{3}y < 6$$

Subtract 2. $7y > 35$ and $-\frac{1}{3}y < 1$ Subtract 5 then multiply both sides by −3.

Divide by 7. $y > 5$ and $y > -3$ Reverse the direction of the inequality symbol.

Step 3:   Graph the solution sets separately so that it is easier to find their intersection.

$y > 5$:
$$\text{⟵}\;|\;|\;|\;|\;|\;|\;|\;|\;|\;|\;|\;\diamond\;|\;\text{⟶}$$
$$-6\;-5\;-4\;-3\;-2\;-1\;\;0\;\;1\;\;2\;\;3\;\;4\;\;5\;\;6$$

$y > -3$:
$$\text{⟵}\;|\;|\;|\;\diamond\;|\;|\;|\;|\;|\;|\;|\;|\;|\;\text{⟶}$$
$$-6\;-5\;-4\;-3\;-2\;-1\;\;0\;\;1\;\;2\;\;3\;\;4\;\;5\;\;6$$

Step 4:   If we were to put the number lines above on top of each other, where would they intersect?

$y > 5$ and $y > -3$: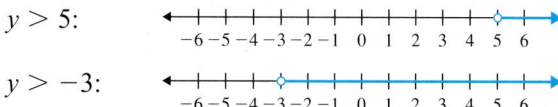
$$-6\;-5\;-4\;-3\;-2\;-1\;\;0\;\;1\;\;2\;\;3\;\;4\;\;5\;\;6$$

Step 5:   The solution, shown in the shaded region in Step 4, is $(5, \infty)$.   ■

You Try 3

Solve each compound inequality and write the answer in interval notation.

a) $4x - 3 > 1$ and $x + 6 < 13$ b) $-\dfrac{4}{5}m > -8$ and $2m + 5 \leq 12$

3. Solve Compound Inequalities Containing the Word *Or*

Recall that the word "*or*" indicates the union of two sets.

Example 4

Solve the compound inequality $6p + 5 \leq -1$ or $p - 3 \geq 1$. Write the answer in interval notation.

In-Class Example 4
Solve the compound inequality
$7n - 4 > 12$ or $3 - 2n \geq 9$.
Write the answer in interval notation.

answer: $(-\infty, -3] \cup \left(\dfrac{16}{7}, \infty\right)$

Solution

Step 1:   These two inequalities are joined by "*or.*" Therefore, the solution set will consist of the values of p that are in the solution set of $6p + 5 \leq -1$ *or* in the solution set of $p - 3 \geq 1$ *or* in *both* solution sets.

Step 2:   Solve each inequality separately.

$$6p + 5 \leq -1 \quad \text{or} \quad p - 3 \geq 1$$
$$6p \leq -6$$
$$p \leq -1 \quad \text{or} \quad p \geq 4$$

Step 3:   Graph the solution sets separately so that it is easier to find the *union* of the sets.

$p \leq -1$:

$p \geq 4$:

Step 4:   The solution set of the compound inequality $6p + 5 \leq -1$ or $p - 3 \geq 1$ consists of the numbers which are solutions to the first inequality *or* the second inequality *or* both. We can visualize it this way: if we put the number lines on top of each other, the solution set of the compound inequality is the **total** (union) of what is shaded.

$p \leq -1$ or $p \geq 4$:

Step 5:   The solution, shown above, is $(-\infty, -1] \cup [4, \infty)$.

Use the *union* symbol for "or."

 **You Try 4**

Solve $t + 8 \geq 14$ or $\dfrac{3}{2}t < 6$ and write the solution in interval notation.

4. Solve Special Compound Inequalities

Example 5

In-Class Example 5
Solve each compound inequality and write the answer in interval notation.

a) $4 + v \leq 3$ and $\frac{1}{2}(v + 3) \geq 4$

b) $3c + 11 > 2$ or $-4c > -8$
answer: a) $\varnothing$ **b)** $(-\infty, \infty)$

Solve each compound inequality and write the answer in interval notation.

a) $k - 5 < -2$ or $4k + 9 > 6$ b) $\frac{1}{2}w \geq 3$ and $1 - w \geq 0$

Solution

a) $k - 5 < -2$ or $4k + 9 > 6$

Step 1: The solution to this "*or*" inequality is the *union* of the solution sets of $k - 5 < -2$ and $4k + 9 > 6$.

Step 2: Solve each inequality separately.

$$k - 5 < -2 \quad \text{or} \quad 4k + 9 > 6$$
$$4k > -3$$
$$k < 3 \quad \text{or} \quad k > -\tfrac{3}{4}$$

Step 3: $k < 3$:

$$\begin{array}{c} \longleftarrow|\!\!+\!\!+\!\!+\!\!+\!\!+\!\!+\!\!\diamond\!\!+\longrightarrow \\ {\scriptstyle -4\ -3\ -2\ -1\ \ 0\ \ 1\ \ 2\ \ 3\ \ 4} \end{array}$$

$k > -\tfrac{3}{4}$:

$$\begin{array}{c} \longleftarrow\!\!+\!\!+\!\!+\!\!\circ\!\!+\!\!+\!\!+\!\!+\!\!+\!\!+\longrightarrow \\ {\scriptstyle -4\ -3\ -2\ -1\ \ 0\ \ 1\ \ 2\ \ 3\ \ 4} \end{array}$$

Step 4: $k < 3$ or $k > -\tfrac{3}{4}$:

$$\begin{array}{c} \longleftarrow\!\!+\!\!+\!\!+\!\!+\!\!+\!\!+\!\!+\!\!+\!\!+\!\!+\longrightarrow \\ {\scriptstyle -4\ -3\ -2\ -1\ \ 0\ \ 1\ \ 2\ \ 3\ \ 4} \end{array}$$

If the number lines in Step 3 were placed on top of each other, the *total* (union) of what would be shaded is the entire number line. This represents all real numbers.

Step 5: The solution set is $(-\infty, \infty)$.

b) $\frac{1}{2}w \geq 3$ and $1 - w \geq 0$

Step 1: The solution to this "*and*" inequality is the *intersection* of the solution sets of $\frac{1}{2}w \geq 3$ and $1 - w \geq 0$.

Step 2: Solve each inequality separately.

$$\frac{1}{2}w \geq 3 \quad \text{and} \quad 1 - w \geq 0$$

Multiply by 2. $\quad w \geq 6$ and $\begin{aligned} 1 &\geq w \qquad &\text{Add } w. \\ w &\leq 1 \qquad &\text{Rewrite } 1 \geq w \text{ as } w \leq 1. \end{aligned}$

Step 3: $w \geq 6$:

$$\begin{array}{c} \longleftarrow\!\!+\!\!+\!\!+\!\!+\!\!+\!\!+\!\!+\!\!\bullet\!\!+\longrightarrow \\ {\scriptstyle -1\ \ 0\ \ 1\ \ 2\ \ 3\ \ 4\ \ 5\ \ 6\ \ 7} \end{array}$$

$w \leq 1$:

$$\begin{array}{c} \longleftarrow\!\!+\!\!+\!\!\bullet\!\!+\!\!+\!\!+\!\!+\!\!+\!\!+\longrightarrow \\ {\scriptstyle -1\ \ 0\ \ 1\ \ 2\ \ 3\ \ 4\ \ 5\ \ 6\ \ 7} \end{array}$$

Step 4: $w \geq 6$ and $w \leq 1$:

$$\begin{array}{c} \longleftarrow\!\!+\!\!+\!\!+\!\!+\!\!+\!\!+\!\!+\!\!+\!\!+\longrightarrow \\ {\scriptstyle -1\ \ 0\ \ 1\ \ 2\ \ 3\ \ 4\ \ 5\ \ 6\ \ 7} \end{array}$$

If the number lines in Step 3 were placed on top of each other, the shaded regions would *not* intersect. Therefore, the solution set is the empty set, $\varnothing$.

Step 5: The solution set of $\frac{1}{2}w \geq 3$ and $1 - w \geq 0$ is $\varnothing$.

You Try 5

Solve the compound inequalities and write the solution in interval notation.

a) $-3w \leq w - 6$ and $5w < 4$ b) $9z - 8 \leq -8$ or $z + 7 \geq 2$

5. Application of Intersection and Union

Example 6

In-Class Example 6
Use the table in Example 6, and list the elements of the set that satisfy the given information.
a) The set of teams with more than 30 play-off appearances and more than 10 championships
b) The set of teams with between 20 and 30 play-off appearances or fewer than 5 championships
answer: a) {Boston Celtics, Los Angeles Lakers}
b) {Chicago Bulls, Cleveland Cavaliers, Detroit Pistons, New York Knicks}

The following table of selected NBA teams contains the number of times they have appeared in the play-offs as well as the number of NBA championships they have won through the 2009–2010 season.

Team	Play-Off Appearances	Championships
Boston Celtics	48	17
Chicago Bulls	29	6
Cleveland Cavaliers	18	0
Detroit Pistons	32	3
Los Angeles Lakers	46	11
New York Knicks	38	2

(www.basketball-reference.com)

List the elements of the set that satisfy the given information.

a)   The set of teams with more than 20 play-off appearances and more than 5 championships

b)   The set of teams with less than 30 play-off appearances or more than 5 championships

Solution

a)   Since the two conditions in this statement are connected by *and*, we must find the team or teams that satisfy *both* conditions. The set of teams is

{Boston Celtics, Chicago Bulls, Los Angeles Lakers}

b)   Since the two conditions in this statement are connected by *or*, we must find the team or teams that satisfy either the first condition, *or* the second condition, *or* both. The set of teams is

{Boston Celtics, Chicago Bulls, Cleveland Cavaliers, Los Angeles Lakers}   ■

You Try 6

Use the table in Example 6, and list the elements of the set that satisfy the given information.

a) The set of teams with less than 40 play-off appearances and at least one championship

b) The set of teams with more than 30 play-off appearances or no championships

---

Answers to You Try Exercises

1) $A \cap B = \{2, 6, 10\}$, $A \cup B = \{1, 2, 4, 5, 6, 8, 9, 10\}$ 2)

$$-5\;-4\;-3\;-2\;-1\quad 0\quad 1\quad 2\quad 3\quad 4\quad 5$$

$(-4, 3]$ 3) a) $(1, 7)$ b) $\left(-\infty, \dfrac{7}{2}\right]$ 4) $(-\infty, 4) \cup [6, \infty)$ 5) a) $\varnothing$ b) $(-\infty, \infty)$

6) a) {Chicago Bulls, Detroit Pistons, New York Knicks}

b) {Boston Celtics, Cleveland Cavaliers, Detroit Pistons, Los Angeles Lakers, New York Knicks}

2.6 Exercises

*Additional answers can be found in the Answers to Exercises appendix.

Objective 1: Find the Intersection and Union of Two Sets

1) Given sets A and B, explain how to find $A \cap B$.

2) Given sets X and Y, explain how to find $X \cup Y$.

Given sets $A = \{2, 4, 6, 8, 10\}$, $B = \{1, 3, 5\}$, $X = \{8, 10, 12, 14\}$, and $Y = \{5, 6, 7, 8, 9\}$ find

3) $X \cap Y$ {8}

4) $A \cap X$ {8, 10}

5) $A \cup Y$ {2, 4, 5, 6, 7, 8, 9, 10}

6) $B \cup Y$ {1, 3, 5, 6, 7, 8, 9}

7) $X \cap B$ ∅

8) $B \cap A$ ∅

9) $A \cup B$ {1, 2, 3, 4, 5, 6, 8, 10}

10) $X \cup Y$ {5, 6, 7, 8, 9, 10, 12, 14}

Each number line represents the solution set of an inequality. Graph the *intersection* of the solution sets and write the intersection in interval notation.

11) $x \geq -3$:
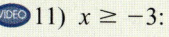

$x \leq 2$:

12) $n \leq 4$:

$n \geq 0$:

13) $t < 3$:

$t > -1$:

14) $y > -4$:

$y < -2$:

15) $c > 1$:

$c \geq 3$:

16) $p < 2$:

$p < -1$:

17) $z \leq 0$:

$z \geq 2$:

18) $g \geq -1$:
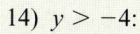

$g < -\dfrac{5}{2}$:
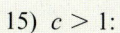

Mixed Exercises: Objectives 2 and 4

Solve each compound inequality. Graph the solution set, and write the answer in interval notation.

19) $a \leq 5$ and $a \geq 2$

20) $k > -3$ and $k < 4$

21) $b - 7 > -9$ and $8b < 24$

22) $3x \leq 1$ and $x + 11 \geq 4$

23) $5w + 9 \leq 29$ and $\dfrac{1}{3}w - 8 > -9$

24) $4y - 11 > -7$ and $\dfrac{3}{2}y + 5 \leq 14$

25) $2m + 15 \geq 19$ and $m + 6 < 5$

26) $d - 1 > 8$ and $3d - 12 < 4$

27) $r - 10 > -10$ and $3r - 1 > 8$

28) $2t - 3 \leq 6$ and $5t + 12 \leq 17$

29) $9 - n \leq 13$ and $n - 8 \leq -7$

30) $c + 5 \geq 6$ and $10 - 3c \geq -5$

Objective 1: Find the Intersection and Union of Two Sets

Each number line represents the solution set of an inequality. Graph the *union* of the solution sets and write the union in interval notation.

31) $p < -1$:

$p > 5$:

32) $z < 2$:

$z > 6$:

33) $a \leq \dfrac{5}{3}$:

$a > 4$:

34) $v \leq -3$:

$v \geq \dfrac{11}{4}$:

35) $y > 1$:

$y > 3$:

36) $x \leq -6$:

$x \leq -2$:

37) $c < \dfrac{7}{2}$:

$c \geq -2$:

38) $q \leq 3$:

$q > -2.7$:

Mixed Exercises: Objectives 3 and 4

Solve each compound inequality. Graph the solution set, and write the answer in interval notation.

39) $z < -1$ or $z > 3$

40) $x \leq -4$ or $x \geq 0$

VIDEO 41) $6m \leq 21$ or $m - 5 > 1$

42) $a + 9 > 7$ or $8a \leq -44$

43) $3t + 4 > -11$ or $t + 19 > 17$

44) $5y + 8 \leq 13$ or $2y \leq -6$

45) $-2v - 5 \leq 1$ or $\frac{7}{3}v < -14$

46) $k - 11 < -4$ or $-\frac{2}{9}k \leq -2$

VIDEO 47) $c + 3 \geq 6$ or $\frac{4}{5}c \leq 10$

48) $\frac{8}{3}g \geq -12$ or $2g + 1 \leq 7$

49) $7 - 6n \geq 19$ or $n + 14 < 11$

50) $d - 4 > -7$ or $-6d \leq 2$

Mixed Exercises: Objectives 2–4

The following exercises contain *and* and *or* inequalities. Solve each inequality, and write the answer in interval notation.

51) $4n + 7 \leq 9$ and $n + 6 \geq 1$ $\left[-5, \frac{1}{2}\right]$

52) $8t - 5 \geq 11$ or $-\frac{2}{5}t \geq 6$ $(-\infty, -15] \cup [2, \infty)$

53) $\frac{4}{3}x + 5 < 2$ or $x + 3 \geq 8$ $\left(-\infty, -\frac{9}{4}\right) \cup [5, \infty)$

54) $p + 10 < -3$ and $8p - 7 > 11$ $\varnothing$

55) $\frac{8}{3}w < -16$ or $w + 9 > -4$ $(-\infty, \infty)$

56) $5y - 2 > 8$ and $\frac{3}{4}y + 2 < 11$ $(2, 12)$

57) $7 - r > 7$ and $0.3r < 6$ $(-\infty, 0)$

58) $2c - 9 \leq -3$ or $10c + 1 \geq 7$ $(-\infty, \infty)$

59) $3 - 2k > 11$ and $\frac{1}{2}k + 5 \geq 1$ $[-8, -4)$

60) $6 - 5a \geq 1$ or $0.8a > 8$ $(-\infty, 1] \cup (10, \infty)$

Objective 5: Application of Intersection and Union

The following table lists the net worth (in billions of dollars) of some of the wealthiest women in the world for the years 2004 and 2008.

Name	Net Worth in 2004	Net Worth in 2008
Liliane Bettencourt	17.2	22.9
Abigail Johnson	12.0	15.0
J.K. Rowling	1.0	1.0
Alice Walton	18.0	19.0
Oprah Winfrey	1.3	2.5

(www.forbes.com)

List the elements of the set that satisfy the given information.

61) The set of women with a net worth more than $15 billion in 2004 and in 2008   {Liliane Bettancourt, Alice Walton}

62) The set of women with a net worth more than $10 billion in 2004 and less than $20 billion in 2008
{Abigail Johnson, Alice Walton}

63) The set of women with a net worth less than $2 billion in 2004 or more than $20 million in 2008
{Liliane Bettancourt, J. K. Rowling, Oprah Winfrey}

64) The set of women with a net worth more than $15 billion in 2004 or more than $2 billion in 2008
{Liliane Bettancourt, Abigail Johnson, Alice Walton. Oprah Winfrey}

Section 2.7 Absolute Value Equations and Inequalities

Objectives

1. **Understand the Meaning of an Absolute Value Equation**
2. **Solve an Equation of the Form** $|ax + b| = k$ **for** $k \geq 0$
3. **Solve an Equation of the Form** $|ax + b| = k$ **for** $k < 0$
4. **Solve an Equation of the Form** $|ax + b| = |cx + d|$
5. **Solve Absolute Value Inequalities Containing** $<$ **or** $\leq$
6. **Solve Absolute Value Inequalities Containing** $>$ **or** $\geq$
7. **Solve Special Cases of Absolute Value Inequalities**
8. **Solve an Applied Problem Using an Absolute Value Inequality**

In Section 1.1 we learned that the absolute value of a number describes its *distance from zero*.

$$|5| = 5 \quad \text{and} \quad |-5| = 5$$

5 units from zero   5 units from zero

$$-7\ -6\ -5\ -4\ -3\ -2\ -1\quad 0\quad 1\quad 2\quad 3\quad 4\quad 5\quad 6\quad 7$$

We use this idea of *distance from zero* to solve absolute value equations and inequalities.

1. Understand the Meaning of an Absolute Value Equation

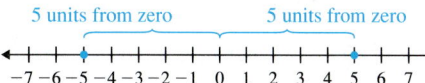 **Example 1**

Solve $|x| = 3$.

Solution

Since the equation contains an absolute value, **solve** $|x| = 3$ means "*Find the number or numbers whose distance from zero is 3.*"

3 units from zero   3 units from zero

$$-6\ -5\ -4\ -3\ -2\ -1\quad 0\quad 1\quad 2\quad 3\quad 4\quad 5\quad 6$$

Those numbers are 3 and -3. Each of them is 3 units from zero. The solution set is $\{-3, 3\}$.

Check: $|3| = 3, |-3| = 3$ ✓

In-Class Example 1
Solve $|v| = 9$.
answer: $\{-9, 9\}$

 **You Try 1**

Solve $|y| = 8$.

Procedure   Solving an Absolute Value Equation

If P represents an expression and k is a positive real number, then to solve $|P| = k$ we rewrite the absolute value equation as the *compound equation*

$$P = k \quad \text{or} \quad P = -k$$

and solve for the variable. P can represent expressions like x, $3a + 2$, $\frac{1}{4}t - 9$, and so on.

2. Solve an Equation of the Form $|ax + b| = k$ for $k \geq 0$

Example 2

Solve each equation.

a) $|m + 1| = 5$ b) $\left|\dfrac{3}{2}t + 7\right| + 5 = 6$

In-Class Example 2
Solve each equation.
a) $|a - 3| = 8$
b) $\left|\dfrac{1}{4}t - 2\right| + 6 = 10$
answers: a) $\{-5, 11\}$
b) $\{-8, 24\}$

Solution

a)   Solving $|m + 1| = 5$ means, "*Find the number or numbers that can be substituted for m so that the quantity m + 1 is 5 units from 0.*"

$m + 1$ will be 5 units from zero if $m + 1 = 5$ or if $m + 1 = -5$, since both 5 and -5 are 5 units from zero. Therefore, we can solve the equation this way:

$$|m + 1| = 5$$

$m + 1 = 5$ or $m + 1 = -5$ Set the quantity inside the absolute
value equal to 5 and -5.

$m = 4$ or $m = -6$ Solve.

Check: $m = 4$: $|4 + 1| \overset{?}{=} 5$ $m = -6$: $|-6 + 1| \overset{?}{=} 5$
$|5| = 5$ ✓ $|-5| = 5$ ✓

The solution set is $\{-6, 4\}$.

b)   Before we rewrite this equation as a compound equation, we must *isolate* the
absolute value (get the absolute value on a side by itself).

$$\left|\frac{3}{2}t + 7\right| + 5 = 6$$

$$\left|\frac{3}{2}t + 7\right| = 1 \qquad \text{Subtract 5 to get the absolute value on a side by itself.}$$

$\frac{3}{2}t + 7 = 1$ or $\frac{3}{2}t + 7 = -1$ Set the quantities inside the absolute
value symbol equal to 1 and -1.

$\frac{3}{2}t = -6$ $\frac{3}{2}t = -8$ Subtract 7.

$\frac{2}{3} \cdot \frac{3}{2}t = \frac{2}{3} \cdot (-6)$ $\frac{2}{3} \cdot \frac{3}{2}t = \frac{2}{3} \cdot (-8)$ Multiply by $\frac{2}{3}$.

$t = -4$ or $t = -\frac{16}{3}$ Solve.

The check is left to the student. The solution set is $\left\{-\dfrac{16}{3}, -4\right\}$. ∎

You Try 2

Solve each equation.

a) $|c - 4| = 3$ b) $\left|\dfrac{1}{4}n - 3\right| + 2 = 5$

3. Solve an Equation of the Form $|ax + b| = k$ for $k < 0$

Example 3

In-Class Example 3
Solve $|5y + 6| = -1$.
answer: ∅

Solve $|4y - 11| = -9$.

Solution

This equation says that the absolute value of the quantity $4y - 11$ equals *negative* 9. Can
an absolute value be negative? No! This equation has *no solution*.

The solution set is ∅. ∎

You Try 3

Solve $|d + 3| = -5$.

4. Solve an Equation of the Form $|ax + b| = |cx + d|$

Another type of absolute value equation involves two absolute values.

Procedure   Solve $|ax + b| = |cx + d|$

If P and Q are expressions, then to solve $|P| = |Q|$ we rewrite the absolute value equation as the compound equation.

$$P = Q \quad \text{or} \quad P = -Q$$

and solve for the variable.

Example 4

Solve $|2w - 3| = |w + 9|$.

In-Class Example 4
Solve $|4z + 1| = |2z - 3|$.

answer: $\left\{-2, \dfrac{1}{6}\right\}$

Solution

This equation is true when the quantities inside the absolute values are the *same* or when they are *negatives* of each other.

$$|2w - 3| = |w + 9|$$

The quantities are the same   or   the quantities are negatives of each other.

$$2w - 3 = w + 9 \qquad\qquad\qquad 2w - 3 = -(w + 9)$$
$$w = 12 \qquad\qquad\qquad\quad 2w - 3 = -w - 9$$
$$3w = -6$$
$$w = -2$$

Check: $w = 12$: $|2(12) - 3| \overset{?}{=} |12 + 9|$ | $w = -2$: $|2(-2) - 3| \overset{?}{=} |-2 + 9|$
$$|24 - 3| \overset{?}{=} |21| \qquad\qquad |-4 - 3| \overset{?}{=} |7|$$
$$|21| = 21 \ \checkmark \qquad\qquad |-7| = 7 \ \checkmark$$

The solution set is $\{-2, 12\}$.

 **BE CAREFUL**   In Example 4 and other examples like it, you *must* put parentheses around the expression with the negative as in $-(w + 9)$.

 You Try 4

Solve $|c + 7| = |3c - 1|$.

Next, we will learn how to solve **absolute value inequalities**. Some examples of absolute value inequalities are

$$|t| < 6, \qquad |n + 2| \leq 5, \qquad |3k - 1| > 11, \qquad \left|5 - \frac{1}{2}y\right| \geq 3$$

5. Solve Absolute Value Inequalities Containing < or ≤

What does it mean to solve $|x| \leq 3$? It means to find the set of all real numbers whose distance from zero is 3 *units or less*.

3 is 3 units from 0.
−3 is 3 units from 0.

Any number *between* 3 and −3 is less than 3 units from zero. For example, if $x = 1$, $|1| \leq 3$. If $x = -2, |-2| \leq 3$. We can represent the solution set on a number line as

We can write the solution set in interval notation as $[-3, 3]$.

> **Procedure**   Solve $|P| \leq k$ or $|P| < k$
>
> Let P be an expression and let k be a positive real number. To solve $|P| \leq k$, solve the three-part inequality $-k \leq P \leq k$. ($<$ may be substituted for $\leq$.)

Example 5

Solve $|t| < 6$. Graph the solution set and write the answer in interval notation.

In-Class Example 5
Solve $|u| < 3$. Graph the solution set and write the answer in interval notation.
answer: $(-3, 3)$

Solution

We must find the set of all real numbers whose distance from zero is less than 6. We can do this by solving the three-part inequality $-6 < t < 6$.

We can represent this on a number line as

We can write the solution set in interval notation as $(-6, 6)$. Any number between −6 and 6 will satisfy the inequality.

You Try 5

Solve. Graph the solution set and write the answer in interval notation.

$$|u| < 9$$

Example 6

Solve each inequality. Graph the solution set and write the answer in interval notation.

a) $|n + 2| \leq 5$ b) $|4 - 5p| < 16$

In-Class Example 6
Solve each inequality.
a) $|a + 4| \leq 10$
b) $|3 - 10m| < 7$
Graph the solution set and write the answer in interval notation.
answer: a) $[-14, 6]$

b) $\left(-\frac{2}{5}, 1\right)$

Solution

a) We must find the set of all real numbers, n, so that $n + 2$ is less than or equal to 5 units from zero. To solve $|n + 2| \leq 5$, we must solve the three-part inequality

$$-5 \leq n + 2 \leq 5$$
$$-7 \leq n \leq 3 \qquad \text{Subtract 2.}$$

The number line representation is

In interval notation, we write the solution as $[-7, 3]$. Any number between −7 and 3 will satisfy the inequality.

b) Solve the three-part inequality.

$$-16 < 4 - 5p < 16$$
$$-20 < -5p < 12 \qquad \text{Subtract 4.}$$
$$4 > p > -\frac{12}{5} \qquad \text{Divide by } -5 \text{ and change the direction of the inequality symbols.}$$

This inequality means *p is less than* 4 *and greater than* $-\dfrac{12}{5}$. We can rewrite it as $-\dfrac{12}{5} < p < 4$.

The number line representation of the solution set is

In interval notation, we write $\left(-\dfrac{12}{5},\, 4\right)$.

You Try 6

Solve. Graph the solution set and write the answer in interval notation.

$$|6k + 5| \le 13$$

6. Solve Absolute Value Inequalities Containing > or ≥

To solve $|x| \ge 4$ means to find the set of all real numbers whose distance from zero is 4 *units or more.*

4 is 4 units from 0.
-4 is 4 units from 0.

Any number greater than 4 *or* less than -4 is more than 4 units from zero.

For example, if $x = 6$, $|6| \ge 4$. If $x = -5$, then $|-5| = 5$ and $5 \ge 4$. We can represent the solution set to $|x| \ge 4$ as

These real numbers are
4 or more units from zero.
These real numbers are
4 or more units from zero.

The solution set consists of two separate regions, so we can write a compound inequality using *or.*

$$x \le -4 \quad \text{or} \quad x \ge 4$$

In interval notation, we write $(-\infty,\, -4] \cup [4,\, \infty)$.

Procedure Solve $|P| \ge k$ or $|P| > k$

Let P be an expression and let k be a positive, real number. To solve $|P| \ge k$ (> may be substituted for ≥), solve the compound inequality $P \ge k$ or $P \le -k$.

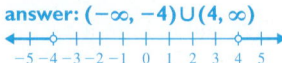

Solve $|r| > 2$. Graph the solution set and write the answer in interval notation.

In-Class Example 7
Solve $|g| > 4$. Graph the solution
set and write the answer in
interval notation.
answer: $(-\infty, -4) \cup (4, \infty)$

Solution

We must find the set of all real numbers whose distance from zero is greater than 2. The solution is the compound inequality $r > 2$ or $r < -2$.

On the number line, we can represent the solution set as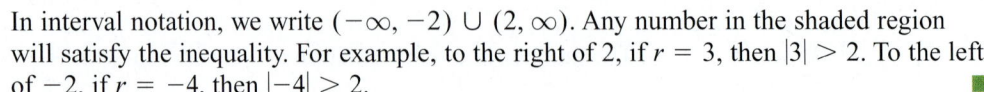

In interval notation, we write $(-\infty, -2) \cup (2, \infty)$. Any number in the shaded region will satisfy the inequality. For example, to the right of 2, if $r = 3$, then $|3| > 2$. To the left of -2, if $r = -4$, then $|-4| > 2$. ■

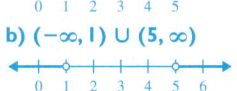

Solve $|d| \geq 5$. Graph the solution set and write the answer in interval notation.

Example 8

Solve each inequality. Graph the solution set and write the answer in interval notation.

a) $|3k - 1| > 11$ b) $|c + 6| + 10 \geq 12$

In-Class Example 8
Solve each inequality. Graph the
solution set and write the
answer in interval notation.
a) $|4c - 7| > 6$
b) $|r - 3| + 12 > 14$
answer:

a) $\left(-\infty, \frac{1}{4}\right) \cup \left(\frac{13}{4}, \infty\right)$

b) $(-\infty, 1) \cup (5, \infty)$

Solution

a)   To solve $|3k - 1| > 11$ means to find the set of all real numbers, k, so that $3k - 1$ is more than 11 units from zero on the number line. We will solve the compound inequality.

$$3k - 1 > 11 \qquad \text{or} \qquad 3k - 1 < -11$$

$3k - 1$ is more than 11 units away
from zero to the *right* of zero. $3k - 1$ is more than 11 units away
from zero to the *left* of zero.

$$3k - 1 > 11 \qquad \text{or} \qquad 3k - 1 < -11$$
$$3k > 12 \qquad\quad \text{or} \qquad\quad 3k < -10 \qquad \text{Add 1.}$$
$$k > 4 \qquad\quad \text{or} \qquad\quad k < -\frac{10}{3} \qquad \text{Divide by 3.}$$

On the number line, we get

From the number line, we can write the interval notation $\left(-\infty, -\frac{10}{3}\right) \cup (4, \infty)$.

Any number in the shaded region will satisfy the inequality.

b)   Begin by getting the absolute value on a side by itself.

$$|c + 6| + 10 \geq 12$$
$$|c + 6| \geq 2 \qquad \text{Subtract 10.}$$

$$c + 6 \geq 2 \quad \text{or} \quad c + 6 \leq -2 \qquad \text{Rewrite as a compound inequality.}$$
$$c \geq -4 \quad \text{or} \qquad c \leq -8 \qquad \text{Subtract 6.}$$

The graph of the solution set is

The interval notation is $(-\infty, -8] \cup [-4, \infty)$. ■

You Try 8

Solve each inequality. Graph the solution set and write the answer in interval notation.

a) $|8q + 9| \geq 7$ b) $|k + 8| - 5 \geq 9$

Example 9 illustrates why it is important to understand what the absolute value inequality means before trying to solve it.

7. Solve Special Cases of Absolute Value Inequalities

Example 9

In-Class Example 9
Solve each inequality.
a) $|h + 2| < -1$
b) $|4q + 9| > 0$
c) $|3t - 5| + 6 \leq 6$
answer: a) $\varnothing$ b) $(-\infty, \infty)$
c) $\left\{ \dfrac{5}{3} \right\}$

Solve each inequality.

a) $|z + 3| < -6$ b) $|2s - 1| \geq 0$ c) $|4d + 7| + 9 \leq 9$

Solution

a) Look carefully at this inequality, $|z + 3| < -6$. It says that the absolute value of a quantity, $z + 3$, is *less than* a negative number. Since the absolute value of a quantity is always zero or positive, this inequality has *no solution*.

The solution set is $\varnothing$.

b) $|2s - 1| \geq 0$ says that the absolute value of a quantity, $2s - 1$, is greater than or equal to zero. An absolute value is *always* greater than or equal to zero, so *any* value of s will make the inequality true.

The solution set consists of all real numbers, which we can write in interval notation as $(-\infty, \infty)$.

c) Begin by isolating the absolute value. $|4d + 7| + 9 \leq 9$

$$|4d + 7| \leq 0 \qquad \text{Subtract 9.}$$

The absolute value of a quantity can *never be less than zero* but it *can equal zero*. To solve this, we must solve $4d + 7 = 0$.

$$4d + 7 = 0$$
$$4d = -7$$
$$d = -\frac{7}{4}$$

The solution set is $\left\{ -\dfrac{7}{4} \right\}$.

You Try 9

Solve each inequality.

a) $|p + 4| \geq 0$ b) $|5n - 7| < -2$ c) $|6y - 1| + 3 \leq 3$

8. Solve an Applied Problem Using an Absolute Value Inequality

Example 10

In-Class Example 10
Refer to Example 10. Assume the can is to be filled with 11 oz of soup with a possible error of ± 0.2 oz.
answer: $|x - 11| \leq 0.2$; the actual amount of soup in the can is between 10.8 oz and 11.2 oz.

On an assembly line, a machine is supposed to fill a can with 19 oz of soup. However, the possibility for error is ± 0.25 oz. Let x represent the range of values for the amount of soup in the can. Write an absolute value inequality to represent the range for the number of ounces of soup in the can, then solve the inequality and explain the meaning of the answer.

Solution

If the *actual* amount of soup in the can is x and there is supposed to be 19 oz in the can, then the error in the amount of soup in the can is $|x - 19|$. If the possible error is ± 0.25 oz, then we can write the inequality

$$|x - 19| \le 0.25$$

$$-0.25 \le x - 19 \le 0.25 \qquad \text{Solve.}$$
$$18.75 \le x \le 19.25$$

The actual amount of soup in the can is between 18.75 and 19.25 oz.   ■

Using Technology

We can use a graphing calculator to solve an equation by entering one side of the equation as Y_1 and the other side as Y_2. Then graph the equations. Remember that absolute value equations like the ones found in this section can have 0, 1, or 2 solutions. *The x-coordinates of their points of intersection are the solutions to the equation.*

We will solve $|3x - 1| = 5$ algebraically and by using a graphing calculator, and then compare the results.

First, use algebra to solve $|3x - 1| = 5$. You should get $\left\{-\dfrac{4}{3}, 2\right\}$.

Next, use a graphing calculator to solve $|3x - 1| = 5$.
We will enter $|3x - 1|$ as Y_1 and 5 as Y_2. To enter $Y_1 = |3x - 1|$,

1) Press the $\boxed{Y=}$ key, so that the cursor is to the right of $\backslash Y_1 =$.

2) Press $\boxed{\text{MATH}}$ and then press the right arrow, to highlight **NUM.** Also highlighted is 1:abs (which stands for *absolute value*).

3) Press $\boxed{\text{ENTER}}$ and you are now back on the $\backslash Y_1 =$ screen. Enter $3x - 1$ with a closing parentheses so that you have now entered $Y_1 = \text{abs}(3x - 1)$.

4) Press the down arrow to enter $\backslash Y_2 = 5$.

5) Press $\boxed{\text{GRAPH}}$.

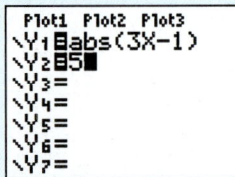 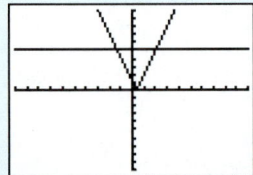

The graphs intersect at two points because there are two solutions to this equation. *Remember that the solutions to the equation are the x-coordinates of the points of intersection.*

To find these x-coordinates we will use the INTERSECT feature.

To find the left-hand intersection point, press $\boxed{\text{2nd}}$ $\boxed{\text{TRACE}}$ and select 5:intersect. Press $\boxed{\text{ENTER}}$. Move the cursor close to the point on the left and press $\boxed{\text{ENTER}}$ three times. You get the result in the screen below on the left.

To find the right-hand intersect point, press $\boxed{\text{2nd}}$ $\boxed{\text{TRACE}}$, select 5:intersect, and press $\boxed{\text{ENTER}}$. Move the cursor close to the point, and press $\boxed{\text{ENTER}}$ three times. You will see the screen that is below on the right.

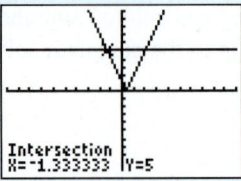 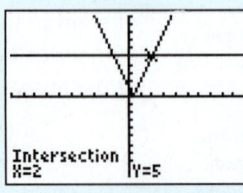

The screen on the left shows $x = -1.333333$. This is the calculator's approximation of $x = -1.\overline{3}$, the decimal equivalent of $x = -\dfrac{4}{3}$, one of the solutions found using algebra.

The screen on the right shows $x = 2$ as a solution, the same solution we obtained algebraically. The calculator gives us a solution set of $\{-1.333333, 2\}$, while the solution set found using algebra is $\left\{-\dfrac{4}{3}, 2\right\}$.

Solve each equation algebraically, then verify your answer using a graphing calculator.

1) $|x - 1| = 2$

2) $|x + 4| = 6$

3) $|2x + 3| = 3$

4) $|4x - 5| = 1$

5) $|3x + 7| - 6 = -8$

6) $|6 - x| + 3 = 3$

Answers to You Try Exercises

1) $\{-8, 8\}$ 2) a) $\{1, 7\}$ b) $\{0, 24\}$ 3) $\varnothing$ 4) $\left\{-\dfrac{3}{2}, 4\right\}$

5) $(-9, 9)$

6) $\left[-3, \dfrac{4}{3}\right]$

7) $(-\infty, -5] \cup [5, \infty)$

8) a) $(-\infty, -2] \cup \left[-\dfrac{1}{4}, \infty\right)$

   b) $(-\infty, -22] \cup [6, \infty)$

9) a) $(-\infty, \infty)$ b) $\varnothing$ c) $\left\{\dfrac{1}{6}\right\}$

Answers to Technology Exercises

1) $\{-1, 3\}$ 2) $\{-10, 2\}$ 3) $\{-3, 0\}$ 4) $\{1, 1.5\}$ 5) $\varnothing$ 6) $\{6\}$

2.7 Exercises

*Additional answers can be found in the Answers to Exercises appendix.

Objective 1: Understand the Meaning of an Absolute Value Equation

1) In your own words, explain the *meaning* of the absolute value of a number.   Answers may vary.

2) Does $|x| = -8$ have a solution? Why or why not?
No; the absolute value of a quantity cannot be negative.

Mixed Exercises: Objectives 2 and 3

Solve.

3) $|q| = 6$ $\{-6, 6\}$

4) $|z| = 7$ $\{-7, 7\}$

5) $|q - 5| = 3$ $\{2, 8\}$

6) $|a + 2| = 13$ $\{-15, 11\}$

7) $|4t - 5| = 7$ $\left\{-\dfrac{1}{2}, 3\right\}$

8) $|9x - 8| = 10$ $\left\{-\dfrac{2}{9}, 2\right\}$

9) $|12c + 5| = 1$ $\left\{-\dfrac{1}{2}, -\dfrac{1}{3}\right\}$

10) $|4 - 5k| = 11$ $\left\{-\dfrac{7}{5}, 3\right\}$

11) $\left|\dfrac{2}{3}b + 3\right| = 13$ $\{-24, 15\}$

12) $\left|\dfrac{3}{4}h + 8\right| = 7$ $\left\{-\dfrac{4}{3}, -20\right\}$

13) $\left| 4 - \dfrac{3}{5}d \right| = 6$ $\left\{ \dfrac{10}{3}, \dfrac{50}{3} \right\}$

14) $\left| \dfrac{3}{2}r + 5 \right| = \dfrac{3}{4}$ $\left\{ -\dfrac{23}{6}, -\dfrac{17}{6} \right\}$

15) $|m - 5| = -3$ $\varnothing$

16) $|2k + 7| = -15$ $\varnothing$

17) $|z - 6| + 4 = 20$ $\{-10, 22\}$

18) $|q + 3| - 1 = 14$ $\{-18, 12\}$

19) $|2a + 5| + 8 = 13$ $\{-5, 0\}$

20) $|6t - 11| + 5 = 10$ $\left\{ 1, \dfrac{8}{3} \right\}$

21) $|w + 14| = 0$ $\{-14\}$

22) $|5h + 7| = -5$ $\varnothing$

23) $|8n + 11| = -1$ $\varnothing$

24) $|4p - 3| = 0$ $\left\{ \dfrac{3}{4} \right\}$

VIDEO 25) $|5b + 3| + 6 = 19$ $\left\{ -\dfrac{16}{5}, 2 \right\}$

26) $1 = |7 - 8x| - 4$ $\left\{ \dfrac{1}{4}, \dfrac{3}{2} \right\}$

27) $|3m - 1| + 5 = 2$ $\varnothing$

28) $\left| \dfrac{5}{4}k + 2 \right| + 9 = 7$ $\varnothing$

Objective 4: Solve an Equation of the Form $|ax + b| = |cx + d|$

Solve the following equations containing two absolute values.

VIDEO 29) $|s + 9| = |2s + 5|$ $\left\{ -\dfrac{14}{3}, 4 \right\}$

30) $|j - 8| = |4j - 7|$ $\left\{ -\dfrac{1}{3}, 3 \right\}$

31) $|3z + 2| = |6 - 5z|$ $\left\{ \dfrac{1}{2}, 4 \right\}$

32) $|1 - 2a| = |10a + 3|$ $\left\{ -\dfrac{1}{2}, -\dfrac{1}{6} \right\}$

33) $\left| \dfrac{3}{2}x - 1 \right| = |x|$ $\left\{ \dfrac{2}{5}, 2 \right\}$

34) $|y| = \left| \dfrac{4}{7}y + 12 \right|$ $\left\{ -\dfrac{84}{11}, 28 \right\}$

35) $\left| \dfrac{1}{4}t - \dfrac{5}{2} \right| = \left| 5 - \dfrac{1}{2}t \right|$ $\{10\}$

36) $\left| k + \dfrac{1}{6} \right| = \left| \dfrac{2}{3}k + \dfrac{1}{2} \right|$ $\left\{ -\dfrac{2}{5}, 1 \right\}$

VIDEO 37) Write an absolute value equation that means *x is 9 units from zero.* $|x| = 9$, may vary

38) Write an absolute value equation that means *y is 6 units from zero.* $|y| = 6$, may vary

39) Write an absolute value equation that has a solution set of $\left\{ -\dfrac{1}{2}, \dfrac{1}{2} \right\}$. $|x| = \dfrac{1}{2}$, may vary

40) Write an absolute value equation that has a solution set of $\{-1.4, 1.4\}$. $|x| = 1.4$, may vary

Mixed Exercises: Objectives 5 and 6

Graph each inequality on a number line and represent the sets of numbers using interval notation.

41) $-1 \le p \le 5$

42) $7 < t < 11$

43) $y < 2$ or $y > 9$

44) $a \le -8$ or $a \ge \dfrac{1}{2}$

45) $n \le -\dfrac{9}{2}$ or $n \ge \dfrac{3}{5}$

46) $-\dfrac{1}{4} \le q \le \dfrac{11}{4}$

Mixed Exercises: Objectives 5 and 7

Solve each inequality. Graph the solution set and write the answer in interval notation.

47) $|m| \le 7$

48) $|c| < 1$

49) $|3k| < 12$

50) $\left| \dfrac{5}{4}z \right| \le 30$

51) $|w - 2| < 4$

52) $|k - 6| \le 2$

VIDEO 53) $|3r + 10| \le 4$

54) $|4a + 1| \le 12$

55) $|7 - 6p| \le 3$

56) $|17 - 9d| < 8$

57) $|5q + 11| < 0$

58) $|6t + 16| < 0$

59) $|2x + 7| \le -12$

60) $|8m - 15| \le -5$

61) $|8c - 3| + 15 < 20$

62) $|2v + 5| + 3 < 14$

63) $\left| \dfrac{3}{2}h + 6 \right| - 2 \le 10$

64) $7 + \left| \dfrac{8}{3}u - 9 \right| < 12$

Mixed Exercises: Objectives 6 and 7

Solve each inequality. Graph the solution set and write the answer in interval notation.

65) $|t| \ge 7$

66) $|p| > 3$

67) $|d + 10| \ge 4$

68) $|q - 7| > 12$

69) $|4v - 3| \ge 9$

70) $|6a + 19| > 11$

VIDEO 71) $|17 - 6x| > 5$

72) $|1 - 4g| \ge 10$

73) $|8k + 5| \ge 0$

74) $|5b - 6| \ge 0$

75) $|z - 3| \ge -5$

76) $|3r + 10| > -11$

77) $|2m - 1| + 4 > 5$

78) $|w + 6| - 4 \ge 2$

79) $-3 + \left| \dfrac{5}{6}n + \dfrac{1}{2} \right| \ge 1$ 80) $\left| \dfrac{3}{2}y - \dfrac{5}{4} \right| + 9 \ge 11$

81) Explain why $|3t - 7| < 0$ has no solution.

82) Explain why $|4l + 9| \le -10$ has no solution.

83) Explain why the solution to $|2x + 1| \ge -3$ is $(-\infty, \infty)$.

84) Explain why the solution to $|7y - 3| \ge 0$ is $(-\infty, \infty)$.

Mixed Exercises: Objectives 1–7

The following exercises contain absolute value equations, linear inequalities, and both types of absolute value inequalities. Solve each. Write the solution set for equations in set notation and use interval notation for inequalities.

85) $|2v + 9| > 3$
$(-\infty, -6) \cup (-3, \infty)$

86) $\left|\dfrac{5}{3}a + 2\right| = 8$ $\left\{-6, \dfrac{18}{5}\right\}$

87) $3 = |4t + 5|$ $\left\{-2, -\dfrac{1}{2}\right\}$

88) $|4k + 9| \le 5$ $\left[-\dfrac{7}{2}, -1\right]$

89) $9 \le |7 - 8q|$
$\left(-\infty, -\dfrac{1}{4}\right] \cup [2, \infty)$

90) $|2p - 5| - 12 = 11$
$\{-9, 14\}$

91) $2(x - 8) + 10 < 4x$
$(-3, \infty)$

92) $\dfrac{1}{2}n + 11 < 8$ $(-\infty, -6)$

93) $|6y + 5| \le -9$ $\varnothing$

94) $8 \le |5v + 2|$
$(-\infty, -2] \cup \left[\dfrac{6}{5}, \infty\right)$

95) $\left|\dfrac{4}{3}x + 1\right| = \left|\dfrac{5}{3}x + 8\right|$
$\{-21, -3\}$

96) $|7z - 8| \le 0$ $\left\{\dfrac{8}{7}\right\}$

97) $|4 - 9t| + 2 = 1$ $\varnothing$

98) $|5b - 11| - 18 < -10$
$\left(\dfrac{3}{5}, \dfrac{19}{5}\right)$

99) $-\dfrac{3}{5} \ge \dfrac{5}{2}a - \dfrac{1}{2}$ $\left(-\infty, -\dfrac{1}{25}\right]$

100) $4 + 3(2r - 5) > 9 - 4r$ $(2, \infty)$

101) $|6k + 17| > -4$
$(-\infty, \infty)$

102) $|5 - w| \ge 3$
$(-\infty, 2] \cup [8, \infty)$

103) $5 \ge |c + 8| - 2$
$[-15, -1]$

104) $0 \le |4a + 1|$
$(-\infty, \infty)$

105) $|5h - 8| > 7$ $\left(-\infty, \dfrac{1}{5}\right) \cup (3, \infty)$

106) $\left|\dfrac{2}{3}y - 1\right| = \left|\dfrac{3}{2}y + 4\right|$ $\left\{-6, -\dfrac{18}{13}\right\}$

Objective 8: Solve an Applied Problem Using an Absolute Value Inequality

107) A gallon of milk should contain 128 oz. The possible error in this measurement, however, is ±0.75 oz. Let a represent the range of values for the amount of milk in the container. Write an absolute value inequality to represent the range for the number of ounces of milk in the container, then solve the inequality and explain the meaning of the answer.

108) Dawn buys a 27-oz box of cereal. The possible error in this amount, however, is ±0.5 oz. Let c represent the range of values for the amount of cereal in the box. Write an absolute value inequality to represent the range for the number of ounces of cereal in the box, then solve the inequality and explain the meaning of the answer.

$|c - 27| \le 0.5$; $26.5 \le c \le 27.5$; there is between 26.5 oz and 27.5 oz of cereal in the box.

109) Emmanuel spent $38 on a birthday gift for his son. He plans on spending within $5 of that amount on his daughter's birthday gift. Let b represent the range of values for the amount he will spend on his daughter's gift. Write an absolute value inequality to represent the range for the amount of money Emmanuel will spend on his daughter's birthday gift, then solve the inequality and explain the meaning of the answer.

110) An employee at a home-improvement store is cutting a window shade for a customer. The customer wants the shade to be 32 in. wide. If the machine's possible error in cutting the shade is $\pm\dfrac{1}{16}$ in., write an absolute value inequality to represent the range for the width of the window shade, and solve the inequality. Explain the meaning of the answer. Let w represent the range of values for the width of the shade.

$|w - 32| \le \dfrac{1}{16}$; $31\dfrac{15}{16} \le w \le 32\dfrac{1}{16}$; the window shade is between $31\dfrac{15}{16}$ in. and $32\dfrac{1}{16}$ in. wide.

Chapter 2: Summary

Definition/Procedure	Example

2.1 Linear Equations in One Variable

The Properties of Equality

1) If $a = b$, then $a + c = b + c$
2) If $a = b$, then $a - c = b - c$
3) If $a = b$, then $ac = bc$
4) If $a = b$, then $\dfrac{a}{c} = \dfrac{b}{c}$ $(c \neq 0)$. **(p. 32)**

Solve $a + 4 = 19$.

$a + 4 - 4 = 19 - 4$ Subtract 4 from
$a = 15$ each side.

The solution set is $\{15\}$.

Solve. $\dfrac{3}{2}t = -30$

$\dfrac{2}{3} \cdot \dfrac{3}{2}t = -30 \cdot \dfrac{2}{3}$ Multiply each side by $\dfrac{2}{3}$.

$t = -20$

The solution set is $\{-20\}$.

How to Solve a Linear Equation

Step 1: **Clear parentheses** and **combine like terms** on each side of the equation.

Step 2: **Get the variable on one side of the equal sign and the constant on the other side of the equal sign** (isolate the variable) using the addition or subtraction property of equality.

Step 3: **Solve for the variable** using the multiplication or division property of equality.

Step 4: **Check the solution** in the original equation. **(p. 34)**

Solve $4(c + 1) + 7 = 2c + 9$.

$4c + 4 + 7 = 2c + 9$ Distribute.
$4c + 11 = 2c + 9$ Combine like terms.
$4c - 2c + 11 = 2c - 2c + 9$ Get variable terms on one side.
$2c + 11 - 11 = 9 - 11$ Get constants on one side.
$2c = -2$
$\dfrac{2c}{2} = \dfrac{-2}{2}$ Division property of equality
$c = -1$

The solution set is $\{-1\}$.

Steps for Solving Applied Problems

Step 1: **Read** the problem carefully, more than once if necessary, until you understand it. Draw a picture, if applicable. Identify what you are being asked to find.

Step 2: **Choose a variable** to represent an unknown quantity. If there are any other unknowns, define them in terms of the variable.

Step 3: **Translate** the problem from English into an equation using the chosen variable.

Step 4: **Solve** the equation.

Step 5: **Check** the answer in the original problem, and **interpret** the solution as it relates to the problem. Be sure your answer makes sense in the context of the problem. **(p. 38)**

The sum of a number and fifteen is eight. Find the number.

Step 1: **Read** the problem carefully.

Step 2: **Choose** a variable.

$x =$ the number

Step 3: "The sum of a number and fifteen is eight" means

The number plus fifteen equals eight.
$x \quad + \quad 15 \quad = \quad 8$

Equation: $x + 15 = 8$

Step 4: **Solve** the equation.

$x + 15 = 8$
$x + 15 - 15 = 8 - 15$
$x = -7$

The number is -7.

Step 5: The **check** is left to the student.

2.2 Applications of Linear Equations

Apply the "**Steps for Solving Applied Problems**" to solve this application involving consecutive odd integers. **(p. 45)**

The sum of three consecutive odd integers is 87. Find the integers.

Step 1: **Read** the problem carefully.

Step 2: **Choose** a variable to represent an unknown, and define the other unknowns.

$x =$ the first odd integer
$x + 2 =$ the second odd integer
$x + 4 =$ the third odd integer

Step 3: "The sum of three consecutive odd integers is 87" means

First odd + Second odd + Third odd = 87
$x \quad + \quad (x + 2) \quad + \quad (x + 4) \quad = 87$

Equation: $x + (x + 2) + (x + 4) = 87$

Definition/Procedure	Example
	Step 4: **Solve** $x + (x + 2) + (x + 4) = 87.$
	$$3x + 6 = 87$$
	$$3x + 6 - 6 = 87 - 6$$
	$$3x = 81$$
	$$\frac{3x}{3} = \frac{81}{3}$$
	$$x = 27$$
	Step 5: Find the values of all of the unknowns.
	$$x = 27, x + 2 = 29, x + 4 = 31$$
	The numbers are 27, 29, and 31.
	The **check** is left to the student.

2.3 Geometry Applications and Solving Formulas

Definition/Procedure	Example
Formulas from geometry can be used to solve applications. **(p. 55)**	A rectangular bulletin board has an area of 180 in². It is 12 in. wide. Find its length.
	Use $A = lw$. Formula for the area of a rectangle
	$A = 180$ in², $w = 12$ in. Find l.
	$$A = lw$$
	$$180 = l(12) \quad \text{Substitute values into } A = lw.$$
	$$\frac{180}{12} = \frac{l(12)}{12}$$
	$$15 = l$$
	The length is 15 in.
To **solve a formula for a specific variable,** think about the steps involved in solving a linear equation in one variable. **(p. 58)**	Solve $C = kr - w$ for r.
	$$C + w = k\boxed{r} - w + w \quad \text{Add } w \text{ to each side.}$$
	$$C + w = k\boxed{r}$$
	$$\frac{C + w}{k} = \frac{k\boxed{r}}{k} \quad \text{Divide each side by } k.$$
	$$\frac{C + w}{k} = r$$

2.4 More Applications of Linear Equations

Definition/Procedure	Example
A **proportion** is a statement that two ratios are equivalent. **(p. 66)**	If Malik can read 2 books in 3 weeks, how long will it take him to read 7 books?
	Step 1: **Read** the problem carefully.
	Step 2: **Choose** a variable.
	$x =$ number of weeks to read 7 books.
	Step 3: Set up a proportion. $\dfrac{2 \text{ books}}{3 \text{ weeks}} = \dfrac{7 \text{ books}}{x \text{ weeks}}$
	Equation: $\dfrac{2}{3} = \dfrac{7}{x}$
	Step 4: **Solve** $\dfrac{2}{3} = \dfrac{7}{x}.$
	$$2x = 3(7) \quad \text{Set cross-products equal.}$$
	$$\frac{2x}{2} = \frac{21}{2}$$
	$$x = \frac{21}{2} = 10\frac{1}{2}$$
	Step 5: It will take Malik $10\frac{1}{2}$ weeks to read 7 books. The **check** is left to the student.

Definition/Procedure	Example

2.5 Linear Inequalities in One Variable

We solve linear inequalities in very much the same way we solve linear equations *except when we multiply or divide by a negative number we must reverse the direction of the inequality symbol.*

We can graph the solution set, write the solution in set notation, or write the solution in interval notation. **(p. 78)**

Solve $x - 5 \leq -3$. Graph the solution set and write the answer in both set notation and interval notation.

$$x - 5 \leq -3$$
$$x - 5 + 5 \leq -3 + 5$$
$$x \leq 2$$

$\{x \mid x \leq 2\}$ Set notation
$(-\infty, 2]$ Interval notation

2.6 Compound Inequalities in One Variable

The solution set of a compound inequality joined by "**and**" will be the **intersection** of the solution sets of the individual inequalities. **(p. 88)**

Solve the compound inequality $5x - 2 \geq -17$ and $x + 8 \leq 9$.
$$5x - 2 \geq -17 \quad \text{and} \quad x + 8 \leq 9$$
$$5x \geq -15$$
$$x \geq -3 \quad \text{and} \quad x \leq 1$$

Solution in interval notation: $[-3, 1]$

The solution set of a compound inequality joined by "**or**" will be the **union** of the solution sets of the individual inequalities. **(p. 89)**

Solve the compound inequality $x - 3 < -1$ or $7x > 42$.
$$x - 3 < -1 \quad \text{or} \quad 7x > 42$$
$$x < 2 \quad \text{or} \quad x > 6$$

Solution in interval notation: $(-\infty, 2) \cup (6, \infty)$

2.7 Absolute Value Equations and Inequalities

Absolute Value Equations

If P represents an expression and k is a positive, real number, then to solve $|P| = k$ we rewrite the absolute value equation as the *compound equation* $P = k$ or $P = -k$ and solve for the variable. **(p. 95)**

Solve $|4a + 10| = 18$.
$$|4a + 10| = 18$$
$$4a + 10 = 18 \quad \text{or} \quad 4a + 10 = -18$$
$$4a = 8 \qquad\qquad 4a = -28$$
$$a = 2 \quad \text{or} \qquad a = -7$$

Check the solutions in the original equation.
The solution set is $\{-7, 2\}$.

Inequalities Containing $<$ or $\leq$

Let P be an expression and let k be a positive, real number. To solve $|P| \leq k$, solve the three-part inequality

$$-k \leq P \leq k$$

($<$ may be substituted for $\leq$.) **(p. 98)**

Solve $|x - 3| \leq 2$. Graph the solution set and write the answer in interval notation.
$$-2 \leq x - 3 \leq 2$$
$$1 \leq x \leq 5$$

In interval notation, we write $[1, 5]$.

Inequalities Containing $>$ or $\geq$

Let P be an expression and let k be a positive, real number. To solve $|P| \geq k$, ($>$ may be substituted for $\geq$) solve the compound inequality

$$P \geq k \text{ or } P \leq -k \qquad\qquad \textbf{(p. 99)}$$

Solve $|2n - 5| > 1$. Graph the solution set and write the answer in interval notation.
$$2n - 5 > 1 \quad \text{or} \quad 2n - 5 < -1 \qquad \text{Solve.}$$
$$2n > 6 \quad \text{or} \qquad 2n < 4 \qquad \text{Add 5.}$$
$$n > 3 \quad \text{or} \qquad n < 2 \qquad \text{Divide by 2.}$$

In interval notation, we write $(\infty, 2) \cup (3, \infty)$.

*Additional answers can be found in the Answers to Exercises appendix.

(2.1) Determine if the given value is a solution to the equation.

1) $2n + 13 = 10$; $n = -\dfrac{3}{2}$ yes

2) $5 + t = 3t - 1$; $t = 4$ no

Solve each equation.

3) $-9z = 30$ $\left\{-\dfrac{10}{3}\right\}$ 4) $p - 11 = -14$ $\{-3\}$

5) $21 = k + 2$ $\{19\}$ 6) $56 = \dfrac{8}{5}m$ $\{35\}$

7) $-\dfrac{4}{9}w = -\dfrac{10}{7}$ $\left\{\dfrac{45}{14}\right\}$ 8) $-c = 4$ $\{-4\}$

9) $21 = 0.6q$ $\{35\}$ 10) $8b - 7 = 57$ $\{8\}$

11) $6 = 15 + \dfrac{9}{2}v$ $\{-2\}$ 12) $2.3a + 1.5 = 10.7$ $\{4\}$

13) $\dfrac{2}{7} - \dfrac{3}{4}k = -\dfrac{17}{14}$ $\{2\}$ 14) $5(2z + 3) - (11z - 4) = 3$ $\{16\}$

15) $11x + 13 = 2x - 5$ $\{-2\}$

16) $5(c + 3) - 2c = 4 + 3(2c + 1)$ $\left\{\dfrac{8}{3}\right\}$

17) $6 - 5(4d - 3) = 7(3 - 4d) + 8d$ {all real numbers}

18) $4k + 19 + 8k = 2(6k - 11)$ $\varnothing$

19) $0.05m + 0.11(6 - m) = 0.08(6)$ $\{3\}$

20) $1 - \dfrac{1}{6}(t + 5) = \dfrac{1}{2}$ $\{-2\}$

Solve using the five "Steps for Solving Applied Problems."

21) Twelve less than a number is five. Find the number.   17

22) A number increased by nine is one less than twice the number. Find the number.   10

(2.2)

23) Mr. Morrissey has 26 children in his kindergarten class, and c children attended preschool. Write an expression for the number of children who did not attend preschool. $26 - c$

24) In a parking lot, there were f foreign cars. Write an expression for the number of American cars in the lot if there were 14 more than the number of foreign cars. $f + 14$

Solve using the five "Steps for Solving Applied Problems."

25) In its first week, American Idol finalist Clay Aiken's debut CD sold 316,000 more copies than Idol winner Kelly Clarkson's debut CD. Together their debut CDs sold 910,000 copies. How many CDs did each of them sell during the first week? (www.top40.com)
Kelly Clarkson: 297,000 copies, Clay Aiken: 613,000 copies

26) The sum of three consecutive even integers is 258. Find the integers.   84, 86, 88

27) The road leading to the Sutter family's farmhouse is 500 ft long. Some of it is paved, and the rest is gravel. If the paved portion is three times as long as the gravel part of the road, determine the length of the gravel portion of the road.   125 ft

28) Before road construction began in front of her ice cream store, Imelda's average monthly revenue was about $8200. In the month since construction began, revenue has decreased by 18%. What was the revenue during the first month of construction? $6724

29) Americans spent about $11.0 billion on veterinary care for their pets in 2008. This is 8.5% more than they spent in 2007. How much did Americans spend on their pets' veterinary care in 2007? (Round to the nearest tenths place.) (http://articles.moneycentral.msn.com.) $10.1 billion

30) Jerome invested some money in an account earning 7% simple interest and $3000 more than that at 8% simple interest. After 1 yr, he earned $915 in interest. How much money did he invest in each account? $4500 at 7% and $7500 at 8%

(2.3)

31) *Use a formula from geometry to solve:* The base of a triangle measures 9 cm. If the area of the triangle is 54 cm², find the height.   12 cm

32) Find the missing angle measures.

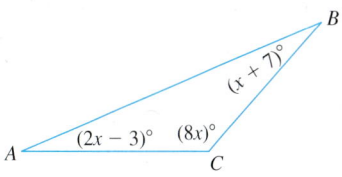

$m\angle A = 29°$, $m\angle B = 23°$, $m\angle C = 128°$

Find the measure of each indicated angle.

33)

$(2x - 1)°$ $(3x - 19)°$ 79°, 101°

34)

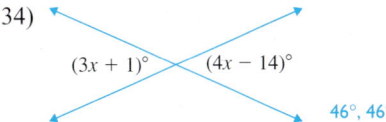

$(3x + 1)°$ $(4x - 14)°$ 46°, 46°

35) *Solve:* Three times the measure of an angle is 12° more than the measure of its supplement. Find the measure of the angle.   48°

Solve for the indicated variable.

36) $y = mx + b$ for m $m = \dfrac{y - b}{x}$

37) $pV = nRT$ for R $R = \dfrac{pV}{nT}$

38) $C = \dfrac{1}{3}n(t + T)$ for t $t = \dfrac{3C}{n} - T$ or $t = \dfrac{3C - Tn}{n}$

(2.4) Solve each proportion.

39) $\dfrac{k}{12} = \dfrac{15}{9}$ {20}

40) $\dfrac{3a + 5}{6} = \dfrac{a - 1}{18}$ {−2}

Set up a proportion and solve.

41) In 2005, a survey found that approximately 1 out of 4 high school students can be classified as heavy drinkers. In a high school of 1768 students, how many would be expected to be heavy drinkers? (www.cdc.gov.)   442

42) Given these two similar triangles, find x.

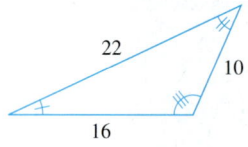
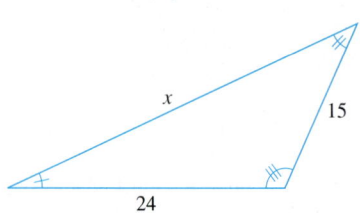

$x = 33$

Solve.

43) A collection of coins contains 91 coins, all dimes and quarters. If the value of the coins is \$14.05, determine the number of each type of coin in the collection.   58 dimes, 33 quarters

44) To make a 6% acid solution, a chemist mixes some 4% acid solution with 12 L of 10% acid solution. How much of the 4% solution must be added to the 10% solution to make the 6% acid solution?   24 L

45) Peter leaves Mitchell's house traveling 30 mph. Mitchell leaves 15 min later, trying to catch up to Peter, going 40 mph. If they drive along the same route, how long will it take Mitchell to catch Peter?   45 min

(2.5) Solve each inequality. Graph the solution set and write the answer in interval notation.

46) $z + 6 \geq 14$

47) $-10y + 7 > 32$

48) $0.03c + 0.09(6 - c) > 0.06(6)$

49) $-15 < 4p - 7 \leq 5$

50) $-1 \leq \dfrac{5 - 3x}{2} \leq 0$

51) *Write an inequality and solve:* Renee's scores on her first three Chemistry tests were 95, 91, and 86. What does she need to make on her fourth test to maintain an average of at least 90?   88 or higher

(2.6)

52) $A = \{10, 20, 30, 40, 50\}$ $B = \{20, 25, 30, 35\}$
   a) Find $A \cup B$.   {10, 20, 25, 30, 35, 40, 50}
   b) Find $A \cap B$.   {20, 30}

Solve each compound inequality. Graph the solution set and write the answer in interval notation.

53) $a + 6 \leq 9$ and $7a - 2 \geq 5$

54) $3r - 1 > 5$ or $-2r \geq 8$

55) $8 - y < 9$ or $\dfrac{1}{10}y > \dfrac{3}{5}$

56) $x + 12 \leq 9$ and $0.2x \geq 3$

The following table lists the number of hybrid vehicles sold in the United States by certain manufacturers in June and July of 2008. (www.hybridcars.com)

Manufacturer	Number Sold in June	Number Sold in July
Toyota	16,330	18,801
Honda	2,717	3,443
Ford	1,910	1,265
Lexus	1,476	1,562
Nissan	1,333	715

List the elements of the set that satisfy the given information.

57) The set of manufacturers who sold more than 3000 hybrid vehicles in June and July   {Toyota}

58) The set of manufacturers who sold more than 5000 hybrid vehicles in June or fewer than 1500 hybrids in July   {Toyota, Ford, Nissan}

(2.7) Solve.

59) $|m| = 9$ {−9, 9}

60) $\left|\dfrac{1}{2}c\right| = 5$ {−10, 10}

61) $|7t + 3| = 4$ $\left\{-1, \dfrac{1}{7}\right\}$

62) $|4 - 3y| = 12$ $\left\{-\dfrac{8}{3}, \dfrac{16}{3}\right\}$

63) $|8p + 11| - 7 = -3$

64) $|5k + 3| - 8 = 4$ $\left\{-3, \dfrac{9}{5}\right\}$

65) $\left|4 - \dfrac{5}{3}x\right| = \dfrac{1}{3}$ $\left\{\dfrac{11}{5}, \dfrac{13}{5}\right\}$

66) $\left|\dfrac{2}{3}w + 6\right| = \dfrac{5}{2}$ $\left\{-\dfrac{51}{4}, -\dfrac{21}{4}\right\}$

67) $|7r - 6| = |8r + 2|$

68) $|3z - 4| = |5z - 6|$ $\left\{1, \dfrac{5}{4}\right\}$

69) $|2a - 5| = -10$ ∅

70) $|h + 6| - 12 = -20$ ∅

71) $|9d + 4| = 0$ $\left\{-\dfrac{4}{9}\right\}$

72) $|6q - 7| = 0$ $\left\{\dfrac{7}{6}\right\}$

73) Write an absolute value equation which means a is 4 *units from zero*. $|a| = 4$, may vary

74) Write an absolute value equation which means t is 7 *units from zero*. $|t| = 7$, may vary

Solve each inequality. Graph the solution set and write the answer in interval notation.

75) $|c| \leq 3$

76) $|w + 1| < 11$

77) $|4t| > 8$

78) $|2v - 7| \geq 15$

79) $|12r + 5| \geq 7$

80) $|3k - 11| < 4$

81) $|4 - a| < 9$

82) $|2 - 5q| > 6$

83) $|4c + 9| - 8 \leq -2$

84) $|3m + 5| + 2 \geq 7$

85) $|5y + 12| - 15 \geq -8$

86) $3 + |z - 6| \leq 13$

87) $|k + 5| > -3$

88) $|4q - 9| < 0$

89) $|12s + 1| \leq 0$

90) A radar gun indicated that a pitcher threw a 93 mph fastball. The radar gun's possible error in measuring the speed of a pitch is ± 1 mph. Write an absolute value inequality to represent the range for the speed of the pitch, and solve the inequality. Explain the meaning of the answer. Let s represent the range of values for the speed of the pitch.

$|s - 93| \leq 1; 92 \leq s \leq 94;$ the speed of the pitch is between 92 mph and 94 mph.

Additional answers can be found in the Answers to Exercises appendix.

Solve.

1) $7p + 16 = 30$ {2}

2) $5 - 3(2c + 7) = 4c - 1 - 5c$ {−3}

3) $\dfrac{5}{8}(3k + 1) - \dfrac{1}{4}(7k + 2) = 1$ {7}

4) $6(4t - 7) = 3(8t + 5)$ $\varnothing$

5) $\dfrac{3 + n}{5} = \dfrac{2n - 3}{8}$ $\left\{\dfrac{39}{2}\right\}$

6) $0.06x + 0.14(x - 5) = 0.10(23)$ {15}

Set up an equation and solve.

7) Nine less than twice a number is 33. Find the number.   21

8) The dosage for a certain medication is $\dfrac{1}{2}$ tsp for every 20 lb of body weight. What is the dosage for a child who weighs 90 lb? $2\dfrac{1}{4}$ tsp

9) Motor oil is available in three types: regular, synthetic, and synthetic blend, which is a mixture of regular oil with synthetic. Bob decides to make 5 qt of his own synthetic blend. How many quarts of regular oil costing $1.20 per quart and how many quarts of synthetic oil costing $3.40 per quart should he mix so that the blend is worth $1.86 per quart?   3.5 qt of regular oil, 1.5 qt of synthetic oil

10) In 2004, Wisconsin had 9 drive-in theaters. This is 82% less than the number of drive-ins in 1967. Determine the number of drive-in theaters in Wisconsin in 1967. (www.drive-ins.com)   50

11) A contractor has 460 ft of fencing to enclose a rectangular construction site. If the length of the site is 160 ft, what is the width?   70 ft

Solve for the indicated variable.

12) $R = \dfrac{kt}{5}$ for t $t = \dfrac{5R}{k}$

13) $S = 2\pi r^2 + 2\pi rh$ for h $h = \dfrac{S - 2\pi r^2}{2\pi r}$

Solve. Graph the solution set, and write the answer in interval notation.

14) $r + 7 \le 2$ $(-\infty, -5]$

15) $9 - 3(2x - 1) < 4x + 5(x + 2) - 8$ $\left(\dfrac{2}{3}, \infty\right)$

16) $-1 < \dfrac{w - 5}{4} \le \dfrac{1}{2}$ (1, 7]

17) *Write an inequality and solve:* Rawlings Builders will rent a forklift for $46.00 per day plus $9.00 per hour. If they have at most $100.00 allotted for a one-day rental, for how long can they keep the forklift and remain within budget?   at most 6 hr

18) Given sets $A = \{1, 2, 3, 6, 12\}$ and $B = \{1, 2, 9, 12\}$, find each of the following.

   a) $A \cup B$ {1, 2, 3, 6, 9, 12}   b) $A \cap B$ {1, 2, 12}

Solve each compound inequality. Write the answer in interval notation.

19) $3n + 5 > 12$ or $\dfrac{1}{4}n < -2$ $(-\infty, -8) \cup \left(\dfrac{7}{3}, \infty\right)$

20) $y - 8 \le -5$ and $2y \ge 0$ [0, 3]

21) $6 - p < 10$ or $p - 7 < 2$ $(-\infty, \infty)$

Solve.

22) $|4y - 9| = 11$ $\left\{-\dfrac{1}{2}, 5\right\}$

23) $|d + 6| - 3 = 7$ {−16, 4}

24) $|3k + 5| = |k - 11|$ $\left\{-8, \dfrac{3}{2}\right\}$

25) $\left|\dfrac{1}{2}n - 1\right| = -8$ $\varnothing$

26) Write an absolute value equation that means x *is 8 units from zero.* $|x| = 8$, may vary

Solve each inequality. Graph the solution set and write the answer in interval notation.

27) $|2z - 7| \le 9$ [−1, 8]

28) $|4m + 9| - 8 \ge 5$ $\left(-\infty, -\dfrac{11}{2}\right] \cup [1, \infty)$

29) $|10 - 3w| < -2$ $\varnothing$

30) A scale in a doctor's office has a possible error of ± 0.75 lb. If Thanh's weight is measured as 168 lb, write an absolute value inequality to represent the range for his weight, and solve the inequality. Let w represent the range of values for Thanh's weight. Explain the meaning of the answer.

   $|w - 168| \le 0.75$; $167.25 \le w \le 168.75$; Thanh's weight is between 167.25 lb and 168.75 lb.

*Additional answers can be found in the Answers to Exercises appendix.

Perform the operations and simplify.

1) $\dfrac{5}{12} - \dfrac{7}{9}$ $-\dfrac{13}{36}$

2) $\dfrac{8}{15} \div 12$ $\dfrac{2}{45}$

3) $52 - 12 \div 4 + 3 \cdot 5$ 64

4) $2^3 - 10(-9 + 4) + (-2)$ 56

5) -3^4 -81

Given the set of numbers $\left\{-13.7, \dfrac{19}{7}, 0, 8, \sqrt{17}, 0.\overline{61}, \sqrt{81}, -2\right\}$ **identify**

6) the rational numbers $\left\{-13.7, \dfrac{19}{7}, 0, 8, 0.\overline{61}, \sqrt{81}, -2\right\}$

7) the integers $\{0, 8 \sqrt{81}, -2\}$

Determine which property is illustrated by each statement.

8) $9 + (4 + 1) = (9 + 4) + 1$ Associative Property

9) $3(n + 7) = 3n + 21$ Distributive Property

10) Rewrite the expression $8 + 5$ using the commutative property. $5 + 8$

Solve each equation.

11) $-31 = \dfrac{4}{7}z + 9$ $\{-70\}$

12) $12 - 4(2a - 5) = 7 + 3(a + 2) - 7a$ $\left\{\dfrac{19}{4}\right\}$

13) $\dfrac{1}{2}(t + 1) - \dfrac{1}{3} = \dfrac{17}{12} + \dfrac{1}{4}(2t - 5)$ {all real numbers}

14) $|9 - 4c| + 3 = 12$ $\left\{0, \dfrac{9}{2}\right\}$

15) Solve for R: $A = P + PRT$. $R = \dfrac{A - P}{PT}$

16) Find the supplement of $59°$. $121°$

17) Find the area and perimeter of the rectangle.

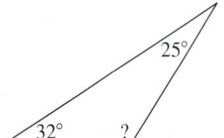

7 cm

15 cm

Area = $105\ \text{cm}^2$; Perimeter = 44 cm

18) Find the missing angle and identify the triangle as acute, obtuse, or right.

25°
32°   ?

$123°$; obtuse

Write an equation and solve.

19) On Thursday, a pharmacist fills twice as many prescriptions with generic drugs as name brand drugs. If he filled 72 prescriptions, how many were with generic drugs, and how many were with name brand drugs?   Generic: 48; Name brand: 24

20) Two friends start biking toward each other from opposite ends of a bike trail. One averages 8 mph, and the other averages 10 mph. If the distance between them when they begin is 9 miles, how long will it take them to meet? $\dfrac{1}{2}$ hour

Solve each inequality. Write the answer in interval notation.

21) $-10w + 3(2w - 5) \le 7 + 2(w - 9)$ $\left[-\dfrac{2}{3}, \infty\right)$

22) $-14 < 6y + 10 < 3$ $\left(-4, -\dfrac{7}{6}\right)$

23) $8x \le -24$ or $4x - 5 \ge 6$ $(-\infty, -3] \cup \left[\dfrac{11}{4}, \infty\right)$

24) $\left|\dfrac{1}{2}m + 7\right| \le 11$ $[-36, 8]$

25) $|3 - r| > 4$ $(-\infty, -1) \cup (7, \infty)$

CHAPTER 3

Linear Equations in Two Variables and Functions

Algebra at Work: Landscape Architecture

We will take a final look at how mathematics is used in landscape architecture.

A landscape architect uses slope in many different ways. David explains that one important application of slope is in designing driveways after a new house has been built. Towns often have building codes that restrict the slope or steepness of a driveway. In this case, the rise of the land is the difference in height between the top and the bottom of the driveway. The run is the linear horizontal distance between those two points. By finding $\dfrac{rise}{run}$, a landscape architect knows how to design the driveway so that it meets the town's building code. This is especially important in cold weather climates, where if a driveway is too steep, a car will too easily slide into the street. If it doesn't meet the code, the driveway may have to be removed and rebuilt, or coils that radiate heat might have to be installed under the driveway to melt the snow in the wintertime. Either way, a mistake in calculating slope could cost the landscape company or the client a lot of extra money.

In Chapter 3, we will learn about slope, its meaning, and different ways to use it.

Section 3.1 Introduction to Linear Equations in Two Variables

Objectives

1. **Plot Ordered Pairs**
2. **Decide Whether an Ordered Pair is a Solution of an Equation**
3. **Graph a Linear Equation in Two Variables by Plotting Points**
4. **Graph a Linear Equation in Two Variables by Finding the Intercepts**
5. **Graph Linear Equations of the Form $x = c$ and $y = d$**
6. **Solve Applications**

We encounter graphs everywhere—in newspapers, books, and on the Internet. The graph in Figure 1 shows how many billions of dollars consumers spent shopping online for consumer electronics during the years 2001–2007.

Graphs like this one are based on the **Cartesian coordinate system,** or the **rectangular coordinate system.** Let's begin this section by reviewing how to work with a Cartesian coordinate system.

1. Plot Ordered Pairs

The Cartesian coordinate system has a horizontal number line, called the **x-axis,** and a vertical number line, called the **y-axis.** The x-axis and y-axis determine a flat surface called a **plane.** The axes divide this plane into four **quadrants.** The point at which the x-axis and y-axis intersect is called the **origin.** The arrow at one end of the x-axis and one end of the y-axis indicates the positive direction on each axis.

Each point in the plane is represented by an **ordered pair** of real numbers (x, y). The first

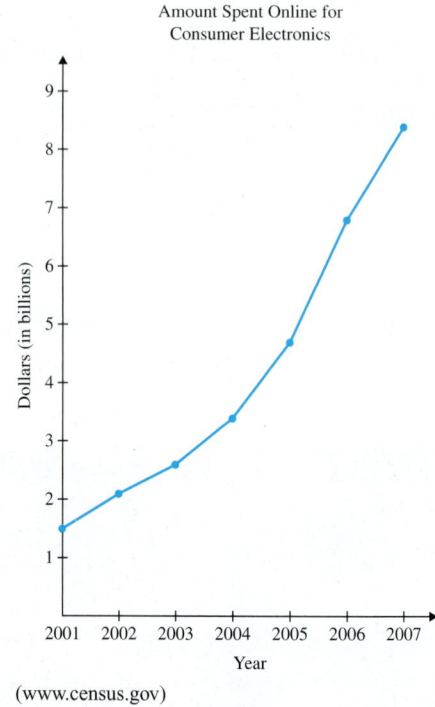

Amount Spent Online for Consumer Electronics

Dollars (in billions)

Year

(www.census.gov)

Figure 1

number in the ordered pair is called the **x-coordinate** (also called the **abscissa**), and it tells us the distance and direction of the point from the origin along the x-axis. The second number in the ordered pair is the **y-coordinate** (also called the **ordinate**), and it tells us the distance and direction of the point from the origin either on the y-axis or parallel to the y-axis. The origin has coordinates $(0, 0)$. See Figure 2.

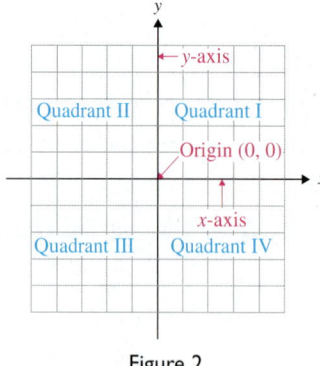

Figure 2

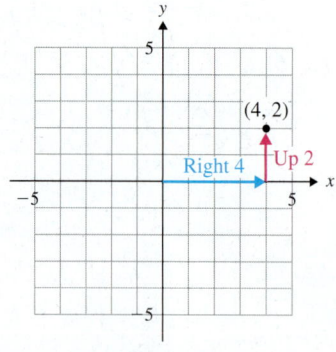

Figure 3

We have graphed the ordered pair $(4, 2)$ on the coordinate system in Figure 3. This is also called **plotting the point** $(4, 2)$. The x-coordinate of the point is 4. It tells us to move 4 units to the right from the origin along the x-axis. The y-coordinate of the point is 2. From the current position, move 2 units up, parallel to the y-axis.

We can use ordered pairs to represent the points on the line graph in Figure 1 just like we use them to represent points in the Cartesian coordinate system. In Figure 1, the years are on the horizontal axis, and the amount of money spent online on consumer electronics is on the vertical axis. If we move along the horizontal axis to the year 2005 and then move up, parallel to the vertical axis, we find the billions of dollars spent shopping online for consumer electronics during that year. This point can be represented by the ordered pair (2005, 4.7), and it tells us that in 2005 shoppers spent approximately $4.7 billion online on consumer electronics.

Example 1

In-Class Example 1
Plot the points.
a) (0, −3) b) (3, 4) c) (−2, −4)
d) (4, 0) e) (1, −2) f) (−5, 1)
answer:

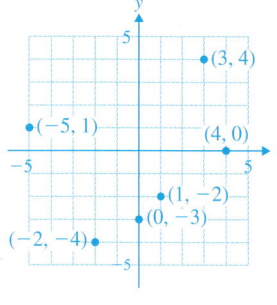

Plot the points.

a) $(-2, 5)$ 　　　　b) $(1, -4)$ 　　　　c) $\left(\dfrac{7}{2}, 3\right)$

d) $(-5, -2)$ 　　　　e) $(0, 1.5)$ 　　　　f) $(-4, 0)$

Solution

To plot each point, move from the origin in the following ways:

a) $(-2, 5)$: Move left 2 units then up 5 units. This point is in quadrant II.

b) $(1, -4)$: Move right 1 unit then down 4 units. This point is in quadrant IV.

c) $\left(\dfrac{7}{2}, 3\right)$: Think of $\dfrac{7}{2}$ as $3\dfrac{1}{2}$. Move right $3\dfrac{1}{2}$ units then up 3 units. This point is in quadrant I.

d) $(-5, -2)$: Move left 5 units then down 2 units. This point is in quadrant III.

e) $(0, 1.5)$: The x-coordinate of 0 means that we don't move in the x-direction (horizontally). From the origin, move up 1.5 units on the y-axis. This point is not in any quadrant.

f) $(-4, 0)$: From the origin, move left 4 units. Since the y-coordinate is zero, we do not move in the y-direction (vertically). This point is not in any quadrant.

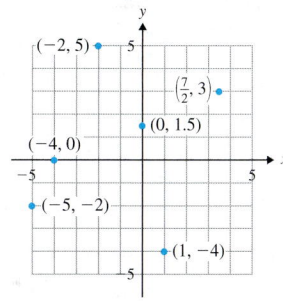

You Try 1

Plot the points.

a) $(3, 1)$ 　　b) $(-2, 4)$ 　　c) $(0, -5)$ 　　d) $(2, 0)$ 　　e) $(-4, -3)$ 　　f) $\left(1, \dfrac{7}{2}\right)$

Note

The coordinate system should always be labeled to indicate how many units each mark represents.

A **solution of an equation in two variables** is written as an ordered pair so that when the values are substituted for the appropriate variables we obtain a true statement.

2. Decide Whether an Ordered Pair Is a Solution of an Equation

Example 2

Determine whether each ordered pair is a solution of $4x + 5y = 11$.

In-Class Example 2
Determine whether each ordered pair is a solution of $3x + 4y = -2$.
a) $(2, -2)$ b) $\left(1, \frac{5}{4}\right)$

answer: a) yes   b) no

a) $(-1, 3)$ b) $\left(\frac{3}{2}, 5\right)$

Solution

a)   Solutions to the equation $4x + 5y = 11$ are written in the form (x, y) where (x, y) is called an *ordered pair*. Therefore, the ordered pair $(-1, 3)$ means that $x = -1$ and $y = 3$.

$$(-1, 3)$$
$$\nearrow \qquad \nwarrow$$
$$x\text{-coordinate} \qquad y\text{-coordinate}$$

To determine if $(-1, 3)$ is a solution of $4x + 5y = 11$, we substitute -1 for x and 3 for y. Remember to put these values in parentheses.

$$4x + 5y = 11$$
$$4(-1) + 5(3) = 11 \qquad \text{Substitute } x = -1 \text{ and } y = 3.$$
$$-4 + 15 = 11 \qquad \text{Multiply.}$$
$$11 = 11 \qquad \text{True}$$

Since substituting $x = -1$ and $y = 3$ into the equation gives the true statement $11 = 11$, $(-1, 3)$ *is a solution* of $4x + 5y = 11$. We say that $(-1, 3)$ *satisfies* $4x + 5y = 11$.

b)   The ordered pair $\left(\frac{3}{2}, 5\right)$ tells us that $x = \frac{3}{2}$ and $y = 5$.

$$4x + 5y = 11$$
$$4\left(\frac{3}{2}\right) + 5(5) = 11 \qquad \text{Substitute } \frac{3}{2} \text{ for } x \text{ and } 5 \text{ for } y.$$
$$6 + 25 = 11 \qquad \text{Multiply.}$$
$$31 = 11 \qquad \text{False}$$

Since substituting $\left(\frac{3}{2}, 5\right)$ into the equation gives the false statement $31 = 11$, the ordered pair is *not* a solution to the equation. ∎

You Try 2

Determine if each ordered pair is a solution of the equation $y = -\frac{3}{4}x + 5$.

a) $(12, -4)$ b) $(0, 7)$ c) $(-8, 11)$

If the variables in the equation are not x and y, then the variables in the ordered pairs are written in alphabetical order. For example, solutions to $2a + b = 7$ are ordered pairs of the form (a, b).

3. Graph a Linear Equation in Two Variables by Plotting Points

We saw in Example 2 that the ordered pair $(-1, 3)$ was a solution of the equation $4x + 5y = 11$. This is just one solution, however. The equation has an infinite number of solutions of the form (x, y), and we can represent these solutions with a graph on the Cartesian coordinate system.

The following table of values contains the solution we just verified as well as other solutions of $4x + 5y = 11$. Plot the points and connect them with a straight line. The line represents all solutions of the equation.

x	y
-1	3
4	-1
9	-5
0	$\frac{11}{5}$
$\frac{11}{4}$	0

$$4x + 5y = 11$$
$$4(-1) + 5(3) = -4 + 15 = 11$$
$$4(4) + 5(-1) = 16 + (-5) = 11$$
$$4(9) + 5(-5) = 36 + (-25) = 11$$
$$4(0) + 5\left(\frac{11}{5}\right) = 0 + 11 = 11$$
$$4\left(\frac{11}{4}\right) + 5(0) = 11 + 0 = 11$$

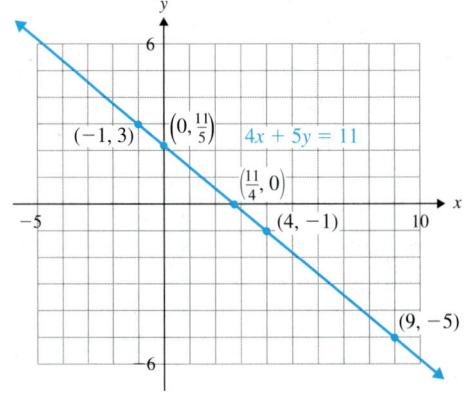

While equations like $x + 3 = 5$ and $2(t - 7) = 3t + 10$ are linear equations in one variable, the equation $4x + 5y = 11$ is an example of a linear equation in *two* variables. **The graph of a linear equation in two variables is a line, and every point on the line is a solution of the equation.**

Definition

A linear equation in two variables can be written in the form $Ax + By = C$ where A, B, and C are real numbers and where both A and B do not equal zero.

Other examples of linear equations in two variables are

$$3x - 5y = 10 \qquad y = \frac{1}{2}x + 7 \qquad -9s + 2t = 4 \qquad x = -8$$

(We can write $x = -8$ as $x + 0y = -8$, therefore it is a linear equation in two variables.) A solution to a linear equation in two variables is written as an ordered pair.

Example 3

Graph $-x + 2y = 4$.

In-Class Example 3
Graph $x - 3y = 4$.
answer:

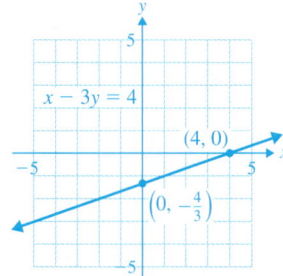

Solution

We will find three ordered pairs which satisfy the equation. Let's complete a table of values for $x = 0$, $x = 2$, and $x = -4$.

$x = 0$:
$$-x + 2y = 4$$
$$-(0) + 2y = 4$$
$$2y = 4$$
$$y = 2$$

$x = 2$:
$$-x + 2y = 4$$
$$-(2) + 2y = 4$$
$$-2 + 2y = 4$$
$$2y = 6$$
$$y = 3$$

$x = -4$:
$$-x + 2y = 4$$
$$-(-4) + 2y = 4$$
$$4 + 2y = 4$$
$$2y = 0$$
$$y = 0$$

We get the table of values

x	y
0	2
2	3
-4	0

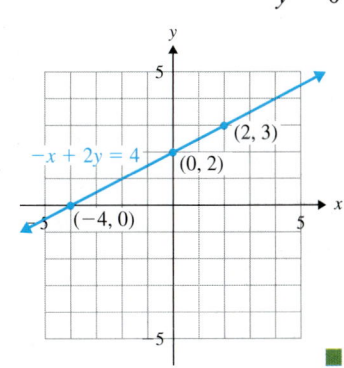

Plot the points $(0, 2)$, $(2, 3)$, and $(-4, 0)$, and draw the line through them.

You Try 3

Graph each line.

a) $3x + 2y = 6$ b) $y = 4x - 8$

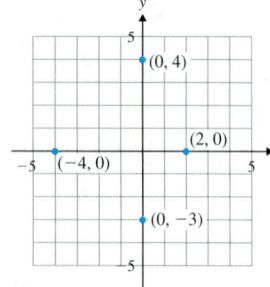

4. Graph a Linear Equation in Two Variables by Finding the Intercepts

In Example 3, the line crosses the x-axis at -4 and crosses the y-axis at 2. These points are called **intercepts**. What is the y-coordinate of any point on the x-axis? It is zero. Likewise, the x-coordinate of any point on the y-axis is zero.

> ### Definition
> The **x-intercept** of the graph of an equation is the point where the graph intersects the x-axis.
> The **y-intercept** of the graph of an equation is the point where the graph intersects the y-axis.

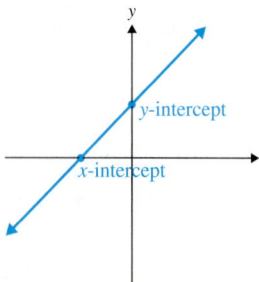

> ### Procedure   Finding Intercepts
> To find the *x-intercept* of the graph of an equation, let $y = 0$ and solve for x.
> To find the *y-intercept* of the graph of an equation, let $x = 0$ and solve for y.

Finding intercepts is very helpful for graphing linear equations in two variables.

Example 4

In-Class Example 4

Graph $y = \frac{2}{3}x + 1$ by finding the intercepts and one other point.
answer:

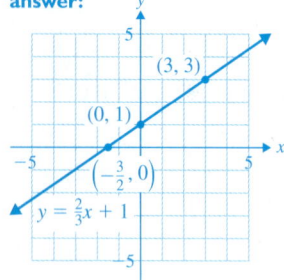

Graph $y = -\dfrac{1}{2}x - 1$ by finding the intercepts and one other point.

Solution

We will begin by finding the intercepts.

x-intercept: Let $y = 0$, and solve for x.
$$0 = -\tfrac{1}{2}x - 1$$
$$1 = -\tfrac{1}{2}x$$
$$-2 = x \qquad \text{Multiply both sides by } -2 \text{ to solve for } x.$$

The x-intercept is $(-2, 0)$.

y-intercept: Let $x = 0$, and solve for y.
$$y = -\tfrac{1}{2}(0) - 1$$
$$y = 0 - 1 = -1$$

The y-intercept is $(0, -1)$.

We must find another point. Let's look closely at the equation $y = -\tfrac{1}{2}x - 1$. The coefficient of x is $-\tfrac{1}{2}$. If we choose a value for x that is a multiple of 2 (the denominator of the fraction), then $-\tfrac{1}{2}x$ will not be a fraction.

Let $x = 2$.
$$y = -\tfrac{1}{2}x - 1$$
$$y = -\tfrac{1}{2}(2) - 1$$
$$y = -1 - 1$$
$$y = -2$$

The third point is $(2, -2)$.

Plot the points, and draw the line through them.

You Try 4

Graph $y = 3x + 6$ by finding the intercepts and one other point.

5. Graph Linear Equations of the Forms $x = c$ and $y = d$

The equation $x = c$ is a linear equation in two variables since it can be written in the form $x + 0y = c$. The same is true for $y = d$. It can be written as $0x + y = d$. Let's see how we can graph these equations.

Example 5

Graph $x = 3$.

In-Class Example 5
Graph $x = -4$.
answer:

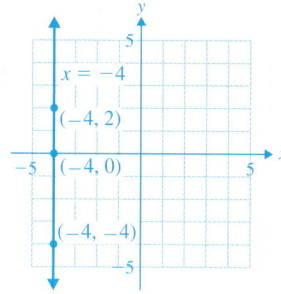

Solution

The equation is $x = 3$. (This is the same as $x + 0y = 3$.) $x = 3$ *means that no matter the value of y, x always equals* 3. We can make a table of values where we choose any value for y, but x is always 3.

x	y
3	0
3	1
3	−2

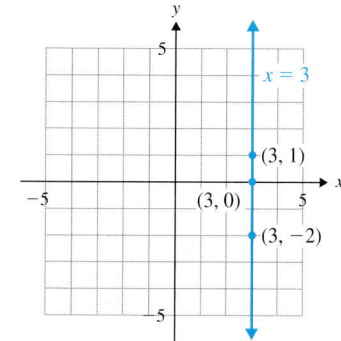

Plot the points, and draw a line through them. The graph of $x = 3$ is a *vertical line*.

We can generalize the result as follows:

Property   The Graph of $x = c$

If c is a constant, then the graph of $x = c$ is a *vertical line* going through the point $(c, 0)$.

You Try 5

Graph $x = -4$.

Example 6

Graph $y = -2$.

Solution

The equation $y = -2$ is the same as $0x + y = -2$, therefore it is linear. $y = -2$ *means that no matter the value of x, y always equals* -2. Make a table of values where we choose any value for x, but y is always -2.

In-Class Example 6
Graph $y = 3$.
answer:

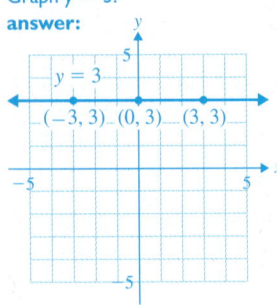

x	y
0	−2
2	−2
−2	−2

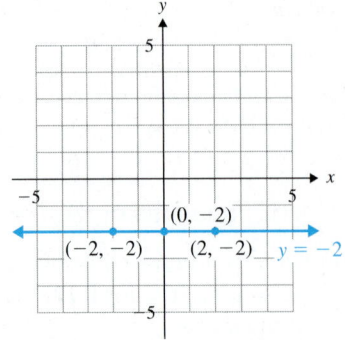

Plot the points, and draw a line through them. The graph of $y = -2$ is a *horizontal line.*

We can generalize the result as follows:

Property The Graph of $y = d$

If d is a constant, then the graph of $y = d$ is a *horizontal line* going through the point $(0, d)$.

You Try 6

Graph $y = 1$.

6. Solve Applications

Next we will see how a linear equation in two variables can be used to solve an application and how its graph can help us visualize the information that we get from the equation.

Example 7

The length of an 18-year-old female's hair is measured to be 250 millimeters (mm) (almost 10 in.). The length of her hair after x days can be approximated by

$$y = 0.30x + 250$$

where y is the length of her hair in millimeters.

a) Find the length of her hair (i) 10 days, (ii) 60 days, and (iii) 90 days after the initial measurement and write the results as ordered pairs.
b) Graph the equation.
c) How long would it take for her hair to reach a length of 274 mm (almost 11 in.)?

In-Class Example 7
Refer to Example 7.
a) Find the length of her hair
(i) 20 days, (ii) 30 days, and
(iii) 70 days after the initial
measurement and write the
results as ordered pairs.
b) Graph the equation.
c) How long would it take for her
hair to reach a length of 262 mm?
answer: a) i) (20, 256);
ii) (30, 259);   iii) (70, 271)
b)

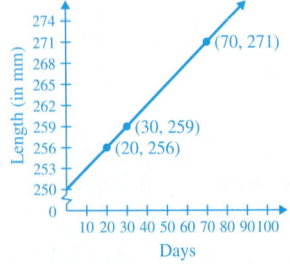

c) It will take 40 days for her
hair to grow to a length of
262 mm.

Solution

a) The problem states that in the equation $y = 0.30x + 250$,

$$x = \text{number of days after the hair was measured}$$
$$y = \text{length of the hair (in millimeters)}$$

We must determine the length of her hair after 10 days, 60 days, and 90 days. We can organize the information in a table of values, shown below.

x	y
10	
60	
90	

i) $x = 10$: $y = 0.30x + 250$

$y = 0.30(10) + 250$ Substitute 10 for x.

$y = 3 + 250 = 253$

After 10 days, her hair is 253 mm long. We can write this as the ordered pair (10, 253).

ii) $x = 60$: $y = 0.30x + 250$

$y = 0.30(60) + 250$ Substitute 60 for x.

$y = 18 + 250 = 268$

After 60 days, her hair is 268 mm long. (Or, after about 2 months, her hair is about 10.5 in. long. It has grown about half of an inch.) We can write this as the ordered pair (60, 268).

iii) $x = 90$: $y = 0.30x + 250$

$y = 0.30(90) + 250$ Substitute 90 for x.

$y = 27 + 250 = 277$

After 90 days, her hair is 277 mm long. Write this as the ordered pair (90, 277).

We can complete the table of values:

x	y
10	253
60	268
90	277

The ordered pairs are (10, 253), (60, 268), and (90, 277).

b) Graph the equation.

The x-axis represents the number of days after the hair was measured. Since it does not make sense to talk about a negative number of days, we will not continue the x-axis in the negative direction.

The y-axis represents the length of the female's hair. Likewise, a negative number does not make sense in this situation, so we will not continue the y-axis in the negative direction.

The scales on the x-axis and y-axis are different. This is because the size of the numbers they represent are quite different.

Here are the ordered pairs we must graph: (10, 253), (60, 268), and (90, 277).

The x-values are 10, 60, and 90, so we will let each mark in the x-direction represent 10 units.

The y-values are 253, 268, and 277. While the numbers are rather large, they do not actually differ by much. We will begin labeling the y-axis at 250, but each mark in the y-direction will represent 3 units. Because there is a large jump in values from 0 to 250 on the y-axis, we indicate this with "⌇" on the axis between the 0 and 250.

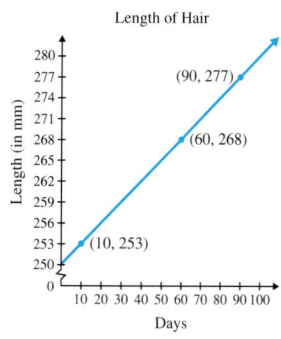

Notice, also, that we have labeled both axes. The ordered pairs are plotted on the graph shown at the left, and the line is drawn through the points.

c)   We must determine how many days it would take for the hair to grow to a length of 274 mm. The length, 274 mm, is the y-value. We must find the value of x that corresponds to $y = 274$ since x represents the number of days.

The equation relating x and y is $y = 0.30x + 250$. We will substitute 274 for y and solve for x.

$$y = 0.30x + 250$$
$$274 = 0.30x + 250$$
$$24 = 0.30x$$
$$80 = x$$

It will take 80 days for her hair to grow to a length of 274 mm. This corresponds to the ordered pair (80, 274) and is consistent with what is illustrated by the graph.

Answers to You Try Exercises

1) a)   b)   c)   d)   e)   f)   2) a) yes   b) no   c) yes

3) a)   b)

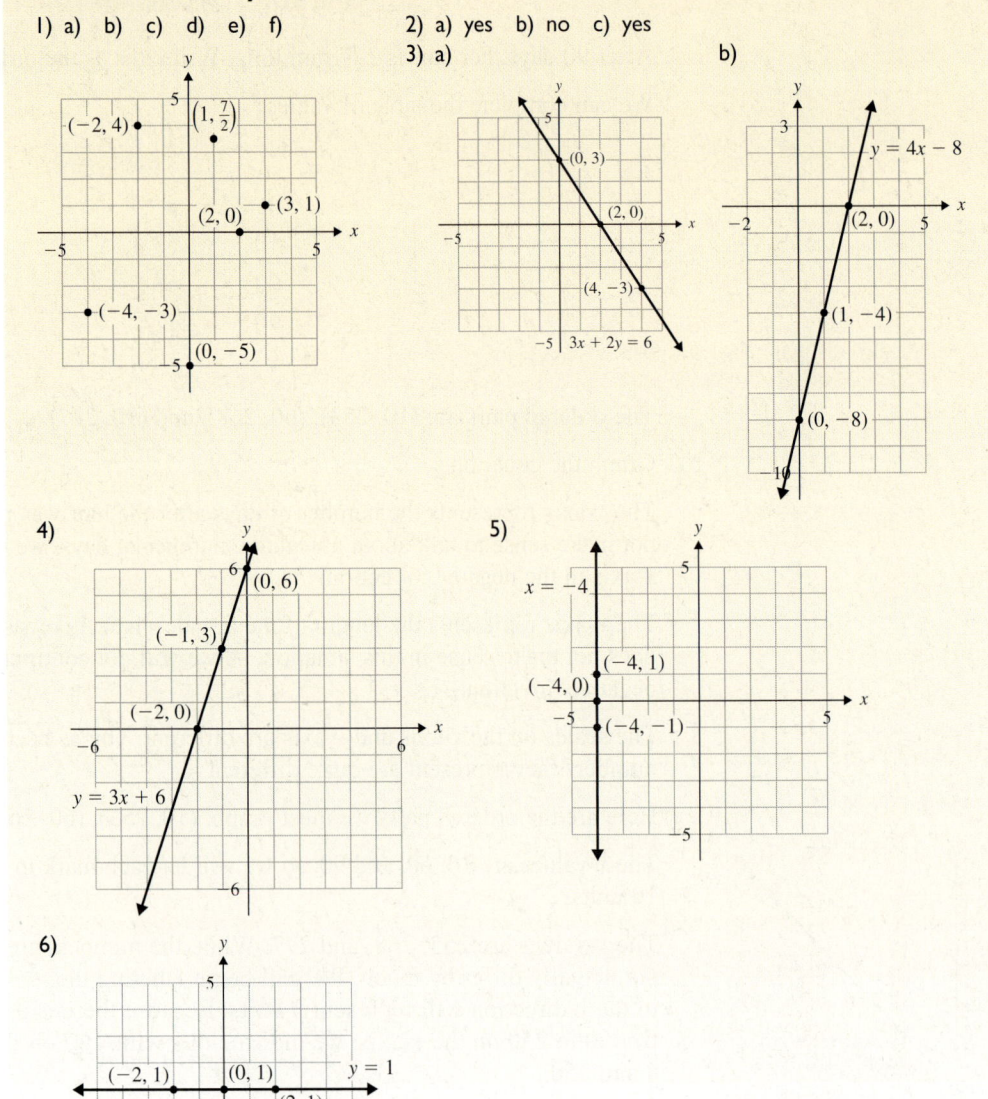

4)

5)

6)

3.1 Exercises

Additional answers can be found in the Answers to Exercises appendix.

Objective 1: Plot Ordered Pairs

1)

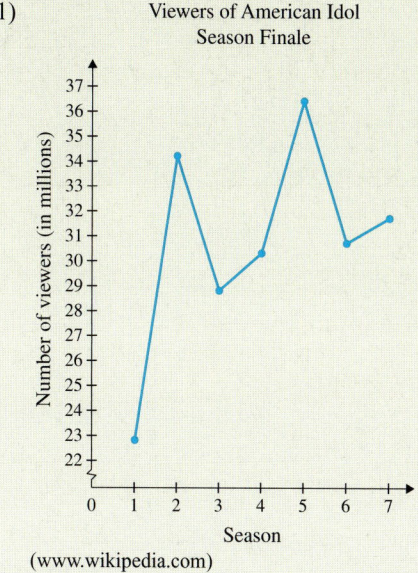

Viewers of American Idol
Season Finale

Number of viewers (in millions)

Season

(www.wikipedia.com)

The graph shows the number of people who watched the season finale of American Idol from Season 1 in 2002 when Kelly Clarkson was the winner through Season 7 in 2008 when David Cook took the crown.

a) If a point on the graph is represented by the ordered pair (x, y), then what do x and y represent?

b) What does the ordered pair (3, 28.8) mean in the context of this problem?   28.8 million people watched the Season 3 finale.

c) Approximately how many people watched the finale in Season 5?   36.4 million

d) Which season finale had approximately 30.7 million viewers?   Season 6

e) Write an ordered pair to represent the number of people who watched Kelly Clarkson win the title of American Idol.   (1, 22.8)

2)

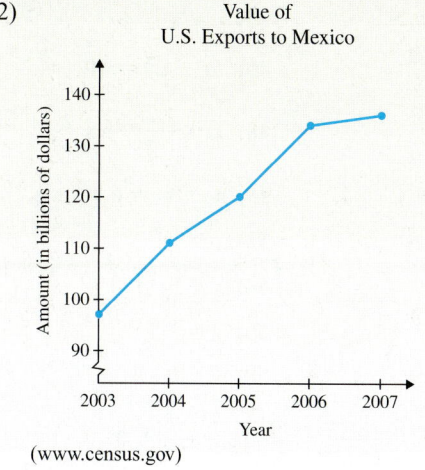

Value of
U.S. Exports to Mexico

Amount (in billions of dollars)

Year

(www.census.gov)

The graph shows the value of U.S. exports to Mexico for the years 2003–2007.

a) If a point on the graph is represented by the ordered pair (x, y), then what do x and y represent?

b) What does the ordered pair (2004, 111) mean in the context of this problem?   In 2004, the value of U.S. exports to Mexico was $111 billion.

c) During which year did exports total about $136 billion?   2007

d) What was the value of exports to Mexico in 2005?   about $120 billion

e) Write an ordered pair to represent the value of U.S. exports to Mexico in 2003.   (2003, 97)

Name each point with an ordered pair, and identify the quadrant in which each point lies.

3)

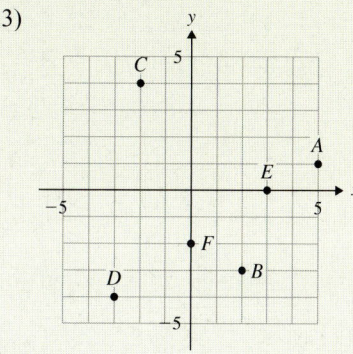

A: (5, 1); quadrant I
B: (2, −3); quadrant IV
C: (−2, 4); quadrant II
D: (−3, −4); quadrant III
E: (3, 0); no quadrant
F: (0, −2); no quadrant

4)

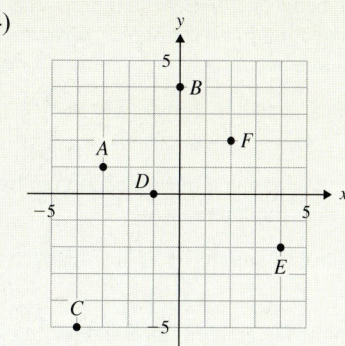

A: (−3, 1); quadrant II
B: (0, 4); no quadrant
C: (−4, −5); quadrant III
D: (−1, 0); no quadrant
E: (4, −2); quadrant IV
F: (2, 2); quadrant I

Graph each ordered pair and explain how you plotted the points.

5) (6, 2)

6) (3, 4)

7) (−1, 4)

8) (−2, −3)

Graph the ordered pairs, labeling each point with the corresponding letter.

9) $A(-3, -5), B(-4, 5), C(2, 0), D(0, 3), E\left(\frac{7}{2}, 1\right),$ $F\left(\frac{5}{2}, -\frac{5}{8}\right), G(-0.7, 2.5)$

10) $A(2, -1), B(5, 2), C(0, 4), D(-3, 0), E\left(4, \frac{8}{3}\right),$ $F\left(-\frac{9}{2}, \frac{5}{4}\right), G(-1.5, -4.2)$

Fill in the blank with *positive, negative,* or *zero.*

11) The x-coordinate of every point in quadrant I is_____. positive

12) The x-coordinate of every point in quadrant II is_____. negative

13) The y-coordinate of every point in quadrant IV is_____. negative

14) The y-coordinate of every point in quadrant II is_____. positive

15) The x-coordinate of every point on the y-axis is _____. zero

16) The y-coordinate of every point on the x-axis is _____. zero

Objective 2: Decide Whether an Ordered Pair is a Solution of an Equation

Determine if each ordered pair is a solution of the given equation.

17) $7x + 2y = 4; (2, -5)$
yes

18) $3x - 5y = 1; (-2, 1)$
no

19) $-2x - y = 13; (-8, 3)$
yes

20) $-4x + 7y = -4; (8, 4)$
yes

21) $y = 5x - 6; (3, 11)$
no

22) $x = -y + 9; (4, -5)$
no

23) $y = -\dfrac{3}{2}x - 19; (-10, -4)$
yes

24) $5y = \dfrac{2}{3}x + 2; (12, 2)$
yes

25) $x = 13; (5, 13)$
no

26) $y = -6; (-7, -6)$
yes

Mixed Exercises: Objectives 3–5

27) Explain the difference between a linear equation in one variable and a linear equation in two variables. Give an example of each. Answers may vary.

28) Is $x^2 + 6y = -5$ a linear equation in two variables? Explain your answer. No. It contains an x^2 term.

29) What do you get when you graph a linear equation in two variables? What does that graph represent?

30) Every linear equation in two variables has how many solutions? an infinite number

31) What is the y-intercept of the graph of an equation? Explain how to find it.

32) What is the x-intercept of the graph of an equation? Explain how to find it.

Complete the table of values and graph each equation.

33) $y = 3x - 1$

x	y
0	−1
1	2
2	5
−1	−4

34) $y = -2x + 5$

x	y
0	5
−1	7
2	1
3	−1

35) $y = -\dfrac{2}{3}x + 4$

x	y
0	4
−3	6
3	2
6	0

36) $y = \dfrac{5}{2}x + 6$

x	y
0	6
2	11
−2	1
−4	−4

37) $-3x + 6y = 9$

x	y
0	$\frac{3}{2}$
−3	0
5	4
−1	1

38) $4x = 1 - y$

x	y
$\frac{1}{4}$	0
0	1
$\frac{5}{2}$	−9
−1	5

39) $y + 4 = 0$

x	y
0	−4
−3	−4
−1	−4
2	−4

40) $x = -\dfrac{3}{2}$

x	y
$-\frac{3}{2}$	5
$-\frac{3}{2}$	0
$-\frac{3}{2}$	−1
$-\frac{3}{2}$	−2

VIDEO 41) For $y = \dfrac{2}{3}x - 7$,

a) find y when $x = 3$, $x = 6$, and $x = -3$. Write the results as ordered pairs. $(3, -5), (6, -3), (-3, -9)$

b) find y when $x = 1$, $x = 5$, and $x = -2$. Write the results as ordered pairs. $\left(1, -\dfrac{19}{3}\right), \left(5, -\dfrac{11}{3}\right), \left(-2, -\dfrac{25}{3}\right)$

c) why is it easier to find the y-values in part a) than in part b)?

42) What ordered pair is a solution to every linear equation of the form $y = mx$, where m is a real number? $(0, 0)$

Graph each equation by finding the intercepts and at least one other point.

43) $y = -2x + 6$

44) $y = x - 3$

VIDEO 45) $3x - 4y = 12$

46) $5x + 2y = 10$

47) $x = \dfrac{1}{4}y - 1$

48) $x = -\dfrac{2}{3}y - 8$

49) $2x + 3y = -6$

50) $x - 4y = 6$

VIDEO 51) $y = -x$

52) $y = x$

53) $5y - 2x = 0$

54) $x + 3y = 0$

55) $x = 5$

56) $y = -1$

57) $y = 0$

58) $x = 0$

59) $y + 3 = 0$

60) $x - \dfrac{5}{2} = 0$

61) $x + 3y = 8$ 62) $6x - y = 7$

63) a) What is the equation of the x-axis?   y = 0

b) What is the equation of the y-axis?   x = 0

64) Let a and b be constants. If the lines $x = a$ and $y = b$ are graphed on the same axes, at what point will they intersect?   (a, b)

Objective 6: Solve Applications

Solve each application.

65) The number of drivers involved in fatal vehicle accidents in 2006 is given in the table. (www.census.gov)

Age	Number of Drivers
16	800
17	1300
18	1800
19	1900

a) Write the information as ordered pairs, (x, y), where x represents the age of the driver and y represents the number of drivers involved in fatal motor vehicle accidents.   (16, 800), (17, 1300), (18, 1800), (19, 1900)

b) Label a coordinate system, choose an appropriate scale, and graph the ordered pairs.

c) Explain the meaning of the ordered pair (18, 1800) in the context of the problem.

66) The number of pounds of potato chips consumed per person is given in the table. (U.S. Dept. of Agriculture)

Year	Pounds of Chips
2003	17.3
2004	16.6
2005	16.0
2006	16.2
2007	15.9

a) Write the information as ordered pairs, (x, y), where x represents the year and y represents the number of pounds of potato chips consumed per person.   (2003, 17.3), (2004, 16.6), (2005, 16.0), (2006, 16.2), (2007, 15.9)

b) Label a coordinate system, choose an appropriate scale, and graph the ordered pairs.

c) Explain the meaning of the ordered pair (2004, 16.6) in the context of the problem.   The consumption of potato chips in 2004 was 16.6 lb per person.

67) Horton's Party Supplies rents a "moon jump" for $100 plus $20 per hour. This can be described by the equation

$$y = 20x + 100$$

where x represents the number of hours and y represents the cost.

a) Complete the table of values, and write the information as ordered pairs.

x	y
1	120
3	160
4	180
6	220

(1, 120), (3, 160), (4, 180), (6, 220)

b) Label a coordinate system, choose an appropriate scale, and graph the equation.

c) Explain the meaning of the ordered pair (4, 180) in the context of the problem.   The cost of renting the moon jump for 4 hours is $180.

d) For how many hours could a customer rent the moon jump if she had $280?   9 hours

68) Kelvin is driving from Los Angeles to Chicago to go to college. His distance from L.A. is given by the equation

$$y = 62x$$

where x represents the number of hours driven, and y represents his distance from Los Angeles.

a) Complete the table of values and write the information as ordered pairs.

x	y
3	186
8	496
15	930
20	1240

(3, 186), (8, 496), (15, 930), (120, 1240)

b) Label a coordinate system, choose an appropriate scale, and graph the equation.

c) Explain the meaning of the ordered pair (20, 1240) in the context of the problem.   After driving 20 hours, Kelvin is 1240 miles from Los Angeles.

d) What does the 62 in $y = 62x$ represent?   Kelvin's average speed

e) How many hours of driving time will it take for Kelvin to get to Chicago if the distance between L.A. and Chicago is about 2034 miles? (Round to the nearest hour.)   about 33 hours

69) Concern about the Leaning Tower of Pisa in Italy led engineers to begin reinforcing the structure in the 1990s. The number of millimeters the Tower was straightened can be described by

$$y = 1.5x$$

where x represents the number of days engineers worked on straightening the Tower and y represents the number of millimeters (mm) the Tower was moved toward vertical. (In the end, the Tower will keep some of its famous lean.) (Reuters-9/7/2000)

a) Make a table of values using $x = 0, 10, 20,$ and $60,$ and write the information as ordered pairs.

b) Explain the meaning of each ordered pair in the context of the problem.

c) Graph the equation using the information in a). Use an appropriate scale.

d) Engineers straightened the Leaning Tower of Pisa a total of about 450 mm. How long did this take?   300 days

70) The blood alcohol percentage of a 180 lb male can be modeled by

$$y = 0.02x$$

where x represents the number of drinks consumed (1 drink = 12 oz of beer, for example) and y represents the blood alcohol percentage. (Taken from data from the U.S. Dept. of Health and Human Services)

a) Make a table of values using $x = 0, 1, 2,$ and 4, and write the information as ordered pairs.

b) Explain the meaning of each ordered pair in the context of the problem.

c) Graph the equation using the information in a). Use an appropriate scale.

d) If a 180 lb male had a blood alcohol percentage of 0.12, how many drinks did he have?   6

71) The relationship between altitude (in feet) and barometric pressure (in inches of mercury) can be modeled by

$$y = -0.001x + 29.86$$

for altitudes between sea level (0 feet) and 5000 feet, where x represents the altitude and y represents the barometric pressure.
(From data taken from www.engineeringtoolbox.com)

Altitude and pressure

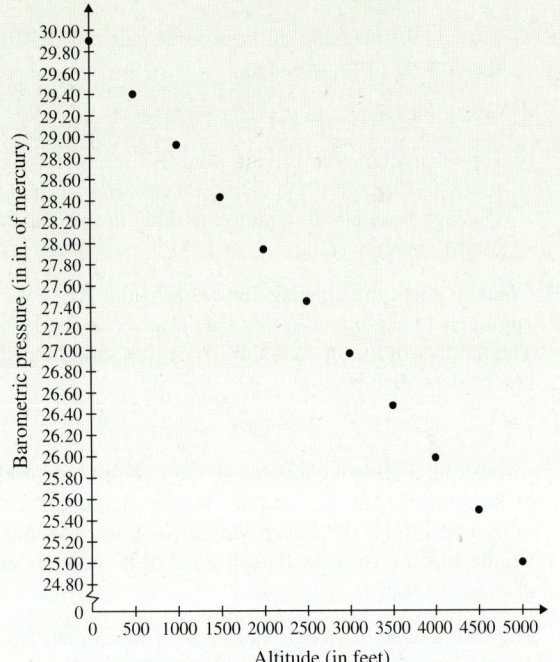

a) From the graph, estimate the pressure at the following altitudes: 0 ft (sea level), 1000 ft, 3500 ft, and 5000 ft.   Answers will vary.

b) Determine the barometric pressures at the same altitudes in a) using the equation. Are the numbers close?   29.86 in.; 28.86 in.; 26.36 in.; 24.86 in.

c) Graph the line that models the data given on the original graph. Use an appropriate scale.

d) Can we use the equation $y = -0.001x + 29.86$ to determine the pressure at 10,000 ft? Why or why not?

72) The graph shows the actual per-pupil spending on education in the state of Washington for the school years 1994–2003.

Per-pupil spending on education in the state of Washington

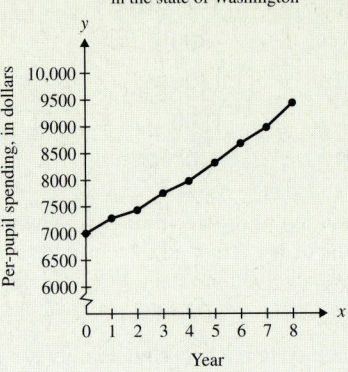

Algebraically, this can be modeled by

$$y = 298.3x + 6878.95$$

where x represents the number of years after the 1994–1995 school year. (So, $x = 0$ represents the 1994–1995 school year, $x = 1$ represents the 1995–1996 school year, etc.). y represents the per-pupil spending on education, in dollars. (www.effwa.org)

a) From the graph, estimate the amount spent per pupil during the following school years: 1994–1995, 1997–1998, and 2001–2002.   Answers may vary.

b) Determine the amount spent during the same school years as in a) using the equation. How do these numbers compare with the estimates from the graph?   1994–1995: $6878.95; 1997–1998: $7773.85; 2001–2002: $8967.05

Section 3.2 Slope of a Line and Slope-Intercept Form

Objectives

1. Understand the Concept of Slope
2. Find the Slope of a Line Given Two Points on the Line
3. Use Slope to Solve Applied Problems
4. Find the Slope of Horizontal and Vertical Lines
5. Use Slope and One Point on a Line to Graph the Line
6. Graph a Line Using Slope and y-Intercept

1. Understand the Concept of Slope

In Section 3.1, we learned to graph lines by plotting points. You may have noticed that some lines are steeper than others. Their "slants" are different too.

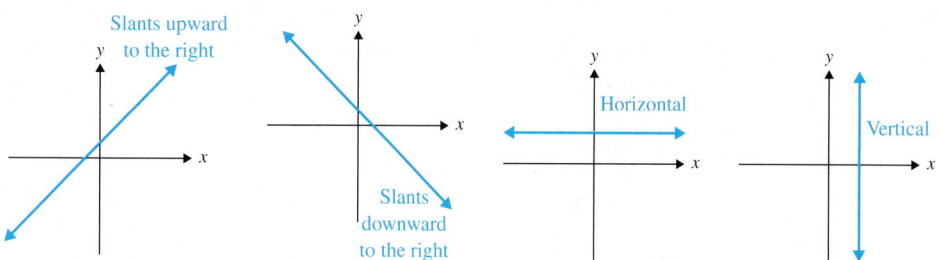

We can describe the steepness of a line with its *slope*.

> **Property** Slope of a Line
>
> The **slope** of a line measures its steepness. It is the ratio of the vertical change in y to the horizontal change in x. Slope is denoted by m.

We can also think of slope as a rate of change. *Slope* is the rate of change between two points. More specifically, it describes the rate of change in y to the change in x.

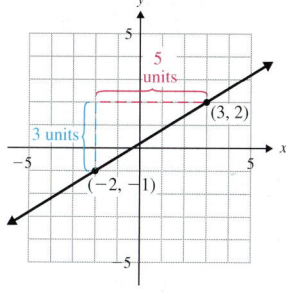 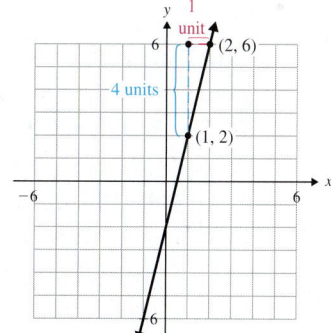

$$\text{Slope} = \frac{3}{5} \quad \begin{array}{l} \leftarrow \text{vertical change} \\ \leftarrow \text{horizontal change} \end{array} \qquad \text{Slope} = 4 \text{ or } \frac{4}{1} \quad \begin{array}{l} \leftarrow \text{vertical change} \\ \leftarrow \text{horizontal change} \end{array}$$

For example in the graph on the left, the line changes 3 units vertically for every 5 units it changes horizontally. Its slope is $\frac{3}{5}$. The line on the right changes 4 units vertically for every 1 unit of horizontal change. It has a slope of $\frac{4}{1}$ or 4.

Notice that the line with slope 4 is steeper than the line that has a slope of $\frac{3}{5}$.

Note

As the magnitude of the slope gets larger the line gets steeper.

Here is an application of slope.

Example 1

In-Class Example 1
Use Example 1.

A sign along a highway through the Rocky Mountains is shown on the left. What does this mean?

Solution

Percent means "out of 100." Therefore, we can write 7% as $\dfrac{7}{100}$. We can interpret $\dfrac{7}{100}$ as the ratio of the vertical change in the road to horizontal change in the road.

The slope of the road is $\dfrac{7}{100}$. $\leftarrow$ vertical change
                                             $\leftarrow$ horizontal change

The highway rises 7 ft for every 100 horizontal feet. ∎

7%

You Try 1

The slope of a skateboard ramp is $\dfrac{7}{16}$ where the dimensions of the ramp are in inches. What does this mean?

2. Find the Slope of a Line Given Two Points on the Line

Here is line *L*. The points (x_1, y_1) and (x_2, y_2) are two points on line *L. We will find the ratio of the vertical change in y to the horizontal change in x between the points (x_1, y_1) and (x_2, y_2).*

To get from (x_1, y_1) to (x_2, y_2), we move *vertically* to point *P* then *horizontally* to (x_2, y_2). The *x*-coordinate of point *P* is x_1 and the *y*-coordinate of *P* is y_2.

When we moved *vertically* from (x_1, y_1) to point $P(x_1, y_2)$, how far did we go? We moved a vertical distance $y_2 - y_1$.

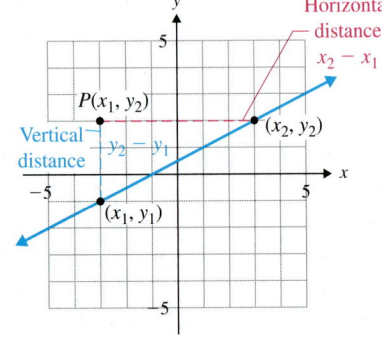

Note

The vertical change is $y_2 - y_1$ and is called the **rise**.

Then we moved *horizontally* from point $P(x_1, y_2)$ to (x_2, y_2). How far did we go? We moved a horizontal distance $x_2 - x_1$.

Note

The horizontal change is $x_2 - x_1$ and is called the **run**.

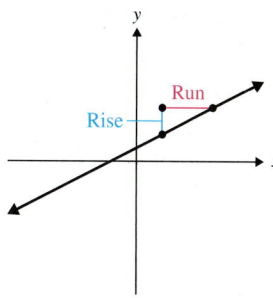

We said that the slope of a line is the ratio of the vertical change (rise) to the horizontal change (run). Therefore,

---

Formula   The Slope of a Line

The **slope** (m) of a line containing the points (x_1, y_1) and (x_2, y_2) is given by

$$m = \frac{\text{vertical change}}{\text{horizontal change}} = \frac{y_2 - y_1}{x_2 - x_1}$$

We can also think of slope as:

$$\frac{\text{rise}}{\text{run}} \quad \text{or} \quad \frac{\text{change in } y}{\text{change in } x}$$

Let's look at some different ways to determine the slope of a line.

Example 2

Determine the slope of each line.

In-Class Example 2
Determine the slope of each line.
a)

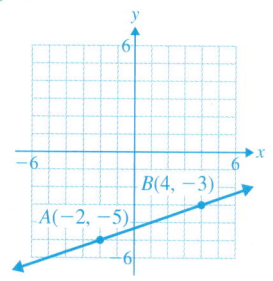

b)

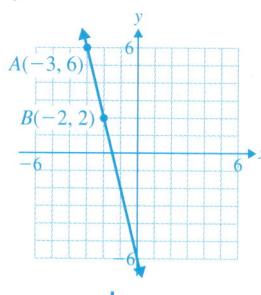

answer: a) $\dfrac{1}{3}$ b) -4

a)

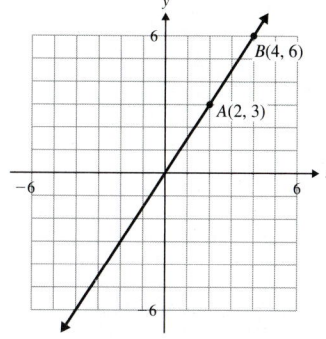

b)

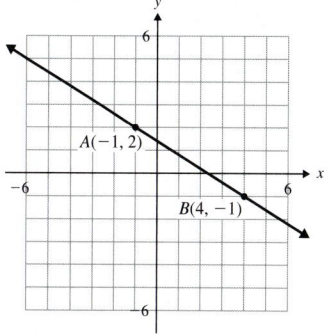

Solution

a) We will find the slope in two ways.

i) First, we will find the vertical change and the horizontal change by counting these changes as we go from A to B.

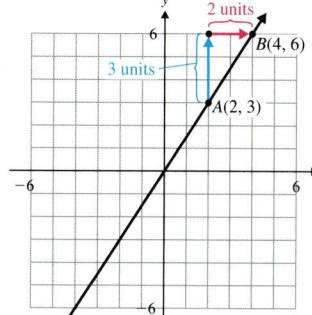

Vertical change (change in y) from A to B:
   3 units

Horizontal change (change in x) from A to B:
   2 units

$$\text{Slope} = \frac{\text{change in } y}{\text{change in } x} = \frac{3}{2} \quad \text{or} \quad m = \frac{3}{2}$$

ii) We can also find the slope using the formula.

Let $(x_1, y_1) = (2, 3)$ and $(x_2, y_2) = (4, 6)$.

$$m = \frac{y_2 - y_1}{x_2 - x_1} = \frac{6 - 3}{4 - 2} = \frac{3}{2}$$

You can see that we get the same result either way we find the slope.

b)   i) First, find the slope by counting the vertical change and horizontal change as we go from A to B.

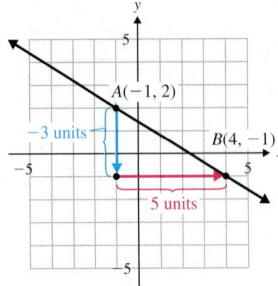

Vertical change (change in y) from A to B:
-3 units

Horizontal change (change in x) from A to B:
5 units

$$\text{Slope} = \frac{\text{change in } y}{\text{change in } x} = \frac{-3}{5} = -\frac{3}{5}$$

$$\text{or} \quad m = -\frac{3}{5}$$

ii) We can also find the slope using the formula.
Let $(x_1, y_1) = (-1, 2)$ and $(x_2, y_2) = (4, -1)$.

$$m = \frac{y_2 - y_1}{x_2 - x_1} = \frac{-1 - 2}{4 - (-1)} = \frac{-3}{5} = -\frac{3}{5}$$

Again, we obtain the same result using either method for finding the slope.   ■

Note

The slope of $-\dfrac{3}{5}$ can be thought of as $\dfrac{-3}{5}$, $\dfrac{3}{-5}$, or $-\dfrac{3}{5}$.

You Try 2

Determine the slope of each line by

a) counting the vertical change and horizontal change.   b) using the formula.

a)

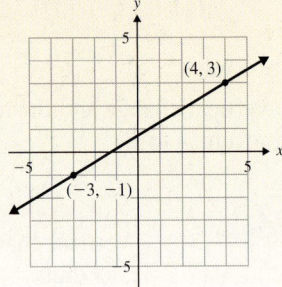

(4, 3)

(−3, −1)

b)

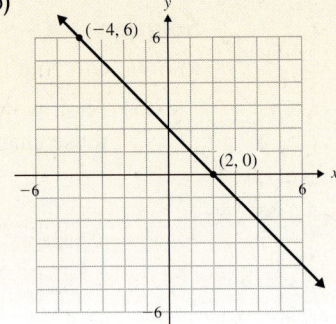

(−4, 6)

(2, 0)

Notice that in Example 2a, the line has a positive slope and slants upward from left to right. As the value of x increases, the value of y increases as well. The line in 2b has a negative slope and slants downward from left to right. Notice, in this case, that as the line goes from left to right, the value of x increases while the value of y decreases. We can summarize these results with the following general statements.

Property   Positive and Negative Slopes

A line with a **positive slope** slants upward from left to right. As the value of x increases, the value of y increases as well.

A line with a **negative slope** slants downward from left to right. As the value of x increases, the value of y decreases.

3. Use Slope to Solve Applied Problems

Example 3

In-Class Example 3
The graph models the number of students at Marquette High School from 2000–2005.

Number of Students at Marquette High School

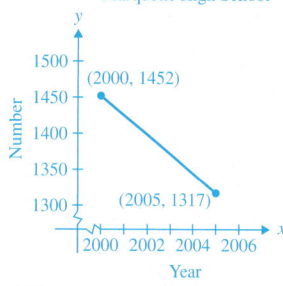

a) How many students attended the school in 2000? In 2005?
answer: 1452; 1317

Refer to Example 3 and answer parts b) and c).
**answer: b) The negative slope means the number of students was decreasing.
c) slope = −27. The number of students at Marquette High School between 2000 and 2005 decreased by 27 per year.**

The graph models the number of students at DeWitt High School from 2004 to 2010.

a) How many students attended the school in 2004? in 2010?

b) What does the sign of the slope of the line segment mean in the context of the problem?

c) Find the slope of the line segment, and explain what it means in the context of the problem.

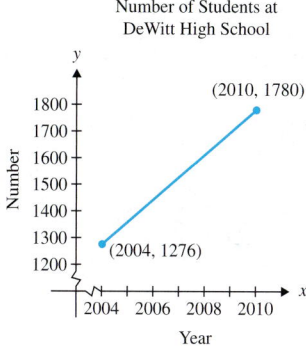

Solution

a) We can determine the number of students by reading the graph. In 2004, there were 1276 students, and in 2010 there were 1780 students.

b) The positive slope tells us that from 2004 to 2010 the number of students was increasing.

c) Let $(x_1, y_1) = (2004, 1276)$ and $(x_2, y_2) = (2010, 1780)$.

$$\text{Slope} = \frac{y_2 - y_1}{x_2 - x_1} = \frac{1780 - 1276}{2010 - 2004} = \frac{504}{6} = 84$$

The slope of the line is 84. Therefore, the number of students attending DeWitt High School between 2004 and 2010 increased by 84 per year.

4. Find the Slope of Horizontal and Vertical Lines

Example 4

In-Class Example 4
Find the slope of the line containing each pair of points.
a) (3, 7) and (−9, 7)
b) (4, −1) and (4, 3)
answer: a) 0   b) undefined

Find the slope of the line containing each pair of points.

a) $(-1, 2)$ and $(3, 2)$ b) $(-3, 4)$ and $(-3, -1)$

Solution

a) Let $(x_1, y_1) = (-1, 2)$ and $(x_2, y_2) = (3, 2)$.

$$m = \frac{y_2 - y_1}{x_2 - x_1} = \frac{2 - 2}{3 - (-1)} = \frac{0}{4} = 0$$

If we plot the points, we see that they lie on a horizontal line. Each point on the line has a y-coordinate of 2, so $y_2 - y_1$ *always* equals zero.
The slope of every horizontal line is zero.

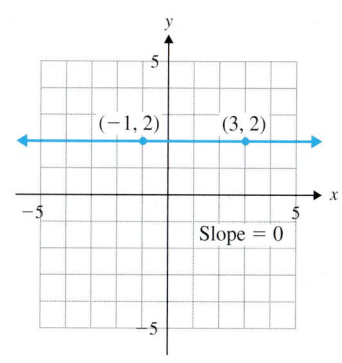

b) Let $(x_1, y_1) = (-3, 4)$ and $(x_2, y_2) = (-3, -1)$.

$$m = \frac{y_2 - y_1}{x_2 - x_1} = \frac{-1 - 4}{-3 - (-3)} = \frac{-5}{0} \text{ undefined}$$

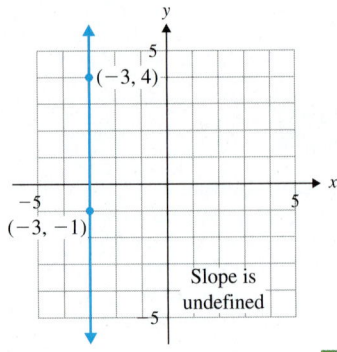

We say that the slope is undefined. Plotting these points gives us a vertical line. Each point on the line has an x-coordinate of -3, so $x_2 - x_1$ *always* equals zero.
    The slope of every vertical line is undefined.

You Try 3

Find the slope of the line containing each pair of points.

a) $(5, 8)$ and $(-2, 8)$ b) $(4, 6)$ and $(4, 1)$

Property   Slopes of Horizontal and Vertical Lines

The slope of a horizontal line, $y = d$, is **zero**. The slope of a vertical line, $x = c$, is **undefined**. (c and d are constants.)

5. Use Slope and One Point on a Line to Graph the Line

We have seen how we can find the slope of a line given two points on the line. Now, we will see how we can use the slope and *one* point on the line to graph the line.

Example 5

In-Class Example 5
Graph the line containing the point
a) $(-1, 4)$ with a slope of -2.
b) $(2, 0)$ with a slope of 4/3.
answer:
a)
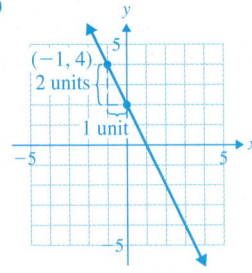
b)
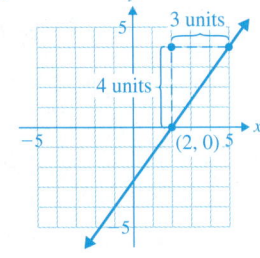

Graph the line containing the point

a) $(2, -5)$ with a slope of $\frac{7}{2}$.   b) $(0, 4)$ with a slope of -3.

Solution

a) Plot the point $(2, -5)$.

Use the slope to find another point on the line.

$$m = \frac{7}{2} = \frac{\text{change in } y}{\text{change in } x}$$

To get from the point $(2, -5)$ to another point on the line, move up 7 units in the y-direction and right 2 units in the x-direction.

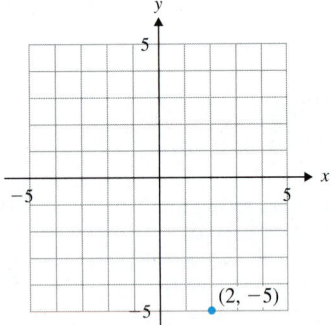

Plot this point, and draw a line through the two points.

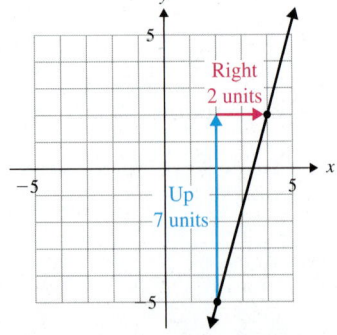

b) Plot the point $(0, 4)$.
    What does the slope, $m = -3$, mean?

$$m = -3 = \frac{-3}{1} = \frac{\text{change in } y}{\text{change in } x}$$

To get from $(0, 4)$ to another point on the line, we will move *down* 3 units in the y-direction and *right* 1 unit in the positive x-direction. We end up at $(1, 1)$.

Plot this point, and draw a line through $(0, 4)$ and $(1, 1)$.

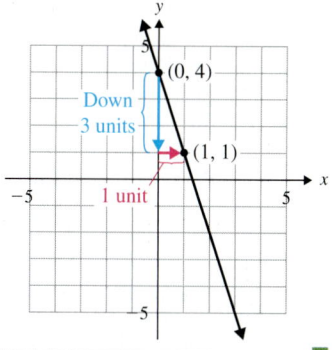

In part b) we could have written $m = -3$ as $m = \dfrac{3}{-1}$. This would have given us a different point on the same line.

 **You Try 4**

Graph the line containing the point

a) $(1, 1)$ with a slope of $-\dfrac{2}{3}$.   b) $(0, -5)$ with a slope of 3.

c) $(-2, 2)$ with an undefined slope.

We have already learned that a linear equation in two variables can be written in the form $Ax + By = C$ (this is called **standard form**), where A, B, and C are real numbers and where both A and B do not equal zero. Equations of lines can take other forms, too, and now we will learn about one of those forms.

In Example 5b, we graphed a line given its y-intercept, $(0, 4)$, and its slope, -3. This leads us into another good method for graphing a line: we can express a line in **slope-intercept** form and then graph it using its slope and y-intercept.

6. Graph a Line Using Slope and y-Intercept

We know that if (x_1, y_1) and (x_2, y_2) are points on a line, then the slope of the line is

$$m = \frac{y_2 - y_1}{x_2 - x_1}$$

Recall that to find the y-intercept of a line, we let $x = 0$ and solve for y. Let one of the points on a line be the y-intercept $(0, b)$, where b is a number. Let another point on the line be (x, y). See the graph on the left.

Substitute the points $(0, b)$ and (x, y) into the slope formula.

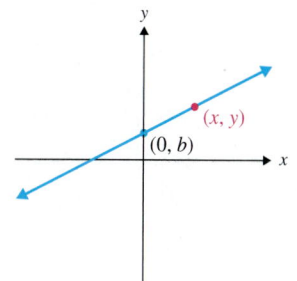

Subtract y-coordinates
$$\downarrow$$
$$m = \frac{y_2 - y_1}{x_2 - x_1} = \frac{y - b}{x - 0} = \frac{y - b}{x}$$
$$\uparrow$$
Subtract x-coordinates

Solve $m = \dfrac{y - b}{x}$ for y.

$$mx = \dfrac{y - b}{x} \cdot x \qquad \text{\color{blue}Multiply by } x \text{ \color{blue}to eliminate the fraction.}$$

$$mx = y - b$$

$$mx + b = y - b + b \qquad \text{\color{blue}Add } b \text{ \color{blue}to each side to solve for } y.$$

$$mx + b = y$$

$$\text{or}$$

$$y = mx + b \qquad \text{\color{blue}Slope-intercept form}$$

> ## Definition
> The **slope-intercept form of a line** is $y = mx + b$, where m is the slope and $(0, b)$ is the y-intercept.

When an equation is in the form $y = mx + b$, we can quickly recognize the y-intercept and slope to graph the line.

Example 6

In-Class Example 6
Graph each line using its slope and y-intercept.

a) $y = -2x + 6$ b) $y = \dfrac{1}{4}x$

c) $7x - 2y = 8$
answer: a)

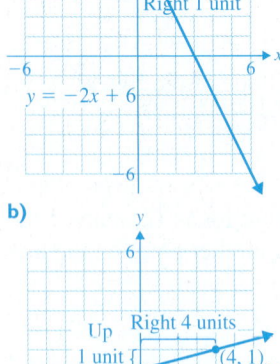

b)

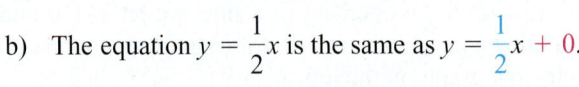

Graph each line using its slope and y-intercept.

a) $y = 3x - 4$ b) $y = \dfrac{1}{2}x$ c) $5x + 3y = 12$

Solution

a) Graph $y = 3x - 4$.

Identify the slope and y-intercept.

$$y = 3x - 4$$

Slope $= 3$, y-intercept is $(0, -4)$.

Plot the y-intercept first, then use the slope to locate another point on the line. Since the slope is 3, think of it as $\dfrac{3}{1} \begin{matrix} \leftarrow \text{change in } y \\ \leftarrow \text{change in } x \end{matrix}$.

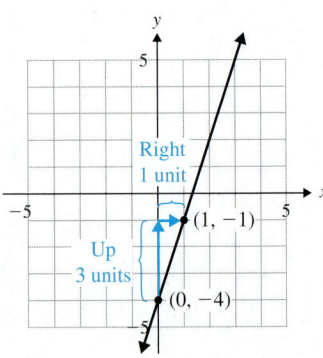

b) The equation $y = \dfrac{1}{2}x$ is the same as $y = \dfrac{1}{2}x + 0$.

Identify the slope and y-intercept.

$$\text{Slope} = \dfrac{1}{2}, \quad y\text{-intercept is } (0, 0).$$

Plot the y-intercept, then use the slope to locate another point on the line.

Since $\dfrac{1}{2}$ is equivalent to $\dfrac{-1}{-2}$, we can use $\dfrac{-1}{-2}$ as the slope to locate yet another point on the line.

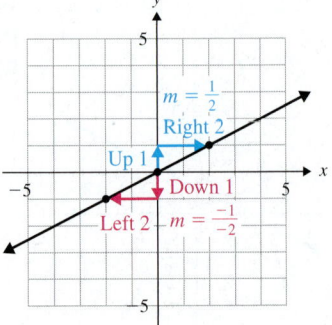

c)

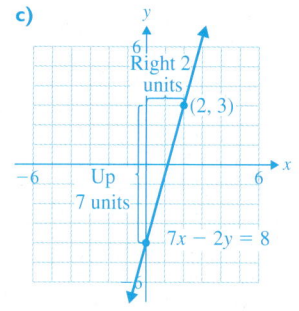

c) Rewrite $5x + 3y = 12$ in slope-intercept form, $y = mx + b$.

$$5x + 3y = 12$$
$$3y = -5x + 12 \qquad \text{Add } -5x \text{ to each side.}$$
$$y = -\frac{5}{3}x + 4 \qquad \text{Divide each side by 3.}$$

Slope $= -\dfrac{5}{3}$, y-intercept is $(0, 4)$.

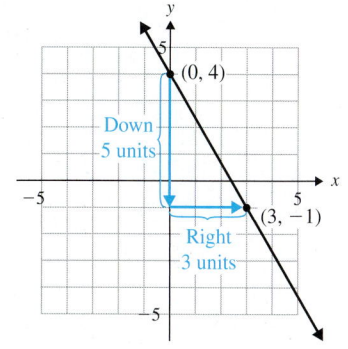

We can interpret the slope of $-\dfrac{5}{3}$ as either $\dfrac{-5}{3}$ or $\dfrac{5}{-3}$. Here we use $\dfrac{-5}{3}$. Using either form will give us a point on the line.

You Try 5

Graph each line using its slope and y-intercept.

a) $y = \dfrac{2}{5}x - 3$ b) $y = -x$ c) $8x - 4y = 12$

Summary   Methods for Graphing a Line

We have learned that we can use different methods for graphing lines. Given the equation of a line we can

1) make a table of values, plot the points, and draw the line through the points.

2) find the x-intercept by letting $y = 0$ and solving for x, and find the y-intercept by letting $x = 0$ and solving for y. Plot the points, then draw the line through the points.

3) put the equation into slope-intercept form, $y = mx + b$, identify the slope and y-intercept, then graph the line.

Using Technology

When we look at the graph of a linear equation, we should be able to estimate its slope. Use the equation $y = x$ as a guideline.

Step 1:   Graph the equation $y = x$.

We can make the graph a thick line (so we can tell it apart from the others) by moving the arrow all the way to the left and hitting ⟨ENTER⟩:

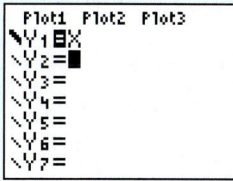   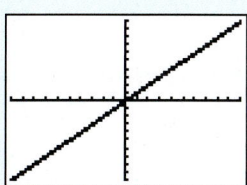

Step 2: Keeping this equation, graph the equation $y = 2x$:

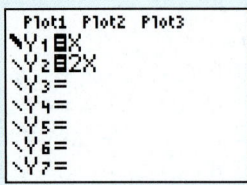 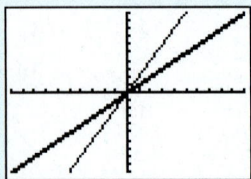

     a. Is the graph steeper or flatter than the graph of $y = x$?

     b. Make a guess as to whether $y = 3x$ will be steeper or flatter than $y = x$. Test your guess by graphing $y = 3x$.

Step 3: Clear the equation $y = 2x$ and graph the equation $y = 0.5x$:

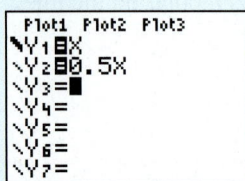 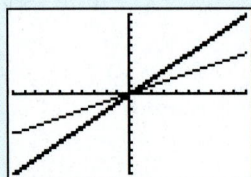

     a. Is the graph steeper or flatter than the graph of $y = x$?

     b. Make a guess as to whether $y = 0.65x$ will be steeper or flatter than $y = x$. Test your guess by graphing $y = 0.65x$.

Step 4: Test similar situations, except with negative slopes: $y = -x$.

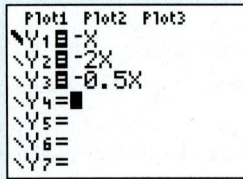 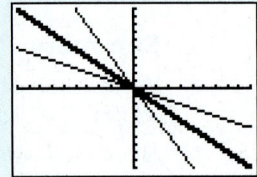

    Did you notice that we have the same relationship, except in the opposite direction? That is, $y = 2x$ is steeper than $y = x$ in the positive direction, and $y = -2x$ is steeper than $y = -x$, but in the negative direction. And $y = 0.5x$ is flatter than $y = x$ in the positive direction, and $y = -0.5x$ is flatter than $y = -x$, but in the negative direction.

Answers to You Try Exercises

1) The ramp rises 7 in. for every 16 horizontal inches.   2) a) $m = \dfrac{4}{7}$ b) $m = -1$

3) a) $m = 0$ b) undefined

4) a)

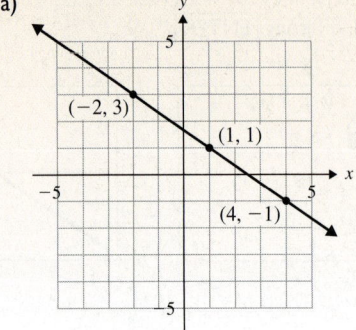

b)

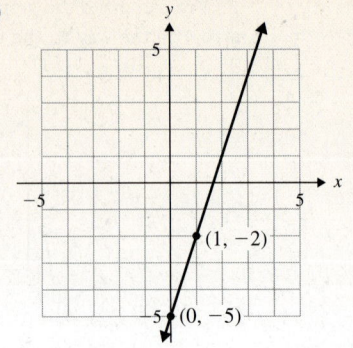

c)

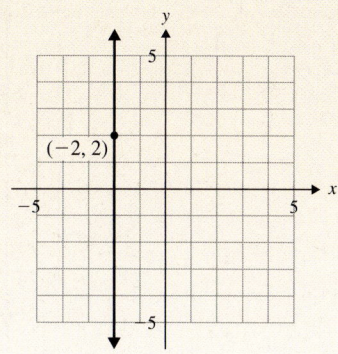

5 a)

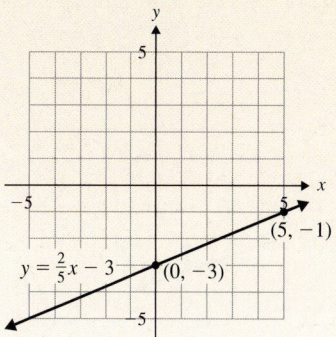

b)

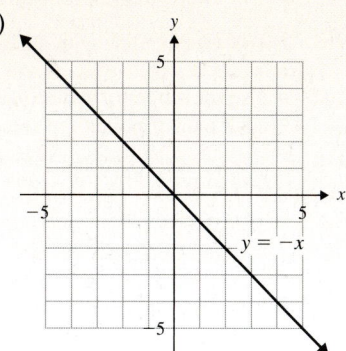

c)

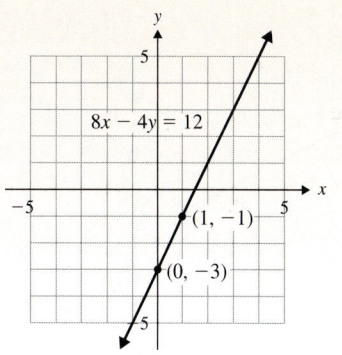

3.2 Exercises

*Additional answers can be found in the Answers to Exercises appendix.

Objective 1: Understand the Concept of Slope

1) Explain the meaning of slope.

2) Describe the slant of a line with a positive slope.
   It slants upward from left to right.

3) Describe the slant of a line with a negative slope.
   It slants downward from left to right.

4) The slope of a horizontal line is _____.   zero.

5) The slope of a vertical line is _____.   undefined

6) If a line contains the points (x_1, y_1) and (x_2, y_2), write the formula for the slope of the line. $m = \dfrac{y_2 - y_1}{x_2 - x_1}$

Mixed Exercises: Objectives 2 and 4

Determine the slope of each line by

   a) counting the vertical change and the horizontal change as you move from one point to the other on the line

   and

   b) using the slope formula. (See Example 2.)

VIDEO 7)

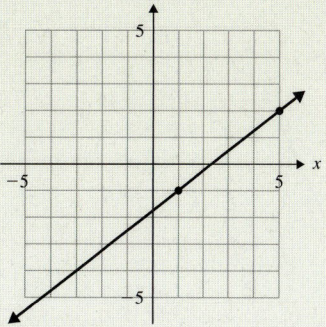

$m = \dfrac{3}{4}$

8)

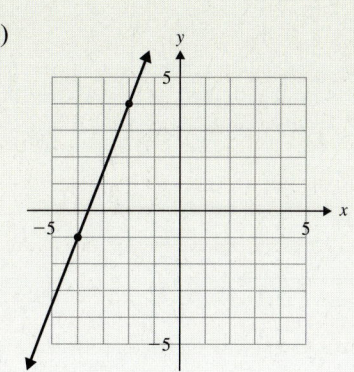

$m = \dfrac{5}{2}$

9)
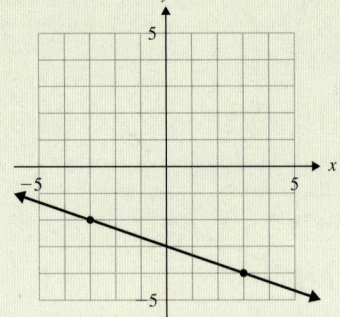
$m = -\dfrac{1}{3}$

10)
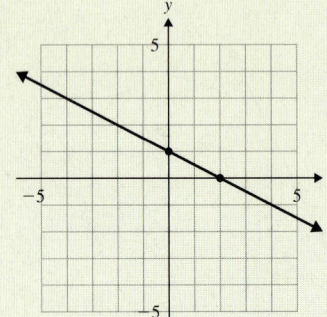
$m = -\dfrac{1}{2}$

11)
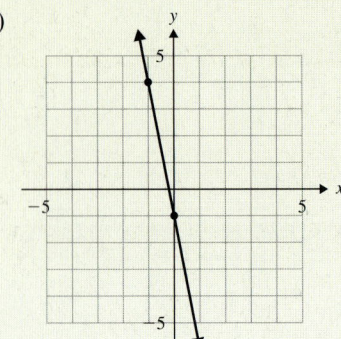
$m = -5$

12)
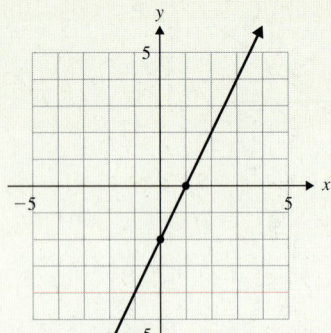
$m = 2$

13)
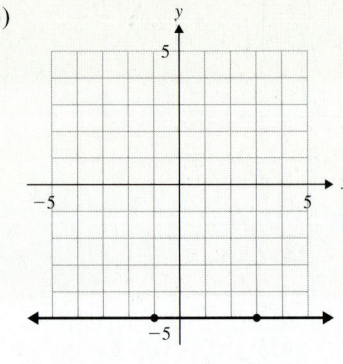
$m = 0$

14)
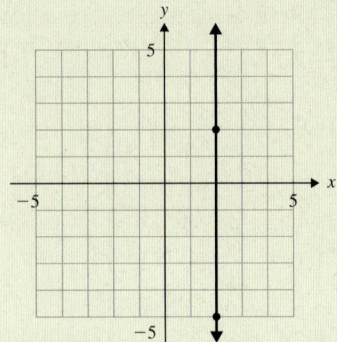
Slope is undefined.

15) Graph a line with a positive slope and a negative x-intercept.

16) Graph a line with a negative slope and a positive y-intercept.

Use the slope formula to find the slope of the line containing each pair of points.

17) $(3, 2)$ and $(9, 5)$ $\dfrac{1}{2}$

18) $(4, 1)$ and $(0, -5)$ $\dfrac{3}{2}$

19) $(-2, 8)$ and $(2, 4)$ -1

20) $(3, -2)$ and $(-1, 6)$ -2

21) $(9, 2)$ and $(0, 4)$ $-\dfrac{2}{9}$

22) $(-5, 1)$ and $(2, -4)$ $-\dfrac{5}{7}$

23) $(3, 5)$ and $(-1, 5)$ 0

24) $(-4, -4)$ and $(-4, 10)$ undefined

25) $(3, 2)$ and $(3, -1)$ undefined

26) $(0, -7)$ and $(-2, -7)$ 0

27) $\left(\dfrac{3}{8}, -\dfrac{1}{3}\right)$ and $\left(\dfrac{1}{2}, \dfrac{1}{4}\right)$ $\dfrac{14}{3}$

28) $\left(\dfrac{3}{2}, \dfrac{7}{3}\right)$ and $\left(\dfrac{1}{3}, 4\right)$ $-\dfrac{10}{7}$

29) $(-1.7, -1.2)$ and $(2.8, -10.2)$ -2

30) $(4.8, -1.6)$ and $(6, 1.4)$ 2.5

Objective 3: Use Slope to Solve Applied Problems

31) The slope of a roof is sometimes referred to as a *pitch*. A garage roof might have a 10-12 *pitch*. The first number refers to the rise of the roof, and the second number refers to how far over you must go to attain that rise (the run).

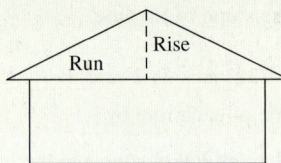

a) Find the slope of a roof with a 10-12 pitch. $\dfrac{5}{6}$

b) Find the slope of a roof with an 8-12 pitch. $\dfrac{2}{3}$

c) If a roof rises 8 in. in a 2-ft run, what is the slope of the roof? How do you write the slope in x-12 *pitch* form? $m = \dfrac{1}{3}$; 4-12 pitch

32) To make all buildings wheelchair accessible, the federal government has mandatory specifications for the slope of wheelchair ramps in new construction. A wheelchair ramp can have a maximum slope of $\frac{1}{12}$. Does the following ramp meet this requirement?

(http://www.access-board.gov/adaag/html/adaag.htm#4.8)

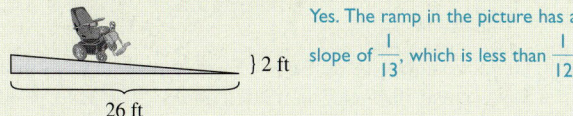

Yes. The ramp in the picture has a slope of $\frac{1}{13}$, which is less than $\frac{1}{12}$.

} 2 ft

26 ft

Use this information for Exercises 33 and 34.

Like many U.S. cities Evanston, Illinois, has building codes to regulate the steepness of driveways and parking garage ramps. A residential driveway can have a maximum slope of 12%, and the maximum slope of a ramp in an indoor parking garage is 15%. (http://www.cityofevanston.org/departments/publicworks/pdf/driveway_regulations.pdf)

33) A homeowner calls a landscape architect to tear out his old driveway and design a new one. The new driveway will have a different slope and will be constructed of brick pavers. In the new plans, the difference in height between the top and the bottom of the driveway will be 4 ft, and the linear distance between those two points will be 40 ft. Does the driveway meet the building code? Explain your answer.   Yes. The slope of the driveway will be 10%.

34) The firm hired to build a parking garage in downtown Evanston wants to use plans from a garage it built in another city. In that structure, the ramps rose 1 ft for every 6.25 ft of horizontal distance. Can the construction firm use the same ramp plans in Evanston, or will it have to redesign them? Explain your answer.   The company will have to redesign the ramps because the slope of the ramps in the existing garage is 16%.

VIDEO 35) Melissa purchased a new car in 2006. The graph shows the value of the car from 2006–2010.

Value of the Car

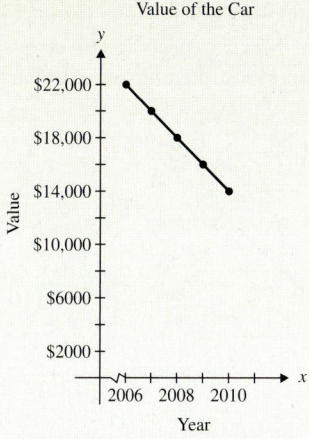

a) What was the value of the car in 2006? $22,000

b) Without computing the slope, determine whether it is positive or negative.   negative

c) What does the sign of the slope mean in the context of this problem?   The value of the car is decreasing over time.

d) Find the slope of the line segment, and explain what it means in the context of the problem.
$m = -2000$; the value of the car is decreasing $2000 per year.

36) The graph shows the approximate number of babies (in thousands) born to teenage girls from 1997–2001. (U.S. Census Bureau)

Number of Births

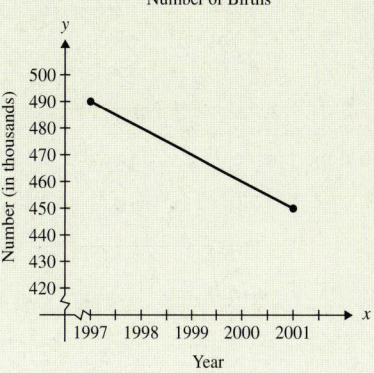

a) Approximately how many babies were born to teen mothers in 1997? in 1998? in 2001?
490,000; 480,000; 450,000

b) Without computing the slope, determine whether it is positive or negative.   negative

c) What does the sign of the slope mean in the context of the problem?   The number of births to teenage girls has been decreasing from 1997 to 2001.

d) Find the slope of the line segment, and explain what it means in the context of the problem. $m = -10$ thousand or $-10,000$; the number of births to teen girls is decreasing by 10,000 per year.

Objective 5: Use Slope and One Point on a Line to Graph a Line

Graph the line containing the given point and with the given slope.

37) $(-3, -2); m = \frac{5}{2}$

38) $(1, 3); m = \frac{1}{4}$

39) $(1, -4); m = \frac{1}{3}$

40) $(-4, 2); m = \frac{2}{7}$

41) $(4, 5); m = -\frac{2}{3}$

42) $(2, -1); m = -\frac{3}{2}$

43) $(-5, 1); m = 3$

44) $(-3, -5); m = 2$

45) $(0, -3); m = -1$

46) $(0, 4); m = -3$

VIDEO 47) $(6, 2); m = -4$

48) $(2, 5); m = -1$

49) $(-2, -1); m = 0$

50) $(-3, 3); m = 0$

51) $(4, 0);$ slope is undefined

52) $(1, 6);$ slope is undefined

53) $(0, 0); m = 1$

54) $(0, 0); m = -1$

Objective 6: Graph a Line Using Slope and y-Intercept

55) The slope-intercept form of a line is $y = mx + b$. What is the slope? What is the y-intercept?
The slope is m, and the y-intercept is $(0, b)$.

56) How do you put an equation that is in standard form, $Ax + By = C$, into slope-intercept form?
Solve the equation for y.

Each of the following equations is in slope-intercept form. Identify the slope and the y-intercept, then graph each line using this information.

VIDEO 57) $y = \frac{2}{5}x - 6$

58) $y = \frac{7}{4}x - 2$

59) $y = -\frac{5}{3}x + 4$

60) $y = -\dfrac{1}{2}x + 5$ 61) $y = \dfrac{3}{4}x + 1$ 62) $y = \dfrac{2}{3}x + 3$

63) $y = 4x - 2$ 64) $y = -3x - 1$ 65) $y = -x + 5$

66) $y = x$ 67) $y = \dfrac{3}{2}x + \dfrac{1}{2}$ 68) $y = -\dfrac{3}{4}x - \dfrac{5}{2}$

69) $y = -2$ 70) $y = 4$

Put each equation into slope-intercept form, if possible, and graph.

VIDEO 71) $x + 3y = -6$ 72) $5x + 2y = 2$

73) $12x - 8y = 32$ 74) $y - x = 1$

75) $x + 9 = 2$ 76) $5 = x + 2$

77) $18 = 6y - 15x$ 78) $20x = 48 - 12y$

79) $y = 0$ 80) $y + 6 = 1$

VIDEO 81) Dave works in sales, and his income is a combination of salary and commission. He earns \$34,000 per year plus 5% of his total sales. The equation $I = 0.05s + 34,000$ represents his total annual income, I, in dollars, when his sales total s dollars.

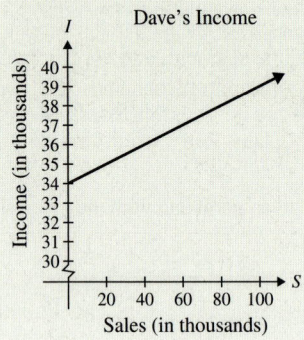

Dave's Income

a) What is the I-intercept? What does it mean in the context of this problem?
(0, 34,000); if Dave has \$0 in sales, his income is \$34,000.

b) What is the slope? What does it mean in the context of the problem? $m = 0.05$; Dave earns \$0.05 for every \$1 in sales.

c) Use the graph to find Dave's income if his total sales are \$80,000. Confirm your answer using the equation.   \$38,000

82) Li Mei gets paid hourly in her after-school job. Her income is given by $I = 7.50h$, where I represents her income in dollars and h represents the number of hours worked.

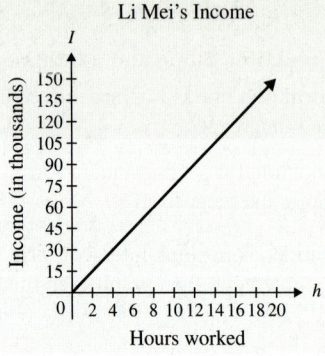

Li Mei's Income

a) What is the I-intercept? What does it mean in the context of the problem?   (0, 0); if Li Mei works 0 hr, she earns \$0.

b) What is the slope? What does it mean in the context of the problem? $m = 7.50$; Li Mei earns \$7.50 per hr.

c) Use the graph to find how much Li Mei earns when she works 14 hr. Confirm your answer using the equation.   \$105

83) The per capita consumption of whole milk in the United States since 1945 can be modeled by $y = -0.59x + 40.53$ where x represents the number of years after 1945, and y represents the per capita consumption of whole milk in gallons. (U.S. Department of Agriculture)

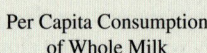

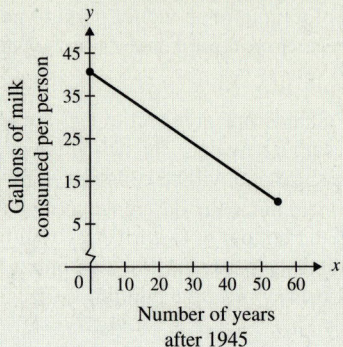

Per Capita Consumption of Whole Milk

a) What is the y-intercept? What does it mean in the context of the problem?

b) What is the slope? What does it mean in the context of the problem?

c) Use the graph to estimate the per capita consumption of whole milk in the year 2000. Then, use the equation to determine this number.

84) On a certain day in 2004, the exchange rate between the American dollar and the Mexican peso was given by $p = 11.40d$, where d represents the number of dollars and p represents the number of pesos.

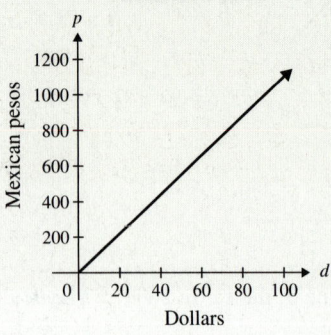

Dollars and Pesos

a) What is the p-intercept? What does it mean in the context of the problem?   (0, 0); \$0 = 0 pesos

b) What is the slope? What does it mean in the context of the problem? $m = 11.40$; each American dollar is worth 11.40 pesos.

c) Use the graph to estimate the value of 600 pesos in dollars. Then, use the equation to determine this number.   *estimate from the graph: $52; from the equation: $52.63*

85) According to information taken from the U.S. Bureau of Labor Statistics, the average annual salary y, in dollars, of a pharmacist from 2000 to 2006 can be approximated by $y = 3986x + 68,613$ where x is the number of years after 2000. (www.bls.gov/oes)

 a) What is the y-intercept? What does it mean in the context of this problem?   *(0, 68,613); in 2000, the average annual salary of a pharmacist was $68,613.*

 b) What is the slope? What does it mean in the context of this problem?   *m = 3986; the average salary of a pharmacist is increasing by $3986 per year.*

 c) What was the average annual salary of a pharmacist in 2004?   *$84,557*

 d) Find the average annual salary of a pharmacist in 2014 if this equation continues to accurately model a pharmacist's average annual salary.   *$124,417*

 e) Using an appropriate scale, graph the equation for the years 2000–2006.

86) According to information taken from the U.S. Bureau of Labor Statistics, the percentage of men, y, ages 20–24 in the civilian labor force from 1986 to 2016 (projected) can be approximated by $y = -0.4x + 85.8$, where x is the

number of years after 1986. (http//www.bls.gov/opub/mlr/2007/11/art3full.pdf)

 a) What is the y-intercept? What does it mean in the context of this problem?   *(0, 85.8); in 1986, 85.8% of men ages 20–24 were in the civilian labor force.*

 b) What is the slope? What does it mean in the context of this problem?   *m = −0.4; the percentage of 20- to 24-year-old men in the civilian labor force is decreasing by 0.4% per year.*

 c) What percentage of 20- to 24-year-old males were in the workforce in 2010?   *76.2%*

 d) What percentage of 20- to 24-year-old males are expected to be in the workforce in 2016?   *73.8%*

 e) Using an appropriate scale, graph the equation for the years 1986–2016.

Write the slope-intercept form for the equation of a line with the given slope and y-intercept.

(VIDEO) 87) $m = -4$; y-int: $(0, 7)$
  y = −4x + 7

88) $m = 5$; y-int: $(0, 3)$
  y = 5x + 3

89) $m = \dfrac{8}{5}$; y-int: $(0, -6)$
  y = $\frac{8}{5}$x − 6

90) $m = \dfrac{4}{9}$; y-int: $(0, -1)$
  y = $\frac{4}{9}$x − 1

91) $m = \dfrac{1}{3}$; y-int: $(0, 5)$
  y = $\frac{1}{3}$x + 5

92) $m = -\dfrac{1}{2}$; y-int: $(0, -3)$
  y = −$\frac{1}{2}$x − 3

93) $m = -1$; y-int: $(0, 0)$
  y = −x

94) $m = 1$; y-int: $(0, 4)$
  y = x + 4

95) $m = 0$; y-int: $(0, -2)$
  y = −2

96) $m = 0$; y-int: $(0, 0)$
  y = 0

Section 3.3 Writing an Equation of a Line

Objectives

1. **Write an Equation of a Line Given Its Slope and y-Intercept**
2. **Use the Point-Slope Formula to Write an Equation of a Line Given Its Slope and a Point on the Line**
3. **Use the Point-Slope Formula to Write an Equation of a Line Given Two Points on the Line**
4. **Write Equations of Horizontal and Vertical Lines**
5. **Use Slope to Determine if Two Lines are Parallel or Perpendicular**
6. **Write an Equation of a Line That is Parallel or Perpendicular to a Given Line**
7. **Write a Linear Equation to Model Real-World Data**

The focus of Chapter 3, thus far, has been on graphing lines given their equations. In this section, we will switch gears. Given information about a line, we will write an equation of that line.

Recall the forms of lines we have discussed so far.

1) **Standard Form:** $Ax + By = C$, where A, B, and C are real numbers and where both A and B do not equal zero.

   We will now set an additional condition for when we write equations of lines in standard form:

   A, B, and C must be integers and A must be positive.

2) **Slope-Intercept Form:** The slope-intercept form of a line is $y = mx + b$, where m is the slope, and the y-intercept is $(0, b)$.

It is common to express the equation of a line in one of these two forms. In the rest of this section, we will learn how to write equations of lines given information about their graphs.

1. Write an Equation of a Line Given Its Slope and y-Intercept

> **Procedure**   Write an Equation of a Line Given Its Slope and y-Intercept
>
> If we are given the slope and y-intercept of a line, use $y = mx + b$ and substitute those values into the equation.

Example 1

Find an equation of the line with slope $= -5$ and y-intercept $(0, 9)$.

In-Class Example 1
Find an equation of the line with
slope $= -3$ and y-intercept $(0, 2)$.
answer: $y = -3x + 2$

Solution

Since we are told the slope and y-intercept, use $y = mx + b$.

$$m = -5 \quad \text{and} \quad b = 9$$

Substitute these values into $y = mx + b$ to get $y = -5x + 9$. ■

You Try 1

Find an equation of the line with slope $= \dfrac{2}{3}$ and y-intercept $(0, -6)$.

2. Use the Point-Slope Formula to Write an Equation of a Line Given Its Slope and a Point on the Line

When we are given the slope of a line and a point on that line, we can use another method to find its equation. This method comes from the formula for the slope of a line.

Let (x_1, y_1) be a given point on a line, and let (x, y) be any other point on the same line. See Figure 4. The slope of that line is

$$m = \frac{y - y_1}{x - x_1} \qquad \text{Definition of slope}$$
$$m(x - x_1) = y - y_1 \qquad \text{Multiply each side by } x - x_1.$$
$$y - y_1 = m(x - x_1) \qquad \text{Rewrite the equation.}$$

We have found the *point-slope form* of the equation of a line.

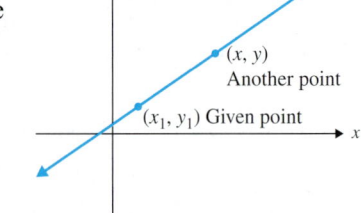

Figure 4

Formula   Point-Slope Formula

Point-Slope Form of a Line. The **point-slope form of a line** is

$$y - y_1 = m(x - x_1)$$

where (x_1, y_1) is a point on the line and m is its slope.

Procedure   Write an Equation of a Line Given Its Slope and a Point on the Line

If we are given the slope of the line and a point on the line, we can use the point-slope formula to find an equation of the line.

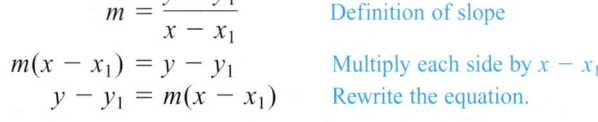
Note

The point-slope formula will help us write an equation of a line. We will not express our final answer in this form. We will write our answer in either slope-intercept form or in standard form.

Example 2

A line has slope -3 and contains the point $(2, 1)$. Find the standard form for the equation of the line.

In-Class Example 2
A line has slope -1 and contains
the point $(-1, -4)$. Find the
standard form for the equation
of the line.
answer: $x + y = -5$

Solution

Although we are told to find the *standard form* for the equation of the line, we do not try to immediately "jump" to standard form.

First, ask yourself, *"What information am I given?"*

We are given the slope and a point on the line. Therefore, we will begin by using the point-slope formula. Our *last* step will be to put it in standard form.

Use $y - y_1 = m(x - x_1)$. Substitute -3 for m. Substitute $(2, 1)$ for (x_1, y_1).

$$y - y_1 = m(x - x_1)$$
$$y - 1 = -3(x - 2) \qquad \text{Substitute 2 for } x_1 \text{ and 1 for } y_1.$$
$$y - 1 = -3x + 6 \qquad \text{Distribute.}$$

Since we are asked to express the answer in standard form, we must get the x- and y-terms on the same side of the equation.

$$3x + y - 1 = 6 \qquad \text{Add } 3x \text{ to each side.}$$
$$3x + y = 7 \qquad \text{Add 1 to each side.}$$

The equation is $3x + y = 7$.

You Try 2

A line has slope -5 and contains the point $(-1, 3)$. Find the standard form for the equation of the line.

3. Use the Point-Slope Formula to Write an Equation of a Line Given Two Points on the Line

We are now ready to discuss how to write an equation of a line when we are given two points on a line.

> **Procedure**   Write an Equation of a Line Given Two Points on the Line
>
> To write an equation of a line given two points on the line,
>
> a) use the points to find the slope of line
>
>   then
>
> b) use the slope and *either one* of the points in the point-slope formula.

Example 3

Write an equation of the line containing the points $(6, 4)$ and $(3, 2)$. Express the answer in slope-intercept form.

In-Class Example 3

Write an equation of the line containing the points $(5, -2)$ and $(-3, 4)$. Express the answer in slope-intercept form.

answer: $y = -\dfrac{3}{4}x + \dfrac{7}{4}$

Solution

We are given two points on the line, so first we will find the slope.

$$m = \frac{2 - 4}{3 - 6} = \frac{-2}{-3} = \frac{2}{3}$$

We will use the slope and *either one* of the points in the point-slope formula. (Each point will give the same result.) We will use $(6, 4)$.

Substitute $\dfrac{2}{3}$ for m. Substitute $(6, 4)$ for (x_1, y_1).

$$y - y_1 = m(x - x_1)$$

$$y - 4 = \frac{2}{3}(x - 6) \qquad \text{Substitute 6 for } x_1 \text{ and 4 for } y_1.$$

$$y - 4 = \frac{2}{3}x - 4 \qquad \text{Distribute.}$$

$$y = \frac{2}{3}x \qquad \text{Add 4 to each side to solve for } y.$$

The equation is $y = \dfrac{2}{3}x$.

You Try 3

Find the standard form for the equation of the line containing the points $(-2, 5)$ and $(1, 1)$.

4. Write Equations of Horizontal and Vertical Lines

In Section 3.2, we learned that the slope of a horizontal line is zero and that it has equation $y = d$, where d is a constant. The slope of a vertical line is undefined, and its equation is $x = c$, where c is a constant.

Formula   Equations of Horizontal and Vertical Lines

Equation of a Horizontal Line: The equation of a horizontal line containing the point (c, d) is $y = d$.

Equation of a Vertical Line: The equation of a vertical line containing the point (c, d) is $x = c$.

Example 4

Write an equation of the horizontal line containing the point $(5, -4)$.

In-Class Example 4
Write an equation of the horizontal line containing the point $(-8, 6)$.
answer: $y = 6$

Solution

The equation of a horizontal line has the form $y = d$, where d is the y-coordinate of the point. The equation of the line is $y = -4$. ∎

You Try 4

Write an equation of the horizontal line containing the point $(1, 7)$.

Summary   Writing Equations of Lines

If you are given

1) **the slope and y-intercept of the line,** use $y = mx + b$ and substitute those values into the equation.

2) **the slope of the line and a point on the line,** use the point-slope formula:

$$y - y_1 = m(x - x_1)$$

Substitute the slope for m and the point you are given for (x_1, y_1). Write your answer in slope-intercept or standard form.

3) **two points on the line,** find the slope of the line and then use the slope and *either* one of the points in the point-slope formula. Write your answer in slope-intercept or standard form.

The equation of a **horizontal line** containing the point (c, d) is $y = d$.

The equation of a **vertical line** containing the point (c, d) is $x = c$.

---

5. Use Slope to Determine if Two Lines are Parallel or Perpendicular

Recall that two lines in a plane are *parallel* if they do not intersect. If we are given the equations of two lines, how can we determine if they are parallel?

Here are the equations of two lines, L_1 and L_2.

$$L_1: y = \frac{2}{3}x + 1 \qquad\qquad L_2: y = \frac{2}{3}x - 5$$

We will graph each line.

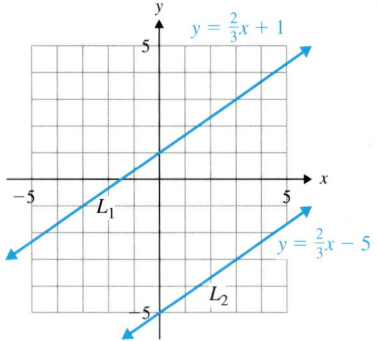

These lines are parallel. Their slopes are the same, but they have different y-intercepts. This is how we determine if two (nonvertical) lines are parallel. They have the same slope, but different y-intercepts.

> **Property**   Parallel Lines
>
> Parallel lines have the same slope. If two lines are vertical they are parallel. However, their slopes are undefined.

The slopes of two lines can tell us about another relationship between the lines. The slopes can tell us if two lines are *perpendicular*. Recall that two lines are **perpendicular** if they intersect at 90° angles.

Here are the graphs of two perpendicular lines and their equations. We will see how their slopes are related.

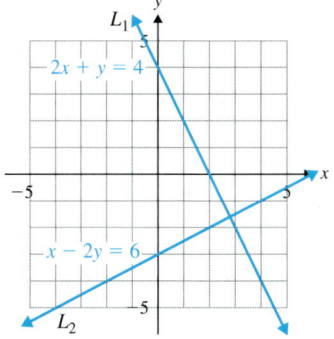

L_1 has equation $2x + y = 4$. L_2 has equation $x - 2y = 6$. Find the slopes of the lines by writing them in slope-intercept form.

$L_1: 2x + y = 4$

$$y = -2x + 4$$

$$m = -2$$

$L_2: x - 2y = 6$

$$-2y = -x + 6$$

$$y = \frac{-x}{-2} + \frac{6}{-2}$$

$$y = \frac{1}{2}x - 3$$

$$m = \frac{1}{2}$$

How are the slopes related? They are **negative reciprocals.** That is, if the slope of one line is a, then the slope of a line perpendicular to it is $-\dfrac{1}{a}$. This is how we determine if two lines are perpendicular (where neither one is vertical).

> **Property**   Perpendicular Lines
>
> When neither line is vertical, perpendicular lines have slopes that are negative reciprocals of each other.

Example 5

In-Class Example 5
Determine whether each pair of lines is parallel, perpendicular, or neither.
a) $-2x + 6y = 1$
   $4x - 12y = 5$
b) $8x + 6y = 5$
   $y = \dfrac{3}{4}x - 11$
c) $2x - y = 5$
   $6x + 3y = 7$

answer: a) parallel
b) perpendicular c) neither

Determine whether each pair of lines is parallel, perpendicular, or neither.

a) $3x + 6y = 10$ b) $2x - 7y = 7$ c) $y = -7x + 4$
   $x + 2y = -12$ $21x + 6y = -2$ $7x - y = 9$

Solution

a) To determine if the lines are parallel or perpendicular, we must find the slope of each line.

Write each equation in slope-intercept form.

$$3x + 6y = 10 \qquad\qquad x + 2y = -12$$
$$6y = -3x + 10 \qquad\qquad 2y = -x - 12$$
$$y = -\frac{3}{6}x + \frac{10}{6} \qquad\qquad y = -\frac{x}{2} - \frac{12}{2}$$
$$y = -\frac{1}{2}x + \frac{5}{3} \qquad\qquad y = -\frac{1}{2}x - 6$$
$$m = -\frac{1}{2} \qquad\qquad m = -\frac{1}{2}$$

Each line has a slope of $-\dfrac{1}{2}$. Their y-intercepts are different. Therefore,

$3x + 6y = 10$ and $x + 2y = -12$ are parallel lines.

b) Begin by writing each equation in slope-intercept form so that we can find their slopes.

$$2x - 7y = 7 \qquad\qquad 21x + 6y = -2$$
$$-7y = -2x + 7 \qquad\qquad 6y = -21x - 2$$
$$y = \frac{-2}{-7}x + \frac{7}{-7} \qquad\qquad y = -\frac{21}{6}x - \frac{2}{6}$$
$$y = \frac{2}{7}x - 1 \qquad\qquad y = -\frac{7}{2}x - \frac{1}{3}$$
$$m = \frac{2}{7} \qquad\qquad m = -\frac{7}{2}$$

The slopes are negative reciprocals, therefore the lines are perpendicular.

c) Again, we must find the slope of each line. $y = -7x + 4$ is already in slope intercept form. Its slope is -7.

Write $7x - y = 9$ in slope-intercept form.

$$-y = -7x + 9 \qquad\qquad \text{Add } -7x \text{ to each side.}$$
$$y = \frac{-7}{-1}x + \frac{9}{-1} \qquad\qquad \text{Divide by } -1.$$
$$y = 7x - 9 \qquad\qquad \text{Simplify.}$$
$$m = 7$$

The slope of $y = -7x + 4$ is -7. The slope of $7x - y = 9$ is 7.

Since the slopes are different and not negative reciprocals, these lines are not parallel, and they are not perpendicular.

You Try 5

Determine if each pair of lines is parallel, perpendicular, or neither.

a) $x + 6y = 12$
   $y = 6x - 9$

b) $-4x + y = 1$
   $2x - 3y = 6$

c) $3x - 2y = -8$
   $15x - 10y = 2$

d) $x = 3$
   $y = 2$

Writing Equations of Parallel and Perpendicular Lines

We have learned that

1) parallel lines have the same slope.

and

2) perpendicular lines have slopes that are negative reciprocals of each other.

We will use this information to write the equation of a line that is parallel or perpendicular to a given line.

6. Write an Equation of a Line That Is Parallel or Perpendicular to a Given Line

Example 6

In-Class Example 6
A line contains the point
$(5, -1)$ and is parallel to the line
$y = 2x - \dfrac{1}{2}$. Write the equation
of the line in slope-intercept
form.
answer: $y = 2x - 11$

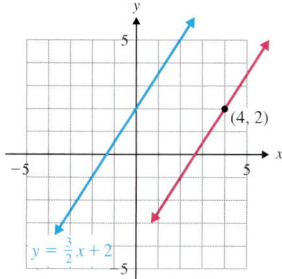

A line contains the point (4, 2) and is parallel to the line $y = \dfrac{3}{2}x + 2$. Write the equation of the line in slope-intercept form.

Solution

Let's look at the graph on the left to help us understand what is happening in this example. We must find the equation of the line in red. It is the line containing the point (4, 2) that is parallel to $y = \dfrac{3}{2}x + 2$.

The line $y = \dfrac{3}{2}x + 2$ has $m = \dfrac{3}{2}$. Therefore, the red line will have $m = \dfrac{3}{2}$ as well.

We know the slope, $\dfrac{3}{2}$, and a point on the line, (4, 2), so we use the point-slope formula to find its equation.

Substitute $\dfrac{3}{2}$ for m. Substitute (4, 2) for (x_1, y_1).

$$y - y_1 = m(x - x_1)$$
$$y - 2 = \dfrac{3}{2}(x - 4) \qquad \text{Substitute 4 for } x_1 \text{ and 2 for } y_1.$$
$$y - 2 = \dfrac{3}{2}x - 6 \qquad \text{Distribute.}$$
$$y = \dfrac{3}{2}x - 4 \qquad \text{Add 2 to each side.}$$

The equation is $y = \dfrac{3}{2}x - 4$.

You Try 6

A line contains the point (8, −5) and is parallel to the line $y = \dfrac{3}{4}x + \dfrac{2}{3}$. Write the equation of the line in slope-intercept form.

Example 7

In-Class Example 7
Find the standard form for the
equation of the line that
contains the point $(-4, -2)$
and that is perpendicular to
$3x + y = 7$.
answer: $x - 3y = 2$

Find the standard form for the equation of the line that contains the point $(-5, 3)$ and that is perpendicular to $5x - 2y = 6$.

Solution

Begin by finding the slope of $5x - 2y = 6$ by putting it into slope-intercept form.

$$5x - 2y = 6$$
$$-2y = -5x + 6 \qquad \text{Add } -5x \text{ to each side.}$$
$$y = \frac{-5}{-2}x + \frac{6}{-2} \qquad \text{Divide by } -2.$$
$$y = \frac{5}{2}x - 3 \qquad \text{Simplify.}$$
$$m = \frac{5}{2}$$

Then, determine the slope of the line containing $(-5, 3)$ by finding the *negative reciprocal* of the slope of the given line.

$$m_{\text{perpendicular}} = -\frac{2}{5}$$

The line for which we need to write an equation has $m = -\frac{2}{5}$ and contains the point $(-5, 3)$.

Use the point-slope formula to find an equation of the line.

Substitute $-\frac{2}{5}$ for m. Substitute $(-5, 3)$ for (x_1, y_1).

$$y - y_1 = m(x - x_1)$$
$$y - 3 = -\frac{2}{5}[x - (-5)] \qquad \text{Substitute } -5 \text{ for } x_1 \text{ and } 3 \text{ for } y_1.$$
$$y - 3 = -\frac{2}{5}(x + 5)$$
$$y - 3 = -\frac{2}{5}x - 2 \qquad \text{Distribute.}$$

Since we are asked to write the equation in standard form, eliminate the fraction by multiplying each side by 5.

$$5(y - 3) = 5\left(-\frac{2}{5}x - 2\right)$$
$$5y - 15 = -2x - 10 \qquad \text{Distribute.}$$
$$5y = -2x + 5 \qquad \text{Add 15 to each side.}$$
$$2x + 5y = 5 \qquad \text{Add } 2x \text{ to each side.}$$

The equation is $2x + 5y = 5$. ∎

You Try 7

Find the equation of the line perpendicular to $-3x + y = 2$ containing the point $(9, 4)$. Write the equation in standard form.

7. Write a Linear Equation to Model Real-World Data

As seen in previous sections of this chapter, equations of lines are used to describe many kinds of real-world situations. We will look at an example in which we must find the equation of a line given some data.

Example 8

In-Class Example 8
Refer to Example 8.
How much sulfur dioxide was released into the air in 2002?
answer: 14,475 thousand tons

Since 1998, sulfur dioxide emissions in the United States have been decreasing by about 1080.5 thousand tons per year. In 2000, approximately 16,636 thousand tons of the pollutant were released into the air. (*Statistical Abstract of the United States*)

a) Write a linear equation to model this data. Let x represent the number of years after 1998, and let y represent the amount of sulfur dioxide (in thousands of tons) released into the air.

b) How much sulfur dioxide was released into the air in 1998? in 2004?

Solution

a) Ask yourself, "What information is given in the problem?"

i) ". . . emissions in the United States have been decreasing by about 1080.5 thousand tons per year" tells us the rate of change of emissions with respect to time. Therefore, this is the *slope*. It will be *negative* since emissions are decreasing.

$$m = -1080.5$$

ii) "In 2000, approximately 16,636 thousand tons . . . were released into the air" gives us a point on the line.

Let x = the number of years after 1998.

The year 2000 corresponds to $x = 2$.

y = amount of sulfur dioxide (in thousands of tons) released into the air.

Then, 16,636 thousand tons corresponds to $y = 16,636$.

A point on the line is **(2, 16,636)**.

Now that we know the slope and a point on the line, we can write an equation of the line using the point-slope formula:

$$y - y_1 = m(x - x_1)$$

Substitute -1080.5 for m. Substitute $(2, 16,636)$ for (x_1, y_1).

$$y - y_1 = m(x - x_1)$$
$$y - 16{,}636 = -1080.5(x - 2) \qquad \text{Substitute 2 for } x_1 \text{ and 16,636 for } y_1.$$
$$y - 16{,}636 = -1080.5x + 2161 \qquad \text{Distribute.}$$
$$y = -1080.5x + 18{,}797 \qquad \text{Add 16,636 to each side.}$$

The equation is $y = -1080.5x + 18{,}797$.

b) To determine the amount of sulfur dioxide released into the air in 1998, let $x = 0$ since x = the number of years *after* 1998.

$$y = -1080.5(0) + 18{,}797 \qquad \text{Substitute } x = 0.$$
$$y = 18{,}797$$

In 1998, 18,797 thousand tons of sulfur dioxide were released. Notice, the equation is in slope-intercept form, $y = mx + b$, and our result is b. That is because to find the y-intercept, let $x = 0$.

To determine how much sulfur dioxide was released in 2004, let $x = 6$ since 2004 is 6 years after 1998.

$$y = -1080.5(6) + 18,797 \qquad \text{Substitute } x = 6.$$
$$y = -6483 + 18,797 \qquad \text{Multiply.}$$
$$y = 12,314$$

In 2004, 12,314 thousand tons of sulfur dioxide were released into the air.   ■

Using Technology

We can use a graphing calculator to explore what we have learned about perpendicular lines.

1) Graph the line $y = -2x + 4$. What is its slope?

2) Find the slope of the line perpendicular to the graph of $y = -2x + 4$.

3) Find the equation of the line perpendicular to $y = -2x + 4$ that passes through the point $(6, 0)$. Express the equation in slope-intercept form.

4) Graph both the original equation and the equation of the perpendicular line:

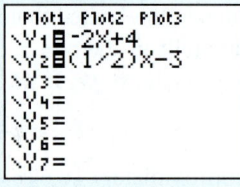

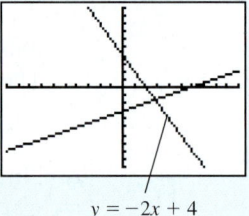

$y = -2x + 4$

5) Do the lines above appear to be perpendicular?

6) Press ZOOM and choose 5:Zsquare:

7) Do the graphs look perpendicular now? Because the viewing window on a graphing calculator is a rectangle, *squaring* the window will give a more accurate picture of the graphs of the equations.

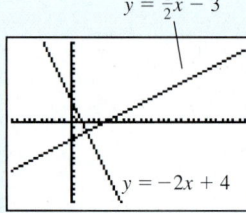

$y = \frac{1}{2}x - 3$

$y = -2x + 4$

Answers to You Try Exercises

1) $y = \dfrac{2}{3}x - 6$ 2) $5x + y = -2$ 3) $4x + 3y = 7$ 4) $y = 7$

5) a) perpendicular   b) neither   c) parallel   d) perpendicular

6) $y = \dfrac{3}{4}x - 11$ 7) $x + 3y = 21$

Answers to Technology Exercises

1) -2 2) $\dfrac{1}{2}$ 3) $y = \dfrac{1}{2}x - 3$ 5) No because they do not meet at 90° angles.

7) Yes because they meet at 90° angles.

3.3 Exercises

Additional answers can be found in the Answers to Exercises appendix.

Objective 1: Write an Equation of a Line Given Its Slope and y-Intercept

1) Explain how to find an equation of a line when you are given the slope and y-intercept of the line.
   Substitute the slope and y-intercept into $y = mx + b$.

2) Is $\frac{2}{3}x - y = 5$ in standard form? Explain your answer.
   No. The coefficient of x must be a positive integer.

Find an equation of the line with the given slope and y-intercept. Express your answer in the indicated form.

3) $m = -7$, y-int: $(0, 2)$; slope-intercept form $y = -7x + 2$

4) $m = 4$, y-int: $(0, -5)$; slope-intercept form $y = 4x - 5$

5) $m = 1$, y-int: $(0, -3)$; standard form $x - y = 3$

6) $m = -2$, y-int: $(0, 8)$; standard form $2x + y = 8$

7) $m = -\frac{1}{3}$, y-int: $(0, -4)$; standard form $x + 3y = -12$

8) $m = \frac{5}{2}$, y-int: $(0, -1)$; standard form $5x - 2y = 2$

9) $m = 1$, y-int: $(0, 0)$; slope-intercept form $y = x$

10) $m = \frac{4}{9}$, y-int: $\left(0, -\frac{1}{6}\right)$; slope-intercept form $y = \frac{4}{9}x - \frac{1}{6}$

Objective 2: Use the Point-Slope Formula to Write an Equation of a Line Given Its Slope and a Point on the Line

11) a) If (x_1, y_1) is a point on a line with slope m, then the point-slope formula is _____. $y - y_1 = m(x - x_1)$

   b) Explain how to find an equation of a line when you are given the slope and a point on the line.
   Substitute the slope and point into the point-slope formula.

Find an equation of the line containing the given point with the given slope. Express your answer in the indicated form.

12) $(5, 8)$, $m = 3$; slope-intercept form $y = 3x - 7$

13) $(1, 6)$, $m = 5$; slope-intercept form $y = 5x + 1$

14) $(3, -2)$, $m = -2$; slope-intercept form $y = -2x + 4$

15) $(-9, 4)$, $m = -1$; slope-intercept form $y = -x - 5$

16) $(-3, -7)$, $m = 1$; standard form $x - y = 4$

17) $(-2, -1)$, $m = 4$; standard form $4x - y = -7$

18) $(2, 3)$, $m = -\frac{4}{5}$; slope-intercept form $y = -\frac{4}{5}x + \frac{23}{5}$

19) $(-4, -5)$, $m = \frac{1}{6}$; slope-intercept form $y = \frac{1}{6}x - \frac{13}{3}$

20) $(-2, 0)$, $m = \frac{5}{8}$; standard form $5x - 8y = -10$

21) $(6, 0)$, $m = -\frac{5}{9}$; standard form $5x + 9y = 30$

22) $\left(\frac{1}{6}, -1\right)$, $m = 4$; slope-intercept form $y = 4x - \frac{5}{3}$

Objective 3: Use the Point-Slope Formula to Write an Equation of a Line Given Two Points on the Line

23) Explain how to find an equation of a line when you are given two points on the line.   Find the slope and use it and one of the points in the point-slope formula.

Find an equation of the line containing the two given points. Express your answer in the indicated form.

24) $(2, 5)$ and $(4, 1)$; slope-intercept form $y = -2x + 9$

25) $(3, 4)$ and $(7, 8)$; slope-intercept form $y = x + 1$

26) $(3, 2)$ and $(4, 5)$; slope-intercept form $y = 3x - 7$

27) $(-2, 4)$ and $(1, 3)$; slope-intercept form $y = -\frac{1}{3}x + \frac{10}{3}$

28) $(5, -2)$ and $(2, -4)$; slope-intercept form $y = \frac{2}{3}x - \frac{16}{3}$

29) $(-1, -5)$ and $(3, -2)$; standard form $3x - 4y = 17$

30) $(-3, 0)$ and $(-5, 1)$; standard form $x + 2y = -3$

31) $(4, 1)$ and $(6, -3)$; standard form $2x + y = 9$

32) $(2, 1)$ and $(4, 6)$; standard form $5x - 2y = 8$

33) $(2.5, 4.2)$ and $(3.1, 7.2)$; slope-intercept form $y = 5.0x - 8.3$

34) $(-3.4, 5.8)$ and $(-1.8, 3.4)$; slope-intercept form $y = -1.5x + 0.7$

Mixed Exercises: Objectives 1–4

Write the slope-intercept form of the equation of each line, if possible.

35)

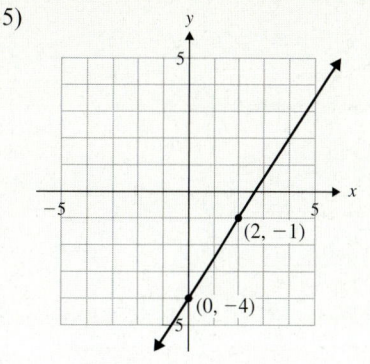

$y = \frac{3}{2}x - 4$

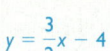

36)

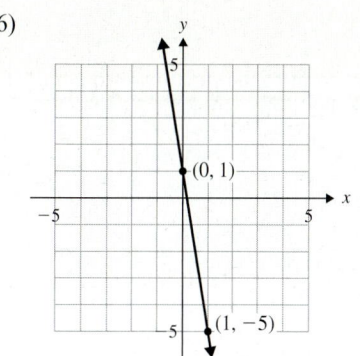

$y = -6x + 1$

37)

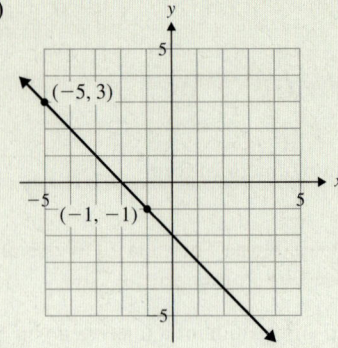

$y = -x - 2$

38)

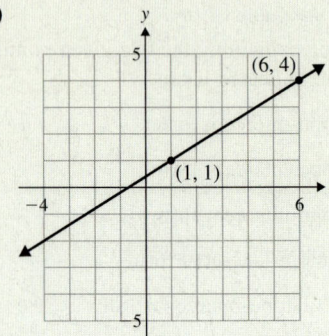

$y = \frac{3}{5}x + \frac{2}{5}$

39)

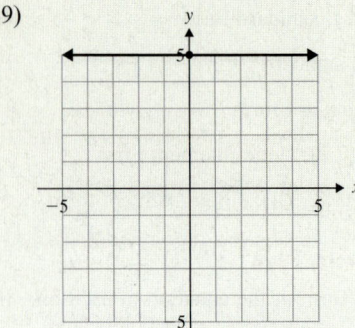

$y = 5$

40)

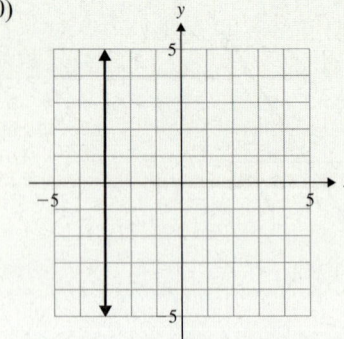

$x = -3$

Write the slope-intercept form of the equation of the line, if possible, given the following information.

41) $m = 4$ and contains $(-4, -1)$ $y = 4x + 15$

42) contains $(3, 0)$ and $(7, -2)$ $y = -\frac{1}{2}x + \frac{3}{2}$

43) $m = \frac{8}{3}$ and y-intercept $(0, -9)$ $y = \frac{8}{3}x - 9$

44) $m = 3$ and contains $(4, 5)$ $y = 3x - 7$

45) contains $(-1, -2)$ and $(-5, 1)$ $y = -\frac{3}{4}x - \frac{11}{4}$

46) y-intercept $(0, 6)$ and $m = 7$ $y = 7x + 6$

47) vertical line containing $(3, 5)$ $x = 3$

48) vertical line containing $\left(-\frac{3}{4}, 2\right)$ $x = -\frac{3}{4}$

49) horizontal line containing $(0, -8)$ $y = -8$

50) horizontal line containing $(7, 4)$ $y = 4$

51) contains $(0, 2)$ and $(6, 0)$ $y = -\frac{1}{3}x + 2$

52) $m = 2$ and y-intercept $(0, -11)$ $y = 2x - 11$

53) $m = 1$ and y-intercept $(0, 0)$ $y = x$

54) contains $(-3, -1)$ and $(2, -3)$ $y = -\frac{2}{5}x - \frac{11}{5}$

Objective 5: Use Slope to Determine if Two Lines are Parallel or Perpendicular

55) How do you know if two lines are perpendicular?
Answers may vary.

56) How do you know if two lines are parallel?
Answers may vary.

Determine if each pair of lines is parallel, perpendicular, or neither.

57) $y = -8x - 6$

$y = \frac{1}{8}x + 3$
perpendicular

58) $y = \frac{6}{11}x + 14$ parallel

$y = \frac{6}{11}x - 2$

59) $y = \frac{2}{9}x + 4$

$4x - 18y = 9$
parallel

60) $y = -\frac{5}{4}x - \frac{1}{3}$

$-4x + 5y = 10$
perpendicular

61) $-3x + 2y = -10$
$3x - 4y = -2$
neither

62) $x - 4y = -12$
$2x - 6y = 9$
neither

63) $y = x$
$x + y = 7$
perpendicular

64) $-x + 5y = -35$
$y = 5x + 2$
neither

65) $4x - 3y = 18$
$-8x + 6y = 5$
parallel

66) $x + y = 4$
$y = -x$
parallel

67) $x = 5$
$x = -2$
parallel

68) $y = 4$
$x = 3$
perpendicular

Lines L_1 and L_2 contain the given points. Determine if lines L_1 and L_2 are parallel, perpendicular, or neither.

69) L_1: $(1, 2)$, $(6, -13)$
L_2: $(-2, 5)$, $(3, -10)$
parallel

70) L_1: $(3, -3)$, $(1, 5)$
L_2: $(4, 3)$, $(-12, -1)$
perpendicular

71) L_1: $(-1, -7)$, $(2, 8)$
L_2: $(10, 2)$, $(0, 4)$
perpendicular

72) L_1: $(1, 7)$, $(-2, -11)$
L_2: $(0, -8)$, $(1, -5)$
neither

73) L_1: $(5, -1)$, $(7, 3)$
L_2: $(-6, 0)$, $(4, 5)$
neither

74) L_1: $(-3, 9)$, $(4, 2)$
L_2: $(6, -8)$, $(-10, 8)$
parallel

Objective 6: Write an Equation of a Line That is Parallel or Perpendicular to a Given Line

Write an equation of the line *parallel* to the given line and containing the given point. Write the answer in slope-intercept form or in standard form, as indicated.

75) $y = 4x + 9$; $(0, 2)$; slope-intercept form $y = 4x + 2$

76) $y = -3x - 1$; $(0, 5)$; slope-intercept form $y = -3x + 5$

77) $y = \dfrac{1}{2}x - 5$; $(4, 5)$; standard form $x - 2y = -6$

78) $y = 2x + 1$; $(-2, -7)$; standard form $2x - y = 3$

79) $4x + 3y = -6$; $(-9, 4)$; standard form $4x + 3y = -24$

80) $x + 4y = 32$; $(-8, 5)$; standard form $x + 4y = 12$

81) $x + 5y = 10$; $(15, 7)$; slope-intercept form $y = -\dfrac{1}{5}x + 10$

82) $18x - 3y = 9$; $(2, -2)$; slope-intercept form $y = 6x - 14$

Write an equation of the line *perpendicular* to the given line and containing the given point. Write the answer in slope-intercept form or in standard form, as indicated.

83) $y = \dfrac{2}{3}x + 4$; $(6, -3)$; slope-intercept form $y = -\dfrac{3}{2}x + 6$

84) $y = -\dfrac{4}{3}x + 2$; $(8, 1)$; slope-intercept form $y = \dfrac{3}{4}x - 5$

85) $y = -5x + 1$; $(10, 0)$; standard form $x - 5y = 10$

86) $y = \dfrac{1}{4}x - 7$; $(-2, 7)$; standard form $4x + y = -1$

87) $x + y = 9$; $(-5, -5)$; slope-intercept form $y = x$

88) $y = x$; $(10, -4)$; slope-intercept form $y = -x + 6$

89) $24x - 15y = 10$; $(16, -7)$; standard form $5x + 8y = 24$

90) $2x + 5y = 11$; $(4, 2)$; standard form $5x - 2y = 16$

Write the slope-intercept form (if possible) of the equation of the line meeting the given conditions.

91) perpendicular to $2x - 6y = -3$ containing $(2, 2)$ $y = -3x + 8$

92) parallel to $6x + y = 4$ containing $(-2, 0)$ $y = -6x - 12$

93) parallel to $y = 2x + 1$ containing $(1, -3)$ $y = 2x - 5$

94) perpendicular to $y = -x - 8$ containing $(4, 11)$ $y = x + 7$

95) parallel to $x = -4$ containing $(-1, -5)$ $x = -1$

96) parallel to $y = 2$ containing $(4, -3)$ $y = -3$

97) perpendicular to $y = 3$ containing $(2, 1)$ $x = 2$

98) perpendicular to $x = 0$ containing $(5, 1)$ $y = 1$

99) perpendicular to $21x - 6y = 2$ containing $(4, -1)$ $y = -\dfrac{2}{7}x + \dfrac{1}{7}$

100) parallel to $-3x + 4y = 8$ containing $(7, 4)$ $y = \dfrac{3}{4}x - \dfrac{5}{4}$

101) parallel to $y = 0$ containing $\left(-3, -\dfrac{5}{2}\right)$ $y = -\dfrac{5}{2}$

102) perpendicular to $y = \dfrac{3}{4}$ containing $(-2, 5)$ $x = -2$

Objective 7: Write a Linear Equation to Model Real-World Data

103) If a man's foot is 10 in. long, his U.S. shoe size is 8. A man's foot length of 10.5 in. corresponds to a shoe size of 9.5. Let L represent the length of a man's foot, and let S represent his shoe size.

   a) Write a linear equation that describes the relationship between the length of a man's foot in terms of his shoe size. $L = \dfrac{1}{3}S + \dfrac{22}{3}$

   b) If a man's foot is 11.5 in. long, what shoe size does he wear?   12.5

104) When a company charges customers $2000 to advertise on a billboard, 30% of its billboards are rented. If it cuts its rental fee to $1000, then 80% of its billboards are rented. Let C represent the cost of renting a billboard, and let P represent the percentage of the company's billboards that are being rented.

   a) Write a linear equation that describes the relationship of the percentage of billboards rented in terms of the cost. $P = -\dfrac{1}{20}C + 130$

   b) What percentage of its billboards would be rented if the company charges $1200?   70%

105) Since 1998 the population of Maine has been increasing by about 8700 people per year. In 2001, the population of Maine was about 1,284,000. (*Statistical Abstract of the United States*)

   a) Write a linear equation to model this data. Let x represent the number of years after 1998, and let y represent the population of Maine. $y = 8700x + 1,257,900$

   b) Explain the meaning of the slope in the context of the problem.   The population of Maine is increasing by 8700 people per year.

   c) According to the equation, how many people lived in Maine in 1998? in 2002?   1,257,900; 1,292,700

   d) If the current trend continues, in what year would the population be 1,431,900?   2018

106) Since 1997, the population of North Dakota has been decreasing by about 3290 people per year. The population was about 650,000 in 1997. (*Statistical Abstract of the United States*)

   a) Write a linear equation to model this data. Let x represent the number of years after 1997, and let y represent the population of North Dakota. $y = -3290x + 650,000$

   b) Explain the meaning of the slope in the context of the problem.   The population of North Dakota is decreasing by 3290 people per year.

   c) According to the equation, how many people lived in North Dakota in 1999? in 2002?   643,420; 633,550

   d) If the current trend holds, in what year would the population be 600,650?   2012

107) The graph here shows the number of farms (in thousands) with milk cows between 1997 and 2002. x represents the number of years after 1997 so that $x = 0$ represents 1997, $x = 1$ represents 1998, and so on. Let y represent the number of farms (in thousands) with milk cows. (USDA, National Agricultural Statistics Service)

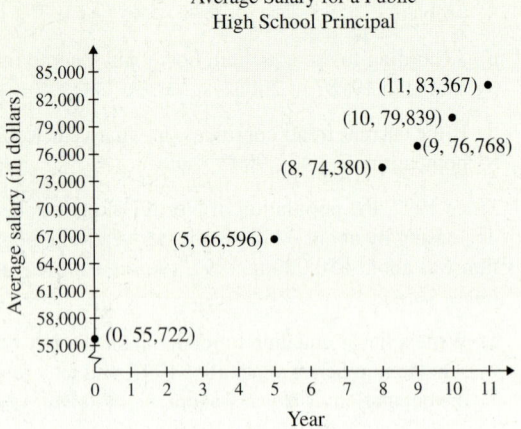

Number of Farms With Milk Cows

a) Write a linear equation to model this data. Use the data points for 1997 and 2002. $y = -6.4x + 124$

b) Explain the meaning of the slope in the context of the problem. The number of farms with milk cows is decreasing by 6.4 thousand (6400) per year.

c) If the current trend continues, find the number of farms with milk cows in 2004.   79.2 thousand (79,200)

108) The graph here shows the average salary for a public school high school principal for several years beginning with 1990. x represents the number of years after 1990 so that $x = 0$ represents 1990, $x = 1$ represents 1991, and so on. Let y represent the average salary of a high school principal. (*Statistical Abstract of the United States*)

Average Salary for a Public High School Principal

a) Write a linear equation to model this data. Use the data points for 1990 and 2001. (Round the slope to the nearest whole number.) $y = 2513x + 55{,}722$

b) Explain the meaning of the slope in the context of the problem.   A principal's average salary is increasing by $2513 per year.

c) Use the equation to estimate the average salary in 1993. $63,261

d) If the current trend continues, find the average salary in 2005. $93,417

109) The chart shows the number of hybrid vehicles registered in the United States from 2000 to 2003. (*American Demographics,* Sept 2004, Vol 26, Issue 7)

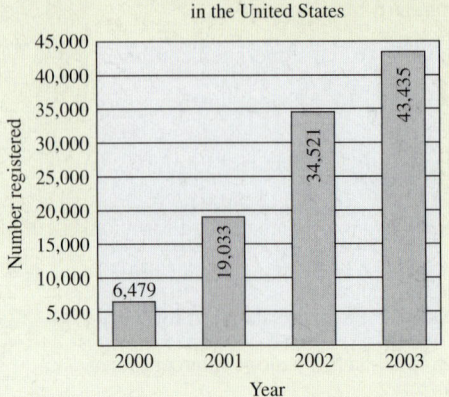

Hybrid Vehicles Registered in the United States

a) Write a linear equation to model this data. Use the data points for 2000 and 2003. Let x represent the number of years after 2000, and let y represent the number of hybrid vehicles registered in the United States. (Round the slope to the tenths place.) $y = 12{,}318.7x + 6479$

b) Explain the meaning of the slope in the context of the problem.   The number of registered hybrid vehicles is increasing by 12,318.7 per year.

c) Use the equation to determine the number of vehicles registered in 2002. How does it compare to the actual number on the chart?
31,116.4; this is slightly lower than the actual value.

d) If the current trend continues, approximately how many hybrid vehicles will be registered in 2010?
129,666

110) The chart shows the percentage of females in Belgium (15–64 yr old) in the workforce between 1980 and 2000. (*Statistical Abstract of the United States*)

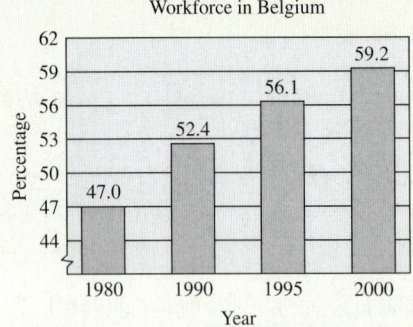

Percentage of Females in the Workforce in Belgium

a) Write a linear equation to model this data. Use the data points for 1980 and 2000. Let x represent the number of years after 1980, and let y represent the percent of females in Belgium in the workforce.
$y = 0.61x + 47.0$

b) Explain the meaning of the slope in the context of the problem.   The percentage of women in Belgium in the workforce is increasing by 0.61 per year.

c) Use the equation to determine the percentage of Belgian women in the workforce in 1990. How does it compare to the actual number on the chart?
53.1%; it is 0.7% higher than the actual value.

d) Do the same as part c) for the year 1995.
56.15%; it is 0.05% higher than the actual value.

e) In what year were 58% of Belgian women working?
1998

Section 3.4 Linear and Compound Linear Inequalities in Two Variables

Objectives

1. Define a Linear Inequality in Two Variables
2. Graph a Linear Inequality in Two Variables
3. Graph a Compound Linear Inequality in Two Variables
4. Solve a Linear Programming Problem

In Chapter 2, we learned how to solve linear inequalities in *one variable* such as $2x - 3 \geq 5$.

We will begin this section by learning how to graph the solution set of linear inequalities in *two variables*. Then we will learn how to graph the solution set of *systems* of linear inequalities in two variables.

1. Define a Linear Inequality in Two Variables

> **Definition**
>
> A **linear equality in two variables** is an inequality that can be written in the form $Ax + By \geq C$ or $Ax + By \leq C$ where A, B, and C are real numbers and where A and B are not both zero. ($>$ and $<$ may be substituted for $\geq$ and $\leq$.)

Here are some examples of linear inequalities in two variables.

$$5x - 3y \geq 6, \qquad y < \frac{1}{4}x + 3, \qquad x \leq 2, \qquad y > -4$$

Note

We can call $x \leq 2$ a linear inequality in two variables because we can write it as $x + 0y \leq 2$. Likewise, we can write $y > -4$ as $0x + y > -4$.

The solutions to linear inequalities in two variables, such as $x + y \geq 3$, are *ordered pairs* of the form (x, y) that make the inequality true. We graph a linear inequality in two variables on a rectangular coordinate system.

Example 1

In-Class Example 1
Shown here is the graph of $x - y \leq 2$. Find three points that solve $x - y \leq 2$ and three points that are not in the solution set.

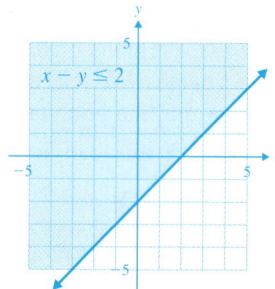

answer: in solution set (0, 4), (0, −2), (−1, 3); not in solution set (0, −3), (4, −1), (−2, −5)

Shown here is the graph of $x + y \geq 3$. Find three points that solve $x + y \geq 3$, and find three points that are not in the solution set.

Solution

The solution set of $x + y \geq 3$ consists of all points either on the line or in the shaded region. *Any* point on the line or in the shaded region will make $x + y \geq 3$ true.

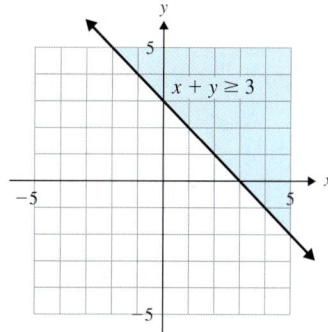

Are These Points Solutions?	Check by Substituting into $x + y \geq 3$	
$(5, 2)$	$5 + 2 \geq 3$	True
$(1, 4)$	$1 + 4 \geq 3$	True
$(3, 0)$ (on the line)	$3 + 0 \geq 3$	True
$(0, 0)$	$0 + 0 \geq 3$	False
$(-4, 1)$	$-4 + 1 \geq 3$	False
$(2, -3)$	$2 + (-3) \geq 3$	False

The points $(5, 2)$, $(1, 4)$, and $(3, 0)$ are some of the points that satisfy $x + y \geq 3$. There are infinitely many solutions. The points $(0, 0)$, $(-4, 1)$, and $(2, -3)$ are three of the points that do not satisfy $x + y \geq 3$. There are infinitely points that are not solutions.

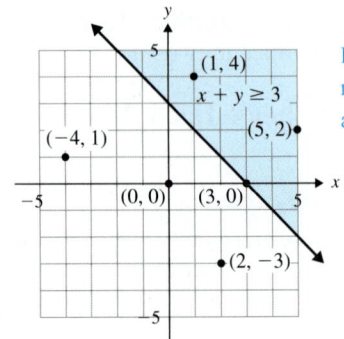

Points in the shaded region and on the line are in the solution set.

The points in the unshaded region are *not* in the solution set.

Note

If the inequality in Example 1 had been $x + y > 3$, then the line would have been drawn as a *dotted line* and all points on the line would *not* be part of the solution set.

You Try 1

Shown here is the graph of $5x - 3y \geq -15$. Find three points that solve $5x - 3y \geq -15$, and find three points that are not in the solution set.

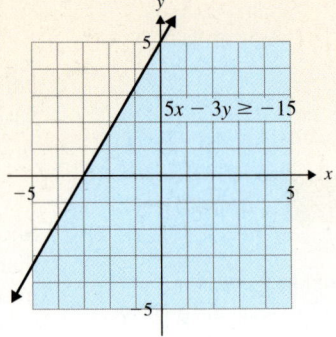

2. Graph a Linear Inequality in Two Variables

As we saw in the graph in Example 1, the line divides the *plane* into two regions or **half planes**. The line $x + y = 3$ is the **boundary line** between the two half planes. We will use this boundary line to graph a linear inequality in two variables. Notice that the boundary line is written as an equation; it uses an equal sign.

Procedure   Graph a Linear Inequality in Two Variables Using a Test Point

1) **Graph the boundary line.** If the inequality contains $\geq$ or $\leq$, make this boundary line *solid*. If the inequality contains $>$ or $<$, make it *dotted*.

2) **Choose a test point not on the line, and shade the appropriate region.** Substitute the test point into the inequality. If $(0, 0)$ is not on the line, it is an easy point to test in the inequality.

   a) If it *makes the inequality true*, shade the region *containing* the test point. All points in the shaded region are part of the solution set.

   b) If the test point *does not satisfy the inequality*, shade the region on the *other* side of the line. All points in the shaded region are part of the solution set.

Example 2

Graph $3x + 4y \leq -8$.

In-Class Example 2
Graph $2x + 3y \leq -9$.
answer:

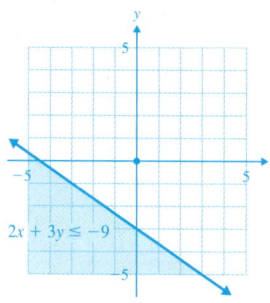

$2x + 3y \leq -9$

Solution

1) Graph the boundary line $3x + 4y = -8$ as a solid line.

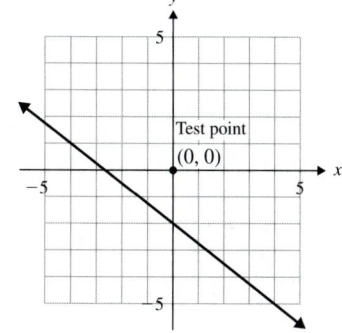

Test point
$(0, 0)$

2) Choose a test point not on the line and substitute it into the inequality to determine whether it makes the inequality true.

Test Point	Substitute into $3x + 4y \leq -8$
$(0, 0)$	$3(0) + 4(0) \leq -8$
	$0 \leq -8$ False

Since the test point $(0, 0)$ does *not* satisfy the inequality we will shade the region that does *not* contain the point $(0, 0)$.

All points on the line and in the shaded region satisfy the inequality $3x + 4y \leq -8$.

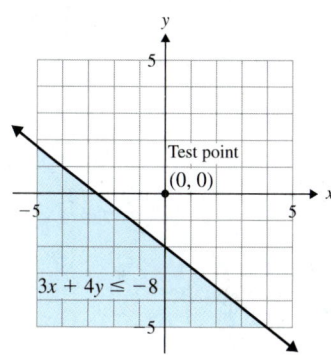

Test point
$(0, 0)$

$3x + 4y \leq -8$

Example 3

Graph $-x + 2y > -4$.

In-Class Example 3
Graph $-2x + y > 1$.
answer:

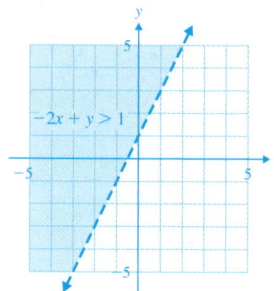

$-2x + y > 1$

Solution

1) Since the inequality symbol is $>$, graph a *dotted* boundary line, $-x + 2y = -4$. (This means that the points *on* the line are not part of the solution set.)

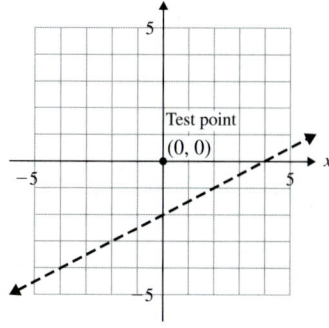

Test point
$(0, 0)$

2) Choose a test point not on the line and substitute it into the inequality to determine whether it makes the inequality true.

Test Point	Substitute into $-x + 2y > -4$
$(0, 0)$	$-(0) + 2(0) > -4$
	$0 > -4$ True

Since the test point $(0, 0)$ satisfies the inequality, shade the region containing that point.

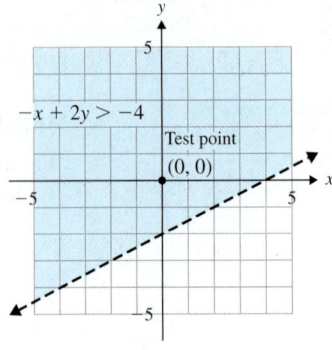

All points in the shaded region satisfy the inequality $-x + 2y > -4$.

You Try 2

Graph each inequality.

a) $2x + y \leq 4$ b) $x + 4y > 12$

If we write the inequality in *slope-intercept form,* we can decide which region to shade without using test points.

Procedure   Graph a Linear Inequality in Two Variables Using the Slope-Intercept Method

1) Write the inequality in the form $y \geq mx + b$ ($y > mx + b$) or $y \leq mx + b$ ($y < mx + b$), and graph the boundary line $y = mx + b$.

2) If the inequality is in the form $y \geq mx + b$ or $y > mx + b$, shade *above* the line.

3) If the inequality is in the form $y \leq mx + b$ or $y < mx + b$, shade *below* the line.

Example 4

Graph each inequality using the slope-intercept method.

a) $y < -\dfrac{1}{3}x + 5$ b) $2x - y \leq -2$

In-Class Example 4
Graph each inequality using the slope-intercept method.

a) $y < \dfrac{3}{4}x - 2$ b) $3x - y \leq 4$

answer:

a)

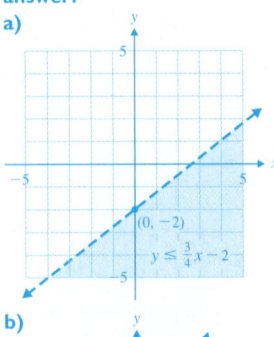

b)

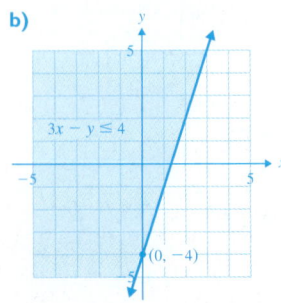

Solution

a) The inequality $y < -\dfrac{1}{3}x + 5$ is already in slope-intercept form.

Graph the boundary line $y = -\dfrac{1}{3}x + 5$ as a *dotted line*.

Since $y < -\dfrac{1}{3}x + 5$ has a *less than* symbol, shade *below* the line. All points in the shaded region satisfy $y < -\dfrac{1}{3}x + 5$.

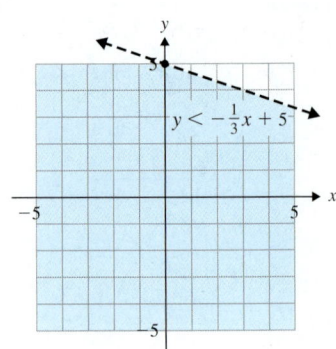

We can choose a point such as $(0, 0)$ in the shaded region as a check. Substituting this point into $y < -\dfrac{1}{3}x + 5$ gives us $0 < -\dfrac{1}{3}(0) + 5$, or $0 < 5$, which is true.

b)   Solve $2x - y \leq -2$ for y.

$$2x - y \leq -2$$
$$-y \leq -2x - 2 \qquad \text{Subtract } 2x.$$
$$y \geq 2x + 2 \qquad \text{Divide by } -1 \text{ and change the direction of the inequality symbol.}$$

Graph $y = 2x + 2$ as a *solid line*.

Since $y \geq 2x + 2$ has a *greater than or equal to* symbol, shade *above* the line.

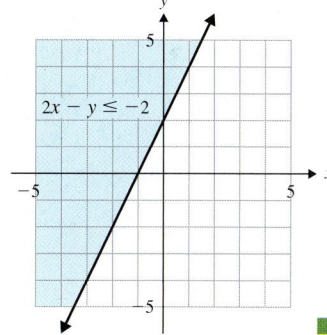

All points on the line and in the shaded region satisfy $2x - y \leq -2$.

You Try 3

Graph each inequality using the slope-intercept method.

a) $y \geq -\dfrac{3}{4}x - 6$ b) $5x - 2y > 4$

3. Graph a Compound Linear Inequality in Two Variables

Linear inequalities in two variables are called *compound linear inequalities* if they are connected by the words *and* or *or*.

The solution set of a compound inequality containing *and* is the *intersection* of the solution sets of the inequalities. The solution set of a compound inequality containing *or* is the *union* of the solution sets of the inequalities.

> **Procedure**   Graphing Compound Linear Inequalities in Two Variables
>
> 1) Graph each inequality separately on the same axes. Shade lightly.
> 2) If the inequality contains *and*, the solution set is the *intersection* of the shaded regions. Heavily shade this region.
> 3) If the inequality contains *or*, the solution set is the *union* (total) of the shaded regions. Heavily shade this region.

Example 5

Graph $x \leq 2$ and $2x + 3y > 3$.

In-Class Example 5
Graph $x < -3$ and $5y - 2x \geq -15$.
answer:

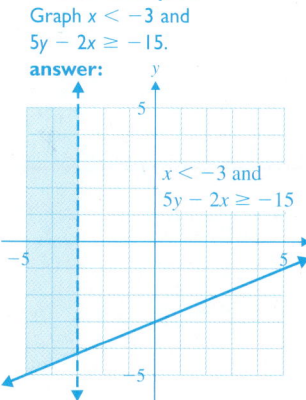

x < −3 and
5*y* − 2*x* ≥ −15

Solution

To graph $x \leq 2$, graph the boundary line $x = 2$ as a solid line. The x-values are *less than* 2 to the *left* of 2, so shade the region to the left of the line $x = 2$.

Graph $2x + 3y > 3$. Use a dotted boundary line.

The region shaded blue in the third graph is the *intersection* of the shaded regions and the solution set of the compound inequality. The part of the line $x = 2$ that is above the line $2x + 3y = 3$ is included in the solution set.

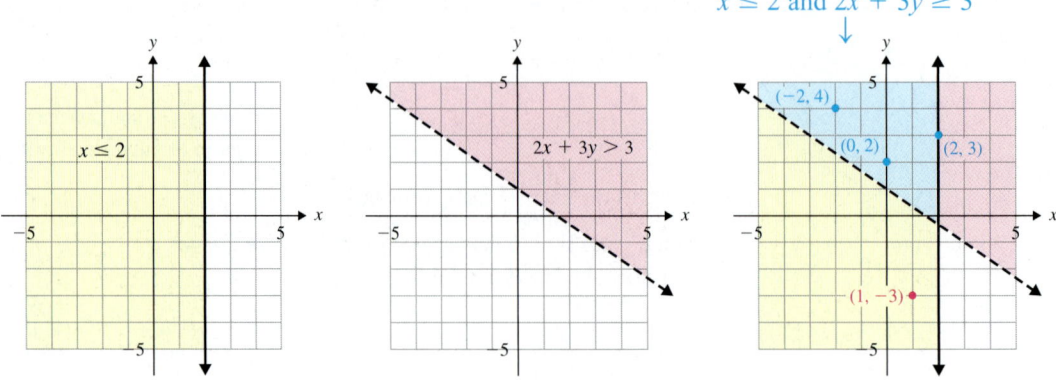

$x \le 2$ and $2x + 3y \ge 3$

Any point in the solution set must satisfy *both* inequalities, and any point *not* in the solution set will not satisfy *both* inequalities. We check three test points next. (See the graph.)

Test Point	Substitute into $x \le 2$	Substitute into $2x + 3y > 3$	Solution?
$(-2, 4)$	$-2 \le 2$ True	$2(-2) + 3(4) > 3$ $8 > 3$ True	Yes
$(0, 2)$	$0 \le 2$ True	$2(0) + 3(2) > 3$ $6 > 3$ True	Yes
$(1, -3)$	$1 \le 2$ True	$2(1) + 3(-3) > 3$ $-7 > 3$ False	No

Although we show three separate graphs in Example 5, it is customary to graph everything on the same axes, shading lightly at first, then to heavily shade the region that is the graph of the compound inequality.

You Try 4

Graph the compound inequality $y \le 3x - 1$ and $y + 2x \le 4$.

Example 6

Graph $y \le \dfrac{1}{2}x$ or $2x + y \ge 2$.

In-Class Example 6

Graph $y \ge \dfrac{3}{2}x$ or $3x + y < 3$.

answer:

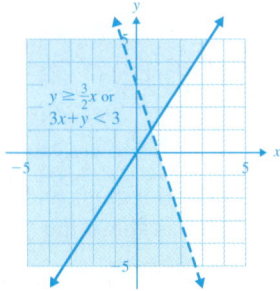

$y \ge \frac{3}{2}x$ or
$3x+y < 3$

Solution

Graph each inequality separately. The solution set of the compound inequality will be the *union* (total) of the shaded regions.

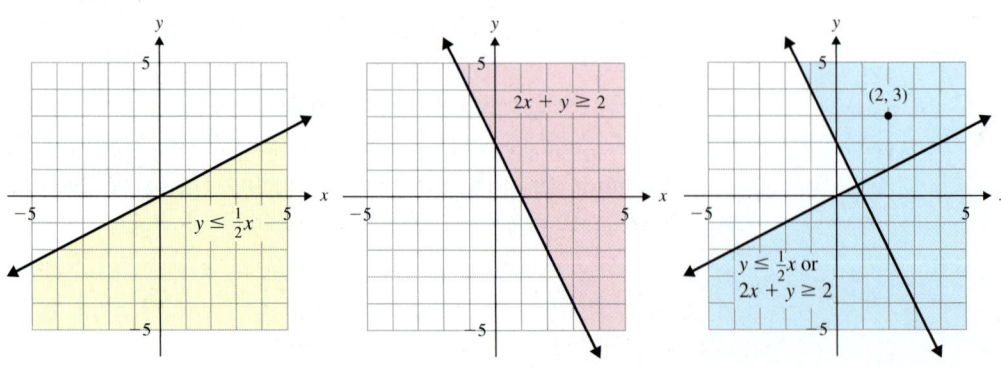

Any point in the shaded region of the third graph will be a solution to the compound inequality $y \leq \frac{1}{2}x$ or $2x + y \geq 2$. This means the point must satisfy $y \leq \frac{1}{2}x$ *or* $2x + y \geq 2$ *or* both. One point in the shaded region is $(2, 3)$.

Test Point	Substitute into $y \leq \frac{1}{2}x$	Substitute into $2x + y \geq 2$	Solution?
$(2, 3)$	$3 \leq \frac{1}{2}(2)$ $3 \leq 1$ False	$2(2) + 3 \geq 2$ $7 \geq 2$ True	Yes

Although $(2, 3)$ does not satisfy $y \leq \frac{1}{2}x$, it *does* satisfy $2x + y \geq 2$, so it *is* a solution of the compound inequality.

Choose a point in the region that is *not* shaded to verify that it does not satisfy either inequality.

You Try 5

Graph the compound inequality $x \geq -4$ or $x - 3y \leq -3$.

4. Solve a Linear Programming Problem

A practical application of linear inequalities in two variables is a process called **linear programming.** Companies use linear programming to determine the best way to use their machinery, employees, and other resources.

A linear programming problem may consist of several inequalities called **constraints.** Constraints describe the conditions that the variables must meet. The graph of the *intersection* of these inequalities is called the **feasible region**—the ordered pairs that are the possible solutions to the problem.

Example 7

In-Class Example 7
Use Example 7.

During a particular week, a company wants Harvey and Amy to work at most 40 hours between them.

Let x = the number of hours Harvey works
     y = the number of hours Amy works

a)   Write the linear inequalities that describe the constraints on the number of hours available to work.

b)   Graph the feasible region (solution set of the intersection of the inequalities), which describes the possible number of hours each person can work.

c)   Find a point in the feasible region and discuss its meaning.

d)   Find a point outside the feasible region and discuss its meaning.

Solution

a)   Since x and y represent the number of hours worked, x and y cannot be negative. We can write $x \geq 0$ and $y \geq 0$.

Together they can work at most 40 hours. We can write $x + y \leq 40$.

The inequalities that describe the constraints on the number of hours available are $x \geq 0$ *and* $y \geq 0$ *and* $x + y \leq 40$. We want to find the *intersection* of these inequalities.

b) The graphs of $x \geq 0$ *and* $y \geq 0$ give us the set of points in the first quadrant since x and y are both positive here.

Graph $x + y \leq 40$. This will be the region *below and including* the line $x + y = 40$ in quadrant I.

The feasible region is shown here.

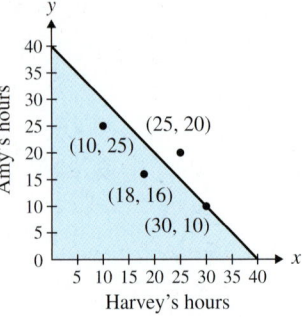

c) One point in the feasible region is (10, 25).

It represents Harvey working 10 hours and Amy working 25 hours. It satisfies all three inequalities.

Test Point	Substitute into $x \geq 0$	Substitute into $y \geq 0$	Substitute into $x + y \leq 40$
(10, 25)	$10 \geq 0$ True	$25 \geq 0$ True	$10 + 25 \leq 40$ $35 \leq 40$ True

d) One point outside the feasible region is (25, 20). It represents Harvey working 25 hours and Amy working 20 hours. This is not possible since it does not satisfy the inequality $x + y \leq 40$.

Test Point	Substitute into $x \geq 0$	Substitute into $y \geq 0$	Substitute into $x + y \leq 40$
(25, 20)	$25 \geq 0$ True	$20 \geq 0$ True	$25 + 20 \leq 40$ $45 \leq 40$ False

Using Technology

To graph a linear inequality in two variables using a graphing calculator, first solve the inequality for y. Then graph the boundary line found by replacing the inequality symbol with an = symbol. For example, to graph the inequality $2x - y \leq 5$, solve it for $y \geq 2x - 5$. Graph the boundary equation $y = 2x - 5$ using a solid line since the inequality symbol is $\leq$. Press Y =, enter $2x - 5$ in Y_1, press ZOOM, and select 6:ZStandard to graph the equation as shown.

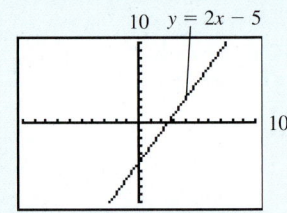

If the inequality symbol is $\leq$, shade below the boundary line. If the inequality symbol is $\geq$, shade above it. To shade above the line, press Y = and move the cursor to the left of Y_1 using the left arrow key. Press ENTER twice and then move the cursor to the next line as shown below left. Press GRAPH to graph the inequality as shown below right.

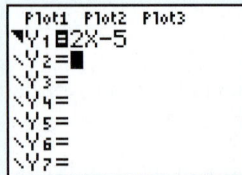

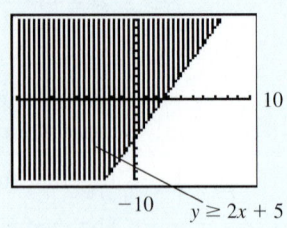

To shade below the line, press $\boxed{Y=}$ and move the cursor to the left of Y₁ using the left arrow key. Keep pressing $\boxed{ENTER}$ until you see ▲ next to Y₁, then move the cursor to the next line as shown below left. Press $\boxed{GRAPH}$ to graph the inequality $y \leq 2x - 5$ as shown below right.

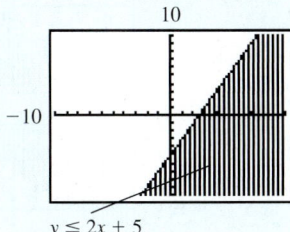

$y \leq 2x + 5$

Graph the linear inequalities in two variables.

1) $y \leq 5x - 2$ 2) $y \geq x - 4$ 3) $x - 2y \leq 6$ 4) $y - x \geq 5$

5) $y \leq -4x + 1$ 6) $y \geq 3x - 6$

Answers to You Try Exercises

1) Answers may vary.   2) a)

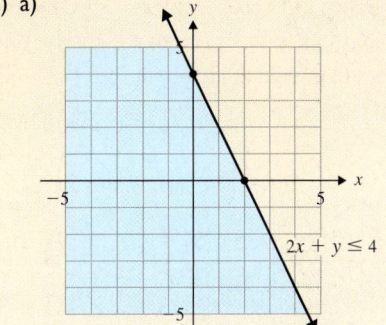

$2x + y \leq 4$

b)

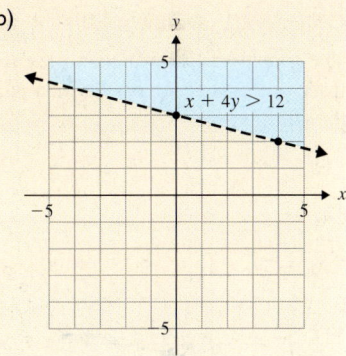

$x + 4y > 12$

3) a)

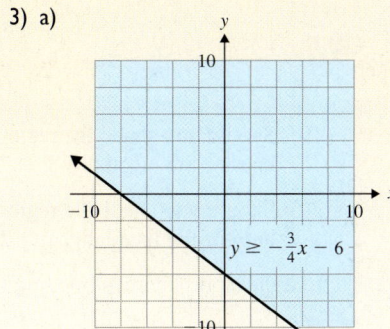

$y \geq -\frac{3}{4}x - 6$

b)

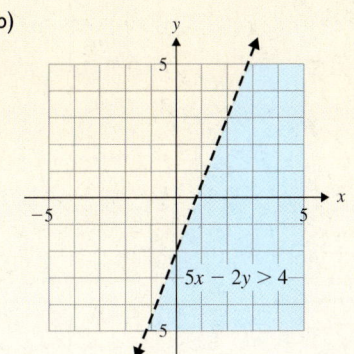

$5x - 2y > 4$

4)

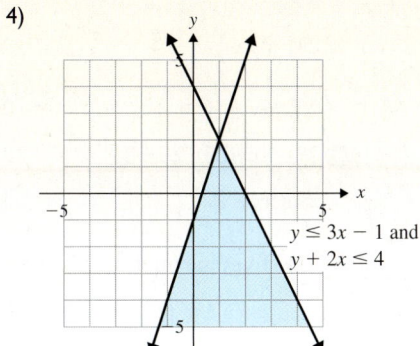

$y \leq 3x - 1$ and $y + 2x \leq 4$

5)

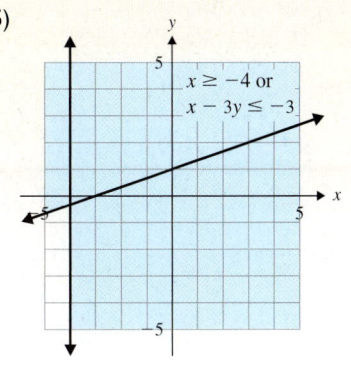

$x \geq -4$ or $x - 3y \leq -3$

Answers to Technology Exercises

1) 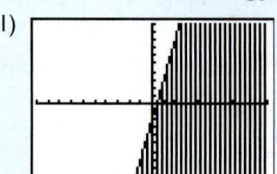   2) 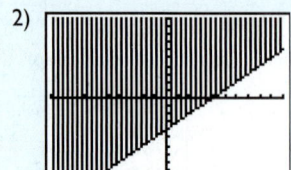   3)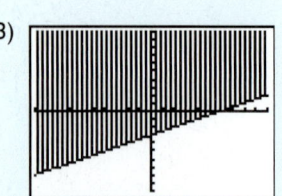

4) 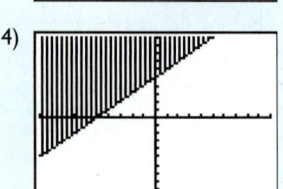   5) 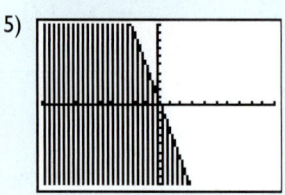   6)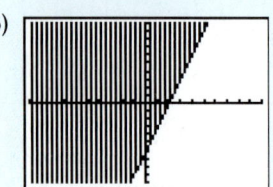

3.4 Exercises

*Additional answers can be found in the Answers to Exercises appendix.

Objective 1: Define a Linear Inequality in Two Variables

The graphs of linear inequalities are given next. For each, find three points that satisfy the inequality and three that are not in the solution set.

1)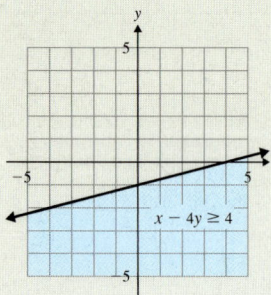

2)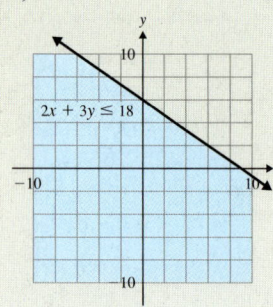

3)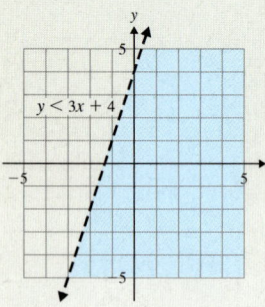

4)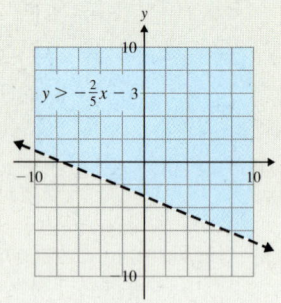

5)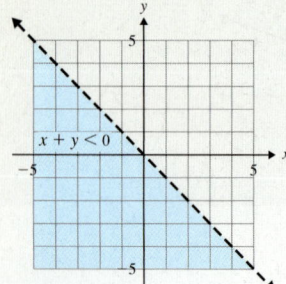

6)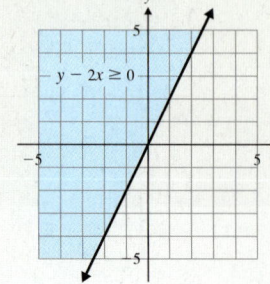

Objective 2: Graph a Linear Inequality in Two Variables

7) Will the boundary line you draw to graph $3x - 4y < 5$ be solid or dotted?   dotted

8) Are points on solid boundary lines included in the inequality's solution set?   yes

Graph the inequalities. Use a test point.

9) $2x + y \geq 6$ 10) $4x + y \leq 3$

11) $y < x + 2$ 12) $y > \frac{1}{2}x - 1$

13) $2x - 7y \leq 14$ 14) $4x + 3y < 15$

15) $y < x$ 16) $y \geq 3x$

17) $y \geq -5$ 18) $x < 1$

19) Should you shade the region above or below the boundary line for the inequality $y \leq 7x + 2$?   below

20) Should you shade the region above or below the boundary line for the inequality $y > 2x + 4$?   above

Use the slope-intercept method to graph each inequality.

21) $y \leq 4x - 3$ 22) $y \geq \frac{5}{2}x - 8$

23) $y > \frac{2}{5}x - 4$ 24) $y < \frac{1}{4}x + 1$

25) $6x + y > 3$ 26) $2x + y > -5$

27) $9x - 3y \leq -21$ 28) $3x + 5y < -20$

29) $x > 2y$ 30) $x - y \leq 0$

31) To graph an inequality like $y \geq \frac{1}{3}x + 2$, would you rather use a test point or use the slope intercept form?   Answers may vary.

32) To graph an inequality like $7x + 2y < 10$, would you rather use a test point or the slope intercept form? Why?   Answers may vary.

Graph using either a test point or the slope-intercept method.

33) $y > -\dfrac{3}{4}x + 1$

34) $y \le \dfrac{1}{3}x - 6$

35) $5x + 2y < -8$

36) $4x + y < 7$

37) $9x - 3y \le 21$

38) $5x - 3y \ge -9$

VIDEO 39) $x > 2$

40) $y \le 4$

41) $3x - 4y > 12$

42) $6x - y \le 2$

Objective 3: Graph a Compound Linear Inequality in Two Variables

43) Is $(3, 5)$ in the solution set of the compound inequality $x - y \ge -6$ and $2x + y < 7$? Why or why not?
    No; it does not satisfy $2x + y < 7$.

44) Is $(3, 5)$ in the solution set of the compound inequality $x - y \ge -6$ or $2x + y < 7$? Why or why not?
    Yes; it satisfies one of the inequalities, $x - y \ge -6$.

Graph each compound inequality.

VIDEO 45) $x \le 4$ and $y \ge -\dfrac{3}{2}x + 3$

46) $y \le \dfrac{1}{4}x + 2$ and $y \ge -1$

47) $y < x + 4$ and $y \ge -3$

48) $x < 3$ and $y > \dfrac{2}{3}x - 1$

49) $2x - 3y < -9$ and $x + 6y < 12$

50) $5x - 3y > 9$ and $2x + 3y \le 12$

51) $y \le -x - 1$ or $x \ge 6$

52) $y \le 2$ or $y \le \dfrac{4}{5}x + 2$

53) $y \le 4$ or $4y - 3x \ge -8$

54) $x + 3y \ge 3$ or $x \ge -2$

55) $y > -\dfrac{2}{3}x + 1$ or $-2x + 5y \le 0$

56) $y > x - 4$ or $3x + 2y \ge 12$

57) $x \ge 5$ and $y \le -3$ 58) $x \le 6$ and $y \ge 1$

59) $y < 4$ or $x \ge -3$ 60) $x \ge 2$ or $y \ge -6$

VIDEO 61) $2x + 5y < 15$ or $y \le \dfrac{3}{4}x - 1$

62) $y - 2x \le 1$ and $y \ge -\dfrac{1}{5}x - 2$

63) $y \ge \dfrac{2}{3}x - 4$ and $4x + y \le 3$

64) $y < 5x + 2$ or $x + 4y < 12$

Objective 4: Solve a Linear Programming Problem

65) During the school year, Tazia earns money by babysitting and tutoring. She can work at most 15 hr per week.

Let x = number of hours Tazia babysits
    y = number of hours Tazia tutors

a) Write the linear inequalities that describe the constraints on the number of hours Tazia can work per week. $x \ge 0$ and $y \ge 0$ and $x + y \le 15$

b) Graph the feasible region that describes how her hours can be distributed between babysitting and tutoring.

c) Find three points in the feasible region and discuss their meanings.

d) Find one point outside the feasible region and discuss its meaning.

66) A machine in a factory can be calibrated to fill either large or small bags of potato chips. The machine will run at most 12 hr per day.

Let x = number of hours the machine fills large bags
    y = number of hours the machine fills small bags

a) Write the linear inequalities that describe the constraints on the number of hours the machine fills the bags each day. $x \ge 0$ and $y \ge 0$ and $x + y \le 12$

b) Graph the feasible region that describes how the hours can be distributed between filling the large and small bags of chips.

c) Find three points in the feasible region and discuss their meanings.

d) Find one point outside the feasible region and discuss its meaning.

67) A lawn mower company produces a push mower and a riding mower. Company analysts predict that, for next spring, the company will need to produce at least 150 push mowers and 100 riding mowers per day, but they can produce at most 250 push mowers and 200 riding mowers per day. To satisfy demand, they will have to ship a total of at least 300 mowers per day.

Let p = number of push mowers produced per day
    r = number of riding mowers produced per day

a) Write the inequalities that describe the constraints on the number of mowers that can be produced per day.
    $150 \le p \le 250$ and $100 \le r \le 200$ and $p + r \ge 300$

b) Graph the feasible region that describes how production can be distributed between the riding mowers and the push mowers. Let p be the horizontal axis and r be the vertical axis.

c) What does the point $(175, 110)$ represent? Will this level of production meet the needs of the company?

d) Find three points in the feasible region and discuss their meanings.   Answers may vary.

e) Find one point outside the feasible region and discuss its meaning.   Answers may vary.

68) A dog food company produces adult dog food and puppy food. The company estimates that for next month it will need to produce at least 12,000 pounds of adult dog food and 8000 pounds of puppy food per day. The factory can produce at most 18,000 pounds of adult dog food and 14,000 pounds of puppy food per day. The company will need to ship a total of at least 25,000 pounds of dog food per day to its customers.

Let a = pounds of adult dog food produced per day
   p = pounds of puppy food produced per day

a) Write the linear inequalities that describe the constraints on the number of pounds of dog food that can be produced per day.
   $12,000 \le a \le 18,000$ and $8000 \le p \le 14,000$ and $a + p \ge 25,000$

b) Graph the feasible region that describes how production can be distributed between the adult dog food and the puppy food. Let a be on the horizontal axis and p be on the vertical axis.

c) What does the point (17,000, 9000) represent? Will this level of production meet the needs of the company?

d) Find three points in the feasible region and discuss their meanings.   Answers may vary.

e) Find one point outside the feasible region and discuss its meaning.   Answers may vary.

Section 3.5 Introduction to Functions

Objectives

1. **Define and Identify Relations, Functions, Domain, and Range**
2. **Given an Equation, Determine Whether y Is a Function of x and Find the Domain**
3. **Use Function Notation**
4. **Find Function Values for Real-Number Values of a Variable**
5. **Evaluate a Function for Variables or Expressions**
6. **Define and Graph a Linear Function**
7. **Solve Problems Using Linear Functions**

If you are driving on a highway at a constant speed of 60 miles per hour, the distance you travel depends on the amount of time spent driving.

Driving Time	Distance Traveled
1 hr	60 mi
2 hr	120 mi
2.5 hr	150 mi
3 hr	180 mi

We can express these relationships with the ordered pairs

(1, 60)   (2, 120)   (2.5, 150)   (3, 180)

where the first coordinate represents the driving time (in hours), and the second coordinate represents the distance traveled (in miles).

We can also describe this relationship with the equation

$$y = 60x$$

where y is the distance traveled, in miles, and x is the number of hours spent driving.

The distance traveled *depends on* the amount of time spent driving. Therefore, the distance traveled is the **dependent variable,** and the driving time is the **independent variable.** In terms of x and y, since the value of y *depends on* the value of x, y is the *dependent variable* and x is the *independent variable.*

1. Define and Identify Relations, Functions, Domain, and Range

Relations and Functions

If we form a set of ordered pairs from the ones listed above

$$\{(1, 60), (2, 120), (2.5, 150), (3, 180)\}$$

we get a *relation*.

> **Definition**
>
> A **relation** is any set of ordered pairs.

> **Definition**
>
> The **domain** of a relation is the set of all values of the independent variable (the first coordinates in the set of ordered pairs). The **range** of a relation is the set of all values of the dependent variable (the second coordinates in the set of ordered pairs).

The domain of the given relation is {1, 2, 2.5, 3}. The range of the relation is {60, 120, 150, 180}.

The relation {(1, 60), (2, 120), (2.5, 150), (3, 180)} is also a *function* because every first coordinate corresponds to *exactly one* second coordinate. A function is a very important concept in mathematics.

> **Definition**
>
> A **function** is a special type of relation. If each element of the domain corresponds to *exactly one* element of the range, then the relation is a function.

Relations and functions can be represented in another way—as a *correspondence* or a *mapping* from one set, the domain, to another, the range.

In this representation, the domain is the set of all values in the first set, and the range is the set of all values in the second set. Our previously stated definition of a function still holds.

Example 1

In-Class Example 1
Identify the domain and range of each relation, and determine whether each relation is a function. a) {(3, 1), (−4, 2), (3, −7), (6, 4)}   b) {(4, −2), (11, 6), $\left(\frac{2}{3}, 4\right), \left(0, -\frac{3}{4}\right)$}

c)

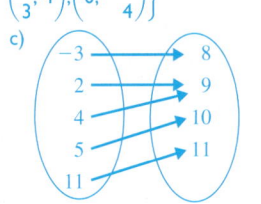

answer: a) domain: {3, −4, 6}; range: {1, 2, −7, 4}; not a function

b) domain: $\left\{4, 11, \frac{2}{3}, 0\right\}$;

range: $\left\{-2, 6, 4, -\frac{3}{4}\right\}$;

function c) domain: {−3, 2, 4, 5, 11}; range: {8, 9, 10, 11}; function

Identify the domain and range of each relation, and determine whether each relation is a function.

a)   {(2, 0), (3, 1), (6, 2), (6, −2)}

b) $\left\{ (-2, -6), (0, -5), \left(1, -\frac{9}{2}\right), (4, -3), \left(5, -\frac{5}{2}\right) \right\}$

c)

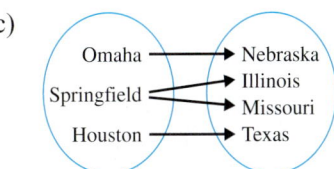

Solution

a)   The *domain* is the set of first coordinates, {2, 3, 6}. (We write the 6 in the set only once even though it appears in two ordered pairs.) The *range* is the set of second coordinates, {0, 1, 2, −2}.

To determine whether or not this relation is a function ask yourself, *"Does every first coordinate correspond to exactly one second coordinate?"* No. In the ordered pairs (6, 2) and (6, −2), the same first coordinate, 6, corresponds to two different second coordinates, 2 and −2. Therefore, this relation is *not* a function.

b)   The *domain* is {−2, 0, 1, 4, 5}. The *range* is $\left\{ -6, -5, -\frac{9}{2}, -3, -\frac{5}{2} \right\}$.

Ask yourself, "Does every first coordinate correspond to *exactly one* second coordinate?" *Yes.* This relation *is* a function.

c)   The *domain* is {Omaha, Springfield, Houston}. The *range* is {Nebraska, Illinois, Missouri, Texas}.

One of the elements in the domain, Springfield, corresponds to *two* elements in the range, Illinois and Missouri. Therefore, this relation is *not* a function.

You Try 1

Identify the domain and range of each relation, and determine whether each relation is a function.

a) $\{(-1, -3), (1, 1), (2, 3), (4, 7)\}$ b) $\{(-12, -6), (-12, 6), (-1, \sqrt{3}), (0, 0)\}$

c)

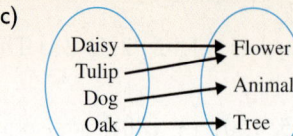

We stated earlier that a relation is a function if each element of the domain corresponds to *exactly one* element of the range.

If the ordered pairs of a relation are such that the first coordinates represent *x*-values and the second coordinates represent *y*-values (the ordered pairs are in the form (x, y)), then we can think of the definition of a function in this way:

Definition

A relation is a **function** if each *x*-value corresponds to exactly one *y*-value.

What does a function look like when it is graphed? Following are the graphs of the ordered pairs in the relations of Example 1a) and 1b).

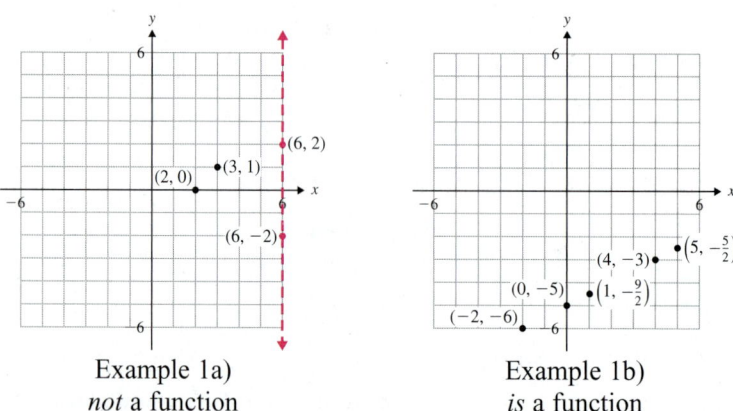

Example 1a)
not a function

Example 1b)
is a function

The relation in Example 1a) is *not* a function since the *x*-value of 6 corresponds to *two different y*-values, 2 and −2. Notice that we can draw a vertical line that intersects the graph in more than one point—the line through (6, 2) and (6, −2).

The relation in Example 1b), however, *is* a function—each *x*-value corresponds to only one *y*-value. Anywhere we draw a vertical line through the points on the graph of this relation, the line intersects the graph in *exactly one point*.

This leads us to the **vertical line test** for a function.

Procedure   The Vertical Line Test

If there is no vertical line that can be drawn through a graph so that it intersects the graph more than once, then the graph represents a function.

If a vertical line *can* be drawn through a graph so that it intersects the graph more than once, then the graph does *not* represent a function.

Example 2

Use the vertical line test to determine whether each graph, in blue, represents a function. Identify the domain and range.

In-Class Example 2
Use the vertical line test to determine whether each graph represents a function. Identify the domain and range.

a)
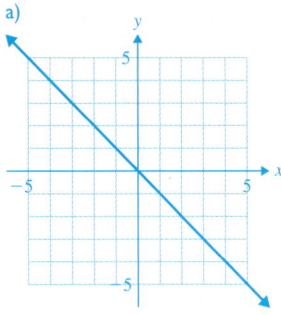

b)
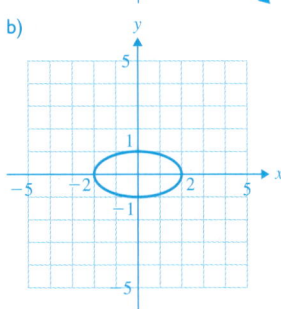

answers: a) function;
domain: $(-\infty, \infty)$; range:
$(-\infty, \infty)$ b) not a function;
domain: $[-2, 2]$; range:
$[-1, 1]$

a)
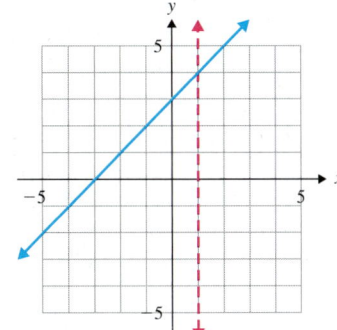

b)
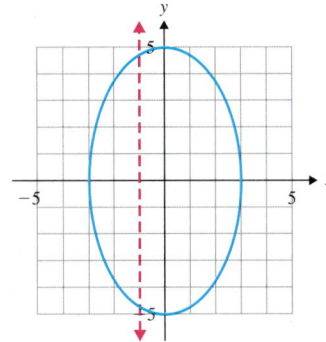

Solution

a) Anywhere a vertical line is drawn through the graph, the line will intersect the graph only once. *This graph represents a function.*

The arrows on the graph indicate that the graph continues without bound.

The domain of this function is the set of x-values on the graph. Since the graph continues indefinitely in the x-direction, the domain is the set of all real numbers. *The domain is $(-\infty, \infty)$.*

The range of this function is the set of y-values on the graph. Since the graph continues indefinitely in the y-direction, the range is the set of all real numbers. *The range is $(-\infty, \infty)$.*

b) This graph fails the vertical line test because we can draw a vertical line through the graph that intersects it more than once. *This graph does not represent a function.*

The set of x-values on the graph includes all real numbers from -3 to 3. *The domain is $[-3, 3]$.*

The set of y-values on the graph includes all real numbers from -5 to 5. *The range is $[-5, 5]$.* ■

You Try 2

Use the vertical line test to determine whether each relation is also a function. Then, identify the domain and range.

a)
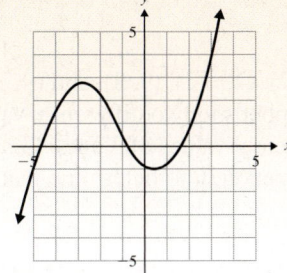

b)
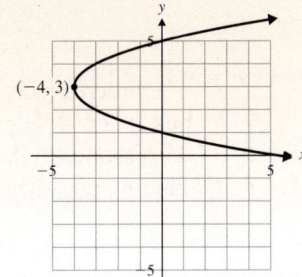

(−4, 3)

2. Given an Equation, Determine Whether y Is a Function of x and Find the Domain

We can also represent relations and functions with equations. The example given at the beginning of the section illustrates this.

The equation $y = 60x$ describes the distance traveled (y, in miles) after x hours of driving at 60 mph. If $x = 2$, $y = 60(2) = 120$. If $x = 3$, $y = 60(3) = 180$, and so on. For *every* value of x that could be substituted into $y = 60x$ there is *exactly one* corresponding value of y. Therefore, $y = 60x$ is a function.

Furthermore, we can say that *y is a function of x*. In the function described by $y = 60x$, the value of y *depends on* the value of x. That is, x is the independent variable and y is the dependent variable.

Property   *y* Is a Function of *x*

If a function describes the relationship between x and y so that x is the independent variable and y is the dependent variable, then we say that *y is a function of x*.

Example 3

In-Class Example 3
Determine whether each relation describes y as a function of x.
a) $y = 3x + 3$ b) $y^2 = 4x$
answer: a) function
b) not a function

Determine whether each relation describes y as a function of x.

a) $y = x + 2$ b) $y^2 = x$

Solution

a)   To begin, substitute a couple of values for x and solve for y to get an idea of what is happening in this relation.

$x = 0$	$x = 3$	$x = -4$
$y = x + 2$	$y = x + 2$	$y = x + 2$
$y = 0 + 2$	$y = 3 + 2$	$y = -4 + 2$
$y = 2$	$y = 5$	$y = -2$

The ordered pairs $(0, 2)$, $(3, 5)$, and $(-4, -2)$ satisfy $y = x + 2$. Each of the values substituted for x has *one* corresponding y-value. In this case, when *any* value is substituted for x there will be *exactly* one corresponding value of y. Therefore, $y = x + 2$ *is a function*.

b)   Substitute a couple of values for x and solve for y to get an idea of what is happening in this relation.

$x = 0$	$x = 4$	$x = 9$
$y^2 = x$	$y^2 = x$	$y^2 = x$
$y^2 = 0$	$y^2 = 4$	$y^2 = 9$
$y = 0$	$y = \pm 2$	$y = \pm 3$

The ordered pairs $(0, 0)$, $(4, 2)$, $(4, -2)$, $(9, 3)$, and $(9, -3)$ satisfy $y^2 = x$. Since $2^2 = 4$ and $(-2)^2 = 4$, $x = 4$ corresponds to two different y-values, 2 and -2. Likewise, $x = 9$ corresponds to the two different y-values of 3 and -3 since $3^2 = 9$ and $(-3)^2 = 9$. Finding one such example is enough to determine that $y^2 = x$ is *not* a function. ∎

You Try 3

Determine whether each relation describes y as a function of x.

a) $y = 3x - 5$ b) $y^2 = x + 1$

Next, we will discuss how to determine the domain of a relation written as an equation.

Sometimes, it is helpful to ask yourself, "Is there any number that *cannot* be substituted for x?"

Example 4

In-Class Example 4
Determine the domain of each relation, and determine whether each relation describes y as a function of x.
a) $y = 4x + 11$
b) $y = \dfrac{5}{12 - x}$ c) $y = \dfrac{4}{5x}$
answer: a) $(-\infty, \infty)$; **function**
b) $(-\infty, 12) \cup (12, \infty)$;
function
c) $(-\infty, 0) \cup (0, \infty)$; **function**

Determine the domain of each relation, and determine whether each relation describes y as a function of x.

a) $y = \dfrac{1}{x}$ b) $y = \dfrac{7}{x - 3}$ c) $y = -2x + 6$

Solution

a) To determine the domain of $y = \dfrac{1}{x}$ ask yourself, "Is there any number that *cannot* be substituted for x?" Yes. *x cannot equal zero because a fraction is undefined if its denominator equals zero.*

The domain contains all real numbers *except* 0. We can write the domain in interval notation as $(-\infty, 0) \cup (0, \infty)$.

$y = \dfrac{1}{x}$ *is a function* since each value of x in the domain will have only one corresponding value of y.

b) Ask yourself, "Is there any number that *cannot* be substituted for x in $y = \dfrac{7}{x - 3}$?"

Look at the denominator. When will it equal 0? Set the denominator equal to 0 and solve for x.

$$x - 3 = 0 \qquad \text{Set the denominator} = 0.$$
$$x = 3 \qquad \text{Solve.}$$

When $x = 3$, the denominator of $y = \dfrac{7}{x - 3}$ equals zero. The domain contains all real numbers *except* 3. Write the domain in interval notation as $(-\infty, 3) \cup (3, \infty)$.

$y = \dfrac{7}{x - 3}$ *is a function.* For every value that can be substituted for x there is only one corresponding value of y.

c) *Is there any number that cannot be substituted for x in $y = -2x + 6$? No.* Any real number can be substituted for x, and $y = -2x + 6$ will be defined.

The domain consists of all real numbers which can be written as $(-\infty, \infty)$.

Every value substituted for x will have exactly one corresponding y-value. $y = -2x + 6$ *is a function.* ∎

Procedure   Finding the Domain of a Relation

The domain of a relation that is written as an equation, where y is in terms of x, is the set of all real numbers that can be substituted for the independent variable, x. When determining the domain of a relation, it can be helpful to keep these tips in mind.

1) Ask yourself, "Is there any number that *cannot* be substituted for x?"

2) If x is in the denominator of a fraction, determine what value of x will make the denominator equal 0 by setting the expression equal to zero. Solve for x. This x-value is *not* in the domain.

The domain consists of all real numbers that can be substituted for x.

 You Try 4

Determine the domain of each relation, and determine whether each relation describes y as a function of x.

a) $y = x - 9$ b) $y = -x^2 + 6$ c) $y = \dfrac{4}{x + 1}$

3. Use Function Notation

We can use *function notation* to name functions. If a relation is a function, then $f(x)$ can be used in place of y. In this case, $f(x)$ *is the same as y.*

For example, $y = x + 3$ is a function. We can also write $y = x + 3$ as $f(x) = x + 3$. *They mean the same thing.*

Definition

$y = f(x)$ is called **function notation**, and it is read as "y equals f of x." $y = f(x)$ means that y is a function of x (y depends on x).

Example 5

In-Class Example 5
a) Evaluate $y = 4x - 1$ for $x = 3$.
b) If $f(x) = 4x - 1$, find $f(3)$.
answer: a) 11   b) 11

a)   Evaluate $y = x + 3$ for $x = 2$.   b)   If $f(x) = x + 3$, find $f(2)$.

Solution

a)   To evaluate $y = x + 3$ for $x = 2$ means to substitute 2 for x and find the corresponding value of y.

$$y = x + 3$$
$$y = 2 + 3 \quad \text{Substitute 2 for } x.$$
$$y = 5$$

When $x = 2$, $y = 5$. We can also say that the ordered pair $(2, 5)$ satisfies $y = x + 3$.

b)   To find $f(2)$ (read as "f of 2") means to find the value of the function when $x = 2$.

$$f(x) = x + 3$$
$$f(2) = 2 + 3 \quad \text{Substitute 2 for } x.$$
$$f(2) = 5$$

We can also say that the ordered pair $(2, 5)$ satisfies $f(x) = x + 3$ where the ordered pair represents $(x, f(x))$.

Note

Example 5 illustrates that evaluating $y = x + 3$ for $x = 2$ and finding $f(2)$ when $f(x) = x + 3$ is *exactly* the same thing. Remember, $f(x)$ is another name for y.

You Try 5

a)   Evaluate $y = -2x + 4$ for $x = 1$.   b)   If $f(x) = -2x + 4$, find $f(1)$.

Different letters can be used to name functions. $g(x)$ is read as "g of x," $h(x)$ is read as "h of x," and so on. Also, the function notation does *not* indicate multiplication; $f(x)$ does *not* mean f times x.

BE CAREFUL

$f(x)$ does *not* mean f times x.

4. Find Function Values for Real-Number Values of the Variable

Sometimes, we call evaluating a function for a certain value *finding a function value*.

 Example 6

Let $f(x) = 6x - 5$ and $g(x) = x^2 - 8x + 3$. Find the following function values.

a) $f(3)$ b) $f(0)$ c) $g(-1)$

In-Class Example 6
Let $f(x) = 2x + 10$ and $g(x) = x^2 + 3x - 11$. Find the following function values.
a) $f(2)$ b) $f(-6)$ c) $g(8)$
answer: a) **14**   b) **-2**   c) **77**

Solution

a) "Find $f(3)$" means to find the value of the function when $x = 3$. Substitute 3 for x.

$$f(x) = 6x - 5$$
$$f(3) = 6(3) - 5 = 18 - 5 = 13$$
$$f(3) = 13$$

We can also say that the ordered pair $(3, 13)$ satisfies $f(x) = 6x - 5$.

b) To find $f(0)$, substitute 0 for x in the function $f(x)$.

$$f(x) = 6x - 5$$
$$f(0) = 6(0) - 5 = 0 - 5 = -5$$
$$f(0) = -5$$

The ordered pair $(0, -5)$ satisfies $f(x) = 6x - 5$.

c) To find $g(-1)$, substitute -1 for every x in the function $g(x)$.

$$g(x) = x^2 - 8x + 3$$
$$g(-1) = (-1)^2 - 8(-1) + 3 = 1 + 8 + 3 = 12$$
$$g(-1) = 12$$

The ordered pair $(-1, 12)$ satisfies $g(x) = x^2 - 8x + 3$.

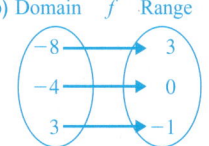 **You Try 6**

Let $f(x) = -4x + 1$ and $h(x) = 2x^2 + 3x - 7$. Find the following function values.

a) $f(5)$ b) $f(-2)$ c) $h(0)$ d) $h(3)$

We can also find function values for functions represented by a set of ordered pairs, a correspondence, or a graph.

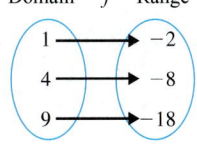 **Example 7**

Find $f(4)$ for each function.

a) $f = \{(-2, -11), (0, -5), (3, 4), (4, 7)\}$

In-Class Example 7
Find $f(3)$ for each function.
a) $f = \{(-7, -1), (-2, 3), (0, 8), (3, 12)\}$
b) Domain f Range

b) Domain f Range   c)

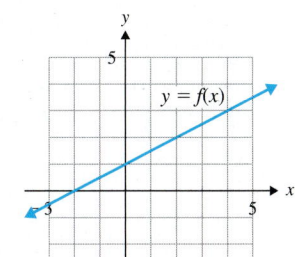

c)

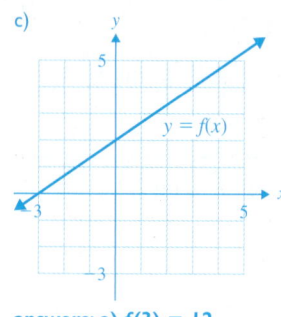

answers: a) $f(3) = 12$
b) $f(3) = -1$ c) $f(3) = 4$

Solution

a) Since this function is expressed as a set of ordered pairs, finding $f(4)$ means finding the y-coordinate of the ordered pair with x-coordinate 4. The ordered pair with x-coordinate 4 is $(4, 7)$, so $f(4) = 7$.

b) In this function, the element 4 in the domain corresponds to the element -8 in the range. Therefore, $f(4) = -8$.

c) To find $f(4)$ from the graph of this function means to find the y-coordinate of the point on the line that has an x-coordinate of 4. Find 4 on the x-axis. Then, go straight up to the graph and move to the left to read the y-coordinate of the point on the graph where the x-coordinate is 4. That y-coordinate is 3. So, $f(4) = 3$. ∎

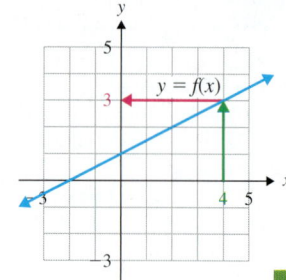

 You Try 7

Find $f(2)$ for each function.

a) $f = \{(-5, 8), (-1, 2), (2, -3), (6, -9)\}$

b) Domain f Range   c)

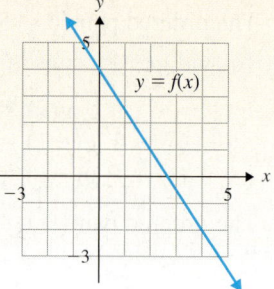

---

5. Evaluate a Function for Variables or Expressions

Functions can be evaluated for variables or expressions.

Example 8

Let $h(x) = 5x + 3$. Find each of the following and simplify.

a) $h(c)$ b) $h(t - 4)$

In-Class Example 8
Let $h(x) = -3x + 1$. Find each
of the following and simplify.
a) $h(m)$ b) $h(w + 5)$
answer: a) $h(m) = -3m + 1$
b) $h(w + 5) = -3w - 14$

Solution

a) Finding $h(c)$ (read as *h of c*) means to substitute c for x in the function h, and simplify the expression as much as possible.

$$h(x) - 5x + 3$$
$$h(c) = 5c + 3 \qquad \text{Substitute } c \text{ for } x.$$

b) Finding $h(t - 4)$ (read as *h of t minus 4*) means to substitute $t - 4$ for x in function h, and simplify the expression as much as possible. *Since $t - 4$ contains two terms, we must put it in parentheses.*

$$h(x) = 5x + 3$$
$$h(t - 4) = 5(t - 4) + 3 \qquad \text{Substitute } t - 4 \text{ for } x.$$
$$h(t - 4) = 5t - 20 + 3 \qquad \text{Distribute.}$$
$$h(t - 4) = 5t - 17 \qquad \text{Combine like terms.}$$

∎

You Try 8

Let $f(x) = 2x - 7$. Find each of the following and simplify.

a) $f(k)$ b) $f(p + 3)$

6. Define and Graph a Linear Function

Earlier in this chapter, we learned that a linear equation can have the form $y = mx + b$. Similarly, a *linear function* has the form $f(x) = mx + b$.

> **Definition**
>
> A **linear function** has the form $f(x) = mx + b$, where m and b are real numbers, m is the *slope* of the line, and $(0, b)$ is the *y-intercept*.

Example 9

Graph $f(x) = -\dfrac{1}{3}x - 1$ using the slope and y-intercept.

In-Class Example 9

Graph $f(x) = \dfrac{1}{3}x - 2$ using the slope and y-intercept.

answer:

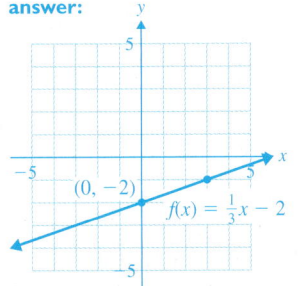

Solution

$$f(x) = -\frac{1}{3}x - 1$$

$$m = -\frac{1}{3} \qquad y\text{-int: } (0, -1)$$

To graph this function, first plot the y-intercept, $(0, -1)$, then use the slope to locate another point on the line.

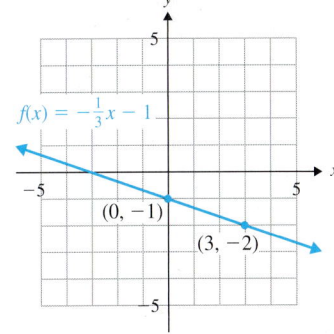

You Try 9

Graph $f(x) = \dfrac{3}{4}x - 2$ using the slope and y-intercept.

7. Solve Problems Using Linear Functions

The independent variable of a function does not have to be x. When using functions to model real-life problems, we often choose a more "meaningful" letter to represent a quantity. For example, if the independent variable represents time, we may use the letter t instead of x. The same is true for naming the function.

No matter what letter is chosen for the independent variable, *the horizontal axis is used to represent the values of the independent variable, and the vertical axis represents the function values.*

Example 10

In-Class Example 10
See Example 10. Change the numbers in a)–c) to the following:
a) 30 sec   b) 2 min
c) 3,087,000 samples of sound
answer: a) 1,323,000
b) 5,292,000   c) 70 sec

A compact disc is read at 44.1 kHz (kilohertz). This means that a CD player scans 44,100 samples of sound per second on a CD to produce the sound that we hear. The function

$$S(t) = 44.1t$$

tells us how many samples of sound, $S(t)$, in *thousands* of samples, are read after t seconds. (www.mediatechnics.com)

a)   How many samples of sound are read after 20 sec?

b)   How many samples of sound are read after 1.5 min?

c)   How long would it take the CD player to scan 1,764,000 samples of sound?

d)   What is the smallest value t could equal in the context of this problem?

e)   Graph the function.

Solution

a)   To determine how much sound is read after 20 sec, let $t = 20$ and find $S(20)$.

$$S(t) = 44.1t$$
$$S(20) = 44.1(20) \qquad \text{Substitute 20 for } t.$$
$$S(20) = 882 \qquad \text{Multiply.}$$

$S(t)$ is in thousands, so the number of samples read is $882 \cdot 1000 = 882{,}000$ samples of sound.

b)   To determine how much sound is read after 1.5 min, do we let $t = 1.5$ and find $S(1.5)$? *No.* Recall that t is in *seconds*. Change 1.5 min to seconds before substituting for t. We must use the correct units in the function.

$$1.5 \text{ min} = 90 \text{ sec}$$

Let $t = 90$ and find $S(90)$.

$$S(t) = 44.1t$$
$$S(90) = 44.1(90)$$
$$S(90) = 3969$$

$S(t)$ is in thousands, so the number of samples read is $3969 \cdot 1000 = 3{,}969{,}000$ samples of sound.

c)   Since we are asked to determine how *long* it would take a CD player to scan 1,764,000 samples of sound, we will be solving for t. What do we substitute for $S(t)$? $S(t)$ is in *thousands*, so substitute $1{,}764{,}000 \div 1000 = 1764$ for $S(t)$. Find t when $S(t) = 1764$.

$$S(t) = 44.1t$$
$$1764 = 44.1t \qquad \text{Substitute 1764 for } S(t).$$
$$40 = t \qquad \text{Divide by 44.1.}$$

It will take 40 sec for the CD player to scan 1,764,000 samples of sound.

d)   Since t represents the number of seconds a CD has been playing, the smallest value that makes sense for t is 0.

e)   Since $S(t)$ is in thousands of samples, the information we obtained in parts a), b), and c) can be written as the ordered pairs (20, 882), (90, 3969), and (40, 1764). In addition, when $t = 0$ (from part d) we obtain $S(0) = 44.1(0) = 0$. (0, 0) is an additional ordered pair on the graph of the function.

Number of Samples of Sound
Scanned by a CD Player in t sec

Answers to You Try Exercises

1) a) domain: $\{-1, 1, 2, 4\}$; range: $\{-3, 1, 3, 7\}$; yes b) domain: $\{-12, -1, 0\}$; range: $\{-6, 6, \sqrt{3}, 0\}$; no
c) domain: {Daisy, Tulip, Dog, Oak}; range: {Flower, Animal, Tree}; yes   2) a) function; domain: $(-\infty, \infty)$;
range: $(-\infty, \infty)$ b) not a function; domain: $[-4, \infty)$; range: $(-\infty, \infty)$ 3) a) yes b) no
4) a) $(-\infty, \infty)$; function b) $(-\infty, \infty)$; function c) $(-\infty, -1) \cup (-1, \infty)$; function
5) a) 2 b) 2   6) a) -19 b) 9 c) -7 d) 20   7) a) -3 b) 5 c) 1

8) a) $f(k) = 2k - 7$ b) $f(p + 3) = 2p - 1$ 9) $m = \dfrac{3}{4}$, y-int: $(0, -2)$

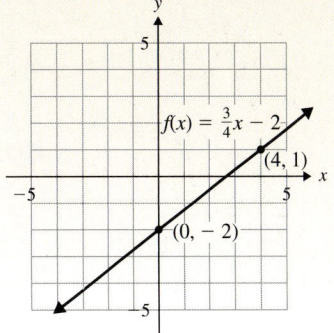

3.5 Exercises

*Additional answers can be found in the Answers to Exercises appendix.

Objective 1: Define and Identify Relations, Functions, Domain, and Range

1) a) What is a relation?   any set of ordered pairs

   b) What is a function?   Answers may vary.

   c) Give an example of a relation that is also a function.
      Answers may vary.

2) Give an example of a relation that is *not* a function.
   Answers may vary.

Identify the domain and range of each relation, and determine whether each relation is a function.

3) $\{(5, 13), (-2, 6), (1, 4), (-8, -3)\}$

4) $\{(0, -3), (1, -4), (1, -2), (16, -5), (16, -1)\}$

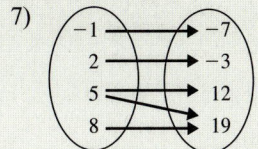 5) $\{(9, -1), (25, -3), (1, 1), (9, 5), (25, 7)\}$

6) $\left\{(-4, -2), \left(-3, -\dfrac{1}{2}\right), \left(-1, -\dfrac{1}{2}\right), (0, -2)\right\}$

7)
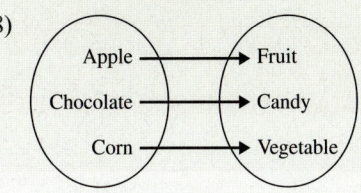
domain: $\{-1, 2, 5, 8\}$;
range: $\{-7, -3, 12, 19\}$;
not a function

8)

Apple ⟶ Fruit
Chocolate ⟶ Candy
Corn ⟶ Vegetable

domain: {Apple, Chocolate, Corn};
range: {Fruit, Candy, Vegetable}; function

9)
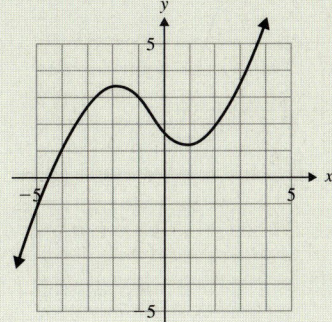
domain: $(-\infty, \infty)$;
range: $(-\infty, \infty)$;
function

10)
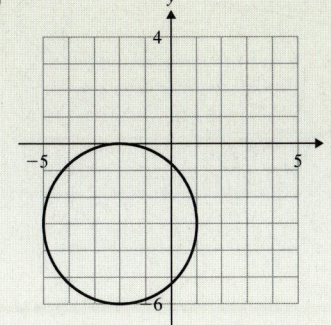
domain: $[-5, 1]$;
range: $[-6, 0]$;
not a function

11)

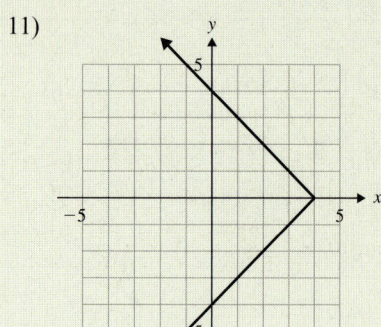

domain: $(-\infty, 4]$;
range: $(-\infty, \infty)$;
not a function

12)

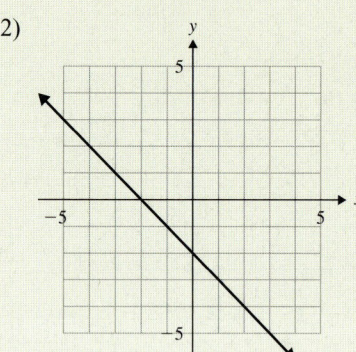

domain: $(-\infty, \infty)$;
range: $(-\infty, \infty)$;
function

13)

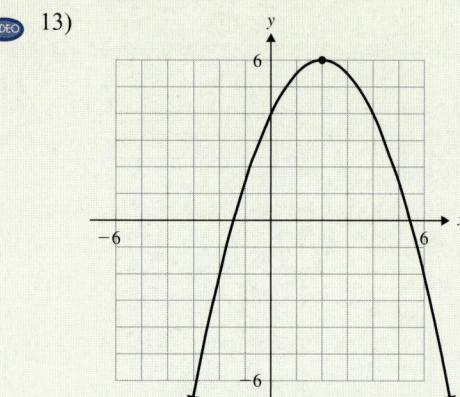

domain: $(-\infty, \infty)$;
range: $(-\infty, 6]$;
function

14)

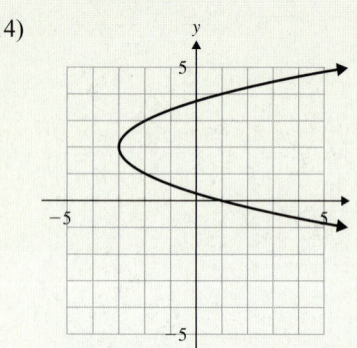

domain: $[-3, \infty)$;
range: $(-\infty, \infty)$;
not a function

Objective 2: Given an Equation, Determine Whether y Is a Function of x and Find the Domain

Determine whether each relation describes y as a function of x.

15) $y = x - 9$ yes   16) $y = x + 4$ yes

17) $y = 2x + 7$ yes   18) $y = \dfrac{2}{3}x + 1$ yes

19) $x = y^4$ no   20) $x = y^2 - 3$ no

21) $y^2 = x - 4$ no   22) $y^2 = x + 9$ no

Determine the domain of each relation, and determine whether each relation describes y as a function of x.

23) $y = x - 5$
$(-\infty, \infty)$; function

24) $y = 2x + 1$
$(-\infty, \infty)$; function

25) $y = x^3 + 2$
$(-\infty, \infty)$; function

26) $y = -x^3 + 4$
$(-\infty, \infty)$; function

27) $x = y^4$
$[0, \infty)$; not a function

28) $x = |y|$
$[0, \infty)$; not a function

29) $y = -\dfrac{8}{x}$
$(-\infty, 0) \cup (0, \infty)$; function

30) $y = \dfrac{5}{x}$
$(-\infty, 0) \cup (0, \infty)$; function

31) $y = \dfrac{9}{x + 4}$
$(-\infty, -4) \cup (-4, \infty)$; function

32) $y = \dfrac{2}{x - 7}$
$(-\infty, 7) \cup (7, \infty)$; function

33) $y = \dfrac{3}{x - 5}$
$(-\infty, 5) \cup (5, \infty)$; function

34) $y = \dfrac{1}{x + 10}$
$(-\infty, -10) \cup (-10, \infty)$; function

35) $y = \dfrac{6}{5x - 3}$

36) $y = -\dfrac{4}{9x + 8}$

37) $y = \dfrac{15}{3x + 4}$

38) $y = \dfrac{5}{6x - 1}$

39) $y = -\dfrac{5}{9 - 3x}$
$(-\infty, 3) \cup (3, \infty)$; function

40) $y = \dfrac{1}{-6 + 4x}$

41) $y = \dfrac{x}{12}$
$(-\infty, \infty)$; function

42) $y = \dfrac{x + 8}{7}$
$(-\infty, \infty)$; function

Mixed Exercises: Objectives 3 and 4

43) Explain what it means when an equation is written in the form $y = f(x)$. y is a function, and y is a function of x.

44) Does $y = f(x)$ mean "$y = f$ times x"? Explain.
No; $f(x)$ is read as "f of x" and $y = f(x)$ means that y is a function of x.

45) a) Evaluate $y = 5x - 8$ for $x = 3$. $y = 7$

    b) If $f(x) = 5x - 8$, find $f(3)$. $f(3) = 7$

46) a) Evaluate $y = -3x - 2$ for $x = -4$. $y = 10$

    b) If $f(x) = -3x - 2$, find $f(-4)$. $f(-4) = 10$

Let $f(x) = -4x + 7$ and $g(x) = x^2 + 9x - 2$. Find the following function values.

47) $f(5)$ -13 48) $f(2)$ -1

49) $f(0)$ 7 50) $f\left(-\dfrac{3}{2}\right)$ 13

51) $g(4)$ 50 52) $g(1)$ 8

53) $g(-1)$ -10 54) $g(0)$ -2

55) $g\left(-\dfrac{1}{2}\right)$ $-\dfrac{25}{4}$ 56) $g\left(\dfrac{1}{3}\right)$ $\dfrac{10}{9}$

57) $f(6) - g(6)$ -105 58) $f(-4) - g(-4)$ 45

For each function f in Exercises 59–64, find $f(-1)$ and $f(4)$.

59) $f = \{(-3, 16), (-1, 10), (0, 7), (1, 4), (4, -5)\}$
$f(-1) = 10, f(4) = -5$

60) $f = \left\{(-8, -1), \left(-1, \dfrac{5}{2}\right), (4, 5), (10, 8)\right\}$
$f(-1) = \dfrac{5}{2}, f(4) = 5$

 61)

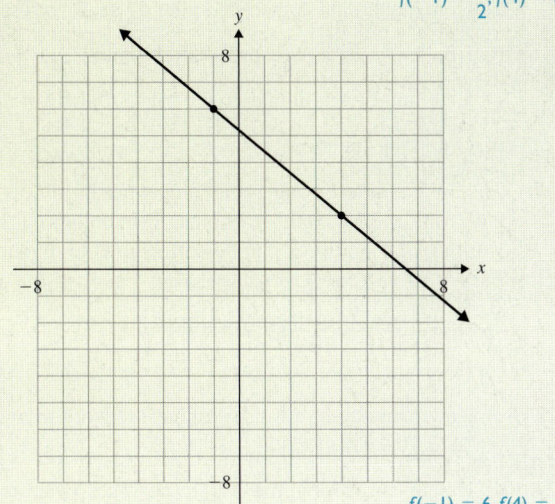

$f(-1) = 6, f(4) = 2$

62)

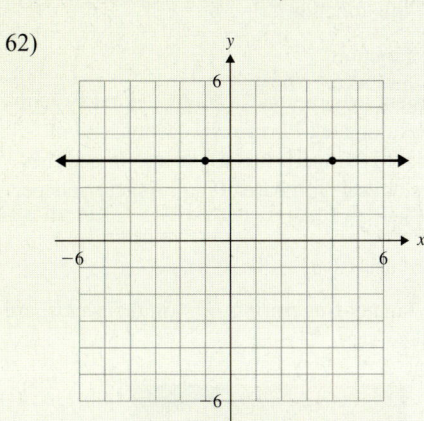

$f(-1) = 3, f(4) = 3$

63)   Domain f Range   64)   Domain f Range

$\begin{array}{l}-3 \rightarrow 10\\ -1 \rightarrow 7\\ 4 \rightarrow 3\\ 9 \rightarrow -1\end{array}$ $\begin{array}{l}-1 \rightarrow -8\\ 0 \rightarrow -5\\ 3 \rightarrow -1\\ 4 \rightarrow 6\end{array}$

$f(-1) = 7, f(4) = 3$ $f(-1) = -8, f(4) = 6$

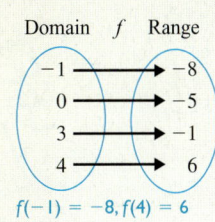 65) $f(x) = -3x - 2$. Find x so that $f(x) = 10$. -4

66) $f(x) = 5x + 4$. Find x so that $f(x) = 9$. 1

67) $g(x) = \dfrac{2}{3}x + 1$. Find x so that $g(x) = 5$. 6

68) $h(x) = -\dfrac{1}{2}x - 6$. Find x so that $h(x) = -2$. -8

Objective 5: Evaluate a Function for Variables or Expressions

Fill It In

Fill in the blanks with either the missing mathematical step or reason for the given step.

69) Let $f(x) = 4x - 5$. Find $f(k + 6)$.

$f(k + 6) = 4(k + 6) - 5$ Substitute $k + 6$ for x.
$= 4k + 24 - 5$ Distribute.
$= 4k + 19$ Simplify.

70) Let $f(x) = -9x + 2$. Find $f(n - 3)$.

$f(n - 3) = -9(n - 3) + 2$ Substitute $n - 3$ for x.
$= -9n + 27 + 2$ Distribute.
$= -9n + 29$ Simplify.

71) $f(x) = -7x + 2$ and $g(x) = x^2 - 5x + 12$. Find each of the following and simplify.

a) $f(c)$ $f(c) = -7c + 2$ b) $f(t)$ $f(t) = -7t + 2$

c) $f(a + 4)$ d) $f(z - 9)$
$f(a + 4) = -7a - 26$ $f(z - 9) = -7z + 65$

e) $g(k)$ $g(k) = k^2 - 5k + 12$ f) $g(m)$ $g(m) = m^2 - 5m + 12$

g) $f(x + h)$ $-7x - 7h + 2$ h) $f(x + h) - f(x)$ $-7h$

72) $f(x) = 5x + 6$ and $g(x) = x^2 - 3x - 11$. Find each of the following and simplify.

a) $f(n)$ $f(n) = 5n + 6$ b) $f(p)$ $f(p) = 5p + 6$

c) $f(w + 8)$ d) $f(r - 7)$
$f(w + 8) = 5w + 46$ $f(r - 7) = 5r - 29$

e) $g(b)$ $g(b) = b^2 - 3b - 11$ f) $g(s)$ $g(s) = s^2 - 3s - 11$

g) $f(x + h)$ $5x + 5h + 6$ h) $f(x + h) - f(x)$ $5h$

Objective 6: Define and Graph a Linear Function

Graph each function by making a table of values and plotting points.

73) $f(x) = x - 4$ 74) $f(x) = x + 2$

75) $f(x) = \dfrac{2}{3}x + 2$ 76) $g(x) = -\dfrac{3}{5}x + 2$

77) $h(x) = -3$ 78) $g(x) = 1$

Graph each function by finding the x- and y-intercepts and one other point.

79) $g(x) = 3x + 3$ 80) $k(x) = -2x + 6$

81) $f(x) = -\dfrac{1}{2}x + 2$ 82) $f(x) = \dfrac{1}{3}x + 1$

83) $h(x) = x$ 84) $f(x) = -x$

Graph each function using the slope and y-intercept.

85) $f(x) = -4x - 1$ 86) $f(x) = -x + 5$

87) $h(x) = \dfrac{3}{5}x - 2$ 88) $g(x) = -\dfrac{1}{4}x - 2$

89) $g(x) = 2x + \dfrac{1}{2}$ 90) $h(x) = 3x + 1$

Graph each function

91) $s(t) = -\dfrac{1}{3}t - 2$ 92) $k(d) = d - 1$

93) $A(r) = -3r$ 94) $N(t) = 3.5t + 1$

Objective 7: Solve Problems Using Linear Functions

95) A truck on the highway travels at a constant speed of 54 mph. The distance, D (in miles), that the truck travels after t hr can be defined by the function

$$D(t) = 54t$$

a) How far will the truck travel after 2 hr?   108 mi

b) How far will the truck travel after 4 hr?   216 mi

c) How long does it take the truck to travel 135 mi?   2.5 hr

d) Graph the function.

96) The velocity of an object, v (in feet per second), of an object during free-fall t sec after being dropped can be defined by the function

$$v(t) = 32t$$

a) Find the velocity of an object 1 sec after being dropped.
32 ft/sec

b) Find the velocity of an object 3 sec after being dropped.
96 ft/sec

c) When will the object be traveling at 256 ft/sec?
after 8 sec

d) Graph the function.

97) Jenelle earns $7.50 per hour at her part-time job. Her total earnings, E (in dollars), for working t hr can be defined by the function

$$E(t) = 7.50t$$

a) Find $E(10)$, and explain what this means in the context of the problem.
$E(10) = 75$; when Jenelle works for 10 hr, she earns $75.00.

b) Find $E(15)$, and explain what this means in the context of the problem.
$E(15) = 112.5$; when Jenelle works 15 hr, she earns $112.50.

c) Find t so that $E(t) = 210$, and explain what this means in the context of the problem.
$t = 28$; for Jenelle to earn $210.00, she must work 28 hr.

98) If gasoline costs $2.50 per gallon, then the cost, C (in dollars), of filling a gas tank with g gal of gas is defined by

$$C(g) = 2.50g$$

a) Find $C(8)$, and explain what this means in the context of the problem. $C(8) = 20$; 8 gal of gas cost $20.00.

b) Find $C(15)$, and explain what this means in the context of the problem. $C(15) = 37.5$; 15 gal of gas cost $37.50.

c) Find g so that $C(g) = 30$, and explain what this means in the context of the problem.
$g = 12$; 12 gal of gas can be purchased for $30.00.

99) A 16× DVD recorder can transfer 21.13 MB (megabytes) of data per second onto a recordable DVD. The function $D(t) = 21.13t$ describes how much data, D (in megabytes), is recorded on a DVD in t sec. (www.osta.org)

a) How much data is recorded after 12 sec?   253.56 MB

b) How much data is recorded after 1 min?   1267.80 MB

c) How long would it take to record 422.6 MB of data?
20 sec

d) Graph the function.

100) The median hourly wage of an embalmer in Illinois in 2002 was $17.82. Seth's earnings, E (in dollars), for working t hr in a week can be defined by the function $E(t) = 17.82t$. (www.igpa.uillinois.edu)

a) How much does Seth earn if he works 30 hr? $534.60

b) How much does Seth earn if he works 27 hr? $481.14

c) How many hours would Seth have to work to make $623.70?   35 hr

d) If Seth can work at most 40 hr per week, what is the domain of this function?   [0, 40]

e) Graph the function.

101) Law enforcement agencies use a computerized system called AFIS (Automated Fingerprint Identification System) to identify fingerprints found at crime scenes. One AFIS system can compare 30,000 fingerprints per second. The function

$$F(s) = 30s$$

describes how many fingerprints, $F(s)$ in thousands, are compared after s sec.

a) How many fingerprints can be compared in 2 sec?
60,000

b) How long would it take AFIS to search through 105,000 fingerprints?   3.5 sec

102) Refer to the function in Exercise 101 to answer the following questions.

a) How many fingerprints can be compared in 3 sec?
90,000

b) How long would it take AFIS to search through 45,000 fingerprints?   1.5 sec

103) Refer to the function in Example 10 on p. 178 to determine the following.

a) Find $S(50)$, and explain what this means in the context of the problem.

b) Find $S(180)$, and explain what this means in the context of the problem.

c) Find t so that $S(t) = 2646$, and explain what this means in the context of the problem.

104) Refer to the function in Exercise 99 to determine the following.

a) Find $D(10)$, and explain what this means in the context of the problem.

b) Find $D(120)$, and explain what this means in the context of the problem.

c) Find t so that $D(t) = 633.9$, and explain what this means in the context of the problem.

105) The graph shows the amount, A, of ibuprofen in Sasha's bloodstream t hr after she takes two tablets for a headache.

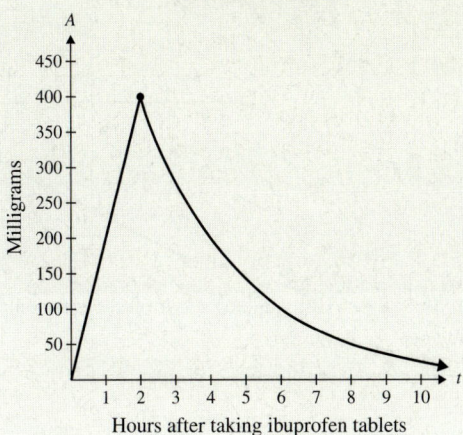

Amount of Ibuprofen in
Sasha's Bloodstream

a) How long after taking the tablets will the amount of ibuprofen in her bloodstream be the greatest? How much ibuprofen is in her bloodstream at this time?

2 hr; 400 mg

b) When will there be 100 mg of ibuprofen in Sasha's bloodstream?   after about 30 min and after 6 hr

c) How much of the drug is in her bloodstream after 4 hr?   200 mg

d) Call this function A. Find $A(8)$, and explain what it means in the context of the problem.

$A(8) = 50$. After 8 hr there are 50 mg of ibuprofen in Sasha's bloodstream.

106) The graph shows the number of gallons (in millions), G, of water entering a water treatment plant t hr after midnight on a certain day.

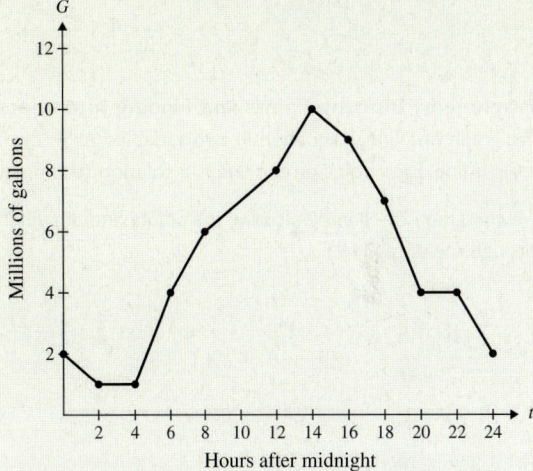

Amount of Water Entering a
Water Treatment Plant

a) Identify the domain and range of this function.

domain: [0, 24]; range: [1,000,000, 10,000,000]

b) How many gallons of water enter the facility at noon? At 10 P.M.?   8 million gallons; 4 million gallons

c) At what time did the most water enter the treatment plant? How much water entered the treatment plant at this time?   2 P.M.; 10 million gallons

d) At what time did the least amount of water enter the treatment plant?   from 2 to 4 A.M.

e) Call this function G, find $G(18)$, and explain what it means in the context of the problem.

$G(18) = 7$. At 6 P.M., 7 gal of water entered the treatment plant.

Definition/Procedure	Example

3.1 Introduction to Linear Equations in Two Variables

A **linear equation in two variables** can be written in the form $Ax + By = C$ where A, B, and C are real numbers and where both A and B do not equal zero.

To determine if an ordered pair is a solution of an equation, substitute the values for the variables. **(p. 119)**

Is $(5, -3)$ a solution of $2x - 7y = 31$?
Substitute 5 for x and -3 for y.

$$2x - 7y = 31$$
$$2(5) - 7(-3) = 31$$
$$10 - (-21) = 31$$
$$10 + 21 = 31$$
$$31 = 31$$

Yes, $(5, -3)$ is a solution.

Graphing by Plotting Points and Finding Intercepts
The graph of a linear equation in two variables, $Ax + By = C$, is a straight line. Each point on the line is a solution to the equation.

We can graph the line by plotting the points and drawing the line through them. **(p. 119)**

Graph $y = \frac{1}{2}x - 4$ by plotting points.

Make a table of values. Plot the points, and draw a line through them.

x	y
0	−4
2	−3
4	−2

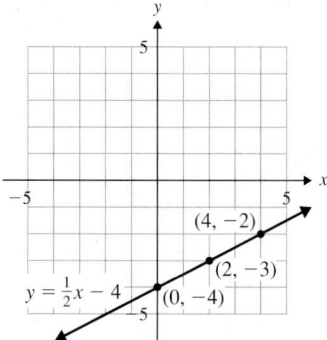

The **x-intercept** of an equation is the point where the graph intersects the x-axis. To find the *x-intercept* of the graph of an equation, let $y = 0$ and solve for x.

The **y-intercept** of an equation is the point where the graph intersects the y-axis. To find the *y-intercept* of the graph of an equation, let $x = 0$ and solve for y. **(p. 120)**

Graph $5x + 2y = 10$ by finding the intercepts and another point on the line.

x-intercept: Let $y = 0$, and solve for x.

$$5x + 2(0) = 10$$
$$5x = 10$$
$$x = 2$$

The *x-intercept* is $(2, 0)$.

y-intercept: Let $x = 0$, and solve for y.

$$5x + 2y = 10$$
$$5(0) + 2y = 10$$
$$2y = 10$$
$$y = 5$$

The *y-intercept* is $(0, 5)$.
Another point on the line is $(4, -5)$.
Plot the points, and draw the line through them.

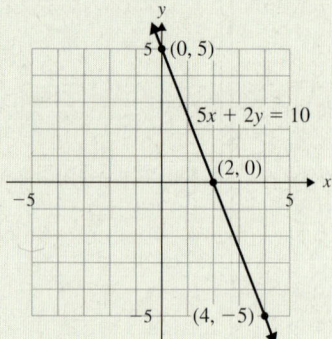

Definition/Procedure	Example

If c is a constant, then the graph of $x = c$ is a *vertical line* going through the point $(c, 0)$.

If d is a constant, then the graph of $y = d$ is a *horizontal line* going through the point $(0, d)$. **(p. 121)**

Graph $x = 2$.   Graph $y = 3$.

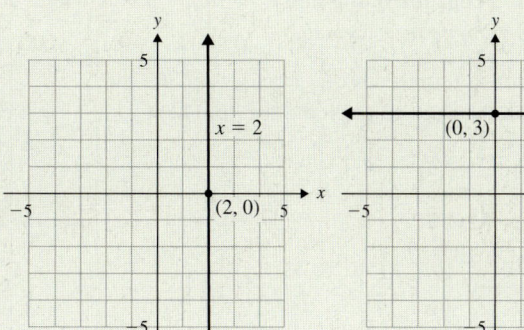

3.2 Slope of a Line and Slope-Intercept Form

The **slope** of a line is the ratio of the vertical change in y to the horizontal change in x. Slope is denoted by m.

The slope of a line containing the points (x_1, y_1) and (x_2, y_2) is

$$m = \frac{y_2 - y_1}{x_2 - x_1}$$

The slope of a horizontal line is zero.
The slope of a vertical line is undefined. **(p. 129)**

Find the slope of the line containing the points $(6, 9)$ and $(-2, 12)$.

$$m = \frac{y_2 - y_1}{x_2 - x_1}$$
$$= \frac{12 - 9}{-2 - 6} = \frac{3}{-8} = -\frac{3}{8}$$

The slope of the line is $-\dfrac{3}{8}$.

If we know the slope of a line and a point on the line, we can graph the line. **(p. 134)**

Graph the line containing the point $(-5, -3)$ with a slope of $\dfrac{4}{7}$.

Start with the point $(-5, -3)$, and use the slope to plot another point on the line.

$$m = \frac{4}{7} = \frac{\text{change in } y}{\text{change in } x}$$

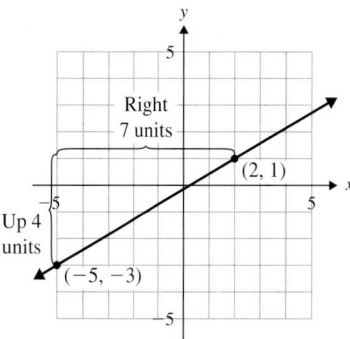

Definition/Procedure	Example

The Slope-Intercept Form of a Line
The **slope-intercept form of a line** is $y = mx + b$, where m is the slope and $(0, b)$ is the y-intercept.

If a line is written in slope-intercept form, we can use the y-intercept and the slope to graph the line. **(p. 136)**

Write the equation in slope-intercept form and graph it.

$$6x + 4y = 16$$
$$4y = -6x + 16$$
$$y = -\frac{6}{4}x + \frac{16}{4}$$
$$y = -\frac{3}{2}x + 4 \qquad \text{Slope-intercept form}$$

$m = -\frac{3}{2}$, y-intercept $(0, 4)$

Plot $(0, 4)$, then use the slope to locate another point on the line.

We will think of the slope as $m = \dfrac{-3}{2} = \dfrac{\text{change in } y}{\text{change in } x}$.

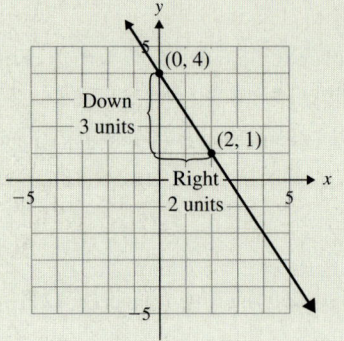

3.3 Writing an Equation of a Line

To write the equation of a line given its slope and y-intercept, use $y = mx + b$ and substitute those values into the equation. **(p. 143)**

Find an equation of the line with slope $= 3$ and y-intercept $(0, -8)$.

$$y = mx + b \qquad \text{Substitute 3 for}$$
$$y = 3x - 8 \qquad m \text{ and } -8 \text{ for } b.$$

Point-Slope Form of a Line: The **point-slope form of a line** is $y - y_1 = m(x - x_1)$, where (x_1, y_1) is a point on the line and m is its slope.

Given the slope of the line and a point on the line, we can use the point-slope formula to find an equation of the line. **(p. 144)**

Find an equation of the line containing the point $(1, -4)$ with slope $= 2$. Express the answer in standard form.

Use $y - y_1 = m(x - x_1)$.

Substitute 2 for m. Substitute $(1, -4)$ for (x_1, y_1).

$$y - (-4) = 2(x - 1)$$
$$y + 4 = 2x - 2$$
$$-2x + y = -6$$
$$2x - y = 6 \qquad \text{Standard form}$$

To write an equation of a line given two points on the line,

a) use the points to find the slope of the line

then

b) use the slope and *either* one of the points in the point-slope formula. **(p. 145)**

Find an equation of the line containing the points $(-2, 6)$ and $(4, 2)$. Express the answer in slope-intercept form.

$$m = \frac{2 - 6}{4 - (-2)} = \frac{-4}{6} = -\frac{2}{3}$$

We will use $m = -\dfrac{2}{3}$ and the point $(-2, 6)$ in the point-slope formula.

$$y - y = m(x - x_1)$$

Definition/Procedure	Example

Substitute $-\dfrac{2}{3}$ for m. Substitute $(-2, 6)$ for (x_1, y_1).

$$y - 6 = -\frac{2}{3}[x - (-2)] \qquad \text{Substitute.}$$

$$y - 6 = -\frac{2}{3}(x + 2)$$

$$y - 6 = -\frac{2}{3}x - \frac{4}{3} \qquad \text{Distribute.}$$

$$y = -\frac{2}{3}x + \frac{14}{3} \qquad \text{Slope-intercept form}$$

The equation of a *horizontal line* containing the point (c, d) is $y = d$.

The equation of a *vertical line* containing the point (c, d) is $x = c$. **(p. 146)**

The equation of a horizontal line containing the point $(7, -4)$ is $y = -4$.
The equation of a vertical line containing the point $(9, 1)$ is $x = 9$.

Parallel and Perpendicular Lines
Parallel lines have the same slope.

Perpendicular lines have slopes that are negative reciprocals of each other. **(p. 147)**

Determine if the lines $5x + y = 3$ and $x - 5y = 20$ are parallel, perpendicular, or neither.

Put each line into slope-intercept form to find their slopes.

$$\begin{array}{c|c}
5x + y = 3 & x - 5y = 20 \\
y = -5x + 3 & -5y = -x + 20 \\
& y = \dfrac{1}{5}x - 4 \\[2mm]
m = -5 & m = \dfrac{1}{5}
\end{array}$$

The lines are *perpendicular* since their slopes are negative reciprocals of each other.

To write an equation of the line parallel or perpendicular to a given line, we must find the slope of the given line first. **(p. 149)**

Write an equation of the line parallel to $2x - 3y = 21$ containing the point $(-6, -3)$. Express the answer in slope-intercept form.

First, find the slope of $2x - 3y = 21$.

$$2x - 3y = 21$$
$$-3y = -2x + 21$$
$$y = \frac{2}{3}x - 7$$
$$m = \frac{2}{3}$$

The slope of the parallel line is also $\dfrac{2}{3}$.

Since this line contains $(-6, -3)$, use the point-slope formula to write its equation.

$$y - y_1 = m(x - x_1)$$

$$y - (-3) = \frac{2}{3}[x - (-6)] \qquad \text{Substitute values.}$$

$$y + 3 = \frac{2}{3}(x + 6)$$

$$y + 3 = \frac{2}{3}x + 4 \qquad \text{Distribute.}$$

$$y = \frac{2}{3}x + 1 \qquad \text{Slope-intercept form}$$

Definition/Procedure	**Example**

3.4 Linear and Compound Linear Inequalities in Two Variables

A **linear inequality in two variables** is an inequality that can be written in the form $Ax + By \geq C$ or $Ax + By \leq C$, where A, B, and C are real numbers and where A and B are not both zero. ($>$ and $<$ may be substituted for $\geq$ and $\leq$.) **(p. 157)**

Some examples of linear inequalities in two variables are

$$x + 3y \leq 2, \quad y > -\frac{2}{3}x + 5, \quad y \geq -1, \quad x < 4$$

Graph a Linear Inequality in Two Variables Using a Test Point

1) *Graph the boundary line.*
  a) If the inequality contains $\geq$ or $\leq$, make the boundary line *solid*.
  b) If the inequality contains $>$ or $<$, make the boundary line *dotted*.
2) *Choose a test point not on the line, and shade the appropriate region.* Substitute the test point into the inequality. If $(0, 0)$ is not on the line, it is an easy point to test in the inequality.
  a) If it *makes the inequality true*, shade the region *containing* the test point. All points in the shaded region are part of the solution set.
  b) If the test point *does not satisfy the inequality*, shade the region on the *other* side of the line. All points in the shaded region are part of the solution set. **(p. 158)**

Graph $2x + y > -3$.

1) Graph the boundary line as a *dotted* line.
2) Choose a test point not on the line and substitute it into the inequality to determine whether it makes the inequality true.

Test Point	Substitute into $2x + y > -3$
$(0, 0)$	$2(0) + (0) > -3$
	$0 > -3$ True

Since the test point satisfies the inequality, shade the region containing $(0, 0)$.

All points in the shaded region satisfy $2x + y > -3$.

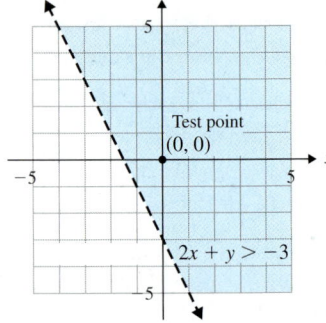

Graph a Linear Inequality in Two Variables Using the Slope-Intercept Method

1) Write the inequality in the form $y \geq mx + b$ ($y > mx + b$) or $y \leq mx + b$ ($y < mx + b$), and graph the boundary line $y = mx + b$.

Graph $-x + 3y \leq 6$ using the slope-intercept method.

Write the inequality in slope-intercept form by solving $-x + 3y \leq 6$ for y.

$$-x + 3y \leq 6$$
$$3y \leq x + 6$$
$$y \leq \frac{1}{3}x + 2$$

2) If the inequality is in the form $y \geq mx + b$ or $y > mx + b$, shade *above* the line.
3) If the inequality is in the form $y \leq mx + b$ or $y < mx + b$, shade *below* the line. **(p. 160)**

Graph $y = \frac{1}{3}x + 2$ as a *solid line.*

Since $y \leq \frac{1}{3}x + 2$

has a $\leq$ symbol, shade *below* the line.

All points on the line and in the shaded region satisfy $-x + 3y \leq 6$.

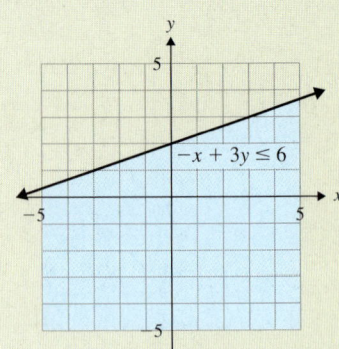

Definition/Procedure	Example

Graphing Compound Linear Inequalities in Two Variables

1) Graph each inequality separately on the same axes. Shade lightly.
2) If the inequality contains *and*, the solution set is the *intersection* of the shaded regions. (Heavily shade this region.)
3) If the inequality contains *or*, the solution set is the *union* (total) of the shaded regions. Heavily shade this region. **(p. 161)**

Graph the compound inequality $y \geq -4x + 3$ and $y \geq 1$.

Since the inequality contains *and*, the solution set is the *intersection* of the shaded regions.

Any point in the shaded area will satisfy *both* inequalities.

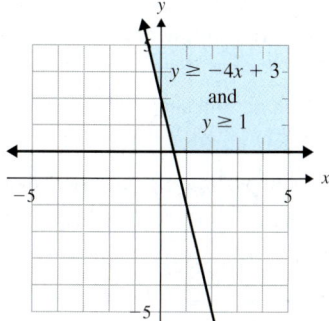

3.5 Introduction to Functions

A **relation** is any set of ordered pairs. A relation can also be represented as a correspondence or mapping from one set to another. **(p. 168)**

Relations:

a) $\{(-4, -12), (-1, -3), (3, 9), (5, 15)\}$

b)
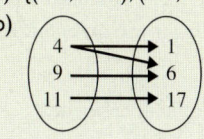

The **domain** of a relation is the set of values of the independent variable (the first coordinates in the set of ordered pairs).

The **range** of a relation is the set of all values of the dependent variable (the second coordinates in the set of ordered pairs). **(p. 169)**

In a) above, the domain is $\{-4, -1, 3, 5\}$, and the range is $\{-12, -3, 9, 15\}$.

In b) above, the domain is $\{4, 9, 11\}$, and the range is $\{1, 6, 17\}$.

A **function** is a relation in which each element of the domain corresponds to *exactly one* element of the range. **(p. 169)**

The relation above in a) *is a function.*

The relation above in b) *is not a function.*

The Vertical Line Test **(p. 170)**

This graph represents a function. Anywhere a vertical line is drawn, it will intersect the graph only once.

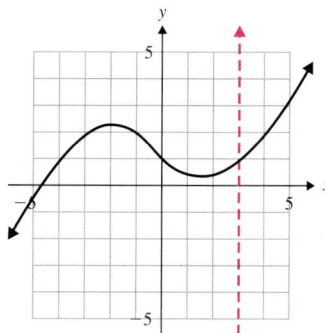

This is *not* the graph of a function. A vertical line can be drawn so that it intersects the graph more than once.

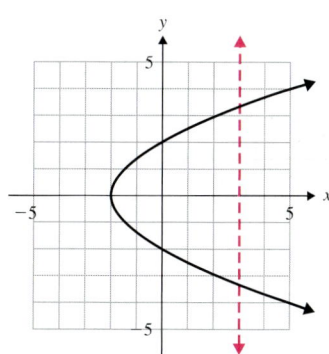

Definition/Procedure	**Example**

The **domain** of a relation that is written as an equation, where y is in terms of x, is the set of all real numbers that can be substituted for the independent variable, x.

When determining the domain of a relation, it can be helpful to keep these tips in mind.

1) Ask yourself, "Is there any number that *cannot* be substituted for x?"

2) If x is in the denominator of a fraction, determine what value of x will make the denominator equal 0 by setting the denominator equal to zero. Solve for x. This x-value is *not* in the domain. **(p. 173)**

Determine the domain of $f(x) = \dfrac{9}{x + 8}$.

$x + 8 = 0$ Set the denominator $= 0$.
$\quad x = -8$ Solve.

When $x = -8$, the denominator of $f(x) = \dfrac{9}{x + 8}$ equals zero.

The domain contains all real numbers *except* -8.

The domain of the function is $(-\infty, -8) \cup (-8, \infty)$.

Function Notation

If a function describes the relationship between x and y so that x is the independent variable and y is the dependent variable, then y is a function of x. $y = f(x)$ is called **function notation** and it is read as "y equals f of x."

Finding a function value means evaluating the function for the given value of the variable. **(p. 174)**

If $f(x) = 9x - 4$, find $f(2)$.

Substitute 2 for x and evaluate.

$f(2) = 9(2) - 4 = 18 - 4 = 14$
$f(2) = 14$

A **linear function** has the form

$$f(x) = mx + b$$

where m and b are real numbers, m is the *slope* of the line, and $(0, b)$ is the *y-intercept*. **(p. 177)**

Graph $f(x) = -3x + 4$ using the slope and y-intercept.

The slope is -3 and the y-intercept is $(0, 4)$. Plot the y-intercept and use the slope to locate another point on the line.

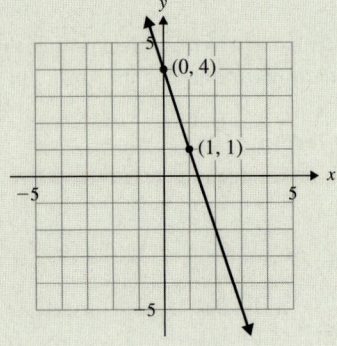

*Additional answers can be found in the Answers to Exercises appendix.

(3.1) Determine if each ordered pair is a solution of the given equation.

1) $4x - y = 9$; $(1, -5)$ yes   2) $3x + 2y = 20$; $(-4, 2)$ no

3) $y = \dfrac{5}{4}x + \dfrac{1}{2}$; $(2, 3)$ yes   4) $x = 7$; $(7, -9)$ yes

Complete the ordered pair for each equation.

5) $y = -6x + 10$; $(-3, \ )$ 6) $y = \dfrac{2}{3}x + 5$; $(12, \ )$
   28   13

7) $y = -8$; $(5, \ )$ -8 8) $5x - 9y = 3$; $(\ , -2)$ -3

Complete the table of values for each equation.

9) $y = x - 11$

x	y
0	-11
3	-8
-1	-12
-5	-16

10) $4x - 6y = 8$

x	y
2	0
0	$-\frac{4}{3}$
3	$\frac{2}{3}$
-4	-4

Plot the ordered pairs on the same coordinate system.

11) a) $(5, 2)$ b) $(-3, 0)$

    c) $(-4, 3)$ d) $(6, -2)$

12) a) $(0, 1)$ b) $(-2, -5)$

    c) $\left(\dfrac{5}{2}, 1\right)$ d) $\left(4, -\dfrac{1}{3}\right)$

13) The fine for an overdue book at the Hinsdale Public Library is given by

$$y = 0.10x$$

where x represents the number of days a book is overdue and y represents the amount of the fine, in dollars.

a) Complete the table of values, and write the information as ordered pairs.

x	y
1	0.10
2	0.20
7	0.70
10	1.00

b) Label a coordinate system, choose an appropriate scale, and graph the ordered pairs.

c) Explain the meaning of the ordered pair $(14, 1.40)$ in the context of the problem.

14) Fill in the blank with positive, negative, or zero.

a) The y-coordinate of every point in quadrant III is _____. negative

b) The x-coordinate of every point in quadrant IV is _____. positive

Complete the table of values and graph each equation.

15) $y = -2x + 3$

x	y
0	3
1	1
2	-1
-2	7

16) $3x + 2y = 4$

x	y
0	2
-2	5
1	$\frac{1}{2}$
4	-4

Graph each equation by finding the intercepts and at least one other point.

17) $x - 2y = 6$ 18) $5x + y = 10$

19) $y = -\dfrac{1}{6}x + 4$ 20) $y = \dfrac{3}{4}x - 7$

21) $x = 5$ 22) $y = -3$

(3.2) Determine the slope of each line.

23)

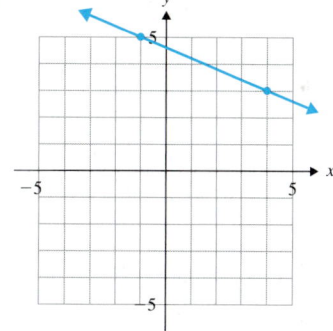

$-\dfrac{2}{5}$

24)

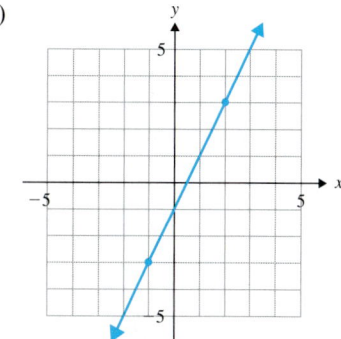

2

Use the slope formula to find the slope of the line containing each pair of points.

25) $(1, 7)$ and $(-4, 2)$ 1

26) $(-2, -3)$ and $(3, -1)$ $\frac{2}{5}$

27) $(-2, 5)$ and $(3, -8)$ $-\frac{13}{5}$

28) $(0, 4)$ and $(8, -2)$ $-\frac{3}{4}$

29) $\left(\frac{3}{2}, -1\right)$ and $\left(-\frac{5}{2}, 7\right)$ -2

30) $(2.5, 5.3)$ and $(-3.5, -1.9)$ 1.2

31) $(9, 0)$ and $(9, 4)$ undefined

32) $(-7, 4)$ and $(1, 4)$ 0

33) Paul purchased some shares of stock in 2004. The graph shows the value of one share of the stock from 2004–2008.

Value of One Share of Stock

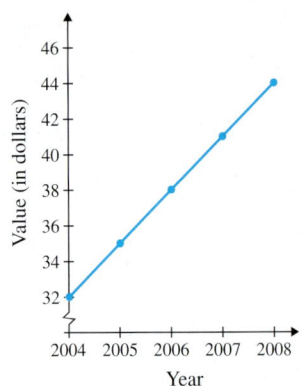

a) What was the value of one share of stock the year that Paul made his purchase?
In 2004, one share of the stock was worth $32.

b) Is the slope of the line segment positive or negative? What does the sign of the slope mean in the context of this problem?
The slope is positive, so the value of the stock is increasing over time.

c) Find the slope. What does it mean in the context of this problem?
$m = 3$; the value of one share of stock is increasing by $3.00 per year.

Graph the line containing the given point and with the given slope.

34) $(-3, -2)$; $m = 4$

35) $(1, 5)$; $m = -3$

36) $(-2, 6)$; $m = -\frac{5}{2}$

37) $(-3, 2)$; slope undefined

38) $(5, 2)$; $m = 0$

Identify the slope and y-intercept, then graph the line.

39) $y = x - 3$

40) $y = -2x + 7$

41) $y = -\frac{3}{4}x + 1$

42) $y = \frac{1}{4}x - 2$

43) $x - 3y = -6$

44) $2x - 7y = 35$

45) $x + y = 0$

46) $y + 3 = 4$

47) Personal consumption expenditures in the United States since 1998 can be modeled by $y = 371.5x + 5920.1$, where x represents the number of years after 1998, and y represents the personal consumption expenditure in billions of dollars. (Bureau of Economic Analysis)

Personal Consumption Expenditures

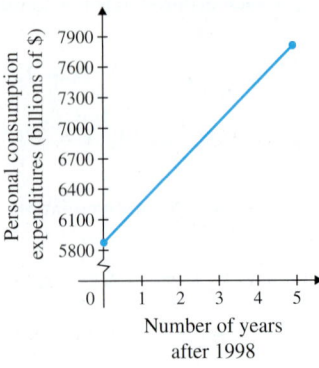

a) What is the y-intercept? What does it mean in the context of the problem? $(0, 5920.1)$; in 1998 the amount of money spent for personal consumption was $5920.1 billion.

b) Has the personal consumption expenditure been increasing or decreasing since 1998? By how much per year? It has been increasing by $371.5 billion per year.

c) Use the graph to estimate the personal consumption expenditure in the year 2002. Then, use the equation to determine this number. estimate from the graph: $7400 billion; number from the equation: $7406.1 billion

(3.3)

48) Write the point-slope formula for the equation of a line with slope m and which contains the point (x_1, y_1).
$y - y_1 = m(x - x_1)$

Write the *slope-intercept form* of the equation of the line, if possible, given the following information.

49) $m = 7$ and contains $(2, 5)$ $y = 7x - 9$

50) $m = -8$ and y-intercept $(0, -1)$ $y = -8x - 1$

51) $m = -\frac{4}{9}$ and y-intercept $(0, 2)$ $y = -\frac{4}{9}x + 2$

52) contains $(-6, -5)$ and $(4, 10)$ $y = \frac{3}{2}x + 4$

53) contains $(3, -6)$ and $(-9, -2)$ $y = -\frac{1}{3}x - 5$

54) $m = \frac{1}{2}$ and contains $(8, -3)$ $y = \frac{1}{2}x - 7$

55) horizontal line containing $(1, 9)$ $y = 9$

56) vertical line containing $(4, 0)$ $x = 4$

Write the *standard form* of the equation of the line given the following information.

57) contains $(-2, 2)$ and $(8, 7)$ $x - 2y = -6$

58) $m = -1$ and contains $(4, -7)$ $x + y = -3$

59) $m = -3$ and contains $\left(\frac{4}{3}, 1\right)$ $3x + y = 5$

60) contains $(15, -2)$ and $(-5, -10)$ $2x - 5y = 40$

61) $m = 6$ and y-intercept $(0, 0)$ $6x - y = 0$

62) $m = -\frac{5}{3}$ and y-intercept $(0, 2)$ $5x + 3y = 6$

63) contains $(1, 1)$ and $(-7, -5)$ $3x - 4y = -1$

64) $m = \dfrac{1}{6}$ and contains $(17, 2)$ $x - 6y = 5$

65) The chart shows the number of wireless communication subscribers worldwide (in millions) from 2001–2004. (Dell'Oro Group and Standard and Poor's Industry Surveys)

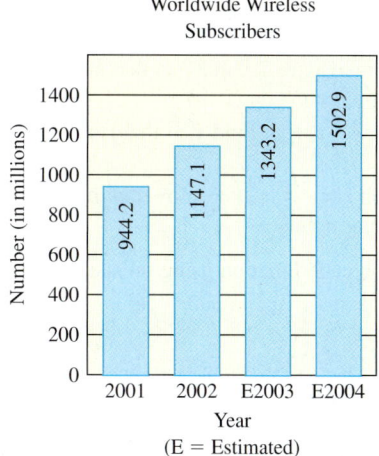

Worldwide Wireless Subscribers

944.2   1147.1   1343.2   1502.9

Number (in millions)

2001   2002   E2003   E2004
Year
(E = Estimated)

a) Write a linear equation to model this data. Use the data points for 2001 and 2004. Let x represent the number of years after 2001, and let y represent the number of worldwide wireless subscribers, in millions. Round to the nearest tenth. $y = 186.2x + 944.2$

b) Explain the meaning of the slope in the context of the problem.   The number of worldwide wireless subscribers is increasing by 186.2 million per year.

c) Use the equation to determine the number of subscribers in 2003. How does it compare to the value given on the chart?   1316.6 million; this is slightly less than the number given on the chart.

Determine if each pair of lines is parallel, perpendicular or neither.

66) $y = -\dfrac{1}{2}x + 7$
    $5x + 10y = 8$ parallel

67) $9x - 4y = -1$
    $-27x + 12y = 2$ parallel

68) $4x - 6y = -3$
    $-3x + 2y = -2$ neither

69) $y = 6$
    $x = -2$ perpendicular

70) $x = 3$
    $x = 1$ parallel

71) $y = 6x - 7$
    $4x + y = 9$ neither

72) $x + 2y = 22$
    $2x - y = 0$ perpendicular

Write an equation of the line *parallel* to the given line and containing the given point. Write the answer in slope-intercept form or in standard form, as indicated.

73) $y = 5x + 14$; $(-2, -4)$; slope-intercept form $y = 5x + 6$

74) $y = -3x + 1$; $(5, -19)$; slope-intercept form $y = -3x - 4$

75) $x - 4y = 9$; $(5, 3)$; standard form $x - 4y = -7$

76) $5x - 3y = 7$; $(4, 8)$; slope-intercept form $y = \dfrac{5}{3}x + \dfrac{4}{3}$

Write an equation of the line *perpendicular* to the given line and containing the given point. Write the answer in slope-intercept form or in standard form, as indicated.

77) $y = -\dfrac{1}{2}x + 9$; $(6, 5)$; slope-intercept form $y = 2x - 7$

78) $y = -x + 11$; $(-10, -8)$; slope-intercept form $y = x + 2$

79) $2x - 11y = 11$; $(2, -7)$; slope-intercept form $y = -\dfrac{11}{2}x + 4$

80) $-2x + 3y = 15$; $(-5, 7)$; standard form $3x + 2y = -1$

81) Write an equation of the line parallel to $x = 7$ containing $(2, 3)$. $x = 2$

82) Write an equation of the line parallel to $y = -5$ containing $(-4, 9)$. $y = 9$

83) Write an equation of the line perpendicular to $y = 6$ containing $(-1, -3)$. $x = -1$

84) Write an equation of the line perpendicular to $x = 7$ containing $(6, 0)$. $y = 0$

(3.4) Graph each linear inequality in two variables.

85) $y \le -2x + 7$

86) $y \ge -\dfrac{3}{2}x + 2$

87) $-3x + 4y > 12$

88) $5x - 2y \ge 8$

89) $y < x$

90) $x \ge 4$

Graph each compound inequality.

91) $y \ge \dfrac{3}{4}x - 4$ and $y \le -5$

92) $y < \dfrac{5}{4}x - 5$ or $y < -3$

93) $4x - y < -1$ or $y > \dfrac{1}{2}x + 5$

94) $2x + 5y \le 10$ and $y \ge \dfrac{1}{3}x + 4$

95) $4x + 2y \ge -6$ and $y \le 2$

96) $2x + y \le 3$ or $6x + y > 4$

(3.5) Identify the domain and range of each relation, and determine whether each relation is a function.

97) $\{(-3, 1), (5, 3), (5, -3), (12, 4)\}$

98)

$\begin{array}{c} 2 \\ -6 \\ 5 \end{array} \rightarrow \begin{array}{c} 0 \\ 1 \\ 8 \\ 13 \end{array}$

domain: $\{-6, 2, 5\}$;
range: $\{0, 1, 8, 13\}$;
not a function

99)

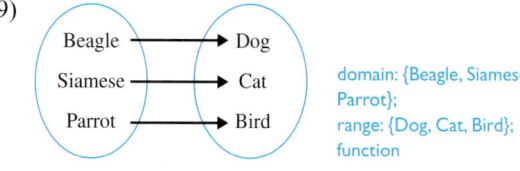

Beagle → Dog
Siamese → Cat
Parrot → Bird

domain: {Beagle, Siamese, Parrot};
range: {Dog, Cat, Bird};
function

100)

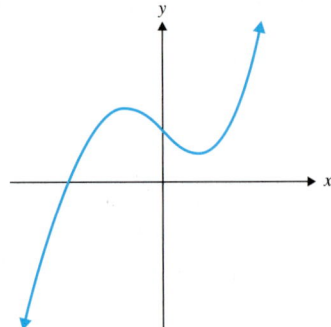

domain: $(-\infty, -\infty)$;
range: $(-\infty, \infty)$;
function

101)

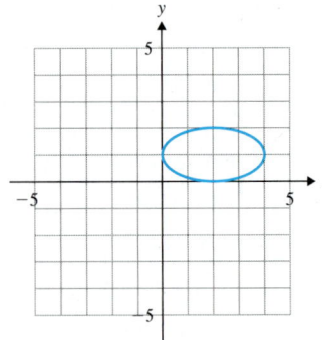

domain: $[0, 4]$;
range: $[0, 2]$;
not a function

Determine the domain of each relation, and determine whether each relation describes y as a function of x.

102) $y = 4x - 7$

$(-\infty, \infty)$; function

103) $y = \dfrac{8}{x + 3}$

$(-\infty, -3) \cup (-3, \infty)$; function

104) $y = \dfrac{15}{x}$

$(-\infty, 0) \cup (0, \infty)$; function

105) $y^2 = x$

$[0, \infty)$; not a function

106) $y = x^2 - 6$

$(-\infty, \infty)$; function

107) $y = \dfrac{5}{7x - 2}$

For each function, f, find $f(3)$ and $f(-2)$.

108) $f = \{(-7, -2), (-2, -5), (1, -10), (3, -14)\}$

$f(3) = -14, f(-2) = -5$

109)

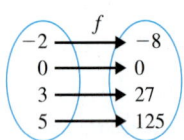

$f(3) = 27, f(-2) = -8$

110)

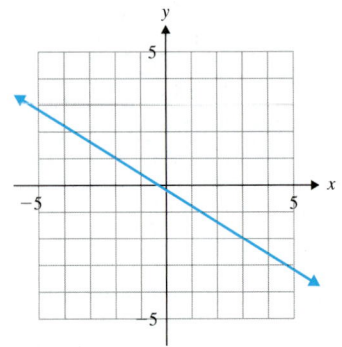

$f(3) = -2, f(-2) = 1$

111) Let $f(x) = 5x - 12$, $g(x) = x^2 + 6x + 5$. Find each of the following and simplify.

a) $f(4)$ 8

b) $f(-3)$ -27

c) $g(3)$ 32

d) $g(0)$ 5

e) $f(a)$ $5a - 12$

f) $g(t)$ $t^2 + 6t + 5$

g) $f(k + 8)$ $5k + 28$

h) $f(c - 2)$ $5c - 22$

i) $f(x + h)$ $5x + 5h - 12$

j) $f(x + h) - f(x)$ $5h$

112) $h(x) = -3x + 7$. Find x so that $h(x) = 19$. -4

113) $f(x) = \dfrac{3}{2}x + 5$. Find x so that $f(x) = \dfrac{11}{2}$. $\dfrac{1}{3}$

114) Graph $f(x) = -2x + 6$ by making a table of values and plotting points.

115) Graph each function using the slope and y-intercept.

a) $f(x) = \dfrac{2}{3}x - 1$

b) $f(x) = -3x + 2$

116) Graph $g(x) = \dfrac{3}{2}x + 3$ by finding the x- and y-intercepts and one other point.

Graph each function.

117) $h(c) = -\dfrac{5}{2}c + 4$

118) $D(t) = 3t$

119) A USB 2.0 device can transfer data at a rate of 480 MB/sec (megabytes/second). Let $f(t) = 480t$ represent the number of megabytes of data that can be transferred in t sec. (www.usb.org)

a) How many megabytes of a file can be transferred in 2 sec? in 6 sec?   960 MB; 2880 MB

b) How long would it take to transfer a 1200 MB file?

2.5 sec

120) A jet travels at a constant speed of 420 mph. The distance D (in miles) that the jet travels after t hr can be defined by the function

$$D(t) = 420t$$

a) Find $D(2)$, and explain what this means in the context of the problem. $D(2) = 840$; after 2 hr, the jet has traveled 840 mi.

b) Find t so that $D(t) = 2100$, and explain what this means in the context of the problem.

$t = 5$; in 5 hr the jet can travel 2100 mi.

Additional answers can be found in the Answers to Exercises appendix.

1) Is $(9, -13)$ a solution of $5x + 3y = 6$?   yes

2) Complete the table of values and graph $y = -2x + 4$.

x	y
0	4
3	-2
-1	6
2	0

3) Fill in the blanks with *positive* or *negative*. In quadrant II, the x-coordinate of every point is _____ and the y-coordinate is _____.   negative; positive

4) For $2x - 3y = 12$,

   a) find the x-intercept. $(6, 0)$

   b) find the y-intercept. $(0, -4)$

   c) find one other point on the line.   Answers may vary.

   d) graph the line.

5) Graph $x = -4$.

6) Find the slope of the line containing the points

   a) $(-8, -5)$ and $(4, -14)$ $-\dfrac{3}{4}$

   b) $(9, 2)$ and $(3, 2)$ 0

7) Graph the line containing the point $(-3, 6)$ with slope $= -\dfrac{2}{5}$.

8) Graph the line containing the point $(4, 1)$ with an undefined slope.

9) Put $6x - 2y = 8$ into slope-intercept form. Then, graph the line.

10) Write the slope-intercept form for the equation of the line with slope -4 and y-intercept $(0, 5)$. $y = -4x + 5$

11) Write the standard form for the equation of a line with slope $\dfrac{1}{2}$ containing the point $(3, 8)$. $x - 2y = -13$

12) Determine if $4x - 7y = -7$ and $14x + 8y = 3$ are parallel, perpendicular, or neither.   perpendicular

13) Find the slope-intercept form of the equation of the line

   a) perpendicular to $y = -3x + 11$ containing $(12, -5)$. $y = \dfrac{1}{3}x - 9$

   b) parallel to $5x - 2y = 2$ containing $(8, 14)$. $y = \dfrac{5}{2}x - 6$

14) Mr. Kumar owns a computer repair business. The graph shows the annual profit since 2004. Let x represent the number of years after 2004, and let y represent the annual profit, in thousands.

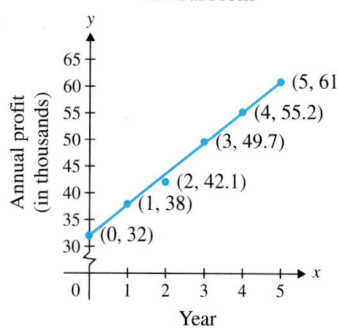

Annual Profit

a) What was the profit in 2008? $55,200

b) Write a linear equation (in slope-intercept form) to model this data. Use the data points for 2004 and 2009. $y = 5.8x + 32$

c) What is the slope of the line? What does it mean in the context of the problem?
   $m = 5.8$; the profit is increasing by $5.8 thousand or $5800 per year.

d) What is the y-intercept? What does it mean in the context of the problem?
   y-int: $(0, 32)$; the profit in 2004 was $32,000.

e) If the profit continues to follow this trend, in what year can Mr. Kumar expect a profit of $90,000?   2014

Graph each inequality.

15) $y \geq 3x + 1$

16) $2x - 5y > 10$

Graph the compound inequality.

17) $-2x + 3y \geq -12$ and $x \leq 3$

18) $y < -x$ or $2x - y > 1$

Identify the domain and range of each relation, and determine whether each relation is a function.

19) $\{(-2, -5), (1, -1), (3, 1), (8, 4)\}$ domain: $\{-2, 1, 3, 8\}$; range: $\{-5, -1, 1, 4\}$; function

20)

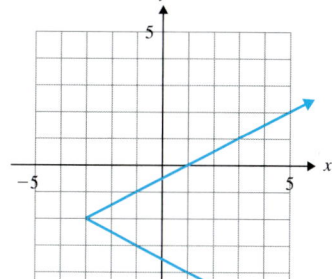

domain: $[-3, \infty)$; range: $(-\infty, \infty)$; not a function

For each function, (a) determine the domain. (b) Is y a function of x?

21) $y = \dfrac{7}{3}x - 5$
   a) $(-\infty, \infty)$ b) yes

22) $y = \dfrac{8}{2x - 5}$
   a) $\left(-\infty, \dfrac{5}{2}\right) \cup \left(\dfrac{5}{2}, \infty\right)$ b) yes

For each function, f, find f(2).

23) $f = \{(-3, -8), (0, -5), (2, -3), (7, 2)\}$ −3

24)

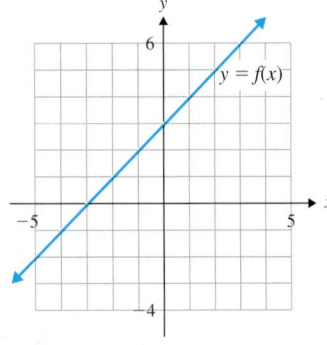

5

Let $f(x) = -4x + 2$ and $g(x) = x^2 - 3x + 7$. Find each of the following and simplify.

25) $f(6)$ −22

26) $g(2)$ 5

27) $g(t)$ $t^2 - 3t + 7$

28) $f(h - 7)$ −4h + 30

Graph the function.

29) $h(x) = -\dfrac{3}{4}x + 5$

30) A USB 1.1 device can transfer data at a rate of 12 MB/sec (megabytes/second). Let $f(t) = 12t$ represent the number of megabytes of data that can be transferred in t sec. (www.usb.org)

   a) How many megabytes of a file can be transferred in 3 sec?   36 MB

   b) How long would it take to transfer 132 MB?   11 sec

*Additional answers can be found in the Answers to Exercises appendix.

1) Write $\dfrac{252}{840}$ in lowest terms. $\dfrac{3}{10}$

2) A rectangular picture frame measures 8 in. by 10.5 in. Find the perimeter of the frame.   37 in.

Evaluate.

3) -2^6 -64

4) $\dfrac{21}{40} \cdot \dfrac{25}{63}$ $\dfrac{5}{24}$

5) $3 - \dfrac{2}{5}$ $\dfrac{13}{5}$

6) Write an expression for "53 less than twice eleven" and simplify. $2(11) - 53;\ -31$

Solve each equation

7) $12 - 5(2n + 9) = 3n + 2(n + 6)$ $\{-3\}$

8) $\dfrac{1}{3}p + \dfrac{1}{4} = \dfrac{2}{3}p - \dfrac{1}{6}(2p + 5)$ $\varnothing$

9) Solve for w: $t + zw = r$ $w = \dfrac{r - t}{z}$

10) $|7y + 2| = 16$ $\left\{-\dfrac{18}{7}, 2\right\}$

Solve. Write the solution in interval notation.

11) $19 - 4x \geq 25$ $\left(-\infty, -\dfrac{3}{2}\right]$

12) $3c - 10 < 2$ and $\dfrac{5}{2}c + 4 > -1$ $(-2, 4)$

For 13–15, write an equation and solve.

13) One serving of Ben and Jerry's Chocolate Chip Cookie Dough Ice Cream has 10% fewer calories than one serving of their Chunky Monkey® Ice Cream. If one serving of Chocolate Chip Cookie Dough has 270 calories, how many calories are in one serving of Chunky Monkey? (www.benjerry.com)   300 calories

14) Find the missing angle measures.

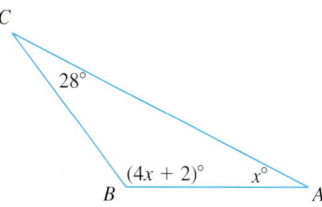

$m\angle A = 30°, m\angle B = 122°$

15) Lynette's age is 7 yr less than three times her daughter's age. If the sum of their ages is 57, how old is Lynette, and how old is her daughter?   Lynette's age = 41; daughter's age = 16

16) Find the slope of the line containing the points $(-7, 8)$ and $(2, 17)$. $m = 1$

17) Graph $4x + y = 5$.

18) Write an equation of the line with slope $-\dfrac{5}{4}$ containing the point $(-8, 1)$. Express the answer in standard form. $5x + 4y = -36$

19) Write an equation of the line perpendicular to $y = \dfrac{1}{3}x + 11$ containing the point $(4, -12)$. Express the answer in slope-intercept form. $y = -3x$

20) Given $y = \dfrac{3}{x + 7}$,

   a) is y a function of x?   yes

   b) determine the domain. $(-\infty, -7) \cup (-7, \infty)$

Let $f(x) = 8x + 3$. Find each of the following and simplify.

21) $f(-5)$ -37

22) $f(a)$ $8a + 3$

23) $f(t + 2)$ $8t + 19$

Graph each function.

24) $f(x) = 2$

25) $h(x) = -\dfrac{1}{4}x + 2$

Solving Systems of Linear Equations

Algebra at Work: Custom Motorcycles

Did you know that algebra can be used in a custom motorcycle shop?

Tanya took apart a transmission to make repairs when she realized that she had mixed up the gears. She was able to replace the shafts onto the bearings, but she could not remember which gear went on which shaft. Tanya measured the distance (in inches) between the shafts, sketched the layout on a piece of paper, and came up with a system of equations to determine which gear goes on which shaft.

If x = the radius of the gear on the left, y = the radius of the gear on the right, and z = the radius of the gear on the bottom, then the system of equations Tanya must solve to determine where to put each gear is

$$x + y = 2.650$$
$$x + z = 2.275$$
$$y + z = 1.530$$

Solving this system, Tanya determines that $x = 1.698$ in., $y = 0.952$ in., and $z = 0.578$ in. Now she knows on which shaft to place each gear.

In this chapter, we will learn how to solve systems of linear equations using graphing, substitution, elimination, and matrices.

Section 4.1 Solving Systems of Linear Equations in Two Variables

Objectives

1. **Determine Whether an Ordered Pair Is a Solution of a System**
2. **Solve a Linear System by Graphing**
3. **Solve a Linear System by Substitution**
4. **Solve a Linear System by Elimination**
5. **Solve Special Systems**

What is a system of linear equations in two variables? A **system of linear equations** consists of two or more linear equations with the same variables. Three examples of such systems are

$$2x + 5y = 5$$
$$x + 4y = -1$$

$$y = \frac{1}{3}x - 8$$
$$5x - 6y = 10$$

$$-3x + y = 1$$
$$x = -2$$

In the third system, we can think of $x = -2$ as an equation in two variables by writing it as $x + 0y = -2$. In this section we will learn how to solve a system of two equations in two variables.

1. Determine Whether an Ordered Pair Is a Solution of a System

We will begin our work with systems of equations by determining whether an ordered pair is a solution of the system.

> **Definition**
>
> A **solution of a system** of two equations in two variables is an ordered pair that is a solution of each equation in the system.

Example 1

In-Class Example 1
Determine if $(-2, 1)$ is a solution of each system of equations.
a) $3x - 7y = -13$
     $-5x + 2y = 9$
b) $x + 10y = 8$
    $2x - 8y = -12$
answer: a) not a solution b) solution

Determine if $(-5, -1)$ is a solution of each system of equations.

a) $x - 3y = -2$
     $-4x + y = 19$

b) $2x - 9y = -1$
    $y = x + 8$

Solution

a)   If $(-5, -1)$ is a solution of $\begin{array}{l} x - 3y = -2 \\ -4x + y = 19 \end{array}$ then when we substitute -5 for x and -1 for y, the ordered pair will make each equation true.

$x - 3y = -2$		$-4x + y = 19$	
$-5 - 3(-1) \stackrel{?}{=} -2$	Substitute.	$-4(-5) + (-1) \stackrel{?}{=} 19$	Substitute.
$-5 + 3 \stackrel{?}{=} -2$		$20 + (-1) \stackrel{?}{=} 19$	
$-2 = -2$	True	$19 = 19$	True

Since $(-5, -1)$ is a solution of each equation, it is a solution of the system.

b)   We will substitute -5 for x and -1 for y to see if $(-5, -1)$ satisfies (is a solution of) each equation.

$2x - 9y = -1$		$y = x + 8$	
$2(-5) - 9(-1) \stackrel{?}{=} -1$	Substitute.	$-1 \stackrel{?}{=} (-5) + 8$	Substitute.
$-10 + 9 \stackrel{?}{=} -1$		$-1 = 3$	False
$-1 = -1$	True		

Although $(-5, -1)$ is a solution of the first equation, it does *not* satisfy $y = x + 8$. Therefore, $(-5, -1)$ is *not* a solution of the system. ■

You Try 1

Determine if $(6, 2)$ is a solution of each system.

a) $-x + 4y = 2$
    $3x - 5y = 6$

b) $y = x - 4$
    $2x - 9y = -6$

Now we will learn how to *find* solutions to systems of linear equations using three methods: graphing, substitution, and elimination.

2. Solve a Linear System by Graphing

To **solve a system of equations in two variables** means to find the ordered pair or pairs that satisfy each equation in the system.

Recall from Chapter 3 that the graph of $Ax + By = C$ is a line. This line represents all solutions of the equation. If two lines intersect at a point, then that point is a solution of each equation.

> ### Definition
> When solving a system of equations by graphing, the point of intersection is the solution of the system. If a system has at least one solution, we say that the system is **consistent.** The equations are **independent** if the system has one solution.

Example 2

In-Class Example 2
Solve the system by graphing.
$$y = -\frac{1}{3}x + \frac{4}{3}$$
$$3x + 2y = -2$$
answer: (−2, 2)

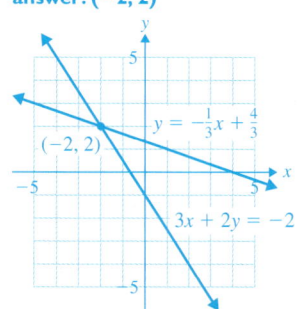

Solve the system by graphing.

$$y = x + 1$$
$$x + 2y = 8$$

Solution

Graph each line on the same axes. The first equation is in slope-intercept form. Graph the second equation by writing it in slope-intercept form or by making a table of values and plotting points.

The point of intersection is (2, 3). Therefore, the solution to the system is (2, 3).

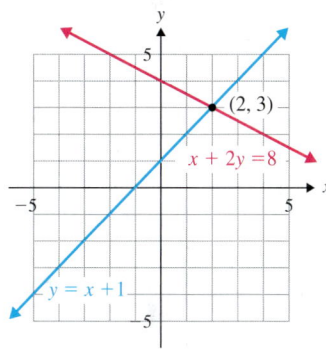

> ### Note
> Always use a straightedge to graph the lines. If the solution of a system contains numbers that are not integers, it may be impossible to accurately read the point of intersection. This is one reason why solving a system by graphing is not always the best way to find the solution. But it can be a useful method, and it is one that is used to solve problems not only in mathematics, but also in areas such as business, economics, and chemistry.

You Try 2

Solve the system by graphing. $2x + 3y = 8$
$y = -4x - 4$

Do two lines *always* intersect? No! Let's see what that tells us about the solution to a system of equations.

Example 3

Solve the system by graphing.

$$2y - x = 2$$
$$-x + 2y = -6$$

In-Class Example 3
Solve the system by graphing.
$-x + 5y = -15$
$2x - 10y = -10$
answer: $\varnothing$

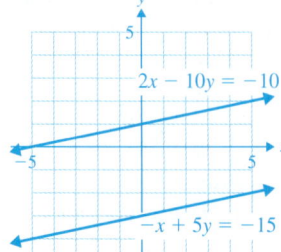

Solution

Graph each line on the same axes. In slope-intercept form, the first equation is $y = \dfrac{1}{2}x + 1$.

The second is $y = \dfrac{1}{2}x - 3$. Both lines have a slope of $\dfrac{1}{2}$. Their y-intercepts, $(0, 1)$ and $(0, -3)$, are different. They are parallel.

Because the lines do not intersect, the system has *no solution*. We write the solution set as $\varnothing$.

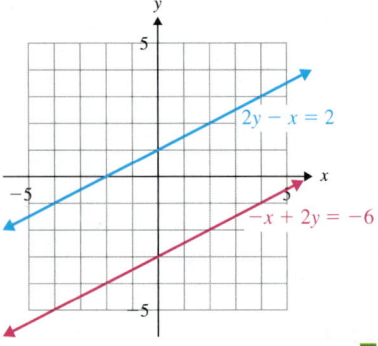

Definition

When solving a system of equations by graphing, if the lines are parallel, then the system has **no solution**. We write this as $\varnothing$. Furthermore, we say that a system that has no solution is **inconsistent**.

Example 4

Solve the system by graphing.

$$y = -\frac{3}{2}x + 2$$
$$6x + 4y = 8$$

In-Class Example 4
Solve the system by graphing.
$y = \dfrac{4}{5}x - 3$
$8x - 10y = 30$
answer: infinite number of
solutions of the form
$\left\{ (x, y) \,\middle|\, y = \dfrac{4}{5}x - 3 \right\}$

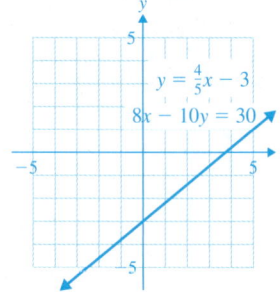

Solution

Graph each line on the same axes. In slope-intercept form, the second equation is $y = -\dfrac{3}{2}x + 2$. The equations are equivalent because the second equation is a multiple of the first. This means that the graph of each equation is the same line. Therefore, each point on the line satisfies each equation.

The system has an *infinite number of solutions* of the form $y = -\dfrac{3}{2}x + 2$. The solution set is

$\left\{ (x, y) \,\middle|\, y = -\dfrac{3}{2}x + 2 \right\}$, which we read as "the set of all ordered pairs (x, y) such that $y = -\dfrac{3}{2}x + 2$."

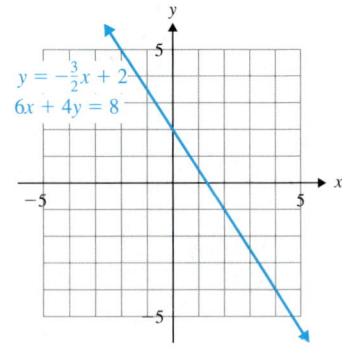

In Example 4, we could have used either equation to write the solution set. However, we will use an equation written in slope-intercept form or an equation written in standard form with integer coefficients that have no common factor other than 1.

> ### Definition
> When solving a system of equations by graphing, if the graph of each equation is the same line, then the system has an **infinite number of solutions.** Because the system has at least one solution, it is **consistent.**

The system in Example 4 contains equations that are equivalent, so the graphs of their lines are identical. We call such equations *dependent*. The systems in Examples 2 and 3, however, do not contain pairs of equivalent equations. The graphs of their lines intersect or are parallel. We call such equations *independent*.

> ### Definition
> The graphs of **dependent** linear equations are identical. The graphs of **independent** linear equations either intersect or are parallel.

> ### Procedure   Solving a System by Graphing
> To solve a system by graphing, graph each line on the same axes.
> 1)   If the lines intersect at a single point, then the point of intersection is the solution of the system. The system is *consistent* and the equations are *independent*. (See Figure 4.1a.)
> 2)   If the lines are parallel, then the system has *no solution*. We write the solution set as ∅. The system is *inconsistent* and the equations are *independent*. (See Figure 4.1b.)
> 3)   If the graphs are the same line, then the system has an *infinite number of solutions*. The system is *consistent* and the equations are *dependent.* (See Figure 4.1c.)

Figure 4.1

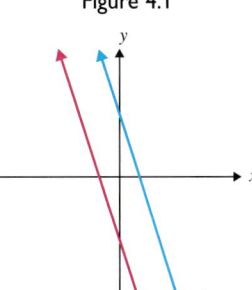

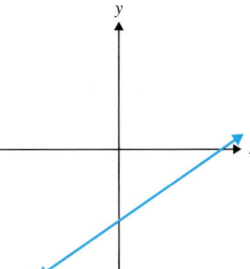

a) One solution
    Consistent system
    Independent
    equations

b) No solution
    Inconsistent system
    Independent
    equations

c) Infinite number
    of solutions
    Consistent system
    Dependent equations

 You Try 3

Solve each system by graphing.

a) $6x - 8y = 8$ b) $x + 2y = 8$
     $3x - 4y = -12$ $-6y - 3x = -24$

The graphs of lines can lead us to the solution of a system. But we can also learn something about the solution by looking at the equations of the lines *without* graphing them.

We saw in Example 4 that if a system's equations have the same slope and the same *y*-intercept, then they are the same line, so the system has an *infinite number of solutions*. Example 3 showed that if a system's equations have the same slope and different *y*-intercepts, then the lines are parallel and the system has *no solution*. We learned in Example 2 that if a system's equations have different slopes, then they will intersect and the system has *one solution*. In Example 5 we will use this information to determine whether a system has no solution, one solution, or an infinite number of solutions.

Example 5

In-Class Example 5
Without graphing, determine whether each system has no solution, one solution, or an infinite number of solutions.
a) $3x - 5y = 6$
$-6x + 10y = -12$
b) $2x + 3y = 18$
$2y = -7x - 16$
c) $5x - y = 1$
$2y = 10x + 14$
answer: a) infinite number of solutions b) one solution c) no solution

Without graphing, determine whether each system has no solution, one solution, or an infinite number of solutions.

a) $y = \dfrac{2}{3}x + 5$
$-4x + 3y = 6$

b) $8x - 12y = 4$
$-6x + 9y = -3$

c) $10x + 4y = -9$
$5x + 2y = 14$

Solution

We will write each equation in slope-intercept form.

a) The first equation is already in slope-intercept form. Write the second equation, $-4x + 3y = 6$, as $y = \dfrac{4}{3}x + 2$. The slope in the first equation is $\dfrac{2}{3}$, and the second is $\dfrac{4}{3}$. The slopes are different, so this system has *one solution*.

b) Write each equation in slope-intercept form.

$$8x - 12y = 4$$
$$-12y = -8x + 4$$
$$y = \dfrac{-8}{-12}x + \dfrac{4}{-12}$$
$$y = \dfrac{2}{3}x - \dfrac{1}{3}$$

$$-6x + 9y = -3$$
$$9y = 6x - 3$$
$$y = \dfrac{6}{9}x - \dfrac{3}{9}$$
$$y = \dfrac{2}{3}x - \dfrac{1}{3}$$

These equations are the same. Therefore, this system has an *infinite number of solutions*.

c) Write each equation in slope-intercept form.

$$10x + 4y = -9$$
$$4y = -10x - 9$$
$$y = \dfrac{-10}{4}x - \dfrac{9}{4}$$
$$y = -\dfrac{5}{2}x - \dfrac{9}{4}$$

$$5x + 2y = 14$$
$$2y = -5x + 14$$
$$y = \dfrac{-5}{2}x + \dfrac{14}{2}$$
$$y = -\dfrac{5}{2}x + 7$$

The equations have the same slope but different y-intercepts. Their lines are parallel, so this system has *no solution*.

You Try 4

Without graphing, determine whether each system has no solution, one solution, or an infinite number of solutions.

a) $10x + y = 4$
$20x + 2y = -7$

b) $y = -\dfrac{2}{3}x + \dfrac{1}{2}$
$4x + 6y = 3$

Using Technology

So far, we have learned that the solution of a system of equations is the point at which their graphs intersect. Let's see how we can solve a system by graphing using a graphing calculator. On the calculator we will solve the following system by graphing:

$$x + y = 5$$
$$y = 2x - 3$$

Begin by entering each equation using the $\boxed{Y=}$ key. Before entering the first equation, we must solve for y.

$$x + y = 5$$
$$y = -x + 5$$

Enter $-x + 5$ in Y_1 and $2x - 3$ in Y_2, press $\boxed{ZOOM}$, and select 6:ZStandard to graph the equations.

Since the lines intersect, the system has a solution. How can we find that solution? Once you see from the graph that the lines intersect, press $\boxed{2nd}$ $\boxed{TRACE}$. Select 5: intersect and then press $\boxed{ENTER}$ three times. The screen will move the cursor to the point of intersection and display the solution to the system on the bottom of the screen.

To obtain the exact solution to the system of equations, first return to the home screen by pressing $\boxed{2nd}$ $\boxed{MODE}$. To display the x-coordinate of the solution, press $\boxed{X,T,\theta,n}$ $\boxed{MATH}$ $\boxed{ENTER}$ $\boxed{ENTER}$ and to display the y-coordinate of the solution press $\boxed{ALPHA}$ $\boxed{1}$ $\boxed{MATH}$ $\boxed{ENTER}$ $\boxed{ENTER}$. The solution to the system is $\left(\dfrac{8}{3}, \dfrac{7}{3}\right)$.

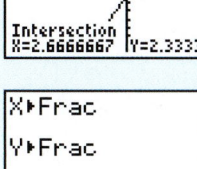

Use a graphing calculator to solve each system.

1) $y = -2x + 5$
    $y = x - 4$

2) $y = -3x - 4$
    $y = x$

3) $2x - y = 3$
    $y - 4x = -8$

4) $x + 2y = -1$
    $x + 4y = 2$

5) $3x + 2y = 4$
    $4x + 3y = 7$

6) $6x - 3y = 10$
    $-2x + y = 4$

3. Solve a Linear System by Substitution

Another way to solve a system of equations is to use the *substitution method. This method is especially good when one of the variables has a coefficient of 1 or -1.*

Example 6

In-Class Example 6
Solve the system by substitution.
$4x - 7y = 8$
$y = 6x + 7$
answer: $\left(-\dfrac{3}{2}, -2\right)$

Solve the system using substitution.

$$2x + 3y = 14$$
$$y = 3x + 1$$

Solution

The second equation is already solved for y; it tells us that y *equals* $3x + 1$. Therefore, we can substitute $3x + 1$ for y in the first equation, then solve for x.

$$
\begin{aligned}
2x + 3y &= 14 && \text{First equation} \\
2x + 3(3x + 1) &= 14 && \text{Substitute.} \\
2x + 9x + 3 &= 14 && \text{Distribute.} \\
11x + 3 &= 14 \\
x &= 1 && \text{Solve for } x.
\end{aligned}
$$

We still need to find y. Substitute $x = 1$ in *either* equation, and solve for y. In this case, we will use the second equation since it is already solved for y.

$$
\begin{aligned}
y &= 3x + 1 && \text{Second equation} \\
y &= 3(1) + 1 && \text{Substitute.} \\
y &= 4
\end{aligned}
$$

Check $x = 1$, $y = 4$ in *both* equations.

$$2x + 3y = 14 \qquad\qquad\qquad\qquad y = 3x + 1$$
$$2(1) + 3(4) \overset{?}{=} 14 \qquad \text{Substitute.} \qquad 4 \overset{?}{=} 3(1) + 1 \qquad \text{Substitute.}$$
$$2 + 12 \overset{?}{=} 14 \qquad \text{True} \qquad\qquad 4 \overset{?}{=} 3 + 1 \qquad \text{True}$$

We write the solution as an ordered pair. The solution to the system is $(1, 4)$.

When we use the **substitution method**, we solve one of the equations for one of the variables in terms of the other. Then we substitute that expression into the other equation. We can do this because solving a system means finding the ordered pair (or pairs) that satisfies *both* equations.

Procedure   Solving a System by Substitution

1) Solve one of the equations for one of the variables. If possible, solve for a variable that has a coefficient of 1 or −1.

2) Substitute the expression found in *step 1* into the *other* equation. The equation you obtain should contain only one variable.

3) Solve the equation in one variable from *step 2*.

4) Substitute the value found in *step 3* into either of the equations to obtain the value of the other variable.

5) Check the values in each of the original equations, and write the solution as an ordered pair.

Example 7

In-Class Example 7
Solve the system by substitution.
$x + 2y = -9$
$5x + 2y = 11$
answer: $(5, -7)$

Solve the system by substitution.

$$x + 4y = 3 \qquad (1)$$
$$2x + 3y = -4 \qquad (2)$$

Solution

We will follow the steps listed above.

1) For which variable should we solve? The x in the first equation is the only variable with a coefficient of 1 or −1. Therefore, we will solve the first equation for x.

$$x + 4y = 3 \qquad \text{Equation (1)}$$
$$x = 3 - 4y \qquad \text{Subtract } 4y.$$

2) Substitute $3 - 4y$ for x in equation (2).

$$2x + 3y = -4 \qquad \text{Equation (2)}$$
$$2(3 - 4y) + 3y = -4 \qquad \text{Substitute.}$$

3) Solve the last equation for y.

$$2(3 - 4y) + 3y = -4$$
$$6 - 8y + 3y = -4 \qquad \text{Distribute.}$$
$$-5y = -10 \qquad \text{Subtract 6; combine like terms.}$$
$$y = 2$$

4) Find x by substituting 2 for y in either equation. We will use equation (1).

$$x + 4y = 3 \qquad (1)$$
$$x + 4(2) = 3 \qquad \text{Substitute 2 for } y.$$
$$x = -5$$

5) The check is left to the reader. The solution of the system is $(-5, 2)$.

If no variable in the system has a coefficient of 1 or -1, solve for any variable.

You Try 5

Solve the system by substitution.

$$10x + 3y = -4$$
$$8x + y = 1$$

If a system contains an equation with fractions, first multiply the equation by the least common denominator to eliminate the fractions. Likewise, if an equation in the system contains decimals, begin by multiplying the equation by the power of 10 that will eliminate the decimals.

Example 8

In-Class Example 8
Solve the system by substitution.
$$\frac{1}{2}x - \frac{2}{3}y = \frac{7}{6}$$
$$\frac{1}{4}x + \frac{1}{2}y = -\frac{3}{8}$$

answer: $\left(-2, \frac{1}{4}\right)$

Solve the system by substitution.

$$\frac{2}{5}x - \frac{1}{3}y = 2 \qquad (1)$$

$$-\frac{1}{6}x + \frac{1}{2}y = \frac{4}{3} \qquad (2)$$

Solution

First, eliminate the fractions in each equation.

$\frac{2}{5}x - \frac{1}{3}y = 2$	Equation (1)
$15\left(\frac{2}{5}x - \frac{1}{3}y\right) = 15 \cdot 2$	Multiply by the LCD: 15.
$6x - 5y = 30 \qquad (3)$	Distribute.

$-\frac{1}{6}x + \frac{1}{2}y = \frac{4}{3}$	Equation (2)
$6\left(-\frac{1}{6}x + \frac{1}{2}y\right) = 6 \cdot \frac{4}{3}$	Multiply by the LCD: 6.
$-x + 3y = 8 \qquad (4)$	Distribute.

We now have an equivalent system of equations.

$$6x - 5y = 30 \qquad (3)$$
$$-x + 3y = 8 \qquad (4)$$

1) The x in equation (4) has a coefficient of -1. Solve this equation for x.

$-x + 3y = 8$	Equation (4)
$-x = 8 - 3y$	Subtract $3y$.
$x = -8 + 3y$	Divide by -1.

2) and 3) Substitute $-8 + 3y$ for x in equation (3) and solve for y.

$6x - 5y = 30$	Equation (3)
$6(-8 + 3y) - 5y = 30$	Substitute.
$-48 + 18y - 5y = 30$	Distribute.
$13y = 78$	Add 48; combine like terms.
$y = 6$	Divide by 13.

4)   Find x by substituting 6 for y in equation (3) or (4). Let's use equation (4) since it has smaller coefficients.

$$\begin{aligned}
-x + 3y &= 8 \qquad &&\text{Equation (4)}\\
-x + 3(6) &= 8 \qquad &&\text{Substitute.}\\
-x + 18 &= 8\\
-x &= -10\\
x &= 10 \qquad &&\text{Divide by } -1.
\end{aligned}$$

5)   Check $x = 10$ and $y = 6$ in the original equations. The solution is $(10, 6)$.   ■

You Try 6

Solve each system by substitution.

a) $\dfrac{2}{3}x - \dfrac{1}{9}y = -3$ b) $0.01x + 0.04y = 0.15$

  $-\dfrac{5}{6}x + \dfrac{1}{2}y = \dfrac{1}{2}$ $0.1x - 0.1y = 0.5$

4. Solve a Linear System by Elimination

The next technique we will learn for solving a system of equations is the **elimination method**. (This is also called the **addition method**.) It is based on the addition property of equality, which says that we can add the *same* quantity to each side of an equation and preserve the equality.

$$\text{If } a = b, \text{ then } a + c = b + c.$$

We can extend this idea by saying that we can add *equal* quantities to each side of an equation and still preserve the equality.

$$\text{If } a = b \text{ and } c = d, \text{ then } a + c = b + d.$$

The objective of the elimination method is to add the equations (or multiples of one or both of the equations) so that one variable is eliminated. Then, we can solve for the remaining variable.

Example 9

In-Class Example 9
Solve the system using the elimination method.
$-x + y = 4$
$x + 3y = 8$
answer: $(-1, 3)$

Solve the system using the elimination method.

$$\begin{aligned}
3x - 2y &= 2 \qquad &&(1)\\
-5x + 2y &= -10 \qquad &&(2)
\end{aligned}$$

Solution

The left side of each equation is equal to the right side of each equation. Therefore, if we add the left sides together and add the right sides together, we can set them equal. We will add these equations vertically, eliminate the y-terms, and solve for x.

$$\begin{aligned}
3x - 2y &= 2 \qquad &&(1)\\
+\ \underline{-5x + 2y} &= \underline{-10} \qquad &&(2)\\
-2x + 0y &= -8 \qquad &&\text{Add equations (1) and (2).}\\
-2x &= -8 \qquad &&\text{Simplify.}\\
x &= 4 \qquad &&\text{Divide by } -2.
\end{aligned}$$

To find y, we can substitute $x = 4$ into either equation. Here, we will use equation (1).

$$3x - 2y = 2 \qquad (1)$$
$$3(4) - 2y = 2$$
$$12 - 2y = 2$$
$$-2y = -10 \qquad \text{Subtract 12.}$$
$$y = 5 \qquad \text{Divide by } -2.$$

Check $x = 4$ and $y = 5$ in *both* equations.

$$3x - 2y = 2 \qquad (1) \qquad\qquad -5x + 2y = -10 \qquad (2)$$
$$3(4) - 2(5) \overset{?}{=} 2 \quad \text{Substitute.} \qquad -5(4) + 2(5) = -10 \quad \text{Substitute.}$$
$$12 - 10 = 2 \quad \text{True} \qquad\qquad -20 + 10 = -10 \quad \text{True}$$

The solution is (4, 5). ■

You Try 7

Solve the system using the elimination method.

$$x + 2y = -6$$
$$-x - 3y = 13$$

In Example 9, we eliminated a variable by adding the equations. Sometimes eliminating a variable requires more steps, however. These steps are listed next.

Procedure   Solving a System of Two Linear Equations by the Elimination Method

1) Write each equation in the form $Ax + By = C$.

2) Determine which variable to eliminate. If necessary, multiply one or both of the equations by a number so that the coefficients of the variable to be eliminated are negatives of one another.

3) Add the equations and solve for the remaining variable.

4) Substitute the value found in *step 3* into either of the original equations to find the value of the other variable.

5) Check the solution in each of the original equations.

Example 10

In-Class Example 10
Solve the system using the elimination method.
$2y = 2 - 5x$
$2x + 3y = -8$
answer: **(2, −4)**

Solve the system using the elimination method.

$$3x = 7y + 5 \qquad (1)$$
$$2x - 3 = 5y \qquad (2)$$

Solution

1) **Write each equation in the form $Ax + By = C$.** We will number these new equations (3) and (4).

$$3x = 7y + 5 \quad (1) \qquad\qquad 2x - 3 = 5y \quad (2)$$
$$3x - 7y = 5 \quad \text{Subtract } 7y. \ (3) \qquad 2x - 5y = 3 \quad \text{Subtract } 5y \text{ and add 3. (4)}$$

2) **Determine which variable to eliminate from equations (3) and (4).** Often, it is easier to eliminate the variable with the smaller coefficients, so we will eliminate x.

The least common multiple of 3 and 2 (the x-coefficients) is 6. We want one x-coefficient to be 6 and the other to be -6. Multiply equation (3) by 2 and equation (4) by -3.

Rewrite the System

$$2(3x - 7y) = 2(5) \qquad \text{2 times (3)}$$
$$-3(2x - 5y) = -3(3) \qquad \text{-3 times (4)}$$

$\longrightarrow$

$$6x - 14y = 10$$
$$-6x + 15y = -9$$

3) **Add the resulting equations to eliminate x. Solve for y.**

$$6x - 14y = 10$$
$$+\ \underline{-6x + 15y = -9}$$
$$y = 1$$

4) **Substitute $y = 1$ into equation (1) and solve for x.**

$$3x = 7y + 5 \qquad (1)$$
$$3x = 7(1) + 5 \qquad \text{Substitute.}$$
$$3x = 12$$
$$x = 4$$

5) **Check** to verify that (4, 1) satisfies each of the original equations. The solution is (4, 1). ■

You Try 8

Solve the system using the elimination method.

$$4x - 10 = -3y$$
$$5x = 4y - 3$$

Sometimes we may want to use the elimination method *twice* to solve a system.

Example 11

Solve using the elimination method.

$$9x - 4y = 7 \qquad (1)$$
$$2x + 3y = -8 \qquad (2)$$

In-Class Example 11
Solve using the elimination
method.
$11x + 5y = 8$
$\ \ x + 3y = 3$
answer: $\left(\dfrac{9}{28}, \dfrac{25}{28} \right)$

Solution

Each equation is written in the form $Ax + By = C$, so we begin with step 2.

2) We will eliminate y from equations (1) and (2).

Rewrite the System

$$3(9x - 4y) = 3(7) \qquad \text{3 times (1)}$$
$$4(2x + 3y) = 4(-8) \qquad \text{4 times (2)}$$

$\longrightarrow$

$$27x - 12y = 21$$
$$8x + 12y = -32$$

3) Add the resulting equations to eliminate y. Solve for x.

$$27x - 12y = \ \ \ 21$$
$$+\ \underline{\ \ 8x + 12y = -32}$$
$$35x = -11$$
$$x = -\frac{11}{35} \qquad \text{Solve for } x.$$

Normally, we would substitute $x = -\frac{11}{35}$ into equation (1) or (2) and solve for y. This time, however, working with a number like $-\frac{11}{35}$ would be difficult, so *we will use the elimination method a second time.*

Go back to the original equations, (1) and (2), and use the elimination method again but eliminate the other variable, x. Then, solve for y.

Rewrite the System

$$-2(9x - 4y) = -2(7) \qquad \text{-2 times (1)} \qquad \longrightarrow \qquad -18x + 8y = -14$$
$$9(2x + 3y) = 9(-8) \qquad \text{9 times (2)} \qquad \qquad \qquad 18x + 27y = -72$$

Add the equations.

$$\begin{array}{r} -18x + 8y = -14 \\ + \quad 18x + 27y = -72 \\ \hline 35y = -86 \end{array}$$

$$y = -\frac{86}{35} \qquad \text{Solve for } y.$$

Check to verify that the solution is $\left(-\dfrac{11}{35}, -\dfrac{86}{35}\right)$. ■

You Try 9

Solve using the elimination method.

$$5x + 7y = 20$$
$$2x - 3y = -4$$

Procedure   Choosing Between the Substitution and Elimination Methods

If an equation contains fractions or decimals, begin by eliminating them. Then use the following guidelines.

1)   If a variable has a coefficient of 1 or −1, solve for that variable and *use the substitution method.*

2)   If none of the variables has a coefficient of 1 or −1, *use the elimination method.*

---

5. Solve Special Systems

As we saw in Examples 3 and 4, some systems have no solution, and some have an infinite number of solutions. How do the substitution and elimination methods illustrate these results?

Example 12

Solve the system.

$$9y = 12x + 5 \qquad (1)$$
$$6y - 8x = -11 \qquad (2)$$

In-Class Example 12
Solve the system.
$8y + 5x = 4$
$-5x = 9 + 8y$
answer: $\varnothing$

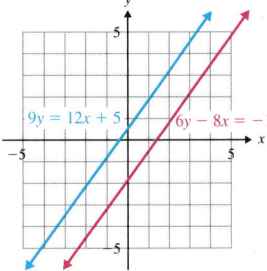

Solution

We will use the elimination method to solve this system.

Step 1:   Write each equation in the form $Ax + By = C$.

$$9y = 12x + 5 \qquad \longrightarrow \qquad -12x + 9y = 5 \qquad (3)$$
$$6y - 8x = -11 \qquad \qquad \qquad -8x + 6y = -11 \qquad (4)$$

Steps 2 and 3:   Determine which variable to eliminate from equations (3) and (4). Eliminate y. The least common multiple of 9 and 6 is 18. One y-coefficient must be 18, and the other must be -18. Then add the equations.

$$-2(-12x + 9y) = -2(5) \qquad \longrightarrow \qquad 24x - 18y = -10$$
$$3(-8x + 6y) = 3(-11) \qquad \qquad \qquad + \quad -24x + 18y = -33$$
$$\qquad \qquad \qquad \qquad \qquad \qquad \qquad \qquad \qquad \overline{\qquad 0 = -43 \qquad} \text{False}$$

Both variables are eliminated, and we get a false statement. The system is inconsistent, and the solution set is ∅. The graph of the system supports our work. The lines are parallel, so the system has no solution. ■

You Try 10

Solve the system.

$$24x + 6y = -7$$
$$4y + 3 = -16x$$

Example 13

In-Class Example 13
Solve the system by substitution.
$$5y - 6 = x$$
$$-4x + 20y = 24$$
answer: infinite number of solutions of the form
$$\{(x, y)\,|\,5y - 6 = x\}$$

Solve the system.

$$2x - 8y = 20 \qquad (1)$$
$$x = 4y + 10 \qquad (2)$$

Solution

Step 1: Equation (2) is already solved for x, so we will use the substitution method and begin with step 2.

Steps 2 and 3: Substitute $4y + 10$ for x in equation (1) and solve for y.

$$2x - 8y = 20 \qquad \text{Equation (1)}$$
$$2(4y + 10) - 8y = 20 \qquad \text{Substitute.}$$
$$8y + 20 - 8y = 20 \qquad \text{Distribute.}$$
$$20 = 20 \qquad \text{True}$$

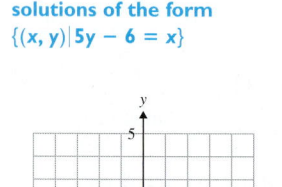
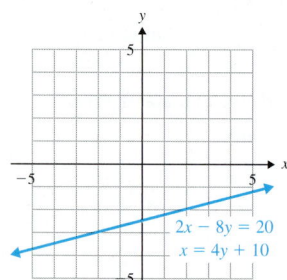

Since both variables are eliminated and we get a true statement, there are an infinite number of solutions to the system. The equations are dependent, and the solution set is $\{(x, y)\,|\,x = 4y + 10\}$, *that is, all the points on the line formed by the equation* $2x - 8y = 20$.
     The graph to the left shows that the equations in the system are the same line, therefore the system has an infinite number of solutions. ■

You Try 11

Solve each system by substitution.

a) $6x + y = -8$ b) $x - 3y = 5$
     $12x + 2y = -9$ $4x - 12y = 20$

Note

When you are solving a system of equations and both variables are eliminated,

1) if you get a *false statement*, like $3 = 5$, then the system has *no solution* and is *inconsistent*.

2) if you get a *true statement*, like $-4 = -4$, then the system has an *infinite number of solutions*. The equations are *dependent*.

Answers to You Try Exercises

1) a) no b) yes

2) $(-2, 4)$

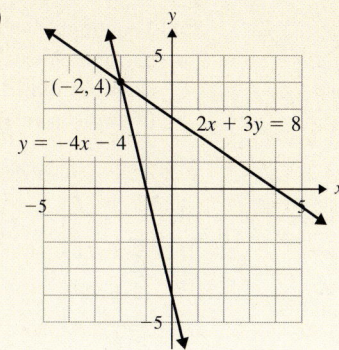

3) a) $\varnothing$

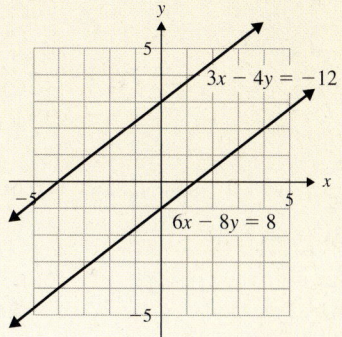

b) $\{(x, y) \mid x + 2y = 8\}$

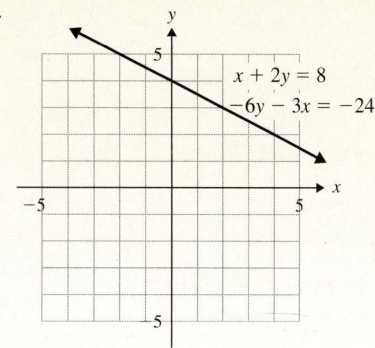

4) a) no solution   b) infinite number of solutions   5) $\left(\dfrac{1}{2}, -3\right)$ 6) a) $(-6, -9)$ b) $(7, 2)$

7) $(8, -7)$ 8) $(1, 2)$ 9) $\left(\dfrac{32}{29}, \dfrac{60}{29}\right)$ 10) $\varnothing$ 11) a) $\varnothing$ b) $\{(x, y) \mid x - 3y = 5\}$

Answers to Technology Exercises

1) $(3, -1)$ 2) $(-1, -1)$ 3) $\left(\dfrac{5}{2}, 2\right)$ 4) $\left(-4, \dfrac{3}{2}\right)$ 5. $(-2, 5)$ 6. $\varnothing$

4.1 Exercises

*Additional answers can be found in the Answers to Exercises appendix.

Objective 1: Determine Whether an Ordered Pair is a Solution of a System

Determine if the ordered pair is a solution of the system of equations.

1) $3x + 2y = 4$ no
   $4x - y = -3$
   $(-2, 5)$

2) $10x + 7y = -13$ yes
   $-6x - 5y = 11$
   $\left(\dfrac{3}{2}, -4\right)$

3) $y = 5x - 7$ no
   $3x + 9 = y$
   $(-1, -2)$

4) $x - 5y = 7$ yes
   $y = 2x + 13$
   $(-8, -3)$

Objective 2: Solve a Linear System by Graphing

5) If you are solving a system of equations by graphing, how do you know if the system has no solution?
   The lines are parallel.

6) If you are solving a system of equations by graphing, how do you know if the system has an infinite number of solutions?   The graphs are the same line.

7) A system with _____ equations has an infinite number of solutions.   dependent

8) If a system of equations has no solution, then the system is said to be _____.   inconsistent

Solve each system of equations by graphing. Identify any inconsistent systems or dependent equations.

VIDEO 9) $y = -\dfrac{2}{3}x + 3$ (3, 1)

$y = x - 2$

10) $y = \dfrac{3}{2}x - 4$ (2, -1)

$y = -2x + 3$

11) $y - 4x = 1$ (-1, -3)

$y = -3$

12) $y = -3x + 1$ (1, -2)

$x = 1$

VIDEO 13) $\dfrac{3}{4}x - y = 0$ ∅; inconsistent

$3x - 4y = 20$

14) $6x - 3y = 12$

$-2x + y = -4$

15) $3x + y = 2$ (1, -1)

$y = 2x - 3$

16) $x = 2y - 6$ (4, 5)

$3x - 2y = 2$

VIDEO 17) $y = -3x + 1$

$12x + 4y = 4$

18) $2x - y = 1$ ∅; inconsistent

$-2x + y = -3$

Write a system of equations so that the given ordered pair is a solution of the system.   For 19–22, answers may vary.

19) (5, 1)

20) (-1, -4)

21) $\left(-\dfrac{1}{2}, 3\right)$

22) $\left(0, \dfrac{2}{3}\right)$

For Exercises 23 and 24, determine which ordered pair could not be a solution to the system of equations that is graphed. Explain why you chose that ordered pair.

23)

24)

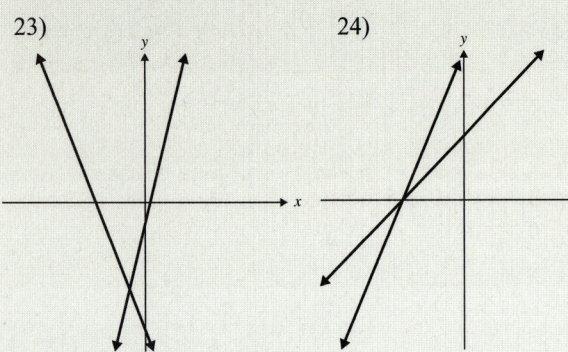

A. (-3, -3)   C. (-1, -8)

B. (-2, 1)   D. $\left(-\dfrac{3}{2}, -\dfrac{9}{2}\right)$

A. (-6, 0)   C. (0, -5)

B. $\left(-\dfrac{1}{2}, 0\right)$ D. (-8.3, 0)

B; (-2, 1) is in quadrant II.   C; (0, -5) is on the y-axis not the x-axis.

25) How do you determine, *without graphing*, that a system of equations has exactly one solution?
The slopes are different.

26) How do you determine, *without graphing*, that a system of equations has no solution?
The slopes are the same, but the y-intercepts are different.

Without graphing, determine whether each system has no solution, one solution, or an infinite number of solutions. Do not solve the system.

27) $y = \dfrac{3}{2}x + \dfrac{7}{2}$

$-9x + 6y = 21$
infinite number of solutions

28) $y = 4x + 6$

$8x - y = -7$
one solution

29) $5x - 2y = -11$

$x + 6y = 18$
one solution

30) $5x - 8y = 24$

$10x - 16y = -9$
no solution

31) $x + y = 10$

$-9x - 9y = 2$
no solution

32) $x - 4y = 2$

$3x + y = 1$
one solution

33) $5x - y = -2$

$x + 6y = 2$
one solution

34) $9x - 6y = 5$

$y = \dfrac{3}{2}x - \dfrac{5}{6}$
infinite number of solutions

35) The graph shows the number of people, seven years of age and older, who have participated more than once in snowboarding and ice/figure skating from 1997–2003. (National Sporting Goods Association)

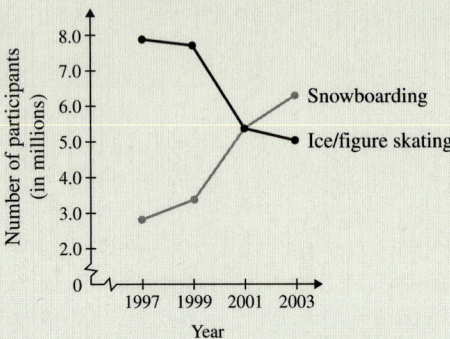

Snowboarders and Ice Skaters

a) When were there more snowboarders than ice/figure skaters?   after 2001

b) When did the number of snowboarders equal the number of skaters? How many people participated in each?   2001; 5.3 million

c) During which years did snowboarding see its greatest increase in participation?   1999–2001

d) During which years did skating see its greatest decrease in participation?   1999–2001

36) The graph shows the percentage of households in Delaware and Nevada with Internet access in various years from 1998 to 2003. (www.census.gov)

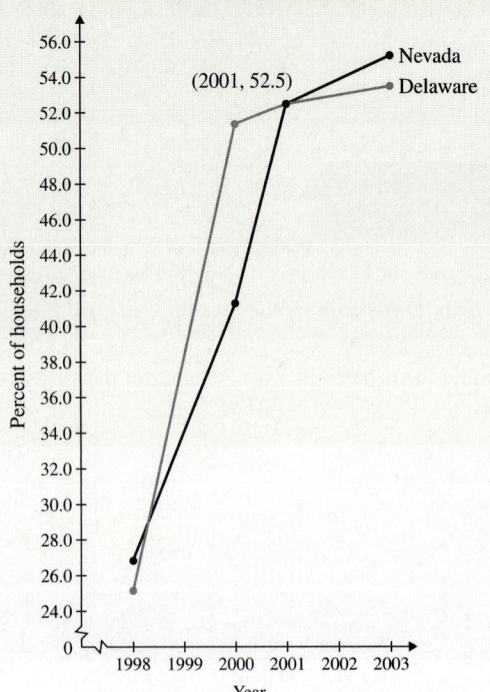

Households with Internet Access

a) In 1998, which state had a greater percentage of households with Internet access? Approximately how many more were there?

b) In what year did both states have the same percentage of households with Internet access? What percentage of households had Internet access that year?

c) Between 2000 and 2001, which state had the greatest increase in the percentage of households with Internet access? How can this information be related to *slope*?

Solve each system using a graphing calculator.

37) $y = -2x + 4$
$y = x - 5$ $(3, -2)$

38) $y = -x + 5$
$y = 3x + 1$ $(1, 4)$

39) $-2x + y = 9$
$x - 3y = -2$ $(-5, -1)$

40) $4x + y = -6$
$x + 3y = 4$ $(-2, 2)$

41) $6x - 5y = 3.25$
$3x + y = -2.75$
$(-0.5, -1.25)$

42) $2x + 5y = 7.5$
$-3x + y = -11.25$
$(3.75, 0)$

Objective 3: Solve a Linear System by Substitution

43) If you were asked to solve this system by substitution, why would it be easiest to begin by solving for y in the second equation?

$$6x - 2y = -5$$
$$3x + y = 4$$

It is the only variable with a coefficient of 1.

Solve each system by substitution. Identify any inconsistent systems or dependent equations.

44) $y = 3x + 4$
$-6x + y = -2$ $(2, 10)$

45) $2x - y = 5$
$x = y + 6$ $(-1, -7)$

46) $x = y - 8$
$-3x - y = 12$ $(-5, 3)$

47) $2x + 5y = 8$
$x - 6y = 4$ $(4, 0)$

48) $x + 3y = -12$
$3x + 4y = -6$ $(6, -6)$

49) $9y - 18x = 5$
$2x - y = 3$
$\varnothing$; inconsistent

50) $2x + 30y = 9$
$x = 6 - 15y$ $\varnothing$; inconsistent

51) $5x - y = 8$
$4y = 10 - x$ $(2, 2)$

52) $y - 4x = -1$ $\left(\frac{1}{4}, 0\right)$
$8x + y = 2$

53) $10x + y = -5$ $\left(-\frac{4}{5}, 3\right)$
$-5x + 2y = 10$

54) $x + 2y = 6$
$x + 20y = -12$ $(8, -1)$

55) $6x + y = -6$
$-12x - 2y = 12$
$\{(x, y)|6x + y = -6\}$; dependent

56) $x - 2y = 10$
$3x - 6y = 30$
$\{(x, y)|x - 2y = 10\}$; dependent

57) $2x - y = 6$
$3y = -18 - x$ $(0, -6)$

58) $2x - 9y = -2$ $\left(-4, -\frac{2}{3}\right)$
$6y - x = 0$

59) $2x - 5y = -4$
$8x - 9y = 6$ $(3, 2)$

60) $2x + 3y = 6$
$5x + 2y = -7$ $(-3, 4)$

61) If an equation in a system contains fractions, what should you do first to make the system easier to solve?
Multiply the equation by the LCD of the fractions to eliminate the fractions.

62) If an equation in a system contains decimals, what should you do first to make the system easier to solve?
Multiply the equation by the power of 10 that will eliminate the decimals.

Solve each system by substitution. Identify any inconsistent systems or dependent equations.

63) $\frac{1}{4}x - \frac{1}{2}y = 1$
$\frac{2}{3}x + \frac{1}{6}y = \frac{25}{6}$ $(6, 1)$

64) $\frac{2}{3}x + \frac{2}{3}y = 6$
$\frac{3}{2}x - \frac{1}{4}y = \frac{13}{2}$ $(5, 4)$

65) $-\frac{2}{15}x - \frac{1}{3}y = \frac{2}{3}$
$\frac{2}{3}x + \frac{5}{3}y = \frac{1}{2}$ $\varnothing$; inconsistent

66) $y - \frac{5}{2}x = -2$
$\frac{3}{4}x - \frac{3}{10}y = \frac{3}{5}$

67) $\frac{3}{4}x + \frac{1}{2}y = 6$
$x = 3y + 8$ $(8, 0)$

68) $0.2x - 0.1y = 0.1$
$0.01x + 0.04y = 0.23$
$(3, 5)$

69) $0.01x + 0.10y = -0.11$
$0.02x - 0.01y = 0.20$
$(9, -2)$

70) $0.3x - 0.1y = 5$
$0.15x + 0.1y = 4$
$(20, 10)$

71) $0.2x - 0.1y = 1$
$0.1x - 0.13y = 0.02$ $(8, 6)$

Mixed Exercises: Objectives 4 and 5

Solve each system using elimination. Identify any inconsistent systems or dependent equations.

72) $3x + 5y = -10$
$7x - 5y = 10$ $(0, -2)$

73) $4x - 3y = -5$
$-4x + 5y = 11$ $(1, 3)$

74) $7x + 6y = 3$
$3x + 2y = -1$ $(-3, 4)$

75) $-8x + 5y = -6$
$4x - 7y = 3$ $\left(\frac{3}{4}, 0\right)$

76) $3x - y = 4$
$-6x + 2y = -8$
$\{(x, y)|3x - y = 4\}$; dependent

77) $5x - 6y = -2$
$10x - 12y = 7$
$\varnothing$; inconsistent

78) $2x - 9 = 8y$
$20y - 5x = 6$
$\varnothing$; inconsistent

79) $x - 6y = -5$
$-24y + 4x = -20$
$\{(x, y)|x - 6y = -5\}$; dependent

80) $8x = 6y - 1$
$10y - 6 = -4x$ $\left(\frac{1}{4}, \frac{1}{2}\right)$

81) $9x - 7y = -14$
$4x + 3y = 6$ $(0, 2)$

82) $6x + 5y = 13$
$5x + 3y = 5$ $(-2, 5)$

83) $7x + 2y = 12$
$24 - 14x = 4y$
$\{(x, y)|7x + 2y = 12\}$; dependent

84) $4 + 9y = -21x$
$14x + 6y = -1$ $\varnothing$; inconsistent

85) $\frac{x}{4} + \frac{y}{2} = -1$
$\frac{3}{8}x + \frac{5}{3}y = -\frac{7}{12}$ $(-6, 1)$

86) $\frac{1}{2}x + \frac{2}{3}y = -\frac{29}{6}$
$-\frac{1}{3}x + y = -4$ $(-3, -5)$

87) $\frac{x}{2} - \frac{y}{5} = \frac{1}{10}$
$\frac{x}{3} + \frac{y}{4} = \frac{5}{6}$ $(1, 2)$

88) $x + \frac{y}{4} = \frac{7}{2}$
$\frac{2}{5}x + \frac{1}{2}y = -1$ $(5, -6)$

89) $0.1x + 2y = -0.8$
$0.03x + 0.10y = 0.26$ $(12, -1)$

90) $0.6x - 0.1y = 0.5$
$0.10x - 0.03y = -0.01$ $(2, 7)$

91) $-0.4x + 0.2y = 0.1$
    $0.6x - 0.3y = 1.5$ $\varnothing$; inconsistent

92) $x - 0.5y = 0.2$
    $-0.3x + 0.15y = -0.06$ $\{(x, y)|x - 0.5y = 0.2\}$; dependent

Solve each system using the elimination method twice. See Example 11.

VIDEO 93) $4x + 5y = -6$ $\left(-\dfrac{123}{17}, \dfrac{78}{17}\right)$ 94) $2x - 11y = 3$ $\left(-\dfrac{65}{63}, -\dfrac{29}{63}\right)$
    $3x + 8y = 15$ $5x + 4y = -7$

95) $2x - 7y = -10$ $\left(\dfrac{85}{46}, \dfrac{45}{23}\right)$ 96) $6x - 9y = -5$ $\left(\dfrac{49}{18}, \dfrac{64}{27}\right)$
    $6x + 2y = 15$ $4x + 3y = 18$

97) Noor needs to rent a car for one day while hers is being repaired. Rent-for-Less charges $0.40 per mile while Frugal Rentals charges $12 per day plus $0.30 per mile. Let $x =$ the number of miles driven, and let $y =$ the cost of the rental. The cost of renting a car from each company can be expressed with the following equations:

Rent-for-Less: $y = 0.40x$
Frugal Rentals: $y = 0.30x + 12$

a) How much would it cost Noor to rent a car from each company if she planned to drive 60 mi?

b) How much would it cost Noor to rent a car from each company if she planned to drive 160 mi?

c) Solve the system of equations using the substitution method, and explain the meaning of the solution.

d) Graph the system of equations, and explain when it is cheaper to rent a car from Rent-for-Less and when it is cheaper to rent a car from Frugal Rentals. When is the cost the same?

98) To rent a moving truck, Discount Van Lines charges $1.20 per mile while Comfort Ride Company charges $60 plus $1.00 per mile. Let $x =$ the number of miles driven, and let $y =$ the cost of the rental. The cost of renting a moving truck from each company can be expressed with the following equations:

Discount Van Lines: $y = 1.20x$
Comfort Ride: $y = 1.00x + 60$

a) How much would it cost to rent a truck from each company if the truck would be driven 100 mi?

b) How much would it cost to rent a truck from each company if the truck would be driven 400 mi?

c) Solve the system of equations using the substitution method, and explain the meaning of the solution.

d) Graph the system of equations, and explain when it is cheaper to rent a truck from Discount Van Lines and when it is cheaper to rent a truck from Comfort Ride. When is the cost the same?

Mixed Exercises: Objectives 3 and 4

Solve each system using *your choice* of the substitution or the elimination method. Identify any inconsistant systems

or dependent equations. See the box about choosing between the substitution and elimination methods on page 211.

99) $-2x - y = -1$ 100) $3x + y = 15$
    $4x + y = -5$ $(-3, 7)$ $y = 4x - 6$ $(3, 6)$

101) $3x + 4y = 9$ $\left(5, -\dfrac{3}{2}\right)$ 102) $0.02x + 0.07y = -0.24$
     $5x + 6y = 16$ $0.05y - 0.04x = 0.10$
                                          $(-5, -2)$

103) $\dfrac{x}{12} - \dfrac{y}{6} = \dfrac{2}{3}$ 104) $y = \dfrac{1}{3}x + 4$
     $\dfrac{x}{4} + \dfrac{y}{3} = 2$ $(8, 0)$ $3y - x = 12$
                                                          $\left\{(x, y) \middle| y = \dfrac{1}{3}x + 4\right\}$; dependent

105) $y = -3x - 20$ 106) $\dfrac{5}{3}x - \dfrac{4}{3}y = -\dfrac{4}{3}$
     $6x + 2y = 5$ $y = 2x + 4$ $(-4, -4)$
     $\varnothing$; inconsistent

107) $0.1x + 0.5y = 0.4$ 108) $6x - 3y = -11$ $\left(\dfrac{5}{3}, 7\right)$
     $-0.03x + 0.01y = 0.2$ $9x - 2y = 1$
     $(-6, 2)$

Solve each system using *your choice* of the substitution or elimination method. Begin by combining like terms.

109) $5(2x - 3) + y - 6x = -24$
     $8y - 3(2y + 3) + x = -6$ $(-3, 3)$

110) $8 + 2(3x - 5) - 7x + 6y = 2$
     $9(y - 2) + 5x - 13y = -12$ $(2, 1)$

111) $6(3x + 4) - 8(x + 2) = 5 - 3y$
     $6x - 2(5y + 2) = -7(2y - 1) - 4$ $\left(-\dfrac{3}{2}, 4\right)$

112) $2(y - 6) = 3y + 4(x - 5)$
     $2(4x + 3) - 5 = 2(1 - y) + 5x$ $(3, -4)$

113) $7x + 3(y - 2) = 7y + 6x - 1$
     $18 + 2(x - y) = 4(x + 2) - 5y$ $(5, 0)$

114) $20 + 3(2y - 3) = 4(2y - 1) - 9x$ $\left(\dfrac{1}{3}, 9\right)$
     $5(3x - 4) + 8y = 3x + 7(y - 1)$

Objective 5: Solve Special Systems

115) When solving a system of linear equations, how do you know if the system has

a) no solution?   The variables are eliminated, and you get a false statement.

b) an infinite number of solutions?   The variables are eliminated, and you get a true statement.

116) Given the following system of equations,

$$x + y = 8$$
$$x + y = c$$

find c so that the system has

a) an infinite number of solutions.   8

b) no solution. c can be any real number except 8.

117) Given the following system of equations,

$$x - y = 3$$
$$x - y = c$$

find c so that the system has

a) an infinite number of solutions.   3

b) no solution. c can be any real number except 3.

118) Given the following system of equations,

$$2x - 3y = 5$$
$$ax - 6y = 10$$

find a so that the system has

a) an infinite number of solutions.   4

b) exactly one solution.   *a can be any real number except 4.*

VIDEO 119) Find b so that $(2, -1)$ is a solution to the system

$$3x - 4y = 10$$
$$-x + by = -7$$ 5

120) Find a so that $(4, 3)$ is a solution to the system

$$ax + 5y = 3$$
$$2x - 3y = -1$$ −3

Extension

Let a, b, and c represent nonzero constants. Solve each system for x and y.

121) $ax + by = 5$
     $2ax - by = 1$ $\left(\dfrac{2}{a}, \dfrac{3}{b}\right)$

122) $-3ax + 2by = 9$
     $3ax - 4by = 1$ $\left(-\dfrac{19}{3a}, -\dfrac{5}{b}\right)$

123) $x - 3by = 2$
     $3x + by = -4$ $\left(-1, -\dfrac{1}{b}\right)$

124) $ax + 4y = 6$
     $2ax + y = 5$ $\left(\dfrac{2}{a}, 1\right)$

125) $ax + by = c$
     $-ax + by = c$ $\left(0, \dfrac{c}{b}\right)$

126) $2ax - by = c$
     $ax + 2by = 8c$ $\left(\dfrac{2c}{a}, \dfrac{3c}{b}\right)$

Systems like those in Exercises 127–130 can be solved by substituting one variable for $\dfrac{1}{x}$, another variable for $\dfrac{1}{y}$, and then using the elimination method. In these systems, substitute

u for $\dfrac{1}{x}$, v for $\dfrac{1}{y}$, and solve for u and v using elimination. Since the original system is in terms of x and y, the last step is to solve for x and y.

In the first equation of Exercise 127, for example, $\dfrac{1}{x} = u$ and $\dfrac{2}{y} = 2 \cdot \dfrac{1}{y} = 2v$. Substitute u for $\dfrac{1}{x}$ and $2v$ for $\dfrac{2}{y}$ so that the equation becomes $u + 2v = -\dfrac{7}{8}$. Use this approach to solve each system.

127) $\dfrac{1}{x} + \dfrac{2}{y} = -\dfrac{7}{8}$
     $\dfrac{1}{x} - \dfrac{1}{y} = \dfrac{5}{8}$ (8, −2)

128) $\dfrac{1}{x} + \dfrac{1}{y} = \dfrac{4}{3}$
     $\dfrac{4}{x} + \dfrac{1}{y} = \dfrac{13}{3}$ (1, 3)

129) $-\dfrac{1}{x} + \dfrac{2}{y} = -\dfrac{7}{3}$
     $\dfrac{2}{x} - \dfrac{3}{y} = 5$ $\left(\dfrac{1}{3}, 3\right)$

130) $\dfrac{6}{x} + \dfrac{1}{y} = 0$
     $-\dfrac{2}{x} + \dfrac{3}{y} = \dfrac{20}{3}$ $\left(-3, \dfrac{1}{2}\right)$

Solve by graphing. Given the functions $f(x)$ and $g(x)$, determine the value of x for which $f(x) = g(x)$.

131) $f(x) = x - 3$, $g(x) = -\dfrac{1}{4}x + 2$ 4

132) $f(x) = x + 4$, $g(x) = -2x - 2$ −2

133) $f(x) = 3x + 3$, $g(x) = x + 1$ −1

134) $f(x) = -\dfrac{2}{3}x + 1$, $g(x) = x - 4$ 3

Section 4.2 Solving Systems of Linear Equations in Three Variables

Objectives

1. Understand Systems of Three Equations in Three Variables
2. Solve Systems of Linear Equations in Three Variables
3. Solve Special Systems in Three Variables
4. Solve a System with Missing Terms

In this section, we will learn how to solve a system of linear equations in *three* variables.

1. Understand Systems of Three Equations in Three Variables

Definition

A **linear equation in three variables** is an equation of the form $Ax + By + Cz = D$ where A, B, and C are not all zero and where A, B, C, and D are real numbers. Solutions to this type of an equation are **ordered triples** of the form (x, y, z).

An example of a linear equation in three variables is

$$2x - y + 3z = 12.$$

This equation has infinitely many solutions. Here are a few:

$(5, 1, 1)$ since $2(5) - (1) + 3(1) = 12$
$(3, 0, 2)$ since $2(3) - (0) + 3(2) = 12$
$(6, -3, -1)$ since $2(6) - (-3) + 3(-1) = 12$

Ordered triples, like $(1, 2, 3)$ and $(3, 0, 2)$, are graphed on a three-dimensional coordinate system, as shown to the right. Notice that the ordered triples are *points*.

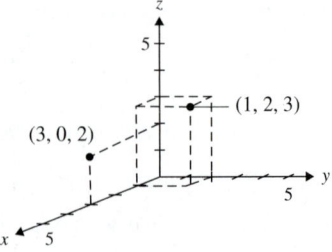

The graph of a linear equation in three variables is a *plane*.

A **solution to a system of linear equations in three variables** is an *ordered triple* that satisfies each equation in the system. Like systems of linear equations in two variables, systems of linear equations in *three* variables can have *one* solution, *no* solution, or *infinitely many* solutions.

Here is an example of a system of linear equations in three variables:

$$x + 4y + 2z = 10$$
$$3x - y + z = 6$$
$$2x + 3y - z = -4$$

In Section 4.1, we solved systems of linear equations in *two* variables by graphing. Since the graph of an equation like $x + 4y + 2z = 10$ is a *plane*, however, solving a system in three variables by graphing would not be practical. But let's look at the graphs of systems of linear equations in three variables that have one solution, no solution, or an infinite number of solutions.

One solution:

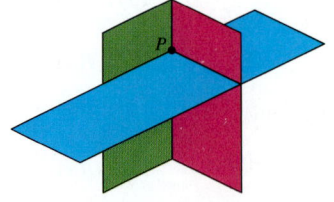

Intersection is at point P.

All three planes intersect at one point; this is the solution of the system.

No solution:

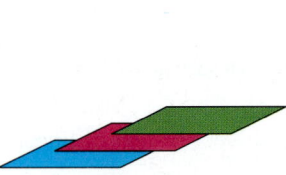

None of the planes may intersect or *two* of the planes may intersect, but if there is no solution to the system, *all three planes* do not have a common point of intersection.

Infinite number of solutions:

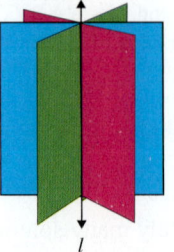

Intersection is the set of points on line l.

Intersection is the set of points on the plane.

The three planes may intersect so that they have a line or a plane in common. The solution to the system is the infinite set of points on the line or the plane, respectively.

2. Solve Systems of Linear Equations in Three Variables

First we will learn how to solve a system in which each equation has three variables.

> **Procedure**   Solving a System of Linear Equations in Three Variables
>
> 1) **Label** the equations ①, ②, and ③.
>
> 2) **Choose a variable to eliminate. Eliminate** this variable from *two* sets of *two* equations using the elimination method. You will obtain two equations containing the same two variables. Label one of these new equations $\boxed{A}$ and the other $\boxed{B}$.
>
> 3) Use the elimination method to **eliminate a variable from equations** $\boxed{A}$ **and** $\boxed{B}$. You have now found the value of one variable.
>
> 4) **Find the value of another variable** by substituting the value found in *Step 3* into equation $\boxed{A}$ or $\boxed{B}$ and solving for the second variable.
>
> 5) **Find the value of the third variable** by substituting the values of the two variables found in *Steps 3* and *4* into equation ①, ②, or ③.
>
> 6) **Check** the solution in each of the original equations, and **write the solution as an ordered triple.**

Example 1

In-Class Example 1
Solve
$x - 3y + z = 5$
$2x + y - 3z = -7$
$x + 2y - z = -4$
answer: (0, −1, 2)

Solve ① $x + 2y - 2z = 3$
         ② $2x + y + 3z = 1$
         ③ $x - 2y + z = -10$

Solution

Steps 1) and 2) We have already **labeled** the equations. We'll **choose** to eliminate the variable y from *two* sets of *two* equations:

a) Add equations ① and ③ to eliminate y. Label the resulting equation $\boxed{A}$.

$$
\begin{array}{ll}
① & \quad x + 2y - 2z = 3 \\
③ \;+ & \quad x - 2y + \;\; z = -10 \\
\hline
\boxed{A} & \quad 2x \qquad\;\; - \;\; z = -7
\end{array}
$$

b) Multiply equation ② by 2 and add it to equation ③ to eliminate y. Label the resulting equation $\boxed{B}$.

$$
\begin{array}{ll}
2 \times ② & \quad 4x + 2y + 6z = 2 \\
③ \;+ & \quad x - 2y + \;\; z = -10 \\
\hline
\boxed{B} & \quad 5x \qquad\quad + 7z = -8
\end{array}
$$

> **Note**
> Equations $\boxed{A}$ and $\boxed{B}$ contain only *two* variables and they are the same variables, x and z.

3) Use the elimination method to **eliminate a variable from equations** $\boxed{A}$ **and** $\boxed{B}$. We will eliminate z from $\boxed{A}$ and $\boxed{B}$. Multiply $\boxed{A}$ by 7 and add it to $\boxed{B}$.

$$
\begin{array}{ll}
7 \times \boxed{A} & \quad 14x - 7z = -49 \\
\boxed{B} \;+ & \quad\;\; 5x + 7z = -8 \\
\hline
& \quad 19x \qquad\;\; = -57 \\
& \quad \boxed{x = -3} \qquad \text{Divide by 19.}
\end{array}
$$

4) **Find the value of another variable** by substituting $x = -3$ into equation $\boxed{A}$ or $\boxed{B}$. We will use $\boxed{A}$ since it has smaller coefficients.

$$\boxed{A} \quad 2x - z = -7$$
$$2(-3) - z = -7 \qquad \text{Substitute } -3 \text{ for } x.$$
$$-6 - z = -7 \qquad \text{Multiply.}$$
$$\boxed{z = 1} \qquad \text{Add 6 and divide by } -1.$$

5) **Find the value of the third variable** by substituting $x = -3$ and $z = 1$ into equation ①, ②, or ③. We will use equation ①.

$$① \quad x + 2y - 2z = 3$$
$$-3 + 2y - 2(1) = 3 \qquad \text{Substitute } -3 \text{ for } x \text{ and 1 for } z.$$
$$-3 + 2y - 2 = 3 \qquad \text{Multiply.}$$
$$2y - 5 = 3 \qquad \text{Combine like terms.}$$
$$\boxed{y = 4} \qquad \text{Add 5 and divide by 2.}$$

6) **Check** the solution, $(-3, 4, 1)$, in each of the original equations, and **write the solution.**

①	②	③
$x + 2y - 2z = 3$	$2x + y + 3z = 1$	$x - 2y + z = -10$
$-3 + 2(4) - 2(1) \overset{?}{=} 3$	$2(-3) + 4 + 3(1) \overset{?}{=} 1$	$-3 - 2(4) + 1 \overset{?}{=} -10$
$-3 + 8 - 2 \overset{?}{=} 3$	$-6 + 4 + 3 \overset{?}{=} 1$	$-3 - 8 + 1 \overset{?}{=} -10$
$3 = 3$	$1 = 1$	$-10 = -10$
True	True	True

The solution is $(-3, 4, 1)$.   ■

You Try 1

Solve $x + 2y + 3z = -11$
 $3x - y + z = 0$
 $-2x + 3y - z = 4$

3. Solve Special Systems in Three Variables

Some systems in three variables have no solution and some have an infinite number of solutions.

Example 2

Solve   ① $-3x + 2y - z = 5$
 ② $x + 4y + z = -4$
 ③ $9x - 6y + 3z = -2$

In-Class Example 2
Solve
$x + 2y + 6z = 0$
$3x - 2y - 3z = 4$
$4x + 8y + 24z = -7$
answer: $\varnothing$

Solution

Steps 1) and 2) We have already *labeled* the equations. The *variable we choose to eliminate* is z, the easiest.

a)   Add equations ① and ② to eliminate z. Label the resulting equation $\boxed{A}$.

$$① \quad -3x + 2y - z = 5$$
$$② + \quad \underline{x + 4y + z = -4}$$
$$\boxed{A} \quad -2x + 6y = 1$$

b) Multiply equation ① by 3 and add it to equation ③ to eliminate z. Label the resulting equation B .

$$
\begin{array}{rl}
① \quad -3x + 2y - z = 5 \longrightarrow 3 \times ① & -9x + 6y - 3z = 15 \\
③ \quad + & 9x - 6y + 3z = -2 \\
\hline
\text{B} & 0 = 13 \quad \text{False}
\end{array}
$$

Since the variables are eliminated and we get the false statement $0 = 13$, equations ① and ③ have no ordered triple that satisfies each equation. The system is inconsistent, so there is no solution. The solution set is $\varnothing$. ■

Note

If the variables are eliminated and you get a false statement, there is *no solution* to the system. The system is inconsistent, so the solution set is $\varnothing$.

Example 3

Solve ① $\quad -4x - 2y + 8z = -12$
② $\qquad 2x + y - 4z = 6$
③ $\quad 6x + 3y - 12z = 18$

In-Class Example 3
Solve $4x - 8y + 12z = 4$
$\qquad x - 2y + 3z = 1$
$\quad -2x + 4y - 6z = -2$

answer:
$\{(x, y, z)\,|\,x - 2y + 3z = 1\}$

Solution

Steps 1) and 2) We label the equations and choose a variable, y, to eliminate.

a) Multiply equation ② by 2 and add it to equation ①. Label the resulting equation A .

$$
\begin{array}{rl}
2 \times ② & 4x + 2y - 8z = \quad 12 \\
① \quad + & -4x - 2y + 8z = -12 \\
\hline
\text{A} & 0 = 0 \quad \text{True}
\end{array}
$$

The variables were eliminated and we obtained the true statement $0 = 0$. This is because equation ① is a multiple of equation ②.

Notice, also, that equation ③ is a multiple of equation ②.

The equations in this system are dependent. There are an infinite number of solutions, and we write the solution set as $\{(x, y, z)\,|\,2x + y - 4z = 6\}$. The equations all have the same graph. ■

You Try 2

Solve each system of equations.

a) $\quad 8x + 20y - 4z = -16$
$\quad -6x - 15y + 3z = 12$
$\quad\quad 2x + 5y - z = -4$

b) $\quad x + 4y - 3z = 2$
$\quad 2x - 5y + 2z = -8$
$\quad -3x - 12y + 9z = 7$

4. Solve a System with Missing Terms

Example 4

In-Class Example 4
Solve $3x + 4y = -1$
$\quad\quad y - 2z = -13$
$\quad\quad 5x - z = -1$
answer: $(1, -1, 6)$

Solve ① $5x - 2y = 6$
     ② $\quad y + 2z = 1$
     ③ $3x - 4z = -8$

Solution

First, notice that while this *is* a system of three equations in three variables, none of the equations contains three variables. Furthermore, each equation is "missing" a different variable.

> **Note**
>
> We will use many of the *ideas* outlined in the steps for solving a system of three equations, but we will use *substitution* rather than the elimination method.

1) Label the equations ①, ②, ③.

2) The goal of step 2 is to obtain two equations that contain the same two variables. We will modify this step from the way it was outlined on p. 219.

    In order to obtain *two* equations with the same *two* variables, we will use *substitution*.

    Since y in equation ② is the only variable in the system with a coefficient of 1, we will solve equation ② for y.

    $$② \; y + 2z = 1$$
    $$y = 1 - 2z \quad\quad \text{Subtract } 2z.$$

    Substitute $y = 1 - 2z$ into equation ① to obtain an equation containing the variables x and z. Simplify. Label the resulting equation $\boxed{A}$.

    $$① \quad\quad\quad 5x - 2y = 6$$
    $$5x - 2(1 - 2z) = 6 \quad\quad \text{Substitute } 1 - 2z \text{ for } y.$$
    $$5x - 2 + 4z = 6 \quad\quad \text{Distribute.}$$
    $$\boxed{A} \quad\quad\quad 5x + 4z = 8 \quad\quad \text{Add 2.}$$

3) The goal of step 3 is to solve for one of the variables. Equations $\boxed{A}$ and ③ contain only x and z.

    We will eliminate z from $\boxed{A}$ and ③. Add the two equations to eliminate z, then solve for x.

    $$\boxed{A} \quad\quad 5x + 4z = 8$$
    $$③ \;\; + \; 3x - 4z = -8$$
    $$\overline{\quad\quad 8x \quad\quad\quad = 0}$$
    $$\boxed{x = 0} \quad\quad \text{Divide by 8.}$$

4) Find the value of another variable by substituting $x = 0$ into either $\boxed{A}$, ①, or ③.

    $$\boxed{A} \quad\quad 5x + 4z = 8$$
    $$5(0) + 4z = 8 \quad\quad \text{Substitute 0 for } x.$$
    $$4z = 8$$
    $$\boxed{z = 2} \quad\quad \text{Divide by 4.}$$

5)   Find the value of the third variable by substituting $x = 0$ into ① or $z = 2$ into ②.

① $5x - 2y = 6$
$5(0) - 2y = 6$ Substitute 0 for x.
$-2y = 6$
$\boxed{y = -3}$ Divide by -2.

6)   Check the solution $(0, -3, 2)$ in each of the original equations. The check is left to the student. The solution is $(0, -3, 2)$.

You Try 3

Solve $x + 2y = 8$
$2y + 3z = 1$
$3x - z = -3$

Answers to You Try Exercises

1) $(2, 1, -5)$ 2) a) $\{(x, y, z)|2x + 5y - z = -4\}$ b) $\varnothing$ c) $(-2, 5, -3)$

4.2 Exercises

*Additional answers can be found in the Answers to Exercises appendix.

Objective 1: Understand Systems of Three Equations in Three Variables

Determine whether the ordered triple is a solution of the system.

1) $4x + 3y - 7z = -6$
    $x - 2y + 5z = -3$
    $-x + y + 2z = 7$
    $(-2, 3, 1)$ yes

2) $3x + y + 2z = 2$
    $-2x - y + z = 5$
    $x + 2y - z = -11$
    $(1, -5, 2)$ yes

3) $-x + y - 2z = 2$
    $3x - y + 5z = 4$
    $2x + 3y - z = 7$
    $(0, 6, 2)$ no

4) $6x - y + 4z = 4$
    $-2x + y - z = 5$
    $2x - 3y + z = 2$
    $\left(-\dfrac{1}{2}, -3, 1\right)$ no

5)   Write a system of equations in x, y, and z so that the ordered triple $(4, -1, 2)$ is a solution of the system.
    Answers may vary.

6)   Find the value of c so that $(6, 0, 5)$ is a solution of the system $2x - 5y - 3z = -3$ 4
    $-x + y + 2z = 4$
    $-2x + 3y + cz = 8$

Objective 2: Solve Systems of Linear Equations in Three Variables

Solve each system. See Example 1.

7) $x + 3y + z = 3$
    $4x - 2y + 3z = 7$
    $-2x + y - z = -1$
    $(-2, 0, 5)$

8) $x - y + 2z = -7$
    $-3x - 2y + z = -10$
    $5x + 4y + 3z = 4$
    $(1, 2, -3)$

9) $5x + 3y - z = -2$
    $-2x + 3y + 2z = 3$
    $x + 6y + z = -1$
    $(1, -1, 4)$

10) $-2x - 2y + 3z = 2$
    $3x + 3y - 5z = -3$
    $-x + y - z = 9$
    $(-5, 4, 0)$

11) $3a + 5b - 3c = -4$
    $a - 3b + c = 6$
    $-4a + 6b + 2c = -6$
    $\left(2, -\dfrac{1}{2}, \dfrac{5}{2}\right)$

12) $a - 4b + 2c = -7$
    $3a - 8b + c = 7$
    $6a - 12b + 3c = 12$
    $\left(6, \dfrac{3}{4}, -5\right)$

Objective 3: Solve Special Systems in Three Variables

Solve each system. Identify any systems that are inconsistent or that have dependent equations. See Examples 2 and 3.

13) $a - 5b + c = -4$ $\varnothing$; inconsistent
    $3a + 2b - 4c = -3$
    $6a + 4b - 8c = 9$

14) $-a + 2b - 12c = 8$ $\varnothing$; inconsistent
    $-6a + 2b - 8c = -3$
    $3a - b + 4c = 4$

15) $-15x - 3y + 9z = 3$
    $5x + y - 3z = -1$ $\{(x, y, z)|5x + y - 3z = -1\}$;
    $10x + 2y - 6z = -2$ dependent equations

16) $-4x + 10y - 16z = -6$
    $-6x + 15y - 24z = -9$ $\{(x, y, z)|2x - 5y + 8z = 3\}$;
    $2x - 5y + 8z = 3$ dependent equations

17) $-3a + 12b - 9c = -3$ $\{(a, b, c)|-a + 4b - 3c = -1\}$;
    $5a - 20b + 15c = 5$ dependent equations
    $-a + 4b - 3c = -1$

18) $3x - 12y + 6z = 4$ $\varnothing$; inconsistent
    $-x + 4y - 2z = 7$
    $5x + 3y + z = -2$

Objective 4: Solve a System with Missing Terms

Solve each system. See Example 4.

19) $5x - 2y + z = -5$ $(2, 5, -5)$
    $x - y - 2z = 7$
    $4y + 3z = 5$

20) $-x + z = 9$ $(-3, 1, 6)$
    $-2x + 4y - z = 4$
    $7x + 2y + 3z = -1$

21) $\begin{aligned} a + 15b &= 5 \\ 4a + 10b + c &= -6 \\ -2a - 5b - 2c &= -3 \end{aligned}$ $\left(-4, \dfrac{3}{5}, 4\right)$

22) $\begin{aligned} 2x - 6y - 3z &= 4 \\ -3y + 2z &= -6 \\ -x + 3y + z &= -1 \end{aligned}$ $\left(1, \dfrac{2}{3}, -2\right)$

23) $\begin{aligned} x + 2y + 3z &= 4 \\ -3x + y &= -7 \\ 4y + 3z &= -10 \end{aligned}$ $(0, -7, 6)$

24) $\begin{aligned} -3a + 5b + c &= -4 \\ a + 5b &= 3 \\ 4a - 3c &= -11 \end{aligned}$ $\left(4, -\dfrac{1}{5}, 9\right)$

25) $\begin{aligned} -5x + z &= -3 \\ 4x - y &= -1 \\ 3y - 7z &= 1 \end{aligned}$ $(1, 5, 2)$

26) $\begin{aligned} a + b &= 1 \\ a - 5c &= 2 \\ b + 2c &= -4 \end{aligned}$ $(7, -6, 1)$

VIDEO 27) $\begin{aligned} 4a + 2b &= -11 \\ -8a - 3c &= -7 \\ b + 2c &= 1 \end{aligned}$ $\left(-\dfrac{1}{4}, -5, 3\right)$

28) $\begin{aligned} 3x + 4y &= -6 \\ -x + 3z &= 1 \\ 2y + 3z &= -1 \end{aligned}$ $\left(-2, 0, -\dfrac{1}{3}\right)$

Mixed Exercises: Objectives 2–4

Solve each system. Identify any systems that are inconsistent or that have dependent equations.

29) $\begin{aligned} 6x + 3y - 3z &= -1 \\ 10x + 5y - 5z &= 4 \\ x - 3y + 4z &= 6 \end{aligned}$ $\varnothing$; inconsistent

30) $\begin{aligned} 2x + 3y - z &= 0 \\ x - 4y - 2z &= -5 \\ -4x + 5y + 3z &= -4 \end{aligned}$ $(5, -1, 7)$

31) $\begin{aligned} 7x + 8y - z &= 16 \\ -\dfrac{1}{2}x - 2y + \dfrac{3}{2}z &= 1 \\ \dfrac{4}{3}x + 4y - 3z &= -\dfrac{2}{3} \end{aligned}$ $\left(4, -\dfrac{3}{2}, 0\right)$

32) $\begin{aligned} 3a + b - 2c &= -3 \\ 9a + 3b - 6c &= -9 \\ -6a - 2b + 4c &= 6 \end{aligned}$ $\{(a, b, c) \mid 3a + b - 2c = -3\};$ dependent equations

33) $\begin{aligned} 2a - 3b &= -4 \\ 3b - c &= 8 \\ -5a + 4c &= -4 \end{aligned}$ $(4, 4, 4)$

34) $\begin{aligned} 5x + y - 2z &= -2 \\ -\dfrac{1}{2}x - \dfrac{3}{4}y + 2z &= \dfrac{5}{4} \\ x - 6z &= 3 \end{aligned}$ $\left(0, -3, -\dfrac{1}{2}\right)$

35) $\begin{aligned} -4x + 6y + 3z &= 3 \\ -\dfrac{2}{3}x + y + \dfrac{1}{2}z &= \dfrac{1}{2} \\ 12x - 18y - 9z &= -9 \end{aligned}$ $\{(x, y, z) \mid -4x + 6y + 3z = 3\};$ dependent equations

36) $\begin{aligned} x - \dfrac{5}{2}y + \dfrac{1}{2}z &= \dfrac{5}{4} \\ x + 3y - z &= 4 \\ -6x + 15y - 3z &= -1 \end{aligned}$ $\varnothing$; inconsistent

37) $\begin{aligned} a + b + 9c &= -3 \\ -5a - 2b + 3c &= 10 \\ 4a + 3b + 6c &= -15 \end{aligned}$ $\left(1, -7, \dfrac{1}{3}\right)$

38) $\begin{aligned} 2x + 3y &= 2 \\ -3x + 4z &= 0 \\ y - 5z &= -17 \end{aligned}$ $(4, -2, 3)$

39) $\begin{aligned} x + 5z &= 10 \\ 4y + z &= -2 \\ 3x - 2y &= 2 \end{aligned}$ $(0, -1, 2)$

40) $\begin{aligned} a + 3b - 8c &= 2 \\ -2a - 5b + 4c &= -1 \\ 4a + b + 16c &= -4 \end{aligned}$ $\left(0, 0, -\dfrac{1}{4}\right)$

VIDEO 41) $\begin{aligned} 2x - y + 4z &= -1 \\ x + 3y + z &= -5 \\ -3x + 2y &= 7 \end{aligned}$ $(-3, -1, 1)$

42) $\begin{aligned} -2a + 3b &= 3 \\ a + 5c &= -1 \\ b - 2c &= -5 \end{aligned}$ $(-6, -3, 1)$

43) Given the following two equations, write a third equation to obtain a system of three equations in x, y, and z so that the system has no solution.   Answers may vary.

$$x + 3y - 2z = -9$$
$$2x - 5y + z = 1$$

44) Given the following two equations, write a third equation to obtain a system of three equations in x, y, and z so that the system has an infinite number of solutions.
Answers may vary.

$$9x - 12y + 3z = 21$$
$$-3x + 4y - z = -7$$

Extension

Extend the concepts of this section to solve each system. Write the solution in the form (a, b, c, d)

45) $\begin{aligned} a - 2b - c + d &= 0 \\ -a + 2b + 3c + d &= 6 \\ 2a + b + c - d &= 8 \\ a - b + 2c + 3d &= 7 \end{aligned}$ $(3, 1, 2, 1)$

46) $\begin{aligned} -a + 4b + 3c - d &= 4 \\ 2a + b - 3c + d &= -6 \\ a + b + c + d &= 0 \\ a - b + 2c - d &= -1 \end{aligned}$ $(-2, 0, 1, 1)$

47) $\begin{aligned} 3a + 4b + c - d &= -7 \\ -3a - 2b - c + d &= 1 \\ a + 2b + 3c - 2d &= 5 \\ 2a + b + c - d &= 2 \end{aligned}$ $(0, -3, 1, -4)$

48) $\begin{aligned} 3a - 4b + c + d &= 12 \\ -3a + 2b - c + 3d &= -4 \\ a - 2b + 2c - d &= 2 \\ -a + 4b + c + d &= 8 \end{aligned}$ $(5, 2, 2, 3)$

Section 4.3 Applications of Systems of Linear Equations

Objectives

1. Solve General Two-Variable Problems
2. Solve Problems Involving Geometry
3. Solve Problems Involving Cost
4. Solve Mixture Problems
5. Solve Problems Involving Distance, Rate, and Time
6. Solve Problems Involving Three Variables

1. Solve General Two-Variable Problems

In Section 2.1, we introduced the five-step problem-solving method. Throughout most of this section, we will modify the method for problems with *two* unknowns and *two* equations.

> **Procedure**   Solving an Applied Problem Using a System of Equations
>
> **Step 1:** **Read** the problem carefully, more than once if necessary. Draw a picture, if applicable. Identify what you are being asked to find.
>
> **Step 2:** **Choose variables** to represent the unknown quantities. Label any pictures with the variables.
>
> **Step 3:** **Write a system of equations using two variables.** It may be helpful to begin by writing the equations in words.
>
> **Step 4:** **Solve** the system.
>
> **Step 5:** **Check** the answer in the original problem, and **interpret** the meaning of the solution as it relates to the problem.

Example 1

In-Class Example 1
Write a system of equations and solve.
   In May 2008 MySpace and Facebook had a total of 96 billion page views. If MySpace had 5.2 billion fewer page views, how many page views did each website have that month? (www.washingtonpost.com)
answer: MySpace: 45.4 billion; Facebook: 50.6 billion

Write a system of equations and solve.

In 2010, the prime-time TV shows of HBO received 53 more Emmy Award nominations than NBC. Together, their shows received a total of 149 nominations. How many Emmy nominations did HBO and NBC each receive in 2010? (www.emmys.org)

Solution

Step 1: **Read** the problem carefully, and identify what we are being asked to find. We must find the number of Emmy nominations received by HBO and by NBC.

Step 2: **Choose variables** to represent the unknown quantities.

x = number of nominations for HBO
y = number of nominations for NBC

Step 3: **Write a system of equations using two variables.** Let's write the equations in English first. Then, we can translate them to algebraic equations.

To get one equation, use the information that says HBO and NBC received a total of 149 nominations.

$$\underset{x}{\text{Number of HBO nominations}} + \underset{y}{\text{Number of NBC nominations}} = \underset{149}{\text{Total number of nominations}} \qquad \text{Equation (1)}$$

To get the second equation, use the information that says that HBO received 53 more nominations than NBC.

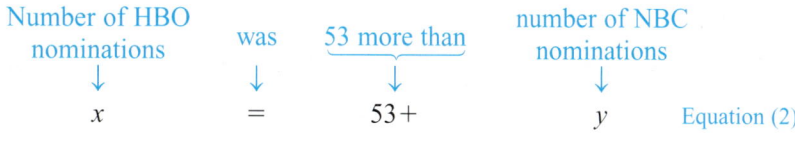

$$\underset{x}{\text{Number of HBO nominations}} \quad \underset{=}{\text{was}} \quad \underset{53+}{\underline{\text{53 more than}}} \quad \underset{y}{\text{number of NBC nominations}} \qquad \text{Equation (2)}$$

The system of equations is
$$x + y = 149$$
$$x = 53 + y.$$

Step 4:   **Solve** the system. Let's use substitution.

$$x + y = 149$$
$$(53 + y) + y = 149 \qquad \text{Substitute.}$$
$$53 + 2y = 149$$
$$2y = 96$$
$$y = 48$$

Find x by substituting $y = 48$ into $x = 53 + y$.

$$x = 53 + 48$$
$$x = 101$$

The solution to the system is $(101, 48)$.

Step 5:   **Check** the answer in the original problem and **interpret** the meaning of the solution as it relates to the problem. *HBO received 101 nominations, and NBC received 48.* The total number of nominations was $101 + 48 = 149$. HBO received 53 more nominations than NBC and $53 + 48 = 101$. The answer is correct. ∎

You Try 1

Write a system of equations and solve.

The Turquoise Bay Motel has 51 rooms. Some of them have two double beds and half as many rooms have one queen-size bed. Determine how many rooms have two beds and how many rooms have one bed.

In Chapter 2 we solved problems involving geometry by defining the unknowns in terms of one variable. Now, we will use a system of equations to solve these applications.

2. Solve Problems Involving Geometry

Example 2

In-Class Example 2
The Herold family purchased a rectangular lot on which they will build a new house. The lot is 40 ft wider than it is long, and the perimeter is 400 ft. Find the length and width of the lot.
answer: width: 120 ft
length: 80 ft

Write a system of equations and solve.

The Alvarez family purchased a rectangular lot on which they will build a new house. The lot is 70 ft longer than it is wide, and the perimeter is 460 ft. Find the length and width of the lot.

Solution

Step 1:   **Read** the problem carefully. Draw a picture. We must find the length and width of the lot.

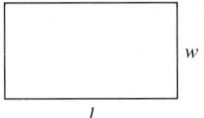

Step 2:   **Choose variables** to represent each unknown quantity and label the picture.

$$l = \text{length of the lot}$$
$$w = \text{width of the lot}$$

Step 3:   **Write a system of equations using two variables.**

We know the lot is 70 ft longer than it is wide, so

Length	is	70 ft more than	the width.
↓	↓	↓	↓
l	$=$	$70+$	w Equation (1)

The perimeter of the lot is 460 feet. Write the second equation using the formula for the perimeter of a rectangle.

Perimeter $=$ 2(length) $+$ 2(width)

$$460 \quad = \quad 2l \quad + \quad 2w \qquad \text{Equation (2)}$$

The system of equations is $\begin{aligned} l &= 70 + w \\ 2l + 2w &= 460. \end{aligned}$

Step 4:   **Solve** the system. Let's use substitution.

$$2l + 2w = 460$$
$$2(70 + w) + 2w = 460 \qquad \text{Substitute.}$$
$$140 + 2w + 2w = 460 \qquad \text{Distribute.}$$
$$4w + 140 = 460$$
$$4w = 320$$
$$w = 80$$

Find l by substituting $w = 80$ into $l = 70 + w$.

$$l = 70 + 80 = 150$$

The solution to the system is (150, 80). (The ordered pair is written as (l, w), in alphabetical order.)

Step 5:   **Check** the answer in the original problem.

The length of the lot is 150 ft, and the width of the lot is 80 ft. Check the solution to verify that the numbers make sense. ■

You Try 2

Write a system of equations and solve.

A rectangular mouse pad is 1.5 in. longer than it is wide. The perimeter of the mouse pad is 29 in. Find its dimensions.

Another common type of problem involves the cost of items and the number of items.

3. Solve Problems Involving Cost

Example 3

In-Class Example 3
Write a system of equations and solve.
Tim bought four tickets to a Kenny Chesney concert and three tickets for a Jimmy Buffett concert for $334.00. Faith bought two tickets to see Kenny Chesney and five tickets to see Jimmy Buffett for $293.00. Find the cost of a ticket to each concert.
answer: Kenny Chesney: $56.50
Jimmy Buffett: $36.00

Write a system of equations and solve.

   In 2004, Usher and Alicia Keys each had concerts at the Allstate Arena near Chicago. Kayla and Levon sat in the same section for each performance. Kayla bought four tickets to see Usher and four to see Alicia Keys for $360. Levon spent $220.50 on two Usher tickets and three Alicia Keys tickets. Find the cost of a ticket to each concert. (www.pollstaronline.com)

Solution

Step 1:   **Read** the problem carefully. We must find the cost of an Usher ticket and the cost of an Alicia Keys ticket.

Step 2:   **Choose variables** to represent each unknown quantity.

$$x = \text{cost of one Usher ticket}$$
$$y = \text{cost of one Alicia Keys ticket}$$

Step 3: **Write a system of equations using two variables.**

Use the information about Kayla's purchase.

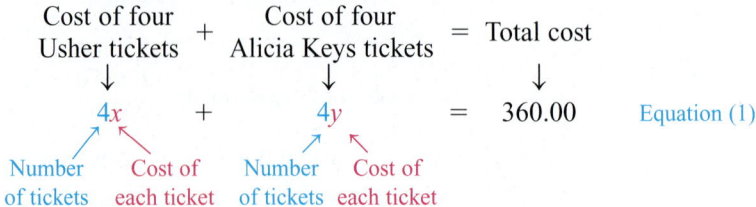

$$4x + 4y = 360.00 \qquad \text{Equation (1)}$$

Use the information about Levon's purchase.

Cost of two Usher tickets	+	Cost of three Alicia Keys tickets	=	Total cost
↓		↓		↓
2x	+	3y	=	220.50

Equation (2)

Number of tickets — Cost of each ticket — Number of tickets — Cost of each ticket

The system of equations is
$$4x + 4y = 360.00$$
$$2x + 3y = 220.50.$$

Step 4: **Solve** the system. Let's use the elimination method. Multiply equation (2) by -2 to eliminate x.

$$
\begin{array}{rl}
4x + 4y = & 360.00 \\
+\ \underline{-4x - 6y = -441.00} \\
-2y = & -81.00 \qquad \text{Add the equations.}\\
y = & 40.50
\end{array}
$$

Find x. We will substitute $y = 40.50$ into $4x + 4y = 360.00$.

$$
\begin{aligned}
4x + 4(40.50) &= 360.00 \qquad \text{Substitute.}\\
4x + 162 &= 360.00 \\
4x &= 198.00 \\
x &= 49.50
\end{aligned}
$$

The solution to the system is (49.50, 40.50).

Step 5: **Check** and interpret the solution.

The cost of a ticket to see Usher in concert was $49.50, and the cost of a ticket to see Alicia Keys was $40.50. Check the numbers to verify that they are correct.

You Try 3

Write a system of equations and solve.

At Julie's Jewelry Box, all necklaces sell for one price and all pairs of earrings sell for one price. Cailen buys three pairs of earrings and a necklace for $19.00, while Marcella buys one pair of earrings and two necklaces for $18.00. Find the cost of a pair of earrings and the cost of a necklace.

Next, we will discuss how to solve mixture problems using two variables and a system of equations.

4. Solve Mixture Problems

Example 4

In-Class Example 4
Write a system of equations and solve.
   How many ounces of a 6% acid solution and how many ounces of a 4% acid solution must be mixed to get 200 oz of a 5% acid solution?
answer: 100 oz of 4% acid solution and 100 oz of 6% acid solution

Write a system of equations and solve.
   How many milliliters of an 8% hydrogen peroxide solution and how many milliliters of a 2% hydrogen peroxide solution should a pharmacist mix to get 300 ml of a 4% hydrogen peroxide solution?

Solution

Step 1: **Read** the problem carefully. We must find the amount of the 8% solution and the amount of the 2% solution she should use.

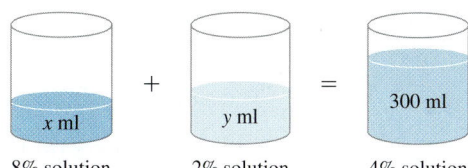

Step 2: **Choose variables** to represent each unknown quantity.

$$x = \text{amount of } 8\% \text{ solution}$$
$$y = \text{amount of } 2\% \text{ solution}$$

Step 3: **Write a system of equations using two variables.**

Make a table to organize the information. To obtain the expression in the last column, multiply the percent of hydrogen peroxide in the solution by the amount of solution to get the amount of hydrogen peroxide in the solution.

	Percent of Hydrogen Peroxide in Solution (as a decimal)	Amount of Solution	Amount of Hydrogen Peroxide in Solution
Mix these	0.08	x	$0.08x$
	0.02	y	$0.02y$
to make →	0.04	300	$0.04(300)$

To get one equation, use the information in the *second column:*

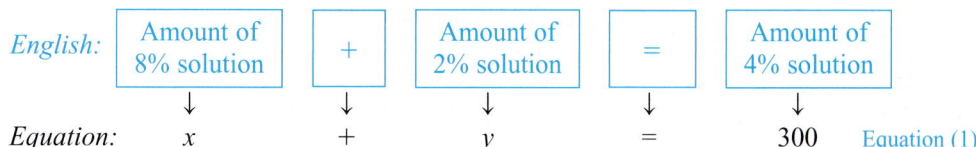

To get the second equation, use the information in the *third column:*

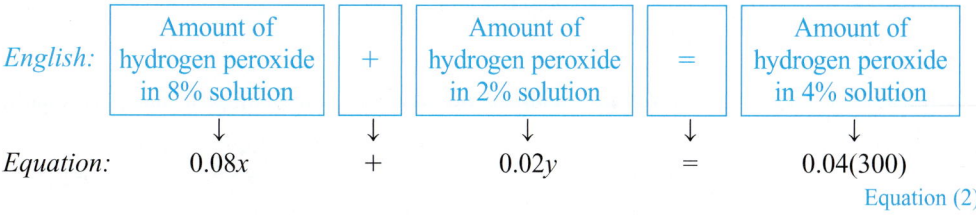

The system of equations is
$$x + y = 300$$
$$0.08x + 0.02y = 0.04(300).$$

Step 4:   **Solve** the system. Multiply the second equation by 100 to eliminate the decimals. We get $8x + 2y = 4(300)$.

$$x + y = 300$$
$$8x + 2y = 1200 \qquad \textcolor{blue}{4(300) = 1200}$$

Use the elimination method. Multiply the first equation by -2 to eliminate y.

$$\begin{aligned} -2x - 2y &= -600 \\ + \quad 8x + 2y &= 1200 \\ \hline 6x &= 600 \qquad \textcolor{red}{\text{Add the equations.}} \\ x &= 100 \end{aligned}$$

To find y, substitute $x = 100$ into $x + y = 300$.

$$100 + y = 300 \qquad \textcolor{blue}{\text{Substitute.}}$$
$$y = 200$$

The solution to the system is (100, 200).

Step 5:   **Check** and **interpret** the solution.

The pharmacist needs 100 ml of an 8% hydrogen peroxide solution and 200 ml of a 2% solution to make 300 ml of a 4% hydrogen peroxide solution. Check the answers in the equations to verify that they are correct. ∎

You Try 4

Write a system of equations and solve.
How many milliliters of a 3% acid solution must be added to 60 ml of an 11% acid solution to make a 9% acid solution?

5. Solve Problems Involving Distance, Rate, and Time

Example 5

Write a system of equations and solve.
Julia and Katherine start at the same point and begin biking in opposite directions. Julia rides 2 mph faster than Katherine. After 2 hr, the girls are 44 mi apart. How fast was each of them riding?

In-Class Example 5
Write a system of equations and solve.
   Jill and Stacey start at the same point and begin running in opposite directions. Jill runs 3 mph slower than Stacey. After 1 hr the two are 9 mi apart. How fast was each one of them running?
**answer: Jill: 3 mph
Stacey: 6 mph**

Solution

Step 1:   **Read** the problem carefully. Draw a picture. We must find the speed at which each girl was riding.

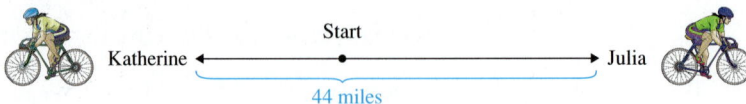

Katherine ← Start → Julia

44 miles

Step 2:   **Choose variables** to represent each unknown quantity.

$$x = \text{Julia's speed}$$
$$y = \text{Katherine's speed}$$

Step 3:   Write a system of equations using two variables.

Let's make a table using the equation $d = rt$. Fill in their speeds (rates) x and y, and fill in 2 for the time since each rode for 2 hr. Multiply these together to fill in the values for the distance.

	d	r	t
Julia	$2x$	x	2
Katherine	$2y$	y	2

Label the picture with expressions for the distances.

To get one equation, look at the picture and think about the distance between the girls after 2 hr.

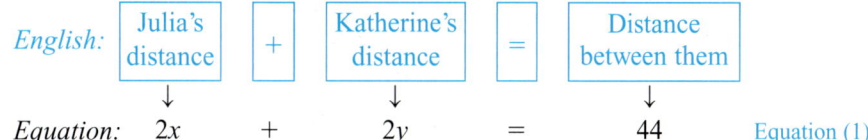

English: | Julia's distance | + | Katherine's distance | = | Distance between them |

Equation: $2x$ + $2y$ =   44   Equation (1)

To get the second equation, use the information that says Julia rides 2 mph faster than Katherine.

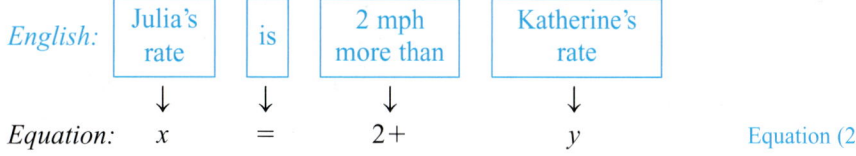

English: | Julia's rate | is | 2 mph more than | Katherine's rate |

Equation: x = $2+$ y Equation (2)

The system of equations is $\begin{aligned} 2x + 2y &= 44 \\ x &= 2 + y. \end{aligned}$

Step 4:   Solve the system. Use substitution.

$$2x + 2y = 44$$
$$2(2 + y) + 2y = 44 \qquad \text{Substitute } 2 + y \text{ for } x.$$
$$4 + 2y + 2y = 44 \qquad \text{Distribute.}$$
$$4y + 4 = 44$$
$$4y = 40$$
$$y = 10$$

Find x by substituting $y = 10$ into $x = 2 + y$.

$$x = 2 + 10 = 12$$

The solution to the system is (12, 10).

Step 5:   Check and **interpret** the solution.

Julia rides her bike at 12 mph, and Katherine rides hers at 10 mph. Julia's speed is 2 mph faster than Katherine's, as stated in the problem. And, since each girl rode for 2 hr, the distance between them is

2(12)   +   2(10) = 24 + 20 = 44 mi. The answer is correct.

Julia's distance   Katherine's distance

You Try 5

Write a system of equations and solve.

Kenny and Kyle start at the same point and begin biking in opposite directions. Kenny rides 1 mph faster than Kyle. After 3 hr, the boys are 51 mi apart. How fast was each of them riding?

6. Solve Problems Involving Three Variables

To solve applications involving a system of three equations in three variables, we will extend the method used for two equations in two variables.

Example 6

In-Class Example 6
Write a system of equations and solve.
   An accounting firm has 41 workers. There are twice as many full-time employees as part-time employees, and the number of temporary workers is seven less than the number of part-timers. How many full-time, part-time, and temporary workers are employed by this company?
answer: 24 full-time, 12 part-time, and 5 temporary

Write a system of equations and solve.

   The top three gold-producing nations in 2002 were South Africa, the United States, and Australia. Together, these three countries produced 37% of the gold during that year. Australia's share was 2% less than that of the United States, while South Africa's percentage was 1.5 times Australia's percentage of world gold production. Determine what percentage of the world's gold supply was produced by each country in 2002. (Market Share Reporter—2005, Vol. 1, "Mine Product" http://www.gold.org/value/market/supply-demand/min_production.html from World Gold Council)

Solution

Step 1: **Read** the problem carefully. We must determine the percentage of the world's gold produced by South Africa, the United States, and Australia in 2002.

Step 2: **Choose variables** to represent the unknown quantities.

x = percentage of world's gold supply produced by South Africa
y = percentage of world's gold supply produced by the United States
z = percentage of world's gold supply produced by Australia

Step 3: **Write a system of equations using the variables.**

To write one equation, we will use the information that says *together* the three countries produced 37% of the gold.

$$x + y + z = 37 \qquad \text{Equation (1)}$$

To write a second equation, we will use the information that says Australia's share was 2% less than that of the United States.

$$z = y - 2 \qquad \text{Equation (2)}$$

To write the third equation, we will use the statement that says South Africa's percentage was 1.5 times Australia's percentage.

$$x = 1.5z \qquad \text{Equation (3)}$$

The system is ① $x + y + z = 37$
② $z = y - 2$
③ $x = 1.5z$

Step 4: **Solve** the system. Since two of the equations contain only two variables, we will modify our steps to solve the system.

Our plan is to rewrite equation ① in terms of a single variable, z, and solve for z.

Solve equation ② for y.

② $z = y - 2$
$z + 2 = y$ \qquad Solve for y.

Now rewrite equation ① using the value for y from equation ② and the value for x from equation ③.

① $x + y + z = 37$ \qquad Equation (1)
$(1.5z) + (z + 2) + z = 37$ \qquad Substitute $1.5z$ for x and $z + 2$ for y.
$3.5z + 2 = 37$ \qquad Combine like terms.
$3.5z = 35$ \qquad Subtract 2.
$z = 10$ \qquad Divide by 3.5.

To solve for x, we can substitute $z = 10$ into equation ③.

③ $x = 1.5z$
     $x = 1.5(10)$ Substitute 10 for z.
     $\boxed{x = 15}$ Multiply.

To solve for y, we substitute $z = 10$ into equation ②.

② $z = y - 2$
     $10 = y - 2$ Substitute 10 for z.
     $\boxed{12 = y}$ Solve for y.

The solution of the system is (15, 12, 10).

Step 5:   **Check** and **interpret** the solution.

In 2002, South Africa produced 15% of the world's gold, the United States produced 12%, and Australia produced 10%. The check is left to the student. ■

You Try 6

Write a system of equations and solve.

Amelia, Bella, and Carmen are sisters. Bella is 5 yr older than Carmen, and Amelia's age is 5 yr less than twice Carmen's age. The sum of their ages is 48. How old is each girl?

Answers to You Try Exercises

1) 34 rooms have two beds, 17 rooms have one bed.   2) 6.5 in. by 8 in.
3) A pair of earrings costs $4.00 and a necklace costs $7.00.   4) 20 ml
5) Kyle: 8 mph; Kenny: 9 mph   6) Amelia: 19; Bella: 17; Carmen: 12

4.3 Exercises

*Additional answers can be found in the Answers to Exercises appendix.

Objective 1: Solve General Two-Variable Problems

Write a system of equations and solve.

1) The sum of two numbers is 36, and one number is two more than the other. Find the numbers.   17 and 19

2) One number is half another number. The sum of the two numbers is 108. Find the numbers.   72 and 36

3) In 2005, *The Aviator* was nominated for four more Academy Awards than the movie *Finding Neverland*. Together they received 18 nominations. How many Academy Award nominations did each movie receive? (*Chicago Tribune*, 1/26/05)   The Aviator: 11; Finding Neverland: 7

4) Through 2009, the University of Southern California football team played in 13 more Rose Bowl games than the University of Michigan. All together, these teams have appeared in 53 Rose Bowls. How many appearances did each team make in the Rose Bowl? (www.tournamentofroses.com)   USC: 33; Michigan: 20

5) There were a total of 2626 IHOP and Waffle House restaurants across the United States at the end of 2004. There were 314 fewer IHOPs than Waffle Houses. Determine the number of IHOP and Waffle House restaurants in the United States. (www.ihop.com, www.ajc.com)   IHOP: 1156; Waffle House: 1470

6) Through 2008, the University of Kentucky men's basketball team won four fewer NCAA championships than UCLA. The two teams won a total of 18 championship titles. How many championships did each team win? (www.ncaasports.com)   UCLA: 11; Kentucky: 7

7) The first NCAA Men's Basketball Championship game was played in 1939 in Evanston, Illinois. In 2004, 38,968 more people attended the final game in San Antonio than attended the first game in Evanston. The total number of people who attended these two games was 49,968. How many people saw the first NCAA championship in 1939, and how many were at the game in 2004? (www.ncaasports.com)   1939: 5500; 2004: 44,468

8) George Strait won 6 more Country Music Awards than Alan Jackson through 2008. They won a total of 38 awards. How many awards did each man win? (www.cmaawards.com)   George Strait: 22; Alan Jackson: 16

VIDEO 9) Annual per capita consumption of chicken in the United States in 2001 was 6.3 lb less than that of beef. Together, each American consumed, on average, 120.5 lb of beef and chicken in 2001. Find the amount of beef and the amount of chicken consumed, per person, in 2001. (U.S. Department of Agriculture)   beef: 63.4 lb; chicken: 57.1 lb

10) Mr. Chen has 27 students in his American History class. For their assignment on the Civil War, twice as many students chose to give a speech as chose to write a paper. How many students will be giving speeches, and how many will be writing papers?   speeches: 18; papers: 9

Objective 2: Solve Problems Involving Geometry

11) The length of a rectangle is twice its width. Find the length and width of the rectangle if its perimeter is 78 in.   length: 26 in.; width: 13 in.

12) The width of a rectangle is 5 cm less than the length. If the perimeter is 46 cm, what are the dimensions of the rectangle?   length: 14 cm.; width: 9 cm

VIDEO 13) Find the dimensions of a rectangular door that has a perimeter of 220 in. if the width is 50 in. less than the height of the door.   width: 30 in.; height: 80 in.

14) Yuki has a rectangular picture she wants to frame. Its perimeter is 42 in., and it is 4 in. longer than it is wide. Find the length and width of the picture.   length: 12.5 in.; width: 8.5 in.

15) An iPod Mini is rectangular in shape and has a perimeter of 28 cm. Its length is 4 cm more than its width. What are the dimensions of the iPod Mini?   length: 9 cm; width: 5 cm

16) Tiny Tots Day Care needs to put a new fence around their rectangular playground. They have determined they will need 220 ft of fencing. If the width of the playground is 30 ft less than the length, what are the playground's dimensions?   width: 40 ft; length: 70 ft

17) Find the measures of angles x and y if the measure of angle x is two-thirds the measure of angle y and if the angles are related according to the figure.

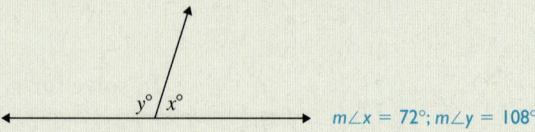

$m\angle x = 72°$; $m\angle y = 108°$

18) Find the measures of angles x and y if the measure of angle y is half the measure of angle x and if the angles are related according to the figure.

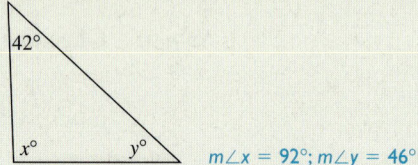

$m\angle x = 92°$; $m\angle y = 46°$

Objective 3: Solve Problems Involving Cost

VIDEO 19) Jennifer and Carlos attended two concerts with their friends at the American Airlines Arena in Miami. Jennifer and her friends bought five tickets to see Marc Anthony and two tickets to the Santana concert for $563. Carlos' group purchased three Marc Anthony tickets and six for Santana for $657. Find the cost of a ticket to each concert. (www.pollstaronline.com)   Marc Anthony: $86; Santana: $66.50

20) Both of the pop groups Train and Maroon 5 played at the House of Blues in North Myrtle Beach, SC in 2003. Three Maroon 5 tickets and two Train tickets would have cost $64, while four Maroon 5 tickets and four Train tickets would have cost $114. Find the cost of a ticket to each concert. (www.pollstaronline.com)   Maroon 5: $7.00; Train: $21.50

21) Every Tuesday, Stacey goes to Panda Express to buy lunch for herself and her colleagues. One day she buys 3 two-item meals and 1 three-item meal for $21.96. The next Tuesday she spends $23.16 on 2 two-item meals and 2 three-item meals. What is the cost of a two-item meal and of a three-item meal? (Panda Express menu)   two-item: $5.19; three-item: $6.39

22) Assume two families pay the average ticket price to attend a Green Bay Packers game in 2004. One family buys four tickets and a parking pass for $242.60. The other family buys six tickets and a parking pass for $351.40. Find the average ticket price and the cost of a parking pass to a Packers game in 2004. (www.teammarketing.com)   average ticket price: $54.40; parking pass: $25.00

23) On vacation, Wendell buys three key chains and five postcards for $10.00, and his sister buys two key chains and three postcards for $6.50. Find the cost of each souvenir.   key chain: $2.50; postcard: $0.50

24) At Sparkle Car Wash, two deluxe car washes and three regular car washes would cost $26.00. Four regular washes and one deluxe wash would cost $23.00. What is the cost of a deluxe car wash and of a regular wash?
deluxe: $7; regular: $4

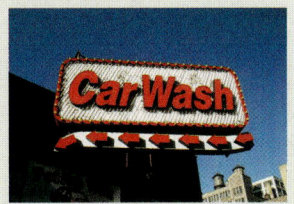

25) Ella spends $7.50 on three cantaloupe and one watermelon at a farmers' market. Two cantaloupe and two watermelon would have cost $9.00. What is the price of a cantaloupe and the price of a watermelon?
cantaloupe: $1.50; watermelon: $3.00

26) One 12-oz serving of Coke and two 12-oz servings of Mountain Dew contain 31.3 tsp of sugar while three servings of Coke and one Mountain Dew contain 38.9 tsp of sugar. How much sugar is in a 12-oz serving of each drink? (www.dentalgentlecare.com)
Coke: 9.3 tsp; Mountain Dew: 11 tsp

27) Carol orders six White Castle hamburgers and a small order of fries for $5.05, and Momar orders eight hamburgers and two small fries for $7.66. Find the cost of a hamburger and the cost of an order of french fries at White Castle. (White Castle menu)
hamburger: $0.61; small fries: $1.39

28) Six White Castle hamburgers and one small order of french fries contain 1150 calories. Eight hamburgers and two orders of fries contain 1740. Determine how many calories are in a White Castle hamburger and in a small order of french fries. (www.whitecastle.com)
hamburger: 140; small fries: 310

Objective 4: Solve Mixture Problems

29) How many ounces of a 9% alcohol solution and how many ounces of a 17% alcohol solution must be mixed to get 12 oz of a 15% alcohol solution?   9%: 3oz; 17%: 9 oz

30) How many milliliters of a 4% acid solution and how many milliliters of a 10% acid solution must be mixed to obtain 54 ml of a 6% acid solution?   4%: 36 ml; 10%: 18 ml

31) How many pounds of peanuts that sell for $1.80 per pound should be mixed with cashews that sell for $4.50 per pound so that a 10-lb mixture is obtained that will sell for $2.61 per pound?   peanuts: 7 lb; cashews: 3 lb

32) Raheem purchases 20 stamps. He buys some $0.44 stamps and some $0.28 stamps and spends $7.52. How many of each type of stamp did he buy? $0.44: 12; $0.28: 8

33) Sally invested $4000 in two accounts, some of it at 3% simple interest, the rest in an account earning 5% simple interest. How much did she invest in each account if she earned $144 in interest after one year?   3%: $2800; 5%: $1200

34) Diego inherited $20,000 and puts some of it into an account earning 4% simple interest and the rest in an account earning 7% simple interest. He earns a total of $1130 in interest after a year. How much did he deposit into each account?   4%: $9000; 7%: $11,000

35) Josh saves all of his quarters and dimes in a bank. When he opens it, he has 110 coins worth a total of $18.80. How many quarters and how many dimes does he have?
52 quarters; 58 dimes

36) Mrs. Kowalski bought nine packages of batteries when they were on sale. The AA batteries cost $1.00 per package and the C batteries cost $1.50 per package. If she spent $11.50, how many packages of each type of battery did she buy?   4 AA batteries; 5 C batteries

37) How much pure acid and how many liters of a 10% acid solution should be mixed to get 12 L of a 40% acid solution?   4 L of pure acid; 8 L of 10% solution

38) How many ounces of pure orange juice and how many ounces of a citrus fruit drink containing 5% fruit juice should be mixed to get 76 oz of a fruit drink that is 25% fruit juice?   16 oz of orange juice, 60 oz of the fruit drink

Objective 5: Solve Problems Involving Distance, Rate, and Time

39) A car and a truck leave the same location, the car headed east and the truck headed west. The truck's speed is 10 mph less than the speed of the car. After 3 hr, the car and truck are 330 mi apart. Find the speed of each vehicle.
car: 60 mph; truck: 50 mph

40) A passenger train and a freight train leave cities 400 mi apart and travel toward each other. The passenger train is traveling 20 mph faster than the freight train. Find the speed of each train if they pass each other after 5 hr.
passenger train: 50 mph; freight train: 30 mph

41) Olivia can walk 8 mi in the same amount of time she can bike 22 mi. She bikes 7 mph faster than she walks. Find her walking and biking speeds.   walking: 4 mph; biking: 11 mph

42) A small, private plane can fly 400 mi in the same amount of time a jet can fly 1000 mi. If the jet's speed is 300 mph more than the speed of the small plane, find the speeds of both planes.   small plane: 200 mph; jet: 500 mph

43) Nick and Scott leave opposite ends of a bike trail 13 mi apart and travel toward each other. Scott is traveling 2 mph slower than Nick. Find each of their speeds if they meet after 30 min.   Nick: 14 mph; Scott: 12 mph

44) Vashon can travel 120 mi by car in the same amount of time he can take the train 150 mi. If the train travels 10 mph faster than the car, find the speed of the car and the speed of the train.   car: 40 mph; train: 50 mph

Objective 6: Solve Problems Involving Three Variables

Write a system of equations and solve.

45) Moe buys two hot dogs, two orders of fries, and a large soda for $9.00. Larry buys two hot dogs, one order of fries, and two large sodas for $9.50, and Curly spends $11.00 on three hot dogs, two orders of fries, and a large soda. Find the price of a hot dog, an order of fries, and a large soda.   hot dog: $2.00; fries: $1.50; soda: $2.00

46) A movie theater charges $9.00 for an adult's ticket, $7.00 for a ticket for seniors 60 and over, and $6.00 for a child's ticket. For a particular movie, the theater sold a total of 290 tickets, which brought in $2400. The number of seniors' tickets sold was twice the number of children's tickets sold. Determine the number of adults', seniors', and children's tickets sold.   adults: 200; seniors: 60; children's: 30

47) A Chocolate Chip Peanut Crunch Clif Bar contains 4 fewer grams of protein than a Chocolate Peanut Butter Balance Bar Plus. A Chocolate Peanut Butter Protein Plus PowerBar contains 9 more grams of protein than the Balance Bar Plus. All three bars contain a total of 50 g of protein. How many grams of protein are in each type of bar? (www.clifbar.com, www.balance.com, www.powerbar.com)   Clif Bar: 11g; Balance Bar: 15 g; PowerBar: 24 g

48) A 1-tablespoon serving size of Hellman's Real Mayonnaise has 55 more calories than the same serving size of Hellman's Light Mayonnaise. Miracle Whip and Hellman's Light have the same number of calories in a 1-tablespoon serving size. If the three spreads have a total of 160 calories in one serving, determine the number of calories in one serving of each. (product labels)   Hellman's Real Mayonnaise: 90; Hellman's Light Mayonnaise: 35; Miracle Whip: 35

49) The three NBA teams with the highest revenues in 2002–2003 were the New York Knicks, the Los Angeles Lakers, and the Chicago Bulls. Their revenues totaled $428 million. The Lakers took in $30 million more than the Bulls, and the Knicks took in $11 million more than the Lakers. Determine the revenue of each team during the 2002–2003 season. (Forbes, Feb. 16, 2004, p. 66)   Knicks: $160 million; Lakers: $149 million; Bulls: $119 million

50) The best-selling paper towel brands in 2002 were Bounty, Brawny, and Scott. Together they accounted for 59% of the market. Bounty's market share was 25% more than Brawny's, and Scott's market share was 2% less than Brawny's. What percentage of the market did each brand hold in 2002? (USA Today, Oct. 23, 2003, p. 3B from Information Resources, Inc.)   Bounty: 37%; Brawny: 12%; Scott: 10%

51) Ticket prices to a Cubs game at Wrigley Field vary depending on whether they are on a value date, a regular date, or a prime date. At the beginning of the 2008 season, Bill, Corrinne, and Jason bought tickets in the bleachers for several games. Bill spent $367 on four value dates, four regular dates, and three prime dates. Corrinne bought tickets for four value dates, three regular dates, and two prime dates for $286. Jason spent $219 on three value dates, three regular dates, and one prime date. How much did it cost to sit in the bleachers at Wrigley Field on a value date, regular date, and prime date in 2008? (http://chicago.cubs.mlb.com)   value: $22; regular: $36; prime: $45

52) To see the Boston Red Sox play at Fenway Park in 2009, two field box seats, three infield grandstand seats, and five bleacher seats cost $530. The cost of four field box seats, two infield grandstand seats, and three bleacher seats was $678. The total cost of buying one of each type of ticket was $201. What was the cost of each type of ticket during the 2009 season? (http://boston.redsox.mlb.com)   field box: $125; infield grandstand: $50; bleacher: $26

53) The measure of the largest angle of a triangle is twice the middle angle. The smallest angle measures 28° less than the middle angle. Find the measures of the angles of the triangle. (Hint: Recall that the sum of the measures of the angles of a triangle is 180°.)   104°, 52°, 24°

54) The measure of the smallest angle of a triangle is one-third the measure of the largest angle. The middle angle measures 30° less than the largest angle. Find the measures of the angles of the triangle. (Hint: Recall that the sum of the measures of the angles of a triangle is 180°.)   90°, 60°, 30°

55) The smallest angle of a triangle measures 44° less than the largest angle. The sum of the two smaller angles is 20° more than the measure of the largest angle. Find the measures of the angles of the triangle.   80°, 64°, 36°

56) The sum of the measures of the two smaller angles of a triangle is 40° less than the largest angle. The measure of the largest angle is twice the measure of the middle angle. Find the measures of the angles of the triangle.   110°, 55°, 15°

57) The perimeter of a triangle is 29 cm. The longest side is 5 cm longer than the shortest side, and the sum of the two smaller sides is 5 cm more than the longest side. Find the lengths of the sides of the triangle.   12 cm, 10 cm, 7 cm

58) The smallest side of a triangle is half the length of the longest side. The sum of the two smaller sides is 2 in. more than the longest side. Find the lengths of the sides if the perimeter is 58 in.   28 in., 16 in., 14 in.

Section 4.4 Solving Systems of Linear Equations Using Matrices

Objectives

1. **Learn the Vocabulary Associated with Gaussian Elimination**
2. **Solve a System Using Gaussian Elimination**

We have learned how to solve systems of linear equations by graphing, substitution, and the elimination method. In this section, we will learn how to use *row operations* and *Gaussian elimination* to solve systems of linear equations. We begin by defining some terms.

1. Learn the Vocabulary Associated with Gaussian Elimination

A **matrix** is a rectangular array of numbers. (The plural of *matrix* is *matrices*.) Each number in the matrix is an **element** of the matrix. An example of a matrix is

$$\begin{array}{ccc} \text{Column 1} & \text{Column 2} & \text{Column 3} \\ \downarrow & \downarrow & \downarrow \end{array}$$

$$\begin{array}{c} \text{Row 1} \rightarrow \\ \text{Row 2} \rightarrow \end{array} \begin{bmatrix} 3 & -1 & 4 \\ 0 & 2 & -5 \end{bmatrix}$$

We can represent a system of equations in an *augmented matrix*. An **augmented matrix** has a vertical line to distinguish between different parts of the equation. For example, we can represent the system below with the augmented matrix shown here:

$$\begin{array}{ll} 5x + 4y = 1 & \text{Equation (1)} \\ x - 3y = 6 & \text{Equation (2)} \end{array} \qquad \left[\begin{array}{cc|c} 5 & 4 & 1 \\ 1 & -3 & 6 \end{array}\right] \qquad \begin{array}{l} \text{Row 1} \\ \text{Row 2} \end{array}$$

Notice that the vertical line separates the system's coefficients from its constants on the other side of the $=$ sign. The system needs to be in standard form, so the first column in the matrix represents the x-coefficients. The second column represents the y-coefficients, and the column on the right represents the constants.

Gaussian elimination is the process of using row operations on an augmented matrix to solve the corresponding system of linear equations. It is a variation of the elimination method and can be very efficient. Computers often use augmented matrices and row operations to solve systems.

The goal of Gaussian elimination is to obtain a matrix of the form $\left[\begin{array}{cc|c} 1 & a & b \\ 0 & 1 & c \end{array}\right]$ or $\left[\begin{array}{ccc|c} 1 & a & b & d \\ 0 & 1 & c & e \\ 0 & 0 & 1 & f \end{array}\right]$ when solving a system of two or three equations, respectively. Notice the 1's along the **diagonal** of the matrix and zeros below the diagonal. We say a matrix is in **row echelon form** when it has 1's along the diagonal and 0's below the diagonal. We get matrices in row echelon form by performing row operations. When we rewrite a matrix that is in row echelon form back into a system, its solution is easy to find.

2. Solve a System Using Gaussian Elimination

The row operations we can perform on augmented matrices are similar to the operations we use to solve a system of equations using the elimination method.

Definition Matrix Row Operations

Performing the following row operations on a matrix produces an equivalent matrix.

1) Interchanging two rows

2) Multiplying every element in a row by a nonzero real number

3) Replacing a row by the sum of it and the multiple of another row

---

Let's use these operations to solve a system using Gaussian elimination. Notice the similarities between this method and the elimination method.

Example 1

In-Class Example 1
Solve using Gaussian elimination.
$x - 3y = 7$
$4x + y = 15$
answer: (4, −1)

Solve using Gaussian elimination. $x + 5y = -1$
$2x - y = 9$

Solution

Begin by writing the system as an augmented matrix. $\begin{bmatrix} 1 & 5 & | & -1 \\ 2 & -1 & | & 9 \end{bmatrix}$

We will use the 1 in Row 1 to make the element below it a zero. If we multiply the 1 by −2 (to get −2) and add it to the 2, we get zero. We must do this operation to the entire row. Denote this as $-2R_1 + R_2 \to R_2$. (Read as, "−2 times Row 1 plus Row 2 makes the new Row 2.") We get a new Row 2.

Use this $\searrow$ $\begin{bmatrix} \textcircled{1} & 5 & | & -1 \\ \boxed{2} & -1 & | & 9 \end{bmatrix}$ to make $\nearrow$ this 0. $-2R_1 + R_2 \to R_2$ $\begin{bmatrix} 1 & 5 & | & -1 \\ -2(1)+2 & -2(5)+(-1) & | & -2(-1)+9 \end{bmatrix}$

$= \begin{bmatrix} 1 & 5 & | & -1 \\ 0 & -11 & | & 11 \end{bmatrix}$ Multiply each element of Row 1 by −2 and add it to the corresponding element of Row 2.

Note

We are not *making* a new Row 1, so it stays the same.

We have obtained the first 1 on the diagonal with a 0 below it. Next we need a 1 on the diagonal in Row 2.

This column is in the correct form. $\downarrow$

$\begin{bmatrix} 1 & 5 & | & -1 \\ 0 & \boxed{-11} & | & 11 \end{bmatrix}$ $\uparrow$ Make this 1. $-\frac{1}{11}R_2 \to R_2$ $\begin{bmatrix} 1 & 5 & | & -1 \\ 0 & 1 & | & -1 \end{bmatrix}$ Multiply each element of Row 2 by $-\frac{1}{11}$ to get a 1 on the diagonal.

We have obtained the final matrix because there are 1's on the diagonal and a 0 below. The matrix is in row echelon form. From this matrix, write a system of equations. The last row gives us the value of y.

$\begin{bmatrix} 1 & 5 & | & -1 \\ 0 & 1 & | & -1 \end{bmatrix}$ $\begin{aligned} 1x + 5y &= -1 \\ 0x + 1y &= -1 \end{aligned}$ or $\begin{aligned} x + 5y &= -1 &\quad \text{Equation (1)} \\ y &= -1 &\quad \text{Equation (2)} \end{aligned}$

$x + 5(-1) = -1$ Substitute −1 for y in equation (1).
$x - 5 = -1$ Multiply.
$x = 4$ Add 5.

The solution is $(4, -1)$. Check by substituting $(4, -1)$ into both equations of the original system.

Here are the steps for using Gaussian elimination to solve a system of any number of equations. Our goal is to obtain a matrix with 1's along the diagonal and 0's below—row echelon form.

> **Procedure** How to Solve a System of Equations Using Gaussian Elimination
>
> **Step 1:** Write the system as an *augmented matrix.*
>
> **Step 2:** Use row operations to make the *first entry in column 1* be a 1.
>
> **Step 3:** Use row operations to make *all entries below the 1 in column 1* be 0's.
>
> **Step 4:** Use row operations to make the *second entry in column 2* be a 1.
>
> **Step 5:** Use row operations to make *all entries below the 1 in column 2* be 0's.
>
> **Step 6:** Continue this procedure until the matrix is in *row echelon form*—1's along the diagonal and 0's below.
>
> **Step 7:** Write the matrix in step 6 as a *system of equations.*
>
> **Step 8:** *Solve the system* from step 7. The last equation in the system will give you the value of one of the variables; find the values of the other variables by using substitution.
>
> **Step 9:** *Check the solution* in each equation of the original system.

You Try 1

Solve the system using Gaussian elimination.

$$x - y = -1$$
$$-3x + 5y = 9$$

Next we will solve a system of three equations using Gaussian elimination.

Example 2

Solve using Gaussian elimination.

$$2x + y - z = -3$$
$$x + 2y - 3z = 1$$
$$-x - y + 2z = 2$$

In-Class Example 2
Solve using Gaussian elimination.
$$3x + 2y + z = 0$$
$$x - y + 2z = 5$$
$$-x + 4y - 4z = -15$$
answer: $(3, -4, -1)$

Solution

Step 1: Write the system as an *augmented matrix.*

$$\begin{bmatrix} 2 & 1 & -1 & -3 \\ 1 & 2 & -3 & 1 \\ -1 & -1 & 2 & 2 \end{bmatrix}$$

Step 2: To make the *first entry in column 1* be a 1, we *could* multiply Row 1 by $\frac{1}{2}$, but this would make the rest of the entries in the first row fractions. Instead, recall that we can interchange two rows. If we interchange Row 1 and Row 2, the first entry in column 1 will be 1.

$$\begin{array}{c} R_1 \leftrightarrow R_2 \\ \text{Interchange} \\ \text{Row 1 and Row 2.} \end{array} \qquad \begin{bmatrix} ① & 2 & -3 & 1 \\ 2 & 1 & -1 & -3 \\ -1 & -1 & 2 & 2 \end{bmatrix}$$

Step 3: We want to make *all the entries below the 1 in column 1* be 0's. To obtain a 0 in place of the 2 in column 1, multiply the 1 by -2 (to get -2) and add it to the 2. Perform that same operation on the entire row to obtain the new Row 2.

$$\begin{array}{c} \text{Use this} \rightarrow \\ \text{to make} \rightarrow \\ \text{this zero.} \end{array} \begin{bmatrix} ① & 2 & -3 & 1 \\ \boxed{2} & 1 & -1 & -3 \\ -1 & -1 & 2 & 2 \end{bmatrix} \begin{array}{c} -2R_1 + R_2 \rightarrow R_2 \\ \text{-2 times Row 1 +} \\ \text{Row 2 = new Row 2} \end{array} \begin{bmatrix} 1 & 2 & -3 & 1 \\ 0 & -3 & 5 & -5 \\ -1 & -1 & 2 & 2 \end{bmatrix}$$

To obtain a 0 in place of the -1 in column 1, add the 1 and the -1. Perform that same operation on the entire row to obtain a new Row 3.

Use this→
to make→
this zero.→
$$\begin{bmatrix} ① & 2 & -3 & | & 1 \\ 0 & -3 & 5 & | & -5 \\ \boxed{-1} & -1 & 2 & | & 2 \end{bmatrix}$$
$R_1 + R_3 \rightarrow R_3$
Row 1 + Row 3 = new Row 3
$$\begin{bmatrix} 1 & 2 & -3 & | & 1 \\ 0 & -3 & 5 & | & -5 \\ 0 & 1 & -1 & | & 3 \end{bmatrix}$$

Step 4:   Next, we want the *second entry in column 2* to be a 1. We *could* multiply Row 2 by $-\dfrac{1}{3}$

to get the 1, but the other entries would be fractions. Instead, interchanging Row 2 and Row 3 will give us a 1 on the diagonal and keep 0's in column 1. (Sometimes, though, fractions are unavoidable.)

$R_2 \leftrightarrow R_3$
Interchange Rows 2 and 3.
$$\begin{bmatrix} 1 & 2 & -3 & | & 1 \\ 0 & ① & -1 & | & 3 \\ 0 & -3 & 5 & | & -5 \end{bmatrix}$$

Step 5:   We want to make *all the entries below the 1 in column 2* be 0's. To obtain a 0 in place of -3 in column 2, multiply the 1 above it by 3 (to get 3) and add it to -3. Perform that same operation on the entire row to obtain a new Row 3.

Use this—
to make—
this zero.
$$\begin{bmatrix} 1 & 2 & -3 & | & 1 \\ 0 & ① & -1 & | & 3 \\ 0 & \boxed{-3} & 5 & | & -5 \end{bmatrix}$$
$3R_2 + R_3 \rightarrow R_3$
3 times Row 2 + Row 3 = new Row 3
$$\begin{bmatrix} 1 & 2 & -3 & | & 1 \\ 0 & 1 & -1 & | & 3 \\ 0 & 0 & 2 & | & 4 \end{bmatrix}$$

We have completed step 5 because there is only one entry below the 1 in column 2.

Step 6:   *Continue this procedure.* The last entry in column 3 needs to be a 1. (This is the last 1 we need along the diagonal.) Multiply Row 3 by $\dfrac{1}{2}$ to obtain the last 1.

$\dfrac{1}{2}R_3 \rightarrow R_3$
Multiply Row 3 by $\dfrac{1}{2}$.
$$\begin{bmatrix} ① & 2 & -3 & | & 1 \\ 0 & ① & -1 & | & 3 \\ 0 & 0 & ① & | & 2 \end{bmatrix}$$

We are done performing row operations because there are 1's on the diagonal and zeros below.

Step 7:   Write the matrix in step 6 as a *system of equations.*

$$\begin{array}{lll} 1x + 2y - 3z = 1 & & x + 2y - 3z = 1 \\ 0x + 1y - 1z = 3 & \text{or} & \quad\quad y - \ z = 3 \\ 0x + 0y + 1z = 2 & & \quad\quad\quad\quad z = 2 \end{array}$$

Step 8:   *Solve the system* in step 7. The last row tells us that $z = 2$. Substitute $z = 2$ into the equation above it ($y - z = 3$) to get the value of y: $y - 2 = 3$, so $y = 5$.

Substitute $y = 5$ and $z = 2$ into $x + 2y - 3z = 1$ to solve for x.

$$\begin{array}{ll} x + 2y - 3z = 1 & \\ x + 2(5) - 3(2) = 1 & \text{Substitute values.} \\ x + 10 - 6 = 1 & \text{Multiply.} \\ x + 4 = 1 & \text{Subtract.} \\ x = -3 & \end{array}$$

The solution of the system is $(-3, 5, 2)$.

Step 9:   *Check the solution* in each equation of the original system. The check is left to the student.

This procedure may seem long and complicated at first, but as you practice and become more comfortable with the steps, you will see that it is actually quite efficient.

You Try 2

Solve the system using Gaussian elimination.
$$x + 3y - 2z = 10$$
$$3x + 2y + z = 9$$
$$-x + 4y - z = -1$$

If we are performing Gaussian elimination and obtain a matrix that produces a false equation as shown, then the system has *no solution*. The system is *inconsistent*.

$$\begin{bmatrix} 1 & -6 & | & 9 \\ 0 & 0 & | & 8 \end{bmatrix} \quad 0x + 0y = 8 \quad \text{False}$$

If, however, we obtain a matrix that produces a row of zeros as shown, then the system has an *infinite number of solutions*. The system is *consistent* with *dependent* equations. We write its solution as we did in previous sections.

$$\begin{bmatrix} 1 & 5 & | & -1 \\ 0 & 0 & | & 0 \end{bmatrix} \quad 0x + 0y = 0 \quad \text{True}$$

Using Technology

In this section, we have learned how to solve a system of three equations using Gaussian elimination. The row operations used to convert an augmented matrix to row echelon form can be performed on a graphing calculator.

Follow the nine-step method given in the text to solve the system using Gaussian elimination:

$$x + 2y - 3z = 1$$
$$y - z = 3$$
$$-2y + 4z = -4$$

Step 1: Write the system as an augmented matrix:

$$\begin{bmatrix} 1 & 2 & -3 & | & 1 \\ 0 & 1 & -1 & | & 3 \\ 0 & -2 & 4 & | & -4 \end{bmatrix}$$

Store the matrix in matrix [A] using a graphing calculator. Press 2nd x^{-1} to select [A]. Press the right arrow key two times and press ENTER to select EDIT. Press 3 ENTER then 4 ENTER to enter the number of rows and number of columns in the augmented matrix. Enter the coefficients one row at a time as follows: 1 ENTER 2 ENTER (−) 3 ENTER 1 ENTER 0 ENTER 1 ENTER (−) 1 ENTER 3 ENTER 0 ENTER (−) 2 ENTER 4 ENTER (−) 4 ENTER.
Press 2nd MODE to return to the home screen. Press 2nd x^{-1} ENTER ENTER to display matrix [A].

```
[A]
 [[1  2  -3  1 ]
  [0  1  -1  3 ]
  [0  -2 4  -4]]
```

Notice that we can omit steps 2–4 because we already have two 1's on the diagonal and 0's below the first 1.

Step 5: Get the element in row 3, column 2 to be 0. Multiply row 2 by the opposite of the number in row 3, column 2 and add to row 3. The graphing calculator row operation used to multiply a row by a nonzero number and add to another row is ***row+(nonzero number, matrix name, first row, second row)**.

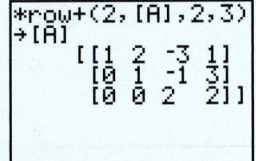

```
*row+(2,[A],2,3)
→[A]
 [[1  2  -3  1 ]
  [0  1  -1  3 ]
  [0  0  2  2]]
```

In this case, we have *row+(2, [A], 2, 3). To enter this row operation on your calculator, press 2nd x^{-1}, then press the right arrow to access the MATH menu. Scroll down to option F and press ENTER to display *row+(then enter 2 , 2nd x^{-1} ENTER , 2 , 3) as shown. Store the result back in matrix [A] by pressing STO> 2nd x^{-1} ENTER ENTER.

Step 6: To make the last number on the diagonal be 1, multiply row 3 by $\frac{1}{2}$. The graphing calculator row operation used to multiply a row by a nonzero number is ***row(nonzero number, matrix name, row)**. In this case, we have *row(1/2, [A], 3). On your calculator, press 2nd x^{-1}, then press the right arrow to access the MATH menu. Scroll down to option E and press ENTER to display *row(then enter 1 ÷ 2 , 2nd x^{-1} ENTER , 3) as shown.

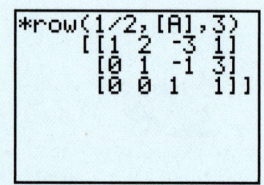

Step 7: Write the matrix from step 6 as:

$$\begin{bmatrix} 1 & 2 & -3 & | & 1 \\ 0 & 1 & -1 & | & 3 \\ 0 & 0 & 1 & | & 1 \end{bmatrix} \quad \begin{array}{l} 1x + 2y - 3z = 1 \\ 0x + 1y - 1z = 3 \quad \text{or} \\ 0x + 0y + 1z = 1 \end{array} \quad \begin{array}{l} x + 2y - 3z = 1 \\ y - z = 3 \\ z = 1 \end{array}$$

Step 8: Solve the system using substitution to obtain the solution $x = -4, y = 4, z = 1$ or $(-4, 4, 1)$.

Step 9: Check the solution.

Using Row Echelon Form

The row echelon form shown above is not unique. Another row echelon form can be obtained *in one step* using a graphing calculator. Given the original augmented matrix stored in [A], press 2nd x^{-1}, then press the right arrow, scroll down to option A, and press ENTER to display ref(which stands for row echelon form, and press ENTER. Press 2nd x^{-1} ENTER) ENTER to show the matrix in row echelon form.

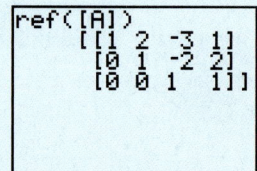

Using Reduced Row Echelon Form

The **reduced row echelon form** of an augmented matrix contains 1's on the diagonal and 0's *above* and *below* the 1's. We can find this using row operations as shown in the 9-step process, or directly in one step. Given the original augmented matrix stored in [A], press 2nd x^{-1}, then press the right arrow, scroll down to option B, and press ENTER to display ref(which stands for reduced row-echelon form, and press ENTER.
Press 2nd x^{-1} ENTER) ENTER to show the matrix in reduced row echelon form.

Write a system of equations from the matrix that is in reduced row echelon form.

$$\begin{bmatrix} 1 & 0 & 0 & | & -4 \\ 0 & 1 & 0 & | & 4 \\ 0 & 0 & 1 & | & 1 \end{bmatrix} \quad \begin{array}{l} 1x + 0y + 0z = -4 \\ 0x + 1y + 0z = 4 \quad \text{or} \\ 0x + 0y + 1z = 1 \end{array} \quad \begin{array}{l} x = -4 \\ y = 4 \\ z = 1 \end{array}$$

Use a graphing calculator to solve each system using Gaussian elimination.

1) $x + 2y = 1$
   $3x - y = 17$

2) $x - 5y = -3$
   $2x - 7y = 3$

3) $-5x + 2y = -4$
   $3x - y = 8$

4) $3x - 5y - 3z = 6$
   $-x + 3y + 2z = 1$
   $-2x + 7y + 5z = 6$

5) $3x + 2y + z = 9$
   $-5x - 2y - z = -7$
   $4x + y + z = 3$

6) $2x - y + 2z = -4$
   $-x + y - 2z = 7$
   $-3x + y - z = -1$

Answers to You Try Exercises

1) $(2, 3)$ 2) $(4, 0, -3)$

Answers to Technology Exercises

1) $(5, -2)$ 2) $(12, 3)$ 3) $(12, 28)$ 4) $(4, -3, 7)$ 5) $(-1, 5, 2)$ 6) $(3, 6, -2)$

4.4 Exercises

Additional answers can be found in the Answers to Exercises appendix.

Objective 1: Learn the Vocabulary Associated with Gaussian Elimination

Write each system in an augmented matrix.

1) $x - 7y = 15$ $\quad\begin{bmatrix} 1 & -7 & | & 15 \\ 4 & 3 & | & -1 \end{bmatrix}$
    $4x + 3y = -1$

2) $x + 6y = 4$ $\quad\begin{bmatrix} 1 & 6 & | & 4 \\ -5 & 1 & | & -3 \end{bmatrix}$
    $-5x + y = -3$

3) $x + 6y - z = -2$
    $3x + y + 4z = 7$
    $-x - 2y + 3z = 8$
$$\begin{bmatrix} 1 & 6 & -1 & | & -2 \\ 3 & 1 & 4 & | & 7 \\ -1 & -2 & 3 & | & 8 \end{bmatrix}$$

4) $x + 2y - 7z = 3$
    $3x - 5y = -1$
    $-x + 2z = -4$
$$\begin{bmatrix} 1 & 2 & -7 & | & 3 \\ 3 & -5 & 0 & | & -1 \\ -1 & 0 & 2 & | & -4 \end{bmatrix}$$

Write a system of linear equations in x and y represented by each augmented matrix.

5) $\begin{bmatrix} 3 & 10 & | & -4 \\ 1 & -2 & | & 5 \end{bmatrix}$ $\quad 3x + 10y = -4$
    $\quad x - 2y = 5$

6) $\begin{bmatrix} 1 & -1 & | & 6 \\ -4 & 7 & | & 2 \end{bmatrix}$ $\quad x - y = 6$
    $\quad -4x + 7y = 2$

7) $\begin{bmatrix} 1 & -6 & | & 8 \\ 0 & 1 & | & -2 \end{bmatrix}$ $\quad x - 6y = 8$
    $\quad y = -2$

8) $\begin{bmatrix} 1 & 2 & | & 11 \\ 0 & 1 & | & 3 \end{bmatrix}$ $\quad x + 2y = 11$
    $\quad y = 3$

Write a system of linear equations in x, y, and z represented by each augmented matrix.

9) $\begin{bmatrix} 1 & -3 & 2 & | & 7 \\ 4 & -1 & 3 & | & 0 \\ -2 & 2 & -3 & | & -9 \end{bmatrix}$

10) $\begin{bmatrix} 1 & 4 & -3 & | & -5 \\ -1 & 2 & 5 & | & 8 \\ 6 & -2 & -1 & | & 3 \end{bmatrix}$

11) $\begin{bmatrix} 1 & 5 & 2 & | & 14 \\ 0 & 1 & -8 & | & 2 \\ 0 & 0 & 1 & | & -3 \end{bmatrix}$

12) $\begin{bmatrix} 1 & 4 & -7 & | & -11 \\ 0 & 1 & 3 & | & -1 \\ 0 & 0 & 1 & | & 6 \end{bmatrix}$

Objective 2: Solve a System Using Gaussian Elimination

Solve each system using Gaussian elimination. Identify any inconsistent systems or dependent equations.

13) $x + 4y = -1$
     $3x + 5y = 4$ $\quad (3, -1)$

14) $x - 3y = 1$
     $-3x + 7y = 3$ $\quad (-8, -3)$

15) $x - 3y = 9$
     $-6x + 5y = 11$ $\quad (-6, -5)$

16) $x + 4y = -6$
     $2x + 5y = 0$ $\quad (10, -4)$

17) $4x - 3y = 6$
     $x + y = -2$ $\quad (0, -2)$

18) $-4x + 5y = -3$
     $x - 8y = -6$ $\quad (2, 1)$

19) $x + y - z = -5$
     $4x + 5y - 2z = 0$
     $8x - 3y + 2z = -4$ $\quad (-1, 4, 8)$

20) $x - 2y + 2z = 3$
     $2x - 3y + z = 13$
     $-4x - 5y - 6z = 8$ $\quad (5, -2, -3)$

21) $x - 3y + 2z = -1$
     $3x - 8y + 4z = 6$
     $-2x - 3y - 6z = 1$ $\quad (10, 1, -4)$

22) $x - 2y + z = -2$
     $2x - 3y + z = 3$
     $3x - 6y + 2z = 1$ $\quad (5, 0, -7)$

23) $-4x - 3y + z = 5$
     $x + y - z = -7$
     $6x + 4y + z = 12$ $\quad (0, 1, 8)$

24) $6x - 9y - 2z = 7$
     $-3x + 4y + z = -4$
     $x - y - z = 1$ $\quad (3, 1, 1)$

25) $x - 3y + z = -4$
     $4x + 5y - z = 0$
     $2x - 6y + 2z = 1$ $\quad \emptyset$; inconsistent

26) $x - y + 3z = 1$
     $5x - 5y + 15z = 5$
     $-4x + 4y - 12z = -4$ $\quad \{(x, y, z) | x - y + 3z = 1\}$; dependent equations

Extension

Extend the concepts of this section to solve these systems using Gaussian elimination.

27) $a + b + 3c + d = -1$
     $-a + c - d = 7$
     $2a + 3b + 9c - 2d = 7$
     $a - 2b + c + 3d = -11$ $\quad (-5, 2, 1, -1)$

28) $a - 2b - c + 3d = 15$
     $2a - 3b + c + 4d = 22$
     $-a + 4b + 6c + 7d = -3$
     $3a + 2b - c - d = -7$ $\quad (1, -4, 0, 2)$

29) $w - 3x + 2y - z = -2$
     $-3w + 8x - 5y + z = 2$
     $2w - x + y + 3z = 7$
     $w - 2x + y + 2z = 3$ $\quad (3, 0, -2, 1)$

30) $w + x - 4y + 2z = -21$
     $3w + 2x + y - z = 6$
     $-2w - x - 2y + 6z = -30$
     $-w + 3x + 4y + z = 1$ $\quad (1, -2, 3, -4)$

Chapter 4: Summary

Definition/Procedure	Example

4.1 Solving Systems of Linear Equations in Two Variables

A **system of linear equations** consists of two or more linear equations with the same variables. A **solution of a system** of two equations in two variables is an ordered pair that is a solution of each equation in the system. **(p. 200)**

Determine if (6, 1) is a solution of the system $\begin{array}{l} x + 3y = 9 \\ -2x + 7y = -5. \end{array}$

$$x + 3y = 9 \qquad\qquad -2x + 7y = -5$$
$$6 + 3(1) \overset{?}{=} 9 \quad \text{Substitute.} \qquad -2(6) + 7(1) \overset{?}{=} -5 \quad \text{Substitute.}$$
$$6 + 3 \overset{?}{=} 9 \qquad\qquad -12 + 7 \overset{?}{=} -5$$
$$9 = 9 \quad \text{True} \qquad\qquad -5 = -5 \quad \text{True}$$

Since (6, 1) is a solution of each equation in the system, yes, it is a solution of the system.

To **solve a system by graphing**, graph each line on the same coordinate axes.

a) If the lines intersect at a single point, then this point is the solution of the system. The system is **consistent** and the equations are **independent.**

b) If the lines are parallel, then the system has **no solution.** We write the solution set as $\varnothing$. The system is **inconsistent** and the equations are **independent.**

c) If the graphs are the same line, then the system has an **infinite number of solutions** consisting of the points on that line. The system is **consistent** and the equations are **dependent. (p. 201)**

Solve by graphing.
$$y = \frac{1}{4}x - 3$$
$$3x + 4y = 4$$

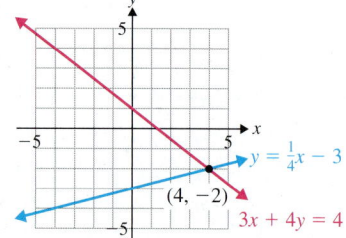

The solution of the system is (4, −2). The system is consistent.

Steps for Solving a System by Substitution

1) Solve one of the equations for one of the variables. If possible, solve for a variable that has a coefficient of 1 or −1.

2) Substitute the expression found in *step 1* into the *other* equation. The equation you obtain should contain only one variable.

3) Solve the equation obtained in *step 2*.

4) Substitute the value found in *step 3* into one of the equations to obtain the value of the other variable.

5) Check the values in the original equations. **(p. 205)**

Solve by substitution.
$$2x - 7y = 2$$
$$x - 2y = -2$$

1) Solve for x in the second equation since its coefficient is 1.
$$x = 2y - 2$$

2) Substitute $2y - 2$ for the x in the first equation.
$$2(2y - 2) - 7y = 2$$

3) Solve the equation in step 2 for y.
$$2(2y - 2) - 7y = 2$$
$$4y - 4 - 7y = 2 \qquad \text{Distribute.}$$
$$-3y - 4 = 2 \qquad \text{Combine like terms.}$$
$$-3y = 6 \qquad \text{Add 4.}$$
$$y = -2 \qquad \text{Divide by } -3.$$

4) Substitute $y = -2$ into the equation in step 1 to find x.
$$x = 2y - 2$$
$$x = 2(-2) - 2 \qquad \text{Substitute } -2 \text{ for } y.$$
$$x = -4 - 2 \qquad \text{Multiply.}$$
$$x = -6$$

5) The solution is (−6, −2). Verify this by substituting (−6, −2) into each of the original equations.

Definition/Procedure	Example

Steps for Solving a System by the Elimination Method

1) Write each equation in the form $Ax + By = C$.
2) Determine which variable to eliminate. If necessary, multiply one or both of the equations by a number so that the coefficients of the variable to be eliminated are opposites.
3) Add the equations, and solve for the remaining variable.
4) Substitute the value found in *Step 3* into either of the original equations to find the value of the other variable.
5) Check the solution in each of the original equations. **(p. 208)**

Solve using the elimination method.
$$7x - 4y = 1$$
$$-4x + 3y = 3$$

Eliminate y. Multiply the first equation by 3, and multiply the second equation by 4 to rewrite the system with equivalent equations.

Rewrite the system:
$$\begin{array}{l} 3(7x - 4y) = 3(1) \\ 4(-4x + 3y) = 4(3) \end{array} \rightarrow \begin{array}{l} 21x - 12y = 3 \\ -16x + 12y = 12 \end{array}$$

Add the equations:
$$\begin{array}{r} 21x - 12y = 3 \\ + \underline{-16x + 12y = 12} \\ 5x = 15 \\ x = 3 \end{array}$$

Substitute $x = 3$ into either of the original equations and solve for y.

$$7x - 4y = 1$$
$$7(3) - 4y = 1$$
$$21 - 4y = 1$$
$$-4y = -20$$
$$y = 5$$

The solution is $(3, 5)$. Verify this by substituting $(3, 5)$ into each of the original equations.

If the variables are eliminated and a false equation is obtained, the system has **no solution. The system is inconsistent, and the solution set is $\emptyset$. (p. 211)**

Solve the system.
$$4x - 12y = 7$$
$$x = 3y - 1$$

We will solve by substitution.

1) The second equation is solved for x.

2) Substitute $3y - 1$ for x in the first equation. $4(3y - 1) - 12y = 7$

3) Solve the equation in step 2 for y.
$$4(3y - 1) - 12y = 7$$
$$12y - 4 - 12y = 7 \quad \text{Distribute.}$$
$$-4 = 7 \quad \text{False}$$

4) The system has no solution. The solution set is $\emptyset$.

If the variables are eliminated and a true equation is obtained, the system has an **infinite number of solutions.** The equations are **dependent. (p. 212)**

Solve the system.
$$y = x + 4$$
$$2x - 2y = -8$$

1) The first equation is solved for y.

2) Substitute $x + 4$ for y in the second equation.

$$2x - 2(x + 4) = -8$$

3) Solve the equation in step 2 for x.
$$2x - 2(x + 4) = -8$$
$$2x - 2x - 8 = -8 \quad \text{Distribute.}$$
$$-8 = -8 \quad \text{True}$$

4) The equations are dependent, and the solution set is $\{(x, y) \mid y = x + 4\}$.

Definition/Procedure	**Example**

4.2 Solving Systems of Linear Equations in Three Variables

A **linear equation in three variables** is an equation of the form $Ax + By + Cz = D$, where A, B, and C are not all zero and where A, B, C, and D are real numbers. Solutions to this type of an equation are **ordered triples** of the form (x, y, z). **(p. 217)**

$5x + 3y + 9z = -2$

One solution of this equation is $(-1, -2, 1)$, since substituting the values for x, y, and z satisfies the equation.

$$5x + 3y + 9z = -2$$
$$5(-1) + 3(-2) + 9(1) \stackrel{?}{=} -2$$
$$-5 - 6 + 9 \stackrel{?}{=} -2$$
$$-2 = -2 \quad \text{True}$$

Solving a System of Three Linear Equations in Three Variables

1) **Label** the equations ①, ②, and ③.
2) **Choose a variable to eliminate. Eliminate** this variable from *two* sets of *two* equations using the elimination method. You will obtain two equations containing the same two variables. Label one of these new equations $\boxed{A}$ and the other $\boxed{B}$.
3) Use the elimination method to **eliminate a variable from equations $\boxed{A}$ and $\boxed{B}$.** You have now found the value of one variable.
4) **Find the value of another variable** by substituting the value found in *Step 3* into equation $\boxed{A}$ or $\boxed{B}$ and solving for the second variable.
5) **Find the value of the third variable** by substituting the values of the two variables found in Steps 3 and 4 into equation ①, ②, or ③.
6) **Check** the solution in each of the original equations, and **write the solution as an ordered triple. (p. 219)**

Solve
① $x + 2y + 3z = 5$
② $4x - 2y - z = -1$
③ $-3x + y + 4z = -12$

1) Label the equations ①, ②, and ③.
2) We will eliminate y from *two* sets of *two* equations.
a) Add equations ① *and* ② to eliminate y. Label the resulting equation $\boxed{A}$.

$$\begin{array}{rl} ① & x + 2y + 3z = 5 \\ ② + & 4x - 2y - z = -1 \\ \hline \boxed{A} & 5x + 2z = 4 \end{array}$$

b) To eliminate y again, multiply equation ③ by 2 and add it to equation ②. Label the resulting equation $\boxed{B}$.

$$\begin{array}{rl} 2 \times ③ & -6x + 2y + 8z = -24 \\ ② + & 4x - 2y - z = -1 \\ \hline \boxed{B} & -2x + 7z = -25 \end{array}$$

3) Eliminate x from $\boxed{A}$ and $\boxed{B}$. Multiply $\boxed{A}$ by 2 and $\boxed{B}$ by 5. Add the resulting equations.

$$\begin{array}{rl} 2 \times \boxed{A} & 10x + 4z = 8 \\ 5 \times \boxed{B} + & -10x + 35z = -125 \\ \hline & 39z = -117 \\ & \boxed{z = -3} \end{array}$$

4) Substitute $z = -3$ into either $\boxed{A}$ or $\boxed{B}$.

$$\begin{array}{ll} \boxed{A} \quad 5x + 2z = 4 & \\ 5x + 2(-3) = 4 & \text{Substitute } -3 \text{ for } z. \\ 5x - 6 = 4 & \text{Multiply.} \\ 5x = 10 & \text{Add 6.} \\ \boxed{x = 2} & \text{Divide by 5.} \end{array}$$

5) Substitute $x = 2$ and $z = -3$ into ① to solve for y.

$$\begin{array}{ll} ① \quad x + 2y + 3z = 5 & \\ 2 + 2y + 3(-3) = 5 & \text{Substitute.} \\ 2 + 2y - 9 = 5 & \text{Multiply.} \\ 2y - 7 = 5 & \text{Combine like terms.} \\ 2y = 12 & \text{Add 7.} \\ \boxed{y = 6} & \text{Divide by 2.} \end{array}$$

6) The solution is $(2, 6, -3)$. The check is left to the student.

Definition/Procedure	**Example**

4.3 Applications of Systems of Linear Equations

Use the **five steps for solving applied problems** outlined in the section to solve an applied problem.

1) **Read** the problem carefully. Draw a picture, if applicable. Identify what you are being asked to find. Label any pictures with the variables.
2) **Choose variables** to represent unknown quantities.
3) **Write a system of equations using the variables.** It may be helpful to begin by writing an equation in words.
4) **Solve** the system.
5) **Check** the answer in the original problem, and **interpret** the meaning of the solution as it relates to the problem. **(p. 225)**

Natalia spent $23.80 at an office supply store when she purchased boxes of pens and paper clips. The pens cost $3.50 per box, and the paper clips cost $0.70 per box. How many boxes of each did she buy if she purchased 10 items all together?

Steps 1) and 2) Define the variables.

x = number of boxes of pens she bought
y = number of boxes of paper clips she bought

3) One equation involves the *cost* of the items:

$$\text{Cost of pens} + \text{Cost of paper clips} = \text{Total cost}$$
$$3.50x \quad + \quad 0.70y \quad = \quad 23.80$$

The other equation involves the number of items:

$$\underset{\text{pens}}{\text{Number of}} + \underset{\text{paper clips}}{\text{Number of}} = \underset{\text{of items}}{\text{Total number}}$$
$$x \quad + \quad y \quad = \quad 10$$

The system is $\quad 3.50x + 0.70y = 23.80$
$$x + y = 10$$

4) Multiply by 10 to eliminate the decimals in the first equation, and then solve the system using substitution.

$$10(3.50x + 0.70y) = 10(23.80) \quad \text{Eliminate decimals.}$$
$$35x + 7y = 238$$

Solve the system $\begin{array}{l} 35x + 7y = 238 \\ x + y = 10 \end{array}$ to determine that the solution is (6, 4).

5) Natalia bought six boxes of pens and four boxes of paper clips. Verify the solution.

4.4 Solving Systems of Linear Equations Using Matrices

An **augmented matrix** contains a vertical line to separate different parts of the matrix. **(p. 237)**

An example of an augmented matrix is $\begin{bmatrix} 1 & 4 & | & -9 \\ 2 & -3 & | & 8 \end{bmatrix}$.

Matrix Row Operations
Performing the following row operations on a matrix produces an equivalent matrix.

1) Interchanging two rows
2) Multiplying every element in a row by a nonzero real number
3) Replacing a row by the sum of it and the multiple of another row

Gaussian elimination is the process of performing row operations on a matrix to put it into *row echelon* form.

A matrix is in **row echelon form** when it has 1's along the diagonal and 0's below.

$$\begin{bmatrix} 1 & a & | & b \\ 0 & 1 & | & c \end{bmatrix} \qquad \begin{bmatrix} 1 & a & b & | & d \\ 0 & 1 & c & | & e \\ 0 & 0 & 1 & | & f \end{bmatrix} \text{ (p. 237)}$$

Solve using Gaussian elimination.

$$x - y = 5$$
$$2x + 7y = 1$$

Write the system in an augmented matrix. Then, perform row operations to get it into row echelon form.

$$\begin{bmatrix} 1 & -1 & | & 5 \\ 2 & 7 & | & 1 \end{bmatrix} \xrightarrow{-2R_1 + R_2 \to R_2} \begin{bmatrix} 1 & -1 & | & 5 \\ 0 & 9 & | & -9 \end{bmatrix} \xrightarrow{\frac{1}{9}R_2 \to R_2} \begin{bmatrix} 1 & -1 & | & 5 \\ 0 & 1 & | & -1 \end{bmatrix}$$

The matrix is in row echelon form since it has 1's on the diagonal and a 0 below.

Write a system of equations from the matrix that is in row echelon form.

$$\begin{bmatrix} 1 & -1 & | & 5 \\ 0 & 1 & | & -1 \end{bmatrix} \quad \begin{array}{l} 1x - 1y = 5 \\ 0x + 1y = -1 \end{array} \quad \text{or} \quad \begin{array}{l} x - y = 5 \\ y = -1 \end{array}$$

Solving the system, we obtain the solution (4, −1).

*Additional answers can be found in the Answers to Exercises appendix.

(4.1) Determine if the ordered pair is a solution of the system of equations.

1) $-2x + y = 3$ no
   $3x - y = -17$
   $(-4, -5)$

2) $3x + 4y = 0$ yes
   $9x + 2y = 5$
   $\left(\dfrac{2}{3}, -\dfrac{1}{2}\right)$

Solve each system by graphing.

3) $y = \dfrac{1}{2}x - 2$ $(-2, -3)$
   $y = 2x + 1$

4) $3x - y = 2$ $(1, 1)$
   $x + y = 2$

5) $-2x + 3y = 15$
   $2x - 3y = -3$
   $\varnothing$; inconsistent

6) $2x + 3y = 5$ $(-2, 3)$
   $y = \dfrac{1}{2}x + 4$

7) $4x + y = -4$ $\{(x, y) | 4x + y = -4\}$; dependent equations
   $-2x - \dfrac{1}{2}y = 2$

Without graphing, determine whether each system has no solution, one solution, or an infinite number of solutions.

8) $x + 7y = -3$ one solution
   $4x - 9y = 1$

9) $y = -\dfrac{2}{3}x - 3$ no solution
   $4x + 6y = 5$

10) $5x - 4y = 2$ infinite number of solutions
    $y = \dfrac{5}{4}x - \dfrac{1}{2}$

11) $15x - 10y = 4$ no solution
    $-9x + 6y = 1$

12) The graph shows the on-time departure percentages during the four quarters of 2003 at San Diego (Lindbergh) and Denver International Airports.

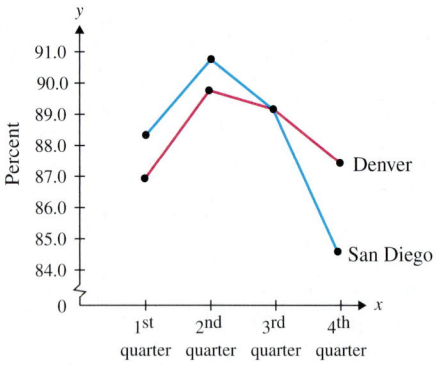

On-time Departures in 2003

(www.census.gov)

a) When was Denver's on-time departure percentage better than San Diego's?   during the 4th quarter of 2003

b) When were their percentages the same and, approximately, what percentage of flights left these airports on time?
   During the 3rd quarter of 2003, approximately 89.1% of flights left on time.

Solve each system by substitution. Identify any inconsistent systems or dependent equations

13) $x + 8y = -2$ $(-10, 1)$
    $2x + 11y = -9$

14) $y = \dfrac{5}{6}x - 2$
    $6y - 5x = -12$

15) $-2x + y = -18$ $(9, 0)$
    $x + 7y = 9$

16) $6x - y = -3$ $\left(\dfrac{1}{3}, 5\right)$
    $15x + 2y = 15$

17) $\dfrac{5}{2}x + \dfrac{9}{2}y = \dfrac{1}{2}$ $(2, -1)$
    $\dfrac{1}{6}x + \dfrac{2}{3}y = -\dfrac{1}{3}$

18) $x + 8y = 2$ $\varnothing$; inconsistent
    $x = 20 - 8y$

Solve each system using the elimination method. Identify any inconsistent systems or dependent equations.

19) $x - y = -8$ $(-5, 3)$
    $3x + y = -12$

20) $-5x + 3y = 17$ $(2, 9)$
    $x + 2y = 20$

21) $4x - 5y = -16$
    $-3x + 4y = 13$
    $(1, 4)$

22) $6x + 8y = 13$
    $9x + 12y = -5$
    $\varnothing$; inconsistent

23) $0.12x + 0.01y = 0.06$
    $0.5x + 0.2y = -0.7$
    $(1, -6)$

24) $3(8 - y) = 5x + 3$
    $x + 2(y - 3) = 2(4 - x)$
    $(0, 7)$

Solve each system by your choice of the substitution or elimination method. Identify any inconsistent systems or dependent equations.

25) $2x - 3y = 3$ $(-3, -3)$
    $3x + 4y = -21$

26) $2(2x - 3) = y + 3$ $(4, 7)$
    $8 + 5(y - 6) = 8(x + 3) - 9x - y$

27) $6x - 4y = 12$ $\{(x, y) | 6x - 4y = 12\}$; dependent
    $15x - 10y = 30$

28) $\dfrac{3}{4}x - y = \dfrac{1}{2}$ $(2, 1)$
    $-\dfrac{x}{3} + \dfrac{y}{2} = -\dfrac{1}{6}$

29) $15 - y = y + 2(4x + 5)$ $\varnothing$; inconsistent
    $2(x + 1) + 3 = 2(y + 4) + 10x$

30) $7x - y = -12$
    $-2x + 3y = 17$ $(-1, 5)$

Solve each system using the elimination method twice.

31) $2x + 7y = -8$ $\left(\dfrac{83}{23}, -\dfrac{50}{23}\right)$
    $3x - y = 13$

32) $x + 4y = 11$ $\left(\dfrac{37}{13}, \dfrac{53}{26}\right)$
    $5x - 6y = 2$

(4.2) Determine if the ordered triple is a solution of the system.

33) $x - 6y + 4z = 13$
    $5x + y + 7z = 8$
    $2x + 3y - z = -5$
    $(-3, -2, 1)$ no

34) $-4x + y + 2z = 1$
    $x - 3y - 4z = 3$
    $-x + 2y + z = -7$
    $(0, -5, 3)$ yes

Solve each system using one of the methods of section 4.2. Identify any inconsistent systems or dependent equations.

35) $2x - 5y - 2z = 3$
    $x + 2y + z = 5$
    $-3x - y + 2z = 0$
    $(3, -1, 4)$

36) $x - 2y + 2z = 6$
    $x + 4y - z = 0$
    $5x + 3y + z = -3$
    $(-4, 3, 8)$

37) $5a - b + 2c = -6$
    $-2a - 3b + 4c = -2$
    $a + 6b - 2c = 10$

38) $2x + 3y - 15z = 5$
    $-3x - y + 5z = 3$
    $-x + 6y - 10z = 12$

39) $4x - 9y + 8z = 2$
$\quad x + 3y = 5$
$\quad 6y + 10z = -1$

40) $-a + 5b - 2c = -3$
$\quad 3a + 2c = -3$
$\quad 2a + 10b = -2$

41) $\quad x + 3y - z = 0$
$\quad 11x - 4y + 3z = 8$
$\quad 5x + 15y - 5z = 1$
$\quad$ ∅; inconsistent

42) $4x + 2y + z = 0$
$\quad 8x + 4y + 2z = 0$
$\quad 16x + 8y + 4z = 0$
$\quad$ {$(x, y, z)|4x + 2y + z = 0$}; dependent

43) $12a - 8b + 4c = 8$
$\quad 3a - 2b + c = 2$
$\quad -6a + 4b - 2c = -4$
$\quad$ {$(a, b, c)|3a - 2b + c = 2$}; dependent

44) $3x - 12y - 6z = -8$
$\quad x + y - z = 5$
$\quad -4x + 16y + 8z = 10$
$\quad$ ∅; inconsistent

45) $5y + 2z = 6$ $\quad$ $(1, 0, 3)$
$\quad -x + 2y = -1$
$\quad 4x - z = 1$

46) $2a - b = 4$ $\quad$ $(3, 2, 2)$
$\quad 3b + c = 8$
$\quad -3a + 2c = -5$

47) $8x + z = 7$ $\quad \left(\dfrac{3}{4}, -2, 1\right)$
$\quad 3y + 2z = -4$
$\quad 4x - y = 5$

48) $6y - z = -2$ $\quad \left(0, \dfrac{1}{3}, 4\right)$
$\quad x + 3y = 1$
$\quad -3x + 2z = 8$

(4.3) Write a system of equations and solve.

49) One day at the Village Veterinary Clinic, the doctors treated twice as many dogs as cats. If they treated a total of 51 cats and dogs, how many of each did the doctors see?
34 dogs, 17 cats

50) At Aurora High School, 183 sophomores study either Spanish or French. If there are 37 fewer French students than Spanish students, how many study each language?
Spanish: 110; French: 73

51) At a Houston Texans football game in 2008, four hot dogs and two sodas cost \$26.50, while three hot dogs and four sodas cost \$28.00. Find the price of a hot dog and the price of a soda at a Texans game.
(www.teammarketing.com) hot dog: \$5.00; soda: \$3.25

52) The perimeter of a rectangular computer monitor is 66 in. The length is 3 in. more than the width. What are the dimensions of the monitor? width: 15 in.; length: 18 in.

53) Find the measures of angles x and y if the measure of x is twice the measure of y.

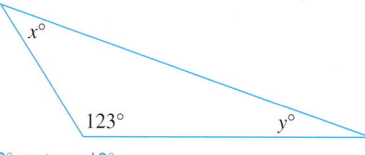

$m\angle x = 38°$; $m\angle y = 19°$

54) A store owner plans to make 10 pounds of a candy mix worth \$1.92/lb. How many pounds of gummi bears worth \$2.40/lb, and how many pounds of jelly beans worth \$1.60/lb must be combined to make the candy mix?
gummi bears: 4 lb; jelly bean: 6 lb

55) How many milliliters of pure alcohol and how many milliliters of a 4% alcohol solution must be combined to make 480 milliliters of an 8% alcohol solution?
pure alcohol: 20 ml; 4% solution: 460 ml

56) A car and a tour bus leave the same location and travel in opposite directions. The car's speed is 12 mph more than the speed of the bus. If they are 270 mi apart after $2\dfrac{1}{2}$ hr, how fast is each vehicle traveling?
car: 60 mph; bus: 48 mph

57) One serving (8 fl oz) of Powerade has 17 mg more sodium than one serving of Propel. One serving of Gatorade has 58 mg more sodium than the same serving size of Powerade. Together the three drinks have 197 mg of sodium. How much sodium is in one serving of each drink?
(Product labels) Propel: 35 mg; Powerade: 52 mg; Gatorade: 110 mg

58) In 2003, the top highway truck tire makers were Goodyear, Michelin, and Bridgestone. Together, they held 53% of the market. Goodyear's market share was 3% more than Bridgestone's, and Michelin's share was 1% less than Goodyear's. What percent of this tire market did each company hold in 2003? (*Market Share Reporter*, Vol. I, 2005, p. 361: From: *Tire Business*, Feb. 2, 2004, p. 9)
Goodyear: 19%; Michelin: 18%; Bridgestone: 16%

59) One Friday, Serena, Blair, and Chuck were busy texting their friends. Together, they sent a total of 140 text messages. Blair sent 15 more texts then Serena, while Chuck sent half as many as Serena. How many texts did each person send that day? Blair: 65; Serena: 50; Chuck: 25

60) Digital downloading of albums has been on the rise. The Recording Industry Association of America reports that in 2005 there were 14 million fewer downloads than in 2006, and in 2007 there were 14.9 million more downloads than the previous year. During all three years, 83.7 million albums were downloaded. How many albums were downloaded in 2005, 2006, and 2007? (www.riaa.com)
2005: 13.6 million; 2006: 27.6 million; 2007: 42.5 million

61) A family of six people goes to an ice cream store every Sunday after dinner. One week, they order two ice cream cones, three shakes, and one sundae for $13.50. The next week they get three cones, one shake, and two sundaes for $13.00. The week after that they spend $11.50 on one shake, one sundae, and four ice cream cones. Find the price of an ice cream cone, a shake, and a sundae.
ice cream cone: $1.50; shake: $2.50; sundae: $3.00

62) An outdoor music theater sells three types of seats—reserved, behind-the-stage, and lawn seats. Two reserved, three behind-the-stage, and four lawn seats cost $360.

Four reserved, two behind-the-stage, and five lawn seats cost $470. One of each type of seat would total $130. Determine the cost of a reserved seat, a behind-the-stage seat, and a lawn seat.
reserved: $60; behind-the-stage: $40; and lawn: $30

63) The measure of the smallest angle of a triangle is one-third the measure of the middle angle. The measure of the largest angle is 70° more than the measure of the smallest angle. Find the measures of the angles of the triangle. $92°, 66°, 22°$

64) The perimeter of a triangle is 40 in. The longest side is twice the length of the shortest side, and the sum of the two smaller sides is four inches longer than the longest side. Find the lengths of the sides of the triangles. 18 in., 13 in., 9 in.

(4.4) Solve each system using Gaussian elimination.

65) $x - y = -11$ $\quad(-9, 2)$
$2x + 9y = 0$

66) $x - 8y = -13$ $\quad(-5, 1)$
$4x + 9y = -11$

67) $5x + 3y = 5$ $\quad(1, 0)$
$-x + 8y = -1$

68) $3x + 5y = 5$ $\quad(10, -5)$
$-4x - 9y = 5$

69) $x - 3y - 3z = -7$
$2x - 5y - 3z = 2$
$-3x + 5y + 4z = -1$
$(5, -2, 6)$

70) $x - 3y + 5z = 3$
$2x - 5y + 6z = -3$
$3x + 2y + 2z = 3$
$(-3, 3, 3)$

Chapter 4: Test

Additional answers can be found in the Answers to Exercises appendix.

1) Determine if $\left(\dfrac{3}{4}, -5\right)$ is a solution of the system.

$$8x + y = 1$$
$$-12x - 4y = 11 \quad \text{yes}$$

In Exercises 2–11, when solving the systems, identify any inconsistent systems or dependent equations.

Solve each system by graphing.

2) $y = 2x - 3$ (2, 1)
   $3x + 2y = 8$

3) $x + y = 3$ $\varnothing$; inconsistent
   $2x + 2y = -2$

Solve each system by substitution.

4) $5x + 9y = 3$ (−3, 2)
   $x - 4y = -11$

5) $-9x + 12y = 21$
   $y = \dfrac{3}{4}x + \dfrac{7}{4}$

Solve each system by elimination.

6) $4x - 3y = -14$ (1, 6)
   $x + 3y = 19$

7) $7x + 8y = 28$ (4, 0)
   $-5x + 6y = -20$

8) $-x + 4y + 3z = 6$ (0, 3, −2)
   $3x - 2y + 6z = -18$
   $x + y + 2z = -1$

Solve each system using any method.

9) $x - 8y = 1$ (9, 1)
   $-2x + 9y = -9$

10) $\dfrac{x}{6} + y = -\dfrac{1}{3}$ $\left(-4, \dfrac{1}{3}\right)$
    $-\dfrac{5}{8}x + \dfrac{3}{4}y = \dfrac{11}{4}$

11) $5(y + 4) - 9 = -3(x - 4) + 4y$ (−2, 7)
    $13x - 2(3x + 2) = 3(1 - y)$

Write a system of equations and solve.

12) A 6-in. Turkey Breast Sandwich from Subway has 400 fewer calories than a Burger King Whopper. If the two sandwiches have a total of 960 calories, determine the number of calories in the Subway sandwich and the number of calories in a Whopper. (www.subway.com, www.bk.com)
Subway: 280 cal; Whopper: 680 cal

13) At a hardware store, three boxes of screws and two boxes of nails sell for $18 while one box of screws and four boxes of nails sell for $16. Find the price of a box of screws and the price of a box of nails.   screws: $4; nails: $3

14) The measure of the smallest angle of a triangle is 9° less than the measure of the middle angle. The largest angle is 30° more than the sum of the two smaller angles. Find the measures of the angles of the triangle.   105°, 42°, 33°

Solve using Gaussian elimination.

15) $x + 5y = -4$ (6, −2)
    $3x + 2y = 14$

16) $-3x + 5y + 8z = 0$
    $x - 3y + 4z = 8$
    $2x - 4y - 3z = 3$ (1, −1, 1)

*Additional answers can be found in the Answers to Exercises appendix.

Perform the operations and simplify.

1) $\dfrac{3}{10} - \dfrac{7}{15}$ $-\dfrac{1}{6}$

2) $5\dfrac{5}{6} \div 1\dfrac{13}{15}$ $3\dfrac{1}{8}$

3) $(5-8)^3 + 40 \div 10 - 6$ -29

4) Find the area of the triangle. $30\ \text{in}^2$

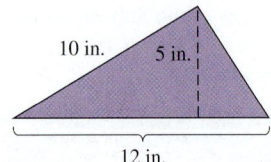

10 in.   5 in.

12 in.

5) Simplify $-8(3x^2 - x - 7)$. $-24x^2 + 8x + 56$

6) Solve $11 - 3(2k - 1) = 2(6 - k)$. $\left\{\dfrac{1}{2}\right\}$

7) Solve $0.04(3p - 2) - 0.02p = 0.1(p + 3)$. $\varnothing$

8) Solve $-47 \le 7t - 5 \le 6$. Write the answer in interval notation. $\left[-6, \dfrac{11}{7}\right]$

9) Write an equation and solve.

The number of plastic surgery procedures performed in the United States in 2003 was 293% more than the number performed in 1997. If approximately 8,253,000 cosmetic procedures were performed in 2003, how many took place in 1997? (American Society for Aesthetic Plastic Surgery)   2,100,000

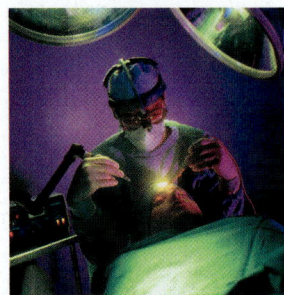

10) Solve $|8w - 3| > 7$. Write the answer in interval notation. $\left(-\infty, -\dfrac{1}{2}\right) \cup \left(\dfrac{5}{4}, \infty\right)$

11) The area, A, of a trapezoid is $A = \dfrac{1}{2}h(b_1 + b_2)$

where h = height of the trapezoid,
b_1 = length of one base of the trapezoid, and
b_2 = length of the second base of the trapezoid.

a) Solve the equation for h. $h = \dfrac{2A}{b_1 + b_2}$

b) Find the height of the trapezoid that has an area of $39\ \text{cm}^2$ and bases of length 8 cm and 5 cm.   6 cm

12) Graph $2x + 3y = 5$.

13) Graph $f(x) = 3x - 2$.

14) Find the x- and y-intercepts of the graph of $4x - 5y = 10$.   x-int: $\left(\dfrac{5}{2}, 0\right)$; y-int: $(0, -2)$

15) Write the slope-intercept form of the equation of the line containing $(-7, 4)$ and $(1, -3)$. $y = -\dfrac{7}{8}x - \dfrac{17}{8}$

16) Determine whether the lines are parallel, perpendicular, or neither.

$$10x + 18y = 9 \qquad \text{perpendicular}$$
$$9x - 5y = 17$$

Determine the domain of each function, and write it in interval notation.

17) $f(x) = -6x + 11$ $(-\infty, \infty)$

18) $g(x) = \dfrac{4}{7 - 2x}$ $\left(-\infty, \dfrac{7}{2}\right) \cup \left(\dfrac{7}{2}, \infty\right)$

19) Let $f(x) = 4x + 9$. Find each of the following.

   a) $f(-5)$ -11
   b) $f(p)$ $4p + 9$
   c) $f(n+2)$ $4n + 17$

Solve each system of equations. Identify any inconsistent systems or dependent equations.

20) $9x + 7y = 7$ $(7, -8)$
    $3x + 4y = -11$

21) $3(2x - 1) - (y + 10) = 2(2x - 3) - 2y$ $(3, 1)$
    $3x + 13 = 4x - 5(y - 3)$

22) $\dfrac{5}{6}x - \dfrac{1}{2}y = \dfrac{2}{3}$ $\varnothing$; inconsistent
    $-\dfrac{5}{4}x + \dfrac{3}{4}y = \dfrac{1}{2}$

23) $4a - 3b = -5$ $(-2, -1, 0)$
    $-a + 5c = 2$
    $2b + c = -2$

Write a system of equations and solve.

24) Dhaval used twice as many 6-ft boards as 4-ft boards when he made a playhouse for his children. If he used a total of 48 boards, how many of each size did he use?
    4-ft boards: 16; 6-ft boards: 32

25) Through 2008, Juanes had won three more Latin Grammy Awards than Alejandro Sanz, while Sanz had won twice as many as Shakira. Together, these three performers had won 38 Latin Grammy Awards. How many did each person win? (www.grammy.com/latin)
    Juanes: 17; Alejandro Sanz: 14; Shakira: 7

Polynomials and Polynomial Functions

Algebra at Work: Custom Motorcycle Shop

The people who build custom motorcycles use a lot of mathematics to do their jobs. Mark is building a chopper frame and needs to make the supports for the axle. He has to punch holes in the plates that will be welded to the frame.

Mark has to punch holes with a diameter of 1 in. in mild steel that is $\frac{3}{8}$-in. thick. The press punches two holes at a time. He must determine how much power is needed to do this job, so he uses a formula containing an exponent, $P = \dfrac{t^2 dN}{3.78}$. After substituting the numbers into the expression, he calculates that the power needed to punch these holes is 0.07 hp.

In this chapter, we will learn more about working with expressions containing exponents.

Section 5.1 The Rules of Exponents

Objectives

1. **Use the Product Rule for Exponents**
2. **Use the Power Rules for Exponents**
3. **Use 0 and Negative Exponents**
4. **Use the Quotient Rule for Exponents**

Recall from Section 1.3 that exponential notation is used as a shorthand way to represent a multiplication problem. For example, $3 \cdot 3 \cdot 3 \cdot 3$ can be written as 3^4, so $3^4 = 81$. The *base* is 3 and the *exponent,* or *power,* is 4. We read 3^4 as "3 to the fourth power" or as "3 to the fourth."

> **Definition**
>
> An **exponential expression** of the form a^n is $a^n = \underbrace{a \cdot a \cdot a \cdots \cdot a}_{n \text{ factors of } a}$, where a is any real number and n is a positive integer. a is the **base**, and n is the **exponent**.

1. Use the Product Rule for Exponents

Is there a rule we can use to *multiply* exponential expressions? Let's rewrite each of the following products as a single power of the base using what we already know:

1) $2^3 \cdot 2^4 = \overbrace{2 \cdot 2 \cdot 2}^{\substack{3 \text{ factors} \\ \text{of } 2}} \cdot \overbrace{2 \cdot 2 \cdot 2 \cdot 2}^{\substack{4 \text{ factors} \\ \text{of } 2}}$

$= \underbrace{2 \cdot 2 \cdot 2 \cdot 2 \cdot 2 \cdot 2 \cdot 2}_{7 \text{ factors of } 2}$

$2^3 \cdot 2^4 = 2^7$

2) $9^2 \cdot 9^3 = \overbrace{9 \cdot 9}^{\substack{2 \text{ factors} \\ \text{of } 9}} \cdot \overbrace{9 \cdot 9 \cdot 9}^{\substack{3 \text{ factors} \\ \text{of } 9}}$

$= \underbrace{9 \cdot 9 \cdot 9 \cdot 9 \cdot 9}_{5 \text{ factors of } 9}$

$9^2 \cdot 9^3 = 9^5$

Do you notice a pattern? *When we multiply expressions with the same base, keep the same base and add the exponents.* This is called the **product rule** for exponents.

> **Property**   Product Rule
>
> Let a be any real number and let m and n be positive integers. Then,
>
> $$a^m \cdot a^n = a^{m+n}$$

Example 1

Find each product.

a) $2^2 \cdot 2^4$
b) $(-x)^8 \cdot (-x)^5$
c) $(-4p^7)(9p^3)$
d) $m^5 n^2$
e) $c^4 \cdot c \cdot c^{10}$
f) $(4a^2 b)(5a^3 b^6)$

In-Class Example 1

Find each product.
a) $5^2 \cdot 5$ b) $y^4 \cdot y^9$
c) $4x^5 \cdot 10x^8$ d) $d \cdot d^7 \cdot d^4$
e) $x^6 y^3$ f) $(-7r^4 t^2)(3rt^9)$

answer: a) 125   **b)** y^{13}
c) $40x^{13}$ **d)** d^{12} **e)** $x^6 y^3$
f) $-21r^5 t^{11}$

Solution

a) $2^2 \cdot 2^4 = 2^{2+4} = 2^6 = 64$ 　　　Since the bases are the same, add the exponents.

b) $(-x)^8 \cdot (x)^5 = (-x)^{8+5} = (-x)^{13} = -x^{13}$

c) $(-4p^7)(9p^3) = (-4 \cdot 9)(p^7 \cdot p^3)$
$= -36p^{10}$ 　　　Associative and commutative properties

d) $m^5 n^2$ 　　　This expression is in simplest form. We cannot use the product rule because the bases are different.

e) $c^4 \cdot c \cdot c^{10} = c^{4+1+10} = c^{15}$

f) $(4a^2 b)(5a^3 b^6) = (4 \cdot 5)(a^2 \cdot a^3)(b^1 \cdot b^6)$
$= 20a^5 b^7$

You Try 1

Find each product.

a) $3^2 \cdot 3^2$ b) $q^6 \cdot q^3$ c) $(-8d^4)(3d^5)$

d) $w^8 \cdot w^5 \cdot w$ e) $w^3 k^5$ f) $(8xy^6)(3x^2y^8)$

BE CAREFUL

Can the product rule be applied to $4^3 \cdot 5^2$? **No!** The bases are not the same, so we cannot simply add the exponents. To evaluate $4^3 \cdot 5^2$, we would evaluate $4^3 = 64$ and $5^2 = 25$, then multiply:

$$4^3 \cdot 5^2 = 64 \cdot 25 = 1600$$

2. Use the Power Rules for Exponents

What does $(2^2)^3$ mean? We can rewrite $(2^2)^3$ as $2^2 \cdot 2^2 \cdot 2^2$. Then, using the product rule for exponents, we get $2^2 \cdot 2^2 \cdot 2^2 = 2^{2+2+2} = 2^6$.

Therefore $(2^2)^3 = 2^{2+2+2}$, or $2^{2\cdot3} = 2^6$. This result suggests the *basic power rule*. Two more power rules are derived from the basic power rule.

Property   The Power Rules for Exponents

Let a be any real number and m and n be integers. Then,

1) **Basic Power Rule** $(a^m)^n = a^{mn}$
    To raise a power to another power, multiply the exponents.

2) **Power Rule for a Product** $(ab)^n = a^n b^n$
    To raise a product to a power, raise each factor to that power.

3) **Power Rule for a Quotient** $\left(\dfrac{a}{b}\right)^n = \dfrac{a^n}{b^n}$, where $b \neq 0$

    To raise a quotient to a power, raise the numerator and denominator to that power.

Example 2

In-Class Example 2

Simplify using the power rules.

a) $(7^6)^2$ b) $(m^2)^5$ c) $(4x)^3$

d) $(2w^4)^5$ e) $\left(\dfrac{4}{5}\right)^3$ f) $\left(\dfrac{-3a}{b^8}\right)^3$

answer: a) 7^{12} **b)** m^{10}

c) $64x^3$ **d)** $32w^{20}$ **e)** $\dfrac{64}{125}$

f) $-\dfrac{27a^3}{b^{24}}$

Simplify using the power rules.

a) $(3^8)^4$ b) $(n^3)^7$ c) $(-9y)^2$ d) $(5c^2)^3$ e) $\left(\dfrac{3}{8}\right)^2$ f) $\left(\dfrac{-2t}{u^4}\right)^5$

Solution

a) $(3^8)^4 = 3^{8\cdot4} = 3^{32}$ Basic power rule

b) $(n^3)^7 = n^{3\cdot7} = n^{21}$ Basic power rule

c) $(-9y)^2 = (-9)^2 y^2 = 81y^2$ Power rule for a product

d) $(5c^2)^3 = 5^3 \cdot (c^2)^3 = 125c^{2\cdot3} = 125c^6$ Begin with the power rule for a product.

e) $\left(\dfrac{3}{8}\right)^2 = \dfrac{3^2}{8^2} = \dfrac{9}{64}$ Power rule for a quotient

f) $\left(\dfrac{-2t}{u^4}\right)^5 = \dfrac{(-2t)^5}{(u^4)^5}$ Power rule for a quotient

$= \dfrac{(-2)^5 \cdot t^5}{u^{4\cdot5}}$ Power rule for a product

Basic power rule

$= \dfrac{-32t^5}{u^{20}}$ Simplify.

$= -\dfrac{32t^5}{u^{20}}, \quad u \neq 0$

You Try 2

Simplify using the power rules.

a) $(5^4)^3$ b) $(j^6)^5$ c) $(10p)^3$ d) $(2k^7)^4$ e) $\left(\dfrac{5}{12}\right)^2$ f) $\left(\dfrac{-4w}{y^9}\right)^3$

Now that we have learned the product and power rules for exponents, let's use the rules together. Remember to follow the order of operations (see Section 1.3).

Example 3

In-Class Example 3
Simplify.
a) $(4f)^2(2f)^3$ b) $-3(4a^2b^3)^2$
c) $\dfrac{(12y^3)^2}{(4z^{10})^3}$

answer: a) $128f^5$
b) $-48a^4b^6$ c) $\dfrac{9y^6}{4z^{30}}$

Simplify.

a) $(2n)^3(3n^6)^2$ b) $2(4a^5b^3)^3$ c) $\dfrac{(6x^9)^2}{(2y^4)^3}$

Solution

a) Remember to perform the exponential operations before multiplication.

$$(2n)^3(3n^6)^2 = (2^3n^3)(3)^2(n^6)^2 \qquad \text{Power rule}$$
$$= (8n^3)(9n^{12}) \qquad \text{Power rule and evaluate exponents}$$
$$= 72n^{15} \qquad \text{Product rule}$$

b) The order of operations tells us to perform exponential operations before multiplication.

$$2(4a^5b^3)^3 = 2 \cdot (4)^3(a^5)^3(b^3)^3 \qquad \text{Order of operations and power rule}$$
$$= 2 \cdot 64a^{15}b^9 \qquad \text{Power rule}$$
$$= 128a^{15}b^9 \qquad \text{Multiply.}$$

c) $\dfrac{(6x^9)^2}{(2y^4)^3}$

Here, the operations are division and exponents. What comes first in the order of operations? **Exponents.**

$$\frac{(6x^9)^2}{(2y^4)^3} = \frac{36x^{18}}{8y^{12}} \qquad \text{Power rule}$$

$$= \frac{\overset{9}{\cancel{36}}x^{18}}{\underset{2}{\cancel{8}}y^{12}} \qquad \text{Divide out the common factor of 4.}$$

$$= \frac{9x^{18}}{2y^{12}}$$

BE CAREFUL

When simplifying the expression $\dfrac{(6x^9)^2}{(2y^4)^3}$ in Example 3, you may want to simplify before applying the product rule, like this:

$$\frac{\overset{3}{\cancel{6}}(6x^9)^2}{\underset{1}{\cancel{(2}}y^4)^3} \neq \frac{(3x^9)^2}{(y^4)^3} = \frac{9x^{18}}{y^{12}} \qquad \text{Incorrect!}$$

This method does not follow the rules for the order of operations, so it does **not** give us the correct answer.

You Try 3

Simplify.

a) $-5(2c^5d^3)^4$ b) $(11r^8t)^2(-r^{10}t^6)^3$ c) $\dfrac{(10a^7b^5)^3}{(8c^3)^2}$

Is it possible to have an exponent of zero or a negative exponent? If so, what do they mean?

3. Use 0 and Negative Exponents

> **Definition**
> **Zero as an Exponent:** If $a \neq 0$, then $a^0 = 1$.

Why does $a^0 = 1$? Let's evaluate $2^0 \cdot 2^3$. Using the product rule we get:

$$2^0 \cdot 2^3 = 2^{0+3} = 2^3 = 8$$

Since $2^0 \cdot 2^3 = 2^3$, it must be true that $2^0 = 1$. This is one way to understand that $a^0 = 1$.

> **Note**
> 0^0 is undefined.

Example 4

In-Class Example 4
Evaluate.
a) 3^0 b) -6^0 c) $(-4)^0$
d) $(9k)^0$
answer: a) 1   b) -1   c) 1
d) 1

Evaluate each expression.

a) 4^0 b) -9^0 c) $(-3)^0$ d) $(8t)^0$

Solution

a) $4^0 = 1$ b) $-9^0 = -1 \cdot 9^0 = -1 \cdot 1 = -1$

c) $(-3)^0 = 1$ d) $(8t)^0 = 1, t \neq 0$ ■

You Try 4

Evaluate.

a) 5^0 b) -7^0 c) $(-6)^0$ d) $(2w)^0$

So far we have worked with exponents that are zero or positive. What does a negative exponent mean? Let's use the product rule to find $2^3 \cdot 2^{-3}$.

$$2^3 \cdot 2^{-3} = 2^{3+(-3)} = 2^0 = 1.$$

Remember that a number multiplied by its reciprocal is 1, and here we have that a quantity, 2^3, times another quantity, 2^{-3}, is 1. Therefore, 2^3 and 2^{-3} are reciprocals!

This leads to the definition of a negative exponent.

> **Property** Negative Exponent
>
> If n is any integer and a and b are not equal to zero, then $a^{-n} = \left(\dfrac{1}{a}\right)^n = \dfrac{1}{a^n}$ and $\left(\dfrac{a}{b}\right)^{-n} = \left(\dfrac{b}{a}\right)^n$.

Therefore, to rewrite an expression of the form a^{-n} or $\left(\dfrac{a}{b}\right)^{-n}$ with a positive exponent, *take the reciprocal of the base and make the exponent positive.*

Example 5

In-Class Example 5
Simplify each expression so that the result contains only positive exponents. Assume all variables do not equal zero.
a) 3^{-3} b) $\left(\dfrac{4}{5}\right)^{-3}$ c) $\left(\dfrac{1}{3}\right)^{-2}$
d) $(-2)^{-2}$ e) n^{-8} f) $\left(\dfrac{2}{m}\right)^{-5}$
g) $7y^{-2}$
answer: a) $\dfrac{1}{27}$ b) $\dfrac{125}{64}$ c) 9
d) $\dfrac{1}{4}$ e) $\dfrac{1}{n^8}$ f) $\dfrac{m^5}{32}$ g) $\dfrac{7}{y^2}$

Simplify each expression so that the result contains only positive exponents. Assume all variables do not equal zero.

a) 5^{-3} b) $\left(\dfrac{8}{7}\right)^{-2}$ c) $\left(\dfrac{1}{2}\right)^{-6}$ d) $(-6)^{-2}$

e) k^{-5} f) $\left(\dfrac{3}{w}\right)^{-4}$ g) $4c^{-2}$

Solution

a) 5^{-3}:   The reciprocal of 5 is $\dfrac{1}{5}$, so $5^{-3} = \left(\dfrac{1}{5}\right)^3 = \dfrac{1^3}{5^3} = \dfrac{1}{125}$.

b) $\left(\dfrac{8}{7}\right)^{-2}$:   The reciprocal of $\dfrac{8}{7}$ is $\dfrac{7}{8}$, so $\left(\dfrac{8}{7}\right)^{-2} = \left(\dfrac{7}{8}\right)^2 = \dfrac{7^2}{8^2} = \dfrac{49}{64}$.

 **BE CAREFUL**   Notice that a negative exponent does not make the answer negative!

c) $\left(\dfrac{1}{2}\right)^{-6}$:   The reciprocal of $\dfrac{1}{2}$ is 2, so $\left(\dfrac{1}{2}\right)^{-6} = 2^6 = 64$.

d) $(-6)^{-2} = \left(-\dfrac{1}{6}\right)^2 = \left(-1 \cdot \dfrac{1}{6}\right)^2 = (-1)^2\left(\dfrac{1}{6}\right)^2 = 1 \cdot \dfrac{1^2}{6^2} = \dfrac{1}{36}$.

e) $k^{-5} = \left(\dfrac{1}{k}\right)^5 = \dfrac{1^5}{k^5} = \dfrac{1}{k^5}$

f) $\left(\dfrac{3}{w}\right)^{-4} = \left(\dfrac{w}{3}\right)^4 = \dfrac{w^4}{3^4} = \dfrac{w^4}{81}$

g) $4c^{-2} = 4 \cdot \left(\dfrac{1}{c}\right)^2$ Note that the base is c, *not* $4c$.

$\qquad = 4 \cdot \dfrac{1}{c^2} = \dfrac{4}{c^2}$

You Try 5

Simplify each expression so that the result contains only positive exponents. Assume all variables do not equal zero.

a) $(11)^{-2}$ b) $\left(\dfrac{1}{10}\right)^{-3}$ c) $\left(\dfrac{2}{3}\right)^{-3}$ d) -5^{-3}

e) $\left(\dfrac{3}{t}\right)^{-4}$ f) z^{-6} g) $9n^{-5}$

How could we rewrite $\dfrac{x^{-2}}{y^{-2}}$ with only positive exponents? One way would be to apply the power rule for exponents: $\dfrac{x^{-2}}{y^{-2}} = \left(\dfrac{x}{y}\right)^{-2} = \left(\dfrac{y}{x}\right)^{2} = \dfrac{y^2}{x^2}.$

Let's do the same for $\dfrac{a^{-5}}{b^{-5}}$: $\dfrac{a^{-5}}{b^{-5}} = \left(\dfrac{a}{b}\right)^{-5} = \left(\dfrac{b}{a}\right)^{5} = \dfrac{b^5}{a^5}.$

Notice that to rewrite the original expression with only positive exponents, the terms with the negative exponents "switch" their positions in the fraction. We can generalize this way:

Definition

If m and n are any integers and a and b are real numbers not equal to zero, then

$$\dfrac{a^{-m}}{b^{-n}} = \dfrac{b^n}{a^m}$$

Example 6

Rewrite the expression with positive exponents. Assume the variables do not equal zero.

a) $\dfrac{m^{-9}}{n^{-4}}$ b) $\dfrac{7r^{-5}}{t^3}$ c) $h^{-2}k^{-1}$ d) $\dfrac{8ab^{-6}}{5c^{-3}}$ e) $\left(\dfrac{xy}{3z}\right)^{-4}$

In-Class Example 6
Rewrite the expression with positive exponents.

a) $\dfrac{f^{-1}}{g^{-5}}$ b) $\dfrac{9s^{-10}}{t^4}$ c) $c^{-2}d^{-7}$

d) $\dfrac{7x^3y^{-8}}{10z^{-4}}$ e) $\left(\dfrac{ab}{4c}\right)^{-3}$

answer: a) $\dfrac{g^5}{f}$ b) $\dfrac{9}{s^{10}t^4}$

c) $\dfrac{1}{c^2d^7}$ d) $\dfrac{7x^3z^4}{10y^8}$ e) $\dfrac{64c^3}{a^3b^3}$

Solution

a) $\dfrac{m^{-9}}{n^{-4}} = \dfrac{n^4}{m^9}$

To make the exponents positive, "switch" the positions of the terms in the fraction.

b) $\dfrac{7r^{-5}}{t^3} = \dfrac{7}{r^5t^3}$

Since the exponent on t is positive, do not change its position in the expression.

c) $h^{-2}k^{-1} = \dfrac{h^{-2}k^{-1}}{1}$

$= \dfrac{1}{h^2k}$

Move $h^{-2}k^{-1}$ to the denominator to write with positive exponents.

d) $\dfrac{8ab^{-6}}{5c^{-3}} = \dfrac{8ac^3}{5b^6}$

To make the exponents positive, switch the positions of the terms with the negative exponents.

e) $\left(\dfrac{xy}{3z}\right)^{-4} = \left(\dfrac{3z}{xy}\right)^{4}$

To make the exponent positive, take the reciprocal of the base.

$= \dfrac{3^4 \cdot z^4}{x^4y^4}$

Use the power rule.

$= \dfrac{81z^4}{x^4y^4}$

Simplify.

You Try 6

Rewrite the expression with positive exponents. Assume the variables do not equal zero.

a) $\dfrac{a^{-2}}{b^{-7}}$ b) $\dfrac{d^{-8}}{2c^4}$ c) $6p^{-5}q^{-1}$ d) $\dfrac{10x^2y^{-9}}{3z^{-8}}$ e) $\left(\dfrac{ab}{5c}\right)^{-3}$

4. Use the Quotient Rule for Exponents

Next, we will discuss how to simplify the quotient of two exponential expressions with the same base. Let's begin by simplifying $\dfrac{9^6}{9^4}$. One way we can do this is by writing the numerator and denominator without exponents:

$$\dfrac{9^6}{9^4} = \dfrac{\cancel{9}\cdot\cancel{9}\cdot\cancel{9}\cdot\cancel{9}\cdot 9\cdot 9}{\cancel{9}\cdot\cancel{9}\cdot\cancel{9}\cdot\cancel{9}}$$ Divide out common factors.

$$= 9\cdot 9 = 9^2 = 81$$

So, $\dfrac{9^6}{9^4} = 9^2$. Do you notice a relationship among the exponents? That's right. We *subtracted* the exponents: $6 - 4 = 2$.

Property   Quotient Rule for Exponents

If m and n are any integers and $a \neq 0$, then

$$\dfrac{a^m}{a^n} = a^{m-n}$$

Notice that the base in the numerator and denominator is a. *In order to apply the quotient rule, the bases must be the same. Subtract the exponent of the denominator from the exponent of the numerator.*

Example 7

Simplify and write with positive exponents.

a) $\dfrac{2^{10}}{2^4}$ b) $\dfrac{w^{13}}{w^5}$ c) $\dfrac{3}{3^{-2}}$ d) $\dfrac{k^3}{k^{12}}$ e) $\dfrac{7^2}{2^3}$ f) $\dfrac{25m^{-8}n^7}{15m^{-5}n^2}$

In-Class Example 7
Simplify and write with positive exponents.
a) $\dfrac{5^8}{5^6}$ b) $\dfrac{d^{10}}{d^4}$ c) $\dfrac{2}{2^{-5}}$ d) $\dfrac{z^2}{z^5}$
e) $\dfrac{3^3}{8^2}$ f) $\dfrac{42r^{-6}t^{10}}{54r^{-2}t^3}$
answer: a) 25 b) d^6 c) 64
d) $\dfrac{1}{z^3}$ e) $\dfrac{27}{64}$ f) $\dfrac{7t^7}{9r^4}$

Solution

a) $\dfrac{2^{10}}{2^4} = 2^{10-4} = 2^6 = 64$ Since the bases are the same, subtract the exponents.

b) $\dfrac{w^{13}}{w^5} = w^{13-5} = w^8$ Since the bases are the same, subtract the exponents.

c) $\dfrac{3}{3^{-2}} = \dfrac{3^1}{3^{-2}} = 3^{1-(-2)}$ Since the bases are the same, subtract the exponents.

$\qquad = 3^3 = 27$ Be careful when subtracting the negative exponent!

d) $\dfrac{k^3}{k^{12}} = k^{3-12} = k^{-9}$ Same base, subtract the exponents.

$\qquad = \left(\dfrac{1}{k}\right)^9 = \dfrac{1}{k^9}$ Write with a positive exponent.

e) $\dfrac{7^2}{2^3} = \dfrac{49}{8}$

Since the bases are not the same, we cannot apply the quotient rule. Evaluate the numerator and denominator separately.

f) $\dfrac{25m^{-8}n^7}{15m^{-5}n^2} = \dfrac{\overset{5}{\cancel{25}}m^{-8}n^7}{\underset{3}{\cancel{15}}m^{-5}n^2}$

Reduce $\dfrac{25}{15}$.

$\qquad = \dfrac{5}{3}m^{-8-(-5)}n^{7-2}$

Apply the quotient rule.

$\qquad = \dfrac{5}{3}m^{-3}n^5$

Subtract.

$\qquad = \dfrac{5n^5}{3m^3}$

Write with only positive exponents.

You Try 7

Simplify and write with positive exponents.

a) $\dfrac{10^9}{10^6}$
b) $\dfrac{v^6}{v^{-2}}$
c) $\dfrac{x^3}{x^7}$
d) $\dfrac{2^8}{2^{11}}$
e) $\dfrac{16a^{11}b^{-5}}{40a^3b^{-2}}$

In Section 5.2 we will learn more about combining the rules of exponents.

Answers to You Try Exercises

1) a) 81 b) q^9 c) $-24d^9$ d) w^{14} e) w^3k^5 f) $24x^3y^{14}$ 2) a) 5^{12} b) j^{30} c) $1000p^3$

d) $16k^{28}$ e) $\dfrac{25}{144}$ f) $-\dfrac{64w^3}{y^{27}}$ 3) a) $-80c^{20}d^{12}$ b) $-121r^{46}t^{20}$ c) $\dfrac{125a^{21}b^{15}}{c^6}$

4) a) 1 b) -1 c) 1 d) 1   5) a) $\dfrac{1}{121}$ b) 1000 c) $\dfrac{27}{8}$ d) $-\dfrac{1}{125}$ e) $\dfrac{t^4}{81}$ f) $\dfrac{1}{z^6}$ g) $\dfrac{9}{n^5}$

6) a) $\dfrac{b^7}{a^2}$ b) $\dfrac{1}{2c^4d^8}$ c) $\dfrac{6}{p^5q}$ d) $\dfrac{10x^2z^8}{3y^9}$ e) $\dfrac{125c^3}{a^3b^3}$ 7) a) 1000 b) v^8 c) $\dfrac{1}{x^4}$ d) $\dfrac{1}{8}$ e) $\dfrac{2a^8}{5b^3}$

5.1 Exercises

*Additional answers can be found in the Answers to Exercises appendix.

Objective 1: Use the Product Rule for Exponents

Identify the base and the exponent in each.

1) $(-7r)^5$ base: $-7r$; exponent: 5 2) $-\dfrac{4}{5}h^3$ base: h; exponent: 3

Rewrite each expression using exponents.

3) $9 \cdot 9 \cdot 9 \cdot 9 \cdot 9 \cdot 9$ 9^6

4) $\left(-\dfrac{2}{7}m\right)\left(-\dfrac{2}{7}m\right)\left(-\dfrac{2}{7}m\right)\left(-\dfrac{2}{7}m\right)$ $\left(-\dfrac{2}{7}m\right)^4$

5) Evaluate $(3 + 4)^2$ and $3^2 + 4^2$. Are they equivalent? Why or why not? $(3 + 4)^2 = 49, 3^2 + 4^2 = 25.$ They are not equivalent because when evaluating $(3 + 4)^2$, first add $3 + 4$ to get 7, then square the 7.

6) For any values of a and b, does $(a + b)^2 = a^2 + b^2$? Why or why not? Answers may vary.

7) Does $-3^4 = (-3)^4$? Why or why not? No. $-3^4 = -1 \cdot 3^4 = -1 \cdot 81 = -81; (-3)^4 = (-3) \cdot (-3) \cdot (-3) \cdot (-3) = 81$

8) Are $2k^4$ and $(2k)^4$ equivalent? Why or why not? No. $2k^4 = 2 \cdot k^4; (2k)^4 = 2^4 \cdot k^4 = 16k^4$

Evaluate the expression using the product rule, where applicable.

9) $2^2 \cdot 2^4$ 64 10) $10^2 \cdot 10$ 1000

11) $3^3 \cdot 5^2$ 675 12) $\left(\dfrac{1}{2}\right)^4 \cdot \left(\dfrac{1}{2}\right)$ $\dfrac{1}{32}$

Simplify the expression using the product rule. Leave your answer in exponential form.

13) $(-4)^2 \cdot (-4)^3 \cdot (-4)^2$ $(-4)^7$

14) $(8) \cdot (8)^7 \cdot (8)^2$ 8^{10}

15) $a^2 \cdot a^3$ a^5

16) $9w^4 \cdot w^3$ $9w^7$

17) $k \cdot k^2 \cdot k^3$ k^6

18) $-6z^5 \cdot z^6 \cdot z$ $-6z^{12}$

19) $(7h^3)(8h^{12})$ $56h^{15}$

20) $(2p^2)(-12p^5)$ $-24p^7$

21) $(5n^3)(-6n^7)(2n^2)$ $-60n^{12}$

22) $\left(\frac{8}{5}y^3\right)(35y)(-2y^4)$ $-112y^8$

23) $\left(\frac{49}{24}t^5\right)(-4t^7)\left(-\frac{12}{7}t\right)$ $14t^{13}$

24) $(6v^9)\left(\frac{5}{8}v^3\right)\left(\frac{4}{15}v^4\right)$ v^{16}

Objective 2: Use the Power Rules for Exponents

Simplify the expression using one of the power rules.

25) $(a^9)^4$ a^{36}

26) $(w^5)^9$ w^{45}

27) $(2^3)^2$ 64

28) $(3^2)^2$ 81

29) $\left(\frac{1}{2}\right)^5$ $\frac{1}{32}$

30) $\left(\frac{3}{2}\right)^4$ $\frac{81}{16}$

31) $\left(\frac{4}{y}\right)^3$ $\frac{64}{y^3}$

32) $\left(\frac{w}{k}\right)^6$ $\frac{w^6}{k^6}$

33) $(-10r)^4$ $10,000r^4$

34) $(2g)^5$ $32g^5$

35) $(-4ab)^3$ $-64a^3b^3$

36) $(-7mn)^2$ $49m^2n^2$

Mixed Exercises: Objectives 1 and 2

Simplify using the product and power rules.

37) $(k^9)^2(k^3)^2$ k^{24}

38) $(p^4)^2(p^7)^3$ p^{29}

39) $(5+3)^2$ 64

40) $(11-8)^2$ 9

41) $8(6k^7l^2)^2$ $288k^{14}l^4$

42) $2(-6a^5b)^2$ $72a^{10}b^2$

43) $-m^4(-5m^2)^3(-m^6)^2$ $125m^{22}$

44) $4h^9(11h^5)^2(-h^4)^3$ $-484h^{31}$

45) $\frac{(4d^9)^2}{(-2c^5)^6}$ $\frac{d^{18}}{4c^{30}}$

46) $\frac{(-5m^7)^3}{(5n^{12})^2}$ $-\frac{5m^{21}}{n^{24}}$

47) $\frac{6(a^8b^3)^5}{(2c)^3}$ $\frac{3a^{40}b^{15}}{4c^3}$

48) $\frac{(3x^8)^3}{15(y^2z^3)^4}$ $\frac{9x^{24}}{5y^8z^{12}}$

49) $(8u^3v^8)^2\left(-\frac{13}{4}uv^5\right)^2$ $676u^8v^{26}$

50) $\left(-\frac{3}{4}h^7k^2\right)^3\left(\frac{2}{9}h^4k\right)^2$ $-\frac{1}{48}h^{29}k^8$

51) $\left(\frac{4r^6s^2}{t^3}\right)^3$ $\frac{64r^{18}s^6}{t^9}$

52) $\left(-\frac{5x^3z}{12y^8}\right)^2$ $\frac{25x^6z^2}{144y^{16}}$

53) $\left(\frac{36n^4}{4m^8p^3}\right)^2$ $\frac{81n^8}{m^{16}p^6}$

54) $\left(\frac{21b^7}{14a^5c^2}\right)^4$ $\frac{81b^{28}}{16a^{20}c^8}$

Objective 3: Use 0 and Negative Exponents

In Exercises 55–58, decide whether each statement is true or false.

55) Raising a positive base to a negative exponent will give a negative result. (Example: 5^{-2}) False

56) $9^0 = 0$ False

57) The reciprocal of 8 is $\frac{1}{8}$. True

58) $4^{-3} + 2^{-3} = 6^{-3}$ False

Evaluate. Assume the variable does not equal zero.

59) 6^0 1

60) -12^0 -1

61) $-(-7)^0$ -1

62) 0^4 0

63) $(11)^0 + (-11)^0$ 2

64) w^0 1

65) $-6y^0$ -6

66) $x^0 - (9x)^0$ 0

67) 6^{-2} $\frac{1}{36}$

68) 2^{-4} $\frac{1}{16}$

69) $\left(\frac{1}{7}\right)^{-2}$ 49

70) $\left(\frac{1}{5}\right)^{-3}$ 125

71) $\left(\frac{4}{3}\right)^{-3}$ $\frac{27}{64}$

72) $\left(\frac{3}{10}\right)^{-3}$ $\frac{1000}{27}$

73) $\left(-\frac{1}{2}\right)^{-5}$ -32

74) $\left(-\frac{6}{11}\right)^{-2}$ $\frac{121}{36}$

75) -2^{-6} $-\frac{1}{64}$

76) -4^{-3} $-\frac{1}{64}$

77) $2^{-3} - 4^{-2}$ $\frac{1}{16}$

78) $5^{-2} + 2^{-2}$ $\frac{29}{100}$

79) $-9^{-2} + 3^{-3} + (-7)^0$ $\frac{83}{81}$

80) $6^0 - 9^{-1} + 4^0 + 3^{-2}$ 2

Rewrite each expression with only positive exponents. Assume the variables do not equal zero.

81) y^{-4} $\frac{1}{y^4}$

82) c^{-1} $\frac{1}{c}$

83) $\frac{a^{-10}}{b^{-3}}$ $\frac{b^3}{a^{10}}$

84) $\frac{u^{-6}}{v^{-2}}$ $\frac{v^2}{u^6}$

85) $\frac{x^4}{10y^{-5}}$ $\frac{x^4y^5}{10}$

86) $\frac{7t^{-3}}{u^6}$ $\frac{7}{t^3u^6}$

87) $8x^3y^{-7}$ $\frac{8x^3}{y^7}$

88) $\frac{1}{2}r^{-5}t^3$ $\frac{t^3}{2r^5}$

89) $\frac{8a^6b^{-1}}{5c^{-10}d}$ $\frac{8a^6c^{10}}{5bd}$

90) $\frac{1}{h^{-7}k^{-4}}$ h^7k^4

91) $\left(\frac{a}{6}\right)^{-2}$ $\frac{36}{a^2}$

92) $\left(\frac{3}{q}\right)^{-4}$ $\frac{q^4}{81}$

93) $\left(\frac{12b}{cd}\right)^{-2}$ $\frac{c^2d^2}{144b^2}$

94) $\left(\frac{2xy}{z}\right)^{-6}$ $\frac{z^6}{64x^6y^6}$

95) $-6r^{-2}$ $-\frac{6}{r^2}$

96) $2w^{-5}$ $\frac{2}{w^5}$

97) $-p^{-8}$ $-\frac{1}{p^8}$

98) $-k^{-7}$ $-\frac{1}{k^7}$

99) $\left(\frac{1}{x}\right)^{-1}$ x

100) $a^4\left(\frac{1}{b}\right)^{-9}$ a^4b^9

Objective 4: Use the Quotient Rule for Exponents

Simplify using the quotient rule. Assume the variables do not equal zero.

101) $\frac{n^9}{n^4}$ n^5

102) $\frac{h^8}{h}$ h^7

103) $\frac{8^{11}}{8^9}$ 64

104) $\frac{2^{15}}{2^{10}}$ 32

105) $\frac{5^6}{5^9}$ $\frac{1}{125}$

106) $\frac{3^7}{3^{11}}$ $\frac{1}{81}$

107) $\dfrac{9d^6}{d}$ $9d^5$

108) $\dfrac{40w^8}{72w^3}$ $\dfrac{5}{9}w^5$

115) $\dfrac{6v^{-1}w}{54v^2w^{-5}}$ $\dfrac{w^6}{9v^3}$

116) $\dfrac{54ab^{-2}}{36a^5b^{-7}}$ $\dfrac{3b^5}{2a^4}$

109) $\dfrac{t^3}{t^5}$ $\dfrac{1}{t^2}$

110) $\dfrac{a^2}{a^9}$ $\dfrac{1}{a^7}$

117) $\dfrac{20m^{-3}n^4}{4m^8n^6}$ $\dfrac{5}{m^{11}n^2}$

118) $\dfrac{3c^{-4}d}{7c^{-1}d^4}$ $\dfrac{3}{7c^3d^3}$

111) $\dfrac{x^{-3}}{x^6}$ $\dfrac{1}{x^9}$

112) $\dfrac{h^{-10}}{h^{-2}}$ $\dfrac{1}{h^8}$

119) $\dfrac{(x+y)^8}{(x+y)^3}$ $(x+y)^5$

120) $\dfrac{(a-3b)^{-4}}{(a-3b)^{-5}}$ $a-3b$

113) $\dfrac{-6k}{k^4}$ $-\dfrac{6}{k^3}$

114) $\dfrac{15m^3}{m^{-1}}$ $15m^4$

Section 5.2 More on Exponents and Scientific Notation

Objectives

1. Combine the Rules of Exponents
2. Understand Scientific Notation
3. Convert a Number from Scientific Notation to Standard Form
4. Convert a Number from Standard Form to Scientific Notation
5. Perform Operations on Numbers in Scientific Notation

1. Combine the Rules of Exponents

Let's review all the rules of exponents and then see how we can combine the rules to simplify expressions.

Summary   Rules of Exponents

In the rules below, a and b are any real numbers and m and n are positive integers.

Product rule	$a^m \cdot a^n = a^{m+n}$
Basic power rule	$(a^m)^n = a^{mn}$
Power rule for a product	$(ab)^n = a^n b^n$
Power rule for a quotient	$\left(\dfrac{a}{b}\right)^n = \dfrac{a^n}{b^n}, (b \neq 0)$
Quotient rule	$\dfrac{a^m}{a^n} = a^{m-n}, (a \neq 0)$

Change from negative to positive exponents, where $a \neq 0$, $b \neq 0$, and m and n are any integers.

$$\dfrac{a^{-m}}{b^{-n}} = \dfrac{b^n}{a^m}, \quad \left(\dfrac{a}{b}\right)^{-n} = \left(\dfrac{b}{a}\right)^n$$

In the definitions below, $a \neq 0$ and n is any integer.

Zero as an exponent	$a^0 = 1$
Negative number as an exponent	$a^{-n} = \left(\dfrac{1}{a}\right)^n = \dfrac{1}{a^n}$

Example 1

Simplify using the rules of exponents. Assume all variables represent nonzero real numbers. Write the answers with positive exponents.

a) $(5k^{-7})^3(2k^6)^2$ b) $\left(\dfrac{9a^8b^5}{a^3b^2}\right)^2$ c) $\dfrac{h^{-4} \cdot h^5}{h^6}$ d) $\left(\dfrac{8x^{-3}y^{10}}{20xy^{-3}}\right)^{-3}$

Solution

a) We must follow the order of operations. Therefore, evaluate the exponents first.

$$
\begin{aligned}
(5k^{-7})^3 \cdot (2k^6)^2 &= 5^3k^{(-7)(3)} \cdot 2^2k^{(6)(2)} &&\text{Apply the power rule.} \\
&= 125k^{-21} \cdot 4k^{12} &&\text{Simplify.} \\
&= 500k^{-21+12} &&\text{Multiply } 125 \cdot 4\text{, and add the exponents.} \\
&= 500k^{-9} &&\text{Add the exponents.} \\
&= \dfrac{500}{k^9} &&\text{Write the answer using a positive exponent.}
\end{aligned}
$$

In-Class Example 1

Simplify using the rules of exponents. Assume all variables represent nonzero real numbers. Write the answers with positive exponents.

a) $(4p^{-8})^2(2p^4)^3$ b) $\left(\dfrac{2c^9d^{11}}{c^2d^5}\right)^4$
c) $\dfrac{k^{12} \cdot k^{-9}}{k^7}$ d) $\left(\dfrac{60a^7b^{-1}}{25a^{-4}b^3}\right)^{-2}$

answer: a) $\dfrac{128}{p^4}$ b) $16c^{28}d^{24}$
c) $\dfrac{1}{k^4}$ d) $\dfrac{25b^8}{144a^{22}}$

b) How can we begin this problem? We can use the quotient rule to simplify the expression before squaring it.

$$\left(\frac{9a^8b^5}{a^3b^2}\right)^2 = (9a^{8-3}b^{5-2})^2 \qquad \text{Apply the quotient rule inside the parentheses.}$$

$$= (9a^5b^3)^2 \qquad\qquad \text{Simplify.}$$

$$= 9^2a^{(5)(2)}b^{(3)(2)} \qquad \text{Apply the product rule.}$$

$$= 81a^{10}b^6$$

c) Let's begin by simplifying the numerator.

$$\frac{h^{-4} \cdot h^5}{h^6} = \frac{h^{-4+5}}{h^6} \qquad \text{Add the exponents in the numerator.}$$

$$= \frac{h^1}{h^6}$$

Now, we can apply the quotient rule:

$$= h^{1-6} = h^{-5} \qquad \text{Subtract the exponents.}$$

$$= \frac{1}{h^5} \qquad\qquad \text{Write the answer using a positive exponent.}$$

d) Eliminate the negative exponent **outside** of the parentheses by taking the reciprocal of the base. Notice that we have **not** eliminated the negatives on the exponents **inside** the parentheses.

$$\left(\frac{8x^{-3}y^{10}}{20xy^{-3}}\right)^{-3} = \left(\frac{20xy^{-3}}{8x^{-3}y^{10}}\right)^3$$

We could apply the exponent of 3 to the quantity inside of the parentheses, but we could also reduce $\dfrac{20}{8}$ first and apply the quotient rule before cubing the quantity.

$$= \left(\frac{5}{2}x^{1-(-3)}y^{-3-10}\right)^3 \qquad \text{Reduce } \frac{20}{8} \text{ and subtract the exponents.}$$

$$= \left(\frac{5}{2}x^4y^{-13}\right)^3$$

$$= \frac{125}{8}x^{12}y^{-39} \qquad\qquad \text{Apply the power rule.}$$

$$= \frac{125x^{12}}{8y^{39}} \qquad\qquad \text{Write the answer using positive exponents.}$$

You Try 1

Simplify using the rules of exponents. Assume all variables represent nonzero real numbers. Write the answers with positive exponents.

a) $\left(\dfrac{2r^{11}t^6}{r^5t}\right)^5$ b) $(-w^{-3})^6(11w^4)^2$ c) $\dfrac{z^8 \cdot z^{-10}}{z^2}$ d) $\left(\dfrac{24a^{12}b^{-4}}{21a^7b}\right)^{-2}$

It is possible for variables to appear in exponents. The same rules apply.

In-Class Example 2
Simplify using the rules of
exponents. Assume all variables
represent nonzero integers.
Write your final answer so that
the exponents have positive
coefficients.

a) $h^{2a} \cdot h^{4a}$ b) $\dfrac{w^{3x}}{w^{8x}}$

answer: a) h^{6a} b) $\dfrac{1}{w^{5x}}$

Example 2

Simplify using the rules of exponents. Assume all variables represent nonzero integers. Write your final answer so that the exponents have positive coefficients.

a) $n^{3x} \cdot n^{2x}$ b) $\dfrac{a^{4b}}{a^{8b}}$

Solution

a) $n^{3x} \cdot n^{2x} = n^{3x+2x} = n^{5x}$ The bases are the same, so apply the product rule. Add the exponents.

b) $\dfrac{a^{4b}}{a^{8b}} = a^{4b-8b}$ The bases are the same, so apply the quotient rule. Subtract the exponents.

$= a^{-4b}$

$= \dfrac{1}{a^{4b}}$ Write the answer with a positive exponent.

You Try 2

Simplify using the rules of exponents. Assume all variables represent nonzero integers. Write your final answer so that the exponents have positive coefficients.

a) $6^{5p} \cdot 6^{p} \cdot 6^{8p}$ b) $(d^4)^{-5k}$

2. Understand Scientific Notation

The distance from the Earth to the Sun is approximately 150,000,000 km.

The gross domestic product of the United States in 2008 was $14,280,700,000,000. (Bureau of Economic Analysis)

A single rhinovirus (cause of the common cold) measures 0.00002 mm across.

Each of these is an example of a very large or very small number containing many zeros. Sometimes, performing operations with so many zeros can be difficult. This is why scientists and economists, for example, often work with such numbers in a different form called *scientific notation*.

Scientific notation is a short-hand method for writing very large and very small numbers. Writing numbers in scientific notation together with applying the rules of exponents can simplify calculations with very large and very small numbers.

Before discussing scientific notation further, we will review some principles behind the notation. Let's look at multiplying numbers by positive powers of 10.

Example 3

In-Class Example 3
Multiply.
a) 0.009152×10^5 b) 48×10^3
answer: a) 915.2 b) $48,000$

Multiply.

a) 0.0538×10^3 b) 76×10^2

Solution

a) $0.0538 \times 10^3 = 0.0538 \times 1000 = 53.8$

b) $76 \times 10^2 = 76 \times 100 = 7600$

Notice that when we multiply a positive number by a positive power of 10, the result is *larger* than the original number. In fact, the exponent determines how many places to the *right* the decimal point is moved.

$$0.0538 \times 10^3 = 53.8 \qquad\qquad 76 \times 10^2 = 76.00 \times 10^2 = 7600$$

3 places to right   2 places to right

You Try 3

Multiply by moving the decimal point the appropriate number of places.

a) 6.44×10^5 b) 0.000937×10^4

What happens to a positive number when we multiply by a *negative* power of 10?

Example 4

In-Class Example 4
Multiply.
a) 32×10^{-3} b) 2.49×10^{-5}
**answer: a) 0.032
b) 0.0000249**

Multiply.

a) 59×10^{-2} b) 138×10^{-4}

Solution

a) $59 \times 10^{-2} = 59 \times \dfrac{1}{100} = \dfrac{59}{100} = 0.59$

b) $138 \times 10^{-4} = 138 \times \dfrac{1}{10,000} = \dfrac{138}{10,000} = 0.0138$

Is there a pattern? When we multiply a positive number by a negative power of 10, the result is *smaller* than the original number. The exponent determines how many places to the *left* the decimal point is moved:

$$59 \times 10^{-2} = 59. \times 10^{-2} = 0.59 \qquad 138 \times 10^{-4} = 0138. \times 10^{-4} = 0.0138$$

2 places to the left   4 places to the left

You Try 4

Multiply.

a) 61×10^{-3} b) 4.9×10^{-4}

Definition

A number is in **scientific notation** if it is written in the form $a \times 10^n$ where $1 \leq |a| < 10$ and n is an integer.

Multiplying $|a|$ by a *positive* power of 10 will result in a number that is *larger* than $|a|$. Multiplying $|a|$ by a *negative* power of 10 will result in a number that is *smaller* than $|a|$. $1 \leq |a| < 10$ means that a is a number that has *one* nonzero digit to the left of the decimal point.

Some examples of numbers written in scientific notation are

$$9.15 \times 10^{-5}, \quad 8.1 \times 10^3, \quad 6 \times 10^{-2}.$$

The following numbers are *not* in scientific notation:

$$26.31 \times 10^5 \qquad 0.87 \times 10^{-4} \qquad 400 \times 10^2$$

↑   ↑   ↑
2 digits to left   Zero is to left   3 digits to left
of decimal point   of decimal point.   of decimal point

3. Convert a Number from Scientific Notation to Standard Form

We will continue our discussion by converting from scientific notation to standard form, that is, a number without exponents.

Example 5	

In-Class Example 5
Rewrite in standard form.
a) 5.281×10^3 b) 4.37×10^{-2}
c) 9.566×10^4
**answer: a) 5281   b) 0.0437
c) 95,660**

Rewrite in standard form.

a) 7.094×10^4 b) 2.6×10^{-3} c) 8.1163×10^3

Solution

a) $7.094 \times 10^4 \rightarrow 7.0940 = 70{,}940$ Remember, multiplying a positive number by a positive
                                                     power of 10 will make the result *larger* than 7.094

         4 places to the right

b) $2.6 \times 10^{-3} \rightarrow 002.6 = 0.0026$ Multiplying 2.6 by a negative power of 10 will make the
                                                     result *smaller* than 2.6.

         3 places to the left

c) $8.1163 \times 10^3 \rightarrow 8.1163 = 8116.3$

         3 places to the right

You Try 5

Rewrite in standard form.

a) 6.09×10^4 b) 1.4×10^{-5} c) 5.02147×10^3

4. Convert a Number from Standard Form to Scientific Notation

To write the number 25,000 in scientific notation, first locate its decimal point.

25,000.
↑
Decimal point is here.

Next, determine where the decimal point will be when the number is in scientific notation.

2500.
∧
Decimal point will be here.

In scientific notation, 25,000 must be 2.5×10^n, where n is an integer. Will n be positive or negative? The number 2.5 must be multiplied by a *positive* power of 10 to make it larger.

Count the number of places between the original and the final decimal place locations.

25000
4 3 2 1

Use the number of places, 4, as the exponent of 10.

$$25{,}000 = 2.5 \times 10^4$$

Example 6	

In-Class Example 6
Write each number in scientific
notation.
a) In 2004, New York City's
population was approximately
8,100,000 people.   b) The speed
of a garden snail is 0.03 mph.
**answer: a) 8.1×10^6 people
b) 3×10^{-2} mph**

Write each number in scientific notation.

Solution

a)   The distance from the Earth to the Sun is approximately 150,000,000 km.

         150,000,000.   150,000,000   Move decimal point eight places.
              ∧   ↖

     Decimal point   Decimal point
     will be here.   is here.

     $150{,}000{,}000 \text{ km} = 1.5 \times 10^8 \text{ km}$

b)   A single rhinovirus measures 0.00002 mm across.

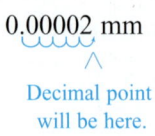

0.00002 mm

Decimal point
will be here.

$$0.00002 \text{ mm} = 2 \times 10^{-5} \text{ mm}$$

Procedure   How to Write a Number in Scientific Notation

1)   Locate the decimal point in the original number.

2)   Determine where the decimal point needs to be when converting to scientific notation. Remember, there will be one nonzero digit to the left of the decimal point.

3)   Count how many places you must move the decimal point to take it from its original place to its position for scientific notation.

4)   If the absolute value of the resulting number is *smaller* than the absolute value of the original number, you will multiply the result by a *positive* power of 10. Example: $350.9 = 3.509 \times 10^2$.

If the absolute value of the resulting number is *larger* than the absolute value of the original number, you will multiply the result by a *negative* power of 10. Example: $0.0000068 = 6.8 \times 10^{-6}$.

You Try 6

Write each number in scientific notation.

a)   The gross domestic product of the United States in 2008 was approximately $14,280,700,000.

b)   The diameter of a human hair is approximately 0.001 in.

---

5. Perform Operations on Numbers in Scientific Notation

We use the rules of exponents to perform operations with numbers in scientific notation.

Example 7

In-Class Example 7
Perform the operations and simplify.
a) $(3 \times 10^2)(-6 \times 10^3)$
b) $\dfrac{7 \times 10^4}{2 \times 10^9}$

answer: a) $-1,800,000$
b) 0.000035

Perform the operations and simplify.

a) $(-2 \times 10^3)(3 \times 10^2)$ b) $\dfrac{9 \times 10^3}{2 \times 10^5}$

Solution

a) $(-2 \times 10^3)(3 \times 10^2) = (-2 \times 3)(10^3 \times 10^2)$ Commutative property
$= -6 \times 10^5$ Add the exponents.
$= -600,000$

b) $\dfrac{9 \times 10^3}{2 \times 10^5} = \dfrac{9}{2} \times \dfrac{10^3}{10^5}$

$= 4.5 \times 10^{-2} = 0.045$ Subtract the exponents.

Write $\dfrac{9}{2}$ in decimal form.

You Try 7

Perform the operations and simplify.

a) $(4.1 \times 10^2)(3 \times 10^4)$ b) $\dfrac{5.2 \times 10^{-10}}{4 \times 10^{-6}}$

Using Technology

We can use a graphing calculator to convert a very large or very small number to scientific notation, or to convert a number in scientific notation to a number written without an exponent. Suppose we are given a very large number such as 35,000,000,000. If you enter any number with more than 10 digits on the home screen of your calculator and press ENTER the number will automatically be displayed in scientific notation as shown on the screen below. A small number with more than two zeros to the right of the decimal point (such as .000123) will automatically be displayed in scientific notation as shown below.

The E shown in the screen refers to a power of 10, so 3.5 E 10 is the number 3.5×10^{10} in scientific notation. 1.23 E-4 is the number 1.23×10^{-4} in scientific notation.

```
35000000000
              3.5E10
.000123
           1.23E -4
```

If a large number has 10 or fewer digits, or if a small number has fewer than three zeros to the right of the decimal point, then the number will not automatically be displayed in scientific notation. To display the number using scientific notation press MODE, select SCI, and press ENTER. When you return to the home screen, all numbers will be displayed in scientific notation as shown below.

```
NORMAL  SCI  ENG
FLOAT   0123456789
RADIAN  DEGREE
FUNC  PAR  POL  SEQ
CONNECTED  DOT
SEQUENTIAL  SIMUL
REAL  a+bi  re^θi
FULL  HORIZ  G-T
SET CLOCK 03/14/01 2:28AM
```

```
5000
              5E3
.0123
           1.23E -2
```

A number written in scientific notation can be entered directly into your calculator. For example, the number 2.38×10^7 can be entered directly on the home screen by typing 2.38 followed by 2nd , 7 ENTER as shown to the right. If you wish to display this number without an exponent, change the mode back to NORMAL and enter the number on the home screen as shown.

```
2.38E7
            2.38E7
2.38E7
          23800000
```

Write each number without an exponent using a graphing calculator.

1) 3.14×10^5 2) 9.3×10^7 3) 1.38×10^{-3}

Write each number in scientific notation using a graphing calculator.

4) 186,000   5) 5280   6) 0.0469

Answers to You Try Exercises

1) a) $32r^{30}t^{25}$ b) $\dfrac{121}{w^{10}}$ c) $\dfrac{1}{z^4}$ d) $\dfrac{49b^{10}}{64a^{10}}$ 2) a) 6^{14p} b) $\dfrac{1}{d^{20k}}$ 3) a) 644,000   b) 9.37

4) a) 0.061   b) 0.00049   5) a) 60,900   b) 0.000014   c) 5021.47

6) a) 1.42807×10^{10} dollars   b) 1.0×10^{-3} in.   7) a) 12,300,000   b) 0.00013

Answers to Technology Exercises

1) 314,000   2) 93,000,000   3) 0.00138   4) 1.86×10^5 5) 5.28×10^3
6) 4.69×10^{-2}

5.2 Exercises

*Additional answers can be found in the Answers to Exercises appendix.

Objective 1: Combine the Rules of Exponents

Simplify. Assume all variables represent nonzero real numbers. Write the answers with positive exponents.

1) $-10(-3g^4)^3$ $\quad 270g^{12}$

2) $6(2d^3)^3$ $\quad 48d^9$

3) $\dfrac{23t}{t^{11}}$ $\quad \dfrac{23}{t^{10}}$

4) $\dfrac{r^{-6}}{r^{-2}}$ $\quad \dfrac{1}{r^4}$

5) $\left(\dfrac{2xy^4}{3x^{-9}y^{-2}}\right)^3$ $\quad \dfrac{8}{27}x^{30}y^{18}$

6) $\left(\dfrac{a^8b^3}{10a^5}\right)^3$ $\quad \dfrac{a^9b^9}{1000}$

7) $\left(\dfrac{7r^3}{s^8}\right)^{-2}$ $\quad \dfrac{s^{16}}{49r^6}$

8) $\left(\dfrac{3n^{-6}}{m^2}\right)^{-4}$ $\quad \dfrac{m^8n^{24}}{81}$

9) $(-k^4)^3$ $\quad -k^{12}$

10) $(t^7)^8$ $\quad t^{56}$

11) $(-2m^5n^2)^5$ $\quad -32m^{25}n^{10}$

12) $(13yz^6)^2$ $\quad 169y^2z^{12}$

13) $\left(-\dfrac{9}{4}z^5\right)\left(\dfrac{2}{3}z^{-1}\right)$ $\quad -\dfrac{3}{2}z^4$

14) $(20w^2)\left(-\dfrac{3}{5}w^6\right)$ $\quad -12w^8$

15) $\left(\dfrac{a^5}{b^4}\right)^{-6}$ $\quad \dfrac{b^{24}}{a^{30}}$

16) $\dfrac{x^{-4}}{y^{10}}$ $\quad \dfrac{1}{x^4y^{10}}$

17) $(-ab^3c^5)^2\left(\dfrac{a^4}{bc}\right)^3$ $\quad a^{14}b^3c^7$

18) $\dfrac{(4v^3)^2}{(6v^8)^2}$ $\quad \dfrac{4}{9v^{10}}$

19) $\left(\dfrac{48u^{-7}v^2}{36u^3v^{-5}}\right)^{-3}$ $\quad \dfrac{27u^{30}}{64v^{21}}$

20) $\left(\dfrac{xy^5}{5x^{-2}y}\right)^{-3}$ $\quad \dfrac{125}{x^9y^{12}}$

21) $\left(\dfrac{-3t^4u}{t^2u^{-4}}\right)^4$ $\quad 81t^8u^{20}$

22) $\left(\dfrac{k^7m^7}{12k^{-1}m^6}\right)^2$ $\quad \dfrac{1}{144}k^{16}m^2$

23) $(h^{-3})^7$ $\quad \dfrac{1}{h^{21}}$

24) $(-n^4)^{-5}$ $\quad -\dfrac{1}{n^{20}}$

25) $\left(\dfrac{h}{2}\right)^5$ $\quad \dfrac{h^5}{32}$

26) $17m^{-2}$ $\quad \dfrac{17}{m^2}$

27) $-7c^8(-2c^2)^3$ $\quad 56c^{14}$

28) $5p^3(4p^7)^2$ $\quad 80p^{17}$

29) $(12a^7)^{-1}(6a)^2$ $\quad \dfrac{3}{a^5}$

30) $(9c^2d)^{-1}$ $\quad \dfrac{1}{9c^2d}$

31) $\left(\dfrac{9}{20}d^5\right)(2d^{-3})\left(\dfrac{4}{33}d^9\right)$

32) $\left(\dfrac{f^8 \cdot f^{-3}}{f^2 \cdot f^9}\right)^5$ $\quad \dfrac{1}{f^{30}}$

33) $\left(\dfrac{56m^4n^8}{21m^4n^5}\right)^{-2}$ $\quad \dfrac{9}{64n^6}$

34) $\dfrac{(2x^4y)^{-2}}{(5xy^3)^2}$ $\quad \dfrac{1}{100x^{10}y^8}$

35) $\dfrac{(-r^{-1}t^7)^{-5}}{(2r^7t^{-1})^{-5}}$ $\quad -\dfrac{32r^{40}}{t^{15}}$

36) $(-10p^{-14})\left(\dfrac{2}{3}p^5\right)\left(\dfrac{9}{20}p^3\right)$ $\quad -\dfrac{3}{p^6}$

37) $\left(\dfrac{c^5}{6a^4b^{-1}}\right)^{-2}\left(\dfrac{a^{-8}c}{2b^3}\right)^5\left(\dfrac{3b^5c^{-2}}{a}\right)^{-3}$ $\quad \dfrac{c}{24a^{29}b^{32}}$

38) $\left(\dfrac{y^{-3}}{5xz^{-1}}\right)^{-1}\left(\dfrac{8x^{-3}}{z^5}\right)^{-2}\left(-\dfrac{2z^{-6}}{x^{-4}y}\right)^3$ $\quad -\dfrac{5x^{19}}{8z^9}$

39) $\dfrac{(x+y)^4}{(x+y)^9}$ $\quad \dfrac{1}{(x+y)^5}$

40) $\dfrac{(s-t)^{10}}{(s-t)^3}$ $\quad (s-t)^7$

Simplify. Assume that the variables represent nonzero integers. Write your final answer so that the exponents have positive coefficients.

41) $(p^{2n})^5$ $\quad p^{10n}$

42) $(3d^{4t})^2$ $\quad 9d^{8t}$

43) $y^m \cdot y^{10m}$ $\quad y^{11m}$

44) $t^{-6c} \cdot t^{9c}$ $\quad t^{3c}$

45) $x^{5a} \cdot x^{-8a}$ $\quad \dfrac{1}{x^{3a}}$

46) $b^{-2y} \cdot b^{-3y}$ $\quad \dfrac{1}{b^{5y}}$

47) $\dfrac{21c^{2x}}{35c^{8x}}$ $\quad \dfrac{3}{5c^{6x}}$

48) $-\dfrac{5y^{-13a}}{8y^{-2a}}$ $\quad -\dfrac{5}{8y^{11a}}$

49) Find the area.

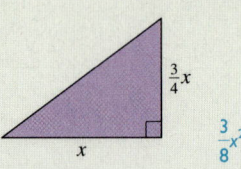

$\frac{3}{8}x^2$ sq units

50) The area of a rectangle is $24n^5$ in². Its width is $6n^3$ in. Find the length. $\quad 4n^2$ in.

Objective 2: Understand Scientific Notation

Determine if each number is in scientific notation.

51) 4.73×10^3 $\quad$ yes

52) -93×10^{-5} $\quad$ no

53) 36.0×10^{-2} $\quad$ no

54) 0.14×10^{-6} $\quad$ no

55) -4.3×10^7 $\quad$ yes

56) -2×10^5 $\quad$ yes

57) 3.5×2^{-8} $\quad$ no

58) Explain, in your own words, how to determine if a number is expressed in scientific notation. $\quad$ Answers may vary.

59) Explain, in your own words, how to write 5.3×10^{-2} in standard form. $\quad$ Answers may vary.

60) Explain, in your own words, how to write 8.76×10^6 in standard form. $\quad$ Answers may vary.

Objective 3: Convert a Number from Scientific Notation to Standard Form

Write each number without an exponent.

61) -6.8×10^{-5} $\quad -0.000068$

62) 2.91×10^4 $\quad 29{,}100$

63) 3.45029×10^4 $\quad 34{,}502.9$

64) -3.57×10^3 $\quad -3570$

65) -5×10^{-5} $\quad -0.00005$

66) 9×10^{-7} $\quad 0.0000009$

67) 8.1×10^{-4} $\quad 0.00081$

68) 2.645067×10^4 $\quad 26{,}450.67$

69) 3×10^6 $\quad 3{,}000{,}000$

70) 7×10^2 $\quad 700$

71) -3.921×10^{-2} $\quad -0.03921$

72) 4.1×10^{-6} $\quad 0.0000041$

Objective 4: Convert a Number from Standard Form to Scientific Notation

Write each number in scientific notation.

73) 2110.5 $\quad 2.1105 \times 10^3$

74) 382.275 $\quad 3.82275 \times 10^2$

75) 0.0048 $\quad 4.8 \times 10^{-3}$

76) 0.000321 $\quad 3.21 \times 10^{-4}$

77) $-400,000$ -4×10^5 78) $92,600$ 9.26×10^4

79) $11,000$ 1.1×10^4 80) $-308,000$ -3.08×10^5

VIDEO 81) 0.0008 8×10^{-4} 82) -0.00000089 -8.9×10^{-7}

83) -0.054 -5.4×10^{-2} 84) 9990 9.99×10^3

85) 6500 6.5×10^3 86) 0.00000002 2×10^{-8}

Write each number in scientific notation.

87) In 1883 it cost approximately $15,000,000 to build the Brooklyn Bridge. (www.nycroads.com) $\$1.5 \times 10^7$

88) A typical music player may hold approximately 16,384,000 bytes of data. 1.6384×10^7 bytes

89) The diameter of an atom is about 0.00000001 cm. 1×10^{-8} cm

90) The oxygen-hydrogen bond-length in a water molecule is 0.000000001 mm. 1×10^{-9} mm

Objective 5: Perform Operations on Numbers in Scientific Notation

Perform the operation as indicated. Write the final answer without an exponent.

91) $\dfrac{8 \times 10^6}{4 \times 10^2}$ $20,000$ 92) $(9 \times 10^5)(3 \times 10^4)$ $27,000,000,000$

VIDEO 93) $(2.3 \times 10^3)(3 \times 10^2)$ $690,000$ 94) $\dfrac{12 \times 10^7}{4 \times 10^4}$ 3000

95) $\dfrac{9.6 \times 10^{11}}{-4 \times 10^4}$ $-24,000,000$ 96) $\dfrac{-8.8 \times 10^{-3}}{-2.2 \times 10^{-8}}$ $400,000$

97) $(-1.5 \times 10^{-6})(6 \times 10^2)$ -0.0009

98) $(-4 \times 10^{-3})(-3.5 \times 10^{-1})$ 0.0014

99) $\dfrac{(0.004)(600,000)}{0.0003}$ $8,000,000$ 100) $\dfrac{360}{(0.006)(2,000,000)}$ 0.03

For each exercise, express each number in scientific notation, then solve.

101) When one of the U.S. space shuttles enters orbit, it travels at about 7800 m/s. How far does it travel in 24 hours?

(Hint: Change hours to seconds, and write all numbers in scientific notation before doing the computations.) (hypertextbook.com) 6.7392×10^8 m

102) According to Nielsen Media Research, over 92,000,000 people watched Super Bowl XLIII in 2009 between the Pittsburgh Steelers and the Arizona Cardinals. The California Avocado Commission estimates that about 46,000,000 pounds of avocados were eaten during that Super Bowl, mostly in the form of guacamole. On average, how many pounds of guacamole did each viewer eat during the Super Bowl?   0.5 pounds per person

103) In 2007, the United States produced about 6×10^9 metric tons of carbon emissions. The U.S. population that year was about 300 million. Find the amount of carbon emissions produced per person that year. (www.eia.doe.gov and U.S. Census Bureau)   20 metric tons

104) A photo printer delivers approximately 1.1×10^6 droplets of ink per square inch. How many droplets of ink would a 4 in. $\times$ 6 in. photo contain?   26,400,000 droplets

105) The Hoover Dam generates approximately 4×10^9 kW-hours (kW-hr) of power per year. How much would it generate in 6 yr? (http://hooverdam.travelnevada.com/greenpower.aspx)   24,000,000,000 kW-hr

106) A three-toed sloth moves at a rate of about 7.38×10^{-2} mph on the ground. How far could it go after 5 hr? (http://hypertextbook.com)   0.369 mi

Section 5.3 Addition and Subtraction of Polynomials and Polynomial Functions

Objectives

1. Learn Polynomial Vocabulary
2. Add and Subtract Polynomials in One Variable
3. Add and Subtract Polynomials in More Than One Variable
4. Define and Evaluate a Polynomial Function
5. Add and Subtract Polynomial Functions

1. Learn Polynomial Vocabulary

In Section 1.3 we defined an *algebraic expression* as a collection of numbers, variables, and grouping symbols connected by operation symbols such as $+$, $-$, $\times$, and $\div$. An example of an algebraic expression is $4x^3 + 6x^2 - x + \frac{5}{2}$. The *terms* of this algebraic expression are $4x^3$, $6x^2$, $-x$, and $\frac{5}{2}$. A **term** is a number or a variable or a product or quotient of numbers and variables. Not only is $4x^3 + 6x^2 - x + \frac{5}{2}$ an expression, it is also a *polynomial*.

> **Definition**
>
> A **polynomial in x** is the sum of a finite number of terms of the form ax^n, where n is a whole number and a is a real number.

Let's look more closely at the polynomial $4x^3 + 6x^2 - x + \dfrac{5}{2}$.

1) The polynomial is written in **descending powers of x** since the powers of x decrease from left to right. Generally, we write polynomials in descending powers of the variable.

2) Recall that the term without a variable is called a **constant.** The constant is 5/2. The **degree of a term** equals the exponent on its variable. (If a term has more than one variable, the degree equals the *sum* of the exponents on the variables.) We will list each term, its coefficient, and its degree.

Term	Coefficient	Degree
$4x^3$	4	3
$6x^2$	6	2
$-x$	-1	1
$\dfrac{5}{2}$	$\dfrac{5}{2}$	$0 \quad \left(\dfrac{5}{2} = \dfrac{5}{2}x^0\right)$

3) The **degree of the polynomial** equals the highest degree of any nonzero term. The polynomial above has degree 3. Or, we say that this is a **third-degree polynomial.**

Example 1

In-Class Example 1
Decide if each expression is or is not a polynomial. If it is a polynomial, identify each term and the degree of each term. Then, find the degree of the polynomial.
a) $2w + 3 + 4w^{-2}$
b) $-18b^4 - 0.02b^3 + 11$
c) $4x^{1/2} + 3x - 1$
d) $4x^3y^2$
answer:
a) not a polynomial
b) polynomial

Term	Degree
$-18b^4$	4
$-0.02b^3$	3
11	0

Degree: 4
c) not a polynomial
d) polynomial

Term	Degree
$4x^3y^2$	5

Degree: 5

Decide if each expression *is* or *is not* a polynomial. If it is a polynomial, identify each term and the degree of each term. Then, find the degree of the polynomial.

a) $-7k^4 + 3.2k^3 - 8k^2 - 11$ b) $5n^2 - \dfrac{4}{3}n + 6 + \dfrac{9}{n^2}$

c) $x^3y^3 + 3x^2y + 3xy + 1$ d) $12t^5$

Solution

a) The expression $-7k^4 + 3.2k^3 - 8k^2 - 11$ is a polynomial in k. Its terms have whole number exponents and real coefficients. Its terms and degrees are in the table. *The degree of this polynomial is 4.*

Term	Degree
$-7k^4$	4
$3.2k^3$	3
$-8k^2$	2
-11	0

b) The expression $5n^2 - \dfrac{4}{3}n + 6 + \dfrac{9}{n^2}$ is *not* a polynomial because one of its terms has a variable in the denominator.

c) The expression $x^3y^3 + 3x^2y + 3xy + 1$ *is* a polynomial because the variables have whole number exponents and the coefficients are real numbers. Since this is a polynomial in two variables, we find the degree of each term by adding the exponents. *The degree of this polynomial is 6.*

Term	Degree
x^3y^3	6
$3x^2y$	3
$3xy$	2
1	0

d) The expression $12t^5$ is a polynomial even though it has only one term. The degree of the term is 5, and that is the degree of the polynomial as well.

You Try 1

Decide if each expression *is* or *is not* a polynomial. If it is a polynomial, identify each term and the degree of each term. Then, find the degree of the polynomial.

a) $g^3 + 8g^2 - \dfrac{5}{g}$ b) $6t^5 - \dfrac{4}{9}t^4 + t + 2$

c) $a^4b^3 - 9a^3b^3 - a^2b + 4a - 7$ d) $x + 5x^{1/2} + 6$

A polynomial with one term, like $12t^5$, is called a **monomial** ("mono" means one). Other examples of monomials are x^2, $-9j^4$, y, x^2y^2, and -2.

A **binomial** is a polynomial that consists of exactly two terms ("bi" means two). Some examples are $n + 6$, $5b^2 - 7$, $c^4 - d^4$, and $-12u^4v^3 + 8u^2v^2$.

A **trinomial** is a polynomial that consists of exactly three terms ("tri" means three). Here are some examples: $p^2 - 5p - 36$, $3k^5 + 24k^2 - 6k$, and $8r^4 + 10r^2s + 3s^2$.

It is important that you understand the meaning of these terms. We will use them throughout our study of algebra.

Recall from beginning algebra that **like terms** contain the same variables with the same exponents. For example, $4y^2$ and $\dfrac{2}{3}y^2$ are like terms, but $5x^6$ and $3x^4$ are not because their exponents are different. We add or subtract like terms by adding or subtracting the coefficients and leaving the variable(s) and exponent(s) the same. We use the same idea for adding and subtracting polynomials.

2. Add and Subtract Polynomials in One Variable

> **Procedure**   Adding Polynomials
>
> To add polynomials, add like terms.

Example 2

Add the polynomials.

a) $11c^2 + 3c - 9$ and $2c^2 + 5c + 1$ b) $(7y^3 - 10y^2 + y - 2) + (3y^3 + 4y^2 - 3)$

Solution

a) The addition problem can be set up horizontally or vertically. We will add these horizontally. Put the polynomials in parentheses since each contains more than one term. Use the associative and commutative properties to rewrite like terms together.

$$(11c^2 + 3c - 9) + (2c^2 + 5c + 1) = (11c^2 + 2c^2) + (3c + 5c) + (-9 + 1)$$
$$= 13c^2 + 8c - 8 \qquad \text{Combine like terms.}$$

b) We will add these polynomials vertically. Line up like terms in columns and add.

$$\begin{array}{r} 7y^3 - 10y^2 + y - 2 \\ + \ \underline{3y^3 + \ 4y^2 \quad\ - 3} \\ 10y^3 - \ 6y^2 + y - 5 \end{array}$$

In-Class Example 2
a) Add $8w^2 - 5w + 16$ and $-4w^2 + 11w + 4$.
b) Add $(4m^4 + 7m^3 + m - 6) + (3m^3 - 2m^2 + m + 4)$.

answer:
a) $4w^2 + 6w + 20$
b) $4m^4 + 10m^3 - 2m^2 + 2m - 2$

You Try 2

Add $(t^3 + 2t^2 - 10t + 1) + (3t^3 - 9t^2 - t + 6)$.

To subtract two polynomials such as $(6p^2 + 2p - 7) - (4p^2 - 9p + 3)$ we will use the distributive property to clear the parentheses in the second polynomial.

Example 3

In-Class Example 3
Subtract $(11h^2 - 3h + 8) - (6h^2 + 2h - 4)$.
answer:
$5h^2 - 5h + 12$

Subtract $(6p^2 + 2p - 7) - (4p^2 - 9p + 3)$.

Solution

$(6p^2 + 2p - 7) - (4p^2 - 9p + 3)$
$\quad = (6p^2 + 2p - 7) \; - 1(4p^2 - 9p + 3)$
$\quad = (6p^2 + 2p - 7) + (-1)(4p^2 - 9p + 3)$ Change -1 to $+(-1)$.
$\quad = (6p^2 + 2p - 7) + (-4p^2 + 9p - 3)$ Distribute.
$\quad = 2p^2 + 11p - 10$ Combine like terms.

In Example 3, notice that we changed the sign of each term in the second polynomial and then added it to the first.

> **Procedure   Subtracting Polynomials**
>
> To subtract two polynomials, change the sign of each term in the second polynomial. Then, add the polynomials.

Let's see how we apply this rule to subtracting polynomials both horizontally and vertically.

Example 4

In-Class Example 4
Subtract
$(3f^3 + f^2 - 2f - 4) - (7f^3 - 8f^2 + 9f - 1)$.
answer:
$-4f^3 + 9f^2 - 11f - 3$

Subtract $(-8k^3 - k^2 + 5k + 7) - (6k^3 - 3k^2 + k - 2)$.

a)   horizontally   b)   vertically

Solution

a) $(-8k^3 - k^2 + 5k + 7) - (6k^3 - 3k^2 + k - 2)$ Change the signs in the
$\quad = (-8k^3 - k^2 + 5k + 7) + (-6k^3 + 3k^2 - k + 2)$ second polynomial and add.
$\quad = -14k^3 + 2k^2 + 4k + 9$ Combine like terms.

b)   To subtract vertically, line up like terms in columns.

$$\begin{array}{r} -8k^3 - k^2 + 5k + 7 \\ -(6k^3 - 3k^2 + k - 2) \end{array}$$ Change the signs in the second polynomial and add the polynomials. $$\begin{array}{r} -8k^3 - k^2 + 5k + 7 \\ +\underline{(-6k^3 + 3k^2 - k + 2)} \\ -14k^3 + 2k^2 + 4k + 9 \end{array}$$

You can see that adding and subtracting polynomials horizontally or vertically gives the same result. ∎

You Try 3

Subtract $(9m^2 - 4m + 2) - (-m^2 + m - 6)$.

3. Add and Subtract Polynomials in More Than One Variable

To add and subtract polynomials in more than one variable, remember that like terms contain the same variables with the same exponents.

Example 5

In-Class Example 5
Perform the indicated operation.
a) $(m^3n^2 + 3mn^2 - 4mn) +$
   $(9m^3n^2 + 3mn^2 + 9mn - 11)$
b) $(2ab - 3b + 6) -$
   $(10ab + a + 9b - 6)$
answer:
a) $10m^3n^2 + 6mn^2 + 5mn - 11$
b) $-8ab - a - 12b + 12$

Perform the indicated operation.

a) $(x^2y^2 + 5x^2y - 8xy - 7) + (5x^2y^2 - 4x^2y - xy + 3)$

b) $(9cd - c + 3d + 1) - (6cd + 5c - 8)$

Solution

a) $(x^2y^2 + 5x^2y - 8xy - 7) + (5x^2y^2 - 4x^2y - xy + 3)$
    $= 6x^2y^2 + x^2y - 9xy - 4$ Combine like terms.

b) $(9cd - c + 3d + 1) - (6cd + 5c - 8) = (9cd - c + 3d + 1) - 6cd - 5c + 8$
    $= 3cd - 6c + 3d + 9$ Combine like terms. ■

You Try 4

Perform the indicated operation.

a) $(-10a^2b^2 + ab - 3b + 4) - (-9a^2b^2 - 6ab + 2b + 4)$

b) $(5.8t^3u^2 + 2.1tu - 7u) + (4.1t^3u^2 - 7.8tu - 1.6)$

4. Define and Evaluate a Polynomial Function

In Chapter 3 we learned about linear functions that have the form $f(x) = mx + b$. A linear function is a special type of polynomial function, which we will study now.

> **Definition**
>
> A **polynomial function of degree n** is given by $f(x) = a_nx^n + a_{n-1}x^{n-1} + \cdots + a_1x + a_0$, where $a_n, a_{n-1}, \cdots, a_1$, and a_0 are real numbers, $a_n \neq 0$, and n is a whole number.

Look at the polynomial $2x^2 - 5x + 7$. If we substitute 3 for x, the *only* value of the expression is 10:

$$2(3)^2 - 5(3) + 7 = 2(9) - 15 + 7 = 18 - 15 + 7 = 10$$

It is true that polynomials have different values depending on what value is substituted for the variable. It is also true that for any value we substitute for x in a polynomial like $2x^2 - 5x + 7$ there will be *only one value* of the expression. Since each value substituted for the variable produces *only one value* of the expression, we can use function notation to represent a polynomial like $2x^2 - 5x + 7$.

$f(x) = 2x^2 - 5x + 7$ is a *polynomial function* since $2x^2 - 5x + 7$ is a polynomial. Therefore, finding $f(3)$ when $f(x) = 2x^2 - 5x + 7$ is the same as evaluating $2x^2 - 5x + 7$ when $x = 3$.

Example 6

In-Class Example 6
If $f(x) = x^3 - 8x^2 - 5x + 9$, find $f(-3)$.
answer: −75

If $f(x) = x^3 - 4x^2 + 3x + 1$, find $f(-2)$.

Solution

$$f(x) = x^3 - 4x^2 + 3x + 1$$
$$f(-2) = (-2)^3 - 4(-2)^2 + 3(-2) + 1 \qquad \text{Substitute } -2 \text{ for } x.$$
$$f(-2) = -8 - 4(4) - 6 + 1$$
$$f(-2) = -8 - 16 - 6 + 1$$
$$f(-2) = -29$$

■

You Try 5

If $g(t) = 2t^4 + t^3 - 7t^2 + 12$, find $g(-1)$.

5. Add and Subtract Polynomial Functions

We have learned that we can add and subtract polynomials. These same operations can be performed with functions.

> **Definition**
>
> Given the functions $f(x)$ and $g(x)$, the **sum and difference of f and g** are defined by
>
> 1) $(f + g)(x) = f(x) + g(x)$
>
> 2) $(f - g)(x) = f(x) - g(x)$
>
> The domain of $(f + g)(x)$ and $(f - g)(x)$ is the intersection of the domains of $f(x)$ and $g(x)$.

Example 7

In-Class Example 7
Let $f(x) = x^2 + 4x - 3$ and $g(x) = x - 1$. Find a) and b) in Example 7.
answer:
a) $x^2 + 5x - 4$
b) $x^2 + 3x - 2$; −4

Let $f(x) = x^2 - 2x + 7$ and $g(x) = 4x - 3$. Find each of the following.

a) $(f + g)(x)$ b) $(f - g)(x)$ and $(f - g)(-1)$

Solution

a) $(f + g)(x) = f(x) + g(x)$
$\qquad\qquad = (x^2 - 2x + 7) + (4x - 3) \qquad$ Substitute the functions.
$\qquad\qquad = x^2 + 2x + 4 \qquad\qquad\qquad$ Combine like terms.

b) $(f - g)(x) = f(x) - g(x)$
$\qquad\qquad = (x^2 - 2x + 7) - (4x - 3) \qquad$ Substitute the functions.
$\qquad\qquad = x^2 - 2x + 7 - 4x + 3 \qquad\quad$ Distribute.
$\qquad\qquad = x^2 - 6x + 10 \qquad\qquad\qquad$ Combine like terms.

Use the result above to find $(f - g)(-1)$.

$$(f - g)(x) = x^2 - 6x + 10$$
$$(f - g)(-1) = (-1)^2 - 6(-1) + 10 \qquad \text{Substitute } -1 \text{ for } x.$$
$$= 1 + 6 + 10$$
$$= 17$$

We can also find $(f - g)(-1)$ using the rule this way:

$$(f - g)(-1) = f(-1) - g(-1)$$
$$= [(-1)^2 - 2(-1) + 7] - [4(-1) - 3] \qquad \text{Substitute } -1 \text{ for } x \text{ in}$$
$$= (1 + 2 + 7) - (-4 - 3) \qquad\qquad\qquad f(x) \text{ and } g(x).$$
$$= 10 - (-7)$$
$$= 17$$

■

You Try 6

Let $f(x) = 3x^2 - 8$ and $g(x) = 2x + 1$. Find each of the following.

a) $(f + g)(x)$ and $(f + g)(-2)$ b) $(f - g)(x)$

We can use polynomial functions to solve real-world problems.

Example 8

In-Class Example 8
Refer to Example 8. Assume that the publisher sells each book for $3.00 so that the revenue is $R(x) = 3x$. Let the cost be $C(x) = 2x + 1000$. Answer parts a) and b).
answer: a) $P(x) = x - 1000$
b) $9000

A publisher sells paperback romance novels to a large bookstore chain for $4.00 per book. Therefore, the publisher's revenue, in dollars, is defined by the function

$$R(x) = 4x$$

where x is the number of books sold to the retailer. The publisher's cost, in dollars, to produce x books is

$$C(x) = 2.5x + 1200$$

In business, profit is defined as revenue − cost. In terms of functions, this is written as $P(x) = R(x) - C(x)$, where $P(x)$ is the profit function.

a) Find the profit function, $P(x)$, that describes the publisher's profit from the sale of x books.

b) If the publisher sells 10,000 books to this chain of bookstores, what is the publisher's profit?

Solution

a) $P(x) = R(x) - C(x)$
$\qquad = 4x - (2.5x + 1200)$ Substitute the functions.
$\qquad = 1.5x - 1200$
$\quad P(x) = 1.5x - 1200$

b) Find $P(10,000)$.

$$P(10,000) = 1.5(10,000) - 1200$$
$$= 15,000 - 1200$$
$$= 13,800$$

The publisher's profit is $13,800.

You Try 7

A candy company sells its Valentine's Day candy to a grocery store retailer for $6.00 per box. The candy company's revenue, in dollars, is defined by $R(x) = 6x$, where x is the number of boxes sold to the retailer. The company's cost, in dollars, to produce x boxes of candy is $C(x) = 4x + 900$.

a) Find the profit function, $P(x)$, that defines the company's profit from the sale of x boxes of candy.

b) Find the candy company's profit from the sale of 2000 boxes of candy.

Answers to You Try Exercises

1) a) not a polynomial   b) It is a polynomial of degree 5.   c) It is a polynomial of degree 7.

Term	Degree
$6t^5$	5
$-\dfrac{4}{9}t^4$	4
t	1
2	0

Term	Degree
a^4b^3	7
$-9a^3b^3$	6
$-a^2b$	3
$4a$	1
-7	0

d) not a polynomial   2) $4t^3 - 7t^2 - 11t + 7$ 3) $10m^2 - 5m + 8$

4) a) $-a^2b^2 + 7ab - 5b$ b) $9.9t^3u^2 - 5.7tu - 7u - 1.6$ 5) 6

6) a) $3x^2 + 2x - 7; 1$ b) $3x^2 - 2x - 9$ 7) a) $P(x) = 2x - 900$ b) $3100

5.3 Exercises

*Additional answers can be found in the Answers to Exercises appendix.

Objective 1: Learn Polynomial Vocabulary

Is the given expression a polynomial? Why or why not?

1) $-5z^2 - 4z + 12$

2) $9t^3 + t^2 - t + \dfrac{3}{8}$

3) $g^3 + 3g^2 + 2g^{-1} - 5$ 4) $6y^4$

5) $m^{2/3} + 4m^{1/3} + 4$ 6) $8c - 5 + \dfrac{2}{c}$

Determine whether each is a monomial, a binomial, or a trinomial.

7) $3x - 7$ binomial

8) $-w^3$ monomial

9) $a^2b^2 + 10ab - 6$ trinomial

10) $16r^2 + 9r$ binomial

11) 1   monomial

12) $v^4 + 7v^2 + 6$ trinomial

13) How do you determine the degree of a polynomial in one variable?

14) Write a fourth-degree polynomial in one variable.
Answers may vary.

15) How do you determine the degree of a term in a polynomial in more than one variable?
Add the exponents on the variables.

16) Write a fifth-degree monomial in x and y.
Answers may vary.

For each polynomial, identify each term in the polynomial, the coefficient and degree of each term, and the degree of the polynomial.

17) $7y^3 + 10y^2 - y + 2$ 18) $4d^2 + 12d - 9$

19) $-9r^3s^2 - r^2s^2 + \dfrac{1}{2}rs + 6s$

20) $8m^2n^2 + 0.5m^2n - mn + 3$

Objective 2: Add and Subtract Polynomials in One Variable

Add.

21) $(11w^2 + 2w - 13) + (-6w^2 + 5w + 7)$ $5w^2 + 7w - 6$

22) $(4f^4 - 3f^2 + 8) + (2f^4 - f^2 + 1)$ $6f^4 - 4f^2 + 9$

23) $(-p + 16) + (-7p - 9)$ $-8p + 7$

24) $(y^3 + 8y^2) + (y^3 - 11y^2)$ $2y^3 - 3y^2$

25) $\left(-7a^4 - \dfrac{3}{2}a + 1\right) + \left(2a^4 + 9a^3 - a^2 - \dfrac{3}{8}\right)$

26) $\left(2d^5 + \dfrac{1}{3}d^4 - 11\right) + (10d^5 - 2d^4 + 9d^2 + 4)$

27) $\left(\dfrac{11}{4}x^3 - \dfrac{5}{6}\right) + \left(\dfrac{3}{8}x^3 + \dfrac{11}{12}\right)$ $\dfrac{25}{8}x^3 + \dfrac{1}{12}$

28) $\left(\dfrac{3}{4}c + \dfrac{1}{8}\right) + \left(\dfrac{3}{2}c - \dfrac{5}{6}\right)$ $\dfrac{9}{4}c - \dfrac{17}{24}$

29) $(6.8k^3 + 3.5k^2 - 10k - 3.3)$ $2.6k^3 + 8.7k^2 - 7.3k - 4.4$
$+ (-4.2k^3 + 5.2k^2 + 2.7k - 1.1)$

30) $(0.6t^4 - 7.3t + 2.2) + (-1.8t^4 + 4.9t^3 + 8.1t + 7.1)$
$-1.2t^4 + 4.9t^3 + 0.8t + 9.3$

Add.

31) $\begin{array}{r} 12x - 11 \\ + \ 5x + \ 3 \end{array}$ $17x - 8$

32) $\begin{array}{r} -6n^3 + 1 \\ + \ 4n^3 - 8 \end{array}$ $-2n^3 - 7$

33) $\begin{array}{r} 9r^2 + 16r + 2 \\ + \ 3r^2 - 10r + 9 \end{array}$ $12r^2 + 6r + 11$

34) $\begin{array}{r} z^2 - 4z \\ + 3z^2 + 9z + 4 \end{array}$ $4z^2 + 5z + 4$

35) $\begin{array}{r} -2.6q^3 - q^2 + 6.9q - \ 1 \\ + \ \ 4.1q^3 \ \ \ \ \ \ \ \ \ \ \ - 2.3q + 16 \end{array}$ $1.5q^3 - q^2 + 4.6q + 15$

36) $\begin{array}{l} \quad 9a^4 + 5.3a^3 - 7a^2 - 1.2a + 6 \\ + \underline{-8a^4 - 2.8a^3 + 4a^2 - 3.9a + 5} \end{array}$

$a^4 + 2.5a^3 - 3a^2 - 5.1a + 11$

Subtract.

37) $(8a^4 - 9a^2 + 17) - (15a^4 + 3a^2 + 3)$ $-7a^4 - 12a^2 + 14$

38) $(16w^3 + 9w - 7) - (27w^3 - 3w - 4)$ $-11w^3 + 12w - 3$

39) $(j^2 + 18j + 2) - (-7j^2 + 6j + 2)$ $8j^2 + 12j$

40) $(-2m^2 + m + 5) - (3m^2 + m + 1)$ $-5m^2 + 4$

41) $(19s^5 - 11s^2) - (10s^5 + 3s^4 - 8s^2 - 2)$
$9s^5 - 3s^4 - 3s^2 + 2$

42) $(h^5 + 7h^3 - 8h) - (-9h^5 + h^4 + 7h^3 - 8h - 6)$
$10h^5 - h^4 + 6$

43) $(-3b^4 - 5b^2 + b + 2) - (-2b^4 + 10b^3 - 5b^2 - 18)$
$-b^4 - 10b^3 + b + 20$

44) $(4t^3 - t^2 + 6) - (t^2 + 7t + 1)$ $4t^3 - 2t^2 - 7t + 5$

45) $\left(-\dfrac{5}{7}r^2 + \dfrac{4}{9}r + \dfrac{2}{3}\right) - \left(-\dfrac{5}{14}r^2 - \dfrac{5}{9}r + \dfrac{11}{6}\right)$

46) $\left(\dfrac{5}{6}y^3 + \dfrac{1}{2}y + 3\right) - \left(-\dfrac{1}{6}y^3 + y^2 - 3y\right)$

Subtract.

47) $\begin{array}{l} \quad 17v + 3 \\ - \underline{2v + 9} \end{array}$ $15v - 6$

48) $\begin{array}{l} \quad 10q - 7 \\ - \underline{4q + 8} \end{array}$ $6q - 15$

49) $\begin{array}{l} \quad 2b^2 - 7b + 4 \\ - \underline{3b^2 + 5b - 3} \end{array}$

50) $\begin{array}{l} \quad -3d^2 + 16d + 2 \\ - \underline{5d^2 + 7d - 3} \end{array}$

51) $\begin{array}{l} \quad a^4 - 2a^3 + 6a^2 - 7a + 11 \\ - \underline{-2a^4 + 9a^3 - a^2 + 3} \end{array}$

52) $\begin{array}{l} \quad 7y^4 + y^3 - 10y^2 + 6y - 2 \\ - \underline{-2y^4 + y^3 - 4y + 1} \end{array}$

53) Explain, in your own words, how to subtract two polynomials.   Answers may vary.

54) Do you prefer adding and subtracting polynomials vertically or horizontally? Why?   Answers may vary.

55) Will the sum of two trinomials always be a trinomial? Why or why not? Give an example.

56) Write a fourth-degree polynomial in x that does not contain a second-degree term.   Answers may vary.

Perform the indicated operations.

57) $(-3b^4 + 4b^2 - 6) + (2b^4 - 18b^2 + 4)$
$+ (b^4 + 5b^2 - 2)$ $-9b^2 - 4$

58) $(-7m^2 - 14m + 56) + (3m^2 + 7m - 6)$
$+ (9m^2 - 10)$ $5m^2 - 7m + 40$

59) $\left(n^3 - \dfrac{1}{2}n^2 - 4n + \dfrac{5}{8}\right) + \left(\dfrac{1}{4}n^3 - n^2 + 7n - \dfrac{3}{4}\right)$

60) $\left(\dfrac{2}{3}z^4 + z^3 - \dfrac{3}{2}z^2 + 1\right)$

$+ \left(z^4 - 2z^3 - \dfrac{1}{6}z^2 + 8z - 1\right)$

61) $(u^3 + 2u^2 + 1) - (4u^3 - 7u^2 + u + 9)$
$+ (8u^3 - 19u^2 + 2)$ $5u^3 - 10u^2 - u - 6$

62) $(21r^3 - 8r^2 + 3r + 2) + (-4r^2 + 5)$
$- (6r^3 - r^2 - 4r)$ $15r^3 - 11r^2 + 7r + 7$

63) $\left(\dfrac{3}{8}k^2 + k - \dfrac{1}{5}\right) - \left(2k^2 + k - \dfrac{7}{10}\right) + (k^2 - 9k)$

64) $\left(y + \dfrac{1}{4}\right) + \left(\dfrac{1}{2}y^2 - 3y + \dfrac{3}{4}\right) - \left(\dfrac{3}{4}y^2 - \dfrac{1}{2}y + 1\right)$

65) $(2t^3 - 8t^2 + t + 10)$
$- [(5t^3 + 3t^2 - t + 8) + (-6t^3 - 4t^2 + 3t + 5)]$
$3t^3 - 7t^2 - t - 3$

66) $(x^2 - 10x - 6)$
$- [(-8x^2 + 11x - 1) + (5x^2 - 9x - 3)]$ $4x^2 - 12x - 2$

67) $(-12a^2 + 9) - (-9a^3 + 7a + 6) + (12a^2 - a + 10)$
$9a^3 - 8a + 13$

68) $(5c + 7) - (c^2 + 4c - 2) - (-7c^3 - c + 4)$
$7c^3 - c^2 + 2c + 5$

Objective 3: Add and Subtract Polynomials in More than One Variable

Each of the polynomials below is a polynomial in two variables. Perform the indicated operation(s).

69) $(4a + 13b) - (a + 5b)$ $3a + 8b$

70) $(-2g - 3h) + (6g + h)$ $4g - 2h$

71) $\left(5m + \dfrac{5}{6}n + \dfrac{1}{2}\right) + \left(-6m + n - \dfrac{3}{4}\right)$ $-m + \dfrac{11}{6}n - \dfrac{1}{4}$

72) $\left(-2c - \dfrac{2}{3}d + 1\right) - \left(2c + \dfrac{1}{9}d - \dfrac{4}{7}\right)$ $-4c - \dfrac{7}{9}d + \dfrac{11}{7}$

73) $(-12y^2z^2 + 5y^2z - 25yz^2 + 16)$
$+ (17y^2z^2 + 2y^2z - 15)$ $5y^2z^2 + 7y^2z - 25yz^2 + 1$

74) $(-8u^2v^2 + 2uv + 3) - (-9u^2v^2 - 14uv + 18)$
$u^2v^2 + 16uv - 15$

75) $(8x^3y^2 - 7x^2y^2 + 7x^2y - 3) + (2x^3y^2 + x^2y - 1)$
$- (4x^2y^2 + 2x^2y + 8)$ $10x^3y^2 - 11x^2y^2 + 6x^2y - 12$

76) $(r^3s^2 + r^2s^2 + 4) - (6r^3s^2 + 14r^2s^2 - 6)$
$+ (8r^3s^2 - 6r^2s^2 - 4)$ $3r^3s^2 - 19r^2s^2 + 6$

Write an expression for each and perform the indicated operation(s).

77) Find the sum of $v^2 - 9$ and $4v^2 + 3v + 1$. $5v^2 + 3v - 8$

78) Add $11d - 12$ to $2d + 3$. $13d - 9$

79) Subtract $g^2 - 7g + 16$ from $5g^2 + 3g + 6$.
$4g^2 + 10g - 10$

80) Subtract $-9y^2 + 4y + 6$ from $2y^2 + y$. $11y^2 - 3y - 6$

81) Subtract the sum of $4n^2 + 1$ and $6n^2 - 10n + 3$ from $2n^2 + n + 4$. $-8n^2 + 11n$

82) Subtract $19x^3 + 4x - 12$ from the sum of $6x^3 + x^2 + x$ and $4x^3 - 3x - 8$. $-9x^3 + x^2 - 6x + 4$

Find the polynomial that represents the perimeter of each rectangle.

83)

$3x + 8$

$x - 1$ 8x + 14 units

84)

$a^2 + 5a - 3$

$a^2 - 2a + 2$ 4a² + 6a − 2 units

85)

$3w^2 - 2w + 4$

$w - 7$ 6w² − 2w − 6 units

86) $\frac{3}{4}t + 2$

$\frac{3}{4}t + 2$ 3t + 8 units

Objective 4: Define and Evaluate a Polynomial Function

87) If $f(x) = 5x^2 + 7x - 8$, find
a) $f(-3)$ 16   b) $f(1)$ 4

88) If $h(a) = -a^2 - 3a + 10$, find
a) $h(5)$ −30   b) $h(-4)$ 6

89) If $P(t) = t^3 - 3t^2 + 2t + 5$, find
a) $P(4)$ 29   b) $P(0)$ 5

90) If $G(c) = 3c^4 + c^2 - 9c - 4$, find
a) $G(0)$ −4   b) $G(-1)$ 9

91) If $H(z) = -4z + 9$, find z so that $H(z) = 11$. $-\frac{1}{2}$

92) If $f(x) = \frac{1}{3}x + 5$, find x so that $f(x) = 7$.   6

93) If $r(k) = \frac{2}{5}k - 3$, find k so that $r(k) = 13$.   40

94) If $Q(a) = 6a - 1$, find a so that $Q(a) = -9$. $-\frac{4}{3}$

Objective 5: Add and Subtract Polynomial Functions

For each pair of functions, find (a) $(f + g)(x)$, (b) $(f + g)(5)$, (c) $(f - g)(x)$, and (d) $(f - g)(2)$.

95) $f(x) = -3x + 1$, $g(x) = 2x - 11$
a) −x − 10 b) −15 c) −5x + 12 d) 2

96) $f(x) = 5x - 9$, $g(x) = x + 4$
a) 6x − 5 b) 25 c) 4x − 13 d) −5

97) $f(x) = 4x^2 - 7x - 1$, $g(x) = x^2 + 3x - 6$
a) 5x² − 4x − 7 b) 98 c) 3x² − 10x + 5 d) −3

98) $f(x) = -2x^2 + x + 8$, $g(x) = 3x^2 - 4x - 6$
a) x² − 3x + 2 b) 12 c) −5x² + 5x + 14 d) 4

Let $f(t) = 4t - 1$, $g(t) = -t^2 + 6$, and $h(t) = 3t^2 - 4t$. Find each of the following.

99) $(g - h)(t)$ −4t² + 4t + 6   100) $(g + h)(t)$ 2t² − 4t + 6

101) $(g - h)(5)$ −74   102) $(g + h)(0)$ 6

103) $(f + g)(-1)$ 0   104) $(f - g)(-2)$ −11

105) $(h - f)\left(\frac{1}{2}\right)$ $-\frac{9}{4}$ 106) $(f - h)\left(\frac{2}{3}\right)$ 3

107) Find two polynomial functions $f(x)$ and $g(x)$ so that
$(f + g)(x) = 5x^2 + 8x - 2$.   Answers may vary.

108) Let $f(x) = 6x^3 - 9x^2 - 4x + 10$. Find $g(x)$ so that
$(f - g)(x) = x^3 + 3x^2 + 8$.   g(x) = 5x³ − 12x² − 4x + 2

109) A manufacturer's revenue, $R(x)$ in dollars, from the sale of x calculators is given by $R(x) = 12x$. The company's cost, $C(x)$ in dollars, to produce x calculators is $C(x) = 8x + 2000$.

a) Find the profit function, $P(x)$, that defines the manufacturer's profit from the sale of x calculators.   P(x) = 4x − 2000

b) What is the profit from the sale of 1500 calculators? $4000

110) $R(x) = 80x$ is the revenue function for the sale of x bicycles, in dollars. The cost to manufacture x bikes, in dollars, is $C(x) = 60x + 7000$.

a) Find the profit function, $P(x)$, that describes the manufacturer's profit from the sale of x bicycles.   P(x) = 20x − 7000

b) What is the profit from the sale of 500 bicycles? $3000

111) $R(x) = 18x$ is the revenue function for the sale of x toasters, in dollars. The cost to manufacture x toasters, in dollars, is $C(x) = 15x + 2400$.

a) Find the profit function, $P(x)$, that describes the profit from the sale of x toasters.   P(x) = 3x − 2400

b) What is the profit from the sale of 800 toasters? $0

112) A company's revenue, $R(x)$ in dollars, from the sale of x dog houses is given by $R(x) = 60x$. The company's cost, $C(x)$ in dollars, to produce x dog houses is $C(x) = 45x + 6000$.

a) Find the profit function, $P(x)$, that describes the company's profit from the sale of x dog houses.   P(x) = 15x − 6000

b) What is the profit from the sale of 300 dog houses?   −$1500; the company loses $1500 if it sells only 300 dog houses.

For Exercises 113 and 114, let x be the number of items sold (in hundreds), and let $R(x)$ and $C(x)$ be in thousands of dollars.

113) A manufacturer's revenue, $R(x)$, from the sale of flat-screen TVs is given by $R(x) = -0.2x^2 + 23x$, while the cost, $C(x)$, is given by $C(x) = 4x + 9$.

a) Find the profit function, $P(x)$, that describes the company's profit from the sale of x hundred flat-screen TVs.   P(x) = −0.2x² + 19x − 9

b) What is the profit from the sale of 2000 TVs? $291,000

114) A manufacturer's revenue, $R(x)$, from the sale of laptop computers is given by $R(x) = -0.4x^2 + 30x$, while the cost, $C(x)$, is given by $C(x) = 3x + 11$.

a) Find the profit function, $P(x)$, that describes the company's profit from the sale of x hundred laptop computers.   P(x) = −0.4x² + 27x − 11

b) What is the profit from the sale of 1500 computers? $304,000

Section 5.4 Multiplication of Polynomials and Polynomial Functions

Objectives

1. Multiply a Monomial and a Polynomial
2. Multiply Two Polynomials
3. Multiply Two Binomials Using FOIL
4. Find the Product of More Than Two Polynomials
5. Find the Product of Binomials of the Form $(a + b)(a - b)$
6. Square a Binomial
7. Multiply Other Binomials
8. Multiply Polynomial Functions

We have already learned that when multiplying two monomials, we multiply the coefficients and add the exponents of the same bases:

$$3x^4 \cdot 5x^2 = 15x^6 \qquad -2a^3b^2 \cdot 6ab^4 = -12a^4b^6$$

In this section we will discuss how to multiply other types of polynomials.

1. Multiply a Monomial and a Polynomial

When multiplying a monomial and a polynomial, we use the distributive property.

 Example 1

Multiply $5n^2(2n^2 + 3n - 4)$.

Solution

$$5n^2(2n^2 + 3n - 4) = (5n^2)(2n^2) + (5n^2)(3n) + (5n^2)(-4) \qquad \text{Distribute.}$$
$$= 10n^4 + 15n^3 - 20n^2 \qquad \text{Multiply.}$$

In-Class Example 1
Multiply $4f^2(-3f^2 + 8f + 1)$.
answer: $-12f^4 + 32f^3 + 4f^2$

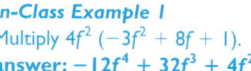 **You Try 1**

Multiply $6a^4(7a^3 - a^2 - 3a + 4)$.

2. Multiply Two Polynomials

To multiply two polynomials, we use the distributive property repeatedly. For example to multiply $(2r - 3)(r^2 + 7r + 9)$, we multiply each term in the second polynomial by $(2r - 3)$.

$$(2r - 3)(r^2 + 7r + 9) = (2r - 3)(r^2) + (2r - 3)(7r) + (2r - 3)(9) \qquad \text{Distribute.}$$

Next, we distribute again.

$$(2r - 3)(r^2) + (2r - 3)(7r) + (2r - 3)(9)$$
$$= (2r)(r^2) - (3)(r^2) + (2r)(7r) - (3)(7r) + (2r)(9) - (3)(9)$$
$$= 2r^3 - 3r^2 + 14r^2 - 21r + 18r - 27 \qquad \text{Multiply.}$$
$$= 2r^3 + 11r^2 - 3r - 27 \qquad \text{Combine like terms.}$$

This process of repeated distribution leads us to the following rule.

Procedure   Multiplying Polynomials

To multiply two polynomials, multiply each term in the second polynomial by each term in the first polynomial. Then, combine like terms. The answer should be written in descending powers.

Let's use this rule to multiply the polynomials in Example 2.

Example 2

Multiply $(c^2 - 3c + 5)(2c^3 + c - 6)$.

In-Class Example 2
Multiply.
a) $(4x - 2)(2x^2 + 3x - 4)$
b) $(2a^2 + a - 3)(a^3 - 2a^2 + 1)$
answer:
a) $8x^3 + 8x^2 - 22x + 8$
b) $2a^5 - 3a^4 - 5a^3 + 8a^2$
$+ a - 3$

Solution

Multiply each term in the second polynomial by each term in the first. Add like terms.

$(c^2 - 3c + 5)(2c^3 + c - 6)$

$= (c^2)(2c^3) + (c^2)(c) + (c^2)(-6) + (-3c)(2c^3) + (-3c)(c) + (-3c)(-6)$
$\quad + (5)(2c^3) + (5)(c) + (5)(-6)$ Distribute.

$= 2c^5 + c^3 - 6c^2 - 6c^4 - 3c^2 + 18c + 10c^3 + 5c - 30$ Multiply.

$= 2c^5 - 6c^4 + 11c^3 - 9c^2 + 23c - 30$ Combine like terms. ■

 **You Try 2**

Multiply $(8z + 1)(7z^2 - z + 2)$.

Polynomials can be multiplied vertically as well. The process is similar to the way we multiply whole numbers. See the example at the right.

$$\begin{array}{r} 458 \\ \times\ 32 \\ \hline 916 \\ 13\ 74 \\ \hline 14{,}656 \end{array}$$

Multiply 458 by 2.
Multiply 458 by 3.
Add.

In the next example, we will find a product of polynomials by multiplying vertically.

Example 3

Multiply vertically. $(3n^2 + 4n - 1)(2n + 7)$

In-Class Example 3
Multiply vertically.
$(5x^2 - 2x + 3)(3x - 1)$
answer:
$15x^3 - 11x^2 + 11x - 3$

Solution

Set up the multiplication problem like you would for whole numbers:

$$\begin{array}{r} 3n^2 + 4n - 1 \\ \times\ \qquad 2n + 7 \\ \hline 21n^2 + 28n - 7 \\ 6n^3 + 8n^2 - 2n \qquad\quad \\ \hline 6n^3 + 29n^2 + 26n - 7 \end{array}$$

Multiply each term in $3n^2 + 4n - 1$ by 7.
Multiply each term in $3n^2 + 4n - 1$ by $2n$.
Line up like terms in the same column. Add. ■

 You Try 3

Multiply vertically.

a) $(5x + 4)(7x^2 - 9x + 2)$ b) $\left(p^2 - \dfrac{3}{2}p - 6\right)(5p^2 + 8p - 4)$

3. Multiply Two Binomials Using FOIL

Multiplying two binomials is one of the most common types of polynomial multiplication used in algebra. A method called **FOIL** is one that is often used to multiply two binomials, and it comes from using the distributive property.

Let's use the distributive property to multiply $(x + 5)(x + 3)$.

$(x + 5)(x + 3) = (x + 5)(x) + (x + 5)(3)$ Distribute.
$\qquad\qquad\quad = x(x) + 5(x) + x(3) + 5(3)$ Distribute.
$\qquad\qquad\quad = x^2 + 5x + 3x + 15$ Multiply.
$\qquad\qquad\quad = x^2 + 8x + 15$ Combine like terms.

To be sure that each term in the first binomial has been multiplied by each term in the second binomial we can use FOIL. **FOIL** stands for **F**irst **O**uter **I**nner **L**ast. Let's see how we can apply FOIL to the binomials we just multiplied.

$$(x + 5)(x + 3) = (x + 5)(x + 3) = x(x) + x(3) + 5(x) + 5(3)$$
$$= x^2 + 3x + 5x + 15 \qquad \text{Multiply.}$$
$$= x^2 + 8x + 15 \qquad \text{Combine like terms.}$$

You can see that we get the same result.

Example 4

Use FOIL to multiply the binomials.

a) $(n + 9)(n - 4)$ b) $(5c - 8)(c - 1)$

c) $(x + 2y)(x - 6y)$ d) $(3a + 2)(b + 7)$

In-Class Example 4
Use FOIL to multiply the binomials.
a) $(k + 10)(k - 7)$
b) $(6p - 5)(p - 3)$
c) $(a + 2b)(a - 8b)$
d) $(5x + 2)(y + 4)$
answer:
a) $k^2 + 3k - 70$
b) $6p^2 - 23p + 15$
c) $a^2 - 6ab - 16b^2$
d) $5xy + 20x + 2y + 8$

Solution

a) $(n + 9)(n - 4) = (n + 9)(n - 4) = n(n) + n(-4) + 9(n) + 9(-4) \qquad$ Use FOIL.
$$= n^2 - 4n + 9n - 36 \qquad \text{Multiply.}$$
$$= n^2 + 5n - 36 \qquad \text{Combine like terms.}$$

Notice that the middle terms are like terms, so we can combine them.

b) $(5c - 8)(c - 1) = 5c(c) + 5c(-1) - 8(c) - 8(-1) \qquad$ Use FOIL.
$$= 5c^2 - 5c - 8c + 8 \qquad \text{Multiply.}$$
$$= 5c^2 - 13c + 8 \qquad \text{Combine like terms.}$$

The middle terms are like terms, so we can combine them.

c) $(x + 2y)(x - 6y) = x(x) + x(-6y) + 2y(x) + 2y(-6y) \qquad$ Use FOIL.
$$= x^2 - 6xy + 2xy - 12y^2 \qquad \text{Multiply.}$$
$$= x^2 - 4xy - 12y^2 \qquad \text{Combine like terms.}$$

Like parts a) and b), we combined the middle terms.

d) $(3a + 2)(b + 7) = 3a(b) + 3a(7) + 2(b) + 2(7) \qquad$ Use FOIL.
$$= 3ab + 21a + 2b + 14 \qquad \text{Multiply.}$$

In this case the middle terms were not like terms, so we could not combine them.   ■

You Try 4

Use FOIL to multiply the binomials.

a) $(a + 9)(a + 2)$ b) $(4t + 3)(t - 5)$
c) $(c - 6d)(c - 3d)$ d) $(5m + 2)(n + 6)$

With practice, you should be able to find the product of two binomials "in your head."

While the polynomial multiplication problems we have seen so far are the most common types we encounter in algebra, there are other products we will see as well.

4. Find the Product of More Than Two Polynomials

Example 5

In-Class Example 5
Multiply a) $12f^2\,(2f-1)(f+2)$
b) $(t+4)(t-1)(t+6)$
answer:
a) $24f^4 + 36f^3 - 24f^2$
b) $t^3 + 9t^2 + 14t - 24$

Multiply $5d^2(4d - 3)(2d - 1)$.

Solution

We can approach this problem a couple of ways.

Method 1

Begin by multiplying the binomials, *then* multiply by the monomial.

$$
\begin{aligned}
5d^2(4d - 3)(2d - 1) &= 5d^2(8d^2 - 4d - 6d + 3) && \text{Use FOIL to multiply the binomials.}\\
&= 5d^2(8d^2 - 10d + 3) && \text{Combine like terms.}\\
&= 40d^4 - 50d^3 + 15d^2 && \text{Distribute.}
\end{aligned}
$$

Method 2

Begin by multiplying $5d^2$ by $(4d - 3)$, then multiply *that* product by $(2d - 1)$.

$$
\begin{aligned}
5d^2(4d - 3)(2d - 1) &= (20d^3 - 15d^2)(2d - 1) && \text{Multiply } 5d^2 \text{ by } (4d - 3).\\
&= 40d^4 - 20d^3 - 30d^3 + 15d^2 && \text{Use FOIL to multiply.}\\
&= 40d^4 - 50d^3 + 15d^2 && \text{Combine like terms.}
\end{aligned}
$$

The result is the same. These may be multiplied by whichever method you prefer. ■

You Try 5

Multiply $-6x^3(x + 5)(3x - 4)$.

Special types of binomial products come up often in algebra. We will look at those next.

5. Find the Product of Binomials of the Form $(a + b)(a - b)$

Let's find the product $(p + 5)(p - 5)$. Using FOIL we get

$$
\begin{aligned}
(p + 5)(p - 5) &= p^2 - 5p + 5p - 25\\
&= p^2 - 25
\end{aligned}
$$

Notice that the "middle terms," the p-terms, drop out. In the result, $p^2 - 25$, the first term (p^2) is the square of p and the last term (25) is the square of 5. They are subtracted. The resulting binomial is a *difference of squares*. This pattern always holds when multiplying two binomials of the form $(a + b)(a - b)$.

> **Formula**   The Product of the Sum and Difference of Two Terms
>
> $$(a + b)(a - b) = a^2 - b^2$$

Example 6

In-Class Example 6
Multiply.
a) $(q + 4)(q - 4)$
b) $(2 - x)(2 + x)$
c) $(5m + 1)(5m - 1)$
answer: a) $q^2 - 16$
b) $4 - x^2$ c) $25m^2 - 1$

Multiply.

a) $(r + 7)(r - 7)$

b) $(3 + y)(3 - y)$

c) $(2z - 5)(2z + 5)$

d) $(4a + 3b)(4a - 3b)$

Solution

a) The product $(r + 7)(r - 7)$ is in the form $(a + b)(a - b)$ because $a = r$ and $b = 7$.

$$
\begin{aligned}
(r + 7)(r - 7) &= r^2 - 7^2\\
&= r^2 - 49
\end{aligned}
$$

b) $(3 + y)(3 - y) = 3^2 - y^2$
$$= 9 - y^2$$

BE CAREFUL Notice that the answer to part (b) is $9 - y^2$ not $y^2 - 9$; subtraction is not commutative.

c)   Since multiplication is commutative (the order in which we multiply does not affect the result), $(2z - 5)(2z + 5)$ is the same as $(2z + 5)(2z - 5)$. This is in the form $(a + b)(a - b)$ where $a = 2z$ and $b = 5$.

$$(2z + 5)(2z - 5) = (2z)^2 - 5^2 \quad \text{Put } 2z \text{ in parentheses.}$$
$$= 4z^2 - 25$$

d) $(4a + 3b)(4a - 3b) = (4a)^2 - (3b)^2 = 16a^2 - 9b^2$

You Try 6

Multiply.

a) $(m + 9)(m - 9)$ b) $(4z + 3)(4z - 3)$

c) $(1 - w)(1 + w)$ d) $(5c + 2d)(5c - 2d)$

6. Square a Binomial

Another type of special binomial product is a **binomial square** such as $(x + 6)^2$. The expression $(x + 6)^2$ means $(x + 6)(x + 6)$. Therefore, we can use FOIL to multiply.

$$(x + 6)^2 = (x + 6)(x + 6) = x^2 + 6x + 6x + 36$$
$$= x^2 + 12x + 36$$

Notice that the outer and inner products, $6x$ and $6x$, are the same. When we add those terms, we see that the middle term of the result is *twice* the product of the terms in each binomial: $12x = 2(x)(6)$.

The *first* term in the result is the square of the *first* term in the binomial, and the *last* term in the result is the square of the *last* term in the binomial. We can express these relationships with these formulas:

Formula   The Square of a Binomial

$$(a + b)^2 = a^2 + 2ab + b^2$$
$$(a - b)^2 = a^2 - 2ab + b^2$$

We can think of the formulas in words as:

To square a binomial, you square the first term, square the second term, then multiply 2 times the first term times the second term and add.

Finding the products $(a + b)^2 = a^2 + 2ab + b^2$ and $(a - b)^2 = a^2 - 2ab + b^2$ is also called *expanding* the binomial squares $(a + b)^2$ and $(a - b)^2$.

BE CAREFUL $(a + b)^2 \neq a^2 + b^2$ and $(a - b)^2 \neq a^2 - b^2$.

Example 7

In-Class Example 7
Expand.
a) $(d - 8)^2$ b) $(a + 5)^2$
c) $(4t - 3)^2$ d) $(5h + 2k)^2$
e) $[(2p - q) + 3]^2$
answer: a) $d^2 - 16d + 64$
b) $a^2 + 10a + 25$
c) $16t^2 - 24t + 9$
d) $25h^2 + 20hk + 4k^2$
e) $4p^2 - 4pq + q^2 +$
$12p - 6q + 9$

Expand.

a) $(q + 4)^2$ b) $(u - 10)^2$ c) $(6n - 5)^2$

d) $(4x + 7y)^2$ e) $[(3x - y) + 1]^2$

Solution

a) $(q + 4)^2 = q^2 \;\; + \;\; 2(q)(4) \;\; + \;\; 4^2$ $a = q, b = 4$

            ↑   ↑   ↑

    Square the   Two times   Square the
    first term   first term   second term
               times second
                   term

        $= q^2 + 8q + 16$

Notice, $(q + 4)^2 \neq q^2 + 16$. Do not "distribute" the power of 2 to each term in the binomial!

b) $(u - 10)^2 = u^2 \;\; - \;\; 2(u)(10) \;\; + \;\; (10)^2$ $a = u, b = 10$

            ↑   ↑   ↑

    Square the   Two times   Square the
    first term   first term   second term
               times second
                   term

        $= u^2 - 20u + 100$

c) $(6n - 5)^2 = (6n)^2 \;\; - \;\; 2(6n)(5) \;\; + \;\; (5)^2$ $a = 6n, b = 5$

            ↑   ↑   ↑

    Square the   Two times   Square the
    first term   first term   second term
               times second
                   term

        $= 36n^2 - 60n + 25$

d) $(4x + 7y)^2 = (4x)^2 + 2(4x)(7y) + (7y)^2$ $a = 4x, b = 7y$
              $= 16x^2 + 56xy + 49y^2$

e) Although this expansion looks complicated, we use the same method. Here, $a = 3x - y$ and $b = 1$.

$[(3x - y) + 1]^2 = (3x - y)^2 + 2(3x - y)(1) + (1)^2$ $(a + b)^2 = a^2 + 2ab + b^2$
               $= (3x)^2 - 2(3x)(y) + (y)^2 + 2(3x - y)(1) + (1)^2$ Expand $(3x - y)^2$.
               $= 9x^2 - 6xy + y^2 + 6x - 2y + 1$ Simplify. ■

You Try 7

Expand.

a) $(t + 7)^2$ b) $(b - 12)^2$ c) $(3v + 4)^2$

d) $(2a + 5b)^2$ e) $[(3c - d) + 4]^2$

7. Multiply Other Binomials

To find other products of binomials, we use techniques we have already discussed.

Example 8

Expand.

a) $(a + 2)^3$

b) $[(3c + d) + 2n][(3c + d) - 2n]$

Solution

a) Just as $x^2 \cdot x = x^3$, $(a + 2)^2 \cdot (a + 2) = (a + 2)^3$. So we can think of $(a + 2)^3$ as $(a + 2)^2(a + 2)$.

$$
\begin{aligned}
(a + 2)^3 &= (a + 2)^2(a + 2) \\
&= (a^2 + 4a + 4)(a + 2) && \text{Square the binomial.} \\
&= a^3 + 2a^2 + 4a^2 + 8a + 4a + 8 && \text{Multiply.} \\
&= a^3 + 6a^2 + 12a + 8 && \text{Combine like terms.}
\end{aligned}
$$

b) Notice that $[(3c + d) + 2n][(3c + d) - 2n]$ has the form $(a + b)(a - b)$, where $a = 3c + d$ and $b = 2n$. So we can use $(a + b)(a - b) = a^2 - b^2$.

$$
\begin{aligned}
[(3c + d) + 2n][(3c + d) - 2n] &= (3c + d)^2 - (2n)^2 \\
&= (3c)^2 + 2(3c)(d) + (d)^2 - (2n)^2 && \text{Expand } (3c + d)^2. \\
&= 9c^2 + 6cd + d^2 - 4n^2 && \text{Simplify.} \quad \blacksquare
\end{aligned}
$$

You Try 8

Expand.

a) $(n - 3)^3$

b) $[(x - 2y) + 3z][(x - 2y) - 3z]$

8. Multiply Polynomial Functions

In Section 5.3 we learned how to add and subtract functions. Next we will learn how to multiply functions.

Definition

Given the functions $f(x)$ and $g(x)$, the **product of f and g** is defined by

$$(fg)(x) = f(x) \cdot g(x)$$

The domain of $(fg)(x)$ is the intersection of the domains of $f(x)$ and $g(x)$.

Example 9

Let $f(x) = 5x - 2$ and $g(x) = 2x^2 + x - 7$. Find

a) $(fg)(x)$ b) $(fg)(-2)$

Solution

a)
$$
\begin{aligned}
(fg)(x) &= f(x) \cdot g(x) \\
&= (5x - 2)(2x^2 + x - 7) && \text{Substitute the functions.} \\
&= 10x^3 + 5x^2 - 35x - 4x^2 - 2x + 14 && \text{Multiply.} \\
&= 10x^3 + x^2 - 37x + 14 && \text{Combine like terms.}
\end{aligned}
$$

In-Class Example 8
Expand.
a) $(g + 4)^3$
b) $[(2a + b) + 4c]$
$\quad\quad\quad [(2a + b) - 4c]$
answer:
a) $g^3 + 12g^2 + 48g + 64$
b) $4a^2 + 4ab + b^2 - 16c^2$

In-Class Example 9
Let $f(x) = 2x - 7$ and
$\quad g(x) = x^2 + 9x + 2$. Find
a) $(fg)(x)$ b) $(fg)(-3)$
answer:
a) $2x^3 + 11x^2 - 59x - 14$
b) 208

b) Since $(fg)(x) = 10x^3 + x^2 - 37x + 14$, substitute -2 for x to find $(fg)(-2)$.

$$(fg)(-2) = 10(-2)^3 + (-2)^2 - 37(-2) + 14 \quad \text{Substitute } -2 \text{ for } x.$$
$$= 10(-8) + 4 + 74 + 14 \quad \text{Simplify.}$$
$$= -80 + 92$$
$$= 12$$

Verify that the result in part b) is the same as $f(-2) \cdot g(-2)$. ■

You Try 9

Let $f(x) = -x + 4$ and $g(x) = 3x^2 - 8x - 6$. Find

a) $(fg)(x)$ b) $(fg)(3)$

Answers to You Try Exercises

1) $42a^7 - 6a^6 - 18a^5 + 24a^4$ 2) $56z^3 - z^2 + 15z + 2$ 3) a) $35x^3 - 17x^2 - 26x + 8$

b) $5p^4 + \dfrac{1}{2}p^3 - 46p^2 - 42p + 24$ 4) a) $a^2 + 11a + 18$ b) $4t^2 - 17t - 15$ c) $c^2 - 9cd + 18d^2$

d) $5mn + 30m + 2n + 12$ 5) $-18x^5 - 66x^4 + 120x^3$ 6) a) $m^2 - 81$ b) $16z^2 - 9$

c) $1 - w^2$ d) $25c^2 - 4d^2$ 7) a) $t^2 + 14t + 49$ b) $b^2 - 24b + 144$ c) $9v^2 + 24v + 16$

d) $4a^2 + 20ab + 25b^2$ e) $9c^2 - 6cd + d^2 + 24c - 8d + 16$ 8) a) $n^3 - 9n^2 + 27n - 27$

b) $x^2 - 4xy + 4y^2 - 9z^2$ 9) a) $-3x^3 + 20x^2 - 26x - 24$ b) -3

5.4 Exercises

*Additional answers can be found in the Answers to Exercises appendix.

Objective 1: Multiply a Monomial and a Polynomial

1) Explain how to multiply two monomials.   Answers may vary.

2) Explain how to multiply a monomial by a trinomial.   Answers may vary.

Multiply.

3) $(7k^4)(2k^2)$ $14k^4$

4) $(8p^6)(-p^4)$ $-8p^{10}$

5) $\left(\dfrac{7}{10}d^9\right)\left(\dfrac{5}{2}d^2\right)$ $\dfrac{7}{4}d^{11}$

6) $\left(-\dfrac{8}{9}c^5\right)\left(\dfrac{3}{10}c^7\right)$ $-\dfrac{4}{15}c^{12}$

7) $7y(4y - 9)$ $28y^2 - 63y$

8) $-12m(11m - 4)$

9) $6v^3(v^2 - 4v - 2)$

10) $3x^4(5x^3 + x - 7)$

11) $-3t^2(9t^3 - 6t^2 - 4t - 7)$ $-27t^5 + 18t^4 + 12t^3 + 21t^2$

12) $-8u^5(9u^4 + 8u^3 + 12u - 1)$

13) $2x^3y(xy^2 + 8xy - 11y + 2)$

14) $5p^5q^2(-5p^2q + 12pq^2 - pq + 2q - 1)$

15) $-\dfrac{3}{4}t^4(20t^3 + 8t^2 - 5t)$ $-15t^7 - 6t^6 + \dfrac{15}{4}t^5$

16) $\dfrac{2}{5}x^4(30x^2 - 15x + 7)$ $12x^6 - 6x^5 + \dfrac{14}{5}x^4$

Objective 2: Multiply Two Polynomials

Perform the indicated operations and simplify.

17) $2(10g^3 + 5g^2 + 4) - (2g^3 - 14g - 20)$

18) $6(7m^2 + 7m + 9) - 11(4m^2 + 7m + 1)$

19) $7(a^3b^3 + 6a^3b^2 + 3)$
    $- 9(6a^3b^3 + 9a^3b^2 - 12a^2b + 8)$

20) $3(x^2y^2 + xy - 2) - 5(x^2y^2 + 6xy - 2)$
    $+ 2(x^2y^2 + 6xy - 2)$ $-15xy$

21) $(q + 3)(5q^2 - 15q + 9)$ $5q^3 - 36q + 27$

22) $(m + 9)(9m^2 + 4m - 7)$ $9m^3 + 85m^2 + 29m - 63$

23) $(p - 6)(2p^2 + 3p - 5)$ $2p^3 - 9p^2 - 23p + 30$

24) $(s - 5)(7s^2 - 3s - 11)$ $7s^3 - 38s^2 + 4s + 55$

25) $(5y^3 - y^2 + 8y + 1)(3y - 4)$ $15y^4 - 23y^3 + 28y^2 - 29y - 4$

26) $(4n^3 + 4n^2 - 5n - 7)(7n - 6)$
    $28n^4 + 4n^3 - 59n^2 - 19n + 42$

27) $\left(\dfrac{1}{2}k^2 + 3\right)(12k^2 + 5k - 10)$ $6k^4 + \dfrac{5}{2}k^3 + 31k^2 + 15k - 30$

28) $\left(\dfrac{2}{3}c^2 - 8\right)(6c^2 - 4c + 9)$ $4c^4 - \dfrac{8}{3}c^3 - 42c^2 + 32c - 72$

29) $(a^2 - a + 3)(a^2 + 4a - 2)$ $a^4 + 3a^3 - 3a^2 + 14a - 6$

30) $(r^2 + 2r + 5)(2r^2 - r - 3)$ $2r^4 + 3r^3 + 5r^2 - 11r - 15$

31) $(3v^2 - v + 2)(-8v^3 + 6v^2 + 5)$

32) $(c^4 + 10c^2 - 7)(c^2 - 2c - 3)$

Multiply each horizontally and vertically. Which method do you prefer and why?

33) $(2x - 3)(4x^2 - 5x + 2)$ $8x^3 - 22x^2 + 19x - 6$

34) $(3n^2 + n - 4)(5n + 2)$ $15n^3 + 11n^2 - 18n - 8$

Objective 3: Multiply Two Binomials Using FOIL

35) What do the letters in the word FOIL represent?
First, Outer, Inner, Last

36) Can FOIL be used to expand $(x + 9)^2$? Explain your answer.   Yes. $(x + 9)^2 = (x + 9)(x + 9)$ so we can use FOIL to find the product.

Use FOIL to multiply.

37) $(w + 8)(w + 7)$ 38) $(k - 5)(k + 9)$

39) $(n - 11)(n - 4)$ 40) $(y - 6)(y - 1)$

41) $(4p + 5)(p - 3)$ 42) $(6t + 1)(t + 7)$

43) $(8n + 3)(3n + 4)$ 44) $(5b - 2)(8b + 5)$

45) $(0.4g - 0.9)(0.1g + 1.1)$ $0.04g^2 + 0.35g - 0.99$

46) $(0.7 + 0.5m)(0.8m - 1.5)$ $0.4m^2 - 0.19m - 1.05$

 47) $(4a - 5b)(3a + 4b)$ 48) $(2x + 3y)(x - 6y)$

49) $(6p + 5q)(10p + 3q)$ 50) $(7m - 3n)(m - n)$

51) $\left(2a - \frac{1}{4}b\right)(2b + a)$ 52) $\left(w + \frac{3}{2}v\right)(2w - 3v)$

Write an expression for a) the perimeter of each figure and b) the area of each figure.

53)
$y - 2$
$y + 6$
a) $4y + 8$ units
b) $y^2 + 4y - 12$ sq units

54)
$4w$
$5w + 7$
a) $18w + 14$ units
b) $20w^2 + 28w$ sq units

55)
$a^2 - a + 8$
$3a$
a) $2a^2 + 4a + 16$ units
b) $3a^3 - 3a^2 + 24a$ sq units

56)
$2x^2 - 3$
$2x^2 - 3$
a) $8x^2 - 12$ units
b) $4x^4 - 12x^2 + 9$ sq units

Objective 4: Find the Product of More than Two Polynomials

57) To find the product $2(n + 6)(n - 1)$, Raman begins by multiplying $2(n + 6)$ and then he multiplies that result by $(n - 1)$. Peggy begins by multiplying $(n + 6)(n - 1)$ and multiplies that result by 2. Who is right?   Both are correct.

58) Find the product $(3a + 2)(a - 4)(a - 2)$

   a) by first multiplying $(3a + 2)(a - 4)$ and then multiplying that result by $(a - 2)$. $3a^3 - 16a^2 + 12a + 16$

   b) by first multiplying $(a - 4)(a - 2)$ and then multiplying that result by $(3a + 2)$. $3a^3 - 16a^2 + 12a + 16$

   c) What do you notice about the results?   They are the same.

Multiply. See Example 5.

59) $3(y + 4)(5y - 2)$ $15y^2 + 54y - 24$

60) $4(7 - 3z)(2z - 1)$ $-24z^2 + 68z - 28$

61) $-7r^2(r - 9)(r - 2)$ $-7r^4 + 77r^3 - 126r^2$

62) $12g^2(2g + 5)(-g + 1)$ $-24g^4 - 36g^3 + 60g^2$

 63) $(c + 3)(c + 4)(c - 1)$ $c^3 + 6c^2 + 5c - 12$

64) $(x - 5)(x - 2)(x + 3)$ $x^3 - 4x^2 - 11x + 30$

65) $10n\left(\frac{1}{2}n^2 + 3\right)(n^2 + 5)$ $5n^5 + 55n^3 + 150n$

66) $12k\left(\frac{1}{4}k^2 - \frac{2}{3}\right)(k^2 + 1)$ $3k^5 - 5k^3 - 8k$

67) $(r + t)(r - 2t)(2r - t)$ $2r^3 - 3r^2t - 3rt^2 + 2t^3$

68) $(x + y)(x - 2y)(x + 3y)$ $x^3 + 2x^2y - 5xy^2 - 6y^3$

Objective 5: Find the Product of Binomials of the Form $(a + b)(a - b)$

Find the following special products. See Example 6.

69) $(3m + 2)(3m - 2)$ 70) $(5y - 4)(5y + 4)$
    $9m^2 - 4$ $25y^2 - 16$

71) $(7a - 8)(7a + 8)$ 72) $(4x - 11)(4x + 11)$
    $49a^2 - 64$ $16x^2 - 121$

73) $(6a - b)(6a + b)$ 74) $(2p + 7q)(2p - 7q)$
    $36a^2 - b^2$ $4p^2 - 49q^2$

75) $\left(n + \frac{1}{2}\right)\left(n - \frac{1}{2}\right)$ $n^2 - \frac{1}{4}$ 76) $\left(b - \frac{1}{5}\right)\left(b + \frac{1}{5}\right)$ $b^2 - \frac{1}{25}$

 77) $\left(\frac{2}{3} - k\right)\left(\frac{2}{3} + k\right)$ $\frac{4}{9} - k^2$ 78) $\left(\frac{4}{3} + z\right)\left(\frac{4}{3} - z\right)$ $\frac{16}{9} - z^2$

79) $(0.3x - 0.4y)(0.3x + 0.4y)$ $0.09x^2 - 0.16y^2$

80) $(1.2a + 0.8b)(1.2a - 0.8b)$ $1.44a^2 - 0.64b^2$

81) $(5x^2 + 4)(5x^2 - 4)$ $25x^4 - 16$

82) $(9k^2 + 3l^2)(9k^2 - 3l^2)$ $81k^4 - 9l^4$

Objective 6: Square a Binomial

Expand.

83) $(y + 8)^2$ $y^2 + 16y + 64$ 84) $(b + 6)^2$ $b^2 + 12b + 36$

85) $(t - 11)^2$ $t^2 - 22t + 121$ 86) $(g - 5)^2$ $g^2 - 10g + 25$

87) $(4w + 1)^2$ $16w^2 + 8w + 1$ 88) $(7n + 2)^2$ $49n^2 + 28n + 4$

89) $(2d - 5)^2$ $4d^2 - 20d + 25$ 90) $(3p - 5)^2$ $9p^2 - 30p + 25$

91) $(6a - 5b)^2$ $36a^2 - 60ab + 25b^2$ 92) $(7x + 6y)^2$ $49x^2 + 84xy + 36y^2$

93) Does $4(t + 3)^2 = (4t + 12)^2$? Why or why not?

94) Explain, in words, how to find the product $3(z - 4)^2$, then find the product.

Find the product.

95) $6(x + 1)^2$ $6x^2 + 12x + 6$ 96) $2(k + 5)^2$ $2k^2 + 20k + 50$

97) $2a(a + 3)^2$ 98) $-3(m - 1)^2$

99) $[(3m + n) + 2]^2$ 100) $[(2c - d) + 7]^2$

101) $[(x - 4) - y]^2$ 102) $[(3r + 2) - t]^2$

Objective 7: Multiply Other Binomials

Expand.

103) $(r + 5)^3$ 104) $(w + 4)^3$

105) $(s - 2)^3$ 106) $(q - 1)^3$

107) $(c^2 - 9)^2$ 108) $\left(\dfrac{3}{8}x + 2\right)^2$

109) $(y + 2)^4$ 110) $(b + 3)^4$

111) $[(v - 5w) + 4][(v - 5w) - 4]$ $v^2 - 10vw + 25w^2 - 16$

112) $[(4p + 3q) + 1][(4p + 3q) - 1]$ $16p^2 + 24pq + 9q^2 - 1$

113) $[(2a + b) + c][(2a + b) - c]$ $4a^2 + 4ab + b^2 - c^2$

114) $[(x - 3y) - 2z][(x - 3y) + 2z]$ $x^2 - 6xy + 9y^2 - 4z^2$

Mixed Exercises: Objectives 1, 3, 6 and 7

115) Does $(x + 5)^2 = x^2 + 25$? Why or why not?
No; $(x + 5)^2 = x^2 + 10x + 25$

116) Does $(y - 3)^3 = y^3 - 27$? Why or why not?
No; $(y - 3)^3 = y^3 - 9y^2 + 27y - 27$

117) Express the volume of the cube as a polynomial.

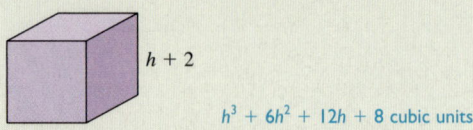

$h + 2$

$h^3 + 6h^2 + 12h + 8$ cubic units

118) Express the area of the square as a polynomial.

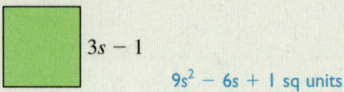

$3s - 1$

$9s^2 - 6s + 1$ sq units

119) Express the area of the shaded region as a polynomial.

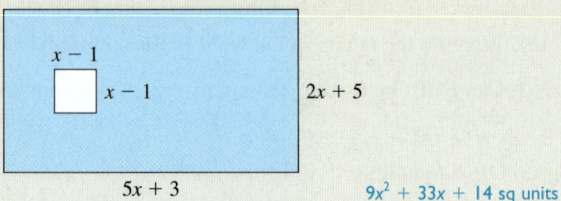

$x - 1$

$x - 1$

$2x + 5$

$5x + 3$

$9x^2 + 33x + 14$ sq units

120) Express the area of the triangle as a polynomial.

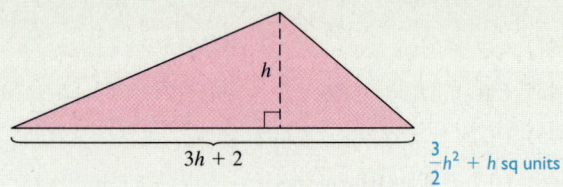

h

$3h + 2$

$\dfrac{3}{2}h^2 + h$ sq units

Objective 8: Multiply Polynomial Functions

Use the following functions for Exercises 121–129.

Let $f(x) = 3x - 5$, $g(x) = x^2$, and $h(x) = 2x^2 - 3x - 1$. Find each of the following.

121) $(fg)(x)$ $3x^3 - 5x^2$ 122) $(gh)(x)$ $2x^4 - 3x^3 - x^2$

123) $(fg)(1)$ -2 124) $(gh)(-2)$ 52

125) $(fh)(x)$ $6x^3 - 19x^2 + 12x + 5$ 126) $(fg)(3)$ 36

127) $(fh)\left(-\dfrac{1}{3}\right)$ $-\dfrac{4}{3}$ 128) $(gh)\left(\dfrac{1}{2}\right)$ $-\dfrac{1}{2}$

129) Find $f(1)$, $g(1)$, and $f(1) \cdot g(1)$. How does this compare to the result in Exercise 123? Why is this true?

130) If $f(x) = x + 6$ and $(fg)(x) = 2x^2 + 12x$, find $g(x)$.
$g(x) = 2x$

Section 5.5 Division of Polynomials and Polynomial Functions

Objectives

1. **Divide a Polynomial by a Monomial**
2. **Divide a Polynomial by a Polynomial**
3. **Divide Polynomial Functions**

The last operation with polynomials we need to discuss is division of polynomials. We will consider this in two parts:

1) Dividing a polynomial by a monomial

Examples: $\dfrac{12a^2 - a + 15}{3}$, $\dfrac{-48m^3 + 30m^2 - 8m}{8m^2}$

and

2)   Dividing a polynomial by a polynomial

$$\text{Examples:}\quad \frac{n^2 + 14n + 48}{n + 6},\quad \frac{27z^3 - 1}{3z - 1}$$

1. Divide a Polynomial by a Monomial

The procedure for dividing a polynomial by a monomial is based on the procedure for adding or subtracting fractions.

To add $\dfrac{4}{15} + \dfrac{7}{15}$, we do the following:

$$\frac{4}{15} + \frac{7}{15} = \frac{4 + 7}{15}\qquad \text{Add numerators, keep the common denominator.}$$

$$= \frac{11}{15}$$

Reversing the process above we can write $\dfrac{11}{15} = \dfrac{4 + 7}{15} = \dfrac{4}{15} + \dfrac{7}{15}$. We can generalize this result and say that $\dfrac{a + b}{c} = \dfrac{a}{c} + \dfrac{b}{c}\quad (c \neq 0)$.

Procedure   Dividing a Polynomial by a Monomial

To divide a polynomial by a monomial, divide *each term* in the polynomial by the monomial and simplify.

Example 1

In-Class Example 1
Divide.
a) $\dfrac{32k^2 + 4k + 12}{4}$
b) $\dfrac{18t^3 - 12t^2 + 6t}{6t}$
answer: a) $8k^2 + k + 3$
b) $3t^2 - 2t + 1$

Divide.

a) $\dfrac{40a^2 - 25a + 10}{5}$ b) $\dfrac{9x^3 + 30x^2 + 3x}{3x}$

Solution

a)   First, note that the polynomial is being divided by a *monomial*. That means we divide each term in the numerator by the monomial 5.

$$\frac{40a^2 - 25a + 10}{5} = \frac{40a^2}{5} - \frac{25a}{5} + \frac{10}{5}$$
$$= 8a^2 - 5a + 2$$

Let's label the components of our division problem the same way as when we divide with integers.

$$\text{Dividend} \rightarrow \frac{40a^2 - 25a + 10}{5} = 8a^2 - 5a + 2 \leftarrow \text{Quotient}$$
$$\text{Divisor} \rightarrow$$

We can check our answer by multiplying the quotient by the divisor. The answer should be the dividend.

$$\text{Check: } 5(8a^2 - 5a + 2) = 40a^2 - 25a + 10 \ \checkmark$$

The quotient $8a^2 - 5a + 2$ is correct.

b) $\dfrac{9x^3 + 30x^2 + 3x}{3x} = \dfrac{9x^3}{3x} + \dfrac{30x^2}{3x} + \dfrac{3x}{3x}$ Divide each term in numerator by $3x$.

$$= 3x^2 + 10x + 1\qquad \text{Apply the quotient rule for exponents.}$$

BE CAREFUL

Students will often incorrectly "cancel out" $\dfrac{3x}{3x}$ and get nothing. But $\dfrac{3x}{3x} = 1$ since a quantity divided by itself equals one.

Check: $3x(3x^2 + 10x + 1) = 9x^3 + 30x^2 + 3x$ ✓   The quotient is correct.   ■

Note

In Example 1b), x cannot equal zero because then the denominator of $\dfrac{9x^3 + 30x^2 + 3x}{3x}$ would equal zero. Remember, a fraction is undefined when its denominator equals zero!

You Try 1

Divide $\dfrac{24t^5 - 6t^4 - 54t^3}{6t^2}$.

Example 2

Divide $(15m + 45m^3 - 4 + 18m^2) \div (9m^2)$.

In-Class Example 2
Divide
$(12 - 6f^3 + 8f + 5f^2) \div (4f^2)$.
answer: $-\dfrac{3}{2}f + \dfrac{5}{4} + \dfrac{2}{f} + \dfrac{3}{f^2}$

Solution

Although this example is written differently, it is the same as the previous examples. Notice, however, the terms in the numerator are not written in descending powers. Rewrite them in descending powers before dividing.

$$\frac{15m + 45m^3 - 4 + 18m^2}{9m^2} = \frac{45m^3 + 18m^2 + 15m - 4}{9m^2}$$

$$= \frac{45m^3}{9m^2} + \frac{18m^2}{9m^2} + \frac{15m}{9m^2} - \frac{4}{9m^2}$$

$$= 5m + 2 + \frac{5}{3m} - \frac{4}{9m^2} \qquad \text{Apply quotient rule and simplify.}$$

The quotient is *not* a polynomial since m and m^2 appear in denominators. The quotient of polynomials is not necessarily a polynomial.   ■

You Try 2

Divide $(6u^2 + 40 - 24u^3 - 8u) \div (8u^2)$.

2. Divide a Polynomial by a Polynomial

When dividing a polynomial by a polynomial containing two or more terms, we use *long division of polynomials*. This method is similar to long division of whole numbers. We will look at a long division problem here so that we can compare the procedure with polynomial long division.

Example 3

In-Class Example 3
Divide 2109 by 9.

answer: $234\dfrac{1}{3}$

Divide 4593 by 8.

Solution

$$\begin{array}{r} 5 \\ 8\overline{)4593} \\ -40\downarrow \\ \hline 59 \end{array}$$

1) How many times does 8 divide into 45 evenly?   5
2) Multiply $5 \times 8 = 40$.
3) Subtract $45 - 40 = 5$.
4) Bring down the 9.

Start the process again.

$$\begin{array}{r} 57 \\ 8\overline{)4593} \\ -40 \\ \hline 59 \\ -56\downarrow \\ \hline 33 \end{array}$$

1) How many times does 8 divide into 59 evenly?   7
2) Multiply $7 \times 8 = 56$.
3) Subtract $59 - 56 = 3$.
4) Bring down the 3.

Do the procedure again.

$$\begin{array}{r} 574 \\ 8\overline{)4593} \\ -40 \\ \hline 59 \\ -56 \\ \hline 33 \\ -32 \\ \hline 1 \end{array}$$

1) How many times does 8 divide into 33 evenly?   4
2) Multiply $4 \times 8 = 32$.
3) Subtract $33 - 32 = 1$.
4) There are no more numbers to bring down, so the remainder is 1.

Write the result.

$$4593 \div 8 = 574\dfrac{1}{8} \begin{array}{l} \leftarrow \text{Remainder} \\ \leftarrow \text{Divisor} \end{array}$$

Check: $(8 \times 574) + 1 = 4592 + 1 = 4593$ ✓

You Try 3

Divide 3827 by 6.

Next we will divide two polynomials using a long division process similar to that of Example 3.

Example 4

In-Class Example 4
Divide $\dfrac{2x^2 + 11x + 12}{x + 4}$.

answer: $2x + 3$

Divide $\dfrac{3x^2 + 19x + 20}{x + 5}$.

Solution

First, notice that we are dividing by more than one term. That tells us to use long division of polynomials.

We will work with the x in $x + 5$ like we worked with the 8 in Example 3.

$$
\begin{array}{r}
3x \\
x + 5 \overline{)\ 3x^2 + 19x + 20} \\
-(3x^2 + 15x) \downarrow \\
\hline
4x + 20
\end{array}
$$

1) By what do we multiply $\underline{x}$ to get $3x^2$? $3x$
   Line up terms in the quotient according to exponents, so write $3x$ above $19x$.

2) Multiply $3x$ by $(x + 5)$: $3x(x + 5) = 3x^2 + 15x$.

3) Subtract $(3x^2 + 19x) - (3x^2 + 15x) = 4x$.

4) Bring down the $+20$.

Start the process again. Remember, work with the x in $x + 5$ like we worked with the 8 in Example 3.

$$
\begin{array}{r}
3x + \ \ 4 \\
x + 5 \overline{)\ 3x^2 + 19x + 20} \\
-(3x^2 + 15x) \\
\hline
4x + 20 \\
-(4x + 20) \\
\hline
0
\end{array}
$$

1) By what do we multiply $\underline{x}$ to get $4x$? 4
   Write $+4$ above $+20$.

2) Multiply 4 by $(x + 5)$: $4(x + 5) = 4x + 20$.

3) Subtract $(4x + 20) - (4x + 20) = 0$.

4) There are no more terms. The remainder is 0.

Write the result.

$$\frac{3x^2 + 19x + 20}{x + 5} = 3x + 4$$

Check: $(x + 5)(3x + 4) = 3x^2 + 4x + 15x + 20 = 3x^2 + 19x + 20$ ✓   ■

You Try 4

Divide.

a) $\dfrac{x^2 + 11x + 24}{x + 8}$ b) $\dfrac{2x^2 + 23x + 45}{x + 9}$

Next, we will look at a division problem with a remainder.

Example 5

Divide $\dfrac{-11c + 16c^3 + 19 - 38c^2}{2c - 5}$.

In-Class Example 5
Divide
$\dfrac{-16k + 12k^2 + 5 + 16k^3}{2k + 3}$.

answer:

$8k^2 - 6k + 1 + \dfrac{2}{2k + 3}$

Solution

When we write our long division problem, the polynomial in the numerator must be rewritten so that the exponents are in descending order. Then, perform the long division.

$$
\begin{array}{r}
8c^2 \\
2c - 5 \overline{)\ 16c^3 - 38c^2 - 11c + 19} \\
-(16c^3 - 40c^2) \downarrow \\
\hline
2c^2 - 11c
\end{array}
$$

1) By what do we multiply $\underline{2c}$ to get $16c^3$? $8c^2$
2) Multiply $8c^2(2c - 5) = 16c^3 - 40c^2$.
3) Subtract.
$$
\begin{aligned}
(16c^3 - 38c^2) &- (16c^3 - 40c^2) \\
&= 16c^3 - 38c^2 - 16c^3 + 40c^2 \\
&= 2c^2
\end{aligned}
$$
4) Bring down the $-11c$.

Repeat the process.

$$\begin{array}{r} 8c^2 + c \\ 2c - 5{\overline{\smash{\big)}\,16c^3 - 38c^2 - 11c + 19}} \\ \underline{-(16c^3 - 40c^2)} \\ 2c^2 - 11c \\ \underline{-(2c^2 - 5c)} \\ -6c + 19 \end{array}$$

1) By what do we multiply $\underline{2c}$ to get $2c^2$? c
2) Multiply $c(2c - 5) = 2c^2 - 5c$.
3) Subtract.
$$(2c^2 - 11c) - (2c^2 - 5c)$$
$$= 2c^2 - 11c - 2c^2 + 5c$$
$$= -6c$$
4) Bring down the $+19$.

Continue.

$$\begin{array}{r} 8c^2 + c - 3 \\ 2c - 5{\overline{\smash{\big)}\,16c^3 - 38c^2 - 11c + 19}} \\ \underline{-(16c^3 - 40c^2)} \\ 2c^2 - 11c \\ \underline{-(2c^2 - 5c)} \\ -6c + 19 \\ \underline{-(-6c + 15)} \\ 4 \end{array}$$

1) By what do we multiply $\underline{2c}$ to get $-6c$? -3
2) Multiply $-3(2c - 5) = -6c + 15$.
3) Subtract.
$$(-6c + 19) - (-6c + 15)$$
$$= -6c + 19 + 6c - 15 = 4$$

We are done with the long division process. How do we know that? Since the degree of 4 (degree zero) is less than the degree of $2c - 5$ (degree one) we cannot divide anymore. *The remainder is 4.*

$$\frac{16c^3 - 38c^2 - 11c + 19}{2c - 5} = 8c^2 + c - 3 + \frac{4}{2c - 5}$$

Check: $(2c - 5)(8c^2 + c - 3) + 4 = 16c^3 + 2c^2 - 6c - 40c^2 - 5c + 15 + 4$
$$= 16c^2 - 38c^2 - 11c + 19 \checkmark$$

■

You Try 5

Divide $-23t^2 - 2 + 20t^3 - 11t$ by $5t + 3$.

As we saw in Example 5, we must write our polynomials so that the exponents are in descending order. We have to watch out for something else as well—missing terms. If a polynomial is missing one or more terms, we put them into the polynomial with coefficients of zero.

Example 6

Divide $x^3 + 125$ by $x + 5$.

In-Class Example 6
Divide $(y^3 + 64)$ by $(y + 4)$.
answer: $y^2 - 4y + 16$

Solution

The degree of the polynomial $x^3 + 125$ is three, but it is missing the x^2-term and the x-term. We will insert these terms into the polynomial by giving them coefficients of zero.

$$x^3 + 125 = x^3 + 0x^2 + 0x + 125$$

Divide.

$$\begin{array}{r} x^2 - 5x + 25 \\ x + 5 \overline{)\,x^3 + 0x^2 + 0x + 125} \\ \underline{-(x^3 + 5x^2)} \\ -5x^2 + 0x \\ \underline{-(-5x^2 - 25x)} \\ 25x + 125 \\ \underline{-(25x + 125)} \\ 0 \end{array}$$

$$(x^3 + 125) \div (x + 5) = x^2 - 5x + 25$$

Check: $(x + 5)(x^2 - 5x + 25) = x^3 - 5x^2 + 25x + 5x^2 - 25x + 125$
$$= x^3 + 125 \checkmark$$

You Try 6

Divide $\dfrac{2k^3 + 5k^2 + 91}{2k + 9}$.

3. Divide Polynomial Functions

Now that we've learned how to add, subtract, and multiply functions, let's learn how to divide functions.

Procedure Dividing Functions

Given the functions $f(x)$ and $g(x)$, the **quotient of f and g** is defined by

$$\left(\frac{f}{g}\right)(x) = \frac{f(x)}{g(x)}, \text{ where } g(x) \neq 0.$$

The domain of $\left(\dfrac{f}{g}\right)(x)$ is the intersection of the domains of $f(x)$ and $g(x)$ except for the values of x for which $g(x) = 0$.

Example 7

Let $f(x) = 3x^2 + 19x + 20$ and $g(x) = x + 5$. Find

a) $\left(\dfrac{f}{g}\right)(x)$, and identify the value of x that is not in its domain.

b) $\left(\dfrac{f}{g}\right)(2)$.

Solution

a) $\left(\dfrac{f}{g}\right)(x) = \dfrac{f(x)}{g(x)}$

$\quad = \dfrac{3x^2 + 19x + 20}{x + 5}$ Substitute the functions.

We can simplify this expression, but first let's look at the denominator of the quotient function. If $x = -5$, the denominator will equal zero. Therefore, -5 is not in the domain of $\left(\dfrac{f}{g}\right)(x)$.

In-Class Example 7
Let $f(x) = 3x^2 - 14x + 8$ and $g(x) = x - 4$. Find

a) $\left(\dfrac{f}{g}\right)(x)$, and identify the value of x that is not in its domain.

b) $\left(\dfrac{f}{g}\right)(5)$.

answer: a) $\left(\dfrac{f}{g}\right)(x) = 3x - 2$, where $x \neq 4$ b) 13

In Example 4 we used long division to find the quotient $\dfrac{3x^2 + 19x + 20}{x + 5} = 3x + 4.$ Therefore,

$$\left(\frac{f}{g}\right)(x) = 3x + 4, \text{ where } x \neq -5.$$

b) Since $\left(\dfrac{f}{g}\right)(x) = 3x + 4$, substitute 2 for x to find $\left(\dfrac{f}{g}\right)(2)$.

$$\left(\frac{f}{g}\right)(2) = 3(2) + 4 \qquad \text{Substitute 2 for } x.$$

$$= 6 + 4 \qquad \text{Multiply.}$$

$$= 10$$

Verify that this result is the same as $\dfrac{f(2)}{g(2)}$.

You Try 7

Let $f(x) = 2x^2 + x - 3$ and $g(x) = x - 1$. Find

a) $\left(\dfrac{f}{g}\right)(x)$, and identify the value of x that is not in its domain.   b) $\left(\dfrac{f}{g}\right)(7)$.

Answers to You Try Exercises

1) $4t^3 - t^2 - 9t$ 2) $-3u + \dfrac{3}{4} - \dfrac{1}{u} + \dfrac{5}{u^2}$ 3) $637\dfrac{5}{6}$ 4) a) $x + 3$ b) $2x + 5$

5) $4t^2 - 7t + 2 - \dfrac{8}{5t + 3}$ 6) $k^2 - 2k + 9 + \dfrac{10}{2k + 9}$ 7) a) $2x + 3$, where $x \neq 1$ b) 17

5.5 Exercises

Additional answers can be found in the Answers to Exercises appendix.

Label the dividend, divisor, and quotient of each division problem.

1) $\dfrac{12c^3 + 20c^2 - 4c}{4c} = 3c^2 + 5c - 1$

2) $2p + 3\overline{)10p^3 + p^2 - 25p - 6}$ $\quad 5p^2 - 7p - 2$

3) Explain, in your own words, how to divide a polynomial by a monomial.   Answers may vary.

4) When do you use long division to divide polynomials?   Use long division when the divisor contains two or more terms.

Objective 1: Divide a Polynomial by a Monomial

Divide.

5) $\dfrac{4a^5 - 10a^4 + 6a^3}{2a^3}$

6) $\dfrac{28k^4 + 8k^3 - 40k^2}{4k^2}$

7) $\dfrac{18u^7 + 18u^5 + 45u^4 - 72u^2}{9u^2}$ $\quad 2u^5 + 2u^3 + 5u^2 - 8$

8) $\dfrac{-15m^6 + 10m^5 + 20m^4 - 35m^3}{5m^3}$ $\quad -3m^3 + 2m^2 + 4m - 7$

9) $(35d^5 - 7d^2) \div (-7d^2)$ $\quad -5d^3 + 1$

10) $(-32q^6 - 8q^3 + 4q^2) \div (-4q^2)$ $\quad 8q^4 + 2q - 1$

11) $\dfrac{9w^5 + 42w^4 - 6w^3 + 3w^2}{6w^3}$ $\quad \dfrac{3}{2}w^2 + 7w - 1 + \dfrac{1}{2w}$

12) $\dfrac{-54j^5 + 30j^3 - 9j^2 + 15}{9j}$ $\quad -6j^4 + \dfrac{10}{3}j^2 - j + \dfrac{5}{3j}$

13) $(10v^7 - 36v^5 - 22v^4 - 5v^2 + 1) \div (4v^4)$

14) $(60z^5 + 3z^4 - 10z) \div (5z^2)$ $\quad 12z^3 + \dfrac{3}{5}z^2 - \dfrac{2}{z}$

Divide.

15) $\dfrac{90a^4b^3 + 60a^3b^3 - 40a^3b^2 + 100a^2b^2}{10ab^2}$ $9a^3b + 6a^2b - 4a^2 + 10a$

16) $\dfrac{24x^6y^6 - 54x^5y^4 - x^3y^3 + 12x^3y^2}{6x^2y}$

17) $(9t^5u^4 - 63t^4u^4 - 108t^3u^4 + t^3u^2) \div (-9tu^2)$

18) $(-45c^8d^6 - 15c^6d^5 + 60c^3d^5 + 30c^3d^3) \div (-15c^3d^2)$
$3c^5d^4 + c^3d^3 - 4d^3 - 2d$

19) Irene divides $16t^3 - 36t^2 + 4t$ by $4t$ and gets a quotient of $4t^2 - 9t$. Is this correct? Why or why not?

20) Kinh divides $\dfrac{15x^2 + 12x}{12x}$ and gets a quotient of $15x^2$.

What was his mistake? What is the correct answer?

Objective 2: Divide a Polynomial by a Polynomial

Divide.

21) $\dfrac{g^2 + 9g + 20}{g + 5}$ $g + 4$

22) $\dfrac{n^2 + 13n + 40}{n + 8}$ $n + 5$

23) $\dfrac{p^2 + 8p + 12}{p + 2}$ $p + 6$

24) $\dfrac{v^2 + 13v + 12}{v + 1}$ $v + 12$

25) $\dfrac{k^2 + 4k - 45}{k + 9}$ $k - 5$

26) $\dfrac{m^2 - 6m - 27}{m + 3}$ $m - 9$

27) $\dfrac{h^2 + 5h - 24}{h - 3}$ $h + 8$

28) $\dfrac{u^2 - 11u + 30}{u - 5}$ $u - 6$

29) $\dfrac{4a^3 - 24a^2 + 29a + 15}{2a - 5}$ $2a^2 - 7a - 3$

30) $\dfrac{28b^3 - 26b^2 + 41b - 15}{7b - 3}$ $4b^2 - 2b + 5$

31) $(p + 45p^2 - 1 + 18p^3) \div (6p + 1)$ $3p^2 + 7p - 1$

32) $(17z^2 - 10 - 12z^3 + 32z) \div (4z + 5)$ $-3z^2 + 8z - 2$

33) $(6t^2 - 7t + 4) \div (t - 5)$ $6t + 23 + \dfrac{119}{t - 5}$

34) $(7d^2 + 57d - 4) \div (d + 9)$ $7d - 6 + \dfrac{50}{d + 9}$

35) $\dfrac{61z + 12z^3 - 37 + 44z^2}{3z + 5}$ $4z^2 + 8z + 7 - \dfrac{72}{3z + 5}$

36) $\dfrac{23k^3 + 22k - 8 + 6k^4 + 44k^2}{6k - 1}$ $k^3 + 4k^2 + 8k + 5 - \dfrac{3}{6k - 1}$

37) $\dfrac{w^3 + 64}{w + 4}$ $w^2 - 4w + 16$

38) $\dfrac{a^3 - 27}{a - 3}$ $a^2 + 3a + 9$

39) $(16r^3 + 58r^2 - 9) \div (8r - 3)$ $2r^2 + 8r + 3$

40) $(50c^3 + 7c + 4) \div (5c + 2)$ $10c^2 - 4c + 3 - \dfrac{2}{5c + 2}$

Mixed Exercises: Objectives 1 and 2

Divide.

41) $\dfrac{6x^4y^4 + 30x^4y^3 - x^2y^2 + 3xy}{6x^2y^2}$ $x^2y^2 + 5x^2y - \dfrac{1}{6} + \dfrac{1}{2xy}$

42) $\dfrac{12v^2 - 23v + 14}{3v - 2}$ $4v - 5 + \dfrac{4}{3v - 2}$

43) $\dfrac{-8g^4 + 49g^2 + 36 - 25g - 2g^3}{4g - 9}$ $-2g^3 - 5g^2 + g - 4$

44) $(12c^2 + 6c - 30c^3 + 48c^4) \div (-6c)$ $-8c^3 + 5c^2 - 2c - 1$

45) $\dfrac{6t^2 - 43t - 20}{t - 8}$ $6t + 5 + \dfrac{20}{t - 8}$

46) $\dfrac{-14u^3v^3 + 7u^2v^3 + 21uv + 56}{7u^2v}$ $-2uv^2 + v^2 + \dfrac{3}{u} + \dfrac{8}{u^2v}$

47) $(8n^3 - 125) \div (2n - 5)$ $4n^2 + 10n + 25$

48) $(12a^4 - 19a^3 + 22a^2 - 9a - 20) \div (3a - 4)$
$4a^3 - a^2 + 6a + 5$

49) $(13x^2 - 7x^3 + 6 + 5x^4 - 14x) \div (x^2 + 2)$ $5x^2 - 7x + 3$

50) $(18m^4 - 66m^3 + 39m^2 + 11m - 7) \div (6m^2 - 1)$
$3m^2 - 11m + 7$

51) $\dfrac{-12a^3 + 9a^2 - 21a}{-3a}$

52) $\dfrac{64r^3 + 27}{4r + 3}$

53) $\dfrac{10h^4 - 6h^3 - 49h^2 + 27h + 19}{2h^2 - 9}$ $5h^2 - 3h - 2 + \dfrac{1}{2h^2 - 9}$

54) $\dfrac{16w^2 - 3 - 7w + 15w^4 - 5w^3}{5w^2 + 7}$ $3w^2 - w - 1 + \dfrac{4}{5w^2 + 7}$

55) $\dfrac{6d^4 + 19d^3 - 8d^2 - 61d - 40}{2d^2 + 7d + 5}$ $3d^2 - d - 8$

56) $\dfrac{8x^4 + 2x^3 - 13x^2 - 53x + 14}{2x^2 + 5x + 7}$ $4x^2 - 9x + 2$

57) $\dfrac{9c^4 - 82c^3 - 41c^2 + 9c + 16}{c^2 - 10c + 4}$ $9c^3 + 8c + 3 + \dfrac{7c + 4}{c^2 - 10c + 4}$

58) $\dfrac{15n^4 - 16n^3 - 31n^2 + 50n - 22}{5n^2 - 7n + 2}$ $3n^2 + n - 6 + \dfrac{6n - 10}{5n^2 - 7n + 2}$

59) $\dfrac{k^4 - 81}{k^2 + 9}$ $k^2 - 9$

60) $\dfrac{b^4 - 16}{b^2 - 4}$ $b^2 + 4$

61) $\dfrac{49a^4 - 15a^2 - 14a^3 + 5a^6}{-7a^3}$ $-\dfrac{5}{7}a^3 - 7a + 2 + \dfrac{15}{7a}$

62) $\dfrac{9q^2 + 26q^4 + 8 - 6q - 4q^3}{2q^2}$ $13q^2 - 2q + \dfrac{9}{2} - \dfrac{3}{q} + \dfrac{4}{q^2}$

63) $\left(x^2 + \dfrac{13}{2}x + 3\right) \div (2x + 1)$ $\dfrac{1}{2}x + 3$

64) $\left(k^2 + \dfrac{11}{3}k + 2\right) \div (3k + 2)$ $\dfrac{1}{3}k + 1$

65) $\left(2w^2 + \dfrac{10}{3}w - 8\right) \div (3w - 4)$ $\dfrac{2}{3}w + 2$

66) $\left(3y^2 - \dfrac{41}{4}y + 9\right) \div (4y - 3)$ $\dfrac{3}{4}y - 2 + \dfrac{3}{4y - 3}$

For each rectangle, find a polynomial that represents the missing side.

67)

$y - 6$

Find the length if the area is given by $4y^2 - 23y - 6$.

$4y + 1$

68)

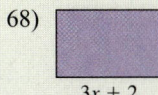

$3x + 2$

Find the width if the area is given by $6x^2 + x - 2$. $2x - 1$

69)

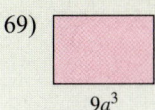

$9a^3$

Find the width if the area is given by $18a^5 - 45a^4 + 9a^3$.

$2a^2 - 5a + 1$

70)

$6w$

Find the length if the area is given by $9w^3 + 6w^2 - 24w$.

$\frac{3}{2}w^2 + w - 4$

71) Find the base of the triangle if the area is given by $6h^3 + 3h^2 + h$.

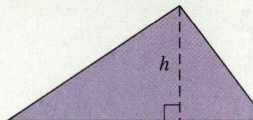

h

$12h^2 + 6h + 2$

72) Find the base of the triangle if the area is given by $6n^3 - 2n^2 + 10n$.

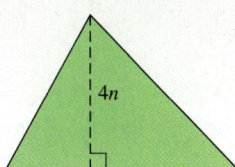

$4n$

$3n^2 - n + 5$

Objective 3: Divide Polynomial Functions

For each pair of functions, find $\left(\dfrac{f}{g}\right)(x)$ and identify any values of x that are not in the domain of the quotient function.

73) $f(x) = 5x^2 + 6x - 27, g(x) = x + 3$

74) $f(x) = 8x^2 - 22x + 15, g(x) = 4x - 5$

75) $f(x) = 12x^3 - 18x^2 + 2x, g(x) = 2x$

76) $f(x) = 24x^4 - 10x^2 + 9x, g(x) = 6x$

77) $f(x) = 3x^4 - 10x^3 + 9x^2 + 2x - 4, g(x) = x - 1$

78) $f(x) = 6x^4 + 11x^3 - 14x^2 - 27x - 4, g(x) = 3x + 4$

Use the following functions for Exercises 79–86.

Let $f(x) = 4x^2 - 1$, $g(x) = 2x + 1$, and $h(x) = 3x$. Find each of the following.

79) $\left(\dfrac{f}{g}\right)(x)$ $2x - 1$, where $x \neq -\dfrac{1}{2}$

80) $\left(\dfrac{h}{g}\right)(x)$ $\dfrac{3x}{2x + 1}$, where $x \neq -\dfrac{1}{2}$

81) $\left(\dfrac{f}{g}\right)(5)$ 9

82) $\left(\dfrac{h}{g}\right)(-1)$ 3

83) $\left(\dfrac{g}{h}\right)(x)$ $\dfrac{2x + 1}{3x}$, where $x \neq 0$

84) $\left(\dfrac{f}{g}\right)\left(\dfrac{1}{4}\right)$ $-\dfrac{1}{2}$

85) $\left(\dfrac{g}{h}\right)\left(-\dfrac{2}{3}\right)$ $\dfrac{1}{6}$

86) $\left(\dfrac{h}{g}\right)(0)$ 0

Chapter 5: Summary

Definition/Procedure	Example

5.1 The Rules of Exponents

Let a and b be real numbers and m and n be integers. The following rules apply:

Product Rule: $a^m \cdot a^n = a^{m+n}$ **(p. 254)**

$x^8 \cdot x^2 = x^{10}$

Basic Power Rule: $(a^m)^n = a^{mn}$ **(p. 255)**

$(t^3)^5 = t^{15}$

Power Rule for a Product: $(ab)^n = a^n b^n$ **(p. 255)**

$(2c)^4 = 2^4 c^4 = 16c^4$

Power Rule for a Quotient: $\left(\dfrac{a}{b}\right)^n = \dfrac{a^n}{b^n}$, where $b \neq 0$. **(p. 255)**

$\left(\dfrac{w}{5}\right)^3 = \dfrac{w^3}{5^3} = \dfrac{w^3}{125}$

Zero Exponent: If $a \neq 0$, then $a^0 = 1$. **(p. 257)**

$(-9)^0 = 1$

Negative Exponent: If n is a natural number and $a \neq 0$ and $b \neq 0$, then $a^{-n} = \left(\dfrac{1}{a}\right)^n = \dfrac{1}{a^n}$, and $\left(\dfrac{a}{b}\right)^{-n} = \left(\dfrac{b}{a}\right)^n$. **(p. 258)**

Evaluate. $\left(\dfrac{5}{2}\right)^{-3} = \left(\dfrac{2}{5}\right)^3 = \dfrac{2^3}{5^3} = \dfrac{8}{125}$

If $a \neq 0$ and $b \neq 0$, then $\dfrac{a^{-m}}{b^{-n}} = \dfrac{b^n}{a^m}$. **(p. 259)**

Rewrite each expression with positive exponents.

a) $\dfrac{x^{-3}}{y^{-7}} = \dfrac{y^7}{x^3}$ b) $\dfrac{14m^{-6}}{n^{-1}} = \dfrac{14n}{m^6}$

Quotient Rule: If $a \neq 0$, then $\dfrac{a^m}{a^n} = a^{m-n}$. **(p. 260)**

Simplify. $\dfrac{4^9}{4^6} = 4^{9-6} = 4^3 = 64$

5.2 More on Exponents and Scientific Notation

Combining the Rules of Exponents (p. 263)

Simplify. $\left(\dfrac{a^4}{2a^7}\right)^{-5} = \left(\dfrac{2a^7}{a^4}\right)^5 = (2a^3)^5 = 32a^{15}$

Scientific Notation

A number is in **scientific notation** if it is written in the form $a \times 10^n$, where $1 \leq |a| < 10$ and n is an integer. That is, a is a number with one nonzero digit to the left of the decimal point. **(p. 266)**

Write in scientific notation.

a) $78,000 \rightarrow 78,000. \rightarrow 7.8 \times 10^4$

b) $0.00293 \rightarrow 0.00293 \rightarrow 2.93 \times 10^{-3}$

Converting from Scientific Notation (p. 268)

Write without exponents.

a) $5 \times 10^{-4} \rightarrow 0005. \rightarrow 0.0005$

b) $1.7 \times 10^6 = 1.700000 \rightarrow 1,700,000$

5.3 Addition and Subtraction of Polynomials and Polynomial Functions

A *polynomial in x* is the sum of a finite number of terms of the form ax^n where n is a whole number and a is a real number. **(p. 272)**

The **degree of a term** equals the exponent on its variable. If a term has more than one variable, the degree equals the *sum of* the exponents on the variables.

The **degree of the polynomial** equals the highest degree of any nonzero term.

Identify each term in the polynomial, the coefficient and degree of each term, and the degree of the polynomial.

$5a^4b^2 - 16a^3b^2 - 4a^2b^3 + ab + 9b$

Term	Coefficient	Degree
$5a^4b^2$	5	6
$-16a^3b^2$	-16	5
$-4a^2b^3$	-4	5
ab	1	2
$9b$	9	1

The degree of the polynomial is 6.

Definition/Procedure	Example
To *add polynomials*, add like terms. Polynomials may be added horizontally or vertically. **(p. 273)**	Add the polynomials. $(6n^2 + 7n - 14) + (-2n^2 + 6n + 3)$ $= [6n^2 + (-2n^2)] + (7n + 6n) + (-14 + 3)$ $= 4n^2 + 13n - 11$
To *subtract two polynomials*, change the sign of each term in the second polynomial. Then, add the polynomials. **(p. 274)**	Subtract. $(3h^3 - 7h^2 + 8h + 4) - (12h^3 - 8h^2 + 3h + 9)$ $= (3h^3 - 7h^2 + 8h + 4) + (-12h^3 + 8h^2 - 3h - 9)$ $= -9h^3 + h^2 + 5h - 5$
$f(x) = 3x^2 + 8x - 4$ is an example of a *polynomial function* since $3x^2 + 8x - 4$ is a polynomial and since each real number that is substituted for x produces only one value for the expression. Finding $f(4)$ is the same as evaluating $3x^2 + 8x - 4$ when $x = 4$. **(p. 275)**	If $f(x) = 3x^2 + 8x - 4$, find $f(4)$. $\quad f(4) = 3(4)^2 + 8(4) - 4$ Substitute 4 for x. $\qquad\quad = 3(16) + 32 - 4$ $\qquad\quad = 48 + 32 - 4$ $\qquad\quad = 76$
Adding and Subtracting Polynomial Functions Given the functions $f(x)$ and $g(x)$, the **sum and difference of f and g** are defined by **(p. 276)** 1) $(f + g)(x) = f(x) + g(x)$ 2) $(f - g)(x) = f(x) - g(x)$ The domain of $(f + g)(x)$ and $(f - g)(x)$ is the intersection of the domains of $f(x)$ and $g(x)$.	Let $f(x) = x^2 - 5x + 3$ and $g(x) = 7x - 9$. Find $(f + g)(x)$ and $(f + g)(-5)$. $\quad (f + g)(x) = f(x) + g(x)$ $\qquad\qquad\quad = (x^2 - 5x + 3) + (7x - 9)$ $\qquad\qquad\quad = x^2 + 2x - 6$ Use $(f + g)(x) = x^2 + 2x - 6$ to find $(f + g)(-5)$. $\quad (f + g)(-5) = (-5)^2 + 2(-5) - 6$ $\qquad\qquad\quad\; = 25 - 10 - 6$ $\qquad\qquad\quad\; = 9$

5.4 Multiplication of Polynomials and Polynomial Functions

When multiplying a *monomial* and a *polynomial*, use the distributive property. **(p. 281)**	Multiply. $4a^3(-3a^2 + 7a - 2)$ $\quad = (4a^3)(-3a^2) + (4a^3)(7a) + (4a^3)(-2)$ Distribute. $\quad = -12a^5 + 28a^4 - 8a^3$
To *multiply two polynomials*, multiply each term in the second polynomial by each term in the first polynomial. Then, combine like terms. **(p. 281)**	Multiply. $(2c + 5)(c^2 - 3c + 6)$ $= (2c)(c^2) + (2c)(-3c) + (2c)(6) + (5)(c^2) + (5)(-3c)$ $\qquad\qquad\qquad\qquad\qquad\qquad\qquad\qquad\quad + (5)(6)$ $= 2c^3 - 6c^2 + 12c + 5c^2 - 15c + 30$ $= 2c^3 - c^2 - 3c + 30$
Multiplying Two Binomials We can use FOIL to multiply two binomials. **FOIL** stands for **F**irst **O**uter **I**nner **L**ast. Multiply the binomials, then, add like terms. **(p. 282)**	Use FOIL to multiply $(3k - 2)(k + 4)$. $\qquad\qquad\qquad\qquad$ F $\qquad$ O $\qquad$ I $\qquad$ L $(3k - 2)(k + 4) = 3k \cdot k + 3k \cdot 4 - 2 \cdot k - 2 \cdot 4$ $\qquad\qquad\qquad\;\; = 3k^2 + 12k - 2k - 8$ $\qquad\qquad\qquad\;\; = 3k^2 + 10k - 8$
Special Products a) $(a + b)(a - b) = a^2 - b^2$ b) $(a + b)^2 = a^2 + 2ab + b^2$ c) $(a - b)^2 = a^2 - 2ab + b^2$ **(p. 284)**	a) Multiply: $(y + 6)(y - 6) = y^2 - 6^2 = y^2 - 36$ b) Expand: $(t + 9)^2 = t^2 + 2(t)(9) + 9^2 = t^2 + 18t + 81$ c) Expand $(4u - 3)^2$. $\quad (4u - 3)^2 = (4u)^2 - 2(4u)(3) + 3^2$ $\qquad\qquad\quad = 16u^2 - 24u + 9$

Definition/Procedure	Example

Multiplying Functions

Given the functions $f(x)$ and $g(x)$, the **product of f and g** is defined by $(fg)(x) = f(x) \cdot g(x)$. The domain of $(fg)(x)$ is the intersection of the domains of $f(x)$ and $g(x)$. **(p. 287)**

Let $f(x) = 3x - 7$ and $g(x) = 2x + 5$. Find $(fg)(x)$.

$$(fg)(x) = f(x) \cdot g(x)$$
$$= (3x - 7)(2x + 5)$$
$$= 6x^2 + x - 35$$

5.5 Division of Polynomials and Polynomial Functions

To *divide a polynomial by a monomial*, divide *each term* in the polynomial by the monomial and simplify. **(p. 291)**

Divide $\dfrac{18r^4 + 2r^3 - 9r^2 + 6r - 10}{2r^2}$.

$$= \frac{18r^4}{2r^2} + \frac{2r^3}{2r^2} - \frac{9r^2}{2r^2} + \frac{6r}{2r^2} - \frac{10}{2r^2}$$

$$= 9r^2 + r - \frac{9}{2} + \frac{3}{r} - \frac{5}{r^2}$$

To *divide a polynomial by another polynomial* containing two or more terms, use *long division*. **(p. 292)**

Divide $\dfrac{12m^3 - 32m^2 - 17m + 25}{6m + 5}$.

$$
\require{enclose}
\begin{array}{r}
2m^2 - 7m + 3 \\
6m + 5 \enclose{longdiv}{12m^3 - 32m^2 - 17m + 25} \\
\underline{-(12m^3 + 10m^2)} \quad\downarrow \\
-42m^2 - 17m \\
\underline{-(-42m^2 - 35m)} \quad\downarrow \\
18m + 25 \\
\underline{-(18m + 15)} \\
\text{Remainder} \leftarrow 10
\end{array}
$$

$$\frac{12m^3 - 32m^2 - 17m + 25}{6m + 5} = 2m^2 - 7m + 3 + \frac{10}{6m + 5}$$

Dividing Polynomial Functions

Given the functions $f(x)$ and $g(x)$, the **quotient of f and g** is defined by $\left(\dfrac{f}{g}\right)(x) = \dfrac{f(x)}{g(x)}$, where $g(x) \neq 0$. The domain of $\left(\dfrac{f}{g}\right)(x)$ is the intersection of the domains of $f(x)$ and $g(x)$ except the values of x for which $g(x) = 0$. **(p. 296)**

Let $f(x) = 5x$ and $g(x) = x - 6$. Find $\left(\dfrac{f}{g}\right)(x)$.

$$\left(\frac{f}{g}\right)(x) = \frac{f(x)}{g(x)}$$
$$= \frac{5x}{x - 6}, \text{ where } x \neq 6$$

Chapter 5: Review Exercises

Additional answers can be found in the Answers to Exercises appendix.

(5.1 and 5.2) Evaluate using the rules of exponents.

1) $\dfrac{3^{10}}{3^6}$ 81

2) 8^{-2} $\dfrac{1}{64}$

3) $\left(\dfrac{5}{4}\right)^{-3}$ $\dfrac{64}{125}$

4) $-4^0 + 7^0$ 0

Simplify. Assume all variables represent nonzero real numbers. Write the answers with positive exponents.

5) $(z^6)^3$ z^{18}

6) $(4p^3)(-3p^7)$ $-12p^{10}$

7) $\dfrac{70r^9}{10r^4}$ $7r^5$

8) $(-5c^4)^2$ $25c^8$

9) $(-9t)(6t^6)$ $-54t^7$

10) $\dfrac{6m^{10}}{24m^6}$ $\dfrac{m^4}{4}$

11) $\dfrac{k^3}{k^{11}}$ $\dfrac{1}{k^8}$

12) $\dfrac{d^{-6}}{d^3}$ $\dfrac{1}{d^9}$

13) $(-2a^2b)^3(5a^{-12}b)$ $-\dfrac{40b^4}{a^6}$

14) $\dfrac{x^5y^{-3}}{x^8y^{-4}}$ $\dfrac{y}{x^3}$

15) $\left(\dfrac{3pq^{-10}}{2p^{-2}q^5}\right)^{-2}$ $\dfrac{4q^{30}}{9p^6}$

16) $(7c^{-8}d^2)(3c^{-2}d)^2$ $\dfrac{63d^4}{c^{12}}$

17) $\left(\dfrac{40}{21}x^{10}\right)(3x^{-12})\left(\dfrac{49}{20}x^2\right)$ 14

18) $\left(\dfrac{4r^{-3}t}{s^2}\right)^{-3}\left(\dfrac{3t^{-5}s}{r^2}\right)^{-2}\left(\dfrac{2rs^2}{t^3}\right)^4$ $\dfrac{r^{17}s^{12}}{36t^5}$

Simplify. Assume that the variables represent nonzero integers. Write the final answer so that the exponents have positive coefficients.

19) $x^{5t} \cdot x^{3t}$ x^{8t}

20) $\dfrac{r^{9a}}{r^{3a}}$ r^{6a}

21) $(y^{2p})^3$ y^{6p}

22) $\dfrac{w^{-12a}}{w^{-3a}}$ $\dfrac{1}{w^{9a}}$

23) True or False: $-x^2 = (-x)^2$ for every real number value of x. Explain your answer.

24) True or False: $(5y)^{-3} = \dfrac{-125}{y^3}$ if $y \neq 0$. Explain your answer.

Write each number without an exponent.

25) 9.38×10^5 938,000

26) -4.185×10^2 -418.5

27) 1.05×10^{-6} 0.00000105

28) 2×10^4 20,000

Write each number in scientific notation.

29) 0.0000575 5.75×10^{-5}

30) 36,940 3.694×10^4

31) 32,000,000 3.2×10^7

32) 0.0000004 4×10^{-7}

Perform the operation as indicated. Write the final answer without an exponent.

33) $\dfrac{8 \times 10^6}{2 \times 10^{13}}$ 0.0000004

34) $(9 \times 10^{-8})(4 \times 10^7)$ 3.6

35) $\dfrac{-3 \times 10^{10}}{-4 \times 10^6}$ 7500

36) $(-4.2 \times 10^2)(3.1 \times 10^3)$ $-1,302,000$

For Exercises 37 and 38, write each of the numbers in scientific notation, then solve the problem. Write the answer without exponents.

37) Eight porcupines have a total of about 2.4×10^5 quills on their bodies. How many quills would one porcupine have?

30,000 quills

38) One molecule of water has a mass of 2.99×10^{-23} g. Find the mass of 100,000,000 molecules. 0.00000000000000299 g

(5.3) Identify each term in the polynomial, the coefficient and degree of each term, and the degree of the polynomial.

39) $4r^3 - 7r^2 + r + 5$

40) $x^3y + 6xy^2 - 8xy + 11y$

41) Evaluate $-x^2y^2 - 7xy + 2x + 5$ for $x = -3$ and $y = 2$. 5

Add or subtract as indicated.

42) $(5t^2 + 11t - 4) - (7t^2 + t - 9)$ $-2t^2 + 10t + 5$

43) $\begin{aligned}5.8p^3 - 1.2p^2 + p - 7.5 \\ + \ 2.1p^3 + 6.3p^2 + 3.8p + 3.9\end{aligned}$ $7.9p^3 + 5.1p^2 + 4.8p - 3.6$

44) $\left(\dfrac{4}{9}w^2 - \dfrac{3}{8}w + \dfrac{2}{5}\right) + \left(\dfrac{2}{9}w^2 + \dfrac{5}{8}w - \dfrac{9}{20}\right)$ $\dfrac{2}{3}w^2 + \dfrac{1}{4}w - \dfrac{1}{20}$

45) Subtract $3a^2b^2 - 10a^2b + ab + 6$ from $a^2b^2 + 7a^2b - 3ab + 11$. $-2a^2b^2 + 17a^2b - 4ab + 5$

46) Find the sum of $6xy + 4x - y - 10$ and $-4xy + 2y + 3$ and subtract it from $-6xy - 7x + y + 2$. $-8xy - 11x + 9$

47) Write a fifth-degree polynomial in x that does not contain a third-degree term. Answers may vary

48) Find the polynomial that represents the perimeter of the rectangle.

$d^2 + 3d + 5$

$d^2 - 5d + 2$

$4d^2 - 4d + 14$ units

49) Let $f(x) = -2x^2 + 5x - 8$. Find each of the following functions values.

    a) $f(-3)$ -41 b) $f\left(\dfrac{1}{2}\right)$ -6

50) Let $f(x) = 3x + 2$ and $g(x) = x^2 + 6x - 10$. Find each of the following.

    a) $(f + g)(x)$ $x^2 + 9x - 8$ b) $(f + g)(0)$ -8

    c) $(f - g)(x)$ $-x^2 - 3x + 12$ d) $(f - g)(-4)$ 8

51) The number of cruise ships, $N(x)$, operating in North America from 2002–2006 can be modeled by the polynomial function $N(x) = 0.5x^3 - 4.357x^2 + 16.429x + 122.686$ where x is the number of years after 2002. (www.census.gov)

    a) Approximately how many cruise ships were operating in 2002? 123

    b) Approximately how many cruise ships were operating in 2006? 151

    c) Find $N(1)$, and explain what it means in the context of the problem. $N(1) = 135$. In 2003, there were approximately 135 cruise ships operating in North America.

52) $R(x) = 20x$ is the revenue function for the sale of x children's soccer uniforms, in dollars. The cost to produce x soccer uniforms, in dollars, is

$$C(x) = 14x + 400$$

    a) Find the profit function, $P(x)$, that describes the profit from the sale of x uniforms. $P(x) = 6x - 400$

    b) What is the profit from the sale of 200 uniforms? $\$800$

(5.4) Find each product.

53) $-6m^3(9m^2 - 3m + 7)$ $-54m^5 + 18m^4 - 42m^3$

54) $7u^4v^2(-8u^2v + 7uv^2 + 12u - 3)$
     $-56u^6v^3 + 49u^5v^4 + 84u^5v^2 - 21u^4v^2$

55) $(2w + 5)(-12w^3 + 6w^2 - 2w + 3)$
     $-24w^4 - 48w^3 + 26w^2 - 4w + 15$

56) $(x^2 + 4x - 11)(12x^4 - 7x^2 + 9)$
     $12x^6 + 48x^5 - 139x^4 - 28x^3 + 86x^2 + 36x - 99$

57) $(y - 7)(y + 8)$ $y^2 + y - 56$

58) $(3n - 7)(2n - 9)$ $6n^2 - 41n + 63$

59) $(ab + 5)(ab + 6)$ $a^2b^2 + 11ab + 30$

60) $(9r + 2s)(r - s)$ $9r^2 - 7rs - 2s^2$

61) $-(4d + 3)(6d + 7)$ $-24d^2 - 46d - 21$

62) $6c^3(4c - 5)(c - 2)$ $24c^5 - 78c^4 + 60c^3$

63) $(p + 3)(p - 6)(p + 2)$

64) $(z + 4)(z + 1)(z + 5)$

65) $\left(\dfrac{3}{5}m + 2\right)\left(\dfrac{1}{3}m - 4\right)$ 66) $\left(\dfrac{2}{9}t - 5\right)\left(\dfrac{1}{10}t - 3\right)$

67) $(z + 9)(z - 9)$ $z^2 - 81$ 68) $\left(\dfrac{1}{5}n - 2\right)\left(\dfrac{1}{5}n + 2\right)$

69) $\left(\dfrac{7}{8} - r^2\right)\left(\dfrac{7}{8} + r^2\right)$ 70) $\left(2a - \dfrac{1}{3}b\right)\left(2a + \dfrac{1}{3}b\right)$

Expand.

71) $(b + 7)^2$ $b^2 + 14b + 49$ 72) $(x - 10)^2$ $x^2 - 20x + 100$

73) $(5q - 2)^2$ $25q^2 - 20q + 4$ 74) $(7 - 3y)^2$ $9y^2 - 42y + 49$

75) $(x - 2)^3$ 76) $(p + 10)^3$
   $x^3 - 6x^2 + 12x - 8$ $p^3 + 30p^2 + 300p + 1000$

77) $-2(3c - 4)^2$ 78) $6w(w + 3)^2$
   $-18c^2 + 48c - 32$ $6w^3 + 36w^2 + 54w$

79) $[(m - 5) + n]^2$ $m^2 - 10m + 25 + 2mn - 10n + n^2$

80) $[(3r + 2t) + 1][(3r + 2t) - 1]$ $9r^2 + 12rt + 4t^2 - 1$

81) Let $f(x) = 2x - 9$ and $g(x) = -5x + 4$. Find each of the following.

    a) $(fg)(x)$ $-10x^2 + 53x - 36$ b) $(fg)(2)$ 30

82) If $f(x) = 3x - 1$ and $(fg)(x) = 1 - 3x$, find $g(x)$. $g(x) = -1$

83) Write an expression for the a) area and b) perimeter of the rectangle.

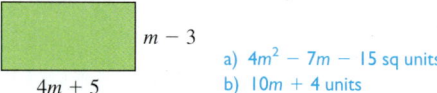
$m - 3$

    a) $4m^2 - 7m - 15$ sq units
$4m + 5$ b) $10m + 4$ units

84) Express the volume of the cube as a polynomial.

$x + 3$

        $x^3 + 9x^2 + 27x + 27$ cubic units

(5.5) Divide

85) $\dfrac{8t^5 - 14t^4 - 20t^3}{2t^3}$ 86) $\dfrac{16p^4 + 56p^3 - 32p^2 + 8p}{-8p}$

87) $\dfrac{c^2 + 8c - 20}{c - 2}$ $c + 10$ 88) $\dfrac{y^2 - 15y + 56}{y - 8}$ $y - 7$

89) $\dfrac{12r^3 - 13r^2 - 5r + 6}{3r + 2}$ $4r^2 - 7r + 3$

90) $\dfrac{-66h^3 - 5h^2 + 45h - 4}{11h - 1}$ $-6h^2 - h + 4$

91) $(15x^4y^4 - 42x^3y^4 - 6x^2y + 10y) \div (-6x^2y)$

92) $(56a^6b^6 + 21a^4b^5 - 4a^3b^4 + a^2b - 7ab) \div (7a^3b^3)$

93) $(6q^2 + 2q - 35) \div (3q + 7)$ $2q - 4 - \dfrac{7}{3q + 7}$

94) $(12r^2 - 16r + 11) \div (6r + 1)$ $2r - 3 + \dfrac{14}{6r + 1}$

95) $\dfrac{23a - 7 + 15a^2}{5a - 4}$ $3a + 7 + \dfrac{21}{5a - 4}$

96) $\dfrac{6m^4 + 2m^3 + 7m^2 + 5m - 20}{2m^2 + 5}$ $3m^2 + m - 4$

97) $\dfrac{8t^4 + 32t^3 - 43t^2 - 44t + 48}{8t^2 - 11}$ $t^2 + 4t - 4 + \dfrac{4}{8t^2 - 11}$

98) $\dfrac{b^3 - 64}{b - 4}$ $b^2 + 4b + 16$

99) $\dfrac{f^3 + 125}{f + 5}$ $f^2 - 5f + 25$

100) $\dfrac{-23 - 46w + 32w^3}{4w + 3}$ $8w^2 - 6w - 7 - \dfrac{2}{4w + 3}$

101) $\dfrac{8k^2 - 8 + 15k^3}{3k - 2}$ $5k^2 + 6k + 4$

102) $(7u^4 - 69u^3 + 15u^2 - 37u + 12) \div (u^2 - 10u + 3)$
$7u^2 + u + 4$

103) $(6c^4 + 13c^3 - 21c^2 - 9c + 10) \div (2c^2 + 5c - 4)$ $3c^2 - c - 2 + \dfrac{-3c + 2}{2c^2 + 5c - 4}$

104) Find the base of the triangle if the area is given by $15y^2 + 12y$.

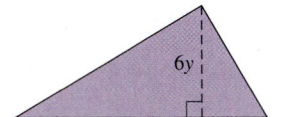

6y

5y + 4

105) Find the length of the rectangle if the area is given by $6x^3 - x^2 + 13x - 10$.

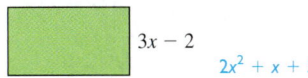

3x − 2

$2x^2 + x + 5$

106) Let $f(x) = x^2 + 6x + 8$, $g(x) = x + 4$, and $h(x) = 7x$. Find each of the following.

a) $\left(\dfrac{f}{g}\right)(x)$ $x + 2$, where $x \neq -4$

b) $\left(\dfrac{f}{g}\right)(-9)$ -7

c) $\left(\dfrac{g}{h}\right)(x)$ $\dfrac{x + 4}{7x}$, where $x \neq 0$

d) $\left(\dfrac{g}{h}\right)(3)$ $\dfrac{1}{3}$

*Additional answers can be found in the Answers to Exercises appendix.

1) Evaluate each expression.

a) -3^4 -81 b) 2^{-5} $\dfrac{1}{32}$ c) $-6^0 - 9^0$ -2

d) $\left(\dfrac{3}{10}\right)^{-3}$ $\dfrac{1000}{27}$ e) $\dfrac{2^{11}}{2^{17}}$ $\dfrac{1}{64}$

Simplify. Assume all variables represent nonzero real numbers. Write the answers with positive exponents.

2) $(-3p^4)(10p^8)$ $-30p^{12}$

3) $\dfrac{a^5 b}{a^9 b^7}$ $\dfrac{1}{a^4 b^6}$

4) $(2y^{-4})^6 \left(\dfrac{1}{2}y^5\right)^3$ $\dfrac{8}{y^9}$

5) $\left(\dfrac{36xy^8}{54x^3 y^{-1}}\right)^{-2}$ $\dfrac{9x^4}{4y^{18}}$

6) $t^{10k} \cdot t^{3k}$ t^{13k}

7) Write 7.283×10^5 without exponents.   728,300

8) Write 0.000165 in scientific notation. 1.65×10^{-4}

9) Divide $\dfrac{-7.5 \times 10^{12}}{1.5 \times 10^8}$. Write the answer without exponents. $-50,000$

10) An electron is a subatomic particle with a mass of 9.1×10^{-28}g. What is the mass of 10,000,000,000 electrons? Write the answer in scientific notation. 9.1×10^{-18} g

11) Given the polynomial $5p^3 - p^2 + 12p + 9$,

a) what is the coefficient of p^2? -1

b) what is the degree of the polynomial?   3

12) Evaluate $-m^2 + 3n$ when $m = -5$ and $n = 8$. -1

In Exercises 13–23, perform the indicated operations.

13) $(10r^3 s^2 + 7r^2 s^2 - 11rs + 5) + (4r^3 s^2 - 9r^2 s^2 + 6rs + 3)$
$14r^3 s^2 - 2r^2 s^2 - 5rs + 8$

14) Subtract $5j^2 - 2j + 9$ from $11j^2 - 10j + 3$. $6j^2 - 8j - 6$

15) $(c - 8)(c - 7)$ $c^2 - 15c + 56$

16) $(3y + 5)(2y + 1)$ $6y^2 + 13y + 5$

17) $\left(u + \dfrac{3}{4}\right)\left(u - \dfrac{3}{4}\right)$ $u^2 - \dfrac{9}{16}$

18) $(2a - 5b)(3a + b)$ $6a^2 - 13ab - 5b^2$

19) $(7m - 5)^2$ $49m^2 - 70m + 25$

20) $6(-n^3 + 4n - 2) - 3(2n^3 + 5n^2 + 8n - 1) + 4n(n^2 - 7n + 3)$
$-8n^3 - 43n^2 + 12n - 9$

21) $(3 - 8m)(2m^2 + 4m - 7)$ $-16m^3 - 26m^2 + 68m - 21$

22) $3x(x + 4)^2$ $3x^3 + 24x^2 + 48x$

23) $[(5a - b) - 3]^2$ $25a^2 - 10ab + b^2 - 30a + 6b + 9$

24) Expand $(s - 4)^3$. $s^3 - 12s^2 + 48s - 64$

Divide.

25) $\dfrac{r^2 + 10r + 21}{r + 7}$ $r + 3$

26) $\dfrac{24t^5 - 60t^4 + 12t^3 - 8t^2}{12t^3}$ $2t^2 - 5t + 1 - \dfrac{2}{3t}$

27) $(38v - 31 + 30v^3 - 51v^2) \div (5v - 6)$ $6v^2 - 3v + 4 - \dfrac{7}{5v - 6}$

28) Write an expression for the base of the triangle if the area is given by $12k^2 + 28k$.

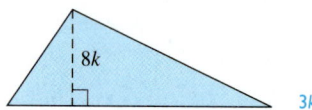
$3k + 7$

29) A company's revenue, $R(x)$ in thousands of dollars, from the sale of x frozen pizzas (in thousands) is given by $R(x) = 5x + 9$. The cost, $C(x)$ in thousands of dollars, of producing these pizzas is given by $C(x) = 3x + 8$.

a) Find the profit function. $P(x)$, that describes the profit from the sale of x pizzas. $P(x) = 2x + 1$

b) What is the profit from the sale of 8000 frozen pizzas? $17,000

30) Let $f(x) = x^2 - 5x - 24$, $g(x) = x - 8$, and $h(x) = x - 2$. Find each of the following.

a) $(f + g)(x)$ $x^2 - 4x - 32$ b) $(f + g)(-2)$ -20

c) $(gh)(x)$ $x^2 - 10x + 16$ d) $(gh)(4)$ -8

e) $\left(\dfrac{f}{g}\right)(x)$ and any values not in its domain $x + 3$ where $x \ne 8$

f) $\left(\dfrac{f}{g}\right)(7)$ 10

Additional answers can be found in the Answers to Exercises appendix.

1) Given the set of numbers
$$\left\{\frac{6}{11}, -14, 2.7, \sqrt{19}, 43, 0.\overline{65}, 0, 8.21079314\ldots\right\}$$
list the

   a) whole numbers. $43, 0$

   b) integers. $-14, 43, 0$

   c) rational numbers. $\frac{6}{11}, -14, 2.7, 43, 0.\overline{65}, 0$

2) Evaluate $-2^4 + 3 \cdot 8 \div (-2)$. -28

3) Divide $2\frac{6}{7} \div 1\frac{4}{21}$. $\frac{12}{5}$ or $2\frac{2}{5}$

Solve.

4) $-\frac{12}{5}c - 7 = 20$ $\left\{-\frac{45}{4}\right\}$

5) $6(w + 4) + 2w = 1 + 8(w - 1)$ $\varnothing$

6) $|9 - 2n| + 5 = 18$ $\{-2, 11\}$

7) $A = \frac{1}{2}h(b_1 + b_2)$ for b_2 $b_2 = \frac{2A}{h} - b_1$

8) Solve the compound inequality and write the answer in interval notation.

$$3y + 16 < 4 \quad \text{or} \quad 8 - y \geq 7 \quad (-\infty, 1]$$

9) *Write an equation in one variable and solve.* How many milliliters of a 12% alcohol solution and how many milliliters of a 4% alcohol solution must be mixed to obtain 60 ml of a 10% alcohol solution?   45 ml of 12% solution, 15 ml of 4% solution

10) Find the x- and y-intercepts of $5x - 2y = 10$ and sketch a graph of the equation.

11) Graph $x = -3$.

12) Write an equation of the line containing the points $(-5, 8)$ and $(1, 2)$. Express the answer in standard form. $x + y = 3$

13) What is the domain of $g(x) = \frac{1}{3}x + 5$? $(-\infty, \infty)$

14) Solve this system using the elimination method.
$$\begin{aligned} -7x + 2y &= 6 \quad (2, 10) \\ 9x - y &= 8 \end{aligned}$$

15) *Write a system of two equations in two variables and solve.* The length of a rectangle is 7 cm less than twice its width. The perimeter of the rectangle is 76 cm. What are the dimensions of the figure?   width: 15 cm; length: 23 cm

Simplify. Assume all variables represent nonzero real numbers. The answers should not contain negative exponents.

16) $-5(3w^4)^2$ $-45w^8$

17) $\left(\frac{2n^{-10}}{n^{-4}}\right)^3$ $\frac{8}{n^{18}}$

18) $p^{10k} \cdot p^{4k}$ p^{14k}

Perform the indicated operations.

19) $(4q^2 + 11q - 2) - 3(6q^2 - 5q + 4) + 2(-10q - 3)$ $-14q^2 + 6q - 20$

20) $(4g - 9)(4g + 9)$ $16g^2 - 81$

21) $\dfrac{8a^4b^4 - 20a^3b^2 + 56ab + 8b}{8a^3b^3}$ $ab - \dfrac{5}{2b} + \dfrac{7}{a^2b^2} + \dfrac{1}{a^3b^2}$

22) $\dfrac{17v^3 - 22v + 7v^4 + 24 - 47v^2}{7v - 4}$ $v^3 + 3v^2 - 5v - 6$

23) $(2p^3 + 5p^2 - 11p + 9) \div (p + 4)$ $2p^2 - 3p + 1 + \dfrac{5}{p + 4}$

24) $(2c - 5)(c - 3)^2$ $2c^3 - 17c^2 + 48c - 45$

25) Let $f(x) = x^2 - 6$ and $g(x) = x^2 + 5x + 4$. Find

   a) $(fg)(x)$. $x^4 + 5x^3 - 2x^2 - 30x - 24$

   b) $(fg)(2)$. -36

Factoring Polynomials

Algebra at Work: Ophthalmology

Mark is an ophthalmologist, a doctor specializing in the treatment of diseases of the eye. He says that he could not do his job without a background in mathematics. While formulas are very important in his work, he says that the thinking skills learned in math courses are the same kinds of thinking skills he uses to treat his patients on a daily basis.

As a physician, Mark says that he must follow a very logical, analytical progression to form an accurate diagnosis and treatment plan. He exam-

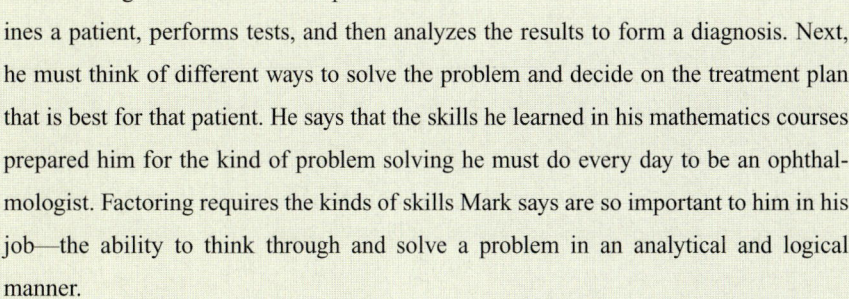

ines a patient, performs tests, and then analyzes the results to form a diagnosis. Next, he must think of different ways to solve the problem and decide on the treatment plan that is best for that patient. He says that the skills he learned in his mathematics courses prepared him for the kind of problem solving he must do every day to be an ophthalmologist. Factoring requires the kinds of skills Mark says are so important to him in his job—the ability to think through and solve a problem in an analytical and logical manner.

In this chapter, we will learn different techniques for factoring polynomials.

Section 6.1 The Greatest Common Factor and Factoring by Grouping

Objectives

1. Find the GCF of a Group of Monomials
2. Factoring vs. Multiplying Polynomials
3. Factor Out the Greatest Common Monomial Factor
4. Factor Out the Greatest Common Binomial Factor
5. Factor by Grouping

Recall that we can write a number as the product of factors:

$$12 = 3 \cdot 4$$
$$\downarrow \quad \downarrow \quad \downarrow$$
$$\text{Product Factor Factor}$$

To **factor** an integer is to write it as the product of two or more integers. Therefore, 12 can also be factored in other ways:

$$12 = 1 \cdot 12 \qquad 12 = 2 \cdot 6 \qquad 12 = -1 \cdot (-12)$$
$$12 = -2 \cdot (-6) \qquad 12 = -3 \cdot (-4) \qquad 12 = 2 \cdot 2 \cdot 3$$

The last **factorization,** $2 \cdot 2 \cdot 3$ or $2^2 \cdot 3$, is called the **prime factorization** of 12 since all of the factors are prime numbers. The factors of 12 are 1, 2, 3, 4, 6, 12, -1, -2, -3, -4, -6, and -12. We can also write the factors as ± 1, ± 2, ± 3, ± 4, ± 6, and ± 12. (Read ± 1 as "plus or minus 1.")

In this chapter, we will learn how to factor polynomials, a skill that is used in many ways throughout algebra.

1. Find the GCF of a Group of Monomials

Definition

The **greatest common factor (GCF)** of a group of two or more integers is the *largest* common factor of the numbers in the group.

For example, if we want to find the GCF of 12 and 20, we can list their positive factors.

$$12: 1, 2, 3, 4, 6, 12$$
$$20: 1, 2, 4, 5, 10, 20$$

The greatest common factor of 12 and 20 is 4. We can also use prime factors.

We begin our study of factoring polynomials by discussing how to find the greatest common factor of a group of monomials.

Example 1

Find the greatest common factor of x^5 and x^3.

In-Class Example 1
Find the greatest common factor of m^8 and m^4.
answer: m^4

Solution

We can write each monomial as the product of its prime factors. To find the GCF use each prime factor the *least* number of times it appears in any of the prime factorizations. Then, multiply.

We can write x^5 and x^3 as

$$x^5 = x \cdot x \cdot x \cdot x \cdot x$$
$$x^3 = x \cdot x \cdot x$$

In x^5, x appears as a factor *five times*. In x^3, x appears as a factor *three times*.

The least number of times x appears as a factor is three. *There will be three factors of x in the GCF.*

$$\text{GCF} = x \cdot x \cdot x = x^3$$

In Example 1, notice that the power of 3 in the GCF is the smallest of the powers when comparing x^5 and x^3. This will always be true.

Note

The exponent on the variable in the GCF will be the *smallest* exponent appearing on the variable in the group of terms.

You Try 1

Find the greatest common factor of y^4 and y^7.

Example 2

In-Class Example 2
Find the greatest common factor for each group of terms.
a) $15c^3, 20c, 5c^7$
b) $35m^3n^4, -84m^6n^3$
answer: a) 5c   b) 7m³n³

Find the greatest common factor for each group of terms.

a) $30k^4, 10k^9, 50k^6$ b) $-12a^8b, 42a^5b^7$ c) $63c^5d^3, 18c^3, 27c^2d^2$

Solution

a) The GCF of the coefficients, 30, 10, and 50, is 10. The smallest exponent on k is 4, so k^4 will be part of the GCF.

$$\text{The GCF of } 30k^4, 10k^9, \text{ and } 50k^6 \text{ is } 10k^4.$$

b) The GCF of the coefficients, -12 and 42, is 6. The smallest exponent on a is 5, so a^5 will be part of the GCF. The smallest exponent on b is 1, so b will be part of the GCF.

$$\text{The GCF of } -12a^8b \text{ and } 42a^5b^7 \text{ is } 6a^5b.$$

c) The GCF of the coefficients is 9. The smallest exponent on c is 2, so c^2 will be part of the GCF. There is no d in the term $18c^3$, so there will be no d in the GCF.

$$\text{The GCF of } 63c^5d^3, 18c^3, \text{ and } 27c^2d^2 \text{ is } 9c^2.$$

You Try 2

Find the greatest common factor for each group of terms.

a) $-16p^7, 8p^5, 40p^8$ b) $r^6s^5, 9r^8s^3, 12r^4s^4$

Factoring Out the Greatest Common Factor

Earlier we said that to **factor an integer** is to write it as the product of two or more integers. To **factor a polynomial** is to write it as a product of two or more polynomials.

Throughout this chapter we will study different factoring techniques. We will begin by discussing how to factor out the greatest common factor.

2. Factoring vs. Multiplying Polynomials

Factoring a polynomial is the opposite of multiplying polynomials. Let's see how these procedures are related.

Example 3

In-Class Example 3
a) Multiply $11v(v + 3)$.
b) Factor out the GCF from
   $11v^2 + 33v$.
**answer: a) $11v^2 + 33v$
b) $11v(v + 3)$**

a) Multiply $2x(x + 5)$.   b) Factor out the GCF from $2x^2 + 10x$.

Solution

a) Use the distributive property to multiply.

$$2x(x + 5) = (2x)x + (2x)(5)$$
$$= 2x^2 + 10x$$

b) Use the distributive property to factor out the greatest common factor from $2x^2 + 10x$.
First, identify the GCF of $2x^2$ and $10x$:   GCF $= 2x$.
Then, rewrite each term as a product of two factors with one factor being $2x$.

$$2x^2 = (2x)(x) \text{ and } 10x = (2x)(5)$$

$$2x^2 + 10x = (2x)(x) + (2x)(5)$$
$$= 2x(x + 5) \qquad \text{Distributive property}$$

When we factor $2x^2 + 10x$, we get $2x(x + 5)$. We can check our result by multiplying.

$$2x(x + 5) = 2x^2 + 10x \checkmark \qquad ■$$

Procedure   Steps for Factoring Out the Greatest Common Factor

1)   Identify the GCF of all of the terms of the polynomial.

2)   Rewrite each term as the product of the GCF and another factor.

3)   Use the distributive property to factor out the GCF from the terms of the polynomial.

4)   Check the answer by multiplying the factors. The result should be the original polynomial.

---

3. Factor Out the Greatest Common Monomial Factor

Example 4

In-Class Example 4
Factor out the GCF.
a) $16t^6 + 32t^4 + 28t^3$
b) $5a^7 + 6a^4$
c) $27x^4y^4 - 81x^3y^5 + 9x^2y^3$
answer:
a) $4t^2(4t^4 + 8t^2 + 7t)$
b) $a^4(5a^3 + 6)$
c) $9x^2y^3(3x^2y - 9xy^2 + 1)$

Factor out the greatest common factor.

a) $12a^5 + 30a^4 + 6a^3$ b) $c^6 - 6c^2$

c) $4x^5y^3 + 12x^5y^2 - 28x^4y^2 - 4x^3y$

Solution

a)   Identify the GCF of all of the terms: GCF $= 6a^3$.

$$12a^5 + 30a^4 + 6a^3 = (6a^3)(2a^2) + (6a^3)(5a) + (6a^3)(1) \quad \text{Rewrite each term using the GCF as one of the factors.}$$

$$= 6a^3(2a^2 + 5a + 1) \qquad \text{Distributive property}$$

Check: $6a^3(2a^2 + 5a + 1) = 12a^5 + 30a^4 + 6a^3 \checkmark$

b)   The GCF of all of the terms is c^2.

$$c^6 - 6c^2 = (c^2)(c^4) - (c^2)(6) \quad \text{Rewrite each term using the GCF as one of the factors.}$$
$$= c^2(c^4 - 6) \qquad \text{Distributive property}$$

Check: $c^2(c^4 - 6) = c^6 - 6c^2 \checkmark$

c)   The GCF of all of the terms is $4x^3y$.

$4x^5y^3 + 12x^5y^2 - 28x^4y^2 - 4x^3y$
$$= (4x^3y)(x^2y^2) + (4x^3y)(3x^2y) - (4x^3y)(7xy) - (4x^3y)(1) \quad \text{Rewrite each term using the GCF as one of the factors.}$$

$$= 4x^3y(x^2y^2 + 3x^2y - 7xy - 1) \qquad \text{Distributive property}$$

Check: $4x^3y(x^2y^2 + 3x^2y - 7xy - 1) = 4x^5y^3 + 12x^5y^2 - 28x^4y^2 - 4x^3y \checkmark \quad ■$

You Try 3

Factor out the greatest common factor.

a) $56k^4 - 24k^3 + 40k^2$ b) $3a^4b^4 - 12a^3b^4 + 18a^2b^4 - 3a^2b^3$

Sometimes we need to take out a negative factor.

Example 5

In-Class Example 5
Factor out $-4f$ from
$-28f^3 + 12f^2 - 16f$.
answer: $-4f(7f^2 - 3f + 4)$

Factor out $-5d$ from $-10d^4 + 45d^3 - 15d^2 + 5d$.

Solution

$$-10d^4 + 45d^3 - 15d^2 + 5d$$
$$= (-5d)(2d^3) + (-5d)(-9d^2) + (-5d)(3d) + (-5d)(-1)$$ Rewrite each term using $-5d$ as one of the factors.

$$= -5d[2d^3 + (-9d^2) + 3d + (-1)]$$ Distributive property
$$= -5d(2d^3 - 9d^2 + 3d - 1)$$ Rewrite $+(-9d^2)$ as $-9d^2$ and $+(-1)$ as -1.

Check: $-5d(2d^3 - 9d^2 + 3d - 1) = -10d^4 + 45d^3 - 15d^2 + 5d$ ✓   ■

BE CAREFUL   When taking out a negative factor, be very careful with the signs!

You Try 4

Factor out $-p^2$ from $-p^5 - 7p^4 + 3p^3 + 11p^2$.

4. Factor Out the Greatest Common Binomial Factor

Until now, all of the GCFs have been monomials. Sometimes, however, the greatest common factor is a *binomial*.

Example 6

In-Class Example 6
Factor out the GCF.
a) $m(n - 6) + 4(n - 6)$
b) $r^2(r + 11) + 9(r + 11)$
c) $u(v - 1) - (v - 1)$
answer: a) $(n - 6)(m + 4)$
b) $(r + 11)(r^2 + 9)$
c) $(v - 1)(u - 1)$

Factor out the greatest common factor.

a) $x(y + 2) + 9(y + 2)$ b) $n^2(n + 6) - 3(n + 6)$ c) $r(s + 4) - (s + 4)$

Solution

a) In the polynomial $\underset{\text{Term}}{\underline{x(y + 2)}} + \underset{\text{Term}}{\underline{9(y + 2)}}$, $x(y + 2)$ is a term and $9(y + 2)$ is a term. What do these terms have in common? $y + 2$

The GCF of $x(y + 2)$ and $9(y + 2)$ is $(y + 2)$. Use the distributive property to factor out $y + 2$.

$$x(y + 2) + 9(y + 2) = (y + 2)(x + 9)$$ Distributive property

Check: $(y + 2)(x + 9) = (y + 2)x + (y + 2)9$ Distribute.

The result $(y + 2)x + (y + 2)9$ is the same as $x(y + 2) + 9(y + 2)$ because multiplication is commutative. ✓

b)   The GCF of the terms in $n^2(n + 6) - 3(n + 6)$ is $n + 6$.

$$n^2(n + 6) - 3(n + 6) = (n + 6)(n^2 - 3)$$ Distributive property

Check: $(n + 6)(n^2 - 3) = (n + 6)n^2 + (n + 6)(-3)$ Distributive property
$$= n^2(n + 6) - 3(n + 6) ✓$$ Commutative property

c)   Let's begin by rewriting $r(s + 4) - (s + 4)$ as $\underbrace{r(s + 4)}_{\text{Term}} - \underbrace{1(s + 4)}_{\text{Term}}$.

The GCF is $s + 4$.

$$r(s + 4) - 1(s + 4) = (s + 4)(r - 1)$$ Distributive property

The check is left to the student.

BE CAREFUL

It is important to write -1 in front of $(s + 4)$. Otherwise, the following mistake is often made:

$$r(s + 4) - (s + 4) = (s + 4)r   \text{THIS IS INCORRECT!}$$

The correct factor is $r - 1$ *not* r.

You Try 5

Factor out the GCF.

a) $t(u - 8) + 5(u - 8)$ b) $z(z^2 + 2) - 6(z^2 + 2)$ c) $2n(m + 7) - (m + 7)$

Taking out a binomial factor leads us to our next method of factoring—factoring by grouping.

5. Factor by Grouping

When we are asked to factor a polynomial containing four terms, we often try to **factor by grouping.**

Example 7

In-Class Example 7
Factor by grouping.
a) $cd - 6c + 4d - 24$
b) $2xz - 8x + 3yz - 12y$
answer: a) $(d - 6)(c + 4)$
b) $(z - 4)(2x + 3y)$

Factor by grouping.

a) $ab + 5a + 3b + 15$ b) $2pr - 5qr + 6p - 15q$ c) $x^3 + 6x^2 - 7x - 42$

Solution

a)   Begin by grouping terms together so that each group has a common factor.

$$\underbrace{ab + 5a}\quad + \quad\underbrace{3b + 15}$$

Factor out a to get $a(b + 5)$. $= a(b + 5) + 3(b + 5)$ Factor out 3 to get $3(b + 5)$.
$$= (b + 5)(a + 3)$$ Factor out $(b + 5)$.

Check: $(b + 5)(a + 3) = ab + 5a + 3b + 15$ ✓

b)   Group terms together so that each group has a common factor.

$$\underbrace{2pr - 5qr}\quad + \quad\underbrace{6p - 15q}$$

Factor out r to get $r(2p - 5q)$. $= r(2p - 5q) + 3(2p - 5q)$ Factor out 3 to get $3(2p - 5q)$.
$$= (2p - 5q)(r + 3)$$ Factor out $(2p - 5q)$.

Check: $(2p - 5q)(r + 3) = 2pr - 5qr + 6p - 15q$ ✓

c)   Group terms together so that each group has a common factor.

$$\underbrace{x^3 + 6x^2}\;\;\underbrace{-\;7x - 42}$$

Factor out x^2 to get $x^2(x + 6)$. $= x^2(x + 6) - 7(x + 6)$ Factor out -7 to get $-7(x + 6)$.
$= (x + 6)(x^2 - 7)$ Factor out $(x + 6)$.

We **must** factor out -7 *not* 7 from the second group so that the binomial factors for both groups are the same! [If we had factored out 7, then the factorization of the second group would have been $7(-x - 6)$.]

Check: $(x + 6)(x^2 - 7) = x^3 + 6x^2 - 7x - 42$ ✓   ■

You Try 6

Factor by grouping.

a) $2cd + 4d + 5c + 10$ b) $4k^2 - 36k + km - 9m$ c) $h^3 + 8h^2 - 5h - 40$

Sometimes we have to rearrange the terms before we can factor.

Example 8

In-Class Example 8
Factor completely.
$36r^2 - 2t + 3r - 24rt$
answer: $(12r + 1)(3r - 2t)$

Factor completely. $24y^2 - 3z + 4y - 18yz$

Solution

Group terms together so that each group has a common factor.

$$\underbrace{24y^2 - 3z}\;\;+\;\underbrace{4y - 18yz}$$

Factor out 3 to get $3(8y^2 - z)$. $= 3(8y^2 - z) + 2y(2 - 9z)$ Factor out $2y$ to get $2y(2 - 9z)$.

The groups do not have common factors! Let's rearrange the terms in the original polynomial and group the terms differently.

$$\underbrace{24y^2 + 4y}\;\;-\;\underbrace{18yz - 3z}$$

Factor out $4y$ to get $4y(6y + 1)$. $= 4y(6y + 1) - 3z(6y + 1)$ Factor out $-3z$ to get $-3z(6y + 1)$.
$= (6y + 1)(4y - 3z)$ Factor out $(6y + 1)$.

Check: $(6y + 1)(4y - 3z) = 24y^2 - 3z + 4y - 18yz$ ✓   ■

Note

Often, there is more than one way that the terms can be rearranged so that the polynomial can be factored by grouping.

You Try 7

Factor completely. $6a^2 - 5b + 15a - 2ab$

Often, we have to combine the two factoring techniques we have learned here. That is, we begin by factoring out the GCF and then we factor by grouping. Let's summarize how to factor a polynomial by grouping and then look at another example.

> **Procedure** Steps for Factoring by Grouping
>
> 1) Before trying to factor by grouping, look at each term in the polynomial and ask yourself, *"Can I factor out a GCF first?"* If so, factor out the GCF from all of the terms.
>
> 2) Make two groups of two terms so that each group has a common factor.
>
> 3) Take out the common factor in each group of terms.
>
> 4) Factor out the common binomial factor using the distributive property.
>
> 5) Check the answer by multiplying the factors.

Example 9

In-Class Example 9
Factor completely.
$2h^4 - 12h^3 + 2h^2 - 12h$
answer: $2h(h - 6)(h^2 + 1)$

Factor completely. $4y^4 + 4y^3 - 20y^2 - 20y$

Solution

Notice that this polynomial has four terms. This is a clue for us to try factoring by grouping. *However,* look at the polynomial carefully and ask yourself, *"Can I factor out a GCF?"* Yes! *Therefore, the first step in factoring this polynomial is to factor out $4y$.*

$$4y^4 + 4y^3 - 20y^2 - 20y = 4y(y^3 + y^2 - 5y - 5) \qquad \text{Factor out the GCF, } 4y.$$

The polynomial in parentheses has 4 terms. Try to factor it by grouping.

$$4y(\underbrace{y^3 + y^2}\ \underbrace{-\ 5y - 5})$$

$$= 4y[y^2(y + 1) - 5(y + 1)] \qquad \text{Take out the common factor in each group.}$$
$$= 4y(y + 1)(y^2 - 5) \qquad \text{Factor out } (y + 1) \text{ using the distributive property.}$$

Check: $4y(y + 1)(y^2 - 5) = 4y(y^3 + y^2 - 5y - 5)$
$$= 4y^4 + 4y^3 - 20y^2 - 20y \ \checkmark$$

You Try 8

Factor completely. $4ab + 14b + 8a + 28$

Remember, seeing a polynomial with four terms is a clue to try factoring by grouping. Not all polynomials will factor this way, however. We will learn other techniques later, and some polynomials must be factored using methods learned in later courses.

Answers to You Try Exercises

1) y^4 2) a) $8p^5$ b) r^4s^3 3) a) $8k^2(7k^2 - 3k + 5)$ b) $3a^2b^3(a^2b - 4ab + 6b - 1)$
4) $-p^2(p^3 + 7p^2 - 3p - 11)$ 5) a) $(u - 8)(t + 5)$ b) $(z^2 + 2)(z - 6)$ c) $(m + 7)(2n - 1)$
6) a) $(c + 2)(2d + 5)$ b) $(k - 9)(4k + m)$ c) $(h + 8)(h^2 - 5)$
7) $(2a + 5)(3a - b)$ 8) $2(b + 2)(2a + 7)$

6.1 Exercises

*Additional answers can be found in the Answers to Exercises appendix.

Objective 1: Find the GCF of a Group of Monomials

Find the greatest common factor of each group of terms.

1) $45m^3, 20m^2$ $5m^2$

2) $18d^6, 21d^2$ $3d^2$

3) $42k^5, 54k^7, 72k^9$ $6k^5$

4) $25t^8, 55t, 30t^3$ $5t$

5) $27x^4y, 45x^2y^3$ $9x^2y$

6) $24r^3s^6, 56r^2s^5$ $8r^2s^5$

7) $28u^2v^5, 20uv^3, -8uv^4$ $4uv^3$

8) $-6a^4b^3, 18a^2b^6, 12a^2b^4$ $6a^2b^3$

9) $21s^2t, 35s^2t^2, s^4t^2$ s^2t

10) $p^4q^4, -p^3q^4, -p^3q$ p^3q

11) $a(n - 7), 4(n - 7)$ $(n - 7)$

12) $x^2(y + 9), z^2(y + 9)$ $(y + 9)$

13) Explain how to find the GCF of a group of terms.   Answers may vary.

14) What does it mean to factor a polynomial?   to write it as a product of two or more polynomials

Mixed Exercises: Objectives 2–4

Factor out the greatest common factor. Be sure to check your answer.

15) $30s + 18$ $6(5s + 3)$ 16) $14a + 24$ $2(7a + 12)$

17) $24z - 4$ $4(6z - 1)$ 18) $63f^2 - 49$ $7(9f^2 - 7)$

19) $3d^2 - 6d$ $3d(d - 2)$ 20) $20m - 5m^2$ $5m(4 - m)$

21) $42y^2 + 35y^3$ $7y^2(6 + 5y)$ 22) $30b^3 - 5b$ $5b(6b^2 - 1)$

23) $t^5 - t^4$ $t^4(t - 1)$ 24) $r^9 + r^2$ $r^2(r^7 + 1)$

25) $\dfrac{1}{2}c^2 + \dfrac{5}{2}c$ $\frac{1}{2}c(c + 5)$ 26) $\dfrac{1}{8}k^2 + \dfrac{7}{8}k$ $\frac{1}{8}k(k + 7)$

27) $10n^5 - 5n^4 + 40n^3$ 28) $18x^7 + 42x^6 - 30x^5$
$5n^3(2n^2 - n + 8)$ $6x^5(3x^2 + 7x - 5)$

29) $2v^8 - 18v^7 - 24v^6 + 2v^5$ $2v^5(v^3 - 9v^2 - 12v + 1)$

30) $12z^6 + 30z^5 - 15z^4 + 3z^3$ $3z^3(4z^3 + 10z^2 - 5z + 1)$

31) $8c^3 + 3d^2$ does not factor 32) $m^5 - 5n^2$ does not factor

33) $a^4b^2 + 4a^3b^3$ 34) $20r^3s^3 - 14rs^4$
$a^3b^2(a + 4b)$ $2rs^3(10r^2 - 7s)$

35) $50x^3y^3 - 70x^3y^2 + 40x^2y$ $10x^2y(5xy^2 - 7xy + 4)$

36) $21b^4d^3 + 15b^3d^3 - 27b^2d^2$ $3b^2d^2(7b^2d + 5bd - 9)$

37) $m(n - 12) + 8(n - 12)$ $(n - 12)(m + 8)$

38) $x(y + 5) + 3(y + 5)$ $(y + 5)(x + 3)$

39) $p(8r - 3) - q(8r - 3)$ $(8r - 3)(p - q)$

40) $a(9c + 4) - b(9c + 4)$ $(9c + 4)(a - b)$

41) $y(z + 11) + (z + 11)$ $(z + 11)(y + 1)$

42) $2u(v - 7) + (v - 7)$ $(v - 7)(2u + 1)$

43) $2k^2(3r + 4) - (3r + 4)$ $(3r + 4)(2k^2 - 1)$

44) $8p(3q + 5) - (3q + 5)$ $(3q + 5)(8p - 1)$

45) Factor out -8 from $-64m - 40$. $-8(8m + 5)$

46) Factor out -7 from $-14k + 21$. $-7(2k - 3)$

47) Factor out $-5t^2$ from $-5t^3 + 10t^2$. $-5t^2(t - 2)$

48) Factor out $-4v^3$ from $-4v^5 - 36v^3$. $-4v^3(v^2 + 9)$

49) Factor out $-a$ from $-3a^3 + 7a^2 - a$. $-a(3a^2 - 7a + 1)$

50) Factor out $-q$ from $-10q^3 - 4q^2 + q$. $-q(10q^2 + 4q - 1)$

51) Factor out -1 from $-b + 8$. $-1(b - 8)$

52) Factor out -1 from $-z - 6$. $-1(z + 6)$

Objective 5: Factor by Grouping

Factor by grouping.

53) $kt + 3k + 8t + 24$ 54) $uv + 5u + 10v + 50$
$(t + 3)(k + 8)$ $(v + 5)(u + 10)$

55) $fg - 7f + 4g - 28$ 56) $cd - 5d + 8c - 40$
$(g - 7)(f + 4)$ $(c - 5)(d + 8)$

57) $2rs - 6r + 5s - 15$ 58) $3jk - 7k + 6j - 14$
$(s - 3)(2r + 5)$ $(3j - 7)(k + 2)$

59) $3xy - 2y + 27x - 18$ 60) $4ab + 32a + 3b + 24$
$(3x - 2)(y + 9)$ $(b + 8)(4a + 3)$

61) $8b^2 + 20bc + 2bc^2 + 5c^3$ $(2b + 5c)(4b + c^2)$

62) $8u^2 - 16uv^2 + 3uv - 6v^3$ $(u - 2v^2)(8u + 3v)$

63) $4a^3 - 12ab + a^2b - 3b^2$ $(a^2 - 3b)(4a + b)$

64) $5x^3 - 30x^2y^2 + xy - 6y^3$ $(x - 6y^2)(5x^2 + y)$

65) $kt + 7t - 5k - 35$ 66) $pq - 2q - 9p + 18$
$(k + 7)(t - 5)$ $(p - 2)(q - 9)$

67) $mn - 8m - 10n + 80$ 68) $hk + 6k - 4h - 24$
$(n - 8)(m - 10)$ $(h + 6)(k - 4)$

69) $dg - d + g - 1$ 70) $qr + 3q - r - 3$
$(g - 1)(d + 1)$ $(r + 3)(q - 1)$

71) $5tu + 6t - 5u - 6$ 72) $4yz + 7z - 20y - 35$
$(5u + 6)(t - 1)$ $(4y + 7)(z - 5)$

73) $36g^4 + 3gh - 96g^3h - 8h^2$ $(12g^3 + h)(3g - 8h)$

74) $40j^3 + 72jk - 55j^2k - 99k^2$ $(5j^2 + 9k)(8j - 11k)$

75) Explain, in your own words, how to factor by grouping.
Answers may vary.

76) What should be the first step in factoring
$3xy + 6x + 15y + 30$?   Factor out 3 from all of the terms.

Factor completely. You may need to begin by taking out the GCF first or by rearranging terms.

Fill It In

Fill in the blanks with either the missing mathematical step or reason for the given step.

77) $5mn + 15m + 10n + 30$

$5mn + 15m + 10n + 30$	
$= 5(mn + 3m + 2n + 6)$	Factor out the GCF.
$= 5[m(n + 3) + 2(n + 3)]$	Group the terms and factor out the GCF from each group.
$= 5(n + 3)(m + 2)$	Take out the binomial factor.

78) $3x^2y - 21x^2 - 6xy + 42x$

$3x^2y - 21x^2 - 6xy + 42x$	
$= 3x(xy - 7x - 2y + 14)$	Factor out the GCF.
$= 3x[x(y - 7) - 2(y - 7)]$	Group the terms and factor out the GCF from each group.
$= 3x(y - 7)(x - 2)$	Take out the binomial factor.

79) $2ab + 8a + 6b + 24$ $2(b + 4)(a + 3)$

80) $7pq + 28q + 14p + 56$ $7(p + 4)(q + 2)$

81) $8s^2t - 40st + 16s^2 - 80s$ $8s(s - 5)(t + 2)$

82) $10hk^3 - 5hk^2 + 30k^3 - 15k^2$ $5k^2(2k - 1)(h + 3)$

83) $7cd + 12 + 28c + 3d$ $(d + 4)(7c + 3)$

84) $9rs + 12 + 2s + 54r$ $(s + 6)(9r + 2)$

85) $42k^3 + 15d^2 - 18k^2d - 35kd$ $(7k - 3d)(6k^2 - 5d)$

86) $12x^3 + 2y^2 - 3x^2y - 8xy$ $(4x - y)(3x^2 - 2y)$

87) $9f^2j^2 + 45fj + 9fj^2 + 45f^2j$ $9fj(f + 1)(j + 5)$

88) $n^3m - 4n^2 + mn^2 - 4n^3$ $n^2(n + 1)(m - 4)$

89) $4x^4y - 14x^3 + 28x^4 - 2x^3y$ $2x^3(2x - 1)(y + 7)$

90) $12a^2c^2 - 20ac - 4ac^2 + 60a^2c$ $4ac(3a - 1)(c + 5)$

94) $18rt^3 + 15t^3 - 18rt^2 - 15t^2$ $3t^2(6r + 5)(t - 1)$

95) $3h^3 - 8k^3 + 12h^2k^2 - 2hk$ $(3h^2 - 2k)(h + 4k^2)$

96) $6yz - 8z - 3y + 4$ $(3y - 4)(2z - 1)$

Mixed Exercises: Objectives 1–5

Factor completely.

97) $2c^4 + 14c^2 + 84c + 12c^3$ $2c(c^2 + 7)(c + 6)$

98) $18w^3 + 45w^2$ $9w^2(2w + 5)$

91) $pq - 8p + 3q - 24$ $(q - 8)(p + 3)$

92) $27d^3 + 36d^2 - 9d$ $9d(3d^2 + 4d - 1)$

93) $a^4b^2 + 2a^3b^3 - 8a^2b^4$ $a^2b^2(a - 2b)(a + 4b)$

99) Factor out $-8v$ from $-16v^3 - 56v^2 + 8v$. $-8v(2v^2 + 7v - 1)$

100) Factor out $-3n^2$ from $-3n^4 + 27n^3 + 33n^2$. $-3n^2(n^2 - 9n - 11)$

Section 6.2 Factoring Trinomials

Objectives

1. Factor a Trinomial of the Form $x^2 + bx + c$
2. More on Factoring a Trinomial of the Form $x^2 + bx + c$
3. Factor a Trinomial Containing Two Variables
4. Factor $ax^2 + bx + c$ $(a \neq 1)$ by Grouping
5. Factor $ax^2 + bx + c$ $(a \neq 1)$ by Trial and Error
6. Factor Using Substitution

One of the factoring problems encountered most often in algebra is the factoring of trinomials. In this section, we will discuss how to factor trinomials like $x^2 + 11x + 18$, $3n^2 - n - 4$, $4a^2 - 11ab + 6b^2$, and many more. Let's begin with trinomials of the form $x^2 + bx + c$, where the coefficient of the squared term is 1.

1. Factor a Trinomial of the Form $x^2 + bx + c$

In Section 6.1 we said that the process of factoring is the opposite of multiplying. Let's see how this will help us understand how to factor a trinomial of the form $x^2 + bx + c$.

Multiply $(x + 4)(x + 7)$ using FOIL.

$$(x + 4)(x + 7) = x^2 + 7x + 4x + 4 \cdot 7 \qquad \text{Multiply using FOIL.}$$
$$= x^2 + (7 + 4)x + 28 \qquad \text{Use the distributive property and multiply } 4 \cdot 7.$$
$$= x^2 + 11x + 28$$

$$(x + 4)(x + 7) = x^2 + 11x + 28$$

11 is the *sum* of 4 and 7.   28 is the *product* of 4 and 7.

So, if we were asked to *factor* $x^2 + 11x + 28$, we need to think of two integers whose *product* is 28 and whose *sum* is 11. Those numbers are 4 and 7. The *factored form* of $x^2 + 11x + 28$ is $(x + 4)(x + 7)$.

> **Procedure** Factoring a Polynomial of the Form $x^2 + bx + c$
>
> To factor a polynomial of the form $x^2 + bx + c$, find two integers m and n whose product is c and whose sum is b. Then,
>
> $$x^2 + bx + c = (x + m)(x + n).$$
>
> 1) If b and c are positive, then both m and n must be positive.
> 2) If c is positive and b is negative, then both m and n must be negative.
> 3) If c is negative, then one integer, m, must be positive and the other integer, n, must be negative.
>
> You can check the answer by multiplying the binomials. The result should be the original polynomial.

Example 1

Factor, if possible.

a) $y^2 - 8y + 15$ b) $k^2 + k - 56$ c) $r^2 + 7r + 9$

In-Class Example 1

Factor, if possible.
a) $d^2 - 15d + 44$
b) $m^2 + 3m - 40$
c) $c^2 + 9c + 12$
answer: a) $(d - 4)(d - 11)$
b) $(m - 5)(m + 8)$ **c) prime**

Solution

a) To factor $y^2 - 8y + 15$, find the two integers whose *product* is 15 and whose *sum* is -8. Since 15 is positive and the coefficient of y is a negative number, -8, both integers will be negative.

Factors of 15	Sum of the Factors
$-1 \cdot (-15) = 15$	$-1 + (-15) = -16$
$-3 \cdot (-5) = 15$	$-3 + (-5) = -8$

The numbers are -3 and -5: $y^2 - 8y + 15 = (y - 3)(y - 5)$.

Check: $(y - 3)(y - 5) = y^2 - 5y - 3y + 15 = y^2 - 8y + 15$ ✓

b) $k^2 + k - 56$

The coefficient of k is 1, so we can think of this trinomial as $k^2 + 1k - 56$.

Find two integers whose *product* is -56 and whose *sum* is 1. Since the last term in the trinomial is negative, one of the integers must be positive and the other must be negative.

Try to find these integers mentally. Two numbers with a product of *positive* 56 are 7 and 8. We need a product of -56, so either the 7 is negative or the 8 is negative.

Factors of -56	Sum of the Factors
$-7 \cdot 8 = -56$	$-7 + 8 = 1$

The numbers are -7 and 8: $k^2 + k - 56 = (k - 7)(k + 8)$.

Check: $(k - 7)(k + 8) = k^2 + 8k - 7k - 56 = k^2 + k - 56$ ✓

c) To factor $r^2 + 7r + 9$, find the two integers whose *product* is 9 and whose *sum* is 7. We are looking for two positive numbers.

Factors of 9	Sum of the Factors
$1 \cdot 9 = 9$	$1 + 9 = 10$
$3 \cdot 3 = 9$	$3 + 3 = 6$

There are no such factors! Therefore, $r^2 + 7r + 9$ does not factor using the methods we have learned here. We say that it is **prime.** ■

Note

We say that trinomials like $r^2 + 7r + 9$ are **prime** if they cannot be factored using the method presented here.

In later mathematics courses, you may learn how to factor such polynomials using other methods so that they are not considered prime.

You Try 1

Factor, if possible.

a) $s^2 + 5s - 66$ b) $d^2 - 6d - 10$ c) $x^2 - 12x + 27$

2. More on Factoring a Trinomial of the Form $x^2 + bx + c$

Sometimes it is necessary to factor out the GCF before applying this method for factoring trinomials.

> **Note**
>
> From this point on, the *first step* in factoring *any* polynomial should be to ask yourself, *"Can I factor out a greatest common factor?"*
>
> And since some polynomials can be factored more than once, after performing one factorization, ask yourself, *"Can I factor again?"* If so, factor again. If not, you know that the polynomial has been completely factored.

Example 2

Factor $5y^3 - 15y^2 - 20y$ completely.

In-Class Example 2
Factor completely.
$8k^2 + 32k - 96$
answer: $8(k - 2)(k + 6)$

Solution

Ask yourself, *"Can I factor out a GCF?"* Yes. The GCF is $5y$.

$$5y^3 - 15y^2 - 20y = 5y(y^2 - 3y - 4)$$

Look at the trinomial and ask yourself, *"Can I factor again?"* Yes. The integers whose product is -4 and whose sum is -3 are -4 and 1. Therefore,

$$5y^3 - 15y^2 - 20y = 5y(y^2 - 3y - 4)$$
$$= 5y(y - 4)(y + 1)$$

We cannot factor again.

$$Check:\ 5y(y - 4)(y + 1) = 5y(y^2 + y - 4y - 4)$$
$$= 5y(y^2 - 3y - 4)$$
$$= 5y^3 - 15y^2 - 20y\ ✓$$

You Try 2

Factor $6g^4 + 42g^3 + 60g^2$ completely.

3. Factor a Trinomial Containing Two Variables

If a trinomial contains two variables and we cannot take out a GCF, the trinomial may still be factored according to the method outlined in this section.

Example 3

Factor $a^2 + 9ab + 18b^2$ completely.

In-Class Example 3
Factor completely.
$y^2 - 14yz + 24z^2$
answer: $(y - 2z)(y - 12z)$

Solution

Ask yourself, *"Can I factor out a GCF?"* No. Notice that the first term is a^2. Let's rewrite the trinomial as

$$a^2 + 9ba + 18b^2$$

so that we can think of $9b$ as the coefficient of a. Find two expressions whose product is $18b^2$ and whose sum is $9b$. They are $3b$ and $6b$ since $3b \cdot 6b = 18b^2$ and $3b + 6b = 9b$.

$$a^2 + 9ab + 18b^2 = (a + 3b)(a + 6b)$$

We cannot factor $(a + 3b)(a + 6b)$ any more, so this is the complete factorization. The check is left to the student.

You Try 3

Factor completely.

a) $x^2 + 15xy + 54y^2$ b) $3k^3 + 18ck^2 - 21c^2k$

If we are asked to factor each of these polynomials, $3x^2 + 18x + 24$ and $3x^2 + 10x + 8$, how do we begin? How do the polynomials differ?

The GCF of $3x^2 + 18x + 24$ is 3. To factor, begin by taking out the 3. We can factor using what we learned earlier.

$$3x^2 + 18x + 24 = 3(x^2 + 6x + 8) \qquad \text{Factor out 3.}$$
$$= 3(x + 4)(x + 2) \qquad \text{Factor the trinomial.}$$

In the second polynomial, $3x^2 + 10x + 8$, we *cannot* factor out the leading coefficient of 3. Next, we will discuss two methods for factoring a trinomial like $3x^2 + 10x + 8$ where we *cannot* factor out the leading coefficient.

4. Factor $ax^2 + bx + c$ ($a \neq 1$) by Grouping

Sum is 10
↓

To factor $3x^2 + 10x + 8$, first find the product of 3 and 8. Then, find two integers

Product: $3 \cdot 8 = 24$

whose *product* is 24 and whose *sum* is 10. The numbers are 6 and 4. Rewrite the middle term, $10x$, as $6x + 4x$, then factor by grouping.

$$3x^2 + 10x + 8 = 3x^2 + 6x + 4x + 8$$
$$= 3x(x + 2) + 4(x + 2) \qquad \text{Take out the common factor from each group.}$$
$$= (x + 2)(3x + 4) \qquad \text{Factor out } (x + 2).$$

$$3x^2 + 10x + 8 = (x + 2)(3x + 4)$$

Check: $(x + 2)(3x + 4) = 3x^2 + 4x + 6x + 8 = 3x^2 + 10x + 8$ ✓

Example 4

Factor completely.

a) $10p^2 - 13p + 4$ b) $5c^2 - 29cd - 6d^2$

In-Class Example 4
Factor completely.
a) $8x^2 - 18x + 9$
b) $18a^2 + 7ab - b^2$
answer: a) $(4x - 3)(2x - 3)$
b) $(9a - b)(2a + b)$

Solution

a)   *Sum* is -13
        ↓
$10p^2 - 13p + 4$

Product: $10 \cdot 4 = 40$

Think of two integers whose *product* is 40 and whose *sum* is -13. (Both numbers will be negative.): -8 *and* -5

Rewrite the middle term, $-13p$, as $-8p - 5p$. Factor by grouping.

$$10p^2 - 13p + 4 = 10p^2 - 8p - 5p + 4$$
$$= 2p(5p - 4) - 1(5p - 4) \qquad \text{Take out the common factor from each group.}$$
$$= (5p - 4)(2p - 1) \qquad \text{Factor out } (5p - 4).$$

Check: $(5p - 4)(2p - 1) = 10p^2 - 13p + 4$ ✓

b)   *Sum is* -29
       ↓
$5c^2 - 29cd - 6d^2$

Product: $5 \cdot (-6) = -30$

The integers whose *product* is -30 and whose *sum* is -29 are -30 *and* 1.

Rewrite the middle term, $-29cd$, as $-30cd + cd$. Factor by grouping.

$$5c^2 - 29cd - 6d^2 = 5c^2 - 30cd + cd - 6d^2$$

$$= 5c(c - 6d) + d(c - 6d) \qquad \text{Take out the common factor from each group.}$$

$$= (c - 6d)(5c + d) \qquad \text{Factor out } (c - 6d).$$

Check: $(c - 6d)(5c + d) = 5c^2 - 29cd - 6d^2$ ✓ ∎

You Try 4

Factor completely.

   a) $2c^2 + 11c + 14$ b) $6n^2 - 23n - 4$ c) $8x^2 - 10xy + 3y^2$

Example 5

Factor $24w^2 - 54w - 15$ completely.

In-Class Example 5
Factor completely.
$8c^2 + 44c + 20$
answer: $4(2c + 1)(c + 5)$

Solution

It is tempting to jump right in and multiply $24 \cdot (-15) = -360$ and try to think of two integers with a product of -360 and a sum of -54. Before doing that, ask yourself, *"Can I factor out a GCF?"* Yes! We can factor out 3.

Sum is -18
↓
$$24w^2 - 54w - 15 = 3(8w^2 - 18w - 5) \qquad \text{Factor out 3.}$$
Product: $8 \cdot (-5) = -40$

Try to factor $8w^2 - 18w - 5$ by finding two integers whose *product* is -40 and whose *sum* is -18. The numbers are -20 and 2.

$$= 3(8w^2 - 20w + 2w - 5)$$

$$= 3[4w(2w - 5) + 1(2w - 5)] \qquad \text{Take out the common factor from each group.}$$

$$= 3(2w - 5)(4w + 1) \qquad \text{Factor out } 2w - 5.$$

Check by multiplying: $3(2w - 5)(4w + 1) = 3(8w^2 - 18w - 5)$

$$= 24w^2 - 54w - 15 \text{ ✓} \quad ∎$$

You Try 5

Factor completely.

   a) $20m^3 - 8m^2 + 4m$ b) $6z^2 + 20z + 16$

---

5. Factor $ax^2 + bx + c$ ($a \neq 1$) by Trial and Error

Earlier, we factored $3x^2 + 10x + 8$ by grouping. Now we will factor it by trial and error, which is just reversing the process of FOIL.

Example 6

Factor $3x^2 + 10x + 8$ completely.

In-Class Example 6
Factor completely.
$7w^2 + 32w + 16$
answer: $(7w + 4)(w + 4)$

Solution

Can we factor out a GCF? No. So try to factor $3x^2 + 10x + 8$ as the product of two binomials. Notice that all terms are positive, so all factors will be positive.

Begin with the squared term, $3x^2$. Which two expressions with integer coefficients can we multiply to get $3x^2$? $3x$ and x. Put these in the binomials.

$$3x^2 + 10x + 8 = (3x \quad)(x \quad) \qquad 3x \cdot x = 3x^2$$

Next, look at the last term, 8. What are the pairs of positive integers that multiply to 8? They are 8 and 1 as well as 4 and 2.

Try these numbers as the last terms of the binomials. The middle term, $10x$, comes from finding the sum of the products of the outer terms and inner terms.

First Try

$$3x^2 + 10x + 8 \overset{?}{=} (3x + 8)(x + 1) \qquad \text{Incorrect!}$$

These must both be $10x$. $8x$ $+$ $3x$ → $11x$

Switch the 8 and 1. $3x^2 + 10x + 8 \overset{?}{=} (3x + 1)(x + 8) \qquad \text{Incorrect!}$

These must both be $10x$. $1x$ $+$ $24x$ → $25x$

Try using 4 and 2. $3x^2 + 10x + 8 \overset{?}{=} (3x + 4)(x + 2) \qquad \text{Correct!}$

These must both be $10x$. $4x$ $+$ $6x$ → $10x$

The factorization of $3x^2 + 10x + 8$ is $(3x + 4)(x + 2)$. Check by multiplying. ■

Example 7

Factor $2r^2 - 13r + 20$ completely.

In-Class Example 7
Factor completely.
$5a^2 + 16a - 45$
answer: $(5a - 9)(a + 5)$

Solution

Can we factor out a GCF? No. To get a product of $2r^2$, we will use $2r$ and r.

$$2r^2 - 13r + 20 = (2r \quad)(r \quad) \qquad 2r \cdot r = 2r^2$$

Since the last term is positive and the middle term is negative, we want pairs of negative integers that multiply to 20. The pairs are -1 and -20, -2 and -10, and -4 and -5. Try these numbers as the last terms of the binomials. The middle term, $-13r$, comes from finding the sum of the products of the outer terms and inner terms.

$$2r^2 - 13r + 20 \overset{?}{=} (2r - 1)(r - 20) \qquad \text{Incorrect!}$$

These must both be $-13r$. $-r$ $+$ $(-40r)$ → $-41r$

Switch the -1 and -20: $2r^2 - 13r + 20 \overset{?}{=} (2r - 20)(r - 1)$

Without multiplying we know that this choice is incorrect. How? In the factor $(2r - 20)$, a 2 can be factored out to get $2(r - 10)$. But, we said that we could not factor out a GCF from the original polynomial, $2r^2 - 13r + 20$. Therefore, it will not be possible to take out a common factor from one of the binomial factors.

Note

If you cannot factor out a GCF from the original polynomial, then you cannot take out a factor from one of the binomial factors either.

For the same reason, the pair -2 and -10 will not give us the correct factorization:

$$2r^2 - 13r + 20 \neq (2r - 2)(r - 10) \qquad \text{2 can be factored out of } (2r - 2).$$
$$2r^2 - 13r + 20 \neq (2r - 10)(r - 2) \qquad \text{2 can be factored out of } (2r - 10).$$

Try using -4 and -5: $2r^2 - 13r + 20 \neq (2r - 4)(r - 5)$ 2 can be factored out of $(2r - 4)$.

Switch the -4 and -5: $2r^2 - 13r + 20 = (2r - 5)(r - 4)$ Correct!

$$-5r$$
These must both be $-13r$. $+ (-8r)$
$\longrightarrow -13r$

The factorization of $2r^2 - 13r + 20$ is $(2r - 5)(r - 4)$. Check by multiplying. ■

You Try 6

Factor completely.

a) $2m^2 + 11m + 12$ b) $3v^2 - 28v + 9$

Example 8

Factor completely.

a) $12d^2 + 46d - 8$ b) $-2h^2 + 9h + 56$

In-Class Example 8
Factor completely.
a) $15y^2 + 51y + 18$
b) $-6t^2 + 13t + 8$
answer: a) $3(5y + 2)(y + 3)$
b) $-(3t - 8)(2t + 1)$

Solution

a)   Ask yourself, *"Can I take out a common factor?"* Yes! The GCF is 2.

$$12d^2 + 46d - 8 = 2(6d^2 + 23d - 4)$$

Now, try to factor $6d^2 + 23d - 4$. To get a product of $6d^2$, we can try either $6d$ *and* d or $3d$ *and* $2d$. Let's start by trying $6d$ *and* d.

$$6d^2 + 23d - 4 = (6d \quad)(d \quad)$$

List pairs of integers that multiply to -4: 4 *and* -1, -4 *and* 1, 2 *and* -2.
   Try 4 and -1. Do not put 4 in the same binomial as $6d$ since then it would be possible to factor out 2. But, a 2 does not factor out of $6d^2 + 23d - 4$. Put the 4 in the same binomial as d.

$$6d^2 + 23d - 4 \overset{?}{=} (6d - 1)(d + 4)$$

$$-d$$
$$+ 24d$$
$$23d \qquad \text{Correct!}$$

Don't forget that the very first step was to factor out a 2. Therefore,

$$12d^2 + 46d - 8 = 2(6d^2 + 23d - 4) = 2(6d - 1)(d + 4).$$

Check by multiplying.

b) Since the coefficient of the squared term is negative, begin by factoring out -1. (There is no other common factor except 1.)

$$-2h^2 + 9h + 56 = -1(2h^2 - 9h - 56)$$

Try to factor $2h^2 - 9h - 56$. To get a product of $2h^2$ we will use $2h$ and h in the binomials.

$$2h^2 - 9h - 56 = (2h \quad)(h \quad)$$

We need pairs of integers so that their product is -56. The ones that come to mind quickly involve 7 and 8 and 1 and 56: -7 and 8, 7 and -8, -1 and 56, 1 and -56.

There are other pairs; if these do not work, we will list others.

Do *not* start with -1 and 56 or 1 and -56 because the middle term, $-9h$, is not very large. Using -1 and 56 or 1 and -56 would likely result in a larger middle term.

Try -7 and 8. Do not put 8 in the same binomial as $2h$ since then it would be possible to factor out 2.

$$2h^2 - 9h - 56 \overset{?}{=} (2h - 7)(h + 8)$$

$$-7h$$
$$+\ 16h$$
$$\overline{\quad 9h \quad}$$ This must equal $-9h$. Incorrect!

Only the sign of the sum is incorrect. *Change the signs in the binomials to get the correct sum.*

$$2h^2 - 9h - 56 \overset{?}{=} (2h + 7)(h - 8)$$

$$7h$$
$$+\ (-16h)$$
$$\overline{\quad -9h \quad}$$ Correct!

Remember that we factored out -1 to begin the problem.

$$-2h^2 + 9h + 56 = -1(2h^2 - 9h - 56) = -(2h + 7)(h - 8)$$

Check by multiplying. ∎

You Try 7

Factor completely.

a) $15b^2 - 55b + 30$ b) $-4p^2 + 3p + 10$

We have seen two methods for factoring $ax^2 + bx + c \ (a \neq 1)$: factoring by grouping and factoring by trial and error. In either case, remember to begin by taking out a common factor from all terms whenever possible.

6. Factor Using Substitution

Some polynomials can be factored using a method called **substitution.** We will illustrate this method in Example 9.

Example 9

Factor $3(x + 4)^2 + 11(x + 4) + 10$ completely.

In-Class Example 9
Factor
$5(x + 2)^2 + 17(x + 2) + 6$
completely.
answer: (5x + 12)(x + 5)

Solution

Notice that the binomial $x + 4$ appears as a *squared quantity* and a *linear quantity*. We will use another letter to represent this quantity. We can use any letter except x. Let's use the letter u.

Let $u = x + 4$. Then, $u^2 = (x + 4)^2$.

Substitute the u and u^2 into the original polynomial and factor.

$3(x + 4)^2 + 11(x + 4) + 10 =$
$3u^2 \qquad + 11u \qquad + 10 = (3u + 5)(u + 2)$
$\qquad\qquad\qquad\qquad = [3(x + 4) + 5][(x + 4) + 2]$ Since the original polynomial was in terms of x, substitute $x + 4$ for u.

$\qquad\qquad\qquad\qquad = (3x + 12 + 5)(x + 4 + 2)$ Distribute.
$\qquad\qquad\qquad\qquad = (3x + 17)(x + 6)$ Combine like terms.

 **BE CAREFUL** The final factorization is **not** the one containing the substitution variable, u. We must go back and replace u with the expression it represented so that the factorization is in terms of the original variable.

 You Try 8

Factor $2(x - 3)^2 + 3(x - 3) - 35$ completely.

 ## Using Technology

We found some ways to narrow down the possibilities when factoring $ax^2 + bx + c$ $(a \neq 1)$ using the trial and error method.

We can also use a graphing calculator to help with the process. Consider the trinomial $2x^2 - 9x - 35$. Enter the trinomial into Y_1 and press [ZOOM] and then enter 6 to display the graph in the standard viewing window.

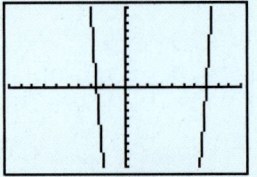

Look on the graph for the x-intercept (if any) that appears to be an integer. It appears that 7 is an x-intercept.

To check if 7 is an x-intercept, press [TRACE] then [7] and press [ENTER]. As shown on the graph, when $x = 7$, $y = 0$, so 7 is an x-intercept.

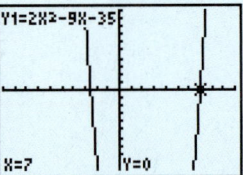

When an x-intercept is an integer, then x minus that x-intercept is a factor of the trinomial. In this case, $x - 7$ is a factor of $2x^2 - 9x - 35$. We can then complete the factoring as $(2x + 5)(x - 7)$, since we must multiply -7 by 5 to obtain -35.

Find an x-intercept using a graphing calculator and factor the trinomial.

1) $3x^2 + 13x - 10$ 2) $2x^2 + x - 36$ 3) $5x^2 + 16x + 3$
4) $2x^2 - 13x + 21$ 5) $3x^2 - 23x - 8$ 6) $7x^2 + 11x - 6$

Answers to You Try Exercises

1) a) $(s + 11)(s - 6)$ b) prime   c) $(x - 9)(x - 3)$ 2) $6g^2(g + 5)(g + 2)$
3) a) $(x + 6y)(x + 9y)$ b) $3k(k + 7c)(k - c)$
4) a) $(2c + 7)(c + 2)$ b) $(6n + 1)(n - 4)$ c) $(4x - 3y)(2x - y)$
5) a) $4m(5m^2 - 2m + 1)$ b) $2(3z + 4)(z + 2)$ 6) a) $(2m + 3)(m + 4)$ b) $(3v - 1)(v - 9)$
7) a) $5(3b - 2)(b - 3)$ b) $-(4p + 5)(p - 2)$ 8) $(2x - 13)(x + 2)$

Answers to Technology Exercises

1) $(3x - 2)(x + 5)$ 2) $(2x + 9)(x - 4)$ 3) $(x + 3)(5x + 1)$ 4) $(x - 3)(2x - 7)$
5) $(x - 8)(3x + 1)$ 6) $(x + 2)(7x - 3)$

6.2 Exercises

*Additional answers can be found in the Answers to Exercises appendix.

Objective 1: Factor a Trinomial of the Form $x^2 + bx + c$

1) If $x^2 + bx + c$ factors to $(x + m)(x + n)$ and if c is positive and b is negative, what do you know about the signs of m and n? They are negative.

2) If $x^2 + bx + c$ factors to $(x + m)(x + n)$ and if b and c are positive, what do you know about the signs of m and n? They are positive.

3) When asked to factor a polynomial, what is the first question you should ask yourself? Can I factor out a GCF?

4) What does it mean to say that a polynomial is prime? It does not factor.

5) After factoring a polynomial, what should you ask yourself to be sure that the polynomial is completely factored? Can I factor again?

6) How do you check the factorization of a polynomial? Multiply the factors. The product should be the original polynomial.

Factor completely, if possible. Check your answer.

7) $g^2 + 8g + 12$
$(g + 6)(g + 2)$

8) $j^2 + 9j + 20$
$(j + 4)(j + 5)$

9) $w^2 + 13w + 42$
$(w + 7)(w + 6)$

10) $t^2 + 15t + 36$
$(t + 12)(t + 3)$

11) $c^2 - 13c + 36$
$(c - 4)(c - 9)$

12) $v^2 - 11v + 24$
$(v - 3)(v - 8)$

13) $b^2 - 2b - 8$
$(b - 4)(b + 2)$

14) $s^2 + 3s - 28$
$(s + 7)(s - 4)$

15) $u^2 + u - 132$
$(u + 12)(u - 11)$

16) $m^2 - m - 110$
$(m - 11)(m + 10)$

17) $q^2 - 8q + 15$
$(q - 5)(q - 3)$

18) $z^2 - 10z + 24$
$(z - 6)(z - 4)$

19) $y^2 + 9y + 10$
prime

20) $a^2 - 16a + 8$
prime

21) $p^2 - 20p + 100$
$(p - 10)(p - 10)$ or $(p - 10)^2$

22) $u^2 + 18u + 81$
$(u + 9)(u + 9)$ or $(u + 9)^2$

Objective 2: More on Factoring a Trinomial of the Form $x^2 + bx + c$

Factor completely, if possible. Check your answer.

23) $3p^2 - 24p - 27$
$3(p - 9)(p + 1)$

24) $5n^2 + 40n + 60$
$5(n + 2)(n + 6)$

25) $2k^3 - 26k^2 + 80k$
$2k(k - 8)(k - 5)$

26) $w^4 - 7w^3 + 12w^2$
$w^2(w - 3)(w - 4)$

27) $a^3b + 9a^2b - 36ab$
$ab(a + 12)(a - 3)$

28) $4x^3y + 32x^2y + 24xy$
$4xy(x^2 + 8x + 6)$

Factor completely by first taking out -1 and then by factoring the trinomial, if possible. Check your answer.

29) $-a^2 - 10a - 16$
$-(a + 8)(a + 2)$

30) $-y^2 - 9y - 18$
$-(y + 6)(y + 3)$

31) $-h^2 + 2h + 15$
$-(h - 5)(h + 3)$

32) $-j^2 - j + 56$
$-(j + 8)(j - 7)$

33) $-k^2 + 11k - 28$
$-(k - 7)(k - 4)$

34) $-b^2 + 17b - 66$
$-(b - 6)(b - 11)$

35) $-n^2 - 14n - 49$ $-(n + 7)(n + 7)$ or $-(n + 7)^2$

36) $-z^2 + 4z - 4$ $-(z - 2)(z - 2)$ or $-(z - 2)^2$

Objective 3: Factor a Trinomial Containing Two Variables

Factor completely. Check your answer.

37) $a^2 + 6ab + 5b^2$
$(a + 5b)(a + b)$

38) $v^2 + 7vw + 6w^2$
$(v + 6w)(v + w)$

39) $m^2 + 4mn - 21n^2$
$(m - 3n)(m + 7n)$

40) $p^2 - 17pq + 72q^2$
$(p - 8q)(p - 9q)$

41) $x^2 - 15xy + 36y^2$
$(x - 12y)(x - 3y)$

42) $r^2 - 9rs + 20s^2$
$(r - 4s)(r - 5s)$

43) $f^2 - 10fg - 11g^2$
$(f + g)(f - 11g)$

44) $u^2 + 2uv - 48v^2$
$(u + 8v)(u - 6v)$

45) $c^2 + 6cd - 55d^2$
$(c - 5d)(c + 11d)$

46) $w^2 + 17wx + 60x^2$
$(w + 5x)(w + 12x)$

Objective 4: Factor $ax^2 + bx + c$ ($a \neq 1$) by Grouping

Factor by grouping.

47) $2r^2 + 11r + 15$
$(2r + 5)(r + 3)$

48) $3a^2 + 10a + 8$
$(3a + 4)(a + 2)$

49) $5p^2 - 21p + 4$
$(5p - 1)(p - 4)$

50) $7j^2 - 30j + 8$
$(7j - 2)(j - 4)$

51) $11m^2 - 18m - 8$
$(11m + 4)(m - 2)$

52) $5b^2 + 9b - 18$
$(5b - 6)(b + 3)$

53) $6v^2 + 11v - 7$
$(3v + 7)(2v - 1)$

54) $8x^2 - 14x + 3$
$(4x - 1)(2x - 3)$

55) $10c^2 + 19c + 6$
$(5c + 2)(2c + 3)$

56) $15n^2 + 22n + 8$
$(5n + 4)(3n + 2)$

57) $9x^2 - 13xy + 4y^2$
$(9x - 4y)(x - y)$

58) $6a^2 + ab - 5b^2$
$(6a - 5b)(a + b)$

Objective 5: Factor $ax^2 + bx + c$ ($a \neq 1$) by Trial and Error

59) How do we know that $(2x - 4)$ cannot be a factor of $2x^2 + 13x - 24$? because 2 can be factored out of $2x - 4$, but 2 cannot be factored out of $2x^2 + 13x - 24$

60) How do we know that $(5c + 10)$ cannot be a factor of $5c^2 + 16c + 30$? because 5 can be factored out of $5c + 10$, but 5 cannot be factored out of $5c^2 + 16c + 30$

Factor by trial and error.

61) $5w^2 + 11w + 6$
$(5w + 6)(w + 1)$

62) $2g^2 + 13g + 18$
$(2g + 9)(g + 2)$

63) $3u^2 - 23u + 30$
$(3u - 5)(u - 6)$

64) $7a^2 - 17a + 6$
$(7a - 3)(a - 2)$

65) $7k^2 + 15k - 18$
$(7k - 6)(k + 3)$

66) $5z^2 - 18z - 35$
$(5z + 7)(z - 5)$

67) $8r^2 + 26r + 15$
$(4r + 3)(2r + 5)$

68) $6t^2 + 23t + 7$
$(2t + 7)(3t + 1)$

69) $6v^2 - 19v + 14$
$(6v - 7)(v - 2)$

70) $10m^2 + 47m - 15$
$(10m - 3)(m + 5)$

71) $10a^2 - 13ab + 4b^2$
$(5a - 4b)(2a - b)$

72) $8x^2 - 19xy + 6y^2$
$(8x - 3y)(x - 2y)$

73) $6c^2 + 31cd + 18d^2$
$(3c + 2d)(2c + 9d)$

74) $12m^2 - 16mn - 35n^2$
$(6m + 7n)(2m - 5n)$

Objective 6: Factor Using Substitution

Use substitution to factor each polynomial.

75) $(n + 5)^2 + 6(n + 5) - 27$ $(n + 14)(n + 2)$

76) $(k - 3)^2 - 9(k - 3) + 8$ $(k - 4)(k - 11)$

77) $(p - 6)^2 + 11(p - 6) + 28$ $(p + 1)(p - 2)$

78) $(t + 4)^2 - 10(t + 4) - 24$ $(t + 6)(t - 8)$

79) $2(w + 1)^2 - 13(w + 1) + 15$ $(2w - 1)(w - 4)$

80) $3(c - 9)^2 + 14(c - 9) + 16$ $(3c - 19)(c - 7)$

81) $6(2y - 1)^2 - 5(2y - 1) - 4$ $(6y - 7)(4y - 1)$

82) $10(3a + 2)^2 - 19(3a + 2) + 6$ $(15a + 8)(6a + 1)$

Mixed Exercises: Objectives 1–6

Factor completely, if possible.

83) $4q^3 - 28q^2 + 48q$ $4q(q - 3)(q - 4)$

84) $2y^2 - 19y + 24$ $(2y - 3)(y - 8)$

85) $6 + 7t + t^2$ $(t + 6)(t + 1)$

86) $m^2n^2 - 5mn - 6$ $(mn - 6)(mn + 1)$

87) $12c^2 + 15c - 18$ $3(4c - 3)(c + 2)$

88) $-h^2 - 3h + 54$ $-(h + 9)(h - 6)$

89) $3(b + 5)^2 + 4(b + 5) - 20$ $(3b + 25)(b + 3)$

90) $a^3b + 10a^2b^2 + 24ab^3$ $ab(a + 6b)(a + 4b)$

91) $7s^2 - 17st + 6t^2$ $(7s - 3t)(s - 2t)$

92) $(x + y)t^2 - 4(x + y)t - 21(x + y)$ $(x + y)(t + 3)(t - 7)$

93) $-10z^2 + 19z - 6$ $-(5z - 2)(2z - 3)$

94) $64p^2 - 112p + 49$ $(8p - 7)^2$

95) $c^2 + 6c - 5$ prime

96) $4(2h + 1)^2 - 3(2h + 1) - 22$ $(8h - 7)(2h + 3)$

97) $r^2 - 11r + 18$ $(r - 9)(r - 2)$

98) $2k^2 + 13k + 21$ $(2k + 7)(k + 3)$

99) $12p^2(q - 1)^2 - 49p(q - 1) + 49(q - 1)^2$ $(q - 1)^2(3p - 7)(4p - 7)$

100) $3w^2 - w - 6$ prime

Section 6.3 Special Factoring Techniques

Objectives

1. Factor a Perfect Square Trinomial
2. Factor the Difference of Two Squares
3. Factor the Sum and Difference of Two Cubes

1. Factor a Perfect Square Trinomial

Recall that we can square a binomial using the formulas

$$(a + b)^2 = a^2 + 2ab + b^2$$
$$(a - b)^2 = a^2 - 2ab + b^2$$

For example, $(x + 5)^2 = x^2 + 2x(5) + 5^2 = x^2 + 10x + 25$.

Since factoring a polynomial means writing the polynomial as a product of its factors, $x^2 + 10x + 25$ factors to $(x + 5)^2$.

The expression $x^2 + 10x + 25$ is a *perfect square trinomial*. A **perfect square trinomial** is a trinomial that results from squaring a binomial.

We can use the factoring method presented in Section 6.2 to factor a perfect square trinomial or we can learn to recognize the special pattern that appears in these trinomials. Above we stated that $x^2 + 10x + 25$ factors to $(x + 5)^2$. How are the terms of the trinomial and binomial related?

Compare $x^2 + 10x + 25$ to $(x + 5)^2$.

x^2 is the square of x, the first term in the binomial.

25 is the square of 5, the last term in the binomial.

We get the term $10x$ by doing the following:

$$10x = 2 \quad \cdot \quad x \quad \cdot \quad 5$$

Two times | First term in binomial | Last term in binomial

This follows directly from how we found $(x + 5)^2$ using the formula.

Formula   Factoring a Perfect Square Trinomial

$$a^2 + 2ab + b^2 = (a + b)^2$$
$$a^2 - 2ab + b^2 = (a - b)^2$$

Note

In order for a trinomial to be a perfect square, two of its terms must be perfect squares.

Example 1

Factor $t^2 + 12t + 36$ completely.

In-Class Example 1
Factor completely.
$r^2 + 10r + 25$
answer: $(r + 5)^2$

Solution

We cannot take out a common factor, so let's see if this follows the pattern of a perfect square trinomial.

$$t^2 + 12t + 36$$

What do you square to get t^2? t $(t)^2$ $(6)^2$ What do you square to get 36?   6

Does the middle term equal $2 \cdot t \cdot 6$? *Yes.*

$$2 \cdot t \cdot 6 = 12t$$

Therefore, $t^2 + 12t + 36 = (t + 6)^2$. Check by multiplying.   ■

Example 2

Factor completely.

In-Class Example 2
Factor completely.
a) $n^2 - 12n + 36$
b) $9x^2 + 24x + 16$
c) $4t^2 + 6t - 28$
d) $9g^3 + 18g^2 + 9g$
answer: a) $(n - 6)^2$
b) $(3x + 4)^2$
c) $2(2t + 7)(t - 2)$
d) $9g(g + 1)^2$

a) $n^2 - 14n + 49$ b) $4p^3 + 24p^2 + 36p$

c) $9k^2 + 30k + 25$ d) $4c^2 + 20c + 9$

Solution

a) We cannot take out a common factor. However, since the middle term is negative and the first and last terms are positive, the sign in the binomial will be a minus $(-)$ sign. Does this fit the pattern of a perfect square trinomial?

$$n^2 - 14n + 49$$

What do you square to get n^2? n $(n)^2$ $(7)^2$ What do you square to get 49?   7

Does the middle term equal $2 \cdot n \cdot 7$? *Yes:* $2 \cdot n \cdot 7 = 14n$

Since there is a minus sign in front of $14n$, $n^2 - 14n + 49$ fits the pattern of $a^2 - 2ab + b^2 = (a - b)^2$ with $a = n$ and $b = 7$.

Therefore, $n^2 - 14n + 49 = (n - 7)^2$. Check by multiplying.

b) From $4p^3 + 24p^2 + 36p$ we *can* begin by taking out the GCF of $4p$.

$$4p^3 + 24p^2 + 36p = 4p(p^2 + 6p + 9)$$

What do you square to get p^2? p $(p)^2$ $(3)^2$ What do you square to get 9?   3

Does the middle term equal $2 \cdot p \cdot 3$? *Yes:* $2 \cdot p \cdot 3 = 6p$.

$$4p^3 + 24p^2 + 36p = 4p(p^2 + 6p + 9) = 4p(p + 3)^2$$

Check by multiplying.

c) We cannot take out a common factor. Since the first and last terms of $9k^2 + 30k + 25$ are perfect squares, let's see if this is a perfect square trinomial.

$$9k^2 + 30k + 25$$

What do you square to get $9k^2$? $3k$ $(3k)^2$ $(5)^2$ What do you square to get 25?   5

Does the middle term equal $2 \cdot 3k \cdot 5$? *Yes*: $2 \cdot 3k \cdot 5 = 30k$.

Therefore, $9k^2 + 30k + 25 = (3k + 5)^2$. Check by multiplying.

d) We cannot take out a common factor. The first and last terms of $4c^2 + 20c + 9$ are perfect squares. Is this a perfect square trinomial?

$$4c^2 + 20c + 9$$

What do you square to get $4c^2$? $2c$ $(2c)^2$ $(3)^2$ What do you square to get 9?   3

Does the middle term equal $2 \cdot 2c \cdot 3$? *No*: $2 \cdot 2c \cdot 3 = 12c$

This is *not* a perfect square trinomial. Applying a method from Section 6.2 we find that the trinomial does factor, however.

$4c^2 + 20c + 9 = (2c + 9)(2c + 1)$. Check by multiplying.   ■

You Try 1

Factor completely.

a) $w^2 + 8w + 16$ b) $a^2 - 20a + 100$ c) $4d^2 - 36d + 81$

2. Factor the Difference of Two Squares

Another common type of factoring problem is a **difference of two squares.** Some examples of these types of binomials are

$$y^2 - 9, \qquad 25m^2 - 16n^2, \qquad 64 - t^2, \qquad \text{and} \qquad h^4 - 16.$$

Notice that in each binomial, the terms are being *subtracted*, and each term is a perfect square.

In Section 5.4, Multiplication of Polynomials, we saw that

$$(a + b)(a - b) = a^2 - b^2.$$

If we reverse the procedure, we get the factorization of the difference of two squares.

Formula   Factoring the Difference of Two Squares

$$a^2 - b^2 = (a + b)(a - b)$$

Don't forget that we can check all factorizations by multiplying.

Example 3

In-Class Example 3
Factor completely.
a) $k^2 - 25$ b) $9y^2 - 121z^2$
c) $r^2 - \dfrac{4}{49}$ d) $t^2 + 100$

answer: a) $(k + 5)(k - 5)$
b) $(3y + 11z)(3y - 11z)$
c) $\left(r + \dfrac{2}{7}\right)\left(r - \dfrac{2}{7}\right)$ **d) prime**

Factor completely.

a) $y^2 - 9$ b) $25m^2 - 16n^2$ c) $w^2 - \dfrac{9}{64}$ d) $c^2 + 36$

Solution

a) First, notice that $y^2 - 9$ is the difference of two terms *and* those terms are perfect squares. We can use the formula $a^2 - b^2 = (a + b)(a - b)$.

Identify a and b.

$$y^2 - 9$$
$$\downarrow \quad \downarrow$$

What do you square to get y^2? y $(y)^2 \ (3)^2$ What do you square to get 9?   3

Then, $a = y$ and $b = 3$. Therefore, $y^2 - 9 = (y + 3)(y - 3)$.

b) Look carefully at $25m^2 - 16n^2$. Each term *is* a perfect square, and they are being subtracted.

Identify a and b.

$$25m^2 - 16n^2$$
$$\downarrow \qquad \downarrow$$

What do you square to get $25m^2$? $5m$ $(5m)^2 \ (4n)^2$ What do you square to get $16n^2$? $4n$

Then, $a = 5m$ and $b = 4n$. So, $25m^2 - 16n^2 = (5m + 4n)(5m - 4n)$.

c) Each term in $w^2 - \dfrac{9}{64}$ is a perfect square, and they are being subtracted.

$$w^2 - \dfrac{9}{64}$$
$$\downarrow \qquad \downarrow$$

What do you square to get w^2? w $(w)^2 \ \left(\dfrac{3}{8}\right)^2$ What do you square to get $\dfrac{9}{64}$? $\dfrac{3}{8}$

So, $a = w$ and $b = \dfrac{3}{8}$. Therefore, $w^2 - \dfrac{9}{64} = \left(w + \dfrac{3}{8}\right)\left(w - \dfrac{3}{8}\right)$.

d) Each term in $c^2 + 36$ is a perfect square, but the expression is the *sum* of two squares. This polynomial does not factor.

$$c^2 + 36 \neq (c + 6)(c - 6) \text{ since } (c + 6)(c - 6) = c^2 - 36.$$
$$c^2 + 36 \neq (c + 6)(c + 6) \text{ since } (c + 6)(c + 6) = c^2 + 12c + 36.$$

So, $c^2 + 36$ is prime. ∎

Note

If the sum of two squares does not contain a common factor, then it cannot be factored.

You Try 2

Factor completely.

a) $r^2 - 25$ b) $49p^2 - 121q^2$ c) $x^2 - \dfrac{25}{144}$ d) $h^2 + 1$

Remember that sometimes we can factor out a GCF first. And, after factoring once, ask yourself, *"Can I factor again?"*

Example 4

Factor completely.

a) $128t - 2t^3$ b) $5x^2 + 45$ c) $h^4 - 16$

In-Class Example 4
Factor completely.
a) $36 - 9m^2$ b) $h^4 - 81$
c) $8p^2 + 800$
answer: a) $9(2 + m)(2 - m)$
b) $(h^2 + 9)(h + 3)(h - 3)$
c) $8(p^2 + 100)$

Solution

a) Ask yourself, *"Can I take out a common factor?"* Yes. Factor out $2t$.
$$128t - 2t^3 = 2t(64 - t^2)$$

Now ask yourself, *"Can I factor again?"* Yes. $64 - t^2$ is the difference of two squares. Identify a and b.
$$64 - t^2$$
$$\downarrow \quad \downarrow$$
$$(8)^2 \quad (t)^2$$

So, $a = 8$ and $b = t$. $64 - t^2 = (8 + t)(8 - t)$.

Therefore, $128t - 2t^3 = 2t(8 + t)(8 - t)$.

 **BE CAREFUL**

$(8 + t)(8 - t)$ is *not* the same as $(t + 8)(t - 8)$ because subtraction is not commutative. While $8 + t = t + 8$, $8 - t$ does not equal $t - 8$. You must write the terms in the correct order.

Another way to see that they are not equivalent is to multiply $(t + 8)(t - 8)$.
$(t + 8)(t - 8) = t^2 - 64$. This is not the same as $64 - t^2$.

b) Ask yourself, *"Can I take out a common factor?"* Yes. Factor out 5.
$$5(x^2 + 9)$$

"Can I factor again?" No; $x^2 + 9$ is the *sum* of two squares.
Therefore, $5x^2 + 45 = 5(x^2 + 9)$.

c) The terms in $h^4 - 16$ have no common factors, but they are perfect squares. Identify a and b.
$$h^4 - 16$$
$$\downarrow \quad \downarrow$$

What do you square to get h^4? h^2 $(h^2)^2 \quad (4)^2$ What do you square to get 16? 4

So, $a = h^2$ and $b = 4$. Therefore, $h^4 - 16 = (h^2 + 4)(h^2 - 4)$.
Can we factor again?

$h^2 + 4$ is the *sum* of two squares. It will not factor.

$h^2 - 4$ is the difference of two squares, so it *will* factor.

$$h^2 - 4$$
$$\downarrow \quad \downarrow \qquad h^2 - 4 = (h + 2)(h - 2)$$
$$(h)^2 \quad (2)^2$$
$$a = h \text{ and } b = 2$$

Therefore, $h^4 - 16 = (h^2 + 4)(h^2 - 4) = (h^2 + 4)(h + 2)(h - 2)$.   ∎

You Try 3

Factor completely.

a) $12p^4 - 27p^2$ b) $y^4 - 1$ c) $2n^2 + 72$

3. Factor the Sum and Difference of Two Cubes

We can understand where we get the formulas for factoring the sum and difference of two cubes by looking at two products.

$$(a + b)(a^2 - ab + b^2) = a(a^2 - ab + b^2) + b(a^2 - ab + b^2) \qquad \text{Distributive property}$$
$$= a^3 - a^2b + ab^2 + a^2b - ab^2 + b^3 \qquad \text{Distribute.}$$
$$= a^3 + b^3 \qquad \text{Combine like terms.}$$

So, $(a + b)(a^2 - ab + b^2) = a^3 + b^3$, the sum of two cubes.

Now, let's multiply $(a - b)(a^2 + ab + b^2)$.

$$(a - b)(a^2 + ab + b^2) = a(a^2 + ab + b^2) - b(a^2 + ab + b^2) \qquad \text{Distributive property}$$
$$= a^3 + a^2b + ab^2 - a^2b - ab^2 - b^3 \qquad \text{Distribute.}$$
$$= a^3 - b^3 \qquad \text{Combine like terms.}$$

So, $(a - b)(a^2 + ab + b^2) = a^3 - b^3$, the difference of two cubes.

The formulas for factoring the sum and difference of two cubes, then, are as follows:

Formula   Factoring the Sum and Difference of Two Cubes

$$a^3 + b^3 = (a + b)(a^2 - ab + b^2)$$
$$a^3 - b^3 = (a - b)(a^2 + ab + b^2)$$

Note

Notice that each factorization is the product of a binomial and a trinomial. To factor the sum and difference of two cubes

Step 1:   Identify a and b.
Step 2:   Place them in the binomial factor and write the trinomial based on a and b.
Step 3:   Simplify.

Example 5

In-Class Example 5
Factor completely.
a) $z^3 + 27$ b) $d^3 - 125$
answer:
a) $(z + 3)(z^2 - 3z + 9)$
b) $(d - 5)(d^2 + 5d + 25)$

Factor completely.

a) $n^3 + 8$ b) $c^3 - 64$ c) $125r^3 + 27s^3$

Solution

a) Use steps 1–3 to factor.

   Step 1: Identify a and b.

$$n^3 + 8$$

What do you cube to get n^3? n $(n)^3$ $(2)^3$ What do you cube to get 8?   2

So, $a = n$ and $b = 2$.

   Step 2: Remember, $a^3 + b^3 = (a + b)(a^2 - ab + b^2)$.
   Write the binomial factor, then write the trinomial.

Square a.   Product of a and b Square b.
Same sign

$$n^3 + 8 = (n + 2)[(n)^2 - (n)(2) + (2)^2]$$

Opposite sign

   Step 3: Simplify: $n^3 + 8 = (n + 2)(n^2 - 2n + 4)$

b) **Step 1:** Identify a and b.

$$c^3 - 64$$

What do you cube to get c^3? c $(c)^3$ $(4)^3$ What do you cube to get 64?   4

So, $a = c$ and $b = 4$.

   Step 2: Write the binomial factor, then write the trinomial. Remember,
   $a^3 - b^3 = (a - b)(a^2 + ab + b^2)$.

Square a.   Product of a and b Square b.
Same sign

$$c^3 - 64 = (c - 4)[(c)^2 + (c)(4) + (4)^2]$$

Opposite sign

   Step 3: Simplify: $c^3 - 64 = (c - 4)(c^2 + 4c + 16)$

c) $125r^3 + 27s^3$

   Step 1: Identify a and b.

$$125r^3 + 27s^3$$

What do you cube to get $125r^3$? $5r$ $(5r)^3$ $(3s)^3$ What do you cube to get $27s^3$? $3s$

So, $a = 5r$ and $b = 3s$.

   Step 2: Write the binomial factor, then write the trinomial. Remember,
   $a^3 + b^3 = (a + b)(a^2 - ab + b^2)$.

$$\overset{\text{Same sign}}{\underbrace{}}\quad \overset{\text{Square } a.}{\downarrow}\quad \overset{\text{Product of } a \text{ and } b}{}\quad \overset{\text{Square } b.}{\downarrow}$$

$$125r^3 + 27s^3 = (5r + 3s)[(5r)^2 - (5r)(3s) + (3s)^2]$$

$$\underset{\text{Opposite sign}}{\underbrace{}}$$

Step 3: Simplify: $125r^3 + 27s^3 = (5r + 3s)(25r^2 - 15rs + 9s^2)$ ■

 You Try 4

Factor completely.

a) $r^3 + 1$ b) $p^3 - 1000$ c) $64x^3 - 125y^3$

Just as in the other factoring problems we've studied so far, the first step in factoring *any* polynomial should be to ask ourselves, *"Can I factor out a GCF?"*

Example 6

Factor $3d^3 - 81$ completely.

In-Class Example 6
Factor completely
$4k^3 + 108$
answer: $4(k + 3)(k^2 - 3k + 9)$

Solution

"Can I factor out a GCF?" Yes. The GCF is 3.

$$3d^3 - 81 = 3(d^3 - 27)$$

Factor $d^3 - 27$. Use $a^3 - b^3 = (a - b)(a^2 + ab + b^2)$.

$$d^3 - 27 = (d - 3)[(d)^2 + (d)(3) + (3)^2]$$
$$(d)^3 - (3)^3 = (d - 3)(d^2 + 3d + 9)$$

$$3d^3 - 81 = 3(d^3 - 27)$$
$$= 3(d - 3)(d^2 + 3d + 9)$$ ■

 You Try 5

Factor completely.

a) $4t^3 + 4$ b) $72a^3 - 9b^6$

As always, the first thing you should do when factoring is ask yourself, *"Can I factor out a GCF?"* and the last thing you should do is ask yourself, *"Can I factor again?"* Now we will summarize the factoring methods discussed in this section.

Summary   Special Factoring Rules

Perfect square trinomials: $a^2 + 2ab + b^2 = (a + b)^2$
                             $a^2 - 2ab + b^2 = (a - b)^2$

Difference of two squares: $a^2 - b^2 = (a + b)(a - b)$

Sum of two cubes: $a^3 + b^3 = (a + b)(a^2 - ab + b^2)$

Difference of two cubes: $a^3 - b^3 = (a - b)(a^2 + ab + b^2)$

Answers to You Try Exercises

1) a) $(w + 4)^2$ b) $(a - 10)^2$ c) $(2d - 9)^2$ 2) a) $(r + 5)(r - 5)$ b) $(7p + 11q)(7p - 11q)$

c) $\left(x - \dfrac{5}{12}\right)\left(x + \dfrac{5}{12}\right)$ d) prime   3) a) $3p^2(2p + 3)(2p - 3)$ b) $(y^2 + 1)(y + 1)(y - 1)$

c) $2(n^2 + 36)$ 4) a) $(r + 1)(r^2 - r + 1)$ b) $(p - 10)(p^2 + 10p + 100)$

c) $(4x - 5y)(16x^2 + 20xy + 25y^2)$ 5) a) $4(t + 1)(t^2 - t + 1)$ b) $9(2a - b^2)(4a^2 + 2ab^2 + b^4)$

6.3 Exercises

*Additional answers can be found in the Answers to Exercises appendix.

Objective 1: Factor a Perfect Square Trinomial

1) Find the following.

a) 6^2 36

b) 10^2 100

c) 4^2 16

d) 11^2 121

e) 3^2 9

f) 8^2 64

g) 12^2 144

h) $\left(\dfrac{1}{2}\right)^2$ $\dfrac{1}{4}$

i) $\left(\dfrac{3}{5}\right)^2$ $\dfrac{9}{25}$

2) What is a perfect square trinomial?
It is a trinomial that results from squaring a binomial.

3) Fill in the blank with a term that has a positive coefficient.

a) $(\_\_)^2 = n^4$ n^2

b) $(\_\_)^2 = 25t^2$ $5t$

c) $(\_\_)^2 = 49k^2$ $7k$

d) $(\_\_)^2 = 16p^4$ $4p^2$

e) $(\_\_)^2 = \dfrac{1}{9}$ $\dfrac{1}{3}$

f) $(\_\_)^2 = \dfrac{25}{4}$ $\dfrac{5}{2}$

4) If x^n is a perfect square, then n is divisible by what number? 2

5) What perfect square trinomial factors to $(z + 9)^2$?
$z^2 + 18z + 81$

6) What perfect square trinomial factors to $(2b - 7)^2$?
$4b^2 - 28b + 49$

7) Why isn't $9c^2 - 12c + 16$ a perfect square trinomial?

8) Why isn't $k^2 + 6k + 8$ a perfect square trinomial?
Only one term, k^2, is a perfect square, so it can't be a perfect square trinomial.

Factor completely.

9) $t^2 + 16t + 64$ $(t + 8)^2$ 10) $x^2 + 12x + 36$ $(x + 6)^2$

11) $g^2 - 18g + 81$ $(g - 9)^2$ 12) $q^2 - 22q + 121$ $(q - 11)^2$

13) $4y^2 + 12y + 9$
$(2y + 3)^2$

14) $49r^2 + 14r + 1$
$(7r + 1)^2$

15) $9k^2 - 24k + 16$
$(3k - 4)^2$

16) $16b^2 - 24b + 9$
$(4b - 3)^2$

17) $a^2 + \dfrac{2}{3}a + \dfrac{1}{9}$ $\left(a + \dfrac{1}{3}\right)^2$ 18) $m^2 + m + \dfrac{1}{4}$ $\left(m + \dfrac{1}{2}\right)^2$

19) $v^2 - 3v + \dfrac{9}{4}$ $\left(v - \dfrac{3}{2}\right)^2$ 20) $h^2 - \dfrac{4}{5}h + \dfrac{4}{25}$ $\left(h - \dfrac{2}{5}\right)^2$

21) $x^2 + 6xy + 9y^2$
$(x + 3y)^2$

22) $9a^2 - 12ab + 4b^2$
$(3a - 2b)^2$

23) $36t^2 - 60tu + 25u^2$
$(6t - 5u)^2$

24) $81k^2 + 18km + m^2$
$(9k + m)^2$

25) $4f^2 + 24f + 36$
$4(f + 3)^2$

26) $9j^2 - 18j + 9$
$9(j - 1)^2$

27) $2p^4 - 24p^3 + 72p^2$
$2p^2(p - 6)^2$

28) $5r^3 + 40r^2 + 80r$
$5r(r + 4)^2$

29) $-18d^2 - 60d - 50$
$-2(3d + 5)^2$

30) $-28z^2 + 28z - 7$
$-7(2z - 1)^2$

31) $12c^3 + 3c^2 + 27c$
$3c(4c^2 + c + 9)$

32) $100n^4 - 8n^3 + 64n^2$
$4n^2(25n^2 - 2n + 16)$

Objective 2: Factor the Difference of Two Squares

33) What binomial factors to

a) $(x + 4)(x - 4)$?
$x^2 - 16$

b) $(4 + x)(4 - x)$?
$16 - x^2$

34) What binomial factors to

a) $(y - 9)(y + 9)$?
$y^2 - 81$

b) $(9 - y)(9 + y)$?
$81 - y^2$

Factor completely.

35) $x^2 - 9$ $(x + 3)(x - 3)$ 36) $q^2 - 49$ $(q + 7)(q - 7)$

37) $n^2 - 121$
$(n + 11)(n - 11)$

38) $d^2 - 81$
$(d + 9)(d - 9)$

39) $m^2 + 64$ prime 40) $q^2 + 9$ prime

41) $y^2 - \dfrac{1}{25}$ $\left(y + \dfrac{1}{5}\right)\left(y - \dfrac{1}{5}\right)$ 42) $t^2 - \dfrac{1}{100}$ $\left(t + \dfrac{1}{10}\right)\left(t - \dfrac{1}{10}\right)$

43) $c^2 - \dfrac{9}{16}$ $\left(c + \dfrac{3}{4}\right)\left(c - \dfrac{3}{4}\right)$ 44) $m^2 - \dfrac{4}{25}$ $\left(m + \dfrac{2}{5}\right)\left(m - \dfrac{2}{5}\right)$

45) $36 - h^2$ $(6 + h)(6 - h)$ 46) $4 - b^2$ $(2 + b)(2 - b)$

47) $169 - a^2$
$(13 + a)(13 - a)$

48) $121 - w^2$
$(11 + w)(11 - w)$

49) $\dfrac{49}{64} - j^2$ $\left(\dfrac{7}{8} + j\right)\left(\dfrac{7}{8} - j\right)$ 50) $\dfrac{144}{49} - r^2$ $\left(\dfrac{12}{7} + r\right)\left(\dfrac{12}{7} - r\right)$

51) $100m^2 - 49$
$(10m + 7)(10m - 7)$

52) $36x^2 - 25$
$(6x + 5)(6x - 5)$

53) $16p^2 - 81$
$(4p + 9)(4p - 9)$

54) $9a^2 - 1$
$(3a + 1)(3a - 1)$

55) $4t^2 + 25$ prime 56) $64z^2 + 9$ prime

57) $\dfrac{1}{4}k^2 - \dfrac{4}{9}$
$\left(\dfrac{1}{2}k + \dfrac{2}{3}\right)\left(\dfrac{1}{2}k - \dfrac{2}{3}\right)$

58) $\dfrac{1}{36}d^2 - \dfrac{4}{49}$ $\left(\dfrac{1}{6}d + \dfrac{2}{7}\right)\left(\dfrac{1}{6}d - \dfrac{2}{7}\right)$

59) $b^4 - 64$
$(b^2 + 8)(b^2 - 8)$

60) $u^4 - 49$
$(u^2 + 7)(u^2 - 7)$

61) $144m^2 - n^4$
$(12m + n^2)(12m - n^2)$

62) $64p^2 - 25q^4$
$(8p + 5q^2)(8p - 5q^2)$

VIDEO 63) $r^4 - 1$
$(r^2 + 1)(r + 1)(r - 1)$

64) $k^4 - 81$
$(k^2 + 9)(k + 3)(k - 3)$

65) $16h^4 - g^4$
$(4h^2 + g^2)(2h + g)(2h - g)$

66) $b^4 - a^4$
$(a^2 + b^2)(b + a)(b - a)$

67) $4a^2 - 100$
$4(a + 5)(a - 5)$

68) $3p^2 - 48$
$3(p + 4)(p - 4)$

69) $2m^2 - 128$
$2(m + 8)(m - 8)$

70) $6j^2 - 6$
$6(j + 1)(j - 1)$

71) $45r^4 - 5r^2$
$5r^2(3r + 1)(3r - 1)$

72) $32n^5 - 200n^3$
$8n^3(2n + 5)(2n - 5)$

Objective 3: Factor the Sum and Difference of Two Cubes

73) Find the following.

a) 4^3 64

b) 1^3 1

c) 10^3 1000

d) 3^3 27

e) 5^3 125

f) 2^3 8

74) If x^n is a perfect cube, then n is divisible by what number?   3

75) Fill in the blank.

a) $(\underline{\quad})^3 = y^3$ y

b) $(\underline{\quad})^3 = 8c^3$ $2c$

c) $(\underline{\quad})^3 = 125r^3$ $5r$

d) $(\underline{\quad})^3 = x^6$ x^2

76) If x^n is a perfect square *and* a perfect cube, then n is divisible by what number?   6

Complete the factorization.

77) $x^3 + 27 = (x + 3)(\quad)$ $x^2 - 3x + 9$

78) $t^3 - 125 = (t - 5)(\quad)$ $t^2 + 5t + 25$

Factor completely.

79) $d^3 + 1$
$(d + 1)(d^2 - d + 1)$

80) $n^3 + 125$
$(n + 5)(n^2 - 5n + 25)$

81) $p^3 - 27$
$(p - 3)(p^2 + 3p + 9)$

82) $g^3 - 8$
$(g - 2)(g^2 + 2g + 4)$

83) $k^3 + 64$
$(k + 4)(k^2 - 4k + 16)$

84) $z^3 - 1000$
$(z - 10)(z^2 + 10z + 100)$

VIDEO 85) $27m^3 - 125$
$(3m - 5)(9m^2 + 15m + 25)$

86) $64c^3 + 1$
$(4c + 1)(16c^2 - 4c + 1)$

87) $125y^3 - 8$
$(5y - 2)(25y^2 + 10y + 4)$

88) $27a^3 + 64$
$(3a + 4)(9a^2 - 12a + 16)$

89) $1000c^3 - d^3$
$(10c - d)(100c^2 + 10cd + d^2)$

90) $125v^3 + w^3$
$(5v + w)(25v^2 - 5vw + w^2)$

91) $8j^3 + 27k^3$
$(2j + 3k)(4j^2 - 6jk + 9k^2)$

92) $125m^3 - 27n^3$
$(5m - 3n)(25m^2 + 15mn + 9n^2)$

93) $64x^3 + 125y^3$
$(4x + 5y)(16x^2 - 20xy + 25y^2)$

94) $27a^3 - 1000b^3$
$(3a - 10b)(9a^2 + 30ab + 100b^2)$

VIDEO 95) $6c^3 + 48$
$6(c + 2)(c^2 - 2c + 4)$

96) $9k^3 - 9$
$9(k - 1)(k^2 + k + 1)$

97) $7v^3 - 7000w^3$
$7(v - 10w)(v^2 + 10vw + 100w^2)$

98) $216a^3 + 64b^3$
$8(3a + 2b)(9a^2 - 6ab + 4b^2)$

99) $h^6 - 64$

100) $p^6 - 1$

Extend the concepts of this section to factor completely.

101) $(x + 5)^2 - (x - 2)^2$ $7(2x + 3)$

102) $(r - 6)^2 - (r + 1)^2$ $-7(2r - 5)$

103) $(2p + 3)^2 - (p + 4)^2$ $(3p + 7)(p - 1)$

104) $(3d - 2)^2 - (d - 5)^2$ $(4d - 7)(2d + 3)$

105) $(t + 5)^3 + 8$ $(t + 7)(t^2 + 8t + 19)$

106) $(c - 2)^3 + 27$ $(c + 1)(c^2 - 7c + 19)$

107) $(k - 9)^3 - 1$ $(k - 10)(k^2 - 17k + 73)$

108) $(y + 3)^3 - 125$ $(y - 2)(y^2 + 11y + 49)$

Putting It All Together

Objective

1. Learn Strategies for Factoring a Given Polynomial

1. Learn Strategies for Factoring a Given Polynomial

In this chapter, we have discussed several different types of factoring problems:

1) Factoring out a GCF (Section 6.1)

2) Factoring by grouping (Section 6.1)

3) Factoring a trinomial of the form $x^2 + bx + c$ (Section 6.2)

4) Factoring a trinomial of the form $ax^2 + bx + c$ (Section 6.2)

5) Factoring a perfect square trinomial (Section 6.3)

6) Factoring the difference of two squares (Section 6.3)

7) Factoring the sum and difference of two cubes (Section 6.3)

We have practiced the factoring methods separately in each section, but how do we know which factoring method to use given many different types of polynomials together?

We will discuss some strategies in this section. First, recall the steps for factoring *any* polynomial:

Summary   To Factor a Polynomial

1) *Always* begin by asking yourself, *"Can I factor out a GCF?"* If so, factor it out.

2) Look at the expression to decide if it will factor further. Apply the appropriate method to factor. If there are

   a) *two terms*, see if it is a difference of two squares or the sum or difference of two cubes as in Section 6.3.

   b) *three terms*, see if it can be factored using the methods of Section 6.2 or determine if it is a perfect square trinomial (Section 6.3).

   c) *four terms*, see if it can be factored by grouping as in Section 6.1.

3) After factoring *always* look carefully at the result and ask yourself, *"Can I factor it again?"* If so, factor again.

Next, we will discuss how to decide which factoring method should be used to factor a particular polynomial.

Example 1

Factor completely.

a) $12a^2 - 27b^2$ b) $y^2 - y - 30$ c) $mn^2 - 4m + 5n^2 - 20$

d) $p^2 - 16p + 64$ e) $8x^2 + 26x + 20$ f) $27k^3 + 8$

g) $t^2 + 36$

In-Class Example 1
Factor completely.
a) $2x^2 - 18y^2$ b) $b^2 + 3b - 28$
c) $d^2 + 144$ d) $2r^3 - 16$
e) $w^3x - x + 4w^3 - 4$
answer: a) $2(x + 3y)(x - 3y)$
b) $(b + 7)(b - 4)$ c) prime
d) $2(r - 2)(r^2 + 2r + 4)$
e) $(x + 4)(w - 1)(w^2 + w + 1)$

Solution

a) *"Can I factor out a GCF?"* is the first thing you should ask yourself. *Yes.* Factor out 3.

$$12a^2 - 27b^2 = 3(4a^2 - 9b^2)$$

Ask yourself, *"Can I factor again?"* Examine $4a^2 - 9b^2$. It has two terms that are being subtracted, and each term is a perfect square. $4a^2 - 9b^2$ is the difference of squares.

$$4a^2 - 9b^2 = (2a + 3b)(2a - 3b)$$
$$\downarrow\qquad\downarrow$$
$$(2a)^2\ (3b)^2$$

$$12a^2 - 27b^2 = 3(4a^2 - 9b^2)$$
$$= 3(2a + 3b)(2a - 3b)$$

"Can I factor again?" No. It is completely factored.

b) *"Can I factor out a GCF?"* No. To factor $y^2 - y - 30$, think of two numbers whose *product* is -30 and *sum* is -1. The numbers are -6 and 5.

$$y^2 - y - 30 = (y - 6)(y + 5)$$

"Can I factor again?" No. It is completely factored.

c) Look at $mn^2 - 4m + 5n^2 - 20$. *"Can I factor out a GCF?"* No. Notice that this polynomial has *four terms*. When a polynomial has *four terms*, think about *factoring by grouping.*

$$\underbrace{mn^2 - 4m}\ \ \underbrace{+\ 5n^2 - 20}$$
$$\downarrow\downarrow$$
$$= m(n^2 - 4) + 5(n^2 - 4)\qquad\text{Take out the common factor from each pair of terms.}$$
$$= (n^2 - 4)(m + 5)\qquad\text{Factor out } (n^2 - 4) \text{ using the distributive property.}$$

Examine $(n^2 - 4)(m + 5)$ and ask yourself, *"Can I factor again?"* Yes! $(n^2 - 4)$ is the difference of two squares. Factor again.

$$(n^2 - 4)(m + 5) = (n + 2)(n - 2)(m + 5)$$

"Can I factor again?" No. It is completely factored.

$$mn^2 - 4m + 5n^2 - 20 = (n + 2)(n - 2)(m + 5)$$

Note

Seeing four terms is a clue to try factoring by grouping.

d) We cannot take out a GCF from $p^2 - 16p + 64$. It is a trinomial, and notice that the first and last terms are perfect squares. *Is this a perfect square trinomial?*

$$p^2 - 16p + 64$$
$$\downarrow \qquad\qquad \downarrow$$
$$(p)^2 \qquad\qquad (8)^2$$

Does the middle term equal $2 \cdot p \cdot (8)$? Yes: $2 \cdot p \cdot (8) = 16p$.
Use $a^2 - 2ab + b^2 = (a - b)^2$ with $a = p$ and $b = 8$.
Then, $p^2 - 16p + 64 = (p - 8)^2$.

"Can I factor again?" No. It is completely factored.

e) It is tempting to jump right in and try to factor $8x^2 + 26x + 20$ as the product of two binomials, but ask yourself, *"Can I take out a GCF?"* Yes! Factor out 2.

$$8x^2 + 26x + 20 = 2(4x^2 + 13x + 10)$$

"Can I factor again?" Yes.

$$2(4x^2 + 13x + 10) = 2(4x + 5)(x + 2)$$

"Can I factor again?" No. So, $8x^2 + 26x + 20 = 2(4x + 5)(x + 2)$.

f) We cannot take out a GCF from $27k^3 + 8$. Notice that $27k^3 + 8$ has two terms, so think about squares and cubes. Neither term is a perfect square *and* the positive terms are being added, so this *cannot* be the difference of squares.

Is each term a perfect cube? *Yes!* $27k^3 + 8$ is the sum of two cubes. We will factor $27k^3 + 8$ using $a^3 + b^3 = (a + b)(a^2 - ab + b^2)$ with $a = 3k$ and $b = 2$.

$$27k^3 + 8 = (3k + 2)[(3k)^2 - (3k)(2) + (2)^2]$$
$$\downarrow \quad \downarrow$$
$$(3k)^3 \ (2)^3$$
$$= (3k + 2)(9k^2 - 6k + 4)$$

"Can I factor again?" No. It is completely factored.

g) Look at $t^2 + 36$ and ask yourself, *"Can I factor out a GCF?"* No. The binomial $t^2 + 36$ is the *sum* of two squares, so it does not factor. This polynomial is prime. ■

You Try 1

Factor completely.

a) $3p^2 + p - 10$ b) $2n^3 - n^2 + 12n - 6$ c) $4k^4 + 36k^3 + 32k^2$

d) $48 - 3y^4$ e) $8r^3 - 125$

Answers to You Try Exercises

1) a) $(3p - 5)(p + 2)$ b) $(n^2 + 6)(2n - 1)$ c) $4k^2(k + 8)(k + 1)$ d) $3(4 + y^2)(2 + y)(2 - y)$
e) $(2r - 5)(4r^2 + 10r + 25)$

Putting It All Together
Exercises

*Additional answers can be found in the Answers to Exercises appendix.

Objective 1: Learn Strategies for Factoring a Given Polynomial

Factor completely.

1) $m^2 + 16m + 60$
   $(m + 10)(m + 6)$

2) $h^2 - 36$
   $(h + 6)(h - 6)$

3) $uv + 6u + 9v + 54$
   $(u + 9)(v + 6)$

4) $2y^2 + 5y - 18$
   $(2y + 9)(y - 2)$

5) $3k^2 - 14k + 8$
   $(3k - 2)(k - 4)$

6) $n^2 - 14n + 49$
   $(n - 7)^2$

7) $16d^6 + 8d^5 + 72d^4$
   $8d^4(2d^2 + d + 9)$

8) $b^2 - 3bc - 4c^2$
   $(b + c)(b - 4c)$

9) $60w^3 + 70w^2 - 50w$
   $10w(3w + 5)(2w - 1)$

10) $7c^3 - 7$
    $7(c - 1)(c^2 + c + 1)$

11) $t^3 + 1000$
    $(t + 10)(t^2 - 10t + 100)$

12) $pq - 6p + 4q - 24$
    $(p + 4)(q - 6)$

13) $49 - p^2$
    $(7 + p)(7 - p)$

14) $h^2 - 15h + 56$
    $(h - 7)(h - 8)$

15) $4x^2 + 4xy + y^2$
    $(2x + y)^2$

16) $27c - 18$
    $9(3c - 2)$

17) $3z^4 - 21z^3 - 24z^2$
    $3z^2(z - 8)(z + 1)$

18) $9a^2 + 6a - 8$
    $(3a - 2)(3a + 4)$

19) $4b^2 + 1$ prime

20) $5abc - 15ac + 10bc - 30c$ $5c(a + 2)(b - 3)$

21) $40x^3 - 135$
    $5(2x - 3)(4x^2 + 6x + 9)$

22) $81z^2 + 36z + 4$
    $(9z + 2)^2$

23) $c^2 - \dfrac{1}{4}$ $\left(c + \dfrac{1}{2}\right)\left(c - \dfrac{1}{2}\right)$

24) $v^2 + 3v + 4$ prime

25) $45s^2t + 4 - 36s^2 - 5t$
    $(3s + 1)(3s - 1)(5t - 4)$

26) $12c^5d - 75cd^3$
    $3cd(2c^2 + 5d)(2c^2 - 5d)$

27) $k^2 + 9km + 18m^2$
    $(k + 3m)(k + 6m)$

28) $64r^3 + 8$
    $8(2r + 1)(4r^2 - 2r + 1)$

29) $z^2 - 3z - 88$
    $(z - 11)(z + 8)$

30) $40f^4g^4 + 8f^3g^3 + 16fg^2$
    $8fg^2(5f^3g^2 + f^2g + 2)$

31) $80y^2 - 40y + 5$
    $5(4y - 1)^2$

32) $4t^2 - t - 5$
    $(4t - 5)(t + 1)$

33) $20c^2 + 26cd + 6d^2$
    $2(10c + 3d)(c + d)$

34) $x^2 - \dfrac{9}{49}$ $\left(x + \dfrac{3}{7}\right)\left(x - \dfrac{3}{7}\right)$

35) $n^4 - 16m^4$
    $(n^2 + 4m^2)(n + 2m)(n - 2m)$

36) $k^2 - 21k + 108$
    $(k - 12)(k - 9)$

37) $2a^2 - 10a - 72$
    $2(a - 9)(a + 4)$

38) $x^2y - 4y + 7x^2 - 28$
    $(x + 2)(x - 2)(y + 7)$

39) $r^2 - r + \dfrac{1}{4}$ $\left(r - \dfrac{1}{2}\right)^2$

40) $v^3 - 125$
    $(v - 5)(v^2 + 5v + 25)$

41) $28gh + 16g - 63h - 36$ $(4g - 9)(7h + 4)$

42) $-24x^3 + 30x^2 - 9x$
    $-3x(2x - 1)(4x - 3)$

43) $8b^2 - 14b - 15$
    $(4b + 3)(2b - 5)$

44) $50u^2 + 60u + 18$ $2(5u + 3)^2$

45) $55a^6b^3 + 35a^5b^3 - 10a^4b - 20a^2b$
    $5a^2b(11a^4b^2 + 7a^3b^2 - 2a^2 - 4)$

46) $64 - u^2$ $(8 + u)(8 - u)$

47) $2d^2 - 9d + 3$ prime

48) $2v^4w + 14v^3w^2 + 12v^2w^3$
    $2v^2w(v + w)(v + 6w)$

49) $9p^2 - 24pq + 16q^2$
    $(3p - 4q)^2$

50) $c^4 - 16$
    $(c^2 + 4)(c + 2)(c - 2)$

51) $30y^2 + 37y - 7$
    $(6y - 1)(5y + 7)$

52) $g^2 + 49$
    prime

53) $80a^3 - 270b^3$
    $10(2a - 3b)(4a^2 + 6ab + 9b^2)$

54) $26n^6 - 39n^4 + 13n^3$
    $13n^3(2n^3 - 3n + 1)$

55) $rt - r - t + 1$
    $(r - 1)(t - 1)$

56) $h^2 + 10h + 25$
    $(h + 5)^2$

57) $4g^2 - 4$
    $4(g + 1)(g - 1)$

58) $25a^2 - 55ab + 24b^2$
    $(5a - 3b)(5a - 8b)$

59) $3c^2 - 24c + 48$
    $3(c - 4)^2$

60) $9t^4 + 64u^2$
    prime

61) $144k^2 - 121$
    $(12k + 11)(12k - 11)$

62) $125p^3 - 64q^3$
    $(5p - 4q)(25p^2 + 20pq + 16q^2)$

63) $-48g^2 - 80g - 12$
    $-4(6g + 1)(2g + 3)$

64) $5d^2 + 60d + 55$
    $5(d + 11)(d + 1)$

65) $q^3 + 1$
    $(q + 1)(q^2 - q + 1)$

66) $9x^2 + 12x + 4$
    $(3x + 2)^2$

67) $81u^4 - v^4$ $(9u^2 + v^2)(3u + v)(3u - v)$

68) $45v^2 + 9vw^2 + 30vw + 6w^3$ $3(5v + w^2)(3v + 2w)$

69) $11f^2 + 36f + 9$
    $(11f + 3)(f + 3)$

70) $4y^3 - 4y^2 - 80y$
    $4y(y - 5)(y + 4)$

71) $2j^{11} - j^3$ $j^3(2j^8 - 1)$

72) $d^2 - \dfrac{169}{100}$ $\left(d + \dfrac{13}{10}\right)\left(d - \dfrac{13}{10}\right)$

73) $w^2 - 2w - 48$
    $(w - 8)(w + 6)$

74) $16a^2 - 40a + 25$
    $(4a - 5)^2$

75) $k^2 + 100$
    prime

76) $24y^3 + 375$
    $3(2y + 5)(4y^2 - 10y + 25)$

77) $m^2 + 4m + 4$ $(m + 2)^2$

78) $r^2 - 15r + 54$ $(r - 9)(r - 6)$

79) $9t^2 - 64$
    $(3t + 8)(3t - 8)$

80) $100c^4 - 36c^2$
    $4c^2(5c + 3)(5c - 3)$

81) $(2z + 1)y^2 + 6(2z + 1)y - 55(2z + 1)$
    $(2z + 1)(y + 11)(y - 5)$

82) $(a + b)c^2 - 5(a + b)c - 24(a + b)$ $(a + b)(c - 8)(c + 3)$

83) $(r - 4)^2 + 11(r - 4) + 28$ $r(r + 3)$

84) $(n + 3)^2 - 2(n + 3) - 35$ $(n - 4)(n + 8)$

85) $(3p - 4)^2 - 5(3p - 4) - 36$ $3p(3p - 13)$

86) $(5w - 2)^2 - 8(5w - 2) + 12$ $(5w - 8)(5w - 4)$

87) $(4k + 1)^2 - (3k + 2)^2$ $(7k + 3)(k - 1)$

88) $(5z + 3)^2 - (3z - 1)^2$ $4(4z + 1)(z + 2)$

89) $(x + y)^2 - (2x - y)^2$ $-3x(x - 2y)$

90) $(3s - t)^2 - (2s + t)^2$ $5s(s - 2t)$

91) $n^2 + 12n + 36 - p^2$ $(n + p + 6)(n - p + 6)$

92) $h^2 - 10h + 25 - k^2$ $(h + k - 5)(h - k - 5)$

93) $x^2 - 2xy + y^2 - z^2$ $(x - y + z)(x - y - z)$

94) $a^2 + 2ab + b^2 - c^2$ $(a + b + c)(a + b - c)$

Section 6.4 Solving Quadratic Equations by Factoring

Objectives

1. **Solve a Quadratic Equation of the Form** $ab = 0$
2. **Solve Quadratic Equations by Factoring**
3. **Solve Higher Degree Equations by Factoring**

In Section 2.1 we began our study of linear equations in one variable. A *linear equation in one variable* is an equation that can be written in the form $ax + b = 0$, where a and b are real numbers and $a \neq 0$.

In this section, we will learn how to solve *quadratic equations*.

> ### Definition
>
> A **quadratic equation** can be written in the form $ax^2 + bx + c = 0$, where a, b, and c are real numbers and $a \neq 0$.

When a quadratic equation is written in the form $ax^2 + bx + c = 0$, we say that it is in **standard form.** But quadratic equations can be written in other forms too.

Some examples of quadratic equations are

$$x^2 + 12x + 27 = 0, \quad 2p(p - 5) = 0, \quad \text{and} \quad (c + 1)(c - 8) = 3.$$

Quadratic equations are also called *second-degree equations* because the highest power on the variable is 2.

There are many different ways to solve quadratic equations. In this section, we will learn how to solve them by factoring; other methods will be discussed later in this book.

Solving a quadratic equation by factoring is based on the *zero product rule*: if the product of two quantities is zero, then one or both of the quantities is zero.

For example, if $5y = 0$, then $y = 0$. If $p \cdot 4 = 0$, then $p = 0$. If $ab = 0$, then either $a = 0$, $b = 0$, or *both a* and *b* equal zero.

> ### Definition
>
> **Zero product rule:** If $ab = 0$, then $a = 0$ or $b = 0$.

We will use this idea to solve quadratic equations by factoring.

1. Solve a Quadratic Equation of the Form $ab = 0$

Example 1

In-Class Example 1
Solve each equation.
a) $h(h - 4) = 0$
b) $(6a - 1)(a + 5) = 0$
answer: a) $\{0, 4\}$
b) $\left\{-5, \dfrac{1}{6}\right\}$

Solve each equation.

a) $x(x + 8) = 0$ b) $(4y - 3)(y + 9) = 0$

Solution

a)   The zero product rule says that at least one of the factors on the left must equal zero in order for the *product* to equal zero.

$$x(x + 8) = 0$$

$$x = 0 \quad \text{or} \quad x + 8 = 0 \qquad \text{Set each factor equal to 0.}$$
$$x = -8 \qquad \text{Solve.}$$

Check the solutions in the original equation:

$$
\begin{array}{l|l}
\text{If } x = 0, & \text{If } x = -8, \\
0(0 + 8) \stackrel{?}{=} 0 & -8(-8 + 8) \stackrel{?}{=} 0 \\
\quad\quad 0(8) = 0 \ \checkmark & \quad\quad -8(0) = 0 \ \checkmark
\end{array}
$$

The solution set is $\{-8, 0\}$.

Note

It is important to remember that the factor x gives us the solution 0.

b) At least one of the factors on the left must equal zero for the *product* to equal zero.

$$(4y - 3)(y + 9) = 0$$

$4y - 3 = 0$ or $y + 9 = 0$ Set each factor equal to 0.
$4y = 3$ $y = -9$ Solve each equation.

$$y = \frac{3}{4}$$

Check in the original equation:

If $y = \frac{3}{4}$,

$$\left[4\left(\frac{3}{4}\right) - 3\right]\left[\frac{3}{4} + 9\right] \stackrel{?}{=} 0$$

$$(3 - 3)\left(\frac{39}{4}\right) \stackrel{?}{=} 0$$

$$0\left(\frac{39}{4}\right) = 0 \ \checkmark$$

If $y = -9$,

$$[4(-9) - 3][-9 + 9] \stackrel{?}{=} 0$$

$$-39(0) = 0 \ \checkmark$$

The solution set is $\left\{-9, \frac{3}{4}\right\}$.

You Try 1

Solve each equation.

a) $c(c - 9) = 0$ b) $(5t + 2)(t - 7) = 0$

2. Solve Quadratic Equations by Factoring

If the equation is in standard form, $ax^2 + bx + c = 0$, begin by factoring the expression.

Example 2

Solve $m^2 - 6m - 40 = 0$.

In-Class Example 2
Solve $g^2 + 10g + 21 = 0$.
answer: $\{-7, -3\}$

Solution

$$m^2 - 6m - 40 = 0$$
$$(m - 10)(m + 4) = 0$$ Factor.

$m - 10 = 0$ or $m + 4 = 0$ Set each factor equal to zero.
$m = 10$ or $m = -4$ Solve.

Check in the original equation:

If $m = 10$,
$$(10)^2 - 6(10) - 40 \stackrel{?}{=} 0$$
$$100 - 60 - 40 = 0 \ \checkmark$$

If $m = -4$,
$$(-4)^2 - 6(-4) - 40 \stackrel{?}{=} 0$$
$$16 + 24 - 40 = 0 \ \checkmark$$

The solution set is $\{-4, 10\}$.

Here are the steps to use to solve a quadratic equation by factoring:

> **Procedure**   Solving a Quadratic Equation by Factoring
>
> 1) Write the equation in the form $ax^2 + bx + c = 0$ (standard form) so that all terms are on one side of the equal sign and zero is on the other side.
>
> 2) Factor the expression.
>
> 3) Set each factor equal to zero, and solve for the variable. (Use the zero product rule.)
>
> 4) Check the answer(s).

You Try 2

Solve $h^2 + 9h + 18 = 0$.

Example 3

In-Class Example 3

Solve each equation by factoring.
a) $8k^2 + 32k = 96$
b) $w^2 = 3(w + 18)$
c) $3g^2 = -18g$
d) $(d - 10)(d - 5) = 6$
e) $6(n^2 + 2n) = -7(n - 1)$
answer: a) $\{-6, 2\}$ b) $\{-6, 9\}$
c) $\{-6, 0\}$ d) $\{4, 11\}$
e) $\left\{-\dfrac{7}{2}, \dfrac{1}{3}\right\}$

Solve each equation by factoring.

a) $2r^2 + 3r = 20$ b) $6d^2 = -42d$ c) $k^2 = -12(k + 3)$

d) $2(x^2 + 5) + 5x = 6x(x - 1) + 16$ e) $(z - 8)(z - 4) = 5$

Solution

a) Begin by writing $2r^2 + 3r = 20$ in standard form, $ar^2 + br + c = 0$.

$$2r^2 + 3r - 20 = 0 \qquad \text{Standard form}$$
$$(2r - 5)(r + 4) = 0 \qquad \text{Factor.}$$

$$2r - 5 = 0 \quad \text{or} \quad r + 4 = 0 \qquad \text{Set each factor equal to zero.}$$
$$2r = 5$$
$$r = \frac{5}{2} \quad \text{or} \qquad r = -4 \qquad \text{Solve.}$$

Check in original equation:

If $r = \dfrac{5}{2}$,

$$2\left(\frac{5}{2}\right)^2 + 3\left(\frac{5}{2}\right) \overset{?}{=} 20$$

$$2\left(\frac{25}{4}\right) + \frac{15}{2} \overset{?}{=} 20$$

$$\frac{25}{2} + \frac{15}{2} \overset{?}{=} 20$$

$$\frac{40}{2} = 20 \checkmark$$

If $r = -4$,

$$2(-4)^2 + 3(-4) \overset{?}{=} 20$$
$$2(16) - 12 \overset{?}{=} 20$$
$$32 - 12 = 20 \checkmark$$

The solution set is $\left\{-4, \dfrac{5}{2}\right\}$.

b) Write $6d^2 = -42d$ in standard form.

$$6d^2 + 42d = 0 \qquad \text{Standard form}$$
$$6d(d + 7) = 0 \qquad \text{Factor.}$$

$$6d = 0 \quad \text{or} \quad d + 7 = 0 \qquad \text{Set each factor equal to zero.}$$
$$d = 0 \quad \text{or} \qquad d = -7 \qquad \text{Solve.}$$

Check. The solution set is $\{-7, 0\}$.

Since both terms in $6d^2 = -42d$ are divisible by 6, we could have started part b) by dividing by 6:

$$\frac{6d^2}{6} = \frac{-42d}{6} \qquad \text{Divide by 6.}$$
$$d^2 = -7d$$
$$d^2 + 7d = 0 \qquad \text{Write in standard form.}$$
$$d(d + 7) = 0 \qquad \text{Factor.}$$

$$d = 0 \quad \text{or} \quad d + 7 = 0 \qquad \text{Set each factor equal to zero.}$$
$$d = -7 \qquad \text{Solve.}$$

The solution set is $\{-7, 0\}$. We get the same result.

 **BE CAREFUL** We cannot divide by d even though each term contains a factor of d. Doing so would eliminate the solution of zero. In general, we can divide an equation by a nonzero real number but we cannot divide an equation by a variable because we may eliminate a solution, and we may be dividing by zero.

c)   To solve $k^2 = -12(k + 3)$, begin by writing the equation in standard form.

$$k^2 = -12k - 36 \qquad \text{Distribute.}$$
$$k^2 + 12k + 36 = 0 \qquad \text{Write in standard form.}$$
$$(k + 6)^2 = 0 \qquad \text{Factor.}$$

Since $(k + 6)^2 = 0$ means $(k + 6)(k + 6) = 0$, setting each factor equal to zero will result in the same value for k.

$$k + 6 = 0 \qquad \text{Set } k + 6 = 0.$$
$$k = -6 \qquad \text{Solve.}$$

Check. The solution set is $\{-6\}$.

d)   We will have to perform several steps to write the equation in standard form.

$$2(x^2 + 5) + 5x = 6x(x - 1) + 16$$
$$2x^2 + 10 + 5x = 6x^2 - 6x + 16 \qquad \text{Distribute.}$$

Move the terms on the left side of the equation to the right side so that the coefficient of x^2 is positive.

$$0 = 4x^2 - 11x + 6 \qquad \text{Write in standard form.}$$
$$0 = (4x - 3)(x - 2) \qquad \text{Factor.}$$

$$4x - 3 = 0 \quad \text{or} \quad x - 2 = 0 \qquad \text{Set each factor equal to zero.}$$
$$4x = 3$$
$$x = \frac{3}{4} \quad \text{or} \qquad x = 2 \qquad \text{Solve.}$$

The check is left to the student. The solution set is $\left\{ \frac{3}{4}, 2 \right\}$.

 **BE CAREFUL** e)   It is tempting to solve $(z - 8)(z - 4) = 5$ like this:

$$(z - 8)(z - 4) = 5$$

$$z - 8 = 5 \quad \text{or} \quad z - 4 = 5 \qquad \text{This is incorrect!}$$

One side of the equation must equal zero in order to set each factor equal to zero.
Begin by multiplying on the left.

$$(z - 8)(z - 4) = 5$$
$$z^2 - 12z + 32 = 5 \qquad \text{Multiply using FOIL.}$$
$$z^2 - 12z + 27 = 0 \qquad \text{Standard form}$$
$$(z - 9)(z - 3) = 0 \qquad \text{Factor.}$$

$$z - 9 = 0 \quad \text{or} \quad z - 3 = 0 \qquad \text{Set each factor equal to zero.}$$
$$z = 9 \quad \text{or} \qquad z = 3 \qquad \text{Solve.}$$

The check is left to the student. The solution set is $\{3, 9\}$.   ■

You Try 3

Solve.

a) $w^2 + 4w - 5 = 0$ b) $29b = 5(b^2 + 4)$ c) $(a + 6)(a + 4) = 3$

d) $t^2 = 8t$ e) $(2y + 1)^2 + 5 = y^2 + 2(y + 7)$

3. Solve Higher Degree Equations by Factoring

Sometimes, equations that are not quadratics can be solved by factoring as well.

Example 4

In-Class Example 4
Solve each equation.
a) $(4y + 3)(y^2 + 2y - 63) = 0$
b) $3a^3 - 48a = 0$

answer: a) $\left\{ -9, -\dfrac{3}{4}, 7 \right\}$

b) $\{-4, 0, 4\}$

Solve each equation.

a) $(2x - 1)(x^2 - 9x - 22) = 0$ b) $4w^3 - 100w = 0$

Solution

a) This is *not* a quadratic equation because if we multiplied the factors on the left we
would get $2x^3 - 19x^2 - 35x + 22 = 0$. This is a *cubic* equation because the degree
of the polynomial on the left is 3.

The original equation is the product of two factors so we can use the zero product rule.

$$(2x - 1)(x^2 - 9x - 22) = 0$$
$$(2x - 1)(x - 11)(x + 2) = 0 \qquad \text{Factor.}$$

$$2x - 1 = 0 \quad \text{or} \quad x - 11 = 0 \quad \text{or} \quad x + 2 = 0 \qquad \text{Set each factor equal to zero.}$$
$$2x = 1$$

$$x = \frac{1}{2} \quad \text{or} \qquad x = 11 \quad \text{or} \qquad x = -2 \qquad \text{Solve.}$$

The check is left to the student. The solution set is $\left\{ -2, \dfrac{1}{2}, 11 \right\}$.

b) The GCF of the terms in the equation is $4w$. Remember, however, that *we can divide
an equation by a constant but we cannot divide an equation by a variable.* Dividing
by a variable may eliminate a solution and may mean we are dividing by zero. So
let's begin by dividing each term by 4.

$$\frac{4w^3}{4} - \frac{100w}{4} = \frac{0}{4} \qquad \text{Divide by 4.}$$
$$w^3 - 25w = 0 \qquad \text{Simplify.}$$
$$w(w^2 - 25) = 0 \qquad \text{Factor out } w.$$
$$w(w + 5)(w - 5) = 0 \qquad \text{Factor } w^2 - 25.$$

$$w = 0 \quad \text{or} \quad w + 5 = 0 \quad \text{or} \quad w - 5 = 0 \qquad \text{Set each factor equal to zero.}$$
$$w = -5 \qquad\qquad w = 5 \qquad \text{Solve}$$

Check. The solution set is $\{0, -5, 5\}$.   ■

You Try 4

Solve.

a) $(c + 10)(2c^2 + 5c - 7) = 0$ b) $r^4 = 25r^2$

In this section, it was possible to solve all of the equations by factoring. Below we show the relationship between solving a quadratic equation by factoring and solving it using a graphing calculator. In Chapter 9 we will learn other methods for solving quadratic equations.

Using Technology

In this section, we learned how to solve a quadratic equation by factoring. We can use a graphing calculator to solve a quadratic equation as well. Let's see how the two are related by using the equation $x^2 - x - 6 = 0$.

$$x^2 - x - 6 = 0$$
$$(x + 2)(x - 3) = 0$$
$$x + 2 = 0 \quad \text{or} \quad x - 3 = 0$$
$$x = -2 \qquad\qquad x = 3$$

The solution set is $\{-2, 3\}$.

Next, solve $x^2 - x - 6 = 0$ using a graphing calculator. Recall from Chapter 3 that to find the x-intercepts of the graph of an equation we let $y = 0$ and solve the equation for x. If we let $y = x^2 - x - 6$, then solving $x^2 - x - 6 = 0$ is the same as finding the x-intercepts of the graph of $y = x^2 - x - 6$. X-intercepts are also called zeros of the equation since they are the values of x that make $y = 0$.

Use $\boxed{\text{Y=}}$ to enter $y = x^2 - x - 6$ into the calculator, press $\boxed{\text{ZOOM}}$ and then enter 6 to display the graph using the standard viewing window as shown at right. We obtain a graph called a parabola, and we can see that it has two x-intercepts. If the scale for each tick mark on the graph is 1, then it appears that the x-intercepts are -2 and 3. To verify press $\boxed{\text{TRACE}}$, type -2, and press $\boxed{\text{ENTER}}$. Since $x = -2$ and $y = 0$, $x = -2$ is an x-intercept.

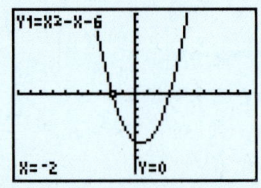

While still in "Trace mode," type 3 and press $\boxed{\text{ENTER}}$. Since $x = 3$ and $y = 0$, $x = 3$ is an x-intercept.

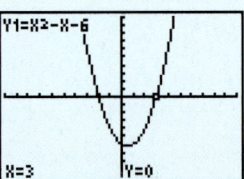

Sometimes an x-intercept is not an integer.

Solve $2x^2 + x - 15 = 0$ using a graphing calculator. Enter $2x^2 + x - 15$ into the calculator and press $\boxed{\text{GRAPH}}$. The x-intercept on the right side of the graph is between two tick marks, so it is not an integer. To find the x-intercept, press $\boxed{\text{2nd}}$ $\boxed{\text{TRACE}}$ and select 2:zero. Move the cursor to the left of one of the intercepts and press $\boxed{\text{ENTER}}$, then move the cursor again, so that it is to the right of the same intercept and press $\boxed{\text{ENTER}}$. Press $\boxed{\text{ENTER}}$ one more time, and the calculator will reveal the intercept and, therefore one solution to the equation.

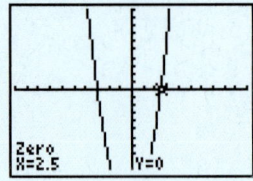

Press **2nd** **MODE** to return to the home screen. Press **X,T,Θ,n**
MATH **ENTER** **ENTER** to display the x-intercept in fraction form:

$x = \dfrac{5}{2}$. Since the other x-intercept appears to be -3, press **TRACE**

-3 **ENTER** to reveal that $x = -3$ and $y = 0$.

```
X▶Frac
              5/2
```

Solve using a graphing calculator.

1) $x^2 - 3x - 4 = 0$
2) $2x^2 + 5x - 3 = 0$
3) $x^2 - 5x - 14 = 0$
4) $3x^2 - 17x + 10 = 0$
5) $x^2 - 7x + 12 = 0$
6) $2x^2 - 9x + 10 = 0$

Answers to You Try Exercises

1) a) $\{0, 9\}$ b) $\left\{-\dfrac{2}{5}, 7\right\}$ 2) $\{-6, -3\}$ 3) a) $\{-5, 1\}$ b) $\left\{\dfrac{4}{5}, 5\right\}$ c) $\{-7, -3\}$

d) $\{0, 8\}$ e) $\left\{-2, \dfrac{4}{3}\right\}$ 4) a) $\left\{-10, -\dfrac{7}{2}, 1\right\}$ b) $\{0, 5, -5\}$

Answers to Technology Exercises

1) $\{-1, 4\}$ 2) $\left\{\dfrac{1}{2}, -3\right\}$ 3) $\{7, -2\}$ 4) $\left\{\dfrac{2}{3}, 5\right\}$ 5) $\{4, 3\}$ 6) $\left\{2, \dfrac{5}{2}\right\}$

6.4 Exercises

Additional answers can be found in the Answers to Exercises appendix.

Objective 1: Solve a Quadratic Equation of the Form $ab = 0$

1) Explain the zero product rule.

2) When Ivan solves $c(c + 7) = 0$, he gets a solution set of $\{-7\}$. Is this correct? Why or why not?
   No. The solution set is $\{-7, 0\}$.

Solve each equation.

3) $(m + 9)(m - 8) = 0$ $\{-9, 8\}$
4) $(a + 10)(a + 4) = 0$ $\{-10, -4\}$
5) $(q - 4)(q - 7) = 0$ $\{4, 7\}$
6) $(x - 5)(x + 2) = 0$ $\{-2, 5\}$
7) $(4z + 3)(z - 9) = 0$ $\left\{-\dfrac{3}{4}, 9\right\}$
8) $(2n + 1)(n - 13) = 0$ $\left\{-\dfrac{1}{2}, 13\right\}$
9) $11s(s + 15) = 0$ $\{-15, 0\}$
10) $-5r(r - 8) = 0$ $\{0, 8\}$
11) $(6x - 5)^2 = 0$ $\left\{\dfrac{5}{6}\right\}$
12) $(d + 7)^2 = 0$ $\{-7\}$
13) $(4h + 7)(h + 3) = 0$ $\left\{-3, -\dfrac{7}{4}\right\}$
14) $(8p - 5)(3p - 11) = 0$ $\left\{\dfrac{5}{8}, \dfrac{11}{3}\right\}$

15) $\left(y + \dfrac{3}{2}\right)\left(y - \dfrac{1}{4}\right) = 0$ $\left\{-\dfrac{3}{2}, \dfrac{1}{4}\right\}$
16) $\left(t - \dfrac{9}{8}\right)\left(t + \dfrac{5}{6}\right) = 0$ $\left\{-\dfrac{5}{6}, \dfrac{9}{8}\right\}$
17) $q(q - 2.5) = 0$ $\{0, 2.5\}$
18) $w(w + 0.8) = 0$ $\{-0.8, 0\}$

Objective 2: Solve Quadratic Equations by Factoring

19) Can we solve $(y + 6)(y - 11) = 8$ by setting each factor equal to 8 like this: $y + 6 = 8$ or $y - 11 = 8$? Why or why not?   No; the product of the factors must equal zero.

20) Explain two ways you could begin to solve $5n^2 - 10n - 40 = 0$. 1) Divide by 5 to get $n^2 - 2n - 8 = 0$ or 2) Factor out 5 to get $5(n^2 - 2n - 8) = 0$.

Solve each equation.

21) $v^2 + 15v + 56 = 0$ $\{-8, -7\}$
22) $y^2 + 2y - 35 = 0$ $\{-7, 5\}$
23) $k^2 + 12k - 45 = 0$ $\{-15, 3\}$
24) $z^2 - 12z + 11 = 0$ $\{1, 11\}$
25) $3y^2 - y - 10 = 0$ $\left\{-\dfrac{5}{3}, 2\right\}$

26) $4f^2 - 15f + 14 = 0$ $\left\{\frac{7}{4}, 2\right\}$

27) $14w^2 + 8w = 0$ $\left\{-\frac{4}{7}, 0\right\}$

28) $10a^2 + 20a = 0$ $\{-2, 0\}$

VIDEO 29) $d^2 - 15d = -54$ $\{6, 9\}$

30) $j^2 + 11j = -28$ $\{-7, -4\}$

31) $t^2 - 49 = 0$ $\{-7, 7\}$

32) $k^2 - 100 = 0$ $\{-10, 10\}$

33) $36 = 25n^2$ $\left\{-\frac{6}{5}, \frac{6}{5}\right\}$

34) $16 = 169p^2$ $\left\{-\frac{4}{13}, \frac{4}{13}\right\}$

35) $m^2 = 60 - 7m$ $\{-12, 5\}$

36) $g^2 + 20 = 12g$ $\{2, 10\}$

37) $55w = -20w^2 - 30$ $\left\{-2, -\frac{3}{4}\right\}$

38) $4v = 14v^2 - 48$ $\left\{-\frac{12}{7}, 2\right\}$

39) $p^2 = 11p$ $\{0, 11\}$

40) $d^2 = d$ $\{0, 1\}$

41) $45k + 27 = 18k^2$ $\left\{-\frac{1}{2}, 3\right\}$

42) $104r + 36 = 12r^2$ $\left\{-\frac{1}{3}, 9\right\}$

43) $b(b - 4) = 96$ $\{-8, 12\}$

44) $54 = w(15 - w)$ $\{6, 9\}$

45) $-63 = 4j(j - 8)$ $\left\{\frac{7}{2}, \frac{9}{2}\right\}$

46) $g(3g + 11) = 70$ $\left\{-7, \frac{10}{3}\right\}$

47) $10x(x + 1) - 6x = 9(x^2 + 5)$ $\{-9, 5\}$

48) $5r(3r + 7) = 2(4r^2 - 21)$ $\{-3, -2\}$

49) $3(h^2 - 4) = 5h(h - 1) - 9h$ $\{1, 6\}$

50) $5(5 + u^2) + 10 = 3u(2u + 1) - u$ $\{-7, 5\}$

VIDEO 51) $\frac{1}{2}(m + 1)^2 = -\frac{3}{4}m(m + 5) - \frac{5}{2}$ $\left\{-3, -\frac{4}{5}\right\}$

52) $(2y - 3)^2 + y = (y - 5)^2 - 6$ $\left\{-\frac{5}{3}, 2\right\}$

53) $3t(t - 5) + 14 = 5 - t(t + 3)$ $\left\{\frac{3}{2}\right\}$

54) $\frac{1}{2}c(2 - c) - \frac{3}{2} = \frac{2}{5}c(c + 1) - \frac{7}{5}$ $\left\{\frac{1}{3}\right\}$

55) $33 = -m(14 + m)$ $\{-11, -3\}$

56) $-84 = s(s + 19)$ $\{-12, -7\}$

57) $(3w + 2)^2 - (w - 5)^2 = 0$ $\left\{-\frac{7}{2}, \frac{3}{4}\right\}$

58) $(2j - 7)^2 - (j + 3)^2 = 0$ $\left\{\frac{4}{3}, 10\right\}$

59) $(q + 3)^2 - (2q - 5)^2 = 0$ $\left\{\frac{2}{3}, 8\right\}$

60) $(6n + 5)^2 - (3n + 4)^2 = 0$ $\left\{-1, -\frac{1}{3}\right\}$

Objective 3: Solve Higher Degree Equations by Factoring

The following equations are not quadratic but can be solved by factoring and applying the zero product rule. Solve each equation.

61) $8y(y + 4)(2y - 1) = 0$ $\left\{-4, 0, \frac{1}{2}\right\}$

62) $-13b(12b + 7)(b - 11) = 0$ $\left\{-\frac{7}{12}, 0, 11\right\}$

63) $(9p - 2)(p^2 - 10p - 11) = 0$ $\left\{-1, \frac{2}{9}, 11\right\}$

64) $(4f + 5)(f^2 - 3f - 18) = 0$ $\left\{-3, -\frac{5}{4}, 6\right\}$

65) $(2r - 5)(r^2 - 6r + 9) = 0$ $\left\{\frac{5}{2}, 3\right\}$

66) $(3x - 1)(x^2 - 16x + 64) = 0$ $\left\{\frac{1}{3}, 8\right\}$

67) $m^3 = 64m$ $\{0, -8, 8\}$

68) $r^3 = 81r$ $\{0, -9, 9\}$

VIDEO 69) $5w^2 + 36w = w^3$ $\{-4, 0, 9\}$

70) $14a^2 - 49a = a^3$ $\{0, 7\}$

71) $2g^3 = 120g - 14g^2$ $\{-12, 0, 5\}$

72) $36z - 24z^2 = -3z^3$ $\{0, 2, 6\}$

73) $45h = 20h^3$ $\left\{0, -\frac{3}{2}, \frac{3}{2}\right\}$

74) $64d^3 = 100d$ $\left\{0, -\frac{5}{4}, \frac{5}{4}\right\}$

75) $2s^2(3s + 2) + 3s(3s + 2) - 35(3s + 2) = 0$ $\left\{-5, -\frac{2}{3}, \frac{7}{2}\right\}$

76) $10n^2(n - 8) + n(n - 8) - 2(n - 8) = 0$ $\left\{-\frac{1}{2}, \frac{2}{5}, 8\right\}$

77) $10a^2(4a + 3) + 2(4a + 3) = 9a(4a + 3)$ $\left\{-\frac{3}{4}, \frac{2}{5}, \frac{1}{2}\right\}$

78) $12d^2(7d - 3) = 5d(7d - 3) + 2(7d - 3)$ $\left\{-\frac{1}{4}, \frac{3}{7}, \frac{2}{3}\right\}$

79) $t^3 + 6t^2 - 4t - 24 = 0$ $\{-6, -2, 2\}$

80) $k^3 - 8k^2 - 9k + 72 = 0$ $\{-3, 3, 8\}$

Find the indicated values for the following polynomial functions.

81) $f(x) = x^2 + 10x + 21$. Find x so that $f(x) = 0$. $-7, -3$

82) $h(t) = t^2 - 6t - 16$. Find t so that $h(t) = 0$. $-2, 8$

VIDEO 83) $g(a) = 2a^2 - 13a + 24$. Find a so that $g(a) = 4$. $\frac{5}{2}, 4$

84) $Q(x) = 4x^2 - 4x + 9$. Find x so that $Q(x) = 8$. $\frac{1}{2}$

85) $H(b) = b^2 + 3$. Find b so that $H(b) = 19$. $-4, 4$

86) $f(z) = z^3 + 3z^2 - 54z + 5$. Find z so that $f(z) = 5$. $-9, 0, 6$

87) $h(k) = 5k^3 - 25k^2 + 20k$. Find k so that $h(k) = 0$. $0, 1, 4$

88) $g(x) = 9x^2 - 10$. Find x so that $g(x) = -6$. $-\frac{2}{3}, \frac{2}{3}$

Section 6.5 Applications of Quadratic Equations

Objectives

1. **Solve Problems Involving Geometry**
2. **Solve Problems Involving Consecutive Integers**
3. **Solve Problems Using the Pythagorean Theorem**
4. **Solve an Applied Problem Using a Given Quadratic Equation**

In Chapters 2 and 4 we explored applications of linear equations. In this section we will look at applications involving quadratic equations. Let's begin by restating the five steps for solving applied problems.

Procedure   Steps for Solving Applied Problems

Step 1: **Read** the problem carefully, more than once if necessary, until you understand it. Draw a picture, if applicable. Identify what you are being asked to find.

Step 2: **Choose a variable** to represent an unknown quantity. If there are any other unknowns, define them in terms of the variable.

Step 3: **Translate** the problem from English into an equation using the chosen variable.

Step 4: **Solve** the equation.

Step 5: **Check** the answer in the original problem, and **interpret** the solution as it relates to the problem. Be sure your answer makes sense in the context of the problem.

---

1. Solve Problems Involving Geometry

Example 1

In-Class Example 1

Solve. A rectangular vegetable garden is 4 ft longer than it is wide. What are the dimensions if it covers 60 ft²?

answer: The length is 10 ft and the width is 6 ft.

Solve. A rectangular vegetable garden is 7 ft longer than it is wide. What are the dimensions of the garden if it covers 60 ft²?

Solution

Step 1: **Read** the problem carefully. Draw a picture.

Step 2: **Choose a variable** to represent the unknown.

$$\text{Let} \qquad w = \text{the width}$$
$$w + 7 = \text{the length}$$

Step 3: **Translate** the information that appears in English into an algebraic equation. We must find the length and width of the garden. We are told that the *area* is 60 ft², so let's use the formula for the area of a rectangle. Then, substitute the expressions above for the length and width and 60 for the area.

$$(length)(width) = Area$$
$$(w + 7)(w) = 60 \qquad \text{length} = w + 7, \text{ width} = w, \text{ area} = 60$$

Step 4: **Solve** the equation.

$$w^2 + 7w = 60 \qquad \text{Distribute.}$$
$$w^2 + 7w - 60 = 0 \qquad \text{Write the equation in standard form.}$$
$$(w + 12)(w - 5) = 0 \qquad \text{Factor.}$$

$$w + 12 = 0 \qquad \text{or} \qquad w - 5 = 0 \qquad \text{Set each factor equal to zero.}$$
$$w = -12 \qquad \text{or} \qquad w = 5 \qquad \text{Solve.}$$

Step 5: **Check** the answer and **interpret** the solution as it relates to the problem. Since w represents the width of the garden, it cannot be a negative number. So, $w = -12$ cannot be the solution. Therefore, the width is 5 ft, which will make the height $5 + 7 = 12$ ft. The area, then, is $(12 \text{ ft}) \cdot (5 \text{ ft}) = 60 \text{ ft}^2$. ■

You Try 1

Solve. The area of the surface of a desk is 8 ft^2. Find the dimensions of the desktop if the width is 2 ft less than the length.

2. Solve Problems Involving Consecutive Integers

In Chapter 2 we solved problems involving consecutive integers. Some applications involving consecutive integers lead to quadratic equations.

Example 2

Solve. Twice the sum of three consecutive odd integers is 9 less than the product of the smaller two. Find the integers.

In-Class Example 2
Solve. Four times the sum of three consecutive even integers is twenty-four less than the product of the larger two. Find the integers.

answer: The even integers are 10, 12, 14 or −4, −2, 0.

Solution

Step 1: **Read** the problem carefully, and identify what we are being asked to find.

We must find three consecutive odd integers.

Step 2: **Choose a variable** to represent an unknown, and define the other unknowns in terms of this variable.

$$x = \text{the first odd integer}$$
$$x + 2 = \text{the second odd integer}$$
$$x + 4 = \text{the third odd integer}$$

Step 3: **Translate** the information that appears in English into an algebraic equation. Read the problem slowly and carefully, breaking it into small parts.

Statement: [Twice the sum of three consecutive odd integers] [is] [9 less than] [the product of the smaller two.]

Equation: $2[x + (x + 2) + (x + 4)] = x(x + 2) \quad -9$

Step 4: **Solve** the equation.

$$2[x + (x + 2) + (x + 4)] = x(x + 2) - 9$$
$$2(3x + 6) = x^2 + 2x - 9 \qquad \text{Combine like terms; distribute.}$$
$$6x + 12 = x^2 + 2x - 9 \qquad \text{Distribute.}$$
$$0 = x^2 - 4x - 21 \qquad \text{Write in standard form.}$$
$$0 = (x + 3)(x - 7) \qquad \text{Factor.}$$

$$x + 3 = 0 \quad \text{or} \quad x - 7 = 0 \qquad \text{Set each factor equal to zero.}$$
$$x = -3 \qquad\qquad x = 7 \qquad \text{Solve.}$$

Step 5: **Check** the answer and **interpret** the solution as it relates to the problem.

We get two sets of solutions. If $x = -3$, then the other odd integers are -1 and 1. If $x = 7$, the other odd integers are 9 and 11.

Check these numbers in the original statement of the problem.

$$2[-3 + (-1) + 1] = (-3)(-1) - 9 \qquad\qquad 2[7 + 9 + 11] = (7)(9) - 9$$
$$2(-3) = 3 - 9 \qquad\qquad\qquad\qquad 2(27) = 63 - 9$$
$$-6 = -6 \qquad\qquad\qquad\qquad\qquad 54 = 54$$

You Try 2

Solve. Find three consecutive even integers such that the product of the two smaller numbers is the same as twice the sum of the integers.

3. Solve Problems Using the Pythagorean Theorem

A **right triangle** is a triangle that contains a 90° **(right)** angle. We can label a right triangle as follows.

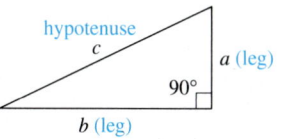

The side opposite the 90° angle is the longest side of the triangle and is called the **hypotenuse.** The other two sides are called the **legs.** The Pythagorean theorem states a relationship between the lengths of the sides of a right triangle. This is a very important relationship in mathematics and is one which is used in many different ways.

Definition   Pythagorean Theorem

Given a right triangle with legs of length a and b and hypotenuse of length c,

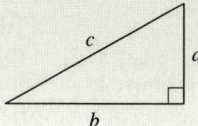

the Pythagorean theorem states that $a^2 + b^2 = c^2$ [or $(\text{leg})^2 + (\text{leg})^2 = (\text{hypotenuse})^2$].

The Pythagorean theorem is true *only* for right triangles.

Example 3

Find the length of the missing side.

In-Class Example 3
Find the length of the missing side.

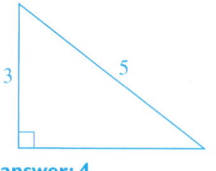

answer: 4

Solution

Since this is a right triangle, we can use the Pythagorean theorem to find the length of the side. Let a represent its length, and label the triangle.

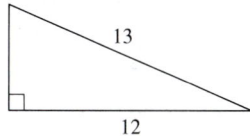

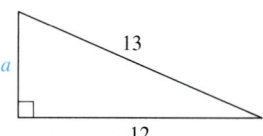

The length of the hypotenuse is 13, so $c = 13$. a and 12 are legs. Let $b = 12$.

$$\begin{aligned}
a^2 + b^2 &= c^2 & &\text{Pythagorean theorem} \\
a^2 + (12)^2 &= (13)^2 & &\text{Substitute values.} \\
a^2 + 144 &= 169 & & \\
a^2 - 25 &= 0 & &\text{Write the equation in standard form.} \\
(a + 5)(a - 5) &= 0 & &\text{Factor.}
\end{aligned}$$

$$a + 5 = 0 \quad \text{or} \quad a - 5 = 0 \qquad \text{Set each factor equal to 0.}$$
$$a = -5 \quad \text{or} \qquad a = 5 \qquad \text{Solve.}$$

$a = -5$ does not make sense as an answer because the length of a side of a triangle cannot be negative. Therefore, $a = 5$.

Check: $\quad 5^2 + (12)^2 \overset{?}{=} (13)^2$
$$25 + 144 = 169 \quad \checkmark$$

You Try 3

Find the length of the missing side.

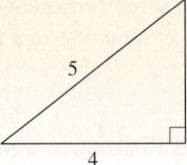

Example 4

In-Class Example 4
See Example 4.

x $x + 5$

$x + 10$

answer: The length of the
fence is 15 ft.

Solve. An animal holding pen situated between two buildings at a right angle with each other will have walls as two of its sides and a fence on the longest side. The side with the fence is 20 ft longer than the shortest side, while the third side is 10 ft longer than the shortest side. Find the length of the fence.

Solution

Step 1: **Read** the problem carefully, and identify what we are being asked to find. Draw a picture.

> We must find the length of the fence.

Step 2: **Choose a variable** to represent an unknown, and define the other unknowns in terms of this variable. Draw and label the picture.

$x + 20$

x

$x + 10$

$x = $ length of the shortest side (a leg)
$x + 10 = $ length of the side along other building (a leg)
$x + 20 = $ length of the fence (hypotenuse)

Step 3: **Translate** the information that appears in English into an algebraic equation. We will use the Pythagorean theorem.

$$a^2 + b^2 = c^2 \qquad \text{Pythagorean theorem}$$
$$x^2 + (x + 10)^2 = (x + 20)^2 \qquad \text{Substitute.}$$

Step 4: **Solve** the equation.

$$x^2 + (x + 10)^2 = (x + 20)^2$$
$$x^2 + x^2 + 20x + 100 = x^2 + 40x + 400 \qquad \text{Multiply using FOIL.}$$
$$2x^2 + 20x + 100 = x^2 + 40x + 400$$
$$x^2 - 20x - 300 = 0 \qquad \text{Write in standard form.}$$
$$(x - 30)(x + 10) = 0 \qquad \text{Factor.}$$

$x - 30 = 0$ or $x + 10 = 0$ Set each factor equal to 0.
$x = 30$ or $x = -10$ Solve.

Step 5: **Check** the answer and **interpret** the solution as it relates to the problem.

The length of the shortest side, x, cannot be a negative number, so x cannot equal -10. Therefore, the length of the shortest side must be 30 ft.

The length of the side along the other building is $x + 10$, so $30 + 10 = 40$ ft.

The length of the fence is $x + 20$, so $30 + 20 = 50$ ft.

Do these lengths satisfy the Pythagorean theorem? Yes.

$$a^2 + b^2 = c^2$$
$$(30)^2 + (40)^2 \overset{?}{=} (50)^2$$
$$900 + 1600 = 2500 \;\checkmark$$

Therefore, the length of the fence is 50 ft.

You Try 4

Solve. A wire is attached to the top of a pole. The wire is 4 ft longer than the pole, and the distance from the wire on the ground to the bottom of the pole is 4 ft less than the height of the pole. Find the length of the wire and the height of the pole.

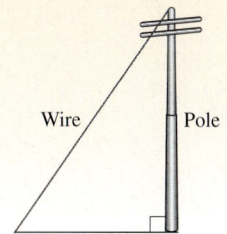

Wire   Pole

Next we will see how to use quadratic equations that model real-life situations.

4. Solve an Applied Problem Using a Given Quadratic Equation

Example 5

In-Class Example 5
Use Example 5.

A Little League baseball player throws a ball upward. The height h of the ball (in feet) t sec after the ball is released is given by the quadratic equation

$$h = -16t^2 + 30t + 4$$

a)   What is the initial height of the ball?

b)   How long does it take the ball to reach a height of 18 ft?

c)   How long does it take for the ball to hit the ground?

Solution

a)   We are asked to find the height at which the ball is released. Since t represents the number of seconds after the ball is thrown, $t = 0$ at the time of release.

Let $t = 0$ and solve for h.

$$h = -16(0)^2 + 30(0) + 4 \qquad \text{Substitute 0 for } t.$$
$$= 0 + 0 + 4$$
$$= 4$$

The initial height of the ball is 4 ft.

b)   We must find the *time* it takes for the ball to reach a height of 18 ft.

Find t when $h = 18$.

$$h = -16t^2 + 30t + 4$$
$$18 = -16t^2 + 30t + 4 \qquad \text{Substitute 18 for } h.$$
$$0 = -16t^2 + 30t - 14 \qquad \text{Write in standard form.}$$
$$0 = 8t^2 - 15t + 7 \qquad \text{Divide by } -2.$$
$$0 = (8t - 7)(t - 1) \qquad \text{Factor.}$$

$$8t - 7 = 0 \quad \text{or} \quad t - 1 = 0 \qquad \text{Set each factor equal to 0.}$$
$$8t = 7$$

$$t = \frac{7}{8} \quad \text{or} \qquad t = 1 \qquad \text{Solve.}$$

How can two answers be possible? After $\dfrac{7}{8}$ sec the ball is 18 ft above the ground *on its way up,* and after 1 sec, the ball is 18 ft above the ground *on its way down.*

The ball reaches a height of 18 ft after $\dfrac{7}{8}$ sec *and* after 1 sec.

c) We must determine the amount of time it takes for the ball to hit the ground. When the ball hits the ground, how high off of the ground is it? *It is 0 ft high.* Find t when $h = 0$.

$$h = -16t^2 + 30t + 4$$
$$0 = -16t^2 + 30t + 4 \qquad \text{Substitute 0 for } h.$$
$$0 = 8t^2 - 15t - 2 \qquad \text{Divide by } -2.$$
$$0 = (8t + 1)(t - 2) \qquad \text{Factor.}$$

$$8t + 1 = 0 \quad \text{or} \quad t - 2 = 0 \qquad \text{Set each factor equal to 0.}$$
$$8t = -1$$
$$t = -\dfrac{1}{8} \quad \text{or} \qquad t = 2 \qquad \text{Solve.}$$

Since t represents time, t cannot equal $-\dfrac{1}{8}$. We reject that as a solution. Therefore, $t = 2$. The ball will hit the ground after 2 sec. ■

Note

In Example 5, the equation can also be written using function notation $h(t) = -16t^2 + 30t + 4$ since the expression $-16t^2 + 30t + 4$ is a polynomial. Furthermore, $h(t) = -16t^2 + 30t + 4$ is a *quadratic function,* and we say that the height, h, is a function of the time, t. We will study quadratic functions in more detail in Chapter 9.

You Try 5

An object is thrown upward from a building. The height h of the object (in feet) t sec after the object is released is given by the quadratic equation

$$h = -16t^2 + 36t + 36.$$

a) What is the initial height of the object?

b) How long does it take the object to reach a height of 44 ft?

c) How long does it take for the object to hit the ground?

Answers to You Try Exercises

1) width = 2 ft; length = 4 ft   2) 6, 8, 10 or −2, 0, 2   3) 3

4) length of wire = 20 ft; height of pole = 16 ft

5) a) 36 ft   b) 0.25 sec and 2 sec   c) 3 sec

6.5 Exercises

*Additional answers can be found in the Answers to Exercises appendix.

Objective 1: Solve Problems Involving Geometry

Find the length and width of each rectangle.

1) Area = 36 in^2

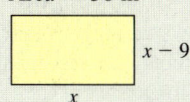
x − 9
x

length = 12 in.; width = 3 in.

2) Area = 40 cm^2

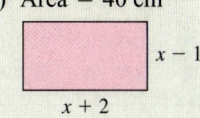
x − 1
x + 2

length = 8 cm; width = 5 cm

Find the base and height of each triangle.

3) Area = 12 cm^2

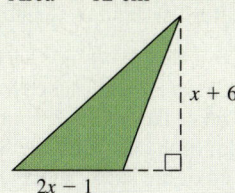
x + 6
2x − 1

base = 3 cm; height = 8 cm

4) Area = 42 in^2

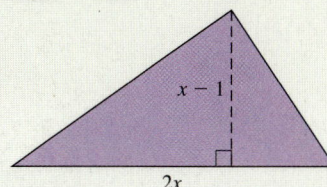
x − 1
2x

base = 14 in.; height = 6 in.

Find the base and height of each parallelogram.

5) Area = 18 in^2

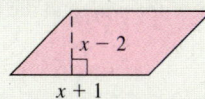
x − 2
x + 1

base = 6 in.; height = 3 in.

6) Area = 50 cm^2

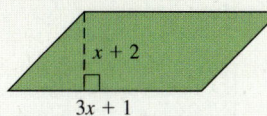
x + 2
3x + 1

base = 10 cm; height = 5 cm

7) The volume of the box is 240 in^3. Find its length and width.

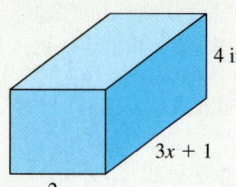
4 in.
3x + 1
2x

length = 10 in.; width = 6 in.

8) The volume of the box is 120 in^3. Find its width and height.

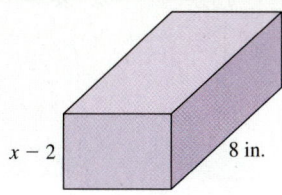
x − 2
8 in.
x

width = 5 in.; height = 3 in.

Write an equation and solve.

9) A rectangular rug is 4 ft longer than it is wide. If its area is 45 ft^2, what is its length and width?

length = 9 ft; width = 5 ft

10) The surface of a rectangular bulletin board has an area of 300 in^2. Find its dimensions if it is 5 in. longer than it is wide.   length = 20 in.; width = 15 in.

11) Judy makes stained glass windows. She needs to cut a rectangular piece of glass with an area of 54 in^2 so that its width is 3 in. less than its length. Find the dimensions of the glass she must cut.

length = 9 in.; width = 6 in.

12) A rectangular painting is twice as long as it is wide. Find its dimensions if it has an area of 12.5 ft^2.

length = 5 ft; width = 2.5 ft

13) The volume of a rectangular storage box is 1440 in^3. It is 20 in. long, and it is half as tall as it is wide. Find the width and height of the box.   width = 12 in.; height = 6 in.

14) A rectangular aquarium is 15 in. high, and its length is 8 in. more than its width. Find the length and width if the volume of the aquarium is 3600 in^3.

length = 20 in.; width = 12 in.

15) The height of a triangle is 3 cm more than its base. Find the height and base if its area is 35 cm^2.

height = 10 cm; base = 7 cm

16) The area of a triangle is 16 cm^2. Find the height and base if its height is half the length of the base.

height = 4 cm; base = 8 cm

Objective 2: Solve Problems Involving Consecutive Integers

Write an equation and solve.

17) The product of two consecutive odd integers is 1 less than three times their sum. Find the integers.   5 and 7 or −1 and 1

18) The product of two consecutive integers is 19 more than their sum. Find the integers.   5 and 6 or −4 and −3

19) Find three consecutive even integers such that the sum of the smaller two is one-fourth the product of the second and third integers.   0, 2, 4 or 2, 4, 6

20) Find three consecutive integers such that the square of the smallest is 29 less than the product of the larger two.   9, 10, 11

21) Find three consecutive integers such that the square of the largest is 22 more than the product of the smaller two.   6, 7, 8

22) Find three consecutive odd integers such that the product of the smaller two is 15 more than four times the sum of the three integers.   13, 15, 17 or −3, −1, 1

Objective 3: Solve Problems Using the Pythagorean Theorem

23) In your own words, explain the Pythagorean theorem.   Answers may vary.

24) Can the Pythagorean theorem be used to find *a* in this triangle? Why or why not?

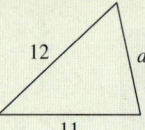

No; it is not a right triangle.

Use the Pythagorean theorem to find the length of the missing side.

25)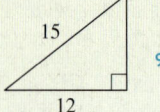

26)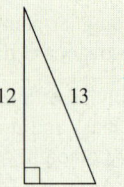

27) 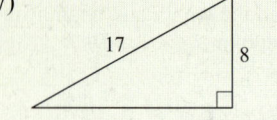   15

28) 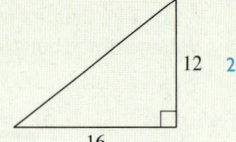   20

29)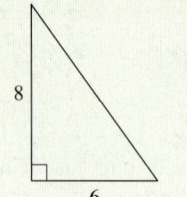

30)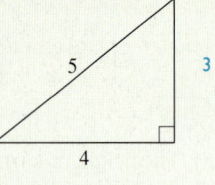

Find the lengths of the sides of each right triangle.

31) 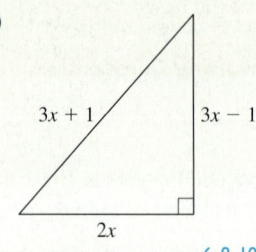   6, 8, 10

32) 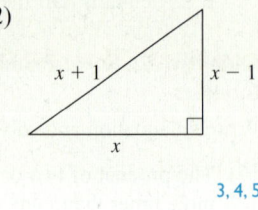   3, 4, 5

33) 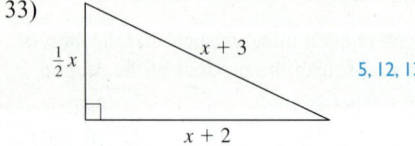   5, 12, 13

34) 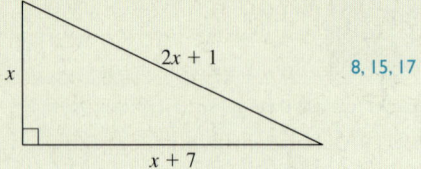   8, 15, 17

Write an equation and solve.

35) The hypotenuse of a right triangle is 2 in. longer than the longer leg. The shorter leg measures 2 in. less than the longer leg. Find the measure of the longer leg of the triangle.   8 in.

36) The longer leg of a right triangle is 7 cm more than the shorter leg. The length of the hypotenuse is 3 cm more than twice the length of the shorter leg. Find the length of the hypotenuse.   13 cm

37) A 13-ft ladder is leaning against a wall. The distance from the top of the ladder to the bottom of the wall is 7 ft more than the distance from the bottom of the ladder to the wall. Find the distance from the bottom of the ladder to the wall.

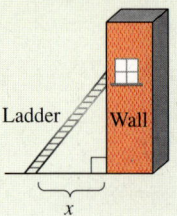

38) A wire is attached to the top of a pole. The pole is 2 ft shorter than the wire, and the distance from the wire on the ground to the bottom of the pole is 9 ft less than the length of the wire. Find the length of the wire and the height of the pole.

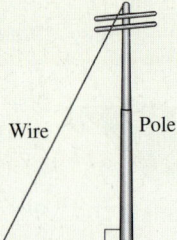

length of wire = 17 ft
length of pole = 15 ft

39) From a bike shop, Rana pedals due north while Yasmeen rides due west. When Yasmeen is 4 mi from the shop, the distance between her and Rana is two miles more than Rana's distance from the bike shop. Find the distance between Rana and Yasmeen.

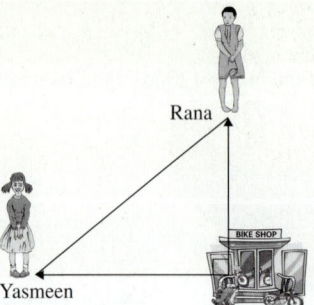

40) Henry and Allison leave home to go to work. Henry drives due west while his wife drives due south. At 8:30 am, Allison is 3 mi farther from home than Henry, and the distance between them is 6 mi more than Henry's distance from home. Find Henry's distance from his house.

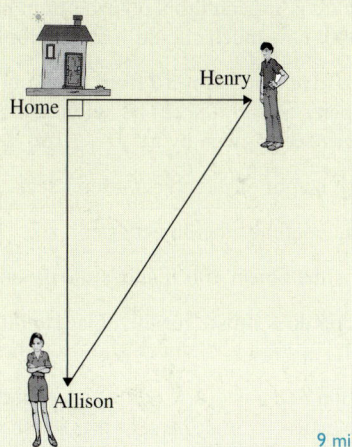

9 mi

Objective 4: Solve Applied Problems Using Given Quadratic Equations

Solve.

41) A rock is dropped from a cliff and into the ocean. The height h (in feet) of the rock after t sec is given by $h = -16t^2 + 144$.

a) What is the initial height of the rock?   144 ft

b) When is the rock 80 ft above the water?   after 2 sec

c) How long does it take the rock to hit the water?   3 sec

42) An object is launched from a platform with an initial velocity of 32 ft/sec. The height h (in feet) of the object after t sec is given by $h = -16t^2 + 32t + 20$.

a) What is the initial height of the object?   20 ft

b) When is the object 32 ft above the ground? $\frac{1}{2}$ sec, $1\frac{1}{2}$ sec

c) How long does it take for the object to hit the ground? $2\frac{1}{2}$ sec

Organizers of fireworks shows use quadratic and linear equations to help them design their programs. *Shells* contain the chemicals that produce the bursts we see in the sky. At a fireworks show the shells are shot from *mortars* and when the chemicals inside the shells ignite they explode, producing the brilliant bursts we see in the night sky.

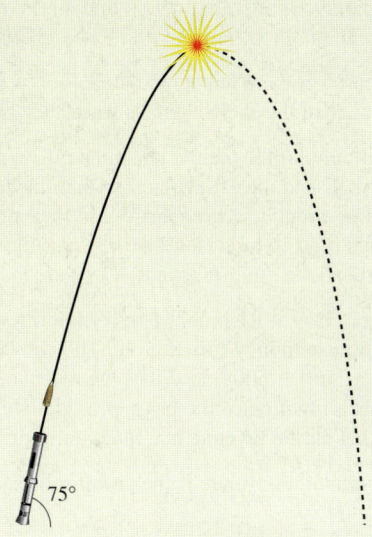

75°

43) At a fireworks show, a 3-in. shell is shot from a mortar at an angle of 75°. The height, y (in feet), of the shell t sec after being shot from the mortar is given by the quadratic equation

$$y = -16t^2 + 144t$$

and the horizontal distance of the shell from the mortar, x (in feet), is given by the linear equation

$$x = 39t$$

(http://library.thinkquest.org/15384/physics/physics.html)

a) How high is the shell after 3 sec?   288 ft

b) What is the shell's horizontal distance from the mortar after 3 sec?   117 ft

c) The maximum height is reached when the shell explodes. How high is the shell when it bursts after 4.5 sec?   324 ft

d) What is the shell's horizontal distance from its launching point when it explodes? (Round to the nearest foot.)   176 ft

44) When a 10-in. shell is shot from a mortar at an angle of 75°, the height, y (in feet), of the shell t sec after being shot from the mortar is given by

$$y = -16t^2 + 264t$$

and the horizontal distance of the shell from the mortar, x (in feet), is given by

$$x = 71t$$

a) How high is the shell after 3 sec?   648 ft

b) Find the shell's horizontal distance from the mortar after 3 sec.   213 ft

c) The shell explodes after 8.25 sec. What is its height when it bursts?   1089 ft

d) What is the shell's horizontal distance from its launching point when it explodes? (Round to the nearest foot.)   586 ft

e) Compare your answers to 43a) and 44a). What is the difference in their heights after 3 sec?   360 ft

f) Compare your answers to 43c) and 44c). What is the difference in the shells' heights when they burst?   765 ft

g) Assuming that the technicians timed the firings of the 3-in. shell and the 10-in. shell so that they exploded at the same time, how far apart would their respective mortars need to be so that the 10-in. shell would burst directly above the 3-in. shell?

45) The senior class at Richmont High School is selling t-shirts to raise money for its prom. The equation $R(p) = -25p^2 + 600p$ describes the revenue, R, in dollars, as a function of the price, p, in dollars, of a t-shirt. That is, the revenue is a function of price.

a) Determine the revenue if the group sells each shirt for $10. $3500

b) Determine the revenue if the group sells each shirt for $15. $3375

c) If the senior class hopes to have a revenue of $3600, how much should it charge for each t-shirt? $12

46) A famous comedian will appear at a comedy club for one performance. The equation $R(p) = -5p^2 + 300p$ describes the relationship between the price of a ticket, p, in dollars, and the revenue, R, in dollars, from ticket sales. That is, the revenue is a function of price.

a) Determine the club's revenue from ticket sales if the price of a ticket is $40. $4000

b) Determine the club's revenue from ticket sales if the price of a ticket is $25. $4375

c) If the club is expecting its revenue from ticket sales to be $4500, how much should it charge for each ticket? $30

47) An object is launched upward from the ground with an initial velocity of 200 ft/sec. The height h (in feet) of the object after t sec is given by $h(t) = -16t^2 + 200t$.

a) Find the height of the object after 1 sec.   184 ft

b) Find the height of the object after 4 sec.   544 ft

c) When is the object 400 ft above the ground?

d) How long does it take for the object to hit the ground? $12\frac{1}{2}$ sec

48) The equation $R(p) = -7p^2 + 700p$ describes the revenue from ticket sales, R, in dollars, as a function of the price, p, in dollars, of a ticket to a fundraising dinner. That is, the revenue is a function of price.

a) Determine the revenue if the ticket price is $40. $16,800

b) Determine the revenue if the group sells each ticket for $70. $14,700

c) If the goal of the organizers is to have ticket revenue of $17,500, how much should it charge for each ticket? $50

Chapter 6: Summary

Definition/Procedure	Example

6.1 The Greatest Common Factor and Factoring by Grouping

To **factor a polynomial** is to write it as a product of two or more polynomials:

To factor out a greatest common factor (GCF),

1) Identify the GCF of all of the terms of the polynomial.
2) Rewrite each term as the product of the GCF and another factor.
3) Use the distributive property to factor out the GCF from the terms of the polynomial.
4) Check the answer by multiplying the factors. **(p. 312)**

Factor out the greatest common factor.

$$16d^6 - 40d^5 + 72d^4$$

The GCF is $8d^4$.

$$16d^6 - 40d^5 + 72d^4 = (8d^4)(2d^2) - (8d^4)(5d) + (8d^4)(9)$$
$$= 8d^4(2d^2 - 5d + 9)$$

Check: $8d^4(2d^2 - 5d + 9) = 16d^6 - 40d^5 + 72d^4$ ✓

The first step in factoring any polynomial is to ask yourself, "*Can I factor out a GCF?*"

The last step in factoring any polynomial is to ask yourself, "*Can I factor again?*"

Try to **factor by grouping** when you are asked to factor a polynomial containing four terms.

1) Make two groups of two terms so that each group has a common factor.
2) Take out the common factor from each group of terms.
3) Factor out the common factor using the distributive property.
4) Check the answer by multiplying the factors. **(p. 316)**

Factor completely. $45tu + 27t + 20u + 12$

Since the four terms have a GCF of 1, we will not factor out a GCF. Begin by grouping two terms together so that each group has a common factor.

$$\underbrace{45tu + 27t}_{\downarrow} + \underbrace{20u + 12}_{\downarrow}$$

$= 9t(5u + 3) + 4(5u + 3)$ Take out the common factor.
$= (5u + 3)(9t + 4)$ Factor out $(5u + 3)$.

Check: $(5u + 3)(9t + 4) = 45tu + 27t + 20u + 12$ ✓

6.2 Factoring Trinomials

Factoring $x^2 + bx + c$

If $x^2 + bx + c = (x + m)(x + n)$, then

1) if b and c are positive, then both m and n must be positive.
2) if c is positive and b is negative, then both m and n must be negative.
3) if c is negative, then one integer, m, must be positive and the other integer, n, must be negative. **(p. 318)**

Factor completely.
a) $y^2 + 7y + 12$

Think of two numbers whose *product* is 12 and whose *sum* is 7. 3 *and* 4. Then,

$$y^2 + 7y + 12 = (y + 3)(y + 4)$$

b) $2r^3 - 26r^2 + 60r$

Begin by factoring out the GCF of $2r$.

$$2r^3 - 26r^2 + 60r = 2r(r^2 - 13r + 30)$$
$$= 2r(r - 10)(r - 3)$$

Factoring $ax^2 + bx + c$ by *Grouping* (p. 321)

Factor completely. $5n^2 + 18n - 8$

$$\overset{\text{Sum is 18}}{\underset{\downarrow}{5n^2 + 18n - 8}}$$

Product: $5 \cdot (-8) = -40$

Think of two integers whose *product* is -40 and whose *sum* is 18.   20 *and* -2

Definition/Procedure	Example
	Factor by grouping.
	$5n^2 + 18n - 8 = \underbrace{5n^2 + 20n}\; \underbrace{- 2n - 8}$ Write $18n$ as $20n - 2n$. $= 5n(n + 4) - 2(n + 4)$ $= (n + 4)(5n - 2)$
Factoring $ax^2 + bx + c$ by _Trial and Error_ When approaching a problem in this way, we must keep in mind that we are reversing the FOIL process. **(p. 322)**	Factor completely. $4x^2 - 16x + 15$ $4x^2 - 16x + 15 = (2x - 3)(2x - 5)$ $\begin{array}{l} -6x \\ + -10x \\ \hline -16x \end{array}$ $4x^2 - 16x + 15 = (2x - 3)(2x - 5)$

6.3 Special Factoring Techniques

A **perfect square trinomial** is a trinomial that results from squaring a binomial.
 	
Factoring a Perfect Square Trinomial:
 	
$a^2 + 2ab + b^2 = (a + b)^2$
$a^2 - 2ab + b^2 = (a - b)^2$ **(p. 328)**
	Factor completely.
	
	a) $c^2 + 24c + 144 = (c + 12)^2$
	   $a = c$ $b = 12$
	
	b) $49p^2 - 56p + 16 = (7p - 4)^2$
	   $a = 7p$ $b = 4$

Factoring the Difference of Two Squares: $a^2 - b^2 = (a + b)(a - b)$ **(p. 330)**	Factor completely. $d^2 - 16 = (d + 4)(d - 4)$ $\downarrow$ $\downarrow$ $(d)^2$ $(4)^2$ $a = d,$ $b = 4$
Factoring the Sum and Difference of Two Cubes: $a^3 + b^3 = (a + b)(a^2 - ab + b^2)$ $a^3 - b^3 = (a - b)(a^2 + ab + b^2)$ **(p. 333)**	Factor completely. $w^3 + 27 = (w + 3)[(w)^2 - (w)(3) + (3)^2]$ $\downarrow$ $\downarrow$ $(w)^3$ $(3)^3$ $a = w,$ $b = 3$ $w^3 + 27 = (w + 3)(w^2 - 3w + 9)$

6.4 Solving Quadratic Equations by Factoring

A **quadratic equation** can be written in the form $ax^2 + bx + c = 0$, where $a, b,$ and c are real numbers and $a \neq 0$. **(p. 341)**	Some examples of quadratic equations are $3x^2 - 5x + 9 = 0,$ $t^2 = 4t + 21,$ and $2(p - 3)^2 = 8 - 7p.$
To solve a quadratic equation by factoring, use the **zero product rule:** If $ab = 0$, then $a = 0$ or $b = 0$. **(p. 341)**	Solve $(y + 9)(y - 4) = 0$ $\swarrow$ $\searrow$ $y + 9 = 0$ or $y - 4 = 0$ Set each factor equal to zero. $y = -9$ or $y = 4$ Solve. The solution set is $\{-9, 4\}$.

Definition/Procedure	Example

Steps for Solving a Quadratic Equation by Factoring

1) Write the equation in the form $ax^2 + bx + c = 0$.
2) Factor the expression.
3) Set each factor equal to zero, and solve for the variable.
4) Check the answer(s). **(p. 343)**

Solve $4m^2 - 11 = m^2 + 2m - 3$.

$$3m^2 - 2m - 8 = 0 \qquad \text{Standard form}$$
$$(3m + 4)(m - 2) = 0 \qquad \text{Factor.}$$

$$3m + 4 = 0 \qquad \text{or} \qquad m - 2 = 0$$
$$3m = -4$$
$$m = -\frac{4}{3} \quad \text{or} \qquad m = 2$$

The solution set is $\left\{-\dfrac{4}{3}, 2\right\}$. Check the answers.

6.5 Applications of Quadratic Equations

Pythagorean Theorem

Given a right triangle with legs of length a and b and hypotenuse of length c,

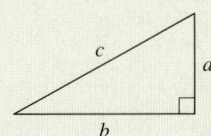

the Pythagorean theorem states that

$$a^2 + b^2 = c^2 \quad \text{(p. 351)}$$

Find the length of side a.

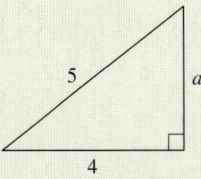

Let $b = 4$ and $c = 5$ in $a^2 + b^2 = c^2$.

$$a^2 + (4)^2 = (5)^2$$
$$a^2 + 16 = 25$$
$$a^2 - 9 = 0$$
$$(a + 3)(a - 3) = 0$$

$$a + 3 = 0 \qquad \text{or} \quad a - 3 = 0$$
$$a = -3 \quad \text{or} \qquad a = 3$$

Reject -3 as a solution since the length of a side cannot be negative.
Therefore, the length of side a is 3.

Additional answers can be found in the Answers to Exercises appendix.

(6.1) Find the greatest common factor of each group of terms.

1) $18, 27$ 9

2) $56, 80, 24$ 8

3) $33p^5q^3, 121p^4q^3, 44p^7q^4$ $11p^4q^3$

4) $42r^4s^3, 35r^2s^6, 49r^2s^4$ $7r^2s^3$

Factor out the greatest common factor.

5) $48y + 84$ $12(4y + 7)$

6) $30a^4 - 9a$ $3a(10a^3 - 3)$

7) $7n^5 - 21n^4 + 7n^3$ $7n^3(n^2 - 3n + 1)$

8) $72u^3v^3 - 42u^3v^2 - 24uv$ $6uv(12u^2v^2 - 7u^2v - 4)$

9) $a(b + 6) - 2(b + 6)$ $(b + 6)(a - 2)$

10) $u(13w - 9) + v(13w - 9)$ $(13w - 9)(u + v)$

Factor by grouping.

11) $mn + 2m + 5n + 10$ $(n + 2)(m + 5)$

12) $jk + 7j - 5k - 35$ $(k + 7)(j - 5)$

13) $5qr - 10q - 6r + 12$ $(r - 2)(5q - 6)$

14) $cd^2 + 5c - d^2 - 5$ $(d^2 + 5)(c - 1)$

15) Factor out $-4x$ from $-8x^3 - 12x^2 + 4x$. $-4x(2x^2 + 3x - 1)$

16) Factor out -1 from $-r^2 + 6r - 2$. $-(r^2 - 6r + 2)$

(6.2) Factor completely.

17) $p^2 + 13p + 40$ $(p + 8)(p + 5)$

18) $f^2 - 17f + 60$ $(f - 12)(f - 5)$

19) $x^2 + xy - 20y^2$ $(x + 5y)(x - 4y)$

20) $t^2 - 2tu - 63u^2$ $(t + 7u)(t - 9u)$

21) $3c^2 - 24c + 36$ $3(c - 6)(c - 2)$

22) $4m^3n + 8m^2n^2 - 60mn^3$ $4mn(m - 3n)(m + 5n)$

23) $5y^2 + 11y + 6$ $(5y + 6)(y + 1)$

24) $3g^2 + g - 44$ $(3g - 11)(g + 4)$

25) $4m^2 - 16m + 15$ $(2m - 5)(2m - 3)$

26) $6t^2 - 49t + 8$ $(6t - 1)(t - 8)$

27) $56a^3 + 4a^2 - 16a$ $4a(7a + 4)(2a - 1)$

28) $18n^2 + 98n + 40$ $2(9n + 4)(n + 5)$

29) $3s^2 + 11st - 4t^2$ $(3s - t)(s + 4t)$

30) $8f^2(g - 11)^3 - 6f(g - 11)^3 - 35(g - 11)^3$ $(g - 11)^3(2f - 5)(4f + 7)$

31) $(3c - 5)^2 + 10(3c - 5) + 24$ $(3c + 1)(3c - 1)$

32) $2(k + 1)^2 - 15(k + 1) + 28$ $(2k - 5)(k - 3)$

(6.3) Factor completely.

33) $n^2 - 25$ $(n + 5)(n - 5)$

34) $49a^2 - 4b^2$ $(7a + 2b)(7a - 2b)$

35) $9t^2 + 16u^2$ prime

36) $z^4 - 1$ $(z^2 + 1)(z + 1)(z - 1)$

37) $10q^2 - 810$ $10(q + 9)(q - 9)$

38) $48v - 27v^3$ $3v(4 + 3v)(4 - 3v)$

39) $a^2 + 16a + 64$ $(a + 8)^2$

40) $4x^2 - 20x + 25$ $(2x - 5)^2$

41) $h^3 + 8$ $(h + 2)(h^2 - 2h + 4)$

42) $q^3 - 1$ $(q - 1)(q^2 + q + 1)$

43) $27p^3 - 64q^3$ $(3p - 4q)(9p^2 + 12pq + 16q^2)$

44) $16c^3 + 250d^3$ $2(2c + 5d)(4c^2 - 10cd + 25d^2)$

Mixed Exercises: (6.1–6.3) Factor completely.

45) $7r^2 + 8r - 12$ $(7r - 6)(r + 2)$

46) $3y^2 + 60y + 300$ $3(y + 10)^2$

47) $\dfrac{9}{25} - x^2$ $\left(\dfrac{3}{5} + x\right)\left(\dfrac{3}{5} - x\right)$

48) $81v^6 + 36v^5 - 9v^4$ $9v^4(9v^2 + 4v - 1)$

49) $st - 5s - 8t + 40$ $(s - 8)(t - 5)$

50) $n^2 - 11n + 30$ $(n - 6)(n - 5)$

51) $w^5 - w^2$ $w^2(w - 1)(w^2 + w + 1)$

52) $gh + 8g - 11h - 88$ $(g - 11)(h + 8)$

53) $a^2 + 3a - 14$ prime

54) $49k^2 - 144$ $(7k + 12)(7k - 12)$

55) $(a - b)^2 - (a + b)^2$ $-4ab$

56) $1000a^3 + 27b^3$ $(10a + 3b)(100a^2 - 30ab + 9b^2)$

57) $6(y - 2)^2 - 13(y - 2) - 8$ $(3y - 14)(2y - 3)$

58) $5a^2 + 22ab + 8b^2$ $(5a + 2b)(a + 4b)$

(6.4) Solve each equation.

59) $c(2c - 1) = 0$ $\left\{0, \dfrac{1}{2}\right\}$

60) $(4z + 7)^2 = 0$ $\left\{-\dfrac{7}{4}\right\}$

61) $3x^2 + x = 2$ $\left\{-1, \dfrac{2}{3}\right\}$

62) $f^2 - 1 = 0$ $\{-1, 1\}$

63) $n^2 = 12n + 45$ $\{-3, 15\}$

64) $10j^2 - 8 + 11j = 0$ $\left\{-\dfrac{8}{5}, \dfrac{1}{2}\right\}$

65) $36 = 49d^2$ $\left\{-\dfrac{6}{7}, \dfrac{6}{7}\right\}$

66) $-13w = w^2$ $\{-13, 0\}$

67) $8b + 64 = 2b^2$ $\{-4, 8\}$

68) $18 = a(9 - a)$ $\{3, 6\}$

69) $y(5y - 9) = -4$ $\left\{\dfrac{4}{5}, 1\right\}$

70) $(z + 2)^2 = -z(3z + 4) + 9$ $\left\{-\dfrac{5}{2}, \dfrac{1}{2}\right\}$

71) $6a^3 - 3a^2 - 18a = 0$

72) $48 = 6r^2 + 12r$ $\{-4, 2\}$

73) $c(5c - 1) + 8 = 4(20 + c^2)$ $\{-8, 9\}$

74) $15t^3 + 40t = 70t^2$ $\left\{0, \dfrac{2}{3}, 4\right\}$

75) $p^2(6p - 1) - 10p(6p - 1) + 21(6p - 1) = 0$ $\left\{\dfrac{1}{6}, 3, 7\right\}$

76) $k^2(4k - 3) - 3k(4k - 3) - 54(4k - 3) = 0$ $\left\{-6, \dfrac{3}{4}, 9\right\}$

(6.5)

77) Find the base and height if the area of the triangle is 15 in².

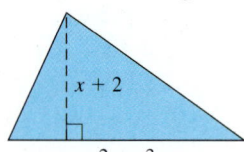

base = 5 in.; height = 6 in.

78) Find the length and width of the rectangle if its area is 28 cm².

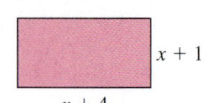

length = 7 cm; width = 4 cm

79) Find the height and length of the box if its volume is 96 in³.

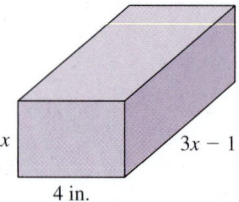

height = 3 in.; length = 8 in.

80) Find the length and width of the box if its volume is 360 in³.

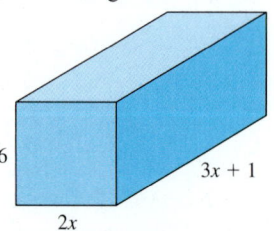

length = 10 in.; width = 6 in.

Use the Pythagorean theorem to find the length of the missing side.

81)
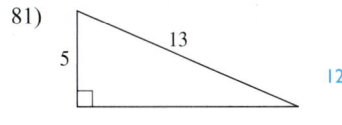
5

13

12

82)
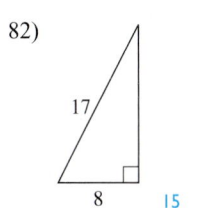
17

8

15

Write an equation and solve.

83) A rectangular countertop has an area of 15 ft². If the width is 3.5 ft shorter than the length, what are the dimensions of the countertop?   length = 6 ft; width = 2.5 ft

84) Kelsey cuts a piece of fabric into a triangle to make a bandana for her dog. The base of the triangle is twice its height. Find the base and height if there is 144 in² of fabric.

base = 24 in.; height = 12 in.

85) The sum of three consecutive integers is one-third the square of the middle number. Find the integers.   −1, 0, 1 or 8, 9, 10

86) Find two consecutive even integers such that their product is 6 more than 3 times their sum.   6, 8 or −2, 0

87) Seth builds a bike ramp in the shape of a right triangle. One leg is one inch shorter than the "ramp" while the other leg, the height of the ramp, is 8 in. shorter than the ramp. What is the height of the ramp?

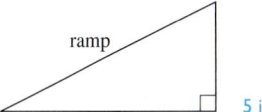
ramp

5 in.

88) A car heads east from an intersection while a motorcycle travels south. After 20 min, the car is 2 mi farther from the intersection than the motorcycle. The distance between the two vehicles is 4 mi more than the motorcycle's distance from the intersection. What is the distance between the car and the motorcycle?

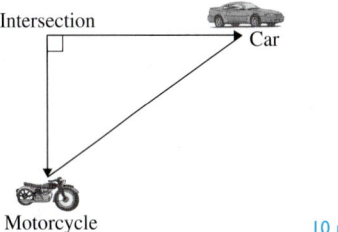

Intersection

Car

Motorcycle   10 mi

89) An object is launched with an initial velocity of 95 ft/sec. The height h (in feet) of the object after t sec is given by $h = -16t^2 + 96t$.

a) From what height is the object launched?   0 ft

b) When does the object reach a height of 128 ft?
after 2 sec and after 4 sec

c) How high is the object after 3 sec?   144 ft

d) When does the object hit the ground?   after 6 sec

Additional answers can be found in the Answers to Exercises appendix.

1) What is the first thing you should do when you are asked to factor a polynomial?   *See if you can factor out a GCF.*

Factor completely.

2) $n^2 - 11n + 30$ $(n - 6)(n - 5)$

3) $16 - b^2$ $(4 + b)(4 - b)$

4) $5a^2 - 13a - 6$ $(5a + 2)(a - 3)$

5) $56p^6q^6 - 77p^4q^4 + 7p^2q^3$ $7p^2q^3(8p^4q^3 - 11p^2q + 1)$

6) $y^3 - 8z^3$ $(y - 2z)(y^2 + 2yz + 4z^2)$

7) $2d^3 + 14d^2 - 36d$ $2d(d + 9)(d - 2)$

8) $r^2 + 25$ *prime*

9) $9h^2 + 24h + 16$ $(3h + 4)^2$

10) $24xy - 36x + 22y - 33$ $(2y - 3)(12x + 11)$

11) $s^2 - 3st - 28t^2$ $(s - 7t)(s + 4t)$

12) $16s^4 - 81t^4$ $(4s^2 + 9t^2)(2s + 3t)(2s - 3t)$

13) $4(3p + 2)^2 + 17(3p + 2) - 15$ $(12p + 5)(3p + 7)$

14) $12b^2 - 44b + 35$ $(2b - 5)(6b - 7)$

15) $m^{12} + m^9$ $m^9(m + 1)(m^2 - m + 1)$

Solve each equation.

16) $b^2 + 7b + 12 = 0$ $\{-4, -3\}$

17) $25k = k^3$ $\{0, -5, 5\}$

18) $144m^2 = 25$ $\left\{-\dfrac{5}{12}, \dfrac{5}{12}\right\}$

19) $(c - 5)(c + 2) = 18$ $\{-4, 7\}$

20) $4q(q - 5) + 14 = 11(2 + q)$ $\left\{-\dfrac{1}{4}, 8\right\}$

21) $24y^2 + 80 = 88y$ $\left\{\dfrac{5}{3}, 2\right\}$

Write an equation and solve.

22) Find the width and height of the storage locker pictured below if its volume is 120 ft³.

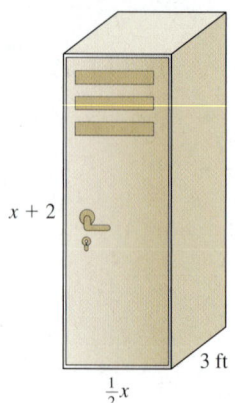

$x + 2$

3 ft

$\dfrac{1}{2}x$

height = 10 ft; width = 4 ft

23) Find three consecutive odd integers such that the sum of the three numbers is 60 less than the square of the largest integer.   *5, 7, 9*

24) Cory and Isaac leave an intersection with Cory jogging north and Isaac jogging west. When Isaac is 1 mi farther from the intersection than Cory, the distance between them is 2 mi more than Cory's distance from the intersection. How far is Cory from the intersection?   *3 mi*

25) The length of a rectangular dog run is 4 ft more than twice its width. Find the dimensions of the run if it covers 96 ft².   *height = 16 ft; width = 6 ft*

26) An object is thrown upward with an initial velocity of 68 ft/sec. The height h (in feet) of the object t sec after it is thrown is given by

$$h = -16t^2 + 68t + 60$$

a) How long does it take for the object to reach a height of 120 ft? $\dfrac{5}{4}$ *sec and 3 sec*

b) What is the initial height of the object?   *60 ft*

c) What is the height of the object after 2 sec?   *132 ft*

d) How long does it take the object to hit the ground?   *5 sec*

*Additional answers can be found in the Answers to Exercises appendix.

Perform the indicated operation(s) and simplify.

1) $\dfrac{3}{8} - \dfrac{5}{6} + \dfrac{7}{12}$ $\quad \dfrac{1}{8}$

2) $-\dfrac{15}{32} \cdot \dfrac{12}{25}$ $\quad -\dfrac{9}{40}$

Simplify. The answer should not contain any negative exponents.

3) $\dfrac{54t^5u^2}{36tu^8}$ $\quad \dfrac{3t^4}{2u^6}$

4) $(8k^6)(-3k^4)$ $\quad -24k^{10}$

5) Write 4.813×10^5 without exponents. $\quad 481{,}300$

6) Solve $\dfrac{1}{3}(n - 2) + \dfrac{1}{4} = \dfrac{5}{12} + \dfrac{1}{6}n$ $\quad \{5\}$

7) Solve for R.
$A = P + PRT$ $\quad R = \dfrac{A - P}{PT}$

8) *Write an equation and solve.*
A Twix candy bar is half the length of a Toblerone candy bar. Together, they are 12 in. long. Find the length of each candy bar. $\quad$ Twix: 4 in.; Toblerone: 8 in.

9) Solve. Write the answer in interval notation.
$$2 + |9 - 5n| \geq 31 \quad (-\infty, -4] \cup \left[\dfrac{38}{5}, \infty\right)$$

10) Graph $y = -\dfrac{3}{5}x + 7$.

11) Write the equation of the line perpendicular to $3x + y = 4$ containing the point $(-6, -1)$. Express the answer in slope-intercept form. $\quad y = \dfrac{1}{3}x + 1$

12) Use any method to solve this system of equations.
$$6(x + 2) + y = x - y - 2$$
$$5(2x - y + 1) = 2(x - y) - 5 \quad (-2, -2)$$

Multiply and simplify.

13) $(6y + 5)(2y - 3)$ $\quad 12y^2 - 8y - 15$

14) $(4p - 7)(2p^2 - 9p + 8)$ $\quad 8p^3 - 50p^2 + 95p - 56$

15) $(c + 8)^2$ $\quad c^2 + 16c + 64$

16) Add $(4a^2b^2 - 17a^2b + 12ab - 11)$
$$+(-a^2b^2 + 10a^2b - 5ab^2 + 7ab + 3).$$
$$3a^2b^2 - 7a^2b - 5ab^2 + 19ab - 8$$

Divide.

17) $\dfrac{12x^4 - 30x^3 - 14x^2 + 27x + 20}{2x - 5}$ $\quad 6x^3 - 7x - 4$

18) $\dfrac{12r^3 + 4r^2 - 10r + 3}{4r^2}$ $\quad 3r + 1 - \dfrac{5}{2r} + \dfrac{3}{4r^2}$

Factor completely, if possible.

19) $bc + 8b - 7c - 56$
$(b - 7)(c + 8)$

20) $54q^2 - 144q + 42$
$6(3q - 7)(3q - 1)$

21) $y^2 + 1$
prime

22) $t^4 - 81$
$(t^2 + 9)(t + 3)(t - 3)$

23) $x^3 - 125$
$(x - 5)(x^2 + 5x + 25)$

Solve.

24) $z^2 + 3z = 40$ $\quad \{-8, 5\}$

25) $-12j(1 - 2j) = 16(5 + j)$ $\quad \left\{-\dfrac{4}{3}, \dfrac{5}{2}\right\}$

Rational Expressions, Equations, and Functions

Algebra at Work: Ophthalmology

At the beginning of Chapter 6 we saw how an ophthalmologist, a doctor specializing in diseases of the eye, uses mathematics every day to treat his patients. Here we will see another example of how math is used in this branch of medicine.

Some formulas in optics involve rational expressions. If Calvin determines that one of his patients needs glasses, he would use the following formula to figure out the proper prescription:

$$P = \frac{1}{f}$$

where f is the focal length, in meters, and P is the power of the lens, in diopters.

While computers now aid in these calculations, physicians believe that it is still important to double-check the calculations by hand.

In this chapter, we will learn how to perform operations with rational expressions and how to solve equations, like the one above, for a specific variable.

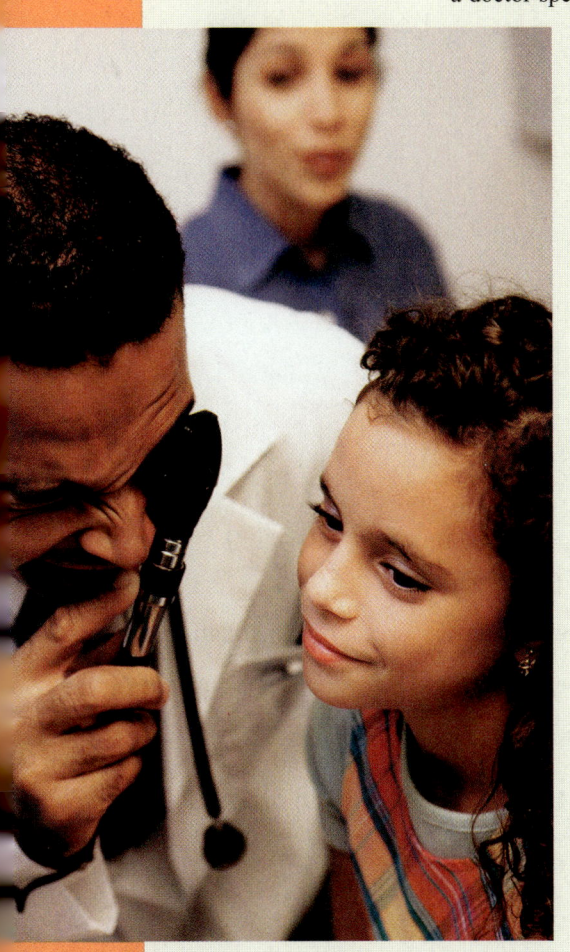

Section 7.1 Simplifying, Multiplying, and Dividing Rational Expressions and Functions

Objectives

1. **Define a Rational Function and Determine the Domain**
2. **Write a Rational Expression in Lowest Terms**
3. **Simplify** $\dfrac{a - b}{b - a}$
4. **Write Equivalent Forms of a Rational Expression**
5. **Multiply Rational Expressions**
6. **Divide Rational Expressions**

What Is a Rational Expression?

In Section 1.1 we defined a **rational number** as the quotient of two integers provided that the denominator does not equal zero. Some examples of rational numbers are

$$\frac{5}{12}, \qquad -\frac{3}{8}, \qquad \text{and} \qquad 23 \left(\text{since } 23 = \frac{23}{1} \right).$$

We can define a rational expression in a similar way. A rational expression is a quotient of two polynomials provided that the denominator does not equal zero. We state the definition formally next.

Definition

A **rational expression** is an expression of the form $\dfrac{P}{Q}$, where P and Q are polynomials and where $Q \neq 0$.

Some examples of rational expressions are

$$\frac{9n^4}{2}, \qquad \frac{3x - 5}{x + 8}, \qquad \frac{6}{c^2 - 2c - 63}, \qquad \text{and} \qquad -\frac{3a + 2b}{a^2 + b^2}.$$

It is important to understand the following facts about rational expressions.

Note

1) A fraction (rational expression) equals zero when its *numerator* equals zero.

2) A fraction (rational expression) is *undefined* when its denominator equals zero.

1. Define a Rational Function and Determine the Domain

Some functions are described by rational expressions. Such functions are called **rational functions.** $f(x) = \dfrac{x + 3}{x - 8}$ is an example of a **rational function** since $\dfrac{x + 3}{x - 8}$ is a rational expression and since each value that can be substituted for x will produce *only one* value for the expression.

Recall from Chapter 3 that the domain of a function $f(x)$ is the set of all real numbers that can be substituted for x. Since a rational expression is undefined when its denominator equals zero, we define the domain of a rational function as follows.

The **domain of a rational function** consists of all real numbers except the value(s) of the variable that make(s) the denominator equal zero.

Therefore, to determine the domain of a rational function we set the denominator equal to zero and solve for the variable. Any value that makes the denominator equal to zero is *not* in the domain of the function.

To determine the domain of a rational function, sometimes it is helpful to ask yourself, "Is there any number that *cannot* be substituted for the variable?"

Example 1

In-Class Example 1

For $f(x) = \dfrac{x^2 - 16}{x + 3}$,

a) find $f(5)$.
b) find x so that $f(x) = 0$.
c) determine the domain of the function.

answer: a) $\dfrac{9}{8}$
b) $f(x) = 0$ when $x = -4$ or $x = 4$.
c) $(-\infty, -3) \cup (-3, \infty)$

For $f(x) = \dfrac{x^2 - 9}{x + 7}$,

a) find $f(6)$.

b) find x so that $f(x) = 0$.

c) determine the domain of the function.

Solution

a) $f(6) = \dfrac{6^2 - 9}{6 + 7} = \dfrac{36 - 9}{13} = \dfrac{27}{13}$

b) To find the values of x that will make $f(x) = 0$, set the function equal to zero and solve for x.

$$\frac{x^2 - 9}{x + 7} = 0$$

The expression $\dfrac{x^2 - 9}{x + 7} = 0$ when its *numerator* equals zero. Set the numerator equal to zero, and solve for x.

$$x^2 - 9 = 0$$
$$(x + 3)(x - 3) = 0 \qquad \text{Factor.}$$

$$x + 3 = 0 \qquad \text{or} \qquad x - 3 = 0 \qquad \text{Set each factor equal to 0.}$$
$$x = -3 \qquad \text{or} \qquad x = 3 \qquad \text{Solve.}$$

$f(x) = 0$ when $x = -3$ or $x = 3$.

c) To determine the domain of $f(x) = \dfrac{x^2 - 9}{x + 7}$ ask yourself, "Is there any number that *cannot* be substituted for x?" Yes. $f(x)$ is **undefined** when the denominator equals zero. Set the denominator equal to zero, and solve for x.

$$x + 7 = 0 \qquad \text{Set the denominator} = 0.$$
$$x = -7 \qquad \text{Solve.}$$

When $x = -7$, the denominator of $f(x) = \dfrac{x^2 - 9}{x + 7}$ equals zero. The domain contains all real numbers *except* -7. Write the domain in interval notation as $(-\infty, -7) \cup (-7, \infty)$.

You Try 1

For $f(x) = \dfrac{x^2 - 25}{x - 2}$,

a) find $f(4)$.

b) find x so that $f(x) = 0$.

c) determine the domain of the function.

Example 2

Determine the domain of each rational function.

In-Class Example 2
Determine the domain of each rational function.

a) $g(t) = \dfrac{8t + 7}{t^2 - 5t - 36}$

b) $h(k) = \dfrac{9k^2 - 1}{6}$

answer:
a)
$(-\infty, -4) \cup (-4, 9) \cup (9, \infty)$
b) $(-\infty, \infty)$

a) $g(c) = \dfrac{6c + 5}{c^2 + 3c - 4}$ b) $h(n) = \dfrac{4n^2 - 9}{7}$

Solution

a)   To determine the domain of $g(c) = \dfrac{6c + 5}{c^2 + 3c - 4}$, ask yourself, "Is there any number that *cannot* be substituted for *c*?" Yes. $g(c)$ is **undefined** when its *denominator* equals zero. Set the denominator equal to zero and solve for *c*.

$$c^2 + 3c - 4 = 0 \qquad \text{Set the denominator} = 0.$$
$$(c + 4)(c - 1) = 0 \qquad \text{Factor.}$$

$$c + 4 = 0 \quad \text{or} \quad c - 1 = 0 \qquad \text{Set each factor equal to 0.}$$
$$c = -4 \quad \text{or} \quad c = 1 \qquad \text{Solve.}$$

When $c = -4$ or $c = 1$, the denominator of $g(c) = \dfrac{6c + 5}{c^2 + 3c - 4}$ equals zero.

The domain contains all real numbers *except* -4 and 1. Write the domain in interval notation as $(-\infty, -4) \cup (-4, 1) \cup (1, \infty)$.

b)   Ask yourself, "Is there any number that *cannot* be substituted for *n*?" No! Looking at the denominator we see that it will never equal zero. Therefore, there *is no value of n* that makes $h(n) = \dfrac{4n^2 - 9}{7}$ undefined. Any real number may be substituted for *n* and the function will be defined.

The domain of the function is the set of all real numbers. Write the domain in interval notation as $(-\infty, \infty)$. ∎

You Try 2

Determine the domain of each rational function.

a) $f(x) = \dfrac{2x - 3}{x^2 - 8x + 12}$ b) $g(a) = \dfrac{a + 4}{10}$

All of the operations that can be performed with fractions can also be done with rational expressions. We begin our study of these operations with rational expressions by learning how to write a rational expression in lowest terms.

2. Write a Rational Expression in Lowest Terms

One way to think about writing a fraction such as $\dfrac{8}{12}$ in lowest terms is

$$\frac{8}{12} = \frac{2 \cdot 4}{3 \cdot 4} = \frac{2}{3} \cdot \frac{4}{4} = \frac{2}{3} \cdot 1 = \frac{2}{3}$$

Since $\dfrac{4}{4} = 1$, we can also think of reducing $\dfrac{8}{12}$ in the following way: $\dfrac{8}{12} = \dfrac{2 \cdot \cancel{4}}{3 \cdot \cancel{4}} = \dfrac{2}{3}$.

We can *factor* the numerator and denominator, then *divide* the numerator and denominator by the common factor, 4. This is the approach we use to write a rational expression in lowest terms.

> **Definition   Fundamental Property of Rational Expressions**
>
> If P, Q, and C are polynomials such that $Q \neq 0$ and $C \neq 0$, then
>
> $$\frac{PC}{QC} = \frac{P}{Q}.$$

This property mirrors the example of writing $\dfrac{8}{12}$ in lowest terms since

$$\frac{PC}{QC} = \frac{P}{Q} \cdot \frac{C}{C} = \frac{P}{Q} \cdot 1 = \frac{P}{Q}.$$

Or, we can also think of the reducing procedure as dividing the numerator and denominator by the common factor, C.

$$\frac{P\cancel{C}}{Q\cancel{C}} = \frac{P}{Q}$$

> **Procedure   Writing a Rational Expression in Lowest Terms**
>
> 1) Completely **factor** the numerator and denominator.
>
> 2) **Divide** the numerator and denominator by the greatest common factor.

Example 3

In-Class Example 3
Write each rational expression in lowest terms.
a) $\dfrac{48n^8}{4n^4}$ b) $\dfrac{3m + 18}{5m^2 + 30m}$
c) $\dfrac{28x^2 - 7}{6x^2 + 9x - 6}$

answer: a) $12n^4$ b) $\dfrac{3}{5m}$
c) $\dfrac{7(2x + 1)}{3(x + 2)}$

Write each rational expression in lowest terms.

a) $\dfrac{20c^6}{5c^4}$ b) $\dfrac{4m + 12}{7m + 21}$ c) $\dfrac{3x^2 - 3}{x^2 + 9x + 8}$

Solution

a) We can simplify $\dfrac{20c^6}{5c^4}$ using the quotient rule presented in Chapter 5.

$$\frac{20c^6}{5c^4} = 4c^2$$

    Divide 20 by 5 and use the quotient rule: $\dfrac{c^6}{c^4} = c^{6-4} = c^2.$

b) $\dfrac{4m + 12}{7m + 21} = \dfrac{4\cancel{(m + 3)}}{7\cancel{(m + 3)}}$ Factor.

$$= \frac{4}{7}$$ Divide out the common factor, $m + 3$.

c) $\dfrac{3x^2 - 3}{x^2 + 9x + 8} = \dfrac{3(x^2 - 1)}{(x + 1)(x + 8)}$ Factor.

$$= \frac{3\cancel{(x + 1)}(x - 1)}{\cancel{(x + 1)}(x + 8)}$$ Factor completely.

$$= \frac{3(x - 1)}{x + 8}$$ Divide out the common factor, $x + 1$.

Notice that we divided by *factors* not *terms*.

$$\frac{\cancel{x + 5}}{2\cancel{(x + 5)}} = \frac{1}{2}$$

Divide by the *factor* $x + 5$.

$$\frac{\cancel{x}}{\cancel{x} + 5} \neq \frac{1}{5}$$

We cannot divide by x because the x in the denominator is a *term* in a sum.

You Try 3

Write each rational expression in lowest terms.

a) $\dfrac{6t - 48}{t^2 - 8t}$ b) $\dfrac{b - 2}{5b^2 - 6b - 8}$ c) $\dfrac{v^3 + 27}{4v^4 - 12v^3 + 36v^2}$

3. Simplify $\dfrac{a - b}{b - a}$

Do you think that $\dfrac{x - 4}{4 - x}$ is in lowest terms? Let's look at it more closely to understand the answer.

$$\frac{x - 4}{4 - x} = \frac{x - 4}{-1(-4 + x)} \qquad \text{Factor } -1 \text{ out of the denominator.}$$

$$= \frac{1(x - 4)}{-1(x - 4)} \qquad \text{Rewrite } -4 + x \text{ as } x - 4.$$

$$= -1$$

Therefore, $\dfrac{x - 4}{4 - x} = -1$.

We can generalize this result as

Note

1) $b - a = -1(a - b)$ and   2) $\dfrac{a - b}{b - a} = -1$

The terms in the numerator and denominator in 2) differ only in sign. They divide out to -1.

Example 4

Write each rational expression in lowest terms.

a) $\dfrac{6 - d}{d - 6}$ b) $\dfrac{25z^2 - 9}{3 - 5z}$

In-Class Example 4
Write each rational expression in lowest terms.
a) $\dfrac{c - 7}{7 - c}$ b) $\dfrac{4z - 11}{121 - 16z^2}$
c) $\dfrac{54w - 6w^3}{w^2 - 6w + 9}$
answer: a) -1 b) $-\dfrac{1}{4z + 11}$
c) $\dfrac{6w(w + 3)}{w - 3}$

Solution

a) $\dfrac{6 - d}{d - 6} = -1$ since $\dfrac{6 - d}{d - 6} = \dfrac{-1(d - 6)}{d - 6} = -1.$

b) $\dfrac{25z^2 - 9}{3 - 5z} = \dfrac{(5z + 3)(5z - 3)^{-1}}{3 - 5z}$ Factor. $\dfrac{5z - 3}{3 - 5z} = -1$

$= -1(5z + 3)$

$= -5z - 3$ Distribute.

You Try 4

Write each rational expression in lowest terms.

a) $\dfrac{x - y}{y - x}$ b) $\dfrac{15n - 5m}{2m - 6n}$ c) $\dfrac{12 - 3y^2}{y^2 - 10y + 16}$

4. Write Equivalent Forms of a Rational Expression

Often, the same rational expression can be written in several different ways. You should be able to recognize equivalent forms of rational expressions because there isn't always just one way to write the correct answer.

Example 5

In-Class Example 5

Write $-\dfrac{5x(x + 2)}{9 - x}$ in three different ways.

answer: i) $\dfrac{5x(x + 2)}{x - 9}$

ii) $\dfrac{-5x(x + 2)}{9 - x}$

iii) $\dfrac{5x(x + 2)}{-9 + x}$

Write $-\dfrac{8u(u - 1)}{1 + u}$ in three different ways.

Solution

The negative sign in front of a fraction can also be applied to the numerator or to the denominator. For example, $-\dfrac{4}{9} = \dfrac{-4}{9} = \dfrac{4}{-9}$. Applying this concept to rational expressions can result in expressions that look quite different but that are, actually, equivalent.

i) Apply the negative sign to the denominator.

$$-\frac{8u(u - 1)}{1 + u} = \frac{8u(u - 1)}{-1(1 + u)}$$
$$= \frac{8u(u - 1)}{-1 - u} \qquad \text{Distribute.}$$

ii) Apply the negative sign to the numerator.

$$-\frac{8u(u - 1)}{1 + u} = \frac{-8u(u - 1)}{1 + u}$$

iii) Apply the negative sign to the numerator, but distribute the -1.

$$-\frac{8u(u - 1)}{1 + u} = \frac{(8u)(-1)(u - 1)}{1 + u}$$
$$= \frac{8u(-u + 1)}{1 + u} \qquad \text{Distribute.}$$
$$= \frac{8u(1 - u)}{1 + u} \qquad \text{Rewrite } -u + 1 \text{ as } 1 - u.$$

Therefore, $\dfrac{8u(u - 1)}{-1 - u}$, $\dfrac{-8u(u - 1)}{1 + u}$, and $\dfrac{8u(1 - u)}{1 + u}$ are *all* equivalent forms of $-\dfrac{8u(u - 1)}{1 + u}$.

Keep this idea of equivalent forms of rational expressions in mind when checking your answers against the answers in the back of the book. Sometimes students believe their answer is wrong because it "looks different" when, in fact, it is an *equivalent form* of the given answer!

You Try 5

Write $\dfrac{-(2 - p)}{7p - 9}$ in three different ways.

5. Multiply Rational Expressions

We multiply rational expressions the same way we multiply rational numbers. Multiply numerators, multiply denominators, and simplify.

Procedure   Multiplying Rational Expressions

If $\dfrac{P}{Q}$ and $\dfrac{R}{T}$ are rational expressions, then $\dfrac{P}{Q} \cdot \dfrac{R}{T} = \dfrac{PR}{QT}$.

To multiply two rational expressions, multiply their numerators, multiply their denominators, and simplify.

Let's begin by reviewing how we multiply two fractions.

Example 6

In-Class Example 6
Multiply $\dfrac{9}{20} \cdot \dfrac{5}{18}$.

answer: $\dfrac{1}{8}$

Multiply $\dfrac{8}{15} \cdot \dfrac{5}{6}$.

Solution

We can multiply numerators, multiply denominators, then simplify by dividing out common factors *or* we can divide out the common factors before multiplying.

$$\frac{8}{15} \cdot \frac{5}{6} = \frac{2 \cdot 4}{3 \cdot \cancel{5}} \cdot \frac{\cancel{5}}{2 \cdot 3} \qquad \text{Factor and divide out common factors.}$$

$$= \frac{4}{3 \cdot 3} \qquad \text{Multiply.}$$

$$= \frac{4}{9} \qquad \text{Simplify.}$$

You Try 6

Multiply $\dfrac{7}{32} \cdot \dfrac{20}{21}$.

Multiplying rational expressions works the same way.

Procedure   Multiplying Rational Expressions

1) Factor.

2) Reduce and multiply.

All products must be written in lowest terms.

Example 7

Multiply.

a) $\dfrac{12a^4}{b^2} \cdot \dfrac{b^5}{6a^9}$

b) $\dfrac{8m + 48}{10m^6} \cdot \dfrac{m^2}{m^2 - 36}$

c) $\dfrac{3k^2 - 11k - 4}{k^2 + k - 20} \cdot \dfrac{k^2 + 10k + 25}{3k^2 + k}$

In-Class Example 7
Multiply.
a) $\dfrac{6y^3}{11z^4} \cdot \dfrac{z^2}{3y}$
b) $\dfrac{3c-12}{8c^{10}} \cdot \dfrac{2c^4}{c^2-16}$
c)
$\dfrac{2d^2-3d-5}{2d^2+6d+4} \cdot \dfrac{4d^2+16d+16}{4d^2-10d}$

answer: a) $\dfrac{2y^2}{11z^2}$
b) $\dfrac{3}{4c^6(c+4)}$ c) $\dfrac{d+2}{d}$

Solution

a) $\dfrac{12a^4}{b^2} \cdot \dfrac{b^5}{6a^9} = \dfrac{\overset{2}{\cancel{12}}\cancel{a^4}}{\cancel{b^2}} \cdot \dfrac{\cancel{b^2} \cdot b^3}{\cancel{6}\cancel{a^4} \cdot a^5}$ Factor and reduce.

$\qquad\qquad\quad = \dfrac{2b^3}{a^5}$ Multiply.

b) $\dfrac{8m+48}{10m^6} \cdot \dfrac{m^2}{m^2-36} = \dfrac{\overset{4}{\cancel{8}}\cancel{(m+6)}}{\underset{5}{\cancel{10}}\cancel{m^2} \cdot m^4} \cdot \dfrac{\cancel{m^2}}{\cancel{(m+6)}(m-6)}$ Factor and reduce.

$\qquad\qquad\qquad\qquad = \dfrac{4}{5m^4(m-6)}$ Multiply.

c) $\dfrac{3k^2-11k-4}{k^2+k-20} \cdot \dfrac{k^2+10k+25}{3k^2+k} = \dfrac{\cancel{(3k+1)}\cancel{(k-4)}}{\cancel{(k+5)}\cancel{(k-4)}} \cdot \dfrac{\overset{(k+5)}{\cancel{(k+5)^2}}}{k\cancel{(3k+1)}}$ Factor and reduce.

$\qquad\qquad\qquad\qquad\qquad = \dfrac{k+5}{k}$ Multiply. ■

You Try 7

Multiply.

a) $\dfrac{x^3}{14y^6} \cdot \dfrac{7y^2}{x^8}$ b) $\dfrac{r^2-25}{r^2-9r} \cdot \dfrac{r^2+r-90}{10-2r}$

6. Divide Rational Expressions

When we divide rational numbers we multiply by a reciprocal. For example,

$\dfrac{7}{4} \div \dfrac{3}{8} = \dfrac{7}{\underset{1}{\cancel{4}}} \cdot \dfrac{\overset{2}{\cancel{8}}}{3} = \dfrac{14}{3}$. We divide rational expressions the same way. To divide rational expressions we multiply the first rational expression by the reciprocal of the second rational expression.

Procedure   Dividing Rational Expressions

If $\dfrac{P}{Q}$ and $\dfrac{R}{T}$ are rational expressions with $Q, R,$ and T not equal to zero, then

$$\dfrac{P}{Q} \div \dfrac{R}{T} = \dfrac{P}{Q} \cdot \dfrac{T}{R} = \dfrac{PT}{QR}$$

Multiply the first rational expression by the reciprocal of the second rational expression.

Example 8

Divide.

a) $\dfrac{36p^5}{q^4} \div \dfrac{4p^2}{q^{10}}$ b) $\dfrac{s^2+8s+16}{s^2-11s+24} \div \dfrac{s^2+5s+4}{15-5s}$

c) $\dfrac{2k^2+k-15}{k^3} \div (2k-5)^2$

In-Class Example 8
Divide.
a) $\dfrac{42b^6}{c^3} \div \dfrac{2b}{c^5}$

b) $\dfrac{r^2 - 13r + 36}{2r + 10} \div \dfrac{12r - 3r^2}{16}$

c) $\dfrac{3n^2 - 22n - 16}{n^2} \div (3n + 2)^2$

answer: a) $21b^5c^2$
b) $-\dfrac{8(r-9)}{3r(r+5)}$ **c)** $\dfrac{n-8}{n^2(3n+2)}$

Solution

a) $\dfrac{36p^5}{q^4} \div \dfrac{4p^2}{q^{10}} = \dfrac{36\,\overset{9p^3}{\cancel{p^5}}}{\cancel{q^4}} \cdot \dfrac{\overset{q^6}{\cancel{q^{10}}}}{\cancel{4p^2}}$ Multiply by the reciprocal and reduce.

$\qquad\qquad = 9p^3q^6$ Multiply.

Notice that we used the *quotient rule* for exponents to reduce:

$$\dfrac{p^5}{p^2} = p^3, \quad \dfrac{q^{10}}{q^4} = q^6$$

b) $\dfrac{s^2 + 8s + 16}{s^2 - 11s + 24} \div \dfrac{s^2 + 5s + 4}{15 - 5s}$

$\qquad = \dfrac{s^2 + 8s + 16}{s^2 - 11s + 24} \cdot \dfrac{15 - 5s}{s^2 + 5s + 4}$ Multiply by the reciprocal.

$\qquad = \dfrac{\overset{(s+4)}{\cancel{(s+4)^2}}}{\cancel{(s-3)}(s - 8)} \cdot \dfrac{5\overset{-1}{\cancel{(3-s)}}}{\cancel{(s+4)}(s + 1)}$ Factor; $\dfrac{3 - s}{s - 3} = -1$.

$\qquad = \dfrac{-5(s + 4)}{(s - 8)(s + 1)}$ Reduce and multiply.

c) $\dfrac{2k^2 + k - 15}{k^3} \div (2k - 5)^2 = \dfrac{\cancel{(2k-5)}(k + 3)}{k^3} \cdot \dfrac{1}{\underset{(2k-5)}{\cancel{(2k-5)^2}}}$ Since $(2k - 5)^2$ can be written as $\dfrac{(2k - 5)^2}{1}$, its reciprocal is $\dfrac{1}{(2k - 5)^2}$.

$\qquad = \dfrac{(k + 3)}{k^3(2k - 5)}$ Reduce and multiply. ■

You Try 8

Divide.

a) $\dfrac{r^6}{35t^9} \div \dfrac{r^4}{14t^3}$ b) $\dfrac{x^2 - 8x - 48}{7x^2 - 42x} \div \dfrac{x^2 - 16}{x^2 - 10x + 24}$

c) $\dfrac{5a^2 + 34a + 24}{a^2} \div (5a + 4)^2$

Answers to You Try Exercises

1) a) $-\dfrac{9}{2}$ b) -5 or 5 c) $(-\infty, 2) \cup (2, \infty)$ 2) a) $(-\infty, 2) \cup (2, 6) \cup (6, \infty)$ b) $(-\infty, \infty)$

3) a) $\dfrac{6}{t}$ b) $\dfrac{1}{5b + 4}$ c) $\dfrac{v + 3}{4v^2}$ 4) a) -1 b) $-\dfrac{5}{2}$ c) $\dfrac{-3(y + 2)}{y - 8}$

5) Some possibilities are $\dfrac{p - 2}{7p - 9}, \dfrac{2 - p}{9 - 7p}, -\dfrac{2 - p}{7p - 9}, \dfrac{2 - p}{-(7p - 9)}$ 6) $\dfrac{5}{24}$

7) a) $\dfrac{1}{2x^5y^4}$ b) $-\dfrac{(r + 5)(r + 10)}{2r}$ 8) a) $\dfrac{2r^2}{5t^6}$ b) $\dfrac{x - 12}{7x}$ c) $\dfrac{a + 6}{a^2(5a + 4)}$

7.1 Exercises

*Additional answers can be found in the Answers to Exercises appendix.

Objective 1: Define a Rational Function and Determine the Domain

1) When does a fraction or a rational expression equal 0?
   when its numerator equals zero

2) When is a fraction or a rational expression undefined?
   when its denominator equals zero

3) How do you determine the value of the variable for which a rational expression is undefined?

4) If $x^2 + 5$ is the numerator of a rational expression, can that expression equal zero? Give a reason.

For each rational function,

   a) find $f(-2)$, if possible.

   b) find x so that $f(x) = 0$.

   c) determine the domain of the function.

5) $f(x) = \dfrac{x + 8}{x + 6}$ a) $\dfrac{3}{2}$ b) -8 c) $(-\infty, -6) \cup (-6, \infty)$

6) $f(x) = \dfrac{x}{3x - 1}$ a) $\dfrac{2}{7}$ b) 0 c) $\left(-\infty, \dfrac{1}{3}\right) \cup \left(\dfrac{1}{3}, \infty\right)$

7) $f(x) = \dfrac{5x - 3}{x + 2}$ a) undefined   b) $\dfrac{3}{5}$ c) $(-\infty, -2) \cup (-2, \infty)$

8) $f(x) = \dfrac{9}{x - 1}$ a) -3 b) never equals zero   c) $(-\infty, 1) \cup (1, \infty)$

9) $f(x) = \dfrac{6}{x^2 + 6x + 5}$ a) -2 b) never equals zero   c) $(-\infty, -5) \cup (-5, -1) \cup (-1, \infty)$

10) $f(x) = \dfrac{2x - 1}{x^2 + x - 12}$ a) $\dfrac{1}{2}$ b) $\dfrac{1}{2}$ c) $(-\infty, -4) \cup (-4, 3) \cup (3, \infty)$

Determine the domain of each rational function.

11) $f(p) = \dfrac{1}{p - 7}$
    $(-\infty, 7) \cup (7, \infty)$

12) $h(z) = \dfrac{z + 8}{z + 3}$
    $(-\infty, -3) \cup (-3, \infty)$

13) $k(r) = \dfrac{r}{5r + 2}$ $\left(-\infty, -\dfrac{2}{5}\right) \cup \left(-\dfrac{2}{5}, \infty\right)$

14) $f(a) = \dfrac{6a}{7 - 2a}$ $\left(-\infty, \dfrac{7}{2}\right) \cup \left(\dfrac{7}{2}, \infty\right)$

15) $g(t) = \dfrac{3t - 4}{t^2 - 9t + 8}$
    $(-\infty, 1) \cup (1, 8) \cup (8, \infty)$

16) $r(c) = \dfrac{c + 9}{c^2 - c - 42}$
    $(-\infty, -6) \cup (-6, 7) \cup (7, \infty)$

17) $h(w) = \dfrac{w + 7}{w^2 - 81}$
    $(-\infty, -9) \cup (-9, 9) \cup (9, \infty)$

18) $k(t) = \dfrac{t}{t^2 - 14t + 33}$
    $(-\infty, 3) \cup (3, 11) \cup (11, \infty)$

19) $A(c) = \dfrac{8}{c^2 + 6}$
    $(-\infty, \infty)$

20) $C(n) = \dfrac{3n + 1}{2}$
    $(-\infty, \infty)$

21) Write your own example of a rational function, $f(x)$, that has a domain of $(-\infty, -8) \cup (-8, \infty)$. *Answers may vary.*

22) Write your own example of a rational function, $g(x)$, that has a domain of $(-\infty, -5) \cup (-5, 6) \cup (6, \infty)$.
    Answers may vary.

Objective 2: Write a Rational Expression in Lowest Terms

Write each rational expression in lowest terms.

23) $\dfrac{12d^5}{30d^8}$ $\dfrac{2}{5d^3}$

24) $\dfrac{108g^4}{9g}$ $12g^3$

25) $\dfrac{3c - 12}{5c - 20}$ $\dfrac{3}{5}$

26) $\dfrac{10d - 5}{12d - 6}$ $\dfrac{5}{6}$

27) $\dfrac{b^2 + b - 56}{b + 8}$ $b - 7$

28) $\dfrac{g^2 + 9g + 20}{g^2 + 2g - 15}$ $\dfrac{g + 4}{g - 3}$

29) $\dfrac{r - 4}{r^2 - 16}$ $\dfrac{1}{r + 4}$

30) $\dfrac{t + 2}{t^2 - 7t - 18}$ $\dfrac{1}{t - 9}$

31) $\dfrac{3k^2 + 28k + 32}{k^2 + 10k + 16}$ $\dfrac{3k + 4}{k + 2}$

32) $\dfrac{3c^2 - 36c + 96}{c - 8}$ $3(c - 4)$

33) $\dfrac{w^3 + 125}{5w^2 - 25w + 125}$ $\dfrac{w + 5}{5}$

34) $\dfrac{4m^3 - 4}{m^2 + m + 1}$ $4(m - 1)$

35) $\dfrac{4m^2 - 20m + 4mn - 20n}{11m + 11n}$ $\dfrac{4(m - 5)}{11}$

36) $\dfrac{uv + 3u - 4v - 12}{v^2 - 9}$ $\dfrac{u - 4}{v - 3}$

37) $\dfrac{x^2 - y^2}{x^3 - y^3}$ $\dfrac{x + y}{x^2 + xy + y^2}$

38) $\dfrac{a^3 + b^3}{a^2 - b^2}$ $\dfrac{a^2 - ab + b^2}{a - b}$

Objective 3: Simplify $\dfrac{a - b}{b - a}$

39) Any rational expression of the form $\dfrac{a - b}{b - a}$ reduces to what? -1

40) Does $\dfrac{z + 9}{z - 9} = -1$?   No

Write each rational expression in lowest terms.

41) $\dfrac{12 - v}{v - 12}$ -1

42) $\dfrac{q - 11}{11 - q}$ -1

43) $\dfrac{k^2 - 49}{7 - k}$ $-k - 7$

44) $\dfrac{m - 10}{20 - 2m}$ $-\dfrac{1}{2}$

45) $\dfrac{30 - 35x}{7x^2 + 8x - 12}$ $-\dfrac{5}{x + 2}$

46) $\dfrac{a^2 - 8a - 33}{11 - a}$ $-a - 3$

47) $\dfrac{16 - 4b^2}{b - 2}$ $-4(b + 2)$

48) $\dfrac{16 - 2w}{w^2 - 64}$ $-\dfrac{2}{w + 8}$

49) $\dfrac{8t^3 - 27}{9 - 4t^2}$ $-\dfrac{4t^2 + 6t + 9}{2t + 3}$

50) $\dfrac{r^3 - 3r^2 + 2r - 6}{21 - 7r}$ $-\dfrac{r^2 + 2}{7}$

Recall that the area of a rectangle is $A = lw$, where $w = $ width and $l = $ length. Solving for the width we get $w = \dfrac{A}{l}$ and solving for the length gives us $l = \dfrac{A}{w}$.

Find the missing side in each rectangle.

51) Area $= 5x^2 + 13x + 6$

$\boxed{}$ $x + 2$

Find the length. $5x + 3$

52) Area $= 2y^2 - y - 15$

$\boxed{}$

$2y + 5$

Find the width. $y - 3$

53) Area $= c^3 - 2c^2 + 4c - 8$

$\boxed{}$

$c^2 + 4$

Find the width. $c - 2$

54) Area $= 2n^3 - 8n^2 + n - 4$

$\boxed{}$ $n - 4$

Find the length. $2n^2 + 1$

Objective 4: Write Equivalent Forms of a Rational Expression

Find three equivalent forms of each rational expression.

55) $-\dfrac{b + 7}{b - 2}$

56) $-\dfrac{8y - 1}{2y + 5}$

57) $-\dfrac{9 - 5t}{2t - 3}$

58) $\dfrac{-12m}{m^2 - 3}$

Objective 5: Multiply Rational Expressions

Multiply.

59) $\dfrac{9}{14} \cdot \dfrac{7}{6}$ $\dfrac{3}{4}$

60) $\dfrac{4}{15} \cdot \dfrac{25}{36}$ $\dfrac{5}{27}$

61) $\dfrac{14u^5}{15v^2} \cdot \dfrac{20v^6}{7u^8}$ $\dfrac{8v^4}{3u^3}$

62) $\dfrac{15s^3}{21t^2} \cdot \dfrac{42t^4}{5s^{12}}$ $\dfrac{6t^2}{s^9}$

63) $\dfrac{5t^2}{(3t - 2)^2} \cdot \dfrac{3t - 2}{10t^3}$ $\dfrac{1}{2t(3t - 2)}$

64) $\dfrac{4u - 5}{9u^2} \cdot \dfrac{3u^6}{(4u - 5)^3}$ $\dfrac{u^4}{3(4u - 5)^2}$

65) $\dfrac{8}{6p + 3} \cdot \dfrac{4p^2 - 1}{12}$ $\dfrac{2(2p - 1)}{9}$

66) $\dfrac{n^2 + 7n + 12}{n + 3} \cdot \dfrac{4}{n + 4}$ 4

67) $\dfrac{2v^2 + 15v + 18}{3v + 18} \cdot \dfrac{12v - 3}{8v + 12}$ $\dfrac{4v - 1}{4}$

68) $\dfrac{y^2 - 4y - 5}{3y^2 + y - 2} \cdot \dfrac{18y - 12}{4y^2}$ $\dfrac{3(y - 5)}{2y^2}$

69) $(x - 8) \cdot \dfrac{4}{x^2 - 8x}$ $\dfrac{4}{x}$

70) $(h^2 + 5h - 6) \cdot \dfrac{10h^2}{2h^2 + 12h}$ $5h(h - 1)$

71) $\dfrac{r^3 + 27}{4t + 20} \cdot \dfrac{rt + 5r - 2t - 10}{r^2 - 9}$ $\dfrac{(r - 2)(r^2 - 3r + 9)}{4(r - 3)}$

72) $\dfrac{36 - w^2}{wt + 6t - w - 6} \cdot \dfrac{8t - 8}{2w^2 - 11w - 6}$ $-\dfrac{8}{2w + 1}$

Objective 6: Divide Rational Expressions

Divide.

73) $\dfrac{4}{5} \div \dfrac{8}{3}$ $\dfrac{3}{10}$

74) $\dfrac{16}{9} \div \dfrac{10}{3}$ $\dfrac{8}{15}$

75) $\dfrac{c^2}{6b} \div \dfrac{c^8}{b}$ $\dfrac{1}{6c^6}$

76) $-\dfrac{15g^3}{14h} \div \dfrac{40g}{7h^3}$ $-\dfrac{3g^2h^2}{16}$

77) $\dfrac{2a - 1}{8a^3} \div \dfrac{(2a - 1)^2}{24a^5}$ $\dfrac{3a^2}{2a - 1}$

78) $\dfrac{2p^4}{(p + 7)^2} \div \dfrac{12p^5}{p + 7}$ $\dfrac{1}{6p(p + 7)}$

79) $\dfrac{18y - 45}{18} \div \dfrac{4y^2 - 25}{10}$ $\dfrac{5}{2y + 5}$

80) $\dfrac{q^2 + q - 56}{5} \div \dfrac{q - 7}{q}$ $\dfrac{q(q + 8)}{5}$

81) $\dfrac{j^2 - 25}{5j + 25} \div \dfrac{7j - 35}{5}$ $\dfrac{1}{7}$

82) $\dfrac{n^2 + 3n - 18}{5n^2 + 30n} \div \dfrac{4n - 12}{8n}$ $\dfrac{2}{5}$

83) $\dfrac{z^2 + 18z + 80}{2z + 1} \div (z + 8)^2$ $\dfrac{z + 10}{(2z + 1)(z + 8)}$

84) $\dfrac{6w^2 - 30w}{7} \div (w - 5)^2$ $\dfrac{6w}{7(w - 5)}$

85) $\dfrac{36a - 12}{16} \div (9a^2 - 1)$ $\dfrac{3}{4(3a + 1)}$

86) $\dfrac{h^2 - 21h + 108}{4h} \div (144 - h^2)$ $\dfrac{9 - h}{4h(h + 12)}$

87) $\dfrac{8d^2 - 8d + 8}{25 - 4d^2} \div \dfrac{d^3 + 1}{2d^2 - 3d - 5}$ $-\dfrac{8}{2d + 5}$

88) $\dfrac{x^2 + 2xy + y^2}{3y - 12} \div \dfrac{7x + 7y}{xy - 4x + 3y - 12}$ $\dfrac{(x + y)(x + 3)}{21}$

89) In the division problem $\dfrac{12}{x} \div \dfrac{3y}{2}$, can $y = 0$? Explain your answer.

90) Find the polynomial in the second denominator so that $\dfrac{2a + 1}{a + 5} \cdot \dfrac{25 - a^2}{?} = -1$. $2a^2 - 9a - 5$

Mixed Exercises: Objectives 5 and 6

Perform the operations and simplify.

91) $\dfrac{a^2 + 4a}{6a + 54} \cdot \dfrac{a^2 + 5a - 36}{16 - a^2}$ $-\dfrac{a}{6}$

92) $\dfrac{3x + 2}{9x^2 - 4} \div \dfrac{4x}{15x^2 - 7x - 2}$ $\dfrac{5x + 1}{4x}$

93) $\dfrac{r^3 + 8}{r + 2} \cdot \dfrac{7}{3r^2 - 6r + 12}$ $\dfrac{7}{3}$

94) $\dfrac{4a^3}{a^2 + a - 72} \cdot (a^2 - a - 56)$ $\dfrac{4a^3(a + 7)}{a + 9}$

95) $\dfrac{54x^8}{22x^3y^2} \div \dfrac{36xy^5}{11x^2y}$ $\dfrac{3x^6}{4y^6}$

96) $\dfrac{2t^2 - 6t + 18}{5t - 5} \cdot \dfrac{t^2 - 9}{t^3 + 27}$ $\dfrac{2(t + 3)}{5(t - 1)}$

97) $\dfrac{2a^2}{a^2 + a - 20} \cdot \dfrac{a^3 + 5a^2 + 4a + 20}{2a^2 + 8}$ $\dfrac{a^2}{a - 4}$

98) $\dfrac{3m^2 + 8m + 4}{4} \div (12m + 8)$ $\dfrac{m + 2}{16}$

99) $\dfrac{30}{4y^2 - 4x^2} \div \dfrac{10x^2 + 10xy + 10y^2}{x^3 - y^3}$ $-\dfrac{3}{4(x + y)}$

100) $\dfrac{28cd^9}{2c^3d} \cdot \dfrac{5d^2}{84c^{10}d^2}$ $\dfrac{5d^8}{6c^{12}}$

101) $\dfrac{4j^2 - 21j + 5}{j^3} \div \left(\dfrac{3j + 2}{j^3 - j^2} \cdot \dfrac{j^2 - 6j + 5}{j} \right)$ $\dfrac{4j - 1}{3j + 2}$

102) $\dfrac{t^3 - 8}{t - 2} \div \left(\dfrac{3t + 11}{5t + 15} \cdot \dfrac{t^2 + 2t + 4}{3t^2 + 11t} \right)$ $5t(t + 3)$

103) If the area of a rectangle is $\dfrac{3}{2xy^6}$ and the width is $\dfrac{y^2}{12x^5}$, what is the length of the rectangle? $\dfrac{18x^4}{y^8}$

104) If the area of a triangle is $\dfrac{3m}{m^2 + 7m + 10}$ and the height is $\dfrac{m - 5}{m + 2}$, what is the length of the base of the triangle? $\dfrac{6m}{m^2 - 25}$

Section 7.2 Adding and Subtracting Rational Expressions

Objectives

1. **Find the Least Common Denominator for a Group of Rational Expressions**
2. **Rewrite Rational Expressions with the LCD as Their Denominators**
3. **Add and Subtract Rational Expressions with a Common Denominator**
4. **Add and Subtract Rational Expressions with Different Denominators**
5. **Add and Subtract Rational Expressions with Denominators Containing Factors $a - b$ and $b - a$**

1. Find the Least Common Denominator for a Group of Rational Expressions

Recall that to add or subtract fractions, they must have a common denominator. Similarly, rational expressions must have common denominators in order to be added or subtracted. In this section, we will discuss how to find the least common denominator (LCD) of rational expressions.

We begin by looking at the fractions $\dfrac{3}{8}$ and $\dfrac{5}{12}$. By inspection we can see that the LCD = 24. But, *why* is that true? Let's write each of the denominators, 8 and 12, as the product of their prime factors:

$$8 = 2 \cdot 2 \cdot 2 = 2^3$$
$$12 = 2 \cdot 2 \cdot 3 = 2^2 \cdot 3$$

The LCD will contain each factor the *greatest* number of times it appears in any single factorization.

2 appears as a factor *three* times in the factorization of 8 and it appears *twice* in the factorization of 12. *The LCD will contain* 2^3.

3 appears as a factor *one* time in the factorization of 12 but does not appear in the factorization of 8. *The LCD will contain* 3.

The LCD, then, is the product of the factors we have identified.

$$\text{LCD of } \dfrac{3}{8} \text{ and } \dfrac{5}{12} = 2^3 \cdot 3 = 8 \cdot 3 = 24$$

This is the same result as the one we obtained just by inspecting the two denominators.

The procedure we just illustrated is the one we use to find the least common denominator of rational expressions.

> **Procedure** Finding the Least Common Denominator (LCD)
>
> **Step 1:** Factor the denominators.
>
> **Step 2:** The LCD will contain each unique factor the *greatest* number of times it appears in any single factorization.
>
> **Step 3:** The LCD is the *product* of the factors identified in Step 2.

Example 1

In-Class Example 1
Find the LCD of each pair of rational expressions.
a) $\dfrac{3}{14}, \dfrac{5}{20}$ b) $\dfrac{8}{15w^4}, \dfrac{6}{5w^2}$
answer: a) 140   b) $15w^4$

Find the LCD of each pair of rational expressions.

a) $\dfrac{7}{18}, \dfrac{11}{24}$ b) $\dfrac{4}{9t^3}, \dfrac{5}{6t^2}$

Solution

a) Follow the steps for finding the least common denominator.

   Step 1: Factor the denominators.

$$18 = 2 \cdot 3 \cdot 3 = 2 \cdot 3^2$$
$$24 = 2 \cdot 2 \cdot 2 \cdot 3 = 2^3 \cdot 3$$

   Step 2: The LCD will contain each unique factor the *greatest* number of times it appears in any factorization. *The LCD will contain 2^3 and 3^2.*

   Step 3: The LCD is the *product* of the factors in step 2.

$$\text{LCD} = 2^3 \cdot 3^2 = 8 \cdot 9 = 72$$

b) To find the LCD of $\dfrac{4}{9t^3}$ and $\dfrac{5}{6t^2}$,

   Step 1: Factor the denominators.

$$9t^3 = 3 \cdot 3 \cdot t^3 = 3^2 \cdot t^3$$
$$6t^2 = 2 \cdot 3 \cdot t^2$$

   Step 2: The LCD will contain each unique factor the *greatest* number of times it appears in any factorization. *It will contain 2, 3^2, and t^3.*

   Step 3: The LCD is the *product* of the factors in Step 2.

$$\text{LCD} = 2 \cdot 3^2 \cdot t^3 = 18t^3$$

You Try 1

Find the LCD of each pair of rational expressions.

a) $\dfrac{4}{15}, \dfrac{9}{20}$ b) $\dfrac{8}{9k^3}, \dfrac{1}{12k^5}$

Example 2

Find the LCD of each group of rational expressions.

a) $\dfrac{6}{x}, \dfrac{2}{x+5}$ b) $\dfrac{10}{c-8}, \dfrac{4c}{c^2-5c-24}$ c) $\dfrac{9}{w^2+2w+1}, \dfrac{1}{2w^2+2w}$

In-Class Example 2

Find the LCD of each group of rational expressions.

a) $\dfrac{3}{d}, \dfrac{11}{d-4}$

b) $\dfrac{6}{t+4}, \dfrac{3}{t^2+14t+40}$

c) $\dfrac{7}{2p^2-6p}, \dfrac{4}{p^2+3p-18}$

answer: a) $d(d-4)$
b) $(t+4)(t+10)$
c) $2p(p-3)(p+6)$

Solution

a) The denominators of $\dfrac{6}{x}$ and $\dfrac{2}{x+5}$ are already in simplest form. It is important to recognize that *x and x + 5 are different factors.*

The LCD will be the product of x and $x + 5$: LCD $= x(x+5)$.

Usually, we leave the LCD in this form; we do not distribute.

b) **Step 1:** Factor the denominators of $\dfrac{10}{c-8}$ and $\dfrac{4c}{c^2-5c-24}$.

$c - 8$ cannot be factored.
$$c^2 - 5c - 24 = (c-8)(c+3)$$

Step 2: The LCD will contain each unique factor the *greatest* number of times it appears in any factorization. *It will contain c − 8 and c + 3.*

Step 3: The LCD is the *product* of the factors identified in Step 2.
$$\text{LCD} = (c-8)(c+3)$$

c) **Step 1:** Factor the denominators of $\dfrac{9}{w^2+2w+1}$ and $\dfrac{1}{2w^2+2w}$.

$$w^2 + 2w + 1 = (w+1)^2$$
$$2w^2 + 2w = 2w(w+1)$$

Step 2: The unique factors are 2, w, and $w + 1$ with $w + 1$ *appearing at most twice. The factors we will use in the LCD are 2, w, and $(w+1)^2$.*

Step 3: The LCD is the *product* of the factors identified in Step 2.
$$\text{LCD} = 2w(w+1)^2 \qquad\blacksquare$$

 You Try 2

Find the LCD of each group of rational expressions.

a) $\dfrac{5}{k}, \dfrac{8k}{k+2}$

b) $\dfrac{12}{p^2-7p}, \dfrac{6}{p-7}$

c) $\dfrac{m}{m^2-25}, \dfrac{8}{m^2+10m+25}$

At first glance it may appear that the least common denominator of $\dfrac{2}{y-3}$ and $\dfrac{10}{3-y}$ is $(y-3)(3-y)$. This is *not* the case. Recall from Section 7.1 that $a - b = -1(b-a)$. We will use this idea to find the LCD of $\dfrac{2}{y-3}$ and $\dfrac{10}{3-y}$.

Example 3

Find the LCD of $\dfrac{2}{y-3}$ and $\dfrac{10}{3-y}$.

In-Class Example 3

Find the LCD of $\dfrac{9}{f-g}$ and $\dfrac{11}{g-f}$.

answer: $f - g$

Solution

Since $3 - y = -(y-3)$, we can rewrite $\dfrac{10}{3-y}$ as $\dfrac{10}{-(y-3)} = -\dfrac{10}{y-3}$.

Therefore, we can now think of our task as finding the LCD of $\dfrac{2}{y-3}$ and $-\dfrac{10}{y-3}$.

The least common denominator is $y - 3$. $\qquad\blacksquare$

You Try 3

Find the LCD of $\dfrac{9}{r-8}$ and $\dfrac{6}{8-r}$.

2. Rewrite Rational Expressions with the LCD as Their Denominators

In order to add or subtract fractions, they must have a common denominator. That means that sometimes we must rewrite one or both fractions so that their denominators contain the LCD. For example, to rewrite $\dfrac{5}{6}$ and $\dfrac{4}{9}$ as equivalent fractions with the least common denominator we begin by identifying the LCD as 18. Then, for each fraction we ask ourselves,

$\dfrac{5}{6}$: *By what number do we multiply 6 to get 18?* 3

Multiply the numerator and denominator of $\dfrac{5}{6}$ by 3 to obtain an equivalent fraction.

$$\frac{5}{6} \cdot \frac{3}{3} = \frac{15}{18}$$

$\dfrac{4}{9}$: *By what number do we multiply 9 to get 18?* 2

Multiply the numerator and denominator of $\dfrac{4}{9}$ by 2 to obtain an equivalent fraction.

$$\frac{4}{9} \cdot \frac{2}{2} = \frac{8}{18}$$

The procedure for rewriting rational expressions as equivalent expressions with the LCD is very similar to the process we use with fractions.

Procedure   Writing Rational Expressions as Equivalent Expressions with the Least Common Denominator

Step 1: Identify and write down the LCD.

Step 2: Look at each rational expression (with its denominator in factored form) and compare its denominator with the LCD. Ask yourself, "What factors are missing?"

Step 3: Multiply the numerator and denominator by the "missing" factors to obtain an equivalent rational expression with the desired LCD. Multiply the terms in the numerator, but leave the denominator as the product of factors.

Example 4

Identify the LCD of each pair of rational expressions, and rewrite each as an equivalent expression with the LCD as its denominator.

a) $\dfrac{7}{8n}, \dfrac{5}{6n^3}$

b) $\dfrac{t}{t-2}, \dfrac{4}{t+7}$

c) $\dfrac{3}{4a^2 - 24a}, \dfrac{9a}{a^2 - 11a + 30}$

d) $\dfrac{r}{r-6}, \dfrac{2}{6-r}$

In-Class Example 4
Identify the LCD of each pair of
rational expressions, and rewrite
each as an equivalent expression
with the LCD as its denominator.

a) $\dfrac{9}{10x}, \dfrac{6}{25x^3}$ b) $\dfrac{3}{c+4}, \dfrac{c}{c-9}$

c) $\dfrac{5}{4t^2-24t}, \dfrac{5t}{t^2-t-30}$

d) $\dfrac{d}{d-4}, \dfrac{11}{4-d}$

answer: a) $50x^3$; $\dfrac{45x^2}{50x^3}, \dfrac{12}{50x^3}$

b) $(c+4)(c-9)$;
$\dfrac{3c-27}{(c+4)(c-9)}, \dfrac{c^2+4c}{(c+4)(c-9)}$

c) $4t(t-6)(t+5)$;
$\dfrac{5t+25}{4t(t-6)(t+5)}, \dfrac{20t^2}{4t(t-6)(t+5)}$

d) $d-4$; $\dfrac{d}{d-4}, -\dfrac{11}{d-4}$

Solution

a)   Follow the steps.

Step 1:   Identify and write down the LCD of $\dfrac{7}{8n}$ and $\dfrac{5}{6n^3}$:   LCD $= 24n^3$.

Step 2:   Compare the denominators of $\dfrac{7}{8n}$ and $\dfrac{5}{6n^3}$ to the LCD and ask yourself, "What's missing?"

$\dfrac{7}{8n}$: $8n$ is "missing" the | $\dfrac{5}{6n^3}$: $6n^3$ is "missing"

factors of 3 and n^2. | the factor 4.

Step 3:   Multiply the numerator and denominator by $3n^2$. | Multiply the numerator and denominator by 4.

$\dfrac{7}{8n} \cdot \dfrac{3n^2}{3n^2} = \dfrac{21n^2}{24n^3}$ | $\dfrac{5}{6n^3} \cdot \dfrac{4}{4} = \dfrac{20}{24n^3}$

$\dfrac{7}{8n} = \dfrac{21n^2}{24n^3}$ and $\dfrac{5}{6n^3} = \dfrac{20}{24n^3}$

b)   Follow the steps.

Step 1:   Identify and write down the LCD of $\dfrac{t}{t-2}$ and $\dfrac{4}{t+7}$:
LCD $= (t-2)(t+7)$.

Step 2:   Compare the denominators of $\dfrac{t}{t-2}$ and $\dfrac{4}{t+7}$ to the LCD and ask yourself, "What's missing?"

$\dfrac{t}{t-2}$: $t-2$ is "missing" the | $\dfrac{4}{t+7}$: $t+7$ is "missing"

factor $t+7$. | the factor $t-2$.

Step 3:   Multiply the numerator and denominator by $t+7$. | Multiply the numerator and denominator by $t-2$.

$\dfrac{t}{t-2} \cdot \dfrac{t+7}{t+7} = \dfrac{t(t+7)}{(t-2)(t+7)}$ | $\dfrac{4}{t+7} \cdot \dfrac{t-2}{t-2} = \dfrac{4(t-2)}{(t+7)(t-2)}$

$= \dfrac{t^2+7t}{(t-2)(t+7)}$ | $= \dfrac{4t-8}{(t-2)(t+7)}$

Notice that we multiplied the factors in the numerator but left the denominator in factored form.

$\dfrac{t}{t-2} = \dfrac{t^2+7t}{(t-2)(t+7)}$ and $\dfrac{4}{t+7} = \dfrac{4t-8}{(t-2)(t+7)}$

c)   Follow the steps.

Step 1:   Identify and write down the LCD of $\dfrac{3}{4a^2-24a}$ and $\dfrac{9a}{a^2-11a+30}$. First, we must factor the denominators.

$\dfrac{3}{4a^2-24a} = \dfrac{3}{4a(a-6)}$, $\dfrac{9a}{a^2-11a+30} = \dfrac{9a}{(a-6)(a-5)}$

We will work with the factored forms of the expressions.

$$\text{LCD} = 4a(a-6)(a-5)$$

Step 2: Compare the denominators of $\dfrac{3}{4a(a-6)}$ and $\dfrac{9a}{(a-6)(a-5)}$ to the LCD, $4a(a-6)(a-5)$, and ask yourself, "What's missing?"

$\dfrac{3}{4a(a-6)}$: $4a(a-6)$ is "missing" the factor $a-5$.

$\dfrac{9a}{(a-6)(a-5)}$: $(a-6)(a-5)$ is "missing" $4a$.

Step 3: Multiply the numerator and denominator by $a-5$.

$$\frac{3}{4a(a-6)} \cdot \frac{a-5}{a-5} = \frac{3(a-5)}{4a(a-6)(a-5)}$$
$$= \frac{3a-15}{4a(a-6)(a-5)}$$

Multiply the numerator and denominator by $4a$.

$$\frac{9a}{(a-6)(a-5)} \cdot \frac{4a}{4a} = \frac{36a^2}{4a(a-6)(a-5)}$$

$$\frac{3}{4a^2-24a} = \frac{3a-15}{4a(a-6)(a-5)} \quad \text{and} \quad \frac{9a}{a^2-11a+30} = \frac{36a^2}{4a(a-6)(a-5)}$$

d) To find the LCD of $\dfrac{r}{r-6}$ and $\dfrac{2}{6-r}$ recall that $6-r$ can be rewritten as $-(r-6)$. So,

$$\frac{2}{6-r} = \frac{2}{-(r-6)} = -\frac{2}{r-6}$$

Therefore, the LCD of $\dfrac{r}{r-6}$ and $-\dfrac{2}{r-6}$ is $r-6$.

The expression $\dfrac{r}{r-6}$ already has the LCD, while $\dfrac{2}{6-r} = -\dfrac{2}{r-6}$. ∎

You Try 4

Identify the least common denominator of each pair of rational expressions, and rewrite each as an equivalent expression with the LCD as its denominator.

a) $\dfrac{9}{7r^5}, \dfrac{4}{21r^2}$ b) $\dfrac{6}{y+4}, \dfrac{8}{3y-2}$ c) $\dfrac{d-1}{d^2+2d}, \dfrac{3}{d^2+12d+20}$ d) $\dfrac{k}{7-k}, \dfrac{4}{k-7}$

We know that in order to add or subtract fractions, they must have a common denominator. For example, $\dfrac{6}{7} - \dfrac{2}{7} = \dfrac{6-2}{7} = \dfrac{4}{7}$. The same is true for rational expressions.

3. Add and Subtract Rational Expressions with a Common Denominator

Example 5

Add $\dfrac{3a}{2a-5} + \dfrac{4a+1}{2a-5}$.

In-Class Example 5

Subtract

$\dfrac{6c}{4c+3} - \dfrac{3c+2}{4c+3}$.

answer: $\dfrac{3c-2}{4c+3}$

Solution

Since $\dfrac{3a}{2a-5}$ and $\dfrac{4a+1}{2a-5}$ have the same denominator, add the terms in the numerator and keep the common denominator.

$$\frac{3a}{2a-5} + \frac{4a+1}{2a-5} = \frac{3a+(4a+1)}{2a-5} \qquad \text{Add terms in the numerator.}$$

$$= \frac{7a+1}{2a-5} \qquad \text{Combine like terms.}$$

∎

We can generalize the procedure for adding and subtracting rational expressions that have a common denominator as follows.

Procedure   Adding and Subtracting Rational Expressions

If $\dfrac{P}{Q}$ and $\dfrac{R}{Q}$ are rational expressions with $Q \neq 0$, then

1) $\dfrac{P}{Q} + \dfrac{R}{Q} = \dfrac{P + R}{Q}$ and   2) $\dfrac{P}{Q} - \dfrac{R}{Q} = \dfrac{P - R}{Q}$

 You Try 5

Add or subtract, as indicated.

a) $\dfrac{9}{11} - \dfrac{4}{11}$ b) $\dfrac{7c}{3c - 4} + \dfrac{2c + 5}{3c - 4}$

All answers to a sum or difference of rational expressions should be in lowest terms. Sometimes it is necessary to simplify our result to lowest terms by factoring the numerator and dividing the numerator and denominator by the greatest common factor.

Example 6

Add or subtract, as indicated.

a) $\dfrac{11}{12t} + \dfrac{7}{12t}$ b) $\dfrac{n^2 - 8}{n(n + 5)} - \dfrac{7 - 2n}{n(n + 5)}$

In-Class Example 6
Add or subtract, as indicated.
a) $\dfrac{13}{7a} - \dfrac{6}{7a}$
b) $\dfrac{k^2 - 15}{k(k - 3)} - \dfrac{12 - 6k}{k(k - 3)}$
answer: a) $\dfrac{1}{a}$ b) $\dfrac{k + 9}{k}$

Solution

a) $\dfrac{11}{12t} + \dfrac{7}{12t} = \dfrac{11 + 7}{12t}$ Add terms in the numerator.

$= \dfrac{18}{12t}$ Combine terms.

$= \dfrac{3}{2t}$ Reduce to lowest terms.

b) $\dfrac{n^2 - 8}{n(n + 5)} - \dfrac{7 - 2n}{n(n + 5)} = \dfrac{(n^2 - 8) - (7 - 2n)}{n(n + 5)}$ Subtract terms in the numerator.

$= \dfrac{n^2 - 8 - 7 + 2n}{n(n + 5)}$ Distribute.

$= \dfrac{n^2 + 2n - 15}{n(n + 5)}$ Combine like terms.

$= \dfrac{(n + 5)(n - 3)}{n(n + 5)}$ Factor the numerator.

$= \dfrac{n - 3}{n}$ Reduce to lowest terms.

 You Try 6

Add or subtract, as indicated.

a) $\dfrac{3}{20c^2} - \dfrac{9}{20c^2}$ b) $\dfrac{k^2 + 2k + 5}{(k + 4)(k - 1)} + \dfrac{5k + 7}{(k + 4)(k + 1)}$ c) $\dfrac{20d - 9}{4d(3d + 1)} - \dfrac{5d - 14}{4d(3d + 1)}$

BE CAREFUL

After combining like terms in the numerator, ask yourself, *"Can I factor the numerator?"* If so, factor it. Sometimes, the expression can be reduced by dividing the numerator and denominator by the greatest common factor.

4. Add and Subtract Rational Expressions with Different Denominators

If we are asked to add or subtract rational expressions with different denominators, we must begin by rewriting each expression with the least common denominator. Then, add or subtract. Simplify the result.

Procedure   Steps for Adding and Subtracting Rational Expressions with Different Denominators

1) Factor the denominators.
2) Write down the LCD.
3) Rewrite each rational expression as an equivalent rational expression with the LCD.
4) Add or subtract the numerators and keep the common denominator in factored form.
5) After combining like terms in the numerator ask yourself, *"Can I factor it?"* If so, factor.
6) Reduce the rational expression, if possible.

Example 7

Add or subtract, as indicated.

In-Class Example 7
Add or subtract, as indicated.
a) $\dfrac{q-3}{5} + \dfrac{q+6}{10}$

b) $\dfrac{8}{9m} - \dfrac{11}{12m^7}$

c) $\dfrac{3k-20}{k^2-16} + \dfrac{k}{k+4}$

answer: a) $\dfrac{3q}{10}$ b) $\dfrac{32m^6-33}{36m^7}$

c) $\dfrac{k-5}{k-4}$

a) $\dfrac{m+8}{3} + \dfrac{m-1}{6}$

b) $\dfrac{3}{4x} - \dfrac{11}{10x^2}$

c) $\dfrac{4a-6}{a^2-9} + \dfrac{a}{a+3}$

Solution

a) The LCD is 6. $\dfrac{m-1}{6}$ already has the LCD.

Rewrite $\dfrac{m+8}{3}$ with the LCD: $\dfrac{m+8}{3} \cdot \dfrac{2}{2} = \dfrac{2(m+8)}{6}$.

$\dfrac{m+8}{3} + \dfrac{m-1}{6} = \dfrac{2(m+8)}{6} + \dfrac{m-1}{6}$ Write each expression with the LCD.

$= \dfrac{2(m+8) + (m-1)}{6}$ Add the numerators.

$= \dfrac{2m+16+m-1}{6}$ Distribute.

$= \dfrac{3m+15}{6}$ Combine like terms.

Ask yourself, *"Can I factor the numerator?"* Yes.

$= \dfrac{\overset{1}{\cancel{3}}(m+5)}{\underset{2}{\cancel{6}}}$ Factor.

$= \dfrac{m+5}{2}$ Reduce.

b)   The LCD of $\dfrac{3}{4x}$ and $\dfrac{11}{10x^2}$ is $20x^2$. Rewrite each expression with the LCD.

$$\frac{3}{4x} \cdot \frac{5x}{5x} = \frac{15x}{20x^2} \quad \text{and} \quad \frac{11}{10x^2} \cdot \frac{2}{2} = \frac{22}{20x^2}$$

$$\frac{3}{4x} - \frac{11}{10x^2} = \frac{15x}{20x^2} - \frac{22}{20x^2} \qquad \text{Write each expression with the LCD.}$$

$$= \frac{15x - 22}{20x^2} \qquad \text{Subtract the numerators.}$$

"Can I factor the numerator?" No. The expression is in simplest form since the numerator and denominator have no common factors.

c)   Begin by factoring the denominator of $\dfrac{4a - 6}{a^2 - 9}$.

$$\frac{4a - 6}{a^2 - 9} = \frac{4a - 6}{(a + 3)(a - 3)}$$

The LCD of $\dfrac{4a - 6}{(a + 3)(a - 3)}$ and $\dfrac{a}{a + 3}$ is $(a + 3)(a - 3)$.

Rewrite $\dfrac{a}{a + 3}$ with the LCD: $\dfrac{a}{a + 3} \cdot \dfrac{a - 3}{a - 3} = \dfrac{a(a - 3)}{(a + 3)(a - 3)}$.

$$\frac{4a - 6}{a^2 - 9} + \frac{a}{a + 3} = \frac{4a - 6}{(a + 3)(a - 3)} + \frac{a}{a + 3} \qquad \begin{array}{l}\text{Factor the}\\ \text{denominator.}\end{array}$$

$$= \frac{4a - 6}{(a + 3)(a - 3)} + \frac{a(a - 3)}{(a + 3)(a - 3)} \qquad \begin{array}{l}\text{Write each expression}\\ \text{with the LCD.}\end{array}$$

$$= \frac{4a - 6 + a(a - 3)}{(a + 3)(a - 3)} \qquad \text{Add the numerators.}$$

$$= \frac{4a - 6 + a^2 - 3a}{(a + 3)(a - 3)} \qquad \text{Distribute.}$$

$$= \frac{a^2 + a - 6}{(a + 3)(a - 3)} \qquad \text{Combine like terms.}$$

Ask yourself, *"Can I factor the numerator?"* Yes.

$$= \frac{\cancel{(a + 3)}(a - 2)}{\cancel{(a + 3)}(a - 3)} \qquad \text{Factor.}$$

$$= \frac{a - 2}{a - 3} \qquad \text{Reduce.}$$

You Try 7

Add or subtract, as indicated.

a) $\dfrac{7}{12t^3} + \dfrac{4}{9t}$ b) $\dfrac{k - 3}{4} - \dfrac{k + 3}{6}$ c) $\dfrac{6}{r - 5} + \dfrac{r^2 - 17r}{r^2 - 25}$

Example 8

Subtract $\dfrac{4w}{w^2 + 9w + 14} - \dfrac{2w + 5}{w^2 + 3w - 28}$.

In-Class Example 8
Subtract
$\dfrac{3n}{n^2 + n - 12} - \dfrac{n + 6}{n^2 - n - 20}$.
answer:
$\dfrac{2n^2 - 18n + 18}{(n - 3)(n + 4)(n - 5)}$

Solution

Factor the denominators, then write down the LCD.

$$\frac{4w}{w^2 + 9w + 14} = \frac{4w}{(w + 7)(w + 2)}, \qquad \frac{2w + 5}{w^2 + 3w - 28} = \frac{2w + 5}{(w + 7)(w - 4)}$$

Rewrite each expression with the LCD, $(w + 7)(w + 2)(w - 4)$.

$$\frac{4w}{(w + 7)(w + 2)} \cdot \frac{w - 4}{w - 4} = \frac{4w(w - 4)}{(w + 7)(w + 2)(w - 4)}$$

$$\frac{2w + 5}{(w + 7)(w - 4)} \cdot \frac{w + 2}{w + 2} = \frac{(2w + 5)(w + 2)}{(w + 7)(w + 2)(w - 4)}$$

$$\frac{4w}{w^2 + 9w + 14} - \frac{2w + 5}{w^2 + 3w - 28}$$

$$= \frac{4w}{(w + 7)(w + 2)} - \frac{2w + 5}{(w + 7)(w - 4)} \qquad \text{Factor denominators.}$$

$$= \frac{4w(w - 4)}{(w + 7)(w + 2)(w - 4)} - \frac{(2w + 5)(w + 2)}{(w + 7)(w + 2)(w - 4)} \qquad \text{Write each expression with the LCD.}$$

$$= \frac{4w(w - 4) - (2w + 5)(w + 2)}{(w + 7)(w + 2)(w - 4)} \qquad \text{Subtract the numerators.}$$

$$= \frac{4w^2 - 16w - (2w^2 + 9w + 10)}{(w + 7)(w + 2)(w - 4)} \qquad \text{Distribute. You must use parentheses.}$$

$$= \frac{4w^2 - 16w - 2w^2 - 9w - 10}{(w + 7)(w + 2)(w - 4)} \qquad \text{Distribute.}$$

$$= \frac{2w^2 - 25w - 10}{(w + 7)(w + 2)(w - 4)} \qquad \text{Combine like terms.}$$

Ask yourself, *"Can I factor the numerator?"* No. The expression is in simplest form since the numerator and denominator have no common factors. ■

BE CAREFUL

In Example 8, when you move from

$$\frac{4w(w - 4) - (2w + 5)(w + 2)}{(w + 7)(w + 2)(w - 4)} \quad \text{to} \quad \frac{4w^2 - 16w - (2w^2 + 9w + 10)}{(w + 7)(w + 2)(w - 4)}$$

you *must* use parentheses since the entire quantity $2w^2 + 9w + 10$ is being subtracted from $4w^2 - 16w$.

You Try 8

Subtract $\dfrac{3d}{d^2 + 13d + 40} - \dfrac{2d - 3}{d^2 + 7d - 8}$.

5. Add and Subtract Rational Expressions with Denominators Containing Factors $a - b$ and $b - a$

Example 9

In-Class Example 9
Add or subtract, as indicated.

a) $\dfrac{4}{r - 4} - \dfrac{2r}{4 - r}$

b) $\dfrac{4}{4 - r} + \dfrac{9}{r^2 - 16}$

answer: a) $\dfrac{4 + 2r}{r - 4}$

b) $\dfrac{-4r - 7}{(r - 4)(r + 4)}$

Add or subtract, as indicated.

a) $\dfrac{s}{s - 3} - \dfrac{11}{3 - s}$ b) $\dfrac{3}{4 - h} + \dfrac{6}{h^2 - 16}$

Solution

a) Recall that $a - b = -(b - a)$. The least common denominator of $\dfrac{s}{s - 3}$ and $\dfrac{11}{3 - s}$ is $s - 3$ or $3 - s$. We will use LCD $= s - 3$.

Rewrite $\dfrac{11}{3 - s}$ with the LCD: $\dfrac{11}{3 - s} = \dfrac{11}{-(s - 3)} = -\dfrac{11}{s - 3}$.

$$\dfrac{s}{s - 3} - \dfrac{11}{3 - s} = \dfrac{s}{s - 3} - \left(-\dfrac{11}{s - 3}\right) \quad \text{Write each expression with the LCD.}$$

$$= \dfrac{s}{s - 3} + \dfrac{11}{s - 3} \quad \text{Distribute.}$$

$$= \dfrac{s + 11}{s - 3} \quad \text{Add the numerators.}$$

b) Factor the denominator of $\dfrac{6}{h^2 - 16}$: $\dfrac{6}{h^2 - 16} = \dfrac{6}{(h + 4)(h - 4)}$.

Rewrite $\dfrac{3}{4 - h}$ with a denominator of $h - 4$: $\dfrac{3}{4 - h} = \dfrac{3}{-(h - 4)} = -\dfrac{3}{h - 4}$.

Now we must find the LCD of $\dfrac{6}{(h + 4)(h - 4)}$ and $-\dfrac{3}{h - 4}$.

$$\text{LCD} = (h + 4)(h - 4)$$

Rewrite $-\dfrac{3}{h - 4}$ with the LCD.

$$-\dfrac{3}{h - 4} \cdot \dfrac{h + 4}{h + 4} = -\dfrac{3(h + 4)}{(h + 4)(h - 4)} = \dfrac{-3(h + 4)}{(h + 4)(h - 4)}$$

$$\dfrac{3}{4 - h} + \dfrac{6}{h^2 - 16} = -\dfrac{3}{h - 4} + \dfrac{6}{(h + 4)(h - 4)}$$

$$= \dfrac{-3(h + 4)}{(h + 4)(h - 4)} + \dfrac{6}{(h + 4)(h - 4)} \quad \text{Write each expression with the LCD.}$$

$$= \dfrac{-3(h + 4) + 6}{(h + 4)(h - 4)} \quad \text{Add the numerators.}$$

$$= \dfrac{-3h - 12 + 6}{(h + 4)(h - 4)} \quad \text{Distribute.}$$

$$= \dfrac{-3h - 6}{(h + 4)(h - 4)} \quad \text{Combine like terms.}$$

Ask yourself, *"Can I factor the numerator?"* Yes.

$$= \dfrac{-3(h + 2)}{(h + 4)(h - 4)} \quad \text{Factor.}$$

Although the numerator factors, the numerator and denominator do not contain any common factors. The result, $\dfrac{-3(h + 2)}{(h + 4)(h - 4)}$, is in simplest form.

You Try 9

Add or subtract, as indicated.

a) $\dfrac{5}{8-y} + \dfrac{3}{y-8}$ b) $\dfrac{1}{x^2-36} - \dfrac{x+4}{6-x}$

Answers to You Try Exercises

1) a) 60   b) $36k^5$ 2) a) $k(k+2)$ b) $p(p-7)$ c) $(m+5)^2(m-5)$ 3) $r-8$

4) a) LCD $= 21r^5$; $\dfrac{9}{7r^5} = \dfrac{27}{21r^5}$, $\dfrac{4}{21r^2} = \dfrac{4r^3}{21r^5}$

b) LCD $= (y+4)(3y-2)$; $\dfrac{6}{y+4} = \dfrac{18y-12}{(y+4)(3y-2)}$, $\dfrac{8}{3y-2} = \dfrac{8y+32}{(y+4)(3y-2)}$

c) LCD $= d(d+2)(d+10)$; $\dfrac{d-1}{d^2+2d} = \dfrac{d^2+9d-10}{d(d+2)(d+10)}$, $\dfrac{3}{d^2+12d+20} = \dfrac{3d}{d(d+2)(d+10)}$

d) LCD $= k-7$; $\dfrac{k}{7-k} = -\dfrac{k}{k-7}$, $\dfrac{4}{k-7} = \dfrac{4}{k-7}$

5) a) $\dfrac{5}{11}$ b) $\dfrac{9c+5}{3c-4}$ 6) a) $-\dfrac{3}{10c^2}$ b) $\dfrac{k+3}{k-1}$ c) $\dfrac{5}{4d}$ 7) a) $\dfrac{16t^2+21}{36t^3}$ b) $\dfrac{k-15}{12}$ c) $\dfrac{r-6}{r+5}$

8) $\dfrac{d^2-10d+15}{(d+8)(d+5)(d-1)}$ 9) a) $-\dfrac{2}{y-8}$ or $\dfrac{2}{8-y}$ b) $\dfrac{(x+5)^2}{(x+6)(x-6)}$

7.2 Exercises

*Additional answers can be found in the Answers to Exercises appendix.

Objective 1: Find the Least Common Denominator for a Group of Rational Expressions

Find the LCD of each group of rational expressions.

1) $\dfrac{5}{8}, \dfrac{9}{20}$ 40

2) $\dfrac{11}{12}, \dfrac{5}{16}$ 48

3) $\dfrac{6}{c^4}, \dfrac{7}{c^3}$ c^4

4) $\dfrac{4}{n^5}, \dfrac{1}{n^9}$ n^9

5) $\dfrac{8}{9p^3}, \dfrac{5}{12p^8}$ $36p^8$

6) $-\dfrac{2}{9z^4}, \dfrac{16}{45z^6}$ $45z^6$

7) $\dfrac{1}{8a^3b^3}, \dfrac{7}{12ab^4}$ $24a^3b^4$

8) $\dfrac{4}{27x^2y^2}, \dfrac{11}{3x^3y^2}$ $27x^3y^2$

9) $\dfrac{3}{n+4}, \dfrac{1}{2}$ $2(n+4)$

10) $\dfrac{7}{9}, \dfrac{6}{z-8}$ $9(z-8)$

11) $\dfrac{10}{w}, \dfrac{6}{2w+1}$ $w(2w+1)$

12) $\dfrac{2}{y}, -\dfrac{9}{3y+5}$ $y(3y+5)$

13) $\dfrac{1}{12a^2-4a}, \dfrac{15}{6a^4-2a^3}$ $4a^3(3a-1)$

14) $\dfrac{1}{10p^3-15p^2}, \dfrac{6}{2p^5-3p^4}$ $5p^4(2p-3)$

15) $\dfrac{8}{r+7}, \dfrac{4}{r-2}$
$(r+7)(r-2)$

16) $\dfrac{m}{m-6}, \dfrac{3}{m-5}$
$(m-6)(m-5)$

17) $\dfrac{w}{w^2-3w-10}, \dfrac{5}{w^2-2w-15}, \dfrac{9w}{w^2+5w+6}$
$(w-5)(w+2)(w+3)$

18) $\dfrac{10t}{t^2-5t-14}, -\dfrac{8}{t^2-49}, \dfrac{t}{t^2+9t+14}$
$(t+7)(t-7)(t+2)$

19) $\dfrac{6}{b-4}, \dfrac{5}{4-b}$
$b-4$ or $4-b$

20) $\dfrac{u}{v-u}, \dfrac{2}{u-v}$
$u-v$ or $v-u$

Objective 2: Rewrite Rational Expressions with the LCD as Their Denominators

21) Explain, in your own words, how to rewrite $\dfrac{5}{x+8}$ as an equivalent rational expression with a denominator of $(x+8)(x-2)$.   Answers may vary.

22) Explain, in your own words, how to rewrite $\dfrac{6}{3-n}$ as an equivalent rational expression with a denominator of $n-3$.   Answers may vary.

Identify the least common denominator of each group of rational expressions, and rewrite each as an equivalent rational expression with the LCD as its denominator.

23) $\dfrac{3}{t}, \dfrac{8}{t^3}$ $\dfrac{3}{t} = \dfrac{3t^2}{t^3}; \dfrac{8}{t^3} = \dfrac{8}{t^3}$

24) $\dfrac{10}{p^5}, \dfrac{7}{p^2}$ $\dfrac{10}{p^5} = \dfrac{10}{p^5}; \dfrac{7}{p^2} = \dfrac{7p^3}{p^5}$

25) $\dfrac{9}{8n^6}, \dfrac{2}{3n^2}$

26) $\dfrac{5}{6a}, \dfrac{7}{8a^5}$

27) $\dfrac{1}{x^3y}, \dfrac{6}{5xy^5}$

28) $\dfrac{5}{6a^2b^4}, \dfrac{5}{a^4b}$

29) $\dfrac{t}{5t-6}, \dfrac{10}{7}$

30) $\dfrac{8}{d}, \dfrac{2}{d-4}$

31) $\dfrac{a}{24a+36}, \dfrac{1}{18a+27}$

32) $\dfrac{7}{12x-4}, \dfrac{x}{18x-6}$

33) $\dfrac{4}{h+5}, \dfrac{7h}{h-3}$

34) $\dfrac{8}{a+9}, \dfrac{a}{2a+7}$

35) $\dfrac{9y}{y^2-y-42}, \dfrac{3}{2y^2+12y}$

36) $\dfrac{4q}{3q^2+24q}, \dfrac{5}{q^2+q-56}$

37) $\dfrac{z}{z^2-10z+25}, \dfrac{15z}{z^2-2z-15}$

38) $\dfrac{c}{c^2+11c+28}, \dfrac{6}{c^2+14c+49}$

39) $\dfrac{11}{g-3}, \dfrac{4}{9-g^2}$

40) $\dfrac{10}{4k-1}, \dfrac{k}{1-16k^2}$

41) $\dfrac{4}{w^2-4w}, \dfrac{6}{7w^2-28w}, \dfrac{11}{w^2-8w+16}$

42) $\dfrac{t}{t^2-4t-21}, \dfrac{2}{t+3}, \dfrac{4}{t^2-49}$

Objective 3: Add and Subtract Rational Expressions with a Common Denominator

Add or subtract, as indicated.

43) $\dfrac{7}{20} + \dfrac{9}{20}$ $\dfrac{4}{5}$

44) $\dfrac{11}{12} - \dfrac{5}{12}$ $\dfrac{1}{2}$

45) $\dfrac{8}{a} + \dfrac{2}{a}$ $\dfrac{10}{a}$

46) $\dfrac{6}{5c} + \dfrac{14}{5c}$ $\dfrac{4}{c}$

47) $\dfrac{8}{x+4} + \dfrac{2x}{x+4}$ 2

48) $\dfrac{7m}{m+5} + \dfrac{35}{m+5}$ 7

49) $\dfrac{7w-4}{w(3w-4)} - \dfrac{20-11w}{w(3w-4)}$ $\dfrac{6}{w}$

50) $\dfrac{10t+7}{t(2t+1)} - \dfrac{2t+3}{t(2t+1)}$ $\dfrac{4}{t}$

51) $\dfrac{2r+15}{(r-5)(r+2)} + \dfrac{r^2-10r}{(r-5)(r+2)}$ $\dfrac{r-3}{r+2}$

52) $\dfrac{d^2-12}{(d+4)(d+1)} + \dfrac{3-8d}{(d+4)(d+1)}$ $\dfrac{d-9}{d+4}$

Objective 4: Add and Subtract Rational Expressions with Different Denominators

53) For $\dfrac{8}{x-3}$ and $\dfrac{2}{x}$:

a) Find the LCD. $x(x-3)$

b) Explain, in your own words, how to rewrite each expression with the LCD.

c) Rewrite each expression with the LCD.

54) For $\dfrac{4}{9b^2}$ and $\dfrac{5}{6b^4}$:

a) Find the LCD. $18b^4$

b) Explain, in your own words, how to rewrite each expression with the LCD.

c) Rewrite each expression with the LCD.

55) How do you find the least common denominator of two rational expressions when their denominators have no common factors?   Find the product of the denominators.

56) Explain, in your own words, how to add two rational expressions with different denominators.   Answers may vary.

Add or subtract as indicated.

57) $\dfrac{5}{8} + \dfrac{1}{6}$ $\dfrac{19}{24}$

58) $\dfrac{9}{10} - \dfrac{5}{6}$ $\dfrac{1}{15}$

59) $\dfrac{5x}{12} - \dfrac{4x}{15}$ $\dfrac{3x}{20}$

60) $\dfrac{9t}{5} + \dfrac{3}{4}$ $\dfrac{3(12t+5)}{20}$

61) $\dfrac{3}{2a} + \dfrac{6}{7a^2}$ $\dfrac{3(7a+4)}{14a^2}$

62) $\dfrac{3}{2f^2} - \dfrac{7}{f}$ $\dfrac{3-14f}{2f^2}$

63) $\dfrac{15}{d-8} - \dfrac{4}{d}$ $\dfrac{11d+32}{d(d-8)}$

64) $\dfrac{8}{r-7} - \dfrac{4}{r}$ $\dfrac{4(r+7)}{r(r-7)}$

65) $\dfrac{1}{z+6} + \dfrac{4}{z+2}$

66) $\dfrac{6}{c-5} + \dfrac{5}{c+3}$

67) $\dfrac{x}{2x+1} - \dfrac{3}{x+5}$

68) $\dfrac{m}{3m+4} - \dfrac{2}{m-9}$

69) $\dfrac{t}{t+7} + \dfrac{11t-21}{t^2-49}$ $\dfrac{t-3}{t-7}$

70) $\dfrac{-3u-5}{u^2-1} + \dfrac{u}{u+1}$ $\dfrac{u-5}{u-1}$

71) $\dfrac{b}{b^2-16} + \dfrac{10}{b^2-5b-36}$ $\dfrac{b^2+b-40}{(b+4)(b-4)(b-9)}$

72) $\dfrac{7g}{g^2-9g-10} + \dfrac{4}{g^2-100}$ $\dfrac{7g^2+74g+4}{(g+1)(g+10)(g-10)}$

73) $\dfrac{3c}{c^2 + 4c - 12} - \dfrac{2c - 5}{c^2 + 2c - 24}$ $\dfrac{(c - 5)(c + 2)}{(c + 6)(c - 2)(c - 4)}$

74) $\dfrac{4a}{a^2 - 5a - 24} - \dfrac{2a + 3}{a^2 - 10a + 16}$ $\dfrac{(2a + 1)(a - 9)}{(a - 8)(a + 3)(a - 2)}$

VIDEO 75) $\dfrac{4b + 1}{3b - 12} + \dfrac{5b}{b^2 - b - 12}$ $\dfrac{4b^2 + 28b + 3}{3(b - 4)(b + 3)}$

76) $\dfrac{k + 9}{2k - 24} + \dfrac{4k}{k^2 - 15k + 36}$ $\dfrac{k^2 + 14k - 27}{2(k - 12)(k - 3)}$

Objective 5: Add and Subtract Rational Expressions with Denominators Containing Factors $a - b$ and $b - a$

77) Is $(x - 7)(7 - x)$ the LCD for $\dfrac{5}{x - 7} + \dfrac{2}{7 - x}$? Why or why not?

78) What is the LCD of $\dfrac{n}{5 - 3n} - \dfrac{8}{3n - 5}$? $3n - 5$ or $5 - 3n$

Add or subtract, as indicated.

79) $\dfrac{9}{z - 6} + \dfrac{2}{6 - z}$

80) $\dfrac{15}{q - 8} + \dfrac{9}{8 - q}$

81) $\dfrac{2c}{12b - 7c} - \dfrac{13}{7c - 12b}$ $\dfrac{2c + 13}{12b - 7c}$ or $\dfrac{2c + 13}{7c - 12b}$

82) $\dfrac{2}{4u - 3v} - \dfrac{6u}{3v - 4u}$ $\dfrac{2(1 + 3u)}{4u - 3v}$ or $\dfrac{2(1 + 3u)}{3v - 4u}$

83) $\dfrac{5}{8 - t} + \dfrac{10}{t^2 - 64}$

84) $\dfrac{8}{r^2 - 9} + \dfrac{2}{3 - r}$

VIDEO 85) $\dfrac{a}{4a^2 - 9} - \dfrac{4}{3 - 2a}$

86) $\dfrac{2y}{9y^2 - 25} - \dfrac{2}{5 - 3y}$ $\dfrac{2(4y + 5)}{(3y + 5)(3y - 5)}$

Mixed Exercises: Objectives 4 and 5

Perform the indicated operations.

87) $\dfrac{2}{j^2 + 8j} + \dfrac{2j}{j + 8} - \dfrac{1}{3j}$ $\dfrac{6j^2 - j - 2}{3j(j + 8)}$

88) $\dfrac{4}{w^2 - 3w} + \dfrac{9}{w} - \dfrac{10w}{w - 3}$ $\dfrac{-10w^2 + 9w - 23}{w(w - 3)}$

VIDEO 89) $\dfrac{c}{c^2 - 8c + 16} - \dfrac{5}{c^2 - c - 12}$ $\dfrac{c^2 - 2c + 20}{(c - 4)^2(c + 3)}$

90) $\dfrac{n}{n^2 + 11n + 30} - \dfrac{3}{n^2 + 10n + 25}$ $\dfrac{n^2 + 2n - 18}{(n + 5)^2(n + 6)}$

91) $\dfrac{1}{x + y} + \dfrac{x}{x^2 - y^2} - \dfrac{4}{2x - 2y}$ $-\dfrac{3y}{(x + y)(x - y)}$

92) $\dfrac{8}{3a + 3b} + \dfrac{3}{a - b} - \dfrac{3a}{a^2 - b^2}$ $\dfrac{8a + b}{3(a + b)(a - b)}$

93) $\dfrac{n + 5}{4n^2 + 7n - 2} - \dfrac{n - 4}{3n^2 + 7n + 2}$ $\dfrac{-n^2 + 33n + 1}{(4n - 1)(3n + 1)(n + 2)}$

94) $\dfrac{3v - 4}{6v^2 - v - 5} - \dfrac{v - 2}{3v^2 + v - 4}$ $\dfrac{(3v - 2)(v + 3)}{(6v + 5)(3v + 4)(v - 1)}$

95) $\dfrac{y + 6}{y^2 - 4y} + \dfrac{y}{2y^2 - 13y + 20} - \dfrac{1}{2y^2 - 5y}$ $\dfrac{3y^2 + 6y - 26}{y(y - 4)(2y - 5)}$

96) $\dfrac{g - 5}{5g^2 - 30g} + \dfrac{g}{2g^2 - 17g + 30} - \dfrac{6}{2g^2 - 5g}$

For each rectangle, find a rational expression in simplest form to represent its a) area and b) perimeter.

97) [rectangle with dimensions $\dfrac{4}{x - 3}$ and $\dfrac{x + 1}{2}$]   a) $\dfrac{2(x + 1)}{x - 3}$ b) $\dfrac{x^2 - 2x + 5}{x - 3}$

98) [rectangle with dimensions $\dfrac{x - 4}{6}$ and $\dfrac{10}{x + 1}$]   a) $\dfrac{5(x - 4)}{3(x + 1)}$ b) $\dfrac{x^2 - 3x + 56}{3(x + 1)}$

99) [rectangle with dimensions $\dfrac{1}{w^2 - 4}$ and $\dfrac{w}{w + 2}$]   a) $\dfrac{w}{(w + 2)^2(w - 2)}$ b) $\dfrac{2(w - 1)^2}{(w + 2)(w - 2)}$

100) [rectangle with dimensions $\dfrac{2}{t^2 + 9t + 20}$ and $\dfrac{t}{t + 5}$]   a) $\dfrac{2t}{(t + 5)^2(t + 4)}$ b) $\dfrac{2(t^2 + 4t + 2)}{(t + 5)(t + 4)}$

Recall that given functions $f(x)$ and $g(x)$, the domains of $(f + g)(x)$, $(f - g)(x)$, $(f \cdot g)(x)$, and $\left(\dfrac{f}{g}\right)(x)$ consist of all values, x, in the domain of f and in the domain of g as well as in the domain of the function obtained by adding, subtracting, multiplying, or dividing the two functions.

For Exercises 101–104, let $f(x) = \dfrac{6}{x}$ and $g(x) = \dfrac{3x + 12}{x + 5}$.

101) Find $(f + g)(x)$ and its domain.

102) Find $(f - g)(x)$ and its domain.

103) Find $(f \cdot g)(x)$ and its domain.

104) Find $\left(\dfrac{f}{g}\right)(x)$ and its domain.

Section 7.3 Simplifying Complex Fractions

Objectives

1. **Simplify a Complex Fraction with One Term in the Numerator and One Term in the Denominator**
2. **Simplify a Complex Fraction with More Than One Term in the Numerator and/or Denominator by Rewriting It as a Division Problem**
3. **Simplify a Complex Fraction with More Than One Term in the Numerator and/or Denominator by Multiplying by the LCD**
4. **Simplify Rational Expressions Containing Negative Exponents**

In algebra we sometimes encounter fractions that contain fractions in their numerators, denominators, or both. Such fractions are called *complex fractions*. Some examples of complex fractions are

$$\frac{\dfrac{5}{8}}{\dfrac{3}{4}}, \quad \frac{\dfrac{1}{2} + \dfrac{1}{3}}{3 - \dfrac{5}{4}}, \quad \frac{\dfrac{2}{x^2 y}}{\dfrac{1}{y} - \dfrac{y}{x}}, \quad \frac{\dfrac{7k - 28}{3}}{\dfrac{k - 4}{k}}$$

> ### Definition
>
> A **complex fraction** is a rational expression that contains one or more fractions in its numerator, its denominator, or both.

A complex fraction is not considered to be an expression in simplest form. In this section, we will learn how to simplify complex fractions to lowest terms.

We begin by looking at two different types of complex fractions.

1) Complex fractions with *one term* in the numerator and *one term* in the denominator

2) Complex fractions that have *more than one term* in their numerators, their denominators, or both

1. Simplify a Complex Fraction with One Term in the Numerator and One Term in the Denominator

Remember that a fraction represents division. For example, $\dfrac{20}{4} = 20 \div 4 = 5$. We use this fact to simplify complex fractions that have one term in the numerator and one term in the denominator.

Example 1

In-Class Example 1
Simplify each complex fraction.
a) $\dfrac{\dfrac{8}{15}}{\dfrac{16}{55}}$ b) $\dfrac{\dfrac{6m + 66}{m}}{\dfrac{m + 11}{9}}$

answer: a) $\dfrac{11}{6}$ b) $\dfrac{54}{m}$

Simplify each complex fraction.

a) $\dfrac{\dfrac{7}{30}}{\dfrac{14}{25}}$

b) $\dfrac{\dfrac{7k - 28}{3}}{\dfrac{k - 4}{k}}$

Solution

a) There is one term in the numerator: $\dfrac{7}{30}$

There is one term in the denominator: $\dfrac{14}{25}$

$\dfrac{\dfrac{7}{30}}{\dfrac{14}{25}}$ means $\dfrac{7}{30} \div \dfrac{14}{25}$. Then,

$$\frac{7}{30} \div \frac{14}{25} = \frac{7}{30} \cdot \frac{25}{14}$$ Multiply by the reciprocal.

$$= \frac{\overset{1}{\cancel{7}}}{\underset{6}{\cancel{30}}} \cdot \frac{\overset{5}{\cancel{25}}}{\underset{2}{\cancel{14}}}$$ Divide 7 and 14 by 7. Divide 25 and 30 by 5.

$$= \frac{5}{12}$$ Multiply.

b)   There is one term in the numerator: $\dfrac{7k - 28}{3}$

There is one term in the denominator: $\dfrac{k - 4}{k}$

To simplify, rewrite as a division problem then carry out the division.

$$\frac{\dfrac{7k - 28}{3}}{\dfrac{k - 4}{k}} = \frac{7k - 28}{3} \div \frac{k - 4}{k}$$ Rewrite the complex fraction as a division problem.

$$= \frac{7k - 28}{3} \cdot \frac{k}{k - 4}$$ Change division to multiplication by the reciprocal of $\dfrac{k - 4}{k}$.

$$= \frac{7\cancel{(k - 4)}}{3} \cdot \frac{k}{\cancel{k - 4}}$$ Factor and divide the numerator and denominator by $k - 4$ to simplify.

$$= \frac{7k}{3}$$ Multiply. ■

You Try 1

Simplify.

a) $\dfrac{\dfrac{10}{27}}{\dfrac{20}{9}}$ b) $\dfrac{\dfrac{6n}{n^2 - 81}}{\dfrac{n^2}{2n + 18}}$

Procedure   Simplify a Complex Fraction with One Term in the Numerator and Denominator

To simplify a complex fraction containing one term in the numerator and one term in the denominator:

1)   Rewrite the complex fraction as a division problem.

2)   Perform the division by multiplying the first fraction by the reciprocal of the second.

(We are multiplying the numerator of the complex fraction by the reciprocal of the denominator.)

2. Simplify a Complex Fraction with More Than One Term in the Numerator and/or Denominator by Rewriting It as a Division Problem

When a complex fraction has more than one term in the numerator and/or the denominator, we can use one of two methods to simplify.

Procedure   Simplify a Complex Fraction Using Method I

1) Combine the terms in the numerator and combine the terms in the denominator so that each contains only one fraction.

2) Rewrite as a division problem.

3) Perform the division by multiplying the first fraction by the reciprocal of the second.

Example 2

In-Class Example 2

Simplify.

a) $\dfrac{\dfrac{2}{3}+\dfrac{3}{5}}{2-\dfrac{4}{5}}$ b) $\dfrac{\dfrac{3}{a^2b^3}}{\dfrac{2}{a}-\dfrac{a^2}{b}}$

answer: a) $\dfrac{19}{18}$

b) $ab^2(2b-a^3)$

Simplify.

a) $\dfrac{\dfrac{1}{2}+\dfrac{1}{3}}{3-\dfrac{5}{4}}$ b) $\dfrac{\dfrac{2}{x^2y}}{\dfrac{1}{y}-\dfrac{y}{x}}$

Solution

a) The numerator is $\dfrac{1}{2}+\dfrac{1}{3}$, and it contains two terms. The denominator is $3-\dfrac{5}{4}$, and it contains two terms.

We will add the terms in the numerator and subtract the terms in the denominator so that the numerator and denominator will each contain one fraction.

$$\dfrac{\dfrac{1}{2}+\dfrac{1}{3}}{3-\dfrac{5}{4}}=\dfrac{\dfrac{3}{6}+\dfrac{2}{6}}{\dfrac{12}{4}-\dfrac{5}{4}}=\dfrac{\dfrac{5}{6}}{\dfrac{7}{4}}$$ Add the fractions in the numerator.
Subtract the terms in the denominator.

Rewrite as a division problem, multiply by the reciprocal, and simplify.

$$\dfrac{5}{6}\div\dfrac{7}{4}=\dfrac{5}{\overset{}{\underset{3}{6}}}\cdot\dfrac{\overset{2}{4}}{7}=\dfrac{10}{21}$$

b) The numerator, $\dfrac{2}{x^2y}$, contains one term; the denominator, $\dfrac{1}{y}-\dfrac{y}{x}$, contains two terms.

We will subtract the terms in the denominator so that it, like the numerator, will contain only one term. The LCD of the expressions in the denominator is xy.

$$\dfrac{\dfrac{2}{x^2y}}{\dfrac{1}{y}-\dfrac{y}{x}}=\dfrac{\dfrac{2}{x^2y}}{\dfrac{x}{xy}-\dfrac{y^2}{xy}}=\dfrac{\dfrac{2}{x^2y}}{\dfrac{x-y^2}{xy}}$$

Rewrite as a division problem, multiply by the reciprocal, and simplify.

$$\dfrac{2}{x^2y}\div\dfrac{x-y^2}{xy}=\dfrac{2}{\underset{x}{x^2y}}\cdot\dfrac{\overset{}{xy}}{x-y^2}=\dfrac{2}{x(x-y^2)}$$

 You Try 2

Simplify.

a) $\dfrac{\dfrac{9}{8}-\dfrac{3}{4}}{\dfrac{2}{3}+\dfrac{1}{4}}$ b) $\dfrac{\dfrac{5}{a}+\dfrac{3}{ab}}{\dfrac{1}{ab}+2}$

3. Simplify a Complex Fraction with More Than One Term in the Numerator and/or Denominator by Multiplying by the LCD

Another method we can use to simplify complex fractions involves multiplying the numerator and denominator of the complex fraction by the LCD of *all* of the fractions in the expression.

Procedure   Simplify a Complex Fraction Using Method 2

1)   Identify and write down the LCD of *all* of the fractions in the complex fraction.
2)   Multiply the numerator and denominator of the complex fraction by the LCD.
3)   Simplify.

We will simplify the complex fractions we simplified in Example 2 using Method 2.

Example 3

In-Class Example 3
Simplify using Method 2.

a) $\dfrac{\dfrac{2}{3} + \dfrac{3}{5}}{2 - \dfrac{4}{5}}$ b) $\dfrac{\dfrac{3}{a^2 b^3}}{\dfrac{2}{a} - \dfrac{a^2}{b}}$

answer: a) $\dfrac{19}{18}$

b) $\dfrac{3}{ab^2(2b - a^3)}$

Simplify using Method 2.

a) $\dfrac{\dfrac{1}{2} + \dfrac{1}{3}}{3 - \dfrac{5}{4}}$ b) $\dfrac{\dfrac{2}{x^2 y}}{\dfrac{1}{y} - \dfrac{y}{x}}$

Solution

a)   Look at *all* of the fractions in the complex fraction. Write down their LCD:

$$\text{LCD} = 12$$

Multiply the numerator and denominator of the complex fraction by the LCD, 12.

$$\frac{12\left(\dfrac{1}{2} + \dfrac{1}{3}\right)}{12\left(3 - \dfrac{5}{4}\right)}$$ We are multiplying the fraction by $\dfrac{12}{12}$, which equals 1.

$$= \frac{12 \cdot \dfrac{1}{2} + 12 \cdot \dfrac{1}{3}}{12 \cdot 3 - 12 \cdot \dfrac{5}{4}}$$ Distribute.

$$= \frac{6 + 4}{36 - 15} = \frac{10}{21}$$ Simplify.

This is the same result we obtained in Example 2 using Method 1.

Note

In the denominator we multiplied the 3 by 12 even though 3 is not a fraction. Remember, *all* terms, not just the fractions, must be multiplied by the LCD.

b)   Look at *all* of the fractions in the complex fraction. Write down their LCD:
LCD = $x^2 y$.

Multiply the numerator and denominator of the complex fraction by the LCD, x^2y.

$$\frac{x^2y\left(\dfrac{2}{x^2y}\right)}{x^2y\left(\dfrac{1}{y}-\dfrac{y}{x}\right)} \qquad \text{We are multiplying the expression by } \frac{x^2y}{x^2y}, \text{ which equals 1.}$$

$$= \frac{x^2y \cdot \dfrac{2}{x^2y}}{x^2y \cdot \dfrac{1}{y} - x^2y \cdot \dfrac{y}{x}} \qquad \text{Distribute.}$$

$$= \frac{2}{x^2 - xy^2} = \frac{2}{x(x - y^2)} \qquad \text{Simplify.}$$

If the numerator and denominator factor, factor them. Sometimes, you can divide by a common factor to simplify.

Notice that the result is the same as what was obtained in Example 2 using Method 1. ■

You Try 3

Simplify using Method 2.

a) $\dfrac{\dfrac{9}{8} - \dfrac{3}{4}}{\dfrac{2}{3} + \dfrac{1}{4}}$

b) $\dfrac{\dfrac{5}{a} + \dfrac{3}{ab}}{\dfrac{1}{ab} + 2}$

You should be familiar with both methods for simplifying complex fractions containing two terms in the numerator or denominator. After a lot of practice, you will be able to decide which method works best for a particular problem.

Example 4

Determine which method to use to simplify each complex fraction, then simplify.

a) $\dfrac{\dfrac{2}{c} + \dfrac{1}{c+6}}{\dfrac{4}{c+6} - \dfrac{1}{c}}$

b) $\dfrac{\dfrac{a^2-9}{5a+10}}{\dfrac{2a-6}{a^2-4}}$

In-Class Example 4
Determine which method to use to simplify each complex fraction, then simplify.

a) $\dfrac{\dfrac{3}{h} + \dfrac{2}{h+5}}{\dfrac{6}{h+5} - \dfrac{1}{h}}$ b) $\dfrac{\dfrac{w^2-49}{6w+24}}{\dfrac{9w-63}{w^2-16}}$

answer: a) $\dfrac{h+3}{h-1}$

b) $\dfrac{(w+7)(w-4)}{54}$

Solution

a) This complex fraction contains two terms in the numerator and two terms in the denominator. Let's use Method 2: multiply the numerator and denominator by the LCD of all of the fractions.

List all of the fractions in the complex fraction: $\dfrac{2}{c}, \dfrac{1}{c+6}, \dfrac{4}{c+6}, \dfrac{1}{c}$. Write down their LCD: LCD $= c(c+6)$.

Multiply the numerator and denominator of the complex fraction by the LCD, $c(c + 6)$, then simplify.

$$\frac{c(c + 6)\left(\dfrac{2}{c} + \dfrac{1}{c + 6}\right)}{c(c + 6)\left(\dfrac{4}{c + 6} - \dfrac{1}{c}\right)} = \frac{c(c + 6) \cdot \dfrac{2}{c} + c(c + 6) \cdot \dfrac{1}{c + 6}}{c(c + 6) \cdot \dfrac{4}{c + 6} - c(c + 6) \cdot \dfrac{1}{c}}$$

Multiply the numerator and denominator by $c(c + 6)$ and distribute.

$$= \frac{2(c + 6) + c}{4c - (c + 6)}$$ 　Multiply.

$$= \frac{2c + 12 + c}{4c - c - 6}$$ 　Distribute.

$$= \frac{3c + 12}{3c - 6}$$ 　Combine like terms.

$$= \frac{3(c + 4)}{3(c - 2)} = \frac{c + 4}{c - 2}$$ 　Factor and simplify.

b)　This complex fraction contains one term in the numerator, $\dfrac{a^2 - 9}{5a + 10}$, and one term in the denominator, $\dfrac{2a - 6}{a^2 - 4}$. To simplify, rewrite as a division problem, multiply by the reciprocal, and simplify.

$$\frac{\dfrac{a^2 - 9}{5a + 10}}{\dfrac{2a - 6}{a^2 - 4}} = \frac{a^2 - 9}{5a + 10} \div \frac{2a - 6}{a^2 - 4}$$ 　Rewrite as a division problem.

$$= \frac{a^2 - 9}{5a + 10} \cdot \frac{a^2 - 4}{2a - 6}$$ 　Multiply by the reciprocal.

$$= \frac{(a + 3)\cancel{(a - 3)}}{5\cancel{(a + 2)}} \cdot \frac{\cancel{(a + 2)}(a - 2)}{2\cancel{(a - 3)}}$$ 　Factor.

$$= \frac{(a + 3)(a - 2)}{10}$$ 　Reduce and multiply.

You Try 4

Determine which method to use to simplify each complex fraction, then simplify.

a)　$\dfrac{\dfrac{8}{k} - \dfrac{1}{k + 5}}{\dfrac{3}{k + 5} + \dfrac{5}{k}}$ 　　b)　$\dfrac{\dfrac{c^2 - 9}{8c - 56}}{\dfrac{2c + 6}{c^2 - 49}}$

4. Simplify Rational Expressions Containing Negative Exponents

If a rational expression contains a negative exponent, rewrite it with positive exponents and simplify.

Example 5

In-Class Example 5

Simplify $\dfrac{m^{-1} - 8n^{-2}}{1 + m^{-2}}$.

answer: $\dfrac{m(n^2 - 8m)}{n^2(m^2 + 1)}$

Simplify $\dfrac{x^{-2} - 3y^{-1}}{1 + x^{-1}}$.

Solution

First, rewrite the expression with positive exponents: $\dfrac{x^{-2} - 3y^{-1}}{1 + x^{-1}} = \dfrac{\dfrac{1}{x^2} - \dfrac{3}{y}}{1 + \dfrac{1}{x}}$

Next we have to simplify. Identify the LCD of all the fractions: x^2y.

Multiply the numerator and denominator of the complex fraction by x^2y and simplify.

$$\dfrac{x^2y\left(\dfrac{1}{x^2} - \dfrac{3}{y}\right)}{x^2y\left(1 + \dfrac{1}{x}\right)} = \dfrac{x^2y \cdot \dfrac{1}{x^2} - x^2y \cdot \dfrac{3}{y}}{x^2y \cdot 1 + x^2y \cdot \dfrac{1}{x}}$$ Multiply and distribute.

$$= \dfrac{y - 3x^2}{x^2y + xy}$$ Multiply.

$$= \dfrac{y - 3x^2}{xy(x + 1)}$$ Factor.

You Try 5

Simplify $\dfrac{4a^{-2} + b^{-1}}{a^{-3} + b}$.

Answers to You Try Exercises

1) a) $\dfrac{1}{6}$ b) $\dfrac{12}{n(n - 9)}$ 2) a) $\dfrac{9}{22}$ b) $\dfrac{5b + 3}{2ab + 1}$ 3) a) $\dfrac{9}{22}$ b) $\dfrac{5b + 3}{2ab + 1}$

4) a) $\dfrac{7k + 40}{8k + 25}$ b) $\dfrac{(c - 3)(c + 7)}{16}$ 5) $\dfrac{a(4b + a^2)}{b(1 + a^3b)}$

7.3 Exercises

Additional answers can be found in the Answers to Exercises appendix.

1) Explain, in your own words, two ways to simplify $\dfrac{\dfrac{2}{9}}{\dfrac{5}{18}}$. Then, simplify it both ways. Which method do you prefer and why?

2) Explain, in your own words, two ways to simplify $\dfrac{\dfrac{3}{2} - \dfrac{1}{5}}{\dfrac{1}{10} + \dfrac{3}{5}}$. Then, simplify it both ways. Which method do you prefer and why?

Objective 1: Simplify a Complex Fraction with One Term in the Numerator and One Term in the Denominator

Simplify completely.

3) $\dfrac{\dfrac{7}{10}}{\dfrac{5}{4}}$ $\dfrac{14}{25}$

4) $\dfrac{\dfrac{3}{8}}{\dfrac{4}{3}}$ $\dfrac{9}{32}$

5) $\dfrac{\dfrac{a^2}{b}}{\dfrac{a}{b^3}}$ ab^2

6) $\dfrac{\dfrac{u^5}{v^2}}{\dfrac{u^2}{v}}$ $\dfrac{u^3}{v}$

7) $\dfrac{\dfrac{s^3}{t^3}}{\dfrac{s^4}{t}}$ $\dfrac{1}{st^2}$

8) $\dfrac{\dfrac{x^4}{y}}{\dfrac{x^2}{y^3}}$ x^2y^2

9) $\dfrac{\dfrac{14m^5n^4}{9}}{\dfrac{35mn^6}{3}}$ $\dfrac{2m^4}{15n^2}$

10) $\dfrac{\dfrac{11b^4c^2}{4}}{\dfrac{55bc}{8}}$ $\dfrac{2b^3c}{5}$

11) $\dfrac{\dfrac{t-6}{5}}{\dfrac{t-6}{t}}$ $\dfrac{t}{5}$

12) $\dfrac{\dfrac{m-3}{m}}{\dfrac{m-3}{16}}$ $\dfrac{16}{m}$

13) $\dfrac{\dfrac{8}{y^2-64}}{\dfrac{6}{y+8}}$ $\dfrac{4}{3(y-8)}$

14) $\dfrac{\dfrac{g^2-36}{15}}{\dfrac{g-6}{45}}$ $3(g+6)$

15) $\dfrac{\dfrac{25w-35}{w^5}}{\dfrac{30w-42}{w}}$ $\dfrac{5}{6w^4}$

16) $\dfrac{\dfrac{d^3}{16d-24}}{\dfrac{d}{40d-60}}$ $\dfrac{5d^2}{2}$

17) $\dfrac{\dfrac{2x}{x+7}}{\dfrac{2}{x^2+4x-21}}$ $x(x-3)$

18) $\dfrac{\dfrac{c^2-7c-8}{6c}}{\dfrac{c-8}{c}}$ $\dfrac{c+1}{6}$

Mixed Exercises: Objectives 2 and 3

Simplify using Method 1 then by using Method 2. Think about which method you prefer and why.

19) $\dfrac{\dfrac{1}{4}+\dfrac{3}{2}}{\dfrac{2}{3}+\dfrac{1}{2}}$ $\dfrac{3}{2}$

20) $\dfrac{\dfrac{7}{9}-\dfrac{1}{3}}{2+\dfrac{1}{9}}$ $\dfrac{4}{19}$

21) $\dfrac{\dfrac{7}{c}+\dfrac{2}{d}}{1-\dfrac{5}{c}}$ $\dfrac{7d+2c}{d(c-5)}$

22) $\dfrac{\dfrac{r}{s}-2}{\dfrac{1}{s}+\dfrac{3}{r}}$ $\dfrac{r(r-2s)}{r+3s}$

23) $\dfrac{\dfrac{5}{z-2}-\dfrac{1}{z+1}}{\dfrac{1}{z-2}+\dfrac{4}{z+1}}$ $\dfrac{4z+7}{5z-7}$

24) $\dfrac{\dfrac{6}{w+4}+\dfrac{4}{w-1}}{\dfrac{5}{w-1}+\dfrac{3}{w+4}}$ $\dfrac{10(w+1)}{8w+17}$

Simplify using either Method 1 or Method 2.

25) $\dfrac{9+\dfrac{5}{y}}{\dfrac{9y+5}{8}}$ $\dfrac{8}{y}$

26) $\dfrac{4-\dfrac{12}{m}}{\dfrac{4m-12}{9}}$ $\dfrac{9}{m}$

27) $\dfrac{x-\dfrac{7}{x}}{x-\dfrac{11}{x}}$ $\dfrac{x^2-7}{x^2-11}$

28) $\dfrac{\dfrac{4}{c}-c}{3+\dfrac{8}{c}}$ $\dfrac{4-c^2}{3c+8}$

29) $\dfrac{\dfrac{4}{3}+\dfrac{2}{5}}{\dfrac{1}{6}-\dfrac{2}{3}}$ $-\dfrac{52}{15}$

30) $\dfrac{\dfrac{1}{4}-\dfrac{5}{6}}{\dfrac{3}{8}+\dfrac{1}{3}}$ $-\dfrac{14}{17}$

31) $\dfrac{\dfrac{2}{a}-\dfrac{2}{b}}{\dfrac{1}{a^2}-\dfrac{1}{b^2}}$ $\dfrac{2ab}{a+b}$

32) $\dfrac{\dfrac{4}{x}-\dfrac{4}{y}}{\dfrac{3}{x^2}-\dfrac{3}{y^2}}$ $\dfrac{4xy}{3(x+y)}$

33) $\dfrac{\dfrac{r}{s^2}+\dfrac{1}{rs}}{\dfrac{s}{r}+\dfrac{1}{r^2}}$ $\dfrac{r(r^2+s)}{s^2(sr+1)}$

34) $\dfrac{\dfrac{n}{m^3}+\dfrac{m}{n}}{\dfrac{3}{n}-\dfrac{m}{n^4}}$ $\dfrac{n^3(n^2+m^4)}{m^3(3n^3-m)}$

35) $\dfrac{1-\dfrac{4}{t+5}}{\dfrac{4}{t^2-25}+\dfrac{t}{t-5}}$ $\dfrac{t-5}{t+4}$

36) $\dfrac{1+\dfrac{4}{t-3}}{\dfrac{t}{t-3}+\dfrac{2}{t^2-9}}$ $\dfrac{t+3}{t+2}$

Mixed Exercises: Objectives 1–3

Simplify completely.

37) $\dfrac{b+\dfrac{1}{b}}{b-\dfrac{3}{b}}$ $\dfrac{b^2+1}{b^2-3}$

38) $\dfrac{\dfrac{z+6}{4}}{\dfrac{z+6}{z}}$ $\dfrac{z}{4}$

39) $\dfrac{\dfrac{m}{n^2}}{\dfrac{m^4}{n}}$ $\dfrac{1}{m^3n}$

40) $\dfrac{\dfrac{z^2+1}{5}}{z+\dfrac{1}{z}}$ $\dfrac{z}{5}$

41) $\dfrac{\dfrac{h^2-1}{4h-12}}{\dfrac{7h+7}{h^2-9}}$ $\dfrac{(h-1)(h+3)}{28}$

42) $\dfrac{\dfrac{r^2+13r+40}{r^2-6r}}{\dfrac{r^2+2r-48}{3r}}$ $\dfrac{3(r+5)}{(r-6)^2}$

43) $\dfrac{\dfrac{6}{x+3} - \dfrac{4}{x-1}}{\dfrac{2}{x-1} + \dfrac{1}{x+2}}$

44) $\dfrac{\dfrac{c^2}{d} + \dfrac{2}{c^2 d}}{\dfrac{d}{c} - \dfrac{c}{d}}$ $\dfrac{c^4 + 2}{c(d+c)(d-c)}$

45) $\dfrac{\dfrac{r^2 - 6}{20}}{r - \dfrac{6}{r}}$ $\dfrac{r}{20}$

46) $\dfrac{\dfrac{1}{6}}{\dfrac{7}{8}}$ $\dfrac{4}{21}$

47) $\dfrac{\dfrac{a-4}{12}}{\dfrac{a-4}{a}}$ $\dfrac{a}{12}$

48) $\dfrac{\dfrac{8}{w} - w}{1 + \dfrac{6}{w}}$ $\dfrac{8 - w^2}{w+6}$

49) $\dfrac{\dfrac{5}{6}}{\dfrac{9}{15}}$ $\dfrac{25}{18}$

50) $\dfrac{\dfrac{5}{h+2} + \dfrac{7}{2h-3}}{\dfrac{1}{h-3} + \dfrac{3}{2h-3}}$ $\dfrac{(17h-1)(h-3)}{(h+2)(5h-12)}$

51) $\dfrac{\dfrac{5}{2n+1} + 1}{\dfrac{1}{n+3} + \dfrac{2}{2n+1}}$ $\dfrac{2(n+3)^2}{4n+7}$

52) $\dfrac{\dfrac{y^4}{z^3}}{\dfrac{y^6}{z^4}}$ $\dfrac{z}{y^2}$

Objective 4: Simplify Rational Expressions Containing Negative Exponents

Simplify.

53) $\dfrac{w^{-1} - v^{-1}}{2w^{-2} + v^{-1}}$ $\dfrac{w(v-w)}{2v+w^2}$

54) $\dfrac{4p^{-2} + q^{-1}}{p^{-1} + q^{-1}}$ $\dfrac{p^2 + 4q}{p(p+q)}$

55) $\dfrac{8x^{-2}}{x^{-1} - y^{-2}}$ $\dfrac{8y^2}{x(y^2 - x)}$

56) $\dfrac{3d^{-1}}{2c^{-2} - d^{-1}}$ $\dfrac{3c^2}{2d - c^2}$

57) $\dfrac{a^{-3} + b^{-2}}{2b^{-2} - 7}$ $\dfrac{a^3 + b^2}{a^3(2 - 7b^2)}$

58) $\dfrac{r^{-2} - t^{-2}}{5 + 7t^{-3}}$ $\dfrac{t(t+r)(t-r)}{r^2(5t^3 + 7)}$

59) $\dfrac{4m^{-1} - n^{-1}}{n^{-1} + m}$ $\dfrac{4n - m}{m(1 + mn)}$

60) $\dfrac{h^{-3} + 9}{k^{-2} - h}$ $\dfrac{k^2(9h^3 + 1)}{h^3(1 - hk^2)}$

For Exercises 61 and 62, let $f(x) = \dfrac{1}{x}$.

61) Complete the table of values. As x gets larger, the value of $f(x)$ gets closer to what number?   0

x	$f(x)$
1	1
2	$\frac{1}{2}$
3	$\frac{1}{3}$
10	$\frac{1}{10}$
100	$\frac{1}{100}$
1000	$\frac{1}{1000}$

62) Complete the table of values. As x gets smaller, what happens to the value of $f(x)$?   It gets larger.

x	$f(x)$
1	1
$\frac{1}{2}$	2
$\frac{1}{3}$	3
$\frac{1}{10}$	10
$\frac{1}{100}$	100
$\frac{1}{1000}$	1000

Section 7.4 Solving Rational Equations

Objectives

1. Differentiate Between Rational Expressions and Rational Equations
2. Solve Rational Equations
3. Solve a Proportion
4. Solve an Equation for a Specific Variable

A **rational equation** is an equation that contains a rational expression. Some examples of rational equations are

$$\frac{1}{3}x + \frac{3}{4} = \frac{5}{6}x - 2, \qquad \frac{5}{c-3} - \frac{c}{c+1} = 4, \quad \text{and} \quad \frac{2t}{t^2 + 6t + 8} + \frac{7}{t+2} = \frac{4}{t+4}.$$

1. Differentiate Between Rational Expressions and Rational Equations

In Chapter 2 we solved rational equations like the first one above, and we learned how to add and subtract rational expressions in Section 7.2. Let's summarize the difference between the two because this is often a point of confusion for students.

Summary   Expressions vs. Equations

1) **The sum or difference of rational expressions does not contain an = sign.** To add or subtract, rewrite each expression with the LCD, and *keep the denominator* while performing the operations.

2) **An equation contains an = sign.** To solve an equation containing rational expressions, *multiply the equation by the LCD* of all fractions to *eliminate the denominators*, then solve.

Example 1

In-Class Example 1
Determine whether each is an equation or is a sum or difference of expressions. Then solve the equation or find the sum or difference.

a) $\dfrac{3q + 2}{3} - \dfrac{q}{6} = \dfrac{3}{4}$

b) $\dfrac{3q + 2}{3} - \dfrac{q}{6}$

answer: a) $\left\{ \dfrac{1}{10} \right\}$ b) $\dfrac{5q + 4}{6}$

Determine whether each is an equation or is a sum or difference of expressions. Then, solve the equation or find the sum or difference.

a) $\dfrac{k - 2}{5} - \dfrac{k}{2} = \dfrac{7}{5}$ b) $\dfrac{k - 2}{5} - \dfrac{k}{2}$

Solution

a)   This is an *equation* because it contains an = sign. We will *solve* for k using the method we learned in Chapter 2: eliminate the denominators by multiplying by the LCD of all of the expressions. LCD = 10.

$$10\left(\frac{k - 2}{5} - \frac{k}{2} \right) = 10 \cdot \frac{7}{5} \qquad \text{\color{blue} Multiply by LCD of 10 to eliminate the denominators.}$$

$$2(k - 2) - 5k = 14 \qquad \text{\color{blue} Distribute and eliminate denominator.}$$

$$2k - 4 - 5k = 14$$

$$-3k - 4 = 14$$

$$-3k = 18$$

$$k = -6$$

Check to verify that the solution set is $\{-6\}$.

b) $\dfrac{k - 2}{5} - \dfrac{k}{2}$ is *not* an equation to be solved because it does *not* contain an = sign.

It is a *difference of rational expressions.* Rewrite each expression with the LCD, then subtract, *keeping the denominators* while performing the operations.

LCD = 10

$$\frac{(k - 2)}{5} \cdot \frac{2}{2} = \frac{2(k - 2)}{10}, \qquad \frac{k}{2} \cdot \frac{5}{5} = \frac{5k}{10}$$

$$\frac{k - 2}{5} - \frac{k}{2} = \frac{2(k - 2)}{10} - \frac{5k}{10} \qquad \text{\color{blue} Rewrite each expression with a denominator of 10.}$$

$$= \frac{2(k - 2) - 5k}{10} \qquad \text{\color{blue} Subtract the numerators.}$$

$$= \frac{2k - 4 - 5k}{10} \qquad \text{\color{blue} Distribute.}$$

$$= \frac{-3k - 4}{10} \qquad \text{\color{blue} Combine like terms.}$$

You Try 1

Determine whether each is an equation or is a sum or difference of expressions. Then solve the equation or find the sum or difference.

a) $\dfrac{n}{9} + \dfrac{n - 11}{6}$ b) $\dfrac{n}{9} + \dfrac{n - 11}{6} = -1$

Let's look at more examples of how to solve equations containing rational expressions.

2. Solve Rational Equations

> **Procedure   How to Solve a Rational Equation**
>
> 1)   If possible, factor all denominators.
>
> 2)   Write down the LCD of all of the expressions.
>
> 3)   Multiply both sides of the equation by the LCD to *eliminate* the denominators.
>
> 4)   Solve the equation.
>
> 5)   Check the solution(s) in the original equation. If a proposed solution makes a denominator equal 0, then it is rejected as a solution.

Example 2

Solve $\dfrac{a}{3} + \dfrac{4}{a} = \dfrac{13}{3}$.

In-Class Example 2

Solve $\dfrac{5}{x} - 2 = \dfrac{x}{4}$.

answer: $\{-10, 2\}$

Solution

Since this is an equation, we will *eliminate* the denominators by multiplying the equation by the LCD of all of the expressions.

LCD $= 3a$

$$3a\left(\frac{a}{3} + \frac{4}{a}\right) = 3a\left(\frac{13}{3}\right) \qquad \text{Multiply both sides of the equation by the LCD, } 3a.$$

$$\cancel{3}a\left(\frac{a}{\cancel{3}}\right) + 3\cancel{a}\left(\frac{4}{\cancel{a}}\right) = \cancel{3}a\left(\frac{13}{\cancel{3}}\right) \qquad \text{Distribute and divide out common factors.}$$

$$a^2 + 12 = 13a$$
$$a^2 - 13a + 12 = 0 \qquad \text{Subtract } 13a.$$
$$(a - 12)(a - 1) = 0 \qquad \text{Factor.}$$

$$a - 12 = 0 \quad \text{or} \quad a - 1 = 0$$
$$a = 12 \quad \text{or} \qquad a = 1$$

Check:

$$a = 12$$
$$\frac{a}{3} + \frac{4}{a} \overset{?}{=} \frac{13}{3}$$
$$\frac{12}{3} + \frac{4}{12} \overset{?}{=} \frac{13}{3}$$
$$\frac{12}{3} + \frac{1}{3} = \frac{13}{3} \quad \checkmark$$

$$a = 1$$
$$\frac{a}{3} + \frac{4}{a} \overset{?}{=} \frac{13}{3}$$
$$\frac{1}{3} + \frac{4}{1} \overset{?}{=} \frac{13}{3}$$
$$\frac{1}{3} + \frac{12}{3} = \frac{13}{3} \quad \checkmark$$

The solution set is $\{1, 12\}$.   ■

  You Try 2

Solve $\dfrac{c}{2} + 1 = \dfrac{24}{c}$.

It is *very* important to check the proposed solution. Sometimes, what appears to be a solution actually is not.

Example 3

In-Class Example 3

Solve $9 + \dfrac{5}{m-5} = \dfrac{m}{m-5}$.

answer: $\varnothing$

Solve $3 - \dfrac{3}{x+3} = \dfrac{x}{x+3}$.

Solution

Since this is an equation, we will *eliminate* the denominators by multiplying the equation by the LCD of all of the expressions.

LCD $= x + 3$

$$(x+3)\left(3 - \dfrac{3}{x+3}\right) = (x+3)\left(\dfrac{x}{x+3}\right)$$ Multiply both sides of the equation by the LCD, $x + 3$.

$$(x+3)3 - \cancel{(x+3)} \cdot \dfrac{3}{\cancel{x+3}} = \cancel{(x+3)}\left(\dfrac{x}{\cancel{x+3}}\right)$$ Distribute and divide out common factors.

$$3x + 9 - 3 = x$$ Multiply.

$$3x + 6 = x$$

$$6 = -2x$$ Subtract $3x$.

$$-3 = x$$ Divide by -2.

$$Check:\ 3 - \dfrac{3}{(-3)+3} \stackrel{?}{=} \dfrac{-3}{(-3)+3}$$ Substitute -3 for x in the original equation.

$$3 - \dfrac{3}{0} = \dfrac{-3}{0}$$

Since $x = -3$ makes the denominator equal zero, -3 cannot be a solution to the equation. Therefore, this equation has no solution. The solution set is $\varnothing$.   ∎

BE CAREFUL

Always check what *appears* to be the solution or solutions to an equation containing rational expressions. If one of these values makes a denominator zero, then it *cannot* be a solution to the equation.

You Try 3

Solve each equation.

a) $\dfrac{7}{s+1} + \dfrac{2s}{s+1} = 3$ b) $\dfrac{3m}{m-4} - 1 = \dfrac{12}{m-4}$

Example 4

In-Class Example 4

Solve $\dfrac{1}{2} - \dfrac{1}{k+1} = \dfrac{k+5}{2k^2-2}$.

answer: {4}

Solve $\dfrac{1}{3} - \dfrac{1}{t+2} = \dfrac{t+14}{3t^2-12}$.

Solution

This is an equation. *Eliminate* the denominators by multiplying by the LCD. Begin by factoring the denominator of $\dfrac{t+14}{3t^2-12}$.

$$\dfrac{1}{3} - \dfrac{1}{t+2} = \dfrac{t+14}{3(t+2)(t-2)}$$ Factor the denominator.

$$LCD = 3(t+2)(t-2)$$ Write down the LCD of all of the expressions.

$$3(t+2)(t-2)\left(\frac{1}{3}-\frac{1}{t+2}\right)=3(t+2)(t-2)\left(\frac{t+14}{3(t+2)(t-2)}\right)$$

Multiply both sides of the equation by the LCD.

$$\cancel{3}(t+2)(t-2)\left(\frac{1}{\cancel{3}}\right)-3\cancel{(t+2)}(t-2)\left(\frac{1}{\cancel{t+2}}\right)=\cancel{3}\cancel{(t+2)}\cancel{(t-2)}\left(\frac{t+14}{\cancel{3}\cancel{(t+2)}\cancel{(t-2)}}\right)$$

Distribute and divide out common factors.

$$
\begin{array}{ll}
(t+2)(t-2)-3(t-2)=t+14 & \text{Multiply.}\\
t^2-4-3t+6=t+14 & \text{Distribute.}\\
t^2-3t+2=t+14 & \text{Combine like terms.}\\
t^2-4t-12=0 & \text{Subtract } t \text{ and subtract } 14.\\
(t-6)(t+2)=0 & \text{Factor.}\\
t-6=0 \quad \text{or} \quad t+2=0 & \text{Set each factor equal to zero.}\\
t=6 \quad \text{or} \qquad t=-2 & \text{Solve.}
\end{array}
$$

Look at the factored form of the equation. If $t=6$, no denominator will equal zero. *If $t=-2$, however, two of the denominators will equal zero. Therefore, we must reject $t=-2$ as a solution.* Check only $t=6$.

$$
\begin{array}{ll}
Check: \dfrac{1}{3}-\dfrac{1}{6+2}\overset{?}{=}\dfrac{6+14}{3(6)^2-12} & \text{Substitute } t=6 \text{ into the original equation.}\\[2mm]
\dfrac{1}{3}-\dfrac{1}{8}\overset{?}{=}\dfrac{20}{108-12} & \text{Simplify.}\\[2mm]
\dfrac{1}{3}-\dfrac{1}{8}\overset{?}{=}\dfrac{20}{96} & \text{Simplify.}\\[2mm]
\dfrac{8}{24}-\dfrac{3}{24}\overset{?}{=}\dfrac{5}{24} & \text{Get a common denominator and reduce } \dfrac{20}{96}.\\[2mm]
\dfrac{5}{24}=\dfrac{5}{24} \quad \checkmark & \text{Subtract.}
\end{array}
$$

The solution set is $\{6\}$.

The previous problem is a good example of why it is necessary to check all "solutions" to equations containing rational expressions.

You Try 4

Solve $\dfrac{w}{w+6}-\dfrac{3}{w+2}=\dfrac{6-3w}{w^2+8w+12}$.

Example 5

Solve $\dfrac{5}{6n^2+18n+12}=\dfrac{n}{2n+2}+\dfrac{1}{3n+6}$.

In-Class Example 5

Solve $-\dfrac{3}{10a^2+20a-80}$
$=\dfrac{a}{5a-10}+\dfrac{1}{2a+8}$.

answer: $\left\{-7,\dfrac{1}{2}\right\}$

Solution

Since this is an equation, we will *eliminate* the denominators by multiplying by the LCD. Begin by factoring all denominators, then identify the LCD.

$$\frac{5}{6(n + 2)(n + 1)} = \frac{n}{2(n + 1)} + \frac{1}{3(n + 2)} \qquad \text{LCD} = 6(n + 2)(n + 1)$$

$$6(n + 2)(n + 1)\left(\frac{5}{6(n + 2)(n + 1)}\right) = 6(n + 2)(n + 1)\left(\frac{n}{2(n + 1)} + \frac{1}{3(n + 2)}\right) \qquad \text{Multiply by the LCD.}$$

$$\cancel{6(n + 2)(n + 1)}\left(\frac{5}{\cancel{6(n + 2)(n + 1)}}\right) = \overset{3}{\cancel{6}}(n + 2)\cancel{(n + 1)}\left(\frac{n}{2\cancel{(n + 1)}}\right) + \overset{2}{\cancel{6}}(n + 2)(n + 1)\left(\frac{1}{3\cancel{(n + 2)}}\right) \qquad \text{Distribute.}$$

$$5 = 3n(n + 2) + 2(n + 1) \qquad \text{Multiply.}$$
$$5 = 3n^2 + 6n + 2n + 2 \qquad \text{Distribute.}$$
$$5 = 3n^2 + 8n + 2 \qquad \text{Combine like terms.}$$
$$0 = 3n^2 + 8n - 3 \qquad \text{Subtract 5.}$$
$$0 = (3n - 1)(n + 3) \qquad \text{Factor.}$$

$$3n - 1 = 0 \qquad \text{or} \qquad n + 3 = 0$$
$$n = \frac{1}{3} \qquad \text{or} \qquad n = -3 \qquad \text{Solve.}$$

If you look at the factored form of the equation you can see that neither $n = \dfrac{1}{3}$ nor $n = -3$ will make a denominator zero. Check the values in the original equation to verify that the solution set is $\left\{-3, \dfrac{1}{3}\right\}$. ■

You Try 5

Solve $\dfrac{11}{6h^2 + 48h + 90} = \dfrac{h}{3h + 15} + \dfrac{1}{2h + 6}$.

3. Solve a Proportion

Example 6

Solve $\dfrac{20}{d + 6} = \dfrac{8}{d}$.

In-Class Example 6
Solve $\dfrac{18}{v + 5} = \dfrac{3}{v}$.
answer: $\{1\}$

Solution

This equation is a *proportion*. A **proportion** is a statement that two ratios are equal. We can solve this proportion as we have solved the other equations in this section, by multiplying both sides of the equation by the LCD. Or, recall from Section 2.4 that *we can solve a proportion by setting the cross products equal to each other.*

$$\frac{20}{d + 6} \diagup\!\!\!\!\diagdown \frac{8}{d}$$

Multiply.   Multiply.

$$20d = 8(d + 6) \qquad \text{Set the cross products equal to each other.}$$
$$20d = 8d + 48 \qquad \text{Distribute.}$$
$$12d = 48 \qquad \text{Subtract } 8d.$$
$$d = 4 \qquad \text{Solve.}$$

$$\text{Check:}\quad \frac{20}{4 + 6} \overset{?}{=} \frac{8}{4} \qquad \text{Substitute } d = 4 \text{ into the original equation.}$$
$$\frac{20}{10} \overset{?}{=} 2$$
$$2 = 2 \quad \checkmark$$

The solution is $\{4\}$.   ■

You Try 6

Solve $\dfrac{9}{y} = \dfrac{5}{y-2}$.

4. Solve an Equation for a Specific Variable

In Section 2.3, we learned how to solve an equation for a specific variable. For example, to solve $2l + 2w = P$ for w, we do the following:

$$2l + 2\boxed{w} = P \qquad \text{Put a box around } w, \text{ the variable for which we are solving.}$$
$$2\boxed{w} = P - 2l \qquad \text{Subtract } 2l.$$
$$w = \frac{P - 2l}{2} \qquad \text{Divide by 2.}$$

Next we discuss how to solve for a specific variable in a rational expression.

Example 7

In-Class Example 7

Solve $t = \dfrac{n}{r - R}$ for r.

answer: $r = \dfrac{n + Rt}{t}$

Solve $m = \dfrac{x}{a - A}$ for a.

Solution

Note that the equation contains a lowercase a and an uppercase A. These represent different quantities, so students should be sure to write them correctly. Put a in a box.

Since a is in the denominator of the rational expression, multiply both sides of the equation by $a - A$ to eliminate the denominator.

$$m = \frac{x}{\boxed{a} - A} \qquad \text{Put } a \text{ in a box.}$$
$$(\boxed{a} - A)m = (\boxed{a} - A)\left(\frac{x}{\boxed{a} - A}\right) \qquad \begin{array}{l}\text{Multiply both sides by } a - A \text{ to} \\ \text{eliminate the denominator.}\end{array}$$
$$\boxed{a}\,m - Am = x \qquad \text{Distribute.}$$
$$\boxed{a}\,m = x + Am \qquad \text{Add } Am.$$
$$a = \frac{x + Am}{m} \qquad \text{Divide by } m.$$

You Try 7

Solve $y = \dfrac{c}{r + d}$ for d.

Example 8

In-Class Example 8

Solve $\dfrac{1}{a} + \dfrac{1}{b} = \dfrac{1}{c}$ for a.

answer: $a = \dfrac{bc}{b - c}$

Solve $\dfrac{1}{a} + \dfrac{1}{b} = \dfrac{1}{c}$ for b.

Solution

Put the b in a box. The LCD of all of the fractions is abc. Multiply both sides of the equation by abc.

$$\frac{1}{a} + \frac{1}{\boxed{b}} = \frac{1}{c} \qquad \text{Put } b \text{ in a box.}$$
$$a\boxed{b}c\left(\frac{1}{a} + \frac{1}{\boxed{b}}\right) = a\boxed{b}c\left(\frac{1}{c}\right) \qquad \begin{array}{l}\text{Multiply both sides by } abc \text{ to} \\ \text{eliminate the denominator.}\end{array}$$
$$a\boxed{b}c \cdot \frac{1}{a} + a\boxed{b}c \cdot \frac{1}{\boxed{b}} = a\boxed{b}c\left(\frac{1}{c}\right) \qquad \text{Distribute.}$$

$$\boxed{b}c + ac = a\boxed{b} \qquad \text{Divide out common factors.}$$

Since we are solving for b and there are terms containing b on each side of the equation, we must get bc and ab on one side of the equation and ac on the other side.

$$ac = a\boxed{b} - \boxed{b}c \qquad \text{Subtract } bc \text{ from each side.}$$

To isolate b, we will *factor* b out of each term on the right-hand side of the equation.

$$ac = \boxed{b}(a - c) \qquad \text{Factor out } b.$$

$$\frac{ac}{a - c} = b \qquad \text{Divide by } a - c.$$

You Try 8

Solve $\dfrac{1}{a} + \dfrac{1}{b} = \dfrac{1}{c}$ for a.

Using Technology

We can use a graphing calculator to solve a rational equation in one variable. First enter the left side of the equation in Y_1 and the right side of the equation in Y_2. Then enter $Y_1 - Y_2$ in Y_3 and graph the equations. The zero(s) [x-intercept(s)] of the graph are the solutions to the equation.

We will solve $\dfrac{2}{x + 5} - \dfrac{3}{x - 2} = \dfrac{4x}{x^2 + 3x - 10}$ using a graphing calculator.

1) Enter $\dfrac{2}{x + 5} - \dfrac{3}{x - 2}$ in Y_1 by entering $2/(x + 5) - 3/(x - 2)$ in Y_1.

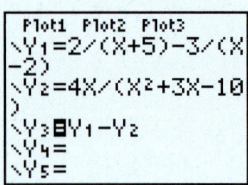

2) Enter $\dfrac{4x}{x^2 + 3x - 10}$ in Y_2 by entering $4x/(x^2 + 3x - 10)$ in Y_2.

3) Enter $Y_1 - Y_2$ in Y_3 as follows: press $\boxed{Y=}$, move the cursor to the right of $\backslash Y_3 =$, press **VARS**, select Y-VARS using the right arrow key, and press $\boxed{\text{ENTER}}$ $\boxed{\text{ENTER}}$ to select Y_1. Press $\boxed{-}$. Press **VARS**, select Y-VARS using the right arrow key, press $\boxed{\text{ENTER}}$ $\boxed{2}$ to select Y_2.

4) Move the cursor to the = right of $\backslash Y_1$ and press $\boxed{\text{ENTER}}$ to deselect Y_1. Move the cursor to the = right of $\backslash Y_2$ and press $\boxed{\text{ENTER}}$ to deselect Y_2. Press $\boxed{\text{GRAPH}}$ to graph $Y_1 - Y_2$ as shown.

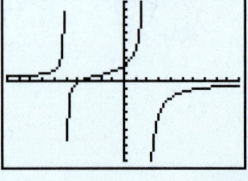

5) Press $\boxed{2^{\text{nd}}}$ $\boxed{\text{TRACE}}$ 2:zero, move the cursor to the left of the zero and press $\boxed{\text{ENTER}}$, move the cursor to the right of the zero and press $\boxed{\text{ENTER}}$, and move the cursor close to the zero and press $\boxed{\text{ENTER}}$ to display the zero as shown on the graph.

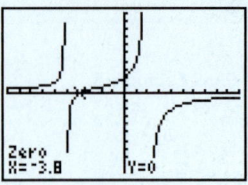

6) Press $\boxed{\text{X,T,}\theta\text{,n}}$ **MATH** $\boxed{\text{ENTER}}$ $\boxed{\text{ENTER}}$ to display the zero $x = -\dfrac{19}{5}$, as shown on the image to the right.

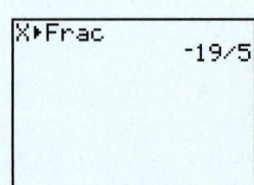

If there is more than one zero, repeat Steps 5 and 6 above for each zero.

Solve each equation using a graphing calculator.

1) $\dfrac{3x}{x-5} + \dfrac{3}{x+2} = \dfrac{5x}{x^2-3x-10}$

2) $\dfrac{5}{x-2} - \dfrac{3}{x+2} = \dfrac{2}{x^2-4}$

3) $\dfrac{3}{x+5} + \dfrac{2}{x-4} = \dfrac{6}{x^2+x-20}$

4) $\dfrac{5}{x-1} + \dfrac{2}{x+3} = \dfrac{4}{x^2+2x-3}$

5) $\dfrac{3}{x+5} + \dfrac{x}{x-2} = \dfrac{5x-2}{x^2+3x-10}$

6) $\dfrac{5}{x+1} - \dfrac{x}{x-4} = \dfrac{-25}{x^2-3x-4}$

Answers to You Try Exercises

1) a) sum; $\dfrac{5n-33}{9}$ b) equation; $\{3\}$ 2) $\{-8, 6\}$ 3) a) $\{4\}$ b) $\varnothing$ 4) $\{4\}$

5) $\left\{-4, -\dfrac{1}{2}\right\}$ 6) $\left\{\dfrac{9}{2}\right\}$ 7) $d = \dfrac{c-ry}{y}$ 8) $a = \dfrac{bc}{b-c}$

Answers to Technology Exercises

1) $\left\{-3, \dfrac{5}{3}\right\}$ 2) $\{-7\}$ 3) $\left\{\dfrac{8}{5}\right\}$ 4) $\left\{-\dfrac{9}{7}\right\}$ 5) $\{-4, 1\}$ 6) $\{5\}$

7.4 Exercises

*Additional answers can be found in the Answers to Exercises appendix.

Mixed Exercises: Objectives 1 and 2

1) When solving an equation containing rational expressions, do you keep the LCD throughout the problem or do you eliminate the denominators?   Eliminate the denominators.

2) When adding or subtracting two rational expressions, do you keep the LCD throughout the problem or do you eliminate the denominators?   Keep the LCD.

Determine whether each is an equation or is a sum or difference of expressions. Then, solve the equation or find the sum or difference.

3) $\dfrac{m}{8} + \dfrac{m-7}{4}$ sum; $\dfrac{3m-14}{8}$

4) $\dfrac{2r+9}{3} - \dfrac{r}{9}$ difference; $\dfrac{5r+27}{9}$

5) $\dfrac{2f-19}{20} = \dfrac{f}{4} + \dfrac{2}{5}$ equation; $\{-9\}$

6) $\dfrac{2h}{5} + \dfrac{2}{3} = \dfrac{h+3}{5}$ equation; $\left\{-\dfrac{1}{3}\right\}$

7) $\dfrac{z}{z-6} - \dfrac{4}{z}$ difference; $\dfrac{z^2-4z+24}{z(z-6)}$

8) $\dfrac{2}{a^2} + \dfrac{1}{a+9}$ sum; $\dfrac{a^2+2a+18}{a^2(a+9)}$

9) $1 + \dfrac{4}{c+2} = \dfrac{9}{c+2}$ equation; $\{3\}$

10) $\dfrac{10}{b-8} - 4 = \dfrac{2}{b-8}$ equation; $\{10\}$

Values that make the denominators equal zero cannot be solutions of an equation. Find *all* of the values that make the denominators zero and that, therefore, cannot be solutions of each equation. Do *not* solve the equation.

11) $\dfrac{t}{t+10} - \dfrac{5}{t} = 3$
    $0, -10$

12) $\dfrac{k+2}{k-3} + 1 = \dfrac{7}{k}$
    $0, 3$

13) $\dfrac{2}{d^2-81} + \dfrac{8}{d} = \dfrac{6}{d+9}$
    $0, 9, -9$

14) $\dfrac{4}{p+2} - \dfrac{5}{p} = \dfrac{p}{p^2-4}$
    $0, 2, -2$

15) $\dfrac{v+7}{v^2-13v+36} - \dfrac{5}{3v-12} = \dfrac{v}{v-9}$
    $4, 9$

16) $\dfrac{3h}{h^2+3h-40} + \dfrac{1}{h+8} = \dfrac{h-1}{4h-20}$
    $-8, 5$

Mixed Exercises: Objectives 2 and 3

Solve each equation.

17) $\dfrac{y}{3} - \dfrac{1}{2} = \dfrac{1}{6}$ $\{2\}$

18) $\dfrac{a}{2} + \dfrac{3}{10} = \dfrac{4}{5}$ $\{1\}$

19) $\dfrac{1}{2}h + h = -3$ $\{-2\}$

20) $\dfrac{1}{6}j - j = -10$ $\{12\}$

21) $\dfrac{7u + 12}{15} = \dfrac{2u}{5} - \dfrac{3}{5}$ $\{-21\}$

22) $\dfrac{8m - 7}{24} = \dfrac{m}{6} - \dfrac{7}{8}$ $\left\{-\dfrac{7}{2}\right\}$

53) $\dfrac{w}{5} = \dfrac{w - 3}{w + 1} + \dfrac{12}{5w + 5}$ $\{1, 3\}$

23) $\dfrac{4}{3t + 2} = \dfrac{2}{2t - 1}$ $\{4\}$

24) $\dfrac{9}{4x + 1} = \dfrac{3}{x + 2}$ $\{5\}$

54) $\dfrac{3y - 2}{y + 2} = \dfrac{y}{4} + \dfrac{1}{4y + 8}$ $\{1, 9\}$

25) $\dfrac{w}{3} = \dfrac{2w - 5}{12}$ $\left\{-\dfrac{5}{2}\right\}$

26) $\dfrac{r - 2}{2} = \dfrac{3r - 5}{4}$ $\{1\}$

VIDEO 55) $\dfrac{8}{p + 2} + \dfrac{p}{p + 1} = \dfrac{5p + 2}{p^2 + 3p + 2}$ $\{-3\}$

27) $\dfrac{12}{a} - 2 = \dfrac{6}{a}$ $\{3\}$

28) $\dfrac{16}{z} + 7 = -\dfrac{12}{z}$ $\{-4\}$

56) $\dfrac{6}{x + 1} + \dfrac{x}{x - 3} = \dfrac{3x + 14}{x^2 - 2x - 3}$ $\{-8, 4\}$

29) $\dfrac{n}{n + 2} + 3 = \dfrac{8}{n + 2}$ $\left\{\dfrac{1}{2}\right\}$

30) $\dfrac{4q}{q + 1} - 2 = \dfrac{3}{q + 1}$ $\left\{\dfrac{5}{2}\right\}$

57) $\dfrac{11}{c + 9} = \dfrac{c}{c - 4} - \dfrac{36 - 8c}{c^2 + 5c - 36}$ $\{-4, -2\}$

VIDEO 31) $\dfrac{2}{s + 6} + 4 = \dfrac{2}{s + 6}$ $\varnothing$

32) $\dfrac{u}{u - 4} + 3 = \dfrac{4}{u - 4}$ $\varnothing$

58) $\dfrac{3}{f + 2} = \dfrac{f}{f + 6} - \dfrac{2}{f^2 + 8f + 12}$ $\{-4, 5\}$

33) $\dfrac{c}{c - 7} - 4 = \dfrac{10}{c - 7}$ $\{6\}$

34) $\dfrac{2b}{b + 9} - 5 = \dfrac{3}{b + 9}$ $\{-16\}$

59) $\dfrac{8}{3g^2 - 7g - 6} + \dfrac{4}{g - 3} = \dfrac{8}{3g + 2}$ $\{-10\}$

35) $\dfrac{32}{g} + 10 = -\dfrac{8}{g}$ $\{-4\}$

36) $\dfrac{9}{r} - 1 = \dfrac{6}{r}$ $\{3\}$

60) $\dfrac{1}{r - 1} + \dfrac{2}{5r - 3} = \dfrac{37}{5r^2 - 8r + 3}$ $\{6\}$

Solve each equation.

37) $\dfrac{1}{m - 1} + \dfrac{2}{m + 3} = \dfrac{4}{m + 3}$ $\{5\}$

61) $\dfrac{h}{h^2 + 2h - 8} + \dfrac{4}{h^2 + 8h - 20} = \dfrac{4}{h^2 + 14h + 40}$ $\{-6\}$

38) $\dfrac{4}{c + 2} + \dfrac{2}{c - 6} = \dfrac{5}{c + 2}$ $\{-10\}$

62) $\dfrac{b}{b^2 + b - 6} + \dfrac{3}{b^2 + 9b + 18} = \dfrac{8}{b^2 + 4b - 12}$ $\{5\}$

39) $\dfrac{4}{w - 8} - \dfrac{10}{w + 8} = \dfrac{40}{w^2 - 64}$ $\{12\}$

63) $\dfrac{u}{8} = \dfrac{2}{10 - u}$ $\{2, 8\}$

64) $-\dfrac{a}{4} = \dfrac{3}{a + 7}$ $\{-4, -3\}$

40) $\dfrac{4}{p - 5} + \dfrac{7}{p + 5} = \dfrac{18}{p^2 - 25}$ $\{3\}$

65) $\dfrac{5}{r + 4} - \dfrac{2}{r} = -1$ $\{-8, 1\}$

66) $\dfrac{6}{c - 5} - \dfrac{2}{c} = 1$ $\{-1, 10\}$

41) $\dfrac{3}{a + 3} + \dfrac{14}{a^2 - 4a - 21} = \dfrac{5}{a - 7}$ $\{-11\}$

67) $\dfrac{q}{q^2 + 4q - 32} + \dfrac{2}{q^2 - 14q + 40} = \dfrac{6}{q^2 - 2q - 80}$ $\varnothing$

42) $\dfrac{5}{k + 4} - \dfrac{3}{k + 2} = \dfrac{8}{k^2 + 6k + 8}$ $\{5\}$

68) $\dfrac{r}{r^2 + 8r + 15} - \dfrac{2}{r^2 + r - 6} = \dfrac{2}{r^2 + 3r - 10}$ $\{-2, 8\}$

43) $\dfrac{9}{t + 4} + \dfrac{8}{t^2 - 16} = \dfrac{1}{t - 4}$ $\varnothing$

69) The average profit a king salmon fisherman receives is given by $A(x) = 6 - \dfrac{1500}{x}$, where x is the number of pounds of king salmon. How many pounds of salmon must he catch so that the profit he earns per pound (the average profit) is $4.00? **750 lb**

44) $\dfrac{12}{g^2 - 9} + \dfrac{2}{g + 3} = \dfrac{7}{g - 3}$ $\varnothing$

45) $\dfrac{4}{x^2 + 2x - 15} = \dfrac{8}{x - 3} + \dfrac{2}{x + 5}$ $\{-3\}$

70) The average profit earned from the sale of x purses is given by $A(x) = 36 - \dfrac{8400}{x}$. How many purses must be sold so that the profit earned per purse (the average profit) is $20.00? **525 purses**

46) $\dfrac{4}{p - 2} - \dfrac{9}{p^2 - 8p + 12} = \dfrac{9}{p - 6}$ $\{-3\}$

47) $\dfrac{k^2}{3} = \dfrac{k^2 + 2k}{4}$ $\{0, 6\}$

48) $\dfrac{x^2}{2} = \dfrac{x^2 - 5x}{3}$ $\{0, -10\}$

71) The formula $P = \dfrac{1}{f}$, where f is the focal length of a lens, in meters, and P is the power of the lens, in diopters, is used to determine an eyeglass prescription. If the power of a lens is 2.5 diopters, what is the focal length of the lens? **0.4 m**

49) $\dfrac{5}{m^2 - 25} = \dfrac{4}{m^2 + 5m}$ $\{-20\}$

50) $\dfrac{3}{t^2} = \dfrac{6}{t^2 + 5t}$ $\{5\}$

51) $\dfrac{10v}{3v - 12} - \dfrac{v + 6}{v - 4} = \dfrac{v}{3}$ $\{2, 9\}$

52) $\dfrac{b - 2}{2b - 12} - \dfrac{b + 2}{b - 6} = \dfrac{b}{2}$ $\{2, 3\}$

72) Magnetic stripe patterns on the ocean floor help geolo-gists study the rates at which oceanic plates are spreading. Geologists use the formula

$$\text{Rate of spreading for stripes} = \frac{\text{width of stripe}}{\text{time duration}}$$

a) If a magnetic stripe on a plate on the ocean floor is 75 miles wide and formed over 2 million years, find the rate at which the plate is spreading.   37.5 miles/million yr

b) If it continues to spread at this rate, how long will it have taken to become 90 miles wide?   2.4 million years

Objective 4: Solve an Equation for a Specific Variable

Solve for the indicated variable.

73) $V = \dfrac{nRT}{P}$ for P $P = \dfrac{nRT}{V}$

74) $W = \dfrac{CA}{m}$ for m $m = \dfrac{CA}{W}$

75) $y = \dfrac{kx}{z}$ for z $z = \dfrac{kx}{y}$

76) $a = \dfrac{rt}{2b}$ for b $b = \dfrac{rt}{2a}$

77) $B = \dfrac{t+u}{3x}$ for x $x = \dfrac{t+u}{3B}$

78) $Q = \dfrac{n-k}{5r}$ for r $r = \dfrac{n-k}{5Q}$

79) $z = \dfrac{a}{b+c}$ for b $b = \dfrac{a-zc}{z}$

80) $d = \dfrac{t}{l-n}$ for n $n = \dfrac{dl-t}{d}$

81) $A = \dfrac{4r}{q-t}$ for t $t = \dfrac{Aq-4r}{A}$

82) $h = \dfrac{3A}{r+s}$ for s $s = \dfrac{3A-hr}{h}$

83) $w = \dfrac{na}{kc+b}$ for c $c = \dfrac{na-wb}{wk}$

84) $r = \dfrac{kx}{y-az}$ for y $y = \dfrac{kx+raz}{r}$

85) $\dfrac{1}{t} = \dfrac{1}{r} - \dfrac{1}{s}$ for r $r = \dfrac{st}{s+t}$

86) $\dfrac{1}{R_1} + \dfrac{1}{R_2} = \dfrac{1}{R_3}$ for R_2 $R_2 = \dfrac{R_1 R_3}{R_1 - R_3}$

87) $\dfrac{2}{A} + \dfrac{1}{C} = \dfrac{3}{B}$ for C $C = \dfrac{AB}{3A - 2B}$

88) $\dfrac{5}{x} = \dfrac{1}{y} - \dfrac{4}{z}$ for z $z = \dfrac{4xy}{x - 5y}$

Putting It All Together

Objective

1. Review the Concepts Presented in Sections 7.1–7.4

1. Review the Concepts Presented in Sections 7.1–7.4

We began Section 7.1 with a discussion of rational functions. Let's review how to find the values of the variable that make a rational function equal zero and how to determine the domain of a rational function.

Example 1

For $g(a) = \dfrac{a+6}{a^2 - 81}$,

In-Class Example 1

Use $g(a) = \dfrac{a-5}{a^2 - 49}$ in Example 1.

answer: a) 5
b)
$(-\infty, -7) \cup (-7, 7) \cup (7, \infty)$

a) find a so that $g(a) = 0$.

b) determine the domain of the function.

Solution

a) To find the value of a that will make $g(a) = 0$, set the function equal to zero and solve for a.

$\dfrac{a+6}{a^2 - 81} = 0$ when its *numerator* equals zero.

Let $a + 6 = 0$, and solve for a.

$$a + 6 = 0$$
$$a = -6$$

$g(a) = 0$ when $a = -6$.

b) Recall that the domain of a rational function consists of all real numbers except the values of the variable that make the denominator equal zero. Set the denominator equal to zero, and solve for a. All values that make the denominator equal zero are *not* in the domain of the function.

$$a^2 - 81 = 0$$
$$(a + 9)(a - 9) = 0 \qquad \text{Factor.}$$
$$a + 9 = 0 \quad \text{or} \quad a - 9 = 0 \qquad \text{Set each factor equal to zero.}$$
$$a = -9 \quad \text{or} \qquad a = 9 \qquad \text{Solve.}$$

When $a = 9$ or $a = -9$, the denominator of $g(a) = \dfrac{a + 6}{a^2 - 81}$ equals zero. The domain contains all real numbers *except* -9 and 9. Write the domain in interval notation as $(-\infty, -9) \cup (-9, 9) \cup (9, \infty)$.

Next let's look at how to write a rational expression in lowest terms.

Example 2

In-Class Example 2
Write in lowest terms:
$$\dfrac{3h^3 + 12h}{h^3 + 4h - 2h^2 - 8}$$
answer: $\dfrac{3h}{h - 2}$

Write in lowest terms: $\dfrac{4x^2 - 28x}{x^2 - 15x + 56}$

Solution

$$\frac{4x^2 - 28x}{x^2 - 15x + 56} = \frac{4x(x - 7)}{(x - 7)(x - 8)} \qquad \text{Factor.}$$

$$= \frac{4x}{x - 8} \qquad \text{Divide by } x - 7.$$

We finished Section 7.1 with multiplying and dividing rational expressions, and then learned how to add and subtract them in Section 7.2. Now we will practice these operations together so that we will learn to recognize the techniques needed to perform these operations.

Example 3

In-Class Example 3
Divide $\dfrac{d^2 + 14d + 45}{d^2 - 25}$
$\div \dfrac{17}{(d - 5)(2d + 1)}.$
answer: $\dfrac{(d + 9)(2d + 1)}{17}$

Divide $\dfrac{y^2 - 6y - 27}{25y^2 - 49} \div \dfrac{y^2 - 9y}{14 - 10y}.$

Solution

Do we need a common denominator to divide? *No.* A common denominator is needed to add or subtract but not to multiply or divide.

To divide, multiply the first rational expression by the reciprocal of the second expression, then factor, reduce, and multiply.

$$\frac{y^2 - 6y - 27}{25y^2 - 49} \div \frac{y^2 - 9y}{14 - 10y} = \frac{y^2 - 6y - 27}{25y^2 - 49} \cdot \frac{14 - 10y}{y^2 - 9y} \qquad \text{Multiply by the reciprocal.}$$

$$= \frac{(y - 9)(y + 3)}{(5y + 7)(5y - 7)} \cdot \frac{\overset{-1}{2(7 - 5y)}}{y(y - 9)} \qquad \text{Factor.}$$

$$= -\frac{2(y + 3)}{y(5y + 7)} \qquad \text{Reduce and multiply.}$$

Recall that $\dfrac{7 - 5y}{5y - 7} = -1.$

Example 4

In-Class Example 4

Add $\dfrac{9}{2w-1} + \dfrac{w}{w+2}$.

answer: $\dfrac{2(w^2+4w+9)}{(2w-1)(w+2)}$

Add $\dfrac{n}{n-6} + \dfrac{2}{n+1}$.

Solution

To add or subtract rational expressions we need a common denominator. We do not need to factor these denominators, so we are ready to identify the LCD.

$$\text{LCD} = (n-6)(n+1)$$

Rewrite each expression with the LCD.

$$\frac{n}{n-6} \cdot \frac{n+1}{n+1} = \frac{n(n+1)}{(n-6)(n+1)}, \qquad \frac{2}{n+1} \cdot \frac{n-6}{n-6} = \frac{2(n-6)}{(n-6)(n+1)}$$

$$\begin{aligned} \frac{n}{n-6} + \frac{2}{n+1} &= \frac{n(n+1)}{(n-6)(n+1)} + \frac{2(n-6)}{(n-6)(n+1)} && \text{Write each expression with the LCD.} \\[2mm] &= \frac{n(n+1) + 2(n-6)}{(n-6)(n+1)} && \text{Add the numerators.} \\[2mm] &= \frac{n^2 + n + 2n - 12}{(n-6)(n+1)} && \text{Distribute.} \\[2mm] &= \frac{n^2 + 3n - 12}{(n-6)(n+1)} && \text{Combine like terms.} \end{aligned}$$

Although this numerator will not factor, remember that sometimes it *is* possible to factor the numerator and simplify the result.

A complex fraction is not considered to be an expression in simplest form. Let's review how to simplify complex fractions.

Example 5

In-Class Example 5

Simplify each complex fraction.

a) $\dfrac{\frac{r-7}{10}}{\frac{3r-21}{5r}}$ b) $\dfrac{\frac{7}{9}-\frac{1}{6}}{1-\frac{2}{3}}$

answer: a) $\dfrac{r}{6}$ b) $\dfrac{11}{6}$

Simplify each complex fraction.

a) $\dfrac{\frac{2a+12}{9}}{\frac{a+6}{5a}}$ b) $\dfrac{1-\frac{3}{4}}{\frac{1}{2}+\frac{1}{3}}$

Solution

a) Since $\dfrac{\frac{2a+12}{9}}{\frac{a+6}{5a}}$ contains one expression in the numerator and one in the denominator, we begin by rewriting it as a division problem.

$$\begin{aligned} \frac{\frac{2a+12}{9}}{\frac{a+6}{5a}} &= \frac{2a+12}{9} \div \frac{a+6}{5a} && \text{Rewrite the complex fraction as a division problem.} \\[2mm] &= \frac{2a+12}{9} \cdot \frac{5a}{a+6} && \text{Change division to multiplication by the reciprocal of } \frac{a+6}{5a}. \\[2mm] &= \frac{2\cancel{(a+6)}}{9} \cdot \frac{5a}{\cancel{a+6}} && \text{Factor and divide numerator and denominator by } a+6 \text{ to simplify.} \\[2mm] &= \frac{10a}{9} && \text{Multiply.} \end{aligned}$$

If a complex fraction contains one term in the numerator and one term in the denominator, it can be simplified by rewriting it as a division problem and then performing the division.

b) We can simplify the complex fraction $\dfrac{1 - \dfrac{3}{4}}{\dfrac{1}{2} + \dfrac{1}{3}}$ in two different ways. We can combine the terms in the numerator, combine the terms in the denominator, and then proceed as in part a). Or, we can follow the steps below.

Step 1: Look at all of the fractions in the complex fraction. They are

$\dfrac{3}{4}, \dfrac{1}{2},$ and $\dfrac{1}{3}.$ Write down their LCD.   LCD = 12

Step 2: Multiply the numerator and denominator of the complex fraction by the LCD, 12.

$$\frac{12\left(1 - \dfrac{3}{4}\right)}{12\left(\dfrac{1}{2} + \dfrac{1}{3}\right)}$$

Step 3: Simplify.

$$\frac{12\left(1 - \dfrac{3}{4}\right)}{12\left(\dfrac{1}{2} + \dfrac{1}{3}\right)} = \frac{12 \cdot 1 - 12 \cdot \dfrac{3}{4}}{12 \cdot \dfrac{1}{2} + 12 \cdot \dfrac{1}{3}} \qquad \text{Distribute.}$$

$$= \frac{12 - 9}{6 + 4} \qquad \text{Multiply.}$$

$$= \frac{3}{10} \qquad \text{Simplify.}$$

If a complex fraction contains more than one term in the numerator and/or denominator, we can multiply the numerator and denominator by the LCD of all of the fractions in the expression and simplify.

Do you remember the difference between an equation and an expression? An equation contains an equal sign and an expression does not. We can solve equations, and we can perform operations with rational expressions.

Example 6

Determine whether each is an equation or a sum or difference of expressions. Then solve the equation, or find the sum or difference.

a) $\dfrac{5w}{w + 1} - 2 = \dfrac{5}{w + 1}$ b) $\dfrac{5w}{w + 1} + \dfrac{5}{w + 1}$

In-Class Example 6
Determine whether each is an
equation or a sum or difference
of expressions. Then solve the
equation, or find the sum or
difference.

a) $\dfrac{3z}{z + 7} - \dfrac{3}{z + 7}$

b) $\dfrac{3z}{z + 7} - 6 = \dfrac{3}{z + 7}$

answer:

a) difference; $\dfrac{3(z - 1)}{z + 7}$

b) equation; $\{-15\}$

Solution

a) This is an *equation* because it contains an $=$ sign. Multiply both sides of the equation
by the LCD of the rational expressions to *eliminate* the denominators: LCD $= w + 1$.

$$(w + 1)\left(\frac{5w}{w + 1} - 2\right) = (w + 1) \cdot \frac{5}{w + 1} \qquad \text{Multiply both sides by } (w + 1).$$

$$5w - 2(w + 1) = 5 \qquad \text{Distribute.}$$
$$5w - 2w - 2 = 5 \qquad \text{Distribute.}$$
$$3w - 2 = 5 \qquad \text{Combine like terms.}$$
$$3w = 7$$
$$w = \frac{7}{3}$$

Verify that the solution set is $\left\{\dfrac{7}{3}\right\}$.

b) $\dfrac{5w}{w + 1} + \dfrac{5}{w + 1}$ is **not** an equation because it does not contain an equal sign. This is
a *sum* of rational expressions.

$$\frac{5w}{w + 1} + \frac{5}{w + 1} = \frac{5w + 5}{w + 1} \qquad \text{Add the expressions.}$$

$$= \frac{5(w + 1)}{w + 1} \qquad \text{Factor.}$$

$$= 5 \qquad \text{Simplify.}$$

 You Try 1

a) Write in lowest terms: $\dfrac{2m^2 - 7m + 5}{1 - m^2}$

b) Subtract $\dfrac{a}{a + 3} - \dfrac{2}{a}$.

c) Multiply $\dfrac{x^3 - 8}{9} \cdot \dfrac{3x + 6}{x^2 - 4}$.

d) Solve $\dfrac{r}{r + 3} + 5 = \dfrac{12}{r + 3}$.

e) For $f(k) = \dfrac{3k - 4}{k^2 + 5k}$,

i) find k so that $f(k) = 0$.

ii) determine the domain of the function.

f) Simplify $\dfrac{\dfrac{w^2 - 25}{8w}}{\dfrac{3w + 15}{6w}}$.

Answers to You Try Exercises

1) a) $\dfrac{5 - 2m}{m + 1}$ b) $\dfrac{a^2 - 2a - 6}{a(a + 3)}$ c) $\dfrac{x^2 + 2x + 4}{3}$ d) $\left\{-\dfrac{1}{2}\right\}$

e) i) $\dfrac{4}{3}$ ii) $(-\infty, -5) \cup (-5, 0) \cup (0, \infty)$ f) $\dfrac{w - 5}{4}$

Putting It All Together
Summary Exercises

*Additional answers can be found in the Answers to Exercises appendix.

Objective 1: Review the Concepts Presented in Sections 7.1–7.4

For each rational function,

a) find $f(2)$, if possible.

b) find x so that $f(x) = 0$.

c) determine the domain of the function.

1) $f(x) = \dfrac{4x}{x^2 - 9}$ a) $-\dfrac{8}{5}$ b) 0 c) $(-\infty, -3) \cup (-3, 3) \cup (3, \infty)$

2) $f(x) = \dfrac{3 - 5x}{x^2 + 2x - 8}$ a) undefined b) $\dfrac{3}{5}$
c) $(-\infty, -4) \cup (-4, 2) \cup (2, \infty)$

3) $f(x) = \dfrac{12}{2x - 1}$ a) 4 b) never equals 0 c) $\left(-\infty, \dfrac{1}{2}\right) \cup \left(\dfrac{1}{2}, \infty\right)$

4) $f(x) = \dfrac{8}{5x}$ a) $\dfrac{4}{5}$ b) never equals 0 c) $(-\infty, 0) \cup (0, \infty)$

5) $f(x) = \dfrac{x - 1}{x^2 + 2}$ a) $\dfrac{1}{6}$ b) 1 c) $(-\infty, \infty)$

6) $f(x) = \dfrac{49 - x^2}{15}$ a) 3 b) -7 and 7 c) $(-\infty, \infty)$

Write each rational expression in lowest terms.

7) $\dfrac{36n^9}{27n^{12}}$ $\dfrac{4}{3n^3}$

8) $\dfrac{8w^{12}}{16w^7}$ $\dfrac{w^5}{2}$

9) $\dfrac{2j + 5}{2j^2 - 3j - 20}$ $\dfrac{1}{j - 4}$

10) $\dfrac{m^2 + 5m - 24}{2m^2 + 2m - 24}$ $\dfrac{m + 8}{2(m + 4)}$

11) $\dfrac{12 - 15n}{5n^2 + 6n - 8}$ $-\dfrac{3}{n + 2}$

12) $\dfrac{-x - y}{xy + y^2 + 3x + 3y}$ $-\dfrac{1}{y + 3}$

Perform the operations and simplify.

13) $\dfrac{5}{f + 8} - \dfrac{2}{f}$ $\dfrac{3f - 16}{f(f + 8)}$

14) $\dfrac{2c^2 - 4c - 6}{c + 1} \div \dfrac{3c - 9}{7}$ $\dfrac{14}{3}$

15) $\dfrac{9a^3}{10b} \cdot \dfrac{40b^2}{81a}$ $\dfrac{4a^2b}{9}$

16) $\dfrac{4j}{j^2 - 81} + \dfrac{3}{j^2 - 3j - 54}$ $\dfrac{4j^2 + 27j + 27}{(j + 9)(j - 9)(j + 6)}$

17) $\dfrac{3}{q^2 - q - 20} + \dfrac{8q}{q^2 + 11q + 28}$ $\dfrac{8q^2 - 37q + 21}{(q - 5)(q + 4)(q + 7)}$

18) $\dfrac{12y^3}{4z} \cdot \dfrac{8z^4}{72y^6}$ $\dfrac{z^3}{3y^3}$

19) $\dfrac{16 - m^2}{m + 4} \div \dfrac{8m - 32}{m + 7}$ $-\dfrac{m + 7}{8}$

20) $\dfrac{12p}{4p^2 + 11p + 6} - \dfrac{5}{p^2 - 4p - 12}$ $\dfrac{12p^2 - 92p - 15}{(4p + 3)(p + 2)(p - 6)}$

21) $\dfrac{13}{r - 8} + \dfrac{4}{8 - r}$ $\dfrac{9}{r - 8}$

22) $\dfrac{1}{4y} + \dfrac{7}{6y^2}$ $\dfrac{3y + 14}{12y^2}$

23) $\dfrac{a^2 - 4}{a^3 + 8} \cdot \dfrac{5a^2 - 10a + 20}{3a - 6}$ $\dfrac{5}{3}$

24) $\dfrac{7}{d + 9} + \dfrac{6}{d^2}$ $\dfrac{7d^2 + 6d + 54}{d^2(d + 9)}$

25) $\dfrac{10}{x - 8} + \dfrac{4}{x + 3}$ $\dfrac{2(7x - 1)}{(x - 8)(x + 3)}$

26) $\dfrac{xy - 4x + 3y - 12}{y^2 - 16} \div \dfrac{x^3 + 27}{6}$ $\dfrac{6}{(y + 4)(x^2 - 3x + 9)}$

27) $\dfrac{13}{5z} - \dfrac{1}{3z}$ $\dfrac{34}{15z}$

28) $\dfrac{6}{k + 2} - \dfrac{1}{k + 5}$ $\dfrac{5k + 28}{(k + 2)(k + 5)}$

29) $\dfrac{10q}{8p - 10q} + \dfrac{8p}{10q - 8p}$ -1

30) $\dfrac{m}{7m - 4n} - \dfrac{20n}{4n - 7m}$ $\dfrac{m + 20n}{7m - 4n}$

31) $\dfrac{6u + 1}{3u^2 - 2u} - \dfrac{u}{3u^2 + u - 2} + \dfrac{10}{u^2 + u}$ $\dfrac{5u^2 + 37u - 19}{u(3u - 2)(u + 1)}$

32) $\dfrac{2p + 3}{p^2 + 7p} - \dfrac{4p}{p^2 - p - 56} + \dfrac{5}{p^2 - 8p}$ $\dfrac{-2p^2 - 8p + 11}{p(p + 7)(p - 8)}$

33) $\dfrac{x}{2x^2 - 7x - 4} - \dfrac{x + 3}{4x^2 + 4x + 1}$ $\dfrac{x^2 + 2x + 12}{(2x + 1)^2(x - 4)}$

34) $\dfrac{f - 12}{f - 7} - \dfrac{5}{7 - f}$ 1

35) $\left(\dfrac{2c}{c + 8} + \dfrac{4}{c - 2}\right) \div \dfrac{6}{5c + 40}$ $\dfrac{5(c^2 + 16)}{3(c - 2)}$

36) $\left(\dfrac{3n}{3n - 1} - \dfrac{4}{n + 4}\right) \cdot \dfrac{9n^2 - 1}{21n^2 + 28}$ $\dfrac{3n + 1}{7(n + 4)}$

37) $\dfrac{3}{w^2 - w} + \dfrac{4}{5w} - \dfrac{3}{w - 1}$ $-\dfrac{11}{5w}$

38) $\dfrac{4}{k^2 + 4k} - \dfrac{3}{2k} + \dfrac{1}{k + 4}$ $-\dfrac{1}{2k}$

For each rectangle, find a rational expression in simplest form to represent its a) area and b) perimeter.

39)

$\dfrac{x}{2}$, $\dfrac{x - 3}{4}$ a) $\dfrac{x(x - 3)}{8}$ b) $\dfrac{3(x - 1)}{2}$

40) [rectangle with width $\dfrac{8}{z+1}$ and height $\dfrac{z}{z+5}$]

a) $\dfrac{8z}{(z+5)(z+1)}$ b) $\dfrac{2(z^2+9z+40)}{(z+5)(z+1)}$

Use $f(x) = \dfrac{x}{3x-1}$ and $g(x) = \dfrac{2x^2}{9x^2-1}$ for Exercises 41 and 42.

41) If $h(x) = f(x) + g(x)$, find $h(x)$ in its simplified form, and determine its domain.

42) If $k(x) = f(x) - g(x)$, find $k(x)$ in its simplified form, and find x so that $k(x) = 0$.

Simplify each complex fraction.

43) $\dfrac{\dfrac{3c^3}{8c+24}}{\dfrac{6c}{c+3}}$ $\dfrac{c^2}{16}$

44) $\dfrac{\dfrac{3}{8}-\dfrac{1}{4}}{\dfrac{1}{6}+\dfrac{2}{3}}$ $\dfrac{3}{20}$

45) $\dfrac{\dfrac{5}{m}+\dfrac{2}{m-3}}{1-\dfrac{4}{m}}$ $\dfrac{7m-15}{(m-3)(m-4)}$

46) $\dfrac{\dfrac{9k^2-1}{14k}}{\dfrac{3k+1}{21k^3}}$ $\dfrac{3k^2(3k-1)}{2}$

47) $\dfrac{\dfrac{25t^2}{6u}}{\dfrac{10t}{9u^4}}$ $\dfrac{15tu^3}{4}$

48) $\dfrac{\dfrac{3}{xy}-\dfrac{y}{x}}{\dfrac{1}{y}+\dfrac{1}{x}}$ $\dfrac{3-y^2}{x+y}$

For Exercises 49 and 50, let $h(x) = \dfrac{f(x)}{g(x)}$. Find a simplified form for $h(x)$, and determine its domain.

49) $f(x) = 6x^2 + 43x - 40$, $g(x) = x + 8$
   $h(x) = 6x - 5$; domain: $(-\infty, -8) \cup (-8, \infty)$

50) $f(x) = x^3 + 7x^2 - 4x - 28$, $g(x) = x^2 - 4$
   $h(x) = x + 7$; domain: $(-\infty, -2) \cup (-2, 2) \cup (2, \infty)$

Solve.

51) $\dfrac{3y}{y+7} - 6 = \dfrac{3}{y+7}$ $\{-15\}$

52) $\dfrac{18}{t} - 2 = \dfrac{10}{t}$ $\{4\}$

53) $k - \dfrac{28}{k} = 3$ $\{-4, 7\}$

54) $\dfrac{6}{m+5} + 1 = \dfrac{2}{m}$ $\{-10, 1\}$

55) $\dfrac{3}{d+2} + \dfrac{10}{d^2-6d-16} = \dfrac{5}{d-8}$ $\{-12\}$

56) $\dfrac{a^2}{2} = \dfrac{a^2-6a}{3}$ $\{0, -12\}$

57) $\dfrac{4}{3x-1} - 7 = \dfrac{4}{3x-1}$ $\varnothing$

58) $\dfrac{3}{n+4} = \dfrac{n}{n+6} - \dfrac{2}{n^2+10n+24}$ $\{-5, 4\}$

Section 7.5 Applications of Rational Equations

Objectives

1. Solve Problems Involving Proportions
2. Solve Problems Involving Distance, Rate, and Time
3. Solve Problems Involving Work

We have studied applications of linear and quadratic equations. Now we turn our attention to applications involving equations with rational expressions. We will continue to use the Steps for Solving Applied Problems outlined in Section 2.1.

1. Solve Problems Involving Proportions

We first solved application problems involving proportions in Section 2.4. We begin this section with a problem involving a proportion.

Example 1

Write an equation and solve.

One morning at a coffee shop, the ratio of the number of customers who ordered regular coffee to the number who ordered decaffeinated coffee was 4 to 1. If the number of people who ordered regular coffee was 126 more than the number who ordered decaf, how many people ordered each type of coffee?

Solution

Step 1: **Read** the problem carefully, and identify what we are being asked to find.

We must find the number of customers who ordered regular coffee and the number who ordered decaffeinated coffee.

In-Class Example 1

Write an equation and solve.
The ratio of the number of students in a fraternity who went on a spring-break trip to the number who did not was 3 to 1. If the number of guys who took a trip was 24 more than the number who did not, how many fraternity members took a trip over spring break?
answer: 36

Step 2: **Choose a variable** to represent the unknown, and define the other unknown in terms of this variable.

$$x = \text{number of people who ordered decaffeinated coffee}$$
$$x + 126 = \text{number of people who ordered regular coffee}$$

Step 3: **Translate** the information that appears in English into an algebraic equation. Write a proportion. We will write our ratios in the form of

$$\frac{\text{number who ordered regular}}{\text{number who ordered decaf}}$$ so that the numerators contain the same quantities

and the denominators contain the same quantities.

number who ordered regular coffee → $\dfrac{4}{1} = \dfrac{x + 126}{x}$ ← number who ordered regular coffee
number who ordered decaf coffee → ← number who ordered decaf coffee

The equation is $\dfrac{4}{1} = \dfrac{x + 126}{x}$.

Step 4: Solve the equation.

$$\frac{4}{1} \begin{matrix} \nearrow \text{Multiply.} \\ \diagdown \\ \nearrow \\ \searrow \text{Multiply.} \end{matrix} \frac{x + 126}{x}$$

$4x = 1(x + 126)$ Set the cross products equal.
$4x = x + 126$
$3x = 126$ Subtract x.
$x = 42$ Divide by 3.

Step 5: **Check** the answer and **interpret** the solution as it relates to the problem.

Therefore, 42 customers ordered decaffeinated coffee and $42 + 126 = 168$ people ordered regular coffee. The check is left to the student. ∎

You Try 1

Write an equation and solve.
During one week at a bookstore, the ratio of the number of romance novels sold to the number of travel books sold was 5 to 3. Determine the number of each type of book sold if customers bought 106 more romance novels than travel books.

2. Solve Problems Involving Distance, Rate, and Time

In Chapter 2 we solved problems involving distance (d), rate (r), and time (t).
The basic formula is $d = rt$. We can solve this formula for r and then for t to obtain

$$r = \frac{d}{t} \qquad \text{and} \qquad t = \frac{d}{r}.$$

In this section, we will encounter problems involving boats going with and against a current, and planes going with and against the wind. Both scenarios use the same idea.

Say a boat's speed is 18 mph in still water. If that same boat had a 4 mph current pushing *against* it, how fast would it be traveling? (The current will cause the boat to slow down.)

$$\text{Speed } against \text{ the current} = 18 \text{ mph} \quad - 4 \text{ mph}$$
$$= 14 \text{ mph}$$

$$\frac{\text{Speed } against}{\text{the current}} = \frac{\text{Speed in}}{\text{still water}} - \frac{\text{Speed of}}{\text{the current}}$$

If the speed of the boat in still water is 18 mph and a 4 mph current is *pushing* the boat, how fast would the boat be traveling *with* the current? (The current will cause the boat to travel faster.)

$$\text{Speed } with \text{ the current} = 18 \text{ mph} \quad + 4 \text{ mph}$$
$$= 22 \text{ mph}$$

$$\frac{\text{Speed } with}{\text{the current}} = \frac{\text{Speed in}}{\text{still water}} + \frac{\text{Speed of}}{\text{the current}}$$

A boat traveling *against* the current is said to be traveling *upstream*. A boat traveling *with* the current is said to be traveling *downstream*. We will use these ideas in Example 2.

Example 2

In-Class Example 2
Write an equation and solve. A boat can travel 9 mi downstream in the same amount of time it can travel 6 mi upstream. If the speed of the current is 3 mph, what is the speed of the boat in still water?
answer: 15 mph

Write an equation and solve.
A boat can travel 8 mi downstream in the same amount of time it can travel 6 mi upstream. If the speed of the current is 2 mph, what is the speed of the boat in still water?

Solution

Step 1: **Read** the problem carefully, and identify what we are being asked to find.

First, we must understand that "8 mi downstream" means 8 *mi with the current,* and "6 mi upstream" means 6 *mi against the current.*

We must find the speed of the boat in still water.

Step 2: **Choose a variable** to represent the unknown, and define the other unknows in terms of this variable.

$$x = \text{the speed of the boat in still water}$$
$$x + 2 = \text{the speed of the boat } with \text{ the current (downstream)}$$
$$x - 2 = \text{the speed of the boat } against \text{ the current (upstream)}$$

Step 3: **Translate** from English to an algebraic equation. Use a table to organize the information.

First, fill in the distances and the rates (or speeds).

	d	*r*	*t*
Downstream	8	$x + 2$	
Upstream	6	$x - 2$	

Next we must write expressions for the time it takes the boat to go downstream and upstream. We know that $d = rt$, so if we solve for t we get $t = \dfrac{d}{r}$.

Substitute the information from the table to get the expressions for the time.

$$\text{Downstream: } t = \frac{d}{r} = \frac{8}{x+2} \qquad \text{Upstream: } t = \frac{d}{r} = \frac{6}{x-2}$$

Put these expressions into the table.

	d	*r*	*t*
Downstream	8	$x + 2$	$\dfrac{8}{x+2}$
Upstream	6	$x - 2$	$\dfrac{6}{x-2}$

The problem states that it takes the boat the *same amount of time* to travel 8 mi downstream as it does to go 6 mi upstream. We can write an equation in English:

$$\text{Time for boat to go 8 mi downstream} = \text{Time for boat to go 6 mi upstream}$$

Looking at the table, we can write the algebraic equation using the expressions for time. The equation is $\dfrac{8}{x+2} = \dfrac{6}{x-2}$.

Step 4:   **Solve** the equation.

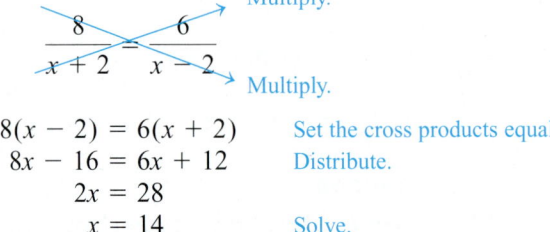

Multiply.

Multiply.

$$8(x-2) = 6(x+2) \qquad \text{Set the cross products equal.}$$
$$8x - 16 = 6x + 12 \qquad \text{Distribute.}$$
$$2x = 28$$
$$x = 14 \qquad \text{Solve.}$$

Step 5:   **Check** the answer and **interpret** the solution as it relates to the problem.

The speed of the boat in still water is 14 mph.

The speed of the boat going downstream is $14 + 2 = 16$ mph, so the time to travel downstream is

$$t = \frac{d}{r} = \frac{8}{16} = \frac{1}{2}\ \text{hr}$$

The speed of the boat going upstream is $14 - 2 = 12$ mph, so the time to travel downstream is

$$t = \frac{d}{r} = \frac{6}{12} = \frac{1}{2}\ \text{hr}$$

So, time upstream = time downstream.   ✓   ■

You Try 2

Write an equation and solve.

It takes a boat the same amount of time to travel 10 mi upstream as it does to travel 15 mi downstream. Find the speed of the boat in still water if the speed of the current is 4 mph.

3. Solve Problems Involving Work

Suppose it takes Brian 5 hr to paint his bedroom. What is the *rate* at which he does the job?

$$\text{rate} = \frac{1\ \text{room}}{5\ \text{hr}} = \frac{1}{5}\ \text{room/hr}$$

Brian works at a rate of $\dfrac{1}{5}$ of a room per hour.

In general, we can say that if it takes t units of time to do a job, then the *rate* at which the job is done is $\dfrac{1}{t}$ job per unit of time.

This idea of *rate* is what we use to determine how long it can take for two or more people or things to do a job.

Let's assume, again, that Brian can paint his room in 5 hr. At this rate, how much of the job can he do in 2 hr?

$$\begin{array}{ccc}
\text{Fractional part} \\
\text{of the job done}
\end{array} = \begin{array}{c}\text{Rate of} \\ \text{work}\end{array} \cdot \begin{array}{c}\text{Amount of} \\ \text{time worked}\end{array}$$

$$= \frac{1}{5} \cdot 2$$

$$= \frac{2}{5}$$

He can paint $\frac{2}{5}$ of the room in 2 hr.

Procedure   Solving Work Problems

The basic equation used to solve work problems is:

$$\begin{array}{c}\text{Fractional part of a job} \\ \text{done by one person or thing}\end{array} + \begin{array}{c}\text{Fractional part of a job} \\ \text{done by another person or thing}\end{array} = 1 \text{ (whole job)}$$

Example 3

In-Class Example 3
Write an equation and solve. If it takes Marshall 12 hr to refinish the hardwood floor in his house while his brother, Nelly, can do the job in 10 hr, how long would it take for there to refinish the floors together?

answer: $5\frac{5}{11}$ hr

Write an equation and solve.

If Brian can paint his bedroom in 5 hr, but his brother, Doug, could paint the room on his own in 4 hr, how long would it take for the two of them to paint the room together?

Solution

Step 1:   **Read** the problem carefully, and identify what we are being asked to find.

We must determine how long it would take Brain and Doug to paint the room together.

Step 2:   **Choose a variable** to represent the unknown.

$$t = \text{the number of hours to paint the room together}$$

Step 3:   **Translate** the information that appears in English into an algebraic equation.

Let's write down their rates:

$$\text{Brian's rate} = \frac{1}{5} \text{ room/hr (since the job takes him 5 hr)}$$

$$\text{Doug's rate} = \frac{1}{4} \text{ room/hr (since the job takes him 4 hr)}$$

It takes them t hr to paint the room together. Recall that

$$\begin{array}{c}\text{Fractional part} \\ \text{of job done}\end{array} = \begin{array}{c}\text{Rate of} \\ \text{work}\end{array} \cdot \begin{array}{c}\text{Amount of} \\ \text{time worked}\end{array}$$

$$\text{Brian's fractional part} = \frac{1}{5} \cdot t = \frac{1}{5}t$$

$$\text{Doug's fractional part} = \frac{1}{4} \cdot t = \frac{1}{4}t$$

The equation we can write comes from

$$\begin{array}{c}\text{Fractional part of the} \\ \text{job done by Brian}\end{array} + \begin{array}{c}\text{Fractional part of the} \\ \text{job done by Doug}\end{array} = 1 \text{ (whole job)}$$

$$\frac{1}{5}t + \frac{1}{4}t = 1$$

Step 4: **Solve** the equation.

$$20\left(\frac{1}{5}t + \frac{1}{4}t\right) = 20(1) \qquad \text{Multiply by the LCD of 20 to eliminate the fractions.}$$

$$20\left(\frac{1}{5}t\right) + 20\left(\frac{1}{4}t\right) = 20(1) \qquad \text{Distribute.}$$

$$4t + 5t = 20 \qquad \text{Multiply.}$$

$$9t = 20 \qquad \text{Combine like terms.}$$

$$t = \frac{20}{9} \qquad \text{Divide by 9.}$$

Step 5: **Check** the answer and **interpret** the solution as it relates to the problem.

Brian and Doug could paint the room together in $\frac{20}{9}$ hr or $2\frac{2}{9}$ hr.

Check:

Fractional part of the job done by Brian	+	Fractional part of the job done by Doug	=	1 whole job
$\frac{1}{5} \cdot \left(\frac{20}{9}\right)$	+	$\frac{1}{4} \cdot \left(\frac{20}{9}\right)$	$\overset{?}{=}$	1
$\frac{4}{9}$	+	$\frac{5}{9}$	=	1

You Try 3

Write an equation and solve.

Krutesh can mow a lawn in 2 hr while it takes Stefan 3 hr to mow the same lawn. How long would it take for them to mow the lawn if they worked together?

Answers to You Try Exercises

1) 159 travel books, 265 romance novels   2) 20 mph   3) $\frac{6}{5}$ hr or $1\frac{1}{5}$ hr

7.5 Exercises

Additional answers can be found in the Answers to Exercises appendix.

Objective 1: Solve Problems Involving Proportion

Solve the following proportions.

1) $\frac{12}{7} = \frac{60}{x}$ {35}

2) $\frac{30}{18} = \frac{45}{y}$ {27}

3) $\frac{6}{13} = \frac{x}{x + 56}$ {48}

4) $\frac{15}{8} = \frac{x}{x - 63}$ {135}

Write an equation for each and solve. See Example 1.

5) At a motocross race, the ratio of male spectators to female spectators was 10 to 3. If there were 370 male spectators, how many females were in the crowd? 111

6) The ratio of students in a history lecture who took notes in pen to those who took notes in pencil was 8 to 3. If 72 students took notes in pen, how many took notes in pencil? 27

7) In a gluten-free flour mixture, the ratio of potato-starch flour to tapioca flour is 2 to 1. If a mixture contains 3 more cups of potato-starch flour than tapioca flour how much of each type of flour is in the mixture? 3 cups of tapioca flour and 6 cups of potato-starch flour

8) Rosa Cruz won an election over her opponent by a ratio of 6 to 5. If her opponent received 372 fewer votes than she did, how many votes did each candidate receive? Rosa: 2232; opponent: 1860

9) The ancient Greeks believed that the rectangle most pleasing to the eye, the golden rectangle, had sides in which the ratio of its length to its width was approximately 8 to 5. They erected many buildings using this golden ratio, including the Parthenon. The marble floor of a museum foyer is to be designed as a golden rectangle. If its width is to be 18 ft less than its length, find the length and width of the foyer.

length: 48 ft; width: 30 ft

10) To obtain a particular color, a painter mixed two colors in a ratio of 7 parts blue to 3 parts yellow. If he used 8 fewer gallons of yellow than blue, how many gallons of blue paint did he use?   14

11) Ms. Hiramoto has invested her money so that the ratio of the amount in bonds to the amount in stocks is 3 to 2. If she has $4000 more invested in bonds than in stocks, how much does she have invested in each?

stocks: $8000; bonds: $12,000

12) At a wildlife refuge, the ratio of deer to rabbits is 4 to 9. Determine the number of deer and rabbits at the refuge if there are 40 more rabbits than deer.   deer: 32; rabbits: 72

13) In a small town, the ratio of households with pets to those without pets is 5 to 4. If 271 more households have pets than do not, how many households have pets?   1355

14) An industrial cleaning solution calls for 5 parts water to 2 parts concentrated cleaner. If a worker uses 15 more quarts of water than concentrated cleaner to make a solution,

   a) how much concentrated cleaner did she use?   10 qt

   b) how much water did she use?   25 qt

   c) how much solution did she make?   35 qt

Objective 2: Solve Problems Involving Distance, Rate, and Time

15) If the speed of a boat in still water is 10 mph,

   a) what is its speed going *against* a 3 mph current?   7 mph

   b) what is its speed *with* a 3 mph current?   13 mph

16) If an airplane travels at a constant rate of 300 mph,

   a) what is its speed going *into* a 25 mph wind?   275 mph

   b) what is its speed going *with* a 25 mph wind?   325 mph

17) If an airplane travels at a constant rate of x mph,

   a) what is its speed going *with* a 30 mph wind? $x + 30$ mph

   b) what is its speed going *against* a 30 mph wind? $x - 30$ mph

18) If the speed of a boat in still water is 13 mph,

   a) what is its speed going *against* a current with a rate of x mph? $13 - x$ mph

   b) what is its speed going *with* a current with a rate of x mph? $13 + x$ mph

Write an equation for each and solve. See Example 2.

19) A current flows at 5 mph. A boat can travel 20 mi downstream in the same amount of time it can go 12 mi upstream. What is the speed of the boat in still water?   20 mph

20) With a current flowing at 4 mph, a boat can travel 32 mi with the current in the same amount of time it can go 24 mi against the current. Find the speed of the boat in still water.   28 mph

21) A boat travels at 16 mph in still water. It takes the same amount of time for the boat to travel 15 mi downstream as to go 9 mi upstream. Find the speed of the current.   4 mph

22) A boat can travel 12 mi downstream in the time it can go 6 mi upstream. If the speed of the boat in still water is 9 mph, what is the speed of the current?

3 mph

23) An airplane flying at constant speed can fly 350 mi with the wind in the same amount of time it can fly 300 mi against the wind. What is the speed of the plane if the wind blows at 20 mph?   260 mph

24) When the wind is blowing at 25 mph, a plane flying at a constant speed can travel 500 mi with the wind in the same amount of time it can fly 400 mi against the wind. Find the speed of the plane.   225 mph

VIDEO 25) In still water the speed of a boat is 10 mph. Against the current it can travel 4 mi in the same amount of time it can travel 6 mi with the current. What is the speed of the current?   2 mph

26) Flying at a constant speed, a plane can travel 800 mi with the wind in the same amount of time it can fly 650 mi against the wind. If the wind blows at 30 mph, what is the speed of the plane?   290 mph

Objective 3: Solve Problems Involving Work

27) Toby can finish a computer programming job in 4 hr. What is the rate at which he does the job? $\frac{1}{4}$ job/hr

28) It takes Crystal 3 hr to paint her backyard fence. What is the rate at which she works? $\frac{1}{3}$ job/hr

29) Eloise can fertilize her lawn in t hr. What is the rate at which she does this job? $\frac{1}{t}$ job/hr

30) It takes Manu twice as long to clean a pool as it takes Anders. If it takes Anders t hr to clean the pool, at what rate does Manu do the job? $\frac{1}{2t}$ job/hr

Write an equation for each and solve. See Example 3.

VIDEO 31) It takes Arlene 2 hr to trim the bushes at a city park while the same job takes Andre 3 hr. How long would it take for them to do the job together? $1\frac{1}{5}$ hr

32) A hot water faucet can fill a sink in 8 min while it takes the cold water faucet only 6 min. How long would it take to fill the sink if both faucets were on? $3\frac{3}{7}$ min

33) Jermaine and Sue must put together notebooks for each person attending a conference. Working alone it would take Jermaine 5 hr while it would take Sue 8 hr. How long would it take for them to assemble the notebooks together? $3\frac{1}{13}$ hr

34) The Williams family has two printers on which they can print out their vacation pictures. The larger printer can print all of the pictures in 3 hr, while it would take 5 hr on the smaller printer. How long would it take to print the vacation pictures using both printers? $1\frac{7}{8}$ hr

35) A faucet can fill a tub in 12 min. The leaky drain can empty the tub in 30 min. If the faucet is on and the drain is leaking, how long would it take to fill the tub?   20 min

36) It takes Deepak 50 min to shovel snow from his sidewalk and driveway. When he works with his brother, Kamal, it takes only 30 min. How long would it take Kamal to do the shoveling himself?   75 min

37) Fatima and Antonio must cut out shapes for an art project at a day-care center. Fatima can do the job twice as fast as Antonio. Together, it takes 2 hr to cut out all of the shapes. How long would it take Fatima to cut out the shapes herself?   3 hr

38) It takes Burt three times as long as Phong to set up a new alarm system. Together they can set it up in 90 min. How long would it take Phong to set up the alarm system by himself?   120 min

39) Working together it takes 2 hr for a new worker and an experienced worker to paint a billboard. If the new employee worked alone, it would take him 6 hr. How long would it take the experienced worker to paint the billboard by himself?   3 hr

40) Audrey can address party invitations in 40 min, while it would take her mom 1 hr. How long would it take for them to address the invitations together?   24 min

41) Homer uses the moving walkway to get to his gate at the airport. He can travel 126 ft when he is walking on the moving walkway in the same amount of time it would take for him to walk only 66 ft on the floor next to it. If the walkway is moving at 2 ft/sec, how fast does Homer walk? $2\frac{1}{5}$ ft/sec

42) Another walkway at the airport moves at $2\frac{1}{2}$ ft/sec. If Bart can travel 140 ft when he is walking on the moving walkway in the same amount of time he can walk 80 ft on the floor next to it, how fast does Bart walk? $3\frac{1}{3}$ ft/sec

Section 7.6 Variation

Objectives

1. Solve Direct Variation Problems
2. Solve Inverse Variation Problems
3. Solve Joint Variation Problems
4. Solve Combined Variation Problems

1. Solve Direct Variation Problems

In Section 3.5 we discussed the following situation:

If you are driving on a highway at a constant speed of 60 mph, the distance you travel depends on the amount of time spent driving.

Let y = the distance traveled, in miles, and let x = the number of hours spent driving. An equation relating x and y is $y = 60x$ and y is a function of x.

We can make a table of values relating x and y.

We can say that the distance traveled, y, is *directly proportional to* the time spent traveling, x. Or y *varies directly as* x.

x	y
1	60
1.5	90
2	120
3	180

Definition

Direct Variation: y varies directly as x (or y is directly proportional to x) means

$$y = kx$$

where k is a nonzero real number. **k is called the constant of variation.**

If two quantities vary directly and $k > 0$, then as one quantity increases the other increases as well. And, as one quantity decreases, the other decreases.

In our example of driving distance, $y = 60x$, 60 *is the constant of variation.* Given information about how variables are related, we can write an equation and solve a variation problem.

Example 1

In-Class Example 1
Suppose *y* varies directly as *x*. If
y = 12 when *x* = 4,
a) find the constant of variation, *k*.
b) write a variation equation relating *x* and *y* using the value of *h* found in *a*).
c) find *y* when *x* = −5.
answer: a) k = 3   b) y = 3x c) y = −15

Suppose *y* varies directly as *x*. If $y = 18$ when $x = 3$,

a)   find the constant of variation, *k*.

b)   write a variation equation relating *x* and *y* using the value of *k* found in a).

c)   find *y* when $x = 11$.

Solution

a)   To find the constant of variation, write a *general* variation equation relating *x* and *y*. *y varies directly as x* means $y = kx$.

We are told that $y = 18$ when $x = 3$. Substitute these values into the equation and solve for *k*.

$$y = kx$$
$$18 = k(3) \qquad \text{Substitute 3 for } x \text{ and 18 for } y.$$
$$6 = k \qquad \text{Divide by 3.}$$

b)   The *specific* variation equation is the equation obtained when we substitute 6 for *k* in $y = kx$: Therefore, $y = 6x$.

c)   To find *y* when $x = 11$, substitute 11 for *x* in $y = 6x$ and evaluate.

$$y = 6x$$
$$= 6(11) \qquad \text{Substitute 11 for } x.$$
$$= 66 \qquad \text{Multiply.}$$

Procedure   Steps for Solving a Variation Problem

Step 1:   Write the *general* variation equation.

Step 2:   Find *k* by substituting the known values into the equation and solving for *k*.

Step 3:   Write the *specific* variation equation by substituting the value of *k* into the *general* variation equation.

Step 4:   Use the specific variation equation to solve the problem.

 You Try 1

Suppose *y* varies directly as *x*. If $y = 40$ when $x = 5$,

a)   find the constant of variation.

b)   write the specific variation equation relating *x* and *y*.

c)   find *y* when $x = 3$.

Example 2

In-Class Example 2
Suppose *p* varies directly as the square root of *z*. If *p* = 20 when *z* = 25, find *p* when *z* = 36.
answer: p = 24

Suppose *p* varies directly as the square of *z*. If $p = 12$ when $z = 2$, find *p* when $z = 5$.

Solution

Step 1:   Write the *general* variation equation.
p varies directly as the *square* of *z* means $p = kz^2$.

Step 2:   Find *k* using the known values: $p = 12$ when $z = 2$.

$$p = kz^2$$
$$12 = k(2)^2 \qquad \text{Substitute 2 for } z \text{ and 12 for } p.$$
$$12 = k(4)$$
$$3 = k$$

Step 3:   Substitute $k = 3$ into $p = kz^2$ to get the *specific* variation equation, $p = 3z^2$.

Step 4:   We are asked to find p when $z = 5$. Substitute $z = 5$ into $p = 3z^2$ to get p.

$$
\begin{aligned}
p &= 3z^2 \\
&= 3(5)^2 \qquad \text{Substitute 5 for } z. \\
&= 3(25) \\
&= 75
\end{aligned}
$$

■

You Try 2

Suppose w varies directly as the cube of n. If $w = 135$ when $n = 3$, find w when $n = 2$.

Example 3

In-Class Example 3
A coffee shop's revenue varies directly as the number of cups sold. If the revenue from the sale of 305 cups is $549, find the revenue from the sale of 400 cups.
answer: $720

A theater's nightly revenue varies directly as the number of tickets sold. If the revenue from the sale of 80 tickets is $3360, find the revenue from the sale of 95 tickets.

Solution

Let n = the number of tickets sold
R = revenue

We will follow the four steps for solving a variation problem.

Step 1:   Write the *general* variation equation: $R = kn$.

Step 2:   Find k using the known values: $R = 3360$ when $n = 80$.

$$
\begin{aligned}
R &= kn \\
3360 &= k(80) \qquad \text{Substitute 80 for } n \text{ and 3360 for } R. \\
42 &= k \qquad\qquad \text{Divide by 80.}
\end{aligned}
$$

Step 3:   Substitute $k = 42$ into $R = 42n$ to get the *specific* variation equation, $R = 42n$.

Step 4:   We must find the revenue from the sale of 95 tickets. Substitute $n = 95$ into $R = 42n$ to find R.

$$
\begin{aligned}
R &= 42n \\
R &= 42(95) \\
R &= 3990
\end{aligned}
$$

The revenue is $3990.

■

You Try 3

The cost to carpet a room varies directly as the area of the room. If it costs $525.00 to carpet a room of area 210 ft², how much would it cost to carpet a room of area 288 ft²?

2. Solve Inverse Variation Problems

If two quantities vary *inversely* (are *inversely* proportional) then as one value increases, the other decreases. Likewise, as one value decreases, the other increases.

Definition

Inverse Variation: y varies inversely as x (or y is inversely proportional to x) means

$$
y = \frac{k}{x}
$$

where k is a nonzero real number. **k is the constant of variation.**

A good example of inverse variation is the relationship between the time, t, it takes to travel a given distance, d, as a function of the rate (or speed), r. We can define this relationship as $t = \dfrac{d}{r}$. As the rate, r, increases, the time, t, that it takes to travel d mi decreases. Likewise, as r decreases, the time, t, that it takes to travel d mi increases. Therefore, t varies *inversely* as r.

Example 4

Suppose q varies inversely as h. If $q = 4$ when $h = 15$, find q when $h = 10$.

In-Class Example 4
Suppose q varies inversely as h. If $q = 9$ when $h = 10$, find q when $h = 15$.
answer: $q = 6$

Solution

Step 1: Write the *general* variation equation, $q = \dfrac{k}{h}$.

Step 2: Find k using the known values: $q = 4$ when $h = 15$.

$$q = \frac{k}{h}$$

$$4 = \frac{k}{15} \qquad \text{Substitute 15 for } h \text{ and 4 for } q.$$

$$60 = k \qquad \text{Multiply by 15.}$$

Step 3: Substitute $k = 60$ into $q = \dfrac{k}{h}$ to get the *specific* variation equation, $q = \dfrac{60}{h}$.

Step 4: Substitute 10 for h in $q = \dfrac{60}{h}$ to find q.

$$q = \frac{60}{10} = 6$$

You Try 4

Suppose m varies inversely as the square of v. If $m = 1.5$ when $v = 4$, find m when $v = 2$.

Example 5

The intensity of light (in lumens) varies inversely as the square of the distance from the source. If the intensity of the light is 40 lumens 5 ft from the source, what is the intensity of the light 4 ft from the source?

In-Class Example 5
See Example 5. Find the intensity of the light 2 ft from the source.
answer: 250 lumens

Solution

Let $d =$ distance from the source (in feet)
$\quad\;\; I =$ intensity of the light (in lumens)

Step 1: Write the *general* variation equation, $I = \dfrac{k}{d^2}$.

Step 2: Find k using the known values: $I = 40$ when $d = 5$.

$$I = \frac{k}{d^2}$$

$$40 = \frac{k}{(5)^2} \qquad \text{Substitute 5 for } d \text{ and 40 for } I.$$

$$40 = \frac{k}{25}$$

$$1000 = k \qquad \text{Multiply by 25.}$$

Step 3: Substitute $k = 1000$ into $I = \dfrac{k}{d^2}$ to get the *specific* variation equation, $I = \dfrac{1000}{d^2}$.

Step 4: Find the intensity, I, of the light 4 ft from the source. Substitute $d = 4$ into $I = \dfrac{1000}{d^2}$ to find I.

$$I = \frac{1000}{(4)^2} = \frac{1000}{16} = 62.5$$

The intensity of the light is 62.5 lumens.   ■

You Try 5

If the voltage in an electrical circuit is held constant (stays the same), then the current in the circuit varies inversely as the resistance. If the current is 40 amps when the resistance is 3 ohms, find the current when the resistance is 8 ohms.

3. Solve Joint Variation Problems

If a variable varies directly as the *product* of two or more other variables, the first variable *varies jointly* as the other variables.

> **Definition**
>
> **Joint Variation:** **y varies jointly as x and z** means $y = kxz$ where k is a nonzero real number.

Example 6

In-Class Example 6
See Example 6. How much interest would the same principal earn if Graham invested it at 3% for 2 yr?
answer: $120

For a given amount invested in a bank account (called the principal), the interest earned varies jointly as the interest rate (expressed as a decimal) and the time the principal is in the account. If Graham earns \$80 in interest when he invests his money for 1 yr at 4%, how much interest would the same principal earn if he invested it at 5% for 2 yr?

Solution

Let r = interest rate (as a decimal)

t = the number of years the principal is invested

I = interest earned

Step 1: Write the *general* variation equation, $I = krt$.

Step 2: Find k using the known values: $I = 80$ when $t = 1$ and $r = 0.04$.

$$
\begin{aligned}
I &= krt \\
80 &= k(0.04)(1) \qquad \text{Substitute the values into } I = krt. \\
80 &= 0.04k \\
2000 &= k \qquad\qquad\ \text{Divide by 0.04.}
\end{aligned}
$$

(The amount he invested, the principal, is \$2000.)

Step 3: Substitute $k = 2000$ into $I = krt$ to get the *specific* variation equation, $I = 2000rt$.

Step 4: Find the interest Graham would earn if he invested \$2000 at 5% interest for 2 yr. Let $r = 0.05$ and $t = 2$. Solve for I.

$$
\begin{aligned}
I &= 2000(0.05)(2) \qquad \text{Substitute 0.05 for } r \text{ and 2 for } t. \\
&= 200 \qquad\qquad\qquad\ \text{Multiply.}
\end{aligned}
$$

Graham would earn \$200.   ■

You Try 6

The volume of a box of constant height varies jointly as its length and width. A box with a volume of 9 ft^3 has a length of 3 ft and a width of 2 ft. Find the volume of a box with the same height, if its length is 4 ft and its width is 3 ft.

4. Solve Combined Variation Problems

A combined variation problem involves both direct and inverse variation.

Example 7

In-Class Example 7
Suppose q varies directly as the square root of r and inversely as p. If $q = 10$ when $r = 25$ and $p = 4$, find q when $r = 49$ and $p = 8$.
answer: 7

Suppose y varies directly as the square root of x and inversely as z. If $y = 12$ when $x = 36$ and $z = 5$, find y when $x = 81$ and $z = 15$.

Solution

Step 1:   Write the *general* variation equation.

$$y = \frac{k\sqrt{x}}{z} \qquad \begin{array}{l} \leftarrow y \text{ varies directly as the square root of } x. \\ \leftarrow y \text{ varies inversely as } z. \end{array}$$

Step 2:   Find k using the known values: $y = 12$ when $x = 36$ and $z = 5$.

$$12 = \frac{k\sqrt{36}}{5} \qquad \text{Substitute the values.}$$
$$60 = 6k \qquad \text{Multiply by 5; } \sqrt{36} = 6.$$
$$10 = k \qquad \text{Divide by 6.}$$

Step 3:   Substitute $k = 10$ into $y = \frac{k\sqrt{x}}{z}$ to get the specific variation equation,

$$y = \frac{10\sqrt{x}}{z}.$$

Step 4:   Find y when $x = 81$ and $z = 15$.

$$y = \frac{10\sqrt{81}}{15} \qquad \text{Substitute 81 for } x \text{ and 15 for } z.$$
$$y = \frac{10 \cdot 9}{15} = \frac{90}{15} = 6$$

You Try 7

Suppose a varies directly as b and inversely as the square of c. If $a = 28$ when $b = 12$ and $c = 3$, find a when $b = 36$ and $c = 4$.

Answers to You Try Exercises

1) a) 8   b) $y = 8x$ c) 24   2) 40   3) \$720.00   4) 6   5) 15 amps   6) 18 ft^3
7) 47.25

7.6 Exercises

Additional answers can be found in the Answers to Exercises appendix.

Mixed Exercises: Objectives 1–4

1) If z varies directly as y, then as y increases, the value of z _____.   increases

2) If a varies inversely as b, then as b increases, the value of a _____.   decreases

Decide whether each equation represents direct, inverse, joint, or combined variation.

3) $y = 6x$ direct

4) $c = 4ab$ joint

5) $f = \dfrac{15}{t}$ inverse

6) $z = 3\sqrt{x}$ direct

7) $p = \dfrac{8q^2}{r}$ combined

8) $w = \dfrac{11}{v^2}$ inverse

Write a general variation equation using k as the constant of variation.

9) M varies directly as n. $M = kn$

10) q varies directly as r. $q = kr$

11) h varies inversely as j. $h = \dfrac{k}{j}$

12) R varies inversely as B. $R = \dfrac{k}{B}$

13) T varies inversely as the square of c. $T = \dfrac{k}{c^2}$

14) b varies directly as the cube of w. $b = kw^3$

15) s varies jointly as r and t. $s = krt$

16) C varies jointly as A and D. $C = kAD$

17) Q varies directly as the square root of z and inversely as m. $Q = \dfrac{k\sqrt{z}}{m}$

18) r varies directly as d and inversely as the square of L.

19) Suppose z varies directly as x. If $z = 63$ when $x = 7$,

   a) find the constant of variation.   9

   b) write the specific variation equation relating z and x. $z = 9x$

   c) find z when $x = 6$.   54

20) Suppose A varies directly as D. If $A = 12$ when $D = 3$,

   a) find the constant of variation.   4

   b) write the specific variation equation relating A and D. $A = 4D$

   c) find A when $D = 11$.   44

21) Suppose N varies inversely as y. If $N = 4$ when $y = 12$,

   a) find the constant of variation.   48

   b) write the specific variation equation relating N and y. $N = \dfrac{48}{y}$

   c) find N when $y = 3$.   16

22) Suppose j varies inversely as m. If $j = 7$ when $m = 9$,

   a) find the constant of variation.   63

   b) write the specific variation equation relating j and m. $j = \dfrac{63}{m}$

   c) find j when $m = 21$.   3

23) Suppose Q varies directly as the square of r and inversely as w. If $Q = 25$ when $r = 10$ and $w = 20$,

   a) find the constant of variation.   5

   b) write the specific variation equation relating Q, r, and w. $Q = \dfrac{5r^2}{w}$

   c) find Q when $r = 6$ and $w = 4$.   45

24) Suppose y varies jointly as a and the square root of b. If $y = 42$ when $a = 3$ and $b = 49$,

   a) find the constant of variation.   2

   b) write the specific variation equation relating y, a, and b. $y = 2a\sqrt{b}$

   c) find y when $a = 4$ and $b = 9$.   24

Solve.

25) If B varies directly as R, and $B = 35$ when $R = 5$, find B when $R = 8$.   56

26) If q varies directly as p, and $q = 10$ when $p = 4$, find q when $p = 10$.   25

27) If L varies inversely as the square of h, and $L = 8$ when $h = 3$, find L when $h = 2$.   18

28) If w varies inversely as d, and $w = 3$ when $d = 10$, find w when $d = 5$.   6

29) If y varies jointly as x and z, and $y = 60$ when $x = 4$ and $z = 3$, find y when $x = 7$ and $z = 2$.   70

30) If R varies directly as P and inversely as the square of Q, and $R = 5$ when $P = 10$ and $Q = 4$, find R when $P = 18$ and $Q = 3$.   16

Solve each problem by writing a variation equation.

31) Kosta is paid hourly at his job. His weekly earnings vary directly as the number of hours worked. If Kosta earned $437.50 when he worked 35 hr, how much would he earn if he worked 40 hr? $500.00

32) The cost of manufacturing a certain brand of spiral note-book is inversely proportional to the number produced. When 16,000 notebooks are produced, the cost per note-book is $0.60. What is the cost of each notebook when 12,000 are produced? $0.80

33) If distance is held constant, the time it takes to travel that distance is inversely proportional to the speed at which one travels. If it takes 14 hr to travel the given distance at 60 mph, how long would it take to travel the same dis-tance at 70 mph?   12 hr

34) The surface area of a cube varies directly as the square of the length of one of its sides. A cube has a surface area of 54 in^2 when the length of each side is 3 in. What is the surface area of a cube with a side of length 6 in.?   216 in^2

35) The power in an electrical system varies jointly as the current and the square of the resistance. If the power is 100 watts when the current is 4 amps and the resistance is 5 ohms, what is the power when the current is 5 amps and the resistance is 6 ohms?   180 watts

36) The force exerted on an object varies jointly as the mass and acceleration of the object. If a 20-newton force is exerted on an object of mass 10 kg and an acceleration of 2 m/sec^2, how much force is exerted on a 50 kg object with an acceleration of 8 m/sec^2?   400 newtons

37) The kinetic energy of an object varies jointly as its mass and the square of its speed. When a roller coaster car with a mass of 1000 kg is traveling at 15 m/sec, its kinetic energy is 112,500 J (joules). What is the kinetic energy of the same car when it travels at 18 m/sec?   162,000 J

38) The volume of a cylinder varies jointly as its height and the square of its radius. The volume of a cylindrical can is 108π cm^3 when its radius is 3 cm and it is 12 cm high. Find the volume of a cylindrical can with a radius of 4 cm and a height of 3 cm.   48π cm^3

39) The frequency of a vibrating string varies inversely as its length. If a 5-ft-long piano string vibrates at 100 cycles/sec, what is the frequency of a piano string that is 2.5 ft long?   200 cycles/sec

40) The amount of pollution produced varies directly as the population. If a city of 500,000 people produces 800,000 tons of pollutants, how many tons of pollutants would be produced by a city of 1,000,000 people?   1,600,000 tons

VIDEO 41) The resistance of a wire varies directly as its length and inversely as its cross-sectional area. A wire of length 40 cm and cross-sectional area 0.05 cm^2 has a resistance of 2 ohms. Find the resistance of 60 cm of the same type of wire.   3 ohms

42) When a rectangular beam is positioned horizontally, the maximum weight that it can support varies jointly as its width and the square of its thickness and inversely as its length. A beam is $\frac{3}{4}$ ft wide, $\frac{1}{3}$ ft thick, and 8 ft long, and it can support 17.5 tons. How much weight can a similar beam support if it is 1 ft wide, $\frac{1}{2}$ ft thick and 12 ft long?   35 tons

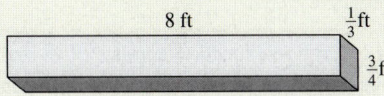
8 ft $\frac{1}{3}$ft $\frac{3}{4}$ft

43) Hooke's law states that the force required to stretch a spring is proportional to the distance that the spring is stretched from its original length. A force of 200 lb is required to stretch a spring 5 in. from its natural length. How much force is needed to stretch the spring 8 in. beyond its natural length?   320 lb

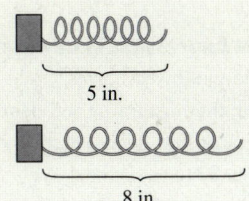
5 in.

8 in.

44) The weight of an object on Earth varies inversely as the square of its distance from the center of the Earth. If an object weighs 300 lb on the surface of the Earth (4000 mi from the center), what is the weight of the object if it is 800 mi above the Earth? (Round to the nearest pound.)   208 lb

Chapter 7: Summary

Definition/Procedure	Example

7.1 Simplifying, Multiplying, and Dividing Rational Expressions and Functions

A **rational expression** is an expression of the form $\frac{P}{Q}$, where P and Q are polynomials and where $Q \neq 0$. **(p. 368)**

Some examples of rational expressions are
$$\frac{4a - 9}{a + 2}, \quad \frac{7w^3}{8}, \quad \frac{12}{k^2 - 9k + 18}, \quad \frac{5x + 3y}{11xy^2}.$$

Rational Functions

$f(x) = \dfrac{x - 9}{x + 2}$ is a rational function since $\dfrac{x - 9}{x + 2}$ is a rational

expression and since each value that can be substituted for x will produce only one value for the expression.

To determine the value of the variable that makes the function equal zero, set the numerator equal to zero and solve for the variable.

The **domain** of a rational function consists of all real numbers except the value(s) of the variable which make the denominator equal zero. **(p. 368)**

$f(x) = \dfrac{x - 9}{x + 2}$

a) Find x so that $f(x) = 0$.
b) Determine the domain of $f(x)$.

a) $\dfrac{x - 9}{x + 2} = 0$ when $x - 9 = 0$.
$$x - 9 = 0$$
$$x = 9$$
When $x = 9$, $f(x) = 0$.

b) $f(x) = \dfrac{x - 9}{x + 2}$
$$x + 2 = 0 \qquad \text{Set the denominator} = 0.$$
$$x = -2 \qquad \text{Solve.}$$

When $x = -2$, the denominator of $f(x) = \dfrac{x - 9}{x + 2}$ equals zero.

The domain contains all real numbers *except* -2. Write the domain in interval notation as $(-\infty, -2) \cup (-2, \infty)$.

Writing a Rational Expression in Lowest Terms
To write an expression in lowest terms,
1) completely **factor** the numerator and denominator.
2) **divide** the numerator and denominator by the greatest common factor. **(p. 371)**

Simplify $\dfrac{2r^2 - 11r + 15}{4r^2 - 36}$.

$$\frac{2r^2 - 11r + 15}{4r^2 - 36} = \frac{(2r - 5)\cancel{(r - 3)}}{4(r + 3)\cancel{(r - 3)}}$$
$$= \frac{2r - 5}{4(r + 3)}$$

Simplifying $\dfrac{a - b}{b - a}$

A rational expression of the form $\dfrac{a - b}{b - a}$ simplifies to -1.

(p. 372)

Simplify $\dfrac{4 - w}{w^2 - 16}$.

$$\frac{4 - w}{w^2 - 16} = \frac{\overset{-1}{\cancel{4 - w}}}{(w + 4)\cancel{(w - 4)}}$$
$$= -\frac{1}{w + 4}$$

Multiplying Rational Expressions
1) Factor the numerators and denominators.
2) Reduce and multiply. **(p. 374)**

Multiply $\dfrac{15v^3}{v^2 + 8v + 12} \cdot \dfrac{2v + 12}{5v}$.

$$\frac{15v^3}{v^2 + 8v + 12} \cdot \frac{2v + 12}{5v} = \frac{\overset{3}{\cancel{15}}v^2 \cdot \cancel{v}}{(v + 2)\cancel{(v + 6)}} \cdot \frac{2\cancel{(v + 6)}}{\cancel{5v}}$$
$$= \frac{6v^2}{v + 2}$$

Definition/Procedure	Example

Dividing Rational Expressions

To divide rational expressions, multiply the first expression by the reciprocal of the second. **(p. 375)**

Divide $\dfrac{3x^2 + 4x}{x + 1} \div \dfrac{9x^2 - 16}{21x - 28}$.

$$\dfrac{3x^2 + 4x}{x + 1} \div \dfrac{9x^2 - 16}{21x - 28} = \dfrac{3x^2 + 4x}{x + 1} \cdot \dfrac{21x - 28}{9x^2 - 16}$$

$$= \dfrac{x(3x + 4)}{x + 1} \cdot \dfrac{7(3x - 4)}{(3x + 4)(3x - 4)}$$

$$= \dfrac{7x}{x + 1}$$

7.2 Adding and Subtracting Rational Expressions

To Find the Least Common Denominator (LCD)

1) Factor the denominators.
2) The LCD will contain each unique factor the greatest number of times it appears in any single factorization.
3) The LCD is the *product* of the factors identified in Step 2. **(p. 380)**

Find the LCD of $\dfrac{5a}{a^2 + 7a}$ and $\dfrac{2}{a^2 + 14a + 49}$.

1) $\quad a^2 + 7a = a(a + 7)$
$\quad a^2 + 14a + 49 = (a + 7)^2$
2) The factors we will use in the LCD are a and $(a + 7)^2$.
3) LCD $= a(a + 7)^2$

Adding and Subtracting Rational Expressions

1) Factor the denominators.
2) Write down the LCD.
3) Rewrite each rational expression as an equivalent expression with the LCD.
4) Add or subtract the numerators and keep the common denominator in factored form.
5) After combining like terms in the numerator ask yourself, "Can I factor it?" If so, factor.
6) Reduce the rational expression, if possible. **(p. 386)**

Add $\dfrac{y}{y + 5} + \dfrac{4y - 30}{y^2 - 25}$.

1) Factor the denominator of $\dfrac{4y - 30}{y^2 - 25}$.

$$\dfrac{4y - 30}{y^2 - 25} = \dfrac{4y - 30}{(y + 5)(y - 5)}$$

2) The LCD is $(y + 5)(y - 5)$.

3) Rewrite $\dfrac{y}{y + 5}$ with the LCD.

$$\dfrac{y}{y + 5} \cdot \dfrac{y - 5}{y - 5} = \dfrac{y(y - 5)}{(y + 5)(y - 5)}$$

4) $\dfrac{y}{y + 5} + \dfrac{4y - 30}{y^2 - 25} = \dfrac{y(y - 5)}{(y + 5)(y - 5)} + \dfrac{4y - 30}{(y + 5)(y - 5)}$

$$= \dfrac{y(y - 5) + 4y - 30}{(y + 5)(y - 5)}$$

$$= \dfrac{y^2 - 5y + 4y - 30}{(y + 5)(y - 5)}$$

$$= \dfrac{y^2 - y - 30}{(y + 5)(y - 5)}$$

5) $\quad = \dfrac{(y + 5)(y - 6)}{(y + 5)(y - 5)}$ Factor.

6) $\quad = \dfrac{y - 6}{y - 5}$ Reduce.

7.3 Simplifying Complex Fractions

A **complex fraction** is a rational expression that contains one or more fractions in its numerator, its denominator, or both. **(p. 393)**

Some examples of complex fractions are

$$\dfrac{\dfrac{9}{10}}{\dfrac{3}{2}}, \quad \dfrac{\dfrac{b + 5}{2}}{\dfrac{4b + 20}{7}}, \quad \text{and} \quad \dfrac{\dfrac{1}{x} - \dfrac{1}{y}}{1 - \dfrac{x}{y}}.$$

Definition/Procedure	Example
To simplify a complex fraction containing one term in the numerator and one term in the denominator, 1) Rewrite the complex fraction as a division problem. 2) Perform the division by multiplying the first fraction by the reciprocal of the second. **(p. 393)**	Simplify $\dfrac{\dfrac{b+5}{2}}{\dfrac{4b+20}{7}}$. $$\dfrac{\dfrac{b+5}{2}}{\dfrac{4b+20}{7}} = \dfrac{b+5}{2} \div \dfrac{4b+20}{7}$$ $$= \dfrac{b+5}{2} \cdot \dfrac{7}{4(b+5)}$$ $$= \dfrac{\cancel{b+5}}{2} \cdot \dfrac{7}{4\cancel{(b+5)}}$$ $$= \dfrac{7}{8}$$
To simplify complex fractions containing more than one term in the numerator and/or the denominator, **Method 1** 1) Combine the terms in the numerator and combine the terms in the denominator so that each contains only one fraction. 2) Rewrite as a division problem. 3) Perform the division. **(p. 395)**	**Method 1** Simplify $\dfrac{\dfrac{1}{x} - \dfrac{1}{y}}{1 - \dfrac{x}{y}}$. 1) $\dfrac{\dfrac{1}{x} - \dfrac{1}{y}}{1 - \dfrac{x}{y}} = \dfrac{\dfrac{y}{xy} - \dfrac{x}{xy}}{\dfrac{y}{y} - \dfrac{x}{y}} = \dfrac{\dfrac{y-x}{xy}}{\dfrac{y-x}{y}}$ 2) $= \dfrac{y-x}{xy} \div \dfrac{y-x}{y}$ 3) $= \dfrac{\cancel{y-x}}{x\cancel{y}} \cdot \dfrac{\cancel{y}}{\cancel{y-x}} = \dfrac{1}{x}$
Method 2 1) Identify and write down the LCD of *all* of the fractions in the complex fraction. 2) Multiply the numerator and denominator of the complex fraction by the LCD. 3) Simplify. **(p. 396)**	**Method 2** Simplify $\dfrac{\dfrac{1}{x} - \dfrac{1}{y}}{1 - \dfrac{x}{y}}$. 1) LCD $= xy$ 2) $\dfrac{xy\left(\dfrac{1}{x} - \dfrac{1}{y}\right)}{xy\left(1 - \dfrac{x}{y}\right)}$ 3) $\dfrac{xy\left(\dfrac{1}{x} - \dfrac{1}{y}\right)}{xy\left(1 - \dfrac{x}{y}\right)} = \dfrac{xy \cdot \dfrac{1}{x} - x\cancel{y} \cdot \dfrac{1}{\cancel{y}}}{xy \cdot 1 - xy \cdot \dfrac{x}{y}}$ Distribute. $= \dfrac{y-x}{xy - x^2}$ Simplify. $= \dfrac{\cancel{y-x}}{x\cancel{(y-x)}}$ $= \dfrac{1}{x}$

Definition/Procedure	Example

7.4 Solving Rational Equations

An equation contains an = sign. To solve a rational equation, **multiply** the equation by the LCD to **eliminate** the denominators, then solve.

Always check the answer to be sure the proposed solution does not make a denominator equal zero. **(p. 402)**

Solve $\dfrac{n}{n+4} + 1 = \dfrac{20}{n+4}$.

This is an equation because it contains an = sign. We must eliminate the denominators. Identify the LCD of all of the expressions in the equation.

$$LCD = (n+4)$$

Multiply both sides of the equation by $(n+4)$.

$$(n+4)\left(\frac{n}{n+4} + 1\right) = (n+4)\left(\frac{20}{n+4}\right)$$

$$(n+4) \cdot \frac{n}{(n+4)} + (n+4) \cdot 1 = (n+4) \cdot \frac{20}{n+4}$$

$$n + n + 4 = 20$$
$$2n + 4 = 20$$
$$2n = 16$$
$$n = 8$$

The solution set is {8}. The check is left to the student.

Solving an Equation for a **Specific Variable (p. 407)**

Solve for n: $x = \dfrac{2a}{n+m}$.

Since we are solving for n, put it in a box.

$$x = \frac{2a}{\boxed{n} + m}$$

$$(\boxed{n} + m)x = (\boxed{n} + m) \cdot \frac{2a}{\boxed{n} + m}$$

$$(\boxed{n} + m)x = 2a$$

$$\boxed{n}\,x + mx = 2a$$

$$\boxed{n}\,x = 2a - mx$$

$$n = \frac{2a - mx}{x}$$

7.5 Applications of Rational Equations

Use the Steps for Solving Word Problems outlined in Section 2.1. **(p. 417)**

Write an equation and solve.

Dimos can put up the backyard pool in 6 hr, but it takes his father only 4 hr to put up the pool. How long would it take the two of them to put up the pool together?

Step 1: **Read** the problem carefully.

Step 2: **Choose** a variable to represent the unknown.

t = number of hours to put up the pool together.

Step 3: **Translate** from English to an algebraic equation.

$$\text{Dimos' rate} = \frac{1}{6}\text{ pool/hr} \quad \text{Father's rate} = \frac{1}{4}\text{ pool/hr}$$

$$\text{Fractional part} = \text{rate} \cdot \text{time}$$

$$\text{Dimos' part} = \frac{1}{6} \cdot t = \frac{1}{6}t$$

$$\text{Father's part} = \frac{1}{4} \cdot t = \frac{1}{4}t$$

Definition/Procedure	**Example**

$$\underset{\text{job by Dimos}}{\text{Fractional}} + \underset{\text{job by his father}}{\text{Fractional}} = 1 \text{ whole job}$$

$$\frac{1}{6}t \quad + \quad \frac{1}{4}t \quad = \quad 1$$

Equation: $\frac{1}{6}t + \frac{1}{4}t = 1.$

Step 4: Solve the equation.

$$12\left(\frac{1}{6}t + \frac{1}{4}t\right) = 12(1) \qquad \text{Multiply by 12, the LCD.}$$

$$12 \cdot \frac{1}{6}t + 12 \cdot \frac{1}{4}t = 12(1) \qquad \text{Distribute.}$$

$$2t + 3t = 12 \qquad \text{Multiply.}$$

$$5t = 12$$

$$t = \frac{12}{5}$$

Step 5: Interpret the solution as it relates to the problem.

Dimos and his father could put up the pool together in $\frac{12}{5}$ hr or $2\frac{2}{5}$ hr. The check is left to the student.

7.6 Variation

Direct Variation

y varies directly as x (or **y is directly proportional to x**) means

$$y = kx$$

where *k* is a nonzero real number. *k* is called the **constant of variation. (p. 424)**

The circumference, *C*, of a circle is given by $C = 2\pi r$. *C* varies directly as *r*, where $k = 2\pi$.

Inverse Variation

y varies inversely as x (or **y is inversely proportional to x**) means

$$y = \frac{k}{x}$$

where *k* is a nonzero real number. **(p. 426)**

The time, *t* (in hours), it takes to drive 600 mi is inversely proportional to the rate, *r*, at which you drive.

$$t = \frac{600}{r}$$

where $k = 600$.

Joint Variation

y varies jointly as x and z means $y = kxz$, where *k* is a nonzero real number. **(p. 428)**

For a given amount, called the principal, deposited in a bank account, the interest earned, *I*, varies jointly as the interest rate, *r*, and the time, *t*, the principal is in the account.

$$I = 1000rt$$

$k = 1000$, the principal.

Definition/Procedure	Example
Combined Variation A **combined variation** problem involves both direct and inverse variation. **(p. 429)**	The resistance of a wire, R, varies directly as its length, L, and inversely as its cross-sectional area, A. $$R = \frac{0.002L}{A}$$ The constant of variation, k, is 0.002. This is the resistivity of the material from which the wire was made.
Steps for Solving a Variation Problem 1) Write the *general* variation equation. 2) Find k by substituting the known values into the equation and solving for k. 3) Write the *specific* variation equation by substituting the value of k into the *general* variation equation. 4) Use the specific variation equation to solve the problem. **(p. 425)**	The cost of manufacturing a certain soccer ball is inversely proportional to the number produced. When 15,000 are made, the cost per ball is $4.00. What is the cost to manufacture each soccer ball when 25,000 are produced? Let n = number of soccer balls produced $\quad C$ = cost of producing each ball 1) *General* variation equation: $C = \dfrac{k}{n}$ 2) Find k using $C = 4$ when $n = 15{,}000$. $$4 = \frac{k}{15{,}000}$$ $$60{,}000 = k$$ 3) *Specific* variation equation: $C = \dfrac{60{,}000}{n}$ 4) Find the cost, C, per ball when $n = 25{,}000$. $$C = \frac{60{,}000}{25{,}000} \qquad \text{Substitute 25,000 for } n.$$ $$C = 2.4$$ The cost per ball is $2.40.

*Additional answers can be found in the Answers to Exercises appendix.

(7.1) Let $f(x) = \dfrac{P(x)}{Q(x)}$ be a rational function.

1) How do we determine the domain of $f(x)$?

2) How do we determine where $f(x) = 0$?
Set $P(x) = 0$, and solve for x.

For each rational function,
a) **find** $f(5)$.
b) **find** x **so that** $f(x) = 0$.
c) **determine the domain of the function.**

3) $f(x) = \dfrac{x + 9}{5x - 1}$ a) $\dfrac{7}{12}$ b) -9 c) $\left(-\infty, \dfrac{1}{5}\right) \cup \left(\dfrac{1}{5}, \infty\right)$

4) $f(x) = \dfrac{8}{x^2 - 100}$ a) $-\dfrac{8}{75}$ b) never equals zero
  c) $(-\infty, -10) \cup (-10, 10) \cup (10, \infty)$

Determine the domain of each rational function.

5) $h(a) = \dfrac{9a}{a^2 - 2a - 24}$ $(-\infty, -4) \cup (-4, 6) \cup (6, \infty)$

6) $k(t) = \dfrac{6t - 1}{t^2 + 7}$ $(-\infty, \infty)$

Write each rational expression in lowest terms.

7) $\dfrac{63a^2}{9a^{11}}$ $\dfrac{7}{a^9}$

8) $\dfrac{15c - 55}{33c - 121}$ $\dfrac{5}{11}$

9) $\dfrac{2z - 7}{6z^2 - 19z - 7}$ $\dfrac{1}{3z + 1}$

10) $\dfrac{10 - x}{x^2 - 100}$ $-\dfrac{1}{x + 10}$

11) $\dfrac{y^2 + 9y - yz - 9z}{yz - 12y - z^2 + 12z}$ $\dfrac{y + 9}{z - 12}$

12) Find three equivalent forms of $-\dfrac{u - 6}{u + 2}$.

Find the missing side in each rectangle.

13) Area $= 2l^2 - 5l - 3$

$l - 3$

Find the length. $2l + 1$

14) Area $= 3b^2 + 17b + 20$

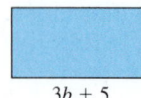

$3b + 5$

Find the width. $b + 4$

Perform the operations and simplify.

15) $\dfrac{16k^4}{3m^2} \div \dfrac{4k^2}{27m}$ $36k^2 \over m$

16) $\dfrac{t + 4}{6} \cdot \dfrac{3(t - 2)}{(t + 4)^2}$ $\dfrac{t - 2}{2(t + 4)}$

17) $\dfrac{6w - 1}{6w^2 + 5w - 1} \cdot \dfrac{3w + 3}{12w}$ $\dfrac{1}{4w}$

18) $\dfrac{3x^2 + 14x + 16}{15x + 40} \div \dfrac{11x + 22}{x - 5}$ $\dfrac{x - 5}{55}$

19) $\dfrac{25 - a^2}{4a^2 + 12a} \div \dfrac{a^3 - 125}{a^2 + 3a} - \dfrac{a + 5}{4(a^2 + 5a + 25)}$

20) $\dfrac{3p^5}{20q^2} \cdot \dfrac{4q^3}{21p^7} \quad \dfrac{q}{35p^2}$

(7.2) Find the LCD of each group of fractions.

21) $\dfrac{2}{k^2}, \dfrac{9}{k} \quad k^2$

22) $\dfrac{4}{9x^2y}, \dfrac{13}{4xy^4} \quad 36x^2y^4$

23) $\dfrac{3}{m}, \dfrac{4}{m + 5} \quad m(m + 5)$

24) $\dfrac{8}{2d^2 - d}, \dfrac{11}{6d - 3} \quad 3d(2d - 1)$

25) $\dfrac{1}{3x + 7}, \dfrac{6x}{x - 9}$
$(3x + 7)(x - 9)$

26) $\dfrac{9}{2 - b}, \dfrac{4b}{b - 2}$
$b - 2$ or $2 - b$

27) $\dfrac{5c - 1}{c^2 + 10c + 24}, \dfrac{9c}{c^2 - 3c - 28}$ $(c + 4)(c + 6)(c - 7)$

28) $\dfrac{6}{x^2 + 8x}, \dfrac{1}{3x^2 + 24x}, \dfrac{13}{x^2 + 16x + 64}$ $3x(x + 8)^2$

29) $\dfrac{3}{a^2 - 13a + 40}, \dfrac{a + 12}{a^2 - 7a - 8}, \dfrac{1}{a^2 - 4a - 5}$
$(a - 5)(a - 8)(a + 1)$

30) $\dfrac{8c}{c^2 - d^2}, \dfrac{d}{d - c}$ $(c + d)(c - d)$ or $(c + d)(d - c)$

Rewrite each rational expression with the indicated denominator.

31) $\dfrac{6}{5r} = \dfrac{24r^2}{20r^3} \quad \dfrac{24r^2}{20r^3}$

32) $\dfrac{8}{3z + 4} = \dfrac{8z}{z(3z + 4)} \quad \dfrac{8z}{z(3z + 4)}$

33) $\dfrac{t - 3}{2t + 1} = \dfrac{t^2 + 2t - 15}{(2t + 1)(t + 5)} \quad \dfrac{t^2 + 2t - 15}{(2t + 1)(t + 5)}$

34) $\dfrac{n}{4 - n} = \dfrac{n}{n - 4} \quad -\dfrac{n}{n - 4}$

Identify the LCD of each group of fractions, and rewrite each as an equivalent fraction with the LCD as its denominator.

35) $\dfrac{3}{8a^3b}, \dfrac{6}{5ab^5}$

36) $\dfrac{8}{p + 7}, \dfrac{2}{p}$

37) $\dfrac{9c}{c^2 + 6c - 16}, \dfrac{4}{c^2 - 4c + 4}$

38) $\dfrac{7}{2r^2 - 12r}, \dfrac{3r}{36 - r^2}, \dfrac{r - 5}{2r^2 + 12r}$

Add or subtract, as indicated.

39) $\dfrac{3}{8c} + \dfrac{7}{8c} \quad \dfrac{5}{4c}$

40) $\dfrac{4m}{m - 3} - \dfrac{5}{m - 3} \quad \dfrac{4m - 5}{m - 3}$

41) $\dfrac{2}{5z^2} + \dfrac{9}{10z} \quad \dfrac{4 + 9z}{10z^2}$

42) $\dfrac{n}{2n - 5} - \dfrac{4}{n} \quad \dfrac{n^2 - 8n + 20}{n(2n - 5)}$

43) $\dfrac{5}{y-2} - \dfrac{6}{y+3}$ $\dfrac{27-y}{(y-2)(y+3)}$

44) $\dfrac{8d-3}{d^2-3d-28} + \dfrac{2d}{5d-35}$ $\dfrac{2d^2+48d-15}{5(d-7)(d+4)}$

45) $\dfrac{10p+3}{4p+4} - \dfrac{8}{p^2-6p-7}$ $\dfrac{10p^2-67p-53}{4(p+1)(p-7)}$

46) $\dfrac{k-3}{k^2+14k+49} - \dfrac{2}{k^2+7k}$ $\dfrac{(k-7)(k+2)}{k(k+7)^2}$

47) $\dfrac{2}{m-11} + \dfrac{19}{11-m}$ $\dfrac{17}{11-m}$ or $-\dfrac{17}{m-11}$

48) $\dfrac{1}{8-r} + \dfrac{16}{r^2-64}$ $-\dfrac{1}{r+8}$

49) $\dfrac{x^2}{x^2-y^2} + \dfrac{x}{y-x}$ $-\dfrac{xy}{(x+y)(x-y)}$

50) $\dfrac{8}{w^2+7w} + \dfrac{3w}{w+7} + \dfrac{2}{5w}$ $\dfrac{15w^2+2w+54}{5w(w+7)}$

51) $\dfrac{3}{g^2-7g} + \dfrac{2g}{5g-35} - \dfrac{6}{5g}$ $\dfrac{2g^2-6g+57}{5g(g-7)}$

52) $\dfrac{d+4}{d^2+2d} + \dfrac{d}{5d^2+7d-6} - \dfrac{10}{5d^2-3d}$ $\dfrac{6d^2+7d-32}{d(d+2)(5d-3)}$

For each rectangle, find a rational expression in simplest form to represent its a) area and b) perimeter.

53) $\dfrac{x}{8}$, $\dfrac{12}{x-4}$
a) $\dfrac{3x}{2(x-4)}$ sq units
b) $\dfrac{x^2-4x+96}{4(x-4)}$ units

54) $\dfrac{2}{x^2}$, $\dfrac{x}{x+1}$
a) $\dfrac{2}{x(x+1)}$ sq units
b) $\dfrac{2x^3+4x+4}{x^2(x+1)}$ units

For Exercises 55 and 56, let $f(x)=\dfrac{5x+3}{x-2}$ and $g(x)=\dfrac{4}{x}$.

55) Find $(f+g)(x)$ and its domain.

56) Find $\left(\dfrac{g}{f}\right)(x)$ and its domain.

(7.3) Simplify completely.

57) $\dfrac{\frac{2}{5}}{\frac{7}{15}}$ $\dfrac{6}{7}$

58) $\dfrac{\frac{f}{g}}{\frac{f^2}{g}}$ $\dfrac{1}{f}$

59) $\dfrac{p+\frac{6}{p}}{\frac{8}{p}+p}$ $\dfrac{p^2+6}{p^2+8}$

60) $\dfrac{\frac{a}{b}-\frac{2a}{b^2}}{\frac{4}{ab}-\frac{a}{b}}$ $\dfrac{a^2(b-2)}{b(2+a)(2-a)}$

61) $\dfrac{\frac{n}{6n+48}}{\frac{n^2}{4n+32}}$ $\dfrac{2}{3n}$

62) $\dfrac{\frac{2}{3}-\frac{4}{5}}{\frac{1}{6}+\frac{1}{2}}$ $-\dfrac{1}{5}$

63) $\dfrac{1-\frac{1}{y-9}}{\frac{2}{y+3}+1}$ $\dfrac{(y+3)(y-10)}{(y-9)(y+5)}$

64) $\dfrac{\frac{4q}{7q+63}}{\frac{q^2}{8q+72}}$ $\dfrac{32}{7q}$

65) $\dfrac{\frac{c}{c+2}+\frac{1}{c^2-4}}{1-\frac{3}{c+2}}$ $\dfrac{c-1}{c-2}$

66) $\dfrac{1+\frac{b}{a-b}}{\frac{b}{a^2-b^2}+\frac{1}{a+b}}$ $a+b$

67) $\dfrac{2x^{-2}+y^{-1}}{x^{-1}-y^{-2}}$ $\dfrac{y(x^2+2y)}{x(y^2-x)}$

68) $\dfrac{12a^{-1}}{4a+b^{-2}}$ $\dfrac{12b^2}{a(4ab^2+1)}$

(7.4) Solve each equation.

69) $\dfrac{5w}{6}-\dfrac{1}{2}=-\dfrac{1}{6}$ $\left\{\dfrac{2}{5}\right\}$

70) $\dfrac{4}{y-6}=\dfrac{12}{y+2}$ $\{10\}$

71) $\dfrac{r}{r+6}+3=\dfrac{10}{r+6}$ $\{-2\}$

72) $\dfrac{3}{x-5}+\dfrac{2}{2x+1}=\dfrac{1}{2x^2-9x-5}$ $\{1\}$

73) $\dfrac{16}{9t-27}+\dfrac{2t-4}{t-3}=\dfrac{t}{9}$ $\{1,20\}$

74) $\dfrac{5}{j+4}+\dfrac{j}{j-2}=\dfrac{2j^2-2j}{j^2+2j-8}$ $\{1,10\}$

75) $\dfrac{3}{b+2}=\dfrac{16}{b^2-4}-\dfrac{4}{b-2}$ $\varnothing$

76) $\dfrac{3k}{k+9}=\dfrac{3}{k+1}$ $\{-3,3\}$

77) $\dfrac{c}{c^2+3c-28}-\dfrac{5}{c^2+15c+56}=\dfrac{5}{c^2+4c-32}$ $\{-3,5\}$

78) $\dfrac{a}{a^2-1}+\dfrac{4}{a^2+9a+8}=\dfrac{8}{a^2+7a-8}$ $\{-6,2\}$

Solve for the indicated variable.

79) $A=\dfrac{2p}{c}$ for c $c=\dfrac{2p}{A}$

80) $R=\dfrac{s+T}{D}$ for D $D=\dfrac{s+T}{R}$

81) $n=\dfrac{t}{a+b}$ for a $a=\dfrac{t-nb}{n}$

82) $w=\dfrac{N}{c-ak}$ for k $k=\dfrac{wc-N}{aw}$

83) $\dfrac{1}{r}=\dfrac{1}{s}+\dfrac{1}{t}$ for s $s=\dfrac{rt}{t-r}$

84) $\dfrac{1}{R_1}+\dfrac{1}{R_2}=\dfrac{1}{R_3}$ for R_1 $R_1=\dfrac{R_2R_3}{R_2-R_3}$

(7.5) Write an equation and solve.

85) The ratio of saturated fat to total fat in a Starbucks tall Caramel Frappuccino is 2 to 3. If there are 4 more grams of total fat in the drink than there are grams of saturated fat, how much total fat is in a Caramel Frappuccino? (Starbucks brochure)   12 g

86) A boat can travel 9 mi downstream in the same amount of time it can travel 6 mi upstream. If the speed of the boat in still water is 10 mph, find the speed of the current.   2 mph

87) When the wind is blowing at 40 mph, a plane flying at a constant speed can travel 800 mi with the wind in the same amount of time it can fly 600 mi against the wind. Find the speed of the plane.   280 mph

88) Wayne can clean the carpets in his house in 4 hr, but it would take his son, Garth, 6 hr to clean them on his own. How long would it take both of them to clean the carpets together? $2\frac{2}{5}$ hr

(7.6)

89) Suppose c varies directly as m. If $c = 56$ when $m = 8$, find c when $m = 3$.   21

90) Suppose A varies jointly as t and r. If $A = 15$ when $t = \dfrac{1}{2}$ and $r = 5$, find A when $t = 3$ and $r = 4$.   72

91) Suppose z varies inversely as the cube of w. If $z = 16$ when $w = 2$, find z when $w = 4$.   2

92) Suppose p varies directly as n and inversely as the square of d. If $p = 42$ when $n = 7$ and $d = 2$, find p when $n = 12$ and $d = 3$.   32

Solve each problem by writing a variation equation.

93) The weight of a ball varies directly as the cube of its radius. If a ball with a radius of 2 in. weighs 0.96 lb, how much would a ball made out of the same material weigh if it had a radius of 3 in.?   3.24 lb

94) If the temperature remains the same, the volume of a gas is inversely proportional to the pressure. If the volume of a gas is 10L (liters) at a pressure of 1.25 atm (atmospheres), what is the volume of the gas at 2 atm?   63.125L

*Additional answers can be found in the Answers to Exercises appendix.

1) $f(x) = \dfrac{x^2 + 4}{x^2 - 2x - 48}$

   a) Find $f(-2)$. $-\dfrac{1}{5}$

   b) Find x so that $f(x) = 0$. $f(x)$ never equals zero

   c) Determine the domain of the function.
   $(-\infty, -8) \cup (-8, 6) \cup (6, \infty)$

2) Determine the domain of $g(x) = \dfrac{x + 9}{2x + 3}$.
  $\left(-\infty, -\dfrac{3}{2}\right) \cup \left(-\dfrac{3}{2}, \infty\right)$

Write each rational expression in lowest terms.

3) $\dfrac{54w^3}{24w^8}$ $\dfrac{9}{4w^5}$

4) $\dfrac{7v^2 + 55v - 8}{v^2 - 64}$ $\dfrac{7v - 1}{v - 8}$

5) Write three equivalent forms of $\dfrac{9 - h}{2h - 3}$.

6) If three rational expressions have denominators of k, $k^2 + 4k + 4$, and $2k^2 + k - 6$, find their least common denominator. $k(2k - 3)(k + 2)^2$

Perform the operations and simplify.

7) $\dfrac{7}{12z} + \dfrac{5}{12z}$ $\dfrac{1}{z}$

8) $\dfrac{21m^4}{n} \div \dfrac{12m^8}{n^3}$ $\dfrac{7n^2}{4m^4}$

9) $\dfrac{r}{2r + 1} + \dfrac{3}{r + 5}$ $\dfrac{r^2 + 11r + 3}{(2r + 1)(r + 5)}$

10) $\dfrac{a^3 - 8}{6a - 66} \cdot \dfrac{a^2 - 9a - 22}{4 - a^2}$ $-\dfrac{a^2 + 2a + 4}{6}$

11) $\dfrac{c - 3}{c - 15} + \dfrac{c + 8}{15 - c}$ $\dfrac{11}{15 - c}$ or $-\dfrac{11}{c - 15}$

12) $\dfrac{x}{x^2 - 49} - \dfrac{3}{x^2 - 2x - 63}$ $\dfrac{x^2 - 12x + 21}{(x + 7)(x - 7)(x - 9)}$

13) Let $f(x) = \dfrac{2}{x}$ and $g(x) = \dfrac{x - 5}{x + 7}$. If $h(x) = f(x) - g(x)$, find $h(x)$ in its simplest form and determine its domain.
  $h(x) = \dfrac{-x^2 + 7x + 14}{x(x + 7)}$; Domain: $(-\infty, -7) \cup (-7, 0) \cup (0, \infty)$

Simplify completely.

14) $\dfrac{1 + \dfrac{2}{d - 3}}{\dfrac{-2d}{d - 3} - d}$ $-\dfrac{1}{d}$

15) $\dfrac{\dfrac{15}{7}}{\dfrac{20}{21}}$ $\dfrac{9}{4}$

16) $\dfrac{\dfrac{1}{x} - \dfrac{1}{y}}{\dfrac{1}{y^2} - \dfrac{1}{x^2}}$ $\dfrac{xy}{x + y}$

17) Write an expression for the base of the triangle if the area is given by $12k^2 + 28k$.

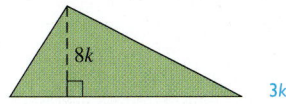

  $8k$ $3k + 7$

18) Find all values that cannot be solutions to the equation $\dfrac{3}{5} - \dfrac{x + 2}{4x - 1} = \dfrac{7}{x}$. Do not solve the equation. $0, \dfrac{1}{4}$

Solve each equation.

19) $\dfrac{7t}{12} + \dfrac{t - 4}{6} = \dfrac{7}{3}$ $\{4\}$

20) $\dfrac{30}{x^2 - 9} = \dfrac{5}{x - 3} - \dfrac{2}{x + 3}$ $\varnothing$

21) $\dfrac{5}{n^2 + 10n + 24} + \dfrac{5}{n^2 + 3n - 18} = \dfrac{n}{n^2 + n - 12}$ $\{-1, 5\}$

22) Solve $y = \dfrac{kxz}{c}$ for c. $c = \dfrac{kxz}{y}$

23) Solve $\dfrac{1}{p} + \dfrac{1}{q} = \dfrac{1}{r}$ for p. $p = \dfrac{qr}{q - r}$

Write an equation for each and solve.

24) A river flows at 3 mph. If a boat can travel 16 mi downstream in the same amount of time it can go 10 mi upstream, find the speed of the boat in still water.   13 mph

25) Suppose n varies jointly as r and the square of s. If $n = 72$ when $r = 2$ and $s = 3$, find n when $r = 3$ and $s = 5$.   300

26) The loudness of a sound is inversely proportional to the square of the distance between the source of the sound and the listener. If the sound level measures 112.5 dB 4 ft from a speaker, how loud is the sound 10 ft from the speaker?   18 dB

Cumulative Review: Chapters 1–7

Additional answers can be found in the Answers to Exercises appendix.

Simplify. Assume all variables represent nonzero real numbers. The answers should not contain negative exponents.

1) $-5(3w^4)^2$ $-45w^8$

2) $\left(\dfrac{2n^{-10}}{n^{-4}}\right)^3$ $\dfrac{8}{n^{18}}$

Solve.

3) $-\dfrac{12}{5}c - 7 = 20$ $\left\{-\dfrac{45}{4}\right\}$

4) *Write an equation in one variable and solve.*
   How many milliliters of a 12% alcohol solution and how many milliliters of a 4% alcohol solution must be mixed to obtain 60 ml of a 10% alcohol solution?
   45 ml of 12% solution; 15 ml of 4% solution

5) Find the x- and y-intercepts of $5x - 2y = 10$ and sketch a graph of the equation. x-int: $(2, 0)$; y-int: $(0, -5)$

6) Graph $f(x) = -\dfrac{1}{3}x + 2$.

7) Write an equation of the line containing the points $(-5, 8)$ and $(1, 2)$. Express the answer in standard form. $x + y = 3$

8) Solve this system using the elimination method. $(2, 10)$
   $-7x + 2y = 6$
   $9x - y = 8$

9) *Write a system of two equations in two variables and solve.*
   The length of a rectangle is 7 cm less than twice its width. The perimeter of the rectangle is 76 cm. What are the dimensions of the figure?   width: 15 cm; length: 23 cm

10) Solve the compound inequality and write the answer in interval notation. $3y + 16 < 4$ or $8 - y \geq 7$ $(-\infty, 1]$

11) Solve $|6p + 13| = 8$. $\left\{-\dfrac{7}{2}, -\dfrac{5}{6}\right\}$

12) Graph $2x - y > 4$.

13) Graph $x \geq -3$ and $y \geq -\dfrac{1}{2}x - 1$.

14) $h(t) = 2t^2 - 11t + 4$
   a) Find $h(3)$. -11
   b) Find t so that $h(t) = -8$. $\dfrac{3}{2}$ or 4

Perform the indicated operation(s).

15) $(4q^2 + 11q - 2) - 3(6q^2 - 5q + 4) + 2(-10q - 3)$
   $-14q^2 + 6q - 20$

16) $(3d^2 - 7)(4d^2 + 6d - 1)$ $12d^4 + 18d^3 - 31d^2 - 42d + 7$

17) $\dfrac{8a^4b^4 - 20a^3b^2 + 56ab + 8b}{8a^3b^3}$ $ab - \dfrac{5}{2b} + \dfrac{7}{a^2b^2} + \dfrac{1}{a^3b^2}$

18) $\dfrac{17v^3 - 22v + 7v^4 + 24 - 47v^2}{7v - 4}$ $v^3 + 3v^2 - 5v - 6$

19) $\dfrac{7}{c - 6} - \dfrac{4}{c}$ $\dfrac{3(c + 8)}{c(c - 6)}$

Factor completely.

20) $25n^2 - 81$ $(5n + 9)(5n - 9)$

21) $3xy^2 + 15xy - 72x$ $3x(y + 8)(y - 3)$

22) $r^2 + 8rt + 16t^2$ $(r + 4t)^2$

23) Determine the domain of $g(a) = \dfrac{a + 3}{8 - a}$. $(-\infty, 8) \cup (8, \infty)$

24) Simplify completely.

   a) $\dfrac{\dfrac{12k^4}{8k^3 - 27}}{\dfrac{9k^5}{3 - 2k}}$ $\dfrac{4}{3k(4k^2 + 6k + 9)}$

   b) $\dfrac{1 + \dfrac{2}{r}}{\dfrac{1}{r} + \dfrac{r}{r + 3}}$ $\dfrac{(r + 3)(r + 2)}{r^2 + r + 3}$

Write an equation and solve.

25) Leticia can assemble a swing set in 3 hr while it takes Betty 5 hr. How long would it take for them to assemble the swing set together? $1\dfrac{7}{8}$ hr

Radicals and Rational Exponents

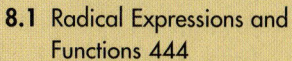

Algebra at Work: Forensics

Forensic scientists use mathematics in many ways to help them analyze evidence and solve crimes. To help him reconstruct an accident scene, Keith can use this formula containing a radical to estimate the minimum speed of a vehicle when the accident occurred:

$$S = \sqrt{30fd}$$

where f = the drag factor, based on the type of road surface

d = the length of the skid, in feet

S = the speed of the vehicle in miles per hour

Keith is investigating an accident in a residential neighborhood where the speed limit is 25 mph. The car involved in the accident left skid marks 60 ft long. Tests showed that the drag factor of the asphalt road was 0.80. Was the driver speeding at the time of the accident?

Substitute the values into the equation and evaluate it to determine the minimum speed of the vehicle at the time of the accident:

$$S = \sqrt{30fd}$$
$$S = \sqrt{30(0.80)(60)}$$
$$S = \sqrt{1440} \approx 38 \text{ mph}$$

The driver was going at least 38 mph when the accident occurred. This is well over the speed limit of 25 mph.

We will learn how to simplify radicals in this chapter as well as how to work with equations like the one given here.

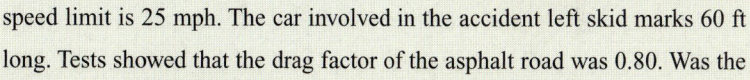

Section 8.1 Radical Expressions and Functions

Objectives

1. Find Square Roots and Principal Square Roots
2. Find Higher Roots
3. Evaluate $\sqrt[n]{a^n}$
4. Determine the Domains of Square Root and Cube Root Functions
5. Graph a Square Root Function
6. Graph a Cube Root Function

Recall that exponential notation represents repeated multiplication. For example,

$$3^2 \text{ means } 3 \cdot 3, \text{ so } 3^2 = 9.$$
$$2^4 \text{ means } 2 \cdot 2 \cdot 2 \cdot 2, \text{ so } 2^4 = 16.$$

In this chapter we will study the opposite, or inverse, procedure, finding **roots** of numbers.

1. Find Square Roots and Principal Square Roots

 **Example 1**

Find all square roots of 25.

Solution

To find a *square* root of 25 ask yourself, "What number do I *square* to get 25?" Or, "What number *multiplied by itself* equals 25?" One number is 5 since $5^2 = 25$. Another number is -5 since $(-5)^2 = 25$. So, 5 and -5 are square roots of 25.

 **You Try 1**

Find all square roots of 64.

The $\sqrt{}$ symbol represents the *positive* square root, or the **principal square root,** of a nonnegative number. For example, $\sqrt{25} = 5$.

BE CAREFUL

Notice that $\sqrt{25} = 5$, but $\sqrt{25} \neq -5$. The $\sqrt{}$ symbol represents *only* the principal square root (positive square root).

To find the **negative square root** of a nonnegative number we must put a $-$ in front of the $\sqrt{}$. For example, $-\sqrt{25} = -5$.

Next we will define some terms associated with the $\sqrt{}$ symbol.

The symbol $\sqrt{}$ is the **square root symbol** or the **radical sign.** The number under the radical sign is the **radicand.** The entire expression, $\sqrt{25}$, is called a **radical.**

$$\text{Radical sign} \rightarrow \underline{\sqrt{25}} \leftarrow \text{Radicand}$$
$$\text{Radical}$$

Example 2

Find each square root, if possible.

a) $\sqrt{100}$ b) $-\sqrt{16}$ c) $\sqrt{\dfrac{4}{25}}$ d) $-\sqrt{\dfrac{81}{49}}$ e) $\sqrt{-9}$

Solution

a) $\sqrt{100} = 10$ since $(10)^2 = 100$.

b) $-\sqrt{16}$ means $-1 \cdot \sqrt{16}$. Therefore, $-\sqrt{16} = -1 \cdot \sqrt{16} = -1 \cdot 4 = -4$.

c) $\sqrt{\dfrac{4}{25}} = \dfrac{2}{5}$ since $\left(\dfrac{2}{5}\right)^2 = \dfrac{4}{25}$.

In-Class Example 1
Find all square roots of 81.
answer: −9, 9

In-Class Example 2
Find each square root, if possible.
a) $\sqrt{16}$ b) $-\sqrt{9}$ c) $\sqrt{\dfrac{36}{49}}$
d) $-\sqrt{\dfrac{100}{9}}$ e) $\sqrt{-64}$
answer: a) 4 b) −3
c) $\dfrac{6}{7}$ d) $-\dfrac{10}{3}$
e) not a real number

d) $-\sqrt{\dfrac{81}{49}}$ means $-1 \cdot \sqrt{\dfrac{81}{49}}$. So, $-\sqrt{\dfrac{81}{49}} = -1 \cdot \sqrt{\dfrac{81}{49}} = -1 \cdot \dfrac{9}{7} = -\dfrac{9}{7}$.

e)   To find $\sqrt{-9}$, ask yourself, "What number do I *square* to get -9?" *There is no such real number* since $3^2 = 9$ and $(-3)^2 = 9$. Therefore, $\sqrt{-9}$ is not a real number.   ■

You Try 2

Find each square root.

a) $\sqrt{9}$ b) $-\sqrt{144}$ c) $\sqrt{\dfrac{25}{36}}$ d) $-\sqrt{\dfrac{1}{64}}$ e) $\sqrt{-49}$

In Example 2, we found the **principal square roots** of 100 and $\dfrac{4}{25}$ and the **negative square roots** of 16 and $\dfrac{81}{49}$.

Let's review what we know about radicands and add a third fact.

Property   Radicands and Square Roots

1)   If the radicand is a *perfect square*, the square root is a *rational* number.

   Example: $\sqrt{16} = 4$ 16 is a perfect square. $\sqrt{\dfrac{100}{49}} = \dfrac{10}{7}$ $\dfrac{100}{49}$ is a perfect square.

2)   If the radicand is a *negative number*, the square root is *not* a real number.

   Example: $\sqrt{-25}$ is *not* a real number.

3)   If the radicand is *positive and not a perfect square*, then the square root is an *irrational* number.

   Example: $\sqrt{13}$ is irrational.   13 is not a perfect square.

The square root of such a number is a real number that is a nonrepeating, nonterminating decimal.

Sometimes, we must plot points containing radicals. For the purposes of graphing, approximating a radical to the nearest tenth is sufficient. A calculator with a $\sqrt{}$ key will give a better approximation of the radical.

---

2. Find Higher Roots

We saw in Example 2a) that $\sqrt{100} = 10$ since $(10)^2 = 100$. We can also find higher roots of numbers like $\sqrt[3]{a}$ (read as "the cube root of a"), $\sqrt[4]{a}$ (read as "the fourth root of a"), $\sqrt[5]{a}$, etc. We will look at a few roots of numbers before learning important rules.

Example 3

In-Class Example 3
Find each root.
a) $\sqrt[3]{27}$ b) $\sqrt[3]{64}$
answer: a) 3   b) 4

Find each root.

a) $\sqrt[3]{125}$ b) $\sqrt[5]{32}$

Solution

a)   To find $\sqrt[3]{125}$ (the cube root of 125) ask yourself, "What number do I *cube* to get 125?" That number is 5.

$$\sqrt[3]{125} = 5 \text{ since } 5^3 = 125$$

Finding the cube root of a number is the *opposite,* or *inverse* procedure, of cubing a number.

b)   To find $\sqrt[5]{32}$ (the fifth root of 32) ask yourself, "What number do I raise to the *fifth power* to get 32?" That number is 2.

$$\sqrt[5]{32} = 2 \text{ since } 2^5 = 32$$

Finding the fifth root of a number and raising a number to the fifth power are *opposite,* or *inverse,* procedures. ∎

You Try 3

Find each root.

a) $\sqrt[3]{27}$ b) $\sqrt[3]{8}$

The symbol $\sqrt[n]{a}$ is read as "the *nth* root of *a*." If $\sqrt[n]{a} = b$, then $b^n = a$.

Index → $\sqrt[n]{a}$ ← Radicand

Radical

We call *n* the **index** of the radical.

Note

When finding *square* roots we do not write $\sqrt[2]{a}$. The square root of *a* is written as $\sqrt{a}$, and the index is understood to be 2.

We know that a positive number, say 36, has a principal square root ($\sqrt{36}$, or 6) and a negative square root ($-\sqrt{36}$, or -6). This is true for all even roots of positive numbers: square roots, fourth roots, sixth roots, and so on. For example, 81 has a principal fourth root ($\sqrt[4]{81}$, or 3) and a negative fourth root ($-\sqrt[4]{81}$, or -3).

Definition

nth Root

For any *positive* number *a* and any *even* index *n*,
　　the **principal nth root** of *a* is $\sqrt[n]{a}$.
　　the **negative nth root** of *a* is $-\sqrt[n]{a}$.

$\overset{even}{\sqrt{}}\,\text{positive} = \text{principal (positive) root}$
$-\overset{even}{\sqrt{}}\,\text{positive} = \text{negative root}$

For any *negative* number *a* and any *even* index *n*,
　　there is **no** real *nth* root of *a*.

$\overset{even}{\sqrt{}}\,\text{negative} = \text{no real root}$

For any number *a* and any *odd* index *n*,
　　there is **one** real *nth* root of *a*, $\sqrt[n]{a}$.

$\overset{odd}{\sqrt{}}\,\text{any number} = \text{exactly one root}$

BE CAREFUL

The definition means that $\sqrt[4]{81}$ cannot be -3 because $\sqrt[4]{81}$ is *defined* as the principal fourth root of 81, which must be positive. $\sqrt[4]{81} = 3$

Example 4

In-Class Example 4
Find each root, if possible.
a) $\sqrt[3]{-27}$ b) $-\sqrt[4]{81}$
c) $\sqrt[4]{-81}$ d) $\sqrt[3]{-125}$
e) $\sqrt[6]{64}$
answer: a) -3 b) -3
c) not a real number
d) -5 e) 2

Find each root, if possible.

a) $\sqrt[4]{16}$ b) $-\sqrt[4]{16}$ c) $\sqrt[4]{-16}$ d) $\sqrt[3]{64}$ e) $\sqrt[3]{-64}$

Solution

a) To find $\sqrt[4]{16}$ ask yourself, "What *positive* number do I raise to the *fourth power* to get 16?" Since $2^4 = 16$ and 2 is positive, $\sqrt[4]{16} = 2$.

b) In part a) we found that $\sqrt[4]{16} = 2$, so $-\sqrt[4]{16} = -(\sqrt[4]{16}) = -2$.

c) To find $\sqrt[4]{-16}$ ask yourself, "What number do I raise to the *fourth power* to get -16?" There is no such real number since $2^4 = 16$ and $(-2)^4 = 16$. Therefore, $\sqrt[4]{-16}$ has *no real root*. (Recall from the definition that $\overset{\text{even}}{\sqrt{}}$ negative has no real root.)

d) To find $\sqrt[3]{64}$ ask yourself, "What number do I *cube* to get 64?" Since $4^3 = 64$ and since we know that $\overset{\text{odd}}{\sqrt{}}$ any number gives exactly one root, $\sqrt[3]{64} = 4$.

e) To find $\sqrt[3]{-64}$ ask yourself, "What number do I *cube* to get -64?" Since $(-4)^3 = -64$ and since we know that $\overset{\text{odd}}{\sqrt{}}$ any number gives exactly one root, $\sqrt[3]{-64} = -4$. ■

You Try 4

Find each root, if possible.

a) $\sqrt[6]{-64}$ b) $\sqrt[3]{-125}$ c) $-\sqrt[4]{81}$ d) $\sqrt[3]{1}$ e) $\sqrt[4]{81}$

3. Evaluate $\sqrt[n]{a^n}$

Earlier we said that the $\sqrt{}$ symbol represents only the *positive* square root of a number. For example, $\sqrt{9} = 3$. It is also true that $\sqrt{(-3)^2} = \sqrt{9} = 3$.

If a variable is in the radicand and we do not know whether the variable represents a positive number, then we must use the absolute value symbol to evaluate the radical. Then we know that the result will be a positive number. For example, $\sqrt{a^2} = |a|$.

What if the index is greater than 2? Let's look at how to find the following roots:

$$\sqrt[4]{(-2)^4} = \sqrt[4]{16} = 2 \qquad \sqrt[3]{(-4)^3} = \sqrt[3]{-64} = -4$$

When the index on the radical is any positive, even integer and we do not know whether the variable in the radicand represents a positive number, we must use the absolute value symbol to write the root. However, when the index is a positive, odd integer, we do not need to use the absolute value symbol.

Procedure   Evaluating $\sqrt[n]{a^n}$

If n is a positive, *even* integer, then $\sqrt[n]{a^n} = |a|$.

If n is a positive, *odd* integer, then $\sqrt[n]{a^n} = a$.

Example 5

In-Class Example 5
Simplify.
a) $\sqrt{(-9)^2}$ b) $\sqrt{r^2}$
c) $\sqrt[3]{(-2)^3}$ d) $\sqrt[7]{p^7}$
e) $\sqrt[4]{(z-6)^4}$ f) $\sqrt[5]{(7m+4)^5}$
answer: a) 9 b) $|r|$ c) -2
d) p e) $|z-6|$ f) $7m+4$

Simplify.

a) $\sqrt{(-7)^2}$ b) $\sqrt{k^2}$ c) $\sqrt[3]{(-5)^3}$ d) $\sqrt[7]{n^7}$

e) $\sqrt[4]{(y-9)^4}$ f) $\sqrt[5]{(8p+1)^5}$

Solution

a) $\sqrt{(-7)^2} = |-7| = 7$ When the index is even, use the absolute value symbol to be certain that the result is not negative.

b) $\sqrt{k^2} = |k|$ When the index is even, use the absolute value symbol to be certain that the result is not negative.

c) $\sqrt[3]{(-5)^3} = -5$ The index is odd, so the absolute value symbol is not necessary.

d) $\sqrt[7]{n^7} = n$ The index is odd, so the absolute value symbol is not necessary.

e) $\sqrt[4]{(y - 9)^4} = |y - 9|$ Even index: use the absolute value symbol to be certain that the result is not negative.

f) $\sqrt[5]{(8p + 1)^5} = 8p + 1$ Odd index: the absolute value symbol is not necessary.   ■

You Try 5

Simplify.

a) $\sqrt{(-12)^2}$ b) $\sqrt{w^2}$ c) $\sqrt[3]{(-3)^3}$ d) $\sqrt[5]{r^5}$

e) $\sqrt[6]{(t + 4)^6}$ f) $\sqrt[7]{(4h - 3)^7}$

4. Determine the Domains of Square Root and Cube Root Functions

An algebraic expression containing a radical is called a **radical expression.** When real numbers are substituted for the variable in radical expressions like $\sqrt{x}$, $\sqrt{4t + 1}$, and $\sqrt[3]{p}$ so that the expression is defined, each value that is substituted will produce *only one* value for the expression. Therefore, function notation can be used to represent radical expressions.

Radical functions are functions of the form $f(x) = \sqrt[n]{x}$. Let's look at some square root and cube root functions.

Two examples of **square root functions** are $f(x) = \sqrt{x}$ and $g(r) = \sqrt{2r - 9}$.

Example 6

In-Class Example 6

Let $f(x) = \sqrt{x}$ and $g(a) = \sqrt{2a - 7}$. Find the function values, if possible.

a) $f(49)$ b) $g(5)$ c) $f(-4)$

d) $g(3)$

answer: a) 7 b) $\sqrt{3}$

c) not a real number

d) not a real number

Let $f(x) = \sqrt{x}$ and $g(r) = \sqrt{2r - 9}$. Find the function values, if possible.

a) $f(64)$ b) $g(7)$ c) $f(-25)$ d) $g(3)$

Solution

a) $f(64) = \sqrt{64} = 8$

b) $g(7) = \sqrt{2(7) - 9} = \sqrt{14 - 9} = \sqrt{5}$

c) $f(-25) = \sqrt{-25}$; not a real number

d) $g(3) = \sqrt{2(3) - 9} = \sqrt{-3}$; not a real number   ■

You Try 6

Let $f(x) = \sqrt{x}$ and $h(t) = \sqrt{3t - 10}$. Find the function values, if possible.

a) $f(25)$ b) $h(9)$ c) $f(-11)$ d) $h(2)$

Parts c) and d) of Example 6 illustrate that when the radicand of a square root function is negative, the function is undefined. Therefore, any value that makes the radicand negative is not in the domain of the function.

> **Definition**
>
> The **domain of a square root function** consists of all of the real numbers that can be substituted for the variable so that radicand is nonnegative.
>
> To determine the domain of a square root function, set up an inequality so that the radicand ≥ 0. Solve for the variable. These are the real numbers in the domain of the function.

Example 7

In-Class Example 7
Determine the domain of each square root function.
a) $f(x) = \sqrt{x}$
b) $g(a) = \sqrt{2a - 7}$
answers:

a) $[0, \infty)$ b) $\left[\dfrac{7}{2}, \infty\right)$

Determine the domain of each square root function.

a) $f(x) = \sqrt{x}$ b) $g(r) = \sqrt{2r - 9}$

Solution

a) The radicand, x, must be greater than or equal to zero. We write that as the inequality $x \geq 0$. In interval notation, we write the domain as $[0, \infty)$.

b) In the square root function $g(r) = \sqrt{2r - 9}$, the radicand, $2r - 9$, must be nonnegative. We write this as $2r - 9 \geq 0$. To determine the domain of the function, solve the inequality $2r - 9 \geq 0$.

$$2r - 9 \geq 0 \qquad \text{The value of the radicand must be} \geq 0.$$
$$2r \geq 9$$
$$r \geq \frac{9}{2} \qquad \text{Solve.}$$

Any value of r that satisfies $r \geq \dfrac{9}{2}$ will make the radicand greater than or equal to zero. The domain of $g(r) = \sqrt{2r - 9}$ is $\left[\dfrac{9}{2}, \infty\right)$.

 You Try 7

Determine the domain of each square root function.

a) $h(x) = \sqrt{x - 9}$ b) $k(t) = \sqrt{7t + 2}$

Two examples of cube root functions are $f(x) = \sqrt[3]{x}$ and $h(a) = \sqrt[3]{a - 5}$. Let's look at these next.

Example 8

In-Class Example 8
Let $f(x) = \sqrt[3]{x}$ and
$h(t) = \sqrt[3]{t} - 9$. Find the
function values, if possible.
a) $f(27)$ b) $f(-10)$ c) $h(1)$
d) $h(13)$
answers: a) 3   **b)** $\sqrt[3]{-10}$
c) -2 **d)** $\sqrt[3]{4}$

Let $f(x) = \sqrt[3]{x}$ and $h(a) = \sqrt[3]{a - 5}$. Find the function values, if possible.

a) $f(125)$ b) $f(-7)$ c) $h(10)$ d) $h(-3)$

Solution

a) $f(125) = \sqrt[3]{125} = 5$

b) $f(-7) = \sqrt[3]{-7}$

c) $h(10) = \sqrt[3]{10 - 5} = \sqrt[3]{5}$

d) $h(-3) = \sqrt[3]{-3 - 5} = \sqrt[3]{-8} = -2$

 You Try 8

Let $f(x) = \sqrt[3]{x}$ and $g(c) = \sqrt[3]{2c + 3}$. Find the function values, if possible.

a) $f(25)$ b) $f(-27)$ c) $g(-2)$ d) $g(2)$

Unlike square root functions, it is possible to evaluate cube root functions when the radicand is negative. Therefore, any real number may be substituted into a cube root function and the function will be defined.

> **Definition**
>
> The **domain of a cube root function** is the set of all real numbers. We can write this in interval notation as $(-\infty, \infty)$.

In fact we can say that when n is an odd, positive number, the domain of $f(x) = \sqrt[n]{x}$ is all real numbers, or $(-\infty, \infty)$. This is because the odd root of any real number is, itself, a real number.

5. Graph a Square Root Function

We need to know the domain of a square root function in order to sketch its graph.

Example 9

In-Class Example 9
Graph each function.
a) $f(x) = \sqrt{x}$
b) $g(x) = \sqrt{x+1}$
answers:
a) See graph in the solution to Example 9a.
b)

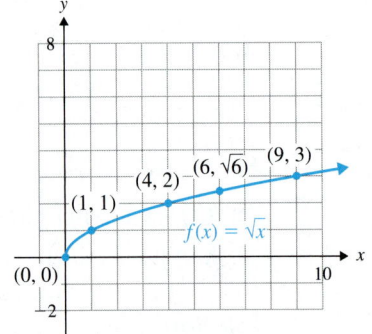

Graph each function.

a) $f(x) = \sqrt{x}$ b) $g(x) = \sqrt{x+4}$

Solution

a) In Example 7 we found that the domain of $f(x) = \sqrt{x}$ is $[0, \infty)$. When we make a table of values, we will start by letting $x = 0$, the smallest number in the domain, and then choose real numbers greater than 0. Usually it is easiest to choose values for x that are perfect squares so that it will be easier to plot the points. We will also plot the point $(6, \sqrt{6})$ so that you can see where it lies on the graph. Connect the points with a smooth curve.

$f(x) = \sqrt{x}$	
x	**f(x)**
0	0
1	1
4	2
6	$\sqrt{6} \approx 2.4$
9	3

The graph reinforces the fact that this is a function. It passes the vertical line test.

b) To graph $g(x) = \sqrt{x+4}$ we will begin by determining its domain. Solve $x + 4 \geq 0$.

$$x + 4 \geq 0 \qquad \text{The value of the radicand must be } \geq 0.$$
$$x \geq -4 \qquad \text{Solve.}$$

The domain of $g(x)$ is $[-4, \infty)$. When we make a table of values, we will start by letting $x = -4$, the smallest number in the domain, and then choose real numbers greater than -4. We will choose values for x so that the radicand will be a perfect square. This will make it easier to plot the points. We will also plot the point $(1, \sqrt{5})$ so that you can see where it lies on the graph. Connect the points with a smooth curve.

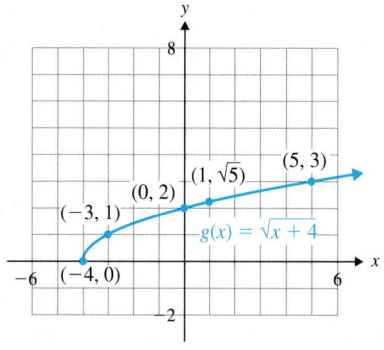

| $g(x) = \sqrt{x + 4}$ | |
x	g(x)
-4	0
-3	1
0	2
1	$\sqrt{5} \approx 2.2$
5	3

Since this graph represents a function, it passes the vertical line test.

 You Try 9

Graph $f(x) = \sqrt{x} + 2$.

6. Graph a Cube Root Function

The domain of a cube root function consists of all real numbers. Therefore, we can substitute any real number into the function and it will be defined. However, we want to choose our numbers carefully. *To make the table of values, pick values in the domain so that the radicand will be a perfect cube, and choose values for the variable that will give us positive numbers, negative numbers, and zero for the value of the radicand.* This will help us to graph the function correctly.

Example 10

In-Class Example 10
Graph each function.
a) $f(x) = \sqrt[3]{x}$
b) $g(x) = \sqrt[3]{x} - 2$
answers:
a) **See graph in the solution to Example 10a.**
b)

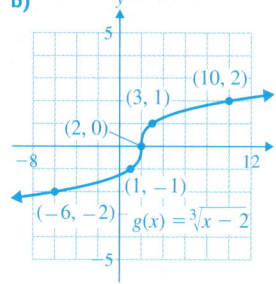

Graph each function.

a) $f(x) = \sqrt[3]{x}$ b) $g(x) = \sqrt[3]{x - 1}$

Solution

a) Make a table of values. Choose x-values that are perfect cubes. Also, remember to choose x-values that are positive, negative, and zero. Plot the points and connect them with a smooth curve.

| $f(x) = \sqrt[3]{x}$ | |
x	f(x)
0	0
1	1
8	2
-1	-1
-8	-2

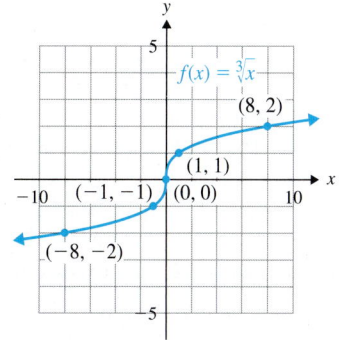

The graph passes the vertical line test for functions.

b) Remember, for the table of values we want to choose values for x that will give us positive numbers, negative numbers, and zero *in the radicand*. First we will determine what value of x will make the radicand in $g(x) = \sqrt[3]{x - 1}$ equal to zero.

$$x - 1 = 0$$
$$x = 1$$

If $x = 1$, the radicand equals zero. Therefore, the first value we will put in the table of values is $x = 1$. Then, choose a couple of numbers *greater than* 1 and a couple that are *less than* 1 so that we get positive and negative numbers in the radicand. Also, we will choose our x-values so that the radicand will be a perfect cube. Plot the points and connect them with a smooth curve.

x	$g(x) = \sqrt[3]{x - 1}$
1	$\sqrt[3]{1 - 1} = \sqrt[3]{0} = 0$
2	$\sqrt[3]{2 - 1} = \sqrt[3]{1} = 1$
9	$\sqrt[3]{9 - 1} = \sqrt[3]{8} = 2$
0	$\sqrt[3]{0 - 1} = \sqrt[3]{-1} = -1$
-7	$\sqrt[3]{-7 - 1} = \sqrt[3]{-8} = -2$

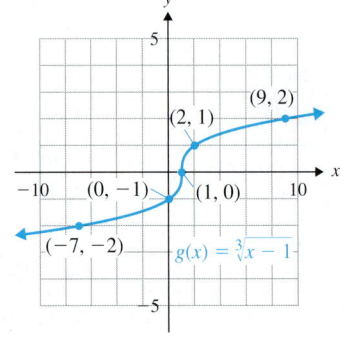

Since this graph represents a function, it passes the vertical line test.

You Try 10

Graph $f(x) = \sqrt[3]{x - 3}$.

Using Technology

We can evaluate square roots, cube roots, or even higher roots using a graphing calculator. A radical sometimes evaluates to an integer and sometimes must be approximated using a decimal.

To evaluate a square root:

For example, to evaluate $\sqrt{9}$ press [2nd] [x^2], enter the radicand [9], and then press [)] [ENTER]. The result is 3 as shown on the screen on the left below. When the radicand is a perfect square such as 9, 16, or 25, then the square root evaluates to a whole number. For example $\sqrt{16}$ evaluates to 4 and $\sqrt{25}$ evaluates to 5 as shown.

If the radicand of a square root is not a perfect square, then the result is a decimal approximation. For example, to evaluate $\sqrt{19}$ press [2nd] [x^2], enter the radicand [1][9], and then press [)] [ENTER]. The result is approximately 4.3589, rounded to four decimal places.

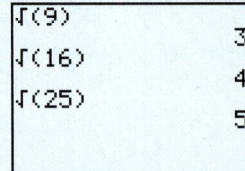

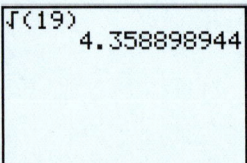

To evaluate a cube root:

For example, to evaluate $\sqrt[3]{27}$ press [MATH] [4], enter the radicand [2][7], and then press [)] [ENTER]. The result is 3 as shown.

If the radicand is a perfect cube such as 27, then the cube root evaluates to an integer. Since 28 is not a perfect cube, the cube root evaluates to approximately 3.0366.

To evaluate radicals with an index greater than 3:

For example, to evaluate $\sqrt[4]{16}$ enter the index $\boxed{4}$, press $\boxed{\text{MATH}}$ $\boxed{5}$, enter the radicand $\boxed{1}\boxed{6}$, and press $\boxed{\text{ENTER}}$. The result is 2.

Since the fifth root of 18 evaluates to a decimal, the result is an approximation of 1.7826 rounded to four decimal places as shown.

```
³√(27)
              3
³√(28)
     3.036588972
```

```
4*√16
              2
5*√18
   1.782602458
```

Evaluate each root using a graphing calculator. If necessary, approximate to the nearest tenth.

1) $\sqrt{25}$ 2) $\sqrt[3]{216}$ 3) $\sqrt{29}$ 4) $\sqrt{324}$ 5) $\sqrt[5]{1024}$ 6) $\sqrt[3]{343}$

Using Technology

We can use a graphing calculator to find the domain of a square root function or cube root function visually. The domain consists of the x-values of the points shown on the graph.

We first consider the basic shape of a square root function. To graph the equation $f(x) = \sqrt{x}$, press $\boxed{\text{2}^{\text{nd}}}$ $\boxed{x^2}$ $\boxed{\text{X,T,}\Theta\text{,n}}$ $\boxed{)}$ to the right of \Y₁ =. Press $\boxed{\text{ZOOM}}$ and select 6:ZStandard to graph the equation.

The left side of the graph begins at the point (0, 0) and the right side of the graph continues up and to the right forever. The x-values of the graph consist of all x-values greater than or equal to 0. In interval notation, the domain is $[0, \infty)$.

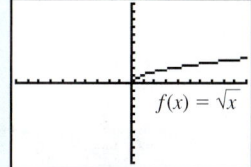

The domain of any square root function can be found using a similar approach. First graph the function and then look at the x-values of the points on the graph. The graph of a square root function will always start at a number and extend to positive or negative infinity.

For example, consider the graph of the function $g(x) = \sqrt{3 - x}$ as shown.

The largest x-value on the graph is 3. The x-values of the graph consist of all x-values less than or equal to 3. In interval notation, the domain is $(-\infty, 3]$.

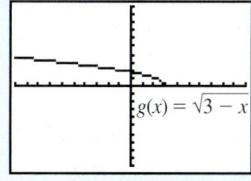

Next consider the basic shape of a cube root function.

To graph the equation $f(x) = \sqrt[3]{x}$ press $\boxed{\text{MATH}}$, select 4: $\sqrt[3]{\ }$ (, and press $\boxed{\text{X,T,}\Theta\text{,n}}$ $\boxed{)}$ to the right of \Y₁=. Press $\boxed{\text{ZOOM}}$ and select 6:ZStandard to graph the equation as shown on the graph at right.

The left side of the graph extends down and to the left forever, and the right side of the graph extends up and to the right forever. In interval notation, the domain is $(-\infty, \infty)$. This is true for any cube root function, so the domain is always $(-\infty, \infty)$.

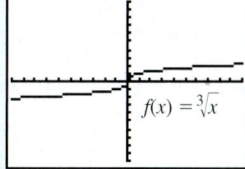

Determine the domain using a graphing calculator. Use interval notation in your answer.

7) $f(x) = \sqrt{x - 2}$ 8) $g(x) = \sqrt{x + 3}$ 9) $h(x) = \sqrt{2 - x}$

10) $f(x) = -\sqrt{x + 1}$ 11) $f(x) = \sqrt[3]{x + 5}$ 12) $g(x) = \sqrt[3]{4 - x}$

Answers to You Try Exercises

1) 8, −8   2) a) 3 b) −12 c) $\dfrac{5}{6}$ d) $-\dfrac{1}{8}$ e) not a real number   3) a) 3 b) 2

4) a) not a real number b) −5 c) −3 d) 1 e) 3   5) a) 12 b) $|w|$ c) −3 d) r
e) $|t + 4|$ f) $4h − 3$ 6) a) 5 b) $\sqrt{17}$ c) not a real number d) not a real number

7) a) $[9, \infty)$ b) $\left[-\dfrac{2}{7}, \infty\right)$ 8) a) $\sqrt[3]{25}$ b) −3 c) −1 d) $\sqrt[3]{7}$

9)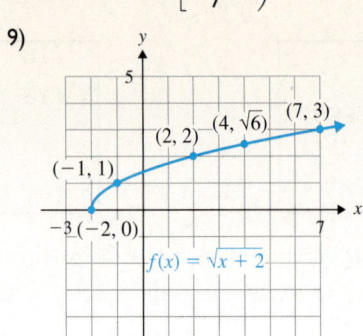

10)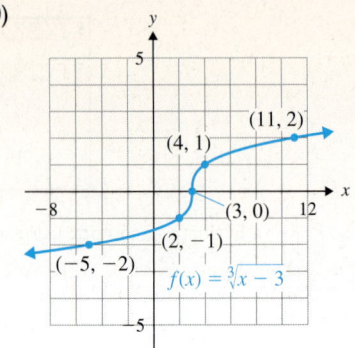

Answers to Technology Exercises

1) 5   2) 6   3) 5.4   4) 18   5) 4   6) 7   7) $[2, \infty)$ 8) $[-3, \infty)$
9) $(-\infty, 2]$ 10) $[-1, \infty)$ 11) $(-\infty, \infty)$ 12) $(-\infty, \infty)$

8.1 Exercises

*Additional answers can be found in the Answers to Exercises appendix.

Objective 1: Find Square Roots and Principal Square Roots

Decide whether each statement is true or false. If it is false, explain why.

1) $\sqrt{121} = 11$ and -11.   False; $\sqrt{121} = 11$ because the $\sqrt{}$ symbol means principal square root.

2) $\sqrt{81} = 9$ True

3) The square root of a negative number is a negative number.   False; the square root of a negative number is not a real number.

4) The even root of a negative number is a negative number.   False; it is not a real number.

Find all square roots of each number.

5) 144   12 and −12

6) 2500   50 and −50

7) $\dfrac{36}{25}$ $\dfrac{6}{5}$ and $-\dfrac{6}{5}$

8) 0.01   0.1 and −0.1

Find each square root, if possible.

9) $\sqrt{49}$ 7

10) $\sqrt{169}$ 13

11) $\sqrt{-4}$ not real

12) $\sqrt{-100}$ not real

13) $\sqrt{\dfrac{81}{25}}$ $\dfrac{9}{5}$

14) $\sqrt{\dfrac{121}{4}}$ $\dfrac{11}{2}$

15) $-\sqrt{36}$ −6

16) $-\sqrt{0.04}$ −0.2

Objective 2: Find Higher Roots

Decide if each statement is true or false. If it is false, explain why.

17) The cube root of a negative number is a negative number.   True

18) The odd root of a negative number is not a real number.   False; the odd root of a negative number is a negative number.

19) Every nonnegative real number has two real, even roots.   False; the only even root of zero is zero.

20) $-\sqrt[4]{10{,}000} = -10$ True

21) Explain how to find $\sqrt[3]{64}$.

22) Explain how to find $\sqrt[4]{16}$.

23) Does $\sqrt[4]{-81} = -3$? Why or why not?

24) Does $\sqrt[3]{-8} = -2$? Why or why not?

Find each root, if possible.

25) $\sqrt[3]{125}$ 5

26) $\sqrt[3]{27}$ 3

27) $\sqrt[3]{-1}$ −1

28) $\sqrt[3]{-8}$ −2

29) $\sqrt[4]{81}$ 3

30) $\sqrt[4]{16}$ 2

31) $\sqrt[4]{-1}$ not real

32) $\sqrt[4]{-81}$ not real

33) $-\sqrt[4]{16}$ −2

34) $-\sqrt[4]{1}$ −1

VIDEO 35) $\sqrt[3]{-32}$ −2

36) $-\sqrt[6]{64}$ −2

37) $-\sqrt[3]{-27}$ 3

38) $-\sqrt[3]{-1000}$ 10

39) $\sqrt[6]{-64}$ not real

40) $\sqrt[4]{-16}$ not real

41) $\sqrt[3]{\dfrac{8}{125}}$ $\dfrac{2}{5}$

42) $\sqrt[4]{\dfrac{81}{16}}$ $\dfrac{3}{2}$

VIDEO 43) $\sqrt{60-11}$ 7

44) $\sqrt{100+21}$ 11

45) $\sqrt[3]{9-36}$ −3

46) $\sqrt{1-9}$ not real

47) $\sqrt{5^2+12^2}$ 13

48) $\sqrt{3^2+4^2}$ 5

Objective 3: Evaluate $\sqrt[n]{a^n}$

49) If n is a positive, even integer and we are not certain that $a \geq 0$, then we must use the absolute value symbol to evaluate $\sqrt[n]{a^n}$. That is, $\sqrt[n]{a^n} = |a|$. Why must we use the absolute value symbol?

50) If n is a positive, odd integer then $\sqrt[n]{a^n} = a$ for any value of a. Why don't we need to use the absolute value symbol?

Simplify.

51) $\sqrt{8^2}$ 8

52) $\sqrt{5^2}$ 5

53) $\sqrt{(-6)^2}$ 6

54) $\sqrt{(-11)^2}$ 11

55) $\sqrt{y^2}$ $|y|$

56) $\sqrt{d^2}$ $|d|$

57) $\sqrt[3]{5^3}$ 5

58) $\sqrt[3]{(-4)^3}$ −4

59) $\sqrt[3]{z^3}$ z

60) $\sqrt[7]{t^7}$ t

61) $\sqrt[4]{h^4}$ $|h|$

62) $\sqrt[6]{m^6}$ $|m|$

63) $\sqrt{(x+7)^2}$ $|x+7|$

64) $\sqrt{(a-9)^2}$ $|a-9|$

65) $\sqrt[3]{(2t-1)^3}$ $2t-1$

66) $\sqrt[5]{(6r+7)^5}$ $6r+7$

67) $\sqrt[4]{(3n+2)^4}$ $|3n+2|$

68) $\sqrt[3]{(x-6)^3}$ $x-6$

69) $\sqrt[7]{(d-8)^7}$ $d-8$

70) $\sqrt[6]{(4y+3)^6}$ $|4y+3|$

Objective 4: Determine the Domains of Square Root and Cube Root Functions

71) Is −1 in the domain of $f(x) = \sqrt{x}$? Explain your answer.   No because $\sqrt{-1}$ is not a real number.

72) Is −1 in the domain of $f(x) = \sqrt[3]{x}$? Explain your answer.   Yes because $\sqrt[3]{-1} = -1$.

73) How do you find the domain of a square root function?

74) What is the domain of a cube root function? $(-\infty, \infty)$

Let $f(x) = \sqrt{x}$ and $g(t) = \sqrt{3t+4}$. Find each of the following, if possible, and simplify.

75) $f(100)$ 10

76) $f(9)$ 3

77) $f(-49)$ not a real number

78) $f(-64)$ not a real number

79) $g(-1)$ 1

80) $g(3)$ $\sqrt{13}$

81) $g(-5)$ not a real number

82) $g(-2)$ not a real number

83) $f(a)$ $\sqrt{a}$

84) $g(w)$ $\sqrt{3w+4}$

85) $f(t+4)$ $\sqrt{t+4}$

86) $f(5p-9)$ $\sqrt{5p-9}$

87) $g(2n-1)$ $\sqrt{6n+1}$

88) $g(m+10)$ $\sqrt{3m+34}$

Let $f(a) = \sqrt[3]{a}$ and $g(x) = \sqrt[3]{4x-1}$. Find each of the following and simplify.

89) $f(64)$ 4

90) $f(-125)$ −5

91) $f(-27)$ −3

92) $g(3)$ $\sqrt[3]{11}$

93) $g(-4)$ $\sqrt[3]{-17}$

94) $g(0)$ −1

95) $g(r)$ $\sqrt[3]{4r-1}$

96) $f(z)$ $\sqrt[3]{z}$

97) $f(c+8)$ $\sqrt[3]{c+8}$

98) $f(5k-2)$ $\sqrt[3]{5k-2}$

99) $g(2a-3)$ $\sqrt[3]{8a-13}$

100) $g(6-w)$ $\sqrt[3]{23-4w}$

Determine the domain of each function.

101) $h(n) = \sqrt{n+2}$ $[-2, \infty)$

102) $g(c) = \sqrt{c+10}$ $[-10, \infty)$

103) $p(a) = \sqrt{a-8}$ $[8, \infty)$

104) $f(a) = \sqrt{a-1}$ $[1, \infty)$

105) $f(a) = \sqrt[3]{a-7}$ $(-\infty, \infty)$

106) $h(t) = \sqrt[3]{t}$ $(-\infty, \infty)$

107) $k(x) = \sqrt{2x-5}$ $\left[\dfrac{5}{2}, \infty\right)$

108) $r(k) = \sqrt{3k+7}$ $\left[-\dfrac{7}{3}, \infty\right)$

109) $g(x) = \sqrt[3]{2x-5}$ $(-\infty, \infty)$

110) $h(c) = \sqrt[3]{-c}$ $(-\infty, \infty)$

111) $g(t) = \sqrt{-t}$ $(-\infty, 0]$

112) $h(x) = \sqrt{3-x}$ $(-\infty, 3]$

113) $r(a) = \sqrt{9-7a}$ $\left(-\infty, \dfrac{9}{7}\right]$

114) $g(c) = \sqrt{8-5c}$ $\left(-\infty, \dfrac{8}{5}\right]$

Objective 5: Graph a Square Root Function

Determine the domain and then graph each function.

115) $f(x) = \sqrt{x-1}$

116) $g(x) = \sqrt{x-4}$

117) $g(x) = \sqrt{x+3}$

118) $h(x) = \sqrt{x+1}$

119) $h(x) = \sqrt{x-2}$

120) $f(x) = \sqrt{x+2}$

121) $f(x) = \sqrt{-x}$

122) $g(x) = \sqrt{-x} - 3$

Objective 6: Graph a Cube Root Function

Determine the domain and then graph each function.

123) $g(x) = \sqrt[3]{x+2}$

124) $f(x) = \sqrt[3]{x+1}$

125) $h(x) = \sqrt[3]{x-2}$

126) $g(x) = \sqrt[3]{-x}$

127) $g(x) = \sqrt[3]{x} + 1$ 128) $h(x) = \sqrt[3]{x} - 1$

129) A stackable storage cube has a storage capacity of 8 ft³. The length of a side of the cube, s, is given by $s = \sqrt[3]{V}$, where V is the volume of the cube. How long is each side of the cube?   2 ft

130) A die (singular of dice) has a surface area of 13.5 cm². Since the die is in the shape of a cube the length of a side of the die, s, is given by $s = \sqrt{\dfrac{A}{6}}$, where A is the surface area of the cube. How long is each side of the cube?

1.5 cm

131) If an object is dropped from the top of a building 160 ft tall, then the formula $t = \sqrt{\dfrac{160 - h}{16}}$ describes how many seconds, t, it takes for the object to reach a height of h ft above the ground. If a piece of ice falls off the top of this building, how long would it take to reach the ground? Give an exact answer and an approximation to two decimal places. $\sqrt{10}$ sec; 3.16 sec

132) A circular flower garden has an area of 51 ft². The radius, r, of a circle in terms of its area is given by $r = \sqrt{\dfrac{A}{\pi}}$, where A is the area of the circle. What is the radius of the garden? Give an exact answer and an approximation to two decimal places. $\sqrt{\dfrac{51}{\pi}}$ ft; 4.03 ft

133) The speed limit on a street in a residential neighborhood is 25 mph. A car involved in an accident on this road left skid marks 40 ft long. Accident investigators determine that the speed of the car can be described by the function $S(d) = \sqrt{22.5d}$, where $S(d)$ is the speed of the car, in miles per hour, at the time of impact and d is the length of the skid marks, in feet. Was the driver speeding at the time of the accident? Show all work to support your answer.   Yes. The car was traveling at 30 mph.

134) A car involved in an accident on a wet highway leaves skid marks 170 ft long. Accident investigators determine that the speed of the car, $S(d)$ in miles per hour, can be described by the function $S(d) = \sqrt{19.8d}$, where d is the length of the skid marks, in feet. The speed limit on the highway is 65 mph. Was the car speeding when the accident occurred? Show all work to support your answer.   No. The car was traveling about 58 mph.

Use the following information for Exercises 135–140.

The period of a pendulum is the time it takes for the pendulum to make one complete swing back and forth. The period, $T(L)$ in seconds, can be described by the function $T(L) = 2\pi\sqrt{\dfrac{L}{32}}$, where L is the length of the pendulum, in feet. For each exercise give an *exact answer and an answer rounded to two decimal places*. Use 3.14 for π.

135) Find the period of the pendulum whose length is 8 ft. π sec; 3.14 sec

136) Find the period of the pendulum whose length is 2 ft. $\dfrac{\pi}{2}$ sec; 1.57 sec

137) Find $T\left(\dfrac{1}{2}\right)$, and explain what it means in the context of the problem. $T\left(\dfrac{1}{2}\right) = \dfrac{\pi}{4}$; A $\dfrac{1}{2}$-ft-long pendulum has a period of $\dfrac{\pi}{4}$ sec. This is approximately 0.79 sec.

138) Find $T(1.5)$, and explain what it means in the context of the problem.

139) Find the period of a 30-inch-long pendulum.

140) Find the period of a 54-inch-long pendulum.

Section 8.2 Rational Exponents

Objectives

1. Evaluate Expressions of the Form $a^{1/n}$
2. Evaluate Expressions of the Form $a^{m/n}$
3. Evaluate Expressions of the Form $a^{-m/n}$
4. Combine the Rules of Exponents
5. Convert a Radical Expression to Exponential Form and Simplify

1. Evaluate Expressions of the Form $a^{1/n}$

In this section, we will explain the relationship between radicals and rational exponents (fractional exponents). Sometimes, converting between these two forms makes it easier to simplify expressions.

Definition

If n is a positive integer greater than 1 and $\sqrt[n]{a}$ is a real number, then

$$a^{1/n} = \sqrt[n]{a}$$

(The denominator of the fractional exponent is the index of the radical.)

Example 1

In-Class Example 1
Write in radical form and evaluate.
a) $9^{1/2}$ b) $27^{1/3}$ c) $32^{1/5}$
d) $(-144)^{1/2}$ e) $-16^{1/4}$
f) $(-32)^{1/5}$
**answer: a) 3   b) 3   c) 2
d) not a real number
e) -2   f) -2**

Write in radical form and evaluate.

a) $8^{1/3}$ b) $49^{1/2}$ c) $81^{1/4}$

d) $-64^{1/6}$ e) $(-16)^{1/4}$ f) $(-125)^{1/3}$

Solution

a) The denominator of the fractional exponent is the index of the radical. Therefore, $8^{1/3} = \sqrt[3]{8} = 2$.

b) The denominator in the exponent of $49^{1/2}$ is 2, so the index on the radical is 2, meaning *square* root.

$$49^{1/2} = \sqrt{49} = 7$$

c) $81^{1/4} = \sqrt[4]{81} = 3$

d) $-64^{1/6} = -(64^{1/6}) = -\sqrt[6]{64} = -2$

e) $(-16)^{1/4} = \sqrt[4]{-16}$, which is not a real number. Remember, the even root of a negative number is not a real number.

f) $(-125)^{1/3} = \sqrt[3]{-125} = -5$ The odd root of a negative number is a negative number. ■

You Try 1

Write in radical form and evaluate.

a) $16^{1/4}$ b) $121^{1/2}$ c) $(-121)^{1/2}$

d) $(-27)^{1/3}$ e) $-81^{1/4}$

2. Evaluate Expressions of the Form $a^{m/n}$

We can add another relationship between rational exponents and radicals.

> **Definition**
>
> If m and n are positive integers and $\dfrac{m}{n}$ is in lowest terms, then
>
> $$a^{m/n} = (a^{1/n})^m = (\sqrt[n]{a})^m$$
>
> if $a^{1/n}$ is a real number.
>
> (The denominator of the fractional exponent is the index of the radical, and the numerator is the power to which we raise the radical expression.) We can also think of $a^{m/n}$ this way:
> $a^{m/n} = (a^m)^{1/n} = \sqrt[n]{a^m}$.

Example 2

In-Class Example 2
Write in radical form and evaluate.
a) $8^{2/3}$ b) $-16^{3/2}$ c) $(-16)^{3/2}$
d) $-32^{4/5}$ e) $(-125)^{2/3}$
**answer: a) 4   b) -64
c) not a real number
d) -16   e) 25**

Write in radical form and evaluate.

a) $25^{3/2}$ b) $-64^{2/3}$ c) $(-81)^{3/2}$ d) $-81^{3/2}$ e) $(-1000)^{2/3}$

Solution

a) The *denominator* of the fractional exponent is the *index* of the radical, and the *numerator* is the *power* to which we raise the radical expression.

$$
\begin{aligned}
25^{3/2} &= (25^{1/2})^3 && \text{Use the definition to rewrite the exponent.}\\
&= (\sqrt{25})^3 && \text{Rewrite as a radical.}\\
&= 5^3 && \sqrt{25} = 5\\
&= 125
\end{aligned}
$$

b)   To evaluate $-64^{2/3}$, *first* evaluate $64^{2/3}$, *then* take the negative of that result.

$$-64^{2/3} = -(64^{2/3}) = -(64^{1/3})^2 \qquad \text{Use the definition to rewrite the exponent.}$$
$$= -(\sqrt[3]{64})^2 \qquad \text{Rewrite as a radical.}$$
$$= -(4)^2 \qquad \sqrt[3]{64} = 4$$
$$= -16$$

c) $(-81)^{3/2} = [(-81)^{1/2}]^3$
$$= (\sqrt{-81})^3 \qquad \text{Not a real number.} \qquad \text{The even root of a negative number is not a real number.}$$

d) $-81^{3/2} = -(81^{1/2})^3 = -(\sqrt{81})^3 = -(9)^3 = -729$

e) $(-1000)^{2/3} = [(-1000)^{1/3}]^2 = (\sqrt[3]{-1000})^2 = (-10)^2 = 100$ ■

You Try 2

Write in radical form and evaluate.

a) $32^{2/5}$ b) $-100^{3/2}$ c) $(-100)^{3/2}$ d) $(-1)^{4/5}$ e) $-1^{5/3}$

BE CAREFUL

In Example 2, notice how the parentheses affect how we evaluate an expression. The base of the expression $(-81)^{3/2}$ is -81, while the base of $-81^{3/2}$ is 81.

3. Evaluate Expressions of the Form $a^{-m/n}$

Recall that if n is any integer and $a \neq 0$, then $a^{-n} = \left(\dfrac{1}{a}\right)^n = \dfrac{1}{a^n}$.

That is, to rewrite the expression with a *positive* exponent, take the reciprocal of the base. For example,

$$2^{-4} = \left(\frac{1}{2}\right)^4 = \frac{1}{16}$$

We can extend this idea to rational exponents.

Definition

If $a^{m/n}$ is a nonzero real number, then

$$a^{-m/n} = \left(\frac{1}{a}\right)^{m/n} = \frac{1}{a^{m/n}}$$

(To rewrite the expression with a *positive* exponent, take the reciprocal of the base.)

Example 3

In-Class Example 3
Rewrite with a positive exponent and evaluate.

a) $9^{-3/2}$ b) $81^{-3/4}$ c) $\left(\dfrac{8}{27}\right)^{-4/3}$

answer: a) $\dfrac{1}{27}$ b) $\dfrac{1}{27}$ c) $\dfrac{81}{16}$

Rewrite with a positive exponent and evaluate.

a) $36^{-1/2}$ b) $32^{-2/5}$ c) $\left(\dfrac{125}{64}\right)^{-2/3}$

Solution

a) To write $36^{-1/2}$ with a positive exponent, take the reciprocal of the base.

$$36^{-1/2} = \left(\frac{1}{36}\right)^{1/2}$$ The reciprocal of 36 is $\dfrac{1}{36}$.

$$= \sqrt{\frac{1}{36}}$$ The denominator of the fractional exponent is the index of the radical.

$$= \frac{1}{6}$$

b) $$32^{-2/5} = \left(\frac{1}{32}\right)^{2/5}$$ The reciprocal of 32 is $\dfrac{1}{32}$.

$$= \left(\sqrt[5]{\frac{1}{32}}\right)^{2}$$ The denominator of the fractional exponent is the index of the radical.

$$= \left(\frac{1}{2}\right)^{2}$$ $\sqrt[5]{\dfrac{1}{32}} = \dfrac{1}{2}$

$$= \frac{1}{4}$$

c) $$\left(\frac{125}{64}\right)^{-2/3} = \left(\frac{64}{125}\right)^{2/3}$$ The reciprocal of $\dfrac{125}{64}$ is $\dfrac{64}{125}$.

$$= \left(\sqrt[3]{\frac{64}{125}}\right)^{2}$$ The denominator of the fractional exponent is the index of the radical.

$$= \left(\frac{4}{5}\right)^{2}$$ $\sqrt[3]{\dfrac{64}{125}} = \dfrac{4}{5}$

$$= \frac{16}{25}$$

 **BE CAREFUL**   The negative exponent does not make the expression negative!

 **You Try 3**

Rewrite with a positive exponent and evaluate.

a) $144^{-1/2}$ b) $16^{-3/4}$ c) $\left(\dfrac{8}{27}\right)^{-2/3}$

4. Combine the Rules of Exponents

We can combine the rules presented in this section with the rules of exponents we learned in Chapter 5 to simplify expressions containing numbers or variables.

Example 4

In-Class Example 4
Simplify completely. The answer should contain only positive exponents.
a) $(3^{2/5})^6$ b) $27^{2/3} \cdot 27^{-1/3}$
c) $\dfrac{4^{13/3}}{4^{7/3}}$

answer: a) $3^{12/5}$ b) 3
c) 16

Simplify completely. The answer should contain only positive exponents.

a) $(6^{1/5})^2$ b) $25^{3/4} \cdot 25^{-1/4}$ c) $\dfrac{8^{2/9}}{8^{11/9}}$

Solution

a) $(6^{1/5})^2 = 6^{2/5}$ Multiply exponents.

b) $25^{3/4} \cdot 25^{-1/4} = 25^{\frac{3}{4}+\left(-\frac{1}{4}\right)}$ Add exponents.

$= 25^{2/4} = 25^{1/2} = 5$

c) $\dfrac{8^{2/9}}{8^{11/9}} = 8^{\frac{2}{9}-\frac{11}{9}}$ Subtract exponents.

$= 8^{-9/9}$ Subtract $\dfrac{2}{9} - \dfrac{11}{9}$.

$= 8^{-1}$ Reduce $-\dfrac{9}{9}$.

$= \left(\dfrac{1}{8}\right)^1 = \dfrac{1}{8}$

You Try 4

Simplify completely. The answer should contain only positive exponents.

a) $49^{3/8} \cdot 49^{1/8}$ b) $(16^{1/12})^3$ c) $\dfrac{7^{2/5}}{7^{4/5}}$

Example 5

In-Class Example 5
Simplify completely. Assume the variables represent positive real numbers. The answer should contain only positive exponents.

a) $x^{1/4} \cdot x^{5/4}$ b) $\left(\dfrac{p^{1/6}}{q^{3/2}}\right)^8$

c) $\dfrac{h^{2/3} \cdot h^{-3/4}}{h^{5/3}}$

d) $\left(\dfrac{m^{1/2}n^{-3}}{m^{3/5}n^{-5}}\right)^{-3/2}$

answer:

a) $x^{3/2}$ b) $\dfrac{p^{4/3}}{q^{12}}$

c) $\dfrac{1}{h^{7/4}}$ d) $\dfrac{m^{3/20}}{n^3}$

Simplify completely. Assume the variables represent positive real numbers. The answer should contain only positive exponents.

a) $r^{1/8} \cdot r^{3/8}$ b) $\left(\dfrac{x^{2/3}}{y^{1/4}}\right)^6$ c) $\dfrac{n^{-5/6} \cdot n^{1/3}}{n^{-1/6}}$ d) $\left(\dfrac{a^{-7}b^{1/2}}{a^5 c^{1/3}}\right)^{-3/4}$

Solution

a) $r^{1/8} \cdot r^{3/8} = r^{\frac{1}{8}+\frac{3}{8}}$ Add exponents.

$= r^{4/8} = r^{1/2}$

b) $\left(\dfrac{x^{2/3}}{y^{1/4}}\right)^6 = \dfrac{x^{\frac{2}{3}\cdot 6}}{y^{\frac{1}{4}\cdot 6}}$ Multiply exponents.

$= \dfrac{x^4}{y^{3/2}}$ Reduce.

c) $\dfrac{n^{-5/6} \cdot n^{1/3}}{n^{-1/6}} = \dfrac{n^{-\frac{5}{6}+\frac{1}{3}}}{n^{-1/6}} = \dfrac{n^{-\frac{5}{6}+\frac{2}{6}}}{n^{-1/6}} = \dfrac{n^{-3/6}}{n^{-1/6}}$ Add exponents.

$= n^{-\frac{3}{6}-\left(-\frac{1}{6}\right)} = n^{-2/6} = n^{-1/3} = \dfrac{1}{n^{1/3}}$ Subtract exponents.

d) $\left(\dfrac{a^{-7}b^{1/2}}{a^5 b^{1/3}}\right)^{-3/4} = \left(\dfrac{a^5 b^{1/3}}{a^{-7}b^{1/2}}\right)^{3/4}$ Eliminate the negative from the outermost exponent by taking the reciprocal of the base.

Simplify the expression inside the parentheses by subtracting the exponents.

$= (a^{5-(-7)}b^{1/3-1/2})^{3/4} = (a^{5+7}b^{2/6-3/6})^{3/4} = (a^{12}b^{-1/6})^{3/4}$

Apply the power rule, and simplify.

$= (a^{12})^{3/4}(b^{-1/6})^{3/4} = a^9 b^{-1/8} = \dfrac{a^9}{b^{1/8}}$

You Try 5

Simplify completely. Assume the variables represent positive real numbers. The answer should contain only positive exponents.

a) $(a^3 b^{1/5})^{10}$ b) $\dfrac{t^{3/10}}{t^{7/10}}$ c) $\dfrac{s^{3/4}}{s^{1/2} \cdot s^{-5/4}}$ d) $\left(\dfrac{x^4 y^{3/8}}{x^9 y^{1/4}}\right)^{-2/5}$

5. Convert a Radical Expression to Exponential Form and Simplify

Some radicals can be simplified by first putting them into rational exponent form and then converting them back to radicals.

Example 6

Rewrite each radical in exponential form, then simplify. Write the answer in simplest (or radical) form. Assume the variable represents a nonnegative real number.

a) $\sqrt[8]{9^4}$ b) $\sqrt[6]{s^4}$

In-Class Example 6
Follow the instructions for Example 6.
a) $\sqrt[10]{36^5}$ b) $\sqrt[8]{x^6}$
answer: a) 6 b) $\sqrt[4]{x^3}$

Solution

a)   Since the index of the radical is the denominator of the exponent and the power is the numerator, we can write

$$\sqrt[8]{9^4} = 9^{4/8} \qquad \text{Write with a rational exponent.}$$
$$= 9^{1/2} = 3$$

b) $\sqrt[6]{s^4} = s^{4/6} \qquad \text{Write with a rational exponent.}$
$$= s^{2/3} = \sqrt[3]{s^2}$$

The expression $\sqrt[6]{s^4}$ is not in simplest form because the 4 and the 6 contain a common factor of 2, but $\sqrt[3]{s^2}$ *is* in simplest form because 2 and 3 do not have any common factors besides 1.

You Try 6

Rewrite each radical in exponential form, then simplify. Write the answer in simplest (or radical) form. Assume the variable represents a nonnegative real number.

a) $\sqrt[6]{125^2}$ b) $\sqrt[10]{p^4}$

In Section 8.1 we said that if a is negative and n is a positive, even number, then $\sqrt[n]{a^n} = |a|$. For example, if we are *not* told that k is positive, then $\sqrt{k^2} = |k|$. However, if we assume that k is positive, then $\sqrt{k^2} = k$. **In the rest of this chapter, we will assume that all variables represent positive, real numbers unless otherwise stated.** When we make this assumption, we do not need to use absolute values when simplifying even roots. And if we consider this together with the relationship between radicals and rational exponents we have another way to explain why $\sqrt[n]{a^n} = a$.

Example 7

In-Class Example 7
Simplify.
a) $\sqrt[3]{11^3}$ b) $(\sqrt[4]{3})^4$ c) $\sqrt{p^2}$
answer: a) 11 b) 3 c) p

Simplify.

a) $\sqrt[3]{5^3}$ b) $(\sqrt[4]{9})^4$ c) $\sqrt{k^2}$

Solution

a) $\sqrt[3]{5^3} = (5^3)^{1/3} = 5^{3 \cdot \frac{1}{3}} = 5^1 = 5$

b) $(\sqrt[4]{9})^4 = (9^{1/4})^4 = 9^{\frac{1}{4} \cdot 4} = 9^1 = 9$

c) $\sqrt{k^2} = (k^2)^{1/2} = k^{2 \cdot \frac{1}{2}} = k^1 = k$

You Try 7

Simplify.

a) $(\sqrt{10})^2$ b) $\sqrt[3]{7^3}$ c) $\sqrt[4]{t^4}$

Using Technology

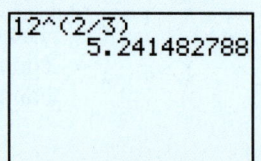

We can evaluate square roots, cube roots, or even higher roots by first rewriting the radical in exponential form and then using a graphing calculator.

For example, to evaluate $\sqrt{49}$, first rewrite the radical as $49^{1/2}$, then enter [4] [9], press [^] [(], enter [1] [÷] [2], and press [)] [ENTER]. The result is 7, as shown on the screen on the left below.

To approximate $\sqrt[3]{12^2}$ rounded to the nearest tenth, first rewrite the radical as $12^{2/3}$, then enter [1] [2], press [^] [(], enter [2] [÷] [3], and press [)] [ENTER]. The result is 5.241482788 as shown on the screen on the right below. The result rounded to the nearest tenth is then 5.2.

```
49^(1/2)
                7
```

```
12^(2/3)
        5.241482788
```

To evaluate radicals with an index greater than 3, follow the same procedure explained above. Evaluate by rewriting in exponential form if necessary and then using a graphing calculator. If necessary, approximate to the nearest tenth.

1) $16^{1/2}$ 2) $\sqrt[3]{512}$ 3) $\sqrt{37}$ 4) $361^{1/2}$ 5) $4096^{2/3}$ 6) $2401^{1/4}$

Answers to You Try Exercises

1) a) 2 b) 11 c) not a real number d) −3 e) −3   2) a) 4 b) −1000

c) not a real number d) 1 e) −1   3) a) $\frac{1}{12}$ b) $\frac{1}{8}$ c) $\frac{9}{4}$ 4) a) 7 b) 2 c) $\frac{1}{7^{2/5}}$

5) a) $a^{30}b^2$ b) $\frac{1}{t^{2/5}}$ c) $s^{3/2}$ d) $\frac{x^2}{y^{1/20}}$ 6) a) 5 b) $\sqrt[5]{p^2}$ 7) a) 10 b) 7 c) t

Answers to Technology Exercises

1) 4   2) 8   3) 6.1   4) 19   5) 256   6) 7

8.2 Exercises

*Additional answers can be found in the Answers to Exercises appendix.

Objective 1: Evaluate Expressions of the Form $a^{1/n}$

1) Explain how to write $25^{1/2}$ in radical form.

2) Explain how to write $1^{1/3}$ in radical form.

Write in radical form and evaluate.

3) $9^{1/2}$ 3

4) $64^{1/2}$ 8

5) $1000^{1/3}$ 10

6) $27^{1/3}$ 3

7) $32^{1/5}$ 2

8) $81^{1/4}$ 3

9) $-125^{1/3}$ -5

10) $-64^{1/6}$ -2

11) $\left(\dfrac{4}{121}\right)^{1/2}$ $\dfrac{2}{11}$

12) $\left(\dfrac{4}{9}\right)^{1/2}$ $\dfrac{2}{3}$

13) $\left(\dfrac{125}{64}\right)^{1/3}$ $\dfrac{5}{4}$

14) $\left(\dfrac{16}{81}\right)^{1/4}$ $\dfrac{2}{3}$

15) $-\left(\dfrac{36}{169}\right)^{1/2}$ $-\dfrac{6}{13}$

16) $-\left(\dfrac{1000}{27}\right)^{1/3}$ $-\dfrac{10}{3}$

17) $(-81)^{1/4}$ not a real number

18) $(-169)^{1/2}$ not a real number

19) $(-1)^{1/7}$ -1

20) $(-8)^{1/3}$ -2

Objective 2: Evaluate Expressions of the Form $a^{m/n}$

21) Explain how to write $16^{3/4}$ in radical form.

22) Explain how to write $100^{3/2}$ in radical form.

Write in radical form and evaluate.

23) $8^{4/3}$ 16

24) $81^{3/4}$ 27

25) $64^{5/6}$ 32

26) $32^{3/5}$ 8

27) $(-125)^{2/3}$ 25

28) $(-1000)^{2/3}$ 100

29) $-36^{3/2}$ -216

30) $-27^{4/3}$ -81

31) $(-81)^{3/4}$ not a real number

32) $(-25)^{3/2}$ not a real number

33) $\left(\dfrac{16}{81}\right)^{3/4}$ $\dfrac{8}{27}$

34) $-16^{5/4}$ -32

35) $-\left(\dfrac{1000}{27}\right)^{2/3}$ $-\dfrac{100}{9}$

36) $-\left(\dfrac{8}{27}\right)^{4/3}$ $-\dfrac{16}{81}$

Objective 3: Evaluate Expressions of the Form $a^{-m/n}$

Decide whether each statement is true or false. Explain your answer.

37) $81^{-1/2} = -9$

38) $\left(\dfrac{1}{100}\right)^{-3/2} = \left(\dfrac{1}{100}\right)^{2/3}$

Rewrite with a positive exponent and evaluate.

Fill It In

Fill in the blanks with either the missing mathematical step or reason for the given step.

39) $64^{-1/2} = \left(\dfrac{1}{64}\right)^{1/2}$ The reciprocal of 64 is $\dfrac{1}{64}$.

$\quad = \sqrt{\dfrac{1}{64}}$ The denominator of the fractional exponent is the index of the radical.

$\quad = \dfrac{1}{8}$ Simplify.

40) $\left(\dfrac{1}{1000}\right)^{-1/3} = (1000)^{1/3}$ The reciprocal of $\dfrac{1}{1000}$ is 1000.

$\quad = \sqrt[3]{1000}$ The denominator of the fractional exponent is the index of the radical.

$\quad = 10$ Simplify.

41) $49^{-1/2}$ $\dfrac{1}{7}$

42) $100^{-1/2}$ $\dfrac{1}{10}$

43) $1000^{-1/3}$ $\dfrac{1}{10}$

44) $27^{-1/3}$ $\dfrac{1}{3}$

45) $\left(\dfrac{1}{81}\right)^{-1/4}$ 3

46) $\left(\dfrac{1}{32}\right)^{-1/5}$ 2

47) $-\left(\dfrac{1}{64}\right)^{-1/3}$ -4

48) $-\left(\dfrac{1}{125}\right)^{-1/3}$ -5

49) $64^{-5/6}$ $\dfrac{1}{32}$

50) $81^{-3/4}$ $\dfrac{1}{27}$

51) $125^{-2/3}$ $\dfrac{1}{25}$

52) $64^{-2/3}$ $\dfrac{1}{16}$

53) $\left(\dfrac{25}{4}\right)^{-3/2}$ $\dfrac{8}{125}$

54) $\left(\dfrac{9}{100}\right)^{-3/2}$ $\dfrac{1000}{27}$

55) $\left(\dfrac{64}{125}\right)^{-2/3}$ $\dfrac{25}{16}$

56) $\left(\dfrac{81}{16}\right)^{-3/4}$ $\dfrac{8}{27}$

Objective 4: Combine the Rules of Exponents

Simplify completely. The answer should contain only positive exponents.

57) $2^{2/3} \cdot 2^{7/3}$ 8

58) $5^{3/4} \cdot 5^{5/4}$ 25

59) $(9^{1/4})^2$ 3

60) $(7^{2/3})^3$ 49

61) $8^{7/5} \cdot 8^{-3/5}$ $8^{4/5}$

62) $6^{-4/3} \cdot 6^{5/3}$ $6^{1/3}$

63) $\dfrac{2^{23/4}}{2^{3/4}}$ 32

64) $\dfrac{5^{3/2}}{5^{9/2}}$ $\dfrac{1}{125}$

65) $\dfrac{4^{2/5}}{4^{6/5} \cdot 4^{3/5}}$ $\dfrac{1}{4^{7/5}}$

66) $\dfrac{6^{-1}}{6^{1/2} \cdot 6^{-5/2}}$ 6

Simplify completely. The answer should contain only positive exponents.

67) $z^{1/6} \cdot z^{5/6}$ z

68) $h^{1/6} \cdot h^{-3/4}$ $\dfrac{1}{h^{7/12}}$

69) $(-9v^{5/8})(8v^{3/4})$ $-72v^{11/8}$

70) $(-3x^{-1/3})(8x^{4/9})$ $-24x^{1/9}$

71) $\dfrac{a^{5/9}}{a^{4/9}}$ $a^{1/9}$

72) $\dfrac{x^{1/6}}{x^{5/6}}$ $\dfrac{1}{x^{2/3}}$

73) $\dfrac{20c^{-2/3}}{72c^{5/6}}$ $\dfrac{5}{18c^{3/2}}$

74) $\dfrac{48w^{3/10}}{10w^{2/5}}$ $\dfrac{24}{5w^{1/10}}$

75) $(x^{-2/9})^3$ $\dfrac{1}{x^{2/3}}$

76) $(n^{-2/7})^3$ $\dfrac{1}{n^{6/7}}$

77) $(z^{1/5})^{2/3}$ $z^{2/15}$

78) $(r^{4/3})^{5/2}$ $r^{10/3}$

79) $(81u^{8/3}v^4)^{3/4}$ $27u^2v^3$

80) $(64x^6y^{12/5})^{5/6}$ $32x^5y^2$

81) $(32r^{1/3}s^{4/9})^{3/5}$ $8r^{1/5}s^{4/15}$

82) $(125a^9b^{1/4})^{2/3}$ $25a^6b^{1/6}$

83) $\left(\dfrac{f^{6/7}}{27g^{-5/3}}\right)^{1/3}$ $\dfrac{f^{2/7}g^{5/9}}{3}$

84) $\left(\dfrac{16c^{-8}}{b^{-11/3}}\right)^{3/4}$ $\dfrac{8b^{11/4}}{c^6}$

85) $\left(\dfrac{x^{-5/3}}{w^{3/2}}\right)^{-6}$ $x^{10}w^9$

86) $\left(\dfrac{t^{-3/2}}{u^{1/4}}\right)^{-4}$ t^6u

87) $\dfrac{y^{1/2} \cdot y^{-1/3}}{y^{5/6}}$ $\dfrac{1}{y^{2/3}}$

88) $\dfrac{t^5}{t^{1/2} \cdot t^{3/4}}$ $t^{15/4}$

89) $\left(\dfrac{a^4b^3}{32a^{-2}b^4}\right)^{2/5}$ $\dfrac{a^{12/5}}{4b^{2/5}}$

90) $\left(\dfrac{16c^{-8}d^3}{c^4d^5}\right)^{3/2}$ $\dfrac{64}{c^{18}d^3}$

91) $\left(\dfrac{r^{4/5}t^{-2}}{r^{2/3}t^5}\right)^{-3/2}$ $\dfrac{t^{21/2}}{r^{1/5}}$

92) $\left(\dfrac{x^{10}y^{1/6}}{x^{-8}y^{2/3}}\right)^{-2/3}$ $\dfrac{y^{1/3}}{x^{12}}$

93) $\left(\dfrac{h^{-2}k^{5/2}}{h^{-8}k^{5/6}}\right)^{-5/6}$ $\dfrac{1}{h^5k^{25/18}}$

94) $\left(\dfrac{c^{1/8}d^{-4}}{c^{3/4}d}\right)^{-8/5}$ cd^8

95) $p^{1/2}(p^{2/3} + p^{1/2})$ $p^{7/6} + p$

96) $w^{4/3}(w^{1/2} - w^3)$ $w^{11/6} - w^{13/3}$

Objective 5: Convert a Radical Expression to Exponential Form and Simplify

Rewrite each radical in exponential form, then simplify. Write the answer in simplest (or radical) form.

Fill It In

Fill in the blanks with either the missing mathematical step or reason for the given step.

97) $\sqrt[12]{25^6} = \underline{25^{6/12}}$ Write with a rational exponent.

  $= 25^{1/2}$ Reduce the exponent.

  $= 5$ <u>Evaluate.</u>

98) $\sqrt[10]{c^4} = \underline{c^{4/10}}$ Write with a rational exponent.

  $= c^{2/5}$ <u>Reduce the exponent.</u>

  $= \underline{\sqrt[5]{c^2}}$ Write in radical form.

99) $\sqrt[6]{49^3}$ 7

100) $\sqrt[9]{8^3}$ 2

101) $\sqrt[4]{81^2}$ 9

102) $\sqrt{3^2}$ 3

103) $(\sqrt{5})^2$ 5

104) $(\sqrt[3]{10})^3$ 10

105) $(\sqrt[3]{12})^3$ 12

106) $(\sqrt[4]{15})^4$ 15

107) $\sqrt[3]{x^{12}}$ x^4

108) $\sqrt[4]{t^8}$ t^2

109) $\sqrt[6]{k^2}$ $\sqrt[3]{k}$

110) $\sqrt[9]{w^6}$ $\sqrt[3]{w^2}$

111) $\sqrt[4]{z^2}$ $\sqrt{z}$

112) $\sqrt[8]{m^4}$ $\sqrt{m}$

113) $\sqrt{d^4}$ d^2

114) $\sqrt{s^6}$ s^3

The wind chill temperature, WC, measures how cold it feels outside (for temperatures under 50 degrees F) when the velocity of the wind, V, is considered along with the air temperature, T. The stronger the wind at a given air temperature, the colder it feels.

The formula for calculating wind chill is

$$WC = 35.74 + 0.6215T - 35.75V^{4/25} + 0.4275TV^{4/25}$$

where WC and T are in degrees Fahrenheit and V is in miles per hour.
(http://www.nws.noaa.gov/om/windchill/windchillglossary.shtml)

Use this information for Exercises 115 and 116, and round all answers to the nearest degree.

115) Determine the wind chill when the air temperature is 20 degrees and the wind is blowing at the given speed.

   a) 5 mph   13 degrees F

   b) 15 mph   6 degrees F

116) Determine the wind chill when the air temperature is 10 degrees and the wind is blowing at the given speed. Round your answer to the nearest degree.

   a) 12 mph -5 degrees F

   b) 20 mph -9 degrees F

Section 8.3 Simplifying Expressions Containing Square Roots

Objectives

1. **Multiply Square Roots**
2. **Simplify the Square Root of a Whole Number**
3. **Use the Quotient Rule for Square Roots**
4. **Simplify Square Root Expressions Containing Variables with Even Exponents**
5. **Simplify Square Root Expressions Containing Variables with Odd Exponents**
6. **Simplify More Square Root Expressions Containing Variables**

In this section, we will introduce rules for finding the product and quotient of square roots as well as for simplifying expressions containing square roots.

1. Multiply Square Roots

Let's begin with the product $\sqrt{4} \cdot \sqrt{9}$. $\sqrt{4} \cdot \sqrt{9} = 2 \cdot 3 = 6$. Also notice that $\sqrt{4} \cdot \sqrt{9} = \sqrt{4 \cdot 9} = \sqrt{36} = 6$.

We obtain the same result. This leads us to the product rule for multiplying expressions containing square roots.

Definition **Product Rule for Square Roots**

Let a and b be nonnegative real numbers. Then,

$$\sqrt{a} \cdot \sqrt{b} = \sqrt{a \cdot b}$$

In other words, the product of two square roots equals the square root of the product.

 Example 1

In-Class Example 1
Multiply.
a) $\sqrt{7} \cdot \sqrt{3}$ b) $\sqrt{5} \cdot \sqrt{g}$
answer: a) $\sqrt{21}$ b) $\sqrt{5g}$

Multiply.   a) $\sqrt{5} \cdot \sqrt{2}$ b) $\sqrt{3} \cdot \sqrt{x}$

Solution

a) $\sqrt{5} \cdot \sqrt{2} = \sqrt{5 \cdot 2} = \sqrt{10}$ b) $\sqrt{3} \cdot \sqrt{x} = \sqrt{3 \cdot x} = \sqrt{3x}$

 **BE CAREFUL**

We can multiply radicals this way *only if* the indices are the same. We will see later how to multiply radicals with different indices such as $\sqrt{5} \cdot \sqrt[3]{t}$.

 You Try 1

Multiply.
a) $\sqrt{6} \cdot \sqrt{5}$ b) $\sqrt{10} \cdot \sqrt{r}$

2. Simplify the Square Root of a Whole Number

Knowing how to simplify radicals is very important in the study of algebra. We begin by discussing how to simplify expressions containing square roots.

How do we know when a square root is simplified?

Property **When Is a Square Root Simplified?**

An expression containing a square root is simplified when all of the following conditions are met:

1) The radicand does not contain any factors (other than 1) that are perfect squares.

2) The radicand does not contain any fractions.

3) There are no radicals in the denominator of a fraction.

Note: Condition 1) implies that the radical cannot contain variables with exponents greater than or equal to 2, the index of the square root.

We will discuss higher roots in Section 8.4.

To simplify expressions containing square roots we reverse the process of multiplying. That is, we use the product rule that says $\sqrt{a \cdot b} = \sqrt{a} \cdot \sqrt{b}$ where a or b are perfect squares.

Example 2

In-Class Example 2
Simplify completely.
a) $\sqrt{45}$ b) $\sqrt{72}$
c) $\sqrt{12}$ d) $\sqrt{15}$
answer: a) $3\sqrt{5}$ b) $6\sqrt{2}$
c) $2\sqrt{3}$ d) $\sqrt{15}$

Simplify completely.

a) $\sqrt{18}$ b) $\sqrt{500}$ c) $\sqrt{21}$ d) $\sqrt{48}$

Solution

a)   The radical $\sqrt{18}$ is not in simplest form since 18 contains a factor (other than 1) that is a perfect square. Think of two numbers that multiply to 18 so that at least one of the numbers is a perfect square: $18 = 9 \cdot 2$.

   (While it is true that $18 = 6 \cdot 3$, neither 6 nor 3 is a perfect square.)
   Rewrite $\sqrt{18}$:

$$\begin{aligned} \sqrt{18} &= \sqrt{9 \cdot 2} & \text{9 is a perfect square.} \\ &= \sqrt{9} \cdot \sqrt{2} & \text{Product rule} \\ &= 3\sqrt{2} & \sqrt{9} = 3 \end{aligned}$$

   $3\sqrt{2}$ is completely simplified because 2 does not have any factors that are perfect squares.

b)   Does 500 have a factor that is a perfect square? Yes! $500 = 100 \cdot 5$. To simplify $\sqrt{500}$, rewrite it as

$$\begin{aligned} \sqrt{500} &= \sqrt{100 \cdot 5} & \text{100 is a perfect square.} \\ &= \sqrt{100} \cdot \sqrt{5} & \text{Product rule} \\ &= 10\sqrt{5} & \sqrt{100} = 10 \end{aligned}$$

   $10\sqrt{5}$ is completely simplified because 5 does not have any factors that are perfect squares.

c) $21 = 3 \cdot 7$ Neither 3 nor 7 is a perfect square.
   $21 = 1 \cdot 21$ Although 1 is a perfect square, it will not help us simplify $\sqrt{21}$.

   $\sqrt{21}$ is in simplest form.

d)   There are different ways to simplify $\sqrt{48}$. We will look at two of them.

   i)   Two numbers that multiply to 48 are 16 and 3 with 16 being a perfect square. We can write

$$\sqrt{48} = \sqrt{16 \cdot 3} = \sqrt{16} \cdot \sqrt{3} = 4\sqrt{3}$$

   ii)   We can also think of 48 as $4 \cdot 12$ since 4 is a perfect square. We can write

$$\sqrt{48} = \sqrt{4 \cdot 12} = \sqrt{4} \cdot \sqrt{12} = 2\sqrt{12}$$

   Therefore, $\sqrt{48} = 2\sqrt{12}$. Is $\sqrt{12}$ in simplest form? *No, because $12 = 4 \cdot 3$ and 4 is a perfect square.* We must continue to simplify.

$$\begin{aligned} \sqrt{48} &= 2\sqrt{12} \\ &= 2\sqrt{4 \cdot 3} = 2\sqrt{4} \cdot \sqrt{3} = 2 \cdot 2 \cdot \sqrt{3} = 4\sqrt{3} \end{aligned}$$

   $4\sqrt{3}$ is completely simplified because 3 does not have any factors that are perfect squares.

Example 2(d) shows that using either $\sqrt{48} = \sqrt{16 \cdot 3}$ or $\sqrt{48} = \sqrt{4 \cdot 12}$ leads us to the same result. Furthermore, this example illustrates that a radical is not always *completely* simplified after just one iteration of the simplification process. It is necessary to always examine the radical to determine whether or not it can be simplified more. ∎

Note

After simplifying a radical, look at the result and ask yourself, "*Is the radical in simplest form?*" If it is not, simplify again. Asking yourself this question will help you to be sure that the radical *is* completely simplified.

You Try 2

Simplify completely.

a) $\sqrt{28}$ b) $\sqrt{75}$ c) $\sqrt{72}$

3. Use the Quotient Rule for Square Roots

Let's simplify $\dfrac{\sqrt{36}}{\sqrt{9}}$. We can say $\dfrac{\sqrt{36}}{\sqrt{9}} = \dfrac{6}{3} = 2$. It is also true that $\dfrac{\sqrt{36}}{\sqrt{9}} = \sqrt{\dfrac{36}{9}} = \sqrt{4} = 2$.

This leads us to the quotient rule for dividing expressions containing square roots.

Definition   Quotient Rule for Square Roots

Let a and b be nonnegative real numbers such that $b \neq 0$. Then,

$$\sqrt{\dfrac{a}{b}} = \dfrac{\sqrt{a}}{\sqrt{b}}$$

The square root of a quotient equals the quotient of the square roots.

Example 3

Simplify completely.

a) $\sqrt{\dfrac{9}{49}}$ b) $\sqrt{\dfrac{200}{2}}$ c) $\dfrac{\sqrt{72}}{\sqrt{6}}$ d) $\sqrt{\dfrac{5}{81}}$

In-Class Example 3
Simplify completely.
a) $\sqrt{\dfrac{8}{49}}$ b) $\sqrt{\dfrac{25}{5}}$
c) $\dfrac{\sqrt{120}}{\sqrt{10}}$ d) $\sqrt{\dfrac{5}{36}}$
answer: a) $\dfrac{2\sqrt{2}}{7}$ **b)** $\sqrt{5}$
c) $2\sqrt{3}$ **d)** $\dfrac{\sqrt{5}}{6}$

Solution

a) Since 9 and 49 are each perfect squares, find the square root of each separately.

$$\sqrt{\dfrac{9}{49}} = \dfrac{\sqrt{9}}{\sqrt{49}} \qquad \text{Quotient rule}$$

$$= \dfrac{3}{7} \qquad \sqrt{9} = 3 \text{ and } \sqrt{49} = 7$$

b) Neither 200 nor 2 is a perfect square, but if we simplify $\dfrac{200}{2}$ we get 100, which *is* a perfect square.

$$\sqrt{\dfrac{200}{2}} = \sqrt{100} \qquad \text{Simplify } \dfrac{200}{2}.$$

$$= 10$$

c)   We can simplify $\dfrac{\sqrt{72}}{\sqrt{6}}$ using two different methods.

   i)   Begin by applying the quotient rule to obtain a fraction under *one* radical and simplify the fraction.

$$\frac{\sqrt{72}}{\sqrt{6}} = \sqrt{\frac{72}{6}} \qquad \text{Quotient rule}$$

$$= \sqrt{12} \; = \; \sqrt{4 \cdot 3} \; = \; \sqrt{4} \cdot \sqrt{3} \; = \; 2\sqrt{3}$$

   ii)   We can apply the product rule to rewrite $\sqrt{72}$ then simplify the fraction.

$$\frac{\sqrt{72}}{\sqrt{6}} = \frac{\sqrt{6} \cdot \sqrt{12}}{\sqrt{6}} \qquad \text{Product rule}$$

$$= \frac{\overset{1}{\cancel{\sqrt{6}}} \cdot \sqrt{12}}{\underset{1}{\cancel{\sqrt{6}}}} \qquad \text{Divide out the common factor.}$$

$$= \sqrt{12} \; = \; \sqrt{4 \cdot 3} \; = \; \sqrt{4} \cdot \sqrt{3} \; = \; 2\sqrt{3}$$

Either method will produce the same result.

d)   The fraction $\dfrac{5}{81}$ does not reduce and 81 *is* a perfect square. Begin by applying the quotient rule.

$$\sqrt{\frac{5}{81}} = \frac{\sqrt{5}}{\sqrt{81}} \qquad \text{Quotient rule}$$

$$= \frac{\sqrt{5}}{9} \qquad \sqrt{81} = 9$$

You Try 3

Simplify completely.

a) $\sqrt{\dfrac{100}{169}}$ b) $\sqrt{\dfrac{27}{3}}$ c) $\dfrac{\sqrt{250}}{\sqrt{5}}$ d) $\sqrt{\dfrac{11}{36}}$

4. Simplify Square Root Expressions Containing Variables with Even Exponents

Recall that a square root is not simplified if it contains any factors that are perfect squares.
   This means that a square root containing variables is simplified if the power on each variable is less than 2. For example, $\sqrt{r^6}$ is not in simplified form. If r represents a nonnegative real number, then we can use rational exponents to simplify $\sqrt{r^6}$.

$$\sqrt{r^6} = (r^6)^{1/2} = r^{6 \cdot \frac{1}{2}} = r^{6/2} = r^3$$

Multiplying $6 \cdot \frac{1}{2}$ is the same as dividing 6 by 2. We can generalize this result with the following statement.

Property $\sqrt{a^m}$

If a is a nonnegative real number and m is an integer, then

$$\sqrt{a^m} = a^{m/2}$$

We can combine this property with the product and quotient rules to simplify radical expressions.

Example 4

In-Class Example 4
Simplify completely.
a) $\sqrt{b^2}$ b) $\sqrt{100a^4}$
c) $\sqrt{24p^8}$ d) $\sqrt{\dfrac{27}{g^6}}$
answer: a) b b) $10a^2$
c) $2p^4\sqrt{6}$ d) $\dfrac{3\sqrt{3}}{g^3}$

Simplify completely.

a) $\sqrt{z^2}$ 　　b) $\sqrt{49t^2}$ 　　c) $\sqrt{18b^{14}}$ 　　d) $\sqrt{\dfrac{32}{n^{20}}}$

Solution

a) $\sqrt{z^2} = z^{2/2} = z^1 = z$

b) $\sqrt{49t^2} = \sqrt{49} \cdot \sqrt{t^2} = 7 \cdot t^{2/2} = 7t$

c) $\sqrt{18b^{14}} = \sqrt{18} \cdot \sqrt{b^{14}}$ 　　Product rule

　　　　 $= \sqrt{9} \cdot \sqrt{2} \cdot b^{14/2}$ 　　9 is a perfect square.

　　　　 $= 3\sqrt{2} \cdot b^7$ 　　Simplify.

　　　　 $= 3b^7\sqrt{2}$ 　　Rewrite using the commutative property.

d)　We begin by using the quotient rule.

$$\sqrt{\frac{32}{n^{20}}} = \frac{\sqrt{32}}{\sqrt{n^{20}}} = \frac{\sqrt{16} \cdot \sqrt{2}}{n^{20/2}} = \frac{4\sqrt{2}}{n^{10}}$$

You Try 4

Simplify completely.

a) $\sqrt{y^{10}}$ 　　b) $\sqrt{144p^{16}}$ 　　c) $\sqrt{\dfrac{45}{w^4}}$

5. Simplify Square Root Expressions Containing Variables with Odd Exponents

How do we simplify an expression containing a square root if the power under the square root is odd? We can use the product rule for radicals and fractional exponents to help us understand how to simplify such expressions.

Example 5

In-Class Example 5
Simplify completely.
a) $\sqrt{m^9}$ b) $\sqrt{t^{13}}$
answer: a) $m^4\sqrt{m}$ b) $t^6\sqrt{t}$

Simplify completely.

a) $\sqrt{x^7}$ 　　b) $\sqrt{c^{11}}$

Solution

a)　To simplify $\sqrt{x^7}$, write x^7 as the product of two factors so that the exponent of one of the factors is the *largest* number less than 7 that is divisible by 2 (the index of the radical).

　　　 $\sqrt{x^7} = \sqrt{x^6 \cdot x^1}$ 　　6 is the largest number less than 7 that is divisible by 2.

　　　　　 $= \sqrt{x^6} \cdot \sqrt{x}$ 　　Product rule

　　　　　 $= x^{6/2} \cdot \sqrt{x}$ 　　Use a fractional exponent to simplify.

　　　　　 $= x^3\sqrt{x}$ 　　$6 \div 2 = 3$

b)　To simplify $\sqrt{c^{11}}$, write c^{11} as the product of two factors so that the exponent of one of the factors is the *largest* number less than 11 that is divisible by 2 (the index of the radical).

　　　 $\sqrt{c^{11}} = \sqrt{c^{10} \cdot c^1}$ 　　10 is the largest number less than 11 that is divisible by 2.

　　　　　 $= \sqrt{c^{10}} \cdot \sqrt{c}$ 　　Product rule

　　　　　 $= c^{10/2} \cdot \sqrt{c}$ 　　Use a fractional exponent to simplify.

　　　　　 $= c^5\sqrt{c}$ 　　$10 \div 2 = 5$

You Try 5

Simplify completely.

a) $\sqrt{m^5}$ b) $\sqrt{z^{19}}$

We used the product rule to simplify each radical in Example 5. During the simplification, however, we always divided an exponent by 2. This idea of division gives us another way to simplify radical expressions. Once again, let's look at the radicals and their simplified forms in Example 5 to see how we can simplify radical expressions using division.

$$\sqrt{x^7} = x^3\sqrt{x^1} = x^3\sqrt{x} \qquad \sqrt{c^{11}} = c^5\sqrt{c^1} = c^5\sqrt{c}$$

Index of radical → 2) 7   3 → Quotient
                        −6
                        1 → Remainder

Index of radical → 2) 11   5 → Quotient
                        −10
                        1 → Remainder

Procedure Simplifying a Radical Containing Variables

To simplify a radical expression containing variables:

1) Divide the original exponent in the radicand by the index of the radical.

2) The exponent on the variable *outside* of the radical will be the *quotient* of the division problem.

3) The exponent on the variable *inside* of the radical will be the *remainder* of the division problem.

Example 6

In-Class Example 6
Simplify completely.
a) $\sqrt{k^5}$ b) $\sqrt{49v^3}$
c) $\sqrt{32m^{19}}$
answer: a) $k^2\sqrt{k}$
b) $7v\sqrt{v}$ c) $4m^9\sqrt{2m}$

Simplify completely.

a) $\sqrt{t^9}$ b) $\sqrt{16b^5}$ c) $\sqrt{45y^{21}}$

Solution

a) To simplify $\sqrt{t^9}$, divide: 2) 9   4 → Quotient
                                                −8
                                                1 → Remainder

$$\sqrt{t^9} = t^4\sqrt{t^1} = t^4\sqrt{t}$$

b) $\sqrt{16b^5} = \sqrt{16} \cdot \sqrt{b^5}$ Product rule
    $\quad\quad\quad = 4 \cdot b^2\sqrt{b^1}$ $5 \div 2$ gives a quotient of 2 and a remainder of 1.
    $\quad\quad\quad = 4b^2\sqrt{b}$

c) $\sqrt{45y^{21}} = \sqrt{45} \cdot \sqrt{y^{21}}$ Product rule
    $\quad\quad\quad = \sqrt{9} \cdot \sqrt{5} \cdot y^{10}\sqrt{y^1}$
    $\quad\quad\quad\quad\quad \uparrow \quad\quad\quad\quad \uparrow$
    $\quad\quad\quad\quad$ Product $21 \div 2$ gives a quotient of 10
    $\quad\quad\quad\quad$ rule   and a remainder of 1.
    $\quad\quad\quad = 3\sqrt{5} \cdot y^{10}\sqrt{y}$ $\sqrt{9} = 3$
    $\quad\quad\quad = 3y^{10} \cdot \sqrt{5} \cdot \sqrt{y}$ Use the commutative property to rewrite the expression.
    $\quad\quad\quad = 3y^{10}\sqrt{5y}$ Use the product rule to write the expression with one radical.

You Try 6

Simplify completely.

a) $\sqrt{m^{13}}$ b) $\sqrt{100v^7}$ c) $\sqrt{32a^3}$

If a radical contains more than one variable, apply the product or quotient rule.

Example 7

In-Class Example 7
Simplify completely.

a) $\sqrt{18m^7n^9}$ b) $\sqrt{\dfrac{3f^9}{g^{12}}}$

answer: a) $3m^3n^4\sqrt{2mn}$

b) $\dfrac{f^4\sqrt{3f}}{g^6}$

Simplify completely.

a) $\sqrt{8a^{15}b^3}$ b) $\sqrt{\dfrac{5r^{27}}{s^8}}$

Solution

a) $\sqrt{8a^{15}b^3} = \sqrt{8} \quad \cdot \quad \sqrt{a^{15}} \quad \cdot \quad \sqrt{b^3}$

$= \sqrt{4} \cdot \sqrt{2} \cdot a^7\sqrt{a^1} \quad \cdot \quad b^1\sqrt{b^1}$

     Product rule   15 ÷ 2 gives a quotient   3 ÷ 2 gives a quotient
                of 7 and a remainder of 1.   of 1 and a remainder of 1.

$= 2\sqrt{2} \cdot a^7\sqrt{a} \cdot b\sqrt{b} \quad \sqrt{4} = 2$

$= 2a^7b \cdot \sqrt{2} \cdot \sqrt{a} \cdot \sqrt{b}$ Use the commutative property to rewrite the expression.

$= 2a^7b\sqrt{2ab}$ Use the product rule to write the expression with one radical.

b) $\sqrt{\dfrac{5r^{27}}{s^8}} = \dfrac{\sqrt{5r^{27}}}{\sqrt{s^8}}$ Quotient rule

$= \dfrac{\sqrt{5} \cdot \sqrt{r^{27}}}{s^4}$ ⟵ Product rule
                     ⟵ 8 ÷ 2 = 4

$= \dfrac{\sqrt{5} \cdot r^{13}\sqrt{r^1}}{s^4}$ 27 ÷ 2 gives a quotient of 13 and a remainder of 1.

$= \dfrac{r^{13} \cdot \sqrt{5} \cdot \sqrt{r}}{s^4}$ Use the commutative property to rewrite the expression.

$= \dfrac{r^{13}\sqrt{5r}}{s^4}$ Use the product rule to write the expression with one radical.

You Try 7

Simplify completely.

a) $\sqrt{c^5d^{12}}$ b) $\sqrt{27x^{10}y^9}$ c) $\sqrt{\dfrac{40u^{13}}{v^{20}}}$

6. Simplify More Square Root Expressions Containing Variables

Next we will look at some examples of multiplying and dividing radical expressions that contain variables. Remember to always look at the result and ask yourself, "*Is the radical in simplest form?*" If it is not, simplify completely.

Example 8

Perform the indicated operation and simplify completely.

a) $\sqrt{6t} \cdot \sqrt{3t}$ b) $\sqrt{2a^3b} \cdot \sqrt{8a^2b^5}$ c) $\dfrac{\sqrt{20x^5}}{\sqrt{5x}}$

In-Class Example 8
Perform the indicated operation and simplify completely.

a) $\sqrt{8b} \cdot \sqrt{2b}$
b) $\sqrt{3x^5y} \cdot \sqrt{6xy^4}$
c) $\dfrac{\sqrt{36r^7}}{\sqrt{2r^3}}$
answer: a) $4b$ b) $3x^3y^2\sqrt{2y}$
c) $3r^2\sqrt{2}$

Solution

a) $\sqrt{6t} \cdot \sqrt{3t} = \sqrt{6t \cdot 3t}$ Product rule

$= \sqrt{18t^2}$

$= \sqrt{18} \cdot \sqrt{t^2}$ Product rule

$= \sqrt{9 \cdot 2} \cdot t = \sqrt{9} \cdot \sqrt{2} \cdot t = 3\sqrt{2} \cdot t = 3t\sqrt{2}$

b) $\sqrt{2a^3b} \cdot \sqrt{8a^2b^5}$

There are two good methods for multiplying these radicals.

i) Multiply the radicands to obtain one radical.

$\sqrt{2a^3b} \cdot \sqrt{8a^2b^5} = \sqrt{2a^3b \cdot 8a^2b^5}$ Product rule

$= \sqrt{16a^5b^6}$ Multiply.

Is the radical in simplest form? *No.*

$= \sqrt{16} \cdot \sqrt{a^5} \cdot \sqrt{b^6}$ Product rule

$= 4 \cdot a^2\sqrt{a} \cdot b^3$ Evaluate.

$= 4a^2b^3\sqrt{a}$ Commutative property

ii) Simplify each radical, then multiply.

$\sqrt{2a^3b} = \sqrt{2} \cdot \sqrt{a^3} \cdot \sqrt{b}$ $\sqrt{8a^2b^5} = \sqrt{8} \cdot \sqrt{a^2} \cdot \sqrt{b^5}$

$= \sqrt{2} \cdot a\sqrt{a} \cdot \sqrt{b}$ $= 2\sqrt{2} \cdot a \cdot b^2\sqrt{b}$

$= a\sqrt{2ab}$ $= 2ab^2\sqrt{2b}$

Then, $\sqrt{2a^3b} \cdot \sqrt{8a^2b^5} = a\sqrt{2ab} \cdot 2ab^2\sqrt{2b}$

$= a \cdot 2ab^2 \cdot \sqrt{2ab} \cdot \sqrt{2b}$ Commutative property

$= 2a^2b^2\sqrt{4ab^2}$ Multiply.

$= 2a^2b^2 \cdot 2 \cdot b \cdot \sqrt{a}$ $\sqrt{4ab^2} = 2b\sqrt{a}$

$= 4a^2b^3\sqrt{a}$ Multiply.

Both methods give the same result.

c) We can use the quotient rule first or simplify first.

i) $\dfrac{\sqrt{20x^5}}{\sqrt{5x}} = \sqrt{\dfrac{20x^5}{5x}}$ Use the quotient rule first.

$= \sqrt{4x^4} = \sqrt{4} \cdot \sqrt{x^4} = 2x^2$

ii) $\dfrac{\sqrt{20x^5}}{\sqrt{5x}} = \dfrac{\sqrt{20} \cdot \sqrt{x^5}}{\sqrt{5x}}$ Simplify first by using the product rule.

$= \dfrac{\sqrt{4} \cdot \sqrt{5} \cdot x^2\sqrt{x}}{\sqrt{5x}}$ Product rule; simplify $\sqrt{x^5}$.

$= \dfrac{2\sqrt{5} \cdot x^2\sqrt{x}}{\sqrt{5x}}$ $\sqrt{4} = 2$

$= \dfrac{2x^2\sqrt{5x}}{\sqrt{5x}}$ Product rule

$= 2x^2$ Divide out the common factor.

Both methods give the same result. In this case, the second method was longer. Sometimes, however, this method *can* be more efficient.

You Try 8

Perform the indicated operation and simplify completely.

a) $\sqrt{2n^3} \cdot \sqrt{6n}$ b) $\sqrt{15cd^5} \cdot \sqrt{3c^2d}$ c) $\dfrac{\sqrt{128k^9}}{\sqrt{2k}}$

Answers to You Try Exercises

1) a) $\sqrt{30}$ b) $\sqrt{10r}$ 2) a) $2\sqrt{7}$ b) $5\sqrt{3}$ c) $6\sqrt{2}$ 3) a) $\dfrac{10}{13}$ b) 3 c) $5\sqrt{2}$ d) $\dfrac{\sqrt{11}}{6}$

4) a) y^5 b) $12p^8$ c) $\dfrac{3\sqrt{5}}{w^2}$ 5) a) $m^2\sqrt{m}$ b) $z^9\sqrt{z}$ 6) a) $m^6\sqrt{m}$ b) $10v^3\sqrt{v}$ c) $4a\sqrt{2a}$

7) a) $c^2d^6\sqrt{c}$ b) $3x^5y^4\sqrt{3y}$ c) $\dfrac{2u^6\sqrt{10u}}{v^{10}}$ 8) a) $2n^2\sqrt{3}$ b) $3cd^3\sqrt{5c}$ c) $8k^4$

8.3 Exercises

*Additional answers can be found in the Answers to Exercises appendix.

Unless otherwise stated, assume all variables represent nonnegative real numbers.

Objective 1: Multiply Square Roots

Multiply and simplify.

1) $\sqrt{3} \cdot \sqrt{7}$ $\sqrt{21}$ 2) $\sqrt{11} \cdot \sqrt{5}$ $\sqrt{55}$

3) $\sqrt{10} \cdot \sqrt{3}$ $\sqrt{30}$ 4) $\sqrt{7} \cdot \sqrt{2}$ $\sqrt{14}$

5) $\sqrt{6} \cdot \sqrt{y}$ $\sqrt{6y}$ 6) $\sqrt{5} \cdot \sqrt{p}$ $\sqrt{5p}$

Objective 2: Simplify the Square Root of a Whole Number

Label each statement as true or false. Give a reason for your answer.

7) $\sqrt{20}$ is in simplest form.
  False; 20 contains a factor of 4 which is a perfect square.

8) $\sqrt{35}$ is in simplest form.

9) $\sqrt{42}$ is in simplest form.

10) $\sqrt{63}$ is in simplest form.
  False; 63 contains a factor of 9 which is a perfect square.

Simplify completely.

Fill It In

Fill in the blanks with either the missing mathematical step or reason for the given step.

11) $\sqrt{60} = \sqrt{4 \cdot 15}$ Factor.
  $= \sqrt{4} \cdot \sqrt{15}$ Product rule
  $= 2\sqrt{15}$ Simplify.

12) $\sqrt{200} = \sqrt{100 \cdot 2}$ Factor.
  $= \sqrt{100} \cdot \sqrt{2}$ Product rule
  $= 10\sqrt{2}$ Simplify.

Simplify completely. If the radical is already simplified, then say so.

13) $\sqrt{20}$ $2\sqrt{5}$ 14) $\sqrt{12}$ $2\sqrt{3}$

15) $\sqrt{54}$ $3\sqrt{6}$ 16) $\sqrt{63}$ $3\sqrt{7}$

17) $\sqrt{33}$ simplified   18) $\sqrt{15}$ simplified

19) $\sqrt{80}$ $4\sqrt{5}$ 20) $\sqrt{108}$ $6\sqrt{3}$

21) $\sqrt{98}$ $7\sqrt{2}$ 22) $\sqrt{96}$ $4\sqrt{6}$

23) $\sqrt{38}$ simplified   24) $\sqrt{46}$ simplified

25) $\sqrt{400}$ 20 26) $\sqrt{900}$ 30

27) $\sqrt{750}$ $5\sqrt{30}$ 28) $\sqrt{420}$ $2\sqrt{105}$

Objective 3: Use the Quotient Rule for Square Roots

Simplify completely.

29) $\sqrt{\dfrac{144}{25}}$ $\dfrac{12}{5}$ 30) $\sqrt{\dfrac{16}{81}}$ $\dfrac{4}{9}$

31) $\dfrac{\sqrt{4}}{\sqrt{49}}$ $\dfrac{2}{7}$ 32) $\dfrac{\sqrt{64}}{\sqrt{121}}$ $\dfrac{8}{11}$

33) $\dfrac{\sqrt{54}}{\sqrt{6}}$ 3 34) $\dfrac{\sqrt{48}}{\sqrt{3}}$ 4

35) $\sqrt{\dfrac{60}{5}}$ $2\sqrt{3}$ 36) $\sqrt{\dfrac{40}{5}}$ $2\sqrt{2}$

37) $\dfrac{\sqrt{120}}{\sqrt{6}}$ $2\sqrt{5}$ 38) $\dfrac{\sqrt{54}}{\sqrt{3}}$ $3\sqrt{2}$

39) $\dfrac{\sqrt{30}}{\sqrt{2}}$ $\sqrt{15}$ 40) $\dfrac{\sqrt{35}}{\sqrt{5}}$ $\sqrt{7}$

41) $\sqrt{\dfrac{6}{49}}$ $\dfrac{\sqrt{6}}{7}$ 42) $\sqrt{\dfrac{2}{81}}$ $\dfrac{\sqrt{2}}{9}$

43) $\sqrt{\dfrac{45}{16}}$ $\dfrac{3\sqrt{5}}{4}$ 44) $\sqrt{\dfrac{60}{49}}$ $\dfrac{2\sqrt{15}}{7}$

Objective 4: Simplify Square Root Expressions Containing Variables with Even Exponents

Simplify completely.

45) $\sqrt{x^8}$ x^4 46) $\sqrt{q^6}$ q^3

47) $\sqrt{w^{14}}$ w^7 48) $\sqrt{t^{16}}$ t^8

49) $\sqrt{100c^2}$ $10c$ 50) $\sqrt{9z^8}$ $3z^4$

51) $\sqrt{64k^6m^{10}}$ $8k^3m^5$ 52) $\sqrt{25p^{20}q^{14}}$ $5p^{10}q^7$

53) $\sqrt{28r^4}$ $2r^2\sqrt{7}$ 54) $\sqrt{27z^{12}}$ $3z^6\sqrt{3}$

55) $\sqrt{300q^{22}t^{16}}$ $10q^{11}t^8\sqrt{3}$ 56) $\sqrt{50n^4y^4}$ $5n^2y^2\sqrt{2}$

57) $\sqrt{\dfrac{81}{c^6}}$ $\dfrac{9}{c^3}$ 58) $\sqrt{\dfrac{h^2}{169}}$ $\dfrac{h}{13}$

59) $\dfrac{\sqrt{40}}{\sqrt{t^8}}$ $\dfrac{2\sqrt{10}}{t^4}$ 60) $\dfrac{\sqrt{18}}{\sqrt{m^{30}}}$ $\dfrac{3\sqrt{2}}{m^{15}}$

61) $\sqrt{\dfrac{75x^2}{y^{12}}}$ $\dfrac{5x\sqrt{3}}{y^6}$ 62) $\sqrt{\dfrac{44}{w^2z^{18}}}$ $\dfrac{2\sqrt{11}}{wz^9}$

Objective 5: Simplify Square Root Expressions Containing Variables with Odd Exponents

Simplify completely.

Fill It In

Fill in the blanks with either the missing mathematical step or reason for the given step.

63) $\sqrt{w^9} = \sqrt{w^8 \cdot w^1}$ Factor.

   $= \sqrt{w^8} \cdot \sqrt{w^1}$ Product rule

   $= w^4\sqrt{w}$ Simplify.

64) $\sqrt{z^{19}} = \sqrt{z^{18} \cdot z^1}$ Factor.

   $= \sqrt{z^{18}} \cdot \sqrt{z^1}$ Product rule

   $= z^9\sqrt{z}$ Simplify.

65) $\sqrt{a^5}$ $a^2\sqrt{a}$ 66) $\sqrt{c^7}$ $c^3\sqrt{c}$

67) $\sqrt{g^{13}}$ $g^6\sqrt{g}$ 68) $\sqrt{k^{15}}$ $k^7\sqrt{k}$

69) $\sqrt{b^{25}}$ $b^{12}\sqrt{b}$ 70) $\sqrt{h^{31}}$ $h^{15}\sqrt{h}$

71) $\sqrt{72x^3}$ $6x\sqrt{2x}$ 72) $\sqrt{100a^5}$ $10a^2\sqrt{a}$

73) $\sqrt{13q^7}$ $q^3\sqrt{13q}$ 74) $\sqrt{20c^9}$ $2c^4\sqrt{5c}$

75) $\sqrt{75t^{11}}$ $5t^5\sqrt{3t}$ 76) $\sqrt{45p^{17}}$ $3p^8\sqrt{5p}$

77) $\sqrt{c^8d^2}$ c^4d 78) $\sqrt{r^4s^{12}}$ r^2s^6

79) $\sqrt{a^4b^3}$ $a^2b\sqrt{b}$ 80) $\sqrt{x^2y^9}$ $xy^4\sqrt{y}$

81) $\sqrt{u^5v^7}$ $u^2v^3\sqrt{uv}$ 82) $\sqrt{f^3g^9}$ $fg^4\sqrt{fg}$

83) $\sqrt{36m^9n^4}$ $6m^4n^2\sqrt{m}$ 84) $\sqrt{4t^6u^5}$ $2t^3u^2\sqrt{u}$

85) $\sqrt{44x^{12}y^5}$ $2x^6y^2\sqrt{11y}$ 86) $\sqrt{63c^7d^4}$ $3c^3d^2\sqrt{7c}$

87) $\sqrt{32t^5u^7}$ $4t^2u^3\sqrt{2tu}$ 88) $\sqrt{125k^3l^9}$ $5kl^4\sqrt{5kl}$

89) $\sqrt{\dfrac{a^7}{81b^6}}$ $\dfrac{a^3\sqrt{a}}{9b^3}$ 90) $\sqrt{\dfrac{x^5}{49y^6}}$ $\dfrac{x^2\sqrt{x}}{7y^3}$

91) $\sqrt{\dfrac{3r^9}{s^2}}$ $\dfrac{r^4\sqrt{3r}}{s}$ 92) $\sqrt{\dfrac{17h^{11}}{k^8}}$ $\dfrac{h^5\sqrt{17h}}{k^4}$

Objective 6: Simplify More Square Root Expressions Containing Variables

Perform the indicated operation and simplify. Assume all variables represent positive real numbers.

93) $\sqrt{5} \cdot \sqrt{10}$ $5\sqrt{2}$ 94) $\sqrt{8} \cdot \sqrt{6}$ $4\sqrt{3}$

95) $\sqrt{21} \cdot \sqrt{3}$ $3\sqrt{7}$ 96) $\sqrt{2} \cdot \sqrt{14}$ $2\sqrt{7}$

97) $\sqrt{w} \cdot \sqrt{w^5}$ w^3 98) $\sqrt{d^3} \cdot \sqrt{d^{11}}$ d^7

99) $\sqrt{n^3} \cdot \sqrt{n^4}$ $n^3\sqrt{n}$ 100) $\sqrt{a^{10}} \cdot \sqrt{a^3}$ $a^6\sqrt{a}$

101) $\sqrt{2k} \cdot \sqrt{8k^5}$ $4k^3$ 102) $\sqrt{5z^9} \cdot \sqrt{5z^3}$ $5z^6$

103) $\sqrt{6x^4y^3} \cdot \sqrt{2x^5y^2}$ 104) $\sqrt{5a^6b^5} \cdot \sqrt{10ab^4}$

105) $\sqrt{8c^9d^2} \cdot \sqrt{5cd^7}$ 106) $\sqrt{6t^3u^3} \cdot \sqrt{3t^7u^4}$

107) $\dfrac{\sqrt{18k^{11}}}{\sqrt{2k^3}}$ $3k^4$ 108) $\dfrac{\sqrt{48m^{15}}}{\sqrt{3m^9}}$ $4m^3$

109) $\dfrac{\sqrt{120h^8}}{\sqrt{3h^2}}$ $2h^3\sqrt{10}$ 110) $\dfrac{\sqrt{72c^{10}}}{\sqrt{6c^2}}$ $2c^4\sqrt{3}$

111) $\dfrac{\sqrt{50a^{16}b^9}}{\sqrt{5a^7b^4}}$ $a^4b^2\sqrt{10ab}$ 112) $\dfrac{\sqrt{21y^8z^{18}}}{\sqrt{3yz^{13}}}$ $y^3z^2\sqrt{7yz}$

113) The velocity v of a moving object can be determined from its mass m and its kinetic energy KE using the formula $v = \sqrt{\dfrac{2KE}{m}}$, where the velocity is in meters/second, the mass is in kilograms, and the KE is measured in joules. A 600-kg roller coaster car is moving along a track and has kinetic energy of 120,000 joules. What is the velocity of the car? 20 m/s

114) The length of a side s of an equilateral triangle is a function of its area A and can be described by $s(A) = \sqrt{\dfrac{4\sqrt{3}A}{3}}$. If an equilateral triangle has an area of $6\sqrt{3}$ cm^2, how long is each side of the triangle? $2\sqrt{6}$ cm

Section 8.4 Simplifying Expressions Containing Higher Roots

Objectives

1. **Multiply Higher Roots**
2. **Simplify Higher Roots of Integers**
3. **Use the Quotient Rule for Higher Roots**
4. **Simplify Radicals Containing Variables**
5. **Multiply and Divide Radicals with Different Indices**

In Section 8.1 we first discussed finding higher roots like $\sqrt[4]{16} = 2$ and $\sqrt[3]{-27} = -3$. In this section, we will extend what we learned about multiplying, dividing, and simplifying *square* roots to doing the same with higher roots.

1. Multiply Higher Roots

Definition   Product Rule for Higher Roots

If $\sqrt[n]{a}$ and $\sqrt[n]{b}$ are real numbers, then

$$\sqrt[n]{a} \cdot \sqrt[n]{b} = \sqrt[n]{a \cdot b}$$

This rule enables us to multiply and simplify radicals with any index in a way that is similar to multiplying and simplifying square roots.

Example 1

In-Class Example 1
Multiply.
a) $\sqrt[3]{3} \cdot \sqrt[3]{8}$ b) $\sqrt[3]{n} \cdot \sqrt[3]{21}$
answer: a) $\sqrt[3]{24}$ **b)** $\sqrt[3]{21n}$

Multiply.

a) $\sqrt[3]{2} \cdot \sqrt[3]{7}$ b) $\sqrt[4]{t} \cdot \sqrt[4]{10}$

Solution

a) $\sqrt[3]{2} \cdot \sqrt[3]{7} = \sqrt[3]{2 \cdot 7} = \sqrt[3]{14}$ b) $\sqrt[4]{t} \cdot \sqrt[4]{10} = \sqrt[4]{t \cdot 10} = \sqrt[4]{10t}$ ■

You Try 1

Multiply.

a) $\sqrt[4]{6} \cdot \sqrt[4]{5}$ b) $\sqrt[5]{8} \cdot \sqrt[5]{k^2}$

BE CAREFUL

Remember that we can apply the product rule in this way *only* if the indices of the radicals are the same. Later in this section we will discuss how to multiply radicals with different indices.

2. Simplify Higher Roots of Integers

In Section 8.3 we said that a simplified *square root* cannot contain any *perfect squares*. Next we list the conditions that determine when a radical with *any* index is in simplest form.

Property   When Is a Radical Simplified?

Let P be an expression and let n be an integer greater than 1. Then $\sqrt[n]{P}$ is completely simplified when all of the following conditions are met:

1) The radicand does not contain any factors (other than 1) that are perfect *n*th powers.
2) The exponents in the radicand and the index of the radical do not have any common factors (other than 1).
3) The radicand does not contain any fractions.
4) There are no radicals in the denominator of a fraction.

Note

Condition 1) implies that the radical cannot contain variables with exponents greater than or equal to n, the index of the radical.

To simplify radicals with any index, use the product rule $\sqrt[n]{a \cdot b} = \sqrt[n]{a} \cdot \sqrt[n]{b}$, where a or b is an nth power.

Remember, to be certain that a radical is simplified completely, always look at the radical carefully and ask yourself, "*Is the radical in simplest form?*"

Example 2

In-Class Example 2
Simplify completely.
a) $\sqrt[3]{81}$ b) $\sqrt[4]{144}$
answer: a) $3\sqrt[3]{3}$ b) $2\sqrt[4]{9}$

Simplify completely.

a) $\sqrt[3]{250}$ b) $\sqrt[4]{48}$

Solution

a)   We will look at two methods for simplifying $\sqrt[3]{250}$.

   i)   Since we must simplify the *cube* root of 250, think of two numbers that multiply to 250 so that at least one of the numbers is a *perfect cube*.

$$250 = 125 \cdot 2$$
$$\sqrt[3]{250} = \sqrt[3]{125 \cdot 2} \qquad \text{125 is a perfect cube.}$$
$$= \sqrt[3]{125} \cdot \sqrt[3]{2} \qquad \text{Product rule}$$
$$= 5\sqrt[3]{2} \qquad \sqrt[3]{125} = 5$$

   Is $5\sqrt[3]{2}$ in simplest form? Yes, because 2 does not have any factors that are perfect cubes.

   ii)   Use a factor tree to find the prime factorization of 250: $250 = 2 \cdot 5^3$.

$$\sqrt[3]{250} = \sqrt[3]{2 \cdot 5^3} \qquad \text{$2 \cdot 5^3$ is the prime factorization of 250.}$$
$$= \sqrt[3]{2} \cdot \sqrt[3]{5^3} \qquad \text{Product rule}$$
$$= \sqrt[3]{2} \cdot 5 \qquad \sqrt[3]{5^3} = 5$$
$$= 5\sqrt[3]{2} \qquad \text{Commutative property}$$

   We obtain the same result using either method.

b)   We will use two methods for simplifying $\sqrt[4]{48}$.

   i)   Since we must simplify the *fourth* root of 48, think of two numbers that multiply to 48 so that at least one of the numbers is a *perfect fourth power*.

$$48 = 16 \cdot 3$$
$$\sqrt[4]{48} = \sqrt[4]{16 \cdot 3} \qquad \text{16 is a perfect fourth power.}$$
$$= \sqrt[4]{16} \cdot \sqrt[4]{3} \qquad \text{Product rule}$$
$$= 2\sqrt[4]{3} \qquad \sqrt[4]{16} = 2$$

   Is $2\sqrt[4]{3}$ in simplest form? Yes, because 3 does not have any factors that are perfect fourth powers.

   ii)   Use a factor tree to find the prime factorization of 48: $48 = 2^4 \cdot 3$.

$$\sqrt[4]{48} = \sqrt[4]{2^4 \cdot 3} \qquad \text{$2^4 \cdot 3$ is the prime factorization of 48.}$$
$$= \sqrt[4]{2^4} \cdot \sqrt[4]{3} \qquad \text{Product rule}$$
$$= 2\sqrt[4]{3} \qquad \sqrt[4]{2^4} = 2$$

Once again, both methods give us the same result.   ■

You Try 2

Simplify completely.

a) $\sqrt[3]{40}$ b) $\sqrt[5]{63}$

3. Use the Quotient Rule for Higher Roots

> **Definition**   Quotient Rule for Higher Roots
>
> If $\sqrt[n]{a}$ and $\sqrt[n]{b}$ are real numbers, $b \neq 0$, and n is a natural number then
>
> $$\sqrt[n]{\frac{a}{b}} = \frac{\sqrt[n]{a}}{\sqrt[n]{b}}$$

We apply the quotient rule when working with nth roots the same way we apply it when working with square roots.

Example 3

In-Class Example 3
Simplify completely.
a) $\sqrt[4]{\dfrac{48}{3}}$ b) $\dfrac{\sqrt[3]{56}}{\sqrt[3]{2}}$

answer: a) 2   b) $\sqrt[3]{28}$

Simplify completely.

a) $\sqrt[3]{-\dfrac{81}{3}}$ b) $\dfrac{\sqrt[3]{96}}{\sqrt[3]{2}}$

Solution

a)   We can think of $-\dfrac{81}{3}$ as $\dfrac{-81}{3}$ or $\dfrac{81}{-3}$. Let's think of it as $\dfrac{-81}{3}$.

Neither -81 nor 3 is a perfect cube, but if we simplify $\dfrac{-81}{3}$ we get -27, which *is* a perfect cube.

$$\sqrt[3]{-\frac{81}{3}} = \sqrt[3]{-27} = -3$$

b)   Let's begin by applying the quotient rule to obtain a fraction under *one* radical, then simplify the fraction.

$$\frac{\sqrt[3]{96}}{\sqrt[3]{2}} = \sqrt[3]{\frac{96}{2}} \qquad \text{Quotient rule}$$
$$= \sqrt[3]{48} \qquad \text{Simplify } \frac{96}{2}.$$
$$= \sqrt[3]{8 \cdot 6} \qquad \text{8 is a perfect cube.}$$
$$= \sqrt[3]{8} \cdot \sqrt[3]{6} \qquad \text{Product rule}$$
$$= 2\sqrt[3]{6} \qquad \sqrt[3]{8} = 2$$

Is $2\sqrt[3]{6}$ in simplest form? Yes, because 6 does not have any factors that are perfect cubes. ■

You Try 3

Simplify completely.

a) $\sqrt[4]{\dfrac{1}{81}}$ b) $\dfrac{\sqrt[3]{162}}{\sqrt[3]{3}}$

4. Simplify Radicals Containing Variables

In Section 8.2 we discussed the relationship between radical notation and fractional exponents. Recall that

> **Property** $\sqrt[n]{a^m}$
>
> If a is a nonnegative number and m and n are integers such that $n > 1$, then
>
> $$\sqrt[n]{a^m} = a^{m/n}.$$
>
> That is, the index of the radical becomes the denominator of the fractional exponent, and the power in the radicand becomes the numerator of the fractional exponent.

This is the principle we use to simplify radicals with indices greater than 2.

Example 4

In-Class Example 4
Simplify completely.
a) $\sqrt[4]{x^{12}}$ b) $\sqrt[3]{8m^9n^6}$
c) $\sqrt[6]{\dfrac{f^{12}}{g^{24}}}$

answer: a) x^3 b) $2m^3n^2$
c) $\dfrac{f^2}{g^4}$

Simplify completely.

a) $\sqrt[3]{y^{15}}$ b) $\sqrt[4]{16t^{24}u^8}$ c) $\sqrt[5]{\dfrac{c^{10}}{d^{30}}}$

Solution

a) $\sqrt[3]{y^{15}} = y^{15/3} = y^5$

b) $\sqrt[4]{16t^{24}u^8} = \sqrt[4]{16} \cdot \sqrt[4]{t^{24}} \cdot \sqrt[4]{u^8}$ Product rule
$\qquad\qquad = 2 \cdot t^{24/4} \cdot u^{8/4}$ Write with rational exponents.
$\qquad\qquad = 2t^6u^2$ Simplify exponents.

c) $\sqrt[5]{\dfrac{c^{10}}{d^{30}}} = \dfrac{\sqrt[5]{c^{10}}}{\sqrt[5]{d^{30}}} = \dfrac{c^{10/5}}{d^{30/5}} = \dfrac{c^2}{d^6}$ Quotient rule

You Try 4

Simplify completely.

a) $\sqrt[3]{a^3b^{21}}$ b) $\sqrt[4]{\dfrac{m^{12}}{16n^{20}}}$

To simplify a radical expression if the power in the radicand does not divide evenly by the index, we use the same methods we used in Section 8.3 for simplifying similar expressions with square roots. We can use the product rule or we can use the idea of quotient and remainder in a division problem.

Example 5

In-Class Example 5
Simplify $\sqrt[4]{d^{31}}$ completely in two ways.
answer: $d^7\sqrt[4]{d^3}$

Simplify $\sqrt[4]{x^{23}}$ completely in two ways: i) use the product rule and ii) divide the exponent by the index and use the quotient and remainder.

Solution

i) Using the product rule:
To simplify $\sqrt[4]{x^{23}}$, write x^{23} as the product of two factors so that the exponent of one of the factors is the *largest* number less than 23 that is divisible by 4 (the index).

$\sqrt[4]{x^{23}} = \sqrt[4]{x^{20} \cdot x^3}$ 20 is the largest number less than 23 that is divisible by 4.
$\qquad = \sqrt[4]{x^{20}} \cdot \sqrt[4]{x^3}$ Product rule
$\qquad = x^{20/4} \cdot \sqrt[4]{x^3}$ Use a fractional exponent to simplify.
$\qquad = x^5\sqrt[4]{x^3}$ $20 \div 4 = 5$

ii) Using the quotient and remainder:

To simplify $\sqrt[4]{x^{23}}$, divide $4\overline{)\,23}$

$$\begin{array}{r} 5 \leftarrow \text{Quotient} \\ 4\overline{)\,23} \\ -20 \\ \hline 3 \leftarrow \text{Remainder} \end{array}$$

Recall from our work with square roots in Section 8.3 that

i) the exponent on the variable *outside* of the radical will be the *quotient* of the division problem,

and

ii) the exponent on the variable *inside* of the radical will be the *remainder* of the division problem.

$$\sqrt[4]{x^{23}} = x^5\sqrt[4]{x^3}$$

Is $x^5\sqrt[4]{x^3}$ in simplest form? Yes, because the exponent inside of the radical is less than the index, and they contain no common factors other than 1.   ■

You Try 5

Simplify $\sqrt[5]{r^{32}}$ completely using both methods shown in Example 5.

We can apply the product and quotient rules together with the methods in Example 5 to simplify certain radical expressions.

Example 6

Completely simplify $\sqrt[3]{56a^{16}b^8}$.

In-Class Example 6
Simplify $\sqrt[4]{32p^5q^8}$ completely.
answer: $2pq^2\sqrt[4]{2p}$

Solution

$$\sqrt[3]{56a^{16}b^8} = \sqrt[3]{56} \cdot \sqrt[3]{a^{16}} \cdot \sqrt[3]{b^8} \qquad \text{Product rule}$$
$$= \sqrt[3]{8} \cdot \sqrt[3]{7} \cdot a^5\sqrt[3]{a^1} \cdot b^2\sqrt[3]{b^2}$$

Product rule   16 ÷ 3 gives a quotient   8 ÷ 3 gives a quotient
                      of 5 and a remainder of 1. of 2 and a remainder of 2.

$$= 2\sqrt[3]{7} \cdot a^5\sqrt[3]{a} \cdot b^2\sqrt[3]{b^2} \qquad \text{Simplify } \sqrt[3]{8}.$$
$$= 2a^5b^2 \cdot \sqrt[3]{7} \cdot \sqrt[3]{a} \cdot \sqrt[3]{b^2} \qquad \text{Use the commutative property to rewrite the expression.}$$
$$= 2a^5b^2\sqrt[3]{7ab^2} \qquad \text{Product rule}$$ ■

You Try 6

Simplify completely.

a) $\sqrt[4]{48x^{15}y^{22}}$ b) $\sqrt[3]{\dfrac{r^{19}}{27s^{12}}}$

5. Multiply and Divide Radicals with Different Indices

The product and quotient rules for radicals apply only when the radicals have the *same* indices. To multiply or divide radicals with *different* indices, we first change the radical expressions to rational exponent form.

Example 7

Multiply the expressions, and write the answer in simplest radical form.

$$\sqrt[3]{x^2} \cdot \sqrt{x}$$

In-Class Example 7
Perform the indicated operation, and write the answer in simplest radical form.

a) $\sqrt[4]{m^5} \cdot \sqrt{m}$ b) $\dfrac{\sqrt[4]{b^3}}{\sqrt[3]{b}}$

answer: a) $m\sqrt[4]{m^3}$
b) $\sqrt[12]{b^5}$

Solution

The indices of $\sqrt[3]{x^2}$ and $\sqrt{x}$ are different, so we *cannot* use the product rule right now. Rewrite each radical as a fractional exponent, use the product rule for *exponents*, then convert the answer back to radical form.

$$\sqrt[3]{x^2} \cdot \sqrt{x} = x^{2/3} \cdot x^{1/2} \qquad \text{Change radicals to fractional exponents.}$$
$$= x^{4/6} \cdot x^{3/6} \qquad \text{Get a common denominator to add exponents.}$$
$$= x^{\frac{4}{6} + \frac{3}{6}} = x^{7/6} \qquad \text{Add exponents.}$$
$$= \sqrt[6]{x^7} = x\sqrt[6]{x} \qquad \text{Rewrite in radical form and simplify.}$$

 You Try 7

Perform the indicated operation, and write the answer in simplest radical form.

a) $\sqrt[4]{y} \cdot \sqrt[6]{y}$ b) $\dfrac{\sqrt[3]{c^2}}{\sqrt{c}}$

Answers to You Try Exercises

1) a) $\sqrt[4]{30}$ b) $\sqrt[5]{8k^2}$ 2) a) $2\sqrt[3]{5}$ b) simplified   3) a) $\dfrac{1}{3}$ b) $3\sqrt[3]{2}$ 4) a) ab^7 b) $\dfrac{m^3}{2n^5}$

5) $r^6\sqrt[5]{r^2}$ 6) a) $2x^3y^5\sqrt[4]{3x^3y^2}$ b) $\dfrac{r^6\sqrt[3]{r}}{3s^4}$ 7) a) $\sqrt[12]{y^5}$ b) $\sqrt[6]{c}$

8.4 Exercises

*Additional answers can be found in the Answers to Exercises appendix.

Mixed Exercises: Objectives 1–3

1) In your own words, explain the product rule for radicals.
Answers may vary.

2) In your own words, explain the quotient rule for radicals.
Answers may vary.

3) How do you know that a radical expression containing a cube root is completely simplified?

4) How do you know that a radical expression containing a fourth root is completely simplified?

Assume all variables represent positive real numbers.

Objective 1: Multiply Higher Roots

Multiply.

5) $\sqrt[3]{5} \cdot \sqrt[3]{4}$ $\sqrt[3]{20}$ 6) $\sqrt[5]{6} \cdot \sqrt[5]{2}$ $\sqrt[5]{12}$

7) $\sqrt[5]{9} \cdot \sqrt[5]{m^2}$ $\sqrt[5]{9m^2}$ 8) $\sqrt[4]{11} \cdot \sqrt[4]{h^3}$ $\sqrt[4]{11h^3}$

9) $\sqrt[3]{a^2} \cdot \sqrt[3]{b}$ $\sqrt[3]{a^2b}$ 10) $\sqrt[5]{t^2} \cdot \sqrt[5]{u^4}$ $\sqrt[5]{t^2u^4}$

Mixed Exercises: Objectives 2 and 3

Simplify completely.

Fill It In

Fill in the blanks with either the missing mathematical step or reason for the given step.

11) $\sqrt[3]{56} = \sqrt[3]{8 \cdot 7}$ Factor.
$= \sqrt[3]{8} \cdot \sqrt[3]{7}$ Product rule
$= 2\sqrt[3]{7}$ Simplify.

12) $\sqrt[4]{80} = \sqrt[4]{16 \cdot 5}$ Factor.
$= \sqrt[4]{16} \cdot \sqrt[4]{5}$ Product rule
$= 2\sqrt[4]{5}$ Simplify.

13) $\sqrt[3]{24}$ $2\sqrt[3]{3}$ 14) $\sqrt[3]{48}$ $2\sqrt[3]{6}$

15) $\sqrt[4]{64}$ $2\sqrt[4]{4}$ 16) $\sqrt[4]{32}$ $2\sqrt[4]{2}$

17) $\sqrt[3]{54}$ $3\sqrt[3]{2}$

18) $\sqrt[3]{88}$ $2\sqrt[3]{11}$

19) $\sqrt[3]{2000}$ $10\sqrt[3]{2}$

20) $\sqrt[3]{108}$ $3\sqrt[3]{4}$

21) $\sqrt[5]{64}$ $2\sqrt[5]{2}$

22) $\sqrt[4]{162}$ $3\sqrt[4]{2}$

23) $\sqrt[4]{\dfrac{1}{16}}$ $\dfrac{1}{2}$

24) $\sqrt[3]{\dfrac{1}{125}}$ $\dfrac{1}{5}$

25) $\sqrt[3]{-\dfrac{54}{2}}$ -3

26) $\sqrt[4]{\dfrac{48}{3}}$ 2

27) $\dfrac{\sqrt[3]{48}}{\sqrt[3]{2}}$ $2\sqrt[3]{3}$

28) $\dfrac{\sqrt[3]{500}}{\sqrt[3]{2}}$ $5\sqrt[3]{2}$

29) $\dfrac{\sqrt[4]{240}}{\sqrt[4]{3}}$ $2\sqrt[4]{5}$

30) $\dfrac{\sqrt[3]{8000}}{\sqrt[3]{4}}$ $10\sqrt[3]{2}$

Objective 4: Simplify Radicals Containing Variables

Simplify completely.

31) $\sqrt[3]{d^6}$ d^2

32) $\sqrt[3]{g^9}$ g^3

33) $\sqrt[4]{n^{20}}$ n^5

34) $\sqrt[4]{t^{36}}$ t^9

35) $\sqrt[5]{x^5y^{15}}$ xy^3

36) $\sqrt[6]{a^{12}b^6}$ a^2b

37) $\sqrt[3]{w^{14}}$ $w^4\sqrt[3]{w^2}$

38) $\sqrt[3]{b^{19}}$ $b^6\sqrt[3]{b}$

39) $\sqrt[4]{y^9}$ $y^2\sqrt[4]{y}$

40) $\sqrt[4]{m^7}$ $m\sqrt[4]{m^3}$

41) $\sqrt[3]{d^5}$ $d\sqrt[3]{d^2}$

42) $\sqrt[3]{c^{29}}$ $c^9\sqrt[3]{c^2}$

43) $\sqrt[3]{u^{10}v^{15}}$ $u^3v^5\sqrt[3]{u}$

44) $\sqrt[3]{x^9y^{16}}$ $x^3y^5\sqrt[3]{y}$

45) $\sqrt[3]{b^{16}c^5}$ $b^5c\sqrt[3]{bc^2}$

46) $\sqrt[4]{r^{15}s^9}$ $r^3s^2\sqrt[4]{r^3s}$

47) $\sqrt[4]{m^3n^{18}}$ $n^4\sqrt[4]{m^3n^2}$

48) $\sqrt[3]{a^{11}b}$ $a^3\sqrt[3]{a^2b}$

49) $\sqrt[3]{24x^{10}y^{12}}$ $2x^3y^4\sqrt[3]{3x}$

50) $\sqrt[3]{54y^{10}z^{24}}$ $3y^3z^8\sqrt[3]{2y}$

51) $\sqrt[3]{250w^4x^{16}}$ $5wx^5\sqrt[3]{2wx}$

52) $\sqrt[3]{72t^{17}u^7}$ $2t^5u^2\sqrt[3]{9t^2u}$

53) $\sqrt[4]{\dfrac{m^8}{81}}$ $\dfrac{m^2}{3}$

54) $\sqrt[4]{\dfrac{16}{x^{12}}}$ $\dfrac{2}{x^3}$

55) $\sqrt[5]{\dfrac{32a^{23}}{b^{15}}}$ $\dfrac{2a^4\sqrt[5]{a^3}}{b^3}$

56) $\sqrt[3]{\dfrac{h^{17}}{125k^{21}}}$ $\dfrac{h^5\sqrt[3]{h^2}}{5k^7}$

57) $\sqrt[4]{\dfrac{t^9}{81s^{24}}}$ $\dfrac{t^2\sqrt[4]{t}}{3s^6}$

58) $\sqrt[5]{\dfrac{32c^9}{d^{20}}}$ $\dfrac{2c\sqrt[5]{c^4}}{d^4}$

59) $\sqrt[3]{\dfrac{u^{28}}{v^3}}$ $\dfrac{u^9\sqrt[3]{u}}{v}$

60) $\sqrt[4]{\dfrac{m^{13}}{n^8}}$ $\dfrac{m^3\sqrt[4]{m}}{n^2}$

Perform the indicated operation and simplify.

61) $\sqrt[3]{6}\cdot\sqrt[3]{4}$ $2\sqrt[3]{3}$

62) $\sqrt[3]{4}\cdot\sqrt[3]{10}$ $2\sqrt[3]{5}$

63) $\sqrt[3]{9}\cdot\sqrt[3]{12}$ $3\sqrt[3]{4}$

64) $\sqrt[3]{9}\cdot\sqrt[3]{6}$ $3\sqrt[3]{2}$

65) $\sqrt[3]{20}\cdot\sqrt[3]{4}$ $2\sqrt[3]{10}$

66) $\sqrt[3]{28}\cdot\sqrt[3]{2}$ $2\sqrt[3]{7}$

67) $\sqrt[3]{m^4}\cdot\sqrt[3]{m^5}$ m^3

68) $\sqrt[3]{t^5}\cdot\sqrt[3]{t}$ t^2

69) $\sqrt[4]{k^7}\cdot\sqrt[4]{k^9}$ k^4

70) $\sqrt[4]{a^9}\cdot\sqrt[4]{a^{11}}$ a^5

71) $\sqrt[3]{r^7}\cdot\sqrt[3]{r^4}$ $r^3\sqrt[3]{r^2}$

72) $\sqrt[3]{y^2}\cdot\sqrt[3]{y^{17}}$ $y^6\sqrt[3]{y}$

73) $\sqrt[5]{p^{14}}\cdot\sqrt[5]{p^9}$ $p^4\sqrt[5]{p^3}$

74) $\sqrt[5]{c^{17}}\cdot\sqrt[5]{c^9}$ $c^5\sqrt[5]{c}$

75) $\sqrt[3]{9z^{11}}\cdot\sqrt[3]{3z^8}$ $3z^6\sqrt[3]{z}$

76) $\sqrt[3]{2h^4}\cdot\sqrt[4]{4h^{16}}$ $2h^6\sqrt[3]{h^2}$

77) $\sqrt[3]{\dfrac{h^{14}}{h^2}}$ h^4

78) $\sqrt[3]{\dfrac{a^{20}}{a^{14}}}$ a^2

79) $\sqrt[3]{\dfrac{c^{11}}{c^4}}$ $c^2\sqrt[3]{c}$

80) $\sqrt[3]{\dfrac{z^{16}}{z^5}}$ $z^3\sqrt[3]{z^2}$

81) $\sqrt[4]{\dfrac{162d^{21}}{2d^2}}$ $3d^4\sqrt[4]{d^3}$

82) $\sqrt[4]{\dfrac{48t^{11}}{3t^6}}$ $2t\sqrt[4]{t}$

Objective 5: Multiply and Divide Radicals with Different Indices

The following radical expressions do not have the same indices. Perform the indicated operation, and write the answer in simplest radical form.

Fill It In

Fill in the blanks with either the missing mathematical step or reason for the given step.

83) $\sqrt{a}\cdot\sqrt[4]{a^3} = a^{1/2}\cdot a^{3/4}$ Change radicals to fractional exponents.

$= a^{2/4}\cdot a^{3/4}$ Rewrite exponents with a common denominator.

$= a^{5/4}$ Add exponents.

$= \sqrt[4]{a^5}$ Rewrite in radical form.

$= a\sqrt[4]{a}$ Simplify.

84) $\sqrt[5]{r^4}\cdot\sqrt[3]{r^2} = r^{4/5}\cdot r^{2/3}$ Change radicals to fractional exponents.

$= r^{12/15}\cdot r^{10/15}$ Rewrite exponents with a common denominator.

$= r^{22/15}$ Add exponents.

$= \sqrt[15]{r^{22}}$ Rewrite in radical form.

$= r\sqrt[15]{r^7}$ Simplify.

85) $\sqrt{p}\cdot\sqrt[3]{p}$ $\sqrt[6]{p^5}$

86) $\sqrt[3]{y^2}\cdot\sqrt[4]{y}$ $\sqrt[12]{y^{11}}$

87) $\sqrt[4]{n^3}\cdot\sqrt{n}$ $n\sqrt[4]{n}$

88) $\sqrt[5]{k^4}\cdot\sqrt{k}$ $k\sqrt[10]{k^3}$

89) $\sqrt[5]{c^3}\cdot\sqrt[3]{c^2}$ $c\sqrt[15]{c^4}$

90) $\sqrt[3]{a^2}\cdot\sqrt[5]{a^2}$ $a\sqrt[15]{a}$

91) $\dfrac{\sqrt{w}}{\sqrt[4]{w}}$ $\sqrt[4]{w}$

92) $\dfrac{\sqrt[4]{m^3}}{\sqrt{m}}$ $\sqrt[4]{m}$

93) $\dfrac{\sqrt[5]{t^4}}{\sqrt[3]{t^2}}$ $\sqrt[15]{t^2}$

94) $\dfrac{\sqrt[4]{h^3}}{\sqrt[3]{h^2}}$ $\sqrt[12]{h}$

95) A block of candle wax in the shape of a cube has a volume of 64 in^3. The length of a side of the block, s, is given by $s = \sqrt[3]{V}$, where V is the volume of the block of candle wax. How long is each side of the block?   4 in.

96) The radius $r(V)$ of a sphere is a function of its volume V and can be described by the function $r(V) = \sqrt[3]{\dfrac{3V}{4\pi}}$. If a spherical water tank has a volume of $\dfrac{256\pi}{3}$ ft^3, what is the radius of the tank?   4 ft

Section 8.5 Adding, Subtracting, and Multiplying Radicals

Objectives

1. **Add and Subtract Radical Expressions**
2. **Simplify Before Adding and Subtracting**
3. **Multiply a Binomial Containing Radical Expressions by a Monomial**
4. **Multiply Radical Expressions Using FOIL**
5. **Square a Binomial Containing Radical Expressions**
6. **Multiply Radical Expressions of the Form** $(a + b)(a - b)$

Just as we can add and subtract like terms such as $4x + 6x = 10x$, we can add and subtract *like radicals* such as $4\sqrt{3} + 6\sqrt{3}$.

> **Definition**
>
> **Like radicals** have the same index and the same radicand.

Some examples of like radicals are

$$4\sqrt{3} \text{ and } 6\sqrt{3}, \qquad -\sqrt[3]{5} \text{ and } 8\sqrt[3]{5}, \qquad \sqrt{x} \text{ and } 7\sqrt{x}, \qquad 2\sqrt[3]{a^2 b} \text{ and } \sqrt[3]{a^2 b}$$

In this section, assume all variables represent nonnegative real numbers.

1. Add and Subtract Radical Expressions

> **Property** Adding and Subtracting Radicals
>
> In order to add or subtract radicals, they must be *like* radicals.

We add and subtract like radicals in the same way we add and subtract like terms—add or subtract the "coefficients" of the radicals and multiply that result by the radical. We are using the distributive property when we combine like terms in this way.

Example 1

Perform the operations and simplify.

a) $4x + 6x$ b) $4\sqrt{3} + 6\sqrt{3}$ c) $\sqrt[4]{5} - 9\sqrt[4]{5}$ d) $3\sqrt{2} + 4\sqrt{3}$

In-Class Example 1
Perform the operations and simplify.
a) $9x + 2x$ b) $9\sqrt{5} + 2\sqrt{5}$
c) $\sqrt[3]{2} - 4\sqrt[3]{2}$
d) $8\sqrt{3} + 2\sqrt{5}$
answer: a) $11x$ b) $11\sqrt{5}$
c) $-3\sqrt[3]{2}$ d) $8\sqrt{3} + 2\sqrt{5}$

Solution

a) First notice that $4x$ and $6x$ are like terms. Therefore, they can be added.

$$4x + 6x = (4 + 6)x \qquad \text{Distributive property}$$
$$= 10x \qquad \text{Simplify.}$$

Or, we can say that by just adding the coefficients, $4x + 6x = 10x$.

b) Before attempting to add $4\sqrt{3}$ and $6\sqrt{3}$, we must be certain that they are like radicals. Since they *are* like, they can be added.

$$4\sqrt{3} + 6\sqrt{3} = (4 + 6)\sqrt{3} \qquad \text{Distributive property}$$
$$= 10\sqrt{3} \qquad \text{Simplify.}$$

Or, we can say that by just adding the coefficients of $\sqrt{3}$, we get $4\sqrt{3} + 6\sqrt{3} = 10\sqrt{3}$.

c) $\sqrt[4]{5} - 9\sqrt[4]{5} = 1\sqrt[4]{5} - 9\sqrt[4]{5} = (1 - 9)\sqrt[4]{5} = -8\sqrt[4]{5}$

d) The radicands in $3\sqrt{2} + 4\sqrt{3}$ are different, so these expressions cannot be combined.

You Try 1

Perform the operations and simplify.

a) $9c + 7c$ b) $9\sqrt{10} + 7\sqrt{10}$ c) $2\sqrt[3]{4} - 8\sqrt[3]{4}$
d) $5\sqrt{6} - 2\sqrt{3}$

Example 2

Perform the operations and simplify. $6\sqrt{x} + 11\sqrt[3]{x} + 2\sqrt{x} - 6\sqrt[3]{x}$

In-Class Example 2
Perform the operations and simplify.
a) $\sqrt{2} - 4 + 16\sqrt{2} + 19$
b) $9\sqrt[3]{y} + 2\sqrt[3]{y} - 3\sqrt{y} + 2\sqrt[3]{y}$
answer: a) $15 + 17\sqrt{2}$
b) $6\sqrt{y} + 4\sqrt[3]{y}$

Solution

Begin by noticing that there are *two* different types of radicals: $\sqrt{x}$ and $\sqrt[3]{x}$. Write the like radicals together.

$$6\sqrt{x} + 11\sqrt[3]{x} + 2\sqrt{x} - 6\sqrt[3]{x} = 6\sqrt{x} + 2\sqrt{x} + 11\sqrt[3]{x} - 6\sqrt[3]{x} \quad \text{Commutative property}$$
$$= (6 + 2)\sqrt{x} + (11 - 6)\sqrt[3]{x} \quad \text{Distributive property}$$
$$= 8\sqrt{x} + 5\sqrt[3]{x}$$

Is $8\sqrt{x} + 5\sqrt[3]{x}$ in simplest form? *Yes.* The radicals are not like (they have different indices) so they cannot be combined further. Also, each radical, $\sqrt{x}$ and $\sqrt[3]{x}$, is in simplest form. ■

You Try 2

Perform the operations and simplify. $8\sqrt[3]{2n} - 3\sqrt{2n} + 5\sqrt{2n} + 5\sqrt[3]{2n}$

2. Simplify Before Adding and Subtracting

Sometimes it looks like two radicals cannot be added or subtracted. But if the radicals can be *simplified* and they turn out to be *like* radicals, then we can add or subtract them.

> **Procedure** Adding and Subtracting Radicals
>
> 1) Write each radical expression in simplest form.
> 2) Combine like radicals.

Example 3

Perform the operations and simplify.

In-Class Example 3
Perform the operations and simplify.
a) $3\sqrt{20} + 2\sqrt{12} + 6\sqrt{5}$
b) $\sqrt[4]{6} + 4\sqrt[4]{96}$
c) $4\sqrt{50n} - 6\sqrt{2n}$
d) $\sqrt[3]{x^3y} + \sqrt[3]{y}$
answer: a) $12\sqrt{5} + 4\sqrt{3}$
b) $9\sqrt[4]{6}$
c) $14\sqrt{2n}$
d) $(x + 1)\sqrt[3]{y}$

a) $8\sqrt{2} + 3\sqrt{50} - \sqrt{45}$ b) $-7\sqrt[3]{40} + \sqrt[3]{5}$

c) $10\sqrt{8t} - 9\sqrt{2t}$ d) $\sqrt[3]{xy^6} + \sqrt[3]{x^7}$

Solution

a) The radicals $8\sqrt{2}$, $3\sqrt{50}$, and $\sqrt{45}$ are not like. The first radical is in simplest form, but $3\sqrt{50}$ and $\sqrt{45}$ should be simplified to determine if any of the radicals can be combined.

$$8\sqrt{2} + 3\sqrt{50} - \sqrt{45} = 8\sqrt{2} + 3\sqrt{25 \cdot 2} - \sqrt{9 \cdot 5} \quad \text{Factor.}$$
$$= 8\sqrt{2} + 3\sqrt{25} \cdot \sqrt{2} - \sqrt{9} \cdot \sqrt{5} \quad \text{Product rule}$$
$$= 8\sqrt{2} + 3 \cdot 5 \cdot \sqrt{2} - 3\sqrt{5} \quad \text{Simplify radicals.}$$
$$= 8\sqrt{2} + 15\sqrt{2} - 3\sqrt{5} \quad \text{Multiply.}$$
$$= 23\sqrt{2} - 3\sqrt{5} \quad \text{Add like radicals.}$$

b) $-7\sqrt[3]{40} + \sqrt[3]{5} = -7\sqrt[3]{8 \cdot 5} + \sqrt[3]{5} \quad$ 8 is a perfect cube.
$$= -7\sqrt[3]{8} \cdot \sqrt[3]{5} + \sqrt[3]{5} \quad \text{Product rule}$$
$$= -7 \cdot 2 \cdot \sqrt[3]{5} + \sqrt[3]{5} \quad \sqrt[3]{8} = 2$$
$$= -14\sqrt[3]{5} + \sqrt[3]{5} \quad \text{Multiply.}$$
$$= -13\sqrt[3]{5} \quad \text{Add like radicals.}$$

c) The radical $\sqrt{2t}$ is simplified, but $\sqrt{8t}$ is not. We must simplify $\sqrt{8t}$:

$$\sqrt{8t} = \sqrt{8} \cdot \sqrt{t} = \sqrt{4} \cdot \sqrt{2} \cdot \sqrt{t} = 2\sqrt{2} \cdot \sqrt{t} = 2\sqrt{2t}$$

Substitute $2\sqrt{2t}$ for $\sqrt{8t}$ in the original expression.

$$\begin{aligned}
10\sqrt{8t} - 9\sqrt{2t} &= 10(2\sqrt{2t}) - 9\sqrt{2t} &&\text{Substitute } 2\sqrt{2t} \text{ for } \sqrt{8t}.\\
&= 20\sqrt{2t} - 9\sqrt{2t} &&\text{Multiply.}\\
&= 11\sqrt{2t} &&\text{Subtract.}
\end{aligned}$$

d) Each radical in the expression $\sqrt[3]{xy^6} + \sqrt[3]{x^7}$ must be simplified.

$$\sqrt[3]{xy^6} = \sqrt[3]{x} \cdot \sqrt[3]{y^6} = \sqrt[3]{x} \cdot y^2 = y^2\sqrt[3]{x} \qquad \bigg| \qquad \sqrt[3]{x^7} = x^2\sqrt[3]{x^1} \quad \begin{array}{l}7 \div 3 \text{ gives a}\\ \text{quotient of 2 and a remainder of 1.}\end{array}$$

$$\begin{aligned}
\sqrt[3]{xy^6} + \sqrt[3]{x^7} &= y^2\sqrt[3]{x} + x^2\sqrt[3]{x} &&\text{Substitute the simplified radicals in the original expression.}\\
&= (y^2 + x^2)\sqrt[3]{x} &&\text{Factor out } \sqrt[3]{x} \text{ from each term.}
\end{aligned}$$

In this problem we cannot *add* $y^2\sqrt[3]{x} + x^2\sqrt[3]{x}$ like we added radicals in previous examples, but we *can* factor out $\sqrt[3]{x}$.

$(y^2 + x^2)\sqrt[3]{x}$ is the completely simplified form of the sum.   ■

You Try 3

Perform the operations and simplify.

a) $7\sqrt{3} - \sqrt{12}$ b) $2\sqrt{63} - 11\sqrt{28} + 2\sqrt{21}$ c) $\sqrt[3]{54} + 5\sqrt[3]{16}$

d) $2\sqrt{6k} + 4\sqrt{54k}$ e) $\sqrt[4]{mn^{11}} + \sqrt[4]{81mn^3}$

In the rest of this section, we will learn how to simplify expressions that combine multiplication, addition, and subtraction of radicals.

3. Multiply a Binomial Containing Radical Expressions by a Monomial

Example 4

Multiply and simplify.

a) $4(\sqrt{5} - \sqrt{20})$ b) $\sqrt{2}(\sqrt{10} + \sqrt{15})$ c) $\sqrt{x}(\sqrt{x} + \sqrt{32y})$

In-Class Example 4
Multiply and simplify.
a) $3(\sqrt{7} + \sqrt{63})$
b) $\sqrt{3}(\sqrt{6} + \sqrt{5})$
c) $\sqrt{m}(3\sqrt{m} + \sqrt{18n})$
answer: a) $12\sqrt{7}$
b) $3\sqrt{2} + \sqrt{15}$
c) $3m + 3\sqrt{2mn}$

Solution

a) Since $\sqrt{20}$ can be simplified, we will do that first.

$$\sqrt{20} = \sqrt{4 \cdot 5} = \sqrt{4} \cdot \sqrt{5} = 2\sqrt{5}$$

Substitute $2\sqrt{5}$ for $\sqrt{20}$ in the original expression.

$$\begin{aligned}
4(\sqrt{5} - \sqrt{20}) &= 4(\sqrt{5} - 2\sqrt{5}) &&\text{Substitute } 2\sqrt{5} \text{ for } \sqrt{20}.\\
&= 4(-\sqrt{5}) &&\text{Subtract.}\\
&= -4\sqrt{5} &&\text{Multiply.}
\end{aligned}$$

b) Neither $\sqrt{10}$ nor $\sqrt{15}$ can be simplified. Begin by applying the distributive property.

$$\begin{aligned}
\sqrt{2}(\sqrt{10} + \sqrt{15}) &= \sqrt{2} \cdot \sqrt{10} + \sqrt{2} \cdot \sqrt{15} &&\text{Distribute.}\\
&= \sqrt{20} + \sqrt{30} &&\text{Product rule}
\end{aligned}$$

Is $\sqrt{20} + \sqrt{30}$ in simplest form? *No.* $\sqrt{20}$ can be simplified.

$$= \sqrt{4 \cdot 5} + \sqrt{30} = \sqrt{4} \cdot \sqrt{5} + \sqrt{30} = 2\sqrt{5} + \sqrt{30}$$

c)   Since $\sqrt{32y}$ can be simplified, we will do that first.

$$\sqrt{32y} = \sqrt{32} \cdot \sqrt{y} = \sqrt{16 \cdot 2} \cdot \sqrt{y} = \sqrt{16} \cdot \sqrt{2} \cdot \sqrt{y} = 4\sqrt{2y}$$

Substitute $4\sqrt{2y}$ for $\sqrt{32y}$ in the original expression.

$$\begin{aligned}
\sqrt{x}(\sqrt{x} + \sqrt{32y}) &= \sqrt{x}(\sqrt{x} + 4\sqrt{2y}) &&\text{Substitute } 4\sqrt{2y} \text{ for } \sqrt{32y}.\\
&= \sqrt{x} \cdot \sqrt{x} + \sqrt{x} \cdot 4\sqrt{2y} &&\text{Distribute.}\\
&= x + 4\sqrt{2xy} &&\text{Multiply.}
\end{aligned}$$

■

You Try 4

Multiply and simplify.

a) $6(\sqrt{75} + 2\sqrt{3})$ b) $\sqrt{3}(\sqrt{3} + \sqrt{21})$ c) $\sqrt{c}(\sqrt{c^3} - \sqrt{100d})$

4. Multiply Radical Expressions Using FOIL

In Chapter 5, we first multiplied binomials using **FOIL** (**F**irst **O**uter **I**nner **L**ast).

$$(2x + 3)(x + 4) = 2x \cdot x + 2x \cdot 4 + 3 \cdot x + 3 \cdot 4$$

$$\ \ \text{F}\qquad\ \ \text{O}\qquad\ \ \text{I}\qquad\ \ \text{L}$$

$$= 2x^2 + 8x + 3x + 12$$

$$= 2x^2 + 11x + 12$$

We can multiply binomials containing radicals the same way.

Example 5

In-Class Example 5

Multiply and simplify.
a) $(3 + \sqrt{3})(6 + \sqrt{3})$
b) $(4\sqrt{2} + \sqrt{5})(2\sqrt{2} - 2\sqrt{5})$
c) $(\sqrt{5} + \sqrt{2t})(\sqrt{5} + 3\sqrt{2t})$
answer: a) 21 + 9√3
b) 6 − 6√10
c) 5 + 4√10t + 6t

Multiply and simplify.

a) $(2 + \sqrt{5})(4 + \sqrt{5})$ b) $(2\sqrt{3} + \sqrt{2})(\sqrt{3} - 5\sqrt{2})$
c) $(\sqrt{r} + \sqrt{3s})(\sqrt{r} + 8\sqrt{3s})$

Solution

a)   Since we must multiply two binomials, we will use FOIL.

$$(2 + \sqrt{5})(4 + \sqrt{5}) = 2 \cdot 4 + 2 \cdot \sqrt{5} + 4 \cdot \sqrt{5} + \sqrt{5} \cdot \sqrt{5}$$

$$\phantom{(2 + \sqrt{5})(4 + \sqrt{5}) =}\ \ \text{F}\qquad\ \text{O}\qquad\ \ \text{I}\qquad\ \ \text{L}$$

$$\begin{aligned}
&= 8 + 2\sqrt{5} + 4\sqrt{5} + 5 &&\text{Multiply.}\\
&= 13 + 6\sqrt{5} &&\text{Combine like terms.}
\end{aligned}$$

b) $(2\sqrt{3} + \sqrt{2})(\sqrt{3} - 5\sqrt{2})$

$$\ \ \ \ \ \ \ \ \ \ \text{F}\qquad\qquad\ \text{O}\qquad\qquad\ \ \text{I}\qquad\qquad\ \ \text{L}$$

$$\begin{aligned}
&= 2\sqrt{3} \cdot \sqrt{3} + 2\sqrt{3} \cdot (-5\sqrt{2}) + \sqrt{2} \cdot \sqrt{3} + \sqrt{2} \cdot (-5\sqrt{2})\\
&= 2 \cdot 3 + (-10\sqrt{6}) + \sqrt{6} + (-5 \cdot 2) &&\text{Multiply.}\\
&= 6 - 10\sqrt{6} + \sqrt{6} - 10 &&\text{Multiply.}\\
&= -4 - 9\sqrt{6} &&\text{Combine like terms.}
\end{aligned}$$

c) $(\sqrt{r} + \sqrt{3s})(\sqrt{r} + 8\sqrt{3s})$

$$\ \ \ \ \ \ \ \ \ \text{F}\qquad\qquad\ \text{O}\qquad\qquad\ \ \text{I}\qquad\qquad\ \ \text{L}$$

$$\begin{aligned}
&= \sqrt{r} \cdot \sqrt{r} + \sqrt{r} \cdot 8\sqrt{3s} + \sqrt{3s} \cdot \sqrt{r} + \sqrt{3s} \cdot 8\sqrt{3s}\\
&= r + 8\sqrt{3rs} + \sqrt{3rs} + 8 \cdot 3s &&\text{Multiply.}\\
&= r + 8\sqrt{3rs} + \sqrt{3rs} + 24s &&\text{Multiply.}\\
&= r + 9\sqrt{3rs} + 24s &&\text{Combine like terms.}
\end{aligned}$$

■

You Try 5

Multiply and simplify.

a) $(6 - \sqrt{7})(5 + \sqrt{7})$ b) $(\sqrt{2} + 4\sqrt{5})(3\sqrt{2} + \sqrt{5})$

c) $(\sqrt{6p} - \sqrt{2q})(\sqrt{6p} - 3\sqrt{2q})$

5. Square a Binomial Containing Radical Expressions

Recall, again, from Chapter 5, that we can use FOIL to square a binomial or we can use these special formulas:

$$(a + b)^2 = a^2 + 2ab + b^2 \qquad (a - b)^2 = a^2 - 2ab + b^2$$

For example,

$$(k + 7)^2 = (k)^2 + 2(k)(7) + (7)^2 \qquad \text{and} \qquad (2p - 5)^2 = (2p)^2 - 2(2p)(5) + (5)^2$$
$$= k^2 + 14k + 49 \qquad\qquad\qquad\qquad\qquad = 4p^2 - 20p + 25$$

To square a binomial containing radicals, we can either use FOIL or we can use the formulas above. Understanding how to use the formulas to square a binomial will make it easier to solve radical equations in Section 8.7.

Example 6

In-Class Example 6
Multiply and simplify.
a) $(\sqrt{5} + 2)^2$
b) $(2\sqrt{t} - 7)^2$
answer: a) $9 + 4\sqrt{5}$
b) $4t - 28\sqrt{t} + 49$

Multiply and simplify.

a) $(\sqrt{10} + 3)^2$ b) $(2\sqrt{x} - 6)^2$

Solution

a) Use $(a + b)^2 = a^2 + 2ab + b^2$.

$$(\sqrt{10} + 3)^2 = (\sqrt{10})^2 + 2(\sqrt{10})(3) + (3)^2 \qquad \text{Substitute } \sqrt{10} \text{ for } a \text{ and } 3 \text{ for } b.$$
$$= 10 + 6\sqrt{10} + 9 \qquad\qquad\qquad \text{Multiply.}$$
$$= 19 + 6\sqrt{10} \qquad\qquad\qquad\quad \text{Combine like terms.}$$

b) Use $(a - b)^2 = a^2 - 2ab + b^2$.

$$(2\sqrt{x} - 6)^2 = (2\sqrt{x})^2 - 2(2\sqrt{x})(6) + (6)^2 \qquad \text{Substitute } 2\sqrt{x} \text{ for } a \text{ and } 6 \text{ for } b.$$
$$= (4 \cdot x) - (4\sqrt{x})(6) + 36 \qquad\qquad \text{Multiply.}$$
$$= 4x - 24\sqrt{x} + 36 \qquad\qquad\qquad \text{Multiply.}$$ ∎

You Try 6

Multiply and simplify.

a) $(\sqrt{6} + 5)^2$ b) $(3\sqrt{2} - 4)^2$ c) $(\sqrt{w} + \sqrt{11})^2$

6. Multiply Radical Expressions of the Form (a + b)(a − b)

We will review one last rule from Chapter 5 on multiplying binomials. We will use this in Section 8.6 when we divide radicals.

$$(a + b)(a - b) = a^2 - b^2$$

For example, $(t + 8)(t - 8) = (t)^2 - (8)^2 = t^2 - 64$.

The same rule applies when we multiply binomials containing radicals.

Example 7

Multiply and simplify $(2\sqrt{x} + \sqrt{y})(2\sqrt{x} - \sqrt{y})$.

In-Class Example 7
Multiply and simplify.
a) $(3 + \sqrt{7})(3 - \sqrt{7})$
b) $(\sqrt{5} - \sqrt{2})(\sqrt{5} + \sqrt{2})$
c) $(\sqrt{r} + \sqrt{5})(\sqrt{r} - \sqrt{5})$
answer: a) 2   b) 3
c) $r - 5$

Solution

Use $(a + b)(a - b) = a^2 - b^2$.

$$(2\sqrt{x} + \sqrt{y})(2\sqrt{x} - \sqrt{y}) = (2\sqrt{x})^2 - (\sqrt{y})^2 \qquad \text{Substitute } 2\sqrt{x} \text{ for } a \text{ and } \sqrt{y} \text{ for } b.$$
$$= 4(x) - y \qquad \text{Square each term.}$$
$$= 4x - y \qquad \text{Simplify.}$$

Note

When we multiply expressions of the form $(a + b)(a - b)$ containing square roots, the radicals are eliminated. *This will always be true.*

You Try 7

Multiply and simplify.

a) $(4 + \sqrt{10})(4 - \sqrt{10})$ b) $(\sqrt{5h} + \sqrt{k})(\sqrt{5h} - \sqrt{k})$

Answers to You Try Exercises

1) a) $16c$ b) $16\sqrt{10}$ c) $-6\sqrt[3]{4}$ d) $5\sqrt{6} - 2\sqrt{3}$ 2) $13\sqrt[3]{2n} + 2\sqrt{2n}$ 3) a) $5\sqrt{3}$
b) $-16\sqrt{7} + 2\sqrt{21}$ c) $13\sqrt[3]{2}$ d) $14\sqrt{6k}$ e) $(n^2 + 3)\sqrt[4]{mn^3}$ 4) a) $42\sqrt{3}$ b) $3 + 3\sqrt{7}$
c) $c^2 - 10\sqrt{cd}$ 5) a) $23 + \sqrt{7}$ b) $26 + 13\sqrt{10}$ c) $6p - 8\sqrt{3pq} + 6q$
6) a) $31 + 10\sqrt{6}$ b) $34 - 24\sqrt{2}$ c) $w + 2\sqrt{11w} + 11$ 7) a) 6 b) $5h - k$

8.5 Exercises

*Additional answers can be found in the Answers to Exercises appendix.

Assume all variables represent nonnegative real numbers.

Objective 1: Add and Subtract Radical Expressions

1) How do you know if two radicals are *like* radicals?
   They have the same index and the same radicand.

2) Are $5\sqrt{3}$ and $7\sqrt[3]{3}$ like radicals? Why or why not?
   No. The indices are different.

Perform the operations and simplify.

3) $5\sqrt{2} + 9\sqrt{2}$ $14\sqrt{2}$ 4) $11\sqrt{7} + 7\sqrt{7}$ $18\sqrt{7}$

5) $7\sqrt[3]{4} + 8\sqrt[3]{4}$ $15\sqrt[3]{4}$ 6) $10\sqrt[3]{5} - 2\sqrt[3]{5}$ $8\sqrt[3]{5}$

7) $6 - \sqrt{13} + 5 - 2\sqrt{13}$ $11 - 3\sqrt{13}$

8) $-8 + 3\sqrt{6} - 4\sqrt{6} + 9$ $1 - \sqrt{6}$

9) $15\sqrt[3]{z^2} - 20\sqrt[3]{z^2}$ $-5\sqrt[3]{z^2}$

10) $7\sqrt[3]{p} - 4\sqrt[3]{p}$ $3\sqrt[3]{p}$

11) $2\sqrt[3]{n^2} + 9\sqrt[5]{n^2} - 11\sqrt[3]{n^2} + \sqrt[5]{n^2}$ $-9\sqrt[3]{n^2} + 10\sqrt[5]{n^2}$

12) $5\sqrt[4]{s} - 3\sqrt[3]{s} + 2\sqrt[3]{s} + 4\sqrt[4]{s}$ $9\sqrt[4]{s} - \sqrt[3]{s}$

13) $\sqrt{5c} - 8\sqrt{6c} + \sqrt{5c} + 6\sqrt{6c}$ $2\sqrt{5c} - 2\sqrt{6c}$

14) $10\sqrt{2m} + 6\sqrt{3m} - \sqrt{2m} + 8\sqrt{3m}$ $9\sqrt{2m} + 14\sqrt{3m}$

Objective 2: Simplify Before Adding and Subtracting

15) What are the steps for adding or subtracting radicals?

16) Is $6\sqrt{2} + \sqrt{10}$ in simplified form? Explain.
    Yes. The radicals are simplified and they are unlike.

Perform the operations and simplify.

Fill It In

Fill in the blanks with either the missing mathematical step or reason for the given step.

17) $\sqrt{48} + \sqrt{3}$
    $= \sqrt{16 \cdot 3} + \sqrt{3}$ .... Factor.
    $= \sqrt{16} \cdot \sqrt{3} + \sqrt{3}$ .... Product rule
    $= 4\sqrt{3} + \sqrt{3}$ .... Simplify.
    $= 5\sqrt{3}$ .... Add like radicals.

18) $\sqrt{44} - 8\sqrt{11}$
    $= \sqrt{4 \cdot 11} - 8\sqrt{11}$ .... Factor.
    $= \sqrt{4} \cdot \sqrt{11} - 8\sqrt{11}$ .... Product rule
    $= 2\sqrt{11} - 8\sqrt{11}$ .... Simplify.
    $= -6\sqrt{11}$ .... Subtract like radicals.

19) $6\sqrt{3} - \sqrt{12}$ $4\sqrt{3}$

20) $\sqrt{45} + 4\sqrt{5}$ $7\sqrt{5}$

21) $\sqrt{32} - 3\sqrt{8}$ $-2\sqrt{2}$

22) $3\sqrt{24} + \sqrt{96}$ $10\sqrt{6}$

23) $\sqrt{12} + \sqrt{75} - \sqrt{3}$ $6\sqrt{3}$

24) $\sqrt{96} + \sqrt{24} - 5\sqrt{54}$ $-9\sqrt{6}$

25) $8\sqrt[3]{9} + \sqrt[3]{72}$ $10\sqrt[3]{9}$

26) $5\sqrt[3]{88} + 2\sqrt[3]{11}$ $12\sqrt[3]{11}$

27) $\sqrt[3]{6} - \sqrt[3]{48}$ $-\sqrt[3]{6}$

28) $11\sqrt[3]{16} + 7\sqrt[3]{2}$ $29\sqrt[3]{2}$

29) $6q\sqrt{q} + 7\sqrt{q^3}$ $13q\sqrt{q}$

30) $11\sqrt{m^3} + 8m\sqrt{m}$ $19m\sqrt{m}$

31) $4d^2\sqrt{d} - 24\sqrt{d^5}$ $-20d^2\sqrt{d}$

32) $16k^4\sqrt{k} - 13\sqrt{k^9}$ $3k^4\sqrt{k}$

33) $9t^3\sqrt[3]{t} - 5\sqrt[3]{t^{10}}$ $4t^3\sqrt[3]{t}$

34) $8r^4\sqrt[3]{r} - 16\sqrt[3]{r^{13}}$ $-8r^4\sqrt[3]{r}$

35) $5a\sqrt[4]{a^7} + \sqrt[4]{a^{11}}$ $6a^2\sqrt[4]{a^3}$

36) $-3\sqrt[4]{c^{11}} + 6c^2\sqrt[4]{c^3}$ $3c^2\sqrt[4]{c^3}$

37) $2\sqrt{8p} - 6\sqrt{2p}$ $-2\sqrt{2p}$

38) $4\sqrt{63t} + 6\sqrt{7t}$ $18\sqrt{7t}$

39) $7\sqrt[3]{81a^5} + 4a\sqrt[3]{3a^2}$ $25a\sqrt[3]{3a^2}$

40) $3\sqrt[3]{40x} - 12\sqrt[3]{5x}$ $-6\sqrt[3]{5x}$

41) $\sqrt{xy^3} + 3y\sqrt{xy}$ $4y\sqrt{xy}$

42) $5a\sqrt{ab} + 2\sqrt{a^3b}$ $7a\sqrt{ab}$

43) $6c^2\sqrt{8d^3} - 9d\sqrt{2c^4d}$ $3c^2d\sqrt{2d}$

44) $11v\sqrt{5u^3} - 2u\sqrt{45uv^2}$ $5uv\sqrt{5u}$

45) $18a^5\sqrt[3]{7a^2b} + 2a^3\sqrt[3]{7a^8b}$ $20a^5\sqrt[3]{7a^2b}$

46) $8p^2q\sqrt[3]{11pq^2} + 3p^2\sqrt[3]{88pq^5}$ $14p^2q\sqrt[3]{11pq^2}$

47) $15cd\sqrt[4]{9cd} - \sqrt[4]{9c^5d^5}$ $14cd\sqrt[4]{9cd}$

48) $7yz^2\sqrt[4]{11y^4z} + 3z\sqrt[4]{11y^8z^5}$ $10y^2z^2\sqrt[4]{11z}$

49) $\sqrt[3]{a^9b} - \sqrt[3]{b^7}$ $\sqrt[3]{b}(a^3 - b^2)$

50) $\sqrt[3]{c^8} + \sqrt[3]{c^2d^3}$ $\sqrt[3]{c^2}(c^2 + d)$

Objective 3: Multiply a Binomial Containing Radical Expressions by a Monomial

Multiply and simplify.

51) $3(x + 5)$ $3x + 15$

52) $8(k + 3)$ $8k + 24$

53) $7(\sqrt{6} + 2)$ $7\sqrt{6} + 14$

54) $5(4 - \sqrt{7})$ $20 - 5\sqrt{7}$

55) $\sqrt{10}(\sqrt{3} - 1)$ $\sqrt{30} - \sqrt{10}$

56) $\sqrt{2}(9 + \sqrt{11})$ $9\sqrt{2} + \sqrt{22}$

57) $-6(\sqrt{32} + \sqrt{2})$ $-30\sqrt{2}$

58) $10(\sqrt{12} - \sqrt{3})$ $10\sqrt{3}$

59) $4(\sqrt{45} - \sqrt{20})$ $4\sqrt{5}$

60) $-3(\sqrt{18} + \sqrt{50})$ $-24\sqrt{2}$

61) $\sqrt{5}(\sqrt{24} - \sqrt{54})$ $-\sqrt{30}$

62) $\sqrt{2}(\sqrt{20} + \sqrt{45})$ $5\sqrt{10}$

63) $\sqrt[4]{3}(5 - \sqrt[4]{27})$ $5\sqrt[4]{3} - 3$

64) $\sqrt[3]{4}(2\sqrt[3]{5} + 7\sqrt[3]{4})$ $2\sqrt[3]{20} + 14\sqrt[3]{2}$

65) $\sqrt{t}(\sqrt{t} - \sqrt{81u})$ $t - 9\sqrt{tu}$

66) $\sqrt{s}(\sqrt{12r} + \sqrt{7s})$ $2\sqrt{3rs} + s\sqrt{7}$

67) $\sqrt{ab}(\sqrt{5a} + \sqrt{27b})$ $a\sqrt{5b} + 3b\sqrt{3a}$

68) $\sqrt{2xy}(\sqrt{2y} - y\sqrt{x})$ $2y\sqrt{x} - xy\sqrt{2y}$

69) $\sqrt[3]{c^2}(\sqrt[3]{c^2} + \sqrt[3]{125cd})$ $c\sqrt[3]{c} + 5c\sqrt[3]{d}$

70) $\sqrt[5]{mn^3}(\sqrt[5]{2m^2n} - n\sqrt[5]{mn^2})$ $\sqrt[5]{2m^3n^4} - n^2\sqrt[5]{m^2}$

Mixed Exercises: Objectives 4–6

71) How are the problems *Multiply* $(x + 8)(x + 3)$ and *Multiply* $(3 + \sqrt{2})(1 + \sqrt{2})$ similar? What method can be used to multiply each of them?

72) How are the problems *Multiply* $(y - 5)^2$ and *Multiply* $(\sqrt{7} - 2)^2$ similar? What method can be used to multiply each of them?

73) What formula can be used to multiply $(5 + \sqrt{6})(5 - \sqrt{6})$? $(a + b)(a - b) = a^2 - b^2$

74) What happens to the radical terms whenever we multiply $(a + b)(a - b)$ where the binomials contain square roots? The radicals are eliminated.

Objective 4: Multiply Radical Expressions Using FOIL

Multiply and simplify.

75) $(p + 7)(p + 6)$ $p^2 + 13p + 42$

76) $(z - 8)(z + 2)$ $z^2 - 6z - 16$

Fill It In

Fill in the blanks with either the missing mathematical step or reason for the given step.

77) $(6 + \sqrt{7})(2 + \sqrt{7})$

$= \underline{6 \cdot 2 + 6\sqrt{7} + 2\sqrt{7} + \sqrt{7} \cdot \sqrt{7}}$ Use FOIL.

$= 12 + 6\sqrt{7} + 2\sqrt{7} + 7$ Multiply.

$= \underline{19 + 8\sqrt{7}}$ Combine like terms.

78) $(3 + \sqrt{5})(1 + \sqrt{5})$

$= 3 \cdot 1 + 3\sqrt{5} + 1\sqrt{5} + \sqrt{5} \cdot \sqrt{5}$ Use FOIL

$= \underline{3 + 3\sqrt{5} + \sqrt{5} + 5}$ Multiply.

$= 8 + 4\sqrt{5}$ Combine like terms.

79) $(\sqrt{2} + 8)(\sqrt{2} - 3)$ $-22 + 5\sqrt{2}$

80) $(\sqrt{6} - 7)(\sqrt{6} + 2)$ $-8 - 5\sqrt{6}$

81) $(\sqrt{5} - 4\sqrt{3})(2\sqrt{5} - \sqrt{3})$ $22 - 9\sqrt{15}$

82) $(5\sqrt{2} - \sqrt{3})(2\sqrt{3} - \sqrt{2})$ $-16 + 11\sqrt{6}$

83) $(5 + 2\sqrt{3})(\sqrt{7} + \sqrt{2})$ $5\sqrt{7} + 5\sqrt{2} + 2\sqrt{21} + 2\sqrt{6}$

84) $(\sqrt{5} + 4)(\sqrt{3} - 6\sqrt{2})$ $\sqrt{15} - 6\sqrt{10} + 4\sqrt{3} - 24\sqrt{2}$

85) $(\sqrt[3]{25} - 3)(\sqrt[3]{5} - \sqrt[3]{6})$ $5 - \sqrt[3]{150} - 3\sqrt[3]{5} + 3\sqrt[3]{6}$

86) $(\sqrt[4]{8} - \sqrt[4]{3})(\sqrt[4]{6} + \sqrt[4]{2})$ $2\sqrt[4]{3} + 2 - \sqrt[4]{18} - \sqrt[4]{6}$

87) $(\sqrt{6p} - 2\sqrt{q})(8\sqrt{q} + 5\sqrt{6p})$ $-2\sqrt{6pq} + 30p - 16q$

88) $(4\sqrt{3r} + \sqrt{s})(3\sqrt{s} - 2\sqrt{3r})$ $10\sqrt{3rs} + 3s - 24r$

Objective 5: Square a Binomial Containing Radical Expressions

89) $(\sqrt{3} + 1)^2$ $4 + 2\sqrt{3}$

90) $(2 + \sqrt{5})^2$ $9 + 4\sqrt{5}$

91) $(\sqrt{11} - \sqrt{5})^2$ $16 - 2\sqrt{55}$

92) $(\sqrt{3} + \sqrt{13})^2$ $16 + 2\sqrt{39}$

93) $(\sqrt{h} + \sqrt{7})^2$ $h + 2\sqrt{7h} + 7$

94) $(\sqrt{m} + \sqrt{3})^2$ $m + 2\sqrt{3m} + 3$

95) $(\sqrt{x} - \sqrt{y})^2$ $x - 2\sqrt{xy} + y$

96) $(\sqrt{b} - \sqrt{a})^2$ $a - 2\sqrt{ab} + b$

Objective 6: Multiply Radical Expressions of the Form $(a + b)(a - b)$

97) $(c + 9)(c - 9)$ $c^2 - 81$ 98) $(g - 7)(g + 7)$ $g^2 - 49$

99) $(6 - \sqrt{5})(6 + \sqrt{5})$ 31   100) $(4 - \sqrt{7})(4 + \sqrt{7})$ 9

101) $(4\sqrt{3} + \sqrt{2})(4\sqrt{3} - \sqrt{2})$ 46

102) $(2\sqrt{2} - 2\sqrt{7})(2\sqrt{2} + 2\sqrt{7})$ -20

103) $(\sqrt[3]{2} - 3)(\sqrt[3]{2} + 3)$ $\sqrt[3]{4} - 9$

104) $(1 + \sqrt[3]{6})(1 - \sqrt[3]{6})$ $1 - \sqrt[3]{36}$

105) $(\sqrt{c} + \sqrt{d})(\sqrt{c} - \sqrt{d})$ $c - d$

106) $(\sqrt{2y} + \sqrt{z})(\sqrt{2y} - \sqrt{z})$ $2y - z$

107) $(8\sqrt{f} - \sqrt{g})(8\sqrt{f} + \sqrt{g})$ $64f - g$

108) $(\sqrt{a} + 3\sqrt{4b})(\sqrt{a} - 3\sqrt{4b})$ $a - 36b$

Extension

Multiply and simplify.

109) $(1 + 2\sqrt[3]{5})(1 - 2\sqrt[3]{5} + 4\sqrt[3]{25})$ 41

110) $(3 + \sqrt[3]{2})(9 - 3\sqrt[3]{2} + \sqrt[3]{4})$ 29

Let $f(x) = x^2$. Find each function value.

111) $f(\sqrt{7} + 2)$ $11 + 4\sqrt{7}$ 112) $f(5 - \sqrt{6})$ $31 - 10\sqrt{6}$

113) $f(1 - 2\sqrt{3})$ $13 - 4\sqrt{3}$ 114) $f(3\sqrt{2} + 4)$ $34 + 24\sqrt{2}$

Section 8.6 Dividing Radicals

Objectives

1. Rationalize a Denominator: One Square Root
2. Rationalize a Denominator: One Higher Root
3. Rationalize a Denominator Containing Two Terms
4. Rationalize a Numerator
5. Divide Out Common Factors from the Numerator and Denominator

It is generally agreed that a radical expression is *not* in simplest form if its denominator contains a radical. For example, $\dfrac{1}{\sqrt{3}}$ is not simplified, but an equivalent form, $\dfrac{\sqrt{3}}{3}$, is simplified.

Later we will show that $\dfrac{1}{\sqrt{3}} = \dfrac{\sqrt{3}}{3}$. The process of eliminating radicals from the denominator of an expression is called **rationalizing the denominator.** We will look at two types of rationalizing problems.

1) Rationalizing a denominator containing one term.
2) Rationalizing a denominator containing two terms.

To rationalize a denominator, we will use the fact that multiplying the numerator and denominator of a fraction by the same quantity results in an equivalent fraction:

$$\frac{2}{3} \cdot \frac{4}{4} = \frac{8}{12} \qquad \frac{2}{3} \text{ and } \frac{8}{12} \text{ are equivalent because } \frac{4}{4} = 1$$

We use the same idea to rationalize the denominator of a radical expression.

1. Rationalize a Denominator: One Square Root

The goal of rationalizing a denominator is to eliminate the radical from the denominator. With regard to square roots, recall that $\sqrt{a} \cdot \sqrt{a} = \sqrt{a^2} = a$ for $a \geq 0$. For example,

$$\sqrt{2} \cdot \sqrt{2} = \sqrt{2^2} = 2, \quad \sqrt{19} \cdot \sqrt{19} = \sqrt{(19)^2} = 19, \quad \sqrt{t} \cdot \sqrt{t} = \sqrt{t^2} = t \ (t \geq 0)$$

We will use this property to rationalize the denominators of the following expressions.

Example 1

In-Class Example 1

Rationalize the denominator of each expression.

a) $\dfrac{2}{\sqrt{5}}$ b) $\dfrac{4}{\sqrt{20}}$ c) $\dfrac{7\sqrt{11}}{\sqrt{6}}$

answer: a) $\dfrac{2\sqrt{5}}{5}$ b) $\dfrac{2\sqrt{5}}{5}$

c) $\dfrac{7\sqrt{66}}{6}$

Rationalize the denominator of each expression.

a) $\dfrac{1}{\sqrt{3}}$ b) $\dfrac{36}{\sqrt{18}}$ c) $\dfrac{5\sqrt{3}}{\sqrt{2}}$

Solution

a) To eliminate the square root from the denominator of $\dfrac{1}{\sqrt{3}}$, ask yourself, "By what do I multiply $\sqrt{3}$ to get a *perfect square* under the square root?" The answer is $\sqrt{3}$ since $\sqrt{3} \cdot \sqrt{3} = \sqrt{3^2} = \sqrt{9} = 3$. Multiply by $\sqrt{3}$ in the numerator *and* denominator. (We are actually multiplying by 1.)

$$\frac{1}{\sqrt{3}} = \frac{1}{\sqrt{3}} \cdot \underset{\uparrow}{\frac{\sqrt{3}}{\sqrt{3}}} = \frac{\sqrt{3}}{\sqrt{3^2}} = \frac{\sqrt{3}}{\sqrt{9}} = \frac{\sqrt{3}}{3}$$

Rationalize the denominator.

 **BE CAREFUL**

$\dfrac{\sqrt{3}}{3}$ is in simplest form. We cannot reduce terms inside and outside of the radical.

$$\frac{\sqrt{3}}{3} = \frac{\sqrt{3}^{1}}{3_1} = \sqrt{1} = 1 \qquad \textbf{\textcolor{red}{Incorrect!}}$$

b) First, simplify the denominator of $\dfrac{36}{\sqrt{18}}$.

$$\frac{36}{\underset{\uparrow}{\sqrt{18}}} = \frac{36}{\underset{\uparrow}{3\sqrt{2}}} = \frac{12}{\sqrt{2}} = \frac{12}{\sqrt{2}} \cdot \underset{\uparrow}{\frac{\sqrt{2}}{\sqrt{2}}} = \frac{12\sqrt{2}}{2} = 6\sqrt{2}$$

Simplify   Simplify.   Rationalize
$\sqrt{18}$.   the denominator.

c) To rationalize $\dfrac{5\sqrt{3}}{\sqrt{2}}$, multiply the numerator and denominator by $\sqrt{2}$.

$$\frac{5\sqrt{3}}{\sqrt{2}} = \frac{5\sqrt{3}}{\sqrt{2}} \cdot \frac{\sqrt{2}}{\sqrt{2}} = \frac{5\sqrt{6}}{2}$$

 You Try 1

Rationalize the denominator of each expression.

a) $\dfrac{1}{\sqrt{7}}$ b) $\dfrac{15}{\sqrt{27}}$ c) $\dfrac{9\sqrt{6}}{\sqrt{5}}$

Sometimes we will apply the quotient or product rule before rationalizing.

Example 2

Simplify completely.

a) $\sqrt{\dfrac{3}{24}}$ b) $\sqrt{\dfrac{5}{14}} \cdot \sqrt{\dfrac{7}{3}}$

In-Class Example 2
Simplify completely.

a) $\sqrt{\dfrac{6}{20}}$ b) $\sqrt{\dfrac{3}{11}} \cdot \sqrt{\dfrac{22}{9}}$

answer: a) $\dfrac{\sqrt{30}}{10}$ b) $\dfrac{\sqrt{6}}{3}$

Solution

a) Begin by simplifying the fraction $\dfrac{3}{24}$ under the radical.

$$\sqrt{\dfrac{3}{24}} = \sqrt{\dfrac{1}{8}} \quad \text{Simplify.}$$

$$= \dfrac{\sqrt{1}}{\sqrt{8}} = \dfrac{1}{\sqrt{4}\cdot\sqrt{2}} = \dfrac{1}{2\sqrt{2}} = \dfrac{1}{2\sqrt{2}}\cdot\dfrac{\sqrt{2}}{\sqrt{2}} = \dfrac{\sqrt{2}}{2\cdot 2} = \dfrac{\sqrt{2}}{4}$$

b) Begin by using the product rule to multiply the radicands.

$$\sqrt{\dfrac{5}{14}}\cdot\sqrt{\dfrac{7}{3}} = \sqrt{\dfrac{5}{14}\cdot\dfrac{7}{3}} \quad \text{Product rule}$$

Multiply the fractions under the radical.

$$= \sqrt{\dfrac{5}{\underset{2}{\cancel{14}}}\cdot\dfrac{\cancel{7}^{\,1}}{3}} = \sqrt{\dfrac{5}{6}} \quad \text{Multiply.}$$

$$= \dfrac{\sqrt{5}}{\sqrt{6}} = \dfrac{\sqrt{5}}{\sqrt{6}}\cdot\dfrac{\sqrt{6}}{\sqrt{6}} = \dfrac{\sqrt{30}}{6}$$

You Try 2

Simplify completely.

a) $\sqrt{\dfrac{10}{35}}$ b) $\sqrt{\dfrac{21}{10}}\cdot\sqrt{\dfrac{2}{7}}$

We work with radical expressions containing variables the same way. **In the rest of this section, we will assume that all variables represent positive real numbers.**

Example 3

In-Class Example 3
Simplify completely.
a) $\dfrac{7}{\sqrt{m}}$ b) $\sqrt{\dfrac{27p^4}{q^3}}$ c) $\sqrt{\dfrac{10xy}{x^4y^2}}$

answer: a) $\dfrac{7\sqrt{m}}{m}$

b) $\dfrac{3p^2\sqrt{3q}}{q^2}$ c) $\dfrac{\sqrt{10xy}}{x^2y}$

Simplify completely.

a) $\dfrac{2}{\sqrt{x}}$ b) $\sqrt{\dfrac{12m^3}{7n}}$ c) $\sqrt{\dfrac{6cd^2}{cd^3}}$

Solution

a) Ask yourself, "By what do I multiply $\sqrt{x}$ to get a *perfect square* under the square root?" The perfect square we want to get is $\sqrt{x^2}$.

$$\sqrt{x}\cdot\sqrt{?} = \sqrt{x^2} = x$$
$$\sqrt{x}\cdot\sqrt{x} = \sqrt{x^2} = x$$

$$\dfrac{2}{\sqrt{x}} = \dfrac{2}{\sqrt{x}}\cdot\underset{\uparrow}{\dfrac{\sqrt{x}}{\sqrt{x}}} = \dfrac{2\sqrt{x}}{\sqrt{x^2}} = \dfrac{2\sqrt{x}}{x}$$

Rationalize the denominator.

b) Before rationalizing, apply the quotient rule and simplify the numerator.

$$\sqrt{\dfrac{12m^3}{7n}} = \dfrac{\sqrt{12m^3}}{\sqrt{7n}} = \dfrac{2m\sqrt{3m}}{\sqrt{7n}}$$

Rationalize the denominator. "By what do I multiply $\sqrt{7n}$ to get a *perfect square* under the square root?" The perfect square we want to get is $\sqrt{7^2 n^2}$ or $\sqrt{49n^2}$.

$$\sqrt{7n} \cdot \sqrt{?} = \sqrt{7^2 n^2} = 7n$$
$$\sqrt{7n} \cdot \sqrt{7n} = \sqrt{7^2 n^2} = 7n$$

$$\sqrt{\frac{12m^3}{7n}} = \frac{2m\sqrt{3m}}{\sqrt{7n}}$$
$$= \frac{2m\sqrt{3m}}{\sqrt{7n}} \cdot \frac{\sqrt{7n}}{\sqrt{7n}} = \frac{2m\sqrt{21mn}}{7n}$$

↑
Rationalize the denominator.

c) $\sqrt{\dfrac{6cd^2}{cd^3}} = \sqrt{\dfrac{6}{d}}$ Simplify the radicand using the quotient rule for exponents.

$$= \frac{\sqrt{6}}{\sqrt{d}} = \frac{\sqrt{6}}{\sqrt{d}} \cdot \frac{\sqrt{d}}{\sqrt{d}} = \frac{\sqrt{6d}}{d}$$

■

You Try 3

Simplify completely.

a) $\dfrac{5}{\sqrt{p}}$ b) $\sqrt{\dfrac{18k^5}{10m}}$ c) $\sqrt{\dfrac{20r^3 s}{s^2}}$

2. Rationalize a Denominator: One Higher Root

Many students assume that to rationalize *all* denominators we simply multiply the numerator and denominator of the expression by the denominator as in $\dfrac{4}{\sqrt{3}} = \dfrac{4}{\sqrt{3}} \cdot \dfrac{\sqrt{3}}{\sqrt{3}} = \dfrac{4\sqrt{3}}{3}$.

We will see, however, why this reasoning is incorrect.

To rationalize an expression like $\dfrac{4}{\sqrt{3}}$ we asked ourselves, "By what do I multiply $\sqrt{3}$ to get a *perfect square* under the *square root*?"

To rationalize an expression like $\dfrac{5}{\sqrt[3]{2}}$ we must ask ourselves, "By what do I multiply $\sqrt[3]{2}$ to get a *perfect cube* under the *cube root*?" The perfect cube we want is 2^3 (since we began with 2) so that $\sqrt[3]{2} \cdot \sqrt[3]{2^2} = \sqrt[3]{2^3} = 2$.

We will practice some fill-in-the-blank problems to eliminate radicals before we move on to rationalizing.

Example 4

In-Class Example 4

Fill in the blank.
a) $\sqrt[3]{6} \cdot \sqrt[3]{?} = \sqrt[3]{6^3} = 6$
b) $\sqrt[3]{49} \cdot \sqrt[3]{?} = \sqrt[3]{7^3} = 7$
c) $\sqrt[3]{?} \cdot \sqrt[3]{m} = \sqrt[3]{m^3} = m$
d) $\sqrt[4]{4} \cdot \sqrt[4]{?} = \sqrt[4]{2^4} = 2$
e) $\sqrt[5]{4} \cdot \sqrt[5]{?} = \sqrt[5]{2^5} = 2$
answer: a) 6^2 b) 7   c) m^2
d) 2^2 e) 2^3

Fill in the blank.

a) $\sqrt[3]{5} \cdot \sqrt[3]{?} = \sqrt[3]{5^3} = 5$ b) $\sqrt[3]{3} \cdot \sqrt[3]{?} = \sqrt[3]{3^3} = 3$

c) $\sqrt[3]{x^2} \cdot \sqrt[3]{?} = \sqrt[3]{x^3} = x$ d) $\sqrt[5]{8} \cdot \sqrt[5]{?} = \sqrt[5]{2^5} = 2$

e) $\sqrt[4]{27} \cdot \sqrt[4]{?} = \sqrt[4]{3^4} = 3$

Solution

a) Ask yourself, "By what do I multiply $\sqrt[3]{5}$ to get $\sqrt[3]{5^3}$?" The answer is $\sqrt[3]{5^2}$.

$$\sqrt[3]{5} \cdot \sqrt[3]{?} = \sqrt[3]{5^3} = 5$$
$$\sqrt[3]{5} \cdot \sqrt[3]{5^2} = \sqrt[3]{5^3} = 5$$

b) "By what do I multiply $\sqrt[3]{3}$ to get $\sqrt[3]{3^3}$?" $\sqrt[3]{3^2}$

$$\sqrt[3]{3} \cdot \sqrt[3]{?} = \sqrt[3]{3^3} = 3$$
$$\sqrt[3]{3} \cdot \sqrt[3]{3^2} = \sqrt[3]{3^3} = 3$$

c) "By what do I multiply $\sqrt[3]{x^2}$ to get $\sqrt[3]{x^3}$?" $\sqrt[3]{x}$

$$\sqrt[3]{x^2} \cdot \sqrt[3]{?} = \sqrt[3]{x^3} = x$$
$$\sqrt[3]{x^2} \cdot \sqrt[3]{x} = \sqrt[3]{x^3} = x$$

d) In this example, $\sqrt[5]{8} \cdot \sqrt[5]{?} = \sqrt[5]{2^5} = 2$, why are we trying to obtain $\sqrt[5]{2^5}$ instead of $\sqrt[5]{8^5}$? Because in the first radical, $\sqrt[5]{8}$, *8 is a power of 2.* Before attempting to fill in the blank, rewrite 8 as 2^3.

$$\sqrt[5]{8} \cdot \sqrt[5]{?} = \sqrt[5]{2^5} = 2$$
$$\sqrt[5]{2^3} \cdot \sqrt[5]{?} = \sqrt[5]{2^5} = 2$$
$$\sqrt[5]{2^3} \cdot \sqrt[5]{2^2} = \sqrt[5]{2^5} = 2$$

e)
$$\sqrt[4]{27} \cdot \sqrt[4]{?} = \sqrt[4]{3^4} = 3$$
$$\sqrt[4]{3^3} \cdot \sqrt[4]{?} = \sqrt[4]{3^4} = 3 \qquad \text{\color{blue}Since 27 is a power of 3, rewrite } \sqrt[4]{27} \text{ as } \sqrt[4]{3^3}.$$
$$\sqrt[4]{3^3} \cdot \sqrt[4]{3} = \sqrt[4]{3^4} = 3$$

You Try 4

Fill in the blank.

a) $\sqrt[3]{2} \cdot \sqrt[3]{?} = \sqrt[3]{2^3} = 2$ b) $\sqrt[5]{t^2} \cdot \sqrt[5]{?} = \sqrt[5]{t^5} = t$ c) $\sqrt[4]{125} \cdot \sqrt[4]{?} = \sqrt[4]{5^4} = 5$

We will use the technique presented in Example 4 to rationalize denominators with indices higher than 2.

Example 5

In-Class Example 5
Rationalize the denominator.
a) $\dfrac{8}{\sqrt[3]{3}}$ b) $\sqrt[4]{\dfrac{2}{9}}$ c) $\dfrac{9}{\sqrt[5]{t}}$
answer: a) $\dfrac{8\sqrt[3]{9}}{3}$ b) $\dfrac{\sqrt[4]{18}}{3}$
c) $\dfrac{9\sqrt[5]{t^4}}{t}$

Rationalize the denominator.

a) $\dfrac{7}{\sqrt[3]{3}}$ b) $\sqrt[5]{\dfrac{3}{4}}$ c) $\dfrac{7}{\sqrt[4]{n}}$

Solution

a) *First* identify what we want the denominator to be *after* multiplying. **We want to obtain $\sqrt[3]{3^3}$ since $\sqrt[3]{3^3} = 3$.**

$$\frac{7}{\sqrt[3]{3}} \cdot \frac{}{} = \frac{}{\sqrt[3]{3^3}} \qquad \text{\color{blue}\leftarrow This is what we want to get.}$$

$$\underset{\color{blue}\text{What is needed here?}}{\uparrow}$$

Ask yourself, "By what do I multiply $\sqrt[3]{3}$ to get $\sqrt[3]{3^3}$?" $\sqrt[3]{3^2}$

$$\frac{7}{\sqrt[3]{3}} \cdot \frac{\sqrt[3]{3^2}}{\sqrt[3]{3^2}} = \frac{7\sqrt[3]{3^2}}{\sqrt[3]{3^3}} \qquad \text{\color{blue}Multiply.}$$

$$= \frac{7\sqrt[3]{9}}{3} \qquad \text{\color{blue}Simplify.}$$

b) Use the quotient rule for radicals to rewrite $\sqrt[5]{\dfrac{3}{4}}$ as $\dfrac{\sqrt[5]{3}}{\sqrt[5]{4}}$. Then, write 4 as 2^2 to get

$$\frac{\sqrt[5]{3}}{\sqrt[5]{4}} = \frac{\sqrt[5]{3}}{\sqrt[5]{2^2}}$$

What denominator do we want to get *after* multiplying? **We want to obtain $\sqrt[5]{2^5}$ since $\sqrt[5]{2^5} = 2$.**

$$\frac{\sqrt[5]{3}}{\sqrt[5]{2^2}} \cdot \frac{\quad}{\quad} = \frac{\quad}{\sqrt[5]{2^5}} \quad \leftarrow \text{This is what we want to get.}$$

$$\uparrow$$
What is needed here?

"By what do I multiply $\sqrt[5]{2^2}$ to get $\sqrt[5]{2^5}$?" $\sqrt[5]{2^3}$

$$\frac{\sqrt[5]{3}}{\sqrt[5]{2^2}} \cdot \frac{\sqrt[5]{2^3}}{\sqrt[5]{2^3}} = \frac{\sqrt[5]{3} \cdot \sqrt[5]{2^3}}{\sqrt[5]{2^5}} \qquad \text{Multiply.}$$

$$= \frac{\sqrt[5]{3} \cdot \sqrt[5]{8}}{2} = \frac{\sqrt[5]{24}}{2} \qquad \text{Multiply.}$$

c) What denominator do we want to get *after* multiplying? We want to obtain $\sqrt[4]{n^4}$ since $\sqrt[4]{n^4} = n$.

$$\frac{7}{\sqrt[4]{n}} \cdot \frac{\quad}{\quad} = \frac{\quad}{\sqrt[4]{n^4}} \qquad \leftarrow \text{This is what we want to get.}$$

$$\uparrow$$
What is needed here?

Ask yourself, "By what do I multiply $\sqrt[4]{n}$ to get $\sqrt[4]{n^4}$?" $\sqrt[4]{n^3}$

$$\frac{7}{\sqrt[4]{n}} \cdot \frac{\sqrt[4]{n^3}}{\sqrt[4]{n^3}} = \frac{7\sqrt[4]{n^3}}{\sqrt[4]{n^4}} \qquad \text{Multiply.}$$

$$= \frac{7\sqrt[4]{n^3}}{n} \qquad \text{Simplify.}$$

You Try 5

Rationalize the denominator.

a) $\dfrac{4}{\sqrt[3]{7}}$ b) $\sqrt[4]{\dfrac{2}{27}}$ c) $\sqrt[5]{\dfrac{8}{w^3}}$

3. Rationalize a Denominator Containing Two Terms

To rationalize the denominator of an expression like $\dfrac{1}{5 + \sqrt{3}}$, we multiply the numerator and the denominator of the expression by the *conjugate* of $5 + \sqrt{3}$.

Definition
The **conjugate** of a binomial is the binomial obtained by changing the sign between the two terms.

Expression	Conjugate
$\sqrt{7} - 2\sqrt{5}$	$\sqrt{7} + 2\sqrt{5}$
$\sqrt{a} + \sqrt{b}$	$\sqrt{a} - \sqrt{b}$

In Section 8.5 we applied the formula $(a + b)(a - b) = a^2 - b^2$ to multiply binomials containing square roots. Recall that the terms containing the square roots were eliminated.

Example 6

Multiply $8 - \sqrt{6}$ by its conjugate.

In-Class Example 6
Multiply $4 + \sqrt{3}$ by its conjugate.
answer: 13

Solution

The conjugate of $8 - \sqrt{6}$ is $8 + \sqrt{6}$. We will first multiply using FOIL to show *why* the radical drops out, then we will multiply using the formula

$$(a + b)(a - b) = a^2 - b^2$$

i) Use FOIL to multiply.

$$(8 - \sqrt{6})(8 + \sqrt{6}) = 8 \cdot 8 + 8 \cdot \sqrt{6} - 8 \cdot \sqrt{6} - \sqrt{6} \cdot \sqrt{6}$$

$$\qquad\qquad\qquad\qquad\quad \text{F} \qquad \text{O} \qquad\quad \text{I} \qquad\quad \text{L}$$

$$= 64 - 6$$
$$= 58$$

ii) Use $(a + b)(a - b) = a^2 - b^2$.

$$(8 - \sqrt{6})(8 + \sqrt{6}) = (8)^2 - (\sqrt{6})^2 \qquad \text{Substitute 8 for } a \text{ and } \sqrt{6} \text{ for } b.$$
$$= 64 - 6 = 58$$

Each method gives the same result. ■

You Try 6

Multiply $2 + \sqrt{11}$ by its conjugate.

Procedure

If the denominator of an expression contains two terms, including one or two square roots, then to rationalize the denominator we multiply the numerator and denominator of the expression by the conjugate of the denominator.

Example 7

Rationalize the denominator and simplify completely.

a) $\dfrac{3}{5 + \sqrt{3}}$ b) $\dfrac{\sqrt{a} + b}{\sqrt{b} - a}$

In-Class Example 7
Rationalize the denominator and simplify completely.
a) $\dfrac{3}{6 + \sqrt{2}}$ b) $\dfrac{\sqrt{c} - d}{\sqrt{d} + c}$
answer: a) $\dfrac{18 - 3\sqrt{2}}{34}$
b) $\dfrac{\sqrt{cd} - c\sqrt{c} - d\sqrt{d} + cd}{d - c^2}$

Solution

a) The denominator of $\dfrac{3}{5 + \sqrt{3}}$ has two terms, so we multiply the numerator and denominator by $5 - \sqrt{3}$, the conjugate of the denominator.

$$\frac{3}{5 + \sqrt{3}} \cdot \frac{5 - \sqrt{3}}{5 - \sqrt{3}} \qquad \text{Multiply by the conjugate.}$$

$$= \frac{3(5 - \sqrt{3})}{(5)^2 - (\sqrt{3})^2} \qquad (a + b)(a - b) = a^2 - b^2$$

$$= \frac{15 - 3\sqrt{3}}{25 - 3} \qquad \text{Simplify.}$$

$$= \frac{15 - 3\sqrt{3}}{22} \qquad \text{Subtract.}$$

b) $$\frac{\sqrt{a} + b}{\sqrt{b} - a} \cdot \frac{\sqrt{b} + a}{\sqrt{b} + a}$$ Multiply by the conjugate.

In the numerator we must multiply $(\sqrt{a} + b)(\sqrt{b} + a)$. We will use FOIL.

$$\frac{\sqrt{a} + b}{\sqrt{b} - a} \cdot \frac{\sqrt{b} + a}{\sqrt{b} + a} = \frac{\sqrt{ab} + a\sqrt{a} + b\sqrt{b} + ab}{(\sqrt{b})^2 - (a)^2}$$ ← Use FOIL in the numerator.
                                            ← $(a + b)(a - b) = a^2 - b^2$

$$= \frac{\sqrt{ab} + a\sqrt{a} + b\sqrt{b} + ab}{b - a^2}$$ Square the terms.   ■

You Try 7

Rationalize the denominator and simplify completely.

a) $\dfrac{1}{\sqrt{7} - 2}$ b) $\dfrac{c + \sqrt{d}}{c - \sqrt{d}}$

4. Rationalize a Numerator

In higher-level math courses sometimes it is necessary to rationlize the *numerator* of a radical expression so that the numerator does not contain a radical.

Example 8

In-Class Example 8
Rationalize the numerator and simplify completely.
a) $\dfrac{\sqrt{11}}{\sqrt{7}}$ b) $\dfrac{4 - \sqrt{6}}{3}$

answer: a) $\dfrac{11}{\sqrt{77}}$

b) $\dfrac{10}{12 + 3\sqrt{6}}$

Rationalize the numerator and simplify completely.

a) $\dfrac{\sqrt{7}}{\sqrt{2}}$ b) $\dfrac{8 - \sqrt{5}}{3}$

Solution

a) Rationalizing the numerator of $\dfrac{\sqrt{7}}{\sqrt{2}}$ means eliminating the square root from the numerator. Multiply the numerator and denominator by $\sqrt{7}$.

$$\frac{\sqrt{7}}{\sqrt{2}} = \frac{\sqrt{7}}{\sqrt{2}} \cdot \frac{\sqrt{7}}{\sqrt{7}} = \frac{7}{\sqrt{14}}$$

b) To rationalize the numerator we must multiply the numerator and denominator by $8 + \sqrt{5}$, the conjugate of the numerator.

$$\frac{8 - \sqrt{5}}{3} \cdot \frac{8 + \sqrt{5}}{8 + \sqrt{5}}$$ Multiply by the conjugate.

$$= \frac{8^2 - (\sqrt{5})^2}{3(8 + \sqrt{5})}$$ ← $(a + b)(a - b) = a^2 - b^2$
                                ← Multiply.

$$= \frac{64 - 5}{24 + 3\sqrt{5}} - \frac{59}{24 + 3\sqrt{5}}$$ ■

You Try 8

Rationalize the numerator and simplify completely.

a) $\dfrac{\sqrt{3}}{\sqrt{5}}$ b) $\dfrac{6 + \sqrt{7}}{4}$

5. Divide Out Common Factors from the Numerator and Denominator

Sometimes it is necessary to simplify a radical expression by dividing out common factors from the numerator and denominator. This is a skill we will need in Chapter 9 to solve quadratic equations, so we will look at an example here.

 **Example 9**

Simplify completely: $\dfrac{4\sqrt{5}+12}{4}$.

In-Class Example 9

Simplify completely: $\dfrac{9\sqrt{3}-18}{9}$.

answer: $\sqrt{3}-2$

Solution

BE CAREFUL

It is tempting to do one of the following:

$$\dfrac{\cancel{4}\sqrt{5}+12}{\cancel{4}} = \sqrt{5}+12 \qquad \textbf{\textcolor{red}{Incorrect!}}$$

or

$$\dfrac{4\sqrt{5}+\overset{3}{\cancel{12}}}{\cancel{4}} = 4\sqrt{5}+3 \qquad \textbf{\textcolor{red}{Incorrect!}}$$

Each is incorrect because $4\sqrt{5}$ is a *term* in a sum and 12 is a *term* in a sum.

The correct way to simplify $\dfrac{4\sqrt{5}+12}{4}$ is to begin by factoring out a 4 in the numerator and *then* divide the numerator and denominator by any common factors.

$$\dfrac{4\sqrt{5}+12}{4} = \dfrac{4(\sqrt{5}+3)}{4} \qquad \textcolor{teal}{\text{Factor out 4 from the numerator.}}$$

$$= \dfrac{\overset{1}{\cancel{4}}(\sqrt{5}+3)}{\underset{1}{\cancel{4}}} \qquad \textcolor{teal}{\text{Divide by 4.}}$$

$$= \sqrt{5}+3 \qquad \textcolor{teal}{\text{Simplify.}}$$

We can divide numerator and denominator by 4 in $\dfrac{4(\sqrt{5}+3)}{4}$ because the 4 in the numerator is part of a *product* not a sum or difference.

 You Try 9

Simplify completely.

a) $\dfrac{5\sqrt{7}-40}{5}$ b) $\dfrac{20+6\sqrt{2}}{4}$

Answers to You Try Exercises

1) a) $\dfrac{\sqrt{7}}{7}$ b) $\dfrac{5\sqrt{3}}{3}$ c) $\dfrac{9\sqrt{30}}{5}$ 2) a) $\dfrac{\sqrt{14}}{7}$ b) $\dfrac{\sqrt{15}}{5}$ 3) a) $\dfrac{5\sqrt{p}}{p}$ b) $\dfrac{3k^2\sqrt{5km}}{5m}$

c) $\dfrac{2r\sqrt{5rs}}{s}$ 4) a) 2^2 or 4   b) t^3 c) 5   5) a) $\dfrac{4\sqrt[3]{49}}{7}$ b) $\dfrac{\sqrt[4]{6}}{3}$ c) $\dfrac{\sqrt[5]{8w^2}}{w}$ 6) -7

7) a) $\dfrac{\sqrt{7}+2}{3}$ b) $\dfrac{c^2+2c\sqrt{d}+d}{c^2-d}$ 8) a) $\dfrac{3}{\sqrt{15}}$ b) $\dfrac{29}{24-4\sqrt{7}}$ 9) a) $\sqrt{7}-8$ b) $\dfrac{10+3\sqrt{2}}{2}$

8.6 Exercises

*Additional answers can be found in the Answers to Exercises appendix.

Assume all variables represent positive real numbers.

Objective 1: Rationalize a Denominator: One Square Root

1) What does it mean to rationalize the denominator of a radical expression?   Eliminate the radical from the denominator.

2) In your own words, explain how to rationalize the denominator of an expression containing one term in the denominator.   Answers may vary.

Rationalize the denominator of each expression.

3) $\dfrac{1}{\sqrt{5}}$ $\dfrac{\sqrt{5}}{5}$

4) $\dfrac{1}{\sqrt{6}}$ $\dfrac{\sqrt{6}}{6}$

5) $\dfrac{9}{\sqrt{6}}$ $\dfrac{3\sqrt{6}}{2}$

6) $\dfrac{25}{\sqrt{10}}$ $\dfrac{5\sqrt{10}}{2}$

7) $-\dfrac{20}{\sqrt{8}}$ $-5\sqrt{2}$

8) $-\dfrac{18}{\sqrt{45}}$ $-\dfrac{6\sqrt{5}}{5}$

9) $\dfrac{\sqrt{3}}{\sqrt{28}}$ $\dfrac{\sqrt{21}}{14}$

10) $\dfrac{\sqrt{8}}{\sqrt{27}}$ $\dfrac{2\sqrt{6}}{9}$

11) $\sqrt{\dfrac{20}{60}}$ $\dfrac{\sqrt{3}}{3}$

12) $\sqrt{\dfrac{12}{80}}$ $\dfrac{\sqrt{15}}{10}$

13) $\dfrac{\sqrt{56}}{\sqrt{48}}$ $\dfrac{\sqrt{42}}{6}$

14) $\dfrac{\sqrt{66}}{\sqrt{12}}$ $\dfrac{\sqrt{22}}{2}$

Multiply and simplify.

15) $\sqrt{\dfrac{10}{7}}\cdot\sqrt{\dfrac{7}{3}}$ $\dfrac{\sqrt{30}}{3}$

16) $\sqrt{\dfrac{11}{5}}\cdot\sqrt{\dfrac{5}{2}}$ $\dfrac{\sqrt{22}}{2}$

17) $\sqrt{\dfrac{6}{5}}\cdot\sqrt{\dfrac{1}{8}}$ $\dfrac{\sqrt{15}}{10}$

18) $\sqrt{\dfrac{11}{10}}\cdot\sqrt{\dfrac{8}{11}}$ $\dfrac{2\sqrt{5}}{5}$

Simplify completely.

19) $\dfrac{8}{\sqrt{y}}$ $\dfrac{8\sqrt{y}}{y}$

20) $\dfrac{4}{\sqrt{w}}$ $\dfrac{4\sqrt{w}}{w}$

21) $\dfrac{\sqrt{5}}{\sqrt{t}}$ $\dfrac{\sqrt{5t}}{t}$

22) $\dfrac{\sqrt{2}}{\sqrt{m}}$ $\dfrac{\sqrt{2m}}{m}$

23) $\sqrt{\dfrac{64v^7}{5w}}$ $\dfrac{8v^3\sqrt{5vw}}{5w}$

24) $\sqrt{\dfrac{81c^5}{2d}}$ $\dfrac{9c^2\sqrt{2cd}}{2d}$

25) $\sqrt{\dfrac{a^3b^3}{3ab^4}}$ $\dfrac{a\sqrt{3b}}{3b}$

26) $\sqrt{\dfrac{m^2n^5}{7m^3n}}$ $\dfrac{n^2\sqrt{7m}}{7m}$

27) $-\dfrac{\sqrt{75}}{\sqrt{b^3}}$ $-\dfrac{5\sqrt{3b}}{b^2}$

28) $-\dfrac{\sqrt{24}}{\sqrt{v^3}}$ $-\dfrac{2\sqrt{6v}}{v^2}$

29) $\dfrac{\sqrt{13}}{\sqrt{j^5}}$ $\dfrac{\sqrt{13j}}{j^3}$

30) $\dfrac{\sqrt{22}}{\sqrt{w^7}}$ $\dfrac{\sqrt{22w}}{w^4}$

Objective 2: Rationalize a Denominator: One Higher Root

Fill in the blank.

31) $\sqrt[3]{2}\cdot\sqrt[3]{?}=\sqrt[3]{2^3}=2$ 2^2 or 4

32) $\sqrt[3]{5}\cdot\sqrt[3]{?}=\sqrt[3]{5^3}=5$ 5^2 or 25

33) $\sqrt[3]{9}\cdot\sqrt[3]{?}=\sqrt[3]{3^3}=3$ 3

34) $\sqrt[3]{4}\cdot\sqrt[3]{?}=\sqrt[3]{2^3}=2$ 2

35) $\sqrt[3]{c}\cdot\sqrt[3]{?}=\sqrt[3]{c^3}=c$ c^2

36) $\sqrt[3]{p}\cdot\sqrt[3]{?}=\sqrt[3]{p^3}=p$ p^2

37) $\sqrt[5]{4}\cdot\sqrt[5]{?}=\sqrt[5]{2^5}=2$ 2^3 or 8

38) $\sqrt[5]{16}\cdot\sqrt[5]{?}=\sqrt[5]{2^5}=2$ 2

39) $\sqrt[4]{m^3}\cdot\sqrt[4]{?}=\sqrt[4]{m^4}=m$ m

40) $\sqrt[4]{k}\cdot\sqrt[4]{?}=\sqrt[4]{k^4}=k$ k^3

Rationalize the denominator of each expression.

41) $\dfrac{4}{\sqrt[3]{3}}$ $\dfrac{4\sqrt[3]{9}}{3}$

42) $\dfrac{26}{\sqrt[3]{5}}$ $\dfrac{26\sqrt[3]{25}}{5}$

43) $\dfrac{12}{\sqrt[3]{2}}$ $6\sqrt[3]{4}$

44) $\dfrac{21}{\sqrt[3]{3}}$ $7\sqrt[3]{9}$

45) $\dfrac{9}{\sqrt[3]{25}}$ $\dfrac{9\sqrt[3]{5}}{5}$

46) $\dfrac{6}{\sqrt[3]{4}}$ $3\sqrt[3]{2}$

47) $\sqrt[4]{\dfrac{5}{9}}$ $\dfrac{\sqrt[4]{45}}{3}$

48) $\sqrt[4]{\dfrac{2}{25}}$ $\dfrac{\sqrt[4]{50}}{5}$

49) $\sqrt[5]{\dfrac{3}{8}}$ $\dfrac{\sqrt[5]{12}}{2}$

50) $\sqrt[5]{\dfrac{7}{4}}$ $\dfrac{\sqrt[5]{56}}{2}$

51) $\dfrac{10}{\sqrt[3]{z}}$ $\dfrac{10\sqrt[3]{z^2}}{z}$

52) $\dfrac{6}{\sqrt[3]{u}}$ $\dfrac{6\sqrt[3]{u^2}}{u}$

53) $\sqrt[3]{\dfrac{3}{n^2}}$ $\dfrac{\sqrt[3]{3n}}{n}$

54) $\sqrt[3]{\dfrac{5}{x^2}}$ $\dfrac{\sqrt[3]{5x}}{x}$

55) $\dfrac{\sqrt[3]{7}}{\sqrt[3]{2k^2}}$ $\dfrac{\sqrt[3]{28k}}{2k}$

56) $\dfrac{\sqrt[3]{2}}{\sqrt[3]{25t}}$ $\dfrac{\sqrt[3]{10t^2}}{5t}$

57) $\dfrac{9}{\sqrt[5]{a^3}}$ $\dfrac{9\sqrt[5]{a^2}}{a}$

58) $\dfrac{8}{\sqrt[5]{h^2}}$ $\dfrac{8\sqrt[5]{h^3}}{h}$

59) $\sqrt[4]{\dfrac{5}{2m}}$ $\dfrac{\sqrt[4]{40m^3}}{2m}$

60) $\sqrt[4]{\dfrac{2}{3t^2}}$ $\dfrac{\sqrt[4]{54t^2}}{3t}$

Objective 3: Rationalize a Denominator Containing Two Terms

61) How do you find the conjugate of an expression with two radical terms?   Change the sign between the two terms.

62) When you multiply a binomial containing a square root by its conjugate, what happens to the radical?   The radical is eliminated.

Find the conjugate of each expression. Then, multiply the expression by its conjugate.

63) $(5+\sqrt{2})$ $(5-\sqrt{2}); 23$

64) $(\sqrt{5}-4)$ $(\sqrt{5}+4); -11$

65) $(\sqrt{2}+\sqrt{6})$ $(\sqrt{2}-\sqrt{6}); -4$

66) $(\sqrt{3}-\sqrt{10})$ $(\sqrt{3}+\sqrt{10}); -7$

67) $(\sqrt{t}-8)$ $(\sqrt{t}+8); t-64$

68) $(\sqrt{p}+5)$ $(\sqrt{p}-5); p-25$

Rationalize the denominator and simplify completely.

Fill It In

Fill in the blanks with either the missing mathematical step or reason for the given step.

69) $\dfrac{6}{4 - \sqrt{5}} = \dfrac{6}{4 - \sqrt{5}} \cdot \dfrac{4 + \sqrt{5}}{4 + \sqrt{5}}$ Multiply by the conjugate.

$= \dfrac{6(4 + \sqrt{5})}{(4)^2 - (\sqrt{5})^2}$ $(a + b)(a - b)$ $= a^2 - b^2$

$= \dfrac{24 + 6\sqrt{5}}{16 - 5}$ Multiply terms in numerator, square terms in denominator.

$= \dfrac{24 + 6\sqrt{5}}{11}$ Simplify.

70) $\dfrac{\sqrt{6}}{\sqrt{7} + \sqrt{2}} = \dfrac{\sqrt{6}}{\sqrt{7} + \sqrt{2}} \cdot \dfrac{\sqrt{7} - \sqrt{2}}{\sqrt{7} - \sqrt{2}}$ Multiply by the conjugate.

$= \dfrac{\sqrt{6}(\sqrt{7} - \sqrt{2})}{(\sqrt{7})^2 - (\sqrt{2})^2}$ $(a + b)(a - b)$ $= a^2 - b^2$

$= \dfrac{\sqrt{42} - \sqrt{12}}{7 - 2}$ Multiply terms in numerator, square terms in denominator.

$= \dfrac{\sqrt{42} - 2\sqrt{3}}{5}$ Simplify.

71) $\dfrac{3}{2 + \sqrt{3}}$ $6 - 3\sqrt{3}$

72) $\dfrac{8}{6 - \sqrt{5}}$ $\dfrac{48 + 8\sqrt{5}}{31}$

73) $\dfrac{10}{9 - \sqrt{2}}$ $\dfrac{90 + 10\sqrt{2}}{79}$

74) $\dfrac{5}{4 + \sqrt{6}}$ $\dfrac{4 - \sqrt{6}}{2}$

75) $\dfrac{\sqrt{8}}{\sqrt{3} + \sqrt{2}}$ $2\sqrt{6} - 4$

76) $\dfrac{\sqrt{32}}{\sqrt{5} - \sqrt{7}}$ $-2\sqrt{10} - 2\sqrt{14}$

77) $\dfrac{\sqrt{3} - \sqrt{5}}{\sqrt{10} - \sqrt{3}}$

78) $\dfrac{\sqrt{3} + \sqrt{6}}{\sqrt{2} + \sqrt{5}}$

VIDEO 79) $\dfrac{\sqrt{m}}{\sqrt{m} + \sqrt{n}}$ $\dfrac{m - \sqrt{mn}}{m - n}$

80) $\dfrac{\sqrt{u}}{\sqrt{u} - \sqrt{v}}$ $\dfrac{u + \sqrt{uv}}{u - v}$

81) $\dfrac{b - 25}{\sqrt{b} - 5}$ $\sqrt{b} + 5$

82) $\dfrac{d - 9}{\sqrt{d} + 3}$ $\sqrt{d} - 3$

83) $\dfrac{\sqrt{x} + \sqrt{y}}{\sqrt{x} - \sqrt{y}}$

84) $\dfrac{\sqrt{f} - \sqrt{g}}{\sqrt{f} + \sqrt{g}}$ $\dfrac{f - 2\sqrt{fg} + g}{f - g}$

Objective 4: Rationalize a Numerator

Rationalize the numerator of each expression and simplify.

85) $\dfrac{\sqrt{5}}{3}$ $\dfrac{5}{3\sqrt{5}}$

86) $\dfrac{\sqrt{2}}{9}$ $\dfrac{2}{9\sqrt{2}}$

87) $\dfrac{\sqrt{x}}{\sqrt{7}}$ $\dfrac{x}{\sqrt{7x}}$

88) $\dfrac{\sqrt{8a}}{\sqrt{b}}$ $\dfrac{4a}{\sqrt{2ab}}$

89) $\dfrac{2 + \sqrt{3}}{6}$ $\dfrac{1}{12 - 6\sqrt{3}}$

90) $\dfrac{1 + \sqrt{7}}{3}$ $\dfrac{2}{\sqrt{7} - 1}$

91) $\dfrac{\sqrt{x} - 2}{x - 4}$ $\dfrac{1}{\sqrt{x} + 2}$

92) $\dfrac{3 - \sqrt{n}}{n - 9}$ $-\dfrac{1}{3 + \sqrt{n}}$

93) $\dfrac{4 - \sqrt{c + 11}}{c - 5}$

94) $\dfrac{\sqrt{x + h} - \sqrt{x}}{h}$

95) Does rationalizing the denominator of an expression change the value of the original expression? Explain your answer.

96) Does rationalizing the numerator of an expression change the value of the original expression? Explain your answer.

Objective 5: Divide Out Common Factors from the Numerator and Denominator

Simplify completely.

97) $\dfrac{5 + 10\sqrt{3}}{5}$ $1 + 2\sqrt{3}$

98) $\dfrac{18 - 6\sqrt{7}}{6}$ $3 - \sqrt{7}$

99) $\dfrac{30 - 18\sqrt{5}}{4}$ $\dfrac{15 - 9\sqrt{5}}{2}$

100) $\dfrac{36 + 20\sqrt{2}}{12}$ $\dfrac{9 + 5\sqrt{2}}{3}$

VIDEO 101) $\dfrac{\sqrt{45} + 6}{9}$ $\dfrac{\sqrt{5} + 2}{3}$

102) $\dfrac{\sqrt{48} + 28}{4}$ $\sqrt{3} + 7$

103) $\dfrac{-10 - \sqrt{50}}{5}$ $-2 - \sqrt{2}$

104) $\dfrac{-35 + \sqrt{200}}{15}$ $\dfrac{-7 + 2\sqrt{2}}{3}$

105) The function $r(A) = \sqrt{\dfrac{A}{\pi}}$ describes the radius of a circle, $r(A)$, in terms of its area, A.

a) If the area of a circle is measured in square inches, find $r(8\pi)$ and explain what it means in the context of the problem.

b) If the area of a circle is measured in square inches, find $r(7)$ and rationalize the denominator. Explain the meaning of $r(7)$ in the context of the problem.

c) Obtain an equivalent form of the function by rationalizing the denominator. $r(A) = \dfrac{\sqrt{A\pi}}{\pi}$

106) The function $r(V) = \sqrt[3]{\dfrac{3V}{4\pi}}$ describes the radius of a sphere, $r(V)$, in terms of its volume, V.

a) If the volume of a sphere is measured in cubic centimeters, find $r(36\pi)$ and explain what it means in the context of the problem.

b) If the volume of a sphere is measured in cubic centimeters, find $r(11)$ and rationalize the denominator. Explain the meaning of $r(11)$ in the context of the problem.

c) Obtain an equivalent form of the function by rationalizing the denominator. $r(V) = \dfrac{\sqrt[3]{6\pi^2 V}}{2\pi}$

Putting It All Together

Objective

1. Review the Concepts Presented in 8.1–8.6

1. Review the Concepts Presented in 8.1–8.6

In 8.1, we learned how to find roots of numbers.

Example 1

In-Class Example 1
Find each root.
a) $\sqrt{36}$ b) $-\sqrt{36}$
c) $\sqrt{-36}$ d) $\sqrt[3]{-1}$
**answer: a) 6   b) −6
c) not a real number
d) −1**

Find each root.

a) $\sqrt{64}$ b) $-\sqrt{64}$ c) $\sqrt{-64}$ d) $\sqrt[3]{-64}$

Solution

a) $\sqrt{64} = 8$ since $8^2 = 64$. Remember that the square root symbol, $\sqrt{\ }$, represents the principal square root (or positive square root) of a number.

b) $-\sqrt{64}$ means $-1 \cdot \sqrt{64}$. Therefore, $-\sqrt{64} = -1 \cdot \sqrt{64} = -1 \cdot 8 = -8$.

c) Recall that the *even* root of a negative number is not a real number. Therefore, $\sqrt{-64}$ is not a real number.

d) The *odd* root of a negative number is a negative number. So, $\sqrt[3]{-64} = -4$ since $(-4)^3 = -64$.

In 8.2, we learned about the relationship between rational exponents and radicals. Recall that if m and n are positive integers and $\dfrac{m}{n}$ is in lowest terms, then $a^{m/n} = (a^{1/n})^m = (\sqrt[n]{a})^m$ provided that $a^{1/n}$ is a real number. For the rest of this section we will assume that all variables represent positive real numbers.

Example 2

In-Class Example 2
Simplify completely. The answer should contain only positive exponents.

a) $(125)^{2/3}$ b) $\left(\dfrac{4c^9 d^{11/6}}{c^{-1}d^2}\right)^{-5/2}$

answer: a) 25   b) $\dfrac{d^{5/12}}{32c^{25}}$

Simplify completely. The answer should contain only positive exponents.

a) $(32)^{4/5}$ b) $\left(\dfrac{a^7 b^{9/8}}{25a^9 b^{3/4}}\right)^{-3/2}$

Solution

a) The denominator of the fractional exponent is the index of the radical, and the numerator is the power to which we raise the radical expression.

$$32^{4/5} = (\sqrt[5]{32})^4 \qquad \text{Write in radical form.}$$
$$= (2)^4 \qquad \sqrt[5]{32} = 2$$
$$= 16$$

b) $\left(\dfrac{a^7 b^{9/8}}{25a^9 b^{3/4}}\right)^{-3/2} = \left(\dfrac{25a^9 b^{3/4}}{a^7 b^{9/8}}\right)^{3/2}$ Eliminate the negative from the outermost exponent by taking the reciprocal of the base.

Simplify the expression inside the parentheses by subtracting the exponents.

$$= (25a^{9-7}b^{3/4-9/8})^{3/2} = (25a^2 b^{6/8-9/8})^{3/2} = (25a^2 b^{-3/8})^{3/2}$$

Apply the power rule, and simplify.

$$= (25)^{3/2}(a^2)^{3/2}(b^{-3/8})^{3/2} = (\sqrt{25})^3 a^3 b^{-9/16} = 5^3 a^3 b^{-9/16} = \dfrac{125a^3}{b^{9/16}}$$

In Sections 8.3–8.6 we learned how to simplify, multiply, divide, add, and subtract radicals. Let's look at these operations together so that we will learn to recognize the techniques needed to perform these operations.

Example 3

In-Class Example 3
Perform the operations and simplify.
a) $\sqrt{5} + 7\sqrt{3} - 11\sqrt{5}$
b) $\sqrt{6}(4\sqrt{3} + 2\sqrt{6})$
answer: a) $7\sqrt{3} - 10\sqrt{5}$
b) $12\sqrt{2} + 12$

Perform the operations and simplify.

a) $\sqrt{3} + 10\sqrt{6} - 4\sqrt{3}$ b) $\sqrt{3}(10\sqrt{6} - 4\sqrt{3})$

Solution

a) This is the *sum and difference* of radicals. Remember that we can only add and subtract radicals that are like radicals.

$$\sqrt{3} + 10\sqrt{6} - 4\sqrt{3} = \sqrt{3} - 4\sqrt{3} + 10\sqrt{6} \quad \text{Write like radicals together.}$$
$$= -3\sqrt{3} + 10\sqrt{6} \quad \text{Subtract.}$$

b) This is the *product* of radical expressions. We must multiply the binomial $10\sqrt{6} - 4\sqrt{3}$ by $\sqrt{3}$ using the distributive property.

$$\sqrt{3}(10\sqrt{6} - 4\sqrt{3}) = \sqrt{3} \cdot 10\sqrt{6} - \sqrt{3} \cdot 4\sqrt{3} \quad \text{Distribute.}$$
$$= 10\sqrt{18} - 4 \cdot 3 \quad \text{Product rule; } \sqrt{3} \cdot \sqrt{3} = 3.$$
$$= 10\sqrt{18} - 12 \quad \text{Multiply.}$$

Ask yourself, "Is $10\sqrt{18} - 12$ in simplest form?" *No.* $\sqrt{18}$ can be simplified.

$$= 10\sqrt{9 \cdot 2} - 12 \quad \text{9 is a perfect square.}$$
$$= 10\sqrt{9} \cdot \sqrt{2} - 12 \quad \text{Product rule}$$
$$= 10 \cdot 3\sqrt{2} - 12 \quad \sqrt{9} = 3$$
$$= 30\sqrt{2} - 12 \quad \text{Multiply.}$$

The expression is now in simplest form.

Next we will look at multiplication problems involving binomials. Remember that the rules we used to multiply binomials like $(x + 4)(x - 9)$ are the same rules we use to multiply binomials containing radicals.

Example 4

In-Class Example 4
Multiply and simplify.
a) $(1 + \sqrt{3})(8 - \sqrt{7})$
b) $(\sqrt{k} + \sqrt{11})(\sqrt{k} - \sqrt{11})$
c) $(4\sqrt{3} - 5)^2$
answer:
a) $8 - \sqrt{7} + 8\sqrt{3} - \sqrt{21}$
b) $k - 11$ **c)** $23 - 40\sqrt{3}$

Multiply and simplify.

a) $(8 + \sqrt{2})(9 - \sqrt{11})$ b) $(\sqrt{n} + \sqrt{7})(\sqrt{n} - \sqrt{7})$
c) $(2\sqrt{5} - 3)^2$

Solution

a) Since we must multiply two binomials, we will use FOIL.

$$\overset{F\qquad O\qquad I\qquad L}{(8 + \sqrt{2})(9 - \sqrt{11}) = 8 \cdot 9 - 8 \cdot \sqrt{11} + 9 \cdot \sqrt{2} - \sqrt{2} \cdot \sqrt{11}} \quad \text{Use FOIL.}$$
$$= 72 - 8\sqrt{11} + 9\sqrt{2} - \sqrt{22} \quad \text{Multiply.}$$

All radicals are simplified and none of them are like radicals, so this expression is in simplest form.

b)   We can multiply $(\sqrt{n} + \sqrt{7})(\sqrt{n} - \sqrt{7})$ using FOIL or, if we notice that this product is in the form $(a + b)(a - b)$ we can apply the rule $(a + b)(a - b) = a^2 - b^2$. Either method will give us the correct answer. We will use the second method.

$$(a + b)(a - b) = a^2 - b^2$$
$$(\sqrt{n} + \sqrt{7})(\sqrt{n} - \sqrt{7}) = (\sqrt{n})^2 - (\sqrt{7})^2 = n - 7 \qquad \text{Substitute } \sqrt{n} \text{ for } a \text{ and } \sqrt{7} \text{ for } b.$$

c)   Once again we can either use FOIL to expand $(2\sqrt{5} - 3)^2$ or we can use the special formula we learned for squaring a binomial.

We will use $(a - b)^2 = a^2 - 2ab + b^2$.

$$\begin{aligned}
(2\sqrt{5} - 3)^2 &= (2\sqrt{5})^2 - 2(2\sqrt{5})(3) + (3)^2 && \text{Substitute } 2\sqrt{5} \text{ for } a \text{ and } 3 \text{ for } b. \\
&= (4 \cdot 5) - 4\sqrt{5}(3) + 9 && \text{Multiply.} \\
&= 20 - 12\sqrt{5} + 9 && \text{Multiply.} \\
&= 29 - 12\sqrt{5} && \text{Combine like terms.}
\end{aligned}$$

Remember that an expression is not considered to be in simplest form if it contains a radical in its denominator. To rationalize the denominator of a radical expression, we must keep in mind the index on the radical and the number of terms in the denominator.

Example 5

In-Class Example 5
Rationalize the denominator of each expression.
a) $\dfrac{14}{\sqrt{2a}}$ b) $\dfrac{14}{\sqrt[3]{2a}}$ c) $\dfrac{\sqrt{14}}{\sqrt{2} + 3}$

answer: a) $\dfrac{7\sqrt{2a}}{a}$

b) $\dfrac{7\sqrt[3]{4a^2}}{a}$ c) $\dfrac{3\sqrt{14} - 2\sqrt{7}}{7}$

Rationalize the denominator of each expression.

a) $\dfrac{10}{\sqrt{2x}}$ b) $\dfrac{10}{\sqrt[3]{2x}}$ c) $\dfrac{\sqrt{10}}{\sqrt{2} - 1}$

Solution

a)   First, notice that the denominator of $\dfrac{10}{\sqrt{2x}}$ contains only one term and it is a *square* root. Ask yourself, "By what do I multiply $\sqrt{2x}$ to get a perfect *square* under the radical?" The answer is $\sqrt{2x}$ since $\sqrt{2x} \cdot \sqrt{2x} = \sqrt{4x^2} = 2x$. Multiply the numerator and denominator by $\sqrt{2x}$, and simplify.

$$\dfrac{10}{\sqrt{2x}} = \dfrac{10}{\sqrt{2x}} \cdot \dfrac{\sqrt{2x}}{\sqrt{2x}} \qquad \text{Rationalize the denominator.}$$
$$= \dfrac{10\sqrt{2x}}{\sqrt{4x^2}} = \dfrac{10\sqrt{2x}}{2x} = \dfrac{5\sqrt{2x}}{x}$$

b)   The denominator of $\dfrac{10}{\sqrt[3]{2x}}$ contains only one term, but it is a *cube* root. Ask yourself, "By what do I multiply $\sqrt[3]{2x}$ to get a radicand that is a perfect *cube*?" The answer is $\sqrt[3]{4x^2}$ since $\sqrt[3]{2x} \cdot \sqrt[3]{4x^2} = \sqrt[3]{8x^3} = 2x$. Multiply the numerator and denominator by $\sqrt[3]{4x^2}$, and simplify.

$$\dfrac{10}{\sqrt[3]{2x}} = \dfrac{10}{\sqrt[3]{2x}} \cdot \dfrac{\sqrt[3]{4x^2}}{\sqrt[3]{4x^2}} \qquad \text{Rationalize the denominator.}$$
$$= \dfrac{10\sqrt[3]{4x^2}}{\sqrt[3]{8x^3}} = \dfrac{10\sqrt[3]{4x^2}}{2x} = \dfrac{5\sqrt[3]{4x^2}}{x}$$

c) The denominator of $\dfrac{\sqrt{10}}{\sqrt{2}-1}$ contains two terms, so how do we rationalize the

denominator of this expression? We multiply the numerator and denominator by the *conjugate* of the denominator.

$$\frac{\sqrt{10}}{\sqrt{2}-1} = \frac{\sqrt{10}}{\sqrt{2}-1} \cdot \frac{\sqrt{2}+1}{\sqrt{2}+1} \qquad \text{Multiply by the conjugate.}$$

$$= \frac{\sqrt{10}(\sqrt{2}+1)}{(\sqrt{2})^2-(1)^2} \qquad \begin{array}{l}\text{Multiply.}\\ (a+b)(a-b)=a^2-b^2\end{array}$$

$$= \frac{\sqrt{20}+\sqrt{10}}{1} \qquad \begin{array}{l}\text{Distribute.}\\ \text{Simplify.}\end{array}$$

$$= 2\sqrt{5}+\sqrt{10} \qquad \sqrt{20}=2\sqrt{5};\ \text{simplify.} \qquad \blacksquare$$

You Try 1

a) Perform the operations and simplify.

  i) $(\sqrt{w}+8)^2$ ii) $(3-\sqrt{5a})(4+\sqrt{5a})$ iii) $\sqrt{2}(9\sqrt{10}-\sqrt{2})$
  iv) $\sqrt{2}+9\sqrt{10}-5\sqrt{2}$ v) $(2\sqrt{3}+y)(2\sqrt{3}-y)$

b) Find each root.

  i) $-\sqrt{\dfrac{121}{16}}$ ii) $\sqrt[3]{-1000}$ iii) $\sqrt{0.09}$ iv) $\sqrt{-49}$

c) Simplify completely. The answer should contain only positive exponents.

  i) $(-64)^{2/3}$ ii) $\left(\dfrac{81x^3y^{1/2}}{x^{-5}y^6}\right)^{-3/4}$

d) Rationalize the denominator of each expression.

  i) $\dfrac{24}{\sqrt[3]{9h}}$ ii) $\dfrac{7+\sqrt{6}}{4+\sqrt{6}}$ iii) $\dfrac{56}{\sqrt{7}}$

Answers to You Try Exercises

1) a) i) $w+16\sqrt{w}+64$ ii) $12-\sqrt{5a}-5a$ iii) $18\sqrt{5}-2$ iv) $-4\sqrt{2}+9\sqrt{10}$ v) $12-y^2$

   b) i) $-\dfrac{11}{4}$ ii) -10 iii) 0.3 iv) not a real number

   c) i) 16 ii) $\dfrac{y^{33/8}}{27x^6}$ d) i) $\dfrac{8\sqrt[3]{3h^2}}{h}$ ii) $\dfrac{22-3\sqrt{6}}{10}$ iii) $8\sqrt{7}$

Putting It All Together
Summary Exercises

*Additional answers can be found in the Answers to Exercises appendix.

Objective 1: Review the Concepts Presented in 8.1–8.6

Assume all variables represent positive real numbers.

Find each root, if possible.

1) $\sqrt[4]{81}$ 3

2) $\sqrt[3]{-1000}$ -10

3) $-\sqrt[6]{64}$ -2

4) $\sqrt{121}$ 11

5) $\sqrt{-169}$ not a real number

6) $\sqrt{\dfrac{144}{49}}$ $\dfrac{12}{7}$

Simplify completely. The answer should contain only positive exponents.

7) $(144)^{1/2}$ 12

8) $(-32)^{4/5}$ 16

9) $-1000^{2/3}$ -100

10) $\left(-\dfrac{16}{81}\right)^{3/4}$ not a real number

11) $125^{-1/3}$ $\dfrac{1}{5}$

12) $\left(\dfrac{100}{9}\right)^{-3/2}$ $\dfrac{27}{1000}$

13) $k^{-3/5}\cdot k^{3/10}$ $\dfrac{1}{k^{3/10}}$

14) $(t^{3/8})^{16}$ t^6

15) $\left(\dfrac{27a^{-8}}{b^9}\right)^{2/3}$ $\dfrac{9}{a^{16/3}b^6}$

16) $\left(\dfrac{18x^{-9}y^{4/3}}{2x^3y}\right)^{-5/2}$ $\dfrac{x^{30}}{243y^{5/6}}$

Simplify completely.

17) $\sqrt{24}$ $2\sqrt{6}$

18) $\sqrt[4]{32}$ $2\sqrt[4]{2}$

19) $\sqrt[3]{72}$ $2\sqrt[3]{9}$

20) $\sqrt[3]{\dfrac{500}{2}}$ $5\sqrt[3]{2}$

21) $\sqrt[4]{243}$ $3\sqrt[4]{3}$

22) $\sqrt{45c^{11}}$ $3c^5\sqrt{5c}$

23) $\sqrt[3]{96m^7n^{15}}$ $2m^2n^5\sqrt[3]{12m}$

24) $\sqrt[5]{\dfrac{64x^{19}}{y^{20}}}$ $\dfrac{2x^3\sqrt[5]{2x^4}}{y^4}$

Perform the operations and simplify.

25) $\sqrt[3]{12}\cdot\sqrt[3]{2}$ $2\sqrt[3]{3}$

26) $\sqrt[4]{\dfrac{96k^{11}}{2k^3}}$ $2k^2\sqrt[4]{3}$

27) $(6+\sqrt{7})(2+\sqrt{7})$ $19+8\sqrt{7}$

28) $4c^2\sqrt[3]{108c}-15\sqrt[3]{32c^7}$ $-18c^2\sqrt[3]{4c}$

29) $\dfrac{18}{\sqrt{6}}$ $3\sqrt{6}$

30) $\dfrac{5}{\sqrt{3}-\sqrt{2}}$ $5\sqrt{3}+5\sqrt{2}$

31) $3\sqrt{75m^3n}+m\sqrt{12mn}$ $17m\sqrt{3mn}$

32) $\sqrt{6p^7q^3}\cdot\sqrt{15pq^2}$ $3p^4q^2\sqrt{10q}$

33) $\dfrac{\sqrt{60t^8u^3}}{\sqrt{5t^2u}}$ $2t^3u\sqrt{3}$

34) $\dfrac{9}{\sqrt[3]{2}}$ $\dfrac{9\sqrt[3]{4}}{2}$

35) $(2\sqrt{3}+10)^2$ $112+40\sqrt{3}$

36) $(\sqrt{2}+3)(\sqrt{2}-3)$ -7

37) $\dfrac{\sqrt{2}}{4+\sqrt{10}}$ $\dfrac{2\sqrt{2}-\sqrt{5}}{3}$

38) $\sqrt[3]{r^2}\cdot\sqrt{r}$ $r\sqrt[6]{r}$

39) $\sqrt[3]{\dfrac{b^2}{9c}}$ $\dfrac{\sqrt[3]{3b^2c^2}}{3c}$

40) $\dfrac{\sqrt[4]{32}}{\sqrt[4]{w^{11}}}$ $\dfrac{2\sqrt[4]{2w}}{w^3}$

For each function, find the domain and graph the function.

41) $f(x)=\sqrt{x-2}$

42) $g(x)=\sqrt[3]{x}$

43) $h(x)=\sqrt[3]{x}+1$

44) $k(x)=\sqrt{x+1}$

45) $g(x)=\sqrt{-x}$

46) $f(x)=\sqrt{x}-2$

Section 8.7 Solving Radical Equations

Objectives

1. **Understand the Steps for Solving a Radical Equation**
2. **Solve an Equation Containing One Square Root**
3. **Solve an Equation Containing Two Square Roots**
4. **Solve an Equation Containing a Cube Root**

In this section, we will learn how to solve *radical equations.*

An equation containing a variable in the radicand is a **radical equation.** Some examples of radical equations are

$$\sqrt{p}=7,\qquad \sqrt[3]{n}=2,\qquad \sqrt{2x+1}+1=x,\qquad \sqrt{5w+6}-\sqrt{4w+1}=1$$

1. Understand the Steps for Solving a Radical Equation

Let's review what happens when we square a square root expression: If $x\ge 0$, then $(\sqrt{x})^2=x$. That is, to eliminate the radical from $\sqrt{x}$, we *square* the expression. Therefore to solve equations like those above containing *square roots,* we *square* both sides of the equation to obtain new equations. The solutions of the new equations contain all of the solutions of the original equation and may also contain *extraneous solutions.*

An **extraneous solution** is a value that satisfies one of the new equations but does not satisfy the original equation. Extraneous solutions occur frequently when solving radical equations, so we *must* check all possible solutions in the original equation and discard any that are extraneous.

Procedure   Solving Radical Equations Containing Square Roots

Step 1:   Get a radical on a side by itself.

Step 2:   Square both sides of the equation to eliminate a radical.

Step 3:   Combine like terms on each side of the equation.

Step 4:   If the equation still contains a radical, repeat Steps 1–3.

Step 5:   Solve the equation.

Step 6:   Check the proposed solutions *in the original equation* and discard extraneous solutions.

2. Solve an Equation Containing One Square Root

Example 1

In-Class Example 1
Solve.
a) $\sqrt{b-8}=2$
b) $\sqrt{w+3}+7=3$
answer: a) {12}   b) $\varnothing$

Solve.

a) $\sqrt{c-2}=3$ b) $\sqrt{t+5}+6=0$

Solution

a) **Step 1:** The radical *is* on a side by itself: $\sqrt{c-2}=3$

   Step 2: *Square* both sides to eliminate the *square root.*

$$(\sqrt{c-2})^2 = 3^2 \qquad \text{Square both sides.}$$
$$c - 2 = 9$$

   Steps 3 and 4 do not apply because there are no like terms to combine and no radicals remain.

   Step 5: Solve the equation.

$$c = 11 \qquad \text{Add 2 to each side.}$$

   Step 6: Check $c = 11$ in the *original* equation.

$$\sqrt{c-2}=3$$
$$\sqrt{11-2} \overset{?}{=} 3$$
$$\sqrt{9}=3 \quad \checkmark$$

   The solution set is {11}.

b) The first step is to get the radical on a side by itself.

$$\sqrt{t+5}+6=0$$
$$\sqrt{t+5} = -6 \qquad \text{Subtract 6 from each side.}$$
$$(\sqrt{t+5})^2 = (-6)^2 \qquad \text{Square both sides to eliminate the radical.}$$
$$t + 5 = 36 \qquad \text{The square root has been eliminated.}$$
$$t = 31 \qquad \text{Solve the equation.}$$

   Check $t = 31$ in the *original* equation.

$$\sqrt{t+5}+6=0$$
$$\sqrt{31+5}+6 \overset{?}{=} 0$$
$$6 + 6 \overset{?}{=} 0 \qquad \text{FALSE}$$

   Because $t = 31$ gives us a false statement, it is an extraneous solution. The equation has no real solution. The solution set is $\varnothing$.   ■

You Try 1

Solve.

a) $\sqrt{a+4}=7$ b) $\sqrt{m-7}+12=9$

Sometimes, we have to square a binomial in order to solve a radical equation. Don't forget that when we square a binomial we can either use FOIL or one of the following formulas:
$(a+b)^2 = a^2 + 2ab + b^2$ or $(a-b)^2 = a^2 - 2ab + b^2$.

Example 2

Solve $\sqrt{2x+1}+1=x$.

In-Class Example 2
Solve $\sqrt{2y+4}-2=y$.
answer: $\{-2, 0\}$

Solution

Start by getting the radical on a side by itself.

$$\sqrt{2x+1}=x-1 \qquad \text{Subtract 1 from each side.}$$
$$(\sqrt{2x+1})^2=(x-1)^2 \qquad \text{Square both sides to eliminate the radical.}$$
$$2x+1=x^2-2x+1 \qquad \text{Simplify; square the binomial.}$$
$$0=x^2-4x \qquad \text{Subtract } 2x; \text{ subtract 1.}$$
$$0=x(x-4) \qquad \text{Factor.}$$

$$x=0 \quad \text{or} \quad x-4=0 \qquad \text{Set each factor equal to zero.}$$
$$x=0 \quad \text{or} \quad x=4 \qquad \text{Solve.}$$

Check $x=0$ and $x=4$ in the *original* equation.

$x=0$:
$$\sqrt{2x+1}+1=x$$
$$\sqrt{2(0)+1}+1 \stackrel{?}{=} 0$$
$$\sqrt{1}+1 \stackrel{?}{=} 0$$
$$2 \stackrel{?}{=} 1 \quad \textbf{FALSE}$$

$x=4$:
$$\sqrt{2x+1}+1=x$$
$$\sqrt{2(4)+1}+1 \stackrel{?}{=} 4$$
$$\sqrt{9}+1 \stackrel{?}{=} 4$$
$$3+1=4 \quad \textbf{TRUE}$$

$x=4$ *is* a solution but $x=0$ is ***not*** because $x=0$ does not satisfy the original equation. The solution set is $\{4\}$. ■

You Try 2

Solve.

a) $\sqrt{3p+10}-4=p$ b) $\sqrt{4h-3}-h=-2$

3. Solve an Equation Containing Two Square Roots

Next, we will take our first look at solving an equation containing two square roots.

Example 3

Solve $\sqrt{2a+4}-3\sqrt{a-5}=0$.

In-Class Example 3
Solve
$\sqrt{3b+4}-2\sqrt{b-1}=0$.
answer: $\{8\}$

Solution

Begin by getting a radical on a side by itself.

$$\sqrt{2a+4}=3\sqrt{a-5} \qquad \text{Add } 3\sqrt{a-5} \text{ to each side.}$$
$$(\sqrt{2a+4})^2=(3\sqrt{a-5})^2 \qquad \text{Square both sides to eliminate the radicals.}$$
$$2a+4=9(a-5) \qquad 3^2=9$$
$$2a+4=9a-45 \qquad \text{Distribute.}$$
$$-7a=-49$$
$$a=7 \qquad \text{Solve.}$$

Check $a=7$ in the original equation.

$$\sqrt{2a+4}-3\sqrt{a-5}=0$$
$$\sqrt{2(7)+4}-3\sqrt{7-5}=0$$
$$\sqrt{14+4}-3\sqrt{2} \stackrel{?}{=} 0$$
$$\sqrt{18}-3\sqrt{2} \stackrel{?}{=} 0$$
$$3\sqrt{2}-3\sqrt{2}=0 \quad \checkmark$$

The solution set is $\{7\}$. ■

You Try 3

Solve $4\sqrt{r-3} - \sqrt{6r+2} = 0$.

Recall from Section 8.5 that we can square binomials containing radical expressions just like we squared $(x-1)^2$ in Example 2. We can use FOIL or the formulas

$$(a+b)^2 = a^2 + 2ab + b^2 \quad \text{or} \quad (a-b)^2 = a^2 - 2ab + b^2$$

Example 4

In-Class Example 4
Square and simplify
$(6 + \sqrt{r+1})^2$.
answer: $37 + 12\sqrt{r+1} + r$

Square and simplify $(3 - \sqrt{m+2})^2$.

Solution

Use the formula $(a-b)^2 = a^2 - 2ab + b^2$.

$$(3 - \sqrt{m+2})^2 = (3)^2 - 2(3)(\sqrt{m+2}) + (\sqrt{m+2})^2 \qquad \text{Substitute 3 for } a \text{ and } \sqrt{m+2} \text{ for } b.$$

$$= 9 - 6\sqrt{m+2} + (m+2)$$

$$= m + 11 - 6\sqrt{m+2} \qquad \text{Combine like terms.} \quad \blacksquare$$

You Try 4

Square and simplify each expression.

a) $(\sqrt{z} - 4)^2$ b) $(5 + \sqrt{3d-1})^2$

To solve the next two equations, we will have to square both sides of the equation twice to eliminate the radicals. Be very careful when you are squaring the binomials that contain a radical.

Example 5

In-Class Example 5
Solve each equation.
a) $\sqrt{c} - \sqrt{c-9} = 1$
b) $\sqrt{7k+8} - \sqrt{3k+4} = 2$
answer: a) {25}   b) {4}

Solve each equation.

a) $\sqrt{x+5} + \sqrt{x} = 5$ b) $\sqrt{5w+6} - \sqrt{4w+1} = 1$

Solution

a) This equation contains two radicals *and* a constant. Get one of the radicals on a side by itself, then square both sides.

$$\sqrt{x+5} = 5 - \sqrt{x} \qquad \text{Subtract } \sqrt{x} \text{ from each side.}$$

$$(\sqrt{x+5})^2 = (5 - \sqrt{x})^2 \qquad \text{Square both sides.}$$

$$x + 5 = (5)^2 - 2(5)(\sqrt{x}) + (\sqrt{x})^2 \qquad \text{Use the formula } (a-b)^2 = a^2 - 2ab + b^2.$$

$$x + 5 = 25 - 10\sqrt{x} + x \qquad \text{Simplify.}$$

The equation still contains a radical. Therefore, repeat Steps 1–3. Begin by getting the radical on a side by itself.

$$5 = 25 - 10\sqrt{x} \qquad \text{Subtract } x \text{ from each side.}$$

$$-20 = -10\sqrt{x} \qquad \text{Subtract 25 from each side.}$$

$$2 = \sqrt{x} \qquad \text{Divide by } -10.$$

$$2^2 = (\sqrt{x})^2 \qquad \text{Square both sides.}$$

$$4 = x \qquad \text{Solve.}$$

The check is left to the student. The solution set is {4}.

b) **Step 1:** Get a radical on a side by itself.

$$\sqrt{5w+6} - \sqrt{4w+1} = 1$$

$$\sqrt{5w+6} = 1 + \sqrt{4w+1} \qquad \text{Add } \sqrt{4w+1} \text{ to each side.}$$

Step 2:   Square both sides of the equation to eliminate a radical.

$$(\sqrt{5w + 6})^2 = (1 + \sqrt{4w + 1})^2$$ Square both sides.

$$5w + 6 = (1)^2 + 2(1)(\sqrt{4w + 1}) + (\sqrt{4w + 1})^2$$ Use the formula $(a + b)^2 = a^2 + 2ab + b^2$.

$$5w + 6 = 1 + 2\sqrt{4w + 1} + 4w + 1$$

Step 3:   Combine like terms on the right side.

$$5w + 6 = 4w + 2 + 2\sqrt{4w + 1}$$ Combine like terms.

Step 4:   The equation still contains a radical, so repeat Steps 1–3.

Step 1:   Get the radical on a side by itself.

$$5w + 6 = 4w + 2 + 2\sqrt{4w + 1}$$

$$w + 4 = 2\sqrt{4w + 1}$$ Subtract $4w$ and subtract 2.

We do not need to eliminate the 2 from in front of the radical before squaring both sides. The radical must not be a part of a *sum* or *difference* when we square.

Step 2:   Square both sides of the equation to eliminate the radical.

$$(w + 4)^2 = (2\sqrt{4w + 1})^2$$ Square both sides.

$$w^2 + 8w + 16 = 4(4w + 1)$$ Square the binomial; $2^2 = 4$.

Steps 3 and 4 no longer apply.

Step 5:   Solve the equation.

$$w^2 + 8w + 16 = 16w + 4$$ Distribute.

$$w^2 - 8w + 12 = 0$$ Subtract $16w$ and subtract 4.

$$(w - 2)(w - 6) = 0$$ Factor.

$$w - 2 = 0 \quad \text{or} \quad w - 6 = 0$$ Set each factor equal to zero.

$$w = 2 \quad \text{or} \quad w = 6$$ Solve.

Step 6:   The check is left to the student. Verify that $w = 2$ and $w = 6$ each satisfy the original equation. The solution set is $\{2, 6\}$.   ■

You Try 5

Solve each equation.

a) $\sqrt{2y + 1} - \sqrt{y} = 1$ b) $\sqrt{3t + 4} + \sqrt{t + 2} = 2$

BE CAREFUL

Watch out for two common mistakes that students make when solving an equation like the one in Example 5b.

1)   Do not square both sides before getting a radical on a side by itself.

   This is incorrect: $(\sqrt{5w + 6} - \sqrt{4w + 1})^2 = 1^2$

$$5w + 6 - (4w + 1) = 1$$

2)   The *second* time we perform Step 2, watch out for this common error.

   This is incorrect: $(w + 4)^2 = (2\sqrt{4w + 1})^2$

$$w^2 + 16 = 2(4w + 1)$$

   On the left we must multiply using FOIL or the formula $(a + b)^2 = a^2 + 2ab + b^2$ and on the right we must remember to square the 2.

4. Solve an Equation Containing a Cube Root

We can solve many equations containing cube roots the same way we solve equations containing square roots except, to eliminate a *cube root,* we *cube* both sides of the equation.

Example 6

In-Class Example 6
Solve
$2\sqrt[3]{5k + 2} - 3\sqrt[3]{k + 3} = 0$.
answer: {5}

Solve $\sqrt[3]{7a + 1} - 2\sqrt[3]{a - 1} = 0$.

Solution

Begin by getting a radical on a side by itself.

$$\sqrt[3]{7a + 1} = 2\sqrt[3]{a - 1} \qquad \text{Add } 2\sqrt[3]{a - 1} \text{ to each side.}$$
$$(\sqrt[3]{7a + 1})^3 = (2\sqrt[3]{a - 1})^3 \qquad \text{Cube both sides to eliminate the radicals.}$$
$$7a + 1 = 8(a - 1) \qquad \text{Simplify; } 2^3 = 8.$$
$$7a + 1 = 8a - 8 \qquad \text{Distribute.}$$
$$9 = a \qquad \text{Subtract } 7a; \text{ add } 8.$$

Check $a = 9$ in the original equation.

$$\sqrt[3]{7a + 1} - 2\sqrt[3]{a - 1} = 0$$
$$\sqrt[3]{7(9) + 1} - 2\sqrt[3]{9 - 1} \stackrel{?}{=} 0$$
$$\sqrt[3]{64} - 2\sqrt[3]{8} \stackrel{?}{=} 0$$
$$4 - 2(2) \stackrel{?}{=} 0$$
$$4 - 4 = 0 \ \checkmark$$

The solution set is $\{9\}$.

You Try 6

Solve $3\sqrt[3]{r - 4} - \sqrt[3]{5r + 2} = 0$.

Using Technology

We can use a graphing calculator to solve a radical equation in one variable. First subtract every term on the right side of the equation from both sides of the equation and enter the result in Y_1. Graph the equation in Y_1. The zeros or *x*-intercepts of the graph are the solutions to the equation.

We will solve $\sqrt{x + 3} = 2$ using a graphing calculator.

1) Enter $\sqrt{x + 3} - 2$ in Y_1.

2) Press ZOOM 6 to graph the function in Y_1 as shown.

3) Press 2nd TRACE 2:zero, move the cursor to the left of the zero and press ENTER, move the cursor to the right of the zero and press ENTER, and move the cursor close to the zero and press ENTER to display the zero. The solution to the equation is $x = 1$.

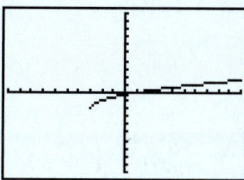
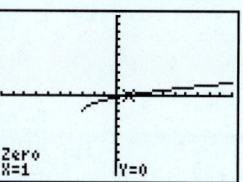

Solve each equation using a graphing calculator.

1) $\sqrt{x - 2} = 1$

2) $\sqrt{3x - 2} = 5$

3) $\sqrt{3x - 2} = \sqrt{x + 2}$

4) $\sqrt{4x - 5} = \sqrt{x + 4}$

5) $\sqrt{2x - 7} = \sqrt{x - 1}$

6) $\sqrt{\sqrt{x} - 1} = 1$

Answers to You Try Exercises

1) a) $\{45\}$ b) $\varnothing$ 2) a) $\{-3, -2\}$ b) $\{7\}$ 3) $\{5\}$ 4) a) $z - 8\sqrt{z} + 16$
b) $3d + 24 + 10\sqrt{3d - 1}$ 5) a) $\{0, 4\}$ b) $\{-1\}$ 6) $\{5\}$

Answers to Technology Exercises

1) $\{3\}$ 2) $\{9\}$ 3) $\{9\}$ 4) $\{3\}$ 5) $\{4\}$ 6) $\{4\}$

8.7 Exercises

*Additional answers can be found in the Answers to Exercises appendix.

Objective 1: Understand the Steps for Solving a Radical Equation

1) Why is it necessary to check the proposed solutions to a radical equation in the original equation?
Sometimes there are extraneous solutions.

2) How do you know, without actually solving and checking the solution, that $\sqrt{y} = -3$ has no solution?
The principle square root of a number cannot equal a negative number.

Objective 2: Solve an Equation Containing One Square Root

Solve.

3) $\sqrt{q} = 7$ $\{49\}$

4) $\sqrt{z} = 10$ $\{100\}$

5) $\sqrt{w} - \dfrac{2}{3} = 0$ $\left\{\dfrac{4}{9}\right\}$

6) $\sqrt{r} - \dfrac{3}{5} = 0$ $\left\{\dfrac{9}{25}\right\}$

7) $\sqrt{a} + 5 = 3$ $\varnothing$

8) $\sqrt{k} + 8 = 2$ $\varnothing$

 9) $\sqrt{b - 11} - 3 = 0$ $\{20\}$

10) $\sqrt{d + 3} - 5 = 0$ $\{22\}$

11) $\sqrt{4g - 1} + 7 = 1$ $\varnothing$

12) $\sqrt{3v + 4} + 10 = 6$ $\varnothing$

13) $\sqrt{3f + 2} + 9 = 11$ $\left\{\dfrac{2}{3}\right\}$

14) $\sqrt{5u - 4} + 12 = 17$ $\left\{\dfrac{29}{5}\right\}$

15) $m = \sqrt{m^2 - 3m + 6}$ $\{2\}$

16) $b = \sqrt{b^2 + 4b - 24}$ $\{6\}$

17) $\sqrt{9r^2 - 2r + 10} = 3r$ $\{5\}$

18) $\sqrt{4p^2 - 3p + 6} = 2p$ $\{2\}$

Square each binomial and simplify.

19) $(n + 5)^2$ $n^2 + 10n + 25$

20) $(z - 3)^2$ $z^2 - 6z + 9$

21) $(c - 6)^2$ $c^2 - 12c + 36$

22) $(2k + 1)^2$ $4k^2 + 4k + 1$

Solve.

23) $p + 6 = \sqrt{12 + p}$ $\{-3\}$

24) $c - 7 = \sqrt{2c + 1}$ $\{12\}$

25) $6 + \sqrt{c^2 + 3c - 9} = c$ $\varnothing$

26) $-4 + \sqrt{z^2 + 5z - 8} = z$ $\varnothing$

27) $w - \sqrt{10w + 6} = -3$ $\{1, 3\}$

28) $3 - \sqrt{8t + 9} = -t$ $\{0, 2\}$

29) $3v = 8 + \sqrt{3v + 4}$ $\{4\}$

30) $4k = 3 + \sqrt{10k + 5}$ $\{2\}$

31) $m + 4 = 5\sqrt{m}$ $\{1, 16\}$

32) $b + 5 = 6\sqrt{b}$ $\{1, 25\}$

33) $y + 2\sqrt{6 - y} = 3$ $\{-3\}$

34) $r - 3\sqrt{r + 2} = 2$ $\{14\}$

35) $\sqrt{r^2 - 8r - 19} = r - 9$ $\{10\}$

36) $\sqrt{x^2 + x + 4} = x + 8$ $\{-4\}$

Objective 3: Solve an Equation Containing Two Square Roots

Solve.

37) $5\sqrt{1 - 5h} = 4\sqrt{1 - 8h}$ $\{-3\}$

38) $3\sqrt{6a - 2} - 4\sqrt{3a + 3} = 0$ $\{11\}$

39) $3\sqrt{3x + 6} - 2\sqrt{9x - 9} = 0$ $\{10\}$

40) $5\sqrt{q + 11} = 2\sqrt{8q + 25}$ $\{25\}$

41) $\sqrt{m} = 3\sqrt{7}$ $\{63\}$

42) $4\sqrt{3} = \sqrt{p}$ $\{48\}$

43) $\sqrt{2w - 1} + 2\sqrt{w + 4} = 0$ $\varnothing$

44) $2\sqrt{3t + 4} + \sqrt{t - 6} = 0$ $\varnothing$

Square each expression and simplify.

45) $(\sqrt{x} + 5)^2$ $x + 10\sqrt{x} + 25$

46) $(\sqrt{y} - 8)^2$ $y - 16\sqrt{y} + 64$

47) $(9 - \sqrt{a + 4})^2$ $85 - 18\sqrt{a + 4} + a$

48) $(4 + \sqrt{p + 5})^2$ $21 + 8\sqrt{p + 5} + p$

49) $(2\sqrt{3n - 1} + 7)^2$ $12n + 28\sqrt{3n - 1} + 45$

50) $(5 - 3\sqrt{2k - 3})^2$ $18k - 30\sqrt{2k - 3} - 2$

Solve.

51) $\sqrt{2y - 1} = 2 + \sqrt{y - 4}$ $\{5, 13\}$

52) $\sqrt{3n + 4} = \sqrt{2n + 1} + 1$ $\{0, 4\}$

53) $1 + \sqrt{3s - 2} = \sqrt{2s + 5}$ $\{2\}$

54) $\sqrt{4p + 12} - 1 = \sqrt{6p - 11}$ $\{6\}$

55) $\sqrt{5a + 19} - \sqrt{a + 12} = 1$ $\left\{\dfrac{1}{4}\right\}$

56) $\sqrt{2u + 3} - \sqrt{5u + 1} = -1$ $\{3\}$

57) $\sqrt{3k + 1} - \sqrt{k - 1} = 2$ $\{1, 5\}$

58) $\sqrt{4z - 3} - \sqrt{5z + 1} = -1$ $\{3, 7\}$

59) $\sqrt{3x + 4} - 5 = \sqrt{3x - 11}$ $\varnothing$

60) $\sqrt{4c - 7} = \sqrt{4c + 1} - 4$ $\varnothing$

61) $\sqrt{3v + 3} - \sqrt{v - 2} = 3$ $\{2, 11\}$

62) $\sqrt{2y + 1} - \sqrt{y} = 1$ $\{0, 4\}$

Objective 4: Solve an Equation Containing a Cube Root

63) How do you eliminate the radical from an equation like $\sqrt[3]{x} = 2$?   Raise both sides of the equation to the third power.

64) Give a reason why $\sqrt[3]{h} = -3$ has no extraneous solutions.   When you solve the equation you get $h = -27$, and the cube root of -27 is -3.

Solve.

65) $\sqrt[3]{y} = 5$ {125}

66) $\sqrt[3]{c} = 3$ {27}

67) $\sqrt[3]{m} = -4$ {−64}

68) $\sqrt[3]{t} = -2$ {−8}

69) $\sqrt[3]{2x - 5} + 3 = 1$ $\left\{-\dfrac{3}{2}\right\}$

70) $\sqrt[3]{4a + 1} + 7 = 4$ {−7}

71) $\sqrt[3]{6j - 2} = \sqrt[3]{j - 7}$ {−1}

72) $\sqrt[3]{w + 3} = \sqrt[3]{2w - 11}$ {14}

73) $\sqrt[3]{3y - 1} - \sqrt[3]{2y - 3} = 0$ {−2}

74) $\sqrt[3]{2 - 2b} + \sqrt[3]{b - 5} = 0$ {−3}

75) $\sqrt[3]{2n^2} = \sqrt[3]{7n + 4}$ $\left\{-\dfrac{1}{2}, 4\right\}$

76) $\sqrt[3]{4c^2 - 5c + 11} = \sqrt[3]{c^2 + 9}$ $\left\{\dfrac{2}{3}, 1\right\}$

Extension

Solve.

77) $p^{1/2} = 6$ {36}

78) $\dfrac{2}{3} = t^{1/2}$ $\left\{\dfrac{4}{9}\right\}$

79) $7 = (2z - 3)^{1/2}$ {26}

80) $(3k + 1)^{1/2} = 4$ {5}

81) $(y + 4)^{1/3} = 3$ {23}

82) $-5 = (a - 2)^{1/3}$ {−123}

83) $\sqrt[4]{n + 7} = 2$ {9}

84) $\sqrt[4]{x - 3} = -1$ ∅

85) $\sqrt{13 + \sqrt{r}} = \sqrt{r + 7}$ {9}

86) $\sqrt{m - 1} = \sqrt{m} - \sqrt{m - 4}$ {5}

87) $\sqrt{y} + \sqrt{y + 5} = \sqrt{y + 2}$ {−1}

88) $\sqrt{2d - \sqrt{d + 6}} = \sqrt{d + 6}$ {10}

Mixed Exercises: Objectives 2 and 4

Solve for the indicated variable.

89) $v = \sqrt{\dfrac{2E}{m}}$ for E $E = \dfrac{mv^2}{2}$

90) $V = \sqrt{\dfrac{300VP}{m}}$ for P $P = \dfrac{mV}{300}$

91) $c = \sqrt{a^2 + b^2}$ for b^2 $b^2 = c^2 - a^2$

92) $r = \sqrt{\dfrac{A}{\pi}}$ for A $A = \pi r^2$

93) $T = \sqrt[4]{\dfrac{E}{\sigma}}$ for σ $\sigma = \dfrac{E}{T^4}$

94) $r = \sqrt[3]{\dfrac{3V}{4\pi}}$ for V $V = \dfrac{4}{3}\pi r^3$

95) The speed of sound is proportional to the square root of the air temperature in still air. The speed of sound is given by the formula.

$$V_S = 20\sqrt{T + 273}$$

where V_S is the speed of sound in meters/second and T is the temperature of the air in °Celsius.

a) What is the speed of sound when the temperature is -17°C (about 1°F)?   320 m/sec

b) What is the speed of sound when the temperature is 16°C (about 61°F)?   340 m/sec

c) What happens to the speed of sound as the temperature increases?   The speed of sound increases.

d) Solve the equation for T. $T = \dfrac{V_s^2}{400} - 273$

96) If the area of a square is A and each side has length l, then the length of a side is given by

$$l = \sqrt{A}$$

A square rug has an area of 25 ft².

a) Find the dimensions of the rug.   5 ft × 5 ft

b) Solve the equation for A. $A = l^2$

97) Let V represent the volume of a cylinder, h represent its height, and r represent its radius. V, h, and r are related according to the formula

$$r = \sqrt{\dfrac{V}{\pi h}}$$

a) A cylindrical soup can has a volume of 28π in³. It is 7 in. high. What is the radius of the can?   2 in.

b) Solve the equation for V. $V = \pi r^2 h$

98) For shallow water waves, the wave velocity is given by

$$c = \sqrt{gH}$$

where g is the acceleration due to gravity (32 ft/sec²) and H is the depth of the water (in feet).

a) Find the velocity of a wave in 8 ft of water.   16 ft/sec

b) Solve the equation for H. $H = \dfrac{c^2}{g}$

99) Refer to the formula given in Problem 98.

The catastrophic Indian Ocean tsunami that hit Banda Aceh, Sumatra, Indonesia, on December 26, 2004 was caused by an earthquake whose epicenter was off the coast of northern Sumatra. The tsunami originated in about 14,400 ft of water.

a) Find the velocity of the wave near the epicenter, in miles per hour. Round the answer to the nearest unit. (Hint: 1 mile = 5280 ft.)   463 mph

b) Banda Aceh, the area hardest hit by the tsunami, was about 60 mi from the tsunami's origin. Approximately how many minutes after the earthquake occurred did the tsunami hit Banda Aceh?
(*Exploring Geology*, McGraw-Hill, 2008.)   about 8 min.

100) The radius r of a cone with height h and volume V is given by $r = \sqrt{\dfrac{3V}{\pi h}}$.

A hanging glass vase in the shape of a cone is 8 in. tall and the radius of the top of the cone is 2 in. How much water will the vase hold? Give an exact answer and an approximation to the tenths place. $\dfrac{32}{3}\pi$ in³; 33.5 in³

Use the following information for Exercises 101 and 102.

The distance a person can see to the horizon is approximated by the function $D(h) = 1.2\sqrt{h}$, where D is the number of miles a person can see to the horizon from a height of h ft.

101) Sig is the captain of an Alaskan crab fishing boat and can see 4.8 mi to the horizon when he is sailing his ship. Find his height above the sea. 16 ft

102) Phil is standing on the deck of a boat and can see 3.6 mi to the horizon. What is his height above the water? 9 ft

Use the following information for Problems 103 and 104.

When the air temperature is 0°F, the wind chill temperature, W, in degrees Fahrenheit is a function of the velocity of the wind, V, in miles per hour and is given by the formula

$$W(V) = 35.74 - 35.75V^{4/25}$$

103) Calculate the wind speed when the wind chill temperature is −10°F. Round to the nearest whole number. 5 mph

104) Find V so that $W(V) = -20$. Round to the nearest whole number. Explain your result in the context of the problem. (http://www.nws.noaa.gov/om/windchill/ windchillglossary.shtml)
When the wind chill temperature is −20°F, the speed of the wind is 16 mph.

Section 8.8 Complex Numbers

Objectives

1. Find the Square Root of a Negative Number
2. Multiply and Divide Square Roots Containing Negative Numbers
3. Add and Subtract Complex Numbers
4. Multiply Complex Numbers
5. Multiply a Complex Number by Its Conjugate
6. Divide Complex Numbers
7. Simplify Powers of i

1. Find the Square Root of a Negative Number

We have seen throughout this chapter that the square root of a negative number does not exist in the real number system because there is no real number that, when squared, will result in a negative number. For example, $\sqrt{-4}$ is not a real number because there is no real number whose square is −4.

The square roots of negative numbers do exist, however, under another system of numbers called *complex numbers*. Before we define a complex number, we must define the number i. The number i is called an *imaginary number*.

Definition

The **imaginary number i** is defined as

$$i = \sqrt{-1}$$

Therefore, squaring both sides gives us

$$i^2 = -1$$

Note

$i = \sqrt{-1}$ and $i^2 = -1$ are two *very* important facts to remember. We will be using them often!

Definition

A **complex number** is a number of the form $a + bi$, where a and b are real numbers; a is called the **real part** and b is called the **imaginary part**.

The following table lists some examples of complex numbers and their real and imaginary parts.

Complex Number	Real Part	Imaginary Part
$5 + 2i$	5	2
$\dfrac{1}{3} - 7i$	$\dfrac{1}{3}$	-7
$8i$	0	8
4	4	0

Note

The complex number $8i$ can be written in the form $a + bi$ as $0 + 8i$. Likewise, besides being a real number, 4 is a complex number since it can be written as $4 + 0i$.

Since all real numbers, a, can be written in the form $a + 0i$, all real numbers are also complex numbers.

Property Real Numbers and Complex Numbers

The set of real numbers is a subset of the set of complex numbers.

Since we defined i as $i = \sqrt{-1}$, we can now evaluate square roots of negative numbers.

 **Example 1**

In-Class Example 1
Simplify.
a) $\sqrt{-4}$ b) $\sqrt{-11}$ c) $\sqrt{-54}$
**answer: a) $2i$ b) $i\sqrt{11}$
c) $3i\sqrt{6}$**

Simplify.

a) $\sqrt{-9}$ b) $\sqrt{-7}$ c) $\sqrt{-12}$

Solution

a) $\sqrt{-9} = \sqrt{-1 \cdot 9} = \sqrt{-1} \cdot \sqrt{9} = i \cdot 3 = 3i$

b) $\sqrt{-7} = \sqrt{-1 \cdot 7} = \sqrt{-1} \cdot \sqrt{7} = i\sqrt{7}$

c) $\sqrt{-12} = \sqrt{-1 \cdot 12} = \sqrt{-1} \cdot \sqrt{12} = i\sqrt{4}\sqrt{3} = i \cdot 2\sqrt{3} = 2i\sqrt{3}$ ■

Note

In Example 1b) we wrote $i\sqrt{7}$ instead of $\sqrt{7}i$, and in Example 1c) we wrote $2i\sqrt{3}$ instead of $2\sqrt{3}i$. We do this to be clear that the i is not under the radical. It is good practice to write the i before the radical.

 **You Try 1**

Simplify.

a) $\sqrt{-36}$ b) $\sqrt{-13}$ c) $\sqrt{-20}$

2. Multiply and Divide Square Roots Containing Negative Numbers

When multiplying or dividing radicals with negative radicands, write each radical in terms of i first. Remember, also, that since $i = \sqrt{-1}$ it follows that $i^2 = -1$. We must keep this in mind when simplifying expressions.

> **Note**
>
> Whenever an i^2 appears in an expression, replace it with -1.

Example 2

In-Class Example 2
Multiply and simplify.
$\sqrt{-12} \cdot \sqrt{-3}$
answer: -6

Multiply and simplify. $\sqrt{-8} \cdot \sqrt{-2}$

Solution

$$
\begin{aligned}
\sqrt{-8} \cdot \sqrt{-2} &= i\sqrt{8} \cdot i\sqrt{2} \quad &&\text{Write each radical in terms of } i \text{ before multiplying.}\\
&= i^2\sqrt{16} \quad &&\text{Multiply.}\\
&= (-1)(4) \quad &&\text{Replace } i^2 \text{ with } -1.\\
&= -4
\end{aligned}
$$

You Try 2

Perform the operation and simplify.

a) $\sqrt{-6} \cdot \sqrt{-3}$

b) $\dfrac{\sqrt{-72}}{\sqrt{-2}}$

3. Add and Subtract Complex Numbers

Just as we can add, subtract, multiply, and divide real numbers, we can perform all of these operations with complex numbers.

> **Procedure   Adding and Subtracting Complex Numbers**
>
> 1) To add complex numbers, add the real parts and add the imaginary parts.
> 2) To subtract complex numbers, apply the distributive property and combine the real parts and combine the imaginary parts.

Example 3

In-Class Example 3
Add or subtract.
a) $(7 + 5i) + (3 + 6i)$
b) $(4 - 3i) - (10 + 2i)$
answer: a) $10 + 11i$
b) $-6 - 5i$

Add or subtract.

a) $(8 + 3i) + (4 + 2i)$ b) $(7 + i) - (3 - 4i)$

Solution

a) $\begin{aligned}[t] (8 + 3i) + (4 + 2i) &= (8 + 4) + (3 + 2)i \quad &&\text{Add real parts; add imaginary parts.}\\ &= 12 + 5i \end{aligned}$

b) $\begin{aligned}[t] (7 + i) - (3 - 4i) &= 7 + i - 3 + 4i \quad &&\text{Distributive property}\\ &= (7 - 3) + (1 + 4)i \quad &&\text{Add real parts; add imaginary parts.}\\ &= 4 + 5i \end{aligned}$

You Try 3

Add or subtract.

a) $(-10 + 6i) + (1 + 8i)$ b) $(2 - 5i) - (-1 + 6i)$

4. Multiply Complex Numbers

We multiply complex numbers just like we would multiply polynomials. There may be an additional step, however. Remember to replace i^2 with -1.

Example 4

In-Class Example 4
Multiply and simplify.
a) $7(4 - 11i)$
b) $(9 - 5i)(-2 + 3i)$
c) $(9 + i)(9 - i)$
answer: a) $28 - 77i$
b) $-3 + 37i$ c) 82

Multiply and simplify.

a) $5(-2 + 3i)$ b) $(8 + 3i)(-1 + 4i)$ c) $(6 + 2i)(6 - 2i)$

Solution

a) $5(-2 + 3i) = -10 + 15i$ Distributive property

b) Look carefully at $(8 + 3i)(-1 + 4i)$. Each complex number has two terms, similar to, say, $(x + 3)(x + 4)$. How can we multiply these two binomials? We can use FOIL.

$$\begin{array}{cccc} \text{F} & \text{O} & \text{I} & \text{L} \end{array}$$
$$(8 + 3i)(-1 + 4i) = (8)(-1) + (8)(4i) + (3i)(-1) + (3i)(4i)$$
$$= \quad -8 \quad + \quad 32i \quad - \quad 3i \quad + \quad 12i^2$$
$$= -8 + 29i + 12(-1) \qquad \text{Replace } i^2 \text{ with } -1.$$
$$= -8 + 29i - 12$$
$$= -20 + 29i$$

c) Use FOIL to find the product $(6 + 2i)(6 - 2i)$.

$$\begin{array}{cccc} \text{F} & \text{O} & \text{I} & \text{L} \end{array}$$
$$(6 + 2i)(6 - 2i) = (6)(6) + (6)(-2i) + (2i)(6) + (2i)(-2i)$$
$$= \quad 36 \quad - \quad 12i \quad + \quad 12i \quad - \quad 4i^2$$
$$= 36 - 4(-1) \qquad \text{Replace } i^2 \text{ with } -1.$$
$$= 36 + 4$$
$$= 40$$

You Try 4

Multiply and simplify.

a) $-3(6 - 7i)$ b) $(5 - i)(4 + 8i)$ c) $(-2 - 9i)(-2 + 9i)$

5. Multiply a Complex Number by Its Conjugate

In Section 8.6, we learned about conjugates of radical expressions. For example, the conjugate of $3 + \sqrt{5}$ is $3 - \sqrt{5}$.

The complex numbers in Example 4c, $6 + 2i$ and $6 - 2i$, are **complex conjugates.**

Definition

The **conjugate** of $a + bi$ is $a - bi$.

We found that $(6 + 2i)(6 - 2i) = 40$ which is a real number. The product of a complex number and its conjugate is *always* a real number, as illustrated next.

$$
\begin{array}{cccc}
\text{F} & \text{O} & \text{I} & \text{L}
\end{array}
$$
$$
\begin{aligned}
(a + bi)(a - bi) &= (a)(a) + (a)(-bi) + (bi)(a) + (bi)(-bi) \\
&= a^2 - abi + abi - b^2 i^2 \\
&= a^2 - b^2(-1) \qquad \text{Replace } i^2 \text{ with } -1. \\
&= a^2 + b^2
\end{aligned}
$$

We can summarize these facts about complex numbers and their conjugates as follows:

Summary   Complex Conjugates

1)   The conjugate of $a + bi$ is $a - bi$.

2)   The product of $a + bi$ and $a - bi$ is a real number.

3)   We can find the product $(a + bi)(a - bi)$ by using FOIL or by using $(a + bi)(a - bi) = a^2 + b^2$.

Example 5

Multiply $-3 + 4i$ by its conjugate using the formula $(a + bi)(a - bi) = a^2 + b^2$.

In-Class Example 5
Multiply $-7 + 2i$ by its conjugate using the formula $(a + bi)(a - bi) = a^2 + b^2$.
answer: 53

Solution

The conjugate of $-3 + 4i$ is $-3 - 4i$.

$$
\begin{aligned}
(-3 + 4i)(-3 - 4i) &= (-3)^2 + (4)^2 \qquad a = -3, b = 4 \\
&= 9 + 16 \\
&= 25
\end{aligned}
$$

You Try 5

Multiply $2 - 9i$ by its conjugate using the formula $(a + bi)(a - bi) = a^2 + b^2$.

6. Divide Complex Numbers

To rationalize the denominator of a radical expression like $\dfrac{2}{3 + \sqrt{5}}$, we multiply the numerator and denominator by $3 - \sqrt{5}$, the conjugate of the denominator. We divide complex numbers in the same way.

Procedure   Dividing Complex Numbers

To divide complex numbers, multiply the numerator and denominator by the *conjugate of the denominator*. Write the quotient in the form $a + bi$.

Example 6

In-Class Example 6
Divide. Write the result in the form $a + bi$.

a) $\dfrac{9}{6 + 3i}$ b) $\dfrac{-3 + i}{4 - 11i}$

answer:

a) $\dfrac{6}{5} - \dfrac{3}{5}i$ b) $-\dfrac{23}{137} - \dfrac{29}{137}i$

Divide. Write the quotient in the form $a + bi$.

a) $\dfrac{3}{4 - 5i}$ b) $\dfrac{6 - 2i}{-7 + i}$

Solution

a) $\dfrac{3}{4 - 5i} = \dfrac{3}{(4 - 5i)} \cdot \dfrac{(4 + 5i)}{(4 + 5i)}$ Multiply the numerator and denominator by the conjugate of the denominator.

$= \dfrac{12 + 15i}{4^2 + 5^2}$ Multiply numerators.

$(a + bi)(a - bi) = a^2 + b^2$

$= \dfrac{12 + 15i}{16 + 25}$

$= \dfrac{12 + 15i}{41}$

$= \dfrac{12}{41} + \dfrac{15}{41}i$ Write the quotient in the form $a + bi$.

Recall that we can find the product $(4 - 5i)(4 + 5i)$ using FOIL *or* by using the formula $(a + bi)(a - bi) = a^2 + b^2$.

b) $\dfrac{6 - 2i}{-7 + i} = \dfrac{(6 - 2i)}{(-7 + i)} \cdot \dfrac{(-7 - i)}{(-7 - i)}$ Multiply the numerator and denominator by the conjugate of the denominator.

$= \dfrac{-42 - 6i + 14i + 2i^2}{(-7)^2 + (1)^2}$ Multiply using FOIL.

$(a + bi)(a - bi) = a^2 + b^2$

$= \dfrac{-42 + 8i - 2}{49 + 1} = \dfrac{-44 + 8i}{50} = -\dfrac{44}{50} + \dfrac{8}{50}i = -\dfrac{22}{25} + \dfrac{4}{25}i$ ■

You Try 6

Divide. Write the result in the form $a + bi$.

a) $\dfrac{6}{-2 + i}$ b) $\dfrac{5 + 3i}{-6 - 4i}$

7. Simplify Powers of i

All powers of i larger than i^1 (or just i) can be simplified. We use the fact that $i^2 = -1$ to simplify powers of i.

Let's write i through i^4 in their simplest forms.

i is in simplest form.
$i^2 = -1$
$i^3 = i^2 \cdot i = -1 \cdot i = -i$
$i^4 = (i^2)^2 = (-1)^2 = 1$

Let's continue by simplifying i^5 and i^6.

$i^5 = i^4 \cdot i$ $i^6 = (i^2)^3$
$\quad = (i^2)^2 \cdot i$ $\quad = (-1)^3$
$\quad = (-1)^2 \cdot i$ $\quad = -1$
$\quad = 1i$
$\quad = i$

The pattern repeats so that all powers of i can be simplified to i, -1, $-i$, or 1.

Example 7

In-Class Example 7
Simplify each power of i.
a) i^{12} b) i^{22} c) i^{15} d) i^{41}
answer: a) 1   b) −1   c) −i
d) i

Simplify each power of i.

a) i^8 b) i^{14} c) i^{11} d) i^{37}

Solution

a) Use the power rule for exponents to simplify i^8. Since the exponent is even we can rewrite it in terms of i^2.

$$i^8 = (i^2)^4 \qquad \text{Power rule}$$
$$= (-1)^4 \qquad i^2 = -1$$
$$= 1 \qquad \text{Simplify.}$$

b) As in Example 7a), the exponent is even. Rewrite i^{14} in terms of i^2.

$$i^{14} = (i^2)^7 \qquad \text{Power rule}$$
$$= (-1)^7 \qquad i^2 = -1$$
$$= -1 \qquad \text{Simplify.}$$

c) The exponent of i^{11} is odd, so first use the product rule to write i^{11} as a product of i and i^{11-1} or i^{10}.

$$i^{11} = i^{10} \cdot i \qquad \text{Product rule}$$
$$= (i^2)^5 \cdot i \qquad \text{10 is even; write } i^{10} \text{ in terms of } i^2.$$
$$= (-1)^5 \cdot i \qquad i^2 = -1$$
$$= -1 \cdot i \qquad \text{Simplify.}$$
$$= -i \qquad \text{Multiply.}$$

d) The exponent of i^{37} is odd. Use the product rule to write i^{37} as a product of i and i^{37-1} or i^{36}.

$$i^{37} = i^{36} \cdot i \qquad \text{Product rule}$$
$$= (i^2)^{18} \cdot i \qquad \text{36 is even; write } i^{36} \text{ in terms of } i^2.$$
$$= (-1)^{18} \cdot i \qquad i^2 = -1$$
$$= 1 \cdot i \qquad \text{Simplify.}$$
$$= i \qquad \text{Multiply.} \qquad \blacksquare$$

You Try 7

Simplify each power of i.

a) i^{18} b) i^{32} c) i^7 d) i^{25}

Using Technology

We can use a graphing calculator to perform operations on complex numbers or to evaluate square roots of negative numbers.

If the calculator is in the default REAL mode the result is an error message "ERR: NONREAL ANS," which indicates that $\sqrt{-4}$ is a complex number rather than a real number. Before evaluating $\sqrt{-4}$ on the home screen of your calculator, check the mode by pressing MODE and looking at row 7 change the mode to complex numbers by selecting $a + bi$, as shown at the left below.

Now evaluating $\sqrt{-4}$ on the home screen results in the correct answer $2i$, as shown on the right below.

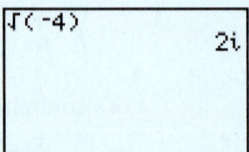

Operations can be performed on complex numbers with the calculator in either REAL or $a + bi$ mode. Simply use the arithmetic operators on the right column on your calculator. To enter the imaginary number i, press 2nd [.]. To add $2 - 5i$ and $4 + 3i$, enter $(2 - 5i) + (4 + 3i)$ on the home screen and press ENTER as shown on the left screen below. To subtract $8 + 6i$ from $7 - 2i$, enter $(7 - 2i) - (8 + 6i)$ on the home screen and press ENTER as shown.

To multiply $3 - 5i$ and $7 + 4i$, enter $(3 - 5i) \cdot (7 + 4i)$ on the home screen and press ENTER as shown on the middle screen below. To divide $2 + 9i$ by $2 - i$, enter $(2 + 9i)/(2 - i)$ on the home screen and press ENTER as shown.

To raise $3 - 4i$ to the fifth power, enter $(3 - 4i)^5$ on the home screen and press ENTER as shown.

Consider the quotient $(5 + 3i)/(4 - 7i)$. The exact answer is $-\dfrac{1}{65} + \dfrac{47}{65}i$. The calculator automatically displays the decimal result. Press MATH 1 ENTER to convert the decimal result to the exact fractional result, as shown on the right screen below.

Perform the indicated operation using a graphing calculator.

1) Simplify $\sqrt{-36}$
2) $(3 + 7i) + (5 - 8i)$
3) $(10 - 3i) - (4 + 8i)$
4) $(3 + 2i)(6 - 3i)$
5) $(4 + 3i) \div (1 - i)$
6) $(5 - 3i)^3$

Answers to You Try Exercises

1) a) $6i$ b) $i\sqrt{13}$ c) $2i\sqrt{5}$ 2) a) $-3\sqrt{2}$ b) 6   3) a) $-9 + 14i$ b) $3 - 11i$

4) a) $-18 + 21i$ b) $28 + 36i$ c) 85   5) 85   6) a) $-\dfrac{12}{5} - \dfrac{6}{5}i$ b) $-\dfrac{21}{26} + \dfrac{1}{26}i$

7) a) -1 b) 1 c) $-i$ d) i

Answers to Technology Exercises

1) $6i$ 2) $8 - i$ 3) $6 - 11i$ 4) $24 + 3i$ 5) $\dfrac{1}{2} + \dfrac{7}{2}i$ 6) $-10 - 198i$

8.8 Exercises

*Additional answers can be found in the Answers to Exercises appendix.

Objective 1: Find the Square Root of a Negative Number

Determine if each statement is true or false.

1) Every complex number is a real number.   False

2) Every real number is a complex number.   True

3) Since $i = \sqrt{-1}$, it follows that $i^2 = -1$.   True

4) In the complex number $-6 + 5i$, -6 is the real part and $5i$ is the imaginary part.   False

Simplify.

5) $\sqrt{-81}$ $9i$

6) $\sqrt{-16}$ $4i$

7) $\sqrt{-25}$ $5i$

8) $\sqrt{-169}$ $13i$

9) $\sqrt{-6}$ $i\sqrt{6}$

10) $\sqrt{-30}$ $i\sqrt{30}$

VIDEO 11) $\sqrt{-27}$ $3i\sqrt{3}$

12) $\sqrt{-75}$ $5i\sqrt{3}$

13) $\sqrt{-60}$ $2i\sqrt{15}$

14) $\sqrt{-28}$ $2i\sqrt{7}$

Objective 2: Multiply and Divide Square Roots Containing Negative Numbers

Find the error in each of the following exercises, then find the correct answer.

15) $\sqrt{-5} \cdot \sqrt{-10} = \sqrt{-5 \cdot (-10)}$
$= \sqrt{50}$
$= \sqrt{25} \cdot \sqrt{2}$
$= 5\sqrt{2}$

16) $(\sqrt{-7})^2 = \sqrt{(-7)^2}$
$= \sqrt{49}$
$= 7$

Perform the indicated operation and simplify.

17) $\sqrt{-1} \cdot \sqrt{-5}$ $-\sqrt{5}$

18) $\sqrt{-5} \cdot \sqrt{-15}$ $-5\sqrt{3}$

19) $\sqrt{-12} \cdot \sqrt{-3}$ -6

20) $\sqrt{-20} \cdot \sqrt{-5}$ -10

21) $\dfrac{\sqrt{-60}}{\sqrt{-15}}$ 2

22) $\dfrac{\sqrt{-2}}{\sqrt{-128}}$ $\dfrac{1}{8}$

23) $(\sqrt{-13})^2$ -13

24) $(\sqrt{-1})^2$ -1

Mixed Exercises: Objectives 3–6

25) Explain how to add complex numbers.
Add the real parts and add the imaginary parts.

26) How is multiplying $(1 + 3i)(2 - 7i)$ similar to multiplying $(x + 3)(2x - 7)$?
Both are products of binomials, so we can multiply both using FOIL.

27) When i^2 appears in an expression, it should be replaced with what? -1

28) Explain how to divide complex numbers.
Multiply the numerator and denominator by the conjugate of the denominator.

Objective 3: Add and Subtract Complex Numbers

Perform the indicated operations.

29) $(-4 + 9i) + (7 + 2i)$ $3 + 11i$

30) $(6 + i) + (8 - 5i)$ $14 - 4i$

31) $(13 - 8i) - (9 + i)$ $4 - 9i$

32) $(-12 + 3i) - (-7 - 6i)$ $-5 + 9i$

33) $\left(-\dfrac{3}{4} - \dfrac{1}{6}i\right) - \left(-\dfrac{1}{2} + \dfrac{2}{3}i\right)$ $-\dfrac{1}{4} - \dfrac{5}{6}i$

34) $\left(\dfrac{1}{2} + \dfrac{7}{9}i\right) - \left(\dfrac{7}{8} - \dfrac{1}{6}i\right)$ $-\dfrac{3}{8} + \dfrac{17}{18}i$

35) $16i - (3 + 10i) + (3 + i)$ $7i$

36) $(-6 - 5i) + (2 + 6i) - (-4 + i)$ 0

Objective 4: Multiply Complex Numbers

Multiply and simplify.

37) $3(8 - 5i)$ $24 - 15i$

38) $-6(8 - i)$ $-48 + 6i$

39) $\dfrac{2}{3}(-9 + 2i)$ $-6 + \dfrac{4}{3}i$

40) $\dfrac{1}{2}(18 + 7i)$ $9 + \dfrac{7}{2}i$

41) $6i(5 + 6i)$ $-36 + 30i$

42) $-4i(6 + 11i)$ $44 - 24i$

43) $(2 + 5i)(1 + 6i)$ $-28 + 17i$

44) $(2 + i)(10 + 5i)$ $15 + 20i$

45) $(-1 + 3i)(4 - 6i)$ $14 + 18i$

46) $(-4 - 9i)(3 - i)$ $-21 - 23i$

47) $(5 - 3i)(9 - 3i)$ $36 - 42i$

48) $(3 - 4i)(6 + 7i)$ $46 - 3i$

49) $\left(\dfrac{3}{4} + \dfrac{3}{4}i\right)\left(\dfrac{2}{5} + \dfrac{1}{5}i\right)$

50) $\left(\dfrac{1}{3} - \dfrac{4}{3}i\right)\left(\dfrac{3}{4} + \dfrac{2}{3}i\right)$

Objective 5: Multiply a Complex Number by Its Conjugate

Identify the conjugate of each complex number, then multiply the number and its conjugate.

51) $11 + 4i$
conjugate: $11 - 4i$; product: 137

52) $-1 - 2i$
conjugate: $-1 + 2i$; product: 5

53) $-3 - 7i$
conjugate: $-3 + 7i$; product: 58

54) $4 + 9i$
conjugate: $4 - 9i$; product: 97

55) $-6 + 4i$
conjugate: $-6 - 4i$; product: 52

56) $6 - 5i$
conjugate: $6 + 5i$; product: 61

57) How are conjugates of complex numbers like conjugates of expressions containing real numbers and radicals?
Answers may vary.

58) Is the product of two complex numbers always a complex number? Explain your answer.
True. For example, a complex number times its conjugate is a real number.

Objective 6: Divide Complex Numbers

Divide. Write the result in the form $a + bi$.

59) $\dfrac{4}{2 - 3i}$ $\dfrac{8}{13} + \dfrac{12}{13}i$

60) $\dfrac{-10}{8 - 9i}$ $-\dfrac{16}{29} - \dfrac{18}{29}i$

61) $\dfrac{8i}{4 + i}$ $\dfrac{8}{17} + \dfrac{32}{17}i$

62) $\dfrac{i}{6 - 5i}$ $-\dfrac{5}{61} + \dfrac{6}{61}i$

63) $\dfrac{2i}{-3 + 7i}$ $\dfrac{7}{29} - \dfrac{3}{29}i$

64) $\dfrac{9i}{-4 + 10i}$ $\dfrac{45}{58} - \dfrac{9}{29}i$

65) $\dfrac{3 - 8i}{-6 + 7i}$ $-\dfrac{74}{85} + \dfrac{27}{85}i$

66) $\dfrac{-5 + 2i}{4 - i}$ $-\dfrac{22}{17} + \dfrac{3}{17}i$

67) $\dfrac{2 + 3i}{5 - 6i}$ $-\dfrac{8}{61} + \dfrac{27}{61}i$

68) $\dfrac{1 + 6i}{5 + 2i}$ $\dfrac{17}{29} + \dfrac{28}{29}i$

69) $\dfrac{9}{i}$ $-9i$

70) $\dfrac{16 + 3i}{-i}$ $-3 + 16i$

Objective 7: Simplify Powers of i

Simplify each power of i.

Fill It In

Fill in the blanks with either the missing mathematical step or reason for the given step.

71) $i^{24} = \underline{(i^2)^{12}}$ Rewrite i^{24} in terms of i^2 using the power rule.
$= \underline{(-1)^{12}}$ $\underline{i^2 = -1}$
$= \underline{1}$ Simplify.

72) $i^{31} = i^{30} \cdot i$ Product rule
$= \underline{(i^2)^{15} \cdot i}$ Rewrite i^{30} in terms of i^2 using the power rule.
$= \underline{(-1)^{15} \cdot i}$ $i^2 = -1$
$= \underline{-1 \cdot i}$ Simplify.
$= \underline{-i}$ Multiply.

73) i^{24} 1

74) i^{16} 1

75) i^{28} 1

76) i^{30} -1

77) i^{9} i

78) i^{19} $-i$

79) i^{35} $-i$

80) i^{29} i

81) i^{23} $-i$

82) i^{40} 1

83) i^{42} -1

84) i^{33} i

85) $(2i)^{5}$ $32i$

86) $(2i)^{6}$ -64

87) $(-i)^{14}$ -1

88) $(-i)^{15}$ i

Expand.

89) $(-2 + 5i)^{3}$ $142 - 65i$

90) $(3 - 4i)^{3}$ $-117 - 44i$

Simplify each expression. Write the result in the form $a + bi$.

91) $1 + \sqrt{-8}$ $1 + 2i\sqrt{2}$

92) $-7 - \sqrt{-48}$ $-7 - 4i\sqrt{3}$

93) $8 - \sqrt{-45}$ $8 - 3i\sqrt{5}$

94) $3 + \sqrt{-20}$ $3 + 2i\sqrt{5}$

95) $\dfrac{-12 + \sqrt{-32}}{4}$ $-3 + i\sqrt{2}$

96) $\dfrac{21 - \sqrt{-18}}{3}$ $7 - i\sqrt{2}$

Used in the field of electronics, the **impedance**, Z, is the total opposition to the current flow of an alternating current (AC) within an electronic component, circuit, or system. It is expressed as a complex number $\boldsymbol{Z = R + Xj,}$ where the i used to represent an imaginary number in most areas of mathematics is replaced by j in electronics. $\boldsymbol{R}$ represents the resistance of a substance, and $\boldsymbol{X}$ represents the reactance.

The **total impedance, Z,** of components connected in series is the *sum* of the individual impedances of each component.

Each exercise contains the impedance of individual circuits. Find the total impedance of a system formed by connecting the circuits in series by finding the sum of the individual impedances.

97) $Z_1 = 3 + 2j$
    $Z_2 = 7 + 4j$
    $Z = 10 + 6j$

98) $Z_1 = 5 + 3j$
    $Z_2 = 9 + 6j$
    $Z = 14 + 9j$

99) $Z_1 = 5 - 2j$
    $Z_2 = 11 + 6j$
    $Z = 16 + 4j$

100) $Z_1 = 4 - 1.5j$
     $Z_2 = 3 + 0.5j$
     $Z = 7 - j$

Chapter 8: Summary

Definition/Procedure	Example

8.1 Radical Expressions and Functions

If the radicand is a perfect square, then the square root is a *rational* number. **(p. 445)**

$\sqrt{49} = 7$ since $7^2 = 49$.

If the radicand is a negative number, then the square root is *not* a real number. **(p. 445)**

$\sqrt{-36}$ is not a real number.

If the radicand is positive and not a perfect square, then the square root is an *irrational* number. **(p. 445)**

$\sqrt{7}$ is irrational because 7 is not a perfect square.

The symbol $\sqrt[n]{a}$ is read as "the *n*th root of *a*." If $\sqrt[n]{a} = b$, then $b^n = a$. We call *n* the **index** of the radical.

For any *positive* number *a* and any *even* index *n*, the **principal nth root** of *a* is $\sqrt[n]{a}$, and the **negative nth root** of *a* is $-\sqrt[n]{a}$. **(p. 446)**

$\sqrt[5]{32} = 2$ since $2^5 = 32$.

$\sqrt[4]{16} = 2$
$-\sqrt[4]{16} = -2$

The *odd root* of a negative number is a negative number. **(p. 446)**

$\sqrt[3]{-125} = -5$ since $(-5)^3 = -125$.

The *even root* of a negative number is not a real number. **(p. 446)**

$\sqrt[4]{-16}$ is not a real number.

If *n* is a positive, even integer, then $\sqrt[n]{a^n} = |a|$.
If *n* is a positive, odd integer, then $\sqrt[n]{a^n} = a$. **(p. 447)**

$\sqrt[4]{(-2)^4} = |-2| = 2$
$\sqrt[3]{(-2)^3} = -2$

The **domain of a square root function** consists of all of the real numbers that can be substituted for the variable so that the radicand is nonnegative. **(p. 448)**

Determine the domain of the square root function.
$$f(x) = \sqrt{6x - 7}$$
$6x - 7 \geq 0$ The value of the radicand must be ≥ 0.
$6x \geq 7$
$x \geq \dfrac{7}{6}$ Solve.

The domain of $f(x) = \sqrt{6x - 7}$ is $\left[\dfrac{7}{6}, \infty\right)$.

The **domain of a cube root function** is the set of all real numbers. We can write this in interval notation as $(-\infty, \infty)$. **(p. 450)**

The domain of $g(x) = \sqrt[3]{x}$ is $(-\infty, \infty)$.

To graph a square root function, find the domain, make a table of values, and graph. **(p. 450)**

Graph $g(x) = \sqrt{x}$.

x	g(x)
0	0
1	1
4	2
6	$\sqrt{6}$
9	3

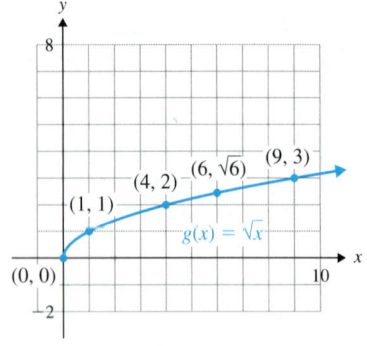

8.2 Rational Exponents

If *n* is a positive integer greater than 1 and $\sqrt[n]{a}$ is a real number, then $a^{1/n} = \sqrt[n]{a}$. **(p. 456)**

$8^{1/3} = \sqrt[3]{8} = 2$

If *m* and *n* are positive integers and $\frac{m}{n}$ is in lowest terms, then $a^{m/n} = (a^{1/n})^m = (\sqrt[n]{a})^m$ if $a^{1/n}$ is a real number. **(p. 457)**

$16^{3/4} = (\sqrt[4]{16})^3 = 2^3 = 8$

Definition/Procedure	Example

If $a^{m/n}$ is a nonzero real number, then

$$a^{-m/n} = \left(\frac{1}{a}\right)^{m/n} = \frac{1}{a^{m/n}}. \text{ (p. 458)}$$

$$25^{-3/2} = \left(\frac{1}{25}\right)^{3/2} = \left(\sqrt{\frac{1}{25}}\right)^{3} = \left(\frac{1}{5}\right)^{3} = \frac{1}{125}$$

 **BE CAREFUL** The negative exponent does not make the expression negative.

8.3 Simplifying Expressions Containing Square Roots

Product Rule for Square Roots

Let a and b be nonnegative real numbers. Then,
$\sqrt{a} \cdot \sqrt{b} = \sqrt{a \cdot b}$. **(p. 465)**

$\sqrt{5} \cdot \sqrt{7} = \sqrt{5 \cdot 7} = \sqrt{35}$

An expression containing a square root is simplified when all of the following conditions are met:

1) The radicand does not contain any factors (other than 1) that are perfect squares.
2) The radicand does not contain any fractions.
3) There are no radicals in the denominator of a fraction.

To *simplify square roots,* rewrite using the product rule as $\sqrt{a \cdot b} = \sqrt{a} \cdot \sqrt{b}$, where a or b is a perfect square.

After simplifying a radical, look at the result and ask yourself, "*Is the radical in simplest form?*" If it is not, simplify again. **(p. 465)**

Simplify $\sqrt{24}$.

$$\begin{aligned}\sqrt{24} &= \sqrt{4 \cdot 6} &&\text{4 is a perfect square.}\\ &= \sqrt{4} \cdot \sqrt{6} &&\text{Product rule}\\ &= 2\sqrt{6} &&\sqrt{4} = 2\end{aligned}$$

Quotient Rule for Square Roots

Let a and b be nonnegative real numbers such that $b \neq 0$.

Then, $\sqrt{\dfrac{a}{b}} = \dfrac{\sqrt{a}}{\sqrt{b}}$. **(p. 467)**

$$\begin{aligned}\sqrt{\frac{72}{25}} &= \frac{\sqrt{72}}{\sqrt{25}} &&\text{Quotient rule}\\ &= \frac{\sqrt{36} \cdot \sqrt{2}}{5} &&\text{Product rule; } \sqrt{25} = 5\\ &= \frac{6\sqrt{2}}{5} &&\sqrt{36} = 6\end{aligned}$$

If a is a nonnegative real number and m is an integer, then $\sqrt{a^{m}} = a^{m/2}$. **(p. 468)**

$\sqrt{k^{18}} = k^{18/2} = k^{9}$
(provided k represents a nonnegative real number)

Two Approaches to Simplifying Radical Expressions Containing Variables

Let a represent a nonnegative real number. To simplify $\sqrt{a^{n}}$ where n is *odd and positive,*

i) Method 1:
Write a^{n} as the product of two factors so that the exponent of one of the factors is the *largest* number less than n that is divisible by 2 (the index of the radical). **(p. 469)**

i) Simplify $\sqrt{x^{9}}$.

$$\begin{aligned}\sqrt{x^{9}} &= \sqrt{x^{8} \cdot x^{1}} &&\text{8 is the largest number less than 9 that is divisible by 2.}\\ &= \sqrt{x^{8}} \cdot \sqrt{x} &&\text{Product rule}\\ &= x^{8/2}\sqrt{x}\\ &= x^{4}\sqrt{x} &&8 \div 2 = 4\end{aligned}$$

Definition/Procedure	Example

ii) Method 2:
1) Divide the exponent in the radicand by the index of the radical.
2) The exponent on the variable *outside* of the radical will be the *quotient* of the division problem.
3) The exponent on the variable *inside* of the radical will be the *remainder* of the division problem. **(p. 470)**

ii) Simplify $\sqrt{p^{15}}$.

$$\sqrt{p^{15}} = p^7\sqrt{p^1}$$
$$= p^7\sqrt{p}$$

15 ÷ 2 gives a quotient of 7 and a remainder of 1.

8.4 Simplifying Expressions Containing Higher Roots

Product Rule for Higher Roots

If $\sqrt[n]{a}$ and $\sqrt[n]{b}$ are real numbers such that the roots exist, then $\sqrt[n]{a} \cdot \sqrt[n]{b} = \sqrt[n]{a \cdot b}$. **(p. 475)**

$$\sqrt[3]{3} \cdot \sqrt[3]{5} = \sqrt[3]{15}$$

Let P be an expression and let n be a positive integer greater than 1. Then $\sqrt[n]{P}$ **is completely simplified when all of the following conditions are met:**

1) The radicand does not contain any factors (other than 1) that are perfect nth powers.
2) The exponents in the radicand and the index of the radical do not have any common factors (other than 1).
3) The radicand does not contain any fractions.
4) There are no radicals in the denominator of a fraction.

To *simplify radicals with any index*, reverse the process of multiplying radicals, where a or b is an nth power.

$$\sqrt[n]{a \cdot b} = \sqrt[n]{a} \cdot \sqrt[n]{b} \quad \text{(p. 475)}$$

Simplify $\sqrt[3]{40}$.

Method 1:
Think of two numbers that multiply to 40 so that one of them is a *perfect cube*.

$$40 = 8 \cdot 5 \qquad \text{8 is a perfect cube.}$$

Then, $\sqrt[3]{40} = \sqrt[3]{8 \cdot 5}$
$= \sqrt[3]{8} \cdot \sqrt[3]{5}$ Product rule
$= 2\sqrt[3]{5}$ $\sqrt[3]{8} = 2$

Method 2:
Begin by using a factor tree to find the prime factorization of 40.

$$40 = 2^3 \cdot 5$$
$$\sqrt[3]{40} = \sqrt[3]{2^3 \cdot 5}$$
$$= \sqrt[3]{2^3} \cdot \sqrt[3]{5} \qquad \text{Product rule}$$
$$= 2\sqrt[3]{5} \qquad \sqrt[3]{2^3} = 2$$

Quotient Rule for Higher Roots

If $\sqrt[n]{a}$ and $\sqrt[n]{b}$ are real numbers, $b \neq 0$ and n is a natural number, then $\sqrt[n]{\dfrac{a}{b}} = \dfrac{\sqrt[n]{a}}{\sqrt[n]{b}}$. **(p. 477)**

$$\sqrt[4]{\frac{32}{81}} = \frac{\sqrt[4]{32}}{\sqrt[4]{81}} = \frac{\sqrt[4]{16} \cdot \sqrt[4]{2}}{3} = \frac{2\sqrt[4]{2}}{3}$$

Simplifying Higher Roots with Variables in the Radicand

If a is a nonnegative number and m and n are integers such that $n > 1$, then $\sqrt[n]{a^m} = a^{m/n}$. **(p. 478)**

Simplify $\sqrt[4]{a^{12}}$.

$$\sqrt[4]{a^{12}} = a^{12/4} = a^3$$

If the exponent does not divide evenly by the index, we can use two methods for simplifying the radical expression. If a is a nonnegative number and m and n are integers such that $n > 1$, then

i) Method 1: Use the product rule.
To simplify $\sqrt[n]{a^m}$, write a^m as the product of two factors so that the exponent of one of the factors is the *largest* number less than m that is divisible by n (the index).

ii) Method 2: Use the quotient and remainder (presented in Section 8.3). **(p. 478)**

i) Simplify $\sqrt[5]{c^{17}}$.

$$\sqrt[5]{c^{17}} = \sqrt[5]{c^{15} \cdot c^2}$$

15 is the largest number less than 17 that is divisible by 5.

$$= \sqrt[5]{c^{15}} \cdot \sqrt[5]{c^2} \qquad \text{Product rule}$$
$$= c^{15/5} \cdot \sqrt[5]{c^2}$$
$$= c^3\sqrt[5]{c^2} \qquad 15 \div 5 = 3$$

ii) Simplify $\sqrt[4]{m^{11}}$.

$$\sqrt[4]{m^{11}} = m^2\sqrt[4]{m^3}$$

11 ÷ 4 gives a quotient of 2 and a remainder of 3.

Definition/Procedure	Example

8.5 Adding, Subtracting, and Multiplying Radicals

Like radicals have the same index and the same radicand. In order to add or subtract radicals, they must be like radicals.

Steps for Adding and Subtracting Radicals
1) Write each radical expression in simplest form.
2) Combine like radicals. **(p. 483)**

Perform the operations and simplify.

a) $5\sqrt{2} + 9\sqrt{7} - 3\sqrt{2} + 4\sqrt{7}$
$= 2\sqrt{2} + 13\sqrt{7}$

b) $\sqrt{72} + \sqrt{18} - \sqrt{45}$
$= \sqrt{36} \cdot \sqrt{2} + \sqrt{9} \cdot \sqrt{2} - \sqrt{9} \cdot \sqrt{5}$
$= 6\sqrt{2} + 3\sqrt{2} - 3\sqrt{5}$
$= 9\sqrt{2} - 3\sqrt{5}$

Combining Multiplication, Addition, and Subtraction of Radicals

Multiply expressions containing radicals using the same techniques that are used for multiplying polynomials. **(p. 484)**

Multiply and simplify.

a) $\sqrt{m}(\sqrt{2m} + \sqrt{n})$
$= \sqrt{m} \cdot \sqrt{2m} + \sqrt{m} \cdot \sqrt{n}$ Distribute.
$= \sqrt{2m^2} + \sqrt{mn}$ Multiply.
$= m\sqrt{2} + \sqrt{mn}$ Simplify.

b) $(\sqrt{k} + \sqrt{6})(\sqrt{k} - \sqrt{2})$

Since we are multiplying two binomials, multiply using FOIL.

$$(\sqrt{k} + \sqrt{6})(\sqrt{k} - \sqrt{2})$$

$$\qquad \text{F} \qquad \text{O} \qquad \text{I} \qquad \text{L}$$
$$= \sqrt{k} \cdot \sqrt{k} - \sqrt{2} \cdot \sqrt{k} + \sqrt{6} \cdot \sqrt{k} - \sqrt{6} \cdot \sqrt{2}$$
$$= k^2 - \sqrt{2k} + \sqrt{6k} - \sqrt{12} \quad \text{Product rule}$$
$$= k^2 - \sqrt{2k} + \sqrt{6k} - 2\sqrt{3} \quad \sqrt{12} = 2\sqrt{3}$$

Squaring a Radical Expression with Two Terms
To square a binomial we can either use FOIL or one of the special formulas from Chapter 5:

$$(a + b)^2 = a^2 + 2ab + b^2$$
$$(a - b)^2 = a^2 - 2ab + b^2 \quad \textbf{(p. 486)}$$

$(\sqrt{7} + 5)^2 = (\sqrt{7})^2 + 2(\sqrt{7})(5) + (5)^2$
$= 7 + 10\sqrt{7} + 25$
$= 32 + 10\sqrt{7}$

Multiply $(a + b)(a - b)$

To multiply binomials of the form $(a + b)(a - b)$ use the formula $(a + b)(a - b) = a^2 - b^2$. **(p. 486)**

$(3 + \sqrt{10})(3 - \sqrt{10}) = (3)^2 - (\sqrt{10})^2$
$= 9 - 10$
$= -1$

8.6 Dividing Radicals

The process of eliminating radicals from the denominator of an expression is called **rationalizing the denominator.**

First, we give examples of rationalizing denominators containing one term. **(p. 489)**

Rationalize the denominator of each expression.

a) $\dfrac{9}{\sqrt{2}} = \dfrac{9}{\sqrt{2}} \cdot \dfrac{\sqrt{2}}{\sqrt{2}} = \dfrac{9\sqrt{2}}{2}$

b) $\dfrac{5}{\sqrt[3]{2}} = \dfrac{5}{\sqrt[3]{2}} \cdot \dfrac{\sqrt[3]{2^2}}{\sqrt[3]{2^2}} = \dfrac{5\sqrt[3]{2^2}}{\sqrt[3]{2^3}} = \dfrac{5\sqrt[3]{4}}{2}$

The *conjugate* of an expression of the form $a + b$ is $a - b$.
(p. 494)

$\sqrt{11} - 4$ conjugate: $\sqrt{11} + 4$
$-8 + \sqrt{5}$ conjugate: $-8 - \sqrt{5}$

Definition/Procedure	Example

Rationalizing a Denominator with Two Terms
If the denominator of an expression contains two terms, including one or two square roots, then to rationalize the denominator, we multiply the numerator and denominator of the expression by the conjugate of the denominator. **(p. 495)**

Rationalize the denominator of $\dfrac{4}{\sqrt{2} - 3}$.

$$\frac{4}{\sqrt{2} - 3} = \frac{4}{\sqrt{2} - 3} \cdot \frac{\sqrt{2} + 3}{\sqrt{2} + 3} \quad \text{Multiply by the conjugate}$$
of the denominator.

$$= \frac{4(\sqrt{2} + 3)}{(\sqrt{2})^2 - (3)^2} \quad (a + b)(a - b) = a^2 - b^2$$

$$= \frac{4(\sqrt{2} + 3)}{2 - 9} \quad \text{Square the terms.}$$

$$= \frac{4(\sqrt{2} + 3)}{-7} = -\frac{4\sqrt{2} + 12}{7}$$

8.7 Solving Radical Equations

Solving Radical Equations Containing Square Roots

Step 1: Get a radical on a side by itself.
Step 2: Square both sides of the equation to eliminate a radical.
Step 3: Combine like terms on each side of the equation.
Step 4: If the equation still contains a radical, repeat Steps 1–3.
Step 5: Solve the equation.
Step 6: Check the proposed solutions *in the original equation* and discard extraneous solutions. **(p. 504)**

Solve $t = 2 + \sqrt{2t - 1}$.

$$t - 2 = \sqrt{2t - 1} \quad \text{Get the radical by itself.}$$
$$(t - 2)^2 = (\sqrt{2t - 1})^2 \quad \text{Square both sides.}$$
$$t^2 - 4t + 4 = 2t - 1$$
$$t^2 - 6t + 5 = 0 \quad \text{Get all terms on the same side.}$$
$$(t - 5)(t - 1) = 0 \quad \text{Factor.}$$
$$t - 5 = 0 \quad \text{or} \quad t - 1 = 0$$
$$t = 5 \quad \text{or} \quad t = 1$$

Check $t = 5$ and $t = 1$ in the *original* equation.

$t = 5$: $t = 2 + \sqrt{2t - 1}$
$5 \overset{?}{=} 2 + \sqrt{2(5) - 1}$
$5 \overset{?}{=} 2 + \sqrt{9}$
$5 = 2 + 3$
True

$t = 1$: $t = 2 + \sqrt{2t - 1}$
$1 \overset{?}{=} 2 + \sqrt{2(1) - 1}$
$1 \overset{?}{=} 2 + 1$
$1 = 3$
False

$t = 5$ is a solution, but $t = 1$ is *not* because $t = 1$ does not satisfy the original equation.

The solution set is $\{5\}$.

8.8 Complex Numbers

Definition of *i*:

$$i = \sqrt{-1}$$

Therefore,

$$i^2 = -1$$

A **complex number** is a number of the form $a + bi$, where a and b are real numbers. a is called the **real part** and b is called the **imaginary part.** The set of real numbers is a subset of the set of complex numbers. **(p. 512)**

Examples of complex numbers:

$-2 + 7i$

5 (since it can be written $5 + 0i$)

$4i$ (since it can be written $0 + 4i$)

Simplifying Complex Numbers
Use the product rule and $i = \sqrt{-1}$. **(p. 513)**

Simplify $\sqrt{-25}$.

$$\sqrt{-25} = \sqrt{-1} \cdot \sqrt{25}$$
$$= i \cdot 5$$
$$= 5i$$

When multiplying or dividing radicals with negative radicands, write each radical in terms of *i* first. **(p. 514)**

Multiply $\sqrt{-12} \cdot \sqrt{-3}$.

$$\sqrt{-12} \cdot \sqrt{-3} = i\sqrt{12} \cdot i\sqrt{3} = i^2\sqrt{36}$$
$$= -1 \cdot 6 = -6$$

Definition/Procedure	Example
Adding and Subtracting Complex Numbers To add and subtract complex numbers, combine the real parts and combine the imaginary parts. **(p. 514)**	Subtract $(10 + 7i) - (-2 + 4i)$. $$(10 + 7i) - (-2 + 4i) = 10 + 7i + 2 - 4i$$ $$= 12 + 3i$$
Multiply complex numbers like we multiply polynomials. Remember to replace i^2 with -1. **(p. 515)**	Multiply and simplify. a) $4(9 + 5i) = 36 + 20i$ $\quad\quad\quad\quad\quad\quad$ F $\quad$ O $\quad$ I $\quad$ L b) $(-3 + i)(2 - 7i) = -6 + 21i + 2i - 7i^2$ $\quad\quad\quad\quad\quad\quad\quad = -6 + 23i - 7(-1)$ $\quad\quad\quad\quad\quad\quad\quad = -6 + 23i + 7$ $\quad\quad\quad\quad\quad\quad\quad = 1 + 23i$
Complex Conjugates 1) The conjugate of $a + bi$ is $a - bi$. 2) The product of $a + bi$ and $a - bi$ is a real number. 3) Find the product $(a + bi)(a - bi)$ using FOIL or recall that $(a + bi)(a - bi) = a^2 + b^2$. **(p. 516)**	Multiply $-5 - 3i$ by its conjugate. The conjugate of $-5 - 3i$ is $-5 + 3i$. Use $(a + bi)(a - bi) = a^2 + b^2$. $$(-5 - 3i)(-5 + 3i) = (-5)^2 + (3)^2$$ $$= 25 + 9$$ $$= 34$$
Dividing Complex Numbers To divide complex numbers, multiply the numerator and denominator by the *conjugate of the denominator*. Write the quotient in the form $a + bi$. **(p. 516)**	Divide $\dfrac{6i}{2 + 5i}$. Write the result in the form $a + bi$. $$\frac{6i}{2 + 5i} = \frac{6i}{2 + 5i} \cdot \frac{(2 - 5i)}{(2 - 5i)}$$ $$= \frac{12i - 30i^2}{2^2 + 5^2}$$ $$= \frac{12i - 30(-1)}{29}$$ $$= \frac{30}{29} + \frac{12}{29}i$$
Simplify Powers of i We can simplify powers of i using $i^2 = -1$. **(p. 517)**	Simplify i^{14}. $\quad i^{14} = (i^2)^7 \quad$ Power rule $\quad\quad = (-1)^7 \quad i^2 = -1$ $\quad\quad = -1 \quad\quad$ Simplify.

Additional answers can be found in the Answers to Exercises appendix.

(8.1) Find each root, if possible.

1) $\sqrt{\dfrac{169}{4}}$ $\dfrac{13}{2}$

2) $\sqrt{-16}$ not real

3) $-\sqrt{81}$ -9

4) $\sqrt[5]{32}$ 2

5) $\sqrt[3]{-1}$ -1

6) $-\sqrt[4]{81}$ -3

7) $\sqrt[6]{-64}$ not real

8) $\sqrt{9-16}$ not real

Simplify. Use absolute values when necessary.

9) $\sqrt{(-13)^2}$ 13

10) $\sqrt[5]{(-8)^5}$ -8

11) $\sqrt{p^2}$ $|p|$

12) $\sqrt[6]{c^6}$ $|c|$

13) $\sqrt[3]{h^3}$ h

14) $\sqrt[4]{(y+7)^4}$ $|y+7|$

15) $f(x) = \sqrt{5x+3}$

    a) Find $f(4)$. $\sqrt{23}$

    b) Find $f(p)$. $\sqrt{5p+3}$

    c) Find the domain of f. $\left[-\dfrac{3}{5}, \infty\right)$

16) $g(x) = \sqrt[3]{x-12}$

    a) Find $g(4)$. -2

    b) Find $g(t+7)$. $\sqrt[3]{t-5}$

    c) Find the domain of g. $(-\infty, \infty)$

17) Graph $k(x) = \sqrt{x+4}$.

18) Graph $h(x) = -\sqrt[3]{x}$.

(8.2)

✏ 19) Explain how to write $8^{2/3}$ in radical form.

✏ 20) Explain how to eliminate the negative from the exponent in an expression like $9^{-1/2}$.
  Take the reciprocal of the base. $9^{-1/2} = \left(\dfrac{1}{9}\right)^{1/2}$

Evaluate.

21) $36^{1/2}$ 6

22) $32^{1/5}$ 2

23) $\left(\dfrac{27}{125}\right)^{1/3}$ $\dfrac{3}{5}$

24) $-16^{1/4}$ -2

25) $32^{3/5}$ 8

26) $\left(\dfrac{64}{27}\right)^{2/3}$ $\dfrac{16}{9}$

27) $81^{-1/2}$ $\dfrac{1}{9}$

28) $\left(\dfrac{1}{27}\right)^{-1/3}$ 3

29) $81^{-3/4}$ $\dfrac{1}{27}$

30) $1000^{-2/3}$ $\dfrac{1}{100}$

31) $\left(\dfrac{27}{1000}\right)^{-2/3}$ $\dfrac{100}{9}$

32) $\left(\dfrac{25}{16}\right)^{-3/2}$ $\dfrac{64}{125}$

From this point forward, assume all variables represent positive real numbers.

Simplify completely. The answer should contain only positive exponents.

33) $3^{6/7} \cdot 3^{8/7}$ 9

34) $(169^4)^{1/8}$ 13

35) $(8^{1/5})^{10}$ 64

36) $\dfrac{8^2}{8^{11/3}}$ $\dfrac{1}{32}$

37) $\dfrac{7^2}{7^{5/3} \cdot 7^{1/3}}$ 1

38) $(2k^{-5/6})(3k^{1/2})$ $\dfrac{6}{k^{1/3}}$

39) $(64a^4b^{12})^{5/6}$ $32a^{10/3}b^{10}$

40) $\left(\dfrac{t^4u^3}{7t^7u^5}\right)^{-2}$ $49t^6u^4$

41) $\left(\dfrac{81c^{-5}d^9}{16c^{-1}d^2}\right)^{-1/4}$ $\dfrac{2c}{3d^{7/4}}$

Rewrite each radical in exponential form, then simplify. Write the answer in simplest (or radical) form.

42) $\sqrt[4]{36^2}$ 6

43) $\sqrt[12]{27^4}$ 3

44) $(\sqrt{17})^2$ 17

45) $\sqrt[3]{7^3}$ 7

46) $\sqrt[5]{t^{20}}$ t^4

47) $\sqrt[4]{k^{28}}$ k^7

48) $\sqrt{x^{10}}$ x^5

49) $\sqrt{w^6}$ w^3

(8.3) Simplify completely.

50) $\sqrt{28}$ $2\sqrt{7}$

51) $\sqrt{1000}$ $10\sqrt{10}$

52) $\dfrac{\sqrt{63}}{\sqrt{7}}$ 3

53) $\sqrt{\dfrac{18}{49}}$ $\dfrac{3\sqrt{2}}{7}$

54) $\dfrac{\sqrt{48}}{\sqrt{121}}$ $\dfrac{4\sqrt{3}}{11}$

55) $\sqrt{k^{12}}$ k^6

56) $\sqrt{\dfrac{40}{m^4}}$ $\dfrac{2\sqrt{10}}{m^2}$

57) $\sqrt{x^9}$ $x^4\sqrt{x}$

58) $\sqrt{y^5}$ $y^2\sqrt{y}$

59) $\sqrt{45t^2}$ $3t\sqrt{5}$

60) $\sqrt{80n^{21}}$ $4n^{10}\sqrt{5n}$

61) $\sqrt{72x^7y^{13}}$ $6x^3y^6\sqrt{2xy}$

62) $\sqrt{\dfrac{m^{11}}{36n^2}}$ $\dfrac{m^5\sqrt{m}}{6n}$

Perform the indicated operation and simplify.

63) $\sqrt{5} \cdot \sqrt{3}$ $\sqrt{15}$

64) $\sqrt{6} \cdot \sqrt{15}$ $3\sqrt{10}$

65) $\sqrt{2} \cdot \sqrt{12}$ $2\sqrt{6}$

66) $\sqrt{b^7} \cdot \sqrt{b^3}$ b^5

67) $\sqrt{11x^5} \cdot \sqrt{11x^8}$ $11x^6\sqrt{x}$

68) $\sqrt{5a^2b} \cdot \sqrt{15a^6b^4}$ $5a^4b^2\sqrt{3b}$

69) $\dfrac{\sqrt{200k^{21}}}{\sqrt{2k^5}}$ $10k^8$

70) $\dfrac{\sqrt{63c^{17}}}{\sqrt{7c^9}}$ $3c^4$

(8.4) Simplify completely.

71) $\sqrt[3]{16}$ $2\sqrt[3]{2}$

72) $\sqrt[3]{250}$ $5\sqrt[3]{2}$

73) $\sqrt[4]{48}$ $2\sqrt[4]{3}$

74) $\sqrt[3]{\dfrac{81}{3}}$ 3

75) $\sqrt[4]{z^{24}}$ z^6

76) $\sqrt[5]{p^{40}}$ p^8

77) $\sqrt[3]{a^{20}}$ $a^6\sqrt[3]{a^2}$

78) $\sqrt[5]{x^{14}y^7}$ $x^2y\sqrt[5]{x^4y^2}$

79) $\sqrt[3]{16z^{15}}$ $2z^5\sqrt[3]{2}$

80) $\sqrt[3]{80m^{17}n^{10}}$ $2m^5n^3\sqrt[3]{10m^2n}$

81) $\sqrt[4]{\dfrac{h^{12}}{81}}$ $\dfrac{h^3}{3}$

82) $\sqrt[5]{\dfrac{c^{22}}{32d^{10}}}$ $\dfrac{c^4\sqrt[5]{c^2}}{2d^2}$

Perform the indicated operation and simplify.

83) $\sqrt[3]{3} \cdot \sqrt[3]{7}$ $\sqrt[3]{21}$

84) $\sqrt[3]{25} \cdot \sqrt[3]{10}$ $5\sqrt[3]{2}$

85) $\sqrt[4]{4t^7} \cdot \sqrt[4]{8t^{10}}$ $2t^4\sqrt[4]{2t}$

86) $\sqrt[5]{\dfrac{x^{21}}{x^{16}}}$ x

87) $\sqrt[3]{n} \cdot \sqrt{n}$ $\sqrt[6]{n^5}$

88) $\dfrac{\sqrt[4]{a^3}}{\sqrt[3]{a}}$ $\sqrt[12]{a^5}$

(8.5) Perform the operations and simplify.

89) $8\sqrt{5} + 3\sqrt{5}$ $11\sqrt{5}$

90) $\sqrt{125} + \sqrt{80}$ $9\sqrt{5}$

91) $\sqrt{80} - \sqrt{48} + \sqrt{20}$ $6\sqrt{5} - 4\sqrt{3}$

92) $9\sqrt[3]{72} - 8\sqrt[3]{9}$ $10\sqrt[3]{9}$

93) $3p\sqrt{p} - 7\sqrt{p^3}$ $-4p\sqrt{p}$

94) $9n\sqrt{n} - 4\sqrt{n^3}$ $5n\sqrt{n}$

95) $10d^2\sqrt{8d} - 32d\sqrt{2d^3}$ $-12d^2\sqrt{2d}$

96) $\sqrt{6}(\sqrt{7} - \sqrt{6})$ $\sqrt{42} - 6$

97) $3\sqrt{k}(\sqrt{20k} + \sqrt{2})$ $6k\sqrt{5} + 3\sqrt{2k}$

98) $(5 - \sqrt{3})(2 + \sqrt{3})$ $7 + 3\sqrt{3}$

99) $(\sqrt{2r} + 5\sqrt{s})(3\sqrt{s} + 4\sqrt{2r})$ $23\sqrt{2rs} + 8r + 15s$

100) $(2\sqrt{5} - 4)^2$ $36 - 16\sqrt{5}$

101) $(1 + \sqrt{y + 1})^2$ $2 + 2\sqrt{y + 1} + y$

102) $(\sqrt{6} - \sqrt{5})(\sqrt{6} + \sqrt{5})$ 1

(8.6) Rationalize the denominator of each expression.

103) $\dfrac{14}{\sqrt{3}}$ $\dfrac{14\sqrt{3}}{3}$

104) $\dfrac{20}{\sqrt{6}}$ $\dfrac{10\sqrt{6}}{3}$

105) $\dfrac{\sqrt{18k}}{\sqrt{n}}$ $\dfrac{3\sqrt{2kn}}{n}$

106) $\dfrac{\sqrt{45}}{\sqrt{m^5}}$ $\dfrac{3\sqrt{5m}}{m^3}$

107) $\dfrac{7}{\sqrt[3]{2}}$ $\dfrac{7\sqrt[3]{4}}{2}$

108) $-\dfrac{15}{\sqrt[3]{9}}$ $-5\sqrt[3]{3}$

109) $\dfrac{\sqrt[3]{x^2}}{\sqrt[3]{y}}$ $\dfrac{\sqrt[3]{x^2y^2}}{y}$

110) $\sqrt[4]{\dfrac{3}{4k^2}}$ $\dfrac{\sqrt[4]{12k^2}}{2k}$

111) $\dfrac{2}{3 + \sqrt{3}}$ $\dfrac{3 - \sqrt{3}}{3}$

112) $\dfrac{z - 4}{\sqrt{z} + 2}$ $\sqrt{z} - 2$

Simplify completely.

113) $\dfrac{8 - 24\sqrt{2}}{8}$ $1 - 3\sqrt{2}$

114) $\dfrac{-\sqrt{48} - 6}{10}$ $\dfrac{-2\sqrt{3} - 3}{5}$

(8.7) Solve.

115) $\sqrt{x + 8} = 3$ $\{1\}$

116) $10 - \sqrt{3r - 5} = 2$ $\{23\}$

117) $\sqrt{3j + 4} = -\sqrt{4j - 1}$ $\varnothing$

118) $\sqrt[3]{6d - 14} = -2$ $\{1\}$

119) $a = \sqrt{a + 8} - 6$ $\{-4\}$

120) $1 + \sqrt{6m + 7} = 2m$ $\{3\}$

121) $\sqrt{4a + 1} - \sqrt{a - 2} = 3$ $\{2, 6\}$

122) $\sqrt{6x + 9} - \sqrt{2x + 1} = 4$ $\{12\}$

123) Solve for V: $r = \sqrt{\dfrac{3V}{\pi h}}$ $V = \frac{1}{3}\pi r^2 h$

124) The velocity of a wave in shallow water is given by $c = \sqrt{gH}$, where g is the acceleration due to gravity (32 ft/sec^2) and H is the depth of the water (in feet). Find the velocity of a wave in 10 ft of water. $8\sqrt{5}$ ft/sec or about 17.9 ft/sec

(8.8) Simplify.

125) $\sqrt{-49}$ $7i$

126) $\sqrt{-8}$ $2i\sqrt{2}$

127) $\sqrt{-2} \cdot \sqrt{-8}$ -4

128) $\sqrt{-6} \cdot \sqrt{-3}$ $-3\sqrt{2}$

Perform the indicated operations.

129) $(2 + i) + (10 - 4i)$ $12 - 3i$

130) $(4 + 3i) - (11 - 4i)$ $-7 + 7i$

131) $\left(\dfrac{4}{5} - \dfrac{1}{3}i\right) - \left(\dfrac{1}{2} + i\right)$ $\dfrac{3}{10} - \dfrac{4}{3}i$

132) $\left(-\dfrac{3}{8} - 2i\right) + \left(\dfrac{5}{8} + \dfrac{3}{2}i\right) - \left(\dfrac{1}{4} - \dfrac{1}{2}i\right)$ 0

Multiply and simplify.

133) $5(-6 + 7i)$ $-30 + 35i$

134) $-8i(4 + 3i)$ $24 - 32i$

135) $3i(-7 + 12i)$ $-36 - 21i$

136) $(3 - 4i)(2 + i)$ $10 - 5i$

137) $(4 - 6i)(3 - 6i)$ $-24 - 42i$

138) $\left(\dfrac{1}{5} - \dfrac{2}{3}i\right)\left(\dfrac{3}{2} - \dfrac{2}{3}i\right)$

Identify the conjugate of each complex number, then multiply the number and its conjugate.

139) $2 - 7i$ conjugate: $2 + 7i$; product: 53

140) $-2 + 3i$ conjugate: $-2 - 3i$; product: 13

Divide. Write the quotient in the form $a + bi$.

141) $\dfrac{6}{2 + 5i}$ $\dfrac{12}{29} - \dfrac{30}{29}i$

142) $\dfrac{-12}{4 - 3i}$ $-\dfrac{48}{25} - \dfrac{36}{25}i$

143) $\dfrac{8}{i}$ $-8i$

144) $\dfrac{4i}{1 - 3i}$ $-\dfrac{6}{5} + \dfrac{2}{5}i$

145) $\dfrac{9 - 4i}{6 - i}$ $\dfrac{58}{37} - \dfrac{15}{37}i$

146) $\dfrac{5 - i}{-2 + 6i}$ $-\dfrac{2}{5} - \dfrac{7}{10}i$

Simplify.

147) i^{10} -1

148) i^{51} $-i$

149) i^{33} i

150) i^{24} 1

*Additional answers can be found in the Answers to Exercises appendix.

Find each real root, if possible.

1) $\sqrt{144}$ 12

2) $\sqrt[3]{-27}$ −3

3) $\sqrt{-16}$ not real

Simplify. Use absolute values when necessary.

4) $\sqrt[4]{w^4}$ $|w|$

5) $\sqrt[5]{(-19)^5}$ −19

6) Let $h(c) = \sqrt{3c + 7}$.

  a) Find $h(-2)$. 1

  b) Find $h(a - 4)$. $\sqrt{3a - 5}$

  c) Determine the domain of h. $\left[-\frac{7}{3}, \infty\right)$

7) Determine the domain of $f(x) = \sqrt{x - 2}$, and graph the function.

Evaluate.

8) $16^{1/4}$ 2

9) $27^{4/3}$ 81

10) $(49)^{-1/2}$ $\frac{1}{7}$

11) $\left(\frac{8}{125}\right)^{-2/3}$ $\frac{25}{4}$

From this point forward, assume all variables represent positive real numbers.

Simplify completely. The answer should contain only positive exponents.

12) $m^{3/8} \cdot m^{1/4}$ $m^{5/8}$

13) $\frac{35a^{1/6}}{14a^{5/6}}$ $\frac{5}{2a^{2/3}}$

14) $(2x^{3/10}y^{-2/5})^{-5}$ $\frac{y^2}{32x^{3/2}}$

Simplify completely.

15) $\sqrt{75}$ $5\sqrt{3}$

16) $\sqrt[3]{48}$ $2\sqrt[3]{6}$

17) $\sqrt{\frac{24}{2}}$ $2\sqrt{3}$

Simplify completely.

18) $\sqrt{y^6}$ y^3

19) $\sqrt[4]{p^{24}}$ p^6

20) $\sqrt{t^9}$ $t^4\sqrt{t}$

21) $\sqrt{63m^5n^8}$ $3m^2n^4\sqrt{7m}$

22) $\sqrt[3]{c^{23}}$ $c^7\sqrt[3]{c^2}$

23) $\sqrt[3]{\frac{a^{14}b^7}{27}}$ $\frac{a^4b^2\sqrt[3]{a^2b}}{3}$

Perform the operations and simplify.

24) $\sqrt{3} \cdot \sqrt{12}$ 6

25) $\sqrt[3]{z^4} \cdot \sqrt[3]{z^6}$ $z^3\sqrt[3]{z}$

26) $\frac{\sqrt{120w^{15}}}{\sqrt{2w^4}}$ $2w^5\sqrt{15w}$

27) $9\sqrt{7} - 3\sqrt{7}$ $6\sqrt{7}$

28) $\sqrt{12} - \sqrt{108} + \sqrt{18}$ $3\sqrt{2} - 4\sqrt{3}$

29) $2h^3\sqrt[4]{h} - 16\sqrt[4]{h^{13}}$ $-14h^3\sqrt[4]{h}$

Multiply and simplify.

30) $\sqrt{6}(\sqrt{2} - 5)$ $2\sqrt{3} - 5\sqrt{6}$

31) $(3 - 2\sqrt{5})(\sqrt{2} + 1)$ $3\sqrt{2} + 3 - 2\sqrt{10} - 2\sqrt{5}$

32) $(\sqrt{7} + \sqrt{3})(\sqrt{7} - \sqrt{3})$ 4

33) $(\sqrt{2p + 1} + 2)^2$ $2p + 5 + 4\sqrt{2p + 1}$

34) $2\sqrt{t}(\sqrt{t} - \sqrt{3u})$ $2t - 2\sqrt{3tu}$

Rationalize the denominator of each expression.

35) $\frac{2}{\sqrt{5}}$ $\frac{2\sqrt{5}}{5}$

36) $\frac{8}{\sqrt{7} + 3}$ $12 - 4\sqrt{7}$

37) $\frac{\sqrt{6}}{\sqrt{a}}$ $\frac{\sqrt{6a}}{a}$

38) $\frac{5}{\sqrt[3]{9}}$ $\frac{5\sqrt[3]{3}}{3}$

39) Simplify completely. $\frac{2 - \sqrt{48}}{2}$ $1 - 2\sqrt{3}$

Solve.

40) $\sqrt{5h + 4} = 3$ {1}

41) $z = \sqrt{1 - 4z} - 5$ {−2}

42) $\sqrt[3]{n - 5} - \sqrt[3]{2n - 18} = 0$ {13}

43) $\sqrt{3k + 1} - \sqrt{2k - 1} = 1$ {1, 5}

44) In the formula $r = \sqrt{\dfrac{V}{\pi h}}$, V represents the volume of a cylinder, h represents the height of the cylinder, and r represents the radius.

  a) A cylindrical container has a volume of 72π in^3. It is 8 in. high. What is the radius of the container? 3 in.

  b) Solve the formula for V. $V = \pi r^2 h$

Simplify.

45) $\sqrt{-64}$ $8i$

46) $\sqrt{-45}$ $3i\sqrt{5}$

47) i^{19} $-i$

Perform the indicated operation and simplify. Write the answer in the form $a + bi$.

48) $(-10 + 3i) - (6 + i)$ $-16 + 2i$

49) $(2 - 7i)(-1 + 3i)$ $19 + 13i$

50) $\frac{8 + i}{2 - 3i}$ $1 + 2i$

Additional answers can be found in the Answers to Exercises appendix.

1) Combine like terms.

$$4x - 3y + 9 - \frac{2}{3}x + y - 1 \qquad \frac{10}{3}x - 2y + 8$$

2) Write in scientific notation.

8,723,000 $\qquad 8.723 \times 10^6$

3) Solve $3(2c - 1) + 7 = 9c + 5(c + 2)$. $\qquad \left\{-\frac{3}{4}\right\}$

4) Graph $3x + 2y = 12$.

5) Write the equation of the line containing the points $(5, 3)$ and $(1, -2)$. Write the equation in slope-intercept form. $\quad y = \frac{5}{4}x - \frac{13}{4}$

6) Solve by substitution.

$$\begin{aligned} 2x + 7y &= -12 \\ x - 4y &= -6 \end{aligned} \qquad (-6, 0)$$

7) Multiply.

$$(5p^2 - 2)(3p^2 - 4p - 1) \qquad 15p^4 - 20p^3 - 11p^2 + 8p + 2$$

8) Divide.

$$\frac{8n^3 - 1}{2n - 1} \qquad 4n^2 + 2n + 1$$

Factor completely.

9) $4w^2 + 5w - 6 \qquad (4w - 3)(w + 2)$

10) $8 - 18t^2 \qquad 2(2 + 3t)(2 - 3t)$

11) Solve $6y^2 - 4 = 5y$. $\qquad \left\{-\frac{1}{2}, \frac{4}{3}\right\}$

12) Solve $3(k^2 + 20) - 4k = 2k^2 + 11k + 6$. $\qquad \{6, 9\}$

13) *Write an equation and solve.* The width of a rectangle is 5 in. less than its length. The area is 84 in². Find the dimensions of the rectangle. $\qquad$ length = 12 in., width = 7 in.

Perform the operations and simplify.

14) $\dfrac{5a^2 + 3}{a^2 + 4a} - \dfrac{3a - 2}{a + 4} \qquad \dfrac{2a^2 + 2a + 3}{a(a + 4)}$

15) $\dfrac{10m^2}{9n} \cdot \dfrac{6n^2}{35m^5} \qquad \dfrac{4n}{21m^3}$

16) Solve $\dfrac{3}{r^2 + 8r + 15} - \dfrac{4}{r + 3} = 1$. $\qquad \{-8, -4\}$

17) Solve $|6g + 1| \geq 11$. Write the answer in interval notation. $\quad (-\infty, -2] \cup \left[\frac{5}{3}, \infty\right)$

18) Solve using Gaussian elimination.

$$\begin{aligned} x + 3y + z &= 3 \\ 2x - y - 5z &= -1 \qquad (4, -1, 2) \\ -x + 2y + 3z &= 0 \end{aligned}$$

19) Simplify. Assume all variables represent nonnegative real numbers.

a) $\sqrt{500} \qquad 10\sqrt{5}$

b) $\sqrt[3]{56} \qquad 2\sqrt[3]{7}$

c) $\sqrt{p^{10}q^7} \qquad p^5q^3\sqrt{q}$

d) $\sqrt[4]{32a^{15}} \qquad 2a^3\sqrt[4]{2a^3}$

20) Evaluate.

a) $81^{1/2} \qquad 9$

b) $8^{4/3} \qquad 16$

c) $(27)^{-1/3} \qquad \dfrac{1}{3}$

21) Multiply and simplify $2\sqrt{3}(5 - \sqrt{3})$. $\qquad 10\sqrt{3} - 6$

22) Rationalize the denominator. Assume the variables represent positive real numbers.

a) $\sqrt{\dfrac{20}{50}} \qquad \dfrac{\sqrt{10}}{5}$

b) $\dfrac{6}{\sqrt[3]{2}} \qquad 3\sqrt[3]{4}$

c) $\dfrac{x}{\sqrt[3]{y^2}} \qquad \dfrac{x\sqrt[3]{y}}{y}$

d) $\dfrac{\sqrt{a} - 2}{1 - \sqrt{a}} \qquad \dfrac{a - 2 - \sqrt{a}}{1 - a}$

23) Solve.

a) $\sqrt{2b - 1} + 7 = 6 \qquad \varnothing$

b) $\sqrt{3z + 10} = 2 - \sqrt{z + 4} \qquad \{-3\}$

24) Simplify.

a) $\sqrt{-49} \qquad 7i$

b) $\sqrt{-56} \qquad 2i\sqrt{14}$

c) $i^8 \qquad 1$

25) Perform the indicated operation and simplify. Write the answer in the form $a + bi$.

a) $(-3 + 4i) + (5 + 3i) \qquad 2 + 7i$

b) $(3 + 6i)(-2 + 7i) \qquad -48 + 9i$

c) $\dfrac{2 - i}{-4 + 3i} \qquad -\dfrac{11}{25} - \dfrac{2}{25}i$

Quadratic Equations and Functions

Algebra at Work: Ophthalmology

We have already seen two applications of mathematics to ophthalmology, and here we have a third. An ophthalmologist can use a quadratic equation to convert between a prescription for glasses and a prescription for contact lenses.

After having reexamined her patient for contact lens use, Sarah can use the following quadratic equation to double-check the prescription for the contact lenses based on the prescription her patient currently has for her glasses.

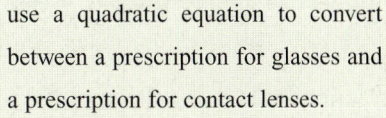

$$D_c = s(D_g)^2 + D_g$$

where D_g = power of the glasses, in diopters

s = distance of the glasses to the eye, in meters

D_c = power of the contact lenses, in diopters

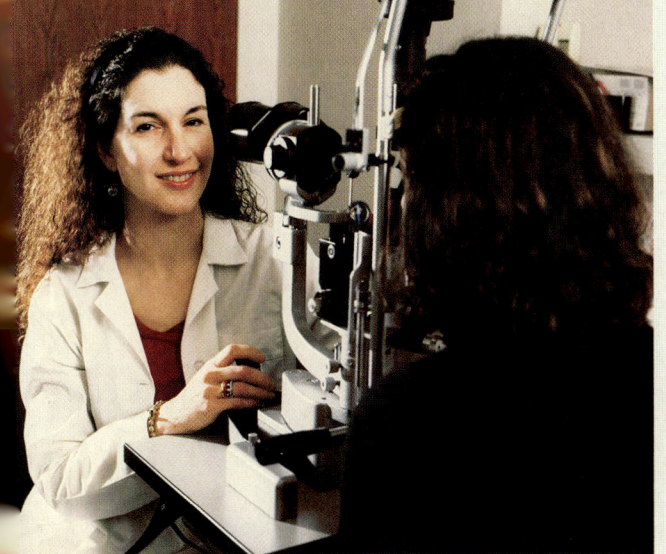

For example, if the power of a patient's eyeglasses is +9.00 diopters and the glasses rest 1 cm or 0.01 m from the eye, the power the patient would need in her contact lenses would be

$$D_c = 0.01(9)^2 + 9$$
$$D_c = 0.01(81) + 9$$
$$D_c = 0.81 + 9$$
$$D_c = 9.81 \text{ diopters}$$

An eyeglass power of +9.00 diopters would convert to a contact lens power of +9.81 diopters.

In this chapter, we will learn different ways to solve quadratic equations.

Section 9.1 The Square Root Property and Completing the Square

We defined a quadratic equation in Chapter 6. Let's restate the definition:

> **Definition**
>
> A **quadratic equation** can be written in the form $ax^2 + bx + c = 0$, where a, b, and c are real numbers and $a \neq 0$.

In Section 6.4 we learned how to solve quadratic equations by factoring. For example, we can use the zero product rule to solve $x^2 - 3x - 40 = 0$.

$$x^2 - 3x - 40 = 0$$
$$(x - 8)(x + 5) = 0 \qquad \text{Factor.}$$

$$x - 8 = 0 \quad \text{or} \quad x + 5 = 0 \qquad \text{Set each factor equal to zero.}$$
$$x = 8 \quad \text{or} \qquad x = -5 \qquad \text{Solve.}$$

The solution set is $\{-5, 8\}$.

It is not easy to solve all quadratic equations by factoring, however. Therefore, we need to learn other methods. In this chapter, we will discuss three more methods for solving quadratic equations. Let's begin with the square root property.

1. Solve an Equation of the Form $x^2 = k$

Look at the equation $x^2 = 9$. We can solve this equation by factoring, like this:

$$x^2 = 9$$
$$x^2 - 9 = 0 \qquad \text{Get all terms on the same side.}$$
$$(x + 3)(x - 3) = 0 \qquad \text{Factor.}$$

$$x + 3 = 0 \quad \text{or} \quad x - 3 = 0 \qquad \text{Set each factor equal to zero.}$$
$$x = -3 \quad \text{or} \qquad x = 3 \qquad \text{Solve.}$$

The solution set is $\{-3, 3\}$.

Or, we can solve an equation like $x^2 = 9$ using the **square root property** as we will see in Example 1a.

> **Definition   The Square Root Property**
>
> Let k be a constant. If $x^2 = k$, then $x = \sqrt{k}$ or $x = -\sqrt{k}$.
>
> (The solution is often written as $x = \pm\sqrt{k}$, read as "x equals plus or minus the square root of k.")

Note

We can use the square root property to solve an equation containing a squared quantity and a constant. To do so we will get the squared quantity containing the variable on one side of the equal sign and the constant on the other side.

Example 1

In-Class Example 1
Solve using the square root property.
a) $g^2 = 36$ b) $t^2 - 60 = 0$
c) $2h^2 + 59 = 9$
answer: a) $\{-6, 6\}$
b) $\{-2\sqrt{15}, 2\sqrt{15}\}$
c) $\{-5i, 5i\}$

Solve using the square root property.

a) $x^2 = 9$ b) $t^2 - 20 = 0$ c) $2a^2 + 21 = 3$

Solution

a)
$$x^2 = 9$$
$$x = \sqrt{9} \quad \text{or} \quad x = -\sqrt{9} \qquad \text{Square root property}$$
$$x = 3 \quad \text{or} \quad x = -3$$

The solution set is $\{-3, 3\}$. The check is left to the student.

An equivalent way to solve $x^2 = 9$ is to write it as

$$x^2 = 9$$
$$x = \pm\sqrt{9} \qquad \text{Square root property}$$
$$x = \pm 3$$

The solution set is $\{-3, 3\}$. We will use this approach when solving equations using the square root property.

b) To solve $t^2 - 20 = 0$, begin by getting t^2 on a side by itself.

$$t^2 - 20 = 0$$
$$t^2 = 20 \qquad \text{Add 20 to each side.}$$
$$t = \pm\sqrt{20} \qquad \text{Square root property}$$
$$t = \pm\sqrt{4} \cdot \sqrt{5} \qquad \text{Product rule for radicals}$$
$$t = \pm 2\sqrt{5} \qquad \sqrt{4} = 2$$

Check:

$t = 2\sqrt{5}$:
$$t^2 - 20 = 0$$
$$(2\sqrt{5})^2 - 20 \stackrel{?}{=} 0$$
$$(4 \cdot 5) - 20 \stackrel{?}{=} 0$$
$$20 - 20 = 0 \checkmark$$

$t = -2\sqrt{5}$:
$$t^2 - 20 = 0$$
$$(-2\sqrt{5})^2 - 20 \stackrel{?}{=} 0$$
$$(4 \cdot 5) - 20 \stackrel{?}{=} 0$$
$$20 - 20 = 0 \checkmark$$

The solution set is $\{-2\sqrt{5}, 2\sqrt{5}\}$.

c) $2a^2 + 21 = 3$
$$2a^2 = -18 \qquad \text{Subtract 21.}$$
$$a^2 = -9 \qquad \text{Divide by 2.}$$
$$a = \pm\sqrt{-9} \qquad \text{Square root property}$$
$$a = \pm 3i$$

Check:

$a = 3i$:
$$2a^2 + 21 = 3$$
$$2(3i)^2 + 21 \stackrel{?}{=} 3$$
$$2(9i^2) + 21 \stackrel{?}{=} 3$$
$$2(9)(-1) + 21 \stackrel{?}{=} 3$$
$$-18 + 21 = 3 \checkmark$$

$a = -3i$:
$$2a^2 + 21 = 3$$
$$2(-3i)^2 + 21 \stackrel{?}{=} 3$$
$$2(9i^2) + 21 \stackrel{?}{=} 3$$
$$2(9)(-1) + 21 \stackrel{?}{=} 3$$
$$-18 + 21 = 3 \checkmark$$

The solution set is $\{-3i, 3i\}$.

You Try 1

Solve using the square root property.

a) $p^2 = 100$ b) $w^2 - 32 = 0$ c) $3m^2 + 19 = 7$

Can we solve $(w - 4)^2 = 25$ using the square root property? Yes. The equation has a *squared quantity* and a *constant*.

2. Solve an Equation of the Form $(ax + b)^2 = k$

Example 2

Solve $x^2 = 25$ and $(w - 4)^2 = 25$ using the square root property.

In-Class Example 2
Solve $n^2 = 49$ and
$(p - 11)^2 = 49$ using the square root property.
answer: $\{-7, 7\}$ and $\{4, 18\}$

Solution

While the equation $(w - 4)^2 = 25$ has a *binomial* that is being squared, the two equations are actually in the same form.

$$x^2 = 25 \qquad\qquad (w - 4)^2 = 25$$

x squared = constant $\qquad$ $(w - 4)$ squared = constant

Solve $x^2 = 25$:

$$x^2 = 25$$
$$x = \pm\sqrt{25} \qquad \text{Square root property}$$
$$x = \pm 5$$

The solution set is $\{-5, 5\}$.

We solve $(w - 4)^2 = 25$ in the same way with some additional steps.

$$(w - 4)^2 = 25$$
$$w - 4 = \pm\sqrt{25} \qquad \text{Square root property}$$
$$w - 4 = \pm 5$$

This means $w - 4 = 5$ or $w - 4 = -5$. Solve both equations.

$$w - 4 = 5 \quad \text{or} \quad w - 4 = -5$$
$$w = 9 \quad \text{or} \qquad w = -1 \qquad \text{Add 4 to each side.}$$

Check:

$w = 9$: $\quad (w - 4)^2 = 25$ $\qquad\qquad$ $w = -1$: $\quad (w - 4)^2 = 25$
$\qquad\qquad (9 - 4)^2 \overset{?}{=} 25$ $\qquad\qquad\qquad\qquad (-1 - 4)^2 \overset{?}{=} 25$
$\qquad\qquad\qquad 5^2 = 25 \ \checkmark$ $\qquad\qquad\qquad\qquad\qquad (-5)^2 = 25 \ \checkmark$

The solution set is $\{-1, 9\}$.   ∎

You Try 2

Solve $(c + 6)^2 = 81$ using the square root property.

Example 3

Solve.

In-Class Example 3
Solve.
a) $(2f + 5)^2 = 16$
b) $(4d - 7)^2 = 5$
c) $(m + 3)^2 + 20 = 11$
d) $(2w - 5)^2 + 32 = 0$

a) $(3t + 4)^2 = 9$ $\qquad$ b) $(2m - 5)^2 = 12$ $\qquad$ c) $(z + 8)^2 + 11 = 7$
d) $(6k - 5)^2 + 20 = 0$

Solution

a) $(3t + 4)^2 = 9$
$$3t + 4 = \pm\sqrt{9} \qquad \text{Square root property}$$
$$3t + 4 = \pm 3$$

answer:

a) $\left\{-\dfrac{9}{2}, -\dfrac{1}{2}\right\}$

b) $\left\{\dfrac{7 - \sqrt{5}}{4}, \dfrac{7 + \sqrt{5}}{4}\right\}$

c) $\{-3 - 3i, -3 + 3i\}$

d) $\left\{\dfrac{5 - 4i\sqrt{2}}{2}, \dfrac{5 + 4i\sqrt{2}}{2}\right\}$

This means $3t + 4 = 3$ or $3t + 4 = -3$. Solve both equations.

$$3t + 4 = 3 \qquad\text{or}\qquad 3t + 4 = -3$$
$$3t = -1 \qquad\qquad 3t = -7 \qquad \text{Subtract 4 from each side.}$$
$$t = -\dfrac{1}{3} \quad\text{or}\quad t = -\dfrac{7}{3} \qquad \text{Divide by 3.}$$

The solution set is $\left\{-\dfrac{7}{3}, -\dfrac{1}{3}\right\}$.

b) $(2m - 5)^2 = 12$

$$2m - 5 = \pm\sqrt{12} \qquad \text{Square root property}$$
$$2m - 5 = \pm2\sqrt{3} \qquad \text{Simplify } \sqrt{12}.$$
$$2m = 5 \pm 2\sqrt{3} \qquad \text{Add 5 to each side.}$$
$$m = \dfrac{5 \pm 2\sqrt{3}}{2} \qquad \text{Divide by 2.}$$

One solution is $\dfrac{5 + 2\sqrt{3}}{2}$, and the other is $\dfrac{5 - 2\sqrt{3}}{2}$.

The solution set, $\left\{\dfrac{5 - 2\sqrt{3}}{2}, \dfrac{5 + 2\sqrt{3}}{2}\right\}$, can also be written as $\left\{\dfrac{5 \pm 2\sqrt{3}}{2}\right\}$.

c) $(z + 8)^2 + 11 = 7$

$$(z + 8)^2 = -4 \qquad \text{Subtract 11 from each side.}$$
$$z + 8 = \pm\sqrt{-4} \qquad \text{Square root property}$$
$$z + 8 = \pm2i \qquad \text{Simplify } \sqrt{-4}.$$
$$z = -8 \pm 2i \qquad \text{Subtract 8 from each side.}$$

The check is left to the student. The solution set is $\{-8 - 2i, -8 + 2i\}$.

d) $(6k - 5)^2 + 20 = 0$

$$(6k - 5)^2 = -20 \qquad \text{Subtract 20 from each side.}$$
$$6k - 5 = \pm\sqrt{-20} \qquad \text{Square root property}$$
$$6k - 5 = \pm2i\sqrt{5} \qquad \text{Simplify } \sqrt{-20}.$$
$$6k = 5 \pm 2i\sqrt{5} \qquad \text{Add 5 to each side.}$$
$$k = \dfrac{5 \pm 2i\sqrt{5}}{6} \qquad \text{Divide by 6.}$$

The check is left to the student. The solution set is $\left\{\dfrac{5 - 2i\sqrt{5}}{6}, \dfrac{5 + 2i\sqrt{5}}{6}\right\}$. ■

You Try 3

Solve.

a) $(7q + 1)^2 = 36$

b) $(5a - 3)^2 = 24$

c) $(c - 7)^2 + 100 = 0$

d) $(2y + 3)^2 - 5 = -23$

Did you notice in Examples 1c), 3c), and 3d) that a complex number *and* its conjugate were the solutions to the equations? This will always be true provided that the variables in the equation have real number coefficients.

Note

If $a + bi$ is a solution of a quadratic equation having only real coefficients, then $a - bi$ is also a solution.

3. Use the Distance Formula

In mathematics, we sometimes need to find the distance between two points in a plane. The **distance formula** enables us to do that. We can use the Pythagorean theorem and the square root property to develop the distance formula.

Suppose we want to find the distance between any two points with coordinates (x_1, y_1) and (x_2, y_2) as pictured here. [We also include the point (x_2, y_1) in our drawing so that we get a right triangle.]

The lengths of the legs are a and b. The length of the hypotenuse is c. Our goal is to find the *distance* between (x_1, y_1) and (x_2, y_2), *which is the same as* finding the length of c.

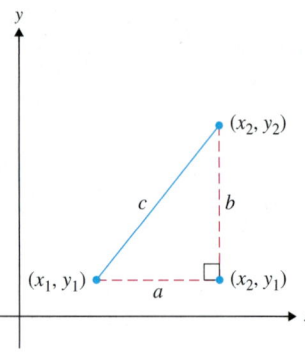

How long is side a? $|x_2 - x_1|$

How long is side b? $|y_2 - y_1|$

The Pythagorean theorem states that $a^2 + b^2 = c^2$. Substitute $|x_2 - x_1|$ for a and $|y_2 - y_1|$ for b, then solve for c.

$$a^2 + b^2 = c^2 \qquad \text{Pythagorean theorem}$$
$$|x_2 - x_1|^2 + |y_2 - y_1|^2 = c^2 \qquad \text{Substitute values.}$$
$$\pm\sqrt{(x_2 - x_1)^2 + (y_2 - y_1)^2} = c \qquad \text{Solve for } c \text{ using the square root property.}$$

The distance between the points (x_1, y_1) and (x_2, y_2) is $c = \sqrt{(x_2 - x_1)^2 + (y_2 - y_1)^2}$. We want only the positive square root since c is a length.

Since this formula represents the *distance* between two points, we usually use the letter d instead of c.

Definition   The Distance Formula

The distance, d, between two points with coordinates (x_1, y_1) and (x_2, y_2) is given by

$$d = \sqrt{(x_2 - x_1)^2 + (y_2 - y_1)^2}.$$

Example 4

Find the distance between the points $(-4, 1)$ and $(2, 5)$.

In-Class Example 4
Find the distance between the points $(-5, 3)$ and $(1, 8)$.
answer: $\sqrt{61}$

Solution

Begin by labeling the points: $\overset{x_1, y_1}{(-4, 1)}, \overset{x_2, y_2}{(2, 5)}$.

Substitute the values into the distance formula.

$$
\begin{aligned}
d &= \sqrt{(x_2 - x_1)^2 + (y_2 - y_1)^2} \\
&= \sqrt{[2 - (-4)]^2 + (5 - 1)^2} \qquad \text{Substitute values.} \\
&= \sqrt{(2 + 4)^2 + (4)^2} \\
&= \sqrt{(6)^2 + (4)^2} = \sqrt{36 + 16} = \sqrt{52} = 2\sqrt{13}
\end{aligned}
$$

You Try 4

Find the distance between the points $(1, 2)$ and $(7, -3)$.

The next method we will learn for solving a quadratic equation is *completing the square*. We need to review an idea first presented in Section 6.3.

A **perfect square trinomial** is a trinomial whose factored form is the square of a binomial. Some examples of perfect square trinomials are

Perfect Square Trinomials	Factored Form
$x^2 + 10x + 25$	$(x + 5)^2$
$d^2 - 8d + 16$	$(d - 4)^2$

In the trinomial $x^2 + 10x + 25$, x^2 is called the *quadratic term*, $10x$ is called the *linear term*, and 25 is called the *constant*.

4. Complete the Square for an Expression of the Form $x^2 + bx$

In a perfect square trinomial where the coefficient of the quadratic term is 1, the constant term is related to the coefficient of the linear term in the following way: *if you find half of the linear coefficient and square the result, you will get the constant term.*

$x^2 + 10x + 25$: The constant, 25, is obtained by

1) finding half of the coefficient of x;   then   2) squaring the result.

$$\frac{1}{2}(10) = 5 \qquad\qquad 5^2 = 25 \text{ (the constant)}$$

$d^2 - 8d + 16$: The constant, 16, is obtained by

1) finding half of the coefficient of d;   then   2) squaring the result.

$$\frac{1}{2}(-8) = -4 \qquad\qquad (-4)^2 = 16 \text{ (the constant)}$$

We can generalize this procedure so that we can find the constant needed to obtain the perfect square trinomial for any quadratic expression of the form $x^2 + bx$. Finding this perfect square trinomial is called **completing the square** because the trinomial will factor to the square of a binomial.

Procedure   Completing the Square for $x^2 + bx$

To find the constant needed to complete the square for $x^2 + bx$:

Step 1: Find half of the coefficient of x: $\frac{1}{2}b$.

Step 2: Square the result: $\left(\frac{1}{2}b\right)^2$.

Step 3: Then add it to $x^2 + bx$ to get $x^2 + bx + \left(\frac{1}{2}b\right)^2$. The factored form is $\left(x + \frac{1}{2}b\right)^2$.

The coefficient of the squared term *must* be 1 before you complete the square!

Example 5

In-Class Example 5
Complete the square for each
expression to obtain a perfect
square trinomial. Then, factor.
a) $f^2 + 10f$ b) $h^2 - 8h$
answer:
a) $f^2 + 10f + 25$; $(f + 5)^2$
b) $h^2 - 8h + 16$; $(h - 4)^2$

Complete the square for each expression to obtain a perfect square trinomial. Then, factor.

a) $y^2 + 6y$ b) $t^2 - 14t$

Solution

a) Find the constant needed to complete the square for $y^2 + 6y$.

  Step 1:   Find half of the coefficient of y:

$$\frac{1}{2}(6) = 3$$

  Step 2:   Square the result:

$$3^2 = 9$$

  Step 3:   Add 9 to $y^2 + 6y$:

$$y^2 + 6y + 9$$

The perfect square trinomial is $y^2 + 6y + 9$. The factored form is $(y + 3)^2$.

b) Find the constant needed to complete the square for $t^2 - 14t$.

  Step 1:   Find half of the coefficient of t:

$$\frac{1}{2}(-14) = -7$$

  Step 2:   Square the result:

$$(-7)^2 = 49$$

  Step 3:   Add 49 to $t^2 - 14t$:

$$t^2 - 14t + 49$$

The perfect square trinomial is $t^2 - 14t + 49$. The factored form is $(t - 7)^2$. ∎

You Try 5

Complete the square for each expression to obtain a perfect square trinomial. Then, factor.

a) $w^2 + 2w$ b) $z^2 - 16z$

We've seen the following perfect square trinomials and their factored forms. We will look at the relationship between the constant in the factored form and the coefficient of the linear term.

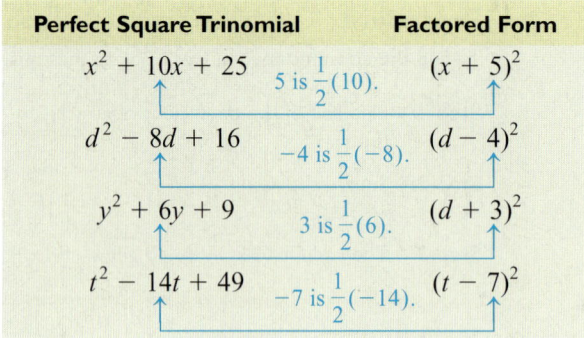

Perfect Square Trinomial	Factored Form
$x^2 + 10x + 25$ 5 is $\frac{1}{2}(10)$.	$(x + 5)^2$
$d^2 - 8d + 16$ -4 is $\frac{1}{2}(-8)$.	$(d - 4)^2$
$y^2 + 6y + 9$ 3 is $\frac{1}{2}(6)$.	$(d + 3)^2$
$t^2 - 14t + 49$ -7 is $\frac{1}{2}(-14)$.	$(t - 7)^2$

This pattern will always hold true and can be helpful in factoring some perfect square trinomials.

Example 6

In-Class Example 6
Complete the square for $p^2 + 11p$ to obtain a perfect square trinomial. Then, factor.
answer:
$p^2 + 11p + \dfrac{121}{4}$; $\left(p + \dfrac{11}{2}\right)^2$

Complete the square for $n^2 + 5n$ to obtain a perfect square trinomial. Then, factor.

Solution

Find the constant needed to complete the square for $n^2 + 5n$.

Step 1: Find half of the coefficient of n: $\dfrac{1}{2}(5) = \dfrac{5}{2}$

Step 2: Square the result: $\left(\dfrac{5}{2}\right)^2 = \dfrac{25}{4}$

Step 3: Add $\dfrac{25}{4}$ to $n^2 + 5n$. The perfect square trinomial is $n^2 + 5n + \dfrac{25}{4}$.

The factored form is $\left(n + \dfrac{5}{2}\right)^2$.

$\dfrac{5}{2}$ is $\dfrac{1}{2}(5)$, the coefficient of n.

$$Check: \left(n + \dfrac{5}{2}\right)^2 = n^2 + 2n\left(\dfrac{5}{2}\right) + \left(\dfrac{5}{2}\right)^2 = n^2 + 5n + \dfrac{25}{4}$$

You Try 6

Complete the square for $p^2 - 3p$ to obtain a perfect square trinomial. Then, factor.

5. Solve a Quadratic Equation by Completing the Square

Any quadratic equation of the form $ax^2 + bx + c = 0$ $(a \neq 0)$ can be written in the form $(x - h)^2 = k$ by completing the square. Once an equation is in this form, we can use the square root property to solve for the variable.

> **Procedure**   Solve a Quadratic Equation $(ax^2 + bx + c = 0)$ by Completing the Square
>
> **Step 1:** **The coefficient of the squared term must be 1.** If it is not 1, divide both sides of the equation by a to obtain a leading coefficient of 1.
>
> **Step 2:** **Get the variables on one side of the equal sign and the constant on the other side.**
>
> **Step 3:** **Complete the square.** Find half of the linear coefficient, then square the result. Add that quantity to *both* sides of the equation.
>
> **Step 4:** **Factor.**
>
> **Step 5:** **Solve using the square root property.**

Example 7

In-Class Example 7
Solve by completing the square.
a) $v^2 + 4v + 9 = 0$
b) $q^2 - 10q + 12 = 0$
c) $15t + 9t^2 = 3$
answer:
a) $\{-2 - i\sqrt{5}, -2 + i\sqrt{5}\}$
b) $\{5 - \sqrt{13}, 5 + \sqrt{13}\}$
c) $\left\{-\dfrac{5}{6} - \dfrac{\sqrt{37}}{6}, -\dfrac{5}{6} + \dfrac{\sqrt{37}}{6}\right\}$

Solve by completing the square.

a) $x^2 + 6x + 8 = 0$ b) $12h + 4h^2 = -24$

Solution

a) $x^2 + 6x + 8 = 0$

Step 1: The coefficient of x^2 is already 1.

Step 2: Get the variables on one side of the equal sign and the constant on the other side: $x^2 + 6x = -8$

Step 3: Complete the square: $\dfrac{1}{2}(6) = 3$

$$3^2 = 9$$

Add 9 to both sides of the equation: $x^2 + 6x + 9 = -8 + 9$
$$x^2 + 6x + 9 = 1$$

Step 4: Factor: $(x + 3)^2 = 1$

Step 5: Solve using the square root property.

$$(x + 3)^2 = 1$$
$$x + 3 = \pm\sqrt{1}$$
$$x + 3 = \pm 1$$

$x + 3 = 1$ or $x + 3 = -1$
$x = -2$ or $x = -4$

The check is left to the student. The solution set is $\{-4, -2\}$.

Note

We would have obtained the same result if we had solved the equation by factoring.

$$x^2 + 6x + 8 = 0$$
$$(x + 4)(x + 2) = 0$$

$x + 4 = 0$ or $x + 2 = 0$
$x = -4$ or $x = -2$

b) $12h + 4h^2 = -24$

Step 1: Since the coefficient of h^2 is *not* 1, divide the whole equation by 4.

$$\frac{12h}{4} + \frac{4h^2}{4} = \frac{-24}{4}$$
$$3h + h^2 = -6$$

Step 2: The constant is on a side by itself. Rewrite the left side of the equation.

$$h^2 + 3h = -6$$

Step 3: Complete the square: $\dfrac{1}{2}(3) = \dfrac{3}{2}$

$$\left(\frac{3}{2}\right)^2 = \frac{9}{4}$$

Add $\dfrac{9}{4}$ to both sides of the equation.

$$h^2 + 3h + \frac{9}{4} = -6 + \frac{9}{4}$$

$$h^2 + 3h + \frac{9}{4} = -\frac{24}{4} + \frac{9}{4} \qquad \text{\color{blue}{Get a common denominator.}}$$

$$h^2 + 3h + \frac{9}{4} = -\frac{15}{4}$$

Step 4:   Factor.

$$\left(h + \frac{3}{2}\right)^2 = -\frac{15}{4}$$

$\uparrow$

$\dfrac{3}{2}$ is $\dfrac{1}{2}(3)$, the coefficient of h.

Step 5:   Solve using the square root property.

$$\left(h + \frac{3}{2}\right)^2 = -\frac{15}{4}$$

$$h + \frac{3}{2} = \pm\sqrt{-\frac{15}{4}}$$

$$h + \frac{3}{2} = \pm\frac{\sqrt{15}}{2}i \qquad \text{\color{blue}{Simplify the radical.}}$$

$$h = -\frac{3}{2} \pm \frac{\sqrt{15}}{2}i \qquad \text{\color{blue}{Subtract }} \frac{3}{2}.$$

The check is left to the student. The solution set is

$$\left\{-\frac{3}{2} - \frac{\sqrt{15}}{2}i, \ -\frac{3}{2} + \frac{\sqrt{15}}{2}i\right\}.$$

You Try 7

Solve by completing the square.

a) $q^2 + 10q - 24 = 0$ b) $2m^2 + 16 = 10m$

Answers to You Try Exercises

1) a) $\{-10, 10\}$ b) $\{-4\sqrt{2}, 4\sqrt{2}\}$ c) $\{-2i, 2i\}$ 2) $\{-15, 3\}$

3) a) $\left\{-1, \dfrac{5}{7}\right\}$ b) $\left\{\dfrac{3 - 2\sqrt{6}}{5}, \dfrac{3 + 2\sqrt{6}}{5}\right\}$ c) $\{7 - 10i, 7 + 10i\}$

d) $\left\{-\dfrac{3}{2} - \dfrac{3\sqrt{2}}{2}i, \ -\dfrac{3}{2} + \dfrac{3\sqrt{2}}{2}i\right\}$ 4) $\sqrt{61}$

5) a) $w^2 + 2w + 1; (w + 1)^2$ b) $z^2 - 16z + 64; (z - 8)^2$

6) $p^2 - 3p + \dfrac{9}{4}; \left(p - \dfrac{3}{2}\right)^2$ 7) a) $\{-12, 2\}$ b) $\left\{\dfrac{5}{2} - \dfrac{\sqrt{7}}{2}i, \dfrac{5}{2} + \dfrac{\sqrt{7}}{2}i\right\}$

9.1 Exercises

*Additional answers can be found in the Answers to Exercises appendix.

Objective 1: Solve an Equation of the Form $x^2 = k$

1) Choose two methods to solve $y^2 - 16 = 0$. Solve the equation using both methods.   Methods may vary; $\{-4, 4\}$

2) If k is a negative number and $x^2 = k$, what can you conclude about the solution to the equation?   The solution is not a real number.

Solve using the square root property.

3) $b^2 = 36$ $\{-6, 6\}$

4) $h^2 = 64$ $\{-8, 8\}$

5) $r^2 - 27 = 0$ $\{-3\sqrt{3}, 3\sqrt{3}\}$

6) $a^2 - 30 = 0$ $\{-\sqrt{30}, \sqrt{30}\}$

7) $n^2 = \dfrac{4}{9}$ $\left\{-\dfrac{2}{3}, \dfrac{2}{3}\right\}$

8) $v^2 = \dfrac{121}{16}$ $\left\{-\dfrac{11}{4}, \dfrac{11}{4}\right\}$

9) $q^2 = -4$ $\{-2i, 2i\}$

10) $w^2 = -121$ $\{-11i, 11i\}$

11) $z^2 + 3 = 0$ $\{-i\sqrt{3}, i\sqrt{3}\}$

12) $h^2 + 14 = -23$ $\{-i\sqrt{37}, i\sqrt{37}\}$

13) $z^2 + 5 = 19$ $\{-\sqrt{14}, \sqrt{14}\}$

14) $q^2 - 3 = 15$ $\{-3\sqrt{2}, 3\sqrt{2}\}$

15) $2d^2 + 5 = 55$ $\{-5, 5\}$

16) $4m^2 + 1 = 37$ $\{-3, 3\}$

17) $5f^2 + 39 = -21$
$\{-2i\sqrt{3}, 2i\sqrt{3}\}$

18) $2y^2 + 56 = 0$
$\{-2i\sqrt{7}, 2i\sqrt{7}\}$

Objective 2: Solve an Equation of the Form $(ax + b)^2 = k$

Solve using the square root property.

19) $(r + 10)^2 = 4$ $\{-12, -8\}$

20) $(x - 5)^2 = 81$ $\{-4, 14\}$

21) $(q - 7)^2 = 1$ $\{6, 8\}$

22) $(c + 12)^2 = 25$ $\{-17, -7\}$

23) $(p + 4)^2 - 18 = 0$
$\{-4 - 3\sqrt{2}, -4 + 3\sqrt{2}\}$

24) $(d + 2)^2 - 7 = 13$
$\{-2 - 2\sqrt{5}, -2 + 2\sqrt{5}\}$

25) $(c + 3)^2 - 4 = -29$
$\{-3 - 5i, -3 + 5i\}$

26) $(u - 15)^2 - 4 = -8$
$\{15 - 2i, 15 + 2i\}$

27) $1 = 15 + (k - 2)^2$
$\{2 - i\sqrt{14}, 2 + i\sqrt{14}\}$

28) $2 = 14 + (g + 4)^2$
$\{-4 - 2i\sqrt{3}, -4 + 2i\sqrt{3}\}$

29) $20 = (2w + 1)^2$

30) $(5b - 6)^2 = 11$

31) $8 = (3q - 10)^2 - 6$

32) $22 = (6x + 11)^2 + 4$

33) $36 + (4p - 5)^2 = 6$

34) $(3k - 1)^2 + 20 = 4$

35) $(6g + 11)^2 + 50 = 1$

36) $9 = 38 + (9s - 4)^2$

37) $\left(\dfrac{3}{4}n - 8\right)^2 = 4$ $\left\{8, \dfrac{40}{3}\right\}$

38) $\left(\dfrac{2}{3}j + 10\right)^2 = 16$ $\{-21, -9\}$

39) $(5y - 2)^2 + 6 = 22$
$\left\{-\dfrac{2}{5}, \dfrac{6}{5}\right\}$

40) $-6 = 3 - (2q - 9)^2$ $\{3, 6\}$

Objective 3: Use the Distance Formula

Find the distance between the given points.

41) $(7, -1)$ and $(3, 2)$ 5

42) $(3, 10)$ and $(12, 6)$ $\sqrt{97}$

43) $(-5, -6)$ and $(-2, -8)$ $\sqrt{13}$

44) $(5, -2)$ and $(-3, 4)$ 10

45) $(0, 13)$ and $(0, 7)$ 6

46) $(-8, 3)$ and $(2, 1)$ $2\sqrt{26}$

47) $(-4, 11)$ and $(2, 6)$ $\sqrt{61}$

48) $(0, 3)$ and $(3, -1)$ 5

49) $(3, -3)$ and $(5, -7)$ $2\sqrt{5}$

50) $(-5, -6)$ and $(-1, 2)$ $4\sqrt{5}$

Objective 4: Complete the Square for an Expression of the Form $x^2 + bx$

51) What is a perfect square trinomial? Give an example.
A trinomial whose factored form is the square of a binomial; examples will vary.

52) Can you complete the square on $3y^2 + 15y$ as it is given? Why or why not?   No, because the coefficient of y^2 is not 1.

Complete the square for each expression to obtain a perfect square trinomial. Then, factor.

Fill It In

Fill in the blanks with either the missing mathematical step or reason for the given step.

53) $w^2 + 8w$

$\dfrac{1}{2}(8) = 4$ Find half of the coefficient of w.

$4^2 = 16$ Square the result.

$w^2 + 8w + 16$ Add the constant to the expression.

The perfect square trinomial is $w^2 + 8w + 16$.

The factored form of the trinomial is $(w + 4)^2$.

54) $n^2 - n$

$\dfrac{1}{2}(-1) = -\dfrac{1}{2}$ Find half of the coefficient of n.

$\left(-\dfrac{1}{2}\right)^2 = \dfrac{1}{4}$ Square the result.

$n^2 - n + \dfrac{1}{4}$ Add the constant to the expression.

The perfect square trinomial is $n^2 - n + \dfrac{1}{4}$.

The factored form of the trinomial is $\left(n - \dfrac{1}{2}\right)^2$.

55) $a^2 + 12a$
$a^2 + 12a + 36; (a + 6)^2$

56) $g^2 + 4g$ $g^2 + 4g + 4; (g + 2)^2$

57) $c^2 - 18c$
$c^2 - 18c + 81; (c - 9)^2$

58) $k^2 - 16k$
$k^2 - 16k + 64; (k - 8)^2$

59) $r^2 + 3r$

60) $z^2 - 7z$

61) $b^2 - 9b$
$b^2 - 9b + \dfrac{81}{4}; \left(b - \dfrac{9}{2}\right)^2$

62) $t^2 + 5t$ $t^2 + 5t + \dfrac{25}{4}; \left(t + \dfrac{5}{2}\right)^2$

63) $x^2 + \dfrac{1}{3}x$

$x^2 + \dfrac{1}{3}x + \dfrac{1}{36}; \left(x + \dfrac{1}{6}\right)^2$

64) $y^2 - \dfrac{3}{5}y$

$y^2 - \dfrac{3}{5}y + \dfrac{9}{100}; \left(y - \dfrac{3}{10}\right)^2$

Objective 5: Solve a Quadratic Equation by Completing the Square

65) What is the first thing you should do if you want to solve $2p^2 - 7p = 8$ by completing the square?
Divide both sides of the equation by 2.

66) Can $x^3 + 10x - 3 = 0$ be solved by completing the square? Give a reason for your answer.
No, because the equation is not quadratic.

Solve by completing the square.

67) $x^2 + 6x + 8 = 0$
$\{-4, -2\}$

68) $t^2 + 12t - 13 = 0$ $\{-13, 1\}$

69) $k^2 - 8k + 15 = 0$ $\{3, 5\}$

70) $v^2 - 6v - 27 = 0$ $\{-3, 9\}$

71) $s^2 + 10 = -10s$
$\{-5 - \sqrt{15}, -5 + \sqrt{15}\}$

72) $u^2 - 9 = 2u$
$\{1 - \sqrt{10}, 1 + \sqrt{10}\}$

73) $t^2 = 2t - 9$
$\{1 - 2i\sqrt{2}, 1 + 2i\sqrt{2}\}$

74) $p^2 = -10p - 26$
$\{-5 - i, -5 + i\}$

75) $v^2 + 4v + 8 = 0$
$\{-2 - 2i, -2 + 2i\}$

76) $a^2 + 19 = 8a$
$\{4 - i\sqrt{3}, 4 + i\sqrt{3}\}$

77) $m^2 + 3m - 40 = 0$
$\{-8, 5\}$

78) $p^2 + 5p + 4 = 0$ $\{-4, -1\}$

79) $x^2 - 7x + 12 = 0$
$\{3, 4\}$

80) $d^2 + d - 72 = 0$ $\{-9, 8\}$

81) $r^2 - r = 3$

82) $y^2 - 3y = 7$

83) $c^2 + 5c + 7 = 0$

84) $b^2 + 14 = 7b$

85) $3k^2 - 6k + 12 = 0$
$\{1 - i\sqrt{3}, 1 + i\sqrt{3}\}$

86) $4f^2 + 16f + 48 = 0$
$\{-2 - 2i\sqrt{2}, -2 + 2i\sqrt{2}\}$

87) $4r^2 + 24r = 8$
$\{-3 - \sqrt{11}, -3 + \sqrt{11}\}$

88) $3h^2 + 6h = 15$
$\{-1 - \sqrt{6}, -1 + \sqrt{6}\}$

89) $10d = 2d^2 + 12$ $\{2, 3\}$

90) $54x - 6x^2 = 48$ $\{1, 8\}$

91) $2n^2 + 8 = 5n$

92) $2t^2 + 3t + 4 = 0$

93) $4a^2 - 7a + 3 = 0$ $\left\{\frac{3}{4}, 1\right\}$

94) $n + 2 = 3n^2$ $\left\{-\frac{2}{3}, 1\right\}$

95) $(y + 5)(y - 3) = 5$
$\{-1 - \sqrt{21}, -1 + \sqrt{21}\}$

96) $(b - 4)(b + 10) = -17$
$\{-3 - 4\sqrt{2}, -3 + 4\sqrt{2}\}$

97) $(2m + 1)(m - 3) = -7$

98) $(3c + 4)(c + 2) = 3$

Use the Pythagorean theorem and the square root property to find the length of the missing side.

99)

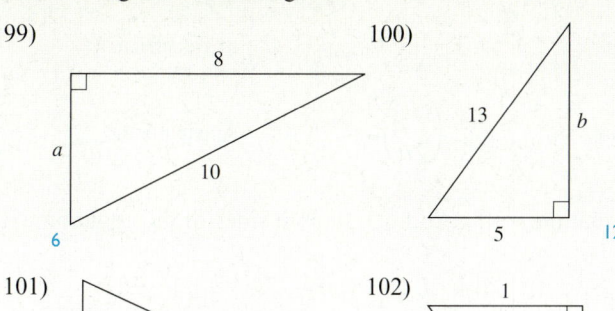

100)

101)

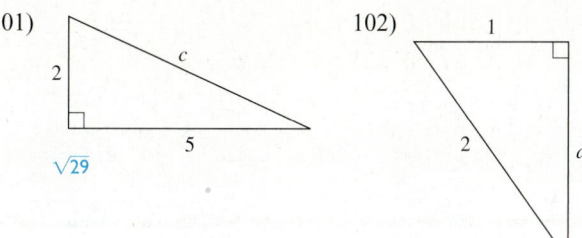

102)

Write an equation and solve. (Hint: Draw a picture.)

103) The width of a rectangle is 4 in., and its diagonal is $2\sqrt{13}$ in. long. What is the length of the rectangle?
6 in.

104) Find the length of the diagonal of a rectangle if it has a width of 5 cm and a length of $4\sqrt{2}$ cm. $\sqrt{57}$ cm

Write an equation and solve.

105) A 13-ft ladder is leaning against a wall so that the base of the ladder is 5 ft away from the wall. How high on the wall does the ladder reach?

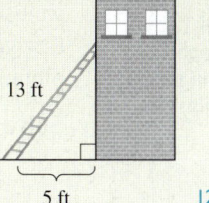

106) Salma is flying a kite. It is 30 ft from her horizontally, and it is 40 ft above her hand. How long is the kite string?

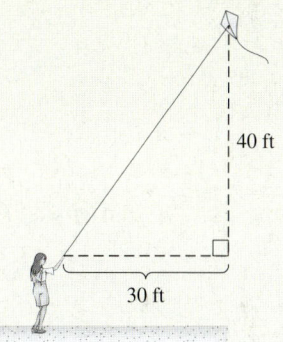

107) Let $f(x) = (x + 3)^2$. Find x so that $f(x) = 49$. $-10, 4$

108) Let $g(t) = (t - 5)^2$. Find t so that $g(t) = 12$.
$5 - 2\sqrt{3}, 5 + 2\sqrt{3}$

Solve each problem by writing an equation and solving it by completing the square.

109) The length of a rectangular garden is 8 ft more than its width. Find the dimensions of the garden if it has an area of 153 ft^2.   width = 9 ft, length = 17 ft

110) The rectangular screen on a laptop has an area of 375 cm^2. Its width is 10 cm less than its length. What are the dimensions of the screen?   width = 15 cm, length = 25 cm

Section 9.2 The Quadratic Formula

1. Derive the Quadratic Formula

In Section 9.1, we saw that any quadratic equation of the form $ax^2 + bx + c = 0$ $(a \neq 0)$ can be solved by completing the square. Therefore, we can solve equations like $x^2 - 8x + 5 = 0$ and $2x^2 + 3x - 1 = 0$ using this method.

We can develop another method for solving quadratic equations by completing the square on the general quadratic equation $ax^2 + bx + c = 0$ $(a \neq 0)$. This will let us derive the *quadratic formula*.

The steps we use to complete the square on $ax^2 + bx + c = 0$ are *exactly* the same steps we use to solve an equation like $2x^2 + 3x - 1 = 0$. We will do these steps side by side so that you can more easily understand how we are solving $ax^2 + bx + c = 0$ for x by completing the square.

Solve by Completing the Square

$$2x^2 + 3x - 1 = 0 \qquad\qquad ax^2 + bx + c = 0$$

Step 1: **The coefficient of the squared term must be 1.**

$$2x^2 + 3x - 1 = 0 \qquad\qquad ax^2 + bx + c = 0$$

$$\frac{2x^2}{2} + \frac{3x}{2} - \frac{1}{2} = \frac{0}{2} \quad \text{Divide by 2.} \qquad \frac{ax^2}{a} + \frac{bx}{a} + \frac{c}{a} = \frac{0}{a} \quad \text{Divide by } a.$$

$$x^2 + \frac{3}{2}x - \frac{1}{2} = 0 \quad \text{Simplify.} \qquad x^2 + \frac{b}{a}x + \frac{c}{a} = 0 \quad \text{Simplify.}$$

Step 2: **Get the constant on the other side of the equal sign.**

$$x^2 + \frac{3}{2}x = \frac{1}{2} \quad \text{Add } \frac{1}{2}. \qquad\qquad x^2 + \frac{b}{a}x = -\frac{c}{a} \quad \text{Subtract } \frac{c}{a}.$$

Step 3: **Complete the square.**

$$\frac{1}{2}\left(\frac{3}{2}\right) = \frac{3}{4} \quad \tfrac{1}{2} \text{ of } x\text{-coefficient} \qquad \frac{1}{2}\left(\frac{b}{a}\right) = \frac{b}{2a} \quad \tfrac{1}{2} \text{ of } x\text{-coefficient}$$

$$\left(\frac{3}{4}\right)^2 = \frac{9}{16} \quad \text{Square the result.} \qquad \left(\frac{b}{2a}\right)^2 = \frac{b^2}{4a^2} \quad \text{Square the result.}$$

$$\text{Add } \frac{9}{16} \text{ to both sides of the equation.} \qquad \text{Add } \frac{b^2}{4a^2} \text{ to both sides of the equation.}$$

$$x^2 + \frac{3}{2}x + \frac{9}{16} = \frac{1}{2} + \frac{9}{16} \qquad x^2 + \frac{b}{a}x + \frac{b^2}{4a^2} = -\frac{c}{a} + \frac{b^2}{4a^2}$$

$$x^2 + \frac{3}{2}x + \frac{9}{16} = \frac{8}{16} + \frac{9}{16} \quad \substack{\text{Get a}\\\text{common}\\\text{denominator.}} \qquad x^2 + \frac{b}{a}x + \frac{b^2}{4a^2} = -\frac{4ac}{4a^2} + \frac{b^2}{4a^2} \quad \substack{\text{Get a}\\\text{common}\\\text{denominator.}}$$

$$x^2 + \frac{3}{2}x + \frac{9}{16} = \frac{17}{16} \quad \text{Add.} \qquad x^2 + \frac{b}{a}x + \frac{b^2}{4a^2} = \frac{b^2 - 4ac}{4a^2} \quad \text{Add.}$$

Step 4: **Factor.**

$$\left(x + \frac{3}{4}\right)^2 = \frac{17}{16}$$

$\frac{3}{4}$ is $\frac{1}{2}\left(\frac{3}{2}\right)$, the coefficient of x.

$$\left(x + \frac{b}{2a}\right)^2 = \frac{b^2 - 4ac}{4a^2}$$

$\frac{b}{2a}$ is $\frac{1}{2}\left(\frac{b}{a}\right)$, the coefficient of x.

Step 5: **Solve using the square root property.**

$$\left(x + \frac{3}{4}\right)^2 = \frac{17}{16}$$

$$x + \frac{3}{4} = \pm\sqrt{\frac{17}{16}}$$

$$x + \frac{3}{4} = \frac{\pm\sqrt{17}}{4} \qquad \sqrt{16} = 4$$

$$x = -\frac{3}{4} \pm \frac{\sqrt{17}}{4} \qquad \text{Subtract } \frac{3}{4}.$$

$$x = \frac{-3 \pm \sqrt{17}}{4} \qquad \begin{array}{l}\text{Same denomi-}\\\text{nators, add}\\\text{numerators.}\end{array}$$

$$\left(x + \frac{b}{2a}\right)^2 = \frac{b^2 - 4ac}{4a^2}$$

$$x + \frac{b}{2a} = \pm\sqrt{\frac{b^2 - 4ac}{4a^2}}$$

$$x + \frac{b}{2a} = \frac{\pm\sqrt{b^2 - 4ac}}{2a} \qquad \sqrt{4a^2} = 2a$$

$$x = -\frac{b}{2a} \pm \frac{\sqrt{b^2 - 4ac}}{2a} \qquad \text{Subtract } \frac{b}{2a}.$$

$$x = \frac{-b \pm \sqrt{b^2 - 4ac}}{2a} \qquad \begin{array}{l}\text{Same denomi-}\\\text{nators, add}\\\text{numerators.}\end{array}$$

The result on the right is called the *quadratic formula*.

Definition   **The Quadratic Formula**

The solutions of any quadratic equation of the form $ax^2 + bx + c = 0$ ($a \neq 0$) are

$$x = \frac{-b \pm \sqrt{b^2 - 4ac}}{2a}$$

Note

1) Write the equation to be solved in the form $ax^2 + bx + c = 0$ so that $a, b,$ and c can be identified correctly.

2) $x = \dfrac{-b \pm \sqrt{b^2 - 4ac}}{2a}$ represents the two solutions $x = \dfrac{-b + \sqrt{b^2 - 4ac}}{2a}$ and

$x = \dfrac{-b - \sqrt{b^2 - 4ac}}{2a}$.

3) Notice that the fraction bar runs under $-b$ *and* under the radical.

$$x = \frac{-b \pm \sqrt{b^2 - 4ac}}{2a} \qquad\qquad x = -b \pm \frac{\sqrt{b^2 - 4ac}}{2a}$$

Correct   Incorrect

4) When deriving the quadratic formula, using the $\pm$ allows us to say that $\sqrt{4a^2} = 2a$.

5) The quadratic formula is a *very* important result and one that we will use often. *It should be memorized!*

2. Solve a Quadratic Equation Using the Quadratic Formula

Example 1

Solve using the quadratic formula.

In-Class Example 1
Solve using the quadratic formula.
a) $4g^2 - 2g - 7 = 0$
b) $y^2 = 6y - 25$
answer:
a) $\left\{\dfrac{1-\sqrt{29}}{4}, \dfrac{1+\sqrt{29}}{4}\right\}$
b) $\{3 - 4i, 3 + 4i\}$

a) $2x^2 + 3x - 1 = 0$ b) $k^2 = 10k - 29$

Solution

a) Is $2x^2 + 3x - 1 = 0$ in the form $ax^2 + bx + c = 0$? Yes. Identify the values of a, b, and c, and substitute them into the quadratic formula.

$$a = 2 \qquad b = 3 \qquad c = -1$$

$$x = \frac{-b \pm \sqrt{b^2 - 4ac}}{2a} \qquad \text{Quadratic formula}$$

$$= \frac{-(3) \pm \sqrt{(3)^2 - 4(2)(-1)}}{2(2)} \qquad \text{Substitute } a = 2, b = 3, \text{ and } c = -1.$$

$$= \frac{-3 \pm \sqrt{9 - (-8)}}{4} \qquad \text{Perform the operations.}$$

$$= \frac{-3 \pm \sqrt{17}}{4}$$

The solution set is $\left\{\dfrac{-3 - \sqrt{17}}{4}, \dfrac{-3 + \sqrt{17}}{4}\right\}$. This is the same result we obtained when we solved this equation by completing the square at the beginning of the section.

b) Is $k^2 = 10k - 29$ in the form $ax^2 + bx + c = 0$? *No.* Begin by writing the equation in the correct form.

$$k^2 - 10k + 29 = 0 \qquad \text{Subtract } 10k \text{ and add } 29 \text{ to both sides.}$$

$$a = 1 \qquad b = -10 \qquad c = 29 \qquad \text{Identify } a, b, \text{ and } c.$$

$$k = \frac{-b \pm \sqrt{b^2 - 4ac}}{2a} \qquad \text{Quadratic formula}$$

$$= \frac{-(-10) \pm \sqrt{(-10)^2 - 4(1)(29)}}{2(1)} \qquad \text{Substitute } a = 1, b = -10, \text{ and } c = 29.$$

$$= \frac{10 \pm \sqrt{100 - 116}}{2} \qquad \text{Perform the operations.}$$

$$= \frac{10 \pm \sqrt{-16}}{2} \qquad 100 - 116 = -16$$

$$= \frac{10 \pm 4i}{2} \qquad \sqrt{-16} = 4i$$

$$= \frac{10}{2} \pm \frac{4}{2}i = 5 \pm 2i$$

The solution set is $\{5 - 2i, 5 + 2i\}$.

You Try 1

Solve using the quadratic formula.

a) $n^2 + 9n + 18 = 0$ b) $5t^2 + t - 2 = 0$

Equations in various forms may be solved using the quadratic formula.

Example 2

Solve using the quadratic formula.

a) $(3p - 1)(3p + 4) = 3p - 5$

b) $\dfrac{1}{2}w^2 + \dfrac{2}{3}w - \dfrac{1}{3} = 0$

In-Class Example 2
Solve using the quadratic formula.
a) $(2d + 5)(2d - 3) = 2d + 1$
b) $\dfrac{1}{4}a^2 - \dfrac{1}{3}a + \dfrac{3}{4} = 0$

answer:

a) $\left\{ \dfrac{-1 - \sqrt{65}}{4}, \dfrac{-1 + \sqrt{65}}{4} \right\}$

b) $\left\{ \dfrac{2}{3} - \dfrac{\sqrt{23}}{3}i, \dfrac{2}{3} + \dfrac{\sqrt{23}}{3}i \right\}$

Solution

a)   Is $(3p - 1)(3p + 4) = 3p - 5$ in the form $ax^2 + bx + c = 0$? *No*. Before we can apply the quadratic formula, we must write it in that form.

$$(3p - 1)(3p + 4) = 3p - 5$$
$$9p^2 + 9p - 4 = 3p - 5 \qquad \text{Multiply using FOIL.}$$
$$9p^2 + 6p + 1 = 0 \qquad \text{Subtract } 3p \text{ and add 5 to both sides.}$$

The equation is in the correct form. Identify a, b, and c. $a = 9 \qquad b = 6 \qquad c = 1$

$$p = \frac{-b \pm \sqrt{b^2 - 4ac}}{2a} \qquad \text{Quadratic formula}$$

$$= \frac{-(6) \pm \sqrt{(6)^2 - 4(9)(1)}}{2(9)} \qquad \text{Substitute } a = 9, b = 6, \text{ and } c = 1.$$

$$= \frac{-6 \pm \sqrt{36 - 36}}{18} \qquad \text{Perform the operations.}$$

$$= \frac{-6 \pm \sqrt{0}}{18}$$

$$= \frac{-6 \pm 0}{18} = \frac{-6}{18} = -\frac{1}{3}$$

The solution set is $\left\{ -\dfrac{1}{3} \right\}$.

b)   Is $\dfrac{1}{2}w^2 + \dfrac{2}{3}w - \dfrac{1}{3} = 0$ in the form $ax^2 + bx + c = 0$? *Yes*. However, working with fractions in the quadratic formula would be difficult. *Eliminate the fractions by multiplying the equation by 6, the least common denominator of the fractions.*

$$6\left(\frac{1}{2}w^2 + \frac{2}{3}w - \frac{1}{3} \right) = 6 \cdot 0 \qquad \text{Multiply by 6 to eliminate the fractions.}$$
$$3w^2 + 4w - 2 = 0$$

Identify a, b, and c from this form of the equation: $a = 3$, $b = 4$, $c = -2$.

$$w = \frac{-b \pm \sqrt{b^2 - 4ac}}{2a} \qquad \text{Quadratic formula}$$

$$= \frac{-(4) \pm \sqrt{(4)^2 - 4(3)(-2)}}{2(3)} \qquad \text{Substitute } a = 3, b = 4, \text{ and } c = -2.$$

$$= \frac{-4 \pm \sqrt{16 - (-24)}}{6} \qquad \text{Perform the operations.}$$

$$= \frac{-4 \pm \sqrt{40}}{6} \qquad 16 - (-24) = 16 + 24 = 40$$

$$= \frac{-4 \pm 2\sqrt{10}}{6} \qquad \sqrt{40} = 2\sqrt{10}$$

$$= \frac{2(-2 \pm \sqrt{10})}{6} \qquad \text{Factor out 2 in the numerator.}$$

$$= \frac{-2 \pm \sqrt{10}}{3} \qquad \text{Divide numerator and denominator by 2 to simplify.}$$

The solution set is $\left\{ \dfrac{-2 - \sqrt{10}}{3}, \dfrac{-2 + \sqrt{10}}{3} \right\}$.

You Try 2

Solve using the quadratic formula.

a) $3 - 2z = -2z^2$ b) $(d + 6)(d - 2) = -10$ c) $\dfrac{5}{4}r^2 + \dfrac{1}{5} = r$

3. Determine the Number and Type of Solutions to a Quadratic Equation Using the Discriminant

We can find the solutions of any quadratic equation of the form $ax^2 + bx + c = 0$ $(a \neq 0)$ using the quadratic formula.

$$x = \frac{-b \pm \sqrt{b^2 - 4ac}}{2a}$$

The radicand in the quadratic formula determines the type of solution a quadratic equation has.

Property   The Discriminant and Solutions

The expression under the radical, $b^2 - 4ac$, is called the **discriminant.** The discriminant tells us what kind of solution a quadratic equation has. If $a, b,$ and c are integers, then

1) if $b^2 - 4ac$ is positive and the square of an integer, the equation has *two rational solutions.*

2) if $b^2 - 4ac$ is positive but not a perfect square, the equation has *two irrational solutions.*

3) if $b^2 - 4ac$ is negative, the equation has *two nonreal, complex solutions of the form $a + bi$ and $a - bi$.*

4) if $b^2 - 4ac = 0$, the equation has *one rational solution.*

Example 3

Find the value of the discriminant. Then, determine the number and type of solutions of each equation.

a) $z^2 + 6z - 4 = 0$ b) $5h^2 = 6h - 2$

In-Class Example 3
Find the value of the discriminant. Then, determine the number and type of solutions of each equation.
a) $y^2 + 3y - 9 = 0$
b) $11c^2 = 9c - 6$
**answer: a) 45; two irrational solutions
b) −183; two nonreal, complex solutions of the form $a + bi$ and $a - bi$**

Solution

a) Is $z^2 + 6z - 4 = 0$ in the form $ax^2 + bx + c = 0$? Yes. Identify $a, b,$ and c.

$$a = 1 \qquad b = 6 \qquad c = -4$$

Discriminant $= b^2 - 4ac = (6)^2 - 4(1)(-4) = 36 + 16 = 52$

Since 52 is positive but *not* a perfect square, the equation will have *two irrational solutions.* ($\sqrt{52}$, or $2\sqrt{13}$, will appear in the solution, and $2\sqrt{13}$ is irrational.)

b) Is $5h^2 = 6h - 2$ in the form $ax^2 + bx + c = 0$? No. Rewrite the equation in that form, and identify $a, b,$ and c.

$$5h^2 - 6h + 2 = 0$$

$$a = 5 \qquad b = -6 \qquad c = 2$$

Discriminant $= b^2 - 4ac = (-6)^2 - 4(5)(2) = 36 - 40 = -4$

Since the discriminant is −4, the equation will have *two nonreal, complex solutions of the form $a + bi$ and $a - bi$,* where $b \neq 0$. ∎

BE CAREFUL

The discriminant is $b^2 - 4ac$ *not* $\sqrt{b^2 - 4ac}$.

You Try 3

Find the value of the discriminant. Then, determine the number and type of solutions of each equation.

a) $2x^2 + x + 5 = 0$ b) $m^2 + 5m = 24$ c) $-3v^2 = 4v - 1$

d) $4r(2r - 3) = -1 - 6r - r^2$

4. Solve an Applied Problem Using the Quadratic Formula

Example 4

In-Class Example 4
Refer to Example 4. Use the
equation $h = -16t^2 + 4t + 20$
to answer a) and b).
answer: a) 1 sec   b) 1.25 sec

A ball is thrown upward from a height of 20 ft. The height h of the ball (in feet) t sec after the ball is released is given by

$$h = -16t^2 + 16t + 20$$

a) How long does it take the ball to reach a height of 8 ft?

b) How long does it take the ball to hit the ground?

Solution

a) Find the *time* it takes for the ball to reach a height of 8 ft.

Find t when $h = 8$.

$$h = -16t^2 + 16t + 20$$
$$8 = -16t^2 + 16t + 20 \qquad \text{Substitute 8 for } h.$$
$$0 = -16t^2 + 16t + 12 \qquad \text{Write in standard form.}$$
$$0 = 4t^2 - 4t - 3 \qquad \text{Divide by } -4.$$

$$t = \frac{-b \pm \sqrt{b^2 - 4ac}}{2a} \qquad \text{Quadratic formula}$$

$$= \frac{-(-4) \pm \sqrt{(-4)^2 - 4(4)(-3)}}{2(4)} \qquad \text{Substitute } a = 4, b = -4, \text{ and } c = -3.$$

$$= \frac{4 \pm \sqrt{16 + 48}}{8} \qquad \text{Perform the operations.}$$

$$= \frac{4 \pm \sqrt{64}}{8} = \frac{4 \pm 8}{8}$$

$$t = \frac{4 + 8}{8} \qquad \text{or} \qquad t = \frac{4 - 8}{8} \qquad \text{The equation has two rational solutions.}$$

$$t = \frac{12}{8} = \frac{3}{2} \qquad \text{or} \qquad t = \frac{-4}{8} = -\frac{1}{2}$$

Since t represents time, t cannot equal $-\dfrac{1}{2}$. We reject that as a solution.

Therefore, $t = \dfrac{3}{2}$ sec or 1.5 sec. The ball will be 8 ft above the ground after 1.5 sec.

b) When the ball hits the ground, it is 0 ft above the ground.

Find t when $h = 0$.

$$h = -16t^2 + 16t + 20$$
$$0 = -16t^2 + 16t + 20 \qquad \text{Substitute 0 for } h.$$
$$0 = 4t^2 - 4t - 5 \qquad \text{Divide by } -4.$$

$$t = \frac{-(-4) \pm \sqrt{(-4)^2 - 4(4)(-5)}}{2(4)} \qquad \text{Substitute } a = 4, b = -4, \text{ and } c = -5.$$

$$= \frac{4 \pm \sqrt{16 + 80}}{8} \qquad \text{Perform the operations.}$$

$$= \frac{4 \pm \sqrt{96}}{8}$$

$$= \frac{4 \pm 4\sqrt{6}}{8} \qquad \sqrt{96} = \sqrt{16} \cdot \sqrt{6} = 4\sqrt{6}$$

$$= \frac{4(1 \pm \sqrt{6})}{8} \qquad \text{Factor out 4 in the numerator.}$$

$$t = \frac{1 \pm \sqrt{6}}{2} \qquad \text{Divide numerator and denominator by 4 to simplify.}$$

$$t = \frac{1 + \sqrt{6}}{2} \quad \text{or} \quad t = \frac{1 - \sqrt{6}}{2} \qquad \text{The equation has two irrational solutions.}$$

$$t \approx \frac{1 + 2.4}{2} \quad \text{or} \quad t \approx \frac{1 - 2.4}{2} \qquad \sqrt{6} \approx 2.4$$

$$t \approx \frac{3.4}{2} = 1.7 \quad \text{or} \quad t \approx -0.7$$

Since t represents time, t cannot equal $\dfrac{1 - \sqrt{6}}{2}$. We reject this as a solution.

Therefore, $t = \dfrac{1 + \sqrt{6}}{2}$ sec or $t \approx 1.7$ sec. The ball will hit the ground after about 1.7 sec.

You Try 4

An object is thrown upward from a height of 12 ft. The height h of the object (in feet) t sec after the object is thrown is given by

$$h = -16t^2 + 56t + 12$$

a) How long does it take the object to reach a height of 36 ft?

b) How long does it take the object to hit the ground?

Answers to You Try Exercises

1) a) $\{-6, -3\}$ b) $\left\{ \dfrac{-1 - \sqrt{41}}{10}, \dfrac{-1 + \sqrt{41}}{10} \right\}$ 2) a) $\left\{ \dfrac{1}{2} - \dfrac{\sqrt{5}}{2}i, \dfrac{1}{2} + \dfrac{\sqrt{5}}{2}i \right\}$

b) $\{-2 - \sqrt{6}, -2 + \sqrt{6}\}$ c) $\left\{ \dfrac{2}{5} \right\}$ 3) a) -39; two nonreal, complex solutions

b) 121; two rational solutions c) 28; two irrational solutions d) 0; one rational solution

4) a) It takes $\dfrac{1}{2}$ sec to reach 36 ft on its way up and 3 sec to reach 36 ft on its way down.

b) $\dfrac{7 + \sqrt{61}}{4}$ sec or approximately 3.7 sec

9.2 Exercises

*Additional answers can be found in the Answers to Exercises appendix.

Mixed Exercises: Objectives 2 and 3

Find the error in each, and correct the mistake.

1) The solution to $ax^2 + bx + c = 0 \ (a \neq 0)$ can be found using the quadratic formula.

$$x = -b \pm \frac{\sqrt{b^2 - 4ac}}{2a}$$

2) In order to solve $5n^2 - 3n = 1$ using the quadratic formula, a student substitutes a, b, and c into the formula in this way: $a = 5$, $b = -3$, $c = 1$.

$$n = \frac{-(-3) \pm \sqrt{(-3)^2 - 4(5)(1)}}{2(5)}$$

3) $\dfrac{-2 \pm 6\sqrt{11}}{2} = -1 \pm 6\sqrt{11}$

4) The discriminant of $3z^2 - 4z + 1 = 0$ is

$$\sqrt{b^2 - 4ac} = \sqrt{(-4)^2 - 4(3)(1)}$$
$$= \sqrt{16 - 12}$$
$$= \sqrt{4}$$
$$= 2.$$

Objective 2: Solve a Quadratic Equation Using the Quadratic Formula

Solve using the quadratic formula.

5) $x^2 + 4x + 3 = 0 \quad \{-3, -1\}$

6) $v^2 - 8v + 7 = 0 \quad \{1, 7\}$

7) $3t^2 + t - 10 = 0 \quad \left\{-2, \dfrac{5}{3}\right\}$

8) $6q^2 + 11q + 3 = 0$

VIDEO 9) $k^2 + 2 = 5k$

10) $n^2 = 5 - 3n$

11) $y^2 = 8y - 25 \quad \{4 - 3i, 4 + 3i\}$

12) $-4x + 5 = -x^2 \quad \{2 - i, 2 + i\}$

13) $3 - 2w = -5w^2$

14) $2d^2 = -4 - 5d$

15) $r^2 + 7r = 0 \quad \{-7, 0\}$

16) $p^2 - 10p = 0 \quad \{0, 10\}$

17) $3v(v + 3) = 7v + 4$

18) $2k(k - 3) = -3$

19) $(2c - 5)(c - 5) = -3 \quad \left\{\dfrac{7}{2}, 4\right\}$

20) $-11 = (3z - 1)(z - 5) \quad \left\{\dfrac{4}{3}, 4\right\}$

21) $\dfrac{1}{6}u^2 + \dfrac{4}{3}u = \dfrac{5}{2}$

22) $\dfrac{5}{2}r^2 + 3r + 2 = 0$

VIDEO 23) $m^2 + \dfrac{4}{3}m + \dfrac{5}{9} = 0$

24) $\dfrac{1}{6}h + \dfrac{1}{2} = \dfrac{3}{4}h^2$

25) $2(p + 10) = (p + 10)(p - 2) \quad \{-10, 4\}$

26) $(t - 8)(t - 3) = 3(3 - t) \quad \{3, 5\}$

27) $4g^2 + 9 = 0 \quad \left\{-\dfrac{3}{2}i, \dfrac{3}{2}i\right\}$

28) $25q^2 - 1 = 0 \quad \left\{-\dfrac{1}{5}, \dfrac{1}{5}\right\}$

29) $x(x + 6) = -34 \quad \{-3 - 5i, -3 + 5i\}$

30) $c(c - 4) = -22 \quad \{2 - 3i\sqrt{2}, 2 + 3i\sqrt{2}\}$

VIDEO 31) $(2s + 3)(s - 1) = s^2 - s + 6 \quad \{-1 - \sqrt{10}, -1 + \sqrt{10}\}$

32) $(3m + 1)(m - 2) = (2m - 3)(m + 2) \quad \{3 - \sqrt{5}, 3 + \sqrt{5}\}$

33) $3(3 - 4y) = -4y^2 \quad \left\{\dfrac{3}{2}\right\}$

34) $5a(5a + 2) = -1 \quad \left\{-\dfrac{1}{5}\right\}$

35) $-\dfrac{1}{6} = \dfrac{2}{3}p^2 + \dfrac{1}{2}p$

36) $\dfrac{1}{2}n = \dfrac{3}{4}n^2 + 2$

37) $4q^2 + 6 = 20q$

38) $4w^2 = 6w + 16$

39) Let $f(x) = x^2 + 6x - 2$. Find x so that $f(x) = 0$.
$-3 - \sqrt{11}, -3 + \sqrt{11}$

40) Let $g(x) = 3x^2 - 4x - 1$. Find x so that $g(x) = 0$.

41) Let $h(t) = 2t^2 - t + 7$. Find t so that $h(t) = 12$.

42) Let $P(a) = a^2 + 8a + 9$. Find a so that $P(a) = -3$.
$-6, -2$

43) Let $f(x) = 5x^2 + 21x - 1$ and $g(x) = 2x + 3$. Find all values of x such that $f(x) = g(x)$. $\quad -4, \dfrac{1}{5}$

44) Let $F(x) = -x^2 + 3x - 2$ and $G(x) = x^2 + 12x + 6$. Find all values of x such that $F(x) = G(x)$.
$\dfrac{-9 - \sqrt{17}}{4}, \dfrac{-9 + \sqrt{17}}{4}$

Objective 3: Determine the Number and Type of Solutions to a Quadratic Equation Using the Discriminant

45) If the discriminant of a quadratic equation is zero, what do you know about the solutions of the equation?
There is one rational solution.

46) If the discriminant of a quadratic equation is negative, what do you know about the solutions of the equation?
There are two nonreal, complex solutions.

Find the value of the discriminant. Then, determine the number and type of solutions of each equation. *Do not solve.*

47) $10d^2 - 9d + 3 = 0$
-39; two nonreal, complex solutions

48) $3j^2 + 8j + 2 = 0$
40; two irrational solutions

49) $4y^2 + 49 = -28y$
0; one rational solution

50) $3q = 1 + 5q^2$
-11; two nonreal, complex solutions

51) $-5 = u(u + 6)$
16; two rational solutions

52) $g^2 + 4 = 4g$
0; one rational solution

VIDEO 53) $2w^2 - 4w - 5 = 0$
56; two irrational solutions

54) $3 + 2p^2 - 7p = 0$
25; two rational solutions

Find the value of a, b, or c so that each equation has only one rational solution.

55) $z^2 + bz + 16 = 0$
-8 or 8

56) $k^2 + bk + 49 = 0$
-14 or 14

57) $4y^2 - 12y + c = 0 \quad 9$

58) $25t^2 - 20t + c = 0 \quad 4$

59) $ap^2 + 12p + 9 = 0 \quad 4$

60) $ax^2 - 6x + 1 = 0 \quad 9$

Objective 4: Solve an Applied Problem Using the Quadratic Formula

Write an equation and solve.

61) One leg of a right triangle is 1 in. more than twice the other leg. The hypotenuse is $\sqrt{29}$ in. long. Find the lengths of the legs. $\quad$ 2 in., 5 in.

62) The hypotenuse of a right triangle is $\sqrt{34}$ in. long. The length of one leg is 1 in. less than twice the other leg. Find the lengths of the legs. $\quad$ 3 in., 5 in.

Solve.

63) An object is thrown upward from a height of 24 ft. The height h of the object (in feet) t sec after the object is released is given by $h = -16t^2 + 24t + 24$.

    a) How long does it take the object to reach a height of 8 ft?

       2 sec

    b) How long does it take the object to hit the ground?

64) A ball is thrown upward from a height of 6 ft. The height h of the ball (in feet) t sec after the ball is released is given by $h = -16t^2 + 44t + 6$.

    a) How long does it take the ball to reach a height of 16 ft?

       0.25 sec on the way up, 2.5 sec on the way down

    b) How long does it take the object to hit the ground?

       $\dfrac{11 + \sqrt{145}}{8}$ sec or about 2.9 sec

Putting It All Together

Objective

1. **Decide Which Method to Use to Solve a Quadratic Equation**

We have learned four methods for solving quadratic equations.

Methods for Solving Quadratic Equations

1) Factoring

2) Square root property

3) Completing the square

4) Quadratic formula

While it is true that the quadratic formula can be used to solve *every* quadratic equation of the form $ax^2 + bx + c = 0$ $(a \neq 0)$, it is not always the most *efficient* method. In this section we will discuss how to decide which method to use to solve a quadratic equation.

1. Decide Which Method to Use to Solve a Quadratic Equation

Example 1

Solve.

    a) $p^2 - 6p = 16$ b) $m^2 - 8m + 13 = 0$

    c) $3t^2 + 8t + 7 = 0$ d) $(2z - 7)^2 - 6 = 0$

In-Class Example 1

Solve.

a) $z^2 + 11z = -30$

b) $k^2 + 6k + 7 = 0$

c) $5r^2 + 6r + 3 = 0$

d) $(3m - 1)^2 + 5 = 0$

answer: a) $\{-6, -5\}$

b) $\{-3 \pm \sqrt{2}\}$

c) $\left\{ -\dfrac{3}{5} \pm \dfrac{\sqrt{6}}{5}i \right\}$

d) $\left\{ \dfrac{1}{3} \pm \dfrac{\sqrt{5}}{3}i \right\}$

Solution

a) Write $p^2 - 6p = 16$ in standard form: $p^2 - 6p - 16 = 0$

    Does $p^2 - 6p - 16$ factor? Yes. *Solve by factoring.*

$$(p - 8)(p + 2) = 0$$

$$p - 8 = 0 \quad \text{or} \quad p + 2 = 0 \qquad \text{Set each factor equal to 0.}$$
$$p = 8 \quad \text{or} \qquad\quad p = -2 \qquad \text{Solve.}$$

    The solution set is $\{-2, 8\}$.

b) To solve $m^2 - 8m + 13 = 0$ ask yourself, "Can I factor $m^2 - 8m + 13$?" No, it does not factor. We could solve this using the quadratic formula, but *completing the square* is also a good method for solving this equation. Why?

    Completing the square is a desirable method for solving a quadratic equation when the coefficient of the squared term is 1 or -1 and when the coefficient of the linear term is even.

    We will solve $m^2 - 8m + 13 = 0$ by completing the square.

    Step 1: The coefficient of m^2 is 1.

    Step 2: Get the variables on one side of the equal sign and the constant on the other side.

$$m^2 - 8m = -13$$

Step 3: Complete the square: $\dfrac{1}{2}(-8) = -4$

$$(-4)^2 = 16$$

Add 16 to both sides of the equation.

$$m^2 - 8m + 16 = -13 + 16$$
$$m^2 - 8m + 16 = 3$$

Step 4: Factor: $(m - 4)^2 = 3$

Step 5: Solve using the square root property:

$$(m - 4)^2 = 3$$
$$m - 4 = \pm\sqrt{3}$$
$$m = 4 \pm \sqrt{3}$$

The solution set is $\{4 - \sqrt{3}, 4 + \sqrt{3}\}$.

Note

Completing the square works well when the coefficient of the squared term is 1 or −1 and when the coefficient of the linear term is *even* because when we complete the square in Step 3, we will not obtain a fraction. (Half of an even number is an integer.)

c) Ask yourself, "Can I solve $3t^2 + 8t + 7 = 0$ by factoring?" No, $3t^2 + 8t + 7$ does not factor. Completing the square would not be a very efficient way to solve the equation because the coefficient of t^2 is 3, and dividing the equation by 3 would give us $t^2 + \dfrac{8}{3}t + \dfrac{7}{3} = 0$.

We will solve $3t^2 + 8t + 7 = 0$ using the quadratic formula.

Identify a, b, and c: $a = 3$ $b = 8$ $c = 7$

$$t = \frac{-b \pm \sqrt{b^2 - 4ac}}{2a} \qquad \text{Quadratic formula}$$

$$= \frac{-(8) \pm \sqrt{(8)^2 - 4(3)(7)}}{2(3)} \qquad \text{Substitute } a = 3, b = 8, \text{ and } c = 7.$$

$$= \frac{-8 \pm \sqrt{64 - 84}}{6} \qquad \text{Perform the operations.}$$

$$= \frac{-8 \pm \sqrt{-20}}{6}$$

$$= \frac{-8 \pm 2i\sqrt{5}}{6} \qquad \sqrt{-20} = i\sqrt{4}\sqrt{5} = 2i\sqrt{5}$$

$$= \frac{2(-4 \pm i\sqrt{5})}{6} \qquad \text{Factor out 2 in the numerator.}$$

$$= \frac{-4 \pm i\sqrt{5}}{3} \qquad \begin{array}{l}\text{Divide numerator and} \\ \text{denominator by 2 to simplify.}\end{array}$$

$$= -\frac{4}{3} \pm \frac{\sqrt{5}}{3}i \qquad \text{Write in the form } a + bi.$$

The solution set is $\left\{ -\dfrac{4}{3} - \dfrac{\sqrt{5}}{3}i, -\dfrac{4}{3} + \dfrac{\sqrt{5}}{3}i \right\}$.

d)   Which method should we use to solve $(2z - 7)^2 - 6 = 0$?

We *could* square the binomial, combine like terms, then solve, possibly, by factoring or using the quadratic formula. However, this would be very inefficient. The equation contains a squared quantity and a constant.

We will solve $(2z - 7)^2 - 6 = 0$ using the square root property.

$$(2z - 7)^2 - 6 = 0$$
$$(2z - 7)^2 = 6 \qquad \text{Add 6 to each side.}$$
$$2z - 7 = \pm\sqrt{6} \qquad \text{Square root property}$$
$$2z = 7 \pm \sqrt{6} \qquad \text{Add 7 to each side.}$$
$$z = \frac{7 \pm \sqrt{6}}{2} \qquad \text{Divide by 2.}$$

The solution set is $\left\{ \dfrac{7 - \sqrt{6}}{2}, \dfrac{7 + \sqrt{6}}{2} \right\}$.

You Try 1

Solve.

a) $2k^2 + 3 = 9k$ b) $2r^2 + 3r - 2 = 0$ c) $(n - 8)^2 + 9 = 0$ d) $y^2 + 4y = -10$

Answers to You Try Exercises

1)   a) $\left\{ \dfrac{9 - \sqrt{57}}{4}, \dfrac{9 + \sqrt{57}}{4} \right\}$ b) $\left\{ -2, \dfrac{1}{2} \right\}$ c) $\{8 - 3i, 8 + 3i\}$ d) $\{-2 - i\sqrt{6}, -2 + i\sqrt{6}\}$

Putting It All Together
Summary Exercises

*Additional answers can be found in the Answers to Exercises appendix.

Objective 1: Decide Which Method to Use to Solve a Quadratic Equation

Keep in mind the four methods we have learned for solving quadratic equations: *factoring, the square root property, completing the square, and the quadratic formula.* Solve the equations using one of these methods.

1) $z^2 - 50 = 0$ $\{-5\sqrt{2}, 5\sqrt{2}\}$

2) $j^2 - 6j = 8$ $\{3 - \sqrt{17}, 3 + \sqrt{17}\}$

3) $a(a + 1) = 20$ $\{-5, 4\}$

4) $2x^2 + 6 = 3x$

5) $u^2 + 7u + 9 = 0$

6) $3p^2 - p - 4 = 0$ $\left\{ -1, \dfrac{4}{3} \right\}$

7) $2k(2k + 7) - 3(k + 1)$

8) $2 = (w + 3)^2 + 8$ $\{-3 - i\sqrt{6}, -3 + i\sqrt{6}\}$

9) $m^2 + 14m + 60 = 0$ $\{-7 - i\sqrt{11}, -7 + i\sqrt{11}\}$

10) $\dfrac{1}{2}y^2 = \dfrac{3}{4} - \dfrac{1}{2}y$

11) $10 + (3b - 1)^2 = 4$

12) $c^2 + 8c + 25 = 0$ $\{-4 - 3i, -4 + 3i\}$

13) $1 = \dfrac{x^2}{12} - \dfrac{x}{3}$ $\{-2, 6\}$

14) $100 = 4d^2$ $\{-5, 5\}$

15) $r^2 - 4r = 3$ $\{2 - \sqrt{7}, 2 + \sqrt{7}\}$

16) $2t^3 + 108t = -30t^2$ $\{-9, -6, 0\}$

17) $p(p + 8) = 3(p^2 + 2) + p$ $\left\{ \dfrac{3}{2}, 2 \right\}$

18) $h^2 = h$ $\{0, 1\}$

19) $\dfrac{10}{z} = 1 + \dfrac{21}{z^2}$ $\{3, 7\}$

20) $2s(2s + 3) = 4s + 5$ $\left\{ \dfrac{-1 - \sqrt{21}}{4}, \dfrac{-1 + \sqrt{21}}{4} \right\}$

21) $(3v + 4)(v - 2) = -9$ $\left\{ \dfrac{1}{3} - \dfrac{\sqrt{2}}{3}i, \dfrac{1}{3} + \dfrac{\sqrt{2}}{3}i \right\}$

22) $34 = 6y - y^2$ $\{3 - 5i, 3 + 5i\}$

23) $(c - 5)^2 + 16 = 0$ $\{5 - 4i, 5 + 4i\}$

24) $(2b + 1)(b + 5) = -7$ $\left\{ -4, -\dfrac{3}{2} \right\}$

25) $3g = g^2$ $\{0, 3\}$

26) $5z^2 + 15z + 30 = 0$

27) $4m^3 = 9m$ $\left\{ -\dfrac{3}{2}, 0, \dfrac{3}{2} \right\}$

28) $\dfrac{9}{2a^2} = \dfrac{1}{6} + \dfrac{1}{a}$ $\{-9, 3\}$

29) $\dfrac{1}{3}q^2 + \dfrac{5}{6}q + \dfrac{4}{3} = 0$ $\left\{ -\dfrac{5}{4} - \dfrac{\sqrt{39}}{4}i, -\dfrac{5}{4} + \dfrac{\sqrt{39}}{4}i \right\}$

30) $-3 = (12d + 5)^2 + 6$ $\left\{ -\dfrac{5}{12} - \dfrac{1}{4}i, -\dfrac{5}{12} + \dfrac{1}{4}i \right\}$

Section 9.3 Equations in Quadratic Form

Objectives

1. **Solve Quadratic Equations Resulting from Equations Containing Fractions or Radicals**
2. **Solve an Equation in Quadratic Form by Factoring**
3. **Solve an Equation in Quadratic Form Using Substitution**
4. **Use Substitution for a Binomial to Solve a Quadratic Equation**

In Chapters 7 and 8, we solved some equations that were *not* quadratic but could be rewritten in the form of a quadratic equation, $ax^2 + bx + c = 0$. Two such examples are:

$$\frac{10}{x} - \frac{7}{x+1} = \frac{2}{3} \qquad \text{and} \qquad r + \sqrt{r} = 12$$

Rational equation (Ch. 7)   Radical equation (Ch. 8)

We will review how to solve each type of equation.

1. Solve Quadratic Equations Resulting from Equations Containing Fractions or Radicals

Example 1

Solve $\dfrac{10}{x} - \dfrac{7}{x+1} = \dfrac{2}{3}$.

In-Class Example 1

Solve $\dfrac{4}{x} - \dfrac{11}{x+1} = \dfrac{8}{3}$.

answer: $\left\{-4, \dfrac{3}{8}\right\}$

Solution

To solve an equation containing rational expressions, *multiply the equation by the LCD of all of the fractions to eliminate the denominators,* then solve.

$$LCD = 3x(x+1)$$

$$3x(x+1)\left(\frac{10}{x} - \frac{7}{x+1}\right) = 3x(x+1)\left(\frac{2}{3}\right)$$

Multiply both sides of the equation by the LCD of the fractions.

$$3x(x+1) \cdot \frac{10}{x} - 3x(x+1) \cdot \frac{7}{x+1} = 3x(x+1) \cdot \left(\frac{2}{3}\right)$$

Distribute and divide out common factors.

$$30(x+1) - 3x(7) = 2x(x+1)$$
$$30x + 30 - 21x = 2x^2 + 2x \qquad \text{Distribute.}$$
$$9x + 30 = 2x^2 + 2x \qquad \text{Combine like terms.}$$
$$0 = 2x^2 - 7x - 30 \qquad \begin{array}{l}\text{Write in the form}\\ ax^2 + bx + c = 0.\end{array}$$
$$0 = (2x+5)(x-6) \qquad \text{Factor.}$$

$$2x + 5 = 0 \quad \text{or} \quad x - 6 = 0 \qquad \begin{array}{l}\text{Set each factor}\\ \text{equal to zero.}\end{array}$$

$$2x = -5$$
$$x = -\frac{5}{2} \quad \text{or} \qquad x = 6 \qquad \text{Solve.}$$

Recall that you *must* check the proposed solutions in the original equation to be certain they do not make a denominator equal zero. The solution set is $\left\{-\dfrac{5}{2}, 6\right\}$.

You Try 1

Solve $\dfrac{1}{m} = \dfrac{1}{2} + \dfrac{m}{m+4}$.

Example 2

Solve $r + \sqrt{r} = 12$.

In-Class Example 2
Solve $d + \sqrt{d} = 20$.
answer: {16}

Solution

The first step in solving a radical equation is getting a radical on a side by itself.

$$r + \sqrt{r} = 12$$
$$\sqrt{r} = 12 - r \qquad \text{Subtract } r \text{ from each side.}$$
$$(\sqrt{r})^2 = (12 - r)^2 \qquad \text{Square both sides.}$$
$$r = 144 - 24r + r^2$$
$$0 = r^2 - 25r + 144 \qquad \text{Write in the form } ax^2 + bx + c = 0.$$
$$0 = (r - 16)(r - 9) \qquad \text{Factor.}$$

$$r - 16 = 0 \quad \text{or} \quad r - 9 = 0 \qquad \text{Set each factor equal to zero.}$$
$$r = 16 \quad \text{or} \qquad r = 9 \qquad \text{Solve.}$$

Recall that you *must* check the proposed solutions in the original equation.

Check $r = 16$:
$$r + \sqrt{r} = 12$$
$$16 + \sqrt{16} \overset{?}{=} 12$$
$$16 + 4 = 12 \qquad \text{False}$$

Check $r = 9$:
$$r + \sqrt{r} = 12$$
$$9 + \sqrt{9} \overset{?}{=} 12$$
$$9 + 3 = 12 \qquad \text{True}$$

16 is an extraneous solution. The solution set is {9}. ■

You Try 2

Solve $y + 3\sqrt{y} = 10$.

2. Solve an Equation in Quadratic Form by Factoring

Some equations that are not quadratic can be solved using the same methods that can be used to solve quadratic equations. These are called **equations in quadratic form.** Some examples of equations in quadratic form are:

$$x^4 - 10x^2 + 9 = 0, \qquad t^{2/3} + t^{1/3} - 6 = 0, \qquad 2n^4 - 5n^2 = -1$$

Let's compare the equations above to *quadratic equations* to understand why they are said to be in quadratic form.

Note

	COMPARE	
An Equation in Quadratic Form	*to*	**A Quadratic Equation**

This exponent is *twice* this exponent.
$$x^4 - 10x^2 + 9 = 0$$

This exponent is *twice* this exponent.
$$x^2 - 10x^1 + 9 = 0$$

This exponent is *twice* this exponent.
$$t^{2/3} + t^{1/3} - 6 = 0$$

This exponent is *twice* this exponent.
$$t^2 + t^1 - 6 = 0$$

This exponent is *twice* this exponent.
$$2n^4 - 5n^2 = -1$$

This exponent is *twice* this exponent.
$$2n^2 - 5n^1 = -1$$

This pattern enables us to work with equations in quadratic form like we can work with quadratic equations.

Example 3

In-Class Example 3
Solve.
a) $x^4 - 29x^2 + 100 = 0$
b) $f^{2/3} + f^{1/3} - 12 = 0$
answer: a) $\{-5, -2, 2, 5\}$
b) $\{-64, 27\}$

Solve.

a) $x^4 - 10x^2 + 9 = 0$ b) $t^{2/3} + t^{1/3} - 6 = 0$

Solution

a) Let's compare $x^4 - 10x^2 + 9 = 0$ to $x^2 - 10x + 9 = 0$.

We can factor $x^2 - 10x + 9$:

$$(x - 9)(x - 1)$$

Confirm by multiplying using FOIL:

$$(x - 9)(x - 1) = x^2 - x - 9x + 9$$
$$= x^2 - 10x + 9$$

Factor $x^4 - 10x^2 + 9$ in a similar way since the exponent, 4, of the first term is twice the exponent, 2, of the second term:

$$x^4 - 10x^2 + 9 = (x^2 - 9)(x^2 - 1)$$

Confirm by multiplying using FOIL:

$$(x^2 - 9)(x^2 - 1) = x^4 - x^2 - 9x^2 + 9$$
$$= x^4 - 10x^2 + 9$$

We can solve $x^4 - 10x^2 + 9 = 0$ by factoring.

$$x^4 - 10x^2 + 9 = 0$$
$$(x^2 - 9)(x^2 - 1) = 0 \qquad \text{Factor.}$$

$$x^2 - 9 = 0 \quad \text{or} \quad x^2 - 1 = 0 \qquad \text{Set each factor equal to 0.}$$
$$x^2 = 9 \qquad\qquad x^2 = 1 \qquad \text{Square root property}$$
$$x = \pm 3 \qquad\qquad x = \pm 1$$

The check is left to the student. The solution set is $\{-3, -1, 1, 3\}$.

b) Compare $t^{2/3} + t^{1/3} - 6 = 0$ to $t^2 + t - 6 = 0$.

We can factor $t^2 + t - 6$:

$$(t + 3)(t - 2)$$

Confirm by multiplying using FOIL:

$$(t + 3)(t - 2) = t^2 - 2t + 3t - 6$$
$$= t^2 + t - 6$$

Factor $t^{2/3} + t^{1/3} - 6$ in a similar way since the exponent, $\dfrac{2}{3}$, of the first term is twice the exponent, $\dfrac{1}{3}$, of the second term:

$$t^{2/3} + t^{1/3} - 6 = (t^{1/3} + 3)(t^{1/3} - 2)$$

Confirm by multiplying using FOIL:

$$(t^{1/3} + 3)(t^{1/3} - 2) = t^{2/3} - 2t^{1/3} + 3t^{1/3} - 6$$
$$= t^{2/3} + t^{1/3} - 6$$

We can solve $t^{2/3} + t^{1/3} - 6 = 0$ by factoring.

$$t^{2/3} + t^{1/3} - 6 = 0$$
$$(t^{1/3} + 3)(t^{1/3} - 2) = 0 \qquad \text{Factor.}$$

$$t^{1/3} + 3 = 0 \quad \text{or} \quad t^{1/3} - 2 = 0 \qquad \text{Set each factor equal to 0.}$$
$$t^{1/3} = -3 \qquad\qquad t^{1/3} = 2 \qquad \text{Isolate the constant.}$$
$$\sqrt[3]{t} = -3 \qquad\qquad \sqrt[3]{t} = 2 \qquad t^{1/3} = \sqrt[3]{t}$$
$$(\sqrt[3]{t})^3 = (-3)^3 \qquad (\sqrt[3]{t})^3 = 2^3 \qquad \text{Cube both sides.}$$
$$t = -27 \quad \text{or} \quad t = 8 \qquad \text{Solve.}$$

The check is left to the student. The solution set is $\{-27, 8\}$.

You Try 3

Solve.

a) $r^4 - 13r^2 + 36 = 0$ b) $c^{2/3} + 4c^{1/3} - 5 = 0$

3. Solve an Equation in Quadratic Form Using Substitution

The equations in Example 3 can also be solved using a method called **substitution.** We will illustrate the method in Example 4.

Example 4

Solve $x^4 - 10x^2 + 9 = 0$ using substitution.

In-Class Example 4
Solve $x^4 - 29x^2 + 100 = 0$
using substitution.
answer: $\{-5, -2, 2, 5\}$

Solution

$$x^4 - 10x^2 + 9 = 0$$
$$\downarrow$$
$$x^4 = (x^2)^2$$

To rewrite $x^4 - 10x^2 + 9 = 0$ in quadratic form, let $u = x^2$.

$$\text{If } u = x^2, \text{ then}$$
$$u^2 = x^4.$$

$$\begin{array}{ll}
x^4 - 10x^2 + 9 = 0 & \\
u^2 - 10u + 9 = 0 & \text{Substitute } u^2 \text{ for } x^4 \text{ and } u \text{ for } x^2. \\
(u - 9)(u - 1) = 0 & \text{Solve by factoring.}
\end{array}$$

$$\begin{array}{lll}
u - 9 = 0 & \text{or} \quad u - 1 = 0 & \text{Set each factor equal to 0.} \\
u = 9 & \text{or} \qquad u = 1 & \text{Solve for } u.
\end{array}$$

Be careful: $u = 9$ and $u = 1$ are *not* the solutions to $x^4 - 10x^2 + 9 = 0$. We still need to solve for x. Above we let $u = x^2$. *To solve for x, substitute 9 for u and solve for x and then substitute 1 for u and solve for x.*

$$\begin{array}{llll}
 & u = x^2 & u = x^2 & \\
\text{Substitute 9 for } u. & 9 = x^2 & 1 = x^2 & \text{Substitute 1 for } u. \\
\text{Square root property} & \pm 3 = x & \pm 1 = x & \text{Square root property}
\end{array}$$

The solution set is $\{-3, -1, 1, 3\}$. This is the same as the result we obtained in Example 3a).   ∎

You Try 4

Solve by substitution.

a) $r^4 - 13r^2 + 36 = 0$ b) $c^{2/3} + 4c^{1/3} - 5 = 0$

If, after substitution, an equation cannot be solved by factoring, we can use the quadratic formula.

Example 5

Solve $2n^4 - 5n^2 = -1$.

In-Class Example 5
Solve $8v^4 + 1 = 7v^2$.

answer: $\left\{ \dfrac{\pm\sqrt{7 \pm \sqrt{17}}}{4} \right\}$

Solution

Write the equation in standard form: $2n^4 - 5n^2 + 1 = 0$.

Can we solve the equation by factoring? *No.*

We will solve $2n^4 - 5n^2 + 1 = 0$ using the quadratic formula. Begin with substitution.

$$\text{If } u = n^2, \text{ then}$$
$$u^2 = n^4.$$

$$2n^4 - 5n^2 + 1 = 0$$
$$2u^2 - 5u + 1 = 0 \qquad \text{Substitute } u^2 \text{ for } n^4 \text{ and } u \text{ for } n^2.$$

$$u = \frac{-(-5) \pm \sqrt{(-5)^2 - 4(2)(1)}}{2(2)} \qquad a = 2, b = -5, c = 1$$

$$u = \frac{5 \pm \sqrt{25 - 8}}{4} = \frac{5 \pm \sqrt{17}}{4}$$

Note that $u = \dfrac{5 \pm \sqrt{17}}{4}$ does not solve the *original* equation. We must solve for x using the fact that $u = x^2$. Since $u = \dfrac{5 \pm \sqrt{17}}{4}$ means $u = \dfrac{5 + \sqrt{17}}{4}$ or $u = \dfrac{5 - \sqrt{17}}{4}$, we get

$$u = x^2 \qquad\qquad\qquad u = x^2$$

$$\frac{5 + \sqrt{17}}{4} = x^2 \qquad\qquad \frac{5 - \sqrt{17}}{4} = x^2$$

$$\pm\sqrt{\frac{5 + \sqrt{17}}{4}} = x \qquad\qquad \pm\sqrt{\frac{5 - \sqrt{17}}{4}} = x \qquad \text{Square root property}$$

$$\frac{\pm\sqrt{5 + \sqrt{17}}}{2} = x \qquad\qquad \frac{\pm\sqrt{5 - \sqrt{17}}}{2} = x \qquad \sqrt{4} = 2$$

The solution set is $\left\{ \dfrac{\sqrt{5 + \sqrt{17}}}{2}, -\dfrac{\sqrt{5 + \sqrt{17}}}{2}, \dfrac{\sqrt{5 - \sqrt{17}}}{2}, -\dfrac{\sqrt{5 - \sqrt{17}}}{2} \right\}$. ■

You Try 5

Solve $2k^4 + 3 = 9k^2$.

4. Use Substitution for a Binomial to Solve a Quadratic Equation

We can use substitution to solve an equation like the one in Example 6.

Example 6

Solve $2(3a + 1)^2 - 7(3a + 1) - 4 = 0$.

In-Class Example 6
Solve
$3(4h + 1)^2 + 5(4h + 1) + 2 = 0$.

answer: $\left\{ -\dfrac{1}{2}, -\dfrac{5}{12} \right\}$

Solution

The binomial $3a + 1$ appears as a *squared quantity* and as a *linear quantity*. Begin by using substitution.

$$\text{Let} \quad u = 3a + 1. \quad \text{Then,} \quad u^2 = (3a + 1)^2.$$

Substitute: $2(3a + 1)^2 - 7(3a + 1) - 4 = 0$

$\qquad\qquad\ 2u^2 \qquad\quad - 7u \qquad\quad - 4 = 0$

Does $2u^2 - 7u - 4 = 0$ factor? *Yes*. Solve by factoring.

$$(2u + 1)(u - 4) = 0 \qquad \text{\color{blue}Factor } 2u^2 - 7u - 4 = 0.$$

$$2u + 1 = 0 \quad \text{or} \quad u - 4 = 0 \qquad \text{\color{blue}Set each factor equal to 0.}$$

$$u = -\frac{1}{2} \quad \text{or} \qquad u = 4 \qquad \text{\color{blue}Solve for } u.$$

Solve for a using $u = 3a + 1$.

When $u = -\dfrac{1}{2}$: $\qquad\qquad\qquad$ When $u = 4$:

$\qquad\qquad\qquad u = 3a + 1 \qquad\qquad\qquad u = 3a + 1$

$\qquad\qquad\qquad -\dfrac{1}{2} = 3a + 1 \qquad\qquad\qquad 4 = 3a + 1$

$\text{\color{blue}Subtract 1.} \qquad -\dfrac{3}{2} = 3a \qquad\qquad\qquad 3 = 3a \qquad \text{\color{blue}Subtract 1.}$

$\text{\color{blue}Multiply by } \dfrac{1}{3}. \qquad -\dfrac{1}{2} = a \qquad\qquad\qquad 1 = a \qquad \text{\color{blue}Divide by 3.}$

The solution set is $\left\{ -\dfrac{1}{2}, 1 \right\}$.

You Try 6

Solve $3(2p - 1)^2 - 11(2p - 1) + 10 = 0$.

BE CAREFUL

Don't forget to solve for the variable in the *original* equation.

Answers to You Try Exercises

1) $\left\{ -2, \dfrac{4}{3} \right\}$ 2) $\{4\}$ 3) a) $\{-3, -2, 2, 3\}$ b) $\{-125, 1\}$ 4) a) $\{-3, -2, 2, 3\}$

b) $\{-125, 1\}$ 5) $\left\{ \dfrac{\sqrt{9 + \sqrt{57}}}{2}, -\dfrac{\sqrt{9 + \sqrt{57}}}{2}, \dfrac{\sqrt{9 - \sqrt{57}}}{2}, -\dfrac{\sqrt{9 - \sqrt{57}}}{2} \right\}$ 6) $\left\{ \dfrac{4}{3}, \dfrac{3}{2} \right\}$

9.3 Exercises

*Additional answers can be found in the Answers to Exercises appendix.

Objective 1: Solve Quadratic Equations Resulting from Equations Containing Fractions or Radicals

Solve.

1) $t - \dfrac{48}{t} = 8$ $\{-4, 12\}$

2) $z + 11 = -\dfrac{24}{z}$ $\{-8, -3\}$

3) $\dfrac{2}{x} + \dfrac{6}{x - 2} = -\dfrac{5}{2}$ $\left\{-2, \dfrac{4}{5}\right\}$

4) $\dfrac{3}{y} - \dfrac{6}{y - 1} = \dfrac{1}{2}$ $\{-3, -2\}$

5) $1 = \dfrac{2}{c} + \dfrac{1}{c - 5}$
   $\{4 - \sqrt{6}, 4 + \sqrt{6}\}$

6) $\dfrac{2}{g} = 1 + \dfrac{g}{g + 5}$

7) $\dfrac{3}{2v + 2} + \dfrac{1}{v} = \dfrac{3}{2}$

8) $\dfrac{1}{b + 3} + \dfrac{1}{b} = \dfrac{1}{3}$

9) $\dfrac{9}{n^2} = 5 + \dfrac{4}{n}$ $\left\{-\dfrac{9}{5}, 1\right\}$

10) $3 - \dfrac{16}{a^2} = \dfrac{8}{a}$ $\left\{-\dfrac{4}{3}, 4\right\}$

11) $\dfrac{5}{6r} = 1 - \dfrac{r}{6r - 6}$

12) $\dfrac{7}{4} - \dfrac{x}{4x + 4} = \dfrac{1}{x}$

13) $g = \sqrt{g + 20}$ $\{5\}$

14) $c = \sqrt{7c - 6}$ $\{1, 6\}$

15) $a = \sqrt{\dfrac{14a - 8}{5}}$ $\left\{\dfrac{4}{5}, 2\right\}$

16) $k = \sqrt{\dfrac{6 - 11k}{2}}$ $\left\{\dfrac{1}{2}\right\}$

17) $p - \sqrt{p} = 6$ $\{9\}$

18) $v + \sqrt{v} = 2$ $\{1\}$

19) $x = 5\sqrt{x} - 4$ $\{1, 16\}$

20) $10 = m - 3\sqrt{m}$ $\{25\}$

21) $2 + \sqrt{2y - 1} = y$ $\{5\}$

22) $1 - \sqrt{5t + 1} = -t$ $\{0, 3\}$

23) $2 = \sqrt{6k + 4} - k$ $\{0, 2\}$

24) $\sqrt{10 - 3q} - 6 = q$ $\{-2\}$

Mixed Exercises: Objectives 2–3

Determine whether each is an equation in quadratic form. Do *not* solve.

25) $n^4 - 12n^2 + 32 = 0$ yes

26) $p^6 + 8p^3 - 9 = 0$ yes

27) $2t^6 + 3t^3 - 5 = 0$ yes

28) $a^4 - 4a - 3 = 0$ no

29) $c^{2/3} - 4c - 6 = 0$ no

30) $3z^{2/3} + 2z^{1/3} + 1 = 0$ yes

31) $m + 9m^{1/2} = 4$ yes

32) $2x^{1/2} - 5x^{1/4} = 2$ yes

33) $5k^4 + 6k - 7 = 0$ no

34) $r^{-2} = 10 - 4r^{-1}$ yes

Solve.

35) $x^4 - 10x^2 + 9 = 0$
    $\{-3, -1, 1, 3\}$

36) $d^4 - 29d^2 + 100 = 0$
    $\{-5, -2, 2, 5\}$

37) $p^4 - 11p^2 + 28 = 0$
    $\{-\sqrt{7}, -2, 2, \sqrt{7}\}$

38) $k^4 - 9k^2 + 8 = 0$
    $\{-2\sqrt{2}, -1, 1, 2\sqrt{2}\}$

39) $a^4 + 12a^2 = -35$
    $\{-i\sqrt{7}, -i\sqrt{5}, i\sqrt{5}, i\sqrt{7}\}$

40) $c^4 + 9c^2 = -18$
    $\{-i\sqrt{6}, -i\sqrt{3}, i\sqrt{3}, i\sqrt{6}\}$

41) $b^{2/3} + 3b^{1/3} + 2 = 0$
    $\{-8, -1\}$

42) $z^{2/3} + z^{1/3} - 12 = 0$
    $\{-64, 27\}$

43) $t^{2/3} - 6t^{1/3} = 40$ $\{-64, 1000\}$

44) $p^{2/3} - p^{1/3} = 6$ $\{-8, 27\}$

45) $2n^{2/3} = 7n^{1/3} + 15$

46) $10k^{1/3} + 8 = -3k^{2/3}$

47) $v - 8v^{1/2} + 12 = 0$ $\{4, 36\}$

48) $j - 6j^{1/2} + 5 = 0$ $\{1, 25\}$

49) $4h^{1/2} + 21 = h$ $\{49\}$

50) $s + 12 = -7s^{1/2}$ $\varnothing$

51) $2a - 5a^{1/2} - 12 = 0$ $\{16\}$

52) $2w = 9w^{1/2} + 18$ $\{36\}$

53) $9n^4 = -15n^2 - 4$

54) $4h^4 + 19h^2 + 12 = 0$

55) $z^4 - 2z^2 = 15$ $\{-\sqrt{5}, \sqrt{5}, -i\sqrt{3}, i\sqrt{3}\}$

56) $a^4 + 2a^2 = 24$ $\{-2, 2, -i\sqrt{6}, i\sqrt{6}\}$

57) $w^4 - 6w^2 + 2 = 0$

58) $p^4 - 8p^2 + 3 = 0$

59) $2m^4 + 1 = 7m^2$

60) $8x^4 + 2 = 9x^2$

61) $t^{-2} - 4t^{-1} - 12 = 0$

62) $d^{-2} + d^{-1} - 6 = 0$

63) $4 = 13y^{-1} - 3y^{-2}$

64) $14h^{-1} + 3 = 5h^{-2}$

Objective 4: Use Substitution for a Binomial to Solve a Quadratic Equation

Solve.

65) $(x - 2)^2 + 11(x - 2) + 24 = 0$ $\{-6, -1\}$

66) $(r + 1)^2 - 3(r + 1) - 10 = 0$ $\{-3, 4\}$

67) $2(3q + 4)^2 - 13(3q + 4) + 20 = 0$ $\left\{-\dfrac{1}{2}, 0\right\}$

68) $4(2b - 3)^2 - 9(2b - 3) - 9 = 0$ $\left\{\dfrac{9}{8}, 3\right\}$

69) $(5a - 3)^2 + 6(5a - 3) = -5$ $\left\{\dfrac{2}{5}, \dfrac{2}{5}\right\}$

70) $(3z - 2)^2 - 8(3z - 2) = 20$ $\{0, 4\}$

71) $3(k + 8)^2 + 5(k + 8) = 12$ $\left\{-11, -\dfrac{20}{3}\right\}$

72) $5(t + 9)^2 + 37(t + 9) + 14 = 0$ $\left\{-\dfrac{47}{5}, -16\right\}$

73) $1 - \dfrac{8}{2w + 1} = -\dfrac{16}{(2w + 1)^2}$ $\left\{\dfrac{3}{2}\right\}$

74) $1 - \dfrac{8}{4p + 3} = -\dfrac{12}{(4p + 3)^2}$ $\left\{-\dfrac{1}{4}, \dfrac{3}{4}\right\}$

75) $1 + \dfrac{2}{h - 3} = \dfrac{1}{(h - 3)^2}$ $\{2 - \sqrt{2}, 2 + \sqrt{2}\}$

76) $\dfrac{2}{(c + 6)^2} + \dfrac{2}{(c + 6)} = 1$ $\{-5 - \sqrt{3}, -5 + \sqrt{3}\}$

Write an equation and solve.

77) It takes Kevin 3 hr longer than Walter to build a tree house. Together they can do the job in 2 hr. How long would it take each man to build the tree house on his own?
Walter: 3 hr; Kevin: 6 hr

78) It takes one pipe 4 hours more to empty a pool than it takes another pipe to fill a pool. If both pipes are accidentally left open, it takes 24 hr to fill the pool. How long does it take the single pipe to fill the pool?   8 hr

79) A boat can travel 9 mi downstream and then 6 mi back upstream in 1 hr. If the speed of the current is 3 mph, what is the speed of the boat in still water?   15 mph

80) A plane can travel 800 mi with the wind and then 650 mi back against the wind in 5 hr. If the wind blows at 30 mph, what is the speed of the plane?   290 mph

81) A large fish tank at an aquarium needs to be emptied so that it can be cleaned. When its large and small drains are opened together, the tank can be emptied in 2 hr. By

itself, it takes the small drain 3 hr longer to empty the tank than it takes the large drain to empty the tank on its own. How much time would it take for each drain to empty the pool on its own?   large drain: 3 hr; small drain: 6 hr

82) Working together, a professor and her teaching assistant can grade a set of exams in 1.2 hr. On her own, the professor can grade the tests 1 hr faster than the teaching assistant can grade them on her own. How long would it take for each person to grade the test by herself?
professor: 2 hr; teaching assistant: 3 hr

83) Miguel took his son to college in Boulder, Colorado, 600 mi from their hometown. On his way home, he was slowed by a snowstorm so that his speed was 10 mph less than when he was driving to Boulder. His total driving time was 22 hr. How fast did Miguel drive on each leg of the trip?   to Boulder: 60 mph; going home: 50 mph

84) Nariko was training for a race and went out for a run. Her speed was 2 mph faster during the first 6 mi than it was for the last 3 mi. If her total running time was $1\frac{3}{4}$ hr, what was her speed on each part of the run?
first 6 mi: 6 mph; last 3 mi: 4 mph

Section 9.4 Formulas and Applications

Objectives

1. **Solve a Formula for a Variable**
2. **Solve an Applied Problem Involving Volume**
3. **Solve an Applied Problem Involving Area**
4. **Solve an Applied Problem Using a Quadratic Equation**

Sometimes, solving a formula for a variable involves using one of the techniques we've learned for solving a quadratic equation or for solving an equation containing a radical.

1. Solve a Formula for a Variable

Example 1

Solve $v = \sqrt{\dfrac{300VP}{m}}$ for m.

Solution

Put a box around the m. The goal is to get m on a side by itself.

$$v = \sqrt{\frac{300VP}{\boxed{m}}}$$

$$v^2 = \frac{300VP}{\boxed{m}} \qquad \text{Square both sides.}$$

In-Class Example 1

Solve $b = \sqrt{\dfrac{40ac}{d}}$ for d.

answer: $d = \dfrac{40ac}{b^2}$

Since we are solving for m and it is in the denominator, multiply both sides by m to eliminate the denominator.

$$\boxed{m}\, v^2 = 300VP \qquad \text{Multiply both sides by } m.$$

$$m = \frac{300VP}{v^2} \qquad \text{Divide both sides by } v^2.$$

You Try 1

Solve $v = \sqrt{\dfrac{2E}{m}}$ for m.

We may need to use the quadratic formula to solve a formula for a variable. Compare the following equations. Each equation is *quadratic in x* because each is written in the form $ax^2 + bx + c = 0$.

$$8x^2 + 3x - 2 = 0 \qquad \text{and} \qquad 8x^2 + tx - z = 0$$
$$a = 8 \quad b = 3 \quad c = -2 \qquad\qquad a = 8 \quad b = t \quad c = -z$$

To solve the equations for x, we can use the quadratic formula.

Example 2

In-Class Example 2
Solve for d.
a) $3d^2 + d - 5 = 0$
b) $3d^2 + cd - b = 0$
answer:
a) $\left\{\dfrac{-1 \pm \sqrt{61}}{6}\right\}$
b) $\left\{\dfrac{-c \pm \sqrt{c^2 + 12b}}{6}\right\}$

Solve for x.

a) $8x^2 + 3x - 2 = 0$ \qquad b) $8x^2 + tx - z = 0$

Solution

a) $8x^2 + 3x - 2$ does not factor, so we will solve using the quadratic formula.

$$8x^2 + 3x - 2 = 0$$

$$a = 8 \quad b = 3 \quad c = -2$$

$$x = \frac{-3 \pm \sqrt{(3)^2 - 4(8)(-2)}}{2(8)} = \frac{-3 \pm \sqrt{9 + 64}}{16} = \frac{-3 \pm \sqrt{73}}{16}$$

The solution set is $\left\{\dfrac{-3 - \sqrt{73}}{16}, \dfrac{-3 + \sqrt{73}}{16}\right\}$.

b) Solve $8x^2 + tx - z = 0$ for x using the quadratic formula.

$$a = 8 \qquad b = t \qquad c = -z$$

$$x = \frac{-t \pm \sqrt{t^2 - 4(8)(-z)}}{2(8)} \qquad\qquad x = \frac{-b \pm \sqrt{b^2 - 4ac}}{2a}$$

$$= \frac{-t \pm \sqrt{t^2 + 32z}}{16} \qquad\qquad \text{Perform the operations.}$$

The solution set is $\left\{\dfrac{-t - \sqrt{t^2 + 32z}}{16}, \dfrac{-t + \sqrt{t^2 + 32z}}{16}\right\}$.

You Try 2

Solve for n.

a) $3n^2 + 5n - 1 = 0$ \qquad b) $3n^2 + pn - r = 0$

2. Solve an Applied Problem Involving Volume

Example 3

In-Class Example 3
A rectangular piece of cardboard is 7 in. longer than it is wide. A square piece that measures 2 in. on each side is cut from each corner, then the sides are turned up to make an uncovered box with volume 396 in³. Find the length and width of the original piece of cardboard.
answer: $w = 15$ in., $l = 22$ in.

A rectangular piece of cardboard is 5 in. longer than it is wide. A square piece that measures 2 in. on each side is cut from each corner, then the sides are turned up to make an uncovered box with volume 252 in³. Find the length and width of the original piece of cardboard.

Solution

Step 1: **Read** the problem carefully. Draw a picture.

Step 2: **Choose a variable** to represent the unknown, and define the other unknown in terms of this variable.

$$\text{Let} \quad x = \text{the width of the cardboard}$$
$$x + 5 = \text{the length of the cardboard}$$

Step 3: **Translate** the information that appears in English into an algebraic equation.

The volume of a box is (length)(width)(height). We will use the formula (length)(width)(height) = 252.

Original Cardboard   **Box**

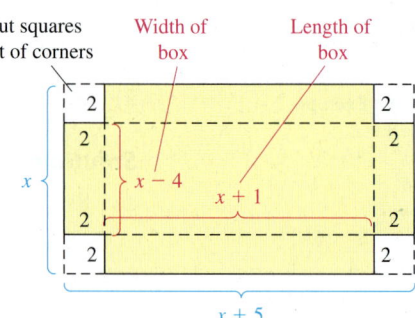

The figure on the left shows the original piece of cardboard with the sides labeled. The figure on the right illustrates how to label the box when the squares are cut out of the corners. When the sides are folded along the dotted lines, we must label the length, width, and height of the box.

$$\text{Length of box} = \begin{matrix}\text{Length of original}\\ \text{cardboard}\end{matrix} - \begin{matrix}\text{Length of}\\ \text{side cut out}\\ \text{on the left}\end{matrix} - \begin{matrix}\text{Length of}\\ \text{side cut out}\\ \text{on the right}\end{matrix}$$

$$= \quad x + 5 \quad - \quad 2 \quad - \quad 2$$
$$= \quad x + 1$$

$$\text{Width of box} = \begin{matrix}\text{Width of original}\\ \text{cardboard}\end{matrix} - \begin{matrix}\text{Length of}\\ \text{side cut out}\\ \text{on top}\end{matrix} - \begin{matrix}\text{Length of}\\ \text{side cut out}\\ \text{on bottom}\end{matrix}$$

$$= \quad x \quad - \quad 2 \quad - \quad 2$$
$$= \quad x - 4$$

$$\text{Height of box} = \text{Length of side cut out}$$
$$= \quad 2$$

Statement: Volume of box = (length)(width)(height)

Equation: $\quad 252 = (x + 1)(x - 4)(2)$

Step 4: **Solve** the equation.

$$252 = (x + 1)(x - 4)(2)$$
$$126 = (x + 1)(x - 4) \qquad \text{Divide both sides by 2.}$$
$$126 = x^2 - 3x - 4 \qquad \text{Multiply.}$$
$$0 = x^2 - 3x - 130 \qquad \text{Write in standard form.}$$
$$0 = (x + 10)(x - 13) \qquad \text{Factor.}$$
$$x + 10 = 0 \quad \text{or} \quad x - 13 = 0 \qquad \text{Set each factor equal to zero.}$$
$$x = -10 \quad \text{or} \qquad x = 13 \qquad \text{Solve.}$$

Step 5: **Check** the answer and **interpret** the solution as it relates to the problem.

Because x represents the width, it cannot be negative. Therefore, the width of the original piece of cardboard is 13 in.

The length of the cardboard is $x + 5$, so $13 + 5 = 18$ in.

Width of cardboard = 13 in.   Length of cardboard = 18 in.

Check:

Width of box = $13 - 4 = 9$ in.; Length of box = $13 + 1 = 14$ in.;
Height of box = 2 in.
Volume of box = $9(14)(2) = 252$ in^3. ■

You Try 3

The width of a rectangular piece of cardboard is 2 in. less than its length. A square piece that measures 3 in. on each side is cut from each corner, then the sides are turned up to make a box with volume 504 in^3. Find the length and width of the original piece of cardboard.

3. Solve an Applied Problem Involving Area

Example 4

A rectangular pond is 20 ft long and 12 ft wide. The pond is bordered by a strip of grass of uniform (the same) width. The area of the grass is 320 ft^2. How wide is the border of grass around the pond?

In-Class Example 4
A rectangular pond is 32 ft long and 21 ft wide. The pond is bordered by a sidewalk of uniform (the same) width. The area of the sidewalk is 228 ft^2. How wide is the sidewalk around the pond?
answer: 2 ft

Solution

Step 1: **Read** the problem carefully. Draw a picture.

Step 2: **Choose a variable** to represent the unknown, and define the other unknowns in terms of this variable.

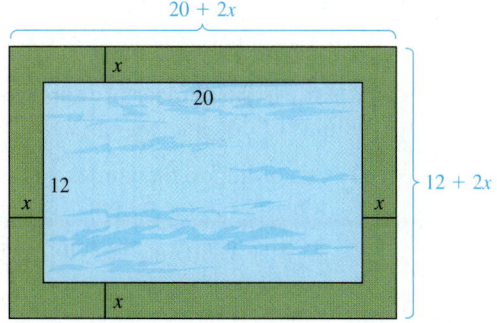

x = width of the strip of grass
$20 + 2x$ = length of pond plus two strips of grass
$12 + 2x$ = width of pond plus two strips of grass

Step 3: **Translate** from English into an algebraic equation.

We know that the area of the grass border is 320 ft². We can calculate the area of the pond since we know its length and width. The pond plus grass border forms a large rectangle of length $20 + 2x$ and width $12 + 2x$. The equation will come from the following relationship:

$$\textit{Statement:} \quad \begin{array}{c} \text{Area of pond} \\ \text{plus grass} \end{array} - \begin{array}{c} \text{Area of} \\ \text{pond} \end{array} = \begin{array}{c} \text{Area of} \\ \text{grass border} \end{array}$$

Equation: $(20 + 2x)(12 + 2x) - 20(12) = 320$

Step 4: **Solve** the equation.

$$
\begin{array}{ll}
(20 + 2x)(12 + 2x) - 20(12) = 320 & \\
240 + 64x + 4x^2 - 240 = 320 & \text{Multiply.} \\
4x^2 + 64x = 320 & \text{Combine like terms.} \\
x^2 + 16x = 80 & \text{Divide by 4.} \\
x^2 + 16x - 80 = 0 & \text{Write in standard form.} \\
(x + 20)(x - 4) = 0 & \text{Factor.} \\
x + 20 = 0 \quad \text{or} \quad x - 4 = 0 & \text{Set each factor equal to 0.} \\
x = -20 \quad \text{or} \quad x = 4 & \text{Solve.}
\end{array}
$$

Step 5: **Check** the answer and **interpret** the solution as it relates to the problem.

x represents the width of the strip of grass, so x cannot equal -20.

The width of the strip of grass is 4 ft.

Check: Substitute $x = 4$ into the equation written in step 3.

$$
\begin{array}{c}
[20 + 2(4)][12 + 2(4)] - 20(12) \overset{?}{=} 320 \\
(28)(20) - 240 \overset{?}{=} 320 \\
560 - 240 = 320 \ \checkmark
\end{array}
$$

You Try 4

A rectangular pond is 6 ft wide and 10 ft long and is surrounded by a concrete border of uniform width. The area of the border is 80 ft². Find the width of the border.

4. Solve an Applied Problem Using a Quadratic Equation

Example 5

The total tourism-related output in the United States from 2000 to 2004 can be modeled by

$$y = 16.4x^2 - 50.6x + 896$$

where x is the number of years since 2000 and y is the total tourism-related output in billions of dollars. (www.bea.gov)

a) According to the model, how much money was generated in 2002 due to tourism-related output?

b) In what year was the total tourism-related output about $955 billion?

In-Class Example 5
Use existing problem from Example 5.
a) According to the model, how much money was generated in 2005 due to tourism-related output?
b) In what year was the total tourism-related output about $891.8 billion?
answer: a) $1053 billion
b) 2003

Solution

a) Since x is the number of years *after* 2000, the year 2002 corresponds to $x = 2$.

$$y = 16.4x^2 - 50.6x + 896$$
$$y = 16.4(2)^2 - 50.6(2) + 896 \qquad \text{Substitute 2 for } x.$$
$$y = 860.4$$

The total tourism-related output in 2002 was approximately $860.4 billion.

b) Since y represents the total tourism-related output (in billions), substitute 955 for y and solve for x.

$$y = 16.4x^2 - 50.6x + 896$$
$$955 = 16.4x^2 - 50.6x + 896 \qquad \text{Substitute 955 for } y.$$
$$0 = 16.4x^2 - 50.6x - 59 \qquad \text{Write in standard form.}$$

Use the quadratic formula to solve for x.

$$a = 16.4 \quad b = -50.6 \quad c = -59$$

$$x = \frac{50.6 \pm \sqrt{(-50.6)^2 - 4(16.4)(-59)}}{2(16.4)} \qquad \begin{array}{l}\text{Substitute the values into}\\ \text{the quadratic formula.}\end{array}$$

$$x \approx 3.99 \approx 4 \text{ or } x \approx -0.90$$

The negative value of x does not make sense in the context of the problem. Use $x \approx 4$, which corresponds to the year 2004. The total tourism-related output was about $955 billion in 2004.

Answers to You Try Exercises

1) $m = \dfrac{2E}{v^2}$
2) a) $\left\{ \dfrac{-5 - \sqrt{37}}{6}, \dfrac{-5 + \sqrt{37}}{6} \right\}$
b) $\left\{ \dfrac{-p - \sqrt{p^2 + 12r}}{6}, \dfrac{-p + \sqrt{p^2 + 12r}}{6} \right\}$

3) length = 20 in., width = 18 in.
4) 2 ft

9.4 Exercises

*Additional answers can be found in the Answers to Exercises appendix.

Objective 1: Solve a Formula for a Variable

Solve for the indicated variable.

1) $A = \pi r^2$ for $r \qquad r = \dfrac{\pm\sqrt{A\pi}}{\pi}$

2) $V = \dfrac{1}{3}\pi r^2 h$ for $r \qquad r = \dfrac{\pm\sqrt{3\pi V h}}{\pi h}$

3) $a = \dfrac{v^2}{r}$ for $v \qquad v = \pm\sqrt{ar}$

4) $K = \dfrac{1}{2}Iw^2$ for $w \qquad w = \dfrac{\pm\sqrt{2KI}}{I}$

5) $E = \dfrac{I}{d^2}$ for $d \qquad d = \dfrac{\pm\sqrt{IE}}{E}$

6) $L = \dfrac{2U}{I^2}$ for $I \qquad I = \dfrac{\pm\sqrt{2UL}}{L}$

7) $F = \dfrac{kq_1q_2}{r^2}$ for $r \qquad r = \dfrac{\pm\sqrt{kq_1q_2F}}{F}$

8) $E = \dfrac{kq}{r^2}$ for $r \qquad r = \dfrac{\pm\sqrt{kqE}}{E}$

9) $d = \sqrt{\dfrac{4A}{\pi}}$ for A

10) $d = \sqrt{\dfrac{12V}{\pi h}}$ for V

11) $T_p = 2\pi\sqrt{\dfrac{l}{g}}$ for l

12) $V = \sqrt{\dfrac{3RT}{M}}$ for T

13) $T_p = 2\pi\sqrt{\dfrac{l}{g}}$ for g

14) $V = \sqrt{\dfrac{3RT}{M}}$ for M

15) Compare the equations $3x^2 - 5x + 4 = 0$ and $rx^2 + 5x + s = 0$.

    a) How are the equations alike?

    b) How can both equations be solved for x?

16) What method could be used to solve $2t^2 + 7t + 1 = 0$ and $kt^2 + mt + n = 0$ for t? Why?

Solve for the indicated variable.

17) $rx^2 - 5x + s = 0$ for $x \qquad x = \dfrac{5 \pm \sqrt{25 - 4rs}}{2r}$

18) $cx^2 + dx - 3 = 0$ for $x \qquad x = \dfrac{-d \pm \sqrt{d^2 + 12c}}{2c}$

19) $pz^2 + rz - q = 0$ for $z \qquad z = \dfrac{-r \pm \sqrt{r^2 + 4pq}}{2p}$

20) $hr^2 - kr + j = 0$ for $r \qquad r = \dfrac{k \pm \sqrt{k^2 - 4hj}}{2h}$

21) $da^2 - ha = k$ for a $a = \dfrac{h \pm \sqrt{h^2 + 4dk}}{2d}$

22) $kt^2 + mt = -n$ for t $t = \dfrac{-m \pm \sqrt{m^2 - 4kn}}{2k}$

23) $s = \dfrac{1}{2}gt^2 + vt$ for t $t = \dfrac{-v \pm \sqrt{v^2 + 2gs}}{g}$

24) $s = 2\pi rh + \pi r^2$ for r $r = \dfrac{-\pi h \pm \sqrt{\pi(\pi h^2 + s)}}{\pi}$

Mixed Exercises: Objectives 2 and 3

Write an equation and solve.

25) The length of a rectangular piece of sheet metal is 3 in. longer than its width. A square piece that measures 1 in. on each side is cut from each corner, then the sides are turned up to make a box with volume 70 in³. Find the length and width of the original piece of sheet metal.   length = 12 in., width = 9 in.

26) The width of a rectangular piece of cardboard is 8 in. less than its length. A square piece that measures 2 in. on each side is cut from each corner, then the sides are turned up to make a box with volume 480 in³. Find the length and width of the original piece of cardboard.   length = 24 in., width = 16 in.

27) A rectangular swimming pool is 60 ft wide and 80 ft long. A nonskid surface of uniform width is to be installed around the pool. If there is 576 ft² of the nonskid material, how wide can the strip of the nonskid surface be?   2 ft

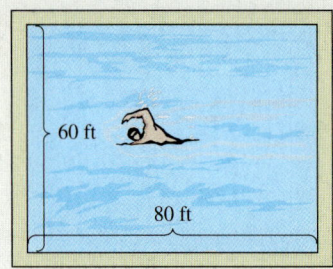

28) A picture measures 10 in. by 12 in. Emilio will get it framed with a border around it so that the total area of the picture plus the frame of uniform width is 168 in². How wide is the border?   1 in.

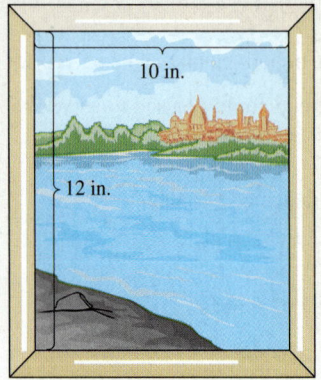

29) The height of a triangular sail is 1 ft less than twice the base of the sail. Find its height and the length of its base if the area of the sail is 60 ft².   base = 8 ft, height = 15 ft

30) Chandra cuts fabric into isosceles triangles for a quilt. The height of each triangle is 1 in. less than the length of the base. The area of each triangle is 15 in². Find the height and base of each triangle.   base = 6 in., height = 5 in.

31) Valerie makes a bike ramp in the shape of a right triangle. The base of the ramp is 4 in. more than twice its height, and the length of the incline is 4 in. less than three times its height. How high is the ramp?   10 in.

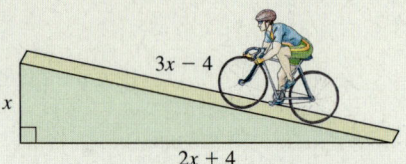

32) The width of a widescreen TV is 10 in. less than its length. The diagonal of the rectangular screen is 10 in. more than the length. Find the length and width of the screen.   length = 40 in., width = 30 in.

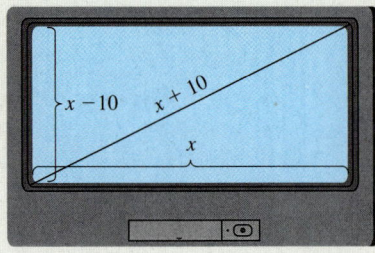

Objective 4: Solve an Applied Problem Using a Quadratic Equation

Solve.

33) An object is propelled upward from a height of 4 ft. The height h of the object (in feet) t sec after the object is released is given by

$$h = -16t^2 + 60t + 4$$

a) How long does it take the object to reach a height of 40 ft?   0.75 sec on the way up, 3 sec on the way down

b) How long does it take the object to hit the ground? $\dfrac{15 + \sqrt{241}}{8}$ sec or about 3.8 sec

34) An object is launched from the ground. The height h of the object (in feet) t sec after the object is released is given by

$$h = -16t^2 + 64t$$

When will the object be 48 ft in the air?

after 1 sec and 3 sec

35) Attendance at Broadway plays from 1996 to 2000 can be modeled by

$$y = -0.25x^2 + 1.5x + 9.5$$

where x represents the number of years after 1996 and y represents the number of people who attended a Broadway play (in millions).
(*Statistical Abstracts of the United States*)

a) Approximately how many people saw a Broadway play in 1996?   9.5 million

b) In what year did approximately 11.75 million people see a Broadway play?   1999

36) The illuminance E (measure of the light emitted, in lux) of a light source is given by

$$E = \frac{I}{d^2}$$

where I is the luminous intensity (measured in candela) and d is the distance, in meters, from the light source. The luminous intensity of a lamp is 2700 candela at a distance of 3 m from the lamp. Find the illuminance, E, in lux.   300 lux

37) A sandwich shop has determined that the demand for its turkey sandwich is $\dfrac{65}{P}$ per day, where P is the price of the sandwich in dollars. The daily supply is given by $10P + 3$. Find the price at which the demand for the sandwich equals the supply. $2.40

38) A hardware store determined that the demand for shovels one winter was $\dfrac{2800}{P}$, where P is the price of the shovel in dollars. The supply was given by $12P + 32$. Find the price at which demand for the shovels equals the supply. $14.00

Use the following formula for Exercises 39 and 40.

A wire is stretched between two poles separated by a distance d, and a weight is in the center of the wire of length L so that the wire is pulled taut as pictured here. The vertical distance, D, between the weight on the wire and the top of the poles is given by $D = \dfrac{\sqrt{L^2 - d^2}}{2}$.

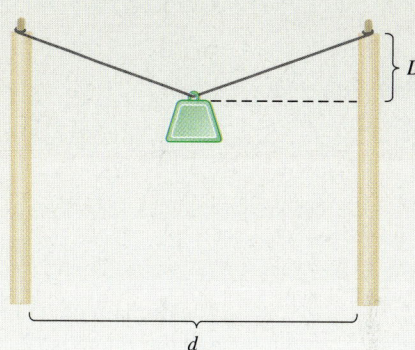

39) A 12.5-ft clothesline is attached to the top of two poles that are 12 ft apart. A shirt is hanging in the middle of the clothesline. Find the distance, D, that the shirt is hanging down.   1.75 ft

40) An 11-ft wire is attached to a ceiling in a loft apartment by hooks that are 10 ft apart. A light fixture is hanging in the middle of the wire. Find the distance, D, between the ceiling and the top of the light fixture. Round the answer to the nearest tenths place.   2.3 ft

Section 9.5 Quadratic Functions and Their Graphs

Objectives

1. Graph a Quadratic Function by Shifting the Graph of $f(x) = x^2$
2. Graph $f(x) = a(x - h)^2 + k$ Using Characteristics of a Parabola
3. Graph $f(x) = ax^2 + bx + c$ by Completing the Square
4. Graph $f(x) = ax^2 + bx + c$ Using $\left(-\dfrac{b}{2a}, f\left(-\dfrac{b}{2a}\right)\right)$

1. Graph a Quadratic Function by Shifting the Graph of $f(x) = x^2$

We were introduced to quadratic functions in Chapter 6, and in this chapter we have learned different methods for solving quadratic equations. In this section, we will learn how to graph quadratic functions. Let's begin with the definition of a quadratic function.

> **Definition**
>
> A **quadratic function** is a function that can be written in the form
>
> $$f(x) = ax^2 + bx + c$$
>
> where a, b, and c are real numbers and $a \neq 0$. An example is $f(x) = x^2 + 6x + 10$. The domain of a quadratic function is $(-\infty, \infty)$.

The simplest form of a quadratic function is $f(x) = x^2$. Let's graph this function as well as a similar function, $g(x) = x^2 + 2$.

Example 1

Graph $f(x) = x^2$ and $g(x) = x^2 + 2$ on the same axes.

In-Class Example 1
Graph $f(x) = x^2$ and
$g(x) = x^2 + 4$ on the same axes.
answer:
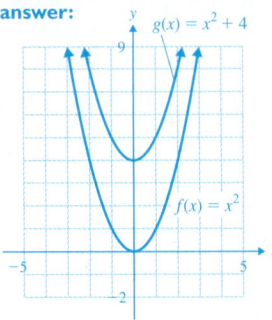

Solution

We will make a table of values for each function, then plot the points.

$f(x) = x^2$	
x	$f(x)$
0	0
1	1
2	4
−1	1
−2	4

$g(x) = x^2 + 2$	
x	$g(x)$
0	2
1	3
2	6
−1	3
−2	6

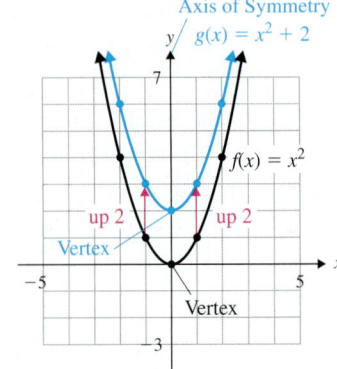

The domain of $f(x)$ is $(-\infty, \infty)$, and the range is $[0, \infty)$. The domain of $g(x)$ is $(-\infty, \infty)$, and the range is $[2, \infty)$.

Definition

The graph of a quadratic function is called a **parabola.** The lowest point on a parabola that opens upward or the highest point on a parabola that opens downward is called the **vertex.**

The vertex of the graph of $f(x)$ in Example 1 is $(0, 0)$, and the vertex of the graph of $g(x)$ is $(0, 2)$.

Every parabola has symmetry. Let's look at the graph of $f(x) = x^2$. If we were to fold the paper along the y-axis, one half of the graph of $f(x) = x^2$ would fall exactly on the other half. The y-axis, or the line $x = 0$, is the **axis of symmetry** of $f(x) = x^2$. [It is also true that the line $x = 0$ is the axis of symmetry of $g(x) = x^2 + 2$.]

We can see from the tables of values in Example 1 that although the x-values are the same in each table, the corresponding y-values in the table for $g(x)$ are *2 more than* the y-values in the first table. In other words, if $f(x) = x^2$, then $g(x) = x^2 + 2$, so $g(x) = f(x) + 2$.

The y-coordinates of the ordered pairs of $g(x)$ are *2 more than* the y-coordinates of the ordered pairs of $f(x)$ when the ordered pairs of f and g have the same x-coordinates. This means that **the graph of g is the same shape as the graph of f, but g is shifted up 2 units.**

We can make the following general statement about shifting the graph of a function vertically.

Property   Vertical Shifts

Given the graph of $f(x)$, if $g(x) = f(x) + k$, where k is a constant, then the graph of $g(x)$ is the same shape as the graph of $f(x)$ but g is shifted **vertically** k units.

 **You Try 1**

Graph $g(x) = x^2 + 1.$

Now let's look at how we can shift the parabola $f(x) = x^2$ horizontally.

Example 2

Graph $f(x) = x^2$ and $g(x) = (x + 3)^2$ on the same axes.

In-Class Example 2
Graph $f(x) = x^2$ and $g(x) = (x - 2)^2$ on the same axes.
answer:

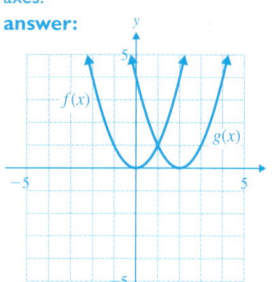

Solution

$f(x) = x^2$	
x	$f(x)$
0	0
1	1
2	4
−1	1
−2	4

$g(x) = (x + 3)^2$	
x	$g(x)$
−3	0
−2	1
−1	4
−4	1
−5	4

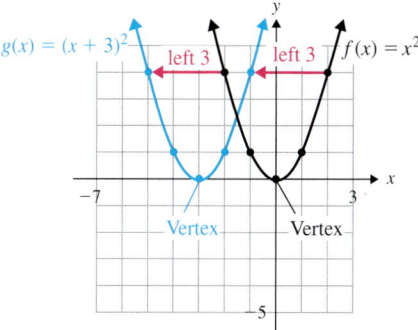

The functions $f(x)$ and $g(x)$ each have a domain of $(-\infty, \infty)$. Each has a range of $[0, \infty)$.

Notice that the y-values are the same in each table. The corresponding x-values in the table for $g(x)$, however, are *3 less than* the x-values in the first table.

The x-coordinates of the ordered pairs of $g(x)$ are *3 less than* the x-coordinates of the ordered pairs of $f(x)$ when the ordered pairs of f and g have the same y-coordinates. This means that **the graph of g is the same shape as the graph of f, but g is shifted left 3 units.** ■

Property   Horizontal Shifts

Given the graph of $f(x)$, if $g(x) = f(x - h)$, where h is a constant, then the graph of $g(x)$ is the same shape as the graph of $f(x)$ but g is shifted **horizontally** h units.

We can think of Example 2 in terms of this horizontal shift. Since $f(x) = x^2$ and $g(x) = (x + 3)^2$, $h = -3$ in $g(x)$ since we can think of $g(x)$ as $g(x) = (x - (-3))^2$.

The graph of g is the same shape as the graph of f but g is shifted -3 units horizontally or 3 units to the **left.**

Note

This vertical and horizontal shifting works for any function, not just quadratic functions.

You Try 2

Graph $g(x) = (x + 4)^2$.

Next we will learn about reflecting the graph of $f(x) = x^2$ about the x-axis.

Example 3

In-Class Example 3
Use given example.

Graph $f(x) = x^2$ and $g(x) = -x^2$ on the same axes.

Solution

$f(x) = x^2$	
x	**f(x)**
0	0
1	1
2	4
−1	1
−2	4

$g(x) = -x^2$	
x	**g(x)**
0	0
1	−1
2	−4
−1	−1
−2	−4

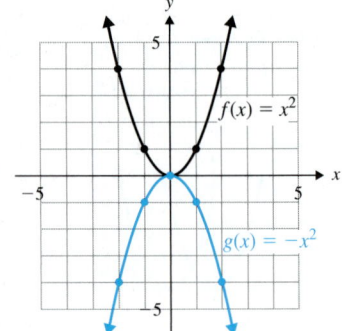

The tables of values show us that although the x-values are the same in each table, the corresponding y-values in the table for $g(x)$ are the *negatives* of the y-values in the first table. With the exception of the vertex, all of the y-coordinates of the points on the graph of g are negative. That is why the graph of $g(x) = -x^2$ is below the x-axis.

Each function has a domain of $(-\infty, \infty)$. The range of $f(x)$ is $[0, \infty)$, and the range of $g(x)$ is $(-\infty, 0]$.

We say that *the graph of g is the reflection of the graph of f about the x-axis.* (The graph of g is the mirror image of the graph of f.)

Property   Reflection about the *x*-axis

Given the graph of any function $f(x)$, if $g(x) = -f(x)$ then the graph of $g(x)$ will be the **reflection of the graph of f about the x-axis.** That is, obtain the graph of g by keeping the x-coordinate of each point on f the same but take the negative of the y-coordinate.

You Try 3

Graph $g(x) = -(x + 2)^2$.

Example 4

In-Class Example 4
Graph $g(x) = (x + 3)^2 + 3$.
answer:

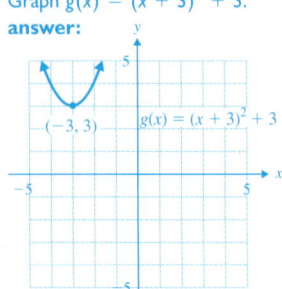

Graph $g(x) = (x - 2)^2 - 1$.

Solution

If we compare $g(x)$ to $f(x) = x^2$, what do the constants in $g(x)$ tell us about transforming the graph of $f(x)$?

$$g(x) = (x - 2)^2 - 1$$

Shift $f(x)$ right 2.   Shift $f(x)$ down 1.

Sketch the graph of $f(x) = x^2$, then move every point on the graph of f right 2 and down 1 to obtain the graph of $g(x)$. This moves the vertex from $(0, 0)$ to $(2, -1)$. Notice that the axis of symmetry of $g(x)$ moves 2 units to the right also. Its equation is $x = 2$. The domain of $g(x)$ is $(-\infty, \infty)$; the range is $[-1, \infty)$.

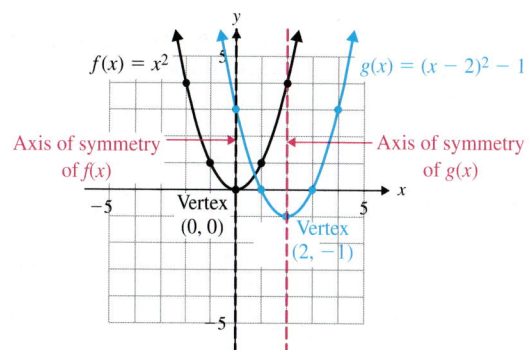

2. Graph $f(x) = a(x - h)^2 + k$ Using Characteristics of a Parabola

When a quadratic function is in the form $f(x) = a(x - h)^2 + k$, we can read the vertex directly from the equation. Furthermore, the value of a tells us if the parabola opens upward or downward and whether the graph is narrower, wider, or the same width as $y = x^2$.

Procedure   Graphing a Quadratic Function of the Form $f(x) = a(x - h)^2 + k$

1) The vertex of the parabola is (h, k).

2) The axis of symmetry is the vertical line with equation $x = h$.

3) If a is positive, the parabola opens upward.

   If a is negative, the parabola opens downward.

4) If $|a| < 1$, then the graph of $f(x) = a(x - h)^2 + k$ is *wider* than the graph of $y = x^2$.

   If $|a| > 1$, then the graph of $f(x) = a(x - h)^2 + k$ is *narrower* than the graph of $y = x^2$.

   If $a = 1$ or $a = -1$, the graph is the *same* width as $y = x^2$.

Example 5

In-Class Example 5

Graph $y = \dfrac{2}{3}(x - 4)^2 + 1$. Also, find the x- and y-intercepts.

answer: no x-ints; y-int:

$\left(0, \dfrac{35}{3}\right)$

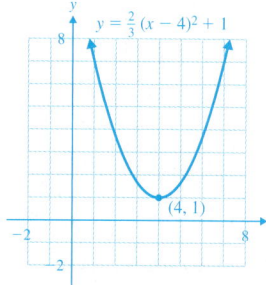

Graph $f(x) = 2(x + 1)^2 - 4$. Also find the x- and y-intercepts.

Solution

Here is the information we can get from the equation:

1) $h = -1$ and $k = -4$. The vertex is $(-1, -4)$.

2) The axis of symmetry is $x = -1$.

3) $a = 2$. Since a is positive, the parabola opens upward.

4) Since $|a| > 1$, the graph of $f(x) = 2(x + 1)^2 - 4$ is *narrower* than the graph of $f(x) = x^2$.

To graph the function, start by putting the vertex on the axes. Then, choose a couple of values of x to the left or right of the vertex to plot more points. Use the axis of symmetry to find the points $(-2, -2)$ and $(-3, 4)$ on the graph of $f(x) = 2(x + 1)^2 - 4$.

x	y
0	-2
1	4

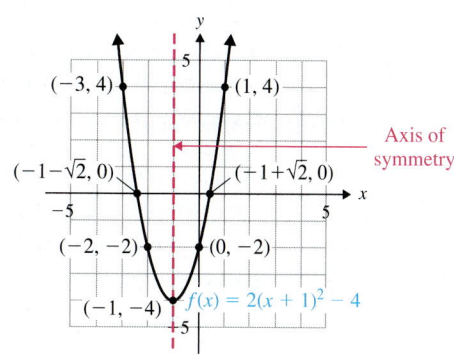

We can read the *y-intercept* from the graph: $(0, -2)$. To find the *x-intercepts*, let $f(x) = 0$ and solve for *x*.

$$f(x) = 2(x + 1)^2 - 4$$
$$0 = 2(x + 1)^2 - 4 \qquad \text{Substitute 0 for } f(x).$$
$$4 = 2(x + 1)^2 \qquad \text{Add 4.}$$
$$2 = (x + 1)^2 \qquad \text{Divide by 2.}$$
$$\pm\sqrt{2} = x + 1 \qquad \text{Square root property}$$
$$-1 \pm \sqrt{2} = x \qquad \text{Add } -1.$$

The *x*-intercepts are $(-1 - \sqrt{2}, 0)$ and $(-1 + \sqrt{2}, 0)$. The domain is $(-\infty, \infty)$; the range is $[-4, \infty)$.

You Try 4

Graph $f(x) = 2(x - 1)^2 - 2$. Also find the *x*- and *y*-intercepts.

When a quadratic function is written in the form $f(x) = ax^2 + bx + c$, there are two methods we can use to graph the function.

Procedure   Graphing Parabolas from the Form $f(x) = ax^2 + bx + c$

There are two methods we can use to graph the function $f(x) = ax^2 + bx + c$.

Method 1:   Rewrite $f(x) = ax^2 + bx + c$ in the form $f(x) = a(x - h)^2 + k$ by *completing the square*.

Method 2:   Use the formula $x = -\dfrac{b}{2a}$ to find the *x*-coordinate of the vertex. Then, the vertex has coordinates $\left(-\dfrac{b}{2a}, f\left(-\dfrac{b}{2a} \right) \right)$.

We will begin with Method 1. We will modify the steps we used in Section 9.1 to solve quadratic equations by completing the square.

3. Graph $f(x) = ax^2 + bx + c$ by Completing the Square

Procedure   Rewriting $f(x) = ax^2 + bx + c$ in the Form $f(x) = a(x - h)^2 + k$ by Completing the Square

Step 1:   **The coefficient of the square term must be 1.** If it is not 1, multiply or divide both sides of the equation (*including f(x)*) by the appropriate value to obtain a leading coefficient of 1.

Step 2:   **Separate the constant from the terms containing the variables by grouping the variable terms with parentheses.**

Step 3:   **Complete the square for the quantity in the parentheses.** Find half of the linear coefficient, then square the result. *Add* that quantity inside the parentheses and *subtract* the quantity from the constant. (Adding and subtracting the same number on the same side of an equation is like adding 0 to the equation.)

Step 4:   **Factor the expression inside the parentheses.**

Step 5:   **Solve for $f(x)$.**

Example 6

In-Class Example 6
Graph each function. Begin by completing the square to rewrite each function in the form $f(x) = a(x - h)^2 + k$.

a) $f(x) = x^2 - 4x + 10$

b) $f(x) = \dfrac{1}{2}x^2 - 8x + 29$

answer:

a) $f(x) = (x - 2)^2 + 6$

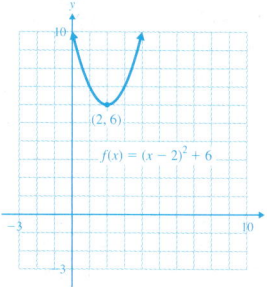

b) $f(x) = \dfrac{1}{2}(x - 8)^2 - 3$

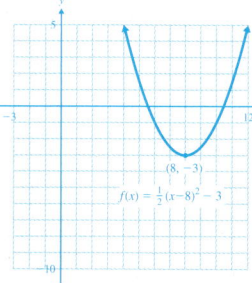

Graph each function. Begin by completing the square to rewrite each function in the form $f(x) = a(x - h)^2 + k$. Include the intercepts.

a) $f(x) = x^2 + 6x + 10$ b) $g(x) = -\dfrac{1}{2}x^2 + 4x - 6$

Solution

a) **Step 1:** The coefficient of x^2 is 1.

Step 2: Separate the constant from the variable terms using parentheses.

$$f(x) = (x^2 + 6x) + 10$$

Step 3: Complete the square for the quantity in the parentheses.

$$\frac{1}{2}(6) = 3$$

$$3^2 = 9$$

Add 9 inside the parentheses and subtract 9 from the 10. This is like adding 0 to the equation.

$$f(x) = (x^2 + 6x + 9) + 10 - 9$$
$$f(x) = (x^2 + 6x + 9) + 1$$

Step 4: Factor the expression inside the parentheses.

$$f(x) = (x + 3)^2 + 1$$

Step 5: The equation *is* solved for $f(x)$.

From the equation $f(x) = (x + 3)^2 + 1$ we can see that

i)   The vertex is $(-3, 1)$.

ii) The axis of symmetry is $x = -3$.

iii) $a = 1$ so the parabola opens upward.

iv) Since $a = 1$, the graph is the same width as $y = x^2$.

Find some other points on the parabola. Use the axis of symmetry.

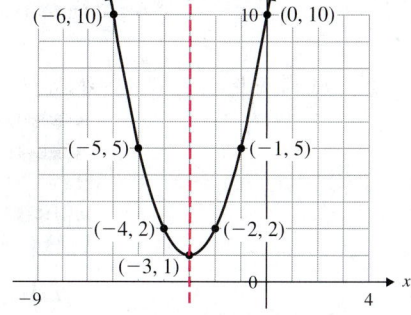

To find the x-intercepts, let $f(x) = 0$ and solve for x. Use *either* form of the equation. We will use $f(x) = (x + 3)^2 + 1$.

x	f(x)
-2	2
-1	5

$$0 = (x + 3)^2 + 1 \qquad \text{Let } f(x) = 0.$$
$$-1 = (x + 3)^2 \qquad \text{Subtract 1.}$$
$$\pm\sqrt{-1} = x + 3 \qquad \text{Square root property}$$
$$-3 \pm i = x \qquad \sqrt{-1} = i; \text{ subtract 3.}$$

Since the solutions to $f(x) = 0$ are *not* real numbers, *there are no x-intercepts.*
To find the y-intercept, let $x = 0$ and solve for $f(0)$.

$$f(x) = (x + 3)^2 + 1$$
$$f(0) = (0 + 3)^2 + 1$$
$$f(0) = 9 + 1 = 10$$

The y-intercept is $(0, 10)$. The domain is $(-\infty, \infty)$, and the range is $[1, \infty)$.

b) ***Step 1:***   The coefficient of x^2 is $-\dfrac{1}{2}$. Multiply both sides of the equation [including the $g(x)$] by -2 so that the coefficient of x^2 will be 1.

$$g(x) = -\frac{1}{2}x^2 + 4x - 6$$

$$-2g(x) = -2\left(-\frac{1}{2}x^2 + 4x - 6\right) \qquad \text{Multiply by } -2.$$

$$-2g(x) = x^2 - 8x + 12 \qquad \text{Distribute.}$$

Step 2:   Separate the constant from the variable terms using parentheses.

$$-2g(x) = (x^2 - 8x) + 12$$

Step 3:   Complete the square for the quantity in parentheses.

$$\frac{1}{2}(-8) = -4$$

$$(-4)^2 = 16$$

Add 16 inside the parentheses and subtract 16 from the 12.

$$-2g(x) = (x^2 - 8x + 16) + 12 - 16$$
$$-2g(x) = (x^2 - 8x + 16) - 4$$

Step 4:   Factor the expression inside the parentheses.

$$-2g(x) = (x - 4)^2 - 4$$

Step 5:   Solve the equation for $g(x)$ by dividing by -2.

$$\frac{-2g(x)}{-2} = \frac{(x-4)^2}{-2} - \frac{4}{-2}$$

$$g(x) = -\frac{1}{2}(x - 4)^2 + 2$$

From $g(x) = -\dfrac{1}{2}(x - 4)^2 + 2$ we can see that

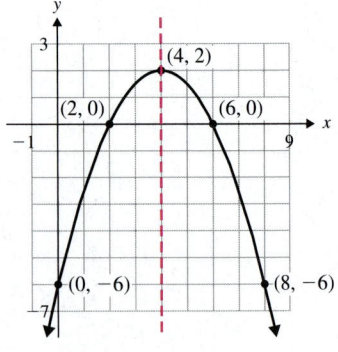

i)   The vertex is $(4, 2)$.

ii) The axis of symmetry is $x = 4$.

iii) $a = -\dfrac{1}{2}$ [the same as in the form

$g(x) = -\dfrac{1}{2}x^2 + 4x - 6$] so the parabola

opens downward.

iv) Since $a = -\dfrac{1}{2}$, the graph of $g(x)$ will be

wider than $y = x^2$.

Find some other points on the parabola.
Use the axis of symmetry.

x	g(x)
6	0
8	-6

Using the axis of symmetry we can see that the x-intercepts are $(6, 0)$ and $(2, 0)$ and that the y-intercept is $(0, -6)$. The domain is $(-\infty, \infty)$; the range is $(-\infty, 2]$.   ■

   You Try 5

Graph each function. Begin by completing the square to rewrite each function in the form $f(x) = a(x - h)^2 + k$. Include the intercepts.

a) $f(x) = x^2 + 4x + 3$ b) $g(x) = -2x^2 + 12x - 8$

4. Graph $f(x) = ax^2 + bx + c$ Using $\left(-\dfrac{b}{2a}, f\left(-\dfrac{b}{2a}\right)\right)$

We can also graph quadratic functions of the form $f(x) = ax^2 + bx + c$ by using the formula $h = -\dfrac{b}{2a}$ to find the x-coordinate of the vertex. This formula comes from completing the square on $f(x) = ax^2 + bx + c$.

Although there is a formula for k, it is only necessary to remember the formula for h. The y-coordinate of the vertex, then, is $k = f\left(-\dfrac{b}{2a}\right)$. The axis of symmetry is $x = h$.

Property   The Vertex Formula

The **vertex** of the graph of $f(x) = ax^2 + bx + c$ $(a \neq 0)$ has coordinates $\left(-\dfrac{b}{2a}, f\left(-\dfrac{b}{2a}\right)\right)$.

Example 7

Graph $f(x) = x^2 - 6x + 3$ using the vertex formula. Include the intercepts.

In-Class Example 7
Graph $f(x) = x^2 - 4x + 4$ using the vertex formula. Include the intercepts.
answer:

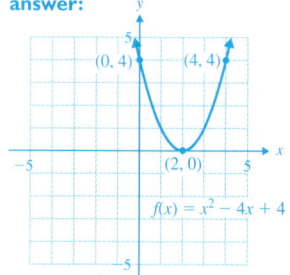

Solution

$a = 1, b = -6, c = 3$. Since $a = +1$, the graph opens upward. The x-coordinate, h, of the vertex is

$$h = -\frac{b}{2a} = -\frac{(-6)}{2(1)} = \frac{6}{2} = 3$$

$h = 3$. Then, the y-coordinate, k, of the vertex is $k = f(3)$.

$$f(x) = x^2 - 6x + 3$$
$$f(3) = 3^2 - 6(3) + 3$$
$$= 9 - 18 + 3 = -6$$

The vertex is $(3, -6)$. The axis of symmetry is $x = 3$.

Find more points on the graph of $f(x) = x^2 - 6x + 3$, then use the axis of symmetry to find other points on the parabola.

x	f(x)
4	−5
5	−2
6	3

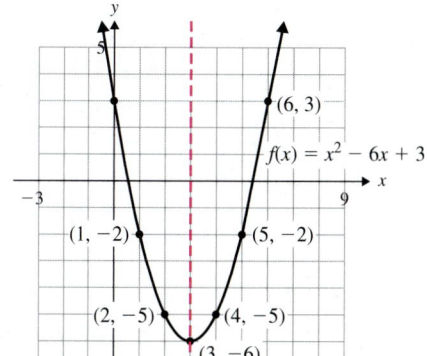

To find the x-intercepts, let $f(x) = 0$ and solve for x.

$$0 = x^2 - 6x + 3$$
$$x = \frac{-(-6) \pm \sqrt{(-6)^2 - 4(1)(3)}}{2(1)} \qquad \text{Solve using the quadratic formula.}$$
$$x = \frac{6 \pm \sqrt{24}}{2} = \frac{6 \pm 2\sqrt{6}}{2} \qquad \text{Simplify.}$$
$$x = 3 \pm \sqrt{6}$$

The x-intercepts are $(3 + \sqrt{6}, 0)$ and $(3 - \sqrt{6}, 0)$.

We can see from the graph that the y-intercept is $(0, 3)$. The domain is $(-\infty, \infty)$; the range is $[-6, \infty)$.

You Try 6

Graph $f(x) = -x^2 - 8x - 13$ using the vertex formula. Include the intercepts.

Using Technology

In Section 6.4 we said that the solutions of the equation $x^2 - x - 6 = 0$ are the x-intercepts of the graph of $y = x^2 - x - 6$.

The x-intercepts are also called the zeros of the equation since they are the values of x that make $y = 0$. Enter $x^2 - x - 6$ in Y_1 then find the x-intercepts shown on the graph by pressing 2nd TRACE and then selecting 2:zero. Move the cursor to the left of an x-intercept using the right arrow key and press ENTER. Move the cursor to the right of the x-intercept using the right arrow key and press ENTER. Move the cursor close to the x-intercept using the left arrow key and press ENTER. Repeat these steps for each x-intercept. The x-intercepts are $(-2, 0)$ and $(3, 0)$ as shown in the graphs to the right.

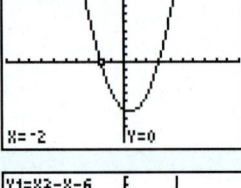

The y-intercept is found by graphing the function and pressing TRACE 0 ENTER. As shown on the graph, the y-intercept for $y = x^2 - x - 6$ is $(0, -6)$.

The x-value of the vertex can be found using the vertex formula. In this case, $a = 1$ and $b = -1$, so $-\dfrac{b}{2a} = \dfrac{1}{2}$. To find the vertex on the graph, press TRACE, type 1/2, and press ENTER. The vertex is shown as $(0.5, -6.25)$ on the graph.

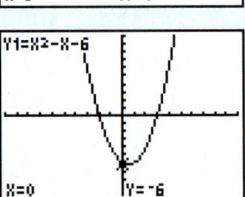

Remember, you can convert the coordinates of the vertex to fractions. Go to the home screen by pressing 2nd MODE. To display the x-value of the vertex, press X,T,θ,n MATH ENTER ENTER. To display the y-value of the vertex, press ALPHA 1 MATH ENTER ENTER. The vertex is then $\left(\dfrac{1}{2}, -\dfrac{25}{4}\right)$.

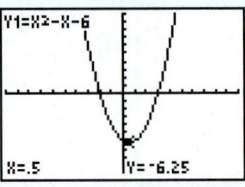

Find the x-intercepts, y-intercept, and vertex using a graphing calculator.

1) $y = x^2 - 2x + 2$ 2) $y = x^2 - 4x - 5$ 3) $y = -(x + 1)^2 + 4$
4) $y = x^2 - 4$ 5) $y = x^2 - 6x + 9$ 6) $y = -x^2 - 8x - 19$

Answers to You Try Exercises

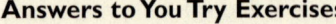

1)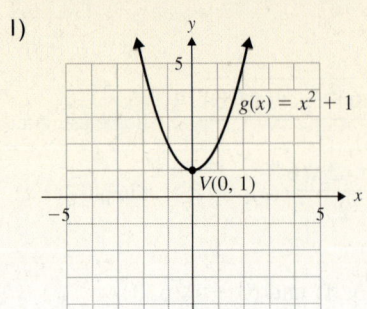

2)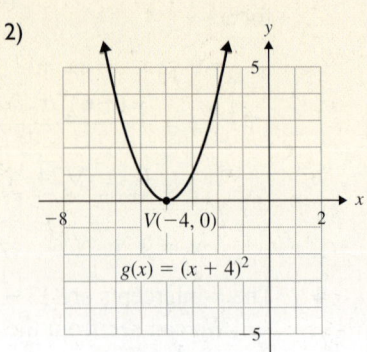

3)

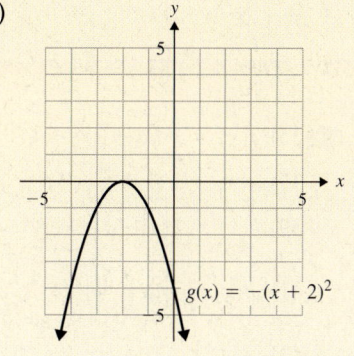

$g(x) = -(x + 2)^2$

4) x-ints: (0, 0), (2, 0); y-int: (0, 0)

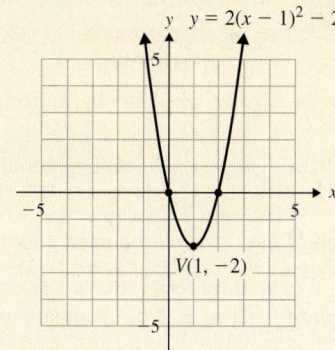

$y = 2(x - 1)^2 - 2$

$V(1, -2)$

5) a) x-ints: $(-3, 0), (-1, 0)$; y-int: $(0, 3)$ b) x-ints: $(3 + \sqrt{5}, 0), (3 - \sqrt{5}, 0)$; y-int: $(0, -8)$

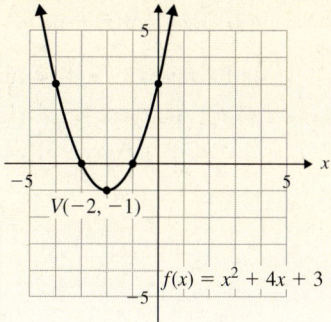

$V(-2, -1)$

$f(x) = x^2 + 4x + 3$

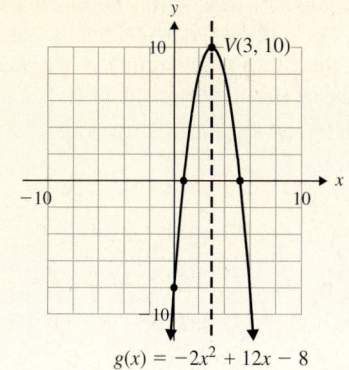

$V(3, 10)$

$g(x) = -2x^2 + 12x - 8$

6) x-ints: $(-4 + \sqrt{3}, 0), (-4 - \sqrt{3}, 0)$; y-int: $(0, -13)$

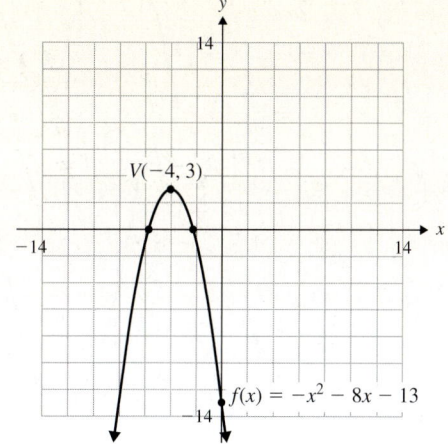

$V(-4, 3)$

$f(x) = -x^2 - 8x - 13$

Answers to Technology Exercises

1) no x-intercepts; y-intercept: (0, 2); vertex: (1, 1)

2) x-intercepts: $(-1, 0), (5, 0)$; y-intercept: $(0, -5)$; vertex: $(2, -9)$

3) x-intercepts: $(-3, 0), (1, 0)$; y-intercept: $(0, 3)$; vertex: $(-1, 4)$

4) x-intercepts: $(-2, 0), (2, 0)$; y-intercept: $(0, -4)$; vertex: $(0, -4)$

5) x-intercept: $(3, 0)$; y-intercept: $(0, 9)$; vertex: $(3, 0)$

6) no x-intercepts; y-intercept: $(0, -19)$; vertex: $(-4, -3)$

9.5 Exercises

*Additional answers can be found in the Answers to Exercises appendix.

Objective 1: Graph a Quadratic Function by Shifting the Graph of $f(x) = x^2$

1) How does the graph of $g(x) = x^2 + 6$ compare to the graph of $f(x) = x^2$? The graph of $g(x)$ is the same shape as the graph of $f(x)$, but $g(x)$ is shifted up 6 units.

2) How does the graph of $h(x) = x^2 - 5$ compare to the graph of $f(x) = x^2$? The graph of $h(x)$ is the same shape as the graph of $f(x)$, but $h(x)$ is shifted down 5 units.

3) How does the graph of $h(x) = (x + 5)^2$ compare to the graph of $f(x) = x^2$? The graph of $h(x)$ is the same shape as the graph of $f(x)$, but $h(x)$ is shifted left 5 units.

4) How does the graph of $g(x) = (x - 4)^2$ compare to the graph of $f(x) = x^2$? The graph of $g(x)$ is the same shape as the graph of $f(x)$, but $g(x)$ is shifted right 4 units.

For Exercises 5–16, sketch the graph of $f(x) = x^2$. Then graph $g(x)$ on the same axes by shifting the graph of $f(x)$.

5) $g(x) = x^2 + 3$ 6) $g(x) = x^2 + 5$

7) $g(x) = x^2 - 4$ 8) $g(x) = x^2 - 1$

9) $g(x) = x^2 - 5$ 10) $g(x) = x^2 + 2$

11) $g(x) = (x + 2)^2$ 12) $g(x) = (x + 1)^2$

13) $g(x) = (x + 5)^2$ 14) $g(x) = (x + 4)^2$

15) $g(x) = (x - 3)^2$ 16) $g(x) = (x - 4)^2$

For Exercises 17–22, graph $f(x) = x^2$ and then graph $g(x)$ on the same axes.

17) $g(x) = -x^2$ 18) $g(x) = -(x - 1)^2$

19) $g(x) = \frac{1}{2}x^2$ 20) $g(x) = 2x^2$

21) $g(x) = (x - 1)^2 - 3$ 22) $g(x) = (x + 2)^2 + 1$

Objective 2: Graph $f(x) = a(x - h)^2 + k$ Using Characteristics of a Parabola

Given a quadratic function of the form $f(x) = a(x - h)^2 + k$,

23) what is the vertex? (h, k)

24) what is the equation of the axis of symmetry? $x = h$

25) how do you know if the parabola opens upward? a is positive.

26) how do you know if the parabola opens downward? a is negative.

27) how do you know if the parabola is narrower than the graph of $y = x^2$? $a > 1$ or $a < -1$

28) how do you know if the parabola is wider than the graph of $y = x^2$? $0 < a < 1$ or $-1 < a < 0$

For each quadratic function, identify the vertex, axis of symmetry, and x- and y-intercepts. Then, graph the function. Determine the domain and range.

29) $f(x) = (x + 1)^2 - 4$ 30) $g(x) = (x - 3)^2 - 1$

31) $g(x) = (x - 2)^2 + 3$ 32) $h(x) = (x + 2)^2 + 7$

33) $y = (x - 4)^2 - 2$ 34) $y = (x + 1)^2 - 5$

35) $f(x) = -(x + 3)^2 + 6$ 36) $g(x) = -(x - 3)^2 + 2$

37) $y = -(x + 1)^2 - 5$ 38) $f(x) = -(x - 2)^2 - 4$

39) $f(x) = 2(x - 1)^2 - 8$ 40) $y = 2(x + 1)^2 - 2$

41) $h(x) = \frac{1}{2}(x + 4)^2$ 42) $g(x) = \frac{1}{4}x^2 - 1$

43) $y = -x^2 + 5$ 44) $h(x) = -(x - 3)^2$

45) $f(x) = -\frac{1}{3}(x + 4)^2 + 3$ 46) $y = -\frac{1}{2}(x - 4)^2 + 2$

47) $g(x) = 3(x + 2)^2 + 5$ 48) $f(x) = 2(x - 3)^2 + 3$

In Exercises 49 and 50, match each function to its graph.

49) $f(x) = x^2 - 3$, $g(x) = (x - 3)^2$, $h(x) = -(x + 3)^2$, $k(x) = -x^2 + 3$

a)
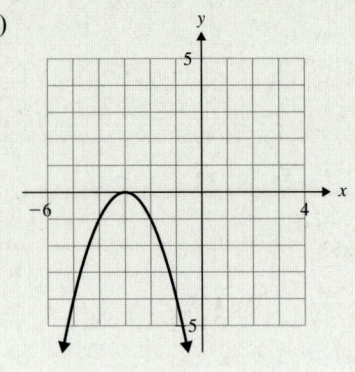
h(x)

b)
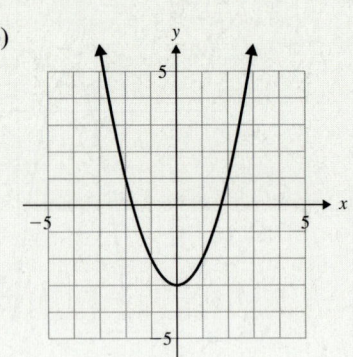
f(x)

c)
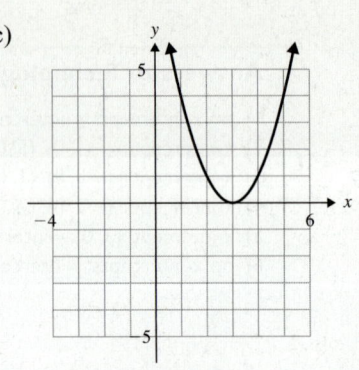
g(x)

d)

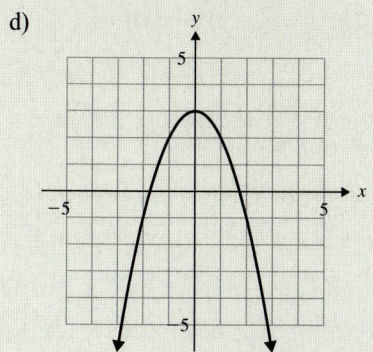

k(x)

50) $f(x) = (x + 1)^2, g(x) = x^2 + 1,$
$h(x) = -(x + 1)^2, k(x) = -x^2 + 1$

a)

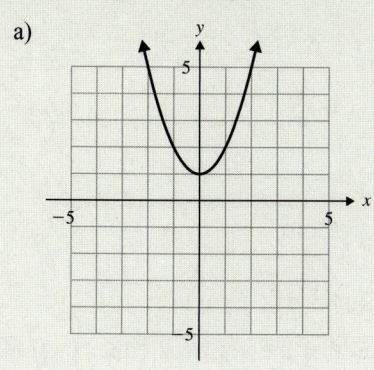

g(x)

b)

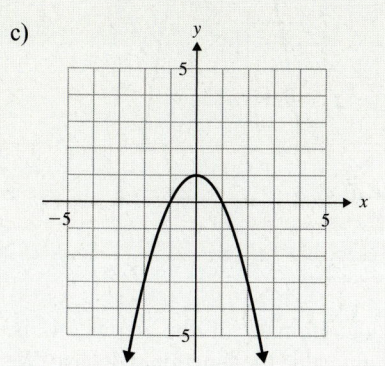

h(x)

c)

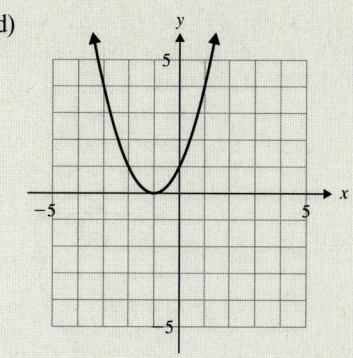

k(x)

d)

f(x)

Exercises 51–56 each present a shift that is performed on the
graph of $f(x) = x^2$ to obtain the graph of $g(x)$. Write the
equation of $g(x)$.

51) $f(x)$ is shifted 8 units to the right. $g(x) = (x - 8)^2$

52) $f(x)$ is shifted down 9 units. $g(x) = x^2 - 9$

53) $f(x)$ is shifted up 3.5 units. $g(x) = x^2 + 3.5$

54) $f(x)$ is shifted left 1.2 units. $g(x) = (x + 1.2)^2$

55) $f(x)$ is shifted left 4 units and down 7 units.
$g(x) = (x + 4)^2 - 7$

56) $f(x)$ is shifted up 2 units and right 5 units.
$g(x) = (x - 5)^2 + 2$

Objective 3: Graph $f(x) = ax^2 + bx + c$ by Completing the Square

Rewrite each function in the form $f(x) = a(x - h)^2 + k$.

Fill It In

Fill in the blanks with either the missing mathematical
step or the reason for the given step.

57) $f(x) = x^2 + 8x + 11$

$f(x) = (x^2 + 8x) + 11$ Group the variable terms
together using parentheses.

$\left[\dfrac{1}{2}(8)\right]^2 = (4)^2 = 16$ Find the number that
completes the square in the
parentheses.

$f(x) = (x^2 + 8x + 16) + 11 - 16$

Add and subtract the number
above to the same side of the
equation.

$f(x) = (x + 4)^2 - 5$ Factor and simplify.

58) $f(x) = x^2 - 4x - 7$
$f(x) = (x^2 - 4x) - 7$ Group the variable terms
together using parentheses.

$\left[\dfrac{1}{2}(-4)\right]^2 = (-2)^2 = 4$ Find the number that completes the square in the
parentheses.

$f(x) = (x^2 - 4x + 4) - 7 - 4$ Add and subtract the
number above to the
same side of the
equation.

$f(x) = (x - 2)^2 - 11$ Factor and simplify.

Rewrite each function in the form $f(x) = a(x - h)^2 + k$ by completing the square. Then, graph the function. Include the intercepts. Determine the domain and range.

(VIDEO) 59) $f(x) = x^2 - 2x - 3$ 60) $g(x) = x^2 + 6x + 8$

61) $y = x^2 + 6x + 7$ 62) $h(x) = x^2 - 4x + 1$

63) $g(x) = x^2 + 4x$ 64) $y = x^2 - 8x + 18$

65) $h(x) = -x^2 - 4x + 5$ 66) $f(x) = -x^2 - 2x + 3$

(VIDEO) 67) $y = -x^2 + 6x - 10$ 68) $g(x) = -x^2 - 4x - 6$

69) $f(x) = 2x^2 - 8x + 4$ 70) $y = 2x^2 - 8x + 2$

71) $g(x) = -\dfrac{1}{3}x^2 - 2x - 9$ 72) $h(x) = -\dfrac{1}{2}x^2 - 3x - \dfrac{19}{2}$

73) $y = x^2 - 3x + 2$ 74) $f(x) = x^2 + 5x + \dfrac{21}{4}$

Objective 4: Graph $f(x) = ax^2 + bx + c$ Using $\left(-\dfrac{b}{2a}, f\left(-\dfrac{b}{2a}\right)\right)$

Graph each function using the vertex formula. Include the intercepts. Determine the domain and range.

75) $y = x^2 + 2x - 3$ 76) $g(x) = x^2 - 6x + 8$

77) $f(x) = -x^2 - 8x - 13$ 78) $y = -x^2 + 2x + 2$

(VIDEO) 79) $g(x) = 2x^2 - 4x + 4$ 80) $f(x) = -4x^2 - 8x - 6$

81) $y = -3x^2 + 6x + 1$ 82) $h(x) = 2x^2 - 12x + 9$

83) $f(x) = \dfrac{1}{2}x^2 - 4x + 5$ 84) $y = \dfrac{1}{2}x^2 + 2x - 3$

85) $h(x) = -\dfrac{1}{3}x^2 - 2x - 5$ 86) $g(x) = \dfrac{1}{5}x^2 - 2x + 8$

Section 9.6 Applications of Quadratic Functions and Graphing Other Parabolas

Objectives

1. Find the Maximum or Minimum Value of a Quadratic Function
2. Given a Quadratic Function, Solve an Applied Problem Involving a Maximum or Minimum Value
3. Write a Quadratic Function to Solve an Applied Problem Involving a Maximum or Minimum Value
4. Graph Parabolas of the Form $x = a(y - k)^2 + h$
5. Rewrite $x = ay^2 + by + c$ as $x = a(y - k)^2 + h$ by Completing the Square
6. Find the Vertex of the Graph of $x = ay^2 + by + c$ Using $y = -\dfrac{b}{2a}$, and Graph the Equation

1. Find the Maximum or Minimum Value of a Quadratic Function

From our work with quadratic functions, we have seen that the vertex is either the lowest point or the highest point on the graph depending on whether the parabola opens upward or downward.

If the parabola opens upward, the vertex is the *lowest* point on the parabola.

If the parabola opens downward, the vertex is the *highest* point on the parabola.

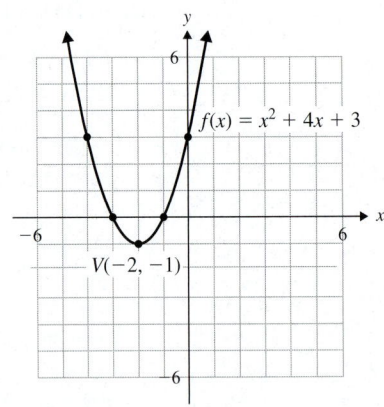

$f(x) = x^2 + 4x + 3$

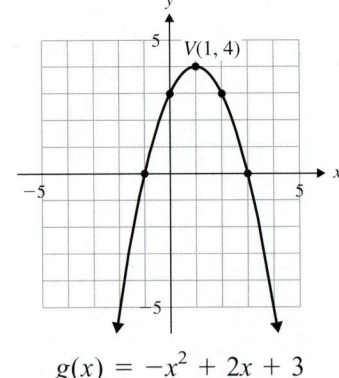

$g(x) = -x^2 + 2x + 3$

The *y*-coordinate of the vertex, -1, is the *smallest* *y*-value the function will have. We say that **-1 is the minimum value of the function.** $f(x)$ has no maximum because the graph continues upward indefinitely—the *y*-values get larger without bound.

The *y*-coordinate of the vertex, 4, is the *largest* *y*-value the function will have. We say that **4 is the maximum value of the function.** $g(x)$ has no minimum because the graph continues downward indefinitely—the *y*-values get smaller without bound.

> **Property**   Maximum and Minimum Values of a Quadratic Function
>
> Let $f(x) = ax^2 + bx + c$.
>
> 1) If a is **positive,** the graph of $f(x)$ opens upward, so the vertex is the lowest point on the parabola. The y-coordinate of the vertex is the **minimum** value of the function $f(x)$.
>
> 2) If a is **negative,** the graph of $f(x)$ opens downward, so the vertex is the highest point on the parabola. The y-coordinate of the vertex is the **maximum** value of the function $f(x)$.

We can use this information about the vertex to help us solve problems.

Example 1

In-Class Example 1
Let $f(x) = -x^2 + 4x + 3$. Answer parts a)–d) as in Example 1.
answer: a) maximum
b) (2, 7)   c) 7
d)

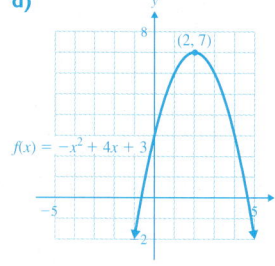

Let $f(x) = -x^2 + 4x + 2$.

a)   Does the function attain a minimum or maximum value at its vertex?

b)   Find the vertex of the graph of $f(x)$.

c)   What is the minimum or maximum value of the function?

d)   Graph the function to verify parts a)−c).

Solution

a)   Since $a = -1$, the graph of $f(x)$ will open downward. Therefore, the vertex will be the *highest* point on the parabola. The function will attain its *maximum* value at the vertex.

b)   Use $x = -\dfrac{b}{2a}$ to find the x-coordinate of the vertex. For $f(x) = -x^2 + 4x + 2$,

$$x = -\frac{b}{2a} = -\frac{(4)}{2(-1)} = 2$$

The y-coordinate of the vertex is $f(2)$.

$$f(2) = -(2)^2 + 4(2) + 2$$
$$= -4 + 8 + 2 = 6$$

The vertex is $(2, 6)$.

c) $f(x)$ has no minimum value. The *maximum* value of the function is 6, the y-coordinate of the vertex. (The largest y-value of the function is 6.)
    We say that the maximum value of the function is 6 and that it occurs at $x = 2$ (the x-coordinate of the vertex).

d)   From the graph of $f(x)$, we can see that our conclusions in parts a)−c) make sense.

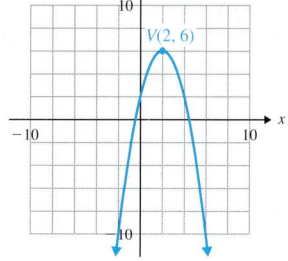

You Try 1

Let $f(x) = x^2 + 6x + 7$. Repeat parts a)–d) from Example 1.

2. Given a Quadratic Function, Solve an Applied Problem Involving a Maximum or Minimum Value

Example 2

In-Class Example 2
A ball is thrown upward from a height of 18 ft. The height h of the ball (in feet) t sec after the ball is released is given by

$$h(t) = -16t^2 + 16t + 18$$

a) How long does it take the ball to reach its maximum height?
b) What is the maximum height attained by the ball?

answer:
a) $t = 0.5$ sec   b) 22 ft

A ball is thrown upward from a height of 24 ft. The height h of the ball (in feet) t sec after the ball is released is given by

$$h(t) = -16t^2 + 16t + 24.$$

a) How long does it take the ball to reach its maximum height?

b) What is the maximum height attained by the ball?

Solution

a) Begin by understanding what the function $h(t)$ tells us: $a = -16$, so the graph of h would open downward. Therefore, the vertex is the highest point on the parabola. The maximum value of the function occurs at the vertex. The ordered pairs that satisfy $h(t)$ are of the form $(t, h(t))$.

To determine how long it takes the ball to reach its maximum height, we must find the t-coordinate of the vertex.

$$t = -\frac{b}{2a} = -\frac{16}{2(-16)} = \frac{1}{2}$$

The ball will reach its maximum height after $\frac{1}{2}$ sec.

b) The maximum height the ball reaches is the y-coordinate (or $h(t)$-coordinate) of the vertex. Since the ball attains its maximum height when $t = \frac{1}{2}$, find $h\left(\frac{1}{2}\right)$.

$$h\left(\frac{1}{2}\right) = -16\left(\frac{1}{2}\right)^2 + 16\left(\frac{1}{2}\right) + 24$$

$$= -16\left(\frac{1}{4}\right) + 8 + 24$$

$$= -4 + 32 = 28$$

The ball reaches a maximum height of 28 ft. ■

You Try 2

An object is propelled upward from a height of 10 ft. The height h of the object (in feet) t sec after the ball is released is given by

$$h(t) = -16t^2 + 32t + 10$$

a) How long does it take the object to reach its maximum height?

b) What is the maximum height attained by the object?

3. Write a Quadratic Function to Solve an Applied Problem Involving a Maximum or Minimum Value

Example 3

In-Class Example 3
Evan plans to put a fence around his rectangular garden. If he has 56 ft of fencing, what is the maximum area he can enclose?
answer: maximum area is 196 ft²

Ayesha plans to put a fence around her rectangular garden. If she has 32 ft of fencing, what is the maximum area she can enclose?

Solution

Begin by drawing a picture.

Let x = the width of the garden

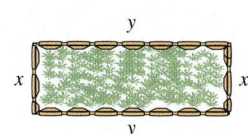

Let y = the length of the garden

Label the picture.

We will write two equations for a problem like this:

1) *The maximize or minimize equation;* this equation describes what we are trying to maximize or minimize.

2) *The constraint equation;* this equation describes the restrictions on the variables or the conditions the variables must meet.

Here is how we will get the equations.

1) We will write a *maximize* equation because we are trying to find the *maximum area* of the garden.

$$\text{Let } A = \text{area of the garden}$$

The area of the rectangle above is xy. Our equation is

$$\text{Maximize: } A = xy$$

2) To write the *constraint* equation, think about the restriction put on the variables. We cannot choose *any* two numbers for x and y. Since Ayesha has 32 ft of fencing, the distance around the garden is 32 ft. This is the perimeter of the rectangular garden. The perimeter of the rectangle drawn above is $2x + 2y$, and it must equal 32 ft.
  The constraint equation is

$$\text{Constraint: } 2x + 2y = 32$$

Set up this maximization problem as

$$\text{Maximize: } A = xy$$
$$\text{Constraint: } 2x + 2y = 32$$

Solve the constraint for a variable, and then substitute the expression into the maximize equation.

$$2x + 2y = 32$$
$$2y = 32 - 2x$$
$$y = 16 - x \qquad \text{Solve the constraint for } y.$$

Substitute $y = 16 - x$ into $A = xy$.

$$A = x(16 - x)$$
$$A = 16x - x^2 \qquad \text{Distribute.}$$
$$A = -x^2 + 16x \qquad \text{Write in descending powers.}$$

Look carefully at $A = -x^2 + 16x$. This is a quadratic function! Its graph is a parabola that opens downward (since $a = -1$). At the vertex, the function attains its maximum. The ordered pairs that satisfy this function are of the form $(x, A(x))$, where x represents the width and $A(x)$ represents the area of the rectangular garden. *The second coordinate of the vertex is the maximum area we are looking for.*

$$A = -x^2 + 16x$$

Use $x = -\dfrac{b}{2a}$ with $a = -1$ and $b = 16$ to find the x-coordinate of the vertex (the width of the rectangle that produces the maximum area).

$$x = -\dfrac{16}{2(-1)} = 8$$

Substitute $x = 8$ into $A = -x^2 + 16x$ to find the maximum area.

$$A = -(8)^2 + 16(8)$$
$$A = -64 + 128$$
$$A = 64$$

The graph of $A = -x^2 + 16x$ is a parabola that opens downward with vertex $(8, 64)$. The maximum area of the garden is 64 ft², and this will occur when the width of the garden is 8 ft. (The length will be 8 ft as well.) ■

Procedure   Steps for Solving a Max/Min Problem Like Example 3

1) Draw a picture, if applicable.

2) Define the unknowns. Label the picture.

3) Write the max/min equation.

4) Write the constraint equation.

5) Solve the constraint for a variable. Substitute the expression into the max/min equation to obtain a quadratic function.

6) Find the vertex of the parabola using the vertex formula, $x = -\dfrac{b}{2a}$.

7) Answer the question being asked.

You Try 3

Find the maximum area of a rectangle that has a perimeter of 28 in.

---

4. Graph Parabolas of the Form $x = a(y - k)^2 + h$

Not all parabolas are functions. Parabolas can open in the x-direction as illustrated below. Clearly, these fail the vertical line test for functions.

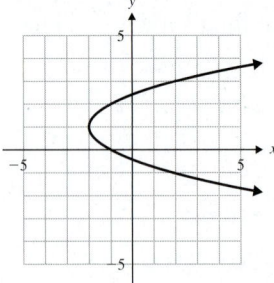   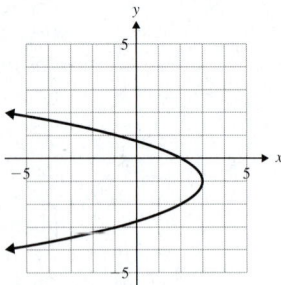

Parabolas that open in the y-direction, or vertically, result from the functions

$$y = a(x - h)^2 + k \qquad \text{or} \qquad y = ax^2 + bx + c.$$

If we interchange the x and y, we obtain the equations

$$x = a(y - k)^2 + h \qquad \text{or} \qquad x = ay^2 + by + c.$$

The graphs of these equations are parabolas that open in the x-direction, or horizontally.

Procedure   Graphing an Equation of the Form $x = a(y - k)^2 + h$

1)   The vertex of the parabola is (h, k). (Notice, however, that h and k have changed their positions in the equation when compared to a quadratic function.)

2)   The axis of symmetry is the horizontal line $y = k$.

3)   If a is positive, the graph opens to the right.
     If a is negative, the graph opens to the left.

Example 4

In-Class Example 4
Graph the equation. Find the
x- and y-intercepts and the
domain and range.
a) $x = (y + 3)^2 + 2$
b) $x = -\dfrac{1}{2}(y - 1)^2 + 4$
answer: a) x-int: (11, 0);
no y-int; domain: $[2, \infty)$;
range: $(-\infty, \infty)$

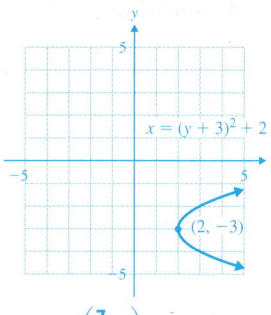

b) x-int: $\left(\dfrac{7}{2}, 0\right)$; y-ints:
$(0, 1 \pm 2\sqrt{2})$; domain:
$(-\infty, 4]$; range: $(-\infty, \infty)$

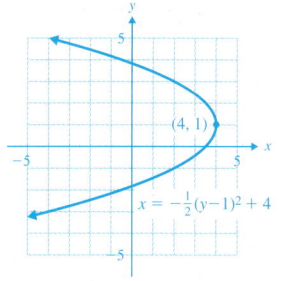

Graph each equation. Find the x- and y-intercepts and the domain and range.

a) $x = (y + 2)^2 - 1$ b) $x = -2(y - 2)^2 + 4$

Solution

a)   1) $h = -1$ and $k = -2$. The vertex is $(-1, -2)$.

2)   The axis of symmetry is $y = -2$.

3) $a = +1$, so the parabola opens to the right. It is the same width as $y = x^2$.

To find the x-intercept, let $y = 0$ and solve for x.

$$x = (y + 2)^2 - 1$$
$$x = (0 + 2)^2 - 1$$
$$x = 4 - 1 = 3$$

The x-intercept is $(3, 0)$.

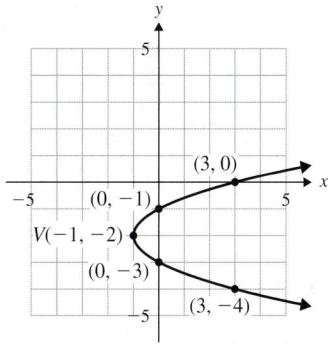

Find the y-intercepts by substituting 0 for x and solving for y.

$$x = (y + 2)^2 - 1$$
$$0 = (y + 2)^2 - 1 \qquad \text{Substitute 0 for } x.$$
$$1 = (y + 2)^2 \qquad \text{Add 1.}$$
$$\pm 1 = y + 2 \qquad \text{Square root property}$$

$$1 = y + 2 \qquad \text{or} \qquad -1 = y + 2$$
$$-1 = y \qquad\qquad\qquad -3 = y \qquad \text{Solve.}$$

The y-intercepts are $(0, -3)$ and $(0, -1)$. Use the axis of symmetry to locate the point $(3, -4)$ on the graph. The domain is $[-1, \infty)$, and the range is $(-\infty, \infty)$.

b) $x = -2(y - 2)^2 + 4$

1) $h = 4$ and $k = 2$. The vertex is $(4, 2)$.

2)   The axis of symmetry is $y = 2$.

3) $a = -2$, so the parabola opens to the left. It is narrower than $y = x^2$.

To find the x-intercept, let $y = 0$ and solve for x.

$$x = -2(y - 2)^2 + 4$$
$$x = -2(0 - 2)^2 + 4$$
$$x = -2(4) + 4 = -4$$

The x-intercept is $(-4, 0)$.

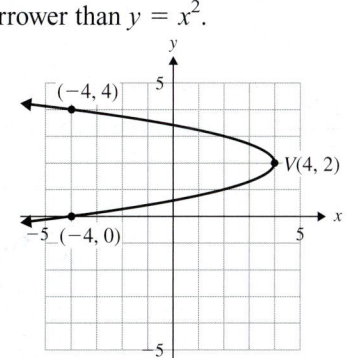

Find the *y*-intercepts by substituting 0 for *x* and solving for *y*.

$$x = -2(y - 2)^2 + 4$$
$$0 = -2(y - 2)^2 + 4 \qquad \text{Substitute 0 for } x.$$
$$-4 = -2(y - 2)^2 \qquad \text{Subtract 4.}$$
$$2 = (y - 2)^2 \qquad \text{Divide by } -2.$$
$$\pm\sqrt{2} = y - 2 \qquad \text{Square root property}$$
$$2 \pm \sqrt{2} = y \qquad \text{Add 2.}$$

The *y*-intercepts are $(0, 2 - \sqrt{2})$ and $(0, 2 + \sqrt{2})$. Use the axis of symmetry to locate the point $(-4, 4)$ on the graph. The domain is $(-\infty, 4]$; the range is $(-\infty, \infty)$.

You Try 4

Graph $x = -(y + 1)^2 - 3$. Find the *x*- and *y*-intercepts and the domain and range.

Procedure   Graphing Parabolas from the Form $x = ay^2 + by + c$

We can use two methods to graph $x = ay^2 + by + c$.

Method 1:   Rewrite $x = ay^2 + by + c$ in the form $x = a(y - k)^2 + h$ by *completing the square.*

Method 2:   Use the formula $y = -\dfrac{b}{2a}$ to find the *y-coordinate* of the vertex. Find the *x-coordinate* by substituting the *y*-value into the equation $x = ay^2 + by + c$.

---

5. Rewrite $x = ay^2 + by + c$ as $x = a(y - k)^2 + h$ by Completing the Square

Example 5

Rewrite $x = 2y^2 - 4y + 8$ in the form $x = a(y - k)^2 + h$ by completing the square.

In-Class Example 5
Rewrite $x = 3y^2 + 18y + 21$ in the form $x = a(y - k)^2 + h$ by completing the square.
answer: $x = 3(y + 3)^2 - 6$

Solution

To complete the square, follow the same procedure used for quadratic functions. (This is outlined on p. 576 in Section 9.5.)

Step 1:   Divide the equation by 2 so that the coefficient of y^2 is 1.

$$\frac{x}{2} = y^2 - 2y + 4$$

Step 2:   Separate the constant from the variable terms using parentheses.

$$\frac{x}{2} = (y^2 - 2y) + 4$$

Step 3:   Complete the square for the quantity in parentheses. Add 1 *inside* the parentheses and *subtract* 1 from the 4.

$$\frac{x}{2} = (y^2 - 2y + 1) + 4 - 1$$

$$\frac{x}{2} = (y^2 - 2y + 1) + 3$$

Step 4:   Factor the expression inside the parentheses.

$$\frac{x}{2} = (y - 1)^2 + 3$$

Step 5: Solve the equation for x by multiplying by 2.

$$2\left(\frac{x}{2}\right) = 2[(y-1)^2 + 3]$$
$$x = 2(y-1)^2 + 6$$

You Try 5

Rewrite $x = -y^2 - 6y - 1$ in the form $x = a(y-k)^2 + h$ by completing the square.

6. Find the Vertex of the Graph of $x = ay^2 + by + c$ Using $y = -\dfrac{b}{2a}$, and Graph the Equation

Example 6

Graph $x = y^2 - 2y + 5$. Find the vertex using the vertex formula. Find the x- and y-intercepts and the domain and range.

In-Class Example 6
Graph $x = y^2 - 4y + 8$. Find the vertex using the vertex formula. Find the x- and y-intercepts and the domain and range.
**answer: vertex: (4, 2);
x-int: (8, 0); no y-int;
domain: [4, ∞); range: (−∞, ∞)**

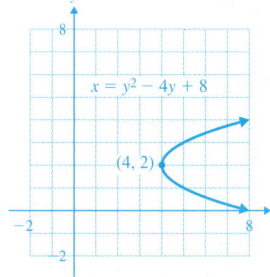

Solution

Since this equation is solved for x and is quadratic in y, it opens in the x-direction. $a = 1$, so it opens to the right. Use the vertex formula to find the *y-coordinate* of the vertex.

$$y = -\frac{b}{2a}$$
$$y = -\frac{-2}{2(1)} = 1 \qquad a = 1, b = -2$$

Substitute $y = 1$ into $x = y^2 - 2y + 5$ to find the x-coordinate of the vertex.

$$x = (1)^2 - 2(1) + 5$$
$$x = 1 - 2 + 5 = 4$$

The vertex is (4, 1). Since the vertex is (4, 1) and the parabola opens to the right, the graph has *no y-intercepts*.

To find the x-intercept, let $y = 0$ and solve for x.

$$x = y^2 - 2y + 5$$
$$x = 0^2 - 2(0) + 5$$
$$x = 5$$

The x-intercept is (5, 0).

Find another point on the parabola by choosing a value for y that is close to the y-coordinate of the vertex. Let $y = -1$. Find x.

$$x = (-1)^2 - 2(-1) + 5$$
$$x = 1 + 2 + 5 = 8$$

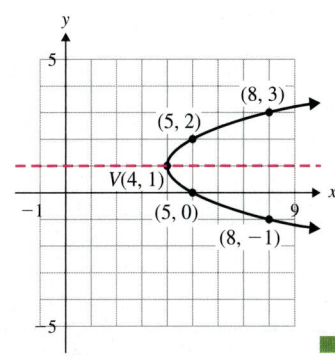

Another point on the parabola is (8, −1). Use the axis of symmetry to locate the additional points (5, 2) and (8, 3). The domain is $[4, \infty)$, and the range is $(-\infty, \infty)$

You Try 6

Graph $x = y^2 + 6y + 3$. Find the vertex using the vertex formula. Find the x- and y-intercepts and the domain and range.

Using Technology

To graph a parabola that is a function, just enter the equation and press GRAPH.

Example 1:   Graph $f(x) = -x^2 + 2$.

Enter $Y_1 = -x^2 + 2$ to graph the function on a calculator.

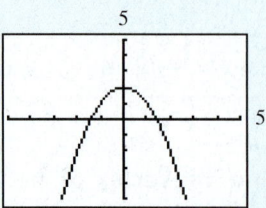

To graph an equation on a calculator, it must be entered so that y is a function of x. Since a parabola that opens horizontally is not a function, we must solve for y in terms of x so that the equation is represented by two different functions.

Example 2:   Graph $x = y^2 - 4$ on a calculator.

Solve for y.

$$x = y^2 - 4$$
$$x + 4 = y^2$$
$$\pm\sqrt{x + 4} = y$$

Now the equation $x = y^2 - 4$ is rewritten so that y is in terms of x. In the graphing calculator, enter $y = \sqrt{x + 4}$ as Y_1. This represents the top half of the parabola since the y-values are positive above the x-axis. Enter $y = -\sqrt{x + 4}$ as Y_2. This represents the bottom half of the parabola since the y-values are negative below the x-axis. Set an appropriate window and press GRAPH.

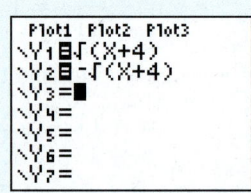   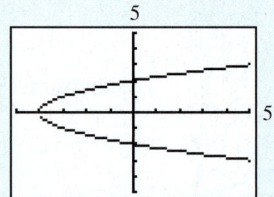

Graph each parabola on a graphing calculator. Where appropriate, rewrite the equation for y in terms of x. These problems come from the homework exercises so that the graphs can be found in the Answers to Exercises appendix.

1) $f(x) = x^2 + 6x + 9$; Exercise 9

2) $x = y^2 + 2$; Exercise 33

3) $x = \dfrac{1}{4}(y + 2)^2$; Exercise 39

4) $f(x) = -\dfrac{1}{2}x^2 + 4x - 6$; Exercise 11

5) $x = -(y - 4)^2 + 5$; Exercise 35

6) $x = y^2 - 4y + 5$; Exercise 41

Answers to You Try Exercises

1) a) minimum value   b) vertex $(-3, -2)$
c) The minimum value of the function is -2.
d)

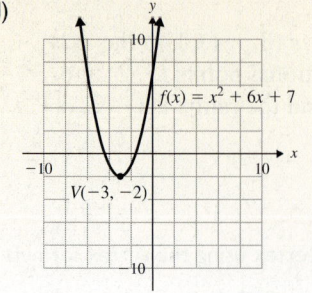

2) a) 1 sec   b) 26 ft   3) 49 in^2
4) $V(-3, -1)$; x-int: $(-4, 0)$; y-int: none; domain: $(-\infty, -1]$; range: $(-\infty, \infty)$

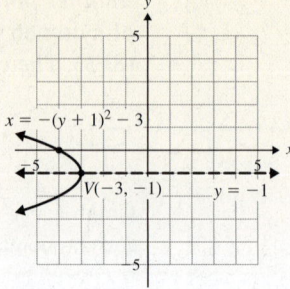

5) $x = -(y + 3)^2 + 8$

6) $V(-6, -3)$; x-int: $(3, 0)$; y-int: $(0, -3 - \sqrt{6})$, $(0, -3 + \sqrt{6})$; domain: $[-6, \infty)$; range: $(-\infty, \infty)$

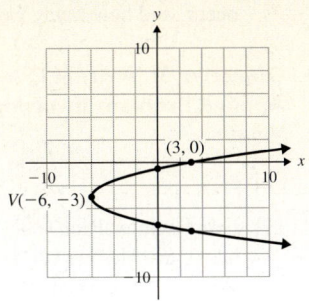

Answers to Technology Exercises

1) The equation can be entered as it is.

2) $Y_1 = \sqrt{x - 2}$; $Y_2 = -\sqrt{x - 2}$

3) $Y_1 = -2 + \sqrt{4x}$; $Y_2 = -2 - \sqrt{4x}$

4) The equation can be entered as it is.

5) $Y_1 = 4 + \sqrt{5 - x}$; $Y_2 = 4 - \sqrt{5 - x}$

6) $Y_1 = 2 + \sqrt{x - 1}$; $Y_2 = 2 - \sqrt{x - 1}$

9.6 Exercises

*Additional answers can be found in the Answers to Exercises appendix.

Objective 1: Find the Maximum or Minimum Value of a Quadratic Function

For Exercises 1–6, determine whether the function has a maximum value, minimum value, or neither.

1)

(maximum)

2)

(neither)

3)

(neither)

4)

(minimum)

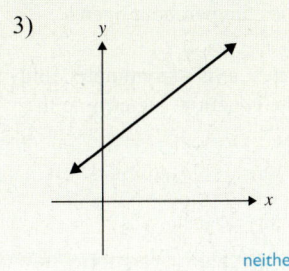

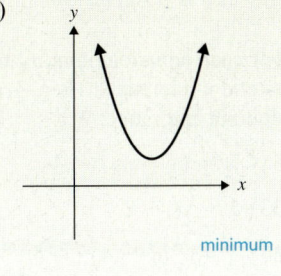

5)

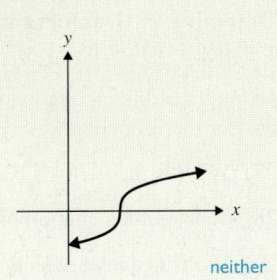

minimum

6)

neither

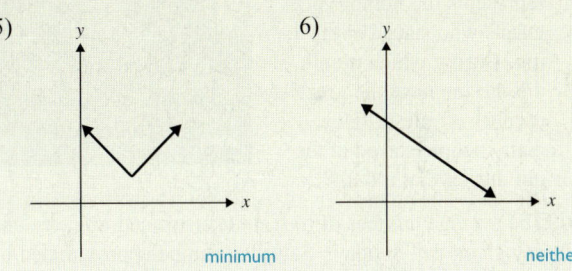

7) Let $f(x) = ax^2 + bx + c$. How do you know whether the function has a maximum or minimum value at the vertex?

8) Is there a maximum value of the function $y = 2x^2 + 12x + 11$? Explain your answer.

For Problems 9–12, answer parts a)–d) for each function, $f(x)$.

   a) Does the function attain a minimum or maximum value at its vertex?

   b) Find the vertex of the graph of $f(x)$.

   c) What is the minimum or maximum value of the function?

   d) Graph the function to verify parts a)–c).

9) $f(x) = x^2 + 6x + 9$

10) $f(x) = -x^2 + 2x + 4$

11) $f(x) = -\dfrac{1}{2}x^2 + 4x - 6$

12) $f(x) = 2x^2 + 4x$

Objective 2: Given a Quadratic Function, Solve an Applied Problem Involving a Maximum or Minimum Value

Solve.

13) An object is fired upward from the ground so that its height h (in feet) t sec after being fired is given by

$$h(t) = -16t^2 + 320t$$

   a) How long does it take the object to reach its maximum height?   10 sec

   b) What is the maximum height attained by the object?   1600 ft

   c) How long does it take the object to hit the ground?   20 sec

14) An object is thrown upward from a height of 64 ft so that its height h (in feet) t sec after being thrown is given by

$$h(t) = -16t^2 + 48t + 64$$

   a) How long does it take the object to reach its maximum height?   1.5 sec

   b) What is the maximum height attained by the object?   100 ft

   c) How long does it take the object to hit the ground?   4 sec

15) The number of guests staying at the Cozy Inn from January to December 2010 can be approximated by

$$N(x) = -10x^2 + 120x + 120$$

   where x represents the number of months after January 2010 ($x = 0$ represents January, $x = 1$ represents February, etc.), and $N(x)$ represents the number of guests who stayed at the inn. During which month did the inn have the greatest number of guests? How many people stayed at the inn during that month?   July; 480 people

16) The average number of traffic tickets issued in a city on any given day Sunday–Saturday can be approximated by

$$T(x) = -7x^2 + 70x + 43$$

   where x represents the number of days after Sunday ($x = 0$ represents Sunday, $x = 1$ represents Monday, etc.), and $T(x)$ represents the number of traffic tickets issued. On which day are the most tickets written? How many tickets are issued on that day?   Friday; 218 tickets

17) The number of babies born to teenage mothers from 1989 to 2002 can be approximated by

$$N(t) = -0.721t^2 + 2.75t + 528$$

   where t represents the number of years after 1989 and $N(t)$ represents the number of babies born (in thousands). According to this model, in what year was the number of babies born to teen mothers the greatest? How many babies were born that year? (U.S. Census Bureau)   1991; 531,000

18) The number of violent crimes in the United States from 1985 to 1999 can be modeled by

$$C(x) = -49.2x^2 + 636x + 12,468$$

where x represents the number of years after 1985 and $C(x)$ represents the number of violent crimes (in thousands). During what year did the greatest number of violent crimes occur, and how many were there? (U.S. Census Bureau)   1991; approximately 14,523,300

Objective 3: Write a Quadratic Function to Solve an Applied Problem Involving a Maximum or Minimum Value

Solve.

19) Every winter Rich makes a rectangular ice rink in his backyard. He has 100 ft of material to use as the border. What is the maximum area of the ice rink?   625 ft²

20) Find the dimensions of the rectangular garden of greatest area that can be enclosed with 40 ft of fencing.   10 ft × 10 ft

21) The Soo family wants to fence in a rectangular area to hold their dogs. One side of the pen will be their barn. Find the dimensions of the pen of greatest area that can be enclosed with 48 ft of fencing.   12 ft × 24 ft

22) A farmer wants to enclose a rectangular area with 120 ft of fencing. One side is a river and will not require a fence. What is the maximum area that can be enclosed?   1800 ft²

23) Find two integers whose sum is 18 and whose product is a maximum.   9 and 9

24) Find two integers whose sum is 26 and whose product is a maximum.   13 and 13

25) Find two integers whose difference is 12 and whose product is a minimum.   6 and −6

26) Find two integers whose difference is 30 and whose product is a minimum.   15 and −15

Objective 4: Graph Parabolas of the Form $x = a(y - k)^2 + h$

Given a quadratic equation of the form $x = a(y - k)^2 + h$, answer the following.

27) What is the vertex? (h, k)

28) What is the equation of the axis of symmetry? $y = k$

29) If a is negative, which way does the parabola open?   to the left

30) If a is positive, which way does the parabola open?   to the right

For each equation, identify the vertex, axis of symmetry, and x- and y-intercepts. Then, graph the equation. Determine the domain and range.

31) $x = (y - 1)^2 - 4$ 32) $x = (y + 3)^2 - 1$

33) $x = y^2 + 2$ 34) $x = (y - 4)^2$

35) $x = -(y - 4)^2 + 5$ 36) $x = -(y + 1)^2 - 7$

37) $x = -2(y - 2)^2 - 9$ 38) $x = -\dfrac{1}{2}(y - 4)^2 + 7$

39) $x = \dfrac{1}{4}(y + 2)^2$ 40) $x = 2y^2 + 3$

Objective 5: Rewrite $x = ay^2 + by + c$ as $x = a(y - k)^2 + h$ VIDEO
by Completing the Square

Rewrite each equation in the form $x = a(y - k)^2 + h$ by completing the square and graph it. Determine the domain and range.

41) $x = y^2 - 4y + 5$

42) $x = y^2 + 4y - 6$

43) $x = -y^2 + 6y + 6$

44) $x = -y^2 - 2y - 5$

VIDEO 45) $x = \dfrac{1}{3}y^2 + \dfrac{8}{3}y - \dfrac{5}{3}$

46) $x = 2y^2 - 4y + 5$

47) $x = -4y^2 - 8y - 10$

48) $x = \dfrac{1}{2}y^2 + 4y - 1$

Objective 6: Find the Vertex of $x = ay^2 + by + c$ Using $y = -\dfrac{b}{2a}$, and Graph the Equation

Graph each equation using the vertex formula. Find the x- and y-intercepts. Determine the domain and range.

49) $x = y^2 - 4y + 3$

50) $x = -y^2 + 4y$

51) $x = -y^2 + 2y + 2$

52) $x = y^2 + 6y - 4$

VIDEO 53) $x = -2y^2 + 4y - 6$

54) $x = 3y^2 + 6y - 1$

55) $x = 4y^2 - 16y + 13$

56) $x = 2y^2 + 4y + 8$

57) $x = \dfrac{1}{4}y^2 - \dfrac{1}{2}y + \dfrac{25}{4}$

58) $x = -\dfrac{3}{4}y^2 + \dfrac{3}{2}y - \dfrac{11}{4}$

Mixed Exercises

Exercises 59–68 contain parabolas that open either horizontally or vertically. Graph each equation. Determine the domain and range.

59) $h(x) = -x^2 + 6$

60) $y = x^2 - 6x - 1$

61) $x = y^2$

62) $f(x) = -3x^2 + 12x - 8$

63) $x = -\dfrac{1}{2}y^2 - 4y - 5$

64) $x = (y - 4)^2 + 3$

65) $y = x^2 + 2x - 3$

66) $x = -3(y + 2)^2 + 11$

67) $f(x) = -2(x - 4)^2 + 3$

68) $g(x) = \dfrac{3}{2}x^2 - 12x + 20$

Section 9.7 Quadratic and Rational Inequalities

Objectives

1. **Solve a Quadratic Inequality by Graphing**
2. **Solve a Quadratic Inequality Using Test Points**
3. **Solve Quadratic Inequalities with Special Solutions**
4. **Solve an Inequality of Higher Degree**
5. **Solve a Rational Inequality**

In Chapter 2, we learned how to solve *linear* inequalities such as $3x - 5 \leq 16$. In this section, we will discuss how to solve *quadratic* and *rational* inequalities.

Definition

A **quadratic inequality** can be written in the form

$$ax^2 + bx + c \leq 0 \quad \text{or} \quad ax^2 + bx + c \geq 0$$

where a, b, and c are real numbers and $a \neq 0$. ($<$ and $>$ may be substituted for $\leq$ and $\geq$.)

1. Solve a Quadratic Inequality by Graphing

To understand how to solve a quadratic inequality, let's look at the graph of a quadratic function.

Example 1

In-Class Example 1
a) Graph $y = x^2 + 6x + 8$.
b) Solve $x^2 + 6x + 8 < 0$.
c) Solve $x^2 + 6x + 8 \geq 0$.
answer: a)

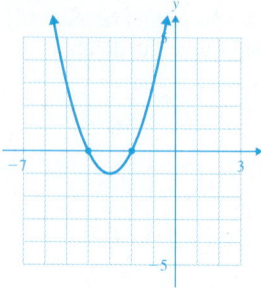

b) $(-4, -2)$
c) $(-\infty, -4] \cup [-2, \infty)$

a) Graph $y = x^2 - 2x - 3$.

b) Solve $x^2 - 2x - 3 < 0$.

c) Solve $x^2 - 2x - 3 \geq 0$.

Solution

a) The graph of the quadratic function $y = x^2 - 2x - 3$ is a parabola that opens upward. Use the vertex formula to confirm that the vertex is $(1, -4)$.
    To find the y-intercept, let $x = 0$ and solve for y.

$$y = 0^2 - 2(0) - 3$$
$$y = -3$$

The y-intercept is $(0, -3)$.
To find the x-intercepts, let $y = 0$ and solve for x.

$0 = x^2 - 2x - 3$
$0 = (x - 3)(x + 1)$ — Factor.
$x - 3 = 0$ or $x + 1 = 0$ — Set each factor equal to 0.
$x = 3$ or $x = -1$ — Solve.

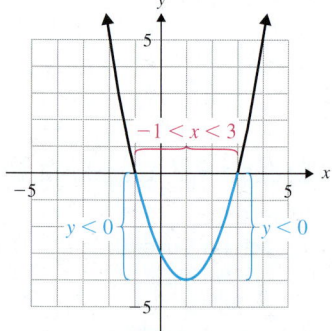

b) We will use the graph of $y = x^2 - 2x - 3$ to solve the inequality $x^2 - 2x - 3 < 0$. That is, to solve $x^2 - 2x - 3 < 0$ we must ask ourselves, "Where are the y-values of the function *less than* zero?"
    The y-values of the function are less than zero when the x-values are greater than -1 and less than 3, as shown to the right.

The solution set of $x^2 - 2x - 3 < 0$ (in interval notation) is $(-1, 3)$.

c) To solve $x^2 - 2x - 3 \geq 0$ means to find the x-values for which the y-values of the function $y = x^2 - 2x - 3$ are *greater than or equal to* zero. (Recall that the x-intercepts are where the function equals zero.)
    The y-values of the function are greater than or equal to zero when $x \leq -1$ or when $x \geq 3$.
    The solution set of $x^2 - 2x - 3 \geq 0$ is $(-\infty, -1] \cup [3, \infty)$.

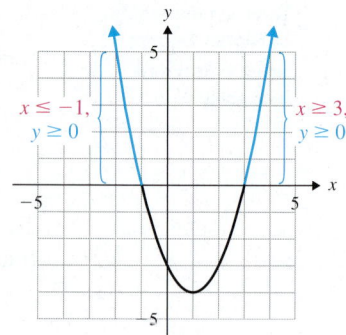

When $x \leq -1$ or $x \geq 3$, the y-values are greater than or equal to 0. ■

You Try 1

a) Graph $y = x^2 + 6x + 5$.   b) Solve $x^2 + 6x + 5 \leq 0$.   c) Solve $x^2 + 6x + 5 > 0$.

2. Solve a Quadratic Inequality Using Test Points

Example 1 illustrates how the x-intercepts of $y = x^2 - 2x - 3$ break up the x-axis into the three separate intervals: $x < -1, -1 < x < 3$, and $x > 3$. We can use this idea of intervals to solve a quadratic inequality without graphing.

 **Example 2**

Solve $x^2 - 2x - 3 < 0$.

In-Class Example 2
Solve $x^2 - 4x - 5 < 0$.
answer: $(-1, 5)$

Solution

Begin by solving the equation $x^2 - 2x - 3 = 0$.

$$x^2 - 2x - 3 = 0$$
$$(x - 3)(x + 1) = 0 \qquad \text{Factor.}$$
$$x - 3 = 0 \quad \text{or} \quad x + 1 = 0 \qquad \text{Set each factor equal to 0.}$$
$$x = 3 \quad \text{or} \qquad x = -1 \qquad \text{Solve.}$$

(These are the x-intercepts of $y = x^2 - 2x - 3$.)

> **Note**
>
> The $<$ indicates that we want to find the values of x that will make $x^2 - 2x - 3 < 0$; that is, find the values of x that make $x^2 - 2x - 3$ a *negative* number.

Put $x = 3$ and $x = -1$ on a number line with the smaller number on the left. This breaks up the number line into three intervals: $x < -1, -1 < x < 3$, and $x > 3$.

Choose a test number in each interval and substitute it into $x^2 - 2x - 3$ to determine whether that value makes $x^2 - 2x - 3$ positive or negative. (If one number in the interval makes $x^2 - 2x - 3$ positive, then *all* numbers in that interval will make $x^2 - 2x - 3$ positive.) Indicate the result on the number line.

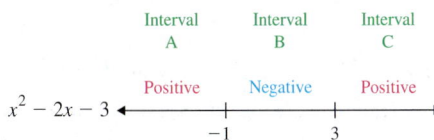

Interval A: $(x < -1)$ As a test number, choose any number less than -1. We will choose -2. Evaluate $x^2 - 2x - 3$ for $x = -2$.

$$x^2 - 2x - 3 = (-2)^2 - 2(-2) - 3 \qquad \text{Substitute } -2 \text{ for } x.$$
$$= 4 + 4 - 3$$
$$= 8 - 3 = 5$$

When $x = -2$, $x^2 - 2x - 3$ is *positive*. Therefore, $x^2 - 2x - 3$ will be positive for all values of x in this interval. Indicate this on the number line as seen above.

Interval B: $(-1 < x < 3)$ As a test number, choose any number between -1 and 3. We will choose 0. Evaluate $x^2 - 2x - 3$ for $x = 0$.

$$x^2 - 2x - 3 = (0)^2 - 2(0) - 3 \qquad \text{Substitute } 0 \text{ for } x.$$
$$= 0 - 0 - 3 = -3$$

When $x = 0$, $x^2 - 2x - 3$ is *negative*. Therefore, $x^2 - 2x - 3$ will be negative for all values of x in this interval. Indicate this on the number line above.

Interval C: $(x > 3)$ As a test number, choose any number greater than 3. We will choose 4. Evaluate $x^2 - 2x - 3$ for $x = 4$.

$$x^2 - 2x - 3 = (4)^2 - 2(4) - 3 \qquad \text{Substitute 4 for } x.$$
$$= 16 - 8 - 3$$
$$= 8 - 3 = 5$$

When $x = 4$, $x^2 - 2x - 3$ is *positive*. Therefore, $x^2 - 2x - 3$ will be positive for all values of x in this interval. Indicate this on the number line.

Look at the number line. The solution set of $x^2 - 2x - 3 < 0$ consists of the interval(s) where $x^2 - 2x - 3$ is *negative*. This is in interval B, $(-1, 3)$.

The graph of the solution set is
$$-5\ -4\ -3\ -2\ -1\ \ 0\ \ 1\ \ 2\ \ 3\ \ 4\ \ 5$$

The solution set is $(-1, 3)$. This is the same as the result we obtained in Example 1 by graphing.   ■

You Try 2

Solve $x^2 + 5x + 4 \le 0$. Graph the solution set and write the solution in interval notation.

Next we will summarize how to solve a quadratic inequality.

Procedure   How to Solve a Quadratic Inequality

Step 1:   **Write the inequality in the form $ax^2 + bx + c \le 0$ or $ax^2 + bx + c \ge 0$.** ($<$ and $>$ may be substituted for $\le$ and $\ge$ 0.) If the inequality symbol is $<$ or $\le$, we are looking for a *negative* quantity in the interval on the number line. If the inequality symbol is $>$ or $\ge$, we are looking for a *positive* quantity in the interval.

Step 2:   **Solve the equation $ax^2 + bx + c = 0$.**

Step 3:   **Put the solutions of $ax^2 + bx + c = 0$ on a number line.** These values break up the number line into intervals.

Step 4:   **Choose a test number in each interval** to determine whether $ax^2 + bx + c$ is positive or negative in each interval. Indicate this on the number line.

Step 5:   **If the inequality is in the form $ax^2 + bx + c \le 0$ or $ax^2 + bx + c < 0$, then the solution set contains the numbers in the interval where $ax^2 + bx + c$ is *negative*.
If the inequality is in the form $ax^2 + bx + c \ge 0$ or $ax^2 + bx + c > 0$, then the solution set contains the numbers in the interval where $ax^2 + bx + c$ is *positive*.**

Step 6:   **If the inequality symbol is $\le$ or $\ge$, then the endpoints of the interval(s) (the numbers found in Step 3) are included in the solution set.** Indicate this with brackets in the interval notation.
If the inequality symbol is $<$ or $>$, then the endpoints of the interval(s) are not included in the solution set. Indicate this with parentheses in interval notation.

---

3. Solve Quadratic Inequalities with Special Solutions

We should look carefully at the inequality before trying to solve it. Sometimes, it is not necessary to go through all of the steps.

Example 3

In-Class Example 3
Solve.
a) $(z + 1)^2 > -1$
b) $(n - 15)^2 \leq -1$
answer: a) $(-\infty, \infty)$ b) $\varnothing$

Solve.

a) $(y + 4)^2 \geq -5$ b) $(t - 8)^2 < -3$

Solution

a) The inequality $(y + 4)^2 \geq -5$ says that a squared quantity, $(y + 4)^2$, is greater than or equal to a *negative* number, -5. *This is always true.* (A squared quantity will *always* be greater than or equal to zero.) Any real number, y, will satisfy the inequality.

   The solution set is (∞, ∞).

b) The inequality $(t - 8)^2 < -3$ says that a squared quantity, $(t - 8)^2$, is less than a *negative* number, -3. *There is no real number value for t so that $(t - 8)^2 < -3$.*

   The solution set is $\varnothing$.

You Try 3

Solve.

a) $(k + 2)^2 \leq -4$ b) $(z - 9)^2 > -1$

4. Solve an Inequality of Higher Degree

Other polynomial inequalities in factored form can be solved in the same way that we solve quadratic inequalities.

Example 4

In-Class Example 4
Solve
$(d + 3)(d - 6)(d - 2) < 0$.
answer: $(-\infty, -3) \cup (2, 6)$

Solve $(c - 2)(c + 5)(c - 4) < 0$.

Solution

This is the factored form of a third-degree polynomial. Since the inequality is $<$, the solution set will contain the intervals where $(c - 2)(c + 5)(c - 4)$ is *negative*.

Solve $(c - 2)(c + 5)(c - 4) = 0$.

$c - 2 = 0$ or	$c + 5 = 0$ or	$c - 4 = 0$	Set each factor equal to 0.
$c = 2$ or	$c = -5$ or	$c = 4$	Solve.

Put $c = 2$, $c = -5$, and $c = 4$ on a number line, and test a number in each interval.

Interval	$c < -5$	$-5 < c < 2$	$2 < c < 4$	$c > 4$
Test number	$c = -6$	$c = 0$	$c = 3$	$c = 5$
Evaluate $(c - 2)(c + 5)(c - 4)$	$(-6 - 2)(-6 + 5)(-6 - 4)$ $= (-8)(-1)(-10)$ $= -80$	$(0 - 2)(0 + 5)(0 - 4)$ $= (-2)(5)(-4)$ $= 40$	$(3 - 2)(3 + 5)(3 - 4)$ $= (1)(8)(-1)$ $= -8$	$(5 - 2)(5 + 5)(5 - 4)$ $= (3)(10)(1)$ $= 30$
Sign	Negative	Positive	Negative	Positive

Negative   Positive   Negative   Positive

$$(c - 2)(c + 5)(c - 4) \longleftarrow \underset{-5}{\quad} \underset{2}{\quad} \underset{4}{\quad} \longrightarrow$$

We can see that the intervals where $(c - 2)(c + 5)(c - 4)$ is negative are $(-\infty, -5)$ and $(2, 4)$. The endpoints are not included since the inequality is $<$.

The graph of the solution set is $\longleftarrow \!\!\!\underset{-8\,-7\,-6\,-5\,-4\,-3\,-2\,-1\ \ 0\ \ 1\ \ 2\ \ 3\ \ 4\ \ 5\ \ 6\ \ 7\ \ 8}{\diamond}\!\!\!\longrightarrow$

The solution set of $(c - 2)(c + 5)(c - 4) < 0$ is $(-\infty, -5) \cup (2, 4)$.   ■

You Try 4

Solve $(y + 3)(y - 1)(y + 1) \geq 0$. Graph the solution set and write the solution in interval notation.

5. Solve a Rational Inequality

An inequality containing a rational expression, $\dfrac{p}{q}$, where p and q are polynomials, is called a **rational inequality.** The way we solve rational inequalities is very similar to the way we solve quadratic inequalities.

Procedure   How to Solve a Rational Inequality

Step 1:   Write the inequality so that there is a 0 on one side and only one rational expression on the other side. If the inequality symbol is $<$ or $\leq$, we are looking for a *negative* quantity in the interval on the number line. If the inequality symbol is $>$ or $\geq$, we are looking for a *positive* quantity in the interval.

Step 2:   Find the numbers that make the numerator equal 0 and any numbers that make the denominator equal 0.

Step 3:   Put the numbers found in Step 2 on a number line. These values break up the number line into intervals.

Step 4:   Choose a test number in each interval to determine whether the rational inequality is positive or negative in each interval. Indicate this on the number line.

Step 5:   If the inequality is in the form $\dfrac{p}{q} \leq 0$ or $\dfrac{p}{q} < 0$, then the solution set contains

the numbers in the interval where $\dfrac{p}{q}$ is *negative.*

If the inequality is in the form $\dfrac{p}{q} \geq 0$ or $\dfrac{p}{q} > 0$, then the solution set contains

the numbers in the interval where $\dfrac{p}{q}$ is *positive.*

Step 6:   Determine whether the endpoints of the intervals are included in or excluded from the solution set. Do not include any values that make the denominator equal 0.

Example 5

In-Class Example 5

Solve $\dfrac{9}{2x+5} > 0$. Graph the solution set and write the solution in interval notation.

answer: $\left(-\dfrac{5}{2}, \infty\right)$

Solve $\dfrac{5}{x+3} > 0$. Graph the solution set and write the solution in interval notation.

Solution

Step 1: The inequality is in the correct form—zero on one side and only one rational expression on the other side. Since the inequality symbol is > 0, the solution set will contain the interval(s) where $\dfrac{5}{x+3}$ is *positive*.

Step 2: Find the numbers that make the numerator equal 0 and any numbers that make the denominator equal 0.

Numerator: 5	Denominator: $x + 3$
The numerator is a constant, 5, so it cannot equal 0.	Set $x + 3 = 0$ and solve for x. $x + 3 = 0$ $x = -3$

Step 3: Put -3 on a number line to break it up into intervals.

$\dfrac{5}{x+3}$ ←————————+————————→
                           -3

Step 4: Choose a test number in each interval to determine whether $\dfrac{5}{x+3}$ is positive or negative in each interval.

Interval	$x < -3$	$x > -3$
Test number	$x = -4$	$x = 0$
Evaluate $\dfrac{5}{x+3}$	$\dfrac{5}{-4+3} = \dfrac{5}{-1} = -5$	$\dfrac{5}{0+3} = \dfrac{5}{3}$
Sign	Negative	Positive

Step 5: The solution set of $\dfrac{5}{x+3} > 0$ contains the numbers in the interval where $\dfrac{5}{x+3}$ is positive. This interval is $(-3, \infty)$.

$\dfrac{5}{x+3}$ ←——— Negative ———+——— Positive ———→
                                  -3

Step 6: Since the inequality symbol is $>$, the endpoint of the interval, -3, is not included in the solution set.

The graph of the solution set is ←—+—+—◇—+—+—+—+—+—+—+—→
                                 $-5\ -4\ -3\ -2\ -1\ \ 0\ \ 1\ \ 2\ \ 3\ \ 4\ \ 5$

The solution set is $(-3, \infty)$. ■

You Try 5

Solve $\dfrac{2}{y-6} < 0$. Graph the solution set and write the solution in interval notation.

Example 6

In-Class Example 6

Solve $\dfrac{5}{x+3} \le 1$. Graph the

solution set and write the
solution in interval notation.

answer: $(-\infty, -3) \cup [2, \infty)$

Solve $\dfrac{7}{a+2} \le 3$. Graph the solution set and write the solution in interval notation.

Solution

Step 1: Get a zero on one side and only one rational expression on the other side.

$$\frac{7}{a+2} \le 3$$

$$\frac{7}{a+2} - 3 \le 0 \qquad \text{Subtract 3.}$$

$$\frac{7}{a+2} - \frac{3(a+2)}{a+2} \le 0 \qquad \text{Get a common denominator.}$$

$$\frac{7}{a+2} - \frac{3a+6}{a+2} \le 0 \qquad \text{Distribute.}$$

$$\frac{1-3a}{a+2} \le 0 \qquad \text{Combine numerators and combine like terms.}$$

From this point forward, we will work with the inequality $\dfrac{1-3a}{a+2} \le 0$. It is equivalent to the original inequality. Since the inequality symbol is $\le$, the solution set contains the interval(s) where $\dfrac{1-3a}{a+2}$ is *negative*.

Step 2: Find the numbers that make the numerator equal 0 and any numbers that make the denominator equal 0.

Numerator	Denominator
$1 - 3a = 0$	$a + 2 = 0$
$-3a = -1$	$a = -2$
$a = \dfrac{1}{3}$	

Step 3: Put $\dfrac{1}{3}$ and -2 on a number line to break it up into intervals.

Step 4: Choose a test number in each interval.

Interval	$a < -2$	$-2 < a < \dfrac{1}{3}$	$a > \dfrac{1}{3}$
Test number	$a = -3$	$a = 0$	$a = 1$
Evaluate $\dfrac{1-3a}{a+2}$	$\dfrac{1-3(-3)}{-3+2} = \dfrac{10}{-1} = -10$	$\dfrac{1-3(0)}{0+2} = \dfrac{1}{2}$	$\dfrac{1-3(1)}{1+2} = -\dfrac{2}{3}$
Sign	Negative	Positive	Negative

Step 5: The solution set of $\dfrac{1-3a}{a+2} \le 0$ $\left(\text{and therefore } \dfrac{7}{a+2} \le 3\right)$ will contain the numbers in the intervals where $\dfrac{1-3a}{a+2}$ is negative. These are the first and last intervals.

Step 6:   Determine whether the endpoints of the intervals, -2 and $\frac{1}{3}$, are included in the solution set. The endpoint $\frac{1}{3}$ is included since it does not make the denominator equal 0. *But -2 is not included because it makes the denominator equal 0.*

The graph of the solution set of $\dfrac{7}{a+2} \le 3$ is

The solution set is $(-\infty, -2) \cup \left[\dfrac{1}{3}, \infty\right)$.

BE CAREFUL

Although an inequality symbol may be $\le$ or $\ge$, an endpoint cannot be included in the solution set if it makes the denominator equal 0.

You Try 6

Solve $\dfrac{3}{z+4} \ge 2$. Graph the solution set and write the solution in interval notation.

Answers to You Try Exercises

1) a)

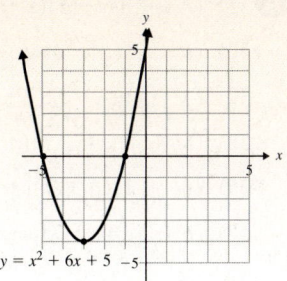

$y = x^2 + 6x + 5$

b) $[-5, -1]$ c) $(-\infty, -5) \cup (-1, \infty)$

2) ; $[-4, -1]$

3) a) $\varnothing$ b) $(-\infty, \infty)$

4) ;

$[-3, -1] \cup [1, \infty)$

5) ; $(-\infty, 6)$

6) ; $\left(-4, -\dfrac{5}{2}\right]$

9.7 Exercises

*Additional answers can be found in the Answers to Exercises appendix.

1) When solving a quadratic inequality, how do you know when to include and when to exclude the endpoints in the solution set?

2) If a rational inequality contains a $\leq$ or $\geq$ symbol, will the endpoints of the solution set always be included? Explain your answer.

Objective 1: Solve a Quadratic Inequality by Graphing

For Exercises 3–6, use the graph of the function to solve each inequality.

3) $y = x^2 + 4x - 5$

4) $y = x^2 - 6x + 8$

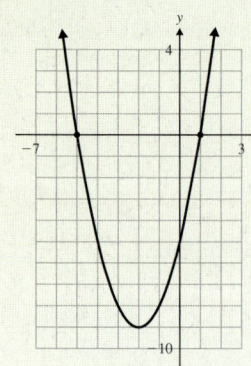

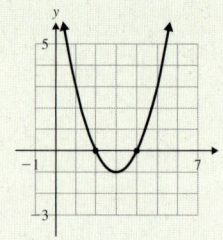

a) $x^2 - 6x + 8 > 0$
   $(-\infty, 2) \cup (4, \infty)$
b) $x^2 - 6x + 8 \leq 0$
   $[2, 4]$

a) $x^2 + 4x - 5 \leq 0$ $[-5, 1]$

b) $x^2 + 4x - 5 > 0$
   $(-\infty, -5) \cup (1, \infty)$

5) $y = -\dfrac{1}{2}x^2 + x + \dfrac{3}{2}$

6) $y = -x^2 - 8x - 12$

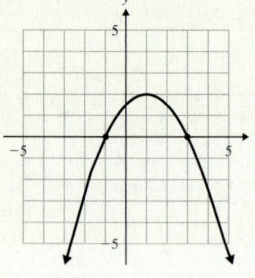

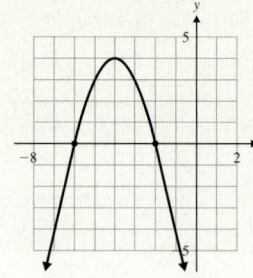

a) $-\dfrac{1}{2}x^2 + x + \dfrac{3}{2} \geq 0$
   $[-1, 3]$
b) $-\dfrac{1}{2}x^2 + x + \dfrac{3}{2} < 0$
   $(-\infty, -1) \cup (3, \infty)$

a) $-x^2 - 8x - 12 < 0$
   $(-\infty, -6) \cup (-2, \infty)$
b) $-x^2 - 8x - 12 \geq 0$
   $[-6, -2]$

Objective 2: Solve a Quadratic Inequality Using Test Points

Solve each quadratic inequality. Graph the solution set and write the solution in interval notation.

7) $x^2 + 6x - 7 \geq 0$

8) $m^2 - 2m - 24 > 0$

9) $c^2 + 5c < 36$

10) $t^2 + 36 \leq 15t$

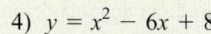 11) $r^2 - 13r > -42$

12) $v^2 + 10v < -16$

13) $3z^2 + 14z - 24 \leq 0$

14) $5k^2 + 36k + 7 \geq 0$

15) $7p^2 - 4 > 12p$

16) $4w^2 - 19w < 30$

17) $b^2 - 9b > 0$

18) $c^2 + 12c \leq 0$

19) $4y^2 \leq -5y$

20) $2a^2 \geq 7a$

21) $m^2 - 64 < 0$

22) $p^2 - 144 > 0$

23) $121 - h^2 \leq 0$

24) $1 - d^2 > 0$

25) $144 \geq 9s^2$

26) $81 \leq 25q^2$

Objective 3: Solve Quadratic Inequalities with Special Solutions

Solve each inequality.

27) $(k + 7)^2 \geq -9$ $(-\infty, \infty)$

28) $(h + 5)^2 \geq -2$ $(-\infty, \infty)$

29) $(3v - 11)^2 > -20$ $(-\infty, \infty)$

30) $(r + 4)^2 < -3$ $\varnothing$

31) $(2y - 1)^2 < -8$ $\varnothing$

32) $(4d - 3)^2 > -1$ $(-\infty, \infty)$

33) $(n + 3)^2 \leq -10$ $\varnothing$

34) $(5s - 2)^2 \leq -9$ $\varnothing$

Objective 4: Solve an Inequality of Higher Degree

Solve each inequality. Graph the solution set and write the solution in interval notation.

35) $(r + 2)(r - 5)(r - 1) \leq 0$

36) $(b + 2)(b - 3)(b - 12) > 0$

37) $(j - 7)(j - 5)(j + 9) \geq 0$

38) $(m + 4)(m - 7)(m + 1) \leq 0$

39) $(6c + 1)(c + 7)(4c - 3) < 0$

40) $(t + 2)(4t - 7)(5t - 1) \geq 0$

Objective 5: Solve a Rational Inequality

Solve each rational inequality. Graph the solution set and write the solution in interval notation.

41) $\dfrac{7}{p + 6} > 0$

42) $\dfrac{3}{v - 2} < 0$

43) $\dfrac{5}{z + 3} \leq 0$

44) $\dfrac{9}{m - 4} \geq 0$

45) $\dfrac{x - 4}{x - 3} > 0$

46) $\dfrac{a - 2}{a + 1} < 0$

47) $\dfrac{h - 9}{3h + 1} \leq 0$

48) $\dfrac{2c + 1}{c + 4} \geq 0$

49) $\dfrac{k}{k + 3} \leq 0$

50) $\dfrac{r}{r - 7} \geq 0$

51) $\dfrac{7}{t+6} < 3$

52) $\dfrac{3}{x+7} < -2$

53) $\dfrac{3}{a+7} \geq 1$

54) $\dfrac{5}{w-3} \leq 1$

55) $\dfrac{2y}{y-6} \leq -3$

56) $\dfrac{3z}{z+4} \geq 2$

57) $\dfrac{3w}{w+2} > -4$

58) $\dfrac{4h}{h+3} < 1$

59) $\dfrac{(6d+1)^2}{d-2} \leq 0$

60) $\dfrac{(x+2)^2}{x+7} \geq 0$

61) $\dfrac{(4t-3)^2}{t-5} > 0$

62) $\dfrac{(2y+3)^2}{y+3} < 0$

63) $\dfrac{n+6}{n^2+4} < 0$

64) $\dfrac{b-3}{b^2+2} > 0$

65) $\dfrac{m+1}{m^2+3} \geq 0$

66) $\dfrac{w-7}{w^2+8} \leq 0$

67) $\dfrac{s^2+2}{s-4} \leq 0$

68) $\dfrac{z^2+10}{z+6} \leq 0$

Mixed Exercises: Objectives 2 and 5

Write an inequality and solve.

69) Compu Corp. estimates that its total profit function, $P(x)$, for producing x thousand units is given by $P(x) = -2x^2 + 32x - 96$.

a) At what level of production does the company make a profit?   between 4000 and 12,000 units

b) At what level of production does the company lose money?
when it produces less than 4000 units or more than 12,000 units

70) A model rocket is launched from the ground with an initial velocity of 128 ft/s. The height $s(t)$, in ft, of the rocket t seconds after liftoff is given by the function $s(t) = -16t^2 + 128t$.

a) When is the rocket more than 192 ft above the ground?   between 2 and 6 sec after it is launched

b) When does the rocket hit the ground?
8 seconds after it is launched

71) A designer purse company has found that the average cost, $\overline{C}(x)$, of producing x purses per month can be described by the function $\overline{C}(x) = \dfrac{10x + 100,000}{x}$.
How many purses must the company produce each month so that the average cost of producing each purse is no more than $20?   10,000 or more

72) A company that produces clay pigeons for target shooting has determined that the average cost, $\overline{C}(x)$, of producing x cases of clay pigeons per month can be described by the function $\overline{C}(x) = \dfrac{2x + 15,000}{x}$. How many cases of clay pigeons must the company produce each month so that the average cost of producing each case is no more than $3?   15,000 or more

Chapter 9: Summary

Definition/Procedure	Example

9.1 The Square Root Property and Completing the Square

The Square Root Property

Let k be a constant. If $x^2 = k$, then
$x = \sqrt{k}$ or $x = -\sqrt{k}$. **(p. 534)**

Solve $6p^2 = 54$.

$$\begin{aligned} p^2 &= 9 && \text{Divide by 6.} \\ p &= \pm\sqrt{9} && \text{Square root property} \\ p &= \pm 3 && \sqrt{9} = 3 \end{aligned}$$

The solution set is $\{-3, 3\}$.

The Distance Formula

The **distance,** d, between two points with coordinates (x_1, y_1) and (x_2, y_2) is given by $d = \sqrt{(x_2 - x_1)^2 + (y_2 - y_1)^2}$. **(p. 538)**

Find the distance between the points $(6, -2)$ and $(0, 2)$.

Label the points: $\overset{x_1\ \ y_1}{(6, -2)}\ \overset{x_2\ y_2}{(0, 2)}$

Substitute the values into the distance formula.

$$\begin{aligned} d &= \sqrt{(0 - 6)^2 + (2 - (-2))^2} \\ &= \sqrt{(-6)^2 + (4)^2} \\ &= \sqrt{36 + 16} = \sqrt{52} = 2\sqrt{13} \end{aligned}$$

A **perfect square trinomial** is a trinomial whose factored form is the square of a binomial. **(p. 539)**

Perfect Square Trinomial	Factored Form
$y^2 + 8y + 16$	$(y + 4)^2$
$9t^2 - 30t + 25$	$(3t - 5)^2$

Complete the Square for $x^2 + bx$

To find the constant needed to complete the square for $x^2 + bx$,

Step 1: Find half of the coefficient of x: $\dfrac{1}{2}b$

Step 2: Square the result: $\left(\dfrac{1}{2}b\right)^2$

Step 3: Add it to $x^2 + bx$: $x^2 + bx + \left(\dfrac{1}{2}b\right)^2$. The factored form is $\left(x + \dfrac{1}{2}b\right)^2$. **(p. 539)**

Complete the square for $x^2 + 12x$ to obtain a perfect square trinomial. Then, factor.

Step 1: Find half of the coefficient of x: $\dfrac{1}{2}(12) = 6$

Step 2: Square the result: $6^2 = 36$

Step 3: Add 36 to $x^2 + 12x$: $x^2 + 12x + 36$

The perfect square trinomial is $x^2 + 12x + 36$.
The factored form is $(x + 6)^2$.

Solve a Quadratic Equation ($ax^2 + bx + c = 0$) by Completing the Square

Step 1: The coefficient of the squared term must be 1. If it is not 1, divide both sides of the equation by a to obtain a leading coefficient of 1.

Step 2: Get the variables on one side of the equal sign and the constant on the other side.

Step 3: Complete the square. Find half of the linear coefficient, then square the result. Add that quantity to both sides of the equation.

Step 4: Factor.

Step 5: Solve using the square root property. **(p. 541)**

Solve $x^2 + 6x + 7 = 0$ by completing the square.

$$\begin{aligned} x^2 + 6x + 7 &= 0 && \text{The coefficient of } x^2 \text{ is 1.} \\ x^2 + 6x &= -7 && \text{Get the constant on the other side of the equal sign.} \end{aligned}$$

Complete the square: $\dfrac{1}{2}(6) = 3$

$$(3)^2 = 9$$

Add 9 to both sides of the equation.

$$\begin{aligned} x^2 + 6x + 9 &= -7 + 9 \\ (x + 3)^2 &= 2 && \text{Factor.} \\ x + 3 &= \pm\sqrt{2} && \text{Square root property} \\ x &= -3 \pm \sqrt{2} \end{aligned}$$

The solution set is $\{-3 - \sqrt{2}, -3 + \sqrt{2}\}$.

606   Chapter 9   Quadratic Equations and Functions

Definition/Procedure	Example

9.2 The Quadratic Formula

The Quadratic Formula
The solutions of any quadratic equation of the form
$ax^2 + bx + c = 0 \ (a \neq 0)$ are

$$x = \frac{-b \pm \sqrt{b^2 - 4ac}}{2a} \quad \textbf{(p. 547)}$$

Solve $2x^2 - 5x - 2 = 0$ using the quadratic formula.

$$a = 2 \quad b = -5 \quad c = -2$$

Substitute the values into the quadratic formula, and simplify.

$$x = \frac{-(-5) \pm \sqrt{(-5)^2 - 4(2)(-2)}}{2(2)}$$

$$x = \frac{5 \pm \sqrt{25 + 16}}{4} = \frac{5 \pm \sqrt{41}}{4}$$

The solution set is $\left\{ \dfrac{5 - \sqrt{41}}{4}, \dfrac{5 + \sqrt{41}}{4} \right\}$.

The expression under the radical, $b^2 - 4ac$ is called the **discriminant.**

1) If $b^2 - 4ac$ is **positive and the square of an integer,** the equation has **two rational solutions.**
2) If $b^2 - 4ac$ is **positive but not a perfect square,** the equation has **two irrational solutions.**
3) If $b^2 - 4ac$ is **negative,** the equation has **two nonreal, complex solutions of the form $a + bi$ and $a - bi$.**
4) If $b^2 - 4ac = 0$, the equation has **one rational solution.**
   (p. 550)

Find the value of the discriminant for $3m^2 + 4m + 5 = 0$ and determine the number and type of solutions of the equation.

$$a = 3 \quad b = 4 \quad c = 5$$

$$b^2 - 4ac = (4)^2 - 4(3)(5) = 16 - 60 = -44$$

Discriminant $= -44$. The equation has two nonreal, complex solutions of the form $a + bi$ and $a - bi$.

9.3 Equations in Quadratic Form

Some equations that are not quadratic can be solved using the same methods that can be used to solve quadratic equations. These are called **equations in quadratic form. (p. 558)**

Solve $\quad r^4 + 2r^2 - 24 = 0.$

$$(r^2 - 4)(r^2 + 6) = 0 \quad \text{Factor.}$$

$$r^2 - 4 = 0 \qquad \text{or} \qquad r^2 + 6 = 0$$
$$r^2 = 4 \qquad\qquad\qquad r^2 = -6$$
$$r = \pm\sqrt{4} \qquad\qquad\quad r = \pm\sqrt{-6}$$
$$r = \pm 2 \qquad\qquad\qquad r = \pm i\sqrt{6}$$

The solution set is $\{-i\sqrt{6}, i\sqrt{6}, -2, 2\}$.

9.4 Formulas and Applications

Solve a Formula for a Variable. (p. 564)

Solve for s: $\quad g = \dfrac{10}{s^2}$

$$s^2 g = 10 \qquad \text{Multiply both sides by } s^2.$$

$$s^2 = \frac{10}{g} \qquad \text{Divide both sides by } g.$$

$$s = \pm\sqrt{\frac{10}{g}} \qquad \text{Square root property}$$

$$s = \frac{\pm\sqrt{10}}{\sqrt{g}} \cdot \frac{\sqrt{g}}{\sqrt{g}} \qquad \text{Rationalize the denominator.}$$

$$s = \frac{\pm\sqrt{10g}}{g}$$

Definition/Procedure	Example

Solving Application Problems Using a Quadratic Equation. (p. 568)

A woman dives off of a cliff 49 m above the ocean. Her height, $h(t)$, in meters, above the water is given by

$$h(t) = -9.8t^2 + 49$$

where t is the time, in seconds, after she leaves the cliff. When will she hit the water?

Let $h(t) = 0$ and solve for t.

$$h(t) = -9.8t^2 + 49$$
$$0 = -9.8t^2 + 49 \qquad \text{Substitute 0 for } h.$$
$$9.8t^2 = 49 \qquad \text{Add } 9.8t^2 \text{ to each side.}$$
$$t^2 = 5 \qquad \text{Divide by 9.8.}$$
$$t = \pm\sqrt{5} \qquad \text{Square root property}$$

Since t represents time, we discard $-\sqrt{5}$. She will hit the water in $\sqrt{5}$, or about 2.2, sec.

9.5 Quadratic Functions and Their Graphs

A **quadratic function** is a function that can be written in the form $f(x) = ax^2 + bx + c$, where a, b, and c are real numbers and $a \neq 0$. The graph of a quadratic function is called a **parabola.**

The lowest point on an upward-opening parabola or the highest point on a downward-opening parabola is called the **vertex.** (pp. 571–572)

$f(x) = 5x^2 + 7x - 9$ is a quadratic function.

A quadratic function can also be written in the form $f(x) = a(x - h)^2 + k$:

1) The vertex of the parabola is (h, k).
2) The axis of symmetry is the vertical line with equation $x = h$.
3) If a is positive, the parabola opens upward. If a is negative, the parabola opens downward.
4) If $|a| < 1$, then the graph of $f(x) = a(x - h)^2 + k$ is *wider* than the graph of $y = x^2$.

   If $|a| > 1$, then the graph of $f(x) = a(x - h)^2 + k$ is *narrower* than the graph of $y = x^2$. (p. 575)

Graph $f(x) = -(x + 3)^2 + 4$.

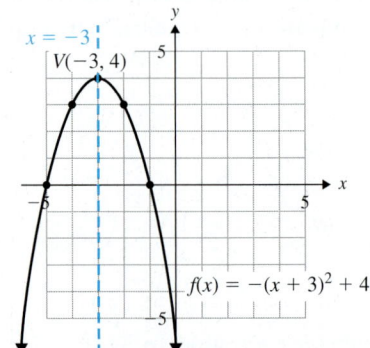

Vertex: $(-3, 4)$

Axis of symmetry: $x = -3$

$a = -1$, so graph opens downward. The domain is $(-\infty, \infty)$; the range is $(-\infty, 4]$.

Definition/Procedure	Example

When a quadratic function is written in the form $f(x) = ax^2 + bx + c$, there are two methods we can use to graph the function.

Method 1: Rewrite $f(x) = ax^2 + bx + c$ in the form $f(x) = a(x - h)^2 + k$ by *completing the square.*

Method 2: Use the formula $h = -\dfrac{b}{2a}$ to find the *x*-coordinate of the vertex. The vertex has coordinates $\left(-\dfrac{b}{2a}, f\left(-\dfrac{b}{2a}\right)\right)$. **(p. 576)**

Graph $f(x) = x^2 + 4x + 5$.

Method 1: Complete the square.

$$f(x) = x^2 + 4x + 5$$
$$f(x) = (x^2 + 4x + 2^2) + 5 - 2^2$$
$$f(x) = (x^2 + 4x + 4) + 5 - 4$$
$$f(x) = (x + 2)^2 + 1$$

The vertex of the parabola is $(-2, 1)$. The axis of symmetry is $x = -2$. The parabola opens upward and has the same shape as $f(x) = x^2$. The graph is shown.

Method 2: Use the formula $h = -\dfrac{b}{2a}$.

$$h = -\frac{4}{2(1)} = -2. \text{ Then, } f(-2) = 1.$$

The vertex of the parabola is $(-2, 1)$. The axis of symmetry is $x = -2$. The domain is $(-\infty, \infty)$; the range is $[-2, \infty)$.

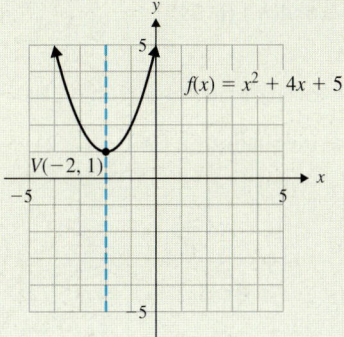

9.6 Applications of Quadratic Functions and Graphing Other Parabolas

Maximum and Minimum Values of a Quadratic Function

Let $f(x) = ax^2 + bx + c$.

1) If *a* is *positive*, the *y*-coordinate of the vertex is the **minimum** value of the function $f(x)$.

2) If *a* is *negative*, the *y*-coordinate of the vertex is the **maximum** value of the function $f(x)$. **(p. 585)**

Find the minimum value of the function $f(x) = 2x^2 + 12x + 7$.

Since *a* is positive ($a = 2$), the function's *minimum value* is at the vertex.

The *x*-coordinate of the vertex is $h = -\dfrac{b}{2a} = -\dfrac{12}{2(2)} = -3$.

The *y*-coordinate of the vertex is
$$f(-3) = 2(-3)^2 + 12(-3) + 7$$
$$= 18 - 36 + 7 = -11$$

The minimum value of the function is -11.

Definition/Procedure	Example

The graph of the quadratic equation $x = ay^2 + by + c$ is a parabola that opens in the x-direction, or horizontally.

The quadratic equation $x = ay^2 + by + c$ can also be written in the form $x = a(y - k)^2 + h$. When it is written in this form we can find the following.

1) The vertex of the parabola is (h, k).

2) The axis of symmetry is the horizontal line $y = k$.

3) If a is positive, the graph opens to the right.

   If a is negative, the graph opens to the left. **(p. 589)**

Graph $x = \dfrac{1}{2}(y + 4)^2 - 2$.

Vertex: $(-2, -4)$

Axis of symmetry: $y = -4$

$a = \dfrac{1}{2}$, so the graph opens to the right. The domain is $[-2, \infty)$; the range is $(-\infty, \infty)$.

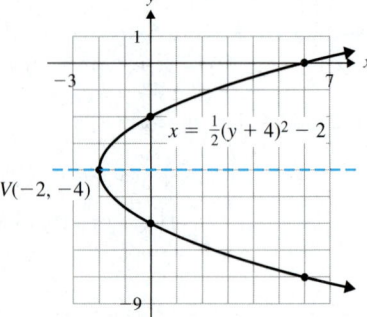

$x = \frac{1}{2}(y + 4)^2 - 2$

$V(-2, -4)$

9.7 Quadratic and Rational Inequalities

A **quadratic inequality** can be written in the form

$$ax^2 + bx + c \le 0 \quad \text{or} \quad ax^2 + bx + c \ge 0$$

where a, b, and c are real numbers and $a \ne 0$. ($<$ and $>$ may be substituted for $\le$ and $\ge$.) **(p. 595)**

An inequality containing a rational expression, like $\dfrac{c - 5}{c + 1} \le 0$, is called a **rational inequality. (p. 600)**

Solve $r^2 - 4r \ge 12$.

Step 1: $r^2 - 4r - 12 \ge 0$ Subtract 12.

Since the inequality symbol is $\ge$, the solution set contains the interval(s) where the quantity $r^2 - 4r - 12$ is *positive.*

Step 2: Solve $r^2 - 4r - 12 = 0$.

$$(r - 6)(r + 2) = 0 \quad \text{Factor.}$$

$$r - 6 = 0 \quad \text{or} \quad r + 2 = 0$$
$$r = 6 \quad \text{or} \quad r = -2$$

Step 3: Put $r = 6$ and $r = -2$ on a number line.

$r^2 - 4r - 12$ Positive   Negative   Positive
-2 6

Step 4: Choose a test number in each interval to determine the sign of $r^2 - 4r - 12$.

Step 5: The solution set will contain the numbers in the intervals where $r^2 - 4r - 12$ is positive.

Step 6: The endpoints of the intervals are included since the inequality is $\ge$. The graph of the solution set is

$-5\ -4\ -3\ -2\ -1\ \ 0\ \ 1\ \ 2\ \ 3\ \ 4\ \ 5\ \ 6\ \ 7\ \ 8\ \ 9\ \ 10$

The solution set of $r^2 - 4r - 12$ is $(-\infty, -2] \cup [6, \infty)$.

Additional answers can be found in the Answers to Exercises appendix.

(9.1) Solve using the square root property.

1) $d^2 = 144$ $\{-12, 12\}$

2) $m^2 = 75$ $\{-5\sqrt{3}, 5\sqrt{3}\}$

3) $v^2 + 4 = 0$ $\{-2i, 2i\}$

4) $2c^2 - 11 = 25$ $\{-3\sqrt{2}, 3\sqrt{2}\}$

5) $(b - 3)^2 = 49$ $\{-4, 10\}$

6) $(6y + 7)^2 - 15 = 0$ $\left\{\dfrac{-7 - \sqrt{15}}{6}, \dfrac{-7 + \sqrt{15}}{6}\right\}$

7) $27k^2 - 30 = 0$ $\left\{\dfrac{-\sqrt{10}}{3}, \dfrac{\sqrt{10}}{3}\right\}$

8) $(j - 14)^2 + 5 = 0$ $\{14 - i\sqrt{5}, 14 + i\sqrt{5}\}$

9) Find the distance between the points $(-8, 3)$ and $(-12, 5)$. $2\sqrt{5}$

10) A rectangle has a length of $5\sqrt{2}$ in. and a width of 4 in. How long is its diagonal? $\sqrt{66}$ in.

Complete the square for each expression to obtain a perfect square trinomial. Then, factor.

11) $r^2 + 10r$
$r^2 + 10r + 25; (r + 5)^2$

12) $z^2 - 12z$
$z^2 - 12z + 36; (z - 6)^2$

13) $c^2 - 5c$
$c^2 - 5c + \dfrac{25}{4}; \left(c - \dfrac{5}{2}\right)^2$

14) $x^2 + x$
$x^2 + x + \dfrac{1}{4}; \left(x + \dfrac{1}{2}\right)^2$

15) $a^2 + \dfrac{2}{3}a$
$a^2 + \dfrac{2}{3}a + \dfrac{1}{9}; \left(a + \dfrac{1}{3}\right)^2$

16) $d^2 - \dfrac{5}{2}d$
$d^2 - \dfrac{5}{2}d + \dfrac{25}{16}; \left(d - \dfrac{5}{4}\right)^2$

Solve by completing the square.

17) $p^2 - 6p - 16 = 0$ $\{-2, 8\}$

18) $w^2 - 2w - 35 = 0$ $\{-5, 7\}$

19) $n^2 + 10n = 6$
$\{-5 - \sqrt{31}, -5 + \sqrt{31}\}$

20) $t^2 + 9 = -4t$
$\{-2 - i\sqrt{5}, -2 + i\sqrt{5}\}$

21) $f^2 + 3f + 1 = 0$ $\left\{-\dfrac{3}{2} - \dfrac{\sqrt{5}}{2}, -\dfrac{3}{2} + \dfrac{\sqrt{5}}{2}\right\}$

22) $j^2 - 7j = 4$ $\left\{\dfrac{7}{2} - \dfrac{\sqrt{65}}{2}, \dfrac{7}{2} + \dfrac{\sqrt{65}}{2}\right\}$

23) $-3q^2 + 7q = 12$ $\left\{\dfrac{7}{6} - \dfrac{\sqrt{95}}{6}i, \dfrac{7}{6} + \dfrac{\sqrt{95}}{6}i\right\}$

24) $6v^2 - 15v + 3 = 0$ $\left\{\dfrac{5}{4} - \dfrac{\sqrt{17}}{4}, \dfrac{5}{4} + \dfrac{\sqrt{17}}{4}\right\}$

(9.2) Solve using the quadratic formula.

25) $m^2 + 4m - 12 = 0$ $\{-6, 2\}$

26) $3y^2 = 10y - 8$ $\left\{\dfrac{4}{3}, 2\right\}$

27) $10g - 5 = 2g^2$ $\left\{\dfrac{5 - \sqrt{15}}{2}, \dfrac{5 + \sqrt{15}}{2}\right\}$

28) $20 = 4x - 5x^2$ $\left\{\dfrac{2}{5} - \dfrac{4\sqrt{6}}{5}i, \dfrac{2}{5} + \dfrac{4\sqrt{6}}{5}i\right\}$

29) $\dfrac{1}{6}t^2 - \dfrac{1}{3}t + \dfrac{2}{3} = 0$ $\{1 - i\sqrt{3}, 1 + i\sqrt{3}\}$

30) $(s - 3)(s - 5) = 9$ $\{4 - \sqrt{10}, 4 + \sqrt{10}\}$

31) $(6r + 1)(r - 4) = -2(12r + 1)$ $\left\{-\dfrac{2}{3}, \dfrac{1}{2}\right\}$

32) $z^2 - \dfrac{3}{2}z + \dfrac{13}{16} = 0$ $\left\{\dfrac{3}{4} - \dfrac{1}{2}i, \dfrac{3}{4} + \dfrac{1}{2}i\right\}$

Find the value of the discriminant. Then, determine the number and type of solutions of each equation. Do not solve.

33) $3n^2 - 2n - 5 = 0$
64; two rational solutions

34) $t^2 = -3(t + 2)$
-15; two nonreal, complex solutions

35) Find the value of b so that $4k^2 + bk + 9 = 0$ has only one rational solution. -12 or 12

36) A ball is thrown upward from a height of 4 ft. The height, $h(t)$, of the ball (in feet) t sec after the ball is released is given by $h(t) = -16t^2 + 52t + 4$.

a) How long does it take the ball to reach a height of 16 ft? 0.25 sec on the way up, 3 sec on the way down

b) How long does it take the ball to hit the ground? $\dfrac{13 + \sqrt{185}}{8}$ sec or about 3.3 sec

(9.1–9.2) Keep in mind the four methods we have learned for solving quadratic equations: *factoring, the square root property, completing the square, and the quadratic formula.* Solve the equations using one of these methods.

37) $3k^2 + 4 = 7k$ $\left\{1, \dfrac{4}{3}\right\}$

38) $n^2 - 6n + 11 = 0$ $\{3 - i\sqrt{2}, 3 + i\sqrt{2}\}$

39) $15 = 3 + (y + 8)^2$ $\{-8 - 2\sqrt{3}, -8 + 2\sqrt{3}\}$

40) $(2a + 1)(a + 2) = 14$ $\left\{-4, \dfrac{3}{2}\right\}$

41) $\dfrac{1}{3}w^2 + w = -\dfrac{5}{6}$ $\left\{-\dfrac{3}{2} - \dfrac{1}{2}i, -\dfrac{3}{2} + \dfrac{1}{2}i\right\}$

42) $4t^2 + 5 = 7$ $\left\{-\dfrac{\sqrt{2}}{2}, \dfrac{\sqrt{2}}{2}\right\}$

43) $6 + p(p - 10) = 2(4p - 15)$ $\{9 - 3\sqrt{5}, 9 + 3\sqrt{5}\}$

44) $6 = 2m - 3m^2$ $\left\{\dfrac{1}{3} - \dfrac{\sqrt{17}}{3}i, \dfrac{1}{3} + \dfrac{\sqrt{17}}{3}i\right\}$

45) $x^3 = x$ $\{-1, 0, 1\}$

46) $\dfrac{1}{12}b^2 - \dfrac{9}{2} = \dfrac{1}{4}b$ $\{-6, 9\}$

47) Let $f(x) = (2x - 1)^2$. Find all values of x so that $f(x) = 25$. $-2, 3$

48) Let $f(x) = \dfrac{1}{10}x^2 + 3x$ and $g(x) = 4x - \dfrac{11}{5}$. Find all values of x such that $f(x) = g(x)$. $5 - \sqrt{3}, 5 + \sqrt{3}$

(9.3) Solve.

49) $\dfrac{5k}{k + 1} = 3k - 4$ $\left\{\dfrac{3 - \sqrt{21}}{3}, \dfrac{3 + \sqrt{21}}{3}\right\}$

50) $\dfrac{10}{m} = 3 + \dfrac{8}{m^2}$ $\left\{\dfrac{4}{3}, 2\right\}$

51) $f = \sqrt{7f - 12}$ $\{3, 4\}$

52) $x - 4\sqrt{x} = 5$ $\{25\}$

53) $n^4 - 17n^2 + 16 = 0$ $\{-4, -1, 1, 4\}$

54) $b^4 + 5b^2 - 14 = 0$ $\{-\sqrt{2}, \sqrt{2}, -i\sqrt{7}, i\sqrt{7}\}$

55) $q^{2/3} + 2q^{1/3} - 3 = 0$ $\{-27, 1\}$

56) $y + 2 = 3y^{1/2}$ $\{1, 4\}$

57) $2r^4 = 7r^2 - 2$

58) $2(v + 2)^2 + (v + 2) - 3 = 0$ $\left\{-\dfrac{7}{2}, -1\right\}$

59) $(2k - 5)^2 - 5(2k - 5) - 6 = 0$ $\left\{2, \dfrac{11}{2}\right\}$

Write an equation and solve.

60) At the end of the day, the employees at Forever Young have to put all clothes left in the dressing room back to their proper places. Working together, Lorena and Erica can put away the clothes in 1 hr 12 min. On her own, it takes Lorena 1 hr longer to put away the clothes than it takes Erica to do it by herself. How long does it take each girl to put away the clothes by herself? *Erica: 2 hr; Lorena: 3 hr*

(9.4) Solve for the indicated variable.

61) $F = \dfrac{mv^2}{r}$ for v $v = \dfrac{\pm\sqrt{Frm}}{m}$

62) $U = \dfrac{1}{2}kx^2$ for x $x = \dfrac{\pm\sqrt{2Uk}}{k}$

63) $r = \sqrt{\dfrac{A}{\pi}}$ for A $A = \pi r^2$

64) $r = \sqrt{\dfrac{V}{\pi l}}$ for V $V = \pi r^2 l$

65) $kn^2 - ln - m = 0$ for n $n = \dfrac{l \pm \sqrt{l^2 + 4km}}{2k}$

66) $2p^2 + t = rp$ for p $p = \dfrac{r \pm \sqrt{r^2 - 8t}}{4}$

Write an equation and solve.

67) Ayesha is making a pillow sham by sewing a border onto an old pillow case. The rectangular pillow case measures 18 in. by 27 in. When she sews a border of uniform width around the pillowcase, the total area of the surface of the pillow sham will be 792 in². How wide is the border? *3 in.*

68) The width of a rectangular piece of cardboard is 4 in. less than its length. A square piece that measures 2 in. on each side is cut from each corner, then the sides are turned up to make a box with volume 280 in³. Find the length and width of the original piece of cardboard. *length = 18 in., width = 14 in.*

69) A flower shop determined that the demand, $D(P)$, for its tulip bouquet is $D(P) = \dfrac{240}{P}$ per week, where P is the price of the bouquet in dollars. The weekly supply, $S(P)$, is given by $S(P) = 4p - 2$. Find the price at which demand for the tulips equals the supply. *$8.00*

70) U.S. sales of a certain brand of wine can be modeled by

$$y = -0.20x^2 + 4.0x + 8.4$$

for the years 1995–2010. x is the number of years after 1995 and y is the number of bottles sold, in millions.

a) How many bottles were sold in 1995? *8.4 million*

b) How many bottles were sold in 2008? *26.6 million*

c) In what year did sales reach 28.4 million bottles? *2005*

(9.5)

71) Given a quadratic function in the form

$$f(x) = a(x - h)^2 + k,$$

a) what is the vertex? *(h, k)*

b) what is the equation of the axis of symmetry? *x = h*

c) what does the sign of a tell us about the graph of f?

72) Match each function to its graph.

$f(x) = (x - 2)^2, g(x) = x^2 - 2, h(x) = -x^2 + 2,$
$k(x) = (x + 2)^2$

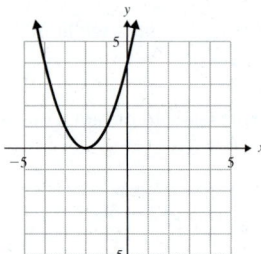

k(x)

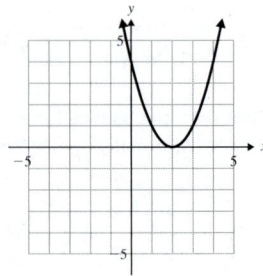

f(x)

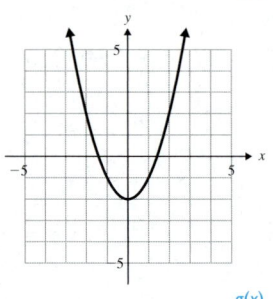

g(x)

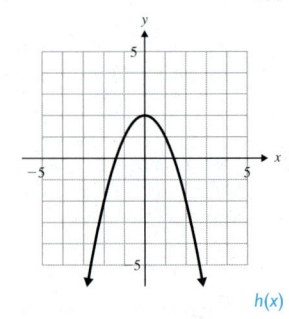

h(x)

For each quadratic function, identify the vertex, axis of symmetry, and x- and y-intercepts. Then, graph the function. Determine the domain and range.

73) $f(x) = x^2 - 4$

74) $h(x) = -(x + 1)^2$

75) $f(x) = (x + 2)^2 - 1$

76) $y = 2x^2$

77) $y = (x - 4)^2 + 2$

78) $g(x) = -\dfrac{1}{2}(x - 3)^2 - 2$

79) If the graph of $f(x) = x^2$ is shifted 6 units to the right to obtain the graph of $g(x)$, what is the equation of $g(x)$? *g(x) = (x - 6)²*

80) What are two ways to find the vertex of the graph of $f(x) = ax^2 + bx + c$?

Rewrite each function in the form $f(x) = a(x - h)^2 + k$ by completing the square. Then, graph the function. Include the intercepts. Determine the domain and range.

81) $f(x) = x^2 - 2x + 3$

82) $y = x^2 + 4x - 1$

83) $y = \dfrac{1}{2}x^2 - 4x + 9$

84) $f(x) = -2x^2 - 8x + 2$

Graph each equation using the vertex formula. Include the intercepts. Determine the domain and range.

85) $y = -x^2 - 6x - 10$ 86) $f(x) = x^2 - 2x - 4$

(9.6) Solve.

87) An object is thrown upward from a height of 240 ft so that its height h (in feet) t sec after being thrown is given by

$$h(t) = -16t^2 + 32t + 240.$$

  a) How long does it take the object to reach its maximum height?   1 sec

  b) What is the maximum height attained by the object?
     256 ft

  c) How long does it take the object to hit the ground?
     5 sec

88) A restaurant wants to add outdoor seating to its inside service. It has 56 ft of fencing to enclose a rectangular, outdoor café. Find the dimensions of the outdoor café of maximum area if the building will serve as one side of the café.

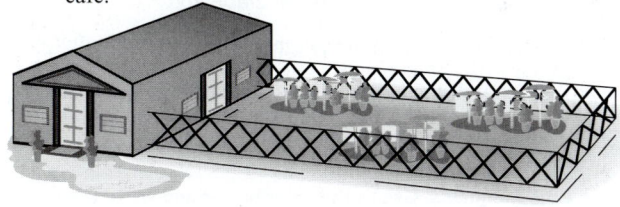

14 ft × 28 ft

For each quadratic equation, identify the vertex, axis of symmetry, and x- and y-intercepts. Then, graph the equation. Determine the domain and range.

89) $x = -(y - 3)^2 + 11$ 90) $x = (y + 1)^2 - 5$

Rewrite each equation in the form $x = a(y - k)^2 + h$ by completing the square. Then, graph the equation. Include the intercepts. Determine the domain and range.

91) $x = y^2 + 8y + 7$ 92) $x = -y^2 + 4y - 4$

Graph each equation using the vertex formula. Include the intercepts. Determine the domain and range.

93) $x = -\dfrac{1}{2}y^2 - 3y - \dfrac{5}{2}$ 94) $x = 3y^2 - 12y$

(9.7) Solve each inequality. Graph the solution set and write the solution in interval notation.

95) $a^2 + 2a - 3 < 0$ $(-3, 1)$

96) $4m^2 + 8m \geq 21$ $\left(-\infty, -\dfrac{7}{2}\right] \cup \left[\dfrac{3}{2}, \infty\right)$

97) $64v^2 \geq 25$ $\left(-\infty, -\dfrac{5}{8}\right] \cup \left[\dfrac{5}{8}, \infty\right)$

98) $36 - r^2 > 0$ $(-6, 6)$

99) $(5c + 2)(c - 4)(3c + 1) < 0$ $\left(-\infty, -\dfrac{2}{5}\right) \cup \left(-\dfrac{1}{3}, 4\right)$

100) $(p - 6)^2 \leq -5$ $\varnothing$

101) $\dfrac{t + 7}{2t - 3} > 0$ $(-\infty, -7) \cup \left(\dfrac{3}{2}, \infty\right)$

102) $\dfrac{6}{g - 7} \leq 0$ $(-\infty, 7)$

103) $\dfrac{z}{z - 2} \leq 3$ $(-\infty, 2) \cup [3, \infty)$

104) $\dfrac{1}{n - 4} > -3$ $\left(-\infty, \dfrac{11}{3}\right) \cup (4, \infty)$

105) $\dfrac{r^2 + 4}{r - 7} \geq 0$ $(7, \infty)$

Solve.

106) Custom Bikes, Inc., estimates that its total profit function, $P(x)$, for producing x thousand units is given by $P(x) = -2x^2 + 32x - 110$. At what level of production does the company make a profit?
     between 5000 and 11,000 units

*Additional answers can be found in the Answers to Exercises appendix.

1) Solve $b^2 + 4b - 7 = 0$ by completing the square.
$$\{-2 - \sqrt{11}, -2 + \sqrt{11}\}$$

2) Solve $x^2 - 8x + 17 = 0$ using the quadratic formula.
$$\{4 - i, 4 + i\}$$

Solve using any method.

3) $(c + 5)^2 + 8 = 2$ $\{-5 - i\sqrt{6}, -5 + i\sqrt{6}\}$

4) $3q^2 + 2q = 8$ $\left\{-2, \dfrac{4}{3}\right\}$

5) $(4n + 1)^2 + 9(4n + 1) + 18 = 0$ $\left\{-\dfrac{7}{4}, -1\right\}$

6) $(2t - 3)(t - 2) = 2$ $\left\{\dfrac{7 - \sqrt{17}}{4}, \dfrac{7 + \sqrt{17}}{4}\right\}$

7) $p^4 + p^2 - 72 = 0$ $\{-2\sqrt{2}, 2\sqrt{2}, -3i, 3i\}$

8) $\dfrac{3}{10x} = \dfrac{x}{x - 1} - \dfrac{4}{5}$ $\left\{-\dfrac{3}{2}, -1\right\}$

9) Find the value of the discriminant. Then, determine the number and type of solutions of the equation. *Do not solve.*
$$5z^2 - 6z - 1 = 0$$
56: two irrational solutions

10) Find the length of the missing side.

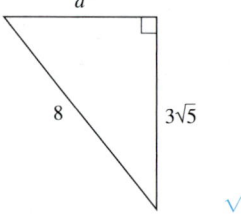

11) Let $P(x) = 5x^2$ and $Q(x) = 2x$. Find all values of x so that $P(x) = Q(x)$. $0, \dfrac{2}{5}$

12) Find the distance between the points $(7, -4)$ and $(5, 6)$. $2\sqrt{26}$

Write an equation and solve.

13) A rectangular piece of sheet metal is 6 in. longer than it is wide. A square piece that measures 3 in. on each side is cut from each corner, then the sides are turned up to make a box with volume 273 in³. Find the length and width of the original piece of sheet metal.   width = 13 in., length = 19 in.

14) Solve for V. $r = \sqrt{\dfrac{3V}{\pi h}}$ $V = \dfrac{1}{3}\pi r^2 h$

15) Solve for t. $rt^2 - st = 6$ $t = \dfrac{s \pm \sqrt{s^2 + 24r}}{2r}$

16) Graph $f(x) = x^2$ and $g(x) = x^2 + 2$.

17) If the graph of $f(x) = x^2$ is shifted 3 units to the left to obtain the graph of $g(x)$, what is the equation of $g(x)$?
$$g(x) = (x + 3)^2$$

Graph each equation. Identify the vertex, axis of symmetry, and intercepts. Determine the domain and range.

18) $f(x) = -(x + 2)^2 + 4$

19) $x = y^2 - 3$

20) $x = 3y^2 - 6y + 5$

21) $g(x) = x^2 - 6x + 8$

22) A ball is projected upward from the top of a 200-ft tall building. The height $h(t)$ of the ball above the ground (in feet) t sec after the ball is released is given by
$$h(t) = -16t^2 + 24t + 200.$$

a) What is the maximum height attained by the ball? 209 ft

b) When will the ball be 40 ft above the ground? after 4 sec

c) When will the ball hit the ground?   after $\dfrac{3 + \sqrt{209}}{4}$ sec or about 4.4 sec

Solve each inequality. Graph the solution set and write the solution in interval notation.

23) $y^2 + 4y - 45 \geq 0$ $(-\infty, -9] \cup [5, \infty)$

24) $\dfrac{m - 5}{m + 3} \geq 0$ $(-\infty, -3) \cup [5, \infty)$

25) A company has determined that the average cost, $\overline{C}(x)$, of producing x backpacks per month can be described by the function $\overline{C}(x) = \dfrac{5x + 80,000}{x}$. How many backpacks must the company produce each month so that the average cost of producing each backpack is no more than $15? 8000 or more

Additional answers can be found in the Answers to Exercises appendix.

1) Simplify $\dfrac{\frac{12}{35}}{\frac{24}{49}}$. $\frac{7}{10}$

Simplify. The final answer should contain only positive exponents.

2) $(-2d^5)^3$ $-8d^{15}$

3) $(5x^4y^{-10})(3xy^3)^2$ $\dfrac{45x^6}{y^4}$

4) *Write an equation and solve.*
In December 2010, an electronics store sold 108 digital cameras. This is a 20% increase over their sales in December 2009. How many digital cameras did they sell in December 2009? 90

5) Solve for m. $y = mx + b$ $m = \dfrac{y-b}{x}$

6) Given the relation
$\{(4, 0), (3, 1), (3, -1), (0, 2)\}$,

   a) what is the domain? $\{0, 3, 4\}$

   b) what is the range? $\{-1, 0, 1, 2\}$

   c) is the relation a function? no

7) Find the x- and y-intercepts of $2x - 5y = 8$ and graph.

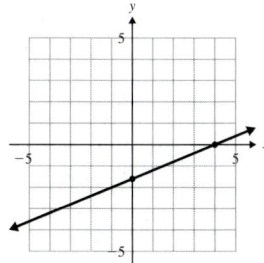

x-int: (4, 0); y-int: $\left(0, -\dfrac{8}{5}\right)$

8) Let $f(x) = \sqrt{x + 3}$.

   a) Find $f(1)$. 2

   b) Find the domain of f. $[-3, \infty)$

   c) Graph the function.

9) *Write a system of two equations in two variables and solve.*
Two bags of chips and three cans of soda cost $3.85, while one bag of chips and two cans of soda cost $2.30. Find the cost of a bag of chips and a can of soda. chips: $0.80; soda: $0.75

10) Subtract
$(4x^2y^2 - 11x^2y + xy + 2) - (x^2y^2 - 6x^2y + 3xy^2 + 10xy - 6)$
$3x^2y^2 - 5x^2y - 3xy^2 - 9xy + 8$

11) Multiply and simplify $3(r - 5)^2$. $3r^2 - 30r + 75$

Factor completely.

12) $4p^3 + 14p^2 - 8p$ $2p(2p - 1)(p + 4)$

13) $a^3 + 125$ $(a + 5)(a^2 - 5a + 25)$

14) Add $\dfrac{z-8}{z+4} + \dfrac{3}{z}$ $\dfrac{z^2 - 5z + 12}{z(z+4)}$

15) Simplify $\dfrac{2 + \frac{6}{c}}{\frac{2}{c^2} - \frac{8}{c}}$ $\dfrac{c(c+3)}{1 - 4c}$

16) Solve $|4k - 3| = 9$. $\left\{-\dfrac{3}{2}, 3\right\}$

17) Solve this system: $4x - 2y + z = -7$
$-3x + y - 2z = 5$
$2x + 3y + 5z = 4$ $(0, 3, -1)$

Simplify. Assume all variables represent nonnegative real numbers.

18) $\sqrt{75}$ $5\sqrt{3}$

19) $\sqrt[3]{40}$ $2\sqrt[3]{5}$

20) $\sqrt{63x^7y^4}$ $3x^3y^2\sqrt{7x}$

21) Simplify $64^{2/3}$. 16

22) Rationalize the denominator: $\dfrac{5}{2 + \sqrt{3}}$. $10 - 5\sqrt{3}$

23) Multiply and simplify $(10 + 3i)(1 - 8i)$. $34 - 77i$

Solve.

24) $3k^2 - 4 = 20$ $\{-2\sqrt{2}, 2\sqrt{2}\}$

25) $\dfrac{3}{5}x^2 + \dfrac{1}{5} = \dfrac{1}{5}x$ $\left\{\dfrac{1}{6} - \dfrac{\sqrt{11}}{6}i, \dfrac{1}{6} + \dfrac{\sqrt{11}}{6}i\right\}$

26) $1 - \dfrac{1}{3h - 2} = \dfrac{20}{(3h - 2)^2}$ $\left\{-\dfrac{2}{3}, \dfrac{7}{3}\right\}$

27) $p^2 + 6p = 27$ $\{-9, 3\}$

28) Solve for V: $r = \sqrt{\dfrac{V}{\pi h}}$. $V = \pi r^2 h$

29) Graph $f(x) = (x - 1)^2 - 4$.

30) Solve $25p^2 \le 144$. $\left[-\dfrac{12}{5}, \dfrac{12}{5}\right]$

Exponential and Logarithmic Functions

Algebra at Work: Finances

Financial planners use mathematical formulas involving exponential functions every day to help clients invest their money.

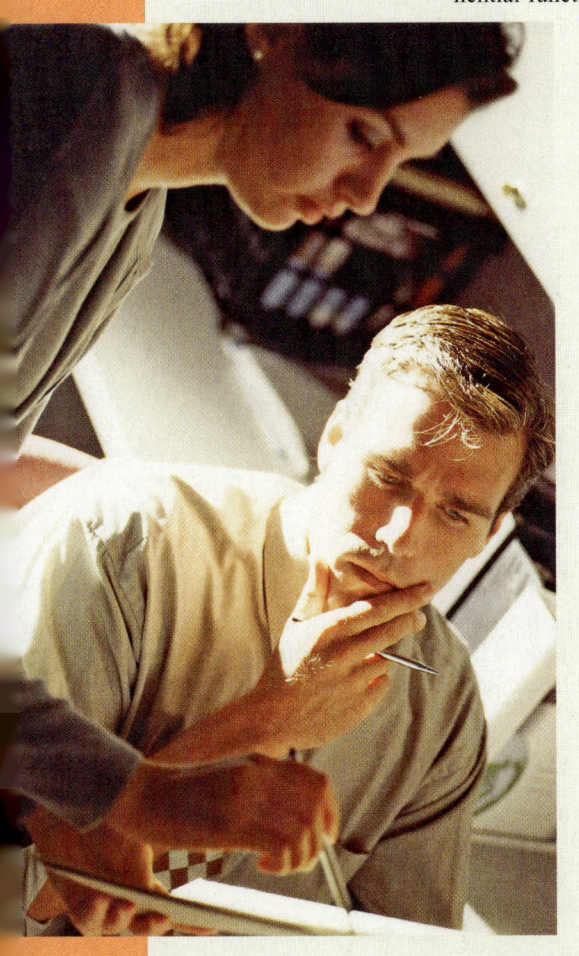

Interest on investments can be paid in various ways—monthly, weekly, continuously, etc.

Isabella has a client who wants to invest $5000 for 4 yr. He would like to know if it is better to invest it at 4.8% compounded continuously or at 5% compounded monthly.

To determine how much the investment will be worth after 4 yr if it is put in the account with interest compounded continuously, Isabella uses the formula $A = Pe^{rt}$ ($e \approx 2.71828$). To determine how much money her client will have after 4 yr if he puts his $5000 into the account earning 5% interest compounded monthly, Isabella uses the formula $A = P\left(1 + \dfrac{r}{n}\right)^{nt}$.

Compounded Continuously

$$A = Pe^{rt}$$

$$A = 5000e^{(0.048)(4)}$$

$$A = \$6058.35$$

Compounded Monthly

$$A = P\left(1 + \frac{r}{n}\right)^{nt}$$

$$A = (5000)\left(1 + \frac{0.05}{12}\right)^{(12)(4)}$$

$$A = \$6104.48$$

If her client invests $5000 in the account paying 4.8% interest compounded continuously, his investment will grow to $6058.35 after 4 yr. If he puts the money in the account paying 5% compounded monthly, he will have $6104.48. Isabella advises him to invest his money in the 5% account.

We will work with these and other exponential functions in this chapter.

Section 10.1 Composite and Inverse Functions

Objectives

1. Find the Composition of Functions
2. Use Function Composition
3. Decide if a Function is One-to-One
4. Find the Inverse of a Function
5. Given the Graph of $f(x)$, Graph $f^{-1}(x)$
6. Show that $(f^{-1} \circ f)(x) = x$ and $(f \circ f^{-1})(x) = x$

Later in this chapter we will learn about exponential and logarithmic functions. These functions are important not only in mathematics but also in areas such as economics, finance, chemistry, and biology. Before we can begin our study of these functions, we must learn about composite, one-to-one, and inverse functions. This is because exponential and logarithmic functions are related in a special way: they are inverses of each other.

1. Find the Composition of Functions

Earlier we learned how to add, subtract, multiply, and divide functions. Now we will combine functions in a new way, using *function composition*. We use these *composite functions* when we are given certain two-step processes and want to combine them into a single step.

For example, if you work x hours per week earning $8 per hour, your earnings before taxes and other deductions can be described by the function $f(x) = 8x$. Your take-home pay is different, however, because of taxes and other deductions. So, if your take-home pay is 75% of your earnings before taxes, then $g(x) = 0.75x$ can be used to compute your take-home pay when x is your earnings before taxes.

We can describe what is happening with two tables of values.

$f(x) = 8x$	
Hours Worked x	Earnings Before Deductions $f(x)$
6	48
10	80
20	160
40	320
x	$f(x)$

$g(x) = 0.75x$	
Earnings Before Deductions x	Take-Home Pay $g(x)$
48	36
80	60
160	120
320	240
x	$g(x)$

One function, $f(x)$, describes total earnings before deductions in terms of the number of hours worked. Another function, $g(x)$, describes take-home pay in terms of the total earnings before deductions. It would be convenient to have a function that would allow us to compute, directly, the take-home pay in terms of the number of hours worked.

$f(x) = 8x$		$g(x) = 0.75x$
Hours Worked	Earnings Before Deductions	Take-Home Pay
6	48	36
10	80	60
20	160	120
40	320	240
x	$f(x)$	$h(x) = g(f(x))$

$$h(x) = (g \circ f)(x) = g(f(x))$$

If we substitute the function $f(x)$ for x in the function $g(x)$, we will get a new function, $h(x)$, where $h(x) = g(f(x))$. The take-home pay in terms of the number of hours worked, $h(x)$, is given by the composition function $g(f(x))$, read as "g of f of x" and is given by

$$h(x) = g(f(x)) = g(8x)$$
$$= 0.75(8x)$$
$$= 6x$$

Therefore, $h(x) = 6x$ allows us to directly compute the take-home pay from the number of hours worked. To find out your take-home pay when you work 20 hr in a week, find $h(20)$.

$$h(x) = 6x$$
$$h(20) = 6(20) = 120$$

Working 20 hr will result in take-home pay of \$120. Notice that this is the same as the take-home pay computed in the tables.

Another way to write $g(f(x))$ is $(g \circ f)(x)$, and both can be read as "g of f of x," or "g composed with f," or "the composition of g and f." Likewise, $f(g(x)) = (f \circ g)(x)$, and these can be read as "f of g of x," or "f composed with g," or "the composition of f and g."

Definition

Given the function $f(x)$ and $g(x)$, the **composition function $f \circ g$** (read "f of g") is defined as

$$(f \circ g)(x) = f(g(x))$$

where $g(x)$ is in the domain of f.

Example 1

In-Class Example 1
Let $f(x) = 3x + 11$ and $g(x) = x - 5$. Find
a) $g(4)$ b) $(f \circ g)(4)$
c) $(f \circ g)(x)$
answer: a) -1 **b)** 8
c) $3x - 4$

Let $f(x) = 2x - 5$ and $g(x) = x + 8$. Find

a) $g(3)$ b) $(f \circ g)(3)$ c) $(f \circ g)(x)$

Solution

a) $g(x) = x + 8$
    $g(3) = 3 + 8 = 11$

b) $(f \circ g)(3) = f(g(3))$ In part a) we found $g(3) = 11$.
    $\qquad\qquad = f(11)$
    $\qquad\qquad = 2(11) - 5$ Substitute 11 for x in $f(x) = 2x - 5$.
    $\qquad\qquad = 17$

c) $(f \circ g)(x) = f(g(x))$
    $\qquad\qquad = f(x + 8)$ Substitute $x + 8$ for $g(x)$.
    $\qquad\qquad = 2(x + 8) - 5$ Substitute $x + 8$ for x in $f(x)$.
    $\qquad\qquad = 2x + 11$

We can also find $(f \circ g)(3)$, the question for part b), by substituting 3 for x in $(f \circ g)(x)$ found in part c).

$$(f \circ g)(x) = 2x + 11$$
$$(f \circ g)(3) = 2(3) + 11 = 17$$

Notice that this is the same as the result we obtained in b). ∎

 You Try 1

Let $f(x) = 3x + 4$ and $g(x) = x - 10$. Find

a) $g(-2)$ b) $(f \circ g)(-2)$ c) $(f \circ g)(x)$

BE CAREFUL

The notation $(f \circ g)(x)$ represents the *composition* of functions, $f(g(x))$; the notation $(f \cdot g)(x)$ represents the *product* of functions, $f(x) \cdot g(x)$.

Example 2

In-Class Example 2
Let $f(x) = 2x + 3$, $g(x) = \sqrt{x}$, and $h(x) = x^2 - 4x + 1$. Find
a) $(f \circ g)(x)$ b) $(g \circ f)(x)$
c) $(h \circ f)(x)$
answer:
a) $(f \circ g)(x) = 2\sqrt{x} + 3$
b) $(g \circ f)(x) = \sqrt{2x + 3}$
c) $(h \circ f)(x) = 4x^2 + 4x - 2$

Let $f(x) = 4x - 1$, $g(x) = x^2$, and $h(x) = x^2 + 5x - 2$. Find

a) $(f \circ g)(x)$ b) $(g \circ f)(x)$ c) $(h \circ f)(x)$

Solution

a) $(f \circ g)(x) = f(g(x))$

   $= f(x^2)$ Substitute x^2 for $g(x)$.

   $= 4(x^2) - 1$ Substitute x^2 for x in $f(x)$.

   $= 4x^2 - 1$

b) $(g \circ f)(x) = g(f(x))$

   $= g(4x - 1)$ Substitute $4x - 1$ for $f(x)$.

   $= (4x - 1)^2$ Substitute $4x - 1$ for x in $g(x)$.

   $= 16x^2 - 8x + 1$ Expand the binomial.

In general, $(f \circ g)(x) \neq (g \circ f)(x)$.

c) $(h \circ f)(x) = h(f(x))$

   $= h(4x - 1)$ Substitute $4x - 1$ for $f(x)$.

   $= (4x - 1)^2 + 5(4x - 1) - 2$ Substitute $4x - 1$ for x in $h(x)$.

   $= 16x^2 - 8x + 1 + 20x - 5 - 2$ Distribute.

   $= 16x^2 + 12x - 6$ Combine like terms.

You Try 2

Let $f(x) = x^2 + 6$, $g(x) = 2x - 3$, and $h(x) = x^2 - 4x + 9$. Find

a) $(g \circ f)(x)$ b) $(f \circ g)(x)$ c) $(h \circ g)(x)$

Sometimes, it is necessary to rewrite a single function in terms of the composition of two other functions. This is called the **decomposition** of functions.

Example 3

In-Class Example 3
Let $h(x) = \sqrt{x^2 + 8}$. Find f and g such that $h(x) = (f \circ g)(x)$.
answer: $f(x) = \sqrt{x}$,
$g(x) = x^2 + 8$

Let $h(x) = \sqrt{x^2 + 5}$. Find f and g such that $h(x) = (f \circ g)(x)$.

Solution

Think about what is happening in the function $h(x)$. We can "build" $h(x)$ in the following way: first find $x^2 + 5$, then take the square root of that quantity. So if we let $g(x) = x^2 + 5$ and $f(x) = \sqrt{x}$, we will get $h(x) = (f \circ g)(x)$. Let's check by finding the composition function, $(f \circ g)(x)$.

$$g(x) = x^2 + 5 \qquad f(x) = \sqrt{x}$$

$(f \circ g)(x) = f(g(x))$

$= f(x^2 + 5)$ Substitute $x^2 + 5$ for $g(x)$.

$= \sqrt{x^2 + 5}$ Substitute $x^2 + 5$ for x in $f(x) = \sqrt{x}$.

Our result is $h(x) = \sqrt{x^2 + 5}$.

In Example 3, forming $h(x)$ using $f(x) = \sqrt{x}$ and $g(x) = x^2 + 5$ is probably the easiest decomposition to "see." However, there is more than one way to decompose a function, $h(x)$, into two functions $f(x)$ and $g(x)$ so that $h(x) = (f \circ g)(x)$.

For example, if $f(x) = \sqrt{x + 5}$ and $g(x) = x^2$, we get

$$(f \circ g)(x) = f(g(x))$$
$$= f(x^2) \qquad \text{Substitute } x^2 \text{ for } g(x).$$
$$= \sqrt{x^2 + 5} \qquad \text{Substitute } x^2 \text{ for } x \text{ in } f(x) = \sqrt{x + 5}.$$

This is another way to obtain $h(x) = \sqrt{x^2 + 5}$.

You Try 3

Let $h(x) = \sqrt{2x^2 + 1}$. Find f and g such that $h(x) = (f \circ g)(x)$.

2. Use Function Composition

Example 4

In-Class Example 4
Use the given example but in
b) find $(A \circ P)(5)$.
answer: b) $(A \circ P)(5) = 25$;
a square that has a side of
length 5 units has an area of
25 square units.

The area, A, of a square expressed in terms of its perimeter, P, is defined by the function

$$A(P) = \frac{1}{16}P^2$$

The perimeter of a square with a side of length x, is defined by the function

$$P(x) = 4x$$

a) Find $(A \circ P)(x)$ and explain what it represents.

b) Find $(A \circ P)(3)$ and explain what it represents.

Solution

a) $(A \circ P)(x) = A(P(x))$

$$= A(4x) \qquad \text{Substitute } 4x \text{ for } P(x).$$
$$= \frac{1}{16}(4x)^2 \qquad \text{Substitute } 4x \text{ for } P \text{ in } A(P) = \frac{1}{16}P^2.$$
$$= \frac{1}{16}(16x^2)$$
$$= x^2$$

$(A \circ P)(x) = x^2$. This is the formula for the area of a square in terms of the length of a side, x.

b) To find $(A \circ P)(3)$, use the result obtained in a).

$$(A \circ P)(x) = x^2$$
$$(A \circ P)(3) = 3^2 = 9$$

A square that has a side of length 3 units has an area of 9 square units.

You Try 4

Let $f(x) = 100x$ represent the number of centimeters in x meters. Let $g(y) = 1000y$ represent the number of meters in y kilometers.

a) Find $(f \circ g)(y)$ and explain what it represents.

b) Find $(f \circ g)(4)$ and explain what it represents.

One-to-One Functions

3. Decide if a Function Is One-to-One

Recall from Section 3.5 that a relation is a *function* if each *x*-value corresponds to exactly one *y*-value. Let's look at two functions, *f* and *g*.

$$f = \{(1, -3), (2, -1), (4, 3), (7, 9)\} \quad g = \{(0, 3), (1, 4), (-1, 4), (2, 7)\}$$

In functions *f* and *g*, each *x*-value corresponds to exactly one *y*-value. That is why they are functions. In function *f*, each *y-value also corresponds to exactly one x-value.* Therefore, *f* is a *one-to-one function*. In function *g*, however, each *y*-value does *not* correspond to exactly one *x*-value. (The *y*-value of 4 corresponds to $x = 1$ and $x = -1$.) Therefore, *g* is *not* a one-to-one function.

Definition

In order for a function to be a **one-to-one function,** each *x*-value corresponds to exactly one *y*-value, and each *y*-value corresponds to exactly one *x*-value.

Alternatively we can say that a function is one-to-one if each value in its domain corresponds to exactly one value in its range *and* if each value in its range corresponds to exactly one value in its domain.

Example 5

In-Class Example 5
Determine whether each function is one-to-one.
a) $f = \{(3, -2), (4, 6), (1, 9), (-2, -4), (0, 0)\}$
b) $m = \{(1, 1), (2, 6), (3, 6), (4, -11)\}$
answer: a) one-to-one
b) not one-to-one

Determine whether each function is one-to-one.

a) $f = \{(-1, 9), (1, -3), (2, -6), (4, -6)\}$

b) $g = \{(-3, 13), (-1, 5), (5, -19), (8, -31)\}$

c)

State	Number of Representatives in U.S. House of Representatives (2010)
Alaska	1
California	53
Connecticut	5
Delaware	1
Ohio	18

d)

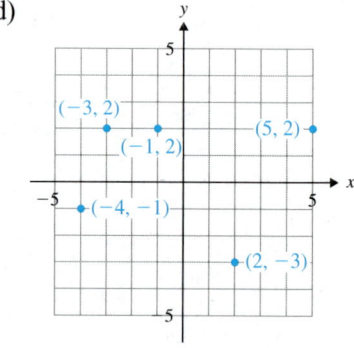

Solution

a) *f* is *not* a one-to-one function since the *y*-value -6 corresponds to two different *x*-values: $(2, -6)$ and $(4, -6)$.

b) *g is* a one-to-one function since each *y*-value corresponds to exactly one *x*-value.

c) The information in the table does *not* represent a one-to-one function since the value 1 in the range corresponds to two different values in the domain, Alaska and Delaware.

d) The graph does *not* represent a one-to-one function since three points have the same *y*-value: $(-3, 2)$, $(-1, 2)$, and $(5, 2)$. ∎

You Try 5

Determine whether each function is one-to-one.

a) $f = \{(-2, -13), (0, -7), (4, 5), (5, 8)\}$ b) $g = \{(-4, 2), (-1, 1), (0, 2), (3, 5)\}$

c)

Element	Atomic Mass (in amu)
Hydrogen	1.00794
Lithium	6.941
Sulfur	32.066
Lead	207.2

d)

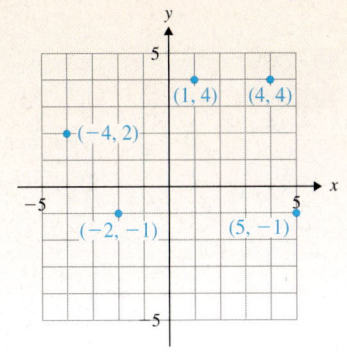

The Horizontal Line Test

Just as we can use the vertical line test to determine if a graph represents a function, we can use the *horizontal line test* to determine if a function is one-to-one.

> **Definition**
>
> **Horizontal Line Test:** If every horizontal line that could be drawn through a function would intersect the graph at most once, then the function is one-to-one.

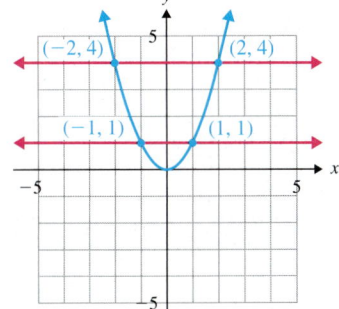

Look at the graph of the given function. We can see that if a horizontal line intersects the graph more than once, then one y-value corresponds to more than one x-value. This means that the function is not one-to-one. For example, the y-value of 1 corresponds to $x = 1$ and $x = -1$.

In-Class Example 6
Determine whether each graph represents a one-to-one function.
a)

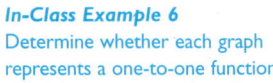
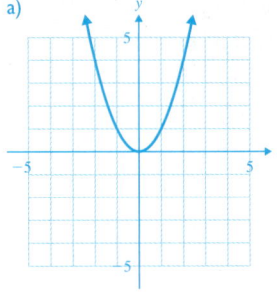

b)

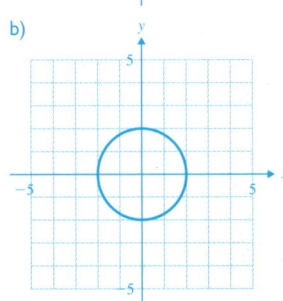

answer: a) not one-to-one
b) not one-to-one

Example 6

Determine whether each graph represents a one-to-one function.

a)

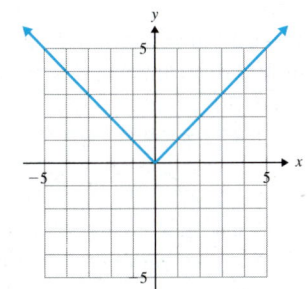

b)

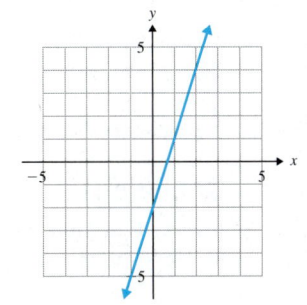

Solution

a) *Not* one-to-one. It is possible to draw a horizontal line through the graph so that it intersects the graph more than once.

b) *Is* one-to-one. Every horizontal line that could be drawn through the graph would intersect the graph at most once. ■

You Try 6

Determine whether each graph represents a one-to-one function.

a)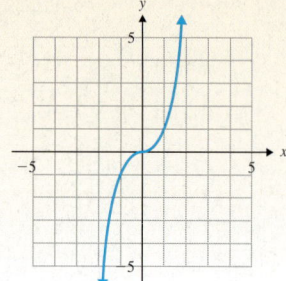

b)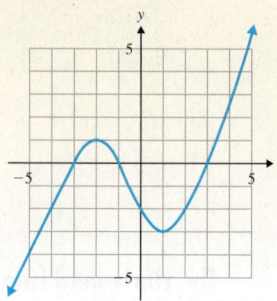

Inverse Functions

4. Find the Inverse of a Function

One-to-one functions lead to other special functions—inverse functions. *A one-to-one function has an inverse function.*

To find the inverse of a one-to-one function, we interchange the coordinates of the ordered pairs.

Example 7

In-Class Example 7
Find the inverse function of
$f = \{(3, 4), (-5, -6), (7, 8), (-9, 10)\}$.
answer: $f^{-1} = \{(4, 3), (-6, -5), (8, 7), (10, -9)\}$

Find the inverse function of $f = \{(4, 2), (9, 3), (36, 6)\}$.

Solution

To find the inverse of f, switch the x- and y-coordinates of each ordered pair. The inverse of f is $\{(2, 4), (3, 9), (6, 36)\}$. ■

You Try 7

Find the inverse function of $f = \{(-5, -1), (-3, 2), (0, 7), (4, 13)\}$.

We use special notation to represent the inverse of a function. If f is a one-to-one function, then f^{-1} (read "f inverse") represents the inverse of f. For Example 7, we can write the inverse as $f^{-1} = \{(2, 4), (3, 9), (6, 36)\}$.

Definition

Inverse Function: Let f be a one-to-one function. The **inverse** of f, denoted by f^{-1}, is a one-to-one function that contains the set of all ordered pairs (y, x), where (x, y) belongs to f.

BE CAREFUL

1) f^{-1} is read "f inverse" *not* "f to the negative one."

2) f^{-1} does *not* mean $\dfrac{1}{f}$.

3) If a function is not one-to-one, it does not have an inverse.

We said that if (x, y) belongs to the one-to-one function $f(x)$, then (y, x) belongs to its inverse, $f^{-1}(x)$ (read as *f inverse of x*). We use this idea to find the equation for the inverse of $f(x)$.

Procedure   How to Find an Equation of the Inverse of $y = f(x)$

Step 1:   Replace $f(x)$ with y.

Step 2:   Interchange x and y.

Step 3:   Solve for y.

Step 4:   Replace y with the inverse notation, $f^{-1}(x)$.

Example 8

Find an equation of the inverse of $f(x) = 3x + 4$.

In-Class Example 8
Find an equation of the inverse of $f(x) = -5x - 7$.

answer: $f^{-1}(x) = -\dfrac{1}{5}x - \dfrac{7}{5}$

Solution

$$f(x) = 3x + 4$$
$$y = 3x + 4 \qquad \text{Replace } f(x) \text{ with } y.$$
$$x = 3y + 4 \qquad \text{Interchange } x \text{ and } y.$$

Solve for y.

$$x - 4 = 3y \qquad \text{Subtract 4.}$$
$$\frac{x - 4}{3} = y \qquad \text{Divide by 3.}$$
$$\frac{1}{3}x - \frac{4}{3} = y \qquad \text{Simplify.}$$
$$f^{-1}(x) = \frac{1}{3}x - \frac{4}{3} \qquad \text{Replace } y \text{ with } f^{-1}(x).$$

You Try 8

Find an equation of the inverse of $f(x) = -5x + 10$.

In Example 9, we will look more closely at the relationship between a function and its inverse.

Example 9

Find the equation of the inverse of $f(x) = 2x - 4$. Then, graph $f(x)$ and $f^{-1}(x)$ on the same axes.

In-Class Example 9
Find the equation of the inverse of $f(x) = 3x + 8$. Then graph $f(x)$ and $f^{-1}(x)$ on the same axes.

answer: $f^{-1}(x) = \dfrac{1}{3}x - \dfrac{8}{3}$

Solution

$$f(x) = 2x - 4$$
$$y = 2x - 4 \qquad \text{Replace } f(x) \text{ with } y.$$
$$x = 2y - 4 \qquad \text{Interchange } x \text{ and } y.$$

Solve for y.

$$x + 4 = 2y \qquad \text{Add 4.}$$

$$\frac{x + 4}{2} = y \qquad \text{Divide by 2.}$$

$$\frac{1}{2}x + 2 = y \qquad \text{Simplify.}$$

$$f^{-1}(x) = \frac{1}{2}x + 2 \qquad \text{Replace } y \text{ with } f^{-1}(x).$$

We will graph $f(x)$ and $f^{-1}(x)$ by making a table of values for each. Then we can see another relationship between the two functions.

$f(x) = 2x - 4$	
x	$y = f(x)$
0	-4
1	-2
2	0
5	6

$f^{-1}(x) = \frac{1}{2}x + 2$	
x	$y = f^{-1}(x)$
-4	0
-2	1
0	2
6	5

Notice that the x- and y-coordinates have switched when we compare the tables of values. Graph $f(x)$ and $f^{-1}(x)$.

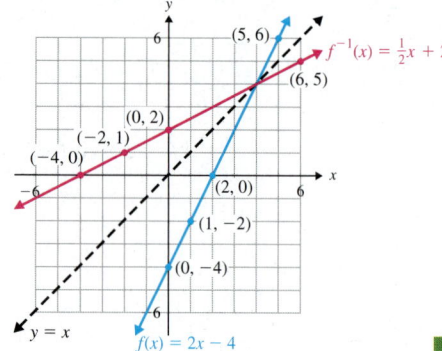

You Try 9

Find the equation of the inverse of $f(x) = -3x + 1$. Then, graph $f(x)$ and $f^{-1}(x)$ on the same axes.

5. Given the Graph of f(x), Graph f⁻¹(x)

Look again at the tables in Example 9. The x-values for $f(x)$ become the y-values of $f^{-1}(x)$, and the y-values of $f(x)$ become the x-values of $f^{-1}(x)$. This is true not only for the values in the tables but for *all* values of x and y. That is, for all ordered pairs (x, y) that belong to $f(x)$, (y, x) belongs to $f^{-1}(x)$. Another way to say this is *the domain of f becomes the range of f^{-1}, and the range of f becomes the domain of f^{-1}.*

Let's turn our attention to the graph in Example 9. The graphs of $f(x)$ and $f^{-1}(x)$ are mirror images of one another with respect to the line $y = x$. We say that *the graphs of $f(x)$ and $f^{-1}(x)$ are symmetric with respect to the line $y = x$.* This is true for every function $f(x)$ and its inverse, $f^{-1}(x)$.

Note

The graphs of $f(x)$ and $f^{-1}(x)$ are symmetric with respect to the line $y = x$.

Example 10

Given the graph of $f(x)$, graph $f^{-1}(x)$.

In-Class Example 10
Given the graph $f(x)$, graph $f^{-1}(x)$.

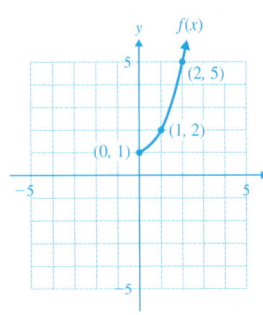

Solution

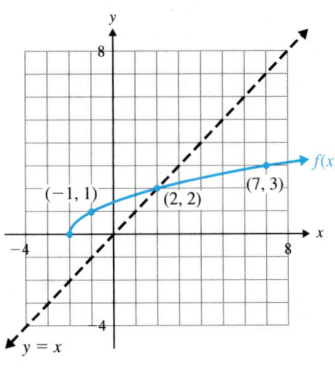

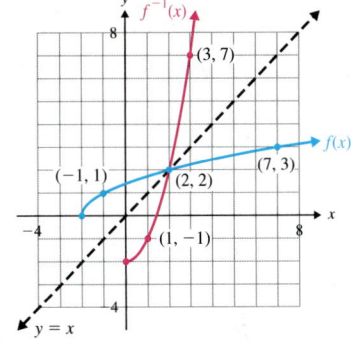

answer:

Some points on the graph of $f(x)$ are $(-2, 0)$, $(-1, 1)$, $(2, 2)$, and $(7, 3)$. We can obtain points on the graph of $f^{-1}(x)$ by interchanging the x- and y-values.

Some points on the graph of $f^{-1}(x)$ are $(0, -2)$, $(1, -1)$, $(2, 2)$, and $(3, 7)$. Plot these points to get the graph of $f^{-1}(x)$. Notice that the graphs are symmetric with respect to the line $y = x$.

You Try 10

Given the graph of $f(x)$, graph $f^{-1}(x)$.

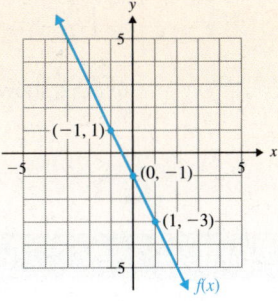

6. Show That $(f^{-1} \circ f)(x) = x$ and $(f \circ f^{-1})(x) = x$

Going back to the tables in Example 9, we see from the first table that $f(0) = -4$ and from the second table that $f^{-1}(-4) = 0$. The first table also shows that $f(1) = -2$ while the second table shows that $f^{-1}(-2) = 1$. That is, putting x into the function f produces $f(x)$. And putting $f(x)$ into $f^{-1}(x)$ produces x.

$$f(0) = -4 \quad \text{and} \quad f^{-1}(-4) = 0$$
$$f(1) = -2 \quad \text{and} \quad f^{-1}(-2) = 1$$
$$\underset{x}{\uparrow} \quad \underset{f(x)}{\uparrow} \qquad \underset{f^{-1}(f(x)) = x}{\uparrow \qquad \uparrow}$$

This leads us to another fact about functions and their inverses.

Note

Let f be a one-to-one function. Then f^{-1} is the inverse of f such that $(f^{-1} \circ f)(x) = x$ and $(f \circ f^{-1})(x) = x$.

Example 11

If $f(x) = 4x + 3$, show that $f^{-1}(x) = \dfrac{1}{4}x - \dfrac{3}{4}$.

In-Class Example 11
If $f(x) = 9x - 1$, show that
$f^{-1}(x) = \dfrac{1}{9}x + \dfrac{1}{9}$.

answer:
$(f^{-1} \circ f)(x) =$
$\dfrac{1}{9}(9x - 1) + \dfrac{1}{9}$

$= x - \dfrac{1}{9} + \dfrac{1}{9} = x$
$(f \circ f^{-1})(x)$
$= 9\left(\dfrac{1}{9}x + \dfrac{1}{9}\right) - 1$
$= x + 1 - 1 = x$

Solution

Show that $(f^{-1} \circ f)(x) = x$ and $(f \circ f^{-1})(x) = x$.

$$(f^{-1} \circ f)(x) = f^{-1}(f(x))$$
$$= f^{-1}(4x + 3) \qquad \text{Substitute } 4x + 3 \text{ for } f(x).$$
$$= \frac{1}{4}(4x + 3) - \frac{3}{4} \qquad \text{Evaluate.}$$
$$= x + \frac{3}{4} - \frac{3}{4} \qquad \text{Distribute.}$$
$$= x$$

$$(f \circ f^{-1})(x) = f(f^{-1}(x))$$
$$= f\left(\frac{1}{4}x - \frac{3}{4}\right) \qquad \text{Substitute } \frac{1}{4}x - \frac{3}{4} \text{ for } f^{-1}(x).$$
$$= 4\left(\frac{1}{4}x - \frac{3}{4}\right) + 3 \qquad \text{Evaluate.}$$
$$= x - 3 + 3 \qquad \text{Distribute.}$$
$$= x$$

You Try 11

If $f(x) = -6x + 2$, show that $f^{-1}(x) = -\dfrac{1}{6}x + \dfrac{1}{3}$.

Using Technology

The composition of two functions can be evaluated analytically, numerically, and graphically using a graphing calculator.

Consider the composition $h(x) = f(g(x))$ given the functions $f(x) = x^2$ and $g(x) = 3x - 2$. The function $h(x)$ is determined analytically by substituting $g(x)$ into the function f.

We can evaluate $h(2) = f(g(2))$ by substituting 2 in for x.

$$h(x) = f(g(x))$$
$$= f(3x - 2)$$
$$= (3x - 2)^2$$
$$= 9x^2 - 12x + 4$$

$$9(2)^2 - 12(2) + 4 = 16$$

Using a graphing calculator, enter $f(x) = x^2$ into Y_1, enter $g(x) = 3x - 2$ into Y_2, and enter $Y_1(Y_2)$ into Y_3 to represent the composition h, as shown. Recall that Y_1 and Y_2 are found by pressing `VARS`, pressing the right arrow key, and pressing `ENTER`.

To evaluate $h(2) = f(g(2))$ using the calculator, press `2nd` `MODE` to return to the home screen and $Y_3(2)$ as shown.

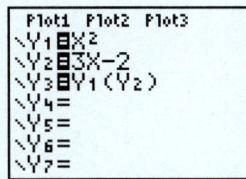
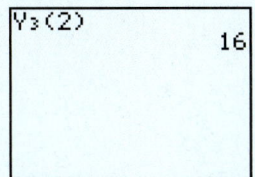

Start with the function Y_1, Y_2 as shown above. $h(2) = f(g(2))$ can be evaluated numerically by setting up a table showing x, Y_1, and Y_2 near $x = 2$. Press `2nd` `WINDOW` and enter 0 after TblStart =. Then press `2nd` `ENTER` to display the table. First evaluate $g(2) = Y_2(2)$ by moving the cursor down to $x = 2$ and then across to the column under Y_2 as shown.

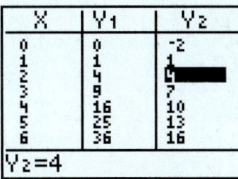

Then substitute the result $g(2) = Y_2(2) = 4$ into the function f to evaluate $f(4) = Y_1(4) = 16$ by moving the cursor down to $x = 4$ and then across to the column under Y_1 as shown. The result is 16 as desired.

To illustrate this composition graphically, change the window by increasing Ymax to 20. $h(2) = f(g(2))$ can be evaluated graphically using the following approach.

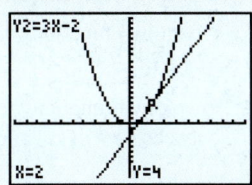

First evaluate $g(2) = Y_2(2)$ by pressing `TRACE`, pressing the down arrow key to switch to Y_2, and pressing `2` `ENTER` resulting in the point $(2, 4)$ on the graph of g, as shown.

Next evaluate $f(4) = Y_1(4)$ by pressing the down arrow key to switch to Y_1, and pressing `4` `ENTER` resulting in the point $(4, 16)$ on the graph of f as shown on the graph. The result is 16 as desired.

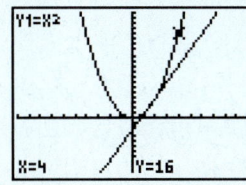

Given the functions $f(x) = x^2 - 5x$ and $g(x) = 2x + 3$, evaluate the following function values using a graphing calculator.

1) $(f \circ g)(3)$ 2) $(f \circ g)(-2)$ 3) $(g \circ f)(2)$
4) $(g \circ f)(-2)$ 5) $(f \circ f)(1)$ 6) $(g \circ g)(2)$

Answers to You Try Exercises

1) a) -12 b) -32 c) $3x - 26$ 2) a) $2x^2 + 9$ b) $4x^2 - 12x + 15$ c) $4x^2 - 20x + 30$

3) $f(x) = \sqrt{x}, g(x) = 2x^2 + 1$; answers may vary.

4) a) $(f \circ g)(y) = 100{,}000y$. This tells us the number of centimeters in y kilometers.
   b) $(f \circ g)(4) = 400{,}000$. There are 400,000 cm in 4 kilometers.

5) a) yes   b) no   c) yes   d) no   6) a) yes   b) no   7) $\{(-1, -5), (2, -3), (7, 0), (13, 4)\}$

8) a) $f^{-1}(x) = -\dfrac{1}{5}x + 2$ 9) $f^{-1}(x) = -\dfrac{1}{3}x + \dfrac{1}{3}$

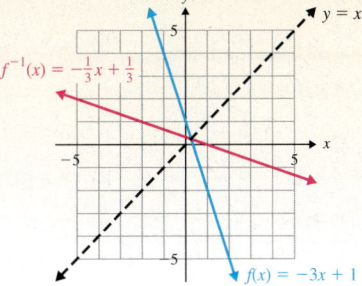

10)

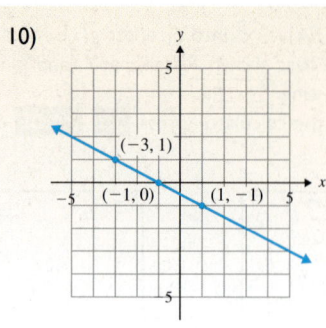

(−3, 1)

(−1, 0)   (1, −1)

11) Show that $(f^{-1} \circ f)(x) = x$ and $(f \circ f^{-1})(x) = x$

Answers to Technology Exercises

1) 36   2) 6   3) −9   4) 31   5) 36   6) 17

10.1 Exercises

*Additional answers can be found in the Answers to Exercises appendix.

Objective 1: Find the Composition of Functions

1) Given two functions $f(x)$ and $g(x)$, explain how to find $(f \circ g)(x)$. $(f \circ g)(x) = f(g(x))$ so substitute the function $g(x)$ into the function $f(x)$ and simplify.

2) Given two functions $f(x)$ and $g(x)$, explain the difference between $(f \circ g)(x)$ and $(f \cdot g)(x)$. $(f \circ g)(x)$ is the composition of functions f and g, $f(g(x))$, and $(f \cdot g)(x) = f(x) \cdot g(x)$, the product of functions f and g.

For Exercises 3–6, find

a) $g(4)$

b) $(f \circ g)(4)$ using the result in part a)

c) $(f \circ g)(x)$

d) $(f \circ g)(4)$ using the result in part c)

3) $f(x) = 3x + 1, g(x) = 2x - 9$
a) −1   b) −2   c) 6x − 26   d) −2

4) $f(x) = -x + 5, g(x) = x + 7$
a) 11   b) −6   c) −x − 2   d) −6

5) $f(x) = x^2 - 5, g(x) = x + 3$
a) 7   b) 44   c) x² + 6x + 4   d) 44

6) $f(x) = x^2 + 2, g(x) = x - 1$
a) 3   b) 11   c) x² − 2x + 3   d) 11

7) Let $f(x) = 5x - 4$ and $g(x) = x + 7$. Find

a) $(f \circ g)(x)$ 5x + 31   b) $(g \circ f)(x)$ 5x + 3

c) $(f \circ g)(3)$ 46

8) Let $f(x) = x - 10$ and $g(x) = 4x + 3$. Find

a) $(f \circ g)(x)$ 4x − 7   b) $(g \circ f)(x)$ 4x − 37

c) $(f \circ g)(-6)$ −31

9) Let $h(x) = -2x + 9$ and $k(x) = 3x - 1$. Find

a) $(k \circ h)(x)$ −6x + 26   b) $(h \circ k)(x)$ −6x + 11

c) $(k \circ h)(-1)$ 32

10) Let $r(x) = 6x + 2$ and $v(x) = -7x - 5$. Find

a) $(v \circ r)(x)$ −42x − 19   b) $(r \circ v)(x)$ −42x − 28

c) $(r \circ v)(2)$ −112

11) Let $g(x) = x^2 - 6x + 11$ and $h(x) = x - 4$. Find

a) $(h \circ g)(x)$ x² − 6x + 7   b) $(g \circ h)(x)$ x² − 14x + 51

c) $(g \circ h)(4)$ 11

12) Let $f(x) = x^2 + 7x - 9$ and $g(x) = x + 2$. Find

a) $(g \circ f)(x)$ x² + 7x − 7   b) $(f \circ g)(x)$ x² + 11x + 9

c) $(g \circ f)(3)$ 23

13) Let $m(x) = x + 8$ and $n(x) = -x^2 + 3x - 8$. Find

a) $(n \circ m)(x)$ −x² − 13x − 48   b) $(m \circ n)(x)$ −x² + 3x

c) $(m \circ n)(0)$ 0

14) Let $f(x) = -x^2 + 10x + 4$ and $g(x) = x + 1$. Find

a) $(g \circ f)(x)$ −x² + 10x + 5   b) $(f \circ g)(x)$ −x² + 8x + 13

c) $(f \circ g)(-2)$ −7

15) Let $f(x) = \sqrt{x + 10}, g(x) = x^2 - 6$. Find

a) $(f \circ g)(x)$ $\sqrt{x^2 + 4}$ b) $(g \circ f)(x)$ x + 4

c) $(f \circ g)(-3)$ $\sqrt{13}$

16) Let $h(x) = x^2 + 7, k(x) = \sqrt{x - 1}$. Find

a) $(h \circ k)(x)$ x + 6   b) $(k \circ h)(x)$ $\sqrt{x^2 + 6}$

c) $(k \circ h)(0)$ $\sqrt{6}$

17) Let $P(t) = \dfrac{1}{t + 8}$, $Q(t) = t^2$. Find

a) $(P \circ Q)(t)$ $\dfrac{1}{t^2 + 8}$

b) $(Q \circ P)(t)$ $\dfrac{1}{(t + 8)^2}$

c) $(Q \circ P)(-5)$ $\dfrac{1}{9}$

18) Let $F(a) = \dfrac{1}{5a}$, $G(a) = a^2$. Find

a) $(G \circ F)(a)$ $\dfrac{1}{25a^2}$

b) $(F \circ G)(a)$ $\dfrac{1}{5a^2}$

c) $(G \circ F)(-2)$ $\dfrac{1}{100}$

For Exercises 19–24, find $f(x)$ and $g(x)$ such that $h(x) = (f \circ g)(x)$.

19) $h(x) = \sqrt{x^2 + 13}$ $f(x) = \sqrt{x}, g(x) = x^2 + 13$; answers may vary.

20) $h(x) = \sqrt{2x^2 + 7}$ $f(x) = \sqrt{x}, g(x) = 2x^2 + 7$; answers may vary.

21) $h(x) = (8x - 3)^2$ $f(x) = x^2, g(x) = 8x - 3$; answers may vary.

22) $h(x) = (4x + 9)^2$ $f(x) = x^2, g(x) = 4x + 9$; answers may vary.

23) $h(x) = \dfrac{1}{6x + 5}$ $f(x) = \dfrac{1}{x}, g(x) = 6x + 5$; answers may vary.

24) $h(x) = \dfrac{2}{x - 10}$ $f(x) = \dfrac{2}{x}, g(x) = x - 10$; answers may vary.

Objective 2: Use Function Composition

25) The sales tax on goods in a major metropolitan area is 7% so that the final cost of an item, $f(x)$, is given by $f(x) = 1.07x$, where x is the cost of the item. A women's clothing store is having a sale so that all of its merchandise is 20% off. If the regular price of an item is x dollars then the sale price, $s(x)$, is given by $s(x) = 0.80x$. Find each of the following and explain their meanings.

a) $s(40)$ $s(40) = 32$. When the regular price of an item is $40, the sale price is $32.

b) $f(32)$ $f(32) = 34.24$. When the cost of an item is $32, the final cost after sale tax is $34.24.

c) $(f \circ s)(x)$ $(f \circ s)(x) = 0.856x$. This is the final cost of the item after the discount and sales tax.

d) $(f \circ s)(40)$ $(f \circ s)(40) = 34.24$. When the original cost of an item is $40, the final cost after the discount and sales tax is $34.24.

26) The function $C(F) = \dfrac{5}{9}(F - 32)$ can be used to convert a temperature from degrees Fahrenheit, F, to degrees Celsius, C. The relationship between the Celsius scale, C, and the Kelvin scale, K, is given by $K(C) = C + 273$. Find each of the following and explain their meanings.

a) $C(59)$ $C(59) = 15$. A temperature of 59°F is equivalent to 15°C.

b) $K(15)$ $K(15) = 288$. A temperature of 15°C is equivalent to 288 K.

c) $K(C(F))$ $K(C(F)) = \dfrac{5}{9}F + \dfrac{2297}{9}$. This is the temperature on the Kelvin scale in terms of the Fahrenheit temperature.

d) $K(C(59))$
$K(C(59)) = 288$. A temperature of 59°F is equivalent to 288 K.

VIDEO 27) Oil spilled from a ship off the coast of Alaska with the oil spreading out in a circle across the surface of the water. The radius of the oil spill is given by $r(t) = 4t$ where t is the number of minutes after the leak began and $r(t)$ is in feet. The area of the spill is given by $A(r) = \pi r^2$ where r represents the radius of the oil slick. Find each of the following and explain their meanings.

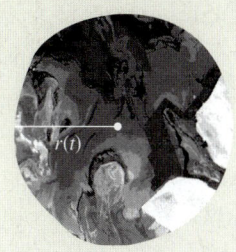

a) $r(5)$ b) $A(20)$ c) $A(r(t))$ d) $A(r(5))$

28) The radius of a circle is half its diameter. We can express this with the function $r(d) = \dfrac{1}{2}d$, where d is the diameter of a circle and r is the radius. The area of a circle in terms of its radius is $A(r) = \pi r^2$. Find each of the following and explain their meanings.

a) $r(6)$ b) $A(3)$ c) $A(r(d))$ d) $A(r(6))$

Objective 3: Decide if a Function is One-to-One

Determine whether each function is one-to-one. If it is one-to-one, find its inverse.

29) $f = \{(-4, 3), (-2, -3), (2, -3), (6, 13)\}$ no

30) $g = \{(0, -7), (1, -6), (4, -5), (25, -2)\}$ yes

31) $h = \{(-5, -16), (-1, -4), (3, 8)\}$ yes

32) $f = \{(-6, 3), (-1, 8), (4, 3)\}$ no

33) $g = \{(2, 1), (5, 2), (7, 14), (10, 19)\}$ yes

34) $h = \{(-1, 4), (0, -2), (5, 1), (9, 4)\}$ no

Determine whether each function is one-to-one.

35) The table shows the average temperature during selected months in Tulsa, Oklahoma. The function matches each month with the average temperature, in °F. Is it one-to-one? (www.noaa.gov)

Month	Average Temp. (°F)
Jan.	36.4
Apr.	60.8
July	83.5
Oct.	62.6

yes

36) The table shows some NCAA conferences and the number of schools in the conference in 2009. The function matches each conference with the number of schools it contains. Is it one-to-one?

Conference	Number of Member Schools
ACC	12
Big 10	11
Big 12	12
MVC	10
Pac10	10

no

Mixed Exercises: Objectives 3–6

37) Do all functions have inverses? Explain your answer.
No; only one-to-one functions have inverses.

38) What test can be used to determine whether the graph of a function has an inverse?   horizontal line test

Determine whether each statement is true or false. If it is false, rewrite the statement so that it is true.

39) $f^{-1}(x)$ is read as "f to the negative one of x."
False; it is read "f inverse of x."

40) If f^{-1} is the inverse of f, then $(f^{-1} \circ f)(x) = x$ and $(f \circ f^{-1})(x) = x$.   true

41) The domain of f is the range of f^{-1}.   true

42) If f is one-to-one and $(5, 9)$ is on the graph of f, then $(-5, -9)$ is on the graph of f^{-1}.
False; $(9, 5)$ is on the graph of f^{-1}.

43) The graphs of $f(x)$ and $f^{-1}(x)$ are symmetric with respect to the x-axis.   False; they are symmetric with respect to $y = x$.

44) Let $f(x)$ be one-to-one. If $f(7) = 2$, then $f^{-1}(2) = 7$.   true

For each function graphed here, answer the following.

a) Determine whether it is one-to-one.

b) If it is one-to-one, graph its inverse.

VIDEO 45)
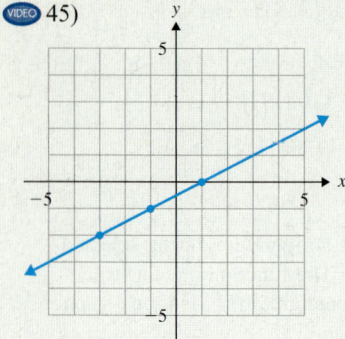

46)
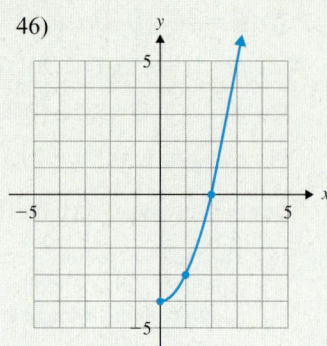

47)
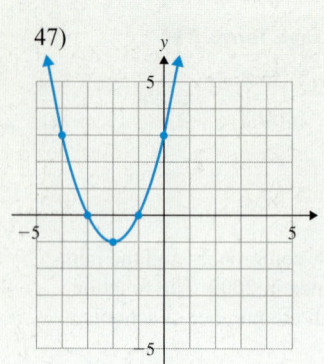
no

48)
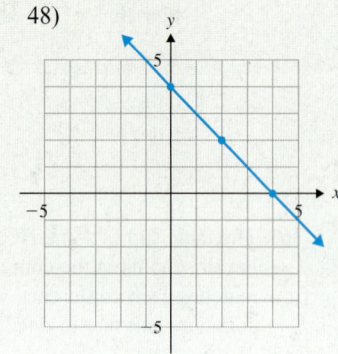

49)   50)
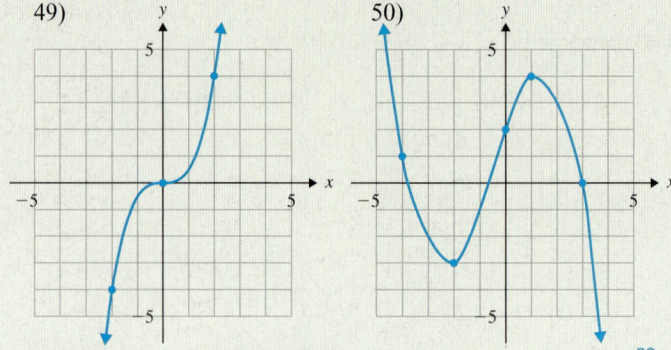
no

Objective 4: Find the Inverse of a Function

Find the inverse of each one-to-one function.

Fill It In

Fill in the blanks with either the missing mathematical step or reason for the given step.

51) $f(x) = 2x - 10$

$y = 2x - 10$ Replace $f(x)$ with y.

$\underline{x = 2y - 10}$ Interchange x and y.

Solve for y.

$x + 10 = 2y$ Add 10.

$\dfrac{1}{2}x + 5 = y$ Divide by 2 and simplify.

$f^{-1}(x) = \dfrac{1}{2}x + 5$ Replace y with $f^{-1}(x)$.

52) $g(x) = \dfrac{1}{3}x + 4$

$y = \dfrac{1}{3}x + 4$ Replace $g(x)$ with y.

$x = \dfrac{1}{3}y + 4$ Interchange x and y.

Solve for y.

$\underline{x - 4 = \dfrac{1}{3}y}$ Subtract 4.

$3x - 12 = y$ Multiply by 3 and simplify.

$\underline{g^{-1}(x) = 3x - 12}$ Replace y with $g^{-1}(x)$.

Find the inverse of each one-to-one function. Then, graph the function and its inverse on the same axes.

53) $g(x) = x - 6$ 54) $h(x) = x + 3$

55) $f(x) = -2x + 5$ 56) $g(x) = 4x - 9$

57) $g(x) = \dfrac{1}{2}x$ 58) $h(x) = -\dfrac{1}{3}x$

59) $f(x) = x^3$ 60) $g(x) = \sqrt[3]{x} + 4$

Find the inverse of each one-to-one function.

61) $f(x) = 2x - 6$ 62) $g(x) = -4x + 8$

VIDEO 63) $h(x) = -\dfrac{3}{2}x + 4$

64) $f(x) = \dfrac{2}{5}x + 1$

74) $f(x) = -\dfrac{5}{4}x + 2$

65) $g(x) = \sqrt[3]{x} + 2$

66) $h(x) = \sqrt[3]{x} - 7$
$h^{-1}(x) = x^3 + 7$

    a) $f(8)$ -8   b) $f^{-1}(-8)$ 8

67) $f(x) = \sqrt{x},\ x \ge 0$
$f^{-1}(x) = x^2,\ x \ge 0$

68) $g(x) = \sqrt{x + 3},\ x \ge -3$
$g^{-1}(x) = x^2 - 3,\ x \ge 0$

75) $f(x) = 2^x$

    a) $f(3)$ 8   b) $f^{-1}(8)$ 3

Objective 6: Show that $(f^{-1} \circ f)(x) = x$ and $(f \circ f^{-1})(x) = x$

Given the one-to-one function $f(x)$, find the function values *without* finding the equation of $f^{-1}(x)$. Find the value in a) before b).

76) $f(x) = 3^x$

    a) $f(-2)$ $\dfrac{1}{9}$ b) $f^{-1}\!\left(\dfrac{1}{9}\right)$ -2

VIDEO 69) $f(x) = 5x - 2$

    a) $f(1)$ 3   b) $f^{-1}(3)$ 1

77) If $f(x) = x + 9$, show that $f^{-1}(x) = x - 9$.

78) If $f(x) = x - 12$, show that $f^{-1}(x) = x + 12$.

70) $f(x) = 3x + 7$

    a) $f(-4)$ -5   b) $f^{-1}(-5)$ -4

VIDEO 79) If $f(x) = -6x + 4$, show that $f^{-1}(x) = -\dfrac{1}{6}x + \dfrac{2}{3}$.

71) $f(x) = -\dfrac{1}{3}x + 5$

    a) $f(9)$ 2   b) $f^{-1}(2)$ 9

80) If $f(x) = -\dfrac{1}{7}x + \dfrac{2}{7}$, show that $f^{-1}(x) = -7x + 2$.

72) $f(x) = \dfrac{1}{2}x - 1$

    a) $f(6)$ 2   b) $f^{-1}(2)$ 6

81) If $f(x) = \dfrac{3}{2}x - 9$, show that $f^{-1}(x) = \dfrac{2}{3}x + 6$.

82) If $f(x) = -\dfrac{5}{8}x + 10$, show that $f^{-1}(x) = -\dfrac{8}{5}x + 16$.

73) $f(x) = -x + 3$

    a) $f(-7)$ 10   b) $f^{-1}(10)$ -7

83) If $f(x) = \sqrt[3]{x - 10}$, show that $f^{-1}(x) = x^3 + 10$.

84) If $f(x) = x^3 - 1$, show that $f^{-1}(x) = \sqrt[3]{x + 1}$.

Section 10.2 Exponential Functions

Objectives

1. Define an Exponential Function
2. Graph $f(x) = a^x$
3. Graph $f(x) = a^{x+c}$
4. Define the Number e and Graph $f(x) = e^x$
5. Solve an Exponential Equation
6. Solve an Applied Problem Using a Given Exponential Function

We have already studied the following types of functions:

Linear functions like $f(x) = 2x + 5$
Quadratic functions like $g(x) = x^2 - 6x + 8$
Absolute value functions like $h(x) = |x|$
Square root functions like $k(x) = \sqrt{x - 3}$

1. Define an Exponential Function

In this section, we will learn about *exponential functions*.

> **Definition**
>
> An **exponential function** is a function of the form
>
> $$f(x) = a^x$$
>
> where $a > 0$, $a \neq 1$, and x is a real number.

Note

1) We stipulate that a is a positive number ($a > 0$) because if a were a negative number, some expressions would not be real numbers.

    Example: If $a = -2$ and $x = \dfrac{1}{2}$, we get $f(x) = (-2)^{1/2} = \sqrt{-2}$ (not real).

    Therefore, a *must* be a positive number.

2) We add the condition that $a \neq 1$ because if $a = 1$, the function would be linear, not exponential.

    Example: If $a = 1$, then $f(x) = 1^x$. This is equivalent to $f(x) = 1$, which is a linear function.

2. Graph $f(x) = a^x$

We can graph exponential functions by plotting points. *It is important to choose many values for the variable so that we obtain positive numbers, negative numbers, and zero in the exponent.*

Example 1

In-Class Example 1
Use the given example.

Graph $f(x) = 2^x$ and $g(x) = 3^x$ on the same axes. Determine the domain and range.

Solution

Make a table of values for each function. Be sure to choose values for x that will give us *positive numbers, negative numbers, and zero* in the exponent.

$f(x) = 2^x$	
x	$f(x)$
0	1
1	2
2	4
3	8
-1	$\dfrac{1}{2}$
-2	$\dfrac{1}{4}$

$g(x) = 3^x$	
x	$g(x)$
0	1
1	3
2	9
3	27
-1	$\dfrac{1}{3}$
-2	$\dfrac{1}{9}$

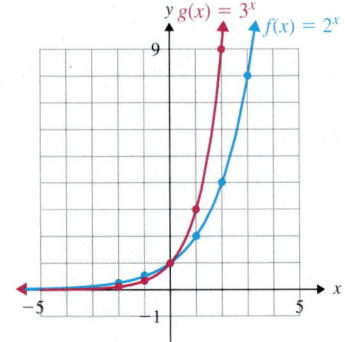

Plot each set of points and connect them with a smooth curve. Note that the larger the value of a, the more rapidly the y-values increase. Additionally, as x increases, the value of y also increases. Here are some other interesting facts to note about the graphs of these functions.

1) Each graph passes the vertical line test so the graphs *do* represent functions.
2) Each graph passes the horizontal line test, so the functions are one-to-one.
3) The y-intercept of each function is $(0, 1)$.
4) The domain of each function is $(-\infty, \infty)$, and the range is $(0, \infty)$.   ∎

You Try 1

Graph $f(x) = 4^x$. Determine the domain and range.

Example 2

Graph $f(x) = \left(\dfrac{1}{2}\right)^x$. Determine the domain and range.

Solution

Make a table of values and be sure to choose values for x that will give us *positive numbers, negative numbers, and zero* in the exponent.

In-Class Example 2

Graph $f(x) = \left(\dfrac{1}{4}\right)^x$. Determine

the domain and range.
**answer: domain: $(-\infty, \infty)$;
range: $(0, \infty)$**

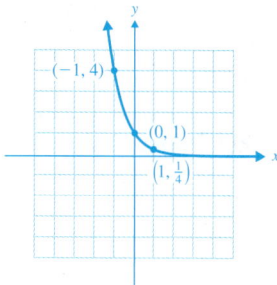

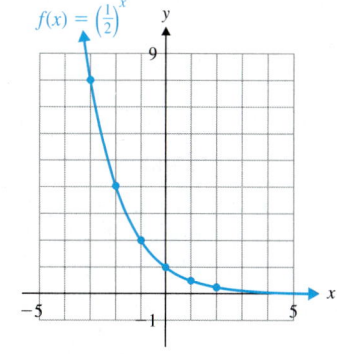

$f(x) = \left(\dfrac{1}{2}\right)^x$	
x	$f(x)$
0	1
1	$\dfrac{1}{2}$
2	$\dfrac{1}{4}$
−1	2
−2	4
−3	8

Like the graphs of $f(x) = 2^x$ and $g(x) = 3^x$ in Example 1, the graph of $f(x) = \left(\dfrac{1}{2}\right)^x$

passes both the vertical and horizontal line tests, making it a one-to-one function.

The y-intercept is $(0, 1)$. The domain is $(-\infty, \infty)$, and the range is $(0, \infty)$.

In the case of $f(x) = \left(\dfrac{1}{2}\right)^x$, however, as the value of x increases, the value of

y *decreases*. This is because $0 < a < 1$. ■

 **You Try 2**

Graph $g(x) = \left(\dfrac{1}{3}\right)^x$. Determine the domain and range.

We can summarize what we have learned so far about exponential functions:

Summary Characteristics of $f(x) = a^x$, where $a > 0$ and $a \neq 1$

1) If $f(x) = a^x$ where $a > 1$, the value of y increases as the value of x increases.

2) If $f(x) = a^x$, where $0 < a < 1$, the value of y decreases as the value of x increases.

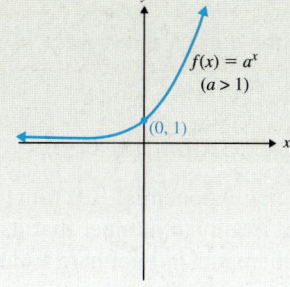

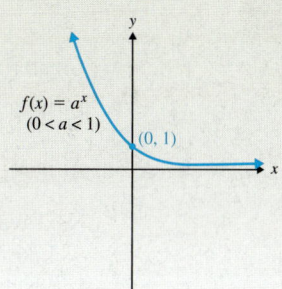

3) The function is one-to-one.

4) The y-intercept is $(0, 1)$.

5) The domain is $(-\infty, \infty)$, and the range is $(0, \infty)$.

3. Graph $f(x) = a^{x+c}$

Next we will graph an exponential function with an expression other than x as its exponent.

Example 3

In-Class Example 3
Graph $f(x) = 2^{x+1}$. Determine the domain and range.
answer: domain: $(-\infty, \infty)$; range: $(0, \infty)$
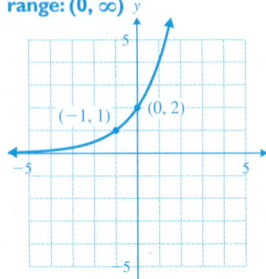

Graph $f(x) = 3^{x-2}$. Determine the domain and range.

Solution

Remember, for the table of values we want to choose values of x that will give us positive numbers, negative numbers, and zero *in the exponent*. First we will determine what value of x will make the exponent equal zero.

$$x - 2 = 0$$
$$x = 2$$

If $x = 2$, the exponent equals zero. Choose a couple of numbers *greater than* 2 and a couple that are *less than* 2 to get positive and negative numbers in the exponent.

	x	$x - 2$	$f(x) = 3^{x-2}$	Plot
	2	0	$3^0 = 1$	$(2, 1)$
Values greater than 2	3	1	$3^1 = 3$	$(3, 3)$
	4	2	$3^2 = 9$	$(4, 9)$
Values less than 2	1	-1	$3^{-1} = \frac{1}{3}$	$\left(1, \frac{1}{3}\right)$
	0	-2	$3^{-2} = \frac{1}{9}$	$\left(0, \frac{1}{9}\right)$

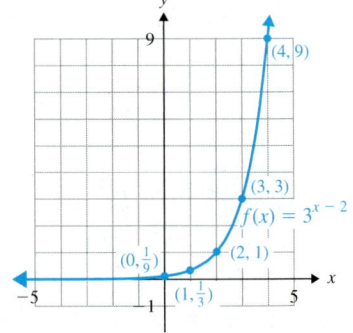

Note that the y-intercept is not $(0, 1)$ because the exponent is $x - 2$, *not x*, as in $f(x) = a^x$. The graph of $f(x) = 3^{x-2}$ is the same shape as the graph of $g(x) = 3^x$ except that the graph of f is shifted 2 units to the right. This is because $f(x) = g(x - 2)$. The domain of f is $(-\infty, \infty)$, and the range is $(0, \infty)$.

 You Try 3

Graph $f(x) = 2^{x+4}$. Determine the domain and range.

4. Define the Number e and Graph $f(x) = e^x$

Next we will introduce a special exponential function, one with a base of e.

Like the number π, e is an irrational number that has many uses in mathematics. In the 1700s, the work of Swiss mathematician Leonhard Euler led him to the approximation of e.

Definition
Approximation of e

$$e \approx 2.718281828459045235$$

One of the questions Euler set out to answer was, what happens to the value of $\left(1 + \dfrac{1}{n}\right)^n$ as n gets larger and larger? He found that as n gets larger, $\left(1 + \dfrac{1}{n}\right)^n$ gets closer to a fixed number. This number is e. Euler approximated e to the 18 decimal places in the definition, and the letter e was chosen to represent this number in his honor. It should be noted that there are other ways to generate e. Finding the value that $\left(1 + \dfrac{1}{n}\right)^n$ approaches as n gets larger and larger is just one way. Also, since e is irrational, it is a nonterminating, nonrepeating decimal.

Example 4

In-Class Example 4
Use the given example.

Graph $f(x) = e^x$. Determine the domain and range.

Solution

A calculator is needed to generate a table of values. We will use either the $\boxed{e^x}$ key or the two keys $\boxed{\text{INV}}$ (or $\boxed{\text{2ND}}$) and $\boxed{\ln x}$ to find powers of e. (Calculators will approximate powers of e to a few decimal places.)

For example, if a calculator has an $\boxed{e^x}$ key, find e^2 by pressing the following keys:

$$\boxed{2}\,\boxed{e^x} \qquad \text{or} \qquad \boxed{e^x}\,\boxed{2}\,\boxed{\text{ENTER}}$$

To four decimal places, $e^2 \approx 7.3891$.

If a calculator has an $\boxed{\ln x}$ key with e^x written above it, find e^2 by pressing the following keys:

$$\boxed{2}\,\boxed{\text{INV}}\,\boxed{\ln x} \qquad \text{or} \qquad \boxed{\text{INV}}\,\boxed{\ln x}\,\boxed{2}\,\boxed{\text{ENTER}}$$

The same approximation for e^2 is obtained.

Remember to choose positive numbers, negative numbers, and zero for x when making the table of values. We will approximate the values of e^x to four decimal places.

$f(x) = e^x$	
x	$f(x)$
0	1
1	2.7183
2	7.3891
3	20.0855
-1	0.3679
-2	0.1353

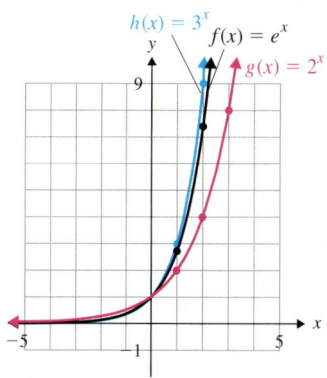

Notice that the graph of $f(x) = e^x$ is between the graphs of $g(x) = 2^x$ and $h(x) = 3^x$. This is because $2 < e < 3$, so e^x grows more quickly than 2^x, but e^x grows more slowly than 3^x. The domain of $f(x) = e^x$ is $(-\infty, \infty)$, and the range is $(0, \infty)$.

We will study e^x and its special properties in more detail later in the chapter.

5. Solve an Exponential Equation

An **exponential equation** is an equation that has a variable in the exponent. Some examples of exponential equations are

$$2^x = 8, \qquad 3^{a-5} = \frac{1}{9}, \qquad e^t = 14, \qquad 5^{2y-1} = 6^{y+4}$$

In this section, we will learn how to solve exponential equations like the first two examples. We can solve those equations by getting the same base.

We know that the exponential function $f(x) = a^x$ ($a > 0$, $a \neq 1$) is one-to-one. This leads to the following property that enables us to solve many exponential equations.

$$\text{If } a^x = a^y, \text{ then } x = y. \ (a > 0, a \neq 1)$$

This property says that if two sides of an equation have the same base, set the exponents equal and solve for the unknown variable.

Procedure   Solving an Exponential Equation

Step 1: **If possible, express each side of the equation with the same base.** If it is *not* possible to get the same base, a different method must be used. (This is presented in Section 10.6.)

Step 2: **Use the rules of exponents to simplify the exponents.**

Step 3: **Set the exponents equal and solve for the variable.**

Example 5

In-Class Example 5
Solve each equation.
a) $3^x = 9$ b) $64^{x+1} = 8^{3x}$
c) $25^{a-3} = 125^{2a}$
d) $6^{2k+1} = \dfrac{1}{36}$

answer:
a) {2}   b) {2}
c) $\left\{ -\dfrac{3}{2} \right\}$ d) $\left\{ -\dfrac{3}{2} \right\}$

Solve each equation.

a) $2^x = 8$ b) $49^{c+3} = 7^{3c}$ c) $9^{6n} = 27^{n-4}$ d) $3^{a-5} = \dfrac{1}{9}$

Solution

a) ***Step 1:*** Express each side of the equation with the same base.

$$2^x = 8$$
$$2^x = 2^3 \qquad \text{Rewrite 8 with a base of 2: } 8 = 2^3.$$

   Step 2: The exponents are simplified.

   Step 3: Since the bases are the same, set the exponents equal and solve.

$$x = 3$$

   The solution set is $\{3\}$.

b) ***Step 1:*** Express each side of the equation with the same base.

$$49^{c+3} = 7^{3c}$$
$$(7^2)^{c+3} = 7^{3c} \qquad \text{Both sides are powers of 7; } 49 = 7^2.$$

   Step 2: Use the rules of exponents to simplify the exponents.

$$7^{2(c+3)} = 7^{3c} \qquad \text{Power rule for exponents}$$
$$7^{2c+6} = 7^{3c} \qquad \text{Distribute.}$$

   Step 3: Since the bases are the same, set the exponents equal and solve.

$$2c + 6 = 3c \qquad \text{Set the exponents equal.}$$
$$6 = c \qquad \text{Subtract } 2c.$$

   The solution set is $\{6\}$.

c)   **Step 1:**   Express each side of the equation with the same base. 9 *and* 27 *are each powers of* 3.

$$9^{6n} = 27^{n-4}$$
$$(3^2)^{6n} = (3^3)^{n-4} \qquad 9 = 3^2;\ 27 = 3^3$$

   Step 2:   Use the rules of exponents to simplify the exponents.

$$3^{2(6n)} = 3^{3(n-4)} \qquad \text{Power rule for exponents}$$
$$3^{12n} = 3^{3n-12} \qquad \text{Multiply.}$$

   Step 3:   Since the bases are the same, set the exponents equal and solve.

$$12n = 3n - 12 \qquad \text{Set the exponents equal.}$$
$$9n = -12 \qquad \text{Subtract } 3n.$$
$$n = -\frac{12}{9} = -\frac{4}{3} \qquad \text{Divide by 9; simplify.}$$

The solution set is $\left\{ -\dfrac{4}{3} \right\}$.

d)   **Step 1:**   Express each side of the equation $3^{a-5} = \dfrac{1}{9}$ with the same base. $\dfrac{1}{9}$ *can be expressed with a base of* 3: $\dfrac{1}{9} = \left(\dfrac{1}{3}\right)^2 = 3^{-2}$.

$$3^{a-5} = \frac{1}{9}$$
$$3^{a-5} = 3^{-2} \qquad \text{Rewrite } \frac{1}{9} \text{ with a base of 3.}$$

   Step 2:   The exponents are simplified.

   Step 3:   Set the exponents equal and solve.

$$a - 5 = -2 \qquad \text{Set the exponents equal.}$$
$$a = 3 \qquad \text{Add 5.}$$

The solution set is $\{3\}$.

You Try 4

Solve each equation.

a) $(12)^x = 144$ b) $6^{t-5} = 36^{t+4}$ c) $32^{2w} = 8^{4w-1}$ d) $8^k = \dfrac{1}{64}$

---

6. Solve an Applied Problem Using a Given Exponential Function

Example 6

The value of a car depreciates (decreases) over time. The value, $V(t)$, in dollars, of a sedan t yr after it is purchased is given by

$$V(t) = 18{,}200(0.794)^t$$

a)   What was the purchase price of the car?

b)   What will the car be worth 5 yr after purchase?

In-Class Example 6

The value of a car depreciates over time. The value, $V(t)$, in dollars, of a car t yr after it is purchased is given by $V(t) = 12{,}500(0.794)^t$.

a) What was the purchase price of the car?

b) What will the car be worth 6 yr after purchase?

answer: a) \$12,500

b) \$3132.08

Solution

a) To find the purchase price of the car, let $t = 0$.
   Evaluate $V(0)$ given that $V(t) = 18{,}200(0.794)^t$.

$$V(0) = 18{,}200(0.794)^0$$
$$= 18{,}200(1)$$
$$= 18{,}200$$

The purchase price of the car was $18,200.

b) To find the value of the car after 5 yr, let $t = 5$. Use a calculator to find $V(5)$.

$$V(5) = 18{,}200(0.794)^5$$
$$= 5743.46$$

The car will be worth about $5743.46.

You Try 5

The value, $V(t)$, in dollars, of a pickup truck t yr after it is purchased is given by

$$V(t) = 23{,}500(0.785)^t$$

a) What was the purchase price of the pickup?

b) What will the pickup truck be worth 4 yr after purchase?

Answers to You Try Exercises

1)

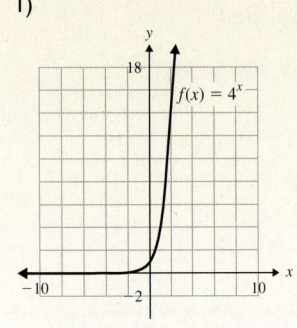

domain: $(-\infty, \infty)$;
range: $(0, \infty)$

2)

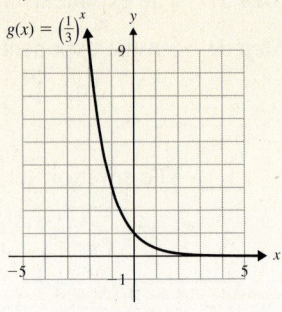

domain: $(-\infty, \infty)$;
range: $(0, \infty)$

3)

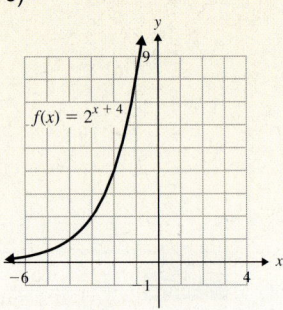

domain: $(-\infty, \infty)$;
range: $(0, \infty)$

4) a) $\{2\}$ b) $\{-13\}$ c) $\left\{\dfrac{3}{2}\right\}$ d) $\{-2\}$ 5) a) $23,500 b) $8923.73

10.2 Exercises

*Additional answers can be found in the Answers to Exercises appendix.

Mixed Exercises: Objectives 1 and 2

1) When making a table of values to graph an exponential function, what kind of values should be chosen for the variable?   Choose values for the variable that will give positive numbers, negative numbers, and zero in the exponent.

2) What is the y-intercept of the graph of $f(x) = a^x$ where $a > 0$ and $a \neq 1$?   (0, 1)

Graph each exponential function. Determine the domain and range.

3) $f(x) = 5^x$
   domain: $(-\infty, \infty)$; range: $(0, \infty)$

4) $g(x) = 4^x$
   domain: $(-\infty, \infty)$; range: $(0, \infty)$

5) $y = 2^x$
   domain: $(-\infty, \infty)$; range: $(0, \infty)$

6) $f(x) = 3^x$
   domain: $(-\infty, \infty)$; range: $(0, \infty)$

7) $h(x) = \left(\dfrac{1}{3}\right)^x$
   domain: $(-\infty, \infty)$; range: $(0, \infty)$

8) $y = \left(\dfrac{1}{4}\right)^x$
   domain: $(-\infty, \infty)$; range: $(0, \infty)$

For an exponential function of the form $f(x) = a^x$ ($a > 0$, $a \neq 1$), answer the following.

9) What is the domain? $(-\infty, \infty)$

10) What is the range? $(0, \infty)$

Objective 3: Graph $f(x) = a^{x+c}$

Graph each exponential function. State the domain and range.

11) $g(x) = 2^{x+1}$
domain: $(-\infty, \infty)$; range: $(0, \infty)$

12) $y = 3^{x+2}$
domain: $(-\infty, \infty)$; range: $(0, \infty)$

13) $f(x) = 3^{x-4}$
domain: $(-\infty, \infty)$; range: $(0, \infty)$

14) $h(x) = 2^{x-3}$
domain: $(-\infty, \infty)$; range: $(0, \infty)$

15) $y = 4^{x+3}$
domain: $(-\infty, \infty)$; range: $(0, \infty)$

16) $g(x) = 4^{x-1}$
domain: $(-\infty, \infty)$; range: $(0, \infty)$

17) $f(x) = 2^{2x}$
domain: $(-\infty, \infty)$; range: $(0, \infty)$

18) $h(x) = 3^{\frac{1}{2}x}$
domain: $(-\infty, \infty)$; range: $(0, \infty)$

19) $y = 2^x + 1$
domain: $(-\infty, \infty)$; range: $(1, \infty)$

20) $f(x) = 2^x - 3$
domain: $(-\infty, \infty)$; range: $(-3, \infty)$

21) $g(x) = 3^x - 2$
domain: $(-\infty, \infty)$; range: $(-2, \infty)$

22) $h(x) = 3^x + 1$
domain: $(-\infty, \infty)$; range: $(1, \infty)$

23) $y = -2^x$
domain: $(-\infty, \infty)$; range: $(-\infty, 0)$

24) $f(x) = -\left(\dfrac{1}{3}\right)^x$
domain: $(-\infty, \infty)$; range: $(-\infty, 0)$

25) As the value of x gets larger, would you expect $f(x) = 2x$ or $g(x) = 2^x$ to grow faster? Why? $g(x) = 2^x$ would grow faster because for values of $x > 2$, $2^x > 2x$.

26) Let $f(x) = \left(\dfrac{1}{5}\right)^x$. The graph of $f(x)$ gets very close to the line $y = 0$ (the x-axis) as the value of x gets larger. Why?

27) If you are given the graph of $f(x) = a^x$, where $a > 0$ and $a \neq 1$, how would you obtain the graph of $g(x) = a^x - 2$?   Shift the graph of $f(x)$ down 2 units.

28) If you are given the graph of $f(x) = a^x$, where $a > 0$ and $a \neq 1$, how would you obtain the graph of $g(x) = a^{x-3}$?   Shift the graph of $f(x)$ right 3 units.

Objective 4: Define the Number e and Graph $f(x) = e^x$

29) What is the approximate value of e to four decimal places?   2.7183

30) Is e a rational or an irrational number? Explain your answer. e is irrational because it is a nonterminating, nonrepeating decimal.

For Exercises 31–34, match each exponential function with its graph.

A)
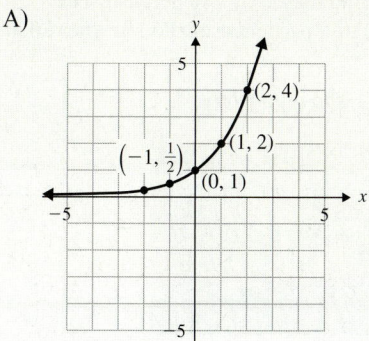

B)
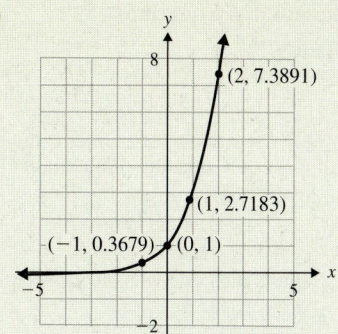

C)
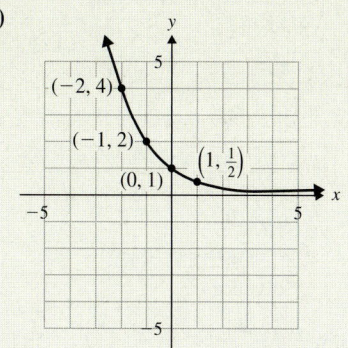

D)
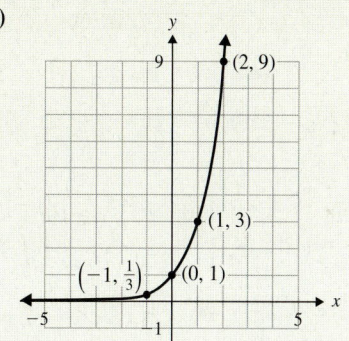

31) $f(x) = e^x$ B

32) $g(x) = 2^x$ A

33) $h(x) = 3^x$ D

34) $k(x) = \left(\dfrac{1}{2}\right)^x$ C

Graph each function. State the domain and range.

35) $f(x) = e^x - 2$
domain: $(-\infty, \infty)$; range: $(-2, \infty)$

36) $g(x) = e^x + 1$
domain: $(-\infty, \infty)$; range: $(1, \infty)$

37) $y = e^{x+1}$
domain: $(-\infty, \infty)$; range: $(0, \infty)$

38) $h(x) = e^{x-3}$
domain: $(-\infty, \infty)$; range: $(0, \infty)$

39) $g(x) = \dfrac{1}{2}e^x$
domain: $(-\infty, \infty)$; range: $(0, \infty)$

40) $y = 2e^x$
domain: $(-\infty, \infty)$; range: $(0, \infty)$

41) $h(x) = -e^x$
domain: $(-\infty, \infty)$; range: $(-\infty, 0)$

42) $f(x) = e^{-x}$
domain: $(-\infty, \infty)$; range: $(0, \infty)$

43) Graph $y = e^x$, and compare it with the graph of $h(x) = -e^x$ in Exercise 41. What can you say about these graphs?   They are symmetric with respect to the x-axis.

44) Graph $y = e^x$, and compare it with the graph of $f(x) = e^{-x}$ in Exercise 42. What can you say about these graphs?   They are symmetric with respect to the y-axis.

Objective 5: Solve an Exponential Equation

Solve each exponential equation.

Fill It In

Fill in the blanks with either the missing mathematical step or reason for the given step.

45) $6^{3n} = 36^{n-4}$

$6^{3n} = (6^2)^{n-4}$	Express each side with the same base.
$6^{3n} = 6^{2(n-4)}$	Power rule for exponents
$6^{3n} = 6^{2n-8}$	Distribute.
$3n = 2n - 8$	Set the exponents equal.
$n = -8$	Solve for n.

The solution set is $\{-8\}$.

46) $125^{2w} = 5^{w+2}$

$(5^3)^{2w} = 5^{w+2}$	Express each side with the same base.
$5^{3(2w)} = 5^{w+2}$	Power rule for exponents
$5^{6w} = 5^{w+2}$	Distribute.
$6w = w + 2$	Set the exponents equal.
$w = \dfrac{2}{5}$	Solve for w.

The solution set is $\left\{\dfrac{2}{5}\right\}$.

47) $9^x = 81$ {2}

48) $4^y = 16$ {2}

49) $5^{4d} = 125$ $\left\{\dfrac{3}{4}\right\}$

50) $4^{3a} = 64$ {1}

51) $16^{m-2} = 2^{3m}$ {8}

52) $3^{5t} = 9^{t+4}$ $\left\{\dfrac{8}{3}\right\}$

53) $7^{2k-6} = 49^{3k+1}$ {-2}

54) $(1000)^{2p-3} = 10^{4p+1}$ {5}

55) $32^{3c} = 8^{c+4}$ {1}

56) $(125)^{2x-9} = 25^{x-3}$ $\left\{\dfrac{21}{4}\right\}$

VIDEO 57) $100^{5z-1} = (1000)^{2z+7}$ $\left\{\dfrac{23}{4}\right\}$

58) $32^{y+1} = 64^{y+2}$ {-7}

59) $81^{3n+9} = 27^{2n+6}$ {-3}

60) $27^{5v} = 9^{v+4}$ $\left\{\dfrac{8}{13}\right\}$

61) $6^x = \dfrac{1}{36}$ {-2}

62) $11^t = \dfrac{1}{121}$ {-2}

63) $2^a = \dfrac{1}{8}$ {-3}

64) $3^z = \dfrac{1}{81}$ {-4}

65) $9^r = \dfrac{1}{27}$ $\left\{-\dfrac{3}{2}\right\}$

66) $16^c = \dfrac{1}{8}$ $\left\{-\dfrac{3}{4}\right\}$

67) $\left(\dfrac{3}{4}\right)^{5k} = \left(\dfrac{27}{64}\right)^{k+1}$ $\left\{\dfrac{3}{2}\right\}$

68) $\left(\dfrac{3}{2}\right)^{y+4} = \left(\dfrac{81}{16}\right)^{y-2}$ {4}

VIDEO 69) $\left(\dfrac{5}{6}\right)^{3x+7} = \left(\dfrac{36}{25}\right)^{2x}$ {-1}

70) $\left(\dfrac{7}{2}\right)^{5w} = \left(\dfrac{4}{49}\right)^{4w+3}$ $\left\{-\dfrac{6}{13}\right\}$

Objective 6: Solve an Applied Problem Using a Given Exponential Function

Solve each application.

71) The value of a car depreciates (decreases) over time. The value, $V(t)$, in dollars, of an SUV t yr after it is purchased is given by

$$V(t) = 32{,}700(0.812)^t$$

a) What was the purchase price of the SUV? $32,700

b) What will the SUV be worth 3 yr after purchase? $17,507.17

72) The value, $V(t)$, in dollars, of a compact car t yr after it is purchased is given by

$$V(t) = 10{,}150(0.784)^t$$

a) What was the purchase price of the car? $10,150

b) What will the car be worth 5 yr after purchase? $3006.40

73) The value, $V(t)$, in dollars, of a minivan t yr after it is purchased is given by

$$V(t) = 16{,}800(0.803)^t$$

a) What was the purchase price of the minivan? $16,800

b) What will the minivan be worth 6 yr after purchase? $4504.04

74) The value, $V(t)$, in dollars, of a sports car t yr after it is purchased is given by

$$V(t) = 48{,}600(0.820)^t$$

a) What was the purchase price of the sports car? $48,600

b) What will the sports car be worth 4 yr after purchase? $21,973.12

75) From 1995 to 2005, the value of homes in a suburb increased by 3% per year. The value, $V(t)$, in dollars, of a particular house t yr after 1995 is given by

$$V(t) = 185{,}200(1.03)^t$$

a) How much was the house worth in 1995? $185,200

b) How much was the house worth in 2002? $227,772.64

76) From 2000 to 2010, the value of condominiums in a big city high-rise building increased by 2% per year. The value, $V(t)$, in dollars, of a particular condo t yr after 2000 is given by

$$V(t) = 420{,}000(1.02)^t$$

a) How much was the condominium worth in 2000? $420,000

b) How much was the condominium worth in 2010? $511,977.66

An *annuity* is an account into which money is deposited every year. The amount of money, A in dollars, in the account after t yr of depositing c dollars at the beginning of every year earning an interest rate r (as a decimal) is

$$A = c\left[\frac{(1+r)^t - 1}{r}\right](1+r)$$

Use the formula for Exercises 77–80.

77) After Fernando's daughter is born, he decides to begin saving for her college education. He will deposit $2000 every year in an annuity for 18 yr at a rate of 9%. How much will be in the account after 18 yr? $90,036.92

78) To save for retirement, Susan plans to deposit $6000 per year in an annuity for 30 yr at a rate of 8.5%. How much will be in the account after 30 yr? $808,637.86

79) Patrice will deposit $4000 every year in an annuity for 10 yr at a rate of 7%. How much will be in the account after 10 yr? $59,134.40

80) Haeshin will deposit $3000 every year in an annuity for 15 yr at a rate of 8%. How much will be in the account after 15 yr? $87,972.85

 81) After taking a certain antibiotic, the amount of amoxicillin $A(t)$, in milligrams, remaining in the patient's system t hr after taking 1000 mg of amoxicillin is

$$A(t) = 1000e^{-0.5332t}$$

How much amoxicillin is in the patient's system 6 hr after taking the medication?   40.8 mg

82) Some cockroaches can reproduce according to the formula

$$y = 2(1.65)^t$$

where y is the number of cockroaches resulting from the mating of two cockroaches and their offspring t months after the first two cockroaches mate.

If Morris finds two cockroaches in his kitchen (assuming one is male and one is female) how large can the cockroach population become after 12 months?   814

Section 10.3 Logarithmic Functions

Objectives

1. Define a Logarithm
2. Convert from Logarithmic Form to Exponential Form
3. Convert from Exponential Form to Logarithmic Form
4. Solve an Equation of the Form $\log_a b = c$
5. Evaluate a Logarithm
6. Evaluate Common Logarithms, and Solve Equations of the Form $\log b = c$
7. Use the Properties $\log_a a = 1$ and $\log_a 1 = 0$
8. Define and Graph a Logarithmic Function
9. Solve an Applied Problem Using a Logarithmic Equation

1. Define a Logarithm

In Section 10.2, we graphed $f(x) = 2^x$ by making a table of values and plotting the points. The graph passes the horizontal line test, making the function one-to-one. Recall that if (x, y) is on the graph of a function, then (y, x) is on the graph of its inverse. We can graph the inverse of $f(x) = 2^x$, $f^{-1}(x)$, by switching the x- and y-coordinates in the table of values and plotting the points.

$f(x) = 2^x$	
x	$y = f(x)$
0	1
1	2
2	4
3	8
-1	$\frac{1}{2}$
-2	$\frac{1}{4}$

$f^{-1}(x)$	
x	$y = f^{-1}(x)$
1	0
2	1
4	2
8	3
$\frac{1}{2}$	-1
$\frac{1}{4}$	-2

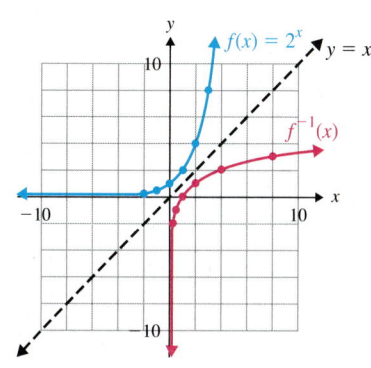

Above is the graph of $f(x) = 2^x$ and its inverse. Notice that, like the graphs of all functions and their inverses, they are symmetric with respect to the line $y = x$.

What is the equation of $f^{-1}(x)$ if $f(x) = 2^x$? We will use the procedure outlined in Section 10.1 to find the equation of $f^{-1}(x)$.

If $f(x) = 2^x$, then find the equation of $f^{-1}(x)$ as follows.

Step 1:   Replace $f(x)$ with y.

$$y = 2^x$$

Step 2:   Interchange x and y.

$$x = 2^y$$

Step 3:   Solve for y.

How do we solve $x = 2^y$ for y? To answer this question, we must introduce another concept called *logarithms*.

Definition

Definition of Logarithm: If $a > 0$, $a \neq 1$, and $x > 0$, then for every real number y,

$$y = \log_a x \text{ means } x = a^y$$

The word **log** is an abbreviation for **logarithm**. We read $\log_a x$ as "log of x to the base a" or "log to the base a of x." *This definition of a logarithm should be memorized!*

Note

It is very important to note that the base of the logarithm must be positive and not equal to 1, and that x must be positive as well.

The relationship between the logarithmic form of an equation ($y = \log_a x$) and the exponential form of an equation ($x = a^y$) is one that has many uses. Notice the relationship between the two forms.

Logarithmic Form		Exponential Form
Value of the logarithm		Exponent
$\downarrow$		$\downarrow$
$y = \log_a x$		$x = a^y$
$\uparrow$		$\uparrow$
Base		Base

From the above, you can see that *a logarithm is an exponent*. $\log_a x$ is the power to which we raise a to get x.

2. Convert from Logarithmic Form to Exponential Form

Much of our work with logarithms involves converting between logarithmic and exponential notation. After working with logs and exponential form, we will come back to the question of how to solve $x = 2^y$ for y.

Example 1

Write in exponential form.

a) $\log_6 36 = 2$ b) $\log_4 \dfrac{1}{64} = -3$ c) $\log_7 1 = 0$

In-Class Example 1
Write in exponential form.
a) $\log_5 25 = 2$ b) $\log_3 27 = 3$
c) $\log_4 1 = 0$
answer: a) $5^2 = 25$
b) $3^3 = 27$ c) $4^0 = 1$

Solution

a) $\log_6 36 = 2$ means that 2 is the power to which we raise 6 to get 36. The exponential form is $6^2 = 36$.

$$\log_6 36 = 2 \text{ means } 6^2 = 36.$$

b) $\log_4 \dfrac{1}{64} = -3$ means $4^{-3} = \dfrac{1}{64}$.

c) $\log_7 1 = 0$ means $7^0 = 1$.

You Try 1

Write in exponential form.

a) $\log_3 81 = 4$ b) $\log_5 \dfrac{1}{25} = -2$ c) $\log_{64} 8 = \dfrac{1}{2}$ d) $\log_{13} 13 = 1$

3. Convert from Exponential Form to Logarithmic Form

Example 2

Write in logarithmic form.

a) $10^4 = 10,000$ b) $9^{-2} = \dfrac{1}{81}$ c) $8^1 = 8$

d) $\sqrt{25} = 5$

In-Class Example 2
Write in logarithmic form.
a) $9^2 = 81$ b) $4^3 = 64$
c) $8^{-2} = \dfrac{1}{64}$ d) $\sqrt[3]{125} = 5$

answer: a) $\log_9 81 = 2$
b) $\log_4 64 = 3$
c) $\log_8 \dfrac{1}{64} = -2$
d) $\log_{125} 5 = \dfrac{1}{3}$

Solution

a) $10^4 = 10,000$ means $\log_{10} 10,000 = 4$.

b) $9^{-2} = \dfrac{1}{81}$ means $\log_9 \dfrac{1}{81} = -2$.

c) $8^1 = 8$ means $\log_8 8 = 1$.

d) To write $\sqrt{25} = 5$ in logarithmic form, rewrite $\sqrt{25}$ as $25^{1/2}$.

$$\sqrt{25} = 5 \text{ is the same as } 25^{1/2} = 5$$

$$25^{1/2} = 5 \text{ means } \log_{25} 5 = \dfrac{1}{2}.$$

Note

When working with logarithms, we will often change radical notation to the equivalent fractional exponent. This is because a logarithm *is* an exponent.

You Try 2

Write in logarithmic form.

a) $7^2 = 49$ b) $5^{-4} = \dfrac{1}{625}$ c) $19^0 = 1$ d) $\sqrt{144} = 12$

Solving Logarithmic Equations

4. Solve an Equation of the Form $\log_a b = c$

A **logarithmic equation** is an equation in which at least one term contains a logarithm. In this section, we will learn how to solve a logarithmic equation of the form $\log_a b = c$. We will learn how to solve other types of logarithmic equations in Sections 10.5 and 10.6.

Procedure Solve an Equation of the Form $\log_a b = c$

To solve a logarithmic equation of the form $\log_a b = c$, write the equation in exponential form $(a^c = b)$ and solve for the variable.

Example 3

In-Class Example 3

Solve each logarithmic equation.
a) $\log_4 h = 2$
b) $\log_2 (2q - 1) = 4$
c) $\log_t 49 = 2$
d) $\log_9 81 = m$
e) $\log_{100} \sqrt[3]{10} = a$

answer: a) $\{16\}$ b) $\left\{\dfrac{17}{2}\right\}$

c) $\{7\}$ d) $\{2\}$ e) $\left\{\dfrac{1}{6}\right\}$

Solve each logarithmic equation.

a) $\log_{10} r = 3$ b) $\log_3(7a + 18) = 4$ c) $\log_w 25 = 2$

d) $\log_2 16 = c$ e) $\log_{36} \sqrt[4]{6} = x$

Solution

a) Write the equation in exponential form and solve for r.

$$\log_{10} r = 3 \quad \text{means} \quad 10^3 = r$$
$$1000 = r$$

The solution set is $\{1000\}$.

b) Write $\log_3(7a + 18) = 4$ in exponential form and solve for a.

$$\log_3(7a + 18) = 4 \quad \text{means} \quad 3^4 = 7a + 18$$
$$81 = 7a + 18$$
$$63 = 7a \qquad \text{Subtract 18.}$$
$$9 = a \qquad \text{Divide by 7.}$$

The solution set is $\{9\}$.

c) Write $\log_w 25 = 2$ in exponential form and solve for w.

$$\log_w 25 = 2 \quad \text{means} \quad w^2 = 25$$
$$w = \pm 5 \qquad \text{Square root property}$$

Although we get $w = 5$ or $w = -5$ when we solve $w^2 = 25$, recall that the base of a logarithm must be a positive number. Therefore, $w = -5$ is *not* a solution.

The solution set is $\{5\}$.

d) Write $\log_2 16 = c$ in exponential form and solve for c.

$$\log_2 16 = c \quad \text{means} \quad 2^c = 16$$
$$c = 4$$

The solution set is $\{4\}$.

e) $\log_{36}\sqrt[4]{6} = x$ means $36^x = \sqrt[4]{6}$

$$(6^2)^x = 6^{1/4}$$ Express each side with the same base;
rewrite the radical as a fractional exponent.

$$6^{2x} = 6^{1/4}$$ Power rule for exponents

$$2x = \frac{1}{4}$$ Set the exponents equal.

$$x = \frac{1}{8}$$ Divide by 2.

The solution set is $\left\{\dfrac{1}{8}\right\}$.

You Try 3

Solve each logarithmic equation.

a) $\log_2 y = 5$ b) $\log_5(3p + 11) = 3$ c) $\log_x 169 = 2$

d) $\log_6 36 = n$ e) $\log_{64}\sqrt[5]{8} = k$

5. Evaluate a Logarithm

Often when working with logarithms, we are asked to *evaluate* them or to find the value of a log.

Example 4

In-Class Example 4
Evaluate.
a) $\log_{11} 121$ b) $\log_4 16$
c) $\log_5 \dfrac{1}{25}$ d) $\log_{64} 8$
answer: a) 2   b) 2   c) −2
d) $\dfrac{1}{2}$

Evaluate.

a) $\log_3 9$ b) $\log_2 8$ c) $\log_{10}\dfrac{1}{10}$ d) $\log_{25} 5$

Solution

a) To evaluate (or find the value of) $\log_3 9$ means to find the power to which we raise 3 to get 9. That power is **2**.

$$\log_3 9 = 2 \quad \text{since} \quad 3^2 = 9$$

b) To evaluate $\log_2 8$ means to find the power to which we raise 2 to get 8. That power is **3**.

$$\log_2 8 = 3 \quad \text{since} \quad 2^3 = 8$$

c) To evaluate $\log_{10}\dfrac{1}{10}$ means to find the power to which we raise 10 to get $\dfrac{1}{10}$. That power is **−1**.

If you don't see that this is the answer, set the expression $\log_{10}\dfrac{1}{10}$ equal to x,

write the equation in exponential form, and solve for x as in Example 3.

$$\log_{10}\dfrac{1}{10} = x \quad \text{means} \quad 10^x = \dfrac{1}{10}$$
$$10^x = 10^{-1} \qquad \dfrac{1}{10} = 10^{-1}$$
$$x = -1$$

Then, $\log_{10}\dfrac{1}{10} = -1$.

d) To evaluate $\log_{25} 5$ means to find the power to which we raise 25 to get 5. That power is $\dfrac{1}{2}$.

Once again, we can also find the value of $\log_{25} 5$ by setting it equal to x, writing the equation in exponential form, and solving for x.

$$\log_{25} 5 = x \quad \text{means} \quad 25^x = 5$$

$$(5^2)^x = 5 \qquad \text{Express each side with the same base.}$$
$$5^{2x} = 5^1 \qquad \text{Power rule; } 5 = 5^1$$
$$2x = 1 \qquad \text{Set the exponents equal.}$$
$$x = \frac{1}{2} \qquad \text{Divide by 2.}$$

Therefore, $\log_{25} 5 = \dfrac{1}{2}$.   ■

You Try 4

Evaluate.

a) $\log_{10} 100$ b) $\log_3 81$ c) $\log_8 \dfrac{1}{8}$ d) $\log_{144} 12$

6. Evaluate Common Logarithms, and Solve Equations of the Form log $b = c$

Logarithms have many applications not only in mathematics but also in other areas such as chemistry, biology, engineering, and economics.

Since our number system is a base 10 system, logarithms to the base 10 are very widely used and are called **common logarithms** or **common logs**. A base 10 log has a special notation—$\log_{10} x$ is written as $\log x$. *When a log is written in this way, the base is assumed to be 10.*

$$\log x \text{ means } \log_{10} x$$

We must keep this in mind when evaluating logarithms and when solving logarithmic equations.

Example 5

In-Class Example 5
Evaluate log 100.
answer: 2

Evaluate log 100.

Solution

log 100 is equivalent to $\log_{10} 100$. To evaluate log 100 means to find the power to which we raise 10 to get 100. That power is **2.**

$$\log 100 = 2$$ ■

You Try 5

Evaluate log 1000.

Example 6

Solve $\log(3x - 8) = 1$.

In-Class Example 6
Solve $\log(c + 2) = 2$.
answer: {98}

Solution

$\log(3x - 8) = 1$ is equivalent to $\log_{10}(3x - 8) = 1$. Write the equation in exponential form and solve for x.

$$\log(3x - 8) = 1 \text{ means } 10^1 = 3x - 8$$
$$10 = 3x - 8$$
$$18 = 3x \qquad \text{Add 8.}$$
$$6 = x \qquad \text{Divide by 3.}$$

The solution set is {6}.

You Try 6

Solve $\log(12q + 16) = 2$.

We will study common logs in more depth in Section 10.5.

7. Use the Properties $\log_a a = 1$ and $\log_a 1 = 0$

There are a couple of properties of logarithms that can simplify our work.

If a is any real number, then $a^1 = a$. Furthermore, if $a \neq 0$, then $a^0 = 1$. Write $a^1 = a$ and $a^0 = 1$ in logarithmic form to obtain these two properties of logarithms:

Properties of Logarithms

If $a > 0$ and $a \neq 1$,

1) $\log_a a = 1$

2) $\log_a 1 = 0$

Example 7

Use the properties of logarithms to evaluate each.

a) $\log_{12} 1$ b) $\log_3 3$ c) $\log 10$ d) $\log_{\sqrt{5}} 1$

In-Class Example 7
Use the properties of logarithms
to evaluate each.
a) $\log_4 4$ b) $\log_6 1$ c) $\log 1$
d) $\log_{1/2} 1$
answer: a) 1   b) 0   c) 0
d) 0

Solution

a) By property 2, $\log_{12} 1 = 0$.

b) By property 1, $\log_3 3 = 1$.

c) The base of $\log 10$ is 10. Therefore, $\log 10 = \log_{10} 10$. By property 1, $\log 10 = 1$.

d) By property 2, $\log_{\sqrt{5}} 1 = 0$.

You Try 7

Use the properties of logarithms to evaluate each.

a) $\log_{16} 16$ b) $\log_{1/3} 1$ c) $\log_{\sqrt{11}} \sqrt{11}$

8. Define and Graph a Logarithmic Function

Next we define a logarithmic function.

> ### Definition
>
> For $a > 0$, $a \neq 1$, and $x > 0$, $f(x) = \log_a x$ is the **logarithmic function with base a.**

> ### Note
>
> $f(x) = \log_a x$ can also be written as $y = \log_a x$. Changing $y = \log_a x$ to exponential form, we get $a^y = x$. Remembering that a is a positive number not equal to 1, it follows that
>
> 1) any real number may be substituted for y. Therefore, **the range of $y = \log_a x$ is $(-\infty, \infty)$.**
>
> 2) x must be a positive number. So, **the domain of $y = \log_a x$ is $(0, \infty)$.**

Let's return to the problem of finding the equation of the inverse of $f(x) = 2^x$ that was first introduced on p. 643.

Example 8

Find the equation of the inverse of $f(x) = 2^x$.

In-Class Example 8
Find the equation of the inverse of $f(x) = 3^x$.
answer: $f^{-1}(x) = \log_3 x$

Solution

Step 1: Replace $f(x)$ with y: $\qquad y = 2^x$

Step 2: Interchange x and y: $\qquad x = 2^y$

Step 3: Solve for y.

To solve $x = 2^y$ for y, write the equation in logarithmic form.

$$x = 2^y \quad \text{means} \quad y = \log_2 x$$

Step 4: Replace y with $f^{-1}(x)$.

$$f^{-1}(x) = \log_2 x$$

The inverse of the exponential function $f(x) = 2^x$ is $f^{-1}(x) = \log_2 x$. ∎

You Try 8

Find the equation of the inverse of $f(x) = 6^x$.

> ### Note
>
> The inverse of the exponential function $f(x) = a^x$ (where $a > 0$, $a \neq 1$, and x is any real number) is $f^{-1}(x) = \log_a x$. Furthermore,
>
> 1) the domain of $f(x)$ is the range of $f^{-1}(x)$.
>
> 2) the range of $f(x)$ is the domain of $f^{-1}(x)$.
>
> Their graphs are symmetric with respect to $y = x$.

To graph a logarithmic function, write it in exponential form first. Then make a table of values, plot the points, and draw the curve through the points.

Example 9

In-Class Example 9
Graph $f(x) = \log_3 x$.
answer:

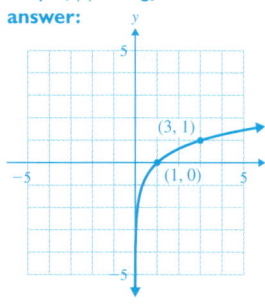

Graph $f(x) = \log_2 x$.

Solution

Substitute y for $f(x)$ and write the equation in exponential form.

$$y = \log_2 x \quad \text{means} \quad 2^y = x$$

To make a table of values, it will be easier to *choose values for y* and compute the corresponding values of x. Remember to choose values of y that will give positive numbers, negative numbers, and zero in the exponent.

$2^y = x$	
x	**y**
1	0
2	1
4	2
8	3
$\dfrac{1}{2}$	−1
$\dfrac{1}{4}$	−2

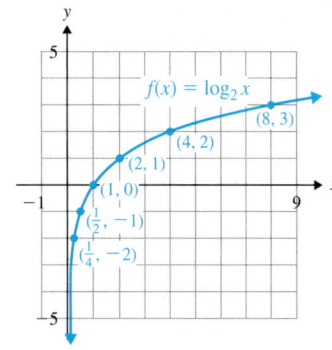

From the graph, we can see that the domain of f is $(0, \infty)$, and the range of f is $(-\infty, \infty)$. ■

You Try 9

Graph $f(x) = \log_4 x$.

Example 10

In-Class Example 10
Graph $f(x) = \log_{1/2} x$.
answer:

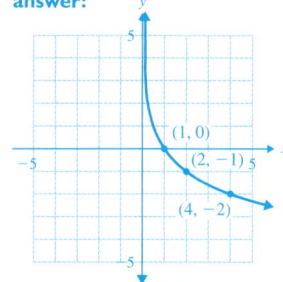

Graph $f(x) = \log_{1/3} x$.

Solution

Substitute y for $f(x)$ and write the equation in exponential form.

$$y = \log_{1/3} x \quad \text{means} \quad \left(\frac{1}{3}\right)^y = x$$

For the table of values, *choose values for y* and compute the corresponding values of x.

$\left(\dfrac{1}{3}\right)^y = x$	
x	**y**
1	0
$\dfrac{1}{3}$	1
$\dfrac{1}{9}$	2
3	−1
9	−2

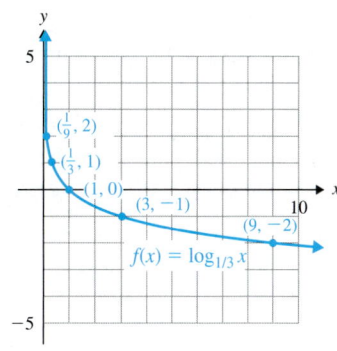

The domain of f is $(0, \infty)$, and the range is $(-\infty, \infty)$. ■

You Try 10

Graph $f(x) = \log_{1/4} x$.

The graphs in Examples 9 and 10 are typical of the graphs of logarithmic functions—Example 9 for functions where $a > 1$ and Example 10 for functions where $0 < a < 1$.

Next is a summary of some characteristics of logarithmic functions.

Summary   Characteristics of a Logarithmic Function $f(x) = \log_a x$, where $a > 0$ and $a \neq 1$

1) If $f(x) = \log_a x$ where $a > 1$, the value of y increases as the value of x increases.

2) If $f(x) = \log_a x$ where $0 < a < 1$, the value of y decreases as the value of x increases.

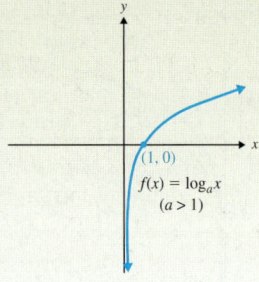

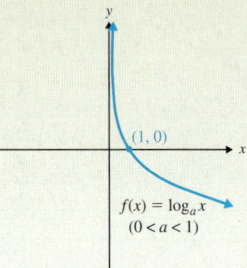

3) The function is one-to-one.

4) The x-intercept is $(1, 0)$.

5) The domain is $(0, \infty)$, and the range is $(-\infty, \infty)$.

6) The inverse of $f(x) = \log_a x$ is $f^{-1}(x) = a^x$.

Compare these characteristics of logarithmic functions to the characteristics of exponential functions on p. 635 in Section 10.2. The domain and range of logarithmic and exponential functions are interchanged since they are inverse functions.

9. Solve an Applied Problem Using a Logarithmic Equation

Example 11

In-Class Example 11
Use Example 11.

A hospital has found that the function $A(t) = 50 + 8 \log_2(t + 2)$ approximates the number of people treated each year since 1995 for severe allergic reactions to peanuts. If $t = 0$ represents the year 1995, answer the following.

a)   How many people were treated in 1995?

b)   How many people were treated in 2001?

c)   In what year were approximately 82 people treated for allergic reactions to peanuts?

Solution

a)   The year 1995 corresponds to $t = 0$. Let $t = 0$, and find $A(0)$.

$$
\begin{aligned}
A(0) &= 50 + 8 \log_2(0 + 2) \qquad &&\text{Substitute 0 for } t. \\
&= 50 + 8 \log_2 2 \\
&= 50 + 8(1) \qquad &&\log_2 2 = 1 \\
&= 58
\end{aligned}
$$

In 1995, 58 people were treated for peanut allergies.

b)   The year 2001 corresponds to $t = 6$. Let $t = 6$, and find $A(6)$.

$$A(6) = 50 + 8 \log_2(6 + 2) \qquad \text{Substitute 6 for } t.$$
$$= 50 + 8 \log_2 8$$
$$= 50 + 8(3) \qquad\qquad \log_2 8 = 3$$
$$= 50 + 24$$
$$= 74$$

In 2001, 74 people were treated for peanut allergies.

c)   To determine in what year 82 people were treated, let $A(t) = 82$ and solve for t.

$$82 = 50 + 8 \log_2(t + 2) \qquad \text{Substitute 82 for } A(t).$$

To solve for t, we first need to get the term containing the logarithm on a side by itself. Subtract 50 from each side.

$$32 = 8 \log_2(t + 2) \qquad \text{Subtract 50.}$$
$$4 = \log_2(t + 2) \qquad \text{Divide by 8.}$$
$$2^4 = t + 2 \qquad\qquad \text{Write in exponential form.}$$
$$16 = t + 2$$
$$14 = t$$

$t = 14$ corresponds to the year 2009. (Add 14 to the year 1995.)

82 people were treated for peanut allergies in 2009.   ∎

You Try 11

The amount of garbage (in millions of pounds) collected in a certain town each year since 1990 can be approximated by $G(t) = 6 + \log_2(t + 1)$, where $t = 0$ represents the year 1990.

a)   How much garbage was collected in 1990?

b)   How much garbage was collected in 1997?

c)   In what year would it be expected that 11,000,000 pounds of garbage will be collected? [Hint: Let $G(t) = 11$.]

Answers to You Try Exercises

1)   a) $3^4 = 81$ b) $5^{-2} = \dfrac{1}{25}$ c) $64^{1/2} = 8$ d) $13^1 = 13$

2)   a) $\log_7 49 = 2$ b) $\log_5 \dfrac{1}{625} = -4$ c) $\log_{19} 1 = 0$ d) $\log_{144} 12 = \dfrac{1}{2}$

3)   a) $\{32\}$ b) $\{38\}$ c) $\{13\}$ d) $\{2\}$ e) $\left\{\dfrac{1}{10}\right\}$

4)   a) 2   b) 4   c) -1 d) $\dfrac{1}{2}$ 5) 3   6) $\{7\}$ 7)   a) 1   b) 0   c) 1   8) $f^{-1}(x) = \log_6 x$

9)
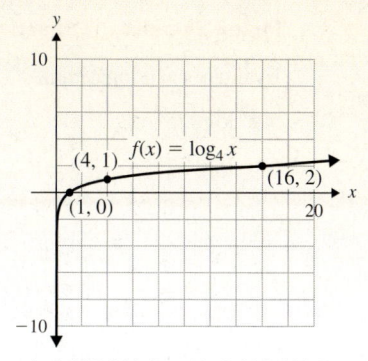

10)
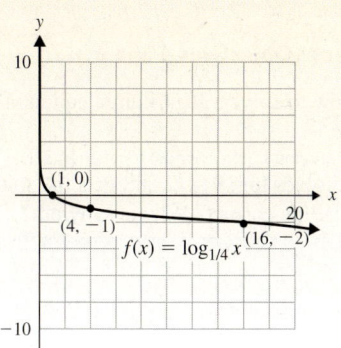

11)   a) 6,000,000 lb   b) 9,000,000 lb   c) 2021

10.3 Exercises

Additional answers can be found in the Answers to Exercises appendix.

Mixed Exercises: Objectives 1 and 2

1) In the equation $y = \log_a x$, a must be what kind of number?
   a must be a positive real number that is not equal to 1.

2) In the equation $y = \log_a x$, x must be what kind of number?
   x must be a positive real number.

3) What is the base of $y = \log x$?   10

4) A base 10 logarithm is called a _____common_____ logarithm.

Write in exponential form.

5) $\log_7 49 = 2$ $7^2 = 49$

6) $\log_{11} 121 = 2$ $11^2 = 121$

7) $\log_2 8 = 3$ $2^3 = 8$

8) $\log_2 32 = 5$ $2^5 = 32$

9) $\log_9 \dfrac{1}{81} = -2$ $9^{-2} = \dfrac{1}{81}$

10) $\log_8 \dfrac{1}{64} = -2$ $8^{-2} = \dfrac{1}{64}$

11) $\log 1{,}000{,}000 = 6$
    $10^6 = 1{,}000{,}000$

12) $\log 10{,}000 = 4$
    $10^4 = 10{,}000$

13) $\log_{25} 5 = \dfrac{1}{2}$ $25^{1/2} = 5$

14) $\log_{64} 4 = \dfrac{1}{3}$ $64^{1/3} = 4$

15) $\log_{13} 13 = 1$ $13^1 = 13$

16) $\log_9 1 = 0$ $9^0 = 1$

Objective 3: Convert from Exponential Form to Logarithmic Form

Write in logarithmic form.

17) $9^2 = 81$ $\log_9 81 = 2$

18) $12^2 = 144$ $\log_{12} 144 = 2$

19) $10^2 = 100$ $\log_{10} 100 = 2$

20) $10^3 = 1000$ $\log_{10} 1000 = 3$

21) $3^{-4} = \dfrac{1}{81}$ $\log_3 \dfrac{1}{81} = -4$

22) $2^{-5} = \dfrac{1}{32}$ $\log_2 \dfrac{1}{32} = -5$

23) $10^0 = 1$ $\log_{10} 1 = 0$

24) $10^1 = 10$ $\log_{10} 10 = 1$

25) $169^{1/2} = 13$ $\log_{169} 13 = \dfrac{1}{2}$

26) $27^{1/3} = 3$ $\log_{27} 3 = \dfrac{1}{3}$

27) $\sqrt{9} = 3$ $\log_9 3 = \dfrac{1}{2}$

28) $\sqrt{64} = 8$ $\log_{64} 8 = \dfrac{1}{2}$

29) $\sqrt[3]{64} = 4$ $\log_{64} 4 = \dfrac{1}{3}$

30) $\sqrt[4]{81} = 3$ $\log_{81} 3 = \dfrac{1}{4}$

Mixed Exercises: Objectives 4 and 6

31) Explain how to solve a logarithmic equation of the form $\log_a b = c$.
    Write the equation in exponential form, then solve for the variable.

32) A student solves $\log_x 9 = 2$ and gets the solution set $\{-3, 3\}$. Is this correct? Why or why not?   *This is incorrect because the base of the logarithm must be positive. The solution set is {3}.*

Solve each logarithmic equation.

Fill It In

Fill in the blanks with either the missing mathematical step or reason for the given step.

33) $\log_2 x = 6$
    $2^6 = x$ Rewrite in exponential form.
    $\underline{64 = x}$ Solve for x.
    The solution set is $\underline{\{64\}}$.

34) $\log_5 t = -3$
    $\underline{5^{-3} = t}$ Rewrite in exponential form.
    $\dfrac{1}{125} = t$ Solve for t.
    The solution set is $\left\{\dfrac{1}{125}\right\}$.

Solve each logarithmic equation.

35) $\log_{11} x = 2$ {121}

36) $\log_5 k = 3$ {125}

37) $\log_4 r = 3$ {64}

38) $\log_2 y = 4$ {16}

39) $\log p = 5$ {100,000}

40) $\log w = 2$ {100}

41) $\log_m 49 = 2$ {7}

42) $\log_x 4 = 2$ {2}

43) $\log_6 h = -2$ $\left\{\dfrac{1}{36}\right\}$

44) $\log_4 b = -3$ $\left\{\dfrac{1}{64}\right\}$

45) $\log_2(a + 2) = 4$ {14}

46) $\log_6(5y + 1) = 2$ {7}

47) $\log_3(4t - 3) = 3$ $\left\{\dfrac{15}{2}\right\}$

48) $\log_2(3n + 7) = 5$ $\left\{\dfrac{25}{3}\right\}$

49) $\log_{81} \sqrt[4]{9} = x$ $\left\{\dfrac{1}{8}\right\}$

50) $\log_{49} \sqrt[3]{7} = d$ $\left\{\dfrac{1}{6}\right\}$

51) $\log_{125} \sqrt{5} = c$ $\left\{\dfrac{1}{6}\right\}$

52) $\log_{16} \sqrt[5]{4} = k$ $\left\{\dfrac{1}{10}\right\}$

53) $\log_{144} w = \dfrac{1}{2}$ {12}

54) $\log_{64} p = \dfrac{1}{3}$ {4}

55) $\log_8 x = \dfrac{2}{3}$ {4}

56) $\log_{16} t = \dfrac{3}{4}$ {8}

57) $\log_{(3m-1)} 25 = 2$ {2}

58) $\log_{(y-1)} 4 = 2$ {3}

Mixed Exercises: Objectives 5–7

Evaluate each logarithm.

59) $\log_5 25$ 2

60) $\log_9 81$ 2

61) $\log_2 32$ 5

62) $\log_4 64$ 3

63) log 100   2

64) log 1000   3

65) $\log_{49} 7$ $\frac{1}{2}$

66) $\log_{36} 6$ $\frac{1}{2}$

67) $\log_8 \frac{1}{8}$ -1

68) $\log_3 \frac{1}{3}$ -1

69) $\log_5 5$ 1

70) $\log_2 1$ 0

VIDEO 71) $\log_{1/4} 16$ -2

72) $\log_{1/3} 27$ -3

Objective 8: Define and Graph a Logarithmic Function

73) Explain how to graph a logarithmic function of the form $f(x) = \log_a x$.

74) What are the domain and range of $f(x) = \log_a x$?
domain: $(0, \infty)$; range $(-\infty, \infty)$

Graph each logarithmic function.

75) $f(x) = \log_3 x$

76) $f(x) = \log_4 x$

77) $f(x) = \log_2 x$

78) $f(x) = \log_5 x$

VIDEO 79) $f(x) = \log_{1/2} x$

80) $f(x) = \log_{1/3} x$

81) $f(x) = \log_{1/4} x$

82) $f(x) = \log_{1/5} x$

Find the inverse of each function.

83) $f(x) = 3^x$ $f^{-1}(x) = \log_3 x$

84) $f(x) = 4^x$ $f^{-1}(x) = \log_4 x$

85) $f(x) = \log_2 x$ $f^{-1}(x) = 2^x$

86) $f(x) = \log_5 x$
$f^{-1}(x) = 5^x$

Objective 9: Solve an Applied Problem Using a Logarithmic Equation

Solve each problem.

VIDEO 87) The function $L(t) = 1800 + 68 \log_3(t + 3)$ approximates the number of dog licenses issued by a city each year since 1980. If $t = 0$ represents the year 1980, answer the following.

a) How many dog licenses were issued in 1980?   1868

b) How many were issued in 2004?   2004

c) In what year would it be expected that 2072 dog licenses will be issued?   2058

88) Until the 1990s, Rock Glen was a rural community outside of a large city. In 1994, subdivisions of homes began to be built. The number of houses in Rock Glen t years after 1994 can be approximated by

$$H(t) = 142 + 58 \log_2(t + 1)$$

where $t = 0$ represents 1994.

a) Determine the number of homes in Rock Glen in 1994.   142

b) Determine the number of homes in Rock Glen in 1997.   258

c) In what year were there approximately 374 homes?   2009

89) A company plans to introduce a new type of cookie to the market. The company predicts that its sales over the next 24 months can be approximated by

$$S(t) = 14 \log_3(2t + 1)$$

where t is the number of months after the product is introduced, and $S(t)$ is in thousands of boxes of cookies.

a) How many boxes of cookies were sold after they were on the market for 1 month?   14,000

b) How many boxes were sold after they were on the market for 4 months?   28,000

c) After 13 months, sales were approximately 43,000. Does this number fall short of, meet, or exceed the number of sales predicted by the formula?
It is 1000 more than what was predicted by the formula.

90) Based on previous data, city planners have calculated that the number of tourists (in millions) to their city each year can be approximated by

$$N(t) = 10 + 1.2 \log_2(t + 2)$$

where t is the number of years after 1995.

a) How many tourists visited the city in 1995?   11,200,000

b) How many tourists visited the city in 2001?   13,600,000

c) In 2009, actual data put the number of tourists at 14,720,000. How does this number compare to the number predicted by the formula?   It is 80,000 less than what would have been expected according to the formula.

Section 10.4 Properties of Logarithms

Logarithms have properties that are very useful in applications and in higher mathematics.

In this section, we will learn more properties of logarithms, and we will practice using them because they can make some very difficult mathematical calculations much easier. *The properties of logarithms come from the properties of exponents.*

1. Use the Product Rule for Logarithms

The product rule for logarithms can be derived from the product rule for exponents.

> **Property** The Product Rule for Logarithms
>
> Let x, y, and a be positive real numbers where $a \neq 1$. Then,
>
> $$\log_a xy = \log_a x + \log_a y$$
>
> The logarithm of a product, xy, is the same as the sum of the logarithms of each factor, x and y.

 **BE CAREFUL** $\log_a xy \neq (\log_a x)(\log_a y)$

Example 1

Rewrite as the sum of logarithms and simplify, if possible. Assume the variables represent positive real numbers.

a) $\log_6(4 \cdot 7)$ b) $\log_4 16t$ c) $\log_8 y^3$ d) $\log 10pq$

Solution

a) The logarithm of a product equals the *sum* of the logs of the factors. Therefore,

$$\log_6(4 \cdot 7) = \log_6 4 + \log_6 7 \qquad \text{Product rule}$$

b) $\log_4 16t = \log_4 16 + \log_4 t \qquad \text{Product rule}$
    $ = 2 + \log_4 t \qquad\qquad \log_4 16 = 2$

Evaluate logarithms, like $\log_4 16$, when possible.

c) $\log_8 y^3 = \log_8(y \cdot y \cdot y) \qquad\qquad$ Write y^3 as $y \cdot y \cdot y$.
    $ = \log_8 y + \log_8 y + \log_8 y \qquad$ Product rule
    $ = 3 \log_8 y$

d) Recall that if no base is written, then it is assumed to be 10.

$$\log 10pq = \log 10 + \log p + \log q \qquad \text{Product rule}$$
$$ = 1 + \log p + \log q \qquad\qquad \log 10 = 1$$

 **You Try 1**

Rewrite as the sum of logarithms and simplify, if possible. Assume the variables represent positive real numbers.

a) $\log_9(2 \cdot 5)$ b) $\log_2 32k$ c) $\log_6 c^4$ d) $\log 100yz$

We can use the product rule for exponents in the "opposite" direction, too. That is, given the sum of logarithms we can write a single logarithm.

Example 2

In-Class Example 2
Write as a single logarithm. Assume the variables represent positive real numbers.
a) $\log_2 12 + \log_2 4$
b) $\log 18 + \log w$
c) $\log_4 p + \log_4 (2p + 3)$
answer: a) $\log_2 48$
b) $\log 18w$
c) $\log_4 (2p^2 + 3p)$

Write as a single logarithm. Assume the variables represent positive real numbers.

a) $\log_8 5 + \log_8 3$ b) $\log 7 + \log r$ c) $\log_3 x + \log_3(x + 4)$

Solution

a) $\log_8 5 + \log_8 3 = \log_8(5 \cdot 3)$ Product rule
$\qquad\qquad\qquad\quad = \log_8 15$ $5 \cdot 3 = 15$

b) $\log 7 + \log r = \log 7r$ Product rule

c) $\log_3 x + \log_3(x + 4) = \log_3 x(x + 4)$ Product rule
$\qquad\qquad\qquad\qquad\quad = \log_3(x^2 + 4x)$ Distribute.

BE CAREFUL $\log_a(x + y) \neq \log_a x + \log_a y$. Therefore, $\log_3(x^2 + 4x)$ does *not* equal $\log_3 x^2 + \log_3 4x$.

You Try 2

Write as a single logarithm. Assume the variables represent positive real numbers.

a) $\log_5 9 + \log_5 4$ b) $\log_6 13 + \log_6 c$ c) $\log y + \log(y - 6)$

2. Use the Quotient Rule for Logarithms

The quotient rule for logarithms can be derived from the quotient rule for exponents.

Property   The Quotient Rule for Logarithms

Let x, y, and a be positive real numbers where $a \neq 1$. Then,

$$\log_a \frac{x}{y} = \log_a x - \log_a y$$

The logarithm of a quotient, $\frac{x}{y}$, is the same as the logarithm of the numerator *minus* the logarithm of the denominator.

BE CAREFUL $\log_a \frac{x}{y} \neq \dfrac{\log_a x}{\log_a y}$.

Example 3

In-Class Example 3
Write as the difference of
logarithms and simplify, if
possible. Assume $r > 0$.

a) $\log_5 \dfrac{8}{11}$ b) $\log_8 \dfrac{16}{r}$

answer: a) $\log_5 8 - \log_5 11$

b) $\dfrac{4}{3} - \log_8 r$

Write as the difference of logarithms and simplify, if possible. Assume $w > 0$.

a) $\log_7 \dfrac{3}{10}$ b) $\log_3 \dfrac{81}{w}$

Solution

a) $\log_7 \dfrac{3}{10} = \log_7 3 - \log_7 10$ Quotient rule

b) $\log_3 \dfrac{81}{w} = \log_3 81 - \log_3 w$ Quotient rule

$= 4 - \log_3 w$ $\log_3 81 = 4$ ∎

You Try 3

Write as the difference of logarithms and simplify, if possible. Assume $n > 0$.

a) $\log_6 \dfrac{2}{9}$ b) $\log_5 \dfrac{n}{25}$

Example 4

In-Class Example 4
Write as a single logarithm.
Assume the variable is defined
so that the expressions are
positive.

a) $\log_3 32 - \log_3 8$
b) $\log_7 (x + 5) - \log_7 (x + 1)$

answer: a) $\log_3 4$

b) $\log_7 \left(\dfrac{x + 5}{x + 1} \right)$

Write as a single logarithm. Assume the variable is defined so that the expressions are positive.

a) $\log_2 18 - \log_2 6$ b) $\log_4(z - 5) - \log_4(z^2 + 9)$

Solution

a) $\log_2 18 - \log_2 6 = \log_2 \dfrac{18}{6}$ Quotient rule

$= \log_2 3$ $\dfrac{18}{6} = 3$

b) $\log_4(z - 5) - \log_4(z^2 + 9) = \log_4 \dfrac{z - 5}{z^2 + 9}$ Quotient rule ∎

BE CAREFUL

$\log_a(x - y) \neq \log_a x - \log_a y$

You Try 4

Write as a single logarithm. Assume the variable is defined so that the expressions are positive.

a) $\log_4 36 - \log_4 3$ b) $\log_5(c^2 - 2) - \log_5(c + 1)$

3. Use the Power Rule for Logarithms

In Example 1c), we saw that $\log_8 y^3 = 3 \log_8 y$ since

$$\log_8 y^3 = \log_8(y \cdot y \cdot y)$$
$$= \log_8 y + \log_8 y + \log_8 y$$
$$= 3 \log_8 y$$

This result can be generalized as the next property and comes from the power rule for exponents.

Property   The Power Rule for Logarithms

Let x and a be positive real numbers, where $a \neq 1$, and let r be any real number. Then,

$$\log_a x^r = r \log_a x$$

BE CAREFUL

The rule applies to $\log_a x^r$ not $(\log_a x)^r$. Be sure you can distinguish between the two expressions.

Example 5

In-Class Example 5
Follow the directions for
Example 5.
a) $\log_x 9^3$ b) $\log_3 81^5$
c) $\log_x \sqrt{6}$ d) $\log_x \dfrac{1}{x^3}$
answer: a) $3 \log_x 9$ b) 20
c) $\dfrac{1}{2} \log_x 6$ d) -3

Rewrite each expression using the power rule and simplify, if possible. Assume the variables represent positive real numbers and that the variable bases are positive real numbers not equal to 1.

a) $\log_9 y^4$ b) $\log_2 8^5$ c) $\log_a \sqrt{3}$ d) $\log_w \dfrac{1}{w}$

Solution

a) $\log_9 y^4 = 4 \log_9 y$ Power rule

b) $\log_2 8^5 = 5 \log_2 8$ Power rule
    $\qquad\quad = 5(3)$ $\log_2 8 = 3$
    $\qquad\quad = 15$ Multiply.

c) *It is common practice to rewrite radicals as fractional exponents when applying the properties of logarithms.* This will be our first step.

$$\log_a \sqrt{3} = \log_a 3^{1/2} \quad \text{Rewrite as a fractional exponent.}$$
$$\qquad\quad = \frac{1}{2} \log_a 3 \quad \text{Power rule}$$

d) Rewrite $\dfrac{1}{w}$ as w^{-1}: $\log_w \dfrac{1}{w} = \log_w w^{-1}$ $\dfrac{1}{w} = w^{-1}$
$$\qquad\qquad\qquad\qquad\qquad\quad = -1 \log_w w \quad \text{Power rule}$$
$$\qquad\qquad\qquad\qquad\qquad\quad = -1(1) \qquad\quad \log_w w = 1$$
$$\qquad\qquad\qquad\qquad\qquad\quad = -1 \qquad\qquad\; \text{Multiply.}$$

You Try 5

Rewrite each expression using the power rule and simplify, if possible. Assume the variables represent positive real numbers and that the variable bases are positive real numbers not equal to 1.

a) $\log_8 t^9$ b) $\log_3 9^7$ c) $\log_a \sqrt[3]{5}$ d) $\log_m \dfrac{1}{m^8}$

The next properties we will look at can be derived from the power rule and from the fact that $f(x) = a^x$ and $g(x) = \log_a x$ are inverse functions.

4. Use the Properties $\log_a a^x = x$ and $a^{\log_a x} = x$

Other Properties of Logarithms

Let a be a positive real number such that $a \neq 1$. Then,

1) $\log_a a^x = x$ for any real number x.

2) $a^{\log_a x} = x$ for $x > 0$.

Example 6

In-Class Example 6
Evaluate each expression.
a) $\log_9 9^3$ b) $\log 10^7$ c) $3^{\log 3^4}$
answer: a) 3   b) 7   c) 4

Evaluate each expression.

a) $\log_6 6^7$ b) $\log 10^8$ c) $5^{\log_5 3}$

Solution

a) $\log_6 6^7 = 7$ $\log_a a^x = x$

b) $\log 10^8 = 8$ The base of the log is 10.

c) $5^{\log_5 3} = 3$ $a^{\log_a x} = x$

You Try 6

Evaluate each expression.

a) $\log_3 3^{10}$ b) $\log 10^{-6}$ c) $7^{\log_7 9}$

Next is a summary of the properties of logarithms. The properties presented in Section 10.3 are included as well.

Summary   Properties of Logarithms

Let x, y, and a be positive real numbers where $a \neq 1$, and let r be any real number. Then,

1) $\log_a a = 1$

2) $\log_a 1 = 0$

3) $\log_a xy = \log_a x + \log_a y$ Product rule

4) $\log_a \dfrac{x}{y} = \log_a x - \log_a y$ Quotient rule

5) $\log_a x^r = r \log_a x$ Power rule

6) $\log_a a^x = x$ for any real number x

7) $a^{\log_a x} = x$

Many students make the same mistakes when working with logarithms. Keep in mind the following to avoid these common errors.

BE CAREFUL

1) $\log_a xy \neq (\log_a x)(\log_a y)$ 4) $\log_a(x - y) \neq \log_a x - \log_a y$

2) $\log_a(x + y) \neq \log_a x + \log_a y$ 5) $(\log_a x)^r \neq r \log_a x$

3) $\log_a \dfrac{x}{y} \neq \dfrac{\log_a x}{\log_a y}$

---

5. Combine the Properties of Logarithms

Not only can the properties of logarithms simplify some very complicated computations, they are also needed for solving some types of logarithmic equations. The properties of logarithms are also used in calculus and many areas of science.

Next, we will see how to use different properties of logarithms together to rewrite logarithmic expressions.

Example 7

In-Class Example 7
Follow the directions for
Example 7.

a) $\log_4 x^3 y$ b) $\log_9 \dfrac{81}{ab^2}$

c) $\log_5 \sqrt{5t}$ d) $\log_v (2v + 3)$
answer: a) $3 \log_4 x + \log_4 y$
b) $2 - \log_9 a - 2 \log_9 b$

c) $\dfrac{1}{2} + \dfrac{1}{2} \log_5 t$

d) $\log_v (2v + 3)$

Write each expression as the sum or difference of logarithms in simplest form. Assume all variables represent positive real numbers and that the variable bases are positive real numbers not equal to 1.

a) $\log_8 r^5 t$ b) $\log_3 \dfrac{27}{ab^2}$ c) $\log_7 \sqrt{7p}$ d) $\log_a (4a + 5)$

Solution

a) $\log_8 r^5 t = \log_8 r^5 + \log_8 t$ Product rule
    $ = 5 \log_8 r + \log_8 t$ Power rule

b) $\log_3 \dfrac{27}{ab^2} = \log_3 27 - \log_3 ab^2$ Quotient rule

$\phantom{\log_3 \dfrac{27}{ab^2}} = 3 - (\log_3 a + \log_3 b^2)$ $\log_3 27 = 3$; product rule
$\phantom{\log_3 \dfrac{27}{ab^2}} = 3 - (\log_3 a + 2 \log_3 b)$ Power rule
$\phantom{\log_3 \dfrac{27}{ab^2}} = 3 - \log_3 a - 2 \log_3 b$ Distribute.

c) $\log_7 \sqrt{7p} = \log_7 (7p)^{1/2}$ Rewrite radical as fractional exponent.

$\phantom{\log_7 \sqrt{7p}} = \dfrac{1}{2} \log_7 (7p)$ Power rule

$\phantom{\log_7 \sqrt{7p}} = \dfrac{1}{2}(\log_7 7 + \log_7 p)$ Product rule

$\phantom{\log_7 \sqrt{7p}} = \dfrac{1}{2}(1 + \log_7 p)$ $\log_7 7 = 1$

$\phantom{\log_7 \sqrt{7p}} = \dfrac{1}{2} + \dfrac{1}{2} \log_7 p$ Distribute.

d) $\log_a (4a + 5)$ is in simplest form and cannot be rewritten using any properties of logarithms. [Recall that $\log_a (x + y) \ne \log_a x + \log_a y$.]

You Try 7

Write each expression as the sum or difference of logarithms in simplest form. Assume all variables represent positive real numbers and that the variable bases are positive real numbers not equal to 1.

a) $\log_2 8s^2 t^5$ b) $\log_a \dfrac{4c^2}{b^3}$ c) $\log_5 \sqrt[3]{\dfrac{25}{n}}$ d) $\dfrac{\log_4 k}{\log_4 m}$

Example 8

In-Class Example 8
Follow the directions for
Example 8.

a) $2 \log_6 9 + \log_6 2$

b) $\dfrac{1}{3} \log_3 m + 4 \log_3 (m + 1)$

answer:
a) $\log_6 162$
b) $\log_3 \sqrt[3]{m} (m + 1)^4$

Write each as a single logarithm in simplest form. Assume the variable represents a positive real number.

a) $2 \log_7 5 + 3 \log_7 2$ b) $\dfrac{1}{2} \log_6 s - 3 \log_6 (s^2 + 1)$

Solution

a) $2 \log_7 5 + 3 \log_7 2 = \log_7 5^2 + \log_7 2^3$ Power rule
    $ = \log_7 25 + \log_7 8$ $5^2 = 25$; $2^3 = 8$
    $ = \log_7 (25 \cdot 8)$ Product rule
    $ = \log_7 200$ Multiply.

b) $\dfrac{1}{2}\log_6 s - 3\log_6(s^2+1) = \log_6 s^{1/2} - \log_6(s^2+1)^3$ Power rule

$$= \log_6 \sqrt{s} - \log_6(s^2+1)^3$$ Write in radical form.

$$= \log_6 \dfrac{\sqrt{s}}{(s^2+1)^3}$$ Quotient rule   ■

You Try 8

Write each as a single logarithm in simplest form. Assume the variables are defined so that the expressions are positive.

a) $2\log 4 + \log 5$

b) $\dfrac{2}{3}\log_5 c + \dfrac{1}{3}\log_5 d - 2\log_5(c-6)$

Given the values of logarithms, we can compute the values of other logarithms using the properties we have learned in this section.

Example 9

In-Class Example 9
Given that $\log 2 \approx 0.3010$ and $\log 3 \approx 0.4771$, use the properties of logarithms to approximate the following.
a) $\log 12$ b) $\log \sqrt{2}$
answer:
a) 1.0791   b) 0.1505

Given that $\log 6 \approx 0.7782$ and $\log 4 \approx 0.6021$, use the properties of logarithms to approximate the following.

a) $\log 24$ b) $\log \sqrt{6}$

Solution

a)   To find the value of $\log 24$, we must determine how to write 24 in terms of 6 or 4 or some combination of the two. Since $24 = 6 \cdot 4$, we can write

$$\log 24 = \log(6 \cdot 4) \qquad\qquad 24 = 6\cdot 4$$
$$= \log 6 + \log 4 \qquad\quad \text{Product rule}$$
$$\approx 0.7782 + 0.6021 \qquad \text{Substitute.}$$
$$= 1.3803 \qquad\qquad\quad \text{Add.}$$

b)   We can write $\sqrt{6}$ as $6^{1/2}$.

$$\log \sqrt{6} = \log 6^{1/2} \qquad\qquad \sqrt{6} = 6^{1/2}$$
$$= \dfrac{1}{2}\log 6 \qquad\qquad \text{Power rule}$$
$$\approx \dfrac{1}{2}(0.7782) \qquad\quad \log 6 \approx 0.7782$$
$$= 0.3891 \qquad\qquad\quad \text{Multiply.}$$ ■

You Try 9

Using the values given in Example 9, use the properties of logarithms to approximate the following.

a) $\log 16$ b) $\log \dfrac{6}{4}$ c) $\log \sqrt[3]{4}$ d) $\log \dfrac{1}{6}$

Answers to You Try Exercises

1) a) $\log_9 2 + \log_9 5$ b) $5 + \log_2 k$ c) $4 \log_6 c$ d) $2 + \log y + \log z$ 2) a) $\log_5 36$
b) $\log_6 13c$ c) $\log(y^2 - 6y)$ 3) a) $\log_6 2 - \log_6 9$ b) $\log_5 n - 2$ 4) a) $\log_4 12$
b) $\log_5 \dfrac{c^2 - 2}{c + 1}$ 5) a) $9 \log_8 t$ b) 14 c) $\dfrac{1}{3}\log_a 5$ d) -8 6) a) 10 b) -6 c) 9

7) a) $3 + 2\log_2 s + 5\log_2 t$ b) $\log_a 4 + 2\log_a c - 3\log_a b$ c) $\dfrac{2}{3} - \dfrac{1}{3}\log_5 n$

d) cannot be simplified   8) a) $\log 80$ b) $\log_5 \dfrac{\sqrt[3]{c^2 d}}{(c - 6)^2}$ 9) a) 1.2042 b) 0.1761
c) 0.2007 d) -0.7782

10.4 Exercises

Additional answers can be found in the Answers to Exercises appendix.

Mixed Exercises: Objectives 1–5

Decide whether each statement is true or false.

1) $\log_6 8c = \log_6 8 + \log_6 c$ true

2) $\log_5 \dfrac{m}{3} = \log_5 m - \log_5 3$ true

3) $\log_9 \dfrac{7}{2} = \dfrac{\log_9 7}{\log_9 2}$ false

4) $\log 1000 = 3$ true

5) $(\log_4 k)^2 = 2 \log_4 k$ false

6) $\log_2(x^2 + 8) = \log_2 x^2 + \log_2 8$ false

7) $5^{\log_5 4} = 4$ true

8) $\log_3 4^5 = 5 \log_3 4$ true

Write as the sum or difference of logarithms and simplify, if possible. Assume all variables represent positive real numbers.

Fill It In

Fill in the blanks with either the missing mathematical step or reason for the given step.

9) $\log_5 25y$

$\log_5 25y = \log_5 25 + \log_5 y$ Product rule

$= \underline{2 + \log_5 y}$ Evaluate $\log_5 25$.

10) $\log_3 \dfrac{81}{n^2}$

$\log_3 \dfrac{81}{n^2} = \log_3 81 - \log_3 n^2$ $\underline{\text{Quotient rule}}$

$= \underline{4 - 2\log_3 n}$ Evaluate $\log_3 81$; use power rule.

11) $\log_8(3 \cdot 10)$ $\log_8 3 + \log_8 10$

12) $\log_2(6 \cdot 5)$ $\log_2 6 + \log_2 5$

13) $\log_7 5d$ $\log_7 5 + \log_7 d$

14) $\log_4 6w$ $\log_4 6 + \log_4 w$

15) $\log_9 \dfrac{4}{7}$ $\log_9 4 - \log_9 7$

16) $\log_5 \dfrac{20}{17}$ $\log_5 20 - \log_5 17$

17) $\log_5 2^3$ $3 \log_5 2$

18) $\log_8 10^4$ $4 \log_8 10$

19) $\log p^8$ $8 \log p$

20) $\log_3 z^5$ $5 \log_3 z$

(VIDEO) 21) $\log_3 \sqrt{7}$ $\dfrac{1}{2}\log_3 7$

22) $\log_7 \sqrt[3]{4}$ $\dfrac{1}{3}\log_7 4$

23) $\log_5 25t$ $2 + \log_5 t$

24) $\log_2 16p$ $4 + \log_2 p$

25) $\log_2 \dfrac{8}{k}$ $3 - \log_2 k$

26) $\log_3 \dfrac{x}{9}$ $(\log_3 x) - 2$

27) $\log_7 49^3$ 6

28) $\log_8 64^{12}$ 24

29) $\log 1000b$ $3 + \log b$

30) $\log_3 27m$ $3 + \log_3 m$

31) $\log_2 32^7$ 35

32) $\log_2 2^9$ 9

33) $\log_5 \sqrt{5}$ $\dfrac{1}{2}$

34) $\log \sqrt[3]{10}$ $\dfrac{1}{3}$

35) $\log \sqrt[3]{100}$ $\dfrac{2}{3}$

36) $\log_2 \sqrt{8}$ $\dfrac{3}{2}$

37) $\log_6 w^4 z^3$ $4 \log_6 w + 3 \log_6 z$

38) $\log_5 x^2 y$ $2 \log_5 x + \log_5 y$

39) $\log_7 \dfrac{a^2}{b^5}$ $2 \log_7 a - 5 \log_7 b$

40) $\log_4 \dfrac{s^4}{t^6}$ $4 \log_4 s - 6 \log_4 t$

41) $\log \dfrac{\sqrt[5]{11}}{y^2}$ $\dfrac{1}{5}\log 11 - 2 \log y$

42) $\log_3 \dfrac{\sqrt{x}}{y^4}$ $\dfrac{1}{2}\log_3 x - 4 \log_3 y$

43) $\log_2 \dfrac{4\sqrt{n}}{m^3}$ $2 + \dfrac{1}{2}\log_2 n - 3\log_2 m$

44) $\log_9 \dfrac{gf^2}{h^3}$ $\log_9 g + 2 \log_9 f - 3 \log_9 h$

(VIDEO) 45) $\log_4 \dfrac{x^3}{yz^2}$

$3 \log_4 x - \log_4 y - 2 \log_4 z$

46) $\log \dfrac{3}{ab^2}$ $\log 3 - \log a - 2 \log b$

47) $\log_5 \sqrt{5c}$ $\dfrac{1}{2} + \dfrac{1}{2} \log_5 c$

48) $\log_8 \sqrt[3]{\dfrac{z}{8}}$ $\left(\dfrac{1}{3}\log_8 z\right) - \dfrac{1}{3}$

49) $\log k(k-6)$ $\log k + \log(k-6)$

50) $\log_2 \dfrac{m^5}{m^2+3}$ $5\log_2 m - \log_2(m^2+3)$

Write as a single logarithm. Assume the variables are defined so that the variable expressions are positive and so that the bases are positive real numbers not equal to 1.

Fill It In

Fill in the blanks with either the missing mathematical step or reason for the given step.

51) $2\log_6 x + \log_6 y$

$2\log_6 x + \log_6 y = \log_6 x^2 + \log_6 y$ Power rule

$= \log_6 x^2 y$ Product rule

52) $5\log 2 + \log c - 3\log d$

$5\log 2 + \log c - 3\log d$

$= \log 2^5 + \log c - \log d^3$ Power rule

$= \log 32 + \log c - \log d^3$ $2^5 = 32$

$= \log 32c - \log d^3$ Product rule

$= \log \dfrac{32c}{d^3}$ Quotient rule

53) $\log_a m + \log_a n$ $\log_a mn$

54) $\log_4 7 + \log_4 x$ $\log_4 7x$

55) $\log_7 d - \log_7 3$ $\log_7 \dfrac{d}{3}$

56) $\log_p r - \log_p s$ $\log_p \dfrac{r}{s}$

57) $4\log_3 f + \log_3 g$ $\log_3 f^4 g$

58) $5\log_y m + 2\log_y n$ $\log_y m^5 n^2$

59) $\log_8 t + 2\log_8 u - 3\log_8 v$ $\log_8 \dfrac{tu^2}{v^3}$

60) $3\log a + 4\log c - 6\log b$ $\log \dfrac{a^3 c^4}{b^6}$

61) $\log(r^2+3) - 2\log(r^2-3)$ $\log \dfrac{r^2+3}{(r^2-3)^2}$

62) $2\log_2 t - 3\log_2(5t+1)$ $\log_2 \dfrac{t^2}{(5t+1)^3}$

63) $3\log_n 2 + \dfrac{1}{2}\log_n k$ $\log_n 8\sqrt{k}$

64) $2\log_z 9 + \dfrac{1}{3}\log_z w$ $\log_z 81\sqrt[3]{w}$

65) $\dfrac{1}{3}\log_d 5 - 2\log_d z$ $\log_d \dfrac{\sqrt[3]{5}}{z^2}$

66) $\dfrac{1}{2}\log_5 a - 4\log_5 b$ $\log_5 \dfrac{\sqrt{a}}{b^4}$

VIDEO 67) $\log_6 y - \log_6 3 - 3\log_6 z$ $\log_6 \dfrac{y}{3z^3}$

68) $\log_7 8 - 4\log_7 x - \log_7 y$ $\log_7 \dfrac{8}{x^4 y}$

69) $4\log_3 t - 2\log_3 6 - 2\log_3 u$ $\log_3 \dfrac{t^4}{36u^2}$

70) $2\log_9 m - 4\log_9 2 - 4\log_9 n$ $\log_9 \dfrac{m^2}{16n^4}$

71) $\dfrac{1}{2}\log_b(c+4) - 2\log_b(c+3)$ $\log_b \dfrac{\sqrt{c+4}}{(c+3)^2}$

72) $\dfrac{1}{2}\log_a r + \dfrac{1}{2}\log_a(r-2) - \log_a(r+2)$ $\log_a \dfrac{\sqrt{r(r-2)}}{r+2}$

73) $\log(a^2+b^2) - \log(a^4-b^4)$ $-\log(a^2-b^2)$

74) $\log_n(x^3-y^3) - \log_n(x-y)$ $\log_n(x^2+xy+y^2)$

Given that $\log 5 \approx 0.6990$ and $\log 9 \approx 0.9542$, use the properties of logarithms to approximate the following. **Do not use a calculator.**

VIDEO 75) $\log 45$ 1.6532

76) $\log 25$ 1.3980

77) $\log 81$ 1.9084

78) $\log \dfrac{9}{5}$ 0.2552

79) $\log \dfrac{5}{9}$ -0.2552

80) $\log \sqrt{5}$ 0.3495

81) $\log 3$ 0.4771

82) $\log \dfrac{1}{9}$ -0.9542

83) $\log \dfrac{1}{5}$ -0.6990

84) $\log 5^8$ 5.5920

85) $\log \dfrac{1}{81}$ -1.9084

86) $\log 90$ 1.9542

87) $\log 50$ 1.6990

88) $\log \dfrac{25}{9}$ 0.4438

89) Since $8 = (-4)(-2)$, can we use the properties of logarithms in the following way? Explain.

$$\log_2 8 = \log_2(-4)(-2)$$
$$= \log_2(-4) + \log_2(-2)$$

No. $\log_a xy$ is defined only if x and y are positive.

90) **Derive the product rule for logarithms from the product rule for exponents.** Assume a, x, and y are positive real numbers with $a \neq 1$. Let $a^m = x$ so that $\log_a x = m$, and let $a^n = y$ so that $\log_a y = n$. Since $a^m \cdot a^n = xy$, show that $\log_a xy = \log_a x + \log_a y$.

$a^m \cdot a^n = xy$

$a^{m+n} = xy$

Change to logarithmic form.

$\log_a xy = m + n$

$\log_a xy = \log_a x + \log_a y$

Section 10.5 Common and Natural Logarithms and Change of Base

Objectives

1. **Evaluate Common Logarithms Without a Calculator**
2. **Evaluate Common Logarithms Using a Calculator**
3. **Solve an Equation Containing a Common Logarithm**
4. **Solve an Applied Problem Given an Equation Containing a Common Logarithm**
5. **Define and Evaluate a Natural Logarithm**
6. **Graph a Natural Logarithm Function**
7. **Solve an Equation Containing a Natural Logarithm**
8. **Solve Applied Problems Using Exponential Functions**
9. **Use the Change-of-Base Formula**

In this section, we will focus our attention on two widely used logarithmic bases—base 10 and base e.

Common Logarithms

1. Evaluate Common Logarithms Without a Calculator

In Section 10.3, we said that a base 10 logarithm is called a **common logarithm**. It is often written as $\log x$.

$$\log x \text{ means } \log_{10} x$$

We can evaluate many logarithms without the use of a calculator because we can write them in terms of a base of 10.

Example 1

Evaluate.

a) $\log 1000$ b) $\log \dfrac{1}{100}$

Solution

a) $\log 1000 = 3$ since $10^3 = 1000$ b) $\log \dfrac{1}{100} = \log 10^{-2}$ $\dfrac{1}{100} = 10^{-2}$

$$= \log_{10} 10^{-2}$$
$$= -2 \qquad \log_a a^x = x$$

In-Class Example 1
Evaluate.
a) log 100   b) log $\dfrac{1}{1000}$
answer: a) 2   b) −3

You Try 1

Evaluate.

a) $\log 100{,}000$ b) $\log \dfrac{1}{10}$

2. Evaluate Common Logarithms Using a Calculator

Common logarithms are used throughout mathematics and other fields to make calculations easier to solve in applications. Often, however, we need a calculator to evaluate the logarithms. Next we will learn how to use a calculator to find the value of a base 10 logarithm. *We will approximate the value to four decimal places.*

Example 2

Find $\log 12$.

In-Class Example 2
Find log 5.
answer: 0.6990

Solution

Enter 12 $\boxed{\text{LOG}}$ or $\boxed{\text{LOG}}$ 12 $\boxed{\text{ENTER}}$ into your calculator.

$$\log 12 \approx 1.0792$$

(Note that $10^{1.0792} \approx 12$. Press 10 $\boxed{y^x}$ 1.0792 $\boxed{=}$ to evaluate $10^{1.0792}$.)

You Try 2

Find log 3.

We can solve logarithmic equations with or without the use of a calculator.

3. Solve an Equation Containing a Common Logarithm

Example 3

Solve $\log x = -3$.

In-Class Example 3
Solve $\log x = -5$.
answer: $\left\{ \dfrac{1}{100,000} \right\}$

Solution

Change to exponential form, and solve for x.

$$\log x = -3 \quad \text{means} \quad \log_{10} x = -3$$
$$10^{-3} = x \qquad \text{Exponential form}$$
$$\frac{1}{1000} = x$$

The solution set is $\left\{ \dfrac{1}{1000} \right\}$. This is the *exact* solution. ■

You Try 3

Solve $\log x = 2$.

For the equation in Example 4, we will give an exact solution *and* a solution that is approximated to four decimal places. This will give us an idea of the size of the exact solution.

Example 4

Solve $\log x = 2.4$. Give an exact solution and a solution that is approximated to four decimal places.

In-Class Example 4
Solve $\log x = 3.3$. Give an exact solution and a solution that is approximated to four decimal places.
answer: $\{10^{3.3}\}$; $\{1995.2623\}$

Solution

Change to exponential form, and solve for x.

$$\log x = 2.4 \quad \text{means} \quad \log_{10} x = 2.4$$
$$10^{2.4} = x \qquad \text{Exponential form}$$
$$251.1886 \approx x \qquad \text{Approximation}$$

The exact solution is $\{10^{2.4}\}$. This is approximately $\{251.1886\}$. ■

You Try 4

Solve $\log x = 0.7$. Give an exact solution and a solution that is approximated to four decimal places.

4. Solve an Applied Problem Given an Equation Containing a Common Logarithm

Example 5

In-Class Example 5
Use Example 5.

The loudness of sound, $L(I)$ in decibels (dB), is given by

$$L(I) = 10 \log \frac{I}{10^{-12}}$$

where I is the intensity of sound in watts per square meter (W/m^2). Fifty meters from the stage at a concert, the intensity of sound is 0.01 W/m^2. Find the loudness of the music at the concert 50 m from the stage.

Solution

Substitute 0.01 for I and find $L(0.01)$.

$$\begin{aligned}
L(0.01) &= 10 \log \frac{0.01}{10^{-12}} \\
&= 10 \log \frac{10^{-2}}{10^{-12}} \qquad 0.01 = 10^{-2} \\
&= 10 \log 10^{10} \qquad \text{Quotient rule for exponents} \\
&= 10(10) \qquad \log 10^{10} = 10 \\
&= 100
\end{aligned}$$

The sound level of the music 50 m from the stage is 100 dB. (To put this in perspective, a normal conversation has a loudness of about 50 dB.)

You Try 5

The intensity of sound from a thunderstorm is about 0.001 W/m^2. Find the loudness of the storm, in decibels.

Natural Logarithms

5. Define and Evaluate a Natural Logarithm

Another base that is often used for logarithms is the number e. In Section 10.2, we said that e, like π, is an irrational number. To four decimal places, $e \approx 2.7183$.

A base e logarithm is called a **natural logarithm** or **natural log.** The notation used for a base e logarithm is ln x (read as "*the natural log of x*" or "*ln of x*"). Since it is a base e logarithm, it is important to remember that

$$\ln x \text{ means } \log_e x$$

Using the properties $\log_a a^x = x$ and $\log_a a = 1$, we can find the value of some natural logarithms without using a calculator.

Example 6

In-Class Example 6
Evaluate.
a) 3 ln e   b) ln e^4
answer: a) 3   b) 4

Evaluate.

a) ln e b) ln e^2

Solution

a) To evaluate $\ln e$, remember that $\ln e = \log_e e = 1$ since $\log_a a = 1$.

$$\ln e = 1$$

This is a value you should remember. We will use this in Section 10.6 to solve exponential equations with base e.

b) $\ln e^2 = \log_e e^2 = 2$ $\log_a a^x = x$ ■

You Try 6

Evaluate.

a) $5 \ln e$ b) $\ln e^8$

We can use a calculator to *approximate natural logarithms to four decimal places* if the properties do not give us an exact value.

Example 7

Find $\ln 5$.

In-Class Example 7
Find $\ln 4$.
answer: 1.3863

Solution

Enter $\boxed{5}\ \boxed{\text{LN}}$ or $\boxed{\text{LN}}\ \boxed{5}\ \boxed{\text{ENTER}}$ into your calculator.

$$\ln 5 \approx 1.6094$$ ■

You Try 7

Find $\ln 9$.

6. Graph a Natural Logarithm Function

We can graph $y = \ln x$ by substituting values for x and using a calculator to approximate the values of y.

Example 8

Graph $y = \ln x$. Determine the domain and range.

In-Class Example 8
Use Example 8.

Solution

Choose values for x, and use a calculator to approximate the corresponding values of y. Remember that $\ln e = 1$, so e is a good choice for x.

x	y
1	0
$e \approx 2.72$	1
6	1.79
0.5	-0.69
0.25	-1.39

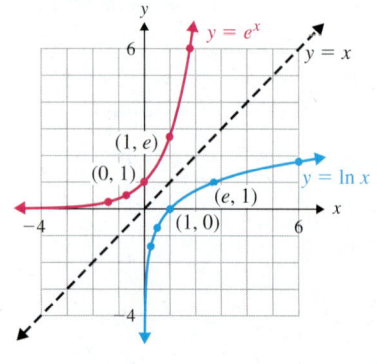

The domain of $y = \ln x$ is $(0, \infty)$, and the range is $(-\infty, \infty)$.

The graph of the inverse of $y = \ln x$ is also shown. We can obtain the graph of the inverse of $y = \ln x$ by reflecting the graph about the line $y = x$. The inverse of $y = \ln x$ is $y = e^x$.

Notice that the domain of $y = e^x$ is $(-\infty, \infty)$, while the range is $(0, \infty)$, the opposite of the domain and range of $y = \ln x$. This is a direct result of the relationship between a function and its inverse. ■

You Try 8

Graph $y = \ln(x + 4)$. Determine the domain and range.

It is important to remember that $y = \ln x$ means $y = \log_e x$. Understanding this relationship allows us to make the following connections:

$$y = \ln x \text{ is equivalent to } y = \log_e x$$

and

$$y = \log_e x \text{ can be written in exponential form as } e^y = x.$$

Therefore, $y = \ln x$ can be written in exponential form as $e^y = x$.

We can use this relationship to show that the inverse of $y = \ln x$ is $y = e^x$ (this is Exercise 97) and also to verify the result in Example 7 when we found that $\ln 5 \approx 1.6094$. We can express $\ln 5 \approx 1.6094$ in exponential form as $e^{1.6094} \approx 5$. Evaluate $e^{1.6094}$ on a calculator to confirm that $e^{1.6094} \approx 5$.

7. Solve an Equation Containing a Natural Logarithm

Note

To solve an equation containing a natural logarithm, like $\ln x = 4$, we change to exponential form and solve for the variable. We can give an exact solution and a solution that is approximated to four decimal places.

Example 9

Solve each equation. Give an exact solution and a solution that is approximated to four decimal places.

a) $\ln x = 4$ b) $\ln(2x + 5) = 3.8$

In-Class Example 9
Solve each equation. Give an exact solution and a solution that is approximated to four decimal places.
a) $\ln d = 6$ b) $\ln(3d - 1) = 2$
answer: a) $\{e^6\}$; **(403.4288)**
b) $\left\{\dfrac{e^2 + 1}{3}\right\}$; **{2.7964}**

Solution

a) $\ln x = 4$ means $\log_e x = 4$

$$e^4 = x \qquad \text{Exponential form}$$
$$54.5982 \approx x \qquad \text{Approximation}$$

The exact solution is $\{e^4\}$. This is approximately $\{54.5982\}$.

b) $\ln(2x + 5) = 3.8$ means $\log_e(2x + 5) = 3.8$

$$e^{3.8} = 2x + 5 \qquad \text{Exponential form}$$
$$e^{3.8} - 5 = 2x \qquad \text{Subtract 5.}$$
$$\frac{e^{3.8} - 5}{2} = x \qquad \text{Divide by 2.}$$
$$19.8506 \approx x \qquad \text{Approximation}$$

The exact solution is $\left\{\dfrac{e^{3.8} - 5}{2}\right\}$. This is approximately $\{19.8506\}$.

You Try 9

Solve each equation. Give an exact solution and a solution that is approximated to four decimal places.

a) $\ln y = 2.7$ b) $\ln(3a - 1) = 0.5$

8. Solve Applied Problems Using Exponential Functions

One of the most practical applications of exponential functions is for compound interest.

Definition

Compound Interest: The amount of money, A, in dollars, in an account after t years is given by

$$A = P\left(1 + \frac{r}{n}\right)^{nt}$$

where P (the principal) is the amount of money (in dollars) deposited in the account, r is the annual interest rate, and n is the number of times the interest is compounded (paid) per year.

Note

We can also think of this formula in terms of the amount of money owed, A, after t yr when P is the amount of money loaned.

Example 10

In-Class Example 10
If $1000 is deposited in an account paying 5% per year, find the total amount in the account after 3 yr if the interest is compounded
a) quarterly.   b) monthly.
**answer: a) $1160.75
b) $1161.47**

If $2000 is deposited in an account paying 4% per year, find the total amount in the account after 5 yr if the interest is compounded

a)   quarterly.   b)   monthly.

(We assume no withdrawals or additional deposits are made.)

Solution

a)   If interest compounds quarterly, then interest is paid four times per year. Use

$$A = P\left(1 + \frac{r}{n}\right)^{nt}$$

with $P = 2000$, $r = 0.04$, $t = 5$, $n = 4$.

$$A = 2000\left(1 + \frac{0.04}{4}\right)^{4(5)}$$
$$= 2000(1.01)^{20}$$
$$\approx 2440.3801$$

Since A is an amount of money, round to the nearest cent. The account will contain $2440.38 after 5 yr.

b) If interest is compounded monthly, then interest is paid 12 times per year. Use

$$A = P\left(1 + \frac{r}{n}\right)^{nt}$$

with $P = 2000$, $r = 0.04$, $t = 5$, $n = 12$.

$$A = 2000\left(1 + \frac{0.04}{12}\right)^{12(5)}$$
$$\approx 2441.9932$$

Round A to the nearest cent. The account will contain $2441.99 after 5 yr. ■

You Try 10

If $1500 is deposited in an account paying 5% per year, find the total amount in the account after 8 yr if the interest is compounded

a)   monthly.   b)   weekly.

In Example 10 we saw that the account contained more money after 5 yr when the interest compounded monthly (12 times per year) versus quarterly (four times per year). This will always be true. The more often interest is compounded each year, the more money that accumulates in the account.

If interest *compounds continuously*, we obtain the formula for *continuous compounding*, $A = Pe^{rt}$.

Definition

Continuous Compounding: If P dollars is deposited in an account earning interest rate r compounded continuously, then the amount of money, A (in dollars), in the account after t years is given by

$$A = Pe^{rt}$$

Example 11

In-Class Example 11
Determine the amount of money in an account after 4 yr if $3000 was initially invested at 6% compounded continuously.
answer: $A = \$3813.75$

Determine the amount of money in an account after 5 yr if $2000 was initially invested at 4% compounded continuously.

Solution

Use $A = Pe^{rt}$ with $P = 2000$, $r = 0.04$, and $t = 5$.

$$A = 2000e^{0.04(5)} \quad \text{Substitute values.}$$
$$= 2000e^{0.20} \quad \text{Multiply } (0.04)(5).$$
$$\approx 2442.8055 \quad \text{Evaluate using a calculator.}$$

Round A to the nearest cent.

The account will contain $2442.81 after 5 yr. Note that, as expected, this is more than the amounts obtained in Example 10 when the same amount was deposited for 5 yr at 4% but the interest was compounded quarterly and monthly. ■

You Try 11

Determine the amount of money in an account after 8 yr if $1500 was initially invested at 5% compounded continuously.

9. Use the Change-of-Base Formula

Sometimes, we need to find the value of a logarithm with a base other than 10 or e—like $\log_3 7$. Some calculators, however, do not calculate logarithms other than common logs (base 10) and natural logs (base e). In such cases, we can use the change-of-base formula to evaluate logarithms with bases other than 10 or e.

Definition

Change-of-Base Formula: If $a, b,$ and x are positive real numbers and $a \neq 1$ and $b \neq 1$, then

$$\log_a x = \frac{\log_b x}{\log_b a}$$

Note

We can choose any positive real number not equal to 1 for b, but it is most convenient to choose 10 or e since these will give us common logarithms and natural logarithms, respectively.

Example 12

In-Class Example 12
Find the value of $\log_4 9$ to four decimal places using
a) common logarithms.
b) natural logarithms.
answer: a) 1.5850   b) 1.5850

Find the value of $\log_3 7$ to four decimal places using

a) common logarithms.   b) natural logarithms.

Solution

a) The base we will use to evaluate $\log_3 7$ is 10—this is the base of a common logarithm. Then

$$\log_3 7 = \frac{\log_{10} 7}{\log_{10} 3} \qquad \text{Change-of-base formula}$$
$$\approx 1.7712 \qquad \text{Use a calculator.}$$

b) The base of a natural logarithm is e. Then

$$\log_3 7 = \frac{\log_e 7}{\log_e 3}$$
$$= \frac{\ln 7}{\ln 3}$$
$$\approx 1.7712 \qquad \text{Use a calculator.}$$

Using either base 10 or base e gives us the same result.   ■

 **You Try 12**

Find the value of $\log_5 38$ to four decimal places using

a) common logarithms.   b) natural logarithms.

Using Technology

Graphing calculators will graph common logarithmic functions and natural logarithmic functions directly using the **log** or **LN** keys.

For example, let's graph $f(x) = \ln x$.

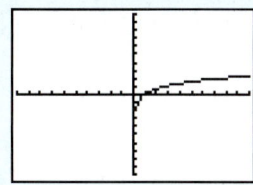

To graph a logarithmic function with a base other than 10 or e, it is necessary to use the change-of-base formula. For example, to graph the function $f(x) = \log_2 x$, first rewrite the function as a quotient of natural logarithms or common logarithms: $f(x) = \log_2 x = \dfrac{\ln x}{\ln 2}$ or $\dfrac{\log x}{\log 2}$. Enter one of these quotients in Y_1 and press GRAPH to graph as shown below. To illustrate that the same graph results in either case, trace to the point where $x = 3$.

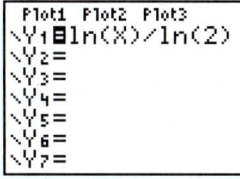
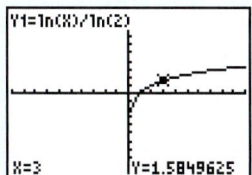

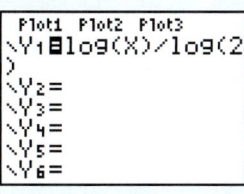
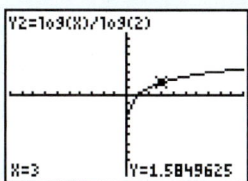

Graph the following functions using a graphing calculator.

1) $f(x) = \log_3 x$ 2) $f(x) = \log_5 x$ 3) $f(x) = 4 \log_2 x + 1$

4) $f(x) = \log_2 (x - 3)$ 5) $f(x) = 2 - \log_4 x$ 6) $f(x) = 3 - \log_2 (x + 1)$

Answers to You Try Exercises

1) a) 5 b) -1 2) 0.4771   3) $\{100\}$ 4) $\{10^{0.7}\}$; $\{5.0119\}$ 5) 90 dB

6) a) 5 b) 8   7) 2.1972

8) domain: $(-4, \infty)$; range: $(-\infty, \infty)$

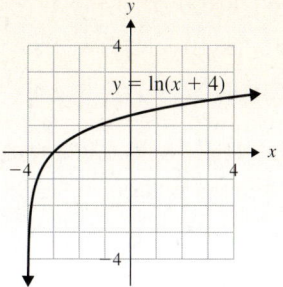

9) a) $\{e^{2.7}\}$; $\{14.8797\}$ b) $\left\{\dfrac{e^{0.5} + 1}{3}\right\}$; $\{0.8829\}$

10) a) \$2235.88 b) \$2237.31 11) \$2237.74

12) a) 2.2602 b) 2.2602

Answers to Technology Exercises

1)
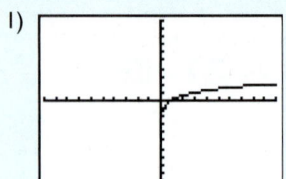

2)
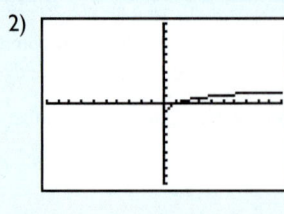

3)
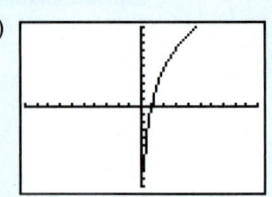

4)
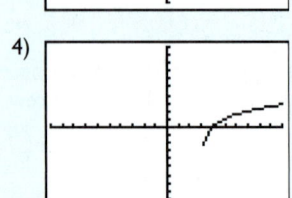

5)
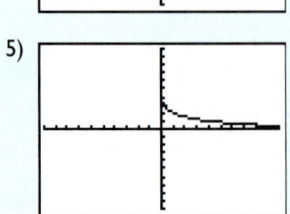

6)
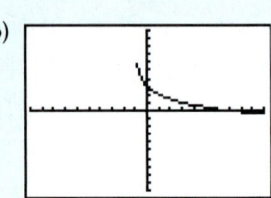

10.5 Exercises

*Additional answers can be found in the Answers to Exercises appendix.

Mixed Exercises: Objectives 1 and 5

1) What is the base of $\ln x$?   e

2) What is the base of $\log x$?   10

Evaluate each logarithm. Do *not* use a calculator.

3) $\log 100$ 2

4) $\log 10,000$ 4

5) $\log \dfrac{1}{1000}$ -3

6) $\log \dfrac{1}{100,000}$ -5

7) $\log 0.1$ -1

8) $\log 0.01$ -2

9) $\log 10^9$ 9

10) $\log 10^7$ 7

VIDEO 11) $\log \sqrt[4]{10}$ $\dfrac{1}{4}$

12) $\log \sqrt[5]{10}$ $\dfrac{1}{5}$

13) $\ln e^6$ 6

14) $\ln e^{10}$ 10

15) $\ln \sqrt{e}$ $\dfrac{1}{2}$

16) $\ln \sqrt[3]{e}$ $\dfrac{1}{3}$

VIDEO 17) $\ln \dfrac{1}{e^5}$ -5

18) $\ln \dfrac{1}{e^2}$ -2

19) $\ln 1$ 0

20) $\log 1$ 0

Mixed Exercises: Objectives 2 and 5

Use a calculator to find the approximate value of each logarithm to four decimal places.

21) $\log 16$ 1.2041

22) $\log 23$ 1.3617

23) $\log 0.5$ -0.3010

24) $\log 627$ 2.7973

25) $\ln 3$ 1.0986

26) $\ln 6$ 1.7918

27) $\ln 1.31$ 0.2700

28) $\ln 0.218$ -1.5233

Objective 3: Solve an Equation Containing a Common Logarithm

Solve each equation. Do *not* use a calculator.

29) $\log x = 3$ {1000}

30) $\log z = 5$ {100,000}

31) $\log k = -1$ $\left\{\dfrac{1}{10}\right\}$

32) $\log c = -2$ $\left\{\dfrac{1}{100}\right\}$

VIDEO 33) $\log(4a) = 2$ {25}

34) $\log(5w) = 1$ {2}

35) $\log(3t + 4) = 1$ {2}

36) $\log(2p + 12) = 2$ {44}

Mixed Exercises: Objectives 3 and 7

Solve each equation. Give an exact solution and a solution that is approximated to four decimal places.

37) $\log a = 1.5$ $\{10^{1.5}\}$; {31.6228}

38) $\log y = 1.8$ $\{10^{1.8}\}$; {63.0957}

39) $\log r = 0.8$ $\{10^{0.8}\}$; {6.3096}

40) $\log c = 0.3$ $\{10^{0.3}\}$; {1.9953}

41) $\ln x = 1.6$ $\{e^{1.6}\}$; {4.9530}

42) $\ln p = 1.1$ $\{e^{1.1}\}$; {3.0042}

VIDEO 43) $\ln t = -2$ $\left\{\dfrac{1}{e^2}\right\}$; {0.1353}

44) $\ln z = 0.25$ $\{e^{0.25}\}$; {1.2840}

45) $\ln(3q) = 2.1$ $\left\{\dfrac{e^{2.1}}{3}\right\}$; {2.7221}

46) $\ln\left(\dfrac{1}{4}m\right) = 3$ $\{4e^3\}$; {80.3421}

47) $\log\left(\dfrac{1}{2}c\right) = 0.47$

48) $\log(6k) = -1$ $\left\{\dfrac{1}{60}\right\}$

49) $\log(5y - 3) = 3.8$

50) $\log(8x + 15) = 2.7$

51) $\ln(10w + 19) = 1.85$

52) $\ln(7a - 4) = 0.6$

53) $\ln(2d - 5) = 0$ {3}

54) $\log(3t + 14) = 2.4$

Objective 6: Graph a Natural Logarithm Function

Graph each function. State the domain and range.

55) $y = \ln x + 2$

56) $f(x) = \ln x - 3$

57) $h(x) = \ln x - 1$

58) $g(x) = \ln x + 1$

59) $f(x) = \ln(x - 2)$

60) $y = \ln(x - 1)$

61) $g(x) = \ln(x + 3)$

62) $h(x) = \ln(x + 2)$

63) $y = -\ln x$

64) $f(x) = \ln(-x)$

65) $h(x) = \log x$

66) $k(x) = \log(x + 4)$

67) If you are given the graph of $f(x) = \ln x$, how could you obtain the graph of $g(x) = \ln(x + 5)$ without making a table of values and plotting points?
Shift the graph of $f(x)$ left 5 units.

68) If you are given the graph of $f(x) = \ln x$, how could you obtain the graph of $h(x) = \ln x + 4$ without making a table of values and plotting points?
Shift the graph of $f(x)$ up 4 units.

Objective 9: Use the Change-of-Base Formula

Use the change-of-base formula with either base 10 or base e to approximate each logarithm to four decimal places.

69) $\log_2 13$ 3.7004

70) $\log_6 25$ 1.7965

71) $\log_9 70$ 1.9336

72) $\log_3 52$ 3.5966

73) $\log_{1/3} 16$ -2.5237

74) $\log_{1/2} 23$ -4.5236

75) $\log_5 3$ 0.6826

76) $\log_7 4$ 0.7124

Mixed Exercises: Objectives 4 and 8

For Exercises 77–80, use the formula

$$L(I) = 10 \log \frac{I}{10^{-12}}$$

where I is the intensity of sound, in watts per square meter, and $L(I)$ is the loudness of sound in decibels. Do *not* use a calculator.

77) The intensity of sound from fireworks is about 0.1 W/m^2. Find the loudness of the fireworks, in decibels.   110 dB

78) The intensity of sound from a dishwasher is about 0.000001 W/m^2. Find the loudness of the dishwasher, in decibels.   60 dB

79) The intensity of sound from a refrigerator is about 0.00000001 W/m^2. Find the loudness of the refrigerator, in decibels.   40 dB

80) The intensity of sound from the takeoff of a space shuttle is $1,000,000$ W/m^2. Find the loudness of the sound made by the space shuttle at takeoff, in decibels.   180 dB

Use the formula $A = P\left(1 + \dfrac{r}{n}\right)^{nt}$ to solve each problem. See Example 10.

81) Isabel deposits $3000 in an account earning 5% per year compounded monthly. How much will be in the account after 3 yr? $3484.42

82) How much money will Pavel have in his account after 8 yr if he initially deposited $6000 at 4% interest compounded quarterly? $8249.64

83) Find the amount Christopher owes at the end of 5 yr if he borrows $4000 at a rate of 6.5% compounded quarterly. $5521.68

84) How much will Anna owe at the end of 4 yr if she borrows $5000 at a rate of 7.2% compounded weekly? $6667.46

Use the formula $A = Pe^{rt}$ to solve each problem. See Example 11.

85) If $3000 is deposited in an account earning 5% compounded continuously, how much will be in the account after 3 yr? $3485.50

86) If $6000 is deposited in an account earning 4% compounded continuously, how much will be in the account after 8 yr? $8262.77

87) How much will Cyrus owe at the end of 6 yr if he borrows $10,000 at a rate of 7.5% compounded continuously? $15,683.12

88) Find the amount Nadia owes at the end of 5 yr if she borrows $4500 at a rate of 6.8% compounded continuously. $6322.26

89) The number of bacteria, $N(t)$, in a culture t hr after the bacteria are placed in a dish is given by

$$N(t) = 5000e^{0.0617t}$$

a) How many bacteria were originally in the culture?   5000

b) How many bacteria are present after 8 hr?   8191

90) The number of bacteria, $N(t)$, in a culture t hr after the bacteria are placed in a dish is given by

$$N(t) = 8000e^{0.0342t}$$

a) How many bacteria were originally in the culture?   8000

b) How many bacteria are present after 10 hr?   11,262

91) The function $N(t) = 10,000e^{0.0492t}$ describes the number of bacteria in a culture t hr after 10,000 bacteria were placed in the culture. How many bacteria are in the culture after 1 day?   32,570

92) How many bacteria are present 2 days after 6000 bacteria are placed in a culture if the number of bacteria in the culture is

$$N(t) = 6000e^{0.0285t}$$

t hr after the bacteria are placed in a dish?   23,565

In chemistry, the pH of a solution is given by

$$pH = -\log[H^+]$$

where $[H^+]$ is the molar concentration of the hydronium ion. A neutral solution has pH $= 7$. *Acidic solutions* have pH < 7, and *basic solutions* have pH > 7.

For Exercises 93–96, the hydronium ion concentrations, $[H^+]$, are given for some common substances. Find the pH of each substance (to the tenths place), and determine whether each substance is acidic or basic.

93) Cola: $[H^+] = 2 \times 10^{-3}$ 2.7; acidic

94) Tomatoes: $[H^+] = 1 \times 10^{-4}$ 4.0; acidic

95) Ammonia: $[H^+] = 6 \times 10^{-12}$ 11.2; basic

96) Egg white: $[H^+] = 2 \times 10^{-8}$ 7.7; basic

Extension

97) Show that the inverse of $y = \ln x$ is $y = e^x$.

Section 10.6 Solving Exponential and Logarithmic Equations

Objectives

1. **Solve an Exponential Equation**
2. **Solve Logarithmic Equations Using the Properties of Logarithms**
3. **Solve Applied Problems Involving Exponential Functions Using a Calculator**
4. **Solve an Applied Problem Involving Exponential Growth or Decay**

In this section, we will learn another property of logarithms that will allow us to solve additional types of exponential and logarithmic equations.

Properties for Solving Exponential and Logarithmic Equations

Let a, x, and y be positive, real numbers, where $a \neq 1$.

1) If $x = y$, then $\log_a x = \log_a y$.

2) If $\log_a x = \log_a y$, then $x = y$.

For example, 1) tells us that if $x = 3$, then $\log_a x = \log_a 3$. Likewise, 2) tells us that if $\log_a 5 = \log_a y$, then $5 = y$. We can use the properties above to solve exponential and logarithmic equations that we could not solve previously.

1. Solve an Exponential Equation

We will look at two types of exponential equations—equations where both sides *can* be expressed with the same base and equations where both sides *cannot* be expressed with the same base. If the two sides of an exponential equation *cannot* be expressed with the same base, we will use logarithms to solve the equation.

Example 1

In-Class Example 1
Solve.
a) $3^x = 9$ b) $3^x = 18$
answer:
a) $\{2\}$ b) $\left\{\dfrac{\ln 18}{\ln 3}\right\}$; the
approximation is $\{2.6309\}$.

Solve.

a) $2^x = 8$ b) $2^x = 12$

Solution

a) Since 8 is a power of 2, we can solve $2^x = 8$ by expressing each side of the equation with the same base and setting the exponents equal to each other.

$$2^x = 8$$
$$2^x = 2^3 \qquad 8 = 2^3$$
$$x = 3 \qquad \text{Set the exponents equal.}$$

The solution set is $\{3\}$.

b) Can we express both sides of $2^x = 12$ with the same base? *No. We will use property* 1) *to solve* $2^x = 12$ *by taking the logarithm of each side.*

We can use a logarithm of *any* base. It is most convenient to use base 10 (common logarithm) or base e (natural logarithm) because this is what we can find most easily on our calculators. *We will take the natural log of both sides.*

$$2^x = 12$$
$$\ln 2^x = \ln 12 \qquad \text{Take the natural log of each side.}$$
$$x \ln 2 = \ln 12 \qquad \log_a x^r = r \log_a x$$
$$x = \frac{\ln 12}{\ln 2} \qquad \text{Divide by } \ln 2.$$

The exact solution is $\left\{\dfrac{\ln 12}{\ln 2}\right\}$. Use a calculator to get an approximation to four decimal places: $x \approx 3.5850$.

The approximation is $\{3.5850\}$. We can verify the solution by substituting it for x in $2^x = 12$: $2^{3.5850} \approx 12$.

BE CAREFUL

$\dfrac{\ln 12}{\ln 2} \neq \ln 6$

Procedure Solving an Exponential Equation

Begin by asking yourself, "*Can I express each side with the same base?*"

1) If the answer is **yes,** then write each side of the equation with the same base, set the exponents equal, and solve for the variable.

2) If the answer is **no,** then take the natural logarithm of each side, use the properties of logarithms, and solve for the variable.

  You Try 1

Solve.

a) $3^{a-5} = 9$ b) $3^t = 24$

Example 2

Solve $5^{x-2} = 16$.

In-Class Example 2
Solve $4^{z+1} = 22$.

answer: $\left\{ \dfrac{\ln 22 - \ln 4}{\ln 4} \right\}$;

$\{1.2297\}$

Solution

Ask yourself, "*Can I express each side with the same base?*" **No.** Therefore, take the natural log of each side.

$$5^{x-2} = 16$$
$$\ln 5^{x-2} = \ln 16 \qquad \text{Take the natural log of each side.}$$
$$(x - 2)\ln 5 = \ln 16 \qquad \log_a x^r = r \log_a x$$

$(x - 2)$ *must* be in parentheses since it contains two terms.

$$x \ln 5 - 2 \ln 5 = \ln 16 \qquad \text{Distribute.}$$
$$x \ln 5 = \ln 16 + 2 \ln 5 \qquad \text{Add } 2 \ln 5 \text{ to get the } x\text{-term by itself.}$$
$$x = \frac{\ln 16 + 2 \ln 5}{\ln 5} \qquad \text{Divide by } \ln 5.$$

The exact solution is $\left\{ \dfrac{\ln 16 + 2 \ln 5}{\ln 5} \right\}$. This is approximately $\{3.7227\}$. ■

 **You Try 2**

Solve $9^{k+4} = 2$.

Recall that $\ln e = 1$. This property is the reason it is convenient to take the *natural logarithm* of both sides of an equation when a base is *e*.

Example 3

Solve $e^{5n} = 4$.

In-Class Example 3
Solve $e^{3x} = 11$.

answer: $\left\{\dfrac{\ln 11}{3}\right\}$; {0.7993}

Solution

Begin by taking the natural log of each side.

$$e^{5n} = 4$$
$$\ln e^{5n} = \ln 4 \qquad \text{Take the natural log of each side.}$$
$$5n \ln e = \ln 4 \qquad \log_a x^r = r \log_a x$$
$$5n(1) = \ln 4 \qquad \ln e = 1$$
$$5n = \ln 4$$
$$n = \frac{\ln 4}{5} \qquad \text{Divide by 5.}$$

The exact solution is $\left\{\dfrac{\ln 4}{5}\right\}$. The approximation is {0.2773}. ■

You Try 3

Solve $e^{6c} = 2$.

2. Solve Logarithmic Equations Using the Properties of Logarithms

We learned earlier that to solve a logarithmic equation like $\log_2(t + 5) = 4$, we write the equation in exponential form and solve for the variable.

$$\log_2(t + 5) = 4$$
$$2^4 = t + 5 \qquad \text{Write in exponential form.}$$
$$16 = t + 5 \qquad 2^4 = 16$$
$$11 = t \qquad \text{Subtract 5.}$$

In this section, we will learn how to solve other types of logarithmic equations as well. We will look at equations where

1) each term in the equation contains a logarithm.

2) one term in the equation does *not* contain a logarithm.

> **Procedure**   How to Solve an Equation Where Each Term Contains a Logarithm
>
> 1) Use the properties of logarithms to write the equation in the form $\log_a x = \log_a y$.
>
> 2) Set $x = y$ and solve for the variable.
>
> 3) Check the proposed solution(s) in the original equation to be sure the values satisfy the equation.

Example 4

Solve.

a) $\log_5(m - 4) = \log_5 9$ b) $\log x + \log(x + 6) = \log 16$

In-Class Example 4
Solve.
a) $\log_3 (v - 2) = \log_3 10$
b) $\log c + \log (c + 3) = \log 28$
answer: a) {12}   b) {4}

Solution

a) To solve $\log_5(m - 4) = \log_5 9$, use the property that states if $\log_a x = \log_a y$, then $x = y$.

$$\log_5(m - 4) = \log_5 9$$
$$m - 4 = 9$$
$$m = 13 \qquad \text{Add 4.}$$

Check to be sure that $m = 13$ satisfies the original equation.

$$\log_5(13 - 4) \overset{?}{=} \log_5 9$$
$$\log_5 9 = \log_5 9 \quad \checkmark$$

The solution set is $\{13\}$.

b)   To solve $\log x + \log(x + 6) = \log 16$, we must begin by using the product rule for logarithms to obtain one logarithm on the left side.

$$
\begin{aligned}
\log x + \log(x + 6) &= \log 16 \\
\log x(x + 6) &= \log 16 && \text{Product rule} \\
x(x + 6) &= 16 && \text{If } \log_a x = \log_a y, \text{ then } x = y. \\
x^2 + 6x &= 16 && \text{Distribute.} \\
x^2 + 6x - 16 &= 0 && \text{Subtract 16.} \\
(x + 8)(x - 2) &= 0 && \text{Factor.} \\
x + 8 = 0 \quad \text{or} \quad x - 2 &= 0 && \text{Set each factor equal to 0.} \\
x = -8 \quad \text{or} \qquad x &= 2 && \text{Solve.}
\end{aligned}
$$

Check to be sure that $x = -8$ and $x = 2$ satisfy the original equation.

Check $x = -8$:

$$\log x + \log(x + 6) = \log 16$$
$$\log(-8) + \log(-8 + 6) \overset{?}{=} \log 16$$
$$\text{FALSE}$$

We reject $x = -8$ as a solution because it leads to $\log(-8)$, which is undefined.

Check $x = 2$:

$$\log x + \log(x + 6) = \log 16$$
$$\log 2 + \log(2 + 6) \overset{?}{=} \log 16$$
$$\log 2 + \log 8 \overset{?}{=} \log 16$$
$$\log(2 \cdot 8) \overset{?}{=} \log 16$$
$$\log 16 = \log 16 \quad \checkmark$$

$x = 2$ satisfies the original equation.

The solution set is $\{2\}$.

BE CAREFUL   Just because a proposed solution is a negative number does *not* mean it should be rejected. You *must* check it in the original equation; it may satisfy the equation.

You Try 4

Solve.

a) $\log_8(z + 3) = \log_8 5$

b) $\log_3 c + \log_3(c - 1) = \log_3 12$

Procedure   How to Solve an Equation Where One Term Does *Not* Contain a Logarithm

1)   Use the properties of logarithms to get one logarithm on one side of the equation and a constant on the other side. That is, write the equation in the form $\log_a x = y$.

2)   Write $\log_a x = y$ in exponential form, $a^y = x$, and solve for the variable.

3)   Check the proposed solution(s) in the original equation to be sure the values satisfy the equation.

Example 5

Solve $\log_2 3w - \log_2(w - 5) = 3$.

In-Class Example 5
Solve
$\log_4 16m - \log_4(m - 3) = 3$.
answer: {4}

Solution

Notice that one term in the equation $\log_2 3w - \log_2(w - 5) = 3$ does *not* contain a logarithm. Therefore, we want to use the properties of logarithms to get *one* logarithm on the left. Then, write the equation in exponential form and solve.

$$\log_2 3w - \log_2(w - 5) = 3$$

$\log_2 \dfrac{3w}{w - 5} = 3$	Quotient rule
$2^3 = \dfrac{3w}{w - 5}$	Write in exponential form.
$8 = \dfrac{3w}{w - 5}$	$2^3 = 8$
$8(w - 5) = 3w$	Multiply by $w - 5$.
$8w - 40 = 3w$	Distribute.
$-40 = -5w$	Subtract $8w$.
$8 = w$	Divide by -5.

Verify that $w = 8$ satisfies the original equation. The solution set is $\{8\}$.

You Try 5

Solve.

a) $\log_4(7p + 1) = 3$ b) $\log_3 2x - \log_3(x - 14) = 2$

Let's look at the two types of equations we have discussed side by side. Notice the difference between them.

Solve each equation

1) $\log_3 x + \log_3(2x + 5) = \log_3 12$

Use the properties of logarithms to get one log on the left.

$$\log_3 x(2x + 5) = \log_3 12$$

Since both terms contain logarithms, use the property that states if $\log_a x = \log_a y$, then $x = y$.

$$x(2x + 5) = 12$$
$$2x^2 + 5x = 12$$
$$2x^2 + 5x - 12 = 0$$
$$(2x - 3)(x + 4) = 0$$
$$2x - 3 = 0 \quad \text{or} \quad x + 4 = 0$$
$$x = \frac{3}{2} \quad \text{or} \quad x = -4$$

Reject -4 as a solution. The solution set is $\left\{\dfrac{3}{2}\right\}$.

2) $\log_3 x + \log_3(2x + 5) = 1$

Use the properties of logarithms to get one log on the left.

$$\log_3 x(2x + 5) = 1$$

The term on the right does *not* contain a logarithm. Write the equation in exponential form and solve.

$$3^1 = x(2x + 5)$$
$$3 = 2x^2 + 5x$$
$$0 = 2x^2 + 5x - 3$$
$$0 = (2x - 1)(x + 3)$$
$$2x - 1 = 0 \quad \text{or} \quad x + 3 = 0$$
$$x = \frac{1}{2} \quad \text{or} \quad x = -3$$

Reject $x = -3$ as a solution. The solution set is $\left\{\dfrac{1}{2}\right\}$.

3. Solve Applied Problems Involving Exponential Functions Using a Calculator

Recall that $A = Pe^{rt}$ is the formula for continuous compound interest where P (the principal) is the amount invested, r is the interest rate, and A is the amount (in dollars) in the account after t yr. Here we will look at how we can use the formula to solve a different problem from the type we solved in Section 10.5.

Example 6

In-Class Example 6
If $1500 is invested at 4% interest compounded continuously, how long would it take for the investment to grow to $2000?
answer: 7.19 yr

If $3000 is invested at 5% interest compounded continuously, how long would it take for the investment to grow to $4000?

Solution

In this problem, we are asked to find t, the amount of *time* it will take for $3000 to grow to $4000 when invested at 5% compounded continuously.

Use $A = Pe^{rt}$ with $P = 3000$, $A = 4000$, and $r = 0.05$.

$$A = Pe^{rt}$$

$$4000 = 3000e^{0.05t} \qquad \text{Substitute the values.}$$

$$\frac{4}{3} = e^{0.05t} \qquad \text{Divide by 3000.}$$

$$\ln \frac{4}{3} = \ln e^{0.05t} \qquad \text{Take the natural log of both sides.}$$

$$\ln \frac{4}{3} = 0.05t \ln e \qquad \log_a x^r = r \log_a x$$

$$\ln \frac{4}{3} = 0.05t(1) \qquad \ln e = 1$$

$$\ln \frac{4}{3} = 0.05t$$

$$\frac{\ln \dfrac{4}{3}}{0.05} = t \qquad \text{Divide by 0.05.}$$

$$5.75 \approx t \qquad \text{Use a calculator to get the approximation.}$$

It would take about 5.75 yr for $3000 to grow to $4000.

 You Try 6

If $4500 is invested at 6% interest compounded continuously, how long would it take for the investment to grow to $5000?

The amount of time it takes for a quantity to double in size are called the *doubling time*. We can use this in many types of applications.

Example 7

In-Class Example 7
Use the given example with 3000 bacteria initially present and $N(t) = 3000e^{0.0462t}$
answer: about 15 hr

The number of bacteria, $N(t)$, in a culture t hr after the bacteria are placed in a dish is given by

$$N(t) = 5000e^{0.0462t}$$

where 5000 bacteria are initially present. How long will it take for the number of bacteria to double?

Solution

If there are 5000 bacteria present initially, there will be $2(5000) = 10{,}000$ bacteria when the number doubles.

Find t when $N(t) = 10{,}000$.

$$N(t) = 5000e^{0.0462t}$$

$$10{,}000 = 5000e^{0.0462t} \qquad \text{Substitute 10,000 for } N(t).$$
$$2 = e^{0.0462t} \qquad\qquad \text{Divide by 5000.}$$
$$\ln 2 = \ln e^{0.0462t} \qquad \text{Take the natural log of both sides.}$$
$$\ln 2 = 0.0462t \ln e \qquad \log_a x^r = r \log_a x$$
$$\ln 2 = 0.0462t(1) \qquad \ln e = 1$$
$$\ln 2 = 0.0462t$$
$$\frac{\ln 2}{0.0462} = t \qquad\qquad\quad \text{Divide by 0.0462.}$$
$$15 \approx t$$

It will take about 15 hr for the number of bacteria to double. ■

You Try 7

The number of bacteria, $N(t)$, in a culture t hr after the bacteria are placed in a dish is given by

$$N(t) = 12{,}000e^{0.0385t}$$

where 12,000 bacteria are initially present. How long will it take for the number of bacteria to double?

4. Solve an Applied Problem Involving Exponential Growth or Decay

We can generalize the formulas used in Examples 6 and 7 with a formula widely used to model situations that grow or decay exponentially. That formula is

$$y = y_0 e^{kt}$$

where y_0 is the initial amount or quantity at time $t = 0$, y is the amount present after time t, and k is a constant. If k is positive, it is called a *growth constant* because the quantity will *increase* over time. If k is negative, it is called a *decay constant* because the quantity will *decrease* over time.

Example 8

In-Class Example 8
Use the given example.

In April 1986, an accident at the Chernobyl nuclear power plant released many radioactive substances into the environment. One such substance was cesium-137. Cesium-137 decays according to the equation

$$y = y_0 e^{-0.0230t}$$

where y_0 is the initial amount present at time $t = 0$ and y is the amount present after t yr. If a sample of soil contains 10 g of cesium-137 immediately after the accident,

a) how many grams will remain after 15 yr?

b) how long would it take for the initial amount of cesium-137 to decay to 2 g?

c) the **half-life** of a substance is the amount of time it takes for a substance to decay to half its original amount. What is the half-life of cesium-137?

Solution

a) The initial amount of cesium-137 is 10 g, so $y_0 = 10$. We must find y when $y_0 = 10$ and $t = 15$.

$$y = y_0 e^{-0.0230t}$$
$$= 10e^{-0.0230(15)} \qquad \text{Substitute the values.}$$
$$\approx 7.08 \qquad \text{Use a calculator to get the approximation.}$$

There will be about 7.08 g of cesium-137 remaining after 15 yr.

b) The initial amount of cesium-137 is $y_0 = 10$. To determine how long it will take to decay to 2 g, let $y = 2$ and solve for t.

$$y = y_0 e^{-0.0230t}$$
$$2 = 10e^{-0.0230t} \qquad \text{Substitute 2 for } y \text{ and 10 for } y.$$
$$0.2 = e^{-0.0230t} \qquad \text{Divide by 10.}$$
$$\ln 0.2 = \ln e^{-0.0230t} \qquad \text{Take the natural log of both sides.}$$
$$\ln 0.2 = -0.0230t \ln e \qquad \log_a x^r = r \log_a x$$
$$\ln 0.2 = -0.0230t \qquad \ln e = 1$$
$$\frac{\ln 0.2}{-0.0230} = t \qquad \text{Divide by } -0.0230.$$
$$69.98 \approx t \qquad \text{Use a calculator to get the approximation.}$$

It will take about 69.98 yr for 10 g of cesium-137 to decay to 2 g.

c) Since there are 10 g of cesium-137 in the original sample, to determine the half-life we will determine how long it will take for the 10 g to decay to 5 g because $\frac{1}{2}(10) = 5$.

Let $y_0 = 10$, $y = 5$, and solve for t.

$$y = y_0 e^{-0.0230t}$$
$$5 = 10e^{-0.0230t} \qquad \text{Substitute the values.}$$
$$0.5 = e^{-0.0230t} \qquad \text{Divide by 10.}$$
$$\ln 0.5 = \ln e^{-0.0230t} \qquad \text{Take the natural log of both sides.}$$
$$\ln 0.5 = -0.0230t \ln e \qquad \log_a x^r = r \log_a x$$
$$\ln 0.5 = -0.0230t \qquad \ln e = 1$$
$$\frac{\ln 0.5}{-0.0230} = t \qquad \text{Divide by } -0.0230.$$
$$30.14 \approx t \qquad \text{Use a calculator to get the approximation.}$$

The half-life of cesium-137 is about 30.14 yr. This means that it would take about 30.14 yr for any quantity of cesium-137 to decay to half of its original amount. ■

You Try 8

Radioactive strontium-90 decays according to the equation
$$y = y_0 e^{-0.0244t}$$
where t is in years. If a sample contains 40 g of strontium-90,

a) how many grams will remain after 8 yr?

b) how long would it take for the initial amount of strontium-90 to decay to 30 g?

c) what is the half-life of strontium-90?

Using Technology

We can solve exponential and logarithmic equations in the same way that we solved other equations—by graphing both sides of the equation and finding where the graphs intersect.

In Example 2 of this section, we learned how to solve $5^{x-2} = 16$. Because the right side of the equation is 16, the graph will have to go at least as high as 16. So set the Y_{max} to be 20, enter the left side of the equation as Y_1 and the right side as Y_2, and press GRAPH :

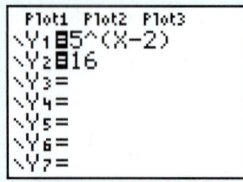

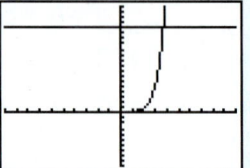

Recall that the x-coordinate of the point of intersection is the solution to the equation. To find the point of intersection, press 2nd TRACE and then highlight 5:intersect and press ENTER . Press ENTER three more times to see that the x-coordinate of the point of intersection is approximately 3.723.

Remember, while the calculator can sometimes save you time, it will often give an approximate answer and not an exact solution.

Use a graphing calculator to solve each equation. Round your answer to the nearest thousandth.

1) $7^x = 49$

2) $6^{2b+1} = 13$

3) $5^{4a+7} = 8^{2a}$

4) $\ln x = 1.2$

5) $\log(k + 9) = \log 11$

6) $\ln(x + 3) = \ln(x - 2)$

Answers to You Try Exercises

1) a) $\{7\}$ b) $\left\{\dfrac{\ln 24}{\ln 3}\right\}$; $\{2.8928\}$ 2) $\left\{\dfrac{\ln 2 - 4\ln 9}{\ln 9}\right\}$; $\{-3.6845\}$ 3) $\left\{\dfrac{\ln 2}{6}\right\}$; $\{0.1155\}$

4) a) $\{2\}$ b) $\{4\}$ 5) a) $\{9\}$ b) $\{18\}$ 6) 1.76 yr   7) 18 hr

8) a) 32.91 g b) 11.79 yr c) 28.41 yr

Answers to Technology Exercises

1) $\{2\}$ 2) $\{0.216\}$ 3) $\{-4.944\}$ 4) $\{3.320\}$ 5) $\{2\}$ 6) $\varnothing$

10.6 Exercises

*Additional answers can be found in the Answers to Exercises appendix.

Objective 1: Solve an Exponential Equation

Solve each equation. Give the exact solution. If the answer contains a logarithm, approximate the solution to four decimal places.

1) $7^x = 49$ $\{2\}$

2) $5^c = 125$ $\{3\}$

3) $7^n = 15$ $\left\{\dfrac{\ln 15}{\ln 7}\right\}$; $\{1.3917\}$

4) $5^a = 38$ $\left\{\dfrac{\ln 38}{\ln 5}\right\}$; $\{2.2602\}$

5) $8^z = 3$ $\left\{\dfrac{\ln 3}{\ln 8}\right\}$; $\{0.5283\}$

6) $4^y = 9$ $\left\{\dfrac{\ln 9}{\ln 4}\right\}$; $\{1.5850\}$

7) $6^{5p} = 36$ $\left\{\dfrac{2}{5}\right\}$

8) $2^{3t} = 32$ $\left\{\dfrac{5}{3}\right\}$

9) $4^{6k} = 2.7$

10) $3^{2x} = 7.8$

11) $2^{4n+1} = 5$

12) $6^{2b+1} = 13$

13) $5^{3a-2} = 8$

14) $3^{2x-3} = 14$

15) $4^{2c+7} = 64^{3c-1}$ $\left\{\dfrac{10}{7}\right\}$

16) $27^{5m-2} = 3^{m+6}$ $\left\{\dfrac{6}{7}\right\}$

17) $9^{5d-2} = 4^{3d}$

18) $5^{4a+7} = 8^{2a}$

Solve each equation. Give the exact solution and the approximation to four decimal places.

19) $e^y = 12.5$ $\{\ln 12.5\}$; $\{2.5257\}$

20) $e^t = 0.36$ $\{\ln 0.36\}$; $\{-1.0217\}$

21) $e^{-4x} = 9$

22) $e^{3p} = 4$

23) $e^{0.01r} = 2$

24) $e^{-0.08k} = 10$

25) $e^{0.006t} = 3$

26) $e^{0.04a} = 12$

27) $e^{-0.4y} = 5$

28) $e^{-0.005c} = 16$

Objective 2: Solve Logarithmic Equations Using the Properties of Logarithms

Solve each equation.

29) $\log_6(k + 9) = \log_6 11$ {2}   30) $\log_5(d - 4) = \log_5 2$ {6}

31) $\log_7(3p - 1) = \log_7 9$ $\left\{ \dfrac{10}{3} \right\}$

32) $\log_4(5y + 2) = \log_4 10$ $\left\{ \dfrac{8}{5} \right\}$

33) $\log x + \log(x - 2) = \log 15$ {5}

34) $\log_9 r + \log_9(r + 7) = \log_9 18$ {2}

35) $\log_3 n + \log_3(12 - n) = \log_3 20$ {2, 10}

36) $\log m + \log(11 - m) = \log 24$ {3, 8}

37) $\log_2(-z) + \log_2(z - 8) = \log_2 15$ $\varnothing$

38) $\log_5 8y - \log_5(3y - 4) = \log_5 2$ $\varnothing$

39) $\log_6(5b - 4) = 2$ {8}   40) $\log_3(4c + 5) = 3$ $\left\{ \dfrac{11}{2} \right\}$

41) $\log(3p + 4) = 1$ {2}   42) $\log(7n - 11) = 1$ {3}

43) $\log_3 y + \log_3(y - 8) = 2$ {9}

44) $\log_4 k + \log_4(k - 6) = 2$ {8}

45) $\log_2 r + \log_2(r + 2) = 3$ {2}

46) $\log_9(z + 8) + \log_9 z = 1$ {1}

47) $\log_4 20c - \log_4(c + 1) = 2$ {4}

48) $\log_6 40x - \log_6(1 + x) = 2$ {9}

49) $\log_2 8d - \log_2(2d - 1) = 4$ $\left\{ \dfrac{2}{3} \right\}$

50) $\log_6(13 - x) + \log_6 x = 2$ {4, 9}

Mixed Exercises: Objectives 3 and 4

Use the formula $A = Pe^{rt}$ to solve Exercises 51–58.

51) If $2000 is invested at 6% interest compounded continuously, how long would it take

    a) for the investment to grow to $2500?   3.72 yr

    b) for the initial investment to double?   11.55 yr

52) If $5000 is invested at 7% interest compounded continuously, how long would it take

    a) for the investment to grow to $6000?   2.60 yr

    b) for the initial investment to double?   9.90 yr

53) How long would it take for an investment of $7000 to earn $800 in interest if it is invested at 7.5% compounded continuously?   1.44 yr

54) How long would it take for an investment of $4000 to earn $600 in interest if it is invested at 6.8% compounded continuously?   2.06 yr

55) Cynthia wants to invest some money now so that she will have $5000 in the account in 10 yr. How much should she invest in an account earning 8% compounded continuously? $2246.64

56) How much should Leroy invest now at 7.2% compounded continuously so that the account contains $8000 in 12 yr? $3371.78

57) Raj wants to invest $3000 now so that it grows to $4000 in 4 yr. What interest rate should he look for? (Round to the nearest tenth of a percent.)   7.2%

58) Marisol wants to invest $12,000 now so that it grows to $20,000 in 7 yr. What interest rate should she look for? (Round to the nearest tenth of a percent.)   7.3%

59) The number of bacteria, $N(t)$, in a culture t hr after the bacteria are placed in a dish is given by

$$N(t) = 4000e^{0.0374t}$$

where 4000 bacteria are initially present.

    a) After how many hours will there be 5000 bacteria in the culture?   6 hr

    b) How long will it take for the number of bacteria to double?   18.5 hr

60) The number of bacteria, $N(t)$, in a culture t hr after the bacteria are placed in a dish is given by

$$N(t) = 10,000e^{0.0418t}$$

where 10,000 bacteria are initially present.

    a) After how many hours will there be 15,000 bacteria in the culture?   9.7 hr

    b) How long will it take for the number of bacteria to double?   16.6 hr

61) The population of an Atlanta suburb is growing at a rate of 3.6% per year. If 21,000 people lived in the suburb in 2004, determine how many people will live in the town in 2012. Use $y = y_0e^{0.036t}$.   28,009

62) The population of a Seattle suburb is growing at a rate of 3.2% per year. If 30,000 people lived in the suburb in 2008, determine how many people will live in the town in 2015. Use $y = y_0e^{0.032t}$.   37,532

63) A rural town in South Dakota is losing residents at a rate of 1.3% per year. The population of the town was 2470 in 1990. Use $y = y_0e^{-0.013t}$ to answer the following questions.

    a) What was the population of the town in 2005?   2032

    b) In what year would it be expected that the population of the town is 1600?   2023

64) In 1995, the population of a rural town in Kansas was 1682. The population is decreasing at a rate of 0.8% per year. Use $y = y_0e^{-0.008t}$ to answer the following questions.

    a) What was the population of the town in 2000?   1616

    b) In what year would it be expected that the population of the town is 1000?   2060

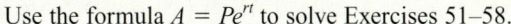

65) Radioactive carbon-14 is a substance found in all living organisms. After the organism dies, the carbon-14 decays according to the equation

$$y = y_0 e^{-0.000121t}$$

where t is in years, y_0 is the initial amount present at time $t = 0$, and y is the amount present after t yr.

a) If a sample initially contains 15 g of carbon-14, how many grams will be present after 2000 yr?   11.78 g

b) How long would it take for the initial amount to decay to 10 g?   3351 yr

c) What is the half-life of carbon-14?   5728 yr

 66) Plutonium-239 decays according to the equation

$$y = y_0 e^{-0.0000287t}$$

where t is in years, y_0 is the initial amount present at time $t = 0$, and y is the amount present after t yr.

a) If a sample initially contains 8 g of plutonium-239, how many grams will be present after 5000 yr?   6.93 g

b) How long would it take for the initial amount to decay to 5 g?   16,376 yr

c) What is the half-life of plutonium-239?   24,151 yr

 67) Radioactive iodine-131 is used in the diagnosis and treatment of some thyroid-related illnesses. The concentration of the iodine in a patient's system is given by

$$y = 0.4 e^{-0.086t}$$

where t is in days and y is in the appropriate units.

a) How much iodine-131 is given to the patient?   0.4 units

b) How much iodine-131 remains in the patient's system 7 days after treatment?   0.22 units

68) The amount of cobalt-60 in a sample is given by

$$y = 30 e^{-0.131t}$$

where t is in years and y is in grams.

a) How much cobalt-60 is originally in the sample?   30 g

b) How long would it take for the initial amount to decay to 10 g?   8.4 yr

Extension

Solve. Where appropriate, give the exact solution and the approximation to four decimal places.

69) $\log_2 (\log_2 x) = 2$ {16}

70) $\log_3 (\log y) = 1$ {1000}

71) $\log_3 \sqrt{n^2 + 5} = 1$ {-2, 2}

72) $\log (p - 7)^2 = 4$ {107}

73) $e^{|t|} = 13$
{-ln 13, ln 13}; {-2.5649, 2.5649}

74) $e^{r^2 - 25} = 1$ {-5, 5}

75) $e^{2y} + 3e^y - 4 = 0$ {0}

76) $e^{2x} - 9e^x + 8 = 0$
{0, ln 8}; {0, 2.0794}

77) $5^{2c} - 4 \cdot 5^c - 21 = 0$
$\left\{ \dfrac{\ln 7}{\ln 5} \right\}$; {1.2091}

78) $9^{2a} + 5 \cdot 9^a - 24 = 0$
$\left\{ \dfrac{1}{2} \right\}$

79) $(\log x)^2 = \log x^3$ {1, 1000}

80) $\log 6^y = y^2$
{0, log 6}; {0, 0.7782}

Definition/Procedure	Example

10.1 Composite and Inverse Functions

Composition of Functions

Given the functions $f(x)$ and $g(x)$, the **composition function** $f \circ g$ (read "f of g") is defined as

$$(f \circ g)(x) = f(g(x))$$

where $g(x)$ is in the domain of f. **(p. 619)**

$f(x) = 4x - 10$ and $g(x) = -3x + 2$.

$$
\begin{aligned}
(f \circ g)(x) &= f(g(x)) \\
&= f(-3x + 2) \\
&= 4(-3x + 2) - 10 \quad \text{Substitute } -3x + 2 \text{ for } x \text{ in } f(x). \\
&= -12x + 8 - 10 \\
&= -12x - 2
\end{aligned}
$$

One-to-One Function

In order for a function to be a **one-to-one function**, each x-value corresponds to exactly one y-value, and each y-value corresponds to exactly one x-value.

The **horizontal line test** tells us how we can determine whether a graph represents a one-to-one function:

If every horizontal line that could be drawn through a function would intersect the graph at most once, then the function is one-to-one. **(pp. 622–623)**

Determine whether each function is one-to-one.

a) $f = \{(-2, 9), (1, 3), (3, -1), (7, -9)\}$ is one-to-one.

b) $g = \{(0, 9), (2, 1), (4, 1), (5, 4)\}$ is *not* one-to-one since the y-value 1 corresponds to two different x-values.

c)

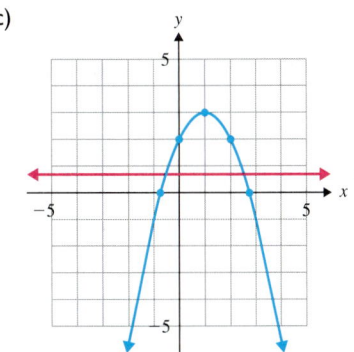

No. It fails the horizontal line test.

Inverse Function

Let f be a one-to-one function. The **inverse** of f, denoted by f^{-1}, is a one-to-one function that contains the set of all ordered pairs (y, x) where (x, y) belongs to f.

How to Find an Equation of the Inverse of $y = f(x)$

Step 1: Replace $f(x)$ with y.

Step 2: Interchange x and y.

Step 3: Solve for y.

Step 4: Replace y with the inverse notation, $f^{-1}(x)$.

The graphs of $f(x)$ and $f^{-1}(x)$ are symmetric with respect to the line $y = x$. **(p. 625)**

Find an equation of the inverse of $f(x) = 2x - 4$.

Step 1: $y = 2x - 4$ Replace $f(x)$ with y.

Step 2: $x = 2y - 4$ Interchange x and y.

Step 3: Solve for y.

$$x + 4 = 2y \quad \text{Add 4.}$$

$$\frac{x + 4}{2} = y \quad \text{Divide by 2.}$$

$$\frac{1}{2}x + 2 = y \quad \text{Simplify.}$$

Step 4: $f^{-1}(x) = \frac{1}{2}x + 2$ Replace y with $f^{-1}(x)$.

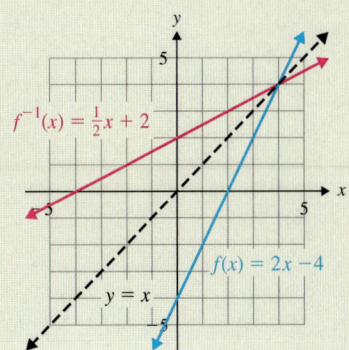

Definition/Procedure	Example

10.2 Exponential Functions

An **exponential function** is a function of the form

$$f(x) = a^x$$

where $a > 0$, $a \neq 1$, and x is a real number. **(p. 633)**

$f(x) = 3^x$

Characteristics of an Exponential Function

$$f(x) = a^x$$

1) If $f(x) = a^x$, where $a > 1$, the value of y increases as the value of x increases.

2) If $f(x) = a^x$ where $0 < a < 1$, the value of y decreases as the value of x increases.

3) The function is one-to-one.

4) The y-intercept is $(0, 1)$.

5) The domain is $(-\infty, \infty)$, and the range is $(0, \infty)$. **(p. 635)**

1)

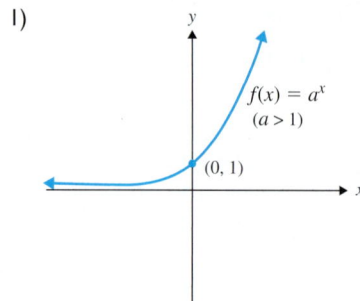

$f(x) = e^x$ is a special exponential function that has many uses in mathematics. Like the number π, e is an irrational number. **(p. 636)**

$$e \approx 2.7183$$

2)

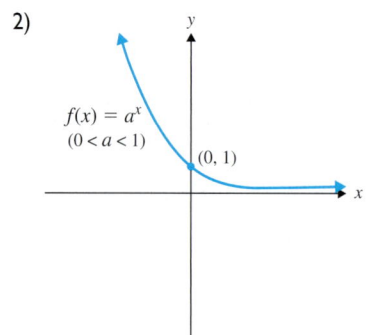

Solving an Exponential Equation

Step 1: **If possible, express each side of the equation with the same base.** If it is not possible to get the same base, a method in Section 10.6 can be used.

Step 2: **Use the rules of exponents to simplify the exponents.**

Step 3: **Set the exponents equal and solve for the variable. (p. 638)**

Solve $5^{4x-1} = 25^{3x+4}$.

Step 1: $5^{4x-1} = (5^2)^{(3x+4)}$ Both sides are powers of 5.

Step 2: $5^{4x-1} = 5^{2(3x+4)}$ Power rule for exponents
$5^{4x-1} = 5^{6x+8}$ Distribute.

Step 3: $4x - 1 = 6x + 8$ The bases are the same. Set the exponents equal.

$-2x = 9$ Subtract $6x$; add 1.

$x = -\dfrac{9}{2}$ Divide by -2.

The solution set is $\left\{ -\dfrac{9}{2} \right\}$.

Definition/Procedure	Example

10.3 Logarithmic Functions

Definition of Logarithm

If $a > 0$, $a \neq 1$, and $x > 0$, then for every real number y, $y = \log_a x$ means $x = a^y$. **(p. 644)**

Write $\log_5 125 = 3$ in exponential form.

$$\log_5 125 = 3 \text{ means } 5^3 = 125$$

A **logarithmic equation** is an equation in which at least one term contains a logarithm.

To solve a logarithmic equation of the form

$$\log_a b = c$$

write the equation in exponential form ($a^c = b$) and solve for the variable. **(p. 646)**

Solve $\log_2 k = 3$.
Write the equation in exponential form and solve for k.

$$\log_2 k = 3 \text{ means } 2^3 = k.$$
$$8 = k$$

The solution set is $\{8\}$.

To evaluate $\log_a b$ means *to find the power to which we raise a to get b.* **(p. 647)**

Evaluate $\log_7 49$.

$$\log_7 49 = 2 \text{ since } 7^2 = 49$$

A base 10 logarithm is called a **common logarithm.** A base 10 logarithm is often written without the base. **(p. 648)**

$\log x$ means $\log_{10} x$.

Characteristics of a Logarithmic Function
$$f(x) = \log_a x, \text{ where } a > 0 \text{ and } a \neq 1$$

1) If $f(x) = \log_a x$, where $a > 1$, the value of y increases as the value of x increases.

2) If $f(x) = \log_a x$, where $0 < a < 1$, the value of y decreases as the value of x increases.

3) The function is one-to-one.

4) The x-intercept is $(1, 0)$.

5) The domain is $(0, \infty)$, and the range is $(-\infty, \infty)$.

6) The inverse of $f(x) = \log_a x$ is $f^{-1}(x) = a^x$. **(p. 652)**

1)

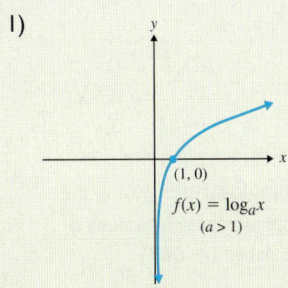

2)

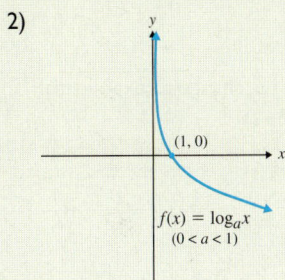

10.4 Properties of Logarithms

Let x, y, and a be positive real numbers where $a \neq 1$ and let r be any real number. Then,

1) $\log_a a = 1$

2) $\log_a 1 = 0$

3) $\log_a xy = \log_a x + \log_a y$ Product rule

4) $\log_a \dfrac{x}{y} = \log_a x - \log_a y$ Quotient rule

5) $\log_a x^r = r \log_a x$ Power rule

6) $\log_a a^x = x$ for any real number x

7) $a^{\log_a x} = x$ **(p. 660)**

Write $\log_4 \dfrac{c^5}{d^2}$ as the sum or difference of logarithms in simplest form. Assume c and d represent positive real numbers.

$$\log_4 \frac{c^5}{d^2} = \log_4 c^5 - \log_4 d^2 \quad \text{Quotient rule}$$
$$= 5 \log_4 c - 2 \log_4 d \quad \text{Power rule}$$

Definition/Procedure	**Example**

10.5 Common and Natural Logarithms and Change of Base

We can evaluate common logarithms with or without a calculator. **(p. 665)**	Find the value of each. a) log 100 b) log 53 a) $\log 100 = \log_{10} 100 = \log_{10} 10^2 = 2$ b) Using a calculator, we get $\log 53 \approx 1.7243$.
The number e is approximately equal to 2.7183. A base e logarithm is called a **natural logarithm**. The notation used for a natural logarithm is ln x. The domain of $f(x) = \ln x$ is $(0, \infty)$, and the range is $(-\infty, \infty)$. **(p. 667)**	$f(x) = \ln x$ means $f(x) = \log_e x$ The graph of $f(x) = \ln x$ looks like this: 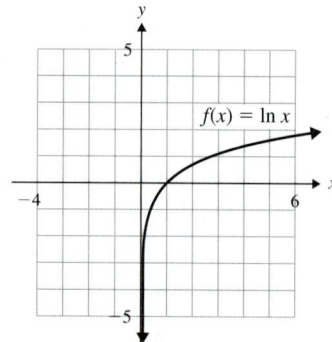

ln e = 1 since ln e = 1 means $\log_e e = 1$.
 	
We can find the values of some natural logarithms using the properties of logarithms. We can approximate the values of other natural logarithms using a calculator. **(p. 667)**
	Find the value of each.
	
	a) $\ln e^{12}$ b) ln 18
	
	a) $\ln e^{12} = 12 \ln e$ Power rule
	           $= 12(1)$ $\ln e = 1$
	           $= 12$
	
	b) Using a calculator, we get $\ln 18 \approx 2.8904$.

To solve an equation such as ln x = 1.6, change to exponential form and solve for the variable. **(p. 669)**	Solve ln x = 1.6.
	
	ln x = 1.6 means $\log_e x = 1.6$.
	
	     $\log_e x = 1.6$
	       $e^{1.6} = x$ Exponential form
	   $4.9530 \approx x$ Approximation
	
	The exact solution is $\{e^{1.6}\}$. The approximation is {4.9530}.

Applications of Exponential Functions
 	
Continuous Compounding
 	
If P dollars is deposited in an account earning interest rate r compounded continuously, then the amount of money, A (in dollars), in the account after t years is given by $A = Pe^{rt}$. **(p. 671)**
	Determine the amount of money in an account after 6 yr if $3000 was initially invested at 5% compounded continuously.
	
	     $A = Pe^{rt}$
	       $= 3000e^{0.05(6)}$ Substitute values.
	       $= 3000e^{0.30}$ Multiply (0.05)(6).
	       ≈ 4049.5764 Evaluate using a calculator.
	       $\approx \$4049.58$ Round to the nearest cent.

Change-of-Base Formula If a, b, and x are positive real numbers and $a \neq 1$ and $b \neq 1$, then $$\log_a x = \frac{\log_b x}{\log_b a}$$ **(p. 672)**	Find $\log_2 75$ to four decimal places. $$\log_2 75 = \frac{\log_{10} 75}{\log_{10} 2} \approx 6.2288$$

Definition/Procedure	Example

10.6 Solving Exponential and Logarithmic Equations

Let a, x, and y be positive real numbers, where $a \neq 1$. 1) If $x = y$, then $\log_a x = \log_a y$. 2) If $\log_a x = \log_a y$, then $x = y$. **Solving an Exponential Equation** Begin by asking yourself, "*Can I express each side with the same base?*" 1) If the answer is **yes,** then write each side of the equation with the same base, set the exponents equal, and solve for the variable. 2) If the answer is **no,** then take the natural logarithm of each side, use the properties of logarithms, and solve for the variable. **(pp. 676–677)**	Solve each equation. a) $4^x = 64$ Ask yourself, "*Can I express both sides with the same base?*" **Yes.** $$4^x = 64$$ $$4^x = 4^3$$ $$x = 3 \quad \text{Set the exponents equal.}$$ The solution set is $\{3\}$. b) $4^x = 9$ Ask yourself, "*Can I express both sides with the same base?*" **No.** Take the natural logarithm of each side. $$4^x = 9$$ $$\ln 4^x = \ln 9 \quad \text{Take the natural log of each side.}$$ $$x \ln 4 = \ln 9 \quad \log_a x^r = r \log_a x$$ $$x = \frac{\ln 9}{\ln 4} \quad \text{Divide by } \ln 4.$$ $$x \approx 1.5850 \quad \text{Use a calculator to get the approximation.}$$ The exact solution is $\left\{\dfrac{\ln 9}{\ln 4}\right\}$. The approximation is $\{1.5850\}$.
Solve an exponential equation with base e by taking the natural logarithm of each side. **(p. 677)**	Solve $e^y = 35.8$. $$\ln e^y = \ln 35.8 \quad \text{Take the natural log of each side.}$$ $$y \ln e = \ln 35.8 \quad \log_a x^r = r \log_a x$$ $$y(1) = \ln 35.8 \quad \ln e = 1$$ $$y = \ln 35.8$$ $$y \approx 3.5779 \quad \text{Approximation}$$ The exact solution is $\{\ln 35.8\}$. The approximation is $\{3.5779\}$.
Solving Logarithmic Equations Sometimes we must use the properties of logarithms to solve logarithmic equations. **(p. 678)**	Solve $\log x + \log(x - 3) = \log 28$. $$\log x + \log(x - 3) = \log 28$$ $$\log x(x - 3) = \log 28 \quad \text{Product rule}$$ $$x(x - 3) = 28 \quad \text{If } \log_a x = \log_a y, \text{ then } x = y.$$ $$x^2 - 3x = 28 \quad \text{Distribute.}$$ $$x^2 - 3x - 28 = 0 \quad \text{Subtract 28.}$$ $$(x - 7)(x + 4) = 0 \quad \text{Factor.}$$ $$x - 7 = 0 \quad \text{or} \quad x + 4 = 0 \quad \text{Set each factor equal to 0.}$$ $$x = 7 \quad \text{or} \quad x = -4 \quad \text{Solve.}$$ Verify that only 7 satisfies the original equation. The solution set is $\{7\}$.

*Additional answers can be found in the Answers to Exercises appendix.

(10.1)

1) Let $f(x) = x + 6$ and $g(x) = 2x - 9$. Find

   a) $(g \circ f)(x)$ $2x + 3$

   b) $(f \circ g)(x)$ $2x - 3$

   c) $(f \circ g)(5)$ 7

2) Let $h(x) = 2x - 1$ and $k(x) = x^2 + 5x - 4$. Find

   a) $(k \circ h)(x)$ $4x^2 + 6x - 8$

   b) $(h \circ k)(x)$ $2x^2 + 10x - 9$

   c) $(h \circ k)(-3)$ -21

3) Antoine's gross weekly pay, G, in terms of the number of hours, h, he worked is given by $G(h) = 12h$. His net weekly pay, N, in terms of his gross pay is given by $N(G) = 0.8G$.

   a) Find $(N \circ G)(h)$ and explain what it represents.

   b) Find $(N \circ G)(30)$ and explain what it represents.

   c) What is his net pay if he works 40 hr in 1 week?

4) Given $h(x)$, find f and g such that $h(x) = (f \circ g)(x)$.

   a) $h(x) = (3x + 10)^2$ b) $h(x) = (8x - 7)^3$

   c) $h(x) = \sqrt{x^2 + 6}$ d) $h(x) = \dfrac{1}{2 - 5x}$

Determine whether each function is one-to-one. If it is one-to-one, find its inverse.

5) $f = \{(-7, -4), (-2, 1), (1, 5), (6, 11)\}$
  yes; $\{(-4, -7), (1, -2), (5, 1), (11, 6)\}$

6) $g = \{(1, 4), (3, 7), (6, 4), (10, 9)\}$ no

Determine whether each function is one-to-one. If it is one-to-one, graph its inverse.

7)
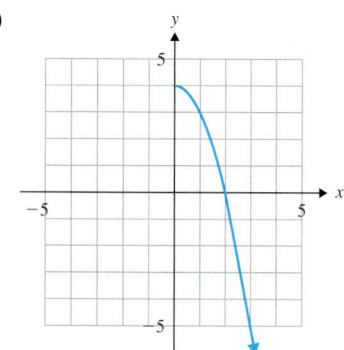

8)
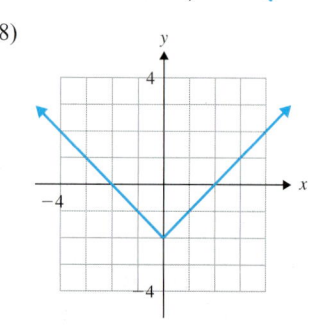

   no

Find the inverse of each one-to-one function. Graph each function and its inverse on the same axes.

9) $f(x) = x + 4$ 10) $g(x) = 2x - 10$

11) $h(x) = \dfrac{1}{3}x - 1$ 12) $f(x) = \sqrt[3]{x} + 2$

Given each one-to-one function $f(x)$, find the following function values *without* finding an equation of $f^{-1}(x)$. Find the value in a) before b).

13) $f(x) = 6x - 1$

   a) $f(2)$ 11 b) $f^{-1}(11)$ 2

14) $f(x) = \sqrt[3]{x + 5}$

   a) $f(-13)$ -2 b) $f^{-1}(-2)$ -13

(10.2) Graph each exponential function. State the domain and range.

15) $f(x) = 2^x$
   domain: $(-\infty, \infty)$; range: $(0, \infty)$

16) $h(x) = \left(\dfrac{1}{3}\right)^x$
   domain: $(-\infty, \infty)$; range: $(0, \infty)$

17) $y = 2^x - 4$
   domain: $(-\infty, \infty)$; range: $(-4, \infty)$

18) $f(x) = 3^{x-2}$
   domain: $(-\infty, \infty)$; range: $(0, \infty)$

19) $f(x) = e^x$
   domain: $(-\infty, \infty)$; range: $(0, \infty)$

20) $g(x) = e^x + 2$
   domain: $(-\infty, \infty)$; range: $(2, \infty)$

Solve each exponential equation.

21) $2^c = 64$ $\{6\}$ 22) $7^{m+5} = 49$ $\{-3\}$

23) $16^{3z} = 32^{2z-1}$ $\left\{-\dfrac{5}{2}\right\}$ 24) $9^y = \dfrac{1}{81}$ $\{-2\}$

25) $\left(\dfrac{3}{2}\right)^{x+4} = \left(\dfrac{4}{9}\right)^{x-3}$ $\left\{\dfrac{2}{3}\right\}$

26) The value, $V(t)$, in dollars, of a luxury car t yr after it is purchased is given by $V(t) = 38{,}200(0.816)^t$.

   a) What was the purchase price of the car? $\$38{,}200$

   b) What will the car be worth 4 yr after purchase?
                             $\$16{,}936.51$

(10.3)

27) What is the domain of $y = \log_a x$? $(0, \infty)$

28) In the equation $y = \log_a x$, a must be what kind of number?
         a must be a positive real number that is not equal to 1.

Write in exponential form.

29) $\log_5 125 = 3$ $5^3 = 125$ 30) $\log_{16}\dfrac{1}{4} = -\dfrac{1}{2}$ $16^{-1/2} = \dfrac{1}{4}$

31) $\log 100 = 2$ $10^2 = 100$ 32) $\log 1 = 0$ $10^0 = 1$

Write in logarithmic form.

33) $3^4 = 81$ $\log_3 81 = 4$ 34) $\left(\dfrac{2}{3}\right)^{-2} = \dfrac{9}{4}$ $\log_{2/3}\dfrac{9}{4} = -2$

35) $10^3 = 1000$ $\log 1000 = 3$ 36) $\sqrt{121} = 11$ $\log_{121} 11 = \dfrac{1}{2}$

Solve.

37) $\log_2 x = 3$ $\{8\}$ 38) $\log_9(4x + 1) = 2$ $\{20\}$

39) $\log_{32} 16 = x$ $\left\{\dfrac{4}{5}\right\}$ 40) $\log(2x + 5) = 1$ $\left\{\dfrac{5}{2}\right\}$

Evaluate.

41) $\log_8 64$ 2

42) $\log_3 27$ 3

43) $\log 1000$ 3

44) $\log 1$ 0

45) $\log_{1/2} 16$ -4

46) $\log_{1/5} \dfrac{1}{25}$ 2

Graph each logarithmic function.

47) $f(x) = \log_2 x$

48) $g(x) = \log_3 x$

49) $h(x) = \log_{1/3} x$

50) $f(x) = \log_{1/4} x$

Find the inverse of each function.

51) $f(x) = 5^x$ $f^{-1}(x) = \log_5 x$

52) $g(x) = 3^x$ $g^{-1}(x) = \log_3 x$

53) $h(x) = \log_6 x$ $h^{-1}(x) = 6^x$

Solve.

54) A company plans to test market its new dog food in a large metropolitan area before taking it nationwide. The company predicts that its sales over the next 12 months can be approximated by

$$S(t) = 10 \log_3(2t + 1)$$

where t is the number of months after the dog food is introduced, and $S(t)$ is in thousands of bags of dog food.

a) How many bags of dog food were sold after 1 month on the market?   10,000

b) How many bags of dog food were sold after 4 months on the market?   20,000

(10.4) Decide whether each statement is true or false.

55) $\log_5(x + 4) = \log_5 x + \log_5 4$ false

56) $\log_2 \dfrac{k}{6} = \log_2 k - \log_2 6$ true

Write as the sum or difference of logarithms and simplify, if possible. Assume all variables represent positive real numbers.

57) $\log_8 3z$ $\log_8 3 + \log_8 z$

58) $\log_7 \dfrac{49}{t}$ $2 - \log_7 t$

59) $\log_4 \sqrt{64}$ $\dfrac{3}{2}$

60) $\log \dfrac{1}{100}$ -2

61) $\log_5 c^4 d^3$ $4\log_5 c + 3\log_5 d$

62) $\log_4 m\sqrt{n}$ $\log_4 m + \dfrac{1}{2}\log_4 n$

63) $\log_a \dfrac{xy}{z^3}$
  $\log_a x + \log_a y - 3\log_a z$

64) $\log_4 \dfrac{a^2}{bc^4}$
  $2\log_4 a - \log_4 b - 4\log_4 c$

65) $\log p(p + 8)$
  $\log p + \log (p + 8)$

66) $\log_6 \dfrac{r^3}{r^2 - 5}$
  $3\log_6 r - \log_6 (r^2 - 5)$

Write as a single logarithm. Assume the variables are defined so that the variable expressions are positive and so that the bases are positive real numbers not equal to 1.

67) $\log c + \log d$ $\log cd$

68) $\log_4 n - \log_4 7$ $\log_4 \dfrac{n}{7}$

69) $9\log_2 a + 3\log_2 b$ $\log_2 a^9 b^3$

70) $\log_5 r - 2\log_5 t$ $\log_5 \dfrac{r}{t^2}$

71) $\log_3 5 + 4\log_3 m - 2\log_3 n$ $\log_3 \dfrac{5m^4}{n^2}$

72) $\dfrac{1}{2}\log_z a - \log_z b$ $\log_z \dfrac{\sqrt{a}}{b}$

73) $3\log_5 c - \log_5 d - 2\log_5 f$ $\log_5 \dfrac{c^3}{df^2}$

74) $2\log_6 x + \dfrac{1}{3}\log_6(x - 4)$ $\log_6 x^2 \sqrt[3]{x - 4}$

Given that $\log 7 \approx 0.8451$ and $\log 9 \approx 0.9542$, use the properties of logarithms to approximate the following. Do NOT use a calculator.

75) $\log 49$ 1.6902

76) $\log 63$ 1.7993

77) $\log \dfrac{7}{9}$ -0.1091

78) $\log \dfrac{1}{7}$ -0.8451

(10.5)

79) What is the base of $\ln x$? e

80) Evaluate $\ln e$.   1

Evaluate each logarithm. Do not use a calculator.

81) $\log 10$ 1

82) $\log 100$ 2

83) $\log \sqrt{10}$ $\dfrac{1}{2}$

84) $\log \dfrac{1}{100}$ -2

85) $\log 0.001$ -3

86) $\ln e^4$ 4

87) $\ln 1$ 0

88) $\ln \sqrt[3]{e}$ $\dfrac{1}{3}$

Use a calculator to find the approximate value of each logarithm to four decimal places.

89) $\log 8$ 0.9031

90) $\log 0.3$ -0.5229

91) $\ln 1.75$ 0.5596

92) $\ln 0.924$ -0.0790

Solve each equation. Do not use a calculator.

93) $\log p = 2$ $\{100\}$

94) $\log(5n) = 3$ $\{200\}$

95) $\log\left(\dfrac{1}{2}c\right) = -1$ $\left\{\dfrac{1}{5}\right\}$

96) $\log(6z - 5) = 1$ $\left\{\dfrac{5}{2}\right\}$

Solve each equation. Give an exact solution and a solution that is approximated to four decimal places.

97) $\log x = 2.1$ $\{10^{2.1}\}; \{125.8925\}$

98) $\log k = -1.4$ $\{10^{-1.4}\}; \{0.0398\}$

99) $\ln y = 2$ $\{e^2\}; \{7.3891\}$

100) $\ln c = -0.5$ $\{e^{-0.5}\}; \{0.6065\}$

101) $\log(4t) = 1.75$ $\left\{\dfrac{10^{1.75}}{4}\right\}; \{14.0585\}$

102) $\ln(2a - 3) = 1$ $\left\{\dfrac{e + 3}{2}\right\}; \{2.8591\}$

Graph each function. State the domain and range.

103) $f(x) = \ln(x - 3)$

104) $g(x) = \ln x - 2$

Use the change-of-base formula with either base 10 or base e to approximate each logarithm to four decimal places.

105) $\log_4 19$ 2.1240

106) $\log_9 42$ 1.7011

107) $\log_{1/2} 38$ −5.2479

108) $\log_6 0.82$ −0.1108

For Exercises 109 and 110, use the formula $L(I) = 10 \log \dfrac{I}{10^{-12}}$, where I is the intensity of sound, in watts per square meter, and $L(I)$ is the loudness of sound in decibels. Do *not* use a calculator.

109) The intensity of sound from the crowd at a college basketball game reached 0.1 W/m². Find the loudness of the crowd, in decibels.   110 dB

110) Find the intensity of the sound of a jet taking off if the noise level can reach 140 dB 25 m from the jet.   100 W/m²

Use the formula $A = P\left(1 + \dfrac{r}{n}\right)^{nt}$ and a calculator to solve Exercises 111 and 112.

111) Pedro deposits $2500 in an account earning 6% interest compounded quarterly. How much will be in the account after 5 yr? $3367.14

112) Find the amount Chelsea owes at the end of 6 yr if she borrows $18,000 at a rate of 7% compounded monthly. $27,361.90

Use the formula $A = Pe^{rt}$ and a calculator to solve Exercises 113 and 114.

113) Find the amount Liang will owe at the end of 4 yr if he borrows $9000 at a rate of 6.2% compounded continuously. $11,533.14

114) If $4000 is deposited in an account earning 5.8% compounded continuously, how much will be in the account after 7 yr? $6003.21

115) The number of bacteria, $N(t)$, in a culture t hr after the bacteria are placed in a dish is given by

$$N(t) = 6000e^{0.0514t}$$

a) How many bacteria were originally in the culture?   6000

b) How many bacteria are present after 12 hr?   11,118

116) The pH of a solution is given by pH = −log[H⁺], where [H⁺] is the molar concentration of the hydronium ion. Find the ideal pH of blood if [H⁺] = 3.98 × 10⁻⁸.   7.4

(10.6) Solve each equation. Give the exact solution. If the answer contains a logarithm, approximate the solution to four decimal places. *Some of these exercises require the use of a calculator to obtain a decimal approximation.*

117) $2^y = 16$ {4}

118) $3^n = 7$ $\left\{\dfrac{\ln 7}{\ln 3}\right\}$; {1.7712}

119) $9^{4k} = 2$ $\left\{\dfrac{\ln 2}{4 \ln 9}\right\}$; {0.0789}

120) $125^{m-4} = 25^{1-m}$ $\left\{\dfrac{14}{5}\right\}$

121) $6^{2c} = 8^{c-5}$ $\left\{\dfrac{5 \ln 8}{\ln 8 - 2 \ln 6}\right\}$; {−6.9127}

122) $e^z = 22$ {ln 22}; {3.0910}

123) $e^{5p} = 8$ $\left\{\dfrac{\ln 8}{5}\right\}$; {0.4159}

124) $e^{0.03t} = 19$ $\left\{\dfrac{\ln 19}{0.03}\right\}$; {98.1480}

Solve each logarithmic equation.

125) $\log_3(5w + 3) = 2$ $\left\{\dfrac{6}{5}\right\}$

126) $\log(3n - 5) = 3$ {335}

127) $\log_2 x + \log_2(x + 2) = \log_2 24$ {4}

128) $\log_7 10p - \log_7(p - 8) = \log_7 6$ ∅

129) $\log_4 k + \log_4(k - 12) = 3$ {16}

130) $\log_3 12m - \log_3(1 + m) = 2$ {3}

Use the formula $A = Pe^{rt}$ to solve Exercises 131 and 132.

131) Jamar wants to invest some money now so that he will have $10,000 in the account in 6 yr. How much should he invest in an account earning 6.5% compounded continuously? $6770.57

132) Samira wants to invest $6000 now so that it grows to $9000 in 5 yr. What interest rate (compounded continuously) should she look for? (Round to the nearest tenth of a percent.)   8.1%

133) The population of a suburb is growing at a rate of 1.6% per year. The population of the suburb was 16,410 in 1990. Use $y = y_0 e^{0.016t}$ to answer the following questions.

a) What was the population of the town in 1995?   17,777

b) In what year would it be expected that the population of the town is 23,000?   2011

134) Radium-226 decays according to the equation

$$y = y_0 e^{-0.000436t}$$

where t is in years, y_0 is the initial amount present at time $t = 0$, and y is the amount present after t yr.

a) If a sample initially contains 80 g of radium-226, how many grams will be present after 500 yr?   64.3 g

b) How long would it take for the initial amount to decay to 20 g?   3180 yr

c) What is the half-life of radium-226?   1590 yr

Additional answers can be found in the Answers to Exercises appendix.

Use a calculator only where indicated.

1) Let $h(x) = 2x + 7$ and $k(x) = x^2 + 5x - 3$. Find

   a) $(h \circ k)(x)$ $2x^2 + 10x + 1$

   b) $(k \circ h)(x)$ $4x^2 + 38x + 81$

   c) $(k \circ h)(-3)$ 3

2) Let $h(x) = (9x - 7)^3$. Find f and g such that $h(x) = (f \circ g)(x)$. $f(x) = x^3, g(x) = 9x - 7$; answers may vary.

Determine whether each function is one-to-one. If it is one-to-one, find its inverse.

3) $f = \{(-4, 5), (-2, 7), (0, 3), (6, 5)\}$ no

4) $g = \left\{(2, 4), (6, 6), \left(9, \dfrac{15}{2}\right), (14, 10)\right\}$

5) Is this function one-to-one? If it is one-to-one, graph its inverse.

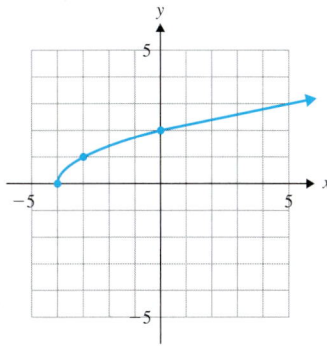

6) Find an equation of the inverse of $f(x) = -3x + 12$. $f^{-1}(x) = -\dfrac{1}{3}x + 4$

Use $f(x) = 2^x$ and $g(x) = \log_2 x$ for Exercises 7–10.

7) Graph $f(x)$.

8) Graph $g(x)$.

9) a) What is the domain of $g(x)$? $(0, \infty)$

   b) What is the range of $g(x)$? $(-\infty, \infty)$

10) How are the functions $f(x)$ and $g(x)$ related?   They are inverses.

11) Write $3^{-2} = \dfrac{1}{9}$ in logarithmic form. $\log_3 \dfrac{1}{9} = -2$

Solve each equation.

12) $9^{4x} = 81$ $\left\{\dfrac{1}{2}\right\}$

13) $125^{2c} = 25^{c-4}$ $\{-2\}$

14) $\log_5 y = 3$ $\{125\}$

15) $\log(3r + 13) = 2$ $\{29\}$

16) $\log_6(2m) + \log_6(2m - 3) = \log_6 40$ $\{4\}$

17) Evaluate.

   a) $\log_2 16$ 4

   b) $\log_7 \sqrt{7}$ $\dfrac{1}{2}$

18) Find $\ln e$. 1

Write as the sum or difference of logarithms and simplify, if possible. Assume all variables represent positive real numbers.

19) $\log_8 5n$ $\log_8 5 + \log_8 n$

20) $\log_3 \dfrac{9a^4}{b^5 c}$ $2 + 4 \log_3 a - 5 \log_3 b - \log_3 c$

21) Write as a single logarithm.

   $2 \log x - 3 \log (x + 1)$ $\log \dfrac{x^2}{(x + 1)^3}$

Use a calculator for the rest of the problems.

Solve each equation. Give an exact solution and a solution that is approximated to four decimal places.

22) $\log w = 0.8$ $\{10^{0.8}\}; \{6.3096\}$

23) $e^{0.3t} = 5$ $\left\{\dfrac{\ln 5}{0.3}\right\}; \{5.3648\}$

24) $\ln x = -0.25$ $\{e^{-0.25}\}; \{0.7788\}$

25) $4^{4a+3} = 9$

Graph each function. State the domain and range.

26) $y = e^x - 4$
   domain: $(-\infty, \infty)$; range: $(-4, \infty)$

27) $f(x) = \ln(x + 1)$
   domain: $(-1, \infty)$; range: $(-\infty, \infty)$

28) Approximate $\log_5 17$ to four decimal places. 1.7604

29) If $6000 is deposited in an account earning 7.4% interest compounded continuously, how much will be in the account after 5 yr? Use $A = Pe^{rt}$. $8686.41

30) Polonium-210 decays according to the equation

   $$y = y_0 e^{-0.00495t}$$

   where t is in days, y_0 is the initial amount present at time $t = 0$, and y is the amount present after t days.

   a) If a sample initially contains 100 g of polonium-210, how many grams will be present after 30 days?   86.2 g

   b) How long would it take for the initial amount to decay to 20 g?   325.1 days

   c) What is the half-life of polonium-210?   140 days

*Additional answers can be found in the Answers to Exercises appendix.

1) Evaluate $40 + 8 \div 2 - 3^2$.   35

2) Evaluate $\dfrac{5}{6} - \dfrac{14}{15} \cdot \dfrac{10}{7}$. $-\dfrac{1}{2}$

Simplify. The answer should not contain any negative exponents.

3) $(-5a^2)(3a^4)$ $-15a^6$

4) $\dfrac{40z^3}{10z^{-5}}$ $4z^8$

5) $\left(\dfrac{2c^{10}}{d^3}\right)^{-3}$ $\dfrac{d^9}{8c^{30}}$

6) Write 0.00009231 in scientific notation. 9.231×10^{-5}

7) *Write an equation and solve.*
A watch is on sale for $38.40. This is 20% off of the regular price. What was the regular price of the watch? $48.00

8) Solve $-4x + 7 < 13$. Graph the solution set and write the answer in interval notation.

9) Solve using the elimination method.

$$x + 4y = -2 \quad (-6, 1)$$
$$-2x + 3y = 15$$

10) Solve using the substitution method.

$$6x + 5y = -8 \quad \left(\dfrac{1}{3}, -2\right)$$
$$3x - y = 3$$

11) Write the equation of a line containing the points $(-2, 5)$ and $(2, -1)$. Write it in slope-intercept form. $y = -\dfrac{3}{2}x + 2$

12) Divide $(6c^3 - 7c^2 - 22c + 5) \div (2c - 5)$. $3c^2 + 4c - 1$

For Exercises 13–15, factor completely.

13) $4w^2 + w - 18$ $(4w + 9)(w - 2)$

14) $3p^3 + 2p^2 - 3p - 2$ $(p + 1)(p - 1)(3p + 2)$

15) $y^2 - 6y + 9$ $(y - 3)^2$

16) Solve $x^2 + 14x = -48$. $\{-8, -6\}$

17) Subtract $\dfrac{r}{r^2 - 49} - \dfrac{3}{r^2 - 2r - 63}$. $\dfrac{r^2 - 12r + 21}{(r + 7)(r - 7)(r - 9)}$

18) Solve $\dfrac{9}{y + 6} + \dfrac{4}{y - 6} = \dfrac{-4}{y^2 - 36}$. $\{2\}$

19) Graph the compound inequality $x + 2y \geq 6$ and $y \quad x \leq 2$.

Simplify. Assume all variables represent positive real numbers.

20) $\sqrt{120}$ $2\sqrt{30}$

21) $\sqrt{45t^9}$ $3t^4\sqrt{5t}$

22) $\sqrt{\dfrac{36a^5}{a^3}}$ $6a$

23) $(27)^{2/3}$ 9

24) Solve $\sqrt{h^2 + 2h - 7} = h - 3$. $\varnothing$

25) Multiply and simplify $(2 - 7i)(3 + i)$. $13 - 19i$

26) Solve by completing the square $k^2 - 8k + 4 = 0$.
$\{4 - 2\sqrt{3}, 4 + 2\sqrt{3}\}$

Solve.

27) $r^2 + 5r = -2$ $\left\{-\dfrac{5}{2} - \dfrac{\sqrt{17}}{2}, -\dfrac{5}{2} + \dfrac{\sqrt{17}}{2}\right\}$

28) $t^2 = 10t - 41$ $\{5 + 4i, 5 - 4i\}$

29) $4m^4 + 4 = 17m^2$ $\left\{-2, -\dfrac{1}{2}, \dfrac{1}{2}, 2\right\}$

30) Find the domain of $f(x) = \dfrac{4}{3x - 2}$. $\left(-\infty, \dfrac{2}{3}\right) \cup \left(\dfrac{2}{3}, \infty\right)$

31) Evaluate $\left(\dfrac{81}{16}\right)^{-5/4}$. $\dfrac{32}{243}$

32) Graph $g(x) = 2x^2 + 4x + 4$.

33) Let $f(x) = x^2 - 6x + 2$ and $g(x) = x - 3$.

   a) Find $f(-1)$.   9

   b) Find $(f \circ g)(x)$. $x^2 - 12x + 29$

   c) Find x so that $g(x) = -7$. -4

34) Graph $f(x) = 2^x - 3$. State the domain and range.
domain: $(-\infty, \infty)$; range: $(-3, \infty)$

35) Solve $16^y = \dfrac{1}{64}$. $\left\{-\dfrac{3}{2}\right\}$

36) Solve $\log_4 (5x + 1) = 2$. $\{3\}$

37) Write as a single logarithm.
$\log a + 2 \log b - 5 \log c$ $\log \dfrac{ab^2}{c^5}$

38) Solve $\log 5r - \log(r + 6) = \log 2$. $\{4\}$

39) Solve $e^{-0.04t} = 6$. Give an exact solution and an approximation to four decimal places. $\left\{-\dfrac{\ln 6}{0.04}\right\}; \{-44.7940\}$

40) Graph $f(x) = -\ln x$. State the domain and range.
domain: $(0, \infty)$; range: $(-\infty, \infty)$

CHAPTER 11

Nonlinear Functions, Conic Sections, and Nonlinear Systems

Algebra at Work: Forensics

We will look at one more application of mathematics to forensic science. Conic sections are used to help solve crimes.

Vanessa is a forensics expert and is called to the scene of a shooting. Blood is spattered everywhere. She uses algebra and

trigonometry to help her analyze the blood stain patterns to determine how far the shooter was standing from the victim and the angle at which the victim was shot. The location and angle help police determine the height of the gun when the shots were fired.

Forensics experts perform blood stain pattern analysis. Pictures are taken of the crime scene, and scientists measure the distance between each of the blood stains and the pool of blood. The individual blood stains are roughly elliptical in shape with tails at the ends. On the pictures, scientists draw a best-fit ellipse into each blood stain, measure the longest axis of the ellipse (the major axis) and the shortest axis of the ellipse (the minor axis) and do calculations that reveal where the shooter was standing at the time of the shooting.

In this chapter, we will learn about the ellipse and other conic sections.

Section 11.1 Graphs of Other Useful Functions

Objectives

1. **Review the Rules for Translating the Graphs of Functions**
2. **Graph Absolute Value Functions**
3. **Graph a Piecewise Function**
4. **Define, Graph, and Apply the Greatest Integer Function**

In-Class Example 1
Graph $g(x) = (x + 2)^2 - 4$ by shifting the graph of $f(x) = x^2$.
answer:

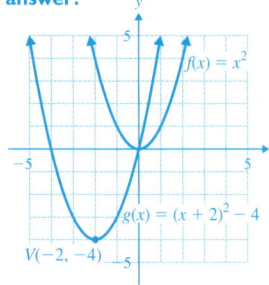

1. Review the Rules for Translating the Graphs of Functions

In Section 9.5 we discussed shifting the graph of $f(x) = x^2$ both horizontally and vertically. We also learned how to reflect the graph about the x-axis. Let's review the rules for translating the graph of $f(x) = x^2$.

Example 1

Graph $g(x) = (x + 3)^2 - 1$ by shifting the graph of $f(x) = x^2$.

Solution

If we compare $g(x)$ to $f(x) = x^2$, what do the constants that have been added to $g(x)$ tell us about translating the graph of $f(x)$?

$$g(x) = (x + 3)^2 - 1$$

Shift $f(x)$ Shift $f(x)$
left 3.   down 1.

Sketch the graph of $f(x) = x^2$, then move every point on f left 3 and down 1 to obtain the graph of $g(x)$. This moves the vertex from $(0, 0)$ to $(-3, -1)$.

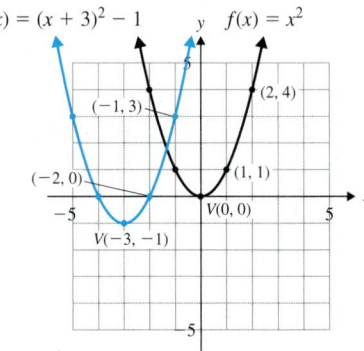

The rules we learned for shifting and reflecting $f(x) = x^2$ works for any function, so we will restate them here and apply them to other functions.

You Try 1

Graph $g(x) = (x - 1)^2 + 2$ by shifting the graph of $f(x) = x^2$.

Property Translating and Reflecting the Graphs of Functions

1) **Vertical Shifts:** Given the graph of any function, $f(x)$, if $g(x) = f(x) + k$ where k is a constant, then the graph of $g(x)$ will be the same shape as the graph of $f(x)$ but g will be shifted **vertically** k units.

2) **Horizontal Shifts:** Given the graph of any function $f(x)$, if $g(x) = f(x - h)$ where h is a constant, then the graph of $g(x)$ will be the same shape as the graph of $f(x)$ but g will be shifted **horizontally** h units.

3) **Reflection about the x-axis:** Given the graph of any function $f(x)$, if $g(x) = -f(x)$, then the graph of $g(x)$ will be the **reflection of the graph of f about the x-axis**. That is, obtain the graph of g by keeping the x-coordinate of each point on f the same but take the negative of the y-coordinate.

Next we will review how to graph the square root functions we learned about in Chapter 8.

Example 2

In-Class Example 2
Use Example 2.

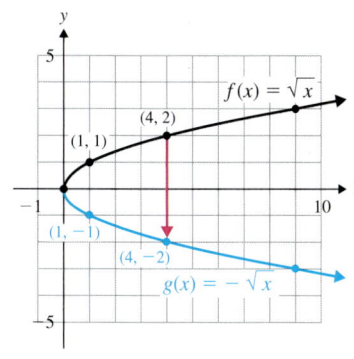

Graph $f(x) = \sqrt{x}$ and $g(x) = -\sqrt{x}$ on the same axes. Identify the domain and range.

Solution

Compare $f(x)$ and $g(x)$.

$$f(x) = \sqrt{x} \qquad g(x) = -\sqrt{x}$$
$$g(x) = -f(x) \qquad \text{Substitute } f(x) \text{ for } \sqrt{x}.$$

Since $g(x) = -f(x)$, the graph of g is the reflection of the graph of f about the x-axis. This means that when the x-coordinates of their ordered pairs are the same, the y-coordinates of the ordered pairs of g are the *negatives* of the y-coordinates in the ordered pairs of f.

The domain of each function is $[0, \infty)$. The range of $f(x)$ is $[0, \infty)$, and the range of $g(x)$ is $(-\infty, 0]$.

You Try 2

Graph $f(x) = \sqrt{x}$ and $g(x) = -\sqrt{x} + 2$ on the same axes. Identify the domain and range.

2. Graph Absolute Value Functions

Example 3

In-Class Example 3
Graph $f(x) = |x|$ and
$g(x) = |x| + 4$ on the same axes.
Identify the domain and range.
answer:
$f(x) = |x|$: The domain is
$(-\infty, \infty)$, and the range is
$[0, \infty)$.

$g(x) = |x| + 4$: The domain is
$(-\infty, \infty)$, and the range is
$[4, \infty)$.

Graph $f(x) = |x|$ and $g(x) = |x| + 2$ on the same axes. Identify the domain and range.

Solution

Compare $f(x)$ and $g(x)$.

$$f(x) = |x| \qquad g(x) = |x| + 2$$
$$g(x) = f(x) + 2 \qquad \text{Substitute } f(x) \text{ for } |x|.$$

The graph of $g(x) = |x| + 2$ is the same shape as the graph of $f(x)$ *except* it will be *shifted up* 2 units. Let's make a table of values for each function so that you can see why this is true.

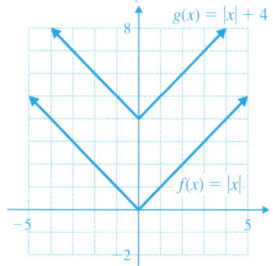

$f(x) = |x|$	
x	$f(x)$
0	0
1	1
2	2
-1	1
-2	2

$g(x) = |x| + 2$	
x	$g(x)$
0	2
1	3
2	4
-1	3
-2	4

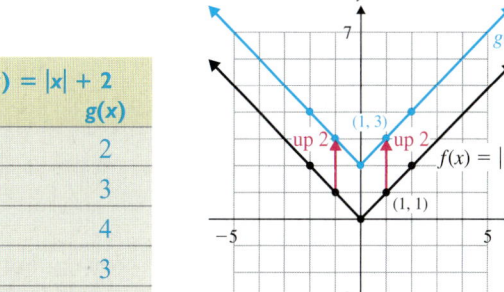

Absolute value functions have **V**-shaped graphs. The domain of $f(x) = |x|$ is $(-\infty, \infty)$, and the range is $[0, \infty)$. The domain of $g(x) = |x| + 2$ is $(-\infty, \infty)$, and the range is $[2, \infty)$.

 **BE CAREFUL** The absolute value functions we will study have V-shaped graphs. The graph of a quadratic function is *not* shaped like a V. It is a parabola.

 You Try 3

Graph $g(x) = |x| - 1$. Identify the domain and range.

We can combine the techniques used in the transformation of the graphs of functions to help us graph more complicated absolute value functions.

Example 4

Graph $h(x) = |x + 2| - 3$.

In-Class Example 4
Graph $h(x) = |x - 3| + 1$.
answer:

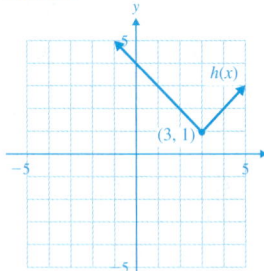

Solution

The graph of $h(x)$ will be the same shape as the graph of $f(x) = |x|$. So, let's see what the constants in $h(x)$ tell us about transforming the graph of $f(x) = |x|$.

$$h(x) = |x + 2| - 3$$

Shift $f(x)$ Shift $f(x)$
left 2.   down 3.

Sketch the graph of $f(x) = |x|$, including some key points, then *move every point on the graph of f left 2 and down 3 to obtain the graph of h*.

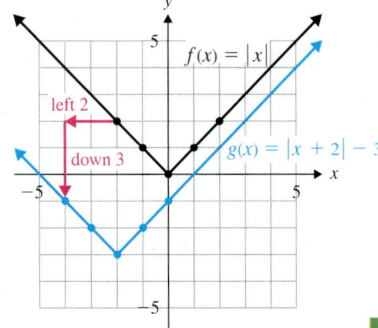

 You Try 4

Graph $h(x) = |x - 2| - 4$.

3. Graph a Piecewise Function

Definition

A **piecewise function** is a single function defined by two or more different rules.

Example 5

In-Class Example 5
Graph the piecewise function
$$f(x) = \begin{cases} 3x + 4, & x \geq 0 \\ -x + 1, & x < 0 \end{cases}$$
answer:

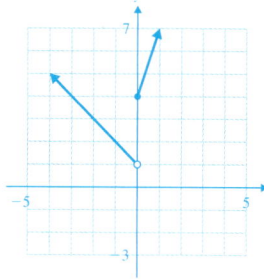

Graph the piecewise function

$$f(x) = \begin{cases} 2x - 4, & x \geq 3 \\ -x + 2, & x < 3 \end{cases}$$

Solution

This is a piecewise function because $f(x)$ is defined by two different rules. *The rule we use to find $f(x)$ depends on what value is substituted for x.*

Graph $f(x)$ by making two separate tables of values, one for each rule.

When $x \geq 3$, use the rule

$$f(x) = 2x - 4$$

The first x-value we will put in the table of values is 3 because it is the smallest number (lower bound) of the domain of $f(x) = 2x - 4$. *The other values we choose for x must be greater than 3 because this is when we use the rule $f(x) = 2x - 4$.* **This part of the graph will not extend to the left of (3, 2).**

$$f(x) = 2x - 4$$
$$(x \geq 3)$$

x	$f(x) = 2x - 4, x \geq 3$
3	2
4	4
5	6
6	8

When $x < 3$, use the rule

$$f(x) = -x + 2$$

The first x-value we will put in the table of values is 3 because it is the upper bound of the domain. *Notice that 3 is not included in the domain (the inequality is $<$, **not** $\leq$) so that the point $(3, f(3))$ will be represented as an open circle on the graph.* The other values we choose for x must be less than 3 because this is when we use the rule $f(x) = -x + 2$. **This part of the graph will not extend to the right of (3, −1).**

$$f(x) = -x + 2$$
$$(x < 3)$$

x	$f(x) = -x + 2, x < 3$
3	−1
2	0
1	1
0	2

$(3, -1)$ is an open circle.

The graph of $f(x)$ is at the left.

$$f(x) = \begin{cases} 2x - 4, x \geq 3 \\ -x + 2, x < 3 \end{cases}$$

You Try 5

Graph the piecewise function

$$f(x) = \begin{cases} -2x + 3, & x \leq -2 \\ \dfrac{3}{2}x - 1, & x > -2 \end{cases}$$

4. Define, Graph, and Apply the Greatest Integer Function

Another function that has many practical applications is the greatest integer function.

Definition

The **greatest integer function**

$$f(x) = [\![x]\!]$$

represents the largest integer less than or equal to x.

Example 6

In-Class Example 6
Let $f(x) = [x]$. Find the
following function values.

a) $f\left(3\frac{1}{4}\right)$ b) $f(19)$ c) $f\left(-4\frac{1}{2}\right)$

answer: a) 3 b) 19 c) −5

Let $f(x) = [x]$. Find the following function values.

a) $f\left(9\frac{1}{2}\right)$ b) $f(6)$ c) $f(-2.3)$

Solution

a) $f\left(9\frac{1}{2}\right) = \left[9\frac{1}{2}\right]$. This is the largest integer *less than or equal to* $9\frac{1}{2}$. That number

is 9. So $f\left(9\frac{1}{2}\right) = \left[9\frac{1}{2}\right] = 9$.

b) $f(x) = [6] = 6$. The largest integer *less than or equal to* 6 is 6.

c) To help us understand how to find this function value we will locate -2.3 on a number line.

The largest integer *less than or equal to* -2.3 is -3, so $f(-2.3) = [-2.3] = -3$. ∎

You Try 6

Let $f(x) = [x]$. Find the following function values.

a) $f(5.1)$ b) $f(0)$ c) $f\left(-5\frac{1}{4}\right)$

Example 7

In-Class Example 7
Use the given function.
answer:

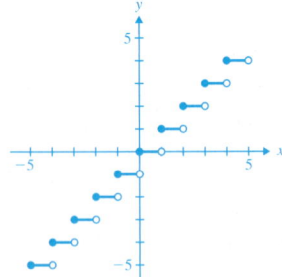

Graph $f(x) = [x]$.

Solution

First, let's look at the part of this function between $x = 0$ and $x = 1$ (when $0 \le x \le 1$).

x	f(x) = [x]	
0	0	
$\frac{1}{4}$	0	For all values of x *greater than or equal to* 0 and
$\frac{1}{2}$	0	*less than* 1, the function value, f(x), equals zero.
$\frac{3}{4}$	0	
⋮	0	
1	1	⟶ When x = 1 the function value changes to 1.

The graph has an open circle at $(1, 0)$ because if $x < 1, f(x) = 0$. That means that x can get *very close* to 1 and the function value will be zero, but $f(1) \ne 0$.

This pattern continues so that for the x-values in the interval $[1, 2)$, the function values are 1. The graph has an open circle at $(2, 1)$.

For the x-values in the interval $[2, 3), f(x) = 2$. The graph has an open circle at $(3, 2)$.

Continuing in this way we get the graph to the right.

The domain of the function is $(-\infty, \infty)$.
The range is the set of all integers
$\{\ldots, -3, -2, -1, 0, 1, 2, 3, \ldots\}$.

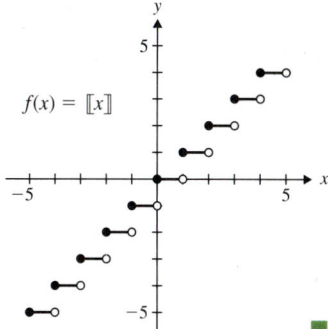

$f(x) = [\![x]\!]$

Because of the appearance of the graph, $f(x) = [\![x]\!]$ is also called a **step function.**   ■

You Try 7

Graph $f(x) = [\![x]\!] - 3$.

Example 8

In-Class Example 8
Use the given example.

To mail a large envelope within the United States in 2009, the U.S. Postal Service charged $0.88 for the first ounce and $0.17 for each additional ounce or fraction of an ounce. Let $C(x)$ represent the cost of mailing a large envelope within the United States, and let x represent the weight of the envelope, in ounces. Graph $C(x)$ for any large envelope weighing up to (and including) 5 ounces. (www.usps.com)

Solution

If a large envelope weighs between 0 and 1 ounce ($0 < x \le 1$), the cost, $C(x)$, is $0.88.

If it weighs more than 1 oz but less than or equal to 2 oz ($1 < x \le 2$), the cost, $C(x)$, is $0.88 + $0.17 = $1.05.

The pattern continues, and we get the graph to the right.

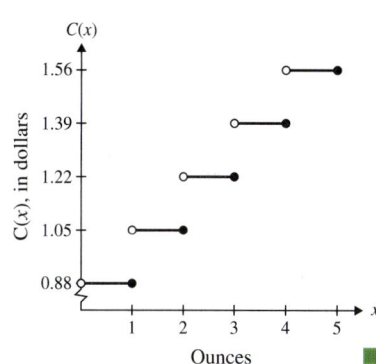

■

You Try 8

To mail a package within the United States at *library rate* in 2009, the U.S. Postal Service charged $2.26 for the first pound and $0.37 for each additional pound or fraction of a pound. Let $C(x)$ represent the cost of mailing a package at library rate and let x represent the weight of the package, in pounds. Graph $C(x)$ for any package weighing up to (and including) 5 pounds. (www.usps.com)

Using Technology

We can graph piecewise functions using a graphing calculator by entering each piece separately.

Suppose that we wish to graph the piecewise function $f(x) = \begin{cases} 2x - 4, & x \geq 3 \\ -x + 2, & x < 3 \end{cases}$.

Enter $2x - 4$ in Y_1 and $-x + 2$ in Y_2. First put parentheses around each function and then put parentheses around the interval of x-values for which that part of the function is defined. In order to enter the inequality symbols, press 2nd MATH and use the arrow keys to scroll to the desired symbol before pressing ENTER .

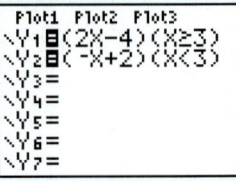

Display the piecewise function using the standard view.

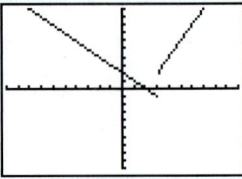

Graph the following functions using a graphing calculator.

1) $f(x) = \begin{cases} 3x - 2, & x \geq 1 \\ -x - 5, & x < 1 \end{cases}$

2) $f(x) = \begin{cases} x + 3, & x \geq 2 \\ -2x + 1, & x < 2 \end{cases}$

3) $f(x) = \begin{cases} -\dfrac{1}{2}x - 1, & x \leq -3 \\ x - 3, & x > -3 \end{cases}$

4) $f(x) = \begin{cases} 4, & x < 4 \\ -x + 2, & x \geq 4 \end{cases}$

5) $f(x) = \begin{cases} -\dfrac{2}{3}x + 1, & x \geq -1 \\ x + 3, & x < -1 \end{cases}$

6) $f(x) = \begin{cases} x, & x \geq 0 \\ 5x - 2, & x < 0 \end{cases}$

Answers to You Try Exercises

1)

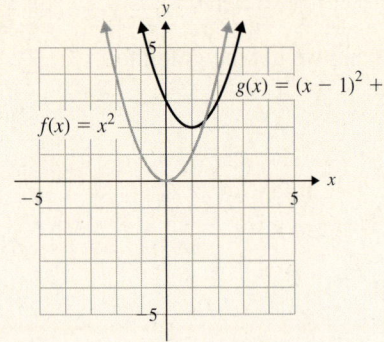

2) $f(x) = \sqrt{x}$; domain: $[0, \infty)$; range: $[0, \infty)$

 $g(x) = -\sqrt{x + 2}$; domain: $[-2, \infty)$; range: $(-\infty, 0]$

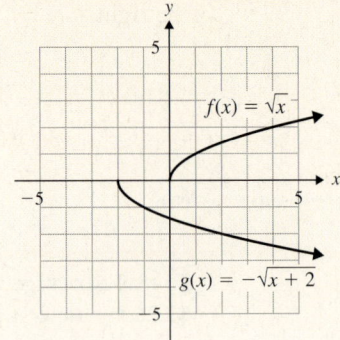

3) domain: $(-\infty, \infty)$, range: $[-1, \infty)$

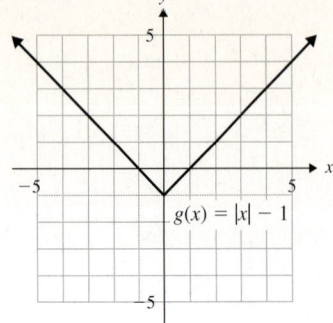

4)

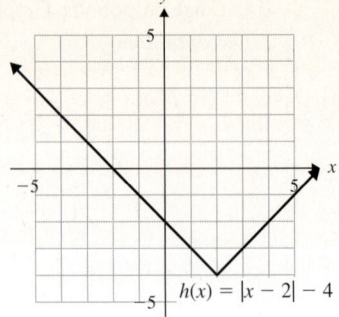

5)

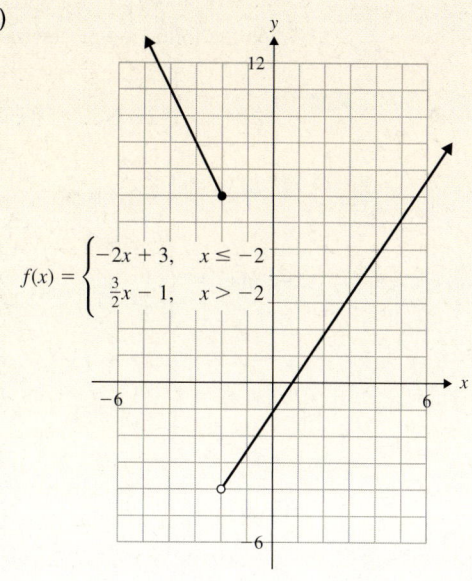

$$f(x) = \begin{cases} -2x + 3, & x \le -2 \\ \frac{3}{2}x - 1, & x > -2 \end{cases}$$

6) a) 5 b) 0 c) −6

7)

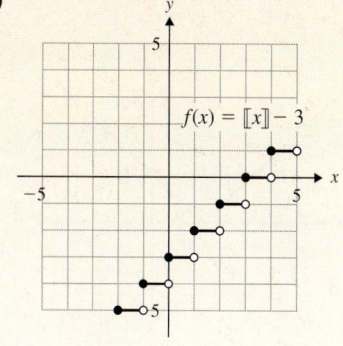

$f(x) = [\![x]\!] - 3$

8)

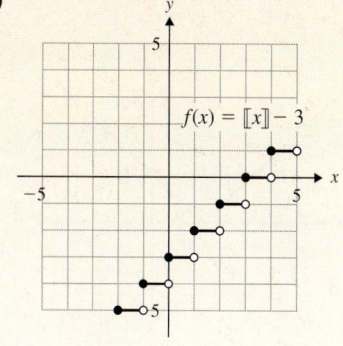

Answers to Technology Exercises

1)
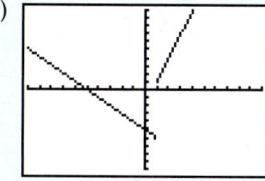

2)
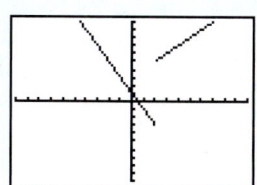

3)
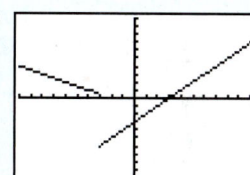

4)
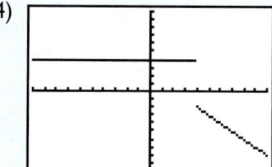

5)
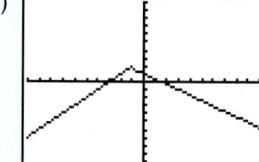

6)
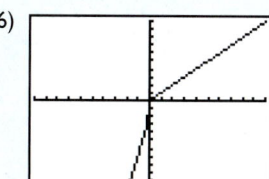

11.1 Exercises

*Additional answers can be found in the Answers to Exercises appendix.

Mixed Exercises: Objectives 1 and 2

Sketch the graph of $f(x)$. Then, graph $g(x)$ on the same axes using the transformation techniques reviewed in this section.

1) $f(x) = |x|$
   $g(x) = |x| - 2$

2) $f(x) = |x|$
   $g(x) = |x| + 1$

3) $f(x) = |x|$
   $g(x) = |x| + 3$

4) $f(x) = |x|$
   $g(x) = |x| - 4$

5) $f(x) = |x|$
   $g(x) = |x - 1|$

6) $f(x) = |x|$
   $g(x) = |x + 3|$

7) $f(x) = |x|$
   $g(x) = |x + 2| + 3$

8) $f(x) = |x|$
   $g(x) = |x - 3| - 4$

9) $f(x) = |x|$
   $g(x) = -|x|$

10) $f(x) = |x|$
   $g(x) = -|x| + 4$

11) $f(x) = |x - 3|$
   $g(x) = -|x - 3|$

12) $f(x) = |x + 4|$
   $g(x) = -|x + 4|$

Match each function to its graph.

13) $f(x) = -|x - 2|, g(x) = |x + 2|,$
$h(x) = -|x| - 2, k(x) = |x| + 2$

a)

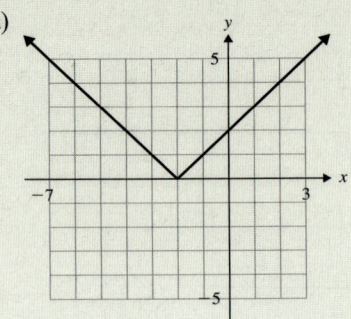

g(x)

b)

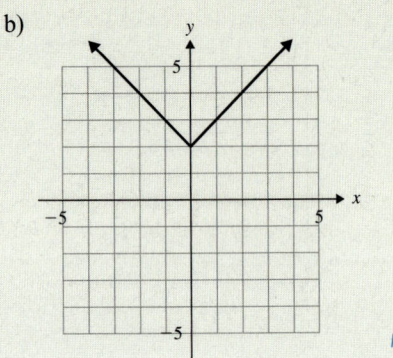

k(x)

c)

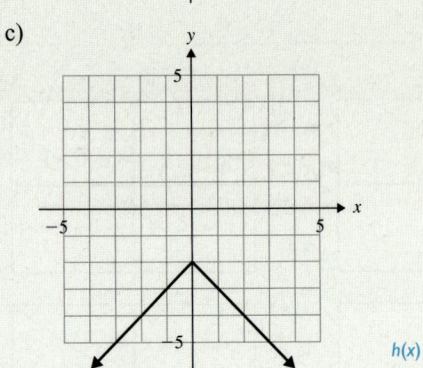

h(x)

d)

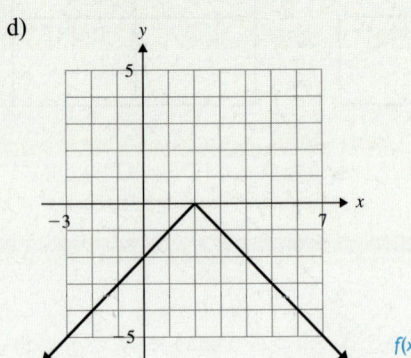

f(x)

If the following transformations are performed on the graph of $f(x)$ to obtain the graph of $g(x)$, write the equation of $g(x)$.

14) $f(x) = |x|$ is shifted left 2 units and down 1 unit.
$g(x) = |x + 2| - 1$

15) $f(x) = |x|$ is shifted right 1 unit and up 4 units.
$g(x) = |x - 1| + 4$

16) $f(x) = |x|$ is reflected about x-axis. $g(x) = -|x|$

Objective 3: Graph a Piecewise Function

Graph the following piecewise functions.

17) $f(x) = \begin{cases} -x - 3, & x \le -1 \\ 2x + 2, & x > -1 \end{cases}$

18) $g(x) = \begin{cases} x - 1, & x \ge 2 \\ -3x + 3, & x < 2 \end{cases}$

19) $h(x) = \begin{cases} -x + 5, & x \ge 3 \\ \dfrac{1}{2}x + 1, & x < 3 \end{cases}$

20) $f(x) = \begin{cases} 2x + 13, & x \le -4 \\ -\dfrac{1}{2}x + 1, & x > -4 \end{cases}$

21) $g(x) = \begin{cases} -\dfrac{3}{2}x - 3, & x < 0 \\ 1, & x \ge 0 \end{cases}$

22) $h(x) = \begin{cases} -\dfrac{2}{3}x - \dfrac{7}{3}, & x \ge -1 \\ 2, & x < -1 \end{cases}$

23) $k(x) = \begin{cases} x + 1, & x \ge -2 \\ 2x + 8, & x < -2 \end{cases}$

24) $g(x) = \begin{cases} x, & x \le 0 \\ 2x + 3, & x > 0 \end{cases}$

25) $f(x) = \begin{cases} 2x - 4, & x > 1 \\ -\dfrac{1}{3}x - \dfrac{5}{3}, & x \le 1 \end{cases}$

26) $k(x) = \begin{cases} \dfrac{1}{2}x + \dfrac{5}{2}, & x < 3 \\ -x + 7, & x \ge 3 \end{cases}$

Objective 4: Define, Graph, and Apply the Greatest Integer Function

Let $f(x) = [\![x]\!]$. Find the following function values.

27) $f\left(3\dfrac{1}{4}\right)$ 3

28) $f\left(10\dfrac{3}{8}\right)$ 10

29) $f(9.2)$ 9

30) $f(7.8)$ 7

31) $f(8)$ 8

32) $f\left(\dfrac{4}{5}\right)$ 0

33) $f\left(-6\dfrac{2}{5}\right)$ −7

34) $f\left(-1\dfrac{3}{4}\right)$ −2

35) $f(-8.1)$ −9

36) $f(-3.6)$ −4

Graph the following greatest integer functions.

37) $f(x) = [\![x]\!] + 1$

38) $g(x) = [\![x]\!] - 2$

39) $h(x) = [\![x]\!] - 4$

40) $k(x) = [\![x]\!] + 3$

41) $g(x) = [\![x + 2]\!]$

42) $h(x) = [\![x - 1]\!]$

43) $k(x) = \left[\!\!\left[\dfrac{1}{2}x\right]\!\!\right]$

44) $f(x) = [\![2x]\!]$

45) To ship small packages within the United States, a shipping company charges \$3.75 for the first pound and \$1.10 for each additional pound or fraction of a pound. Let $C(x)$ represent the cost of shipping a package, and let x represent the weight of the package. Graph $C(x)$ for any package weighing up to (and including) 6 lb.

46) To deliver small packages overnight, an express delivery service charges \$15.40 for the first pound and \$4.50 for each additional pound or fraction of a pound. Let $C(x)$ represent the cost of shipping a package overnight, and let x represent the weight of the package. Graph $C(x)$ for any package weighing up to (and including) 6 lb.

47) Visitors to downtown Hinsdale must pay the parking meters to park their cars. The cost of parking is 5¢ for the first 12 min and 5¢ for each additional 12 min or fraction of this time. Let $P(t)$ represent the cost of parking, and let t represent the number of minutes the car is parked at the meter. Graph $P(t)$ for parking a car for up to (and including) 1 hr.

48) To consult with an attorney costs \$35 for every 10 min or fraction of this time. Let $C(t)$ represent the cost of meeting an attorney, and let t represent the length of the meeting, in minutes. Graph $C(t)$ for meeting with the attorney for up to (and including) 1 hr.

Extension

In earlier chapters, we learned how to graph $f(x) = x^3, f(x) = \sqrt{x}$, and $f(x) = \sqrt[3]{x}$. Use the translation techniques reviewed in this section to graph each function in Exercises 49–56. State the domain and range of each function.

49) a) $f(x) = x^3$ domain: $(-\infty, \infty)$; range: $(-\infty, \infty)$

   b) $f(x) = \sqrt{x}$ domain: $[0, \infty)$; range: $[0, \infty)$

   c) $f(x) = \sqrt[3]{x}$ domain: $(-\infty, \infty)$; range: $(-\infty, \infty)$

50) $g(x) = (x + 2)^3$
domain: $(-\infty, \infty)$; range: $(-\infty, \infty)$

51) $y = \sqrt{x - 1}$
domain: $[1, \infty)$; range: $[0, \infty)$

52) $g(x) = \sqrt[3]{x} + 4$
domain: $(-\infty, \infty)$; range: $(-\infty, \infty)$

53) $h(x) = x^3 - 3$
domain: $(-\infty, \infty)$; range: $(-\infty, \infty)$

54) $h(x) = \sqrt{x - 3}$
domain: $[3, \infty)$; range: $[0, \infty)$

55) $k(x) = \sqrt[3]{x - 2}$
domain: $(-\infty, \infty)$; range: $(-\infty, \infty)$

56) $g(x) = x^3 + 1$ domain: $(-\infty, \infty)$; range: $(-\infty, \infty)$

Mixed Exercises: Objectives 1–3

Graph each function. State the domain and range.

57) $g(x) = |x + 2| + 3$
domain: $(-\infty, \infty)$; range: $[3, \infty)$

58) $h(x) = |x + 1| - 5$
domain: $(-\infty, \infty)$; range: $[-5, \infty)$

59) $k(x) = \dfrac{1}{2}|x|$
domain: $(-\infty, \infty)$; range: $[0, \infty)$

60) $g(x) = 2|x|$
domain: $(-\infty, \infty)$; range: $[0, \infty)$

61) $f(x) = \sqrt{x + 4} - 2$
domain: $[-4, \infty)$; range: $[-2, \infty)$

62) $y = (x - 3)^3 + 1$
domain: $(-\infty, \infty)$; range: $(-\infty, \infty)$

63) $h(x) = -x^3$
domain: $(-\infty, \infty)$; range: $(-\infty, \infty)$

64) $g(x) = -\sqrt[3]{x}$
domain: $(-\infty, \infty)$; range: $(-\infty, \infty)$

If the following transformations are performed on the graph of $f(x)$ to obtain the graph of $g(x)$, write the equation of $g(x)$.

65) $f(x) = \sqrt{x}$ is shifted 5 units to the left. $g(x) = \sqrt{x + 5}$

66) $f(x) = \sqrt[3]{x}$ is shifted down 6 units. $g(x) = \sqrt[3]{x} - 6$

67) $f(x) = x^3$ is shifted left 2 units and down 1 unit.
$g(x) = (x + 2)^3 - 1$

68) $f(x) = \sqrt{x}$ is shifted right 1 unit and up 4 units.
$g(x) = \sqrt{x - 1} + 4$

69) $f(x) = \sqrt[3]{x}$ is reflected about the x-axis. $g(x) = -\sqrt[3]{x}$

70) $f(x) = x^3$ is reflected about the x-axis. $g(x) = -x^3$

Section 11.2 The Circle

Objectives

1. Define a Conic Section
2. Use the Midpoint Formula
3. Graph a Circle Given in the Form $(x - h)^2 + (y - k)^2 = r^2$
4. Graph a Circle of the Form $Ax^2 + Ay^2 + Cx + Dy + E = 0$

1. Define a Conic Section

In this chapter, we will study the *conic sections*. When a right circular cone is intersected by a plane, the result is a **conic section.** The conic sections are parabolas, circles, ellipses, and hyperbolas. The following figures show how each conic section is obtained from the intersection of a cone and a plane.

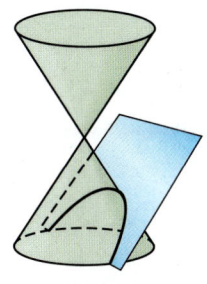
Parabola

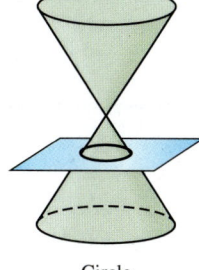
Circle

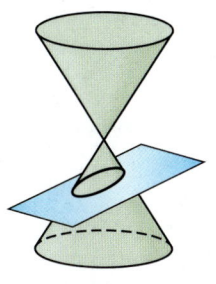
Ellipse

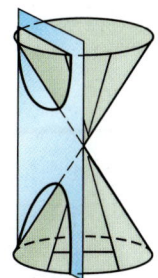
Hyperbola

In Chapter 9 we learned how to graph parabolas. The graph of a quadratic function, $f(x) = ax^2 + bx + c$, is a *parabola* that opens vertically. Another form this function may take is $f(x) = a(x - h)^2 + k$. The graph of a quadratic equation of the form $x = ay^2 + by + c$, or $x = a(y - k)^2 + h$, is a *parabola* that opens horizontally. The next conic section we will discuss is the circle.

We will use the distance formula, presented in Section 9.1, to derive the equation of a circle. But first, let's learn the **midpoint formula.**

2. Use the Midpoint Formula

The **midpoint** of a line segment is the point that is exactly halfway between the endpoints of a line segment. We use the *midpoint formula* to find the midpoint.

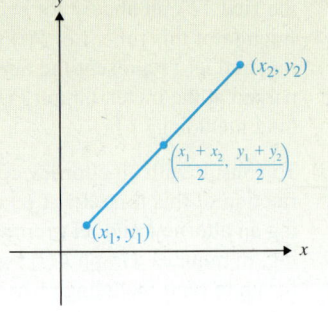

Definition   The Midpoint Formula

If (x_1, y_1) and (x_2, y_2) are the endpoints of a line segment, then the **midpoint** of the segment has coordinates

$$\left(\frac{x_1 + x_2}{2}, \frac{y_1 + y_2}{2} \right)$$

Note

The *x*-coordinate of the midpoint is the *average* of the *x*-coordinates of the endpoints. The *y*-coordinate of the midpoint is the *average* of the *y*-coordinates of the endpoints.

Example 1

Find the midpoint of the line segment with endpoints $(-3, 4)$ and $(1, -2)$.

In-Class Example 1
Find the midpoint of the line segment with endpoints $(-2, -5)$ and $(6, 3)$.
answer: (2, −1)

Solution

Begin by labeling the points: $\overset{(x_1, y_1)}{(-3, 4)}, \overset{(x_2, y_2)}{(1, -2)}$.

Substitute the values into the midpoint formula.

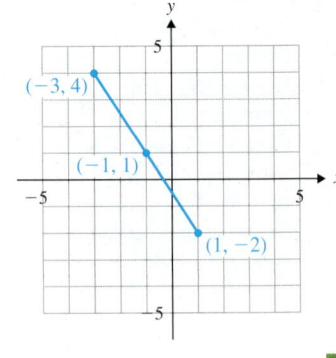

$$\text{Midpoint} = \left(\frac{x_1 + x_2}{2}, \frac{y_1 + y_2}{2} \right)$$

$$= \left(\frac{-3 + 1}{2}, \frac{4 + (-2)}{2} \right) \qquad \text{Substitute values.}$$

$$= \left(\frac{-2}{2}, \frac{2}{2} \right)$$

$$= (-1, 1) \qquad \text{Simplify.}$$

You Try 1

Find the midpoint of the line segment with endpoints $(5, 2)$ and $(1, -3)$.

The midpoint of a diameter of a circle is the **center** of the circle.

3. Graph a Circle Given in the Form $(x - h)^2 + (y - k)^2 = r^2$

A **circle** is defined as the set of all points in a plane equidistant (the same distance) from a fixed point. The fixed point is the **center** of the circle. The distance from the center to a point on the circle is the **radius** of the circle.

Let the center of a circle have coordinates (h, k) and let (x, y) represent any point on the circle. Let r represent the distance between these two points. r is the radius of the circle.

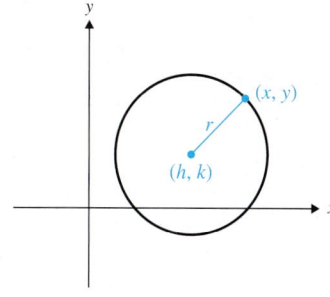

We will use the distance formula to find the distance between the center, (h, k), and the point (x, y) on the circle.

$$d = \sqrt{(x_2 - x_1)^2 + (y_2 - y_1)^2} \qquad \text{Distance formula}$$

Substitute (x, y) for (x_2, y_2), (h, k) for (x_1, y_1), and r for d.

$$r = \sqrt{(x - h)^2 + (y - k)^2}$$
$$r^2 = (x - h)^2 + (y - k)^2 \qquad \text{Square both sides.}$$

This is the **standard form** for the equation of a circle.

Definition

Standard Form for the Equation of a Circle: The standard form for the equation of a circle with center (h, k) and radius r is

$$(x - h)^2 + (y - k)^2 = r^2$$

Example 2

Graph $(x - 2)^2 + (y + 1)^2 = 9$.

In-Class Example 2
Graph
$(x + 1)^2 + (y + 2)^2 = 4$.
answer:

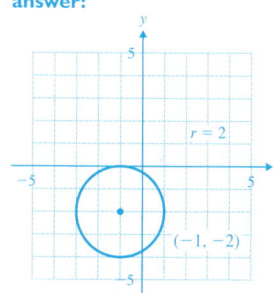

Solution

Standard form is $(x - h)^2 + (y - k)^2 = r^2$.
Our equation is $(x - 2)^2 + (y + 1)^2 = 9$.

$$h = 2 \qquad k = -1 \qquad r = \sqrt{9} = 3$$

The center is $(2, -1)$. The radius is 3.

To graph the circle, first plot the center $(2, -1)$. Use the radius to locate four points on the circle. From the center, move 3 units up, down, left, and right. Draw a circle through the four points.

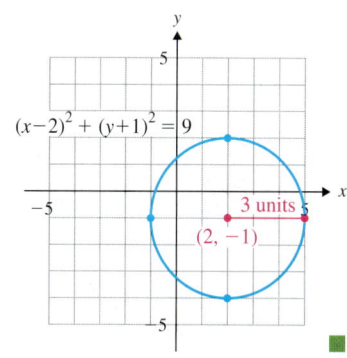

 **You Try 2**

Graph $(x + 3)^2 + (y - 1)^2 = 16$.

Example 3

Graph $x^2 + y^2 = 1$.

In-Class Example 3
Graph $x^2 + y^2 = 25$.
answer:

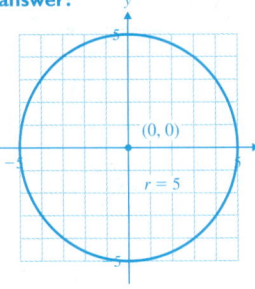

Solution

Standard form is $(x - h)^2 + (y - k)^2 = r^2$.
Our equation is $\quad x^2 \quad + \quad y^2 \quad = 1$.

$$h = 0 \qquad k = 0 \qquad r = \sqrt{1} = 1$$

The center is $(0, 0)$. The radius is 1. Plot $(0, 0)$, then use the radius to locate four points on the circle. From the center, move 1 unit up, down, left, and right. Draw a circle through the four points.

The circle $x^2 + y^2 = 1$ is used often in other areas of mathematics such as trigonometry. $x^2 + y^2 = 1$ is called the **unit circle.**

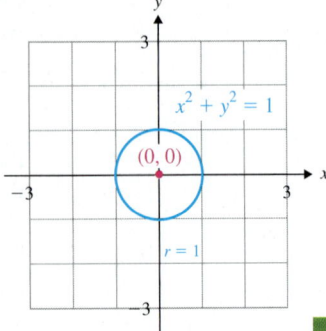

You Try 3

Graph $x^2 + y^2 = 25$.

If we are told the center and radius of a circle, we can write its equation.

Example 4

Find an equation of the circle with center $(-5, 0)$ and radius $\sqrt{7}$.

In-Class Example 4
Find an equation of the circle with center $(3, -2)$ and radius $\sqrt{6}$.
answer:
$(x - 3)^2 + (y + 2)^2 = 6$

Solution

The x-coordinate of the center is h. $\quad h = -5$
The y-coordinate of the center is k. $\quad k = 0$

$$r = \sqrt{7}$$

Substitute these values into $(x - h)^2 + (y - k)^2 = r^2$.

$$[x - (-5)]^2 + (y - 0)^2 = (\sqrt{7})^2 \qquad \text{Substitute } -5 \text{ for } x, 0 \text{ for } k, \text{ and } \sqrt{7} \text{ for } r.$$
$$(x + 5)^2 + y^2 = 7$$

You Try 4

Find an equation of the circle with center $(4, 7)$ and radius 5.

4. Graph a Circle of the Form $Ax^2 + Ay^2 + Cx + Dy + E = 0$

The equation of a circle can take another form—general form.

Definition

General Form for the Equation of a Circle: An equation of the form $Ax^2 + Ay^2 + Cx + Dy + E = 0$, where $A, C, D,$ and E are real numbers, is the **general form** for the equation of a circle.

The coefficients of x^2 and y^2 must be the same in order for this to be the equation of a circle.

To graph a circle given in this form, we complete the square on x and on y to put it into standard form.

After we learn *all* of the conic sections, it is very important that we understand how to identify each one. To do this we will usually look at the coefficients of the square terms.

Example 5

Graph $x^2 + y^2 + 6x + 2y + 6 = 0$.

In-Class Example 5
Graph
$x^2 + y^2 + 8x + 4y + 13 = 0$.
answer:

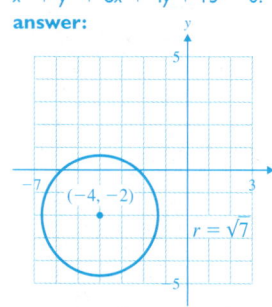

Solution

The coefficients of x^2 and y^2 are each 1. Therefore, this is the equation of a circle.

Our goal is to write the given equation in standard form, $(x - h)^2 + (y - k)^2 = r^2$, so that we can identify its center and radius. To do this we will group x^2 and $6x$ together, group y^2 and $2y$ together, then complete the square on each group of terms.

$$x^2 + y^2 + 6x + 2y + 6 = 0 \qquad \text{Group } x^2 \text{ and } 6x \text{ together.}$$
$$(x^2 + 6x) + (y^2 + 2y) = -6 \qquad \text{Group } y^2 \text{ and } 2y \text{ together.}$$
Move the constant to the other side.

Complete the square for each group of terms.

$$(x^2 + 6x + 9) + (y^2 + 2y + 1) = -6 + 9 + 1$$
$$(x + 3)^2 + (y + 1)^2 = 4$$

Since 9 and 1 are added on the left, they must also be added on the right. Factor; add.

The center of the circle is $(-3, -1)$. The radius is 2.

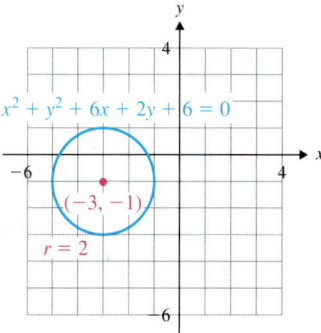

You Try 5

Graph $x^2 + y^2 + 10x - 4y + 20 = 0$.

Note

If we rewrite $Ax^2 + Ay^2 + Cx + Dy + E = 0$ in standard form and get $(x - h)^2 + (y - k)^2 = 0$, then the graph is just the point (h, k). If the constant on the right side of the standard form equation is a negative number then the equation has no graph.

Using Technology

Recall that the equation of a circle is not a function. However, if we want to graph an equation on a graphing calculator, it must be entered as a function or a pair of functions. Therefore, to graph a circle we must solve the equation for y in terms of x.

Let's discuss how to graph $x^2 + y^2 = 4$ on a graphing calculator.

We must solve the equation for y.

$$x^2 + y^2 = 4$$
$$y^2 = 4 - x^2$$
$$y = \pm\sqrt{4 - x^2}$$

Now the equation of the circle $x^2 + y^2 = 4$ is rewritten so that y is in terms of x. In the graphing calculator, enter $y = \sqrt{4 - x^2}$ as Y_1. This represents the top half of the circle since the y-values are positive above the x-axis. Enter $y = -\sqrt{4 - x^2}$ as Y_2. This represents the bottom half of the circle since the y-values are negative below the x-axis. Here we have the window set from -3 to 3 in both the x- and y-directions. Press $\boxed{\text{GRAPH}}$.

The graph is distorted and does not actually look like a circle! This is because the screen is rectangular, and the graph is longer in the x-direction. We can "fix" this by squaring the window.

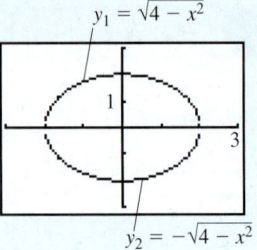

To square the window and get a better representation of the graph of $x^2 + y^2 = 4$, press $\boxed{\text{ZOOM}}$ and choose 5:ZSquare. The graph reappears on a "squared" window and now looks like a circle.

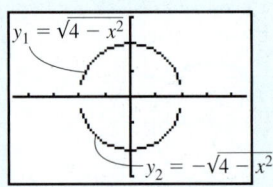

Identify the center and radius of each circle. Then, rewrite each equation for y in terms of x, and graph each circle on a graphing calculator. These problems come from the homework exercises.

1) $x^2 + y^2 = 36$; Exercise 23

2) $x^2 + y^2 = 9$; Exercise 25

3) $(x + 3)^2 + y^2 = 4$; Exercise 19

4) $x^2 + (y - 1)^2 = 25$; Exercise 27

Answers to You Try Exercises

1) $\left(3, -\dfrac{1}{2}\right)$ 2)

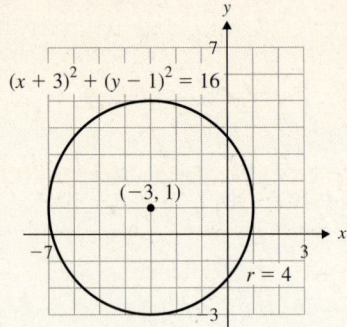

3)

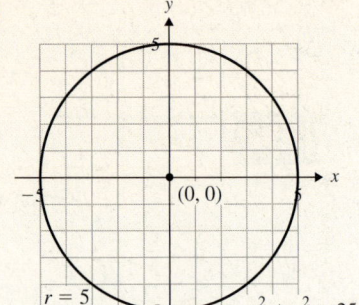

4) $(x - 4)^2 + (y - 7)^2 = 25$ 5)

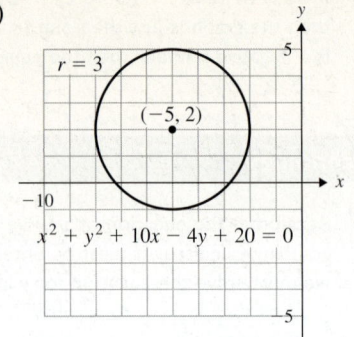

Answers to Technology Exercises

1) Center $(0, 0)$; radius = 6; $Y_1 = \sqrt{36 - x^2}$, $Y_2 = -\sqrt{36 - x^2}$
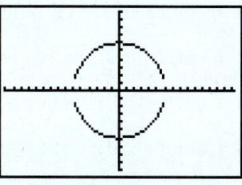

2) Center $(0, 0)$; radius = 3; $Y_1 = \sqrt{9 - x^2}$, $Y_2 = -\sqrt{9 - x^2}$
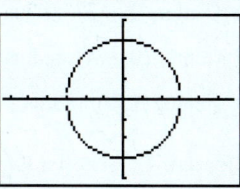

3) Center $(-3, 0)$; radius = 2; $Y_1 = \sqrt{4 - (x + 3)^2}$,
   $Y_2 = -\sqrt{4 - (x + 3)^2}$
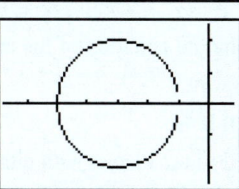

4) Center $(0, 1)$; radius = 5; $Y_1 = 1 + \sqrt{25 - x^2}$, $Y_2 = 1 - \sqrt{25 - x^2}$
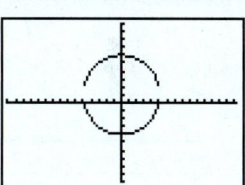

11.2 Exercises

Additional answers can be found in the Answers to Exercises appendix.

Objective 2: Use the Midpoint Formula

1) $(1, 3)$ and $(7, 9)$ $(4, 6)$

2) $(2, 10)$ and $(8, 4)$ $(5, 7)$

3) $(-5, 2)$ and $(-1, -8)$ $(-3, -3)$

4) $(6, -3)$ and $(0, 5)$ $(3, 1)$

5) $(-3, -7)$ and $(1, -2)$

6) $(-1, 3)$ and $(2, -9)$

7) $(4, 0)$ and $(-3, -5)$ $\left(\frac{1}{2}, -\frac{5}{2}\right)$

8) $(-2, 4)$ and $(9, 3)$ $\left(\frac{7}{2}, \frac{7}{2}\right)$

9) $\left(\frac{3}{2}, -1\right)$ and $\left(\frac{5}{2}, \frac{7}{2}\right)$ $\left(2, \frac{5}{4}\right)$

10) $\left(\frac{9}{2}, \frac{3}{2}\right)$ and $\left(-\frac{7}{2}, -5\right)$ $\left(\frac{1}{2}, -\frac{7}{4}\right)$

11) $(-6.2, 1.5)$ and $(4.8, 5.7)$ $(-0.7, 3.6)$

12) $(-3.7, -1.8)$ and $(3.7, -3.6)$ $(0, -2.7)$

Objective 3: Graph a Circle Given in the Form $(x - h)^2 + (y - k)^2 = r^2$

13) Is the equation of a circle a function? Explain your answer.

14) The standard form for the equation of a circle is
$$(x - h)^2 + (y - k)^2 = r^2$$
Identify the center and the radius.   center: (h, k); radius: r

Identify the center and radius of each circle and graph.

VIDEO 15) $(x + 2)^2 + (y - 4)^2 = 9$

16) $(x + 1)^2 + (y + 3)^2 = 25$

17) $(x - 5)^2 + (y - 3)^2 = 1$

18) $x^2 + (y - 5)^2 = 9$

19) $(x + 3)^2 + y^2 = 4$

20) $(x - 2)^2 + (y - 2)^2 = 36$

21) $(x - 6)^2 + (y + 3)^2 = 16$

22) $(x + 8)^2 + (y - 4)^2 = 4$

23) $x^2 + y^2 = 36$

24) $x^2 + y^2 = 16$

25) $x^2 + y^2 = 9$

26) $x^2 + y^2 = 25$

27) $x^2 + (y - 1)^2 = 25$

28) $(x + 3)^2 + y^2 = 1$

Find an equation of the circle with the given center and radius.

VIDEO 29) Center $(4, 1)$; radius = 5 $(x - 4)^2 + (y - 1)^2 = 25$

30) Center $(3, 5)$; radius = 2 $(x - 3)^2 + (y - 5)^2 = 4$

31) Center $(-3, 2)$; radius = 1 $(x + 3)^2 + (y - 2)^2 = 1$

32) Center $(4, -6)$; radius = 3 $(x - 4)^2 + (y + 6)^2 = 9$

33) Center $(-1, -5)$; radius $= \sqrt{3}$ $(x + 1)^2 + (y + 5)^2 = 3$

34) Center $(-2, -1)$; radius $= \sqrt{5}$ $(x + 2)^2 + (y + 1)^2 = 5$

35) Center $(0, 0)$; radius $= \sqrt{10}$ $x^2 + y^2 = 10$

36) Center $(0, 0)$; radius $= \sqrt{6}$ $x^2 + y^2 = 6$

37) Center $(6, 0)$; radius $= 4$ $(x - 6)^2 + y^2 = 16$

38) Center $(0, -3)$; radius $= 5$ $x^2 + (y + 3)^2 = 25$

39) Center $(0, -4)$; radius $= 2\sqrt{2}$ $x^2 + (y + 4)^2 = 8$

40) Center $(1, 0)$; radius $= 3\sqrt{2}$ $(x - 1)^2 + y^2 = 18$

Objective 4: Graph a Circle of the Form $Ax^2 + Ay^2 + Cx + Dy + E = 0$

Write the equation of the circle in standard form.

Fill It In

Fill in the blanks with either the missing mathematical step or reason for the given step.

41) $x^2 + y^2 - 8x + 2y + 8 = 0$

$(x^2 - 8x) + (y^2 + 2y) = -8$ Group x- and y-terms separately.

$(x^2 - 8x + 16) + (y^2 + 2y + 1)$

$\qquad = -8 + 16 + 1$ Complete the square.

$(x - 4)^2 + (y + 1)^2 = 9$ Factor.

42) $x^2 + y^2 + 2x + 10y + 10 = 0$

$(x^2 + 2x) + (y^2 + 10y) = -10$ Group x- and y-terms separately.

$(x^2 + 2x + 1) + (y^2 + 10y + 25)$

$\qquad = -10 + 1 + 25$ Complete the square.

$(x + 1)^2 + (y + 5)^2 = 16$ Factor.

Put the equation of each circle in the form $(x - h)^2 + (y - k)^2 = r^2$, identify the center and the radius, and graph.

[VIDEO] 43) $x^2 + y^2 + 2x + 10y + 17 = 0$

44) $x^2 + y^2 - 4x - 6y + 9 = 0$

45) $x^2 + y^2 + 8x - 2y - 8 = 0$

46) $x^2 + y^2 - 6x + 8y + 24 = 0$

47) $x^2 + y^2 - 10x - 14y + 73 = 0$

48) $x^2 + y^2 + 12x + 12y + 63 = 0$

49) $x^2 + y^2 + 6y + 5 = 0$

50) $x^2 + y^2 + 2x - 24 = 0$

51) $x^2 + y^2 - 4x - 1 = 0$

52) $x^2 + y^2 - 10y + 22 = 0$

53) $x^2 + y^2 - 8x + 8y - 4 = 0$

54) $x^2 + y^2 - 6x + 2y - 6 = 0$

[VIDEO] 55) $4x^2 + 4y^2 - 12x - 4y - 6 = 0$
(Hint: Begin by dividing the equation by 4.)

56) $16x^2 + 16y^2 + 16x - 24y - 3 = 0$
(Hint: Begin by dividing the equation by 16.)

Mixed Exercises: Objectives 3 and 4

57) The London Eye is a Ferris wheel that opened in London in March 2000. It is 135 m high, and the bottom of the wheel is 7 m off the ground.

135 m

7 m

a) What is the diameter of the wheel?   128 m

b) What is the radius of the wheel?   64 m

c) Using the axes in the illustration, what are the coordinates of the center of the wheel? $(0, 71)$

d) Write the equation of the wheel. $x^2 + (y - 71)^2 = 4096$
(www.aviewoncities.com/london/londoneye.htm)

58) The first Ferris wheel was designed and built by George W. Ferris in 1893 for the Chicago World's Fair. It was 264 ft tall, and the wheel had a diameter of 250 ft.

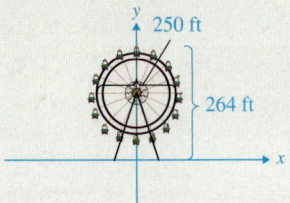

250 ft

264 ft

a) What is the radius of the wheel?   125 ft

b) Using the axes in the illustration, what are the coordinates of the center of the wheel? $(0, 139)$

c) Write the equation of the wheel. $x^2 + (y - 139)^2 = 15,625$

59) A CD is placed on axes as shown in the figure where the units of measurement for x and y are millimeters. Using $\pi = 3.14$, what is the surface area of a CD (to the nearest square millimeter)?   11,127 mm²

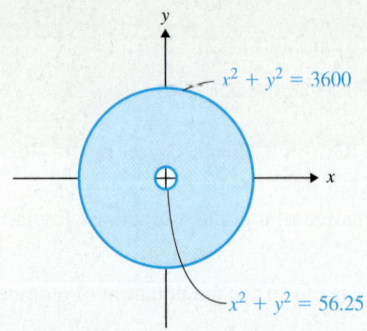

$x^2 + y^2 = 3600$

$x^2 + y^2 = 56.25$

60) A storage container is in the shape of a right circular cylinder. The top of the container may be described by the equation $x^2 + y^2 = 5.76$, as shown in the figure (x and y are in feet). If the container is 3.2 ft tall, what is the storage capacity of the container (to the nearest ft³)?

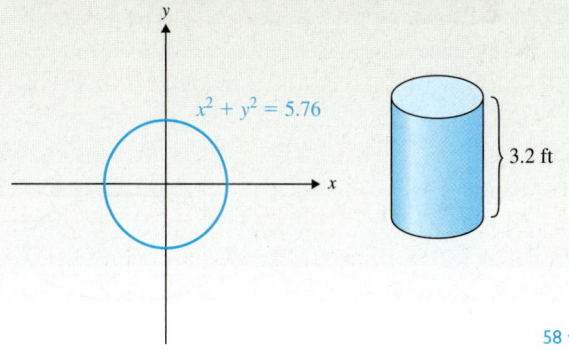

$x^2 + y^2 = 5.76$

3.2 ft

58 ft³

Use a graphing calculator to graph each equation. Square the viewing window.

61) $(x - 5)^2 + (y - 3)^2 = 1$

62) $x^2 + y^2 = 16$

63) $(x + 2)^2 + (y - 4)^2 = 9$

64) $(x + 3)^2 + y^2 = 1$

Section 11.3 The Ellipse

Objectives
1. Graph an Ellipse

1. Graph an Ellipse

The next conic section we will study is the *ellipse*. An **ellipse** is the set of all points in a plane such that the *sum* of the distances from a point on the ellipse to two fixed points is constant. Each fixed point is called a **focus** (plural: **foci**). The point halfway between the foci is the **center** of the ellipse.

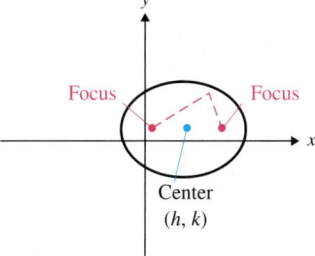

Focus   Focus

Center
(h, k)

The orbits of planets around the sun as well as satellites around the Earth are elliptical. Statuary Hall in the U.S. Capitol building is an ellipse. If a person stands at one focus of this ellipse and whispers, a person standing across the room on the other focus can clearly hear what was said. Properties of the ellipse are used in medicine as well. One procedure for treating kidney stones involves immersing the patient in an elliptical tub of water. The kidney stone is at one focus, while at the other focus, high energy shock waves are produced, which destroy the kidney stone.

Definition
Standard Form for the Equation of an Ellipse: The standard form for the equation of an ellipse is

$$\frac{(x - h)^2}{a^2} + \frac{(y - k)^2}{b^2} = 1$$

The center of the ellipse is (h, k).

It is important to remember that the terms on the left are *both* positive quantities.

Example 1

Graph $\dfrac{(x-3)^2}{16} + \dfrac{(y-1)^2}{4} = 1$.

In-Class Example 1
Graph
$\dfrac{(x+4)^2}{25} + \dfrac{(y-1)^2}{9} = 1$.

answer:

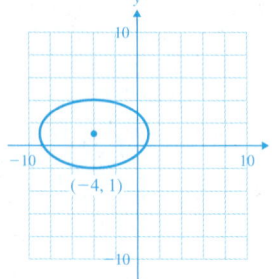

Solution

Standard form is $\dfrac{(x-h)^2}{a^2} + \dfrac{(y-k)^2}{b^2} = 1$.

Our equation is $\dfrac{(x-3)^2}{16} + \dfrac{(y-1)^2}{4} = 1$.

$$h = 3 \qquad k = 1$$
$$a = \sqrt{16} = 4 \qquad b = \sqrt{4} = 2$$

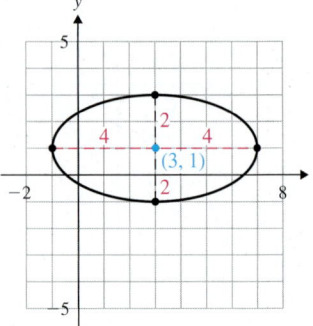

The center is $(3, 1)$.

To graph the ellipse, first plot the center $(3, 1)$. Since $a = 4$ and a^2 is under the squared quantity containing the **x**, move 4 units each way in the x-direction from the center. These are two points on the ellipse.

Since $b = 2$ and b^2 is under the squared quantity containing the **y**, move 2 units each way in the y-direction from the center. These are two more points on the ellipse. Sketch the ellipse through the four points.   ■

You Try 1

Graph $\dfrac{(x+2)^2}{25} + \dfrac{(y-3)^2}{16} = 1$.

Example 2

Graph $\dfrac{x^2}{9} + \dfrac{y^2}{25} = 1$.

In-Class Example 2
Graph $\dfrac{x^2}{16} + \dfrac{y^2}{4} = 1$.

answer:

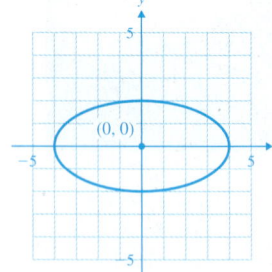

Solution

Standard form is $\dfrac{(x-h)^2}{a^2} + \dfrac{(y-k)^2}{b^2} = 1$.

Our equation is $\dfrac{x^2}{9} + \dfrac{y^2}{25} = 1$.

$$h = 0 \qquad k = 0$$
$$a = \sqrt{9} = 3 \qquad b = \sqrt{25} = 5$$

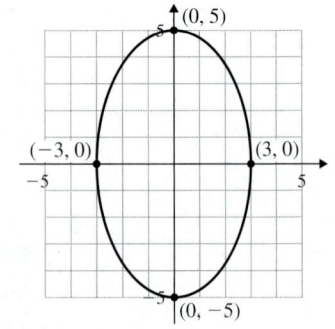

The center is $(0, 0)$.

Plot the center $(0, 0)$. Since $a = 3$ and a^2 is under the x^2, move 3 units each way in the x-direction from the center. These are two points on the ellipse.

Since $b = 5$ and b^2 is under the y^2, move 5 units each way in the y-direction from the center. These are two more points on the ellipse. Sketch the ellipse through the four points.   ■

You Try 2

Graph $\dfrac{x^2}{36} + \dfrac{y^2}{9} = 1$.

In Example 2, note that the *origin*, (0, 0), is the center of the ellipse. Notice that $a = 3$ and the x-intercepts are (3, 0) and (−3, 0); $b = 5$ and the y-intercepts are (0, 5) and (0, −5). We can generalize these relationships as follows.

Definition

Equation of an Ellipse with Center at Origin: The graph of $\dfrac{x^2}{a^2} + \dfrac{y^2}{b^2} = 1$ is an ellipse with center at the origin, x-intercepts (a, 0) and (−a, 0), and y-intercepts (0, b) and (0, −b).

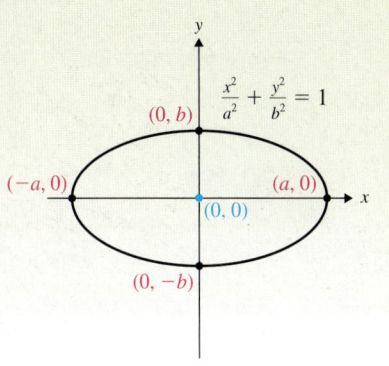

Looking at Examples 1 and 2 we can make another interesting observation.

Example 1	**Example 2**
$\dfrac{(x-3)^2}{16} + \dfrac{(y-1)^2}{4} = 1$	$\dfrac{x^2}{9} + \dfrac{y^2}{25} = 1$
$a^2 = 16 \quad b^2 = 4$	$a^2 = 9 \quad b^2 = 25$
$a^2 > b^2$	$b^2 > a^2$

The number under $(x-3)^2$ is greater than the number under $(y-1)^2$. The ellipse is longer in the x-direction.

The number under y^2 is greater than the number under x^2. The ellipse is longer in the y-direction.

This relationship between a^2 and b^2 will always produce the same result. The equation of an ellipse can take other forms.

Example 3

Graph $4x^2 + 25y^2 = 100$.

Solution

How can we tell if this is a circle or an ellipse? We look at the coefficients of x^2 and y^2. Both of the coefficients are positive, *and* they are different. *This is an ellipse.* (If this were a circle, the coefficients would be the same.)

Since the standard form for the equation of an ellipse has a 1 on one side of the = sign, divide both sides of $4x^2 + 25y^2 = 100$ by 100 to obtain a 1 on the right.

$$4x^2 + 25y^2 = 100$$
$$\frac{4x^2}{100} + \frac{25y^2}{100} = \frac{100}{100} \qquad \text{Divide both sides by 100.}$$
$$\frac{x^2}{25} + \frac{y^2}{4} = 1 \qquad \text{Simplify.}$$

In-Class Example 3
Graph $9x^2 + 16y^2 = 144$.
answer:

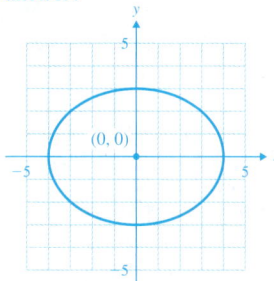

The center is $(0, 0)$. $a = \sqrt{25} = 5$ and $b = \sqrt{4} = 2$. Plot $(0, 0)$. Move 5 units each way from the center in the x-direction. Move 2 units each way from the center in the y-direction.

Notice that the x-intercepts are $(5, 0)$ and $(-5, 0)$. The y-intercepts are $(2, 0)$ and $(-2, 0)$.

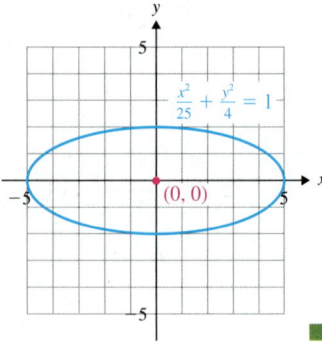

You Try 3

Graph $x^2 + 4y^2 = 4$.

Note

You may have noticed that if $a^2 = b^2$, then the ellipse is a circle.

Just like with the equation of a circle, sometimes we must complete the square to put the equation of an ellipse into standard form.

Example 4

Graph $9x^2 + 4y^2 + 18x - 16y - 11 = 0$.

In-Class Example 4
Graph
$4x^2 + 9y^2 - 8x + 72y + 112 = 0$.
answer:
$\dfrac{(x - 1)^2}{9} + \dfrac{(y + 4)^2}{4} = 1$

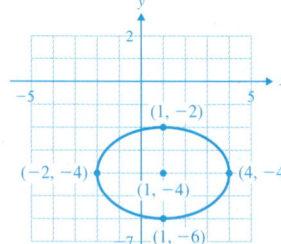

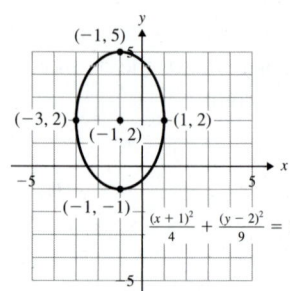

Solution

Our goal is to write the given equation in standard form, $\dfrac{(x - h)^2}{a^2} + \dfrac{(y - k)^2}{b^2} = 1$. To do this we will group the x-terms together, group the y-terms together, then complete the square on each group of terms.

$$9x^2 + 4y^2 + 18x - 16y - 11 = 0 \qquad \text{Group the x-terms together and the y-terms together.}$$
$$(9x^2 + 18x) + (4y^2 - 16y) = 11 \qquad \text{Move the constant to the other side.}$$
$$9(x^2 + 2x) + 4(y^2 - 4y) = 0 \qquad \text{Factor out the coefficients of the squared terms.}$$

Complete the square for each group of terms.

$$9 \cdot 1 = 9 \qquad\qquad 4 \cdot 4 = 16$$
$$9(x^2 + 2x + 1) + 4(y^2 - 4y + 4) = 11 + 9 + 16 \qquad \text{Since 9 and 16 are added on the left, they must also be added on the right.}$$
$$9(x + 1)^2 + 4(y - 2)^2 = 36 \qquad \text{Factor; add.}$$
$$\frac{9(x + 1)^2}{36} + \frac{4(y - 2)^2}{36} = \frac{36}{36} \qquad \text{Divide by 36 to get 1 on the right.}$$
$$\frac{(x + 1)^2}{4} + \frac{(y - 2)^2}{9} = 1 \qquad \text{Standard form}$$

The center of the ellipse is $(-1, 2)$. The graph is at left.

You Try 4

Graph $4x^2 + 25y^2 - 24x - 50y - 39 = 0$.

Using Technology

We graph ellipses on a graphing calculator in the same way that we graphed circles: solve the equation for y in terms of x, enter both values of y, and graph both equations.

Let's graph the ellipse given in Example 3: $4x^2 + 25y^2 = 100$.

Solve for y.

$$25y^2 = 100 - 4x^2$$

$$y^2 = 4 - \frac{4x^2}{25}$$

$$y = \pm\sqrt{4 - \frac{4x^2}{25}}$$

Enter $Y_1 = \sqrt{4 - \frac{4x^2}{25}}$, the top half of the ellipse (the y-values are positive above the x-axis). Enter $Y_2 = -\sqrt{4 - \frac{4x^2}{25}}$, the bottom half of the ellipse (the y-values are negative below the x-axis). Set an appropriate window, and press $\boxed{\text{GRAPH}}$.

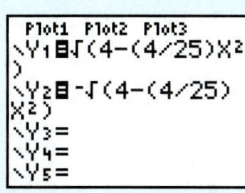
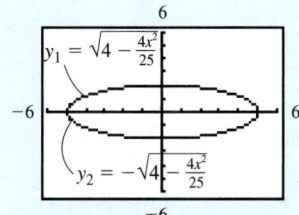

Identify the center of each ellipse, rewrite each equation for y in terms of x, and graph each equation on a graphing calculator.

1) $4x^2 + 9y^2 = 36$ 2) $x^2 + \dfrac{y^2}{4} = 1$ 3) $\dfrac{x^2}{25} + (y + 4)^2 = 1$ 4) $25x^2 + y^2 = 25$

Answers to You Try Exercises

1)

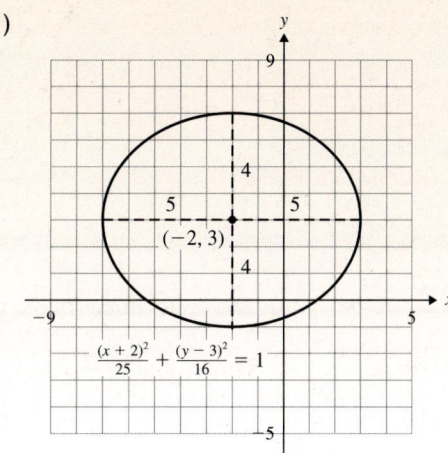

2)

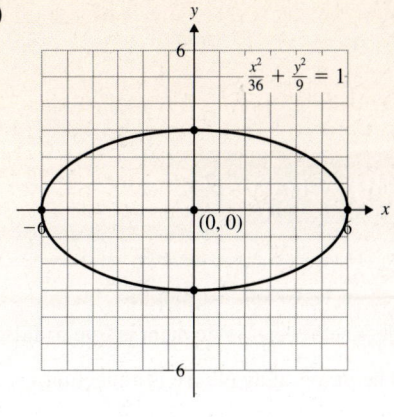

3)

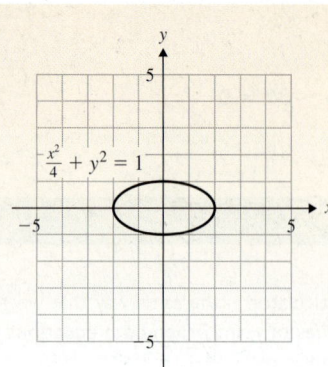

4)

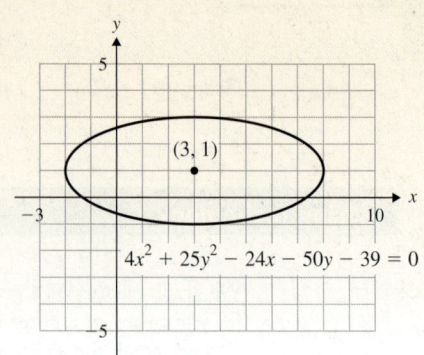

Answers to Technology Exercises

1) Center $(0, 0)$; $Y_1 = \sqrt{4 - \dfrac{4x^2}{9}}$; $Y_2 = -\sqrt{4 - \dfrac{4x^2}{9}}$

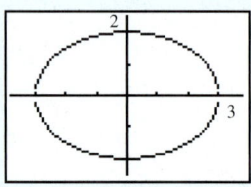

2) Center $(0, 0)$; $Y_1 = \sqrt{4 - 4x^2}$; $Y_2 = -\sqrt{4 - 4x^2}$

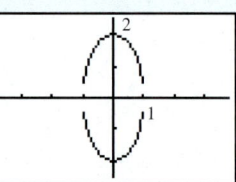

3) Center $(0, -4)$; $Y_1 = -4 + \sqrt{1 - \dfrac{x^2}{25}}$; $Y_2 = -4 - \sqrt{1 - \dfrac{x^2}{25}}$

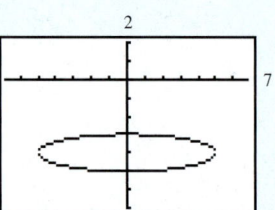

4) Center $(0, 0)$; $Y_1 = \sqrt{25 - 25x^2}$; $Y_2 = -\sqrt{25 - 25x^2}$

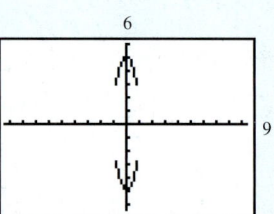

11.3 Exercises

*Additional answers can be found in the Answers to Exercises appendix.

Objective 1: Graph an Ellipse

Decide whether each statement is true or false.

1) The graph of an ellipse is a function.   false

2) The center of the ellipse with equation
$$\frac{(x - 1)^2}{16} + \frac{(y - 5)^2}{9} = 1 \text{ is } (-1, -5). \quad \text{false}$$

3) The center of the ellipse with equation
$$\frac{(x - 8)^2}{9} + \frac{(y + 3)^2}{25} = 1 \text{ is } (8, -3). \quad \text{true}$$

4) The center of the ellipse with equation
$$\frac{x^2}{7} + \frac{y^2}{4} = 1 \text{ is } (0, 0). \quad \text{true}$$

5) The graph of $4x^2 + 25y^2 = 100$ is an ellipse.   true

6) The graph of $\dfrac{(x+3)^2}{4} - \dfrac{(y+2)^2}{9} = 1$ is an ellipse.   false

7) The graph of $9y^2 - x^2 = 9$ is an ellipse.   false

8) The equation of $4x^2 + y^2 + 8x - 10y + 8 = 0$ can be put into the standard form for the equation of an ellipse by completing the square.   true

Identify the center of each ellipse and graph the equation.

9) $\dfrac{(x+2)^2}{9} + \dfrac{(y-1)^2}{4} = 1$

10) $\dfrac{(x-4)^2}{4} + \dfrac{(y-3)^2}{16} = 1$

11) $\dfrac{(x-3)^2}{9} + \dfrac{(y+2)^2}{16} = 1$

12) $\dfrac{(x+4)^2}{25} + \dfrac{(y-5)^2}{16} = 1$ 13) $\dfrac{x^2}{36} + \dfrac{y^2}{16} = 1$

14) $\dfrac{x^2}{36} + \dfrac{y^2}{4} = 1$ 15) $x^2 + \dfrac{y^2}{4} = 1$

16) $\dfrac{x^2}{9} + y^2 = 1$ 17) $\dfrac{x^2}{25} + (y+4)^2 = 1$

18) $(x+3)^2 + \dfrac{(y+4)^2}{9} = 1$

19) $\dfrac{(x+1)^2}{4} + \dfrac{(y+3)^2}{9} = 1$

20) $\dfrac{(x-2)^2}{16} + \dfrac{y^2}{25} = 1$

21) $4x^2 + 9y^2 = 36$ 22) $x^2 + 4y^2 = 16$

23) $25x^2 + y^2 = 25$ 24) $9x^2 + y^2 = 36$

Write the equation of the ellipse in standard form.

Fill It In

Fill in the blanks with either the missing mathematical step or the reason for the given step.

25) $3x^2 + 2y^2 - 6x + 4y - 7 = 0$

$(3x^2 - 6x) + (2y^2 + 4y) = 7$	Group x- and y-terms separately.
$3(x^2 - 2x) + 2(y^2 + 2y) = 7$	Factor out the coefficients of the squared terms.
$3(x^2 - 2x + 1) + 2(y^2 + 2y + 1)$ $= 7 + 3(1) + 2(1)$	Complete the square.
$3(x-1)^2 + 2(y+1)^2 = 12$	Factor.
$\dfrac{(x-1)^2}{4} + \dfrac{(y+1)^2}{6} = 1$	Divide both sides by 12.

26) $4x^2 + 9y^2 + 16x + 54y + 61 = 0$

$(4x^2 + 16x) + (9y^2 + 54y) = -61$	Group x- and y-terms separately.
$4(x^2 + 4x) + 9(y^2 + 6y) = -61$	Factor out the coefficients of the squared terms.
$4(x^2 + 4x + 4) + 9(y^2 + 6y + 9)$ $= -61 + 4(4) + 9(9)$	Complete the square.
$4(x+2)^2 + 9(y+3)^2 = 36$	Factor.
$\dfrac{(x+2)^2}{9} + \dfrac{(y+3)^2}{4} = 1$	Divide both sides by 36.

Put each equation into the standard form for the equation of an ellipse, and graph.

27) $x^2 + 4y^2 - 2x - 24y + 21 = 0$

28) $9x^2 + 4y^2 + 36x - 8y + 4 = 0$

29) $9x^2 + y^2 + 72x + 2y + 136 = 0$

30) $x^2 + 4y^2 - 6x - 40y + 105 = 0$

31) $4x^2 + 9y^2 - 16x - 54y + 61 = 0$

32) $4x^2 + 25y^2 + 8x + 200y + 304 = 0$

33) $25x^2 + 4y^2 + 150x + 125 = 0$

34) $4x^2 + y^2 - 2y - 15 = 0$

Extension

Write an equation of the ellipse containing the following points.

35) $(-3, 0), (3, 0), (0, -5),$ and $(0, 5)$ $\dfrac{x^2}{9} + \dfrac{y^2}{25} = 1$

36) $(-6, 0), (6, 0), (0, -2),$ and $(0, 2)$ $\dfrac{x^2}{36} + \dfrac{y^2}{4} = 1$

37) $(-7, 0), (7, 0), (0, -1),$ and $(0, 1)$ $\dfrac{x^2}{49} + y^2 = 1$

38) $(-1, 0), (1, 0), (0, -9),$ and $(0, 9)$ $x^2 + \dfrac{y^2}{81} = 1$

39) $(3, 5), (3, -3), (1, 1),$ and $(5, 1)$ $\dfrac{(x-3)^2}{4} + \dfrac{(y-1)^2}{16} = 1$

40) $(-1, 1), (-1, -5), (4, -2),$ and $(-6, -2)$ $\dfrac{(x+1)^2}{25} + \dfrac{(y+2)^2}{9} = 1$

41) Is a circle a special type of ellipse? Explain your answer.

42) The Oval Office in the White House is an ellipse about 36 ft long and 29 ft wide. If the center of the room is at the origin of a Cartesian coordinate system and the length of the room is along the x-axis, write an equation of the elliptical room. (www.whitehousehistory.org) $\dfrac{x^2}{324} + \dfrac{y^2}{210.25} = 1$

43) The fuselage of a Boeing 767 jet has an elliptical cross section that is 198 in. wide and 213 in. tall. If the center of this cross section is at the origin and the width is along the x-axis, write an equation of this ellipse. (Jan Roskam, *Airplane Design*, p. 89). $\dfrac{x^2}{9801} + \dfrac{y^2}{11{,}342.25} = 1$

44) The arch of a bridge over a canal in Amsterdam is half of an ellipse. At water level the arch is 14 ft wide, and it is 6 ft tall at its highest point.

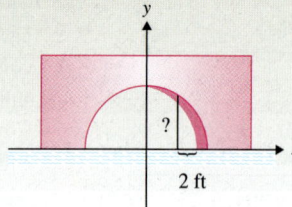

2 ft

a) Write an equation of the arch. $y = \sqrt{36 - \dfrac{36x^2}{49}}$

b) What is the height of the arch (to the nearest foot) 2 ft from the bottom edge?   4 ft

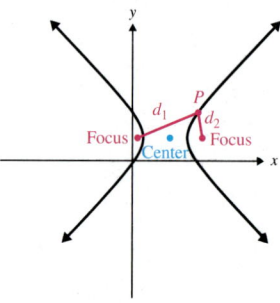 Use a graphing calculator to graph the following. Square the viewing window.

45) $\dfrac{x^2}{36} + \dfrac{y^2}{16} = 1$

46) $9x^2 + y^2 = 36$

Section 11.4 The Hyperbola

Objectives

1. **Graph a Hyperbola in Standard Form**
2. **Graph a Hyperbola in Nonstandard Form**
3. **Graph Other Square Root Functions**

1. Graph a Hyperbola in Standard Form

The last of the conic sections is the *hyperbola*. A **hyperbola** is the set of all points, P, in a plane such that the absolute value of the *difference* of the distances $|d_1 - d_2|$, from two fixed points is constant. Each fixed point is called a **focus**. The point halfway between the foci is the **center** of the hyperbola.

Some navigation systems used by ships are based on the properties of hyperbolas. A lamp casts a hyperbolic shadow on a wall, and many telescopes use hyperbolic lenses.

A hyperbola is a graph consisting of two branches.

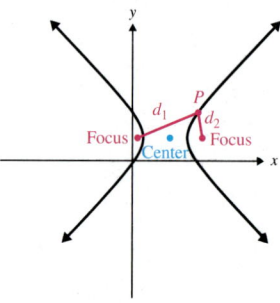

Definition   Standard Form for the Equation of a Hyperbola

1) A hyperbola with center (h, k) and branches that open in the *x-direction* has equation

$$\frac{(x - h)^2}{a^2} - \frac{(y - k)^2}{b^2} = 1.$$

Its graph is to the right.

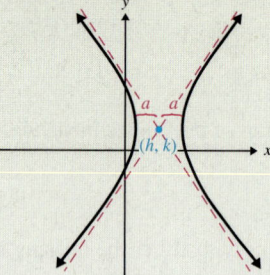

2) A hyperbola with center (h, k) and branches that open in the *y-direction* has equation

$$\frac{(y - k)^2}{b^2} - \frac{(x - h)^2}{a^2} = 1.$$

Its graph is to the right.

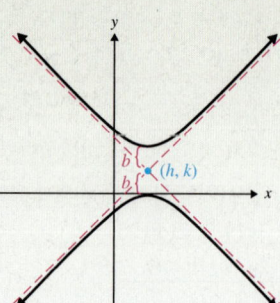

Notice in 1) that $\dfrac{(x - h)^2}{a^2}$ is the positive quantity, and the branches open in the *x-direction*.

In 2), the positive quantity is $\dfrac{(y - k)^2}{b^2}$, and the branches open in the *y-direction*.

In 1) and 2) notice how the branches of the hyperbola get closer to the dotted lines as the branches continue indefinitely. These dotted lines are called **asymptotes.** They are not an actual part of the graph of the hyperbola, but we can use them to help us obtain the hyperbola.

Example 1

In-Class Example 1

Graph
$$\frac{(x-3)^2}{4} - \frac{(y+2)^2}{9} = 1.$$

answer:

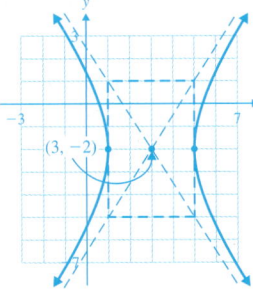

Graph $\dfrac{(x+2)^2}{9} - \dfrac{(y-1)^2}{4} = 1.$

Solution

How do we know that this is a hyperbola and not an ellipse? *It is a hyperbola because there is a subtraction sign between the two quantities on the left.* If it was addition, it would be an ellipse.

Standard form is $\dfrac{(x-h)^2}{a^2} - \dfrac{(y-k)^2}{b^2} = 1.$

Our equation is $\dfrac{(x+2)^2}{9} - \dfrac{(y-1)^2}{4} = 1.$

$$h = -2 \qquad k = 1$$
$$a = \sqrt{9} = 3 \qquad b = \sqrt{4} = 2$$

The center is $(-2, 1)$. *Since the quantity* $\dfrac{(x-h)^2}{a^2}$ *is the positive quantity, the branches of the hyperbola will open in the x-direction.*

We will use the center, $a = 3$, and $b = 2$ to draw a *reference rectangle*. The diagonals of this rectangle are the asymptotes of the hyperbola.

First, plot the center $(-2, 1)$. Since $a = 3$ and a^2 is under the squared quantity containing the x, move 3 units each way in the x-direction from the center. These are two points on the rectangle.

Since $b = 2$ and b^2 is under the squared quantity containing the y, move 2 units each way in the y-direction from the center. These are two more points on the rectangle.

Draw the rectangle containing these four points, then draw the diagonals of the rectangle as dotted lines. These are the asymptotes of the hyperbola.

Sketch the branches of the hyperbola opening in the x-direction with the branches approaching the asymptotes.

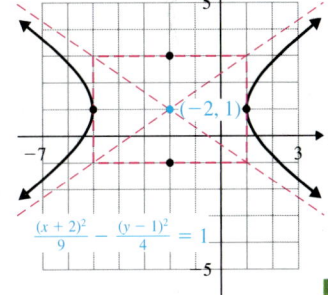

You Try 1

Graph $\dfrac{(x+1)^2}{9} - \dfrac{(y+1)^2}{16} = 1.$

Example 2

In-Class Example 2

Graph $\dfrac{y^2}{9} - \dfrac{x^2}{4} = 1$.

answer:

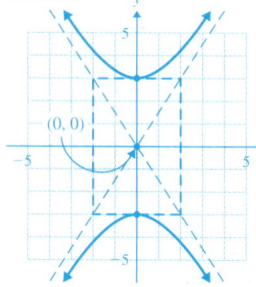

Graph $\dfrac{y^2}{4} - \dfrac{x^2}{25} = 1$.

Solution

Standard form is $\dfrac{(y-k)^2}{b^2} - \dfrac{(x-h)^2}{a^2} = 1$. Our equation is $\dfrac{y^2}{4} - \dfrac{x^2}{25} = 1$.

$$k = 0 \qquad h = 0$$
$$b = \sqrt{4} = 2 \qquad a = \sqrt{25} = 5$$

The center is $(0, 0)$. *Since the quantity* $\dfrac{y^2}{4}$ *is the* positive *quantity, the branches of the hyperbola will open in the y-direction.*

Use the center, $a = 5$, and $b = 2$ to draw the reference rectangle and its diagonals.

Plot the center $(0, 0)$. Since $a = 5$ and a^2 is under the x^2, move 5 units each way in the x-direction from the center to get two points on the rectangle.

Since $b = 2$ and b^2 is under the y^2, move 2 units each way in the y-direction from the center to get two more points on the rectangle.

Draw the rectangle and its diagonals as dotted lines. These are the asymptotes of the hyperbola.

Sketch the branches of the hyperbola opening in the y-direction approaching the asymptotes.

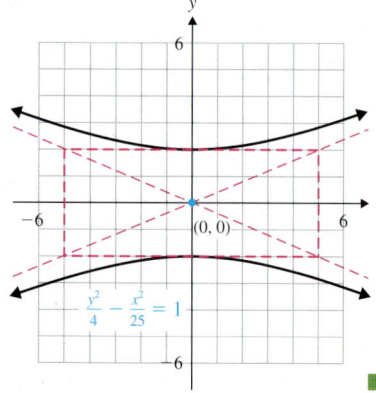

You Try 2

Graph $\dfrac{y^2}{16} - \dfrac{x^2}{16} = 1$.

Definition   Equation of a Hyperbola with Center at the Origin

1) The graph of $\dfrac{x^2}{a^2} - \dfrac{y^2}{b^2} = 1$ is a hyperbola with center $(0, 0)$ and x-intercepts $(a, 0)$ and $(-a, 0)$ as shown below.

2) The graph $\dfrac{y^2}{b^2} - \dfrac{x^2}{a^2} = 1$ is a hyperbola with center $(0, 0)$ and y-intercepts $(0, b)$ and $(0, -b)$ as shown below.

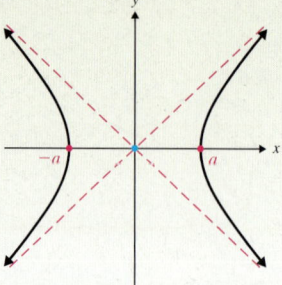

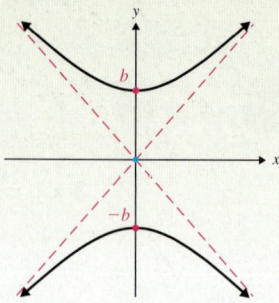

The equations of the asymptotes are $y = \dfrac{b}{a}x$ and $y = -\dfrac{b}{a}x$.

Let's look at another example.

Example 3

In-Class Example 3
Graph $16x^2 - y^2 = 16$.
answer:

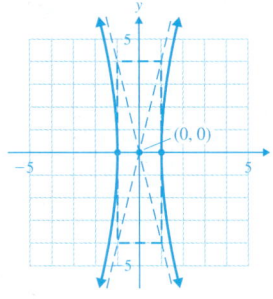

Graph $y^2 - 9x^2 = 9$.

Solution

This is a hyperbola since there is a subtraction sign between the two terms.

Since the standard form for the equation of the hyperbola has a 1 on one side of the = sign, divide both sides of $y^2 - 9x^2 = 9$ by 9 to obtain a 1 on the right.

$$y^2 - 9x^2 = 9$$

$$\frac{y^2}{9} - \frac{9x^2}{9} = \frac{9}{9} \qquad \text{Divide both sides by 9.}$$

$$\frac{y^2}{9} - x^2 = 1 \qquad \text{Simplify.}$$

The center is $(0, 0)$. *The branches of the hyperbola will open in the y-direction since $\frac{y^2}{9}$ is a positive quantity.*

x^2 is the same as $\frac{x^2}{1}$, so $a = \sqrt{1} = 1$ and $b = \sqrt{9} = 3$.

Plot the center at the origin. Move 1 unit each way in the *x*-direction from the center and 3 units each way in the *y*-direction. Draw the rectangle and the asymptotes.

We can find the equations of the asymptotes using $a = 1$ and $b = 3$:

$$y = \frac{3}{1}x \text{ and } y = -\frac{3}{1}x \qquad y = \frac{b}{a}x \text{ and } y = -\frac{b}{a}x.$$

The equations of the asymptotes are $y = 3x$ and $y = -3x$.

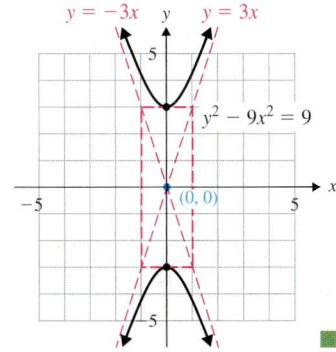

Sketch the branches of the hyperbola opening in the *y*-direction approaching the asymptotes. The *y*-intercepts are $(0, 3)$ and $(0, -3)$. There are no *x*-intercepts.

 You Try 3

Graph $4x^2 - 9y^2 = 36$.

2. Graph a Hyperbola in Nonstandard Form

Equations of hyperbolas can take other forms. We look at one here.

> **Definition**   A Nonstandard Form for the Equation of a Hyperbola
>
> The graph of the equation $xy = c$, where c is a nonzero constant, is a hyperbola whose asymptotes are the *x*- and *y*-axes.

Example 4

In-Class Example 4
Use Example 4.

Graph $xy = 4$.

Solution

Solve for *y*.

$$y = \frac{4}{x} \qquad \text{Divide by } x.$$

Notice that we cannot substitute 0 for x in the equation $y = \dfrac{4}{x}$ because then the denominator would equal zero. Also notice that as $|x|$ gets larger, the value of y gets closer to 0. Likewise, we cannot substitute 0 for y. The x-axis is a horizontal asymptote, and the y-axis is a vertical asymptote.

Make a table of values, plot the points, and sketch the branches of the hyperbola so that they approach the asymptotes.

x	y
1	4
−1	−4
2	2
−2	−2
4	1
−4	−1
8	$\frac{1}{2}$
−8	$-\frac{1}{2}$

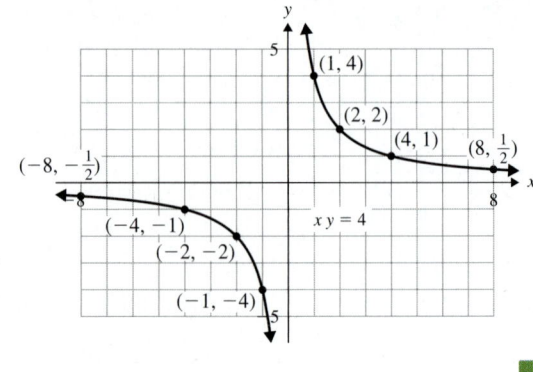

You Try 4

Graph $xy = 8$.

3. Graph Other Square Root Functions

We have already learned how to graph square root functions like $f(x) = \sqrt{x}$ and $g(x) = \sqrt{x - 3}$. Next, we will learn how to graph other square root functions by relating them to the graphs of conic sections.

The vertical line test shows that horizontal parabolas, circles, ellipses, and some hyperbolas are not the graphs of functions. What happens, however, if we look at a portion of the graph of a conic section? Let's start with a circle.

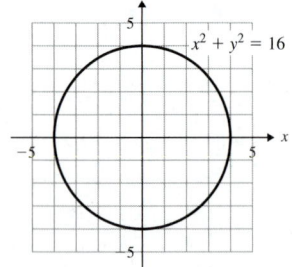

The graph of $x^2 + y^2 = 16$, at the left, is a circle with center $(0, 0)$ and radius 4. If we solve this equation for y we get $y = \pm\sqrt{16 - x^2}$. This represents two equations, $y = \sqrt{16 - x^2}$ and $y = -\sqrt{16 - x^2}$.

The graph of $y = \sqrt{16 - x^2}$ is the **top half of the circle** since the y-coordinates of all points on the graph will be non-negative. The domain is $[-4, 4]$, and the range is $[0, 4]$.

Because of the negative sign out front of the radical, the graph of $y = -\sqrt{16 - x^2}$ is the **bottom half of the circle** since the y-coordinates of all points on the graph will be non-positive. The domain is $[-4, 4]$, and the range is $[-4, 0]$.

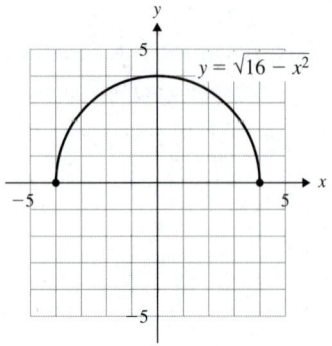

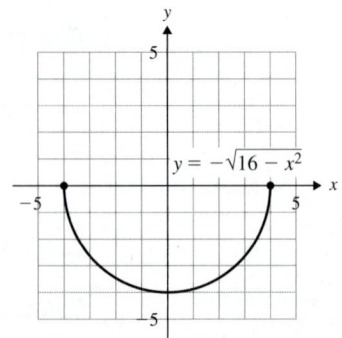

Therefore, to graph $y = \sqrt{16 - x^2}$ it is helpful if we recognize that it is the top half of the circle with equation $x^2 + y^2 = 16$. Likewise, if we are asked to graph $y = -\sqrt{16 - x^2}$, we should recognize that it is the bottom half of the graph of $x^2 + y^2 = 16$.

Let's graph another square root function by first relating it to a conic section.

Example 5

Graph $f(x) = -5\sqrt{1 - \dfrac{x^2}{9}}$. Identify the domain and range.

In-Class Example 5

Graph $f(x) = -2\sqrt{1 - \dfrac{x^2}{36}}$.

Identify the domain and range.
answer: domain $[-6, 6]$;
range: $[-2, 0]$

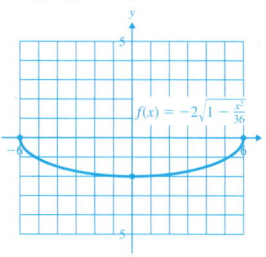

Solution

The graph of this function is half of the graph of a conic section. First, notice that $f(x)$ is always a nonpositive quantity. Since all nonpositive values of y are on or below the x-axis, the graph of this function will only be on or below the x-axis.

Replace $f(x)$ with y, and rearrange the equation into a form we recognize as a conic section.

$$y = -5\sqrt{1 - \frac{x^2}{9}} \qquad \text{Replace } f(x) \text{ with } y.$$

$$-\frac{y}{5} = \sqrt{1 - \frac{x^2}{9}} \qquad \text{Divide by } -5.$$

$$\frac{y^2}{25} = 1 - \frac{x^2}{9} \qquad \text{Square both sides.}$$

$$\frac{x^2}{9} + \frac{y^2}{25} = 1 \qquad \text{Add } \frac{x^2}{9}.$$

The equation $\dfrac{x^2}{9} + \dfrac{y^2}{25} = 1$ represents an ellipse centered at the origin. Its domain is $[-3, 3]$, and its range is $[-5, 5]$. It is not a function.

The graph of $f(x) = -5\sqrt{1 - \dfrac{x^2}{9}}$ is the bottom half of the ellipse, and it is a function. Its domain is $[-3, 3]$, and its range is $[-5, 0]$.

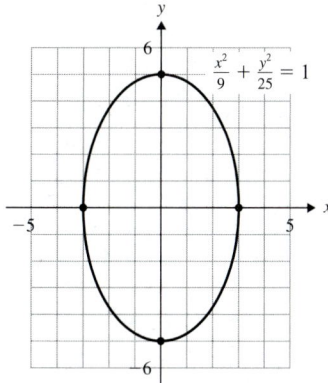

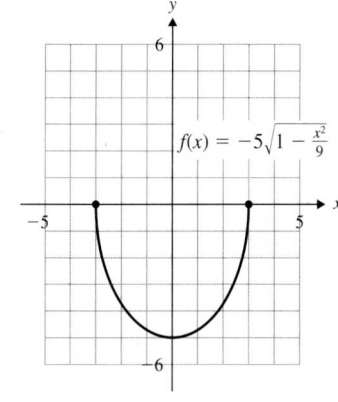

You Try 5

Graph $f(x) = \sqrt{\dfrac{x^2}{4} - 1}$. Identify the domain and range.

More about the conic sections and their characteristics are studied in later mathematics courses.

Using Technology

To sketch the graph of a hyperbola on a graphing calculator, we use the same technique we used for graphing an ellipse: solve the equation for y in terms of x, enter both values of y, and graph both equations. If we are graphing a hyperbola such as $xy = 4$, we will solve for y and enter the single equation.

Graph $9x^2 - y^2 = 16$. First, solve for y.

$$9x^2 - y^2 = 16$$
$$-y^2 = 16 - 9x^2$$
$$y^2 = 9x^2 - 16$$
$$y = \pm\sqrt{9x^2 - 16}$$

Enter $y = \sqrt{9x^2 - 16}$ as Y_1. This represents the portions of the branches of the hyperbola that are on and above the x-axis since the y-values are all non-negative. Enter $y = -\sqrt{9x^2 - 16}$ as Y_2. This represents the portions of the branches of the hyperbola that are on and below the x-axis since the y-values are nonpositive. Set an appropriate window, and press $\boxed{\text{GRAPH}}$.

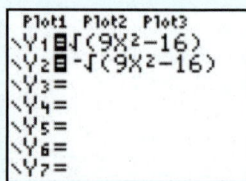

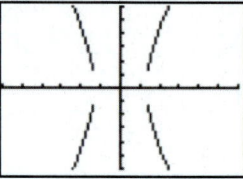

Although it appears that there is a "break" in the graph near the x-axis, the graph does actually continue across the x-axis with the x-intercepts at $\left(\dfrac{4}{3}, 0\right)$ and $\left(-\dfrac{4}{3}, 0\right)$.

Graph each equation on a graphing calculator. These problems come from the homework exercises.

1) $9x^2 - y^2 = 36$; Exercise 17

2) $y^2 - x^2 = 1$; Exercise 19

3) $\dfrac{y^2}{16} - \dfrac{x^2}{4} = 1$; Exercise 7

4) $\dfrac{x^2}{9} - \dfrac{y^2}{25} = 1$; Exercise 5

5) $y^2 - \dfrac{(x-1)^2}{9} = 1$; Exercise 13

6) $xy = 1$; Exercise 29

Answers to You Try Exercises

1)
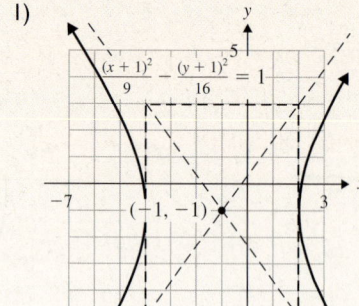

2)
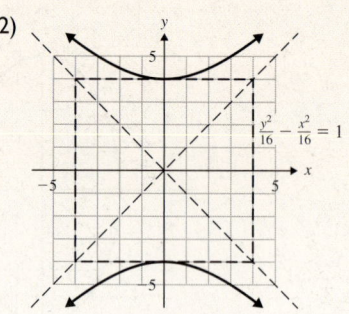

3)
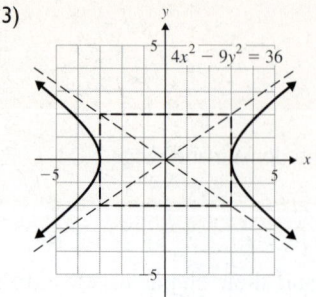

4)

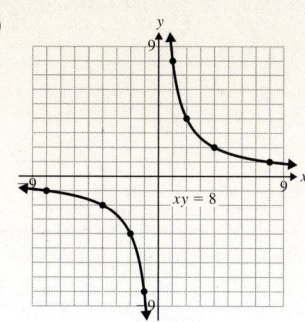

$xy = 8$

5) domain: $(-\infty, -2] \cup [2, \infty)$; range: $[0, \infty)$

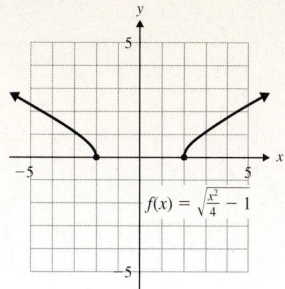

$f(x) = \sqrt{\frac{x^2}{4} - 1}$

Answers to Technology Exercises

1) $9x^2 - y^2 = 36$

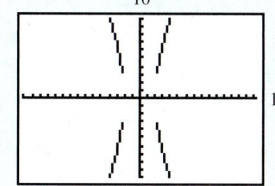

2) $y^2 - x^2 = 1$

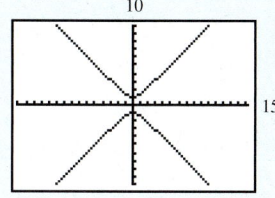

3) $\dfrac{y^2}{16} - \dfrac{x^2}{4} = 1$

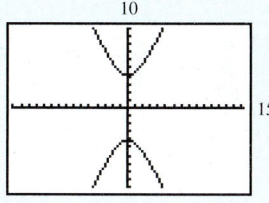

4) $\dfrac{x^2}{9} - \dfrac{y^2}{25} = 1$

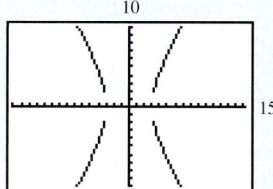

5) $y^2 - \dfrac{(x - 1)^2}{9} = 1$

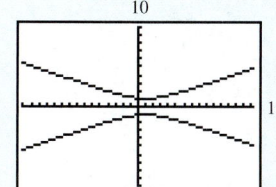

6) $xy = 1$

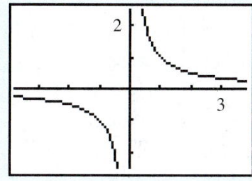

11.4 Exercises

*Additional answers can be found in the Answers to Exercises appendix.

Objective 1: Graph a Hyperbola in Standard Form

Decide whether each statement is true or false.

1) The graph of $\dfrac{(x - 7)^2}{9} + \dfrac{(y - 2)^2}{16} = 1$ is a hyperbola with center $(7, 2)$.   false

2) The center of the hyperbola with equation $\dfrac{(x - 5)^2}{16} - \dfrac{(y + 3)^2}{25} = 1$ is $(5, -3)$.   true

3) The center of the hyperbola with equation $\dfrac{(y - 6)^2}{25} - \dfrac{(x + 4)^2}{4} = 1$ is $(-4, 6)$.   true

4) The graph of $9y^2 - x^2 = 9$ is a hyperbola with center at the origin.   true

Graph each hyperbola. Identify the center and sketch the asymptotes.

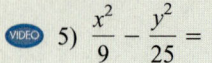 5) $\dfrac{x^2}{9} - \dfrac{y^2}{25} = 1$

6) $\dfrac{x^2}{9} - \dfrac{y^2}{4} = 1$

7) $\dfrac{y^2}{16} - \dfrac{x^2}{4} = 1$

8) $\dfrac{y^2}{4} - \dfrac{x^2}{4} = 1$

9) $\dfrac{(x-2)^2}{9} - \dfrac{(y+3)^2}{16} = 1$

10) $\dfrac{(x+3)^2}{4} - \dfrac{(y+1)^2}{16} = 1$

11) $\dfrac{(y+1)^2}{25} - \dfrac{(x+4)^2}{4} = 1$

12) $\dfrac{(y-1)^2}{36} - \dfrac{(x+1)^2}{9} = 1$

13) $y^2 - \dfrac{(x-1)^2}{9} = 1$

14) $\dfrac{(y+4)^2}{4} - x^2 = 1$

15) $\dfrac{(x-1)^2}{25} - \dfrac{(y-2)^2}{25} = 1$

16) $\dfrac{(x-2)^2}{16} - \dfrac{(y-3)^2}{9} = 1$

17) $9x^2 - y^2 = 36$

18) $4y^2 - x^2 = 16$

19) $y^2 - x^2 = 1$

20) $x^2 - y^2 = 25$

Write the equations of the asymptotes of the graph in each of the following exercises. *Do not graph.*

21) Exercise 5 $y = \frac{5}{3}x$ and $y = -\frac{5}{3}x$

22) Exercise 6 $y = \frac{2}{3}x$ and $y = -\frac{2}{3}x$

23) Exercise 7 $y = 2x$ and $y = -2x$

24) Exercise 8 $y = x$ and $y = -x$

25) Exercise 17 $y = 3x$ and $y = -3x$

26) Exercise 18 $y = \frac{1}{2}x$ and $y = -\frac{1}{2}x$

27) Exercise 19 $y = x$ and $y = -x$

28) Exercise 20 $y = x$ and $y = -x$

Objective 2: Graph a Hyperbola in Nonstandard Form

Graph each equation.

29) $xy = 1$

30) $xy = 6$

31) $xy = 10$

32) $xy = 2$

33) $xy = -4$

34) $xy = -8$

35) $xy = -6$

36) $xy = -1$

Objective 3: Graph Other Square Root Functions

Graph each square root function. Identify the domain and range.

37) $f(x) = \sqrt{9 - x^2}$

38) $g(x) = \sqrt{25 - x^2}$

39) $h(x) = -\sqrt{1 - x^2}$

40) $k(x) = -\sqrt{9 - x^2}$

41) $g(x) = -2\sqrt{1 - \dfrac{x^2}{9}}$

42) $f(x) = 3\sqrt{1 - \dfrac{x^2}{16}}$

43) $h(x) = -3\sqrt{\dfrac{x^2}{4} - 1}$

44) $k(x) = 2\sqrt{\dfrac{x^2}{16} - 1}$

Sketch the graph of each equation.

45) $x = \sqrt{16 - y^2}$

46) $x = -\sqrt{4 - y^2}$

47) $x = -3\sqrt{1 - \dfrac{y^2}{4}}$

48) $x = \sqrt{1 - \dfrac{y^2}{9}}$

Extension

We have learned that sometimes it is necessary to complete the square to put the equations of circles and ellipses into standard form. The same is true for hyperbolas. In Exercises 49 and 50, practice going through the steps of putting the equation of a hyperbola into standard form.

Fill It In

Fill in the blanks with either the missing mathematical step or reason for the given step.

49) $4x^2 - 9y^2 - 8x - 18y - 41 = 0$

$4x^2 - 8x - 9y^2 - 18y = 41$ Group x- and y-terms separately.

$4(x^2 - 2x) - 9(y^2 + 2y) = 41$ Factor out the coefficients of the squared terms.

$4(x^2 - 2x + 1) - 9(y^2 + 2y + 1)$ Complete the square.
$= 41 + 4(1) - 9(1)$

$4(x - 1)^2 - 9(y + 1)^2 = 36$ Factor.

$\dfrac{(x-1)^2}{9} - \dfrac{(y+1)^2}{4} = 1$ Divide both sides by 36.

50) $-x^2 + 4y^2 + 6x - 16y + 3 = 0$

$-x^2 + 6x + 4y^2 - 16y = -3$ Group the x- and y-terms separately.

$-(x^2 - 6x) + 4(y^2 - 4y) = -3$ Factor out the coefficients of the squared terms.

$-(x^2 - 6x + 9) + 4(y^2 - 4y + 4)$ Complete the square.
$= -3 - 1(9) + 4(4)$

$-(x - 3)^2 + 4(y - 2)^2 = 4$ Factor.

$(y - 2)^2 - \dfrac{(x-3)^2}{4} = 1$ Divide both sides by 4.

For Exercises 51–54, put each equation into the standard form for the equation of a hyperbola and graph.

51) $x^2 - 4y^2 - 2x - 24y - 51 = 0$

52) $9x^2 - 4y^2 + 90x - 16y + 173 = 0$

53) $16y^2 - 9x^2 + 18x - 64y - 89 = 0$

54) $y^2 - 4x^2 - 16x - 6y - 23 = 0$

We know that the standard form for the equation of a hyperbola is

$$\dfrac{(x-h)^2}{a^2} - \dfrac{(y-k)^2}{b^2} = 1 \quad \text{or} \quad \dfrac{(y-k)^2}{b^2} - \dfrac{(x-h)^2}{a^2} = 1.$$

The equations for the asymptotes are

$$y - k = \frac{b}{a}(x - h) \text{ and } y - k = -\frac{b}{a}(x - h)$$

Use these formulas to write the equations of the asymptotes of the graph in each of the following exercises. *Do not graph.*

55) Exercise 9

56) Exercise 10

57) Exercise 11

58) Exercise 12

59) Exercise 13

60) Exercise 14

61) Exercise 15

62) Exercise 16

63) A hyperbola centered at the origin opens in the

y-direction and has asymptotes with equations $y = \frac{1}{2}x$ and

$y = -\frac{1}{2}x$. Write an equation of the hyperbola. $y^2 - \frac{x^2}{4} = 1$

64) A hyperbola centered at the origin opens in the

x-direction and has asymptotes with equations $y = \frac{3}{2}x$ and

$y = -\frac{3}{2}x$. Write an equation of the hyperbola. $\frac{x^2}{4} - \frac{y^2}{9} = 1$

Use a graphing calculator to graph the following. Square the viewing window.

65) $\dfrac{(x - 2)^2}{9} - \dfrac{(y + 3)^2}{16} = 1$

66) $\dfrac{(x + 3)^2}{4} - \dfrac{(y + 1)^2}{16} = 1$

67) $\dfrac{(y + 1)^2}{25} - \dfrac{(x + 4)^2}{4} = 1$

68) $\dfrac{(y - 1)^2}{36} - \dfrac{(x + 1)^2}{9} = 1$

Putting It All Together

Objective

1. Identify and Graph Different Types of Conic Sections

1. Identify and Graph Different Types of Conic Sections

Sometimes the most difficult part of graphing a conic section is identifying which type of graph will result from the given equation. In this section, we will discuss how to look at an equation and decide what type of conic section it represents.

Example 1

Graph $x^2 + y^2 + 4x - 6y + 9 = 0$.

In-Class Example 1
Graph
$x^2 + y^2 - 8x - 6y + 9 = 0$.
answer:

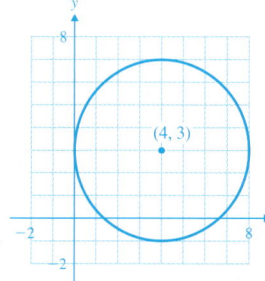

Solution

First, notice that this equation has two squared terms. Therefore, its graph cannot be a parabola since the equation of a parabola contains only one squared term. Next, observe that the coefficients of x^2 and y^2 are each 1. Since the coefficients are the same, *this is the equation of a circle.*

Write the equation in the form $(x - h)^2 + (y - k)^2 = r^2$ by completing the square on the x-terms and on the y-terms.

$$x^2 + y^2 + 4x - 6y + 9 = 0$$
$$(x^2 + 4x) + (y^2 - 6y) = -9$$

Group the x-terms together and group the y-terms together. Move the constant to the other side.

$$(x^2 + 4x + 4) + (y^2 - 6y + 9) = -9 + 4 + 9$$

Complete the square for each group of terms.

$$(x + 2)^2 + (y - 3)^2 = 4$$

Factor; add.

The center of the circle is $(-2, 3)$. The radius is 2.

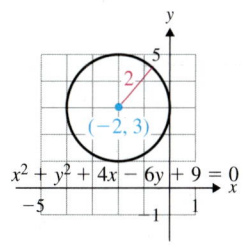

Example 2

In-Class Example 2
Graph $x = y^2 - 8y + 19$.
answer:

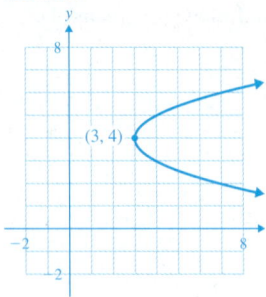

Graph $x = y^2 + 4y + 3$.

Solution

This equation contains only one squared term. Therefore, *this is the equation of a parabola.* Since the squared term is y^2 and $a = 1$, the parabola will open to the right.

Use the formula $y = -\dfrac{b}{2a}$ to find the y-coordinate of the vertex.

$$a = 1 \quad b = 4 \quad c = 3$$

$$y = -\frac{4}{2(1)} = -2$$

$$x = (-2)^2 + 4(-2) + 3 = -1$$

The vertex is $(-1, -2)$. Make a table of values to find other points on the parabola, and use the axis of symmetry to find more points.

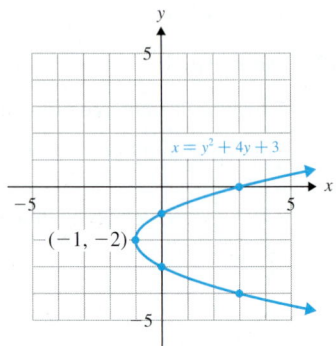

x	y
0	−1
3	0

Plot the points in the table. Locate the points $(0, -3)$ and $(3, -4)$ using the axis of symmetry, $y = -2$.

Example 3

In-Class Example 3
Graph
$\dfrac{(x + 1)^2}{16} - \dfrac{(y - 3)^2}{4} = 1$.
answer:

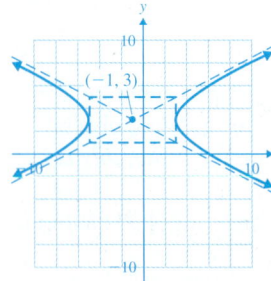

Graph $\dfrac{(y - 1)^2}{9} - \dfrac{(x - 3)^2}{4} = 1$.

Solution

In this equation we see the *difference* of two squares. *The graph of this equation is a hyperbola. The branches of the hyperbola will open in the y-direction since the quantity containing the variable y,* $\dfrac{(y - 1)^2}{9}$, *is the positive, squared quantity.*

The center is $(3, 1)$; $a = \sqrt{4} = 2$ and $b = \sqrt{9} = 3$. Draw the reference rectangle and its diagonals, the asymptotes of the graph.

   The branches of the hyperbola approach the asymptotes.

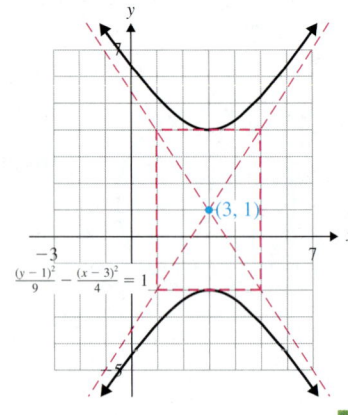

Example 4

In-Class Example 4
Graph $4x^2 + y^2 = 16$.
answer:

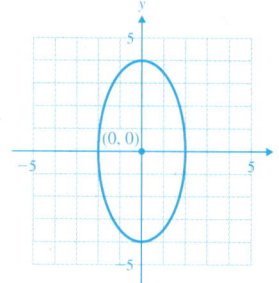

Graph $x^2 + 9y^2 = 36$.

Solution

This equation contains the *sum* of two squares with *different* coefficients. *This is the equation of an ellipse.* (If the coefficients were the same, the graph would be a circle.)

Divide both sides of the equation by 36 to get 1 on the right side of the = sign.

$$x^2 + 9y^2 = 36$$
$$\frac{x^2}{36} + \frac{9y^2}{36} = \frac{36}{36} \qquad \text{Divide both sides by 36.}$$
$$\frac{x^2}{36} + \frac{y^2}{4} = 1 \qquad \text{Simplify.}$$

The center is $(0, 0)$, $a = 6$ and $b = 2$.

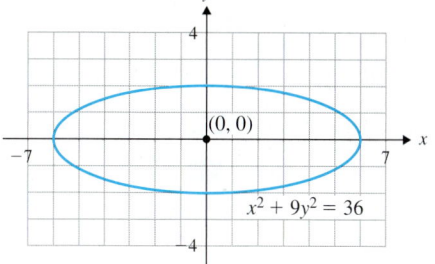

You Try 1

Determine whether the graph of each equation is a parabola, circle, ellipse, or hyperbola. Then, graph each equation.

a) $4x^2 - 25y^2 = 100$

b) $x^2 + y^2 - 6x - 12y + 9 = 0$

c) $y = -x^2 - 2x + 4$

d) $x^2 + \dfrac{(y + 4)^2}{9} = 1$

Answers to You Try Exercises

1) a) hyperbola

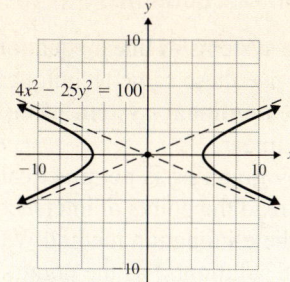

b) circle

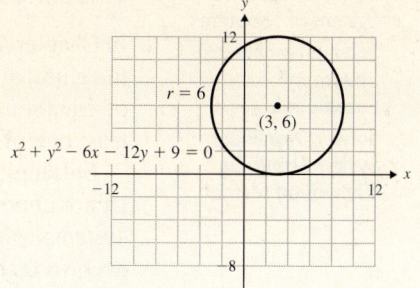

c) parabola

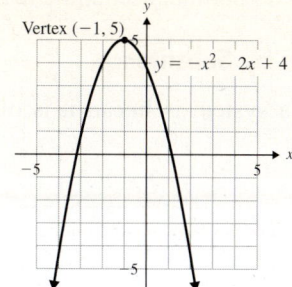

d) ellipse

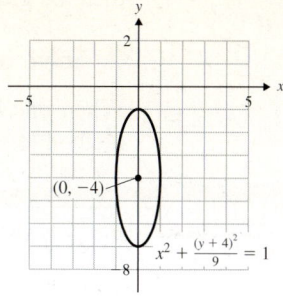

Putting It All Together
Summary Exercises

*Additional answers can be found in the Answers to Exercises appendix.

Objective 1: Identify and Graph Different Types of Conic Sections

Determine whether the graph of each equation is a parabola, circle, ellipse, or hyperbola. Then, graph each equation.

1) $y = x^2 + 4x + 8$

2) $(x + 5)^2 + (y - 3)^2 = 25$

3) $\dfrac{(y + 4)^2}{9} - \dfrac{(x + 1)^2}{4} = 1$

4) $x = (y - 1)^2 + 8$ 5) $16x^2 + 9y^2 = 144$

6) $x^2 - 4y^2 = 36$ 7) $x^2 + y^2 + 8x - 6y - 11 = 0$

8) $\dfrac{(x - 2)^2}{25} + \dfrac{(y - 2)^2}{36} = 1$

9) $(x - 1)^2 + \dfrac{y^2}{16} = 1$ 10) $x^2 + y^2 + 8y + 7 = 0$

11) $x = -(y + 4)^2 - 3$ 12) $\dfrac{(y + 4)^2}{4} - (x + 2)^2 = 1$

13) $25x^2 - 4y^2 = 100$ 14) $4x^2 + y^2 = 16$

15) $(x - 3)^2 + y^2 = 16$ 16) $y = -x^2 + 6x - 7$

17) $x = \dfrac{1}{2}y^2 + 2y + 3$ 18) $\dfrac{(x - 3)^2}{16} + y^2 = 1$

19) $(x - 2)^2 - (y + 1)^2 = 9$

20) $x^2 + y^2 + 6x - 8y + 9 = 0$

Where appropriate, write the equation in standard form. Then, graph each equation.

21) $xy = 5$

22) $25x^2 - 4y^2 + 150x + 125 = 0$

23) $9x^2 + y^2 - 54x + 4y + 76 = 0$

24) $xy = -2$

25) $9y^2 - 4x^2 - 18y + 16x - 43 = 0$

26) $4x^2 + 9y^2 - 8x - 54y + 49 = 0$

Use a graphing calculator to graph Exercises 27–30. Square the viewing window.

27) $x^2 + y^2 + 8x - 6y - 11 = 0$

28) $(x - 1)^2 + \dfrac{y^2}{16} = 1$

29) $x = -(y + 4)^2 - 3$ 30) $25x^2 - 4y^2 = 100$

Section 11.5 Nonlinear Systems of Equations

Objectives

1. Define a Nonlinear System of Equations
2. Solve a Nonlinear System by Substitution
3. Solve a Nonlinear System Using the Elimination Method

1. Define a Nonlinear System of Equations

In Chapter 4, we learned to solve systems of linear equations by graphing, substitution, and the elimination method. We can use these same techniques for solving a *nonlinear* system of equations in two variables. A **nonlinear system of equations** is a system in which at least one of the equations is not linear.

Solving a nonlinear system by graphing is not practical since it would be very difficult (if not impossible) to accurately read the points of intersection. Therefore, we will solve the systems using substitution and the elimination method. We will graph the equations, however, so that we can visualize the solution(s) as the point(s) of intersection of the graphs.

We are interested only in real-number solutions. If a system has imaginary solutions, then the graphs of the equations do not intersect in the real-number plane.

2. Solve a Nonlinear System by Substitution

When one of the equations in a system is linear, it is often best to use the substitution method to solve the system.

Example 1

Solve the system $x^2 - 2y = 2$ (1)
$\qquad\qquad\qquad -x + y = 3$ (2)

In-Class Example 1
Solve the system $x^2 + 3y = 4$
$\qquad\qquad x + y = 2$
answer: {(1, 1), (2, 0)}

Solution

The graph of equation (1) is a parabola, and the graph of equation (2) is a line. Let's begin by thinking about the number of possible points of intersection the graphs can have.

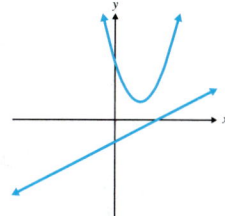
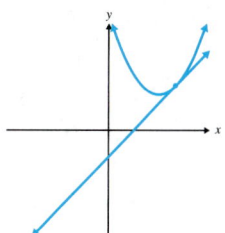
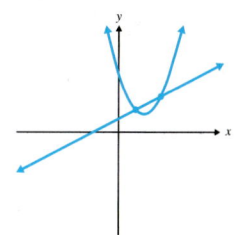

No points of intersection
The system has no solution.

One point of intersection
The system has one solution.

Two points of intersection
The system has two solutions.

Solve the linear equation for one of the variables.

$$-x + y = 3$$
$$y = x + 3 \quad (3) \qquad \text{Solve for } y.$$

Substitute $x + 3$ for y in equation (1).

$$
\begin{array}{ll}
x^2 - 2y = 2 & \text{Equation (1).} \\
x^2 - 2(x + 3) = 2 & \text{Substitute.} \\
x^2 - 2x - 6 = 2 & \text{Distribute.} \\
x^2 - 2x - 8 = 0 & \text{Subtract 2.} \\
(x - 4)(x + 2) = 0 & \text{Factor.} \\
x = 4 \text{ or } x = -2 & \text{Solve.}
\end{array}
$$

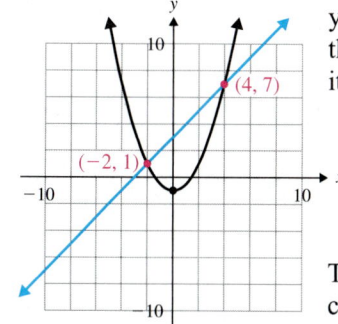

To find the corresponding value of y for each value of x, we can substitute $x = 4$ and then $x = -2$ into *either* equation (1), (2), or (3). No matter which equation you choose, you should always check the solutions in *both* of the original equations. We will substitute the values into equation (3) because this is just an alternative form of equation (2), and it is already solved for y.

Substitute each value into equation (3) to find y.

$$
\begin{array}{ll}
x = 4: \quad y = x + 3 & \qquad x = -2: \quad y = x + 3 \\
 \quad y = 4 + 3 & \qquad \quad y = -2 + 3 \\
 \quad y = 7 & \qquad \quad y = 1
\end{array}
$$

The proposed solutions are $(4, 7)$ and $(-2, 1)$. Verify that they solve the system by checking them in equation (1). The solution set is $\{(4, 7), (-2, 1)\}$. We can see on the graph to the left that these are the points of intersection of the graphs. ■

 **You Try 1**

Solve the system $x^2 + 3y = 6$
$$x + y = 2$$

Example 2

Solve the system $x^2 + y^2 = 1 \quad (1)$
$$x + 2y = -1 \quad (2)$$

In-Class Example 2
Solve the system $x^2 + y^2 = 1$
$$2x - y = 1$$
answer: $\{(0, -1), (\frac{4}{5}, \frac{3}{5})\}$

Solution

The graph of equation (1) is a circle, and the graph of equation (2) is a line. These graphs can intersect at zero, one, or two points. Therefore, this system will have zero, one, or two solutions.

We will not solve equation (1) for a variable because doing so would give us a radical in the expression. It will be easiest to solve equation (2) for x because its coefficient is 1.

$$
\begin{array}{ll}
x + 2y = -1 & \quad (2) \\
x = -2y - 1 & \quad (3) \qquad \text{Solve for } x.
\end{array}
$$

Substitute $-2y - 1$ for x in equation (1).

$$x^2 + y^2 = 1 \quad (1)$$
$$(-2y - 1)^2 + y^2 = 1 \qquad \text{Substitute.}$$
$$4y^2 + 4y + 1 + y^2 = 1 \qquad \text{Expand } (-2y - 1)^2.$$
$$5y^2 + 4y = 0 \qquad \text{Combine like terms; subtract 1.}$$
$$y(5y + 4) = 0 \qquad \text{Factor.}$$
$$y = 0 \quad \text{or} \quad 5y + 4 = 0 \qquad \text{Set each factor equal to zero.}$$
$$y = -\frac{4}{5} \qquad \text{Solve for } y.$$

Substitute $y = 0$ and then $y = -\dfrac{4}{5}$ into equation (3) to find their corresponding values of x.

$$y = 0: x = -2y - 1 \qquad\qquad y = -\frac{4}{5}: x = -2y - 1$$
$$x = -2(0) - 1 \qquad\qquad\qquad x = -2\left(-\frac{4}{5}\right) - 1$$
$$x = -1 \qquad\qquad\qquad\qquad x = \frac{8}{5} - 1 = \frac{3}{5}$$

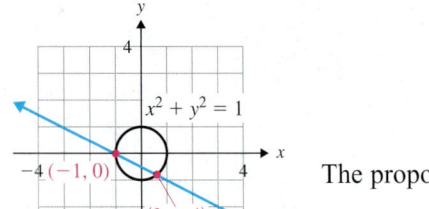

The proposed solutions are $(-1, 0)$ and $\left(\dfrac{3}{5}, -\dfrac{4}{5}\right)$. Check them in equations (1) and (2).

The solution set is $\left\{(-1, 0), \left(\dfrac{3}{5}, -\dfrac{4}{5}\right)\right\}$. The graph at left shows that these are the points of intersection of the graphs.

You Try 2

Solve the system $x^2 + y^2 = 25$
$x - y = 7$

Note

We must always check the proposed solutions in *each* equation in the system.

3.   Solve a Nonlinear System Using the Elimination Method

The elimination method can be used to solve a system when both equations are second-degree equations.

Example 3

Solve the system $5x^2 + 3y^2 = 21 \quad (1)$
$4x^2 - y^2 = 10 \quad (2)$

In-Class Example 3
Solve the system $4x^2 + 3y^2 = 23$
$2x^2 - y^2 = 9$
answer: $\{(\sqrt{5}, 1), (\sqrt{5}, -1),$
$(-\sqrt{5}, 1), (-\sqrt{5}, -1)\}$

Solution

Each equation is a second-degree equation. The first is an ellipse and the second is a hyperbola. They can have zero, one, two, three, or four points of intersection. Multiply equation (2) by 3. Then adding the two equations will eliminate the y^2-terms.

Original System	Rewrite the System
$5x^2 + 3y^2 = 21$	$5x^2 + 3y^2 = 21$
$4x^2 - y^2 = 10$ $\rightarrow$	$12x^2 - 3y^2 = 30$

$$5x^2 + 3y^2 = 21$$
$$+ \ \underline{12x^2 - 3y^2 = 30} \qquad \text{Add the equations to eliminate } y^2.$$
$$17x^2 = 51$$
$$x^2 = 3$$
$$x = \pm\sqrt{3}$$

Find the corresponding values of y for $x = \sqrt{3}$ and $x = -\sqrt{3}$.

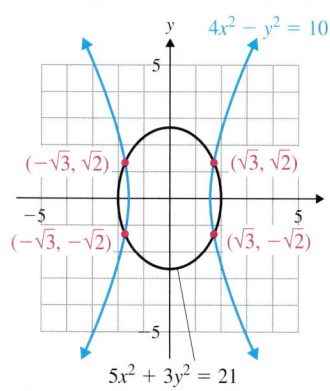

$x = \sqrt{3}$:
$$4x^2 - y^2 = 10 \quad (2)$$
$$4(\sqrt{3})^2 - y^2 = 10$$
$$12 - y^2 = 10$$
$$-y^2 = -2$$
$$y^2 = 2$$
$$y = \pm\sqrt{2}$$

This gives us $(\sqrt{3}, \sqrt{2})$ and $(\sqrt{3}, -\sqrt{2})$.

$x = -\sqrt{3}$:
$$4x^2 - y^2 = 10 \quad (2)$$
$$4(-\sqrt{3})^2 - y^2 = 10$$
$$12 - y^2 = 10$$
$$-y^2 = -2$$
$$y^2 = 2$$
$$y = \pm\sqrt{2}$$

This gives us $(-\sqrt{3}, \sqrt{2})$ and $(-\sqrt{3}, -\sqrt{2})$.

Check the proposed solutions in equation (1) to verify that they satisfy that equation as well.

The solution set is $\{(\sqrt{3}, \sqrt{2}), (\sqrt{3}, -\sqrt{2}), (-\sqrt{3}, \sqrt{2}), (-\sqrt{3}, -\sqrt{2})\}$.

You Try 3

Solve the system $2x^2 - 13y^2 = 20$
$-x^2 + 10y^2 = 4$

For solving some systems, using *either* substitution or the elimination method works well. Look carefully at each system to decide which method to use.

We will see in Example 4 that not all systems have solutions.

Example 4

Solve the system
$$y = \sqrt{x} \quad (1)$$
$$y^2 - 4x^2 = 4 \quad (2)$$

In-Class Example 4
Solve the system $y = 2\sqrt{x}$
$4y^2 - 9x^2 = 36$
answer: $\varnothing$

Solution

The graph of the square root function $y = \sqrt{x}$ is half of a parabola. The graph of equation (2) is a hyperbola. Solve this system by substitution. Replace y in equation (2) with $\sqrt{x}$ from equation (1).

$$y^2 - 4x^2 = 4 \quad (2)$$
$$(\sqrt{x})^2 - 4x^2 = 4 \qquad \text{Substitute } y = \sqrt{x} \text{ into equation (2)}.$$
$$x - 4x^2 = 4$$
$$0 = 4x^2 - x + 4$$

Since the right-hand side does not factor, solve it using the quadratic formula.

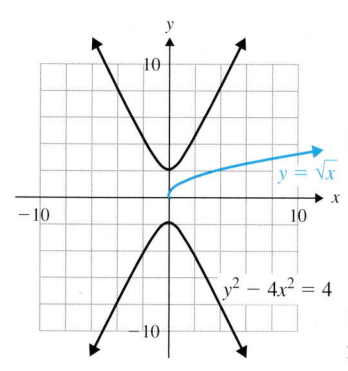

$$4x^2 - x + 4 = 0 \qquad a = 4 \qquad b = -1 \qquad c = 4$$

$$x = \frac{-(-1) \pm \sqrt{(-1)^2 - 4(4)(4)}}{2(4)} = \frac{1 \pm \sqrt{1 - 64}}{8} = \frac{1 \pm \sqrt{-63}}{8}$$

Since $\sqrt{-63}$ is not a real number, there are no real-number values for x. The system has no solution, so the solution set is $\varnothing$. The graph is shown on the left.

You Try 4

Solve the system $4x^2 + y^2 = 4$
$x - y = 3$

Using Technology

We can solve systems of nonlinear equations on the graphing calculator just like we solved systems of linear equations in Chapter 4—graph the equations and find their points of intersection.

Let's look at Example 3:

$$5x^2 + 3y^2 = 21$$
$$4x^2 - y^2 = 10$$

Solve each equation for y and enter them into the calculator.

Solve $5x^2 + 3y^2 = 21$ for y:

$$y = \pm\sqrt{7 - \frac{5}{3}x^2}$$

Enter $\sqrt{7 - \frac{5}{3}x^2}$ as Y_1.

Enter $-\sqrt{7 - \frac{5}{3}x^2}$ as Y_2.

Solve $4x^2 - y^2 = 10$ for y:

$$y = \pm\sqrt{4x^2 - 10}$$

Enter $\sqrt{4x^2 - 10}$ as Y_3.

Enter $-\sqrt{4x^2 - 10}$ as Y_4.

After entering the equations, press GRAPH.

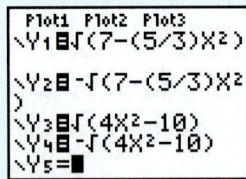

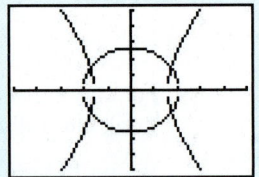

The system has four real solutions since the graphs have four points of intersection. We can use the INTERSECT option to find the solutions. Since we graphed four functions, we must tell the calculator which point of intersection we want to find. Note that the point where the graphs intersect in the first quadrant comes from the intersection of equations Y_1 and Y_3. Press 2nd TRACE and choose 5:intersect and you will see the screen to the right.

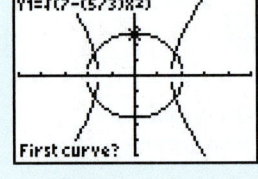

Notice that the top left of the screen to the right displays the function Y_1. Since we want to find the intersection of Y_1 and Y_3, press ENTER when Y_1 is displayed. Now Y_2 appears at the top left, but we do not need this function. Press the down arrow to see the equation for Y_3 and be sure that the cursor is close to the intersection point in quadrant I. Press ENTER twice. You will see the approximate solution (1.732, 1.414), as shown to the right.

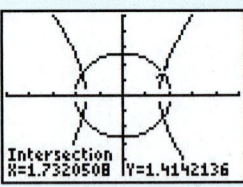

In Example 3 we found the exact solutions algebraically. The calculator solution, (1.732, 1.414), is an approximation of the exact solution $(\sqrt{3}, \sqrt{2})$.

The other solutions of the system can be found in the same way.

Use the graphing calculator to find all real-number solutions of each system. These are taken from the examples in the section and from the Chapter Summary.

1) $x^2 - 2y = 2$
    $-x + y = 3$

2) $x^2 + y^2 = 1$
    $x + 2y = -1$

3) $x - y^2 = 3$
    $x - 2y = 6$

4) $y = \sqrt{x}$
    $y^2 - 4x^2 = 4$

Answers to You Try Exercises

1) $\{(0, 2), (3, -1)\}$ 2) $\{(4, -3), (3, -4)\}$ 3) $\{(6, 2), (6, -2), (-6, 2), (-6, -2)\}$ 4) $\varnothing$

Answers to Technology Exercises

1) $\{(4, 7), (-2, 1)\}$ 2) $\{(-1, 0), (0.6, -0.8)\}$ 3) $\{(12, 3), (4, -1)\}$ 4) $\varnothing$

11.5 Exercises

*Additional answers can be found in the Answers to Exercises appendix.

Objective 1: Define a Nonlinear System of Equations

If a nonlinear system consists of equations with the following graphs,

    a) sketch the different ways in which the graphs can intersect.

    b) make a sketch in which the graphs do not intersect.

    c) how many possible solutions can each system have?

1) circle and line   2) parabola and line

3) parabola and ellipse   4) ellipse and hyperbola

5) parabola and hyperbola   6) circle and ellipse

Mixed Exercises: Objectives 2 and 3

Solve each system using either substitution or the elimination method.

7) $x^2 + 4y = 8$
    $x + 2y = -8$
    $\{(-4, -2), (6, -7)\}$

8) $x^2 + y = 1$
    $-x + y = -5$
    $\{(-3, -8), (2, -3)\}$

9) $x + 2y = 5$
    $x^2 + y^2 = 10$
    $\{(-1, 3), (3, 1)\}$

10) $y = 2$
    $x^2 + y^2 = 8$
    $\{(2, 2), (-2, 2)\}$

11) $y = x^2 - 6x + 10$
    $y = 2x - 6$ $\{(4, 2)\}$

12) $y = x^2 - 10x + 22$
    $y = 4x - 27$ $\{(7, 1)\}$

13) $x^2 + 2y^2 = 11$
    $x^2 - y^2 = 8$

14) $2x^2 - y^2 = 7$
    $2y^2 - 3x^2 = 2$

15) $x^2 + y^2 = 6$
    $2x^2 + 5y^2 = 18$

16) $5x^2 - y^2 = 16$
    $x^2 + y^2 = 14$

17) $3x^2 + 4y = -1$
    $x^2 + 3y = -12$
    $\{(3, -7), (-3, -7)\}$

18) $2x^2 + y = 9$
    $y = 3x^2 + 4$
    $\{(-1, 7), (1, 7)\}$

19) $y = 6x^2 - 1$
    $2x + 5y = -5$ $\{(0, -1)\}$

20) $x^2 + 2y = 5$
    $-3x^2 + 2y = 5$ $\left\{\left(0, \frac{5}{2}\right)\right\}$

21) $x^2 + y^2 = 4$
    $-2x^2 + 3y = 6$ $\{(0, 2)\}$

22) $x^2 + y^2 = 49$
    $x - 2y^2 = 7$ $\{(7, 0)\}$

23) $x^2 + y^2 = 3$
    $x + y = 4$ $\varnothing$

24) $y - x = 1$
    $4y^2 - 16x^2 = 64$ $\varnothing$

25) $x = \sqrt{y}$
    $x^2 - 9y^2 = 9$ $\varnothing$

26) $x = \sqrt{y}$
    $x^2 - y^2 = 4$ $\varnothing$

27) $9x^2 + y^2 = 9$
    $x^2 + y^2 = 5$

28) $x^2 + y^2 = 6$
    $5x^2 + y^2 = 10$

29) $y = -x^2 - 2$
    $x^2 + y^2 = 4$ $\{(0, -2)\}$

30) $x^2 + y^2 = 1$
    $y = x^2 + 1$ $\{(0, 1)\}$

Write a system of equations and solve.

31) Find two numbers whose product is 40 and whose sum is 13.
    8 and 5

32) Find two numbers whose product is 28 and whose sum is 11.
    7 and 4

33) The perimeter of a rectangular computer screen is 38 in. Its area is 88 in^2. Find the dimensions of the screen.

8 in. × 11 in.

34) The area of a rectangular bulletin board is 180 in^2, and its perimeter is 54 in. Find the dimensions of the bulletin board.
    12 in. × 15 in.

35) A sporting goods company estimates that the cost y, in dollars, to manufacture x thousands of basketballs is given by

$$y = 6x^2 + 33x + 12$$

The revenue y, in dollars, from the sale of x thousands of basketballs is given by

$$y = 15x^2$$

The company breaks even on the sale of basketballs when revenue equals cost. The point, (x, y), at which this occurs is called the *break-even point*. Find the break-even point for the manufacture and sale of the basketballs.   4000 basketballs, $240

36) A backpack manufacturer estimates that the cost y, in dollars, to make x thousands of backpacks is given by

$$y = 9x^2 + 30x + 18$$

The revenue y, in dollars, from the sale of x thousands of backpacks is given by

$$y = 21x^2$$

Find the break-even point for the manufacture and sale of the backpacks. (See Exercise 35 for an explanation.)
3000 backpacks, $189

Section 11.6 Second-Degree Inequalities and Systems of Inequalities

Objectives

1. Graph Second-Degree Inequalities
2. Graph Systems of Nonlinear Inequalities

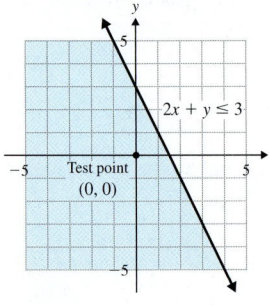

1. Graph Second-Degree Inequalities

In Section 3.4, we learned how to graph linear inequalities in two variables such as $2x + y \le 3$. To graph this inequality, first graph the boundary line $2x + y = 3$. Then we can choose a test point on one side of the line, say $(0, 0)$. Since $(0, 0)$ satisfies the inequality $2x + y \le 3$, we shade the side of the line containing $(0, 0)$. All points in the shaded region satisfy $2x + y \le 3$. (If the test point had *not* satisfied the inequality, we would have shaded the other side of the line.)

The graph of $2x + y \le 3$ is to the left.

A **second-degree inequality** contains at least one squared term and no variable with degree greater than 2. We graph second-degree inequalities the same way we graph linear inequalities in two variables.

Example 1

Graph $x^2 + y^2 < 25$.

In-Class Example 1
Graph $x^2 + y^2 < 1$.
answer:

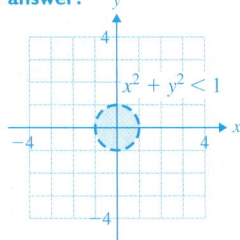

Solution

Begin by graphing the *circle*, $x^2 + y^2 = 25$, as a dotted curve since the inequality is $<$. (Points *on* the circle will not satisfy the inequality.)

Next, choose a test point not on the boundary curve: $(0, 0)$. Does the test point satisfy $x^2 + y^2 < 25$? Yes: $0^2 + 0^2 < 25$ is true. Shade the region inside the circle. All points in the shaded region satisfy $x^2 + y^2 < 25$.

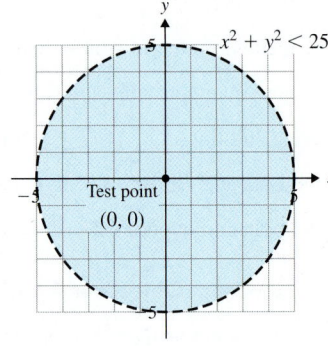

You Try 1

Graph $\dfrac{x^2}{25} + \dfrac{y^2}{9} < 1$.

Example 2

Graph $4x^2 - 9y^2 \ge 36$.

In-Class Example 2
Graph $x^2 - 9y^2 \ge 9$.
answer:

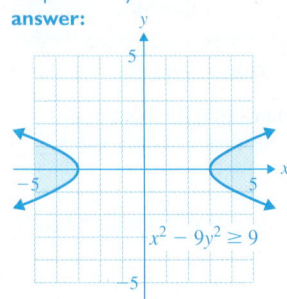

Solution

First, graph the *hyperbola*, $4x^2 - 9y^2 = 36$, as a solid curve since the inequality is $\ge$.

$$4x^2 - 9y^2 = 36$$
$$\frac{4x^2}{36} - \frac{9y^2}{36} = \frac{36}{36} \qquad \text{Divide by 36.}$$
$$\frac{x^2}{9} - \frac{y^2}{4} = 1 \qquad \text{Simplify.}$$

The center is $(0, 0)$, $a = 3$, and $b = 2$.

Next, choose a test point not on the boundary curve: (0, 0). Does (0, 0) satisfy $4x^2 - 9y^2 \geq 36$? No: $4(0)^2 - 9(0)^2 \geq 36$ is false.

Since (0, 0) does not satisfy the inequality, we do not shade the region containing (0, 0). Shade the other side of the branches of the hyperbola.

All points in the shaded region satisfy $4x^2 - 9y^2 \geq 36$.

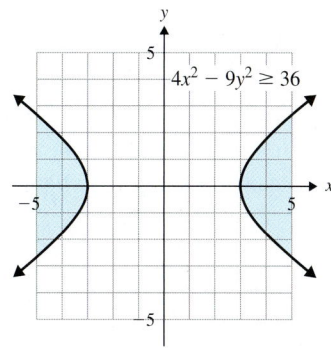

You Try 2

Graph $y \geq -x^2 + 3$.

2. Graph Systems of Nonlinear Inequalities

The **solution set of a system of inequalities** consists of the set of points that satisfy *all* the inequalities in the system.

We first discussed this in Section 3.4 when we graphed the solution set of a system like

$$x \geq -2 \text{ and } y \leq x - 3$$

The solution set of such a system of *linear* inequalities is the intersection of their graphs. The solution set of *second-degree* inequalities is also the intersection of the graphs of the individual inequalities.

Example 3

Graph the solution set of the system.

$$4x^2 + y^2 < 16$$
$$-x + 2y > 2$$

In-Class Example 3

Graph the solution set of the system.

$$4x^2 + 9y^2 < 36$$
$$2x + y > -1$$

answer:

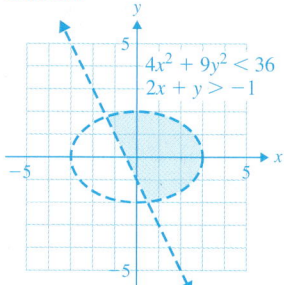

Solution

First, graph the *ellipse*, $4x^2 + y^2 = 16$, as a dotted curve since the inequality is $<$.

$$4x^2 + y^2 = 16$$
$$\frac{4x^2}{16} + \frac{y^2}{16} = \frac{16}{16} \qquad \text{Divide by 16.}$$
$$\frac{x^2}{4} + \frac{y^2}{16} = 1 \qquad \text{Simplify.}$$

The test point (0, 0) satisfies the inequality $4x^2 + y^2 < 16$, so shade inside the dotted curve of the ellipse, as shown to the right.

Graph the *line*, $-x + 2y = 2$, as a dotted line since the inequality is $>$.

$$-x + 2y > 2$$
$$2y > x + 2$$
$$y > \frac{1}{2}x + 1 \qquad \text{Solve for } y.$$

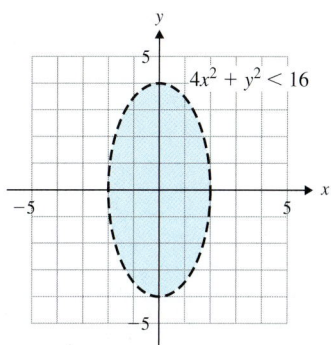

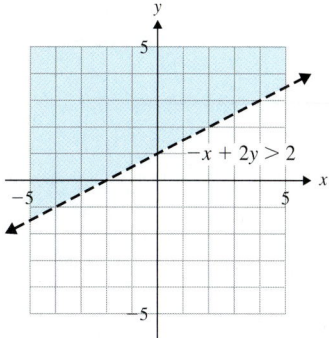

Shade above the line since the test point (0, 0) does *not* satisfy $y > \frac{1}{2}x + 1$.

The solution set of the system is the *intersection* of the two graphs. The shaded regions overlap as shown to the right. This is the solution set of the system.

All points in the shaded region satisfy *both* inequalities.

$$4x^2 + y^2 < 16$$
$$-x + 2y > 2$$

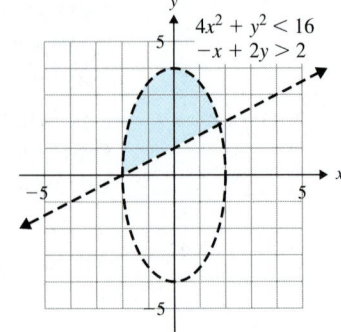

You Try 3

Graph the solution set of the system.

$$y < x - 4$$
$$x^2 + y > 0$$

Example 4

Graph the solution set of the system.

In-Class Example 4
Use Example 4.

$$x \geq 0$$
$$y - x^2 \leq 1$$
$$x^2 + y^2 \leq 4$$

Solution

First, the graph of $x = 0$ is the y-axis. Therefore, the graph of $x \geq 0$ consists of quadrants I and IV. See the figure below left.

Graph the *parabola, $y - x^2 = 1$,* as a solid curve since the inequality is $\leq$. Rewrite the inequality as $y \leq x^2 + 1$ to determine that the vertex is $(0, 1)$. The test point $(0, 0)$ satisfies the inequality $y \leq x^2 + 1$, so shade outside of the parabola. See the figure below center.

The graph of $x^2 + y^2 \leq 4$ is the inside of the circle $x^2 + y^2 = 4$. See the figure below right.

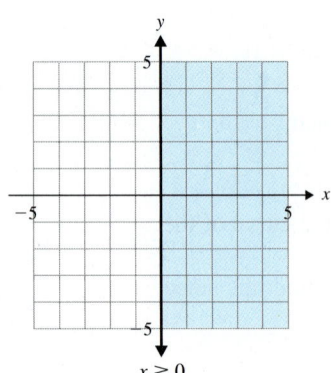

$x \geq 0$

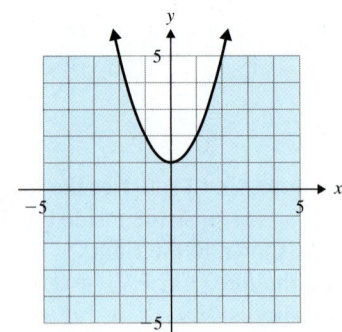

$y - x^2 \leq 1$

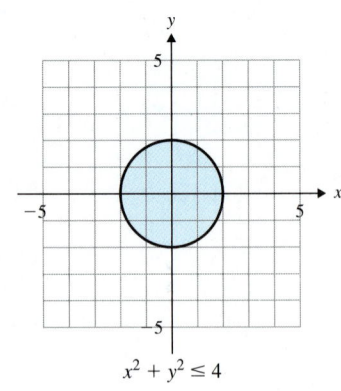

$x^2 + y^2 \leq 4$

Finally, the graph of the solution set of the system is the intersection, or overlap, of these three regions, as shown to the right.

All points in the shaded region satisfy each of the inequalities in the system

$$x \geq 0$$
$$y - x^2 \leq 1$$
$$x^2 + y^2 \leq 4$$

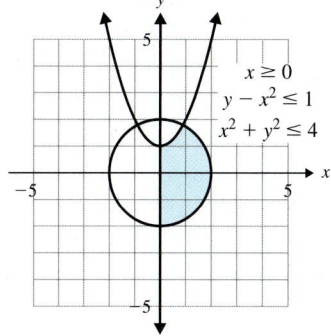

You Try 4

Graph the solution set of the system.

$$y \geq 0$$
$$x^2 + y^2 \leq 9$$
$$y - x^2 \geq -2$$

Answers to You Try Exercises

1)

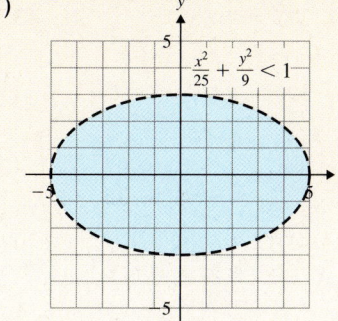

2)

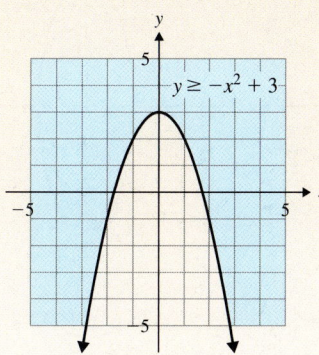

3)

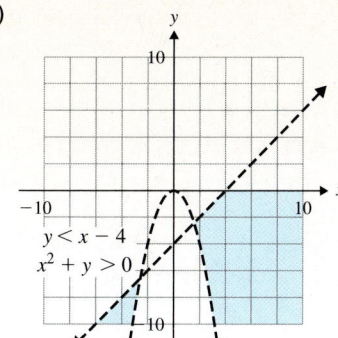

4)

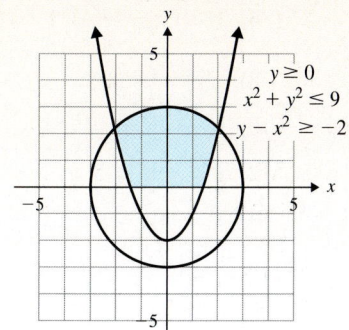

11.6 Exercises

*Additional answers can be found in the Answers to Exercises appendix.

Objective 1: Graph Second-Degree Inequalities

The graphs of second-degree inequalities are given below. For each, find three points which satisfy the inequality and three points that are not in the solution set.

1) $x^2 + y^2 \geq 36$

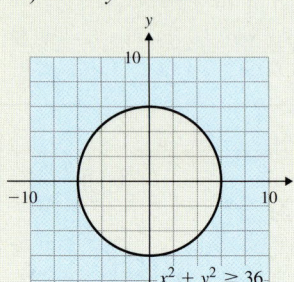

2) $x > y^2 + 2$

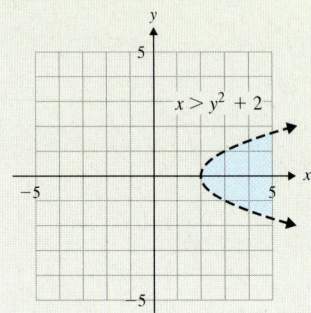

3) $4y^2 - x^2 < 4$

4) $25x^2 + 4y^2 \leq 100$

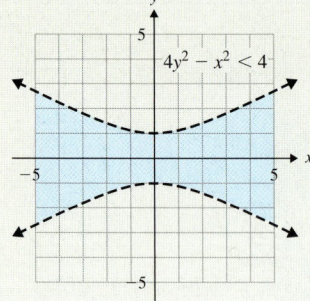

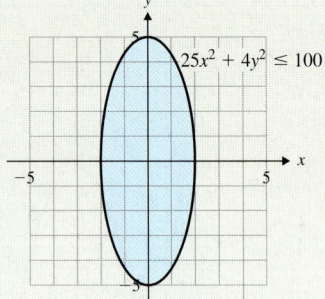

Graph each inequality.

5) $\dfrac{x^2}{9} + \dfrac{y^2}{16} < 1$

6) $y \geq (x - 2)^2 - 1$

7) $y < (x + 1)^2 + 3$

8) $x^2 + y^2 < 16$

9) $25y^2 - 4x^2 \geq 100$

10) $x^2 - 4y^2 \leq 4$

11) $x^2 + y^2 \geq 3$

12) $y \geq -x^2 + 5$

13) $x > y^2 - 2$

14) $x^2 + 4y^2 \leq 4$

15) $\dfrac{x^2}{4} - \dfrac{y^2}{9} \leq 1$

16) $x \geq -y^2 - 2y + 1$

17) $x^2 + (y + 4)^2 < 9$

18) $y^2 - \dfrac{x^2}{9} > 1$

19) $x^2 + 9y^2 \geq 9$

20) $(x + 3)^2 + y^2 \geq 4$

21) $y \leq -x^2 - 2x + 3$

22) $\dfrac{x^2}{36} + \dfrac{y^2}{25} > 1$

Objective 2: Graph Systems of Nonlinear Inequalities

Graph the solution set of each system.

23) $y \geq x - 2$
    $x^2 + y \leq 1$

24) $y + x < 3$
    $x^2 + y^2 < 9$

25) $2y - x > 4$
    $4x^2 + 9y^2 > 36$

26) $y < x - 3$
    $y - x^2 < -5$

27) $x^2 + y^2 \geq 16$
    $25x^2 - 4y^2 \leq 100$

28) $x^2 - y^2 \leq 1$
    $4x^2 + 9y^2 \leq 36$

29) $x^2 + y^2 \leq 9$
    $9x^2 + 4y^2 \geq 36$

30) $x^2 + y^2 \leq 16$
    $y^2 - 4x^2 \geq 4$

31) $y^2 - x^2 < 1$
    $4x^2 + y^2 > 16$

32) $x - y^2 > -4$
    $2y - 3x > 4$

33) $x + y^2 < 0$
    $x^2 + y^2 < 16$

34) $x^2 + y^2 \geq 4$
    $x + y \geq 1$

35) $\dfrac{x^2}{16} + \dfrac{y^2}{9} \leq 1$
    $4x^2 - y^2 \geq 16$

36) $\dfrac{x^2}{25} + \dfrac{y^2}{9} > 1$
    $y > x^2$

37) $y \leq -x$
    $y - 2x^2 \leq -2$

38) $x^2 + y^2 \geq 1$
    $x^2 + 25y^2 \leq 25$

39) $x \geq 0$
    $x^2 + y^2 \leq 25$

40) $y \leq 0$
    $x^2 + 4y^2 \leq 36$

41) $y < 0$
    $4x^2 - 9y^2 < 36$

42) $x > 0$
    $y^2 - 4x^2 < 4$

43) $x \geq 0$
    $y \geq x^2 + 4$
    $x + 2y \leq 12$

44) $y \geq 0$
    $4x^2 + 25y^2 \leq 100$
    $2x + 5y \leq 10$

45) $y < 0$
    $x^2 + y^2 < 9$
    $y > x + 1$

46) $y < 0$
    $y < -x^2 + 4$
    $y < \dfrac{1}{2}x - 1$

47) $y \geq 0$
    $y \leq x^2$
    $\dfrac{x^2}{4} + \dfrac{y^2}{9} \leq 1$

48) $x \geq 0$
    $y \leq x^2 - 1$
    $x^2 + y^2 \leq 16$

49) $y < 0$
    $4x^2 + 9y^2 < 36$
    $x^2 - y^2 > 1$

50) $x < 0$
    $x^2 + y^2 < 9$
    $y < -x^2$

51) $x \geq 0$
    $y \geq 0$
    $x^2 + y^2 \geq 4$
    $x \geq y^2$

52) $x \geq 0$
    $y \geq 0$
    $y \geq x^2$
    $x^2 + 4y^2 \geq 16$

Definition/Procedure	Example

11.1 Graphs of Other Useful Functions

Absolute value functions have V-shaped graphs. **(p. 699)**

Graph $f(x) = |x|$.

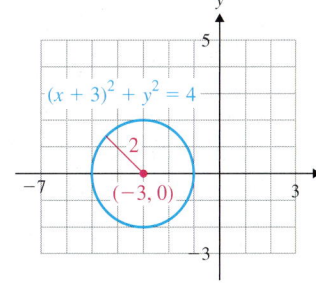

A **piecewise function** is a single function defined by two or more different rules. **(p. 700)**

$$f(x) = \begin{cases} x + 3, & x > -2 \\ -\dfrac{1}{2}x + 2, & x \le -2 \end{cases}$$

The **greatest integer function**, $f(x) = [\![x]\!]$, represents the largest integer less than or equal to x. **(p. 701)**

$[\![8.3]\!] = 8, \quad \left[\!\!\left[-4\dfrac{3}{8}\right]\!\!\right] = -5$

11.2 The Circle

The Midpoint Formula

If (x_1, y_1) and (x_2, y_2) are the endpoints of a line segment, then the **midpoint** of the segments has coordinates

$$\left(\frac{x_1 + x_2}{2}, \frac{y_1 + y_2}{2}\right) \text{ (p. 708)}$$

Find the midpoint of the line segment with endpoints $(-2, 5)$ and $(6, 3)$.

$$\text{Midpoint} = \left(\frac{-2 + 6}{2}, \frac{5 + 3}{2}\right) = \left(\frac{4}{2}, \frac{8}{2}\right) = (2, 4)$$

Parabolas, circles, ellipses, and hyperbolas are called **conic sections.**

The **standard form for the equation of a circle** with center (h, k) and radius r is

$$(x - h)^2 + (y - k)^2 = r^2 \text{ (p. 709)}$$

Graph $(x + 3)^2 + y^2 = 4$.

The center is $(-3, 0)$. The radius is $\sqrt{4} = 2$.

The **general form for the equation of a circle** is

$$Ax^2 + Ay^2 + Cx + Dy + E = 0$$

where A, C, D, and E are real numbers.

To rewrite the equation in the form $(x - h)^2 + (y - k)^2 = r^2$, divide the equation by A so that the coefficient of each squared term is 1, then complete the square on x and on y to put it into standard form. **(p. 710)**

Write $x^2 + y^2 - 16x + 4y + 67 = 0$ in the form $(x - h)^2 + (y - k)^2 = r^2$.

Group the x-terms together and group the y-terms together.

$$(x^2 - 16x) + (y^2 + 4y) = -67$$

Complete the square for each group of terms.

$$(x^2 - 16x + 64) + (y^2 + 4y + 4) = -67 + 64 + 4$$
$$(x - 8)^2 + (y + 2)^2 = 1$$

Definition/Procedure	Example

11.3 The Ellipse

The **standard form for the equation of an ellipse** is

$$\frac{(x - h)^2}{a^2} + \frac{(y - k)^2}{b^2} = 1$$

The center of the ellipse is (h, k). **(p. 715)**

Graph $\dfrac{(x - 1)^2}{9} + \dfrac{(y - 2)^2}{4} = 1$.

The center is $(1, 2)$.
$a = \sqrt{9} = 3$
$b = \sqrt{4} = 2$

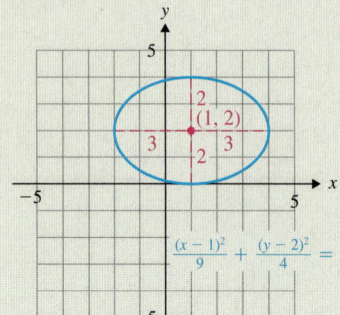

11.4 The Hyperbola

Standard Form for the Equation of a Hyperbola

1) A hyperbola with center (h, k) with branches that open in the *x-direction* has equation

$$\frac{(x - h)^2}{a^2} - \frac{(y - k)^2}{b^2} = 1$$

2) A hyperbola with center (h, k) with branches that open in the *y-direction* has equation

$$\frac{(y - k)^2}{b^2} - \frac{(x - h)^2}{a^2} = 1$$

Notice in 1) that $\dfrac{(x - h)^2}{a^2}$ is the positive quantity, and the branches open in the *x-direction*.

In 2), the positive quantity is $\dfrac{(y - k)^2}{b^2}$, and the branches open in the *y-direction*. **(p. 722)**

Graph $\dfrac{(y - 1)^2}{9} - \dfrac{(x - 4)^2}{4} = 1$.

The center is $(4, 1)$, $a = \sqrt{4} = 2$, and $b = \sqrt{9} = 3$.

Use the center, $a = 2$, and $b = 3$ to draw the reference rectangle. The diagonals of the rectangle are the asymptotes of the hyperbola.

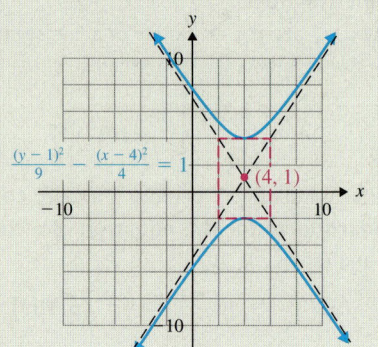

A Nonstandard Form for the Equation of a Hyperbola
The graph of the equation $xy = c$, where c is a nonzero constant, is a hyperbola whose asymptotes are the *x*- and *y*-axes. **(p. 725)**

Graph $xy = 4$.

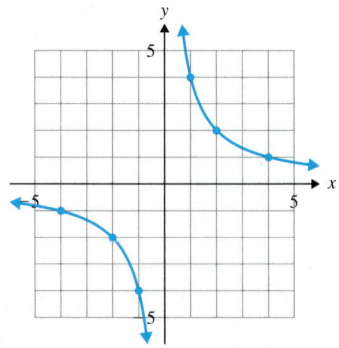

11.5 Nonlinear Systems of Equations

A **nonlinear system of equations** is a system in which at least one of the equations is not linear. We can solve nonlinear systems by substitution or the elimination method. **(p. 734)**

Solve $\quad x - y^2 = 3 \qquad$ (1)
$\qquad\quad x - 2y = 6 \qquad$ (2)

$\qquad\quad x - y^2 = 3 \qquad$ (1) $\qquad$ Solve equation (1) for x.
$\qquad\qquad\quad x = y^2 + 3 \quad$ (3)

Definition/Procedure	Example

	Substitute $x = y^2 + 3$ into equation (2).
	
	$$\begin{aligned} (y^2 + 3) - 2y &= 6 \\ y^2 - 2y - 3 &= 0 \quad \text{Subtract 6.} \\ (y - 3)(y + 1) &= 0 \quad \text{Factor.} \\ y - 3 = 0 \quad \text{or} \quad y + 1 &= 0 \quad \text{Set each factor equal to 0.} \\ y = 3 \quad \text{or} \quad y &= -1 \quad \text{Solve.} \end{aligned}$$
	
	Substitute each value into equation (3).
	
	$y = 3:$ $x = y^2 + 3$ $y = -1:$ $x = y^2 + 3$
	           $x = (3)^2 + 3$ $x = (-1)^2 + 3$
	           $x = 12$ $x = 4$
	
	The proposed solutions are $(12, 3)$ and $(4, -1)$.
	Verify that they also satisfy (2).
	The solution set is $\{(12, 3), (4, -1)\}$.

11.6 Second-Degree Inequalities and Systems of Inequalities

To graph a **second-degree inequality,** graph the boundary curve, choose a test point, and shade the appropriate region. **(p. 740)**	Graph $x^2 + y^2 < 9$. Graph the *circle* $x^2 + y^2 = 9$ as a dotted curve since the inequality is $<$. Choose a test point not on the boundary curve: $(0, 0)$. Since the test point $(0, 0)$ satisfies the inequality, shade the region inside the circle. 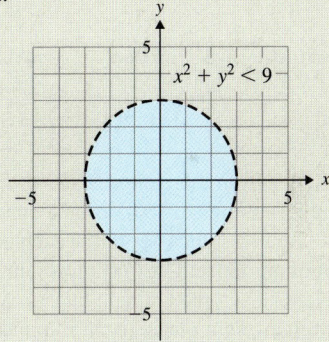 All points in the shaded region satisfy $x^2 + y^2 < 9$.
The **solution set of a system of inequalities** consists of the set of points that satisfy all inequalities in the system. The solution set is the intersection of the graphs of the individual inequalities. **(p. 741)**	Graph the solution set of the system $$\begin{aligned} x^2 + y &\leq 3 \\ x + y &\geq 1 \end{aligned}$$ First, graph the *parabola* $x^2 + y = 3$ as a solid curve since the inequality is $\leq$. The test point $(0, 0)$ satisfies the inequality $x^2 + y \leq 3$, so shade inside the parabola. Graph the *line* $x + y = 1$ as a solid line since the inequality is $\geq$. Shade above the line since the test point $(0, 0)$ does not satisfy $x + y \geq 1$. 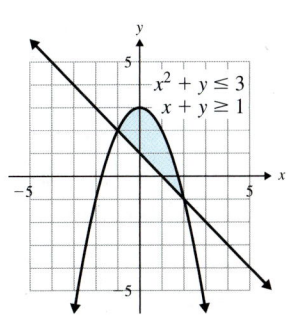 The solution set of the system is the *intersection* of the shaded regions of the two graphs. All points in the shaded region satisfy both inequalities.

*Additional answers can be found in the Answers to Exercises appendix.

(11.1) Graph each function, and identify the domain and range.

1) $g(x) = |x|$

2) $f(x) = |x + 1|$

3) $h(x) = |x| - 4$

4) $h(x) = |x - 1| + 2$

5) $k(x) = -|x| + 5$

6) If the graph of $f(x) = |x|$ is shifted right 5 units to obtain the graph of $g(x)$, write the equation of $g(x)$. $g(x) = |x - 5|$

Graph each piecewise function.

7) $f(x) = \begin{cases} -\dfrac{1}{2}x - 2, & x \le 2 \\ x - 3, & x > 2 \end{cases}$

8) $g(x) = \begin{cases} 1, & x < -3 \\ x + 4, & x \ge -3 \end{cases}$

Let $f(x) = [\![x]\!]$. Find the following function values.

9) $f\left(7\dfrac{2}{3}\right)$ 7

10) $f(2.1)$ 2

11) $f\left(-8\dfrac{1}{2}\right)$ -9

12) $f(-5.8)$ -6

13) $f\left(\dfrac{3}{8}\right)$ 0

Graph each greatest integer function.

14) $f(x) = [\![x]\!]$

15) $g(x) = \left[\!\!\left[\dfrac{1}{2}x\right]\!\!\right]$

16) To mail a letter from the United States to Mexico costs $0.60 for the first ounce, $0.85 over 1 ounce but less than or equal to 2 ounces, then $0.40 for each additional ounce or fraction of an ounce. Let $C(x)$ represent the cost of mailing a letter from the United States to Mexico, and let x represent the weight of the letter, in ounces. Graph $C(x)$ for any letter weighing up to (and including) 5 ounces. (www.usps.com)

(11.2) Find the midpoint of the line segment with the given endpoints.

17) $(3, 8)$ and $(5, 2)$ $(4, 5)$

18) $(-6, 1)$ and $(-2, -1)$ $(-4, 0)$

19) $(7, -3)$ and $(6, -4)$ $\left(\dfrac{13}{2}, -\dfrac{7}{2}\right)$

20) $\left(\dfrac{2}{3}, \dfrac{1}{4}\right)$ and $\left(-\dfrac{1}{6}, \dfrac{5}{8}\right)$ $\left(\dfrac{1}{4}, \dfrac{7}{16}\right)$

Identify the center and radius of each circle and graph.

21) $(x + 3)^2 + (y - 5)^2 = 36$

22) $x^2 + (y + 4)^2 = 9$

23) $x^2 + y^2 - 10x - 4y + 13 = 0$

24) $x^2 + y^2 + 4x + 16y + 52 = 0$

Find an equation of the circle with the given center and radius.

25) Center $(3, 0)$; radius $= 4$ $(x - 3)^2 + y^2 = 16$

26) Center $(1, -5)$; radius $= \sqrt{7}$ $(x - 1)^2 + (y + 5)^2 = 7$

(11.3)

27) When is an ellipse also a circle? when $a = b$ in $\dfrac{(x - h)^2}{a^2} + \dfrac{(y - k)^2}{b^2} = 1$

Identify the center of the ellipse and graph the equation.

28) $\dfrac{x^2}{25} + \dfrac{y^2}{36} = 1$

29) $\dfrac{(x + 3)^2}{9} + \dfrac{(y - 3)^2}{4} = 1$

30) $(x - 4)^2 + \dfrac{(y - 2)^2}{16} = 1$

31) $25x^2 + 4y^2 = 100$

Write each equation in standard form and graph.

32) $4x^2 + 9y^2 + 8x + 36y + 4 = 0$ $\dfrac{(x + 1)^2}{9} + \dfrac{(y + 2)^2}{4} = 1$

33) $25x^2 + 4y^2 - 100x = 0$ $\dfrac{(x - 2)^2}{4} + \dfrac{y^2}{25} = 1$

(11.4)

34) How can you distinguish between the equation of an ellipse and the equation of a hyperbola?

Identify the center of the hyperbola and graph the equation.

35) $\dfrac{y^2}{9} - \dfrac{x^2}{25} = 1$

36) $\dfrac{(y - 3)^2}{4} - \dfrac{(x + 2)^2}{9} = 1$

37) $\dfrac{(x + 1)^2}{4} - \dfrac{(y + 2)^2}{4} = 1$

38) $16x^2 - y^2 = 16$

39) Graph $xy = 6$.

40) Write the equations of the asymptotes of the graph in Exercise 35. $y = \dfrac{3}{5}x$ and $y = -\dfrac{3}{5}x$

Write each equation in standard form and graph.

41) $16y^2 - x^2 + 2x + 96y + 127 = 0$ $(y + 3)^2 - \dfrac{(x - 1)^2}{16} = 1$

42) $x^2 - y^2 - 4x + 6y - 9 = 0$ $\dfrac{(x - 2)^2}{4} - \dfrac{(y - 3)^2}{4} = 1$

Graph each function. Identify the domain and range.

43) $h(x) = 2\sqrt{1 - \dfrac{x^2}{9}}$

44) $f(x) = -\sqrt{4 - x^2}$

(11.2–11.4) Determine whether the graph of each equation is a parabola, circle, ellipse, or hyperbola, then graph each equation.

45) $x^2 + 9y^2 = 9$

46) $x^2 + y^2 = 25$

47) $x = -y^2 + 6y - 5$

48) $x^2 - y = 3$

49) $\dfrac{(x - 3)^2}{16} - \dfrac{(y - 4)^2}{25} = 1$

50) $\dfrac{(x + 3)^2}{16} + \dfrac{(y + 1)^2}{25} = 1$

51) $x^2 + y^2 - 2x + 2y - 2 = 0$

52) $4y^2 - 9x^2 = 36$

53) $y = \dfrac{1}{2}(x + 2)^2 + 1$

54) $x^2 + y^2 - 6x - 8y + 16 = 0$

(11.5)

55) If a nonlinear system of equations consists of an ellipse and a hyperbola, how many possible solutions can the system have? $0, 1, 2, 3,$ or 4

56) If a nonlinear system of equations consists of a line and a circle, how many possible solutions can the system have? $0, 1,$ or 2

Solve each system.

57) $-4x^2 + 3y^2 = 3$ $\{(6, 7), (6, -7), (-6, 7), (-6, -7)\}$
$7x^2 - 5y^2 = 7$

58) $y - x^2 = 7$ $\{(0, 7)\}$
$3x^2 + 4y = 28$

59) $y = 3 - x^2$ $\{(1, 2), (-2, -1)\}$
$x - y = -1$

60) $x^2 + y^2 = 9$ $\{(2, \sqrt{5}), (2, -\sqrt{5})\}$
$8x + y^2 = 21$

61) $4x^2 + 9y^2 = 36$ $\varnothing$
$y = \dfrac{1}{3}x - 5$

62) $4x + 3y = 0$ $\left\{\left(-\dfrac{3}{2}, 2\right), \left(\dfrac{3}{2}, -2\right)\right\}$
$4x^2 + 4y^2 = 25$

Write a system of equations and solve.

63) Find two numbers whose product is 36 and whose sum is 13. 9 and 4

64) The perimeter of a rectangular window is 78 in., and its area is 378 in.^{2}. Find the dimensions of the window. 18 in. $\times$ 21 in.

(11.6) Graph each inequality.

65) $x^2 + y^2 \leq 4$

66) $y > -(x - 3)^2 + 2$

67) $\dfrac{x^2}{9} + \dfrac{y^2}{4} > 1$

68) $\dfrac{x^2}{9} - \dfrac{y^2}{4} \geq 1$

69) $x < -y^2 + 2y - 12$

70) $4x^2 + 25y^2 < 100$

Graph the solution set of each system.

71) $y - x^2 > 2$
$x + y > 5$

72) $x^2 + y^2 \leq 25$
$y - x \leq 2$

73) $\dfrac{x^2}{16} + \dfrac{y^2}{25} < 1$
$y + 3 > x^2$

74) $x^2 - y^2 \leq 4$
$x^2 + y^2 \geq 16$

75) $y \geq 0$
$x^2 + y^2 \leq 36$
$x + y^2 \leq 0$

76) $x \leq 0$
$4x^2 + 25y^2 \geq 100$
$-x + y \leq 1$

*Additional answers can be found in the Answers to Exercises appendix.

1) Determine the midpoint of the line segment with endpoints $(2, 1)$ and $(10, -6)$. $\left(6, -\dfrac{5}{2}\right)$

Graph each function.

2) $g(x) = |x| + 2$

3) $h(x) = -|x - 1|$

4) $f(x) = \begin{cases} x + 3, & x > -1 \\ -2x - 5, & x \leq -1 \end{cases}$

Determine whether the graph of each equation is a parabola, circle, ellipse, or hyperbola, then graph each equation.

5) $\dfrac{(x - 2)^2}{25} + \dfrac{(y + 3)^2}{4} = 1$

6) $y = -2x^2 + 6$

7) $y^2 - 4x^2 = 16$

8) $x^2 + (y - 1)^2 = 9$

9) $xy = 2$

10) Write $x^2 + y^2 + 2x - 6y - 6 = 0$ in the form $(x - h)^2 + (y - k)^2 = r^2$. Identify the center and radius, and graph the equation.

11) Write an equation of the circle with center $(5, 2)$ and radius $\sqrt{11}$. $(x - 5)^2 + (y - 2)^2 = 11$

12) The Colosseum in Rome is an ellipse measuring 188 m long and 156 m wide. If the Colosseum is represented on a Cartesian coordinate system with the center of the ellipse at the origin and the longer axis along the x-axis, write an equation of this elliptical structure. (www.romaviva.ccm/Colosseo/colosseum.htm) $\dfrac{x^2}{8836} + \dfrac{y^2}{6084} = 1$

13) Graph $f(x) = -\sqrt{25 - x^2}$. State the domain and range.

14) Suppose a nonlinear system consists of the equation of a parabola and a circle.

   a) Sketch the different ways in which the graphs can intersect.

   b) Make a sketch in which the graphs do not intersect.

   c) How many possible solutions can the system have? 0, 1, 2, 3, or 4

Solve each system.

15) $\begin{aligned} x - 2y^2 &= -1 \\ x + 4y &= -1 \end{aligned}$ $\{(-1, 0), (7, -2)\}$

16) $\begin{aligned} 2x^2 + 3y^2 &= 21 \\ -x^2 + 12y^2 &= 3 \end{aligned}$ $\{(3, 1), (3, -1), (-3, 1), (-3, -1)\}$

17) The perimeter of a rectangular picture frame is 44 in. The area is 112 in^2. Find the dimensions of the frame. 8 in. × 14 in.

Graph each inequality.

18) $y \geq x^2 - 2$

19) $x^2 - 4y^2 < 36$

Graph the solution set of each system.

20) $\begin{aligned} x^2 + y^2 &> 9 \\ y &< 2x - 1 \end{aligned}$

21) $\begin{aligned} x &\geq 0 \\ y &\leq x^2 \\ 4x^2 + 9y^2 &\leq 36 \end{aligned}$

Additional answers can be found in the Answers to Exercises appendix.

Perform the indicated operations and simplify.

1) $\dfrac{1}{6} - \dfrac{11}{12}$ $-\dfrac{3}{4}$

2) $16 + 20 \div 4 - (5 - 2)^2$ 12

Find the area and perimeter of each figure.

3)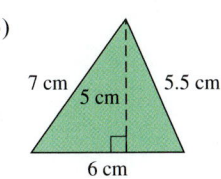

7 cm   5 cm   5.5 cm

6 cm

$A = 15 \text{ cm}^2; P = 18.5 \text{ cm}$

4)

5 in.

4 in.

10 in.

18 in.

$A = 128 \text{ in}^2; P = 56 \text{ in.}$

Evaluate.

5) $(-1)^5$ -1

6) 2^4 16

7) Simplify $\left(\dfrac{2a^8b}{a^2b^{-4}}\right)^{-3}$.

8) Solve $\dfrac{3}{8}k + 11 = -4$. $\{-40\}$

9) Solve for n. $an + z = c$

10) Solve $8 - 5p \le 28$. $[-4, \infty)$

11) *Write an equation and solve.*
The sum of three consecutive odd integers is 13 more than twice the largest integer. Find the numbers. 15, 17, 19

12) Find the slope of the line containing the points $(-6, 4)$ and $(2, -4)$. -1

13) What is the slope of the line with equation $y = 3$? 0

14) Graph $y = -2x + 5$.

15) Write the slope-intercept form of the line containing the points $(-4, 7)$ and $(4, 1)$. $y = -\dfrac{3}{4}x + 4$

16) Solve the system $\begin{aligned} 3x + 4y &= 3 \\ 5x + 6y &= 4 \end{aligned}$ $\left(-1, \dfrac{3}{2}\right)$

17) *Write a system of equations and solve.*
How many milliliters of an 8% alcohol solution and how many milliliters of a 16% alcohol solution must be mixed to make 20 mL of a 14% alcohol solution? 5 mL of 8%, 15 mL of 16%

18) Subtract $5p^2 - 8p + 4$ from $2p^2 - p + 10$. $-3p^2 + 7p + 6$

19) Multiply and simplify. $(4w - 3)(2w^2 + 9w - 5)$ $8w^3 + 30w^2 - 47w + 15$

20) Divide. $(x^3 - 7x - 36) \div (x - 4)$ $x^2 + 4x + 9$

Factor completely.

21) $6c^2 - 14c + 8$ $2(3c - 4)(c - 1)$

22) $m^3 - 8$ $(m - 2)(m^2 + 2m + 4)$

23) Solve $(x + 1)(x + 2) = 2(x + 7) + 5x$. $\{-2, 6\}$

24) Multiply and simplify. $\dfrac{a^2 + 3a - 54}{4a + 36} \cdot \dfrac{10}{36 - a^2}$

25) Simplify $\dfrac{\dfrac{t^2 - 9}{4}}{\dfrac{t - 3}{24}}$. $6(t + 3)$

26) Solve $|3n + 11| = 7$. $\left\{-6, -\dfrac{4}{3}\right\}$

27) Solve $|5r + 3| > 12$. $(-\infty, -3) \cup \left(\dfrac{9}{5}, \infty\right)$

Simplify. Assume all variables represent nonnegative real numbers.

28) $\sqrt{75}$ $5\sqrt{3}$

29) $\sqrt[3]{48}$ $2\sqrt[3]{6}$

30) $\sqrt[3]{27a^5b^{13}}$ $3ab^4\sqrt[3]{a^2b}$

31) $(16)^{-3/4}$ $\dfrac{1}{8}$

32) $\dfrac{18}{\sqrt{12}}$ $3\sqrt{3}$

33) Rationalize the denominator of $\dfrac{5}{\sqrt{3} + 4}$. $\dfrac{20 - 5\sqrt{3}}{13}$

Solve.

34) $(2p - 1)^2 + 16 = 0$ $\left\{\dfrac{1}{2} - 2i, \dfrac{1}{2} + 2i\right\}$

35) $y^2 = -7y - 3$ $\left\{-\dfrac{7}{2} - \dfrac{\sqrt{37}}{2}, -\dfrac{7}{2} + \dfrac{\sqrt{37}}{2}\right\}$

36) Solve $25p^2 \le 144$. $\left[-\dfrac{12}{5}, \dfrac{12}{5}\right]$

37) Solve $\dfrac{t - 3}{2t + 5} > 0$. $\left(-\infty, -\dfrac{5}{2}\right) \cup (3, \infty)$

38) Given the relation $\{(4, 0), (3, 1), (3, -1), (0, 2)\}$,
   a) what is the domain? $\{0, 3, 4\}$
   b) what is the range? $\{-1, 0, 1, 2\}$
   c) is the relation a function? no

39) Graph $f(x) = \sqrt{x}$ and $g(x) = \sqrt{x + 3}$ on the same axes.

40) $f(x) = 3x - 7$ and $g(x) = 2x + 5$.
   a) Find $g(-4)$. -3
   b) Find $(g \circ f)(x)$. $6x - 9$
   c) Find x so that $f(x) = 11$. 6

41) Given the function $f = \{(-5, 9), (-2, 11), (3, 14), (7, 9)\}$,
   a) is f one-to-one? no
   b) does f have an inverse? no

42) Find an equation of the inverse of $f(x) = \dfrac{1}{3}x + 4$. $f^{-1}(x) = 3x - 12$

Solve.

43) $8^{5t} = 4^{t-3}$ $\left\{-\dfrac{6}{13}\right\}$

44) $\log_3 (4n - 11) = 2$ $\{5\}$

45) Evaluate $\log 100$. 2

46) Graph $f(x) = \log_2 x$.

47) Solve $e^{3k} = 8$. Give an exact solution and an approximation to four decimal places. $\left\{\dfrac{\ln 8}{3}\right\}$; $\{0.6931\}$

48) Graph $\dfrac{y^2}{4} - \dfrac{x^2}{9} = 1$.

49) Graph $x^2 + y^2 - 2x + 6y - 6 = 0$.

50) Solve the system. $\begin{aligned} y - 5x^2 &= 3 \\ x^2 + 2y &= 6 \end{aligned}$ $(0, 3)$

Sequences and Series

Algebra at Work: Finance

Many people deposit money in an account on a regular basis to save for their children's college educations. The formula used to determine how much money will be in such an account results from a geometric series since it is found from a sum of deposits and interest payments over a period of time.

Mr. and Mrs. Chesbit consult a financial planner, George, for advice on how much money they will save for their daughter's college education if they deposit $200 per month for 18 yr into an account earning 6% annual interest compounded monthly. Using a formula based on a geometric series, George calculates that they will have saved $77,470.64. Of this amount, $43,200 is money they have invested, and the rest is interest.

We will learn more about geometric series in this chapter.

Section 12.1 Sequences and Series

1. Define Infinite Sequence, Finite Sequence, and General Term of a Sequence

Suppose that a math class is conducting an experiment. On the first day, 2 pennies will be placed in a jar. On the second day, 4 pennies will be placed in the jar. On the third day, students will put 6 pennies in the jar, on the fourth day they will put 8 pennies in the jar, and so on. The number of pennies placed in the jar each day forms a **sequence**.

If the students continue to deposit pennies into the jar in this way indefinitely, we obtain the sequence

$$2, 4, 6, 8, 10, 12, 14, 16, \ldots$$

This sequence has an infinite number of terms and is called an *infinite sequence*.

Suppose, however, that the students will stop putting pennies in the jar after 5 days. Then we obtain the sequence

$$2, 4, 6, 8, 10$$

This sequence has a finite number of terms and is called a *finite sequence*.

The number of pennies placed in the jar on a given day is related to how many days into the experiment the class is.

2	4	6	8	10
↑	↑	↑	↑	↑
Day 1	Day 2	Day 3	Day 4	Day 5

The number of pennies placed in the jar on day n is 2n, where n is a natural number beginning with 1.

Each number in the sequence $2, 4, 6, 8, 10, \ldots$ is called a **term** of the sequence. The terms of a sequence are related to the set of natural numbers, and a sequence is a function.

Definition

An **infinite sequence** is a function whose domain is the set of natural numbers.

A **finite sequence** is a function whose domain is the set of the first n natural numbers.

Previously, we used the notation $f(x)$ to denote a function. When describing sequences, however, we use the notation a_n (read *a sub n*).

To describe the sequence $2, 4, 6, 8, 10, \ldots$, we use the notation $a_n = 2n$.

The first term of the sequence is denoted by a_1, the second term is denoted by a_2, the third term by a_3, and so on. Therefore, using the formula $a_n = 2n$:

$$a_1 = 2(1) = 2 \quad \text{Let } n = 1 \text{ to find the first term.}$$
$$a_2 = 2(2) = 4 \quad \text{Let } n = 2 \text{ to find the second term.}$$
$$a_3 = 2(3) = 6 \quad \text{Let } n = 3 \text{ to find the third term.}$$
$$a_4 = 2(4) = 8 \quad \text{Let } n = 4 \text{ to find the fourth term.}$$

etc.

The *n*th term of the sequence, a_n, is called the **general term** of the sequence. Therefore, the general term of the sequence $2, 4, 6, 8, 10, \ldots$ is $a_n = 2n$.

What is the difference between the two functions $f(x) = 2x$ and $a_n = 2n$? The domain of $f(x) = 2x$ is the set of real numbers, while the domain of $a_n = 2n$ is the set of natural numbers.

We can see how their graphs differ. The graphing calculator boxes illustrate that $f(x) = 2x$ is a line, while $a_n = 2n$ consists of function values found when n is a *natural number* so that the points are not connected.

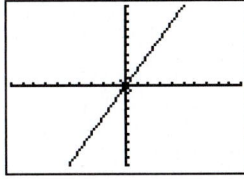

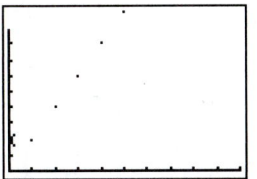

The graph of $f(x) = 2x$ The graph of the first five terms of $a_n = 2n$

2. Write the Terms of a Sequence

| Example 1 |

In-Class Example 1
Write the first five terms of
each sequence with general
term a_n.
a) $a_n = 2n + 7$
b) $a_n = 4 \cdot \left(\dfrac{1}{2}\right)^n$
c) $a_n = (-1)^{n-1} \cdot 4n$
answer: a) 9, 11, 13, 15, 17
b) 2, 1, $\dfrac{1}{2}, \dfrac{1}{4}, \dfrac{1}{8}$
c) 4, −8, 12, −16, 20

Write the first five terms of each sequence with general term a_n.

a) $a_n = 4n - 1$ b) $a_n = 3 \cdot \left(\dfrac{1}{2}\right)^n$ c) $a_n = (-1)^{n+1} \cdot 2n$

Solution

a) Evaluate $a_n = 4n - 1$ for $n = 1, 2, 3, 4,$ and 5.

n	$a_n = 4n - 1$
1	$a_1 = 4(1) - 1 = 3$
2	$a_2 = 4(2) - 1 = 7$
3	$a_3 = 4(3) - 1 = 11$
4	$a_4 = 4(4) - 1 = 15$
5	$a_5 = 4(5) - 1 = 19$

The first five terms of the sequence are 3, 7, 11, 15, 19.

b) Evaluate $a_n = 3 \cdot \left(\dfrac{1}{2}\right)^n$ for $n = 1, 2, 3, 4,$ and 5.

n	$a_n = 3 \cdot \left(\dfrac{1}{2}\right)^n$
1	$a_1 = 3 \cdot \left(\dfrac{1}{2}\right)^1 = \dfrac{3}{2}$
2	$a_2 = 3 \cdot \left(\dfrac{1}{2}\right)^2 = \dfrac{3}{4}$
3	$a_3 = 3 \cdot \left(\dfrac{1}{2}\right)^3 = \dfrac{3}{8}$
4	$a_4 = 3 \cdot \left(\dfrac{1}{2}\right)^4 = \dfrac{3}{16}$
5	$a_5 = 3 \cdot \left(\dfrac{1}{2}\right)^5 = \dfrac{3}{32}$

The first five terms of the sequence are $\dfrac{3}{2}, \dfrac{3}{4}, \dfrac{3}{8}, \dfrac{3}{16}, \dfrac{3}{32}$.

c) Evaluate $a_n = (-1)^{n+1} \cdot 2n$ for $n = 1, 2, 3, 4,$ and 5.

n	$a_n = (-1)^{n+1} \cdot 2n$
1	$a_1 = (-1)^{1+1} \cdot 2(1) = (-1)^2 \cdot 2 = 2$
2	$a_2 = (-1)^{2+1} \cdot 2(2) = (-1)^3 \cdot 4 = -4$
3	$a_3 = (-1)^{3+1} \cdot 2(3) = (-1)^4 \cdot 6 = 6$
4	$a_4 = (-1)^{4+1} \cdot 2(4) = (-1)^5 \cdot 8 = -8$
5	$a_5 = (-1)^{5+1} \cdot 2(5) = (-1)^6 \cdot 10 = 10$

The first five terms of the sequence are $2, -4, 6, -8, 10$. Notice that the terms of this sequence have alternating signs. A sequence in which the signs of the terms alternate is called an **alternating sequence**.

You Try 1

Write the first five terms of each sequence with general term a_n.

a) $a_n = 2n - 5$ b) $a_n = \left(\dfrac{1}{3}\right)^{n-1}$ c) $a_n = (-1)^n \cdot n$

Example 2

In-Class Example 2
The general term of a sequence is given by
$a_n = (-1)^n \cdot (n + 6)$. Find each of the following.
a) the first term of the sequence   b) a_8 c) the 21st term of the sequence
answer: a) -7 b) 14
c) -27

The general term of a sequence is given by $a_n = (-1)^n \cdot (2n + 1)$. Find each of the following.

a) the first term of the sequence

b) a_6

c) the 49th term of the sequence

Solution

a) The first term of the sequence is a_1. To find a_1, let $n = 1$ and evaluate.

$$a_n = (-1)^n \cdot (2n + 1)$$
$$a_1 = (-1)^1 \cdot [2(1) + 1] \qquad \text{Substitute 1 for } n.$$
$$a_1 = -1 \cdot (3)$$
$$a_1 = -3$$

b) a_6 is the sixth term of the sequence. Substitute 6 for n and evaluate.

$$a_n = (-1)^n \cdot (2n + 1)$$
$$a_6 = (-1)^6 \cdot [2(6) + 1] \qquad \text{Substitute 6 for } n.$$
$$a_6 = 1 \cdot (13)$$
$$a_6 = 13$$

c) The 49th term of the sequence is a_{49}. To find a_{49}, let $n = 49$ and evaluate.

$$a_n = (-1)^n \cdot (2n + 1)$$
$$a_{49} = (-1)^{49} \cdot [2(49) + 1] \qquad \text{Substitute 49 for } n.$$
$$a_{49} = -1 \cdot (99)$$
$$a_{49} = -99$$

You Try 2

The general term of a sequence is given by $a_n = \dfrac{(-1)^{n+1}}{5n}$. Find each of the following.

a)   the first term of the sequence   b) a_{10} c)   the 32nd term of the sequence

3. Find the General Term of a Sequence

Next we will look at the opposite procedure. Given the first few terms of a sequence we will find a formula for the general term, a_n. To do this, look for a pattern and try to find a relationship between the term and the term number.

Example 3

In-Class Example 3
Find a formula for the general term, a_n, of each sequence.
a) $4, 8, 12, 16, 20, \ldots$
b) $1, 3, 5, 7, 9, \ldots$
c) $\dfrac{3}{2}, \dfrac{3}{4}, \dfrac{1}{2}, \dfrac{3}{8}, \dfrac{3}{10}, \ldots$
d) $-\dfrac{1}{4}, \dfrac{1}{4}, -\dfrac{1}{4}, \dfrac{1}{4}, -\dfrac{1}{4}, \ldots$
answer: a) $a_n = 4n$
b) $a_n = 2n - 1$
c) $a_n = \dfrac{3}{2n}$ d) $a_n = \dfrac{(-1)^n}{4}$

Find a formula for the general term, a_n, of each sequence.

a) $1, 2, 3, 4, 5, \ldots$ b) $7, 14, 21, 28, 35, \ldots$

c) $\dfrac{1}{1}, \dfrac{1}{4}, \dfrac{1}{9}, \dfrac{1}{16}, \dfrac{1}{25}, \ldots$ d) $-3, 9, -27, 81, -243, \ldots$

Solution

a)   It is helpful to write each term with its term number below. Ask yourself, *"What is the relationship between the term and the term number?"*

Term:	1	2	3	4	5
Term number:	a_1	a_2	a_3	a_4	a_5

Each term is the same as the subscript on its term number. The *n*th term can be written as $a_n = n$.

b)   Write each term with its term number below. *What is the relationship between the term and the term number?*

Term:	7	14	21	28	35
Term number:	a_1	a_2	a_3	a_4	a_5

Each term is 7 times the subscript on its term number. The *n*th term may be written as $a_n = 7n$.

c)   Write each term with its term number below. *What is the relationship between the term and the term number?*

Term:	$\dfrac{1}{1}$	$\dfrac{1}{4}$	$\dfrac{1}{9}$	$\dfrac{1}{16}$	$\dfrac{1}{25}$
Term number:	a_1	a_2	a_3	a_4	a_5

Each numerator is 1. Each denominator is the square of the subscript on its term number. The *n*th term may be written as $a_n = \dfrac{1}{n^2}$.

d)   Write each term with its term number below. *What is the relationship between the term and the term number?*

Term:	-3	9	-27	81	-243
Term number:	a_1	a_2	a_3	a_4	a_5

The terms alternate in sign with the first term being negative. $(-1)^n$ will give us the desired alternating signs.

Disregarding the signs, each term is a power of 3: $3 = 3^1, 9 = 3^2, 27 = 3^3, 81 = 3^4, 243 = 3^5$, with the exponent being the subscript of the term number.

The *n*th term may be written as $a_n = (-1)^n \cdot 3^n$.

BE CAREFUL

There may be more than one general term, a_n, that produces the first few terms of a sequence.

You Try 3

Find a formula for the general term, a_n, of each sequence.

a) $11, 22, 33, 44, 55, \ldots$ b) $5, 6, 7, 8, 9, \ldots$

c) $\dfrac{1}{6}, \dfrac{1}{12}, \dfrac{1}{18}, \dfrac{1}{24}, \dfrac{1}{30}, \ldots$ d) $4, -16, 64, -256, 1024, \ldots$

4. Solve an Applied Problem Using a Sequence

Sequences can be used to model many real-life situations.

Example 4

In-Class Example 4
A country club needs to raise its membership dues by 5% per year for the next several years. If membership currently costs $3000 per year, write a sequence that represents the amount the members will pay for the next 4 yr.
answer: $3150, $3307.50, $3472.88, $3646.52

A university anticipates that it will need to increase its tuition 10% per year for the next several years. If tuition is currently $8000 per year, write a sequence that represents the amount of tuition students will pay each of the next 4 years.

Solution

Next year	Tuition: $8000 + 0.10($8000) = $8800
In 2 years	Tuition: $8800 + 0.10($8800) = $9680
In 3 years	Tuition: $9680 + 0.10($9680) = $10,648
In 4 years	Tuition: $10,648 + 0.10($10,648) = $11,712.80

The amount of tuition students will pay each of the next 4 years is given by the sequence

$8800, $9680, $10,648, $11,712.80

You Try 4

The number of customers paying their utility bills online has been increasing by 5% per year for the last several years. This year, 1200 customers of a particular utility company pay their bills online. Write a sequence that represents the number of customers paying their bills online each of the next 4 years.

Series

5. Understand the Meaning of Series and Summation Notation

If we add the terms of a sequence, we get a *series*.

Definition

A sum of the terms of a sequence is called a **series**.

Just like a sequence can be finite or infinite, a series can be finite or infinite. In this section, we will discuss only finite series.

Suppose the *sequence*

$$\$10, \$10.05, \$10.10, \$10.15, \$10.20, \$10.25$$

represents the amount of interest earned in an account each month over a 6-month period. Then the *series*

$$\$10 + \$10.05 + \$10.10 + \$10.15 + \$10.20 + \$10.25$$

represents the total amount of interest earned during the 6-month period. The total amount of interest earned is $60.75.

We use a shorthand notation, called **summation notation**, to denote a series when we know the general term, a_n, of the sequence. The Greek letter Σ (sigma) is used to mean *sum*. Instead of using the letter n, the letters $i, j,$ or k are usually used with Σ.

For example, $\sum_{i=1}^{4}(3i - 5)$ means "find the sum of the first four terms of the sequence defined by $a_n = 3n - 5$." We read $\sum_{i=1}^{4}(3i - 5)$ as "the sum of $3i - 5$ as i goes from 1 to 4."

The variable i is called the **index of summation**. The $i = 1$ under the Σ tells us that 1 is the first value to substitute for i, and the 4 above the Σ tells us that 4 is the last value to substitute for i. To evaluate $\sum_{i=1}^{4}(3i - 5)$, find the first four terms of the sequence by first substituting 1 for i, then 2, then 3, and finally let $i = 4$. Then add the terms of the sequence.

$$
\begin{array}{ccccccccc}
 & i=1 & & i=2 & & i=3 & & i=4 & \\
\sum_{i=1}^{4}(3i - 5) = & [3(1) - 5] & + & [3(2) - 5] & + & [3(3) - 5] & + & [3(4) - 5] & \\
= & -2 & + & 1 & + & 4 & + & 7 & = 10
\end{array}
$$

 **BE CAREFUL**

The i used in summation notation has no relationship with the complex number i.

6. Evaluate a Series

Example 5

In-Class Example 5
Evaluate each series.
a) $\sum_{i=1}^{5}\dfrac{-i + 6}{2}$
b) $\sum_{i=2}^{7}(-1)^{i} \cdot 2(i + 6)$

answer: a) 7.5 or $\dfrac{15}{2}$
b) -6

Evaluate each series.

a) $\displaystyle\sum_{i=1}^{5}\frac{i + 1}{3}$ b) $\displaystyle\sum_{i=3}^{6}(-1)^{i} \cdot 2^{i}$

Solution

a) i will start with 1 and end with 5. To evaluate the series, find the terms and then find their sum.

$$
\begin{array}{ccccccccccc}
 & & i=1 & & i=2 & & i=3 & & i=4 & & i=5 \\
\sum_{i=1}^{5}\dfrac{i + 1}{3} & = & \dfrac{1 + 1}{3} & + & \dfrac{2 + 1}{3} & + & \dfrac{3 + 1}{3} & + & \dfrac{4 + 1}{3} & + & \dfrac{5 + 1}{3} \\
 & = & \dfrac{2}{3} & + & \dfrac{3}{3} & + & \dfrac{4}{3} & + & \dfrac{5}{3} & + & \dfrac{6}{3} = \dfrac{20}{3}
\end{array}
$$

b) *i* will start with 3 and end with 6. Find the terms, then find their sum.

$$\sum_{i=3}^{6} (-1)^i \cdot 2^i = (-1)^3 \cdot 2^3 + (-1)^4 \cdot 2^4 + (-1)^5 \cdot 2^5 + (-1)^6 \cdot 2^6$$

	$i = 3$	$i = 4$	$i = 5$	$i = 6$	
$=$	-8	$+$ 16	$+$ (-32)	$+$ 64	$= 40$ ■

You Try 5

Evaluate each series.

a) $\displaystyle\sum_{i=1}^{6} (-1)^{i+1} \cdot (2i^2)$ b) $\displaystyle\sum_{i=4}^{8} (10 - i)$

7. Write a Series Using Summation Notation

To write a series using summation notation, we need to find a general term, a_n, that will pro-duce the terms in the sum. Remember, try to find a relationship between the term and the term number. There may be more than one way to represent a series in summation notation.

Example 6

Write each series using summation notation.

a) $\dfrac{1}{2} + \dfrac{2}{3} + \dfrac{3}{4} + \dfrac{4}{5} + \dfrac{5}{6} + \dfrac{6}{7}$ b) $9 + 16 + 25 + 36$

In-Class Example 6
Write each series using sum-mation notation.
a) $\dfrac{2}{3} + \dfrac{4}{5} + \dfrac{6}{7} + \dfrac{8}{9}$
b) $3 + 8 + 15 + 24$

answer: a) $\displaystyle\sum_{i=1}^{4} \dfrac{2i}{2i+1}$

b) $\displaystyle\sum_{i=2}^{5} (i^2 - 1)$

Solution

a) Find a general term, a_n, that will produce the terms in the series. Write each term with its term number below.

Term:	$\dfrac{1}{2}$	$\dfrac{2}{3}$	$\dfrac{3}{4}$	$\dfrac{4}{5}$	$\dfrac{5}{6}$	$\dfrac{6}{7}$
Term number:	a_1	a_2	a_3	a_4	a_5	a_6

The numerator of each term is the *same* as the subscript on its term number. The denominator of each term is *one more than* the subscript on its term number.

Therefore, the terms of this series can be produced if $a_n = \dfrac{n}{n+1}$ for $n = 1$ to 6.

In summation notation, we can write $\displaystyle\sum_{i=1}^{6} \dfrac{i}{i+1}$.

b) We will use this example to illustrate how two different summation notations can represent the same series.

i) Write each term with its term number below. Note that each term is a perfect square.

Term:	$9 = 3^2$	$16 = 4^2$	$25 = 5^2$	$36 = 6^2$
Term number:	a_1	a_2	a_3	a_4

The base of each exponential expression is *two more than* its term number. The terms of this series can be produced if $a_n = (n + 2)^2$ for $n = 1$ to 4.

In summation notation, we can write $\displaystyle\sum_{i=1}^{4} (i + 2)^2$.

ii) Write each term in the series as a perfect square:

$$9 + 16 + 25 + 36 = 3^2 + 4^2 + 5^2 + 6^2$$

If we let *i* begin at 3 and end at 6, we can write $\displaystyle\sum_{i=3}^{6} i^2$.

The series $9 + 16 + 25 + 36$ is a good example of one that can be written in summation notation in more than one way.   ■

You Try 6

Write each series using summation notation.

a) $7 + \dfrac{7}{2} + \dfrac{7}{3} + \dfrac{7}{4} + \dfrac{7}{5}$ b) $8 + 9 + 10 + 11 + 12 + 13$

8. Use Summation Notation to Represent the Average of a Group of Numbers

Summation notation is widely used in statistics. To find the average of a group of numbers, for example, we find the sum of the numbers and divide by the number of numbers in the group.

In statistics, we use *summation notation* to represent the **average** or **arithmetic mean** of a group of numbers. The arithmetic mean is represented by $\bar{x}$, and is given by the formula

$$\bar{x} = \frac{\displaystyle\sum_{i=1}^{n} x_i}{n}$$

where $x_1, x_2, x_3, \ldots, x_n$ are the numbers in the group and n is the number of numbers in the group.

Example 7

In-Class Example 7
Find the average of 16, 20, 24, 28, and 32.
answer: 24

The number of calories in 12-oz cans of five different soft drinks is 150, 120, 170, 150, and 140. What is the average number of calories in a 12-oz can of soda?

Solution

Since we must find the average of *five* numbers, the formula for $\bar{x}$ will be

$$\bar{x} = \frac{\displaystyle\sum_{i=1}^{5} x_i}{5}$$

where $x_1 = 150$, $x_2 = 120$, $x_3 = 170$, $x_4 = 150$, and $x_5 = 140$.

$$\bar{x} = \frac{150 + 120 + 170 + 150 + 140}{5}$$
$$= 146$$

The average number of calories in a 12-oz can of soda is 146. ■

You Try 7

A movie theater complex has eight separate theaters. They had movies starting between 7 P.M. and 8 P.M. on a Friday night. Attendance figures in these theaters were as follows: 138, 58, 79, 178, 170, 68, 115, and 94. Find the average attendance per theater.

Using Technology

We can use a graphing calculator to display the terms of a sequence or to find the sum of a finite number of terms of a sequence, which is the same as evaluating a series. For example, consider the sequence 1, 4, 7, 10, 13, ... A sequence of numbers is represented as a list on a graphing calculator. The list variables are L1, L2, L3, L4, L5, and L6 and can be accessed by pressing [2nd] [1], [2nd] [2], [2nd] [3], [2nd] [4], [2nd] [5], and [2nd] [6], respectively. Store the first five numbers in the sequence 1, 4, 7, 10, 13, . . . in L1 by pressing [STAT] [ENTER] and then entering each number followed by [ENTER], as shown at the left below.

The numbers stored in L1 can be displayed on the home screen by pressing [2nd] [1] [ENTER], as shown on the right screen below.

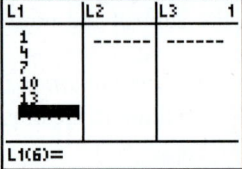

To find the sum of the terms of this sequence stored in L1, press [2nd] [STAT], then press the right arrow twice to MATH, select 5:sum(then enter [2nd] [1] [ )] on the home screen as shown. To find the sum of the 2nd, 3rd, and 4th terms of this sequence, enter "sum(L1, 2, 4)" as shown here.

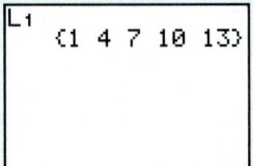

Given the nth term of an arithmetic sequence, a graphing calculator can display a finite number of terms. First change the mode of the calculator to sequence mode by pressing [MODE] and selecting SEQ in row 4, as shown at the left below.

The syntax for displaying terms of an arithmetic sequence is seq(nth term, n, starting index, ending index). For example, to display the first five terms of the arithmetic sequence with nth term $2n + 1$, enter seq($2n + 1, n, 1, 5$) on the home screen. To do this, press [2nd] [STAT], right arrow to OPS menu, and select 5:seq(2 [X,T,θ,n] + 1, [X,T,θ,n], 1, 5). To find the sum if the first five terms of the arithmetic sequence, press [2nd] [STAT] and right arrow twice to MATH menu and select 5:sum(followed by the sequence definition shown above, as shown at the right below.

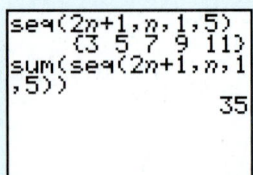

Evaluate each sum using a graphing calculator.

1) $\displaystyle\sum_{i=1}^{6} (3i + 8)$ 2) $\displaystyle\sum_{i=1}^{9} (2i - 14)$ 3) $\displaystyle\sum_{i=1}^{12} (-4i + 9)$

4) $\displaystyle\sum_{i=1}^{215} (i - 10)$ 5) $\displaystyle\sum_{i=1}^{100} \left(\frac{1}{2}i + 6\right)$ 6) $\displaystyle\sum_{i=1}^{20} (4i - 5)$

Answers to You Try Exercises

1) a) $-3, -1, 1, 3, 5$ b) $1, \dfrac{1}{3}, \dfrac{1}{9}, \dfrac{1}{27}, \dfrac{1}{81}$ c) $-1, 2, -3, 4, -5$ 2) a) $a_1 = \dfrac{1}{5}$ b) $a_{10} = -\dfrac{1}{50}$

c) $a_{32} = -\dfrac{1}{160}$ 3) a) $a_n = 11n$ b) $a_n = n + 4$ c) $a_n = \dfrac{1}{6n}$ d) $a_n = (-1)^{n+1} \cdot 4^n$

4) $1260, 1323, 1389, 1458$ 5) a) -42 b) 20 6) a) $\displaystyle\sum_{i=1}^{5} \dfrac{7}{i}$ b) $\displaystyle\sum_{i=8}^{13} i$ or $\displaystyle\sum_{i=1}^{6} (i + 7)$

7) 112.5

Answers to Technology Exercises

1) 111 2) -36 3) -204 4) $21{,}070$ 5) 3125 6) 740

12.1 Exercises

Additional answers can be found in the Answers to Exercises appendix.

Objective 2: Write the Terms of a Sequence

Write out the first five terms of each sequence.

1) $a_n = n + 2$ $3, 4, 5, 6, 7$

2) $a_n = n - 4$ $-3, -2, -1, 0, 1$

3) $a_n = 3n - 4$ $-1, 2, 5, 8, 11$

4) $a_n = 4n + 1$ $5, 9, 13, 17, 21$

5) $a_n = 2n^2 - 1$ $1, 7, 17, 31, 49$

6) $a_n = 2n^2 + 3$ $5, 11, 21, 35, 53$

7) $a_n = 3^{n-1}$ $1, 3, 9, 27, 81$

8) $a_n = 2^n$ $2, 4, 8, 16, 32$

9) $a_n = 5 \cdot \left(\dfrac{1}{2}\right)^n$

10) $a_n = 6 \cdot \left(\dfrac{1}{3}\right)^{n-1}$

VIDEO 11) $a_n = (-1)^{n+1} \cdot 7n$ $7, -14, 21, -28, 35$

12) $a_n = (-1)^n \cdot (n + 1)$ $-2, 3, -4, 5, -6$

13) $a_n = \dfrac{n - 4}{n + 3}$

14) $a_n = \dfrac{n^2 - 1}{n}$

Given the general term of each sequence, find each of the following.

VIDEO 15) $a_n = 3n + 2$

a) the first term of the sequence 5

b) a_5 17

c) the 28th term 86

16) $a_n = 3n - 11$

a) the first term of the sequence -8

b) a_7 10

c) the 32nd term 85

17) $a_n = \dfrac{n - 4}{n + 6}$

a) a_1 $-\dfrac{3}{7}$

b) a_2 $-\dfrac{1}{4}$

c) the 16th term $\dfrac{6}{11}$

18) $a_n = \dfrac{3n - 1}{4n + 5}$

a) a_1 $\dfrac{2}{9}$

b) a_{10} $\dfrac{29}{45}$

c) the 21st term $\dfrac{62}{89}$

19) $a_n = 10 - n^2$

a) the first term of the sequence 9

b) the 6th term -26

c) a_{20} -390

20) $a_n = 4n^2 - 9$

a) the first term of the sequence -5

b) the fourth term 55

c) the 13th term 667

Objective 3: Find the General Term of a Sequence

Find a formula for the general term, a_n, of each sequence.

21) $2, 4, 6, 8, \ldots$ $a_n = 2n$

22) $9, 18, 27, 36, \ldots$ $a_n = 9n$

23) $1, 4, 9, 16, \ldots$ $a_n = n^2$

24) $1, 8, 27, 64, \ldots$ $a_n = n^3$

25) $\dfrac{1}{3}, \dfrac{1}{9}, \dfrac{1}{27}, \dfrac{1}{81}, \ldots$ $a_n = \left(\dfrac{1}{3}\right)^n$

26) $\dfrac{4}{5}, \dfrac{4}{25}, \dfrac{4}{125}, \dfrac{4}{625}, \ldots$ $a_n = \dfrac{4}{5^n}$

VIDEO 27) $\dfrac{1}{2}, \dfrac{2}{3}, \dfrac{3}{4}, \dfrac{4}{5}, \ldots$ $a_n = \dfrac{n}{n + 1}$

28) $1, \dfrac{1}{2}, \dfrac{1}{3}, \dfrac{1}{4}, \ldots$ $a_n = \dfrac{1}{n}$

29) $5, -10, 15, -20, \ldots$ $a_n = (-1)^{n+1} \cdot (5n)$

30) $-2, 4, -6, 8, \ldots$ $a_n = (-1)^n \cdot (2n)$

31) $-\dfrac{1}{2}, \dfrac{1}{4}, -\dfrac{1}{8}, \dfrac{1}{16}, \ldots$ $a_n = (-1)^n \left(\dfrac{1}{2}\right)^n$

32) $\dfrac{1}{4}, -\dfrac{1}{16}, \dfrac{1}{64}, -\dfrac{1}{256}, \ldots$ $a_n = (-1)^{n+1} \left(\dfrac{1}{4}\right)^n$

Objective 4: Solve an Applied Problem Using a Sequence

33) A television's value decreases by $\frac{1}{3}$ each year. If it was purchased for \$2592, write a sequence that represents the value of the TV at the beginning of each of the next 4 years.   \$1728, \$1152, \$768, \$512

34) Due to an increase in costs, Hillcrest Health Club has decided to increase dues by 10% each year for the next 5 years. A membership currently costs \$1000 per year. Write a sequence that represents the cost of a membership each of the next 5 years.   \$1100, \$1210, \$1331, \$1464.10, \$1610.51

35) Carlton wants to improve his bench press. He plans on adding 10 lb to the bar each week. If he can lift 100 lb this week, how much will he lift 6 weeks from now?   160 lb

36) Currently, Sierra earns \$8.80 per hour, and she can get a raise of \$0.50 per hour every 6 months. What will be her hourly wage 18 months from now?   \$10.30

Mixed Exercises: Objectives 5 and 6

37) What is the difference between a sequence and a series?
A sequence is a list of terms in a certain order, and a series is a sum of the terms of a sequence.

38) Explain what $\sum\limits_{i=1}^{5} (7i + 2)$ means.
It means to find the sum of the first five terms of the sequence defined by $a_n = 7n + 2$.

Evaluate each series.

39) $\sum\limits_{i=1}^{6} (2i + 1)$ 48

40) $\sum\limits_{i=1}^{5} (4i + 3)$ 75

41) $\sum\limits_{i=1}^{5} (i - 8)$ −25

42) $\sum\limits_{i=1}^{4} (5 - 2i)$ 0

43) $\sum\limits_{i=1}^{4} (4i^2 - 2i)$ 100

44) $\sum\limits_{i=1}^{6} (3i^2 - 4i)$ 189

45) $\sum\limits_{i=1}^{6} \frac{i}{2}$ $\frac{21}{2}$

46) $\sum\limits_{i=1}^{3} \frac{2i}{i + 3}$ $\frac{23}{10}$

47) $\sum\limits_{i=1}^{5} (-1)^{i+1} \cdot (i)$ 3

48) $\sum\limits_{i=1}^{6} (-1)^{i} \cdot (i)$ 3

49) $\sum\limits_{i=5}^{9} (i - 2)$ 25

50) $\sum\limits_{i=6}^{10} (2i - 3)$ 65

51) $\sum\limits_{i=3}^{6} (i^2)$ 86

52) $\sum\limits_{i=2}^{7} (i - 1)^2$ 91

Objective 7: Write a Series Using Summation Notation

Write each series using summation notation.

53) $1 + \frac{1}{2} + \frac{1}{3} + \frac{1}{4} + \frac{1}{5}$ $\sum\limits_{i=1}^{5} \frac{1}{i}$

54) $11 + \frac{11}{2} + \frac{11}{3} + \frac{11}{4} + \frac{11}{5} + \frac{11}{6}$ $\sum\limits_{i=1}^{6} \frac{11}{i}$

55) $3 + 6 + 9 + 12$ $\sum\limits_{i=1}^{4} (3i)$

56) $4 + 8 + 12 + 16 + 20 + 24 + 28$ $\sum\limits_{i=1}^{7} (4i)$

57) $5 + 6 + 7 + 8 + 9 + 10$ $\sum\limits_{i=1}^{6} (i + 4)$

58) $4 + 5 + 6 + 7$ $\sum\limits_{i=1}^{4} (i + 3)$

59) $-1 + 2 - 3 + 4 - 5 + 6 - 7$ $\sum\limits_{i=1}^{7} (-1)^{i} \cdot (i)$

60) $2 - 4 + 8 - 16 + 32$ $\sum\limits_{i=1}^{5} (-1)^{i+1} \cdot (2^i)$

61) $3 - 9 + 27 - 81$ $\sum\limits_{i=1}^{4} (-1)^{i+1} \cdot (3^i)$

62) $-1 + 4 - 9 + 16 - 25$ $\sum\limits_{i=1}^{5} (-1)^{i} \cdot (i^2)$

Objective 8: Use Summation Notation to Represent the Average of a Group of Numbers

Find the arithmetic mean of each group of numbers.

63) 19, 24, 20, 17, 23, 17   20

64) 38, 31, 43, 40, 33   37

65) 8, 7, 11, 9, 12   9.4

66) 5, 9, 6, 5, 8, 3, 1, 7   5.5

67) Corey's credit card balance each month from January through June of 2011 is given in this table. Find the average monthly credit card balance during this period.   \$1054.09

Month	Balance
January	\$1431.60
February	\$1117.82
March	\$985.43
April	\$1076.22
May	\$900.00
June	\$813.47

68) The annual rainfall amounts (in inches) at Lindbergh Field in San Diego from 1999 to 2005 are listed below. Find the average annual rainfall during this period. (Round the answer to the hundredths place.) (www.sdcwa.org)   8.99 in.

Year	Total Rainfall (inches)
1999	6.51
2000	5.77
2001	8.82
2002	3.44
2003	10.24
2004	5.31
2005	22.81

Section 12.2 Arithmetic Sequences and Series

Objectives

1. Define Arithmetic Sequence and Common Difference
2. Find the Common Difference for an Arithmetic Sequence
3. Write the Terms of a Sequence
4. Find the General Term of an Arithmetic Sequence
5. Find a Specified Term of an Arithmetic Sequence
6. Determine the Number of Terms in an Arithmetic Sequence
7. Solve an Applied Problem Involving an Arithmetic Sequence
8. Find the Sum of Terms of an Arithmetic Sequence
9. Solve an Applied Problem Involving an Arithmetic Series

1. Define Arithmetic Sequence and Common Difference

In this section, we will discuss a special type of sequence called an *arithmetic sequence*.

Definition

An **arithmetic sequence** is a sequence in which each term after the first differs from the preceding term by a constant amount, d. d is called the **common difference**.

Note

An arithmetic sequence is also called an **arithmetic progression**.

For example, 3, 7, 11, 15, 19, . . . is an arithmetic sequence. If we subtract each term from the term that follows it, we see that the common difference d, equals 4.

$$7 - 3 = 4$$
$$11 - 7 = 4$$
$$15 - 11 = 4$$
$$19 - 15 = 4$$
$$\vdots$$
$$a_n - a_{n-1} = 4$$

nth term   Previous term

For this sequence $a_n - a_{n-1} = 4$. That is, choose a term in the sequence (a_n) and subtract the term before it (a_{n-1}) to get 4.

Note

For any arithmetic sequence, $d = a_n - a_{n-1}$.

2. Find the Common Difference for an Arithmetic Sequence

Example 1

In-Class Example 1
Find d for each arithmetic sequence.
a) 18, 24, 30, 36
b) 17, 10, 3, −4, −11
answer: a) 6   b) −7

Find d for each arithmetic sequence.

a)   −4, −1, 2, 5, 8, . . .   b)   35, 27, 19, 11, 3, . . .

Solution

a)   To find d, choose any term and subtract the preceding term: $d = 5 - 2 = 3$.

You can see that choosing a different pair of terms will produce the same result.

$$d = 2 - (-1) = 3$$

b)   Choose any term and subtract the preceding term: $d = 27 - 35 = -8$.   ■

You Try 1

Find d for each arithmetic sequence.

a)   −2, 3, 8, 13, 18, . . .   b)   25, 13, 1, −11, −23, . . .

3. Write the Terms of a Sequence

If we are given the first term, a_1, of an arithmetic sequence and the common difference, d, we can write the terms of the sequence.

Example 2

Write the first five terms of the arithmetic sequence with first term 5 and common difference 6.

In-Class Example 2
Write the first five terms of the arithmetic sequence with first term -4 and common difference -3.
answer:
$-4, -7, -10, -13, -16$

Solution

Since the first term is 5, $a_1 = 5$. Add 6 to get the second term, a_2. Continue adding 6 to get each term.

$$a_1 = 5$$
$$a_2 = 5 + 6 = 11$$
$$a_3 = 11 + 6 = 17$$
$$a_4 = 17 + 6 = 23$$
$$a_5 = 23 + 6 = 29$$

The first five terms of the sequence are 5, 11, 17, 23, 29.

You Try 2

Write the first five terms of the arithmetic sequence with first term -3 and common difference 2.

4. Find the General Term of an Arithmetic Sequence

Given the first term, a_1, of an arithmetic sequence and the common difference, d, we can find the general term of the sequence, a_n.

Consider the sequence in Example 2: 5, 11, 17, 23, 29. We will show that each term can be written in terms of a_1 and d, leading us to a formula for a_n.

5	11	17	23	29
$a_1 = 5$	$a_2 = 5 + 6$	$a_3 = 11 + 6$	$a_4 = 17 + 6$	$a_5 = 23 + 6$
	$a_2 = a_1 + d$	$a_3 = a_2 + d$	$a_4 = a_3 + d$	$a_5 = a_4 + d$
		$a_3 = (a_1 + d) + d$	$a_4 = (a_1 + 2d) + d$	$a_5 = (a_1 + 3d) + d$
		$a_3 = a_1 + 2d$	$a_4 = a_1 + 3d$	$a_5 = a_1 + 4d$

For each term after the first, the coefficient of d is one less than the term number.

This pattern applies for any arithmetic sequence. Therefore, the general term, a_n, of an arithmetic sequence is given by $a_n = a_1 + (n - 1)d$.

Definition

General Term of an Arithmetic Sequence: The general term of an arithmetic sequence with first term a_1 and common difference d is given by

$$a_n = a_1 + (n - 1)d$$

Example 3

In-Class Example 3
Given the arithmetic sequence
5, 7, 9, 11, 13, ... find
a) a_1 and d.
b) a formula for a_n.
c) the 25th term of the sequence.
answer: a) $a_1 = 5; d = 2$
b) $a_n = 2n + 3$ c) 53

Given the arithmetic sequence $9, 13, 17, 21, 25, \ldots$, find

a) a_1 and d.   b) a formula for a_n.   c) the 31st term of the sequence.

Solution

a) a_1 is the first term of the sequence. $a_1 = 9$. To find d, choose any term and subtract the preceding term.

$$d = 13 - 9 = 4$$

Therefore, $a_1 = 9$ and $d = 4$.

b) To find a_n, begin with the formula and substitute 9 for a_1 and 4 for d.

$$\begin{aligned}
a_n &= a_1 + (n - 1)d && \text{Formula for } a_n \\
a_n &= 9 + (n - 1)(4) && \text{Substitute 9 for } a_1 \text{ and 4 for } d. \\
a_n &= 9 + 4n - 4 && \text{Distribute.} \\
a_n &= 4n + 5 && \text{Combine like terms.}
\end{aligned}$$

c) Finding the 31st term of the sequence means finding a_{31}. Let $n = 31$ in the formula for a_n.

$$\begin{aligned}
a_n &= 4n + 5 \\
a_{31} &= 4(31) + 5 \\
a_{31} &= 129
\end{aligned}$$
■

You Try 3

Given the arithmetic sequence $4, 11, 18, 25, 32, \ldots$, find

a) a_1 and d.   b) a formula for a_n.   c) the 25th term of the sequence.

Example 4

In-Class Example 4
Find the general term, a_n, and
the 12th term of the arithmetic
sequence 99, 88, 77, 66, 55, ...
answer:
$a_n = -11n + 110; a_{12} = -22$

Find the general term, a_n, and the 54th term of the arithmetic sequence $49, 41, 33, 25, 17, \ldots$.

Solution

To find a_n, we must use the formula $a_n = a_1 + (n - 1)d$. Therefore, identify a_1 and find d.

$$a_1 = 49 \qquad d = 41 - 49 = -8$$

Substitute 49 for a_1 and -8 for d in the formula for a_n.

$$\begin{aligned}
a_n &= a_1 + (n - 1)d && \text{Formula for } a_n \\
a_n &= 49 + (n - 1)(-8) && \text{Substitute 49 for } a_1 \text{ and } -8 \text{ for } d. \\
a_n &= 49 - 8n + 8 && \text{Distribute.} \\
a_n &= -8n + 57 && \text{Combine like terms.}
\end{aligned}$$

The fifty-fourth term of the sequence is a_{54}. Let $n = 54$ in the formula for a_n.

$$\begin{aligned}
a_n &= -8n + 57 \\
a_{54} &= -8(54) + 57 \\
a_{54} &= -375
\end{aligned}$$
■

You Try 4

Find the general term, a_n, and the 41st term of the arithmetic sequence $-1, -6, -11, -16, -21, \ldots$.

5. Find a Specified Term of an Arithmetic Sequence

If we know a_1 and d for an arithmetic sequence, we can find any term without having the formula for a_n.

Example 5

In-Class Example 5
Find the eighth term of the arithmetic sequence with first term 0 and common difference -12.
answer: -84

Find the 15th term of the arithmetic sequence with first term -17 and common difference 3.

Solution

The first term is -17, so $a_1 = -17$. The common difference is 3, so $d = 3$. Finding the 15th term means finding a_{15}. Use the formula $a_n = a_1 + (n - 1)d$ and substitute 15 for n, -17 for a_1, and 3 for d.

$$a_n = a_1 + (n - 1)d \qquad n = 15, a_1 = -17, d = 3$$
$$a_{15} = -17 + (15 - 1)(3)$$
$$a_{15} = -17 + 14(3) \qquad \text{Subtract.}$$
$$a_{15} = -17 + 42 \qquad \text{Multiply.}$$
$$a_{15} = 25 \qquad \text{Add.}$$

The 15th term is 25.

You Try 5

Find the 23rd term of the arithmetic sequence with first term 18 and common difference 4.

Sometimes, we can write and solve a system of equations to find the general term, a_n.

Example 6

In-Class Example 6
The third term of an arithmetic sequence is 8 and the seventh term is 24. Find
a) the general term, a_n.
b) the 13th term.
answer: a) $a_n = 4n - 4$
b) $a_{13} = 48$

The fourth term of an arithmetic sequence is -10 and the ninth term is 25. Find

a)   the general term, a_n.   b)   the 16th term.

Solution

a)   Since the fourth term is -10, $a_4 = -10$. Since the ninth term is 25, $a_9 = 25$.

Use the formula for a_n along with $a_4 = -10$ and $a_9 = 25$ to obtain two equations containing the two variables a_1 and d.

$$a_4 = -10 \qquad\qquad\qquad\qquad a_9 = 25$$

$$\begin{aligned} a_n &= a_1 + (n - 1)d \\ a_4 &= a_1 + (4 - 1)d \quad &n = 4 \\ -10 &= a_1 + 3d \quad &a_4 = -10 \end{aligned} \qquad \begin{aligned} a_n &= a_1 + (n - 1)d \\ a_9 &= a_1 + (9 - 1)d \quad &n = 9 \\ 25 &= a_1 + 8d \quad &a_9 = 25 \end{aligned}$$

We obtain the system of equations

$$-10 = a_1 + 3d$$
$$25 = a_1 + 8d$$

Multiply the first equation by -1 and add the two equations. This will eliminate a_1 and enable us to solve for d.

$$\begin{aligned} 10 &= -a_1 - 3d \qquad \text{Multiply first equation by } -1. \\ + \; 25 &= a_1 + 8d \\ \hline 35 &= 5d \\ 7 &= d \end{aligned}$$

Substitute $d = 7$ into $-10 = a_1 + 3d$ to solve for a_1.

$$-10 = a_1 + 3(7)$$
$$-10 = a_1 + 21$$
$$-31 = a_1$$

$a_1 = -31$ and $d = 7$. Substitute these values into $a_n = a_1 + (n - 1)d$ to find a formula for a_n.

$$
\begin{aligned}
a_n &= a_1 + (n - 1)d \\
&= -31 + (n - 1)(7) \qquad d = 7 \\
&= -31 + 7n - 7 \qquad \text{Distribute.} \\
&= 7n - 38 \qquad \text{Combine like terms.}
\end{aligned}
$$

The general term is $a_n = 7n - 38$.

b)   The 16th term is a_{16}. Substitute 16 for n in $a_n = 7n - 38$ to find the 16th term.

$$
\begin{aligned}
a_n &= 7n - 38 \\
a_{16} &= 7(16) - 38 \qquad \text{Let } n = 16. \\
&= 74
\end{aligned}
$$

$a_{16} = 74.$

You Try 6

The third term of an arithmetic sequence is 11 and the tenth term is -31. Find

a)   the general term, a_n.   b)   the 22nd term.

6. Determine the Number of Terms in an Arithmetic Sequence

We can determine the number of terms in a sequence using the formula for a_n.

Example 7

In-Class Example 7
Find the number of terms in the arithmetic sequence
$9, 7, 5, 3, \ldots, -7.$
answer: 9

Find the number of terms in the arithmetic sequence $18, 14, 10, 6, \ldots, -54$.

Solution

The first term in the sequence is 18, so $a_1 = 18$.

Let $n = $ the number of terms in the sequence.

Then $a_n = -54$, since -54 is the last (or nth) term.

$d = -4$ since $14 - 18 = -4$.

Substitute 18 for a_1, -4 for d, and -54 for a_n into $a_n = a_1 + (n - 1)d$, and solve for n.

$$
\begin{aligned}
a_n &= a_1 + (n - 1)d \\
-54 &= 18 + (n - 1)(-4) \qquad \text{Let } a_n = -54, a_1 = 18, \text{ and } d = -4. \\
-54 &= 18 - 4n + 4 \qquad \text{Distribute.} \\
-54 &= 22 - 4n \qquad \text{Combine like terms.} \\
-76 &= -4n \qquad \text{Subtract 22.} \\
19 &= n \qquad \text{Divide by } -4.
\end{aligned}
$$

This sequence has 19 terms.

You Try 7

Find the number of terms in the arithmetic sequence $3, 8, 13, 18, \ldots, 48$.

7. Solve an Applied Problem Involving an Arithmetic Sequence

Example 8

In-Class Example 8
Use the given example.

Due to a decrease in sales, a company decides to decrease the number of employees at one of its manufacturing plants by 25 per month for the next 6 months. The plant currently has 483 workers.

a) Find the general term of an arithmetic sequence, a_n, that models the number of employees working at the manufacturing plant.

b) How many employees remain after 6 months?

Solution

a) The first term of the arithmetic sequence is the current number of employees. So $a_1 = 483$.

The number of workers will decrease by 25 per month, so $d = -25$.
Substitute $a_1 = 483$ and $d = -25$ into $a_n = a_1 + (n - 1)d$.

$$a_n = a_1 + (n - 1)d$$
$$a_n = 483 + (n - 1)(-25) \qquad a_1 = 483, d = -25$$
$$a_n = 483 - 25n + 25 \qquad \text{Distribute.}$$
$$a_n = -25n + 508 \qquad \text{Combine like terms.}$$

b) The number of employees remaining after 6 months is the sixth term of the sequence, a_6.

Substitute 6 for n in $a_n = -25n + 508$.

$$a_n = -25n + 508$$
$$a_6 = -25(6) + 508$$
$$a_6 = -150 + 508$$
$$a_6 = 358$$

There will be 358 employees.

You Try 8

Arianna wants to save money for a car. She makes an initial deposit of $2000, and then she will deposit $300 at the beginning of each month.

a) Find the general term for an arithmetic sequence, a_n, that models the amount of money (ignoring interest) that she has saved.

b) How much has she saved (ignoring interest) after 12 months?

Arithmetic Series

8. Find the Sum of Terms of an Arithmetic Sequence

We first defined a series in Section 12.1 as a sum of the terms of a sequence.

Therefore, an **arithmetic series** is a sum of terms of an arithmetic sequence.

It would not be difficult to find the sum of, say, the first 5 terms of an arithmetic sequence. But if we were asked to find the sum of the first 50 terms, using a formula would be more convenient.

Let S_n represent the first n terms of an arithmetic sequence. Then,

$$S_n = a_1 + (a_1 + d) + (a_1 + 2d) + (a_1 + 3d) + \cdots + [a_1 + (n - 1)d]$$

We can write the sum with the terms in reverse order as

$$S_n = a_n + (a_n - d) + (a_n - 2d) + (a_n - 3d) + \cdots + [a_n - (n - 1)d]$$

Next, add the two expressions by adding the corresponding terms. We get

$$2S_n = (a_1 + a_n) + (a_1 + a_n) + (a_1 + a_n) + (a_1 + a_n) + \cdots + (a_1 + a_n)$$
$$2S_n = n(a_1 + a_n) \qquad \text{There are } n \ (a_1 + a_n)\text{-terms.}$$
$$S_n = \frac{n}{2}(a_1 + a_n) \qquad \text{Divide by 2.}$$

$S_n = \frac{n}{2}(a_1 + a_n)$ is one formula for the sum of the first n terms of an arithmetic sequence.

We can derive another formula for the sum if we substitute $a_n = a_1 + (n - 1)d$ for a_n in the formula above.

$$S_n = \frac{n}{2}(a_1 + a_n)$$
$$= \frac{n}{2}[a_1 + (a_1 + (n - 1)d)] \qquad a_n = a_1 + (n - 1)d$$
$$= \frac{n}{2}[2a_1 + (n - 1)d] \qquad \text{Combine like terms.}$$

Another formula for the sum of the first n terms of an arithmetic sequence is

$$S_n = \frac{n}{2}[2a_1 + (n - 1)d]$$

Formula   Sum of the First n Terms of an Arithmetic Sequence

The sum of the first n terms, S_n, of an arithmetic sequence with first term a_1, nth term a_n, and common difference d is given by

1) $S_n = \frac{n}{2}(a_1 + a_n)$

2) $S_n = \frac{n}{2}[2a_1 + (n - 1)d]$

Note

It is convenient to use 1) when the first term, the last term, and the number of terms are known. If the last term is not known, then 2) may be a better choice for finding the sum.

Example 9

Find the sum of the first 17 terms of the arithmetic sequence with first term 41 and last term -23.

In-Class Example 9
Find the sum of the first 13 terms of the arithmetic sequence with first term 11 and last term -33.
answer: -143

Solution

We are given the first term, the last term, and the number of terms. *We will use formula 1)* $S_n = \frac{n}{2}(a_1 + a_n)$ to find S_{17}, the sum of the first 17 terms.

$$a_1 = 41, \quad a_{17} = -23, \quad n = 17$$

$$S_n = \frac{n}{2}(a_1 + a_n) \qquad \text{Formula 1)}$$

$$S_{17} = \frac{17}{2}(41 + a_{17}) \qquad \text{Let } n = 17 \text{ and } a_1 = 41.$$

$$S_{17} = \frac{17}{2}[41 + (-23)] \qquad a_{17} = -23$$

$$S_{17} = \frac{17}{2}(18) \qquad \text{Add.}$$

$$S_{17} = 153 \qquad \text{Multiply.}$$

The sum of the first 17 terms is 153. ■

You Try 9

Find the sum of the first 15 terms of the arithmetic sequence with first term 2 and last term 72.

Example 10

Find the sum of the first 12 terms of the arithmetic sequence with first term 7 and common difference 3.

In-Class Example 10
Find the sum of the first nine terms of the arithmetic sequence with first term −5 and common difference 4.
answer: 99

Solution

We are given the first term ($a_1 = 7$), the common difference ($d = 3$), and the number of terms ($n = 12$). The last term is not known. *We will use formula 2)*

$S_n = \frac{n}{2}[2a_1 + (n - 1)d]$ to find S_{12}, the sum of the first 12 terms.

$$a_1 = 7, d = 3, n = 12$$

$$S_n = \frac{n}{2}[2a_1 + (n - 1)d] \qquad \text{Formula 2)}$$

$$S_{12} = \frac{12}{2}[2(7) + (12 - 1)(3)] \qquad \text{Let } a_1 = 7, d = 3, \text{ and } n = 12.$$

$$S_{12} = 6[14 + (11)(3)]$$
$$S_{12} = 6(47)$$
$$S_{12} = 282$$

The sum of the first 12 terms is 282. ■

You Try 10

Find the sum of the first 10 terms of the arithmetic sequence with first term 6 and common difference 4.

The general term of an arithmetic sequence has the form $a_n = bn + c$, where b and c are constants. Therefore, $\sum\limits_{i=1}^{n} (bn + c)$ represents an arithmetic series or the sum of the first n terms of an arithmetic sequence. We can use formula 1) to evaluate $\sum\limits_{i=1}^{n} (bn + c)$.

Example 11

In-Class Example 11

Evaluate $\sum_{i=1}^{20} (2i + 11)$.

answer: 640

Evaluate $\sum_{i=1}^{16} (3i - 19)$.

Solution

Since i begins at 1 and ends at 16, to evaluate $\sum_{i=1}^{16} (3i - 19)$ means to find the sum of the first 16 terms, S_{16}, of the arithmetic sequence with general term $a_n = 3n - 19$.

Find the first term by substituting 1 for i: $a_1 = 3(1) - 19 = -16$

Find the last (the sixteenth) term by substituting 16 for i: $a_{16} = 3(16) - 19 = 29$.

There are 16 terms, so $n = 16$.

Because we know that $a_1 = -16$, $a_{16} = 29$, and $n = 16$, use formula 1) to evaluate $\sum_{i=1}^{16} (3i - 19)$.

$$S_n = \frac{n}{2}(a_1 + a_n) \qquad \text{Formula 1)}$$

$$S_{16} = \frac{16}{2}(-16 + a_{16}) \qquad \text{Let } n = 16 \text{ and } a_1 = -16.$$

$$S_{16} = 8(-16 + 29) \qquad a_{16} = 29$$
$$S_{16} = 8(13)$$
$$S_{16} = 104$$

Therefore, $\sum_{i=1}^{16} (3i - 19) = 104$.

You Try 11

Evaluate $\sum_{i=1}^{19} (2i + 1)$.

9. Solve an Applied Problem Involving an Arithmetic Series

Example 12

In-Class Example 12
Use the given example.

An acrobatic troupe forms a human pyramid with six people in the bottom row, five people in the second row, four people in the third row, and so on. How many people are in the pyramid if the pyramid has six rows?

Solution

The information in the problem suggests the arithmetic sequence 6, 5, 4, . . . , 1, where each term represents the number of people in a particular row of the pyramid.

We are asked to find the *total* number of people in the pyramid, so we are finding S_6, the sum of the six terms of the sequence.

Since there are six rows, $n = 6$.

There are six people in the first row, so $a_1 = 6$.

There is 1 person in the last row, so $a_6 = 1$. Use formula 1).

$$S_n = \frac{n}{2}(a_1 + a_n)$$

$$S_6 = \frac{6}{2}(6 + a_6) \qquad \text{Let } n = 6 \text{ and } a_1 = 6.$$

$$S_6 = 3(6 + 1) \qquad \text{Let } a_6 = 1.$$

$$S_6 = 3(7) = 21$$

There are 21 people in the pyramid.   ■

You Try 12

A child builds a tower with blocks so that the bottom row contains nine blocks, the next row contains seven blocks, the next row contains five blocks, and so on. If the tower has five rows, how many blocks are in the tower?

Answers to You Try Exercises

1) a) $d = 5$ b) $d = -12$ 2) $-3, -1, 1, 3, 5$ 3) a) $a_1 = 4, d = 7$ b) $a_n = 7n - 3$ c) 172
4) $a_n = -5n + 4; a_{41} = -201$ 5)   106   6) a) $a_n = -6n + 29$ b) -103 7)   10
8)   a) $a_n = 300n + 1700$ b) $5300   9)   555   10)   240   11)   399   12)   25

12.2 Exercises

*Additional answers can be found in the Answers to Exercises appendix.

Mixed Exercises: Objectives 1 and 2

1) What is an arithmetic sequence? Give an example.

2) How do you find the common difference for an arithmetic sequence?   Choose any term and subtract the term that precedes it.

Determine whether each sequence is arithmetic. If it is, find the common difference, d.

3) 3, 11, 19, 27, 35, . . .
yes, $d = 8$

4) 4, 7, 10, 13, 16, . . .
yes, $d = 3$

5) 10, 6, 2, -2, -6, . . .
yes, $d = -4$

6) 27, 20, 13, 6, -1, . . .
yes, $d = -7$

7) 4, 8, 16, 32, 64, . . .   no

8) 1, 3, 6, 10, 15, . . .   no

9) $-17, -14, -11, -8, -5,$ . . .   yes, $d = 3$

10) $-12, -10, -8, -6, -4,$ . . .   yes, $d = 2$

Objective 3: Write the Terms of a Sequence

Write the first five terms of each arithmetic sequence with the given first term and common difference.

11) $a_1 = 7, d = 2$
7, 9, 11, 13, 15

12) $a_1 = 20, d = 4$
20, 24, 28, 32, 36

13) $a_1 = 15, d = -8$
15, 7, -1, -9, -17

14) $a_1 = -3, d = -2$
$-3, -5, -7, -9, -11$

15) $a_1 = -10, d = 3$
$-10, -7, -4, -1, 2$

16) $a_1 = -19, d = 5$
$-19, -14, -9, -4, 1$

Write the first five terms of the arithmetic sequence with general term a_n.

17) $a_n = 6n + 7$
13, 19, 25, 31, 37

18) $a_n = 2n + 7$
9, 11, 13, 15, 17

19) $a_n = 5 - n$
4, 3, 2, 1, 0

20) $a_n = 3 - 4n$
$-1, -5, -9, -13, -17$

Mixed Exercises: Objectives 4 and 5

21) Given the arithmetic sequence
4, 7, 10, 13, 16, . . .

a) Find a_1 and d. $a_1 = 4, d = 3$

b) Find a formula for the general term of the sequence, a_n. $a_n = 3n + 1$

c) Find the 35th term of the sequence.   106

22) Given the arithmetic sequence
$-5, -3, -1, 1, 3,$ . . .

a) Find a_1 and d. $a_1 = -5, d = 2$

b) Find a formula for the general term of the sequence, a_n. $a_n = 2n - 7$

c) Find the 35th term of the sequence.   63

23) Given the arithmetic sequence
4, −1, −6, −11, −16, . . .

   a) Find a_1 and d. $a_1 = 4, d = -5$

   b) Find a formula for the general term of the
   sequence, a_n. $a_n = -5n + 9$

   c) Find a_{19}.   −86

24) Given the arithmetic sequence
−9, −21, −33, −45, −57, . . .

   a) Find a_1 and d. $a_1 = -9, d = -12$

   b) Find a formula for the general term of the
   sequence, a_n. $a_n = -12n + 3$

   c) Find a_{15}.   −177

For each arithmetic sequence, find a_n and then use a_n to find
the indicated term.

25) −7, −5, −3, −1, 1, . . .; a_{25} $a_n = 2n - 9; a_{25} = 41$

26) 13, 19, 25, 31, 37, . . .; a_{30} $a_n = 6n + 7; a_{30} = 187$

27) $1, \dfrac{3}{2}, 2, \dfrac{5}{2}, 3, \ldots; a_{18}$ 28) $\dfrac{1}{3}, \dfrac{2}{3}, 1, \dfrac{4}{3}, \dfrac{5}{3}, \ldots; a_{21}$

29) $a_1 = 0, d = -5; a_{23}$ 30) $a_1 = -5, d = -7; a_{14}$
   $a_n = -5n + 5; a_{23} = -110$ $a_n = -7n + 2; a_{14} = -96$
Find the indicated term for each arithmetic sequence.

31) $a_1 = -5, d = 4; a_{16}$ 55   32) $a_1 = 10, d = 3; a_{29}$ 94

33) $a_1 = -7, d = -5; a_{21}$ 34) $a_1 = 27, d = -4; a_{32}$
   −107   −97

Two terms of an arithmetic sequence are given in each
problem. Find the general term of the sequence, a_n, and find
the indicated term.

35) $a_3 = 11, a_7 = 19; a_{11}$ 36) $a_5 = 13, a_{11} = 31; a_{16}$
   $a_n = 2n + 5; a_{11} = 27$ $a_n = 3n - 2; a_{16} = 46$
37) $a_2 = 7, a_6 = -13; a_{14}$ $a_n = -5n + 17; a_{14} = -53$

38) $a_3 = -9, a_7 = -25; a_{10}$ $a_n = -4n + 3; a_{10} = -37$

39) $a_4 = -5, a_{11} = 16; a_{18}$ 40) $a_4 = -10, a_9 = 0; a_{17}$
   $a_n = 3n - 17; a_{18} = 37$ $a_n = 2n - 18; a_{17} = 16$

**Objective 6: Determine the Number of Terms in an
Arithmetic Sequence**

Find the number of terms in each arithmetic sequence.

41) 8, 13, 18, 23, . . . , 63   12

42) 8, 11, 14, 17, . . . , 50   15

43) 9, 7, 5, 3, . . . , −27   19

44) −7, −11, −15, −19, . . . , −91   22

**Objective 8: Find the Sum of Terms of an
Arithmetic Sequence**

45) For a particular sequence, suppose you are asked to find
S_{15}. What are you finding?
S_{15} is the sum of the first 15 terms of the sequence.

46) Write down the two formulas for S_n, and explain when to
use each formula.

47) Find the sum of the first 10 terms of the arithmetic
sequence with first term 14 and last term 68.   410

48) Find the sum of the first nine terms of the arithmetic
sequence with first term 2 and last term 34.   162

49) Find the sum of the first seven terms of the arithmetic
sequence with first term 3 and last term −9.   −21

50) Find the sum of the first 11 terms of the arithmetic
sequence with first term −8 and last term −58.   −363

Find S_8 for each arithmetic sequence described below.

51) $a_1 = -1, a_8 = -29$ −120   52) $a_1 = -5, a_8 = 9$ 16

53) $a_1 = 3, d = 5$ 164   54) $a_1 = 2, d = 3$ 100

55) $a_1 = 10, d = -6$ −88   56) $a_1 = -1, d = -3$ −92

57) $a_n = -4n - 1$ −152   58) $a_n = 4n + 1$ 152

59) $a_n = 3n + 4$ 140   60) $a_n = -6n + 5$ −176

61) a) Evaluate $\displaystyle\sum_{i=1}^{10} (2i + 7)$ by writing out each term and
   finding the sum.
   $9 + 11 + 13 + 15 + 17 + 19 + 21 + 23 + 25 + 27 = 180$
   b) Evaluate $\displaystyle\sum_{i=1}^{10} (2i + 7)$ using a formula for S_n.   180

   c) Which method do you prefer and why?   Answers may vary.

Evaluate each sum using a formula for S_n.

62) $\displaystyle\sum_{i=1}^{6} (3i + 8)$ 111   63) $\displaystyle\sum_{i=1}^{5} (8i - 5)$ 95

64) $\displaystyle\sum_{i=1}^{9} (2i - 14)$ −36   65) $\displaystyle\sum_{i=1}^{7} (-2i + 7)$ −7

66) $\displaystyle\sum_{i=1}^{12} (-4i + 9)$ −204   67) $\displaystyle\sum_{i=1}^{18} (3i - 11)$ 315

68) $\displaystyle\sum_{i=1}^{215} (i - 10)$ 21,070   69) $\displaystyle\sum_{i=1}^{500} i$ 125,250

70) $\displaystyle\sum_{i=1}^{100} \left(\dfrac{1}{2}i + 6\right)$ 3125

Mixed Exercises: Objectives 7 and 9

Solve each application.

71) This month, Warren deposited $1500 into a bank account.
He will deposit $100 into the account at the beginning of
each month. Disregarding interest, how much money will
Warren have saved after 9 months? $2300

72) When Antoinnette is hired for a job, she signs a contract
for a salary of $34,000 plus a raise of $1800 each year for
the next 4 years. What will be her salary in the last year
of her contract? $41,200

73) Beginning the first week of June, Noor will begin to deposit money in her bank. She will deposit $1 the first week, $2 the second week, $3 the third week, $4 the fourth week, and she will continue to deposit money in this way for 24 weeks. How much money will she have saved after 24 weeks? $300

74) Refer to Exercise 73. If Tracy deposits money weekly but deposits $1 then $3 then $5, etc., how much will Tracy have saved after 24 weeks? $576

75) A stack of logs has 12 logs in the bottom row (the first row), 11 logs in the second row, 10 logs in the third row, and so on, until the last row contains one log.

   a) How many logs are in the eighth row?   5

   b) How many logs are in the stack?   78

76) A landscaper plans to put a pyramid design in a brick patio so that the bottom row of the pyramid contains 9 bricks and every row above it contains two fewer bricks. How many bricks does she need to make the design?   25

77) A lecture hall has 14 rows. The first row has 12 seats, and each row after that has 2 more seats than the previous row. How many seats are in the last row? How many seats are in the lecture hall?   38; 350

78) A theater has 23 rows. The first row contains 10 seats, the next row has 12 seats, the next row has 14 seats, and so on. How many seats are in the last row? How many seats are in the theater?   54; 736

79) The main floor of a concert hall seats 860 people. The first row contains 24 seats, and the last row contains 62 seats. If each row has 2 more seats than the previous row, how many rows of seats are on the main floor of the concert hall?   20

80) A child builds a tower with blocks so that the bottom row contains 9 blocks and the top row contains 1 block. If he uses 45 blocks, how many rows are in the tower?   9

Section 12.3 Geometric Sequences and Series

Objectives

1. Define Geometric Sequence and Common Ratio
2. Find the Common Ratio for a Geometric Sequence
3. Find the Terms of a Geometric Sequence
4. Find the General Term of a Geometric Sequence
5. Find a Specified Term of a Geometric Sequence
6. Solve an Applied Problem Involving a Geometric Sequence
7. Find the Sum of Terms of a Geometric Sequence
8. Find the Sum of Terms of an Infinite Geometric Sequence
9. Solve an Applied Problem Involving an Infinite Series
10. Distinguish Between an Arithmetic and a Geometric Sequence

1. Define Geometric Sequence and Common Ratio

In Section 12.2, we learned that a sequence such as

$$5, 9, 13, 17, 21, \ldots$$

is an *arithmetic sequence* because each term differs from the previous term by a constant amount, called the common difference, d. In this case, $d = 4$.

The terms of the sequence

$$3, 6, 12, 24, 48, \ldots$$

do not differ by a constant amount, but each term after the first is obtained by *multiplying* the preceding term by 2. Such a sequence is called a *geometric sequence*.

Definition

A **geometric sequence** is a sequence in which each term after the first is obtained by multiplying the preceding term by a constant, r. r is called the **common ratio**.

Note

A geometric sequence is also called a **geometric progression**.

In the sequence 3, 6, 12, 24, 48, . . . the common ratio, r, is 2. We can find the value of r by dividing any term after the first by the preceding term. For example,

$$r = \frac{6}{3} = \frac{12}{6} = \frac{24}{12} = \frac{48}{24} = 2$$

2. Find the Common Ratio for a Geometric Sequence

Example 1

In-Class Example 1
Find the common ratio, r, for each geometric sequence.
a) 2, 6, 18, 54, . . .
b) 24, 6, $\frac{3}{2}$, $\frac{3}{8}$, . . .
answer: a) 3   b) $\frac{1}{4}$

Find the common ratio, r, for each geometric sequence.

a) 5, 15, 45, 135, . . .   b) 12, 6, 3, $\frac{3}{2}$, . . .

Solution

a) To find r, choose any term and divide it by the term preceding it: $r = \frac{15}{5} = 3$.

It is important to realize that dividing *any* term by the term immediately before it will give the same result.

b) Choose any term and divide by the term preceding it: $r = \frac{3}{6} = \frac{1}{2}$.

You Try 1

Find the common ratio, r, for each geometric sequence.

a) 2, 12, 72, 432, . . .   b) 15, 5, $\frac{5}{3}$, $\frac{5}{9}$, . . .

3. Find the Terms of a Geometric Sequence

Example 2

In-Class Example 2
Write the first five terms of the geometric sequence with first term 6 and common ratio $\frac{1}{3}$.
answer: 6, 2, $\frac{2}{3}$, $\frac{2}{9}$, $\frac{2}{27}$

Write the first five terms of the geometric sequence with first term 10 and common ratio 2.

Solution

Each term after the first is obtained by multiplying by 2.

$$a_1 = 10$$
$$a_2 = 10(2) = 20$$
$$a_3 = 20(2) = 40$$
$$a_4 = 40(2) = 80$$
$$a_5 = 80(2) = 160$$

The first five terms of the sequence are 10, 20, 40, 80, 160.

You Try 2

Write the first five terms of the geometric sequence with first term 2 and common ratio 2.

4. Find the General Term of a Geometric Sequence

Example 2 suggests a pattern that enables us to find a formula for the general term, a_n, of a geometric sequence. The common ratio, r, is 2. The first five terms of the sequence are

10	20	40	80	160
$a_1 = 10$	$a_2 = 10 \cdot 2$	$a_3 = 10 \cdot 4$	$a_4 = 10 \cdot 8$	$a_5 = 10 \cdot 16$
	$a_2 = a_1 \cdot r$	$a_3 = a_1 \cdot r^2$	$a_4 = a_1 \cdot r^3$	$a_5 = a_1 \cdot r^4$

The exponent on r is one less than the term number.

This pattern applies for any geometric sequence. Therefore, the general term, a_n, of a geometric sequence is given by $a_n = a_1 r^{n-1}$.

Definition

General Term of a Geometric Sequence: The general term of a geometric sequence with first term a_1 and common ratio r is given by

$$a_n = a_1 r^{n-1}$$

Example 3

In-Class Example 3
Find the general term, a_n, and the eighth term of the geometric sequence 4, 16, 64, 256, ...
answer: $a_n = 4(4)^{n-1}$; $a_8 = 65{,}536$

Find the general term, a_n, and the eighth term of the geometric sequence 3, 6, 12, 24, 48, ...

Solution

$a_1 = 3$ and $r = \dfrac{6}{3} = 2$. Substitute these values into $a_n = a_1 r^{n-1}$.

$$a_n = a_1 r^{n-1}$$
$$a_n = 3(2)^{n-1} \qquad \text{Let } a_1 = 3 \text{ and } r = 2.$$

The general term, $a_n = 3(2)^{n-1}$, is in simplest form.

To find the eighth term, a_8, substitute 8 for n and simplify.

$$a_n = 3(2)^{n-1}$$
$$a_8 = 3(2)^{8-1} \qquad \text{Let } n = 8.$$
$$a_8 = 3(2)^7 \qquad \text{Subtract.}$$
$$a_8 = 3(128) \qquad 2^7 = 128$$
$$a_8 = 384 \qquad \text{Multiply.}$$

The eighth term is 384.

BE CAREFUL Remember the order of operations when evaluating $3(2)^7$. We evaluate exponents before we do multiplication to find $2^7 = 128$ *before* multiplying by 3.

You Try 3

Find the general term, a_n, and the fifth term of the geometric sequence 2, 6, 18, ...

5. Find a Specified Term of a Geometric Sequence

If we know a_1 and r for a geometric sequence, we can find any term using $a_n = a_1 r^{n-1}$.

Example 4

In-Class Example 4
Find the sixth term of the
geometric sequence 32, -16,
8, -4, . . .
answer: -1

Find the sixth term of the geometric sequence $12, -4, \dfrac{4}{3}, \ldots$.

Solution

$a_1 = 12$ and $r = -\dfrac{4}{12} = -\dfrac{1}{3}$. Find a_6 using $a_n = a_1 r^{n-1}$.

$$a_n = a_1 r^{n-1}$$
$$a_6 = (12)\left(-\dfrac{1}{3}\right)^{6-1} \qquad \text{Let } n = 6,\, a_1 = 12, \text{ and } r = -\dfrac{1}{3}.$$
$$a_6 = 12\left(-\dfrac{1}{3}\right)^{5}$$
$$a_6 = 12\left(-\dfrac{1}{243}\right)$$
$$a_6 = -\dfrac{4}{81}$$

The sixth term is $-\dfrac{4}{81}$.

You Try 4

Find the seventh term of the geometric sequence $-50, 25, -\dfrac{25}{2}, \ldots$.

6. Solve an Applied Problem Involving a Geometric Sequence

Example 5

In-Class Example 5
Substitute $26,000 for the
$24,000 in Example 5.
answer:
a) $a_n = 26,000(0.75)^{n-1}$
b) $10,968.75

A pickup truck purchased for $24,000 (This is its value at the beginning of year 1.) depreciates 25% each year. That is, its value each year is 75% of its value the previous year.

a) Find the general term, a_n, of the geometric sequence that models the value of the truck at the beginning of each year.

b) What is the pickup truck worth at the beginning of the fourth year?

Solution

a) To find the value of the pickup each year, we *multiply* how much it was worth the previous year by 0.75. Therefore, the value of the truck each year can be modeled by a *geometric* sequence.

$a_1 = 24,000$ since this is the value at the beginning of year 1. $r = 0.75$ since we will *multiply* the value each year by 0.75 to find the value the next year.

Substitute $a_1 = 24,000$ and $r = 0.75$ into $a_n = a_1 r^{n-1}$ to find the general term.

$$a_n = a_1 r^{n-1}$$
$$a_n = 24,000(0.75)^{n-1} \qquad \text{Let } a_1 = 24,000 \text{ and } r = 0.75.$$

b) To find the value of the pickup truck at the beginning of the fourth year, use a_n from a) and let $n = 4$.

$$a_n = 24{,}000(0.75)^{n-1}$$
$$a_4 = 24{,}000(0.75)^{4-1} \qquad \text{Let } n = 4.$$
$$a_4 = 24{,}000(0.75)^3$$
$$a_4 = 10{,}125$$

The truck is worth \$10,125 at the beginning of the fourth year.   ∎

You Try 5

A minivan purchased for \$27,000 depreciates 30% each year. That is, its value each year is 70% of its value the previous year.

a) Find the general term, a_n, of the geometric sequence that models the value of the minivan at the beginning of each year.

b) What is the minivan worth at the beginning of the third year?

Geometric Series

7. Find the Sum of Terms of a Geometric Sequence

A **geometric series** is a sum of terms of a geometric sequence.

Just as we can use a formula to find the sum of the first n terms of an arithmetic sequence, there is a formula to find the sum of the first n terms of a geometric sequence.

Let S_n represent the sum of the first n terms of a geometric sequence. Then,

$$S_n = a_1 + a_1 r + a_1 r^2 + a_1 r^3 + \cdots + a_1 r^{n-1}$$

Multiply both sides of the equation by r.

$$rS_n = a_1 r + a_1 r^2 + a_1 r^3 + a_1 r^4 + \cdots + a_1 r^n$$

Subtract rS_n from S_n.

$$S_n - rS_n = (a_1 - a_1 r) + (a_1 r - a_1 r^2) + (a_1 r^2 - a_1 r^3) + (a_1 r^3 - a_1 r^4) + \cdots$$
$$+ (a_1 r^{n-1} - a_1 r^n)$$

Regrouping the right-hand side gives us

$$S_n - rS_n = a_1 + (a_1 r - a_1 r) + (a_1 r^2 - a_1 r^2) + (a_1 r^3 - a_1 r^3) + \cdots$$
$$+ (a_1 r^{n-1} - a_1 r^{n-1}) - a_1 r^n$$

The differences in parentheses equal zero, and we get $S_n - rS_n = a_1 - a_1 r^n$.

Factor out S_n on the left-hand side and a_1 on the right-hand side.

$$S_n(1 - r) = a_1(1 - r^n)$$
$$S_n = \frac{a_1(1 - r^n)}{1 - r} \qquad \text{Divide by } (1 - r).$$

Definition

Sum of the First n Terms of a Geometric Sequence: The sum of the first n terms, S_n, of a geometric sequence with first term a_1 and common ratio r is given by

$$S_n = \frac{a_1(1 - r^n)}{1 - r}$$

where $r \neq 1$.

Example 6

In-Class Example 6
Find the sum of the first five terms of the geometric sequence with first term 9 and common ratio −3.
answer: 549

Find the sum of the first four terms of the geometric sequence with first term 2 and common ratio 5.

Solution

$a_1 = 2$, $r = 5$, and $n = 4$. We are asked to find S_4, the sum of the first four terms of the geometric sequence. Use the formula

$$S_n = \frac{a_1(1 - r^n)}{1 - r}$$

$$S_4 = \frac{2[1 - (5)^4]}{1 - 5} \qquad \text{Let } n = 4, r = 5, \text{ and } a_1 = 2.$$

$$S_4 = \frac{2(1 - 625)}{-4} \qquad 5^4 = 625$$

$$S_4 = \frac{2(-624)}{-4}$$

$$S_4 = 312$$

We will verify that this result is the same as the result we would obtain by finding the first four terms of the sequence and then finding their sum.

$$a_1 = 2 \qquad a_2 = 2 \cdot 5 = 10 \qquad a_3 = 10 \cdot 5 = 50 \qquad a_4 = 50 \cdot 5 = 250$$

The terms are 2, 10, 50, and 250. Their sum is $2 + 10 + 50 + 250 = 312$. ∎

You Try 6

Find the sum of the first five terms of the geometric sequence with first term 3 and common ratio 4.

Using summation notation, $\displaystyle\sum_{i=1}^{n} a \cdot b^i$ (where a and b are constants) represents a geometric series or the sum of the first n terms of a geometric sequence. Furthermore, the first term is found by substituting 1 for i (so that the first term is ab) and the common ratio is b. We can evaluate $\displaystyle\sum_{i=1}^{n} a \cdot b^i$ using the formula for S_n.

Example 7

In-Class Example 7
Evaluate $\displaystyle\sum_{i=1}^{4} -2(4)^i$.
answer: −680

Evaluate $\displaystyle\sum_{i=1}^{5} 6(2)^i$.

Solution

Use the formula $S_n = \dfrac{a_1(1 - r^n)}{1 - r}$ to find the sum. If we let $i = 1$, we obtain $a_1 = 12$, $r = 2$, and $n = 5$.

Substitute these values into the formula for S_n.

$$S_n = \frac{a_1(1 - r^n)}{1 - r}$$

$$S_5 = \frac{12(1 - 2^5)}{1 - 2} \qquad \text{Let } n = 5, a_1 = 12, \text{ and } r = 2.$$

$$S_5 = \frac{12(1 - 32)}{-1} \qquad 2^5 = 32$$

$$S_5 = -12(-31)$$

$$S_5 = 372$$

$$\sum_{i=1}^{5} 6 \cdot 2^i = 372$$ ∎

You Try 7

Evaluate $\displaystyle\sum_{i=1}^{4} 3(5)^i$.

8. Find the Sum of Terms of an Infinite Geometric Sequence

Until now, we have considered only the sum of the first n terms of a geometric sequence. That is, we have discussed the sum of a *finite* series. Is it possible to find the sum of an *infinite* series?

Consider a geometric series with common ratio $r = \dfrac{1}{2}$. What happens to the value of $\left(\dfrac{1}{2}\right)^n$, or r^n, as n gets larger?

We will make a table of values containing n and $\left(\dfrac{1}{2}\right)^n$.

At the left you can see that as the value of n gets larger, the value of r^n gets smaller. In fact, the value of r^n gets closer and closer to zero. *We say that as n approaches infinity, r^n approaches zero.*

How does this affect the formula for the sum of the first n terms of a geometric sequence? The formula is $S_n = \dfrac{a_1(1 - r^n)}{1 - r}$.

If n approaches infinity and r^n approaches zero, we get $S = \dfrac{a_1(1 - 0)}{1 - r} = \dfrac{a_1}{1 - r}$.

This formula will hold for $|r| < 1$.

n	$r^n = \left(\dfrac{1}{2}\right)^n$
1	$\dfrac{1}{2} = 0.5$
2	$\dfrac{1}{4} = 0.25$
3	$\dfrac{1}{8} = 0.125$
4	$\dfrac{1}{16} = 0.0625$
5	$\dfrac{1}{32} = 0.03125$
6	$\dfrac{1}{64} = 0.015625$
⋮	
15	$\dfrac{1}{32,768} \approx 0.0000305$

Definition

Sum of the Terms of an Infinite Geometric Sequence: The sum of the terms, S, of an infinite geometric sequence with first term a_1 and common ratio r, where $|r| < 1$, is given by

$$S = \frac{a_1}{1 - r}$$

If $|r| \geq 1$, then the sum does not exist.

Example 8

In-Class Example 8
Find the sum of the terms of the infinite geometric sequence
$12, 9, \dfrac{27}{4}, \dfrac{81}{16}, \ldots$

answer: 48

Find the sum of the terms of the infinite geometric sequence $6, 4, \dfrac{8}{3}, \dfrac{16}{9}, \ldots$.

Solution

We will use the formula $S = \dfrac{a_1}{1-r}$, so we must identify a_1 and find r. $a_1 = 6$.

$$r = \frac{4}{6} = \frac{2}{3}$$

Since $|r| = \left|\dfrac{2}{3}\right| < 1$, the sum exists. Substitute $a_1 = 6$ and $r = \dfrac{2}{3}$ into the formula.

$$S = \frac{a_1}{1-r} = \frac{6}{1 - \dfrac{2}{3}} = \frac{6}{\dfrac{1}{3}} = 6 \cdot 3 = 18$$

The sum is 18. ∎

 You Try 8

Find the sum of the terms of the infinite geometric sequence $1, \dfrac{3}{5}, \dfrac{9}{25}, \dfrac{27}{125}, \ldots$

 **BE CAREFUL**

Remember, if $|r| \geq 1$ then the sum does not exist!

9. Solve an Applied Problem Involving an Infinite Series

Example 9

In-Class Example 9
In Example 9, substitute 75% for 80% and 18 in. for 2 ft.

answer: 72 in.

Each time a certain pendulum swings, it travels 80% of the distance it traveled on the previous swing. If it travels 2 ft on its first swing, find the total distance the pendulum travels before coming to rest.

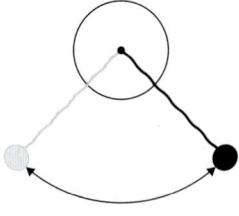

Solution

The geometric series that models this problem is

$$2 + 1.6 + 1.28 + \cdots$$

where

$$2 = \text{number of feet traveled on the first swing}$$
$$2(0.80) = 1.6 = \text{number of feet traveled on the second swing}$$
$$1.6(0.80) = 1.28 = \text{number of feet traveled on the third swing}$$
$$\text{etc.}$$

We can use the formula $S = \dfrac{a_1}{1-r}$ with $a_1 = 2$ and $r = 0.80$ to find the total distance the pendulum travels before coming to rest.

$$S = \frac{a_1}{1-r} = \frac{2}{1 - 0.80} = \frac{2}{0.20} = 10$$

The pendulum travels 10 ft before coming to rest. ∎

You Try 9

Each time a certain pendulum swings, it travels 90% of the distance it traveled on the previous swing. If it travels 20 in. on its first swing, find the total distance the pendulum travels before coming to rest.

10. Distinguish Between an Arithmetic and a Geometric Sequence

Example 10

In-Class Example 10

Determine whether each sequence is arithmetic or geometric. Then find the sum of the first six terms of the sequence.

a) $2, 4, 6, 8, 10, \ldots$
b) $2, 4, 8, 16, \ldots$

**answer: a) arithmetic; 42
b) geometric; 126**

Determine whether each sequence is arithmetic or geometric. Then find the sum of the first six terms of each sequence.

a) $-8, -4, -2, -1, \ldots$ b) $-9, -4, 1, 6, \ldots$

Solution

a) If every term of the sequence is obtained by *adding* the same constant to the previous term, then the sequence is arithmetic. (The sequence has a common difference, d.)

   If every term is obtained by *multiplying* the previous term by the same constant, then the sequence is geometric. (The sequence has a common ratio, r.)

   By inspection we can see that the terms of the sequence

$$-8, -4, -2, -1, \ldots$$

are *not* obtained by adding the same amount to each term. For example, $-4 = -8 + 4$, but $-2 = -4 + 2$.

Is there a common ratio?

$$\frac{-4}{-8} = \frac{1}{2}, \quad \frac{-2}{-4} = \frac{1}{2}, \quad \frac{-1}{-2} = \frac{1}{2}$$

Yes, $r = \dfrac{1}{2}$. *The sequence is geometric.*

Use $S_n = \dfrac{a_1(1 - r^n)}{1 - r}$ with $a_1 = -8$, $r = \dfrac{1}{2}$, and $n = 6$ to find S_6, the sum of the first six terms of this geometric sequence.

$$S_n = \frac{a_1(1 - r^n)}{1 - r}$$

$$S_6 = \frac{-8\left[1 - \left(\dfrac{1}{2}\right)^6\right]}{1 - \left(\dfrac{1}{2}\right)} \qquad \text{Let } a_1 = -8, r = \frac{1}{2}, \text{ and } n = 6.$$

$$S_6 = \frac{-8\left(1 - \dfrac{1}{64}\right)}{\dfrac{1}{2}} \qquad \left(\frac{1}{2}\right)^6 = \frac{1}{64}$$

$$S_6 = \frac{-8\left(\dfrac{63}{64}\right)}{\dfrac{1}{2}} \qquad \text{Subtract.}$$

$$S_6 = \frac{-\dfrac{63}{8}}{\dfrac{1}{2}} = -\frac{63}{8} \cdot 2 = -\frac{63}{4}$$

The sum of the first six terms of the geometric sequence $-8, -4, -2, -1, \ldots$ is $-\dfrac{63}{4}$.

b) Each term in the sequence $-9, -4, 1, 6$, is obtained by adding 5 to the previous term. *This is an arithmetic sequence with $a_1 = -9$ and common difference $d = 5$.* Since we know $a_1, d,$ and n ($n = 6$), we will use the formula

$$S_n = \frac{n}{2}[2a_1 + (n-1)d]$$

to find S_6, the sum of the first six terms of this arithmetic sequence.

$$S_6 = \frac{6}{2}[2(-9) + (6-1)5] \qquad \text{Let } a_1 = -9, d = 5, \text{ and } n = 6.$$
$$S_6 = 3[-18 + (5)(5)]$$
$$S_6 = 3[-18 + 25] = 3(7) = 21$$

The sum of the first six terms of the arithmetic sequence $-9, -4, 1, 6, \dots$ is 21. ■

You Try 10

Determine whether each sequence is arithmetic or geometric. Then, find the sum of the first seven terms of each sequence.

a) $25, 22, 19, 16, \dots$ b) $\frac{1}{6}, \frac{1}{3}, \frac{2}{3}, \frac{4}{3}, \dots$

Answers to You Try Exercises

1) a) 6 b) $\frac{1}{3}$ 2) $2, 4, 8, 16, 32$ 3) $a_n = 2(3)^{n-1}; a_5 = 162$ 4) $a_7 = -\frac{25}{32}$

5) a) $a_n = 27{,}000(0.70)^{n-1}$ b) \$13,230   6) $S_5 = 1023$ 7) 2340   8) $\frac{5}{2}$

9) 200 in.   10) a) arithmetic; $S_7 = 112$ b) geometric; $S_7 = \frac{127}{6}$

12.3 Exercises

Additional answers can be found in the Answers to Exercises appendix.

Mixed Exercises: Objectives 1 and 2

 1) What is the difference between an arithmetic and a geometric series?

2) Give an example of a geometric sequence. Answers may vary.

Find the common ratio, r, for each geometric sequence.

3) $1, 2, 4, 8, \dots$ 2

4) $3, 12, 48, 192, \dots$ 4

5) $9, 3, 1, \frac{1}{3}, \dots$ $\frac{1}{3}$

6) $8, 4, 2, 1, \dots$ $\frac{1}{2}$

7) $-2, \frac{1}{2}, -\frac{1}{8}, \frac{1}{32}, \dots$ $-\frac{1}{4}$

8) $2, -6, 18, -54, \dots$ -3

Objective 3: Find the Terms of a Geometric Sequence

Write the first five terms of the geometric sequence with the given first term and common ratio.

9) $a_1 = 2, r = 5$
2, 10, 50, 250, 1250

10) $a_1 = 3, r = 2$
3, 6, 12, 24, 48

11) $a_1 = \frac{1}{4}, r = -2$

12) $a_1 = 250, r = \frac{1}{5}$

13) $a_1 = 72, r = \frac{2}{3}$

14) $a_1 = -20, r = -\frac{3}{2}$

Mixed Exercises: Objectives 4 and 5

Find the general term, a_n, for each geometric sequence. Then, find the indicated term.

15) $a_1 = 4, r = 7; a_3$
$a_n = 4(7)^{n-1}; 196$

16) $a_1 = 3, r = 8; a_3$
$a_n = 3(8)^{n-1}; 192$

17) $a_1 = -1, r = 3; a_5$
$a_n = -1(3)^{n-1}; -81$

18) $a_1 = -5, r = -\frac{1}{3}; a_4$

19) $a_1 = 2, r = \frac{1}{5}; a_4$

20) $a_1 = 7, r = 3; a_5$
$a_n = 7(3)^{n-1}; 567$

21) $a_1 = -\frac{1}{2}, r = -\frac{3}{2}; a_4$

22) $a_1 = \frac{3}{5}, r = 2; a_6$

Find the general term of each geometric sequence.

23) $5, 10, 20, 40, \dots$
$a_n = 5(2)^{n-1}$

24) $4, 12, 36, 108, \dots$
$a_n = 4(3)^{n-1}$

25) $-3, -\dfrac{3}{5}, -\dfrac{3}{25}, -\dfrac{3}{125}, \ldots$ $a_n = -3\left(\dfrac{1}{5}\right)^{n-1}$

26) $-1, 4, -16, 64, \ldots$ $a_n = -1(-4)^{n-1}$

27) $3, -6, 12, -24, \ldots$ $a_n = 3(-2)^{n-1}$

28) $2, \dfrac{2}{3}, \dfrac{2}{9}, \dfrac{2}{27}, \ldots$ $a_n = 2\left(\dfrac{1}{3}\right)^{n-1}$

VIDEO 29) $\dfrac{1}{3}, \dfrac{1}{12}, \dfrac{1}{48}, \dfrac{1}{192}, \ldots$ $a_n = \dfrac{1}{3}\left(\dfrac{1}{4}\right)^{n-1}$

30) $-\dfrac{1}{5}, -\dfrac{3}{10}, -\dfrac{9}{20}, -\dfrac{27}{40}, \ldots$ $a_n = -\dfrac{1}{5}\left(\dfrac{3}{2}\right)^{n-1}$

Find the indicated term of each geometric sequence.

31) $1, 2, 4, 8, \ldots; a_{12}$ 2048

32) $1, 3, 9, 27, \ldots; a_{10}$ 19,683

33) $27, -9, 3, -1, \ldots; a_8$ $-\dfrac{1}{81}$

34) $-\dfrac{1}{125}, -\dfrac{1}{25}, -\dfrac{1}{5}, -1, \ldots; a_7$ -125

35) $-\dfrac{1}{64}, -\dfrac{1}{32}, -\dfrac{1}{16}, -\dfrac{1}{8}, \ldots; a_{12}$ -32

36) $-5, 10, -20, 40, \ldots; a_8$ 640

Objective 10: Distinguish Between an Arithmetic and a Geometric Sequence

Determine whether each sequence is arithmetic or geometric. Then, find the general term, a_n, of the sequence.

VIDEO 37) $15, 24, 33, 42, 51, \ldots$ arithmetic; $a_n = 9n + 6$

38) $-1, -3, -9, -27, -81, \ldots$ geometric; $a_n = -1(3)^{n-1}$

39) $-2, 6, -18, 54, -162, \ldots$ geometric; $a_n = -2(-3)^{n-1}$

40) $8, 3, -2, -7, -12, \ldots$ arithmetic; $a_n = -5n + 13$

41) $\dfrac{1}{9}, \dfrac{1}{18}, \dfrac{1}{36}, \dfrac{1}{72}, \dfrac{1}{144}, \ldots$ geometric; $a_n = \dfrac{1}{9}\left(\dfrac{1}{2}\right)^{n-1}$

42) $11, 22, 44, 88, 176, \ldots$ geometric; $a_n = 11(2)^{n-1}$

43) $-31, -24, -17, -10, -3, \ldots$ arithmetic; $a_n = 7n - 38$

44) $\dfrac{3}{2}, 2, \dfrac{5}{2}, 3, \dfrac{7}{2}, \ldots$ arithmetic; $a_n = \dfrac{1}{2}n + 1$

Objective 6: Solve an Applied Problem Involving a Geometric Sequence

Solve each application.

45) A sports car purchased for $40,000 depreciates 20% each year.

   a) Find the general term, a_n, of the geometric sequence that models the value of the sports car at the beginning of each year. $a_n = 40{,}000(0.80)^{n-1}$

   b) How much is the sports car worth at the beginning of the fifth year? $16,384

46) A luxury car purchased for $64,000 depreciates 15% each year.

   a) Find the general term, a_n, of the geometric sequence that models the value of the car at the beginning of each year. $a_n = 64{,}000(0.85)^{n-1}$

   b) How much is the luxury car worth at the beginning of the fourth year? $39,304

47) A company's advertising budget is currently $500,000 per year. For the next several years, the company will cut the budget by 10% per year.

   a) Find the general term, a_n, of the geometric sequence that models the company's advertising budget for each of the next several years. $a_n = 500{,}000(0.90)^{n-1}$

   b) What is the advertising budget 3 years from now? $364,500

48) In January 2011, approximately 1000 customers at a grocery store used the self-checkout lane. The owners predict that number will increase by 20% per month for the next year.

   a) Find the general term, a_n, of the geometric sequence that models the number of customers expected to use the self-checkout lane each month for the next year. $a_n = 1000(1.20)^{n-1}$

   b) Predict how many people will use the self-checkout lane in September 2011. Round to the nearest whole number.   4300

49) A home purchased for $160,000 increases in value by 4% per year.

   a) Find the general term of the geometric sequence that models the future value of the house. $a_n = 160{,}000(1.04)^{n-1}$

   b) How much is the home worth 5 years after it is purchased? (Hint: Think carefully about what number to substitute for n.) Round the answer to the nearest dollar. $194,664

50) A home purchased for $140,000 increases in value by 5% per year.

   a) Find the general term of the geometric sequence that models the future value of the house.
     $a_n = 140,000(1.05)^{n-1}$

   b) How much is the home worth 8 years after it is purchased? (Hint: Think carefully about what number to substitute for n.) Round the answer to the nearest dollar. $206,844

Objective 7: Find the Sum of Terms of a Geometric Sequence

51) Find the sum of the first six terms of the geometric sequence with $a_1 = 9$ and $r = 2$.   567

52) Find the sum of the first four terms of the geometric sequence with $a_1 = 6$ and $r = 3$.   240

Use the formula for S_n to find the sum of the terms of each geometric sequence.

53) 7, 28, 112, 448, 1792, 7168, 28672 $S_7 = 38,227$

54) $-5, -30, -180, -1080, -6480$ $S_5 = -7775$

55) $-\dfrac{1}{4}, -\dfrac{1}{2}, -1, -2, -4, -8$ $S_6 = -\dfrac{63}{4}$

56) $\dfrac{3}{8}, \dfrac{3}{2}, 6, 24, 96, 384$ $S_6 = \dfrac{4095}{8}$ 57) $1, \dfrac{1}{3}, \dfrac{1}{9}, \dfrac{1}{27}, \dfrac{1}{81}$ $S_5 = \dfrac{121}{81}$

58) $-3, \dfrac{6}{5}, -\dfrac{12}{25}, \dfrac{24}{125}, -\dfrac{48}{625}$ $S_5 = -\dfrac{1353}{625}$

59) $\displaystyle\sum_{i=1}^{7} 9(2)^i$ 2286

60) $\displaystyle\sum_{i=1}^{8} 5(2)^i$ 2550

61) $\displaystyle\sum_{i=1}^{5} (-4)(3^i)$ -1452

62) $\displaystyle\sum_{i=1}^{6} (-7)(-2)^i$ -294

63) $\displaystyle\sum_{i=1}^{6} 3\left(-\dfrac{1}{2}\right)^i$ $-\dfrac{63}{64}$

64) $\displaystyle\sum_{i=1}^{5} 2\left(\dfrac{1}{3}\right)^i$ $\dfrac{242}{243}$

65) $\displaystyle\sum_{i=1}^{4} (-18)\left(-\dfrac{2}{3}\right)^i$ $\dfrac{52}{9}$

66) $\displaystyle\sum_{i=1}^{4} 10\left(-\dfrac{2}{5}\right)^i$ $-\dfrac{348}{125}$

67) Gemma decides to save some pennies so that she'll put 1¢ in her bank on the first day, 2¢ on the second day, 4¢ on the third day, 8¢ on the fourth day, and so on. If she continued in this way,

   a) how many pennies would she have to put in her bank on the tenth day to continue the pattern?   512

   b) how much money will she have saved after 10 days? $10.23

68) The number of bacteria in a culture doubles every day. If a culture begins with 1000 bacteria, how many bacteria are present after 7 days?   64,000

Objective 8: Find the Sum of Terms of an Infinite Geometric Sequence

Find the sum of the terms of the infinite geometric sequence, if possible.

69) $a_1 = 8, r = \dfrac{1}{4}$ $\dfrac{32}{3}$

70) $a_1 = 18, r = \dfrac{1}{3}$ 27

71) $a_1 = 5, r = -\dfrac{4}{5}$ $\dfrac{25}{9}$

72) $a_1 = 20, r = -\dfrac{3}{4}$ $\dfrac{80}{7}$

73) $a_1 = 9, r = \dfrac{5}{3}$ Sum does not exist.

74) $a_1 = 3, r = \dfrac{3}{2}$ Sum does not exist.

75) $8, \dfrac{16}{3}, \dfrac{32}{9}, \dfrac{64}{27}, \ldots$ 24

76) $\dfrac{7}{2}, \dfrac{7}{4}, \dfrac{7}{8}, \dfrac{7}{16}, \ldots$ 7

77) $-\dfrac{15}{2}, \dfrac{15}{4}, -\dfrac{15}{8}, \dfrac{15}{16}, \ldots$ -5

78) $-12, 8, -\dfrac{16}{3}, \dfrac{32}{9}, \ldots$ $-\dfrac{36}{5}$

79) $\dfrac{1}{25}, \dfrac{1}{5}, 1, 5, \ldots$ Sum does not exist.

80) $36, 6, 1, \dfrac{1}{6}, \ldots$ $\dfrac{216}{5}$

81) $-40, -30, -\dfrac{45}{2}, -\dfrac{135}{8}, \ldots$ -160

82) $4, -12, 36, -108, \ldots$ Sum does not exist.

Objective 9: Solve an Applied Problem Involving an Infinite Series

Solve each application.

83) Each time a certain pendulum swings, it travels 75% of the distance it traveled on the previous swing. If it travels 3 ft on its first swing, find the total distance the pendulum travels before coming to rest.   12 ft

84) Each time a certain pendulum swings, it travels 70% of the distance it traveled on the previous swing. If it travels 42 in. on its first swing, find the total distance the pendulum travels before coming to rest.   140 in.

85) A ball is dropped from a height of 27 ft. Each time the ball bounces, it rebounds to $\dfrac{2}{3}$ of its previous height.

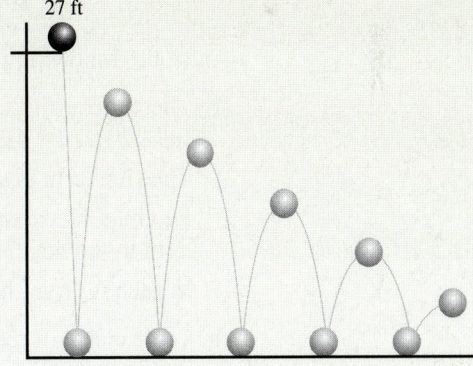

27 ft

   a) Find the height the ball reaches after the fifth bounce. $3\dfrac{5}{9}$ ft

   b) Find the total vertical distance the ball has traveled when it comes to rest.   135 ft

86) A ball is dropped from a height of 16 ft. Each time the ball bounces, it rebounds to $\dfrac{3}{4}$ of its previous height.

   a) Find the height the ball reaches after the fourth bounce. $5\dfrac{1}{16}$ ft

   b) Find the total vertical distance the ball has traveled when it comes to rest.   112 ft

Section 12.4 The Binomial Theorem

Objectives

1. Expand $(a + b)^n$ Using Pascal's Triangle
2. Evaluate Factorials
3. Evaluate $\binom{n}{r}$
4. Expand $(a + b)^n$ Using the Binomial Theorem
5. Find a Specified Term in the Expansion of $(a + b)^n$

In this section, we will learn how to expand a binomial, $(a + b)^n$, where n is a nonnegative integer. We first encountered expansion of binomials in Chapter 5 when we learned to expand binomials such as $(a + b)^2$ and $(a + b)^3$. To expand $(a + b)^2$ means to find the product $(a + b)(a + b)$. Multiplying the binomials using FOIL gives us

$$\begin{aligned}(a + b)^2 &= (a + b)(a + b) \\ &= a^2 + ab + ab + b^2 \qquad \text{Multiply using FOIL.} \\ &= a^2 + 2ab + b^2\end{aligned}$$

To expand $(a + b)^3$ means to find the product $(a + b)(a + b)(a + b)$ or $(a + b)(a + b)^2$. Expanding $(a + b)^3$ gives us

$$\begin{aligned}(a + b)^3 &= (a + b)(a + b)(a + b) \\ &= (a + b)(a + b)^2 \\ &= (a + b)(a^2 + 2ab + b^2) \\ &= a^3 + 2a^2b + ab^2 + a^2b + 2ab^2 + b^3 \qquad \text{Distribute.} \\ &= a^3 + 3a^2b + 3ab^2 + b^3 \qquad \text{Combine like terms.}\end{aligned}$$

Expanding, for example, $(a + b)^5$ in this way would be an extremely long process. There are other ways to expand binomials, and the first one we will discuss is *Pascal's triangle*.

1. Expand $(a + b)^n$ Using Pascal's Triangle

Here are the expansions of $(a + b)^n$ for several values of n:

$$\begin{aligned}(a + b)^0 &= 1 \\ (a + b)^1 &= a + b \\ (a + b)^2 &= a^2 + 2ab + b^2 \\ (a + b)^3 &= a^3 + 3a^2b + 3ab^2 + b^3 \\ (a + b)^4 &= a^4 + 4a^3b + 6a^2b^2 + 4ab^3 + b^4 \\ (a + b)^5 &= a^5 + 5a^4b + 10a^3b^2 + 10a^2b^3 + 5ab^4 + b^5\end{aligned}$$

Notice the following patterns in the expansion of $(a + b)^n$:

1) There are $n + 1$ terms in the expansion of $(a + b)^n$. For example, in the expansion of $(a + b)^4$, $n = 4$ and the expansion contains $4 + 1 = 5$ terms.

2) The first term is a^n and the last term is b^n.

3) Reading the expansion from left to right, the exponents on a **decrease by 1** from one term to the next, while the exponents on b **increase by 1** from one term to the next.

4) In each term in the expansion, the sum of the exponents of the variables is n.

The coefficients of the terms in the expansion follow a pattern too. If we write the coefficients in triangular form, we obtain **Pascal's triangle**, named after seventeenth-century French mathematician Blaise Pascal. *The numbers in the nth row of the triangle tell us the coefficients of the terms in the expansion of $(a + b)^n$.*

Coefficients of the Terms in the Expansion of: **Pascal's Triangle**

$$\begin{array}{cc} (a + b)^0\text{:} & 1 \\ (a + b)^1\text{:} & 1 \quad 1 \\ (a + b)^2\text{:} & 1 \quad 2 \quad 1 \\ (a + b)^3\text{:} & 1 \quad 3 \quad 3 \quad 1 \\ (a + b)^4\text{:} & 1 \quad 4 \quad 6 \quad 4 \quad 1 \\ (a + b)^5\text{:} & 1 \quad 5 \quad 10 \quad 10 \quad 5 \quad 1 \\ & \text{etc.} \end{array}$$

Notice that the first and last numbers of each row in the triangle are 1. The other numbers in the triangle are obtained by adding the two numbers above it. For example, here is how to obtain the sixth row from the fifth:

Fifth row ($n = 4$):   1   4   6   4   1

Sixth row ($n = 5$):   1   5   10   10   5   1

Example 1

In-Class Example 1
Use the given example.

Expand $(a + b)^6$.

Solution

The coefficients of the terms in $(a + b)^5$ are given by the last row of the triangle above. We must find the next row of the triangle to find the coefficients of the terms in the expansion of $(a + b)^6$.

$(a + b)^5$:   1   5   10   10   5   1

$(a + b)^6$:   1   6   15   20   15   6   1

Recall that the first term is a^n and the last term is b^n. Since $n = 6$, the first term will be a^6, and the exponent of a will decrease by 1 for each term. The variable b will appear in the second term and increase by 1 for each term until the last term, b^6.

$$(a + b)^6 = a^6 + 6a^5b + 15a^4b^2 + 20a^3b^3 + 15a^2b^4 + 6ab^5 + b^6$$

You Try 1

Expand $(a + b)^7$.

Although Pascal's triangle is a better way to expand $(a + b)^n$ than doing repeated polynomial multiplication, it can be tedious for large values of n. A more practical way to expand a binomial is by using the **binomial theorem**. Before learning this method, we need to learn about **factorials** and **binomial coefficients**.

2. Evaluate Factorials

The notation $n!$ is read as "n factorial."

Definition

$n! = n(n - 1)(n - 2)(n - 3) \cdots (1)$, where n is a positive integer.

Note

By definition, $0! = 1$.

Example 2

Evaluate.

a) 4!   b) 7!

In-Class Example 2
Evaluate. a) 5! b) 8!
answer: a) 120 b) 40,320

Solution

a) $4! = 4 \cdot 3 \cdot 2 \cdot 1 = 24$ b) $7! = 7 \cdot 6 \cdot 5 \cdot 4 \cdot 3 \cdot 2 \cdot 1 = 5040$ ■

You Try 2

Evaluate.

a) 3!   b) 6!

3. Evaluate $\binom{n}{r}$

Factorials are used to evaluate binomial coefficients. A **binomial coefficient** has the form $\binom{n}{r}$, read as "the number of combinations of n items taken r at a time" or as "n choose r." $\binom{n}{r}$ is used extensively in many areas of mathematics including probability. Another notation for $\binom{n}{r}$ is nCr.

Definition

Binomial Coefficient

$$\binom{n}{r} = \frac{n!}{r!(n-r)!}$$

where n and r are positive integers and $r \le n$.

Example 3

Evaluate.

a) $\binom{5}{3}$ b) $\binom{9}{2}$ c) $\binom{3}{3}$ d) $\binom{4}{0}$

In-Class Example 3
Evaluate.
a) $\binom{6}{4}$ b) $\binom{7}{3}$
c) $\binom{9}{9}$ d) $\binom{3}{0}$
answer: a) 15 b) 35
c) 1 d) 1

Solution

a) To evaluate $\binom{5}{3}$, substitute 5 for n and 3 for r.

$$\binom{n}{r} = \frac{n!}{r!(n-r)!}$$

$$\binom{5}{3} = \frac{5!}{3!(5-3)!} \qquad \text{Let } n = 5 \text{ and } r = 3.$$

$$= \frac{5!}{3!2!} \qquad \text{Subtract.}$$

$$= \frac{5 \cdot 4 \cdot 3 \cdot 2 \cdot 1}{(3 \cdot 2 \cdot 1)(2 \cdot 1)} \qquad \text{Rewrite each factorial as a product.}$$

At this point, do *not* find the products in the numerator and denominator. Instead, divide out common factors.

$$= \frac{5 \cdot 4 \cdot 3 \cdot 2 \cdot 1}{(3 \cdot 2 \cdot 1)(2 \cdot 1)} \qquad \text{Divide out common factors.}$$

$$= \frac{20}{2} \qquad \text{Multiply.}$$

$$= 10 \qquad \text{Simplify.}$$

$$\binom{5}{3} = 10$$

b) To evaluate $\binom{9}{2}$, substitute 9 for n and 2 for r.

$$\binom{9}{2} = \frac{9!}{2!(9-2)!} \qquad \text{Let } n = 9 \text{ and } r = 2.$$

$$= \frac{9!}{2!\,7!} \qquad \text{Subtract.}$$

$$= \frac{9 \cdot 8 \cdot 7 \cdot 6 \cdot 5 \cdot 4 \cdot 3 \cdot 2 \cdot 1}{(2 \cdot 1)(7 \cdot 6 \cdot 5 \cdot 4 \cdot 3 \cdot 2 \cdot 1)} \qquad \text{Rewrite each factorial as a product.}$$

$$= \frac{9 \cdot 8 \cdot 7 \cdot 6 \cdot 5 \cdot 4 \cdot 3 \cdot 2 \cdot 1}{(2 \cdot 1)(7 \cdot 6 \cdot 5 \cdot 4 \cdot 3 \cdot 2 \cdot 1)} \qquad \text{Divide out common factors.}$$

$$= \frac{72}{2}$$

$$= 36$$

$$\binom{9}{2} = 36$$

c) To evaluate $\binom{3}{3}$, substitute 3 for n and for r.

$$\binom{n}{r} = \frac{n!}{r!(n-r)!}$$

$$\binom{3}{3} = \frac{3!}{3!(3-3)!} \qquad \text{Let } n = 3 \text{ and } r = 3.$$

$$= \frac{3!}{3!\,0!} \qquad \text{Subtract.}$$

$$= \frac{3!}{3!(1)} \qquad \text{Divide out common factors; } 0! = 1.$$

$$= \frac{1}{1} = 1 \qquad \text{Simplify.}$$

$$\binom{3}{3} = 1$$

d) To evaluate $\binom{4}{0}$, substitute 4 for n and 0 for r.

$$\binom{4}{0} = \frac{4!}{0!(4-0)!} \qquad \text{Let } n = 4 \text{ and } r = 0.$$

$$= \frac{4!}{0!\,4!} \qquad \text{Subtract.}$$

$$= \frac{4!}{(1)(4!)} \qquad \text{Divide out common factors; } 0! = 1.$$

$$= \frac{1}{1} = 1 \qquad \text{Simplify.}$$

$$\binom{4}{0} = 1$$

Note

We can extend the results of c) and d) and say that for any natural number n,

$$\binom{n}{n} = 1 \quad \text{and} \quad \binom{n}{0} = 1$$

You Try 3

Evaluate.

a) $\binom{4}{1}$ b) $\binom{8}{5}$ c) $\binom{3}{0}$ d) $\binom{6}{6}$

4. Expand $(a + b)^n$ Using the Binomial Theorem

Now that we can evaluate a binomial coefficient, we state the binomial theorem for expanding $(a + b)^n$.

Definition

Binomial Theorem: For any positive integer n,

$$(a + b)^n = a^n + \binom{n}{1}a^{n-1}b + \binom{n}{2}a^{n-2}b^2 + \binom{n}{3}a^{n-3}b^3 + \cdots + \binom{n}{n-1}ab^{n-1} + b^n$$

The same patterns that emerged in the expansion of $(a + b)^n$ using Pascal's triangle appear when using the binomial theorem. Keep in mind that

1) there are $n + 1$ terms in the expansion.

2) the first term in the expansion is a^n and the last term is b^n.

3) after a^n, the exponents on *a decrease by 1* from one term to the next, while b is introduced in the second term and then the exponents on *b increase by 1* from one term to the next.

4) in each term in the expansion, the sum of the exponents of the variables is n.

Example 4

Use the binomial theorem to expand $(a + b)^4$.

In-Class Example 4
Use the binomial theorem to expand $(a + b)^5$.
answer: $a^5 + 5a^4b + 10a^3b^2 + 10a^2b^3 + 5ab^4 + b^5$

Solution

Let $n = 4$ in the binomial theorem.

$$(a + b)^4 = a^4 + \binom{4}{1}a^{4-1}b + \binom{4}{2}a^{4-2}b^2 + \binom{4}{3}a^{4-3}b^3 + b^4$$

$$= a^4 + \binom{4}{1}a^3b + \binom{4}{2}a^2b^2 + \binom{4}{3}ab^3 + b^4$$

Notice that the exponents of a decrease by 1 while the exponents of b increase by 1.

$$= a^4 + \frac{4!}{1!\,3!}a^3b + \frac{4!}{2!\,2!}a^2b^2 + \frac{4!}{3!\,1!}ab^3 + b^4$$

$$= a^4 + 4a^3b + 6a^2b^2 + 4ab^3 + b^4$$

This is the same result as the expansion on p. 788.

You Try 4

Use the binomial theorem to expand $(a + b)^3$.

Example 5

Use the binomial theorem to expand $(x + 6)^3$.

In-Class Example 5
Use the binomial theorem to expand $(x + 2)^3$.
answer: $x^3 + 6x^2 + 12x + 8$

Solution

Substitute x for a, 6 for b, and 3 for n in the binomial theorem to expand $(x + 6)^3$.

$$(x + 6)^3 = (x)^3 + \binom{3}{1}(x)^{3-1}(6) + \binom{3}{2}(x)^{3-2}(6)^2 + (6)^3$$
$$= x^3 + 3(x^2)(6) + (3)x(36) + 216$$
$$= x^3 + 18x^2 + 108x + 216$$

You Try 5

Use the binomial theorem to expand $(y + 5)^4$.

BE CAREFUL

When expanding a binomial containing the *difference* of two terms, rewrite the expression in terms of addition.

Example 6

Use the binomial theorem to expand $(2x - 3y)^5$.

In-Class Example 6
Use the binomial theorem to expand $(2b - c)^4$.
answer: $16b^4 - 32b^3c + 24b^2c^2 - 8bc^3 + c^4$

Solution

Since the binomial theorem applies to the expansion of $(a + b)^n$, rewrite $(2x - 3y)^5$ as $[2x + (-3y)]^5$.

Substitute $2x$ for a, $-3y$ for b, and 5 for n in the binomial theorem. *Be sure to put $2x$ and $-3y$ in parentheses to find the expansion correctly.*

$$[2x + (-3y)]^5 = (2x)^5 + \binom{5}{1}(2x)^{5-1}(-3y) + \binom{5}{2}(2x)^{5-2}(-3y)^2$$
$$+ \binom{5}{3}(2x)^{5-3}(-3y)^3 + \binom{5}{4}(2x)^{5-4}(-3y)^4 + (-3y)^5$$
$$= 32x^5 + (5)(2x)^4(-3y) + (10)(2x)^3(9y^2) + (10)(2x)^2(-27y^3)$$
$$+ (5)(2x)^1(81y^4) + (-243y^5)$$
$$= 32x^5 + (5)(16x^4)(-3y) + (10)(8x^3)(9y^2)$$
$$+ (10)(4x^2)(-27y^3) + (5)(2x)(81y^4) + (-243y^5)$$
$$= 32x^5 - 240x^4y + 720x^3y^2 - 1080x^2y^3 + 810xy^4 - 243y^5$$

You Try 6

Use the binomial theorem to expand $(3x - 4y)^4$.

5. Find a Specified Term in the Expansion of $(a + b)^n$

If we want to find a specific term of a binomial expansion without writing out the entire expansion, we can use the following formula.

> **Definition**
>
> **The kth Term of a Binomial Expansion:** The kth term of the expansion of $(a + b)^n$ is given by
>
> $$\frac{n!}{(n - k + 1)!(k - 1)!} a^{n-k+1} b^{k-1}$$
>
> where $k \leq n + 1$.

Example 7

Find the fifth term in the expansion of $(c^2 + 2d)^8$.

In-Class Example 7
Find the fifth term in the expansion of $(3x^2 + 2y)^8$.
answer: $90{,}720x^8y^4$

Solution

Since we want to find the fifth term, $k = 5$, use the formula above with $a = c^2$, $b = 2d$, $n = 8$, and $k = 5$. The fifth term is

$$\frac{8!}{(8 - 5 + 1)!(5 - 1)!} (c^2)^{8-5+1}(2d)^{5-1} = \frac{8!}{4!\,4!} (c^2)^4 (2d)^4$$
$$= 70c^8(16d^4)$$
$$= 1120c^8d^4$$

You Try 7

Find the sixth term in the expansion of $(2m + n^2)^9$.

Using Technology

We will discuss how to compute factorials and the binomial coefficient on a graphing calculator. Sometimes it is quicker to calculate them by hand, and sometimes a calculator will make our work easier.

Evaluating 3! can be done very easily by multiplying: $3! = 3 \times 2 \times 1 = 6$. To find 10! by hand we would multiply: $10! = 10 \times 9 \times 8 \times 7 \times 6 \times 5 \times 4 \times 3 \times 2 \times 1 = 3{,}628{,}800$. On a graphing calculator, we could find 10! either by performing this multiplication or we can use a special function.

Graphing calculators have a factorial key built in. It is found using the MATH key. When you press MATH, move the arrow over to the PRB column, and you will see this menu:

```
MATH NUM CPX PRB
1:rand
2:nPr
3:nCr
4:!
5:randInt(
6:randNorm(
7:randBin(
```

Notice that choice 4 is the factorial symbol.

To compute 10!, enter 10 and then press MATH. Highlight PRB so that you see the screen at above right. Choose 4:! and press ENTER. The screen displays 10!. Press ENTER to see that $10! = 3{,}628{,}800$.

```
10!
        3628800
```

Because the binomial coefficient, $\binom{n}{r}$, is used so often in mathematical applications, most graphing calculators have a built-in key that performs the calculations for you. It is located on the same menu as the factorial. Refer to the first calculator screen to see that nCr is choice number 3.

To find the value of $\binom{9}{4}$, press $\boxed{9}$ $\boxed{\text{MATH}}$, and then highlight PRB.

Choose 3: nCr and press $\boxed{\text{ENTER}}$. Now enter 4 and press $\boxed{\text{ENTER}}$. The screen will look like the next screen here. The value of $\binom{9}{4}$ is 126.

```
9 nCr 4
            126
```

Although the calculator has functions to evaluate factorials and binomial coefficients, sometimes it is actually quicker to evaluate them by hand. Think about this as you evaluate the following problems.

Evaluate each of the following using the methods discussed in this section. Verify the result using a graphing calculator. Think about which method you prefer for each problem.

1) 4! 2) 6! 3) 9! 4) $\binom{5}{2}$

5) $\binom{7}{4}$ 6) $\binom{15}{8}$ 7) $\binom{18}{14}$ 8) $\binom{25}{24}$

Answers to You Try Exercises

1) $a^7 + 7a^6b + 21a^5b^2 + 35a^4b^3 + 35a^3b^4 + 21a^2b^5 + 7ab^6 + b^7$
2) a) 6 b) 720   3) a) 4 b) 56 c) 1 d) 1   4) $a^3 + 3a^2b + 3ab^2 + b^3$
5) $y^4 + 20y^3 + 150y^2 + 500y + 625$ 6) $81x^4 - 432x^3y + 864x^2y^2 - 768xy^3 + 256y^4$
7) $2016m^4n^{10}$

Answers to Technology Exercises

1) 24   2) 720   3) 362,880   4) 10   5) 35   6) 6435   7) 3060   8) 25

12.4 Exercises

*Additional answers can be found in the Answers to Exercises appendix.

Objective 1: Expand $(a + b)^n$ Using Pascal's Triangle

1) In your own words, explain how to construct Pascal's triangle.   Answers may vary.

2) What are the first and last terms in the expansion of $(a + b)^n$?   first term: a^n; last term: b^n

Use Pascal's triangle to expand each binomial.

3) $(r + s)^3$ 4) $(m + n)^4$
5) $(y + z)^5$ 6) $(c + d)^6$
7) $(x + 5)^4$ 8) $(k + 2)^5$

Objective 2: Evaluate Factorials

9) In your own words, explain how to evaluate $n!$ for any positive integer.   Answers may vary.

10) Evaluate 0!. 1

Evaluate.

11) 2! 2 12) 3! 6
13) 5! 120 14) 6! 720

Objective 3: Evaluate $\binom{n}{r}$

Evaluate each binomial coefficient.

15) $\binom{5}{2}$ 10

16) $\binom{4}{2}$ 6

17) $\binom{7}{3}$ 35

18) $\binom{8}{5}$ 56

19) $\binom{10}{4}$ 210

20) $\binom{9}{3}$ 84

21) $\binom{9}{7}$ 36

22) $\binom{11}{8}$ 165

23) $\binom{4}{4}$ 1

24) $\binom{5}{5}$ 1

25) $\binom{6}{1}$ 6

26) $\binom{3}{1}$ 3

27) $\binom{5}{0}$ 1

28) $\binom{7}{0}$ 1

Objective 4: Expand $(a + b)^n$ Using the Binomial Theorem

29) How many terms are in the expansion of $(a + b)^9$?   10

30) Before expanding $(t - 4)^6$ using the binomial theorem, how should the binomial be rewritten? $[t + (-4)]^6$

Use the binomial theorem to expand each expression.

31) $(f + g)^3$

32) $(c + d)^5$

33) $(w + 2)^4$

34) $(h + 4)^4$

35) $(b + 3)^5$

36) $(t + 9)^3$

37) $(a - 3)^4$

38) $(p - 2)^3$

39) $(u - v)^3$

40) $(p - q)^5$

41) $(3m + 2)^4$

42) $(2k + 1)^4$

43) $(3a - 2b)^5$

44) $(4c - 3d)^4$

45) $(x^2 + 1)^3$

46) $(w^3 + 2)^3$

47) $\left(\dfrac{1}{2}m - 3n\right)^4$

48) $\left(\dfrac{1}{3}a + 2b\right)^5$

49) $\left(\dfrac{1}{3}y + 2z^2\right)^3$

50) $\left(t^2 - \dfrac{1}{2}u\right)^4$

Objective 5: Find a Specified Term in the Expansion of $(a + b)^n$

Find the indicated term of each binomial expansion.

51) $(k + 5)^8$; third term $700k^6$

52) $(y + 4)^7$; fifth term $8960y^3$

53) $(w + 1)^{15}$; tenth term $5005w^6$

54) $(z + 3)^9$; seventh term $61,236z^3$

55) $(q - 3)^9$; second term $-27q^8$

56) $(u - 2)^7$; fourth term $-280u^4$

57) $(3x - 2)^6$; fifth term $2160x^2$

58) $(2w - 1)^9$; seventh term $672w^3$

59) $(2y^2 + z)^{10}$; eighth term $960y^6z^7$

60) $(p + 3q^2)^8$; fifth term $5670p^4q^8$

61) $(c^3 - 3d^2)^7$; third term $189c^{15}d^4$

62) $(2r^3 - s^4)^6$; sixth term $-12r^3s^{20}$

63) $(5u + v^3)^{11}$; last term v^{33}

64) $(4h - k^4)^{12}$; last term k^{48}

65) Show that $\binom{n}{n} = 1$ for any positive integer n.   Answers may vary.

66) Show that $\binom{n}{1} = n$ for any positive integer n.   Answers may vary.

Chapter 12: Summary

Definition/Procedure	Example

12.1 Sequences and Series

Finite Sequence

A **finite sequence** is a function whose domain is the set of the first n natural numbers. **(p. 754)**

$a_n = \{1, 2, 3, \ldots, n\}$ for $n \geq 0$

Infinite Sequence

An **infinite sequence** is a function whose domain is the set of natural numbers. **(p. 754)**

$a = \{1, 2, 3, \ldots\}$

General Term

The nth term of the sequence, a_n, is the **general term** of the sequence. **(p. 754)**

Write the first five terms of the sequence with general term $a_n = 5n - 9$.

Evaluate $a_n = 5n - 9$ for $n = 1, 2, 3, 4,$ and 5.

The first five terms are $-4, 1, 6, 11, 16$.

Given some terms of a sequence, we can find a formula for a_n. **(p. 757)**

Find the general term, a_n, for the sequence
$3, 9, 27, 81, 243, \ldots$

Write each term as a power of 3.

Term: $\quad 3^1, \quad 3^2, \quad 3^3, \quad 3^4, \quad 3^5$

Term number: $a_1 \quad a_2 \quad a_3 \quad a_4 \quad a_5$

The nth term may be written as $a_n = 3^n$.

Series

A sum of the terms of a sequence is called a **series**. A series can be finite or infinite. **(p. 758)**

$S_n = a_1 + a_2 + \cdots + a_n$

Summation Notation

Σ (sigma) is shorthand notation for a series. The letters i, j, and k are often used as variables for the **index of summation**. **(p. 759)**

Evaluate $\displaystyle\sum_{i=1}^{3} (-1)^i \cdot 5^i$.

$$\sum_{i=1}^{3} (-1)^i \cdot 5^i = (-1)^1 \cdot 5^1 + (-1)^2 \cdot 5^2 + (-1)^3 \cdot 5^3$$

$$= -5 \quad + \quad 25 \quad + \quad (-125)$$

$$= -105$$

Arithmetic Mean

The **average** of a group of numbers is represented by $\bar{x}$, and is given by the formula $\bar{x} = \dfrac{\displaystyle\sum_{i=1}^{n} x_i}{n}$, where

$x_1, x_2, x_3, \ldots, x_n$ are the numbers in the group and n is the number of numbers in the group. **(p. 761)**

Jin's grades on his five sociology tests were 86, 91, 83, 78, and 88. What is his test average?

$\bar{x} = \dfrac{\displaystyle\sum_{i=1}^{5} x_i}{5}$, where $x_1 = 86$, $x_2 = 91$, $x_3 = 83$,

$x_4 = 78$, and $x_5 = 88$.

$$\bar{x} = \frac{86 + 91 + 83 + 78 + 88}{5} = 85.2$$

Definition/Procedure	Example

12.2 Arithmetic Sequences and Series

Arithmetic Sequence

An **arithmetic sequence** is a sequence in which each term after the first differs from the preceding term by a constant amount d. d is called the **common difference. (p. 765)**

10, 13, 16, 19, 22, ... is an arithmetic sequence since each term is 3 more than the previous term.

The common difference, d, is 3.

Sum of the First n Terms of an Arithmetic Sequence

An **arithmetic series** is a sum of terms of an arithmetic sequence. The sum of the first n terms, S_n, is given by

$$S_n = \frac{n}{2}(a_1 + a_n) \quad \text{or} \quad S_n = \frac{n}{2}[2a_1 + (n-1)d]. \textbf{ (p. 770)}$$

Find the sum of the first 20 terms of the arithmetic sequence with first term -9 and last term 29.

Since we are given $n = 20$, $a_1 = -9$, and $a_{20} = 29$, find S_{20}.

$$S_n = \frac{n}{2}(a_1 + a_n)$$

$$S_{20} = \frac{20}{2}(-9 + a_{20}) \qquad \text{Let } n = 20 \text{ and } a_1 = -9.$$

$$S_{20} = 10(-9 + 29) \qquad \text{Let } a_{20} = 29.$$

$$S_{20} = 10(20) = 200$$

12.3 Geometric Sequences and Series

A **geometric sequence** is a sequence in which each term after the first is obtained by multiplying the preceding term by a constant, r. r is called the **common ratio. (p. 776)**

$-2, -6, -18, -54, -162, \dots$ is a geometric sequence since each term after the first is obtained by multiplying the previous term by 3. The common ratio, r, is 3.

We can find r by dividing any term by the preceding term:

$$r = \frac{-6}{-2} = \frac{-18}{-6} = \frac{-54}{-18} = \frac{-162}{-54} = 3$$

The **general term of a geometric sequence** with first term a_1 and common ratio r is given by $a_n = a_1 r^{n-1}$. **(p. 778)**

Find a_n for the geometric sequence

$-2, -6, -18, -54, -162, \dots$

$$a_1 = -2 \qquad r = 3$$
$$a_n = a_1 r^{n-1}$$
$$a_n = (-2)(3)^{n-1} \qquad \text{Let } a_1 = -2 \text{ and } r = 3.$$

Sum of the First n Terms of a Geometric Sequence

A **geometric series** is a sum of terms of a geometric sequence.

The sum of the first n terms, S_n, of a geometric sequence with first term a_1 and common ratio r is given by

$$S_n = \frac{a_1(1 - r^n)}{1 - r} \text{ where } r \neq 1 \textbf{ (p. 780)}$$

Find the sum of the first six terms of the geometric sequence with first term 7 and common ratio 2.

$n = 6$, $a_1 = 7$, $r = 2$. Find S_6.

$$S_n = \frac{a_1(1 - r^n)}{1 - r}$$

$$S_6 = \frac{7(1 - 2^6)}{1 - 2} \qquad \text{Let } n = 6, a_1 = 7, r = 2.$$

$$S_6 = \frac{7(-63)}{-1} = 441$$

Sum of the Terms of an Infinite Geometric Sequence

The sum of the terms, S, of an infinite geometric sequence with first term a_1 and common ratio r, where $|r| < 1$, is given by

$S = \dfrac{a_1}{1 - r}$. If $|r| \geq 1$, then the sum does not exist. **(p. 782)**

Find the sum of the terms of the infinite geometric sequence

$75, 15, 3, \dfrac{3}{5}, \dots$

$$a_1 = 75, r = \frac{15}{75} = \frac{1}{5}$$

Since $|r| = \left|\dfrac{1}{5}\right| < 1$, the sum exists.

$$S = \frac{a_1}{1 - r} = \frac{75}{1 - \frac{1}{5}} = \frac{75}{\frac{4}{5}} = 75 \cdot \frac{5}{4} = \frac{375}{4}$$

Definition/Procedure	Example

12.4 The Binomial Theorem

Pascal's Triangle

The numbers in the rows of Pascal's triangle tell us the coefficients of the terms in the expansion of $(a + b)^n$. **(p. 788)**

Expand $(a + b)^4$.

The coefficients of the terms in the expansion are in the fifth row of Pascal's triangle.

$$(a + b)^4 = a^4 + 4a^3b + 6a^2b^2 + 4ab^3 + b^4$$

Binomial Coefficient

$\binom{n}{r} = \dfrac{n!}{r!(n - r)!}$, where n and r are positive integers and $r \leq n$.

$$\binom{n}{n} = 1 \quad \text{and} \quad \binom{n}{0} = 1 \text{ (p. 790)}$$

Evaluate $\binom{6}{2}$.

$$\binom{6}{2} = \frac{6!}{2!(6 - 2)!} = \frac{6!}{2! \, 4!} = \frac{6 \cdot 5 \cdot 4 \cdot 3 \cdot 2 \cdot 1}{(2 \cdot 1)(4 \cdot 3 \cdot 2 \cdot 1)} \quad \text{Divide out}$$
common factors.
$$= \frac{30}{2} = 15$$

Binomial Theorem

For any positive integer n,

$$(a + b)^n = a^n + \binom{n}{1}a^{n-1}b + \binom{n}{2}a^{n-2}b^2$$

$$+ \binom{n}{3}a^{n-3}b^3 + \cdots + \binom{n}{n-1}ab^{n-1} + b^n \text{ (p. 792)}$$

Use the binomial theorem to expand $(2c + 5)^4$.

Let $a = 2c$, $b = 5$, and $n = 4$.

$$(2c + 5)^4 = (2c)^4 + \binom{4}{1}(2c)^3(5)$$

$$+ \binom{4}{2}(2c)^2(5)^2 + \binom{4}{3}(2c)(5)^3 + (5)^4$$

$$= 16c^4 + (4)(8c^3)(5) + 6(4c^2)(25) + (4)(2c)(125) + 625$$

$$= 16c^4 + 160c^3 + 600c^2 + 1000c + 625$$

The *k*th Term of a Binomial Expansion

The *k*th term of the expansion of $(a + b)^n$ is given by

$$\frac{n!}{(n - k + 1)!(k - 1)!} a^{n-k+1}b^{k-1}, \text{ where } k \leq n + 1. \text{ (p. 794)}$$

Find the sixth term in the expansion of $(x - 2y)^9$.

$$n = 9, k = 6, a = x, b = -2y$$

The sixth term is

$$\frac{9!}{(9 - 6 + 1)!(6 - 1)!}(x)^{9-6+1}(-2y)^{6-1}$$

$$= \frac{9!}{4! \, 5!}(x)^4(-2y)^5$$

$$= 126x^4(-32y^5)$$

$$= -4032x^4y^5$$

*Additional answers can be found in the Answers to Exercises appendix.

(12.1) Write out the first five terms of each sequence.

1) $a_n = 7n + 1$
8, 15, 22, 29, 36

2) $a_n = n^2 - 7$
−6, −3, 2, 9, 18

3) $a_n = (-1)^{n+1} \cdot \left(\dfrac{1}{n^2} \right)$

4) $a_n = \dfrac{n+1}{n^2}$

Find a formula for the general term, a_n, of each sequence.

5) 5, 10, 15, 20, . . . $a_n = 5n$

6) $1, \dfrac{1}{8}, \dfrac{1}{27}, \dfrac{1}{64}, \ldots$ $a_n = \dfrac{1}{n^3}$

7) $-2, -\dfrac{3}{2}, -\dfrac{4}{3}, -\dfrac{5}{4}, \ldots$

8) $-1, 2, -3, 4, \ldots$
$a_n = (-1)^n(n)$

9) Currently, Maura earns $8.25 per hour, and she can get a raise of $0.25 per hour every 4 months. Write a sequence that represents her current wage as well as her hourly wage 4, 8, and 12 months from now.
$8.25, $8.50, $8.75, $9.00

10) Dorian wants to increase the number of students at his martial arts school by 20% every 6 months. If he currently has 40 students, how many does he hope to have 18 months from now? Round to the nearest whole number.

69

11) What is the difference between a sequence and a series?

12) Write in summation notation. *Find the sum of the first eight terms in the sequence given by $a_n = 6n - 1$. Do* **not** *evaluate.* $\displaystyle\sum_{i=1}^{8} (6i - 1)$

Write out the terms in each series and evaluate.

13) $\displaystyle\sum_{i=1}^{5} (2i^2 + 1)$
3 + 9 + 19 + 33 + 51 = 115

14) $\displaystyle\sum_{i=1}^{6} (-1)^i(2i)$
−2 + 4 − 6 + 8 − 10 + 12 = 6

Write each series using summation notation.

15) $13 + \dfrac{13}{2} + \dfrac{13}{3} + \dfrac{13}{4}$ $\displaystyle\sum_{i=1}^{4} \dfrac{13}{i}$

16) $2 - 4 + 8 - 16 + 32 - 64$ $\displaystyle\sum_{i=1}^{6} (-1)^{i+1}(2^i)$

17) Find the arithmetic mean of this group of numbers: 18, 25, 26, 20, 22   22.2

18) The Shiu family's heating bills from December 2010 to March 2011 are given in this table. Find the average amount they paid per month to heat their home during this time period. $190.01

Month	Heating Bill
December	$143.88
January	$210.15
February	$227.90
March	$178.11

(12.2) Write the first five terms of each arithmetic sequence with the given first term and common difference.

19) $a_1 = 11, d = 7$
11, 18, 25, 32, 39

20) $a_1 = 0, d = -4$
0, −4, −8, −12, −16

21) $a_1 = -58, d = 8$
−58, −50, −42, −34, −26

22) $a_1 = -1, d = \dfrac{1}{2}$
$-1, -\dfrac{1}{2}, 0, \dfrac{1}{2}, 1$

For each arithmetic sequence, find a) a_1 and d, b) a formula for the general term of the sequence, a_n, and c) a_{20}.

23) 6, 10, 14, 18, 22, . . .

24) −13, −6, 1, 8, 15, . . .

25) −8, −13, −18, −23, −28, . . .

26) 14, 17, 20, 23, 26, . . .

Find the indicated term for each arithmetic sequence.

27) $-15, -9, -3, 3, 9, \ldots; a_{15}$ 69

28) $5, 2, -1, -4, -7, \ldots; a_{22}$ −58

29) $a_1 = -4, d = -\dfrac{3}{2}; a_{21}$ −34

30) $a_1 = -6, d = 7; a_{23}$ 148

Two terms of an arithmetic sequence are given in each problem. Find the general term of the sequence, a_n, and find the indicated term.

31) $a_6 = 24, a_9 = 36; a_{12}$ $a_n = 4n; a_{12} = 48$

32) $a_4 = 1, a_{10} = 13; a_{15}$ $a_n = 2n - 7; a_{15} = 23$

33) $a_5 = -5, a_{10} = -15; a_{17}$ $a_n = -2n + 5; a_{17} = -29$

34) $a_3 = -13, a_8 = -28; a_{20}$ $a_n = -3n - 4; a_{20} = -64$

Find the number of terms in each arithmetic sequence.

35) $-4, -2, 0, 2, \ldots, 34$ 20

36) $2, -3, -8, -13, \ldots, -158$ 33

37) $-8, -7, -6, -5, \ldots, 43$ 52

38) $7, 13, 19, 25, \ldots, 109$ 18

39) Find the sum of the first eight terms of the arithmetic sequence with first term 5 and last term -27. -88

40) Find the sum of the first 15 terms of the arithmetic sequence with the first term -9 and last term 19. 75

Find S_{10} for each arithmetic sequence described below.

41) $a_1 = -6, d = 7$ 255 42) $a_1 = 8, a_{10} = 35$ 215

43) $a_1 = 13, a_{10} = -59$ -230 44) $a_1 = -11, d = -5$ -335

45) $a_n = 2n - 5$ 60 46) $a_n = n - 7$ -15

47) $a_n = -7n + 16$ -225 48) $a_n = 9n - 2$ 475

Evaluate each sum using a formula for S_n.

49) $\sum\limits_{i=1}^{4} (-11i - 2)$ $S_4 = -118$ 50) $\sum\limits_{i=1}^{8} (2i - 7)$ $S_8 = 16$

51) $\sum\limits_{i=1}^{13} (3i + 4)$ $S_{13} = 325$ 52) $\sum\limits_{i=1}^{15} (-5i + 1)$ $S_{15} = -585$

53) $\sum\limits_{i=1}^{11} (-4i + 2)$ $S_{11} = -242$ 54) $\sum\limits_{i=1}^{19} (-6i - 2)$ $S_{19} = -1178$

Solve each application.

55) A theater has 20 rows. The first row contains 15 seats, the next row has 17 seats, the next row has 19 seats, and so on. How many seats are in the last row? How many seats are in the theater? $53; 680$

56) A stack of logs has 16 logs in the bottom row, 15 logs in the second row, 14 logs in the third row, and so on until the last row contains one log.

   a) How many logs are in the 12th row? 5

   b) How many logs are in the stack? 136

57) On March 1, Darshan deposits \$2 into his bank. The next week he deposits \$4, the week after that he saves \$6, the next week he deposits \$8, and so on. If he saves his money in this way for a total of 30 weeks, how much money will he have saved? $\$930$

58) Laura's company currently employs 14 people. She plans to expand her business, so starting next month she will hire three new workers per month for the next 6 months. How many employees will she have after that hiring period? 32

(12.3) Find the common ratio, r, for each geometric sequence.

59) $4, 20, 100, 500, \ldots$ 5

60) $8, 2, \dfrac{1}{2}, \dfrac{1}{8}, \ldots$ $\dfrac{1}{4}$

Write the first five terms of the geometric sequence with the given first term and common ratio.

61) $a_1 = 7, r = 2$
   $7, 14, 28, 56, 112$

62) $a_1 = 5, r = -3$
   $5, -15, 45, -135, 405$

63) $a_1 = 48, r = \dfrac{1}{4}$
   $48, 12, 3, \dfrac{3}{4}, \dfrac{3}{16}$

64) $a_1 = -16, r = \dfrac{3}{2}$
   $-16, -24, -36, -54, -81$

Find the general term, a_n, for each geometric sequence. Then, find the indicated term.

65) $a_1 = 3, r = 2; a_6$
   $a_n = 3(2)^{n-1}; 96$

66) $a_1 = 4, r = 3; a_5$
   $a_n = 4(3)^{n-1}; 324$

67) $a_1 = 8, r = \dfrac{1}{3}; a_4$
   $a_n = 8\left(\dfrac{1}{3}\right)^{n-1}; \dfrac{8}{27}$

68) $a_1 = -\dfrac{1}{2}, r = -3; a_4$
   $a_n = -\dfrac{1}{2}(-3)^{n-1}; \dfrac{27}{2}$

Find the general term of each geometric sequence.

69) $7, 42, 252, 1512, \ldots$ $a_n = 7(6)^{n-1}$

70) $-4, -\dfrac{4}{5}, -\dfrac{4}{25}, -\dfrac{4}{125}, \ldots$ $a_n = (-4)\left(\dfrac{1}{5}\right)^{n-1}$

71) $-15, 45, -135, 405, \ldots$ $a_n = (-15)(-3)^{n-1}$

72) $\dfrac{1}{9}, \dfrac{4}{9}, \dfrac{16}{9}, \dfrac{64}{9}, \ldots$ $a_n = \dfrac{1}{9}(4)^{n-1}$

Find the indicated term of each geometric sequence.

73) $1, 3, 9, 27, \ldots; a_8$ 2187 74) $4, 2, 1, \dfrac{1}{2}, \ldots; a_9$ $\dfrac{1}{64}$

75) Find the sum of the first five terms of the geometric sequence with $a_1 = 8$ and $r = 3$. 968

76) Find the sum of the first four terms of the geometric sequence with $a_1 = -2$ and $r = 5$. -312

Use the formula for S_n to find the sum of each geometric sequence.

77) $8, 40, 200, 1000, 5000$
   $S_5 = 6248$

78) $-\dfrac{1}{3}, 1, -3, 9, -27$
   $S_5 = -\dfrac{61}{3}$

79) $8, 4, 2, 1, \dfrac{1}{2}, \dfrac{1}{4}$ $S_6 = \dfrac{63}{4}$

80) $7, 14, 28, 56, 112, 224$
   $S_6 = 441$

81) $\sum\limits_{i=1}^{5} 7\left(\dfrac{1}{2}\right)^{i}$ $S_5 = \dfrac{217}{32}$ 82) $\sum\limits_{i=1}^{4} 2(3)^{i}$ $S_4 = 240$

Solve each application.

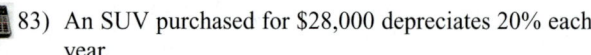

83) An SUV purchased for \$28,000 depreciates 20% each year.

   a) Find the general term, a_n, of the geometric sequence that models the value of the SUV at the beginning of each year. $a_n = 28{,}000(0.80)^{n-1}$

   b) How much is the SUV worth at the beginning of the third year?   \$17,920

84) A home purchased for \$180,000 increases in value by 4% per year.

   a) Find the general term of the geometric sequence that models the future value of the house. $a_n = 180{,}000(1.04)^{n-1}$

   b) How much is the home worth 6 years after it is purchased? Round the answer to the nearest dollar.
   \$227,757

For each sequence,

   a) determine whether it is arithmetic or geometric.

   b) find a_n.

   c) find S_8.

85) 7, 9, 11, 13, ...

86) 33, 28, 23, 18, ...

87) $9, \dfrac{9}{2}, \dfrac{9}{4}, \dfrac{9}{8}, \ldots$

88) $-\dfrac{9}{2}, -4, -\dfrac{7}{2}, -3, \ldots$

89) $-1, -3, -9, -27, \ldots$

90) $5, -10, 20, -40, \ldots$

Solve each application.

91) The number of bacteria in a culture increases by 50% each day. If a culture begins with 2000 bacteria, how many bacteria are present after 5 days?   10,125

92) Trent will put 2¢ in his bank on one day, 4¢ on the second day, 8¢ on the third day, 16¢ on the fourth day, and so on. If he continues in this way,

   a) how much money will Trent have to put in his bank on the 11th day to continue the pattern?   \$20.48

   b) how much money will he have saved after 11 days?
   \$40.94

93) When does the sum of an infinite geometric sequence exist?   when $|r| < 1$

Find the sum of the terms of each infinite geometric sequence, if possible.

94) $a_1 = 24, r = \dfrac{3}{7}$ 42

95) $a_1 = -3, r = \dfrac{1}{8}$ $-\dfrac{24}{7}$

96) $a_1 = 9, r = \dfrac{4}{3}$
Sum does not exist.

97) $-15, 10, -\dfrac{20}{3}, \dfrac{40}{9}, \ldots$
-9

98) $20, -5, \dfrac{5}{4}, -\dfrac{5}{16}, \ldots$ 16

99) $-4, 12, -36, 108, \ldots$
Sum does not exist.

Solve.

100) Each time a certain pendulum swings, it travels 80% of the distance it traveled on the previous swing. If it travels 4 ft on its first swing, find the total distance the pendulum travels before coming to rest.   20 ft

(12.4) Use Pascal's triangle to expand each binomial.

101) $(y + z)^4$
$y^4 + 4y^3z + 6y^2z^2 + 4yz^3 + z^4$

102) $(c + d)^3$
$c^3 + 3c^2d + 3cd^2 + d^3$

Evaluate.

103) 6!   720

104) 8!   40,320

105) $\dbinom{5}{3}$ 10

106) $\dbinom{7}{2}$ 21

107) $\dbinom{9}{1}$ 9

108) $\dbinom{6}{6}$ 1

Use the binomial theorem to expand each expression.

109) $(m + n)^4$

110) $(k + 2)^6$

111) $(h - 9)^3$

112) $(2w - 5)^3$

113) $(2p^2 - 3r)^5$

114) $\left(\dfrac{1}{3}b + c\right)^4$

Find the indicated term of each binomial expansion.

115) $(z + 4)^8$; fifth term $17{,}920z^4$

116) $(y - 6)^8$; sixth term $-435{,}456y^3$

117) $(2k - 1)^{13}$; 11th term $2288k^3$

118) $(3p + q)^9$; seventh term $2268p^3q^6$

*Additional answers can be found in the Answers to Exercises appendix.

Write out the first five terms of each sequence.

1) $a_n = 2n - 3$ $-1, 1, 3, 5, 7$

2) $a_n = (-1)^{n+1}\left(\dfrac{n}{n+2}\right)$ $\dfrac{1}{3}, -\dfrac{1}{2}, \dfrac{3}{5}, -\dfrac{2}{3}, \dfrac{5}{7}$

3) What is the difference between an arithmetic and a geometric sequence?

4) Write the first five terms of the geometric sequence with first term 32 and common ratio $-\dfrac{1}{2}$. $32, -16, 8, -4, 2$

5) Find the common difference for the arithmetic sequence $-17, -11, -5, 1, 7, \ldots$ $d = 6$

Determine whether each sequence is arithmetic or geometric, and find a_n.

6) $4, 12, 36, 108, 324, \ldots$ geometric; $a_n = 4(3)^{n-1}$

7) $5, 2, -1, -4, -7, \ldots$ arithmetic; $a_n = -3n + 8$

8) Find the forty-first term of the arithmetic sequence with $a_6 = -3$ and $a_{11} = 7$. 67

9) Write out each term of $\displaystyle\sum_{i=1}^{4}(5i^2 + 6)$ and find the sum. $11 + 26 + 51 + 86 = 174$

10) Use the formula for S_n to find the sum of the first six terms of the geometric sequence with $a_1 = 9$ and $r = 2$. 567

11) Use a formula for S_n to find the sum of the first 11 terms of the arithmetic sequence with $a_1 = 5$ and $d = 3$. 220

12) Evaluate $\displaystyle\sum_{i=1}^{100}(-4i + 3)$. $-19{,}900$

13) Find the sum of the terms of the infinite geometric sequence with $a_1 = 7$ and $r = \dfrac{3}{10}$. 10

Solve each application.

14) At a construction site, pipes are stacked so that there are 14 in the bottom row, 13 in the next row, 12 in the next row, and so on until the top row contains one pipe. How many pipes are in the stack? 105

15) In 2007, a bank found that 11,000 of its customers used online banking. That number continued to increase by 10% per year. How many customers used online banking in 2010? Round the answer to the nearest whole number. $14{,}641$

16) Each time a certain pendulum swings, it travels 75% of the distance it traveled on the previous swing. If it travels 40 in. on its first swing, find the total distance the pendulum travels before coming to rest. 160 in.

Evaluate.

17) $5!$ 120

18) $\dbinom{6}{2}$ 15

19) Expand using the binomial theorem: $(3x + 1)^4$.
   $81x^4 + 108x^3 + 54x^2 + 12x + 1$

20) Find the fourth term in the expansion of $(k - 2)^9$.
   $-672k^6$

*Additional answers can be found in the Answers to Exercises appendix.

1) Add $\dfrac{5}{8} + \dfrac{1}{6} + \dfrac{3}{4}$. $\dfrac{37}{24}$

2) Find the a) area and b) circumference of the circle. Give the exact answer for each in terms of π, and give an approximation using 3.14 for π. Include the correct units.

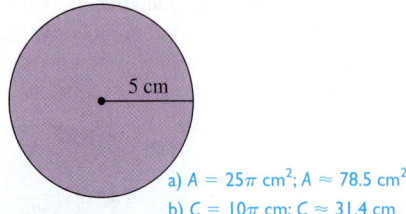

5 cm

a) $A = 25\pi$ cm^2; $A \approx 78.5$ cm^2
b) $C = 10\pi$ cm; $C \approx 31.4$ cm

3) Simplify completely. The answer should contain only positive exponents.

a) $(-9k^4)^2$ $81k^8$

b) $(-7z^3)(8z^{-9})$ $-\dfrac{56}{z^6}$

c) $\left(\dfrac{40a^{-7}b^3}{8ab^{-2}}\right)^{-3}$ $\dfrac{a^{24}}{125b^{15}}$

4) Write 0.00008215 in scientific notation. 8.215×10^{-5}

5) Solve each equation.

a) $\dfrac{4}{3}y - 7 = 13$ $\{15\}$

b) $3(n - 4) + 11 = 5n - 2(3n + 8)$ $\left\{\dfrac{15}{4}\right\}$

6) *Write an equation and solve.*
The width of a rectangle is 5 in. less than its length. If the perimeter of the rectangle is 26 in., find its length and width.   length = 9 in., width = 4 in.

7) *Write an equation and solve.*
Kathleen invests a total of $9000. She invests some of it at 4% simple interest and the rest at 6% simple interest. If she earns $480 in interest after 1 year, how much did she invest at each rate? $3000 at 4% and $6000 at 6%

8) Solve $5 - 8x \geq 9$. Write the answer in interval notation.

9) Solve the compound inequality $2c + 9 < 3$ or $c - 7 > -2$. Write the answer in interval notation.
$(-\infty, -3) \cup (5, \infty)$

10) Graph each line.

a) $3x + y = 4$

b) $y = \dfrac{1}{2}x - 3$

c) $x = 1$

11) Write the equation of the line (in slope-intercept form) containing the points $(-3, 5)$ and $(6, 2)$. $y = -\dfrac{1}{3}x + 4$

12) Write an equation of the line perpendicular to $2x + 3y = 15$ containing the point $(-4, 2)$. Write the equation in standard form. $3x - 2y = -16$

13) Solve using the elimination method.

$$3x + 5y = 12$$
$$2x - 3y = 8$$ $(4, 0)$

14) *Write a system of two equations and solve.*
How many milliliters of a 7% acid solution and how many milliliters of a 15% acid solution must be mixed to make 60 mL of a 9% acid solution?
15 mL of 15% solution, 45 mL of 7% solution

15) Perform the operations and simplify.

a) $(9m^3 - 7m^2 + 3m + 2)$
    $-(4m^3 + 11m^2 - 7m - 1)$
    $5m^3 - 18m^2 + 10m + 3$

b) $5(2p + 3) + \dfrac{2}{3}(4p - 9)$ $\dfrac{38}{3}p + 9$

16) Multiply and simplify.

a) $(2d - 7)(3d + 4)$ $6d^2 - 13d - 28$

b) $(h - 6)(3h^2 - 7h + 2)$ $3h^3 - 25h^2 + 44h - 12$

17) Divide.

a) $\dfrac{12x^3 + 7x^2 - 37x + 18}{3x - 2}$ $4x^2 + 5x - 9$

b) $\dfrac{20a^3b^3 - 45a^2b + 10ab + 60}{10ab}$ $2a^2b^2 - \dfrac{9a}{2} + 1 + \dfrac{6}{ab}$

18) Factor completely.

a) $b^2 + 4b - 12$

b) $20xy + 5x + 4y + 1$

c) $5c^2 + 27cd - 18d^2$

d) $4r^2 - 36$

e) $27a^3 + 125b^3$

f) $w^2 - 16w + 64$

19) Solve each equation.

a) $m^2 - 15m + 54 = 0$ $\{6, 9\}$

b) $x^3 = 3x^2 + 28x$ $\{-4, 0, 7\}$

20) Divide $\dfrac{3z^2 + 22z - 16}{z + 8} \div \dfrac{8 - 12z}{7z + 14}$. $-\dfrac{7(z + 2)}{4}$

21) Add $\dfrac{k}{k^2 - 11k + 18} + \dfrac{k + 2}{2k^2 - 17k - 9}$. $\dfrac{(3k + 4)(k - 1)}{(2k + 1)(k - 2)(k - 9)}$

22) Simplify each complex fraction.

a) $\dfrac{\dfrac{8v^2w}{3}}{\dfrac{16vw^2}{21}}$ $\dfrac{7v}{2w}$

b) $\dfrac{\dfrac{9}{c} + 4c}{5 + \dfrac{6}{c}}$ $\dfrac{9 + 4c^2}{5c + 6}$

23) Solve $\dfrac{1}{a + 1} - \dfrac{a}{6} = \dfrac{a - 4}{a + 1}$. $\{-10, 3\}$

24) Solve $|8y + 3| = 19$. $\left\{-\dfrac{11}{4}, 2\right\}$

25) Solve $\left|\dfrac{1}{4}t - 5\right| \leq 2$. Write the answer in interval notation. $[12, 28]$

26) Graph $x + 3y < 6$.

27) Solve the system.
$$-x + 3y + 3z = 5 \qquad (-2, 2, -1)$$
$$3x - 2y - z = -9$$
$$-3x + y + 3z = 5$$

28) Evaluate.

a) $\sqrt{25}$ 5

b) $\sqrt[3]{27}$ 3

c) $32^{3/5}$ 8

d) $125^{-2/3}$ $\dfrac{1}{25}$

29) Simplify. Assume all variables represent positive real numbers.

a) $\sqrt{63}$ $3\sqrt{7}$

b) $\sqrt[4]{48}$ $2\sqrt[4]{3}$

c) $\sqrt{20x^2y^9}$ $2xy^4\sqrt{5y}$

d) $\sqrt[3]{250c^{17}d^{12}}$ $5c^5d^4\sqrt[3]{2c^2}$

30) Perform the operations and simplify.
$\sqrt{18} + \sqrt{2} - \sqrt{72}$ $-2\sqrt{2}$

31) Rationalize each denominator.

a) $\dfrac{5}{\sqrt{t}}$ $\dfrac{5\sqrt{t}}{t}$

b) $\dfrac{4}{\sqrt{6} - 2}$ $2\sqrt{6} + 4$

c) $\dfrac{n}{\sqrt[3]{4}}$ $\dfrac{n\sqrt[3]{2}}{2}$

32) Solve $\sqrt{3x - 2} + \sqrt{x + 2} = 4$.   {2}

33) Evaluate $\sqrt{-16}$. $4i$

34) Solve by completing the square.
$r^2 - 8r + 11 = 0$ $\{4 - \sqrt{5}, 4 + \sqrt{5}\}$

35) Solve each equation.

a) $9h^2 + 2h + 1 = 0$ $\left\{-\dfrac{1}{9} - \dfrac{2\sqrt{2}}{9}i, -\dfrac{1}{9} + \dfrac{2\sqrt{2}}{9}i\right\}$

b) $(w + 11)^2 + 4 = 0$ $\{-11 - 2i, -11 + 2i\}$

c) $(b + 4)^2 - (b + 4) = 12$ $\{-7, 0\}$

d) $k^4 + 15 = 8k^2$ $\{-\sqrt{5}, -\sqrt{3}, \sqrt{3}, \sqrt{5}\}$

36) Let $f(x) = 3x + 1$ and $g(x) = x^2 - 6x + 2$.

a) Find $f(-4)$.

b) Find $g(5)$.

c) Find $(g \circ f)(x)$.

d) Find $(f \circ g)(x)$.

e) Find the domain of $f(x)$.

f) Graph $f(x)$.

37) Determine the domain of each function.

a) $f(x) = \dfrac{6}{5x - 10}$ $(-\infty, 2) \cup (2, \infty)$

b) $h(x) = \sqrt{2x + 3}$ $\left[-\dfrac{3}{2}, \infty\right)$

38) Graph $f(x) = |x + 2|$.

39) Sketch the graph of each equation. Identify the vertex.

a) $x = (y - 1)^2 - 3$

b) $f(x) = -\dfrac{1}{2}x^2 + 2x + 1$

40) Write an equation of the inverse of $f(x) = -6x + 8$.
$f^{-1}(x) = -\dfrac{1}{6}x + \dfrac{4}{3}$

41) Evaluate.

a) $\log_2 16$ 4

b) $\log 100$ 2

c) $\ln e$ 1

d) $\log_3 \sqrt{3}$ $\dfrac{1}{2}$

42) Write as a single logarithm: $\log_a 5 + 2 \log_a r - 3 \log_a s$
$\log_a \dfrac{5r^2}{s^3}$

43) Solve.

a) $5^{2y} = 125^{y+4}$ $\{-12\}$

b) $4^{x-3} = 3^{2x}$ $\left\{\dfrac{3 \ln 4}{\ln 4 - 2 \ln 3}\right\}$; approximation: $\{-5.13\}$

c) $e^{-6t} = 8$ $\left\{-\dfrac{\ln 8}{6}\right\}$; approximation: $\{-0.35\}$

44) Solve.

a) $\log_3(2x + 1) = 2$ $\{4\}$

b) $\log_6 9m - \log_6(2m + 5) = \log_6 2$ $\{2\}$

45) Graph $(x - 2)^2 + (y + 1)^2 = 9$.

46) Solve the system.
$$x^2 + y^2 = 20$$
$$x + y = -2 \qquad \{(-4, 2), (2, -4)\}$$

47) Solve each inequality. Write the answer in interval notation.

a) $w^2 + 5w > -6$ $(-\infty, -3) \cup (-2, \infty)$

b) $\dfrac{c + 1}{c + 7} \le 0$ $(-7, -1]$

48) Determine whether each sequence is arithmetic or geometric, and find a_n.

a) 7, 11, 15, 19, 23, . . .   arithmetic; $a_n = 4n + 3$

b) 2, -6, 18, -54, 162, . . .   geometric; $a_n = 2(-3)^{n-1}$

49) Use a formula for S_n to find S_6 for each sequence.

a) the geometric sequence with $a_1 = 48$ and $r = \dfrac{1}{2}$ $\dfrac{189}{2}$

b) the arithmetic sequence with $a_1 = -10$ and $a_6 = -25$ -105

50) Expand $(2t + 5)^3$ using the binomial theorem.
$8t^3 + 60t^2 + 150t + 125$

Appendix A

Section A.1 Geometry Review

Objectives

1. Identify Angles and Parallel and Perpendicular Lines
2. Identify Triangles
3. Use Area, Perimeter, and Circumference Formulas
4. Use Volume Formulas

Thousands of years ago, the Egyptians collected taxes based on how much land a person owned. They developed measuring techniques to accomplish such a task. Later the Greeks formalized the process of measurements such as this into a branch of mathematics we call geometry. "Geometry" comes from the Greek words for "earth measurement." In this section, we will review some basic geometric concepts that we will need in the study of algebra.

Let's begin by looking at angles. An angle can be measured in **degrees.** For example, 45° is read as "45 degrees."

1. Identify Angles and Parallel and Perpendicular Lines

Angles

An **acute angle** is an angle whose measure is greater than 0° and less than 90°.

A **right angle** is an angle whose measure is 90°, indicated by the ∟ symbol.

An **obtuse angle** is an angle whose measure is greater than 90° and less than 180°.

A **straight angle** is an angle whose measure is 180°.

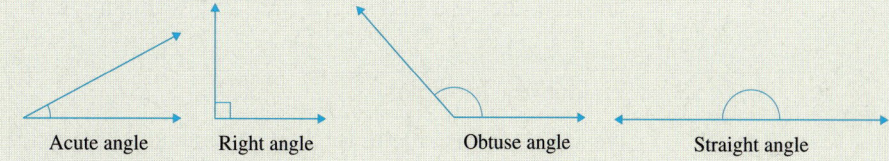

Acute angle   Right angle   Obtuse angle   Straight angle

Two angles are **complementary** if their measures add to 90°.
Two angles are **supplementary** if their measures add to 180°.

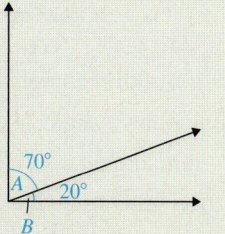

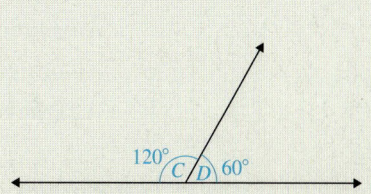

A and B are **complementary angles** since $m\angle A + m\angle B = 70° + 20° = 90°$.

C and D are **supplementary angles** since $m\angle C + m\angle D = 120° + 60° = 180°$.

Note

The measure of angle A is denoted by $m\angle A$.

In-Class Example 1
$m\angle A = 34°$. Find its complement.
answer: 56°

Example 1

$m\angle A = 41°$. Find its complement.

Solution

$$\text{Complement} = 90° - 41° = 49°$$

Since the sum of two complementary angles is 90°, if one angle measures 41°, its complement has a measure of $90° - 41° = 49°$.

You Try 1

$m\angle A = 62°$. Find its supplement.

Next, we will explore some relationships between lines and angles.

Vertical Angles

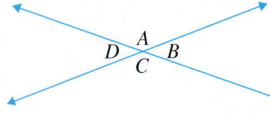

Figure A.1

When two lines intersect, four angles are formed (see Figure A.1). The pair of opposite angles are called **vertical angles**. Angles *A* and *C* are *vertical angles,* and angles *B* and *D* are *vertical angles. The measures of vertical angles are equal.* Therefore, $m\angle A = m\angle C$ and $m\angle B = m\angle D$.

Parallel lines

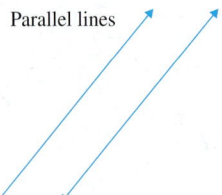

Figure A.2

Parallel and Perpendicular Lines

Parallel lines are lines in the same plane that do not intersect (Figure A.2). **Perpendicular lines** are lines that intersect at right angles (Figure A.3).

2. Identify Triangles

We can classify triangles by their angles and by their sides.

Perpendicular lines

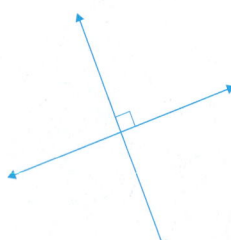

Figure A.3

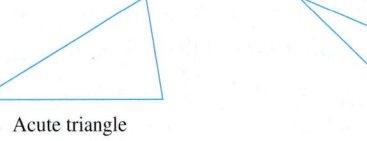
Acute triangle

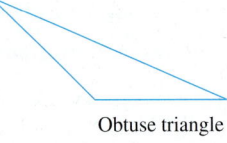
Obtuse triangle

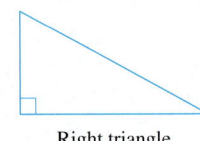
Right triangle

An **acute triangle** is one in which all three angles are acute.

An **obtuse triangle** contains one obtuse angle.

A **right triangle** contains one right angle.

> ### Property
> The sum of the measures of the angles of any triangle is 180°.

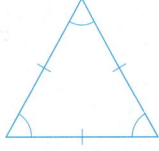
Equilateral triangle

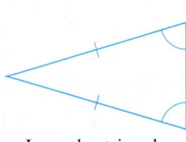
Isosceles triangle

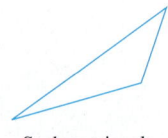
Scalene triangle

If a triangle has three sides of equal length, it is an **equilateral triangle.** (Each angle measure of an equilateral triangle is 60°.)

If a triangle has two sides of equal length, it is an **isosceles triangle.** (The angles opposite the equal sides have the same measure.)

If a triangle has no sides of equal length, it is a **scalene triangle.** (No angles have the same measure.)

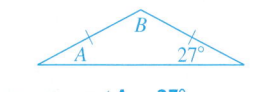
Example 2

In-Class Example 2
Find the measures of angles A and B in this isosceles triangle.

answer: $m\angle A = 27°$;
$m\angle B = 126°$

Find the measures of angles A and B in this isosceles triangle.

Solution

The single hash marks on the two sides of the triangle mean that those sides are of equal length.

$m\angle B = 39°$ Angle measures opposite sides of equal length are the same.
$39° + m\angle B = 39° + 39° = 78°$.

We have found that the sum of two of the angles is 78°. Since all of the angle measures add up to 180°,

$$m\angle A = 180° - 78° = 102°$$

You Try 2

Find the measures of angles A and B in this isosceles triangle.

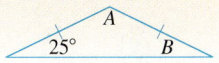

3. Use Area, Perimeter, and Circumference Formulas

The **perimeter** of a figure is the distance around the figure, while the **area** of a figure is the number of square units enclosed within the figure. For some familiar shapes, we have the following formulas:

Figure		Perimeter	Area
Rectangle:		$P = 2l + 2w$	$A = lw$
Square:		$P = 4s$	$A = s^2$
Triangle: h = height		$P = a + b + c$	$A = \dfrac{1}{2}bh$
Parallelogram: h = height		$P = 2a + 2b$	$A = bh$
Trapezoid: h = height		$P = a + c + b_1 + b_2$	$A = \dfrac{1}{2}h(b_1 + b_2)$

The perimeter of a circle is called the **circumference.** The **radius,** r, is the distance from the center of the circle to a point on the circle. A line segment that passes through the center of the circle and has its endpoints on the circle is called a **diameter.**

Pi, π, is the ratio of the circumference of any circle to its diameter. $\pi \approx 3.14159265\ldots$, but we will use 3.14 as an approximation for π. The symbol $\approx$ is read as "approximately equal to."

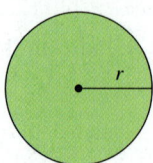

Circumference	Area
$C = 2\pi r$	$A = \pi r^2$

Example 3

Find the perimeter and area of each figure.

In-Class Example 3
Find the perimeter and area of each figure.
a)

5 in.

9 in.

b)
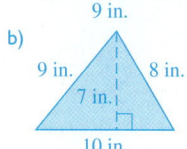
9 in.   8 in.
7 in.
10 in.

answer: a) $P = 28$ in., $A = 45$ in^2
b) $P = 27$ in., $A = 35$ in^2

a)
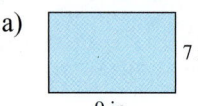
7 in.

9 in.

b)
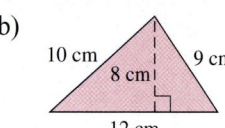
10 cm   9 cm
8 cm
12 cm

Solution

a) This figure is a rectangle.

Perimeter: $P = 2l + 2w$
$P = 2(9 \text{ in.}) + 2(7 \text{ in.})$
$P = 18 \text{ in.} + 14 \text{ in.}$
$P = 32$ in.

Area: $A = lw$
$A = (9 \text{ in.})(7 \text{ in.})$
$A = 63$ in^2 or 63 square inches

b) This figure is a triangle.

Perimeter: $P = a + b + c$
$P = 9 \text{ cm} + 12 \text{ cm} + 10 \text{ cm}$
$P = 31$ cm

Area: $A = \dfrac{1}{2}bh$

$A = \dfrac{1}{2}(12 \text{ cm})(8 \text{ cm})$

$A = 48$ cm^2 or 48 square centimeters

You Try 3

Find the perimeter and area of the figure.

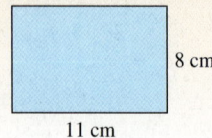
8 cm

11 cm

Example 4

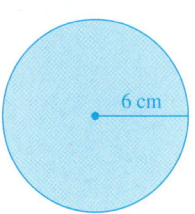

4 cm

In-Class Example 4
Find (a) the circumference and
(b) the area of the circle. Give
an exact answer for each and
give an approximation using
3.14 for π.
answer: a) $C = 12\pi$ cm;
$C \approx 37.68$ cm
b) $A = 36\pi$ cm^2;
$A \approx 113.04$ cm^2

6 cm

Find (a) the circumference and (b) the area of the circle shown at left. Give an exact answer for each and give an approximation using 3.14 for π.

Solution

a)   The formula for the circumference of a circle is $C = 2\pi r$. The radius of the given circle is 4 cm. Replace r with 4 cm.

$$C = 2\pi r$$
$$= 2\pi (4 \text{ cm}) \qquad \text{Replace } r \text{ with 4 cm.}$$
$$= 8\pi \text{ cm} \qquad \text{Multiply.}$$

Leaving the answer in terms of π gives us the exact circumference of the circle, 8π cm.
To find an approximation for the circumference, substitute 3.14 for π and simplify.

$$C = 8\pi \text{ cm}$$
$$\approx 8(3.14) \text{ cm} = 25.12 \text{ cm}$$

b)   The formula for the area of a circle is $A = \pi r^2$. Replace r with 4 cm.

$$A = \pi r^2$$
$$= \pi (4 \text{ cm})^2 \qquad \text{Replace } r \text{ with 4 cm.}$$
$$= 16\pi \text{ cm}^2 \qquad 4^2 = 16$$

Leaving the answer in terms of π gives us the exact area of the circle, 16π cm^2.
To find an approximation for the area, substitute 3.14 for π and simplify.

$$A = 16\pi \text{ cm}^2$$
$$\approx 16(3.14) \text{ cm}^2$$
$$= 50.24 \text{ cm}^2$$

You Try 4

Find (a) the circumference and (b) the area of the circle. Give an exact answer for each and give an approximation using 3.14 for π.

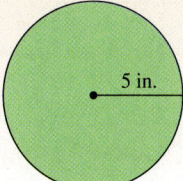

5 in.

A **polygon** is a closed figure consisting of three or more line segments. (See the figure.) We can extend our knowledge of perimeter and area to determine the area and perimeter of a polygon.

   Polygons:

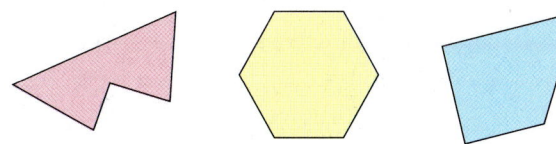

Example 5

Find the perimeter and area of the figure shown here.

In-Class Example 5
Find the perimeter and area of
the figure.
answer: $P = 46$ in.
$A = 128$ in^2

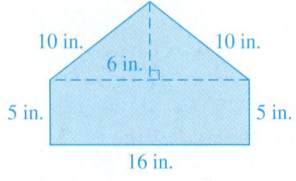

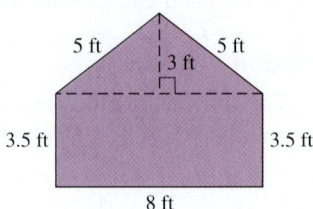

Solution

Perimeter: The perimeter is the distance around the figure.

$$P = 5 \text{ ft} + 5 \text{ ft} + 3.5 \text{ ft} + 8 \text{ ft} + 3.5 \text{ ft}$$
$$P = 25 \text{ ft}$$

Area: To find the area of this figure, think of it as two regions: a triangle and a rectangle.

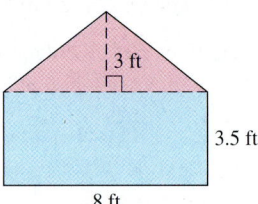

$$\text{Total area} = \text{Area of triangle} + \text{Area of rectangle}$$
$$= \frac{1}{2} bh + lw$$
$$= \frac{1}{2} (8 \text{ ft})(3 \text{ ft}) + (8 \text{ ft})(3.5 \text{ ft})$$
$$= 12 \text{ ft}^2 + 28 \text{ ft}^2$$
$$= 40 \text{ ft}^2$$

You Try 5

Find the perimeter and area of the figure.

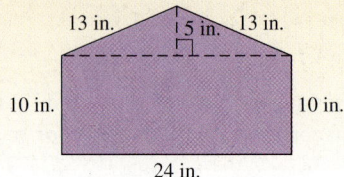

4. Use Volume Formulas

The **volume** of a three-dimensional object is the amount of space occupied by the object. Volume is measured in cubic units such as cubic inches (in^3), cubic centimeters (cm^3), cubic feet (ft^3), and so on. Volume also describes the amount of a substance that can be enclosed within a three-dimensional object. Therefore, volume can also be measured in quarts, liters, gallons, and so on. In the figures, l = length, w = width, h = height, s = length of a side, and r = radius.

Volumes of Three-Dimensional Figures

Rectangular solid		$V = lwh$
Cube		$V = s^3$
Right circular cylinder		$V = \pi r^2 h$
Sphere		$V = \dfrac{4}{3} \pi r^3$
Right circular cone		$V = \dfrac{1}{3} \pi r^2 h$

Example 6

Find the volume of each. In (b) give the answer in terms of π.

In-Class Example 6
Find the volume of each. In (b) give the answer in terms of π.
a)   b)

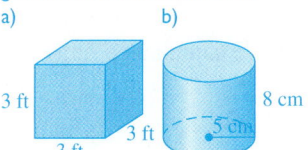

3 ft
3 ft   3 ft
8 cm
5 cm

answer: a) $V = 27 \text{ ft}^3$
b) $V = 200\pi \text{ cm}^3$

a)

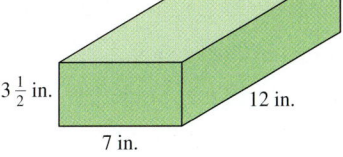

$3\frac{1}{2}$ in.   12 in.
7 in.

b)

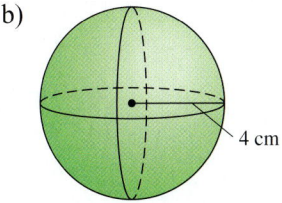

4 cm

Solution

a) $V = lwh$ Volume of a rectangular solid

$= (12 \text{ in.})(7 \text{ in.})\left(3\frac{1}{2} \text{ in.}\right)$ Substitute values.

$= (12 \text{ in.})(7 \text{ in.})\left(\frac{7}{2} \text{ in.}\right)$ Change to an improper fraction.

$= \left(84 \cdot \frac{7}{2}\right) \text{in}^3$ Multiply.

$= 294 \text{ in}^3$ or 294 cubic inches

b) $V = \dfrac{4}{3} \pi r^3$ Volume of a sphere

$= \dfrac{4}{3} \pi (4 \text{ cm})^3$ Replace r with 4 cm.

$= \dfrac{4}{3} \pi (64 \text{ cm}^3)$ $4^3 = 64$

$= \dfrac{256}{3} \pi \text{ cm}^3$ Multiply.

You Try 6

Find the volume of each figure. In (b) give the answer in terms of π.

a)   A box with length = 3 ft, width = 2 ft, and height = 1.5 ft

b)   A sphere with radius = 3 in.

Example 7

Application

A large truck has a fuel tank in the shape of a right circular cylinder. Its radius is 1 ft, and it is 4 ft long.

a)   How many cubic feet of diesel fuel will the tank hold? (Use 3.14 for π.)

b)   How many gallons will it hold? Round to the nearest gallon. (1 ft³ ≈ 7.48 gal)

c)   If diesel fuel costs $1.75 per gallon, how much will it cost to fill the tank?

Solution

a)   We're asked to determine how much fuel the tank will hold. We must find the *volume* of the tank.

$$\text{Volume of a cylinder} = \pi r^2 h$$
$$\approx (3.14)(1 \text{ ft})^2 (4 \text{ ft})$$
$$= 12.56 \text{ ft}^3$$

The tank will hold 12.56 ft³ of diesel fuel.

b)   We must convert 12.56 ft³ to gallons. Since 1 ft³ ≈ 7.48 gal, we can change units by multiplying:

$$12.56 \text{ ft}^3 \cdot \left(\frac{7.48 \text{ gal}}{1 \text{ ft}^3} \right) = 93.9488 \text{ gal}$$
$$\approx 94 \text{ gal}$$

We can divide out units in fractions the same way we can divide out common factors.

The tank will hold approximately 94 gal.

c)   Diesel fuel costs $1.75 per gallon. We can figure out the total cost of the fuel the same way we did in (b).

$1.75 *per* gallon
↓

$$94 \text{ gal} \cdot \left(\frac{\$1.75}{\text{gal}} \right) = \$164.50$$ Divide out the units of gallons.

It will cost about $164.50 to fill the tank.

In-Class Example 7
Application
A small truck has a fuel tank in the shape of a right circular cylinder. Its radius is 1 ft, and it is 2 ft long.
a) How many cubic feet of diesel fuel will the tank hold? (Use 3.14 for π.)
b) How many gallons of fuel will it hold? Round to the nearest gallon. (1 ft³ ≈ 7.48 gal)
c) If diesel fuel costs $1.75 per gallon, how much will it cost to fill the tank?
answer: a) 6.28 ft³
b) 47 gal   c) $82.25

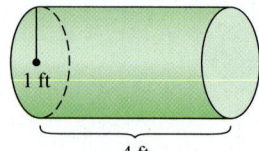

1 ft

4 ft

You Try 7

A large truck has a fuel tank in the shape of a right circular cylinder. Its radius is 1 ft, and it is 3 ft long.

a) How many cubic feet of diesel fuel will the tank hold? (Use 3.14 for π.)

b) How many gallons of fuel will it hold? Round to the nearest gallon. (1 ft^3 $\approx$ 7.48 gal)

c) If diesel fuel costs $1.75 per gallon, how much will it cost to fill the tank?

Answers to You Try Exercises

1) 118°   2) $m\angle A = 130°; m\angle B = 25°$ 3) $P = 38$ cm; $A = 88$ cm^2 4) a) $C = 10\pi$ in.;
$C \approx 31.4$ in.   b) $A = 25\pi$ in^2; $A \approx 78.5$ in^2 5) $P = 70$ in.; $A = 300$ in^2 6) a) 9 ft^3 b) 36π in^3
7) a) 9.42 ft^3 b) 70 gal   c) $122.50

A.1 Exercises

*Additional answers can be found in the Answers to Exercises appendix.

Objective 1: Identify Angles and Parallel and Perpendicular Lines

1) An angle whose measure is between 0° and 90° is a(n) ____acute____ angle.

2) An angle whose measure is 90° is a(n) ____right____ angle.

3) An angle whose measure is 180° is a(n) ____straight____ angle.

4) An angle whose measure is between 90° and 180° is a(n) ____obtuse____ angle.

5) If the sum of two angles is 180°, the angles are _____.
    If the sum of two angles is 90°, the angles are _____.
    supplementary; complementary

6) If two angles are supplementary, can both of them be obtuse? Explain.   No. If both angles are obtuse, then their sum will be greater than 180°, so they cannot be supplementary.

Find the complement of each angle.

7) 59°   31°

8) 84°   6°

9) 12°   78°

10) 40°   50°

Find the supplement of each angle.

11) 143°   37°

12) 62°   118°

13) 38°   142°

14) 155°   25°

Find the measure of the missing angles.

15)

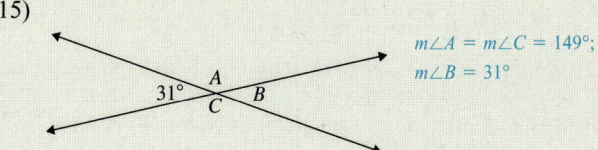

$m\angle A = m\angle C = 149°;$
$m\angle B = 31°$

16)

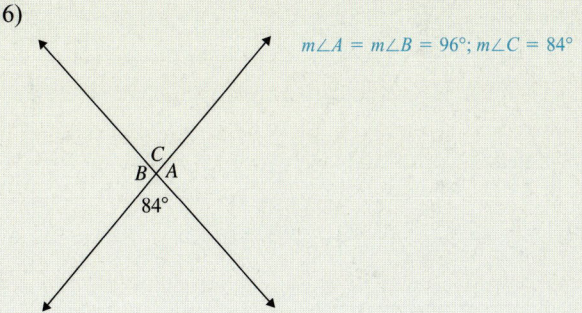

$m\angle A = m\angle B = 96°; m\angle C = 84°$

Objective 2: Identify Triangles

17) The sum of the angles in a triangle is ____180____ degrees.

Find the missing angle and classify each triangle as acute, obtuse, or right.

18)

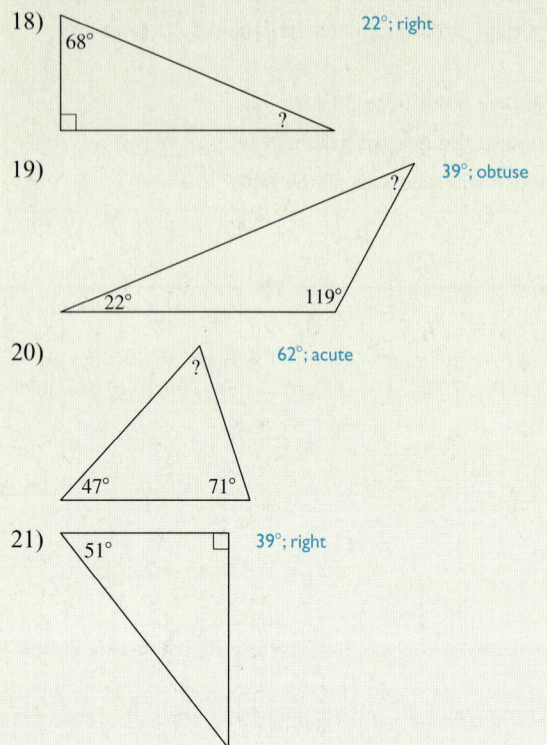

22°; right

19)

39°; obtuse

20)

62°; acute

21)

39°; right

22) Can a triangle contain more than one obtuse angle? Explain.   No. The sum of the angles in a triangle is 180°. If a triangle had two obtuse angles, their sum would be greater than 180°.

Classify each triangle as equilateral, isosceles, or scalene.

23)

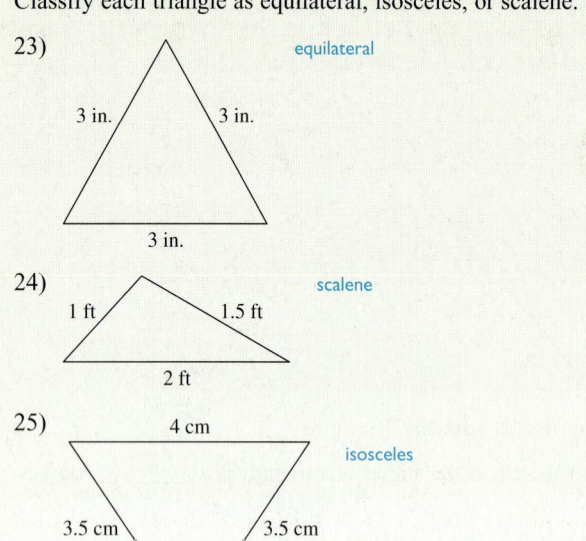

equilateral

24)

scalene

25)

isosceles

26) What can you say about the measures of the angles in an equilateral triangle?   Each of them is a 60° angle.

27) True or False: A right triangle can also be isosceles.   true

28) True or False: If a triangle has two sides of equal length, then the angles opposite these sides are equal.   true

Objective 3: Use Area, Perimeter, and Circumference Formulas

Find the area and perimeter of each figure. Include the correct units.

29)

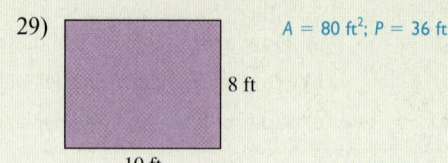

$A = 80$ ft²; $P = 36$ ft

30)

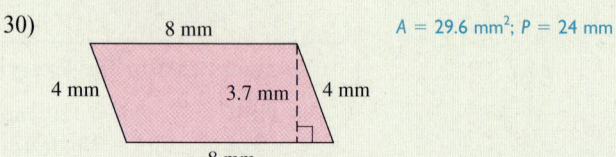

$A = 29.6$ mm²; $P = 24$ mm

31)

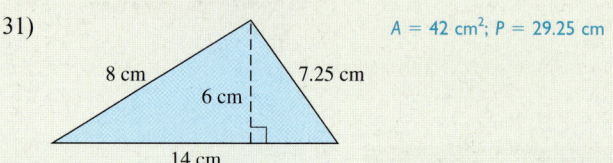

$A = 42$ cm²; $P = 29.25$ cm

32)

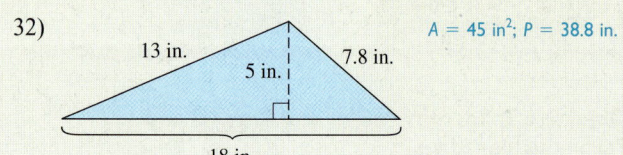

$A = 45$ in²; $P = 38.8$ in.

33)

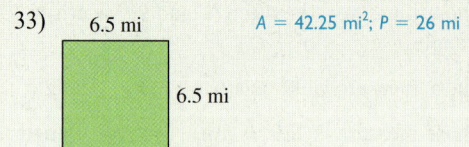

$A = 42.25$ mi²; $P = 26$ mi

34)

$2\frac{1}{2}$ ft $A = 19\frac{1}{6}$ ft²; $P = 20\frac{1}{3}$ ft

$7\frac{2}{3}$ ft

35)

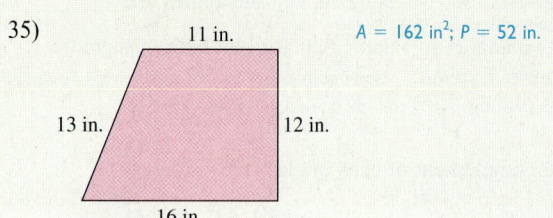

$A = 162$ in²; $P = 52$ in.

36)

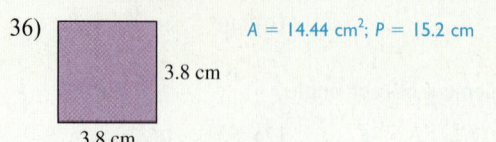

$A = 14.44$ cm²; $P = 15.2$ cm

For 37–40, find (a) the area and (b) the circumference of the circle. Give an exact answer for each and give an approximation using 3.14 for π. Include the correct units.

37)   38)

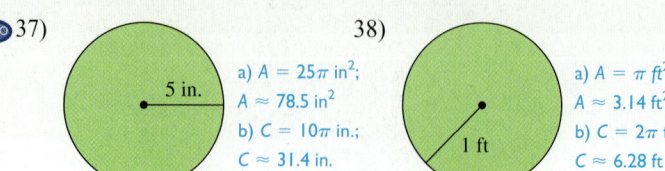

37) a) $A = 25\pi$ in²; $A \approx 78.5$ in² b) $C = 10\pi$ in.; $C \approx 31.4$ in.

38) a) $A = \pi$ ft²; $A \approx 3.14$ ft² b) $C = 2\pi$ ft; $C \approx 6.28$ ft

39)
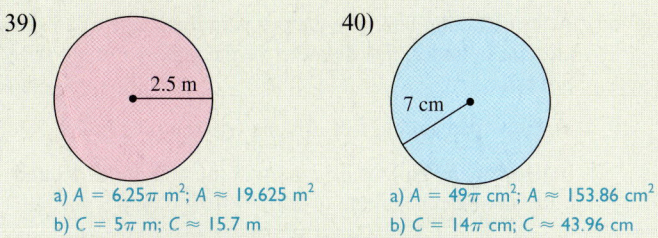

a) $A = 6.25\pi$ m^2; $A \approx 19.625$ m^2
b) $C = 5\pi$ m; $C \approx 15.7$ m

40)
a) $A = 49\pi$ cm^2; $A \approx 153.86$ cm^2
b) $C = 14\pi$ cm; $C \approx 43.96$ cm

For 41–44, find the exact area and circumference of the circle in terms of π. Include the correct units.

41)
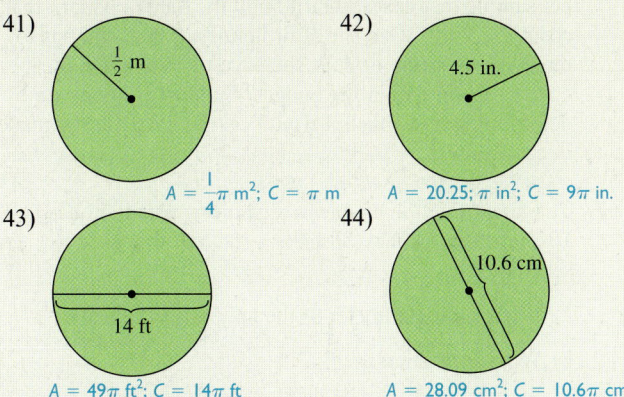

42)
$A = \frac{1}{4}\pi$ m^2; $C = \pi$ m $A = 20.25; \pi$ in^2; $C = 9\pi$ in.

43)

44)
$A = 49\pi$ ft^2; $C = 14\pi$ ft $A = 28.09$ cm^2; $C = 10.6\pi$ cm

Find the area and perimeter of each figure. Include the correct units.

45)
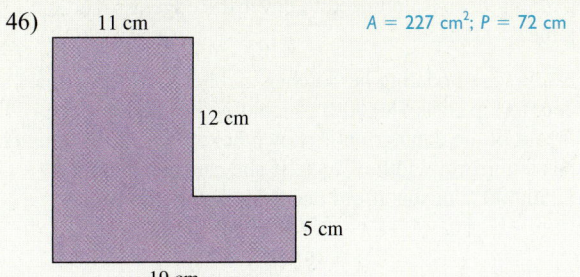
$A = 376$ m^2;
$P = 86$ m

46)
$A = 227$ cm^2; $P = 72$ cm

47)
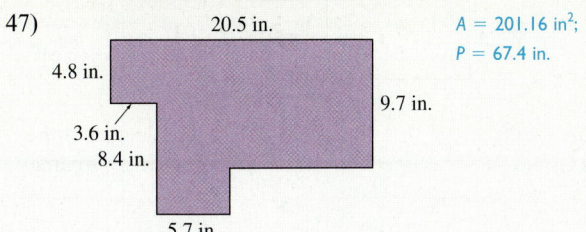
$A = 201.16$ in^2;
$P = 67.4$ in.

48)
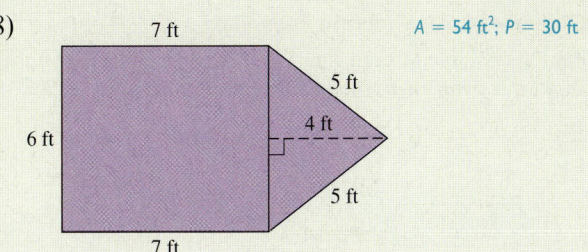
$A = 54$ ft^2; $P = 30$ ft

Find the area of the shaded region. Use 3.14 for π. Include the correct units.

49)
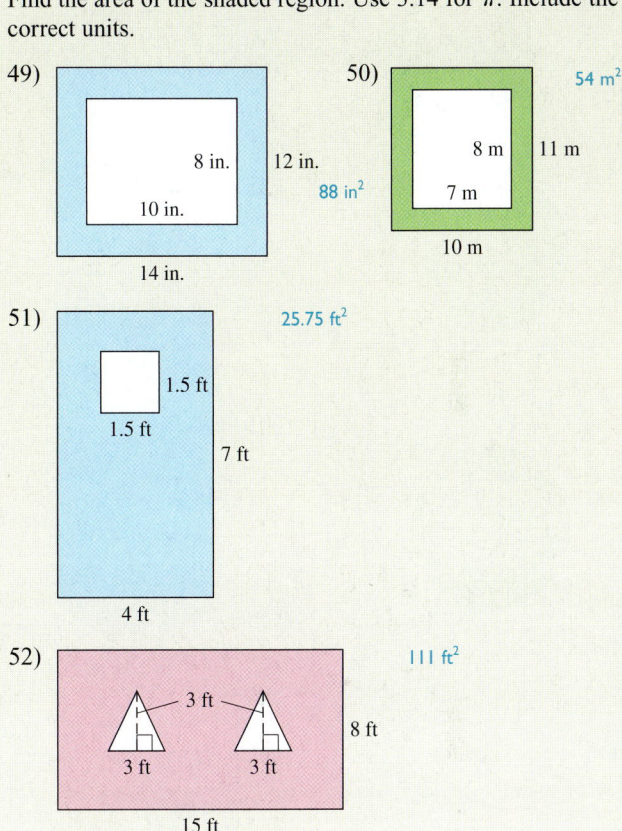
88 in^2

50)
54 m^2

51)
25.75 ft^2

52)
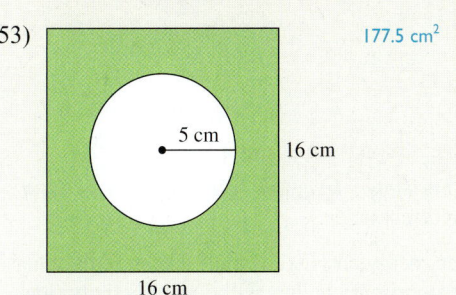
111 ft^2

53)
177.5 cm^2

54)
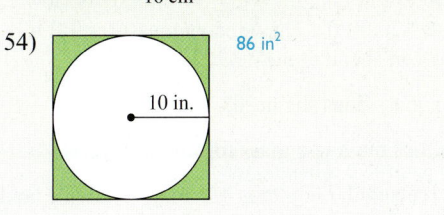
86 in^2

Objective 4: Use Volume Formulas

Find the volume of each figure. Where appropriate, give the answer in terms of π. Include the correct units.

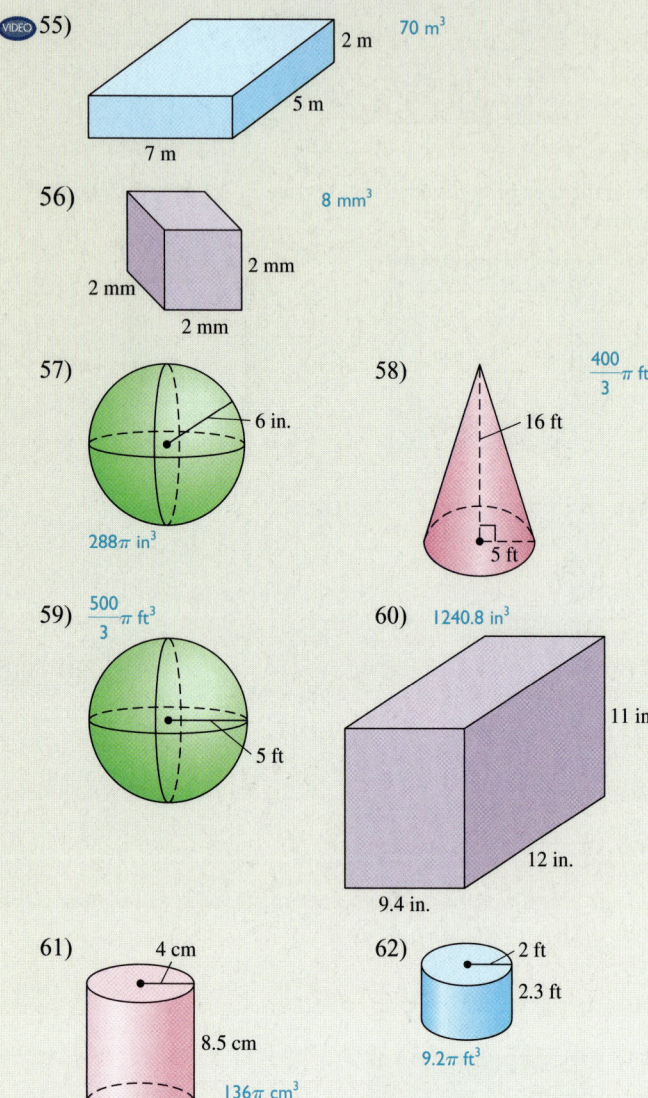

55)   2 m   70 m³   5 m   7 m

56)   8 mm³   2 mm   2 mm   2 mm

57)   6 in.   288π in³

58) $\frac{400}{3}\pi$ ft³   16 ft   5 ft

59) $\frac{500}{3}\pi$ ft³   5 ft

60)   1240.8 in³   11 in.   12 in.   9.4 in.

61)   4 cm   8.5 cm   136π cm³

62)   2 ft   2.3 ft   9.2π ft³

Mixed Exercises: Objectives 3 and 4

Applications of Perimeter, Area, and Volume: Use 3.14 for π and include the correct units.

63) To lower her energy costs, Yun would like to replace her rectangular storefront window with low-emissivity (low-e) glass that costs $20.00/ft². The window measures 9 ft by 6.5 ft, and she can spend at most $900.

   a) How much glass does she need?   58.5 ft²

   b) Can she afford the low-e glass for this window?
      No, it would cost $1170 to use this glass.

64) An insulated rectangular cooler is 15" long, 10" wide, and 13.6" high. What is the volume of the cooler?   2040 in³

65) A fermentation tank at a winery is in the shape of a right circular cylinder. The diameter of the tank is 6 ft, and it is 8 ft tall.

   a) How many cubic feet of wine will the tank hold?
      226.08 ft³

   b) How many gallons of wine will the tank hold? Round to the nearest gallon. (1 ft³ ≈ 7.48 gallons)   1691 gal

66) Yessenia wants a custom-made area rug measuring 5 ft by 8 ft. She has budgeted $500. She likes the Alhambra carpet sample that costs $9.80/ft² and the Sahara pattern that costs $12.20/ft². Can she afford either of these fabrics to make the area rug, or does she have to choose the cheaper one to remain within her budget? Support your answer by determining how much it would cost to have the rug made in each pattern.   The Alhambra would cost $392.00 and the Sahara would cost $488.00, so she could afford either one.

67) The lazy Susan on a table in a Chinese restaurant has a 10-in. radius. (A lazy Susan is a rotating tray used to serve food.)

   a) What is the perimeter of the lazy Susan?   62.8 in.

   b) What is its area?   314 in²

68) Find the perimeter of home plate given the dimensions below.   58 in.

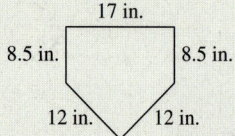

17 in.   8.5 in.   8.5 in.   12 in.   12 in.

69) A rectangular reflecting pool is 30 ft long, 19 ft wide and 1.5 ft deep. How many gallons of water will this pool hold? (1 ft³ ≈ 7.48 gallons)   6395.4 gal

70) Ralph wants to childproof his house now that his daughter has learned to walk. The round, glass-top coffee table in his living room has a diameter of 36 in. How much soft padding does Ralph need to cover the edges around the table?   113.04 in.

71) Nadia is remodeling her kitchen and her favorite granite countertop costs $80.00/ft², including installation. The layout of the countertop is shown below, where the counter has a uniform width of $2\frac{1}{4}$ ft. If she can spend at most $2500.00, can she afford her first-choice granite?
   No. This granite countertop would cost $2970.00.

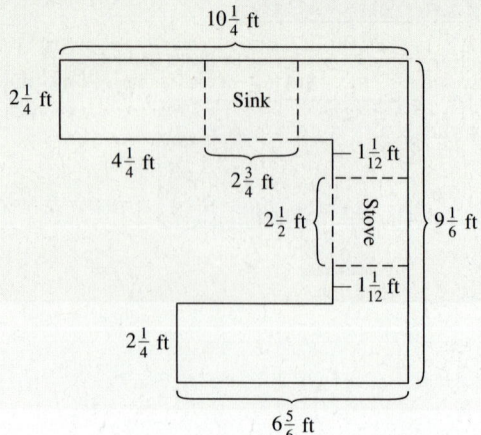

$10\frac{1}{4}$ ft $2\frac{1}{4}$ ft   Sink $4\frac{1}{4}$ ft $2\frac{3}{4}$ ft $1\frac{1}{12}$ ft $2\frac{1}{2}$ ft   Stove $9\frac{1}{6}$ ft $1\frac{1}{12}$ ft $2\frac{1}{4}$ ft $6\frac{5}{6}$ ft

72) A container of lip balm is in the shape of a right circular cylinder with a radius of 1.5 cm and a height of 2 cm. How much lip balm will the container hold?   14.13 cm³

73) The radius of a women's basketball is approximately 4.6 in. Find its circumference to the nearest tenth of an inch.   28.9 in.

74) The chamber of a rectangular laboratory water bath measures $6'' \times 11\frac{3}{4}'' \times 5\frac{1}{2}''$.

   a) How many cubic inches of water will the water bath hold? $387\frac{3}{4}$ in³ or 387.75 in³

   b) How many liters of water will the water bath hold? (1 in³ ≈ 0.016 liter)   6.204 liters

75) A town's public works department will install a flower garden in the shape of a trapezoid. It will be enclosed by decorative fencing that costs $23.50/ft.

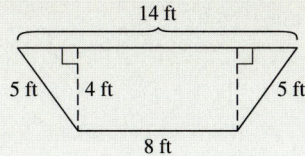

   a) Find the area of the garden.   44 ft²

   b) Find the cost of the fence. $752

76) Jaden is making decorations for the bulletin board in his fifth-grade classroom. Each equilateral triangle has a height of 15.6 in. and sides of length 18 in.

   a) Find the area of each triangle.   140.4 in²

   b) Find the perimeter of each triangle.   54 in.

77) The top of a counter-height pub table is in the shape of an equilateral triangle. Each side has a length of 18 in., and the height of the triangle is 15.6 in. What is the area of the table top?   140.4 in²

78) The dimensions of Riyad's home office are 10′ × 12′. He plans to install laminated hardwood flooring that costs $2.69/ft². How much will the flooring cost? $322.80

79) Salt used to melt road ice in winter is piled in the shape of a right circular cone. The radius of the base is 12 ft, and the pile is 8 ft high. Find the volume of salt in the pile.   1205.76 ft³

80) Find the volume of the ice cream pictured below. Assume that the right circular cone is completely filled and that the scoop on top is half of a sphere.   6.28 in³

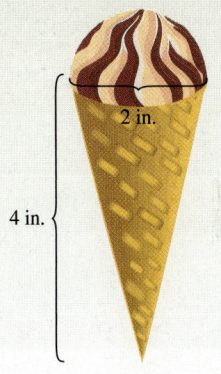

Section A.2 Determinants and Cramer's Rule

Objectives

1. Identify the Order of a Matrix
2. Find the Determinant of a 2 × 2 Matrix
3. Find the Determinant of a 3 × 3 Matrix
4. Use Cramer's Rule to Solve a System of Linear Equations in Two Variables
5. Use Cramer's Rule to Solve a System of Linear Equations in Three Variables

We first learned about matrices in Section 4.4. In this section we begin by learning some new terms associated with matrices.

1. Identify the Order of a Matrix

Let's look at the matrix A. $A = \begin{bmatrix} -6 & 5 \\ 1 & -8 \\ 0 & 3 \end{bmatrix}$

We can describe a matrix by its *order*. The order of a matrix is determined by the number of rows and columns it contains. The order of matrix A is 3 × 2 (read as "three by two") because it has three rows and two columns. The **order** of matrix is written as *number of rows × number of columns*. Another word for order is **dimension.**

   A **square matrix** is a matrix in which the number of rows equals the number of columns. Look at these two matrices.

$$B = \begin{bmatrix} 8 & -2 \\ 5 & 3 \end{bmatrix} \qquad C = \begin{bmatrix} -2 & 3 & 0 \\ 5 & 0 & 6 \\ 1 & -4 & 1 \end{bmatrix}$$

We say that B is a 2×2 matrix and that C is a 3×3 matrix. Matrices B and C are square matrices. Since the order of matrix A is 3×2, it is *not* a square matrix.

2. Find the Determinant of a 2 × 2 Matrix

Every square matrix has a number associated with it called its **determinant.** How a determinant is computed depends on its order. In this section, we will learn how to find the determinant of 2×2 and 3×3 matrices. Finding the determinants of larger matrices will be left for future courses.

Procedure   Evaluating the Determinant of a 2 × 2 Matrix

Given the 2×2 matrix $\begin{bmatrix} a_1 & b_1 \\ a_2 & b_2 \end{bmatrix}$, its **determinant** is denoted by $\begin{vmatrix} a_1 & b_1 \\ a_2 & b_2 \end{vmatrix}$ and is computed as follows:

$$\begin{vmatrix} a_1 & b_1 \\ a_2 & b_2 \end{vmatrix} = a_1 b_2 - a_2 b_1$$

Note

To find the determinant of a 2×2 matrix we find the *difference* of the products along the diagonals:

$$\begin{vmatrix} a_1 & b_1 \\ a_2 & b_2 \end{vmatrix} = a_1 b_2 - a_2 b_1$$

Example 1

Evaluate $\begin{vmatrix} 8 & -2 \\ 5 & 3 \end{vmatrix}$.

In-Class Example 1

Evaluate $\begin{vmatrix} 7 & -3 \\ 2 & 5 \end{vmatrix}$.

answer: 41

Solution

Find the products along the diagonals and subtract.

$$\begin{vmatrix} 8 & -2 \\ 5 & 3 \end{vmatrix} = 8 \cdot 3 - 5(-2) = 24 - (-10) = 24 + 10 = 34$$

You Try 1

Evaluate $\begin{vmatrix} -4 & 6 \\ 3 & 1 \end{vmatrix}$.

BE CAREFUL

A matrix is an *array* of numbers and is enclosed with brackets, but the determinant of a matrix is enclosed with vertical lines and is a *number* associated with the matrix.

Matrix: $\begin{bmatrix} 8 & -2 \\ 5 & 3 \end{bmatrix}$ Determinant of the Matrix: $\begin{vmatrix} 8 & -2 \\ 5 & 3 \end{vmatrix} = 34$

3. Find the Determinant of a 3 × 3 Matrix

We will evaluate the determinant of a 3 × 3 matrix by **expanding by the minors of the first column.**

To evaluate $\begin{vmatrix} a_1 & b_1 & c_1 \\ a_2 & b_2 & c_2 \\ a_3 & b_3 & c_3 \end{vmatrix}$, begin by circling the first column: $\begin{vmatrix} \boxed{a_1} & b_1 & c_1 \\ a_2 & b_2 & c_2 \\ a_3 & b_3 & c_3 \end{vmatrix}$

For each element in column 1, cross off the row and column in which the element appears to get its minor, the 2 × 2 determinant formed by the remaining elements.

$$\begin{vmatrix} a_1 & b_1 & c_1 \\ a_2 & b_2 & c_2 \\ a_3 & b_3 & c_3 \end{vmatrix} \qquad \text{Minor of } a_1: \begin{vmatrix} b_2 & c_2 \\ b_3 & c_3 \end{vmatrix}$$

$$\begin{vmatrix} a_1 & b_1 & c_1 \\ a_2 & b_2 & c_2 \\ a_3 & b_3 & c_3 \end{vmatrix} \qquad \text{Minor of } a_2: \begin{vmatrix} b_1 & c_1 \\ b_3 & c_3 \end{vmatrix}$$

$$\begin{vmatrix} a_1 & b_1 & c_1 \\ a_2 & b_2 & c_2 \\ a_3 & b_3 & c_3 \end{vmatrix} \qquad \text{Minor of } a_3: \begin{vmatrix} b_1 & c_1 \\ b_2 & c_2 \end{vmatrix}$$

We use these minors to evaluate the determinant of a 3 × 3 matrix.

Procedure Evaluating the Determinant of a 3 × 3 matrix

The determinant of a 3 × 3 matrix is computed as follows

$$\begin{vmatrix} a_1 & b_1 & c_1 \\ a_2 & b_2 & c_2 \\ a_3 & b_3 & c_3 \end{vmatrix} = a_1 \begin{vmatrix} b_2 & c_2 \\ b_3 & c_3 \end{vmatrix} - a_2 \begin{vmatrix} b_1 & c_1 \\ b_3 & c_3 \end{vmatrix} + a_3 \begin{vmatrix} b_1 & c_1 \\ b_2 & c_2 \end{vmatrix}$$

where the 2 × 2 determinants are the minors of their corresponding elements in column 1. Notice that the operation in front of the second product is *subtraction*. In this method we are expanding by the minors of the first column.

In other words, to evaluate the determinant of a 3 × 3 matrix by expanding by the minors of the first column, we do the following.

$$\begin{vmatrix} a_1 & b_1 & c_1 \\ a_2 & b_2 & c_2 \\ a_3 & b_3 & c_3 \end{vmatrix} = a_1(\text{minor of } a_1) - a_2(\text{minor of } a_2) + a_3(\text{minor of } a_3)$$

Remember that the operation in front of the second product is subtraction!

Example 2

In-Class Example 2

Evaluate $\begin{vmatrix} 3 & 2 & -5 \\ 4 & 1 & -1 \\ -2 & 0 & 3 \end{vmatrix}$.

answer: −21

Evaluate $\begin{vmatrix} 2 & -5 & 1 \\ -3 & 4 & -2 \\ -1 & 0 & 6 \end{vmatrix}$.

Solution

Circle the first column and find the minor of each element in that column.

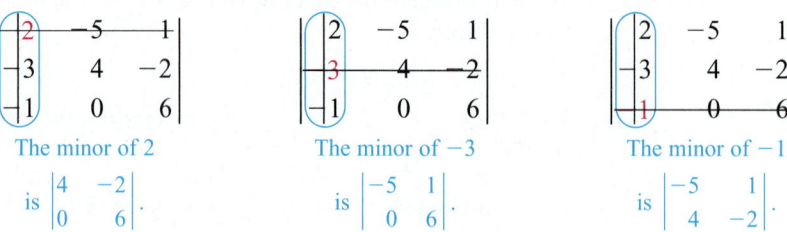

The minor of 2 is $\begin{vmatrix} 4 & -2 \\ 0 & 6 \end{vmatrix}$.

The minor of -3 is $\begin{vmatrix} -5 & 1 \\ 0 & 6 \end{vmatrix}$.

The minor of -1 is $\begin{vmatrix} -5 & 1 \\ 4 & -2 \end{vmatrix}$.

Use the definition to get

$$\begin{vmatrix} 2 & -5 & 1 \\ -3 & 4 & -2 \\ -1 & 0 & 6 \end{vmatrix} = 2\begin{vmatrix} 4 & -2 \\ 0 & 6 \end{vmatrix} - (-3)\begin{vmatrix} -5 & 1 \\ 0 & 6 \end{vmatrix} + (-1)\begin{vmatrix} -5 & 1 \\ 4 & -2 \end{vmatrix}$$

$$= 2[4(6) - 0(-2)] + 3[-5(6) - 0(1)] - 1[-5(-2) - 4(1)] \quad \text{Evaluate the } 2 \times 2 \text{ determinants.}$$

$$= 2(24 - 0) + 3(-30 - 0) - 1(10 - 4) \quad \text{Multiply inside the brackets.}$$

$$= 2(24) + 3(-30) - 1(6) \quad \text{Subtract inside the parentheses.}$$

$$= 48 - 90 - 6$$

$$= -48$$

You Try 2

Evaluate $\begin{vmatrix} -4 & 3 & -2 \\ 2 & -1 & 5 \\ 1 & 8 & 0 \end{vmatrix}$.

4. Use Cramer's Rule to Solve a System of Linear Equations in Two Variables

Cramer's rule is a method that uses determinants to solve a system of linear equations. We can use this method if the system contains the same number of variables as equations and if all of the constants are not zero.

We can see where Cramer's rule comes from if we solve the following system for x and y using the elimination method.

$$a_1x + b_1y = c_1 \qquad (1)$$
$$a_2x + b_2y = c_2 \qquad (2)$$

Let's solve for x by eliminating y. Multiply (1) by b_2, and multiply (2) by $-b_1$.

Original System		Rewrite the system
$a_1x + b_1y = c_1$	b_2 times (1)	$a_1b_2x + b_1b_2y = b_2c_1$
$a_2x + b_2y = c_2$	$-b_1$ times (2)	$-a_2b_1x - b_1b_2y = -b_1c_2$

Add the equations in the rewritten system. The y is eliminated.

$$\begin{array}{r} a_1b_2x + b_1b_2y = b_2c_1 \\ + \; -a_2b_1x - b_1b_2y = -b_1c_2 \\ \hline (a_1b_2 - a_2b_1)x = b_2c_1 - b_1c_2 \end{array}$$

$$x = \frac{b_2c_1 - b_1c_2}{a_1b_2 - a_2b_1}$$

We can rewrite the numerator and denominator as determinants.

Numerator	**Denominator**
$b_2c_1 - b_1c_2 = \begin{vmatrix} c_1 & b_1 \\ c_2 & b_2 \end{vmatrix}$	$a_1b_2 - a_2b_1 = \begin{vmatrix} a_1 & b_1 \\ a_2 & b_2 \end{vmatrix}$

We can write the value of x, then, in terms of determinants. We call the determinant in the numerator D_x and the determinant in the denominator D.

$$x = \frac{\begin{vmatrix} c_1 & b_1 \\ c_2 & b_2 \end{vmatrix}}{\begin{vmatrix} a_1 & b_1 \\ a_2 & b_2 \end{vmatrix}} = \frac{D_x}{D}$$

If we went back to the original system and solved for y by eliminating x we would get

$$y = \frac{\begin{vmatrix} a_1 & c_1 \\ a_2 & c_2 \end{vmatrix}}{\begin{vmatrix} a_1 & b_1 \\ a_2 & b_2 \end{vmatrix}} = \frac{D_y}{D}$$

where the determinant in the numerator is called D_y and the determinant in the denominator is D, the same as the one we obtained in the denominator of x.

Now we will summarize Cramer's rule for solving a system of two linear equations.

Procedure   Cramer's Rule for Solving a System of Two Linear Equations

To solve a system of linear equations of the form $a_1x + b_1y = c_1$
$\qquad\qquad\qquad\qquad\qquad\qquad\qquad\qquad\qquad a_2x + b_2y = c_2$

Step 1:   Set up and evaluate the determinants.

$$D = \begin{vmatrix} a_1 & b_1 \\ a_2 & b_2 \end{vmatrix} \qquad D \text{ contains the coefficients of the variables.}$$

$$D_x = \begin{vmatrix} c_1 & b_1 \\ c_2 & b_2 \end{vmatrix} \qquad \text{To get } D_x, \text{ the constants replace the coefficients of } x \text{ in } D.$$

$$D_y = \begin{vmatrix} a_1 & c_1 \\ a_2 & c_2 \end{vmatrix} \qquad \text{To get } D_y, \text{ the constants replace the coefficients of } y \text{ in } D.$$

Step 2:   Find x and y using the values of the determinants.

$$x = \frac{D_x}{D} \quad \text{and} \quad y = \frac{D_y}{D} \quad \text{if } D \neq 0$$

The value of D gives us information about the solution of the system. If $D \neq 0$, then the system has a single solution. If $D = 0$ and if at least one of the other determinants is not zero, then the system is inconsistent and, therefore, has no solution. If $D = 0$ and if all other determinants are zero, then the equations are dependent and there are an infinite number of solutions.

Example 3

Solve using Cramer's rule. $2x + 5y = -1$
$\qquad\qquad\qquad\qquad\qquad x - 2y = -5$

In-Class Example 3
Solve using Cramer's rule.
$\quad 3x - y = 1$
$-2x + 3y = 11$
answer: (2, 5)

Solution

Step 1:   **Set up and evaluate the determinants.**

D contains the coefficients of the variables.

$$D = \begin{vmatrix} 2 & 5 \\ 1 & -2 \end{vmatrix} = 2(-2) - 1(5) = -4 - 5 = -9$$

To get D_x the constants replace the coefficients of x in D.

$$D_x = \begin{vmatrix} -1 & 5 \\ -5 & -2 \end{vmatrix} = -1(-2) - (-5)(5) = 2 + 25 = 27$$

To get D_y the constants replace the coefficients of y in D.

$$D_y = \begin{vmatrix} 2 & -1 \\ 1 & -5 \end{vmatrix} = 2(-5) - 1(-1) = -10 + 1 = -9$$

Step 2:   **Find x and y.**

$$x = \frac{D_x}{D} = \frac{27}{-9} = -3 \quad \text{and} \quad y = \frac{D_y}{D} = \frac{-9}{-9} = 1$$

The solution is $(-3, 1)$. The check is left to the student.   ■

You Try 3

Solve using Cramer's rule. $3x - 2y = -8$
                                      $2x + y = 11$

5. Use Cramer's Rule to Solve a System of Linear Equations in Three Variables

Cramer's rule can be extended so that we can solve a system of three linear equations in three variables.

Procedure   Cramer's Rule for Solving a System of Three Linear Equations

To solve a system of linear equations of the form $a_1x + b_1y + c_1z = d_1$
                                                 $a_2x + b_2y + c_2z = d_2$
                                                 $a_3x + b_3y + c_3z = d_3$

Step 1:   **Set up and evaluate the determinants.**

$$D = \begin{vmatrix} a_1 & b_1 & c_1 \\ a_2 & b_2 & c_2 \\ a_3 & b_3 & c_3 \end{vmatrix} \qquad D \text{ contains the coefficients of the variables.}$$

$$D_x = \begin{vmatrix} d_1 & b_1 & c_1 \\ d_2 & b_2 & c_2 \\ d_3 & b_3 & c_3 \end{vmatrix} \qquad \text{To get } D_x, \text{ the constants replace the coefficients of } x \text{ in } D.$$

$$D_y = \begin{vmatrix} a_1 & d_1 & c_1 \\ a_2 & d_2 & c_2 \\ a_3 & d_3 & c_3 \end{vmatrix} \qquad \text{To get } D_y, \text{ the constants replace the coefficients of } y \text{ in } D.$$

$$D_z = \begin{vmatrix} a_1 & b_1 & d_1 \\ a_2 & b_2 & d_2 \\ a_3 & b_3 & d_3 \end{vmatrix} \qquad \text{To get } D_z, \text{ the constants replace the coefficients of } z \text{ in } D.$$

Step 2:   **Find x, y, and z using the values of the determinants.**

$$x = \frac{D_x}{D}, \quad y = \frac{D_y}{D}, \quad \text{and} \quad z = \frac{D_z}{D} \quad \text{if } D \neq 0$$

Example 4

In-Class Example 4
Use Example 4.

Solve using Cramer's rule. $\begin{aligned} x - 2y + 3z &= 0 \\ -2x + 6y - z &= -4 \\ x + 4y + 5z &= 1 \end{aligned}$

Solution

Step 1:   Set up and evaluate the determinants.

D contains the coefficients of the variables.

$$D = \begin{vmatrix} 1 & -2 & 3 \\ -2 & 6 & -1 \\ 1 & 4 & 5 \end{vmatrix}$$

Expand by minors about column 1.

$$D = 1\begin{vmatrix} 6 & -1 \\ 4 & 5 \end{vmatrix} - (-2)\begin{vmatrix} -2 & 3 \\ 4 & 5 \end{vmatrix} + 1\begin{vmatrix} -2 & 3 \\ 6 & -1 \end{vmatrix}$$

$$= 1(30 + 4) + 2(-10 - 12) + 1(2 - 18)$$
$$= 1(34) + 2(-22) + 1(-16)$$
$$= 34 - 44 - 16$$
$$= -26$$

To get D_x the constants replace the coefficients of x in D.

$$D_x = \begin{vmatrix} 0 & -2 & 3 \\ -4 & 6 & -1 \\ 1 & 4 & 5 \end{vmatrix}$$

Verify that $D_x = -104$.

To get D_y the constants replace the coefficients of y in D.

$$D_y = \begin{vmatrix} 1 & 0 & 3 \\ -2 & -4 & -1 \\ 1 & 1 & 5 \end{vmatrix}$$

Verify that $D_y = -13$.

To get D_z the constants replace the coefficients of z in D.

$$D_z = \begin{vmatrix} 1 & -2 & 0 \\ -2 & 6 & -4 \\ 1 & 4 & 1 \end{vmatrix}$$

Verify that $D_z = 26$.

Step 2:   Find x, y, and z.

$$x = \frac{D_x}{D} = \frac{-104}{-26} = 4, \quad y = \frac{D_y}{D} = \frac{-13}{-26} = \frac{1}{2}, \quad \text{and} \quad z = \frac{D_z}{D} = \frac{26}{-26} = -1$$

The solution is $\left(4, \dfrac{1}{2}, -1\right)$. The check is left to the student.

You Try 4

Solve using Cramer's rule. $\begin{aligned} 2x + y - 2z &= -4 \\ x - 3y - 6z &= -5 \\ -3x - 2y + 2z &= 6 \end{aligned}$

Answers to You Try Exercises

1) -22 2) 141 3) $(2, 7)$ 4) $\left(-5, 3, -\dfrac{3}{2}\right)$

A.2 Exercises

*Additional answers can be found in the Answers to Exercises appendix.

Mixed Exercises: Objectives 1 and 2

1) What does the order of a matrix tell us about the matrix?

2) What is a square matrix?
   It is a matrix in which the number of rows equals the number of columns.

3) What is the difference between a matrix and the determinant of a matrix?

4) Does $\begin{bmatrix} 6 & 4 \\ 1 & 5 \end{bmatrix} = 26$? Explain your answer.

Mixed Exercises: Objectives 2 and 3

Evaluate.

5) $\begin{vmatrix} 2 & 1 \\ 3 & 9 \end{vmatrix}$ 15

6) $\begin{vmatrix} 5 & 7 \\ -1 & -4 \end{vmatrix}$ -13

7) $\begin{vmatrix} -4 & -3 \\ -8 & 2 \end{vmatrix}$ -32

8) $\begin{vmatrix} 7 & 0 \\ -10 & -9 \end{vmatrix}$ -63

9) $\begin{vmatrix} 0 & 8 \\ -6 & 12 \end{vmatrix}$ 48

10) $\begin{vmatrix} -10 & -3 \\ -2 & -5 \end{vmatrix}$ 44

11) $\begin{vmatrix} \dfrac{2}{3} & 20 \\ -\dfrac{3}{5} & -18 \end{vmatrix}$ 0

12) $\begin{vmatrix} 10 & 24 \\ 3 & 3 \\ 4 & 2 \end{vmatrix}$ -3

13) $\begin{vmatrix} \dfrac{9}{8} & \dfrac{2}{3} \\ \dfrac{5}{2} & \dfrac{1}{3} \end{vmatrix}$ $-\dfrac{31}{24}$

14) $\begin{vmatrix} -\dfrac{1}{2} & -3 \\ \dfrac{7}{4} & \dfrac{3}{5} \end{vmatrix}$ $\dfrac{99}{20}$

15) $\begin{vmatrix} 2 & 0 & 4 \\ 3 & -1 & 1 \\ 1 & 5 & 2 \end{vmatrix}$ 50

16) $\begin{vmatrix} 1 & -3 & -6 \\ 2 & 5 & 1 \\ 4 & 1 & 0 \end{vmatrix}$ 95

17) $\begin{vmatrix} -1 & 7 & 0 \\ -4 & 6 & -2 \\ 5 & 1 & -1 \end{vmatrix}$ -94

18) $\begin{vmatrix} 2 & 1 & -2 \\ -1 & 4 & 0 \\ -3 & -5 & 1 \end{vmatrix}$ -25

19) $\begin{vmatrix} 0 & 8 & 1 \\ 1 & -4 & -3 \\ -6 & 0 & 5 \end{vmatrix}$ 80

20) $\begin{vmatrix} -5 & 0 & 1 \\ 0 & 6 & -4 \\ 2 & 8 & 3 \end{vmatrix}$ -262

21) $\begin{vmatrix} -3 & -1 & 0 \\ 2 & 1 & 5 \\ 0 & -2 & 3 \end{vmatrix}$ -33

22) $\begin{vmatrix} 0 & -5 & -9 \\ -4 & 7 & 1 \\ 0 & 6 & 7 \end{vmatrix}$ 76

23) $\begin{vmatrix} 2 & -8 & 4 \\ 4 & 0 & 1 \\ 3 & -12 & 6 \end{vmatrix}$ 0

24) $\begin{vmatrix} 0 & 1 & 4 \\ 6 & 9 & -3 \\ -2 & -3 & 1 \end{vmatrix}$ 0

Mixed Exercises: Objectives 4 and 5

25) Suppose you are solving a system of two linear equations and that you get $D = -6$, $D_x = -24$, and $D_y = 2$. What is the solution of the system? $\left(4, -\dfrac{1}{3}\right)$

26) Suppose you are solving a system of three linear equations and that you get $D = 10$, $D_x = -50$, $D_y = 20$, and $D_z = 0$. What is the solution of the system? $(-5, 2, 0)$

27) If you are solving a system of linear equations using Cramer's rule, how do you know if the system is inconsistent?
    $D = 0$ and at least one of the other determinants is not zero.

28) If you are solving a system of linear equations using Cramer's rule, how do you know if the system has an infinite number of solutions?
    $D = 0$ and all of the other determinants equal zero.

Objective 4: Use Cramer's Rule to Solve a System of Linear Equations in Two Variables

Solve each system using Cramer's rule.

29) $\begin{aligned} x + 3y &= 1 \\ 2x + 9y &= 8 \end{aligned}$ $(-5, 2)$

30) $\begin{aligned} -x + 2y &= 11 \\ 3x - 2y &= -5 \end{aligned}$ $(3, 7)$

31) $\begin{aligned} -7x + 6y &= 0 \\ x - y &= 1 \end{aligned}$ $(-6, -7)$

32) $\begin{aligned} -3x + y &= -16 \\ x + 3y &= 2 \end{aligned}$ $(5, -1)$

33) $\begin{aligned} 4a - 5b &= -5 \\ -6a + 7b &= 6 \end{aligned}$ $\left(\dfrac{5}{2}, 3\right)$

34) $\begin{aligned} 2x + 9y &= -10 \\ 5x - 3y &= -8 \end{aligned}$ $\left(-2, -\dfrac{2}{3}\right)$

35) $\begin{aligned} 3x &= 8 - 8y \\ 3y - 5x &= 3 \end{aligned}$ $(0, 1)$

36) $\begin{aligned} a + 6 &= 7b \\ 2a + 12 &= -5b \end{aligned}$ $(-6, 0)$

37) $\begin{aligned} -2x + 6y &= -6 \\ 9y &= 3x - 14 \end{aligned}$ $\varnothing$

38) $\begin{aligned} 5 - 8y &= 4x \\ 2x + 4y &= 9 \end{aligned}$ $\varnothing$

39) $\begin{aligned} x + 4y &= -3 \\ -3x - 12y &= 9 \end{aligned}$ $\{(x, y) \mid x + 4y = -3\}$

40) $\begin{aligned} 2x + 10y &= 16 \\ x + 5y &= 8 \end{aligned}$ $\{(x, y) \mid x + 5y = 8\}$

Objective 5: Use Cramer's Rule to Solve a System of Linear Equations in Three Variables

Solve each system using Cramer's rule.

41) $\begin{aligned} x + 5y + 2z &= 9 \\ 3x - y + z &= 0 \\ -4x + y - 2z &= 0 \end{aligned}$ $(1, 2, -1)$

42) $\begin{aligned} 2x + 3y + z &= 1 \\ x + y - z &= -6 \\ x + 5y - 3z &= -10 \end{aligned}$ $(-3, 1, 4)$

43) $\begin{aligned} 2x + 6y + z &= 4 \\ -3x - 3y + z &= -3 \\ -x - 2y - z &= 4 \end{aligned}$ $(-4, 3, -6)$

44) $\begin{aligned} 7x + y + 7z &= 2 \\ -x - y - 2z &= 1 \\ 5x + 2y + 5z &= 4 \end{aligned}$ $(3, 2, -3)$

45) $\begin{aligned} x + 2y - 6z &= 3 \\ -2x + y + 6z &= -8 \\ x - 4y - 3z &= 4 \end{aligned}$ $\left(5, 0, \dfrac{1}{3}\right)$

46) $\begin{aligned} -4x - 3y + z &= 5 \\ -6x + 5y + 2z &= -2 \\ 2x + y - z &= -2 \end{aligned}$ $\left(-\dfrac{1}{2}, -1, 0\right)$

47) $-3a + 5b - c = -8$
    $a - 2b = 2$
    $-a + 7b + 2c = 5$
    $(4, 1, 1)$

48) $a + b - 3c = 4$
    $2a - b + 2c = 1$
    $-a - 3c = -5$
    $(2, 5, 1)$

58) $0.1x + 0.2y + 0.3z = 0.5$ $(-5, -1, 4)$
    $y + 0.2z = -0.2$
    $-0.2x + 0.5y + 0.1z = 0.9$

49) $x - 4y = 1$
    $2x + 5z = 1$ $\left(-2, -\dfrac{3}{4}, 1\right)$
    $8y + 3z = -3$

50) $x + 3y = 2$
    $-6y - z = 0$ $\left(0, \dfrac{2}{3}, -4\right)$
    $x + z = -4$

Extension

Solve for x.

51) $x - 2y + z = 3$ $\varnothing$
    $x + y + 3z = 4$
    $-x + 5y + z = 0$

52) $x + 5y - 3z = 9$ $\varnothing$
    $2x + 8y - 6z = 7$
    $-3x - 12y + 9z = -2$

59) $\begin{vmatrix} 6 & 3 \\ 5 & x \end{vmatrix} = 33$ {8}

60) $\begin{vmatrix} 4 & 5 \\ 2 & x \end{vmatrix} = 18$ {7}

53) $3x + 6y - 3z = -9$
    $x + 2y - z = -3$
    $-2x - 4y + 2z = 6$
    $\{(x, y, z) | x + 2y - z = -3\}$

54) $x - y + 2z = -1$
    $2x - 2y + 4z = -2$
    $3x - 3y + 6z = -3$
    $\{(x, y, z) | x - y + 2z = -1\}$

61) $\begin{vmatrix} 2 & 2 \\ x & 3 \end{vmatrix} = -12$ {9}

62) $\begin{vmatrix} -1 & x \\ 2 & 7 \end{vmatrix} = 1$ {-4}

55) $-\dfrac{1}{2}a + \dfrac{1}{4}b + c = \dfrac{3}{4}$
    $\dfrac{1}{3}b + \dfrac{2}{3}c = -\dfrac{1}{3}$
    $\dfrac{1}{4}a + c = 2$
    $(0, -5, 2)$

56) $\dfrac{1}{4}a + \dfrac{3}{2}b - \dfrac{1}{4}c = \dfrac{5}{2}$
    $\dfrac{1}{2}a + \dfrac{1}{3}c = \dfrac{3}{2}$
    $b - \dfrac{1}{2}c = \dfrac{1}{2}$
    $(1, 2, 3)$

63) $\begin{vmatrix} 2 & -3 & 1 \\ 0 & 4 & 1 \\ x & 0 & 5 \end{vmatrix} = 47$ {-1}

64) $\begin{vmatrix} x & 4 & 0 \\ -2 & 2 & 1 \\ 0 & 1 & -3 \end{vmatrix} = -45$ {3}

57) $x - 0.3y - 0.1z = -0.4$
    $0.1x - 0.2y + 0.5z = 0.3$
    $0.1y - 0.1z = 0.2$ $(1, 4, 2)$

65) $\begin{vmatrix} 0 & x & 1 \\ -1 & 0 & 5 \\ 3 & -1 & 2 \end{vmatrix} = 35$ {2}

66) $\begin{vmatrix} 1 & 3 & 0 \\ 0 & 1 & x \\ -2 & 1 & 2 \end{vmatrix} = 37$ {-5}

Section A.3 Synthetic Division and the Remainder Theorem

Objectives

1. Use Synthetic Division
2. Use the Remainder Theorem

1. Use Synthetic Division

When we divide a polynomial by a binomial of the form $x - c$, we can use long division or another method called *synthetic division*. **Synthetic division** uses only the numerical coefficients of the variables to find the quotient.

Consider the division problem $(3x^3 - 5x^2 - 6x + 13) \div (x - 2)$. On the left, we will illustrate the long division process as we have already presented it. On the right, we will show the process using only the coefficients of the variables.

$$
\begin{array}{r}
3x^2 + x - 4 \\
x - 2 \overline{) 3x^3 - 5x^2 - 6x + 13} \\
\underline{-(3x^3 - 6x^2)} \\
x^2 - 6x \\
\underline{-(x^2 - 2x)} \\
-4x + 13 \\
\underline{-(-4x + 8)} \\
5
\end{array}
$$

$$
\begin{array}{r}
3 + 1 - 4 \\
1 - 2 \overline{) 3 - 5 - 6 + 13} \\
\underline{-(3 - 6)} \\
1 - 6 \\
\underline{-(1 - 2)} \\
-4 + 13 \\
\underline{-(-4 + 8)} \\
5
\end{array}
$$

> **Note**
>
> As long as we keep the like terms lined up in the correct columns, the variables do not affect the numerical coefficients of the quotient. This process of using only the numerical coefficients to divide a polynomial by a binomial of the form $x - c$ is called *synthetic division*. Using synthetic division is often quicker than using the traditional long division process.

We will present the steps for performing synthetic division by looking at the above example again.

Example 1

In-Class Example 1
Use synthetic division to divide
$(2y^3 - 11y^2 + 13y + 8)$ by
$(y - 4)$.
answer:

$2y^2 - 3y + 1 + \dfrac{12}{y - 4}$

Use synthetic division to divide $(3x^3 - 5x^2 - 6x + 13)$ by $(x - 2)$.

Solution

Remember, in order to be able to use synthetic division, the divisor must be in the form $x - c$. $x - 2$ is in the form $x - c$, and $c = 2$.

Procedure   How to Perform Synthetic Division

Step 1: Write the dividend in descending powers of x. If a term of any degree is missing, insert the term into the polynomial with a coefficient of 0.
The dividend in the example is $3x^3 - 5x^2 - 6x + 13$. It is written in descending order, and no terms are missing.

Step 2: Write the value of c in an open box. Next to it, on the right, write the coefficients of the terms of the dividend. Skip a line and draw a horizontal line under the coefficients. Bring down the first coefficient.
In this example, $c = 2$.

$$\underline{2\,|}\ \ 3 \quad -5 \quad -6 \quad 13$$
$$\overline{\ \ 3}$$

Step 3: Multiply the number in the box by the coefficient under the horizontal line. (Here, that is $2 \cdot 3 = 6$.) Write the product under the next coefficient. (Write the 6 under the -5.) Then, *add* the numbers in the second column. (Here, we get 1.)

$$\underline{2\,|}\ \ 3 \quad -5 \quad -6 \quad 13 \qquad\qquad 2 \cdot 3 = 6;\ -5 + 6 = 1$$
$$\phantom{\underline{2\,|}\ \ 3 \quad} 6$$
$$\overline{\ \ 3 \quad\ \ 1}$$

Step 4: Multiply the number in the box by the number under the horizontal line in the second column. (Here, that is $2 \cdot 1 = 2$.) Write the product under the next coefficient. (Write the 2 under the -6.) Then, *add* the numbers in the third column. (Here, we get -4.)

$$\underline{2\,|}\ \ 3 \quad -5 \quad -6 \quad 13 \qquad\qquad 2 \cdot 1 = 2;\ -6 + 2 = -4$$
$$\phantom{\underline{2\,|}\ \ 3 \quad} 6 \quad\ \ 2$$
$$\overline{\ \ 3 \quad\ \ 1 \quad -4}$$

Step 5: Repeat the procedure of Step 4 with subsequent columns until there is a number in each column in the row under the horizontal line.

$$\underline{2\,|}\ \ 3 \quad -5 \quad -6 \quad 13 \qquad\qquad 2 \cdot (-4) = -8;\ 13 + (-8) = 5$$
$$\phantom{\underline{2\,|}\ \ 3 \quad} 6 \quad\ \ 2 \quad -8$$
$$\overline{\ \ 3 \quad\ \ 1 \quad -4 \quad\ \ 5}$$

The numbers in the last row represent the quotient and the remainder. The last number is the remainder. The numbers before it are the coefficients of the quotient. *The degree of the quotient is one less than the degree of the dividend.*

In our example, the dividend is a *third-degree* polynomial. Therefore, the quotient is a *second-degree* polynomial.

Since the 3 in the first row is the coefficient of x^3 in the dividend, the 3 in the last row is the coefficient of x^2 in the quotient, and so on.

Dividend
$$3x^3 - 5x^2 - 6x + 13$$
$$\overbrace{}$$
$$\underline{2\,|}\ \ 3 \quad -5 \quad -6 \quad 13$$
$$\phantom{\underline{2\,|}\ \ 3 \quad} 6 \quad\ \ 2 \quad -8$$
$$\overline{\ \ 3 \quad\ \ 1 \quad -4 \quad\ \ 5} \to \text{Remainder}$$
$$3x^2 + 1x - 4$$
Quotient

$$(3x^3 - 5x^2 - 6x + 13) \div (x - 2) = 3x^2 + x - 4 + \dfrac{5}{x - 2}$$

You Try 1

Use synthetic division to divide $(2x^3 + x^2 - 16x - 7) \div (x - 3)$.

If the divisor is $x + 3$, then it can be written in the form $x - c$ as $x - (-3)$. So, $c = -3$.

 **BE CAREFUL**

Synthetic division can be used only when dividing a polynomial by a binomial of the form $x - c$. If the divisor is not in the form $x - c$, use long division.

Synthetic division can be used to find $(4x^2 - 19x + 16) \div (x - 4)$ because $x - 4$ is in the form $x - c$.

Synthetic division *cannot* be used to find $(4x^2 - 19x + 16) \div (2x - 3)$ because $2x - 3$ is *not* in the form $x - c$. Use long division.

Example 2

In-Class Example 2
$P(t) = 4t^2 + 3t - 10$
a) Find $P(-2)$.
b) Use synthetic division to find the quotient and the remainder when $P(t)$ is divided by $t + 2$.
answer: a) $P(-2) = 0$
b) $\dfrac{4t^2 + 3t - 10}{t + 2} = 4t - 5$;
the remainder is 0.

$P(x) = 3x^2 + 14x + 8$

a) Find $P(-4)$.

b) Use synthetic division to find the quotient and the remainder when $P(x)$ is divided by $x + 4$.

Solution

a) $P(-4) = 3(-4)^2 + 14(-4) + 8$ Substitute -4 for x.
$= 3(16) - 56 + 8$
$= 48 - 56 + 8$
$= 0$

$P(-4) = 0$

b) Remember, in order to be able to use synthetic division, the divisor must be in the form $x - c$. Write $x + 4$ as $x - (-4)$. Therefore, $c = -4$.

Step 1: The dividend, $3x^2 + 14x + 8$, is written in descending powers of x, and it is not missing any terms.

Steps 2–5: Write the value of c in an open box. Here, $c = -4$. Set up the synthetic division problem and perform the steps until there are numbers in every column of the last row.

$$
\begin{array}{r|rrr}
-4 & 3 & 14 & 8 \\
   &   & -12 & -8 \\
\hline
   & 3 & 2 & 0 \rightarrow \text{Remainder}
\end{array}
$$

$$3x + 2$$
Quotient

$$\frac{3x^2 + 14x + 8}{x + 4} = 3x + 2. \text{ The remainder is 0.}$$

 You Try 2

$P(n) = 5n^2 + n - 4$

a) Find $P(-1)$.

b) Use synthetic division to find the quotient and the remainder when $P(n)$ is divided by $n + 1$.

2. Use the Remainder Theorem

Notice in Example 2 that $P(-4) = 0$ and that when $P(x)$ was divided by $x + 4$ the remainder was zero. This is a consequence of the **remainder theorem.** The remainder theorem tells us the relationship between $P(c)$ and the remainder when $P(x)$ is divided by $x - c$.

Procedure   The Remainder Theorem

When $P(x)$ is divided by $x - c$, the remainder is $P(c)$.

Example 3

In-Class Example 3
Let $P(x) = x^4 + 5x^3 + 9x + 7$.
Use the remainder theorem and synthetic division to find $P(-2)$. Then, use direct substitution to verify the result.
answer: -35

Let $P(x) = 2x^4 - 11x^2 - 9x + 5$. Use the remainder theorem and synthetic division to find $P(3)$. Then, use direct substitution to verify the result.

Solution

Since there is no x^3-term in the original polynomial, insert $0x^3$ into $P(x)$.

$$P(x) = 2x^4 + 0x^3 - 11x^2 - 9x + 5$$

The remainder theorem tells us that $P(3)$ is the remainder when $P(x)$ is divided by $x - 3$.

Write the value of c in an open box. Here, $c = 3$. Set up the synthetic division problem and perform the steps until there are numbers in every column of the last row.

$$
\begin{array}{r|rrrrr}
3 & 2 & 0 & -11 & -9 & 5 \\
  &   & 6 & 18 & 21 & 36 \\
\hline
  & 2 & 6 & 7 & 12 & 41 \rightarrow \text{Remainder}
\end{array}
$$

The remainder is 41. Therefore, $P(3) = 41$.
Let's verify this result with direct substitution.

$$P(x) = 2x^4 - 11x^2 - 9x + 5$$
$$P(3) = 2(3)^4 - 11(3)^2 - 9(3) + 5 \qquad \text{Substitute 3 for } x.$$
$$= 2(81) - 11(9) - 27 + 5$$
$$= 162 - 99 - 27 + 5$$
$$= 41$$

You Try 3

Let $P(a) = 3a^4 - 10a^3 + a + 8$. Use the remainder theorem and synthetic division to find $P(2)$. Then, use direction substitution to verify the result.

Because the remainder was zero and $P(-4) = 0$ in Example 2, $x + 4$ is a factor of $3x^2 + 14x + 8$. We can factor $3x^2 + 14x + 8$ as $(x + 4)(3x + 2)$. In Example 3, however, we found that $P(3) = 41$, which tells us that when $P(x) = 2x^4 - 11x^2 - 9x + 5$ is divided by $x - 3$, the remainder is 41. Therefore, $x - 3$ is *not* a factor of $P(x)$.

It can be proven that if $P(c) = 0$, then $x - c$ is a factor of $P(x)$. Likewise, if $P(c) \neq 0$ then $x - c$ is *not* a factor of $P(x)$.

Answers to You Try Exercises

1) $2x^2 + 7x + 5 + \dfrac{8}{x - 3}$ 2) a) 0   b) $\dfrac{5n^2 + n - 4}{n + 1} = 5n - 4$; the remainder is 0.

3) The remainder is -22; $P(2) = -22$.

A.3 Exercises

*Additional answers can be found in the Answers to Exercises appendix.

Objective 1: Use Synthetic Division

1) Explain when synthetic division may be used to divide polynomials.
   Use synthetic division when the divisor is in the form $x - c$.

Can synthetic division be used to divide the polynomials in Exercises 2–4? Why or why not?

2) $(x^2 + 8x^3 - 10x + 3) \div (x - 6)$
   Yes. The divisor, $x - 6$, is in the form $x - c$.

3) $(2x^3 - 7x^2 - 2x + 10) \div (2x + 5)$
   No. The divisor, $2x + 5$, is not in the form $x - c$.

4) $\dfrac{x^3 - 15x^2 + 8x + 12}{x^2 - 2}$
   No. The divisor, $x^2 - 2$, is not in the form $x - c$.

Use synthetic division to divide the polynomials.

5) $(t^2 + 5t - 36) \div (t - 4)$ $t + 9$

6) $(m^2 - 2m - 24) \div (m - 6)$ $m + 4$

7) $\dfrac{5n^2 + 21n + 20}{n + 3}$ $5n + 6 + \dfrac{2}{n + 3}$

8) $\dfrac{6k^2 + 4k - 19}{k + 2}$ $6k - 8 - \dfrac{3}{k + 2}$

9) $(2y^3 + 7y^2 - 10y + 21) \div (y + 5)$ $2y^2 - 3y + 5 - \dfrac{4}{y + 5}$

10) $(4z^3 - 11z^2 - 6z + 10) \div (z - 3)$ $4z^2 + z - 3 + \dfrac{1}{z - 3}$

11) $(4p - 3 - 10p^2 + 3p^3) \div (p - 3)$ $3p^2 - p + 1$

12) $(10c^2 + 3c + 2c^3 - 20) \div (c + 4)$ $2c^2 + 2c - 5$

13) $(2 + 5x^4 - 8x + 7x^3 - x^2) \div (x + 1)$ $5x^3 + 2x^2 - 3x - 5 + \dfrac{7}{x + 1}$

14) $(-4w^3 + w - 8 + w^4 + 7w^2) \div (w - 2)$ $w^3 - 2w^2 + 3w + 7 + \dfrac{6}{w - 2}$

15) $\dfrac{r^3 - 3r^2 + 4}{r - 2}$ $r^2 - r - 2$

16) $\dfrac{m^4 - 81}{m - 3}$ $m^3 + 3m^2 + 9m + 27$

17) $(2c^5 - 3c^4 - 11c) \div (c - 2)$ $2c^4 + c^3 + 2c^2 + 4c - 3 - \dfrac{6}{c - 2}$

18) $(n^5 - 29n^2 - 2n) \div (n - 3)$ $n^4 + 3n^3 + 9n^2 - 2n - 8 - \dfrac{24}{n - 3}$

19) $(2x^3 + 7x^2 - 16x + 6) \div \left(x - \dfrac{1}{2}\right)$ $2x^2 + 8x - 12$

20) $(3t^3 - 25t^2 + 14t - 2) \div \left(t - \dfrac{1}{3}\right)$ $3t^2 - 24t + 6$

21) Do you prefer long division or synthetic division? Why?
    Answers may vary.

22) In addition to performing long division or synthetic division,

   a) what is a third method for dividing $\dfrac{3x^2 + 2x - 8}{x + 2}$?
      Factor the numerator, and divide out common factors.

   b) Which method do you prefer? Why?   Answers may vary.

23) $P(x) = 2x^2 - 19x + 45$

   a) Find $P(5)$.   0

   b) Use synthetic division to find the quotient and remainder when $P(x)$ is divided by $x - 5$.
      quotient: $2x - 9$; remainder: 0

24) $P(t) = 3t^2 - 5t - 2$

   a) Find $P(2)$.   0

   b) Use synthetic division to find the quotient and remainder when $P(t)$ is divided by $t - 2$.
      quotient: $3t + 1$; remainder: 0

25) $P(r) = 5r^3 + 13r^2 - 10r - 12$

   a) Find $P(-3)$.   0

   b) Use synthetic division to find the quotient and remainder when $P(r)$ is divided by $r + 3$.
      quotient: $5r^2 - 2r - 4$; remainder: 0

26) $P(k) = 4k^3 + 5k^2 + 8k + 7$

   a) Find $P(-1)$.   0

   b) Use synthetic division to find the quotient and remainder when $P(k)$ is divided by $k + 1$.
      quotient: $4k^2 + k + 7$; remainder: 0

27) $P(n) = 3n^2 + 7n + 6$

   a) Find $P(-2)$.   4

   b) Use synthetic division to find the quotient and remainder when $P(n)$ is divided by $n + 2$.
      quotient: $3n + 1$; remainder: 4

28) $P(w) = 2w^2 - 9w + 8$

   a) Find $P(3)$. -1

   b) Use synthetic division to find the quotient and remainder when $P(x)$ is divided by $x - 5$.
      quotient: $2w - 3$; remainder: -1

Objective 2: Use the Remainder Theorem

For each polynomial, $P(x)$, and the given value of c, use the remainder theorem and synthetic division to find $P(c)$. Then, use direct substitution to verify the result.

29) $P(x) = x^2 + 6x + 4$; $c = 2$ $P(2) = 20$

30) $P(x) = x^2 + x - 8$; $c = 4$ $P(4) = 12$

31) $P(x) = x^3 - 5x^2 - x + 9$; $c = -3$ $P(-3) = -60$

32) $P(x) = 2x^3 + 3x^2 - 5x + 6$; $c = -2$ $P(-2) = 12$

33) $P(x) = 2x^4 + x^2 - 3$; $c = -1$ $P(-1) = 0$

34) $P(x) = x^4 - 11x^2 + 18$; $c = 3$ $P(3) = 0$

35) $P(x) = x^4 - 3x^3 - 5$; $c = 2$ $P(2) = -13$

36) $P(x) = 3x^4 + 7x^3 + 2$; $c = -1$ $P(-1) = -2$

Use synthetic division to find each function value.

37) $f(x) = 5x^3 - 12x^2 - 7x + 10$; $f(4)$ $f(4) = 110$

38) $g(x) = 6x^3 + 5x^2 - 11x - 8$; $g(-3)$ $g(-3) = -92$

39) $P(x) = 3x^4 + 11x^3 - 14x^2 + 20x - 19; P(-5)$

$P(-5) = 31$

40) $k(x) = 2x^4 - 15x^3 + 19x^2 - 8x + 17; k(6)$ $k(6) = 5$

41) $h(x) = 2x^4 + 7x^3 + 13x - 12; h(-4)$ $h(-4) = 0$

42) $F(x) = 5x^4 - 17x^2 - 8x + 4; F(2)$ $F(2) = 0$

Extension

43) The binomial $x - c$ is a factor of the polynomial function $P(x)$ if $P(c)$ equals what number?   0

44) If $P(x)$ is a polynomial function and $P(2) = 9$, then is $x - 2$ a factor of $P(x)$? Explain your answer.

In Exercises 45–50, use synthetic division to determine if the given binomial is a factor of the given polynomial.

45) $x + 2; x^3 + 5x^2 + x - 10$ yes

46) $x - 5; x^3 - 4x^2 - 6x + 5$ yes

47) $x - 3; 2x^3 - 8x^2 + 5x + 9$ no

48) $x + 4; 3x^3 + 10x^2 - 9x + 7$ no

49) $x + 1; 4x^4 - 7x^2 + 3$ yes

50) $x - 1; 2x^4 + x^3 + 4x - 2$ no

51) Let $P(x) = x^3 - 2x^2 - 11x + 12$.

a) Use synthetic division to find $P(1)$.   0

b) Completely factor $P(x)$. $P(x) = (x - 1)(x + 3)(x - 4)$

c) Solve $P(x) = 0$. $\{-3, 1, 4\}$

52) Let $f(x) = x^3 - x^2 - 14x + 24$.

a) Use synthetic division to find $f(-4)$.   0

b) Completely factor $f(x)$. $f(x) = (x + 4)(x - 2)(x - 3)$

c) Solve $f(x) = 0$. $\{-4, 2, 3\}$

Answers to Exercises

Chapter 1

Section 1.1

1) a) $6, 0$ b) $-14, 6, 0$ c) $\sqrt{19}$ d) 6

   e) $-14, 6, \dfrac{2}{5}, 0, 3.\overline{28}, -1\dfrac{3}{7}, 0.95$

   f) $-14, 6, \dfrac{2}{5}, \sqrt{19}, 0, 3.\overline{28}, -1\dfrac{3}{7}, 0.95$

2) a) $34, -18, 0$ b) 34 c) $5.2, 34, -\dfrac{9}{4}, -18, 0, 0.\overline{7}, \dfrac{5}{6}$

   d) $34, 0$ e) $\sqrt{6}, 4.3811275\ldots$

   f) $5.2, 34, -\dfrac{9}{4}, -18, 0, 0.\overline{7}, \dfrac{5}{6}, \sqrt{6}, 4.3811275\ldots$

3) true 4) false 5) false 6) true 7) true 8) true

9)

$$-4 \quad -1\tfrac{1}{2} \quad 0 \ \tfrac{3}{4} \qquad\qquad 6$$

   $-7\ -6\ -5\ -4\ -3\ -2\ -1\ \ 0\ \ 1\ \ 2\ \ 3\ \ 4\ \ 5\ \ 6\ \ 7$

10)

$$-4\tfrac{5}{6} \ \ -3 \qquad\quad 1\ 2\tfrac{1}{4} \qquad 5\tfrac{2}{3}$$

   $-7\ -6\ -5\ -4\ -3\ -2\ -1\ \ 0\ \ 1\ \ 2\ \ 3\ \ 4\ \ 5\ \ 6\ \ 7$

11)

$$-5 \quad -2\tfrac{5}{7} \qquad\qquad 2\tfrac{3}{4}\ \ 4.3\ \ 6\tfrac{1}{8}$$

   $-7\ -6\ -5\ -4\ -3\ -2\ -1\ \ 0\ \ 1\ \ 2\ \ 3\ \ 4\ \ 5\ \ 6\ \ 7$

12)

$$-5 \ \ -2\tfrac{1}{2}\ -\tfrac{4}{5} \qquad 1.7\ 3\tfrac{1}{5}$$

   $-6\ -5\ -4\ -3\ -2\ -1\ \ 0\ \ 1\ \ 2\ \ 3\ \ 4\ \ 5\ \ 6$

13)

$$-6.8 \ \ -4\tfrac{1}{3} \qquad\quad -\tfrac{3}{8}\ 0.2\ 1\tfrac{8}{9}$$

   $-7\ -6\ -5\ -4\ -3\ -2\ -1\ \ 0\ \ 1\ \ 2\ \ 3\ \ 4\ \ 5\ \ 6\ \ 7$

14)

$$-1\tfrac{2}{3}\ 0.61\ 1\tfrac{7}{10} \qquad 5.9$$

   $-7\ -6\ -5\ -4\ -3\ -2\ -1\ \ 0\ \ 1\ \ 2\ \ 3\ \ 4\ \ 5\ \ 6\ \ 7$

15) 13 16) 8 17) $\dfrac{3}{2}$ 18) 23 19) -10 20) -6 21) -19

22) $-1\dfrac{3}{5}$ 23) -11 24) -5 25) 7 26) $\dfrac{1}{2}$ 27) 4.2

28) -2.9 29) $-10, -2, 0, \dfrac{9}{10}, 3.8, 7$

30) $-7, -6, -1, 5.2, 5.9, 6$

31) $-9, -4\dfrac{1}{2}, -0.3, \dfrac{1}{4}, \dfrac{5}{8}, 1$ 32) $-5\dfrac{2}{3}, -5, \dfrac{6}{7}, 1, 13.6, 14$

33) true 34) false 35) true 36) true 37) false

38) true 39) false 40) false 41) true 42) false

43) -27 44) 2830 45) -0.5% 46) -3.8

47) -4371 48) 18.567 49) -6333 50) -0.6

Section 1.2

1) Answers may vary. 2) Answers may vary.

3) Answers may vary. 4) Answers may vary.

5) -4 6) -12 7) -14 8) 7 9) 13 10) 27 11) 18

12) -34 13) -1451 14) 188 15) $\dfrac{1}{2}$ 16) $-\dfrac{11}{18}$

17) $-\dfrac{31}{24}$ or $-1\dfrac{7}{24}$ 18) $\dfrac{31}{56}$ 19) $-\dfrac{13}{36}$ 20) $-\dfrac{11}{12}$ 21) -3.9

22) 6.09 23) 7.31 24) -10.4 25) -5.2 26) 286.743

27) 6 28) -3 29) 22 30) -12 31) 0.4 32) $\dfrac{1}{2}$

33) true 34) false 35) false 36) false 37) true 38) true

39) $29{,}028 - (-36{,}201) = 65{,}229$. There is a 65,229-ft difference between Mt. Everest and the Mariana Trench.

40) $67{,}630{,}052 - 67{,}859{,}176 = -229{,}124$. Attendance at Major League Baseball games decreased by 229,124 from 2002 to 2003.

41) $51{,}700 - 51{,}801 = -101$. The median income for a male with a bachelor's degree decreased by $101 from 2004 to 2005.

42) $6288 - 6296 = -8$. New Orleans lies 8 ft below sea level.

43) $-79.8 + 213.8 = 134$. The highest temperature on record in the United States is 134°F.

44) $12 - 73 = -61$. The coldest temperature recorded in Colorado was -61°F.

45) $7 + 4 + 1 + 6 - 10 = 8$. The Patriots' net yardage on this offensive drive was 8 yd.

46) a) $-\$2{,}628{,}933$ b) $\$827{,}162$ c) $-\$2{,}023{,}724$

47) a) -5000 b) $10{,}000$ c) $18{,}000$ d) $-21{,}000$

48) a) -2384 b) -1426 c) 1203

49) positive 50) negative 51) -45 52) -33 53) 42

54) 496 55) 30 56) -42 57) $-\dfrac{14}{15}$ 58) $\dfrac{3}{20}$ 59) -0.3

60) 26.98 61) -64 62) 120 63) 0 64) -40

65) negative 66) positive 67) 7 68) -12 69) -8 70) 8

71) -4 72) -24 73) $\dfrac{10}{13}$ 74) $\dfrac{21}{5}$ or $4\dfrac{1}{5}$ 75) 0

76) 0 77) $-\dfrac{9}{7}$ or $-1\dfrac{2}{7}$ 78) $\dfrac{2}{5}$ 79) $\dfrac{4}{3}$ or $1\dfrac{1}{3}$

80) $-\dfrac{4}{5}$ 81) -0.05 82) -100

Section 1.3

1) 9^4 2) 2^8 3) negative   4) positive   5) positive

6) -3^4 means $-1 \cdot 3^4$. So, $-1 \cdot 3^4 = -1 \cdot 81 = -81$.
$(-3)^4$ means $(-3)(-3)(-3)(-3) = 81$.

7) 32   8) 81   9) 121   10) 64   11) 16   12) -125

13) -49 14) -36 15) -8 16) -81 17) $\dfrac{1}{125}$ 18) $\dfrac{81}{16}$

19) 0.25 or $\dfrac{1}{4}$

20) 1 raised to any natural number power equals 1.

21) False; the $\sqrt{}$ symbol means to find only the positive square root of 49.

22) true   23) true

24) False; the square root of a negative number is not a real number.

25) 8 and -8 26) 5 and -5 27) 20 and -20

28) 90 and -90 29) $\dfrac{5}{4}$ and $-\dfrac{5}{4}$ 30) $\dfrac{7}{12}$ and $-\dfrac{7}{12}$ 31) 6

32) 13   33) -1 34) -30 35) not real   36) not real

37) $\dfrac{10}{11}$ 38) $\dfrac{2}{3}$ 39) $-\dfrac{1}{8}$ 40) $-\dfrac{1}{5}$ 41) Answers may vary.

42) 33   43) 10   44) -26 45) 23   46) 26   47) $\dfrac{7}{25}$

48) $\dfrac{11}{16}$ 49) 3   50) 39   51) -27 52) 33   53) -20

54) 45   55) 8   56) 12   57) $\dfrac{5}{6}$ 58) $\dfrac{6}{5}$ 59) a) 37   b) 28

60) a) 33   b) -21 61) -9 62) -24 63) 0   64) -70

65) -1 66) 1   67) $-\dfrac{9}{5}$ 68) 2   69) 0   70) 1   71) $\dfrac{1}{6}$

72) 9   73) associative   74) commutative   75) commutative

76) inverse   77) associative   78) distributive

79) distributive   80) identity   81) identity   82) inverse

83) distributive   84) distributive   85) $7u + 7v$

86) $(12 + 3) + 7$ 87) $4 + k$ 88) $-8c - 40$

89) $-4z$ 90) $11r + 9$ 91) No. Subtraction is not commutative.

92) Yes. Addition is commutative.

93) $5 \cdot 4 + 5 \cdot 3 = 20 + 15 = 35$

94) $8 \cdot 1 + 8 \cdot 5 = 8 + 40 = 48$

95) $(-2) \cdot 5 + (-2) \cdot 7 = -10 + (-14) = -24$

96) $6 \cdot 9 + 6 \cdot (-4) = 54 + (-24) = 30$

97) $(-7) \cdot 2 + (-7) \cdot (-6) = -14 + 42 = 28$

98) $-9 + 5 = -4$ 99) $-6 - 1 = -7$

100) $8 \cdot 4 + (-2) \cdot 4 = 32 + (-8) = 24$

101) $(-10) \cdot 5 + 3 \cdot 5 = -50 + 15 = -35$

102) $2 \cdot (-6) + 2 \cdot 5 + 2 \cdot 3 = -12 + 10 + 6 = 4$

103) $9g + 9 \cdot 6 = 9g + 54$ 104) $4t + 4(-5) = 4t - 20$

105) $-5z + (-5) \cdot 3 = -5z - 15$

106) $-2m + (-2) \cdot 11 = -2m - 22$

107) $-8u + (-8) \cdot (-4) = -8u + 32$

108) $-3h + (-3) \cdot (-9) = -3h + 27$

109) $-v + 6$ 110) $-y + 13$

111) $10m + 10 \cdot 5n + 10 \cdot (-3) = 10m + 50n - 30$

112) $12 \cdot 2a + 12 \cdot (-3b) + 12c = 24a - 36b + 12c$

113) $8c - 9d + 14$ 114) $-x + 4y - 10z$

Chapter 1 Review Exercises

1) a) $0, 2$ b) 2   c) $-6, 0, 2$

d) $-6, 14.38, \dfrac{3}{11}, 2, 5.\overline{7}, 0$ e) $\sqrt{23}, 9.21743819\ldots$

2)

3) 10   4) -14 5) 75   6) $-\dfrac{31}{24}$ or $-1\dfrac{7}{24}$ 7) -5.1

8) $-17°$F   9) 70   10) -10 11) -7.77 12) -30

13) 60   14) -9 15) 2   16) $-\dfrac{19}{22}$ 17) $-\dfrac{25}{18}$ or $-1\dfrac{7}{18}$ 18) $\dfrac{2}{9}$

19) -25 20) 25   21) 81   22) -1 23) -64 24) 4   25) 7

26) -2 27) -6 28) not a real number   29) -2 30) 79

31) -40 32) $-\dfrac{1}{8}$ 33) $\dfrac{15}{28}$ 34) -2 35) $\dfrac{1}{6}$ 36) $-\dfrac{6}{19}$

37) 9   38) -16

39)

Term	Coeff.
c^4	1
$12c^3$	12
$-c^2$	-1
$-3.8c$	-3.8
11	11

40) -32 41) $-\dfrac{29}{9}$ or $-3\dfrac{2}{9}$ 42) identity   43) associative

44) inverse   45) commutative   46) distributive

47) $3 \cdot 10 - 3 \cdot 6 = 30 - 18 = 12$

48) $3 \cdot 2 + 9 \cdot 2 = 6 + 18 = 24$ 49) $-12 - 5 = -17$

50) $(-7) \cdot 2c + (-7)(-d) + (-7) \cdot 4 = -14c + 7d - 28$

Chapter 1 Test

1) a) $41, -8, 0$ b) $\sqrt{75}, 6.37528861\ldots$ c) 41

   d) $41, -8, 0, 2.\overline{83}, 6.5, 4\frac{5}{8}$ e) $41, 0$

2)

3) $\frac{3}{16}$ 4) $\frac{3}{5}$ 5) $3\frac{1}{12}$ 6) $\frac{2}{13}$ 7) $-\frac{11}{42}$ 8) -30 9) 28

10) $-\frac{9}{25}$ 11) 48 12) -6.5 13) 0 14) 1 15) 32

16) -81 17) 92 18) -10 19) -24 20) 23 21) $-\frac{5}{8}$

22) true   23) false   24) false   25) false   26) 14,787 ft

27) 0 28) a) distributive   b) inverse   c) commutative

29) a) $(-2) \cdot 5 + (-2) \cdot 3 = -10 + (-6) = -16$

   b) $5t + 5 \cdot 9u + 5 \cdot 1 = 5t + 45u + 5$

30) No. Subtraction is not commutative.

Chapter 2

Section 2.1

1) equation   2) equation   3) expression   4) expression

5) No, it is an expression.   6) Yes, it is an equation.   7) b, d

8) a, b   9) yes   10) no   11) no   12) yes   13) $\{17\}$

14) $\{-7\}$ 15) $\{-4\}$ 16) $\{7\}$ 17) $\left\{-\frac{1}{8}\right\}$ 18) $\left\{\frac{7}{12}\right\}$

19) $\{10\}$ 20) $\{8\}$ 21) $\{-48\}$ 22) $\{-60\}$ 23) $\{-15\}$

24) $\{24\}$ 25) $\{18\}$ 26) $\left\{\frac{15}{8}\right\}$ 27) $\{12\}$ 28) $\{10\}$

29) $\{8\}$ 30) $\{-3\}$ 31) $\{0\}$ 32) $\left\{\frac{12}{7}\right\}$ 33) $\{27\}$

34) $\{21\}$ 35) $\left\{-\frac{7}{5}\right\}$ 36) $\left\{\frac{50}{3}\right\}$ 37) $\{6\}$ 38) $\{8\}$

39) $\{-3\}$ 40) $\{-13\}$ 41) $\left\{-\frac{5}{4}\right\}$ 42) $\left\{\frac{4}{3}\right\}$ 43) $\{0\}$

44) $\{-5\}$ 45) $\{3\}$ 46) $\{4\}$ 47) $\{-3\}$ 48) $\{-7\}$

49) $\left\{\frac{7}{3}\right\}$ 50) $\left\{-\frac{2}{3}\right\}$ 51) $\{4\}$ 52) $\left\{-\frac{3}{4}\right\}$ 53) $\varnothing$

54) $\varnothing$ 55) {all real numbers}   56) {all real numbers}

57) {all real numbers}   58) $\varnothing$ 59) $\{0\}$ 60) $\{-15\}$

61) $\left\{\frac{11}{2}\right\}$ 62) Multiply the equation by 12.   63) $\{5\}$

64) $\{-1\}$ 65) $\{3\}$ 66) $\{5\}$ 67) $\left\{-\frac{15}{2}\right\}$ 68) $\{-2\}$

69) $\{-8\}$ 70) $\left\{\frac{17}{7}\right\}$ 71) $\{5\}$ 72) $\{4\}$ 73) $\{4\}$

74) $\{24\}$ 75) $\{300\}$ 76) $\{800\}$ 77) $x + 4 = 15; 11$

78) $x + 13 = 8; -5$ 79) $x - 7 = 22; 29$ 80) $x - 9 = 11; 20$

81) $2x = -16; -8$ 82) $6x = 54; 9$ 83) $2x + 7 = 35; 14$

84) $2x + 5 = 53; 24$ 85) $3x - 8 = 40; 16$

86) $2x - 7 = -13; -3$ 87) $\frac{1}{2}x + 10 = 3; -14$

88) $\frac{1}{3}x + 4 = 1; -9$ 89) $2x - 3 = x + 8; 11$

90) $5x - 12 = x + 16; 7$ 91) $\frac{1}{3}x + 10 = x - 2; 18$

92) $x - 9 = \frac{1}{2}x + 7; 32$ 93) $2(x + 5) = 16; 3$

94) $2[x + (-8)] = 4; 10$ 95) $3x = 15 + \frac{1}{2}x; 6$

96) $x + 4 = 2x - 9; 23$ 97) $x - 6 = 5 + 2x; -11$

98) $\frac{x}{4} = x - 9; 12$

Section 2.2

1) $c + 5$ 2) $h + 23$ 3) $p - 31$ 4) $a - 3.8$ 5) $3w$

6) $\frac{1}{3}s$ 7) $14 - x$ 8) $142 - m$ mi

9) Pepsi = 9.8 tsp, Gatorade = 3.3 tsp

10) MI = 70 mi^2, Lich = 62 mi^2

11) Greece: 16, Thailand: 8   12) Golden = 64 lb, BC = 32 lb

13) Columbia: 1240 mi, Ohio: 1310 mi

14) Siegelman: 669,105; Riley: 672,225

15) 11 in. and 25 in.   16) 51 in. and 24 in.   17) 12 ft and 6 ft

18) 33 ft and 22 ft   19) 64, 65, 66   20) 38, 39   21) 12, 14

22) 31, 33, 35   23) $-15, -13, -11$ 24) 48, 50

25) 172, 173   26) 17 Arthur Ave. and 19 Arthur Ave.

27) $63.75   28) $55.20   29) $11.55   30) $102.00

31) $11.60   32) $299.25   33) $140.00   34) $17.80

35) $10.95   36) $499.00   37) $32.50   38) $24.00   39) 80

40) 1800 acres   41) 425   42) $32,500   43) $32   44) $770

45) $6890   46) $3165   47) $380   48) $522

49) $9000 at 6% and $6000 at 7%

50) $3500 at 4% and $4500 at 5%

51) $1400 at 6% and $1600 at 5%

52) $4000 at 7.4% and $800 at 9%

53) $2800 at 9.5% and $4200 at 7%

54) $1300 at 5% and $1700 at 6.5%   55) ride: 4 mi; walk: 3 mi

56) 500,000   57) $6000 at 5% and $14,000 at 9%   58) 186, 187

59) 2500   60) 2002: 9 matches; 2003: 17 matches

61) 12 in., 17 in., and 24 in.   62) IRA: $2500, CD: $6000

63) 72, 74, 76   64) 13 ft, 19 ft, and 26 ft

65) *Shrek*: $267.7 million; *Harry Potter*: $309.7 million

66) $22.50   67) CD: $1500; IRA: $3000; Mutual Fund: $2500

68) Taylor Swift: 3.2 mil; Susan Boyle: 3.1 mil;
Michael Jackson: 2.3 mil

69) $38,600   70) $79.00

Section 2.3

1) 25 ft   2) 12 in.   3) 6 in.   4) 15 in.   5) 415 ft^2

6) 113 ft^2 7) 12 ft   8) a) 90 ft   b) 440 ft   9) 6 in.

10) 8 in.   11) $m\angle A = 26°, m\angle B = 52°$

12) $m\angle B = 43°, m\angle C = 78°$

13) $m\angle A = 37°, m\angle B = 55°, m\angle C = 88°$

14) $m\angle A = 107°, m\angle B = 39°, m\angle C = 34°$

15) 68°, 68°   16) 123°, 123°   17) 150°, 150°   18) 51°, 51°

19) 133°, 47°   20) 162°, 18°   21) 79°, 101°   22) 65°, 115°

23) $180 - x$ 24) $90 - x$ 25) 17°   26) 64°

27) angle: 20°, comp.: 70°, supp.: 160°

28) angle: 12°, comp.: 78°, supp.: 168°

29) 35°   30) 41°   31) 40°   32) 73°   33) 2   34) 0.08

35) 4   36) 20   37) 18   38) 12   39) 2   40) 6   41) 7

42) 8   43) 20   44) 13

45) a) $x = 23$ b) $x = p - n$ c) $x = v - q$

46) a) $t = 22$ b) $t = m + w$ c) $t = j + v$

47) a) $n = 6$ b) $n = \dfrac{c}{y}$ c) $n = \dfrac{d}{w}$

48) a) $y = 9$ b) $y = \dfrac{x}{a}$ c) $y = \dfrac{r}{p}$

49) a) $c = 21$ b) $c = ur$ c) $c = xt$

50) a) $m = 16$ b) $m = zp$ c) $m = qf$

51) a) $d = 3$ b) $d = \dfrac{z + a}{k}$

52) a) $g = -7$ b) $g = \dfrac{\pi - k}{c}$

53) a) $z = -\dfrac{5}{2}$ b) $z = \dfrac{w - t}{y}$

54) a) $p = \dfrac{11}{5}$ b) $p = \dfrac{d + r}{n}$

55) $m = \dfrac{F}{a}$ 56) $r = \dfrac{C}{2\pi}$ 57) $c = nv$ 58) $R = 2f$

59) $\sigma = \dfrac{E}{T^4}$ 60) $\rho = \dfrac{p}{gy}$ 61) $h = \dfrac{3V}{\pi r^2}$ 62) $L = \dfrac{2U}{I^2}$

63) $E = IR$ 64) $b = \dfrac{2A}{h}$ 65) $R = \dfrac{I}{PT}$ 66) $P = \dfrac{I}{RT}$

67) $I = \dfrac{P - 2w}{2}$ or $I = \dfrac{P}{2} - w$

68) $T = \dfrac{A - P}{PR}$ or $T = \dfrac{A}{PR} - \dfrac{1}{R}$ 69) $N = \dfrac{2.5H}{D^2}$ 70) $A = \dfrac{3V}{H}$

71) $b_2 = \dfrac{2A}{h} - b_1$ or $b_2 = \dfrac{2A - hb_1}{h}$

72) $r^2 = R^2 - \dfrac{A}{\pi}$ or $r^2 = \dfrac{\pi R^2 - A}{\pi}$

73) $h^2 = \dfrac{S}{\pi} - \dfrac{c^2}{4}$ or $h^2 = \dfrac{1}{4}\left(\dfrac{4S}{\pi} - c^2\right)$

74) $c^2 = \dfrac{4S}{\pi} - 4h^2$ or $c^2 = \dfrac{4S - 4\pi h^2}{\pi}$

75) a) $w = \dfrac{P - 2l}{2}$ b) 3 cm   76) a) $h = \dfrac{2A}{b}$ b) 5 in.

77) a) $F = \dfrac{9}{5}C + 32$ b) 77°F   78) 23°F

Section 2.4

1) $\dfrac{3}{5}$ 2) $\dfrac{2}{3}$ 3) $\dfrac{4}{3}$ 4) $\dfrac{19}{16}$

5) A ratio is a quotient of two quantities. A proportion is a statement that two ratios are equal.

6) Yes, the ratio "9 to 20" can be written as $\dfrac{9}{20}$. This reduces to 0.45.

7) false   8) true   9) true   10) false   11) {2}   12) {4}

13) $\left\{\dfrac{48}{5}\right\}$ 14) $\left\{\dfrac{8}{3}\right\}$ 15) {-1}   16) {3}   17) {-2}

18) $\left\{\dfrac{1}{2}\right\}$ 19) 30   20) $7.10   21) 82.5 mg   22) 34.5 mg

23) 168   24) 102 krone   25) $x = 10$ 26) $x = 48$

27) $x = 13$ 28) $x = 12$ 29) $x = 63$ 30) $x - \dfrac{39}{2}$

31) a) $0.80   b) 80¢   32) a) $1.60   b) 160¢

33) a) $2.17   b) 217¢   34) a) $3.00   b) 300¢

35) a) $2.95   b) 295¢   36) a) $1.59   b) 159¢

37) a) $0.25q$ b) $25q$ 38) a) $0.01p$ b) p

39) a) $0.10d$ b) $10d$ 40) a) $0.05n$ b) $5n$

41) a) $0.01p + 0.25q$ b) $p + 25q$

42) a) $0.05n + 0.10d$ b) $5n + 10d$

43) 9 nickels, 17 quarters   44) 78 pennies and 53 nickels

45) 11 $5 bills, 14 $1 bills   46) 22 $20 bills, 11 $10 bills

47) 38 adult tickets, 19 children's tickets

48) 20 44¢ stamps, 8 28¢ stamps

49) 2 oz   50) 1.8 mL   51) 7.6 mL   52) 5.2 oz

53) 16 oz of the 4% acid solution, 8 oz of the 10% acid solution

54) $106\frac{2}{3}$ mL   55) $2\frac{1}{2}$ L

56) $133\frac{1}{3}$ mL of the 7% solution, $266\frac{2}{3}$ mL of the 1% solution

57) 2 lb of Aztec and 3 lb of Cinnamon   58) 2 lb

59) eastbound: 65 mph; westbound: 73 mph

60) northbound: 200 mph; southbound: 250 mph

61) $\frac{5}{6}$ hr   62) $1\frac{2}{3}$ hr

63) passenger train: 50 mph; freight train: 30 mph

64) 11:30 A.M.   65) 36 min   66) 12:45 P.M.   67) 4:30 P.M.

68) $2\frac{1}{2}$ hr   69) 48 mph   70) 9:30 A.M.

71) 23 dimes, 16 quarters   72) $466\frac{2}{3}$ g   73) 1560 calories

74) 164   75) jet: 400 mph, small plane: 200 mph

76) 114 adult tickets, 31 children's tickets

77) 8 cc of the 0.08% solution and 12 cc of the 0.03% solution

78) $\frac{1}{4}$ hr or 15 min   79) $1\frac{1}{5}$ gal   80) 3315 yen

Section 2.5

1) You use parentheses when there is a $<$ or $>$ symbol or when you use ∞ or $-\infty$.

2) You use brackets when there is a $\leq$ or $\geq$ symbol.

3)
a) $\{x|x \geq 3\}$ b) $[3, \infty)$

4)
a) $\{t|t \geq -4\}$ b) $[-4, \infty)$

5)
a) $\{c|c < -1\}$ b) $(-\infty, -1)$

6)
a) $\left\{r|r < \frac{5}{2}\right\}$ b) $\left(-\infty, \frac{5}{2}\right)$

7)
a) $\left\{w \middle| w > -\frac{11}{3}\right\}$ b) $\left(-\frac{11}{3}, \infty\right)$

8)
a) $\{p|p \leq 2\}$ b) $(-\infty, 2]$

9)
a) $\{n|1 \leq n \leq 4\}$ b) $[1, 4]$

10)
a) $\{g|-3 \leq g \leq 2\}$ b) $[-3, 2]$

11)
a) $\{a|-2 < a < 1\}$ b) $(-2, 1)$

12)
a) $\{d|-4 < d < 0\}$ b) $(-4, 0)$

13)
a) $\left\{z \middle| \frac{1}{2} < z \leq 3\right\}$ b) $\left(\frac{1}{2}, 3\right]$

14)
a) $\{y|-2 \leq y < 3\}$ b) $[-2, 3)$

15)
a) $\{n|n \leq 5\}$ b) $(-\infty, 5]$

16)
a) $\{p|p \geq -2\}$ b) $[-2, \infty)$

17)
a) $\{y|y \geq -4\}$ b) $[-4, \infty)$

18)
a) $\{r|r \leq 4\}$ b) $(-\infty, 4]$

19)
a) $\{c|c > 4\}$ b) $(4, \infty)$

20)
a) $\{v|v > 3\}$ b) $(3, \infty)$

21)
a) $\left\{k \middle| k < -\frac{11}{3}\right\}$ b) $\left(-\infty, -\frac{11}{3}\right)$

22)
a) $\left\{m \middle| m < -\frac{7}{4}\right\}$ b) $\left(-\infty, -\frac{7}{4}\right)$

23)
a) $\{b|b \geq -8\}$ b) $[-8, \infty)$

24)

-5 -4 -3 -2 -1 0 1 2

a) $\{a|a \le -3\}$ b) $(-\infty, -3]$

25)

-3 -2 -1 0 1 2 3 4

a) $\{w|w < 3\}$ b) $(-\infty, 3)$

26)

-5 -4 -3 -2 -1 0 1 2 3 4 5

a) $\left\{t\middle|t > \dfrac{3}{5}\right\}$ b) $\left(\dfrac{3}{5}, \infty\right)$

27)

-7 -6 -5 -4 -3 -2 -1 0

a) $\{x|x < -6\}$ b) $(-\infty, -6)$

28)

-16   -14   -12   -10

a) $\{z|z \ge -15\}$ b) $[-15, \infty)$

29)

-12 -11 -10 -9 -8 -7 -6

a) $\{p|p \le -10\}$ b) $(-\infty, -10]$

30)

3 4 5 6 7 8 9 10

a) $\{y|y > 8\}$ b) $(8, \infty)$

31)

-3 -2 -1 0 1 2 3

$(-1, \infty)$

32)

-3 -2 -1 0 1 2 3 4 5

$(-\infty, 4]$

33)

-3 -2 -1 0 1 2 3

$\left(-\infty, -\dfrac{3}{7}\right]$

34)

-3 -2 -1 0 1 2 3

$\left(-\dfrac{3}{2}, \infty\right)$

35)

-4 -3 -2 -1 0 1 2 3 4

$(-3, \infty)$

36)

0 1 2 3 4 5 6 7 8

$(-\infty, 7]$

37)

-4 -3 -2 -1 0 1 2

$(-\infty, -2]$

38)

-5 -4 -3 -2 -1 0 1 2 3 4 5

$\left(\dfrac{6}{11}, \infty\right)$

39)

-2 -1 0 1 2 3 4 5

$(-\infty, 4)$

40)

-3 -2 -1 0 1 2 3

$[-1, \infty)$

41)

-3 -2 -1 0 1 2 3

$(0, \infty)$

42)

10 11 12 13 14 15

$\left[\dfrac{27}{2}, \infty\right)$

43)

-1 0 1 2 3 4 5 6

$(-\infty, 5]$

44)

0 1 2 3 4 5 6 7 8 9

$(-\infty, 8]$

45)

-4 -3 -2 -1 0 1 2 3 4

$[-3, 1]$

46)

-5 -4 -3 -2 -1 0 1 2 3 4 5

$[-2, 4]$

47)

-3 -2 -1 0 1 2 3 4

$\left(\dfrac{3}{2}, 3\right)$

48)

-6 -5 -4 -3 -2 -1 0 1 2 3 4 5

$\left(-5, \dfrac{7}{2}\right)$

49)

-6 -5 -4 -3 -2 -1 0 1 2

$[-4, -1]$

50)

0 1 2 3 4 5 6 7

$[2, 5]$

51)

0 1 2 3 4 5 6 7

$\left[\dfrac{7}{4}, 3\right)$

52)

-5 -4 -3 -2 -1 0 1 2 3 4 5

$\left(-2, -\dfrac{2}{3}\right]$

53)

-10 -8 -6 -4 -2 0 2 4 6 8

$(-8, 4)$

54)

0 1 2 3 4 5 6

$[2, 4]$

55)

-5 -4 -3 -2 -1 0 1 2 3 4 5

$\left[-1, -\dfrac{2}{5}\right]$

56)

-4 -3 -2 -1 0 1 2 3 4 5

$\left(-\dfrac{1}{2}, 4\right)$

57)

(1, 3]

58)

[5, 8)

59)

$\left[-1, \dfrac{4}{3}\right]$

60)

(−5, 1)

61) (−7, ∞)   62) (−4, 3)   63) $\left(-\infty, \dfrac{4}{3}\right]$ 64) [3, ∞)

65) (−∞, −12)   66) (−6, ∞)   67) $\left(-15, -\dfrac{15}{4}\right]$

68) (−∞, 6]   69) [−9, ∞)   70) $\left[-\dfrac{1}{3}, 1\right]$

71) [−2, 0)   72) (−∞, −2]

73) He can rent the truck for at most 2 hr 45 min.

74) at most $5\dfrac{1}{2}$ hr   75) at most 8 mi   76) at most $12\dfrac{1}{2}$ mi

77) 89 or higher   78) 81 or higher

Section 2.6

1) $A \cap B$ means "A intersect B." $A \cap B$ is the set of all numbers which are in set A and in set B.

2) $X \cup Y$ means "X union Y." $X \cup Y$ is the set of all numbers that are in set X or set Y or in both.

3) {8}   4) {8, 10}   5) {2, 4, 5, 6, 7, 8, 9, 10}

6) {1, 3, 5, 6, 7, 8, 9}   7) ∅   8) ∅

9) {1, 2, 3, 4, 5, 6, 8, 10}   10) {5, 6, 7, 8, 9, 10, 12, 14}

11)

[−3, 2]

12)

[0, 4]

13)

(−1, 3)

14)

(−4, −2)

15)

[3, ∞)

16)

(−∞, −1)

17)

∅

18)

∅

19)

[2, 5]

20)

(−3, 4)

21)

(−2, 3)

22)

$\left[-7, \dfrac{1}{3}\right]$

23)

(−3, 4]

24)

(1, 6]

25)

∅

26)

∅

27)

(3, ∞)

28)

(−∞, 1]

29)

[−4, 1]

30)

[1, 5]

31)

(−∞, −1) ∪ (5, ∞)

32)

(−∞, 2) ∪ (6, ∞)

33)

$\left(-\infty, \dfrac{5}{3}\right] \cup (4, \infty)$

34)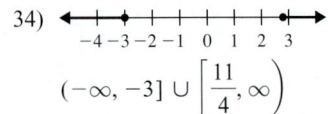

$(-\infty, -3] \cup \left[\dfrac{11}{4}, \infty\right)$

35)

$(1, \infty)$

36)

$(-\infty, -2]$

37)

$(-\infty, \infty)$

38)

$(-\infty, \infty)$

39)

$(-\infty, -1) \cup (3, \infty)$

40)

$(-\infty, -4] \cup [0, \infty)$

41)

$\left(-\infty, \dfrac{7}{2}\right] \cup (6, \infty)$

42)

$\left(-\infty, -\dfrac{11}{2}\right] \cup (-2, \infty)$

43)

$(-5, \infty)$

44)

$(-\infty, 1]$

45)

$(-\infty, -6) \cup [-3, \infty)$

46)

$(-\infty, 7) \cup [9, \infty)$

47)

$(-\infty, \infty)$

48)

$(-\infty, \infty)$

49)

$(-\infty, -2]$

50)

$(-3, \infty)$

51) $\left[-5, \dfrac{1}{2}\right]$ 52) $(-\infty, -15] \cup [2, \infty)$

53) $\left(-\infty, -\dfrac{9}{4}\right) \cup [5, \infty)$ 54) $\varnothing$ 55) $(-\infty, \infty)$

56) $(2, 12)$ 57) $(-\infty, 0)$ 58) $(-\infty, \infty)$ 59) $[-8, -4)$

60) $(-\infty, 1] \cup (10, \infty)$

61) {Liliane Bettancourt, Alice Walton}

62) {Abigail Johnson, Alice Walton}

63) {Liliane Bettancourt, J.K. Rowling, Oprah Winfrey}

64) {Liliane Bettancourt, Abigail Johnson, Alice Walton, Oprah Winfrey}

Section 2.7

1) Answers may vary.

2) No; the absolute value of a quantity cannot be negative.

3) $\{-6, 6\}$ 4) $\{-7, 7\}$ 5) $\{2, 8\}$ 6) $\{-15, 11\}$

7) $\left\{-\dfrac{1}{2}, 3\right\}$ 8) $\left\{-\dfrac{2}{9}, 2\right\}$ 9) $\left\{-\dfrac{1}{2}, -\dfrac{1}{3}\right\}$

10) $\left\{-\dfrac{7}{5}, 3\right\}$ 11) $\{-24, 15\}$ 12) $\left\{-\dfrac{4}{3}, -20\right\}$

13) $\left\{-\dfrac{10}{3}, \dfrac{50}{3}\right\}$ 14) $\left\{-\dfrac{23}{6}, -\dfrac{17}{6}\right\}$ 15) $\varnothing$ 16) $\varnothing$

17) $\{-10, 22\}$ 18) $\{-18, 12\}$ 19) $\{-5, 0\}$ 20) $\left\{1, \dfrac{8}{3}\right\}$

21) $\{-14\}$ 22) $\varnothing$ 23) $\varnothing$ 24) $\left\{\dfrac{3}{4}\right\}$ 25) $\left\{-\dfrac{16}{5}, 2\right\}$

26) $\left\{\dfrac{1}{4}, \dfrac{3}{2}\right\}$ 27) $\varnothing$ 28) $\varnothing$ 29) $\left\{-\dfrac{14}{3}, 4\right\}$

30) $\left\{-\dfrac{1}{3}, 3\right\}$ 31) $\left\{\dfrac{1}{2}, 4\right\}$ 32) $\left\{-\dfrac{1}{2}, -\dfrac{1}{6}\right\}$ 33) $\left\{\dfrac{2}{5}, 2\right\}$

34) $\left\{-\dfrac{84}{11}, 28\right\}$ 35) $\{10\}$ 36) $\left\{-\dfrac{2}{5}, 1\right\}$

37) $|x| = 9$, may vary   38) $|y| = 6$, may vary

39) $|x| = \dfrac{1}{2}$, may vary   40) $|x| = 1.4$, may vary

41) $[-1, 5]$

42) $(7, 11)$

43) $(-\infty, 2) \cup (9, \infty)$

44) $(-\infty, -8] \cup \left[\dfrac{1}{2}, \infty\right)$

45) $\left(-\infty, -\dfrac{9}{2}\right] \cup \left[\dfrac{3}{5}, \infty\right)$

46) $\left[-\dfrac{1}{4}, \dfrac{11}{4}\right]$

47) $[-7, 7]$

48) $(-1, 1)$

49) $(-4, 4)$

50) $[-24, 24]$

51) $(-2, 6)$

52) $[4, 8]$

53) $\left[-\dfrac{14}{3}, -2\right]$

54) $\left[-\dfrac{13}{4}, \dfrac{11}{4}\right]$

55) $\left[\dfrac{2}{3}, \dfrac{5}{3}\right]$

56) $\left(1, \dfrac{25}{9}\right)$

57) $\varnothing$

58) $\varnothing$

59) $\varnothing$

60) $\varnothing$

61) $\left(-\dfrac{1}{4}, 1\right)$

62) $(-8, 3)$

63) $[-12, 4]$

64) $\left(\dfrac{3}{2}, \dfrac{21}{4}\right)$

65) $(-\infty, -7] \cup [7, \infty)$

66) $(-\infty, -3) \cup (3, \infty)$

67) $(-\infty, -14] \cup [-6, \infty)$

68) $(-\infty, -5) \cup (19, \infty)$

69) $\left(-\infty, -\dfrac{3}{2}\right] \cup [3, \infty)$

70) $(-\infty, -5) \cup \left(-\dfrac{4}{3}, \infty\right)$

71) $(-\infty, 2) \cup \left(\dfrac{11}{3}, \infty\right)$

72) $\left(-\infty, -\dfrac{9}{4}\right] \cup \left[\dfrac{11}{4}, \infty\right)$

73) $(-\infty, \infty)$

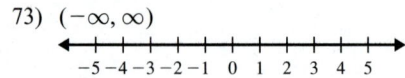

74) $(-\infty, \infty)$

75) $(-\infty, \infty)$

76) $(-\infty, \infty)$

77) $(-\infty, 0) \cup (1, \infty)$

78) $(-\infty, -12] \cup [0, \infty)$

79) $\left(-\infty, -\dfrac{27}{5}\right] \cup \left[\dfrac{21}{5}, \infty\right)$

80) $\left(-\infty, -\dfrac{1}{2}\right] \cup \left[\dfrac{13}{6}, \infty\right)$

81) The absolute value of a quantity is always 0 or positive; it cannot be less than 0.

82) The absolute value of a quantity is always 0 or positive; it cannot be less than or equal to a negative number.

83) The absolute value of a quantity is always 0 or positive, so for any real number, x, the quantity $|2x + 1|$ will be greater than -3.

84) The absolute value of a quantity is always 0 or positive, so for any real number, y, the quantity $|7y - 3|$ will be greater than 0.

85) $(-\infty, -6) \cup (-3, \infty)$

86) $\left\{-6, \dfrac{18}{5}\right\}$ 87) $\left\{-2, -\dfrac{1}{2}\right\}$ 88) $\left[-\dfrac{7}{2}, -1\right]$

89) $\left(-\infty, -\dfrac{1}{4}\right] \cup [2, \infty)$ 90) $\{-9, 14\}$ 91) $(-3, \infty)$

92) $(-\infty, -6)$ 93) $\varnothing$ 94) $(-\infty, -2] \cup \left[\dfrac{6}{5}, \infty\right)$

95) $\{-21, -3\}$ 96) $\left\{\dfrac{8}{7}\right\}$ 97) $\varnothing$ 98) $\left(\dfrac{3}{5}, \dfrac{19}{5}\right)$

99) $\left(-\infty, -\dfrac{1}{25}\right]$ 100) $(2, \infty)$ 101) $(-\infty, \infty)$

102) $(-\infty, 2] \cup [8, \infty)$ 103) $[-15, -1]$ 104) $(-\infty, \infty)$

105) $\left(-\infty, \dfrac{1}{5}\right) \cup (3, \infty)$ 106) $\left\{-6, -\dfrac{18}{13}\right\}$

107) $|a - 128| \le 0.75$; $127.25 \le a \le 128.75$; there is between 127.25 oz and 128.75 oz of milk in the container.

108) $|c - 27| \le 0.5$; $26.5 \le c \le 27.5$; there is between 26.5 oz and 27.5 oz of cereal in the box.

109) $|b - 38| \le 5$; $33 \le b \le 43$; he will spend between \$33 and \$43 on his daughter's gift.

110) $|w - 32| \le \dfrac{1}{16}$; $31\dfrac{15}{16} \le w \le 32\dfrac{1}{16}$; the window shade is between $31\dfrac{15}{16}$ in. and $32\dfrac{1}{16}$ in. wide.

Chapter 2 Review Exercises

1) yes   2) no   3) $\left\{-\dfrac{10}{3}\right\}$ 4) $\{-3\}$ 5) $\{19\}$ 6) $\{35\}$

7) $\left\{\dfrac{45}{14}\right\}$ 8) $\{-4\}$ 9) $\{35\}$ 10) $\{8\}$ 11) $\{-2\}$ 12) $\{4\}$

13) $\{2\}$ 14) $\{16\}$ 15) $\{-2\}$ 16) $\left\{\dfrac{8}{3}\right\}$

17) {all real numbers}   18) $\varnothing$ 19) $\{3\}$ 20) $\{-2\}$

21) 17   22) 10   23) $26 - c$ 24) $f + 14$

25) Kelly Clarkson: 297,000 copies; Clay Aiken: 613,000 copies

26) 84, 86, 88   27) 125 ft   28) \$6724   29) \$10.1 billion

30) \$4500 at 7% and \$7500 at 8%   31) 12 cm

32) $m\angle A = 29°$, $m\angle B = 23°$, $m\angle C = 128°$ 33) $79°, 101°$

34) $46°, 46°$ 35) $48°$ 36) $m = \dfrac{y - b}{x}$ 37) $R = \dfrac{pV}{nT}$

38) $t = \dfrac{3C}{n} - T$ or $t = \dfrac{3C - Tn}{n}$ 39) $\{20\}$

40) $\{-2\}$ 41) 442   42) $x = 33$

43) 58 dimes, 33 quarters   44) 24 L   45) 45 min

46)

$[8, \infty)$

47)

$\left(-\infty, -\dfrac{5}{2}\right)$

48)

$(-\infty, 3)$

49)

$(-2, 3]$

50)

$\left[\dfrac{5}{3}, \dfrac{7}{3}\right]$

51) 88 or higher

52) a) {10, 20, 25, 30, 35, 40, 50}   b) {20, 30}

53)

[1, 3]

54)

$(-\infty, -4] \cup (2, \infty)$

55)

$(-1, \infty)$

56)

∅

57) {Toyota}   58) {Toyota, Ford, Nissan}   59) {−9, 9}

60) {−10, 10}   61) $\left\{-1, \dfrac{1}{7}\right\}$ 62) $\left\{-\dfrac{8}{3}, \dfrac{16}{3}\right\}$

63) $\left\{-\dfrac{15}{8}, -\dfrac{7}{8}\right\}$ 64) $\left\{-3, \dfrac{9}{5}\right\}$ 65) $\left\{\dfrac{11}{5}, \dfrac{13}{5}\right\}$

66) $\left\{-\dfrac{51}{4}, -\dfrac{21}{4}\right\}$ 67) $\left\{-8, \dfrac{4}{15}\right\}$ 68) $\left\{1, \dfrac{5}{4}\right\}$

69) ∅   70) ∅   71) $\left\{-\dfrac{4}{9}\right\}$ 72) $\left\{\dfrac{7}{6}\right\}$

73) $|a| = 4$, may vary   74) $|t| = 7$, may vary

75) [−3, 3]

76) (−12, 10)

77) $(-\infty, -2) \cup (2, \infty)$

78) $(-\infty, -4] \cup [11, \infty)$

79) $(-\infty, -1] \cup \left[\dfrac{1}{6}, \infty\right)$

80) $\left(\dfrac{7}{3}, 5\right)$

81) (−5, 13)

82) $\left(-\infty, -\dfrac{4}{5}\right) \cup \left(\dfrac{8}{5}, \infty\right)$

83) $\left[-\dfrac{15}{4}, -\dfrac{3}{4}\right]$

84) $\left(-\infty, -\dfrac{10}{3}\right] \cup [0, \infty)$

85) $\left(-\infty, -\dfrac{19}{5}\right] \cup [-1, \infty)$

86) [−4, 16]

87) $(-\infty, \infty)$

88) ∅

89) $\left\{-\dfrac{1}{12}\right\}$

90) $|s - 93| \le 1$; $92 \le s \le 94$; the speed of the pitch is between 92 mph and 94 mph.

Chapter 2 Test

1) {2}   2) −3   3) {7}   4) ∅   5) $\left\{\dfrac{39}{2}\right\}$ 6) {15}   7) 21

8) $2\dfrac{1}{4}$ tsp   9) 3.5 qt of regular oil, 1.5 qt of synthetic oil

10) 50   11) 70 ft   12) $t = \dfrac{5R}{k}$ 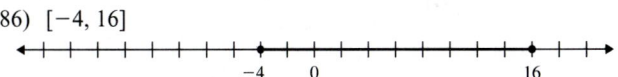13) $h = \dfrac{S - 2\pi r^2}{2\pi r}$

14)

$(-\infty, -5]$

15)

$\left(\dfrac{2}{3}, \infty\right)$

16)

+--+--o--+--+--+--+--+--●--+--+--→
0 1 2 3 4 5 6 7 8 9

(1, 7]

17) at most 6 hr   18) a) {1, 2, 3, 6, 9, 12}   b) {1, 2, 12}

19) $(-\infty, -8) \cup \left(\frac{7}{3}, \infty\right)$ 20) [0, 3]   21) $(-\infty, \infty)$

22) $\left\{-\frac{1}{2}, 5\right\}$ 23) {-16, 4}   24) $\left\{-8, \frac{3}{2}\right\}$

25) ∅   26) |x| = 8, may vary

27) [-1, 8]

+--+--+--+--+--+--+--+--+--●--+--+--+--+--+--+--+--+--●--+--+--→
                          -1 0   8

28) $\left(-\infty, -\frac{11}{2}\right] \cup [1, \infty)$

$-\frac{11}{2}$

●--+--+--+--+--+--+--●--+--+--+--→
-5 -4 -3 -2 -1 0 1 2 3 4 5

29) ∅

30) |w − 168| ≤ 0.75; 167.25 ≤ w ≤ 168.75; Thanh's weight is between 167.25 lb and 168.75 lb.

Cumulative Review for Chapters 1–2

1) $-\frac{13}{36}$ 2) $\frac{2}{45}$ 3) 64   4) 56   5) −81

6) $\left\{-13.7, \frac{19}{7}, 0, 8, 0.\overline{61}, \sqrt{81}, -2\right\}$ 7) {0, 8, $\sqrt{81}$, −2}

8) Associative Property   9) Distributive Property

10) 5 + 8   11) {−70}   12) $\left\{\frac{19}{4}\right\}$ 13) {all real numbers}

14) $\left\{0, \frac{9}{2}\right\}$ 15) $R = \frac{A - P}{PT}$ 16) 121°

17) Area = 105 cm²; Perimeter = 44 cm   18) 123°; obtuse

19) Generic: 48; Name brand: 24   20) $\frac{1}{2}$ hr   21) $\left[-\frac{2}{3}, \infty\right)$

22) $\left(-4, -\frac{7}{6}\right)$ 23) $(-\infty, -3] \cup \left[\frac{11}{4}, \infty\right)$

24) [-36, 8]   25) $(-\infty, -1) \cup (7, \infty)$

Chapter 3

Section 3.1

1) a) x represents the season and y represents the number of people who watched the finale of that season.
b) 28.8 million people watched the Season 3 finale.
c) 36.4 million   d) Season 6   e) (1, 22.8)

2) a) x represents the year and y represents the value (in billions of dollars) of U.S. exports to Mexico.

b) In 2004, the value of U.S. exports to Mexico was $111 billion.
c) 2007   d) about $120 billion   e) (2003, 97)

3) A: (5, 1); quadrant I
B: (2, −3); quadrant IV
C: (−2, 4); quadrant II
D: (−3, −4); quadrant III
E: (3, 0); no quadrant
F: (0, −2); no quadrant

4) A: (−3, 1); quadrant II
B: (0, 4); no quadrant
C: (−4, −5); quadrant III
D: (−1, 0); no quadrant
E: (4, −2); quadrant IV
F: (2, 2); quadrant I

5–8)

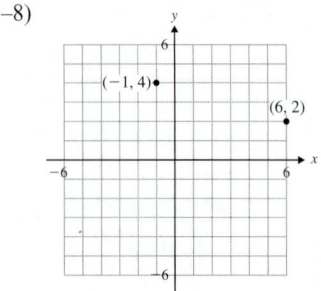

9)

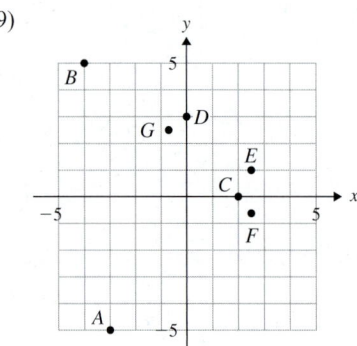

10)

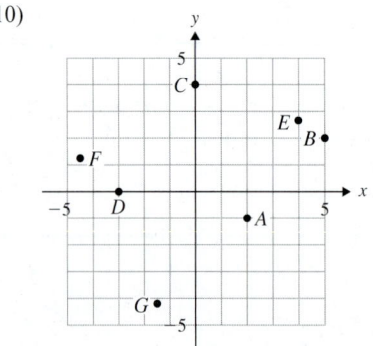

11) positive   12) negative   13) negative   14) positive

15) zero   16) zero   17) yes   18) no   19) yes   20) yes

21) no   22) no   23) yes   24) yes   25) no   26) yes

27) Answers may vary.   28) No. It contains an x² term.

29) A line: The line represents all solutions to the equation. Every point on the line is a solution of the equation.

30) an infinite number

31) It is the point where the graph intersects the y-axis. To find the y-intercept, let x = 0 and solve for y.

32) It is the point where the graph intersects the x-axis. To find the x-intercept, let y = 0 and solve for x.

33)

x	y
0	−1
1	2
2	5
−1	−4

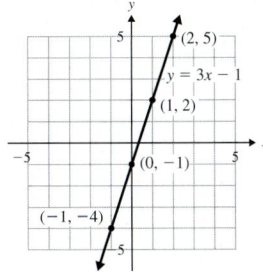

34)

x	y
0	5
−1	7
2	1
3	−1

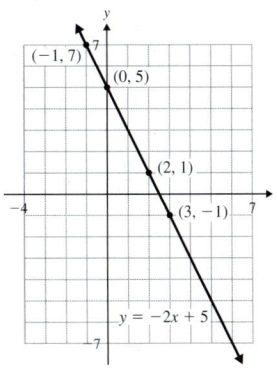

35)

x	y
0	4
−3	6
3	2
6	0

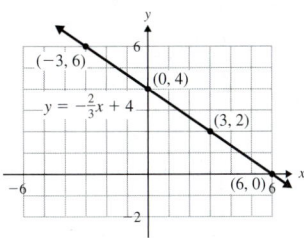

36)

x	y
0	6
2	11
−2	1
−4	−4

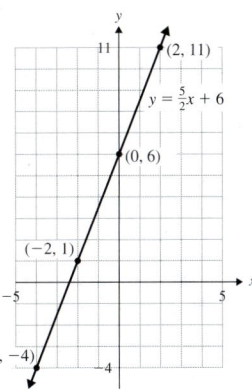

37)

x	y
0	$\frac{3}{2}$
−3	0
5	4
−1	1

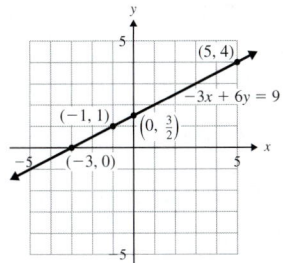

38)

x	y
$\frac{1}{4}$	0
0	1
$\frac{5}{2}$	−9
−1	5

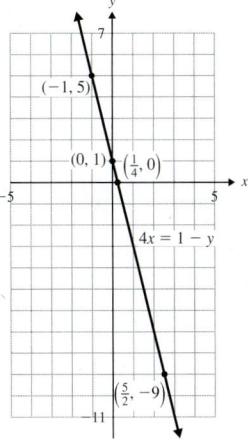

39)

x	y
0	−4
−3	−4
−1	−4
2	−4

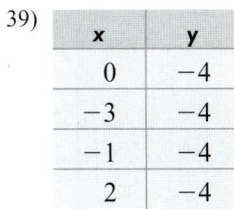
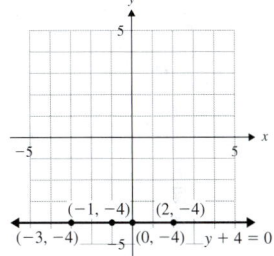

40)

x	y
$-\frac{3}{2}$	5
$-\frac{3}{2}$	0
$-\frac{3}{2}$	−1
$-\frac{3}{2}$	−2

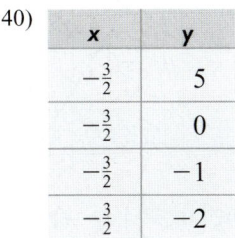
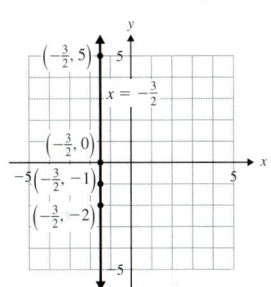

41) a) $(3, -5), (6, -3), (-3, -9)$

b) $\left(1, -\frac{19}{3}\right), \left(5, -\frac{11}{3}\right), \left(-2, -\frac{25}{3}\right)$

c) The x-values in part a) are multiples of the denominator of $\frac{2}{3}$. So, when you multiply $\frac{2}{3}$ by a multiple of 3 the fraction is eliminated.

42) $(0, 0)$

43) $(3, 0), (0, 6), (1, 4)$ 44) $(3, 0), (0, -3), (2, -1)$

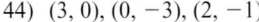
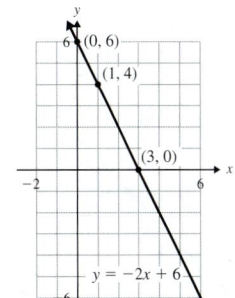
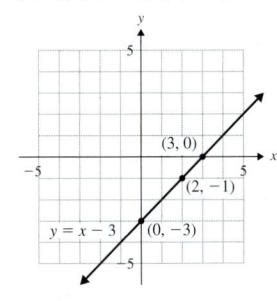

45) $(4, 0), (0, -3), \left(2, -\dfrac{3}{2}\right)$ 46) $(2, 0), (0, 5), \left(-1, \dfrac{15}{2}\right)$

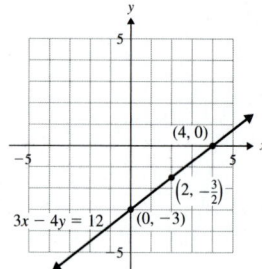   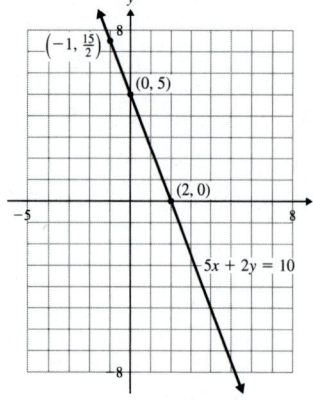
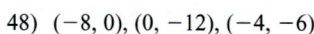

53) $(0, 0), (5, 2), (-5, -2)$ 54) $(0, 0), (3, -1), (-6, 2)$

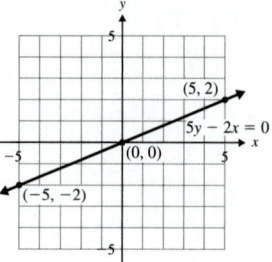   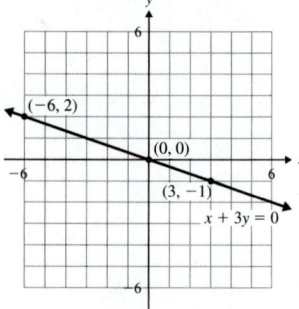
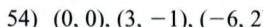

47) $(-1, 0), (0, 4), (1, 8)$ 48) $(-8, 0), (0, -12), (-4, -6)$

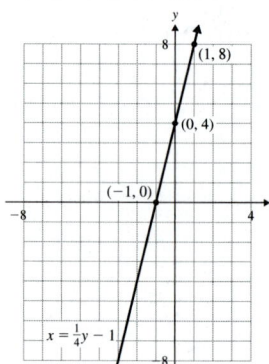   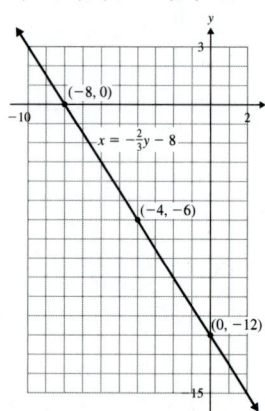

55) $(5, 0), (5, 2), (5, -1)$ 56) $(0, -1), (1, -1), (-3, -1)$
Answers may vary.   Answers may vary.

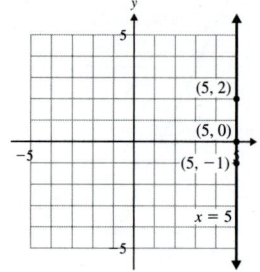   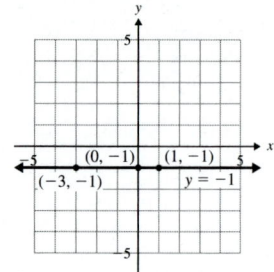

57) $(0, 0), (1, 0), (-2, 0)$ 58) $(0, 0), (0, -1), (0, 2)$
Answers may vary.   Answers may vary.

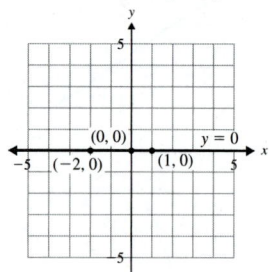   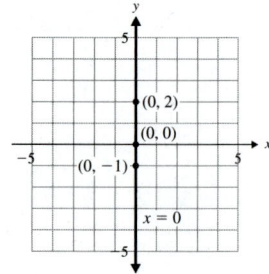

49) $(-3, 0), (0, -2), \left(1, -\dfrac{8}{3}\right)$ 50) $(6, 0), \left(0, -\dfrac{3}{2}\right), (2, -1)$

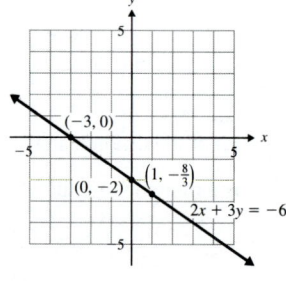   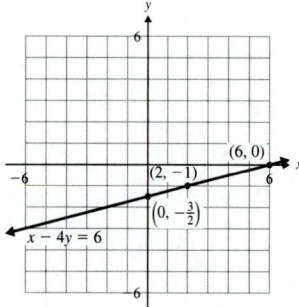
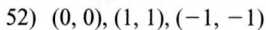

59) $(0, -3), (1, -3), (-3, -3)$ 60) $\left(\dfrac{5}{2}, 0\right), \left(\dfrac{5}{2}, 1\right), \left(\dfrac{5}{2}, -2\right)$
Answers may vary.   Answers may vary.

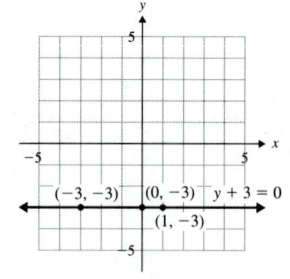   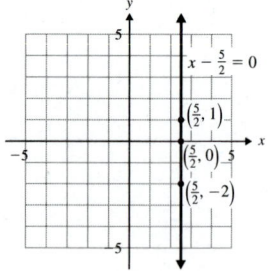
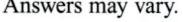

51) $(0, 0), (1, -1), (-1, 1)$ 52) $(0, 0), (1, 1), (-1, -1)$

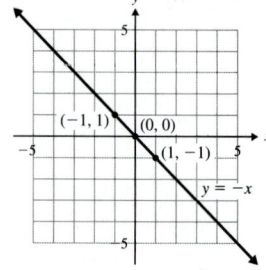   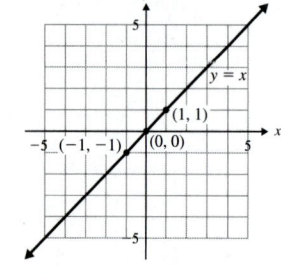

61) $(8, 0), \left(0, \dfrac{8}{3}\right), (2, 2)$

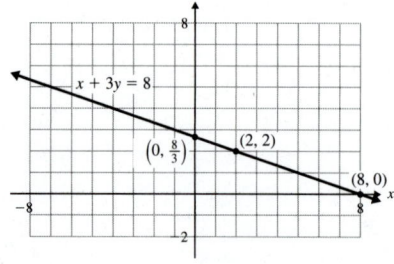

62) $\left(\frac{7}{6}, 0\right)$, $(0, -7)$, $(2, 5)$

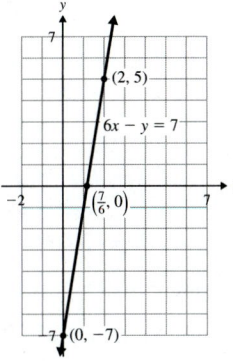

63) a) $y = 0$ b) $x = 0$ 64) (a, b)

65) a) $(16, 800)$, $(17, 1300)$, $(18, 1800)$, $(19, 1900)$

b)

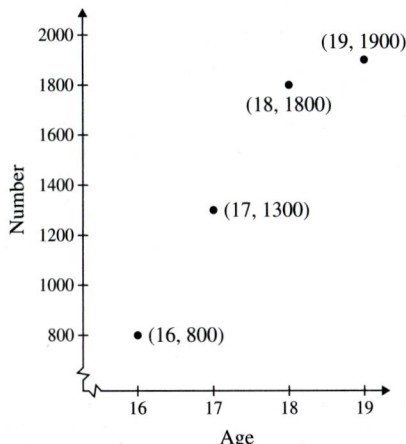

c) There were 1800 18-year old drivers in fatal motor vehicle accidents in 2006.

66) a) $(2003, 17.3)$, $(2004, 16.6)$, $(2005, 16.0)$, $(2006, 16.2)$, $(2007, 15.9)$

b)

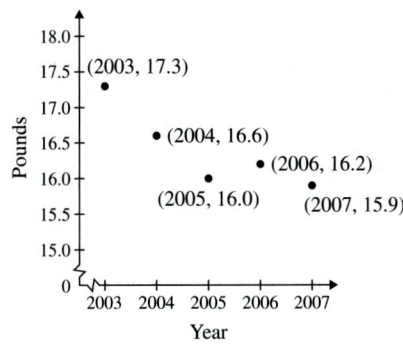

c) The consumption of potato chips in 2004 was 16.6 lb per person.

67) a)

x	y
1	120
3	160
4	180
6	220

$(1, 120)$, $(3, 160)$, $(4, 180)$, $(6, 220)$

b)

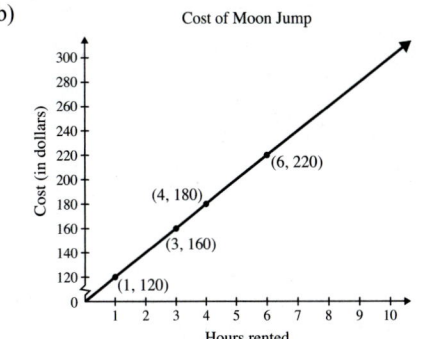

c) The cost of renting the moon jump for 4 hr is $180.

d) 9 hr

68) a)

x	y
3	186
8	496
15	930
20	1240

$(3, 186)$, $(8, 496)$, $(15, 930)$, $(20, 1240)$

b)

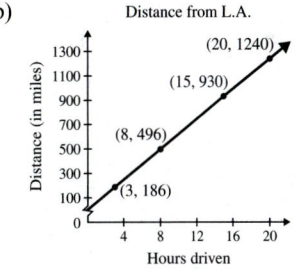

c) After driving 20 hr. Kelvin is 1240 mi from Los Angeles.

d) Kelvin's average speed

e) about 33 hr

69) a)

x	y
0	0
10	15
20	30
60	90

$(0, 0)$, $(10, 15)$, $(20, 30)$, $(60, 90)$

b) $(0, 0)$: Before engineers began working ($x = 0$ days), the tower did not move toward vertical ($y = 0$).

$(10, 15)$: After 10 days of working, the tower was moved 15 mm toward vertical.

$(20, 30)$: After 20 days of working, the tower was moved 30 mm toward vertical.

$(60, 90)$: After 60 days of working, the tower was moved 90 mm toward vertical.

c)

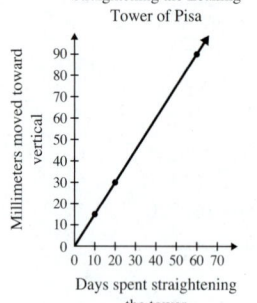

Straightening the Leaning
Tower of Pisa

d) 300 days

70) a)

x	y
0	0
1	0.02
2	0.04
4	0.08

(0, 0), (1, 0.02), (2, 0.04), (4, 0.08)

b) (0, 0): If no drinks are consumed, the blood alcohol percentage is 0.
(1, 0.02): After 1 drink, the blood alcohol percentage is 0.02.
(2, 0.04): After 2 drinks, the blood alcohol percentage is 0.04.
(4, 0.08): After 4 drinks, the blood alcohol percentage is 0.08.

c)

Blood alcohol percentage
of 180-lb male

d) 6

71) a) Answers will vary.
b) 29.86 in.; 28.86 in.; 26.36 in.; 24.86 in.
c)

Altitude and pressure

d) No, because the problem states that the equation applies
to altitudes 0 ft–5000 ft.

72) a) Answers may vary.   b) 1994–1995: $6878.95;
1997–1998: $7773.85;
2001–2002: $8967.05

Section 3.2

1) The slope of a line is the ratio of vertical change to horizontal
change. It is $\dfrac{\text{change in } y}{\text{change in } x}$ or $\dfrac{\text{rise}}{\text{run}}$ or $\dfrac{y_2 - y_1}{x_2 - x_1}$ where (x_1, y_1)
and (x_2, y_2) are points on the line.

2) It slants upward from left to right.

3) It slants downward from left to right.   4) zero

5) undefined   6) $m = \dfrac{y_2 - y_1}{x_2 - x_1}$ 7) $m = \dfrac{3}{4}$ 8) $m = \dfrac{5}{2}$

9) $m = -\dfrac{1}{3}$ 10) $m = -\dfrac{1}{2}$ 11) $m = -5$ 12) $m = 2$

13) $m = 0$ 14) Slope is undefined.

15)

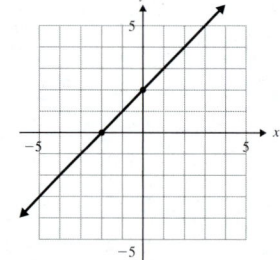

16)

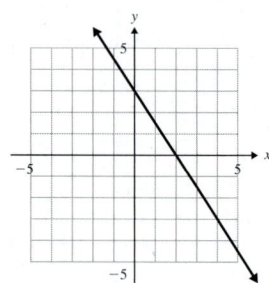

17) $\dfrac{1}{2}$ 18) $\dfrac{3}{2}$ 19) -1 20) -2 21) $-\dfrac{2}{9}$ 22) $-\dfrac{5}{7}$ 23) 0

24) undefined   25) undefined   26) 0 27) $\dfrac{14}{3}$ 28) $-\dfrac{10}{7}$

29) -2 30) 2.5 31) a) $\dfrac{5}{6}$ b) $\dfrac{2}{3}$ c) $m = \dfrac{1}{3}$; 4-12 pitch

32) Yes. The ramp in the picture has a slope of $\frac{1}{13}$, which is less
than $\frac{1}{12}$.

33) Yes. The slope of the driveway will be 10%.

34) The company will have to redesign the ramps because the
slopes of the ramps in the existing garage is 16%.

35) a) $22,000   b) negative
c) The value of the car is decreasing over time.
d) $m = -2000$; the value of the car is decreasing $2000 per
year.

36) a) 490,000; 480,000; 450,000   b) negative
c) The number of births to teenage girls has been decreasing
from 1997 to 2001.
d) $m = -10$ thousand or $-10,000$; the number of births to
teen girls is decreasing by 10,000 per year.

37)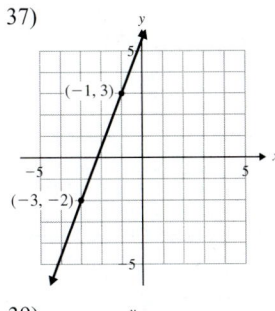

38)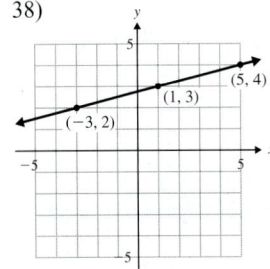

46)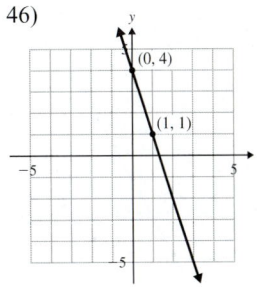

47)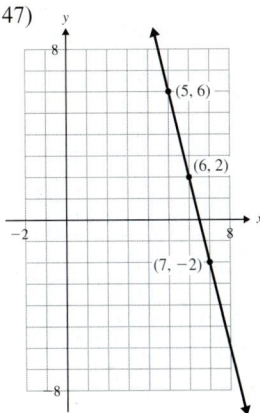

39)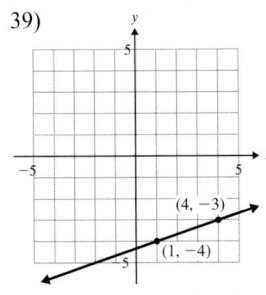

40)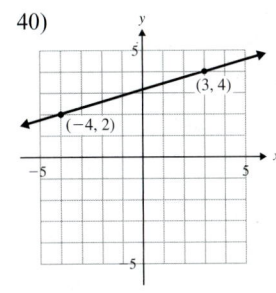

48)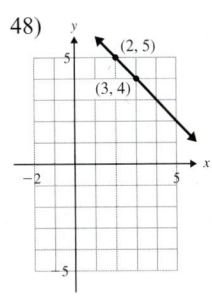

49)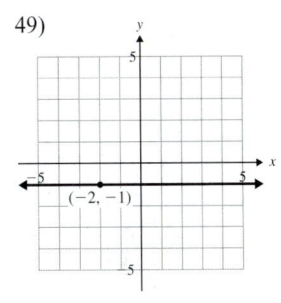

41)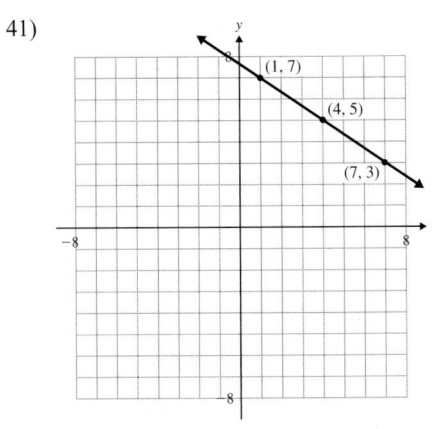

50)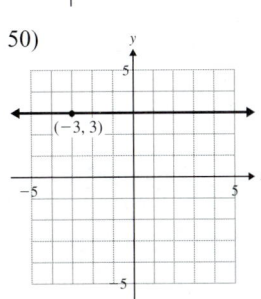

51)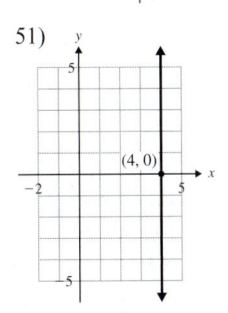

42)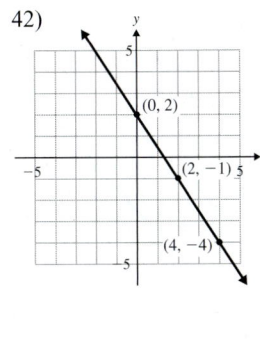

43)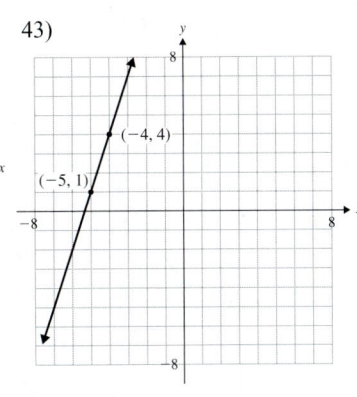

52)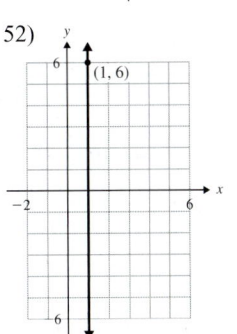

53)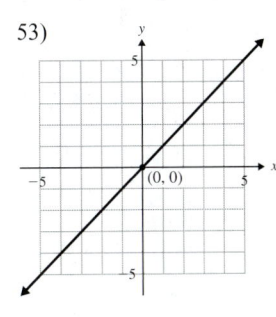

54)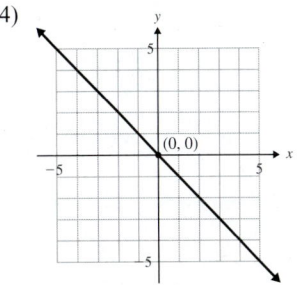

44)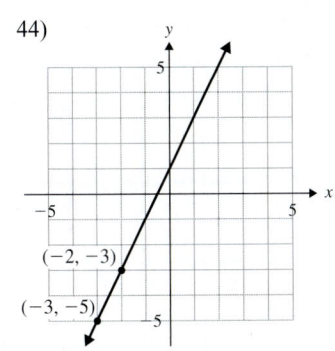

45)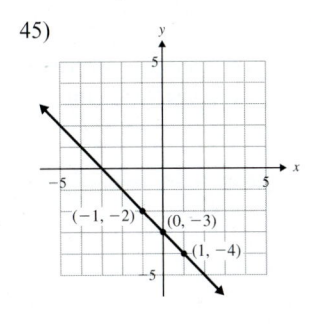

55) The slope is m, and the y-intercept is $(0, b)$.

56) Solve the equation for y.

57) $m = \dfrac{2}{5}$, y-int: $(0, -6)$

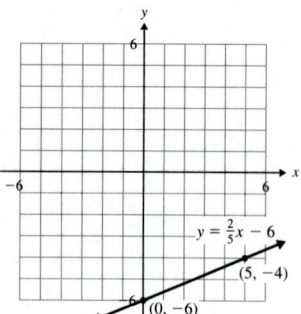

58) $m = \dfrac{7}{4}$, y-int: $(0, -2)$

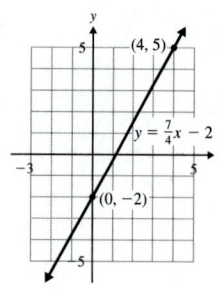

59) $m = -\dfrac{5}{3}$, y-int: $(0, 4)$

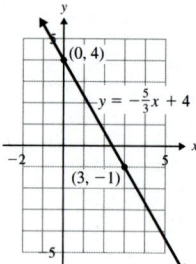

60) $m = -\dfrac{1}{2}$, y-int: $(0, 5)$

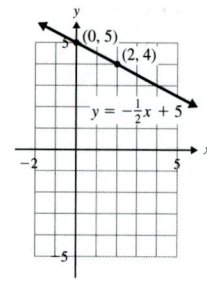

61) $m = \dfrac{3}{4}$, y-int: $(0, 1)$

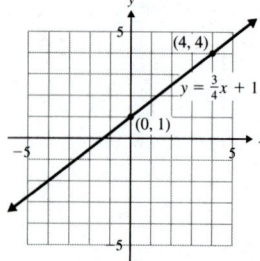

62) $m = \dfrac{2}{3}$, y-int: $(0, 3)$

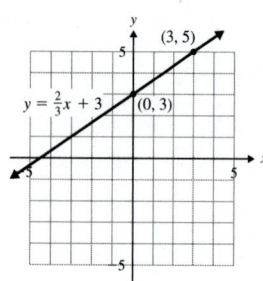

63) $m = 4$, y-int: $(0, -2)$

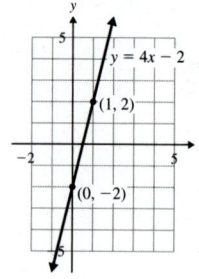

64) $m = -3$, y-int: $(0, -1)$

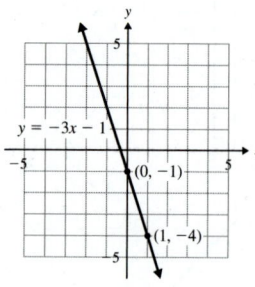

65) $m = -1$, y-int: $(0, 5)$

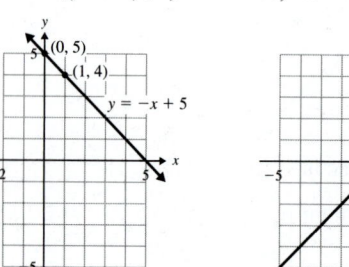

66) $m = 1$, y-int: $(0, 0)$

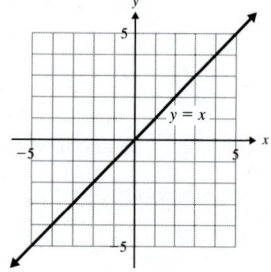

67) $m = \dfrac{3}{2}$, y-int: $\left(0, \dfrac{1}{2}\right)$

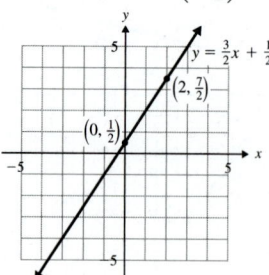

68) $m = -\dfrac{3}{4}$, y-int: $\left(0, -\dfrac{5}{2}\right)$

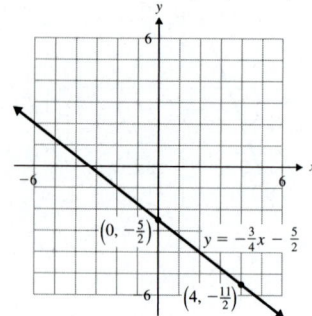

69) $m = 0$, y-int: $(0, -2)$

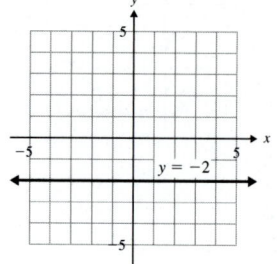

70) $m = 0$, y-int: $(0, 4)$

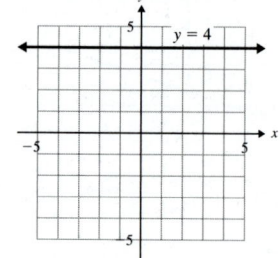

71) $y = -\dfrac{1}{3}x - 2$

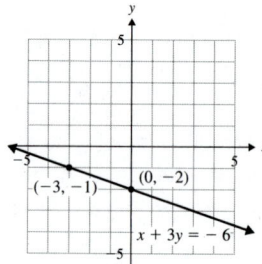

72) $y = -\dfrac{5}{2}x + 1$

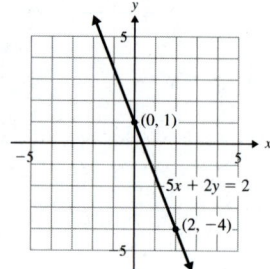

73) $y = \dfrac{3}{2}x - 4$

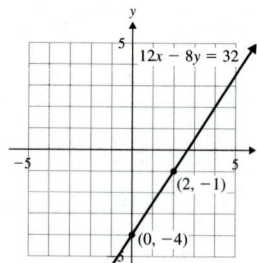

74) $y = x + 1$

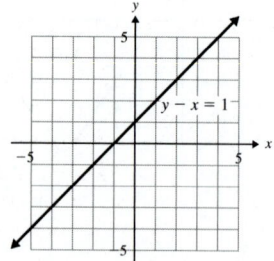

75) This cannot be written in slope-intercept form.

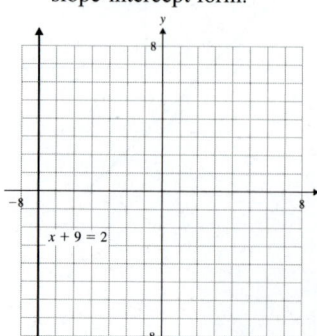

76) This cannot be written in slope-intercept form.

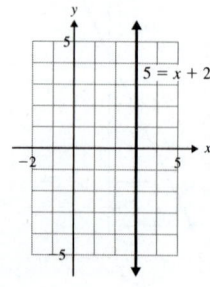

77) $y = \dfrac{5}{2}x + 3$

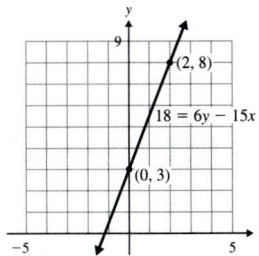

78) $y = -\dfrac{5}{3}x + 4$

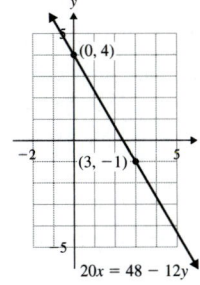

79) $y = 0$

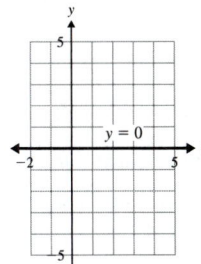

80) $y = -5$

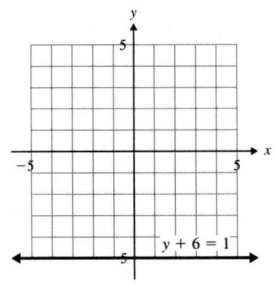

81) a) (0, 34,000); if Dave has $0 in sales, his income is $34,000.
   b) $m = 0.05$; Dave earns $0.05 for every $1 in sales.
   c) $38,000

82) a) (0, 0); if Li Mei works 0 hr, she earns $0.
   b) $m = 7.50$; Li Mei earns $7.50 per hr.   c) $105

83) a) In 1945, 40.53 gal of whole milk were consumed per person, per year.
   b) $m = -0.59$; since 1945, Americans have been consuming 0.59 fewer gallons of whole milk each year.
   c) estimate from the graph: 7.5 gal; consumption from the equation: 8.08 gal.

84) a) (0, 0); $0 = 0 pesos
   b) $m = 11.40$; each American dollar is worth 11.40 pesos.
   c) estimate from the graph: $52; from the equation: $52.63

85) a) (0, 68,613); in 2000, the average annual salary of a pharmacist was $68,613.
   b) $m = 3986$; the average salary of a pharmacist is increasing by $3986 per year.
   c) $84,557   d) $124,417   e) Answers may vary.

Average Annual Salary
of a Pharmacist

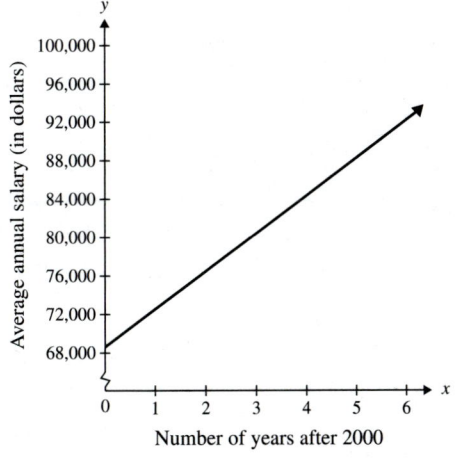

86) a) (0, 85.8); in 1986, 85.8% of men ages 20–24 were in the civilian labor force.
   b) $m = -0.4$; the percentage of men ages 20–24 in the civilian labor force is decreasing by 0.4% per year.
   c) 76.2%   d) 73.8%   e) Answers may vary.

Percentage of Men, Ages 20–24,
in the Civilian Labor Force from
1986 to 2016 (projected)

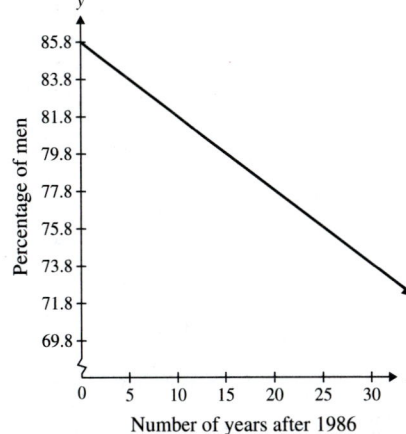

87) $y = -4x + 7$ 88) $y = 5x + 3$ 89) $y = \dfrac{8}{5}x - 6$

90) $y = \dfrac{4}{9}x - 1$ 91) $y = \dfrac{1}{3}x + 5$ 92) $y = -\dfrac{1}{2}x - 3$

93) $y = -x$ 94) $y = x + 4$ 95) $y = -2$ 96) $y = 0$

Section 3.3

1) Substitute the slope and y-intercept into $y = mx + b$.

2) No. The coefficient of x must be a positive integer.

3) $y = -7x + 2$ 4) $y = 4x - 5$ 5) $x - y = 3$

6) $2x + y = 8$ 7) $x + 3y = -12$ 8) $5x - 2y = 2$

9) $y = x$ 10) $y = \dfrac{4}{9}x - \dfrac{1}{6}$

11) a) $y - y_1 = m(x - x_1)$
   b) Substitute the slope and point into the point-slope formula.

12) $y = 3x - 7$ 13) $y = 5x + 1$ 14) $y = -2x + 4$

15) $y = -x - 5$ 16) $x - y = 4$ 17) $4x - y = -7$

18) $y = -\dfrac{4}{5}x + \dfrac{23}{5}$ 19) $y = \dfrac{1}{6}x - \dfrac{13}{3}$ 20) $5x - 8y = -10$

21) $5x + 9y = 30$ 22) $y = 4x - \dfrac{5}{3}$

23) Find the slope and use it and one of the points in the point-slope formula.

24) $y = -2x + 9$ 25) $y = x + 1$ 26) $y = 3x - 7$

27) $y = -\dfrac{1}{3}x + \dfrac{10}{3}$ 28) $y = \dfrac{2}{3}x - \dfrac{16}{3}$ 29) $3x - 4y = 17$

30) $x + 2y = -3$ 31) $2x + y = 9$ 32) $5x - 2y = 8$

33) $y = 5.0x - 8.3$ 34) $y = -1.5x + 0.7$ 35) $y = \frac{3}{2}x - 4$

36) $y = -6x + 1$ 37) $y = -x - 2$ 38) $y = \frac{3}{5}x + \frac{2}{5}$

39) $y = 5$ 40) $x = -3$ 41) $y = 4x + 15$

42) $y = -\frac{1}{2}x + \frac{3}{2}$ 43) $y = \frac{8}{3}x - 9$ 44) $y = 3x - 7$

45) $y = -\frac{3}{4}x - \frac{11}{4}$ 46) $y = 7x + 6$ 47) $x = 3$

48) $x = -\frac{3}{4}$ 49) $y = -8$ 50) $y = 4$ 51) $y = -\frac{1}{3}x + 2$

52) $y = 2x - 11$ 53) $y = x$ 54) $y = -\frac{2}{5}x - \frac{11}{5}$

55) Answers may vary.   56) Answers may vary.

57) perpendicular   58) parallel   59) parallel

60) perpendicular   61) neither   62) neither   63) perpendicular

64) neither   65) parallel   66) parallel   67) parallel

68) perpendicular   69) parallel   70) perpendicular

71) perpendicular   72) neither   73) neither   74) parallel

75) $y = 4x + 2$ 76) $y = -3x + 5$ 77) $x - 2y = -6$

78) $2x - y = 3$ 79) $4x + 3y = -24$ 80) $x + 4y = 12$

81) $y = -\frac{1}{5}x + 10$ 82) $y = 6x - 14$ 83) $y = -\frac{3}{2}x + 6$

84) $y = \frac{3}{4}x - 5$ 85) $x - 5y = 10$ 86) $4x + y = -1$

87) $y = x$ 88) $y = -x + 6$ 89) $5x + 8y = 24$

90) $5x - 2y = 16$ 91) $y = -3x + 8$ 92) $y = -6x - 12$

93) $y = 2x - 5$ 94) $y = x + 7$ 95) $x = -1$ 96) $y = -3$

97) $x = 2$ 98) $y = 1$ 99) $y = -\frac{2}{7}x + \frac{1}{7}$ 100) $y = \frac{3}{4}x - \frac{5}{4}$

101) $y = -\frac{5}{2}$ 102) $x = -2$ 103) a) $L = \frac{1}{3}S + \frac{22}{3}$ b) 12.5

104) a) $P = -\frac{1}{20}C + 130$ b) 70%

105) a) $y = 8700x + 1{,}257{,}900$
   b) The population of Maine is increasing by 8700 people per year.
   c) 1,257,900; 1,292,700   d) 2018

106) a) $y = -3290x + 650{,}000$
   b) The population of North Dakota is decreasing by 3290 people per year.
   c) 643,420; 633,550   d) 2012

107) a) $y = -6.4x + 124$
   b) The number of farms with milk cows is decreasing by 6.4 thousand (6400) per year.
   c) 79.2 thousand (79,200)

108) a) $y = 2513x + 55{,}722$
   b) A principal's average salary is increasing by $2513 per year.
   c) $63,261   d) $93,417

109) a) $y = 12{,}318.7x + 6479$
   b) The number of registered hybrid vehicles is increasing by 12,318.7 per year.
   c) 31,116.4; this is slightly lower than the actual value.
   d) 129,666

110) a) $y = 0.61x + 47.0$
   b) The percentage of women in Belgium in the workforce is increasing by 0.61 per year.
   c) 53.1%; it is 0.7% higher than the actual value.
   d) 56.15%; it is 0.05% higher than the actual value.
   e) 1998

Section 3.4

1) Answers may vary.   2) Answers may vary.

3) Answers may vary.   4) Answers may vary.

5) Answers may vary.   6) Answers may vary.

7) dotted   8) yes

9) 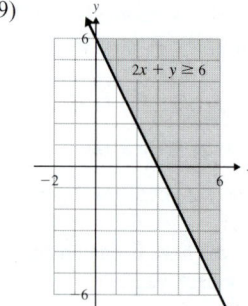   10)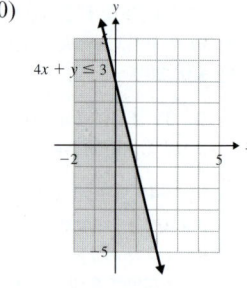

11) 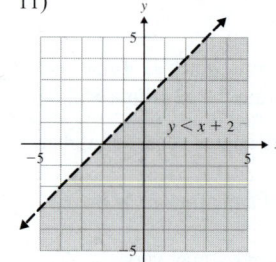   12)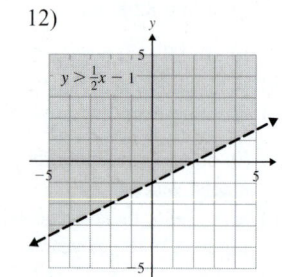

13)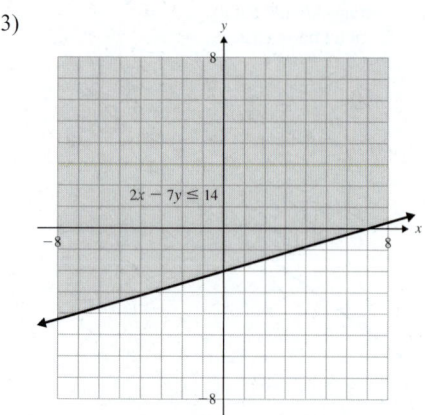

14)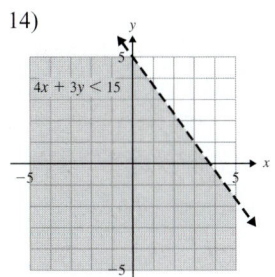

15)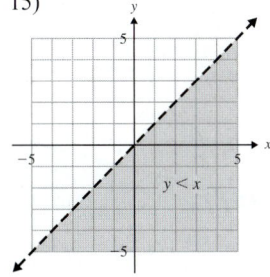

25)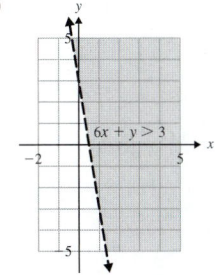

26)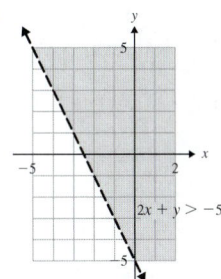

16)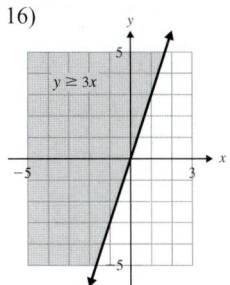

17)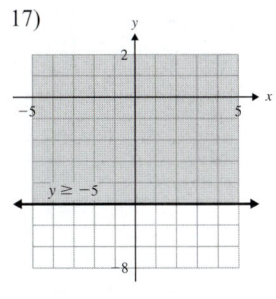

27)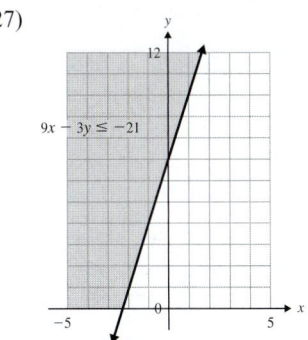

18)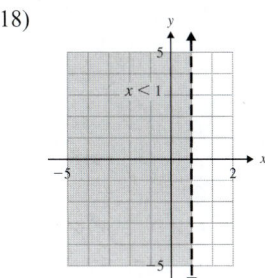

19) below   20) above

28)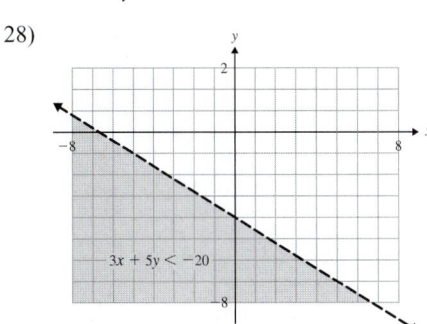

21)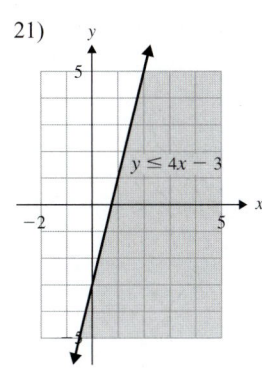

22)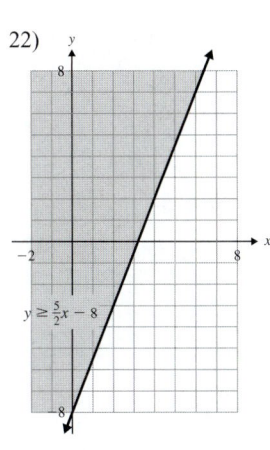

29)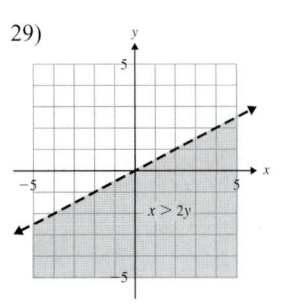

30)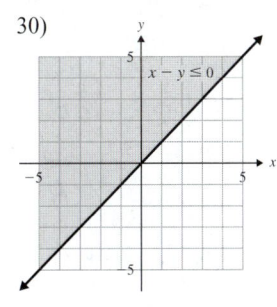

31) Answers may vary.   32) Answers may vary.

23)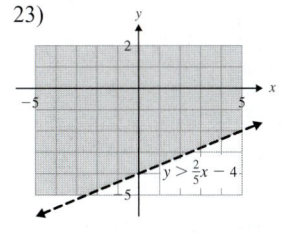

24)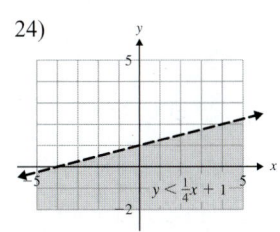

33)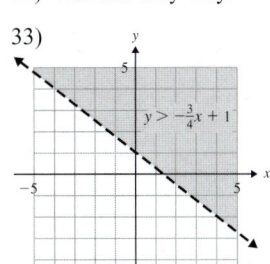

34)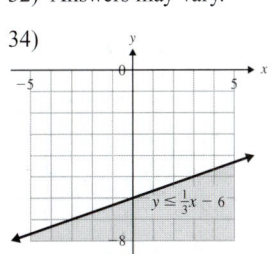

35)
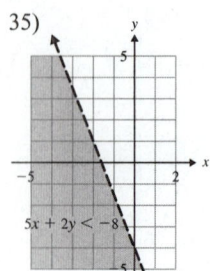
$5x + 2y < -8$

36)
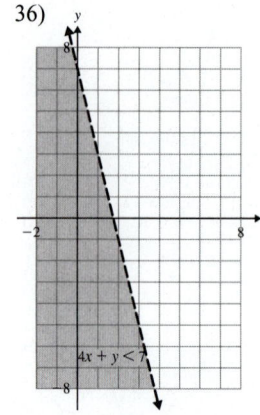
$4x + y < 7$

46)
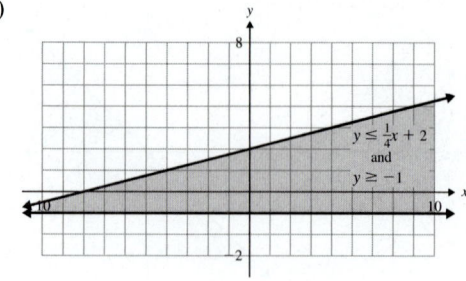
$y \le \frac{1}{4}x + 2$
and
$y \ge -1$

37)
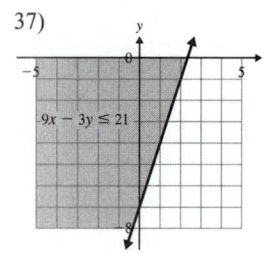
$9x - 3y \le 21$

38)
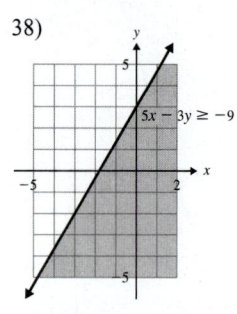
$5x - 3y \ge -9$

47)
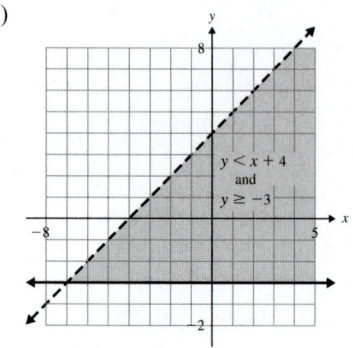
$y < x + 4$
and
$y \ge -3$

39)
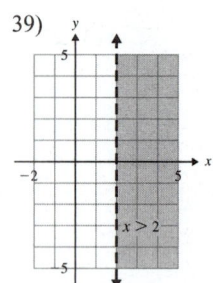
$x > 2$

40)
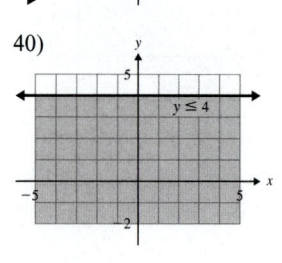
$y \le 4$

48)
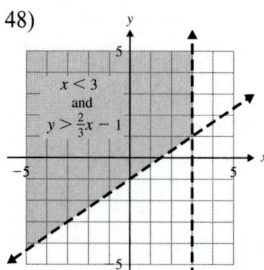
$x < 3$
and
$y > \frac{2}{3}x - 1$

49)
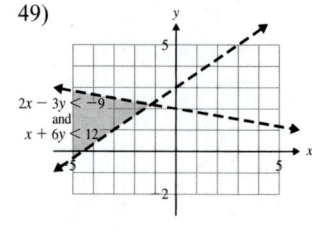
$2x - 3y < -9$
and
$x + 6y < 12$

41)
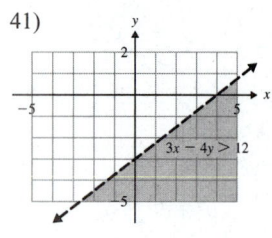
$3x - 4y > 12$

42)
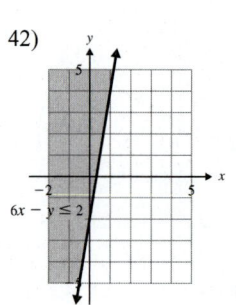
$6x - y \le 2$

50)
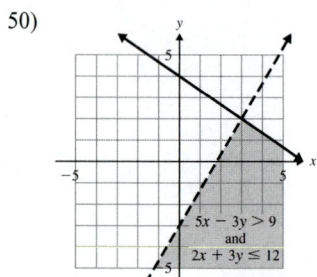
$5x - 3y > 9$
and
$2x + 3y \le 12$

43) No; it does not satisfy $2x + y < 7$.

44) Yes; it satisfies one of the inequalities, $x - y \ge -6$.

45)
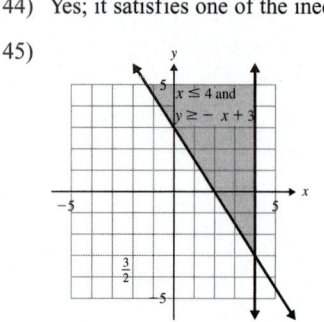
$x \le 4$ and
$y \ge -x + 3$

51)
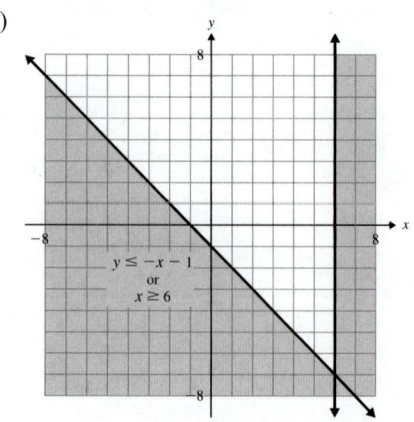
$y \le -x - 1$
or
$x \ge 6$

52)

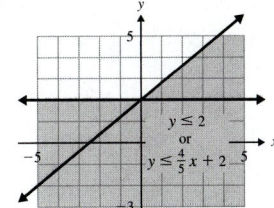

$y \leq 2$
or
$y \leq \frac{4}{5}x + 2$

53)

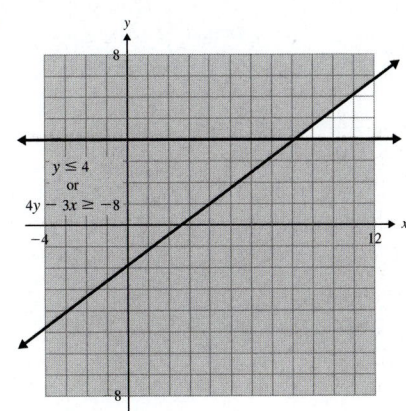

$y \leq 4$
or
$4y - 3x \geq -8$

54)

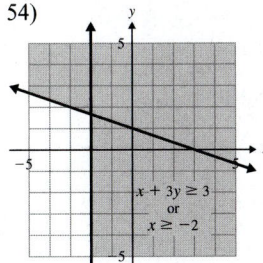

$x + 3y \geq 3$
or
$x \geq -2$

55)

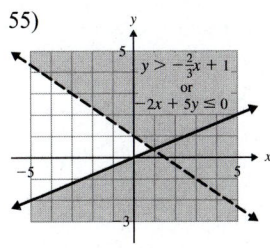

$y > -\frac{2}{3}x + 1$
or
$-2x + 5y \leq 0$

56)

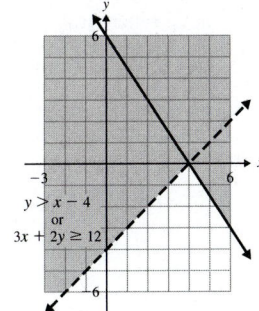

$y > x - 4$
or
$3x + 2y \geq 12$

57)

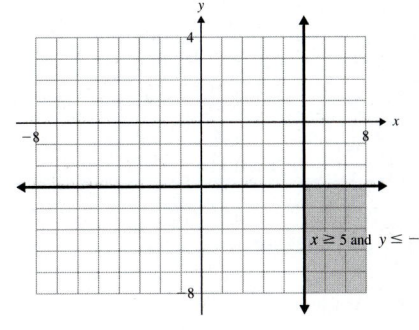

$x \geq 5$ and $y \leq -3$

58)

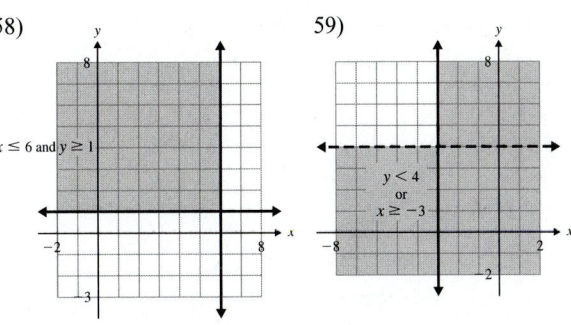

$x \leq 6$ and $y \geq 1$

59)

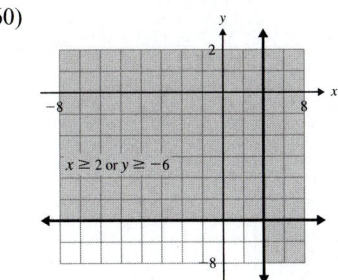

$y < 4$
or
$x \geq -3$

60)

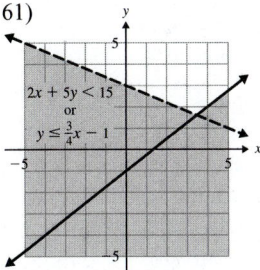

$x \geq 2$ or $y \geq -6$

61)

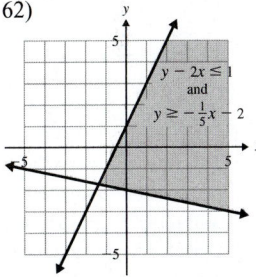

$2x + 5y < 15$
or
$y \leq \frac{3}{4}x - 1$

62)

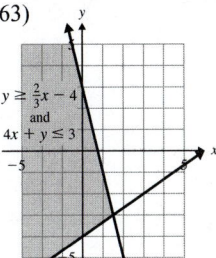

$y - 2x \leq 1$
and
$y \geq -\frac{1}{5}x - 2$

63)

$y \geq \frac{2}{3}x - 4$
and
$4x + y \leq 3$

64)

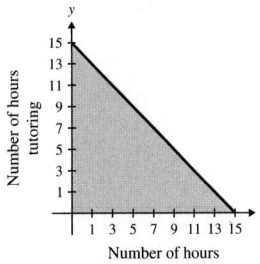

$y < 5x + 2$
or
$x + 4y < 12$

65) a) $x \geq 0$ and $y \geq 0$ and $x + y \leq 15$

b)

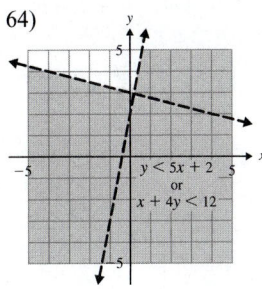

Number of hours tutoring

Number of hours babysitting

c) Answers may vary.   d) Answers may vary.

66) a) $x \geq 0$ and $y \geq 0$ and $x + y \leq 12$
b)

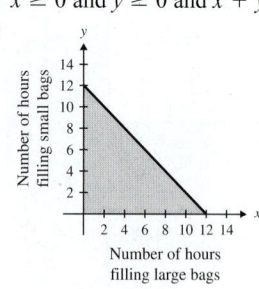

c) Answers may vary.   d) Answers may vary.

67) a) $150 \leq p \leq 250$ and $100 \leq r \leq 200$ and $p + r \geq 300$
b)

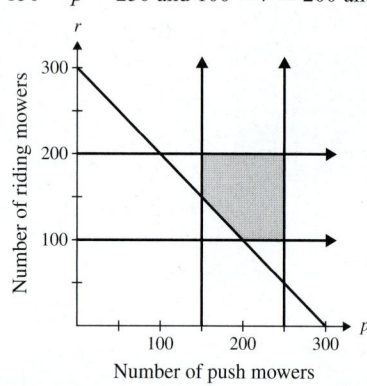

Number of push mowers

c) It represents the production of 175 push mowers and 110 riding mowers per day. This does not meet the level of production needed because it is not a total of at least 300 mowers per day and is not in the feasible region.
d) Answers may vary.   e) Answers may vary.

68) a) $12{,}000 \leq a \leq 18{,}000$ and $8000 \leq p \leq 14{,}000$ and $a + p \geq 25{,}000$
b)

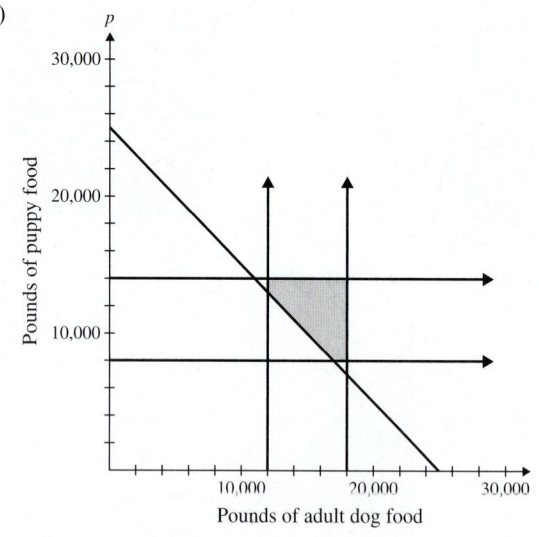

Pounds of adult dog food

c) It represents the production of 17,000 pounds of adult dog food and 9000 pounds of puppy food per day. This meets the level of production needed because it satisfies all of the inequalities and is in the feasible region.
d) Answers may vary.   e) Answers may vary.

Section 3.5

1) a) any set of ordered pairs   2) Answers may vary.
   b) Answers may vary.
   c) Answers may vary.

3) domain: $\{-8, -2, 1, 5\}$ 4) domain: $\{0, 1, 16\}$
   range: $\{-3, 4, 6, 13\}$ range: $\{-5, -4, -3, -2, -1\}$
   function   not a function

5) domain: $\{1, 9, 25\}$ 6) domain: $\{-4, -3, -1, 0\}$
   range: $\{-3, -1, 1, 5, 7\}$ range: $\left\{-2, -\dfrac{1}{2}\right\}$
   not a function   function

7) domain: $\{-1, 2, 5, 8\}$
   range: $\{-7, -3, 12, 19\}$
   not a function

8) domain: $\{$Apple, Chocolate, Corn$\}$
   range: $\{$Fruit, Candy, Vegetable$\}$
   function

9) domain: $(-\infty, \infty)$;   10) domain: $[-5, 1]$;
   range: $(-\infty, \infty)$ range: $[-6, 0]$
   function   not a function

11) domain: $(-\infty, 4]$;   12) domain: $(-\infty, \infty)$;
    range: $(-\infty, \infty)$ range: $(-\infty, \infty)$
    not a function   function

13) domain: $(-\infty, \infty)$;   14) domain: $[-3, \infty)$;
    range: $(-\infty, 6]$ range: $(-\infty, \infty)$
    function   not a function

15) yes   16) yes   17) yes   18) yes   19) no   20) no

21) no   22) no   23) $(-\infty, \infty)$; function

24) $(-\infty, \infty)$; function   25) $(-\infty, \infty)$; function

26) $(-\infty, \infty)$; function   27) $[0, \infty)$; not a function

28) $[0, \infty)$; not a function   29) $(-\infty, 0) \cup (0, \infty)$; function

30) $(-\infty, 0) \cup (0, \infty)$; function

31) $(-\infty, -4) \cup (-4, \infty)$; function

32) $(-\infty, 7) \cup (7, \infty)$; function

33) $(-\infty, 5) \cup (5, \infty)$; function

34) $(-\infty, -10) \cup (-10, \infty)$; function

35) $\left(-\infty, \dfrac{3}{5}\right) \cup \left(\dfrac{3}{5}, \infty\right)$; function

36) $\left(-\infty, -\dfrac{8}{9}\right) \cup \left(-\dfrac{8}{9}, \infty\right)$; function

37) $\left(-\infty, -\dfrac{4}{3}\right) \cup \left(-\dfrac{4}{3}, \infty\right)$; function

38) $\left(-\infty, \dfrac{1}{6}\right) \cup \left(\dfrac{1}{6}, \infty\right)$; function

39) $(-\infty, 3) \cup (3, \infty)$; function

40) $\left(-\infty, \dfrac{3}{2}\right) \cup \left(\dfrac{3}{2}, \infty\right)$; function

41) $(-\infty, \infty)$; function   42) $(-\infty, \infty)$; function

43) y is a function, and y is a function of x.

44) No; $f(x)$ is read as "f of x" and $y = f(x)$ means that y is a function of x.

45) a) $y = 7$ b) $f(3) = 7$ 46) a) $y = 10$ b) $f(-4) = 10$

47) -13 48) -1 49) 7 50) 13 51) 50 52) 8

53) -10 54) -2 55) $-\dfrac{25}{4}$ 56) $\dfrac{10}{9}$ 57) -105 58) 45

59) $f(-1) = 10, f(4) = -5$ 60) $f(-1) = \dfrac{5}{2}, f(4) = 5$

61) $f(-1) = 6, f(4) = 2$ 62) $f(-1) = 3, f(4) = 3$

63) $f(-1) = 7, f(4) = 3$ 64) $f(-1) = -8, f(4) = 6$ 65) -4

66) 1 67) 6 68) -8

69) Substitute $k + 6$ for x; $4k + 24 - 5$; $4k + 19$

70) $f(n - 3) = -9(n - 3) + 2$; Distribute; $-9n + 29$

71) a) $f(c) = -7c + 2$ b) $f(t) = -7t + 2$
    c) $f(a + 4) = -7a - 26$ d) $f(z - 9) = -7z + 65$
    e) $g(k) = k^2 - 5k + 12$ f) $g(m) = m^2 - 5m + 12$
    g) $-7x - 7h + 2$ h) $-7h$

72) a) $f(n) = 5n + 6$ b) $f(p) = 5p + 6$
    c) $f(w + 8) = 5w + 46$ d) $f(r - 7) = 5r - 29$
    e) $g(b) = b^2 - 3b - 11$ f) $g(s) = s^2 - 3s - 11$
    g) $5x + 5h + 6$ h) $5h$

73)
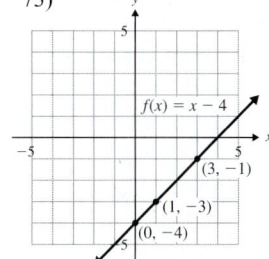

74)
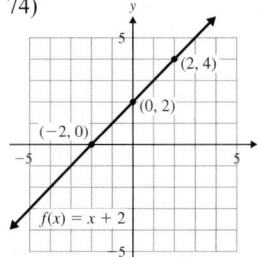

75)
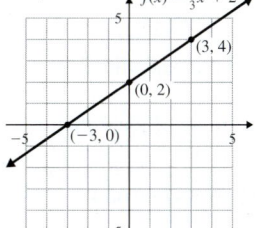

76)
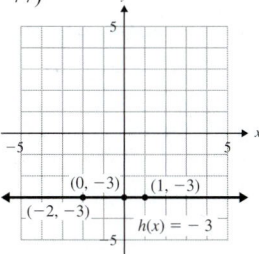

77)

78)
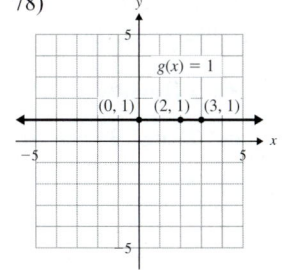

79) x-int: $(-1, 0)$; y-int: $(0, 3)$ 80) x-int: $(3, 0)$; y-int: $(0, 6)$
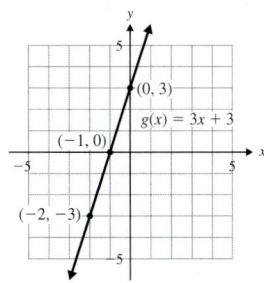
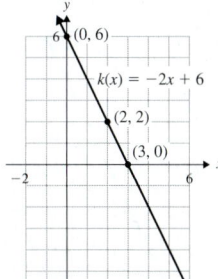

81) x-int: $(4, 0)$; y-int: $(0, 2)$ 82) x-int: $(-3, 0)$; y-int: $(0, 1)$
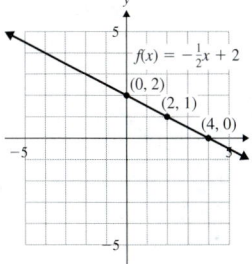
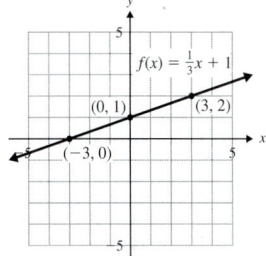

83) intercept: $(0, 0)$ 84) intercept: $(0, 0)$
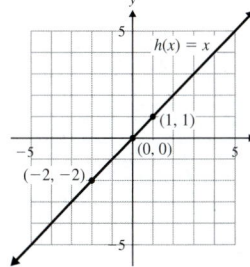
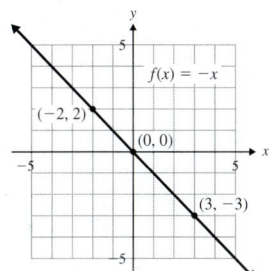

85) $m = -4$; y-int: $(0, -1)$ 86) $m = -1$; y-int: $(0, 5)$
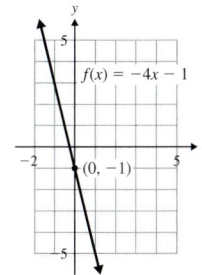
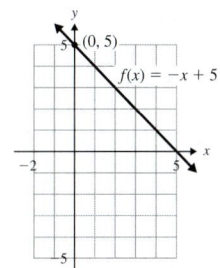

87) $m = \dfrac{3}{5}$; y-int: $(0, -2)$ 88) $m = -\dfrac{1}{4}$; y-int: $(0, -2)$
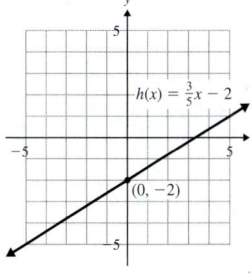
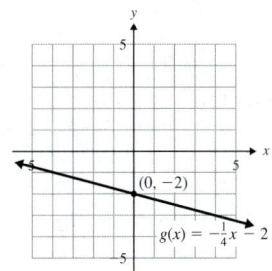

89) $m = 2$; y-int: $\left(0, \dfrac{1}{2}\right)$ 90) $m = 3$; y-int: $(0, 1)$

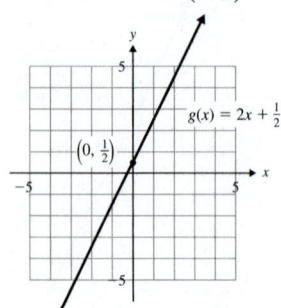

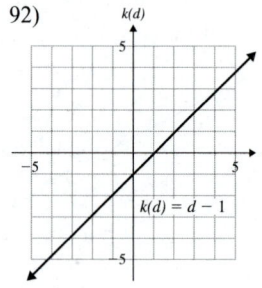

91)

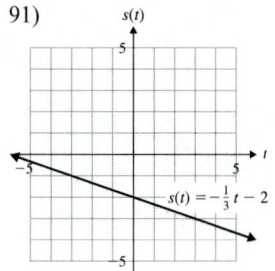

92)

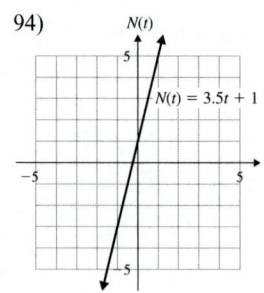

93)

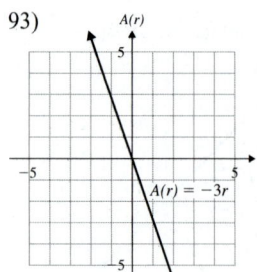

94)

95) a) 108 mi   b) 216 mi   c) 2.5 hr
    d)

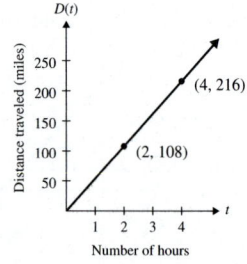

96) a) 32 ft/sec   b) 96 ft/sec   c) after 8 sec
    d)

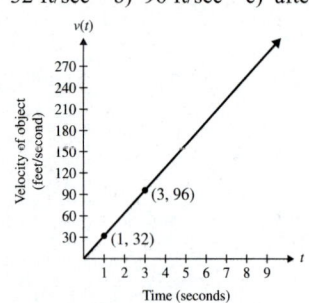

97) a) $E(10) = 75$; when Jenelle works for 10 hr,
       she earns $75.00.

b) $E(15) = 112.5$; when Jenelle works 15 hr,
   she earns $112.50.
c) $t = 28$; for Jenelle to earn $210.00, she must work 28 hr.

98) a) $C(8) = 20$; 8 gal of gas cost $20.00.
    b) $C(15) = 37.5$; 15 gal of gas cost $37.50.
    c) $g = 12$; 12 gal of gas can be purchased for $30.00.

99) a) 253.56 MB   b) 1267.8 MB   c) 20 sec
    d)

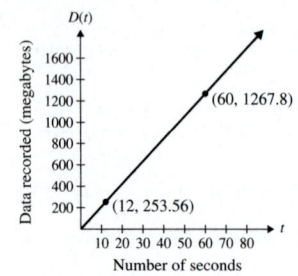

100) a) $534.60   b) $481.14   c) 35 hr   d) [0, 40]
     e)

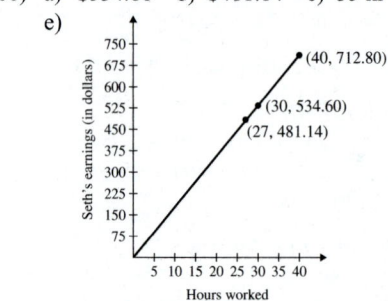

101) a) 60,000   b) 3.5 sec   102) a) 90,000   b) 1.5 sec

103) a) $S(50) = 2205$; after 50 sec, the CD player reads
        2,205,000 samples of sound.
     b) $S(180) = 7938$; after 180 sec (or 3 min), the CD player
        reads 7,938,000 samples of sound.
     c) $t = 60$; the CD player reads 2,646,000 samples of sound
        in 60 seconds (or 1 min).

104) a) $D(10) = 211.3$; after 10 sec, 211.3 MB of data are
        recorded onto a DVD.
     b) $D(120) = 2535.6$; after 120 sec (or 2 min), 2535.6 MB
        of data are recorded onto a DVD.
     c) $t = 30$; a 16 × DVD recorder transfers 633.9 MB of
        data onto a DVD in 30 sec.

105) a) 2 hr; 400 mg   b) after about 30 min and after 6 hr
     c) 200 mg
     d) $A(8) = 50$. After 8 hr there are 50 mg of ibuprofen in
        Sasha's bloodstream.

106) a) domain: [0, 24]; range: [1,000,000, 10,000,000]
     b) 8 million gallons; 4 million gallons
     c) 2 P.M.; 10 million gallons   d) from 2 to 4 A.M.
     e) $G(18) = 7$. At 6 P.M., 7 gal of water entered the
        treatment plant.

Chapter 3 Review Exercises

1) yes   2) no   3) yes   4) yes   5) 28   6) 13

7) -8 8) -3

9)

x	y
0	−11
3	−8
−1	−12
−5	−16

10)

x	y
2	0
0	$-\dfrac{4}{3}$
3	$\dfrac{2}{3}$
−4	−4

16)

x	y
0	2
−2	5
1	$\dfrac{1}{2}$
4	−4

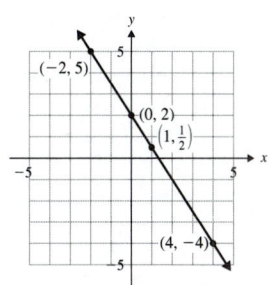

11)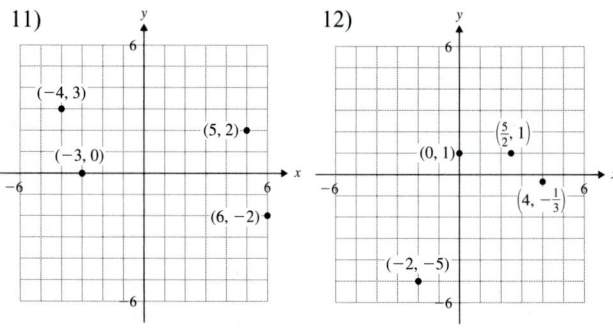

12)

17) (6, 0), (0, −3);
(2, −2) may vary.

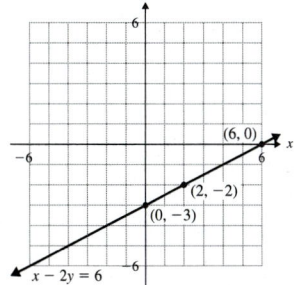

18) (2, 0), (0, 10);
(3, −5) may vary.

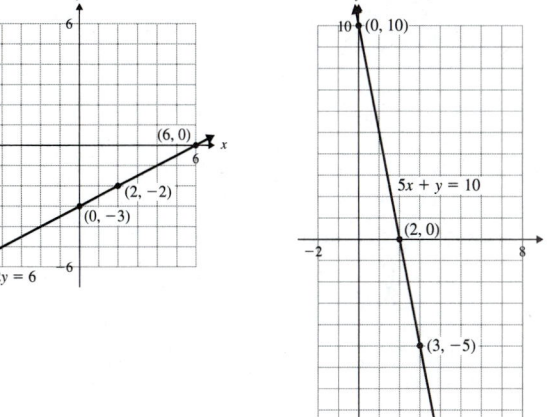

13) a)

x	y
1	0.10
2	0.20
7	0.70
10	1.00

(1, 0.10), (2, 0.20), (7, 0.70), (10, 1.00)

b)

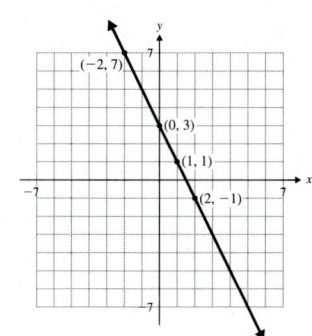

c) If a book is 14 days overdue, the fine is $1.40.

14) a) negative   b) positive

15)

x	y
0	3
1	1
2	−1
−2	7

19) (24, 0), (0, 4);
(12, 2) may vary.

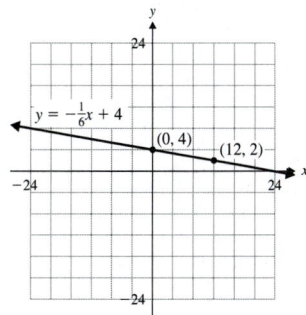

20) $\left(\dfrac{28}{3}, 0\right)$, (0, −7);
(8, −1) may vary.

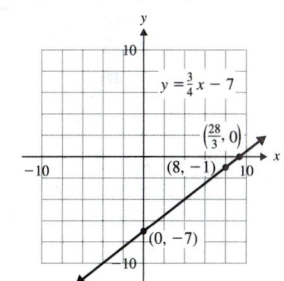

21) (5, 0), (5, 1) may vary,
(5, 2) may vary.

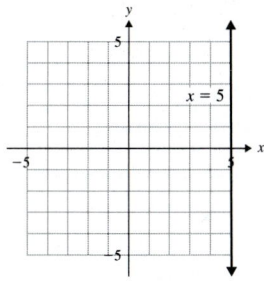

22) (0, −3), (1, −3) may vary,
(2, −3) may vary.

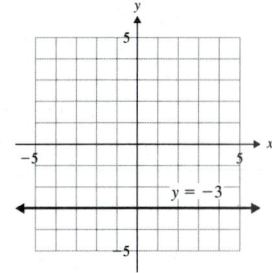

23) $-\dfrac{2}{5}$ 24) 2   25) 1   26) $\dfrac{2}{5}$ 27) $-\dfrac{13}{5}$ 28) $-\dfrac{3}{4}$

29) −2   30) 1.2   31) undefined   32) 0

33) a) In 2004, one share of the stock was worth $32.
    b) The slope is positive, so the value of the stock is increasing over time.
    c) $m = 3$; the value of one share of stock is increasing by $3.00 per year.

34)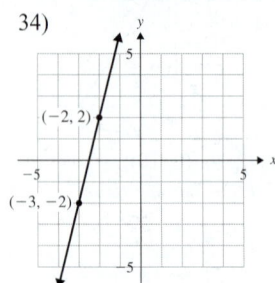

35)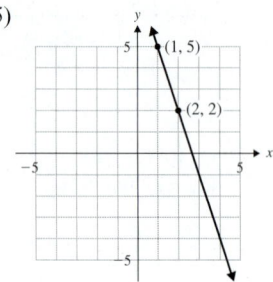

36)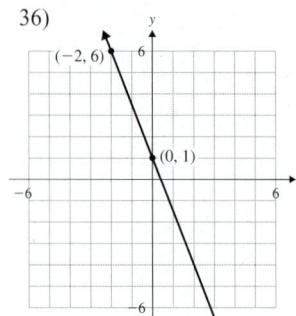

37)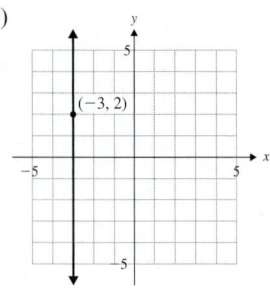

38)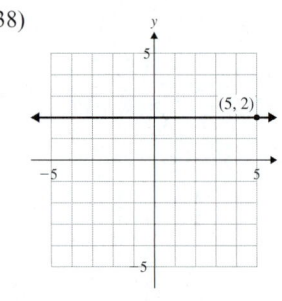

39) $m = 1$, y-int: $(0, -3)$
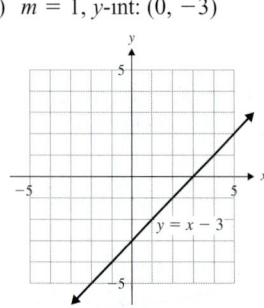

40) $m = -2$, y-int: $(0, 7)$

41) $m = -\dfrac{3}{4}$, y-int: $(0, 1)$
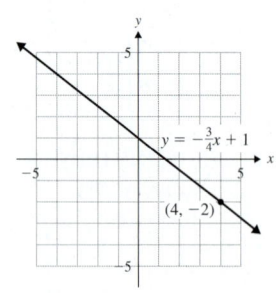

42) $m = \dfrac{1}{4}$, y-int: $(0, -2)$

43) $m = \dfrac{1}{3}$, y-int: $(0, 2)$
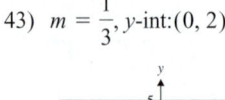
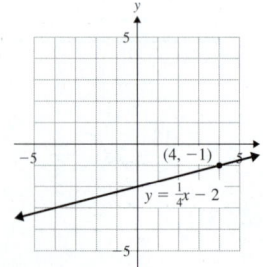
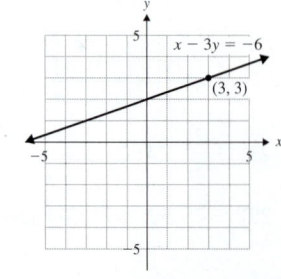

44) $m = \dfrac{2}{7}$, y-int: $(0, -5)$

45) $m = -1$, y-int: $(0, 0)$
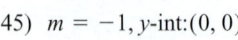
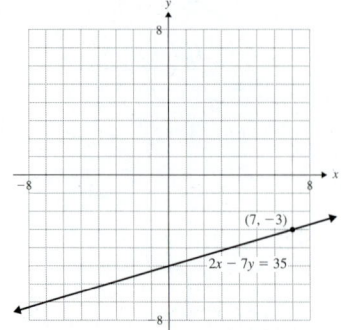
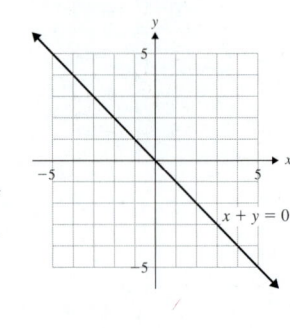

46) $m = 0$, y-int: $(0, 1)$
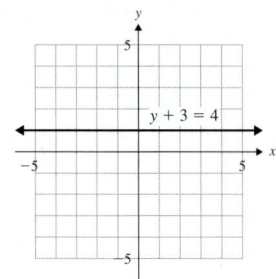

47) a) $(0, 5920.1)$; in 1998 the amount of money spent for personal consumption was $5920.1 billion.
    b) It has been increasing by $371.5 billion per year.
    c) estimate from the graph: $7400 billion; number from the equation: $7406.1 billion.

48) $y - y_1 = m(x - x_1)$ 49) $y = 7x - 9$ 50) $y = -8x - 1$

51) $y = -\dfrac{4}{9}x + 2$ 52) $y = \dfrac{3}{2}x + 4$ 53) $y = -\dfrac{1}{3}x - 5$

54) $y = \dfrac{1}{2}x - 7$ 55) $y = 9$ 56) $x = 4$ 57) $x - 2y = -6$

58) $x + y = -3$ 59) $3x + y = 5$ 60) $2x - 5y = 40$

61) $6x - y = 0$ 62) $5x + 3y = 6$ 63) $3x - 4y = -1$

64) $x - 6y = 5$

65) a) $y = 186.2x + 944.2$
    b) The number of worldwide wireless subscribers is increasing by 186.2 million per year.
    c) 1316.6 million; this is slightly less than the number given on the chart.

66) parallel   67) parallel   68) neither   69) perpendicular

70) parallel   71) neither   72) perpendicular

73) $y = 5x + 6$ 74) $y = -3x - 4$ 75) $x - 4y = -7$

76) $y = \dfrac{5}{3}x + \dfrac{4}{3}$ 77) $y = 2x - 7$ 78) $y = x + 2$

79) $y = -\dfrac{11}{2}x + 4$ 80) $3x + 2y = -1$ 81) $x = 2$

82) $y = 9$ 83) $x = -1$ 84) $y = 0$

85)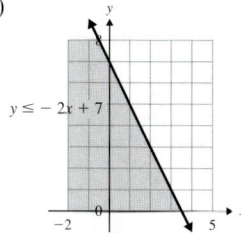

$y \leq -2x + 7$

86)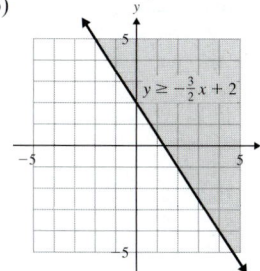

$y \geq -\frac{3}{2}x + 2$

87)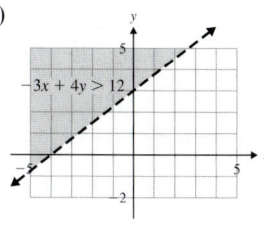

$-3x + 4y > 12$

88)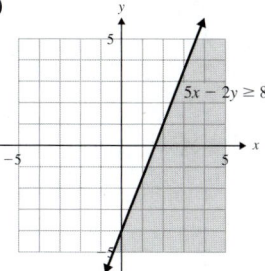

$5x - 2y \geq 8$

89)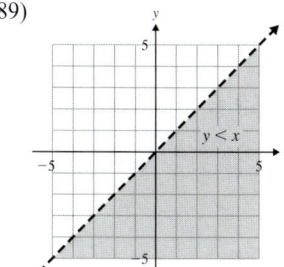

$y < x$

90)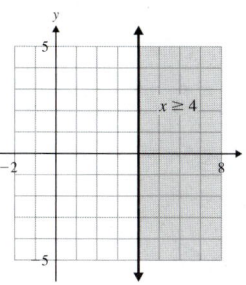

$x \geq 4$

91)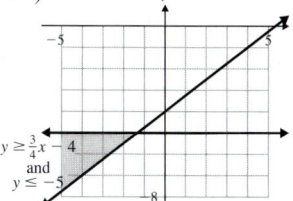

$y \geq \frac{3}{4}x - 4$
and
$y \leq -5$

92)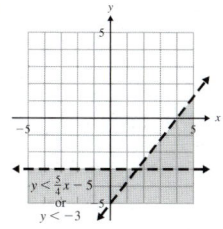

$y < \frac{5}{2}x - 5$
or
$y < -3$

93)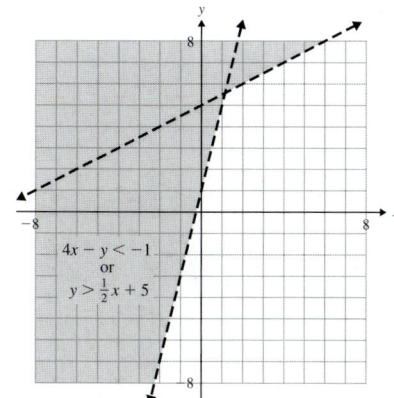

$4x - y < -1$
or
$y > \frac{1}{2}x + 5$

94)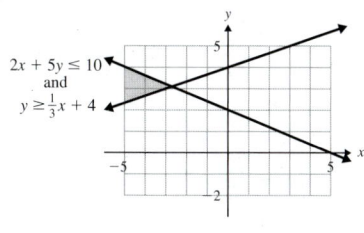

$2x + 5y \leq 10$
and
$y \geq \frac{1}{3}x + 4$

95)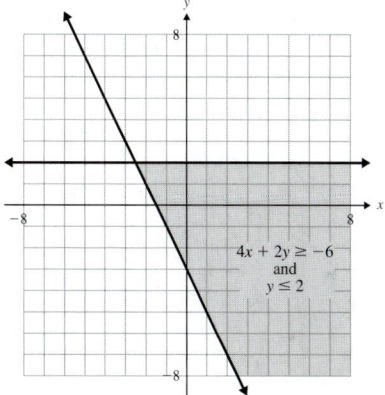

$4x + 2y \geq -6$
and
$y \leq 2$

96)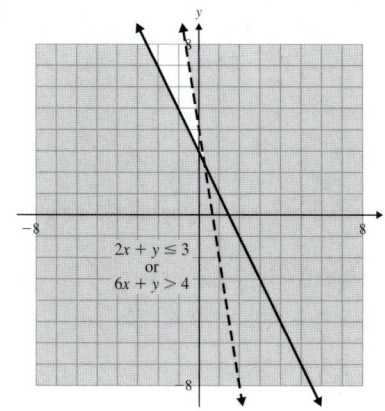

$2x + y \leq 3$
or
$6x + y > 4$

97) domain: $\{-3, 5, 12\}$
range: $\{1, 3, -3, 4\}$
not a function

98) domain: $\{-6, 2, 5\}$
range: $\{0, 1, 8, 13\}$
not a function

99) domain: {Beagle, Siamese, Parrot}
range: {Dog, Cat, Bird}
function

100) domain: $(-\infty, \infty)$
range: $(-\infty, \infty)$
function

101) domain: $[0, 4]$
range: $[0, 2]$
not a function

102) $(-\infty, \infty)$; function   103) $(-\infty, -3) \cup (-3, \infty)$; function

104) $(-\infty, 0) \cup (0, \infty)$; function   105) $[0, \infty)$; not a function

106) $(-\infty, \infty)$; function   107) $\left(-\infty, \frac{2}{7}\right) \cup \left(\frac{2}{7}, \infty\right)$; function

108) $f(3) = -14, f(-2) = -5$ 109) $f(3) = 27, f(-2) = -8$

110) $f(3) = -2, f(-2) = 1$

111) a) 8   b) -27 c) 32   d) 5   e) $5a - 12$
f) $t^2 + 6t + 5$ g) $5k + 28$ h) $5c - 22$
i) $5x + 5h - 12$ j) $5h$

112) -4 113) $\frac{1}{3}$

114)

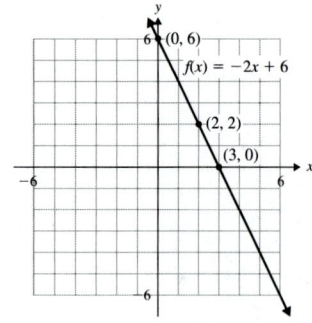

115) a)

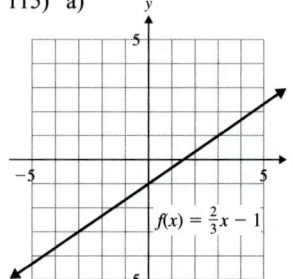

b)

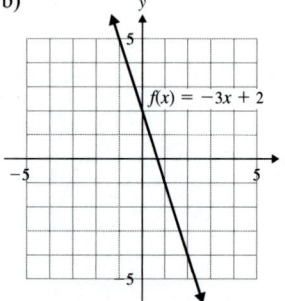

116) x-int: $(-2, 0)$; y-int: $(0, 3)$

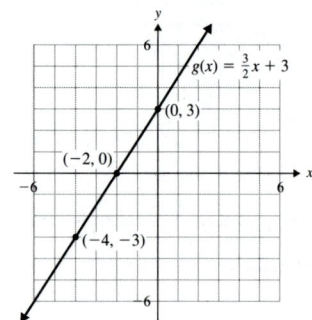

117)

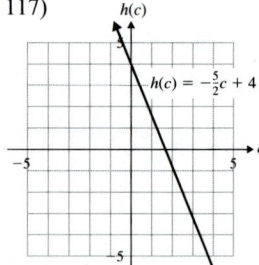

118)

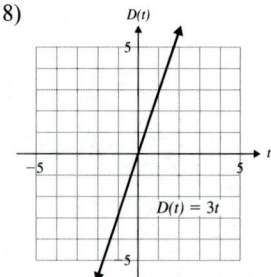

119) a) 960 MB; 2880 MB   b) 2.5 sec

120) a) $D(2) = 840$; after 2 hr, the jet has traveled 840 mi.
       b) $t = 5$; in 5 hr the jet can travel 2100 mi.

Chapter 3 Test

1) yes

2)

x	y
0	4
3	−2
−1	6
2	0

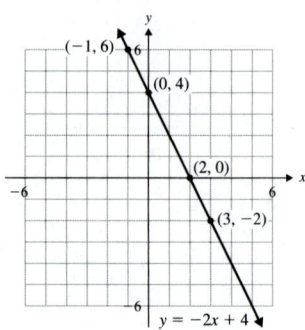

3) negative; positive

4) a) $(6, 0)$ b) $(0, -4)$ c) Answers may vary.

   d)

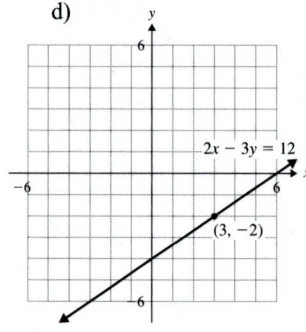

5)

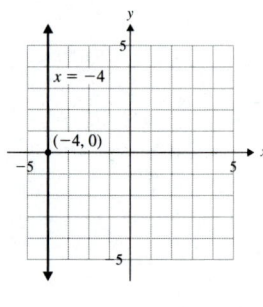

6) a) $-\dfrac{3}{4}$ b) 0

7)

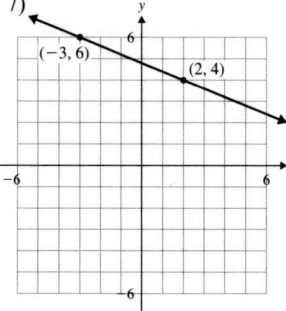

8)

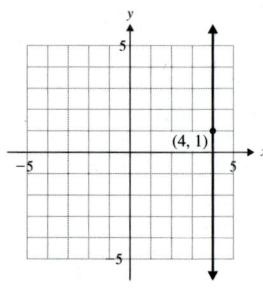

9) $y = 3x - 4$

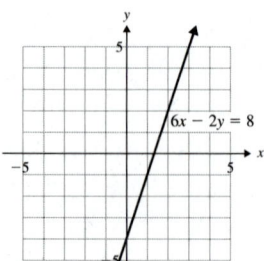

10) $y = -4x + 5$ 11) $x - 2y = -13$ 12) perpendicular

13) a) $y = \dfrac{1}{3}x - 9$ b) $y = \dfrac{5}{2}x - 6$

14) a) $55,200   b) $y = 5.8x + 32$

c) $m = 5.8$; the profit is increasing by $5.8 thousand or
$5800 per year.

d) y-int: $(0, 32)$; the profit in 2004 was $32,000.

e) 2014

15)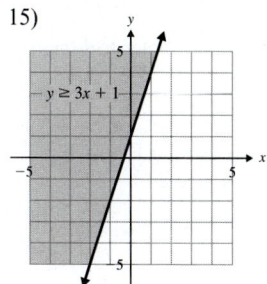

16)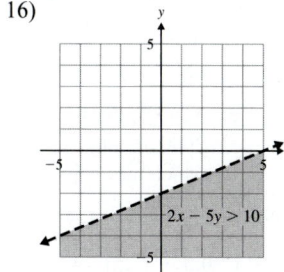

17)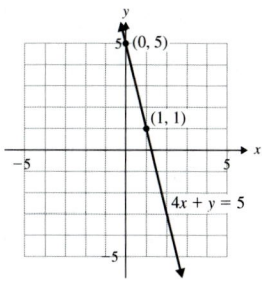

18) $5x + 4y = -36$ 19) $y = -3x$

20) a) yes   b) $(-\infty, -7) \cup (-7, \infty)$ 21) -37

22) $8a + 3$ 23) $8t + 19$

17)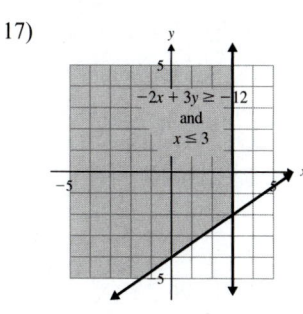

18)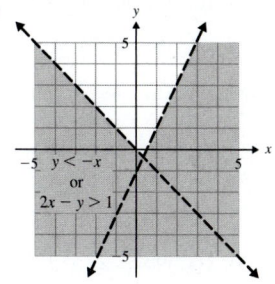

24)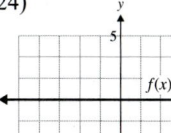

25)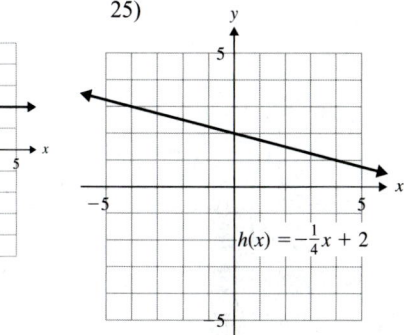

19) domain: $\{-2, 1, 3, 8\}$;
range: $\{-5, -1, 1, 4\}$;
function

20) domain: $[-3, \infty)$; range: $(-\infty, \infty)$; not a function

21) a) $(-\infty, \infty)$ b) yes   22) a) $\left(-\infty, \frac{5}{2}\right) \cup \left(\frac{5}{2}, \infty\right)$ b) yes

23) -3 24) 5 25) -22 26) 5 27) $t^2 - 3t + 7$

28) $-4h + 30$ 29)

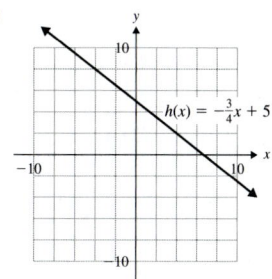

30) a) 36 MB   b) 11 sec

Cumulative Review for Chapters 1–3

1) $\frac{3}{10}$ 2) 37 in.   3) -64 4) $\frac{5}{24}$ 5) $\frac{13}{5}$

6) $2(11) - 53; -31$ 7) $\{-3\}$ 8) $\varnothing$ 9) $w = \frac{r - t}{z}$

10) $\left\{-\frac{18}{7}, 2\right\}$ 11) $\left(-\infty, -\frac{3}{2}\right]$ 12) $(-2, 4)$

13) 300 calories   14) $m\angle A = 30°$, $m\angle B = 122°$

15) Lynette's age = 41; daughter's age = 16   16) $m = 1$

Chapter 4

Section 4.1

1) no   2) yes   3) no   4) yes   5) The lines are parallel.

6) The graphs are the same line.   7) dependent

8) inconsistent

9) $(3, 1)$

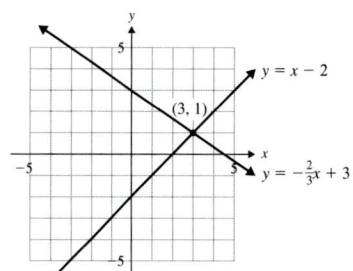

10) $(2, -1)$

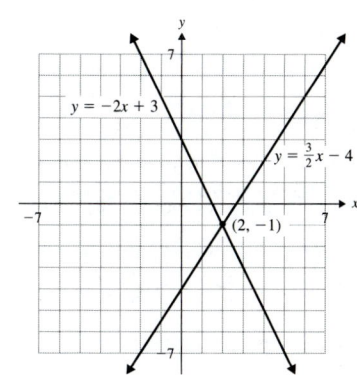

11) $(-1, -3)$

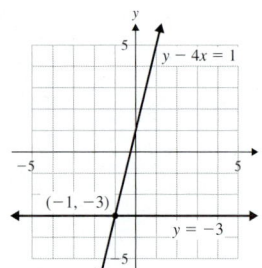

12) $(1, -2)$

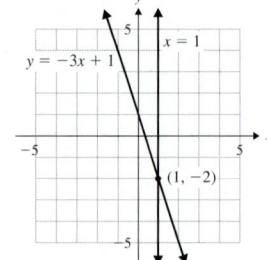

13) $\varnothing$; inconsistent system

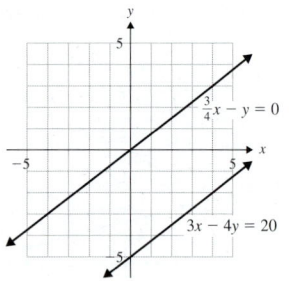

14) $\{(x, y)|-2x + y = -4\}$; dependent equations

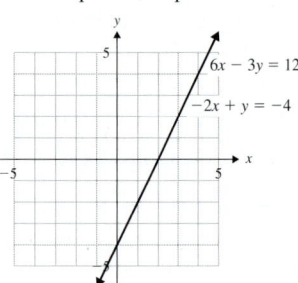

15) $(1, -1)$

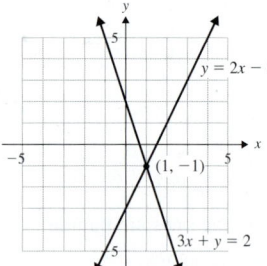

16) $(4, 5)$

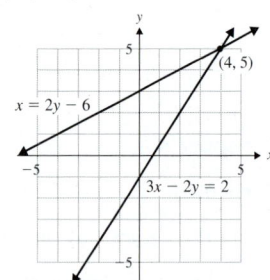

17) $\{(x, y)|y = -3x + 1\}$; dependent equations

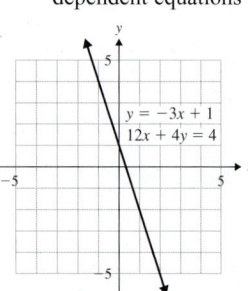

18) $\varnothing$; inconsistent system

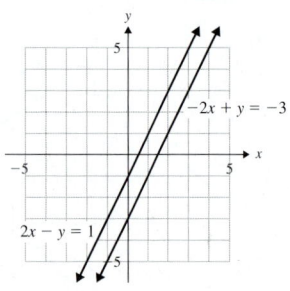

19–22) Answers may vary.   23) B; $(-2, 1)$ is in quadrant II.

24) C; $(0, -5)$ is on the y-axis not the x-axis.

25) The slopes are different.

26) The slopes are the same, but the y-intercepts are different.

27) infinite number of solutions   28) one solution

29) one solution   30) no solution   31) no solution

32) one solution   33) one solution

34) infinite number of solutions

35) a) after 2001   b) 2001: 5.3 million

c) 1999–2001   d) 1999–2001

36) a) In 1998 there were approximately 1% more households with Internet access in Nevada than in Delaware.

b) In 2001, 52.5% of households in both states had Internet access.

c) Nevada; the slope of the line segment for Nevada from 2000 to 2001 is greater (steeper) than for Delaware, which means its rate of change is greater.

37) $(3, -2)$ 38) $(1, 4)$ 39) $(-5, -1)$ 40) $(-2, 2)$

41) $(-0.5, -1.25)$ 42) $(3.75, 0)$

43) It is the only variable with a coefficient of 1.

44) $(2, 10)$ 45) $(-1, -7)$ 46) $(-5, 3)$ 47) $(4, 0)$

48) $(6, -6)$ 49) $\varnothing$; inconsistent   50) $\varnothing$; inconsistent

51) $(2, 2)$ 52) $\left(\dfrac{1}{4}, 0\right)$ 53) $\left(-\dfrac{4}{5}, 3\right)$ 54) $(8, -1)$

55) $\{(x, y)|6x + y = -6\}$; dependent

56) $\{(x, y)|x - 2y = 10\}$; dependent   57) $(0, -6)$

58) $\left(-4, -\dfrac{2}{3}\right)$ 59) $(3, 2)$ 60) $(-3, 4)$

61) Multiply the equation by the LCD of the fractions to eliminate the fractions.

62) Multiply the equation by the power of 10 that will eliminate the decimals.

63) $(6, 1)$ 64) $(5, 4)$ 65) $\varnothing$; inconsistent

66) $\left\{(x, y)\,\middle|\, y - \dfrac{5}{2}x = -2\right\}$; dependent

67) $(8, 0)$ 68) $(3, 5)$ 69) $(9, -2)$ 70) $(20, 10)$

71) $(8, 6)$ 72) $(0, -2)$ 73) $(1, 3)$ 74) $(-3, 4)$

75) $\left(\dfrac{3}{4}, 0\right)$ 76) $\{(x, y)|3x - y = 4\}$; dependent

77) $\varnothing$; inconsistent   78) $\varnothing$; inconsistent

79) $\{(x, y)|x - 6y = -5\}$; dependent   80) $\left(\dfrac{1}{4}, \dfrac{1}{2}\right)$

81) $(0, 2)$ 82) $(-2, 5)$

83) $\{(x, y)|7x + 2y = 12\}$; dependent   84) $\varnothing$; inconsistent

85) $(-6, 1)$ 86) $(-3, -5)$ 87) $(1, 2)$ 88) $(5, -6)$

89) $(12, -1)$ 90) $(2, 7)$ 91) $\varnothing$; inconsistent

92) $\{(x, y)|x - 0.5y = 0.2\}$; dependent

93) $\left(-\dfrac{123}{17}, \dfrac{78}{17}\right)$ 94) $\left(-\dfrac{65}{63}, -\dfrac{29}{63}\right)$ 95) $\left(\dfrac{85}{46}, \dfrac{45}{23}\right)$

96) $\left(\dfrac{49}{18}, \dfrac{64}{27}\right)$

97) a) Rent-for-Less: $24; Frugal: $30

   b) Rent-for-Less: $64; Frugal: $60

   c) (120, 48); if the car is driven 120 mi, the cost would be the same from each company: $48.

   d) If a car is driven less than 120 mi, it is cheaper to rent from Rent-for-Less. If a car is driven more than 120 mi, it is cheaper to rent from Frugal Rentals. If a car is driven exactly 120 mi, the cost is the same from each company.

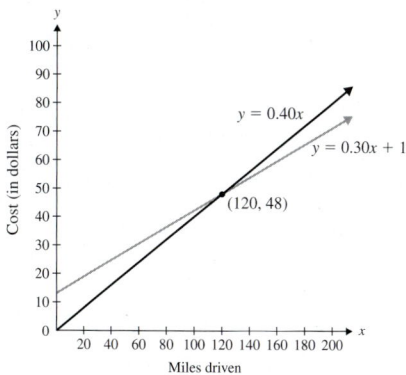

98) a) Discount: $120; Comfort Ride: $160

   b) Discount: $480; Comfort Ride: $460

   c) (300, 360); if the truck is driven 300 mi, the cost would be the same for each company: $360.

   d) If a truck is driven less than 300 mi, it is cheaper to rent from Discount Van Lines. If a truck is driven more than 300 mi, it is cheaper to rent from Comfort Ride. If a truck is driven exactly 300 mi, the cost is the same from each company.

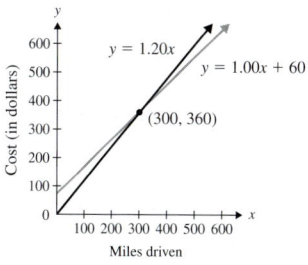

99) $(-3, 7)$ 100) $(3, 6)$ 101) $\left(5, -\dfrac{3}{2}\right)$ 102) $(-5, -2)$

103) $(8, 0)$ 104) $\left\{(x, y) \,\middle|\, y = \dfrac{1}{3}x + 4\right\}$; dependent

105) $\varnothing$; inconsistent   106) $(-4, -4)$ 107) $(-6, 2)$

108) $\left(\dfrac{5}{3}, 7\right)$ 109) $(-3, 3)$ 110) $(2, 1)$ 111) $\left(-\dfrac{3}{2}, 4\right)$

112) $(3, -4)$ 113) $(5, 0)$ 114) $\left(\dfrac{1}{3}, 9\right)$

115) a) The variables are eliminated, and you get a false statement.

   b) The variables are eliminated, and you get a true statement.

116) a) 8   b) c can be any real number except 8.

117) a) 3   b) c can be any real number except 3.

118) a) 4   b) a can be any real number except 4.

119) 5   120) -3

121) $\left(\dfrac{2}{a}, \dfrac{3}{b}\right)$ 122) $\left(-\dfrac{19}{3a}, -\dfrac{5}{b}\right)$ 123) $\left(-1, \dfrac{1}{b}\right)$

124) $\left(\dfrac{2}{a}, 1\right)$ 125) $\left(0, \dfrac{c}{b}\right)$ 126) $\left(\dfrac{2c}{a}, \dfrac{3c}{b}\right)$ 127) $(8, -2)$

128) $(1, 3)$ 129) $\left(\dfrac{1}{3}, 3\right)$ 130) $\left(-3, \dfrac{1}{2}\right)$

131) 4

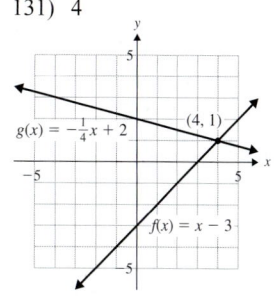

132) -2

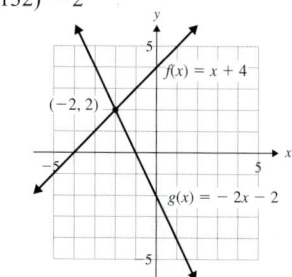

133) -1

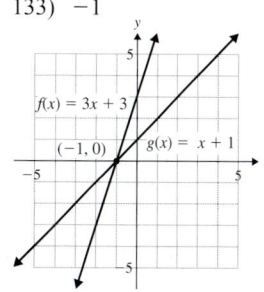

134) 3

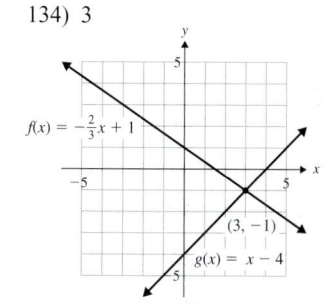

Section 4.2

1) yes   2) yes   3) no   4) no   5) Answers may vary.   6) 4

7) $(-2, 0, 5)$ 8) $(1, 2, -3)$ 9) $(1, -1, 4)$ 10) $(-5, 4, 0)$

11) $\left(2, -\dfrac{1}{2}, \dfrac{5}{2}\right)$ 12) $\left(6, \dfrac{3}{4}, -5\right)$ 13) $\varnothing$; inconsistent

14) $\varnothing$; inconsistent

15) $\{(x, y, z) \,|\, 5x + y - 3z = -1\}$; dependent equations

16) $\{(x, y, z) \,|\, 2x - 5y + 8z = 3\}$; dependent equations

17) $\{(a, b, c) \,|\, -a + 4b - 3c = -1\}$; dependent equations

18) $\varnothing$; inconsistent   19) $(2, 5, -5)$ 20) $(-3, 1, 6)$

21) $\left(-4, \dfrac{3}{5}, 4\right)$ 22) $\left(1, \dfrac{2}{3}, -2\right)$ 23) $(0, -7, 6)$

24) $\left(4, -\dfrac{1}{5}, 9\right)$ 25) $(1, 5, 2)$ 26) $(7, -6, 1)$

27) $\left(-\dfrac{1}{4}, -5, 3\right)$ 28) $\left(-2, 0, -\dfrac{1}{3}\right)$ 29) $\varnothing$; inconsistent

30) $(5, -1, 7)$ 31) $\left(4, -\dfrac{3}{2}, 0\right)$

32) $\{(a, b, c)|3a + b - 2c = -3\}$; dependent equations

33) $(4, 4, 4)$ 34) $\left(0, -3, -\dfrac{1}{2}\right)$

35) $\{(x, y, z)|-4x + 6y + 3z = 3\}$; dependent equations

36) $\varnothing$; inconsistent   37) $\left(1, -7, \dfrac{1}{3}\right)$ 38) $(4, -2, 3)$

39) $(0, -1, 2)$ 40) $\left(0, 0, -\dfrac{1}{4}\right)$ 41) $(-3, -1, 1)$

42) $(-6, -3, 1)$ 43) Answers may vary.

44) Answers may vary.   45) $(3, 1, 2, 1)$ 46) $(-2, 0, 1, 1)$

47) $(0, -3, 1, -4)$ 48) $(5, 2, 2, 3)$

Section 4.3

1) 17 and 19   2) 72 and 36

3) *The Aviator:* 11; *Finding Neverland:* 7

4) USC: 33; Michigan: 20

5) IHOP: 1156; Waffle House: 1470

6) UCLA: 11; Kentucky: 7   7) 1939: 5500; 2004: 44,468

8) George Strait: 22; Alan Jackson: 16

9) beef: 63.4 lb; chicken: 57.1 lb

10) speeches: 18; papers: 9   11) length: 26 in.; width: 13 in.

12) length: 14 cm; width: 9 cm   13) width: 30 in.; height: 80 in.

14) length: 12.5 in.; width: 8.5 in.   15) length: 9 cm; width: 5 cm

16) width: 40 ft; length: 70 ft   17) $m\angle x = 72°$; $m\angle y = 108°$

18) $m\angle x = 92°$; $m\angle y = 46°$

19) Marc Anthony: $86; Santana: $66.50

20) Maroon 5: $7.00; Train: $21.50

21) two-item: $5.19; three-item: $6.39

22) average ticket price: $54.40; parking pass: $25.00

23) key chain: $2.50; postcard: $0.50

24) deluxe: $7; regular: $4

25) cantaloupe: $1.50; watermelon: $3.00

26) Coke: 9.3 tsp; Mountain Dew: 11 tsp

27) hamburger: $0.61; small fries: $1.39

28) hamburger: 140; small fries: 310   29) 9%: 3 oz; 17%: 9 oz

30) 4%: 36 ml; 10%: 18 ml   31) peanuts: 7 lb; cashews: 3 lb

32) $0.44: 12; $0.28: 8   33) 3%: $2800; 5%: $1200

34) 4%: $9000; 7%: $11,000   35) 52 quarters, 58 dimes

36) 4 AA batteries, 5 C batteries

37) 4 L of pure acid, 8 L of 10% solution

38) 16 oz of orange juice, 60 oz of the fruit drink

39) car: 60 mph; truck: 50 mph

40) passenger train: 50 mph; freight train: 30 mph

41) walking: 4 mph; biking: 11 mph

42) small plane: 200 mph; jet: 500 mph

43) Nick: 14 mph; Scott: 12 mph

44) car: 40 mph; train: 50 mph

45) hot dog: $2.00; fries: $1.50; soda: $2.00

46) adults': 200; seniors': 60; children's: 30

47) Clif Bar: 11g; Balance Bar: 15 g; PowerBar: 24 g

48) Hellman's Real Mayonnaise: 90 cal; Hellman's Light Mayonnaise: 35 cal; Miracle Whip: 35 cal

49) Knicks: $160 million; Lakers: $149 million; Bulls: $119 million

50) Bounty: 37%; Brawny: 12%; Scott: 10%

51) value: $22; regular: $36; prime: $45

52) field box: $125; infield grandstand: $50; bleacher: $26

53) $104°, 52°, 24°$ 54) $90°, 60°, 30°$ 55) $80°, 64°, 36°$

56) $110°, 55°, 15°$ 57) 12 cm, 10 cm, 7 cm

58) 28 in., 16 in., 14 in.

Section 4.4

1) $\begin{bmatrix} 1 & -7 & | & 15 \\ 4 & 3 & | & -1 \end{bmatrix}$

2) $\begin{bmatrix} 1 & 6 & | & 4 \\ -5 & 1 & | & -3 \end{bmatrix}$

3) $\begin{bmatrix} 1 & 6 & -1 & | & -2 \\ 3 & 1 & 4 & | & 7 \\ -1 & -2 & 3 & | & 8 \end{bmatrix}$

4) $\begin{bmatrix} 1 & 2 & -7 & | & 3 \\ 3 & -5 & 0 & | & -1 \\ -1 & 0 & 2 & | & -4 \end{bmatrix}$

5) $\begin{aligned} 3x + 10y &= -4 \\ x - 2y &= 5 \end{aligned}$

6) $\begin{aligned} x - y &= 6 \\ -4x + 7y &= 2 \end{aligned}$

7) $\begin{aligned} x - 6y &= 8 \\ y &= -2 \end{aligned}$

8) $\begin{aligned} x + 2y &= 11 \\ y &= 3 \end{aligned}$

9) $\begin{aligned} x - 3y + 2z &= 7 \\ 4x - y + 3z &= 0 \\ -2x + 2y - 3z &= -9 \end{aligned}$

10) $\begin{aligned} x + 4y - 3z &= -5 \\ -x + 2y + 5z &= 8 \\ 6x - 2y - z &= 3 \end{aligned}$

11) $\begin{aligned} x + 5y + 2z &= 14 \\ y - 8z &= 2 \\ z &= -3 \end{aligned}$

12) $\begin{aligned} x + 4y - 7z &= -11 \\ y + 3z &= -1 \\ z &= 6 \end{aligned}$

13) $(3, -1)$ 14) $(-8, -3)$ 15) $(-6, -5)$ 16) $(10, -4)$

17) $(0, -2)$ 18) $(2, 1)$ 19) $(-1, 4, 8)$ 20) $(5, -2, -3)$

21) $(10, 1, -4)$ 22) $(5, 0, -7)$ 23) $(0, 1, 8)$ 24) $(3, 1, 1)$

25) $\varnothing$; inconsistent

26) $\{(x, y, z)|x - y + 3z = 1\}$; dependent equations

27) $(-5, 2, 1, -1)$ 28) $(1, -4, 0, 2)$ 29) $(3, 0, -2, 1)$

30) $(1, -2, 3, -4)$

Chapter 4 Review Exercises

1) no   2) yes

3) $(-2, -3)$ 4) $(1, 1)$

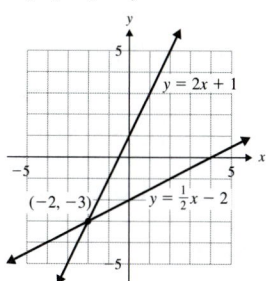
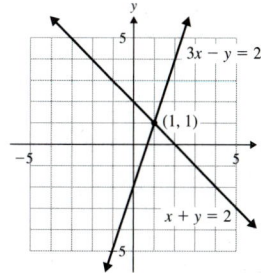

5) $\varnothing$ 6) $(-2, 3)$

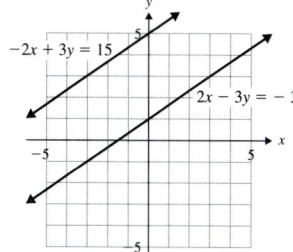
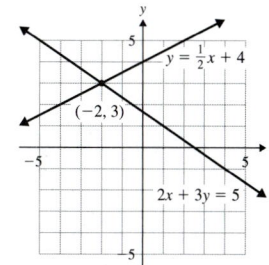

7) $\{(x, y)|4x + y = -4\}$; dependent equations

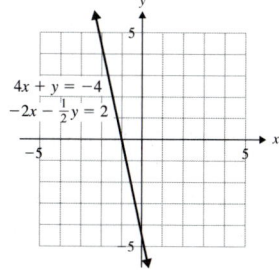

8) one solution   9) no solution

10) infinite number of solutions   11) no solution

12) a) during the 4th quarter of 2003

b) During the 3rd quarter of 2003, approximately 89.1% of flights left on time.

13) $(-10, 1)$ 14) $\left\{(x, y)\left|y = \dfrac{5}{6}x - 2\right.\right\}$; dependent equations

15) $(9, 0)$ 16) $\left(\dfrac{1}{3}, 5\right)$ 17) $(2, -1)$ 18) $\varnothing$; inconsistent

19) $(-5, 3)$ 20) $(2, 9)$ 21) $(1, 4)$ 22) $\varnothing$; inconsistent

23) $(1, -6)$ 24) $(0, 7)$ 25) $(-3, -3)$ 26) $(4, 7)$

27) $\{(x, y)|6x - 4y = 12\}$; dependent equations

28) $(2, 1)$ 29) $\varnothing$; inconsistent   30) $(-1, 5)$

31) $\left(\dfrac{83}{23}, -\dfrac{50}{23}\right)$ 32) $\left(\dfrac{37}{13}, \dfrac{53}{26}\right)$ 33) no   34) yes

35) $(3, -1, 4)$ 36) $(-4, 3, 8)$

37) $\left(-1, 2, \dfrac{1}{2}\right)$ 38) $\left(-2, 1, -\dfrac{2}{5}\right)$ 39) $\left(3, \dfrac{2}{3}, -\dfrac{1}{2}\right)$

40) $\left(-5, \dfrac{4}{5}, 6\right)$ 41) $\varnothing$; inconsistent

42) $\{(x, y, z)|4x + 2y + z = 0\}$; dependent equations

43) $\{(a, b, c)|3a - 2b + c = 2\}$; dependent equations

44) $\varnothing$; inconsistent   45) $(1, 0, 3)$ 46) $(3, 2, 2)$

47) $\left(\dfrac{3}{4}, -2, 1\right)$ 48) $\left(0, \dfrac{1}{3}, 4\right)$ 49) 34 dogs, 17 cats

50) Spanish: 110; French: 73   51) hot dog: $5.00; soda: $3.25

52) width: 15 in; length: 18 in.

53) $m\angle x = 38°$; $m\angle y = 19°$

54) gummi bears: 4 lb; jelly beans: 6 lb

55) pure alcohol: 20 ml; 4% solution: 460 ml

56) car: 60 mph; bus: 48 mph

57) Propel: 35 mg; Powerade: 52 mg; Gatorade: 110 mg

58) Goodyear: 19%; Michelin: 18%; Bridgestone: 16%

59) Blair: 65; Serena: 50; Chuck: 25

60) 2005: 13.6 million; 2006: 27.6 million; 2007: 42.5 million

61) ice cream cone: $1.50; shake: $2.50; sundae: $3.00

62) reserved: $60; behind-the-stage: $40; lawn: $30

63) 92°, 66°, 22°   64) 18 in., 13 in., 9 in.   65) $(-9, 2)$

66) $(-5, 1)$ 67) $(1, 0)$ 68) $(10, -5)$ 69) $(5, -2, 6)$

70) $(-3, 3, 3)$

Chapter 4 Test

1) yes

2) $(2, 1)$ 3) $\varnothing$; inconsistent

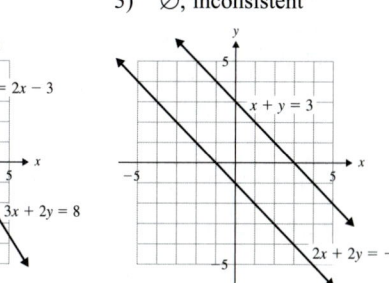

4) $(-3, 2)$ 5) $\left\{(x, y)\left|y = \dfrac{3}{4}x + \dfrac{7}{4}\right.\right\}$; dependent equations

6) $(1, 6)$ 7) $(4, 0)$ 8) $(0, 3, -2)$ 9) $(9, 1)$ 10) $\left(-4, \dfrac{1}{3}\right)$

11) $(-2, 7)$ 12) Subway: 280 cal; Whopper: 680 cal

13) screws: $4; nails: $3   14) 105°, 42°, 33°

15) $(6, -2)$ 16) $(1, -1, 1)$

Cumulative Review for Chapters 1–4

1) $-\dfrac{1}{6}$ 2) $3\dfrac{1}{8}$ 3) -29 4) 30 in^2 5) $-24x^2 + 8x + 56$

6) $\left\{\dfrac{1}{2}\right\}$ 7) $\varnothing$ 8) $\left[-6, \dfrac{11}{7}\right]$ 9) 2,100,000

10) $\left(-\infty, -\dfrac{1}{2}\right) \cup \left(\dfrac{5}{4}, \infty\right)$ 11) a) $h = \dfrac{2A}{b_1 + b_2}$ b) 6 cm

12) 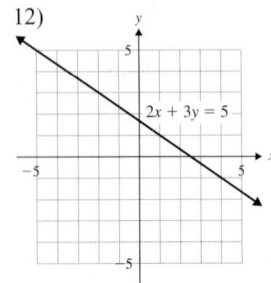 13)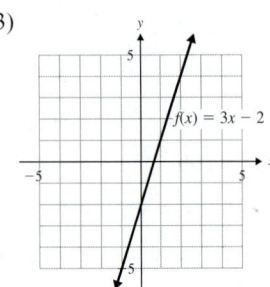

14) x-int: $\left(\dfrac{5}{2}, 0\right)$; y-int: $(0, -2)$ 15) $y = -\dfrac{7}{8}x - \dfrac{17}{8}$

16) perpendicular 17) $(-\infty, \infty)$ 18) $\left(-\infty, \dfrac{7}{2}\right) \cup \left(\dfrac{7}{2}, \infty\right)$

19) a) -11 b) $4p + 9$ c) $4n + 17$ 20) $(7, -8)$

21) $(3, 1)$ 22) $\varnothing$; inconsistent 23) $(-2, -1, 0)$

24) 4-ft boards: 16; 6-ft boards: 32

25) Juanes: 17; Alejandro Sanz: 14; Shakira: 7

Chapter 5

Section 5.1

1) base: $-7r$; exponent: 5 2) base: h; exponent: 3 3) 9^6

4) $\left(-\dfrac{2}{7}m\right)^4$

5) $(3 + 4)^2 = 49, 3^2 + 4^2 = 25$. They are not equivalent because when evaluating $(3 + 4)^2$, we first add $3 + 4$ to get 7, then square the 7.

6) Answers may vary.

7) No. $3^4 = -1 \cdot 3^4 = -1 \cdot 81 = -81$; $(-3)^4 = (-3) \cdot (-3) \cdot (-3) \cdot (-3) = 81$

8) No. $2k^4 = 2 \cdot k^4$; $(2k)^4 = 2^4 \cdot k^4 = 16k^4$ 9) 64

10) 1000 11) 675 12) $\dfrac{1}{32}$ 13) $(-4)^7$ 14) 8^{10} 15) a^5

16) $9w^7$ 17) k^6 18) $-6z^{12}$ 19) $56h^{15}$ 20) $-24p^7$

21) $-60n^{12}$ 22) $-112y^8$ 23) $14t^{13}$ 24) v^{16} 25) a^{36}

26) w^{45} 27) 64 28) 81 29) $\dfrac{1}{32}$ 30) $\dfrac{81}{16}$ 31) $\dfrac{64}{y^3}$

32) $\dfrac{w^6}{k^6}$ 33) $10{,}000r^4$ 34) $32g^5$ 35) $-64a^3b^3$

36) $49m^2n^2$ 37) k^{24} 38) p^{29} 39) 64 40) 9

41) $288k^{14}t^4$ 42) $72a^{10}b^2$ 43) $125m^{22}$ 44) $-484h^{31}$

45) $\dfrac{d^{18}}{4c^{30}}$ 46) $-\dfrac{5m^{21}}{n^{24}}$ 47) $\dfrac{3a^{40}b^{15}}{4c^3}$ 48) $\dfrac{9x^{24}}{5y^8z^{12}}$

49) $676u^8v^{26}$ 50) $-\dfrac{1}{48}h^{29}k^8$ 51) $\dfrac{64r^{18}s^6}{t^9}$ 52) $\dfrac{25x^6z^2}{144y^{16}}$

53) $\dfrac{81n^8}{m^{16}p^6}$ 54) $\dfrac{81b^{28}}{16a^{20}c^8}$ 55) False 56) False 57) True

58) False 59) 1 60) -1 61) -1 62) 0 63) 2 64) 1

65) -6 66) 0 67) $\dfrac{1}{36}$ 68) $\dfrac{1}{16}$ 69) 49 70) 125

71) $\dfrac{27}{64}$ 72) $\dfrac{1000}{27}$ 73) -32 74) $\dfrac{121}{36}$ 75) $-\dfrac{1}{64}$

76) $-\dfrac{1}{64}$ 77) $\dfrac{1}{16}$ 78) $\dfrac{29}{300}$ 79) $\dfrac{83}{81}$ 80) 2 81) $\dfrac{1}{y^4}$

82) $\dfrac{1}{c}$ 83) $\dfrac{b^3}{a^{10}}$ 84) $\dfrac{v^2}{u^6}$ 85) $\dfrac{x^4y^5}{10}$ 86) $\dfrac{7}{t^3u^6}$ 87) $\dfrac{8x^3}{y^7}$

88) $\dfrac{t^3}{2r^5}$ 89) $\dfrac{8a^6c^{10}}{5bd}$ 90) h^7k^4 91) $\dfrac{36}{a^2}$ 92) $\dfrac{q^4}{81}$

93) $\dfrac{c^2d^2}{144b^2}$ 94) $\dfrac{z^6}{64x^6y^6}$ 95) $-\dfrac{6}{r^2}$ 96) $\dfrac{2}{w^5}$ 97) $-\dfrac{1}{p^8}$

98) $-\dfrac{1}{k^7}$ 99) x 100) a^4b^9 101) n^5 102) h^7 103) 64

104) 32 105) $\dfrac{1}{125}$ 106) $\dfrac{1}{81}$ 107) $9d^5$ 108) $\dfrac{5}{9}w^5$ 109) $\dfrac{1}{t^2}$

110) $\dfrac{1}{a^7}$ 111) $\dfrac{1}{x^9}$ 112) $\dfrac{1}{h^8}$ 113) $-\dfrac{6}{k^3}$ 114) $15m^4$

115) $\dfrac{w^6}{9v^3}$ 116) $\dfrac{3b^5}{2a^4}$ 117) $\dfrac{5}{m^{11}n^2}$ 118) $\dfrac{3}{7c^3d^3}$

119) $(x + y)^5$ 120) $a - 3b$

Section 5.2

1) $270g^{12}$ 2) $48d^9$ 3) $\dfrac{23}{t^{10}}$ 4) $\dfrac{1}{r^4}$ 5) $\dfrac{8}{27}x^{30}y^{18}$ 6) $\dfrac{a^9b^9}{1000}$

7) $\dfrac{s^{16}}{49r^6}$ 8) $\dfrac{m^8n^{24}}{81}$ 9) $-k^{12}$ 10) t^{56} 11) $-32m^{25}n^{10}$

12) $169y^2z^{12}$ 13) $-\dfrac{3}{2}z^4$ 14) $-12w^8$ 15) $\dfrac{b^{24}}{a^{30}}$ 16) $\dfrac{1}{x^4y^{10}}$

17) $a^{14}b^3c^7$ 18) $\dfrac{4}{9v^{10}}$ 19) $\dfrac{27u^{30}}{64v^{21}}$ 20) $\dfrac{125}{x^9y^{12}}$ 21) $81t^8u^{20}$

22) $\dfrac{1}{144}k^{16}m^2$ 23) $\dfrac{1}{h^{21}}$ 24) $-\dfrac{1}{n^{20}}$ 25) $\dfrac{h^5}{32}$ 26) $\dfrac{17}{m^2}$

27) $56c^{14}$ 28) $80p^{17}$ 29) $\dfrac{3}{a^5}$ 30) $\dfrac{1}{9c^2d}$ 31) $\dfrac{6}{55}d^{11}$

32) $\dfrac{1}{f^{30}}$ 33) $\dfrac{9}{64n^6}$ 34) $\dfrac{1}{100x^{10}y^8}$ 35) $-\dfrac{32r^{40}}{t^{15}}$ 36) $-\dfrac{3}{p^6}$

37) $\dfrac{c}{24a^{29}b^{32}}$ 38) $-\dfrac{5x^{19}}{8z^9}$ 39) $\dfrac{1}{(x+y)^5}$ 40) $(s-t)^7$

41) p^{10n} 42) $9d^{8t}$ 43) y^{11m} 44) t^{3c} 45) $\dfrac{1}{x^{3a}}$ 46) $\dfrac{1}{b^{5y}}$

47) $\dfrac{3}{5c^{6x}}$ 48) $-\dfrac{5}{8y^{11a}}$ 49) $\dfrac{3}{8}x^2$ sq units 50) $4n^2$ in.

51) yes 52) no 53) no 54) no 55) yes 56) yes

57) no 58) Answers may vary. 59) Answers may vary.

60) Answers may vary. 61) -0.000068 62) 29,100

63) 34,502.9 64) -3570 65) -0.00005 66) 0.0000009

67) 0.00081 68) 26,450.67 69) 3,000,000 70) 700

71) -0.03921 72) 0.0000041 73) 2.1105×10^3

74) 3.82275×10^2 75) 4.8×10^{-3} 76) 3.21×10^{-4}

77) -4×10^5 78) 9.26×10^4 79) 1.1×10^4

80) -3.08×10^5 81) 8×10^{-4} 82) -8.9×10^{-7}

83) -5.4×10^{-2} 84) 9.99×10^3 85) 6.5×10^3

86) 2×10^{-8} 87) $\$1.5 \times 10^7$ 88) 1.6384×10^7 bytes

89) 1×10^{-8} cm 90) 1×10^{-9} mm 91) 20,000

92) 27,000,000,000 93) 690,000 94) 3000

95) $-24,000,000$ 96) 400,000 97) -0.0009

98) 0.0014 99) 8,000,000 100) 0.03 101) 6.7392×10^8

102) 0.5 pounds per person 103) 20 metric tons

104) 26,400,000 droplets 105) 24,000,000,000 kW-hr

106) 0.369 mi

Section 5.3

1) Yes; the coefficients are real numbers and the exponents are whole numbers.

2) Yes; the coefficients are real numbers and the exponents are whole numbers.

3) No; one of the exponents is a negative number.

4) Yes; the coefficient of y is a real number and the exponent is a whole number.

5) No; two of the exponents are fractions.

6) No; there is a variable in the denominator.

7) binomial 8) monomial 9) trinomial 10) binomial

11) monomial 12) trinomial

13) It is the same as the degree of the term in the polynomial with the highest degree.

14) Answers may vary.

15) Add the exponents on the variables.

16) Answers may vary.

17)

Term	Coeff.	Degree
$7y^3$	7	3
$10y^2$	10	2
$-y$	-1	1
2	2	0

Degree is 3.

18)

Term	Coeff.	Degree
$4d^2$	4	2
$12d$	12	1
-9	-9	0

Degree is 2.

19)

Term	Coeff.	Degree
$-9r^3s^2$	-9	5
$-r^2s^2$	-1	4
$\dfrac{1}{2}rs$	$\dfrac{1}{2}$	2
$6s$	6	1

Degree is 5.

20)

Term	Coeff.	Degree
$8m^2n^2$	8	4
$0.5m^2n$	0.5	3
$-mn$	-1	2
3	3	0

Degree is 4.

21) $5w^2 + 7w - 6$ 22) $6f^4 - 4f^2 + 9$ 23) $-8p + 7$

24) $2y^3 - 3y^2$ 25) $-5a^4 + 9a^3 - a^2 - \dfrac{3}{2}a + \dfrac{5}{8}$

26) $12d^5 - \dfrac{5}{3}d^4 + 9d^2 - 7$ 27) $\dfrac{25}{8}x^3 + \dfrac{1}{12}$

28) $\dfrac{9}{4}c - \dfrac{17}{24}$ 29) $2.6k^3 + 8.7k^2 - 7.3k - 4.4$

30) $-1.2t^4 + 4.9t^3 + 0.8t + 9.3$ 31) $17x - 8$

32) $-2n^3 - 7$ 33) $12r^2 + 6r + 11$ 34) $4z^2 + 5z + 4$

35) $1.5q^3 - q^2 + 4.6q + 15$

36) $a^4 + 2.5a^3 - 3a^2 - 5.1a + 11$

37) $-7a^4 - 12a^2 + 14$ 38) $-11w^3 + 12w - 3$

39) $8j^2 + 12j$ 40) $-5m^2 + 4$ 41) $9s^5 - 3s^4 - 3s^2 + 2$

42) $10h^5 - h^4 + 6$ 43) $-b^4 - 10b^3 + b + 20$

44) $4t^3 - 2t^2 - 7t + 5$ 45) $-\dfrac{5}{14}r^2 + r - \dfrac{7}{6}$

46) $y^3 - y^2 + \dfrac{7}{2}y + 3$ 47) $15v - 6$ 48) $6q - 15$

49) $-b^2 - 12b + 7$ 50) $-8d^2 + 9d + 5$

51) $3a^4 - 11a^3 + 7a^2 - 7a + 8$ 52) $9y^4 - 10y^2 + 10y - 3$

53) Answers may vary.   54) Answers may vary.

55) No. If the coefficients of the like terms are opposite in sign, their sum will be zero. Example:

$$(3x^2 + 4x + 5) + (2x^2 - 4x + 1) = 5x^2 + 6$$

56) Answers may vary.   57) $-9b^2 - 4$ 58) $5m^2 - 7m + 40$

59) $\dfrac{5}{4}n^3 - \dfrac{3}{2}n^2 + 3n - \dfrac{1}{8}$ 60) $\dfrac{5}{3}z^4 - z^3 - \dfrac{5}{3}z^2 + 8z$

61) $5u^3 - 10u^2 - u - 6$ 62) $15r^3 - 11r^2 + 7r + 7$

63) $-\dfrac{5}{8}k^2 - 9k + \dfrac{1}{2}$ 64) $-\dfrac{1}{4}y^2 - \dfrac{3}{2}y$

65) $3t^3 - 7t^2 - t - 3$ 66) $4x^2 - 12x - 2$

67) $9a^3 - 8a + 13$ 68) $7c^3 - c^2 + 2c + 5$ 69) $3a + 8b$

70) $4g - 2h$ 71) $-m + \dfrac{11}{6}n - \dfrac{1}{4}$ 72) $-4c - \dfrac{7}{9}d + \dfrac{11}{7}$

73) $5y^2z^2 + 7y^2z - 25yz^2 + 1$ 74) $u^2v^2 + 16uv - 15$

75) $10x^3y^2 - 11x^2y^2 + 6x^2y - 12$

76) $3r^3s^2 - 19r^2s^2 + 6$ 77) $5v^2 + 3v - 8$ 78) $13d - 9$

79) $4g^2 + 10g - 10$ 80) $11y^2 - 3y - 6$

81) $-8n^2 + 11n$ 82) $-9x^3 + x^2 - 6x + 4$

83) $8x + 14$ units   84) $4a^2 + 6a - 2$ units

85) $6w^2 - 2w - 6$ units   86) $3t + 8$ units

87) a) 16   b) 4   88) a) -30 b) 6   89) a) 29   b) 5

90) a) -4 b) 9   91) $-\dfrac{1}{2}$ 92) 6   93) 40   94) $-\dfrac{4}{3}$

95) a) $-x - 10$ b) -15 c) $-5x + 12$ d) 2

96) a) $6x - 5$ b) 25   c) $4x - 13$ d) -5

97) a) $5x^2 - 4x - 7$ b) 98   c) $3x^2 - 10x + 5$ d) -3

98) a) $x^2 - 3x + 2$ b) 12   c) $-5x^2 + 5x + 14$ d) 4

99) $-4t^2 + 4t + 6$ 100) $2t^2 - 4t + 6$ 101) -74 102) 6

103) 0   104) -11 105) $-\dfrac{9}{4}$ 106) 3

107) Answers may vary.   108) $g(x) = 5x^3 - 12x^2 - 4x + 2$

109) a) $P(x) = 4x - 2000$ b) $4000

110) a) $P(x) = 20x - 7000$ b) $3000

111) a) $P(x) = 3x - 2400$ b) $0

112) a) $P(x) = 15x - 6000$ b) $-$1500; the company loses $1500 if it sells only 300 dog houses.

113) a) $P(x) = -0.2x^2 + 19x - 9$ b) $291,000

114) a) $P(x) = -0.4x^2 + 27x - 11$ b) $304,000

Section 5.4

1) Answers may vary.   2) Answers may vary.   3) $14k^6$

4) $-8p^{10}$ 5) $\dfrac{7}{4}d^{11}$ 6) $-\dfrac{4}{15}c^{12}$ 7) $28y^2 - 63y$

8) $-132m^2 + 48m$ 9) $6v^5 - 24v^4 - 12v^3$

10) $15x^7 + 3x^5 - 21x^4$ 11) $-27t^5 + 18t^4 + 12t^3 + 21t^2$

12) $-72u^9 - 64u^8 - 96u^6 + 8u^5$

13) $2x^4y^3 + 16x^4y^2 - 22x^3y^2 + 4x^3y$

14) $-25p^7q^3 + 60p^6q^4 - 5p^6q^3 + 10p^5q^3 - 5p^5q^2$

15) $-15t^7 - 6t^6 + \dfrac{15}{4}t^5$ 16) $12x^6 - 6x^5 + \dfrac{14}{5}x^4$

17) $18g^3 + 10g^2 + 14g + 28$ 18) $-2m^2 - 35m + 43$

19) $-47a^3b^3 - 39a^3b^2 + 108a^2b - 51$

20) $-15xy$ 21) $5q^3 - 36q + 27$

22) $9m^3 + 85m^2 + 29m - 63$ 23) $2p^3 - 9p^2 - 23p + 30$

24) $7s^3 - 38s^2 + 4s + 55$

25) $15y^4 - 23y^3 + 28y^2 - 29y - 4$

26) $28n^4 + 4n^3 - 59n^2 - 19n + 42$

27) $6k^4 + \dfrac{5}{2}k^3 + 31k^2 + 15k - 30$

28) $4c^4 - \dfrac{8}{3}c^3 - 42c^2 + 32c - 72$

29) $a^4 + 3a^3 - 3a^2 + 14a - 6$

30) $2r^4 + 3r^3 + 5r^2 - 11r - 15$

31) $-24v^5 + 26v^4 - 22v^3 + 27v^2 - 5v + 10$

32) $c^6 - 2c^5 + 7c^4 - 20c^3 - 37c^2 + 14c + 21$

33) $8x^3 - 22x^2 + 19x - 6$ 34) $15n^3 + 11n^2 - 18n - 8$

35) First, Outer, Inner, Last

36) Yes. $(x + 9)^2 = (x + 9)(x + 9)$ so we can use FOIL to find the product.

37) $w^2 + 15w + 56$ 38) $k^2 + 4k - 45$

39) $n^2 - 15n + 44$ 40) $y^2 - 7y + 6$ 41) $4p^2 - 7p - 15$

42) $6t^2 + 43t + 7$ 43) $24n^2 + 41n + 12$

44) $40b^2 + 9b - 10$ 45) $0.04g^2 + 0.35g - 0.99$

46) $0.4m^2 - 0.19m - 1.05$ 47) $12a^2 + ab - 20b^2$

48) $2x^2 - 9xy - 18y^2$ 49) $60p^2 + 68pq + 15q^2$

50) $7m^2 - 10mn + 3n^2$ 51) $2a^2 + \dfrac{15}{4}ab - \dfrac{1}{2}b^2$

52) $2w^2 - \dfrac{9}{2}v^2$ 53) a) $4y + 8$ units   b) $y^2 + 4y - 12$ sq units

54) a) $18w + 14$ units   b) $20w^2 + 28w$ sq units

55) a) $2a^2 + 4a + 16$ units   b) $3a^3 - 3a^2 + 24a$ sq units

56) a) $8x^2 - 12$ units   b) $4x^4 - 12x^2 + 9$ sq units

57) Both are correct.

58) a) $3a^3 - 16a^2 + 12a + 16$ b) $3a^3 - 16a^2 + 12a + 16$
    c) They are the same.

59) $15y^2 + 54y - 24$ 60) $-24z^2 + 68z - 28$

61) $-7r^4 + 77r^3 - 126r^2$ 62) $-24g^4 - 36g^3 + 60g^2$

63) $c^3 + 6c^2 + 5c - 12$ 64) $x^3 - 4x^2 - 11x + 30$

65) $5n^5 + 55n^3 + 150n$ 66) $3k^5 - 5k^3 - 8k$

67) $2r^3 - 3r^2t - 3rt^2 + 2t^3$ 68) $x^3 + 2x^2y - 5xy^2 - 6y^3$

69) $9m^2 - 4$ 70) $25y^2 - 16$ 71) $49a^2 - 64$

72) $16x^2 - 121$ 73) $36a^2 - b^2$ 74) $4p^2 - 49q^2$

75) $n^2 - \dfrac{1}{4}$ 76) $b^2 - \dfrac{1}{25}$ 77) $\dfrac{4}{9} - k^2$ 78) $\dfrac{16}{9} - z^2$

79) $0.09x^2 - 0.16y^2$ 80) $1.44a^2 - 0.64b^2$ 81) $25x^4 - 16$

82) $81k^4 - 9l^4$ 83) $y^2 + 16y + 64$ 84) $b^2 + 12b + 36$

85) $t^2 - 22t + 121$ 86) $g^2 - 10g + 25$

87) $16w^2 + 8w + 1$ 88) $49n^2 + 28n + 4$

89) $4d^2 - 20d + 25$ 90) $9p^2 - 30p + 25$

91) $36a^2 - 60ab + 25b^2$ 92) $49x^2 + 84xy + 36y^2$

93) No. The order to operations tells us to perform exponents, $(t + 3)^2$, before multiplying by 4.

94) First find $(z - 4)^2$, then multiply by 3.
    $3(z - 4)^2 = 3z^2 - 24z + 48$

95) $6x^2 + 12x + 6$ 96) $2k^2 + 20k + 50$

97) $2a^3 + 12a^2 + 18a$ 98) $-3m^2 + 6m - 3$

99) $9m^2 + 6mn + n^2 + 12m + 4n + 4$

100) $4c^2 - 4cd + d^2 + 28c - 14d + 49$

101) $x^2 - 8x + 16 - 2xy + 8y + y^2$

102) $9r^2 + 12r + 4 - 6rt - 4t + t^2$

103) $r^3 + 15r^2 + 75r + 125$ 104) $w^3 + 12w^2 + 48w + 64$

105) $s^3 - 6s^2 + 12s - 8$ 106) $q^3 - 3q^2 + 3q - 1$

107) $c^4 - 18c^2 + 81$ 108) $\dfrac{9}{64}x^2 + \dfrac{3}{2}x + 4$

109) $y^4 + 8y^3 + 24y^2 + 32y + 16$

110) $b^4 + 12b^3 + 54b^2 + 108b + 81$

111) $v^2 - 10vw + 25w^2 - 16$ 112) $16p^2 + 24pq + 9q^2 - 1$

113) $4a^2 + 4ab + b^2 - c^2$ 114) $x^2 - 6xy + 9y^2 - 4z^2$

115) No. $(x + 5)^2 = x^2 + 10x + 25$

116) No. $(y - 3)^3 = y^3 - 9y^2 + 27y - 27$

117) $h^3 + 6h^2 + 12h + 8$ cubic units

118) $9s^2 - 6s + 1$ sq units

119) $9x^2 + 33x + 14$ sq units   120) $\dfrac{3}{2}h^2 + h$ sq units

121) $3x^3 - 5x^2$ 122) $2x^4 - 3x^3 - x^2$ 123) -2 124) 52

125) $6x^3 - 19x^2 + 12x + 5$ 126) 36 127) $-\dfrac{4}{3}$ 128) $-\dfrac{1}{2}$

129) It is the same because multiplication of functions is defined as $(fg)(x) = f(x) \cdot g(x)$.

130) $g(x) = 2x$

Section 5.5

1) dividend: $12c^3 + 20c^2 - 4c$; divisor: $4c$;
   quotient: $3c^2 + 5c - 1$

2) dividend: $10p^3 + p^2 - 25p - 6$; divisor: $2p + 3$;
   quotient: $5p^2 - 7p - 2$

3) Answers may vary.

4) Use long division when the divisor contains two or more terms.

5) $2a^2 - 5a + 3$ 6) $7k^2 + 2k - 10$

7) $2u^5 + 2u^3 + 5u^2 - 8$ 8) $-3m^3 + 2m^2 + 4m - 7$

9) $-5d^3 + 1$ 10) $8q^4 + 2q - 1$

11) $\dfrac{3}{2}w^2 + 7w - 1 + \dfrac{1}{2w}$ 12) $-6j^4 + \dfrac{10}{3}j^2 - j + \dfrac{5}{3j}$

13) $\dfrac{5}{2}v^3 - 9v - \dfrac{11}{2} - \dfrac{5}{4v^2} + \dfrac{1}{4v^4}$ 14) $12z^3 + \dfrac{3}{5}z^2 - \dfrac{2}{z}$

15) $9a^3b + 6a^2b - 4a^2 + 10a$

16) $4x^4y^5 - 9x^3y^3 - \dfrac{1}{6}xy^2 + 2xy$

17) $-t^4u^2 + 7t^3u^2 + 12t^2u^2 - \dfrac{1}{9}t^2$

18) $3c^5d^4 + c^3d^3 - 4d^3 - 2d$

19) The answer is incorrect. When you divide $4t$ by $4t$, you get 1. The quotient should be $4t^2 - 9t + 1$.

20) He cancelled out the $12x$ terms so that he was left with $15x^2$.
    The correct answer is $\dfrac{5}{4}x + 1$.

21) $g + 4$ 22) $n + 5$ 23) $p + 6$ 24) $v + 12$ 25) $k - 5$

26) $m - 9$ 27) $h + 8$ 28) $u - 6$ 29) $2a^2 - 7a - 3$

30) $4b^2 - 2b + 5$ 31) $3p^2 + 7p - 1$ 32) $-3z^2 + 8z - 2$

33) $6t + 23 + \dfrac{119}{t - 5}$ 34) $7d - 6 + \dfrac{50}{d + 9}$

35) $4z^2 + 8z + 7 - \dfrac{72}{3z + 5}$ 36) $k^3 + 4k^2 + 8k + 5 - \dfrac{3}{6k - 1}$

37) $w^2 - 4w + 16$ 38) $a^2 + 3a + 9$

39) $2r^2 + 8r + 3$ 40) $10c^2 - 4c + 3 - \dfrac{2}{5c + 2}$

41) $x^2y^2 + 5x^2y - \dfrac{1}{6} + \dfrac{1}{2xy}$ 42) $4v - 5 + \dfrac{4}{3v - 2}$

43) $-2g^3 - 5g^2 + g - 4$ 44) $-8c^3 + 5c^2 - 2c - 1$

45) $6t + 5 + \dfrac{20}{t - 8}$ 46) $-2uv^2 + v^2 + \dfrac{3}{u} + \dfrac{8}{u^2v}$

47) $4n^2 + 10n + 25$ 48) $4a^3 - a^2 + 6a + 5$

49) $5x^2 - 7x + 3$ 50) $3m^2 - 11m + 7$ 51) $4a^2 - 3a + 7$

52) $16r^2 - 12r + 9$ 53) $5h^2 - 3h - 2 + \dfrac{1}{2h^2 - 9}$

54) $3w^2 - w - 1 + \dfrac{4}{5w^2 + 7}$ 55) $3d^2 - d - 8$

56) $4x^2 - 9x + 2$ 57) $9c^2 + 8c + 3 + \dfrac{7c + 4}{c^2 - 10c + 4}$

58) $3n^2 + n - 6 + \dfrac{6n - 10}{5n^2 - 7n + 2}$ 59) $k^2 - 9$

60) $b^2 + 4$ 61) $-\dfrac{5}{7}a^3 - 7a + 2 + \dfrac{15}{7a}$

62) $13q^2 - 2q + \dfrac{9}{2} - \dfrac{3}{q} + \dfrac{4}{q^2}$ 63) $\dfrac{1}{2}x + 3$ 64) $\dfrac{1}{3}k + 1$

65) $\dfrac{2}{3}w + 2$ 66) $\dfrac{3}{4}y - 2 + \dfrac{3}{4y - 3}$ 67) $4y + 1$

68) $2x - 1$ 69) $2a^2 - 5a + 1$ 70) $\dfrac{3}{2}w^2 + w - 4$

71) $12h^2 + 6h + 2$ 72) $3n^2 - n + 5$

73) $\left(\dfrac{f}{g}\right)(x) = 5x - 9$, where $x \neq -3$

74) $\left(\dfrac{f}{g}\right)(x) = 2x - 3$, where $x \neq \dfrac{5}{4}$

75) $\left(\dfrac{f}{g}\right)(x) = 6x^2 - 9x + 1$, where $x \neq 0$

76) $\left(\dfrac{f}{g}\right)(x) = 4x^3 - \dfrac{5}{3}x + \dfrac{3}{2}$, where $x \neq 0$

77) $\left(\dfrac{f}{g}\right)(x) = 3x^3 - 7x^2 + 2x + 4$, where $x \neq 1$

78) $\left(\dfrac{f}{g}\right)(x) = 2x^3 + x^2 - 6x - 1$, where $x \neq -\dfrac{4}{3}$

79) $2x - 1$, where $x \neq -\dfrac{1}{2}$ 80) $\dfrac{3x}{2x + 1}$, where $x \neq -\dfrac{1}{2}$

81) 9   82) 3   83) $\dfrac{2x + 1}{3x}$, where $x \neq 0$ 84) $-\dfrac{1}{2}$ 85) $\dfrac{1}{6}$

86) 0

Chapter 5 Review Exercises

1) 81   2) $\dfrac{1}{64}$ 3) $\dfrac{64}{125}$ 4) 0   5) z^{18} 6) $-12p^{10}$

7) $7r^5$ 8) $25c^8$ 9) $-54t^7$ 10) $\dfrac{m^4}{4}$ 11) $\dfrac{1}{k^8}$ 12) $\dfrac{1}{d^9}$

13) $-\dfrac{40b^4}{a^6}$ 14) $\dfrac{y}{x^3}$ 15) $\dfrac{4q^{30}}{9p^6}$ 16) $\dfrac{63d^4}{c^{12}}$ 17) 14

18) $\dfrac{r^7s^{12}}{36t^5}$ 19) x^{8t} 20) r^{6a} 21) y^{6p} 22) $\dfrac{1}{w^{9a}}$

23) False. $-x^2 = -1 \cdot x \cdot x$. If $x \neq 0$, then $-x^2$ is a negative number. But $(-x)^2 = -x \cdot (-x)$, which is a positive number when $x \neq 0$.

24) False. A negative exponent does not make an expression negative. $(5y)^{-3} = \dfrac{1}{(5y)^3} = \dfrac{1}{125y^3}$

25) 938,000   26) -418.5 27) 0.00000105   28) 20,000

29) 5.75×10^{-5} 30) 3.694×10^4 31) 3.2×10^7

32) 4×10^{-7} 33) 0.0000004   34) 3.6   35) 7500

36) $-1,302,000$ 37) 30,000 quills

38) 0.00000000000000299 g

39)

Term	Coeff.	Degree
$4r^3$	4	3
$-7r^2$	-7	2
r	1	1
5	5	0

Degree is 3.

40)

Term	Coeff.	Degree
x^3y	1	4
$6xy^2$	6	3
$-8xy$	-8	2
$11y$	11	1

Degree is 4.

41) 5   42) $-2t^2 + 10t + 5$ 43) $7.9p^3 + 5.1p^2 + 4.8p - 3.6$

44) $\dfrac{2}{3}w^2 + \dfrac{1}{4}w - \dfrac{1}{20}$ 45) $-2a^2b^2 + 17a^2b - 4ab + 5$

46) $-8xy - 11x + 9$ 47) Answers may vary.

48) $4d^2 - 4d + 14$ units   49) a) -41 b) -6

50) a) $x^2 + 9x - 8$ b) -8 c) $x^2 - 3x + 12$ d) 8

51) a) 123   b) 151   c) $N(1) = 135$. In 2003, there were approximately 135 cruise ships operating in North America.

52) a) $P(x) = 6x - 400$ b) \$800

53) $-54m^5 + 18m^4 - 42m^3$

54) $-56u^6v^3 + 49u^5v^4 + 84u^5v^2 - 21u^4v^2$

55) $-24w^4 - 48w^3 + 26w^2 - 4w + 15$

56) $12x^6 + 48x^5 - 139x^4 - 28x^3 + 86x^2 + 36x - 99$

57) $y^2 + y - 56$ 58) $6n^2 - 41n + 63$

59) $a^2b^2 + 11ab + 30$ 60) $9r^2 - 7rs - 2s^2$

61) $-24d^2 - 46d - 21$ 62) $24c^5 - 78c^4 + 60c^3$

63) $p^3 - p^2 - 24p - 36$ 64) $z^3 + 10z^2 + 29z + 20$

65) $\dfrac{1}{5}m^2 - \dfrac{26}{15}m - 8$ 66) $\dfrac{1}{45}t^2 - \dfrac{7}{6}t + 15$

67) $z^2 - 81$ 68) $\dfrac{1}{25}n^2 - 4$ 69) $\dfrac{49}{64} - r^4$ 70) $4a^2 - \dfrac{1}{9}b^2$

71) $b^2 + 14b + 49$ 72) $x^2 - 20x + 100$

73) $25q^2 - 20q + 4$ 74) $9y^2 - 42y + 49$

75) $x^3 - 6x^2 + 12x - 8$ 76) $p^3 + 30p^2 + 300p + 1000$

77) $-18c^2 + 48c - 32$ 78) $6w^3 + 36w^2 + 54w$

79) $m^2 - 10m + 25 + 2mn - 10n + n^2$

80) $9r^2 + 12rt + 4t^2 - 1$ 81) a) $-10x^2 + 53x - 36$ b) 30

82) $g(x) = -1$

83) a) $4m^2 - 7m - 15$ sq units   b) $10m + 4$ units

84) $x^3 + 9x^2 + 27x + 27$ cubic units

85) $4t^2 - 7t - 10$ 86) $-2p^3 - 7p^2 + 4p - 1$

87) $c + 10$ 88) $y - 7$ 89) $4r^2 - 7r + 3$

90) $-6h^2 - h + 4$ 91) $-\dfrac{5}{2}x^2y^3 + 7xy^3 + 1 - \dfrac{5}{3x^2}$

92) $8a^3b^3 + 3ab^2 - \dfrac{4}{7}b + \dfrac{1}{7ab^2} - \dfrac{1}{a^2b^2}$

93) $2q - 4 - \dfrac{7}{3q + 7}$ 94) $2r - 3 + \dfrac{14}{6r + 1}$

95) $3a + 7 + \dfrac{21}{5a - 4}$ 96) $3m^2 + m - 4$

97) $t^2 + 4t - 4 + \dfrac{4}{8t^2 - 11}$ 98) $b^2 + 4b + 16$

99) $f^2 - 5f + 25$ 100) $8w^2 - 6w - 7 - \dfrac{2}{4w + 3}$

101) $5k^2 + 6k + 4$ 102) $7u^2 + u + 4$

103) $3c^2 - c - 2 + \dfrac{-3c + 2}{2c^2 + 5c - 4}$ 104) $5y + 4$

105) $2x^2 + x + 5$

106) a) $x + 2$, where $x \neq -4$ b) -7

   c) $\dfrac{x + 4}{7x}$, where $x \neq 0$ d) $\dfrac{1}{3}$

Chapter 5 Test

1) a) -81 b) $\dfrac{1}{32}$ c) -2 d) $\dfrac{1000}{27}$ e) $\dfrac{1}{64}$ 2) $-30p^{12}$

3) $\dfrac{1}{a^4b^6}$ 4) $\dfrac{8}{y^9}$ 5) $\dfrac{9x^4}{4y^{18}}$ 6) t^{13k} 7) 728,300   8) 1.65×10^{-4}

9) $-50,000$ 10) 9.1×10^{-18} g   11)   a) -1 b) 3

12) -1 13) $14r^3s^2 - 2r^2s^2 - 5rs + 8$ 14) $6j^2 - 8j - 6$

15) $c^2 - 15c + 56$ 16) $6y^2 + 13y + 5$ 17) $u^2 - \dfrac{9}{16}$

18) $6a^2 - 13ab - 5b^2$ 19) $49m^2 - 70m + 25$

20) $-8n^3 - 43n^2 + 12n - 9$

21) $-16m^3 - 26m^2 + 68m - 21$ 22) $3x^3 + 24x^2 + 48x$

23) $25a^2 - 10ab + b^2 - 30a + 6b + 9$

24) $s^3 - 12s^2 + 48s - 64$ 25) $r + 3$

26) $2t^2 - 5t + 1 - \dfrac{2}{3t}$ 27) $6v^2 - 3v + 4 - \dfrac{7}{5v - 6}$

28) $3k + 7$ 29) a) $P(x) = 2x + 1$ b) \$17,000

30) a) $x^2 - 4x - 32$ b) -20 c) $x^2 - 10x + 16$ d) -8

   e) $x + 3$, where $x \neq 8$ f) 10

Cumulative Review for Chapters 1–5

1) a) 43, 0   b) $-14, 43, 0$ c) $\dfrac{6}{11}, -14, 2.7, 43, 0.\overline{65}, 0$

2) -28 3) $\dfrac{12}{5}$ or $2\dfrac{2}{5}$ 4) $\left\{ -\dfrac{45}{4} \right\}$ 5) $\varnothing$ 6) $\{-2, 11\}$

7) $b_2 = \dfrac{2A}{h} - b_1$ 8) $(-\infty, 1]$

9) 45 ml of 12% solution, 15 ml of 4% solution

10) x-int: (2, 0); y-int: (0, −5)   11)

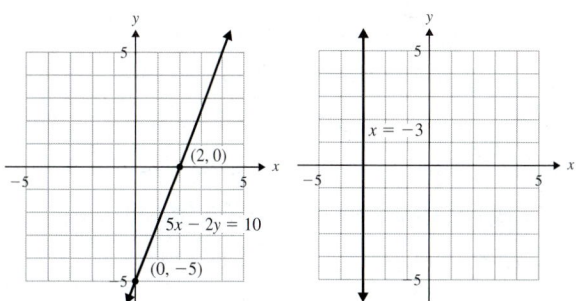

12) $x + y = 3$ 13) $(-\infty, \infty)$ 14) $(2, 10)$

15) width: 15 cm; length: 23 cm   16) $-45w^8$ 17) $\dfrac{8}{n^{18}}$

18) p^{14k} 19) $-14q^2 + 6q - 20$ 20) $16g^2 - 81$

21) $ab - \dfrac{5}{2b} + \dfrac{7}{a^2b^2} + \dfrac{1}{a^3b^2}$ 22) $v^3 + 3v^2 - 5v - 6$

23) $2p^2 - 3p + 1 + \dfrac{5}{p + 4}$ 24) $2c^3 - 17c^2 + 48c - 45$

25) a) $x^4 + 5x^3 - 2x^2 - 30x - 24$ b) -36

Chapter 6

Section 6.1

1) $5m^2$ 2) $3d^2$ 3) $6k^5$ 4) $5t$ 5) $9x^2y$ 6) $8r^2s^5$ 7) $\frac{1}{n}4uv^3$

8) $6a^2b^3$ 9) s^2t 10) p^3q 11) $(n-7)$ 12) $(y+9)$

13) Answers may vary.

14) to write it as a product of two or more polynomials

15) $6(5s+3)$ 16) $2(7a+12)$ 17) $4(6z-1)$

18) $7(9f^2-7)$ 19) $3d(d-2)$ 20) $5m(4-m)$

21) $7y^2(6+5y)$ 22) $5b(6b^2-1)$ 23) $t^4(t-1)$

24) $r^2(r^7+1)$ 25) $\frac{1}{2}c(c+5)$ 26) $\frac{1}{8}k(k+7)$

27) $5n^3(2n^2-n+8)$ 28) $6x^5(3x^2+7x-5)$

29) $2v^5(v^3-9v^2-12v+1)$ 30) $3z^3(4z^3+10z^2-5z+1)$

31) does not factor   32) does not factor   33) $a^3b^2(a+4b)$

34) $2rs^3(10r^2-7s)$ 35) $10x^2y(5xy^2-7xy+4)$

36) $3b^2d^2(7b^2d+5bd-9)$ 37) $(n-12)(m+8)$

38) $(y+5)(x+3)$ 39) $(8r-3)(p-q)$

40) $(9c+4)(a-b)$ 41) $(z+11)(y+1)$

42) $(v-7)(2u+1)$ 43) $(3r+4)(2k^2-1)$

44) $(3q+5)(8p-1)$ 45) $-8(8m+5)$ 46) $-7(2k-3)$

47) $-5t^2(t-2)$ 48) $-4v^3(v^2+9)$ 49) $-a(3a^2-7a+1)$

50) $-q(10q^2+4q-1)$ 51) $-1(b-8)$ 52) $-1(z+6)$

53) $(t+3)(k+8)$ 54) $(v+5)(u+10)$

55) $(g-7)(f+4)$ 56) $(c-5)(d+8)$

57) $(s-3)(2r+5)$ 58) $(3j-7)(k+2)$

59) $(3x-2)(y+9)$ 60) $(b+8)(4a+3)$

61) $(2b+5c)(4b+c^2)$ 62) $(u-2v^2)(8u+3v)$

63) $(a^2-3b)(4a+b)$ 64) $(x-6y^2)(5x^2+y)$

65) $(k+7)(t-5)$ 66) $(p-2)(q-9)$

67) $(n-8)(m-10)$ 68) $(h+6)(k-4)$

69) $(g-1)(d+1)$ 70) $(r+3)(q-1)$

71) $(5u+6)(t-1)$ 72) $(4y+7)(z-5)$

73) $(12g^3+h)(3g-8h)$ 74) $(5j^2+9k)(8j-11k)$

75) Answers may vary.   76) Factor out 3 from all of the terms.

77) $5(mn+3m+2n+6)$; Group the terms and factor out the GCF from each group.
$5(n+3)(m+2)$

78) $3x(xy-7x-2y+14)$; Group the terms and factor out the GCF from each group.
$3x(y-7)(x-2)$

79) $2(b+4)(a+3)$ 80) $7(p+4)(q+2)$

81) $8s(s-5)(t+2)$ 82) $5k^2(2k-1)(h+3)$

83) $(d+4)(7c+3)$ 84) $(s+6)(9r+2)$

85) $(7k-3d)(6k^2-5d)$ 86) $(4x-y)(3x^2-2y)$

87) $9fj(f+1)(j+5)$ 88) $n^2(n+1)(m-4)$

89) $2x^3(2x-1)(y+7)$ 90) $4ac(3a-1)(c+5)$

91) $(q-8)(p+3)$ 92) $9d(3d^2+4d-1)$

93) $a^2b^2(a-2b)(a+4b)$ 94) $3t^2(6r+5)(t-1)$

95) $(3h^2-2k)(h+4k^2)$ 96) $(3y-4)(2z-1)$

97) $2c(c^2+7)(c+6)$ 98) $9w^2(2w+5)$

99) $-8v(2v^2+7v-1)$ 100) $-3n^2(n^2-9n-11)$

Section 6.2

1) They are negative.   2) They are positive.

3) Can I factor out a GCF?   4) It does not factor.

5) Can I factor again?

6) Multiply the factors. The product should be the original polynomial.

7) $(g+6)(g+2)$ 8) $(j+4)(j+5)$ 9) $(w+7)(w+6)$

10) $(t+12)(t+3)$ 11) $(c-4)(c-9)$

12) $(v-3)(v-8)$ 13) $(b-4)(b+2)$

14) $(s+7)(s-4)$ 15) $(u+12)(u-11)$

16) $(m-11)(m+10)$ 17) $(q-5)(q-3)$

18) $(z-6)(z-4)$ 19) prime   20) prime

21) $(p-10)(p-10)$ or $(p-10)^2$

22) $(u+9)(u+9)$ or $(u+9)^2$

23) $3(p-9)(p+1)$ 24) $5(n+2)(n+6)$

25) $2k(k-8)(k-5)$ 26) $w^2(w-3)(w-4)$

27) $ab(a+12)(a-3)$ 28) $4xy(x^2+8x+6)$

29) $-(a+8)(a+2)$ 30) $-(y+6)(y+3)$

31) $-(h-5)(h+3)$ 32) $-(j+8)(j-7)$

33) $-(k-7)(k-4)$ 34) $-(b-6)(b-11)$

35) $-(n+7)(n+7)$ or $-(n+7)^2$

36) $-(z-2)(z-2)$ or $-(z-2)^2$ 37) $(a+5b)(a+b)$

38) $(v+6w)(v+w)$ 39) $(m-3n)(m+7n)$

40) $(p-8q)(p-9q)$ 41) $(x-12y)(x-3y)$

42) $(r-4s)(r-5s)$ 43) $(f+g)(f-11g)$

44) $(u+8v)(u-6v)$ 45) $(c-5d)(c+11d)$

46) $(w+5x)(w+12x)$ 47) $(2r+5)(r+3)$

48) $(3a+4)(a+2)$ 49) $(5p-1)(p-4)$

50) $(7j-2)(j-4)$ 51) $(11m+4)(m-2)$

52) $(5b-6)(b+3)$ 53) $(3v+7)(2v-1)$

54) $(4x - 1)(2x - 3)$ 55) $(5c + 2)(2c + 3)$

56) $(5n + 4)(3n + 2)$ 57) $(9x - 4y)(x - y)$

58) $(6a - 5b)(a + b)$

59) because 2 can be factored out of $2x - 4$, but 2 cannot be factored out of $2x^2 + 13x - 24$

60) because 5 can be factored out of $5c + 10$, but 5 cannot be factored out of $5c^2 + 16c + 30$

61) $(5w + 6)(w + 1)$ 62) $(2g + 9)(g + 2)$

63) $(3u - 5)(u - 6)$ 64) $(7a - 3)(a - 2)$

65) $(7k - 6)(k + 3)$ 66) $(5z + 7)(z - 5)$

67) $(4r + 3)(2r + 5)$ 68) $(2t + 7)(3t + 1)$

69) $(6v - 7)(v - 2)$ 70) $(10m - 3)(m + 5)$

71) $(5a - 4b)(2a - b)$ 72) $(8x - 3y)(x - 2y)$

73) $(3c + 2d)(2c + 9d)$ 74) $(6m + 7n)(2m - 5n)$

75) $(n + 14)(n + 2)$ 76) $(k - 4)(k - 11)$

77) $(p + 1)(p - 2)$ 78) $(t + 6)(t - 8)$

79) $(2w - 1)(w - 4)$ 80) $(3c - 19)(c - 7)$

81) $(6y - 7)(4y - 1)$ 82) $(15a + 8)(6a + 1)$

83) $4q(q - 3)(q - 4)$ 84) $(2y - 3)(y - 8)$

85) $(t + 6)(t + 1)$ 86) $(mn - 6)(mn + 1)$

87) $3(4c - 3)(c + 2)$ 88) $-(h + 9)(h - 6)$

89) $(3b + 25)(b + 3)$ 90) $ab(a + 6b)(a + 4b)$

91) $(7s - 3t)(s - 2t)$ 92) $(x + y)(t + 3)(t - 7)$

93) $-(5z - 2)(2z - 3)$ 94) $(8p - 7)^2$ 95) prime

96) $(8h - 7)(2h + 3)$ 97) $(r - 9)(r - 2)$

98) $(2k + 7)(k + 3)$ 99) $(q - 1)^2(3p - 7)(4p - 7)$

100) prime

Section 6.3

1) a) 36   b) 100   c) 16   d) 121   e) 9   f) 64   g) 144

   h) $\dfrac{1}{4}$ i) $\dfrac{9}{25}$

2) It is a trinomial that results from squaring a binomial.

3) a) n^2 b) $5t$ c) $7k$ d) $4p^2$ e) $\dfrac{1}{3}$ f) $\dfrac{5}{2}$

4) 2   5) $z^2 + 18z + 81$ 6) $4b^2 - 28b + 49$

7) The middle term does not equal $2(3c)(-4)$. It would have to equal $-24c$ to be a perfect square trinomial.

8) Only one term, k^2, is a perfect square, so it can't be a perfect square trinomial.

9) $(t + 8)^2$ 10) $(x + 6)^2$ 11) $(g - 9)^2$ 12) $(q - 11)^2$

13) $(2y + 3)^2$ 14) $(7r + 1)^2$ 15) $(3k - 4)^2$

16) $(4b - 3)^2$ 17) $\left(a + \dfrac{1}{3}\right)^2$ 18) $\left(m + \dfrac{1}{2}\right)^2$

19) $\left(v - \dfrac{3}{2}\right)^2$ 20) $\left(h - \dfrac{2}{5}\right)^2$ 21) $(x + 3y)^2$

22) $(3a - 2b)^2$ 23) $(6t - 5u)^2$ 24) $(9k + m)^2$

25) $4(f + 3)^2$ 26) $9(j - 1)^2$ 27) $2p^2(p - 6)^2$

28) $5r(r + 4)^2$ 29) $-2(3d + 5)^2$ 30) $-7(2z - 1)^2$

31) $3c(4c^2 + c + 9)$ 32) $4n^2(25n^2 - 2n + 16)$

33) a) $x^2 - 16$ b) $16 - x^2$ 34) a) $y^2 - 81$ b) $81 - y^2$

35) $(x + 3)(x - 3)$ 36) $(q + 7)(q - 7)$

37) $(n + 11)(n - 11)$ 38) $(d + 9)(d - 9)$

39) prime   40) prime   41) $\left(y + \dfrac{1}{5}\right)\left(y - \dfrac{1}{5}\right)$

42) $\left(t + \dfrac{1}{10}\right)\left(t - \dfrac{1}{10}\right)$ 43) $\left(c + \dfrac{3}{4}\right)\left(c - \dfrac{3}{4}\right)$

44) $\left(m + \dfrac{2}{5}\right)\left(m - \dfrac{2}{5}\right)$ 45) $(6 + h)(6 - h)$

46) $(2 + b)(2 - b)$ 47) $(13 + a)(13 - a)$

48) $(11 + w)(11 - w)$ 49) $\left(\dfrac{7}{8} + j\right)\left(\dfrac{7}{8} - j\right)$

50) $\left(\dfrac{12}{7} + r\right)\left(\dfrac{12}{7} - r\right)$ 51) $(10m + 7)(10m - 7)$

52) $(6x + 5)(6x - 5)$ 53) $(4p + 9)(4p - 9)$

54) $(3a + 1)(3a - 1)$ 55) prime   56) prime

57) $\left(\dfrac{1}{2}k + \dfrac{2}{3}\right)\left(\dfrac{1}{2}k - \dfrac{2}{3}\right)$ 58) $\left(\dfrac{1}{6}d + \dfrac{2}{7}\right)\left(\dfrac{1}{6}d - \dfrac{2}{7}\right)$

59) $(b^2 + 8)(b^2 - 8)$ 60) $(u^2 + 7)(u^2 - 7)$

61) $(12m + n^2)(12m - n^2)$ 62) $(8p + 5q^2)(8p - 5q^2)$

63) $(r^2 + 1)(r + 1)(r - 1)$ 64) $(k^2 + 9)(k + 3)(k - 3)$

65) $(4h^2 + g^2)(2h + g)(2h - g)$

66) $(a^2 + b^2)(b + a)(b - a)$ 67) $4(a + 5)(a - 5)$

68) $3(p + 4)(p - 4)$ 69) $2(m + 8)(m - 8)$

70) $6(j + 1)(j - 1)$ 71) $5r^2(3r + 1)(3r - 1)$

72) $8n^3(2n + 5)(2n - 5)$

73) a) 64   b) 1   c) 1000   d) 27   e) 125   f) 8

74) 3   75) a) y b) $2c$ c) $5r$ d) x^2 76) 6

77) $x^2 - 3x + 9$ 78) $t^2 + 5t + 25$

79) $(d + 1)(d^2 - d + 1)$ 80) $(n + 5)(n^2 - 5n + 25)$

81) $(p - 3)(p^2 + 3p + 9)$ 82) $(g - 2)(g^2 + 2g + 4)$

83) $(k + 4)(k^2 - 4k + 16)$ 84) $(z - 10)(z^2 + 10z + 100)$

85) $(3m - 5)(9m^2 + 15m + 25)$

86) $(4c + 1)(16c^2 - 4c + 1)$

87) $(5y - 2)(25y^2 + 10y + 4)$

88) $(3a + 4)(9a^2 - 12a + 16)$

89) $(10c - d)(100c^2 + 10cd + d^2)$

90) $(5v + w)(25v^2 - 5vw + w^2)$

91) $(2j + 3k)(4j^2 - 6jk + 9k^2)$

92) $(5m - 3n)(25m^2 + 15mn + 9n^2)$

93) $(4x + 5y)(16x^2 - 20xy + 25y^2)$

94) $(3a - 10b)(9a^2 + 30ab + 100b^2)$

95) $6(c + 2)(c^2 - 2c + 4)$ 96) $9(k - 1)(k^2 + k + 1)$

97) $7(v - 10w)(v^2 + 10vw + 100w^2)$

98) $8(3a + 2b)(9a^2 - 6ab + 4b^2)$

99) $(h + 2)(h - 2)(h^2 - 2h + 4)(h^2 + 2h + 4)$

100) $(p + 1)(p - 1)(p^2 - p + 1)(p^2 + p + 1)$

101) $7(2x + 3)$ 102) $-7(2r - 5)$ 103) $(3p + 7)(p - 1)$

104) $(4d - 7)(2d + 3)$ 105) $(t + 7)(t^2 + 8t + 19)$

106) $(c + 1)(c^2 - 7c + 19)$ 107) $(k - 10)(k^2 - 17k + 73)$

108) $(y - 2)(y^2 + 11y + 49)$

Putting It All Together

1) $(m + 10)(m + 6)$ 2) $(h + 6)(h - 6)$

3) $(u + 9)(v + 6)$ 4) $(2y + 9)(y - 2)$

5) $(3k - 2)(k - 4)$ 6) $(n - 7)^2$ 7) $8d^4(2d^2 + d + 9)$

8) $(b + c)(b - 4c)$ 9) $10w(3w + 5)(2w - 1)$

10) $7(c - 1)(c^2 + c + 1)$ 11) $(t + 10)(t^2 - 10t + 100)$

12) $(p + 4)(q - 6)$ 13) $(7 + p)(7 - p)$

14) $(h - 7)(h - 8)$ 15) $(2x + y)^2$ 16) $9(3c - 2)$

17) $3z^2(z - 8)(z + 1)$ 18) $(3a - 2)(3a + 4)$

19) prime   20) $5c(a + 2)(b - 3)$

21) $5(2x - 3)(4x^2 + 6x + 9)$ 22) $(9z + 2)^2$

23) $\left(c + \dfrac{1}{2}\right)\left(c - \dfrac{1}{2}\right)$ 24) prime

25) $(3s + 1)(3s - 1)(5t - 4)$ 26) $3cd(2c^2 + 5d)(2c^2 - 5d)$

27) $(k + 3m)(k + 6m)$ 28) $8(2r + 1)(4r^2 - 2r + 1)$

29) $(z - 11)(z + 8)$ 30) $8fg^2(5f^3g^2 + f^2g + 2)$

31) $5(4y - 1)^2$ 32) $(4t - 5)(t + 1)$

33) $2(10c + 3d)(c + d)$ 34) $\left(x + \dfrac{3}{7}\right)\left(x - \dfrac{3}{7}\right)$

35) $(n^2 + 4m^2)(n + 2m)(n - 2m)$ 36) $(k - 12)(k - 9)$

37) $2(a - 9)(a + 4)$ 38) $(x + 2)(x - 2)(y + 7)$

39) $\left(r - \dfrac{1}{2}\right)^2$ 40) $(v - 5)(v^2 + 5v + 25)$

41) $(4g - 9)(7h + 4)$ 42) $-3x(2x - 1)(4x - 3)$

43) $(4b + 3)(2b - 5)$ 44) $2(5u + 3)^2$

45) $5a^2b(11a^4b^2 + 7a^3b^2 - 2a^2 - 4)$ 46) $(8 + u)(8 - u)$

47) prime   48) $2v^2w(v + w)(v + 6w)$ 49) $(3p - 4q)^2$

50) $(c^2 + 4)(c + 2)(c - 2)$ 51) $(6y - 1)(5y + 7)$

52) prime   53) $10(2a - 3b)(4a^2 + 6ab + 9b^2)$

54) $13n^3(2n^3 - 3n + 1)$ 55) $(r - 1)(t - 1)$ 56) $(h + 5)^2$

57) $4(g + 1)(g - 1)$ 58) $(5a - 3b)(5a - 8b)$

59) $3(c - 4)^2$ 60) prime   61) $(12k + 11)(12k - 11)$

62) $(5p - 4q)(25p^2 + 20pq + 16q^2)$

63) $-4(6g + 1)(2g + 3)$ 64) $5(d + 11)(d + 1)$

65) $(q + 1)(q^2 - q + 1)$ 66) $(3x + 2)^2$

67) $(9u^2 + v^2)(3u + v)(3u - v)$ 68) $3(5v + w^2)(3v + 2w)$

69) $(11f + 3)(f + 3)$ 70) $4y(y - 5)(y + 4)$

71) $j^3(2j^8 - 1)$ 72) $\left(d + \dfrac{13}{10}\right)\left(d - \dfrac{13}{10}\right)$

73) $(w - 8)(w + 6)$ 74) $(4a - 5)^2$ 75) prime

76) $3(2y + 5)(4y^2 - 10y + 25)$ 77) $(m + 2)^2$

78) $(r - 9)(r - 6)$ 79) $(3t + 8)(3t - 8)$

80) $4c^2(5c + 3)(5c - 3)$ 81) $(2z + 1)(y + 11)(y - 5)$

82) $(a + b)(c - 8)(c + 3)$ 83) $r(r + 3)$

84) $(n - 4)(n + 8)$ 85) $3p(3p - 13)$

86) $(5w - 8)(5w - 4)$ 87) $(7k + 3)(k - 1)$

88) $4(4z + 1)(z + 2)$ 89) $-3x(x - 2y)$

90) $5s(s - 2t)$ 91) $(n + p + 6)(n - p + 6)$

92) $(h + k - 5)(h - k - 5)$ 93) $(x - y + z)(x - y - z)$

94) $(a + b + c)(a + b - c)$

Section 6.4

1) It says that if the product of two quantities equals 0, then one or both of the quantities must be zero.

2) No. The solution set is $\{-7, 0\}$.   3) $\{-9, 8\}$ 4) $\{-10, -4\}$

5) $\{4, 7\}$ 6) $\{-2, 5\}$ 7) $\left\{-\dfrac{3}{4}, 9\right\}$ 8) $\left\{-\dfrac{1}{2}, 13\right\}$

9) $\{-15, 0\}$ 10) $\{0, 8\}$ 11) $\left\{\dfrac{5}{6}\right\}$ 12) $\{-7\}$

13) $\left\{-3, -\dfrac{7}{4}\right\}$ 14) $\left\{\dfrac{5}{8}, \dfrac{11}{3}\right\}$ 15) $\left\{-\dfrac{3}{2}, \dfrac{1}{4}\right\}$

16) $\left\{-\dfrac{5}{6}, \dfrac{9}{8}\right\}$ 17) $\{0, 2.5\}$ 18) $\{-0.8, 0\}$

19) No; the product of the factors must equal zero.

20) 1) Divide by 5 to get $n^2 - 2n - 8 = 0$ or
2) Factor out 5 to get $5(n^2 - 2n - 8)$.

21) $\{-8, -7\}$ 22) $\{-7, 5\}$ 23) $\{-15, 3\}$ 24) $\{1, 11\}$

25) $\left\{-\dfrac{5}{3}, 2\right\}$ 26) $\left\{\dfrac{7}{4}, 2\right\}$ 27) $\left\{-\dfrac{4}{7}, 0\right\}$ 28) $\{-2, 0\}$

29) $\{6, 9\}$ 30) $\{-7, -4\}$ 31) $\{-7, 7\}$ 32) $\{-10, 10\}$

33) $\left\{-\dfrac{6}{5}, \dfrac{6}{5}\right\}$ 34) $\left\{-\dfrac{4}{13}, \dfrac{4}{13}\right\}$ 35) $\{-12, 5\}$ 36) $\{2, 10\}$

37) $\left\{-2, -\dfrac{3}{4}\right\}$ 38) $\left\{-\dfrac{12}{7}, 2\right\}$ 39) $\{0, 11\}$ 40) $\{0, 1\}$

41) $\left\{-\dfrac{1}{2}, 3\right\}$ 42) $\left\{-\dfrac{1}{3}, 9\right\}$ 43) $\{-8, 12\}$ 44) $\{6, 9\}$

45) $\left\{\dfrac{7}{2}, \dfrac{9}{2}\right\}$ 46) $\left\{-7, \dfrac{10}{3}\right\}$ 47) $\{-9, 5\}$ 48) $\{-3, -2\}$

49) $\{1, 6\}$ 50) $\{-7, 5\}$ 51) $\left\{-3, -\dfrac{4}{5}\right\}$ 52) $\left\{-\dfrac{5}{3}, 2\right\}$

53) $\left\{\dfrac{3}{2}\right\}$ 54) $\left\{\dfrac{1}{3}\right\}$ 55) $\{-11, -3\}$ 56) $\{-12, -7\}$

57) $\left\{-\dfrac{7}{2}, \dfrac{3}{4}\right\}$ 58) $\left\{\dfrac{4}{3}, 10\right\}$ 59) $\left\{\dfrac{2}{3}, 8\right\}$ 60) $\left\{-1, -\dfrac{1}{3}\right\}$

61) $\left\{-4, 0, \dfrac{1}{2}\right\}$ 62) $\left\{-\dfrac{7}{12}, 0, 11\right\}$ 63) $\left\{-1, \dfrac{2}{9}, 11\right\}$

64) $\left\{-3, -\dfrac{5}{4}, 6\right\}$ 65) $\left\{\dfrac{5}{2}, 3\right\}$ 66) $\left\{\dfrac{1}{3}, 8\right\}$ 67) $\{0, -8, 8\}$

68) $\{0, -9, 9\}$ 69) $\{-4, 0, 9\}$ 70) $\{0, 7\}$ 71) $\{-12, 0, 5\}$

72) $\{0, 2, 6\}$ 73) $\left\{0, -\dfrac{3}{2}, \dfrac{3}{2}\right\}$ 74) $\left\{0, -\dfrac{5}{4}, \dfrac{5}{4}\right\}$

75) $\left\{-5, -\dfrac{2}{3}, \dfrac{7}{2}\right\}$ 76) $\left\{-\dfrac{1}{2}, \dfrac{2}{5}, 8\right\}$ 77) $\left\{-\dfrac{3}{4}, \dfrac{2}{5}, \dfrac{1}{2}\right\}$

78) $\left\{-\dfrac{1}{4}, \dfrac{3}{7}, \dfrac{2}{3}\right\}$ 79) $\{-6, -2, 2\}$ 80) $\{-3, 3, 8\}$

81) $-7, -3$ 82) $-2, 8$ 83) $\dfrac{5}{2}, 4$ 84) $\dfrac{1}{2}$ 85) $-4, 4$

86) $-9, 0, 6$ 87) $0, 1, 4$ 88) $-\dfrac{2}{3}, \dfrac{2}{3}$

Section 6.5

1) length $=$ 12 in.; width $=$ 3 in.

2) length $=$ 8 cm; width $=$ 5 cm

3) base $=$ 3 cm; height $=$ 8 cm

4) base $=$ 14 in.; height $=$ 6 in.

5) base $=$ 6 in.; height $=$ 3 in.

6) base $=$ 10 cm; height $=$ 5 cm

7) length $=$ 10 in.; width $=$ 6 in.

8) width $=$ 5 in.; height $=$ 3 in.

9) length $=$ 9 ft; width $=$ 5 ft

10) length $=$ 20 in.; width $=$ 15 in.

11) length $=$ 9 in.; width $=$ 6 in.

12) length $=$ 5 ft; width $=$ 2.5 ft

13) width $=$ 12 in.; height $=$ 6 in.

14) length $=$ 20 in.; width $=$ 12 in.

15) height $=$ 10 cm; base $=$ 7 cm

16) height $=$ 4 cm; base $=$ 8 cm

17) 5 and 7 or -1 and 1   18) 5 and 6 or -4 and -3

19) 0, 2, 4 or 2, 4, 6   20) 9, 10, 11

21) 6, 7, 8   22) 13, 15, 17 or -3, -1, 1

23) Answers may vary.   24) No; it is not a right triangle.

25) 9   26) 5   27) 15   28) 20   29) 10   30) 3

31) 6, 8, 10   32) 3, 4, 5   33) 5, 12, 13   34) 8, 15, 17

35) 8 in.   36) 13 cm   37) 5 ft

38) length of wire $=$ 17 ft
    height of pole $=$ 15 ft

39) 5 mi   40) 9 mi   41) a) 144 ft   b) after 2 sec   c) 3 sec

42) a) 20 ft

b) when $t = \dfrac{1}{2}$ sec and when $t = 1\dfrac{1}{2}$ sec   c) $2\dfrac{1}{2}$ sec

43) a) 288 ft   b) 117 ft   c) 324 ft   d) 176 ft

44) a) 648 ft   b) 213 ft   c) 1089 ft   d) 586 ft   e) 360 ft

f) 765 ft

g) The 10-in. shell would need to be 410 ft farther horizontally from the point of explosion than the 3-in. shell.

45) a) $3500   b) $3375   c) $12

46) a) $4000   b) $4375   c) $30

47) a) 184 ft   b) 544 ft

c) when $t = 2\dfrac{1}{2}$ sec and when $t = 10$ sec   d) $12\dfrac{1}{2}$ sec

48) a) $16,800   b) $14,700   c) $50

Chapter 6 Review Exercises

1) 9   2) 8   3) $11p^4q^3$ 4) $7r^2s^3$ 5) $12(4y + 7)$

6) $3a(10a^3 - 3)$ 7) $7n^3(n^2 - 3n + 1)$

8) $6uv(12u^2v^2 - 7u^2v - 4)$ 9) $(b + 6)(a - 2)$

10) $(13w - 9)(u + v)$ 11) $(n + 2)(m + 5)$

12) $(k + 7)(j - 5)$ 13) $(r - 2)(5q - 6)$

14) $(d^2 + 5)(c - 1)$ 15) $-4x(2x^2 + 3x - 1)$

16) $-(r^2 - 6r + 2)$ 17) $(p + 8)(p + 5)$

18) $(f - 12)(f - 5)$ 19) $(x + 5y)(x - 4y)$

20) $(t + 7u)(t - 9u)$ 21) $3(c - 6)(c - 2)$

22) $4mn(m - 3n)(m + 5n)$ 23) $(5y + 6)(y + 1)$

24) $(3g - 11)(g + 4)$ 25) $(2m - 5)(2m - 3)$

26) $(6t - 1)(t - 8)$ 27) $4a(7a + 4)(2a - 1)$

28) $2(9n + 4)(n + 5)$ 29) $(3s - t)(s + 4t)$

30) $(g - 11)^3(2f - 5)(4f + 7)$ 31) $(3c + 1)(3c - 1)$

32) $(2k - 5)(k - 3)$ 33) $(n + 5)(n - 5)$

34) $(7a + 2b)(7a - 2b)$ 35) prime

36) $(z^2 + 1)(z + 1)(z - 1)$ 37) $10(q + 9)(q - 9)$

38) $3v(4 + 3v)(4 - 3v)$ 39) $(a + 8)^2$ 40) $(2x - 5)^2$

41) $(h + 2)(h^2 - 2h + 4)$ 42) $(q - 1)(q^2 + q + 1)$

43) $(3p - 4q)(9p^2 + 12pq + 16q^2)$

44) $2(2c + 5d)(4c^2 - 10cd + 25d^2)$

45) $(7r - 6)(r + 2)$ 46) $3(y + 10)^2$ 47) $\left(\dfrac{3}{5} + x\right)\left(\dfrac{3}{5} - x\right)$

48) $9v^4(9v^2 + 4v - 1)$ 49) $(s - 8)(t - 5)$

50) $(n - 6)(n - 5)$ 51) $w^2(w - 1)(w^2 + w + 1)$

52) $(g - 11)(h + 8)$ 53) prime   54) $(7k + 12)(7k - 12)$

55) $-4ab$ 56) $(10a + 3b)(100a^2 - 30ab + 9b^2)$

57) $(3y - 14)(2y - 3)$ 58) $(5a + 2b)(a + 4b)$

59) $\left\{0, \dfrac{1}{2}\right\}$ 60) $\left\{-\dfrac{7}{4}\right\}$ 61) $\left\{-1, \dfrac{2}{3}\right\}$ 62) $\{-1, 1\}$

63) $\{-3, 15\}$ 64) $\left\{-\dfrac{8}{5}, \dfrac{1}{2}\right\}$ 65) $\left\{-\dfrac{6}{7}, \dfrac{6}{7}\right\}$ 66) $\{-13, 0\}$

67) $\{-4, 8\}$ 68) $\{3, 6\}$ 69) $\left\{\dfrac{4}{5}, 1\right\}$ 70) $\left\{-\dfrac{5}{2}, \dfrac{1}{2}\right\}$

71) $\left\{0, -\dfrac{3}{2}, 2\right\}$ 72) $\{-4, 2\}$ 73) $\{-8, 9\}$ 74) $\left\{0, \dfrac{2}{3}, 4\right\}$

75) $\left\{\dfrac{1}{6}, 3, 7\right\}$ 76) $\left\{-6, \dfrac{3}{4}, 9\right\}$

77) base = 5 in.; height = 6 in.

78) length = 7 cm; width = 4 cm

79) height = 3 in.; length = 8 in.

80) length = 10 in.; width = 6 in.

81) 12   82) 15   83) length = 6 ft; width = 2.5 ft

84) base = 24 in.; height = 12 in.   85) $-1, 0, 1$ or 8, 9, 10

86) 6, 8 or $-2, 0$ 87) 5 in.   88) 10 mi

89) a) 0 ft   b) after 2 sec and after 4 sec   c) 144 ft

   d) after 6 sec

Chapter 6 Test

1) See if you can factor out a GCF.   2) $(n - 6)(n - 5)$

3) $(4 + b)(4 - b)$ 4) $(5a + 2)(a - 3)$

5) $7p^2q^3(8p^4q^3 - 11p^2q + 1)$

6) $(y - 2z)(y^2 + 2yz + 4z^2)$

7) $2d(d + 9)(d - 2)$ 8) prime   9) $(3h + 4)^2$

10) $(2y - 3)(12x + 11)$ 11) $(s - 7t)(s + 4t)$

12) $(4s^2 + 9t^2)(2s + 3t)(2s - 3t)$ 13) $(12p + 5)(3p + 7)$

14) $(2b - 5)(6b - 7)$ 15) $m^9(m + 1)(m^2 - m + 1)$

16) $\{-4, -3\}$ 17) $\{0, -5, 5\}$ 18) $\left\{-\dfrac{5}{12}, \dfrac{5}{12}\right\}$

19) $\{-4, 7\}$ 20) $\left\{-\dfrac{1}{4}, 8\right\}$ 21) $\left\{\dfrac{5}{3}, 2\right\}$

22) height = 10 ft; width = 4 ft   23) 5, 7, 9

24) 3 mi   25) length = 16 ft; width = 6 ft

26) a) $\dfrac{5}{4}$ sec and 3 sec   b) 60 ft   c) 132 ft   d) 5 sec

Cumulative Review Chapters 1–6

1) $\dfrac{1}{8}$ 2) $-\dfrac{9}{40}$ 3) $\dfrac{3t^4}{2u^6}$ 4) $-24k^{10}$ 5) 481,300   6) $\{5\}$

7) $R = \dfrac{A - P}{PT}$ 8) Twix: 4 in.; Toblerone: 8 in.

9) $(-\infty, -4] \cup \left[\dfrac{38}{5}, \infty\right)$

10)

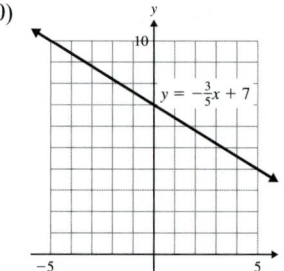

11) $y = \dfrac{1}{3}x + 1$ 12) $(-2, -2)$ 13) $12y^2 - 8y - 15$

14) $8p^3 - 50p^2 + 95p - 56$ 15) $c^2 + 16c + 64$

16) $3a^2b^2 - 7a^2b - 5ab^2 + 19ab - 8$ 17) $6x^3 - 7x - 4$

18) $3r + 1 - \dfrac{5}{2r} + \dfrac{3}{4r^2}$ 19) $(b - 7)(c + 8)$

20) $6(3q - 7)(3q - 1)$ 21) prime

22) $(t^2 + 9)(t + 3)(t - 3)$ 23) $(x - 5)(x^2 + 5x + 25)$

24) $\{-8, 5\}$ 25) $\left\{-\dfrac{4}{3}, \dfrac{5}{2}\right\}$

Chapter 7

Section 7.1

1) when its numerator equals zero

2) when its denominator equals zero

3) Set the denominator equal to zero and solve for the variable. That value cannot be substituted into the expression because it will make the denominator equal to zero.

4) No. A rational expression equals zero when its numerator equals zero. If $x = 0$, then the numerator equals 5. If x is a positive or a negative number, then x^2 is positive and $x^2 + 5$ is positive too. Therefore, $x^2 + 5$ cannot equal zero; it is always positive.

5) a) $\dfrac{3}{2}$ b) -8 c) $(-\infty, -6) \cup (-6, \infty)$

6) a) $\dfrac{2}{7}$ b) 0   c) $\left(-\infty, \dfrac{1}{3}\right) \cup \left(\dfrac{1}{3}, \infty\right)$

7) a) undefined   b) $\dfrac{3}{5}$ c) $(-\infty, -2) \cup (-2, \infty)$

8) a) -3 b) never equals zero   c) $(-\infty, 1) \cup (1, \infty)$

9) a) -2 b) never equals zero

   c) $(-\infty, -5) \cup (-5, -1) \cup (-1, \infty)$

10) a) $\dfrac{1}{2}$ b) $\dfrac{1}{2}$ c) $(-\infty, -4) \cup (-4, 3) \cup (3, \infty)$

11) $(-\infty, 7) \cup (7, \infty)$ 12) $(-\infty, -3) \cup (-3, \infty)$

13) $\left(-\infty, -\dfrac{2}{5}\right) \cup \left(-\dfrac{2}{5}, \infty\right)$ 14) $\left(-\infty, \dfrac{7}{2}\right) \cup \left(\dfrac{7}{2}, \infty\right)$

15) $(-\infty, 1) \cup (1, 8) \cup (8, \infty)$

16) $(-\infty, -6) \cup (-6, 7) \cup (7, \infty)$

17) $(-\infty, -9) \cup (-9, 9) \cup (9, \infty)$

18) $(-\infty, 3) \cup (3, 11) \cup (11, \infty)$

19) $(-\infty, \infty)$ 20) $(-\infty, \infty)$ 21) Answers may vary.

22) Answers may vary.   23) $\dfrac{2}{5d^3}$ 24) $12g^3$ 25) $\dfrac{3}{5}$ 26) $\dfrac{5}{6}$

27) $b - 7$ 28) $\dfrac{g + 4}{g - 3}$ 29) $\dfrac{1}{r + 4}$ 30) $\dfrac{1}{t - 9}$ 31) $\dfrac{3k + 4}{k + 2}$

32) $3(c - 4)$ 33) $\dfrac{w + 5}{5}$ 34) $4(m - 1)$ 35) $\dfrac{4(m - 5)}{11}$

36) $\dfrac{u - 4}{v - 3}$ 37) $\dfrac{x + y}{x^2 + xy + y^2}$ 38) $\dfrac{a^2 - ab + b^2}{a - b}$ 39) -1

40) No   41) -1 42) -1 43) $-k - 7$ 44) $-\dfrac{1}{2}$

45) $-\dfrac{5}{x + 2}$ 46) $-a - 3$ 47) $-4(b + 2)$ 48) $-\dfrac{2}{w + 8}$

49) $-\dfrac{4t^2 + 6t + 9}{2t + 3}$ 50) $-\dfrac{r^2 + 2}{7}$ 51) $5x + 3$ 52) $y - 3$

53) $c - 2$ 54) $2n^2 + 1$

55) Possible answers:
$\dfrac{-b - 7}{b - 2}, \dfrac{-(b + 7)}{b - 2}, \dfrac{b + 7}{2 - b}, \dfrac{b + 7}{-(b - 2)}, \dfrac{b + 7}{-b + 2}$

56) Possible answers:
$\dfrac{-8y + 1}{2y + 5}, \dfrac{1 - 8y}{2y + 5}, \dfrac{-(8y - 1)}{2y + 5}, \dfrac{8y - 1}{-2y - 5}, \dfrac{8y - 1}{-(2y + 5)}$

57) Possible answers:
$\dfrac{-9 + 5t}{2t - 3}, \dfrac{5t - 9}{2t - 3}, \dfrac{-(9 - 5t)}{2t - 3}, \dfrac{9 - 5t}{-2t + 3}, \dfrac{9 - 5t}{3 - 2t}, \dfrac{9 - 5t}{-(2t - 3)}$

58) Possible answers:
$-\dfrac{12m}{m^2 - 3}, \dfrac{12m}{-(m^2 - 3)}, \dfrac{12m}{-m^2 + 3}, \dfrac{12m}{3 - m^2}$

59) $\dfrac{3}{4}$ 60) $\dfrac{5}{27}$ 61) $\dfrac{8v^4}{3u^3}$ 62) $\dfrac{6t^2}{s^9}$ 63) $\dfrac{1}{2t(3t - 2)}$

64) $\dfrac{u^4}{3(4u - 5)^2}$ 65) $\dfrac{2(2p - 1)}{9}$ 66) 4   67) $\dfrac{4v - 1}{4}$

68) $\dfrac{3(y - 5)}{2y^2}$ 69) $\dfrac{4}{x}$ 70) $5h(h - 1)$

71) $\dfrac{(r - 2)(r^2 - 3r + 9)}{4(r - 3)}$ 72) $-\dfrac{8}{2w + 1}$ 73) $\dfrac{3}{10}$ 74) $\dfrac{8}{15}$

75) $\dfrac{1}{6c^6}$ 76) $-\dfrac{3g^2h^2}{16}$ 77) $\dfrac{3a^2}{2a - 1}$ 78) $\dfrac{1}{6p(p + 7)}$

79) $\dfrac{5}{2y + 5}$ 80) $\dfrac{q(q + 8)}{5}$ 81) $\dfrac{1}{7}$ 82) $\dfrac{2}{5}$

83) $\dfrac{z + 10}{(2z + 1)(z + 8)}$ 84) $\dfrac{6w}{7(w - 5)}$ 85) $\dfrac{3}{4(3a + 1)}$

86) $\dfrac{9 - h}{4h(h + 12)}$ 87) $-\dfrac{8}{2d + 5}$ 88) $\dfrac{(x + y)(x + 3)}{21}$

89) No. To divide the rational expressions we rewrite the problem as $\dfrac{12}{x} \cdot \dfrac{2}{3y}$. If $y = 0$, then the denominator of $\dfrac{2}{3y}$ will equal zero and the expression will be undefined.

90) $2a^2 - 9a - 5$ 91) $-\dfrac{a}{6}$ 92) $\dfrac{5x + 1}{4x}$ 93) $\dfrac{7}{3}$

94) $\dfrac{4a^3(a + 7)}{a + 9}$ 95) $\dfrac{3x^6}{4y^6}$ 96) $\dfrac{2(t + 3)}{5(t - 1)}$ 97) $\dfrac{a^2}{a - 4}$

98) $\dfrac{m + 2}{16}$ 99) $-\dfrac{3}{4(x + y)}$ 100) $\dfrac{5d^8}{6c^{12}}$ 101) $\dfrac{4j - 1}{3j + 2}$

102) $5t(t + 3)$ 103) $\dfrac{18x^4}{y^8}$ 104) $\dfrac{6m}{m^2 - 25}$

Section 7.2

1) 40   2) 48   3) c^4 4) n^9 5) $36p^8$ 6) $45z^6$ 7) $24a^3b^4$

8) $27x^3y^2$ 9) $2(n + 4)$ 10) $9(z - 8)$ 11) $w(2w + 1)$

12) $y(3y + 5)$ 13) $4a^3(3a - 1)$ 14) $5p^4(2p - 3)$

15) $(r + 7)(r - 2)$ 16) $(m - 6)(m - 5)$

17) $(w - 5)(w + 2)(w + 3)$ 18) $(t + 7)(t - 7)(t + 2)$

19) $b - 4$ or $4 - b$ 20) $u - v$ or $v - u$

21) Answers may vary.   22) Answers may vary.

23) $\dfrac{3}{t} = \dfrac{3t^2}{t^3}; \dfrac{8}{t^3} = \dfrac{8}{t^3}$ 24) $\dfrac{10}{p^5} = \dfrac{10}{p^5}; \dfrac{7}{p^2} = \dfrac{7p^3}{p^5}$

25) $\dfrac{9}{8n^6} = \dfrac{27}{24n^6}; \dfrac{2}{3n^2} = \dfrac{16n^4}{24n^6}$ 26) $\dfrac{5}{6a} = \dfrac{20a^4}{24a^5}; \dfrac{7}{8a^5} = \dfrac{21}{24a^5}$

27) $\dfrac{1}{x^3y} = \dfrac{5y^4}{5x^3y^5}; \dfrac{6}{5xy^5} = \dfrac{6x^2}{5x^3y^5}$

28) $\dfrac{5}{6a^2b^4} = \dfrac{5a^2}{6a^4b^4}; \dfrac{5}{a^4b} = \dfrac{30b^3}{6a^4b^4}$

29) $\dfrac{t}{5t - 6} = \dfrac{7t}{7(5t - 6)}; \dfrac{10}{7} = \dfrac{50t - 60}{7(5t - 6)}$

30) $\dfrac{8}{d} = \dfrac{8d - 32}{d(d - 4)}; \dfrac{2}{d - 4} = \dfrac{2d}{d(d - 4)}$

31) $\dfrac{a}{24a + 36} = \dfrac{3a}{36(2a + 3)}; \dfrac{1}{18a + 27} = \dfrac{4}{36(2a + 3)}$

32) $\dfrac{7}{12x - 4} = \dfrac{21}{12(3x - 1)}; \dfrac{x}{18x - 6} = \dfrac{2x}{12(3x - 1)}$

33) $\dfrac{4}{h + 5} = \dfrac{4h - 12}{(h + 5)(h - 3)}; \dfrac{7h}{h - 3} = \dfrac{7h^2 + 35h}{(h + 5)(h - 3)}$

34) $\dfrac{8}{a + 9} = \dfrac{16a + 56}{(a + 9)(2a + 7)}; \dfrac{a}{2a + 7} = \dfrac{a^2 + 9a}{(a + 9)(2a + 7)}$

35) $\dfrac{9y}{y^2 - y - 42} = \dfrac{18y^2}{2y(y + 6)(y - 7)};$
$\dfrac{3}{2y^2 + 12y} = \dfrac{3y - 21}{2y(y + 6)(y - 7)}$

36) $\dfrac{4q}{3q^2 + 24q} = \dfrac{4q^2 - 28q}{3q(q + 8)(q - 7)};$
$\dfrac{5}{q^2 + q - 56} = \dfrac{15q}{3q(q + 8)(q - 7)}$

37) $\dfrac{z}{z^2 - 10z + 25} = \dfrac{z^2 + 3z}{(z - 5)^2(z + 3)};$
$\dfrac{15z}{z^2 - 2z - 15} = \dfrac{15z^2 - 75z}{(z - 5)^2(z + 3)}$

38) $\dfrac{c}{c^2 + 11c + 28} = \dfrac{c^2 + 7c}{(c + 7)^2(c + 4)};$
$\dfrac{6}{c^2 + 14c + 49} = \dfrac{6c + 24}{(c + 7)^2(c + 4)}$

39) $\dfrac{11}{g - 3} = \dfrac{11g + 33}{(g + 3)(g - 3)}; \dfrac{4}{9 - g^2} = -\dfrac{4}{(g + 3)(g - 3)}$

40) $\dfrac{10}{4k - 1} = \dfrac{40k + 10}{(4k + 1)(4k - 1)};$
$\dfrac{k}{1 - 16k^2} = -\dfrac{k}{(4k + 1)(4k - 1)}$

41) $\dfrac{4}{w^2 - 4w} = \dfrac{28w - 112}{7w(w - 4)^2}; \dfrac{6}{7w^2 - 28w} = \dfrac{6w - 24}{7w(w - 4)^2};$
$\dfrac{11}{w^2 - 8w + 16} = \dfrac{77w}{7w(w - 4)^2}$

42) $\dfrac{t}{t^2 - 4t - 21} = \dfrac{t^2 + 7t}{(t + 7)(t - 7)(t + 3)};$
$\dfrac{2}{t + 3} = \dfrac{2t^2 - 98}{(t + 7)(t - 7)(t + 3)};$
$\dfrac{4}{t^2 - 49} = \dfrac{4t + 12}{(t + 7)(t - 7)(t + 3)}$

43) $\dfrac{4}{5}$ 44) $\dfrac{1}{2}$ 45) $\dfrac{10}{a}$ 46) $\dfrac{4}{c}$ 47) 2   48) 7

49) $\dfrac{6}{w}$ 50) $\dfrac{4}{t}$ 51) $\dfrac{r - 3}{r + 2}$ 52) $\dfrac{d - 9}{d + 4}$

53) a) $x(x - 3)$

b) Multiply the numerator and denominator of $\dfrac{8}{x - 3}$ by x and the numerator and denominator of $\dfrac{2}{x}$ by $x - 3$.

c) $\dfrac{8}{x - 3} = \dfrac{8x}{x(x - 3)}, \dfrac{2}{x} = \dfrac{2x - 6}{x(x - 3)}$

54) a) $18b^4$

b) Multiply the numerator and denominator of $\dfrac{4}{9b^2}$ by $2b^2$ and the numerator and denominator of $\dfrac{5}{6b^4}$ by 3.

c) $\dfrac{4}{9b^2} = \dfrac{8b^2}{18b^4}, \dfrac{5}{6b^4} = \dfrac{15}{18b^4}$

55) Find the product of the denominators.

56) Answers may vary.   57) $\dfrac{19}{24}$ 58) $\dfrac{1}{15}$ 59) $\dfrac{3x}{20}$

60) $\dfrac{3(12t + 5)}{20}$ 61) $\dfrac{3(7a + 4)}{14a^2}$ 62) $\dfrac{3 - 14f}{2f^2}$

63) $\dfrac{11d + 32}{d(d - 8)}$ 64) $\dfrac{4(r + 7)}{r(r - 7)}$ 65) $\dfrac{5z + 26}{(z + 6)(z + 2)}$

66) $\dfrac{11c - 7}{(c - 5)(c + 3)}$ 67) $\dfrac{x^2 - x - 3}{(2x + 1)(x + 5)}$

68) $\dfrac{m^2 - 15m - 8}{(3m + 4)(m - 9)}$ 69) $\dfrac{t - 3}{t - 7}$ 70) $\dfrac{u - 5}{u - 1}$

71) $\dfrac{b^2 + b - 40}{(b + 4)(b - 4)(b - 9)}$ 72) $\dfrac{7g^2 + 74g + 4}{(g + 1)(g + 10)(g - 10)}$

73) $\dfrac{(c - 5)(c + 2)}{(c + 6)(c - 2)(c - 4)}$ 74) $\dfrac{(2a + 1)(a - 9)}{(a - 8)(a + 3)(a - 2)}$

75) $\dfrac{4b^2 + 28b + 3}{3(b - 4)(b + 3)}$ 76) $\dfrac{k^2 + 14k - 27}{2(k - 12)(k - 3)}$

77) No; if the sum is rewritten as $\dfrac{5}{x - 7} - \dfrac{2}{x - 7}$ then the LCD $= x - 7$. If the sum is rewritten as $\dfrac{-5}{7 - x} + \dfrac{2}{7 - x}$, then the LCD is $7 - x$.

78) $3n - 5$ or $5 - 3n$ 79) $\dfrac{7}{z - 6}$ or $-\dfrac{7}{6 - z}$

80) $\dfrac{6}{q-8}$ or $-\dfrac{6}{8-q}$ 81) $\dfrac{2c+13}{12b-7c}$ or $-\dfrac{2c+13}{7c-12b}$

82) $\dfrac{2(1+3u)}{4u-3v}$ or $-\dfrac{2(1+3u)}{3v-4u}$ 83) $-\dfrac{5(t+6)}{(t+8)(t-8)}$

84) $-\dfrac{2(r-1)}{(r+3)(r-3)}$ 85) $\dfrac{3(3a+4)}{(2a+3)(2a-3)}$

86) $\dfrac{2(4y+5)}{(3y+5)(3y-5)}$ 87) $\dfrac{6j^2-j-2}{3j(j+8)}$

88) $\dfrac{-10w^2+9w-23}{w(w-3)}$ 89) $\dfrac{c^2-2c+20}{(c-4)^2(c+3)}$

90) $\dfrac{n^2+2n-18}{(n+5)^2(n+6)}$ 91) $-\dfrac{3y}{(x+y)(x-y)}$

92) $\dfrac{8a+b}{3(a+b)(a-b)}$ 93) $\dfrac{-n^2+33n+1}{(4n-1)(3n+1)(n+2)}$

94) $\dfrac{(3v-2)(v+3)}{(6v+5)(3v+4)(v-1)}$ 95) $\dfrac{3y^2+6y+26}{y(y-4)(2y-5)}$

96) $\dfrac{7g^2-45g+205}{5g(g-6)(2g-5)}$ 97) a) $\dfrac{2(x+1)}{x-3}$ b) $\dfrac{x^2-2x+5}{x-3}$

98) a) $\dfrac{5(x-4)}{3(x+3)}$ b) $\dfrac{x^2-3x+56}{3(x+1)}$

99) a) $\dfrac{w}{(w+2)^2(w-2)}$ b) $\dfrac{2(w-1)^2}{(w+2)(w-2)}$

100) a) $\dfrac{2t}{(t+5)^2(t+4)}$ b) $\dfrac{2(t^2+4t+2)}{(t+5)(t+4)}$

101) $\dfrac{3(x^2+6x+10)}{x(x+5)}$; Domain: $(-\infty,-5)\cup(-5,0)\cup(0,\infty)$

102) $\dfrac{-3(x^2+2x-10)}{x(x+5)}$;
Domain: $(-\infty,-5)\cup(-5,0)\cup(0,\infty)$

103) $\dfrac{18(x+12)}{x(x+5)}$; Domain: $(-\infty,-5)\cup(-5,0)\cup(0,\infty)$

104) $\dfrac{2(x+5)}{x(x+4)}$;
Domain: $(-\infty,-5)\cup(-5,-4)\cup(-4,0)\cup(0,\infty)$

Section 7.3

1) Method 1: Rewrite it as a division problem, then simplify.

$\dfrac{2}{9}\div\dfrac{5}{18}=\dfrac{2}{\cancel{9}_1}\cdot\dfrac{\cancel{18}^2}{5}=\dfrac{4}{5}$

Method 2: Multiply the numerator and denominator by 18, the LCD of $\dfrac{2}{9}$ and $\dfrac{5}{18}$. Then, simplify.

$\dfrac{\cancel{18}^2\left(\dfrac{2}{9}\right)}{\cancel{18}_1\left(\dfrac{5}{18}\right)}=\dfrac{4}{5}$

2) Method 1: Subtract the fractions in the numerator and add the fractions in the denominator. Then, rewrite the complex fraction as a division problem and simplify.

$\dfrac{\dfrac{3}{2}-\dfrac{1}{5}}{\dfrac{1}{10}+\dfrac{3}{5}}=\dfrac{\dfrac{15}{10}-\dfrac{2}{10}}{\dfrac{1}{10}+\dfrac{6}{10}}$

$=\dfrac{\dfrac{13}{10}}{\dfrac{7}{10}}$

$=\dfrac{13}{10}\div\dfrac{7}{10}$

$=\dfrac{13}{\cancel{10}}\cdot\dfrac{\cancel{10}}{7}$

$=\dfrac{13}{7}$

Method 2: Multiply the numerator and denominator by 10, the LCD of all of the fractions. Simplify.

$\dfrac{10\left(\dfrac{3}{2}-\dfrac{1}{5}\right)}{10\left(\dfrac{1}{10}+\dfrac{3}{5}\right)}=\dfrac{15-2}{1+6}=\dfrac{13}{7}$

3) $\dfrac{14}{25}$ 4) $\dfrac{9}{32}$ 5) ab^2 6) $\dfrac{u^3}{v}$ 7) $\dfrac{1}{st^2}$ 8) x^2y^2 9) $\dfrac{2m^4}{15n^2}$

10) $\dfrac{2b^3c}{5}$ 11) $\dfrac{t}{5}$ 12) $\dfrac{16}{m}$ 13) $\dfrac{4}{3(y-8)}$ 14) $3(g+6)$

15) $\dfrac{5}{6w^4}$ 16) $\dfrac{5d^2}{2}$ 17) $x(x-3)$ 18) $\dfrac{c+1}{6}$ 19) $\dfrac{3}{2}$ 20) $\dfrac{4}{19}$

21) $\dfrac{7d+2c}{d(c-5)}$ 22) $\dfrac{r(r-2s)}{r+3s}$ 23) $\dfrac{4z+7}{5z-7}$ 24) $\dfrac{10(w+1)}{8w+17}$

25) $\dfrac{8}{y}$ 26) $\dfrac{9}{m}$ 27) $\dfrac{x^2-7}{x^2-11}$ 28) $\dfrac{4-c^2}{3c+8}$ 29) $-\dfrac{52}{15}$

30) $-\dfrac{14}{17}$ 31) $\dfrac{2ab}{a+b}$ 32) $\dfrac{4xy}{3(x+y)}$ 33) $\dfrac{r(r^2+s)}{s^2(sr+1)}$

34) $\dfrac{n^3(n^2+m^4)}{m^3(3n^3-m)}$ 35) $\dfrac{t-5}{t+4}$ 36) $\dfrac{t+3}{t+2}$ 37) $\dfrac{b^2+1}{b^2-3}$

38) $\dfrac{z}{4}$ 39) $\dfrac{1}{m^3n}$ 40) $\dfrac{z}{5}$ 41) $\dfrac{(h-1)(h+3)}{28}$ 42) $\dfrac{3(r+5)}{(r-6)^2}$

43) $\dfrac{2(x-9)(x+2)}{3(x+3)(x+1)}$ 44) $\dfrac{c^4+2}{c(d+c)(d-c)}$ 45) $\dfrac{r}{20}$ 46) $\dfrac{4}{21}$

47) $\dfrac{a}{12}$ 48) $\dfrac{8-w^2}{w+6}$ 49) $\dfrac{25}{18}$ 50) $\dfrac{(17h-1)(h-3)}{(h+2)(5h-12)}$

51) $\dfrac{2(n+3)^2}{4n+7}$ 52) $\dfrac{z}{y^2}$ 53) $\dfrac{w(v-w)}{2v+w^2}$ 54) $\dfrac{p^2+4q}{p(p+q)}$

55) $\dfrac{8y^2}{x(y^2-x)}$ 56) $\dfrac{3c^2}{2d-c^2}$ 57) $\dfrac{a^3+b^2}{a^3(2-7b^2)}$

58) $\dfrac{t(t+r)(t-r)}{r^2(5t^3+7)}$ 59) $\dfrac{4n-m}{m(1+mn)}$ 60) $\dfrac{k^2(9h^3+1)}{h^3(1-hk^2)}$

61) 0

x	f(x)
1	1
2	$\dfrac{1}{2}$
3	$\dfrac{1}{3}$
10	$\dfrac{1}{10}$
100	$\dfrac{1}{100}$
1000	$\dfrac{1}{1000}$

62) It gets larger.

x	f(x)
1	1
$\dfrac{1}{2}$	2
$\dfrac{1}{3}$	3
$\dfrac{1}{10}$	10
$\dfrac{1}{100}$	100
$\dfrac{1}{1000}$	1000

Section 7.4

1) Eliminate the denominators.   **2)** Keep the LCD.

3) sum; $\dfrac{3m - 14}{8}$ **4)** difference; $\dfrac{5r + 27}{9}$ **5)** equation; $\{-9\}$

6) equation; $\left\{-\dfrac{1}{3}\right\}$ **7)** difference; $\dfrac{z^2 - 4z + 24}{z(z - 6)}$

8) sum; $\dfrac{a^2 + 2a + 18}{a^2(a + 9)}$ **9)** equation; $\{3\}$ **10)** equation; $\{10\}$

11) $0, -10$ **12)** $0, 3$ **13)** $0, 9, -9$ **14)** $0, 2, -2$ **15)** $4, 9$

16) $-8, 5$ **17)** $\{2\}$ **18)** $\{1\}$ **19)** $\{-2\}$ **20)** $\{12\}$

21) $\{-21\}$ **22)** $\left\{-\dfrac{7}{2}\right\}$ **23)** $\{4\}$ **24)** $\{5\}$ **25)** $\left\{-\dfrac{5}{2}\right\}$

26) $\{1\}$ **27)** $\{3\}$ **28)** $\{-4\}$ **29)** $\left\{\dfrac{1}{2}\right\}$ **30)** $\left\{\dfrac{5}{2}\right\}$ **31)** $\varnothing$

32) $\varnothing$ **33)** $\{6\}$ **34)** $\{-16\}$ **35)** $\{-4\}$ **36)** $\{3\}$ **37)** $\{5\}$

38) $\{-10\}$ **39)** $\{12\}$ **40)** $\{3\}$ **41)** $\{-11\}$ **42)** $\{5\}$

43) $\varnothing$ **44)** $\varnothing$ **45)** $\{-3\}$ **46)** $\{-3\}$ **47)** $\{0, 6\}$

48) $\{0, -10\}$ **49)** $\{-20\}$ **50)** $\{5\}$ **51)** $\{2, 9\}$ **52)** $\{2, 3\}$

53) $\{1, 3\}$ **54)** $\{1, 9\}$ **55)** $\{-3\}$ **56)** $\{-8, 4\}$

57) $\{-4, -2\}$ **58)** $\{-4, 5\}$ **59)** $\{-10\}$ **60)** $\{6\}$

61) $\{-6\}$ **62)** $\{5\}$ **63)** $\{2, 8\}$ **64)** $\{-4, -3\}$

65) $\{-8, 1\}$ **66)** $\{-1, 10\}$ **67)** $\varnothing$ **68)** $\{-2, 8\}$

69) 750 lb   **70)** 525 purses   **71)** 0.4 m

72) a) 37.5 miles/million years   b) 2.4 million years

73) $P = \dfrac{nRT}{V}$ **74)** $m = \dfrac{CA}{W}$ **75)** $z = \dfrac{kx}{y}$ **76)** $b = \dfrac{rt}{2a}$

77) $x = \dfrac{t + u}{3B}$ **78)** $r = \dfrac{n - k}{5Q}$ **79)** $b = \dfrac{a - zc}{z}$

80) $n = \dfrac{dl - t}{d}$ **81)** $t = \dfrac{Aq - 4r}{A}$ **82)** $s = \dfrac{3A - hr}{h}$

83) $c = \dfrac{na - wb}{wk}$ **84)** $y = \dfrac{kx + raz}{r}$ **85)** $r = \dfrac{st}{s + t}$

86) $R_2 = \dfrac{R_1 R_3}{R_1 - R_3}$ **87)** $C = \dfrac{AB}{3A - 2B}$ **88)** $z = \dfrac{4xy}{x - 5y}$

Putting It All Together

1) a) $-\dfrac{8}{5}$ b) 0   c) $(-\infty, -3) \cup (-3, 3) \cup (3, \infty)$

2) a) undefined   b) $\dfrac{3}{5}$ c) $(-\infty, -4) \cup (-4, 2) \cup (2, \infty)$

3) a) 4   b) never equals 0   c) $\left(-\infty, \dfrac{1}{2}\right) \cup \left(\dfrac{1}{2}, \infty\right)$

4) a) $\dfrac{4}{5}$ b) never equals 0   c) $(-\infty, 0) \cup (0, \infty)$

5) a) $\dfrac{1}{6}$ b) 1   c) $(-\infty, \infty)$

6) a) 3   b) -7 and 7   c) $(-\infty, \infty)$

7) $\dfrac{4}{3n^3}$ **8)** $\dfrac{w^5}{2}$ **9)** $\dfrac{1}{j - 4}$ **10)** $\dfrac{m + 8}{2(m + 4)}$ **11)** $-\dfrac{3}{n + 2}$

12) $-\dfrac{1}{y + 3}$ **13)** $\dfrac{3f - 16}{f(f + 8)}$ **14)** $\dfrac{14}{3}$ **15)** $\dfrac{4a^2 b}{9}$

16) $\dfrac{4j^2 + 27j + 27}{(j + 9)(j - 9)(j + 6)}$ **17)** $\dfrac{8q^2 - 37q + 21}{(q - 5)(q + 4)(q + 7)}$

18) $\dfrac{z^3}{3y^3}$ **19)** $-\dfrac{m + 7}{8}$ **20)** $\dfrac{12p^2 - 92p - 15}{(4p + 3)(p + 2)(p - 6)}$

21) $\dfrac{9}{r - 8}$ **22)** $\dfrac{3y + 14}{12y^2}$ **23)** $\dfrac{5}{3}$ **24)** $\dfrac{7d^2 + 6d + 54}{d^2(d + 9)}$

25) $\dfrac{2(7x - 1)}{(x - 8)(x + 3)}$ **26)** $\dfrac{6}{(y + 4)(x^2 - 3x + 9)}$ **27)** $\dfrac{34}{15z}$

28) $\dfrac{5k + 28}{(k + 2)(k + 5)}$ **29)** -1 **30)** $\dfrac{m + 20n}{7m - 4n}$

31) $\dfrac{5u^2 + 37u - 19}{u(3u - 2)(u + 1)}$ **32)** $\dfrac{-2p^2 - 8p + 11}{p(p + 7)(p - 8)}$

33) $\dfrac{x^2 + 2x + 12}{(2x + 1)^2 (x - 4)}$ **34)** 1   **35)** $\dfrac{5(c^2 + 16)}{3(c - 2)}$

36) $\dfrac{3n + 1}{7(n + 4)}$ **37)** $-\dfrac{11}{5w}$ **38)** $-\dfrac{1}{2k}$

39) a) $\dfrac{x(x - 3)}{8}$ b) $\dfrac{3(x - 1)}{2}$

40) a) $\dfrac{8z}{(z + 5)(z + 1)}$ b) $\dfrac{2(z^2 + 9z + 40)}{(z + 5)(z + 1)}$

41) $h(x) = \dfrac{x(5x + 1)}{(3x + 1)(3x - 1)}$;

Domain: $\left(-\infty, -\dfrac{1}{3}\right) \cup \left(-\dfrac{1}{3}, \dfrac{1}{3}\right) \cup \left(\dfrac{1}{3}, \infty\right)$

42) $k(x) = \dfrac{x(x + 1)}{(3x + 1)(3x - 1)}$; $k(x) = 0$ when $x = 0$ or $x = -1$

43) $\dfrac{c^2}{16}$ 44) $\dfrac{3}{20}$ 45) $\dfrac{7m - 15}{(m - 3)(m - 4)}$ 46) $\dfrac{3k^2(3k - 1)}{2}$

47) $\dfrac{15tu^3}{4}$ 48) $\dfrac{3 - y^2}{x + y}$

49) $h(x) = 6x - 5$; Domain: $(-\infty, -8) \cup (-8, \infty)$

50) $h(x) = x + 7$; Domain: $(-\infty, -2) \cup (-2, 2) \cup (2, \infty)$

51) $\{-15\}$ 52) $\{4\}$ 53) $\{-4, 7\}$ 54) $\{-10, 1\}$

55) $\{-12\}$ 56) $\{0, -12\}$ 57) $\varnothing$ 58) $\{-5, 4\}$

Section 7.5

1) $\{35\}$ 2) $\{27\}$ 3) $\{48\}$ 4) $\{135\}$ 5) 111   6) 27

7) 3 cups of tapioca flour and 6 cups of potato-starch flour

8) Rosa: 2232; opponent: 1860   9) length: 48 ft; width: 30 ft

10) 14   11) stocks: $8000; bonds: $12,000

12) deer: 32; rabbits: 72   13) 1355

14) a) 10 qt   b) 25 qt   c) 35 qt

15) a) 7 mph   b) 13 mph   16) a) 275 mph   b) 325 mph

17) a) $x + 30$ mph   b) $x - 30$ mph

18) a) $13 - x$ mph   b) $13 + x$ mph   19) 20 mph

20) 28 mph   21) 4 mph   22) 3 mph   23) 260 mph

24) 225 mph   25) 2 mph   26) 290 mph   27) $\dfrac{1}{4}$ job/hr

28) $\dfrac{1}{3}$ job/hr   29) $\dfrac{1}{t}$ job/hr   30) $\dfrac{1}{2t}$ job/hr   31) $1\dfrac{1}{5}$ hr

32) $3\dfrac{3}{7}$ min   33) $3\dfrac{1}{13}$ hr   34) $1\dfrac{7}{8}$ hr   35) 20 min   36) 75 min

37) 3 hr   38) 120 min   39) 3 hr   40) 24 min

41) $2\dfrac{1}{5}$ ft/sec   42) $3\dfrac{1}{3}$ ft/sec

Section 7.6

1) increases   2) decreases   3) direct   4) joint   5) inverse

6) direct   7) combined   8) inverse   9) $M = kn$

10) $q = kr$ 11) $h = \dfrac{k}{j}$ 12) $R = \dfrac{k}{B}$ 13) $T = \dfrac{k}{c^2}$

14) $b = kw^3$ 15) $s = krt$ 16) $C = kAD$ 17) $Q = \dfrac{k\sqrt{z}}{m}$

18) $r = \dfrac{kd}{L^2}$ 19) a) 9   b) $z = 9x$ c) 54

20) a) 4   b) $A = 4D$ c) 44   21) a) 48   b) $N = \dfrac{48}{y}$ c) 16

22) a) 63   b) $j = \dfrac{63}{m}$ c) 3   23) a) 5   b) $Q = \dfrac{5r^2}{w}$ c) 45

24) a) 2   b) $y = 2a\sqrt{b}$ c) 24   25) 56   26) 25   27) 18

28) 6   29) 70   30) 16   31) $500.00   32) $0.80   33) 12 hr

34) 216 in^2 35) 180 watts   36) 400 newtons   37) 162,000 J

38) 48π cm^3 39) 200 cycles/sec   40) 1,600,000 tons

41) 3 ohms   42) 35 tons   43) 320 lb   44) 208 lb

Chapter 7 Review Exercises

1) Set $Q(x) = 0$ and solve for x. The domain of $f(x)$ contains all real numbers except the values of x that make $Q(x) = 0$.

2) Set $P(x) = 0$ and solve for x.

3) a) $\dfrac{7}{12}$ b) -9 c) $\left(-\infty, \dfrac{1}{5}\right) \cup \left(\dfrac{1}{5}, \infty\right)$

4) a) $-\dfrac{8}{75}$ b) never equals zero

c) $(-\infty, -10) \cup (-10, 10) \cup (10, \infty)$

5) $(-\infty, -4) \cup (-4, 6) \cup (6, \infty)$ 6) $(-\infty, \infty)$

7) $\dfrac{7}{a^9}$ 8) $\dfrac{5}{11}$ 9) $\dfrac{1}{3z + 1}$ 10) $-\dfrac{1}{x + 10}$ 11) $\dfrac{y + 9}{z - 12}$

12) Possible answers:

$\dfrac{-u + 6}{u + 2}, \dfrac{6 - u}{u + 2}, \dfrac{u - 6}{-u - 2}, \dfrac{-(u - 6)}{u + 2}, \dfrac{u - 6}{-(u + 2)}$

13) $2l + 1$ 14) $b + 4$ 15) $\dfrac{36k^2}{m}$ 16) $\dfrac{t - 2}{2(t + 4)}$ 17) $\dfrac{1}{4w}$

18) $\dfrac{x - 5}{55}$ 19) $-\dfrac{a + 5}{4(a^2 + 5a + 25)}$ 20) $\dfrac{q}{35p^2}$ 21) k^2

22) $36x^2y^4$ 23) $m(m + 5)$ 24) $3d(2d - 1)$

25) $(3x + 7)(x - 9)$ 26) $b - 2$ or $2 - b$

27) $(c + 4)(c + 6)(c - 7)$ 28) $3x(x + 8)^2$

29) $(a - 5)(a - 8)(a + 1)$

30) $(c + d)(c - d)$ or $(c + d)(d - c)$

31) $\dfrac{24r^2}{20r^3}$ 32) $\dfrac{8z}{z(3z + 4)}$ 33) $\dfrac{t^2 + 2t - 15}{(2t + 1)(t + 5)}$ 34) $-\dfrac{n}{n - 4}$

35) $\dfrac{3}{8a^3b} = \dfrac{15b^4}{40a^3b^5}; \dfrac{6}{5ab^5} = \dfrac{48a^2}{40a^3b^5}$

36) $\dfrac{8}{p + 7} = \dfrac{8p}{p(p + 7)}; \dfrac{2}{p} = \dfrac{2p + 14}{p(p + 7)}$

37) $\dfrac{9c}{c^2 + 6c - 16} = \dfrac{9c^2 - 18c}{(c - 2)^2(c + 8)};$

$\dfrac{4}{c^2 - 4c + 4} = \dfrac{4c + 32}{(c - 2)^2(c + 8)}$

38) $\dfrac{7}{2r^2 - 12r} = \dfrac{7r + 42}{2r(r + 6)(r - 6)};$

$\dfrac{3r}{36 - r^2} = -\dfrac{6r^2}{2r(r + 6)(r - 6)};$

$\dfrac{r - 5}{2r^2 + 12r} = \dfrac{r^2 - 11r + 30}{2r(r + 6)(r - 6)}$

39) $\dfrac{5}{4c}$ 40) $\dfrac{4m-5}{m-3}$ 41) $\dfrac{4+9z}{10z^2}$ 42) $\dfrac{n^2-8n+20}{n(2n-5)}$

43) $\dfrac{27-y}{(y-2)(y+3)}$ 44) $\dfrac{2d^2+48d-15}{5(d-7)(d+4)}$

45) $\dfrac{10p^2-67p-53}{4(p+1)(p-7)}$ 46) $\dfrac{(k-7)(k+2)}{k(k+7)^2}$

47) $\dfrac{17}{11-m}$ or $-\dfrac{17}{m-11}$ 48) $-\dfrac{1}{r+8}$

49) $-\dfrac{xy}{(x+y)(x-y)}$ 50) $\dfrac{15w^2+2w+54}{5w(w+7)}$

51) $\dfrac{2g^2-6g+57}{5g(g-7)}$ 52) $\dfrac{6d^2+7d-32}{d(d+2)(5d-3)}$

53) a) $\dfrac{3x}{2(x-4)}$ sq units   b) $\dfrac{x^2-4x+96}{4(x-4)}$ units

54) a) $\dfrac{2}{x(x+1)}$ sq units   b) $\dfrac{2x^3+4x+4}{x^2(x+1)}$ units

55) $\dfrac{5x^2+7x-8}{x(x-2)}$; Domain: $(-\infty,0)\cup(0,2)\cup(2,\infty)$

56) $\dfrac{4(x-2)}{x(5x+3)}$;

Domain: $\left(-\infty,-\dfrac{3}{5}\right)\cup\left(-\dfrac{3}{5},0\right)\cup(0,2)\cup(2,\infty)$

57) $\dfrac{6}{7}$ 58) $\dfrac{1}{f}$ 59) $\dfrac{p^2+6}{p^2+8}$ 60) $\dfrac{a^2(b-2)}{b(2+a)(2-a)}$ 61) $\dfrac{2}{3n}$

62) $-\dfrac{1}{5}$ 63) $\dfrac{(y+3)(y-10)}{(y-9)(y+5)}$ 64) $\dfrac{32}{7q}$ 65) $\dfrac{c-1}{c-2}$

66) $a+b$ 67) $\dfrac{y(x^2+2y)}{x(y^2-x)}$ 68) $\dfrac{12b^2}{a(4ab^2+1)}$ 69) $\left\{\dfrac{2}{5}\right\}$

70) $\{10\}$ 71) $\{-2\}$ 72) $\{1\}$ 73) $\{1,20\}$ 74) $\{1,10\}$

75) $\varnothing$ 76) $\{-3,3\}$ 77) $\{-3,5\}$ 78) $\{-6,2\}$

79) $c=\dfrac{2p}{A}$ 80) $D=\dfrac{s+T}{R}$ 81) $a=\dfrac{t-nb}{n}$

82) $k=\dfrac{wc-N}{aw}$ 83) $s=\dfrac{rt}{t-r}$ 84) $R_1=\dfrac{R_2R_3}{R_2-R_3}$

85) 12 g   86) 2 mph   87) 280 mph   88) $2\dfrac{2}{5}$ hr   89) 21

90) 72   91) 2   92) 32   93) 3.24 lb   94) 63.125 liters

Chapter 7 Test

1) a) $-\dfrac{1}{5}$ b) never equals zero

c) $(-\infty,-8)\cup(-8,6)\cup(6,\infty)$

2) $\left(-\infty,-\dfrac{3}{2}\right)\cup\left(-\dfrac{3}{2},\infty\right)$ 3) $\dfrac{9}{4w^5}$ 4) $\dfrac{7v-1}{v-8}$

5) Possible answers: $-\dfrac{h-9}{2h-3},\dfrac{h-9}{3-2h},\dfrac{-h+9}{2h-3},\dfrac{h-9}{-2h+3}$

6) $k(2k-3)(k+2)^2$ 7) $\dfrac{1}{z}$ 8) $\dfrac{7n^2}{4m^4}$

9) $\dfrac{r^2+11r+3}{(2r+1)(r+5)}$ 10) $-\dfrac{a^2+2a+4}{6}$

11) $\dfrac{11}{15-c}$ or $-\dfrac{11}{c-15}$ 12) $\dfrac{x^2-12x+21}{(x+7)(x-7)(x-9)}$

13) $h(x)=\dfrac{-x^2+7x+14}{x(x+7)}$;

Domain: $(-\infty,-7)\cup(-7,0)\cup(0,\infty)$

14) $-\dfrac{1}{d}$ 15) $\dfrac{9}{4}$ 16) $-\dfrac{xy}{x+y}$ 17) $3k+7$ 18) $0,\dfrac{1}{4}$

19) $\{4\}$ 20) $\varnothing$ 21) $\{-1,5\}$ 22) $c=\dfrac{kxz}{y}$ 23) $p=\dfrac{qr}{q-r}$

24) 13 mph   25) 300   26) 18 dB

Cumulative Review Chapters 1–7

1) $-45w^8$ 2) $\dfrac{8}{n^{18}}$ 3) $\left\{-\dfrac{45}{4}\right\}$

4) 45 ml of 12% solution; 15 ml of 4% solution

5) x-int: $(2,0)$; y-int: $(0,-5)$

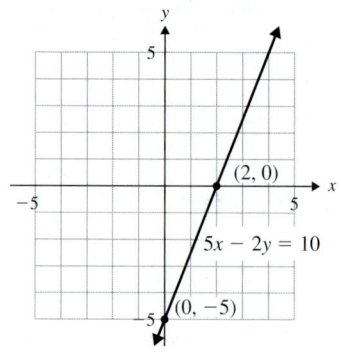

6)

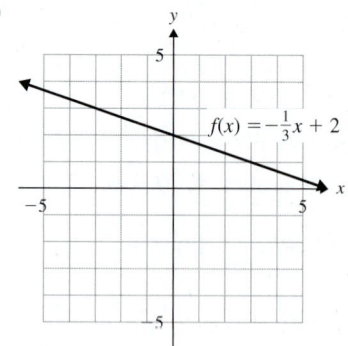

7) $x+y=3$ 8) $(2,10)$ 9) width: 15 cm; length: 23 cm

10) $(-\infty,1]$ 11) $\left\{-\dfrac{7}{2},-\dfrac{5}{6}\right\}$

12)

13)

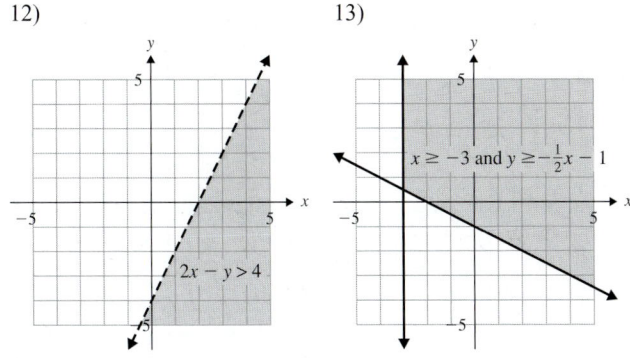

14) a) -11 b) $\dfrac{3}{2}$ or 4   15) $-14q^2 + 6q - 20$

16) $12d^4 + 18d^3 - 31d^2 - 42d + 7$

17) $ab - \dfrac{5}{2b} + \dfrac{7}{a^2b^2} + \dfrac{1}{a^3b^2}$ 18) $v^3 + 3v^2 - 5v - 6$

19) $\dfrac{3(c + 8)}{c(c - 6)}$ 20) $(5n + 9)(5n - 9)$ 21) $3x(y + 8)(y - 3)$

22) $(r + 4t)^2$ 23) $(-\infty, 8) \cup (8, \infty)$

24) a) $-\dfrac{4}{3k(4k^2 + 6k + 9)}$ b) $\dfrac{(r + 3)(r + 2)}{r^2 + r + 3}$ 25) $1\dfrac{7}{8}$ hr

Chapter 8

Section 8.1

1) False; $\sqrt{121} = 11$ because the $\sqrt{}$ symbol means principal square root.

2) True

3) False; the square root of a negative number is not a real number.

4) False; it is not a real number.   5) 12 and -12

6) 50 and -50 7) $\dfrac{6}{5}$ and $-\dfrac{6}{5}$ 8) 0.1 and -0.1 9) 7

10) 13   11) not real   12) not real   13) $\dfrac{9}{5}$ 14) $\dfrac{11}{2}$ 15) -6

16) -0.2 17) True

18) False; the odd root of a negative number is a negative number.

19) False; the only even root of zero is zero.   20) True

21) $\sqrt[3]{64}$ is the number you cube to get 64. $\sqrt[3]{64} = 4$

22) $\sqrt[4]{16}$ is the number you raise to the fourth power to get 16. $\sqrt[4]{16} = 2$

23) No; the even root of a negative number is not a real number.

24) Yes; $\sqrt[3]{-8} = -2$ because $(-2)^3 = -8$.   25) 5   26) 3

27) -1 28) -2 29) 3   30) 2   31) not real   32) not real

33) -2 34) -1 35) -2 36) -2 37) 3   38) 10

39) not real   40) not real   41) $\dfrac{2}{5}$ 42) $\dfrac{3}{2}$ 43) 7   44) 11

45) -3 46) not real   47) 13   48) 5

49) If a is negative and we didn't use the absolute values, the result would be negative. This is incorrect because if a is negative and n is even, then $a^n > 0$ so that $\sqrt[n]{a^n} > 0$. Using absolute values ensures a positive result.

50) If a is negative, then $\sqrt[n]{a^n}$ will simplify to a negative number. So when simplifying $\sqrt[n]{a^n}$, the result is not always nonnegative.

51) 8   52) 5   53) 6   54) 11   55) $|y|$ 56) $|d|$ 57) 5

58) -4 59) z 60) t 61) $|h|$ 62) $|m|$ 63) $|x + 7|$

64) $|a - 9|$ 65) $2t - 1$ 66) $6r + 7$ 67) $|3n + 2|$

68) $x - 6$ 69) $d - 8$ 70) $|4y + 3|$

71) No, because $\sqrt{-1}$ is not a real number.

72) Yes, because $\sqrt[3]{-1} = -1$.

73) Set up an inequality so that the radicand is greater than or equal to 0. Solve for the variable. These are the real numbers in the domain of the function.

74) $(-\infty, \infty)$ 75) 10   76) 3   77) not a real number

78) not a real number   79) 1   80) $\sqrt{13}$

81) not a real number   82) not a real number   83) $\sqrt{a}$

84) $\sqrt{3w + 4}$ 85) $\sqrt{t + 4}$ 86) $\sqrt{5p - 9}$ 87) $\sqrt{6n + 1}$

88) $\sqrt{3m + 34}$ 89) 4   90) -5 91) -3 92) $\sqrt[3]{11}$

93) $\sqrt[3]{-17}$ 94) -1 95) $\sqrt[3]{4r - 1}$ 96) $\sqrt[3]{z}$ 97) $\sqrt[3]{c + 8}$

98) $\sqrt[3]{5k - 2}$ 99) $\sqrt[3]{8a - 13}$ 100) $\sqrt[3]{23 - 4w}$

101) $[-2, \infty)$ 102) $[-10, \infty)$ 103) $[8, \infty)$ 104) $[1, \infty)$

105) $(-\infty, \infty)$ 106) $(-\infty, \infty)$ 107) $\left[\dfrac{5}{2}, \infty\right)$

108) $\left[-\dfrac{7}{3}, \infty\right)$ 109) $(-\infty, \infty)$ 110) $(-\infty, \infty)$

111) $(-\infty, 0]$ 112) $(-\infty, 3)$ 113) $\left(-\infty, \dfrac{9}{7}\right]$

114) $\left(-\infty, \dfrac{8}{5}\right]$

115) Domain: $[1, \infty)$ 116) Domain: $[4, \infty)$

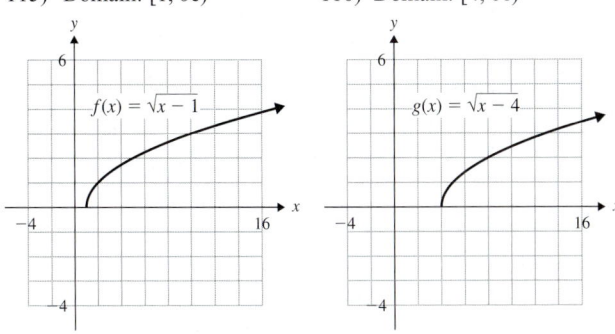

117) Domain: $[-3, \infty)$

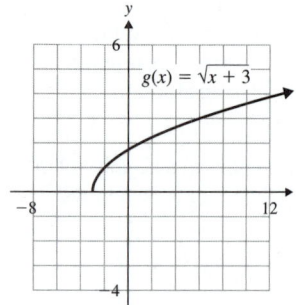

118) Domain: $[-1, \infty)$

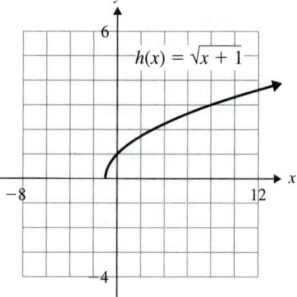

127) Domain: $(-\infty, \infty)$

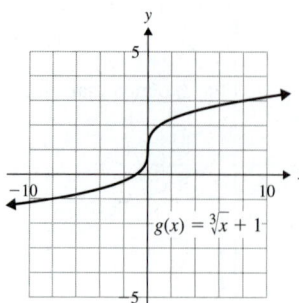

128) Domain: $(-\infty, \infty)$

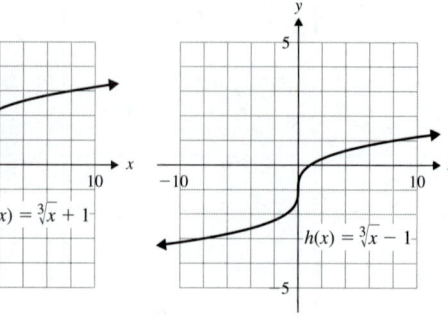

119) Domain: $[0, \infty)$

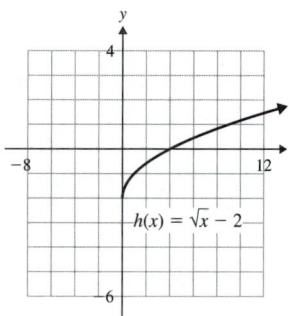

120) Domain: $[0, \infty)$

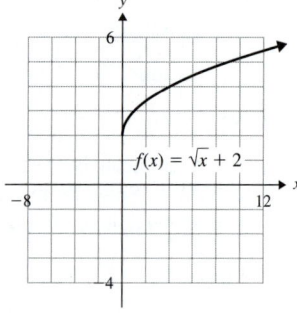

129) 2 ft   130) 1.5 cm   131) $\sqrt{10}$ sec; 3.16 sec

132) $\sqrt{\dfrac{51}{\pi}}$ ft; 4.03 ft   133) Yes. The car was traveling at 30 mph.

134) No. The car was traveling about 58 mph.

135) π sec; 3.14 sec   136) $\dfrac{\pi}{2}$ sec; 1.57 sec

137) $T\left(\dfrac{1}{2}\right) = \dfrac{\pi}{4}$; A $\dfrac{1}{2}$-ft-long pendulum has a period of $\dfrac{\pi}{4}$ sec.
    This is approximately 0.79 sec.

121) Domain: $(-\infty, 0]$

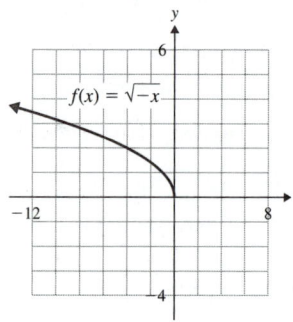

122) Domain: $(-\infty, 0]$

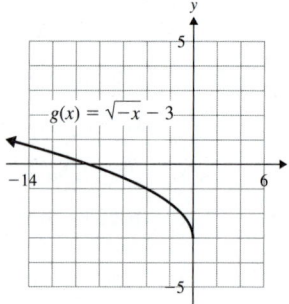

138) $T(1.5) = \dfrac{\sqrt{3}}{4}\pi$; A 1.5-ft-long pendulum has a period of

    $\dfrac{\sqrt{3}}{4}\pi$ sec. This is approximately 1.36 sec.

139) $\dfrac{\sqrt{5}}{4}\pi$ sec; 1.76 sec   140) $\dfrac{3}{4}\pi$ sec; 2.36 sec

Section 8.2

1) The denominator of 2 becomes the index of the radical.
   $25^{1/2} = \sqrt{25}$

2) The denominator of 3 becomes the index of the radical.
   $1^{1/3} = \sqrt[3]{1}$

123) Domain: $(-\infty, \infty)$

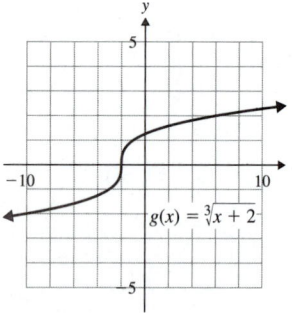

124) Domain: $(-\infty, \infty)$

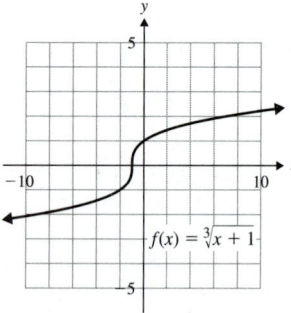

3) 3   4) 8   5) 10   6) 3   7) 2   8) 3   9) −5   10) −2

11) $\dfrac{2}{11}$ 12) $\dfrac{2}{3}$ 13) $\dfrac{5}{4}$ 14) $\dfrac{2}{3}$ 15) $-\dfrac{6}{13}$ 16) $-\dfrac{10}{3}$

17) not a real number   18) not a real number   19) −1   20) −2

21) The denominator of 4 becomes the index of the radical. The
    numerator of 3 is the power to which we raise the radical
    expression. $16^{3/4} = (\sqrt[4]{16})^3$

22) The denominator of 2 becomes the index of the radical.
    The numerator of 3 is the power to which we raise the
    radical expression. $100^{3/2} = (\sqrt{100})^3$

125) Domain: $(-\infty, \infty)$

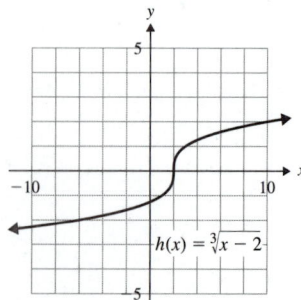

126) Domain: $(-\infty, \infty)$

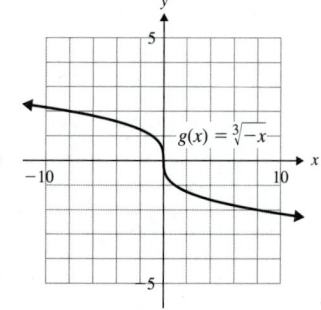

23) 16   24) 27   25) 32   26) 8   27) 25   28) 100

29) −216   30) −81   31) not a real number

32) not a real number   33) $\dfrac{8}{27}$ 34) −32

35) $-\dfrac{100}{9}$ 36) $-\dfrac{16}{81}$

37) False; the negative exponent does not make the result negative.
$81^{-1/2} = \dfrac{1}{9}$

38) False; to eliminate the negative from the exponent, take the
reciprocal of the base. $\left(\dfrac{1}{100}\right)^{-3/2} = (100)^{3/2}$

39) $\dfrac{1}{64}$; $\dfrac{1}{64}$; The denominator of the fractional exponent is
the index of the radical.; $\dfrac{1}{8}$

40) 1000; The reciprocal of 1000 is $\dfrac{1}{1000}$. The denominator of
the fractional exponent is the index of the radical; 10

41) $\dfrac{1}{7}$ 42) $\dfrac{1}{10}$ 43) $\dfrac{1}{10}$ 44) $\dfrac{1}{3}$ 45) 3   46) 2   47) -4

48) -5 49) $\dfrac{1}{32}$ 50) $\dfrac{1}{27}$ 51) $\dfrac{1}{25}$ 52) $\dfrac{1}{16}$ 53) $\dfrac{8}{125}$

54) $\dfrac{1000}{27}$ 55) $\dfrac{25}{16}$ 56) $\dfrac{8}{27}$ 57) 8   58) 25   59) 3   60) 49

61) $8^{4/5}$ 62) $6^{1/3}$ 63) 32   64) $\dfrac{1}{125}$ 65) $\dfrac{1}{4^{7/5}}$ 66) 6

67) z 68) $\dfrac{1}{h^{7/12}}$ 69) $-72v^{11/8}$ 70) $-24x^{1/9}$ 71) $a^{1/9}$

72) $\dfrac{1}{x^{2/3}}$ 73) $\dfrac{5}{18c^{3/2}}$ 74) $\dfrac{24}{5w^{1/10}}$ 75) $\dfrac{1}{x^{2/3}}$ 76) $\dfrac{1}{n^{6/7}}$

77) $z^{2/15}$ 78) $r^{10/3}$ 79) $27u^2v^3$ 80) $32x^5y^2$ 81) $8r^{1/5}s^{4/15}$

82) $25a^6b^{1/6}$ 83) $\dfrac{f^{2/7}g^{5/9}}{3}$ 84) $\dfrac{8b^{11/4}}{c^6}$ 85) $x^{10}w^9$ 86) t^6u

87) $\dfrac{1}{y^{2/3}}$ 88) $t^{15/4}$ 89) $\dfrac{a^{12/5}}{4b^{2/5}}$ 90) $\dfrac{64}{c^{18}d^3}$ 91) $\dfrac{t^{21/2}}{r^{1/5}}$

92) $\dfrac{y^{1/3}}{x^{12}}$ 93) $\dfrac{1}{h^5k^{25/18}}$ 94) cd^8 95) $p^{7/6} + p$

96) $w^{11/6} - w^{13/3}$ 97) $25^{6/12}$; $25^{1/2}$; Evaluate.

98) $c^{4/10}$; Reduce the exponent; $\sqrt{c^2}$ 99) 7   100) 2   101) 9

102) 3   103) 5   104) 10   105) 12   106) 15   107) x^4

108) t^2 109) $\sqrt[3]{k}$ 110) $\sqrt[3]{w^2}$ 111) $\sqrt{z}$ 112) $\sqrt{m}$

113) d^2 114) s^3 115) a) 13 degrees F   b) 6 degrees F

116) a) -5 degrees F   b) -9 degrees F

Section 8.3

1) $\sqrt{21}$ 2) $\sqrt{55}$ 3) $\sqrt{30}$ 4) $\sqrt{14}$ 5) $\sqrt{6y}$ 6) $\sqrt{5p}$

7) False; 20 contains a factor of 4 which is a perfect square.

8) True; 35 does not have any factors (other than 1) that are
perfect squares.

9) True; 42 does not have any factors (other than 1) that are
perfect squares.

10) False; 63 contains a factor of 9 which is a perfect square.

11) Factor; $\sqrt{4} \cdot \sqrt{15}$; $2\sqrt{15}$

12) $\sqrt{100 \cdot 2}$; Product rule; $10\sqrt{2}$

13) $2\sqrt{5}$ 14) $2\sqrt{3}$ 15) $3\sqrt{6}$ 16) $3\sqrt{7}$ 17) simplified

18) simplified   19) $4\sqrt{5}$ 20) $6\sqrt{3}$ 21) $7\sqrt{2}$ 22) $4\sqrt{6}$

23) simplified   24) simplified   25) 20   26) 30   27) $5\sqrt{30}$

28) $2\sqrt{105}$ 29) $\dfrac{12}{5}$ 30) $\dfrac{4}{9}$ 31) $\dfrac{2}{7}$ 32) $\dfrac{8}{11}$ 33) 3   34) 4

35) $2\sqrt{3}$ 36) $2\sqrt{2}$ 37) $2\sqrt{5}$ 38) $3\sqrt{2}$ 39) $\sqrt{15}$

40) $\sqrt{7}$ 41) $\dfrac{\sqrt{6}}{7}$ 42) $\dfrac{\sqrt{2}}{9}$ 43) $\dfrac{3\sqrt{5}}{4}$ 44) $\dfrac{2\sqrt{15}}{7}$ 45) x^4

46) q^3 47) w^7 48) t^8 49) $10c$ 50) $3z^4$ 51) $8k^3m^5$

52) $5p^{10}q^7$ 53) $2r^2\sqrt{7}$ 54) $3z^6\sqrt{3}$ 55) $10q^{11}t^8\sqrt{3}$

56) $5n^2y^2\sqrt{2}$ 57) $\dfrac{9}{c^3}$ 58) $\dfrac{h}{13}$

59) $\dfrac{2\sqrt{10}}{t^4}$ 60) $\dfrac{3\sqrt{2}}{m^{15}}$ 61) $\dfrac{5x\sqrt{3}}{y^6}$ 62) $\dfrac{2\sqrt{11}}{wz^9}$

63) Factor; $\sqrt{w^8} \cdot \sqrt{w^1}$; Simplify.

64) Factor; Product rule; $z^9\sqrt{z}$ 65) $a^2\sqrt{a}$ 66) $c^3\sqrt{c}$

67) $g^6\sqrt{g}$ 68) $k^7\sqrt{k}$ 69) $b^{12}\sqrt{b}$ 70) $h^{15}\sqrt{h}$ 71) $6x\sqrt{2x}$

72) $10a^2\sqrt{a}$ 73) $q^3\sqrt{13q}$ 74) $2c^4\sqrt{5c}$ 75) $5t^5\sqrt{3t}$

76) $3p^8\sqrt{5p}$ 77) c^4d 78) r^2s^6 79) $a^2b\sqrt{b}$ 80) $xy^4\sqrt{y}$

81) $u^2v^3\sqrt{uv}$ 82) $fg^4\sqrt{fg}$ 83) $6m^4n^2\sqrt{m}$ 84) $2t^3u^2\sqrt{u}$

85) $2x^6y^2\sqrt{11y}$ 86) $3c^3d^2\sqrt{7c}$ 87) $4t^2u^3\sqrt{2tu}$

88) $5kl^4\sqrt{5kl}$ 89) $\dfrac{a^3\sqrt{a}}{9b^3}$ 90) $\dfrac{x^2\sqrt{x}}{7y^3}$ 91) $\dfrac{r^4\sqrt{3r}}{s}$

92) $\dfrac{h^5\sqrt{17h}}{k^4}$ 93) $5\sqrt{2}$ 94) $4\sqrt{3}$ 95) $3\sqrt{7}$ 96) $2\sqrt{7}$

97) w^3 98) d^7 99) $n^3\sqrt{n}$ 100) $a^6\sqrt{a}$ 101) $4k^3$

102) $5z^6$ 103) $2x^4y^2\sqrt{3xy}$ 104) $5a^3b^4\sqrt{2ab}$

105) $2c^5d^4\sqrt{10d}$ 106) $3t^5u^3\sqrt{2u}$ 107) $3k^4$ 108) $4m^3$

109) $2h^3\sqrt{10}$ 110) $2c^4\sqrt{3}$ 111) $a^4b^2\sqrt{10ab}$

112) $y^3z^2\sqrt{7yz}$ 113) 20 m/s   114) $2\sqrt{6}$ cm

Section 8.4

1) Answers may vary.   2) Answers may vary.

3) i) Its radicand will not contain any factors that are perfect
cubes.
   ii) The radicand will not contain fractions.
   iii) There will be no radical in the denominator of a fraction.

4) i) Its radicand will not contain any factors that are perfect
fourth powers.
   ii) The radicand will not contain fractions.
   iii) There will be no radical in the denominator of a fraction.

5) $\sqrt[3]{20}$ 6) $\sqrt[5]{12}$ 7) $\sqrt[5]{9m^2}$ 8) $\sqrt[4]{11h^3}$ 9) $\sqrt[3]{a^2b}$

10) $\sqrt[5]{t^2u^4}$ 11) Factor; $\sqrt[3]{8} \cdot \sqrt[3]{7}$; $2\sqrt[3]{7}$

12) Factor; Product rule; $2\sqrt[4]{5}$ 13) $2\sqrt[3]{3}$ 14) $2\sqrt[3]{6}$

15) $2\sqrt[4]{4}$ 16) $2\sqrt[4]{2}$ 17) $3\sqrt[3]{2}$ 18) $2\sqrt[3]{11}$ 19) $10\sqrt[3]{2}$

20) $3\sqrt[3]{4}$ 21) $2\sqrt[5]{2}$ 22) $3\sqrt[4]{2}$ 23) $\dfrac{1}{2}$ 24) $\dfrac{1}{5}$ 25) -3

26) 2 27) $2\sqrt[3]{3}$ 28) $5\sqrt[3]{2}$ 29) $2\sqrt[4]{5}$ 30) $10\sqrt[3]{2}$ 31) d^2

32) g^3 33) n^5 34) t^9 35) xy^3 36) a^2b 37) $w^4\sqrt[3]{w^2}$

38) $b^6\sqrt[3]{b}$ 39) $y^2\sqrt[4]{y}$ 40) $m\sqrt[4]{m^3}$ 41) $d\sqrt[3]{d^2}$ 42) $c^9\sqrt[3]{c^2}$

43) $u^3v^5\sqrt[3]{u}$ 44) $x^3y^5\sqrt[3]{y}$ 45) $b^5c\sqrt[3]{bc^2}$ 46) $r^3s^2\sqrt[4]{r^3s}$

47) $n^4\sqrt[4]{m^3n^2}$ 48) $a^3\sqrt[3]{a^2b}$ 49) $2x^3y^4\sqrt[3]{3x}$ 50) $3y^3z^8\sqrt[3]{2y}$

51) $5wx^5\sqrt[3]{2wx}$ 52) $2t^5u^2\sqrt[3]{9t^2u}$ 53) $\dfrac{m^2}{3}$ 54) $\dfrac{2}{x^3}$

55) $\dfrac{2a^4\sqrt[5]{a^3}}{b^3}$ 56) $\dfrac{h^5\sqrt[3]{h^2}}{5k^7}$ 57) $\dfrac{t^2\sqrt[4]{t}}{3s^6}$ 58) $\dfrac{2c\sqrt[5]{c^4}}{d^4}$

59) $\dfrac{u^9\sqrt[3]{u}}{v}$ 60) $\dfrac{m^3\sqrt[4]{m}}{n^2}$ 61) $2\sqrt[3]{3}$ 62) $2\sqrt[3]{5}$ 63) $3\sqrt[3]{4}$

64) $3\sqrt[3]{2}$ 65) $2\sqrt[3]{10}$ 66) $2\sqrt[3]{7}$ 67) m^3 68) t^2 69) k^4

70) a^5 71) $r^3\sqrt[3]{r^2}$ 72) $y^6\sqrt[3]{y}$ 73) $p^4\sqrt[5]{p^3}$ 74) $c^5\sqrt[5]{c}$

75) $3z^6\sqrt[3]{z}$ 76) $2h^6\sqrt[3]{h^2}$ 77) h^4 78) a^2 79) $c^2\sqrt[3]{c}$

80) $z^3\sqrt[3]{z^2}$ 81) $3d^4\sqrt[4]{d^3}$ 82) $2t\sqrt[4]{t}$

83) Change radicals to fractional exponents.; Rewrite exponents with a common denominator.; $a^{5/4}$; Rewrite in radical form. $a\sqrt[4]{a}$

84) $r^{4/5} \cdot r^{2/3}$; $r^{12/15} \cdot r^{10/15}$; Add exponents.; $\sqrt[15]{r^{22}}$; $r\sqrt[15]{r^7}$

85) $\sqrt[6]{p^5}$ 86) $\sqrt[12]{y^{11}}$ 87) $n\sqrt[4]{n}$ 88) $k\sqrt[10]{k^3}$ 89) $c\sqrt[15]{c^4}$

90) $a\sqrt[15]{a}$ 91) $\sqrt[12]{w}$ 92) $\sqrt[4]{m}$ 93) $\sqrt[15]{t^2}$ 94) $\sqrt[12]{h}$

95) 4 in.   96) 4 ft

Section 8.5

1) They have the same index and the same radicand.

2) No. The indices are different.   3) $14\sqrt{2}$ 4) $18\sqrt{7}$

5) $15\sqrt[3]{4}$ 6) $8\sqrt[3]{5}$ 7) $11 - 3\sqrt{13}$ 8) $1 - \sqrt{6}$

9) $-5\sqrt[3]{z^2}$ 10) $3\sqrt[3]{p}$ 11) $-9\sqrt[3]{n^2} + 10\sqrt[5]{n^2}$

12) $9\sqrt[4]{s} - \sqrt[3]{s}$ 13) $2\sqrt{5c} - 2\sqrt{6c}$ 14) $9\sqrt{2m} + 14\sqrt{3m}$

15)   i) Write each radical expression in simplest form.

   ii) Combine like radicals.

16) Yes. The radicals are simplified and they are unlike.

17) Factor.; $\sqrt{16} \cdot \sqrt{3} + \sqrt{3}$; Simplify.; $5\sqrt{3}$

18) Factor.; Product rule; $2\sqrt{11} - 8\sqrt{11}$; $-6\sqrt{11}$ 19) $4\sqrt{3}$

20) $7\sqrt{5}$ 21) $-2\sqrt{2}$ 22) $10\sqrt{6}$ 23) $6\sqrt{3}$ 24) $-9\sqrt{6}$

25) $10\sqrt[3]{9}$ 26) $12\sqrt[3]{11}$ 27) $-\sqrt[3]{6}$ 28) $29\sqrt[3]{2}$

29) $13q\sqrt{q}$ 30) $19m\sqrt{m}$ 31) $-20d^2\sqrt{d}$ 32) $3k^4\sqrt{k}$

33) $4t^3\sqrt[3]{t}$ 34) $-8r^4\sqrt[3]{r}$ 35) $6a^2\sqrt[4]{a^3}$ 36) $3c^2\sqrt[4]{c^3}$

37) $-2\sqrt{2p}$ 38) $18\sqrt{7t}$ 39) $25a\sqrt[3]{3a^2}$ 40) $-6\sqrt[3]{5x}$

41) $4y\sqrt{xy}$ 42) $7a\sqrt{ab}$ 43) $3c^2d\sqrt{2d}$ 44) $5uv\sqrt{5u}$

45) $20a^5\sqrt[3]{7a^2b}$ 46) $14p^2q\sqrt[3]{11pq^2}$ 47) $14cd\sqrt[4]{9cd}$

48) $10y^2z^2\sqrt[4]{11z}$ 49) $\sqrt[3]{b}(a^3 - b^2)$ 50) $\sqrt[3]{c^2}(c^2 + d)$

51) $3x + 15$ 52) $8k + 24$ 53) $7\sqrt{6} + 14$ 54) $20 - 5\sqrt{7}$

55) $\sqrt{30} - \sqrt{10}$ 56) $9\sqrt{2} + \sqrt{22}$ 57) $-30\sqrt{2}$

58) $10\sqrt{3}$ 59) $4\sqrt{5}$ 60) $-24\sqrt{2}$ 61) $-\sqrt{30}$ 62) $5\sqrt{10}$

63) $5\sqrt[4]{3} - 3$ 64) $2\sqrt[3]{20} + 14\sqrt[3]{2}$ 65) $t - 9\sqrt{tu}$

66) $2\sqrt{3rs} + s\sqrt{7}$ 67) $a\sqrt{5b} + 3b\sqrt{3a}$

68) $2y\sqrt{x} - xy\sqrt{2y}$ 69) $c\sqrt[3]{c} + 5c\sqrt[3]{d}$

70) $\sqrt[5]{2m^3n^4} - n^2\sqrt[5]{m^2}$

71) Both are examples of multiplication of two binomials. They can be multiplied using FOIL.

72) Both are examples of the square of a binomial. We can multiply them using the formula $(a - b)^2 = a^2 - 2ab + b^2$.

73) $(a + b)(a - b) = a^2 - b^2$

74) The radicals are eliminated.

75) $p^2 + 13p + 42$ 76) $z^2 - 6z - 16$

77) $6 \cdot 2 + 6\sqrt{7} + 2\sqrt{7} + \sqrt{7} \cdot \sqrt{7}$; Multiply.; $19 + 8\sqrt{7}$

78) Use FOIL.; $3 + 3\sqrt{5} + \sqrt{5} + 5$; $8 + 4\sqrt{5}$

79) $-22 + 5\sqrt{2}$ 80) $-8 - 5\sqrt{6}$ 81) $22 - 9\sqrt{15}$

82) $-16 + 11\sqrt{6}$ 83) $5\sqrt{7} + 5\sqrt{2} + 2\sqrt{21} + 2\sqrt{6}$

84) $\sqrt{15} - 6\sqrt{10} + 4\sqrt{3} - 24\sqrt{2}$

85) $5 - \sqrt[3]{150} - 3\sqrt[3]{5} + 3\sqrt[3]{6}$

86) $2\sqrt[4]{3} + 2 - \sqrt[4]{18} - \sqrt[4]{6}$

87) $-2\sqrt{6pq} + 30p - 16q$ 88) $10\sqrt{3rs} + 3s - 24r$

89) $4 + 2\sqrt{3}$ 90) $9 + 4\sqrt{5}$ 91) $16 - 2\sqrt{55}$

92) $16 + 2\sqrt{39}$ 93) $h + 2\sqrt{7h} + 7$ 94) $m + 2\sqrt{3m} + 3$

95) $x - 2\sqrt{xy} + y$ 96) $a - 2\sqrt{ab} + b$ 97) $c^2 - 81$

98) $g^2 - 49$ 99) 31 100) 9 101) 46 102) -20

103) $\sqrt[3]{4} - 9$ 104) $1 - \sqrt[3]{36}$ 105) $c - d$ 106) $2y - z$

107) $64f - g$ 108) $a - 36b$ 109) 41 110) 29

111) $11 + 4\sqrt{7}$ 112) $31 - 10\sqrt{6}$ 113) $13 - 4\sqrt{3}$

114) $34 + 24\sqrt{2}$

Section 8.6

1) Eliminate the radical from the denominator.

2) Answers may vary.   3) $\dfrac{\sqrt{5}}{5}$ 4) $\dfrac{\sqrt{6}}{6}$ 5) $\dfrac{3\sqrt{6}}{2}$

6) $\dfrac{5\sqrt{10}}{2}$ 7) $-5\sqrt{2}$ 8) $-\dfrac{6\sqrt{5}}{5}$ 9) $\dfrac{\sqrt{21}}{14}$ 10) $\dfrac{2\sqrt{6}}{9}$

11) $\dfrac{\sqrt{3}}{3}$ 12) $\dfrac{\sqrt{15}}{10}$ 13) $\dfrac{\sqrt{42}}{6}$ 14) $\dfrac{\sqrt{22}}{2}$ 15) $\dfrac{\sqrt{30}}{3}$

16) $\dfrac{\sqrt{22}}{2}$ 17) $\dfrac{\sqrt{15}}{10}$ 18) $\dfrac{2\sqrt{5}}{5}$ 19) $\dfrac{8\sqrt{y}}{y}$ 20) $\dfrac{4\sqrt{w}}{w}$

21) $\dfrac{\sqrt{5t}}{t}$ 22) $\dfrac{\sqrt{2m}}{m}$ 23) $\dfrac{8v^3\sqrt{5vw}}{5w}$ 24) $\dfrac{9c^2\sqrt{2cd}}{2d}$

25) $\dfrac{a\sqrt{3b}}{3b}$ 26) $\dfrac{n^2\sqrt{7m}}{7m}$ 27) $-\dfrac{5\sqrt{3b}}{b^2}$ 28) $-\dfrac{2\sqrt{6v}}{v^2}$

29) $\dfrac{\sqrt{13j}}{j^3}$ 30) $\dfrac{\sqrt{22w}}{w^4}$ 31) 2^2 or 4   32) 5^2 or 25

33) 3   34) 2   35) c^2 36) p^2 37) 2^3 or 8   38) 2   39) m

40) k^3 41) $\dfrac{4\sqrt[3]{9}}{3}$ 42) $\dfrac{26\sqrt[3]{25}}{5}$ 43) $6\sqrt[3]{4}$ 44) $7\sqrt[3]{9}$

45) $\dfrac{9\sqrt[3]{5}}{5}$ 46) $3\sqrt[3]{2}$ 47) $\dfrac{\sqrt[4]{45}}{3}$ 48) $\dfrac{\sqrt[4]{50}}{5}$ 49) $\dfrac{\sqrt[5]{12}}{2}$

50) $\dfrac{\sqrt[5]{56}}{2}$ 51) $\dfrac{10\sqrt[3]{z^2}}{z}$ 52) $\dfrac{6\sqrt[3]{u^2}}{u}$ 53) $\dfrac{\sqrt[3]{3n}}{n}$ 54) $\dfrac{\sqrt[3]{5x}}{x}$

55) $\dfrac{\sqrt[3]{28k}}{2k}$ 56) $\dfrac{\sqrt[3]{10t^2}}{5t}$ 57) $\dfrac{9\sqrt[5]{a^2}}{a}$ 58) $\dfrac{8\sqrt[5]{h^3}}{h}$

59) $\dfrac{\sqrt[4]{40m^3}}{2m}$ 60) $\dfrac{\sqrt[4]{54t^2}}{3t}$

61) Change the sign between the two terms.

62) The radical is eliminated.   63) $(5-\sqrt{2})$; 23

64) $(\sqrt{5}+4)$; -11 65) $(\sqrt{2}-\sqrt{6})$; -4

66) $(\sqrt{3}+\sqrt{10})$; -7 67) $(\sqrt{t}+8)$; $t-64$

68) $(\sqrt{p}-5)$; $p-25$

69) Multiply by the conjugate.; $(a+b)(a-b) = a^2 - b^2$;
$\dfrac{24+6\sqrt{5}}{16-5}$; $\dfrac{24+6\sqrt{5}}{11}$

70) Multiply by the conjugate.; $(a+b)(a-b) = a^2 - b^2$;
$\dfrac{\sqrt{42}-\sqrt{12}}{7-2}$; $\dfrac{\sqrt{42}-2\sqrt{3}}{5}$

71) $6-3\sqrt{3}$ 72) $\dfrac{48+8\sqrt{5}}{31}$ 73) $\dfrac{90+10\sqrt{2}}{79}$

74) $\dfrac{4-\sqrt{6}}{2}$ 75) $2\sqrt{6}-4$ 76) $-2\sqrt{10}-2\sqrt{14}$

77) $\dfrac{\sqrt{30}-5\sqrt{2}+3-\sqrt{15}}{7}$

78) $\dfrac{\sqrt{15}+\sqrt{30}-\sqrt{6}-2\sqrt{3}}{3}$ 79) $\dfrac{m-\sqrt{mn}}{m-n}$

80) $\dfrac{u+\sqrt{uv}}{u-v}$ 81) $\sqrt{b}+5$ 82) $\sqrt{d}-3$

83) $\dfrac{x+2\sqrt{xy}+y}{x-y}$ 84) $\dfrac{f-2\sqrt{fg}+g}{f-g}$ 85) $\dfrac{5}{3\sqrt{5}}$

86) $\dfrac{2}{9\sqrt{2}}$ 87) $\dfrac{x}{\sqrt{7x}}$ 88) $\dfrac{4a}{\sqrt{2ab}}$ 89) $\dfrac{1}{12-6\sqrt{3}}$

90) $\dfrac{2}{\sqrt{7}-1}$ 91) $\dfrac{1}{\sqrt{x}+2}$ 92) $-\dfrac{1}{3+\sqrt{n}}$

93) $-\dfrac{1}{4+\sqrt{c+11}}$ 94) $\dfrac{1}{\sqrt{x+h}+\sqrt{x}}$

95) No, because when we multiply the numerator and denominator by the conjugate of the denominator, we are multiplying the original expression by 1.

96) No, because when we multiply the numerator and denominator by the conjugate of the numerator, we are multiplying the original expression by 1.

97) $1+2\sqrt{3}$ 98) $3-\sqrt{7}$ 99) $\dfrac{15-9\sqrt{5}}{2}$

100) $\dfrac{9+5\sqrt{2}}{3}$ 101) $\dfrac{\sqrt{5}+2}{3}$ 102) $\sqrt{3}+7$

103) $-2-\sqrt{2}$ 104) $\dfrac{-7+2\sqrt{2}}{3}$

105) a) $r(8\pi) = 2\sqrt{2}$; When the area of a circle is 8π in^2, its radius is $2\sqrt{2}$ in.

b) $r(7) = \dfrac{\sqrt{7\pi}}{\pi}$; When the area of a circle is 7 in^2, its radius is $\dfrac{\sqrt{7\pi}}{\pi}$ in. (This is approximately 1.5 in.)

c) $r(A) = \dfrac{\sqrt{A\pi}}{\pi}$

106) a) $r(36\pi) = 3$; When the volume of a sphere is 36π cm^3, its radius is 3 cm.

b) $r(11) = \dfrac{\sqrt[3]{66\pi^2}}{2\pi}$; When the volume of a sphere is 11 cm^3, its radius is $\dfrac{\sqrt[3]{66\pi^2}}{2\pi}$ cm. (This is approximately 1.4 cm.)

c) $r(V) = \dfrac{\sqrt[3]{6\pi^2 V}}{2\pi}$

Putting It All Together

1) 3   2) -10 3) -2 4) 11   5) not a real number   6) $\dfrac{12}{7}$

7) 12   8) 16   9) -100 10) not a real number   11) $\dfrac{1}{5}$

12) $\dfrac{27}{1000}$ 13) $\dfrac{1}{k^{3/10}}$ 14) t^6 15) $\dfrac{9}{a^{16/3}b^6}$ 16) $\dfrac{x^{30}}{243y^{5/6}}$

17) $2\sqrt{6}$ 18) $2\sqrt[4]{2}$ 19) $2\sqrt[3]{9}$ 20) $5\sqrt[3]{2}$ 21) $3\sqrt[4]{3}$

22) $3c^5\sqrt{5c}$ 23) $2m^2n^5\sqrt[3]{12m}$ 24) $\dfrac{2x^3\sqrt[5]{2x^4}}{y^4}$ 25) $2\sqrt[3]{3}$

26) $2k^2\sqrt[4]{3}$ 27) $19+8\sqrt{7}$ 28) $-18c^2\sqrt[3]{4c}$ 29) $3\sqrt{6}$

30) $5\sqrt{3}+5\sqrt{2}$ 31) $17m\sqrt{3mn}$ 32) $3p^4q^2\sqrt{10q}$

33) $2t^3u\sqrt{3}$ 34) $\dfrac{9\sqrt[3]{4}}{2}$ 35) $112+40\sqrt{3}$ 36) -7

37) $\dfrac{2\sqrt{2}-\sqrt{5}}{3}$ 38) $r\sqrt[6]{r}$ 39) $\dfrac{\sqrt[3]{3b^2c^2}}{3c}$ 40) $\dfrac{2\sqrt[4]{2w}}{w^3}$

41) Domain: $[2, \infty)$

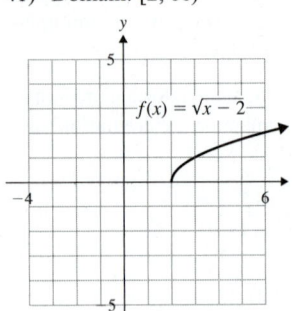

42) Domain: $(-\infty, \infty)$

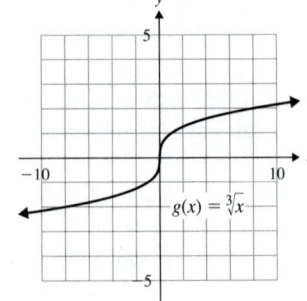

43) Domain: $(-\infty, \infty)$

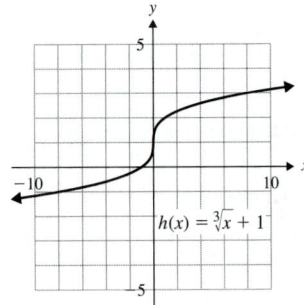

44) Domain: $[-1, \infty)$

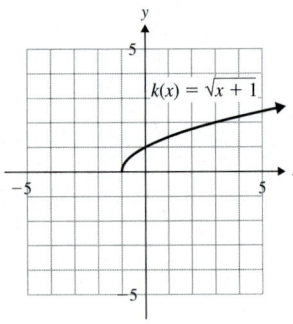

45) Domain: $(-\infty, 0]$

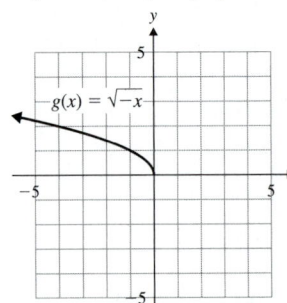

46) Domain: $[0, \infty)$

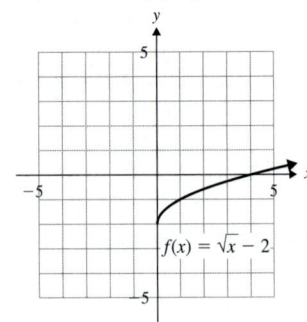

Section 8.7

1) Sometimes there are extraneous solutions.

2) The principal square root of a number cannot equal a negative number.

3) $\{49\}$ 4) $\{100\}$ 5) $\left\{\dfrac{4}{9}\right\}$ 6) $\left\{\dfrac{9}{25}\right\}$ 7) $\varnothing$ 8) $\varnothing$

9) $\{20\}$ 10) $\{22\}$ 11) $\varnothing$ 12) $\varnothing$ 13) $\left\{\dfrac{2}{3}\right\}$ 14) $\left\{\dfrac{29}{5}\right\}$

15) $\{2\}$ 16) $\{6\}$ 17) $\{5\}$ 18) $\{2\}$ 19) $n^2 + 10n + 25$

20) $z^2 - 6z + 9$ 21) $c^2 - 12c + 36$ 22) $4k^2 + 4k + 1$

23) $\{-3\}$ 24) $\{12\}$ 25) $\varnothing$ 26) $\varnothing$ 27) $\{1, 3\}$ 28) $\{0, 2\}$

29) $\{4\}$ 30) $\{2\}$ 31) $\{1, 16\}$ 32) $\{1, 25\}$ 33) $\{-3\}$

34) $\{14\}$ 35) $\{10\}$ 36) $\{-4\}$ 37) $\{-3\}$ 38) $\{11\}$

39) $\{10\}$ 40) $\{25\}$ 41) $\{63\}$ 42) $\{48\}$ 43) $\varnothing$ 44) $\varnothing$

45) $x + 10\sqrt{x} + 25$ 46) $y - 16\sqrt{y} + 64$

47) $85 - 18\sqrt{a + 4} + a$ 48) $21 + 8\sqrt{p + 5} + p$

49) $12n + 28\sqrt{3n - 1} + 45$ 50) $18k - 30\sqrt{2k - 3} - 2$

51) $\{5, 13\}$ 52) $\{0, 4\}$ 53) $\{2\}$ 54) $\{6\}$ 55) $\left\{\dfrac{1}{4}\right\}$

56) $\{3\}$ 57) $\{1, 5\}$ 58) $\{3, 7\}$ 59) $\varnothing$ 60) $\varnothing$

61) $\{2, 11\}$ 62) $\{0, 4\}$

63) Raise both sides of the equation to the third power.

64) When you solve the equation, you get $h = -27$, and the cube root of -27 is -3.

65) $\{125\}$ 66) $\{27\}$ 67) $\{-64\}$ 68) $\{-8\}$ 69) $\left\{-\dfrac{3}{2}\right\}$

70) $\{-7\}$ 71) $\{-1\}$ 72) $\{14\}$ 73) $\{-2\}$ 74) $\{-3\}$

75) $\left\{-\dfrac{1}{2}, 4\right\}$ 76) $\left\{\dfrac{2}{3}, 1\right\}$ 77) $\{36\}$ 78) $\left\{\dfrac{4}{9}\right\}$

79) $\{26\}$ 80) $\{5\}$ 81) $\{23\}$ 82) $\{-123\}$ 83) $\{9\}$

84) $\varnothing$ 85) $\{9\}$ 86) $\{5\}$ 87) $\{-1\}$ 88) $\{10\}$

89) $E = \dfrac{mv^2}{2}$ 90) $P = \dfrac{mV}{300}$ 91) $b^2 = c^2 - a^2$

92) $A = \pi r^2$ 93) $\sigma = \dfrac{E}{T^4}$ 94) $V = \dfrac{4}{3}\pi r^3$

95) a) 320 m/sec   b) 340 m/sec

c) The speed of sound increases.   d) $T = \dfrac{V_s^2}{400} - 273$

96) a) 5 ft $\times$ 5 ft   b) $A = l^2$ 97) a) 2 in.   b) $V = \pi r^2 h$

98) a) 16 ft/sec   b) $H = \dfrac{c^2}{g}$

99) a) 463 mph   b) about 8 min.   100) $\dfrac{32}{3}\pi$ in^3; 33.5 in^3

101) 16 ft   102) 9 ft   103) 5 mph

104) When the wind chill temperature is $-20°F$, the speed of the wind is 16 mph.

Section 8.8

1) False   2) True   3) True   4) False   5) $9i$ 6) $4i$

7) $5i$ 8) $13i$ 9) $i\sqrt{6}$ 10) $i\sqrt{30}$ 11) $3i\sqrt{3}$ 12) $5i\sqrt{3}$

13) $2i\sqrt{15}$ 14) $2i\sqrt{7}$

15) Write each radical in terms of i before multiplying.
$$\sqrt{-5} \cdot \sqrt{-10} = i\sqrt{5} \cdot i\sqrt{10}$$
$$= i^2\sqrt{50}$$
$$= (-1)\sqrt{25} \cdot \sqrt{2}$$
$$= -5\sqrt{2}$$

16) Write the radical in terms of i before squaring.
$$(\sqrt{-7})^2 = (i\sqrt{7})^2$$
$$= i^2 \cdot 7$$
$$= (-1) \cdot 7$$
$$= -7$$

17) $-\sqrt{5}$ 18) $-5\sqrt{3}$ 19) -6 20) -10 21) 2 22) $\dfrac{1}{8}$

23) -13 24) -1

25) Add the real parts and add the imaginary parts.

26) Both are products of binomials, so we can multiply both using FOIL.

27) -1

28) Multiply the numerator and denominator by the conjugate of the denominator.

29) $3 + 11i$ 30) $14 - 4i$ 31) $4 - 9i$ 32) $-5 + 9i$

33) $-\dfrac{1}{4} - \dfrac{5}{6}i$ 34) $-\dfrac{3}{8} + \dfrac{17}{18}i$ 35) $7i$ 36) 0 37) $24 - 15i$

38) $-48 + 6i$ 39) $-6 + \dfrac{4}{3}i$ 40) $9 + \dfrac{7}{2}i$ 41) $-36 + 30i$

42) $44 - 24i$ 43) $-28 + 17i$ 44) $15 + 20i$ 45) $14 + 18i$

46) $-21 - 23i$ 47) $36 - 42i$ 48) $46 - 3i$

49) $\dfrac{3}{20} + \dfrac{9}{20}i$ 50) $\dfrac{41}{36} - \dfrac{7}{9}i$

51) conjugate: $11 - 4i$; 52) conjugate: $-1 + 2i$;
    product: 137   product: 5

53) conjugate: $-3 + 7i$; 54) conjugate: $4 - 9i$;
    product: 58   product: 97

55) conjugate: $-6 - 4i$; 56) conjugate: $6 + 5i$;
    product: 52   product: 61

57) Answers may vary.

58) True. For example, a complex number times its conjugate is a real number.

59) $\dfrac{8}{13} + \dfrac{12}{13}i$ 60) $-\dfrac{16}{29} - \dfrac{18}{29}i$ 61) $\dfrac{8}{17} + \dfrac{32}{17}i$

62) $-\dfrac{5}{61} + \dfrac{6}{61}i$ 63) $\dfrac{7}{29} - \dfrac{3}{29}i$ 64) $\dfrac{45}{58} - \dfrac{9}{29}i$

65) $-\dfrac{74}{85} + \dfrac{27}{85}i$ 66) $-\dfrac{22}{17} + \dfrac{3}{17}i$ 67) $-\dfrac{8}{61} + \dfrac{27}{61}i$

68) $\dfrac{17}{29} + \dfrac{28}{29}i$ 69) $-9i$ 70) $-3 + 16i$ 71) $(i^2)^{12}; i^2 = -1; 1$

72) Product rule; Rewrite i^{30} in terms of i^2 using the power rule; $(-1)^{15} \cdot i; -1 \cdot i; -i$

73) 1 74) 1 75) 1 76) -1 77) i 78) $-i$ 79) $-i$

80) i 81) $-i$ 82) 1 83) -1 84) i 85) $32i$ 86) -64

87) -1 88) i 89) $142 - 65i$ 90) $-117 - 44i$

91) $1 + 2i\sqrt{2}$ 92) $-7 - 4i\sqrt{3}$ 93) $8 - 3i\sqrt{5}$

94) $3 + 2i\sqrt{5}$ 95) $-3 + i\sqrt{2}$ 96) $7 - i\sqrt{2}$

97) $Z = 10 + 6j$ 98) $Z = 14 + 9j$ 99) $Z = 16 + 4j$

100) $Z = 7 - j$

Chapter 8 Review Exercises

1) $\dfrac{13}{2}$ 2) not real 3) -9 4) 2 5) -1 6) -3

7) not real 8) not real 9) 13 10) -8 11) $|p|$ 12) $|c|$

13) h 14) $|y + 7|$

15) a) $\sqrt{23}$ b) $\sqrt{5p + 3}$ c) $\left[-\dfrac{3}{5}, \infty\right)$

16) a) -2 b) $\sqrt[3]{t - 5}$ c) $(-\infty, \infty)$

17)   18)

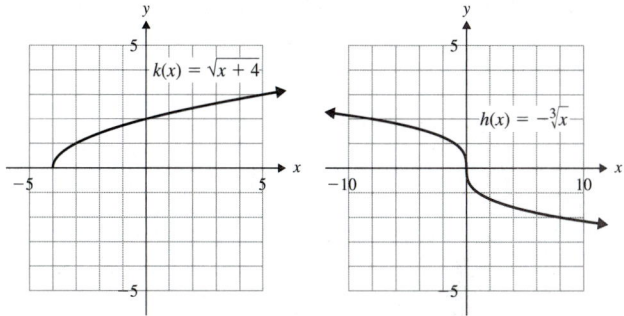

19) The denominator of the fractional exponent becomes the index on the radical. The numerator is the power to which we raise the radical expression. $8^{2/3} = (\sqrt[3]{8})^2$

20) Take the reciprocal of the base. $9^{-1/2} = \left(\dfrac{1}{9}\right)^{1/2}$

21) 6 22) 2 23) $\dfrac{3}{5}$ 24) -2 25) 8 26) $\dfrac{16}{9}$ 27) $\dfrac{1}{9}$

28) 3 29) $\dfrac{1}{27}$ 30) $\dfrac{1}{100}$ 31) $\dfrac{100}{9}$ 32) $\dfrac{64}{125}$ 33) 9

34) 13 35) 64 36) $\dfrac{1}{32}$ 37) 1 38) $\dfrac{6}{k^{1/3}}$ 39) $32a^{10/3}b^{10}$

40) $49t^6u^4$ 41) $\dfrac{2c}{3d^{7/4}}$ 42) 6 43) 3 44) 17 45) 7

46) t^4 47) k^7 48) x^5 49) w^3 50) $2\sqrt{7}$ 51) $10\sqrt{10}$

52) 3 53) $\dfrac{3\sqrt{2}}{7}$ 54) $\dfrac{4\sqrt{3}}{11}$ 55) k^6 56) $\dfrac{2\sqrt{10}}{m^2}$ 57) $x^4\sqrt{x}$

58) $y^2\sqrt{y}$ 59) $3t\sqrt{5}$ 60) $4n^{10}\sqrt{5n}$ 61) $6x^3y^6\sqrt{2xy}$

62) $\dfrac{m^5\sqrt{m}}{6n}$ 63) $\sqrt{15}$ 64) $3\sqrt{10}$ 65) $2\sqrt{6}$ 66) b^5

67) $11x^6\sqrt{x}$ 68) $5a^4b^2\sqrt{3b}$ 69) $10k^8$ 70) $3c^4$ 71) $2\sqrt[3]{2}$

72) $5\sqrt[3]{2}$ 73) $2\sqrt[4]{3}$ 74) 3 75) z^6 76) p^8 77) $a^6\sqrt[3]{a^2}$

78) $x^2y\sqrt[5]{x^4y^2}$ 79) $2z^5\sqrt[3]{2}$ 80) $2m^5n^3\sqrt[3]{10m^2n}$ 81) $\dfrac{h^3}{3}$

82) $\dfrac{c^4\sqrt[5]{c^2}}{2d^2}$ 83) $\sqrt[3]{21}$ 84) $5\sqrt[3]{2}$ 85) $2t^4\sqrt[4]{2t}$ 86) x

87) $\sqrt[6]{n^5}$ 88) $\sqrt[12]{a^5}$ 89) $11\sqrt{5}$ 90) $9\sqrt{5}$

91) $6\sqrt{5} - 4\sqrt{3}$ 92) $10\sqrt[3]{9}$ 93) $-4p\sqrt{p}$ 94) $5n\sqrt{n}$

95) $-12d^2\sqrt{2d}$ 96) $\sqrt{42} - 6$ 97) $6k\sqrt{5} + 3\sqrt{2k}$

98) $7 + 3\sqrt{3}$ 99) $23\sqrt{2rs} + 8r + 15s$ 100) $36 - 16\sqrt{5}$

101) $2 + 2\sqrt{y+1} + y$ 102) 1   103) $\dfrac{14\sqrt{3}}{3}$ 104) $\dfrac{10\sqrt{6}}{3}$

105) $\dfrac{3\sqrt{2kn}}{n}$ 106) $\dfrac{3\sqrt{5m}}{m^3}$ 107) $\dfrac{7\sqrt[3]{4}}{2}$ 108) $-5\sqrt[3]{3}$

109) $\dfrac{\sqrt[3]{x^2 y^2}}{y}$ 110) $\dfrac{\sqrt[4]{12k^2}}{2k}$ 111) $\dfrac{3 - \sqrt{3}}{3}$ 112) $\sqrt{z} - 2$

113) $1 - 3\sqrt{2}$ 114) $\dfrac{-2\sqrt{3} - 3}{5}$ 115) $\{1\}$ 116) $\{23\}$

117) $\varnothing$ 118) $\{1\}$ 119) $\{-4\}$ 120) $\{3\}$ 121) $\{2, 6\}$

122) $\{12\}$ 123) $V = \dfrac{1}{3}\pi r^2 h$

124) $8\sqrt{5}$ ft/sec or about 17.9 ft/sec   125) $7i$ 126) $2i\sqrt{2}$

127) -4 128) $-3\sqrt{2}$ 129) $12 - 3i$ 130) $-7 + 7i$

131) $\dfrac{3}{10} - \dfrac{4}{3}i$ 132) 0   133) $-30 + 35i$ 134) $24 - 32i$

135) $-36 - 21i$ 136) $10 - 5i$ 137) $-24 - 42i$

138) $-\dfrac{13}{90} - \dfrac{17}{15}i$

139) conjugate: $2 + 7i$;   140) conjugate: $-2 - 3i$;
product: 53   product: 13

141) $\dfrac{12}{29} - \dfrac{30}{29}i$ 142) $-\dfrac{48}{25} - \dfrac{36}{25}i$ 143) $-8i$ 144) $-\dfrac{6}{5} + \dfrac{2}{5}i$

145) $\dfrac{58}{37} - \dfrac{15}{37}i$ 146) $-\dfrac{2}{5} - \dfrac{7}{10}i$ 147) -1 148) $-i$

149) i 150) 1

Chapter 8 Test

1) 12   2) -3 3) not real   4) $|w|$ 5) -19

6) a) 1   b) $\sqrt{3a - 5}$ c) $\left[-\dfrac{7}{3}, \infty\right)$

7) Domain: $[2, \infty)$

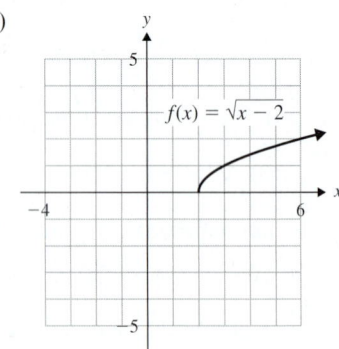

8) 2   9) 81   10) $\dfrac{1}{7}$ 11) $\dfrac{25}{4}$ 12) $m^{5/8}$ 13) $\dfrac{5}{2a^{2/3}}$

14) $\dfrac{y^2}{32x^{3/2}}$ 15) $5\sqrt{3}$ 16) $2\sqrt[3]{6}$ 17) $2\sqrt{3}$ 18) y^3 19) p^6

20) $t^4\sqrt{t}$ 21) $3m^2 n^4\sqrt{7m}$ 22) $c^7\sqrt[3]{c^2}$ 23) $\dfrac{a^4 b^2\sqrt[3]{a^2 b}}{3}$

24) 6   25) $z^3\sqrt[3]{z}$ 26) $2w^5\sqrt{15w}$ 27) $6\sqrt{7}$

28) $3\sqrt{2} - 4\sqrt{3}$ 29) $-14h^3\sqrt[4]{h}$ 30) $2\sqrt{3} - 5\sqrt{6}$

31) $3\sqrt{2} + 3 - 2\sqrt{10} - 2\sqrt{5}$ 32) 4

33) $2p + 5 + 4\sqrt{2p + 1}$ 34) $2t - 2\sqrt{3tu}$ 35) $\dfrac{2\sqrt{5}}{5}$

36) $12 - 4\sqrt{7}$ 37) $\dfrac{\sqrt{6a}}{a}$ 38) $\dfrac{5\sqrt[3]{3}}{3}$ 39) $1 - 2\sqrt{3}$

40) $\{1\}$ 41) $\{-2\}$ 42) $\{13\}$ 43) $\{1, 5\}$

44) a) 3 in.   b) $V = \pi r^2 h$ 45) $8i$ 46) $3i\sqrt{5}$ 47) $-i$

48) $-16 + 2i$ 49) $19 + 13i$ 50) $1 + 2i$

Cumulative Review for Chapters 1–8

1) $\dfrac{10}{3}x - 2y + 8$ 2) 8.723×10^6 3) $\left\{-\dfrac{3}{4}\right\}$

4)

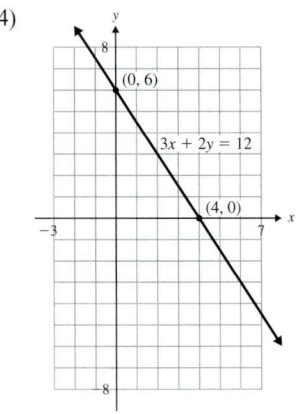

5) $y = \dfrac{5}{4}x - \dfrac{13}{4}$ 6) $(-6, 0)$

7) $15p^4 - 20p^3 - 11p^2 + 8p + 2$ 8) $4n^2 + 2n + 1$

9) $(4w - 3)(w + 2)$ 10) $2(2 + 3t)(2 - 3t)$

11) $\left\{-\dfrac{1}{2}, \dfrac{4}{3}\right\}$ 12) $\{6, 9\}$ 13) length = 12 in., width = 7 in.

14) $\dfrac{2a^2 + 2a + 3}{a(a + 4)}$ 15) $\dfrac{4n}{21m^3}$ 16) $\{-8, -4\}$

17) $(-\infty, -2] \cup \left[\dfrac{5}{3}, \infty\right)$ 18) $(4, -1, 2)$

19) a) $10\sqrt{5}$ b) $2\sqrt[3]{7}$ c) $p^5 q^3\sqrt{q}$ d) $2a^3\sqrt[4]{2a^3}$

20) a) 9   b) 16   c) $\dfrac{1}{3}$ 21) $10\sqrt{3} - 6$

22) a) $\dfrac{\sqrt{10}}{5}$ b) $3\sqrt[3]{4}$ c) $\dfrac{x\sqrt[3]{y}}{y}$ d) $\dfrac{a - 2 - \sqrt{a}}{1 - a}$

23) a) $\varnothing$ b) $\{-3\}$ 24) a) $7i$ b) $2i\sqrt{14}$ c) 1

25) a) $2 + 7i$ b) $-48 + 9i$ c) $-\dfrac{11}{25} - \dfrac{2}{25}i$

Chapter 9
Section 9.1

1) Methods may vary; $\{-4, 4\}$

2) The solution is not a real number.   3) $\{-6, 6\}$

4) $\{-8, 8\}$ 5) $\{-3\sqrt{3}, 3\sqrt{3}\}$ 6) $\{-\sqrt{30}, \sqrt{30}\}$

7) $\left\{-\dfrac{2}{3}, \dfrac{2}{3}\right\}$ 8) $\left\{-\dfrac{11}{4}, \dfrac{11}{4}\right\}$ 9) $\{-2i, 2i\}$

10) $\{-11i, 11i\}$ 11) $\{-i\sqrt{3}, i\sqrt{3}\}$ 12) $\{-i\sqrt{37}, i\sqrt{37}\}$

13) $\{-\sqrt{14}, \sqrt{14}\}$ 14) $\{-3\sqrt{2}, 3\sqrt{2}\}$ 15) $\{-5, 5\}$

16) $\{-3, 3\}$ 17) $\{-2i\sqrt{3}, 2i\sqrt{3}\}$ 18) $\{-2i\sqrt{7}, 2i\sqrt{7}\}$

19) $\{-12, -8\}$ 20) $\{-4, 14\}$ 21) $\{6, 8\}$ 22) $\{-17, -7\}$

23) $\{-4 - 3\sqrt{2}, -4 + 3\sqrt{2}\}$

24) $\{-2 - 2\sqrt{5}, -2 + 2\sqrt{5}\}$ 25) $\{-3 - 5i, -3 + 5i\}$

26) $\{15 - 2i, 15 + 2i\}$ 27) $\{2 - i\sqrt{14}, 2 + i\sqrt{14}\}$

28) $\{-4 - 2i\sqrt{3}, -4 + 2i\sqrt{3}\}$

29) $\left\{\dfrac{-1 - 2\sqrt{5}}{2}, \dfrac{-1 + 2\sqrt{5}}{2}\right\}$

30) $\left\{\dfrac{6 - \sqrt{11}}{5}, \dfrac{6 + \sqrt{11}}{5}\right\}$ 31) $\left\{\dfrac{10 - \sqrt{14}}{3}, \dfrac{10 + \sqrt{14}}{3}\right\}$

32) $\left\{\dfrac{-11 - 3\sqrt{2}}{6}, \dfrac{-11 + 3\sqrt{2}}{6}\right\}$

33) $\left\{\dfrac{5}{4} - \dfrac{\sqrt{30}}{4}i, \dfrac{5}{4} + \dfrac{\sqrt{30}}{4}i\right\}$ 34) $\left\{\dfrac{1}{3} - \dfrac{4}{3}i, \dfrac{1}{3} + \dfrac{4}{3}i\right\}$

35) $\left\{-\dfrac{11}{6} - \dfrac{7}{6}i, -\dfrac{11}{6} + \dfrac{7}{6}i\right\}$ 36) $\left\{\dfrac{4}{9} - \dfrac{\sqrt{29}}{9}i, \dfrac{4}{9} + \dfrac{\sqrt{29}}{9}i\right\}$

37) $\left\{8, \dfrac{40}{3}\right\}$ 38) $\{-21, -9\}$ 39) $\left\{-\dfrac{2}{5}, \dfrac{6}{5}\right\}$ 40) $\{3, 6\}$

41) 5   42) $\sqrt{97}$ 43) $\sqrt{13}$ 44) 10   45) 6   46) $2\sqrt{26}$

47) $\sqrt{61}$ 48) 5   49) $2\sqrt{5}$ 50) $4\sqrt{5}$

51) A trinomial whose factored form is the square of a binomial; examples will vary.

52) No, because the coefficient of y^2 is not 1.

53) $\dfrac{1}{2}(8) = 4; 4^2 = 16; w^2 + 8w + 16;$
$w^2 + 8w + 16; (w + 4)^2$

54) Find half of the coefficient of n.; Square the result.; Add the constant to the expression.; $n^2 - n + \dfrac{1}{4}; \left(n - \dfrac{1}{2}\right)^2$

55) $a^2 + 12a + 36; (a + 6)^2$ 56) $g^2 + 4g + 4; (g + 2)^2$

57) $c^2 - 18c + 81; (c - 9)^2$ 58) $k^2 - 16k + 64; (k - 8)^2$

59) $r^2 + 3r + \dfrac{9}{4}; \left(r + \dfrac{3}{2}\right)^2$ 60) $z^2 - 7z + \dfrac{49}{4}; \left(z - \dfrac{7}{2}\right)^2$

61) $b^2 - 9b + \dfrac{81}{4}; \left(b - \dfrac{9}{2}\right)^2$ 62) $t^2 + 5t + \dfrac{25}{4}; \left(t + \dfrac{5}{2}\right)^2$

63) $x^2 + \dfrac{1}{3}x + \dfrac{1}{36}; \left(x + \dfrac{1}{6}\right)^2$ 64) $y^2 - \dfrac{3}{5}y + \dfrac{9}{100}; \left(y - \dfrac{3}{10}\right)^2$

65) Divide both sides of the equation by 2.

66) No, because the equation is not quadratic.

67) $\{-4, -2\}$ 68) $\{-13, 1\}$ 69) $\{3, 5\}$ 70) $\{-3, 9\}$

71) $\{-5 - \sqrt{15}, -5 + \sqrt{15}\}$ 72) $\{1 - \sqrt{10}, 1 + \sqrt{10}\}$

73) $\{1 - 2i\sqrt{2}, 1 + 2i\sqrt{2}\}$ 74) $\{-5 - i, -5 + i\}$

75) $\{-2 - 2i, -2 + 2i\}$ 76) $\{4 - i\sqrt{3}, 4 + i\sqrt{3}\}$

77) $\{-8, 5\}$ 78) $\{-4, -1\}$ 79) $\{3, 4\}$ 80) $\{-9, 8\}$

81) $\left\{\dfrac{1}{2} - \dfrac{\sqrt{13}}{2}, \dfrac{1}{2} + \dfrac{\sqrt{13}}{2}\right\}$ 82) $\left\{\dfrac{3}{2} - \dfrac{\sqrt{37}}{2}, \dfrac{3}{2} + \dfrac{\sqrt{37}}{2}\right\}$

83) $\left\{-\dfrac{5}{2} - \dfrac{\sqrt{3}}{2}i, -\dfrac{5}{2} + \dfrac{\sqrt{3}}{2}i\right\}$ 84) $\left\{\dfrac{7}{2} - \dfrac{\sqrt{7}}{2}i, \dfrac{7}{2} + \dfrac{\sqrt{7}}{2}i\right\}$

85) $\{1 - i\sqrt{3}, 1 + i\sqrt{3}\}$ 86) $\{-2 - 2i\sqrt{2}, -2 + 2i\sqrt{2}\}$

87) $\{-3 - \sqrt{11}, -3 + \sqrt{11}\}$ 88) $\{-1 - \sqrt{6}, -1 + \sqrt{6}\}$

89) $\{2, 3\}$ 90) $\{1, 8\}$ 91) $\left\{\dfrac{5}{4} - \dfrac{\sqrt{39}}{4}i, \dfrac{5}{4} + \dfrac{\sqrt{39}}{4}i\right\}$

92) $\left\{-\dfrac{3}{4} - \dfrac{\sqrt{23}}{4}i, -\dfrac{3}{4} + \dfrac{\sqrt{23}}{4}i\right\}$ 93) $\left\{\dfrac{3}{4}, 1\right\}$

94) $\left\{-\dfrac{2}{3}, 1\right\}$ 95) $\{-1 - \sqrt{21}, -1 + \sqrt{21}\}$

96) $\{-3 - 4\sqrt{2}, -3 + 4\sqrt{2}\}$ 97) $\left\{\dfrac{5}{4} - \dfrac{\sqrt{7}}{4}i, \dfrac{5}{4} + \dfrac{\sqrt{7}}{4}i\right\}$

98) $\left\{-\dfrac{5}{3} - \dfrac{\sqrt{10}}{3}, -\dfrac{5}{3} + \dfrac{\sqrt{10}}{3}\right\}$ 99) 6   100) 12

101) $\sqrt{29}$ 102) $\sqrt{3}$ 103) 6 in.   104) $\sqrt{57}$ cm

105) 12 ft   106) 50 ft   107) $-10, 4$

108) $5 - 2\sqrt{3}, 5 + 2\sqrt{3}$ 109) width = 9 ft, length = 17 ft

110) width = 15 cm, length = 25 cm

Section 9.2

1) The fraction bar should also be under $-b$:
$$x = \dfrac{-b \pm \sqrt{b^2 - 4ac}}{2a}$$

2) The equation must be written as $5n^2 - 3n - 1 = 0$ before identifying the values of a, b, and c.
$$a = 5, b = -3, c = -1$$
$$n = \dfrac{-(-3) \pm \sqrt{(-3)^2 - 4(5)(-1)}}{2(5)}$$

3) You cannot divide only the -2 by 2.
$$\dfrac{-2 \pm 6\sqrt{11}}{2} = \dfrac{2(-1 \pm 3\sqrt{11})}{2} = -1 \pm 3\sqrt{11}$$

4) The discriminant does *not* include the square root. Discriminant $= b^2 - 4ac = 4$.

5) $\{-3, -1\}$ 6) $\{1, 7\}$ 7) $\left\{-2, \dfrac{5}{3}\right\}$ 8) $\left\{-\dfrac{3}{2}, -\dfrac{1}{3}\right\}$

9) $\left\{\dfrac{5 - \sqrt{17}}{2}, \dfrac{5 + \sqrt{17}}{2}\right\}$ 10) $\left\{\dfrac{-3 - \sqrt{29}}{2}, \dfrac{-3 + \sqrt{29}}{2}\right\}$

11) $\{4 - 3i, 4 + 3i\}$ 12) $\{2 - i, 2 + i\}$

13) $\left\{\dfrac{1}{5} - \dfrac{\sqrt{14}}{5}i, \dfrac{1}{5} + \dfrac{\sqrt{14}}{5}i\right\}$ 14) $\left\{-\dfrac{5}{4} - \dfrac{\sqrt{7}}{4}i, -\dfrac{5}{4} + \dfrac{\sqrt{7}}{4}i\right\}$

15) $\{-7, 0\}$ 16) $\{0, 10\}$ 17) $\left\{\dfrac{-1 - \sqrt{13}}{3}, \dfrac{-1 + \sqrt{13}}{3}\right\}$

18) $\left\{\dfrac{3 - \sqrt{3}}{2}, \dfrac{3 + \sqrt{3}}{2}\right\}$ 19) $\left\{\dfrac{7}{2}, 4\right\}$ 20) $\left\{\dfrac{4}{3}, 4\right\}$

21) $\{-4 - \sqrt{31}, -4 + \sqrt{31}\}$

22) $\left\{-\dfrac{3}{5} - \dfrac{\sqrt{11}}{5}i, -\dfrac{3}{5} + \dfrac{\sqrt{11}}{5}i\right\}$ 23) $\left\{-\dfrac{2}{3} - \dfrac{1}{3}i, -\dfrac{2}{3} + \dfrac{1}{3}i\right\}$

24) $\left\{\dfrac{1 - \sqrt{55}}{9}, \dfrac{1 + \sqrt{55}}{9}\right\}$ 25) $\{-10, 4\}$ 26) $\{3, 5\}$

27) $\left\{-\dfrac{3}{2}i, \dfrac{3}{2}i\right\}$ 28) $\left\{-\dfrac{1}{5}, \dfrac{1}{5}\right\}$ 29) $\{-3 - 5i, -3 + 5i\}$

30) $\{2 - 3i\sqrt{2}, 2 + 3i\sqrt{2}\}$ 31) $\{-1 - \sqrt{10}, -1 + \sqrt{10}\}$

32) $\{3 - \sqrt{5}, 3 + \sqrt{5}\}$ 33) $\left\{\dfrac{3}{2}\right\}$ 34) $\left\{-\dfrac{1}{5}\right\}$

35) $\left\{-\dfrac{3}{8} - \dfrac{\sqrt{7}}{8}i, -\dfrac{3}{8} + \dfrac{\sqrt{7}}{8}i\right\}$ 36) $\left\{\dfrac{1}{3} - \dfrac{\sqrt{23}}{3}i, \dfrac{1}{3} + \dfrac{\sqrt{23}}{3}i\right\}$

37) $\left\{\dfrac{5 - \sqrt{19}}{2}, \dfrac{5 + \sqrt{19}}{2}\right\}$ 38) $\left\{\dfrac{3 - \sqrt{73}}{4}, \dfrac{3 + \sqrt{73}}{4}\right\}$

39) $-3 - \sqrt{11}, -3 + \sqrt{11}$ 40) $\dfrac{2 - \sqrt{7}}{3}, \dfrac{2 + \sqrt{7}}{3}$

41) $\dfrac{1 - \sqrt{41}}{4}, \dfrac{1 + \sqrt{41}}{4}$ 42) $-6, -2$ 43) $-4, \dfrac{1}{5}$

44) $\dfrac{-9 - \sqrt{17}}{4}, \dfrac{-9 + \sqrt{17}}{4}$ 45) There is one rational solution.

46) There are two nonreal, complex solutions.

47) -39; two nonreal, complex solutions

48) 40; two irrational solutions   49) 0; one rational solution

50) -11; two nonreal, complex solutions

51) 16; two rational solutions   52) 0; one rational solution

53) 56; two irrational solutions   54) 25; two rational solutions

55) -8 or 8   56) -14 or 14   57) 9   58) 4   59) 4   60) 9

61) 2 in., 5 in.   62) 3 in., 5 in.

63) a) 2 sec   b) $\dfrac{3 + \sqrt{33}}{4}$ sec or about 2.2 sec

64) a) 0.25 sec on the way up, 2.5 sec on the way down
b) $\dfrac{11 + \sqrt{145}}{8}$ sec or about 2.9 sec

Chapter 9 Putting It All Together

1) $\{-5\sqrt{2}, 5\sqrt{2}\}$ 2) $\{3 - \sqrt{17}, 3 + \sqrt{17}\}$ 3) $\{-5, 4\}$

4) $\left\{\dfrac{3}{4} - \dfrac{\sqrt{39}}{4}i, \dfrac{3}{4} + \dfrac{\sqrt{39}}{4}i\right\}$ 5) $\left\{\dfrac{-7 - \sqrt{13}}{2}, \dfrac{-7 + \sqrt{13}}{2}\right\}$

6) $\left\{-1, \dfrac{4}{3}\right\}$ 7) $\left\{-3, \dfrac{1}{4}\right\}$ 8) $\{-3 - i\sqrt{6}, -3 + i\sqrt{6}\}$

9) $\{-7 - i\sqrt{11}, -7 + i\sqrt{11}\}$ 10) $\left\{\dfrac{-1 - \sqrt{7}}{2}, \dfrac{-1 + \sqrt{7}}{2}\right\}$

11) $\left\{\dfrac{1}{3} - \dfrac{\sqrt{6}}{3}i, \dfrac{1}{3} + \dfrac{\sqrt{6}}{3}i\right\}$ 12) $\{-4 - 3i, -4 + 3i\}$

13) $\{-2, 6\}$ 14) $\{-5, 5\}$ 15) $\{2 - \sqrt{7}, 2 + \sqrt{7}\}$

16) $\{-9, -6, 0\}$ 17) $\left\{\dfrac{3}{2}, 2\right\}$ 18) $\{0, 1\}$ 19) $\{3, 7\}$

20) $\left\{\dfrac{-1 - \sqrt{21}}{4}, \dfrac{-1 + \sqrt{21}}{4}\right\}$ 21) $\left\{\dfrac{1}{3} - \dfrac{\sqrt{2}}{3}i, \dfrac{1}{3} + \dfrac{\sqrt{2}}{3}i\right\}$

22) $\{3 - 5i, 3 + 5i\}$ 23) $\{5 - 4i, 5 + 4i\}$ 24) $\left\{-4, -\dfrac{3}{2}\right\}$

25) $\{0, 3\}$ 26) $\left\{-\dfrac{3}{2} - \dfrac{\sqrt{15}}{2}i, -\dfrac{3}{2} + \dfrac{\sqrt{15}}{2}i\right\}$

27) $\left\{-\dfrac{3}{2}, 0, \dfrac{3}{2}\right\}$ 28) $\{-9, 3\}$

29) $\left\{-\dfrac{5}{4} - \dfrac{\sqrt{39}}{4}i, -\dfrac{5}{4} + \dfrac{\sqrt{39}}{4}i\right\}$

30) $\left\{-\dfrac{5}{12} - \dfrac{1}{4}i, -\dfrac{5}{12} + \dfrac{1}{4}i\right\}$

Section 9.3

1) $\{-4, 12\}$ 2) $\{-8, -3\}$ 3) $\left\{-2, \dfrac{4}{5}\right\}$ 4) $\{-3, -2\}$

5) $\{4 - \sqrt{6}, 4 + \sqrt{6}\}$ 6) $\left\{\dfrac{-3 - \sqrt{89}}{4}, \dfrac{-3 + \sqrt{89}}{4}\right\}$

7) $\left\{\dfrac{1 - \sqrt{7}}{3}, \dfrac{1 + \sqrt{7}}{3}\right\}$ 8) $\left\{\dfrac{3 - 3\sqrt{5}}{2}, \dfrac{3 + 3\sqrt{5}}{2}\right\}$

9) $\left\{-\dfrac{9}{5}, 1\right\}$ 10) $\left\{-\dfrac{4}{3}, 4\right\}$ 11) $\left\{\dfrac{11 - \sqrt{21}}{10}, \dfrac{11 + \sqrt{21}}{10}\right\}$

12) $\left\{\dfrac{-3 - \sqrt{105}}{12}, \dfrac{-3 + \sqrt{105}}{12}\right\}$ 13) $\{5\}$ 14) $\{1, 6\}$

15) $\left\{\dfrac{4}{5}, 2\right\}$ 16) $\left\{\dfrac{1}{2}\right\}$ 17) $\{9\}$ 18) $\{1\}$ 19) $\{1, 16\}$

20) $\{25\}$ 21) $\{5\}$ 22) $\{0, 3\}$ 23) $\{0, 2\}$ 24) $\{-2\}$

25) yes   26) yes   27) yes   28) no   29) no   30) yes

31) yes   32) yes   33) no   34) yes   35) $\{-3, -1, 1, 3\}$

36) $\{-5, -2, 2, 5\}$ 37) $\{-\sqrt{7}, -2, 2, \sqrt{7}\}$

38) $\{-2\sqrt{2}, -1, 1, 2\sqrt{2}\}$ 39) $\{-i\sqrt{7}, -i\sqrt{5}, i\sqrt{5}, i\sqrt{7}\}$

40) $\{-i\sqrt{6}, -i\sqrt{3}, i\sqrt{3}, i\sqrt{6}\}$ 41) $\{-8, -1\}$ 42) $\{-64, 27\}$

43) $\{-64, 1000\}$ 44) $\{-8, 27\}$ 45) $\left\{-\dfrac{27}{8}, 125\right\}$

46) $\left\{-8, -\dfrac{64}{27}\right\}$ 47) $\{4, 36\}$ 48) $\{1, 25\}$ 49) $\{49\}$

50) $\varnothing$ 51) $\{16\}$ 52) $\{36\}$

53) $\left\{-\dfrac{2\sqrt{3}}{3}i, -\dfrac{\sqrt{3}}{3}i, \dfrac{\sqrt{3}}{3}i, \dfrac{2\sqrt{3}}{3}i\right\}$

54) $\left\{-2i, -\dfrac{\sqrt{3}}{2}i, \dfrac{\sqrt{3}}{2}i, 2i\right\}$ 55) $\{-\sqrt{5}, \sqrt{5}, -i\sqrt{3}, i\sqrt{3}\}$

56) $\{-2, 2, -i\sqrt{6}, i\sqrt{6}\}$

57) $\{-\sqrt{3+\sqrt{7}}, \sqrt{3+\sqrt{7}}, -\sqrt{3-\sqrt{7}}, \sqrt{3-\sqrt{7}}\}$

58) $\{-\sqrt{4+\sqrt{13}}, \sqrt{4+\sqrt{13}}, -\sqrt{4-\sqrt{13}}, \sqrt{4-\sqrt{13}}\}$

59) $\left\{-\dfrac{\sqrt{7+\sqrt{41}}}{2}, \dfrac{\sqrt{7+\sqrt{41}}}{2},\right.$

$\left.-\dfrac{\sqrt{7-\sqrt{41}}}{2}, \dfrac{\sqrt{7-\sqrt{41}}}{2}\right\}$

60) $\left\{-\dfrac{\sqrt{9+\sqrt{17}}}{4}, \dfrac{\sqrt{9+\sqrt{17}}}{4},\right.$

$\left.-\dfrac{\sqrt{9-\sqrt{17}}}{4}, \dfrac{\sqrt{9-\sqrt{17}}}{4}\right\}$

61) $\left\{-\dfrac{1}{2}, \dfrac{1}{6}\right\}$ 62) $\left\{-\dfrac{1}{3}, \dfrac{1}{2}\right\}$ 63) $\left\{\dfrac{1}{4}, 3\right\}$ 64) $\left\{-5, \dfrac{1}{3}\right\}$

65) $\{-6, -1\}$ 66) $\{-3, 4\}$ 67) $\left\{-\dfrac{1}{2}, 0\right\}$ 68) $\left\{\dfrac{9}{8}, 3\right\}$

69) $\left\{-\dfrac{2}{5}, \dfrac{2}{5}\right\}$ 70) $\{0, 4\}$ 71) $\left\{-11, -\dfrac{20}{3}\right\}$

72) $\left\{-\dfrac{47}{5}, -16\right\}$ 73) $\left\{\dfrac{3}{2}\right\}$ 74) $\left\{-\dfrac{1}{4}, \dfrac{3}{4}\right\}$

75) $\{2 - \sqrt{2}, 2 + \sqrt{2}\}$ 76) $\{-5 - \sqrt{3}, -5 + \sqrt{3}\}$

77) Walter: 3 hr; Kevin: 6 hr   78) 8 hr   79) 15 mph

80) 290 mph   81) large drain: 3 hr; small drain: 6 hr

82) professor: 2 hr; teaching assistant: 3 hr

83) to Boulder: 60 mph; going home: 50 mph

84) first 6 mi: 6 mph; last 3 mi: 4 mph

Section 9.4

1) $r = \dfrac{\pm\sqrt{A\pi}}{\pi}$ 2) $r = \dfrac{\pm\sqrt{3\pi Vh}}{\pi h}$ 3) $v = \pm\sqrt{ar}$

4) $w = \dfrac{\pm\sqrt{2KI}}{I}$ 5) $d = \dfrac{\pm\sqrt{IE}}{E}$ 6) $I = \dfrac{\pm\sqrt{2UL}}{L}$

7) $r = \dfrac{\pm\sqrt{kq_1q_2F}}{F}$ 8) $r = \dfrac{\pm\sqrt{kqE}}{E}$ 9) $A = \dfrac{1}{4}\pi d^2$

10) $V = \dfrac{1}{12}\pi d^2 h$ 11) $l = \dfrac{gT_P^2}{4\pi^2}$ 12) $T = \dfrac{V^2 M}{3R}$

13) $g = \dfrac{4\pi^2 l}{T_P^2}$ 14) $M = \dfrac{3RT}{V^2}$

15) a) Both are written in the standard form for a quadratic equation, $ax^2 + bx + c = 0$.
   b) Use the quadratic formula.

16) Use the quadratic formula because both are quadratic in t and written in standard form $at^2 + bt + c = 0$.

17) $x = \dfrac{5 \pm \sqrt{25 - 4rs}}{2r}$ 18) $x = \dfrac{-d \pm \sqrt{d^2 + 12c}}{2c}$

19) $z = \dfrac{-r \pm \sqrt{r^2 + 4pq}}{2p}$ 20) $r = \dfrac{k \pm \sqrt{k^2 - 4hj}}{2h}$

21) $a = \dfrac{h \pm \sqrt{h^2 + 4dk}}{2d}$ 22) $t = \dfrac{-m \pm \sqrt{m^2 - 4kn}}{2k}$

23) $t = \dfrac{-v \pm \sqrt{v^2 + 2gs}}{g}$ 24) $r = \dfrac{-\pi h \pm \sqrt{\pi(\pi h^2 + s)}}{\pi}$

25) length = 12 in., width = 9 in.

26) length = 24 in., width = 16 in.   27) 2 ft   28) 1 in.

29) base = 8 ft, height = 15 ft   30) base = 6 in., height = 5 in.

31) 10 in.   32) length = 40 in., width = 30 in.

33) a) 0.75 sec on the way up, 3 sec on the way down
   b) $\dfrac{15 + \sqrt{241}}{8}$ sec or about 3.8 sec

34) after 1 sec and 3 sec   35) a) 9.5 million   b) 1999

36) 300 lux   37) $2.40   38) $14.00   39) 1.75 ft   40) 2.3 ft

Section 9.5

1) The graph of $g(x)$ is the same shape as the graph of $f(x)$, but $g(x)$ is shifted up 6 units.

2) The graph of $h(x)$ is the same shape as the graph of $f(x)$, but $h(x)$ is shifted down 5 units.

3) The graph of $h(x)$ is the same shape as the graph of $f(x)$, but $h(x)$ is shifted left 5 units.

4) The graph of $g(x)$ is the same shape as the graph of $f(x)$, but $g(x)$ is shifted right 4 units.

5)

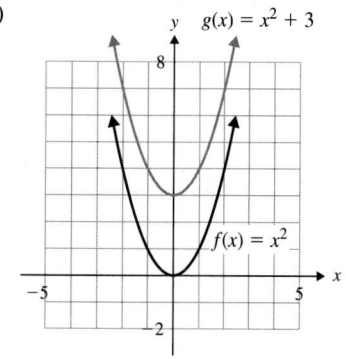

6)

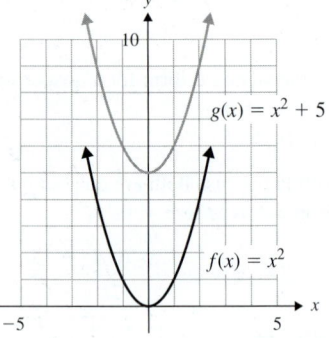

$g(x) = x^2 + 5$

$f(x) = x^2$

7)

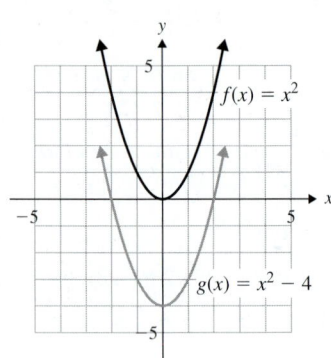

$f(x) = x^2$

$g(x) = x^2 - 4$

8)

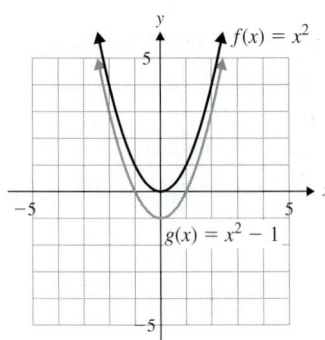

$f(x) = x^2$

$g(x) = x^2 - 1$

9)

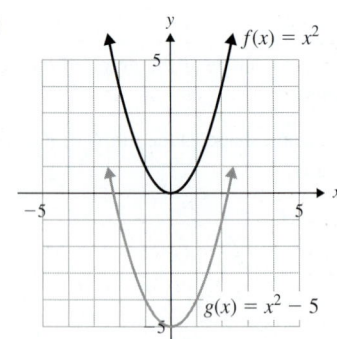

$f(x) = x^2$

$g(x) = x^2 - 5$

10)

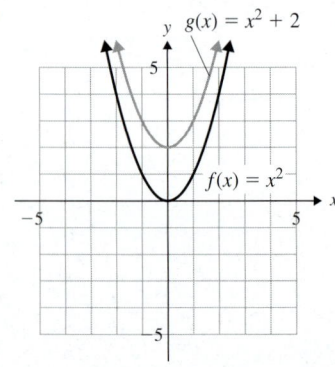

$g(x) = x^2 + 2$

$f(x) = x^2$

11)

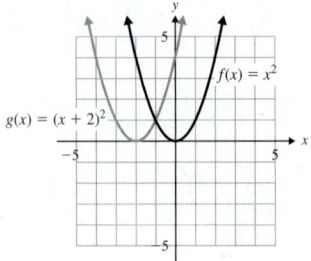

$f(x) = x^2$

$g(x) = (x + 2)^2$

12)

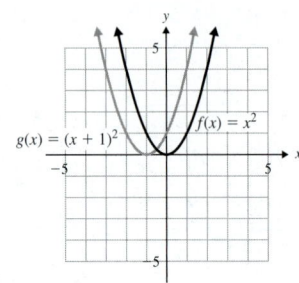

$g(x) = (x + 1)^2$

$f(x) = x^2$

13)

$g(x) = (x + 5)^2$

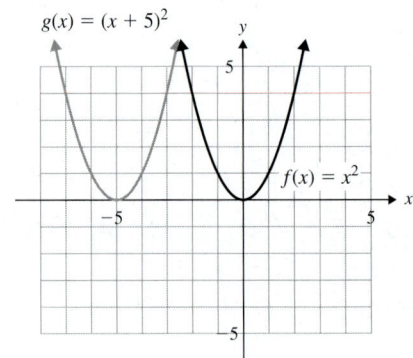

$f(x) = x^2$

14)

$g(x) = (x + 4)^2$

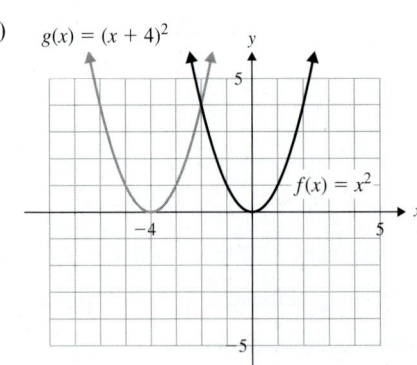

$f(x) = x^2$

15)

$g(x) = (x - 3)^2$

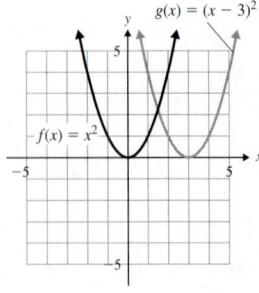

$f(x) = x^2$

16)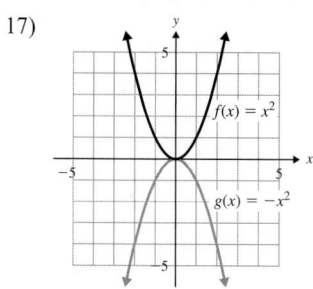

17)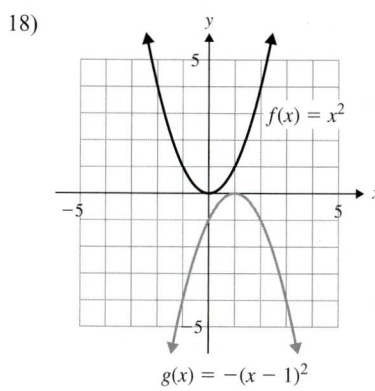

18)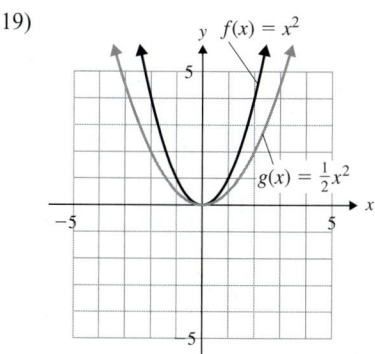

19)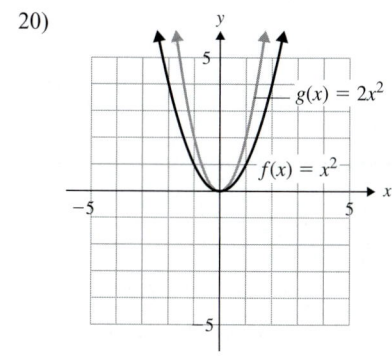

20)

21)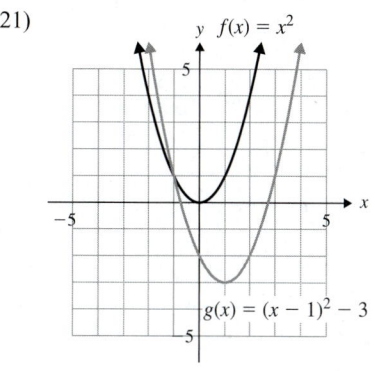

22)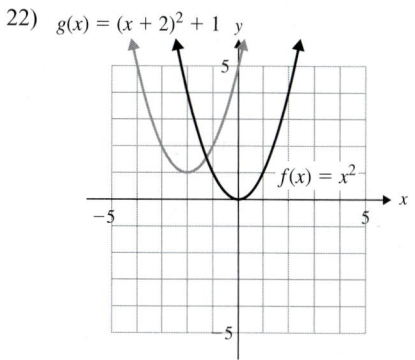

23) (h, k) 24) $x = h$ 25) a is positive.   26) a is negative.

27) $a > 1$ or $a < -1$ 28) $0 < a < 1$ or $-1 < a < 0$

29) $V(-1, -4)$; $x = -1$;
x-ints: $(-3, 0)$, $(1, 0)$;
y-int: $(0, -3)$; domain:
$(-\infty, \infty)$; range: $[-4, \infty)$

30) $V(3, -1)$; $x = 3$;
x-ints: $(2, 0)$, $(4, 0)$;
y-int: $(0, 8)$; domain:
$(-\infty, \infty)$; range: $[-1, \infty)$

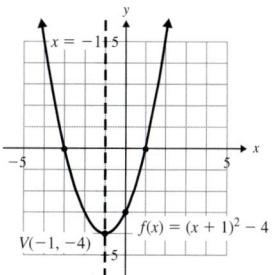

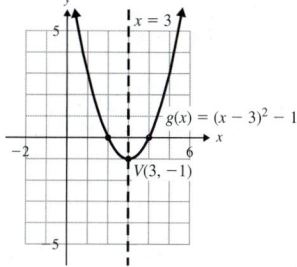

31) $V(2, 3)$; $x = 2$; x-ints:
none; y-int: $(0, 7)$;
domain: $(-\infty, \infty)$;
range: $[3, \infty)$

32) $V(-2, 7)$; $x = -2$;
x-ints: none; y-int:
$(0, 11)$; domain:
$(-\infty, \infty)$; range: $[7, \infty)$

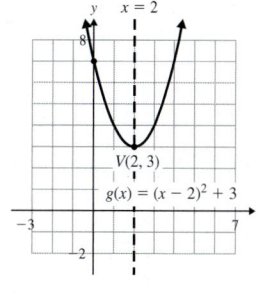

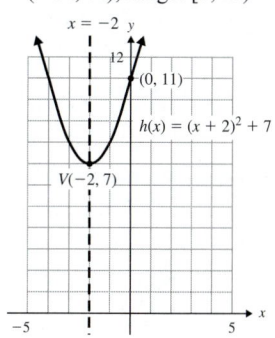

33) $V(4, -2)$; $x = 4$; x-ints:
$(4 - \sqrt{2}, 0)$,
$(4 + \sqrt{2}, 0)$; y-int:
$(0, 14)$; domain:
$(-\infty, \infty)$; range: $[-2, \infty)$

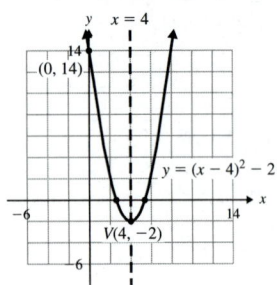

34) $V(-1, -5)$; $x = -1$;
x-ints: $(-1 - \sqrt{5}, 0)$,
$(-1 + \sqrt{5}, 0)$; y-int:
$(0, -4)$; domain: $(-\infty, \infty)$;
range: $[-5, \infty)$

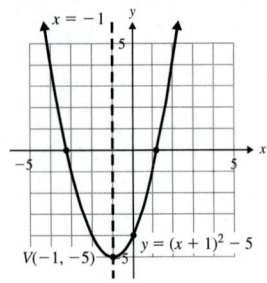

35) $V(-3, 6)$; $x = -3$; x-ints: $(-3 - \sqrt{6}, 0)$, $(-3 + \sqrt{6}, 0)$;
y-int: $(0, -3)$; domain: $(-\infty, \infty)$; range: $(-\infty, 6]$

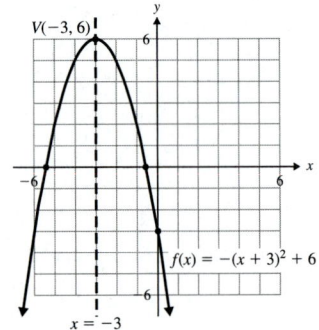

36) $V(3, 2)$; $x = 3$; x-ints: $(3 - \sqrt{2}, 0)$, $(3 + \sqrt{2}, 0)$; y-int:
$(0, -7)$; domain: $(-\infty, \infty)$; range: $(-\infty, 2]$

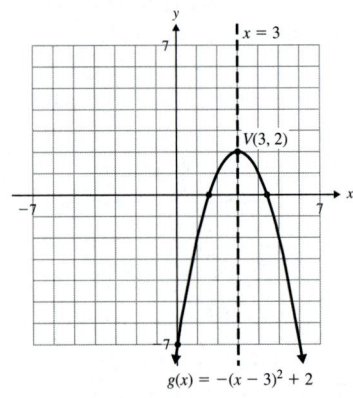

37) $V(-1, -5)$; $x = -1$:
x-ints: none; y-int: $(0, -6)$;
domain: $(-\infty, \infty)$;
range: $(-\infty, -5]$

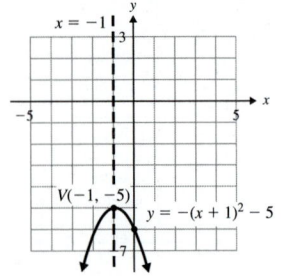

38) $V(2, -4)$; $x = 2$; x-ints:
none; y-int: $(0, -8)$;
domain: $(-\infty, \infty)$;
range: $(-\infty, -4]$

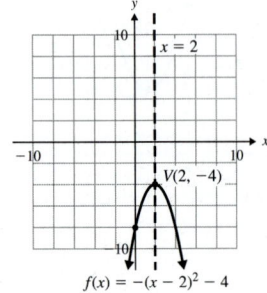

39) $V(1, -8)$; $x = 1$;
x-ints: $(-1, 0)$, $(3, 0)$;
y-int: $(0, -6)$; domain:
$(-\infty, \infty)$; range: $[-8, \infty)$

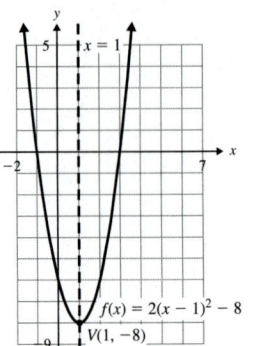

40) $V(-1, -2)$; $x = -1$;
x-ints: $(-2, 0)$, $(0, 0)$;
y-int: $(0, 0)$; domain:
$(-\infty, \infty)$; range: $[-2, \infty)$

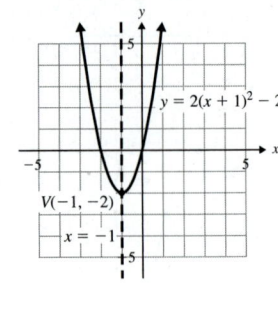

41) $V(-4, 0)$; $x = -4$; x-int: $(-4, 0)$; y-int: $(0, 8)$; domain:
$(-\infty, \infty)$; range $[0, \infty)$

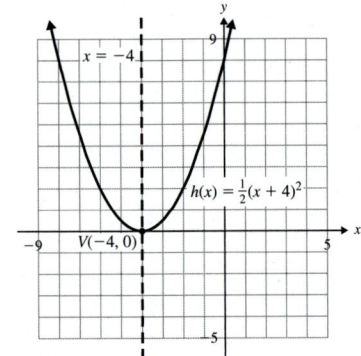

42) $V(0, -1)$; $x = 0$; x-ints:
$(-2, 0)$, $(2, 0)$; y-int:
$(0, -1)$; domain:
$(-\infty, \infty)$; range: $[-1, \infty)$

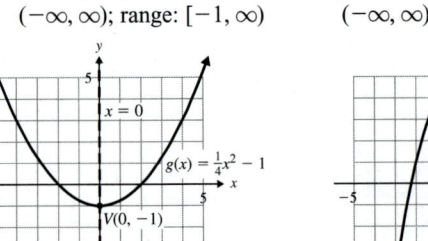

43) $V(0, 5)$; $x = 0$; x-ints:
$(-\sqrt{5}, 0)$, $(\sqrt{5}, 0)$;
y-int: $(0, 5)$; domain:
$(-\infty, \infty)$; range: $(-\infty, 5]$

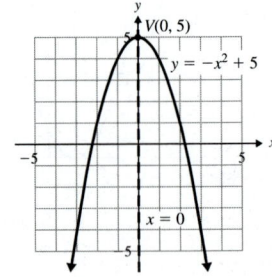

44) $V(3, 0)$; $x = 3$; x-int: $(3, 0)$; y-int: $(0, -9)$; domain:
$(-\infty, \infty)$; range: $(-\infty, 0]$

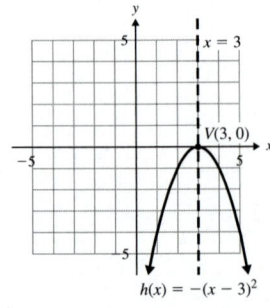

45) $V(-4, 3)$; $x = -4$; x-ints: $(-7, 0)$, $(-1, 0)$; y-int: $\left(0, -\dfrac{7}{3}\right)$;

domain: $(-\infty, \infty)$; range: $(-\infty, 3]$

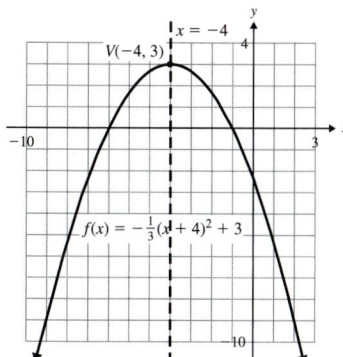

46) $V(4, 2)$; $x = 4$; x-ints: $(2, 0)$, $(6, 0)$; y-int: $(0, -6)$; domain: $(-\infty, \infty)$; range: $(-\infty, 2]$

47) $V(-2, 5)$; $x = -2$; x-int: none; y-int: $(0, 17)$; domain: $(-\infty, \infty)$; range: $[5, \infty)$

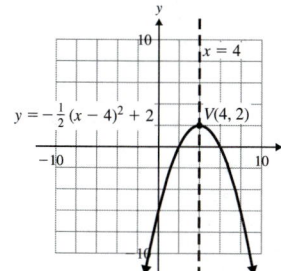
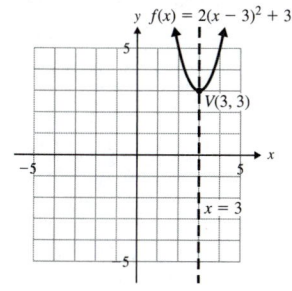

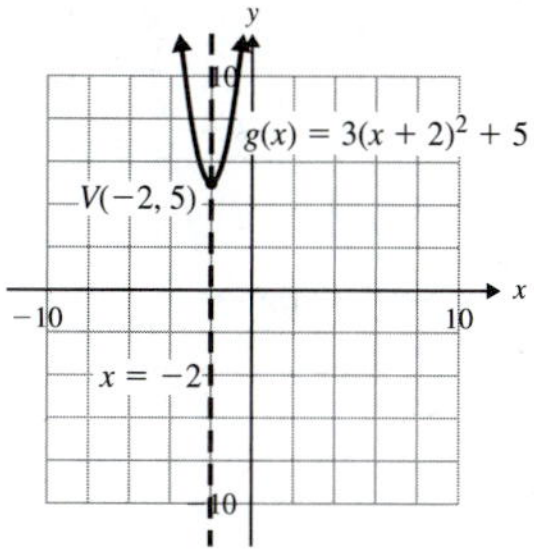

48) $V(3, 3)$; $x = 3$; x-int: none; y-int: $(0, 21)$; domain: $(-\infty, \infty)$; range: $[3, \infty)$

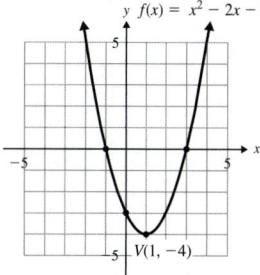

49) a) $h(x)$ b) $f(x)$ c) $g(x)$ d) $k(x)$

50) a) $g(x)$ b) $h(x)$ c) $k(x)$ d) $f(x)$

51) $g(x) = (x - 8)^2$ 52) $g(x) = x^2 - 9$

53) $g(x) = x^2 + 3.5$ 54) $g(x) = (x + 1.2)^2$

55) $g(x) = (x + 4)^2 - 7$ 56) $g(x) = (x - 5)^2 + 2$

57) $f(x) = (x^2 + 8x) + 11$; $\left[\dfrac{1}{2}(8)\right]^2 = (4)^2 = 16$; Add and subtract the number above to the same side of the equation.; $f(x) = (x + 4)^2 - 5$

58) Group the variable terms together using parentheses.; $\left[\dfrac{1}{2}(-4)\right]^2 = (-2)^2 = 4$; $f(x) = (x^2 - 4x + 4) - 7 - 4$; Factor and simplify.

59) $f(x) = (x - 1)^2 - 4$; x-ints: $(-1, 0)$, $(3, 0)$; y-int: $(0, -3)$; domain: $(-\infty, \infty)$; range: $[-4, \infty)$

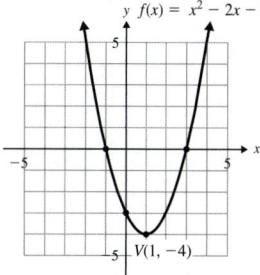

60) $g(x) = (x + 3)^2 - 1$; x-ints: $(-4, 0)$, $(-2, 0)$; y-int: $(0, 8)$; domain: $(-\infty, \infty)$; range: $[-1, \infty)$

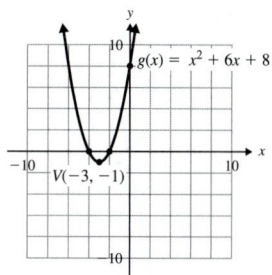

61) $y = (x + 3)^2 - 2$; x-ints: $(-3 - \sqrt{2}, 0)$, $(-3 + \sqrt{2}, 0)$; y-int: $(0, 7)$; domain: $(-\infty, \infty)$; range: $[-2, \infty)$

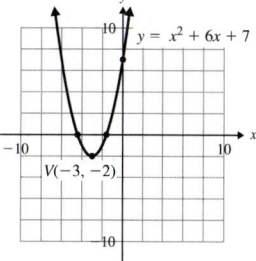

62) $h(x) = (x - 2)^2 - 3$; x-ints: $(2 - \sqrt{3}, 0)$, $(2 + \sqrt{3}, 0)$; y-int: $(0, 1)$; domain: $(-\infty, \infty)$; range: $[-3, \infty)$

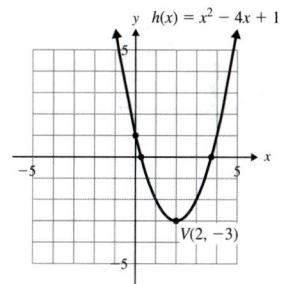

63) $g(x) = (x + 2)^2 - 4$; x-ints: $(-4, 0)$, $(0, 0)$; y-int: $(0, 0)$; domain: $(-\infty, \infty)$; range: $[-4, \infty)$

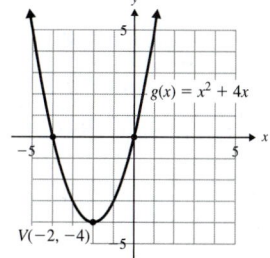

64) $y = (x - 4)^2 + 2$; x-ints: none; y-int: $(0, 18)$; domain: $(-\infty, \infty)$; range: $[2, \infty)$

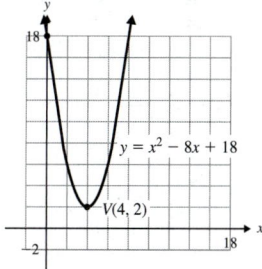

65) $h(x) = -(x + 2)^2 + 9$; x-ints: $(-5, 0)$, $(1, 0)$; y-int: $(0, 5)$; domain: $(-\infty, \infty)$; range: $(-\infty, 9]$

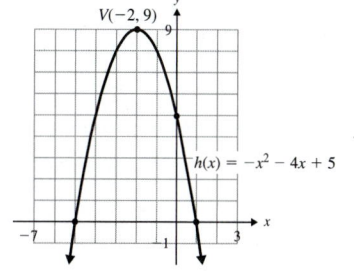

66) $f(x) = -(x + 1)^2 + 4$; x-ints: $(-3, 0)$, $(1, 0)$; y-int: $(0, 3)$; domain: $(-\infty, \infty)$; range: $(-\infty, 4]$

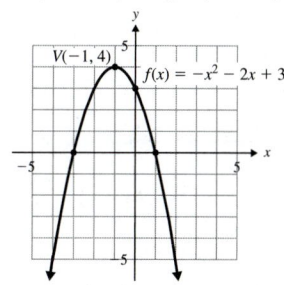

67) $y = -(x - 3)^2 - 1$;
x-ints: none; y-int: $(0, -10)$;
domain: $(-\infty, \infty)$;
range: $(-\infty, -1]$

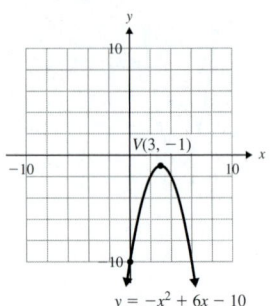

68) $g(x) = -(x + 2)^2 - 2$;
x-ints: none; y-int: $(0, -6)$;
domain: $(-\infty, \infty)$;
range: $(-\infty, -2]$

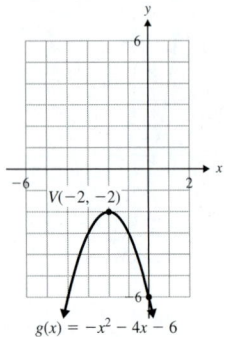

73) $y = \left(x - \dfrac{3}{2}\right)^2 - \dfrac{1}{4}$; x-ints: $(1, 0)$, $(2, 0)$; y-int: $(0, 2)$;
domain: $(-\infty, \infty)$; range: $\left[-\dfrac{1}{4}, \infty\right)$

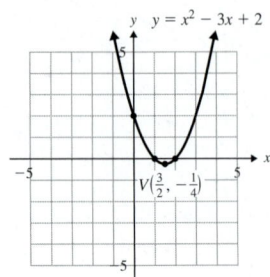

69) $f(x) = 2(x - 2)^2 - 4$;
x-ints: $(2 - \sqrt{2}, 0)$,
$(2 + \sqrt{2}, 0)$; y-int: $(0, 4)$;
domain: $(-\infty, \infty)$;
range: $[-4, \infty)$

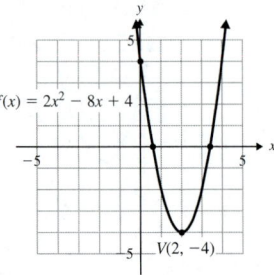

70) $y = 2(x - 2)^2 - 6$;
x-ints: $(2 - \sqrt{3}, 0)$,
$(2 + \sqrt{3}, 0)$; y-int:
$(0, 2)$; domain: $(-\infty, \infty)$;
range: $[-6, \infty)$

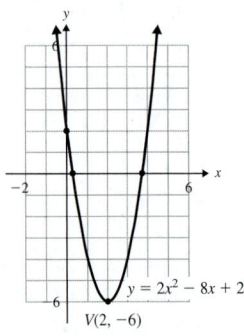

74) $f(x) = \left(x + \dfrac{5}{2}\right)^2 - 1$; x-ints: $\left(-\dfrac{7}{2}, 0\right)$, $\left(-\dfrac{3}{2}, 0\right)$; y-int:
$\left(0, \dfrac{21}{4}\right)$; domain: $(-\infty, \infty)$; range: $[-1, \infty)$

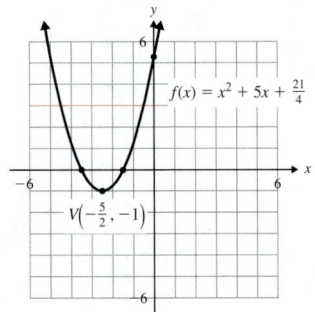

71) $g(x) = -\dfrac{1}{3}(x + 3)^2 - 6$;
x-ints: none; y-int: $(0, -9)$;
domain: $(-\infty, \infty)$; range:
$(-\infty, -6]$

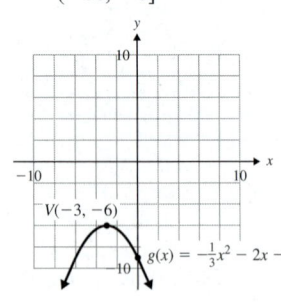

72) $h(x) = -\dfrac{1}{2}(x + 3)^2 - 5$;
x-ints: none; y-int:
$\left(0, -\dfrac{19}{2}\right)$; domain:
$(-\infty, \infty)$; range: $(-\infty, -5]$

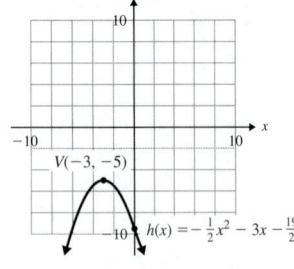

75) $V(-1, -4)$; x-ints: $(-3, 0)$,
$(1, 0)$; y-int: $(0, -3)$; domain:
$(-\infty, \infty)$; range: $[-4, \infty)$

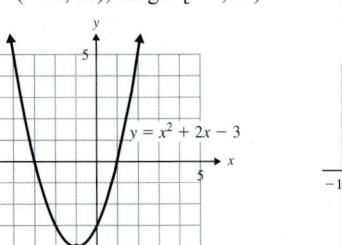

76) $V(3, -1)$; x-ints: $(2, 0)$,
$(4, 0)$; y-int: $(0, 8)$; domain:
$(-\infty, \infty)$; range: $[-1, \infty)$

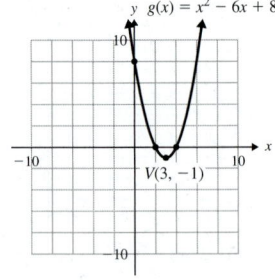

77) $V(-4, 3)$; x-ints:
$(-4 - \sqrt{3}, 0)$,
$(-4 + \sqrt{3}, 0)$; y-int:
$(0, -13)$; domain:
$(-\infty, \infty)$; range: $(-\infty, 3]$

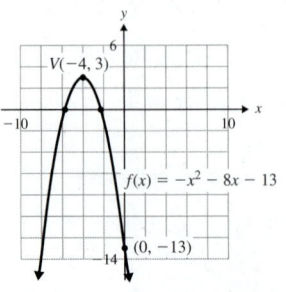

78) $V(1, 3)$; x-ints:
$(1 - \sqrt{3}, 0)$,
$(1 + \sqrt{3}, 0)$; y-int:
$(0, 2)$; domain: $(-\infty, \infty)$;
range: $(-\infty, 3]$

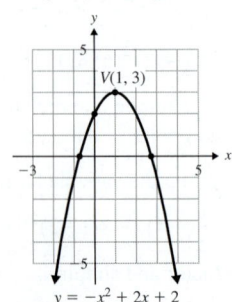

79) $V(1, 2)$; x-int: none; y-int: $(0, 4)$; domain: $(-\infty, \infty)$; range: $[2, \infty)$

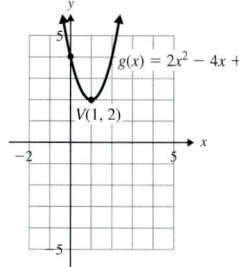

80) $V(-1, -2)$; x-int: none; y-int: $(0, -6)$; domain: $(-\infty, \infty)$; range: $(-\infty, -2]$

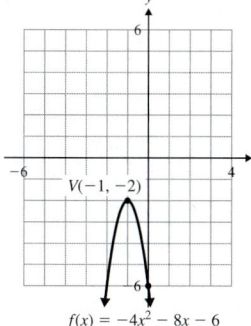

81) $V(1, 4)$; x-ints: $\left(1 + \dfrac{2\sqrt{3}}{3}, 0\right)$, $\left(1 - \dfrac{2\sqrt{3}}{3}, 0\right)$; y-int: $(0, 1)$; domain: $(-\infty, \infty)$; range: $(-\infty, 4]$

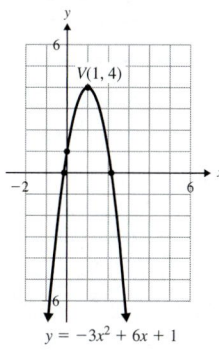

82) $V(3, -9)$; x-ints: $\left(3 - \dfrac{3\sqrt{2}}{2}, 0\right)$, $\left(3 + \dfrac{3\sqrt{2}}{2}, 0\right)$; y-int: $(0, 9)$; domain: $(-\infty, \infty)$; range: $[-9, \infty)$

83) $V(4, -3)$; x-ints: $(4 - \sqrt{6}, 0)$, $(4 + \sqrt{6}, 0)$; y-int: $(0, 5)$; domain: $(-\infty, \infty)$; range: $[-3, \infty)$

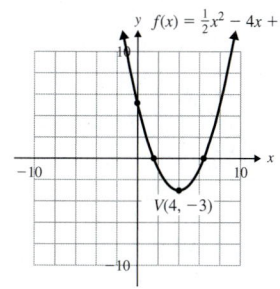

84) $V(-2, -5)$; x-ints: $(-2 - \sqrt{10}, 0)$, $(-2 + \sqrt{10}, 0)$; y-int: $(0, -3)$; domain: $(-\infty, \infty)$; range: $[-5, \infty)$

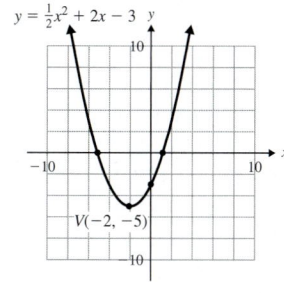

85) $V(-3, -2)$; x-int: none; y-int: $(0, -5)$; domain: $(-\infty, \infty)$; range: $(-\infty, -2]$

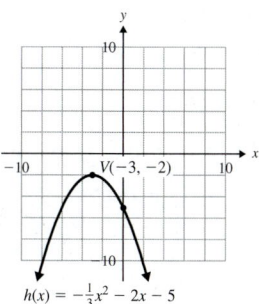

86) $V(5, 3)$; x-int: none; y-int: $(0, 8)$; domain: $(-\infty, \infty)$; range: $[3, \infty)$

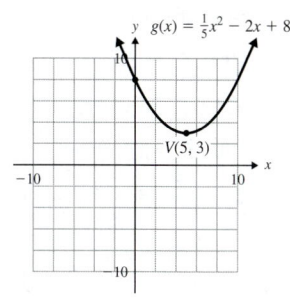

Section 9.6

1) maximum   2) neither   3) neither   4) minimum

5) minimum   6) neither

7) If a is positive the graph opens upward, so the y-coordinate of the vertex is the minimum value of the function. If a is negative the graph opens downward, so the y-coordinate of the vertex is the maximum value of the function.

8) No; the graph of the function opens upward, so the function attains its minimum value at the vertex. But, the graph continues upward indefinitely, so there is no maximum value of y.

9) a) minimum
   b) $(-3, 0)$ c) 0
   d)

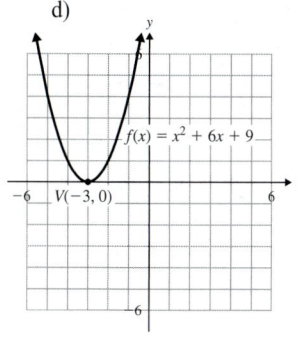

10) a) maximum
    b) $(1, 5)$ c) 5
    d)

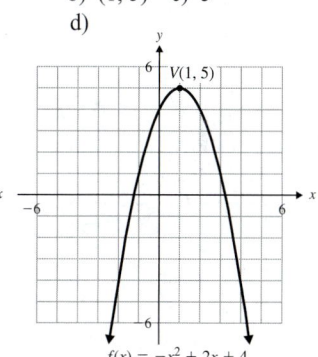

11) a) maximum
    b) $(4, 2)$ c) 2
    d)

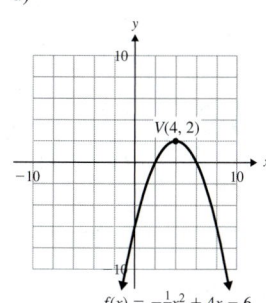

12) a) minimum
    b) $(-1, -2)$ c) -2
    d)

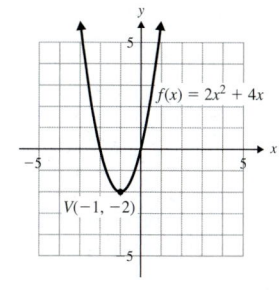

13) a) 10 sec   b) 1600 ft   c) 20 sec

14) a) 1.5 sec   b) 100 ft   c) 4 sec

15) July; 480 people   16) Friday; 218 tickets

17) 1991; 531,000   18) 1991; approximately 14,523,300

19) 625 ft^2 20) 10 ft $\times$ 10 ft   21) 12 ft $\times$ 24 ft

22) 1800 ft^2 23) 9 and 9   24) 13 and 13   25) 6 and -6

26) 15 and -15 27) (h, k) 28) $y = k$ 29) to the left

30) to the right

31) $V(-4, 1)$; $y = 1$;
x-int: $(-3, 0)$;
y-ints: $(0, -1)$, $(0, 3)$;
domain: $[-4, \infty)$;
range: $(-\infty, \infty)$

32) $V(-1, -3)$; $y = -3$;
x-int: $(8, 0)$;
y-ints: $(0, -4)$, $(0, -2)$;
domain: $[-1, \infty)$;
range: $(-\infty, \infty)$

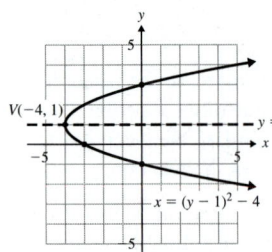

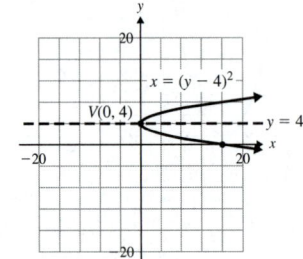

33) $V(2, 0)$; $y = 0$;
x-int: $(2, 0)$; y-int: none;
domain: $[2, \infty)$; range:
$(-\infty, \infty)$

34) $V(0, 4)$; $y = 4$;
x-int: $(16, 0)$; y-int: $(0, 4)$;
domain: $[0, \infty)$;
range: $(-\infty, \infty)$

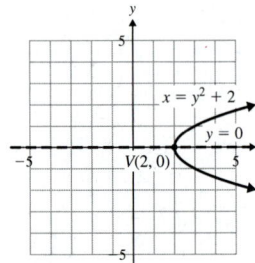

35) $V(5, 4)$; $y = 4$; x-int: $(-11, 0)$; y-ints:
$(0, 4 - \sqrt{5})$, $(0, 4 + \sqrt{5})$; domain: $(-\infty, 5]$; range: $(-\infty, \infty)$

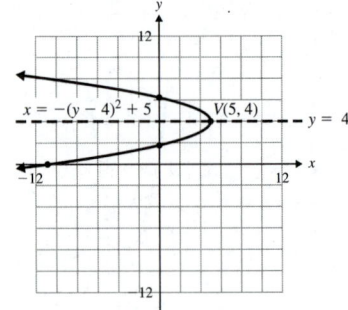

36) $V(-7, -1)$; $y = -1$; x-int:
$(-8, 0)$; y-int: none;
domain: $(-\infty, -7]$;
range: $(-\infty, \infty)$

37) $V(-9, 2)$; $y = 2$; x-int:
$(-17, 0)$; y-int: none;
domain: $(-\infty, -9]$;
range: $(-\infty, \infty)$

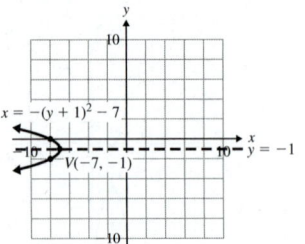

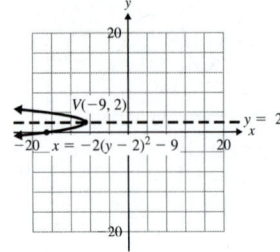

38) $V(7, 4)$; $y = 4$;
x-int: $(-1, 0)$;
y-ints: $(0, 4 - \sqrt{14})$,
$(0, 4 + \sqrt{14})$; domain:
$(-\infty, 7]$; range: $(-\infty, \infty)$

39) $V(0, -2)$; $y = -2$;
x-int: $(1, 0)$;
y-int: $(0, -2)$;
domain: $[0, \infty)$;
range: $(-\infty, \infty)$

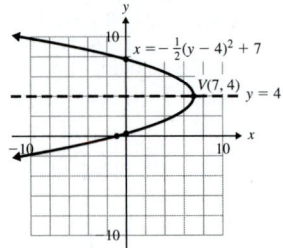

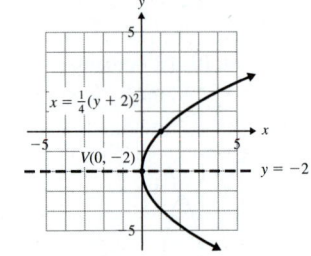

40) $V(3, 0)$; $y = 0$;
x-int: $(3, 0)$; y-int: none;
domain: $[3, \infty)$;
range: $(-\infty, \infty)$

41) $x = (y - 2)^2 + 1$;
domain: $[1, \infty)$;
range: $(-\infty, \infty)$

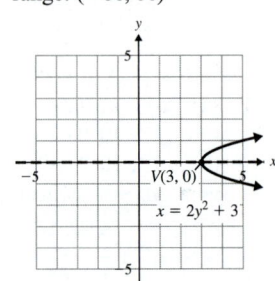

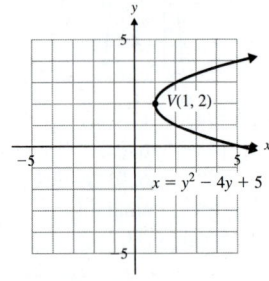

42) $x = (y + 2)^2 - 10$;
domain: $[-10, \infty)$;
range: $(-\infty, \infty)$

43) $x = -(y - 3)^2 + 15$;
domain: $(-\infty, 15]$;
range: $(-\infty, \infty)$

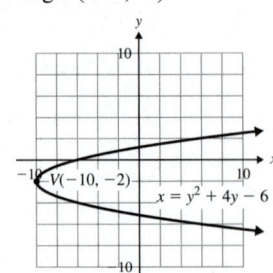

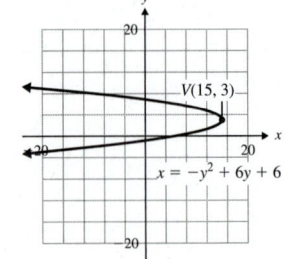

44) $x = -(y + 1)^2 - 4$;
domain: $(-\infty, -4]$;
range: $(-\infty, \infty)$

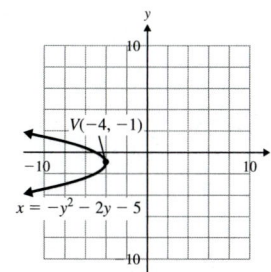

45) $x = \frac{1}{3}(y + 4)^2 - 7$;
domain: $[-7, \infty)$;
range: $(-\infty, \infty)$

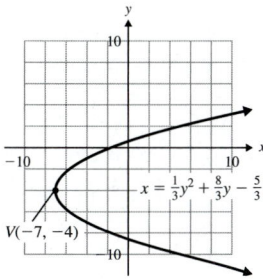

52) $V(-13, -3)$; x-int: $(-4, 0)$;
y-ints: $(0, -3 - \sqrt{13})$,
$(0, -3 + \sqrt{13})$;
domain: $[-13, \infty)$;
range: $(-\infty, \infty)$

53) $V(-4, 1)$;
x-int: $(-6, 0)$;
y-int: none;
domain: $(-\infty, -4]$;
range: $(-\infty, \infty)$

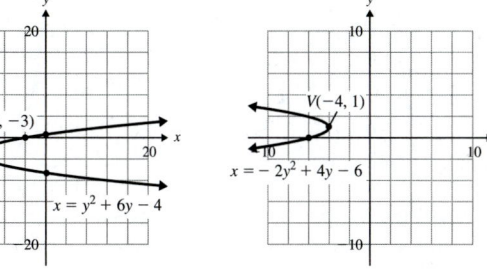

46) $x = 2(y - 1)^2 + 3$;
domain: $[3, \infty)$;
range: $(-\infty, \infty)$

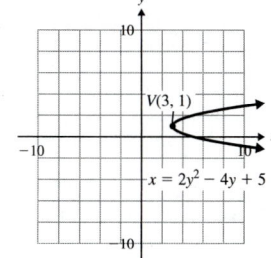

47) $x = -4(y + 1)^2 - 6$;
domain: $(-\infty, -6]$;
range: $(-\infty, \infty)$

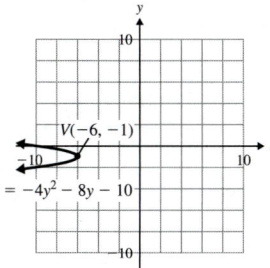

54) $V(-4, -1)$; x-int: $(-1, 0)$;
y-ints: $\left(0, -1 - \frac{2\sqrt{3}}{3}\right)$,
$\left(0, -1 + \frac{2\sqrt{3}}{3}\right)$;
domain: $[-4, \infty)$;
range: $(-\infty, \infty)$

55) $V(-3, 2)$; x-int: $(13, 0)$;
y-ints: $\left(0, 2 - \frac{\sqrt{3}}{2}\right)$,
$\left(0, 2 + \frac{\sqrt{3}}{2}\right)$;
domain: $[-3, \infty)$;
range: $(-\infty, \infty)$

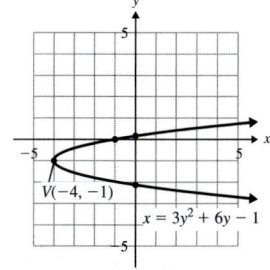

48) $x = \frac{1}{2}(y + 4)^2 - 9$;
domain: $[-9, \infty)$;
range: $(-\infty, \infty)$

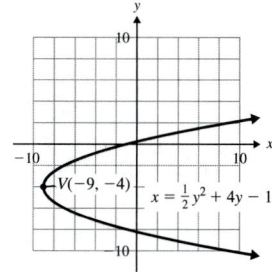

49) $V(-1, 2)$; x-int: $(3, 0)$;
y-ints: $(0, 1)$, $(0, 3)$;
domain: $[-1, \infty)$;
range: $(-\infty, \infty)$

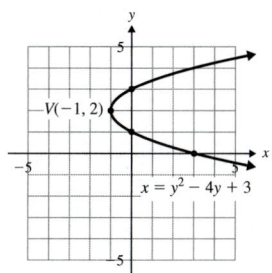

56) $V(6, -1)$; x-int: $(8, 0)$;
y-int: none;
domain: $[6, \infty)$;
range: $(-\infty, \infty)$

57) $V(6, 1)$; x-int: $\left(\frac{25}{4}, 0\right)$;
y-int: none;
domain: $[6, \infty)$;
range: $(-\infty, \infty)$

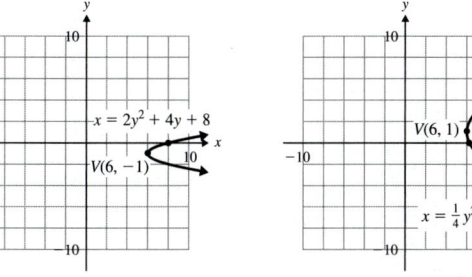

50) $V(4, 2)$; x-int: $(0, 0)$;
y-ints: $(0, 0)$, $(0, 4)$;
domain: $(-\infty, 4]$;
range: $(-\infty, \infty)$

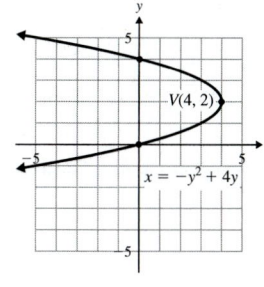

51) $V(3, 1)$; x-int: $(2, 0)$;
y-ints:
$(0, 1 - \sqrt{3})$, $(0, 1 + \sqrt{3})$;
domain: $(-\infty, 3]$;
range: $(-\infty, \infty)$

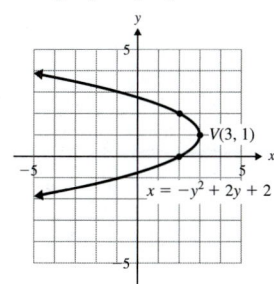

58) $V(-2, 1)$; x-int: $\left(-\frac{11}{4}, 0\right)$; y-int: none; domain: $(-\infty, -2]$;
range: $(-\infty, \infty)$

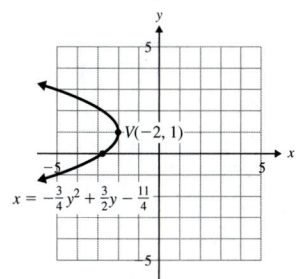

59) domain: $(-\infty, \infty)$; range: $(-\infty, 6]$

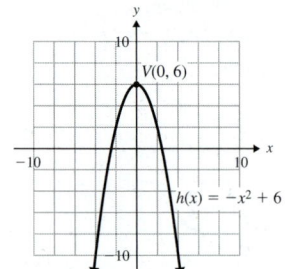

60) domain: $(-\infty, \infty)$; range: $[-10, \infty)$

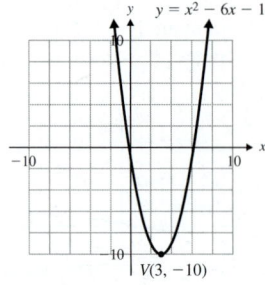

61) domain: $[0, \infty)$; range: $(-\infty, \infty)$

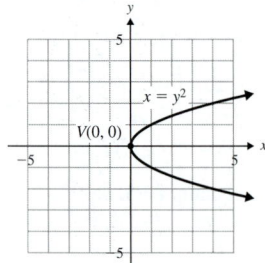

62) domain: $(-\infty, \infty)$; range: $(-\infty, 4]$

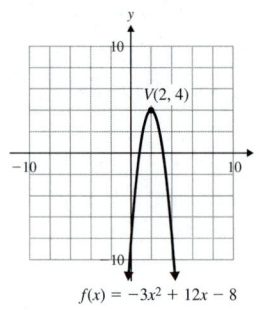

63) domain: $(-\infty, 3]$; range: $(-\infty, \infty)$

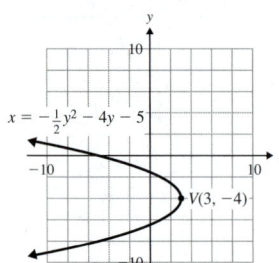

64) domain: $[3, \infty)$; range: $(-\infty, \infty)$

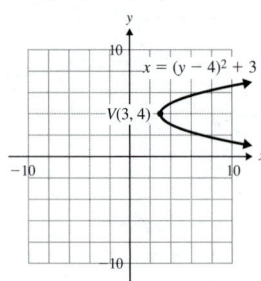

65) domain: $(-\infty, \infty)$; range: $[-4, \infty)$

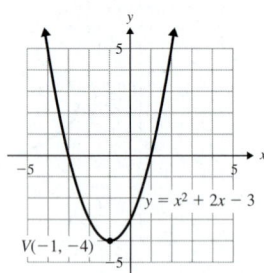

66) domain: $(-\infty, 11]$; range: $(-\infty, \infty)$

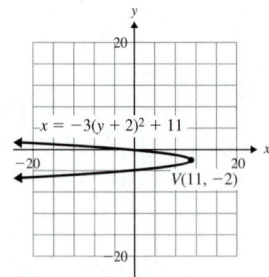

67) domain: $(-\infty, \infty)$; range: $(-\infty, 3]$

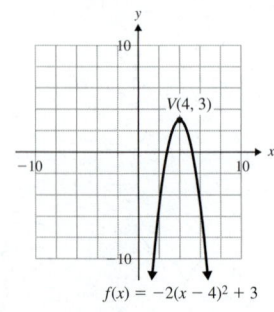

68) domain: $(-\infty, \infty)$; range: $[-4, \infty)$

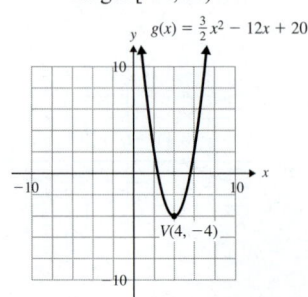

Section 9.7

1) The endpoints are included when the inequality symbol is $\leq$ or $\geq$. The endpoints are not included when the symbol is $<$ or $>$.

2) No; if an endpoint makes a denominator zero, then it will not be part of the solution set.

3) a) $[-5, 1]$ b) $(-\infty, -5) \cup (1, \infty)$

4) a) $(-\infty, 2) \cup (4, \infty)$ b) $[2, 4]$

5) a) $[-1, 3]$ b) $(-\infty, -1) \cup (3, \infty)$

6) a) $(-\infty, -6) \cup (-2, \infty)$ b) $[-6, -2]$

7) $(-\infty, -7] \cup [1, \infty)$

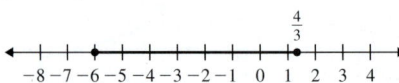

8) $(-\infty, -4) \cup (6, \infty)$

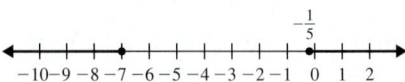

9) $(-9, 4)$

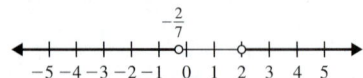

10) $[3, 12]$

11) $(-\infty, 6) \cup (7, \infty)$

12) $(-8, -2)$

13) $\left[-6, \dfrac{4}{3}\right]$

14) $(-\infty, -7] \cup \left[-\dfrac{1}{5}, \infty\right)$

15) $\left(-\infty, -\dfrac{2}{7}\right) \cup (2, \infty)$

16) $\left(-\dfrac{5}{4}, 6\right)$

17) $(-\infty, 0) \cup (9, \infty)$

18) $[-12, 0]$

$-15 \quad -13 \quad -11 \quad -9\,-8\,-7\,-6\,-5\,-4\,-3\,-2\,-1 \quad 0 \quad 1 \quad 2 \quad 3$

19) $\left[-\dfrac{5}{4}, 0\right]$

$-\dfrac{5}{4}$

$-5\,-4\,-3\,-2\,-1 \quad 0 \quad 1 \quad 2 \quad 3 \quad 4 \quad 5$

20) $(-\infty, 0] \cup \left[\dfrac{7}{2}, \infty\right)$

$\dfrac{7}{2}$

$-5\,-4\,-3\,-2\,-1 \quad 0 \quad 1 \quad 2 \quad 3 \quad 4 \quad 5$

21) $(-8, 8)$

$-10\,-9\,-8\,-7\,-6\,-5\,-4\,-3\,-2\,-1 \quad 0 \quad 1 \quad 2 \quad 3 \quad 4 \quad 5 \quad 6 \quad 7 \quad 8 \quad 9 \quad 10$

22) $(-\infty, -12) \cup (12, \infty)$

$-15 \quad -13 \quad -11 \quad -9\,-8\,-7\,-6\,-5\,-4\,-3\,-2\,-1 \quad 0 \quad 1\,2\,3\,4\,5\,6\,7\,8\,9\,10\,11\,12\,13\,14\,15$

23) $(-\infty, -11] \cup [11, \infty)$

$-15 \quad -13 \quad -11 \quad -9\,-8\,-7\,-6\,-5\,-4\,-3\,-2\,-1 \quad 0 \quad 1\,2\,3\,4\,5\,6\,7\,8\,9\,10\,11\,12\,13\,14\,15$

24) $(-1, 1)$

$-5\,-4\,-3\,-2\,-1 \quad 0 \quad 1 \quad 2 \quad 3 \quad 4 \quad 5$

25) $[-4, 4]$

$-5\,-4\,-3\,-2\,-1 \quad 0 \quad 1 \quad 2 \quad 3 \quad 4 \quad 5$

26) $\left(-\infty, -\dfrac{9}{5}\right] \cup \left[\dfrac{9}{5}, \infty\right)$

$-\dfrac{9}{5} \qquad \dfrac{9}{5}$

$-5\,-4\,-3\,-2\,-1 \quad 0 \quad 1 \quad 2 \quad 3 \quad 4 \quad 5$

27) $(-\infty, \infty)$ 28) $(-\infty, \infty)$ 29) $(-\infty, \infty)$ 30) $\varnothing$

31) $\varnothing$ 32) $(-\infty, \infty)$ 33) $\varnothing$ 34) $\varnothing$

35) $(-\infty, -2] \cup [1, 5]$

$-6\,-5\,-4\,-3\,-2\,-1 \quad 0 \quad 1 \quad 2 \quad 3 \quad 4 \quad 5 \quad 6$

36) $(-2, 3) \cup (12, \infty)$

$-5\,-4\,-3\,-2\,-1 \quad 0 \quad 1\,2\,3\,4\,5\,6\,7\,8\,9\,10\,11\,12\,13\,14\,15$

37) $[-9, 5] \cup [7, \infty)$

$-10\,-9\,-8\,-7\,-6\,-5\,-4\,-3\,-2\,-1 \quad 0 \quad 1\,2\,3\,4\,5\,6\,7\,8\,9\,10$

38) $(-\infty, -4] \cup [-1, 7]$

$-8\,-7\,-6\,-5\,-4\,-3\,-2\,-1 \quad 0 \quad 1 \quad 2 \quad 3 \quad 4 \quad 5 \quad 6 \quad 7 \quad 8$

39) $(-\infty, -7) \cup \left(-\dfrac{1}{6}, \dfrac{3}{4}\right)$

$-\dfrac{1}{6} \quad \dfrac{3}{4}$

$-10\,-9\,-8\,-7\,-6\,-5\,-4\,-3\,-2\,-1 \quad 0 \quad 1 \quad 2 \quad 3$

40) $\left[-2, \dfrac{1}{5}\right] \cup \left[\dfrac{7}{4}, \infty\right)$

$\dfrac{1}{5} \quad \dfrac{7}{4}$

$-5\,-4\,-3\,-2\,-1 \quad 0 \quad 1 \quad 2 \quad 3 \quad 4 \quad 5$

41) $(-6, \infty)$

$-8\,-7\,-6\,-5\,-4\,-3\,-2\,-1 \quad 0$

42) $(-\infty, 2)$

$-5\,-4\,-3\,-2\,-1 \quad 0 \quad 1 \quad 2 \quad 3 \quad 4 \quad 5$

43) $(-\infty, -3)$

$-5\,-4\,-3\,-2\,-1 \quad 0 \quad 1 \quad 2 \quad 3 \quad 4 \quad 5$

44) $(4, \infty)$

$-2\,-1 \quad 0 \quad 1 \quad 2 \quad 3 \quad 4 \quad 5 \quad 6 \quad 7$

45) $(-\infty, 3) \cup (4, \infty)$

$-2\,-1 \quad 0 \quad 1 \quad 2 \quad 3 \quad 4 \quad 5 \quad 6 \quad 7$

46) $(-1, 2)$

$-5\,-4\,-3\,-2\,-1 \quad 0 \quad 1 \quad 2 \quad 3 \quad 4 \quad 5$

47) $\left(-\dfrac{1}{3}, 9\right]$

$-\dfrac{1}{3}$

$-3\,-2\,-1 \quad 0 \quad 1 \quad 2 \quad 3 \quad 4 \quad 5 \quad 6 \quad 7 \quad 8 \quad 9 \quad 10 \quad 11 \quad 12$

48) $(-\infty, -4) \cup \left[-\dfrac{1}{2}, \infty\right)$

$-\dfrac{1}{2}$

$-8\,-7\,-6\,-5\,-4\,-3\,-2\,-1 \quad 0 \quad 1 \quad 2$

49) $(-3, 0]$

$-5\,-4\,-3\,-2\,-1 \quad 0 \quad 1 \quad 2 \quad 3 \quad 4 \quad 5$

50) $(-\infty, 0] \cup (7, \infty)$

$-3\,-2\,-1 \quad 0 \quad 1 \quad 2 \quad 3 \quad 4 \quad 5 \quad 6 \quad 7 \quad 8 \quad 9 \quad 10$

51) $(-\infty, -6) \cup \left(-\dfrac{11}{3}, \infty\right)$

$-\dfrac{11}{3}$

$-10\,-9\,-8\,-7\,-6\,-5\,-4\,-3\,-2\,-1 \quad 0$

52) $\left(-\dfrac{17}{2}, -7\right)$

$-\dfrac{17}{2}$

$-10\,-9\,-8\,-7\,-6\,-5\,-4\,-3\,-2\,-1 \quad 0$

53) $(-7, -4]$

$-10\,-9\,-8\,-7\,-6\,-5\,-4\,-3\,-2\,-1 \quad 0$

54) $(-\infty, 3) \cup [8, \infty)$

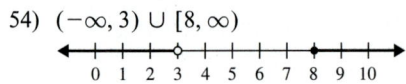

55) $\left[\dfrac{18}{5}, 6\right)$

56) $(-\infty, -4) \cup [8, \infty)$

57) $(-\infty, -2) \cup \left(-\dfrac{8}{7}, \infty\right)$

58) $(-3, 1)$

59) $(-\infty, 2)$

60) $(-7, \infty)$

61) $(5, \infty)$

62) $(-\infty, -3)$

63) $(-\infty, -6)$

64) $(3, \infty)$

65) $[-1, \infty)$

66) $(-\infty, 7]$

67) $(-\infty, 4)$

68) $(-\infty, -6)$

69) a) between 4000 and 12,000 units
    b) when it produces less than 4000 units or more than 12,000 units

70) a) between 2 and 6 sec after it is launched
    b) 8 sec after it is launched

71) 10,000 or more   72) 15,000 or more

Chapter 9 Review Exercises

1) $\{-12, 12\}$ 2) $\{-5\sqrt{3}, 5\sqrt{3}\}$ 3) $\{-2i, 2i\}$

4) $\{-3\sqrt{2}, 3\sqrt{2}\}$ 5) $\{-4, 10\}$

6) $\left\{\dfrac{-7 - \sqrt{15}}{6}, \dfrac{-7 + \sqrt{15}}{6}\right\}$ 7) $\left\{-\dfrac{\sqrt{10}}{3}, \dfrac{\sqrt{10}}{3}\right\}$

8) $\{14 - i\sqrt{5}, 14 + i\sqrt{5}\}$ 9) $2\sqrt{5}$ 10) $\sqrt{66}$ in.

11) $r^2 + 10r + 25; (r + 5)^2$ 12) $z^2 - 12z + 36; (z - 6)^2$

13) $c^2 - 5c + \dfrac{25}{4}; \left(c - \dfrac{5}{2}\right)^2$ 14) $x^2 + x + \dfrac{1}{4}; \left(x + \dfrac{1}{2}\right)^2$

15) $a^2 + \dfrac{2}{3}a + \dfrac{1}{9}; \left(a + \dfrac{1}{3}\right)^2$ 16) $d^2 - \dfrac{5}{2}d + \dfrac{25}{16}; \left(d - \dfrac{5}{4}\right)^2$

17) $\{-2, 8\}$ 18) $\{-5, 7\}$ 19) $\{-5 - \sqrt{31}, -5 + \sqrt{31}\}$

20) $\{-2 - i\sqrt{5}, -2 + i\sqrt{5}\}$ 21) $\left\{-\dfrac{3}{2} - \dfrac{\sqrt{5}}{2}, -\dfrac{3}{2} + \dfrac{\sqrt{5}}{2}\right\}$

22) $\left\{\dfrac{7}{2} - \dfrac{\sqrt{65}}{2}, \dfrac{7}{2} + \dfrac{\sqrt{65}}{2}\right\}$ 23) $\left\{\dfrac{7}{6} - \dfrac{\sqrt{95}}{6}i, \dfrac{7}{6} + \dfrac{\sqrt{95}}{6}i\right\}$

24) $\left\{\dfrac{5}{4} - \dfrac{\sqrt{17}}{4}, \dfrac{5}{4} + \dfrac{\sqrt{17}}{4}\right\}$ 25) $\{-6, 2\}$ 26) $\left\{\dfrac{4}{3}, 2\right\}$

27) $\left\{\dfrac{5 - \sqrt{15}}{2}, \dfrac{5 + \sqrt{15}}{2}\right\}$ 28) $\left\{\dfrac{2}{5} - \dfrac{4\sqrt{6}}{5}i, \dfrac{2}{5} + \dfrac{4\sqrt{6}}{5}i\right\}$

29) $\{1 - i\sqrt{3}, 1 + i\sqrt{3}\}$ 30) $\{4 - \sqrt{10}, 4 + \sqrt{10}\}$

31) $\left\{-\dfrac{2}{3}, \dfrac{1}{2}\right\}$ 32) $\left\{\dfrac{3}{4} - \dfrac{1}{2}i, \dfrac{3}{4} + \dfrac{1}{2}i\right\}$

33) 64; two rational solutions

34) -15; two nonreal, complex solutions   35) -12 or 12

36) a) 0.25 sec on the way up, 3 sec on the way down
    b) $\dfrac{13 + \sqrt{185}}{8}$ sec or about 3.3 sec

37) $\left\{1, \dfrac{4}{3}\right\}$ 38) $\{3 - i\sqrt{2}, 3 + i\sqrt{2}\}$

39) $\{-8 - 2\sqrt{3}, -8 + 2\sqrt{3}\}$ 40) $\left\{-4, \dfrac{3}{2}\right\}$

41) $\left\{-\dfrac{3}{2} - \dfrac{1}{2}i, -\dfrac{3}{2} + \dfrac{1}{2}i\right\}$ 42) $\left\{-\dfrac{\sqrt{2}}{2}, \dfrac{\sqrt{2}}{2}\right\}$

43) $\{9 - 3\sqrt{5}, 9 + 3\sqrt{5}\}$ 44) $\left\{\dfrac{1}{3} - \dfrac{\sqrt{17}}{3}i, \dfrac{1}{3} + \dfrac{\sqrt{17}}{3}i\right\}$

45) $\{-1, 0, 1\}$ 46) $\{-6, 9\}$ 47) $-2, 3$

48) $5 - \sqrt{3}, 5 + \sqrt{3}$ 49) $\left\{\dfrac{3 - \sqrt{21}}{3}, \dfrac{3 + \sqrt{21}}{3}\right\}$

50) $\left\{\dfrac{4}{3}, 2\right\}$ 51) $\{3, 4\}$ 52) $\{25\}$ 53) $\{-4, -1, 1, 4\}$

54) $\{-\sqrt{2}, \sqrt{2}, -i\sqrt{7}, i\sqrt{7}\}$ 55) $\{-27, 1\}$ 56) $\{1, 4\}$

57) $\left\{-\dfrac{\sqrt{7+\sqrt{33}}}{2}, \dfrac{\sqrt{7-\sqrt{33}}}{2},\right.$

$\left.-\dfrac{\sqrt{7-\sqrt{33}}}{2}, \dfrac{\sqrt{7+\sqrt{33}}}{2}\right\}$

58) $\left\{-\dfrac{7}{2}, -1\right\}$ 59) $\left\{2, \dfrac{11}{2}\right\}$ 60) Erica: 2 hr; Lorena: 3 hr

61) $v = \dfrac{\pm\sqrt{Frm}}{m}$ 62) $x = \dfrac{\pm\sqrt{2Uk}}{k}$ 63) $A = \pi r^2$

64) $V = \pi r^2 l$ 65) $n = \dfrac{l \pm \sqrt{l^2 + 4km}}{2k}$

66) $p = \dfrac{r \pm \sqrt{r^2 - 8t}}{4}$ 67) 3 in.

68) length = 18 in., width = 14 in.   69) $8.00

70) a) 8.4 million   b) 26.6 million   c) 2005

71) a) (h, k) b) $x = h$
    c) If a is positive, the parabola opens upward. If a is
    negative, the parabola opens downward.

72) $k(x), f(x), g(x), h(x)$

73) $V(0, -4)$; $x = 0$; x-ints:
    $(-2, 0), (2, 0)$; y-int:
    $(0, -4)$; domain:
    $(-\infty, \infty)$; range: $[-4, \infty)$

74) $V(-1, 0)$; $x = -1$;
    x-int: $(-1, 0)$; y-int:
    $(0, -1)$; domain:
    $(-\infty, \infty)$; range: $(-\infty, 0]$

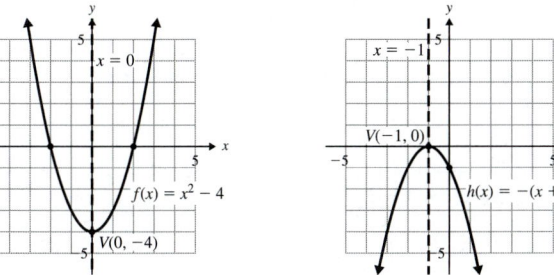

75) $V(-2, -1)$; $x = -2$;
    x-ints: $(-3, 0), (-1, 0)$;
    y-int: $(0, 3)$; domain:
    $(-\infty, \infty)$; range: $[-1, \infty)$

76) $V(0, 0)$; $x = 0$; x-int:
    $(0, 0)$; y-int: $(0, 0)$;
    domain: $(-\infty, \infty)$;
    range: $[0, \infty)$

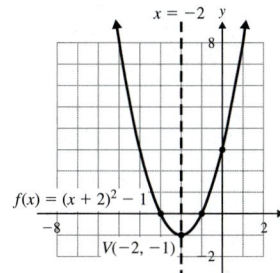
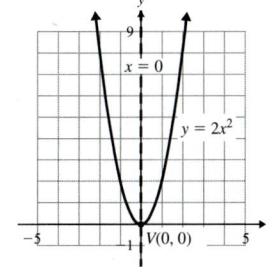

77) $V(4, 2)$; $x = 4$; x-ints:
    none; y-int: $(0, 18)$;
    domain: $(-\infty, \infty)$;
    range: $[2, \infty)$

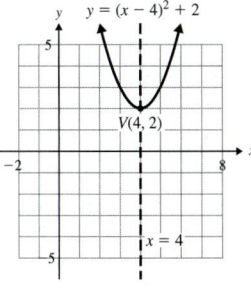

78) $V(3, -2)$; $x = 3$;
    x-ints: none; y-int:
    $\left(0, -\dfrac{13}{2}\right)$; domain:
    $(-\infty, \infty)$; range: $(-\infty, -2]$

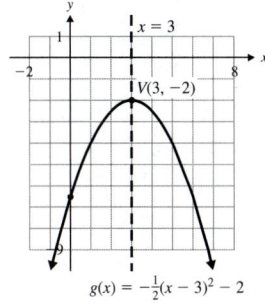

79) $g(x) = (x - 6)^2$

80) i) Rewrite the function in the form $f(x) = a(x - h)^2 + k$.
    The vertex is (h, k).

    ii) Use the vertex formula. The vertex is $\left(-\dfrac{b}{2a}, f\left(-\dfrac{b}{2a}\right)\right)$.

81) $f(x) = (x - 1)^2 + 2$;
    x-ints: none; y-int: $(0, 3)$;
    domain: $(-\infty, \infty)$;
    range: $[2, \infty)$

82) $y = (x + 2)^2 - 5$
    x-ints: $(-2 - \sqrt{5}, 0)$,
    $(-2 + \sqrt{5}, 0)$; y-int:
    $(0, -1)$; domain:
    $(-\infty, \infty)$; range: $[-5, \infty)$

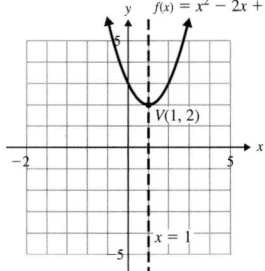
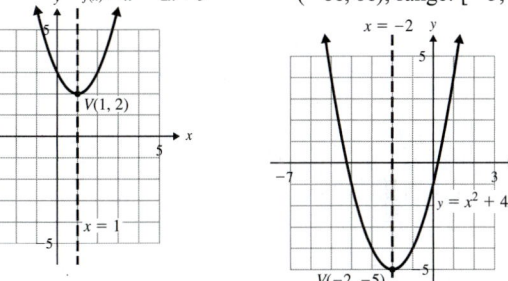

83) $y = \dfrac{1}{2}(x - 4)^2 + 1$;
    x-ints: none; y-int: $(0, 9)$;
    domain: $(-\infty, \infty)$;
    range: $[1, \infty)$

84) $f(x) = -2(x + 2)^2 + 10$;
    x-ints: $(-2 - \sqrt{5}, 0)$,
    $(-2 + \sqrt{5}, 0)$; y-int:
    $(0, 2)$; domain: $(-\infty, \infty)$;
    range: $(-\infty, 10]$

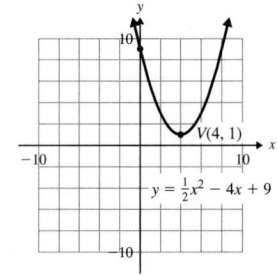
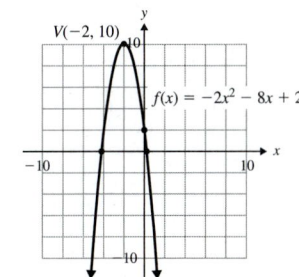

85) $V(-3, -1)$; x-ints: none; y-int: $(0, -10)$; domain: $(-\infty, \infty)$; range: $(-\infty, -1]$

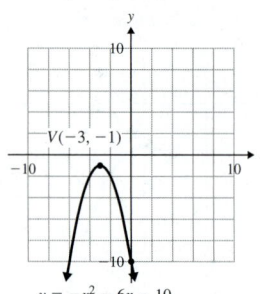

86) $V(1, -5)$; x-ints: $(1 - \sqrt{5}, 0)$, $(1 + \sqrt{5}, 0)$; y-int: $(0, -4)$; domain: $(-\infty, \infty)$; range: $[-5, \infty)$

92) $x = -(y - 2)^2$; x-int: $(-4, 0)$; y-int: $(0, 2)$; domain: $(-\infty, 0)$; range: $(-\infty, \infty)$

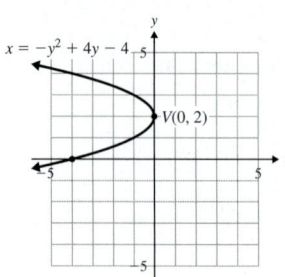

93) $V(2, -3)$; x-int: $\left(-\dfrac{5}{2}, 0\right)$; y-ints: $(0, -5)$, $(0, -1)$; domain: $(-\infty, 2]$; range: $(-\infty, \infty)$

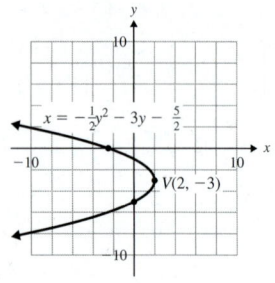

87) a) 1 sec b) 256 ft c) 5 sec   88) 14 ft × 28 ft

89) $V(11, 3)$; $y = 3$; x-int: $(2, 0)$; y-ints: $(0, 3 - \sqrt{11})$, $(0, 3 + \sqrt{11})$; domain: $(-\infty, 11]$; range: $(-\infty, \infty)$

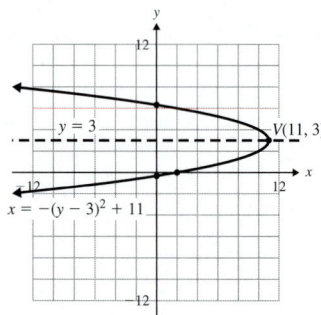

94) $V(-12, 2)$; x-int: $(0, 0)$; y-ints: $(0, 0)$, $(0, 4)$; domain: $[-12, \infty)$; range: $(-\infty, \infty)$

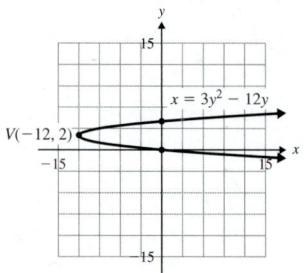

95) $(-3, 1)$

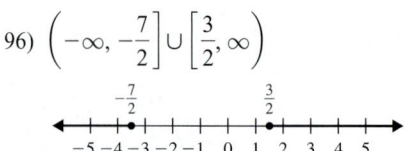

96) $\left(-\infty, -\dfrac{7}{2}\right] \cup \left[\dfrac{3}{2}, \infty\right)$

97) $\left(-\infty, -\dfrac{5}{8}\right] \cup \left[\dfrac{5}{8}, \infty\right)$

90) $V(-5, -1)$; $y = -1$; x-int: $(-4, 0)$; y-ints: $(0, -1 - \sqrt{5})$, $(0, -1 + \sqrt{5})$; domain: $[-5, \infty)$; range: $(-\infty, \infty)$

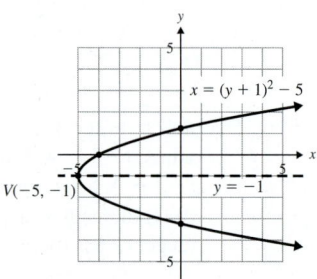

98) $(-6, 6)$

99) $\left(-\infty, -\dfrac{2}{5}\right) \cup \left(-\dfrac{1}{3}, 4\right)$

100) $\varnothing$

91) $x = (y + 4)^2 - 9$; x-int: $(7, 0)$; y-ints: $(0, -1)$, $(0, -7)$; domain: $[-9, \infty)$; range: $(-\infty, \infty)$

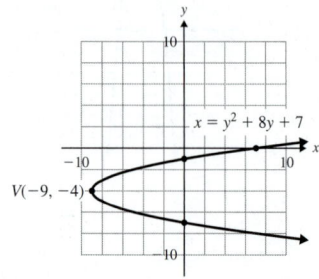

101) $(-\infty, -7) \cup \left(\dfrac{3}{2}, \infty\right)$

102) $(-\infty, 7)$

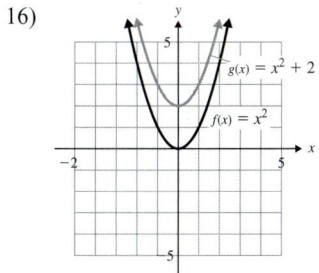

103) $(-\infty, 2) \cup [3, \infty)$

104) $\left(-\infty, \dfrac{11}{3}\right) \cup (4, \infty)$

105) $(7, \infty)$

106) between 5000 and 11,000 units

Chapter 9 Test

1) $\{-2 - \sqrt{11}, -2 + \sqrt{11}\}$ 2) $\{4 - i, 4 + i\}$

3) $\{-5 - i\sqrt{6}, -5 + i\sqrt{6}\}$ 4) $\left\{-2, \dfrac{4}{3}\right\}$ 5) $\left\{-\dfrac{7}{4}, -1\right\}$

6) $\left\{\dfrac{7 - \sqrt{17}}{4}, \dfrac{7 + \sqrt{17}}{4}\right\}$ 7) $\{-2\sqrt{2}, 2\sqrt{2}, -3i, 3i\}$

8) $\left\{-\dfrac{3}{2}, -1\right\}$ 9) 56; two irrational solutions   10) $\sqrt{19}$

11) $0, \dfrac{2}{5}$ 12) $2\sqrt{26}$ 13) width = 13 in., length = 19 in.

14) $V = \dfrac{1}{3}\pi r^2 h$ 15) $t = \dfrac{s \pm \sqrt{s^2 + 24r}}{2r}$

16)

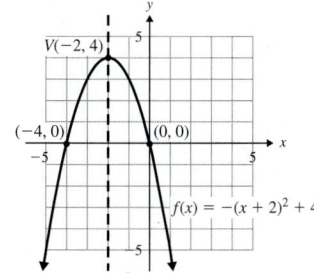

17) $g(x) = (x + 3)^2$

18) domain: $(-\infty, \infty)$;
    range: $(-\infty, 4]$

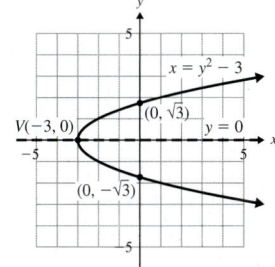

19) domain: $[-3, \infty)$;
    range: $(-\infty, \infty)$

20) domain: $[2, \infty)$;
    range: $(-\infty, \infty)$

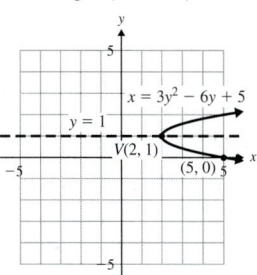

21) domain: $(-\infty, \infty)$;
    range: $[-1, \infty)$

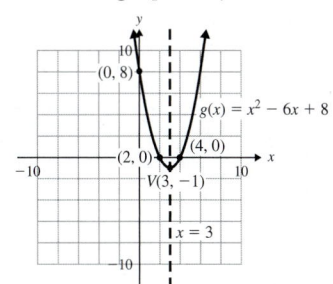

22) a) 209 ft   b) after 4 sec

   c) after $\dfrac{3 + \sqrt{209}}{4}$ sec or about 4.4 sec

23) $(-\infty, -9] \cup [5, \infty)$

24) $(-\infty, -3) \cup [5, \infty)$

25) 8000 or more

Cumulative Review Chapters 1–9

1) $\dfrac{7}{10}$ 2) $-8d^{15}$ 3) $\dfrac{45x^6}{y^4}$ 4) 90   5) $m = \dfrac{y - b}{x}$

6) a) $\{0, 3, 4\}$ b) $\{-1, 0, 1, 2\}$ c) no

7) x-int: $(4, 0)$;
   y-int: $\left(0, -\dfrac{8}{5}\right)$

8) a) 2   b) $[-3, \infty)$
   c)

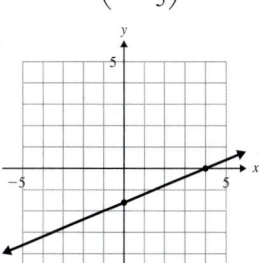

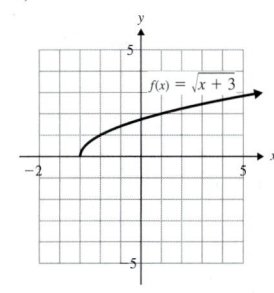

9) chips: $0.80, soda: $0.75

10) $3x^2y^2 - 5x^2y - 3xy^2 - 9xy + 8$ 11) $3r^2 - 30r + 75$

12) $2p(2p - 1)(p + 4)$ 13) $(a + 5)(a^2 - 5a + 25)$

14) $\dfrac{z^2 - 5z + 12}{z(z + 4)}$ 15) $\dfrac{c(c + 3)}{1 - 4c}$ 16) $\left\{-\dfrac{3}{2}, 3\right\}$ 17) $(0, 3, -1)$

18) $5\sqrt{3}$ 19) $2\sqrt[3]{5}$ 20) $3x^3y^2\sqrt{7x}$ 21) 16   22) $10 - 5\sqrt{3}$

23) $34 - 77i$ 24) $\{-2\sqrt{2}, 2\sqrt{2}\}$

25) $\left\{\dfrac{1}{6} - \dfrac{\sqrt{11}}{6}i, \dfrac{1}{6} + \dfrac{\sqrt{11}}{6}i\right\}$ 26) $\left\{-\dfrac{2}{3}, \dfrac{7}{3}\right\}$ 27) $\{-9, 3\}$

28) $V = \pi r^2 h$ 29)

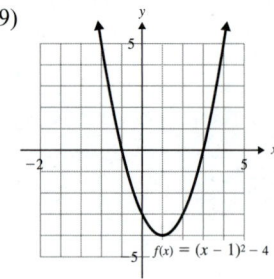

$f(x) = (x - 1)^2 - 4$

30) $\left[-\dfrac{12}{5}, \dfrac{12}{5} \right]$

c) $(f \circ s)(x) = 0.856x$. This is the final cost of the item after the discount and sales tax.

d) $(f \circ s)(40) = \$34.24$. When the original cost of an item is \$40, the final cost after the discount and sales tax is \$34.24.

26) a) $C(59) = 15$. A temperature of $59°F$ is equivalent to $15°C$.

b) $K(15) = 288$. A temperature of $15°C$ is equivalent to 288 K.

c) $K(C(F)) = \dfrac{5}{9}F + \dfrac{2297}{9}$. This is the temperature on the Kelvin scale in terms of the Fahrenheit temperature.

d) $K(C(59)) = 288$. A temperature of $59°F$ is equivalent to 288 K.

27) a) $r(5) = 20$. The radius of the spill 5 min after the ship started leaking was 20 ft.

b) $A(20) = 400\pi$. The area of the oil slick is 400π ft^2 when its radius is 20 ft.

c) $A(r(t)) = 16\pi t^2$. This is the area of the oil slick in terms of t, the number of minutes after the leak began.

d) $A(r(5)) = 400\pi$. The area of the oil slick 5 min after the ship began leaking was 400π ft^2.

28) a) $r(6) = 3$. When the diameter of a circle is 6 units, its radius is 3 units.

b) $A(3) = 9\pi$. When the radius of a circle is 3 units, its area is 9π square units.

c) $A(r(d)) = \dfrac{1}{4}\pi d^2$. This is the area of a circle in terms of its diameter.

d) $A(r(6)) = 9\pi$. When the diameter of a circle is 6 units, its area is 9π square units.

Chapter 10

Section 10.1

1) $(f \circ g)(x) = f(g(x))$, so substitute the function $g(x)$ into the function $f(x)$ and simplify.

2) $(f \circ g)(x)$ is the composition of functions f and g, so $(f \circ g)(x) = f(g(x))$. However, $(f \cdot g)(x)$ is the product of functions f and g, so $(f \cdot g)(x) = f(x) \cdot g(x)$.

3) a) -1 b) -2 c) $6x - 26$ d) -2

4) a) 11 b) -6 c) $-x - 2$ d) -6

5) a) 7 b) 44 c) $x^2 + 6x + 4$ d) 44

6) a) 3 b) 11 c) $x^2 - 2x + 3$ d) 11

7) a) $5x + 31$ b) $5x + 3$ c) 46

8) a) $4x - 7$ b) $4x - 37$ c) -31

9) a) $-6x + 26$ b) $-6x + 11$ c) 32

10) a) $-42x - 19$ b) $-42x - 28$ c) -112

11) a) $x^2 - 6x + 7$ b) $x^2 - 14x + 51$ c) 11

12) a) $x^2 + 7x - 7$ b) $x^2 + 11x + 9$ c) 23

13) a) $-x^2 - 13x - 48$ b) $-x^2 + 3x$ c) 0

14) a) $-x^2 + 10x + 5$ b) $-x^2 + 8x + 13$ c) -7

15) a) $\sqrt{x^2 + 4}$ b) $x + 4$ c) $\sqrt{13}$

16) a) $x + 6$ b) $\sqrt{x^2 + 6}$ c) $\sqrt{6}$

17) a) $\dfrac{1}{t^2 + 8}$ b) $\dfrac{1}{(t + 8)^2}$ c) $\dfrac{1}{9}$

18) a) $\dfrac{1}{25a^2}$ b) $\dfrac{1}{5a^2}$ c) $\dfrac{1}{100}$

19) $f(x) = \sqrt{x}, g(x) = x^2 + 13$; answers may vary.

20) $f(x) = \sqrt{x}, g(x) = 2x^2 + 7$; answers may vary.

21) $f(x) = x^2, g(x) = 8x - 3$; answers may vary.

22) $f(x) = x^2, g(x) = 4x + 9$; answers may vary.

23) $f(x) = \dfrac{1}{x}, g(x) = 6x + 5$; answers may vary.

24) $f(x) = \dfrac{2}{x}, g(x) = x - 10$; answers may vary.

25) a) $s(40) = 32$. When the regular price of an item is \$40, the sale price is \$32.

b) $f(32) = 34.24$. When the cost of an item is \$32, the final cost after sales tax is \$34.24.

29) no   30) yes; $g^{-1} = \{(-7, 0), (-6, 1), (-5, 4), (-2, 25)\}$

31) yes; $h^{-1} = \{(-16, -5), (-4, -1), (8, 3)\}$

32) no   33) yes; $g^{-1} = \{(1, 2), (2, 5), (14, 7), (19, 10)\}$

34) no   35) yes   36) no

37) No; only one-to-one functions have inverses.

38) horizontal line test   39) False; it is read "f inverse of x."

40) true   41) true   42) False; $(9, 5)$ is on the graph of f^{-1}.

43) False; they are symmetric with respect to $y = x$.

44) true

45) a) yes
b)

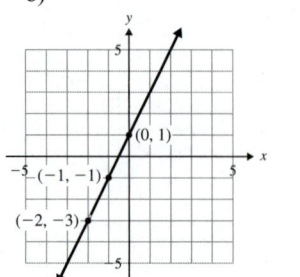

$(0, 1)$
$(-1, -1)$
$(-2, -3)$

46) a) yes
b)

$(0, 2)$
$(-3, 1)$
$(-4, 0)$

47) no

48) a) yes
   b)

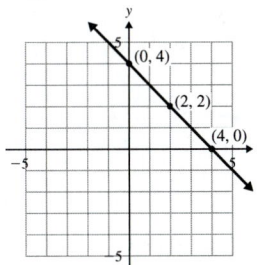

49) a) yes
   b)

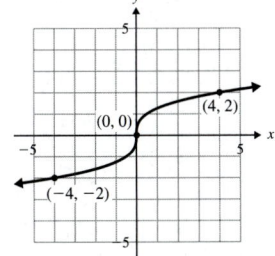

50) no

51) Replace $f(x)$ with y.; $x = 2y - 10$; Add 10.; $\frac{1}{2}x + 5 = y$; Replace y with $f^{-1}(x)$.

52) $y = \frac{1}{3}x + 4$; Interchange x and y.; $x - 4 = \frac{1}{3}y$; Multiply by 3 and simplify.; $g^{-1}(x) = 3x - 12$

53) $g^{-1}(x) = x + 6$

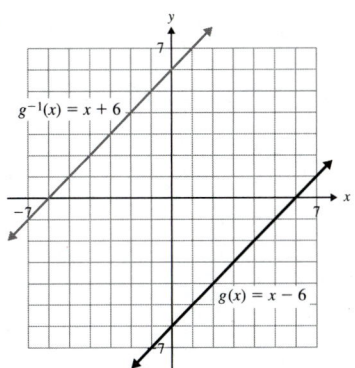

54) $h^{-1}(x) = x - 3$

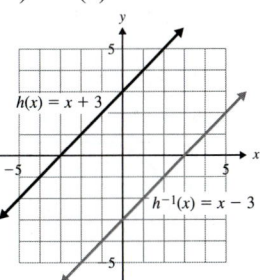

55) $f^{-1}(x) = -\frac{1}{2}x + \frac{5}{2}$

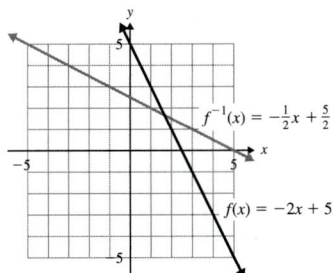

56) $g^{-1}(x) = \frac{1}{4}x + \frac{9}{4}$

57) $g^{-1}(x) = 2x$

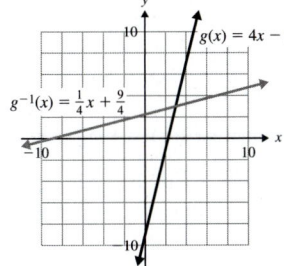

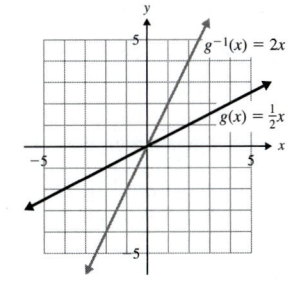

58) $h^{-1}(x) = -3x$

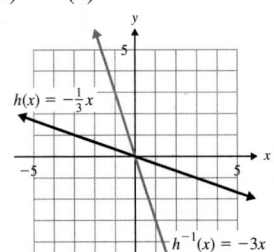

59) $f^{-1}(x) = \sqrt[3]{x}$

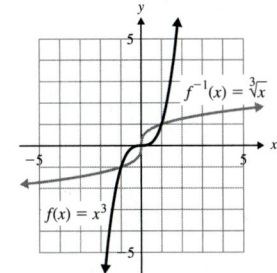

60) $g^{-1}(x) = (x - 4)^3$

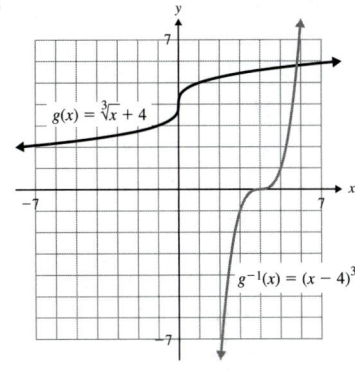

61) $f^{-1}(x) = \frac{1}{2}x + 3$ 62) $g^{-1}(x) = -\frac{1}{4}x + 2$

63) $h^{-1}(x) = -\frac{2}{3}x + \frac{8}{3}$ 64) $f^{-1}(x) = \frac{5}{2}x - \frac{5}{2}$

65) $g^{-1}(x) = x^3 - 2$ 66) $h^{-1}(x) = x^3 + 7$

67) $f^{-1}(x) = x^2, x \geq 0$ 68) $g^{-1}(x) = x^2 - 3, x \geq 0$

69) a) 3 b) 1   70) a) -5 b) -4 71) a) 2 b) 9

72) a) 2 b) 6   73) a) 10 b) -7 74) a) -8 b) 8

75) a) 8 b) 3   76) a) $\frac{1}{9}$ b) -2 77–84) Answers may vary.

Section 10.2

1) Choose values for the variable that will give positive numbers, negative numbers, and zero in the exponent.

2) $(0, 1)$

3)

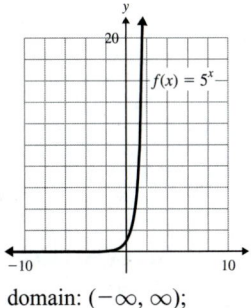

domain: $(-\infty, \infty)$;
range: $(0, \infty)$

4)

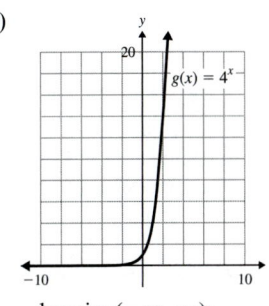

domain: $(-\infty, \infty)$;
range: $(0, \infty)$

5)

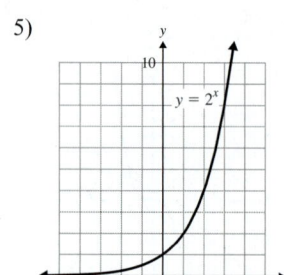

domain: $(-\infty, \infty)$;
range: $(0, \infty)$

6)

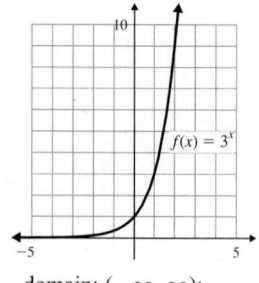

domain: $(-\infty, \infty)$;
range: $(0, \infty)$

17)

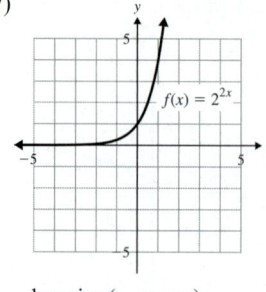

domain: $(-\infty, \infty)$;
range: $(0, \infty)$

18)

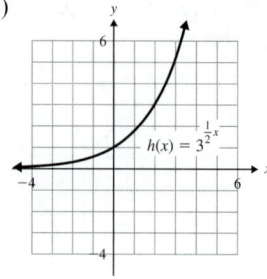

domain: $(-\infty, \infty)$;
range: $(0, \infty)$

7)

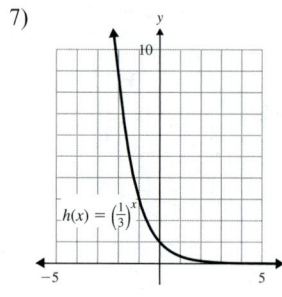

domain: $(-\infty, \infty)$;
range: $(0, \infty)$

8)

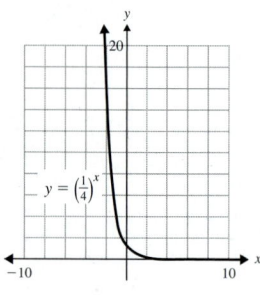

domain: $(-\infty, \infty)$;
range: $(0, \infty)$

19)

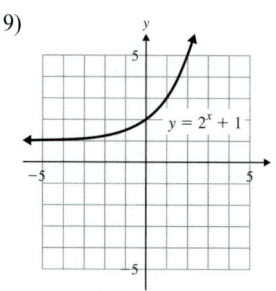

domain: $(-\infty, \infty)$;
range: $(1, \infty)$

20)

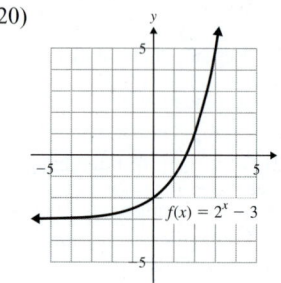

domain: $(-\infty, \infty)$;
range: $(-3, \infty)$

9) $(-\infty, \infty)$ 10) $(0, \infty)$

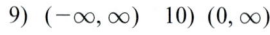

11)

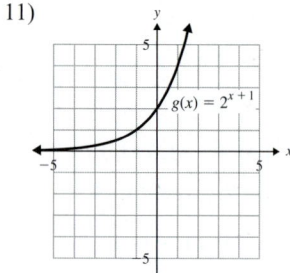

domain: $(-\infty, \infty)$;
range: $(0, \infty)$

12)

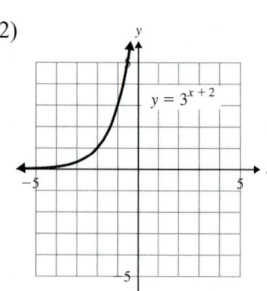

domain: $(-\infty, \infty)$;
range: $(0, \infty)$

21)

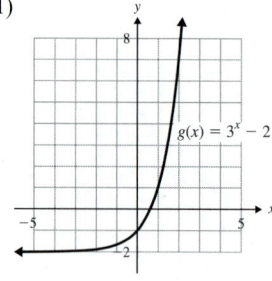

domain: $(-\infty, \infty)$;
range: $(-2, \infty)$

22)

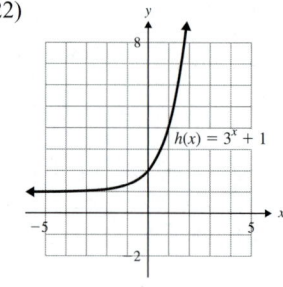

domain: $(-\infty, \infty)$;
range: $(1, \infty)$

13)

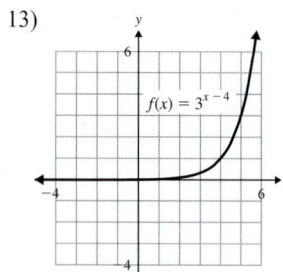

domain: $(-\infty, \infty)$;
range: $(0, \infty)$

14)

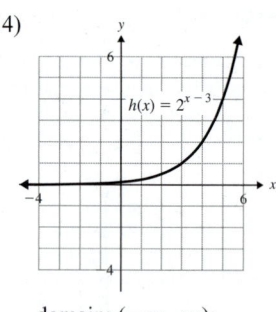

domain: $(-\infty, \infty)$;
range: $(0, \infty)$

23)

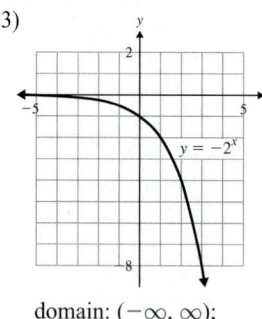

domain: $(-\infty, \infty)$;
range: $(-\infty, 0)$

24)

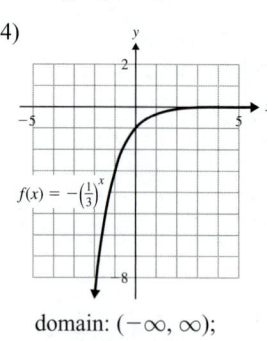

domain: $(-\infty, \infty)$;
range: $(-\infty, 0)$

25) $g(x) = 2^x$ would grow faster because for values of $x > 2$, $2^x > 2x$.

26) As x gets larger, the value of the function $f(x) = \left(\dfrac{1}{5}\right)^x$ gets closer to zero. Therefore, the y-coordinates of the points get closer to zero so that, if we graphed the function, we would see that the graph gets closer to the line $y = 0$.

27) Shift the graph of $f(x)$ down 2 units.

28) Shift the graph of $f(x)$ right 3 units.   29) 2.7183

15)

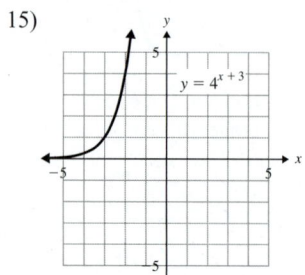

domain: $(-\infty, \infty)$;
range: $(0, \infty)$

16)

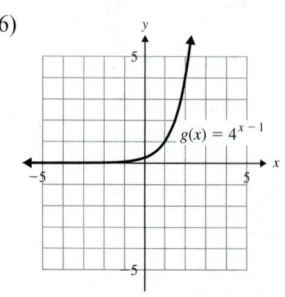

domain: $(-\infty, \infty)$;
range: $(0, \infty)$

30) e is irrational because it is a nonterminating, nonrepeating decimal.

31) B   32) A   33) D   34) C

35)

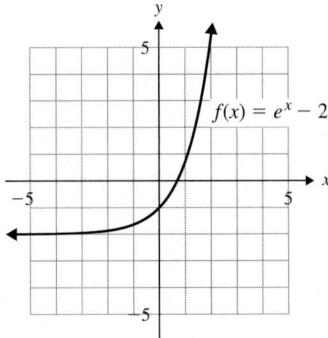

domain: $(-\infty, \infty)$; range: $(-2, \infty)$

36)

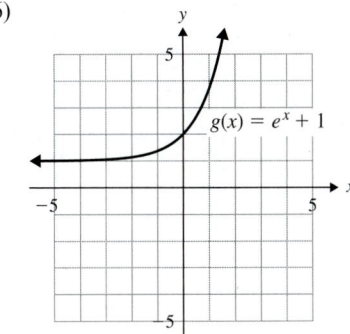

domain: $(-\infty, \infty)$; range: $(1, \infty)$

37)

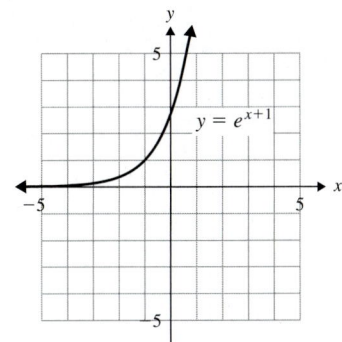

domain: $(-\infty, \infty)$; range: $(0, \infty)$

38)

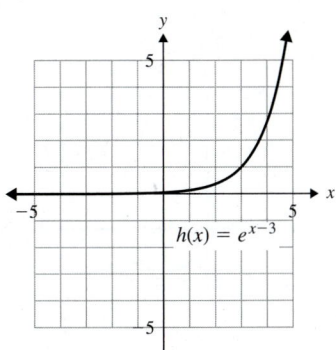

domain: $(-\infty, \infty)$; range: $(0, \infty)$

39)

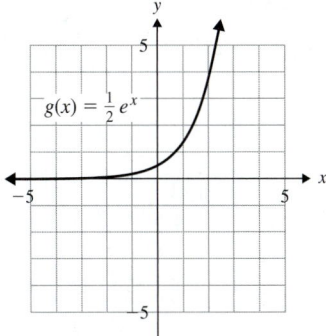

domain: $(-\infty, \infty)$; range: $(0, \infty)$

40)

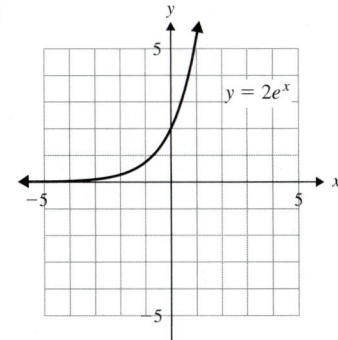

domain: $(-\infty, \infty)$; range: $(0, \infty)$

41)

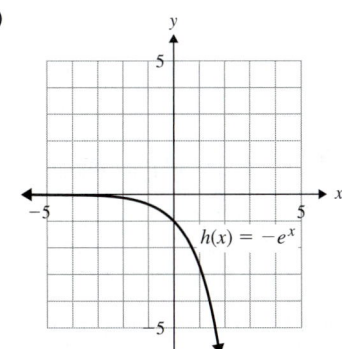

domain: $(-\infty, \infty)$; range: $(-\infty, 0)$

42)

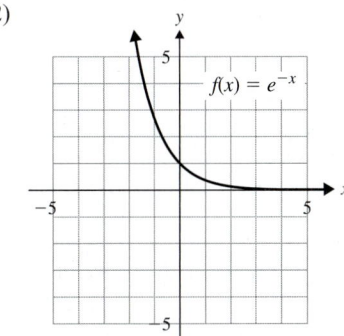

domain: $(-\infty, \infty)$; range: $(0, \infty)$

43) They are symmetric with respect to the x-axis.

44) They are symmetric with respect to the y-axis.

45) $6^{3n} = (6^2)^{n-4}$; Power rule for exponents; Distribute.; $3n = 2n - 8$; $n = -8$; $\{-8\}$

46) Express each side with the same base.; $5^{3(2w)} = 5^{w+2}$; Distribute.; Set the exponents equal.; $w = \dfrac{2}{5}$; $\left\{\dfrac{2}{5}\right\}$

47) $\{2\}$ 48) $\{2\}$ 49) $\left\{\dfrac{3}{4}\right\}$ 50) $\{1\}$ 51) $\{8\}$ 52) $\left\{\dfrac{8}{3}\right\}$

53) $\{-2\}$ 54) $\{5\}$ 55) $\{1\}$ 56) $\left\{\dfrac{21}{4}\right\}$ 57) $\left\{\dfrac{23}{4}\right\}$

58) $\{-7\}$ 59) $\{-3\}$ 60) $\left\{\dfrac{8}{13}\right\}$ 61) $\{-2\}$ 62) $\{-2\}$

63) $\{-3\}$ 64) $\{-4\}$ 65) $\left\{-\dfrac{3}{2}\right\}$ 66) $\left\{-\dfrac{3}{4}\right\}$ 67) $\left\{\dfrac{3}{2}\right\}$

68) $\{4\}$ 69) $\{-1\}$ 70) $\left\{-\dfrac{6}{13}\right\}$

71) a) \$32,700 b) \$17,507.17 72) a) \$10,150 b) \$3006.40

73) a) \$16,800 b) \$4504.04 74) a) \$48,600 b) \$21,973.12

75) a) \$185,200 b) \$227,772.64

76) a) \$420,000 b) \$511,977.66

77) \$90,036.92 78) \$808,637.86 79) \$59,134.40

80) \$87,972.85 81) 40.8 mg 82) 814

Section 10.3

1) a must be a positive real number that is not equal to 1.

2) x must be a positive real number. 3) 10 4) common

5) $7^2 = 49$ 6) $11^2 = 121$ 7) $2^3 = 8$ 8) $2^5 = 32$

9) $9^{-2} = \dfrac{1}{81}$ 10) $8^{-2} = \dfrac{1}{64}$ 11) $10^6 = 1,000,000$

12) $10^4 = 10,000$ 13) $25^{1/2} = 5$ 14) $64^{1/3} = 4$

15) $13^1 = 13$ 16) $9^0 = 1$ 17) $\log_9 81 = 2$

18) $\log_{12} 144 = 2$ 19) $\log_{10} 100 = 2$ 20) $\log_{10} 1000 = 3$

21) $\log_3 \dfrac{1}{81} = -4$ 22) $\log_2 \dfrac{1}{32} = -5$ 23) $\log_{10} 1 = 0$

24) $\log_{10} 10 = 1$ 25) $\log_{169} 13 = \dfrac{1}{2}$ 26) $\log_{27} 3 = \dfrac{1}{3}$

27) $\log_9 3 = \dfrac{1}{2}$ 28) $\log_{64} 8 = \dfrac{1}{2}$ 29) $\log_{64} 4 = \dfrac{1}{3}$

30) $\log_{81} 3 = \dfrac{1}{4}$

31) Write the equation in exponential form, then solve for the variable.

32) This is incorrect because the base of a logarithm must be positive. The solution set is $\{3\}$.

33) Rewrite in exponential form.; $64 = x$; $\{64\}$

34) $5^{-3} = t$; $\dfrac{1}{125} = t$; $\left\{\dfrac{1}{125}\right\}$

35) $\{121\}$ 36) $\{125\}$ 37) $\{64\}$ 38) $\{16\}$ 39) $\{100,000\}$

40) $\{100\}$ 41) $\{7\}$ 42) $\{2\}$ 43) $\left\{\dfrac{1}{36}\right\}$ 44) $\left\{\dfrac{1}{64}\right\}$

45) $\{14\}$ 46) $\{7\}$ 47) $\left\{\dfrac{15}{2}\right\}$ 48) $\left\{\dfrac{25}{3}\right\}$ 49) $\left\{\dfrac{1}{8}\right\}$

50) $\left\{\dfrac{1}{6}\right\}$ 51) $\left\{\dfrac{1}{6}\right\}$ 52) $\left\{\dfrac{1}{10}\right\}$ 53) $\{12\}$ 54) $\{4\}$

55) $\{4\}$ 56) $\{8\}$ 57) $\{2\}$ 58) $\{3\}$ 59) 2 60) 2

61) 5 62) 3 63) 2 64) 3 65) $\dfrac{1}{2}$ 66) $\dfrac{1}{2}$ 67) -1

68) -1 69) 1 70) 0 71) -2 72) -3

73) Replace $f(x)$ with y, write $y = \log_a x$ in exponential form, make a table of values, then plot the points and draw the curve.

74) domain: $(0, \infty)$; range: $(-\infty, \infty)$

75)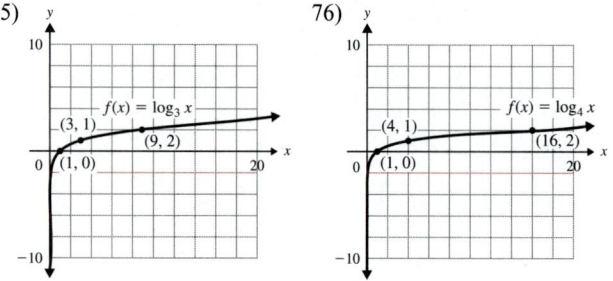

76)

77)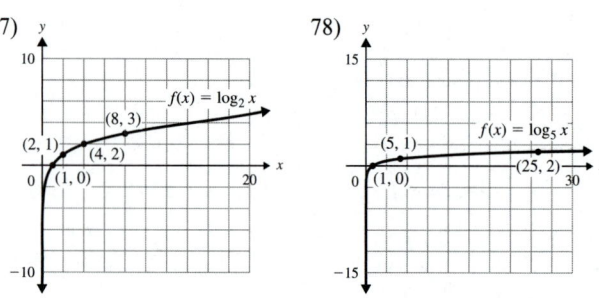

78)

79)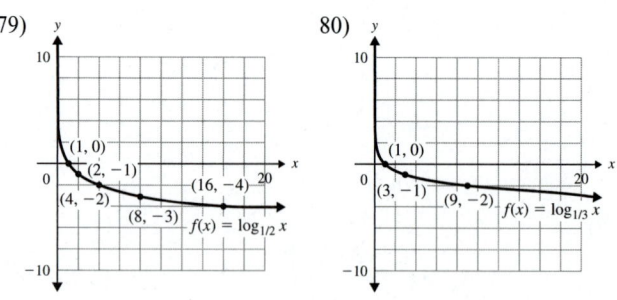

80)

81)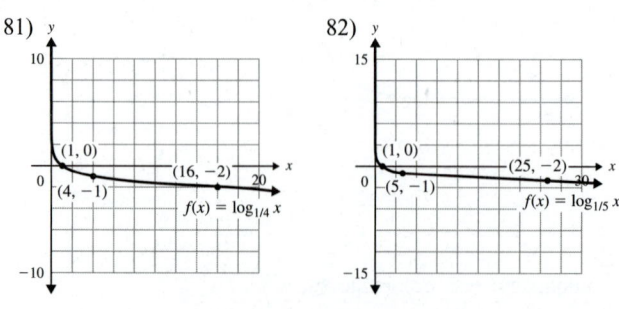

82)

83) $f^{-1}(x) = \log_3 x$ 84) $f^{-1}(x) = \log_4 x$ 85) $f^{-1}(x) = 2^x$

86) $f^{-1}(x) = 5^x$ 87) a) 1868   b) 2004   c) 2058

88) a) 142   b) 258   c) 2009   89) a) 14,000   b) 28,000
 c) It is 1000 more than what was predicted by the formula.

90) a) 11,200,000   b) 13,600,000
 c) It is 80,000 less than what would have been expected
 according to the formula.

Section 10.4

1) true   2) true   3) false   4) true   5) false   6) false

7) true   8) true   9) Product rule; $2 + \log_5 y$

10) Quotient rule; $4 - 2\log_3 n$ 11) $\log_8 3 + \log_8 10$

12) $\log_2 6 + \log_2 5$ 13) $\log_7 5 + \log_7 d$ 14) $\log_4 6 + \log_4 w$

15) $\log_9 4 - \log_9 7$ 16) $\log_5 20 - \log_5 17$ 17) $3\log_5 2$

18) $4\log_8 10$ 19) $8\log p$ 20) $5\log_3 z$ 21) $\dfrac{1}{2}\log_3 7$

22) $\dfrac{1}{3}\log_7 4$ 23) $2 + \log_5 t$ 24) $4 + \log_2 p$ 25) $3 - \log_2 k$

26) $(\log_3 x) - 2$ 27) 6   28) 24   29) $3 + \log b$

30) $3 + \log_3 m$ 31) 35   32) 9   33) $\dfrac{1}{2}$ 34) $\dfrac{1}{3}$ 35) $\dfrac{2}{3}$

36) $\dfrac{3}{2}$ 37) $4\log_6 w + 3\log_6 z$ 38) $2\log_5 x + \log_5 y$

39) $2\log_7 a - 5\log_7 b$ 40) $4\log_4 s - 6\log_4 t$

41) $\dfrac{1}{5}\log 11 - 2\log y$ 42) $\dfrac{1}{2}\log_3 x - 4\log_3 y$

43) $2 + \dfrac{1}{2}\log_2 n - 3\log_2 m$ 44) $\log_9 g + 2\log_9 f - 3\log_9 h$

45) $3\log_4 x - \log_4 y - 2\log_4 z$ 46) $\log 3 - \log a - 2\log b$

47) $\dfrac{1}{2} + \dfrac{1}{2}\log_5 c$ 48) $\left(\dfrac{1}{3}\log_8 z\right) - \dfrac{1}{3}$ 49) $\log k + \log(k-6)$

50) $5\log_2 m - \log_2(m^2 + 3)$ 51) Power rule; $\log_6 x^2 y$

52) $\log 2^5 + \log c - \log d^3$; $2^5 = 32$;
 $\log 32\,c - \log d^3$; Quotient rule

53) $\log_a mn$ 54) $\log_4 7x$ 55) $\log_7 \dfrac{d}{3}$ 56) $\log_p \dfrac{r}{s}$

57) $\log_3 f^4 g$ 58) $\log_y m^5 n^2$ 59) $\log_8 \dfrac{tu^2}{v^3}$ 60) $\log \dfrac{a^3 c^4}{b^6}$

61) $\log \dfrac{r^2 + 3}{(r^2 - 3)^2}$ 62) $\log_2 \dfrac{t^2}{(5t+1)^3}$ 63) $\log_n 8\sqrt{k}$

64) $\log_z 81\sqrt[3]{w}$ 65) $\log_d \dfrac{\sqrt[3]{5}}{z^2}$ 66) $\log_5 \dfrac{\sqrt{a}}{b^4}$ 67) $\log_6 \dfrac{y}{3z^3}$

68) $\log_7 \dfrac{8}{x^4 y}$ 69) $\log_3 \dfrac{t^4}{36u^2}$ 70) $\log_9 \dfrac{m^2}{16n^4}$

71) $\log_b \dfrac{\sqrt{c+4}}{(c+3)^2}$ 72) $\log_a \dfrac{\sqrt{r(r-2)}}{r+2}$ 73) $-\log(a^2 - b^2)$

74) $\log_n(x^2 + xy + y^2)$ 75) 1.6532   76) 1.3980   77) 1.9084

78) 0.2552   79) -0.2552 80) 0.3495   81) 0.4771

82) -0.9542 83) -0.6990 84) 5.5920   85) -1.9084

86) 1.9542   87) 1.6990   88) 0.4438

89) No. $\log_a xy$ is defined only if x and y are positive.

90) $a^m \cdot a^n = xy$; $a^{m+n} = xy$; Change to logarithmic form;
 $\log_a xy = m + n$; $\log_a xy = \log_a x + \log_a y$

Section 10.5

1) e 2) 10   3) 2   4) 4   5) -3 6) -5 7) -1 8) -2

9) 9   10) 7   11) $\dfrac{1}{4}$ 12) $\dfrac{1}{5}$ 13) 6   14) 10   15) $\dfrac{1}{2}$ 16) $\dfrac{1}{3}$

17) -5 18) -2 19) 0   20) 0   21) 1.2041   22) 1.3617

23) -0.3010 24) 2.7973   25) 1.0986   26) 1.7918

27) 0.2700   28) -1.5233 29) $\{1000\}$ 30) $\{100,000\}$

31) $\left\{\dfrac{1}{10}\right\}$ 32) $\left\{\dfrac{1}{100}\right\}$ 33) $\{25\}$ 34) $\{2\}$ 35) $\{2\}$

36) $\{44\}$ 37) $\{10^{1.5}\}$; $\{31.6228\}$ 38) $\{10^{1.8}\}$; $\{63.0957\}$

39) $\{10^{0.8}\}$; $\{6.3096\}$ 40) $\{10^{0.3}\}$; $\{1.9953\}$

41) $\{e^{1.6}\}$; $\{4.9530\}$ 42) $\{e^{1.1}\}$; $\{3.0042\}$

43) $\left\{\dfrac{1}{e^2}\right\}$; $\{0.1353\}$ 44) $\{e^{0.25}\}$; $\{1.2840\}$

45) $\left\{\dfrac{e^{2.1}}{3}\right\}$; $\{2.7221\}$ 46) $\{4e^3\}$; $\{80.3421\}$

47) $\{2 \cdot 10^{0.47}\}$; $\{5.9024\}$ 48) $\left\{\dfrac{1}{60}\right\}$

49) $\left\{\dfrac{3 + 10^{3.8}}{5}\right\}$; $\{1262.5147\}$ 50) $\left\{\dfrac{10^{2.7} - 15}{8}\right\}$; $\{60.7734\}$

51) $\left\{\dfrac{e^{1.85} - 19}{10}\right\}$; $\{-1.2640\}$ 52) $\left\{\dfrac{4 + e^{0.6}}{7}\right\}$; $\{0.8317\}$

53) $\{3\}$ 54) $\left\{\dfrac{10^{2.4} - 14}{3}\right\}$; $\{79.0629\}$

55)

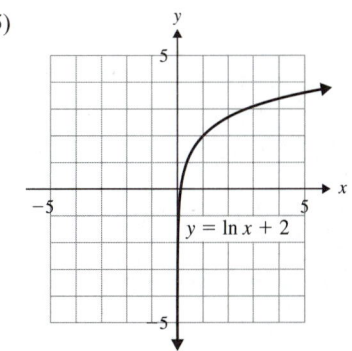

$y = \ln x + 2$

domain: $(0, \infty)$; range: $(-\infty, \infty)$

56)

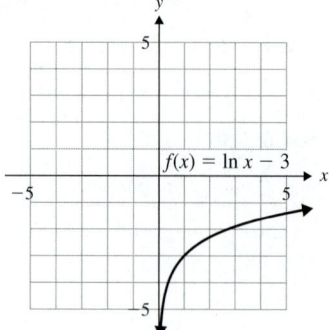

domain: $(0, \infty)$; range: $(-\infty, \infty)$

57)

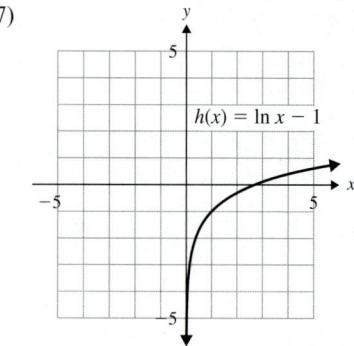

domain: $(0, \infty)$; range: $(-\infty, \infty)$

58)

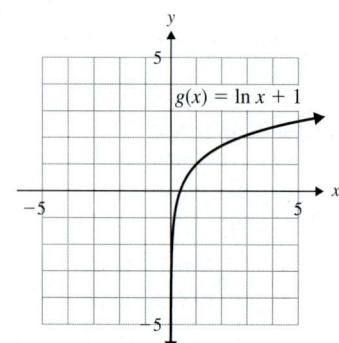

domain: $(0, \infty)$; range: $(-\infty, \infty)$

59)

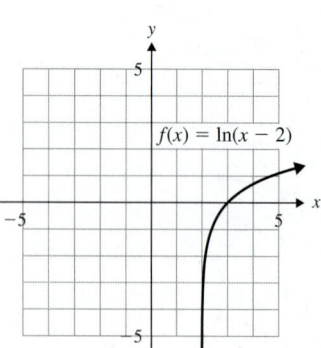

domain: $(2, \infty)$; range: $(-\infty, \infty)$

60)

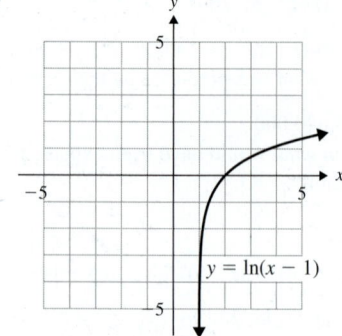

domain: $(1, \infty)$; range: $(-\infty, \infty)$

61)

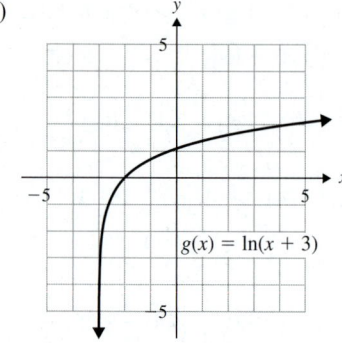

domain: $(-3, \infty)$; range: $(-\infty, \infty)$

62)

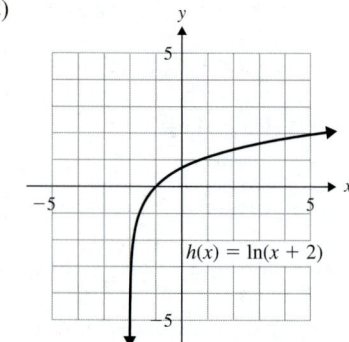

domain: $(-2, \infty)$; range: $(-\infty, \infty)$

63)

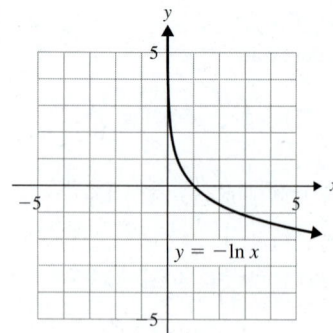

domain: $(0, \infty)$; range: $(-\infty, \infty)$

64)

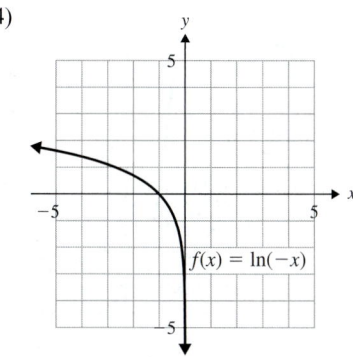

domain: $(-\infty, 0)$; range: $(-\infty, \infty)$

65)

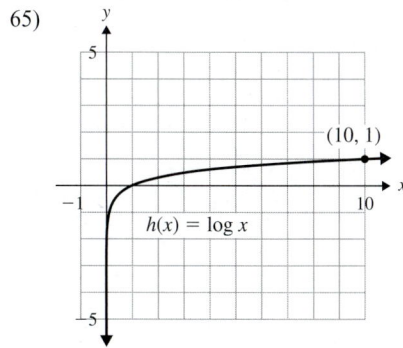

domain: $(0, \infty)$; range: $(-\infty, \infty)$

66)

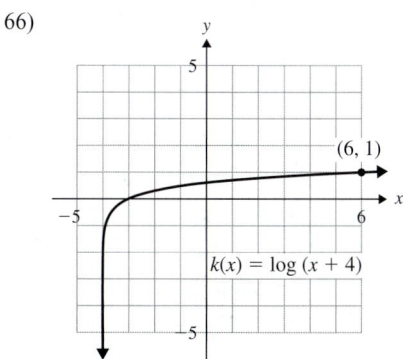

domain: $(-4, \infty)$; range: $(-\infty, \infty)$

67) Shift the graph of $f(x)$ left 5 units.

68) Shift the graph of $f(x)$ up 4 units.   69) 3.7004

70) 1.7965   71) 1.9336   72) 3.5966   73) -2.5237

74) -4.5236 75) 0.6826   76) 0.7124   77) 110 dB

78) 60 dB   79) 40 dB   80) 180 dB   81) $3484.42

82) $8249.64   83) $5521.68   84) $6667.46   85) $3485.50

86) $8262.77   87) $15,683.12   88) $6322.26

89) a) 5000   b) 8191   90) a) 8000   b) 11,262   91) 32,570

92) 23,565   93) 2.7; acidic   94) 4.0; acidic   95) 11.2; basic

96) 7.7; basic   97) Answers may vary.

Section 10.6

1) $\{2\}$ 2) $\{3\}$ 3) $\left\{\dfrac{\ln 15}{\ln 7}\right\}$; $\{1.3917\}$

4) $\left\{\dfrac{\ln 38}{\ln 5}\right\}$; $\{2.2602\}$ 5) $\left\{\dfrac{\ln 3}{\ln 8}\right\}$; $\{0.5283\}$

6) $\left\{\dfrac{\ln 9}{\ln 4}\right\}$; $\{1.5850\}$ 7) $\left\{\dfrac{2}{5}\right\}$ 8) $\left\{\dfrac{5}{3}\right\}$

9) $\left\{\dfrac{\ln 2.7}{6 \ln 4}\right\}$; $\{0.1194\}$ 10) $\left\{\dfrac{\ln 7.8}{2 \ln 3}\right\}$; $\{0.9349\}$

11) $\left\{\dfrac{\ln 5 - \ln 2}{4 \ln 2}\right\}$; $\{0.3305\}$ 12) $\left\{\dfrac{\ln 13 - \ln 6}{2 \ln 6}\right\}$; $\{0.2158\}$

13) $\left\{\dfrac{\ln 8 + 2 \ln 5}{3 \ln 5}\right\}$; $\{1.0973\}$

14) $\left\{\dfrac{\ln 14 + 3 \ln 3}{2 \ln 3}\right\}$; $\{2.7011\}$ 15) $\left\{\dfrac{10}{7}\right\}$ 16) $\left\{\dfrac{6}{7}\right\}$

17) $\left\{\dfrac{2 \ln 9}{5 \ln 9 - 3 \ln 4}\right\}$; $\{0.6437\}$

18) $\left\{\dfrac{7 \ln 5}{2 \ln 8 - 4 \ln 5}\right\}$; $\{-4.9437\}$ 19) $\{\ln 12.5\}$; $\{2.5257\}$

20) $\{\ln 0.36\}$; $\{-1.0217\}$ 21) $\left\{-\dfrac{\ln 9}{4}\right\}$; $\{-0.5493\}$

22) $\left\{\dfrac{\ln 4}{3}\right\}$; $\{0.4621\}$ 23) $\left\{\dfrac{\ln 2}{0.01}\right\}$; $\{69.3147\}$

24) $\left\{-\dfrac{\ln 10}{0.08}\right\}$; $\{-28.7823\}$ 25) $\left\{\dfrac{\ln 3}{0.006}\right\}$; $\{183.1021\}$

26) $\left\{\dfrac{\ln 12}{0.04}\right\}$; $\{62.1227\}$ 27) $\left\{-\dfrac{\ln 5}{0.4}\right\}$; $\{-4.0236\}$

28) $\left\{-\dfrac{\ln 16}{0.005}\right\}$; $\{-554.5177\}$ 29) $\{2\}$ 30) $\{6\}$

31) $\left\{\dfrac{10}{3}\right\}$ 32) $\left\{\dfrac{8}{5}\right\}$ 33) $\{5\}$ 34) $\{2\}$ 35) $\{2, 10\}$

36) $\{3, 8\}$ 37) $\varnothing$ 38) $\varnothing$ 39) $\{8\}$ 40) $\left\{\dfrac{11}{2}\right\}$ 41) $\{2\}$

42) $\{3\}$ 43) $\{9\}$ 44) $\{8\}$ 45) $\{2\}$ 46) $\{1\}$ 47) $\{4\}$

48) $\{9\}$ 49) $\left\{\dfrac{2}{3}\right\}$ 50) $\{4, 9\}$ 51) a) 3.72 yr   b) 11.55 yr

52) a) 2.60 yr   b) 9.90 yr   53) 1.44 yr   54) 2.06 yr

55) $2246.64   56) $3371.78   57) 7.2%   58) 7.3%

59) a) 6 hr   b) 18.5 hr   60) a) 9.7 hr   b) 16.6 hr

61) 28,009   62) 37,532   63) a) 2032   b) 2023

64) a) 1616   b) 2060   65) a) 11.78 g   b) 3351 yr   c) 5728 yr

66) a) 6.93 g   b) 16,376 yr   c) 24,151 yr

67) a) 0.4 units   b) 0.22 units   68) a) 30 g   b) 8.4 yr

69) $\{16\}$ 70) $\{1000\}$ 71) $\{-2, 2\}$ 72) $\{107\}$

73) $\{-\ln 13, \ln 13\}; \{-2.5649, 2.5649\}$ 74) $\{-5, 5\}$

75) $\{0\}$ 76) $\{0, \ln 8\}; \{0, 2.0794\}$ 77) $\left\{\dfrac{\ln 7}{\ln 5}\right\}; \{1.2091\}$

78) $\left\{\dfrac{1}{2}\right\}$ 79) $\{1, 1000\}$ 80) $\{0, \log 6\}; \{0, 0.7782\}$

Chapter 10 Review Exercises

1) a) $2x + 3$ b) $2x - 3$ c) 7

2) a) $4x^2 + 6x - 8$ b) $2x^2 + 10x - 9$ c) -21

3) a) $(N \circ G)(h) = 9.6h$. This is Antoine's net pay in terms of how many hours he has worked.
   b) $(N \circ G)(30) = 288$. When Antoine works 30 hr, his net pay is \$288.
   c) \$384

4) a) $f(x) = x^2, g(x) = 3x + 10$; answers may vary.
   b) $f(x) = x^3, g(x) = 8x - 7$; answers may vary.
   c) $f(x) = \sqrt{x}, g(x) = x^2 + 6$; answers may vary.
   d) $f(x) = \dfrac{1}{x}, g(x) = 2 - 5x$; answers may vary.

5) yes; $\{(-4, -7), (1, -2), (5, 1), (11, 6)\}$ 6) no

7) yes   8) no

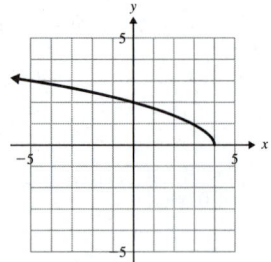

9) $f^{-1}(x) = x - 4$ 10) $g^{-1}(x) = \dfrac{1}{2}x + 5$

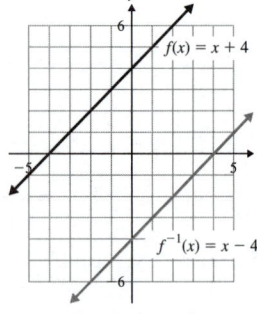   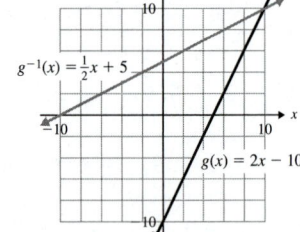

11) $h^{-1}(x) = 3x + 3$ 12) $f^{-1}(x) = (x - 2)^3$

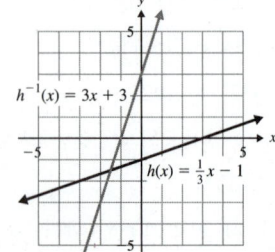   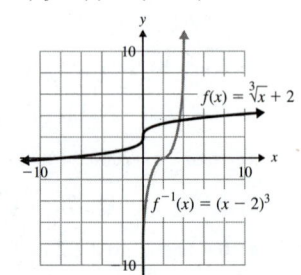

13) a) 11   b) 2   14) a) -2 b) -13

15)   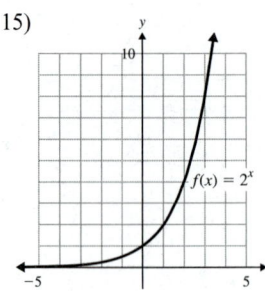   16)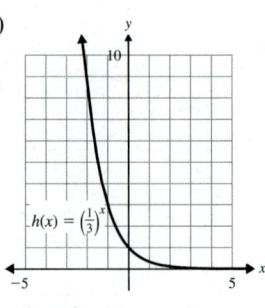

domain: $(-\infty, \infty)$;   domain: $(-\infty, \infty)$;
range: $(0, \infty)$ range: $(0, \infty)$

17)   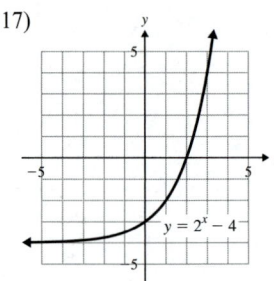   18)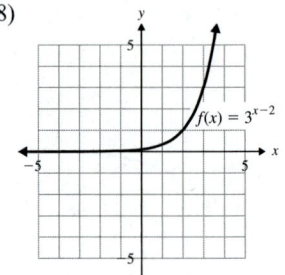

domain: $(-\infty, \infty)$;   domain: $(-\infty, \infty)$;
range: $(-4, \infty)$ range: $(0, \infty)$

19)   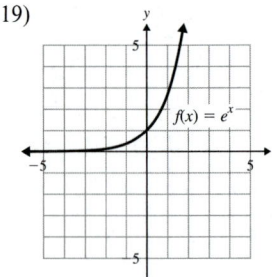   20)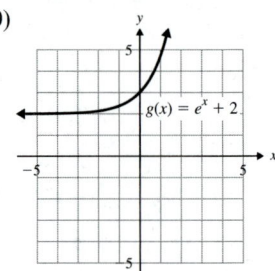

domain: $(-\infty, \infty)$;   domain: $(-\infty, \infty)$;
range: $(0, \infty)$ range: $(2, \infty)$

21) $\{6\}$ 22) $\{-3\}$ 23) $\left\{-\dfrac{5}{2}\right\}$ 24) $\{-2\}$ 25) $\left\{\dfrac{2}{3}\right\}$

26) a) \$38,200   b) \$16,936.51   27) $(0, \infty)$

28) a must be a positive real number that is not equal to 1.

29) $5^3 = 125$ 30) $16^{-1/2} = \dfrac{1}{4}$ 31) $10^2 = 100$ 32) $10^0 = 1$

33) $\log_3 81 = 4$ 34) $\log_{2/3}\dfrac{9}{4} = -2$ 35) $\log 1000 = 3$

36) $\log_{121} 11 = \dfrac{1}{2}$ 37) $\{8\}$ 38) $\{20\}$ 39) $\left\{\dfrac{4}{5}\right\}$

40) $\left\{\dfrac{5}{2}\right\}$ 41) 2   42) 3   43) 3   44) 0   45) -4 46) 2

47)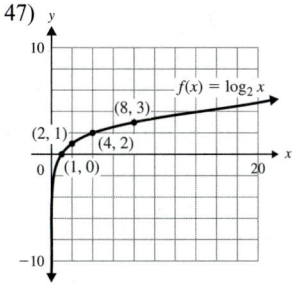

48)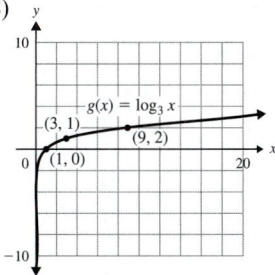

49)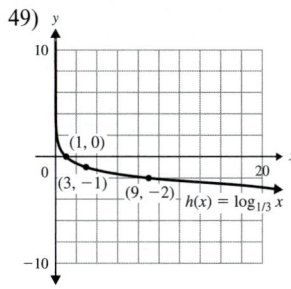

50)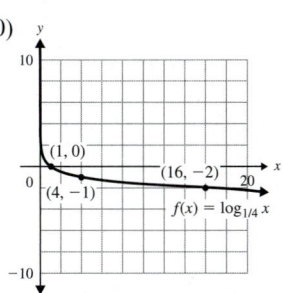

103)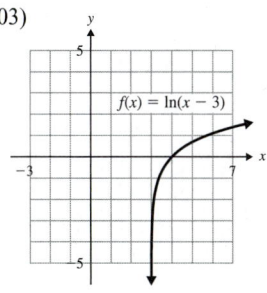

domain: $(3, \infty)$;
range: $(-\infty, \infty)$

104)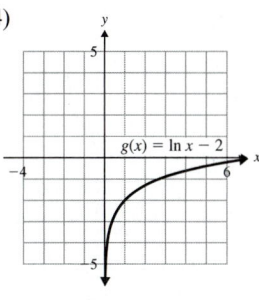

domain: $(0, \infty)$;
range: $(-\infty, \infty)$

105) 2.1240   106) 1.7011   107) -5.2479 108) -0.1108

109) 110 dB   110) 100 W/m^2 111) \$3367.14

112) \$27,361.90   113) \$11,533.14   114) \$6003.21

115) a) 6000   b) 11,118   116) 7.4   117) $\{4\}$

118) $\left\{\dfrac{\ln 7}{\ln 3}\right\}$; $\{1.7712\}$ 119) $\left\{\dfrac{\ln 2}{4 \ln 9}\right\}$; $\{0.0789\}$

120) $\left\{\dfrac{14}{5}\right\}$ 121) $\left\{\dfrac{5 \ln 8}{\ln 8 - 2 \ln 6}\right\}$; $\{-6.9127\}$

122) $\{\ln 22\}$; $\{3.0910\}$ 123) $\left\{\dfrac{\ln 8}{5}\right\}$; $\{0.4159\}$

124) $\left\{\dfrac{\ln 19}{0.03}\right\}$; $\{98.1480\}$ 125) $\left\{\dfrac{6}{5}\right\}$ 126) $\{335\}$

127) $\{4\}$ 128) $\varnothing$ 129) $\{16\}$ 130) $\{3\}$ 131) \$6770.57

132) 8.1%   133) a) 17,777   b) 2011

134) a) 64.3 g   b) 3180 yr   c) 1590 yr

Chapter 10 Test

51) $f^{-1}(x) = \log_5 x$ 52) $g^{-1}(x) = \log_3 x$ 53) $h^{-1}(x) = 6^x$

54) a) 10,000   b) 20,000   55) false   56) true

57) $\log_8 3 + \log_8 z$ 58) $2 - \log_7 t$ 59) $\dfrac{3}{2}$ 60) -2

61) $4 \log_5 c + 3 \log_5 d$ 62) $\log_4 m + \dfrac{1}{2} \log_4 n$

63) $\log_a x + \log_a y - 3 \log_a z$ 64) $2 \log_4 a - \log_4 b - 4 \log_4 c$

65) $\log p + \log(p + 8)$ 66) $3 \log_6 r - \log_6(r^2 - 5)$

67) $\log cd$ 68) $\log_4 \dfrac{n}{7}$ 69) $\log_2 a^9 b^3$ 70) $\log_5 \dfrac{r}{t^2}$

71) $\log_3 \dfrac{5m^4}{n^2}$ 72) $\log_z \dfrac{\sqrt{a}}{b}$ 73) $\log_5 \dfrac{c^3}{df^2}$ 74) $\log_6 x^2 \sqrt[3]{x - 4}$

75) 1.6902   76) 1.7993   77) -0.1091 78) -0.8451

79) e 80) 1   81) 1   82) 2   83) $\dfrac{1}{2}$ 84) -2 85) -3

86) 4   87) 0   88) $\dfrac{1}{3}$ 89) 0.9031   90) -0.5229

91) 0.5596   92) -0.0790 93) $\{100\}$ 94) $\{200\}$ 95) $\left\{\dfrac{1}{5}\right\}$

96) $\left\{\dfrac{5}{2}\right\}$ 97) $\{10^{2.1}\}$; $\{125.8925\}$ 98) $\{10^{-1.4}\}$; $\{0.0398\}$

99) $\{e^2\}$; $\{7.3891\}$ 100) $\{e^{-0.5}\}$; $\{0.6065\}$

101) $\left\{\dfrac{10^{1.75}}{4}\right\}$; $\{14.0585\}$ 102) $\left\{\dfrac{e + 3}{2}\right\}$; $\{2.8591\}$

1) a) $2x^2 + 10x + 1$ b) $4x^2 + 38x + 81$ c) 3

2) $f(x) = x^3, g(x) = 9x - 7$; answers may vary.

3) no   4) yes; $g^{-1} = \left\{(4, 2), (6, 6), \left(\dfrac{15}{2}, 9\right), (10, 14)\right\}$

5) yes   6) $f^{-1}(x) = -\dfrac{1}{3}x + 4$

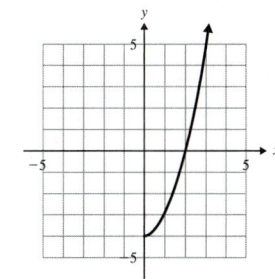

7)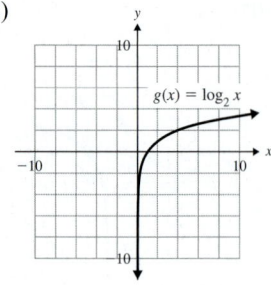

8)

19)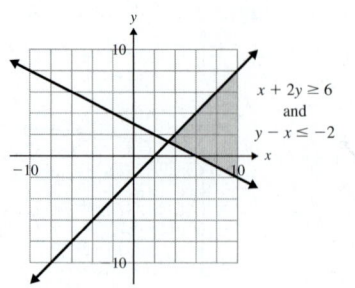

9) a) $(0, \infty)$ b) $(-\infty, \infty)$ 10) They are inverses.

11) $\log_3 \dfrac{1}{9} = -2$ 12) $\left\{\dfrac{1}{2}\right\}$ 13) $\{-2\}$ 14) $\{125\}$

15) $\{29\}$ 16) $\{4\}$ 17) a) 4   b) $\dfrac{1}{2}$

18) $\{1\}$ 19) $\log_8 5 + \log_8 n$

20) $2 + 4\log_3 a - 5\log_3 b - \log_3 c$

21) $\log \dfrac{x^2}{(x+1)^3}$ 22) $\{10^{0.08}\}$; $\{6.3096\}$

23) $\left\{\dfrac{\ln 5}{0.3}\right\}$; $\{5.3648\}$ 24) $\{e^{-0.25}\}$; $\{0.7788\}$

25) $\left\{\dfrac{\ln 9 - 3\ln 4}{4\ln 4}\right\}$; $\{-0.3538\}$

26)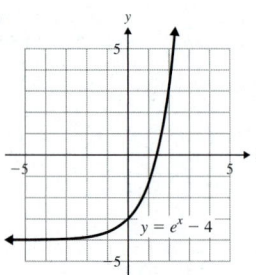

domain: $(-\infty, \infty)$;
range: $(-4, \infty)$

27)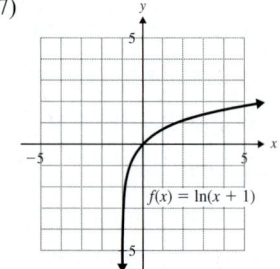

domain: $(-1, \infty)$;
range: $(-\infty, \infty)$

28) 1.7604   29) $8686.41

30) a) 86.2 g   b) 325.1 days   c) 140 days

Cumulative Review for Chapters 1–10

1) 35   2) $-\dfrac{1}{2}$ 3) $-15a^6$ 4) $4z^8$ 5) $\dfrac{d^9}{8c^{30}}$

6) 9.231×10^{-5} 7) $48.00

8) 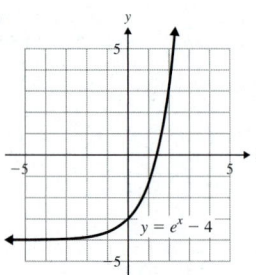 ; $\left(-\dfrac{3}{2}, \infty\right)$

9) $(-6, 1)$ 10) $\left(\dfrac{1}{3}, -2\right)$ 11) $y = -\dfrac{3}{2}x + 2$

12) $3c^2 + 4c - 1$ 13) $(4w + 9)(w - 2)$

14) $(p + 1)(p - 1)(3p + 2)$ 15) $(y - 3)^2$ 16) $\{-8, -6\}$

17) $\dfrac{r^2 - 12r + 21}{(r + 7)(r - 7)(r - 9)}$ 18) $\{2\}$

20) $2\sqrt{30}$ 21) $3t^4\sqrt{5t}$ 22) $6a$ 23) 9   24) $\varnothing$

25) $13 - 19i$ 26) $\{4 - 2\sqrt{3}, 4 + 2\sqrt{3}\}$

27) $\left\{-\dfrac{5}{2} - \dfrac{\sqrt{17}}{2}, -\dfrac{5}{2} + \dfrac{\sqrt{17}}{2}\right\}$ 28) $\{5 + 4i, 5 - 4i\}$

29) $\left\{-2, -\dfrac{1}{2}, \dfrac{1}{2}, 2\right\}$ 30) $\left(-\infty, \dfrac{2}{3}\right) \cup \left(\dfrac{2}{3}, \infty\right)$

31) $\dfrac{32}{243}$ 32)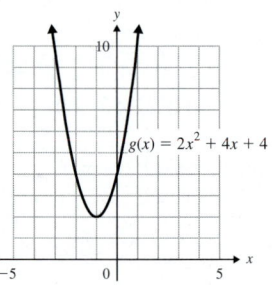

33) a) 9   b) $x^2 - 12x + 29$ c) -4

34)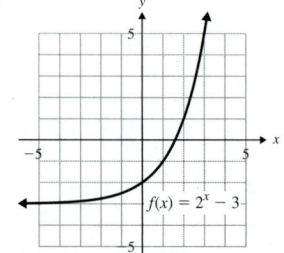

domain: $(-\infty, \infty)$;   range: $(-3, \infty)$

35) $\left\{-\dfrac{3}{2}\right\}$ 36) $\{3\}$ 37) $\log \dfrac{ab^2}{c^5}$ 38) $\{4\}$

39) $\left\{-\dfrac{\ln 6}{0.04}\right\}$; $\{-44.7940\}$

40)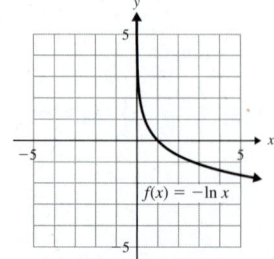

domain: $(0, \infty)$;   range: $(-\infty, \infty)$

Chapter 11

Section 11.1

1)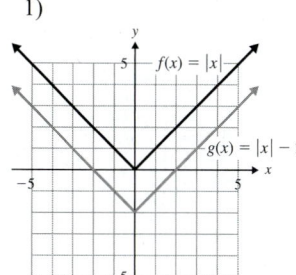

2)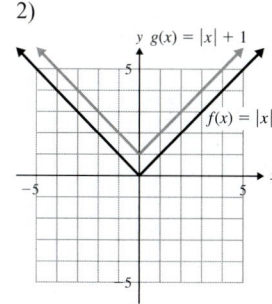

3)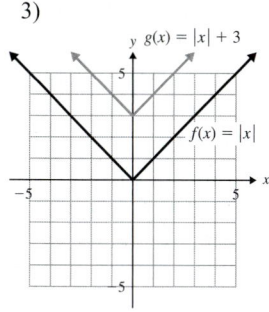

4)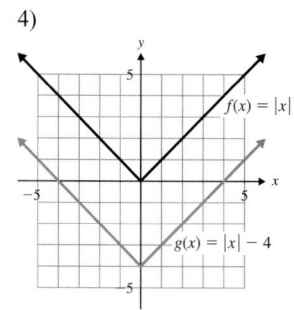

5)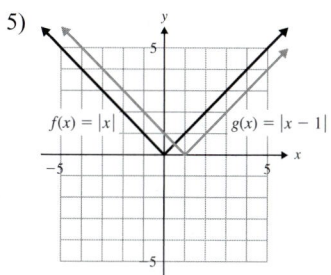

6)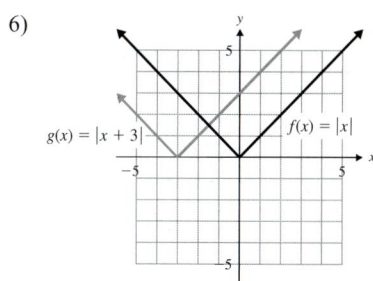

7)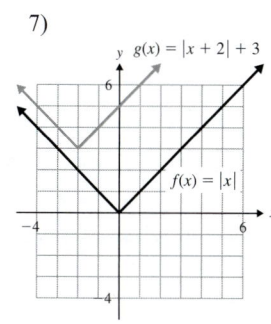

8)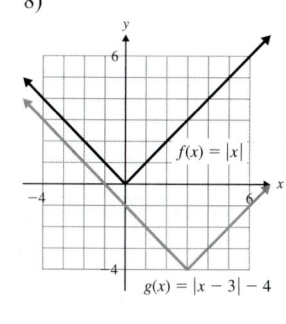

9)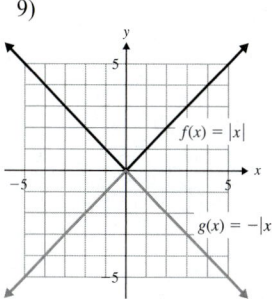

10)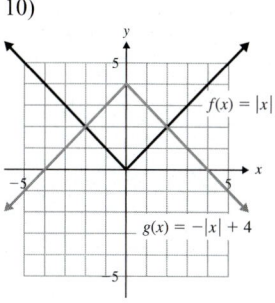

11)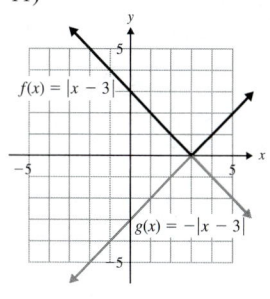

12)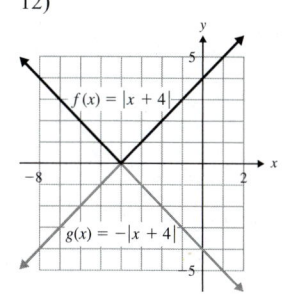

13) a) $g(x)$ b) $k(x)$ c) $h(x)$ d) $f(x)$

14) $g(x) = |x + 2| - 1$ 15) $g(x) = |x - 1| + 4$

16) $g(x) = -|x|$

17)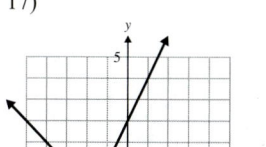

18)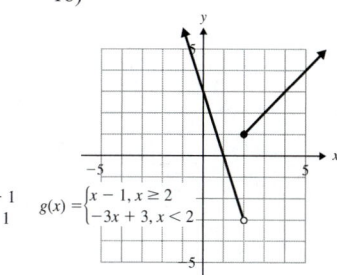

19)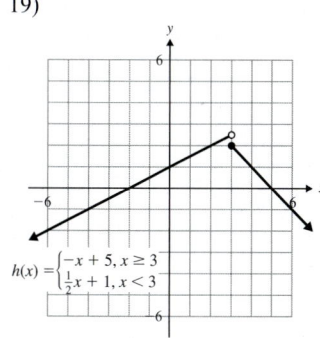

20)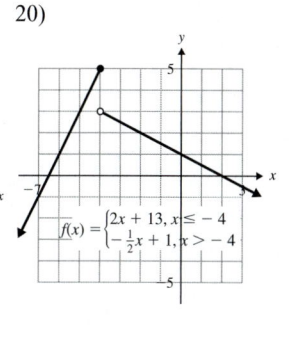

21)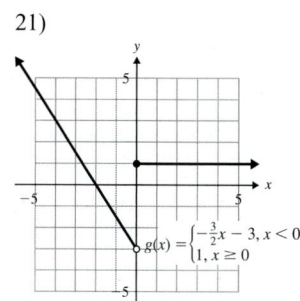

22)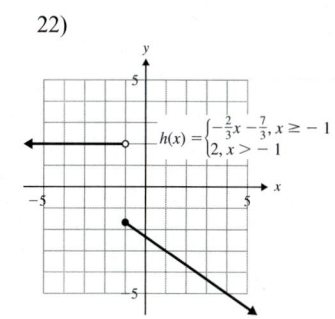

23)

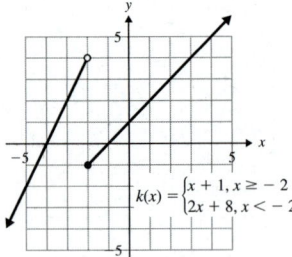

$$k(x) = \begin{cases} x + 1, x \ge -2 \\ 2x + 8, x < -2 \end{cases}$$

24)

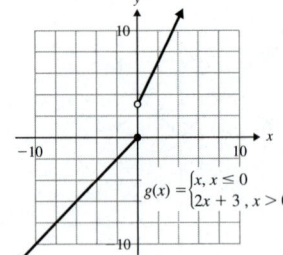

$$g(x) = \begin{cases} x, x \le 0 \\ 2x + 3, x > 0 \end{cases}$$

25)

$$f(x) = \begin{cases} 2x - 4, x > 1 \\ -\frac{1}{3}x - \frac{5}{3}, x \le 1 \end{cases}$$

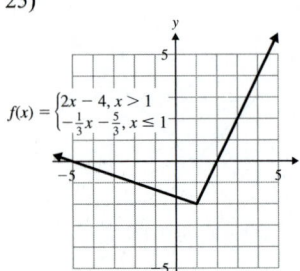

26)

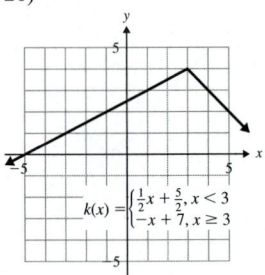

$$k(x) = \begin{cases} \frac{1}{2}x + \frac{5}{2}, x < 3 \\ -x + 7, x \ge 3 \end{cases}$$

27) 3   28) 10   29) 9   30) 7   31) 8   32) 0

33) −7   34) −2   35) −9   36) −4

37)

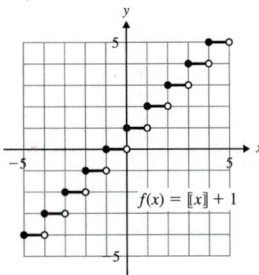

$f(x) = [\![x]\!] + 1$

38)

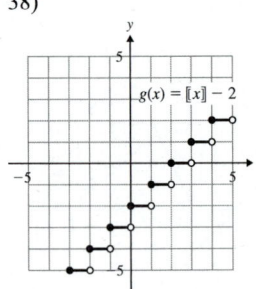

$g(x) = [\![x]\!] - 2$

39)

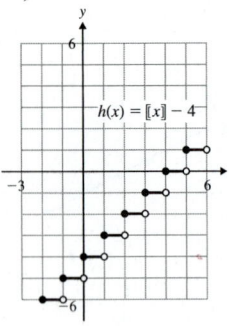

$h(x) = [\![x]\!] - 4$

40)

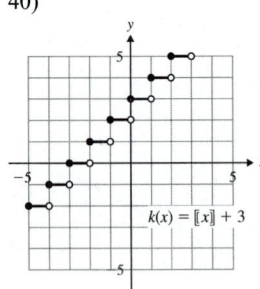

$k(x) = [\![x]\!] + 3$

41)

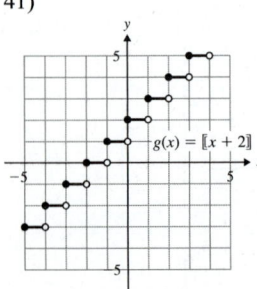

$g(x) = [\![x + 2]\!]$

42)

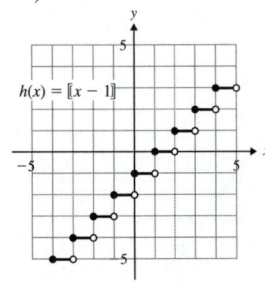

$h(x) = [\![x - 1]\!]$

43)

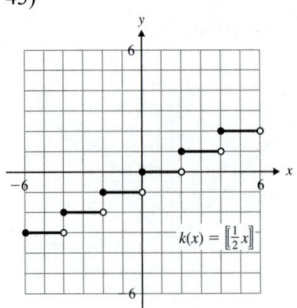

$k(x) = \left[\!\!\left[\frac{1}{2}x\right]\!\!\right]$

44)

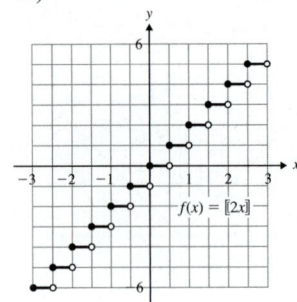

$f(x) = [\![2x]\!]$

45)

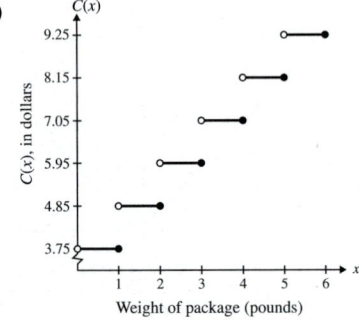

46)

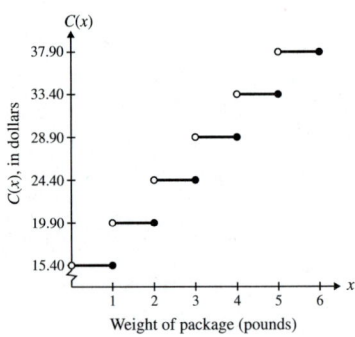

47)

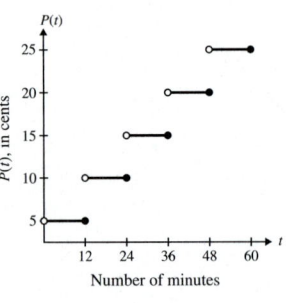

48)

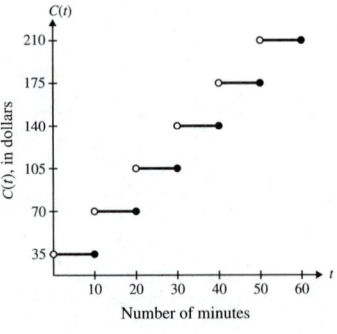

49) a) domain: $(-\infty, \infty)$;
    range: $(-\infty, \infty)$

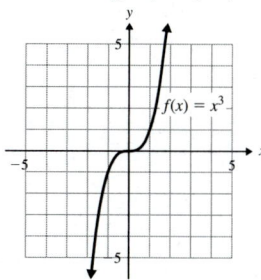

b) domain: $[0, \infty)$;
   range: $[0, \infty)$

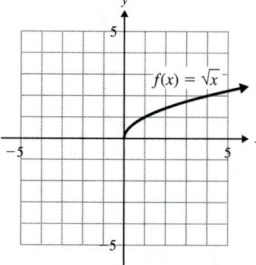

57) domain: $(-\infty, \infty)$;
    range: $[3, \infty)$

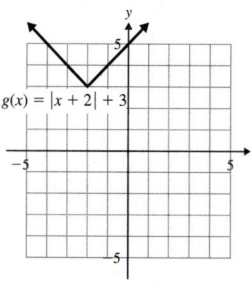

58) domain: $(-\infty, \infty)$;
    range: $[-5, \infty)$

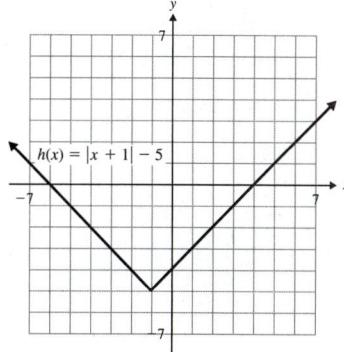

c) domain: $(-\infty, \infty)$;
   range: $(-\infty, \infty)$

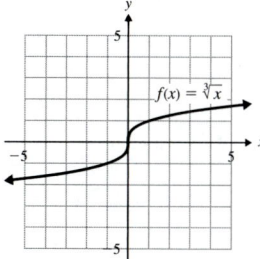

50) domain: $(-\infty, \infty)$;
    range: $(-\infty, \infty)$

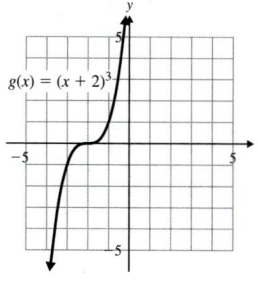

59) domain: $(-\infty, \infty)$;
    range: $[0, \infty)$

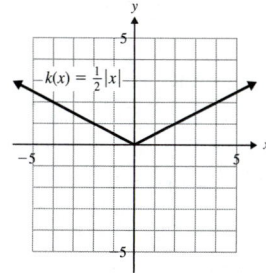

60) domain: $(-\infty, \infty)$;
    range: $[0, \infty)$

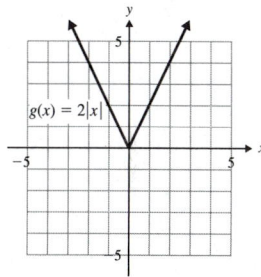

51) domain: $[1, \infty)$;
    range: $[0, \infty)$

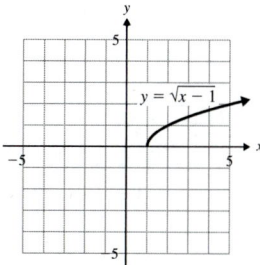

52) domain: $(-\infty, \infty)$;
    range: $(-\infty, \infty)$

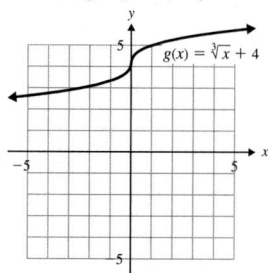

61) domain: $[-4, \infty)$;
    range: $[-2, \infty)$

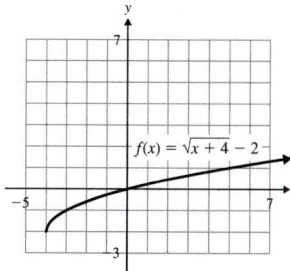

62) domain: $(-\infty, \infty)$;
    range: $(-\infty, \infty)$

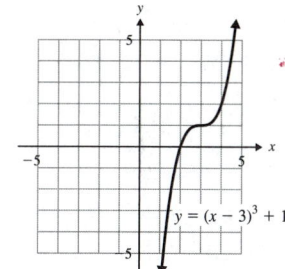

53) domain: $(-\infty, \infty)$;
    range: $(-\infty, \infty)$

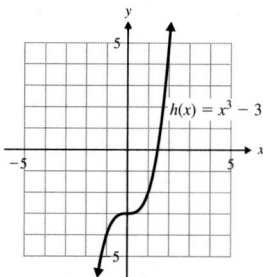

54) domain: $[3, \infty)$;
    range: $[0, \infty)$

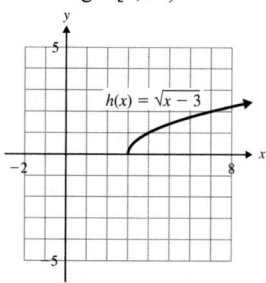

63) domain: $(-\infty, \infty)$;
    range: $(-\infty, \infty)$

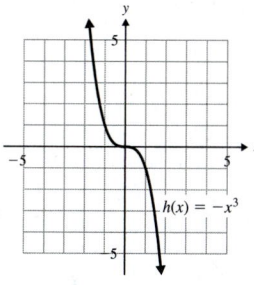

64) domain: $(-\infty, \infty)$;
    range: $(-\infty, \infty)$

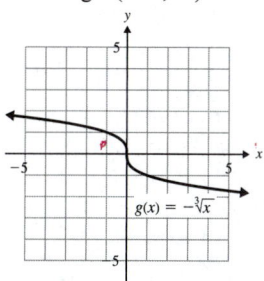

55) domain: $(-\infty, \infty)$;
    range: $(-\infty, \infty)$

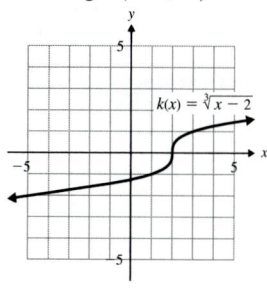

56) domain: $(-\infty, \infty)$;
    range: $(-\infty, \infty)$

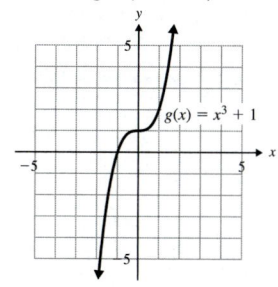

65) $g(x) = \sqrt{x+5}$ 66) $g(x) = \sqrt[3]{x} - 6$

67) $g(x) = (x+2)^3 - 1$ 68) $g(x) = \sqrt{x-1} + 4$

69) $g(x) = -\sqrt[3]{x}$ 70) $g(x) = -x^3$

Section 11.2

1) $(4, 6)$ 2) $(5, 7)$ 3) $(-3, -3)$ 4) $(3, 1)$

5) $\left(-1, -\frac{9}{2}\right)$ 6) $\left(\frac{1}{2}, -3\right)$ 7) $\left(\frac{1}{2}, -\frac{5}{2}\right)$ 8) $\left(\frac{7}{2}, \frac{7}{2}\right)$

9) $\left(2, \dfrac{5}{4}\right)$ 10) $\left(\dfrac{1}{2}, -\dfrac{7}{4}\right)$ 11) $(-0.7, 3.6)$ 12) $(0, -2.7)$

13) No; there are values in the domain that give more than one value in the range. The graph fails the vertical line test.

14) center: (h, k); radius: r

15) center: $(-2, 4)$; $r = 3$

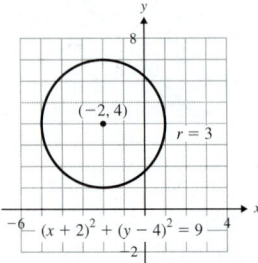

16) center: $(-1, -3)$; $r = 5$

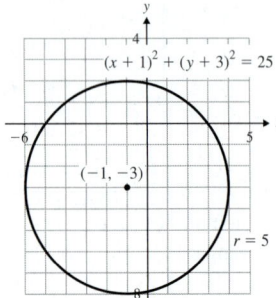

17) center: $(5, 3)$; $r = 1$

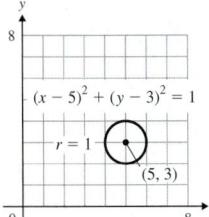

18) center: $(0, 5)$; $r = 3$

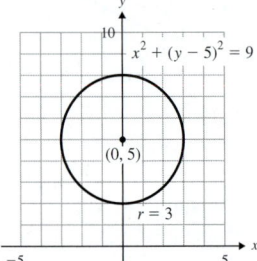

19) center: $(-3, 0)$; $r = 2$

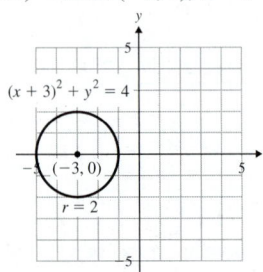

20) center: $(2, 2)$; $r = 6$

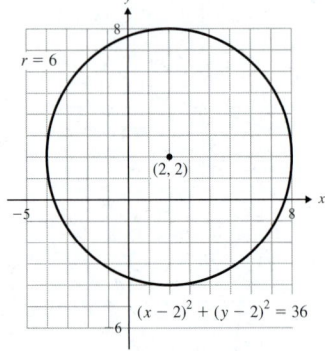

21) center: $(6, -3)$; $r = 4$

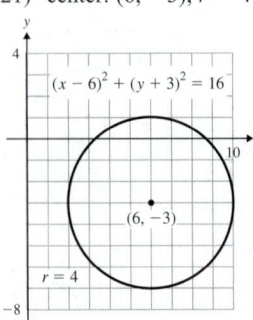

22) center: $(-8, 4)$; $r = 2$

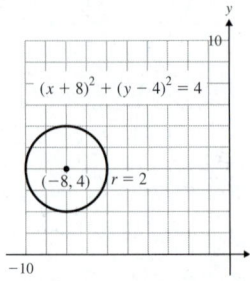

23) center: $(0, 0)$; $r = 6$

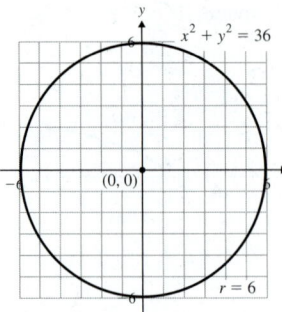

24) center: $(0, 0)$; $r = 4$

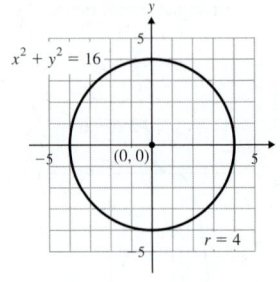

25) center: $(0, 0)$; $r = 3$

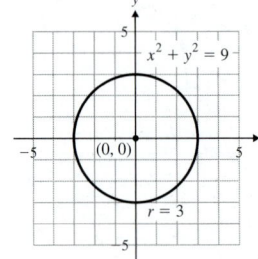

26) center: $(0, 0)$; $r = 5$

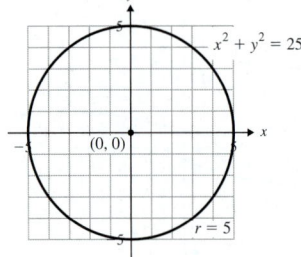

27) center: $(0, 1)$; $r = 5$

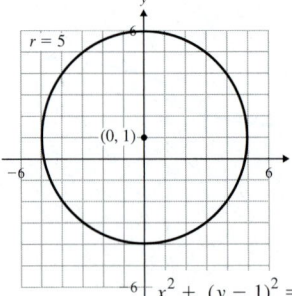

28) center: $(-3, 0)$; $r = 1$

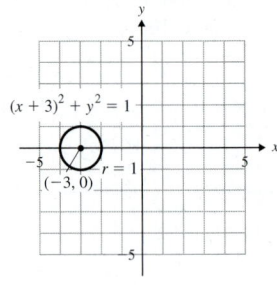

29) $(x - 4)^2 + (y - 1)^2 = 25$ 30) $(x - 3)^2 + (y - 5)^2 = 4$

31) $(x + 3)^2 + (y - 2)^2 = 1$ 32) $(x - 4)^2 + (y + 6)^2 = 9$

33) $(x + 1)^2 + (y + 5)^2 = 3$ 34) $(x + 2)^2 + (y + 1)^2 = 5$

35) $x^2 + y^2 = 10$ 36) $x^2 + y^2 = 6$ 37) $(x - 6)^2 + y^2 = 16$

38) $x^2 + (y + 3)^2 = 25$ 39) $x^2 + (y + 4)^2 = 8$

40) $(x - 1)^2 + y^2 = 18$

41) Group x- and y-terms separately.;
$(x^2 - 8x + 16) + (y^2 + 2y + 1) = -8 + 16 + 1$;
$(x - 4)^2 + (y + 1)^2 = 9$

42) Group x- and y-terms separately.;
$(x^2 + 2x + 1) + (y^2 + 10y + 25) = -10 + 1 + 25$;
$(x + 1)^2 + (y + 5)^2 = 16$

43) $(x + 1)^2 + (y + 5)^2 = 9$;
    center: $(-1, -5)$; $r = 3$

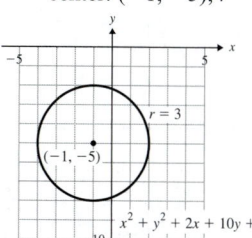

44) $(x - 2)^2 + (y - 3)^2 = 4$;
    center: $(2, 3)$; $r = 2$

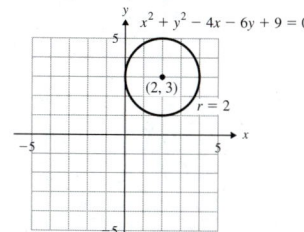

53) $(x - 4)^2 + (y + 4)^2 = 36$;
    center: $(4, -4)$; $r = 6$

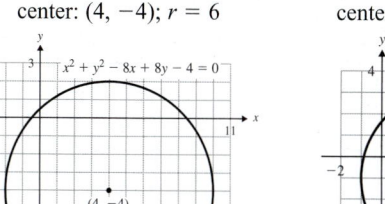

54) $(x - 3)^2 + (y + 1)^2 = 16$;
    center: $(3, -1)$; $r = 4$

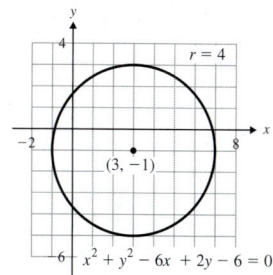

45) $(x + 4)^2 + (y - 1)^2 = 25$;
    center: $(-4, 1)$; $r = 5$

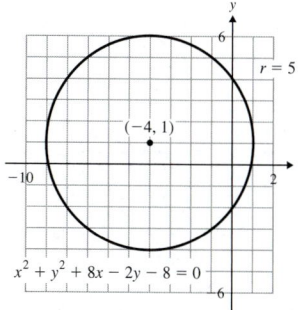

46) $(x - 3)^2 + (y + 4)^2 = 1$;
    center: $(3, -4)$; $r = 1$

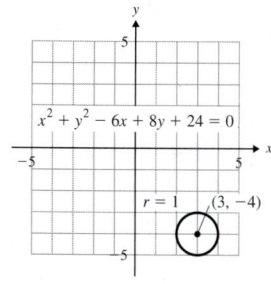

55) $\left(x - \dfrac{3}{2}\right)^2 + \left(y - \dfrac{1}{2}\right)^2 = 4$;
    center: $\left(\dfrac{3}{2}, \dfrac{1}{2}\right)$; $r = 2$

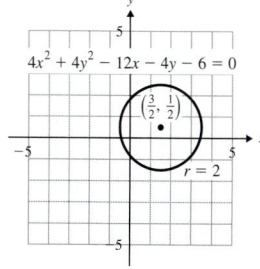

56) $\left(x + \dfrac{1}{2}\right)^2 + \left(y - \dfrac{3}{4}\right)^2 = 1$;
    center: $\left(-\dfrac{1}{2}, \dfrac{3}{4}\right)$; $r = 1$

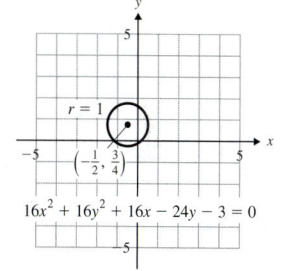

47) $(x - 5)^2 + (y - 7)^2 = 1$;
    center: $(5, 7)$; $r = 1$

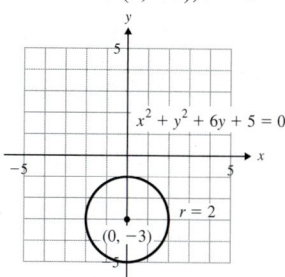

48) $(x + 6)^2 + (y + 6)^2 = 9$;
    center: $(-6, -6)$; $r = 3$

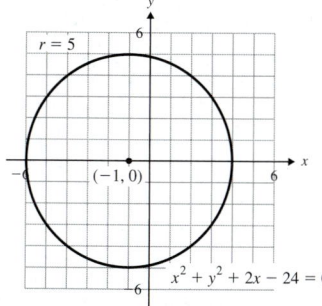

57) a) 128 m   b) 64 m   c) $(0, 71)$
    d) $x^2 + (y - 71)^2 = 4096$

58) a) 125 ft   b) $(0, 139)$ c) $x^2 + (y - 139)^2 = 15,625$

59) 11,127 mm^2 60) 58 ft^3

61) $(x - 5)^2 + (y - 3)^2 = 1$

62) $x^2 + y^2 = 16$

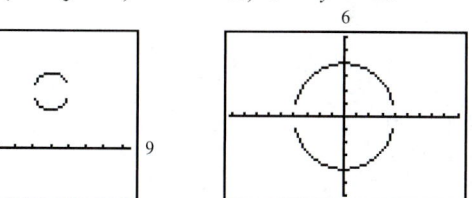

49) $x^2 + (y + 3)^2 = 4$;
    center: $(0, -3)$; $r = 2$

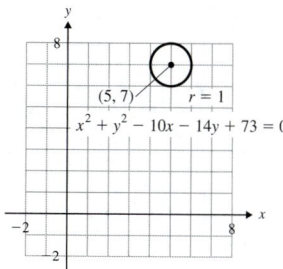

50) $(x + 1)^2 + y^2 = 25$;
    center: $(-1, 0)$; $r = 5$

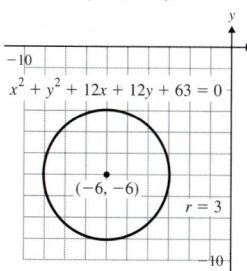

63) $(x + 2)^2 + (y - 4)^2 = 9$

64) $(x + 3)^2 + y^2 = 1$

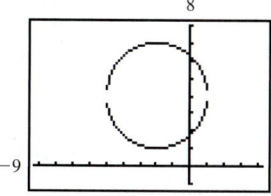
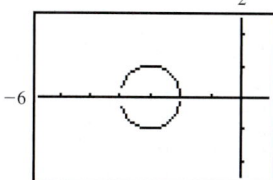

51) $(x - 2)^2 + y^2 = 5$;
    center: $(2, 0)$; $r = \sqrt{5}$

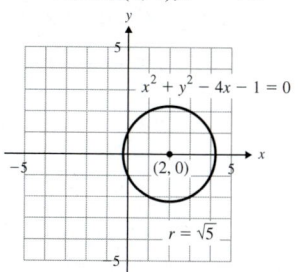

52) $x^2 + (y - 5)^2 = 3$;
    center: $(0, 5)$; $r = \sqrt{3}$

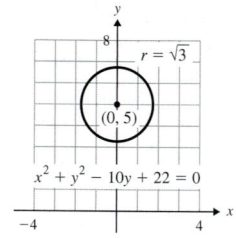

Section 11.3

1) false   2) false   3) true   4) true   5) true   6) false

7) false   8) true

9) center: $(-2, 1)$

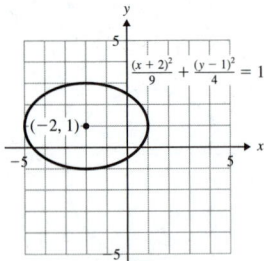

10) center: $(4, 3)$

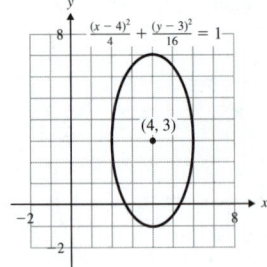

19) center: $(-1, -3)$

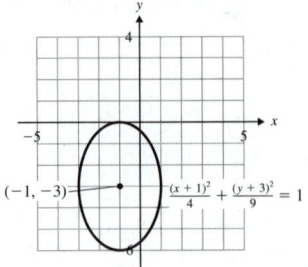

20) center: $(2, 0)$

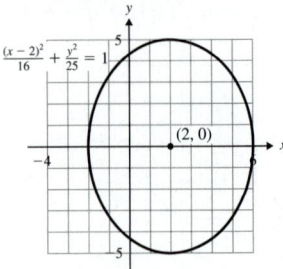

11) center: $(3, -2)$

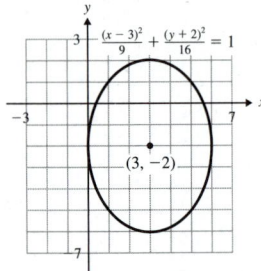

12) center: $(-4, 5)$

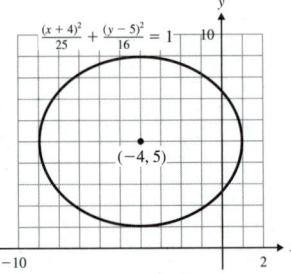

21) center: $(0, 0)$

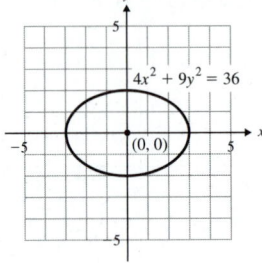

22) center: $(0, 0)$

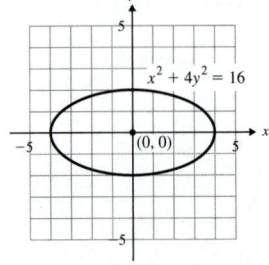

13) center: $(0, 0)$

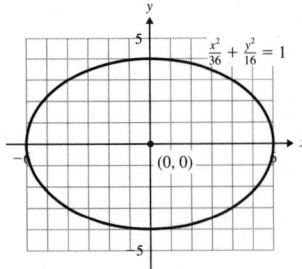

14) center: $(0, 0)$

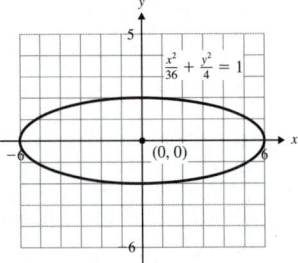

23) center: $(0, 0)$

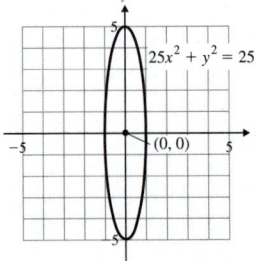

24) center: $(0, 0)$

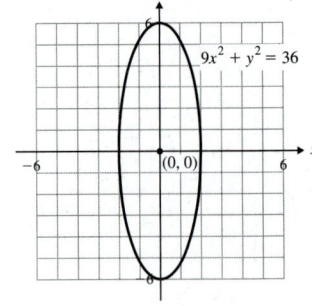

15) center: $(0, 0)$

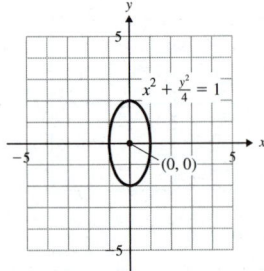

16) center: $(0, 0)$

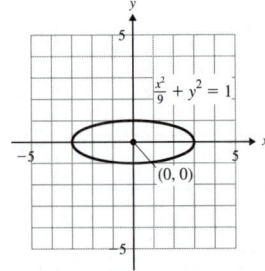

25) Group x- and y-terms separately.;
$3(x^2 - 2x) + 2(y^2 + 2y) = 7$; Complete the square.;
$3(x - 1)^2 + 2(y + 1)^2 = 12$; $\dfrac{(x - 1)^2}{4} + \dfrac{(x + 1)^2}{6} = 1$

26) Group x- and y-terms separately.;
$4(x^2 + 4x) + 9(y^2 + 6y) = -61$;
$4(x^2 + 4x + 4) + 9(y^2 + 6y + 9) = -61 + 4(4) + 9(9)$;
Factor.; $\dfrac{(x + 2)^2}{9} + \dfrac{(x + 3)^2}{4} = 1$

17) center: $(0, -4)$

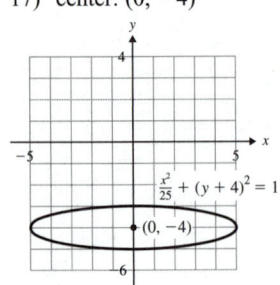

18) center: $(-3, -4)$

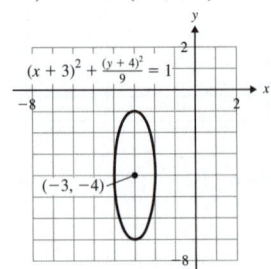

27) $\dfrac{(x - 1)^2}{16} + \dfrac{(y - 3)^2}{4} = 1$

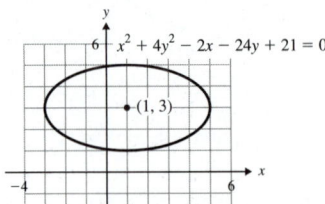

28) $\dfrac{(x + 2)^2}{4} + \dfrac{(y - 1)^2}{9} = 1$

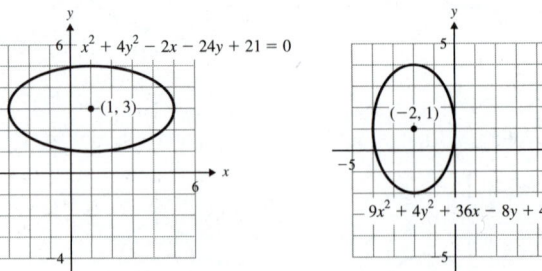

29) $(x + 4)^2 + \dfrac{(y + 1)^2}{9} = 1$ 30) $\dfrac{(x - 3)^2}{4} + (y - 5)^2 = 1$

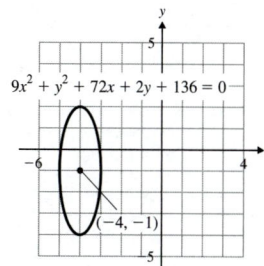

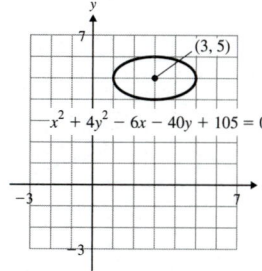

31) $\dfrac{(x - 2)^2}{9} + \dfrac{(y - 3)^2}{4} = 1$

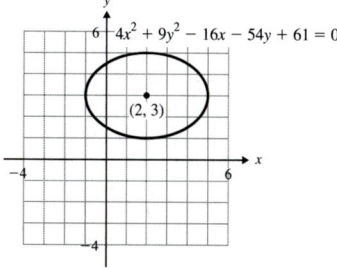

32) $\dfrac{(x + 1)^2}{25} + \dfrac{(y + 4)^2}{4} = 1$

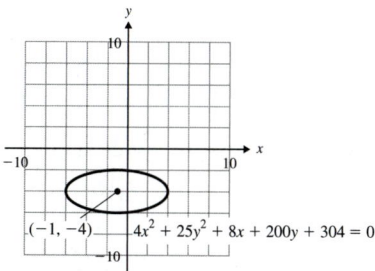

33) $\dfrac{(x + 3)^2}{4} + \dfrac{y^2}{25} = 1$

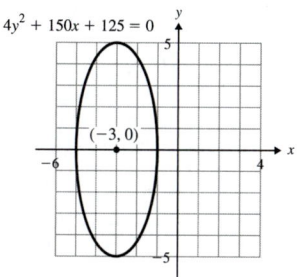

34) $\dfrac{x^2}{4} + \dfrac{(y - 1)^2}{16} = 1$

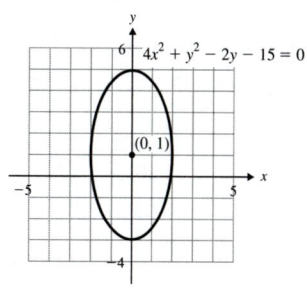

35) $\dfrac{x^2}{9} + \dfrac{y^2}{25} = 1$ 36) $\dfrac{x^2}{36} + \dfrac{y^2}{4} = 1$ 37) $\dfrac{x^2}{49} + y^2 = 1$

38) $x^2 + \dfrac{y^2}{81} = 1$ 39) $\dfrac{(x - 3)^2}{4} + \dfrac{(y - 1)^2}{16} = 1$

40) $\dfrac{(x + 1)^2}{25} + \dfrac{(y + 2)^2}{9} = 1$

41) Yes. If $a = b$ in the equation $\dfrac{(x - h)^2}{a^2} + \dfrac{(y - k)^2}{b^2} = 1$, then the ellipse is a circle.

42) $\dfrac{x^2}{324} + \dfrac{y^2}{210.25} = 1$ 43) $\dfrac{x^2}{9801} + \dfrac{y^2}{11,342.25} = 1$

44) a) $y = \sqrt{36 - \dfrac{36x^2}{49}}$ b) 4 ft

45) $\dfrac{x^2}{36} + \dfrac{y^2}{16} = 1$ 46) $9x^2 + y^2 = 36$

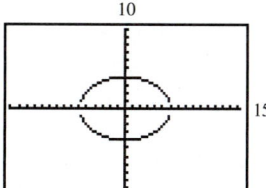

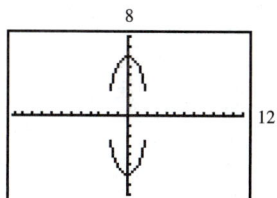

Section 11.4

1) false   2) true   3) true   4) true

5) center: $(0, 0)$ 6) center: $(0, 0)$

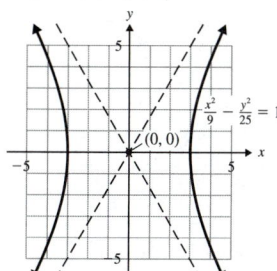

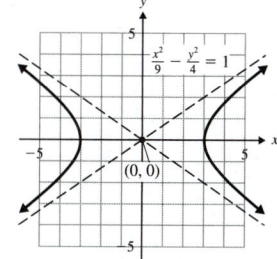

7) center: $(0, 0)$ 8) center: $(0, 0)$

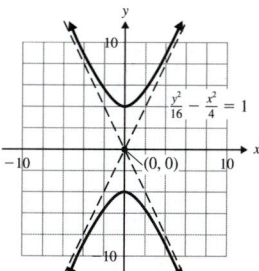

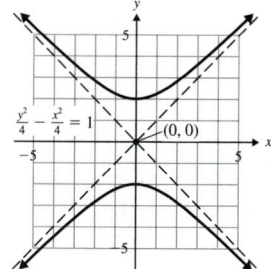

9) center: $(2, -3)$

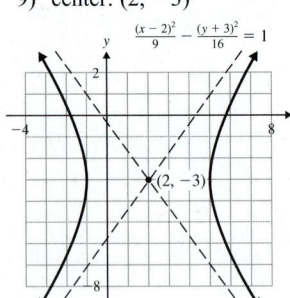

10) center: $(-3, -1)$

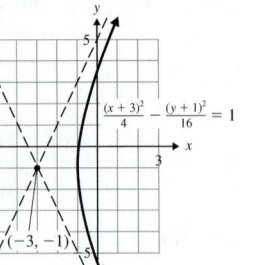

18) center: $(0, 0)$

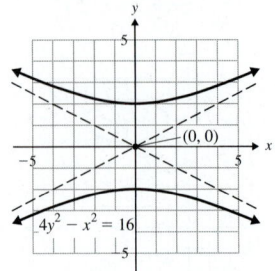

19) center: $(0, 0)$

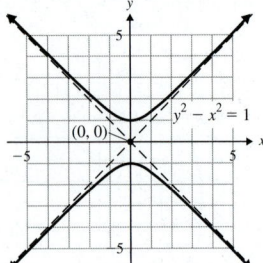

11) center: $(-4, -1)$

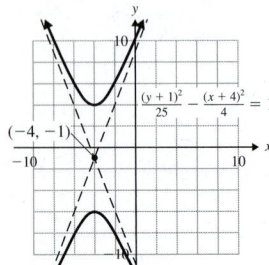

12) center: $(-1, 1)$

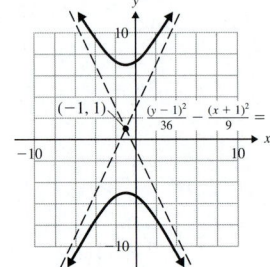

20) center: $(0, 0)$

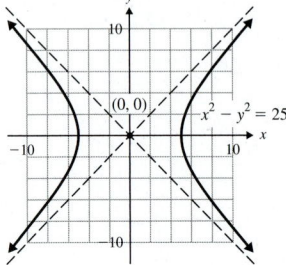

13) center: $(1, 0)$

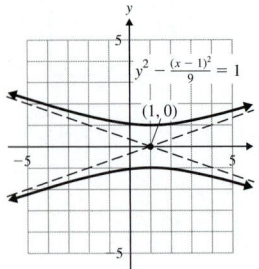

14) center: $(0, -4)$

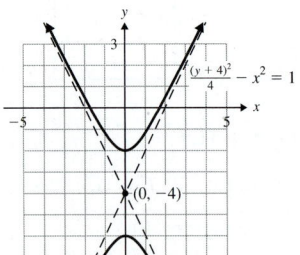

21) $y = \dfrac{5}{3}x$ and $y = -\dfrac{5}{3}x$ 22) $y = \dfrac{2}{3}x$ and $y = -\dfrac{2}{3}x$

23) $y = 2x$ and $y = -2x$ 24) $y = x$ and $y = -x$

25) $y = 3x$ and $y = -3x$ 26) $y = \dfrac{1}{2}x$ and $y = -\dfrac{1}{2}x$

27) $y = x$ and $y = -x$ 28) $y = x$ and $y = -x$

29)

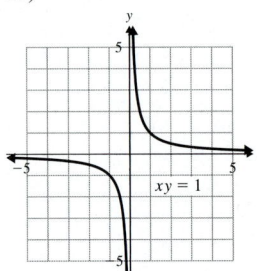

30)

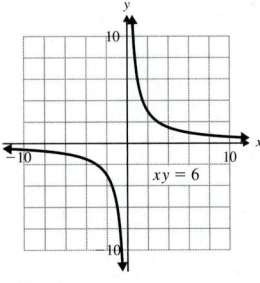

15) center: $(1, 2)$

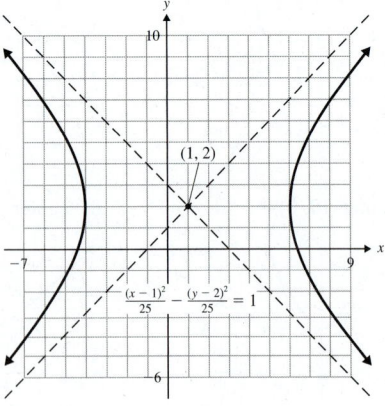

31)

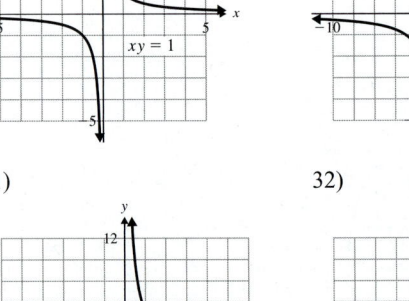

32)

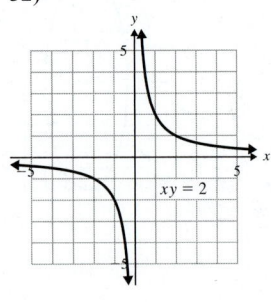

16) center: $(2, 3)$

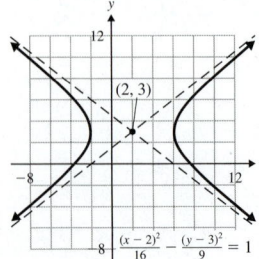

17) center: $(0, 0)$

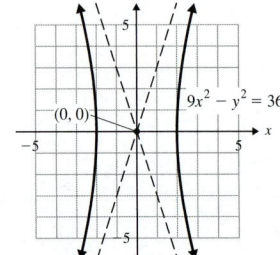

33)

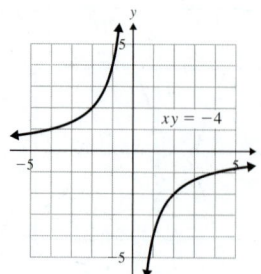

34)

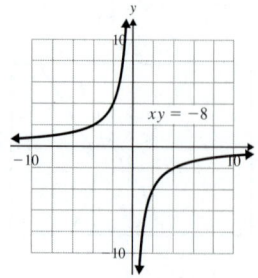

35)

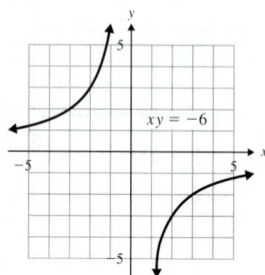

36)

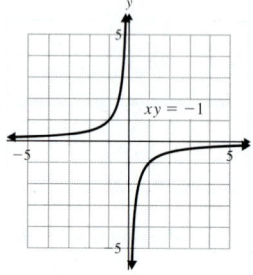

37) domain: $[-3, 3]$;
range: $[0, 3]$

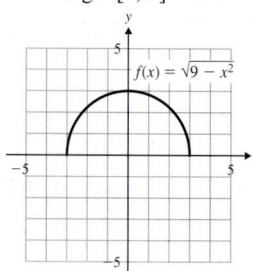

38) domain: $[-5, 5]$;
range: $[0, 5]$

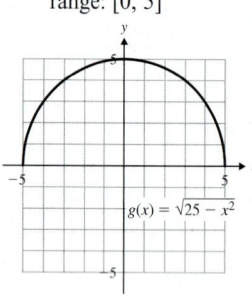

39) domain: $[-1, 1]$;
range: $[-1, 0]$

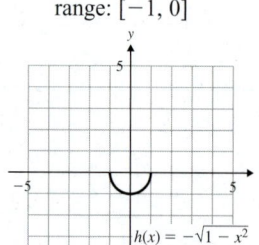

40) domain: $[-3, 3]$;
range: $[-3, 0]$

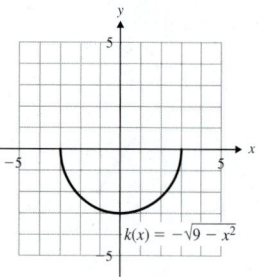

41) domain: $[-3, 3]$;
range: $[-2, 0]$

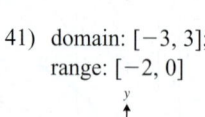

42) domain: $[-4, 4]$;
range: $[0, 3]$

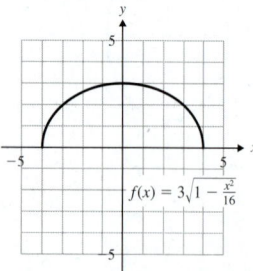

43) domain:
$(-\infty, -2] \cup [2, \infty)$;
range: $(-\infty, 0]$

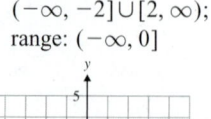

44) domain:
$(-\infty, -4] \cup [4, \infty)$;
range: $[0, \infty)$

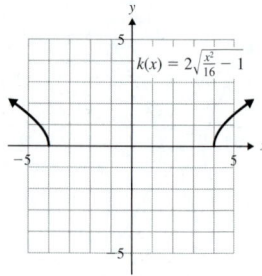

45)

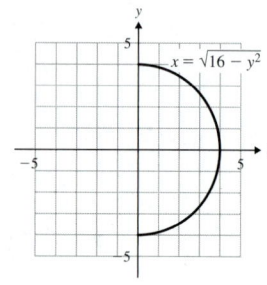

46)

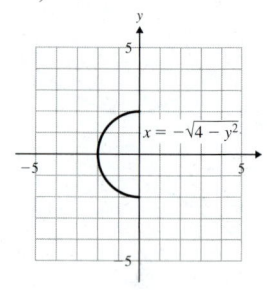

47)

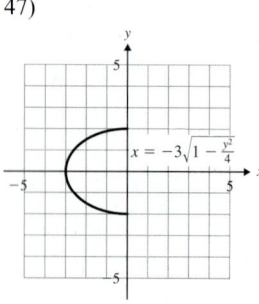

48)

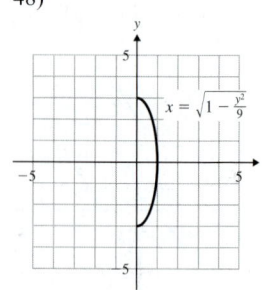

49) Group the x- and y-terms separately.;
$4(x^2 - 2x) - 9(y^2 + 2y) = 41$;
Complete the square.;
$4(x - 1)^2 - 9(y + 1)^2 = 36$; $\dfrac{(x - 1)^2}{9} - \dfrac{(y + 1)^2}{4} = 1$

50) Group the x- and y-terms separately.;
$-(x^2 - 6x) + 4(y^2 - 4y) = -3$;
$-(x^2 - 6x + 9) + 4(y^2 - 4y + 4) = -3 - 1(9) + 4(4)$;
Factor.; $(y - 2)^2 - \dfrac{(x - 3)^2}{4} = 1$

51) $\dfrac{(x - 1)^2}{16} - \dfrac{(y + 3)^2}{4} = 1$

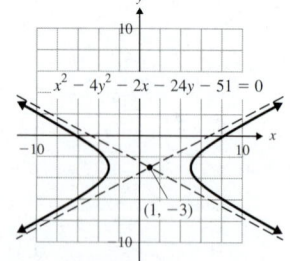

52) $\dfrac{(x+5)^2}{4} - \dfrac{(y+2)^2}{9} = 1$

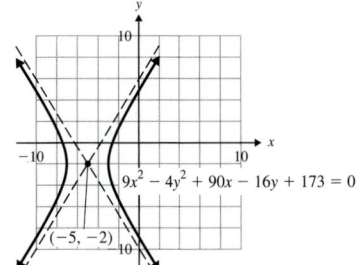

$9x^2 - 4y^2 + 90x - 16y + 173 = 0$

$(-5, -2)$

53) $\dfrac{(y-2)^2}{9} - \dfrac{(x-1)^2}{16} = 1$

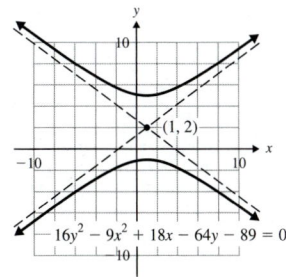

$(1, 2)$

$16y^2 - 9x^2 + 18x - 64y - 89 = 0$

54) $\dfrac{(y-3)^2}{16} - \dfrac{(x+2)^2}{4} = 1$

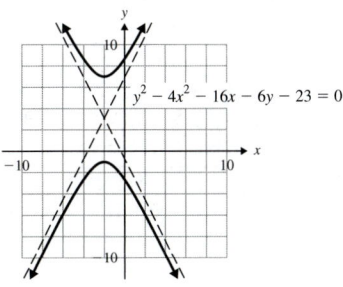

$y^2 - 4x^2 - 16x - 6y - 23 = 0$

55) $y + 3 = \dfrac{4}{3}(x - 2)$ and $y + 3 = -\dfrac{4}{3}(x - 2)$

56) $y + 1 = 2(x + 3)$ and $y + 1 = -2(x + 3)$

57) $y + 1 = \dfrac{5}{2}(x + 4)$ and $y + 1 = -\dfrac{5}{2}(x + 4)$

58) $y - 1 = 2(x + 1)$ and $y - 1 = -2(x + 1)$

59) $y = \dfrac{1}{3}(x - 1)$ and $y = -\dfrac{1}{3}(x - 1)$

60) $y + 4 = 2x$ and $y + 4 = -2x$

61) $y = x + 1$ and $y = -x + 3$

62) $y - 3 = \dfrac{3}{4}(x - 2)$ and $y - 3 = -\dfrac{3}{4}(x - 2)$

63) $y^2 - \dfrac{x^2}{4} = 1$ 64) $\dfrac{x^2}{4} - \dfrac{y^2}{9} = 1$

65) $\dfrac{(x-2)^2}{9} - \dfrac{(y+3)^2}{16} = 1$ 66) $\dfrac{(x+3)^2}{4} - \dfrac{(y+1)^2}{16} = 1$

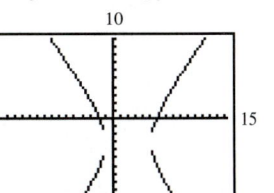 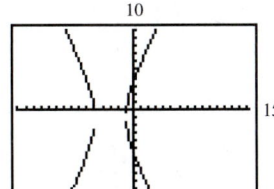

67) $\dfrac{(y+1)^2}{25} - \dfrac{(x+4)^2}{4} = 1$ 68) $\dfrac{(y-1)^2}{36} - \dfrac{(x+1)^2}{9} = 1$

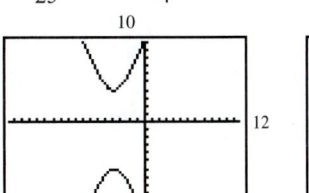 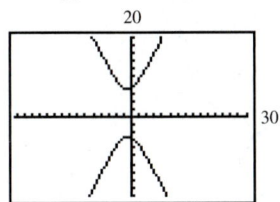

Chapter 11 Putting It All Together

1) parabola

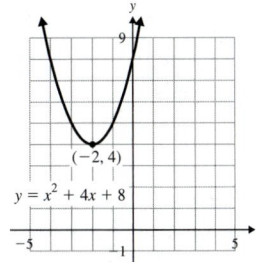

$(-2, 4)$

$y = x^2 + 4x + 8$

2) circle

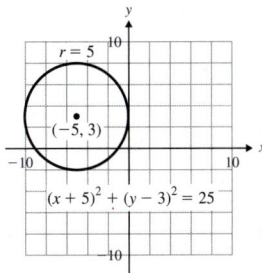

$r = 5$

$(-5, 3)$

$(x + 5)^2 + (y - 3)^2 = 25$

3) hyperbola

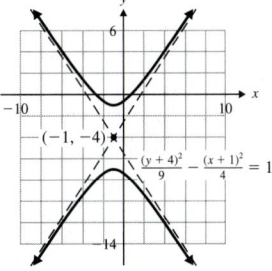

$(-1, -4)$

$\dfrac{(y+4)^2}{9} - \dfrac{(x+1)^2}{4} = 1$

4) parabola

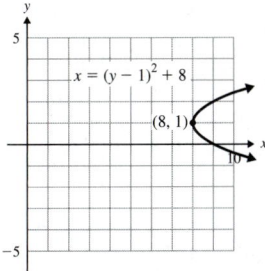

$x = (y - 1)^2 + 8$

$(8, 1)$

5) ellipse

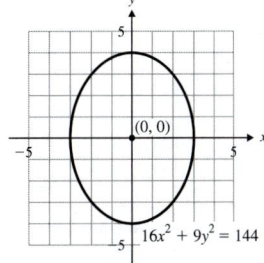

$(0, 0)$

$16x^2 + 9y^2 = 144$

6) hyperbola

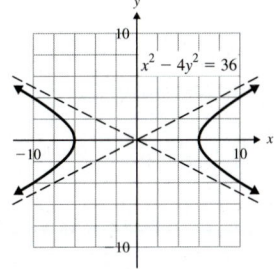

$x^2 - 4y^2 = 36$

7) circle

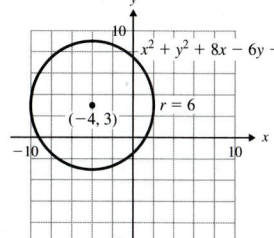

$x^2 + y^2 + 8x - 6y - 11 = 0$
$(-4, 3)$
$r = 6$

8) ellipse

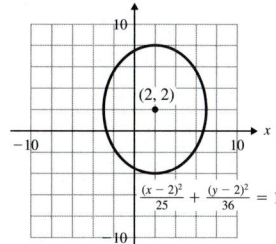

$(2, 2)$
$\frac{(x-2)^2}{25} + \frac{(y-2)^2}{36} = 1$

17) parabola

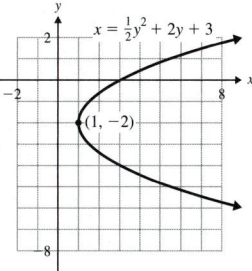

$x = \frac{1}{2}y^2 + 2y + 3$
$(1, -2)$

18) ellipse

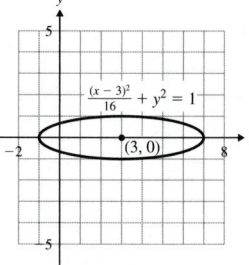

$\frac{(x-3)^2}{16} + y^2 = 1$
$(3, 0)$

9) ellipse

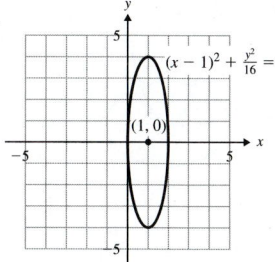

$(x - 1)^2 + \frac{y^2}{16} = 1$
$(1, 0)$

10) circle

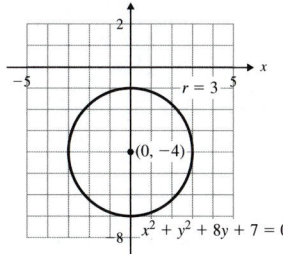

$r = 3$
$(0, -4)$
$x^2 + y^2 + 8y + 7 = 0$

19) hyperbola

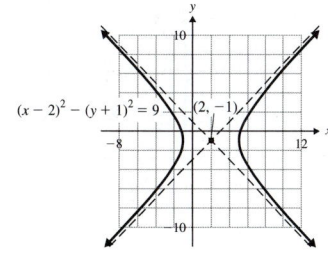

$(x-2)^2 - (y+1)^2 = 9$
$(2, -1)$

20) circle

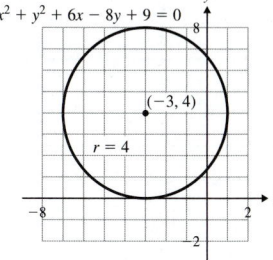

$x^2 + y^2 + 6x - 8y + 9 = 0$
$(-3, 4)$
$r = 4$

11) parabola

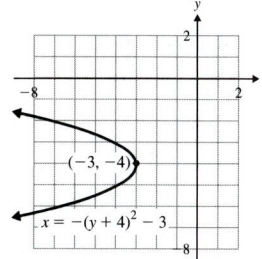

$(-3, -4)$
$x = -(y + 4)^2 - 3$

12) hyperbola

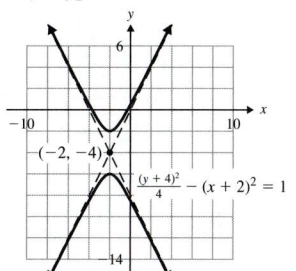

$(-2, -4)$
$\frac{(y+4)^2}{4} - (x+2)^2 = 1$

21)

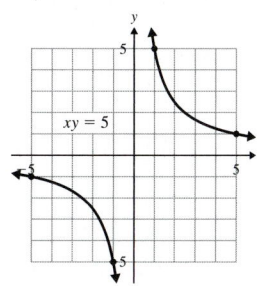

$xy = 5$

22) $\dfrac{(x+3)^2}{4} - \dfrac{y^2}{25} = 1$

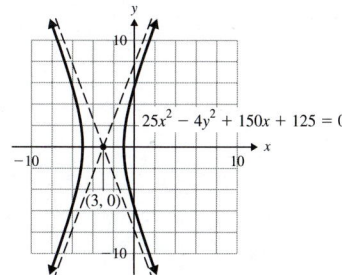

$25x^2 - 4y^2 + 150x + 125 = 0$
$(3, 0)$

13) hyperbola

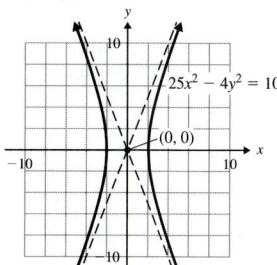

$25x^2 - 4y^2 = 100$
$(0, 0)$

14) ellipse

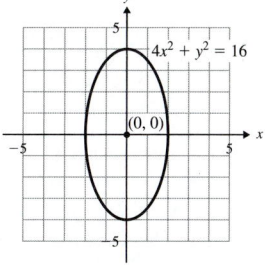

$4x^2 + y^2 = 16$
$(0, 0)$

23) $(x - 3)^2 + \dfrac{(y + 2)^2}{9} = 1$ 24)

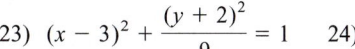

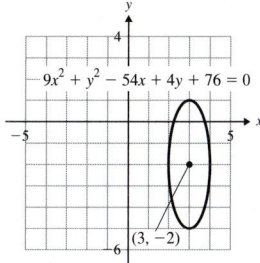

$9x^2 + y^2 - 54x + 4y + 76 = 0$
$(3, -2)$

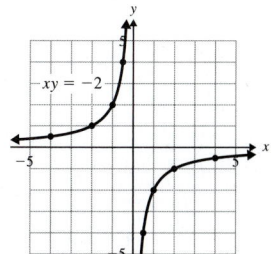

$xy = -2$

15) circle

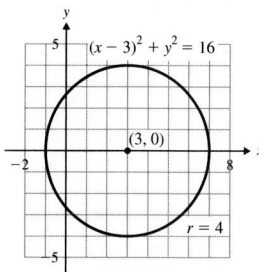

$(x - 3)^2 + y^2 = 16$
$(3, 0)$
$r = 4$

16) parabola

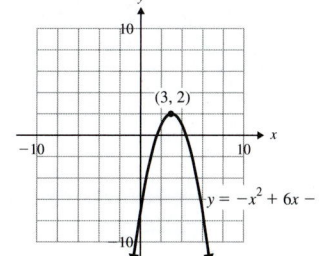

$(3, 2)$
$y = -x^2 + 6x - 7$

25) $\dfrac{(y-1)^2}{4} - \dfrac{(x-2)^2}{9} = 1$ 26) $\dfrac{(x-1)^2}{9} + \dfrac{(y-3)^2}{4} = 1$

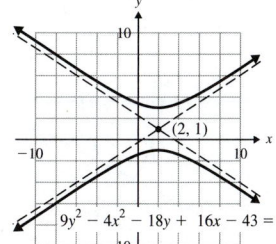

$(2, 1)$
$9y^2 - 4x^2 - 18y + 16x - 43 = 0$

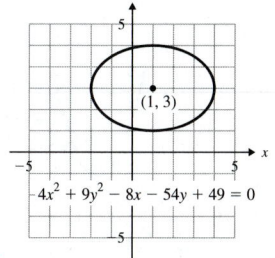

$(1, 3)$
$-4x^2 + 9y^2 - 8x - 54y + 49 = 0$

27) $x^2 + y^2 + 8x - 6y - 11 = 0$ 28) $(x - 1)^2 + \dfrac{y^2}{16} = 1$

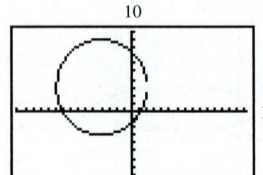   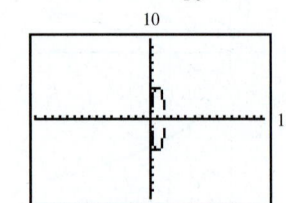

29) $x = -(y + 4)^2 - 3$ 30) $25x^2 - 4y^2 = 100$

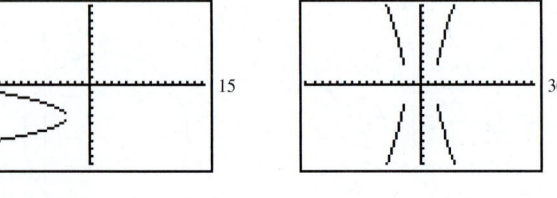

Section 11.5

1) c) 0, 1, or 2

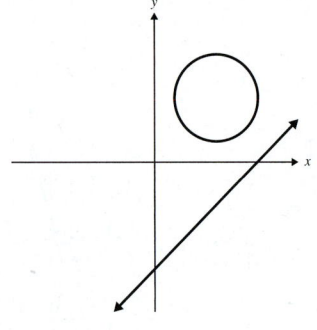

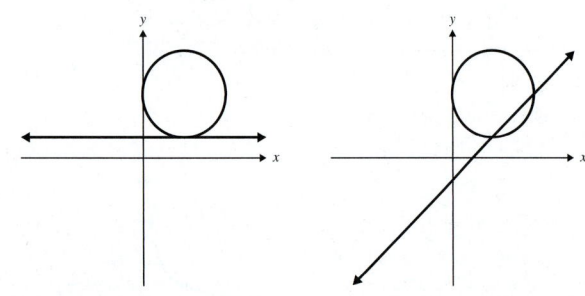

2) c) 0, 1, or 2

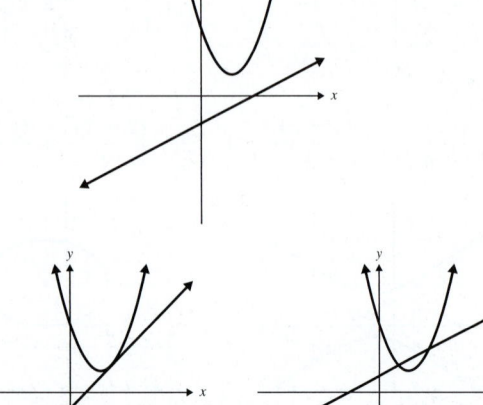

3) c) 0, 1, 2, 3, or 4

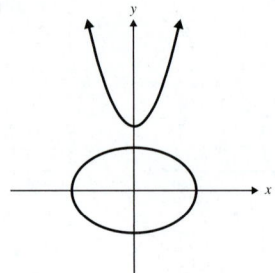

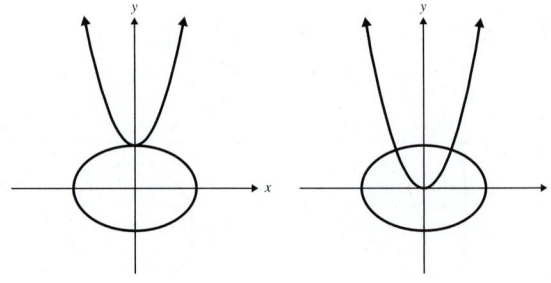

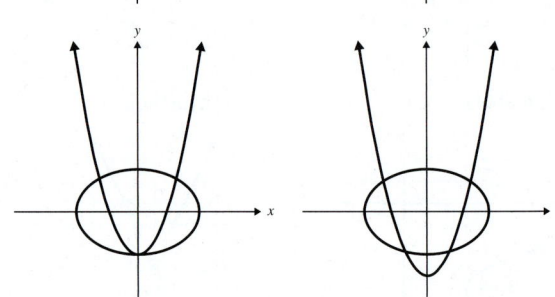

4) c) 0, 1, 2, 3, or 4

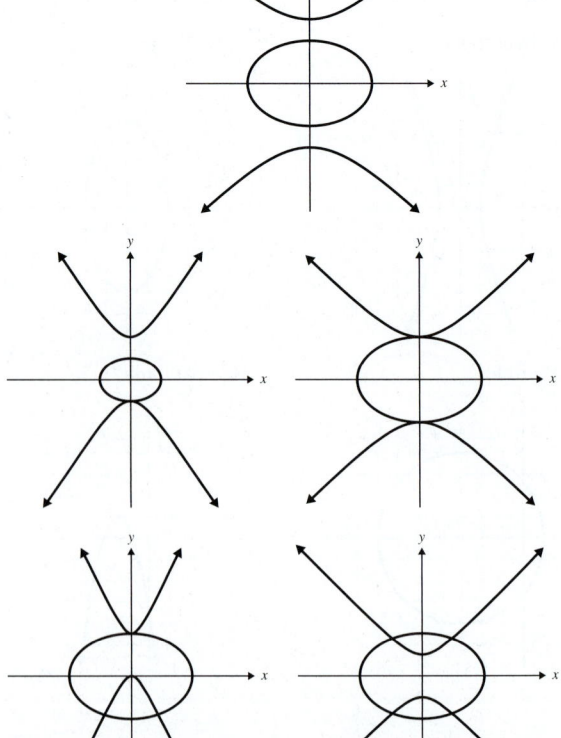

5) c) 0, 1, 2, 3, or 4

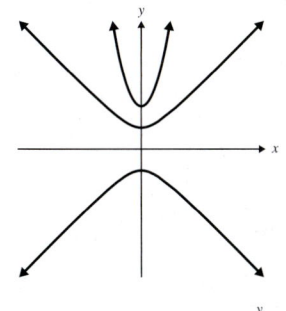

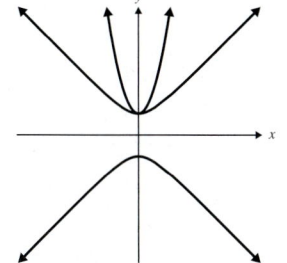

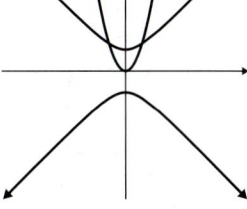

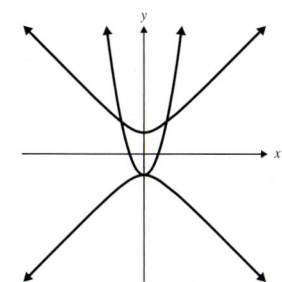

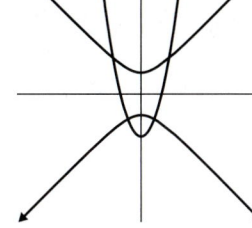

6) c) 0, 1, 2, 3, or 4

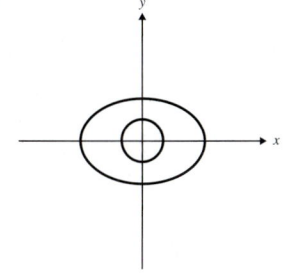

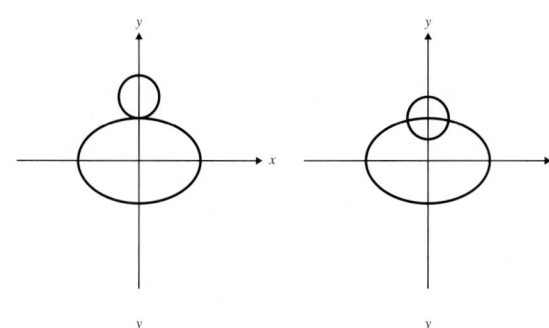

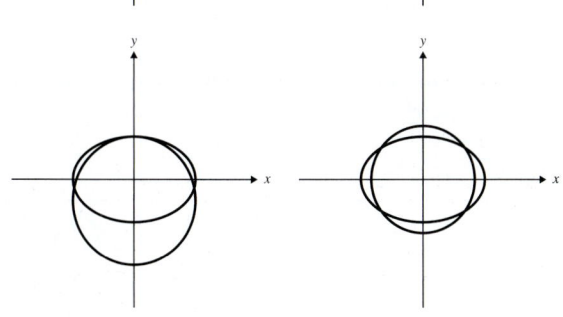

7) $\{(-4, -2), (6, -7)\}$ 8) $\{(-3, -8), (2, -3)\}$

9) $\{(-1, 3), (3, 1)\}$ 10) $\{(2, 2), (-2, 2)\}$ 11) $\{(4, 2)\}$

12) $\{(7, 1)\}$ 13) $\{(3, 1), (3, -1), (-3, 1), (-3, -1)\}$

14) $\{(4, 5), (4, -5), (-4, 5), (-4, -5)\}$

15) $\{(2, \sqrt{2}), (2, -\sqrt{2}), (-2, \sqrt{2}), (-2, -\sqrt{2})\}$

16) $\{(\sqrt{5}, 3), (\sqrt{5}, -3), (-\sqrt{5}, 3), (-\sqrt{5}, -3)\}$

17) $\{(3, -7), (-3, -7)\}$ 18) $\{(-1, 7), (1, 7)\}$ 19) $\{(0, -1)\}$

20) $\left\{\left(0, \dfrac{5}{2}\right)\right\}$ 21) $\{(0, 2)\}$ 22) $\{(7, 0)\}$

23) $\varnothing$ 24) $\varnothing$ 25) $\varnothing$ 26) $\varnothing$

27) $\left\{\left(\dfrac{\sqrt{2}}{2}, \dfrac{3\sqrt{2}}{2}\right), \left(\dfrac{\sqrt{2}}{2}, -\dfrac{3\sqrt{2}}{2}\right),\right.$
$\left.\left(-\dfrac{\sqrt{2}}{2}, \dfrac{3\sqrt{2}}{2}\right), \left(-\dfrac{\sqrt{2}}{2}, -\dfrac{3\sqrt{2}}{2}\right)\right\}$

28) $\{(1, \sqrt{5}), (1, -\sqrt{5}), (-1, \sqrt{5}), (-1, -\sqrt{5})\}$

29) $\{(0, -2)\}$ 30) $\{(0, 1)\}$ 31) 8 and 5   32) 7 and 4

33) 8 in. $\times$ 11 in.   34) 12 in. $\times$ 15 in.

35) 4000 basketballs; $240   36) 3000 backpacks; $189

Section 11.6

1) Three points that satisfy the inequality: (8, 0), (0, −8), and (6, 0); three that do not: (0, 0), (−3, −2), and (1, 1): answers may vary.

2) Three points that satisfy the inequality: (3, 0), (4, 1), and (5, −1); three that do not: (0, 0), (2, 0), and (−1, −1): answers may vary.

3) Three points that satisfy the inequality: (0, 0), (−4, −1), and (5, 2); three that do not: (0, 1), (0, −2), and (3, 3): answers may vary.

4) Three points that satisfy the inequality: (0, 0), (0, 5), and (1, −1); three that do not: (−3, 0), (−4, 4), and (5, 0): answers may vary.

5)   6)

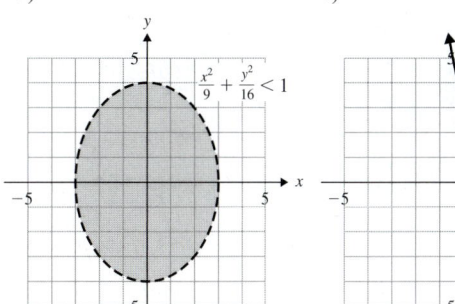

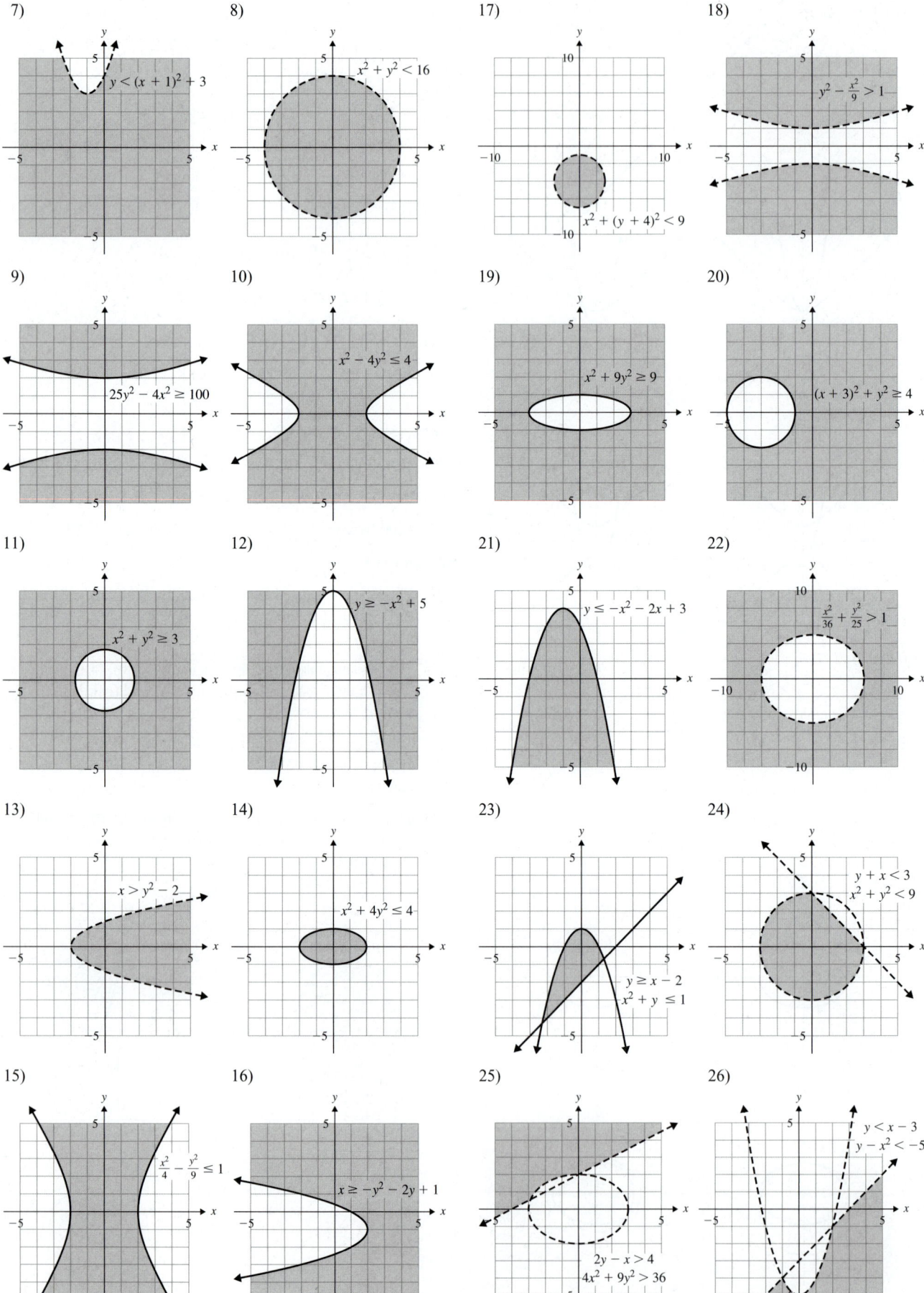

7) $y < (x + 1)^2 + 3$

8) $x^2 + y^2 < 16$

17) $x^2 + (y + 4)^2 < 9$

18) $y^2 - \dfrac{x^2}{9} > 1$

9) $25y^2 - 4x^2 \geq 100$

10) $x^2 - 4y^2 \leq 4$

19) $x^2 + 9y^2 \geq 9$

20) $(x + 3)^2 + y^2 \geq 4$

11) $x^2 + y^2 \geq 3$

12) $y \geq -x^2 + 5$

21) $y \leq -x^2 - 2x + 3$

22) $\dfrac{x^2}{36} + \dfrac{y^2}{25} > 1$

13) $x > y^2 - 2$

14) $x^2 + 4y^2 \leq 4$

23) $y \geq x - 2$
     $x^2 + y \leq 1$

24) $y + x < 3$
     $x^2 + y^2 < 9$

15) $\dfrac{x^2}{4} - \dfrac{y^2}{9} \leq 1$

16) $x \geq -y^2 - 2y + 1$

25) $2y - x > 4$
     $4x^2 + 9y^2 > 36$

26) $y < x - 3$
     $y - x^2 < -5$

27)

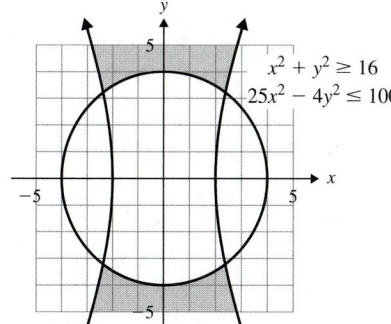

$x^2 + y^2 \geq 16$
$25x^2 - 4y^2 \leq 100$

35)

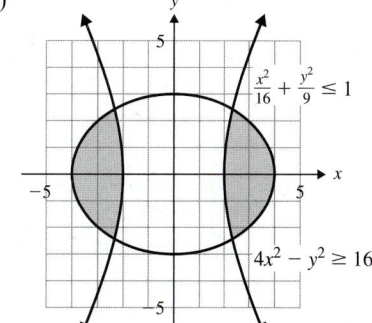

$\frac{x^2}{16} + \frac{y^2}{9} \leq 1$

$4x^2 - y^2 \geq 16$

28)

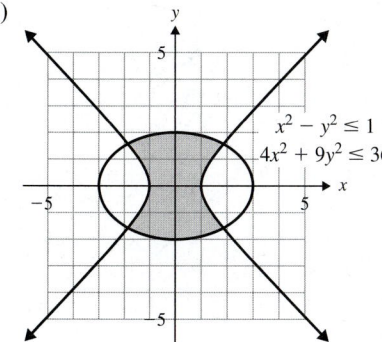

$x^2 - y^2 \leq 1$
$4x^2 + 9y^2 \leq 36$

36)

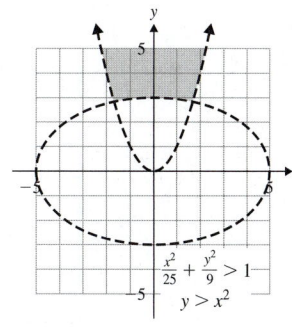

$\frac{x^2}{25} + \frac{y^2}{9} > 1$
$y > x^2$

37)

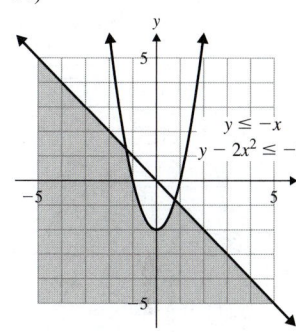

$y \leq -x$
$y - 2x^2 \leq -2$

29)

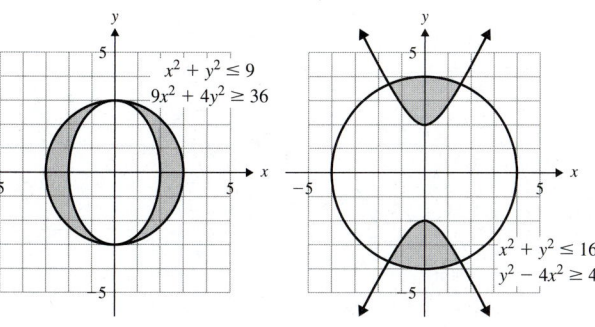

$x^2 + y^2 \leq 9$
$9x^2 + 4y^2 \geq 36$

30)

$x^2 + y^2 \leq 16$
$y^2 - 4x^2 \geq 4$

38)

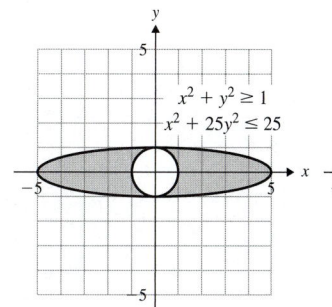

$x^2 + y^2 \geq 1$
$x^2 + 25y^2 \leq 25$

39)

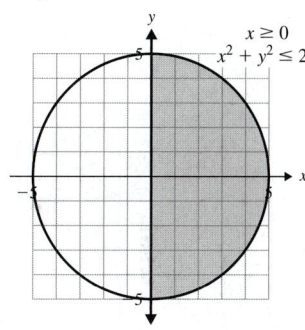

$x \geq 0$
$x^2 + y^2 \leq 25$

31)

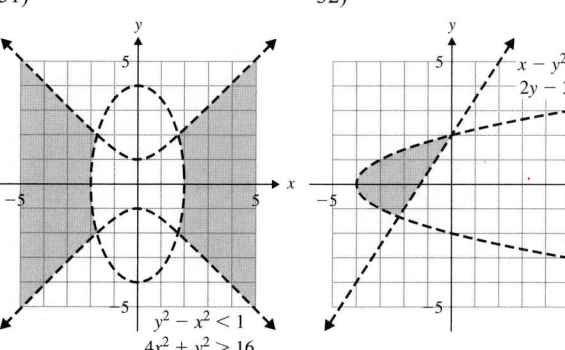

$y^2 - x^2 < 1$
$4x^2 + y^2 > 16$

32)

$x - y^2 > -4$
$2y - 3x > 4$

40)

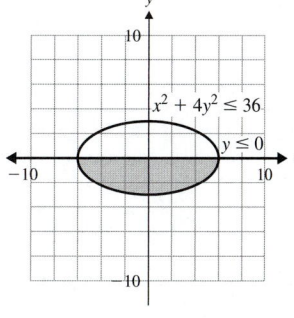

$x^2 + 4y^2 \leq 36$
$y \leq 0$

41)

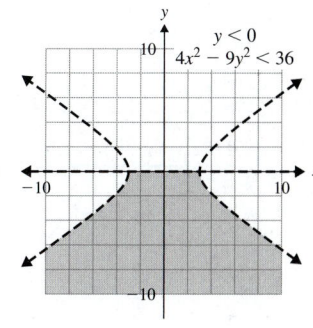

$y < 0$
$4x^2 - 9y^2 < 36$

33)

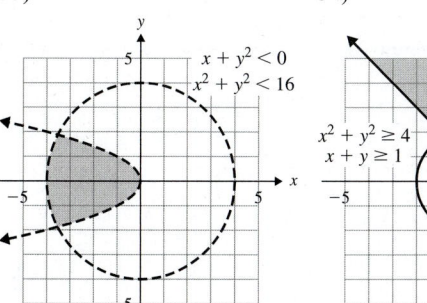

$x + y^2 < 0$
$x^2 + y^2 < 16$

34)

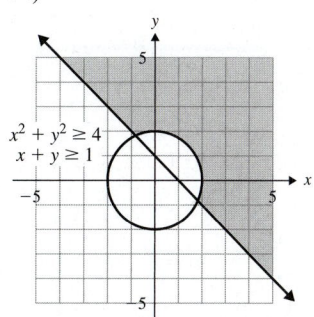

$x^2 + y^2 \geq 4$
$x + y \geq 1$

42)

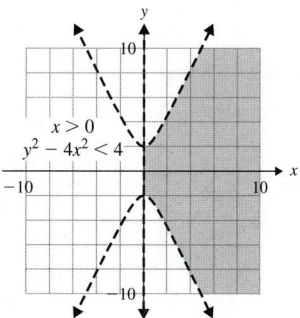

$x > 0$
$y^2 - 4x^2 < 4$

43)

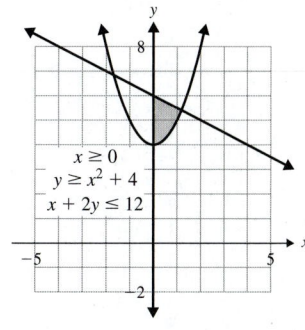

$x \geq 0$
$y \geq x^2 + 4$
$x + 2y \leq 12$

44)

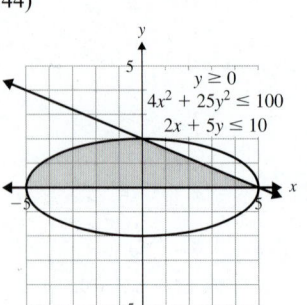

$y \geq 0$
$4x^2 + 25y^2 \leq 100$
$2x + 5y \leq 10$

45)

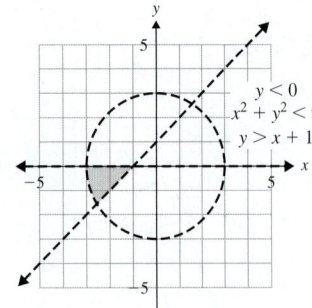

$y < 0$
$x^2 + y^2 < 9$
$y > x + 1$

52)

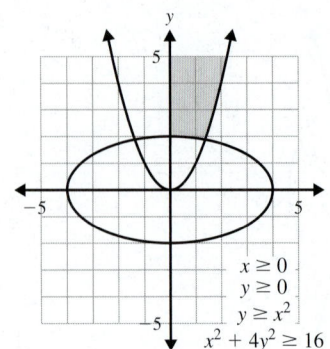

$x \geq 0$
$y \geq 0$
$y \geq x^2$
$x^2 + 4y^2 \geq 16$

46)

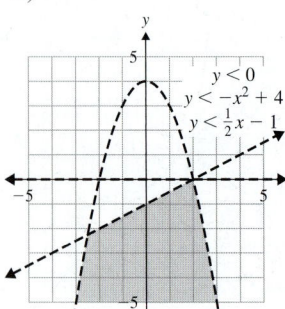

$y < 0$
$y < -x^2 + 4$
$y < \frac{1}{2}x - 1$

47)

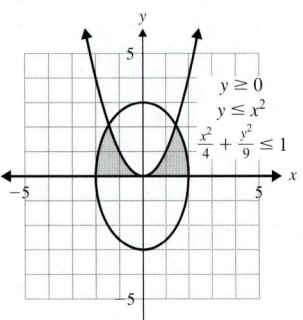

$y \geq 0$
$y \leq x^2$
$\frac{x^2}{4} + \frac{y^2}{9} \leq 1$

Chapter 11 Review Exercises

1) domain: $(-\infty, \infty)$;
   range: $[0, \infty)$

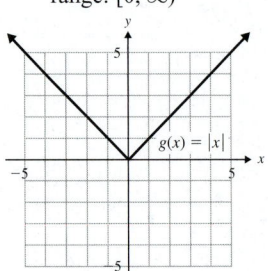

$g(x) = |x|$

2) domain: $(-\infty, \infty)$;
   range: $[0, \infty)$

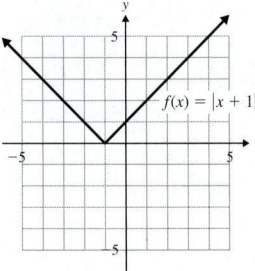

$f(x) = |x + 1|$

48)

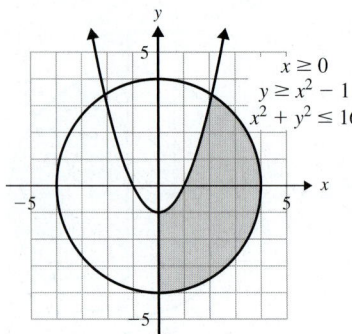

$x \geq 0$
$y \geq x^2 - 1$
$x^2 + y^2 \leq 16$

3) domain: $(-\infty, \infty)$;
   range: $[-4, \infty)$

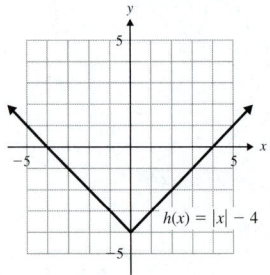

$h(x) = |x| - 4$

4) domain: $(-\infty, \infty)$;
   range: $[2, \infty)$

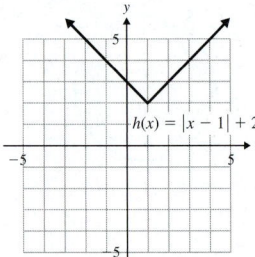

$h(x) = |x - 1| + 2$

49)

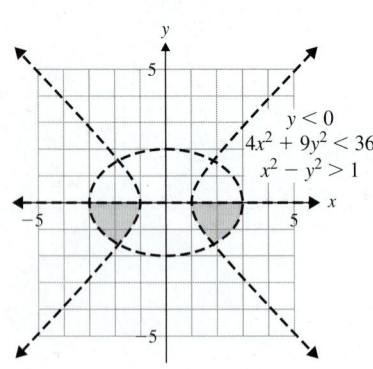

$y < 0$
$4x^2 + 9y^2 < 36$
$x^2 - y^2 > 1$

5) domain: $(-\infty, \infty)$;
   range: $(-\infty, 5]$

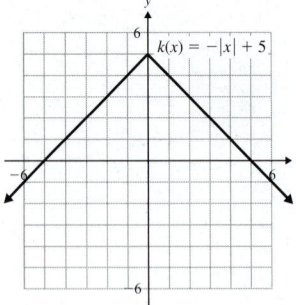

$k(x) = -|x| + 5$

6) $g(x) = |x - 5|$

50)

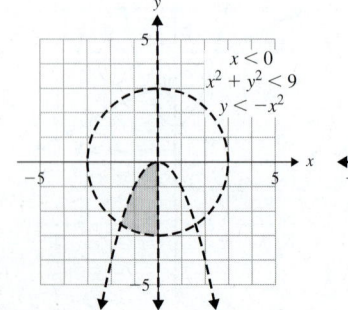

$x < 0$
$x^2 + y^2 < 9$
$y < -x^2$

51)

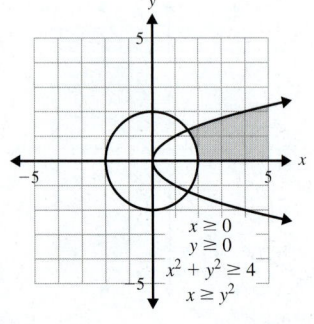

$x \geq 0$
$y \geq 0$
$x^2 + y^2 \geq 4$
$x \geq y^2$

7)

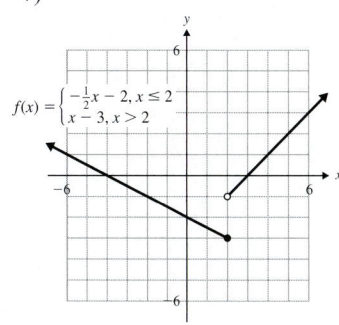

$f(x) = \begin{cases} -\frac{1}{2}x - 2, x \le 2 \\ x - 3, x > 2 \end{cases}$

8)

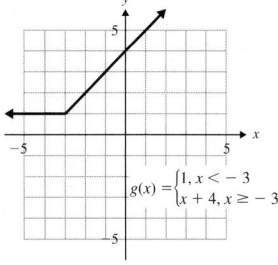

$g(x) = \begin{cases} 1, x < -3 \\ x + 4, x \ge -3 \end{cases}$

9) 7   10) 2   11) −9   12) −6   13) 10

14)

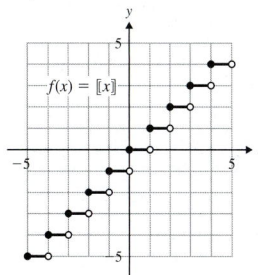

$f(x) = [\![x]\!]$

15)

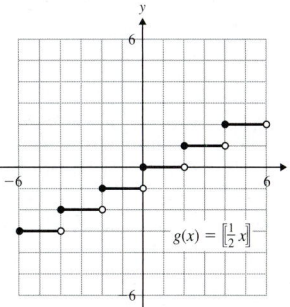

$g(x) = [\![\frac{1}{2}x]\!]$

16)

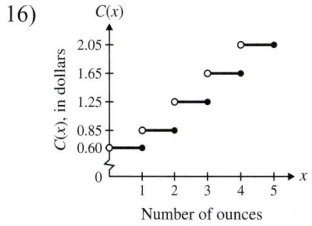

17) (4, 5)   18) (−4, 0)   19) $\left(\dfrac{13}{2}, -\dfrac{7}{2}\right)$ 20) $\left(\dfrac{1}{4}, \dfrac{7}{16}\right)$

21) center: (−3, 5);
    $r = 6$

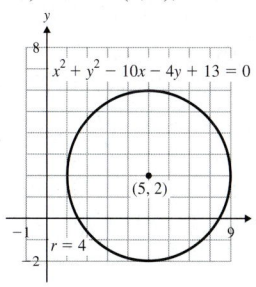

$(x + 3)^2 + (y - 5)^2 = 36$

22) center: (0, −4);
    $r = 3$

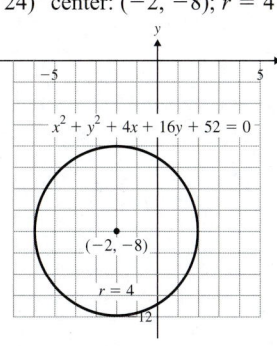

$x^2 + (y + 4)^2 = 9$

23) center: (5, 2); $r = 4$

$x^2 + y^2 - 10x - 4y + 13 = 0$

24) center: (−2, −8); $r = 4$

$x^2 + y^2 + 4x + 16y + 52 = 0$

25) $(x - 3)^2 + y^2 = 16$ 26) $(x - 1)^2 + (y + 5)^2 = 7$

27) when $a = b$ in $\dfrac{(x - h)^2}{a^2} + \dfrac{(y - k)^2}{b^2} = 1$

28) center (0, 0)

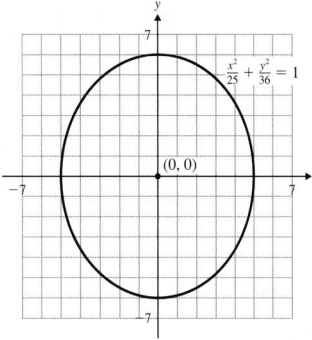

$\dfrac{x^2}{25} + \dfrac{y^2}{36} = 1$

29) center (−3, 3)

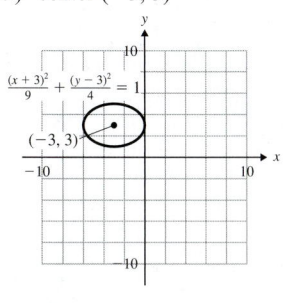

$\dfrac{(x + 3)^2}{9} + \dfrac{(y - 3)^2}{4} = 1$

30) center (4, 2)

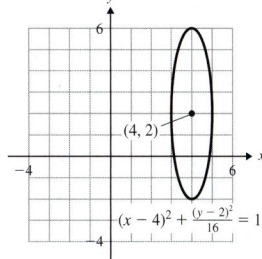

$(x - 4)^2 + \dfrac{(y - 2)^2}{16} = 1$

31) center (0, 0)

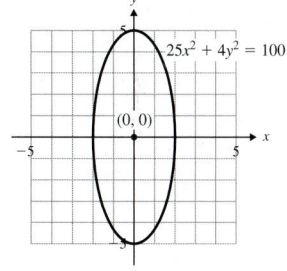

$25x^2 + 4y^2 = 100$

32) $\dfrac{(x + 1)^2}{9} + \dfrac{(y + 2)^2}{4} = 1$

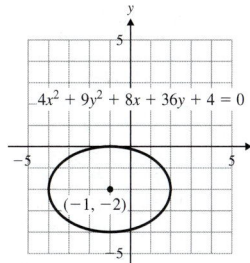

$4x^2 + 9y^2 + 8x + 36y + 4 = 0$

33) $\dfrac{(x - 2)^2}{4} + \dfrac{y^2}{25} = 1$

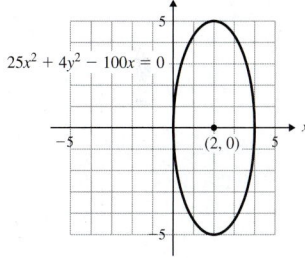

$25x^2 + 4y^2 - 100x = 0$

34) The equation of an ellipse contains the sum of two squared variable terms, but the equation of a hyperbola contains the difference of two squared variable terms.

35) center (0, 0)

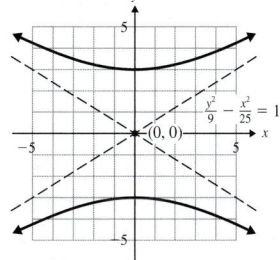

$\dfrac{y^2}{9} - \dfrac{x^2}{25} = 1$

36) center (−2, 3)

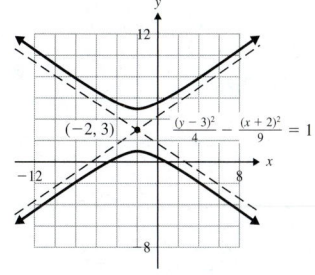

$\dfrac{(y - 3)^2}{4} - \dfrac{(x + 2)^2}{9} = 1$

37) center $(-1, -2)$

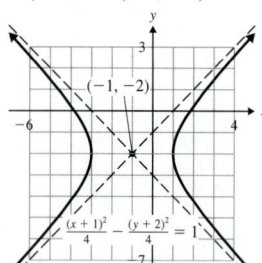

$$\frac{(x+1)^2}{4} - \frac{(y+2)^2}{4} = 1$$

38) center $(0, 0)$

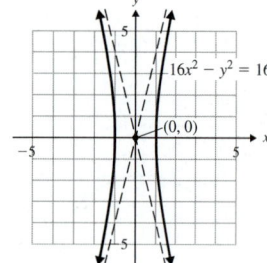

$16x^2 - y^2 = 16$

47) parabola

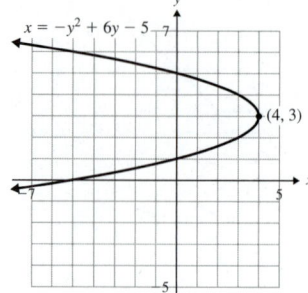

$x = -y^2 + 6y - 5$

48) parabola

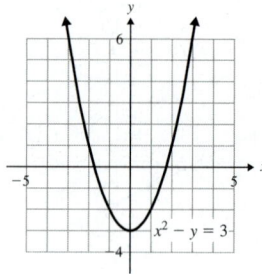

$x^2 - y = 3$

39)

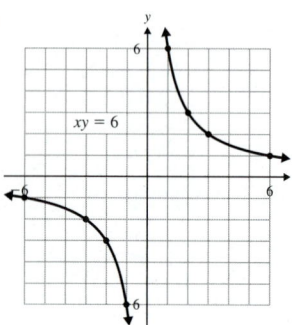

$xy = 6$

40) $y = \frac{3}{5}x$ and $y = -\frac{3}{5}x$

49) hyperbola

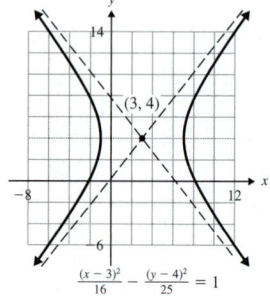

$$\frac{(x-3)^2}{16} - \frac{(y-4)^2}{25} = 1$$

50) ellipse

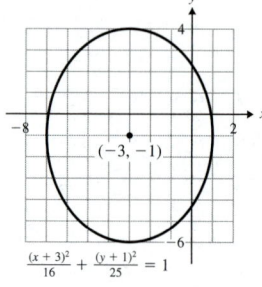

$$\frac{(x+3)^2}{16} + \frac{(y+1)^2}{25} = 1$$

41) $(y + 3)^2 - \dfrac{(x-1)^2}{16} = 1$

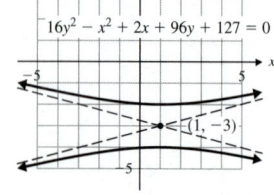

$16y^2 - x^2 + 2x + 96y + 127 = 0$

42) $\dfrac{(x-2)^2}{4} - \dfrac{(y-3)^2}{4} = 1$

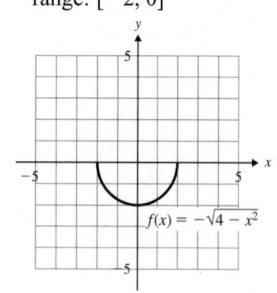

$x^2 - y^2 - 4x + 6y - 9 = 0$

51) circle

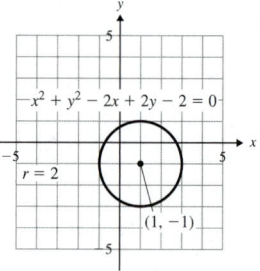

$-x^2 + y^2 - 2x + 2y - 2 = 0$

$r = 2$

52) hyperbola

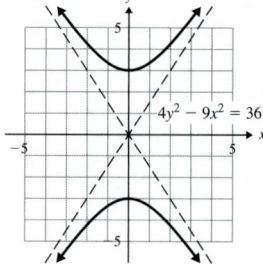

$4y^2 - 9x^2 = 36$

43) domain: $[-3, 3]$; range: $[0, 2]$

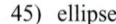

$h(x) = 2\sqrt{1 - \frac{x^2}{9}}$

44) domain: $[-2, 2]$; range: $[-2, 0]$

$f(x) = -\sqrt{4 - x^2}$

53) parabola

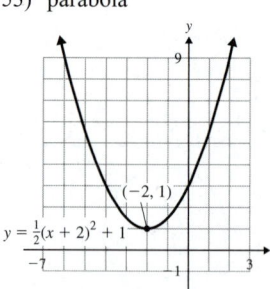

$y = \frac{1}{2}(x+2)^2 + 1$

54) circle

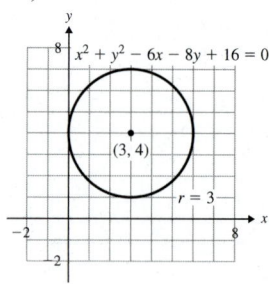

$x^2 + y^2 - 6x - 8y + 16 = 0$

$r = 3$

55) 0, 1, 2, 3, or 4   56) 0, 1, or 2

57) $\{(6, 7), (6, -7), (-6, 7), (-6, -7)\}$ 58) $\{(0, 7)\}$

59) $\{(1, 2), (-2, -1)\}$ 60) $\{(2, \sqrt{5}), (2, -\sqrt{5})\}$ 61) $\varnothing$

62) $\left\{\left(-\dfrac{3}{2}, 2\right), \left(\dfrac{3}{2}, -2\right)\right\}$ 63) 9 and 4   64) 18 in. $\times$ 21 in.

45) ellipse

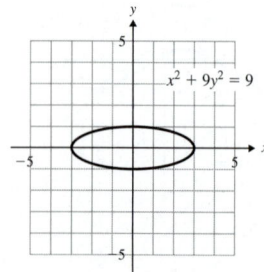

$x^2 + 9y^2 = 9$

46) circle

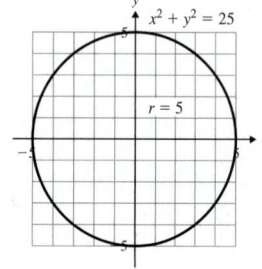

$x^2 + y^2 = 25$

$r = 5$

65)
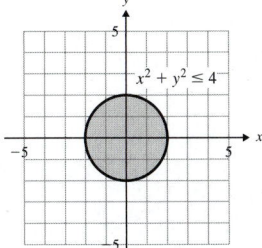

66)
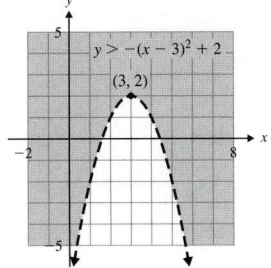

75)
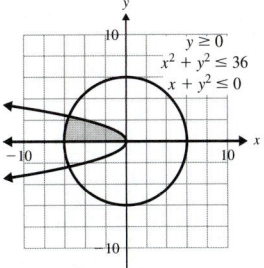

76)
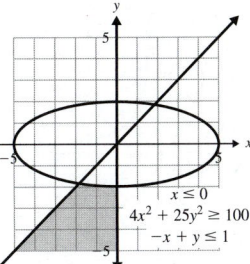

67)
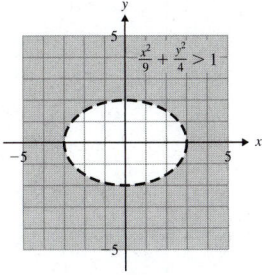

68)
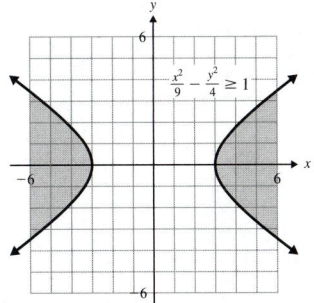

Chapter 11 Test

1) $\left(6, -\dfrac{5}{2}\right)$ 2)
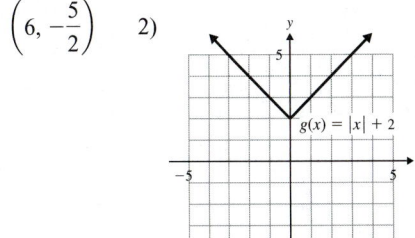

69)
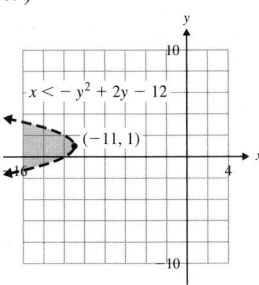

70)
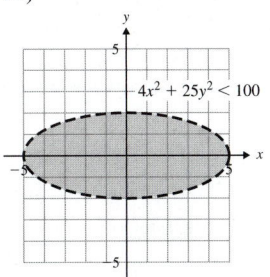

3)
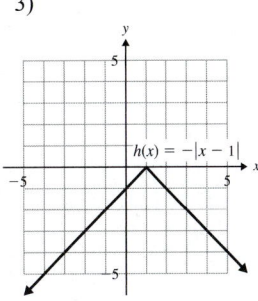

4)
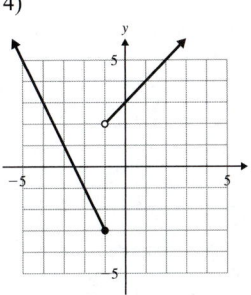

71)
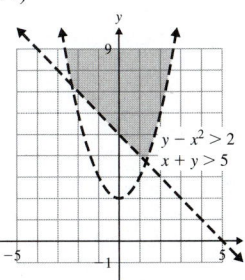

72)
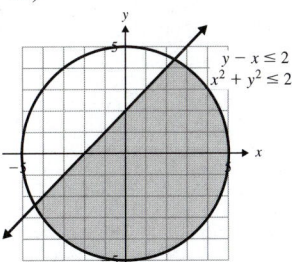

5) ellipse
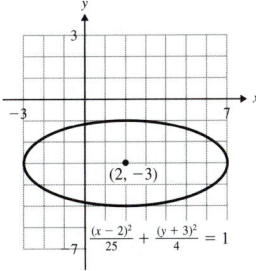

6) parabola
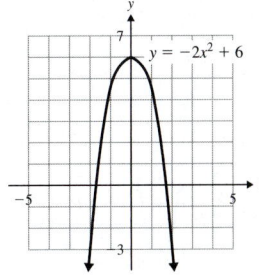

73)
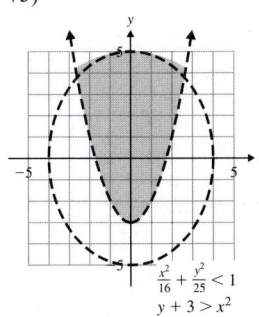

74)
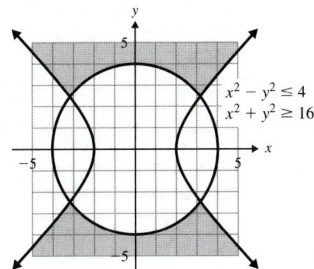

7) hyperbola
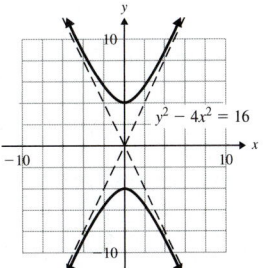

8) circle
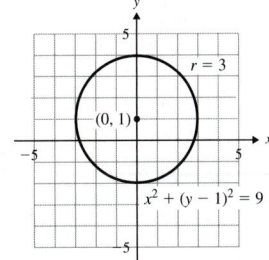

9) hyperbola

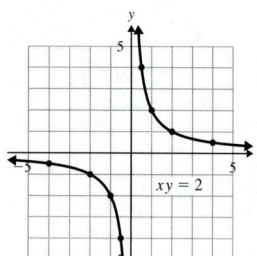

10) $(x + 1)^2 + (y - 3)^2 = 16;$
center $(-1, 3); r = 4$

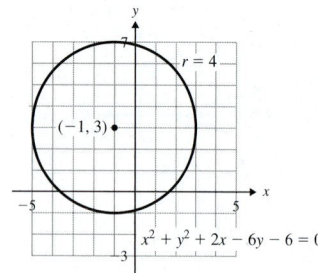

11) $(x - 5)^2 + (y - 2)^2 = 11$ 12) $\dfrac{x^2}{8836} + \dfrac{y^2}{6084} = 1$

13) domain: $[-5, 5]$;
range: $[-5, 0]$

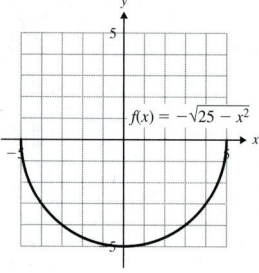

14) a)

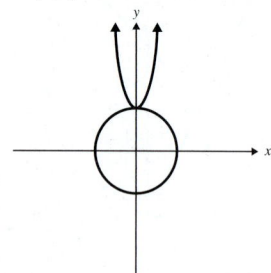

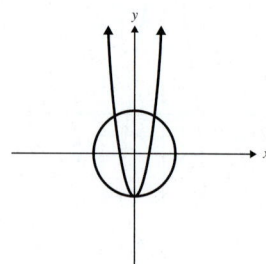

b)

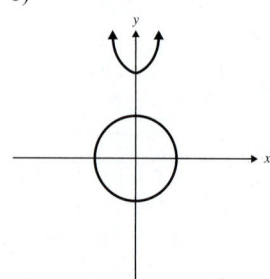

c) 0, 1, 2, 3, or 4

15) $\{(-1, 0), (7, -2)\}$ 16) $\{(3, 1), (3, -1), (-3, 1), (-3, -1)\}$

17) 8 in. $\times$ 14 in.

18)

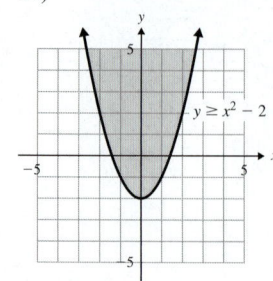

19)

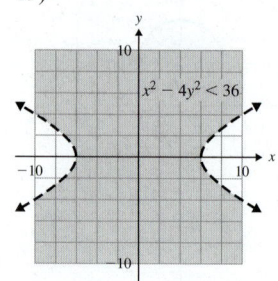

20)

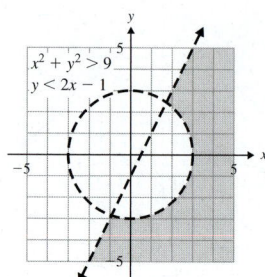

21)

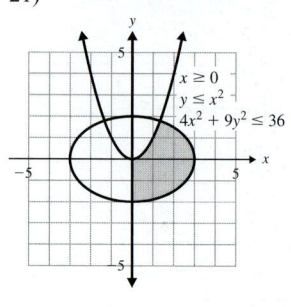

Cumulative Review for Chapters 1–11

1) $-\dfrac{3}{4}$ 2) 12   3) $A = 15$ cm^2; $P = 18.5$ cm

4) $A = 128$ in^2; $P = 56$ in.   5) -1 6) 16   7) $\dfrac{1}{8a^{18}b^{15}}$

8) $\{-40\}$ 9) $n = \dfrac{c - z}{a}$ 10) $[-4, \infty)$ 11) 15, 17, 19

12) -1 13) 0

14)

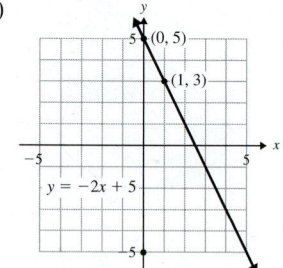

15) $y = -\dfrac{3}{4}x + 4$ 16) $\left(-1, \dfrac{3}{2}\right)$

17) 5 mL of 8%, 15 mL of 16%   18) $-3p^2 + 7p + 6$

19) $8w^3 + 30w^2 - 47w + 15$ 20) $x^2 + 4x + 9$

21) $2(3c - 4)(c - 1)$ 22) $(m - 2)(m^2 + 2m + 4)$

23) $\{-2, 6\}$ 24) $-\dfrac{5}{2(a + 6)}$ 25) $6(t + 3)$ 26) $\left\{-6, -\dfrac{4}{3}\right\}$

27) $(-\infty, -3) \cup \left(\dfrac{9}{5}, \infty\right)$ 28) $5\sqrt{3}$ 29) $2\sqrt[3]{6}$

30) $3ab^4\sqrt[3]{a^2b}$ 31) $\dfrac{1}{8}$ 32) $3\sqrt{3}$ 33) $\dfrac{20-5\sqrt{3}}{13}$

34) $\left\{\dfrac{1}{2}+2i,\dfrac{1}{2}-2i\right\}$ 35) $\left\{-\dfrac{7}{2}+\dfrac{\sqrt{37}}{2},-\dfrac{7}{2}-\dfrac{\sqrt{37}}{2}\right\}$

36) $\left[-\dfrac{12}{5},\dfrac{12}{5}\right]$ 37) $\left(-\infty,-\dfrac{5}{2}\right)\cup(3,\infty)$

38) a) $\{0,3,4\}$ b) $\{-1,0,1,2\}$ c) no

39)

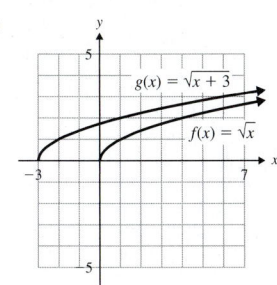

40) a) -3 b) $6x-9$ c) 6   41) a) no   b) no

42) $f^{-1}(x)=3x-12$ 43) $\left\{-\dfrac{6}{13}\right\}$ 44) $\{5\}$ 45) 2

46)

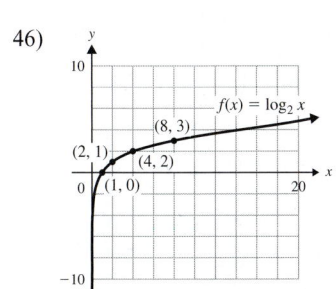

47) $\left\{\dfrac{\ln 8}{3}\right\}$; $\{0.6931\}$

48)   49)

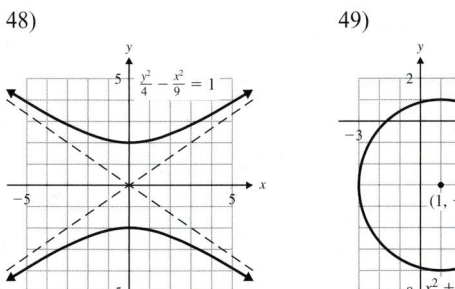

50) $(0,3)$

Chapter 12

Section 12.1

1) $3,4,5,6,7$ 2) $-3,-2,-1,0,1$ 3) $-1,2,5,8,11$

4) $5,9,13,17,21$ 5) $1,7,17,31,49$ 6) $5,11,21,35,53$

7) $1,3,9,27,81$ 8) $2,4,8,16,32$ 9) $\dfrac{5}{2},\dfrac{5}{4},\dfrac{5}{8},\dfrac{5}{16},\dfrac{5}{32}$

10) $6,2,\dfrac{2}{3},\dfrac{2}{9},\dfrac{2}{27}$ 11) $7,-14,21,-28,35$

12) $-2,3,-4,5,-6$ 13) $-\dfrac{3}{4},-\dfrac{2}{5},-\dfrac{1}{6},0,\dfrac{1}{8}$

14) $0,\dfrac{3}{2},\dfrac{8}{3},\dfrac{15}{4},\dfrac{24}{5}$ 15) a) 5   b) 17   c) 86

16) a) -8 b) 10   c) 85   17) a) $-\dfrac{3}{7}$ b) $-\dfrac{1}{4}$ c) $\dfrac{6}{11}$

18) a) $\dfrac{2}{9}$ b) $\dfrac{29}{45}$ c) $\dfrac{62}{89}$ 19) a) 9   b) -26 c) -390

20) a) -5 b) 55   c) 667   21) $a_n=2n$ 22) $a_n=9n$

23) $a_n=n^2$ 24) $a_n=n^3$ 25) $a_n=\left(\dfrac{1}{3}\right)^n$ 26) $a_n=\dfrac{4}{5^n}$

27) $a_n=\dfrac{n}{n+1}$ 28) $a_n=\dfrac{1}{n}$ 29) $a_n=(-1)^{n+1}\cdot(5n)$

30) $a_n=(-1)^n\cdot(2n)$ 31) $a_n=(-1)^n\left(\dfrac{1}{2}\right)^n$

32) $a_n=(-1)^{n+1}\left(\dfrac{1}{4}\right)^n$ 33) $1728, \$1152, \$768, \$512$

34) $\$1100, \$1210, \$1331, \$1464.10, \$1610.51$

35) 160 lb   36) 10.30

37) A sequence is a list of terms in a certain order, and a series is a sum of the terms of a sequence.

38) It means to find the sum of the first five terms of the sequence defined by $a_n=7n+2$.

39) 48   40) 75   41) -25 42) 0   43) 100   44) 189

45) $\dfrac{21}{2}$ 46) $\dfrac{23}{10}$ 47) 3   48) 3   49) 25   50) 65   51) 86

52) 91   53) $\displaystyle\sum_{i=1}^{5}\dfrac{1}{i}$ 54) $\displaystyle\sum_{i=1}^{6}\dfrac{11}{i}$ 55) $\displaystyle\sum_{i=1}^{4}(3i)$ 56) $\displaystyle\sum_{i=1}^{7}(4i)$

57) $\displaystyle\sum_{i=1}^{6}(i+4)$ 58) $\displaystyle\sum_{i=1}^{4}(i+3)$ 59) $\displaystyle\sum_{i=1}^{7}(-1)^i\cdot(i)$

60) $\displaystyle\sum_{i=1}^{5}(-1)^{i+1}\cdot(2^i)$ 61) $\displaystyle\sum_{i=1}^{4}(-1)^{i+1}(3^i)$

62) $\displaystyle\sum_{i=1}^{5}(-1)^i\cdot(i^2)$ 63) 20   64) 37   65) 9.4   66) 5.5

67) $\$1054.09$ 68) 8.99 in.

Section 12.2

1) It is a list of numbers in a specific order such that the difference between any two successive terms is the same (common difference). One example is $1,4,7,10,\ldots$. The common difference is 3.

2) Choose any term and subtract the term that precedes it.

3) yes, $d=8$ 4) yes, $d=3$ 5) yes, $d=-4$ 6) yes, $d=-7$

7) no   8) no   9) yes, $d=3$ 10) yes, $d=2$

11) $7,9,11,13,15$ 12) $20,24,28,32,36$

13) $15,7,-1,-9,-17$ 14) $-3,-5,-7,-9,-11$

15) $-10,-7,-4,-1,2$ 16) $-19,-14,-9,-4,1$

17) 13, 19, 25, 31, 37   18) 9, 11, 13, 15, 17

19) 4, 3, 2, 1, 0   20) $-1, -5, -9, -13, -17$

21) a) $a_1 = 4, d = 3$ b) $a_n = 3n + 1$ c) 106

22) a) $a_1 = -5, d = 2$ b) $a_n = 2n - 7$ c) 63

23) a) $a_1 = 4, d = -5$ b) $a_n = -5n + 9$ c) -86

24) a) $a_1 = -9, d = -12$ b) $a_n = -12n + 3$ c) -177

25) $a_n = 2n - 9; a_{25} = 41$ 26) $a_n = 6n + 7; a_{30} = 187$

27) $a_n = \frac{1}{2}n + \frac{1}{2}; a_{18} = \frac{19}{2}$ 28) $a_n = \frac{1}{3}n; a_{21} = 7$

29) $a_n = -5n + 5; a_{23} = -110$

30) $a_n = -7n + 2; a_{14} = -96$ 31) 55   32) 94   33) -107

34) -97 35) $a_n = 2n + 5; a_{11} = 27$

36) $a_n = 3n - 2; a_{16} = 46$ 37) $a_n = -5n + 17; a_{14} = -53$

38) $a_n = -4n + 3; a_{10} = -37$ 39) $a_n = 3n - 17; a_{18} = 37$

40) $a_n = 2n - 18; a_{17} = 16$ 41) 12   42) 15   43) 19

44) 22   45) S_{15} is the sum of the first 15 terms of the sequence.

46) (i) $S_n = \frac{n}{2}(a_1 + a_n)$; use this formula when the first term

and the last term are known.

(ii) $S_n = \frac{n}{2}[2a_1 + (n - 1)d]$; use this formula when the

first term and the common difference are known.

47) 410   48) 162   49) -21 50) -363 51) -120

52) 16   53) 164   54) 100   55) -88 56) -92 57) -152

58) 152   59) 140   60) -176

61) a) $9 + 11 + 13 + 15 + 17 + 19 + 21 + 23 + 25 + 27 = 180$
b) 180   c) Answers may vary.

62) 111   63) 95   64) -36 65) -7 66) -204

67) 315   68) 21,070   69) 125,250   70) 3125   71) $2300

72) $41,200   73) $300   74) $576   75) a) 5   b) 78

76) 25   77) 38; 350   78) 54; 736   79) 20   80) 9

Section 12.3

1) A sequence is arithmetic if the difference between
consecutive terms is constant, but a sequence is geometric
if each term after the first is obtained by multiplying the
preceding term by a common ratio.

2) Answers may vary.   3) 2   4) 4   5) $\frac{1}{3}$ 6) $\frac{1}{2}$ 7) $-\frac{1}{4}$

8) -3 9) 2, 10, 50, 250, 1250   10) 3, 6, 12, 24, 48

11) $\frac{1}{4}, -\frac{1}{2}, 1, -2, 4$ 12) $250, 50, 10, 2, \frac{2}{5}$

13) $72, 48, 32, \frac{64}{3}, \frac{128}{9}$ 14) $-20, 30, -45, \frac{135}{2}, -\frac{405}{4}$

15) $a_n = 4(7)^{n-1}$; 196   16) $a_n = 3(8)^{n-1}$; 192

17) $a_n = -1(3)^{n-1}$; -81 18) $a_n = -5\left(-\frac{1}{3}\right)^{n-1}; \frac{5}{27}$

19) $a_n = 2\left(\frac{1}{5}\right)^{n-1}; \frac{2}{125}$ 20) $a_n = 7(3)^{n-1}$; 567

21) $a_n = -\frac{1}{2}\left(-\frac{3}{2}\right)^{n-1}; \frac{27}{16}$ 22) $a_n = \frac{3}{5}(2)^{n-1}; \frac{96}{5}$

23) $a_n = 5(2)^{n-1}$ 24) $a_n = 4(3)^{n-1}$ 25) $a_n = -3\left(\frac{1}{5}\right)^{n-1}$

26) $a_n = -1(-4)^{n-1}$ 27) $a_n = 3(-2)^{n-1}$

28) $a_n = 2\left(\frac{1}{3}\right)^{n-1}$ 29) $a_n = \frac{1}{3}\left(\frac{1}{4}\right)^{n-1}$

30) $a_n = -\frac{1}{5}\left(\frac{3}{2}\right)^{n-1}$ 31) 2048   32) 19,683   33) $-\frac{1}{81}$

34) -125 35) -32 36) 640   37) arithmetic; $a_n = 9n + 6$

38) geometric; $a_n = -1(3)^{n-1}$

39) geometric; $a_n = -2(-3)^{n-1}$

40) arithmetic; $a_n = -5n + 13$

41) geometric; $a_n = \frac{1}{9}\left(\frac{1}{2}\right)^{n-1}$ 42) geometric; $a_n = 11(2)^{n-1}$

43) arithmetic; $a_n = 7n - 38$ 44) arithmetic; $a_n = \frac{1}{2}n + 1$

45) a) $a_n = 40,000(0.80)^{n-1}$ b) $16,384

46) a) $a_n = 64,000(0.85)^{n-1}$ b) $39,304

47) a) $a_n = 500,000(0.90)^{n-1}$ b) $364,500

48) a) $a_n = 1000(1.20)^{n-1}$ b) 4300

49) a) $a_n = 160,000(1.04)^{n-1}$ b) $194,664

50) a) $a_n = 140,000(1.05)^{n-1}$ b) $206,844

51) 567   52) 240   53) $S_7 = 38,227$ 54) $S_5 = -7775$

55) $S_6 = -\frac{63}{4}$ 56) $S_6 = \frac{4095}{8}$ 57) $S_5 = \frac{121}{81}$

58) $S_5 = -\frac{1353}{625}$ 59) 2286   60) 2550   61) -1452

62) -294 63) $-\frac{63}{64}$ 64) $\frac{242}{243}$ 65) $\frac{52}{9}$ 66) $-\frac{348}{125}$

67) a) 512   b) $10.23   68) 64,000   69) $\frac{32}{3}$ 70) 27

71) $\frac{25}{9}$ 72) $\frac{80}{7}$ 73) Sum does not exist.

74) Sum does not exist.   75) 24   76) 7   77) -5 78) $-\frac{36}{5}$

79) Sum does not exist.   80) $\frac{216}{5}$ 81) -160

82) Sum does not exist.   83) 12 ft   84) 140 in.

85) a) $3\dfrac{5}{9}$ ft   b) 135 ft   86) a) $5\dfrac{1}{16}$ ft   b) 112 ft

Section 12.4

1) Answers may vary.   2) first term: a^n, last term: b^n

3) $r^3 + 3r^2s + 3rs^2 + s^3$ 4) $m^4 + 4m^3n + 6m^2n^2 + 4mn^3 + n^4$

5) $y^5 + 5y^4z + 10y^3z^2 + 10y^2z^3 + 5yz^4 + z^5$

6) $c^6 + 6c^5d + 15c^4d^2 + 20c^3d^3 + 15c^2d^4 + 6cd^5 + d^6$

7) $x^4 + 20x^3 + 150x^2 + 500x + 625$

8) $k^5 + 10k^4 + 40k^3 + 80k^2 + 80k + 32$

9) Answers may vary.   10) 1   11) 2   12) 6   13) 120

14) 720   15) 10   16) 6   17) 35   18) 56   19) 210   20) 84

21) 36   22) 165   23) 1   24) 1   25) 6   26) 3   27) 1

28) 1   29) 10   30) $[t + (-4)]^6$ 31) $f^3 + 3f^2g + 3fg^2 + g^3$

32) $c^5 + 5c^4d + 10c^3d^2 + 10c^2d^3 + 5cd^4 + d^5$

33) $w^4 + 8w^3 + 24w^2 + 32w + 16$

34) $h^4 + 16h^3 + 96h^2 + 256h + 256$

35) $b^5 + 15b^4 + 90b^3 + 270b^2 + 405b + 243$

36) $t^3 + 27t^2 + 243t + 729$

37) $a^4 - 12a^3 + 54a^2 - 108a + 81$

38) $p^3 - 6p^2 + 12p - 8$ 39) $u^3 - 3u^2v + 3uv^2 - v^3$

40) $p^5 - 5p^4q + 10p^3q^2 - 10p^2q^3 + 5pq^4 - q^5$

41) $81m^4 + 216m^3 + 216m^2 + 96m + 16$

42) $16k^4 + 32k^3 + 24k^2 + 8k + 1$

43) $243a^5 - 810a^4b + 1080a^3b^2 - 720a^2b^3 + 240ab^4 - 32b^5$

44) $256c^4 - 768c^3d + 864c^2d^2 - 432cd^3 + 81d^4$

45) $x^6 + 3x^4 + 3x^2 + 1$ 46) $w^9 + 6w^6 + 12w^3 + 8$

47) $\dfrac{1}{16}m^4 - \dfrac{3}{2}m^3n + \dfrac{27}{2}m^2n^2 - 54mn^3 + 81n^4$

48) $\dfrac{1}{243}a^5 + \dfrac{10}{81}a^4b + \dfrac{40}{27}a^3b^2 + \dfrac{80}{9}a^2b^3 + \dfrac{80}{3}ab^4 + 32b^5$

49) $\dfrac{1}{27}y^3 + \dfrac{2}{3}y^2z^2 + 4yz^4 + 8z^6$

50) $t^8 - 2t^6u + \dfrac{3}{2}t^4u^2 - \dfrac{1}{2}t^2u^3 + \dfrac{1}{16}u^4$ 51) $700k^6$

52) $8960y^3$ 53) $5005w^6$ 54) $61{,}236z^3$ 55) $-27q^8$

56) $-280u^4$ 57) $2160x^2$ 58) $672w^3$ 59) $960y^6z^7$

60) $5670p^4q^8$ 61) $189c^{15}d^4$ 62) $-12r^3s^{20}$ 63) v^{33} 64) k^{48}

65) Answers may vary.   66) Answers may vary.

Chapter 12 Review Exercises

1) 8, 15, 22, 29, 36   2) $-6, -3, 2, 9, 18$

3) $1, -\dfrac{1}{4}, \dfrac{1}{9}, -\dfrac{1}{16}, \dfrac{1}{25}$ 4) $2, \dfrac{3}{4}, \dfrac{4}{9}, \dfrac{5}{16}, \dfrac{6}{25}$ 5) $a_n = 5n$

6) $a_n = \dfrac{1}{n^3}$ 7) $a_n = -\dfrac{n+1}{n}$ 8) $a_n = (-1)^n(n)$

9) \$8.25, \$8.50, \$8.75, \$9.00   10) 69

11) A sequence is a list of terms in a certain order, and a series is the indicated sum of the terms of a sequence.

12) $\displaystyle\sum_{i=1}^{8}(6i - 1)$ 13) $3 + 9 + 19 + 33 + 51 = 115$

14) $-2 + 4 - 6 + 8 - 10 + 12 = 6$ 15) $\displaystyle\sum_{i=1}^{4}\dfrac{13}{i}$

16) $\displaystyle\sum_{i=1}^{6}(-1)^{i+1}(2^i)$ 17) 22.2   18) \$190.01

19) 11, 18, 25, 32, 39   20) $0, -4, -8, -12, -16$

21) $-58, -50, -42, -34, -26$ 22) $-1, -\dfrac{1}{2}, 0, \dfrac{1}{2}, 1$

23) a) $a_1 = 6, d = 4$ b) $a_n = 4n + 2$ c) 82

24) a) $a_1 = -13, d = 7$ b) $a_n = 7n - 20$ c) 120

25) a) $a_1 = -8, d = -5$ b) $a_n = -5n - 3$ c) -103

26) a) $a_1 = 14, d = 3$ b) $a_n = 3n + 11$ c) 71

27) 69   28) -58 29) -34 30) 148   31) $a_n = 4n; a_{12} = 48$

32) $a_n = 2n - 7; a_{15} = 23$ 33) $a_n = -2n + 5; a_{17} = -29$

34) $a_n = -3n - 4; a_{20} = -64$ 35) 20   36) 33   37) 52

38) 18   39) -88 40) 75   41) 255   42) 215   43) -230

44) -335 45) 60   46) -15 47) -225 48) 475

49) $S_4 = -118$ 50) $S_8 = 16$ 51) $S_{13} = 325$ 52) $S_{15} = -585$

53) $S_{11} = -242$ 54) $S_{19} = -1178$ 55) 53; 680

56) a) 5   b) 136   57) \$930   58) 32   59) 5   60) $\dfrac{1}{4}$

61) 7, 14, 28, 56, 112   62) $5, -15, 45, -135, 405$

63) $48, 12, 3, \dfrac{3}{4}, \dfrac{3}{16}$ 64) $-16, -24, -36, -54, -81$

65) $a_n = 3(2)^{n-1}; 96$ 66) $a_n = 4(3)^{n-1}; 324$

67) $a_n = 8\left(\dfrac{1}{3}\right)^{n-1}; \dfrac{8}{27}$ 68) $a_n = \left(-\dfrac{1}{2}\right)(-3)^{n-1}; \dfrac{27}{2}$

69) $a_n = 7(6)^{n-1}$ 70) $a_n = (-4)\left(\dfrac{1}{5}\right)^{n-1}$

71) $a_n = (-15)(-3)^{n-1}$ 72) $a_n = \dfrac{1}{9}(4)^{n-1}$ 73) 2187

74) $\dfrac{1}{64}$ 75) 968   76) -312 77) $S_5 = 6248$

78) $S_5 = -\dfrac{61}{3}$ 79) $S_6 = \dfrac{63}{4}$ 80) $S_6 = 441$ 81) $S_5 = \dfrac{217}{32}$

82) $S_4 = 240$ 83) a) $a_n = 28{,}000(0.80)^{n-1}$ b) $17{,}920

84) a) $a_n = 180{,}000(1.04)^{n-1}$ b) $227{,}757

85) a) arithmetic b) $a_n = 2n + 5$ c) $S_8 = 112$

86) a) arithmetic b) $a_n = -5n + 38$ c) $S_8 = 124$

87) a) geometric b) $a_n = 9\left(\dfrac{1}{2}\right)^{n-1}$ c) $S_8 = \dfrac{2295}{128}$

88) a) arithmetic b) $a_n = \dfrac{1}{2}n - 5$ c) $S_8 = -22$

89) a) geometric b) $a_n = (-1)(3)^{n-1}$ c) $S_8 = -3280$

90) a) geometric b) $a_n = 5(-2)^{n-1}$ c) $S_8 = -425$

91) $10{,}125$ 92) a) $20.48 b) $40.94 93) when $|r| < 1$

94) 42 95) $-\dfrac{24}{7}$ 96) Sum does not exist. 97) -9

98) 16 99) Sum does not exist. 100) 20 ft

101) $y^4 + 4y^3z + 6y^2z^2 + 4yz^3 + z^4$

102) $c^3 + 3c^2d + 3cd^2 + d^3$ 103) 720 104) 40,320

105) 10 106) 21 107) 9 108) 1

109) $m^4 + 4m^3n + 6m^2n^2 + 4mn^3 + n^4$

110) $k^6 + 12k^5 + 60k^4 + 160k^3 + 240k^2 + 192k + 64$

111) $h^3 - 27h^2 + 243h - 729$

112) $8w^3 - 60w^2 + 150w - 125$

113) $32p^{10} - 240p^8r + 720p^6r^2 - 1080p^4r^3 + 810p^2r^4 - 243r^5$

114) $\dfrac{1}{81}b^4 + \dfrac{4}{27}b^3c + \dfrac{2}{3}b^2c^2 + \dfrac{4}{3}bc^3 + c^4$ 115) $17{,}920z^4$

116) $-435{,}456y^3$ 117) $2288k^3$ 118) $2268p^3q^6$

Chapter 12 Test

1) $-1, 1, 3, 5, 7$ 2) $\dfrac{1}{3}, -\dfrac{1}{2}, \dfrac{3}{5}, -\dfrac{2}{3}, \dfrac{5}{7}$

3) An arithmetic sequence is obtained by adding the common difference to each term to obtain the next term, while a geometric sequence is obtained by multiplying a term by the common ratio to get the next term.

4) $32, -16, 8, -4, 2$ 5) $d = 6$ 6) geometric; $a_n = 4(3)^{n-1}$

7) arithmetic; $a_n = -3n + 8$ 8) 67

9) $11 + 26 + 51 + 86 = 174$ 10) 567 11) 220

12) $-19{,}900$ 13) 10 14) 105 15) 14,641 16) 160 in.

17) 120 18) 15 19) $81x^4 + 108x^3 + 54x^2 + 12x + 1$

20) $-672k^6$

Cumulative Review: Chapters 1–12

1) $\dfrac{37}{24}$

2) a) $A = 25\pi$ cm^2; $A \approx 78.5$ cm^2
   b) $C = 10\pi$ cm; $C \approx 31.4$ cm

3) a) $81k^8$ b) $-\dfrac{56}{z^6}$ c) $\dfrac{a^{24}}{125b^{15}}$ 4) 8.215×10^{-5}

5) a) $\{15\}$ b) $\left\{-\dfrac{15}{4}\right\}$ 6) length = 9 in., width = 4 in.

7) $3000 at 4% and $6000 at 6% 8) $\left(-\infty, -\dfrac{1}{2}\right]$

9) $(-\infty, -3) \cup (5, \infty)$

10) a)

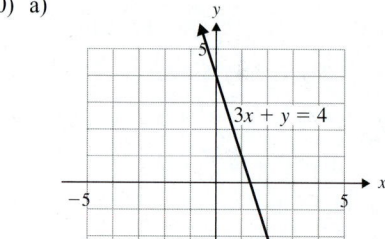

b)

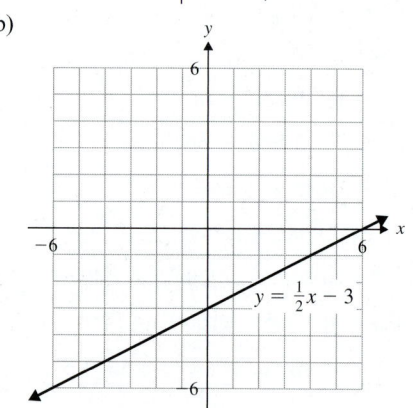

c)

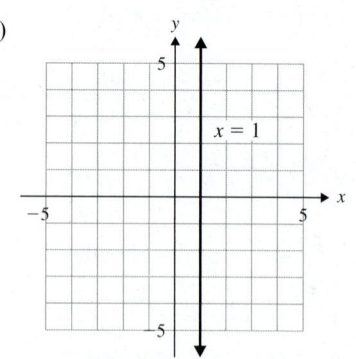

11) $y = -\dfrac{1}{3}x + 4$ 12) $3x - 2y = -16$ 13) $(4, 0)$

14) 15 mL of 15% solution, 45 mL of 7% solution

Photo Credits

Page 1: © The McGraw-Hill Companies, Inc./Lars A. Niki, photographer; **p. 31:** © EP100/PhotoDisc/Getty RF; **p. 43:** © Creatstock Photographic/Alamy RF; **p. 53(top):** © Image Source/PunchStock RF; **p. 53(middle):** © Vol. 25 PhotoDisc/Getty RF; **p. 53(bottom):** © BrandX/Punchstock RF; **p. 54:** © Corbis RF; **p. 62:** © Vol. 5 PhotoDisc/Getty RF; **p. 70:** © StockDisc/Getty RF; **p. 75:** © Corbis RF; **p. 76:** © Comstock/Punchstock RF; **p. 77(left):** © Getty RF; **p. 77(right):** © Corbis RF; **p. 85:** © Vol. 56 PhotoDisc/Getty RF; **p. 87:** © DynamicGraphics/Jupiter Images RF; **p. 102:** © The McGraw-Hill Companies, Inc./Andrew Resek, photographer; **p. 105:** © The McGraw-Hill Companies, Inc./Andrew Resek, photographer; **p. 109:** © Adrian Sherratt/Alamy RF; **p. 111:** © Vol. 10 PhotoDisc/Getty RF; **p. 115:** © The McGraw-Hill Companies, Inc./Jill Braaten, photographer; **p. 151:** © Stockbyte/Punchstock Images RF; **p. 168:** © BrandX 195/Getty RF; **p. 182 (left):** © The McGraw-Hill Companies, Inc./Gary He, photographer; **p. 182 (right):** © SS18 PhotoDisc/Getty RF; **p. 199:** © DT01/Getty RF; **p. 225:** © Bluestone Productions/Superstock RF; **p. 233:** © Royalty-Free/Corbis; **p. 234 (left):** © BrandX/Punchstock RF; **p. 235 (left):** © Corbis RF; **p. 235 (right):** © Stockbyte/Punchstock RF; **p. 236:** © Corbis RF; **p. 249(left):** © RP020/Getty RF; **p. 249(right):** © Ingram Publishing/Alamy RF; **p. 250:** © BrandX Pictures/Punchstock RF; **p. 251:** © The McGraw-Hill Companies, Inc./Andrew Resek, photographer; **p. 252:** © Vol. 54 PhotoDisc/Getty RF; **p. 253:** © Corbis RF; **p. 271:** © BananaStock/Punchstock RF; **p. 277:** © Vol. 21 PhotoDisc/Getty RF; **p. 303:** © Digital Vision RF; **p. 304:** © Stockbyte/PunchStock RF; **p. 309:** © The McGraw-Hill Companies, Inc./Jill Braaten, photographer; **p. 355(top):** © SS14 PhotoDisc/Getty RF; **p. 355(bottom):** © The McGraw-Hill Companies, Inc./Ken Cavanagh, photographer; **p. 363:** BrandX/Punchstock RF; **p. 367:** © EP046 PhotoDisc/Getty RF; **p. 417:** © Vol. 21 PhotoDisc/Getty RF; **p. 423(left):** © Corbis RF; **p. 423(right):** © Corbis RF; **p. 427:** © Comstock/JupiterImages RF; **p. 431(top):** © OS40 PhotoDisc/Getty RF; **p. 431(bottom):** © Creatas/PictureQuest RF; **p. 443:** © PhotoDisc/Getty RF; **p. 456:** The McGraw-Hill Companies, Inc./Mark Dierker, photographer; **p. 464:** © image100/Getty RF; **p. 511:** © Goodshoot/Alamy RF; **p. 533:** © Vol. 86 PhotoDisc/Getty RF; **p. 564(left):** © Corbis/Superstock RF; **p. 564(right):** © Brand X Pictures/JupiterImages RF; **p. 568:** © Getty RF; **p. 594:** © Vol. 21 PhotoDisc/Getty RF; **p. 617:** © PhotoDisc/Getty RF; **p. 642:** © Vol. 1 PhotoDisc/Getty RF; **p. 652:** ©: image100/PunchStock RF; **p. 655:** © Stockbyte/Getty RF; **p. 675:** © Brand X Pictures/PunchStock RF; **p. 682:** © Creatas/PunchStock RF; **p. 693:** © Brand X Pictures/PunchStock RF; **p. 694:** © Vol. 10 PhotoDisc/Getty RF; **p. 697:** © Ingram Publishing/Fotosearch RF; **p. 703:** © OS31 PhotoDisc/Getty RF; **p. 714:** © Ingram Publishing/AGE Fotostock RF; **p. 715:** © Brand X Pictures/Punchstock RF; **p. 739:** © Comstock/PictureQuest RF; **p.753:** © moodboard/Corbis RF; **p. 758:** © BananaStock/JupiterImages RF; **p. 764:** © Imageshop/Punchstock RF; **p. 776:** © Vol. 115 PhotoDisc/Getty RF; **p. 786(top):** © The McGraw-Hill Companies, Inc./Gary He, photographer; **p. 786(bottom):** © Vol. 17 PhotoDisc/Getty RF; **p. 800:** © Corbis RF; **p. 803:** © Vol. 80 PhotoDisc/Getty RF.

Index

Figure		Perimeter	Area
Rectangle:		$P = 2l + 2w$	$A = lw$
Square:		$P = 4s$	$A = s^2$
Triangle: h = height		$P = a + b + c$	$A = \dfrac{1}{2}bh$
Parallelogram: h = height		$P = 2a + 2b$	$A = bh$
Trapezoid: h = height		$P = a + c + b_1 + b_2$	$A = \dfrac{1}{2}h(b_1 + b_2)$
Circle:		**Circumference** $C = 2\pi r$	**Area** $A = \pi r^2$

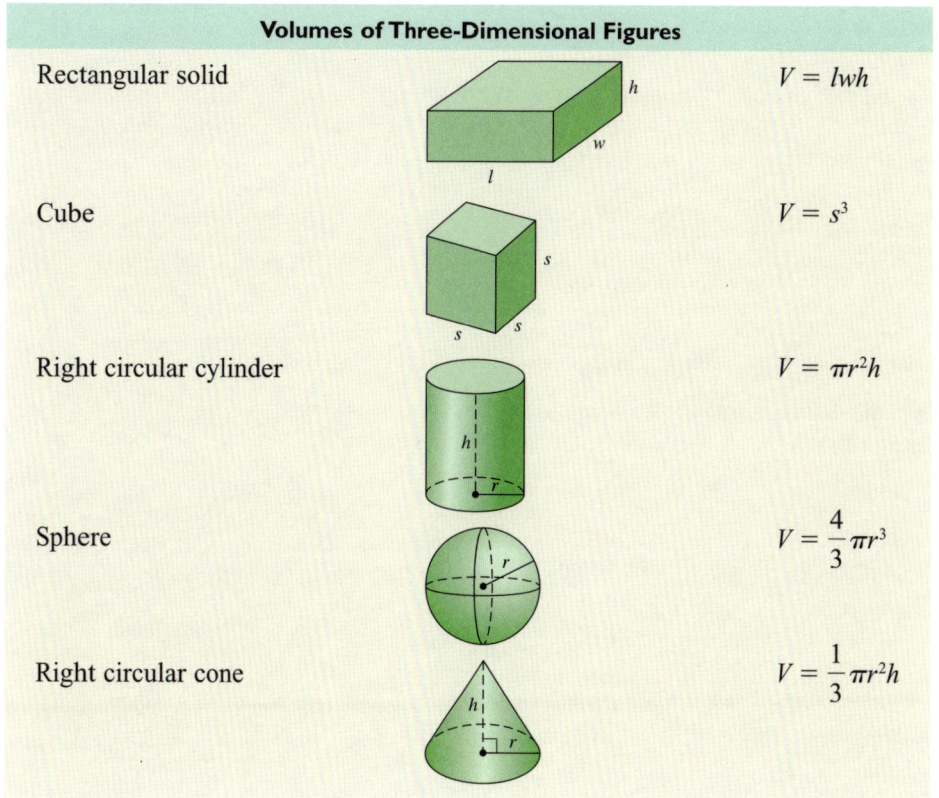

Volumes of Three-Dimensional Figures		
Rectangular solid		$V = lwh$
Cube		$V = s^3$
Right circular cylinder		$V = \pi r^2 h$
Sphere		$V = \dfrac{4}{3}\pi r^3$
Right circular cone		$V = \dfrac{1}{3}\pi r^2 h$

Angles

An **acute angle** is an angle whose measure is greater than 0° and less than 90°.

A **right angle** is an angle whose measure is 90°.

An **obtuse angle** is an angle whose measure is greater than 90° and less than 180°.

A **straight angle** is an angle whose measure is 180°.

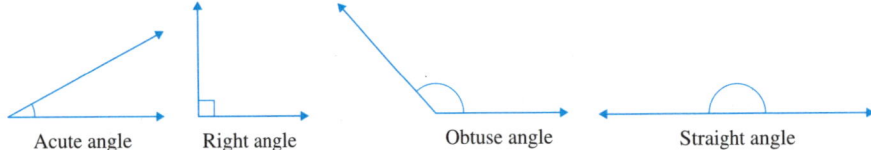

| Acute angle | Right angle | Obtuse angle | Straight angle |

Two angles are **complementary** if their measures add to 90°.

Two angles are **supplementary** if their measures add to 180°.

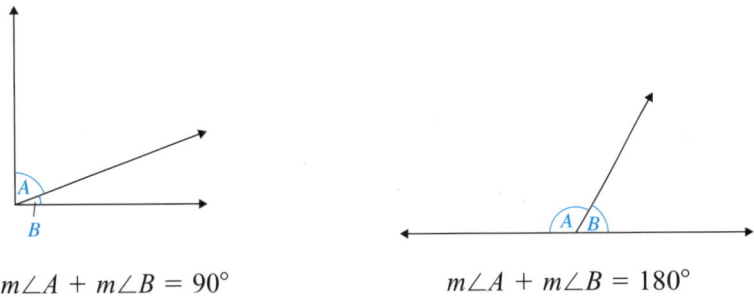

$$m\angle A + m\angle B = 90°$$ $$m\angle A + m\angle B = 180°$$

Vertical Angles

The pair of opposite angles are called **vertical angles**. Angles A and C are *vertical angles,* and angles B and D are *vertical angles. The measures of vertical angles are equal.* Therefore, $m\angle A = m\angle C$ and $m\angle B = m\angle D$.

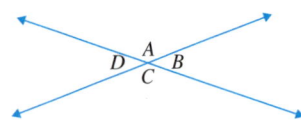

Parallel and Perpendicular Lines

Parallel lines are lines in the same plane that do not intersect.

Perpendicular lines are lines that intersect at right angles.

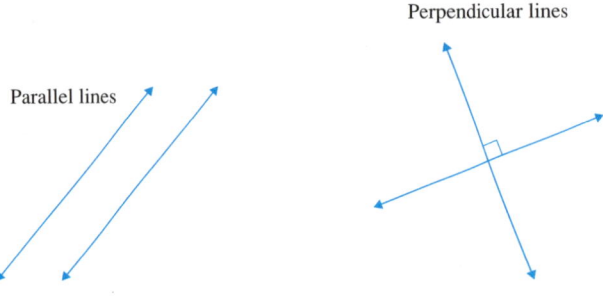

Perpendicular lines

Parallel lines